Deuterostomes

Chordata

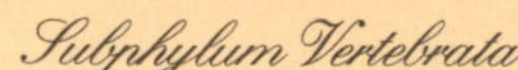

Subphylum Urochordata

Subphylum Cephalochordata

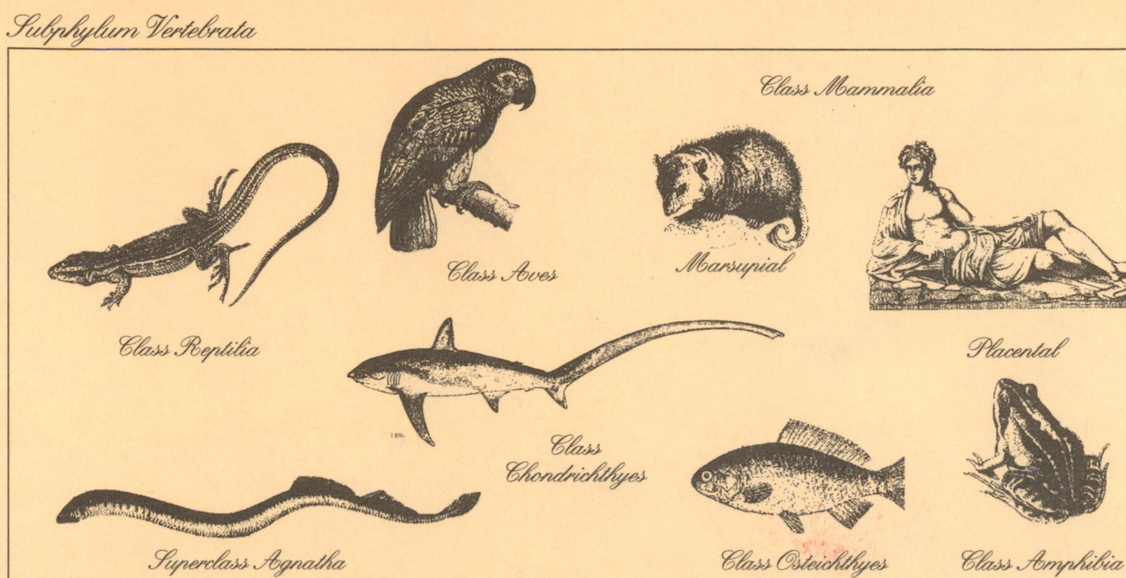

Echinodermata

Chaetognatha

Hemichordata

Loricifera

Entoprocta

Gastrotricha

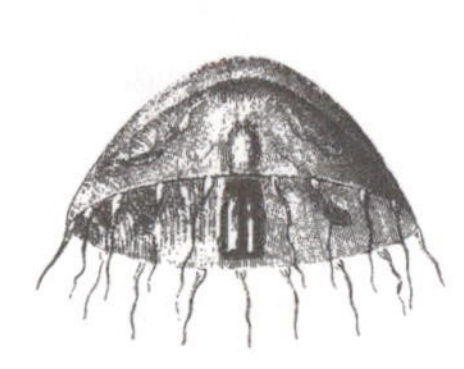

Cnidaria

Ctenophora

Porifera

Placozoa

Mesozoa

CONCEPTS IN ZOOLOGY

Second Edition

CONCEPTS IN

ZOOLOGY

Second Edition

C. LEON HARRIS

State University of New York

Plattsburgh, New York

HarperCollinsCollegePublishers

Sponsoring Editor: Bonnie Roesch
Developmental Editor: Vicki Cohen
Project Coordination: Electronic Publishing Services Inc.
Design Administrator: Jess Schaal
Text Design: Electronic Publishing Services Inc.
Art Studio: Hudson River Studios
Cover Design: Lesiak/Crampton Design Inc.: Lucy Lesiak
Cover Photo: Tony Stone Images: Erik Svenson
Photo Researcher: Mira Schachne
Production Administrator: Randee Wire
Compositor: Electronic Publishing Services Inc.
Printer and Binder: R.R. Donnelley & Sons Company
Cover Printer: Phoenix Design

For permission to use copyrighted material, grateful acknowledgment is made to the copyright holders on pp. C-1–C-4, which are hereby made part of this copyright page.

Concepts in Zoology, Second Edition

Library of Congress Cataloging-in-Publication Data
Harris, C. Leon.
Concepts in zoology / C. Leon Harris.—2nd ed.
p. cm.
Includes bibliographical references and index.
ISBN 0-673-99243-8
1. Zoology. I. Title.
QL47.2.H38 1995
591—dc20 95-24505
CIP

96 97 98 9 8 7 6 5 4 3 2

Leon Harris studied electrical engineering and physics at M. I. T. and Virginia Tech, then earned graduate degrees in biophysics at Penn State. He says that "Having to teach myself zoology has given me an appreciation for what beginning students face and what they need in a textbook." Dr. Harris has taught biology at the State University of New York at Plattsburgh since 1970. In addition to general zoology, he teaches courses in animal physiology, evolution, and writing for biology students. His current research is on the mechanisms of learning, using cockroaches as experimental subjects. When he is not writing and doing research on animals, he likes to watch and photograph them on hikes through the Adirondacks and in Vermont. He also enjoys running, cross-country skiing, and playing classic guitar.

FOR MARY JANE

I think I could turn and live with animals, they are so placid and self-contain'd,
I stand and look at them long and long.

They do not sweat and whine about their condition,
They do not lie awake in the dark and weep for their sins,
They do not make me sick discussing their duty to God,
Not one is dissatisfied, not one is demented with the mania of owning things,
Not one kneels to another, nor to his kind that lived thousands of years ago,
Not one is respectable or unhappy over the whole earth.

So they show their relations to me and I accept them,
They bring me tokens of myself, they evince them plainly in their possession.

I wonder where they get those tokens,
Did I pass that way huge times ago and negligently drop them?

Walt Whitman, from "Song of Myself"

BRIEF CONTENTS

DETAILED CONTENTS

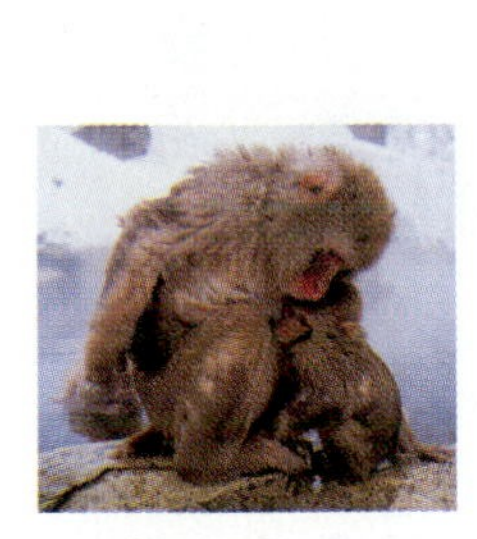

UNIT 4: THE DIVERSITY OF ANIMALS 426

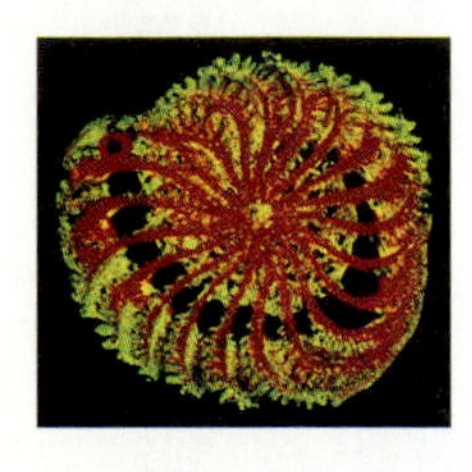

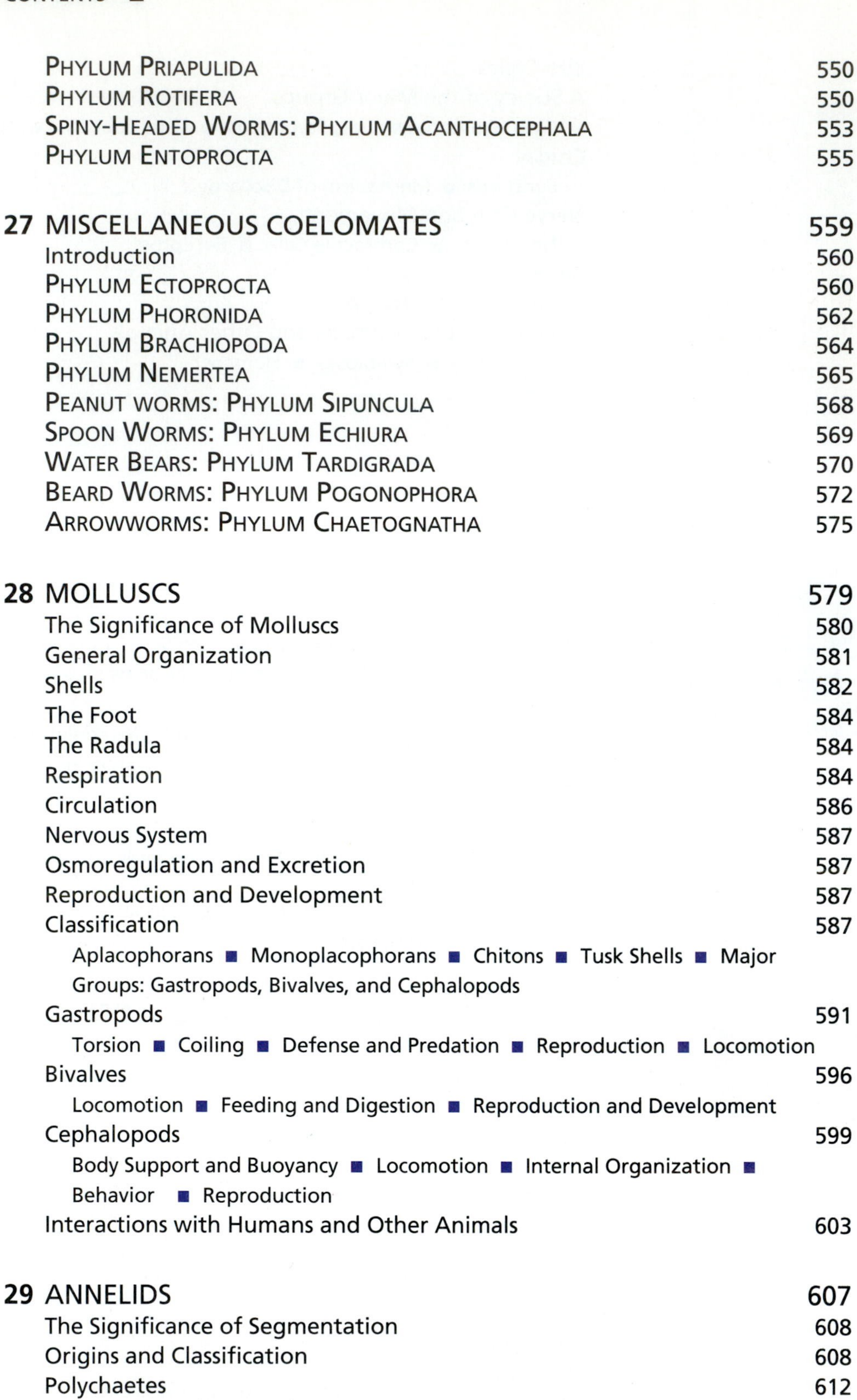

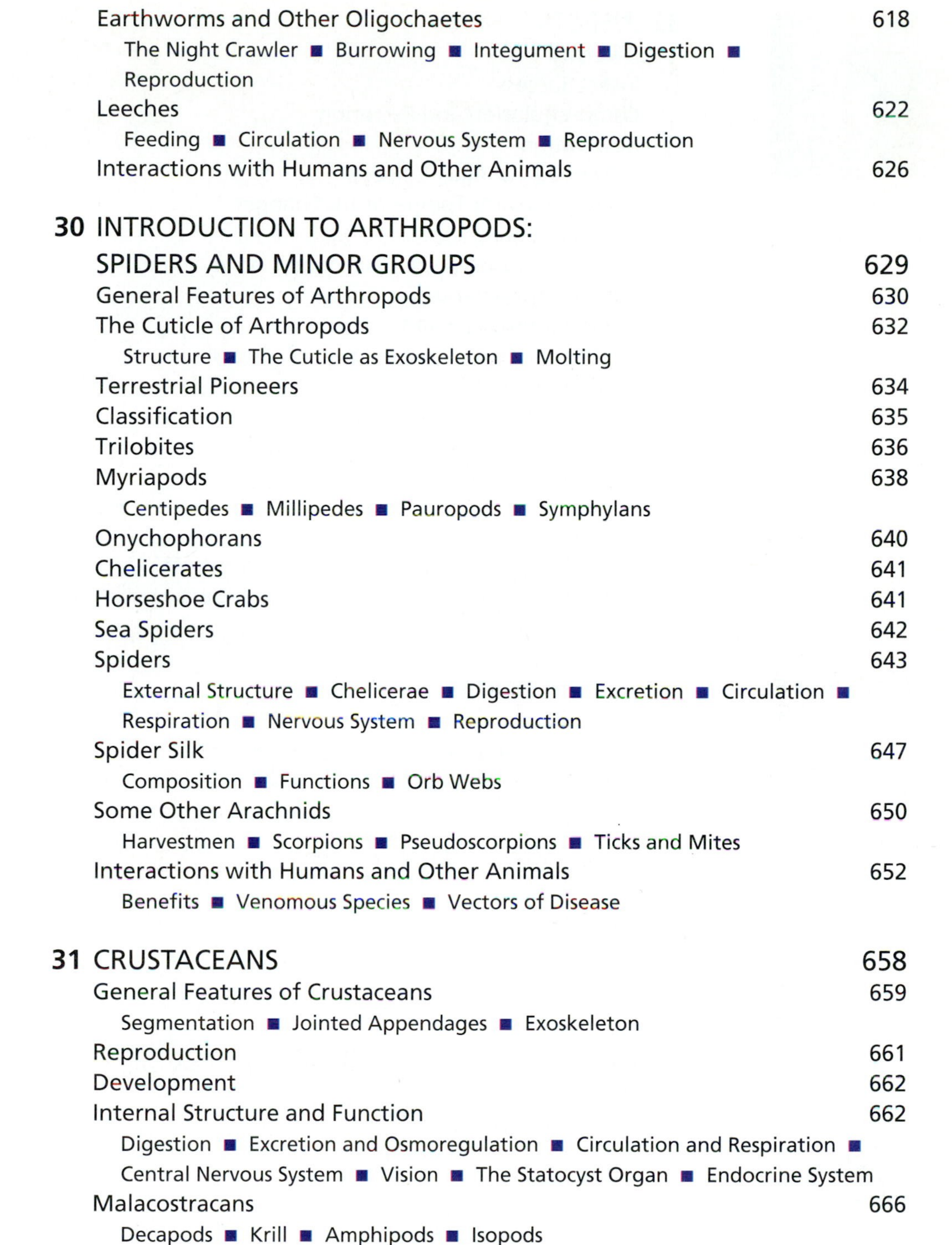

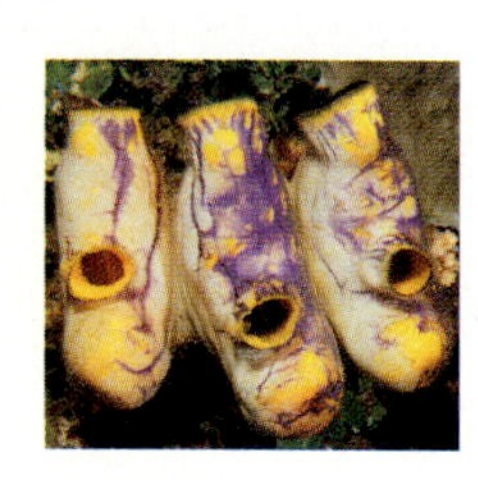

The Conceptual Approach to Zoology

This second edition of *Concepts in Zoology* embodies many changes from the first edition, some of which I describe below. One thing that has not changed, however, is the emphasis on concepts rather than details of zoology. I considered making this emphasis more obvious by highlighting the concepts in boldface, as a growing number of general biology texts are doing. I quickly realized, however, that concepts are by nature and definition "abstract generalizations about facts." They are not themselves facts that can be stated in simple phrases or sentences, and we as scientists do not use them in that way when we are thinking conceptually. Chloroforming concepts and pinning them to the page in bold type would thus give students a misleading picture of how scientists think, and it would change the concepts into little more than items to be memorized for the next exam.

The conceptual approach in this book can best be seen by reading a chapter, perhaps side-by-side with a chapter on the same material from another textbook. Rather than begin immediately with definitions and the naming of parts, I first lay a conceptual framework into which the student can assimilate such details. For example, I introduce the concept of a Cnidarian in Chapter 24 by describing the body plan and its adaptive significance before delving into details of the gastrovascular cavity. I explain the conceptual approach in more detail in the Preface to Students for the benefit of those who may be more accustomed to learning facts than ideas. The Introduction (Chapter 1) also describes the importance of concepts in science and gives some explicit examples.

Organization

If you are familiar with the first edition of *Concepts in Zoology,* you will notice that the book you are holding has defied the tendency of most textbooks (not to mention their authors) to increase their bulk. The first edition was already one of the most concise general zoology texts, and with the guidance of many reviewers and more than 200 of our fellow teachers, I was able to cut quite a lot of material of peripheral importance. The crucial content on animal diversity remains largely uncut, however, and I have added important new information that has become available in the last five years.

The second edition of *Concepts in Zoology* follows the same inverse-hierarchical organization as the first edition, starting with molecular and cellular concepts and building up through animal physiology, ecology, evolution, and behavior before surveying the major groups of animals. As before, each chapter can stand alone, and they can be used in virtually any order. The only change in the order of chapters has been to move the chapter on hormones before the chapters on neurophysiology, which then leads to the discussion of muscle. The chapters on development and on ecology have been reorganized, and discussions of onychophorans and chaetognaths have also been moved in response to recent taxonomic information. Many lesser changes, too numerous to mention here, are indicated in the Instructor's Manual and Test Bank, which is described below. Referring to these changes should make it easy to update your lecture notes and keep abreast of recent changes in the literature.

Pedagogical Aids

As in the first edition, each chapter opens with a list of headings to help students get the big picture of what the chapter is about. As a further aid to students, each chapter has a concise Summary. Although it comes near the end of each chapter, it might not be a bad idea to point out to students that there is no rule against reading the summary before as well as after the rest of the chapter. Also at the end of each chapter is a Chapter Test on the main concepts. You may want to suggest that students test themselves or each other with these questions, and you might find them helpful as models for your own essay questions for exams. New in many of the figure captions are questions intended to encourage students to focus greater attention on the figures, although I hope they will also be fun. Answers are provided at the end of each chapter and in the Instructor's Manual and Test Bank.

Although the focus is on concepts rather than terminology, the beginning zoology student must inevitably be introduced to many technical terms, unless you want them to go around babbling about "the doohickey that snails eat with." Technical terms are written in boldface the first time they occur within a chapter, and the term is defined within or shortly after the sentence where it first occurs. In addition, technical terms are defined in the Glossary. With more than 1800 zoological terms, this is the most extensive glossary in a zoology text, and it should prove valuable in subsequent biology courses that students will take. Where appropriate I have also provided pronunciation guides or derivations of new terms in context, and there is a list of common root words and a general guide to pronunciation on the back endpaper. The most important new words are listed as Key Terms at the end of each chapter in the order in which they first appear. To keep the list manageable, it does not include all the boldfaced terms, but only those that are key to understanding the main concepts.

The thorough Index is another valuable resource for students who want to look up particular topics either in the introductory zoology course or in later courses. You and your students may also find the two lists of references at the end of each chapter useful. The first list is Recommended Readings for students (or even instructors) who want to delve more deeply into particular subjects. The second list includes some of the Additional References I used in preparing the chapter. Instructors may find these more helpful than students do, because many contain recent information that may not be widely known.

Ancillaries

Students may also benefit, at least indirectly through you, from the following supplements:

Laboratory Manual. *The HarperCollins Zoology Laboratory Manual* was written by Bill Tietjen of Bellarmine College and myself. Its 26 well-illustrated exercises are not merely dissection guides, but explorations of animals as living organisms rather than collections of dead parts.

The HarperCollins Biology Encyclopedia Laser Disk. This laser disk provides instant access to transparencies, micrographs, slides, animations, and film clips.

Instructor's Manual and Test Bank. Ken Saladin of Georgia College wrote the original test bank consisting of more than 2300 questions in various formats, and I have reviewed every question and added some new ones for the present edition. In addition, the Instructor's Manual and Test Bank provides a summary of the main concept in each section of the text. It also notes the major changes from the first to the second edition. Another valuable aid is the list of appropriate transparencies for each section of the text.

The HarperCollins Test Master, available for both IBM and Macintosh, offers a computerized version of the test bank questions.

Acknowledgments

Once again I am indebted to so many of my fellow zoologists for suggestions and helpful criticisms that it is impossible to list them all. You know who you are, and I hope this book makes you proud of your contribution. I must give a special thanks to my colleagues in the Department of Biological Sciences and to the administration at SUNY Plattsburgh for their continued encouragement and a sabbatic leave. I am especially grateful to all my talented friends at HarperCollins who turned so much raw information into a book. Glyn Davies and Susan McLaughlin were unfailing in their enthusiasm and guidance. Vicki Cohen shepherded the manuscript through what seemed like endless rounds of review and revision. Susan Posmentier of Hudson River Studios supervised the brilliant artwork, and Jeff Chen and Patricia Andrews at Electronic Publishing Services Inc. guided it and the words onto the page. Lisa De Mol somehow managed to bring it all together as Project Editor.

Finally, I thank all the reviewers listed below. Their staggering diversity of suggestions convinced me that there is an infinite variety of ways to teach zoology. Alas, a book can do it only one way.

Robert C. Anderson, *Idaho State University*
W. Sylvester Allred, *Northern Arizona University*
Joseph A. Arruda, *Pittsburgh State University*
Andrew R. Blaustein, *Oregon State University*
Lee F. Braithwaite, *Brigham Young University*
David F. Brakke, *University of Wisconsin—Eau Claire*
Frank J. Bulow, *Tennessee Technological University*
Bryan R. Burrage, *College of the Desert*
Robert Day, *Ohio State University*
Gerald L. DeMoss, *Morehead State University*
Peter Ducey, *State University of New York at Cortland*
DuWayne C. Englert, *Southern Illinois University—Carbondale*
Bruce Felgenhauer, *University of Southwest Louisiana*
Jeffrey Freeman, *Castleton State College*
Karen Hart, *Peninsula College*
Arthur C. Hulse, *Indiana University of Pennsylvania*
Ronald L. Jenkins, *Samford University*
Eric Larsen, *Villanova University*
Bernard A. Marcus, *Genessee Community College*
Richard N. Mariscal, *Florida State University*
Brian T. Miller, *Middle Tennessee State University*
Thomas C. Moon, *California University of Pennsylvania*
Keith Morrill, *South Dakota State University*
Robert K. Okazaki, *Southeastern Louisiana University*
Karen Olmstead, *University of South Dakota*
Robert S. Prezant, *Indiana University of Pennsylvania*
James A. Raines, *North Harris College*
Barbara A. Ramey, *Eastern Kentucky University*
Fred Searcy, Jr., *Broward Community College*
Walter K. Taylor, *University of Central Florida*
Mary Katherine Wicksten, *Texas A&M University*

Partnership Reviewers

Eastern Michigan University:
Howard D. Booth
Allen Kurta
Peter Reinthal

University of Guelph:
Edward D. Bailey
Roy Danzman
Alex Middleton

Southwest Texas State University:
Richard W. Manning
Thomas R. Simpson
Samuel F. Tarsitano
Donald W. Tuff

C. Leon Harris

I might as well confess at the very beginning that I have written this book with the intention that you enjoy it. If you have been brought up to believe that good medicine has to taste bad, this may seem an unpardonable sin, but I have two excuses. First, you probably learn most easily what you enjoy knowing. Second, animals *are* enjoyable, and a zoology text that obscures that fact is simply wrong from cover to cover. Most successful zoologists would agree because they were first attracted to the field by the shapes and colors of shells, butterflies, beetles, or birds. They remain zoologists because of the pleasure of understanding animals. Zoologists are often enjoyable too, by the way, and many are as interesting as the animals they study. I want to introduce you to some of these zoologists in this text, to help you understand how they think and work. I chose some contemporary zoologists because I hope to discourage the impression one often gets from textbooks—that the only good zoologist is a dead zoologist.

You will find that, contrary to the popular impression, good zoologists are not people gifted with an infallible vision of truth. Usually their work involves more *re*vision than vision. Most of the concepts of zoology presented here have been revised in the past, and many will undoubtedly have to be revised in the future. Often I will tell you that zoologists don't agree or simply don't know something. If your goal is merely to learn facts, this will discourage you. If your goal is to learn how to be a zoologist, however, you will be glad to know that there is plenty left for you to do.

One can really enjoy animals only after understanding many facts about them, and understanding those facts requires hard work. I have tried to reduce the amount of work by avoiding the presentation of facts for their own sake. Even more important, I have tried to present the facts within the context of broader concepts. Zoologists use concepts to handle all the information they must know, and you, too, will find it easier to organize and interrelate facts about animals if you first grasp concepts. Concepts are defined as abstract generalizations derived from facts, but they are not themselves facts. Because they are conceptual they do not lend themselves to easy summarization in simple phrases and sentences, so don't look for them spelled out that way here, and try to avoid thinking of concepts as items to be memorized for the next exam. Instead, try to develop a conceptual understanding about how each animal relates to its natural environment. Try to see the world from the animal's point of view. What are its problems, and how does it solve them? Why, for example, do freshwater polyps, jellyfish, and sea anemones all have hollow bodies and thin tissues? Once that concept is grasped, the gastrovascular cavity becomes more than a strange word to be memorized.

I have tried to reduce the amount of work by introducing only the terminology needed to understand a concept. Too many textbooks assume that they have covered a topic if they have used every term related to it and that defining a term is the same as explaining a concept. This tendency is especially unfortunate in zoology because much of the terminology was established in the days when every zoologist knew Greek and Latin. I know of one student who asked his zoology professor after the first lecture whether the course would ever be offered in English. Unfortunately there is no way to avoid entirely the specialized terminology of zoology unless you want to go around talking about "the big thingy in the middle of a jellyfish." In spite of my best efforts, therefore, this book includes hundreds of terms that are probably new to you. These new terms are printed in bold type where they first occur, and the most important ones are listed as Key Terms at the end of each chapter. The Key Terms are listed in the order in which they occur within the chapter rather than alphabetically, so you can quickly review them by browsing through the chapter. All are defined within the context in which they first appear and also in the glossary at the end of the text. With more than 1800 terms, this is by far the most complete glossary of any general zoology text, and it should prove useful in other courses that you may take in the future.

A good way to learn new terms is to pronounce them aloud. This can be a source of great fun at a party, and it should not be embarrassing if you remember that there is no one correct way to pronounce most zoological terms, since no one knows how ancient Greek and Latin were spoken. Zoologists usually follow the pronunciation of their teachers, which varies widely. British and American zoologists don't even agree on the pronunciation of "zoology": The British say zoo-AH-low-gee, and Americans say zoe-AH-low-gee. The most difficult terms in the text are followed by a simple guide to pronunciation (like this: pro-NUN-see-A-shun), with the spelling simplified and accented syllables capitalized. A table inside the back cover offers general rules for pronunciation. Many of the new terms can be dissected into root words, some of which you already know because they occur in everyday English. On the inside back cover is a list of common word roots. It would save time in the long run to learn as many of these as you can.

In addition to the specialized terminology, you will also find many funny-looking names in italics. These are the scientific names for species. I include these scientific names because the common names often vary from one region to another, and because you should get used to seeing how zoologists refer to animals. In this system the name of a species consists of two italicized words: for example, *Canis familiaris.* The first word is capitalized and refers to the genus group, which usually includes several different species. The second word is not capitalized and differs for each species in a genus. These names may look imposing, but many of them make sense. *Canis familiaris,* for example, is simply the familiar canine. You will see some of these names so often that you will learn them without even trying, and you will soon get into the habit of reading through the others without losing much time.

There are more than a million of these species names for animals. This is far too many for any one person to know, so zoologists generalize about large groups of animals, such as protostomes, vertebrates, and phyla (FY-luh, plural of phylum). The 31 phyla of animals recognized in this text are represented in a figure inside the front cover. Referring to the figure frequently will be a relatively painless way to become familiar with the major animal groups.

The references listed at the end of each chapter will provide further information. There are two kinds of references. Recommended Readings are those that I consider useful and readable background. Additional References are sources of specialized information, especially information that is not widely known. As is the common practice in scientific literature, I cite many of these additional references within the text like this: (Groucho, Harpo, and Karl 1948). You might find these citations distracting at first, but I think you will soon learn to read through the ones that don't interest you. Also at the end of each chapter is a Chapter Test that enables you to see how well you have grasped the main concepts. Many of the figure captions also contain questions of a different kind: little puzzles that can usually be answered by looking at the figure carefully. You can check your answers by referring to the end of each chapter. Try not to feel too guilty if you enjoy solving these puzzles and learning about all the other intricate puzzles of animal life.

The Harris Award for Excellence in Zoology Teaching

Although I have tried to create a book that will awaken your desire to want to learn about animals, I realize that your instructor can have a far greater influence in the role of mentor and good example. Starting with this edition I would like to recognize the importance of such instructors with the Harris Award for Excellence in Zoology Teaching. One awardee each year will be selected by me from nominations made by students. Near the end of your zoology course, if your instructor has been particularly effective in stirring your interest in animals, please send me a letter explaining briefly how she or he did so. In addition to the full name of the instructor, please include the name and number of the course, the name of the school and department, the term in which you took the course, and the approximate number of students who finished the course with you. Please send your nominations to me at the Department of Biological Sciences, SUNY, Plattsburgh, NY 12901.

C. Leon Harris

CHAPTER

1

Introduction

Human and camel (*Camelus bactrianus*)—an ancient relationship.

An Ancient Profession

For as long as there have been humans there have been those who studied animals. The evidence for this is that such slow, frail creatures as ourselves have managed to exist for two million years. That would not have been possible if our ancestors had not acquired an impressive understanding of our prey and predators. In some parts of what is now Europe, primitive zoologists appear to have deliberately attempted to pass on their learning to future generations. With paintings as elegant as they were anatomically detailed, they revealed the strategy of the hunt and pointed out the vulnerable parts of prey. In many cases animals were portrayed not merely as potential food, but as objects of beauty and admiration (Fig. 1.1).

No one would have had to explain to primitive people why they needed to understand animals. Few of us, however, now hunt from necessity, and even fewer have to grapple tooth and claw with predators. Now our struggle to survive is more subtle, but no less real. Legions of insects feast on crops and stored food that could feed the millions of humans who are now starving. Animal parasites still weaken and kill millions of people every year. Our bodies and those of other animals on whom we depend for food are threatened by infections from viruses, bacteria, and fungi, as well as by other disorders. Only through an understanding of the functioning of animals, including ourselves, have we been able to hold some of these threats more or less to a standoff. Because of the continuing increase in the human population and the appearance of new diseases, our survival demands even greater efforts to understand animals. Even if our survival did not depend on it, however, there would undoubtedly still be zoologists, simply because zoology is one of the most exciting and enjoyable endeavors many of us can imagine.

What Zoologists Do

Zoologists are people who use their scientific knowledge of animals in a variety of activities (Figs. 1.2 and 1.3). Many with bachelor's degrees in zoology become technicians, attendants, research associates, conservation officers, rangers, or guides at zoos, museums, universities, game preserves, parks, and forests. Many of the same organizations also employ zoologists with graduate degrees and research experience in teaching and

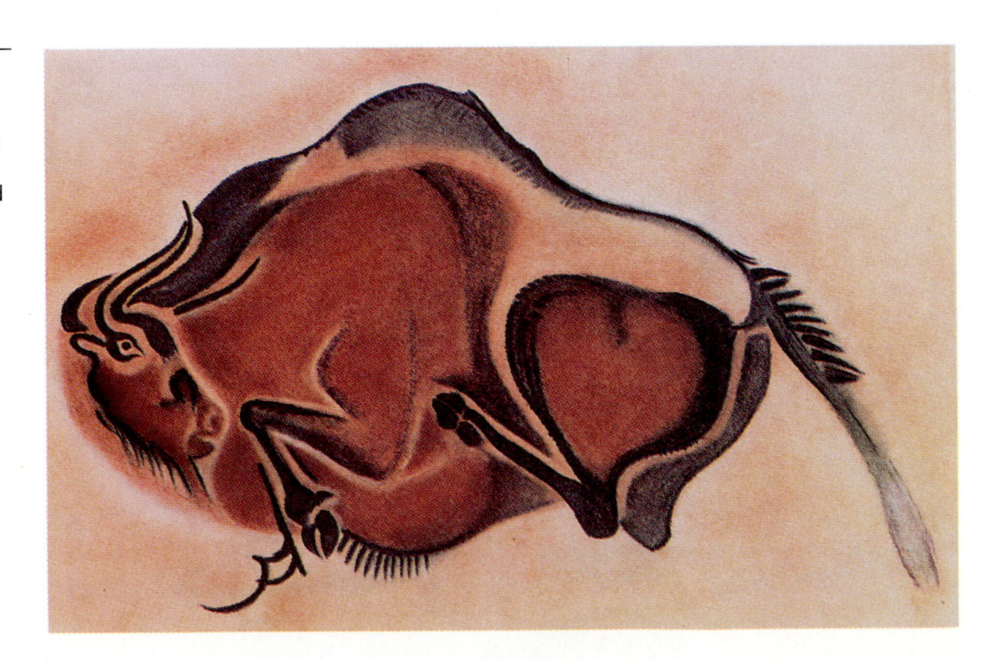

FIGURE 1.1

A charging bull painted 14,000 years ago in a cave at Altamira, Spain by a Cro-Magnon person. Although the bull may have been hunted for food, it appears that the artist was also impressed with the animal's beauty. The detail is remarkably accurate, considering that the artist undoubtedly worked from memory.

research. Zoologists also work in environmental conservation and wildlife management, assessing and maintaining the health and populations of animals. In these days when many people learn about animals through photographs and television, there are also opportunities in photography and cinematography for those whose knowledge allows them to get close enough to animals without disturbing them and without getting themselves killed. Other zoologists find that their interest in animals is mainly physiological, and they may ultimately go into veterinary or human medicine. This is by no means a complete list of the activities engaged in by people who think of themselves as zoologists.

FIGURE 1.2

A zoologist, interacting closely with an animal in its environment. Here the zoologist is measuring the beak of a gull in a study of food preferences. Field research like this is often lonely and physically demanding but provides opportunities for close associations with animals.

Regardless which career a zoologist chooses, it is essential to have not only a good knowledge of subjects directly related to zoology, but also a broad background in the other sciences and in mathematics. Many areas of zoology also require abilities in statistics, photography, drawing, computer use, economics, management, and other subjects. Zoologists also benefit from a broad education in many other ways, simply because they often have to interact with people of various political, economic, social, and aesthetic persuasions. It is also essential that zoologists be able to communicate well, both orally and in writing. No matter how much knowledge a zoologist has, it is useless unless she or he can communicate it to employers, peers, or the public. Most important, zoologists must have a capacity and enthusiasm for a lifetime of learning about animals. Many of the "facts" in this book will prove to be incorrect, irrelevant, or incomplete in the future; but I hope the book will provide a conceptual foundation that will enable you to acquire better knowledge, along with the motivation for wanting to.

If you do have a broad range of skills, you may find a wide choice of careers in zoology. How do you choose one? First, you should remember that you will be engaged in a career for a long time, and that no salary is worth being miserable for the best part of your life. Before choosing, ask yourself whether you prefer working indoors or outdoors. (Which would you enjoy more—sitting at a lab bench examining animal feces for eight hours, or tracking moose in mud while donating blood to mosquitos?) Do you prefer working with whole animals or with their organs? Also consider whether the kinds of animals you enjoy working with includes *Homo sapiens.*

Even if you eventually choose some career unrelated to zoology, you should never consider your knowledge of animals wasted. Many of the most significant contributions to zoology have been made by people who did not earn a living as zoologists. The list includes Charles Darwin and the Emperor of Japan. You may also have noticed that those whose lives are the fullest and most enjoyable also usually enjoy animals.

How Zoologists Think

THE CONCEPTUAL FOUNDATION OF ZOOLOGY. Zoology has changed during the time that textbooks have replaced cave art, but zoologists still ask many of the same questions about animals that our early ancestors did. Indeed, it seems that zoologists know even less than our ancestors did, for the more we learn, the more we realize how little we know about the vast world of animal life. Now it takes more than one elder in a tribe to pass on all the knowledge about animals. The relatively little we know is already too much for one person to master. A taxonomist who can identify beetles is an authority on beetles, but that taxonomist would probably have to seek help from a physiologist to understand the functioning of the beetle's internal organs, from an ecologist to understand its relation to other organisms, and from an embryologist, a geneticist, and an evolutionist to understand why the beetle appears as it does. These would be only a few of the many kinds of zoologists, because any biologist whose primary focus is on animals can be called a zoologist.

FIGURE 1.3

Many zoologists work in teams, taking advantage of the controlled and more comfortable environment of the laboratory.

A zoologist is not someone who knows all the details about every kind of animal, but someone who out of necessity and curiosity has learned a great deal about some of them. More important, a zoologist has organized his learning into a relatively few concepts that are usually also held by other zoologists. New facts add to established concepts and support them, but a concept is not itself a fact. By definition a concept is an abstract generalization about facts. Most concepts defy reduction to a single phrase or sentence, and scientists usually do not use concepts in that way. Some examples of zoological concepts are presented later in this chapter, starting on page 10.

The Myth of Method. New zoological facts are discovered by observation under natural conditions and by experimentation under controlled conditions. Such observations and experimentation are often done to test hypotheses and theories, which are tentative explanations for phenomena. The process of testing hypotheses and theories by observations and experiments is often referred to as "the scientific method," but it is not a method in the same sense that there is a method for repairing a car or changing a diaper. One often sees "the scientific method" laid out in stepwise fashion in the introductions of science textbooks, but that is about the only place where you will find such recipes for doing science. Zoologists and other scientists do not consciously follow such steps, and they seldom think about "the scientific method."

Well, how do zoologists think? Just about like everyone else does, although with perhaps more awareness of the possibility that they might be wrong. They formulate tentative explanations and test them by making observations and conducting experiments to see whether the expected consequences actually occur. You do the same when you figure out why your car won't start or why the baby keeps crying. As Charles Darwin once wrote, "I have often said and thought that the process of scientific discovery was identical with everyday thought, only with more care." Of course, some people, whether zoologists, auto mechanics, or baby sitters, are more careful than others and are more talented at creating and testing hypotheses and theories.

Scientific hypotheses and theories are not inherently more likely to be true than those of other areas of endeavor. In fact, the real difference between scientific and nonscientific hypotheses and theories about natural phenomena is that only the scientific ones can be proved false. That is, a theory is scientific only if one can describe an experiment that might disprove it. This method of distinguishing science from nonscience, called the **Criterion of Demarcation,** was formulated in the 1930s by the influential philosopher of science, Karl Popper (1902–1994). The insistence that hypotheses and theories be potentially falsifiable gives zoologists and other scientists a critical attitude that often makes them annoying to nonscientists who would prefer to believe unquestioningly in quack cures, alien encounters, creation "science," and other pseudoscience.

The Integrity of Science. The critical attitude of scientists is the main safeguard that justifies confidence in science. Scientists usually don't have the chance to do research unless their plans survive the critical judgment of other scientists who are expert in the same field. After research is completed, the results must then pass a similar "peer review" before they can be published in a reputable journal. Even then, if the research is sufficiently important, other scientists will repeat it to see if it is valid. This process does not make science inerrant, but it does help prevent embarrassments like the cold-fusion fiasco that occurred because scientists bypassed peer review and reported their work directly to the press.

Like everyone else, of course, scientists usually find it easier to be critical of others than of themselves. The desire to reap the honors and material rewards of success, or at least to feed themselves and their families, also sometimes makes them overlook evidence that might disprove their ideas. It can even tempt them to take credit for work others have done or for work that was never done. As the competition for jobs, promotions, and grants has increased, more and more "scientists" have been found to have plagiarized, fabricated, or misrepresented research. Dishonesty in science is still rare enough that even actions that would be accepted as routine in business and politics evoke shock and condemnation when they occur in science. Any dishonesty at all, however, is too much, and it takes a lot of the fun out of science for those who are attracted to it because of its integrity.

Perhaps there is a scientific method after all:

1. Plan experiments and observations in such a way that data are as likely to disprove as to support the theory that you hope or believe is valid.
2. Treat data as the sacred and unalterable word of nature.
3. Be as critical of your own ideas as you are of the ideas of others.

The Use of Animals in Experimentation. Scientific concepts are tested by experimentation, so by their very nature zoological concepts require animal experimentation. Such experimentation often comes into conflict with another human trait: empathy—the ability to imagine one's self in the place of another. Inspired by empathy for animals, many individuals and groups have succeeded in improving the welfare of research subjects. Some others are not content with improving animal welfare, however, but insist that animals have rights just as humans do. In the words of a co-founder of People for the Ethical Treatment of Animals (PETA), "There is really no rational reason for saying a human being has special rights. . . . A rat is a pig is a dog is a boy." Some individuals and groups have taken legal and illegal measures toward ending all research on animals, and ultimately all other uses of animals, including as food and pets. Some have damaged laboratories and threatened researchers with violence, and several such groups are on government lists of terrorist organizations. Others have discouraged research by convincing law-makers to require expensive and time-consuming regulations. More and more zoologists are finding themselves compelled to justify to others and to themselves their use of animals in research.

Dishonesty in science will meet with nearly universal condemnation, but it is harder to find a consensus on animal welfare and rights, for it is almost impossible to take a rigid point of view on this subject without self-contradiction. There are many who condemn research on animals, but who nonetheless avail themselves of foods, medicines, and surgical procedures that are based on such research. At the other extreme are physiologists who perform surgery on animals but are sickened by hunting for sport. Each person has to decide individually what uses of animals are proper. The questions in the box on p. 6 may stimulate your own thinking on the subject.

The myth of the scientific method and the necessity of using animals in experimentation sometimes give students and nonscientists the impression that the study of animal life is itself lifeless, cold, and cruelly rational. Most zoologists do not perceive zoology this way, however. It is certainly not the way I present zoology in this book. On the contrary, I hope to show that zoology exalts and celebrates humanity, for without it none of us would know that humanity is the product of billions of years of evolution, that we are brothers and sisters to all other living things and the children of vanished organisms that only science can memorialize, and that we are nevertheless unique in being the only animals capable of understanding all this. Zoology is itself an expression of humanity. It is conducted by human beings who share with artists, theologians, philosophers, and historians a desire to understand their lives and to find out what we

Ethical Considerations in the Use of Live Animals in Research

1. What kinds of animals feel pain? Are only humans truly conscious of pain? Can other animals feel pain too, or do they merely react with defensive or withdrawal reflexes? Can dogs feel pain? Rats? Frogs? Cockroaches? Sponges? Protozoa?
2. Is the ability to perceive pain the only criterion for judging the ethics of experimentation? If so, then is all animal experimentation ethical as long as anesthesia is used? What about experimentation on a comatose human?
3. If the ability to perceive pain is not the only criterion, then what are the other criteria? If those other criteria make it unethical to use animals for experimentation, do they also make it unethical to perform experiments on plants? Do these criteria also rule out any use of animals or plants for food and other purposes?
4. How much pain is inflicted in the name of science compared with that inflicted from carelessness and for sport, agriculture, and development? Each year an estimated 10 million vertebrate animals, mostly rodents, are used for experimental research in the United States. About twice as many cats and dogs are abandoned by their owners to starve to death or to be "put to sleep" by humane societies. Many more pets and farm animals are neutered, often without anesthetics, than are subjected to experimental surgery. The destruction of habitat destroys countless millions of animals. Of all these activities, why is animal experimentation singled out? Could it be because scientists are not as politically powerful as pet owners, farmers, and developers? Will animal rights advocates be satisfied to eliminate research, or will they next try to eliminate pet ownership, animal agriculture, and development?
5. How much pain results from research compared with that which is a natural part of the lives of wild animals, few of whom die a comfortable death in old age? If other animals inflict pain on their prey to meet their needs, why cannot humans inflict pain to fulfill their needs? Or does this commit the **naturalistic fallacy** of asserting that what occurs in nature is morally right for people?
6. Is it possible that the pain inflicted through research is outweighed by the reduction in pain for both humans and animals that is possible because of advances in human and veterinary medicine that result from such research?
7. Does government regulation of animal research make sense considering that it generally protects only warm, furry animals while leaving unprotected the cold, wet ones?
8. Shouldn't scientists use every means to reduce the amount of unnecessary pain, by using statistics to determine the minimum number of animals required and by using isolated organs, cell cultures, mathematical models, and computer simulations when possible?
9. If scientists can inflict pain on other animals, what is to prevent scientists of future generations from adopting a similar attitude toward experimentation on humans? Does inflicting pain on an animal dull the conscience of the experimenter, making her or him less sensitive to pain in other humans? Does an unconcern for pain in animals diminish humans by blurring one distinction, empathy, that separates us from most other animals? Or does a deeper understanding of the structure and functioning of animals make us more respectful of all animals, including humans?

are doing and ought to be doing on Earth. A considerable number of zoologists are motivated by the most humanitarian of all goals, that of improving the well-being of humans and other animals.

What Is an Animal?

This may seem a silly question, but in fact biologists have often debated whether such organisms as sponges and freshwater polyps ought to be considered animals, and recent studies comparing the molecules of such organisms have reopened the debate. Until this century, organisms were considered to be animals if they moved actively and to be plants if they did not. This division of life into only two **kingdoms**, Animalia and Plantae, persisted into this century even though an enormous variety of microorganisms had been seen since the discovery of the microscope. Eventually biologists added a third kingdom, Monera, to accommodate the bacteria. Since the late 1950s biologists have added two more kingdoms, Fungi and Protista, for fungi and protists (protozoans and some algae). Most biologists now accept this division of all living things on this planet into these five kingdoms.

The five-kingdom classification is currently under challenge, partly because bacteria (kingdom Monera) differ so much from organisms in the other four kingdoms. Bacteria lack cell nuclei and are therefore said to be **prokaryotes** (Greek *pro-* before, *karyon* kernel, i.e., nucleus). All other organisms have nuclei in their cells and are called **eukaryotes** (pronounced yew-KARE-ee-oats; Greek *eu-* true). Most eukaryotes are also similar to each other in having other intracellular structures that are not found in prokaryotes. Thus it seems unjustified to divide the eukaryotes into four kingdoms at an equal level with prokaryotes. Comparisons of the structures of certain molecules do, indeed, show that bacteria are vastly different from other organisms. In fact, there are two extremely different kinds of bacteria, the eubacteria and the archaebacteria (pronounced AR-kee-bak-TEER-ee-uh), that differ from each other more than do all the eukaryotes (Fig. 1.4). Many biologists therefore now recognize three rather than five major groups: archaebacteria, eubacteria, and eukaryota.

The **archaebacteria** live in environments that are extremely salty, hot, cold, or anoxic, so people and other animals seldom encounter them. **Eubacteria**, however, asso-

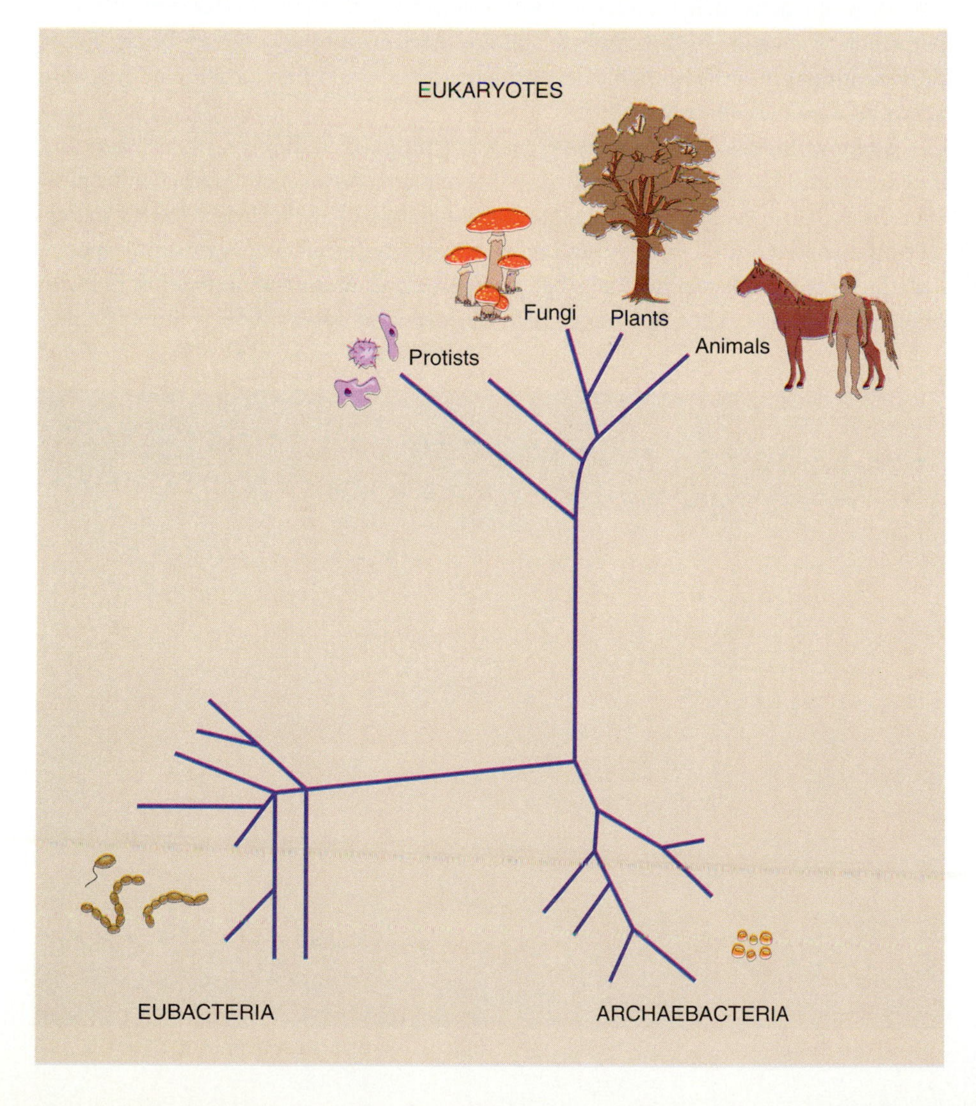

FIGURE 1.4

The evolutionary relationship of different kinds of organisms to each other, based on comparisons of molecular structure. The distance along a set of lines between any two groups of organisms represents the evolutionary distance separating the two groups. Animals are closely related to other eukaryotes but not to the two major groups of bacteria.

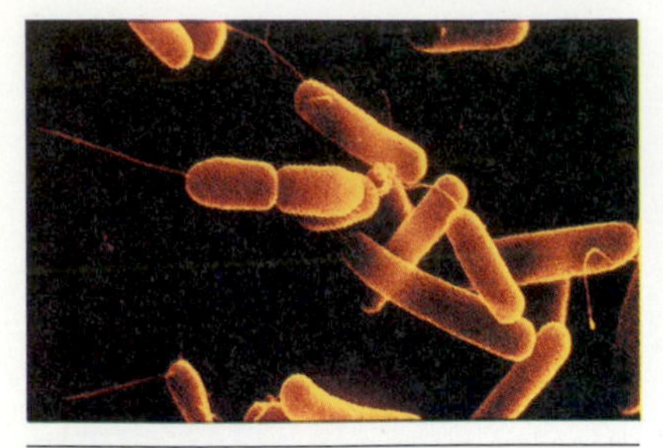

FIGURE 1.5

Scanning electron micrograph of *Escherichia coli* bacteria on tissue from the human cheek. Each bacterium is about 3 micrometers (millionth of a meter) long. This species is found in the digestive tract of almost all mammals. It is usually harmless, but it can cause infections of the urinary tract, and some strains can be lethal. In some species, such as rats and rabbits, *E. coli* contributes useful amounts of vitamins.

ciate intimately with animals. The eubacteria include fewer than 5000 known species, although many times that number undoubtedly remain to be discovered. Actually the term "species" is difficult to apply to these and other organisms that reproduce asexually. We know many bacteria as agents of disease and food spoilage, but probably just as many species of eubacteria are beneficial to animals. Among them are the cyanobacteria (also called blue-green algae), which produce oxygen by photosynthesis just as plants and algae do. In fact, the cyanobacteria may have been the first photosynthesizers and might have created the oxygen-rich atmosphere that enabled animals to evolve. Some bacteria help animals by enabling food plants to grow, others help animals digest food, and still others help eliminate the bodies of animals after they have died. No matter how clean you are, you harbor about a hundred quadrillion (100,000,000,000,000,000) bacteria—about 10 times as many bacterial cells as human cells (Fig. 1.5).

Among the eukaryotes more than 60,000 species are neither fungi, nor plants, nor animals, but are lumped together as **protists.** Protists are usually divided between photosynthesizing forms—algae—and nonphotosynthesizing forms—**protozoa.** In fact, however, they are so diverse that one taxonomist has suggested that they be classified into 20 different kingdoms. Most are single-celled and microscopic, but some consist of large colonies of more or less identical cells. The giant kelp, for example, grows to be hundreds of meters long (Fig. 1.6). Algae are important because their photosynthesis converts solar energy into chemical nutrients that can be used by animals that feed on them. Without algae, oceans and lakes would be devoid of animal life. The protists also include protozoans such as *Amoeba* and *Paramecium,* which were once considered to be primitive animals. Protozoans are still discussed in zoology courses because they provide insight into the probable ancestors of animals, and they provide information about how animal cells function.

There are approximately 69,000 species of **fungi,** including yeasts, mushrooms, and organisms that cause various forms of rot in plant and animal tissues. Unlike most bacteria and protists, fungi reproduce sexually—that is, by the fusion of genetic material from two individuals. In these and other sexually reproducing organisms, a **species** is defined as a group of individuals that can reproduce among themselves but not with other such groups. They also reproduce by means of spores, which are extremely

FIGURE 1.6

Giant kelp *Macrocystis pyrifera.* Even though this species grows to 60 meters long and resembles a plant, its cells identify it as a relative of single-celled protists. Fifty million dollars worth of giant kelp are harvested in California to produce emulsifiers for ice cream and other foods, as well as supplements for cattle feed. Other animals, such as sea urchins, anemones, sea otters, barnacles, snails, worms, crabs, and fishes, use these protists for food and habitat.

FIGURE 1.7

A worker ant (*Atta colombica tonsipes*) on its subterranean fungus garden. (The term *tonsipes* denotes the subspecies.) Because fungi do not photosynthesize, they can live in the darkness of ant burrows.

durable packets of genetic material. Each fungus consists of one type of cell, with each cell enclosed in a rigid wall. Many fungi grow as fibrous networks (mycelia). The fungi neither eat nor photosynthesize, but obtain energy by absorbing organic matter. If the organic matter happens to belong to a human, then the fungus causes an infection such as athlete's foot or jock itch. Other fungi, such as mushrooms, are important food sources for various species of animals (Fig. 1.7), and many are decomposers that allow molecules from dead tissues to be recycled back into living animals.

Approximately a quarter of a million species are now considered to be **plants.** Like fungi, their cells are enclosed in rigid walls, but unlike fungi, each plant consists of different kinds of cells. Because of their ability to convert solar energy into chemical-bond energy through photosynthesis, many plants are essential sources of nutrients for animals. Plants also provide shelter and habitat for many terrestrial animals (Fig. 1.8).

FIGURE 1.8

Mountain gorilla, *Gorilla gorilla beringei,* in the lush vegetation of Rwanda in Africa. An adult male gorilla eats approximately 60 pounds of vegetation each day. Gorillas also use leaves as sleeping nests and shelter from rain. Although plants provide abundant food and shelter, mountain gorillas are endangered by a soaring population of humans competing for their territory.

Finally we come to **animals**, which could be defined as organisms with eukaryotic cells that are neither protists, nor fungi, nor plants. More precisely, animals are organisms with numerous and diverse eukaryotic cells that lack walls or the ability to photosynthesize. Narrow as this definition might seem, it is broad enough to embrace more than a million known species of living animals and many times that number of species that are extinct or not yet discovered.

An Overview of Zoology

CELLS: WHERE LIFE DWELLS. Although our ancestral zoologists had an intimate knowledge of animals, they must have wondered in vain about life itself. Many probably dissected living animals (including humans) in attempts to locate the place where the spirit dwells—the place where life itself lives. Probably some thought they had found it within the lungs or the heart. The Latin word *anima* meant both breath and soul, and even now we speak of being "a heartbeat away from death." What primitive zoologists did not know, and what no one could imagine until the invention of science, is that life is not a thing that can be dissected out and examined. It is a process, or many processes, that occur everywhere within an organism, in every living cell. The cells are where life dwells.

The cell doctrine—the realization that all organisms consist of cells and materials made by cells—was perhaps the single most important concept in the quest to understand life. It tells us that what an animal must do to stay alive is to provide the internal conditions necessary to keep its cells alive. The beautiful part of this concept is that each cell has a reciprocal need to contribute to the maintenance of the conditions required for the survival of other cells. Animals and other multicellular organisms are therefore made of cells, by cells, and for cells. If too many of the cells fail to do their jobs, other cells will die, and the animal will die.

Cells do inevitably lose their ability to function, of course. Many of them are replaced, but there is a limit to how often that can be done. Therefore it is inevitable that each animal will die. In order for the species to survive, some cells must be capable of forming new animals of about the same kind before they die. If they did not, a zoology text would be just a collection of short stories instead of a long saga with many heroes and adventures. All aspects of zoology, whether developmental, physiological, evolutionary, ecological, or behavioral, are ultimately devoted to ensuring that the saga continues—that the cells responsible for reproducing the organism have the requirements of life.

Even though we have "cornered" life within microscopic cells, it is still not easy to identify or define life. About the best we can do is characterize it in terms of the processes associated with it. Each generation of zoologists, and perhaps each individual zoologist, could propose his own list of the characteristics of life. In Unit 1 of this text we shall examine the main concepts of cellular processes, including the following:

1. **Living cells maintain a greater degree of order within themselves than occurs outside.** As we shall see, animal cells contain orderly arrangements of **organelles** ("little organs") made of molecules that are themselves orderly. Immediately following death, this ordered state begins to reverse; the molecules and organelles break down, and the cells decay. In other words, death brings a return to the normal situation found among nonliving things, which decay and crumble to disordered dust.

2. **Living cells transfer energy.** The way in which cells increase order within themselves is by taking many disordered molecules from the environment and converting them into relatively few orderly molecules. They convert simple amino acids in the diet, for example, into complex proteins. This process and many others performed by cells require energy. According to the First Law of Thermodynamics, energy cannot be created. The energy used by cells must therefore come from outside the cell, and ultimately from outside the animal in the form of food.
3. **Living cells have means of ensuring that their genetic information survives after the cell has died.** Since every cell within an animal usually has the same genetic information, not all of them have to reproduce to form a new individual of the same species. A few reproductive cells are capable of passing to the next generation all the genetic information contained in the individual organism. By cooperating to ensure the survival of these reproductive cells, the nonreproductive cells ensure the survival of their own genetic information.
4. **The functions of living cells are regulated.** The processes occurring within cells do not happen randomly. For example, the sequence of amino acids in proteins is determined by genes, and the transfer of energy is controlled by enzymes that are produced in response to gene activation.
5. **Cells become specialized for different functions during embryonic development.** Every cell of an individual contains the same genetic information that was in the single cell from which it developed, but differences in the expression of this genetic information enables each cell to perform a particular function.

THE INTERNAL ENVIRONMENT. Most of an animal's activities are directed toward the maintenance of an internal environment compatible with life in the cells. In the mid-19th century the French physiologist Claude Bernard called attention to this fact in his statement that "all the vital mechanisms, varied as they are, have only one object: that of preserving constant the conditions of life in the *milieu intérieur.*" This quotation neatly summarizes one of the most important concepts of animal physiology, but after more than a century we need to qualify it somewhat. Bernard wrote vaguely of the *milieu intérieur* as being the body fluids, but today we would more precisely say that the internal environment is the **interstitial fluid** that surrounds cells.

Within the interstitial fluid the levels of nutrients, oxygen, ions, and sometimes temperature are maintained within narrow limits. This maintenance, referred to as **homeostasis**, is controlled by organ systems that will be described in Unit 2. Briefly, the feeding and digestive systems are responsible for homeostasis in the concentrations of nutrients, such as sugars used for energy and amino acids needed for protein synthesis. The respiratory system maintains the level of oxygen, which is required to obtain energy from nutrients. Excretory organs such as the kidney regulate concentrations of ions. Various homeostatic systems in birds and mammals keep body temperature constant within a few degrees by regulating the production, gain, and loss of heat. The hormonal, neural, muscular, and circulatory systems are all involved in maintaining these levels of nutrients, oxygen, ions, and temperature. Their own activities often fluctuate radically as they play their roles in homeostasis. The reproductive system is the only physiological system with neither a constant level of activity nor a role in homeostasis.

INTERACTIONS OF ANIMALS WITH THEIR ENVIRONMENTS AND EACH OTHER. The need for animals to provide their cells with nutrients, oxygen, ions, and suitable temperatures inevitably leads to interactions with their environments and with each other. These interactions are studied in the fields of ecology, evolution, and behavior, which we shall survey in Unit 3. Three fundamental concepts characterize the scientific study of these aspects of zoology:

1. The ecological interaction of animals with their external environments is largely a struggle for energy, because energy is required to maintain the internal environment and to reproduce. That struggle manifests itself in all the varieties of ways by which animals obtain food.
2. One of the most striking things about organisms is that they are well adapted to their environments. Yet environments have changed over the 3.5 billion years that life has existed, so the organisms must also have changed. This change is evolution. The record of evolution is preserved in fossils and also in comparisons of the millions of species of organisms still living. This evolution results mainly from the large amount of genetic variability within each group of organisms and from the competition among them to survive and reproduce. Genetic variations that bring success in the competition naturally tend to be found in higher proportions in succeeding generations. In effect, nature tends to select the organisms that are genetically endowed with the ability to leave more offspring. The increasing proportion of individuals with those naturally selected traits is eventually seen as a change in the species.
3. We assume that animal behavior results from physiological mechanisms that are in part under genetic control. We therefore assume that the behavior of animals also evolves in a way that improves the ability to reproduce.

DIVERSITY. Unit 4 will describe some of the million or so species of animals known to be alive. In an attempt to keep track of so many animals, taxonomists traditionally classify them into different taxa (plural of **taxon**). Taxonomists use at least seven levels of taxa for animals:

Kingdom (Animalia)
Phylum (plural, phyla)
Class
Order
Family
Genus (plural, genera)
Species (plural also species)

(You might remember these taxonomic levels by reciting, "Kind Professor, Carefully Observing Fauna, Guiding Students.") Each level of the series includes all the lower levels. That is, each phylum includes one or more classes, each class includes one or more orders, and so on. Phylum Chordata includes class Mammalia, which includes order Primates, which includes family Hominidae, which includes genus Homo, which includes the species *Homo sapiens.* Even these seven levels are sometimes inadequate, so they are often divided into higher and lower levels, such as superclass and subspecies. As the previous discussion of kingdoms indicates, taxa above the level of the species group are not fixed and absolute, but created solely for the convenience of biologists. For sexual organisms, however, species can be defined as groups of interbreeding organisms.

In the past few decades an increasing number of zoologists have adopted an alternative approach to classification called **cladistics**. Instead of grouping organisms into phyla, classes, orders, and so on, they group them in **clades**. Cladistics will be described in detail in Chapter 21, but briefly, a clade consists of all the organisms that are descendants of a given ancestor. For example, mammals make up a large clade that includes many smaller clades, such as the one comprising primates. Within the primate clade there are still smaller ones, such as the one that includes humans and chimpanzees. This is noted here only because you may encounter phrases like "the mammalian clade" before you get to Chapter 21.

Obviously, it will not be possible to cover every species of animal or even every genus, family, or order in this text. However, some attention will be given to each phylum and most classes, and a great deal of attention will be devoted to lesser taxa about which zoologists have been most interested. One chapter each will deal with the vertebrate classes (fishes, amphibians, reptiles, birds, and our fellow mammals) even though they represent less than 5% of animal species.

Throughout this text there is repeated reference to one particular species, because zoologists, like most people, have always been more interested in it than in any other. This species is, of course, the one we flatter with the name *Homo sapiens*—wise man. Discussion of the way we treat our fellow animals will often make us wonder whether the name is deserved. We shall ponder whether some intelligent species coming after us will view the history of humankind as a brief zoological catastrophe, whether the little we have learned through science will prove too much for our own good, and whether it might not have been better for the other animals if we had remained in our caves merely painting and admiring them. On the whole, however, I hope you will come to appreciate how a scientific understanding of animals can enable us to live with more regard for ourselves and for the other species with which we share this planet.

Summary

Zoology, the study of animals, is an endeavor that humans have always engaged in for survival and pleasure. There are numerous kinds of careers in zoology, either in research or in practical applications to improving the well-being of humans and animals. Zoology requires a variety of skills and talents, and an understanding of concepts is more important than absorption of facts supposedly established by "the scientific method." Integrity is another indispensable trait.

Animals differ from bacteria, fungi, protozoans, and plants in having diverse cells with nuclei and organelles but not cell walls or the ability to photosynthesize. The life of animals is carried out by cells, which cooperate to create an "internal environment" in which they can survive and in which reproductive cells can produce new offspring. The cellular processes of life consist of creating more order within the organism than occurs outside, and in animals this requires energy derived from food. These life processes are regulated, ultimately by genetic instructions that are passed from one generation to the next. Animal cells require certain levels of nutrients, oxygen, ions, and sometimes temperature, which are maintained homeostatically by physiological systems.

Animals must obtain nutrients, oxygen, and the other requirements of life from their environments, often by consuming other animals. Interactions of animals with their environments and each other comprise the study of ecology. Because environments have changed during the history of animals, the animals themselves must also have changed, since they continue to interact successfully with their environments. These changes, called evolution, have occurred largely because animals with genes that gave them a greater ability to survive and produce offspring were more likely to pass those genes to following generations. Among the evolutionary changes are those affecting behavior.

Animals have evolved into millions of different species, approximately a million of which survive and have been described. These are traditionally classified by taxonomists into phyla, classes, orders, families, genera, and species, although a growing number of zoologists group them into clades.

Key Terms

prokaryote
eukaryote
protist
protozoa
organelle
interstitial fluid
homeostasis
taxon
kingdom
phylum
class
order
family
genus
species
clade

Chapter Test

1. Think about the following lines:
   What am I, Life? A thing of watery salt
   Held in cohesion by unresting cells,
   Which work they know not why, which never halt,
   Myself unwitting where their Master dwells?
   John Masefield, *Sonnets 14*
   a. To what extent are the lines biologically relevant?
   b. To what degree is scientific zoology capable of answering the questions asked by Masefield? (pp. 10–11)
2. Make up a sentence that is scientific but false. Make up one that is nonscientific but true. Which of the following statements from this chapter is scientific according to the criterion of falsifiability? (*Hint:* Try to imagine experiments potentially capable of proving the statements false.) (p. 4)
   a. Zoology is one of the most exciting and enjoyable endeavors many of us can imagine.
   b. All organisms consist of cells and materials made by cells.
   c. Evolution results from the large amount of genetic variability within each group of organisms and from the competition among them to survive and reproduce.
   d. Zoology is itself an expression of humanity.
3. If you are taking a laboratory with this course and your instructor requires reports, discuss what you should do if you get results that differ from what lectures and the text had led you to expect. Should you report actual observations, or should you change the results to make them consistent with what the book says? Suppose you did not follow instructions correctly: should you "cook" the data to hide your error? Try to formulate a general rule governing when it is ethical to report data that you did not actually observe. (pp. 4–5)
4. For each of five characteristics of living cells (maintenance of internal order, transformation of energy, regulation, reproduction, specialization) name a nonliving thing that also has such a characteristic. Of the nonliving things you named, are there any that have all five characteristics? A virus consists of a protein coat surrounding genetic material, and it can reproduce only with the aid of a cell it infects. Are viruses alive? (pp. 10–11)
5. Explain why no animal can live in isolation from an environment and other organisms. Explain how interactions with the environment and other organisms could lead to ecological, behavioral, and evolutionary differences among animals. (pp. 11–12)
6. Name one kind of bacterium, protist, fungus, and plant, and describe how it interacts with an animal. (pp. 7–9)

Readings

Recommended Readings

Alternatives to Animal Use in Research, Testing, and Education. Washington, DC: Office of Technology Assessment. 1986.

Beveridge, W. I. B. 1957. *The Art of Scientific Investigation.* London: William Heinemann Ltd.

Chubin, D. E. 1985. Research malpractice. *BioScience* 35:80–89.

Fox, M. A. 1986. *The Case for Animal Experimentation.* Berkeley: University of California Press.

Goodfield, J. 1981. *An Imagined World: A Story of Scientific Discovery.* New York: Harper & Row. (*Excellent account of the daily life of an immunological researcher. See especially Chapter 15 on "scientific method" and creativity.*)

Heinrich, B. 1989. *Ravens in Winter.* New York: Summit Books. (*The thrills and chills of field research.*)

Janovy, J., Jr. 1985. *On Becoming a Biologist.* New York: Harper & Row.

Klemm, W. R. (Ed.). 1977. *Discovery Processes in Modern Biology.* Huntington, NY: Robert E. Krieger Publishing Company. (*Autobiographical accounts of discovery.*)

March, B. E. 1984. Bioethical problems: animal welfare, animal rights. *BioScience* 34:615–620.

Regan, T. 1985. *The Case for Animal Rights.* Berkeley: University of California Press.
Ritvo, H. 1984. *Plus ça change:* antivivisection then and now. *BioScience* 34:626–633.
Rodd, R. 1990. *Biology, Ethics and Animals.* New York: Oxford University Press.
Sechzer, J. A. (Ed.). 1983. *The Role of Animals in Biomedical Research. Ann. NY Acad. Sci.* 406.
Sperlinger, D. (Ed.). 1981. *Animals in Research: New Perspectives on Animal Experimentation.* New York: Wiley.
Starr, D. 1984. Equal rights. *Audubon* 86(6):30–35 (Nov). (*The activities of various animal rights groups.*)
Wilson, E. O. 1994. *Naturalist.* Washington, DC: Island Press. (*Autobiography of an eminent zoologist.*)

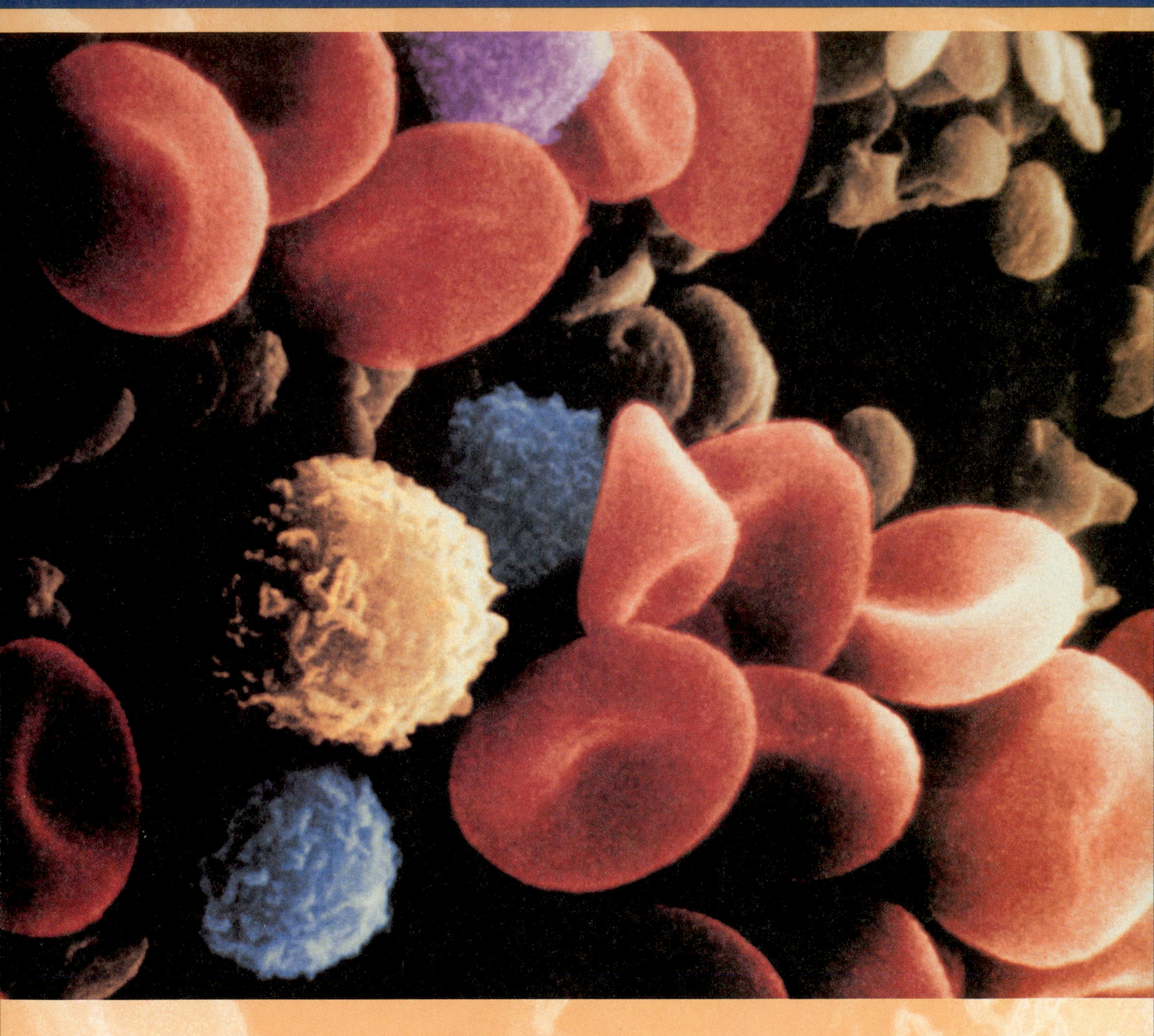

OVERVIEW

UNIT

1

The Cellular and Molecular Bases of Life

These blood cells illustrate some of the concepts to be covered in this unit. One of the major concepts is that animals are composed of numerous cells of a great variety, each specialized for a unique function in the life of the animal. The biconcave disc-shaped cells, artificially colored red in this scanning electron micrograph, are red blood cells also called erythrocytes. Each is about 7.5 millionths of a meter wide (7.5 micrometers—7.5 μm). Erythrocytes are specialized to carry oxygen to every other cell, which depends on the oxygen for the production of energy. The spherical, rough-coated cells are white blood cells, also called leukocytes. There are several kinds of leukocytes, shown here in different colors, and they help destroy damaged cells and invading bacteria and other parasites.

The structures of cells and their abilities to contribute to the life of the animal are determined by genetic instructions encoded in large molecules within the cells. These large molecules, called nucleic acids, make up genes. Genes control structure and functions by instructing the cell in how to make various proteins. These proteins then regulate other activities within the cell and also compose most of the structure of the cell. Genes also have to make copies of themselves and direct the synthesis of new molecules that will enable each cell to divide into two new and identical cells.

Remarkably, all the many kinds of cells in an animal have the same genetic information. At some point a cell in the human embryo acquires the ability to become bone marrow and later to become even more specialized as either red blood cells or white blood cells.

CHAPTER

2

Materials and Mechanisms of Cells

Sockeye salmon *Oncorhynchus nerka* "running" upstream.

Water and Related Matters

It may seem strange to begin the study of such a fascinating subject as life with such an ordinary subject as water. Unless you live in the desert or in one of the increasing areas where the water is undrinkable, you probably seldom think about water. You, yourself, are more than half water, however, and so am I. Some animals, such as jellyfishes, are as much as 90% water. A few animals can survive almost complete drying in an inactive state, but humans and most other animals die if they lose just one-fifth of their body water. Water is the canvas on which life is painted. Like the artist, we must begin by preparing the canvas.

Another reason why it is appropriate to begin with water is that historically that is the way the scientific study of life began. The first known scientist, Thales (pronounced THAY-leez), who lived on the Mediterranean coast around 600 B.C., thought that all matter was made of water. A few decades later Anaximander, perhaps impressed with the perpetual tide of life washed up on the Mediterranean shores, speculated that even humans originated in the sea. Aristotle (384–322 B.C.) and other philosopher-scientists of ancient Greece soon realized that there were many more phenomena of nature than could be accounted for by the properties of one substance—even water. They therefore added three other substances: earth, fire, and air.

Thales, Anaximander, and the Greek philosopher-scientists adhered to a view that still guides modern scientists, at least unconsciously. That view is called materialism. Since biologists are not noted for being greedy for material wealth, perhaps we should use the term **scientific materialism.** Scientific materialism is the belief that all natural phenomena, including those of life, are due to the properties of matter. It is a view that I shall assume throughout this chapter and throughout the text. According to this view, the phenomena of life are due in large part to the properties of the matter composing living things. Since so much of every organism is water, scientific materialism suggests that much of life depends on the properties of water.

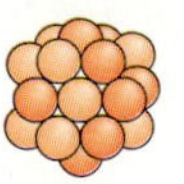

FIGURE 2.1
(A) Models of a hydrogen (H) and an oxygen (O) nucleus. Electrons are not shown since the page would have to be about 30 feet wide to fit them into the figure. (B) Models of hydrogen and oxygen atoms showing the electrons. At the atomic scale the position of particles cannot be defined, so the cloudlike figures represent the probable locations of electrons. A hypothetical camera with its shutter left open might record such blurred images of electrons, which can be thought of as small, negatively charged particles. The nuclei are vanishingly small, somewhere near the center. One million atoms in a row would not quite reach across the period at the end of this sentence.

■ **Question:** How many electrons does oxygen have? Why are only six shown? (Answers to this and other figure questions in this chapter are on p. 49.)

Atoms and Molecules

Some of the Greek philosopher-scientists considered water, earth, fire, and air to be made of particles that could not be divided into smaller particles without destroying their properties. They called such particles **atoms,** meaning "indivisible." We still use the term "atom" for such indivisible particles or elements, but we now know that water, earth, fire, and air are not atoms at all. Each particle of water, for example, is a **molecule:** a collection of two or more atoms that are bonded together. Water is made of one atom of oxygen and two atoms of hydrogen and is therefore represented by the familiar formula H_2O. Like all other atoms, hydrogen and oxygen have nuclei made of protons and usually some neutrons. A proton is relatively massive and carries one positive charge. All atoms of the same type, or element, have the same number of protons in the nucleus. Hydrogen always has one proton in the nucleus, and oxygen always has eight protons in its nucleus. A neutron has about the same mass as a proton, but no charge. Hydrogen and other atoms can exist in several forms, called **isotopes,** which differ in the number of neutrons in the nucleus. The most common isotope of hydrogen has one proton and no neutron in the nucleus. The most common isotope of oxygen has eight protons and eight neutrons in the nucleus. Around the nucleus are electrons, which have little mass and one negative charge each. The number of electrons equals the number of protons, so the net charge on an atom is zero (Fig. 2.1). The bonds that hold atoms together in a molecule are due to electrons.

Each electron is confined to one of several **energy levels** that can be thought of as being different distances from the nucleus. The first energy level, which is closest to the nucleus, can hold no more than two electrons. The second and third levels can hold no more than eight electrons each. If the highest level is not completely filled, the atom can share one or more pairs of electrons with another atom. This sharing of electrons forms a type of chemical bond called a **covalent bond.** Hydrogen has only one electron, so it has one vacancy in the first energy level where it can share a pair of electrons with another atom. For example, a pair of hydrogen atoms can form covalent bonds with each other, making a hydrogen molecule (H_2). Oxygen has eight electrons—two in the first energy level and six in the second—so it has two vacancies in the second energy level for electrons from other atoms to share. Like hydrogen, a pair of oxygen atoms can share electrons with each other, forming an oxygen molecule (O_2).

Water as a Molecule

Water is formed when two hydrogen atoms form covalent bonds with one oxygen atom. The chemical reaction in which hydrogen forms covalent bonds with oxygen to make water can be written as follows:

$$2H + O \longrightarrow H_2O$$

The two hydrogen atoms do not stick out of a water molecule at opposite ends, as one might expect, but actually form a 105° angle with each other (Fig. 2.2). As a result, the positive charges of the hydrogen nuclei are concentrated on one side of the water mol-

FIGURE 2.2

(A) Two hydrogen atoms form covalent bonds with an oxygen atom to make water. (B) The unequal distribution of charge on water molecules causes weak attractions between them, called hydrogen bonds. Water molecules are also weakly attracted to negatively charged parts of other molecules.

■ **Question:** What would be the effect of the attraction of water molecules for a swimming protozoan or tiny animal?

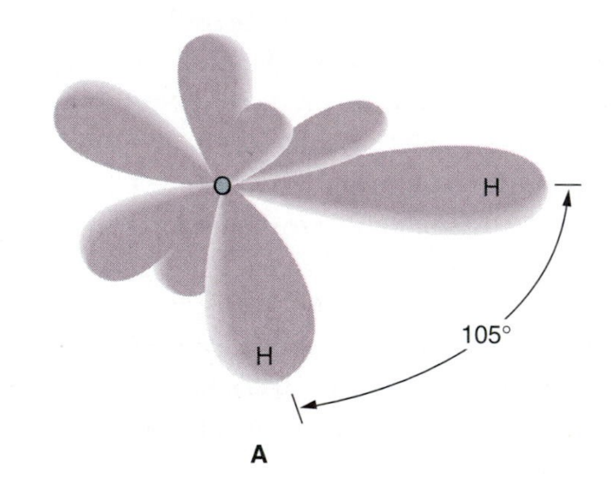

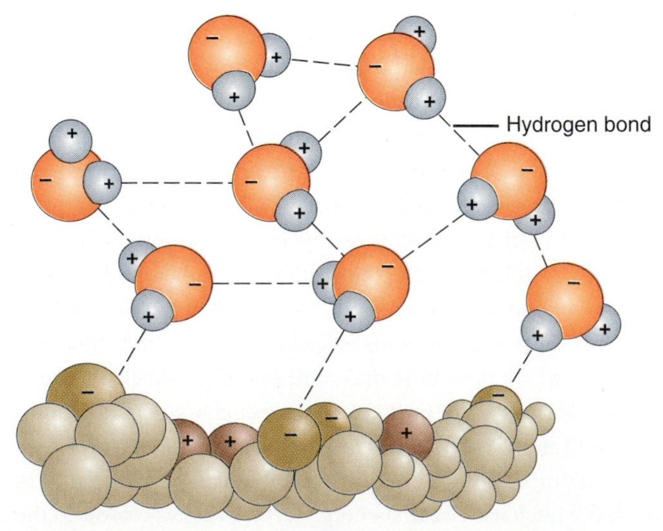

A

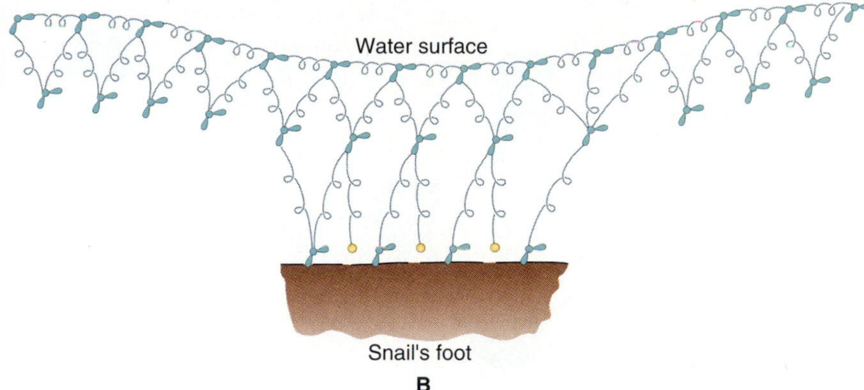

B

FIGURE 2.3

(A) An illustration of cohesion and adhesion. The common pond snail *Physa* hangs from the surface of the water because molecules in its foot adhere to the water molecules. Surface tension, which is due to the cohesion of water molecules, supports the weight of the snail. (B) Cohesion and adhesion are due to hydrogen bonds, represented here as springs.

ecule, and the negative charges of most of the oxygen electrons are concentrated at the opposite end. This separation of charge gives water a **polar** structure, like a bar magnet with opposite poles on each end. Just as the north pole of one bar magnet tends to stick to the south pole of another, the positively charged end of a polar molecule such as water is attracted to the negatively charged region of other polar molecules. Such an attraction between two water molecules is an example of a **hydrogen bond.** A hydrogen bond is the charge attraction between a hydrogen atom that is covalently bonded to an oxygen or nitrogen atom in a polar molecule, and a negative region of the same or a different polar molecule. A single hydrogen bond is quite weak, but in large numbers hydrogen bonds account for many of the life-giving properties of water, as well as proteins and the molecules responsible for heredity.

One of the vital properties of water that is due to hydrogen bonds is **cohesion:** the attraction of identical molecules to each other. Cohesion accounts for the tendency of water to form drops. Were it not for cohesion, cells would lose their shapes, and body fluids would leak out of the tiny crevices in tissues. Conversely, if the cohesion of water molecules were much greater, water would not be a liquid at the temperatures usually encountered on Earth. Equally important is **adhesion,** which is the attraction of different kinds of molecules to each other (Fig. 2.3). Water adheres to other polar molecules because it forms hydrogen bonds with the negatively charged parts of those molecules. Without adhesion, water would not make anything wet. Life without wet cells and organs is inconceivable.

The number of molecules of a substance and the mass of the substance are related to each other by the ***mole.*** *A mole of any substance equals 6.02×10^{23} molecules (Avogadro's number) and has a mass in grams equal to one* ***gram molecular mass*** *(= gram molecular weight). The gram molecular mass of a substance is a mass in grams approximately equal to the total number of protons and neutrons in a molecule of the substance. One mole of H_2 (one gram molecular mass) therefore has a mass of 2 grams, and 1 mole of O_2 has a mass of 32 grams. Concentration can therefore be expressed as the number of moles in a liter of solution, which equals the* ***molarity.*** *A 1 molar solution contains 1 mole of solute per liter of solution, or 6.02×10^{23} molecules per liter, regardless of the identity of the solute.*

Water as a Solvent

The property of adhesion explains why many substances dissolve in water. When water molecules adhere to individual molecules of a substance, the cohesion between molecules of the substance is weakened. The molecules therefore disperse, or dissolve, in the water. The substance that dissolves is called the **solute,** and the substance in which it dissolves is called the **solvent.** The combination of solute and solvent is called a **solution.** Water is one of the best natural solvents for polar molecules. This property is essential for life, because it allows large amounts of biologically important substances to be contained within cells and to circulate within the body fluids of organisms. The number of molecules of solute dissolved in a given volume of solution determines the **concentration** of the solution. Since it would be rather tedious to count the number of molecules in a volume of solution, the number of molecules is usually expressed indirectly as the mass of solute.

Many molecules form chemical bonds in which electrons are not shared as in covalent bonds, but are transferred from one atom to another. These are called **ionic**

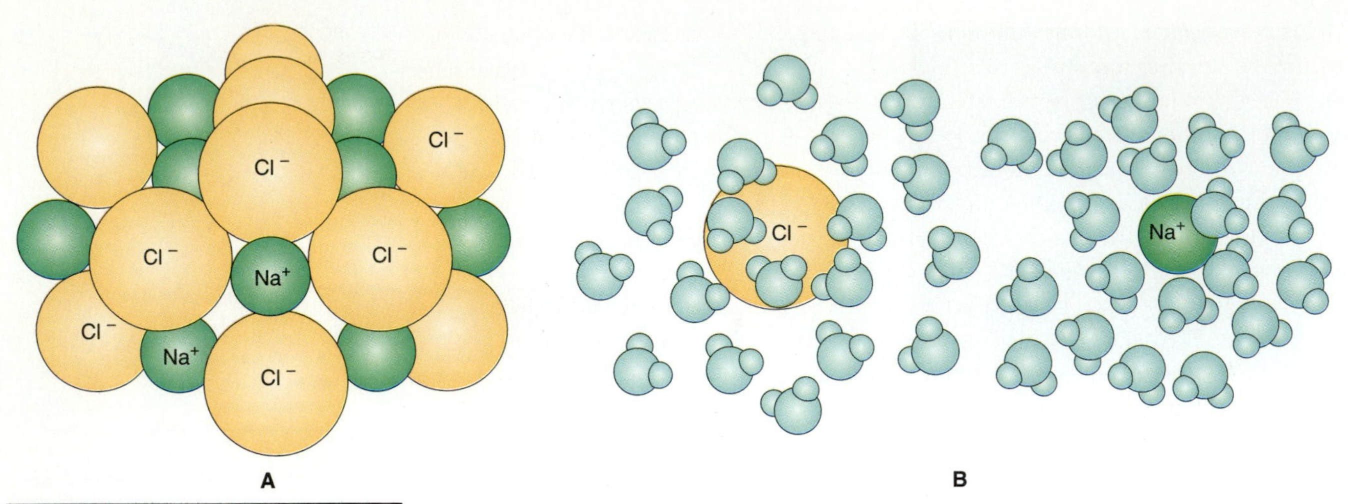

FIGURE 2.4

(A) Atoms of sodium (Na) and chlorine (Cl) linked by ionic bonds form regular arrays in a crystal of ordinary table salt (NaCl). (B) When dissolved in water, NaCl dissociates into sodium ions (Na^+) and chloride ions (Cl^-). Because of charge attraction, water molecules orient in a shell around these ions. This phenomenon is called hydration.

■ **Question:** Why are the water molecules oriented differently toward Na^+ than toward Cl^-?

bonds. Ionic bonds are easily broken when molecules dissolve in water, allowing atoms in the molecules to dissociate. During such dissociation an electron from one atom of the molecule remains with the other part of the molecule, forming one or more **ions.** Ions are atoms or molecules that carry a net charge, because the number of electrons does not equal the number of protons. For example, when crystals of ordinary table salt, sodium chloride (NaCl), dissolve in water, they dissociate into sodium and chloride ions (Fig. 2.4). During this dissociation, sodium loses the single electron in its third energy level and becomes a sodium ion. Since the sodium ion has one more proton than electrons, it is written as Na^+. The lost electron takes up residence in the available place in the third energy level of chlorine, which therefore becomes the chloride ion. The chloride ion then has one extra negative charge, and it is written Cl^-. Positively charged ions such as Na^+ are called **cations.** Negatively charged ions like Cl^- are called **anions.**

Acidity

A small proportion of water molecules also dissociate, forming hydrogen ions (H^+ = protons) and hydroxyl ions (OH^-):

$$H_2O \longrightarrow H^+ + OH^-$$

Other molecules, called **acids,** can release hydrogen ions much more readily, and those ions can have profound effects on biological processes. The reason is that the distribution of charges on proteins and other biological molecules determines their ability to function. If the negatively charged parts of such molecules are covered by hydrogen ions, they may function at a different rate, or perhaps not at all. In some cases molecules that normally adhere to each other by charge attraction split apart if the concentration of H^+ is too high.

The concentration of hydrogen ions in a solution is usually measured by **pH,** which is the negative exponent of the H^+ concentration expressed as a power of 10. For example, the average concentration of H^+ ions in pure water is 10^{-7} molar, so the pH of pure water is 7. This is said to be neutral. By adding an acid, which as any substance from which hydrogen ions dissociate, the concentration might be raised to 10^{-1} molar, for a pH of 1. Such extreme acidity occurs in the human stomach when too much hydrochloric acid (HCl) is secreted (Fig. 2.5). Adding a base, which is a substance that contributes hydroxyl ions, reduces the concentration of H^+ by combining them with

OH^- to form water. Adding sodium hydroxide (NaOH) to water, for example, could reduce the H^+ concentration to 10^{-11} molar, for a pH of 11.

Certain molecules or combinations of molecules have the property of stabilizing, or buffering, the concentration of H^+ in water. Such molecules are called **buffers.** Buffers work by releasing H^+ when the pH begins to rise, and binding H^+ when the pH begins to fall. A buffer often consists of a weak acid and a related salt. In human blood, for example, the weak acid is carbonic acid (H_2CO_3), and the salt is sodium bicarbonate ($NaHCO_3$), which dissociates into Na^+ and bicarbonate (HCO_3^-). If the pH is too low, the excess H^+ tends to bind with bicarbonate, forming carbonic acid. If the pH is too high, the carbonic acid releases H^+. Proteins also have the ability to release or bind hydrogen ions, and they therefore act as buffers in body fluids and within cells. Buffering is extremely important in preventing changes in pH from upsetting the functioning of numerous molecules essential for life. In human blood a deviation of only 0.2 pH units from the normal value of approximately 7.4 can be life-threatening.

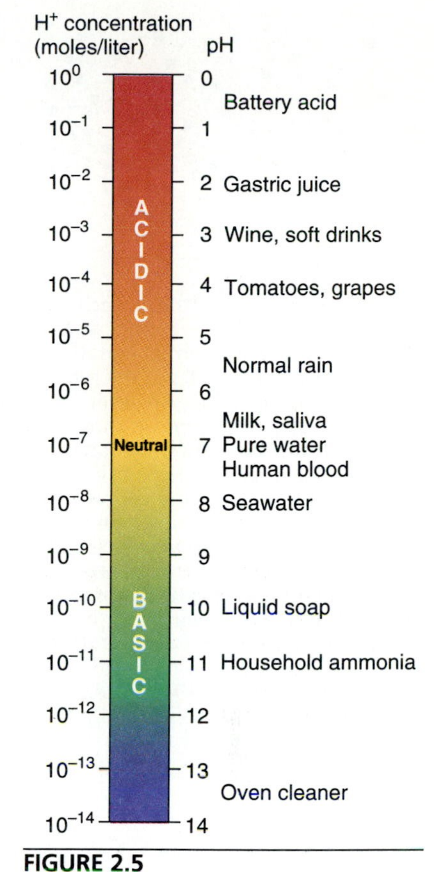

FIGURE 2.5
A scale showing H^+ concentration, the corresponding pH values, and some representative solutions.

Heat

Many of the properties of water and other molecules depend on how much they are agitated by thermal energy. A measure of this thermal energy is **temperature.** The form of the energy that causes this thermal agitation is **heat.** Thus adding heat to a substance increases its temperature. One of the interesting and important properties of water is that it takes more heat to raise the temperature of a certain mass of it than for any other abundant substance. Conversely, water has to lose a lot of heat to cool off. In other words, water has a high **heat capacity.** It takes one **calorie** of heat to raise the temperature of one gram of water by 1°C. The average adult human has from 40,000 to 50,000 grams of water in the body, making a volume of 40 to 50 liters. If you imagine trying to heat more than 10 gallons of water on a stove, you will appreciate why your body temperature does not readily fluctuate. The same resistance to fluctuation in temperature also protects animals that live in lakes and oceans. Although air temperature can change drastically within a few hours, the temperatures of large bodies of water tend to remain constant for days and weeks at a time. Water has several other interesting and vital thermal properties that will be discussed in more detail in Chapter 15.

Diffusion

The jostling of water and other molecules due to thermal agitation gives rise to three phenomena that lie at the heart of any understanding of life. The first such phenomenon is **diffusion.** Diffusion refers to the tendency of molecules to move from a region where they are highly concentrated to a region of lower concentration until the molecules are equally distributed. If one drops a cube of sugar into a cup of coffee, for example, the sugar will dissolve and eventually become evenly distributed throughout the coffee, even without stirring. A great deal could be written about the cause of diffusion, but the concept is fundamentally simple. Thermal agitation causes each molecule to move in a random direction. It is therefore just as likely that a molecule in a region where it is highly concentrated will move to a region of low concentration as it is that a molecule in a region of low concentration will move to one with a high concentration. There are more molecules in the region with high concentration, however, so more molecules will be moving from the high to the low concentration than will be moving in the reverse direction (Fig. 2.6 on page 24). Eventually, so many molecules will have moved from regions of high to low concentration that the concentrations will be equal.

The molecules will then continue to move about in random directions, but there will be no further change in concentration. This state is called **equilibrium.**

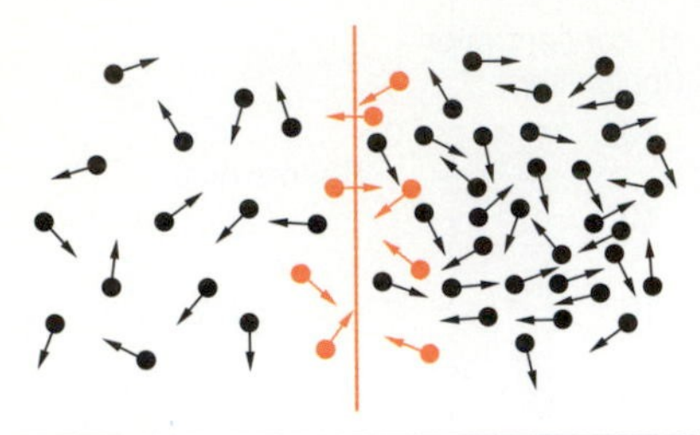

FIGURE 2.6
The explanation for diffusion. Each molecule is moving in a random direction (arrows) due to thermal agitation. There are more molecules in the region with high concentration, however, so more molecules will be moving toward the region with low concentration than in the reverse direction. Here five molecules (color) from the more concentrated solution will cross into the dilute solution, and only three molecules will cross in the reverse direction. Thus, there is a net flow, or diffusion, from high to low concentrations.

■ **Question:** Can you think of a way to reverse this diffusion? How about without using energy?

Osmosis

Water is a molecule that follows the same principles of physics and probability as any other molecule, so it also diffuses. Water diffuses from a dilute solution into one that is more concentrated—a phenomenon known as **osmosis.** Osmosis is usually observed only when solutions having two different concentrations are separated by a membrane or other structure that allows water, but not solute, to pass through (Fig. 2.7). Because of osmosis, the more concentrated solution will accumulate water and therefore develop an increased pressure, called **osmotic pressure.** Osmotic pressure is generally proportional to the difference in concentrations in the two solutions. Animal cells have a definite shape partly because of the osmotic pressure inside them.

A similar pressure can also occur when certain molecules that do not dissolve are added to water to form a **colloid.** Gelatin dessert is an example of a colloid. The attraction of water to the undissolved molecules results in diffusion of water toward the molecules. This water movement results in a **colloid osmotic pressure,** also called **oncotic pressure.** In contrast to osmotic pressure, the magnitude of the colloid osmotic pressure produced by a given amount of substance depends on the identity of the substance.

Entropy

Diffusion and osmosis are two examples of a far more general principle: Any natural process results in an increase in disorder. The reason is that there are many more ways for objects to be disordered than for them to be ordered, so any change in a system is likely to increase its disorder. For example, there are many ways for sugar molecules to be distributed randomly throughout a coffee cup, but many fewer ways for them to be arranged in a cube. Diffusion or any other natural change in the distribution of sugar molecules is therefore likely to increase disorder. Sugar dissolves spontaneously; it does not form cubes spontaneously in water. The same concept applies even to mundane objects. There are countless places to lose one sock, but only a few places where the sock belongs. A sock is therefore much more likely to become misplaced than it is to spontaneously end up neatly folded in a drawer beside its mate.

This concept is so important that it has been elevated to the status of a physical law called the **Second Law of Thermodynamics.** There are various ways of stating the Second Law. One is as follows: No process is possible in which the only result is the complete conversion of heat into work. One cannot, for example, create a steam engine

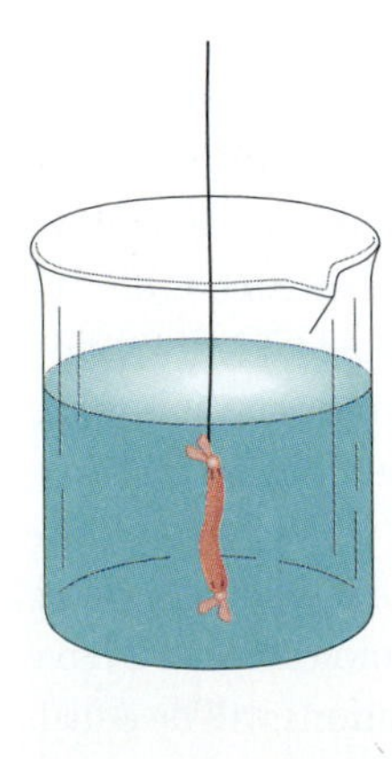

FIGURE 2.7
If a cellophane bag lightly filled with a sugar solution is suspended in water, the bag will accumulate water because of osmosis and will swell, producing osmotic pressure. The same thing can happen to animal cells (see Fig. 14.1).

■ **Question:** Why does your skin become wrinkled after soaking in water?

that would convert all the heat in steam into useful work. This is true even though the conversion of the heat energy into an equivalent amount of mechanical energy would not violate the **First Law of Thermodynamics,** which states that energy can be converted from one form to another but can never be destroyed.

The Second Law can be stated more precisely in terms of a quantity called **entropy,** which is a kind of measure of disorder. It is impossible to observe or measure entropy directly, but one can calculate how much it changes. If an object at temperature T gains a small quantity of heat Q, then the object gains an amount of entropy equal to Q/T. It is a familiar fact that some heat is produced as a byproduct of any natural process. Therefore every natural process, whether it is diffusion, osmosis, operating a steam engine, or losing one's socks, increases the entropy of the universe. This is another way of stating the Second Law.

The Second Law of Thermodynamics is bad news for those who like to think very, very far ahead. It means that the universe is constantly becoming more disordered, and there is nothing we can do about it. In fact, the only way to slow down the increase in entropy is to do absolutely nothing. Inevitably the universe will reach a state of total uniformity, with matter and temperature evenly distributed. That prospect is a long way off, however. The Second Law raises more immediate concerns having to do with our own lives and the lives of every other organism.

Organic Molecules

At first glance, life appears to violate the Second Law of Thermodynamics. Instead of increasing disorder, living organisms increase order within themselves as they grow, reproduce, and metabolize. During photosynthesis, for example, plants convert six molecules of CO_2 into a single, more-ordered molecule of the sugar glucose. Furthermore, plants and animals use the energy in glucose to create carbon-based molecules that are even more highly ordered. Such complex, carbon-based molecules are called **organic molecules,** because they now occur naturally only where there are (or were) organisms. Despite appearances, the synthesis of organic molecules and the other activities of living do not actually violate the Second Law. Any increase in order that results from the synthesis of organic molecules is accompanied by an even greater increase in disorder outside the organism due to the production of heat and metabolic wastes. Consequently, living does increase the total entropy of the universe, as predicted by the Second Law.

Although it is not against the Second Law to make organic molecules, it is nevertheless noteworthy that in nature only living organisms now do it. Many respected scientists once regarded this fact as proof that living organisms have unique powers. The belief in such unique powers of life, called **vitalism,** is the antithesis of scientific materialism. Vitalism was once taken quite seriously by many scientists, but it began to weaken in 1828 after Friedrich Wöhler synthesized the simple organic substance urea from inorganic molecules. Nowadays, of course, plastics and other complex organic molecules are routinely manufactured without any organisms present. Organic molecules apparently also occurred spontaneously on Earth billions of years ago, even before life appeared (see p. 381). Therefore no vitalistic power is needed to account for organic molecules.

Organic molecules are special, nonetheless. What makes them so is their enormous variety of structures and properties. This variety is due to each carbon atom's ability to form four covalent bonds with a variety of other atoms. Carbon has a total of six electrons. Therefore it has four electrons in its second energy level that it can share with other atoms, and four vacancies in that energy level that can accommodate electrons from other atoms (Fig. 2.8A). In organic molecules, carbons (C) commonly bond with each other as well as with hydrogen (H), nitrogen (N), oxygen (O), phosphorus (P), and sulfur (S) (CHNOPS). These other atoms often associate with each other in common **functional groups** that affect the properties of organic substances (Fig. 2.8B). For example, the properties of alcohols are due to one or more hydroxyl (—OH) groups. The carboxyl group (—COOH) makes an organic molecule acidic, because it can lose the H as a hydrogen ion.

The most common organic molecules in living organisms belong to one of four classes: nucleic acids, carbohydrates, lipids, and proteins. The nucleic acids, deoxyribonucleic acid (DNA) and ribonucleic acid (RNA), are involved in heredity. They consist of long sequences of sugars (either deoxyribose or ribose) linked to each other by phosphate groups, with nitrogenous bases attached to the sugars. The sequence of nitrogenous bases controls the production of proteins, thereby determining hereditary traits. The structure and functioning of nucleic acids will be described in detail in Chapter 5.

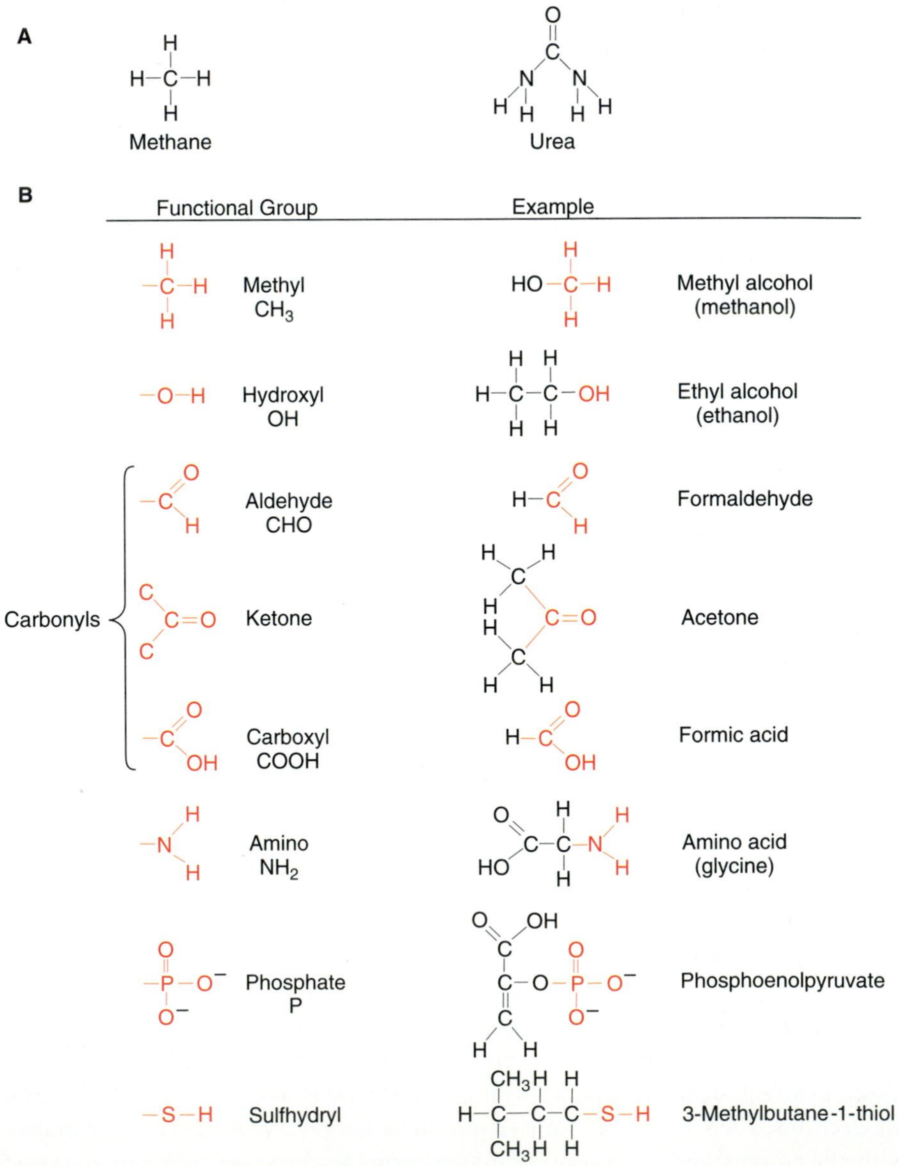

FIGURE 2.8

(A) A carbon atom can covalently bond with up to four other atoms, forming a variety of organic molecules. Two of the simplest organic molecules are methane (CH_4) and urea [$CO(NH_2)_2$]. In methane the carbon and each of the hydrogens share one pair of electrons, forming four single bonds. In urea the carbon and the oxygen share two pairs of electrons, forming a double bond. (B) Carbon also bonds with a variety of functional groups that affect the properties of organic molecules. Some examples are shown. Formic acid is an irritant secreted by ants in self-defense, and 3-methylbutane-1-thiol is the odorant secreted by skunks in self-defense.

■ **Question:** What other functional groups occur in methanol; ethanol; glycine?

A

Methane

Urea

B

Functional Group		Example
Methyl CH_3		Methyl alcohol (methanol)
Hydroxyl OH		Ethyl alcohol (ethanol)
Carbonyls	Aldehyde CHO	Formaldehyde
	Ketone	Acetone
	Carboxyl COOH	Formic acid
Amino NH_2		Amino acid (glycine)
Phosphate P		Phosphoenolpyruvate
Sulfhydryl		3-Methylbutane-1-thiol

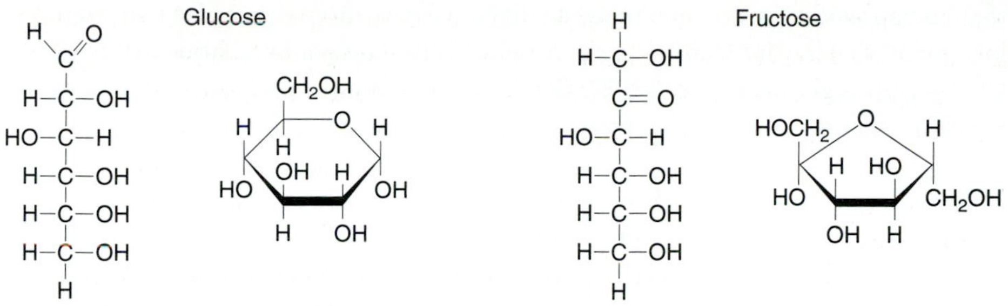

FIGURE 2.9

The structure of two simple sugars represented in linear form (left) and in the ring form that occurs in solution. The Cs in the ring occur at the angles and are usually not shown in diagrams. Glucose circulates in the blood of humans and many other animals and is a major source of energy. Fructose occurs in semen as an energy source for sperm.

CARBOHYDRATES. Carbohydrates include simple sugars such as glucose, as well as more complex sugars such as starch and cellulose. A simple sugar has an aldehyde or ketone group (Fig. 2.8) and two or more hydroxyl groups, and it consists of C, H, and O in the ratio 1:2:1. The formulas for both glucose and fructose, for example, are $C_6H_{12}O_6$. These six-carbon sugars are called **hexoses.** When crystallized, simple sugars take the form of carbon chains. When dissolved in water, however, simple sugars tend to form rings (Fig. 2.9). When the covalent bonds between carbon atoms in simple sugars are broken inside cells, using the controlled processes that will be described in the next chapter, the energy from those bonds can be captured by the cell to perform other functions. Simple sugars are therefore used as immediate sources of energy for cellular functions. In humans and many other animals, glucose circulates in blood and other body fluids as a ready source of energy.

Simple sugars are also called **monosaccharides.** When two simple sugars are linked together they form a **disaccharide.** Ordinary table sugar is the disaccharide sucrose, which forms when a molecule of glucose links to a molecule of fructose (Fig. 2.10A). The linkage, formed by oxygen, is called a **glycosidic link.** Digestion breaks glycosidic links, liberating monosaccharides that can enter the bloodstream as sources of cellular energy. Another important disaccharide in humans and other mammals is lactose, the sugar in milk. Lactose, consisting of glucose linked to a similar hexose called galactose, provides much of the energy needed by infant mammals.

Quite long chains of simple sugars can be joined by glycosidic linkages to form **polysaccharides.** Polysaccharides are much less soluble than mono- and disaccharides and therefore generally do not circulate in blood, milk, or other body fluids. They serve mainly as long-term energy stores or for structural support. Two common polysaccharides, cellulose and starch, are made by plants. Both consist entirely of glucose, but their glycosidic linkages differ. Cellulose, which helps form the walls around plant cells, has glycosidic links that neither humans nor most other animals can digest. We and most other animals that eat plants excrete cellulose as roughage in the feces. Starch, the white material that plants synthesize to store glucose, has glycosidic links

FIGURE 2.10

(A) Common table sugar, sucrose, is a disaccharide consisting of a glucose and a fructose joined by a glycosidic link. (B) The polysaccharide glycogen is a branched chain of glucose subunits. Plant starch is similar except that its chains have fewer branches.

■ **Question:** What atoms are removed from glucose and fructose to form sucrose? What molecule would those atoms form?

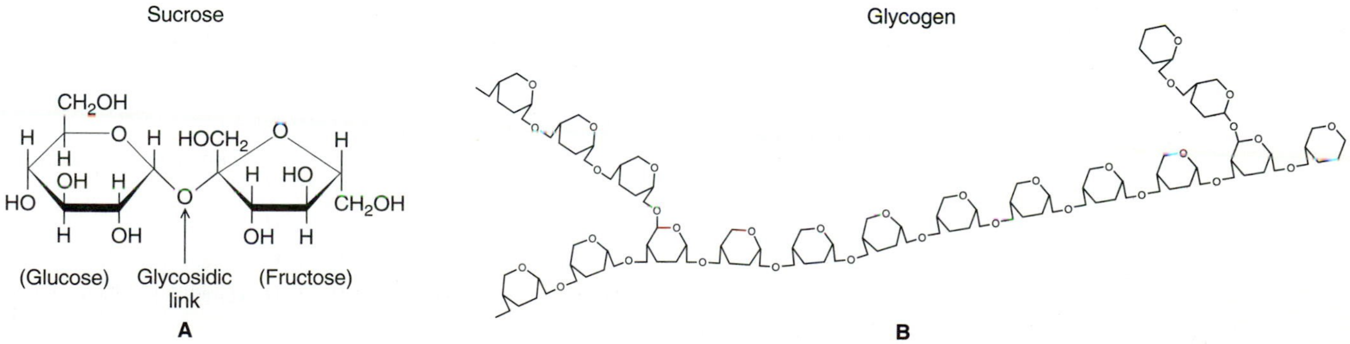

that humans and most other animals can digest. Glucose from digested starch is a major source of energy for many animals. Another important polysaccharide that is similar to starch is **glycogen** (Fig. 2.10B). Certain cells in humans and other animals produce glycogen as between-meal reserves of glucose.

LIPIDS. Fats and several other kinds of organic molecules are called lipids. Fats—**neutral fats,** to be precise—are the most familiar lipids. A neutral fat consists of a three-carbon molecule of **glycerol** to which up to three **fatty acids** are attached (Fig. 2.11). Fatty acids are long chains of carbon with hydrogens attached to the carbons and an acidic carboxyl group at one end. If three fatty acids are attached, the neutral fat is a **triglyceride** (= triacylglycerol). If one or two fatty acids are attached, it is a monoglyceride or diglyceride, respectively. In the same triglyceride any fatty acid can differ from the others in the number of carbons and in the number of double bonds between adjacent carbons. The fewer double bonds a fat has, the more hydrogen can attach to the carbons, and the more **saturated** the fat is said to be. The degree of saturation of a glyceride affects its fluidity at a given temperature. Saturated glycerides are fluid at higher temperatures but become solid, like lard, at room temperature. Saturated fats are more common in birds and mammals ("warm-blooded" animals), as well as in other kinds of animals and plants from warm climates. Unsaturated glycerides are fluid (oils) at room temperature and are more common in plants and animals from cold climates.

Lipids are said to be **hydrophobic** ("water-fearing") because they dissolve in water much less readily than do carbohydrates and many other organic molecules. In fact, lipids and other hydrophobic molecules interact more readily with each other than with water, which explains why fats and oils form globules in dishwater. Glycerides are hydrophobic because of the hydrogen nuclei that surround the fatty acid chains. These positive charges repel the positive charges of the hydrogens in water, preventing the

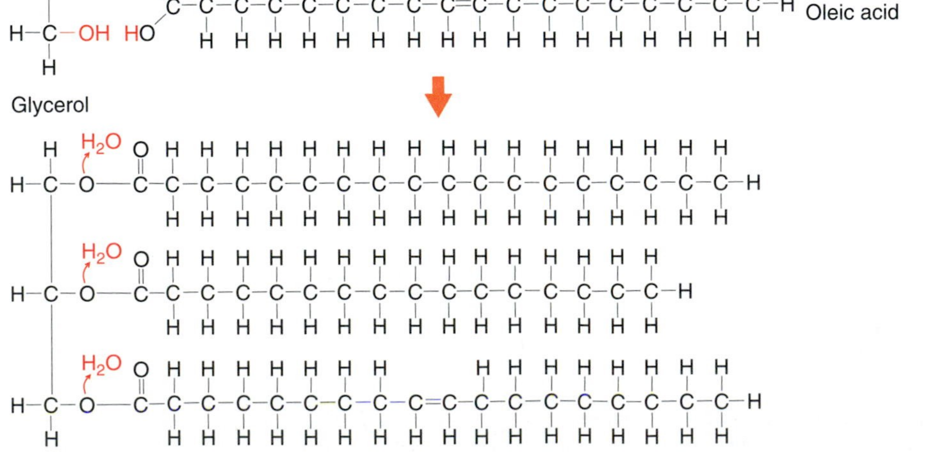

FIGURE 2.11
The formation of a triglyceride. The upper part of the figure shows glycerol on the left and three different fatty acids. Stearic acid and palmitic acid are both saturated but differ in the number of carbon atoms. (The number of carbons is even because fatty acids are synthesized two carbons at a time.) Oleic acid has 18 carbons, like stearic acid, but is unsaturated because of the double bond. During formation of a triglyceride, hydroxyl groups from glycerol and an H from the fatty acid are removed as water, producing a bond through oxygen (an ester bond). Formation of a chemical bond by removal of H_2O is called a dehydration reaction. Breaking the bond by adding H_2O, as occurs when fats are digested, is called a hydrolysis reaction.

Stearic acid
Palmitic acid
Oleic acid
Glycerol
H_2O
H_2O
H_2O

formation of hydrogen bonds. Neutral fats do, however, dissolve in many other lipids and in solvents such as acetone and alcohol.

The low solubility of neutral fats in body fluids and their large number of carbon–carbon bonds make them ideal as long-term energy stores. In times when food is readily available, animals tend to eat more than they immediately need. Fat-storing cells, called **adipose cells,** use the excess energy (**calories**) to synthesize triglycerides. When food is scarce, much of this energy can be recovered by breaking down the triglycerides. Adipose tissues are also effective as insulation against cold. In humans the amount of fat deposited depends on a variety of environmental, genetic, and behavioral factors (see pp. 271–272). The average young American man has a little over 10% of his body weight in the form of triglycerides. This is enough energy to see him through a 40-day fast. American women have an even higher proportion of body fat (approximately 18%), much of it distributed in a layer beneath the skin.

Another important kind of lipid, **phospholipid,** is similar in structure to a triglyceride. The main difference is that in a phospholipid, one fatty acid is replaced by a different kind of substance, called a **head group,** that is linked to the glycerol by a phosphate (Fig. 2.12A). The head group is polar, so it is able to form hydrogen bonds with water. This makes the head group soluble, or **hydrophilic** ("water-loving"). Because phospholipids contain both hydrophobic and hydrophilic regions, they can mix with both lipids and water. One practical application of this property is that lecithin, a phospholipid in egg yolks, binds to both the oil and the lemon juice in mayonnaise and

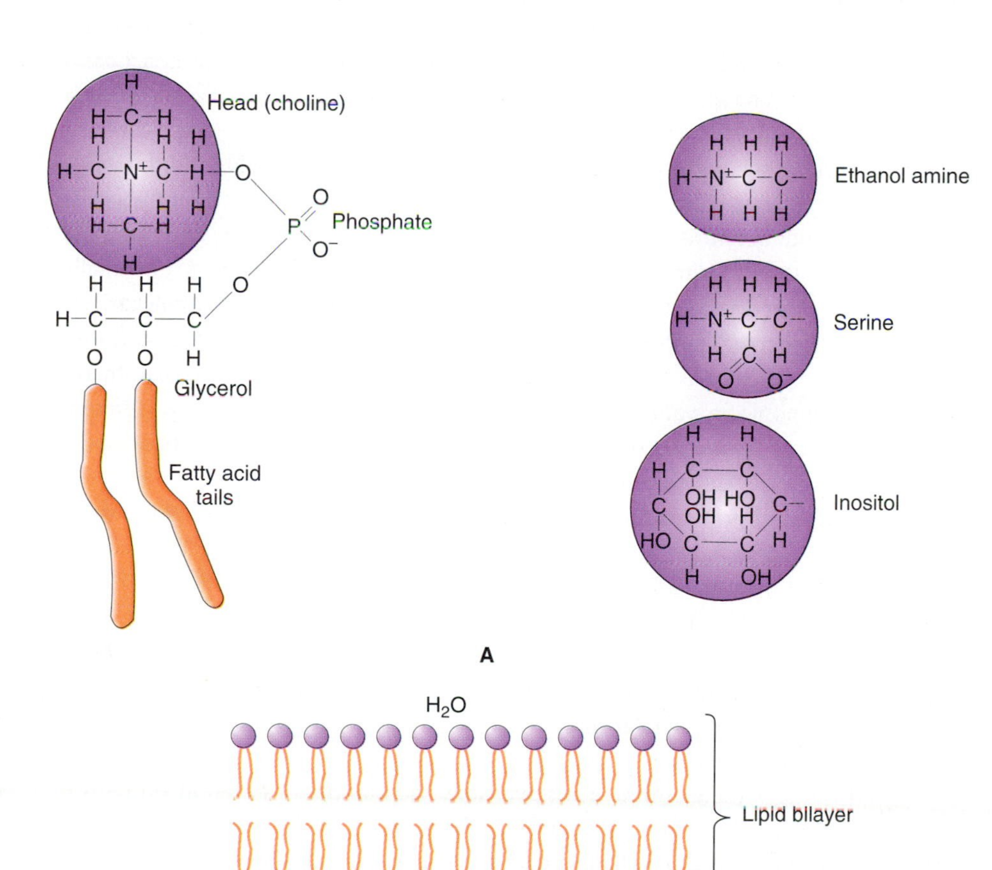

FIGURE 2.12

(A) A phospholipid is essentially a triglyceride with one fatty acid replaced by a phosphate-linked head group. The head group shown here is choline, forming the phospholipid lecithin (= phosphatidylcholine). Three other common head groups are ethanolamine, the amino acid serine, and inositol. (B) Because the head groups are hydrophilic and the fatty acid tails are hydrophobic, phospholipids in water spontaneously form a lipid bilayer.

keeps them from separating. There is another property of phospholipids that is even more useful since it is essential for the formation of cell membranes. When added to water they spontaneously form a double layer called a **lipid bilayer.** The hydrophobic tails avoid the water by pointing toward each other into the middle of the bilayer, and the hydrophilic heads point outward toward the water on each side of the layer (Fig. 2.12B).

Also included among lipids are **steroids,** such as cholesterol and certain hormones (see Fig. 7.2). Steroids differ from other lipids in that they are based on a four-ring structure. Differences in the functional groups attached to the rings determine how the steroids function. For example, the male sex hormone testosterone differs from the female sex hormone estradiol mainly in having a ketone group where estradiol has a hydroxyl group. Cholesterol is synthesized from saturated fatty acids and is an important constituent in membranes. Another group of lipids are **waxes,** which consist of long-chain fatty acids linked to long-chain alcohols or carbon rings. Waxes have a variety of special applications, such as the comb of beehives and as waterproof coverings in animal skins, hair, and feathers.

PROTEINS. The last major class of organic molecules to be described here are the proteins. Proteins consist of **polypeptides,** which are chains of **amino acids.** Polypeptides with only a few amino acids are referred to as **peptides.** As the name implies, an amino acid is a molecule with both an amino group and an acidic carboxyl group. Although an infinite variety of amino acids is possible, only 20 occur in natural proteins. These amino acids all have both the amino and carboxyl groups attached to the same carbon. Also attached to that carbon is an H, along with an **R group** that is different for each of the 20 amino acids (Fig. 2.13).

Differences in the R groups affect the properties of the amino acids and the properties of the peptides and proteins in which they occur. Eight of the natural amino acids are nonpolar and therefore hydrophobic, like fats. The others are polar and therefore hydrophilic. Of the 12 polar amino acids, seven bear no net charge at the pH normally found in cells. Three polar amino acids are positively charged, and two are negatively charged. These charged amino acids can interact electrically with each other and with other charged molecules. A large protein with many different amino acids can therefore be hydrophobic in some parts and hydrophilic in others and have different charges in different areas. These regional differences affect the overall shapes of proteins and the way they interact with other molecules.

In peptides and proteins the amino acids link to each other by **peptide bonds.** A peptide bond is formed when the OH is removed from the carboxyl of one amino acid, an H is removed from the amino group on another amino acid, and the two groups bond to each other (Fig. 2.14 on page 32). Two amino acids linked by a peptide bond form a dipeptide, three linked in this way form a tripeptide, and so on. Many peptides function as hormones. Some proteins function as hormones, and many others are responsible for maintaining cellular shape, for muscle contraction, for regulating chemical reactions within cells, and for such diverse structures as tendons, hair, feathers, horns, and claws. With 20 amino acids available, the number of different proteins is unimaginably large. The number of different dipeptides is $20 \times 20 = 20^2 = 400$, the number of tripeptides is $20^3 = 8000$, and so on. The number of polypeptides with 100 amino acids is 20^{100}, which equals approximately 10^{130}. Since there are only about 10^{27} molecules in a human, and only about 10^{71} atoms in the entire universe, it is obvious

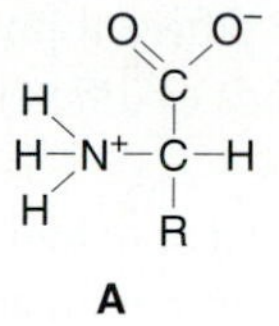

A

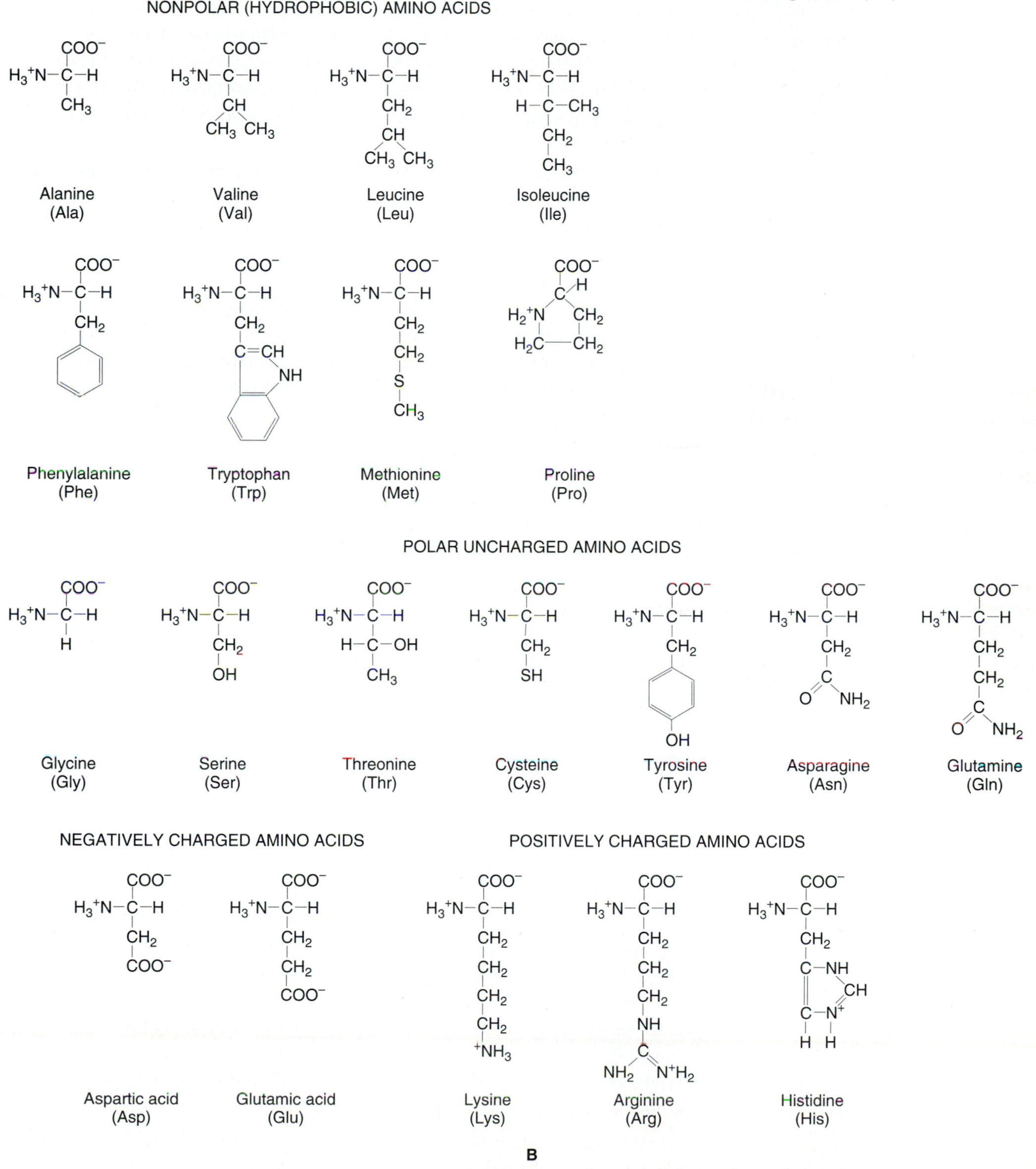

NONPOLAR (HYDROPHOBIC) AMINO ACIDS

Alanine (Ala) | Valine (Val) | Leucine (Leu) | Isoleucine (Ile)

Phenylalanine (Phe) | Tryptophan (Trp) | Methionine (Met) | Proline (Pro)

POLAR UNCHARGED AMINO ACIDS

Glycine (Gly) | Serine (Ser) | Threonine (Thr) | Cysteine (Cys) | Tyrosine (Tyr) | Asparagine (Asn) | Glutamine (Gln)

NEGATIVELY CHARGED AMINO ACIDS

Aspartic acid (Asp) | Glutamic acid (Glu)

POSITIVELY CHARGED AMINO ACIDS

Lysine (Lys) | Arginine (Arg) | Histidine (His)

B

FIGURE 2.13

(A) The general structure of an amino acid. At the normal pH found in cells the amino groups often have an additional H^+ and the carboxyls often lose an H^+, as shown. (B) The 20 amino acids that occur in natural proteins, grouped according to their properties.

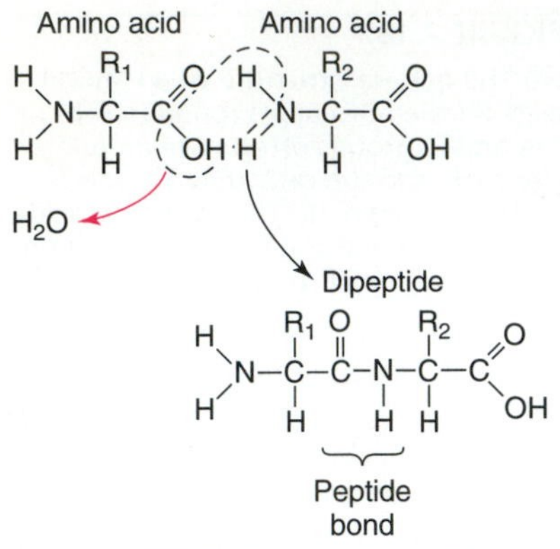

FIGURE 2.14
Formation of a dipeptide from two amino acids.

■ **Question:** What happens when proteins are digested?

that no animal could ever produce all the different possible proteins. However, even the simplest animal must produce thousands of different proteins that perform all the functions required for life.

The different functions of proteins are due to their different structures. One way in which they can differ is in the sequence of amino acids, called the **primary structure.** The primary structure is determined genetically, as will be described in Chapter 5. Superposed on the primary structure are characteristic patterns, called **secondary structure,** that are due to hydrogen bonding between adjacent parts of a protein. Two patterns of secondary structure are quite common: the α-helix and the β-sheet (alpha helix and beta sheet) (Fig. 2.15). Because these secondary structures form hydrogen bonds within themselves rather than with water, they tend to be hydrophobic.

Most proteins also have a **tertiary structure** that is determined by interactions among R groups. These interactions can be due to hydrogen bonding between polar R groups, attraction or repulsion between charged R groups, association of hydrophobic R groups with each other, and covalent bonding between R groups. The latter kind of interaction is especially common between two cysteines, whose sulfhydryls (–SH) tend to form **disulfide bonds** (= S–S bridges). These disulfide bonds are especially strong reinforcements of tertiary structure. They determine the curliness of hair, for example. Finally, many proteins also have a **quaternary structure,** in which two or more polypeptides combine into one protein. The four levels of protein structure are illustrated in Fig. 2.16.

Proteins tend to be either globular or fibrous in structure. **Globular proteins** are usually soluble and are common in body fluids, inside cells, and incorporated into cell membranes. **Fibrous proteins** form long, insoluble strands. Examples of fibrous proteins are **collagen,** which is a reinforcement in skin, bone, and ligaments, and **keratin,** which occurs in silk, horns, beaks, hooves, nails, feathers, and hair.

Mechanism, Measurement, and Microscopy

Carbohydrates, lipids, and proteins do not occur randomly within cells, but generally as parts of precisely shaped structures called **organelles** ("little organs"). Like the workings of a clock, organelles are the **mechanisms** of life within cells. The concept that there must be mechanisms responsible for life is as characteristic of modern science as is materialism. One of the most important organelles is the **plasma membrane,** which encloses every cell and serves as a mechanism that regulates the passage of molecules into and out of the cell. Another important organelle is the nucleus, which contains the mechanisms that control protein synthesis and heredity. Several other organelles are located within the **cytoplasm** (Fig. 2.17 on page 34). One of these, the mitochondrion that provides most of the power for the other mechanisms, will be described in the next chapter. Before describing the other organelles, it is important to understand how we can know anything at all about them. Most are so small, less than one-thousandth of a millimeter (1 micrometer), that it is hard even to imagine them. The first box (Fig. 2.18 on page 35) is intended to help you grasp just how small they are. The second box (pp. 36–38) explains how microscopy is used to study the structure of cells.

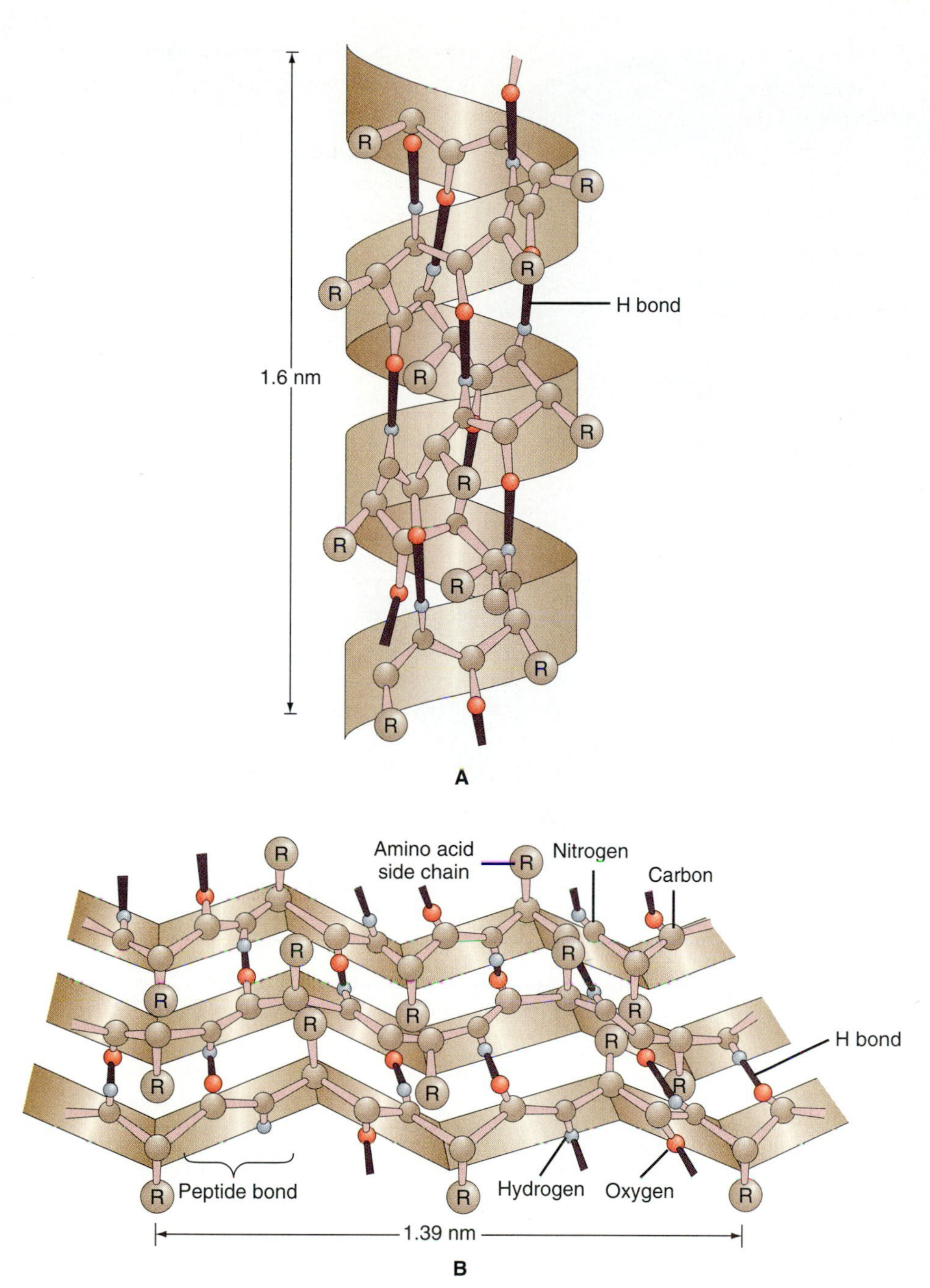

FIGURE 2.15

Two common types of secondary structure are maintained by hydrogen bonds between the hydrogens (green) and the oxygens (red). (A) An α-helix. (B) A β-sheet.

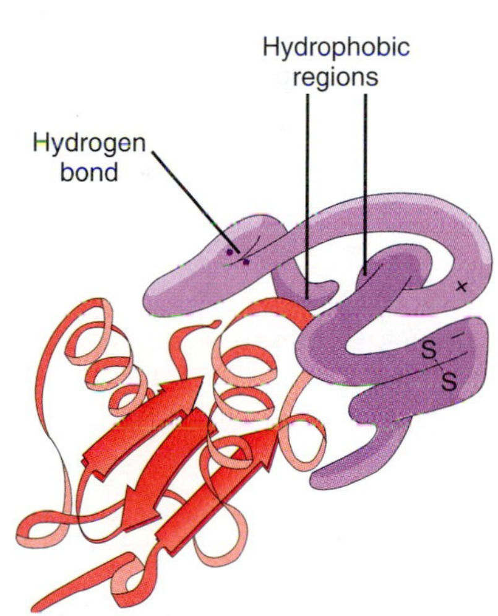

FIGURE 2.16

Levels of structure in a hypothetical protein with two polypeptide subunits. The subunit on the left is shown in a "ribbon model," while the one on the right is shown in a "sausage model." The primary structure is the sequence of amino acids in the polypeptide strands (not represented). The secondary structure consists of the α-helices (spirals) and β-sheets (broad arrows) shown in the ribbon model. The tertiary structure is the overall shape of each polypeptide, which is determined by hydrogen and disulfide bonds and by interactions between charged and hydrophobic regions (sausage model). The quarternary structure is the association of the two polypeptides into a single functional unit.

The Plasma Membrane

STRUCTURE. The plasma membrane is largely responsible for keeping the low-entropy cytoplasm from mixing with the high-entropy environment of cells. Its importance is clear from the fact that damage to the plasma membrane usually kills the cell. Under the electron microscope the plasma membrane has a three-layered appearance, because the metals used to stain the membrane adhere only to the inner and outer surfaces. This simple appearance in the electron microscope belies a much more complicated structure. According to the **fluid mosaic model** that has been widely accepted

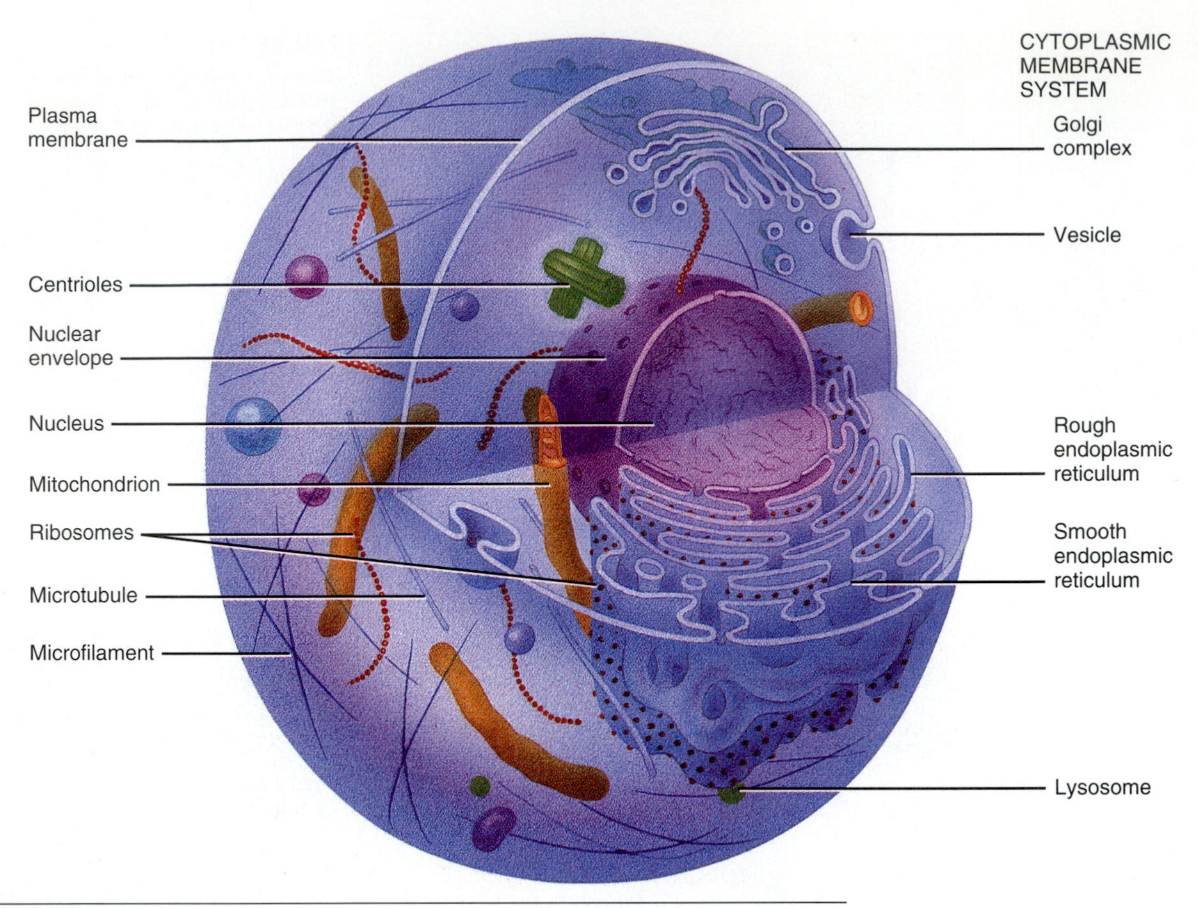

FIGURE 2.17

A generalized animal cell, showing most of the organelles. The cell consists of the plasma membrane, the cytoplasm, and the nucleus. The plasma membrane surrounds every cell and controls the passage of material into and out of it. In the cytoplasm, the cytoplasmic membrane system, which includes rough and smooth endoplasmic reticulum and the Golgi complex, produces vesicles for new plasma membrane, lysosomes, and secretion. Other vesicles bring in certain materials, which are released into the cell after the vesicle is digested by lysosomes. Mitochondria produce most of the energy used by cells, centrioles are involved with the movement of genetic material during cell division, and ribosomes on the rough endoplasmic reticulum produce protein. These three organelles will be described in Chapters 3, 4, and 5, respectively. All these organelles are considered part of the cytoplasm. The nucleus, which is not considered part of the cytoplasm, is enclosed by a porous nuclear envelope that is connected with the cytoplasmic membrane system. The cell shape is maintained by tubules and filaments of the cytoskeleton.

since 1972, the plasma membrane actually consists of a fluid film of lipids with proteins floating within it (Fig. 2.21 on page 38).

In most animal cells the lipids, especially phospholipids, account for about half the mass of the plasma membrane. As explained above, these phospholipids spontaneously form a bilayer that is the main structural component of the membrane. Each square micrometer of membrane contains approximately two million molecules of phospholipid. In most plasma membranes each phospholipid has a cholesterol molecule situated between the hydrophilic tails near the polar head group. This cholesterol stiffens the tails and prevents the bilayer from becoming too fluid, and it also keeps the bilayer from hardening at low temperature. This is an important adaptation for animals that live in the cold. On the outer surface of the plasma membrane are **glycolipids:** lipids with carbohydrates attached.

Dimensions of Life

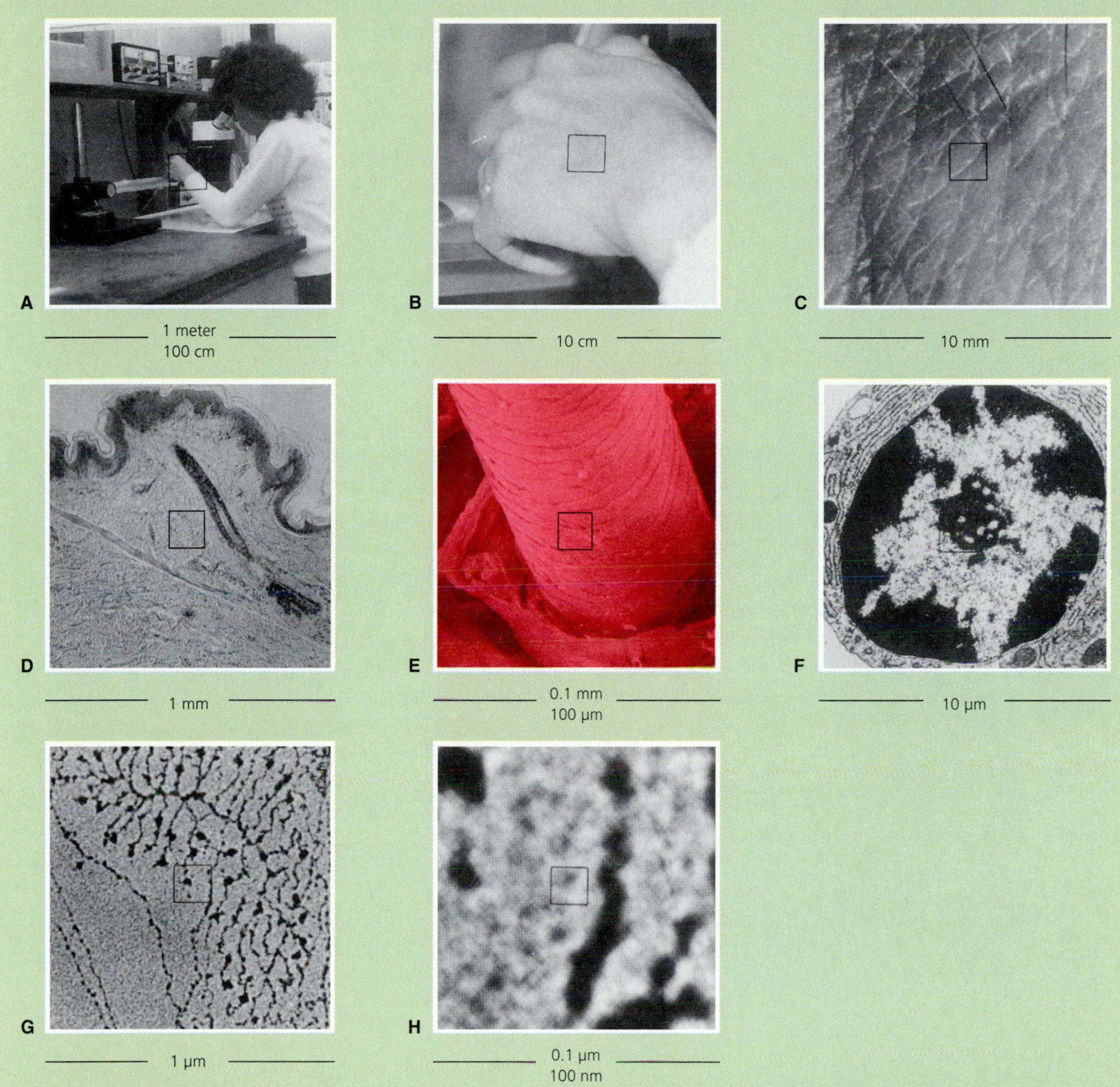

FIGURE 2.18

Relative sizes of structures commonly encountered in zoology. Each figure covers 10 times the width and 100 times the area of the following. (A) Humans and other large animals are conveniently measured in units of meters. (B) The width of the hand is approximately 0.1 meter, or 10 centimeters (cm). (C) With a field of view 1 cm wide, hairs become visible. (D) Objects smaller than 1 millimeter (mm) are not easily seen with the unaided eye. This stained slice of human skin showing a hair follicle was photographed using light microscopy at a magnification of 100. (E) Scanning electron microscopy reveals three-dimensional structure of a hair at a higher magnification. (F) Most animal cells are on the order of 10 micrometers (μm) wide. Their internal structure is best viewed with transmission electron microscopy. (G) Transmission electron microscopy at high power even reveals the molecules responsible for heredity. (H) The fine structure of molecules is beyond the limits of transmission electron microscopy, although computers can enhance the image and enable the artist to represent it. At such small dimensions another unit of measure, the angstrom (Å, = 10^{-10} meter = 0.1 nm), is also common.

Microscopy

Our understanding of cells has been intimately linked to developments in microscopy. Even the name "cell" derives from *cellulae,* the name Robert Hooke gave to the hollow spaces in cork that he observed under the microscope in 1663. Antony van Leeuwenhoek (an-TON-ee van LAY-win-huke) was probably the first to see living cells, using simple microscopes that he made with glass beads as lenses, starting in about 1670. This Dutch cloth merchant had intended to use his microscopes to examine the closeness of weave, but his first sight of a drop of pond water drew him into a new world inhabited by bacteria and "animalcules." Monumental as his observations were, their significance was not grasped for more than a century. Most biologists were more interested in the easily seen structures such as plant and animal fibers, and they regarded cells as byproducts.

It was not until the 19th century that improved techniques in microscopy allowed biologists to realize the significance of cells. First Carl Zeiss and others perfected the **compound microscope,** which incorporates several lenses to increase magnification and to correct the distortions inherent in single lenses. Just as significant was the discovery of dyes and techniques for preserving and slicing biological specimens. By the middle 1800s, observations based on these techniques (as well as philosophical and even political reasons that we can scarcely comprehend today) convinced most biologists that cells were not merely byproducts of living tissues but were essential for all life. In 1839 Theodor Schwann pronounced what we still regard as the cell theory: "that there exists one general principle for the formation of all organic productions, and that this principle is the formation of cells." That is, organisms consist of cells and the products of cells.

By the turn of the century it was firmly established that the essence of life is in cells. All that remained was to discover how the materials and mechanisms of cells accounted for life. As always, our understanding of the internal workings of cells depends on advances in microscopical techniques.

Light Microscopy Light microscopy depends on light passing through different structures of a specimen in different ways so that the structures can be distinguished from the background. After the light has passed through the specimen, it is magnified by lenses (left side of Fig. 2.19). Ordinary light passes through all parts of cells in about the same way, so most cells look virtually transparent and structureless (Fig. 2.20A). To remedy this problem the light can be manipulated by polarizing it, changing its angle of incidence, splitting it into two rays and recombining them

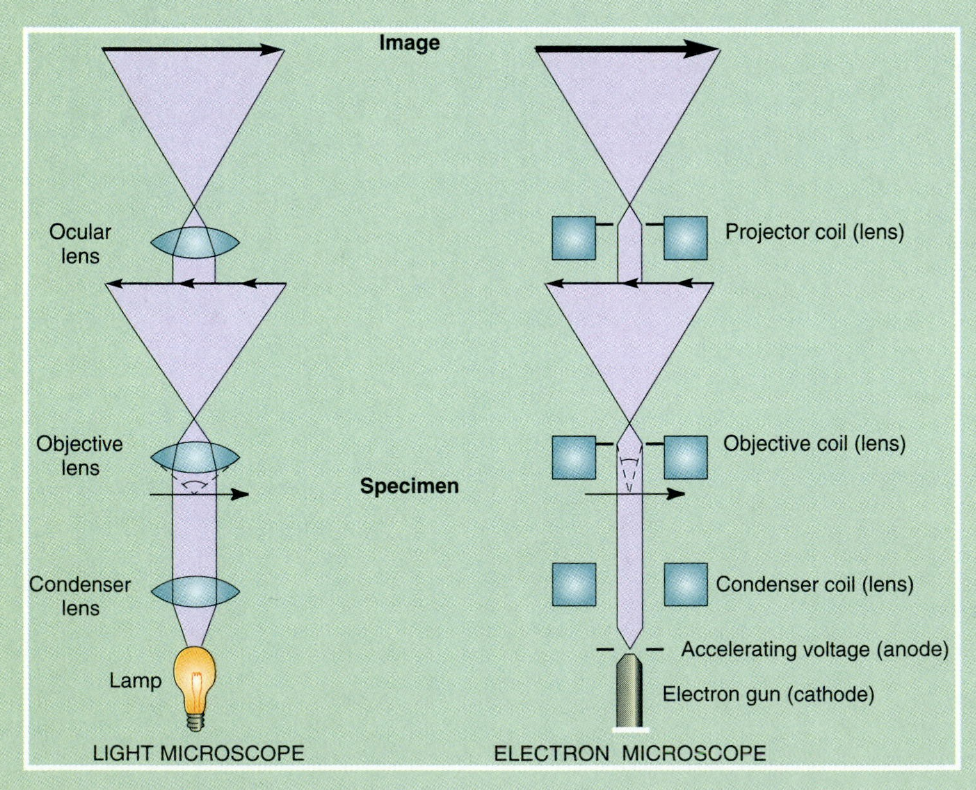

FIGURE 2.19
The pathway of light through a light microscope (left) compared with the pathway of electrons through a transmission electron microscope. The electron microscope is shown upside down for convenience.

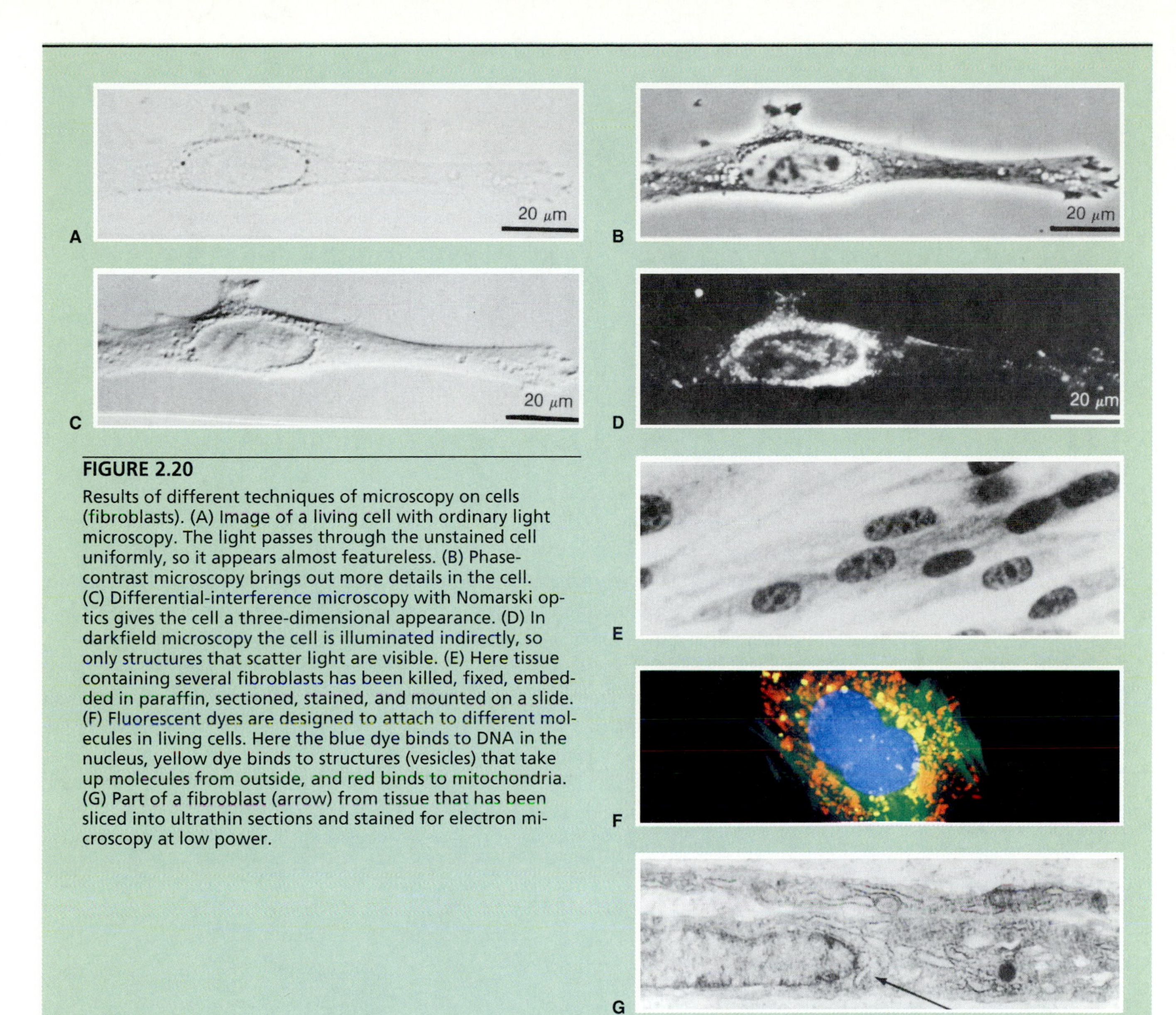

FIGURE 2.20

Results of different techniques of microscopy on cells (fibroblasts). (A) Image of a living cell with ordinary light microscopy. The light passes through the unstained cell uniformly, so it appears almost featureless. (B) Phase-contrast microscopy brings out more details in the cell. (C) Differential-interference microscopy with Nomarski optics gives the cell a three-dimensional appearance. (D) In darkfield microscopy the cell is illuminated indirectly, so only structures that scatter light are visible. (E) Here tissue containing several fibroblasts has been killed, fixed, embedded in paraffin, sectioned, stained, and mounted on a slide. (F) Fluorescent dyes are designed to attach to different molecules in living cells. Here the blue dye binds to DNA in the nucleus, yellow dye binds to structures (vesicles) that take up molecules from outside, and red binds to mitochondria. (G) Part of a fibroblast (arrow) from tissue that has been sliced into ultrathin sections and stained for electron microscopy at low power.

in different ways, or other sophisticated optical techniques. Each technique brings out different features of a cell, often quite dramatically (Fig. 2.20B–D).

Another approach to revealing cellular structure with light microscopy is to use dyes that stain structures differentially. Although there are some **vital stains** that can be used with living tissues, most stains require that the specimen be killed, **fixed** to coagulate proteins and hold the shape of cells, then preserved in alcohol or some other substance. The specimen is then embedded in paraffin and sliced into sections usually 10 to 20 micrometers (μm) thick and mounted on glass slides (Fig. 2.20E and F).

Electron Microscopy Although the techniques of light microscopy have become quite sophisticated, they cannot overcome a fundamental limitation. This limitation is the finite wavelength of light, which prevents any light microscope from distinguishing (**resolving**) an object smaller than 0.2 μm. Many of the most interesting parts of cells are approximately this size and cannot be studied with the light microscope. Under certain circumstances, however, electrons also behave like waves, and they have much shorter wavelengths than light. This fact is the basis for the **electron microscope.** The most powerful electron microscopes have a **resolving power** 1000 times better than that of the light microscope. Electron microscopes can resolve objects as small as 1 nanometer (nm = 0.001 μm), which is about the size of many molecules.

Neither the cellular structures nor the stains used in light microscopy will absorb electrons, so heavy metals

box continues

have to be used as stains in electron microscopy. In **transmission electron microscopy** the specimen first has to be killed, fixed, and embedded. The specimen is then sliced into sections less than 0.1 μm thick (ultrathin) to allow the electrons to pass through (right side of Fig. 2.19). The section is then stained with heavy metals, taking advantage of the fact that the metal adheres to some molecules better than to others (Fig. 2.20G). It is important to keep in mind that what one actually sees with the electron microscope is the metal stain, not the molecules in the specimen. The image is uncolored, although false color is often added to photographed images. With another type of electron microscopy, called **scanning electron microscopy** (SEM), the specimen is usually not sliced. Instead, a metal stain is applied to the surface, allowing biologists to view the external structure of cells, tissues, and even entire animals with lifelike depth. [For examples, see the figure opening this unit (page 16) and Fig. 32.22.]

Proteins account for the other half of the mass of plasma membrane. Most of these proteins have attached carbohydrates that protrude from the outer surface of the membrane only, making them **glycoproteins.** The glycoproteins are dissolved within the lipid bilayer and cannot flip over or pop out, because they are anchored by hydrophobic and hydrophilic interactions with phospholipids in the bilayer.

DIFFUSION. The ability of plasma membranes to prevent the passage of some molecules while allowing the passage of others is a good illustration of the application of scientific materialism and mechanism. Because of the lipid bilayer, only organic molecules that are soluble in lipids (hydrophobic) can diffuse through the membrane. Most molecules that are free to move in and around cells are hydrophilic, however, because they have to be dissolved in water. Therefore the lipid bilayer is impermeable to most of the organic molecules in organisms. Unless they are hydrophobic, organic mole-

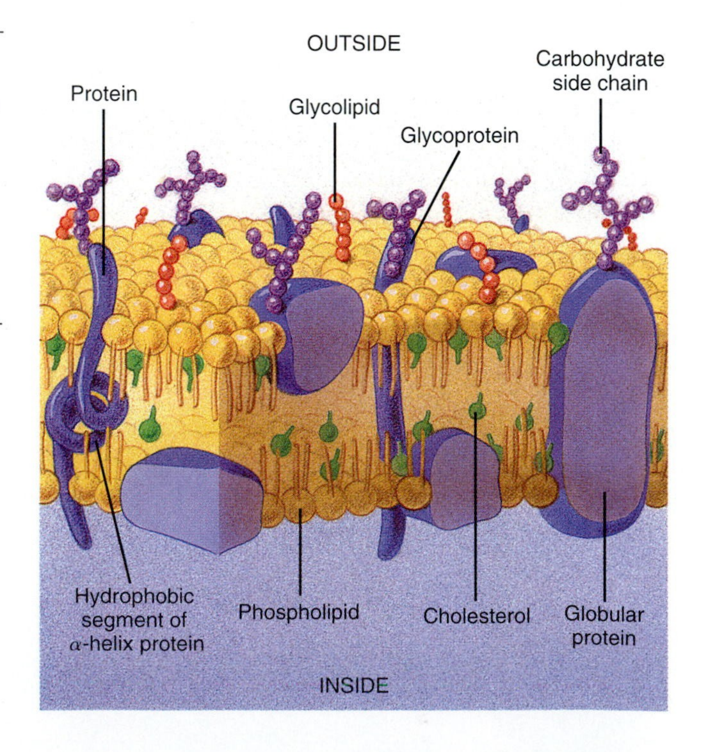

FIGURE 2.21

The plasma membrane consists of a lipid bilayer with proteins embedded in it. Cholesterol helps regulate the fluidity of the membrane. The proteins are either globular or fibrous. A hydrophobic part of the protein is embedded in the hydrophobic layer of the lipid bilayer, and hydrophilic parts of the protein are embedded in the outer layers. Other proteins are attached to the surface. Carbohydrate side chains on glycoproteins and glycolipids are sites of attachment for toxins, viruses, bacteria, hormones, and many other molecules.

cules inside cells tend to stay in the cytoplasm, and those that are outside generally cannot enter unless they are brought in by some specific mechanism. In addition to blocking the diffusion of most organic molecules, the lipid bilayer also keeps out most inorganic molecules. There are only two kinds of inorganic molecules that can diffuse through the lipid bilayer: (1) small, nonpolar molecules, such as O_2; and (2) small, uncharged, polar molecules, including H_2O and CO_2.

In addition to H_2O, O_2, and CO_2, other natural molecules that can diffuse through the lipid bilayer are fats and steroids. There is also an increasing number of manufactured organic molecules that diffuse through plasma membranes, for good or ill. These include synthetic steroids, such as those in oral contraceptives. If they were not able to penetrate plasma membranes, they could not be absorbed across the intestinal lining and would have to be administered by injection. Steroid hormones can also be absorbed through the skin, and some women have had unnecessary hysterectomies because steroids in some skin creams induced symptoms indistinguishable from uterine cancer. Of even more concern are industrial solvents, pesticides, and other manufactured substances that are designed to penetrate lipids. Because they are absorbed through skin, lungs, and intestines, and may also damage plasma membranes, many have a high potential for inducing cancers, mutations, and birth defects.

In addition to molecules that can diffuse through the lipid bilayer, some small ions can diffuse through the plasma membrane through **channels** made of glycoproteins. Because of such channels, plasma membranes are generally permeable to H^+, K^+, and Cl^-. Usually they are less permeable to larger ions such as Na^+. For example, the plasma membranes of red blood cells are 50 times more permeable to K^+ and Cl^- than to Na^+. Most other ions are even larger than Na^+ and are essentially impermeant. Because membranes are more permeable to some ions than to others, they are said to be **semipermeable.** This semipermeability gives rise to a voltage (**membrane potential**) across the plasma membrane that is essential in the functioning of neural and other cells, as will be explained in Chapter 8. This membrane potential is generally between 0.05 and 0.09 volts, negative inside the cell compared with outside.

Under certain circumstances the permeability to sodium and other ions can be increased by opening **gated channels.** The signal opening the gate of such a channel can be a particular chemical or a change in membrane potential. For example, several different kinds of potassium channels are known, some voltage-dependent, some responsive to the levels of calcium in the cytoplasm.

Diffusion through channels is usually limited to small ions, but larger ions and even molecules can diffuse across the plasma membrane by means of **facilitated diffusion.** In facilitated diffusion a molecule diffuses down its concentration gradient by attaching to a **transporter** molecule. This is the route by which glucose enters cells.

ACTIVE TRANSPORT. The diffusion of water, ions, and other molecules, even if facilitated by a transporter, is always from the side of the plasma membrane with the higher concentration to the side with the lower concentration. If diffusion were the only means by which a cell could bring in the substances it requires, it would be severely limited. There are mechanisms, however, that enable a plasma membrane to transport materials in the direction opposite the one they would passively follow. This process is referred to as **active transport,** because it requires energy. The mechanisms of active transport are referred to as **pumps.** The best-known example is the **sodium–potassium pump,** which pumps three sodium ions out of a cell while pumping two potassium ions into it. The sodium–potassium pump is also known as the

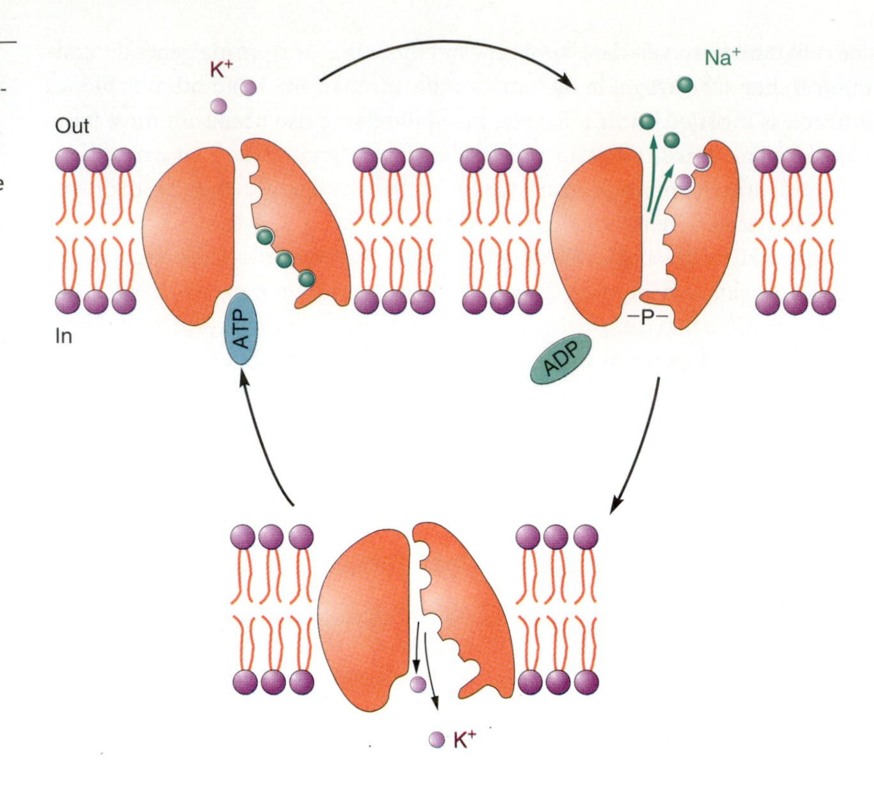

FIGURE 2.22

Schematic representation of the operation of the Na^+–K^+ ATPase that is the sodium–potassium pump. In the upper left the binding of three Na^+ ions (red dots) triggers a change in the structure of the ATPase that moves the Na^+ outside the cell, against diffusion and the negative membrane potential. Afterward the binding of two K^+ ions (green dots) restores the structure of the ATPase, which transports the K^+ into the cell.

■ **Question:** Of what kind of molecule would the pump be made?

Na^+–K^+ ATPase, because it obtains the energy to actively transport Na^+ and K^+ by breaking bonds in a molecule called ATP. It is not yet clear how the Na^+–K^+ ATPase works, but Fig. 2.22 summarizes what it does.

Na^+–K^+ ATPases have been found in the plasma membranes of every organism examined, and by some accounts they use up to 40% of an organism's energy. Evidently they are important. One of the things sodium–potassium pumps do is to produce a lower concentration of Na^+ inside the cell than outside. The pumps also bring some K^+ into the cell, although the negative membrane potential is much more important in this regard and also maintains a low concentration of Cl^- inside the cell (Fig. 2.23). Na^+–K^+ ATPases are also coupled to other kinds of pumps in the plasma membranes and are therefore indirectly responsible for other kinds of active transport. Such coupled pumps transport glucose, amino acids, and other nutrients from the intestine into the bloodstream. In addition, there are also Ca^{2+} pumps, proton (H^+) pumps, and other kinds of pumps that operate independently of sodium–potassium pumps.

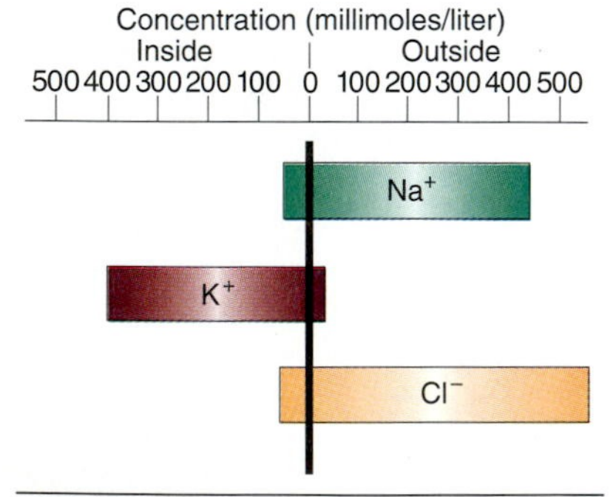

FIGURE 2.23

Typical concentrations of Na^+ K^+, and Cl^- inside and outside a cell, represented by the width of each bar.

■ **Question:** Why is there so much Na^+ outside when there is a negative voltage inside?

Exocytosis and Endocytosis. Two other mechanisms for getting materials out of or into cells are exocytosis and endocytosis. **Exocytosis** (Greek *exo-* out + *kytos* hollow vessel, i.e., a cell) occurs when material in the cytoplasm is packaged within a membrane-bound sphere called a **vesicle.** The vesicle contents are then released externally when the vesicle fuses with the plasma membrane. A similar process is responsible for building new membrane. Besides the plasma membrane, several other organelles are involved in exocytosis, as will be described in more detail later. Among the important materials secreted by exocytosis are hormones, mucus, and substances that trigger muscle contraction or modify neural activity.

Endocytosis (Greek *endo-* in) is essentially the reverse of exocytosis. Material outside the cell binds to particular sites on the plasma membrane, which then invaginates, enclosing the material in a vesicle. The vesicle then pinches off the inner surface of the membrane and enters the cytoplasm. In the cytoplasm the vesicle contents are released when the vesicle is digested by an organelle called the **lysosome.** There are three types of endocytosis: pinocytosis, phagocytosis, and receptor-mediated endocytosis (Fig. 2.24). **Pinocytosis** (Greek *pino* to drink) refers to the uptake of water and dissolved material. **Phagocytosis** (Greek *phagein* to eat) is the uptake of any undissolved material, such as food, viruses, bacteria, and cellular debris. **Receptor-mediated endocytosis** occurs when membrane receptors bind a particular substance and migrate to a depres-

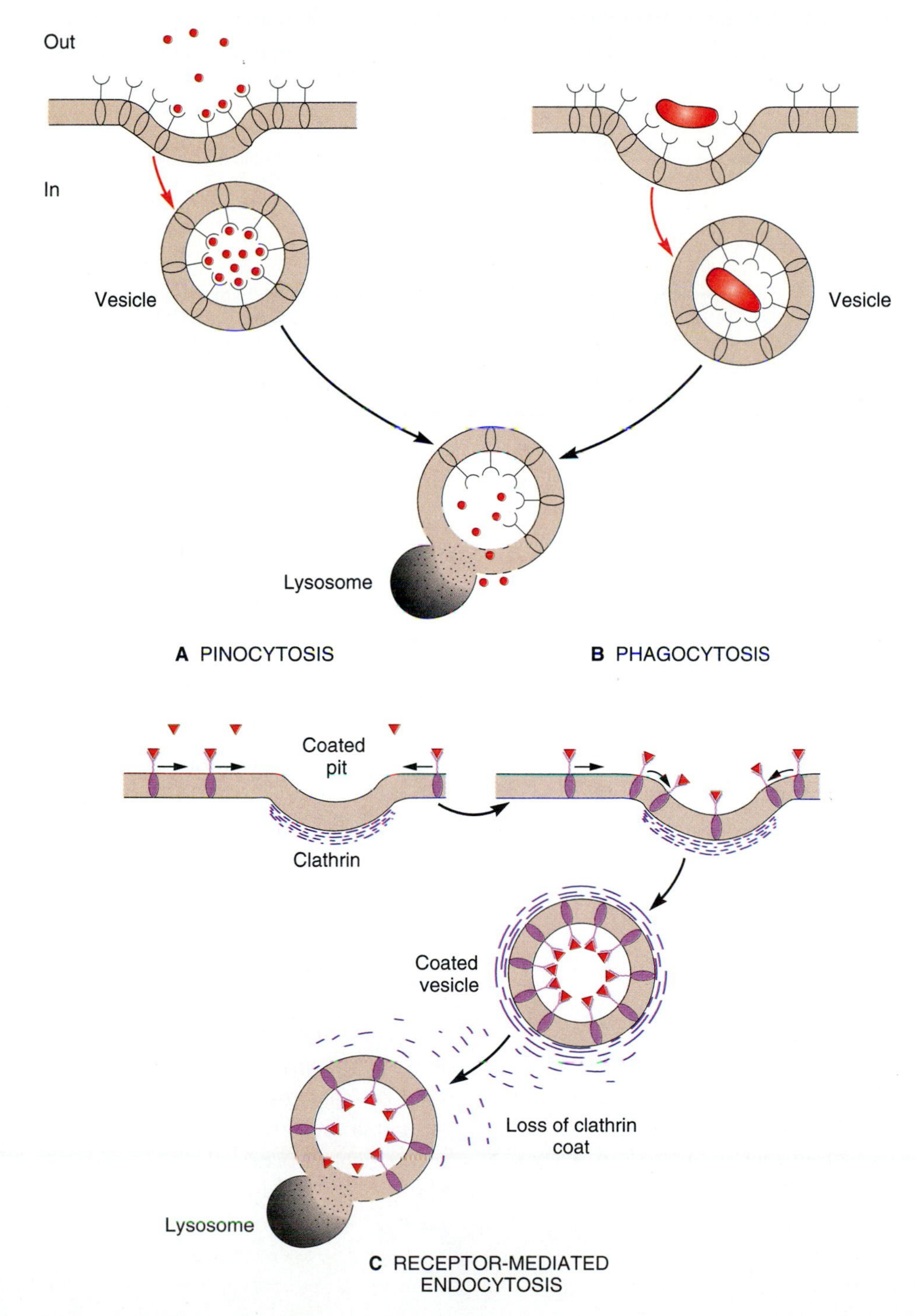

FIGURE 2.24

Three types of endocytosis, followed by digestion by a lysosome. (A) Pinocytosis, the uptake of water along with dissolved materials and small particles in the water. (B) Phagocytosis, the uptake of solids. (C) Receptor-mediated endocytosis. Receptors carrying particular molecules migrate to pits coated on the inner surface with the protein clathrin. Vesicles form from these coated pits.

sion called a **coated pit,** where the endocytosis occurs. Figure 2.25 summarizes the mechanisms by which the plasma membrane controls the movement of material into and out of cells.

Why Do Animals Have More Than One Cell?

The importance of the plasma membrane in the transport of material into and out of the cell may partly explain a fundamental fact of life in animals and other multicellular organisms. These organisms consist of many small cells rather than one or a few large ones. The explanation has to do with the surface-to-volume ratio, a factor that recurs frequently throughout this text. As an object grows in size, its surface area increases

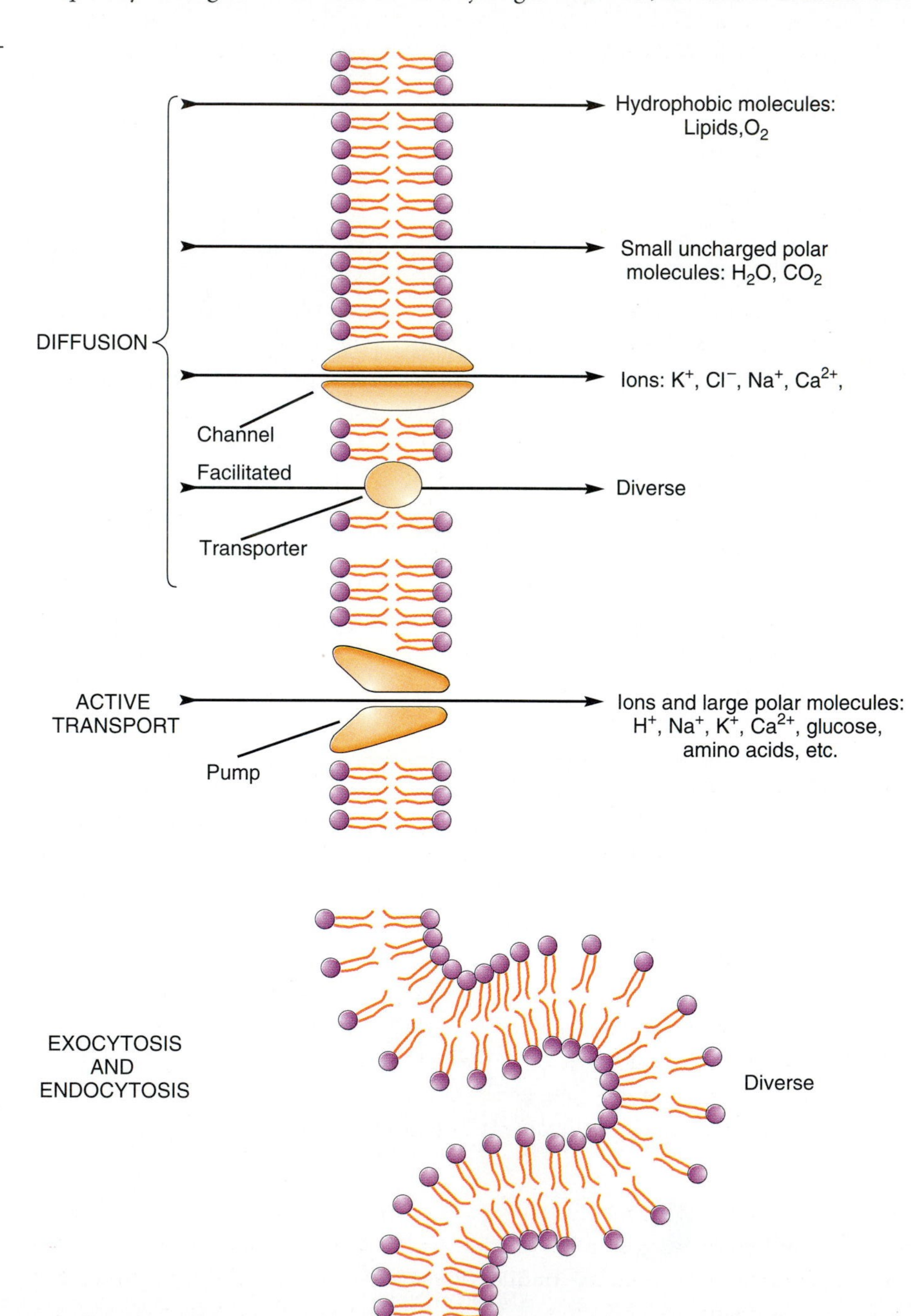

FIGURE 2.25
Summary diagram of the methods by which plasma membrane controls the entry and exit of material.

■ **Question:** Na^+, Ca^{2+}, and glucose can move across the membrane either by facilitated diffusion or active transport. What determines which method is used?

more slowly than its volume. If a microscopic single-celled organism were to become a larger organism by increasing the diameter of its cell tenfold, for example, its volume would increase a thousandfold (10^3), but its surface area would increase by only a hundredfold (10^2). Thus there would be a thousand times as much volume to supply nutrients to and eliminate wastes from, but only a hundred times the plasma membrane to do the job. For this and other reasons discussed elsewhere (p. 476), single-celled organisms evolved into animals by keeping the size of their cells constant, but increasing their number.

The many cells that make up an animal do not function independently, but coordinate their activities, often by means of chemical messengers. Such chemical messengers are secreted by certain cells and often bind to specific **receptors** on the plasma membranes of other cells. In response to binding of the messenger, the receptor molecule triggers a change on the inner surface of the membrane, causing a response in the cytoplasm. Among these chemical messengers are some hormones (Chapter 7) and synaptic transmitters (Chapter 8) that allow nerve cells to communicate with each other.

Cellular Shape, Movement, and Interconnections

The Cytoskeleton and Extracellular Matrix. Although the plasma membrane is quite fluid, the cell is not merely a shapeless bag of cytoplasm. Cells have definite structure. This structure is partly due to a positive oncotic pressure resulting from a greater amount of protein and other molecules in the cytoplasm than outside the cell. In addition, there is an internal network of protein fibers that constitutes a **cytoskeleton.** The cytoskeleton consists of three major components: microtubules, microfilaments, and intermediate filaments. **Microtubules** are the thickest of the three (22 nm), and they radiate from the nucleus of the cell to just under the plasma membrane. They are believed to act as a framework for the internal organization of cells and are also essential in cellular reproduction, as will be described in a later chapter. In nerve cells and perhaps other kinds of cells, microtubules also serve as tracks along which vesicles and organelles move in both directions.

Microfilaments are the thinnest components of the cytoskeleton (6 nm) and are made of **actin,** which is also a major protein in muscle. Microfilaments run just beneath the plasma membrane and function in cell movements and as **stress fibers** that reinforce the plasma membrane (Fig. 2.26). **Intermediate filaments** are between microtubules and microfilaments in diameter (7 to 11 nm) and consist of different proteins in different kinds of cells. The best known of these intermediate filaments is keratin, which was previously mentioned as the fibrous protein in horns, beaks, hooves, nails, hair, and feathers.

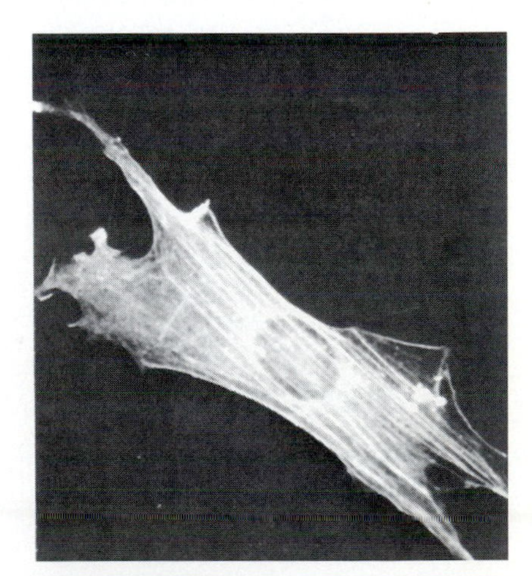

FIGURE 2.26
Bundles of thin filaments revealed by fluorescent antibodies attached to the actin.

In addition to these intracellular proteins, there are also large fibrous proteins and globular glycoproteins that make up an **extracellular matrix** (ECM). The ECM helps maintain cell structure, and it helps cells adhere to each other. The ECM has also attracted considerable interest because of recent evidence that it may transmit messages across the plasma membrane to direct cellular activities.

Motility. Many of the proteins that make up the cytoskeleton are also involved in cell movement. Actin microfilaments, for example, are partly responsible for the contraction of muscle and some other kinds of cells. Actin is also believed to be involved in **ameboid movement,** which is a kind of creeping movement of certain protozoans such as *Amoeba* and of white blood cells (see Figs. 10.16 and 22.6). Microtubules are the main elements in **cilia** and **flagella,** which are hair-shaped structures on the outer surface of the plasma membrane (see Fig. 10.17). Cilia are short and numerous, and they are responsible for locomotion in some protozoans and invertebrates and for moving fluids past tissues in many animals. Flagella are long and sparse and are responsible for the swimming of some protozoans and sperm and for the propulsion of fluid in sponges and some other invertebrates. All these kinds of movement are discussed in detail in Chapter 10. The formation of cilia and flagella is initiated by small, cylindrical organelles called **centrioles,** which are themselves made of microtubules (Fig. 2.27).

INTERCELLULAR CONNECTIONS. Most cells are connected to others in ways that affect their shape and restrict their movements. There are three kinds of **intercellular junctions:** desmosomes, tight junctions, and gap junctions (Fig. 2.28). **Desmosomes** maintain the structural integrity of tissues by fastening together the plasma membranes of adjacent cells. **Tight junctions** block the passage of substances around cells by joining plasma membranes in a band that encircles each cell. **Gap junctions** form protein channels that allow the passage of monosaccharides, amino acids, and ions between neighboring cells.

Cytoplasmic Membranes

The plasma membrane surrounding every cell is closely related to a system of membranes within the cytoplasm. This **cytoplasmic membrane system** is connected to the plasma membrane, exchanges phospholipids and glycoproteins with it, and prepares

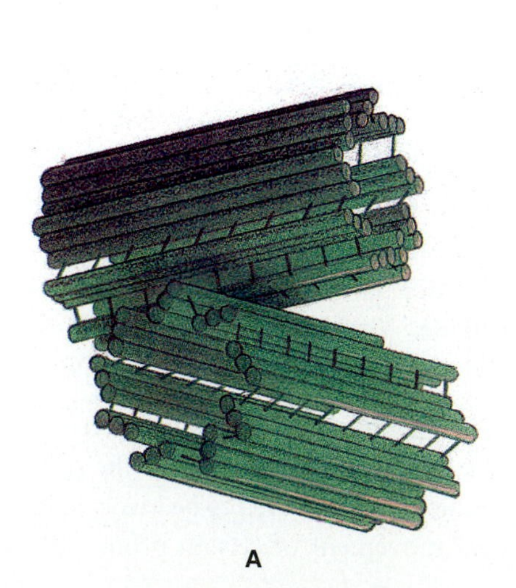

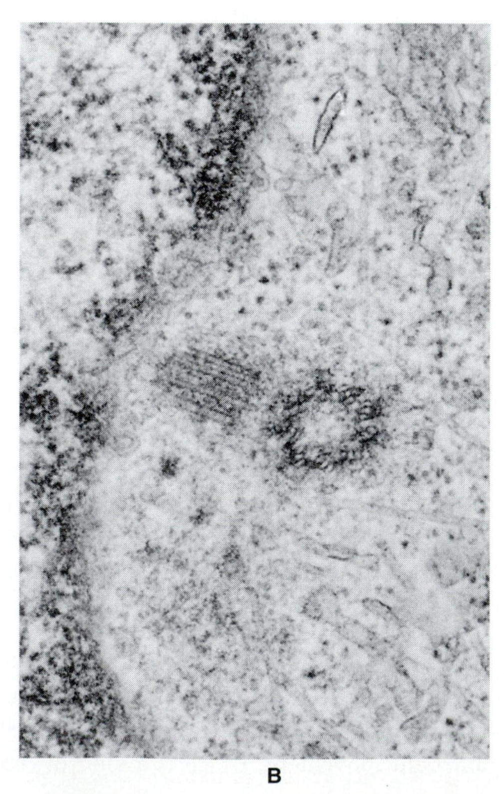

A B

FIGURE 2.27

(A) Schematic diagram of two centrioles. (B) Electron micrograph showing two centrioles, one in cross section and one in longitudinal section. Diameter of each centriole is approximately 0.2 μm.

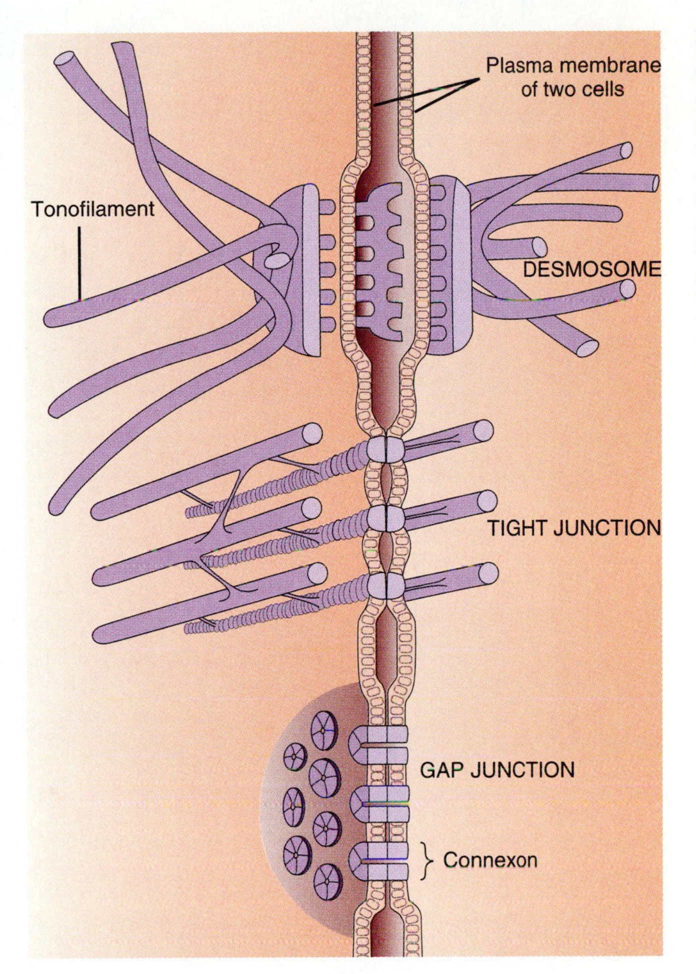

FIGURE 2.28
Types of intercellular junction. Like spot welds, desmosomes join the plasma membranes of two cells for structural integrity. Tonofilaments are intermediate filaments made of keratin that anchor the desmosome to the cytoplasm. A tight junction prevents passage of substances between two cells. A gap junction has central channels wide enough (2 nm) to permit the movement of ions, simple sugars, and amino acids between cells, but not larger molecules. Connexons in the gap junction consist of two sets of six protein subunits.

glycoproteins for exocytosis from the plasma membrane. The cytoplasmic membrane system comprises several distinct organelles, including endoplasmic reticulum, Golgi (GOAL-jee) complexes, and lysosomes. Briefly, proteins are synthesized on endoplasmic reticulum, where many are packaged inside vesicles or as parts of the membranes enclosing the vesicles. The vesicles then enter Golgi complexes, which modify the proteins and repackage them into new vesicles and then send them to their destinations. Some proteins inside the vesicles are incorporated into lysosomes, which are organelles containing enzymes for intracellular digestion. Other proteins are secreted during exocytosis (as in Fig. 8.7B). Proteins in the vesicle membrane become incorporated into the plasma membrane.

The first organelle of the cytoplasmic membrane system to be discussed is the **endoplasmic reticulum** (ER) (Fig. 2.29 on page 46). Most ER is of a type called **rough endoplasmic reticulum,** because small organelles called **ribosomes** give it a grainy appearance in the electron microscope. As will be described in Chapter 5, the ribosomes on rough ER begin the synthesis of proteins according to genetic instructions from the nucleus. Many proteins have particular sequences of amino acids that label them for packaging in vesicles and further processing by the **Golgi complex.** A Golgi complex (= Golgi body or Golgi apparatus) consists of a stack of hollow, disc-shaped compartments. Each compartment absorbs vesicles, modifies the proteins in them, repackages them into new vesicles, then sends these vesicles on to the next compartment. Each compartment sends the protein to its appropriate destination, whether to a lysosome, to a secretory vesicle, or to the plasma membrane (Fig. 2.30 on page 47).

FIGURE 2.29

Two types of endoplasmic reticulum. Rough endoplasmic reticulum bears ribosomes, which are organelles used in protein synthesis. They are not visible in the scanning electron micrograph (A), but appear as tiny, dark spots in the transmission electron micrograph (B). Smooth endoplasmic reticulum (C and D) does not have ribosomes. It is the site of lipid synthesis.

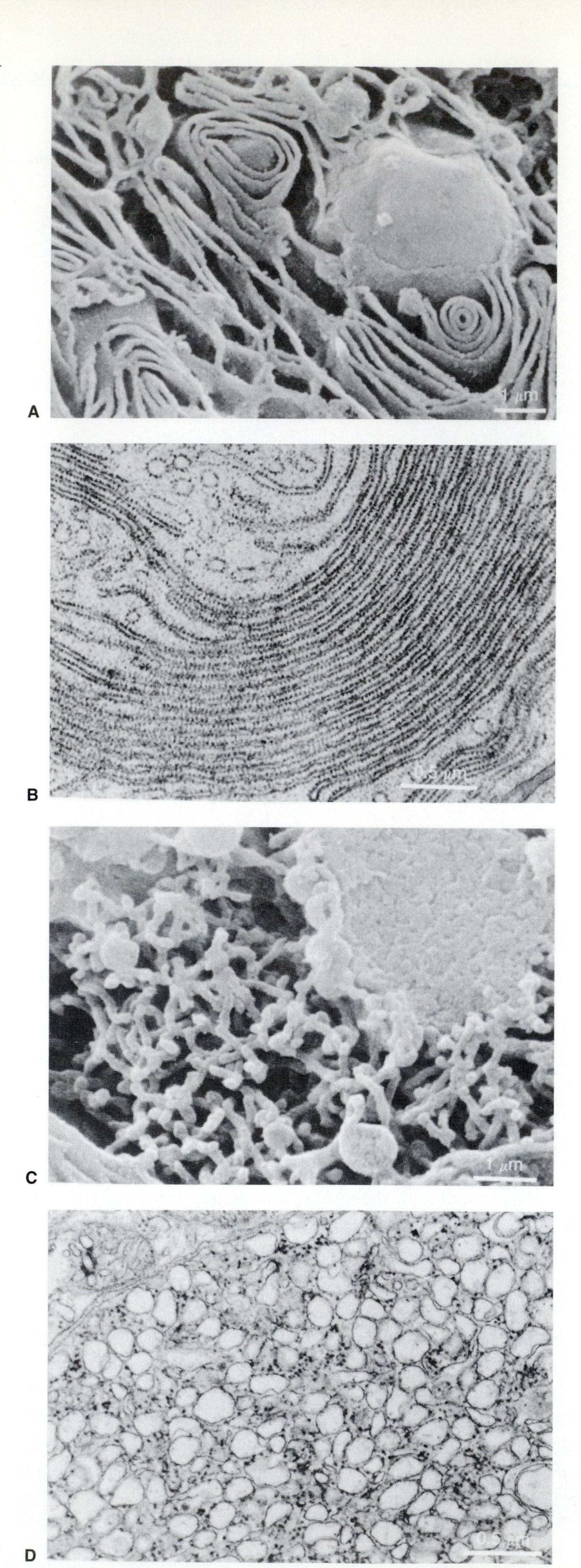

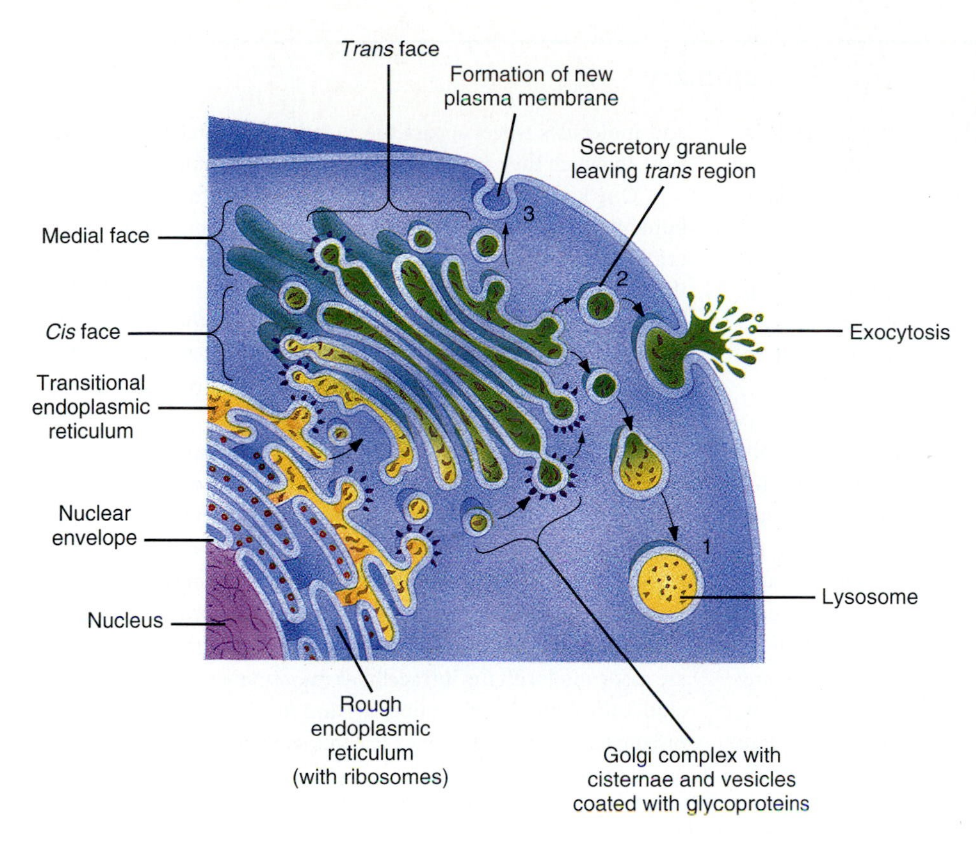

FIGURE 2.30

Functioning of the cytoplasmic membrane system. Proteins from rough endoplasmic reticulum are packaged inside vesicles or their membranes. The vesicles are then absorbed by a Golgi complex, which has three distinct groups of cisternae (compartments): *cis,* medial, and *trans.* The *cis* cisternae label proteins to be incorporated into lysosomes (1). The medial cisternae label proteins that are to be secreted (2). The *trans* cisternae repackage all the proteins into vesicles and send them to their appropriate destinations. Proteins incorporated into vesicle membranes progress through the Golgi complex and eventually become incorporated into plasma membrane (3).

Closely associated with the cytoplasmic membrane system is the **nuclear envelope,** which encloses the nucleus. The nuclear envelope consists of an inner and outer membrane penetrated by several thousand large pores (Fig. 2.31). The pores, held open by a complex of proteins, permit the flow of genetic information from the nucleus to the ribosomes.

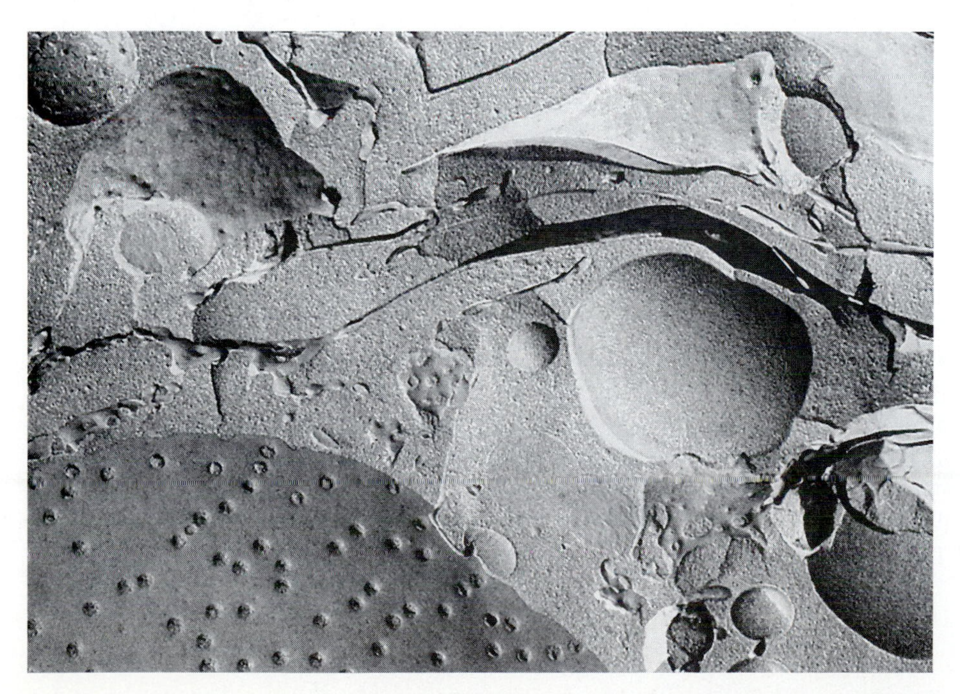

FIGURE 2.31

The nuclear envelope (lower left). Pores are clearly visible. This scanning electron micrograph was prepared by freezing and fracturing the cell, eroding part of the exposed surface, then making a platinum replica (freeze-etch technique).

Summary

Our understanding of life has progressed as far as it has because of the concepts of scientific materialism and mechanism. These two assumptions allow us to appreciate why water is essential to life. The molecular properties of water enable it to serve as a solvent for a wide variety of other essential materials, including ions, such as H^+, and many other inorganic and organic substances. Another important property of water is its large capacity for thermal energy, which means that its temperature tends to remain constant.

Diffusion and osmosis tend to make the concentrations of molecules uniform, increasing entropy in accordance with the Second Law of Thermodynamics. There are numerous mechanisms, however, that maintain lower entropy inside cells than outside. Some of these mechanisms are associated with the plasma membrane, which allows the diffusion of some molecules but not of others, depending on the properties of the molecules. Generally, a molecule can diffuse through the lipid bilayer of the plasma membrane if it can dissolve in lipid, or if it is small and uncharged. Some small ions can diffuse through the plasma membrane through special channels made of protein. Other ions and molecules travel across the membrane by facilitated diffusion, in which they attach to a transporter molecule.

One kind of organic molecule that can diffuse through the lipid bilayer of plasma membrane is lipid itself, including triglyceride. Carbohydrates, such as sugars, are not soluble in lipid and therefore cannot diffuse through plasma membrane. Since they are important sources of energy for the cell, they have to be brought in by another process, called active transport. The mechanism of active transport depends on proteins in the plasma membrane.

Proteins are also unable to diffuse across plasma membrane. Many of them carry out essential functions inside the cells where they are produced. Among these proteins are enzymes, parts of the cytoskeleton that maintains the shape of the cell and enables it to move, and structures that form desmosomes, tight junctions, and gap junctions with other cells. Some proteins are secreted from the cell by exocytosis. The mechanisms of exocytosis are associated with the intracellular membrane system, which includes the endoplasmic reticulum and the Golgi complex. These organelles also produce new plasma membrane.

Key Terms

energy level
covalent bond
hydrogen bond
solute
solvent
concentration
ionic bond
acid
diffusion
osmosis
entropy
organic molecule
carbohydrate
lipid
hydrophobic
hydrophilic
phospholipid
protein
plasma membrane
channel
active transport
exocytosis
endocytosis
cytoskeleton
endoplasmic reticulum
Golgi complex

Chapter Test

1. Describe three properties of water that are important in life. (pp. 20–24)
2. Explain how the structure of water accounts for each of the properties in question 1. (pp. 20–21)
3. Explain how each of the properties in question 1 is important to life. (pp. 20–24)
4. Explain how diffusion and osmosis illustrate the Second Law of Thermodynamics. (pp. 24–25)
5. For each of the following kinds of organic molecules, carbohydrates, lipids, and proteins, (a) describe the general structure, (b) name one example, (c) describe the general role in animals, and (d) explain how the molecular properties contribute to their performing that role. (pp. 27–32)
6. Define the primary, secondary, tertiary, and quaternary structures of proteins. How is each kind of structure determined? Explain the importance of tertiary and quaternary structure in the functioning of proteins. (pp. 30–32)
7. Sketch a portion of plasma membrane showing the main features of the fluid mosaic model. (p. 38)
8. Explain the role of phospholipids in the plasma membrane. (p. 34)
9. What kinds of molecules can diffuse through the lipid bilayer of the plasma membrane? How do ions and large polar molecules penetrate the plasma membrane? (pp. 38–42)
10. Describe the roles of rough endoplasmic reticulum, smooth endoplasmic reticulum, and the Golgi complex in exocytosis. (pp. 44–47)

■ Answers to the Figure Questions

2.1 Atoms have the same number of electrons as protons, so there are eight electrons in oxygen. Only the six electrons at the higher energy level are shown in part B.

2.2 For small organisms in which the exposed surface is large relative to body mass, hydrogen bonds could be significant compared with other forces, and they would tend to resist movement of the organism.

2.4 Water is a polar molecule, and the negatively charged oxygen tends to be attracted to Na^+ and repelled by Cl^-.

2.6 The direction of diffusion could be reversed by increasing the pressure or the thermal agitation (temperature) on the side with the lower concentration. Either method would require energy. A method that might not require energy would be to have a solvent on the side with the higher concentration that would slow the molecular movement.

2.7 One usually expects fruits and other organic matter to become wrinkled when they dry out and swollen when they absorb water. The difference here is that the fingers and other parts of the body absorb water only in the outer layer of the skin, which therefore expands until its area is greater than that of the underlying tissues.

2.8 Methanol, like all alcohols, also has a hydroxyl group; ethanol has a methyl group; glycine (like all amino acids) also has a carboxyl group.

2.10 Comparison with Fig. 2.9 shows that an H and an OH are removed to form a glycosidic link. This H and OH are removed as H_2O.

2.14 During digestion of a protein, peptide bonds are broken (hydrolyzed) by adding H_2O to reverse the process shown in the figure.

2.22 Proteins are the only kinds of molecules in the plasma membrane that could have the stable structures required for such a mechanism as the sodium–potassium pump.

2.23 The sodium–potassium pump actively transports Na^+ out of the cell, and the plasma membrane does not normally allow it to diffuse back in.

2.25 Facilitated diffusion works only in the direction of higher to lower concentration. Active transport can move ions and molecules from lower to higher concentration.

Readings

Recommended Readings

Allen, R. D. 1987. The microtubule as an intracellular engine. *Sci. Am.* 256(2):42–49 (Feb).

Atkins, P. W. 1987. *Molecules.* New York: Scientific American Library.

Boatman, E. S., et al. 1987. Today's microscopy. *BioScience* 37:384–394.

Bretscher, M. S. 1985. The molecules of the cell membrane. *Sci. Am.* 253(4):100–108 (Oct).

Bretscher, M. S. 1987. How animal cells move. *Sci. Am.* 257(6):72–90 (Dec).

Dautry-Varsat, A., and H. F. Lodish. 1984. How receptors bring proteins and particles into cells. *Sci. Am.* 250(5):52–58 (May).

De Duve, C. 1984. *A Guided Tour of the Living Cell,* 2 vols. New York: Scientific American Library.

Doolittle, R. F. 1985. Proteins. *Sci. Am.* 253(4):88–99 (Oct).

Ford, B. J. 1991. *The Leeuwenhoek Legacy.* Bristol and London: Biopress Ltd. and Farrand Press.

Goodsell, D. S. 1992. A look inside the living cell. *Am. Sci.* 80:457–465.

Goodsell, D. S. 1993. *The Machinery of Life.* New York: Springer-Verlag.

Karplus, M., and J. A. McCammon. 1986. The dynamics of proteins. *Sci. Am.* 254(4):42–51 (Apr).

Kessel, R. G., and R. H. Kardon. 1979. *Tissues and Organs: A Text-Atlas of Scanning Electron Microscopy.* San Francisco: W. H. Freeman.

Lienhard, G. E., et al. 1992. How cells absorb glucose. *Sci. Am.* 266(1):86–91 (Jan).

Olson, A. J., and D. S. Goodsell. 1992. Visualizing biological molecules. *Sci. Am.* 267(5):76–81 (Nov).

Orci, L., J.-D. Vassalli, and A. Perrelet. 1988. The insulin factory. *Sci. Am.* 259(3):85–94 (Sept). (*Production and secretion of an important protein.*)

Rothman, J. E. 1985. The compartmental organization of the Golgi apparatus. *Sci. Am.* 253(3):74–89 (Sept).

Sharon, N. 1980. Carbohydrates. *Sci. Am.* 243(5):90–116 (Nov).

Sharon, N., and H. Lis. 1993. Carbohydrates in cell recognition. *Sci. Am.* 268(1):82–89 (Jan).

Staehelin, L. A., and B. E. Hull. 1978. Junctions between living cells. *Sci. Am.* 238(5):140–152 (May).

Taylor, D. L., et al. 1992. The new vision of light microscopy. *Am. Sci.* 80:322–335.

Unwin, N., and R. Henderson. 1984. The structure of proteins in biological membranes. *Sci. Am.* 250(2):78–94 (Feb).

Weber, K., and M. Osborn. 1985. The molecules of the cell matrix. *Sci. Am.* 253(4):110–120 (Oct).

Weinberg, R. A. 1985. The molecules of life. *Sci. Am.* 253(4):48–57 (Oct). (*Introduction to an issue devoted to this topic.*)

CHAPTER

3

Energetics of Cells

Hump-backed whale (*Megaptera novaeangliae*).

Energy

Like machines, the natural mechanisms of life require energy. Many plants, algae, and bacteria obtain their energy from sunlight by the process of photosynthesis, and they store this energy in organic molecules, such as starch. Most other organisms, including all animals, are incapable of obtaining energy from inorganic sources and must get it by consuming the organic molecules of other organisms, usually plants and animals. Through the process of digestion they break these organic molecules into simple nutrients, such as simple sugars, fatty acids, and amino acids, which the cells then absorb. Inside the cells these nutrients are **metabolized.** Metabolism involves the synthesis of new organic molecules from nutrients (**anabolism**) or the breakdown of nutrient molecules to extract energy (**catabolism**).

The energy in organic molecules is released by breaking their chemical bonds. This is also the way energy is released during the burning of coal or gasoline, but obviously cells don't work like stoves or engines. Catabolism has to be more controlled, and some of the energy has to be in a form more usable than heat. During catabolism, therefore, organic molecules are degraded gradually, bond by bond. Energy from some of these bonds then transfers to other organic molecules, which can, in turn, transfer the energy to a variety of mechanism in the cell. The energy freed during catabolism, which is then free to participate in cellular mechanisms, is called Gibbs free energy, or simply **free energy.** Free energy is formally defined as the potentially available energy content of a system, minus the energy that is unavailable due to the increase in entropy. Entropy always increases in a spontaneous reaction, so free energy always decreases in a natural reaction. Therefore free energy transfers from one molecule to another only "downhill," from a high energy level to a lower one.

ATP

Generally, the molecule that ultimately accepts free energy from a nutrient during catabolism and transfers it to mechanisms that use the free energy is **adenosine triphosphate,** better known as ATP. ATP is often called the "energy currency" of cells. Just as money facilitates the selling of products or services in order to purchase other products or services, ATP allows the cell to exchange the free energy from nutrients to obtain work from its mechanisms. ATP is therefore an energy coupler, not an energy reserve. In fact, a day's supply of ATP weighs about twice as much as you do. It is therefore not produced unless it is needed, and it is spent as soon as it is available.

Adenosine triphosphate consists of three phosphate groups attached in series to the five-carbon sugar **ribose,** which is attached to a structure called **adenine** (Fig. 3.1A on page 52). ATP acquires energy when a molecule of adenosine diphosphate (ADP), which has two phosphate groups, picks up a third phosphate (Fig. 3.1B on page 52). When ATP gives up its free energy to perform some function, this last phosphate breaks off, forming ADP once again. Since the terminal phosphate bond absorbs H_2O as it breaks off from ATP, the reaction is an example of hydrolysis. Figure 2.22 illustrated one way that the hydrolysis of ATP is coupled to a cellular mechanism.

The bonds between phosphate groups in ATP are often called "high-energy bonds," but there is actually only a moderate amount of free energy associated with them. Since free energy moves spontaneously only from a higher level to a lower one, a chemical bond that is broken to make ATP must release more free energy than can be released during the hydrolysis of the terminal phosphate bond of the ATP. Likewise the terminal phosphate bond in ATP must release more free energy during hydrolysis than is used by whatever mechanism uses the ATP. The amount of free energy released during the hydrolysis of the terminal phosphate bonds of a mole (505 grams) of ATP is

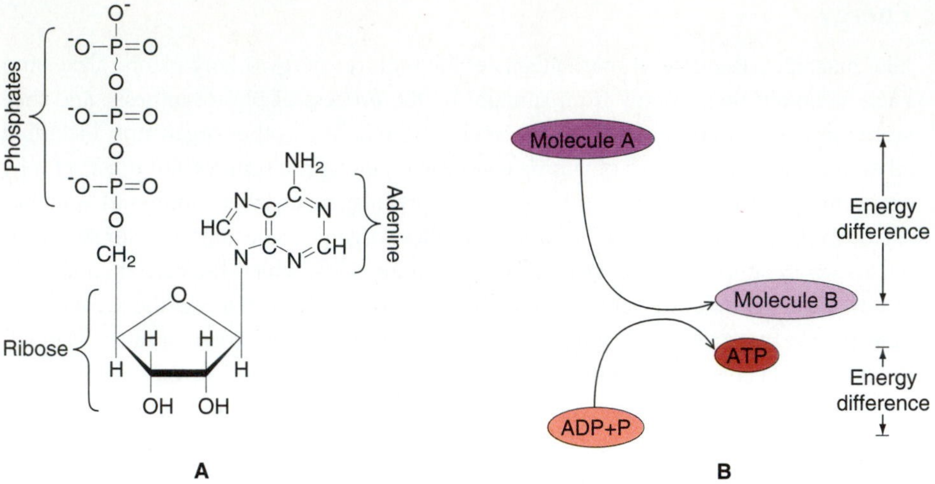

FIGURE 3.1

(A) A molecule of adenosine triphosphate (ATP). Adenosine diphosphate is identical except that it has one less phosphate. The crowding of negative charge in each phosphate gives the phosphate bond a moderate amount of energy. (B) ATP is synthesized when some of the energy released from a chemical reaction (molecule A ⟶ molecule B) is used to attach a phosphate group to ADP. This is referred to as a coupled reaction. The phosphate is either inorganic phosphate or a phosphate group on the original molecule A. When ATP drives a cellular process, essentially the reverse kind of reaction occurs.

generally calculated to be 7.3 kilocalories. (The exact amount depends on the conditions under which ATP is used.) This much energy is enough for about a minute of running, and it is about one-tenth the free energy provided by a slice of white bread.

Activation Energy

One of the major nutrients used by cells to make ATP is the simple sugar glucose (see Fig. 2.9). By a long series of reactions that will be described in the next section, glucose breaks down bond by bond, releasing some of its free energy as heat and transferring some of the free energy to make ATP. Each day more than a kilogram of glucose breaks down in this way inside you. Yet you can walk into a chemistry stockroom and find jars of glucose sitting around not releasing energy as either heat or ATP. Although it contains a lot of free energy, glucose is stable. Otherwise it would break down in plants before animals consumed it, and it would break down in your bloodstream before your cells got it. Other nutrients are also stable. For that matter, the same is true for TNT. Even though these molecules have a lot of free energy to release, they do not do so until activated.

Like TNT, glucose and other nutrients have to be activated by the addition of some form of energy. This added energy, called the **activation energy,** makes the molecule unstable, thereby allowing it to react chemically. Activation energy can be visualized as a shove that starts a molecule rolling (Fig. 3.2A). In cells the only source of activation energy is usually thermal agitation, which, by itself, is seldom enough to make metabolic reactions proceed at an appreciable rate. One way to speed up chemical reactions in cells is to heat them. To some extent many animals, including ourselves, do this by maintaining our body temperatures above the environmental temperature. Obviously, however, there are limits to this approach, not the least of which is that heating the cells uses up most of the additional ATP produced.

Enzymes

A far more practical mechanism for increasing a reaction rate is to reduce the required activation energy by means of an **enzyme** (Fig. 3.2B). An enzyme is an organic catalyst—a molecule that speeds up a chemical reaction, commonly by a factor of a million to a trillion. Enzymes are not changed by the reactions they catalyze, and they can therefore be used over and over again. With few exceptions, enzymes are proteins. The tertiary structure of the enzyme forms an **active site** in which the reacting molecule—the **substrate**—

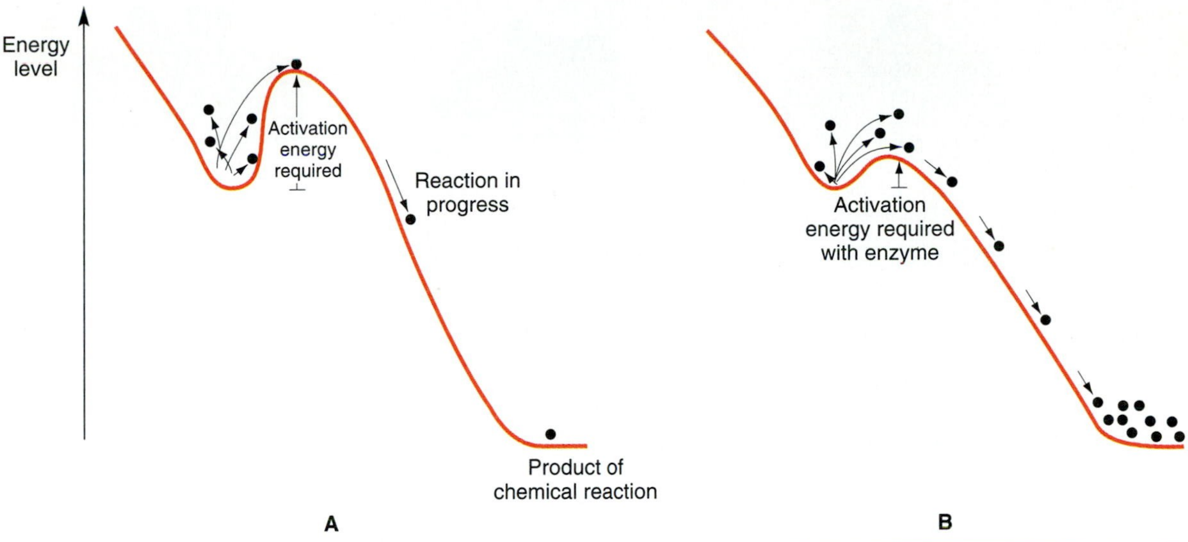

FIGURE 3.2

Activation energy and the role of enzymes. (A) Normally the chemical breakdown of a nutrient molecule requires more activation energy than can be provided by thermal agitation. Therefore even though the chemical reaction is energetically favorable ("downhill"), it seldom occurs. (B) Enzymes lower the required activation energy, accelerating the reaction.

■ **Question:** How did the nutrient molecules get up the hill in the first place?

fits (Fig. 3.3 on page 54). The active site of an enzyme is generally highly specific for a particular substrate. An enzyme increases a reaction rate because the active site briefly holds the substrate in a position that favors the reaction. If the substrate is to be bonded to another molecule, the enzyme holds the substrate and the molecule close to each other while they react. If the reaction involves the breaking of a bond in the substrate, the enzyme applies a strain to the bond.

Many enzymes require **cofactors** in order to function. Some cofactors are **coenzymes,** which are organic molecules that transfer a functional group to or from the substrate. ATP is a coenzyme that transfers a phosphate group, for example. Other cofactors are metal ions, such as magnesium, calcium, or zinc. In addition, many molecules have **allosteric effects** on enzymes. By binding to a particular site on an enzyme they either increase or decrease the ability of the active site to bind to the substrate, thereby increasing or decreasing the rate of the reaction.

Although enzymes work wonders in speeding up certain reactions, they cannot work miracles. They cannot force a reaction to proceed if the reaction is not spontaneous according to the Second Law of Thermodynamics. This means that a metabolic process won't proceed, even with an enzyme, unless it increases the total entropy. Nor can an enzyme make a metabolic process run "uphill," increasing the total free energy. Therefore, the only way to synthesize a molecule such as ATP is to couple the synthesis to a reaction such as the breakdown of glucose, which produces more entropy than ATP loses, and which loses more free energy than the ATP gains.

Glycolysis

The first stage in the breakdown of glucose to form ATP occurs in the cytoplasm. This stage, called glycolysis, involves the attachment of phosphate to sugar molecules, rearrangement of the sugar molecules so that energy in the sugars is concentrated in the phosphate bonds, then transferring the phosphate bonds to ADP to form ATP. Although this sounds simple, it takes more than nine separate reactions (Fig. 3.4 on page 55). These reactions appear mystifying at first glance, but there is actually a good deal of logic in the sequence. The first four reactions accomplish three things.

1. They rearrange glucose into a fructose structure, which is more symmetric. The advantage of this symmetry is that the six-carbon sugar can be split into two

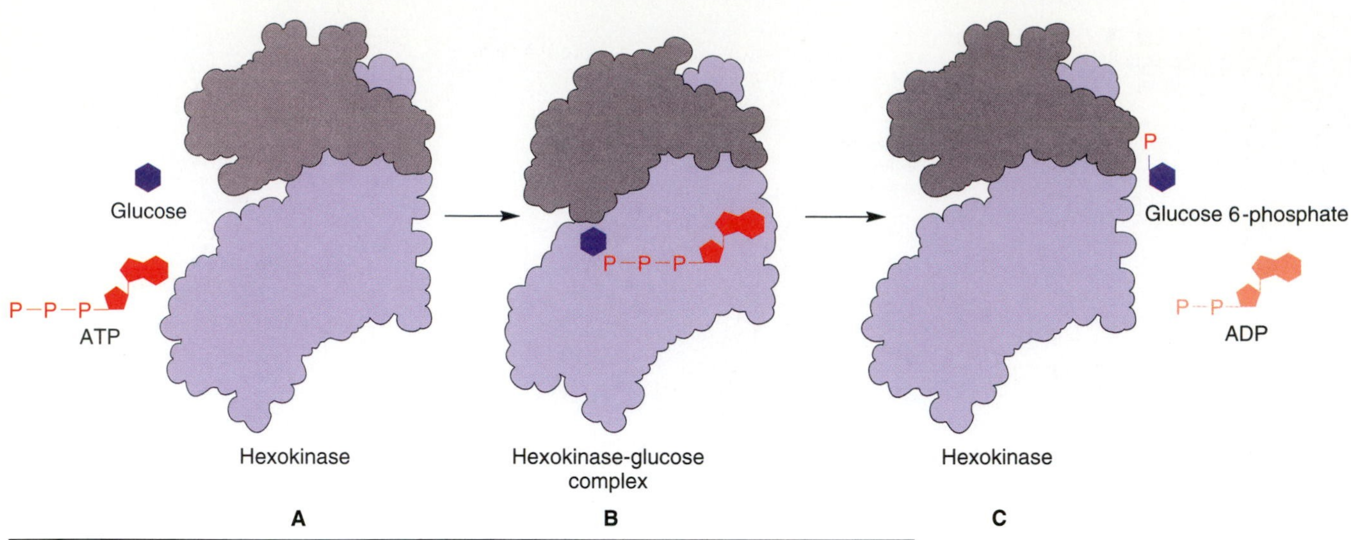

FIGURE 3.3

A model of the action of the enzyme hexokinase, which catalyzes the first reaction in the metabolism of glucose. (A) Glucose and ATP move to the active site. (B) The enzyme changes shape (a process called induced fit). The enzyme and its bound substrate are now called an enzyme–substrate complex. In this configuration the enzyme holds the glucose and an ATP molecule in such a way that a phosphate is transferred to the glucose. (C) After several microseconds the products of the reaction leave the enzyme, which is unaltered and ready to repeat the process.

■ **Question:** Suppose there is an unlimited amount of ATP, 1 molecule of glucose, and 10 molecules of enzyme. How would the reaction rate be affected if you add a second molecule of glucose? Ten molecules of glucose? A hundred? A thousand? Can you show the relationship of reaction rate to substrate concentration on a graph?

identical three-carbon sugars (trioses), which can then be metabolized by the same set of enzymes, rather than have two separate pathways for each triose.

2. Glucose and fructose receive activation energy from two ATPs. This activation energy makes the fructose unstable, so it breaks down into the two trioses.
3. Each of the two trioses ends up with a phosphate group from one of the ATPs.

It might appear strange that the first steps in producing ATP actually use ATP. It is really no stranger, however, than investing money into a business in the expectation of making a profit later. In the remaining five reactions of glycolysis, in fact, the invested two ATPs are recovered, and a profit of two more ATPs is realized. The formation of two additional ATPs occurs because each of the two trioses acquires a second phosphate group (step 5 in Fig. 3.4). In order for the triose and an inorganic phosphate to bind to each other in this reaction, they have to lose a hydrogen atom and an electron (e^-). H and e^- combine to form **hydride** ($H{:}^-$). This hydride is then transferred to a coenzyme that serves as an **electron carrier.**

The electron carrier in glycolysis is **nicotinamide adenine dinucleotide** (NAD+). NAD+ consists of nicotinamide (which is derived from the vitamin niacin), adenine, and two riboses joined by phosphates. NAD+ accepts the hydride and becomes NADH. In this process NAD+ is said to be **reduced** to NADH, while the triose is **oxidized.** NAD+ acquires not only a hydride during reduction, but also a great deal of energy. As we shall see later, that energy can be used to form more ATP.

*Any reaction in which a molecule gains an electron is called a **reduction.** The molecule that gains the electron is said to be **reduced.** Every gain of an electron by one molecule requires the loss of an electron by another. The loss of an electron is called **oxidation,** and the molecule that loses the electron is said to be **oxidized.** Thus reduction and oxidation reactions always occur together and are called **oxidation–reduction reactions,** or **redox reactions.** The term oxidation comes from the fact that oxygen takes electrons from substances when it oxidizes them. However, oxygen is not required for all oxidation reactions, including those in glycolysis.*

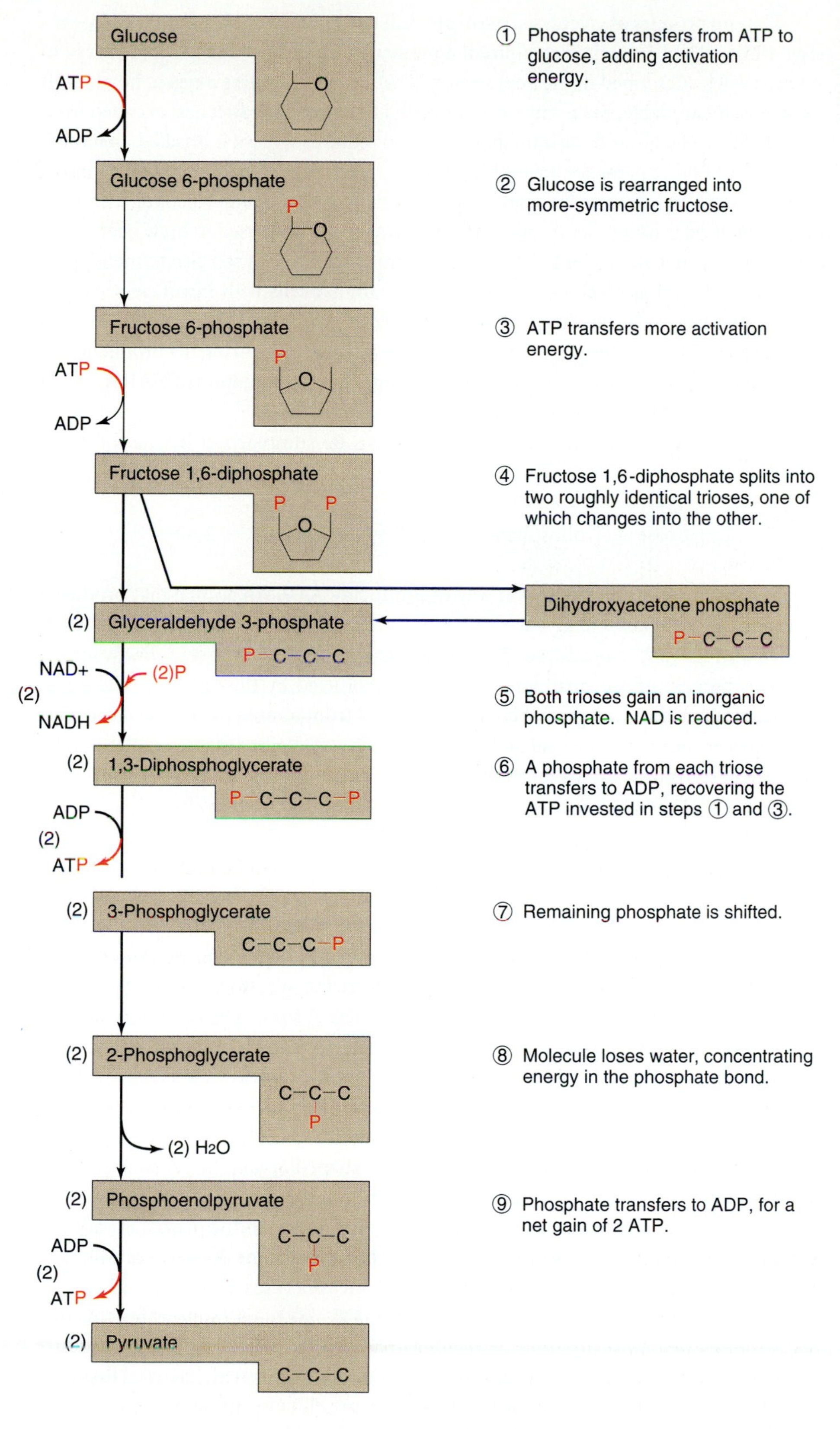

① Phosphate transfers from ATP to glucose, adding activation energy.

② Glucose is rearranged into more-symmetric fructose.

③ ATP transfers more activation energy.

④ Fructose 1,6-diphosphate splits into two roughly identical trioses, one of which changes into the other.

⑤ Both trioses gain an inorganic phosphate. NAD is reduced.

⑥ A phosphate from each triose transfers to ADP, recovering the ATP invested in steps ① and ③.

⑦ Remaining phosphate is shifted.

⑧ Molecule loses water, concentrating energy in the phosphate bond.

⑨ Phosphate transfers to ADP, for a net gain of 2 ATP.

FIGURE 3.4

An outline of glycolysis, emphasizing the transfer of energy with phosphate and reduction of NAD+ (shown in color).

■ **Question:** How would ATP production be affected if dihydroxyacetone phosphate were removed, as it is during synthesis of fat?

The end product of glycolysis, **pyruvate,** still has a considerable amount of free energy. Most cells in animals metabolize the pyruvate further and transfer that energy to ATP, as will be described in the next section. That process requires oxygen, however. If oxygen is not available, the pyruvate is converted to a waste product and excreted from the cell. Since glycolysis by this method does not require oxygen, it is called **anaerobic glycolysis.** Another name for anaerobic glycolysis is **fermentation.** One familiar kind of fermentation involves the conversion of pyruvate into CO_2 and ethyl alcohol. This kind of fermentation is usually associated with yeasts, such as those used to brew beer. If deprived of oxygen, goldfish and certain small worms (nematodes) can also ferment pyruvate into CO_2 and alcohol. Generally, however, animal cells with insufficient oxygen convert pyruvate to **lactate.** Lactate forms in skeletal muscles during exercise, and it is largely responsible for the fatigue and soreness that can result. During the production of lactate, pyruvate is reduced. This reduction is coupled to the oxidation of NADH, which means that the NAD+ is completely recycled.

Anaerobic glycolysis in most animal cells can be summarized by the following equation:

$$\text{glucose} + 2\ \text{phosphate} + 2\ \text{ADP} \longrightarrow 2\ \text{lactate} + 2\ \text{ATP}$$

It is hard to imagine a more elegant way of producing ATP. An animal that produced all its energy by anaerobic glycolysis would not even have to breathe, since no O_2 is needed and no CO_2 is produced. Such an animal would be at a severe disadvantage, however, because of the small amount of ATP produced by this method. The maximum amount of free energy that could be released from a mole of glucose is 686 kcal (kilocalories), but only 14.6 kcal of that energy finds its way into the two moles of ATP produced by glycolysis. This means that the efficiency of anaerobic glycolysis is just a little over 2% (14.6/686). Even your bank offers a better return than 2%.

Cellular Respiration

If oxygen is present in the cytoplasm, the pyruvate produced in glycolysis is not metabolized into lactate or other waste products. Instead, pyruvate undergoes further breakdown in which its free energy is eventually transferred to ATP as C–H bonds are broken and C is eliminated as CO_2. This complete catabolism of glucose under aerobic conditions is called **respiration,** or **cellular respiration** to avoid confusion with breathing. The use of the term respiration for both the aerobic breakdown of glucose and for breathing is not just an unfortunate coincidence. Cellular respiration requires O_2 and produces CO_2—the two main reasons for breathing.

The catabolism of pyruvate occurs in sausage-shaped organelles called mitochondria (singular, **mitochondrion**) (Fig. 3.5). Mitochondria are bounded by two distinct membranes. The **outer membrane** is similar in composition to the plasma membrane and cytoplasmic membranes of the cell. The **inner membrane** is biochemically and functionally more like the plasma membranes of bacteria. This is one line of evidence that mitochondria originated as bacteria (see pp. 385–386). Also supporting this theory is the fact that mitochondria still retain a great deal of autonomy. They reproduce independently of the rest of the cell, and they have their own genetic material that controls the synthesis of some of the enzymes needed for cellular respiration.

Cellular respiration will be described in detail later, but for now it is important merely to see how it occurs in relation to mitochondrial structure (Fig. 3.6 on page 58). Cellular respiration involves two sets of processes: the **Krebs citric acid cycle,** which

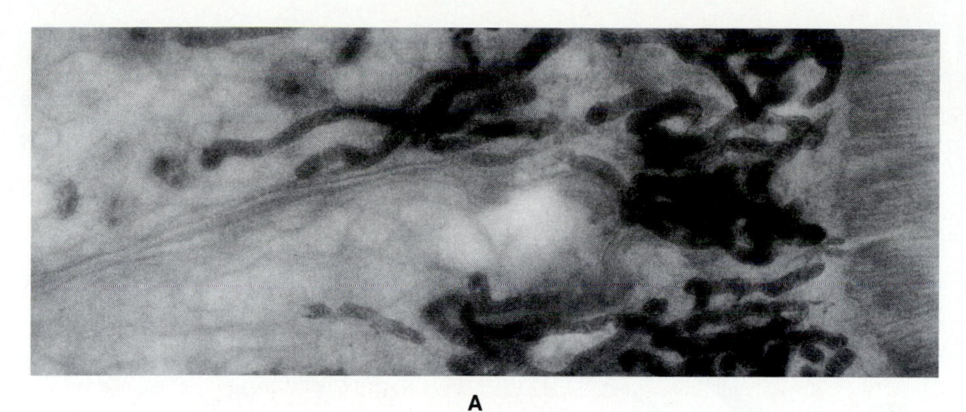

A

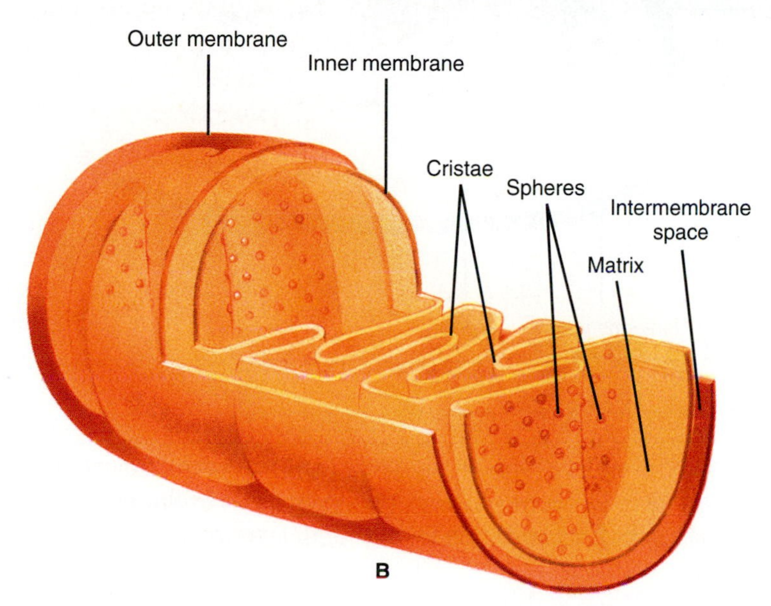

B

FIGURE 3.5

(A) Electron micrograph of mitochondria in skin cells of a snail. Each cell may have from dozens to thousands of mitochondria, each typically 0.2 μm wide and several times as long. (B) Schematic diagram of a mitochondrion.

■ **Question:** Although mitochondria are generally several time longer than they are wide, in electron micrographs they usually appear to be short. Why?

occurs in the central space, or **matrix** of the mitochondrion, and the **respiratory chain,** which occurs on the inner membrane, especially on infolding pockets called **cristae.** The Krebs citric acid cycle, also called the **tricarboxylic acid (TCA) cycle,** catabolizes pyruvate. Most of the free energy released by this catabolism is transferred with hydride to NAD+, which is reduced to NADH. Before it can transfer this energy to ATP, the NADH must enter the respiratory chain. Some of the enzymes that catalyze the production of ATP in the respiratory chain are visible under the electron microscope as **spheres** that dot the cristae.

In the respiratory chain NADH is oxidized back to NAD+ when it loses its hydride. The hydride then splits into one proton (H^+) and two electrons. The energy associated with the electrons drives proton pumps. The proton pumps store the energy temporarily by transporting the H^+ from the hydride and other sources into the **intermembrane space** between the inner and outer membranes. The stored energy is released when H^+ diffuses back into the matrix. During this diffusion the energy drives a chemical reaction in which inorganic phosphate binds to ADP, forming ATP. Having expended their energy, the electrons attach to O_2, forming oxygen atoms that each have two electrons. Each oxygen atom then binds to two H^+ in the matrix, forming water as a waste product. This is the only function of oxygen in animal life: to serve as an **electron acceptor.** In fact, if O_2 were not used in this way it would build up to toxic levels that would oxidize organic molecules.

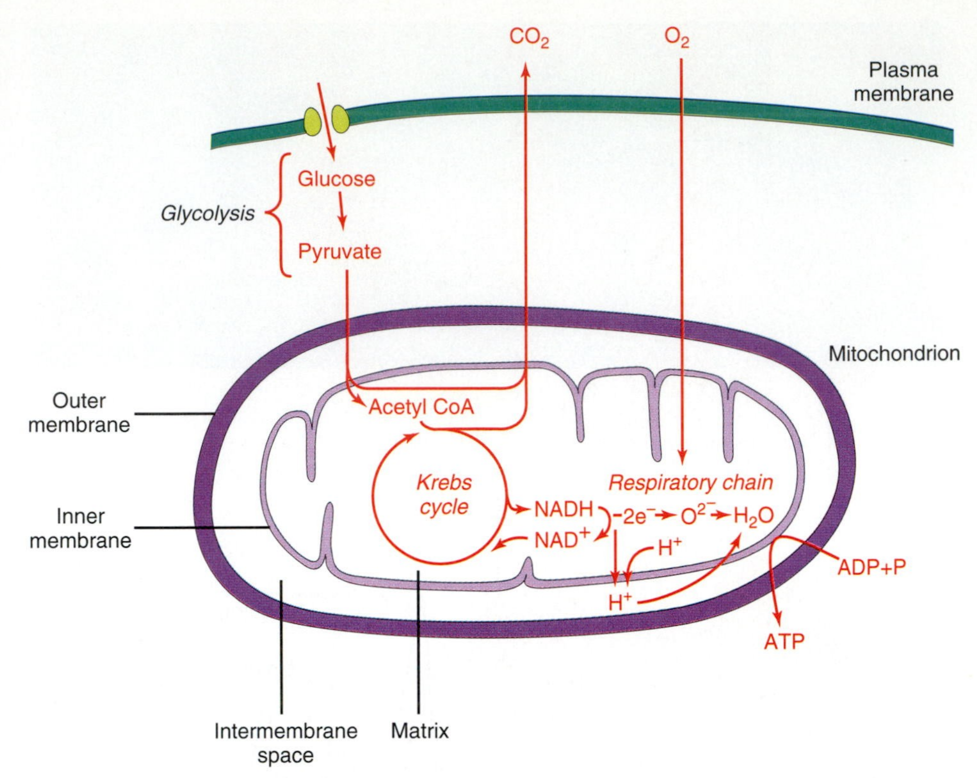

FIGURE 3.6

Outline of how most of the ATP is produced in cellular respiration. Pyruvate produced by glycolysis in the cytoplasm enters the mitochondrion, where it is converted to acetyl-CoA and CO_2. In the Krebs cycle the remaining carbons are lost as CO_2, and NADH is produced. Hydride (H:⁻) from NADH enters the respiratory chain, where proton pumps use energy from the transport of e^- to pump H^+ into the intermembrane space. Energy from the diffusion of H^+ back into the matrix generates ATP. The electrons are eventually accepted by oxygen atoms, which then combine with H^+ in the matrix to form H_2O.

KREBS CYCLE. Now let's consider the Krebs citric acid cycle in more detail. Every molecule of glucose produces two molecules of pyruvate. Each pyruvate starts one turn of the Krebs cycle, which accomplishes four major things:

1. The three carbons in each pyruvate molecule are eliminated as CO_2. One carbon is eliminated as CO_2 immediately (step 1, Fig. 3.7). The two carbons left from pyruvate bond to **coenzyme A** (a derivative of the vitamin pantothenic acid), forming **acetyl-CoA.** The function of coenzyme A is to reserve a bond for the attachment of a four-carbon molecule, oxaloacetate. This bonding forms the six-carbon molecule citrate. Citrate then undergoes a series of reactions in which two carbons are eliminated as CO_2 (steps 4 and 5), once more forming oxaloacetate. The cycle is completed when oxaloacetate attaches to another acetyl-CoA.
2. Free energy in the intermediates in the Krebs cycle is concentrated around hydrogen, which is then transferred as hydride to NAD+. In this way four molecules of NADH are produced for each molecule of pyruvate (steps 1, 4, 5, and 8 in Fig. 3.7). Each of these NADH molecules has enough free energy associated with it to produce approximately three ATPs in the respiratory chain, thereby accounting for most of the ATP produced under aerobic conditions.
3. A molecule of **flavine adenine dinucleotide** (FAD) is reduced to $FADH_2$ (step 6). FAD is similar to NAD+ except that it has a flavine group (derived from the vitamin riboflavin) instead of a nicotinamide. Like NAD+, FAD is a coenzyme that gains free energy when it is reduced. This energy is sufficient to produce approximately two ATPs in the respiratory chain.
4. One molecule of ATP is produced (step 5). [Actually **guanosine triphosphate** (GTP), a molecule similar to ATP, is produced. GTP then transfers its terminal phosphate group to ADP.]

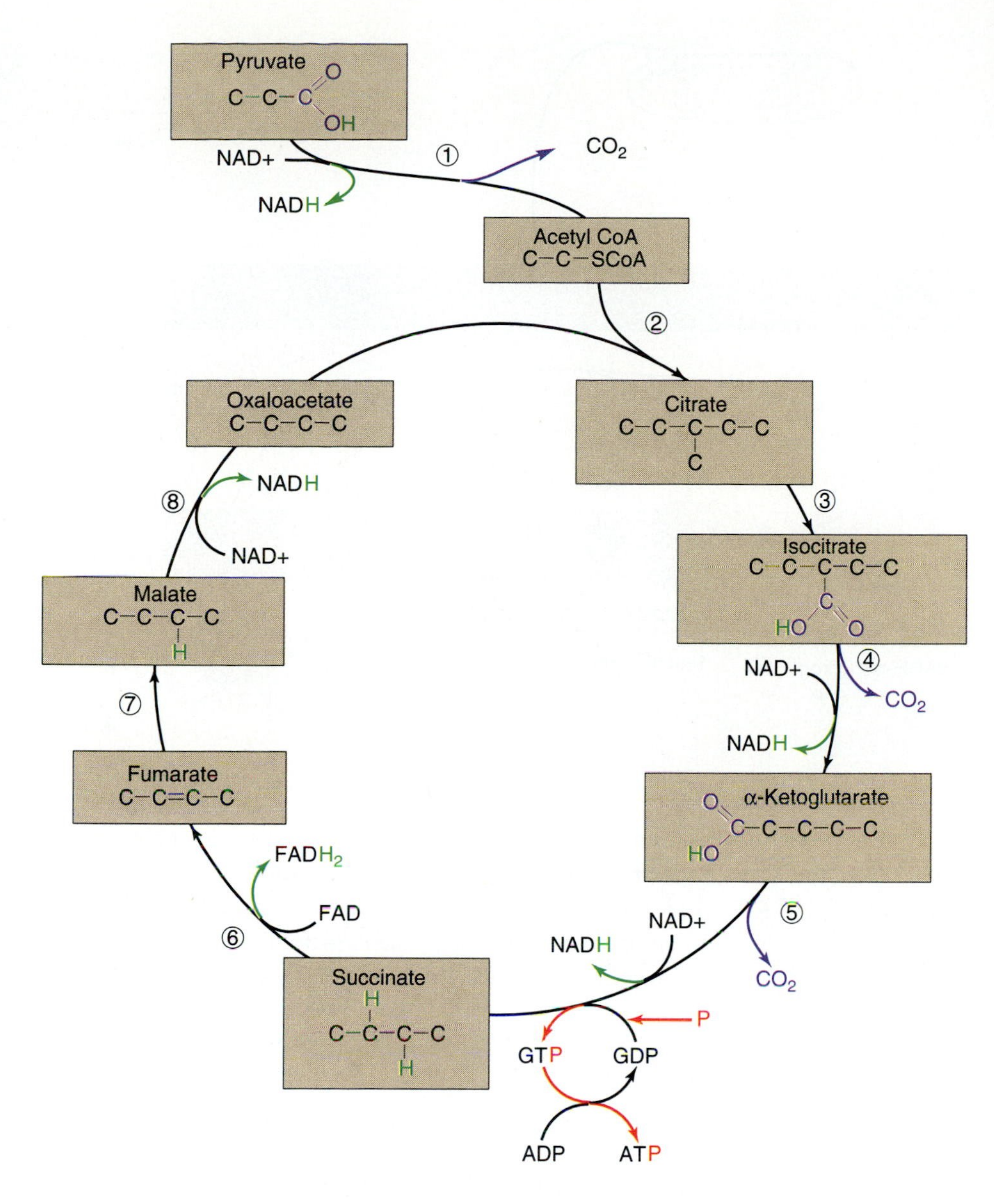

FIGURE 3.7

Outline of the Krebs cycle. The movement of carbons to CO_2, H to NADH, and phosphate to ATP are shown in color.

The Respiratory Chain. Although some ATP is produced in the Krebs cycle, most of the free energy in pyruvate ends up in NADH, with a little in $FADH_2$. This energy is transferred to ATP in the **respiratory chain.** The precise mechanism by which the respiratory chain produces ATP is still being worked out, but the current evidence supports the **chemiosmotic theory** first proposed by Peter Mitchell in 1961. The basic idea of Mitchell's chemiosmotic theory is that the respiratory chain drives **proton pumps** that transport hydrogen ions into the intermembrane space. The energy to pump the protons comes from electrons moving from higher to lower energy levels along an **electron transport chain.** This electron transport chain, along with the proton pumps, is associated with three **enzyme complexes** bound to the inner membrane (Fig. 3.8 on page 60). After being pumped into the intermembrane space, the H^+ then diffuses back into the matrix through another complex that is essentially a proton pump that runs in reverse. Instead of using ATP to transport H^+, this proton pump uses energy from the diffusion of H^+ to produce ATP.

The Efficiency of Cellular Respiration. The amount of ATP produced by cellular respiration depends on how much H^+ is pumped into the intermembrane

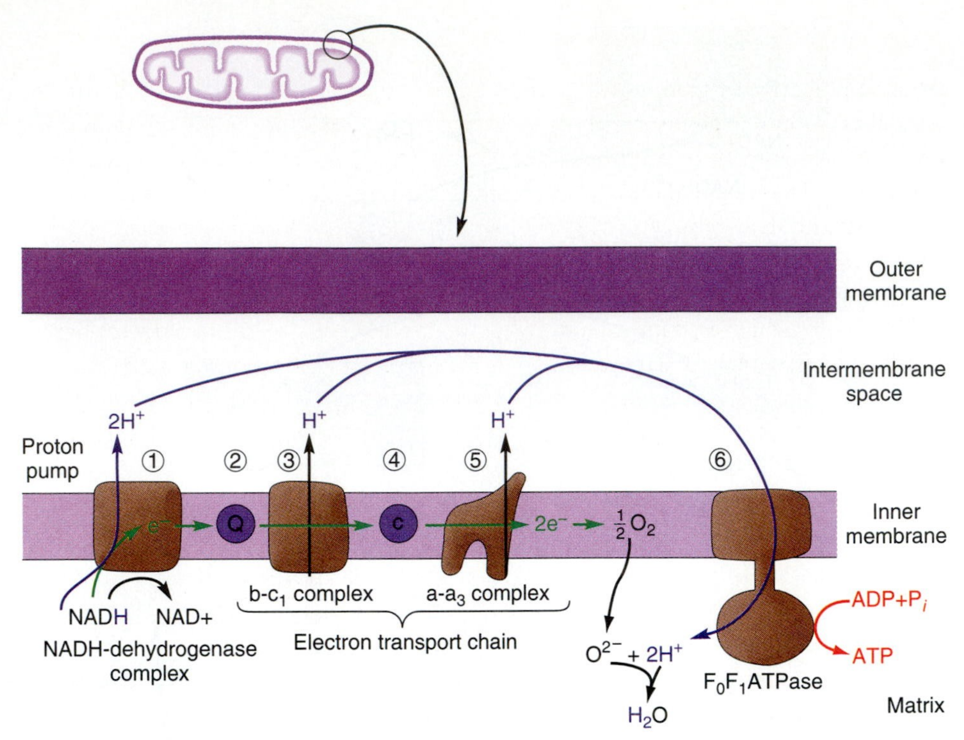

FIGURE 3.8

The respiratory chain. (1) The NADH-dehydrogenase complex is named for an enzyme that catalyzes the oxidation of NADH to NAD+. Hydride from NADH dissociates into a proton (H^+) and two electrons (e-). The proton, together with one from the matrix, is pumped into the intermembrane space. (Alternatively, the two hydrogen atoms can come from $FADH_2$.) (2) The two electrons then transfer to coenzyme Q (Q). Coenzyme Q transfers the electrons to the b–c_1 complex. (3) The b–c_1 complex includes two cytochromes, b and c_1. (Cytochromes are proteins with an iron-bearing heme group that is readily reduced and oxidized. Their name comes from the fact that they give cells a slight color.) Energy transferred from the two electrons drives the b–c_1 complex to pump an undetermined number of H^+ into the intermembrane space. (4) The two electrons then transfer to cytochrome c, which transfers them to the a–a_3 complex. (5) The a–a_3 complex, also known as cytochrome oxidase, includes cytochromes a and a_3. As the two e- pass through, the complex pumps more H+ into the intermembrane space. The two e- then transfer to an oxygen atom ($\frac{1}{2}O_2$). (6) The H^+ pumped into the intermembrane space diffuses back into the matrix through the F_0F_1 ATPase, which is in a sphere on the cristae. This complex is essentially a reverse proton pump coupling the energy from H^+ diffusion to the attachment of a phosphate group to ADP to make ATP. Each pair of protons combines with the oxygen atom, forming water ($2H^+ + \frac{1}{2}O_2 \longrightarrow H_2O$).

space, which varies with conditions in the cell. On average, however, one molecule of NADH from the Krebs cycle provides enough free energy to produce approximately three molecules of ATP. Under aerobic conditions there is also a net production of two molecules of NADH in glycolysis (step 5, Fig. 3.4), which are used to make two molecules of $FADH_2$. These molecules of $FADH_2$, and the two from the Krebs cycle, transfer their H a little farther down the respiratory chain and therefore allow for the production of only about two ATPs each. Table 3.1 is an accounting of all the ATP produced. The "bottom line" is that the aerobic catabolism of one mole of glucose produces approximately 36 moles of ATP. (The exact amount may be as low as 25 moles or as high as 38 moles.)

The equation describing the complete aerobic breakdown of glucose is

$$C_6H_{12}O_6 + 6\,O_2 + 36\,\text{ADP} + 36\,\text{phosphate} \longrightarrow 6\,CO_2 + 6\,H_2O + 36\,\text{ATP}$$

Comparison with the equation for glycolysis (see p. 56) shows that cellular respiration produces much more ATP but requires oxygen and produces carbon dioxide. Is the additional ATP from cellular respiration worth the trouble of having to breathe (not to mention the trouble of having to learn about it)? Assuming that each mole of ATP provides 7.3 kcal of free energy, the total yield is approximately 263 kcal. Since a mole of glucose can release 686 kcal of free energy, the efficiency of cellular respiration is approximately 263/686 = 38%. The remaining energy—62% of the energy in glucose—is lost as heat. This may seem wasteful, but in fact it is not bad compared with the most efficient artificial engine, the gas turbine, which has 40% efficiency. It is conservative compared with a well-tuned gasoline engine, which delivers only about 15% to 20% of the free energy from gasoline to the wheels. Cellular respiration is efficient, indeed, compared with the 2% efficiency of anaerobic glycolysis. If you are still not convinced that cellular respiration is worth the trouble, just hold your breath until you change your mind.

Alternative Metabolic Pathways

Man does not live by glucose alone, nor does any other animal. Other carbohydrates, as well as fats and proteins, make up a significant portion of the diet. In addition, animals convert glucose into other carbohydrates and into fats and proteins. Many of these molecules are metabolized in alternative pathways that intersect the pathways of glycolysis and the Krebs cycle (Fig. 3.9). For example, liver and muscle cells produce glycogen, a starchlike substance (see Fig. 2.10B), whenever glucose is abundant. When glucose levels in the blood start to fall, the glycogen breaks down into glucose 6-phosphate, which is then catabolized in the cell or used to make glucose that is pumped into the blood. Glycogen can thus provide cells with a between-meal snack of glucose. Most adult humans store enough glycogen to provide about a day's worth of energy. Some other carbohydrates, especially ribose and other five-carbon sugars, do not fit so easily into the glycolytic pathway. A complicated series of reactions called the **pentose shunt** produces these pentoses from combinations of three- and six-carbon sugars.

Triglycerides (neutral fats) stored in adipose tissue or ingested with the diet are catabolized by converting them into glycerol and fatty acids. The glycerol is easily changed into dihydroxyacetone phosphate for use in glycolysis, and the fatty acids are (less easily) converted into acetyl-CoA for use in the Krebs cycle. If excess nutrients are available, triglycerides are synthesized by essentially reversing these processes. In other words, if you eat too much you get fat. Adult humans with normal body mass store enough fat to provide energy for a month.

The metabolism of proteins is complicated by the fact that 20 different amino acids are involved. In humans eight of these amino acids, called essential amino acids, cannot be either catabolized or synthesized. Some of the remaining 12 amino acids, however, are similar to intermediates in the Krebs cycle once they have lost their amino

TABLE 3.1

A balance sheet for the production of ATP by aerobic catabolism of one mole of glucose.

The number of ATPs produced by one NADH from Krebs cycle is assumed to be three, and from glycolysis two. The number of ATPs from one $FADH_2$ is assumed to be two. Remember that the Krebs cycle occurs twice for each molecule of glucose.

SOURCE OF ATP	MOLES OF ATP
Glycolosis	2
2 NADH in glycolosis	4
Krebs cycle (2 GTP)	2
8 NADH in Krebs cycle	24
2 FADH2 in Krebs cycle	4
Total	36

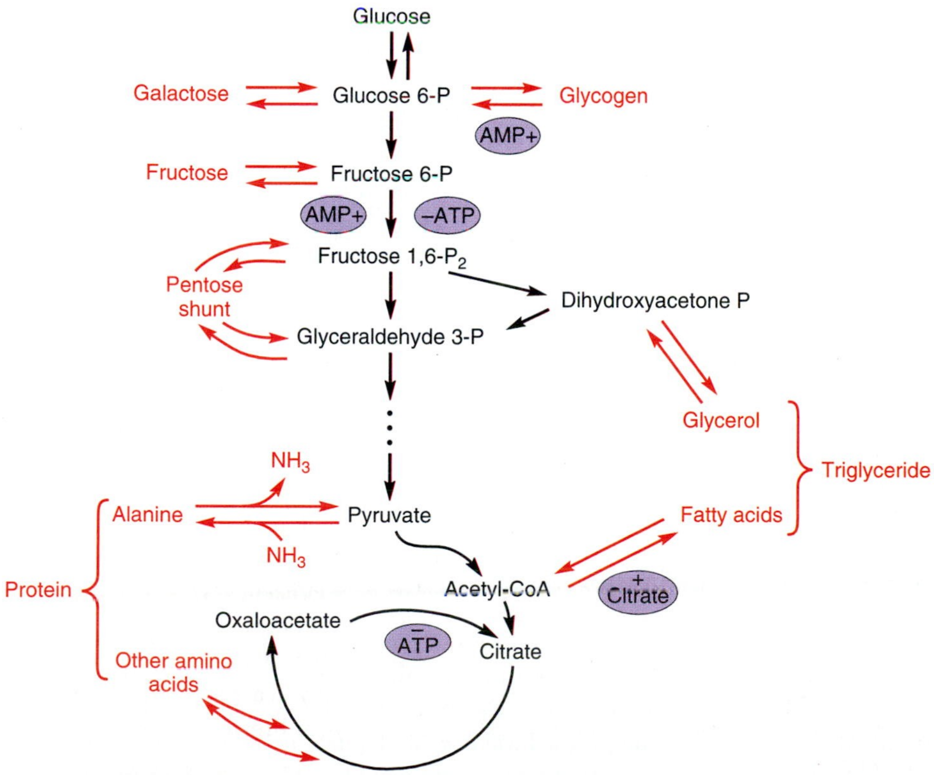

FIGURE 3.9
Alternative metabolic pathways. ATP, AMP, and citrate stimulate (+) or inhibit (–) reactions at the points shown.

■ **Question:** When a fatty acid is catabolized, it is first activated with one ATP, then broken down two carbons at a time, with each breakdown producing one NADH and one $FADH_2$. From Figs. 3.4 and 3.7 and Table 3.1, show that 443 ATP molecules are produced by the complete catabolism of the triglyceride in Fig. 2.11.

groups. The amino acid alanine, for example, is catabolized by deaminating it to form pyruvate, which then enters the Krebs cycle.

All these intersecting metabolic pathways (and these are only a few of them) make one wonder how the cell maintains any order at all. In general, each pathway is regulated by **feedback inhibition** from products of the pathway. For example, surplus ATP inhibits its own production by inhibiting glycolysis and the Krebs cycle (Fig. 3.9). As the Krebs cycle slows down, citrate accumulates, which causes acetyl-CoA to be diverted into the synthesis of fatty acids. In contrast, adenosine monophosphate (AMP), which accumulates from the breakdown of ATP, stimulates glycolysis and the breakdown of glycogen by its allosteric effect on certain enzymes. In addition to feedback control, signals from outside the cell also affect metabolic pathways. The hormone insulin (see pp. 270–271), for example, stimulates the production of glycogen, fat, and muscle, thereby removing excess glucose from the blood. The hormones glucagon (see p. 271) and adrenaline promote the breakdown of glycogen to provide additional glucose.

Metabolic Rates

Trying to follow the metabolism of all the various nutrients in every cell of an animal would be so difficult that no scientist would attempt it. It is relatively easy, however, to measure an animal's overall rate of energy exchange, defined as the metabolic rate (MR). The most straightforward way of measuring MR is **direct calorimetry**—measuring heat production. Since all the energy released by metabolism is ultimately converted into heat, either during the metabolism or while using the ATP it produces, one can determine the rate of energy conversion by measuring the heat produced by an animal over a period of time. In order to measure MR, one could submerge an animal in a known volume of water and measure the rise in temperature per unit of time. Knowing that it takes one calorie of heat to raise a gram of water by 1°C, one could then determine the MR in units of calories per minute, or, more conveniently, kilocalories per hour (kcal/hr). (One kilocalorie is equal to the capitalized Calorie used in diet books.)

A more convenient method is to measure the rate of oxygen consumption or carbon dioxide production. This method, called **indirect calorimetry,** works because a given nutrient requires a fixed amount of O_2 to catabolize it, and it releases a fixed amount of CO_2. MR determined by indirect calorimetry is often expressed as milliliters of O_2 consumed per minute, or it may be converted to kcal/hr. For normal human adults the metabolic rate is from 1000 to 2500 kcal/day. A typical human therefore produces enough heat in a day to heat 2000 liters of water by 1°C or 1 liter of water by 2000°C! This is approximately the same rate of heat production as for a 100-watt light bulb. Exercise, growth, increased thyroid activity, and exposure to cold sharply raise the MR above this level.

The First Law of Thermodynamics states that energy is neither created nor destroyed. Therefore all the energy released as heat, stored in the body as glycogen, fat, and protein, and lost as waste must have come from the calories in nutrients. That is,

$$\text{caloric intake} = \text{MR} \times \text{time} + \text{stored energy} + \text{waste}$$

Usually the energy lost in waste (mainly feces) is negligible. If the caloric intake is less than the amount expended (MR × time), then the stored energy must be negative. That is, if an animal uses more energy than it takes in with food, it will lose weight. Calories absorbed in excess of the amount used will be stored as fat, increasing body weight. If caloric intake equals caloric expenditure, weight will be stable. Therefore an average

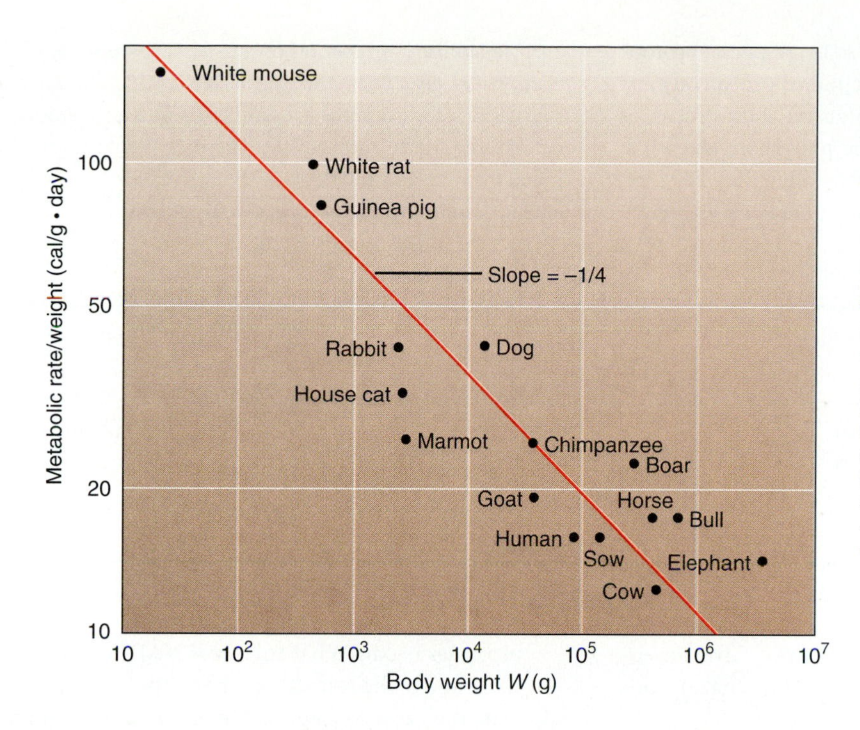

FIGURE 3.10

The "mouse-to-elephant" graph of metabolic rate per gram of body mass versus the body mass. Both axes are plotted as powers of 10 (logarithmic). The relationship is linear with a slope of $-\frac{1}{4}$, which indicates that the metabolic rate (MR) divided by body mass is proportional to the body mass to the $-\frac{1}{4}$ power: $MR = aW^{-1/4}$, where a is constant. Thus an average gram of tissue in a resting mammal has a lower metabolic rate the larger the animal is. The MR of an average gram of tissue in a mouse is approximately 115 calories per gram per day, while that in an elephant is only about 15 calories per gram per day. Similar graphs occur for nonmammals, although the exponent is not always –0.25. It ranges from –0.30 for freshwater fishes to –0.14 for turtles.

adult must consume at least 2000 kcal/day to maintain constant body weight. This is not hard for most of us (see pp. 271–272). About five slices of apple pie provide that much energy (though not all the essential nutrients). Increased activity will increase caloric requirements. Walking briskly for 2 hours, for example, will require another piece of pie.

Body size also affects metabolic rate. This is not surprising, since a large animal has more metabolizing tissue than a smaller one. It is surprising, however, that the metabolic rate does not increase in direct proportion to body weight. An animal that weighs twice as much as another usually has a metabolic rate that is less than twice as high. This means that an average gram of mouse tissue has a higher metabolic rate than an average gram of elephant tissue (Fig. 3.10). One explanation is that smaller animals have relatively larger hearts, kidneys, and other organs with high metabolic activities. Another contributing factor may be that for some reason smaller animals have a higher rate of proton leak across the inner membrane of the mitochondria (Porter and Brand 1993). Whatever the explanation, large animals tend to be more efficient in their metabolism. A 2.5-gram shrew must, in fact, eat almost constantly during its waking hours, while most large animals have time for other things.

Summary

Cells require energy to power their mechanisms, and they derive that energy from the chemical bonds in nutrients, especially glucose. Glucose is a stable molecule, requiring activation energy before it breaks down and releases the energy in its chemical bonds. Cells are able to extract that energy at normal body temperatures only because they produce enzymes that reduce the amount of activation energy needed.

The energy from chemical bonds in glucose is usually transferred to ATP, which can then transfer the energy to cellular mechanisms that require it. A few types of animal cell can produce ATP without oxygen in the series of reactions called glycolysis, in which glucose is converted into pyruvate. Only about 2% of the energy in glucose gets into ATP, however.

Most animal cells increase the efficiency by further metabolizing the pyruvate. Metabolism of pyruvate occurs in mitochondria during cellular respiration. Cellular respiration comprises two major sets of processes: Krebs citric acid cycle and the respiratory chain. In the Krebs cycle the carbons in pyruvate are converted into CO_2, and most of the energy from the chemical bonds is transferred to NAD+ to form NADH. In the respiratory chain the NADH is converted back to NAD+, releasing protons and electrons. Energy from the electrons pumps protons against a concentration gradient. Energy from diffusion of the protons is then used to convert ADP and inorganic phosphate into ATP. Oxygen is required to accept the electrons after they have expended their energy. Cellular respiration is about 40% efficient.

Metabolism of most other nutrients share some of the same reactions as glycolysis and the Krebs cycle. The fatty acids in triglycerides, for example, are converted into acetyl-CoA, which is metabolized in the Krebs cycle. The metabolic rate can be determined from heat production or the exchange of O_2 or CO_2, and it depends on the overall level of activity and body size of the animal.

Key Terms

metabolism
ATP
activation energy
enzyme
substrate
glycolysis
oxidation–reduction reaction
respiration
mitochondrion
Krebs citric acid cycle
respiratory chain
metabolic rate

Chapter Test

1. What are enzymes? How do they work? Give an example of a particular enzyme, and explain what it does. (pp. 52–53)
2. Describe the structure of ATP and relate it to its functioning. (pp. 51–52)
3. The overall equation for glycolysis in most animal cells is

$$\text{glucose} + 2\ \text{ADP} + 2\ \text{phosphate} \longrightarrow 2\ \text{lactate} + 2\ \text{ATP}$$

   Explain each term in the equation. ATP has more free energy than ADP; explain how it acquires that free energy. In what part of the cell does glycolysis occur? (pp. 53–56)
4. Briefly describe the Krebs cycle and the respiratory chain. Diagram a mitochondrion showing where these two processes occur. What role does NADH play in the Krebs Cycle and in the respiratory chain? (pp. 56–58)
5. State the basic concept of Mitchell's chemiosmotic theory. (p. 59)
6. Cellular respiration is described by the following equation:

$$\text{glucose} + 6\ O_2 + 36\ \text{ADP} + 36\ \text{P} \longrightarrow 6\ CO_2 + 6\ H_2O + 36\ \text{ATP}$$

   Which oxygen atoms on the right-hand side of the equation come from the O_2 on the left-hand side? Explain your answer. From where does the carbon in CO_2 come? By what series of reactions? (pp. 56–58)
7. Define metabolic rate, and describe two general methods for measuring it. (p. 62)

■ Answers to Figure Questions

3.2 The energy in nutrient molecules came from the ATP used by the organisms that synthesized the molecules. Ultimately this energy is almost always from photosynthesis, which uses solar energy.

3.3 Increasing the number of molecules of substrate, in this case glucose, will give a proportional increase in the reaction rate as long as the number of enzyme molecules is so large that there are always some available to catalyze the reaction. Beyond that point, adding further substrate gives less of an increase in reaction rate, and eventually no increase at all. Thus doubling the number of glucose molecules would probably double the reaction rate if there are 10 times as many hexokinase molecules as there are glucose molecules. Increasing the number of glucose molecules 10-fold would probably increase the rate more, but not by 10 times, and increasing the number of glucose molecules by 100 or 1000 times might not increase the rate any further. (The dependence of reaction rate on substrate concentration is covered by the Michaelis–Menten equation, which you will eventually encounter in chemistry courses.)

3.4 The direct production of ATP during glycolysis would be zero. However, in the processes to be covered later (Krebs cycle and the respiratory chain), production of ATP from pyruvate and NADH would be reduced only by half.

3.5 Slicing the cells for microscopy produces round or elliptical sections in the same way that slicing pepperoni does.

3.9 The triglyceride in Fig. 2.11 yields a molecule of glycerol plus three fatty acids. The molecule of glycerol yields 1 NADH and 2 ATP in glycolysis (Fig. 3.4) plus 4 NADH, 1 $FADH_2$, and 1 ATP in one turn of Krebs cycle (Fig. 3.7). Assuming 2 ATP for the first NADH and 3 ATP from each of the NADH from Krebs cycle, plus 2 ATP for the $FADH_2$, we find a total of 19 ATP from the glycerol alone. From this we subtract 3 ATP needed to start the breakdown of the three fatty acids; this leaves 16. Turning each of the two 18-carbon fatty acids into acetyl-CoA requires the breaking of eight bonds. (Not nine; this is the tricky part.) Similarly, turning the 16-carbon fatty acid into acetyl-CoA requires breaking seven bonds. These 23 reactions generate 23 NADH and 23 $FADH_2$ and produce 26 molecules of acetyl-CoA, each of which generates 3 NADH, 1 $FADH_2$, and 1 ATP in a turn of Krebs cycle. This gives a total of 101 NADH ($23 + 26 \times 3$), 49 $FADH_2$, and 26 ATP. Assuming 3 ATP per NADH and 2 ATP per $FADH_2$ brings the total number of molecules of ATP from fatty acids to 427. Add to this the 16 remaining from glycerol gives 443.

Readings

Recommended Readings

Dressler, D., and H. Potter. 1991. *Discovering Enzymes.* New York: Scientific American Library.

Hinkle, P. C., and R. E. McCarty. 1978. How cells make ATP. *Sci. Am.* 238(3):104–123 (Mar).

Additional Reference

Porter, R. K., and M. D. Brand. 1993. Body mass dependence of H^+ leak in mitochondria and its relevance to metabolic rate. *Nature* 362:628–630.

CHAPTER

4

Genetic Inheritance

Mother and young mountain zebra, *Equus zebra*.

Early Notions of Inheritance

Anyone who has studied the materials and mechanisms of cells described in the previous chapters should be impressed, and perhaps mystified, by how well they perform their functions. How do cells produce materials and mechanisms that are so well adapted? Where do they get the information that tells them how to make an enzyme and how a microtubule is supposed to work? How is it that we can talk about *the* cell at all? Why are they and the organisms they compose all so similar yet continually varying from one generation to the next?

The ancients had an explanation for these phenomena. It made such good sense that it is still often repeated by those who have not studied biology. According to this commonsense theory, cells and organisms essentially learn from their parents how to function, because the events that affect an organism in its lifetime are passed on to its offspring. We call this concept **the inheritance of acquired characters.** Inheritance of acquired characters has probably been common "knowledge" for thousands of years. *Genesis* 30, for example, records that Jacob produced spotted kids and lambs by placing spotted tree branches in front of the goats and sheep while they were conceiving.

Inheritance of acquired characters implies another commonsense theory: that every part of a parent's body contributes to the characteristics of the offspring. This would permit the visual system of Jacob's goats and sheep to influence their offspring. This concept is called **pangenesis,** from the Greek words for "all" and "origin." In the 4th century B.C., Aristotle easily refuted both inheritance of acquired characters and pangenesis (*On the Generation of Animals I.* 17, 18). He noted, for example, that children are not born with the beards of their fathers, and that children often resemble their grandparents more than their parents.

Unfortunately, not even Aristotle could propose an explanation for heredity as seductive as inheritance of acquired characters and pangenesis. As late as 1809, Jean Baptiste Lamarck used inheritance of acquired characters (now called "Lamarckism") as a cornerstone of his theory of evolution. Even Charles Darwin accepted inheritance of acquired characters. In his book *The Variation of Animals and Plants Under Domestication,* published in 1868, Darwin elaborated a theory of pangenesis that he was just as proud of as his theory of natural selection.

The Genesis of Genetics

Darwin did not know that just two years before publication of his book on pangenesis, evidence against it and the inheritance of acquired characters had been published by an obscure monk, **Gregor Mendel** (Fig. 4.1). Johann Mendel (1822–1884) was the son of poor peasants who scrimped to finance his education in physics and mathematics. At age 21 he became an Augustinian novice, taking the name Gregor. He was ordained a priest in 1847 and was appointed deputy teacher of high-school mathematics in 1849. After failing the biology portion of the teachers' qualifying examination, he was sent to the University of Vienna for two years to study zoology, plant physiology, chemistry, and physics. He applied again for the teachers' exam but had to withdraw because of sickness. At age 34 he entered the Altbrünn Monastery in what is now Brno, The Czech Republic, where he eventually became Abbot.

FIGURE 4.1
Gregor Mendel in 1862, four years before publication of his research on heredity.

Mendel began his studies while engaged in one of the enterprises of the monastery, raising ornamental flowers. Mendel noticed that whenever he allowed pollen from one variety of plant to fertilize the flowers of a different variety of the same species, the resulting hybrids showed a remarkable consistency in bearing the traits of one variety or the other. Darwin and others had already studied hybridization, but they looked at the overall features of hybrids. Since characters from both parents occurred in the hybrid offspring, they had concluded that the hereditary material blended together in a hopelessly complicated manner. Mendel, with the analytical approach of the physicist, decided to look at one character at a time. He chose garden peas, partly because their closed flowers resist random pollination by insects, and he used true-breeding varieties, which consistently produce offspring with the same characteristics as the parents. In 1856 he began systematically cross-pollinating various true-breeding varieties of pea plants that differed in only one trait. This procedure is now called a **monohybrid cross,** because the resulting hybrids come from parents that differ in only one (*mono*) trait. Mendel found, as one example, that a monohybrid cross of a plant with purple flowers and a plant with white flowers produced only hybrids with purple flowers. In Mendel's terminology, which we still use, the purple-flower character was **dominant** over the white-flower character, which was **recessive.**

If heredity were due to substances transmitted from the parents, what happened to the substance determining the recessive character? Did it simply disappear? To find out, Mendel allowed the hybrid flowers to become fertilized with their own pollen (a normal occurrence because of the closed flower structure), to see if any white flowers would reappear in the next generation. Of 929 plants produced from self-fertilized hybrids, 705 bore purple flowers, and 224 had white flowers (Fig. 4.2). The factors determining the recessive white-flower character had evidently survived within the purple-flowered hybrid. This in itself argued against the inheritance of acquired characters and pangenesis, since the white flowers in the second generation did not inherit the characters of their parents.

Mendel found similar results for six other characters (Fig. 4.3). In every case about three-fourths of the offspring of self-fertilized hybrids had the dominant character, and about one-fourth had the recessive character. In other words, **the ratio of dominant to recessive characters in the offspring of hybrid parents is approximately 3:1.** When the plants with the recessive characters were self-fertilized, they invariably pro-

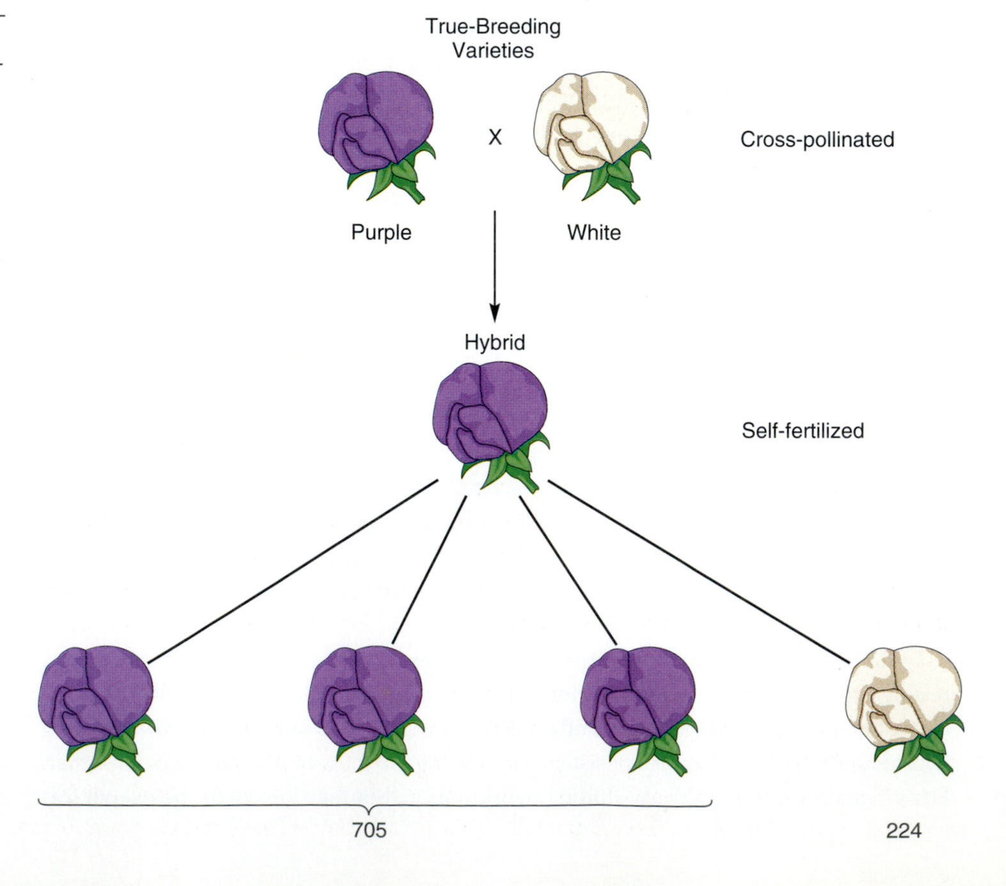

FIGURE 4.2

Mendel's experiment showing that a recessive character (white flowers) is not expressed in a hybrid but emerges in approximately one-fourth of the offspring of self-fertilized hybrids.

■ **Question:** The allele for black hair is dominant over the allele for blonde hair. Why don't one-fourth of Swedes and Chinese have blonde hair?

	Dominant character	Recessive character	Ratio
Form of seed	Round  5474	Wrinkled  1850	2.96:1
Color of albumen	Yellow  6022	Green  2001	3.01:1
Color of seed coat and color of flower	Grey-brown  Purple  705	White  White  224	3.15:1
Form of pod	Inflated  882	Constricted  299	2.95:1
Color of unripe pod	Green  428	Yellow  152	2.82:1
Position of flowers	Axial  651	Terminal  207	3.14:1
Length of stem	Long  787	Short  277	2.84:1
Totals	14,949	5010	2.98:1

FIGURE 4.3
Summary of Mendel's results from self-fertilized hybrids. True-breeding varieties of pea plants differing in the seven traits listed in the left column were cross-pollinated, producing hybrids with the dominant character. After the hybrids were self-fertilized, they produced offspring with the numbers of dominant and recessive characters shown in the boxes. The ratio of dominants to recessives for each character is given in the right column. The bottom row shows the total numbers and the ratio of dominant to recessive characters—2.98:1, which Mendel rounded off to 3:1.

■ **Question:** Mendel got more ratios so close to 3:1 than is statistically likely, suggesting that he "cooked" his data or was biased in analyzing it. But how did he know that the correct answer was 3:1?

duced offspring with the recessive characters. In other words, they were true-breeding. One-third of the plants with dominant characters were also true-breeding, producing only offspring with the dominant character. However, the other two-thirds of the plants with dominant characters produced both dominant and recessive offspring in the ratio of 3:1. Evidently they were hybrids like their parents.

Mendel proposed a simple mathematical model for these findings. He let A denote the dominant character, a the recessive character, and Aa the hybrid. He assumed that each **gamete** (ovum in a flower or sperm cell in pollen) received either A or a with equal likelihood. In a self-fertilizing hybrid, therefore, an A ovum has an equal chance of being fertilized by A pollen or a pollen. Likewise an a ovum has an equal chance of being fertilized by *A* or *a* pollen. All the combinations, *AA, Aa, aA,* and *aa,* should therefore occur with equal frequency. *Aa* and *aA* are identical, so the ratio of *AA* to *Aa* to *aa* is 1:2:1. Since *AA* and *Aa* look alike, however, the ratio of those with the dominant character to those with the recessive character is 3:1. When those with the dominant characters self-fertilize, two out of the three are shown to be *Aa* hybrids.

Segregation and Independent Assortment

This model works only if *A* and *a* do not blend but remain **segregated** from each other during the formation of gametes. This principle of segregation has come to be known as **Mendel's First Law.** Mendel did not say so, but the simplest explanation for segregation is that each plant has a pair of "units" for each character—what we now call **genes.** The different forms of a gene, whether dominant or recessive, we now call **alleles.** There may be many kinds of alleles for each gene, but only one allele for each gene is transmitted from a plant though each gamete.

The principle of segregation states that the allele determining a dominant character behaves independently of a recessive allele. But what about alleles for two different characters? For example, will round versus wrinkled seeds still occur in a 3:1 ratio whether or not the albumen (Fig. 4.3) is yellow or green? If so, then three out of four of the offspring will be dominant in one trait, and three out of four of those plants will also be dominant in the other trait. Thus the proportion of offspring with both characters dominant should be $^3/_4 \times {}^3/_4 = {}^9/_{16}$. Similarly, the proportion with one character dominant and one recessive should be $^3/_4 \times {}^1/_4 = {}^3/_{16}$, and the proportion with both recessive characters should be $^1/_4 \times {}^1/_4 = {}^1/_{16}$. In other words, the ratio of offspring with the four combinations of characters should be 9:3:3:1.

To test this prediction Mendel produced hybrids of parents that differed in two characters, a procedure now called a **dihybrid cross,** as opposed to the monohybrid crosses described previously. The two characters were round seeds versus wrinkled seeds, and yellow versus green albumen. Mendel then permitted the hybrids to self-pollinate. Out of 556 seeds produced by self-fertilized hybrids, the expected numbers were 313 ($= 556 \times 9/16$) round with yellow albumen, 104 round with green albumen, 104 wrinkled with yellow albumen, and 35 wrinkled with green albumen. The actual numbers were 315, 108, 101, and 32. This close agreement (some say suspiciously close) supported the hypothesis that the different characters assort independently of each other. This concept of **independent assortment** is now known as **Mendel's Second Law.** As we shall see, however, it is a "law" that is widely disobeyed.

Mendel's research was published in 1866, but at that time few biologists could understand the mathematical reasoning or the far-reaching implications. It was not until 1900 that three researchers independently obtained some of the same results, then discovered Mendel's paper. By then the inheritance of acquired characteristics and pangenesis were virtually forgotten issues, so the fact that Mendel disproved them was unremarkable. What remained remarkable, however, was that patterns of heredity could be explained using a few relatively simple concepts. The early geneticists had to create a new vocabulary to describe these concepts. Individuals that developed from zygotes with both alleles of the same type (both dominant or both recessive) were said to be **homozygous.** Individuals with two different alleles for the same gene were **heterozygous.** The actual genetic makeup of the individual was called the **genotype.** The appearance or other property of the organism, which depends partly on genotype and partly on environmental influences, was named the **phenotype.** The first generation resulting from a monohybrid or dihybrid cross was named the F_1 **generation,** and the offspring resulting from self-fertilization or interbreeding of the F_1 generation was named the F_2 **generation.**

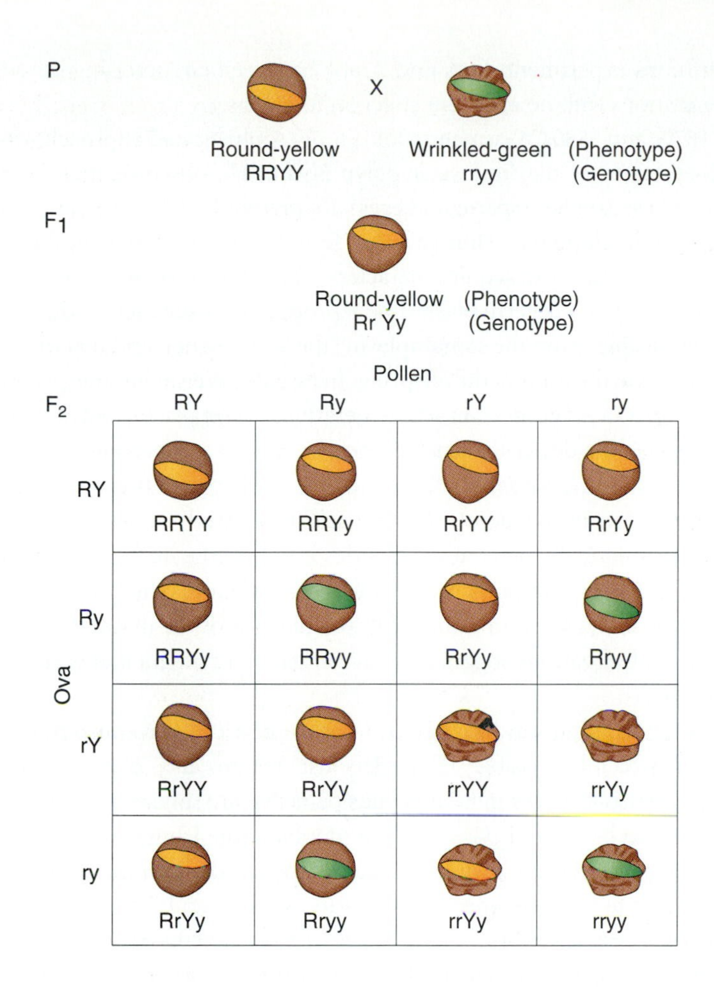

FIGURE 4.4

A Punnett square that predicts the ratios of the four phenotypes resulting from the dihybrid cross of pea plants heterozygous for form of seed and color of albumen, discussed previously in the text. Capital letters denote dominant alleles, while lowercase letters denote recessive alleles. At top, in the parental (P) generation, a plant that is true-breeding (homozygous) for the dominant alleles for round seeds and yellow albumen is crossed with a plant that is true-breeding for the recessive alleles for wrinkled seeds and green albumen. Part of the seed coat is shown torn away to reveal the yellow or green albumen. The phenotypes are, respectively, round-yellow and wrinkled-green, and the genotypes are symbolized RRYY and rryy. The F_1 (first filial) generation has the dominant phenotype (round-yellow), and its genotype is heterozygous (RrYy). This hybrid produces pollen and ova with either the R or r allele, and either the Y or y allele. These recombine with equal probability in the F_2 generation produced by self-fertilized hybrids. Counting the number of squares with each of the four possible phenotypes gives the 9:3:3:1 ratio predicted from independent assortment.

Using this terminology we would say that Mendel's true-breeding plants were either homozygous dominants or homozygous recessives, and that when they were crossed with each other the F_1 generation consisted of heterozygotes that developed into plants with the dominant phenotype. When the heterozygous F_1 plants self-fertilized, they produced F_2 offspring with a genotype ratio of 1:2:1 and a phenotype ratio of 3:1. An equally important contribution of the early geneticists were statistical and mathematical methods to determine whether actual genotype and phenotype ratios are consistent with theoretical predictions. One of these methods is the **Punnett square,** which allows the simple calculation of genotype and phenotype ratios (Fig. 4.4).

Although the concept that genes control inherited traits quickly became established, it also soon became clear that most genes affect organisms in a far more complicated way than the genes that Mendel studied. It is generally the case that a given trait is not determined solely by one pair of alleles, but by many alleles interacting with each other and with the environment at various times during development and later life.

Chromosomal Inheritance

During the 35 years that Mendel's work was being ignored, the mechanism of inheritance was being approached from a different direction by biologists using the improving techniques of microscopy to study cell structure. One of these pioneering cytologists was **August Weismann** (1834–1914). Weismann devoted his early career to disproving the inheritance of acquired characters and pangenesis. At first he took the straightforward approach of cutting off the tails of 40 generations of mice, demonstrating that tails in the 40th generation were just as long as those in the first. (Actually this was a superfluous experiment. Jews and Arabs had been circumcising boys for hundreds of generations with no apparent effect on foreskins.)

In the 1870s and 1880s Weismann took a more sophisticated approach with microscopical observations of the freshwater polyp *Hydra.* He observed that the cells that eventually produce gametes (sperm and eggs) are present in definite locations from the earliest stages of development. Thus gametes do not come from all over the body, and they have little opportunity to acquire characters during the life of the animal. Weismann therefore proposed that the **germ plasm** (the reproductive tissue that produces gametes) is separate and distinct from the **somatoplasm** (the body tissue), and that only the germ plasm contributes to the traits of the offspring. In essence, Weismann was proposing that what is really reproduced generation after generation is germ plasm, and that the body of an organism is merely a device that enables the germ plasm to reproduce.

Weismann and others continually refined the germ-plasm theory using the latest results from microscopical studies. One crucial line of study concerned the **nucleus.** Cytologists observed that when one cell divides into two, the nucleus also divides into two nuclei. In the early 1880s Oskar Hertwig and Hermann Fol also observed that the nucleus of a sea urchin's sperm unites with the nucleus of the ovum during fertilization. Both these observations suggested to Weismann and others that germ plasm was located in nuclei.

Later the germ plasm was localized in what we now call **chromosomes,** which are thread-shaped structures visible in nuclei just before and during cell division. Chromosomes generally occur in homologous pairs that are similar in size and shape, so the cells are said to be **diploid** (Greek *diploos* double). The laboratory organisms commonly used in the 19th century have so many chromosomes that it was hard to follow their movements under the microscope. Edouard van Beneden and Theodor Boveri found, however, that the parasitic worm *Ascaris bivalens* has only two pairs of chromosomes. They were therefore able to see that only one member of each pair of chromosomes enters each newly formed ovum or sperm cell. The gametes are therefore **haploid** (Greek *haploos* single). When a sperm cell and an ovum unite during fertilization, the resulting **zygote** has its number of chromosomes restored to the normal diploid number. Weismann had previously predicted that something like this halving of the genetic material in gametes would have to occur to prevent the amount of hereditary material from doubling with each generation. He and others therefore concluded that the germ plasm was in the chromosomes.

Weismann's distinction between somatoplasm and germ plasm and his conclusion that the chromosomes are responsible for heredity suggest two different processes of heredity. First, there must be some way by which genetic information in chromosomes is passed from a zygote to the somatoplasm to determine the characteristics of the organism. Second, there must be a process by which the gamete is made haploid so that the zygote becomes diploid. The two processes are called, respectively, mitosis and meiosis.

Mitosis

Cytologists in the 19th century found that each zygote divides into two **daughter cells,** each of which divides into two more daughter cells, and so on, through embryonic development and later growth. Just before each cell division the chromosomes become visible within the nuclei. Cytologists could then see that every cell has the same number of chromosomes as the original zygote. They could also see that the chromosomes move about in a quite remarkable pattern called **mitosis.** Mitosis accounts for the number of chromosomes remaining constant after each cell division.

Although mitosis is a continuous process, it is convenient to divide it into four phases (Fig. 4.5). First is **prophase.** During prophase the **nuclear envelope** breaks

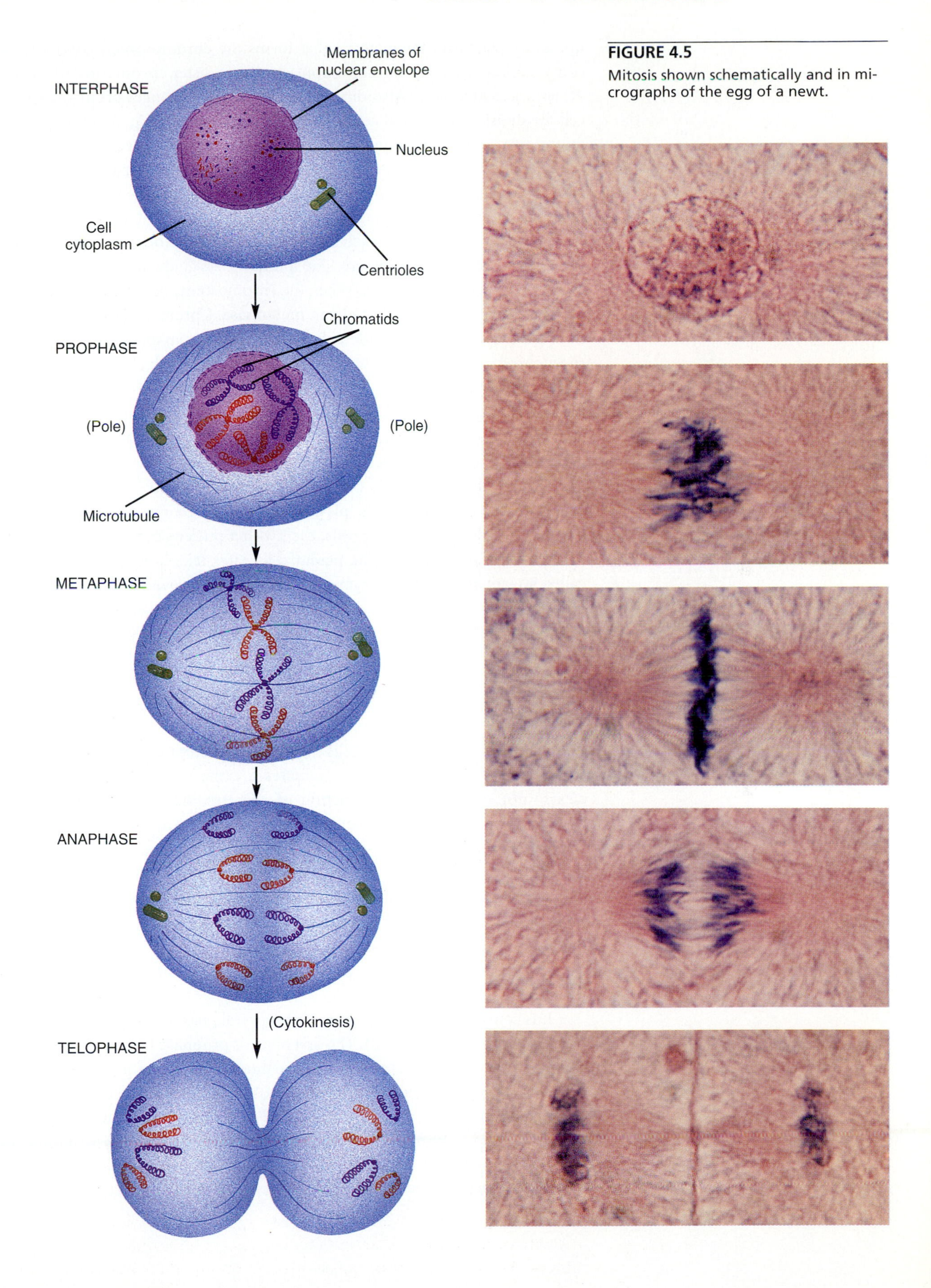

FIGURE 4.5

Mitosis shown schematically and in micrographs of the egg of a newt.

up, and **chromatin,** the material that forms the chromosomes, condenses and becomes visible. Each chromosome consists of two sister **chromatids** attached to each other at a **centromere.** Also during prophase other changes occur that prepare the cell for division. In animal cells a pair of **centrioles** (see Fig. 2.27) in the cytoplasm migrate to each of two opposite **poles.** These centrioles are composed of microtubules, and they apparently organize the assembly of other microtubules into the **spindle apparatus.** The spindle apparatus radiates from the two poles like iron filings around the poles of a magnet.

Spindle fibers attach to each chromosome at a **kinetochore.** The kinetochore is a protein located at the centromere. The spindle fibers guide the chromosomes to a disc-shaped area between the poles. When the chromosomes are lined up midway between the centrioles, the cell is said to be in **metaphase.** Chromosomes are most easily seen during metaphase (Fig. 4.6). Metaphase is followed by **anaphase.** During anaphase the sister chromatids separate toward opposite poles, and each chromatid becomes a new chromosome. Arrival of the daughter chromosomes at each pole signals the onset of the last phase, **telophase.** During telophase the chromosomes begin to disperse into chromatin, and a new nuclear envelope starts to form around them. Mitosis concludes at this point, when two nuclei have formed.

Usually the cytoplasmic components of the cell also start to move apart, in a process called **cytokinesis.** The plasma membrane becomes constricted, dividing the cell into two diploid daughter cells, each with a nucleus containing a set of chromosomes identical to those of the parent cell. After telophase the cells enter the state called **interphase,** in which the chromatin directs the growth and other activities of the cell.

Meiosis

As the result of mitosis, most cells in an animal inherit copies of the same chromosomes that were in the zygote. Gametes are the major exceptions. Mitosis will clearly not serve for the production of gametes, since they have to be haploid. Instead of mitosis, therefore, cells produce gametes through a process called **meiosis.** Meiosis begins in much the same way as mitosis, with each chromosome in a diploid cell appearing as two sister chromatids. The main difference is that there are two cell divisions, called **meiosis I** and **meiosis II.** During meiosis I the two chromatids do not separate, as in mitosis. Instead, each chromosome goes to one of two daughter cells. Each of these cells then divides again in meiosis II. This time the chromatids do separate, and each becomes a chromosome. Thus four haploid cells result from meiosis.

It is convenient to divide meiosis into several phases that are similar to the phases of mitosis (Fig. 4.7 on page 76). The first phase is **prophase I** (prophase of meiosis I), in which the chromosomes appear much like those in prophase of mitosis. In prophase I, however, each pair of homologous chromosomes line up side by side, and the four chromatids join. This association is called **synapsis,** and the interconnected chromatids are said to form a **tetrad.** The chromatids in each tetrad then intertwine in a process called **crossing over** (Fig. 4.8 on page 77). Although it could not be observed in the 19th century, genetic evidence to be described later shows that segments of the two

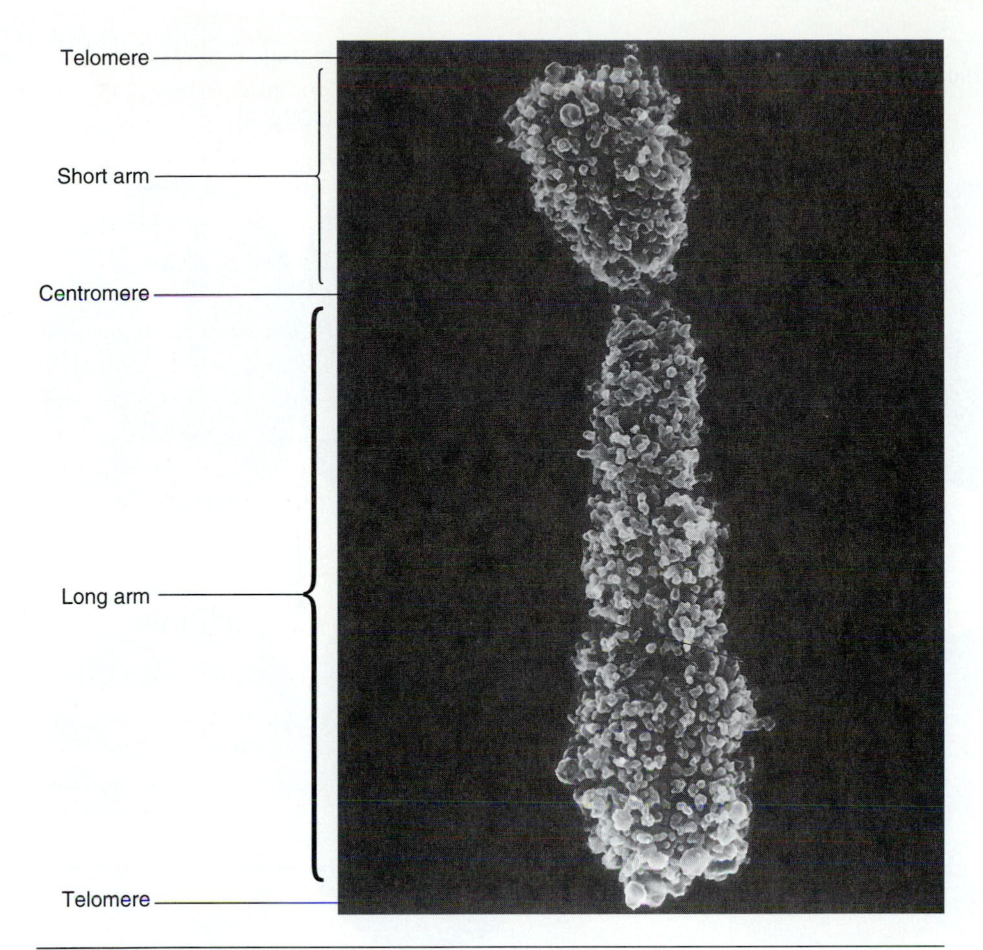

FIGURE 4.6

Scanning electron micrograph of a metaphase chromosome. In metaphase the chromosome consists of tightly condensed strands of chromatin. The constriction shows the location of the centromere, to which the kinetochore (not visible) attaches. The ends are defined by special regions of chromatin called telomeres.

chromatids in one homologous chromosome actually interchange with each other during prophase I.

During prophase I the chromosomes attach to spindles and migrate to an area between the two cellular poles, as in prophase of mitosis. Homologous chromosomes are still joined as tetrads during **metaphase I.** Metaphase I is followed by **anaphase I,** in which the homologous chromosomes migrate toward opposite poles. Meiosis I is completed when the chromosomes arrive at the poles, during **telophase I.** At this point each chromosome still consists of two attached sister chromatids, although parts of each chromatid may have come from the homologous chromosome that is now at the opposite pole. After telophase I comes cytokinesis, followed by division into two daughter cells.

Each of the two daughter cells formed in meiosis I has one set of chromosomes, each consisting of two chromatids. They may then proceed directly to metaphase II, or they may enter a brief period of interphase, during which the chromatin disperses. In that case the chromosomes reappear in **prophase II,** the first phase of meiosis II.

FIGURE 4.7
Meiosis.

■ **Question:** How much genetic material is there during telophase II compared with prophase I?

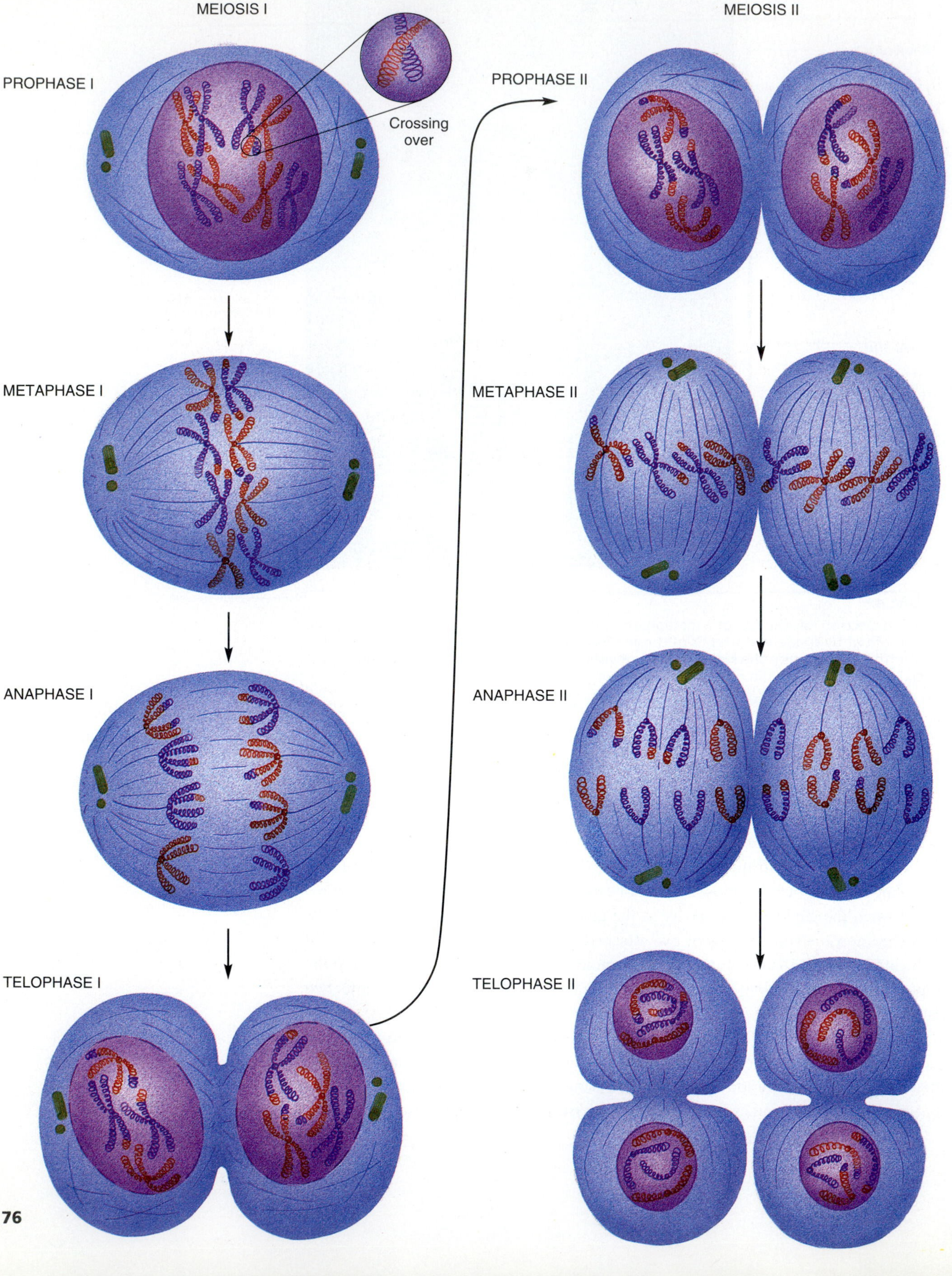

MEIOSIS I

PROPHASE I

Crossing over

METAPHASE I

ANAPHASE I

TELOPHASE I

MEIOSIS II

PROPHASE II

METAPHASE II

ANAPHASE II

TELOPHASE II

During prophase II each chromosome, consisting of two sister chromatids, attaches to spindles and moves to a position between the poles. This is followed by **metaphase II, anaphase II,** and **telophase II,** which resemble the corresponding phases of mitosis, but with a different result. In mitosis there are four chromatids for each pair of homologous chromosomes, and two chromatids—one from each chromosome of a pair—go to each daughter cell and become diploid chromosomes. In meiosis II, on the other hand, there are only two sister chromatids, and these separate to opposite poles and become haploid chromosomes. Cytokinesis and cell division then result in the production of haploid cells. The development of gametes from these cells is described in Chapter 6.

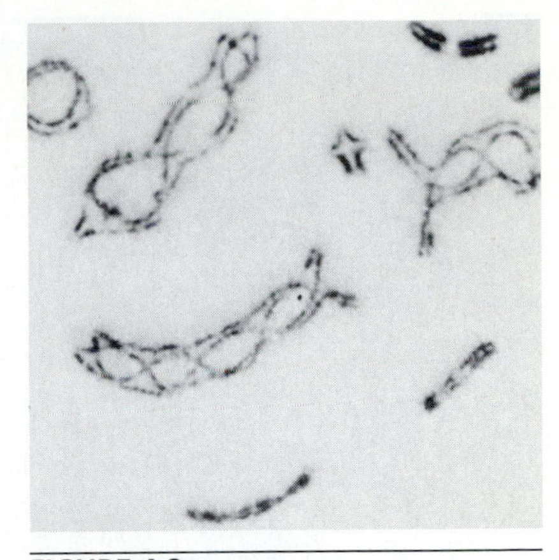

FIGURE 4.8
Synapsis and multiple crossing over during prophase I of meiosis in a grasshopper. Each pair of homologous chromosomes forms a tetrad of intertwining chromatids.

The Linkage of Genes to Chromosomes

Almost immediately after Mendel's work was rediscovered in 1900, some geneticists realized that the movements of chromosomes during mitosis and meiosis were similar to the patterns by which genes are inherited by somatic cells and gametes. One of the first to make this observation was W. S. Sutton. In 1903, while a graduate student, Sutton pointed out that, like genes, chromosomes occur in pairs in somatic cells and are transmitted individually through gametes. During his brief scientific career before he had to go into medicine to make a living, Sutton used the lubber grasshopper *Brachystola* to show that chromosomes also remain distinct from each other, and that the inheritance of one chromosome in a homologous pair does not affect the inheritance of a chromosome in a different pair. In other words, chromosomes, like genes, obey the rules of segregation and independent assortment. Sutton's proposal that genes are parts of chromosomes seems obvious to us now, but it was widely opposed at the time.

One opponent of the concept was Thomas Hunt Morgan. The demonstration that genes are indeed parts of chromosomes was due mainly to T. H. Morgan himself, along with his students at Columbia University and millions of little flies named *Drosophila melanogaster* (Fig. 4.9 on page 78). When he began his work with drosophila in about 1909, Morgan was known mainly as an embryologist interested in evolution. At the time he doubted Mendel's work, as well as Darwin's theory of natural selection. He expected that his research would show that inheritance in animals is not due to genes, and that sudden changes (mutations) were far more important in evolution than was natural selection.

Morgan chose *Drosophila,* commonly called a fruit fly, allegedly because he could not get funding to work with mammals. Whatever the reason, drosophila was a good choice. These flies are inexpensive to raise in large numbers, and they have a short life cycle. Each female can lay several hundred eggs, providing a large sample of offspring for statistical analysis. Within two days each egg hatches into a maggot, which develops into a pupa and finally an adult within 10 days. The adult becomes sexually mature within two days and can live for two more weeks. Another major advantage of drosophila is that it has only four pairs of chromosomes, which facilitates tracking them through mitosis and meiosis.

WHITE-EYED AND WILD-TYPE FRUIT FLIES. Drosophila provided three main lines of evidence that convinced Morgan and almost every other biologist that genes do determine heredity and are parts of chromosomes. The first line of evidence originated

A

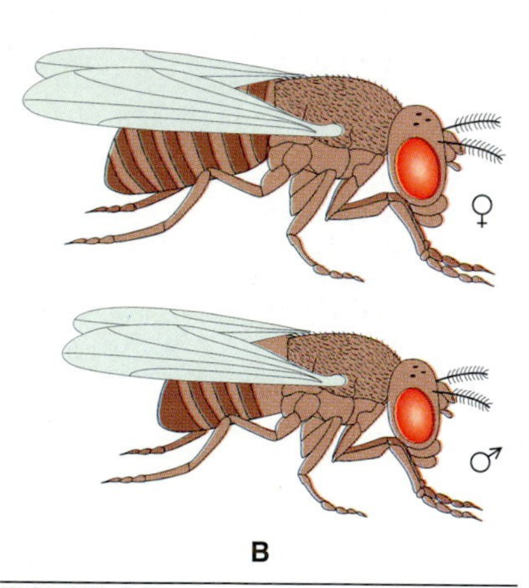

B

FIGURE 4.9

Major contributors to 20th-century genetics. (A) T. H. Morgan in the Fly Room at Columbia University around 1917, surrounded by milk bottles in which fruit flies were raised. (B) Male and female *Drosophila melanogaster.* Length approximately 2 mm. Drosophila is commonly called a fruit fly, but it is more correctly called a vinegar fly or pomace fly (family Drosophilidae) to distinguish it from true fruit flies (family Tephritidae).

in 1909 with a chance discovery by Calvin B. Bridges, who was then an undergraduate hired to wash the milk bottles in which the flies were raised. One day Bridges noticed a male fly with white eyes, in contrast to the normal red eyes of the "wild-type" drosophila. Rather than flick the mutant fly away, Bridges brought it to Morgan. Morgan then set out to study the inheritance of the mutation, expecting that it would disobey the Mendelian laws. Instead, the hybrids resulting from crossing the white-eyed male with a red-eyed female were all red-eyed, as would be expected if the white-eyed character were recessive. When these F_1 hybrids were interbred they produced an F_2 generation consisting of red-eyed and white-eyed flies in a 3:1 ratio, again as Mendel would have predicted. There was one major un-Mendelian feature, however. All the white-eyed flies were males; no female had white eyes. Somehow the inheritance of the white-eye character depended on the sex of the offspring.

A simple explanation for the inheritance of the white-eye character was already at hand in the fact that some chromosomes differ, depending on gender. Of the eight chromosomes in *Drosophila,* six occur in three similar, homologous pairs, regardless of the sex of the fly. Such chromosomes are called **autosomes.** In males the two chromosomes of the remaining pair are dissimilar (Fig. 4.10). One is designated X and the other Y. Females have two X chromosomes. This pair of chromosomes, either XY in males or XX in females, are called **sex chromosomes.** (A similar arrangement of sex chromosomes also occurs in humans and most other animals. In many animals, however, the male lacks a Y chromosome and is designated XO. In moths, butterflies, birds, and many lizards and snakes, it is the females that have different sex chromosomes.) Every ovum of drosophila has one X chromosome from the female parent. Every sperm cell, however, has either the X chromosome or the Y chromosome from the male parent. An ovum fertilized by a sperm cell with the X develops into a female, and an ovum fertilized by a sperm cell with the Y develops into a male.

The explanation for the inheritance of the white-eye character by half the males and by no female in F_2 is that the recessive allele for the white-eye character is on the X chromosome. In other words, the allele is **sex-linked.** The F_1 females were heterozygous for the allele, so half the F_2 males inherited the dominant allele and were therefore red-eyed like the wild type. The other F_2 males, however, inherited the X chromosome with the recessive white-eye allele. They had white eyes, since there was no allele on the Y chromosome to block expression of the recessive allele (Fig. 4.11). This explanation was confirmed by crossing white-eyed F_2 males with the heterozygous F_1 females. Half the offspring, whether male or female, were white-eyed.

At first the fact that the white-eye allele was inherited in exactly the same pattern as the X chromosome did not convince Morgan that genes are parts of chromosomes. Bridges later became a graduate student in Morgan's lab, however, and found that X linkage of the white-eye allele could also explain eye color of abnormal flies that inherited an extra X or Y chromosome. This ability to explain this rare phenomenon convinced even Morgan. It is now established that genes do occur on chromosomes. Moreover, each gene has a particular location, or **locus,** on a particular chromosome.

LINKAGE GROUPS. The evidence that genes are parts of chromosomes raised another question, the answer to which provided the third line of evidence that genes are on chromosomes. Since there are many more genes than chromosomes, many genes must have their loci (LOW-sy, plural of locus) on the same chromosome. How, then, could there be independent assortment? All the genes on a chromosome should be inherited together. In fact, many geneticists had found that there *is* a tendency for certain genes to be inherited together. The opportunity to study whether this tendency was due to genes occurring on the same chromosome arose after Bridges discovered numerous other mutations. Another of Morgan's students, A. H. Sturtevant, also spotted numerous mutants (in spite of being colorblind). Within a few years they had identified 85 mutations and had studied their patterns of inheritance. They found, for example, that if a fly carried both an allele for white eyes and an allele for hairy wings, its offspring tended to inherit either both mutations or neither of them. Since the white-eye allele was known to be on the X chromosome, the logical inference was that the allele for hairy wings was also on the X chromosome. A mutant allele causing bent wings was inherited independently from those two mutations, however, suggesting that it was autosomal (due to an allele located on an autosome). By this kind of reasoning, the 85 mutations were divided among different **linkage groups.** The number of linkage groups was four: the same as the number of pairs of chromosomes.

Two alleles in the same linkage group were not always inherited together, however. Evidently pairs of alleles on a single chromosome could become separated during crossing over in prophase I. In this way alleles on two homologous chromosomes **recombine** in different combinations. The greater the distance between the locus of one allele and the locus of another, the more likely this recombination. Sturtevant and others laboriously worked out the frequencies of recombination of numerous alleles. By this procedure they were able to map the relative distances between loci for the alleles.

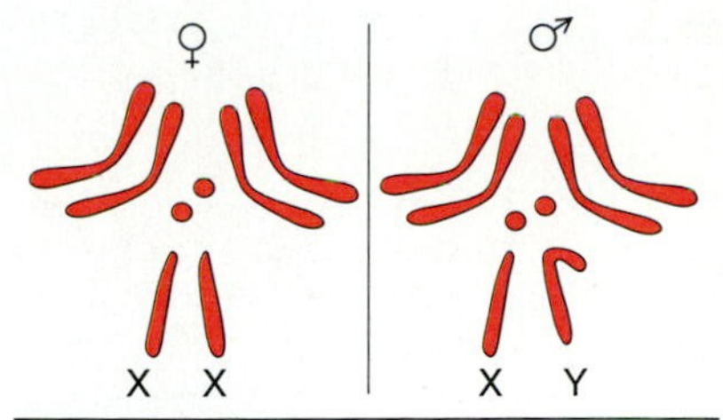

FIGURE 4.10

The four pairs of chromosomes in female and male *Drosophila melanogaster.* The pairs at the top are autosomes; the pairs at the bottom (X and Y) are sex chromosomes.

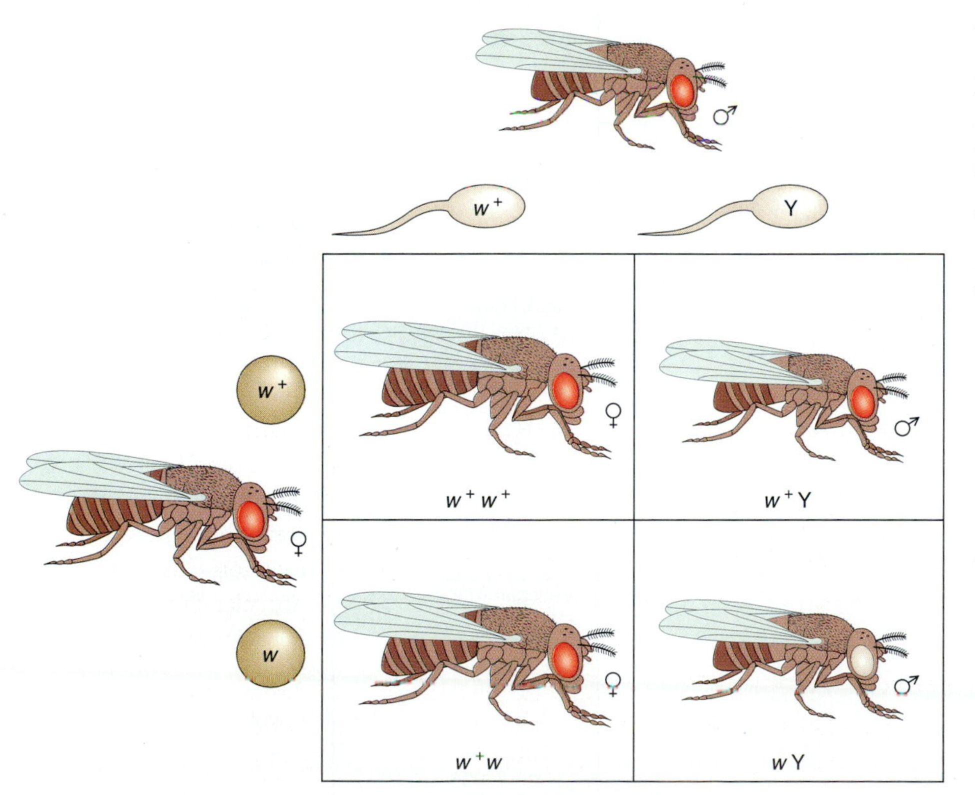

FIGURE 4.11

A Punnett square explaining the inheritance of a sex-linked allele for white eyes in the F_2 generation of drosophila. Above and to the side of the square are shown the two parents and their gametes. w^+ represents the dominant wild-type allele for red eyes, and *w* represents the recessive allele for white eyes. The sperm carrying a Y chromosome have neither of these two alleles.

The ability to produce such **linkage maps** confirmed the physical identity of genes as parts of chromosomes (Fig. 4.12).

That alleles on the same chromosome can become separated accounts for the fact that, in general, genes are independently assorted. Nevertheless, in some cases assortment is not independent. This naturally raises the question whether extraordinary luck or intercession by the saints guided Mendel in choosing the seven traits he studied. In pea plants there are only seven pairs of chromosomes, so it might appear that each of the seven genes must have been on a different chromosome. In fact, however, several of the genes for those traits are on the same chromosomes, but at loci far enough away that the linkage is weak.

FIGURE 4.12

A linkage map showing the locations of genes in drosophila. The Y chromosome is not represented, although it is known to carry at least seven genes affecting male fertility, among others. The numbers represent map distances of the loci from one end of the chromosome, as inferred from the probability of recombination.

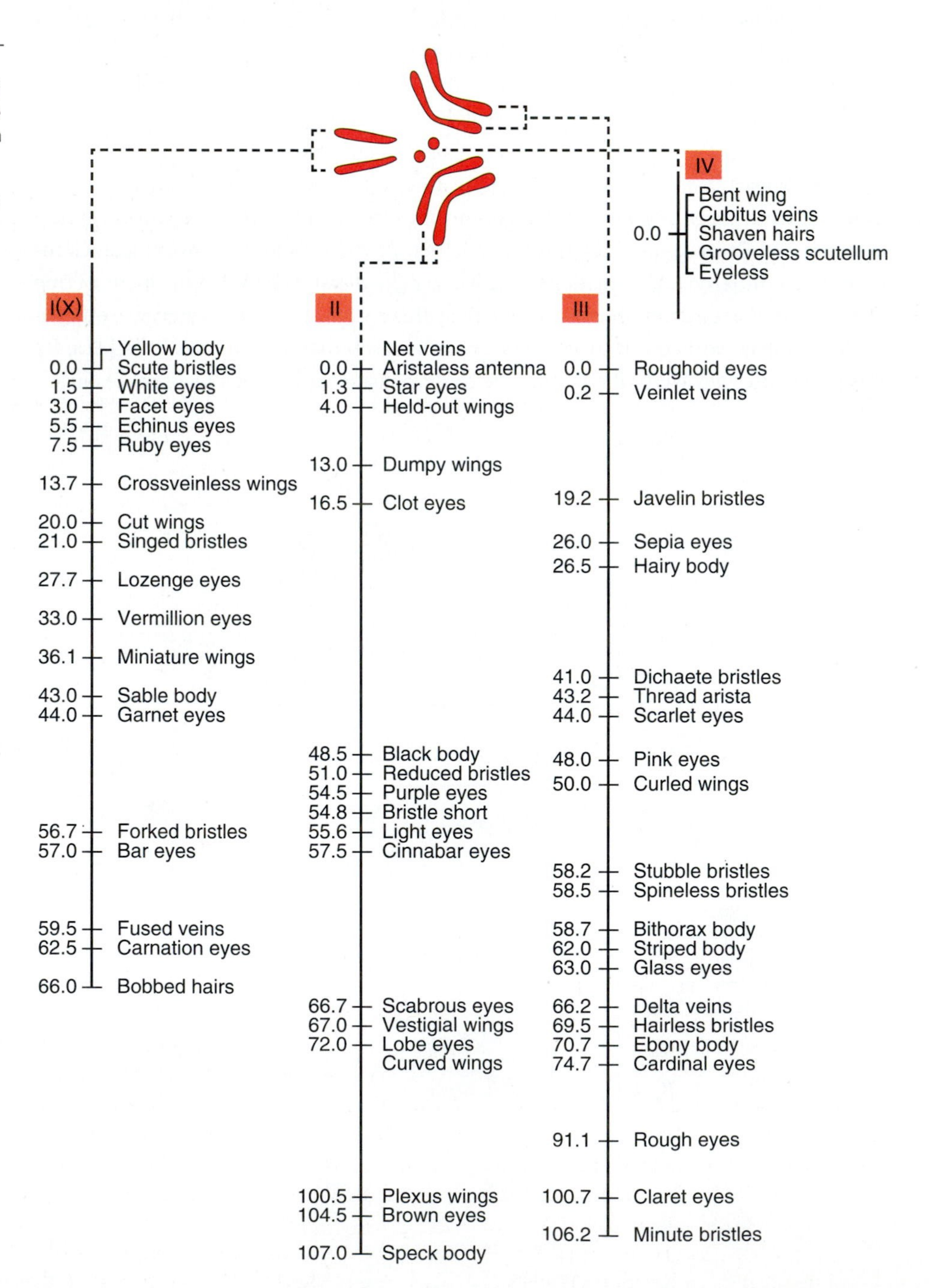

Summary

Gregor Mendel's research with garden peas disproved the age-old assumptions of inheritance of acquired characters and pangenesis and provided the first clues for the existence of genes. Mendel found that the offspring of two true-breeding varieties all had the appearance of only one of the parents. When these hybrids were self-fertilized or cross-pollinated, however, one-fourth of the offspring did not look like the hybrid parents, but like the other true-breeding variety. Mendel's interpretation, in modern terms, was that both true-breeding varieties were homozygous, one for a dominant allele and the other for a recessive. The hybrid offspring of the F_1 generation were heterozygous in genotype but had the phenotype of the dominant parent. The F_2 offspring of these hybrids, however, had a probability of one in four of getting a recessive allele from both parents. Such homozygous recessives therefore had the recessive phenotype. Likewise, one-fourth of the F_2 offspring were homozygous dominant, and half were heterozygous. Since these had the dominant phenotype, the phenotypic ratio of dominant to recessive was 3:1.

This pattern of inheritance implies two concepts: segregation and independent assortment. Segregation means that each allele behaves separately: the effects of a dominant allele do not blend with the effects of a recessive allele. Independent assortment means that the probability of inheriting one allele is not affected by the inheritance of another. Mendel demonstrated the validity of independent assortment by means of a dihybrid cross, but we now know that assortment is not independent for two genes linked on the same chromosome.

Evidence that chromosomes are involved in inheritance was found in the late 19th century by cytologists unaware of Mendel's research. They found that the cells of most organisms are diploid; that is, they have two copies of each chromosome. Gametes, however, are haploid, with only one chromosome of each pair. During fertilization two gametes combine, producing a diploid cell that divides repeatedly to produce all the diploid cells of a new organism. The movement of chromosomes prior to such cell division is called mitosis, and it occurs in four phases: prophase, metaphase, anaphase, and telophase. The movement of chromosomes that produce haploid gametes is called meiosis. It is divided into meiosis I and meiosis II, each of which has prophase, metaphase, anaphase, and telophase.

The unification of Mendel's work with that of the cytologists began early in this century, when it was observed that chromosomes obey the principles of segregation and independent assortment, just as genes do. The proof that genes are parts of chromosomes came first from studies of drosophila, in which particular mutations, such as white eyes, were inherited along with particular chromosomes. By using pairs of mutations and calculating the probability of their being inherited together, the position of mutated genes could be mapped on the chromosomes.

Key Terms

dominant
recessive
gene
allele
segregation
independent assortment
homozygous
heterozygous
genotype
phenotype
chromosome
diploid
haploid
mitosis
chromatin
meiosis
crossing over
linkage
autosome
sex chromosome
sex-linked

Chapter Test

1. Explain how Mendel's research argues against pangenesis and inheritance of acquired characteristics. (pp. 67–68)
2. Explain how a genotypic ratio of 1:2:1 in the F_2 generation of Mendel's studies gave a phenotypic ration of 3:1. (pp. 69–70)
3. "Mendel's First Law" proposes that a dominant allele remains segregated from its recessive counterpart. Describe the experimental evidence for segregation. (pp. 68–70)
4. "Mendel's Second Law" proposes that different genes assort independently. Describe the experimental evidence for independent assortment. Give one example of an exception to independent assortment. (pp. 70–79)
5. Prior to 1900, what kind of evidence suggested that chromosomes were involved in heredity? Describe two kinds of evidence that convinced Morgan and others that genes were parts of chromosomes. (pp. 71–72, 78–79)
6. Construct a Punnett square showing the genotypes and phenotypes of F_1 offspring of a white-eyed (homozygous recessive, sex-linked) female and a red-eyed (wild-type) male drosophila. Use Fig. 4.11 as a model. (p. 79)
7. Explain the relationship of each of the following to each other: chromatin, chromatid, chromosome, centromere, kinetochore. (pp. 72–75)
8. Explain the differences between mitosis and meiosis that make mitosis suitable for transmitting identical genetic information to all daughter cells, and meiosis suitable for transmitting different genetic information to gametes. (pp. 72–77)

Answers to Figure Questions

4.2 The 3:1 ratio applies only if both parents ar hybrids. This would seldom be the case in either population.

4.3 Apparently Mendel already had in mind the theory that traits are inherited through pairs of particles.

4.7 The same amount.

Readings

Recommended Readings

Crow, J. F. 1979. Genes that violate Mendel's rules. *Sci. Am.* 240(2):134–146 (Feb).

Glover, D. M., C. Gonzalez, and J. W. Raff. 1993. The centrosome. *Sci. Am.* 268(6):62–68 (June).

Moore, J. A. 1986. Science as a way of knowing—genetics. *Am. Zool.* 26:583–747. (*Comprehensive treatment of classical genetics in a historical context. See also other papers on genetics in the same issue.*)

CHAPTER

5

Genetic Control

Albino koala *Phasolarctos cinereus.*

FIGURE 5.1
James D. Watson and Francis H. C. Crick at the unveiling of their double-helix model of DNA in 1953.

What Are Genes Made Of?

It took Morgan and his students less than a decade to convince themselves and almost everyone else that heredity is due to genes that are parts of chromosomes. During the following two or three decades, others showed that the concept of genetic change within populations was consistent with Darwin's theory of natural selection (see pp. 366–367). In the meantime there was naturally a good deal of curiosity about the materials and mechanisms of genes. By the early 1940s it was shown that genes direct the production of enzymes and other proteins. The gene is, in fact, loosely defined in the phrase "one gene, one polypeptide chain."

Also by the 1940s there was good evidence that genes consist of **deoxyribonucleic acid** (DNA). The structure of DNA will be described in more detail, but briefly it consists of a chain of sugars (deoxyribose) linked to each other by phosphate groups, with each sugar bearing a nitrogen-containing structure called a **nitrogenous base.** There are only four kinds of nitrogenous bases in DNA: adenine, thymine, guanine, and cytosine. Once the chemistry of DNA was known, there began a deliberate race to learn how these long molecules were arranged in the cell and how they controlled heredity. The race was won in a remarkably short time in 1953 by a young American postdoctoral student, James D. Watson, and his mentor, Francis Crick, at Cambridge University (Fig. 5.1).

Watson and Crick relied on two major clues. First was a discovery by E. Chargaff that there is a pattern to the amounts of the four nitrogenous bases. The number of adenine bases in DNA is approximately equal to the number of thymines, and the number of guanines is similar to the number of cytosines. The second line of evidence was from analysis of the patterns into which x rays are diffracted as they pass through crystals of DNA. The x-ray diffraction studies of DNA by Maurice Wilkins and Rosalind Franklin suggested that the sugar and phosphate backbones of DNA were in the form of a helix, or spiral.

The DNA in bacteria and in mitochondria differ in structure and functioning from the nuclear DNA of eukaryotes that is described here. Instead of strands of DNA organized around histones, mitochondrial DNA occurs as a single closed loop without associated proteins. In this and other ways it is like bacterial DNA, as might be expected from other evidence that mitochondria originated as prokaryotic endosymbionts (see pp. 385–386). The genes on mitochondrial DNA specify some proteins of the respiratory chain, as well as other molecules needed to produce those proteins. The mitochondrial DNA also duplicates itself when the mitochondria reproduce. One of the most interesting properties of mitochondrial genes is that in humans and most other animals they are almost always inherited only from the mother, since the mitochondria of sperm remain outside the ovum during fertilization.

Watson and Crick applied this skimpy evidence in a way that was familiar to physicists but unorthodox to biologists: they tried to guess a structure of DNA that would work. Using simple models with the sugar–phosphate backbones represented by wire and the bases cut out of sheet metal, they tried various arrangements to see if one was consistent with the evidence and also intuitively compelling. One early model had three backbones twisted around each other and the bases sticking out. They abandoned this triple-helix model because it did not have the "ring of truth" that one expects from a correct theory of this importance. As Watson (1968) put it, the model "smelled bad." Finally in 1953 Watson and Crick hit upon the **double-helix** model of DNA, with two backbones outside, like rails of a spiral staircase, and with the bases sticking inward like steps (Fig. 5.2). To their delight, this model worked only if adenine projected inward toward thymine and if guanine projected toward cytosine. The model therefore neatly explained Chargaff's ratios. Variations in the sequence of bases would account for the genetic information, while the uniform sugar–phosphate backbones would preserve the crystalline structure outside. In addition, the duplication of DNA during mitosis and meiosis could be explained by the separation and copying of both strands. The model explained so many properties of genes that the subsequent experimental proof was anticlimactic.

By the time of Watson and Crick the works of Mendel, Morgan, and others were classic. Only later did the relationship of DNA to genes and chromosomes become clear. We now know that each chromosome contains a single molecule of DNA, and that genes occur in series along this molecule. During mitosis the DNA is tightly condensed around proteins (see Fig. 4.6). The degree of condensation is evident from the fact that the average human chromatid is only a few micrometers long but contains a molecule of DNA ap-

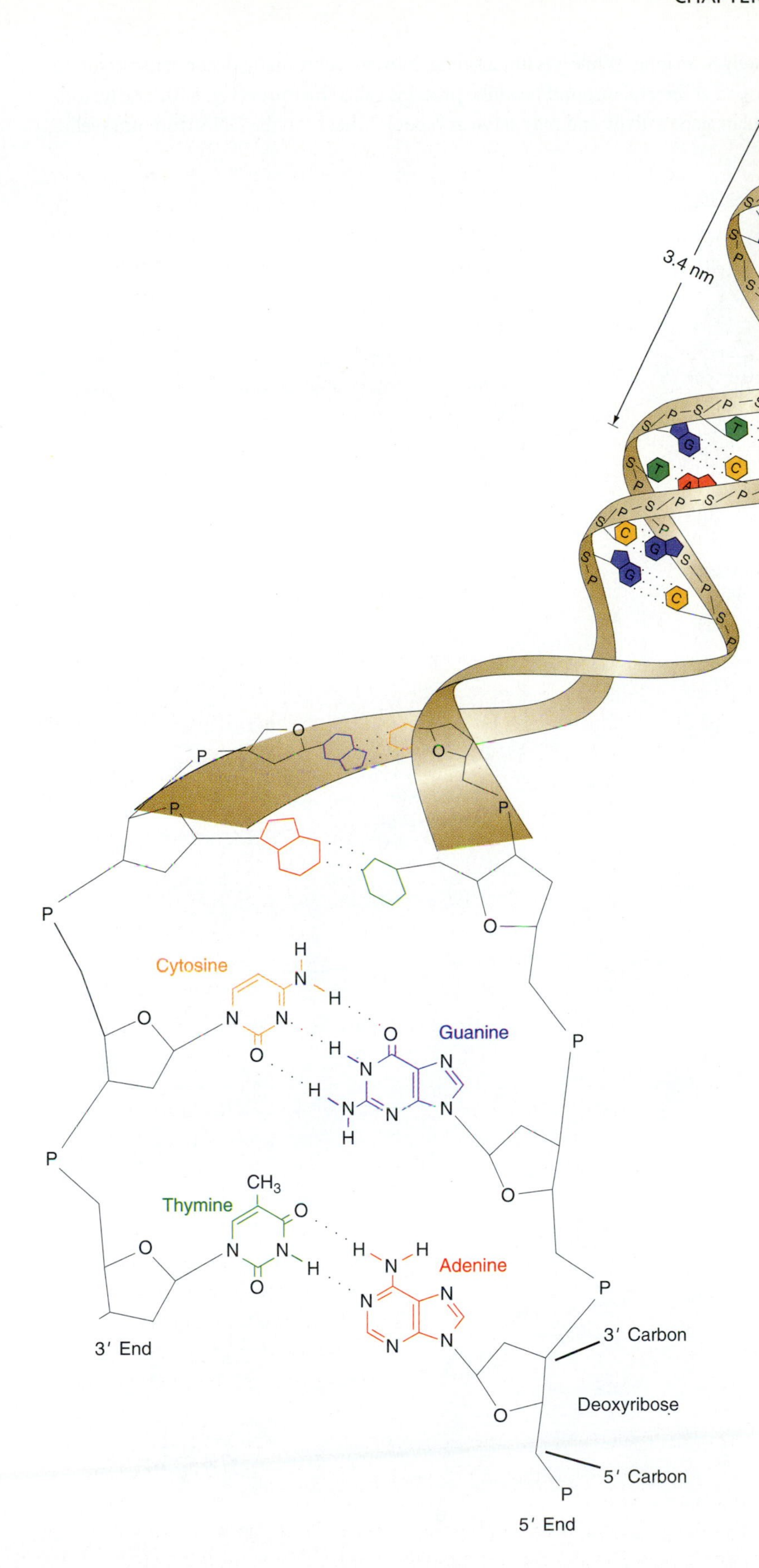

FIGURE 5.2

The double-helix model of DNA proposed by Watson and Crick. The right-hand spiral backbones consist of the sugar deoxyribose (S) and phosphate (P). The two chains are held together by hydrogen bonds (dotted lines) between complementary nitrogenous bases. Adenine (A) binds to thymine (T), and guanine (G) binds to cytosine (C). A and G belong to the class of molecules called purines; T and C are pyrimidines. The structure repeats every 3.4 nm, although the sequence of bases varies. The two DNA strands run in opposite directions, indicated by the numbers assigned to two carbons, 3′ and 5′ (three-prime and five-prime) on the deoxyribose.

proximately 5 cm long. While it is functioning, however, chromatin decondenses, with the DNA looped at intervals around beadlike proteins called **histones** (Fig. 5.3). The histones help regulate gene activity and may serve as "spools" that keep the DNA from unraveling.

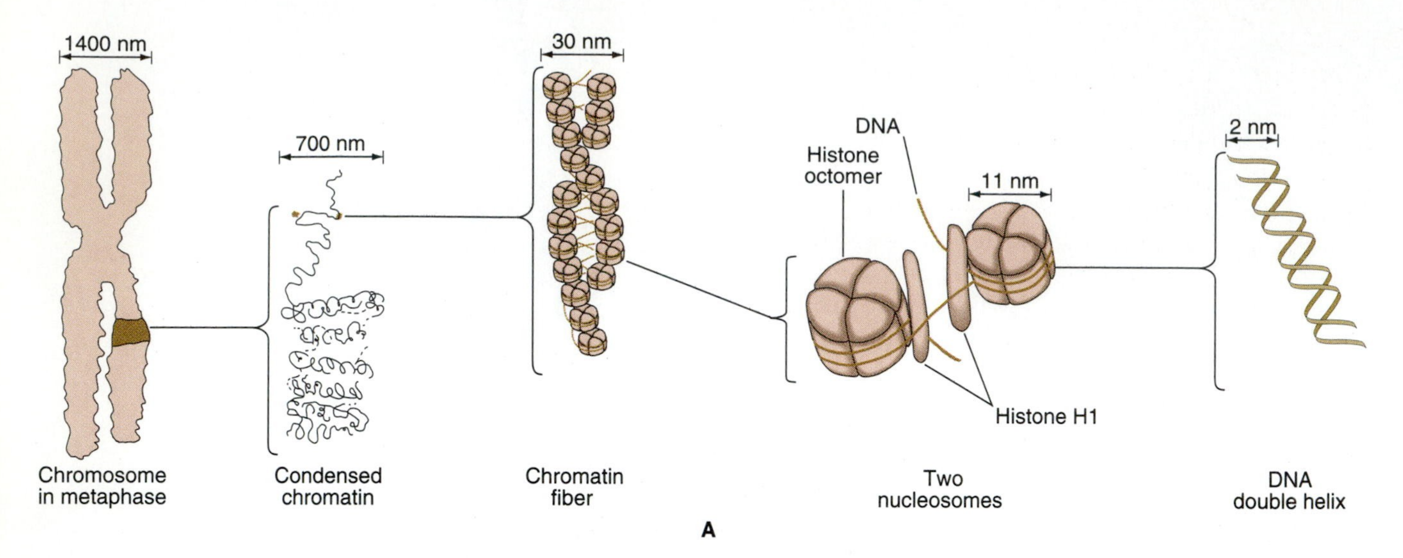

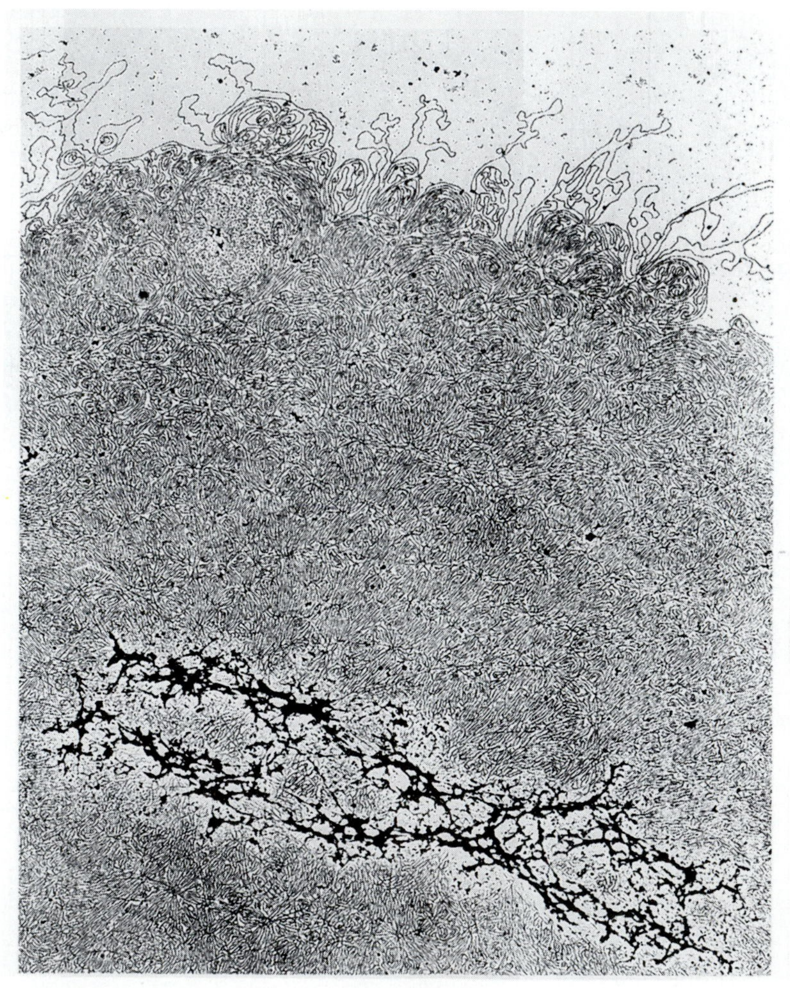

FIGURE 5.3

The organization of DNA in chromosomes. (A) In metaphase the DNA in chromosomes is highly condensed by proteins. During interphase the chromatin is partially unraveled, but is still supercoiled into a chromatin fiber approximately 15 times as thick as a single DNA molecule. Portions of DNA that are functioning apparently unravel further into a series of nucleosomes. Each nucleosome consists of approximately 70 nm of DNA (approximately 200 base pairs), part of which is wrapped 1.8 times around a histone octomer consisting of four pairs of different histones. Another type of histone (H1) apparently keeps the DNA bound to the octomers. (B) Removing the histones from a metaphase chromosome allows the DNA to unravel. Only about half the DNA of this human chromosome is shown. The dark-stained material is the protein scaffolding of the chromosome.

Each histone and the length of DNA associated with it make up the fundamental unit of chromatin structure, called the **nucleosome.**

The Cell Cycle

Genes direct the synthesis of proteins and duplicate themselves for mitosis in a regular pattern called the cell cycle. Each cell cycle starts with interphase (see Fig. 4.5) in a new daughter cell at the end of a previous mitosis, and the cycle continues through the end of the following mitosis. Interphase is divided into three parts: two **gap phases,** G_1 and G_2, and the **S (synthetic) phase** that comes between G_1 and G_2 (Fig. 5.4). During G_1, genes actively direct the production of enzymes and other proteins, especially those needed for the daughter cell to grow to its final size. During S phase the cell synthesizes a replica of its DNA and produces histones and other proteins needed to duplicate chromatids prior to mitosis. During G_2 the genes continue to direct the synthesis of proteins, especially those needed to make cytoplasmic materials for the two new daughter cells. As will be described in more detail later, the production of proteins during all these phases involves (1) the **transcription** of the information in the DNA into a messenger and (2) the **translation** of that messenger into a protein. The synthesis of new DNA during S phase is called **replication.**

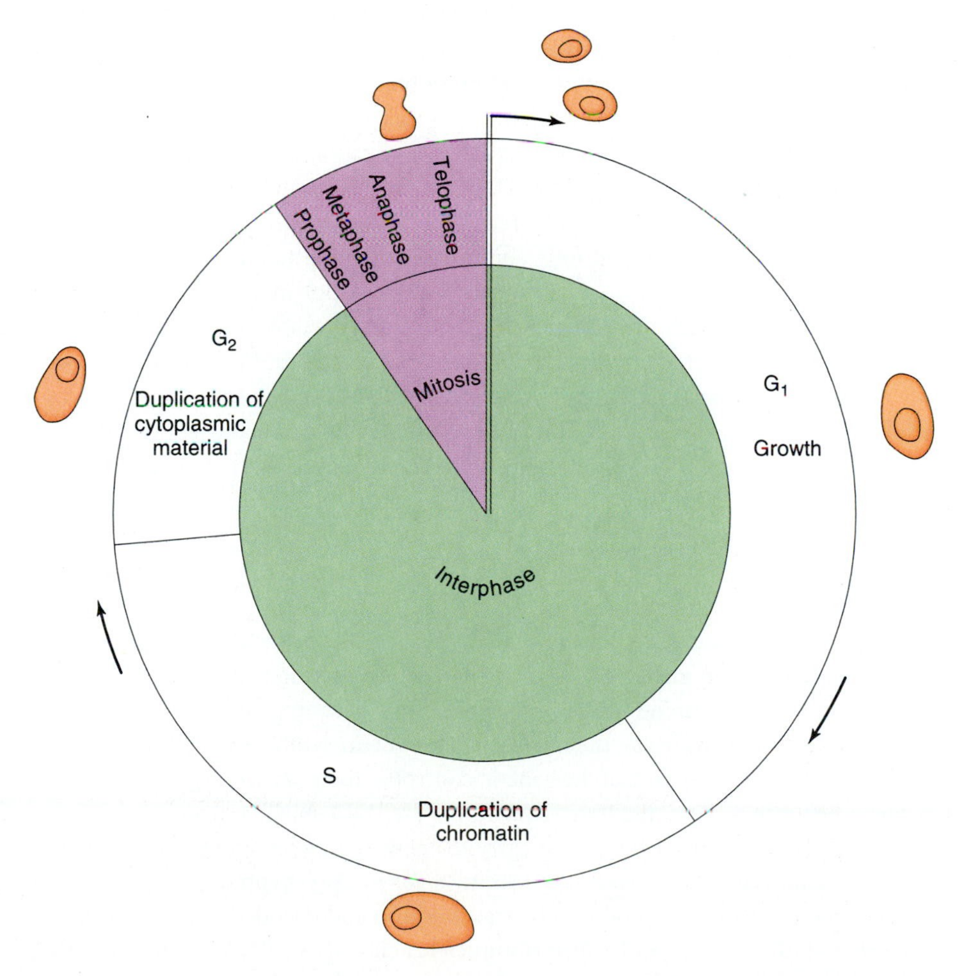

FIGURE 5.4

The cell cycle. In animals and other eukaryotes the cell cycle lasts from several hours to many years. The first gap phase, G_1, is the most variable in duration and depends on how small the daughter cell is compared with its final size. S phase, during which DNA and histones are synthesized, lasts 6 to 10 hours. G_2 lasts 3 to 5 hours, and mitosis takes 1 to 2 hours.

Replication

During S phase the cell has to construct a replica of its chromatin so that both daughter cells will receive a complete stock of genes identical to its own. Since the sequence of bases in DNA contains the genetic information, the problem of replicating the DNA is essentially one of synthesizing a new molecule of DNA with the same sequence of bases. One of the major achievements of the Watson–Crick model of DNA is that the solution to this problem falls out almost immediately. Because of the complementary base pairing of adenine to thymine, and of guanine to cytosine, it is only necessary to unzip the two strands of the double helix and use each one as a **template** for the construction of a complementary strand (Fig. 5.5A). In this way each of the two old strands of DNA gets a new complementary strand. Since half the old DNA is conserved on each new molecule, the process is called **semiconservative replication.**

The base and the energy required to attach it to the new DNA strand come from a molecule consisting of three phosphates attached to deoxyribose, to which is attached the base. An example of such a triphosphate is deoxyadenosine triphosphate (dATP), which has adenine as its base. dATP is identical to adenosine triphosphate (ATP) except that its sugar is deoxyribose, which has one less oxygen atom than ribose (see Fig. 3.1A). Like ATP, dATP transfers energy when the two terminal phosphates break off. The molecule that is left is the **nucleotide** adenosine, which consists of adenine, deoxyribose, and one phosphate. The energy released from dATP is used to bond the adenosine to the growing DNA strand, adding adenine to the sequence of bases (Fig. 5.5B). In the same manner dGTP, dCTP, and dTTP provide the energy to attach the nucleotides guanosine, cytidine, and thymidine where they are called for by the template strand of the DNA. Enzymes called **DNA polymerases** catalyze the attachment of these nucleotides to the new DNA strand.

The enzymes involved in semiconservative replication also help ensure the correctness of the sequence of bases. After DNA polymerases catalyze the attachment of nucleotides during replication, it **proofreads** the sequence. Because of proofreading, the wrong nucleotide is inserted only about one time in a billion, or approximately three times during each mitosis of a human cell. DNA polymerases also help repair damaged DNA. Although the double helix is intrinsically stable, ultraviolet radiation from sunlight, ionizing radiation, and other hazards can alter the bases, turning genetic information into gibberish. DNA polymerases and other repair enzymes usually recognize such abnormal bases, cut them out, and substitute the correct nucleotide called for by the complementary DNA strand. **DNA ligase** then catalyzes the patching of the new nucleotides to the old strand.

Transcription

Throughout interphase, genes direct the synthesis of all the proteins needed by the cell. The Watson–Crick model of DNA immediately suggests that the sequence of nucleotides in DNA determines the sequence of amino acids in proteins. In eukaryotes, however, DNA remains inside the nucleus while protein synthesis occurs in the cytoplasm. It follows, therefore, that the genetic control of the production of protein must occur in two stages. First, the base sequence in the DNA must be **transcribed** into a messenger. Next, the messenger must carry the genetic information to the cytoplasm where it is **translated** into a sequence of amino acids in a polypeptide.

The messenger that is produced by transcription and decoded during translation is another kind of nucleic acid, called **ribonucleic acid** (RNA). RNA is similar to DNA except for the following differences:

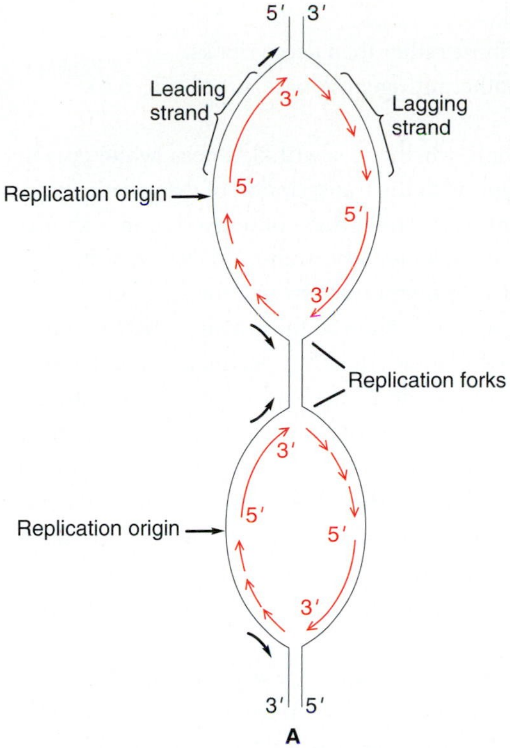

FIGURE 5.5

Schematic representation of semiconservative replication. No attempt is made to represent the helical structure of DNA or histones on the DNA. (A) Replication occurs at a rate of about three bases per minute at numerous replication origins. At each replication origin the two DNA strands are separated by an enzyme (DNA helicase), forming two replication forks that move away from each other. As the replication forks move apart, "unzipping" the double helix, each single strand of DNA serves as the template for the synthesis of a new strand. Nucleotides can be added to a growing new strand (color) only at the 3′ end of DNA, so each new strand can elongate only in its 5′-to-3′ direction indicated by arrows. Therefore only one strand, the leading strand, can grow steadily in the same direction as the movement of the replication fork. The other, lagging strand, must wait for the replication fork to move a certain distance (100 to 200 bases), then grow in the opposite direction. An enzyme, DNA ligase, splices the sections of the lagging strand together. Replication stops when one replication fork hits another. (B) DNA at a replication fork. On the left a deoxyadenosine triphosphate (dATP) has just lost two phosphates, leaving an adenosine nucleotide that is about to join the leading strand opposite a thymidine. On the right, deoxyguanosine triphosphate (dGTP) is about to add a guanosine to the lagging strand. Addition of new nucleotides is catalyzed by DNA polymerase. Note that the newly synthesized double helices will have identical base sequences.

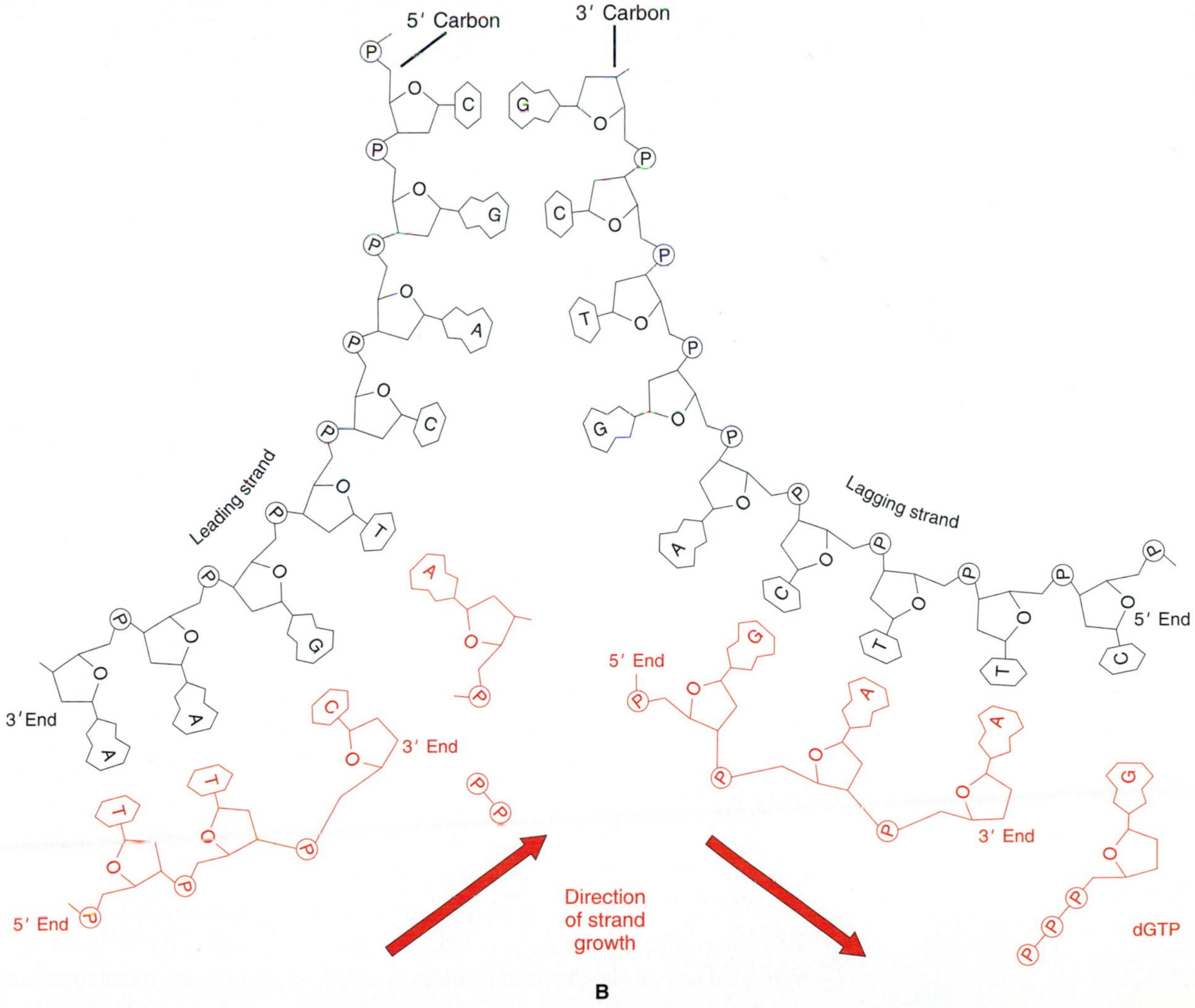

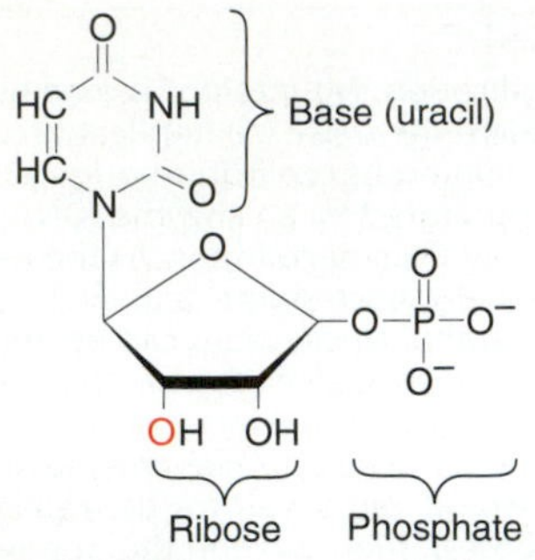

1. RNA is normally single-stranded.
2. The sugar in ribonucleic acid is ribose rather than deoxyribose.
3. In place of thymine, RNA has another nitrogenous base, uracil (Fig. 5.6).

FIGURE 5.6

A nucleotide of RNA consists of a phosphate, attached to the sugar ribose, to which is attached a nitrogenous base. In this example the base is uracil, forming the nucleotide uridine. Three of the RNA bases—adenine, guanine, and cytosine—are identical to those found in DNA nucleotides. Deoxyribose in DNA differs from ribose in that it lacks an oxygen (color).

The expression of genetic information, whether for wrinkled peas, white eyes in drosophila, or skin color in humans, begins with the transcription of the genetic information of DNA into **messenger RNA** (mRNA). The process of transcription is similar in many respects to the replication of a complementary strand of DNA. A section of the DNA double helix unzips, and an **RNA polymerase** enzyme moves along one of the DNA strands attaching complementary nucleotides to the growing mRNA strand. Wherever a thymine (T) occurs on the DNA strand, the RNA polymerase uses a molecule of ATP to add an adenosine (A) nucleotide to the end of the mRNA strand. Similarly, wherever G or C occurs on the DNA, the RNA polymerase adds a cytosine or guanosine, respectively. A major difference between transcription and replication is that when the RNA polymerase encounters an A on the DNA strand, it attaches the nucleotide uracil (U) rather than thymine. Another major difference is that the RNA strand does not remain attached to the DNA strand, but continually peels off as it is synthesized. The double helix zips up again about one turn past the point of transcription.

Transcription begins 20 to 30 bases away from a special sequence of DNA bases called the **promoter.** In order for a gene to become active, **transcription factors,** also called regulatory proteins, must bind to the promoter. The RNA polymerase then also binds to the promoter. RNA polymerase attaches in one orientation only, thereby determining the direction along the DNA in which the polymerase will transcribe. Since transcription can occur in only one direction, the polymerase will transcribe one strand of the DNA and not the other (Fig. 5.7). Transcription stops when the polymerase reaches a **termination sequence.** Typically, the distance between the promoter and the termination sequence is a few thousand bases, requiring several minutes to transcribe.

In eukaryotic cells the resulting RNA strand, called the **primary transcript** or **mRNA precursor,** is still not ready to direct the synthesis of a protein. First the primary transcript must undergo **processing.** The first step in mRNA processing is the attachment of a **5′ cap** of methyl guanosine. Following this capping, the other end of the primary transcript is cut, and a **poly-A tail** consisting of 150 to 200 adenosines is attached. Then, in seeming defiance of the economy of nature, up to 90% of the laboriously transcribed RNA is cut out and discarded. The discarded bits of RNA (as well as the sequences of DNA that code for them) are called intervening sequences: **introns,** for short. The remaining portions of mRNA (as well as the sequences of DNA that code for them) are called **exons.** An average gene consists of 15 to 20 exons. There is further **editing,** by addition or deletion of bases here and there, before the mature mRNA is ready. This mature mRNA, consisting of a series of edited exons with a poly-A tail and a cap at the ends, is then exported through the pores in the nuclear envelope (see Fig. 2.31) into the cytoplasm.

Translation

Messenger RNA directs the assembly of a polypeptide by specifying the amino acids that are to be attached as the polypeptide is produced. These instructions are contained in the sequence of bases in the mRNA, which was transcribed from the DNA of the gene coding for the polypeptide. The process by which the language of genes is translated into the language of proteins is called, appropriately, translation. The key to

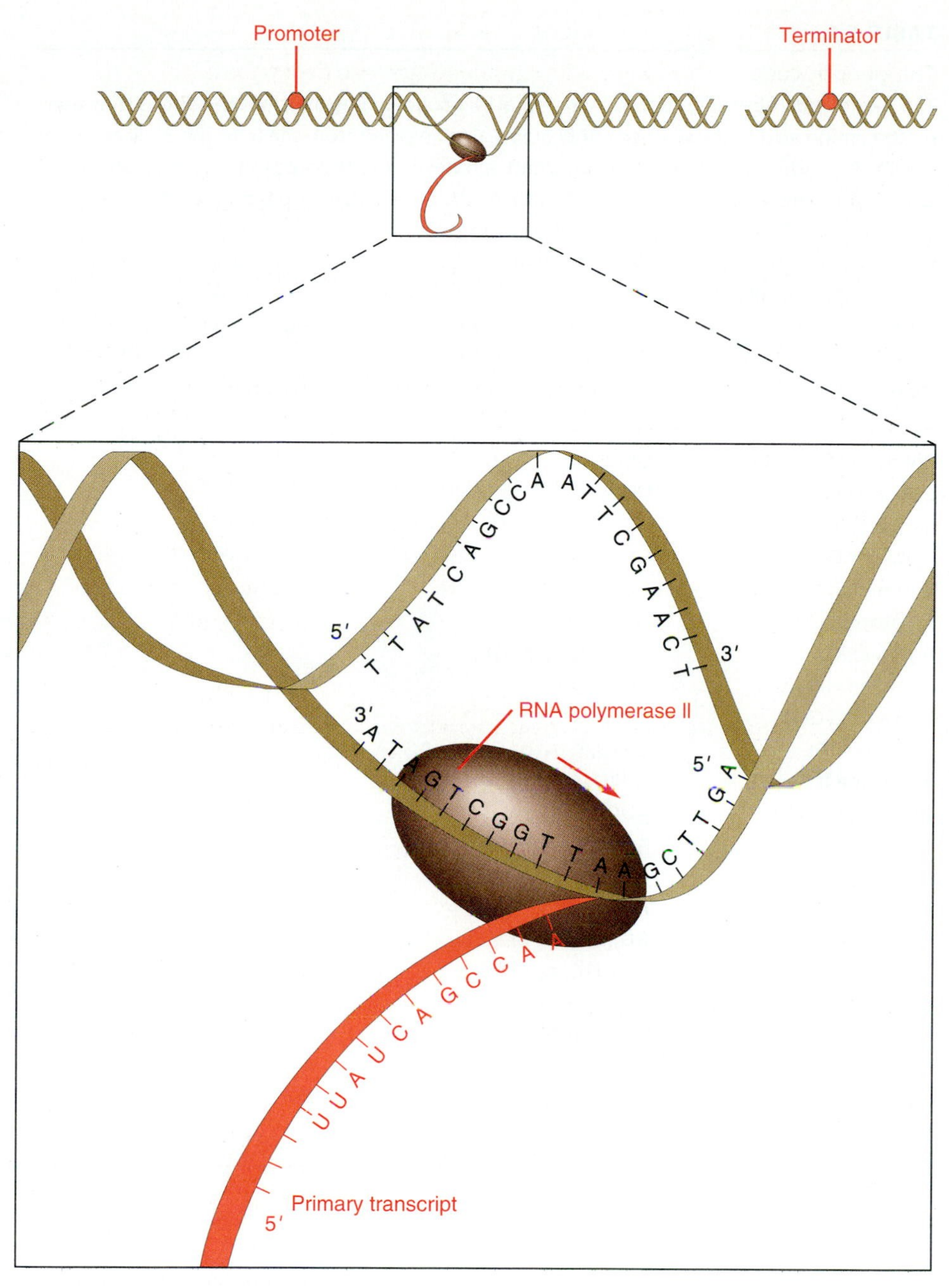

FIGURE 5.7

Transcription. Beginning 20 to 30 nucleotides "downstream" from the promoter sequence on DNA, RNA polymerase II unzips a turn of the double helix and synthesizes a complementary strand of RNA in the 5′-to-3′ direction. Transcription stops at the terminator sequence. Note that the base sequence on the primary RNA transcript is identical to the base sequence on the nontranscribed DNA strand, except that uracil substitutes for thymine.

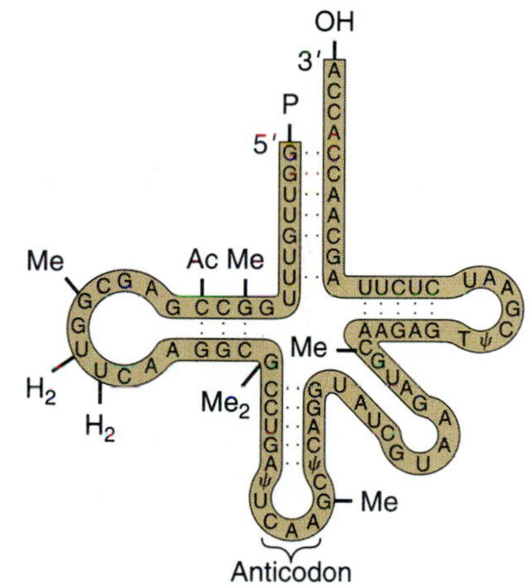

FIGURE 5.8

The structure of transfer ribonucleic acid in yeast. Self-complementary sequences (shown joined by dots) cause the tRNA to fold into this complex shape. Enzymes called aminoacyl-tRNA synthetases catalyze the attachment of the appropriate amino acid to the 3′ attachment site of the tRNA. Letters other than U, C, A, and G refer to modified bases. I, for example, represents inosine.

■ **Question:** What amino acid would be attached at the 3′ end of this tRNA?

translation is the **genetic code.** With only four different bases in mRNA and 20 different amino acids in proteins, the genetic code clearly cannot simply specify one amino acid whenever a particular base occurs. Even two bases would be inadequate to specify a particular amino acid, since the number of combinations of four bases taken two at a time is only 16 (4^2). However, the number of combinations of three bases is 64 (4^3), which is more than adequate to specify 20 amino acids. Intensive research throughout the 1960s established that the genetic code is indeed based on triplets of bases. Each triplet on the mRNA, called a **codon,** specifies an amino acid to be added during the synthesis of a protein, or it signals the stop or start of translation (Table 5.1). Most codons have "synonyms" that code for the same amino acid. The genetic code is therefore said to be **degenerate.**

The actual translation of the genetic code is performed by a different kind of RNA called **transfer RNA** (tRNA) (Fig. 5.8). Transfer RNA is similar to mRNA except that

TABLE 5.1

The genetic code.

The name and abbreviation of each amino acid is followed by a list of the codons that specify the amino acid in the synthesis of proteins. Codons read from the 5′ to the 3′ direction of mRNA. Parentheses enclose all the alternative bases in the third position of the codon that specify the same amino acid. The codons UAA, UAG, and UGA stop synthesis of the protein. The codon AUG codes for methionine, except for the first AUG near the 5′ end of the mRNA, which starts translation. A few minor "dialects" in the genetic code have been discovered. For example, in certain protozoans called ciliates, UAA and UAG are codons for glutamine rather than stop codons. Also, in the mitochondria of mammals, AGA and AGG are stop codons rather than codons for arginine, AUA is a codon for methionine rather than isoleucine, and UGA codes for tryptophan instead of being a stop codon.

Phenylalanine PHE	UU(U or C)
Leucine LEU	UU(A or G) and CU (U, C, A, or G)
Serine SER	UC (U, C, A, or G) and AG (U or C)
Tyrosine TYR	UA (U or C)
Cysteine CYS	UG (U or C)
Tryptophan TRP	UGG
Proline PRO	CC (U, C, A, or G)
Histidine HIS	CA (U or C)
Glutamine GLN	CA (A or G)
Arginine ARG	CG (U, C, A, or G) and AG (A or G)
Isoleucine ILE	AU (U, C, or A)
Methionine MET	AUG
Threonine THR	AC (U, C, A, or G)
Asparagine ASN	AA (U or C)
Lysine LYS	AA (A or G)
Valine VAL	GU (U, C, A, or G)
Alanine ALA	GC (U, C, A, or G)
Aspartic acid ASP	GA (U or C)
Glutamic acid GLU	GA (A or G)
Glycine GLY	GG (U, C, A, or G)
Stop	UAA, UAG, and UGA
Start	First AUG at 5′ end

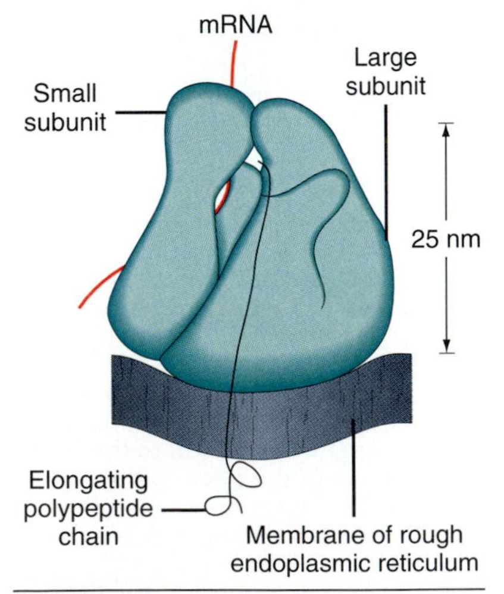

FIGURE 5.9

Model of a ribosome bound to rough endoplasmic reticulum, showing the position of mRNA and the protein during translation.

many of the bases are altered. Each kind of tRNA attaches to a different amino acid. In addition, each kind of tRNA has a different combination of three bases, forming an **anticodon** that recognizes a complementary codon on mRNA during translation. For example, the tRNA that carries the amino acid phenylalanine has the anticodon sequence AAG in the 3′ to 5′ direction. AAG binds to the codon UUC on mRNA in the 5′ to 3′ direction, causing the tRNA to transfer the phenylalanine to the protein being synthesized. Like most tRNAs, this one has some **wobble** to it, so it will also bind to the codon UUU. Wobble partly accounts for the degeneracy of the genetic code.

The actual synthesis of proteins takes place on organelles called **ribosomes,** which are located either on rough endoplasmic reticulum (see p. 45) or free within the cytoplasm. Ribosomes are made within the cell nucleus in a dark-staining region called the **nucleolus.** Each ribosome consists of a large and a small subunit comprising several kinds of **ribosomal RNA** (rRNA) and more than a hundred different proteins (Fig. 5.9). A ribosome begins translation at the start (AUG) codon nearest the 5′ cap, then moves along the mRNA in one direction (Fig. 5.10). It moves three bases at a time, pausing at

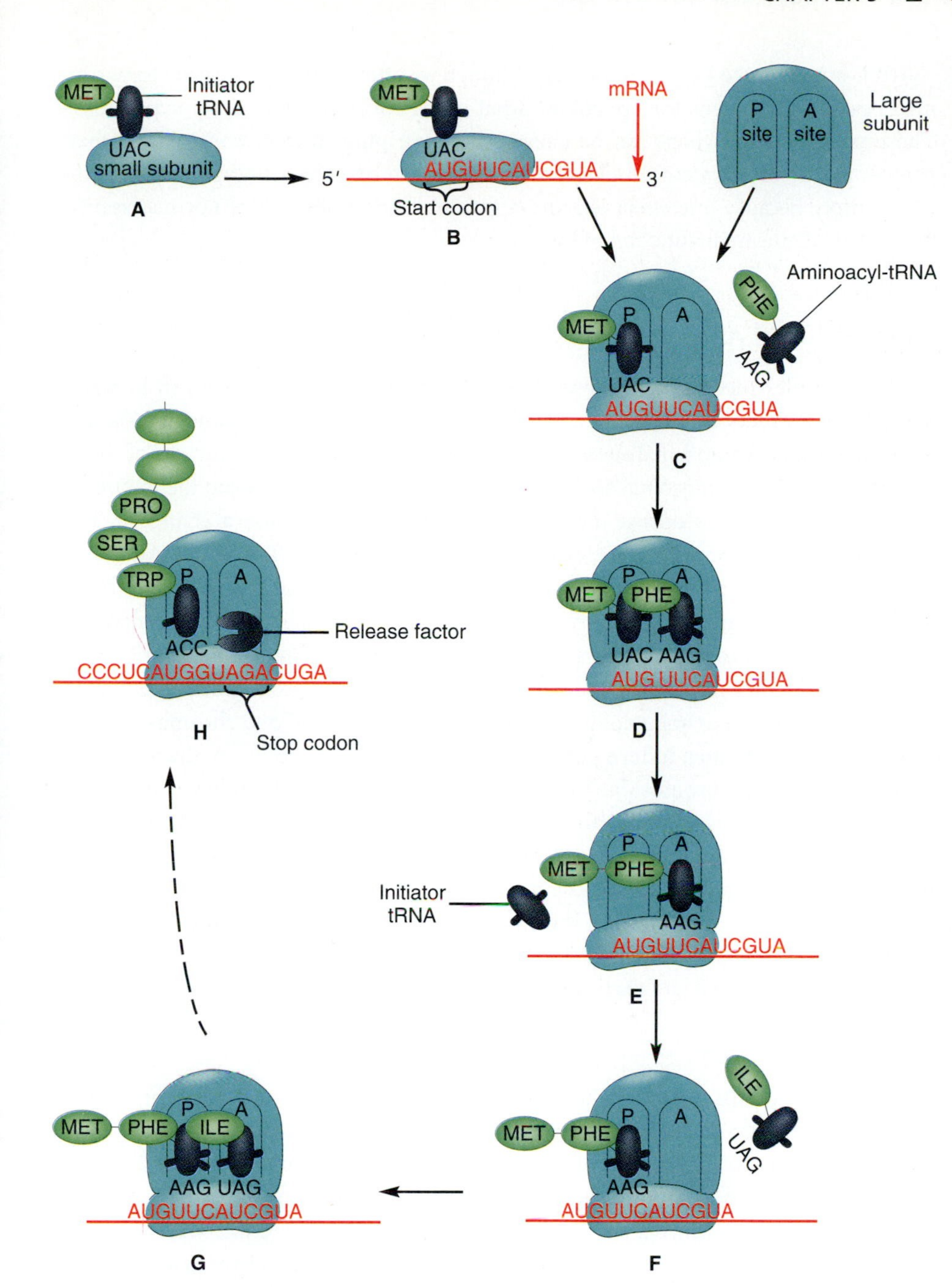

FIGURE 5.10

Schematic representation of translation and protein synthesis. Before translation the two ribosomal subunits are separated. (A) Translation begins when an initiator tRNA carrying methionine, together with certain initiation factors, binds to a small subunit. (B) This initiation complex then binds to the start codon of mRNA. (C) A large ribosomal subunit with P and A sites then attaches to the small subunit in such a way that a P site of the large subunit is next to the initiator tRNA. The codon on the 3′ side of the start codon is near the A site on the large subunit. This is where the appropriate aminoacyl-tRNA (tRNA with its attached amino acid) attaches. (D, E) After the aminoacyl-tRNA attaches, a peptide bond is formed between its amino acid and the methionine (MET). During this step the methionine detaches from the initiator tRNA, forming a dipeptide, and the initiator tRNA then drops off the P site. (F) The ribosome then moves three bases in the 5′-to-3′ direction on the mRNA. The tRNA with the attached dipeptide (called a peptidyl-tRNA) is then at the P site of the ribosome. (G) Another aminoacyl-tRNA then attaches to the A site. A peptide bond joins its amino acid to the dipeptide, the tRNA at the P site drops off, the ribosome shifts again to the next codon, and so on. (H) This process continues until the A site of the ribosome reaches a stop codon. When that occurs, a release factor attaches at the A site, terminating translation. The protein is released, and the two ribosomal subunits separate from the mRNA.

each codon until the tRNA matching the codon attaches. An enzyme then catalyzes the formation of a peptide bond between the tRNA's amino acid and the peptide chain being synthesized. This process continues until the ribosome reaches a stop codon.

Ribosomes spend an average of 50 milliseconds at each codon, attaching amino acids at a rate of about 20 per second. By the time the first ribosome has reached the stop codon, dozens of others may have begun transcribing the same mRNA, resulting in a chain of ribosomes linked to a single mRNA strand, forming a **polysome.** In this way a single mRNA can direct the production of many copies of a polypeptide simultaneously. Actual measurements of the total rate of protein synthesis are impressive. A silkworm caterpillar (*Bombyx mori*) makes 10,000 copies of the mRNA for fibroin, the protein in silk. Over a period of four days, each of these mRNAs is translated into 100,000 fibroin molecules, for a total of 10^9.

It is important to keep in mind that although we often speak of "the gene for wrinkled peas," or "the gene for eye color," what genes are really "for" are polypeptides. These polypeptides, which can be enzymes, transcription factors controlling other genes, or structural proteins, usually have many effects on the phenotype of an organism. Simply because one effect is more apparent than another does not necessarily mean that it is the main function of the gene.

Chromosomal Aberrations

Like all complex mechanisms, those of genetics sometimes malfunction. In humans some 4000 disorders are known to result from genetic malfunctions. Some of these are associated with chromosomal aberrations: changes in the number or structure of chromosomes. Some chromosomal aberrations result from **nondisjunction:** the failure of chromatids to separate during meiosis. Nondisjunction can cause a change in the number of chromosomes in a set, called **aneuploidy** (AN-you-PLOY-dee). Although aneuploidy is often harmful or lethal, some animals, especially salamanders and frogs, normally have a form of aneuploidy called **polyploidy,** in which there are as many as six entire sets of chromosomes in addition to the usual two. That is, instead of being merely diploid, they may be triploid, tetraploid, or as much as octaploid. Polyploidy is often not as harmful as aneuploidy with only one extra copy of one chromosome. In fact, plant breeders often induce polyploidy to improve crop species. A change in the structure of a single chromosome is often more serious than aneuploidy. Such a chromosomal aberration is often due to the **translocation** of a fragment of a chromosome to the end of a different chromosome. The individual therefore receives an extra copy of the genes on the fragment or lacks a copy of those genes.

Aneuploidy and translocation show up in the **karyotype,** which is the set of chromosomes as they appear under the microscope. Karyotypes of human fetuses can be obtained by collecting fetal cells from the amniotic fluid through the process of amniocentesis.

Down Syndrome. Several kinds of abnormal karyotypes are frequent in humans. Probably the most familiar is Down syndrome, also called trisomy 21. Down syndrome is due to inheritance of an extra copy of chromosome 21 (Fig. 5.11) or translocation of part of chromosome 21 to some other chromosome. In the United States, Down syndrome occurs in approximately 1 out of 700 live births. Most cases are thought to be due to some mishap of meiosis in ova, because the chances of having a child with Down syndrome rises sharply if the mother is over age 35. Aneuploidy and translocation affect other chromosomes as well, but they are generally lethal to the fetus. Chromosome 21 is one of the smallest chromosomes, however, and it presumably contains fewer genes.

About 20% of fetuses trisomic for chromosome 21 survive to birth, although often with severe abnormalities. The major consequences of Down syndrome include mental retardation, defects in the heart and immune system, increased risk of leukemia and cataracts, and characteristic anatomical features such as flattened face, short stature, and unusual palm creases. Thanks to improved medical care and support, however, many victims of Down syndrome now live happily past middle age.

Aneuploidy of Sex Chromosomes. Other common chromosomal aberrations in humans involve aneuploidy of X or Y chromosomes. Trisomy XXY results in

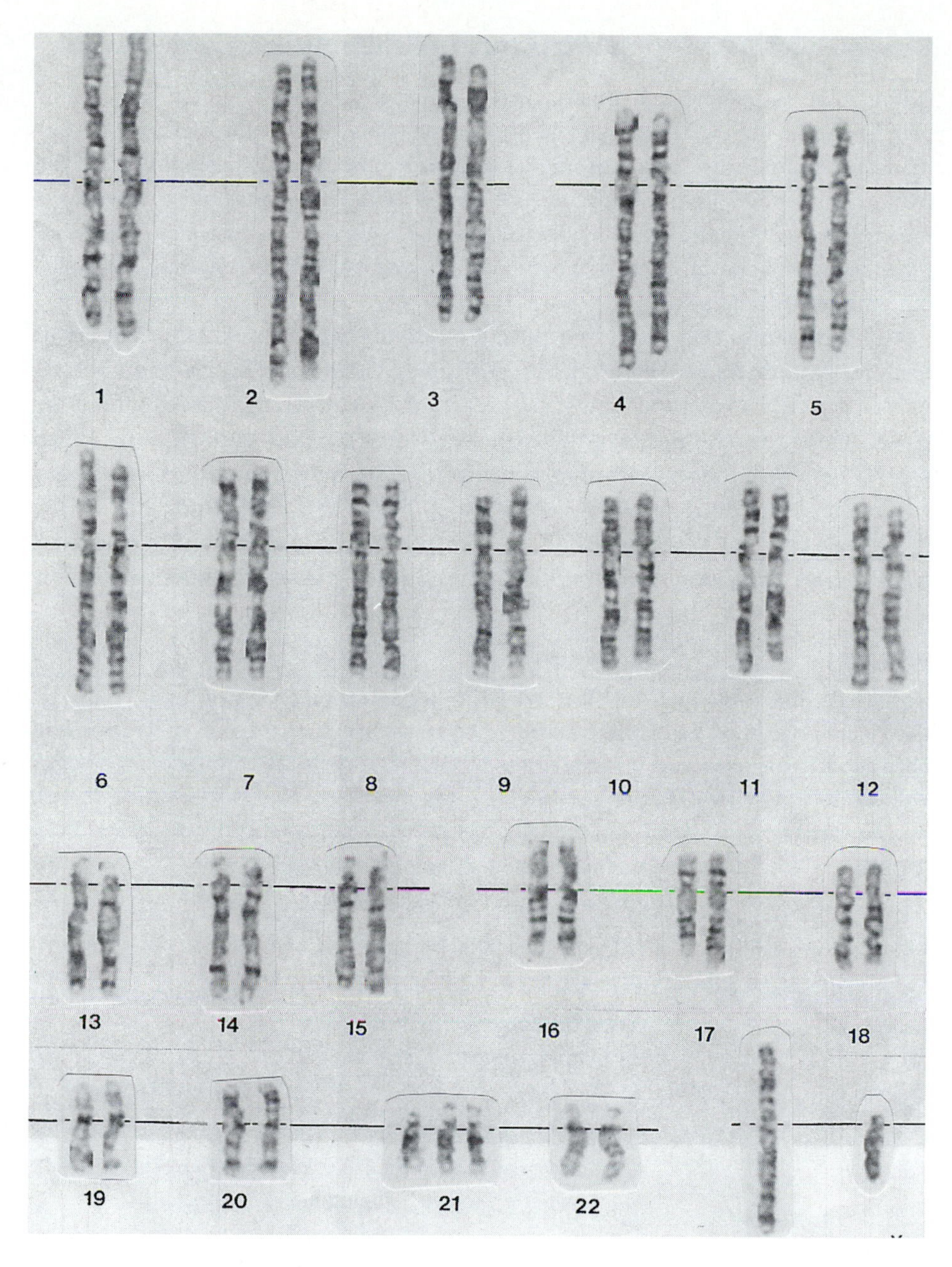

FIGURE 5.11

The karyotype of a person with Down syndrome. Note the extra chromosome numbered 21.

■ **Question:** Is this person male or female?

the development of a male, since a gene on the Y chromosome determines maleness in humans, but the additional X chromosome leads to testicular underdevelopment and feminization (Fig. 5.12 on page 96). An individual with only the X chromosome (XO) develops as a female who lacks ovaries and has other abnormalities. The most common aneuploidy of the sex chromosomes, affecting an average of several males in 10 thousand, is trisomy XYY. Individual XYY males are difficult to pick out of a crowd, but on average they are taller and slightly retarded. They are also more likely to be in prison than are men with normal karyotypes. This has led some to suppose that there was a gene for criminality on the extra Y chromosome. It is just as likely, however, that the intellectual impairment leads them to commit impulsive crimes and to make mistakes that get them convicted. This is a good illustration of how risky it is to jump to conclusions about genetic causes for behavior. (See pp. 406–407 for further discussion of the genetics of human behavior.)

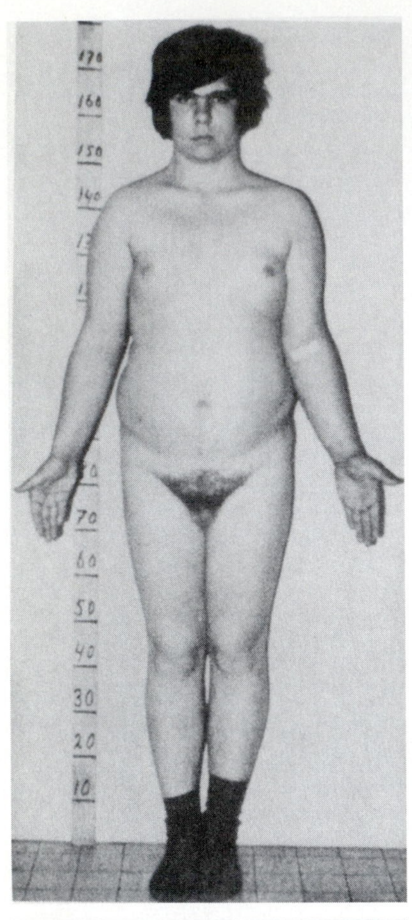

FIGURE 5.12

One or two males in a thousand have Klinefelter's syndrome, trisomy XXY, resulting in atrophied tecticles and enlarged breasts. Treatment with testosterone corrects most of the symptoms.

Changes in Chromosomal Structure in Evolution and Development

Although devastating to individuals, chromosomal aberrations are often harmless or beneficial over the evolutionary scale of time. At some time in the evolution of humans, for example, two chromosomes fused, reducing our haploid number from the 24 found in living apes to our present 23. Apparently this reduction in chromosome number did us no harm, presumably because there was no loss of genes. These and other changes in chromosome number have occurred, apparently randomly, throughout evolution (Table 5.2).

Chromosomes also undergo a variety of lesser changes in structure. Some of these result from **unequal crossing over.** Unequal crossing over occurs in prophase I of meiosis when two homologous chromosomes do not break in exactly the same place. One of the chromosomes therefore loses some genes or parts of genes to the other. Two other causes of chromosomal change are inversion and transposition. **Inversion** occurs during synapsis when a loop forms in a chromatid, then twists, breaks off, and reattaches backward. In **transposition** a segment of DNA called a **transposable element,** or sometimes a "jumping gene," moves to a different location

TABLE 5.2.

Diploid chromosome numbers for animals representative of major taxa.

Even within classes of animals the number can vary widely, suggesting that chromosomal aberrations occur randomly. In many fishes, reptiles, and birds the chromosomes occur in two distinct sizes: macrochromosomes and microchromosomes.

SPONGES	Freshwater sponge, *Spongilla lacustris*	10
	Marine sponge, *Leucosolenia ciliata*	26
NEMATODES (Males)	Soil nematode, *Rhabditis* sp.	13
	Intestinal roundworm, *Ascaris lumbricoides*	43
ANNELIDS	Leech, *Dina lineata*	18
	Earthworm, *Lumbricus terrestris*	36
ARTHROPODS	Crayfish, *Cambarus clarkii*	200
	House fly, *Musca domestica*	12
	Ant, *Formica sanguinea* (females; males are haploid)	48
	Ant, *Myrmecia pilosula* (females; males are haploid)	2
FISHES	Northern pike, *Esox lucius*	18
	Goldfish, *Carassius auratus*	100
AMPHIBIANS	Grass frog, *Rana pipiens*	26
	Clawed frog, *Xenopus laevis*	36
BIRDS	House sparrow, *Passer domesticus*	76
	Pigeon, *Columba livia*	80
MAMMALS	Chinese hamster, *Cricetulus griseus*	22
	Golden hamster, *Mesocricetus auratus*	44
	Dog, *Canis familiaris*	78
	Donkey, *Equus asinus*	63
	Horse, *Equus caballus*	64
	Rhesus monkey, *Macaca mulatta*	42
	Gorilla, *Gorilla gorilla*	48
	Chimpanzee, *Pan troglodytes*	48
	Human, *Homo sapiens*	46

on the same chromosome or another chromosome. Unlike other chromosomal changes, which are almost always random and harmful, many inversions and transpositions are examples of "programmed gene rearrangements" that occur in a systematic and useful manner (Borst and Greaves 1987). They allow for the duplication and deletion of genes at appropriate times in development, and they enable gene segments to recombine in a wide range of combinations. The latter function accounts for our ability to produce millions of different antibodies from just a few genes. There is also evidence that transposable elements have been transferred between different species of *Drosophila* by mites that feed on the blood of both species (Rennie 1993). Some biologists suspect that such interspecific transposition could be important in evolution.

Geneticists now believe that changes in chromosome structure account for an essential process in evolution: the acquisition of new genes. Translocation, aneuploidy, inversion, transposition, and unequal crossing over can give rise to duplicate genes that subsequently mutate and produce new products. Evidence that new genes arise by duplication and subsequent mutation lies in the existence of **protein families.** One example of a protein family is the three pigments in the human eye that are responsible for color vision. All three genes for these proteins are similar and close to each other on the X chromosome, and they are therefore likely to have arisen through unequal crossing over.

Instead of taking on a new role, a duplicated gene can also lose its promoter and become a **pseudogene.** Pseudogenes are never transcribed, but they are recognizable because their base sequences are similar to those of functioning genes. Pseudogenes and introns make up a large proportion of most organisms' **genomes** (the genetic material in a haploid set of chromosomes). In humans about 97% of the chromatin consists of such DNA. Such "junk DNA" does not encode polypeptides and has no known function other than to replicate itself. Perhaps because so much of the genome consists of introns and pseudogenes, there is no consistent relationship between phylogeny and genome size (Fig. 5.13).

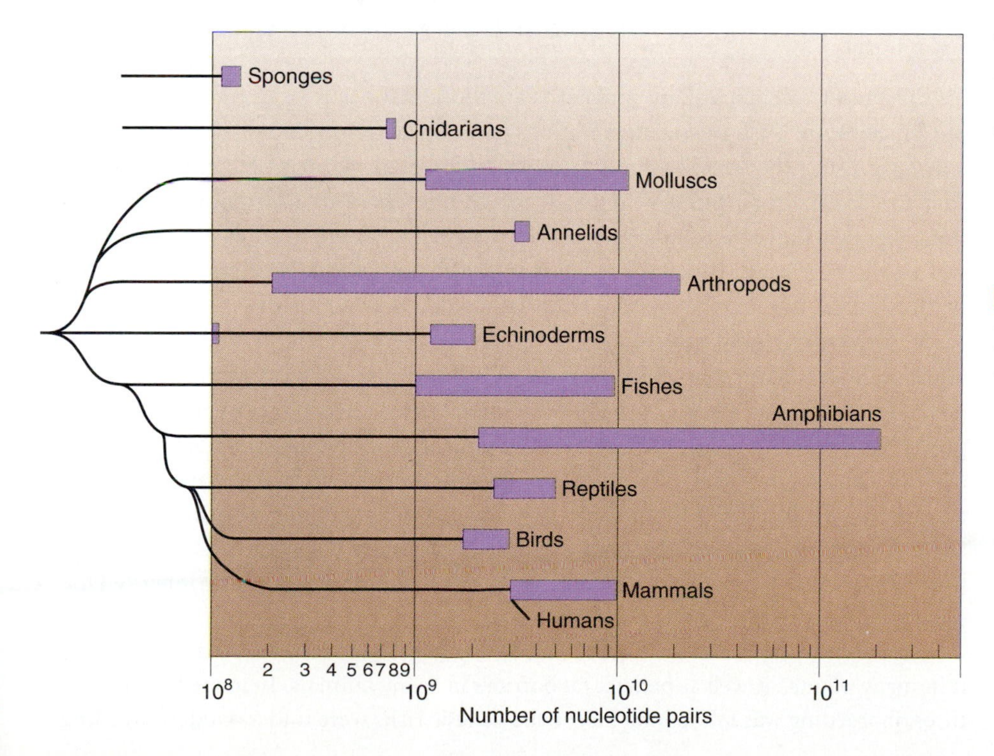

FIGURE 5.13
The range of genome size in a variety of animals. The lack of relationship between DNA content and either phylogenetic position or physiological complexity is called the C-value paradox. Genome size, however, is well correlated with cell size.

■ **Question:** Does physiological complexity necessarily require more genetic information?

Recent evidence suggests that changes in chromosome structure also control cell division, and therefore tissue growth. Each end of a chromosome is normally protected by a combination of special proteins and base sequences called the **telomere** (see Fig. 4.6). During each S phase of the cell cycle, however, parts of the telomere are lost, eventually causing most cells to stop dividing and die. Each cell is capable of producing an enzyme called **telomerase** that repairs the damage to the telomeres and makes the cell immortal, but most cells normally do not do so. While immortality might appear to be a good thing for the cell, it would allow uncontrolled growth of tissue. It appears, in fact, that such repair of telomeres is what enables tumors to grow unchecked once cell division has been stimulated by cancer-causing genes (**oncogenes**).

Point Mutations

Most mutations do not affect the number or structure of chromosomes, but only a single DNA base pair. These are called **point mutations.** Thousands of point mutations occur in each cell every day, but most are promptly repaired. Nevertheless, point mutations accumulate at a level of about one per billion base pairs. **Mutagens,** such as radiation and certain chemicals, increase the rate. Fortunately, most point mutations are **neutral:** they have no effect on the gene product because they occur in introns or at the "wobbly" third base of a codon. Point mutations are more common in these harmless positions, apparently because they do not reduce the ability to reproduce. Some point mutations have no effect on the functioning of a protein even if they do change the amino acid sequence. Bovine (cow) insulin differs from human insulin in three different amino acids out of 51, yet bovine insulin works quite well in human diabetics. Apparently the amino acid differences occur at sites on the insulin that are not crucial for its functioning.

On the other hand, a point mutation that changes the tertiary structure of an enzyme or other crucial protein could have serious consequences. For example, if the + charge that helps determine the configuration of the protein in Fig. 2.16 were due to lysine, and a point mutation changed the first base in its codon from A to G, the entire structure of the protein might be affected. (Refer to Tables 5.1 and Fig. 2.13 to see why.)

The most damaging point mutations are those that occur in a gene for ribosomal or transfer RNA. Such **missense mutations** can stop all translation or affect every protein by causing the wrong amino acids to be inserted. Some point mutations also change a codon for an amino acid into a stop codon. These are called **nonsense mutations,** because they result in useless peptide fragments. There are also mutations that add or delete a base. These are called **frameshift mutations,** because they shift the **reading frame,** altering the message of every triplet following the site of the mutation.

Even harmful point mutations may have no immediate effect if they are recessive, as most are. It is likely that you have three, four, or five such recessive mutations. The normal homologous allele is usually dominant because it produces enough normal product to compensate for the mutant allele. Only if an individual inherits one mutant allele from both parents and becomes homozygous recessive for the mutation will it be expressed. The likelihood of both parents being carriers of a mutated form of the same allele is normally remote. The risks increase appreciably when there is **inbreeding,** however, because there is then a much greater risk that both parents have inherited the same mutation from the same ancestor.

Apparently the harmful effects of inbreeding have led to structural barriers against it in many plants, as well as behavioral barriers in many animals. In many human cultures, inbreeding was taboo long before the genetic risks were understood. Inbreeding is a major concern in rearing captive animals and in managing small nature reserves. For that reason, managers of zoos frequently exchange animals for breeding purposes. Inbreeding also threatens wild endangered species, which may succumb to internal genetic disorders even if they escape external threats.

Even without inbreeding, disorders due to certain point mutations do occur repeatedly in populations. Defects due to recessive point mutations linked to the X chromosome are especially common. One example of such a sex-linked trait is the inability to distinguish red and green, which affects 5% of males of northern European descent. A more serious X-linked mutation is Duchenne muscular dystrophy, which totally paralyzes about one in every 3500 boys. The cause of the disease is a mutation in the gene for a muscle protein called dystrophin. One might think that a gene that restricted its own opportunities for inheritance so severely would soon disappear from a population. Because females have a normal gene on one of their X chromosomes, however, they can act as **carriers,** passing the mutation on for several generations before it occurs in a male and is selected against. Half the male children of carriers will inherit the disorder. Moreover, the gene for dystrophin is one of the largest ever discovered, and it is therefore a large target for spontaneous mutation. Approximately one-third of all cases are due to new mutations.

A similar set of circumstances is responsible for most cases of **hemophilia,** a disorder that impairs blood clotting in about one male out of every 10,000 (Fig. 5.14). In the most common form of hemophilia, a point mutation prevents an allele from producing normal factor VIII, one of the molecules essential for the clotting of blood. Until injections of factor VIII became widely available, hemophiliacs usually died of internal or external bleeding before they reached age 20. Factor VIII consists of 2332 amino acids, so, like the gene for dystrophin, its gene is a relatively large target for spontaneous mutation.

SICKLE-CELL ANEMIA. Hemophilia is an example of a frequent harmful mutation that survives in large numbers because its effects are not immediately selected against. There are also genetic defects that survive in large numbers because selection favors them under certain conditions. The classic example is sickle-cell anemia. The allele for sickle-cell anemia is the mutated gene on chromosome 11 that codes for the beta chain of hemoglobin, the oxygen-carrying molecule that fills red blood cells (see Fig. 12.3).

FIGURE 5.14

A pedigree chart showing the occurrence of hemophilia among the royal families of England, Germany, Russia, and Spain. One of the first steps in studying a genetic mutation is the recording of such pedigrees. This particular example is known because royalty record their pedigrees in unusual detail.

■ **Question:** How do we know that Queen Victoria and some others were carriers, since they had no symptoms? Why is the status of many of the women uncertain?

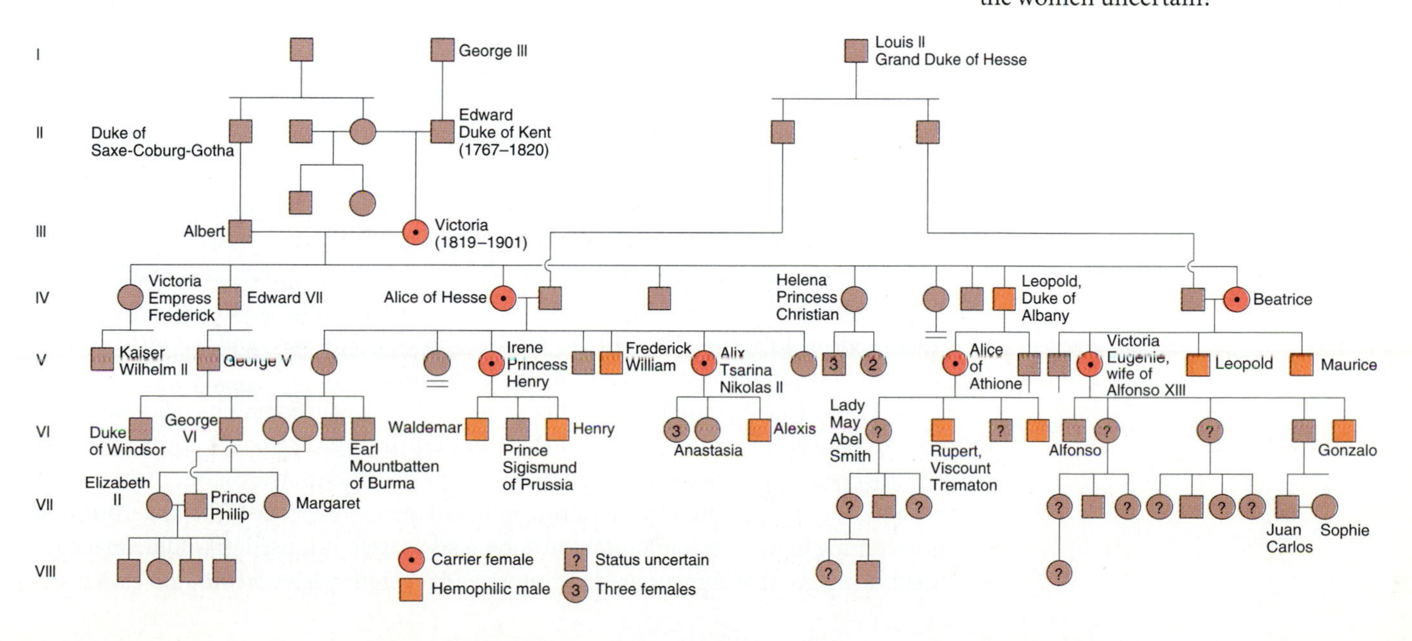

This point mutation consists of the substitution of an adenine for a uracil in one codon. This causes valine to be substituted for glutamic acid at the sixth position from the amino end of the beta chain, which causes the red blood cell to assume a sickle shape when the oxygen level in blood is low. Those who are homozygous for the allele suffer from sickle-cell anemia, and most die from clogged blood vessels. The allele is not recessive, so heterozygotes are also affected, though not as severely. Heterozygotes, who are said to have **sickle-cell trait,** live normal lives under most circumstances. Only when their blood is depleted of oxygen do the abnormal cells become sickle shaped, resulting in severe pain and tissue damage.

In spite of the clear disadvantages of the allele, it is common in much of Africa and southern Asia. Eight percent of Americans of African descent also carry the allele. (A related disorder, **beta thalassemia,** affects some of Mediterranean descent.) Why hasn't the allele disappeared? The first clue to this mystery came from maps showing that the native range for sickle-cell anemia coincides with that for malaria. It happens that the sickling of red blood cells inhibits the reproduction of the malaria parasite *Plasmodium* within them (see p. 472). The debilitating effects of sickle-cell trait are therefore compensated by resistance to the even more debilitating disease of malaria. Homozygous sickle-cell anemia is not compensated, of course, since it is even worse than malaria. Apparently, however, the disadvantages of homozygosity are balanced against the advantages of heterozygosity. Loss of the allele due to the death of homozygotes is offset by the increased reproduction of the allele by heterozygotes.

Genetic Screening

An understanding of genetics at the molecular level has led to a variety of techniques for analyzing the genes carried by individual organisms. Genetic screening requires only a small sample of blood, saliva, or other body fluids. Tiny amounts of DNA from such samples can be multiplied many times by **polymerase chain reaction** (PCR). In PCR the DNA is repeatedly cleaved into single strands, then treated with polymerase to double the amount of DNA. The DNA can then be mixed with radioactive **complementary DNA** (cDNA) probes. These are segments of DNA with a base sequence similar to that of a gene for which an individual is being screened. When mixed with the single-stranded DNA of the individual, the cDNA probe binds with the gene, revealing its presence by the radiation.

Another powerful tool for genetic screening is **restriction enzyme** from bacteria. Restriction enzymes cut DNA at specific sequences of bases, usually where four to six bases on each strand have the same sequence in opposite directions. For example, the restriction enzyme EcoRI cleaves DNA at the sequence GAATTC, whose complementary strand reads the same in reverse. Because of point mutations at these sequences, a restriction enzyme will cut at certain sites on the DNA of one individual, but not in another. Thus the lengths of the fragments will vary from one individual to another. The different DNA fragment lengths are called **restriction fragment length polymorphisms, or RFLPs.** Numerous RFLPs can be isolated in this way, and it is virtually impossible for all of them to be identical in two individuals.

RFLPs make it possible to identify an individual organism and its relationship to others. This process is referred to as **DNA fingerprinting.** The best-known and most controversial applications of DNA fingerprinting are in criminal trials, in which guilt or innocence may be established from samples of semen or blood. DNA fingerprinting is becoming increasingly important to zoologists as well. It is used in wildlife management to ensure that animals bred to restore endangered species are not too closely related to each other. It is being used in zoos to prevent unwanted hybridization by being sure that animals selected for breeding belong to the same subspecies. DNA fingerprinting is also used in wildlife conservation to determine whether animals or their tusks, hides, or other products were obtained illegally. Zoologists who study animal behavior also use DNA fingerprinting to determine the degree to which animals that are thought to be monogamous are faithful to their mates. (Apparently few are.)

RFLPs also serve as **markers** that indicate the presence of a harmful mutation. If a particular RFLP is found to occur consistently in individuals suffering from a mutation, then the presence of the marker can be used to determine whether a person is likely to be carrying the mutant allele. One application is in identifying potential victims of Huntington's disease. Huntington's disease results from an autosomal dominant allele that causes the brain to degenerate, starting in middle age and continuing until death after decades of suffering. Without a genetic marker the disease cannot be detected until its symptoms begin, usually after victims have already had children, each of whom will have a 50-50 chance of having inherited the allele. Using an RFLP marker to detect the presence or absence of the allele would alleviate their uncertainty and would help them to decide whether to have children.

Using RFLPs and linkage mapping, researchers have located the genes for many disorders, including Huntington's disease, Duchenne muscular dystrophy, cystic fibrosis, familial Alzheimer's disease, and neurofibromatosis (so called "Elephant man's disease"). Ultimately, it should be possible to map all of the 100,000 human genes, providing a linkage map similar to that for drosophila chromosomes (see Fig. 4.12). Such a gene map could be the first step toward a brave new world in which humans directly intervene in altering their own genetic makeup and that of other organisms. The first steps in such an undertaking are already underway as part of the Human Genome Project, which aims to determine all three billion base pairs in the human genome.

Genetic Engineering

Genetic screening and mapping may soon make us among the last to grow up believing that genetic defects are as inescapable as an unlucky throw of the dice. We may be among the first to be able to load the genetic dice or take back the throw by repairing such defects through genetic engineering. The main tool enabling genetic engineers to fulfill this goal is **recombinant DNA** technology. Recombinant DNA technology, also called gene-splicing, refers to the artificial transfer of a gene or genes from one organism into another. It is becoming a standard tool in research, wildlife management, and other areas of zoology. It is important, therefore, to understand what the technology involves.

Some bacteria naturally exchange genetic material from one species to another in the form of a circular strand of DNA called a **plasmid.** This natural recombination accounts for the spread of antibiotic resistance among different species of bacteria. *Agrobacterium tumefaciens,* the bacterium responsible for crown gall disease in plants, also transfers plasmids to the plants it infects. In certain cases genetic engineers can substitute desired genes for those on a plasmid, then allow bacteria to transfer the plasmid. Probably the most common way to introduce foreign genetic material is to use viruses. When viruses infect cells, they introduce their own genetic material, which then essentially hijacks the infected cell and forces it to produce more viruses. The genetic material of viruses can be DNA or, in the case of **retroviruses,** RNA. Retroviruses are particularly useful, because the RNA is first transcribed into a DNA copy, which then often becomes permanently incorporated into the DNA of the infected cell. (Normal humans contain numerous copies of such DNA sequences acquired from

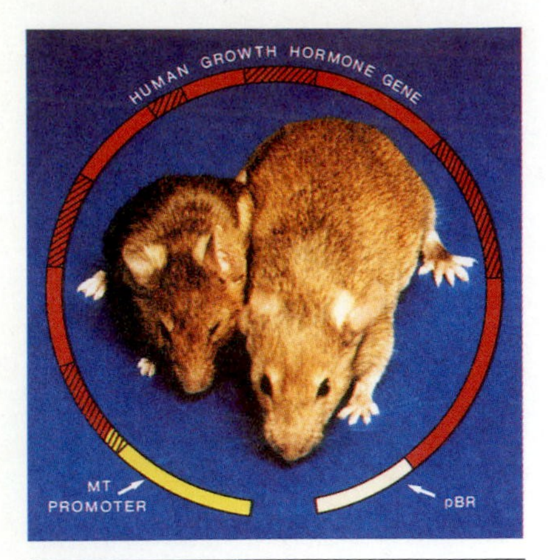

FIGURE 5.15
Ordinary mouse (left) and a "super mouse" bearing recombined genes for human growth hormone. The mouse on the right weighs more than twice as much as the normal mouse. The ovum from which the "super mouse" developed had genes injected into it on circles of DNA called plasmids (pBR), which also incorporated a promoter from a mouse gene. Subsequently, the promoter was activated in a variety of tissues, which synthesized the growth hormone. The growth hormone gene is represented in red, with the exons crosshatched.

retroviruses.) By removing certain genes from the viral genetic material and substituting a desired gene, genetic engineers can use viruses to introduce the desired genes into cells.

Among the potential benefits of recombinant DNA technology could be the production of organisms with unique abilities, such as bacteria that could digest oil spills, plastics, and other pollutants. Another possibility is that crops and endangered species could be given genes that enable them to tolerate environmental stresses so that they could expand their ranges or growing seasons, or tolerate pollution. It is even possible that certain traits of extinct species such as the hairy mammoth, or even of mummified humans, could be studied by recombining their DNA into living cells or organisms. Because ancient DNA continues to accumulate mutations, however, it is virtually impossible that an entire extinct organism, such as a dinosaur, could be recreated from a few of its cells.

Already recombinant DNA technology has been responsible for an impressive list of accomplishments:

1. Children suffering from a lack of the enzyme adenosine deaminase, which is essential for immunity, have almost certainly been spared an early death by having the gene for the enzyme injected in genetically engineered white blood cells. **Gene therapy** is also being tested in animals or humans for a variety of other diseases, including malignant melanoma and cystic fibrosis (Thompson 1992).
2. Human insulin to treat diabetes, growth hormone to prevent dwarfism, and factor VIII to treat hemophilia are being manufactured in large quantities by bacteria with human genes.
3. Genes for growth hormone are being recombined into cows to make them produce more milk, and into fish to make them grow larger. Experiments with mice suggest that farm mammals could also be made larger in this way (Fig. 5.15).

The Recombinant DNA Controversy

Like every new technology, recombinant DNA technology poses potential risks. One can easily imagine a demented genetic engineer using recombinant DNA technology to "improve" the AIDS virus to make it as communicable as the common cold, and even a well-meaning genetic engineer might accidentally start a plague. One researcher was about to concoct a strain of intestinal bacteria that would digest cellulose and thereby provide an additional source of carbohydrate until it occurred to him that such a bacterium might cause intestinal gas or fatal diarrhea. Another potentially unforeseen outcome might be oil-digesting bacteria that would cause a plague in automobiles and other machines. There are government regulations that are supposed to prevent this kind of thing, but the history of nuclear energy does not inspire much confidence in them. In the short history of genetic engineering there have already been several cases in which scientists deliberately bypassed regulations.

It is also unsettling when genetic engineers reveal their ignorance of ecology by claiming that releasing a genetically altered organism into the environment will have only the intended consequence without any side effects. In real life, as Garrett Hardin says, "You can never do just one thing." Introducing natural organisms can wreak havoc upon an ecosystem, so engineered organisms surely have the same potential. Even if an engineered organism has direct effects that are entirely beneficial, there may be indirect effects that are difficult to predict. Who will profit, for example, from using the new technology to increase milk production when many farmers are already going out of business because of surpluses? Will the new technology be affordable by small farmers, or will it accelerate the growth of huge agribusinesses? Will the technology ever relieve hunger in underdeveloped countries? If so, what effect will that have on overpopulation?

Summary

The information that is passed from generation to generation is encoded in the genetic material, deoxyribonucleic acid. DNA consists of two helical chains of deoxyribose alternating with phosphate, with one of four nitrogenous bases attached to each deoxyribose. The sequence of the four bases thymine, cytosine, adenine, and guanine encodes the information. This sequence is copied when DNA is replicated for mitosis and meiosis, thus enabling the genetic information to pass to daughter cells and gametes.

The base sequence also controls the sequence of amino acids in proteins produced by a cell, and it thus controls much of the activity of cells and of the organism. The information is first transcribed to messenger RNA. mRNA is similar to DNA except that it is single-stranded, has ribose instead of deoxyribose, and has uracil instead of thymine. After processing, the mRNA goes to ribosomes where the base sequence is translated into an amino-acid sequence. The bases on the mRNA are divided into triplets, each of which is a codon that specifies an amino acid. Each codon selects a particular transfer RNA that carries a particular amino acid. The tRNA then inserts its amino acid in the growing polypeptide chain as it is synthesized.

Normally, DNA is coiled tightly in chromosomes, and each gene has a certain location in a certain chromosome. Changes in the number and structure of chromosomes will therefore affect the number of genes, with consequences that are often detrimental, but sometimes normal and beneficial. These kinds of changes include aneuploidy, in which the number of chromosomes differs from that of the normal karyotype, and translocation, in which part of a chromosome changes location. Another kind of genetic change results from point mutation, in which one base in DNA is substituted for another. Such changes may have no effect on the protein coded by the gene, or they can lead to changes that may result in disease.

Genetic screening is now being used increasingly to detect mutations, as well as for other purposes. Ultimately, it should be possible to map all human genes, and even to determine the sequence of all bases in human DNA. This knowledge will increase the ability to change the genetic composition of humans and other organisms. Already genes can be transferred from one organism to another by recombinant DNA techniques, leading to improvements in food and other products.

Key Terms

deoxyribonucleic acid
nitrogenous base
double helix
nucleosome
cell cycle
replication
nucleotide
transcription
ribonucleic acid
translation
genetic code
codon
transfer RNA
ribosome
chromosomal aberration
aneuploidy
karyotype
genome
point mutation
restriction enzyme
recombinant DNA

Chapter Test

1. Describe the major activities of genes during each part of the cell cycle: G_1, S, G_2, mitosis. (p. 87)
2. Name and describe the process by which genes are duplicated in mitosis and meiosis. (p. 88)
3. Antidiuretic hormone is a peptide with the amino acid sequence CYS-TYR-PHE-GLN-ASN-CYS-PRO-ARG-GLY. Using the genetic code (Table 5.1), write a sequence of mRNA bases that would code for this peptide. In addition to this sequence, what other bases would the processed mRNA have? Write the sequence of DNA bases from which your mRNA sequence would have been transcribed. In addition to that DNA sequence, what additional bases would have been associated with the gene? (pp. 90–92)
4. The hormone oxytocin is identical to antidiuretic hormone except that ILE is found in the third position instead of PHE, and LEU is found in the next-to-last position in place of ARG. From Table 5.1, explain how each of these two substitutions could have arisen during evolution as a result of single point mutations. (pp. 98–99)
5. Describe how transfer RNA translates the sequence of codons in mRNA to an amino acid sequence. What role do ribosomes play in translation? (pp. 91–93)
6. Describe a chromosomal aberration and describe its consequences in humans. Describe a disease caused by a point mutation. (pp. 94–100)
7. Explain what recombinant DNA is. How can it be produced artificially? Why would one want to? (pp. 101–102)

■ Answers to the Figure Questions

5.8 Leucine, not valine. The anticodon read in the 3′ to 5′ direction is AAC, which complements the codon UUG (5′ to 3′) on mRNA. Note that the 5′ end of the bound mRNA would be to the right in this figure, and the codon would read from right to left.

5.11 The presence of the Y chromosome indicates that the person is male.

5.13 Physiological complexity is difficult to measure. If by complexity one means that the organism is capable of carrying out more functions, then one might well assume that each additional function requires additional genetic information. On the other hand, a bacterial cell carries out all the functions of metabolism, reproduction, and locomotion that many animals do, but with much less genetic information. Moreover, one would be hard put to argue that humans are less complex than most other mammals, not to mention annelids and amphibians, which have more genetic material than we do.

5.14 The large number of her sons who were hemophiliacs makes it highly likely that she was a carrier. The status of a woman would be uncertain if she were the daughter of a hemophiliac or a carrier, but did not have any descendants who were hemophiliacs.

Readings

Recommended Readings

Arnheim, N., T. White, and W. E. Rainey. 1990. Application of PCR: organismal and population biology. *BioScience* 40:174–182.

Capecchi, M. R. 1994. Targeted gene replacement. *Sci. Am.* 270(3):52–59 (Mar).

Cavenee, W. K., and R. L. White. 1995. The genetic basis of cancer. *Sci. Am.*. 272(3):72–79 (Mar).

Chambon, P. 1981. Split genes. *Sci. Am.* 244(5):60–71 (May).

Darnell, J. E., Jr. 1985. RNA. *Sci. Am.* 253(4):68–78 (Oct).

Dickerson, R. E. 1983. The DNA helix and how it is read. *Sci. Am.* 249(6):94–111 (Dec).

Felsenfeld, G. 1985. DNA. *Sci. Am.* 253(4):58–67 (Oct).

Grivell, L. A. 1983. Mitochondrial DNA. *Sci. Am.* 248(3):78–89 (Mar).

Grunstein, M. 1992. Histones as regulators of genes. *Sci. Am.* 267(4):68–74B (Oct).

Hirsch, M. S., and J. C. Kaplan. 1987. Antiviral therapy. *Sci. Am.* 256(4):76–85 (Apr). (*How drugs that interfere with transcription and translation combat viruses.*)

Holliday, R. 1989. A different kind of inheritance. *Sci. Am.* 260(6):60–73 (June). (*How different genes may be activated in different cells.*)

Kornberg, R. D., and A. Klug. 1981. The nucleosome. *Sci. Am.* 244(2):52–64 (Feb).

Lawn, R. M., and G. A. Vehar. 1986. The molecular genetics of hemophilia. *Sci. Am.* 254(3):48–54 (Mar).

McKnight, S. L. 1991. Molecular zippers in gene regulation. *Sci. Am.* 264(4):54–64 (Apr).

Moyzis, R. K. 1991. The human telomere. *Sci. Am.* 265(2):48–55 (Aug).

Murray, A. W., and M. W. Kirschner. 1991. What controls the cell cycle. *Sci. Am.* 264(3):56–63 (Mar).

Nathans, J. 1989. The genes for color vision. *Sci. Am.* 260(2):42–49 (Feb).

Paterson, D. 1987. The causes of Down syndrome. *Sci. Am.* 257(2):52–60 (Aug).

Rennie, J. 1993. DNA's new twists. *Sci. Am.* 266(3):122–132 (Mar).

Rhodes, D., and A. Klug. 1993. Zinc fingers. *Sci. Am.* 265(2):56–65 (Feb). (*Crucial parts of transcription factors.*)

Ross, J. 1989. The turnover of messenger RNA. *Sci. Am.* 260(4):48–55 (Apr).

Sapienza, C. 1990. Parental imprinting of genes. *Sci. Am.* 263(4):52–60 (Oct). (*Identical genes may have different effects, depending on which parent they come from.*)

Stahl, F. W. 1987. Genetic recombination. *Sci. Am.* 256(2):90–101 (Feb).

Strange, C. 1992. Cell cycle advances. *BioScience* 42:252–256. (*How cyclins control the cell cycle.*)

Tjian, R. 1995. Molecular machines that control genes. *Sci. Am.* 272(2):54–61 (Feb).

Todorov, I. N. 1990. How cells maintian stability. *Sci. Am.* 263(6):66–75 (Dec). (*How cells recover after an antibiotic blocks translation.*)

Verma, I. M. 1990. Gene therapy. *Sci. Am.* 263(5):68–81 (Nov).

Watson, J. D. 1968. *The Double Helix: A Personal Account of the Discovery of the Structure of DNA.* New York: Signet.

Weintraub, H. M. 1990. Antisense RNA and DNA. *Sci. Am.* 262(1):40–46 (Jan).

White, R., and J-M. Lalouel. 1988. Chromosome mapping with DNA markers. *Sci. Am.* 258(2):40–48 (Feb).

Additional References

Borst, P., and D. R. Greaves. 1987. Programmed gene rearrangements altering gene expression. *Science* 235:658–667.

DuPuis, E. M., and C. Geisler. 1988. Biotechnology and the small farm. *BioScience* 38:406–411.

Fischetti, M. 1991. A feast of gene-splicing down on the fish farm. *Science* 253:512–513.

Martin, J. B. 1987. Molecular genetics: applications to the clinical neurosciences. *Science* 238:765–772.

Marx, J. 1994. Chromosome ends catch fire. *Science* 265:1656–1658.

Mulligan, R. C. 1993. The basic science of gene therapy. *Science* 260:926–932.

Palmiter, R. D., et al. 1983. Metallothionein–human GH fusion genes stimulate growth of mice. *Science* 222:809–814.

Tangley, L. 1986. Biotechnology on the farm. *BioScience* 36:590–593.

Thompson, L. 1992. At age 2, gene therapy enters a growth phase. *Science* 258:744–746.

CHAPTER

6

Development

Tiger salamander embryo (*Ambystoma tigrinum*).

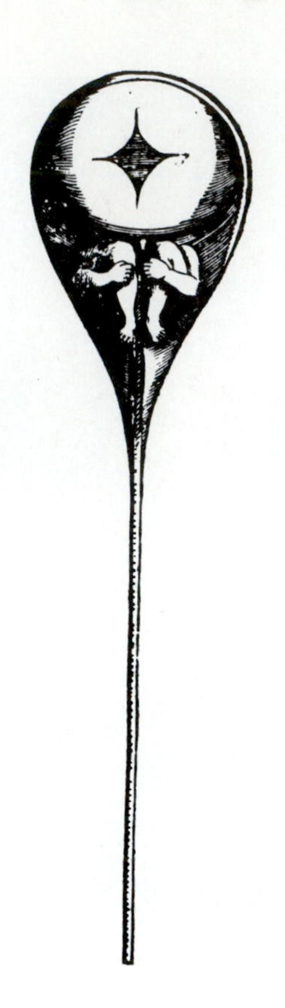

FIGURE 6.1

Hartsoeker's representation of a human infant preformed within the head of a sperm. The diamond shape is the fontanel between the cranial bones of the infant. Hartsoeker never claimed to actually see such sperm, but another preformationist claimed to see tiny chickens, horses, and mules in the semen from these species.

■ **Question:** State five good arguments against preformationism, besides the one mentioned in the text.

Early Concepts of Development

It is quite likely that until the domestication of animals around 10,000 years ago, many people did not associate copulation with reproduction. Until people were able to observe that birth regularly followed mating in their animals, copulation and having babies were two unrelated things, each miraculous in its own way. Discovering that babies develop as a result of fertilization of eggs by sperm in no way lessened the awe and mystery. Even with our knowledge of cells and genes, it is difficult for us not to feel the same way. Nevertheless, we know that underlying the mystery of development are materials and mechanisms that we can potentially understand scientifically.

One of the first to approach development with a scientific perspective was Aristotle in the 4th century B.C. In his book *On the Generation of Animals,* Aristotle says humans develop out of menstrual fluid that has been organized and vitalized by the semen. We are apt to laugh at such an idea now, but it took 2000 years to improve upon it. In the 17th century William Harvey studied development of the chick and concluded that all animals develop from an egg: *ex ovo omnia.* (Of course, Harvey did not believe all animals develop from the kind of things one might serve scrambled with English muffin. The term egg refers to any female gamete.) Harvey's emphasis on eggs led early microscopists to look for them in mammals, and in 1672 they found something that fit their expectations. Actually, however, it was the follicle in which the egg develops. The mammalian egg was not identified until more than a century later.

Several years after the mammalian egg was thought to have been discovered, Antony van Leeuwenhoek observed human sperm through his microscope but thought they were parasites. At about the same time, Nicolas Hartsoeker discovered sperm independently and realized their significance. The discoveries of eggs and sperm helped establish development as a product of materials and mechanisms. Nevertheless, they left unsolved the question of how a fully formed animal emerges from such a tiny beginning. One explanation advocated by many scientists, including Hartsoeker, is referred to as **preformationism.** Preformationism is the belief that each organism was already preformed within either the egg or the sperm (Fig. 6.1). With our knowledge of cells and genetics, it is not hard to come up with half a dozen arguments against preformationism now. Indeed, there were good arguments against it in its own time, one being that each unborn animal would have to contain its own preformed offspring, which would contain their own offspring, and so on for the entire future of the species.

Epigenesis

By the 19th century, preformationism had been replaced by its alternative concept, epigenesis. Epigenesis refers to the gradual development of new structures that did not exist previously as such. In the modern view of epigenesis, an animal begins as a fertilized egg (zygote) that lacks the features that will eventually appear. When the zygote begins to develop, it becomes an **embryo,** which gradually acquires the features characteristic of the animal. Although animals are extremely diverse, most develop through similar stages, called cleavage, gastrulation, and morphogenesis (Fig. 6.2):

1. In **cleavage** the zygote divides repeatedly, resulting in a mass of cells, called a **blastula.**
2. Some outer cells of the blastula then move inward during the process called **gastrulation.**

3. Gastrulation marks the beginning of **morphogenesis,** in which the embryo begins to take on a definite shape, and cells start to become differentiated from each other.

Gametogenesis

The development of an organism from a fertilized egg begins long before fertilization, since the gametes themselves must first develop. Hormonal signals (described in Chapter 16) trigger the development of gametes from **primordial germ cells** by the process called gametogenesis. The primordial germ cells that produce sperm are called **spermatogonia,** and the process of sperm production is called **spermatogenesis.** Spermatogonia first undergo repeated mitosis, producing genetically identical diploid **primary spermatocytes** in enormous numbers (billions in men) (Fig. 6.3 on page 108). Each primary spermatocyte then undergoes meiosis (see pp. 74–77). Meiosis I results in two haploid secondary spermatocytes. Because of crossing over during meiosis I, secondary spermatocytes are all genetically different from each other. Each of these undergoes meiosis II, resulting in a total of four haploid **spermatids.** The spermatids mature (during **spermiogenesis**) into functional sperm.

The earliest stages of egg production, or **oogenesis** (pronounced oh-uh- JEN-uh-sis), are similar to those of spermatogenesis. The germ cells, called **oogonia** (oh-uh-GO-nee-uh), first proliferate mitotically. A few of the oogonia then grow and develop into genetically identical **primary oocytes.** Primary oocytes undergo meiosis I, producing genetically diverse **secondary oocytes** and small, functionless **first polar bodies.** In meiosis II each secondary oocyte divides into a mature haploid egg, or **ovum,** and a **second polar body.** Human females produce up to a million oogonia during the first three months as a fetus. These oogonia begin meiosis but stop after prophase I. Fewer than 500 then resume meiosis, usually one at a time starting at puberty and continuing until menopause. In women and in most other mammals, meiosis II is completed only after fertilization. The first polar body also completes meiosis, bringing the total number of polar bodies to three.

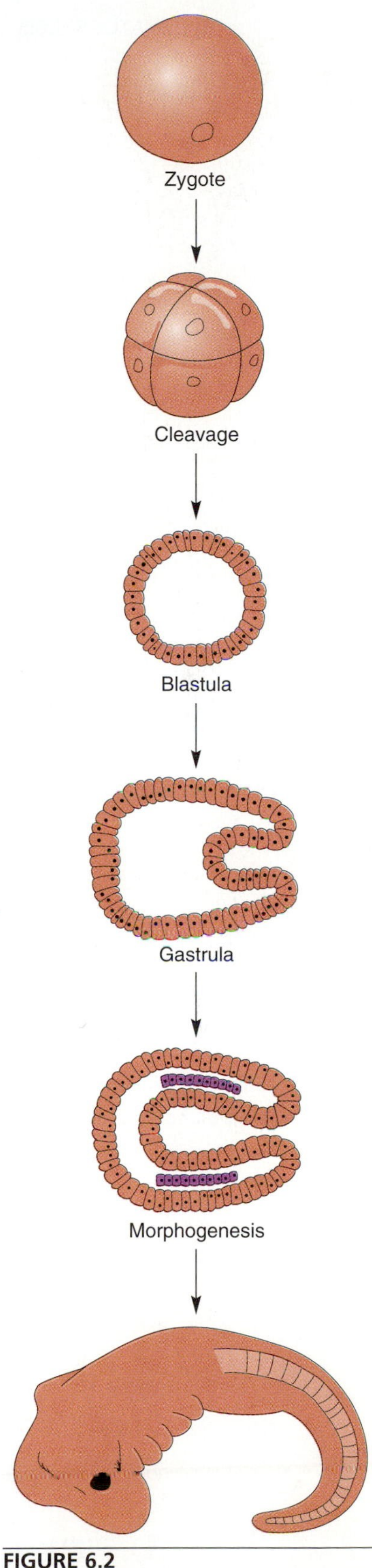

FIGURE 6.2
The modern epigenetic view. Most animals develop through these embryonic stages.

Fertilization

The overall function of fertilization is to combine two haploid gametes into a single zygote with the normal diploid number of chromosomes. This seemingly straightforward objective is complicated by the fact that the egg is usually exposed to numerous sperm simultaneously. In fact, in humans and many other species, numerous sperm are required to achieve fertilization, although only one sperm fertilizes an ovum. The entry of more than one sperm—**polyspermy**—has to be prevented or the number of chromosomes will be abnormal. In species in which fertilization occurs outside the body, the egg also generally has to resist fertilization by sperm from different species. Fertilization must therefore balance the conflicting needs of encouraging fertilization by sperm of the same species while preventing polyspermy and rejecting sperm of another species. There is an enormous variety of adaptations by which different groups of animals achieve these goals, but it has been possible to study fertilization in only a few groups. Echinoderms, especially sea urchins (see Fig. 33.1C), are favorite subjects of fertilization studies. Advances in tissue culture have also made it possible to study fertilization in mammals. Fertilization will be described in both groups.

Before fertilization can occur, sperm must move into the vicinity of the egg. The sperm of most animals propel themselves actively, using long **flagella** (see pp.

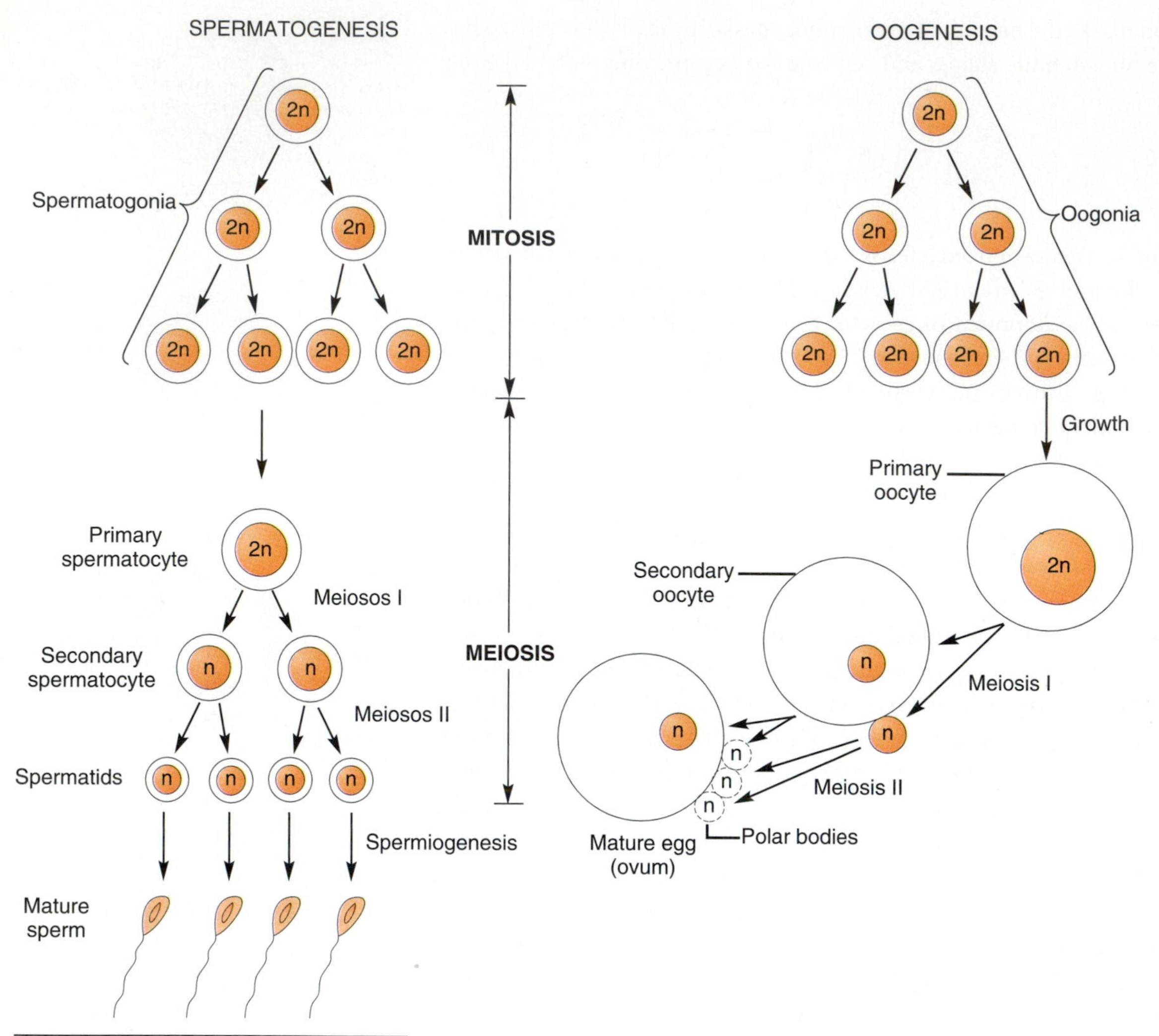

FIGURE 6.3
Spermatogenesis and oogenesis.

■ **Question:** Does it make sense to produce three polar cells that are useless? Why not produce four small ova?

210–211; Fig. 6.4). The flagellum of each sperm is powered by mitochondria densely packed in the **midpiece.** Anterior to the midpiece and separated by a neck is the **head** of the sperm, which contains the "payload" of tightly condensed chromatin in the nucleus. The egg is generally many times larger than the male gamete and incapable of moving under its own power (Fig. 6.5 on page 110). Substances enclosing the egg provide protection but allow penetration by sperm of the same species. Sea urchin eggs are protected by a thick **jelly coat** and a tough **vitelline envelope** made of glycoproteins. In mammals the vitelline envelope is called the **zona pellucida.**

Depending on the species, fertilization occurs at different times in the development of the egg. In sea urchins, meiosis is complete by the time of fertilization, and the egg has matured into an ovum. In mammals, however, the egg is arrested in metaphase II at the time of fertilization and is therefore properly referred to as an oocyte rather than an ovum. As soon as fertilization occurs, the developing zygote may be referred to as an **embryo.** (The term *embryo* is sometimes applied to any developmental stage before birth or hatching, especially in early stages before there is some recognizable form. In humans the term is generally used only up to the eighth week after fertilization, after which the term *fetus* is used.)

Fertilization proceeds by several steps:

1. First there is a loose and nonspecific **attachment** of the sperm to the jelly coat (sea urchins) or zona pellucida (mammals).

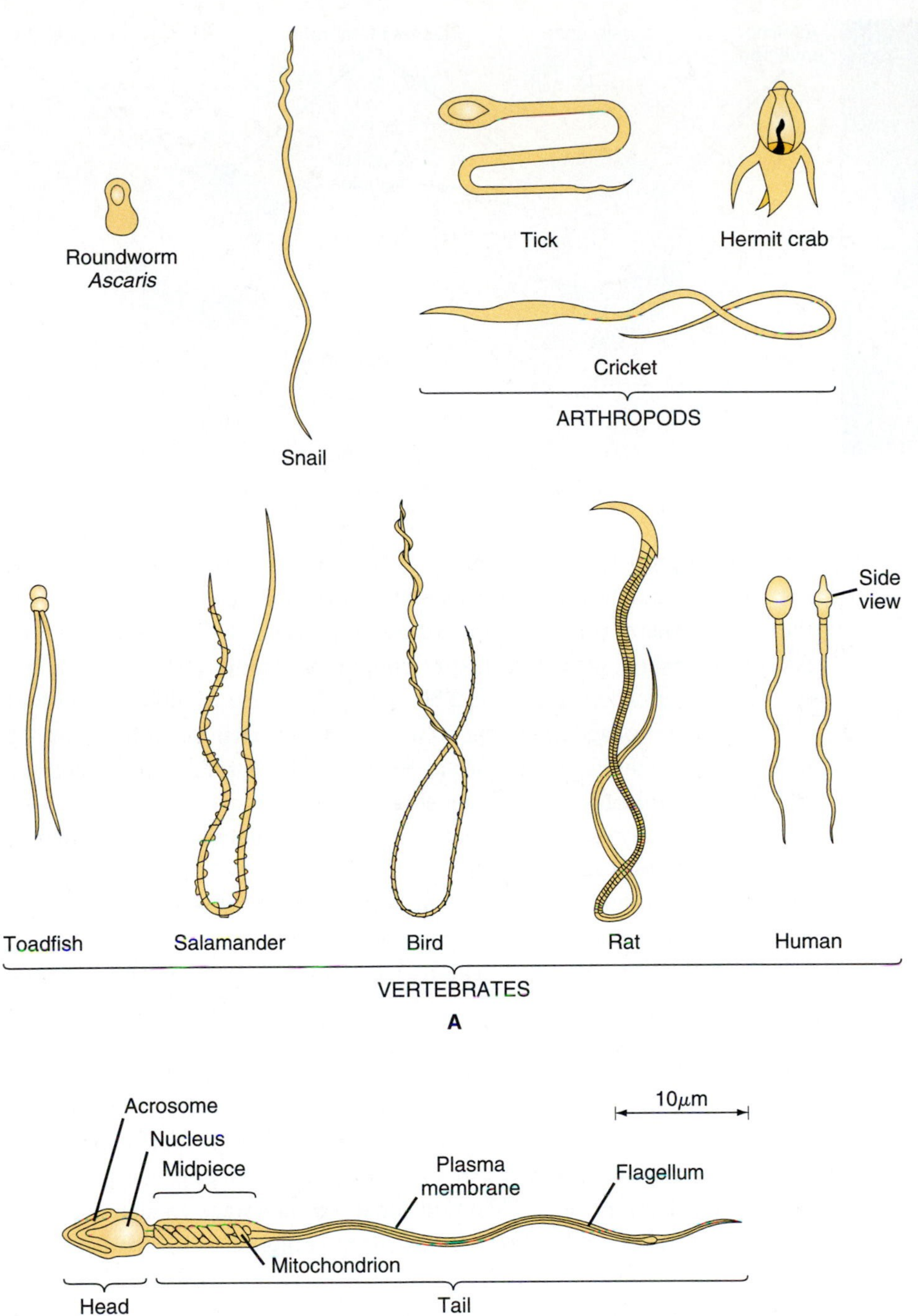

FIGURE 6.4

(A) Sperm from various groups of animals. Some, such as those of crustaceans, are nonmotile. Sperm of round worms and arthropods propel themselves with ameboid motion, but most sperm use flagella. All drawn at the same scale. (B) Detail of a human sperm cell.

2. Attachment is followed by the **acrosome reaction** and strong **binding** of sperm. The acrosome reaction is the rupture of the sperm's **acrosome** (= acrosomal vesicle; Fig. 6.4B). The acrosome reaction releases enzymes that digest a pathway through the jelly coat and exposes molecules called **bindins** on the sperm. Bindins cause a specific binding of sperm to receptor molecules on the vitelline envelope. Generally, the receptor molecules attach only to bindins on sperm of the same species. In sea urchins the acrosome reaction occurs before binding. In mammals, binding precedes the acrosome reaction and is triggered by **ZP3,** one of the three kinds of glycoproteins that compose the zona pellucida.

A

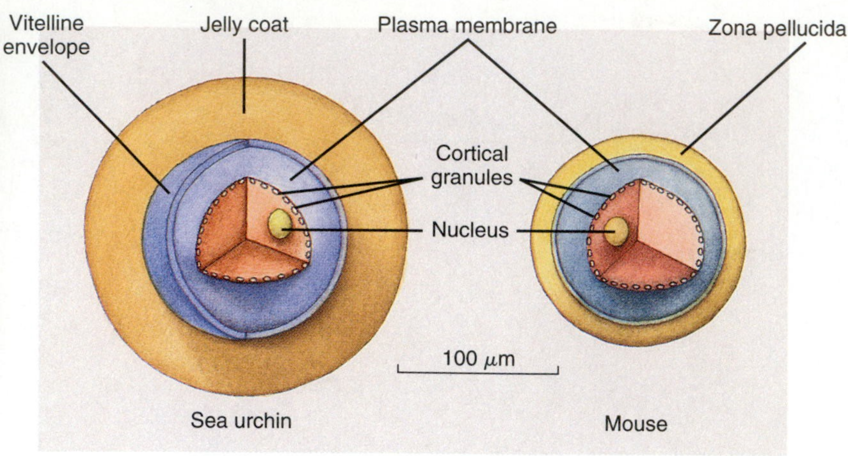

B

FIGURE 6.5

(A) A human oocyte in the oviduct. The halolike ring around the oocyte is the zona pellucida. The oocyte is about the size of the period at the end of this sentence. (B) Schematic comparison of the ovum from a sea urchin and the oocyte from a mammal.

■ **Question:** Why would the sea urchin but not the mammal require a thick jelly coat?

3. Fertilization proper begins when the plasma membranes of the two gametes fuse together, and the head of the sperm is drawn into the egg (Fig. 6.6).
4. Entry of the sperm triggers a series of reactions, each of which is an essential part of fertilization. One of the first reactions blocks fertilization by additional sperm. In sea urchins there is a **fast block** to polyspermy. Within 3 seconds of fertilization the voltage across the plasma membrane changes from about –60 millivolts inside the egg to as much as +20 millivolts. A similar process occurs in some other animals, but not in most mammals.
5. Subsequent changes act as a stronger **slow block** to polyspermy. The first of these is the **cortical reaction,** which is the release of the contents of **cortical granules** just beneath the plasma membrane (Fig. 6.5B). In sea urchins the cortical reaction makes the plasma membrane impenetrable to other sperm (Fig. 6.7). In mammals the cortical reaction blocks polyspermy by triggering a **zona reaction** in which the zona pellucida hardens and ZP3 changes so that it no longer binds sperm.
6. Penetration by sperm triggers **egg activation,** which is a sharp rise in the metabolic activities that are necessary for continued development. Activation does not depend on any molecular contribution from the sperm, but is apparently due only to mechanical stimulation. In many species a pin prick will trigger activation, and development will continue up to a certain stage. In mammals, fertilization also stimulates the oocyte nucleus to complete meiosis and become the female pronucleus.
7. Finally, the genetic material in the two haploid gametes fuses. The genetic material in the male and female **pronuclei** combine into a diploid **zygote nucleus** (Fig. 6.8 on page 112).

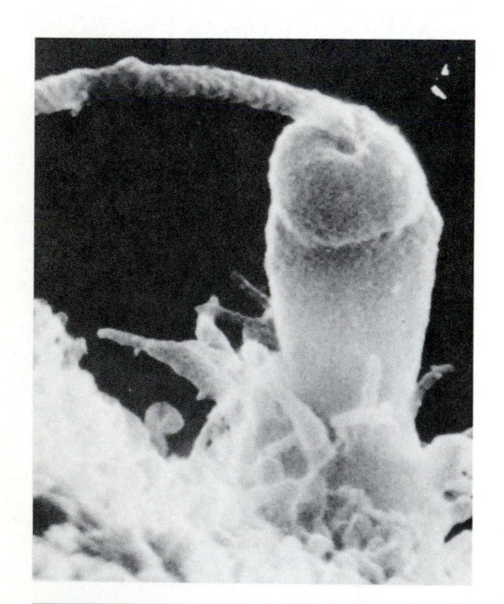

FIGURE 6.6

Scanning electron micrograph showing fertilization of a sea urchin egg. A group of microvilli on the egg's plasma membrane has formed a "fertilization cone" that draws the sperm head inside.

Cleavage

Egg activation initiates the series of cell divisions called cleavage. Unlike most mitotic divisions, those involved in cleavage are not followed by the G_1 phase during which cellular growth occurs (see p. 87). Instead, each mitosis is followed directly by S phase, in which DNA is replicated for the next mitosis. The total mass of cells therefore remains the same as in the zygote, and each cell, or **blastomere,** gets smaller with each division. (That is, the blastomeres become more like normal body cells in size.) Protein synthesis during cleavage depends on a massive amount of messenger RNA (mRNA) and other

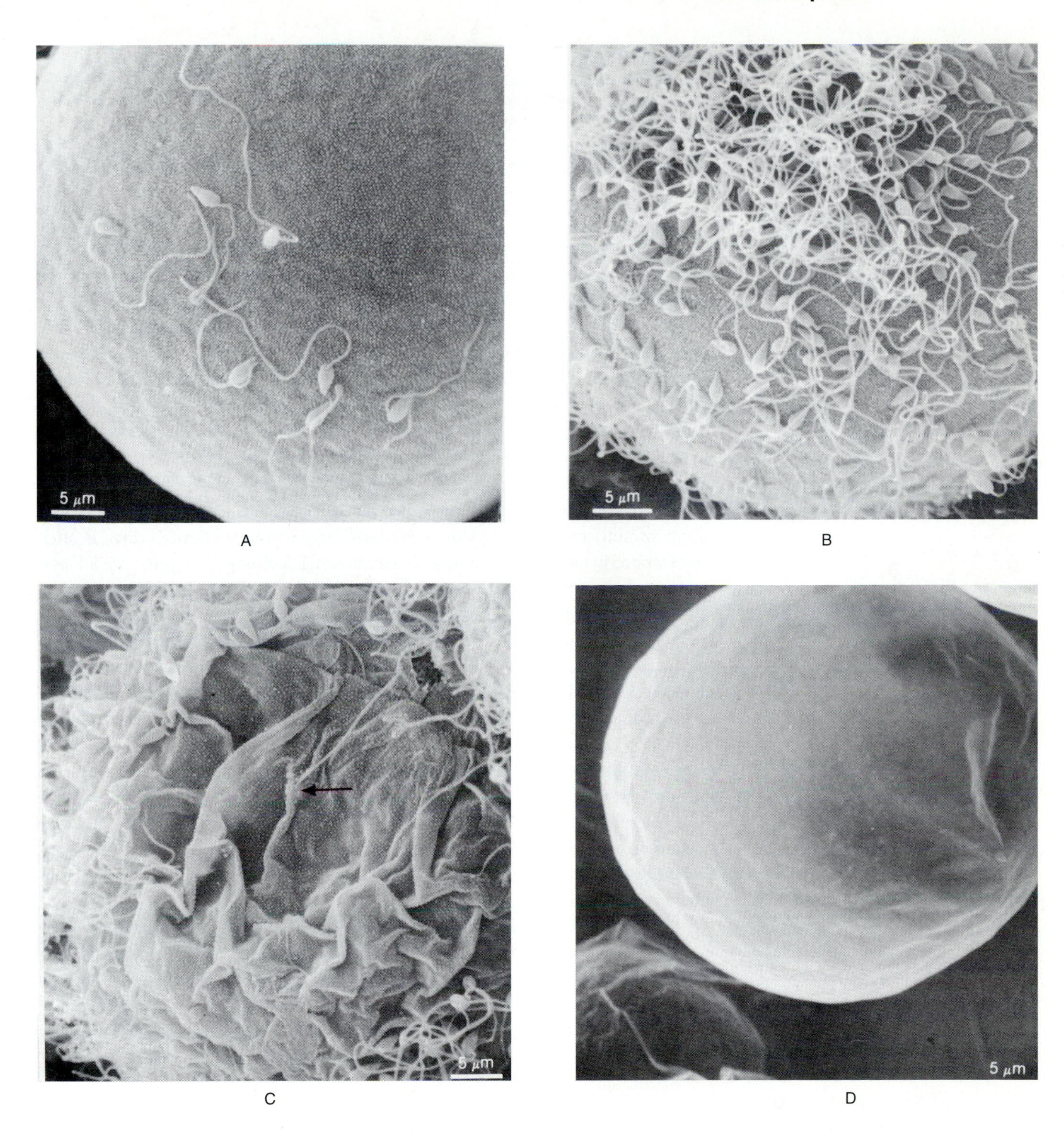

FIGURE 6.7
The cortical reaction in sea urchins. (A) At fertilization, only a few sperm have attached to the egg. (B) Sperm continue to attach, but the fast block prevents polyspermy. (C) Thirty seconds after fertilization the cortical reaction has cleared a zone around the point of fertilization (arrow). (D) Within 3 minutes the vitelline membrane is hardened, and sperm are dispersed.

maternal factors synthesized by the oocyte before fertilization. The main thing achieved during cleavage is multicellularity. Cleavage takes days in frogs and produces tens of thousands of blastomeres. It is much slower in mammals, with humans taking a week to produce a few hundred blastomeres (Fig. 6.9 on page 113). Cleavage results in the formation of a distinctive sphere of blastomeres called the **blastula** (Greek *blastos* a bud). In mammals the blastula is called a **blastocyst.** Generally, the blastula contains a fluid-filled cavity, the **blastocoel** (pronounced BLAST-oh-SEAL).

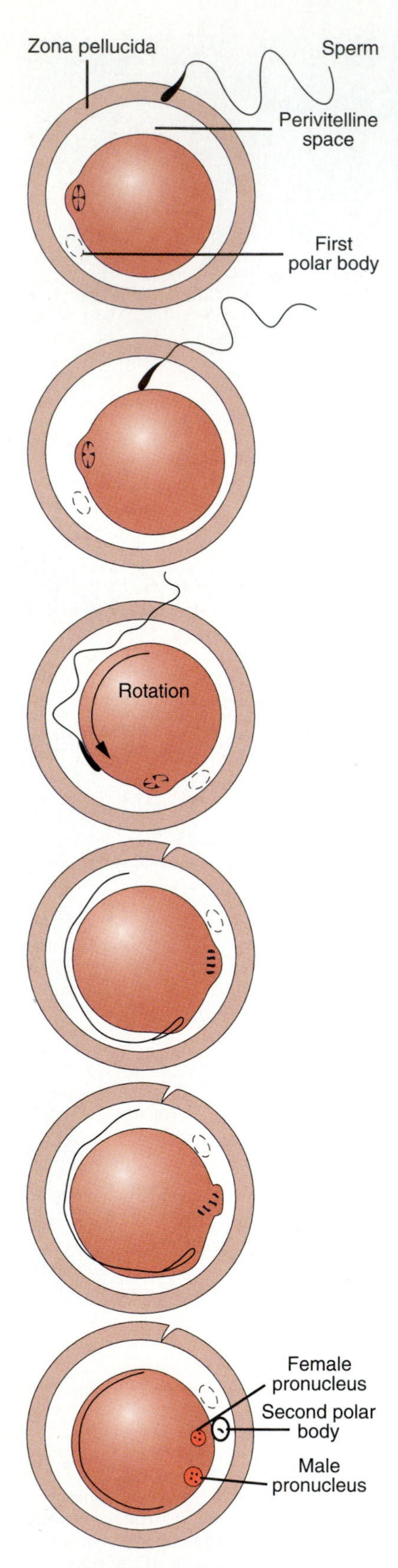

FIGURE 6.8

Fertilization in mammals takes more than a day. The sperm rotates the oocyte and enters sideways. The oocyte then completes meiosis II, forming the second polar body and the female pronucleus.

■ **Question:** What, if anything, does the common phrase "moment of conception" mean?

Cleavage Patterns. The sizes and movements of blastomeres follow what is called the cleavage pattern. Cleavage patterns tend to be conserved among related groups of animals, and they are of considerable interest to zoologists because they are thought to provide clues to the evolutionary relationships among various groups of animals. (See Chapter 21 for further discussion.) One of the factors affecting cleavage pattern is the amount and distribution of reserve nutrients, especially proteins, in the egg. These nutrients make up the **yolk.** Eggs that develop quickly into larval or adult forms that can feed themselves, or that are nourished by the mother by means of a placenta, generally have little yolk. Eggs with sparse yolk that is evenly distributed are said to be **isolecithal** (Fig. 6.10 on page 114). Animals such as reptiles, birds, and many fishes undergo prolonged development outside the mother's body with no external supply of nutrients. Their eggs therefore have large amounts of yolk, which is often concentrated in one place. Such eggs are said to be **telolecithal.** The eggs of cephalopod molluscs, such as squids and octopuses, and those of amphibians and some fishes are moderately telolecithal, or **mesolecithal.** In insects and most other arthropods the eggs are said to be **centrolecithal,** because the yolk is concentrated toward the center.

In isolecithal and mesolecithal eggs the entire egg can divide, and cleavage is therefore termed **holoblastic** (Greek *holo-* whole). Among animals with holoblastic cleavage, the two most common patterns of cleavage are spiral and radial. In **spiral cleavage,** which is common in invertebrates, the first two cleavages are equal, resulting in four blastomeres of the same size (Fig. 6.11 on page 114). The furrows separating the four cells are said to be **meridional,** by analogy with the meridians on a globe. The third cleavage is unequal, dividing the four blastomeres into four **micromeres** and four larger **macromeres.** The micromeres spiral into the furrows between the macromeres, disrupting the meridional pattern of the cleavage furrows. This spiral pattern continues during subsequent cleavages.

In sponges, cnidarians, echinoderms, the invertebrate chordate *Branchiostoma* (see Fig. 34.7), and amphibians there is **radial cleavage.** In radial cleavage, daughter cells tend to remain radially aligned, and the furrows remain meridional (Fig. 6.12 on page 115). As in spiral cleavage, the third cleavage is unequal, resulting in micromeres and macromeres of unequal size. The micromeres occur at what is called the **animal pole** of the embryo. The macromeres, which contain more yolk, occur at the **vegetal pole.**

Besides spiral and radial, there are two other cleavage patterns that occur in animals with holoblastic cleavage. In humans and other mammals that develop with the aid of a placenta (placental mammals), there is a **rotational cleavage pattern.** After the second cleavage, one pair of daughter cells lies at right angles to the other pair (see four-cell stage of Fig. 6.9A and Fig. 21.9). Cleavage in mammals is also unusual in that the blastomeres do not divide simultaneously, so at any one time the number of cells may not be even. Cephalopod molluscs (squids and octopuses) and the invertebrate chordates called tunicates (see Fig. 34.7) have a **bilateral cleavage pattern.** The first cleavage furrow defines a plane of bilateral symmetry. Cleavage thereafter results in blastomeres on one side of the plane being mirror images of blastomeres on the other side.

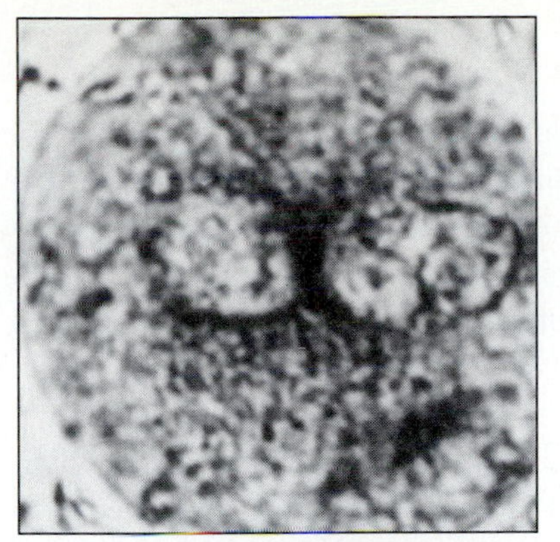

2-cell embryo; 30 hr

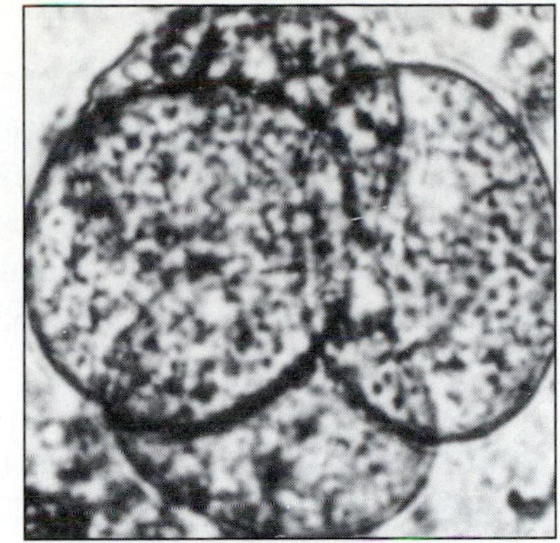

4-cell; 40 hr

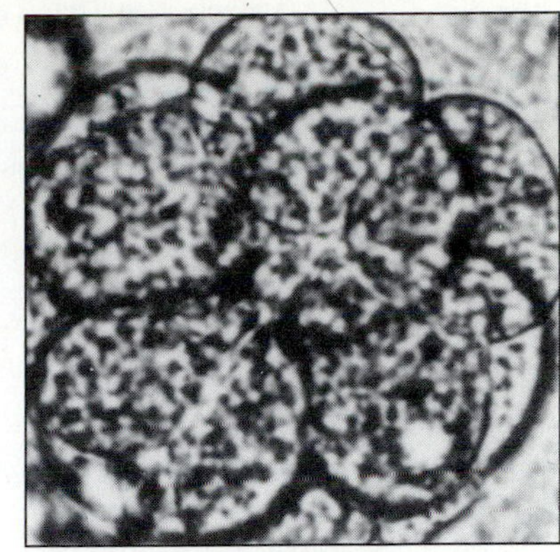

8-cell; 55 hr

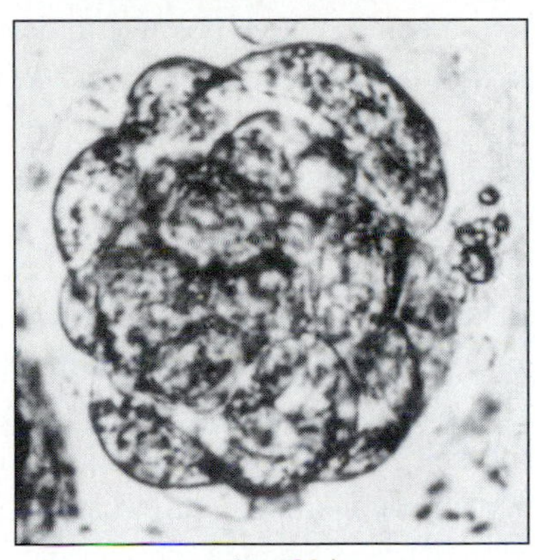

morula; 100 hr

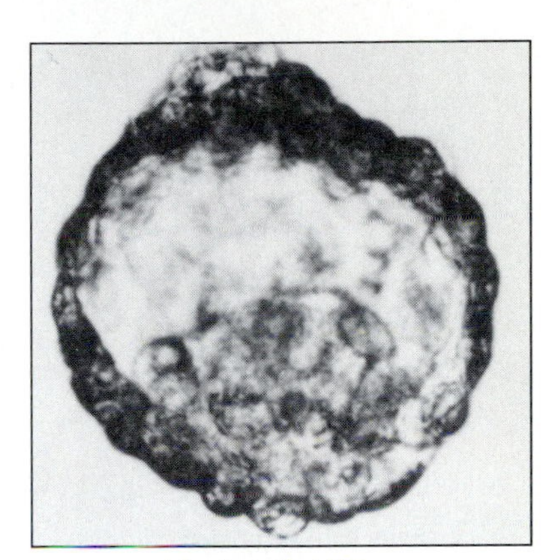

blastocyst; 140 hr

A

FIGURE 6.9

(A) Cleavage in a human, beginning with the two-cell stage following first cleavage and ending with the blastocyst (the preferred term for blastula in mammals). The number of hours indicates time since the completion of fertilization. Note the zona pellucida, which starts to degenerate in the blastocyst. (B) Cleavage in women occurs as the zygote is propelled down the oviduct by cilia. Four cleavages occur during the first four days after fertilization, resulting in a morula of approximately 16 cells. The morula then develops into a free blastocyst that enters the uterus. Approximately five to six days after fertilization the blastocyst implants into the uterine lining.

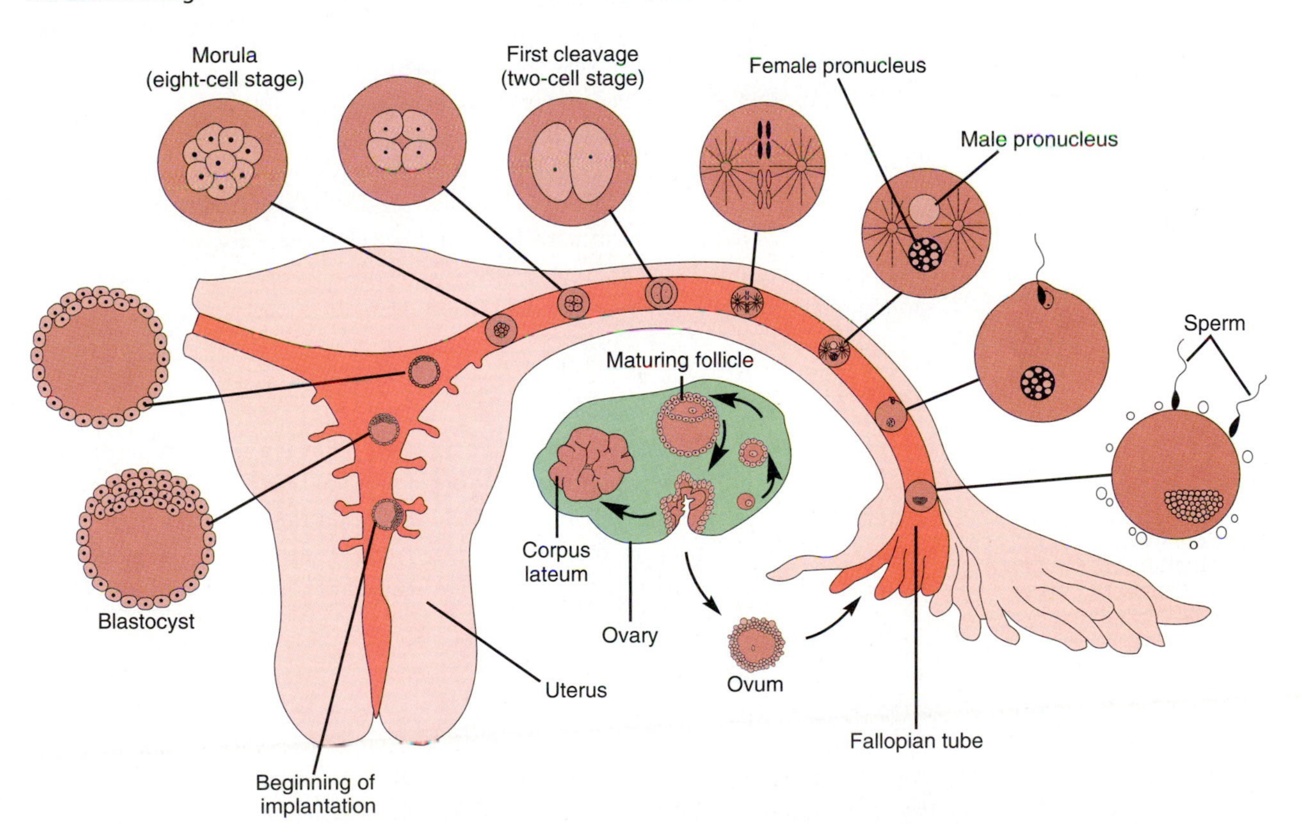

B

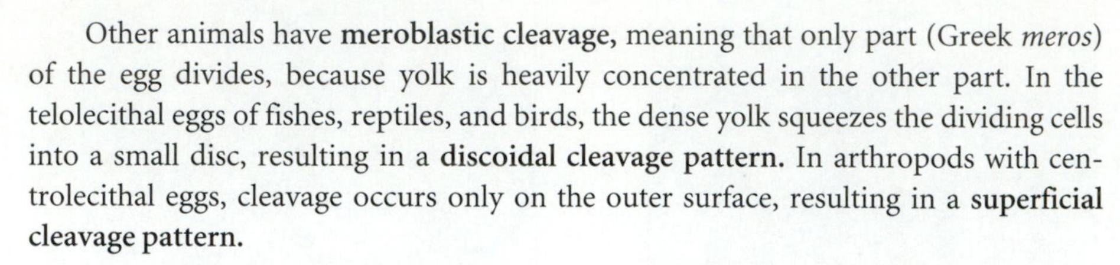

Other animals have **meroblastic cleavage,** meaning that only part (Greek *meros*) of the egg divides, because yolk is heavily concentrated in the other part. In the telolecithal eggs of fishes, reptiles, and birds, the dense yolk squeezes the dividing cells into a small disc, resulting in a **discoidal cleavage pattern.** In arthropods with centrolecithal eggs, cleavage occurs only on the outer surface, resulting in a **superficial cleavage pattern.**

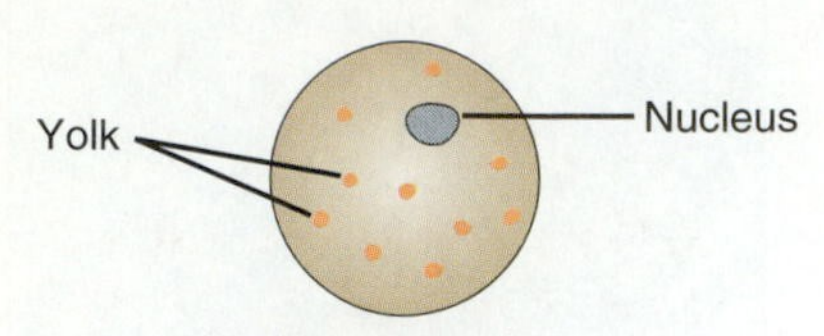

ISOLECITHAL
Sponges, cnidarians, echinoderms, many flatworms, nematodes, rotifers, most molluscs, many annelids, invertebrate chordates, placental mammals

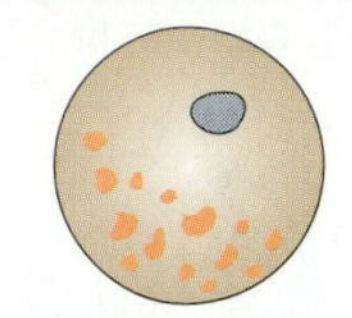

MESOLECITHAL
Cephalopod molluscs, amphibians, some fishes

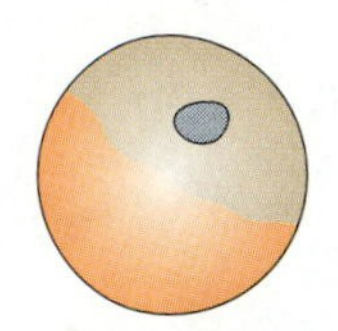

TELOLECITHAL
Many fishes, reptiles, birds

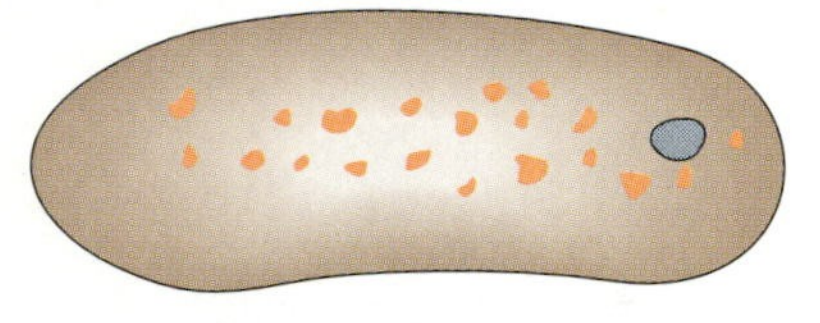

CENTROLECITHAL
Most arthropods

FIGURE 6.10
Eggs classified according to the amount and distribution of yolk.

Determinate Versus Indeterminate Cleavage. In many species, blastomeres begin early in cleavage to "decide" what they are going to be when they grow up. In particular, **germ cells** that will later develop into gametes are determined during cleavage, just as Weismann proposed (p. 72). By destroying blastomeres or injecting dye into them, it is often possible to determine what contribution each one makes to the organism. One can then construct a **fate map** of the blastomeres. In some cases the embryos are transparent enough to allow the construction of fate maps simply by observing blastomeres as they divide and differentiate. One example is the round worm *Caenorhabditis elegans* (pronounced SEE-no-rab-DY-tiss ELL-i-ganz; phylum Nematoda; see pp. 541–542 for details). All 959 of its somatic cells, as well as its gametes, have been traced to just six blastomeres (Fig. 6.13 on page 116).

In *C. elegans* and many other invertebrates the fates of blastomeres are determined from the start and are not affected by position or interactions within the embryo. If two such blastomeres are separated, they will not develop into identical twins; but each will develop for a limited time into whatever part of the animal it would have become had the cells not been separated. This is called **determinate cleavage.** Another name for determinate cleavage is **mosaic development.** Like a mosaic, the entire organism is assembled from pieces (blastomeres) that are not interchangeable. Determinate cleavage probably results from different kinds and amounts of messenger RNA and other maternal factors going to different blastomeres as they divide.

In many other species, blastomeres have the capacity to modify their fate under abnormal conditions. For example, if two blastomeres of an echinoderm or a chordate are separated after the first cleavage, each blastomere can develop into a complete, normal animal. This sometimes happens in humans, producing identical twins. This type of cleavage is said to be **indeterminate.** Indeterminate cleavage is also called **regulative development** because each blastomere seems to regulate its own development in response to position in the embryo and interactions with other blastomeres.

Gastrulation

During cleavage the blastomeres have generally been following a pattern of division and movement controlled by maternal factors that were in the ovum. By the time the blastula forms, this division and movement have slowed down considerably. Then suddenly something triggers the cells to take a more active role in epigenesis. Blastomeres begin to transcribe their own DNA instead of using maternal mRNA in-

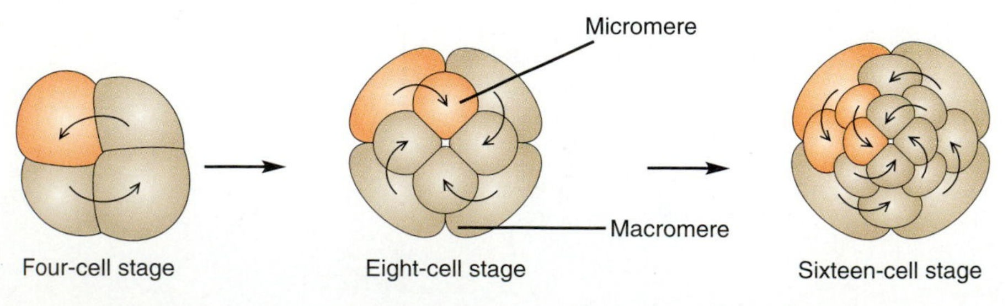

FIGURE 6.11
The spiral cleavage pattern viewed from above one pole. Arrows show movement of daughter cells from each cleavage. In three dimensions the arrowheads would point slightly toward the reader.

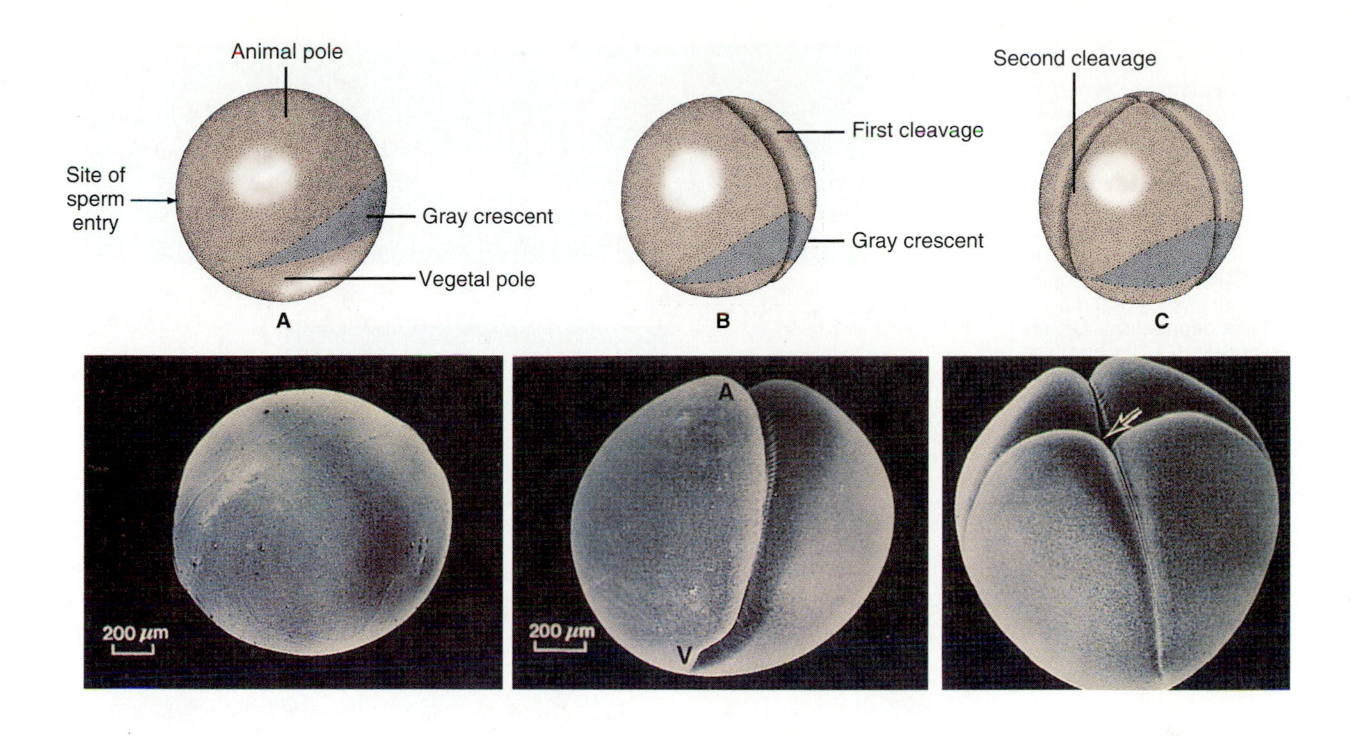

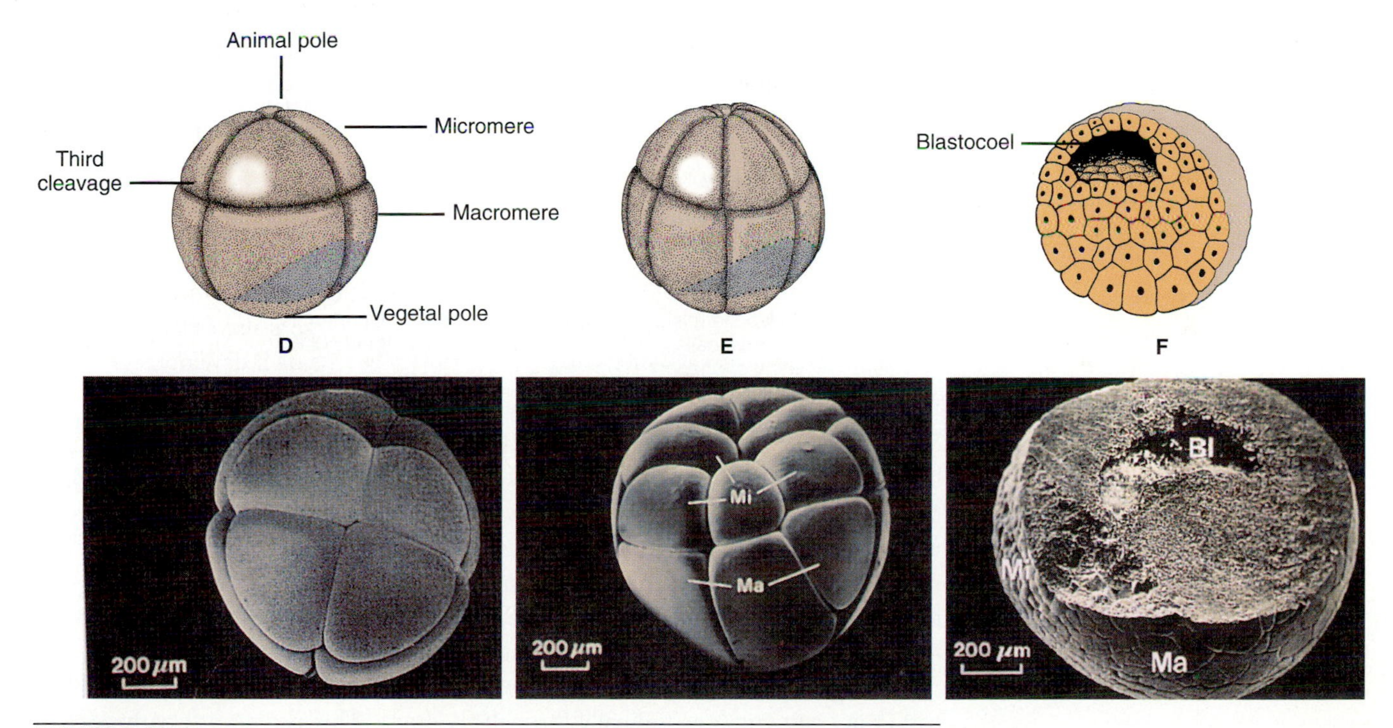

FIGURE 6.12

Schematic representations and scanning electron micrographs of radial cleavage in a frog egg. (A) Zygote before first cleavage. Pigment in the animal pole has migrated toward the site of sperm cell entry, producing a less-pigmented gray crescent. (B) First cleavage is from pole to pole, intersecting the site of fertilization and bisecting the gray crescent. (C) Second cleavage is meridional at right angles forming a cross (arrow) at the animal pole.
(D, E) Third through sixth cleavages produce micromeres (Mi) at the animal pole that tend to remain radially aligned with macromeres (Ma) at the vegetal pole. (F) Late blastula, about 10 hours after fertilization, with the blastocoel (Bl) toward the animal pole.

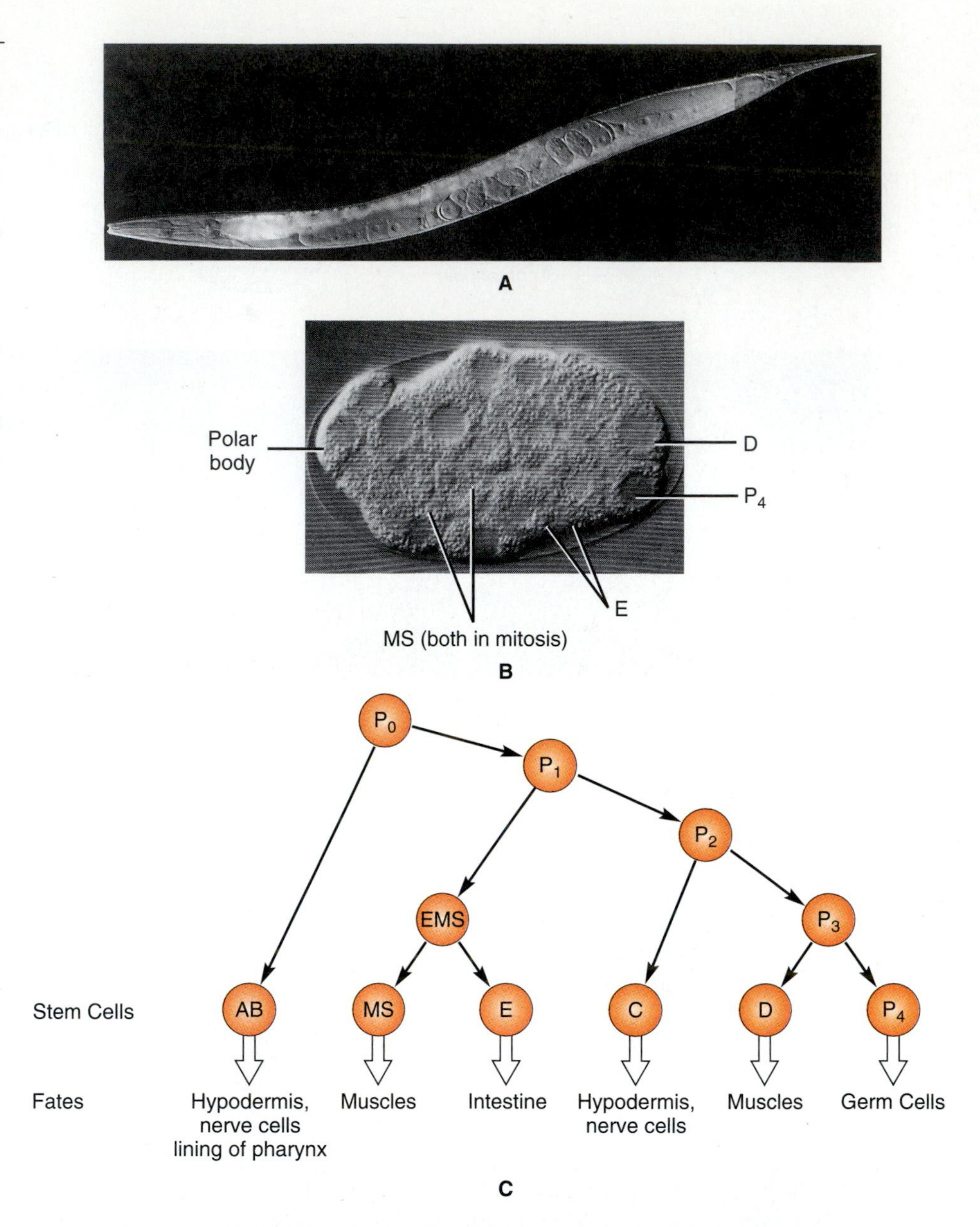

FIGURE 6.13

(A) The adult nematode *Caenorhabditis elegans.* (Approximately 1.5 mm long.) (B) An embryo of *C. elegans* approximately 90 minutes after fertilization, in the 24-cell stage near the end of cleavage. Already it is determined that the anterior of the worm will be to the left, and dorsal will be to the top as shown on the page. A special microscopical technique (Nomarski optics) makes the nuclei look like depressions. Letters identify stem cells. (C) The origin of the six stem cells and the fates of their descendants. The zygote P_0 divides into P_1 and stem cell AB. P_1 then divides into P_2 and EMS, and so on. (Hypodermis, also called epidermis, covers the animal and secretes the outer cuticle.)

herited with the ovum. In each cell a different combination of genes is transcribed, so that each becomes progressively different and interacts with its neighbors in increasingly divergent ways. The cells begin to divide and move around dramatically. Out of a mass of seemingly identical blastomeres an embryo takes shape, and tissues and organs form. This process is called **morphogenesis.** The stage of development that marks the transition from cleavage to morphogenesis is called gastrulation.

In most animals, gastrulation lays down the basic "tube-within-a-tube" body plan, with a gut running longitudinally through the body. This process is clearly seen in echinoderms such as the sea urchin (Fig. 6.14). In the sea urchin, blastomeres at the vegetal pole push inward (invaginate), forming an opening called the **blastopore.** The blastopore leads into the primitive gut cavity, called the **archenteron.** Eventually, the archenteron completely penetrates the embryo and emerges at what will later be the mouth. Since this second opening, not the blastopore, becomes the mouth, echinoderms are said to be **deuterostomes** (Greek *deutero-* second + *stoma* mouth). Chordates, including humans and other vertebrates, are also deuterostomes. In most other animals the blastopore develops into the mouth. Since this is the first opening in

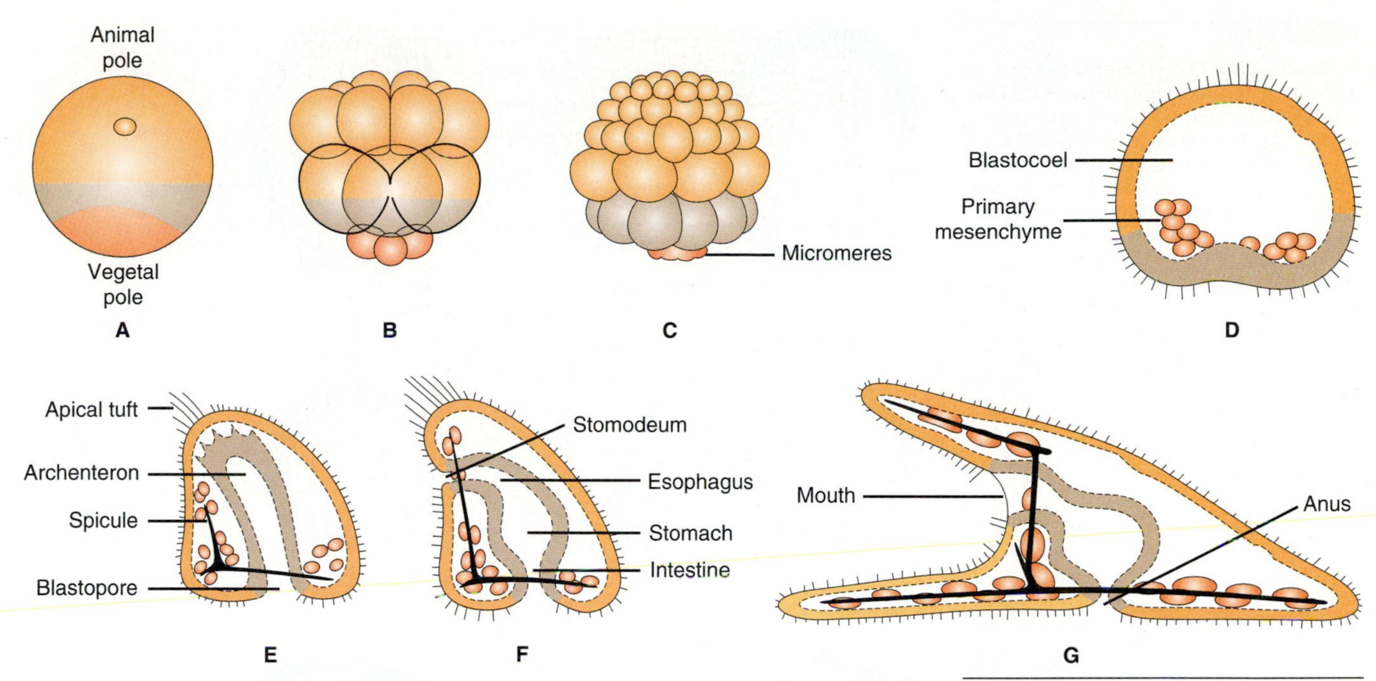

FIGURE 6.14

Gastrulation in a sea urchin. Parts of egg are shaded to show fates. (A–C) Radial cleavage produces micromeres at the vegetal pole. (D) In the blastula the micromeres migrate into the blastocoel and form diffuse connective tissue (primary mesenchyme). (E) Gastrulation. The vegetal plate invaginates, forming the blastopore and archenteron. Primary mesenchyme constructs spicules that compose the skeleton. The apical tuft of cilia will enable the embryo to swim. (F) Continued development of the archenteron forms the mouth and digestive tract. (G) Resulting echinopluteus larva, which swims and feeds on its own. See Fig. 33.16 for external view.

the embryo, these animals are called **protostomes** (Greek *proto-* first). The distinction between protostomes and deuterostomes is significant in classification, as will be discussed in Chapter 21.

In addition to establishing the basic body plan, gastrulation in most animals divides the cells into three **germ layers.** The outer germ layer is **ectoderm,** and cells descending from it will generally become epidermis, nervous tissue, and related cells. The inner germ layer, the **endoderm,** will usually become gut lining and various internal glands and organs. A third germ layer, the **mesoderm,** develops from endoderm or (less often) ectoderm. Mesoderm usually gives rise to blood, bone, and connective tissues, as well as to muscle. Most animals have these three germ layers and are said to be **triploblastic.**

In most animals, one of the most important creations of the mesoderm is a particular type of body cavity called the **coelom** (SEAL-um). Technically, a coelom is a body cavity that develops from mesoderm and is enclosed by a **peritoneum,** which is a membrane consisting of nucleated cells. The coelom generally arises in one of two different ways. In most protostomes and in mammals and some other chordates, the mesoderm splits into two layers, and the coelom forms between them (Fig. 6.15A on page 118). These animals are said to be **schizocoelous** (Greek *schizo-* split). In many deuterostomes (echinoderms and many chordates) the mesoderm forms a pouch that pinches off the archenteron, enclosing the coelom within it (Fig. 6.15B on page 118). These animals are termed **enterocoelous** (Greek *entero-* gut).

In vertebrates the formation of the gastrula is somewhat different from that in echinoderms. In amphibians, for example, the yolk prevents simple invagination of the vegetal pole. Instead, certain cells move into the blastocoel, producing a slit like blastopore into which cells on the dorsal lip migrate and form the archenteron (Fig. 6.16 on page 119). Animals with even more yolk, such as fishes, reptiles, birds, and egg-laying mammals, begin morphogenesis without forming either a blastopore or an archenteron.

The gastrulas of vertebrates also form **extraembryonic membranes** that are essential in maintaining a suitable environment for the embryo (Fig. 6.17 on page 119). In most vertebrates the largest extraembryonic membrane is the **yolk sac,** which is an extension of the primitive gut that encloses the yolk. Reptiles, birds, and mammals have

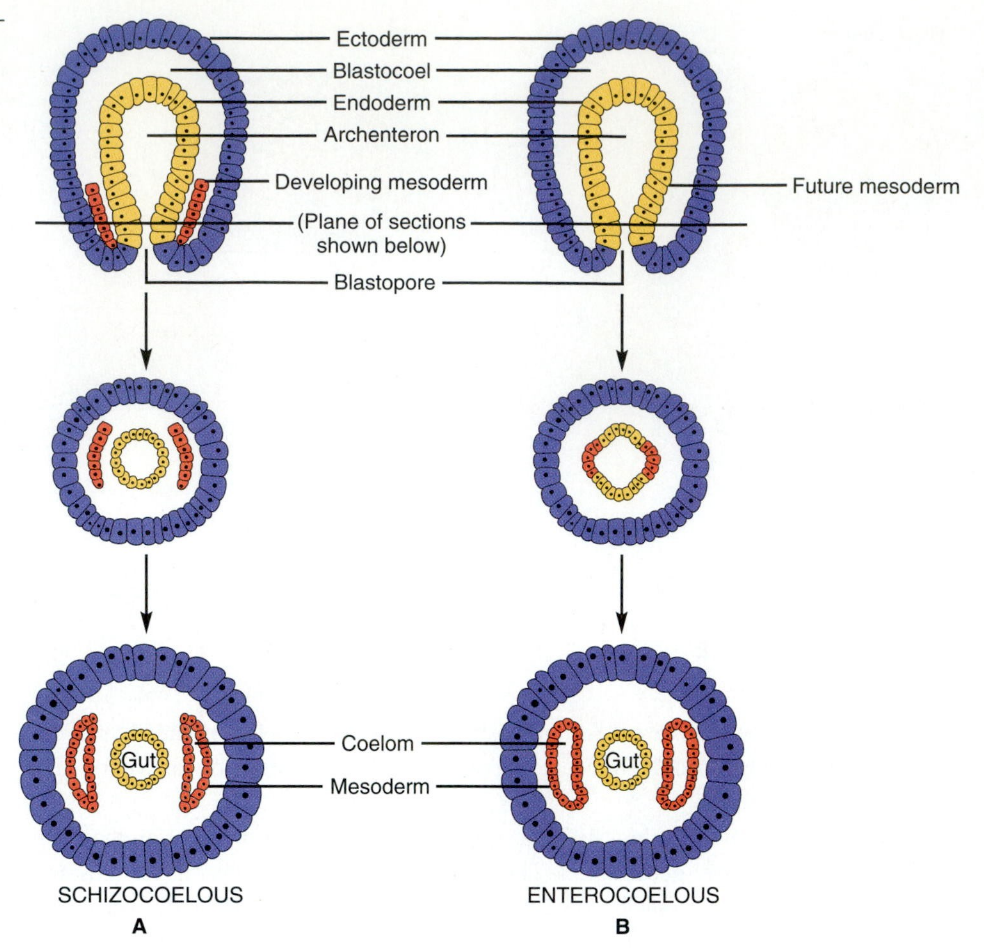

FIGURE 6.15
Representation of two methods of coelom formation from mesoderm. Top row, longitudinal section. Middle and bottom rows, cross sections. (A) Most protostomes and many chordates are schizocoelous: The coelom forms by the splitting of mesoderm. (B) Echinoderms and primitive chordates are enterocoelous: The coelom forms as an outpocketing of the lining of the archenteron.

an additional extraembryonic membrane, the **amnion,** that contains the amniotic fluid that moistens and cushions the embryo. The amnion is considered an evolutionary milestone, since it enabled these vertebrates—the **amniotes**—to develop on land. A third kind of extraembryonic membrane of amniotes is the **allantois,** which stores metabolic wastes and helps exchange oxygen and carbon dioxide. Finally amniotes have a **chorion** that also exchanges respiratory gases.

Even though humans and other **placental mammals** develop within the moist environment of the uterus, their gastrulation is much like that of vertebrates with yolky eggs. Presumably, this similarity reflects our evolution from egg-laying reptiles. The cells in the blastocyst that will later form the fetus are restricted to an **inner cell mass,** as if they were squeezed there by yolk (Fig. 6.18 on page 120). Some of these cells separate into a **hypoblast.** This hypoblast forms a yolk sac like that in fishes, reptiles, and birds, even though in placental mammals there is no yolk for it to enclose. There is also an allantois that stores metabolic wastes, although it is vestigial in humans and many other placental mammals. Later it becomes the lining of the urinary bladder.

The remaining cells of the inner cell mass form the **epiblast** in placental mammals. The epiblast subdivides into two layers, one of which forms the amnion. The other layer is the **embryonic epiblast,** which includes the cells of the embryo. The blastocyst is surrounded by a layer of blastomeres that forms a **trophoblast.** About five or six days after fertilization the trophoblast somehow works its way into the uterine lining during implantation. During the following week the trophoblast will develop into the chorion,

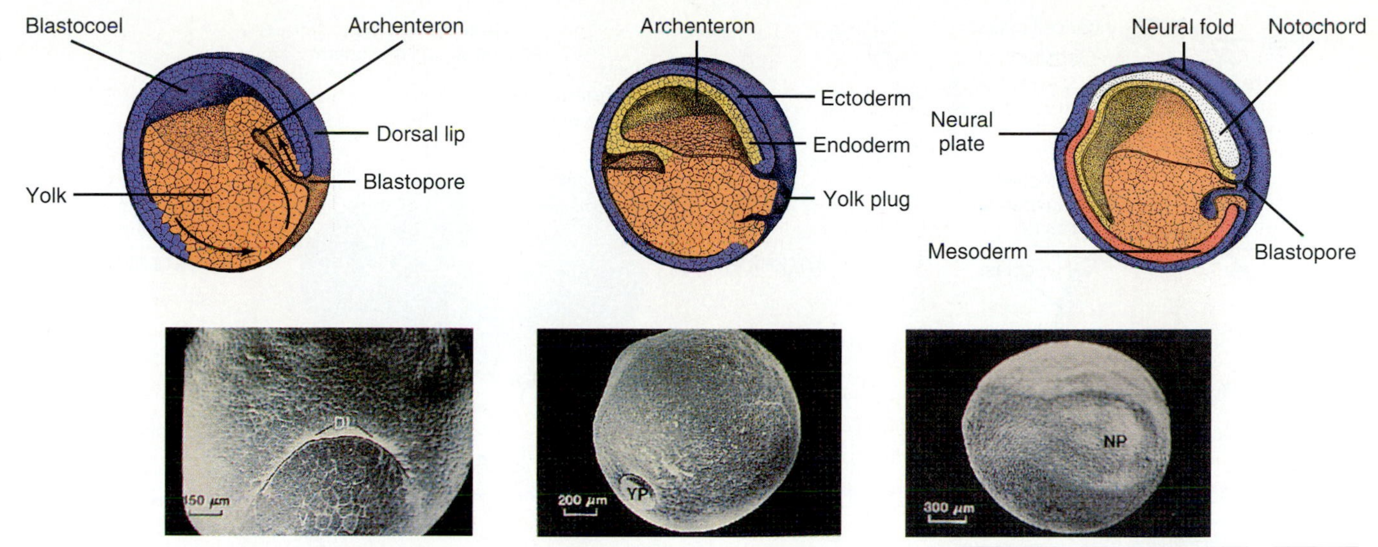

FIGURE 6.16

Gastrulation and subsequent development in the frog, continuing from Fig. 6.12. (Left) Gastrulation, about 10 hours after fertilization. There are around 30,000 cells. Cells in the region of the gray crescent migrate inward, forming a slitlike blastopore that will become the anus. Cells migrating over the dorsal lip (DL) of the blastopore and into the blastocoel form the archenteron.
(Middle) Ectoderm and endoderm form. Migration of ectoderm over the vegetal pole covers the yolk except for a yolk plug (YP) at the blastopore.
(Right) Ectoderm covers the entire gastrula. Some endoderm forms mesoderm. Gastrulation is complete after about 8 hours. The number of cells has more than doubled. The notochord and neural folds are the future location of the brain and spinal cord.

which will extract nutrients from the mother. Later it will develop into the fetal part of the placenta.

Cells accumulate at the posterior end of the embryonic epiblast of amniotes, forming a **primitive streak** into which other cells migrate. Cells entering the anterior end of the primitive streak (**Hensen's node**) and continuing anteriorly form the **notochord.** The notochord is a rod of stiff tissue that is one of the hallmarks of the phylum Chordata, although it disappears before birth in mammals and most other vertebrates. Cells that migrate laterally into the primitive streak will become endoderm and mesoderm (Fig. 6.19 on page 120).

Morphogenesis

DIFFERENTIATION INTO TISSUES. Gastrulation is the first part of morphogenesis, during which each embryonic cell makes a commitment that it and all its progeny will develop into a particular type of cell in the adult. This commitment is called **determination.** Determination is usually not immediately apparent at the time it happens, but gradually cells become visibly different from each other in a process called **differentiation.** Groups of similar cells then form **tissues,** which are collections of cells that are structurally and functionally related to each other. There are four fundamental types of tissue: epithelium, connective tissue, nerve, and muscle. In mammals some 200 subtypes are recognized.

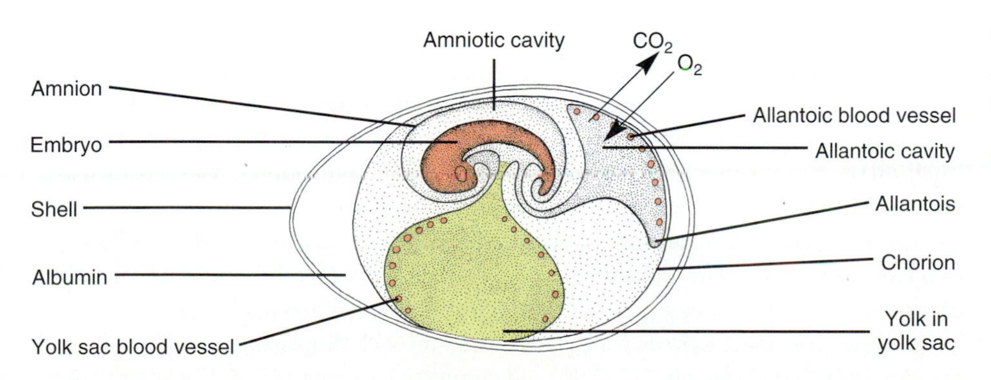

FIGURE 6.17

The embryo of a bird, emphasizing the extraembryonic membranes. Compare Fig. 38.18, which shows a recently laid egg.

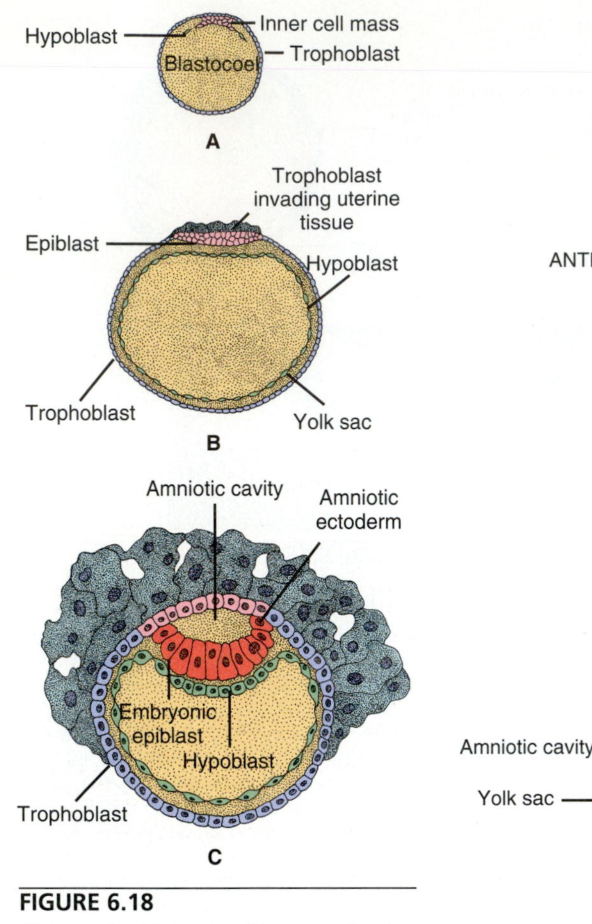

FIGURE 6.18

Changes in the human blastocyst leading up to gastrulation. (A) The blastocyst before implantation. Compare Fig. 6.9. (B) The blastocyst during implantation by the trophoblast. Hypoblast cells from the inner cell mass have separated from the epiblast to form a yolk sac. (C) Approximately 10 days later, just before gastrulation. Trophoblast is proliferating. Amniotic ectoderm from the epiblast is forming an amnion enclosing the amniotic cavity.

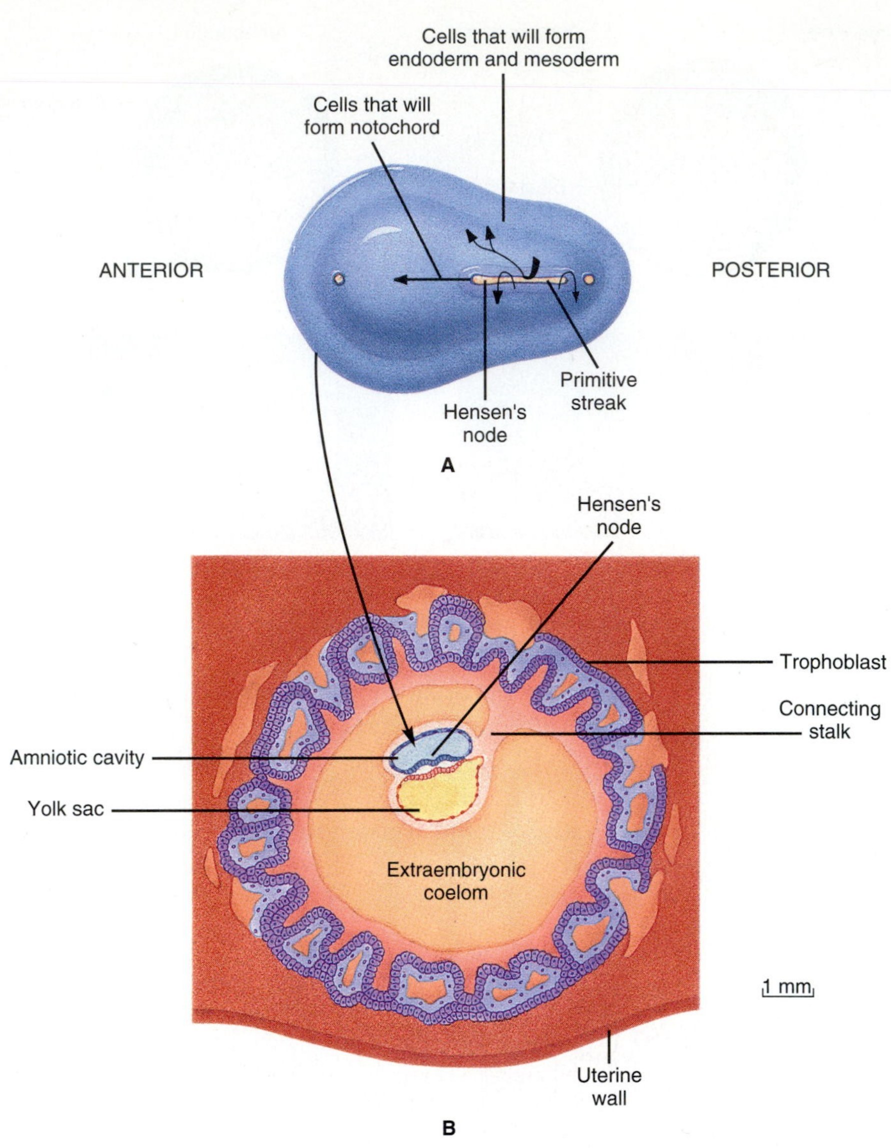

FIGURE 6.19

Gastrulation in the human embryo. (A) Dorsal view of the embryonic disc. Cells migrating into the primitive streak anteriorly will become the notochord; those migrating laterally will become endoderm and mesoderm. (B) Longitudinal section through the embryo at the time of gastrulation, about 17 days after fertilization. The embryonic disc lies between the amniotic cavity and the yolk sac, and all are attached by a connecting stalk to the developing placenta. The anterior of the embryonic disc is to the left.

Epithelial tissue derives from ectoderm and endoderm. It serves as a covering for the body surface and for internal cavities and tubes, and it also forms the secretory portion of glands (Fig. 6.20). Epithelial tissue can consist of one layer of cells (**simple epithelium**) or two or more layers (**stratified epithelium**). Epithelium is also classified as **squamous, cuboidal,** or **columnar,** depending on whether the cells on the outer surface are flattened, cubical, or column-shaped. The inner surface of epithelium is bound by a **basement membrane** (= **basal lamina**). The basement membrane consists mainly of protein fibers that provide mechanical support for the epithelium.

Connective tissue derives from mesoderm and includes cartilage and bone (see pp. 195, 197) , blood (see pp. 217–218), and **connective tissue proper.** The latter tissue

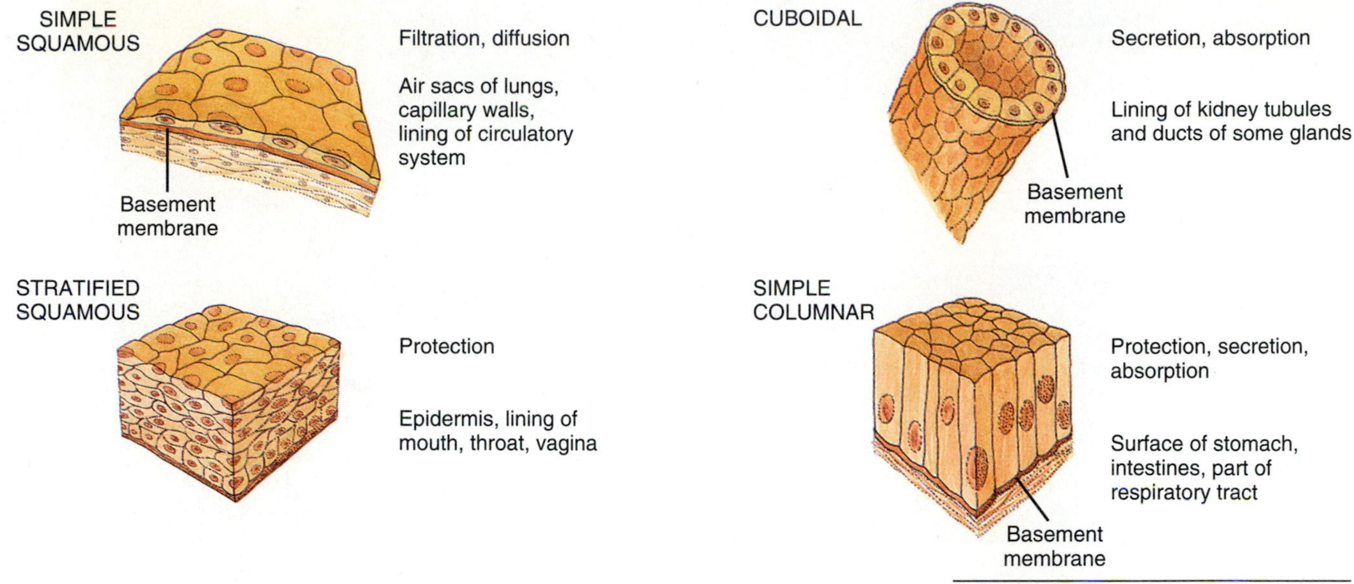

FIGURE 6.20

Some types of epithelial tissue, and their functions and locations.

includes adipose tissue (fat), which serves as an energy reserve and as thermal and mechanical insulation. Connective tissue proper also includes **fibroblast** cells that secrete collagen and other fibers that interconnect other kinds of tissues. Connective tissue proper is categorized either as dense (tendons and ligaments) or as loose. The remaining two types of tissue, neural and muscular, derive from ectoderm and mesoderm, respectively, and are described in later chapters. Figure 6.21 on page 122 traces the lineage of various tissues in the human.

■ **Question:** What is the essential difference between simple and stratified squamous epithelium? What would stratified columnar epithelium look like?

NEURULATION. During morphogenesis, tissues somehow shape themselves into organs. The first organ to form in humans and many other vertebrates is the central nervous system, which includes the brain and spinal cord. The rudiments of the brain and spinal cord take shape as the **neural tube** forms along the notochord in the process called neurulation. The embryo at this stage is called a **neurula** (Fig. 6.22A on page 123). **Neural crest** cells that form within the neurula later migrate and give rise to other parts of the nervous system, as well as to other structures such as the dentin of the teeth. (Refer to Fig. 6.21.)

CONTINUED MORPHOGENESIS IN THE HUMAN. Soon after neurulation, the heart and major blood vessels start to form, and the heart starts to beat a few days later. At about the same time, mesoderm on each side of the neural tube collects into dense structures called **somites.** The number of somites gradually increases, reaching the maximum of 40 at around one month of gestation in humans. These somites will differentiate into skeleton, muscles, and connective tissues of the skin (see Fig. 6.21 on page 122). Approximately 100 primordial germ cells for the next generation are already determined. They originate in the yolk sac but soon migrate to the hindgut.

By the fourth week after fertilization the human embryo has taken on a recognizable form (Fig. 6.23A on page 124). The heart continues to beat as a large bulge outside the chest, and all or most of the somites have formed. The outline of the eye can be made out. On each side of the heart are limb buds that will become arms, and farther back are the limb buds that will become legs. Between the posterior limb buds is a tail—a relic of our evolution. **Pharyngeal slits** in the neck also attest that our ancestors had gills. The embryo is approximately 6 mm long at this stage, and the placenta is fully able to handle the increasing demand for nutrients and oxygen.

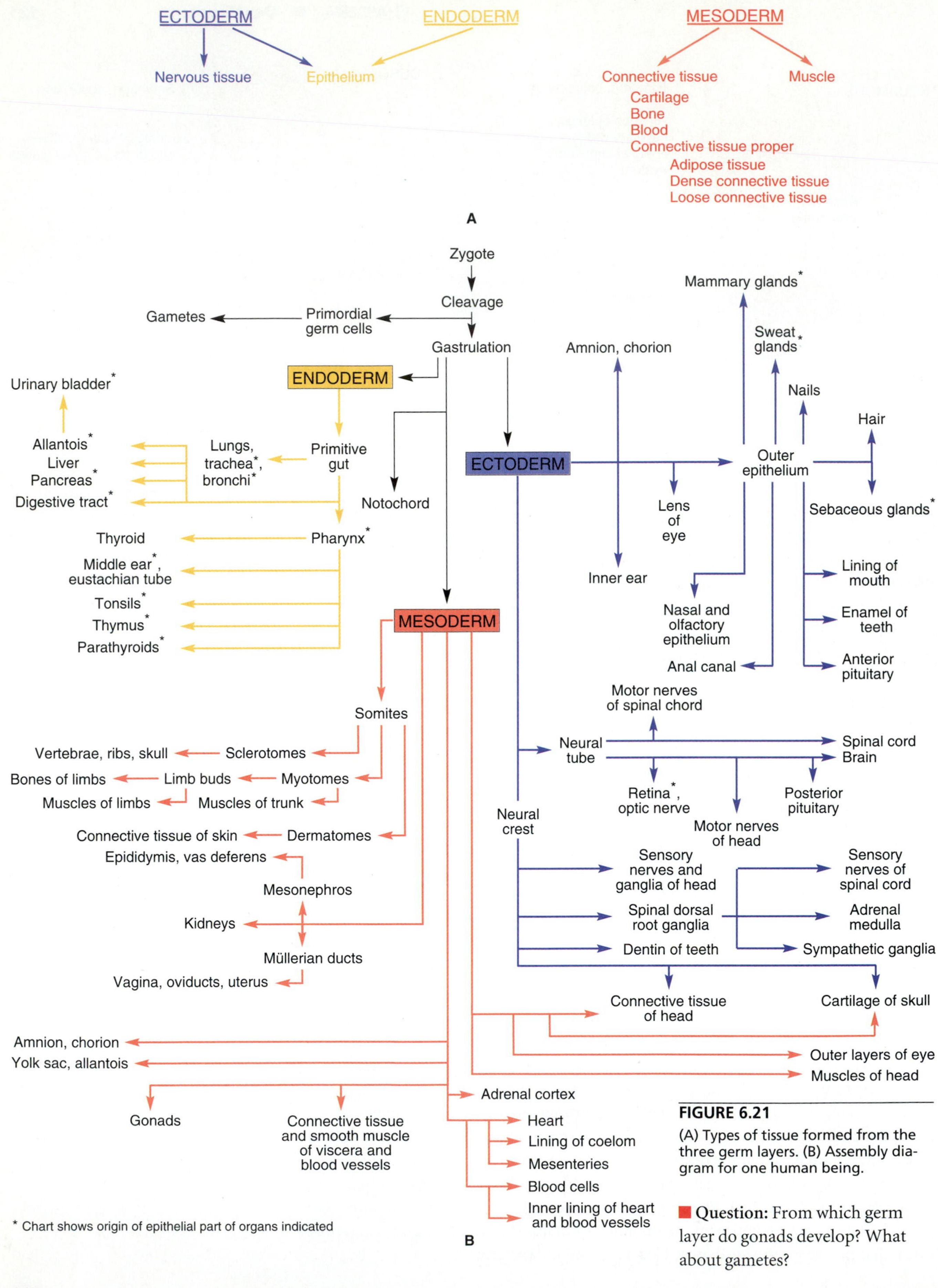

FIGURE 6.21

(A) Types of tissue formed from the three germ layers. (B) Assembly diagram for one human being.

■ **Question:** From which germ layer do gonads develop? What about gametes?

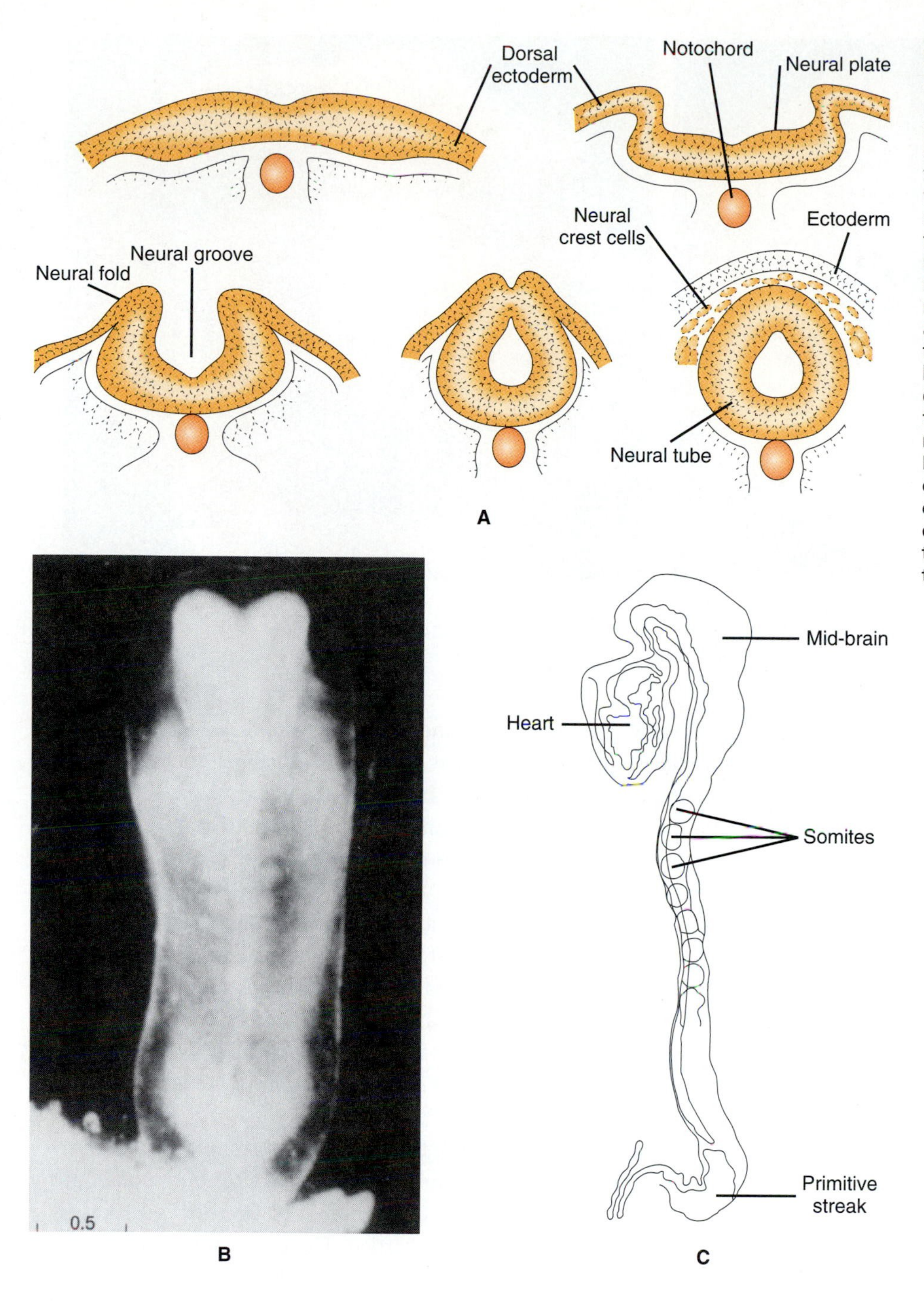

FIGURE 6.22

(A) Cross-sectional representation of neurulation. For an external view of the neural folds and neural plate in a frog see Fig. 6.16. The notochord induces ectoderm on each side to rise up in neural folds that fuse together. This curls the neural plate between the folds, forming the neural tube that will become the brain and spinal cord. Between the neural tube and outer ectoderm is the neural crest, parts of which will migrate and form a variety of neural and other tissues. (B) Dorsal view of a human embryo during neurulation, at about 21 days after fertilization (approximately four days after Fig. 6.18). Neural folds rising over the head will form the brain. Neural folds posteriorly have already closed over the neural tube that will become the spinal cord. Note the amnion enclosing the embryo. (Scale in millimeters.) (C) Drawing showing left side of the embryo in part B.

By about six weeks the embryo has grown to approximately 15 mm long (Fig. 6.23B on page 124). Outlines of fingers and toes are visible, although they are linked by webs of skin. The pharyngeal slits are becoming incorporated into the neck and developing face. At about this time a gene on the Y chromosome of male embryos switches on, producing a **testis determining factor** (TDF). TDF causes the primordial gonads to develop into testes rather than ovaries. The testes then secrete testosterone and other substances that shape the development of male internal reproductive organs and genitals (Fig. 6.24 on page 125). **Anti-Müllerian duct factor** from the testes causes the **Müllerian ducts** to degenerate. Each of the two temporary **mesonephric kidneys** and the **Wolffian duct** that drains it then develop into male reproductive organs associated with one testis. In female embryos the absence of these secretions from the testes causes

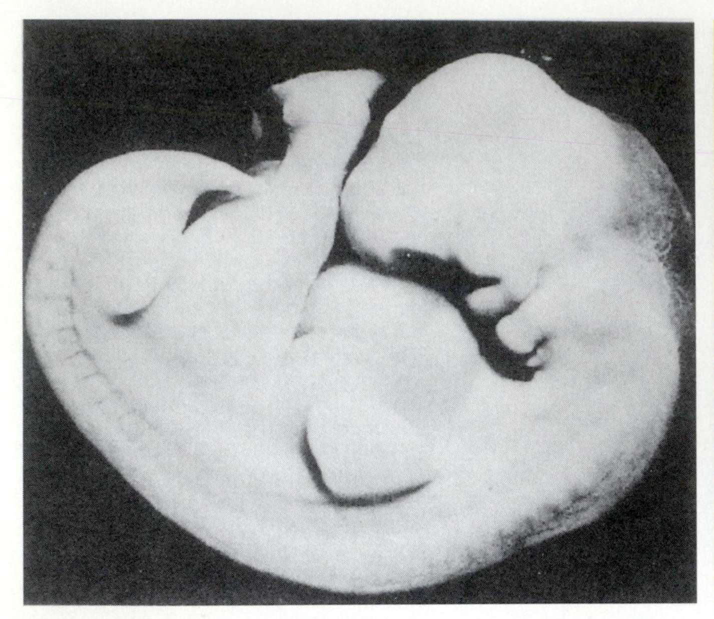

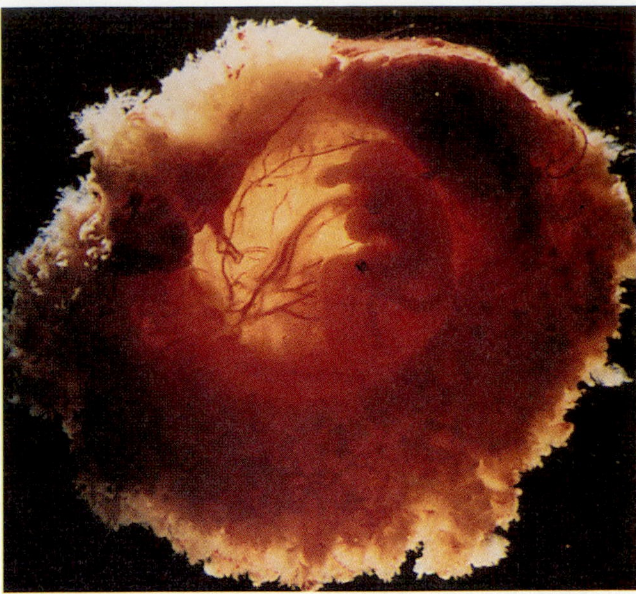

FIGURE 6.23

(A) Human embryo at about one month after fertilization. (Length approximately 6 mm.) (B) At about six weeks. (Length approximately 15 mm.)

the degeneration of the two mesonephric kidneys and the Wolffian ducts. Müllerian ducts then continue to develop into the oviducts, uterus, and upper part of the vagina.

By the time these changes are underway at around the eighth week, the human embryo has essentially reached its final form and is now referred to as a "fetus" (Fig. 6.25 on page 126). Although it is only about 29 mm long, it has approximately 90% of the 4500 named structures that will appear in the adult. The fetus still has about 17 more weeks of development, however, before it has a chance of surviving outside the mother's body, even with support from the best technology available today. The external genitals are developed, although one cannot tell by looking whether they are male or female. The tail and the webs between fingers have degenerated by a process of **controlled cell death.** (Sometimes, however, the tail or webs persist even at birth.) The skeleton is present as cartilage that will later be replaced by bone, a process that is not complete until after puberty.

The subsequent prenatal development of a fetus mainly involves growth of the organs and development of their ability to function. In the human fetus during the third month, the nervous system and muscles begin to function. The fetus kicks, makes facial expressions, sucks, swallows, and jerks as if startled. These are spontaneous or reflex movements, rather than conscious actions. The parts of the brain responsible for consciousness, perception, and voluntary movement will not fully develop until months after birth. The lungs and kidneys are developed, although the placenta continues to do their work for them. By the end of the first trimester, about three months after fertilization, the fetus is about 6 cm long.

During the second trimester (fourth through sixth months) the fetus grows to a sitting height of approximately 35 cm. Bones replace much of the cartilage, and the fetus can kick hard enough to make its presence known to the mother. The heart can be heard beating. The skin is wrinkled and red, covered with downy hair and protected by a cheesy material. The eyes open. Perhaps 10% of fetuses delivered around the end of the second trimester survive, but only if body temperature and breathing are maintained artificially. During the final trimester the fetus grows at such a rapid pace that if

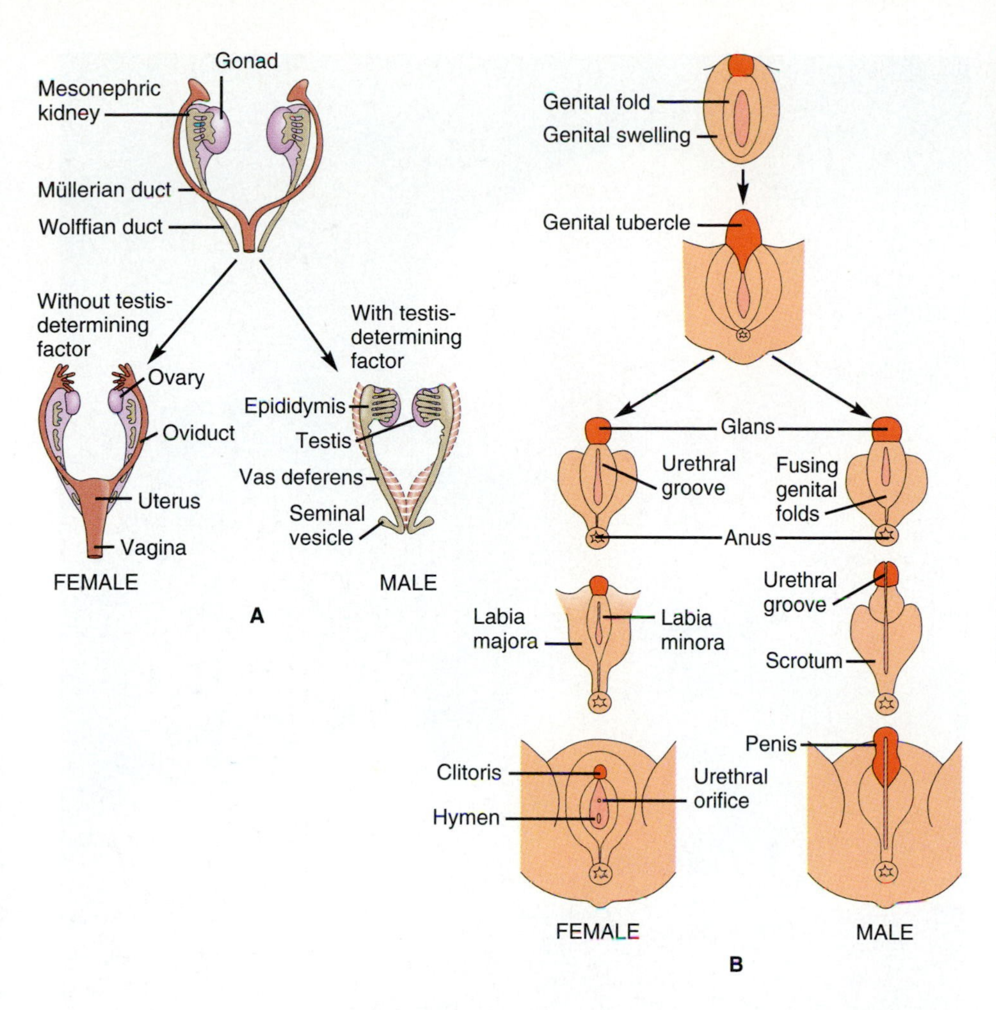

FIGURE 6.24

Changes in internal and external sex organs beginning at about eight weeks after fertilization. (A) TDF causes the gonad to develop into a testis. The testis then secretes anti-Müllerian duct factor, which causes the Müllerian ducts to degenerate, and testosterone, which causes the Wolffian ducts to develop. Each mesonephric kidney and Wolffian duct then becomes the epididymis, vas deferens, and seminal vesicle. (Compare Fig. 16.4.) In females the mesonephric kidneys and Wolffian ducts degenerate, and the Müllerian ducts becomes the oviduct, uterus, and upper part of the vagina. (Compare Fig. 16.6.) (B) Changes in external genitalia. In females the genital tubercle becomes the clitoris, and the genital folds become the labia minora. In males the genital tubercle and genital folds of the embryo become the penis. The genital swelling becomes the labia majora or the scrotum.

it continued, the baby would weigh 200 pounds by its first birthday. A major change during this period is that the lungs and circulatory system prepare for the transition to air-breathing. Before birth the blood is diverted into arteries that bypass the lungs, but within minutes after birth, ducts in major arteries close, forcing blood into the lungs.

Mechanisms of Development

Watching cells in a developing human or other animal is a little like watching traffic from an airplane: There is obviously some kind of orderliness in the overall activity, and individuals seem to move with a purpose, but it is impossible to say what determines the destination of each one. The mystery is especially perplexing since studies of mitosis show that all the cells receive the same genetic information that was in the zygote. This fact was dramatically confirmed in th e early 1960s by John B. Gurdon at Oxford. Gurdon destroyed the nuclei from eggs of African clawed frogs, *Xenopus laevis,* and replaced them with nuclei from the intestine of tadpoles. Some of these eggs went on to develop into normal frogs genetically identical to the "donor" tadpole. Therefore, even though most genes are inactive in the intestinal cells of a tadpole, they must have been present in order to provide all the information to make a complete frog.

These and other studies suggest that development is a process of repeated cell division in which different combinations of genes become inactivated in different cells. These different gene combinations first establish the basic body pattern, then trigger cell differentiation in morphogenesis, and finally stimulate growth of tissues.

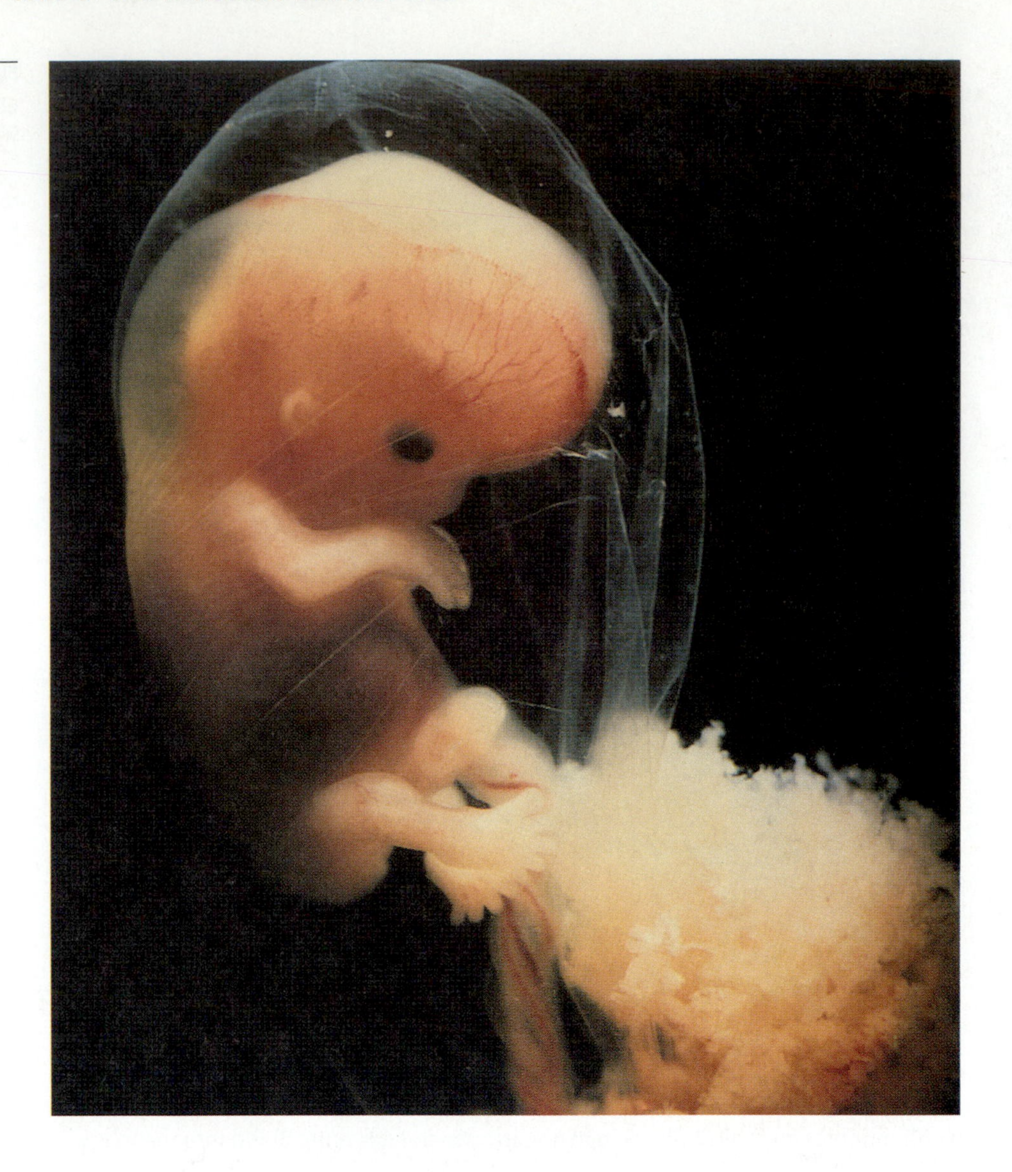

FIGURE 6.25
The human fetus at about the eighth week after fertilization.

PATTERN FORMATION. One of the first steps in development is determining front from rear and top from bottom—that is, determining **anterior–posterior** and **dorsal–ventral axes.** In *Drosophila,* Christiane Nüsslein-Volhard and her colleagues in Germany have shown that the anteroposterior axis is determined before fertilization by RNA secreted into the egg by nurse cells from the mother. RNA transcribed by a gene called *bicoid* (pronounced BICK-oyd) in nurse cells collects at one end of the egg and directs the synthesis of proteins that activate some genes and repress others, thereby determining that that end will be anterior. In similar fashion, RNA from a gene called *nanos* at the posterior pole of the *Drosophila* egg directs production of substances that allow development of the abdomen. The general term for the protein products of the *bicoid* and *nanos* genes and for other substances that control development is **morphogens.** Morphogens are believed to diffuse from their source, setting up chemical gradients that, in effect, tell other cells where they are in the embryo and what they ought to become. Proteins produced by some of the genes activated by *bicoid* subsequently activate other genes that produce different morphogens (Fig. 6.26).

Another aspect of pattern formation that is important in many animals is **segmentation,** also called **metamerism** (see p. 608). Segmented animals, including annelids,

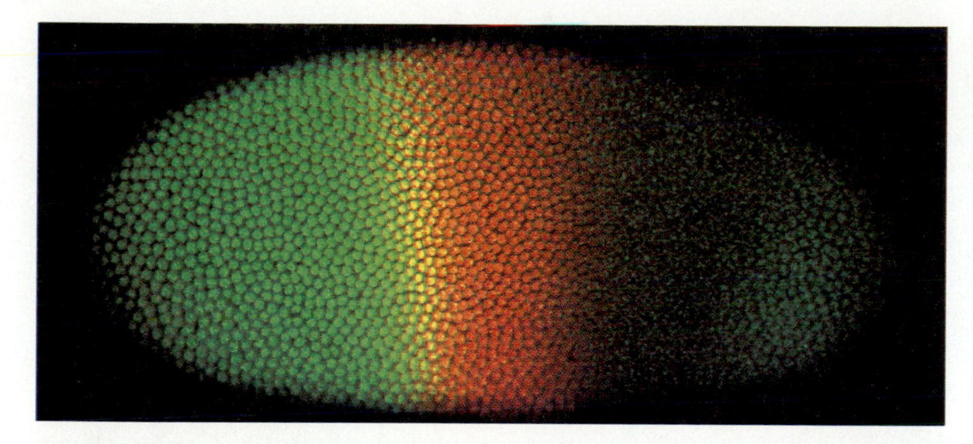

FIGURE 6.26

Products of the *bicoid* gene activate other genes in different parts of the embryo, including two called *Krüppel* and *hunchback*. These two genes produce proteins that can be labeled by fluorescent antibodies. Red shows the region where the *Krüppel* gene is active, and green shows where the *hunchback* gene is active. Yellow shows where the regions overlap. Products of the *nanos* gene at the posterior end inhibit the *hunchback* gene, thereby allowing development of the abdomen.

arthropods, and chordates, develop from a lengthwise series of segments or metameres that are identical early in development. The activation of different combinations of genes in each segment causes the segment to form different structures. For example, the anterior segment of insect embryos will form antennae and mouthparts, while segments farther back will form legs. The antennae, mouthparts, legs, and other segmental structures that have similar embryological origins are said to be **serially homologous.** In chordates, segmentation is apparent in the somites, which give rise to homologous groups of muscles and bones, such as the vertebrae.

The number and orientation of segments is controlled by **segmentation genes,** as was first discovered in *Drosophila* from mutations. For example, a mutation in the segmentation gene *Krüppel* (see Fig. 6.26) results in a loss of most of the segments in the anterior abdomen. The identity of each segment is determined by **homeotic genes,** such as *bicoid.* A mutation in some homeotic genes changes the identity of a segment, causing its appendages to develop into the homologous appendages of a different segment (Fig. 6.27). For example, *Bithorax* mutations cause fruit flies to develop an extra pair of wings on the posterior thoracic segment, where peglike structures would normally occur. One of the intriguing discoveries about homeotic and segmentation genes in drosophila is that most of them include an identical sequence of 180 base pairs, called the **homeo box.** Even more exciting, the same homeo box occurs in certain genes of many other animals, including cnidarians, nematodes, earthworms, sea urchins, frogs, chickens, mice, and humans, and also in plants and yeasts. Even more incredible, in animals the order of these gene clusters on the chromosome is the same order in which they are expressed in the body.

FIGURE 6.27

Scanning electron micrographs of the head of a normal *Drosophila* (left) and of one homozygous for a mutation in the *Antennapedia* complex of genes. In the mutant, the antennae are replaced by tiny legs.

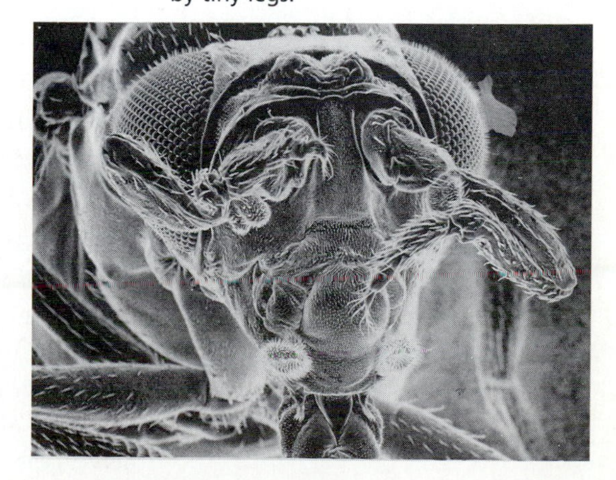

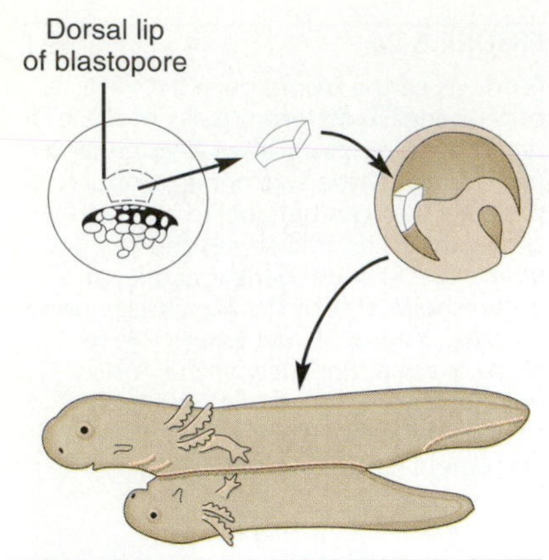

FIGURE 6.28

Spemann and Mangold's demonstration of induction. The dorsal blastopore lip of the newt *Triturus cristatus* was transplanted to the gastrula of *T. taeniatus* in the area fated to become the belly. The fact that *T. taeniatus* is pigmented, while *T. cristatus* is transparent, enabled Spemann and Mangold to determine whether later tissues came from the transplant or the host. The transplant induced the development of Siamese twin newts formed mainly of pigmented host tissue.

In amphibians the animal and vegetal poles and the site of sperm entry establish the first axes for pattern formation. The gray crescent that forms opposite the site of sperm entry becomes the dorsal part of the animal (see Fig. 6.12). During gastrulation the anterior–posterior axis is aligned with the poles. The dorsal lip of the blastopore (see Fig. 6.16) then functions as an **organizer** that determines the rest of the body plan, as was demonstrated in 1924 by Hans Spemann and Hilde Mangold. They grafted the dorsal lip of the blastopore of one newt into the gastrula of another newt in an area that was fated to become the belly. The transplanted dorsal blastopore lip induced the host embryo to develop a second larva joined at the belly as a Siamese twin (Fig. 6.28). This process is referred to as **induction.** No other part of the gastrula has the ability to induce development of a second larva, although other tissues can induce development of smaller organs. Recently it has been found that the dorsal lip of the blastopore in the African clawed frog *Xenopus laevis* produces proteins similar to those produced by *bicoid* and other homeotic genes in *Drosophila.*

MECHANISMS OF MORPHOGENESIS. Once the pattern of an embryo has been established, the processes of morphogenesis direct particular cells to move to appropriate places in accordance with that pattern. These cellular movements are due to **extension, adhesion,** and **contraction** of cells. The migration of micromeres during gastrulation in sea urchins (Fig. 6.14D) and the subsequent movement of the primary mesenchyme illustrate these processes. First the cells extend **pseudopods** with tips that adhere to appropriate cells on the inner surface of the blastocoel. Once the pseudopods are attached, they contract, pulling the rest of the cell in. Like pattern formation, morphogenesis is thought to be guided by morphogens that distinguish anteroposterior and dorsoventral axes of individual organs and limbs. Recently it has been discovered that cells in the posterior part of vertebrate limbs produce a protein that guides the development of fingers and toes in the appropriate order. This protein is the product of a gene called *Sonic hedgehog,* named for the appearance of fruit flies in which the gene is mutated. *Sonic hedgehog* in the notochord and neural plate (Fig 6.22) is also involved in guiding the development of the spinal cord and brain.

What guides each cell within a developing organ remains largely a mystery, especially in the case of the central nervous system, where hundreds of billions of nerve cells must make precise connections with other nerve cells. Human brains have a great deal of research to do before they understand their own development. There is growing evidence, however, that certain chemical factors either promote or inhibit the formation of nerve-cell extensions called **growth cones** (Fig. 6.29). Concentration gradients of the same or different chemical factors attract or repel growth cones. If a growth cone contacts an appropriate target cell, it somehow recognizes it and adheres. Such processes continue in humans even after birth, and they may be influenced by sensory stimulation and learning. In addition to forming new connections, superfluous connections are deleted until human adolescents have only about half as many connections between nerve cells as when they were eight months old.

MECHANISMS OF GROWTH. Later embryonic development and early postnatal life are periods of rapid growth for most animals. This growth is due to rapid cell division in response to polypeptide **growth factors.** The first growth factor was discovered in the 1950s by Rita Levi-Montalcini, Viktor Hamburger, and Stanley Cohen. They found that a substance that happened to be abundant in certain tumors, the venom of snakes, and the saliva of male mice was necessary for the growth and maintenance of particular nerve cells (those derived from the neural crest). They called the substance **nerve**

growth factor (NGF). Afterward came the discovery of several other growth factors, such as **epidermal growth factor** (EGF). EGF stimulates division of epidermal, neural, and some other kinds of cells and also stimulates spermatogenesis. It also occurs in milk and stimulates continued development of the infant's digestive tract. As development proceeds, levels of growth factors decline, cells differentiate into their ultimate specialized forms, and cell division slows down. Growth factors can trigger a resumption of cell division during wound repair, however, and possibly also in the formation of tumors.

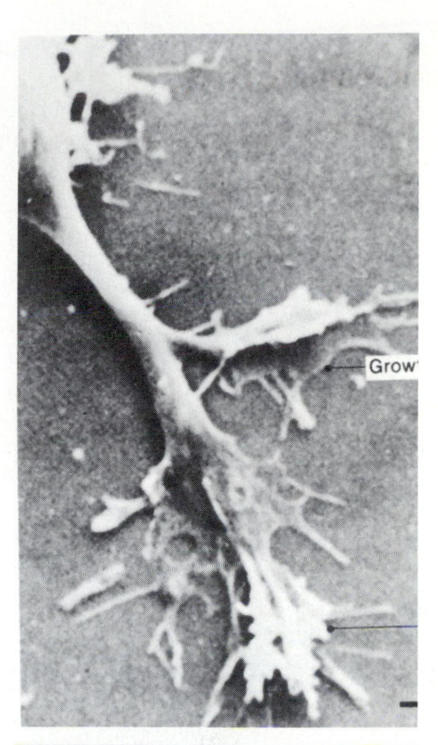

FIGURE 6.29

Scanning electron micrograph of a nerve cell in tissue culture searching for an appropriate target cell. The cell body containing the nucleus is at the upper left. The process running to the lower right divides and expands into two growth cones from which thin filopodia extend.

■ **Question:** How might a growth cone "know" when it has made a correct connection? What happens to it if it does not?

DEVELOPMENTAL DEFECTS. The multitude of events required for the development of a new organism provides ample opportunity for something to go wrong. In fact, approximately one-third of all human embryos miscarry, often without the mother knowing she was pregnant. Other developmental abnormalities are not fatal to the embryo but result in defects in the child. Such **developmental defects** are of two types: **genetic defects** and **congenital defects.** Genetic defects are due to point mutations or chromosomal aberrations, and they result from absent or incorrect gene products during meiosis or development. Down syndrome (see p. 94) is just one of many such genetic defects. Congenital defects are not inherited but result from external factors, called **teratogens,** that interfere with a normal process of development. In humans virtually any substance that can be transferred from maternal to fetal blood through the placenta is a potential teratogen. The list of known and suspected teratogens includes viruses (such as the type that causes German measles), alcohol, and numerous drugs (including aspirin).

Metamorphosis and Regeneration

As if once were not impressive enough, many animals develop twice during their life cycles. They are born as **larvae,** then undergo **metamorphosis** into adults with completely different forms. Familiar examples of larvae include the tadpoles of frogs and the caterpillars of butterflies and moths. Animals that undergo such metamorphosis are said to have **indirect development.** Development directly into a form that resembles the adult is called **direct development.** In some cases, larvae are essentially embryos that continue their development outside the egg, consuming food instead of yolk. The grubs and maggots of certain insects are examples. In most cases, however, larvae are adapted for essential functions that the adult cannot perform. In sponges, sea anemones, and many molluscs, for example, larvae are often responsible for dispersing the species to new habitats, swimming by means of cilia or drifting with currents. The adults, in contrast, are often sedentary and adapted mainly for reproduction.

Just as different parts of the egg, or different blastomeres, are fated to become different parts of the organism, different parts of the larva are fated to be reorganized into adult parts. This is clearly seen in *Drosophila* maggots, where different parts, called **imaginal discs,** are determined to become various parts of the adult (Fig. 6.30 on page 130). Changes in the levels of hormones (especially juvenile hormone) trigger the development of imaginal discs into adult structures (see Fig. 7.14). In frogs and other amphibians, increased levels of thyroid hormones trigger metamorphosis.

Like metamorphosis, **regeneration** is a kind of postnatal development of the adult form. Animals vary widely in their ability to regenerate. Certain flatworms can regenerate a new cerebral ganglion ("brain") if the original one has been removed surgically, but humans and other mammals are incapable of replacing even a single nerve cell that has died. Most insects and amphibians can regenerate a lost leg, but the best humans and other mammals can do is to regenerate much of the liver or a lost fingertip. One of the main reasons for studying regeneration, aside from the intrinsic interest and the

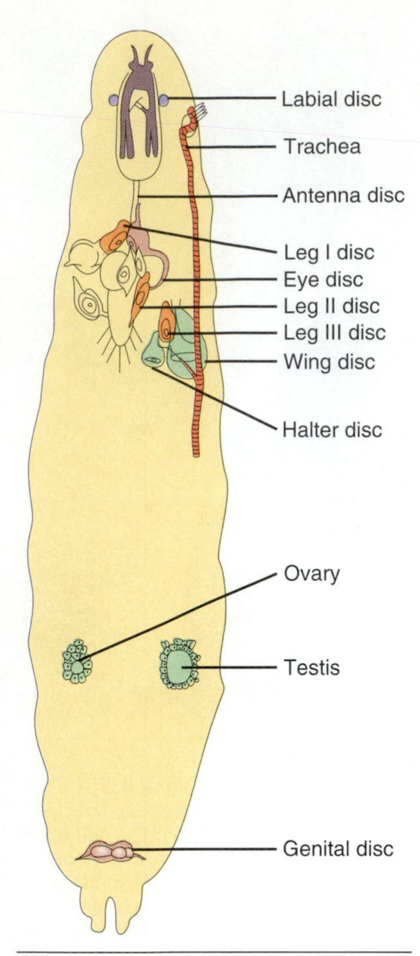

FIGURE 6.30
Imaginal discs in the larva of a drosophila. The term "imaginal" comes from the fact that the adult insect is called the "imago."

■ **Question:** Suggest an experiment by which you could verify the fate of an imaginal disc.

potential for regenerating lost organs in humans, is that it may provide clues to developmental processes. During regeneration, growth factors cause cells at the site of the amputation to revert to an undifferentiated state like that in the embryo and to undergo rapid division. The accumulation of undifferentiated cells is called **blastema.** In salamanders and other tailed amphibians, which are the only vertebrates that can regenerate limbs as adults, blastema forms on the stump a week or so after amputation of a limb. Positional information in the form of chemicals then directs the organization of the blastema into a functional limb.

Development and Evolution

As noted previously, one of the main motivations for zoologists to study development is that it provides insights regarding taxonomic relationships among groups of animals. In the early 19th century the Estonian biologist Karl Ernst von Baer made a number of observations that suggest such a relationship, although he explicitly rejected any evolutionary implications. What is known as **von Baer's law** states that embryonic development in vertebrates goes from general forms common to all vertebrates to increasingly specialized forms characteristic of classes, orders, and lower taxonomic levels. Thus the early embryos of all vertebrates, whether fish, frog, hog, or human, all look alike (Fig. 6.31). Later it is possible to tell the human embryo from the fish but not from the hog, and still later one can see a difference between these two mammals.

In 1866, inspired by Darwin's *Origin of Species,* Ernst Haeckel brought out the evolutionary implications of von Baer's law and extended it to invertebrates. Unfortunately, Haeckel went too far with his **biogenetic law,** which states that "the history of the fetus is a recapitulation of the history of the race [species]." In other words, Haeckel mistakenly thought ontogeny (the development of the individual) is a recapitulation of phylogeny (the evolution of the species). Biologists enthusiastically embraced the idea that ontogeny recapitulates phylogeny until early in this century, and then with even more enthusiasm they rejected it and rejected von Baer's law for good measure.

Von Baer's insight is now respectable once more, although few biologists would call it a law. The explanation for von Baer's "law" is probably similar to the reason why no one tries to design an automobile from the ground up. It is much easier to start with a successful design and tinker with it, even if it means retaining obsolete features. Changing the basic design of the chassis would have serious and unforeseen consequences for the engine, transmission, and suspension that may lead to the car's failure in the marketplace. Early changes in development may likewise have consequences that lead to the failure of the organism in its natural environment. Thus human embryos develop pharyngeal slits, hair, and a tail perhaps because those that have the genetic mutations needed to omit those steps die from other consequences.

The study of developmental biology also provides other insights regarding evolution. One of these is the realization that evolution can result not only from mutations in the genes that are directly responsible for structures, but also in the genes that regulate the activity of other genes during development. Mutations such as those described above that affect segmentation can lead to radical changes in structure, some of which might be beneficial. It is not difficult to imagine, for example, that the *Bithorax* mutant could give rise to an entirely new group of four-winged flies.

Mutations in genes that affect development could also give rise to evolution by changing the timing of developmental events. Differences in developmental timing are, in fact, well known in some groups of animals and are referred to as **heterochrony.** Heterochrony causes certain organisms to develop certain features either earlier or

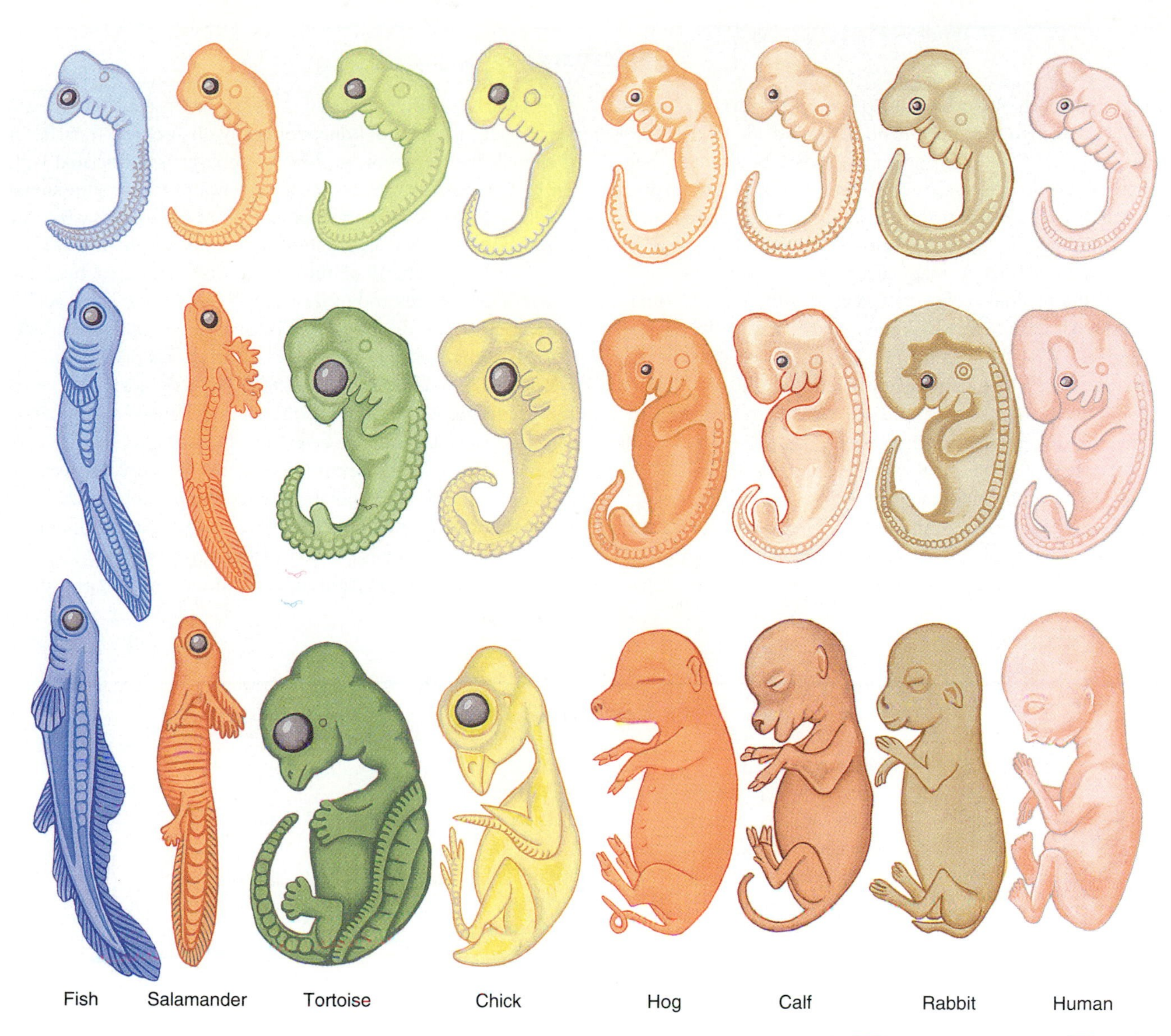

FIGURE 6.31

An illustration of von Baer's discovery that the development of vertebrates proceeds from the general to the specific. The three embryos in each column are successive stages from the same species, and those in each row are at comparable stages of development.

later in life than they would ordinarily. One kind of heterochrony, called **paedomorphosis,** occurs when juvenile features are retained in the adults. Paedomorphosis can occur in two ways: by the retention of juvenile features into adulthood, called **neoteny,** or by the accelerated development of sexual maturity in the juvenile, called **progenesis.** Paedomorphosis can clearly be seen in adult animals that look like the larvae of related species (Figs. 6.32 and 36.16). It seems likely that paedomorphosic species evolved as a result of mutations that affected the timing of developmental events.

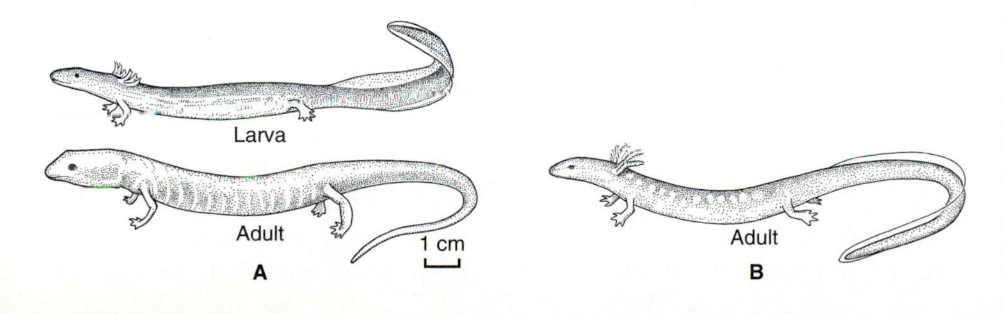

FIGURE 6.32

Paedomorphosis in salamanders. (A) Larvae of salamanders, such as the grotto salamander *Typhlotriton spelaeus,* are generally aquatic and have external gills, which they lose after metamorphosis to adults. (B) In paedomorphic species, such as the Texas salamander *Eurycea neotenes,* the adult retains the external gills. Paedomorphosis in some salamanders is due to the absence of a rise in thyroid hormones that normally triggers metamorphosis. Injecting thyroid hormones into larvae that are normally paedomorphic will cause them to lose their external gills.

Summary

During gametogenesis, primordial germ cells undergo repeated mitosis, producing genetically identical diploid spermatocytes or primary oocytes. These then undergo meiosis, becoming genetically diverse haploid sperm or ova. During fertilization, a sperm cell propels itself toward an ovum, binds to receptor molecules specific for its species, then digests a pathway into the ovum. This triggers a cortical reaction that helps prevent polyspermy. Inside the ovum the haploid male pronucleus and the haploid female pronucleus combine to form a diploid zygote. The zygote then begins to divide repeatedly, in the process called cleavage. The cells follow specific cleavage patterns, depending on the amount and distribution of yolk and on the kind of animal. In some species, development is determinate, with the fate of each blastomere already determined. Cleavage ends with the formation of the blastula.

Beginning with gastrulation, the cells become increasingly specialized in gene expression, and they begin to move about. During gastrulation, three germ layers are established in most animals: ectoderm, which will form nerve and skin; endoderm, which will become gut and other internal organs; and mesoderm, which will form blood, bone, connective tissue, and muscle. In mammals the formation of these germ layers is associated with the primitive streak, which occurs near the future site of the spinal cord and brain. As morphogenesis continues, the cells continue to differentiate and to aggregate as different kinds of tissues.

The body pattern of the embryo is established by morphogens, which are produced by certain genes and which act on other genes. The mechanisms responsible for cell movements during morphogenesis are cellular extension, adhesion, and contraction. Subsequent growth of differentiated tissues is stimulated by various growth factors. Disturbances in these developmental processes can result in birth defects. Metamorphosis and regeneration in some animals appear to be related to normal patterns of development. Developmental patterns appear to have changed gradually during evolutionary history, providing a useful tool for inferring the evolutionary history of animals. Some changes, such as heterochrony, may explain some evolutionary changes.

Key Terms

cleavage
blastula
gastrulation
morphogenesis
gametogenesis
germ cell
fertilization
embryo
yolk
spiral cleavage pattern
radial cleavage pattern
determinate cleavage
indeterminate cleavage
blastopore
deuterostome
protostome
germ layer
ectoderm
endoderm
mesoderm
triploblastic
coelom
amnion
differentiation
tissue
segmentation
induction
growth factor
metamorphosis
regeneration
paedomorphosis

Chapter Test

1. Explain the implications of the experiment by Gurdon, in which he transplanted the nuclei from intestinal cells of tadpoles into frog eggs. (p. 125)
2. Diagram the processes of spermatogenesis and oogenesis. Which stages are diploid, and which are haploid? What are polar bodies? (pp. 107–108)
3. Eggs must encourage fertilization by sperm of the same species while preventing polyspermy and fertilization of a different species. Explain how these conflicting requirements are achieved. (pp. 107–110)
4. Briefly describe what occurs during cleavage, gastrulation, and morphogenesis. (pp. 106–107)
5. How does the distribution of yolk affect cleavage? (pp. 112, 114)
6. The body plan of an animal is said to be established during gastrulation. Explain this statement for a sea urchin, a frog, and a human. (pp. 112, 114)
7. Name the three germ layers. Name three structures in your body that originate from each germ layer. (p. 117)
8. Human embryonic development is apparently inefficient in many ways. For example, a tail is formed only to be eliminated, and there is a yolk sac but no yolk. How do most zoologists explain the occurrence of these structures? (p. 130)
9. Describe one example illustrating how advances in the understanding of cellular, genetic, and molecular biology have contributed to our understanding of the mechanism of development. (pp. 125–129)
10. Regeneration is also often compared with development. In what ways are they similar? How do they differ? (pp. 129–130)

■ Answers to the Figure Questions

6.1 (1) Because each organism would have to contain all the genetic information for a potentially infinite number of offspring, there would have to be an infinite amount of genetic material. (2) Only the parent in which the offspring was preformed would contribute to the offspring's genetic inheritance. (3) There would be no apparent means by which offspring of a given individual could differ from each other. (4) Genetic damage would continually accumulate, resulting in a decline in fitness with each generation. (5) There would be no explanation for the specific effects of agents that cause birth defects. (6) The functioning of organs depends greatly on their size, so it is unlikely that an embryo could be simply a tiny version of the adult.

6.3 In placental mammals that give multiple births, it might make more sense to produce four smaller eggs rather than one large one. In most animals, however, the egg has to be large enough to contain all the material from which the embryo will develop.

6.5 Unlike the embryos of placental mammals, which develop in a uterus, sea urchin embryos have only the jelly coat for protection.

6.8 If conception is defined, as it often is, as the penetration of the egg's plasma membrane by the sperm, then the brief time when that occurs is the "moment of conception." That is only the beginning of a long series of processes necessary for complete conception, however, so the phrase is meaningless and misleading.

6.20 Stratified tissue has more than one layer (stratum) of cells. Stratified columnar epithelium has more than one layer, with columnar cells in the outer layer.

6.21 Gonads develop from mesoderm, but gametes do not develop from any germ layer. They are segregated before the germ layers are defined.

6.29 It is reasonable to suppose that growth cones are guided by chemical gradients and that they recognize targets chemically. They would regress if they do not find an appropriate connection.

6.30 Removing an imaginal disc from a larva should result in an adult that lacks the appropriate organ. Alternatively, one might try growing imaginal discs in tissue culture.

Readings

Recommended Readings

Beams, H. W., and R. G. Kessel. 1976. Cytokinesis: a comparative study of cytoplasmic division in animal cells. *Am. Sci.* 64:279–290.

Beardsley, T. 1991. Smart genes. *Sci. Am.* 265(2):86–95 (Aug). (*Genetic control of development.*)

De Robertis, E. M., G. Oliver, and C. V. E. Wright. 1990. Homeobox genes and the vertebrate body plan. *Sci. Am.* 263(1):46–52 (July).

Edelman, G. M. 1984. Cell-adhesion molecules: a molecular basis for animal form. *Sci. Am.* 250(4):118–129 (Apr).

Edelman, G. M. 1989. Topobiology. *Sci. Am.* 260(5):76–88 (May).

Epel, D. 1977. The program of fertilization. *Sci. Am.* 237(5):128–138 (Nov).

Gerhart, J., et al. 1986. Amphibian early development. *BioScience* 36:541–549.

Goodman, C. S., and M. J. Bastiani. 1984. How embryonic nerve cells recognize one another. *Sci. Am.* 251(6):58–66 (Dec).

Hall, B. K. 1988. The embryonic development of bone. *Am. Sci.* 76:174–181.

Hynes, R. O. 1986. Fibronectins. *Sci. Am.* 254(6):42–51 (June).

Kline, D. 1991. Activation of the egg by the sperm. *BioScience* 41:89–95.

Levi-Montalcini, R., and P. Calissano. 1979. The nerve-growth factor. *Sci. Am.* 240(6):68–77 (June).

McGinnis, W., and M. Kuziora. 1994. The molecular architects of body design. *Sci. Am.* 270(2):58–66 (Feb).

Nilsson, L. 1977. *A Child Is Born.* New York: Delacorte Press. (*Extraordinary photographs of human development.*)

Sachs, L. 1986. Growth, differentiation and the reversal of malignancy. *Sci. Am.* 254(1):40–47 (Jan).

Stent, G. S., and D. A. Weisblat. 1982. The development of a simple nervous system. *Sci. Am.* 246(1):136–146 (Jan). (*The leech.*)

Wassarman, P. M. 1988. Fertilization in mammals. *Sci. Am.* 259(6):78–84 (Dec).

Additional References

Bard, J. B. L. 1994. *Embryos: Color Atlas of Development.* Prescott, AZ: Wolfe. (*Covers 10 species of animal commonly used in research.*)

Bryant, S. V., D. M. Gardiner, and K. Muneoka. 1987. Limb development and regeneration. *Am. Zool.* 27:675–696.

Conn, D. B. 1991. *Atlas of Invertebrate Reproduction and Development.* New York: Wiley–Liss.

England, M. A. 1983. *Color Atlas of Life Before Birth.* Chicago: Yearbook Medical Publishers. (*Detailed color photographs of human development.*)

Foe, V. E., and G. M. Odell. 1989. Mitotic domains partition fly embryos, reflecting early cell biological consequences of determination in progress. *Am. Zool.* 29:617–652.

Ford, N. M. 1988. *When Did I Begin?* New York: Cambridge University Press. (*After reviewing religious and philosophical opinions and embryological evidence, Ford concludes that the embryo becomes distinctly human at gastrulation.*)

Gilbert, S. G. 1989. *Pictorial Human Embryology.* Seattle: University of Washington Press.

Kessler, D. S., and D. A. Meltor. 1994. Vertebrate embryonic induction: mesodermal and neural patterning. *Science* 266:596–604.

Lawrence, P. A. 1992. *The Making of a Fly: The Genetics of Animal Design.* Boston: Blackwell Scientific.

Moore, J. A. 1987. Science as a way of knowing—developmental biology. *Am. Zool.* 27:415–573. (*Key concepts of development in historical context.*)

O'Rahilly, R., and F. Müller. 1987. *Developmental Stages in Human Embryos.* Washington, DC: Carnegie Institution of Washington.

OVERVIEW

UNIT

2

Maintaining the Cellular Environment

Musk oxen were once hunted nearly to extinction, but they are now valued for their fine undercoat of fur, called *qiviut* (KEE-vee-ut). *Qiviut* also saves musk oxen from extinction by insulating them from the hostile environment of the North American Arctic. It enables them to maintain a body temperature of 38°C even when the air temperature is 40°C below freezing.

Qiviut is part of the integument—the body covering that separates the harsh conditions of the external environment from the benign conditions of the internal environment. The internal environment is where the cells live, provided with just the right levels of temperature, water, ions, nutrients, and oxygen. Maintaining these levels in the internal environment also requires various physiological systems. For example, the muscular and nervous systems enable musk oxen cells to obtain sufficient water and nutrients in an external environment that is as dry as the deserts of southwest America and where the only food is scattered tufts of grass.

Maintaining correct internal levels of temperature, water, ions, nutrients, and oxygen also requires the excretory, digestive, circulatory, and respiratory systems. These systems, in turn, require chemical and neural signals to coordinate them. The one physiological system that does not appear to be directly involved in maintaining the internal environment is the reproductive system, which enables musk oxen to survive as a species even though all the individuals inevitably succumb to the harshness of the external environment.

CHAPTER

7

Hormones and Other Molecular Messengers

Hormone-induced sex differences in wood ducks *Aix sponsa*.

Chemical Coordination

Beginning with the French physician Claude Bernard more than a century ago, biologists have realized that proper levels of temperature, nutrients, water, ions, and oxygen tend to be maintained in the internal environment of an animal. The maintenance of conditions suitable for the life of cells in the internal environment is now called **homeostasis.** It seems obvious that homeostasis requires the transfer of information to coordinate the physiological systems that maintain homeostasis. The fact that such information could be transmitted by chemicals was first demonstrated in 1902 by W. M. Bayliss and E. H. Starling. They found that a dog's pancreas increased its rate of secretion after acid was injected into the intestine, even if nerves to the pancreas had been cut. This suggested that a chemical messenger from the intestine circulated to the pancreas and stimulated secretion. In order to test this hypothesis, they crushed some of the lining of the gut after it had been exposed to acid, filtered it, then injected the extract into the blood of another dog. Within about a minute this dog's pancreas increased its rate of secretion. Thus was discovered the first hormone, now called secretin.

We now know of dozens of hormones as well as numerous other chemical messengers. By definition, **hormones are secreted by specific organs called glands, they travel through the blood, and they evoke a response in distant target cells** (Fig. 7.1). Since hormones travel within the circulatory system, they are also called **endocrine** secretions, to distinguish them from exocrine secretions such as sweat, tears, and digestive juices, which stay out of the bloodstream. Many other chemical messengers resemble hormones in their chemical properties or modes of action, but they act locally. These secretions, referred to here as **local messengers,** act on specific target cells nearby, without being transported through the blood.

How Chemical Messengers Work

All chemical messengers work by binding temporarily to other molecules that have specific shapes and charge distributions that complement the messengers. Messengers and other substances that bind specifically to such molecules are called **ligands,** and

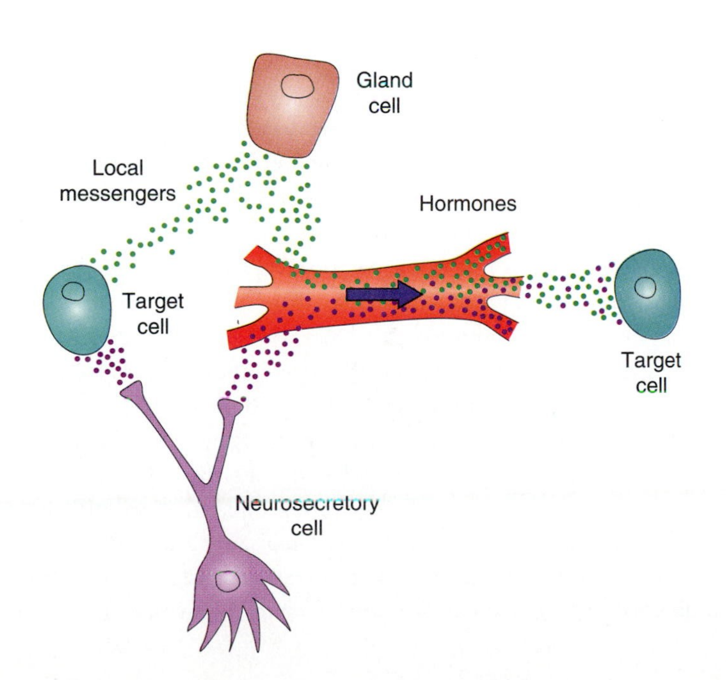

FIGURE 7.1
Chemical messengers are secreted by gland cells or nerve cells. Hormones, but not local messengers, travel through the bloodstream to their target cells.

■ **Question:** What chemical properties of local messengers might keep them from functioning as hormones?

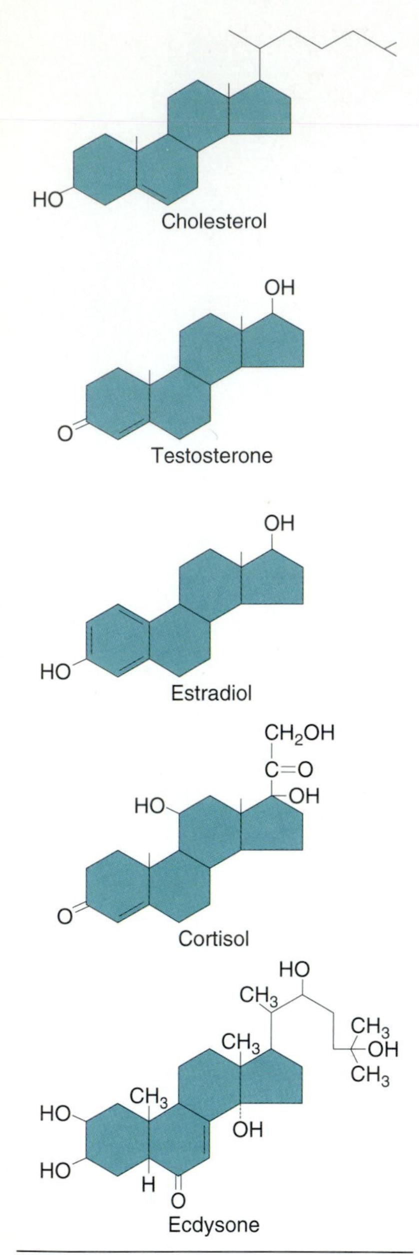

FIGURE 7.2

Cholesterol and some of the steroid hormones derived from it. All the steroids share a similar four-ring structure; the variety of their effects is due entirely to variations in side groups. All these hormones occur in vertebrates, except for ecdysone, which occurs in arthropods.

■ **Question:** Why would minor changes in side groups produce such different effects?

the molecules to which they bind are called **receptors.** It is convenient to divide ligands into two major groups, depending on whether their receptors are inside or outside the target cell.

STEROIDS AND RELATED MESSENGERS. Steroids and a few other ligands are able to penetrate the plasma membranes of cells and bind to receptors inside target cells. Steroids have a four-ring molecular structure derived from cholesterol (Fig. 7.2), and most can diffuse directly across plasma membranes by dissolving in the lipid bilayer (pp. 38–39). Once inside a target cell, the steroid binds to the receptor molecule, and the hormone–receptor complex activates particular genes (Fig. 7.3). Only certain cells have the receptor molecule for a given steroid, so even though steroid hormones penetrate all cells, they normally act only on target cells. Some thyroid hormones (Fig. 7.4) and many local messengers also have a chemical structure that enables them to penetrate cell membranes and bind to intracellular receptor molecules in target cells.

NONSTEROID MESSENGERS. Most nonsteroid ligands are proteins, glycoproteins, peptides, or amines. Because they are so large or bear a charge, they do not diffuse across plasma membranes into cells, but must bind to receptor molecules on the plasma membranes of target cells. **Second messenger** molecules then transmit the signal from the hormone receptor into the cell. The best-known second messenger is **cyclic adenosine monophosphate** (cAMP). Cyclic AMP is like the ordinary adenosine monophosphate of RNA except that its phosphate is attached to the ribose at two places rather than at just one (Fig. 7.5). Cyclic AMP is produced from ATP in response to the activation of a **G protein** after the hormone binds to its receptor (Fig. 7.6 on page 140). Cyclic AMP then activates enzymes that catalyze the target cell's response to the messenger. For example, when the hormone glucagon binds to its receptor on liver cells, the resulting increase in cAMP triggers an increase in the activity of enzymes that convert glycogen into glucose. Other second messengers in other cells include **cyclic guanosine monophosphate** (cGMP), Ca^{2+}, and both **inositol triphosphate** (IP_3) and **diacylglycerol** (DAG), which are produced from a membrane phospholipid (phosphatidyl inositol; see Fig. 2.12).

Local Messengers

There is an enormous variety of chemical messengers that act locally in coordinating cellular activities. Among the most important of these local messengers are the **prostaglandins,** which were first discovered in the 1930s by researchers who noted that something in human semen caused the smooth muscle of the uterus to contract. Thinking that the substance came from the prostate gland, they called it prostaglandin. That particular prostaglandin actually came from the seminal vesicles, and other prostaglandins were later found to come from a wide range of other tissues. The basic prostaglandin structure has 20 carbons with a five-carbon ring (Fig. 7.7 on page 140). Prostaglandins are synthesized from fatty acids, and they retain the lipid solubility that allows them to diffuse through plasma membranes. They break down so quickly that they have only local effects. Besides affecting uterine smooth muscle, various prostaglandins affect blood pressure, blood clotting, inflammation, and numerous other processes. Prostaglandins occur not only in mammals, but also in other vertebrates and in invertebrates. In fact, the richest source of raw materials for the manufacture of prostaglandins is the sea whip coral *Plexaura homomalla* (phylum Cnidaria).

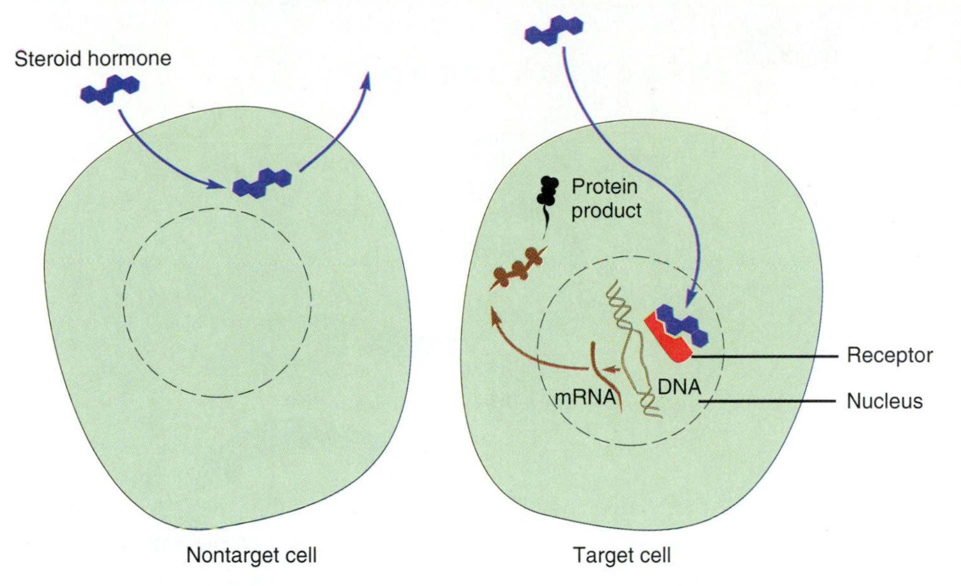

FIGURE 7.3
The action of steroids. An essential feature of steroids is their lipid solubility, which allows them to pass through membranes and evoke their responses within target cells that have receptors for them.

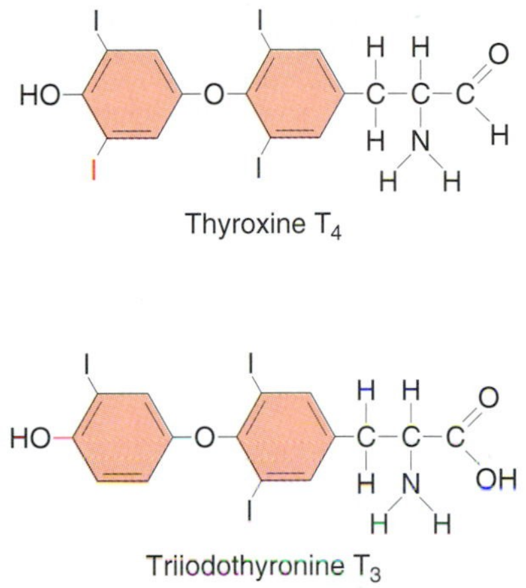

Thyroxine T_4

Triiodothyronine T_3

FIGURE 7.4
The structures of two thyroid hormones, thyroxine (T_4) and triiodothyronine (T_3).

One of the most surprising discoveries of the 1990s is that the simple gases **nitric oxide** (NO) and **carbon monoxide** (CO) are local messengers in a variety of tissues. Both gases are toxic when inhaled, because they displace oxygen in the blood by binding to heme portion of hemoglobin (pp. 241–242). In their normal roles as local messengers, NO and CO also bind to the heme that is part of the enzyme guanylyl cyclase, which catalyzes the production of the second messenger cGMP in a variety of cells. Both NO and CO apparently serve as synaptic transmitters in the nervous system (p. 157), and NO is a major regulator of blood pressure, since it causes arteries to dilate. One of the arteries that NO dilates is essential for erection of the penis in mammals. Since NO breaks down within 10 seconds, it can have only local effects.

Many **growth factors** (pp. 128–129) are hormones, but others are local messengers. Growth factors stimulate cellular reproduction during development, growth, and regeneration. Two other important local messengers are **histamine,** which is similar to the amino acid histidine, and **kinins,** which are polypeptides. In response to tissue damage, histamine and kinins trigger **inflammation** by dilating arteries, increasing the permeability of capillaries, and causing pain, swelling, and immune responses (p. 231).

What Hormones Do

Hormones differ from local messengers in that they act on many different cells, often in many areas of the body, and for periods ranging from minutes to perhaps years. In addition, hormones are usually secreted in response to some overall imbalance in one of the requirements for the lives of cells: either nutrients, oxygen, water, ions, or temperature. Hormones direct particular organs to correct such imbalances, thereby helping to maintain homeostasis. (The major exceptions to this generalization are the reproductive hormones, which do not help maintain homeostasis.)

Other chapters in this unit will present many examples of how hormones help maintain homeostasis, but just one example will be presented briefly here to illustrate the concept involved. The requirement for life that is regulated in this example is the blood sugar glucose, which supplies most of the energy for cells. Two of the hormones

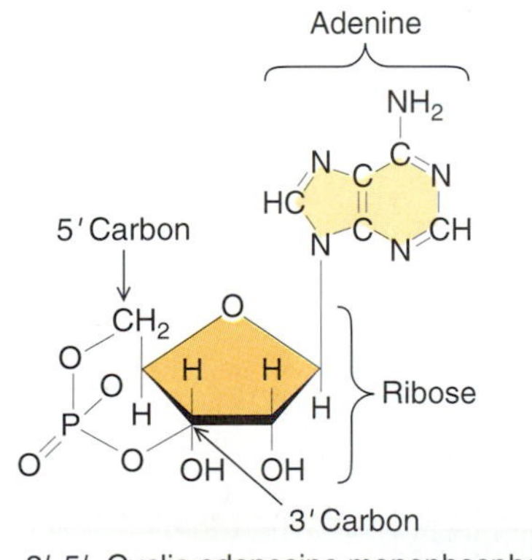

FIGURE 7.5
Cyclic adenosine monophosphate (cAMP).

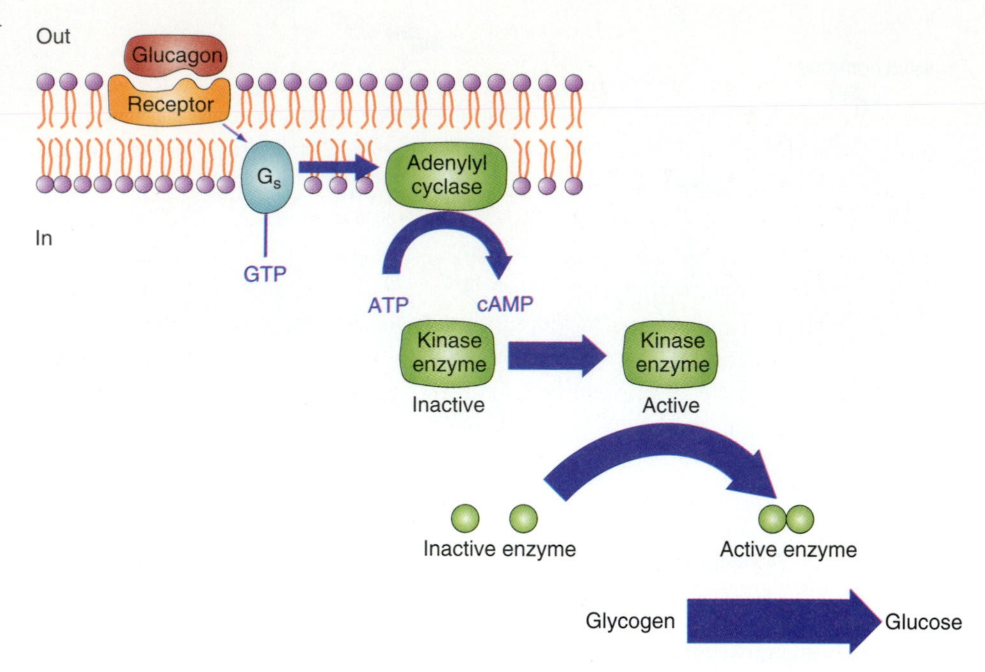

FIGURE 7.6

The action of the hormone glucagon, like many others, is mediated by the second messenger cyclic adenosine monophosphate (cAMP). Glucagon is released when the amount of glucose in the bloodstream declines. The binding of glucagon to its receptor on the plasma membrane of a liver cell activates a G protein (G_s), which binds to guanosine triphosphate (GTP). The activated G protein stimulates the production of cAMP, which then activates a kinase enzyme that attaches a phosphate group to an inactive form of another enzyme. This active enzyme then catalyzes the breakdown of glycogen and the release of glucose by the liver cell. The increasing widths of the arrows indicate that each step in the sequence amplifies the response.

involved in maintaining glucose levels are glucagon and insulin, which are both secreted by the pancreas. As noted previously, glucagon tends to keep the level of glucose from falling, by stimulating the release of glucose from the liver. **Insulin** helps control glucose level in the blood in the opposite way. Insulin is secreted when the blood glucose level rises above normal, as it does following a meal. The increasing level of glucose in the blood directly stimulates cells of the pancreas to release insulin, which then circulates through the bloodstream. Insulin then stimulates cells, especially in the liver and muscles, to take in the excess glucose and convert it to glycogen, fat, and muscle protein. In this way the blood glucose level is prevented from becoming so high that it might cause glucose to be wasted in the urine or affect the functioning of the brain. Thus, glucagon, insulin, and hormones in general are secreted when an organ detects an imbalance in the body fluids, and they help correct the imbalance by directing target cells to take appropriate action.

Some Major Hormones of Vertebrates

There are so many hormones that a single chapter can do little more than survey the best-known ones, which are naturally those of humans and other mammals. Fortunately, such a survey is not as limited in scope as it might seem, because the endocrine organs and hormones that occur in mammals also occur in most other vertebrates, although they sometimes have different functions. For example, prolactin, which stimulates milk production in female mammals, also occurs in amphibians, which don't produce milk. In amphibians, prolactin stimulates the regeneration of lost limbs, as well as some reproductive functions. Figure 7.8 and Table 7.1 summarize the major endocrine organs and best-known hormones of mammals and most other vertebrates. Many hormones of invertebrates are quite different from those of vertebrates and will be surveyed later in this chapter.

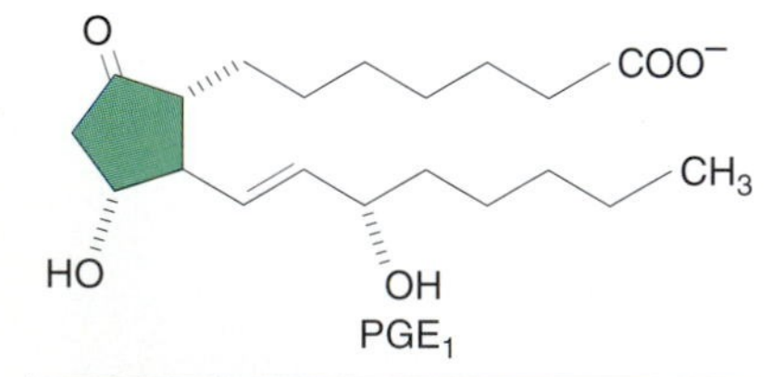

FIGURE 7.7

One of several types of prostaglandin. Prostaglandins are synthesized from fatty acids in cell membranes.

Hormones of the Hypothalamus and Pituitary

The pituitary (also called the **hypophysis**) is only the size of a pea in humans, but it is probably the most important endocrine gland, because it regulates the secretion of so many other hormones. The pituitary is therefore traditionally called "the master gland." In fact, however, the pituitary simply relays instructions from a part of the brain called the **hypothalamus,** so it might be more appropriate to call it the "overseer gland" (see Fig. 7.9 on page 144 and Fig. 9.8). The pituitary is actually two distinct types of tissue that come together during embryonic development. Each part takes its orders from the hypothalamus in a different language. The **anterior pituitary** develops from tissue in the roof of the mouth and secretes hormones in response to **releasing hormones** and **release-inhibiting hormones** from the hypothalamus. All the releasing and release-inhibiting hormones are peptides except one (dopamine). They reach the anterior pituitary by means of a short **portal vessel.**

The **posterior pituitary** (= neurohypophysis) develops as an extension of the hypothalamus, and it releases its hormones in response to action potentials in nerve cells. The hormones of the posterior pituitary are actually produced in **neurosecretory cells** in the hypothalamus. They travel down axons and are released in response to action

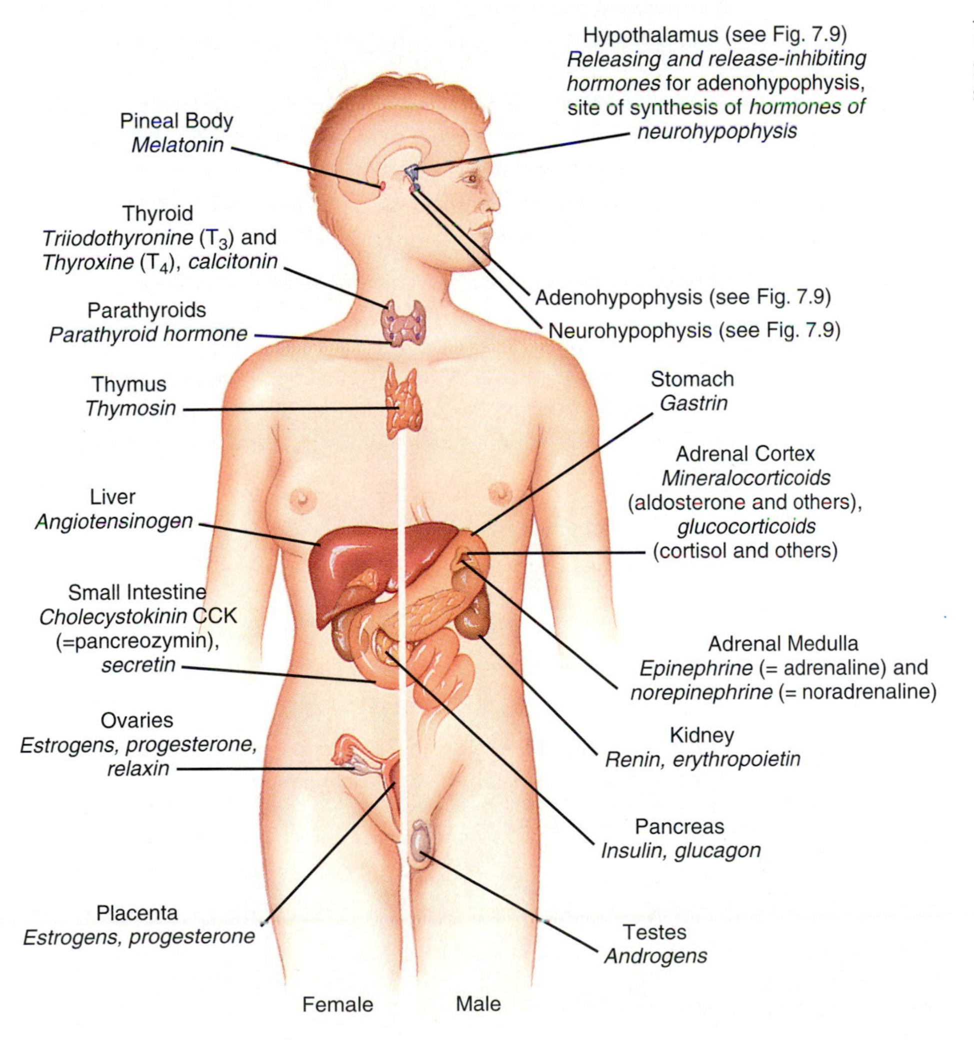

FIGURE 7.8
Some major endocrine organs and their secretions in humans.

TABLE 7.1

Major endocrine organs and their hormones in humans and many other vertebrates.

HYPOTHALAMUS (See Fig. 7.9)		
Releasing and release-inhibiting hormones to the anterior pituitary.		
Hormones of the posterior pituitary.		
PITUITARY (See Fig. 7.9)		
PINEAL GLAND (= PINEAL BODY)		
Melatonin.	Derivative of tryptophan.	Secreted in humans and many other mammals with a daily rhythm. In many vertebrates the level of melatonin appears to help maintain rhythms, such as the seasonal rhythm of reproduction (see pp. 410–412). In amphibians, melatonin lightens skin color by inhibiting secretion of melanocyte-stimulating hormone (MSH) from the anterior pituitary.
THYROID		
Triiodothyronine T_3 and Thyroxine T_4.	Derived from two tyrosine molecules with three or four iodine atoms attached (see Fig. 7.4).	Most T_4 is converted to T_3, the more active form. T_3 is lipid-soluble and diffuses into cells, where it binds to receptor molecules in the nucleus. T_3 and T_4 increase metabolic rates and are essential for development of the nervous system. Thyroxine also stimulates metamorphosis in many animals, such as jellyfish, fishes, and frogs.
Calcitonin.	A peptide.	Reduces Ca^{2+} levels in blood by inhibiting release from bone.
PARATHYROIDS		
Parathyroid hormone.	A peptide.	Increases Ca^{2+} levels in the blood by stimulating removal from bone and the reabsorption of Ca^{2+} by kidney. Stimulates conversion of vitamin D to an active hormone, calcitrol, which increases Ca^{2+} absorption by the intestine.
THYMUS		
Thymosin.	Several peptides.	Stimulate T cells, which are important in immunity.
ADRENAL CORTEX		
Mineralocorticoids (aldosterone and others).	Steroids (Fig. 7.2).	Stimulate Na^+ and H_2O reabsorption by kidney.
Glucocorticoids (cortisol and others).	Steroids.	Increase glucose levels in blood by stimulating conversion of fat and amino acids to glucose. Counteract inflammation.
Estrogens and **Testosterone.**		See below, under testes and ovaries.
ADRENAL MEDULLA (= CHROMAFFIN TISSUE)		
Epinephrine (= adrenaline) and **Norepinephrine** (= noradrenaline).	Amines (Fig. 7.11) derived from tyrosine.	Increase pumping of blood, blood pressure, and oxygen consumption.

HEART **Atrial natriuretic peptide.**		Reduces blood volume and blood pressure.
KIDNEY **Renin** (pronounced REE-nin to avoid confusion with the digestive enzyme rennin).	A protein.	Activates angiotensin II hormone (see under liver) in response to low blood flow in the kidneys.
Erythropoietin.	A glycoprotein.	A growth factor that stimulates production of red blood cells in response to reduced availability of O_2.
LIVER **Angiotensinogen.**	A peptide.	Precursor of the hormone angiotensin II, activated by renin in the blood. Angiotensin II increases blood pressure, stimulates thirst, and stimulates the secretion of aldosterone by the adrenal glands and antidiuretic hormone from the posterior pituitary.
PANCREAS **Insulin.**	A protein.	Decreases concentrations of glucose in blood by stimulating uptake of glucose and conversion to glycogen by liver and muscle.
Glucagon.	A protein.	Increases concentration of glucose in blood by stimulating breakdown of glycogen in liver and muscle.
STOMACH **Gastrin.**	A protein.	Secreted in response to peptides and proteins in stomach. Stimulates secretion of acid and protein-digesting enzyme by the stomach.
SMALL INTESTINE **Cholecystokinin** (CCK) (pronounced ko-lay-SIS-toe-KY-nin).	A peptide.	Secreted in response to fats and amino acids in small intestine. Stimulates release of digestive enzymes by pancreas and release of bile by gallbladder.
Secretin.	A peptide.	Secreted in response to food and acid in intestine. Stimulates release of bicarbonate by pancreas. Inhibits stomach secretion and motility.
TESTES **Androgens,** mainly testosterone.	Steroids.	Promote male behavior, morphology, and reproductive functions.
OVARIES **Estrogens,** such as estradiol.	Steroids.	Promote female behavior, morphology, and reproductive functions.
Progesterone.	A steroid.	Promotes functioning of uterus and breasts.
Relaxin.	A peptide.	Relaxes pelvic ligaments and cervix during delivery of infant.
PLACENTA **Estrogens** and **Progesterone.**		See above under ovaries.
Human chorionic gonadotropin (hCG).	A protein.	Essential for maintenance of pregnancy.

Adenohypophysis: anterior lobe
Adrenocorticotropic hormone (ACTH), a peptide, stimulates secretion of glucocorticoids by adrenal cortex

Thyroid-stimulating hormone (TSH), a glycoprotein, stimulates secretion of thyroxine and triiodothyronine by thyroid gland

Follicle-stimulating hormone (FSH), a glycoprotein, stimulates development of the follicle in ovaries; stimulates sperm production in testes

Luteinizing hormone (LH) (= interstitial-cell-stimulating hormone, ICSH), a glycoprotein, stimulates ovulation in females, testosterone production by testes of males

Growth hormone (GH), a peptide, stimulates protein synthesis and growth

Prolactin, a peptide, stimulates production of milk

Adenohypophysis: intermediate lobe
(= pars intermedia), lacking in adult humans

Malanocyte-stimulating hormone (MSH), a peptide. In many species causes darkening of skin, hair or feathers by stimulating melanin synthesis by melanocytes (birds and mammals) of dispersal of melanophores (other vertebrates)

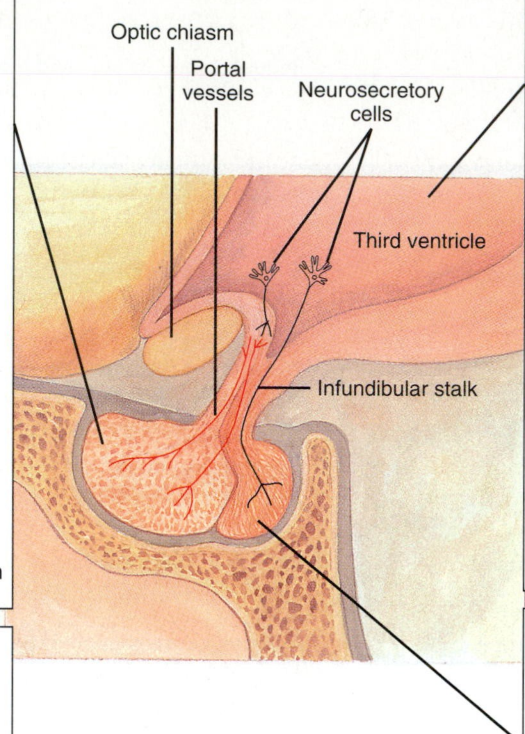

Hypothalamus
Corticotropin-releasing hormone (CRH) stimulates release of adrenocorticotropic hormone (ACTH)

Thyrotropin-releasing hormone (TRH) stimulates release of thyroid-stimulating hormone (TSH)

Gonadotropin-releasing hormone (GnRH) stimulates release of luteinizing hormone (LH) and follicle-stimulating hormone (FSH)

Growth-hormone-releasing hormone (GRH) stimulates release of growth hormone (GH)

Somatostatin (= growth-hormone-release-inhibiting hormone, GIH) inhibits secretion of growth hormone (GH)

Dopamine (= prolactin release-inhibiting hormone, PIH), derived from tyrosine, inhibits secretion of prolactin

Neurohypophysis
Oxytocin, an octapeptide, stimulates milk release

Antidiuretic hormone (ADH) (= vasopressin), an octapeptide, increases reabsorption of water by kidney and causes slight increase in blood pressure

Vasotocin, an octapeptide, occurs in fetal mammals, and in the adults of most other vertebrates. Effects are similar to those of ADH and oxytocin

FIGURE 7.9
Hormones of the hypothalamus and pituitary in humans.

■ **Question:** What disorders in humans would result from failure to produce each of these hormones?

potentials in those cells. These neurosecretions are all octapeptides (peptides with eight amino acids). At least 10 are secreted by the posterior pituitaries of various vertebrates. Oxytocin and antidiuretic hormone (= vasopressin) are the two secreted by the human posterior pituitary (Fig. 7.10).

Stress

As a part of the brain, the hypothalamus coordinates the activity of the endocrine system with that of the central nervous system. Many of its regulatory functions will be described in later chapters in this unit, but for now it will suffice to give an illustration of how the hypothalamus and pituitary integrate neural and hormonal responses. The stimulus in this example is stress, which physiologists define as the response to any stimulus that tends to upset homeostasis. Injury, pain, threat, or some other stressor triggers the hypothalamus to release **corticotropin releasing hormone** (CRH). CRH is a peptide that circulates to the anterior pituitary, where it stimulates the release of **adrenocorticotropic hormone** (ACTH).

ACTH is a peptide that circulates to the cortex of the adrenal gland (see Fig. 14.9), stimulating it to secrete a number of steroid hormones. The most important of these hormones in combatting stress are the **glucocorticoids,** especially **cortisol.** Glucocorticoids stimulate the conversion of amino acids into glucose as an energy source. They also have near-miraculous powers to overcome the pain and swelling of inflammation. For a while after their discovery in the late 1940s, they were used for everything from severe burns to diaper rash. Physicians became more conservative, however, after observing that glucocorticoids also inhibit immune responses and slow healing. Prolonged use can also produce such side effects as loss of hair, brittleness of bones, sterility, and behavioral changes. Normally the endocrine system does not secrete glucocorticoids or other potentially harmful hormones for long periods. ACTH limits the secretion of glucocorticoids by inhibiting its own release by the anterior pituitary, and also by inhibiting the release of CRH from the hypothalamus.

Stress also causes a part of the nervous system (the sympathetic nervous system; p. 178) to release the neurosecretion **norepinephrine** in a variety of tissues (Fig. 7.11). This norepinephrine, also known as noradrenaline, is a local messenger that increases the force and rate of contraction of the heart, increases blood pressure, stimulates conversion of glycogen into glucose by the liver, and has numerous other effects that help the animal deal with the cause of the stress. In addition, this norepinephrine stimulates the release of norepinephrine as a hormone from the medulla of the adrenal gland into the bloodstream. At the same time, the adrenal medulla also releases the related hormone **epinephrine,** which is also known as adrenaline. As hormones, norepinephrine and epinephrine act on the circulatory system and glucose level in much the same way that norepinephrine does as a neurosecretion.

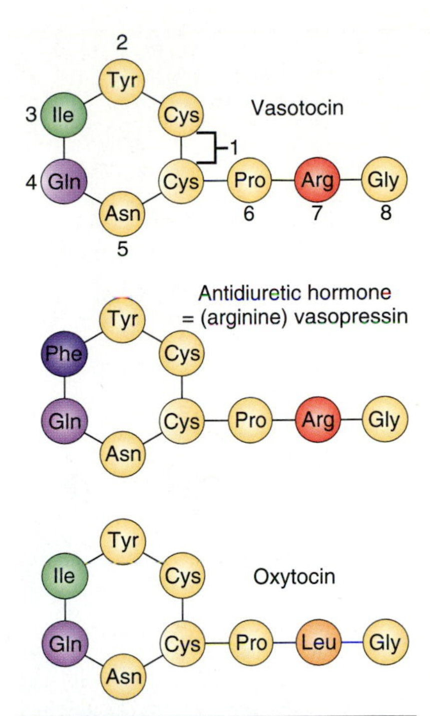

FIGURE 7.10

Three of the 10 or more hormones secreted from the neurohypophyses of various vertebrates. Nine of those 10 octapeptides are identical except at the sites numbered 3, 4, and 7. Arginine vasotocin appears to be the most primitive of those octapeptides and may have been the one from which the others evolved. It is replaced by vasopressin in adult mammals. The vasopressin of most mammals is often called arginine vasopressin to distinguish it from lysine vasopressin in pigs, which differs at amino acid number 7.

Some Hormones of Invertebrates

Virtually every invertebrate examined has been found to produce hormones, many of which differ in chemistry and mode of action from those of vertebrates. Table 7.2 indicates some of the most important hormones of invertebrates and the functions they serve. As an example of hormonal coordination in invertebrates, the following section describes how hormones control molting in insects.

Hormonal Coordination of Insect Molting

Insects have to shed their external skeletons in order to grow and develop into adults, and this molting requires the hormonal coordination of numerous behavioral and metabolic activities (see pp. 633–634). One of the first to study the hormonal coordination of molting was Vincent B. Wigglesworth (1899–1994), who chose the blood-sucking bug *Rhodnius* as his experimental subject. This might seem an unsavory choice, but it made sense because *Rhodnius* transmits Chagas" disease (see p. 468), making the research potentially beneficial to humans. *Rhodnius* normally molts four times into a juvenile of increasing size, and then a fifth time into an adult. Wigglesworth found that if the last **instar** (juvenile stage) of *Rhodnius* were joined to an early instar, the last-instar juvenile molted into an extra juvenile stage rather than into an adult. Evidently a substance in the early instar, now called **juvenile hormone** (JH), suppressed the development of adult features. Subsequent experiments revealed that JH is normally secreted in juveniles until the final molt, and that JH must be absent for development into an adult. Wigglesworth's simple but effective experiments left an impression on an entire generation of insect endocrinologists. Figure 7.12B on page 146 shows a more literal way in which Wigglesworth left his mark.

It occurred to Carroll M. Williams of Harvard University that Wigglesworth's discovery could be used as the basis for a "third-generation insecticide." First-generation insecticides, using such natural toxins as nicotine, had failed because insects quickly learned to avoid their strong tastes or odors. Second-generation insecticides, using synthetic chemicals such as DDT, had failed because the insects had become tolerant to

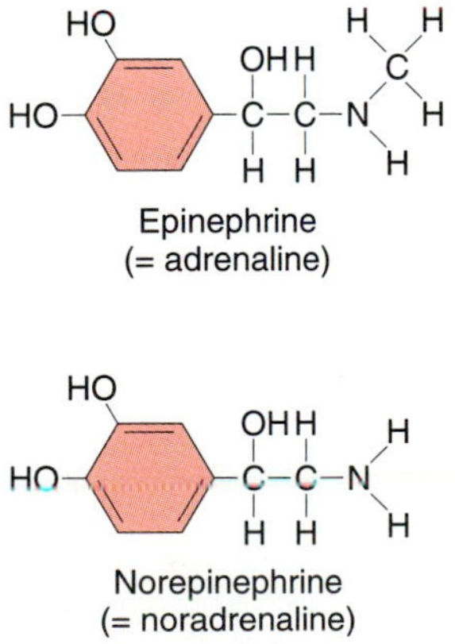

Epinephrine
(= adrenaline)

Norepinephrine
(= noradrenaline)

FIGURE 7.11

The hormones epinephrine and norepinephrine.

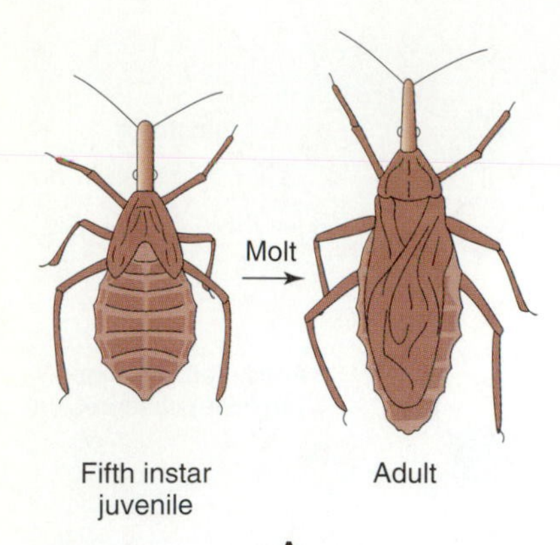

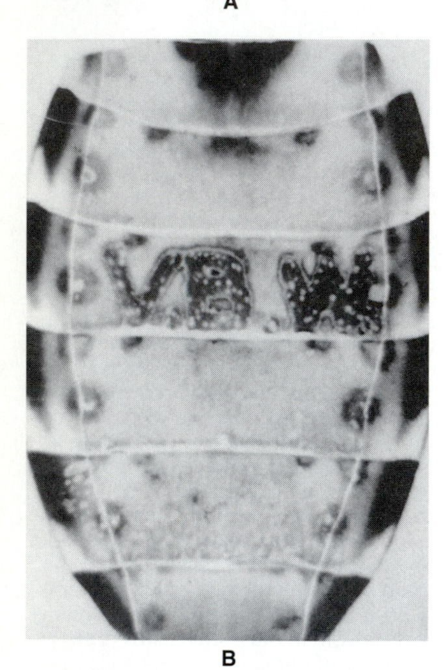

FIGURE 7.12

(A) The last juvenile instar and the adult of *Rhodnius prolixus.* This bug molts after a meal of blood has sufficiently distended the abdomen. In the last molt the adult acquires wings, and the cuticle on the abdomen becomes more pale and less mottled. (B) The initials of Vincent B. Wigglesworth inscribed into the abdominal cuticle of a *Rhodnius* adult. Juvenile hormone was used to write the initials on the juvenile prior to its last molt. The hormone caused the cuticle to retain the juvenile pattern.

TABLE 7.2

The sources and effects of some hormones of invertebrates.

ANNELIDS	
Earthworms	A neurosecretion stimulates formation of gametes and sex characteristics and increases concentrations of sugar in the blood.
Leeches	A neurosecretion stimulates development of gametes and triggers color changes.
MOLLUSCS	
Gastropods	Bag cells in abdominal ganglion of marine slug *Aplysia* secretes **egg-laying hormone**. A neurosecretion in the common land snail *Helix* stimulates spermatogenesis; another hormone stimulates egg development; hormones from ovary and testis stimulate accessory sex organs.
Cephalopods	Optic gland in the eye stalk of octopus, squid, and others produces a hormone that stimulates egg development, proliferation of spermatogonia, and secondary sex characteristics. A secretion from this gland kills the female after she has brooded her young.
ARTHROPODS	
Crustaceans	The Y organ in the head produces **ecdysone**, which triggers molting of old cuticle. The X-organ–sinus-gland (XOSG) system near the optic nerves releases **molt-inhibiting hormone** (MIH), which inhibits release of ecdysone. The XOSG system also releases other hormones that control pigments in the cuticle, increase glucose in the blood, and regulate other functions.
Insects	Corpora cardiaca (plural of corpus cardiacum; see Fig. 7.14) release hormones that regulate water balance, **adipokinetic hormone** that promotes the conversion of fat to blood sugar, and hormones that regulate heart beat. **Diapause hormone** arrests development of the eggs. **Eclosion hormone** evokes unique movements of the pupa that allow the insect to emerge as an adult. **Bursicon** promotes hardening and tanning of the exoskeleton.
ECHINODERMS	Hypodermal cell layer of radial nerve secretes **gonad-stimulating substance**. Follicle cells of ovary secrete **1-methyl-adenine**, which indirectly stimulates maturation of gametes.

them. These problems would not be expected to occur with a synthetic juvenile hormone or anti-JH, however. Chemicals that disrupted the functioning of JH in insects would also be safer than previous insecticides, since they would not affect other animals.

It turned out that plants had come up with the same idea long before Williams. While working in Williams's lab, Karel Slàma found that the insects he had brought from Czechoslovakia would not develop normally. He eventually traced the problem to American paper towels that he had placed in the insect containers. The wood of balsam fir that went into the paper contains a ligand that mimics JH and prevents development into adults (Fig. 7.13). Other plants, such as the common bedding plant *Ageratum,* produce an anti-JH that destroys the secretory cells of JH. Juvenile insects that feed on these plants develop prematurely into small adults.

In another series of studies, Williams found that the isolated abdomens of the pupae of cecropia moths (*Hyalophora cecropia*) would molt into adult abdomens if the

Juvenile hormone III

A

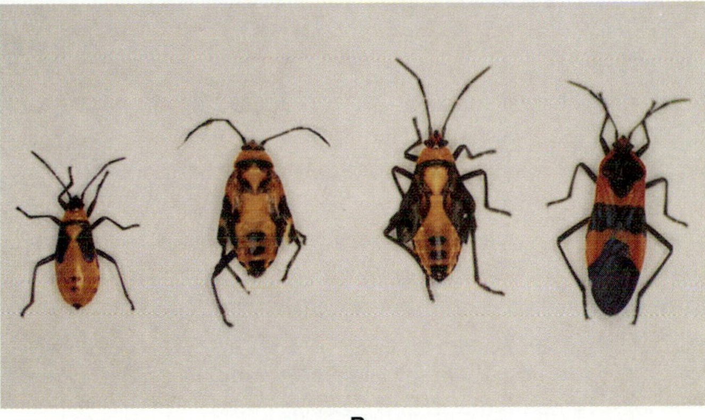

B

FIGURE 7.13

(A) Juvenile hormone (JH), normally secreted in the early stages of insect development, prevents the formation of adult features. (B) Some plants, such as the balsam fir tree and the sweet basil herb, produce substances (juvocimenes) that mimic JH. Application of the JH analog from sweet basil to the last nymphal instar of the large milkweed bug *Oncopeltus fasciatus* promotes the development into additional juvenile stages. The milkweed bug on the extreme left is a normal last-instar juvenile, and that on the far right is a normal adult. The second insect from the left was treated with a large dose of JH analog. Instead of developing into a normal adult, it developed into an extra juvenile instar. A smaller dose of the JH mimic had a smaller effect (third from left).

cerebral ganglia and the **prothoracic gland** were implanted into the abdomens. It has since been found that **prothoracicotropic hormone** (PTTH) from the cerebral ganglia normally stimulates the prothoracic gland to secrete a steroid called **ecdysone,** which triggers molting. Thus the cerebral ganglion and prothoracic gland in the isolated abdomen triggered molting by secreting PTTH and ecdysone as they would prior to a normal molt. Figure 7.14 summarizes how JH and ecdysone normally function in insect development.

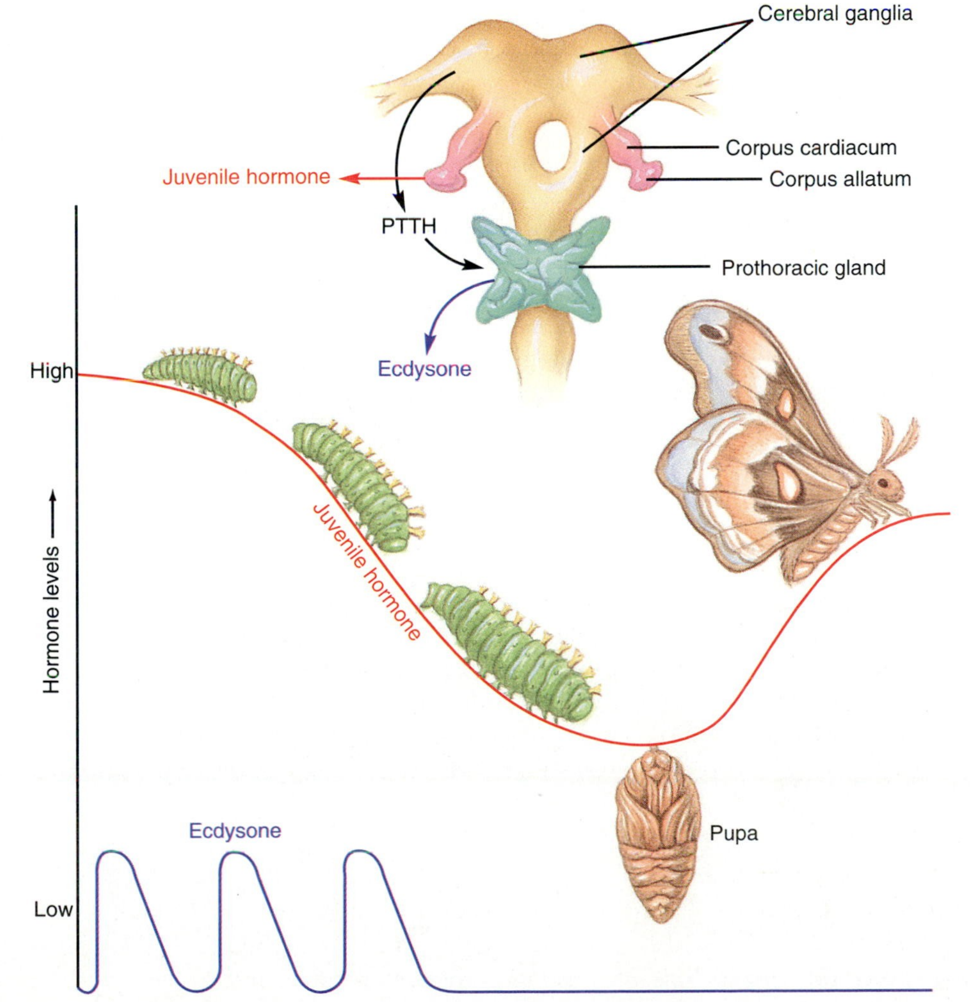

FIGURE 7.14

Some of the hormones that control development in a moth. High levels of juvenile hormone (JH) from the corpora allata during early juvenile stages keep the animal in the caterpillar stage. Ecdysone must be released from the prothoracic gland [in response to prothoracicotropin (PTTH) from the cerebral ganglion] during a critical period prior to each molt. The insect becomes a pupa when JH falls below a certain level. JH levels rise again in the adult and stimulate reproductive functions.

Summary

Many physiological functions are coordinated by chemicals that function as local messengers and hormones. Local messengers act on tissues near the cells that secrete them, while hormones travel in blood to organs that are generally far from the glands that secrete them. Hormones generally help maintain homeostasis. Steroids and some other chemical messengers enter target cells where they bind to receptor molecules and induce gene activity. Nonsteroids and other messengers that are charged or too large to enter the cell bind to a receptor molecule in the plasma membrane, which then evokes a response by means of a second messenger inside the target cell.

In vertebrates the secretion of many hormones is influenced by the pituitary gland, which is in turn controlled by the hypothalamus. In mammals the pituitary gland has an anterior lobe, which secretes hormones in response to releasing hormones from the hypothalamus, and a posterior lobe, which releases neurosecretions. One secretion (ACTH) from the anterior pituitary stimulates the secretion of steroid hormones, norepinephrine, and epinephrine from the adrenal glands, and these hormones help coordinate the stress response.

Different hormones are secreted by various invertebrates. In insects, for example, several hormones coordinate molting and development. Ecdysone triggers molting, and juvenile hormone prevents a juvenile from molting into an adult.

Key Terms

homeostasis
gland
endocrine
steroid
second messenger
insulin
pituitary
hypothalamus
releasing hormone
adrenocorticotropic hormone
juvenile hormone
ecdysone

Chapter Test

1. What conditions must be satisfied before a chemical messenger is considered to be a hormone? Describe some chemical messengers that are not hormones. How do they differ from hormones? (pp. 137–139)
2. Explain how the action of steroid hormones on their target cells differs from the action of nonsteroid hormones on their target cells. (p. 138)
3. List as many endocrine organs of mammals as you can, and compare your list with Fig. 7.8.
4. Which of the endocrine organs in Fig. 7.8 release their hormones by neurosecretion? Which ones secrete steroid hormones? Which ones affect at least one other endocrine gland? (pp. 142–143)
5. Explain how the functioning of the anterior pituitary differs from that of the posterior pituitary. (pp. 141, 144)
6. In what way are the secretions of the posterior pituitary similar to each other? (p. 144)
7. Explain how mimics of juvenile-hormone or anti-juvenile hormones might be used as insecticides. What advantages and disadvantages would such third- and fourth-generation insecticides have over conventional insecticides? (pp. 145–146)

■ Answers to the Figure Questions

7.1 The local messengers might be large molecules (proteins) that do not diffuse into blood capillaries, or they might quickly bind to molecules on the target cells.

7.2 Changes in the side groups would affect which receptor molecules the steroid binds to, which would affect the response.

7.9 CRH and ACTH: inability to respond to a physiological stress. TRH and TSH: decreased metabolic rate. GnRH and LH: infertility and loss of sexual function. FSH: infertility. GRH and GH: inhibition of growth. GIH: excessive growth. PIH: lactation. Prolactin: inability to lactate. MSH: lightening of skin color. Oxytocin: inability to nurse. ADH: excessive urine production, dehydration, reduced blood pressure. Vasotocin: inability to nurse, excessive urine production, dehydration, reduced blood pressure.

Readings

Recommended Readings

Berridge, M. J. 1985. The molecular basis for communication within the cell. *Sci. Am.* 253(4):142–152 (Oct).

Carmichael, S. W., and H. Winkler. 1985. The adrenal chromaffin cell. *Sci. Am.* 253(2):40–49 (Aug).

Evans, H. E. 1968. *Life on a Little-Known Planet.* New York: E. P. Dutton. (*Chapter 9 is an engaging history of research in insect endocrinology.*)

Lancaster, J. R., Jr. 1992. Nitric oxide in cells. *Am. Sci.* 80:248–259.

Linder, M. E., and A. G. Gilman. 1992. G proteins. *Sci. Am.* 267(1):56–65 (July).

Rasmussen, H. 1989. The cycling of calcium as an intracellular messenger. *Sci. Am.* 261(4):66–73 (Oct).

Sapolsky, R. M. 1990. Stress in the wild. *Sci. Am.* 262(1):116–123 (Jan).

Scheller, R. H., and R. Axel. 1984. How genes control an innate behavior. *Sci. Am.* 250(3):54–62 (Mar). (*The egg-laying hormone of a sea slug.*)

Snyder, S. H. 1985. The molecular basis of communication between cells. *Sci. Am.* 253(4):132–140 (Oct).

Snyder, S. H., and D. S. Bredt. 1992. Biological roles of nitric oxide. *Sci. Am.* 266(5):68–77.

Uvnäs-Moberg, K. 1989. The gastrointestinal tract in growth and reproduction. *Sci. Am.* 261(1):78–83 (July).

Additional References

Bowers, W. S., and R. Nishida. 1980. Juvocimenes: potent juvenile hormone mimics from sweet basil. *Science* 209:1030–1032.

Feir, D. 1984. Inhibition of gland development in insects by a naturally occurring antiallatotropin (anti-hormone). In: C. L. Harris (Ed.). *Tested Studies for Laboratory Teaching: Proceedings of the Third Workshop/Conference of the Association for Biology Laboratory Education (ABLE).* Dubuque, IA: Kendall/Hunt, Chapter 6.

Kaltenbach, J. C. 1988. Endocrine aspects of homeostasis. *Am. Zool.* 28:761–773.

Laufer, H., and R. G. H. Downer (Eds.). 1988. *Endocrinology of Selected Invertebrate Types.* New York: Alan R. Liss.

CHAPTER

8

Nerve Cells

The Squid *Loligo*.

The Nature of the Nervous System

As you read these words you may imagine that they are somehow entering your brain, speaking almost audibly to an inner self. A moment's thought tells you, however, that the words stop at the retinas of your eyes and that there must be some way the retinas convert their images into a form that can be used within your dark and silent brain. Millions of your nerve cells are in fact translating these words into an electrical and chemical code. Understanding this code is the ultimate challenge of science, because all that we feel and know—even science itself—results from such processes.

Until a few centuries ago, most people believed that understanding the brain was safely beyond the ability of humans, because our thoughts, sensations, and behavior seemed too miraculous to comprehend. This belief began to fade in the 17th century, thanks largely to the French philosopher-mathematician-scientist René Descartes (pronounced day-cart). Descartes argued that all animals were machines and that humans were animals with souls. It followed, therefore, that humans—even their thoughts, sensations, and behavior—could be understood like any other machine. Perhaps inspired by the water-powered, lifelike robots that were then the rage of Paris, Descartes suggested that nerves activate muscles by squirting fluid into them. Descartes' explanation of how the brain controls behavior was wrong, of course, but it opened the way to studying the mechanisms of the brain without encroaching on religion. Although the concept that our brains are no different in principle from machines or the brains of other animals makes many people uncomfortable, it has paradoxically led to a more humane treatment of behavioral differences. For example, thanks to research inspired by this realization, people with schizophrenia are now treated with drugs rather than tortured to drive out demons, locked up for life, or subjected to futile psychoanalysis.

Another breakthrough in understanding the nervous system came in the 19th century with the realization that, like all other organs, it consists of cells and the products of cells. The most important type of cell in nervous systems is the nerve cell, also called a **neuron.** Nerve cells carry out the same functions of protein synthesis and metabolism as other cells, but they are also specialized in many ways for transmitting electrical and chemical signals among sensory receptors and other nerve cells, as well as for activating muscles and glands. The electrical signals generally travel along a wirelike extension called the **axon.** The chemical signals are released at the end of the axon and elsewhere by **synapses** (Fig. 8.1 on page 152).

Understanding the activities of nerve cells requires an understanding of some fundamental electrical concepts. ***Voltage,*** *also called* ***potential*** *and* ***electromotive force,*** *is a force that tends to move electrons, ions, and other charged matter.* ***Charge*** *is a fundamental property, like mass, length, and time. Just as gravity tends to move mass, voltage tends to move charge. Unlike gravity and mass, however, voltages and charges come in pairs: positive and negative. Positive voltage attracts electrons and negatively charged ions (anions). Negative voltage attracts positively charged ions (cations). The amount of charge moved by a voltage in a unit of time is the* ***current.*** *Increasing the voltage increases the current. The movement of charge is opposed by electrical* ***resistance,*** *which is analogous to friction. The greater the electrical resistance of a substance, the less current there will be through it when a particular voltage is applied.*

A current through a resistance will generate a voltage across the resistance, with the magnitude of the voltage proportional to both the current and the resistance. If the current consists of cations, such as Na^+ or K^+, moving through channels in the plasma membrane, the voltage produced across the membrane will be negative on the surface where the ions enter and positive on the surface where they exit. Voltage can also be produced by the separation of positive charges from negative charges. The voltage produced is directly proportional to the amount of charge and is negative where the negative charges accumulate.

Electrical Signals

MEMBRANE POTENTIALS. Electrical signals in nervous systems result from changes in the voltage across the plasma membranes of nerve cells. Such membrane potentials occur because a plasma membrane is not equally permeable to all ions and also because ions do not have equal concentrations on each side of the membrane. (See pp. 38–39 for a discussion of membrane permeability.) For example, proteins and amino acids (represented by A^- because they are anions) have much higher concentrations inside cells than in the interstitial fluid. Because of the negative charge and large size of these anions, the plasma membrane is not permeable to them. Another important ion that also usually cannot diffuse through the plasma membrane is sodium (Na^+). Na^+ has a higher concentration outside the cell than inside because of active transport by the sodium–potassium pumps (see pp. 39–40).

In contrast to Na^+ and A^-, potassium (K^+) and chloride (Cl^-) ions ordinarily diffuse freely through channels in membranes. You might expect that because of diffusion

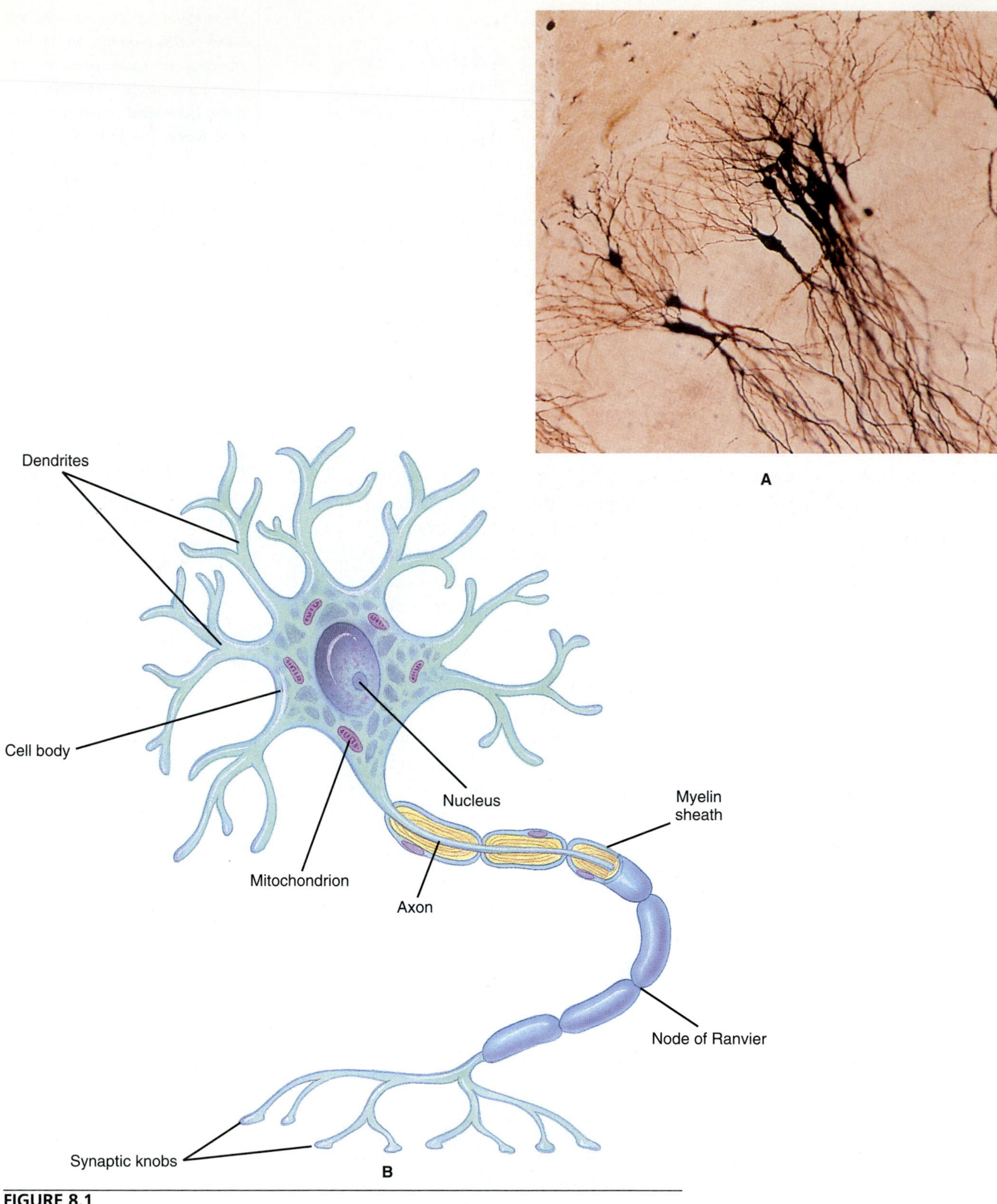

FIGURE 8.1

(A) Nerve cells have an enormous variety of shapes and sizes. These are from the human brain. The dark swelling near the center are cell bodies, which have numerous branches called dendrites spreading to the upper left. Dendrites receive chemical signals from other nerve cells. A single process called the axon runs from each cell body and branches to the lower right, conducting electrical signals. The ends of the branches transmit chemical signals to dendrites of other nerve cells. (B) The structure of a representative nerve cell.

there would be equal concentrations of K^+ and Cl^- on each side of the membrane. This is not the case, however, because A^- attracts K^+ into the cell, and Na^+ attracts Cl^- outside (Fig. 8.2). The positive charge (mainly K^+) inside a cell equals the negative charge (mainly A^-), and the negative charge (mainly Cl^-) in the interstitial fluid equals the positive charge (mainly Na^+).

Because they can diffuse readily across the membrane and have unequal concentrations on each side of the membrane, K^+ tends to diffuse out of a cell, and Cl^- tends to diffuse into the cell. The slight diffusion of K^+ out and Cl^- in produces a charge separation across the plasma membrane, with more negative charge on the inner surface of the membrane and more positive charge on the outer surface. The diffusion of these two ions continues until the voltage produced by the charge separation exactly counterbalances the tendency of the ions to diffuse further. The exact balance of the voltage and diffusion leads to an **equilibrium** in which there is no net diffusion. At equilibrium, the negative voltage inside the cell holds the K^+ in and keeps the Cl^- out. This voltage is called the **resting membrane potential.** The resting membrane potential is generally from 50 to 70 millivolts (0.05 to 0.07 volts) in nerve cells, with the inside negative.

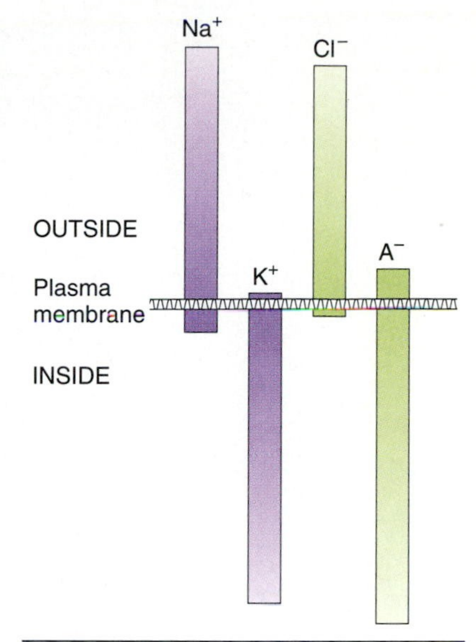

FIGURE 8.2
Relative concentrations of ions inside and outside a typical axon. A- represents anions. Positively and negatively charged ions are shown in different colors. The concentration of positive charge must equal that for negative charge both inside and outside a cell.

■ **Question:** Suppose a living nerve cell is placed in a solution containing the same concentrations of potassium and chloride as occurs in the cell. How will the ions move across the membrane, and why?

How Electrical Signals Are Produced. Neural signals result from changing the permeability of the plasma membrane to one or more ions, chiefly potassium and sodium. Chloride is usually at equilibrium with the resting membrane potential, so changing the permeability of the membrane to Cl^- will usually not generate a neural signal. In fact, increasing the permeability to Cl^- will tend to keep the membrane potential from changing. In contrast, potassium is somewhat out of equilibrium, because the sodium–potassium pumps transport more K^+ into a cell than can be held there by the resting membrane potential alone. As a result, increasing the permeability of the membrane to K^+ will allow K^+ to diffuse out more readily, which will produce additional negative voltage across the membrane. The resulting increase in the polarization of the membrane is called **hyperpolarization.**

Sodium concentrations are far out of equilibrium with the resting membrane potential, because of active transport. This means that an increase in the permeability to Na^+ above its normally low value will allow much more Na^+ to enter the cell, by both diffusion and attraction to the negative voltage. The voltage generated by the inward movement of Na^+ will be positive on the inside and will reduce the magnitude of the membrane potential, or even reverse it to positive on the inside. Such a decrease or reversal of membrane potential is called **depolarization.** All electrical signals in neurons are depolarizations and hyperpolarizations resulting from such changes in permeability (Fig. 8.3 on page 154).

Action Potentials. The type of neuronal signal responsible for carrying information over long distances is the action potential. An action potential is a brief change in membrane potential that travels along the plasma membranes of certain nerve cells and also muscle cells. In nerve cells, action potentials generally occur only in the plasma membranes surrounding axons. Our understanding of the action potential is due largely to studies on **giant axons** of the squid *Loligo* (Fig. 8.4). A. L. Hodgkin, A. F. Huxley, and others who pioneered these studies just before and after World War II chose these axons because they are up to half a millimeter in diameter. The size permits the insertion of fine wires into the axon to measure directly the voltage across the membrane. What they discovered with squid giant axons has since been confirmed in much smaller axons of many other animals, by penetrating the membrane with **microelectrodes** having tip diameters of less than a micrometer.

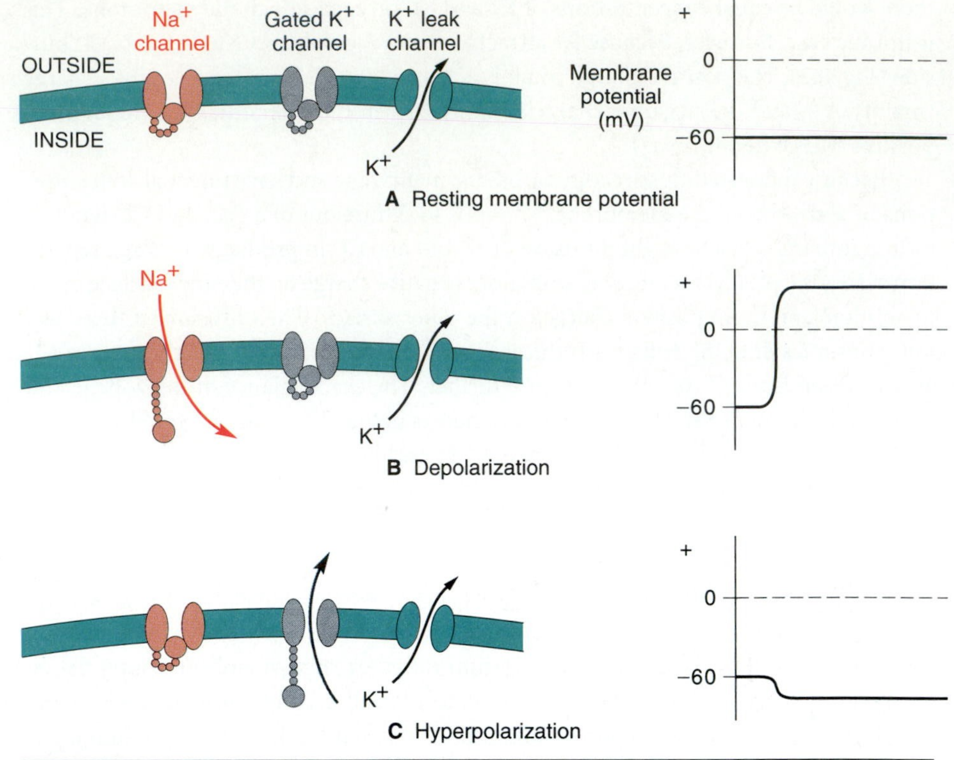

FIGURE 8.3

Schematic representation of how changes in permeability to NA^+ and K^+ change membrane potentials. Changes in ion permeability are due to opening and closing of ion channels, represented here according to the "ball-and-chain" model of ion channels. (A) At rest, sodium channels are closed, so the permeability to sodium is negligible. Potassium permeability is much larger, allowing some of the K^+ pumped in by sodium–potassium pumps to leak out. (B) Opening sodium channels allows a small amount of Na^+ to enter the cell and depolarizes the membrane. (C) Opening the gated potassium channels allows more K^+ to diffuse out and hyperpolarizes the membrane.

Hodgkin and Huxley found that when a giant axon is stimulated with a brief shock, the action potential begins with a rapid depolarization. In fact, there is an overshoot, with the voltage inside the axon becoming positive. The membrane potential then returns to negative inside and hyperpolarizes briefly. The entire action potential lasts only about 3 milliseconds (Fig. 8.5). Hodgkin, Huxley, and others eventually learned that the depolarization, recovery, and hyperpolarization of the action potential are due to changes in the permeability of the membrane to sodium and potassium ions. These permeability changes occur in excitable cells because the channels for Na^+ and K^+ open and close in response to changes in the membrane potential.

This is how the changes in permeability produce an action potential: First the stimulus depolarizes the membrane slightly. This depolarization opens the Na^+ channels somewhat, increasing the permeability to sodium. The resulting inward sodium current depolarizes the membrane further. This depolarization opens the sodium channels further, which increases sodium permeability more, and so on. This sequence accounts for the rapid depolarization during the rising phase of the action potential. When the membrane potential becomes inside-positive, it triggers a separate change that closes the sodium channels. This helps terminate the action potential. In squid giant axons and in many other axons, the depolarization also slowly opens voltage-dependent potassium channels, increasing the permeability to potassium. The resulting

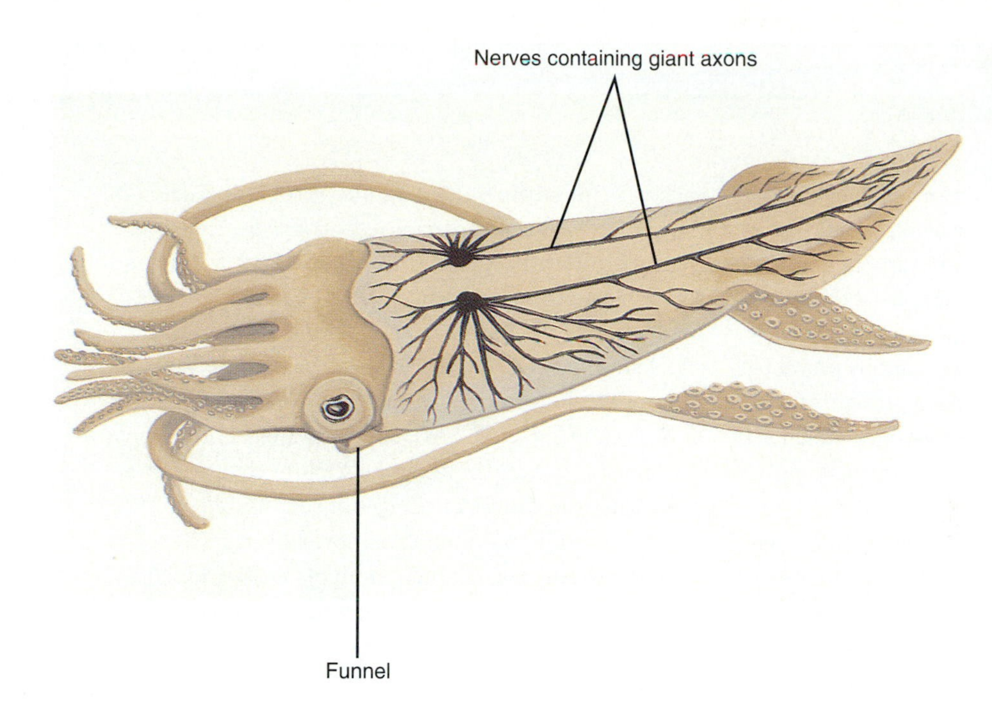

FIGURE 8.4

The squid *Loligo pealei,* showing the location of giant axons. Giant axons trigger jet-propelled escape from predators by causing the body to contract and force water out the movable funnel.

K^+ current helps bring the membrane potential back to the resting level. In fact, it causes the membrane to hyperpolarize momentarily.

PROPERTIES OF ACTION POTENTIALS. Once the action potential is triggered, the changes in permeability and membrane potentials play themselves out automatically. The magnitude and duration of the action potential depend only on the concentrations of ions, which are constant in a healthy animal. Consequently, the action potentials are all the same in each animal, and they either occur completely or do not occur at all. This is called the **all or none law.** As long as the stimulus produces a sufficient triggering depolarization, an action potential will occur. In other words, there exists a **threshold** for an action potential. As long as a stimulus depolarizes the membrane beyond threshold, the size of the action potential will not increase with the size of the stimulus, any more than the bang from a stick of dynamite will increase if you light it with a blowtorch rather than a match.

The energy that drives the action potential does not come from the stimulus, but from the high external concentration of sodium that was previously produced by the sodium–potassium pumps. The numbers of ions moving across the membrane during an action potential are relatively minuscule. For a squid giant axon with a radius of 0.2 mm and a length of 8 cm, only 10^{-12} mole of Na^+ enters per action potential. This is only 1/10,000 as much Na^+ as is already in the cell, and it requires only 3×10^{-13} mole of ATP to pump out. Human brain cells are even more efficient. Your brain consumes only as much power as a 10-watt light bulb, no matter how bright you are.

Although the mechanisms for the action potential were discovered in the squid giant axon, the same principles have been found to apply to other kinds of excitable cells. There are some notable differences in detail, however. In many axons of mammals, for example, high chloride permeability rather than voltage-dependent potassium channels speed the recovery from action potentials. Consequently, most action potentials in humans and other mammals do not end with a hyperpolarization.

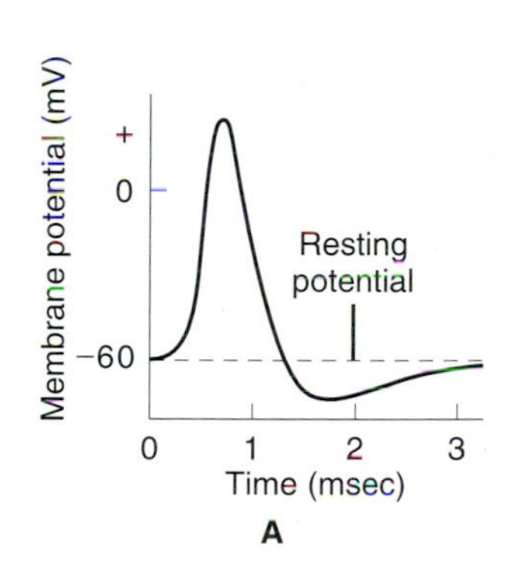

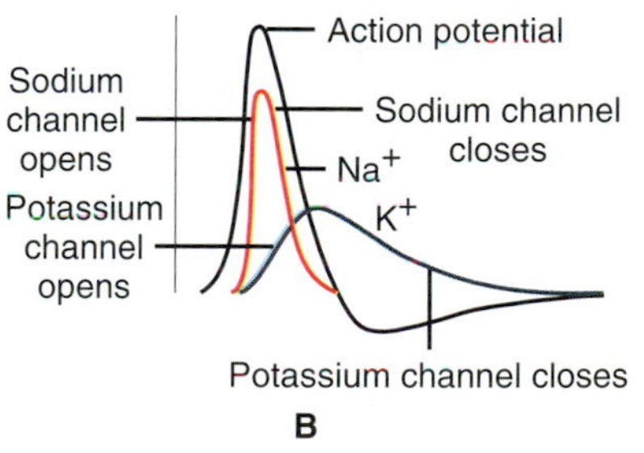

FIGURE 8.5

(A) An action potential from an isolated squid giant axon. (B) The changes in permeability to sodium and potassium that cause the action potential.

■ **Question:** A toxin in the venom of certain scorpions blocks potassium channels. Diagram an action potential in a squid that has had the unlikely misfortune of being stung by such a scorpion.

Neurotoxins

The details in the previous section enable neurophysiologists to understand the effects of many neurotoxins that occur in plants and animals. For example, cocaine and its derivatives, benzocaine, xylocaine, and others, are **local anesthetics** when applied directly to a nerve, because they block action potentials. They work by preventing sodium channels from opening. Another neurotoxin that blocks sodium channels is tetrodotoxin, which is produced by symbiotic bacteria in a variety of animals, including the puffer fish *Fugu.* Eating puffer fish without removing the ovaries, liver, and intestines often causes death by paralysis. (In spite of that, or perhaps because of it, *Fugu* is a delicacy in Japan. Even though chefs must be licensed to prepare *Fugu,* approximately 100 people die each year from eating it.) A similar substance, saxitoxin, is produced by the protozoans responsible for "red tides." Saxitoxin often kills fish that eat the protozoans and sometimes causes "paralytic shellfish poisoning" in people who eat molluscs that have fed on the protozoa.

Batrachotoxin, which occurs in the poison-dart frogs of South America (see p. 789) and in certain birds of New Guinea (see p. 836), exerts its fatal action in just the opposite way, by holding the sodium channels open. The toxin from American scorpions such as *Centruroides* makes the sodium channels abnormally responsive to depolarization. Toxin in the venom of Old World scorpions such as *Leiurus* blocks potassium channels, thereby slowing the recovery of action potentials.

CONDUCTION OF ACTION POTENTIALS. Action potentials would be useless unless they went from one place to another in the nervous system. Axons conduct action potentials automatically, because an action potential at one place on a membrane is a large enough depolarization to excite adjacent areas of membrane. Thus one action potential triggers another nearby on the membrane, and that triggers another, and so on. Consequently, an action potential starting at one end of an axon appears to travel to the other end. In fact, it is not the same action potential that is traveling, but a series of continually regenerating action potentials. Usually an action potential begins at the end of the axon near the cell body and travels in one direction.

Most axons of invertebrates conduct at speeds of less than 1 meter per second (m/sec). At that slow rate it might take a large animal several seconds to find out that it is being eaten by a predator. The speed of conduction increases with the diameter of the axon, however, and many invertebrates have large-diameter giant axons that trigger rapid escape responses from potential predators. Such giant axons occur not only in the squid, but also in cockroaches and some other insects, in many fishes, and in earthworms and other annelids. The thickest known axon diameter, 1.7 mm, pulls the marine annelid *Myxicola* into its protective tube.

Even greater conduction speeds occur in many vertebrate axons even though they have smaller diameters, thanks to **myelin.** Myelin consists of sheaths of membrane that partially surround the axon. During embryonic development, myelin-forming cells line up along the axons, then spiral around them, wrapping them in layers of membrane. Spaces, called **nodes of Ranvier,** remain between each deposit of myelin at intervals of about 1 mm (Figs. 8.1B and 8.6). Action potentials can occur only at these nodes, where the axon membrane is exposed. Instead of traveling along the plasma membrane, the voltage from an action potential at one node passes through the interstitial fluid to the next node, where it evokes an action potential. That action potential then excites the next node, and so on. This apparent jumping of action potentials from one node to the next is called **saltatory conduction,** from the Latin word *saltare,* meaning "to leap." The reason saltatory conduction is so much faster—up to 100 m/sec—is that current passes more rapidly through the interstitial fluid than an action potential conducts along membrane.

Another advantage of myelination is that it conserves energy. Na^+ does not have to be pumped back out all along the axon after each action potential, but only at the nodes.

Chemical Signals

SYNAPSES. Once action potentials reach their destinations, they have to evoke some response in another cell, usually either a muscle cell or another neuron. The structures by which neurons evoke such responses are synapses. It would be reasonable to guess that synapses simply transmit action potentials from the nerve cell to the other cell, but that occurs in only a minority of synapses—the **electrical synapses.** Electrical synapses occur at gap junctions (see Fig. 2.28), often in rapidly conducting pathways, such as the giant fibers that trigger escape in crayfish. The advantage of electrical synapses is that they conduct action potentials very rapidly. Their disadvantage is that rapid conduction of action potentials is all they can do.

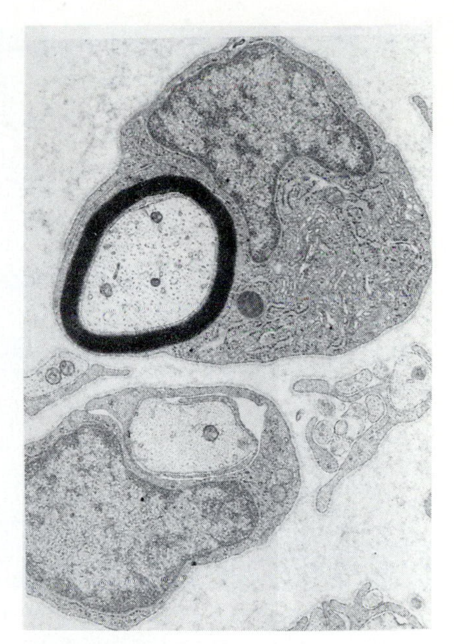

FIGURE 8.6

Electron micrograph showing two Schwann cells forming myelin in a peripheral nerve of a laboratory rat. The Schwann cell in the upper right has already laid down many darkly staining layers of myelin by moving around the axon. The Schwann cell in the lower left has just began encircling an axon with layers of membrane. In the brain and spinal cord, myelin is similar in structure but is formed by different cells, the oligodendrologia. Compare Fig. 8.1B.

Most synapses are **chemical synapses,** which can have varying effects, either evoking or inhibiting action potentials. A chemical synapse works by releasing one or more substances called **synaptic transmitters** (= neurotransmitters) onto another cell. The amino acid glutamate is the most common transmitter in vertebrates, but for historical reasons, acetylcholine is the most familiar. Some of the approximately 60 known transmitters will be described in more detail later. Some synaptic transmitters simply diffuse through membranes. Such transmitters include the gases nitric oxide (NO) and carbon monoxide (CO), which have only recently been discovered to be transmitters. Most known synaptic transmitters, however, are packaged within spherical **vesicles** and are released by exocytosis (see p. 40; Fig. 8.7 on page 158). Often the vesicles are located in bulb-shaped endings called **synaptic knobs,** which are often located near dendrites of other neurons (Fig. 8.1B). The membrane of the synaptic knob is said to be **presynaptic,** and the opposite membrane is **postsynaptic.** Presynaptic and postsynaptic membranes are separated by a **synaptic cleft.**

Although the synaptic cleft is only 0.02 μm wide, action potentials cannot jump it. In fact, they simply vanish at the presynaptic membrane. Before they vanish, however, they cause some of the synaptic vesicles to fuse with the presynaptic membrane. Vesicles then rupture, releasing the synaptic transmitters within them. How action potentials trigger the release of transmitter is not fully understood, but it requires an influx of Ca^{2+} into the synaptic knob. After the molecules of transmitter diffuse across the cleft, they bind to specific **receptor molecules** on the postsynaptic membrane. Postsynaptic membranes have many types of receptor molecules that react in different ways to the same transmitter.

These receptor molecules are ligand-gated ion channels. When the transmitter (ligand) binds to them, they change the permeability of the postsynaptic membrane to one or more ions. If the transmitter increases the permeability to Na^+, the postsynaptic membrane will be depolarized. A depolarizing postsynaptic potential is called an **excitatory postsynaptic potential** (EPSP), because it tends to excite the target cell to produce action potentials. If the transmitter increases the permeability to K^+, the postsynaptic membrane will be hyperpolarized. Such a hyperpolarization is called an **inhibitory postsynaptic potential** (IPSP), because it inhibits the generation of action potentials. A transmitter that increases the permeability to Cl^- can also inhibit action potentials by keeping the resting membrane potential constant, resisting the effects of EPSPs.

EPSPs and IPSPs are examples of **graded potentials.** Unlike action potentials, graded potentials occur in various sizes, and they decrease in size as they travel along membrane. Graded potentials are useless for transferring information over long distances, but they are satisfactory in small nerve cells like those in the retina of the eye, and

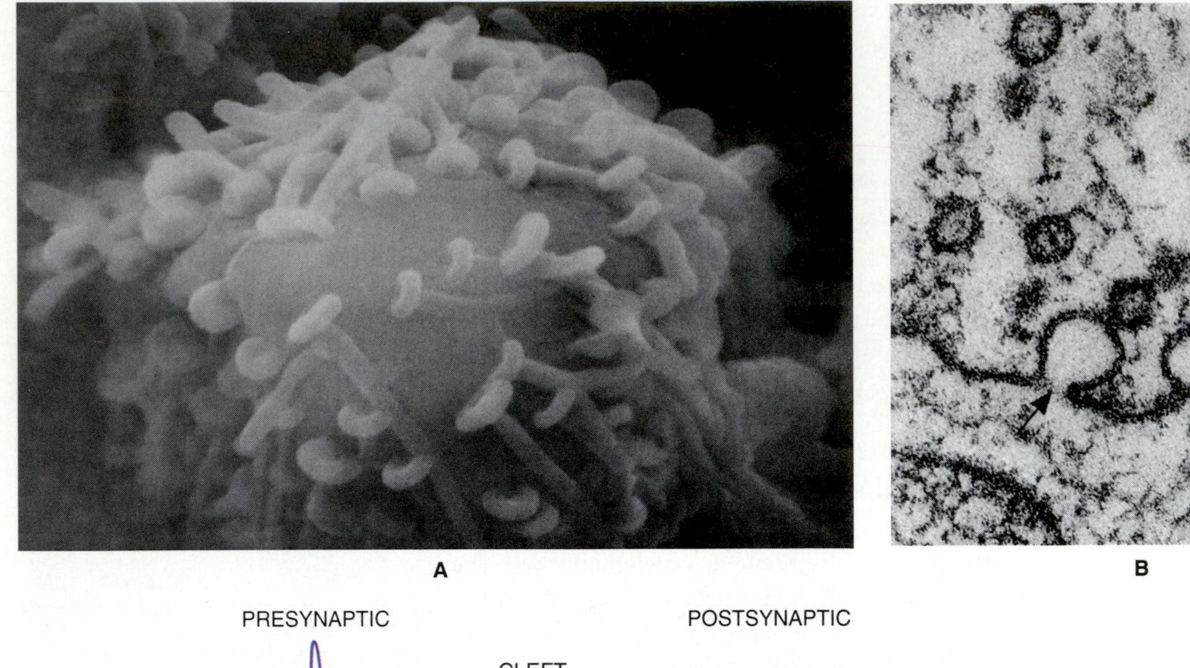

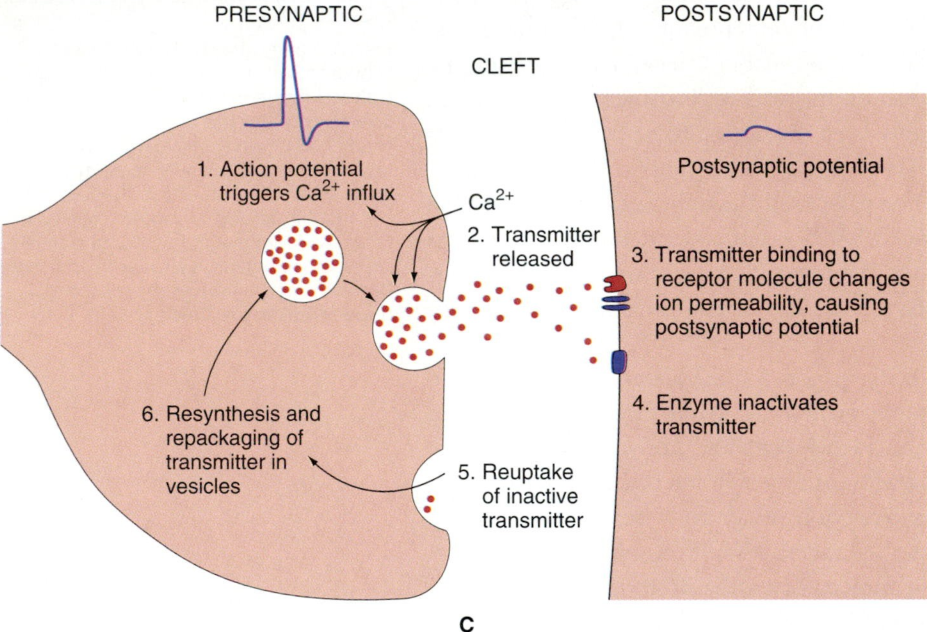

FIGURE 8.7

(A) Scanning electron micrograph of synaptic knobs in the sea hare *Aplysia californica.* (B) Transmission electron micrograph of the neuromuscular synapse of the common grass frog *Rana pipiens.* The nerve and muscle were quickly frozen just after stimulation and were thereby caught in the act of releasing the transmitter, acetylcholine. In this species each vesicle of 40-nm diameter contains approximately 10,000 molecules of acetylcholine. (C) Diagram of transmitter release in a generalized chemical synapse.

they are ideal for transferring postsynaptic responses over a dendrite and cell body to the axon. If the EPSPs are not canceled by IPSPs, and if they produce an above-threshold depolarization when they reach the excitable portion of an axon, they can evoke action potentials. While postsynaptic potentials are being generated, the transmitter has to be removed from the postsynaptic membrane and the cleft in a process called **reuptake.** Often the transmitter is first inactivated by an enzyme. The entire sequence of synaptic events occurs in about 1 msec—much less time than it takes to read about it.

Modification of Synapses. One of the most impressive things about the complex interactions of neurons is that they can be modified by a variety of natural and artificial factors. For example, the functioning of many synapses distributed throughout the nervous system can be altered by a variety of secretions that act as **neuromodulators.** Synaptic functioning is also altered during **learning,** and such modifications appear to be the basis for **memory.** There is good evidence that learning and memory result from strengthening and weakening synapses in a variety of ways: by changing the number of synaptic contacts, the number of vesicles released per action potential, the number of receptor molecules on the postsynaptic membrane, the length or shape of postsynaptic dendrites, the rate of inactivation and reuptake of transmitter, or the permeability of either the presynaptic or postsynaptic membrane.

Not even the Treasurer of the United States can imagine a hundred billion of anything, but if you could imagine a hundred billion nerve cells, and if you could imagine thousands of synapses on each one with varying degrees of excitation or inhibition, then you would have some idea of the complexity of your brain. No computer will ever achieve such elaborate "wiring." The resulting power of nervous systems and the advantages of chemical over electrical synapses are suggested by the simple example in Fig. 8.8.

SYNAPTIC CHEMISTRY. The same features of chemical synapses that make learning possible also make them susceptible to chemical disturbances. Many plant toxins, such as cocaine, heroin, and tetrahydrocannabinol from marijuana, affect synaptic functioning, and most psychologists now realize that many behavioral disorders are due to chemical imbalances affecting synapses. Neurotoxins can affect synapses in a variety of ways, including the following:

1. They can be ligands that bind to the postsynaptic receptor and mimic the action of the transmitter.
2. They can be ligands that bind to the postsynaptic receptor and block the binding of the transmitter.
3. They can affect the breakdown or reuptake of transmitter.
4. They can affect the synthesis of transmitter.

Table 8.1 describes some chemical transmitters, their modes of action, where they occur, and ways in which their actions are affected by drugs.

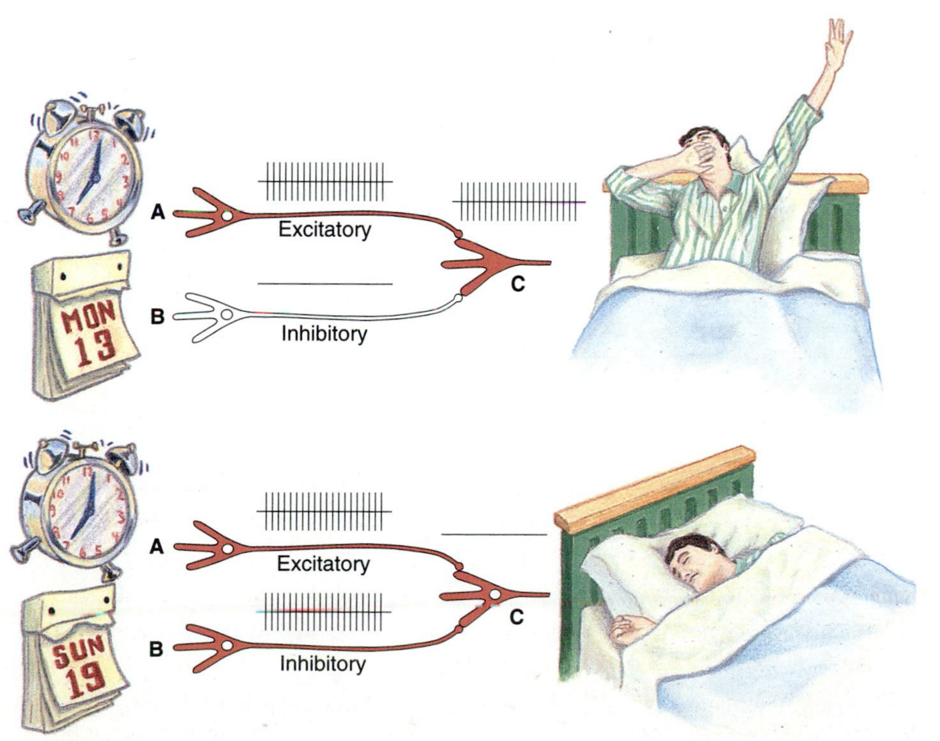

FIGURE 8.8

Representation of a simple neural circuit theoretically capable of performing the following logical operation: "If A and not B, then C." In this case the operation determines whether the individual will arise in the morning. Neuron A fires when the alarm clock rings and tends to excite Neuron C, which causes the individual to wake up. Neuron B is spontaneously active when the individual realizes he has no morning classes; it inhibits C. Such circuits with excitatory and inhibitory synapses are capable of handling any logical proposition that is either true or false.

■ **Question:** What physiological explanation (excuse) does this model provide for sometimes oversleeping?

TABLE 8.1

Synaptic transmitters and drugs that affect their functioning.

Transmitter	Action	Description
ACETYLCHOLINE (ACh)	Excitatory	Released by parasympathetic nervous system (see p. 178), which slows down heart and speeds up digestion, and by synapses that control vertebrate skeletal muscle. Also common in the central nervous systems of both vertebrates and invertebrates.

Nicotine binds to ACh receptors and mimics the action of ACh in the brain and parasympathetic nervous system. Nicotine was the first insecticide, but insects quickly learn to avoid the strong odor. Now its main use is to addict people to tobacco. Dependence on nicotine is more difficult for an addict to break than dependence on cocaine or heroin.

Curare, a substance from certain South American plants, blocks ACh receptors at synapses onto skeletal muscle, causing paralysis. The first use was as an arrow poison; now it is commonly used in surgery to keep muscles from contracting.

Diisopropylfluorophosphate (DFP) and related **organophosphorus compounds** destroy the enzyme (acetylcholinesterase) that normally inactivates ACh in the synaptic cleft. These chemicals were first developed as nerve gas for warfare against humans. They are now used in the war against insects, but they still kill and injure many farm workers.

Transmitter	Action	Description
NOREPINEPHRINE	Excitatory or inhibitory, depending on the site	A transmitter in the central nervous system in various animals; also released by the sympathetic nervous system (see p. 178). High levels of norepinephrine (and serotonin) occur in the brain during the manic phase of bipolar affective (manic–depressive) disorder, and low levels occur during the depressive phase.

Amphetamines stimulate the release of norepinephrine; prolonged use can cause psychosis by depleting the synaptic vesicles.

Transmitter	Action	Description
DOPAMINE	Excitatory or inhibitory	Secreted in central nervous system. Excessive response to dopamine in the brain is thought to be associated with schizophrenia. Low levels in one area of the brain occur in Parkinson's disease, which is characterized by tremors, loss of facial expression, and shuffling gait.

Antischizophrenic (= neuroleptic) drugs, such as clozapine, depress certain dopamine synapses.

Cocaine inhibits the reuptake of dopamine (as well as serotonin and norepinephrine). Cocaine produces a euphoric mood in those who ingest it and also in the criminals who get rich by selling it. A highly purified and addicting form, called "crack," can cause cardiac arrest by constricting blood vessels to the heart.

Amphetamines promote the release of dopamine (as well as norepinephrine). They also inhibit the storage of dopamine in vesicles.

L-DOPA, a metabolic precursor of dopamine, is used to treat Parkinson's disease.

<table>
<tr><td>SEROTONIN</td><td>Excitatory</td><td>In central nervous system. Has a wide range of effects in different parts of the brain.</td></tr>
<tr><td colspan="3">The antidepressant Prozac increases the effect of serotonin by interfering with its reuptake.

Hallucinogenic drugs often affect serotonin synapses. Dimethyltryptamine (DMT), for example, is chemically similar to serotonin and mimics its effects. Lysergic acid diethylamide (LSD) blocks receptors for serotonin.

Tryptophan, an amino acid, is a metabolic precursor to serotonin. Increasing the levels of tryptophan in the brain induces drowsiness.</td></tr>
<tr><td>ENKEPHALIN</td><td>Inhibitory</td><td>In central nervous system of vertebrates.</td></tr>
<tr><td colspan="3">Opiates (heroin, morphine, codeine, etc.) bind to the enkephalin receptors and mimic its action, causing euphoria and blocking pain.</td></tr>
<tr><td>GAMMA-AMINOBUTYRIC ACID (GABA)</td><td>Inhibitory</td><td>The most important inhibitory transmitter in the vertebrate brain and in many invertebrates. Increases chloride permeability.</td></tr>
<tr><td colspan="3">The tranquilizer Xanax increases the action of GABA. Ethyl alcohol may exert its intoxicating effect partly in the same way as Xanax, and partly by disrupting the structure of neuronal membranes.</td></tr>
</table>

Concepts of Sensory Reception

So far we have passed over the question of where the stimuli come from that normally evoke action potentials. Many neurons are spontaneously active, but most neural activity originates at sensory receptors. Sensory receptors convert energy from the environment into graded changes in membrane potential, called **receptor potentials.** In many receptors the receptor potential is a depolarization that triggers action potentials directly in an axon that is part of the receptor cell. In other receptors the receptor potential activates a synapse that then generates action potentials in a separate nerve cell. In either case, the action potentials conduct through **sensory** (= afferent) **neurons** into the central nervous system. This is the only known way that information about the outside real world gets into the nervous system. This book and everything else that is part of what we call reality are nothing more than action potentials, as far as your brain is concerned.

The fact that all sensations are based on action potentials presents a problem, since every action potential is identical to every other one in the nervous system. How can we distinguish among the countless different odors, sounds, colors, tastes, and other sensations using action potentials that are identical, odorless, silent, colorless, and tasteless? The answer must be that there is a kind of sensory code based on the pattern of action potentials, like Morse code, which is based on patterns of dots and dashes. Patterns of action potentials in sensory axons can vary in only three ways: in frequency (number per second), in time of occurrence, and in the nerve cells in which they occur.

1. **The frequency of action potentials conveys information about the intensity of the stimulus.** The stronger the stimulus, the more action potentials evoked

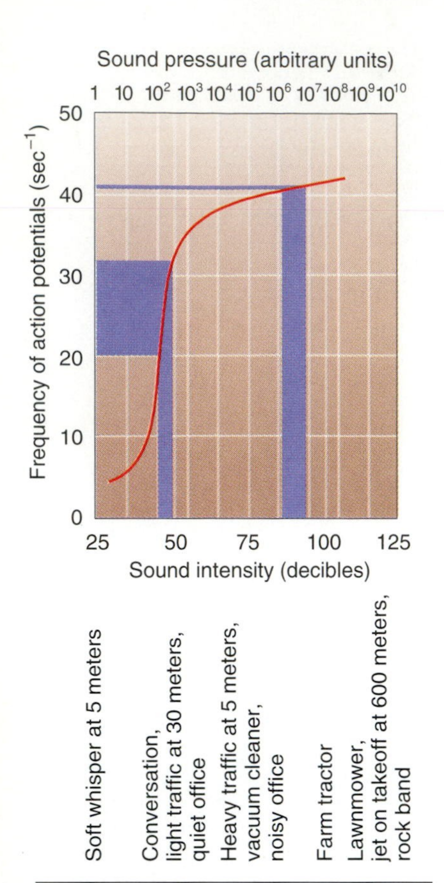

FIGURE 8.9

The frequency of action potentials in an auditory axon of a cat in response to brief tone bursts of different loudness. Sound intensity is given in the decibel (dB) scale, which is proportional to subjective loudness, and also as sound pressure, which reflects the actual force on the eardrum. The threshold for hearing is 0 dB. In a quiet room a person can barely detect a 1-dB sound. Every 10 dB represents a tenfold increase in sound pressure. The colored bars show that a sound that would produce a large frequency change at low background intensity would produce little change at high background levels. Thus, like most sensory systems, the auditory system combines extreme sensitivity with an enormous functional range of intensities.

■ **Question:** Why can you hear a pin drop in a quiet room but not in a noisy one?

in a sensory neuron per second (Fig. 8.9). The brain interprets the frequency of action potentials as intensity—for example, as loudness or brightness. Some receptor cells are especially suited to providing this kind of information. They produce a constant frequency of action potentials as long as the stimulus intensity stays the same. Such receptors are said to be **tonic.**

2. **The time of occurrence of action potentials provides information about the time of occurrence of the stimulus.** This may seem obvious, but some receptors, called **phasic** receptors, are especially adapted for providing this kind of information. In contrast to tonic receptors, phasic receptors initiate action potentials only when there is a change in stimulus intensity. Each sound wave, for example, triggers a brief burst of action potentials from certain receptor cells in the ear, and this brief burst goes to the brain where it indicates precisely when the wave occurred. The system is so precise that you can tell the direction of a sound from the difference in time of the phasic response from the two ears.
3. **Which neurons conduct the action potentials provides information about the modality, quality, and location of the stimulus.**

   Modality refers to the major category of sensation, whether sight, sound, smell, taste, or touch. Humans perceive any stimulus as belonging to the visual modality if the stimulus evokes action potentials in axons of the optic nerve or in the visual areas of the brain. Getting hit in the eye, for example, produces the visual sensation of "seeing stars."

   Quality refers to subcategories within each modality. Color and pitch are qualities within the modalities of vision and hearing, for example. Different receptors and neurons respond to different colors and different pitches.

   Location of a stimulus is often determined by which receptor is stimulated, which determines which neurons are excited. If an itching sensation is detected by receptors on the elbow, then action potentials will occur in neurons in the brain that respond to that elbow.

All these kinds of information about a stimulus are sent to the central nervous system simultaneously for parallel processing. Somehow animals are able to put this meager input of information together into a perception of their environments. Perhaps, like us, other animals use this neural code to create a rich image of the world. Until techniques are available to measure what another animal is thinking, however, we should resist assuming that the world appears to other species as it appears to us. In fact, we cannot even say that every human perceives stimuli in the same way. We have all learned to call light with a wavelength of 650 nanometers "red," but there is no way to tell whether the sensation of redness is the same for everyone.

How Receptors Work

THE STRETCH RECEPTOR OF THE CRAYFISH. The way in which receptor cells convert stimuli into receptor potentials can be explained for a few types of receptors. One of the simplest is the stretch receptor of crayfish and other crustaceans. These receptors evoke action potentials whenever a muscle stretches, thereby indicating the positions of legs and other parts of the animal's body. Crayfish stretch receptors are essentially neurons attached to the muscles (Fig. 8.10). When the muscles are stretched, so are the receptors. The stretch increases the sodium permeability of the receptor membrane, perhaps by mechanically opening the sodium channels. The increase in sodium

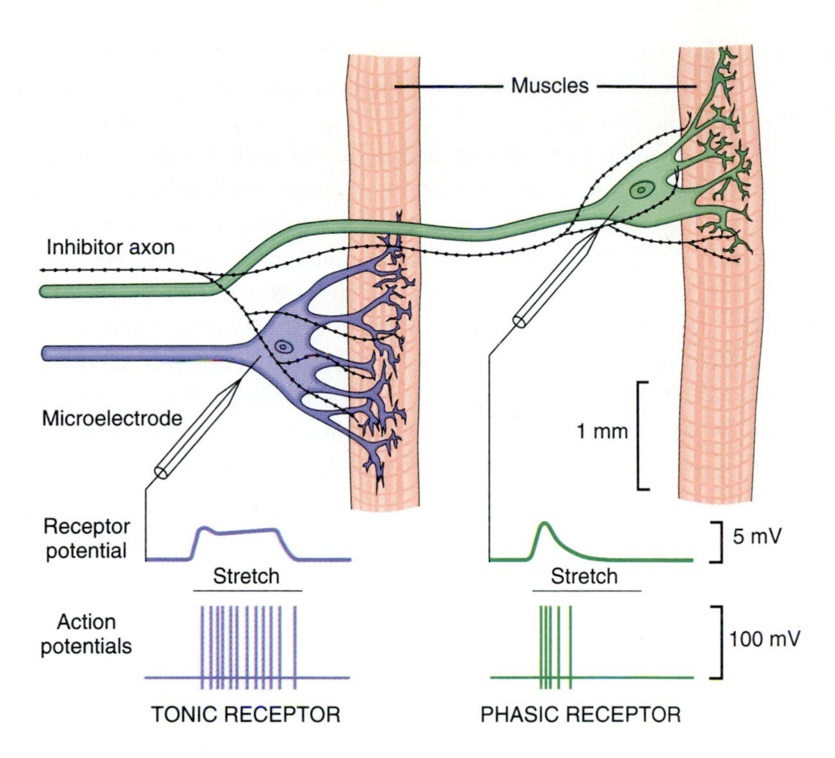

FIGURE 8.10

Two stretch receptors attached to parallel muscle fibers in a crayfish. One is a tonic receptor and the other is a phrasic receptor. The inhibitory neuron to the receptors can block the receptor potential. Receptor potentials and action potentials were recorded by microelectrodes during a sudden stretch that was then held. Although the stimulus to both receptor is the same, the phasic receptor responds only at the start of stimulation.

■ **Question:** What would be the pattern of action potentials in each cell for a steadily increasing stretch?

permeability produces a depolarizing receptor potential that can evoke action potentials. In the tonic stretch receptors, the greater the stretch, the larger the depolarization, and the higher the frequency of action potentials. Generally parallel to the tonic stretch receptor is a phasic stretch receptor that provides information about the time of occurrence of the stretch. As is common in receptors, the sensitivity can be reduced by inhibitory nerve cells conducting action potentials from the central nervous system.

Rods and Cones of the Eye. Among the most complex receptor cells are the photoreceptor cells of vertebrates. These are called rods or cones, depending on their shapes (Fig. 8.1). Rods have cylindrical **outer segments** that contain approximately 2000 disc-shaped membranes bearing light-absorbing pigments. This pigment is **rhodopsin,** a yellow substance that absorbs a broad range of wavelengths. Rhodopsin combines a vitamin-A derivative called **retinal** with a protein called **opsin.** Cones are similar, except that their outer segments are cone-shaped, and the opsins differ. Vertebrates typically have several types of cones with different opsins that enable them to perceive different colors. In addition, lungfishes, many reptiles, and birds have colored drops of oil in the cones, which narrow the color sensitivity. Mammals have cones with three different opsins, each of which absorbs light most effectively at a different wavelength from the other two. In humans these three wavelengths are perceived as red, blue, and green. Although we have only three kinds of cones, any color can be perceived from the combination of cones it stimulates. For example, stimulation of both red-sensitive and blue-sensitive cones would indicate the color purple.

Rods and cones also differ in their sensitivity. Rods are so sensitive that they can respond to individual photons, but they become blinded in bright daylight. They are therefore useful mainly at night and in heavy shadow. Cones are not sensitive enough to work in darkness, but the different color sensitivities enable them to transmit information about color in bright light. Good vision in the dark requires a large number of rods, while color vision in daylight requires numerous cones. The retinas of most mammals, especially nocturnal ones, have primarily rods. The retinas of humans and

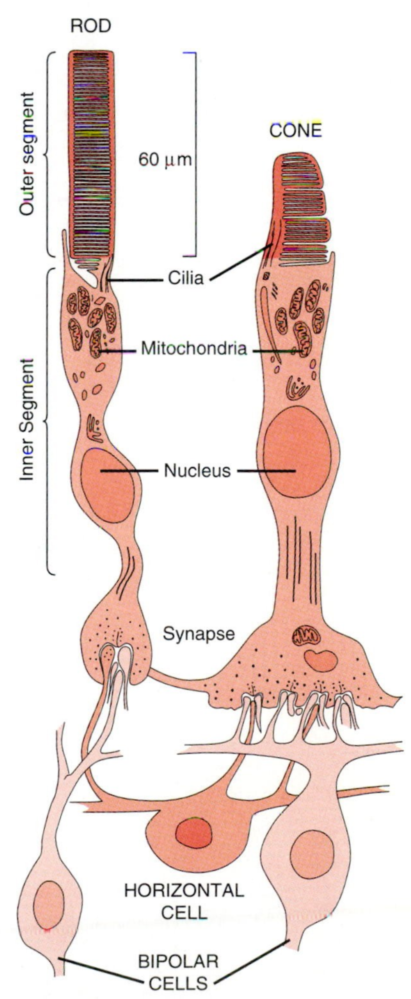

FIGURE 8.11

Structure of the mammalian rod and cone, and their connections to other cells in the retina.

other primates have a mixture of rods and cones in peripheral areas and have only cones in the center of focus, called the **fovea.** Humans have a total of about 100 million rods and 3 million cones.

The mechanism by which rods and cones convert light into electrical signals is still being worked out, but already the results are surprising. Unlike the crayfish stretch receptor, rods and cones depolarize when they are *not* stimulated—that is, in the dark. The reason is that their membranes have a high permeability to Na^+ in the dark, because the sodium channels are kept open by cyclic guanosine monophosphate (cGMP). Light triggers reactions that break down the cGMP, thereby reducing sodium permeability and producing a hyperpolarizing receptor potential. This receptor potential travels to a chemical synapse. In the dark this synapse leaks transmitter, but in the light the hyperpolarizing receptor potential reduces the rate of transmitter release. This change in synaptic activity affects other retinal cells, which then produce action potentials in the optic nerve.

Photoreceptor Organs

Receptor cells usually occur as parts of sensory organs that protect and tune the receptor cells to appropriate stimulation. The survey of animals in the final chapters describes many specialized receptor organs. For now we focus on organs that illustrate general concepts, especially in humans and other vertebrates.

Photoreceptor organs include eyes and other structures that respond to electromagnetic radiation with wavelengths in or near the range that humans perceive as light. The exact range varies with species. Rattlesnakes, pythons, and boa constrictors have not only eyes, but also infrared (IR) receptors that allow them to "see" the warmth given off by prey (see p. 817). The eyes of many insects, birds, fishes, and other animals respond to ultraviolet (UV) wavelengths that are invisible to us.

There are numerous types of photoreceptor organs ranging in complexity from simple light detectors to complex eyes capable of analyzing detailed images. (For examples see Fig. 28.27.) The simplest photoreceptor organ is the **eyespot,** which may be little more than a nerve cell containing pigment. Some animals have eyes with lenses that focus light onto the photoreceptor cells. If there are several photoreceptor cells, the eye may be able to detect the position of light, depending on which cells the lens focuses it on. Only if there are numerous photoreceptor cells at the correct focusing distance from the lens will the eye be able to analyze an image. Image-forming eyes occur in only five phyla: Cnidaria, Mollusca, Annelida, Arthropoda, and Chordata. Crustaceans and insects have **compound eyes** consisting of numerous subunits (see pp. 665–666).

Vertebrate eyes have millions of photoreceptor cells (rods and cones) combined in each **retina** (Fig. 8.12A). A lens focuses light onto the retina, and each rod or cone samples the light from a small area of the visual field. The lenses in fishes, amphibians, and snakes focus on near or distant objects by moving back and forth. In most reptiles, birds, and mammals the lens focuses on nearby objects by becoming more spherical. In terrestrial animals the clear covering of the eye, called the **cornea,** is twice as effective as the lens in focusing light, because of the refraction of light as it goes from air into tissue. The diameter of the **pupil** controls the amount of light entering the eye.

The image on the retina is inverted, although that does not matter as far as the brain is concerned. (If you wore goggles that invert images, the world would look normal after a few days. Then, if you took the goggles off, everything would look upside down.) The light also seems to come in the wrong way, passing through the other retinal cells before reaching the photopigments. This does not affect vision, however, be-

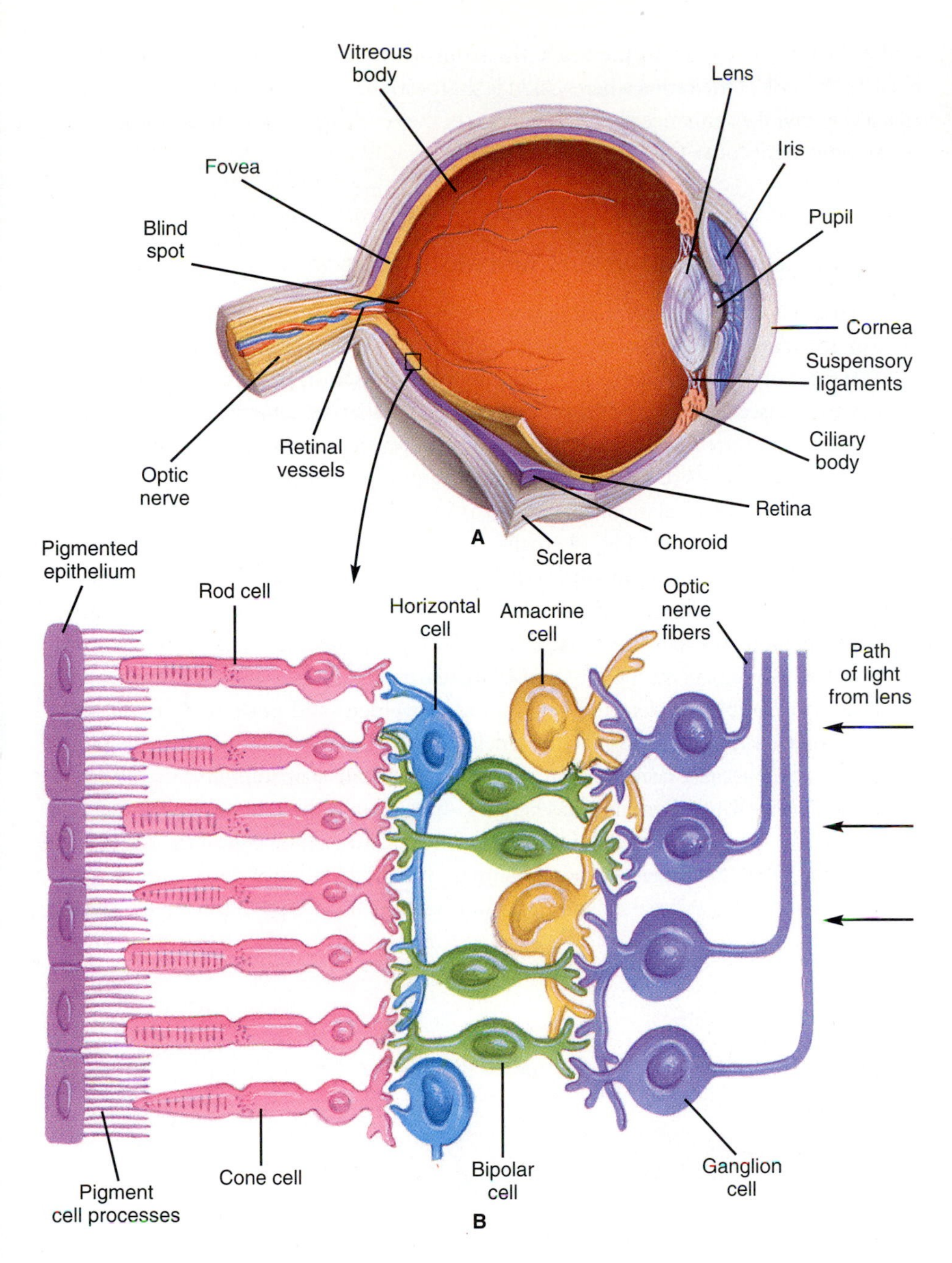

FIGURE 8.12

Structure of the mammalian eye. (A) The human eyeball. Contraction of muscles in the ciliary body causes the suspensory ligaments to relax tension on the lens, enabling the lens to focus nearby objects onto the retina. The iris prevents excessive light from entering. (B) Enlarged sections of the retina. Complex chemical synapses translate excitations of the rods and cones into patterns of action potentials in ganglion cell axons in the optic nerve.

cause these cells are transparent. In most animals, any light not absorbed by the photopigments is absorbed by a layer of **melanin** pigment. Many nocturnal animals, however, have a reflecting layer, called the **tapetum,** that bounces the light back through the rods and gives them a second chance to absorb the light. This tapetum causes the characteristic eye shine of cats and many other nocturnal animals.

Although the vertebrate eye has many fine features, the image it forms on the retina is really quite poor. A camera store that sold human eyes would soon be out of business. The reason is that no simple lens—not even a living one—can form a sharp image over a wide area for all colors. That is why good camera lenses are so expensive; they combine many separate elements, each of which corrects defects in the others. Instead of using multiple lenses, vertebrate eyes correct the fuzzy image neuronally, through interactions of **horizontal, bipolar, amacrine,** and **ganglion** cells in the retina (Fig. 8.12B). These cells also do preliminary processing of visual information before sending action potentials to the brain via axons of the ganglion cells. In the fovea, which is the area of the retina where vision is sharpest, these cells are interconnected in such a way that they enhance contrast. In areas of the retina nearest the nose, these cells detect movement "out of the corner of the eye."

Thermoreceptors, Nociceptors, and Mechanoreceptors

Somesthetic (body sense) receptors respond to thermal and mechanical stimulation of the skin, to pain, and to the position of limbs. They include thermoreceptors that react to either heat or cold, nociceptors that produce the sensation of pain in response to substances released by damaged tissue, and various mechanoreceptors. Little is known about the structure and functioning of thermoreceptors and nociceptors. Mechanoreceptor cells are associated with a variety of organs that respond to many types of mechanical stimulation. These organs include stretch receptors like those in crayfish and vertebrates (see Figs. 8.10 and 9.2), as well as receptors for touch, pressure, acceleration, and position with respect to gravity. Some mechanoreceptors appear to be little more than the ends of nerve cells, while others have elaborate accessory structures that tune the receptor cells to particular types of stimulation (Fig. 8.13).

The most complex mechanoreceptor organs in vertebrates are the **inner ears,** which are responsible for detecting sound and movement and position of the head. Inner ears employ receptor cells called **hair cells.** Hair cells have nothing to do with hair, but are so named because each has a bundle of 30 to 300 hair-shaped projections of various lengths (Fig. 8.14). Bending these "hairs" in a certain direction depolarizes the membrane and causes the release of transmitter from the synaptic end of the hair cell. The resulting EPSP then triggers action potentials. Bending the projections in the opposite direction hyperpolarizes the membranes and reduces the release of transmitter.

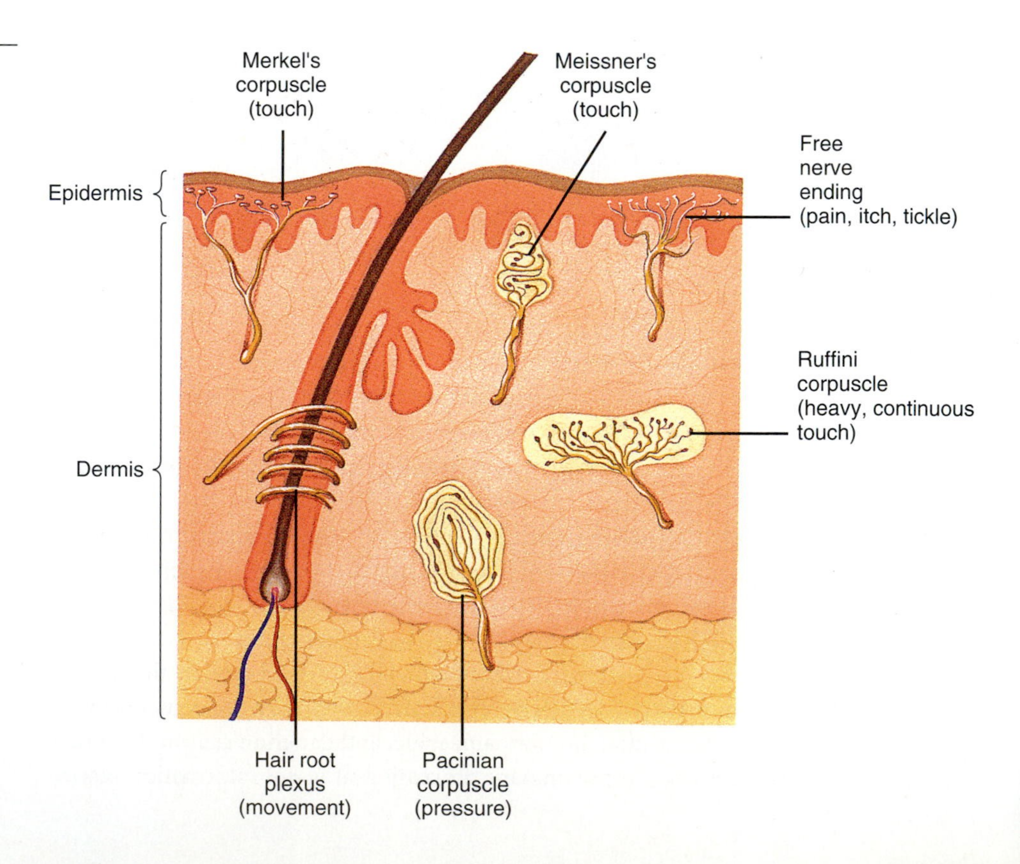

FIGURE 8.13
A few of the dozens of more types of cutaneous receptors in mammals.

■ **Question:** Why are there three types of touch receptors? How does the structure and position of the receptors relate to the stimulus to which each one responds?

The Vestibular Apparatus. The part of the inner ear that responds to movement and position of the head is the vestibular apparatus (Fig. 8.15 on page 168). Information from the vestibular apparatus, combined in the brain with information from stretch receptors in the neck and eye muscles, helps maintain balance and distinguish movement of the body from movement of the surroundings. The vestibular apparatus includes the semicircular canals and the otolith organs. Each of the three **semicircular canals** of each inner ear responds to rotation in a different plane, such as while turning the head, falling forward, or tilting sideways. During such events the fluid **endolymph** that fills the semicircular canals tends not to move while the surrounding tissue is moving. (The same effect occurs when you quickly turn a glass of water.) The motion of endolymph relative to surrounding tissue causes a deflection of hair cells, triggering action potentials that go to the brain (Fig. 8.15B, C).

The **otolith organs**—the **utricle** and **saccule**—are responsible for balance. They are named for otoliths (= otoconia), which are dense, gelatinous structures containing crystals of calcium carbonate (Fig. 8.15D). Projections of hair cells are bent by the weight of the otoliths, generating action potentials that carry information about the direction of gravity.

Hearing. Although most of us take hearing for granted, it is a rare ability in animals, occurring only in some arthropods and vertebrates. Hearing is due to a response by hair cells to waves of increasing and decreasing air pressure rippling from a source of sound. The **frequency** of the sound is the number of cycles of pressure change per second, measured in hertz (Hz). Humans perceive different sound frequencies as pitch. The difference between minimum and maximum pressures during a wave determines the intensity—or loudness—of the sound.

In crocodilians, birds, and mammals, the part of the inner ear responsible for hearing is the **cochlea.** In most mammals the cochleae are snail-shaped, and in humans each one is about the size of a house fly. The cochlea responds to vibration of the **middle ear bones,** which, in turn, respond to vibration of the **eardrum** (= tympanum) (Fig. 8.15). Changing pressure in the endolymph inside the cochlea vibrates the **basilar membrane** inside, which supports the **organ of Corti** (pronounced kor-tee; Fig. 8.16 on page 169). The mechanosensitive hair cells lie in the organ of Corti, with the hairs pointed out toward the **tectorial membrane.** As the organ of Corti vibrates up and down in response to sound, the tectorial membrane rubs against the hair cells, generating receptor potentials. These receptor potentials then trigger the release of synaptic transmitter that generates action potentials in axons of the auditory nerve. Surprisingly, receptor potentials also cause the hair cells to shorten and vibrate the basilar membrane (Mammano and Ashmore 1993). This process is thought to amplify the response, and in some cases it actually *produces* a sound that can be heard by another person.

The louder the sound, the greater the stimulation of the hair cells, and the higher the frequency of action potentials (Fig. 8.9). Thus, as usual, the intensity of stimulation is coded by the frequency of action potentials. The pitch of sound appears to be coded according to which receptor cells are most intensely stimulated. This tuning to different frequencies results in part from mechanical properties of the basilar membrane. For each pitch there is a particular portion of the basilar membrane that vibrates with the greatest amplitude. In mammals the end of the basilar membrane nearest the middle ear bones vibrates maximally in response to high-frequency sound, and the other end responds most to low-frequency sound. You can demonstrate the effect with a fishing rod or similar long, flexible pole. Shaking the rod rapidly produces the most vibration near your hand, and shaking it slowly causes the tip of the rod to vibrate most.

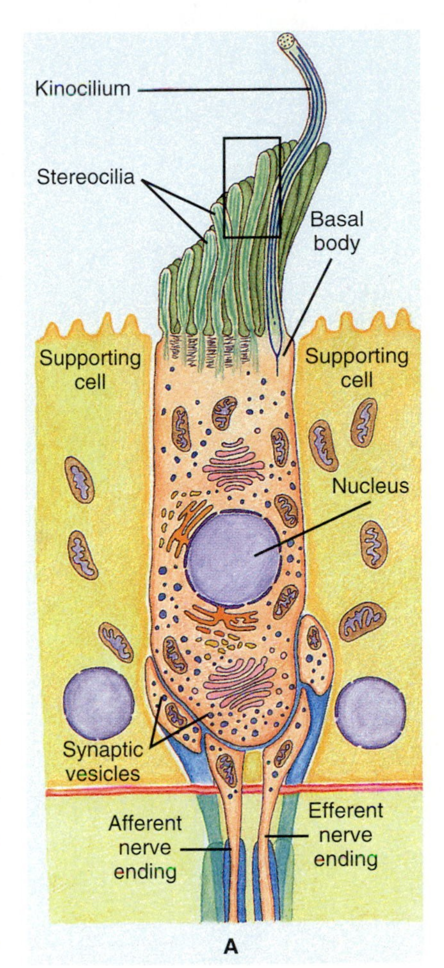

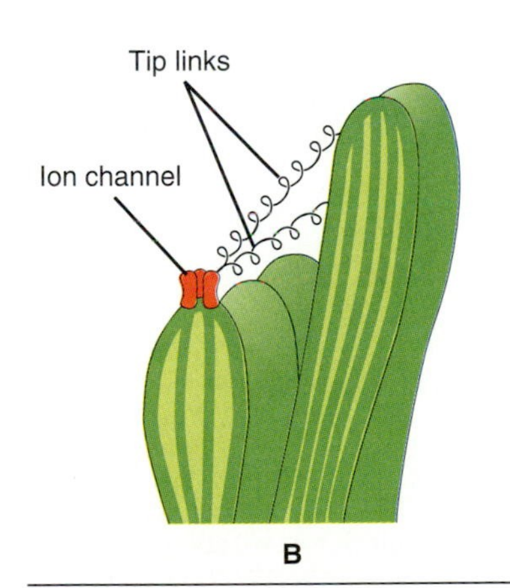

FIGURE 8.14

(A) A human hair cell with rows of stereocilia and one kinocilium. The role of the kinocilium is not clear; it is absent in the cochlea. Bending the stereocilia depolarizes or hyperpolarizes the plasma membrane of the hair cell, producing a receptor potential. The receptor potential regulates the release of excitatory transmitter from the synaptic portion of the cell. Note the efferent neuron that controls the sensitivity of the hair cell. (B) Detail of the area in the box in part (A). Bending a stereocilium pulls or releases a tip link that opens or closes an ion channel in an adjacent stereocilium.

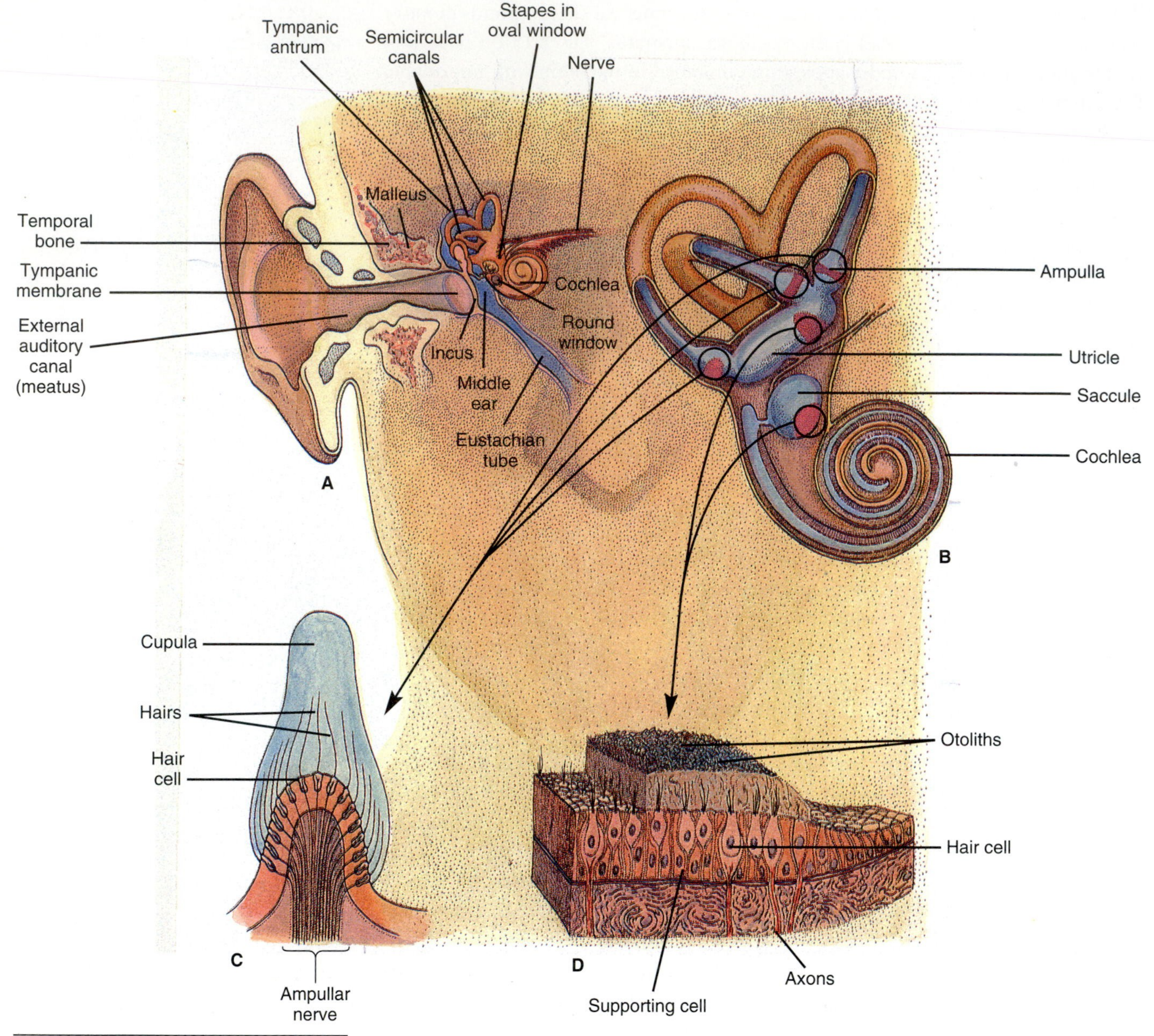

FIGURE 8.15

(A) The auditory organs and vestibular apparatus of a human. (B) The inner ear, consisting of the vestibular apparatus (semicircular canals and otolith organs) and the cochlea. The three semicircular canals detect rotation, the otolith organs (utricle and saccule) detect linear acceleration (gravity), and the cochlea detects sounds. (C) Enlarged section of one ampulla of the semicircular canals, where hair cells are stimulated by the relative movement of endolymph against the cupula. (D) A portion of the utricle or saccule. The weight of otoconia stimulates hair cells.

Chemoreceptors

Chemoreceptors in humans respond to taste (= gustation) and smell (= olfaction), but these two modalities do not begin to describe all the kinds of chemoreception that occur among animals. Fishes and many other aquatic animals, for example, have chemoreceptors on the body surface that respond to substances given off by prey, predators, and members of their own species. Many insects have virtually an unabridged dictionary of chemicals by which they communicate (see pp. 408–409).

In spite of their diversity, most chemoreceptors operate on similar principles. Generally, the outer surface of the chemoreceptor cell bears receptor molecules that bind to particular substances, depending on the molecular shape and charge distribution of the substance. Like the interaction of a synaptic transmitter with the postsynap-

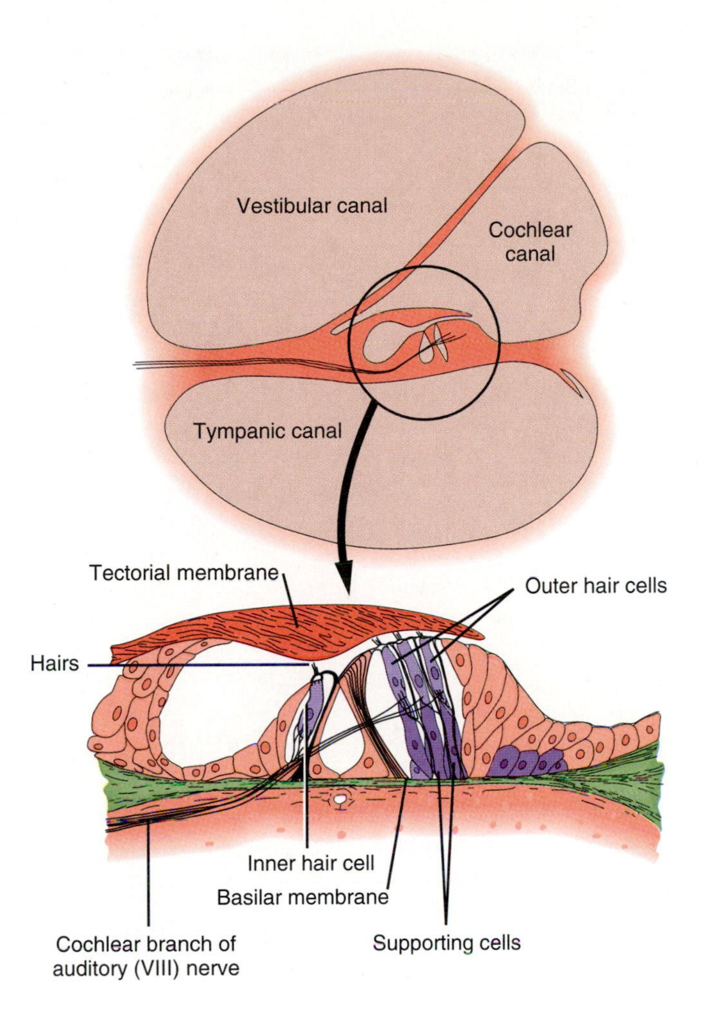

FIGURE 8.16

Cross section of one turn of the cochlea. Pressure bends the basilar membrane and the organ of Corti on it, causing the tectorial membrane to deflect the hair cells. The hair cells then evoke action potentials in axons of the auditory (= vestibulocochlear) nerve. The greater the deflection of the hair cells, the higher the frequency of action potentials and the greater the perceived loudness.

■ **Question:** How does loud sound at a particular pitch cause deafness for that particular pitch?

tic membrane receptor, this interaction triggers action potentials by changing membrane permeability. In order to detect a large variety of substances, several types of chemoreceptor are needed, each specific for a particular type of substance. Different animals have chemoreceptors that respond to different substances. Bees are indifferent to urine, and dogs do not go around sniffing flowers. Animals also differ in the sensitivity of their chemoreceptor organs. Most dogs, for example, have a sense of smell so keen that they trust their noses and ignore visible tracks.

In mammals the detection of airborne chemicals is called **olfaction.** Chemoreceptors for olfaction lie in the **olfactory epithelium** on the roof of the nasal cavity. Short axons conduct action potentials from the olfactory epithelium to the two **olfactory bulbs** in the brain, where the information is analyzed (Fig. 8.17 on page 170). Neural pathways from the olfactory bulbs go to the deeper areas of the brain that are related to emotional and survival reactions, which may explain why odors often evoke strong reactions that are more emotional than cognitive. Most mammals, as well as amphibians and reptiles, also have another set of olfactory receptors in the **vomeronasal organ** (= Jacobson's organ) in the roof of the mouth. The vomeronasal organ responds mainly to pheromones, which are chemicals in urine, sweat, and other secretions that are important in social interactions. The vomeronasal organ of humans is generally thought to be degenerate.

The other type of chemoreception in humans is **taste. Taste cells** are located throughout the mouth, but primarily in **taste buds** clustered on **taste papillae** on the

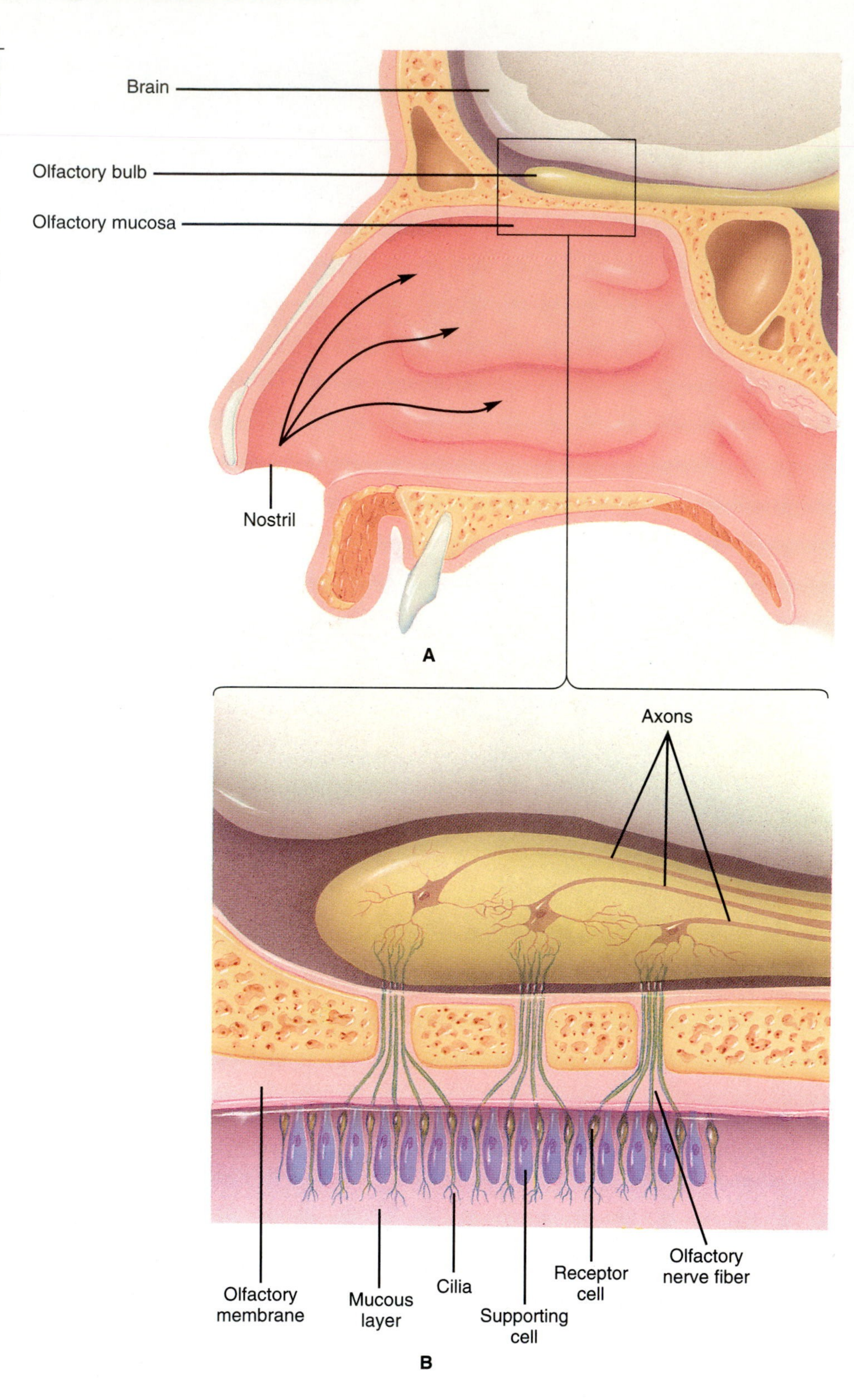

FIGURE 8.17

(A) Location of the olfactory epithelium and the olfactory bulb of the human brain. (B) Structure of the olfactory cells. The numerous cilia are embedded in a mucous layer; their function is unknown. Odorant molecules carried through the mucus by odorant-binding protein interact with plasma membranes, which then generate action potentials that travel the short distance to the olfactory bulb.

tongue (Fig. 8.18). Human taste cells respond to four primary taste qualities—sour, salt, sweet, and bitter. Pain receptors in taste buds also contribute to the taste of onions, pepper, mustard and other "hot" foods. How a substance tastes depends on its molecular properties. Acidic substances release H^+, which enters ion channels on certain taste cells, giving rise to a sour taste. The salt taste appears to arise from the influx

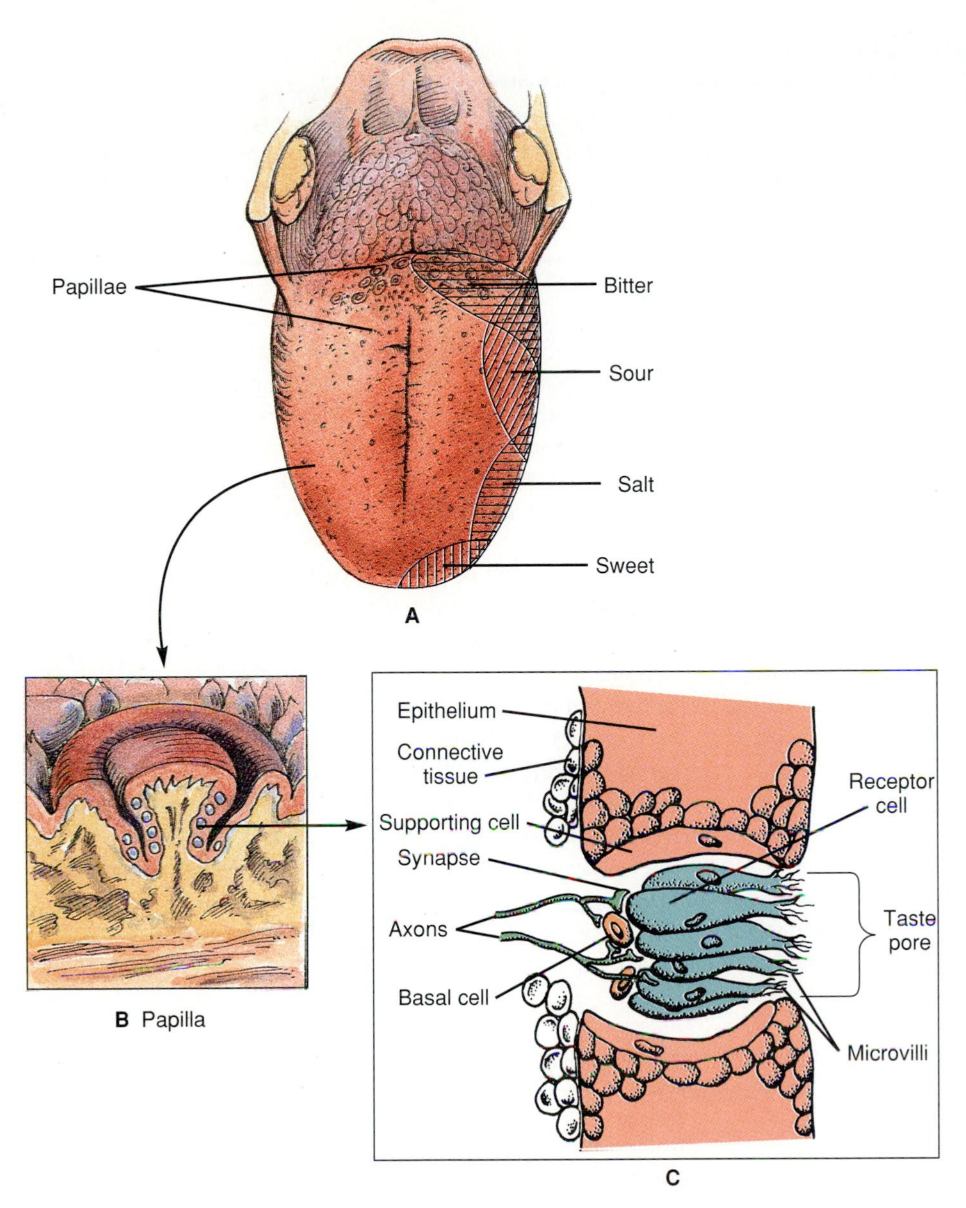

FIGURE 8.18

(A) Taste papillae on the human tongue. Different areas of the tongue differ in sensitivity to different tastes. (B) One papilla, showing several taste buds. (C) One of the 9000 taste buds in the human tongue. Each taste bud contains approximately 100 taste cells (receptor cells and basal cells) that respond to dissolve chemicals. Synaptic interactions among the taste cells generate action potentials in the axons, which signal one of the primary taste qualities.

of Na^+ and other ions through channels. Sugars bind to molecular receptor molecules. Bitter-tasting substances (often plant toxins) work by a variety of mechanisms. The main function of taste may be to inform the animal that what it is about to eat contains needed calories or minerals (sweet or salt), or is acidic or toxic (sour or bitter). In this latter role, taste buds serve a function analogous to that of human tasters who used to sample the food of kings and queens to be sure it had not been poisoned. Like many tasters who served unpopular rulers, taste cells survive for less than two weeks.

Electroreceptors and Magnetoreceptors

The most recently discovered receptor type is the electroreceptor, which is found in all cartilaginous fishes (such as sharks and rays), many bony fishes, some amphibians, and egg-laying mammals (the platypus and echidnas). Electroreceptors enable these animals to locate prey, navigate by means of electric fields generated by ocean currents, avoid solid objects that deflect electric fields, or communicate with members of their own species by means of weak electrical signals (see pp. 776–777). Electroreceptors are so sensitive in some fishes that they can detect the electric field produced by the muscles of prey buried in sand. The organs responsible for sensitivity to magnetic fields are generally unknown, although crystals of magnetite occur in many animals. The main evidence for the existence of magnetoreceptors is that some birds, insects, salamanders, and perhaps mammals use the Earth's magnetic field to navigate during homing and migration (see pp. 412–413).

Summary

Nerve cells rapidly communicate information from sensory receptors to other nerve cells and to muscles and glands. This information is in the form of action potentials: reversals of membrane voltage that last only a few milliseconds. Action potentials are triggered by changes in the permeability to ions, mainly by an increase and then a decrease in the permeability of voltage-sensitive sodium channels. Various chemical substances, such as local anesthetics and neurotoxins, alter the generation and conduction of action potentials. Communication from one neuron to another is most often by release of chemical transmitters from synapses. The transmitters change the permeabilities of postsynaptic membranes, producing postsynaptic potentials that either excite or inhibit action potentials. Synaptic functioning is altered during learning and by a variety of chemicals.

Information from sensory receptors is encoded as patterns of action potentials in sensory neurons. The intensity of stimulation is encoded as frequency of action potentials from tonic receptors, the time of stimulation is signaled by brief bursts of action potentials from phasic receptors, and the modality, quality, and location are encoded by which neurons conduct the action potentials. Sensory receptors vary in complexity. Some, such as the crayfish stretch receptor, produce a receptor potential that directly excites an afferent axon. In other receptor cells, such as rods and cones of the vertebrate retina, the receptor potential first affects release of synaptic transmitter onto other nerve cells, which then interact in a complex way before exciting afferent cells.

Receptor cells are generally parts of receptor organs, which can be classified according to modality. Photoreceptor organs, such as eyes, respond to electromagnetic radiation that is visible to humans, and some respond to infrared or ultraviolet that can be detected in other species. Somesthetic receptors include thermoreceptors that respond to temperature, nociceptors that respond to painful stimuli, and mechanoreceptors that respond to such stimuli as touch, pressure, and sound. Vertebrate mechanoreceptors include the vestibular apparatus, which senses head orientation and movement, and the cochlea, which is responsible for hearing. Chemoreceptors respond to chemical stimulation, and in humans they include the cells responsible for olfaction and taste.

Key Terms

resting membrane potential
depolarization
hyperpolarization
action potential
axon
all or none law
threshold
conduction
myelin
synapse
synaptic transmitter
synaptic knob
postsynaptic potential
receptor potential
tonic
phasic
rhodopsin
vestibular apparatus
cochlea

Chapter Test

1. Explain why there are unequal concentrations of K^+ and Cl^- on each side of the plasma membrane. (p. 153)
2. Explain how the unequal distribution of K^+ and Cl^- across a membrane gives rise to a voltage across the membrane. (p. 153)
3. Which ion is most important in initiating an action potential? Explain. (pp. 154–155)
4. Why is the speed of axonal conduction important? Describe two adaptations that increase the speed of conduction. (p. 156)
5. All stages of neural functioning depend on changing membrane voltages by changing the permeability to one or more ions. Illustrate this statement by explaining excitatory and inhibitory synapses. (p. 157)
6. List all the ways you can imagine by which the function of chemical synapses could be modified by either learning or drugs. (pp. 157–161)
7. Imagine a pin prick on the back of your left hand. Explain in your own words how your brain knows (1) the intensity of

the stimulus, (2) when the stimulus is occurring, and (3) that the stimulus is cutaneous, that it is a sharp pain rather than pressure, and that it is located on the back of your left hand. (pp. 161–162)

8. Explain how the crayfish stretch receptor works. (pp. 162–163)
9. Explain how hair cells can respond differently, depending on whether they are in the semicircular canals, the otolith organs, or the cochlea. (p. 167)

Answers to the Figure Questions

8.2 In a solution with high concentration of K^+ and low concentration of Cl^- like those inside the cell, one might assume that neither K^+ nor Cl^- would diffuse. If the concentrations of Na^+ and A^- are unchanged, however, K^+ must diffuse into the cell, and Cl^- must diffuse out of the cell until the concentration ratio is the same as shown in the figure. That is the nature of equilibrium: It reestablishes itself after disturbance.

8.5 The falling phase of the action potential would be more gradual, and there would be no hyperpolarization.

8.8 It is a familiar fact that the nervous system adapts to a familiar stimulus and eventually ignores it. The equivalent in this figure would be a gradual decline in the responsiveness of neuron A until it no longer evokes activity in C.

8.9 The quiet room is represented by the extreme left of the *x* axis, where a small increase in intensity evokes a large change in the frequency of action potentials. At the right end of the *x* axis, representing a noisy room, the same increase in intensity does not evoke a large change in frequency of action potentials.

8.10 For the tonic receptor there would be a steady increase in frequency. For the phasic receptor the frequency would be constant during the stretch.

8.13 Merkel's corpuscles are unguarded and at the skin surface, and they probably respond to light touch. Meissner's corpuscles are somewhat deeper and are within a capsule, so they probably respond to more intense touch. Ruffini corpuscles are still deeper and within a thicker capsule, which explains why they respond only to heavy continuous touch.

8.16 The hair cells that are most intensely stimulated by a particular pitch may be damaged by the movement of the tectorial membrane. Thereafter they would be unable to respond to that pitch.

Readings

Recommended Readings

Alkon, D. L. 1989. Memory storage and neural systems. *Sci. Am.* 261(1):42–50 (July).

Barondes, S. H. 1993. *Molecules and Mental Illness.* New York: Scientific American Books.

Changeux, J-P. 1993. Chemical signalling in the brain. *Sci. Am.* 269(5):58–62 (Nov).

Cronin, T. W., N. J. Marshall, and M. F. Land. 1994. The unique visual system of the mantis shrimp. *Am. Sci.* 82:356–365.

Dunant, Y., and M. Israël. 1985. The release of acetylcholine. *Sci. Am.* 252(4):58–66 (Apr).

Gottlieb, D. I. 1988. GABAergic neurons. *Sci. Am.* 258(2):82–89 (Feb).

Hawryshyn, C. W. 1992. Polarization vision in fish. *Am. Sci.* 80:164–175.

Hudspeth, A. J. 1983. The hair cells of the inner ear. *Sci. Am.* 248(1):54–64 (Jan).

Jacobs, B. L. 1987. How hallucinogenic drugs work. *Am. Sci.* 75:386–391.

Jacobs, B. L. 1994. Serotonin, motor activity and depression-related disorders. *Am. Sci.* 82:456–463.

Koretz, J. F., and G. H. Handelman. 1988. How the human eye focuses. *Sci. Am.* 259(1):92–99 (July).

Levine, J. S., and E. F. MacNichol, Jr. 1982. Color vision in fishes. *Sci. Am.* 246(2):140–149 (Feb).

Loeb, G. E. 1985. The functional replacement of the ear. *Sci. Am.* 252(2):104–111 (Feb).

Masland, R. H. 1986. The functional architecture of the retina. *Sci. Am.* 255(6):102–111 (Dec).

McLaughlin, S., and R. F. Margolskee. 1994. The sense of taste. *Am. Sci.* 82:538–545.

Morell, P., and W. T. Norton. 1980. Myelin. *Sci. Am.* 242(5):88–116 (May).

Neher, E., and B. Sakmann. 1992. The patch clamp technique. *Sci. Am.* 266(3):44–51 (Mar). (*Studies of the acetylcholine receptor.*)

Newman, E. A., and P. H. Hartline. 1982. The infrared "vision" of snakes. *Sci. Am.* 246(3):116–127 (Mar).

Nilsson, D-E. 1989. Vision optics and evolution. *BioScience* 39:298–307.

Parker, D. E. 1980. The vestibular apparatus. *Sci. Am.* 243(5):118–135 (Nov).

Poggio, T., and C. Koch. 1987. Synapses that compute motion. *Sci. Am.* 256(1):46–52 (May). (*Motion detectors in the retina.*)

Schnapf, J. L., and D. A. Baylor. 1987. How photoreceptor cells respond to light. *Sci. Am.* 256(4):40–47 (Apr).

Snyder, S. H. 1986. *Drugs and the Brain.* New York: Scientific American Library.

Stryer, L. 1987. The molecules of visual excitation. *Sci. Am.* 257(1):42–50 (July).

Van Dyke, C., and R. Byck. 1982. Cocaine. *Sci. Am.* 246(3):128–141 (Mar).

Additional Reference

Mammano, F., and J. F. Ashmore. 1993. Reverse transduction measured in the isolated cochlea by laser Michelson interferometry. *Nature* 365:838—841.

CHAPTER

9

Nerve Systems

Burrowing owl *Athene cunicularia.*

Neural Subsystems

No sensible person would claim to understand an entire culture after having met only one individual from that culture. Likewise, we should not presume that our consideration of individual cells in the previous chapter provides an understanding of an entire nervous system. Scientists have worked out the complete "wiring" diagram of the 302 neurons and 8000 synapses in the nematode *Caenorhabditis elegans,* yet no one pretends to completely understand this little worm's behavior. Within nervous systems, however, there are many subsystems consisting of relatively few nerve cells that can be understood. Examples of such subsystems are reflex mechanisms, servomechanisms, and central pattern generators.

REFLEXES. A reflex is a stereotyped response due to a relatively simple connection from receptors to muscles or glands (Table 9.1). In the simplest reflexes, called **monosynaptic reflexes,** afferent neurons from receptors activate efferent neurons to muscles or glands through only one synaptic connection. The best-known example of a monosynaptic reflex is the **withdrawal reflex,** which pulls a limb away from injury (Fig. 9.1 on page 176). This neural pathway begins with nociceptors (pain receptors), which excite afferent neurons. In vertebrates the axons of the afferent neurons are in a nerve that connects to the spinal cord. Within the spinal cord, branches from the afferent axons have excitatory synapses onto motor neurons that go to **flexor** muscles of the injured limb. These muscles pull the limb toward the body and usually away from the cause of the pain. At the same time, other branches inhibit the motor neurons to the **extensor muscles** of the same limb. Still other branches send action potentials to the opposite side of the spinal cord to excite the extensor muscles and inhibit the flexor muscles, so that the other limb can support the body. All this happens before the brain knows about it. In fact, these reflexes can occur even without a brain.

SERVOMECHANISMS AND MUSCLE CONTROL. Once triggered, reflexes occur without further modification. Servomechanisms, in contrast, are neural subsystems

TABLE 9.1

Some reflexes of vertebrates.

Withdrawal reflex	Withdraws limb from injury. See text for further description.
Grasping reflex	In infant primates the fingers and toes flex in response to pressure on the palm or sole of the foot. This presumably helps infants grasp limbs and the mother's hair.
Rooting and sucking reflexes	In infant mammals a touch on the cheek causes the infant to move its mouth toward the object. This rooting reflex allows the infant to find the mother's nipple. Contact on the lips initiates sucking.
Swallowing reflex	Pressure on the soft palate at the back of the mouth triggers contraction of the muscles involved in swallowing.
Vestibulo-ocular reflex	When the semicircular canals sense rotation of the head, the eyes rotate in the opposite direction. This occurs even in the dark and keeps the gaze fixed on one spot even when the head is moving.

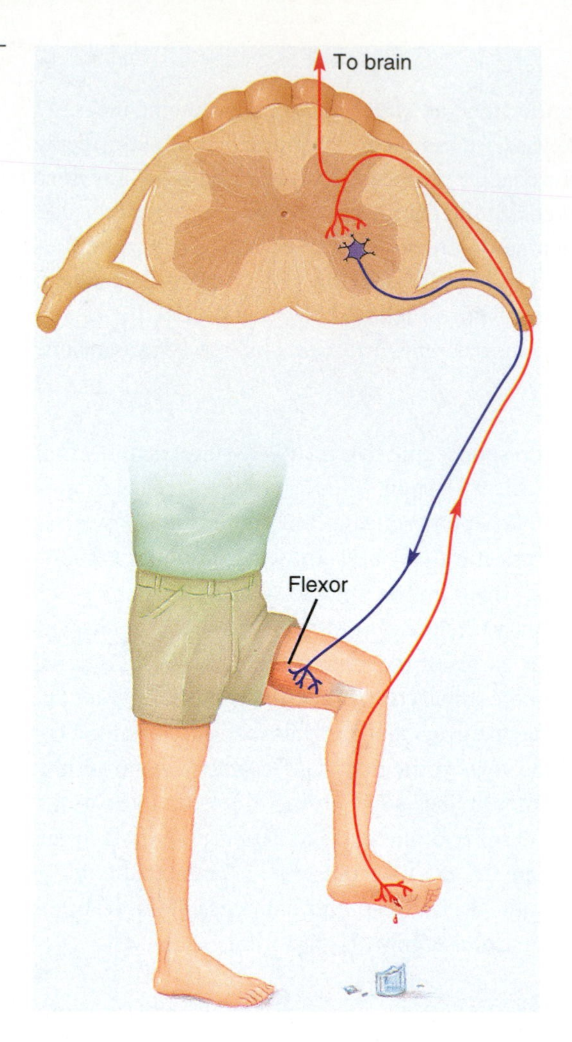

FIGURE 9.1
Neural pathway for the withdrawal reflex in a human.

■ **Question:** Add pathways that would excite the extensors and inhibit the flexors on the right leg.

that continually modify their output. As an example of a servomechanism, consider the fact that you can hold your hand out in a steady position without looking at it or thinking about it. The neural subsystem responsible for this ability begins with stretch receptors called **muscle spindles** in your skeletal muscles (Fig. 9.2). Muscle spindles respond to change in muscle length and therefore to change in posture. Each gram of muscle has from several to hundreds of spindle receptors. Each spindle has muscles of its own at each end, and the receptor portion is in the middle. Stretch on the receptor portion triggers action potentials in afferent axons to the spinal cord. This stretch can be due either to contraction of the muscle in which the spindle occurs or to contraction of the muscles within the spindle due to voluntary activation.

Figure 9.2 shows how the spindle receptor organ enables you to hold out your hand steadily. In the spinal cord the afferents from muscle spindles excite the motor neurons to the muscles in which the spindles occur. If the hand starts to sag under the pull of gravity, the muscle spindles will be stretched. This will increase the frequency of action potentials in the motor neurons, which will cause the muscle to contract with more force, restoring the hand to its original position. This mechanism maintains constant position even if someone unexpectedly drops a weight in your hand.

Central Pattern Generators. A third type of elementary subsystem is the central pattern generator, which consists of neurons that produce patterns of action

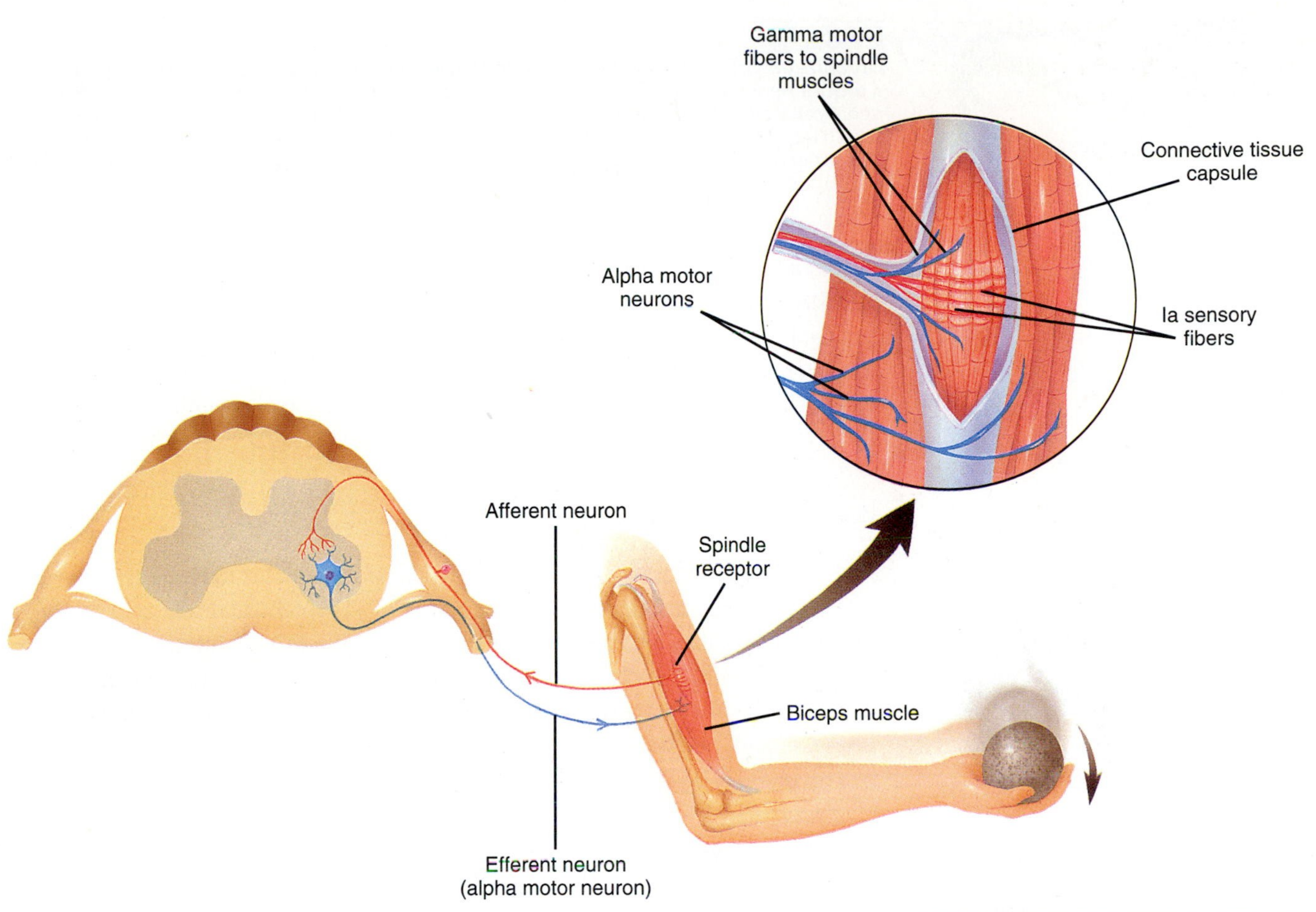

FIGURE 9.2

The spindle receptor organ is part of a servomechanism that maintains limb position. The receptor cells include the spiral endings of Ia afferent axons. Stretch applied to these endings generates action potentials in the Ia afferents. At each end of the spindle are muscle fibers controlled by small motor neurons called gamma efferent fibers. If the biceps muscle starts to stretch because of the weight of the lower arm, the spindle receptor generates action potentials in the Ia afferent, which excites the alpha motor neuron to the biceps.

potentials that do not depend on sensory input. Central pattern generators are often responsible for rhythmic behaviors such as breathing and walking. They include at least one nerve cell that produces action potentials spontaneously. Associated neurons may then shape the pattern of action potentials.

Nervous Systems of Invertebrates

These and other elementary subsystems are combined in a variety of ways in different animals. It is difficult to generalize about nervous systems except to note that the more complex they are, the more complex the behaviors they produce. Sponges, which do not have any nerve cells, spend their adult lives attached to substratum, hardly behaving at all. *Hydra,* jellyfishes, and sea anemones (phylum Cnidaria) can detect prey and actively avoid predators. Their nerve cells form a diffuse **nerve net** in which stimulation at one point sends electrical activity radiating throughout the system (Fig. 9.3 on page 178). They have no collection of nerve cells that would justify the term **central nervous system** (CNS). Only in bilaterally symmetric animals are there concentrations of cell bodies and synaptic connections that constitute a central nervous system. Central nervous systems greatly increase the combinations of interactions of nerve cells and thus allow more complex and varied behaviors. Even the simplest CNS, such as that of a flatworm (phylum Platyhelminthes), can perform complex behaviors such as learning to negotiate a maze.

Often the largest part of the invertebrate CNS is the **cerebral** (= cephalic) **ganglion** in the head, which handles sensory input from eyes, antennae, and other anterior re-

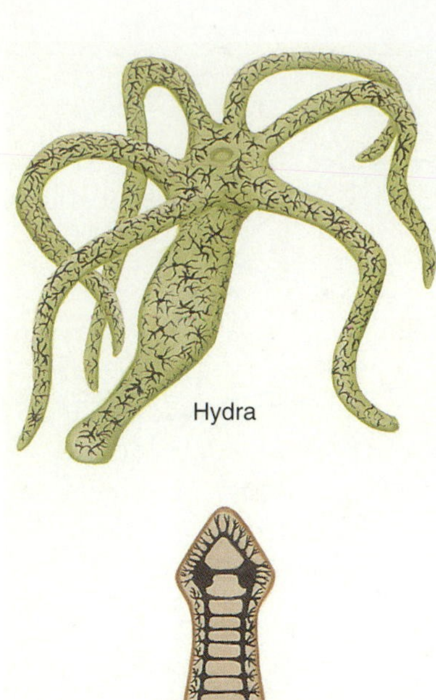

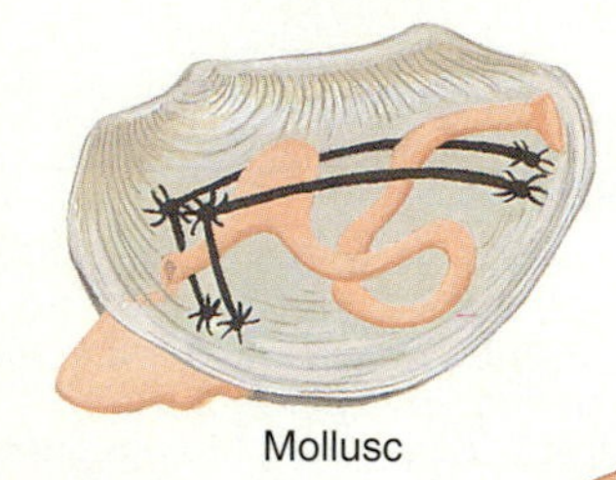

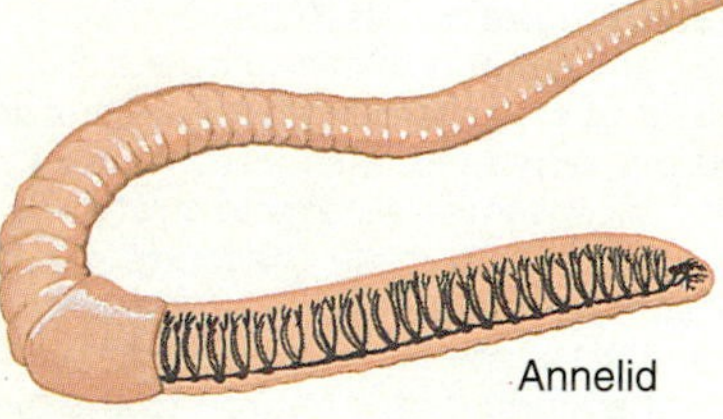

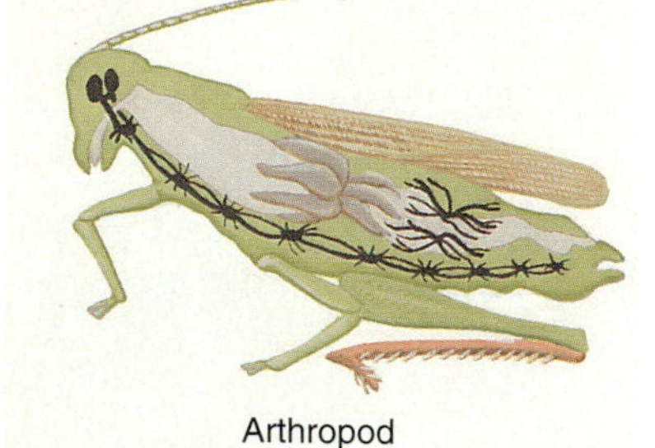

FIGURE 9.3

Organization of nervous systems in various invertebrates. *Hydra* and other cnidarians have a diffuse nerve net. Most invertebrates have a central nervous system consisting of knots of nerve cell bodies and synapses called ganglia, which are joined by nerves called connectives. Flatworms have two cerebral ganglia and two rows of smaller ganglia interconnected by connectives. Molluscs have several ganglia joined by connectives in a loop. Annelids and arthropods have a chain of ganglia along the ventral midline.

ceptors. Cerebral ganglia of invertebrates are often called "brains," but this usage can be misleading, because the cerebral ganglion is generally only one of several important ganglia. In most invertebrates, each ganglion controls one part of the animal, and the ganglia are interconnected by nerves called **connectives.** Connectives are handy for research, since it is easy to record action potentials in the axons within them, and they can be cut to isolate ganglia from each other.

Vertebrate Nervous Systems

THE AUTONOMIC NERVOUS SYSTEM. The most complex nervous systems occur in vertebrates such as ourselves, and the rest of this chapter will be devoted to general concepts of how they function. The central nervous system, consisting of the brain and spinal cord, communicates with the rest of the body by means of afferent and efferent axons in nerves that make up the **peripheral nervous system.** An important part of the vertebrate peripheral nervous system is the autonomic nervous system, which exercises involuntary control over digestion, circulation, breathing, copulation, and some other functions. The autonomic nervous system has two divisions, the sympathetic and the parasympathetic. Each division releases a different **neurosecretion** from synapselike nerve endings. The **sympathetic division** tends to be active during stress ("fight or flight" situations; see pp. 144–145). It secretes norepinephrine, which inhibits digestion and sexual arousal and stimulates circulation and breathing. The sympathetic division originates in a chain of ganglia on each side of the middle portion of the spinal cord (Figs. 9.4 and 9.5). The **parasympathetic division** tends to be active during rest. Its neurosecretion is acetylcholine, which stimulates digestion and sexual arousal and inhibits circulation and breathing. The parasympathetic division originates at both ends of the spinal cord. The functions of these two divisions in regulating organ systems will be detailed in later chapters in this unit.

THE SPINAL CORD. The central nervous system consists of the brain and spinal cord, which is enclosed within the vertebrae and is the origin of most of the nerves in the peripheral nervous system (Fig. 9.5 on page 180). Near the spinal cord each nerve splits into two branches: the **dorsal root** and the **ventral root.** In mammals almost all afferent axons enter the spinal cord via the dorsal root, and all efferent axons leave by the ventral root. Afferent neurons have their cell bodies in the dorsal root ganglia, and efferent neurons have cell bodies in the central **gray matter** of the spinal cord. In the gray matter are also found synaptic connections among neurons (such as those involved in reflexes). Surrounding the gray matter is **white matter,** consisting of myelinated axons connecting various levels of the spinal cord and the brain.

THE BRAIN. The brains of all vertebrates develop from three embryonic divisions: the **forebrain** (= prosencephalon), the **midbrain** (= mesencephalon), and the **hindbrain** (= rhombencephalon). These three divisions begin development with similar pro-

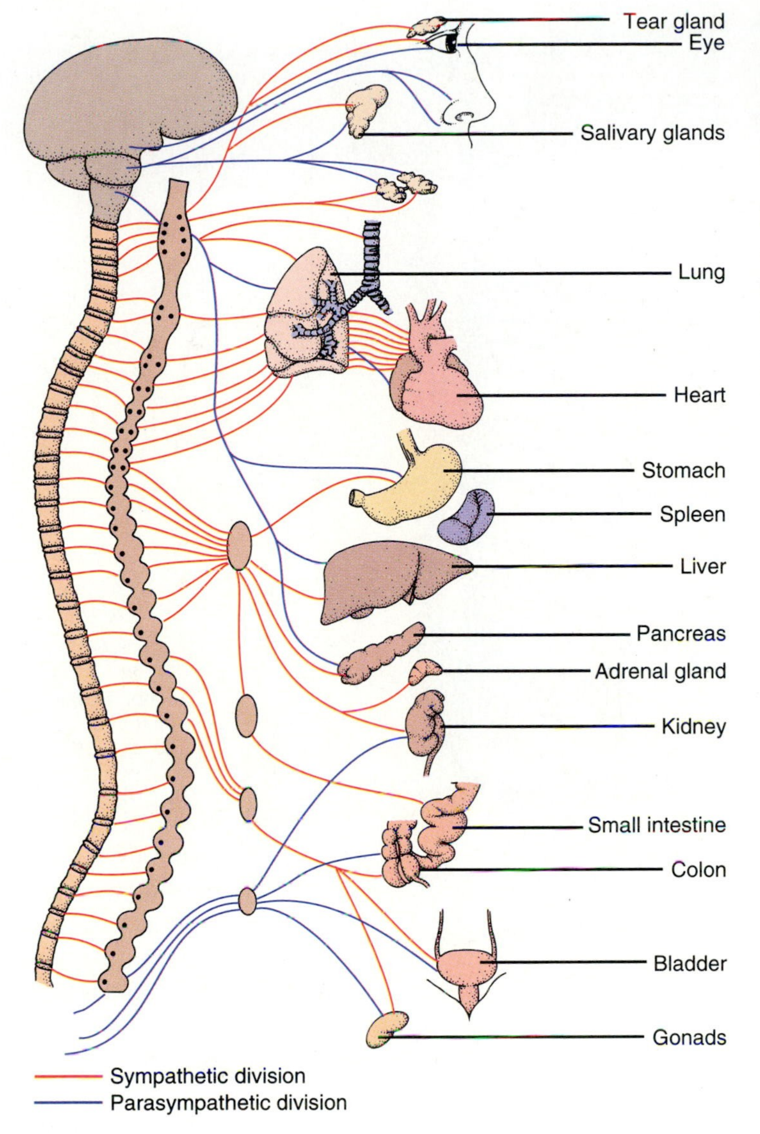

FIGURE 9.4

Schematic diagram of the human autonomic nervous system showing the organs it affects. The spinal cord is shown to the left of the chain of sympathetic ganglia.

■ **Question:** Why would severe fright cause a dry mouth, rapid heart beat, and perhaps urination or defecation? What effect do you think the parasympathetic branch would have on each of the organs shown?

portions in all vertebrates, but the parts grow at different rates in different groups of vertebrates (Fig. 9.6 on page 181). In general, the mass of the brain increases with the mass of the body, and birds and mammals have larger brains than fishes, amphibians, and reptiles of the same body mass (Fig. 9.7 on page 181). Naturally the largest brains occur in the huge mammals, such as whales and elephants. Together with porpoises, however, humans get about seven times as much brain as we deserve based on body mass.

In birds and mammals, the forebrain grows disproportionately large and gives rise to the **cerebrum** (Fig. 9.8 on p. 182). The cerebrum is divided into two cerebral hemispheres. Its outer layer, the **cerebral cortex,** is responsible for perception and conscious behavior in mammals, as will be described later. In mammals, and especially in humans, the cerebrum grows so large that its surface seems to crumple, and it seems to spill over the sides. The hindbrain develops into the **cerebellum,** which coordinates movements. The cerebellum is large in birds relative to other vertebrates, presumably because of the greater demands for coordination in flight. All three embryonic divisions contribute to

FIGURE 9.5

(A) The position of the spinal cord in a human showing the many nerves branching to each side. (B) A cross section of the spinal cord. The spinal nerves split into the dorsal and ventral roots at the spinal cord.

■ **Question:** Describe a spinal injury that could cause loss of sensation, but not paralysis, from the waist down.

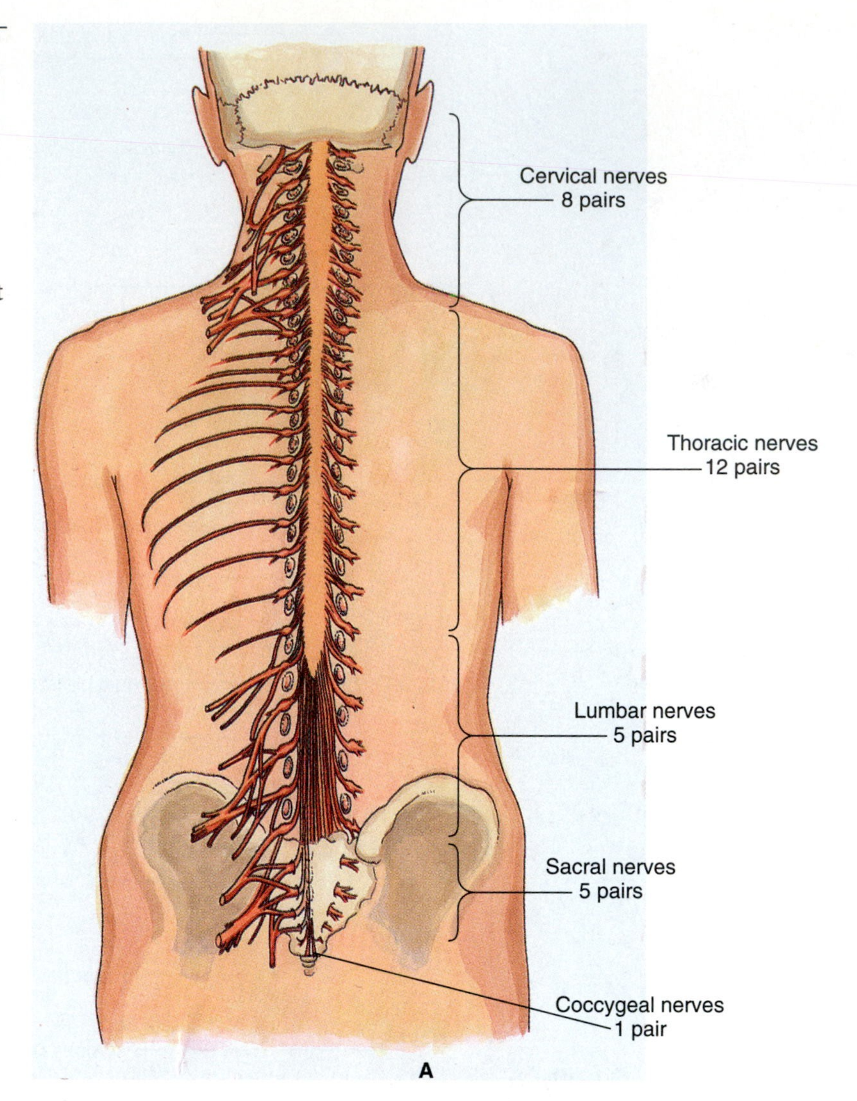

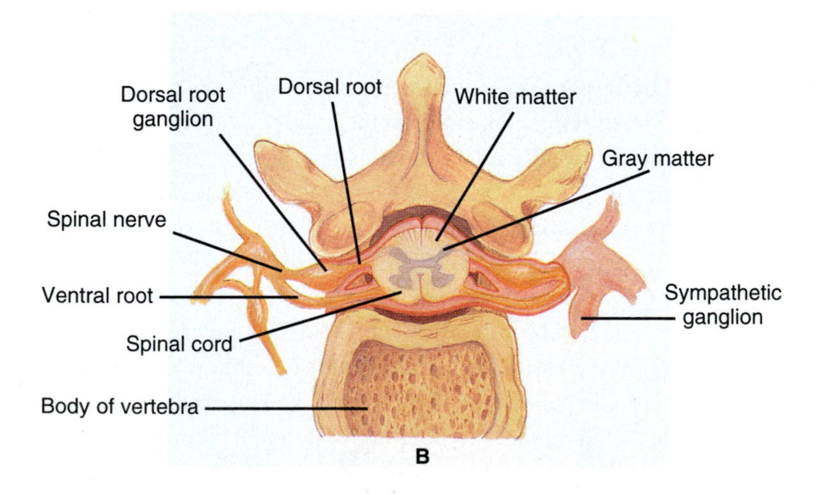

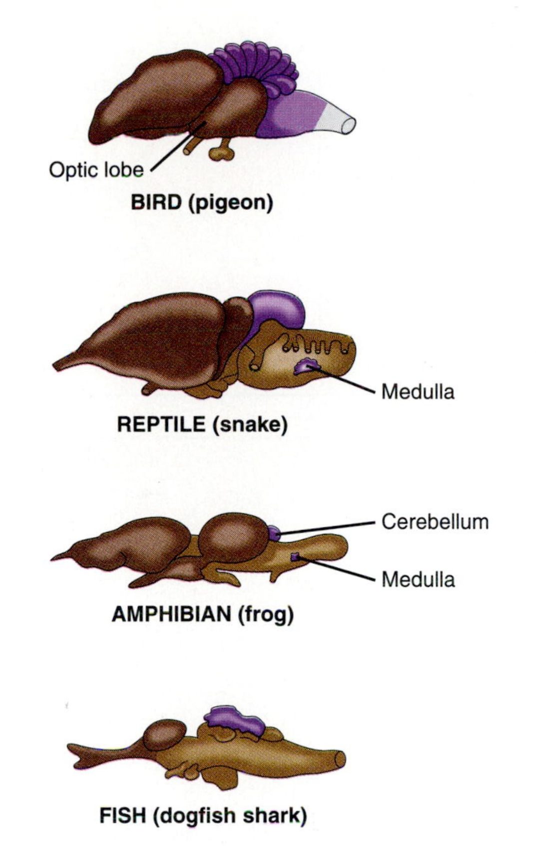

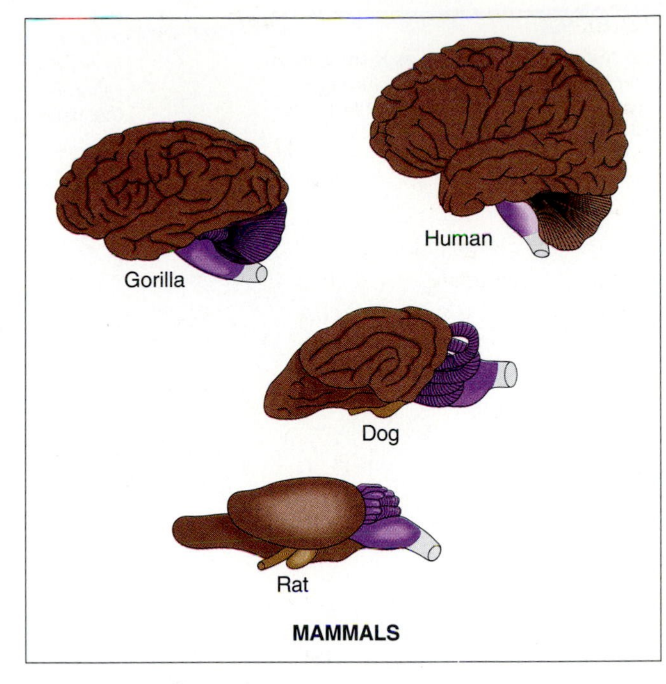

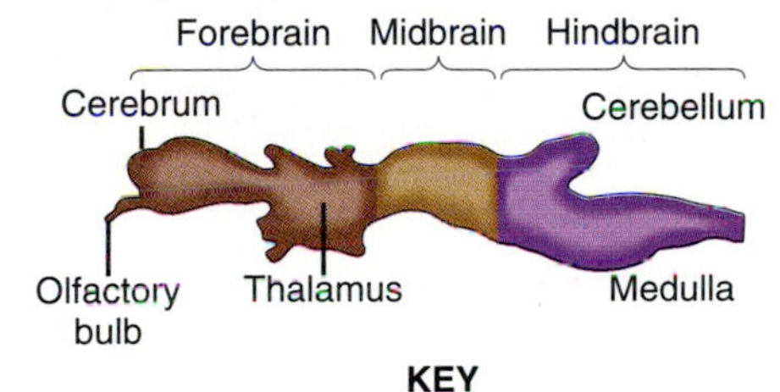

FIGURE 9.6
Representative brains from several vertebrate classes drawn on a scale that emphasizes the relative size of different parts, rather than the actual size.

the **brainstem,** which is the part of the brain that would be left after removing the cerebrum and cerebellum.

Surrounded by the cerebrum at the upper part of the brainstem is the **thalamus.** Much of the thalamus consists of **nuclei:** clusters of cell bodies and synapses that process information to and from the cerebrum. For example, the lateral geniculate nuclei process information from the retinas and forward it to the visual areas of the cerebrum. Just beneath the thalamus is the **hypothalamus.** In addition to regulating secre-

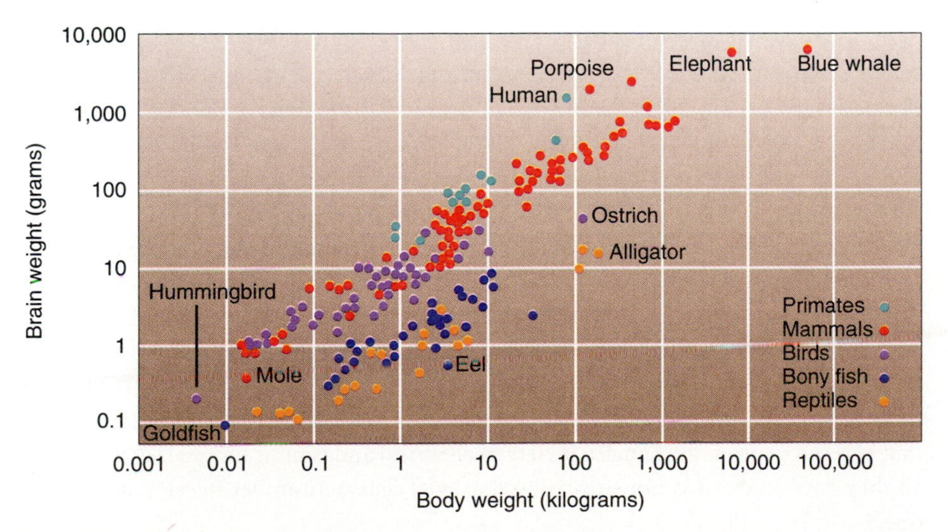

FIGURE 9.7
The relationship of brain mass to body mass in vertebrates. Birds and mammals generally have larger brains than fishes and reptiles of the same body size. The graph of brain mass to body mass for birds and mammals on this double-logarithmic plot is approximately linear, indicating that brain mass is generally proportional to (body mass)^p, where p is a number (⅔).

■ **Question:** Why does brain mass increase with body size? Are large animals necessarily smarter?

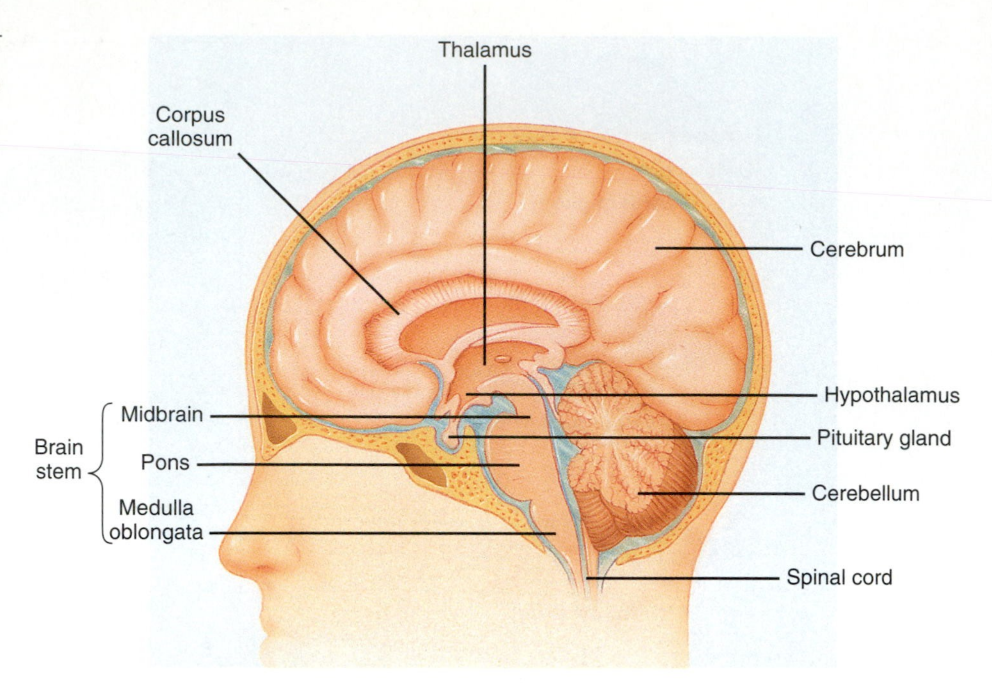

FIGURE 9.8
The human brain shown as if split down the middle. The corpus callosum is a band of axons connecting the two cerebral hemispheres.

tion from the pituitary gland (pp. 141, 144), the hypothalamus helps regulate feeding, water and ion balance, and body temperature.

Farther down the brainstem is the **tectum,** which coordinates reflex movements of the head and eyes toward a sound or a movement detected by the retina. Below it is the **pons** (Latin for bridge) with horizontal axons that bridge the two halves of the brain. The pons helps with orientation toward visual stimuli and also helps regulate breathing. A little farther down is the **medulla oblongata,** which regulates circulation and breathing and which triggers the more-or-less vital functions of coughing, sneezing, swallowing, and hiccupping.

Internal Environment of the Brain and Spinal Cord

Unlike the peripheral nervous system, which is necessarily exposed to all kinds of dangers as it carries messages to and from various bodily outposts, the central nervous system is a well-guarded command center. Nerve cells in the brain and spinal cord are cushioned against mechanical and chemical shocks by cerebrospinal fluid and glial cells. **Cerebrospinal fluid** (CSF) circulates water, ions, nutrients, and other small, soluble molecules around the brain and spinal cord and through large cavities in the brain, called **ventricles.** CSF is a filtrate of blood formed by tissues lining the ventricles. Nonneural **glial cells** help regulate the composition of CSF. The glial cells called **astrocytes** transport glucose and other nutrients from the blood into the CSF. Astrocytes also prevent the entry of many other substances from blood into the CSF, thereby forming a **blood–brain barrier.** They also remove excess ions and transmitters. During development other glial cells (**oligodendroglia**) help axons reach their destinations and form myelin sheaths around axons (see p. 156). A third category of glial cells, called **microglia,** ward off infection and repair injury. Glial cells have generally been thought to serve only in such supporting roles for neurons, but recent research shows that glia and neurons exchange a variety of electrical and chemical signals. Considering all they have to do, it is not surprising that glial cells outnumber nerve cells nine-to-one.

Motor Functions of the Cerebrum and Cerebellum

Even the simplest vertebrate brain is too complex to understand in any detail, but the general role of different areas is known fairly well from studies on nonhuman vertebrates, from the effects of strokes and other brain injuries in humans, and from modern imaging techniques. The cerebral cortex is largely divided into sensory and motor areas (Fig. 9.9). The motor cortex generates commands for voluntary movement, with each area of the cortex controlling a particular part of the body, as will be described later. Patterns of action potentials encoding these commands are apparently modified en route to the efferent neurons by **basal ganglia** in the cerebrum and by the **cerebellum.** Damage to the motor cortex results in complete paralysis of a part of the body, while damage to a basal ganglion (for example, in Parkinson's disease) or the cerebellum affects motor coordination. The basal ganglia are involved in such automatic movements as swinging the arms while walking and maintaining the appropriate muscle tone for posture. The cerebellum appears to combine sensory information from the eyes, vestibular apparatus, and muscle and joint receptors to convert the motor command into a skilled movement. The cerebellum also apparently calculates one's position in the near future. Damage to one area will cause a person to continually bump into walls, even if the walls are clearly visible.

Sensory Functions of the Cerebral Cortex

Only 350,000 efferent axons control all of a person's actions, but it takes two to three million afferent axons to provide enough information for the person to know how to act. Nearly half the sensory input enters through the optic nerves, and most of the other inputs are through the spinal cord. The cerebrum is responsible for analyzing most of the information brought in through the afferent neurons. Virtually all of the rear half of the cerebral cortex (the **occipital lobe**) is involved in visual perception. Perception of sound is localized in the **temporal lobes** near the ears. Just behind the motor cortex and overlapping with it slightly is the somatosensory cortex, which is responsible for perception of taste, pain, touch, and other bodily sensations. The **parietal**

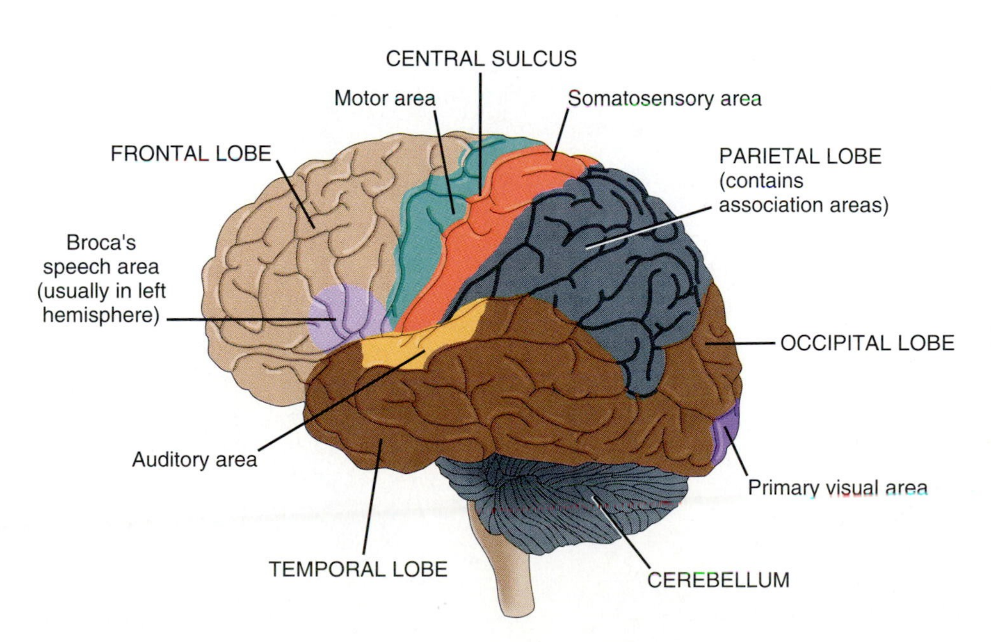

FIGURE 9.9

The location of the primary sensory and motor areas of the human left cerebral cortex. The major anatomical divisions and landmarks are indicated.

■ **Question:** You have no doubt heard that humans use only a tenth of their brains. Does this now seem likely to you? What could have been the experimental basis, if any, for such a statement?

lobe between the occipital lobe and the somatosensory area is an association area, which coordinates information from all the sensory modalities.

Although these various areas of the brain are involved in diverse and enormously complex functions, all operate according to four general concepts of brain function: lateralization, topographical organization, stimulus filtering, and distributed hierarchy.

Lateralization. In general, each cerebral hemisphere is responsible for movements and sensations in the opposite side of the body. For example, the somatosensory cortex of the left hemisphere interprets information mainly from somatic receptors of the right side of the body, and the visual cortex of the right cerebral hemisphere interprets visual stimuli seen on the left (right half of the retina) (Fig. 9.10). The **amytal test** demonstrates this lateralization dramatically. When the barbiturate amytal is injected into one carotid artery in the neck, the opposite half of the body immediately goes limp and numb.

Topographical Organization. A second generalization is that adjacent areas of the cerebral cortex handle information from adjacent groups of receptor cells. As shown in Fig. 9.10, for example, this topographical organization preserves the shape of an arrow across the two areas of primary visual cortex. Each cerebral hemisphere has about 20 other areas that also independently map visual information. In the primary auditory cortex there is a "tonotopic" mapping: Pitches that are close to each other excite areas of the auditory cortex that are close to each other. The somatosensory cortex, and also the motor cortex, is organized like a warped map of the body surface (Fig. 9.11).

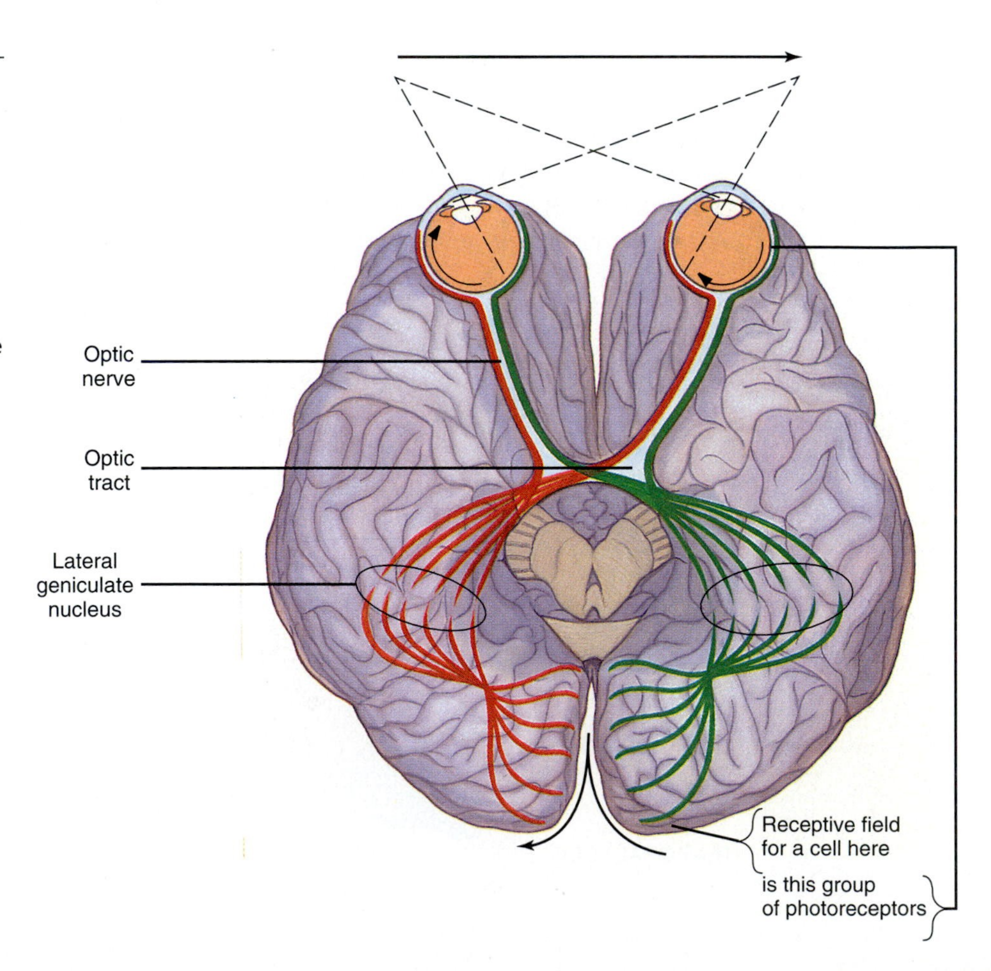

FIGURE 9.10

Lateralization of visual processing in the cerebrum of a cat. Axons in the optic nerve go to the lateral geniculate nuclei in the thalamus, which sends the information to the primary visual cortex. Because axons from the medial halves of each retina cross over to the opposite lateral geniculate nucleus and those from the lateral halves do not, each side of the brain handles visual information from the opposite side of the stimulus. Each neuron in the visual cortex responds to stimulation of a small number of retinal receptor cells, which makes up the receptive field for that neuron.

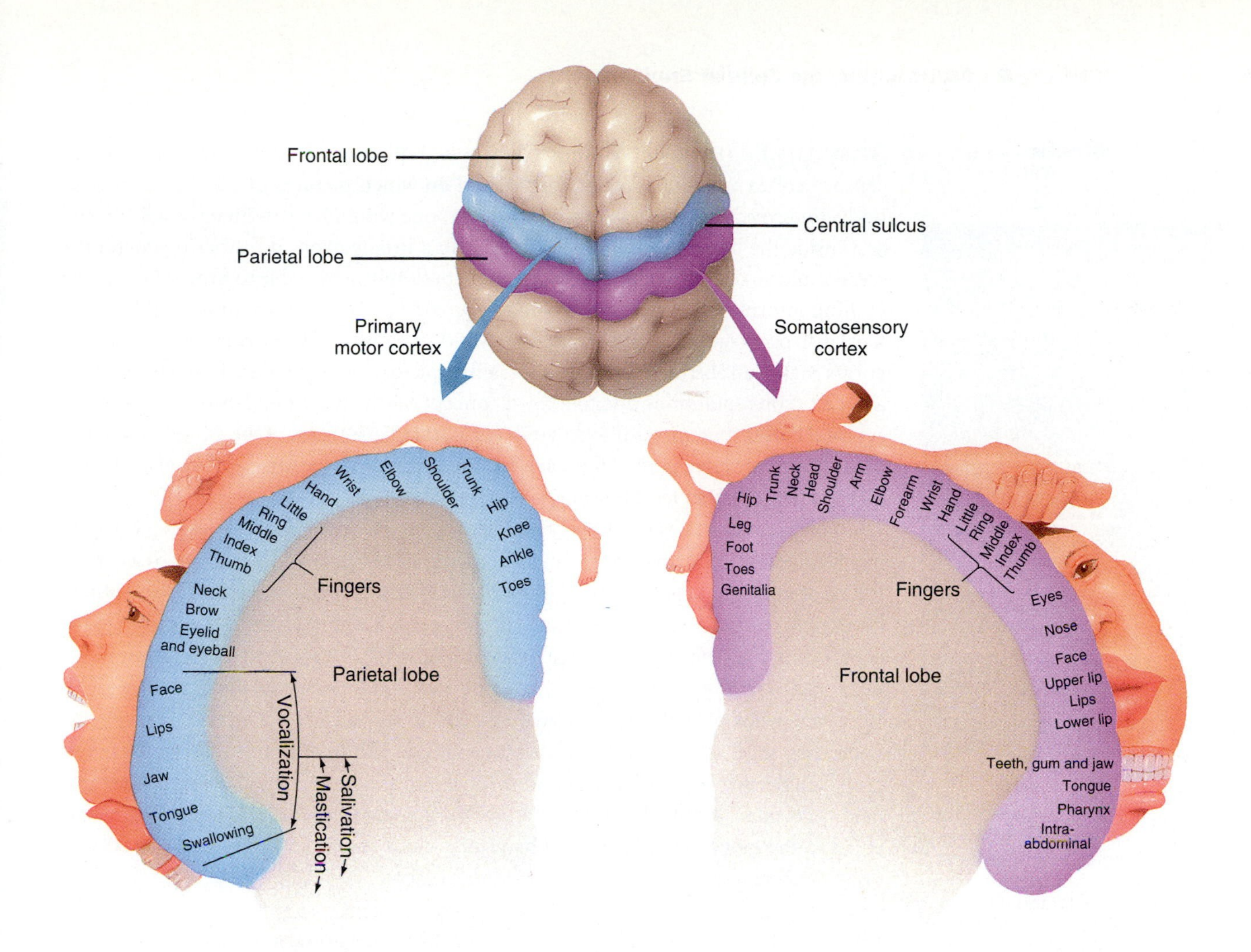

A

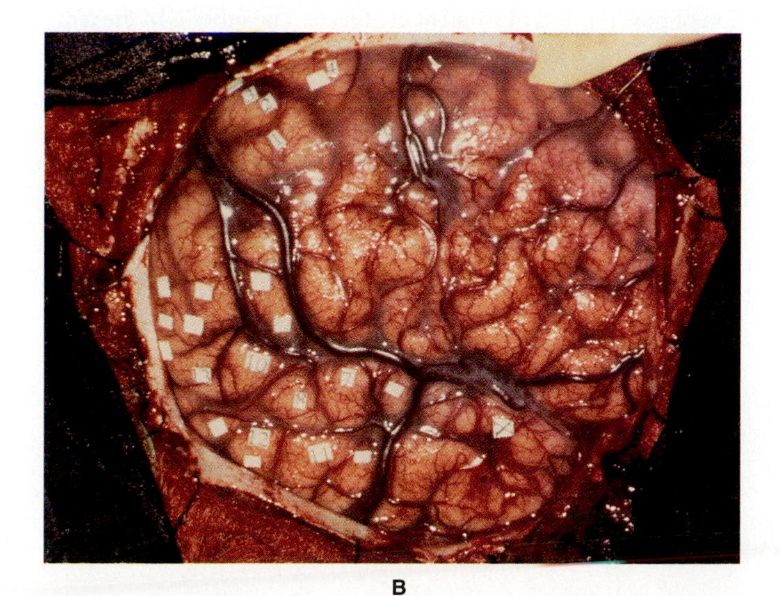

B

FIGURE 9.11

(A) The somatotopic organization of the somatosensory cortex and primary motor cortex. (B) How the somatotopic organization of the brain is determined. Here the right cerebral cortex of an epilepsy patient is exposed under local anesthesia prior to surgery. The patient is facing right, and the temporal lobe, outlined by the large veins, is at the bottom of the picture. To map the motor cortex, different areas, marked by numbered pieces of paper, are electrically stimulated, and the resulting movements are recorded by the surgeons. A similar procedure is used to map sensory areas, except that the fully awake patient reports the sensations he feels as a result of the stimulation. In this patient, stimulation of the following locations evoked the following results: (1) tingling in left thumb; (2) tingling in left ring finger; (3) tingling in left middle finger; (4) flexing of left fingers and wrist; (8 and 13) complex memories.

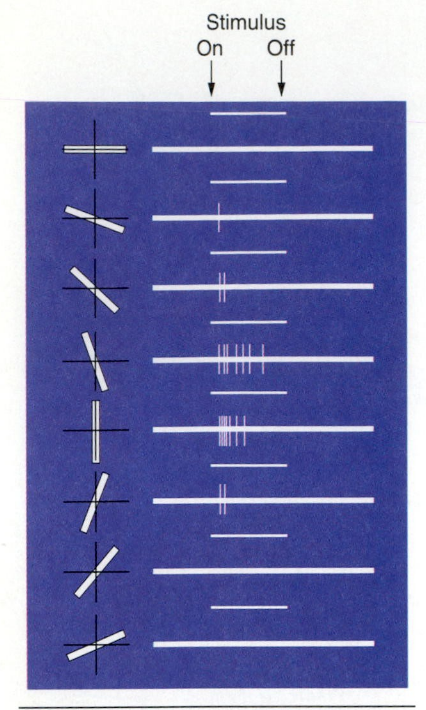

FIGURE 9.12

The response of a single neuron in the primary visual cortex of a cat to a bar of light at various angles. For this simple cell the bar of light must be in the center of the receptive field and must be oriented at a certain angle (vertical in this example) in order to evoke a discharge of action potentials.

Stimulus Filtering. The preceding paragraph might lead you to think that the sensory cortex is a kind of computer screen on which pictures of the world and the body are projected. If that were true, however, one would have to imagine a little person inside the brain watching the screen, with a little person in his brain watching a screen, and so on. Moreover, the brain would probably not be able to process all the incoming information, since the eyes alone provide enough information to fill up the hard disk on a computer in a few seconds. To deal with all this information the sensory cortex makes an abstract rather than a realistic picture of the world and the body. This abstract representation is due to a third concept of organization: Sensory areas of the cerebral cortex function as filters that detect particular features of the stimulus, such as color, form, or movement. After each section analyzes the feature, it forwards the information to a higher level for integration.

Stimulus filtering is most easily understood for vision, thanks to studies pioneered by David Hubel and Torsten Wiesel using electrodes implanted into visual areas of the brains of cats, monkeys, and blind human volunteers. Hubel, Wiesel, and many others have found that each cell in the visual cortex responds only to a restricted group of receptor cells that make up the **receptive field** (Fig. 9.10). Thus the first level of stimulus filtering involves each nerve cell's filtering out stimuli outside the receptive field. Receptive fields overlap, so each area of a visual stimulus is covered by many cells. Each of these cells is further specialized for a different kind of stimulus filtering. For example, in the primary visual cortex some cells respond only if an edge in the receptive field is oriented within about 20° of a preferred angle (Fig. 9.12). (Twenty degrees is equal to the movement the minute hand of a clock makes in 3 minutes and 20 seconds.) Others respond only to color or to movement.

Distributed Hierarchy. Information from the primary visual cortex is then distributed through at least 305 neural pathways that interconnect 32 visual areas. Different features of a stimulus, such as color, form, or movement, are processed at the same time, and the information is finally sent to two higher centers that are no longer selective for receptive field or for color. One of these is in the parietal lobe and is responsible for perception of position and movement of objects. The other is in the temporal lobe and is responsible for recognition of objects. Damage to one of these areas by a stroke or other injury can have bizarre effects on seeing. Some people with damage to the parietal lobes can see and recognize a bus standing at the curb, but they cannot see it at all if it is moving. Others with damage to the temporal lobes can perceive that the object is moving, but they cannot recognize it as a bus. Some, in fact, cannot even recognize their own faces in a mirror. Those of us with reasonably normal brains somehow "bind" the two kinds of visual perception into a single image. In fact, our brains normally add information to fill in gaps in sensory information (Fig. 9.13).

FIGURE 9.13

A demonstration of the brain's ability to fill in missing information. With the left eye closed, focus the right eye on the circle and move the page closer or farther away until you find the distance where the X is no longer visible. At this point the image of the X is focused on the blind spot of the retina, where the optic nerves enter and receptors are lacking. Note that you do not see a black hole there, because the brain fills in the blind spot. It "assumes" that whatever is hidden in the blind spot looks just like the area surrounding it. Now advance the tip of a pencil slowly toward the X. Note that the tip seems to disappear as its image on the retina enters the blind spot. As you continue to advance the pencil, however, it suddenly appears whole as soon as the image of the tip passes out of the blind spot. Apparently the brain fills in the missing portion of the pencil by assuming that it is like the rest of the pencil.

Memory

During the interpretation of sensory information, and during such complex functions as language and reasoning, numerous and diverse bits of information have to be stored temporarily and manipulated in a sort of mental scratchpad called **working memory.** The **frontal lobe** (Fig. 9.9) appears to be important in processing working memory. Destroying the frontal lobes, as was once done in the "treatment" of certain mental disorders, destroys the ability to reason. After information is processed, it can be stored in **long-term memory** by the **hippocampus,** which curves around the brainstem. Damage to the hippocampus due to a vitamin deficiency often associated with alcoholism can prevent the formation of new memories, essentially trapping the victim in time. No matter how often the patient is told the date, he will still think it is just before his hippocampus was damaged. The diverse kinds of information that make up a memory (visual, auditory, etc.) are apparently transferred weeks or months later from the hippocampus to appropriate areas of the cerebral cortex for permanent storage. Such permanent memories might be responsible for recognition of, for example, a bus.

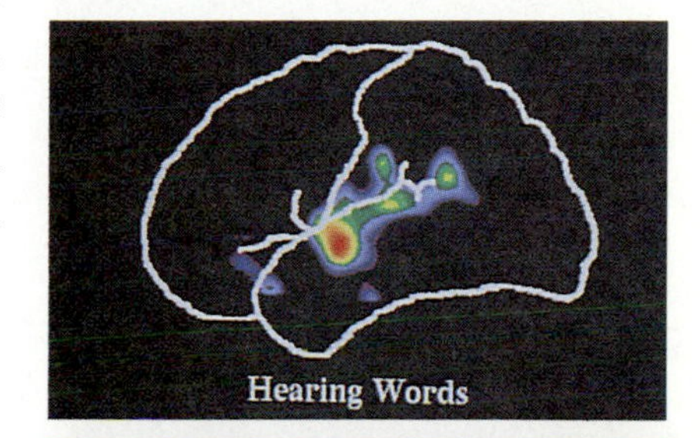

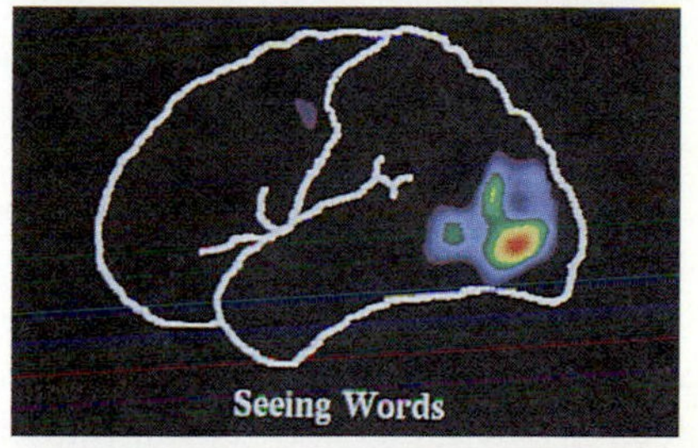

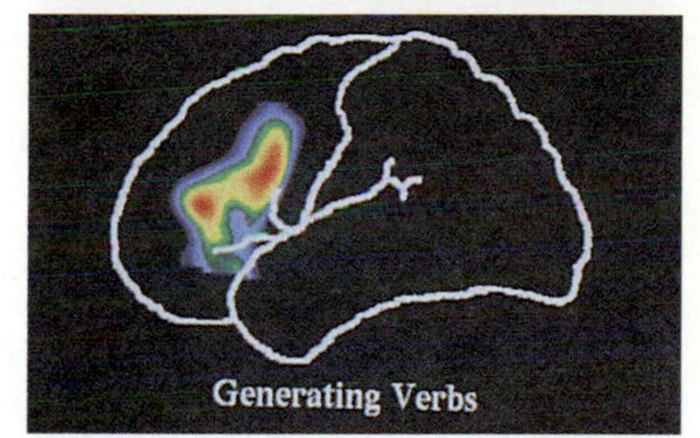

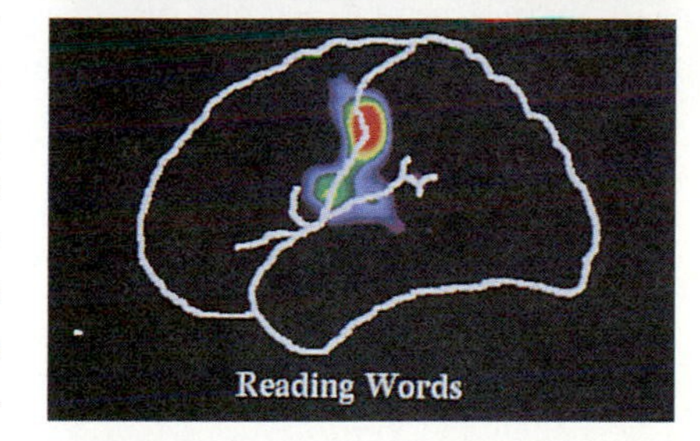

FIGURE 9.14

Positron-emission tomography (PET) scans showing local blood flow, and therefore the level of activity, in different parts of the brain during different language processes.

■ **Question:** Using Fig. 9.9, can you name the areas of cortex involved in each of these four language activities?

Language

Language is one of the most important functions of the brain in humans and perhaps to some extent in other animals. Language functions are carried out mainly in the left cerebral hemisphere in 99% of right-handed people and 70% of left-handers. This lateralization of language function is more pronounced in men than in women (Shaywitz et al. 1995). Reading, listening, formulating speech, and actually speaking are all managed by different areas (Fig. 9.14). The generation of words in correct sequence and context occurs in **Broca's speech area.** The function of this area was discovered by Paul Broca in the previous century during autopsies of former patients who had spoken in a halting, "telegraphic" manner. People with damage to Broca's area have difficulty forming sentences but generally have no trouble understanding speech. Damage a few centimeters posteriorly, however, in **Wernicke's area,** renders them incapable of understanding speech.

The right half of the brain appears to be more specialized for nonverbal information processing, such as mathematics, music, face recognition, and emotion. Normally the two halves of the brain keep each other informed through axons in the **cerebral commissure.** If the cerebral commissure is cut, however, as is sometimes done to control epilepsy, each half of the brain can think and behave independently of the other. Some of the consequences are startling. If a familiar object, like a hammer, is placed in the left hand of a blindfolded commissurotomized ("split-brain") person, he can use the object normally, just as anyone can. He cannot name the object, however. If the object is placed in the right hand, he can name it. If two different images are flashed in the left and right visual fields, the patient can talk about the right image only, although the right half of the brain can still perceive the left image. In one experiment a commissurotomized patient was shown a picture of a flower on the right and a picture of a nude in the left. The patient grinned and said, "Wow! That's some flower!"

Emotion and Consciousness

In humans and other vertebrates, a group of structures that encircle the brainstem and make up the **limbic system** appears to be responsible for producing emotions. The limbic system includes part of the hypothalamus, the hippocampus, an almond-shaped structure called the **amygdala,** and others. The amygdala is responsible for violent behavior. Some violent criminals have voluntarily had their amygdalas removed as a means of controlling their urges. The amygdala is also involved in recognizing emotion in the facial expressions of others (Adolphs et al. 1994). Electrically stimulating the amygdala or certain other areas of the brain can evoke or inhibit a variety of emotions such as anxiety and pleasure. It is perhaps reassuring to note that the patients feel no less human; they clearly understand the difference between the evoked reactions and their own free will.

The topics covered so far in this chapter by no means exhaust all the functions of the brains of humans and many other animals. Some of these functions, such as consciousness, are so mysterious that no one has devised a way of studying them experimentally, or even of defining them. Thus, neuroscientists have no way of knowing whether consciousness is distributed throughout the brain or localized in a particular place that each of us thinks of as "I." Nor can we say whether other animals are conscious. Quite possibly the behaviors of many animals are like reflexes in humans, occurring automatically without any awareness. On the other hand, the behaviors of many vertebrates and the organization of their nervous systems are so similar to our own that it is reasonable to assume some degree of consciousness.

Summary

Nerve systems are complex in their organization and in the behaviors they produce, but parts—subsystems—can be understood. Simple connections between afferent and efferent nerve cells, such as occur in the vertebrate spinal cord, are responsible for reflex responses to stimulation. Another kind of subsystem, the servomechanism, continually monitors and corrects itself to control such behaviors as maintaining posture. Another subsystem is the central pattern generator, which produces patterns of action potentials underlying rhythmic behavior.

The nervous systems of different animals are extremely diverse in organization and complexity. Some, such as those of jellyfishes, are diffuse networks. Bilaterally symmetric animals generally have a central nervous system with several ganglia. In vertebrates the central nervous system consists of the brain and spinal cord with branching nerves in the peripheral nervous system. The peripheral nervous system also includes an autonomic nervous system, with sympathetic and parasympathetic divisions that have opposite effects on digestion, copulation, breathing, circulation, and other functions.

The brain consists of the cerebrum, cerebellum, and brainstem. The cerebellum coordinates learned movements. Much of the cerebral cortex is divided among areas responsible for voluntary movements and for the perception of various kinds of sensation. The cerebrum is lateralized, with each half responsible for movements and sensations of the opposite side of the body. The cerebral cortex has a topographical organization, with adjacent areas of cortex responsible for adjacent areas of the body. Each nerve cell responds to selective features of its receptive field, such as color or pitch. Information from each cell is then distributed through a hierarchy of sensory areas responsible for various aspects of the stimulus.

Memory generally progresses from working memory in the frontal lobes, to long-term memory in the hippocampus, and finally to permanent storage in appropriate areas of the cerebral cortex. In most people the left cerebral cortex is responsible for language, with different areas responsible for reading, hearing, and speaking.

Key Terms

reflex
muscle spindle
servomechanism
central pattern generator
nerve net
central nervous system
ganglion
autonomic nervous system
sympathetic division
parasympathetic division
dorsal root
ventral root
cerebrum
cerebellum
brainstem
cortex
cerebrospinal fluid
glial cell
somatosensory cortex
lateralization
receptive field
Broca's speech area
cerebral commissure

Chapter Test

1. Sketch a simple diagram that could account for the eye-blink reflex when something touches the cornea. (p. 175)
2. Which type of neural subsystem, a servomechanism or a central pattern generator, would be responsible for the coordination of eye and hand movements? Which one could account for the movements of a dog's leg when it scratches? (pp. 175–177)
3. Describe the neural organization found among some invertebrates, proceeding from simplest to most complex. (pp. 177–178)
4. Describe the functions of each of the following divisions of the vertebrate nervous system: central nervous system, peripheral nervous system, autonomic nervous system, sympathetic division, parasympathetic division. (p. 178)
5. What types of function are handled by various parts of the brainstem? (pp. 181–182)
6. How are the functions of the cerebrum distributed anatomically? (pp. 181–182)
7. What parts of the brain are involved in the coordination of movement, and how are they involved? (p. 183)
8. Give an example of stimulus filtering. (p. 186)
9. Trace the flow of information through your brain as you read this question, formulate an answer, and write it. (p. 187)

■ Answers to the Figure Questions

9.1 In the cross section of the spinal cord, draw a branch from the afferent neuron to the other side of the spinal cord with an excitatory synapse onto a motor neuron going to the extensor. The tricky part is getting inhibition of the flexor from this excitatory neuron. To do that, draw another branch from the afferent and a motor neuron going to the flexor muscle, then put a small inhibitory interneuron between the two. Excitation of the inhibitory interneuron will then inhibit the flexor motoneuron.

9.4 A fright would stimulate the sympathetic branch, which has (1) inhibitory output to the salivary glands and sphincter muscles that control urination and defecation and (2) excitatory output to the heart. The parasympathetic branch stimulates tear glands, salivary glands, digestive organs, gonads, and muscles that contract bronchial tubes in the lungs, while slowing the heart.

9.5 Damage to all the dorsal roots of the lumbar, sacral, and coccygeal nerves would have this effect.

9.7 Much of the brain is involved in analyzing sensory information and coordinating motor output, so it is not surprising that brain mass increases with body mass. The fact that brain mass increases in proportion to the surfaces area (body mass 2/3) makes sense if the number of receptor and muscle cells is proportional to surface area. A large animal is not necessarily smarter, because much of its larger brain is occupied with controlling the larger muscles and receiving information from the larger body surface.

9.9 No one has yet discovered an area of the brain that has no activity in a normal person. In fact, this old saying was around long before there were any scientific methods capable of supporting it.

9.14 Hearing words: auditory area of the temporal lobe and Wernicke's area in the parietal lobe. Seeing words: visual area of the occipital lobe: Generating words: Broca's speech area in the left frontal lobe. Speaking words: motor and somatosensory areas of the frontal and parietal lobes.

Readings

Recommended Readings

Aoki, C., and P. Siekevitz. 1988. Plasticity in brain development. *Sci. Am.* 259(6):56–64 (Dec).

Calvin, W. H. 1994. The emergence of intelligence. *Sci. Am.* 271(4):100–107 (Oct).

Fischbach, G. D. 1992. Mind and brain. *Sci. Am.* 267(3):48–57 (Sept). (*Introduction to an entire issue on this subject.*)

Glickstein, M. 1988. The discovery of the visual cortex. *Sci. Am.* 259(3):118–127 (Sept).

Goldstein, G. W., and A. L. Betz. 1986. The blood–brain barrier. *Sci. Am.* 255(3):74–83 (Sept).

Horgan, J. 1994. Can science explain consciousness? *Sci. Am.* 271(1):88–94 (July).

Hubel, D. H. 1988. *Eye, Brain, and Vision.* New York: Scientific American Books.

Jerison, H. J. 1976. Paleoneurology and the evolution of mind. *Sci. Am.* 234(1):90–101 (Jan). (*The relation of brain size to intelligence among different species.*)

Kalil, R. E. 1989. Synapse formation in the developing brain. *Sci. Am.* 261(6):76–85 (Dec).

Kimelberg, H. K., and M. D. Norenberg. 1989. Astrocytes. *Sci. Am.* 260(4):66–76 (Apr). (*The varied functions of a glial cell.*)

Konishi, M. 1993. Listening with two ears. *Sci. Am.* 268(4):66–73 (Apr). (*How the brain of the barn owl interprets sound.*)

LeDoux, J. E. 1994. Emotion, memory and the brain. *Sci. Am.* 270(6):50–57 (June).

Livingstone, M. S. 1988. Art, illusion and the visual system. *Sci. Am.* 258(1):78–85 (Jan).

Melzack, R. 1992. Phantom limbs. *Sci. Am.* 266(4):120–126 (Apr).

(Amputees may still feel sensations that appear to come from missing limbs.)

Mishkin, M., and T. Appenzeller. 1987. The anatomy of memory. *Sci. Am.* 256(6):80–89 (June).

Morrison, A. R. 1983. A window on the sleeping brain. *Sci. Am.* 248(4):94–102 (Apr).

Posner, M. I., and M. E. Raichle. 1994. *Images of Mind.* New York: Scientific American Books.

Raichle, M. E. 1994. Visualizing the mind. *Sci. Am.* 270(4):58–64 (Apr).

Ramachandran, V. S. 1992. Blind spots. *Sci. Am.* 266(5):86–91 (May).

Sacks, O. 1985. *The Man Who Mistook His Wife for a Hat.* New York: Summit Books. *(This and other books by Sacks offer fascinating and instructive accounts of neurological disorders.)*

Spector, R., and C. E. Johanson. 1989. The mammalian choroid plexus. *Sci. Am.* 261(5):68–74 (Nov). *(The formation of cerebrospinal fluid.)*

Additional References

Adolphs, R., D. Tranel, H. Damasio, and A. Damasio. 1994. Impaired recognition of emotion in facial expressions following bilateral damage to the human amygdala. *Nature* 372:669–672.

Shaywitz, B. A., et al. 1995. Sex differences in the functional organization of the brain for language. *Nature* 373:607–609.

CHAPTER

10

Integument, Skeleton, and Muscle

Orangutan *Pongo pygmaeus.*

Integuments

Animals are multicellular organisms that cannot produce their own nutrients, so they have to move about through often inhospitable environments to obtain the necessities of life for their internal environments. Many small, aquatic animals use cilia to get about, but most are so large that they require the much greater forces produced by muscles. These animals generally require some kind of skeleton to prevent their bodies from collapsing under the forces produced by their own muscles or from external forces such as gravity. As will be described later in this chapter, some animals are supported by internal hydrostatic pressure, some other animals, including ourselves, have skeletons enclosed within other tissues, and others have skeletons consisting of the body covering, called the **integument.** The integument also protects the other body tissues, separating the carefully maintained internal environment from the external environment. Integuments vary widely, depending on the amount of support and protection needed.

INVERTEBRATE EPIDERMIS. In many small, marine animals, buoyancy supports the body against the force of gravity, and the composition of the water may differ little from that of the interstitial fluid. In many of these animals the integument need be no more than a single layer of cells, called an **epidermis.** The epidermis may be permeable to nutrients, wastes, and respiratory gases, enabling the animal to do without special digestive, excretory, or respiratory organs. In larger invertebrates, especially if they live in fresh water or on land, the epidermis has additional adaptations for protection and support. In flatworms, earthworms, snails, and many other animals, epidermal cells secrete **mucus,** which is a watery solution of specialized glycoproteins (Fig. 10.1A). The mucus protects against abrasion and can also provide a track on which the animal glides during locomotion.

In many animals the epidermis secretes minerals that protect and support the body. The shells of molluscs and many other invertebrates are made largely of calcium carbonate $CaCO_3$. In other animals, such as nematodes, annelids, and arthropods, the epidermis secretes a tough, water-resistant **cuticle.** In such animals the epidermis is also called a **hypodermis.** A disadvantage of shells and cuticle is that they require some provision to allow the animal to grow within them. Molluscs and some other animals with calcareous shells continually enlarge the shells as they grow. Arthropods periodically shed the old, outgrown cuticle in a process called **molting.**

SKIN. The integument of vertebrates, known to all as skin, affords good protection and also grows as the animal does. Skin has two layers, the outer epidermis and the dermis beneath it (Fig. 10.1B). The epidermis of vertebrates is quite different from the epidermis of invertebrates in that it has several layers of cells. The innermost cells undergo frequent division, which pushes the older cells toward the surface. During this process much of the cytoplasm is replaced by **keratin,** a complex of proteins held together by disulfide bonds. This process of **keratinization** eventually kills the cells, which end up forming a scaly **cornified** layer that is resistant to abrasion and is impermeable to water. Each of us sheds billions of these cornified cells each day, many of them accumulating as house dust.

Beneath the epidermis is the **dermis,** or true skin. Dermis consists mainly of connective tissue with large amounts of the fibrous protein **collagen.** Collagen accounts for the toughness and flexibility of skin. (Contrary to advertisements for many skin lotions, however, there is no way that collagen applied externally can penetrate the epidermis to restore the skin.) The dermis of mammals also includes the hair follicles, cu-

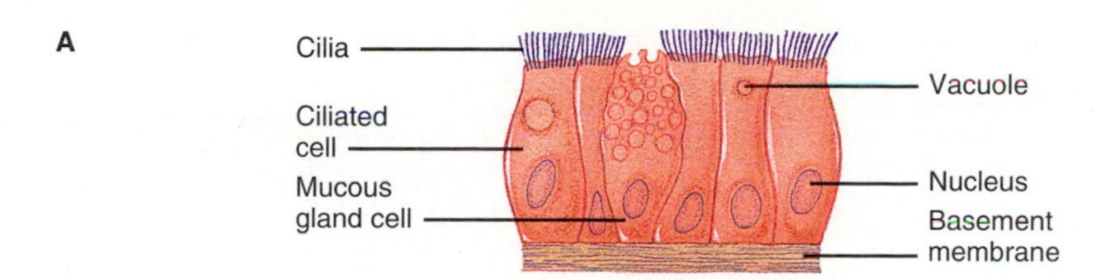

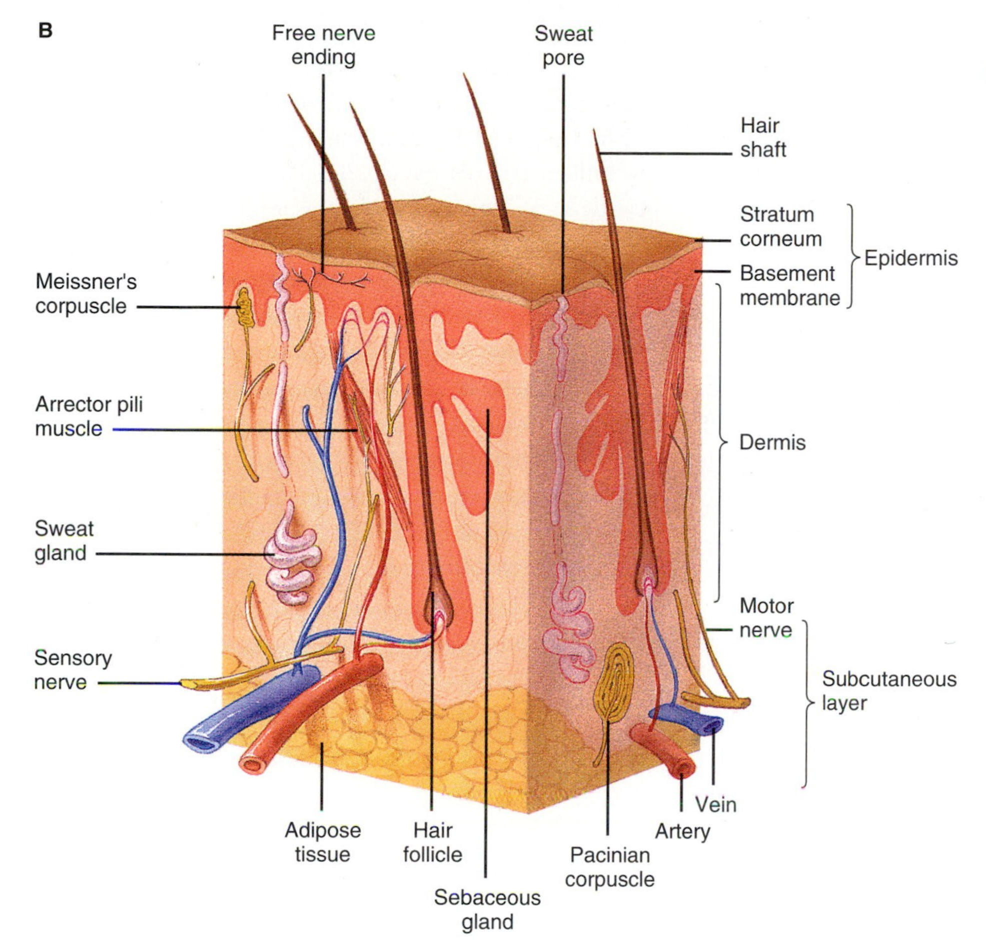

FIGURE 10.1

Representative integuments of an invertebrate and a vertebrate. (A) The simple integument of a flatworm (phylum Platyhelminthes) consists of an epidermis supported by a basement membrane. Flatworms use a combination of mucus and cilia on the epidermal cells in locomotion. (B) The more complex integument of a human consists of a multilayered epidermis and dermis.

■ **Question:** How is each type of integument protected from abrasion and dehydration?

taneous sensory receptors (see Fig. 8.13), and several types of glands that originate in the epidermis. The glands include **sebaceous glands,** which produce a fatty secretion (**sebum**) that is released through hair follicles (Fig. 10.1B). Sebum helps protect the skin from dehydration. In addition, there are two types of sweat glands. **Eccrine sweat glands** produce a watery solution that cools the body by evaporation. **Apocrine sweat glands** produce an odorous secretion that aids in courtship and other kinds of social communication—in most mammals, at least. In humans, the apocrine glands occur mainly in the arm pits and groin.

COLORATION. In addition to providing structural support and protection, many integuments also provide coloration. Colors may serve as camouflage or, for venomous, toxic, or bad-tasting animals, they may warn off potential predators. In addition, colors often serve in social communication, helping members of the same species identify individuals, their gender, reproductive status, social rank, and so on. Some animals are colored simply because their blood or whatever they have eaten shows

FIGURE 10.2
(A) Hair reflects all light and is therefore white unless it contains pigments that absorb particular colors. In many mammals, such as the eastern chipmunk *Tamias striatus,* melanin pigments make the hair black or brown. (B) Almost all the color of the peacock *Pavo cristatus* is structural color due to reflection of particular wavelengths by the feathers.

■ **Question:** Why are laboratory mice and rats white with red eyes? What is the probable source of each of the colors in the peacock?

through their tissues, but most colors are due either to **pigments** or to fine structures that refract or reflect light (**structural colors**) (Table 10.1).

Pigments may be dispersed in epidermis or they may be contained in special cells called **chromatophores.** Under neural and hormonal control, the pigments in chromatophores can condense or expand, reducing or intensifying the color. This control over chromatophores accounts for the chameleon's famous ability to change appearance. Feathers and hair may also acquire structural colors and pigments (Fig. 10.2). Black or brown hair color is due to melanin pigment, which the hair absorbs from nearby **melanocytes** as it grows. Melanocyte stimulating hormone (MSH) from the pituitary gland stimulates melanin production by melanocytes. MSH can be blocked, however, by **agouti peptide** from certain cells in the hair follicle, which causes the melanocyte to release yellowish-red pigment instead. Often a single hair will have bands of both colors. Under the influence of natural or artificial selection, the genetic control of MSH and agouti peptide has been manipulated to produce the enormous variety of coat colors seen in dogs, cats, and other domesticated animals, as well as in wild mammals.

TABLE 10.1
Animal colors and their sources.

White	White is due to the reflection of all wavelengths of light. In insects it is often a structural color—**structural white**—due to dust-sized, transparent particles that scatter all wavelengths. A silvery white, like that of fish scales, is produced by reflection from a purine such as guanine or adenine. These pigments are often found in a type of chromatophore called an **iridophore**.
Iridescence	Interference between incident and reflected light causes some colors to be blocked while others are intensified, depending on the angle of viewing. The luster of pearls and some feathers and butterfly wings is a structural color due to overlapping translucent layers (**interference layers**). The cuticles of some insects and the scales of some fishes are finely grooved and produce iridescent **diffraction colors** like those on a CD. Some birds' feathers contain bundles of thin tubes that produce diffraction colors.
Blue	Blue is generally a structural color called **Tyndall blue**, which is due to the scattering of light by particles about the size of the wavelength of light. Blue light is scattered most, producing a blue color in the same way that scattering of light by air produces a blue sky.
Green	Green is generally produced by Tyndall blue filtered through one of the yellow pigments noted below. Some birds, however, have feathers with green pigments.
Yellow, orange, and red	Yellow, orange, and red are generally due to carotenoids or pteridines, often in a type of chromatophore called a **xanthophore**. Some **melanins** produce yellow or dull red.
Brown, gray, and black	Brown, gray, and black are due to melanins. Chromatophores containing melanins are called **melanophores**.

Skeletal Support

HYDROSTATIC SKELETONS. In many soft-bodied invertebrates (indicated in Fig. 21.11) the integument also serves as a skeleton, even if it is not immediately recognizable as such. The internal pressure of body fluids may stretch the epidermis like an inflated balloon, creating a **hydrostatic skeleton** (= hydroskeleton). Animals with hydrostatic skeletons typically control internal pressure with two sets of muscles: one running longitudinally along the body and the other running circularly around the body. Coordinated contractions of these muscles also allows for burrowing, swimming, and other forms of locomotion (Fig. 10.3).

EXOSKELETONS. In many other invertebrates the integuments provide protection and support as exoskeletons: rigid structures that enclose all or part of the body. As was previously noted, exoskeletons generally take the form of shells or cuticle, usually consisting of hard plates hinged together by flexible connective tissue. Movement is generally produced by contraction of muscles attached to the inner surfaces of the plates. In clams, oysters, scallops, and other bivalve molluscs, for example, the two parts of the shells, called **valves,** are closed by contraction of muscle that attach to both valves. An elastic ligament opens the valves when the muscles relax. In insects and other arthropods the muscles attach to the cuticular plates. Contraction of one set of muscles moves a joint in one direction, and contraction of another set of muscles (antagonist muscles) moves the joint in the opposite direction (Fig. 10.4 on page 196).

ENDOSKELETONS. Endoskeletons are, by definition, enclosed by other body tissues, and they occur in a variety of forms. Sponges have endoskeletons that consist of mineral **spicules** (see Figs. 23.3 through 23.6) and fibers of **spongin** that keep the body from collapsing. Adult sponges remain attached to substratum (they are sessile), so they have no need for muscles attached to the endoskeleton. Echinoderms such as starfishes and sea urchins have endoskeletons made up of small, calcareous plates called **ossicles.**

The most familiar endoskeletons are, of course, your own and those of vertebrates you may eat. Vertebrate skeletons are composed of **cartilage** and bone. Skeletons are

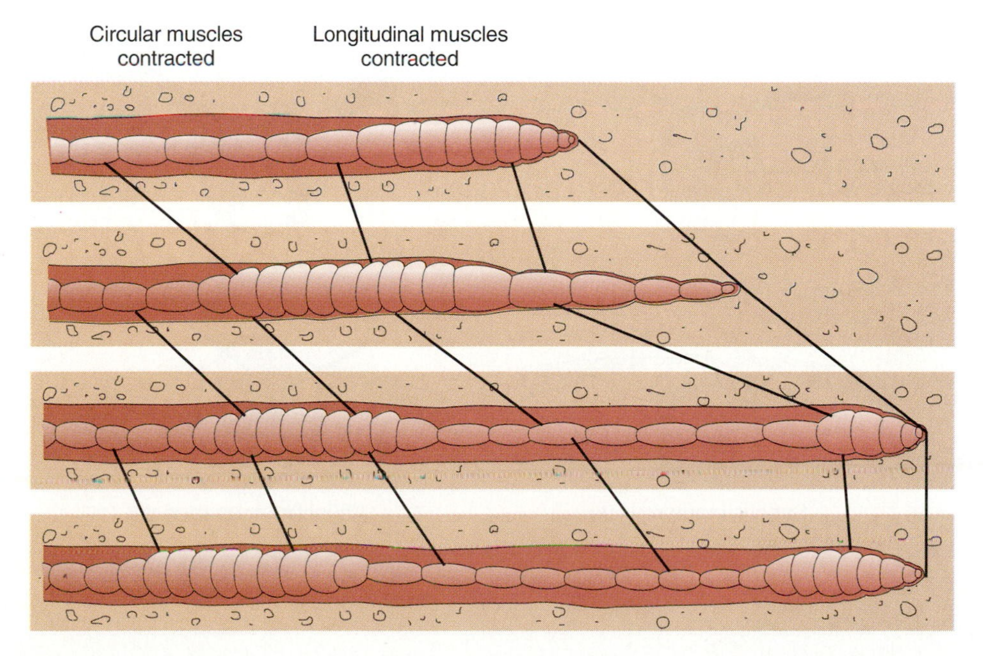

FIGURE 10.3
How an animal that lacks bones can burrow through earth. The earthworm *Lumbricus terrestris* uses peristaltic progression, which is possible only because the hydrostatic skeleton keeps the body from collapsing when its muscles contract. In the top part of the figure, contraction of the longitudinal muscles in front thickens the segments and allows them to make stronger contact with the sides of the burrow. The circular muscles at the anterior end then contract, narrowing the segments and pushing them forward (second part of the sequence). A wave of contraction of the circular muscles moves backward, alternating with contraction of the longitudinal muscles, resulting in forward movement. Progression is shown by the lines connecting every fifth segment.

FIGURE 10.4

Exoskeleton and antagonist muscle attachment in arthropods. (A) The exoskeleton of the lobster *Homarus americanus* with detail of the crusher claw showing the attachment of the opener and closer muscles. (B) Attachment of flight muscles in insects. The arrangement on the left, which occurs in dragonflies, uses elevator and depressor muscles that attach to the wing. Elevators pivot the wings up, and depressors pull them down. The contraction of each muscle is synchronous with a burst of action potentials from the nervous system. The arrangement on the right is typical of flies, bees, beetles, and bugs. The muscles do not attach directly to the wings, but to the exoskeleton of the thorax. Both elevators and depressors are asynchronous muscles, which contract in response to being stretched. The elevators pull the top of the thorax down, causing the wings to spring up. This stretches the depressors, which run longitudinally through the thorax. When they contract they bow the thorax out, which depresses the wings. This, in turn, stretches the elevators, causing them to contract, and so on.

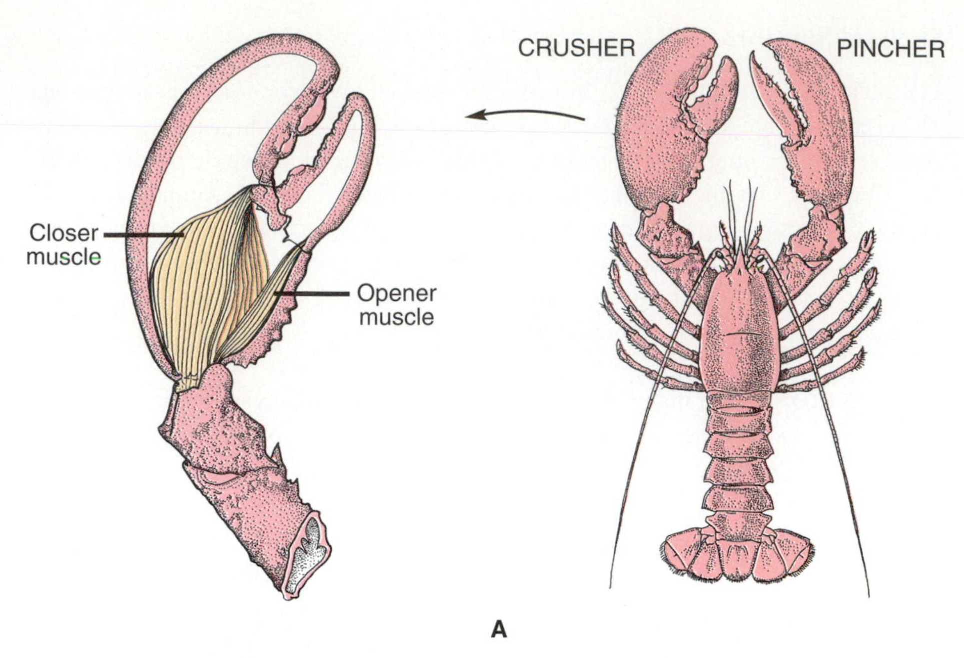

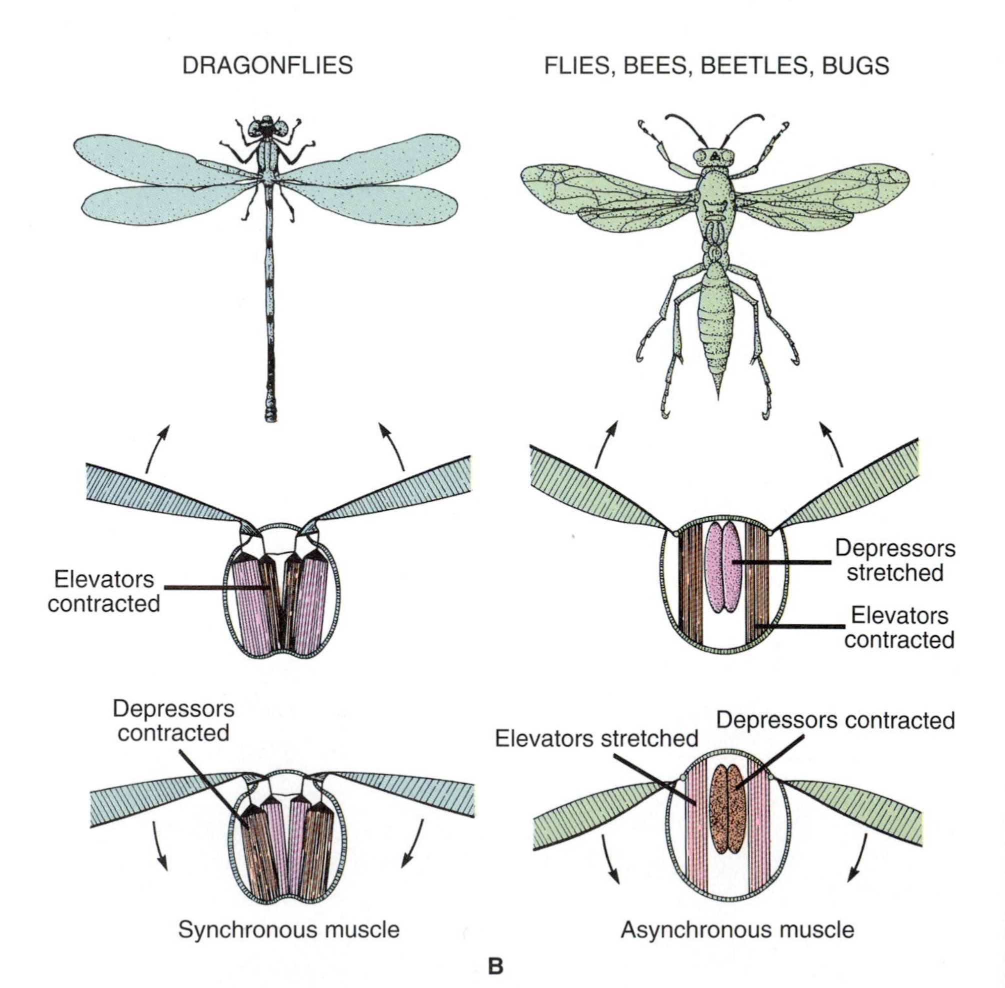

entirely **cartilaginous** in the invertebrate chordate *Branchiostoma* (amphioxus; see Fig. 34.8), in jawless vertebrates (hagfishes and lampreys; see Fig. 35.1), and in sharks, rays, and related fishes (class Chondrichthyes; see Fig. 35.2). In other vertebrates, cartilage forms the skeleton of the embryo, but bone later replaces most of the cartilage. Even in those animals, however, some cartilage may remain as flexible supports in the nose, ears, and trachea and as cushions between joints (Fig. 10.5A). The ability of cartilage to remain flexible, yet retain its shape and resist damage, lies in the fact that it consists largely of water. The water is held in place by a network of collagen and complex molecules called **proteoglycans,** both of which are secreted by cells called **chondrocytes.**

The skeletons of most vertebrates are composed mainly of **bone.** One type of bone, called **dermal bone** (= membranous bone), is produced by layers of embryonic connective tissue rather than by replacement of cartilage. Dermal bone occurs in the skull and face of mammals and in such nonskeletal structures as fish scales, turtle shells, and antlers. Most structural support, however, is provided by **dense bone** (= compact bone) (Fig. 10.5). Dense bone is produced by cells called **osteoblasts** (Greek *osteon* bone + *blasti* a bud), which replace cartilage with calcium carbonate and calcium phosphate. During embryonic development in birds and mammals, osteoblasts deposit these minerals in concentric layers (**lamellae**). The centers of these layers remain as hollow channels, called **central canals** (= Haversian canals), which contain blood vessels and nerves. In bones such as those of the arms and legs, dense bone surrounds a core of **spongy bone** (= trabecular or cancellous bone). The spaces in spongy bone are hollow in birds (see Fig. 38.11A), but in mammals they are filled with marrow, which produces blood cells.

Contrary to appearances, bone is living, dynamic tissue. Osteoblasts continue to work throughout life, depositing new bone in response to mechanical stress and to repair fractures. Other cells, called **osteoclasts** (Greek *klasis* a breaking), can break down the minerals of bone. This is usually in response to hormones, such as parathyroid hormone from the parathyroid glands (see p. 142), which signal a need for the calcium in the body fluids. The lack of mechanical stress, as when bedridden, and the absence of estrogens following menopause can also lead to excessive loss of minerals from bones.

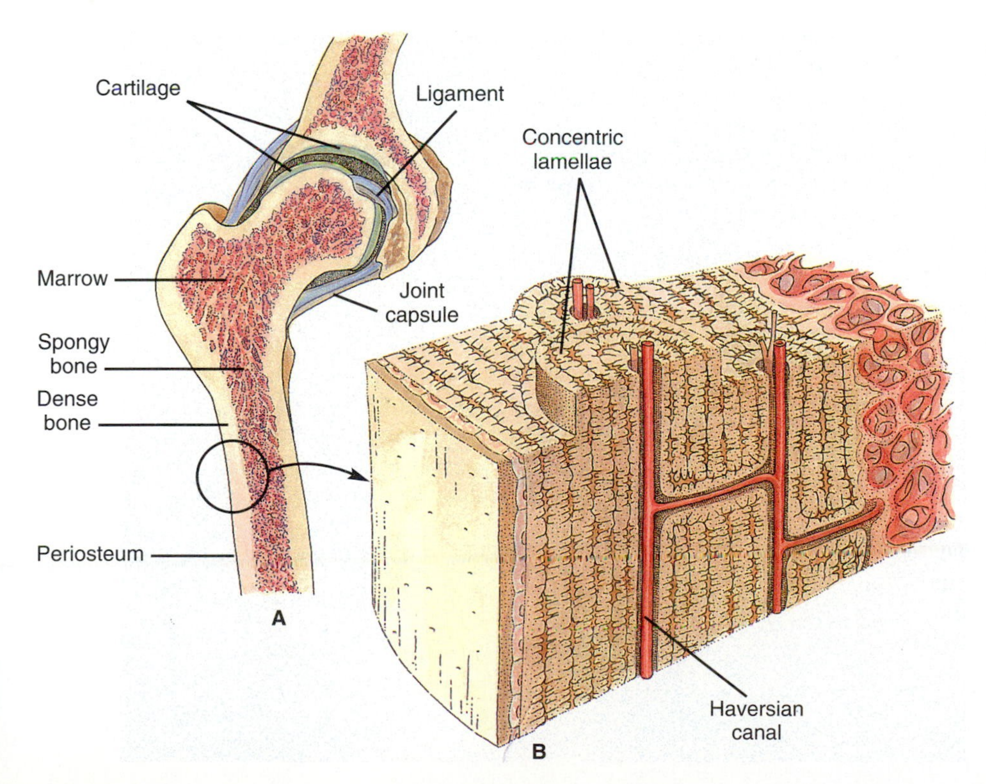

FIGURE 10.5

(A) Bone and cartilage at a human hip joint. The cartilage helps cushion the bones and reduces friction between them. Ligaments hold the bones together. The bones are surrounded by a sheath (the periosteum). The outer dense bone encloses the spongy bone. At the ends of long bones such as these, the spongy bone spaces are filled with red marrow, which produces red blood cells. (B) Details of dense and spongy bone.

Figure 10.6 shows the major bones of the human skeleton. Similar bones occur in other vertebrates, with sizes and shapes adapted to different functions. These bones are divided into those of the **axial skeleton** and of the **appendicular skeleton.** The axial skeleton lies along the body axis and includes the bones of the **skull, vertebral column,** and **thoracic basket** (rib cage). The appendicular skeleton comprises the bones of the arms and legs, which attach to the **pectoral girdle** and the **pelvic girdle,** respectively.

Muscular Anatomy

In addition to supporting the body, a skeleton provides sites for the attachment of skeletal muscles. In vertebrates these attachments are connective tissues called **tendons,** which generally connect each end of a muscle to two different bones that are separated by one or more joints. Contraction of a muscle then causes one bone to move with respect to the other. Muscles can exert force only while contracting, not while

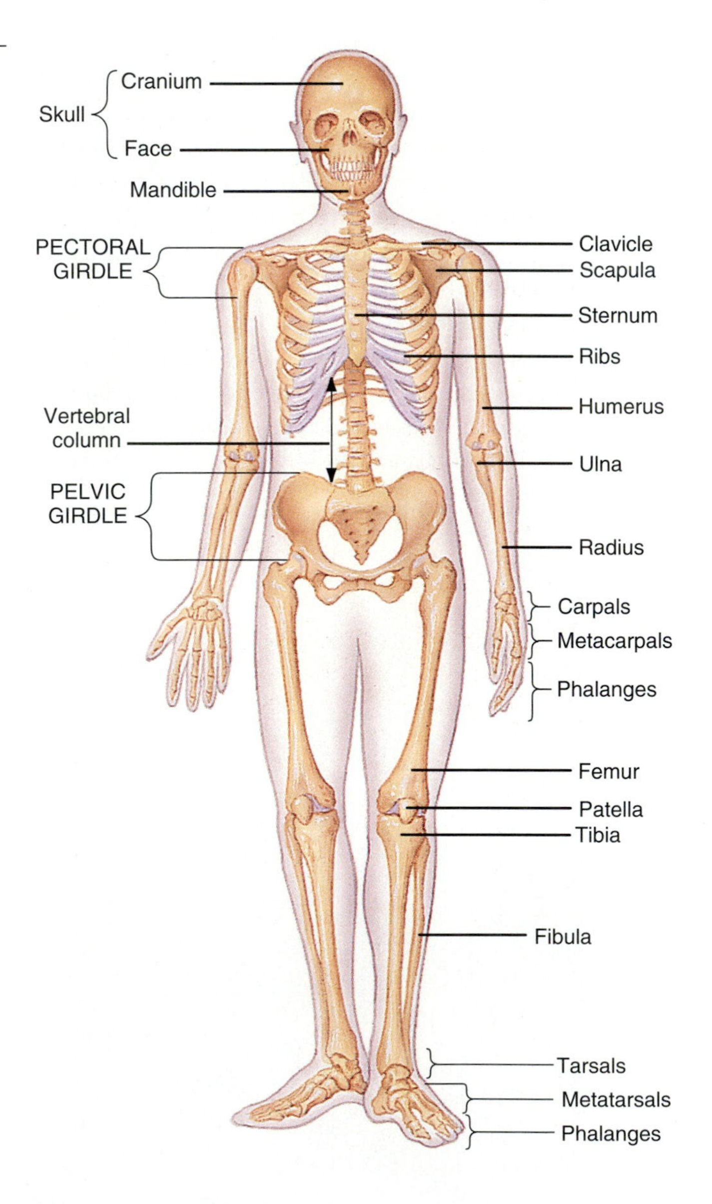

FIGURE 10.6
Major bones of the human skeleton.

■ **Question:** What bones of the leg are homologous with the humerus, radius, ulna, and carpals of the arm?

lengthening, so they have to be arranged in **antagonistic** pairs. For example, the biceps brachii in the front of the upper arm of a human is a **flexor** muscle that helps bring the hand toward the body. The triceps brachii on the opposite side of the arm is an **extensor** that extends the hand away from the body. Figure 10.7 shows some of the most im-

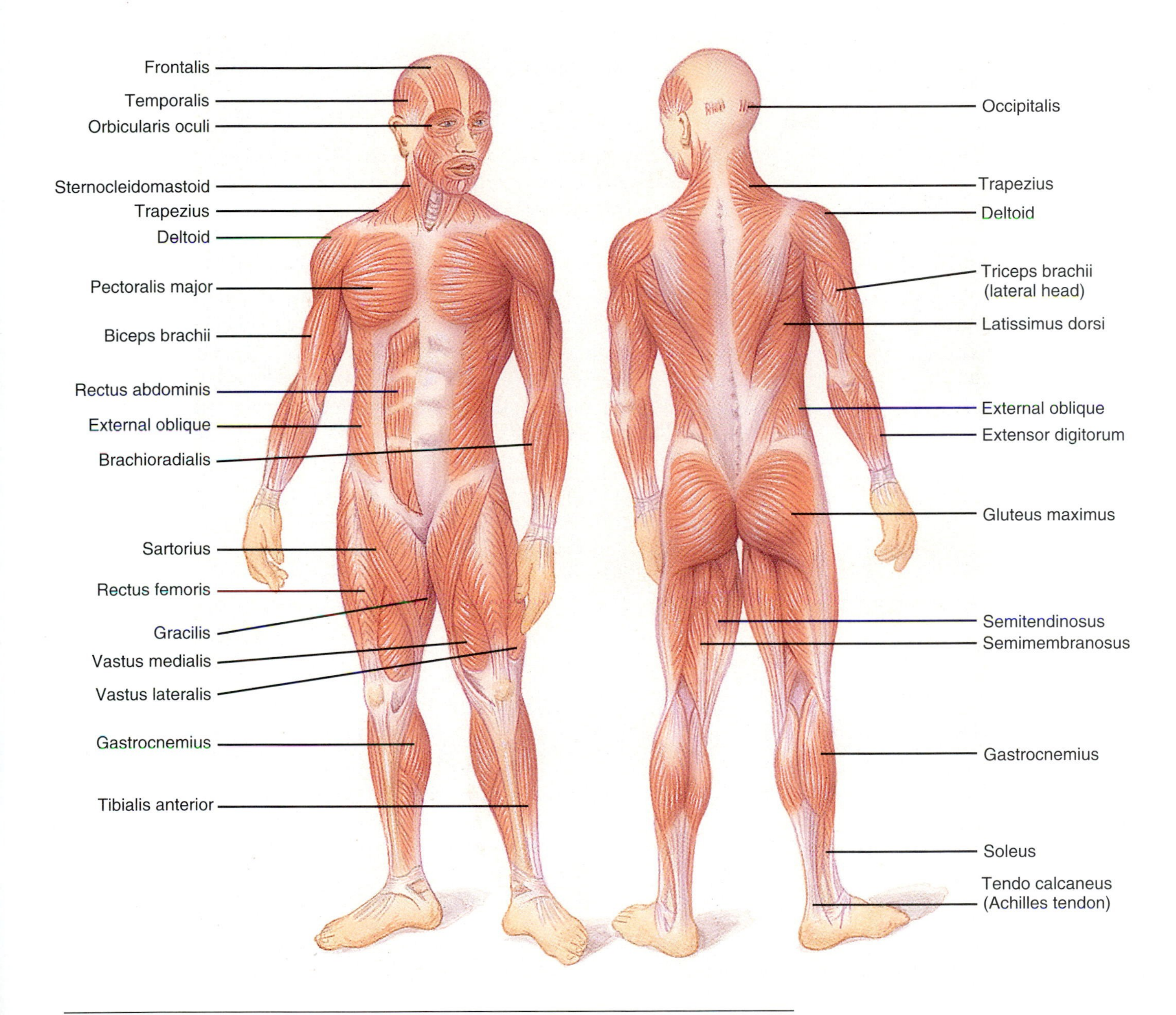

FIGURE 10.7
Some major muscles of the human.

■ **Question:** Based on this figure, what movement would you predict for contraction of each of the following muscles: gastrocnemius, triceps brachii, rectus abdominis. Test your predictions on yourself.

portant skeletal muscles in the human. You should try to locate some of them on yourself and observe the movements they cause.

The way in which a muscle is attached to the skeleton determines not only the type of motion produced when it contracts, but also the speed and force with which the body part moves. This is so because the bones move like levers (Fig. 10.8).

Activation of Skeletal Muscle

Galen in the 2nd century and Descartes in the 17th thought movement was due to the expansion of a neural vapor within muscles. Their concepts were probably influenced by the steam technology of their times, just as our concepts are influenced by computer technology. We now think muscle contractions are controlled by patterns of electrical activity from the brain and spinal cord. This electrical activity, in the form of action potentials in motor nerve cells, interacts with skeletal muscle cells through **neuromuscular synapses.** These neuromuscular synapses, also known as neuromuscular junctions or endplates, work much like the chemical synapses within the central nervous system (CNS). The major differences between neuromuscular synapses and CNS synapses are structural. Neuromuscular synapses in vertebrates are usually elongated extensions of the axon, rather than knob-shaped structures. (Compare Figs. 8.7 and 10.9.) In neuromuscular synapses the vesicles are clustered just opposite **subjunctional folds** in the postsynaptic membrane of the muscle cell. The subjunctional folds contain the receptor molecules that bind the transmitter substance released by the neuromuscular synapse.

Another difference is that neuromuscular synapses do not release as large a variety of excitatory and inhibitory transmitters. In vertebrates, in fact, there are no inhibitory neuromuscular transmitters, and the only excitatory transmitter is acetylcholine (ACh). ACh depolarizes the postsynaptic membrane of the muscle cell, producing an **endplate potential** (EPP) that is similar to an excitatory postsynaptic potential. EPPs travel to the excitable portion of the muscle membrane, where they evoke action poten-

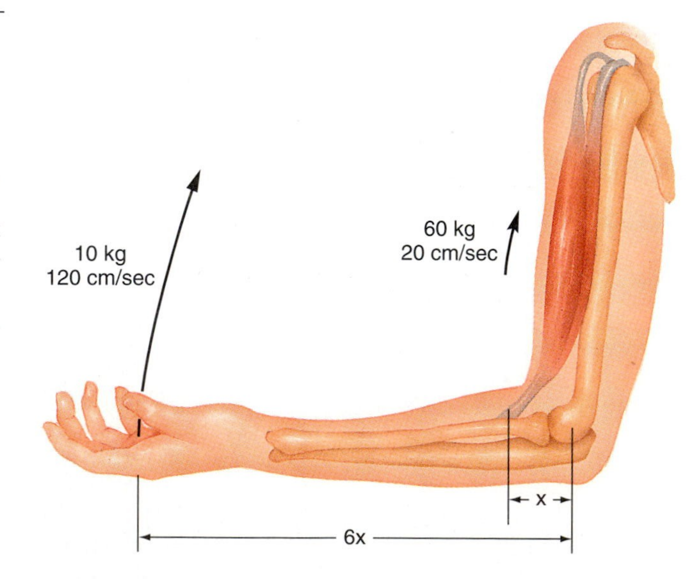

FIGURE 10.8

The human elbow, showing the attachment of the biceps brachii muscle to the radius and scapula. The numbers show the mass (in kilograms) that can be lifted and the speed (in centimeters per second) of movement of the muscle and of the hand. Because of the lever action of the joint, the force and speed are not the same for the hand as for the muscle.

■ **Question:** If arm strength had been more important in human evolution, would the biceps brachii have been attached to the radius closer to the elbow or farther away? What would have been sacrificed for the greater strength?

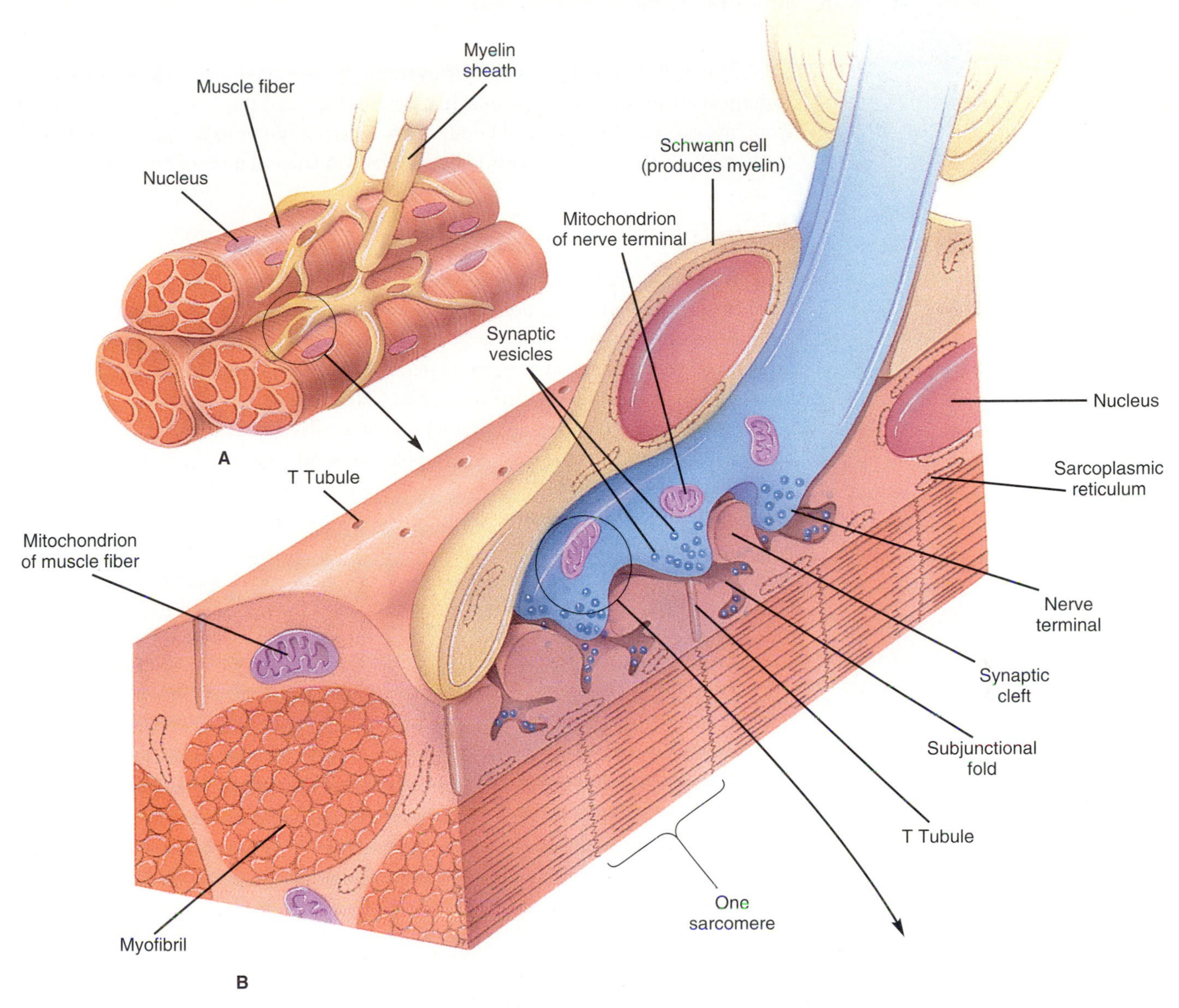

FIGURE 10.9

The structure of neuromuscular synapses in vertebrate skeletal muscle. (A) Neuromuscular junctions. (B) An enlargement of the circled portion of (A), showing a longitudinal section of the neuromuscular junction. (C) Electron micrograph of a cross section of the neuromuscular synapse of the leopard frog *Rana pipiens.* Each of the vesicles (left part of figure) is about 40 nm in diameter. Longer circles on the right are mitochondria.

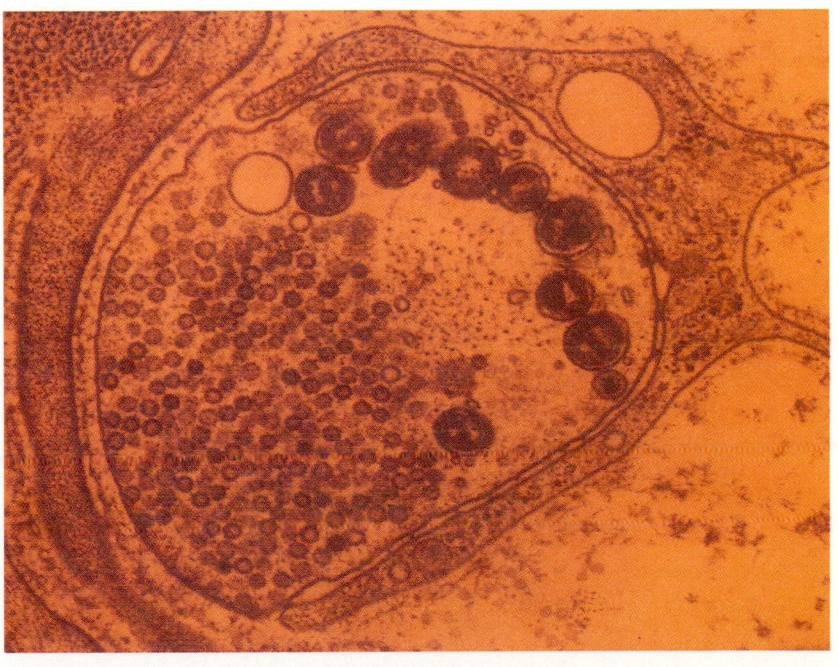

tials. In vertebrates and a few invertebrates the action potentials arise from inward sodium currents like those in axons. In most invertebrates the action potentials result from inward calcium currents. In both cases the action potentials trigger muscle contraction after a sequence of events to be described in later sections of this chapter.

Control of Muscle Contraction

As with the action potentials in axons, those in the membranes of muscles are all identical. How, then, can action potentials trigger so many different kinds of movement? As with action potentials from sensory receptors, the information directing bodily movements must be encoded in the pattern of action potentials to the muscles. One kind of muscle information is obvious: **The CNS controls the type of movement by activating different muscles.** Sending action potentials to a flexor muscle, for example, will cause a movement that is just the opposite of sending action potentials to the antagonistic extensor muscle.

A second principle of the motor code is revealed by the pattern of connections between motor neurons and skeletal muscle fibers (Fig. 10.10). Each branch of a peripheral nerve going to a particular muscle contains many motor axons, and each axon connects to different muscle cells. In vertebrates each muscle cell, also called a muscle fiber, receives synaptic input from only one motor axon, but each motor axon can have synapses onto several muscle fibers. Since an action potential in the motor axon conducts to all its branches and activates all the muscle fibers with which they connect, all those muscle fibers contract as a unit. A **motor unit** thus consists of one motor axon and all the muscle fibers it controls. Thus the second principle of the motor code is as follows: **The strength of muscle contraction increases with the number of motor units activated** (Fig. 10.11). The fewer the muscle fibers in a motor unit, the more precise the control over the muscle. The muscles controlling eye movements have as few as three muscle fibers per motor unit, while the large gastrocnemius muscle in the calf of the leg has 2000 fibers per motor unit.

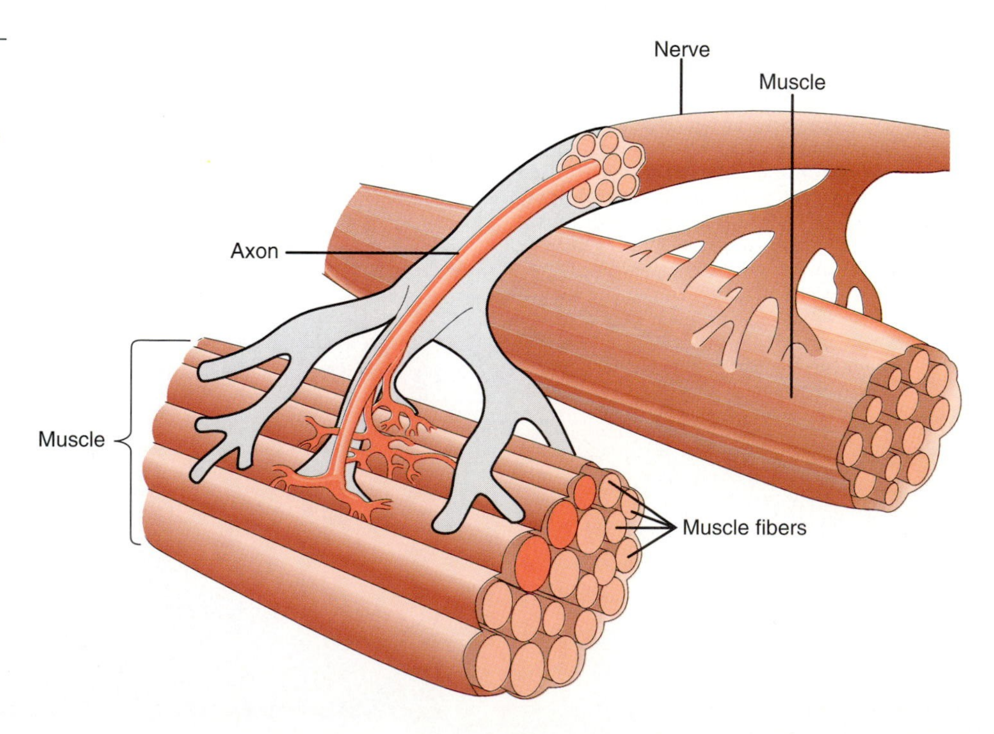

FIGURE 10.10

Two small muscles innervated by branches of one nerve. Each nerve includes numerous axons, with each axon of a motor neuron controlling one motor unit. One motor unit—with three muscle fibers controlled by one axon—is shown in color.

A

FIGURE 10.11

How the force of contraction increases with increasing number of motor units activated. (A) The experimental preparation. A leopard frog was pithed (its brain and spinal cord were destroyed by a probe to prevent voluntary movement and pain). Electrodes were then placed on the sciatic nerve near the spinal cord, and a force transducer was attached by a thread to the Achilles tendon on the gastrocnemius muscle. This muscle helps power the hop in the frog. (B) Brief shocks applied to the sciatic nerve evoked twitches in the gastrocnemius muscle, and the force of the twitches increased with the voltage of the shocks. Note, however, that although the voltage was increased gradually, the force of contraction increased in jumps. A representation of the sciatic nerve in cross section shows that the increasing voltage excited an increasing number of motor units in all-or-none fashion.

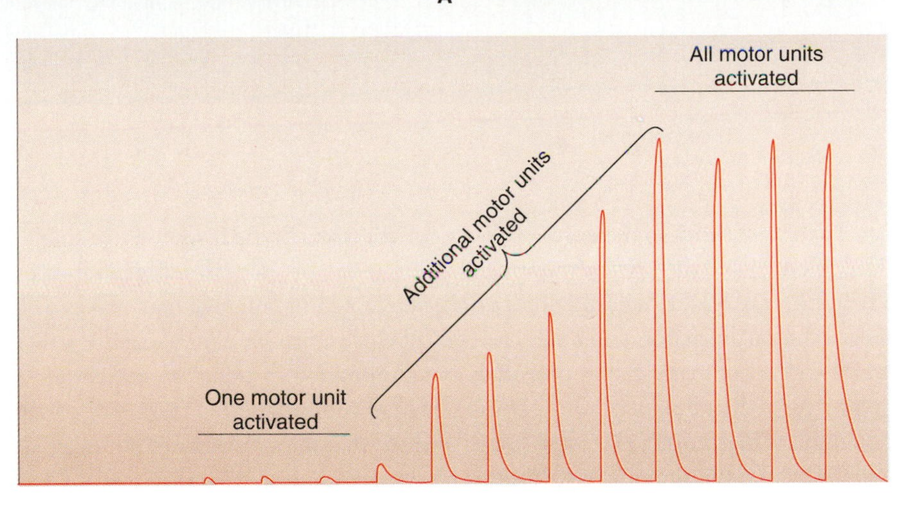

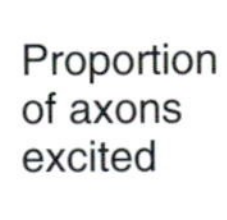

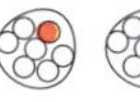
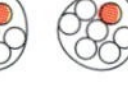

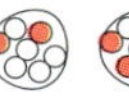

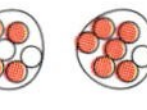

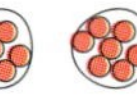

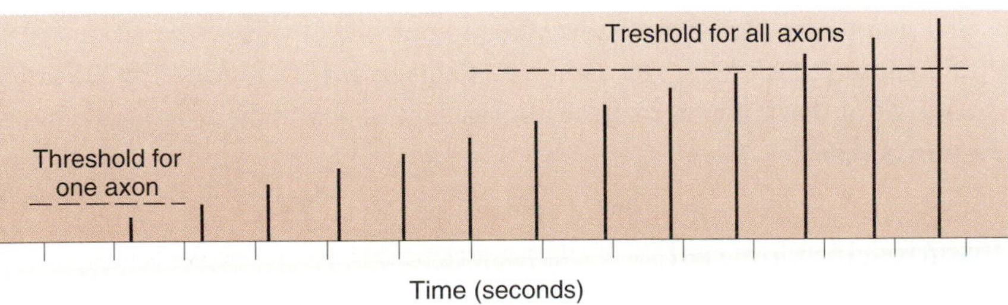

B

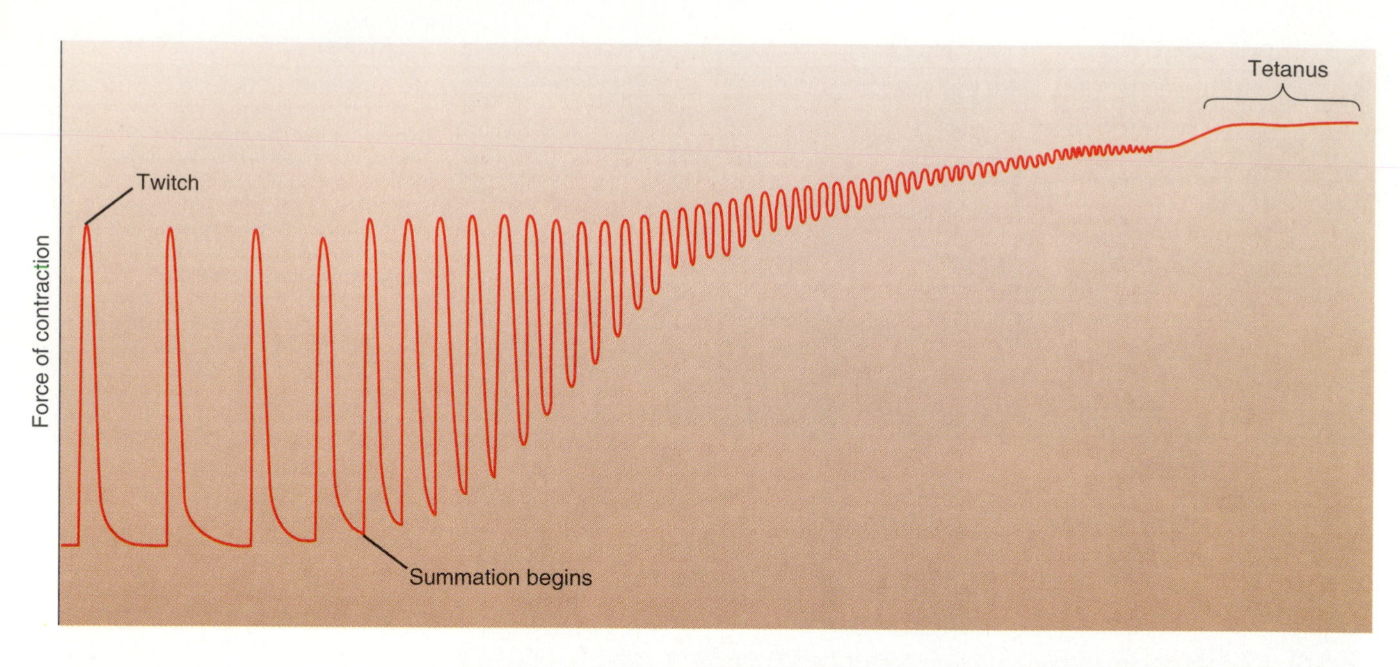

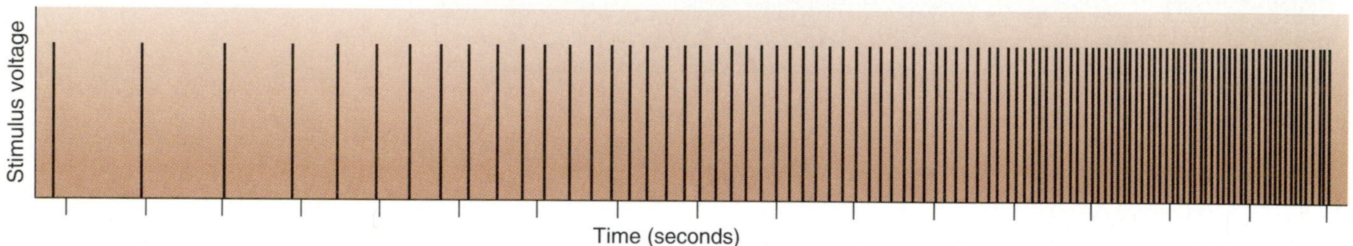

FIGURE 10.12

The effect of stimulus frequency on the force of contraction. The procedure was the same as in Fig. 10.11 except that instead of increasing the voltage of the shocks, the frequency was increased. When the frequency was high enough (about two shocks per second), each twitch began before the previous one was over, and the twitches began to summate. As the frequency reached about 15 shocks per second, individual twitches disappeared, and the contraction became steady. This steady contraction, called tetanus, is the normal mode of muscle contraction.

Each brief shock to the nerve evokes a **twitch,** which is the response to a single action potential and the briefest movement a muscle can produce. Brief as it is, however, a twitch lasts at least a hundred times as long as the action potential that evokes it. This means that many action potentials can occur before a muscle can relax after a twitch, and the effect of many action potentials can **summate** to produce a contraction with greater force than that of a twitch. During normal contraction the frequency of action potentials is high enough (at least 15 per second) that the muscle does not relax at all, and the force is continuous (Fig. 10.12). This normal condition is called **tetanus,** which also happens to be the name of a disease characterized by steady contraction of muscles. Thus, increasing the number of action potentials in a given period increases the force of contraction. This is the third principle of the motor code: **The strength of muscle contraction increases with the frequency of action potentials.**

To summarize, the type of movement is controlled by directing action potentials to different sets of muscles, and the force of movement is controlled by activating the appropriate number of motor units within each muscle with the appropriate frequency of action potentials.

How Muscles Contract

The Structure of Skeletal Muscle. Antony van Leeuwenhoek, the first microscopist, observed that meat, which is skeletal muscle, has an intriguing pattern of bands or striations running through its fibers (Fig. 10.13). These bands are now desig-

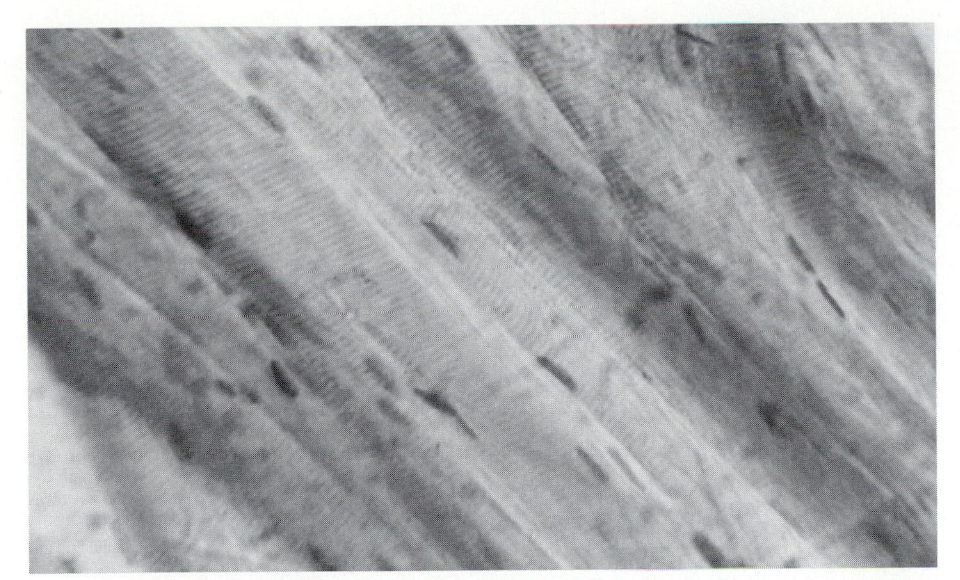

A

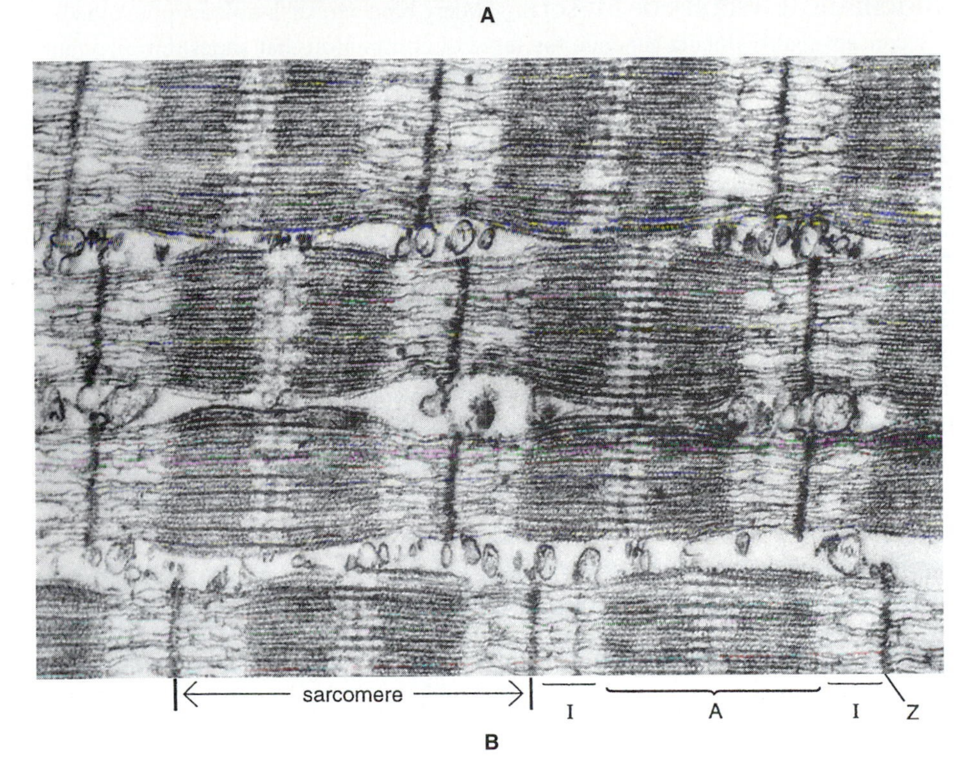

B

FIGURE 10.13

(A) Human striated muscle as seen in the light microscope. The muscle cells run from the upper left to lower right. Note the many nuclei (dark ovals) in each cell. At right angles to the cells are dark A bands alternating with I bands. (B) In electron micrographs of frog muscle the I bands can be seen to consist of thin lengthwise filaments attached to Z lines, and the A band to consist of thick and thin filaments. The distance between Z lines is about 2 μm. Compare Fig. 10.9B.

nated **A bands** and **I bands.** The A and I bands repeat from one end of a fiber to the other, with each A band enclosed by two I bands. Until a few decades ago the only contribution of these striations to the understanding of skeletal muscle was a new name: **striated muscle.** Striations were generally ignored earlier in this century because there was no place for them in the prevailing theory of contraction, which assumed that elastic filaments contracted within the fibers like rubber bands.

Refinements in light and electron microscopy in the early 1950s paved the way for a new theory in which the banding pattern of striated muscle did fit. A. F. Huxley (cited in Chapter 8 for his work on the squid giant axon) invented a new type of microscope that allowed him to see that the A bands of frog muscles did not shorten during contraction. This was contrary to the old theory of elastic filaments. Independently, H. E. Huxley (not directly related to A. F.) helped refine electron microscopy to reveal that muscle fibers have two types of filament in them, neither of which changes length. **Thick filaments** occur only in A bands. **Thin filaments** occur in I bands and extend into the A bands interspersed among the thick filaments. Both types of filament can be seen in Fig. 10.9B, which shows a cross section of an A band. The thin filaments connect to a **Z line,** which is actually a Z disc in three-dimensional life. Thus the sequence ZIAIZ repeats all along the length of the muscle fiber. Each repetition is called a **sarcomere** (Fig. 10.13B). Presumably the function of an entire muscle can be understood from just one sarcomere.

In 1954 both Huxleys jointly proposed a new theory of muscle contraction that is now generally accepted. Their **sliding filament theory** proposes that the thin and thick filaments slide among each other, and that the force of muscle contraction comes from the interaction of the two types of filaments.

BIOCHEMICAL STUDIES OF MUSCLE. Many biochemical studies support the basic concept that filaments generate the force of contraction that slides thin filaments past thick filaments. Each thick filament consists of several hundred long protein strands with globular protein heads at each end. The protein that makes up thick filaments is **myosin.** The thin filaments consist of two twisted strands of globular proteins called **actin.** (This protein is also found in microfilaments of other cells.) One line of support for the sliding filament theory comes from studies of purified actin and myosin, which form a gel-like suspension when mixed. If ATP is added to this suspension, the actin and myosin bind together to form a precipitate, but only if Ca^{2+} is also added. Apparently the force of contraction produced by the interaction of thick and thin filaments uses ATP as an energy source and Ca^{2+} as an activator.

In electron micrographs one can actually see connections, called **cross-bridges,** between actin and myosin in muscle. These cross-bridges are the heads of the thick filaments. Biochemical studies have shown that these heads have ATPase enzyme activity in the presence of actin. It appears, therefore, that the cross-bridges from myosin to actin are the sites where the energy from ATP is converted to the mechanical energy of contraction. How this energy conversion occurs is still uncertain. The ATP somehow causes the cross-bridges at both ends of a thick filament to pull the actin toward the middle of the sarcomere. Since this happens all along the length of a muscle fiber, the fiber shortens.

Calcium activates muscle contraction by first attaching to another protein, **troponin,** which resides on actin. Ca^{2+} triggers contraction by causing a change in troponin, which then acts on **tropomyosin,** another protein found on actin. It is thought that at rest tropomyosin blocks the sites for myosin attachment to actin, and the change in troponin displaces the tropomyosin, allowing the cross-bridges to attach. Once the myosin heads attach to actin, their ATPase activity triggers contraction of the muscle.

As usual, answering these questions has raised others. From where does the Ca^{2+} enter the sarcomere, and how do action potentials trigger its entrance? For invertebrate muscles the answer to this question appears to be that the influx of Ca^{2+} during the action potential also triggers contraction. For vertebrates the answer is more complicated. It has long been known that a type of smooth endoplasmic reticulum (see p. 45) called **sarcoplasmic reticulum** (Fig. 10.9) actively takes up Ca^{2+} from an external solution. It seems reasonable, therefore, that Ca^{2+} is stored in the sarcoplasmic reticulum when the muscle is relaxed and is released when contraction is initiated. Apparently the calcium permeability of the membranes of sarcoplasmic reticulum increases in response to the action potential in the muscle fiber, releasing the Ca^{2+} that triggers contraction. Action potentials travel to the sarcoplasmic reticulum by way of **T tubules** (transverse tubules) that run from the plasma membrane into the fibers.

SUMMARY OF THE SLIDING FILAMENT THEORY. All this information and theory from several decades and many laboratories by many techniques can now be integrated into a coherent idea of the mechanism by which a muscle fiber contracts. Action potentials in motor axons release acetylcholine from a neuromuscular synapse, which evokes a depolarizing endplate potential (EPP) on the plasma membrane of the muscle fiber. This EPP triggers action potentials in the plasma membrane, which travel down the T tubules that encircle the myofibrils (Fig. 10.14). The action potentials trigger an increase in calcium permeability in the membrane of sarcoplasmic reticulum, which allows the Ca^{2+} stored in the sarcoplasmic reticulum to leak out. The Ca^{2+} causes a change in the configuration of troponin on the actin molecules, which displaces tropomyosin from the binding site for the myosin heads. The myosin heads can then attach to the actin, forming cross-bridges. The myosin heads also convert ATP energy into mechanical energy, which causes the cross-bridges to pull on the actin filaments at each end.

Relaxation occurs when the action potentials cease, and the calcium permeability of the sarcoplasmic reticulum returns to its resting level. Calcium pumps then transport the Ca^{2+} from the sarcomeres into the sarcoplasmic reticulum, breaking the cross-bridges. It may seem incredible that something as weak as the bond between two protein filaments could account for the strength in your own muscles, let alone the strength of a professional weight-lifter. There are millions of cross-bridges in each muscle fiber, however, and there are thousands of muscle fibers in a muscle. A skeletal muscle 1 cm^2 in diameter has, in fact, enough cross-bridges to support a mass of up to 3 kg. Thicker muscles generate greater forces in direct proportion to their cross-sectional area.

Because the number of possible cross-bridges depends on the amount of overlap of thick and thin filaments, the maximum force that can be produced by a muscle varies with its length. If the muscle is stretched or compressed, it will not be able to produce as much force as it could at an intermediate length. As one might expect, in a resting leopard frog the lengths of the muscles used for jumping are optimal for producing maximal force for a sudden leap, as are the number of motor units activated and the frequency of action potentials (Lutz and Rome 1994).

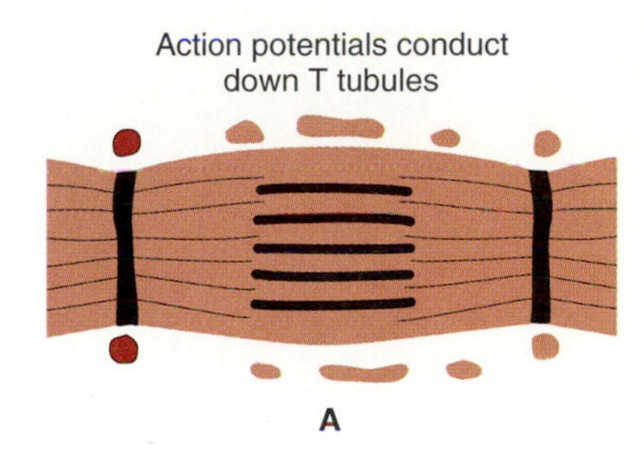

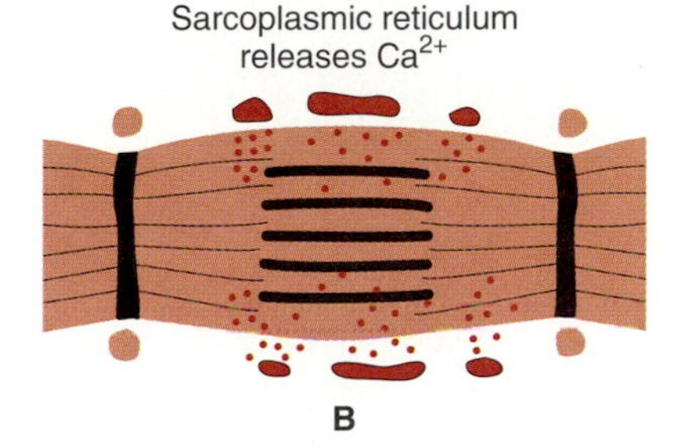

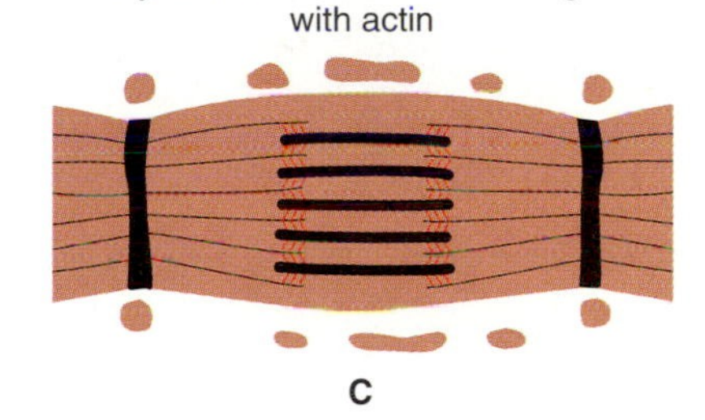

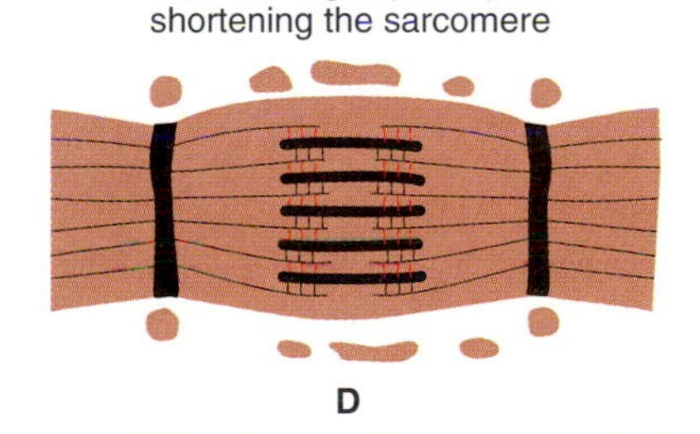

FIGURE 10.14

Activation of vertebrate skeletal muscle. (A) Action potentials in the motor axon evoke action potentials in the muscle fiber membrane. These action potentials travel down the T tubules. (B) The action potentials trigger the release of Ca^{2+} from the sarcoplasmic reticulum. (C) Ca^{2+} causes a change in the structure of troponin, which allows cross-bridges to form. (D) The cross-bridges (myosin heads) pull the actin and attached Z discs toward the center of the sarcomere.

■ **Question:** What is the probable cause of the stiffening of muscle after death or during cramping?

Energetics of Muscle Contraction

SOURCES OF ENERGY. The energy for muscle contraction comes from the conversion of ATP into ADP and inorganic phosphate. Approximately 70% of the available energy is used directly for contraction; the remainder is used in transporting Ca^{2+} back into the sarcoplasmic reticulum during relaxation. There is not enough ATP in a sarcomere to power more than a few twitches. If that small amount is depleted, the result is **cramping,** in which cross-bridges form but there is no contraction. The muscle simply becomes rigid. The same thing occurs after death in **rigor mortis.** Cramping does not often occur, however, because molecules called **phosphagens** usually maintain a steady level of ATP by transferring phosphate groups to ADP. In vertebrates and some invertebrates the phosphagen is creatine phosphate; in most invertebrates it is arginine phosphate.

FATIGUE. Muscles usually fatigue from some other cause before they run out of ATP. Fatigue—the reduction in force that a muscle can generate—is often due to the accumulation of byproducts of the production and breakdown of ATP. Inorganic phosphate released from ATP is one cause of fatigue. Lactic acid from the anaerobic production of ATP (see pp. 53–56) also contributes to fatigue by reducing the pH inside the sarcomeres. In addition, lactic acid is one cause of muscle damage in severe exercise, which is responsible for soreness. Training can increase the resistance to fatigue by increasing the blood supply, amounts of metabolic enzymes, and levels of the protein **myoglobin,** which transports oxygen to mitochondria. In addition, exercise stimulates muscle fibers to produce a larger number of filaments, which increases the diameter of the fibers. Exercise may also increase the number of muscle fibers.

Varieties of Skeletal Muscle

One expects that as the result of evolution, animals come to use their muscles in such a way that they work optimally. For some muscles the optimal performance would occur when they do as much work as they can in the least time. For others, optimal performance would occur when they do as much work as they can using the least energy. In other words, some muscles must produce peak **power,** while others must have peak **efficiency.** The speed of contraction influences both power and efficiency, but the optimal speed for power is generally about 1.5 times that for efficiency. For this reason there are generally at least two categories of skeletal muscle—one specialized for efficiency and one for power. Those specialized for efficiency are called **slow-twitch** or **Type I fibers.** They produce ATP aerobically, and they contract relatively slowly. They have a reddish color due mainly to a generous supply of blood vessels and to the cytochromes in their numerous mitochondria. Type I fibers also have a large amount of the oxygen-binding pigment myoglobin. Muscles specialized for power are **fast-twitch** or **Type II white fibers.** They contract more rapidly but rely on the anaerobic breakdown of glycogen for their ATP. Type II fibers have a light color, few mitochondria, and little myoglobin. They are also generally thicker than Type I fibers.

Both types of fiber can occur within the same muscle, although many muscles have much more of one type of fiber than of the other. Slow-twitch fibers predominate in muscles that must contract for long periods to maintain an erect posture. A familiar example of slow-twitch fibers is the dark meat of chickens, which occurs in the legs and back. Fast-twitch fibers predominate in muscles that contract rapidly but not often. The light meat in the breast (flight muscles) of chickens consists largely of fast-twitch fibers. If required to work for long periods, fast-twitch fibers quickly fatigue and get sore. Intermediate types of muscle also occur. In ducks and most other birds that can fly for long distances, the breast muscles consist largely of an intermediate type of fiber—the **fast-contracting red fiber.** This type, also called the **Type II red fiber,** has many mitochondria but can contract rapidly.

Training may affect the proportions of muscle types in a muscle. Sprinters generally have a greater-than-average proportion of fast-twitch fibers in their muscles, while marathoners usually have a greater proportion of slow-twitch fibers, and middle-distance runners have about average proportions of fast- and slow-twitch fibers. These training effects may result from the fact that sustained stimulation of the motor nerve to a fast-twitch muscle will transform it into a slower muscle.

Cardiac Muscle

Vertebrates have two other types of muscle besides skeletal muscle. Cardiac muscle, of which hearts are made, has the striations and parallel filaments of skeletal muscle but differs from it in many respects (Fig. 10.15). The cells of cardiac muscle do not fuse into cylindrical fibers, but are joined by electrically conducting **intercalated discs.** The cells branch out to form interconnecting networks that ensure that action potentials trigger contraction throughout the entire mass of cardiac muscle. Cardiac muscle also differs from skeletal muscle in not requiring excitation by the nervous system. Instead, some of the cardiac muscle cells generate action potentials spontaneously, and these action potentials conduct through the intercalated discs and trigger contraction in other cardiac muscle cells. Thus vertebrate hearts can go on contracting after they have been entirely removed from the body. Cardiac muscle also differs from skeletal muscle in having fewer T tubules (virtually none in most vertebrate classes), and in getting more of its Ca^{2+} influx across the plasma membrane rather than from the sarcoplasmic reticulum only. The relevance of all this to heart function will be explained in the following chapter.

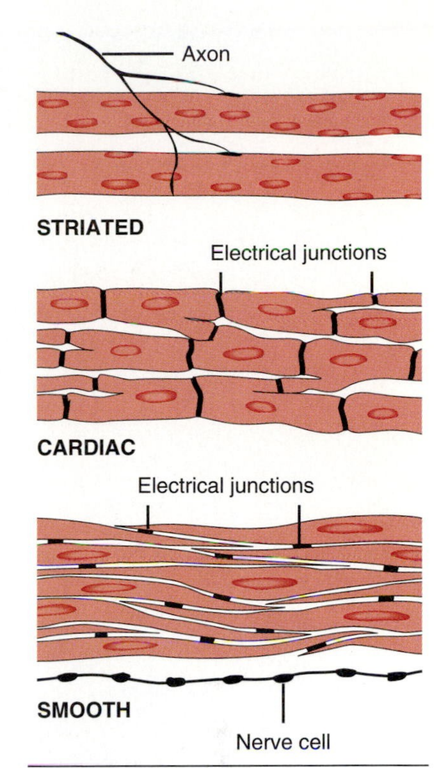

FIGURE 10.15

The three types of vertebrate muscle: skeletal, cardiac, and smooth. While skeletal muscle is multinucleate, cardiac and smooth muscle have only one nucleus per cell. Both skeletal and cardiac muscle cells are striated and fiber shaped, while smooth muscle cells are not striated and are spindle-shaped. Neither cardiac nor smooth muscle cells receive direct innervation through neuromuscular junctions, but smooth muscle can be controlled by nerve cells that release transmitters in their vicinity. Both cardiac and smooth muscle cells are electrically interconnected.

Smooth Muscle

A third type of vertebrate muscle is smooth muscle, so called because it lacks the striations of skeletal and cardiac muscle (Fig. 10.15). Smooth muscle is usually found in hollow organs, such as blood vessels, the digestive tract, and the urinary bladder. Smooth muscle cells are spindle-shaped. Their thick and thin filaments are not arranged in neat array like those of striated muscles, but in such a way that during contraction each cell twists like a cloth being wrung out. Like the cells of cardiac muscle, those of smooth muscle are electrically interconnected so an action potential in one cell sends a wave of contraction throughout the muscle. One example of such a wave of contraction is the **peristalsis** that occurs in the esophagus during swallowing.

Action potentials or graded depolarizations in smooth muscle result from an influx of Ca^{2+} rather than Na^{+}. The same Ca^{2+} triggers the contraction. There is no troponin for the Ca^{2+} to bind to in smooth muscle. Instead, Ca^{2+} binds to **calmodulin,** a troponin-like protein that regulates calcium action in many cells. A Ca–calmodulin complex acts on the myosin heads to activate cross-bridge formation. Contraction is extremely slow in smooth muscle and can be excited by some nerve endings and inhibited by others. These nerve endings are not synapses, and they are not always required to trigger contraction. Smooth muscle of the intestine, for example, will contract in response to direct mechanical stimulation. The excitability of many smooth muscles is controlled by the autonomic nervous system, with the parasympathetic and sympathetic divisions having opposite effects.

Types of Invertebrate Muscle

A few differences among invertebrate muscles should be mentioned to indicate the numerous departures from the vertebrate patterns described above. The control of vertebrate striated muscle is not at all like that of most animals. In the majority of invertebrates, Ca^{2+} triggers contraction not only by its effect on troponin but also by an effect on myosin. Other differences occur in the pattern of innervation of skeletal muscle. In arthropods, for example, a typical muscle fiber is innervated by several neurons. One excitatory nerve fiber produces a large depolarization that triggers a fast contraction, and another produces a small depolarization and a slow contraction. There is also at least one inhibitory neuron that causes hyperpolarization. This multiple innervation seems to be a way of getting fine gradations of force with a limited number of muscle fibers.

Another variation in muscle function occurs in bees and wasps, flies, beetles, and bugs. The flight muscles of these animals do not contract in synchrony with the action potentials in their motor axons, but often at a frequency higher than the maximum frequency of action potentials. These muscles are therefore called **asynchronous muscles** (= fibrillar muscles). Neural stimulation of these asynchronous muscles maintains a constant level of Ca^{2+} in the sarcomeres, but cross-bridges do not form until the muscles are quickly stretched. The movement of the wing upward therefore activates the muscles that produce the following down-stroke (Fig. 10.4B). In a midge this cycle can happen 1000 times a second.

Another interesting variation in muscle function is found in the adductor muscles of bivalve molluscs such as mussels, oysters, scallops, and clams. Once the cross-bridges form in these **catch muscles,** they remain rigid with little further use of ATP. These smooth muscles give bivalves the ability to "clam up" against predators for days at little energetic expense. For their size they are the strongest muscles known.

Movement Without Muscle

AMEBOID MOVEMENT. Many protozoans and many cells within animals, propel themselves by means other than muscle. One such mechanism is ameboid movement, named for the protozoan *Amoeba* (pp. 458–459). The osteoblasts, osteoclasts, and white blood cells of humans and many other vertebrates wander about by ameboid movement. Under the light microscope it appears that ameboid motion results from a flow of **endoplasm,** a fluid form of cytoplasm, in the direction of overall motion (Fig. 10.16). The endoplasm moves into projections of the cell called **pseudopodia** and then turns into a clearer, less-fluid **ectoplasm.** The transformation from endoplasm to ectoplasm involves the assembly of actin filaments under the control of Ca^{2+}.

CILIA AND FLAGELLA. Another type of cellular movement without muscle uses cilia and flagella. Cilia and the flagella of eukaryotes are virtually identical to each other

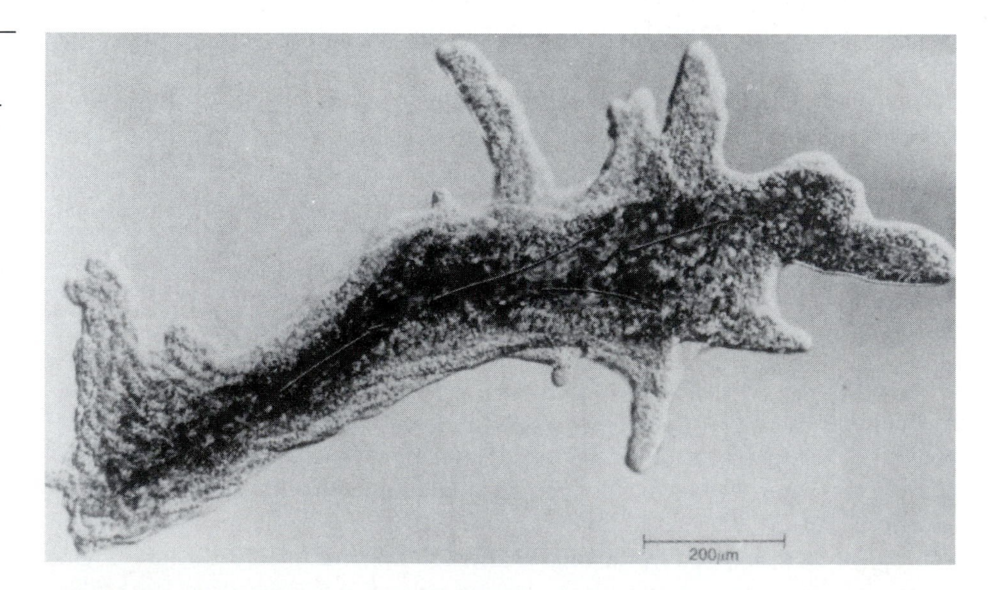

FIGURE 10.16
Light micrograph of the giant ameba *Chaos carolinensis.* Blurring of the central portion of this protozoan resulted from the streaming of cytoplasm responsible for ameboid movement.

in structure, but cilia are usually shorter (less than 15 μm long) and occur in large groups. Flagella are up to 2 mm long (in some insect sperm) and usually occur individually or in small groups. Both cilia and flagella have similar core structures, called **axonemes.** The axonemes usually have a distinctive "9 + 2" arrangement of nine microtubule doublets surrounding two microtubules (Fig. 10.17A). Cilia and flagella attach to a **basal body** (= kinetosome) in the cytoplasm. How they are controlled is a mystery.

Cilia are responsible for the locomotion of many protozoans such as *Paramecium,* some small invertebrates such as flatworms and ribbon worms (phyla Platyhelminthes and Nemertea), some aquatic molluscs, and some invertebrate larvae. In some animals, cilia circulate water during filter feeding and respiration. In mammals, cilia lining the respiratory tract propel debris-trapping mucus away from the lungs, and cilia lining the oviduct propel the oocyte toward the uterus.

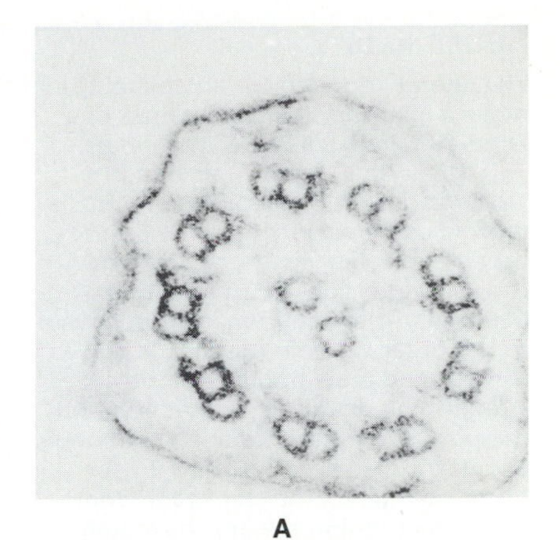

A

B

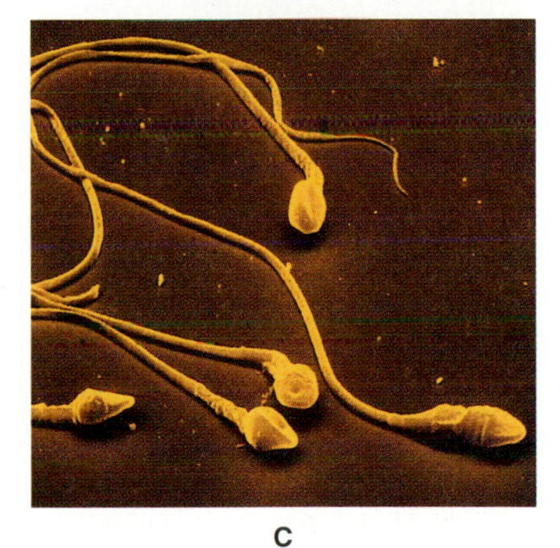

C

FIGURE 10.17

(A) Electron micrograph showing cross section of a flagellum from a sperm cell of the sea urchin *Lytechinus.* The "9 + 2" configuration of microtubules is the axoneme. (B) Cilia of the protozoan *Opalina* beat in waves. The distance between successive waves of cilia is about 8 μm. (C) Human sperm cells, each with a single flagellum about 40 μm long.

Body Size and Locomotion

ALLOMETRY. Among the favorite characters of fairy tales are humans and other animals who have grown or shrunk incredibly. Although such dwarfs and giants are usually portrayed as being ordinary in every way except size, their movements would in fact be far from ordinary. Consider what would happen if you suddenly grew to 10 times your present height and retained the same proportions. Your weight would have increased in proportion to the cube of your height. However, the cross-sectional areas of your body and limbs would have increased only in proportion to the square of your height. Thus you would weigh a thousand (10^3) times as much, but your muscles and bones would be only a hundred (10^2) times stronger. If you now weigh 130 pounds, you would weigh a staggering (literally) 130,000 pounds, and all that weight would have to rest on legs adapted to support only 13,000 pounds. Your bones would probably be crushed, or at the very least you would be pinned to the floor by your own weight. Other problems would arise if you shrank to one-tenth your present height while retaining the same proportions. You would weigh 0.13 pounds (about 2 ounces) but would have the strength appropriate for a person 10 times as massive. You would probably smash yourself against the ceiling the first time you took a step.

The study of the differential effects of changing linear, area, and volume dimensions in organisms is referred to as **allometry** (Greek *allo-* different + *metron* a measure) or **scaling.** Allometric considerations suggest that during evolution there have probably been many compromises among linear dimension, area, and volume. For example, it is probable that in the evolution of squirrels there were times when being larger would have helped preserve them from predators, but the increased weight would have hindered their escaping through the trees. Allometric constraints are evident in several mathematical regularities in animals of various sizes. One of the most studied of these regularities is in the energy an animal uses in locomotion (Fig. 10.18 on page 212). For animals ranging in size from mosquitoes to horses, the energy needed for an animal to transport a unit of its body mass for a given distance is proportional to the animal's mass raised to some power that depends on whether the animal swims, flies, or runs.

SWIMMING. This and other allometric relationships have led a growing number of zoologists to look for physical explanations. The case of swimming is most easily understood, since buoyancy allows us to ignore the effects of gravity. Swimming is possible when an animal can continue its forward movement between swimming strokes. This can happen only if the **inertial force** (the acceleration of the animal with respect to

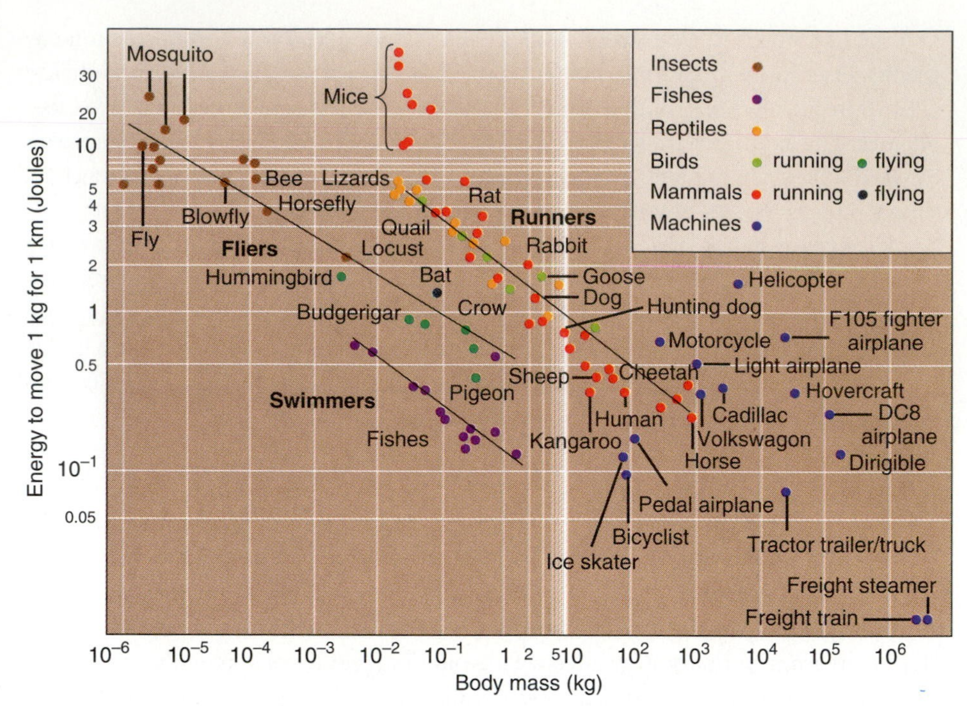

FIGURE 10.18

The energy required for an animal to move a kilogram of its body mass one kilometer, versus its total body mass. This is a double logarithmic plot in which each unit on an axis represents a tenfold change. Plotting data in this way is equivalent to taking the logarithm of both sides of an equation. If one side of the equation is proportional to the other variable raised to a power, this method plots the data along a straight line with a slope that equals the power. This technique is common in allometry. (See Figs. 3.10 and 9.7 for other examples.) The slopes for swimming and running are approximately –⅓. This shows that for swimming and running the energy required to move a kilogram of body mass one kilometer is proportional to one divided by the cube root of the body mass. Since the body mass is proportional to the cube of body length, the energy per kilogram per kilometer is proportional to one divided by body length. The slope for flying is approximately –¼.

the water) exceeds the **viscous force,** which tends to stop the animal. The ratio of inertial to viscous forces is given by the **Reynolds number,** Re:

$$\text{Re} = \frac{(\text{length of body}) \times (\text{speed}) \times (\text{density of water})}{(\text{viscosity of water})}$$
$$= 100 \times (\text{length of body in cm}) \times (\text{speed in cm/sec})$$

All else being equal, therefore, the higher the Reynolds number, the more easily an animal can swim. For similarly shaped animals, the longest and fastest ones will expend less energy in moving a kilogram of their body mass for one kilometer. The higher Reynolds number of larger animals may explain why larger fishes are more efficient as swimmers, as shown in Fig. 10.18.

The blue whale (*Balaenoptera musculus*), with a body length of about 2500 cm and a cruising speed of 1000 cm/sec, has an Re of more than 10^8. It should be able to swim with extreme efficiency, although no one has yet figured out how to demonstrate this. Salmon, which have to be highly efficient to migrate upstream without eating, have an Re of about 10^6. Human swimmers, perhaps surprisingly, have an Re about equal to that of salmon. We are less efficient probably because we are not as streamlined. For much shorter and slower swimmers, Re drops sharply. Viscous forces overcome inertial forces, and water starts to act like syrup or even tar. Swimming for such animals is inefficient, if not impossible. Diving beetles, with an Re of about 10^2, are at about the limit at which real swimming is possible. Many invertebrate larvae and fish hatchlings with lower Reynolds numbers join the plankton, drifting passively with the current. Under a microscope many protozoans with even smaller values of Re often appear to move quite rapidly, but they are not actually swimming. A 0.02-cm paramecium moving through water at about 0.1 cm/sec (Re approximately 0.2) is essentially crawling through the water with its cilia.

FLYING. The limitations imposed by body size also apply to flying if one corrects for the much lower density and viscosity of air. (For the same length and speed, Re for flying is about 1/15 that for swimming.) The ability to glide through the air between wing strokes also depends on having a high Re. The California condor (*Gymnogyps californianus*) has an Re value on the order of 10^5 and can glide for extended periods. Even large dragonflies and butterflies have Re values large enough (more than 10^3) to permit periodic gliding. Birds and insects smaller than that have to flap continually to maintain forward motion.

Unlike swimming, flying is complicated by the fact that the animal must not only move forward but also generate **lift** to overcome gravity. Scientists do not fully understand how flying animals generate lift (see Chapter 32 for discussion of flight in insects and Chapter 38 for flight in birds). There is, however, one fact that is firmly grounded (so to speak) in numerous unsuccessful attempts at human flight: The surface area of wings has to be large enough to generate enough lift to overcome the weight of the body. In other words, the **wing loading** (body mass divided by wing area) must not be too large. Birds with large wing loading, such as ducks, must fly fast to generate additional lift. However, since body mass generally increases as the cube of body length, but wing area increases only as the square of body length, there is an upper limit on body size for flapping flight.

The generation of lift by flapping while in flight is hard to study, for obvious reasons. It is much easier to study how an animal takes off from a dead start by flapping. By adding weights until animals were no longer able to take off, Jim Marden (1987) has measured the maximum lift generated by more than 160 insects, bats, and birds of numerous species. He found that the maximum lift is directly proportional to the mass of the flight muscles (Fig. 10.19). This result is rather surprising for two reasons. First, allometric considerations suggest that lift should be proportional to $(\text{mass})^{2/3}$ rather than directly proportional to mass, since force is proportional to the cross-sectional area of muscle. Second, the result implies that the diverse flight muscles among insects, bats, and birds all generate the same lift per gram.

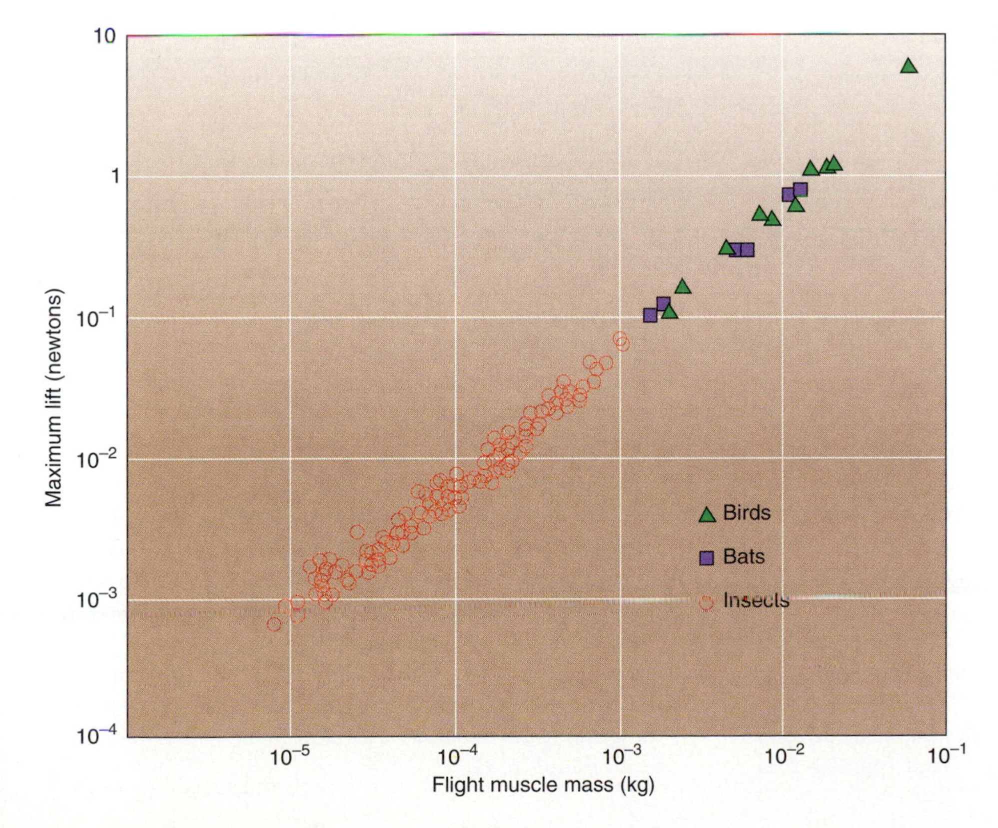

FIGURE 10.19

Double logarithmic plot of the lift produced by flapping in more than 160 insects, birds, and bats versus the mass of their flight muscles. The data plot along a line with slope equal to 1.0, showing that lift is directly proportional to the mass of flight muscle.

Summary

The internal environments of animals are protected by integuments that range in complexity from a thin epidermis to thick shells, cuticles, and skins. The skin of vertebrates consists of a protective epidermis and an underlying dermis. Vertebrate skin also includes many other structures, such as hairs, sensory organs, and glands, as well as pigments and structures that produce color.

The bodies of most animals are supported by skeletons. The skeleton may simply be internal fluid pressure—a hydroskeleton—or it may be a rigid structure. Many invertebrates have rigid exoskeletons that are parts of the integument, while other animals have endoskeletons that are enclosed within other tissues. In vertebrates the endoskeleton consists mainly of bone, which replaces most of the cartilage during embryonic development.

Skeletal muscle attached to the skeleton enables animals to move. Different movements are evoked by activating muscles that attach to different bones, and the force of contraction is controlled by varying the number of motor units activated and the frequency of action potentials. The connection between nerve cells and muscle cells is the neuromuscular synapse, which triggers changes in the voltages across membranes in muscle cells. These voltages trigger a release of calcium, which triggers interaction between thick and thin protein filaments in the muscle. According to the sliding filament theory, muscle contraction is not due to shortening of filaments but to sliding of thick filaments past thin filaments. In response to calcium, thick filaments form cross-bridges with thin filaments, then pull on the thin filaments to shorten the muscle. The energy for contraction comes from ATP.

Skeletal muscle fibers are specialized for either efficiency or power. The former are slow-twitch fibers, and the latter are fast-twitch fibers. In addition to these two types of skeletal muscle cells, vertebrates also have cardiac muscle, which makes up the heart, and smooth muscle, which controls the movement of the digestive tract and the diameter of hollow and tubular organs. Numerous other kinds of muscles occur in invertebrates. Other mechanisms for movement are ameboid motion and cilia or flagella.

Length, surface area, and volume change by different amounts when body size changes. Swimming and flying are most efficient for long animals, but flying is impossible when body mass reaches a certain limit.

Key Terms

integument
epidermis
dermis
hydrostatic skeleton
exoskeleton
endoskeleton
cartilage
flexor
extensor
neuromuscular synapse
motor unit
tetanus
striated muscle
thin and thick filaments
sarcomere
sliding filament theory
myosin
actin
cross-bridge
sarcoplasmic reticulum
T tubules
slow-twitch and fast-twitch fibers
cardiac muscle
smooth muscle
ameboid movement
cilium
flagellum
allometry
Reynolds number

Chapter Test

1. Describe how the epidermis of an invertebrate differs from that of a vertebrate. (pp. 192–193)
2. What functions are served by the vertebrate epidermis? What functions are performed by the dermis? (pp. 192–193)
3. Give an example of an animal with each type of skeleton—hydrostatic, exoskeleton, and endoskeleton—and explain how the contractions of its muscles produce locomotion. (pp. 195–197)
4. Describe three factors that affect how much force you exert when lifting an object with one hand. (pp. 200–204)
5. From memory make a sketch that shows a motor axon, a motor unit, a muscle fiber, and a sarcomere. Compare your sketch with Figs. 10.9 and 10.10.
6. Describe the relationship among the following: actin, myosin, thin filaments, thick filaments, I band, A band. (pp. 204–206)
7. State the essence of the sliding filament theory. Describe a competing theory and explain why it is not as widely accepted as the sliding filament theory. (pp. 204–206)
8. Describe the functions of Ca^{2+} in movement and support. (Don't forget the role of Ca^{2+} in the neuromuscular junction.) (pp. 197, 206–207)
9. For each of the following, state one similarity and one difference between it and vertebrate skeletal muscle: insect asynchronous muscle; the catch muscle of molluscs; ameboid movement; the movement of cilia or flagella. (pp. 209–211)
10. Examine the pictures of very large and very small mammals in Chapter 39. What do you notice about the thickness of the legs in relation to body size? How would you explain this relationship? (p. 216)

11. By filling in the following table, show how the properties of slow-twitch and fast-twitch muscle fibers relate to the greater efficiency of the former and the greater power of the latter. (p. 208)

	Slow-Twitch	Fast-Twitch
Speed of contraction (fast or slow)		
Usage (continuous or intermittent)		
Number of mitochondria (few or many)		
Type of metabolism (aerobic or anaerobic)		
Resistance to fatigue (little or great)		
Color (red or white)		

12. Fill in the following table and explain how each item is appropriate for the type of muscle (pp. 208–209):

	Skeletal Muscle	Cardiac Muscle	Smooth Muscle
Typical location in vertebrates			
Voluntary control?			
Striated?			
Each cell with only one nucleus?			
Cells interconnect electrically?			
Neuromuscular junctions present?			

■ Answers to the Figure Questions

10.1 The flatworm integument is protected by mucus; that of the human is protected by the stratum corneum, which consists of dead, dry cells.

10.2 Like all albinos, laboratory rats and mice lack functional genes for production of melanin in hair and the retina, so the hair is white and the retina appears red from the blood circulating in it. (Albinism is simply a useful indicator of the genetic purity of laboratory rodents.) The colors of the peacock are probably as follows: blue = Tyndall blue; yellow-green due to Tyndall blue through yellow pigment; orange due to pigment; white due to absence of pigment or structural color; black melanin.

10.6 The femur is homologous with the humerus, the tibia is homologous with the radius, the fibula is homologous with the ulna, and the tarsals are homologous with the carpals.

10.7 Gastrocnemius—extension of the foot, as in standing on tiptoes; triceps brachii—straightening of arm; rectus abdominus—tightening of waist.

10.8 Farther away, applying the force closer to the hand. Speed of hand movement would have been reduced.

10.14 Cross-bridges connecting myosin to actin do not break. (Relaxation requires the presence of ATP.)

Readings

Recommended Readings

Alexander, R. M. 1991. How dinosaurs ran. *Sci. Am.* 264(4):130–136 (Apr).

Alexander, R. M. 1992. *Exploring Biomechanics: Animals in Motion.* New York: Scientific American Library.

Caplan, A. I. 1984. Cartilage. *Sci. Am.* 251(4):84–94 (Oct).

Denny, M. W. 1990. Terrestrial *versus* aquatic biology: the medium and its message. *Am. Zool.* 30:111–121. (*Swimming versus flying.*)

Govind, C. K. 1989. Asymmetry in lobster claws. *Am. Sci.* 77:468–474.

McMahon, T. A., and J. T. Bonner. 1983. *On Size and Shape.* New York: Scientific American Books.

Smith, K. K., and W. M. Kier. 1989. Trunks, tongues, and tentacles: moving with skeletons of muscles. *Am. Sci.* 77:28–35.

Stossel, T. P. 1994. The machinery of cell crawling. *Sci. Am.* 271(3):54–63 (Sept).

Tucker, V. A. 1975. The energetic cost of moving about. *Am. Sci.* 63:413–419.

Webb, P. W. 1988. Simple physical principles and vertebrate aquatic locomotion. *Am. Zool.* 28:709–725.

Additional References

Lutz, G. J., and L. C. Rome. 1994. Built for jumping: the design of the frog muscular system. *Science* 263:370–372.

Marden, J. H. 1987. Maximum lift production during takeoff in flying animals. *J. Exp. Biol.* 130:235-258.

CHAPTER

11

Circulation and Immunity

Glass frog (*Centrolenella*) with heart visible through skin.

All animals, including humans, evolved from single-celled marine organisms that used the sea as a source of oxygen, minerals, nutrients, and thermal energy and as a dump for metabolic wastes. In a sense, we have never left our marine origins. Our cells are still bathed in a kind of internal sea—a salty liquid called the **interstitial fluid.** In large animals the interstitial fluid surrounding most cells are far from the sources of oxygen, nutrients, and the other necessities of life, but these requirements come to the interstitial fluid on the ebb and flow of an internal tide of **blood.** Blood flows through a **circulatory** (= vascular) **system,** propelled in most animals by contractions of a **heart.** This chapter presents the general concepts of blood and its circulation, as well as its defenses against bacteria, viruses, and other pirates of the internal sea.

Blood

In many invertebrates, blood is similar to the interstitial fluid. In other invertebrates and in all vertebrates, blood contains specialized cells and molecules that aid in its functions. Removing the cells from vertebrate blood leaves the fluid portion, called **plasma.** Plasma often contains substances that cause **clotting,** which reduces blood loss in case of injury. If plasma is allowed to clot and the clot is removed, the remaining fluid is **serum.** Blood may also contain **respiratory pigments,** which increase the ability of blood to transport oxygen, as will be described in the next chapter. The respiratory pigments may simply be suspended in the blood, or they may be contained in blood cells. In vertebrates the respiratory pigment **hemoglobin** is contained in **red blood cells** (erythrocytes). Still other blood cells and molecules combat viruses, bacteria, and parasites. In vertebrates the **white blood cells** (leukocytes) serve these functions. (See Fig. 11.1 on page 218 and the figure on p. 16.)

Open and Closed Circulatory Systems

Some of the enormous variety of circulatory systems will be described in the chapters in Unit 4. For now we can consider two major types of systems—closed and open (Fig. 11.2 on page 219). In **closed circulatory systems,** blood is always confined within the vascular system (arteries, capillaries, veins, and heart). The interstitial fluid is formed by filtration of blood across the capillary walls, and it therefore resembles serum in being without blood cells and large molecules. Closed circulatory systems occur in vertebrates, such as ourselves, and in annelids such as earthworms (Fig. 11.3A on page 220), cephalopod molluscs such as the octopus, and a few other groups. (See Fig. 21.10 for a summary.) Our own closed system will be examined more closely later in this chapter.

In **open circulatory systems,** blood is not confined within the vascular system. Instead the blood that leaves the heart simply empties into spaces that make up the **hemocoel** (HEEM-oh-seal; Fig. 11.3B). The blood then moves through the hemocoel in a leisurely fashion, usually under low pressure. Open systems occur in most molluscs and arthropods and in some invertebrate chordates. The molluscs with open circulatory systems are clams, snails, and other sluggish forms that don't demand much oxygen from their blood. Insects are able to sustain high levels of activity with open circulatory systems by having air tubes that bring air directly to cells, as will be described in the next chapter. An advantage of open circulatory systems is that blood pressure is usually low, so there is little danger of bleeding to death from an injury.

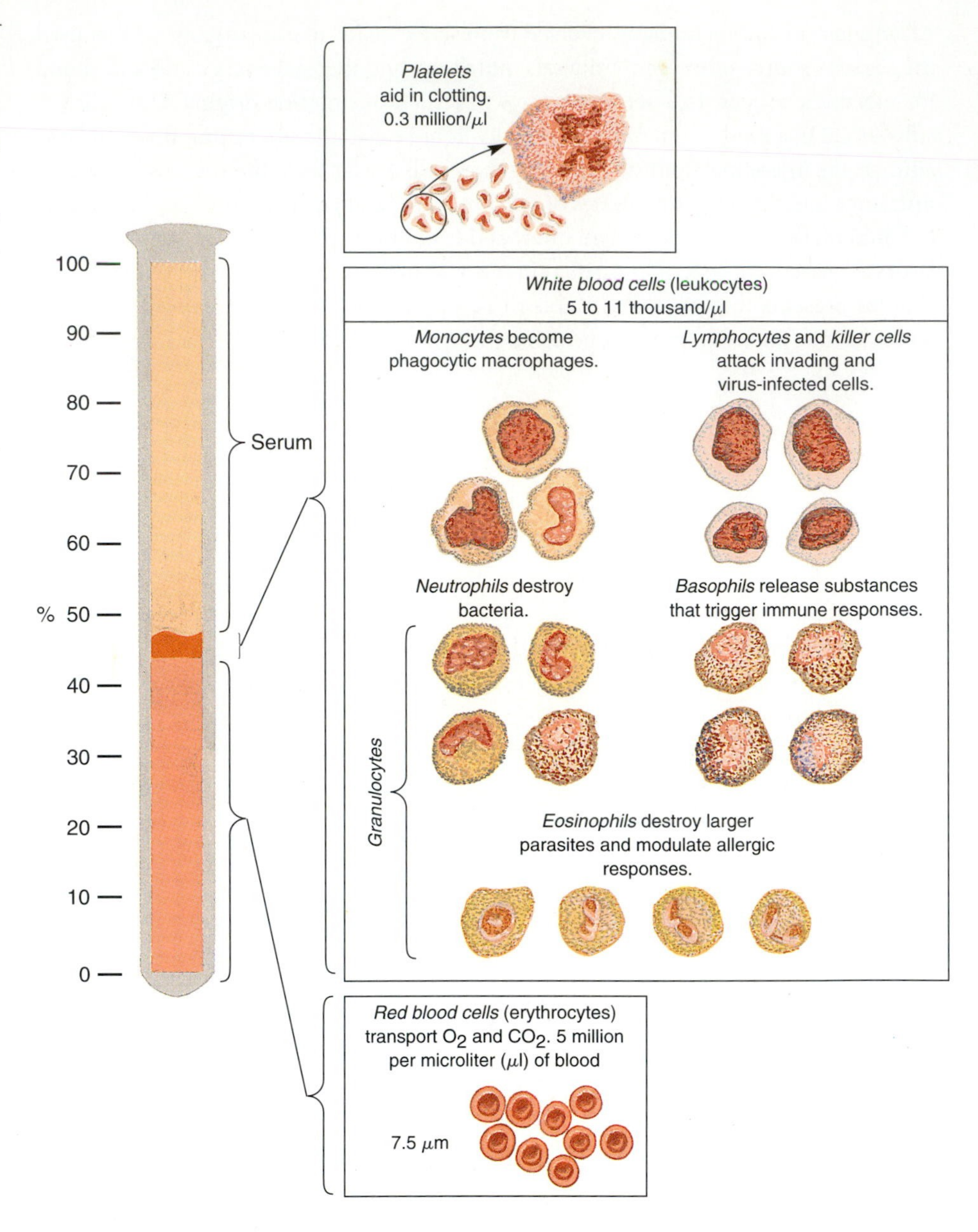

FIGURE 11.1

The composition of human blood. The figure at left shows blood from a normal person after removing the clotting proteins and centrifuging it to make the cells settle. A little over half the volume is the fluid serum. Approximately 45% (a little less in women) is red blood cells. Between serum and red blood cells is a "buffy coat" containing platelets and white blood cells. Platelets, red blood cells, and white blood cells make up the formed elements, all of which originate as stem cells in bone marrow. Platelets are fragments of white blood cells called megakaryocytes. All the white blood cells are colorless under the light microscope; special stains are used to reveal structural differences such as intracellular granules.

Comparison of Vertebrate Hearts

In both closed and open systems, blood is pumped by contractions of the heart, aided in some species by contractions of blood vessels. The term "heart" is applied to a wide variety of hollow structures, but almost all consist of muscle, and most have valves that allow only one-way flow of blood. In vertebrates the heart is divided into two or more chambers. The two major kinds of chambers are the **atrium** and the **ventricle.** Atria pump blood from veins into the ventricles, and the ventricles pump the blood into arteries. In addition, fishes, amphibians, and most reptiles have a **sinus venosus** that receives venous blood and serves as a reservoir for the atrium. (In birds and mammals the sinus venosus occurs only in the embryo.) Fishes also have a **bulbus arteriosus** that receives blood from the ventricle and smoothes the pulsations. Vertebrate hearts vary primarily in the number of atria and ventricles, following an evolutionary trend toward increasing the number by dividing the atria and ventricles into halves. This division allows one half of a chamber to pump oxygenated blood from the respiratory organs to

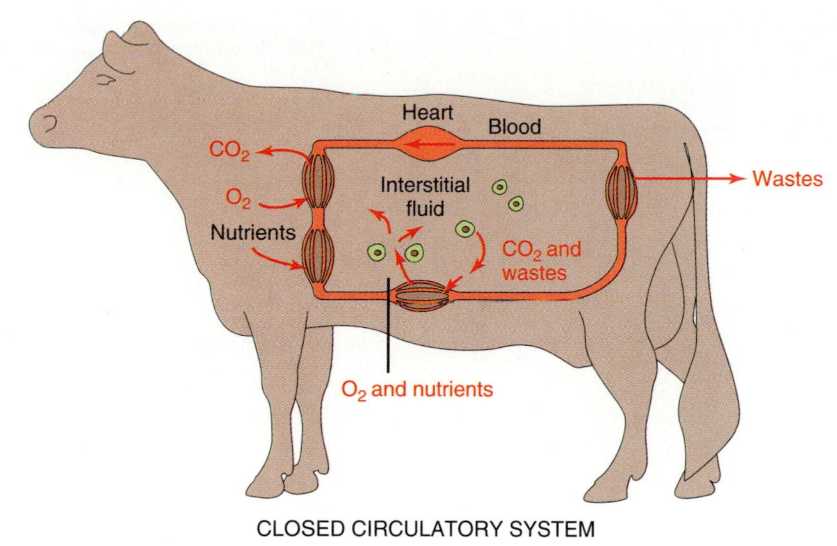

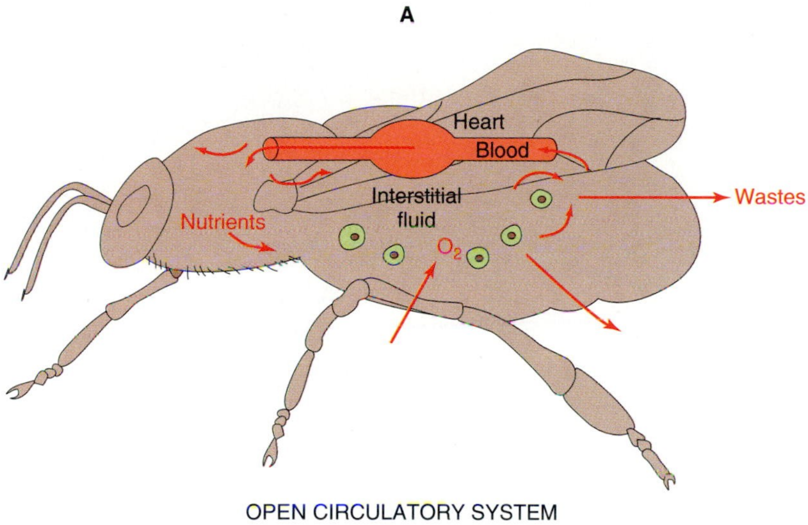

FIGURE 11.2

The relationships among the external environment, blood, and the interstitial fluid. (A) In closed circulatory systems, blood is confined within a system of vessels. As it circulates through capillaries in the respiratory and digestive systems, it absorbs oxygen and nutrients for delivery through other capillaries to metabolizing cells. The blood absorbs carbon dioxide and metabolic wastes from those cells and delivers them to the respiratory and excretory systems for elimination. (B) In open circulatory systems the blood is not confined to the vessels, and there is no clear distinction between blood and the interstitial fluid.

the body, and the other half to pump deoxygenated blood from the body back to the respiratory organs.

In fishes the chambers are all in series, so there is no separation of oxygenated from deoxygenated blood (Fig. 11.4A on page 221). Oxygen-depleted venous blood enters the sinus venosus, and the atrium and then the ventricle pump it through the bulbus artriosus to the gills, where oxygen is absorbed and carbon dioxide eliminated. The blood loses much of its pressure in the gills before it continues through arteries to the systemic circulation.

In amphibians, two atria empty into a single ventricle. Deoxygenated blood from the body tissues enters the sinus venosus from the major vein, the **vena cava,** and then goes into the right atrium (Fig. 11.4B). At the same time, blood that has been oxygenated in the lungs or skin passes through the **pulmonary vein** to the left atrium. The deoxygenated and oxygenated blood from the two atria enter the single ventricle, but structures within the ventricle reduce mixing (see Fig. 36.8). Most of the oxygenated blood leaves the ventricle through the **aorta,** while the deoxygenated blood goes to the lungs and skin through the **pulmocutaneous artery.** Deoxygenated blood becomes oxygenated in the lungs if the amphibian is breathing air, and it becomes oxygenated in the skin if the animal is submerged.

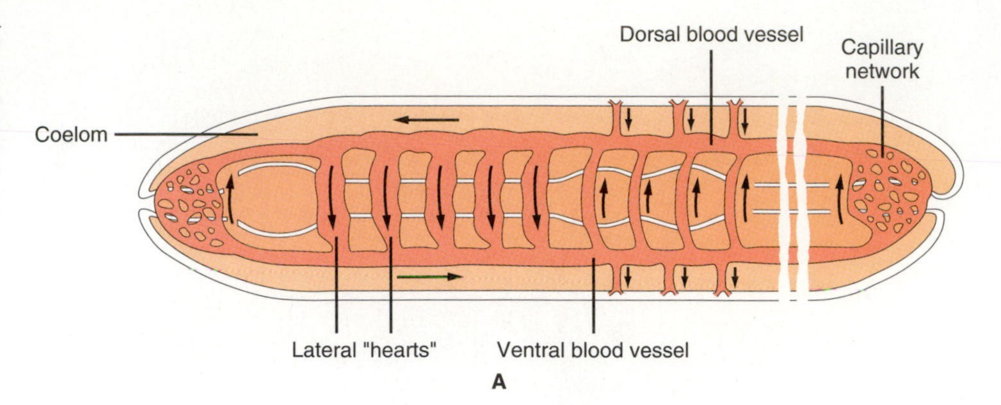

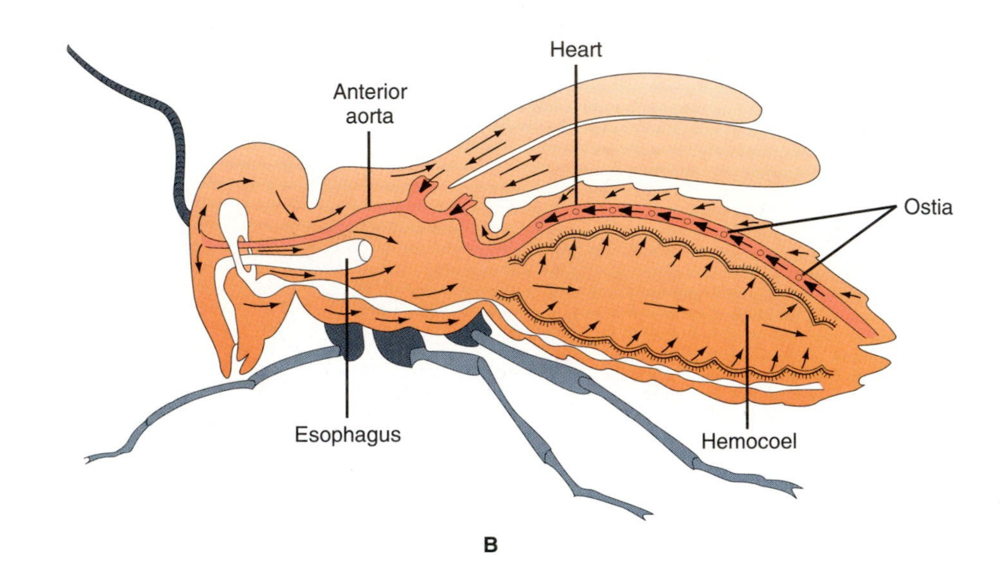

FIGURE 11.3

(A) Schematic diagram of the closed circulatory system of the giant earthworm *Glossoscolex giganteus.* Because of the large size of this Brazilian species (up to 0.5 kg), it has been possible to measure its blood pressure directly. In the ventral blood vessel the pressure varies between 35 and 70 mmHg. (B) Schematic diagram of the open circulatory system of a generalized insect. Blood is drawn into the heart through one-way openings called ostia and is then pumped through the anterior aorta. From there it percolates among the organs in the body cavity (hemocoel). Open systems generally have low pressures. In the grasshopper *Locusta,* for example, blood pressure varies from about 2 to 6 mmHg.

The hearts of most reptiles are similar to those of amphibians except that the sinus venosus is fused with the right atrium, and the ventricle is more nearly divided into two chambers (Fig. 11.4C). In crocodiles, alligators, and related reptiles (order Crocodilia) there is complete division of the ventricle into two chambers. Oxygenated blood therefore does not mix with deoxygenated blood while the crocodilian is breathing. If the animal holds its breath under water, however, a valve in the pulmonary artery shunts deoxygenated blood into the aorta, bypassing the lungs.

Like crocodilians, birds and mammals have four-chambered hearts, with complete separation between the **pulmonary circulation,** which pumps oxygen-depleted blood through the lungs, and the **systemic circulation,** which pumps the oxygenated blood to metabolizing tissues (Fig. 11.4D). The complete separation of systemic and pulmonary circulation is necessary, because these animals have much higher levels of activity than do reptiles. The occasional child born with a hole in the septum between the two ventricles will die within a few years unless the hole is surgically repaired. The separation of circulation also enables the left ventricle to produce high pressure to get blood up to the brain, while the right ventricle produces a lower pressure to avoid flooding the lungs with fluid. This is not a trivial matter for giraffes.

The existence of a separate pulmonary circulatory system raises an interesting question. Does it function in a mammalian embryo, where lungs are of little use and are, in fact, collapsed? The answer is no. High resistance to flow through the collapsed lungs forces the blood from the pulmonary artery into the aorta through a bypass

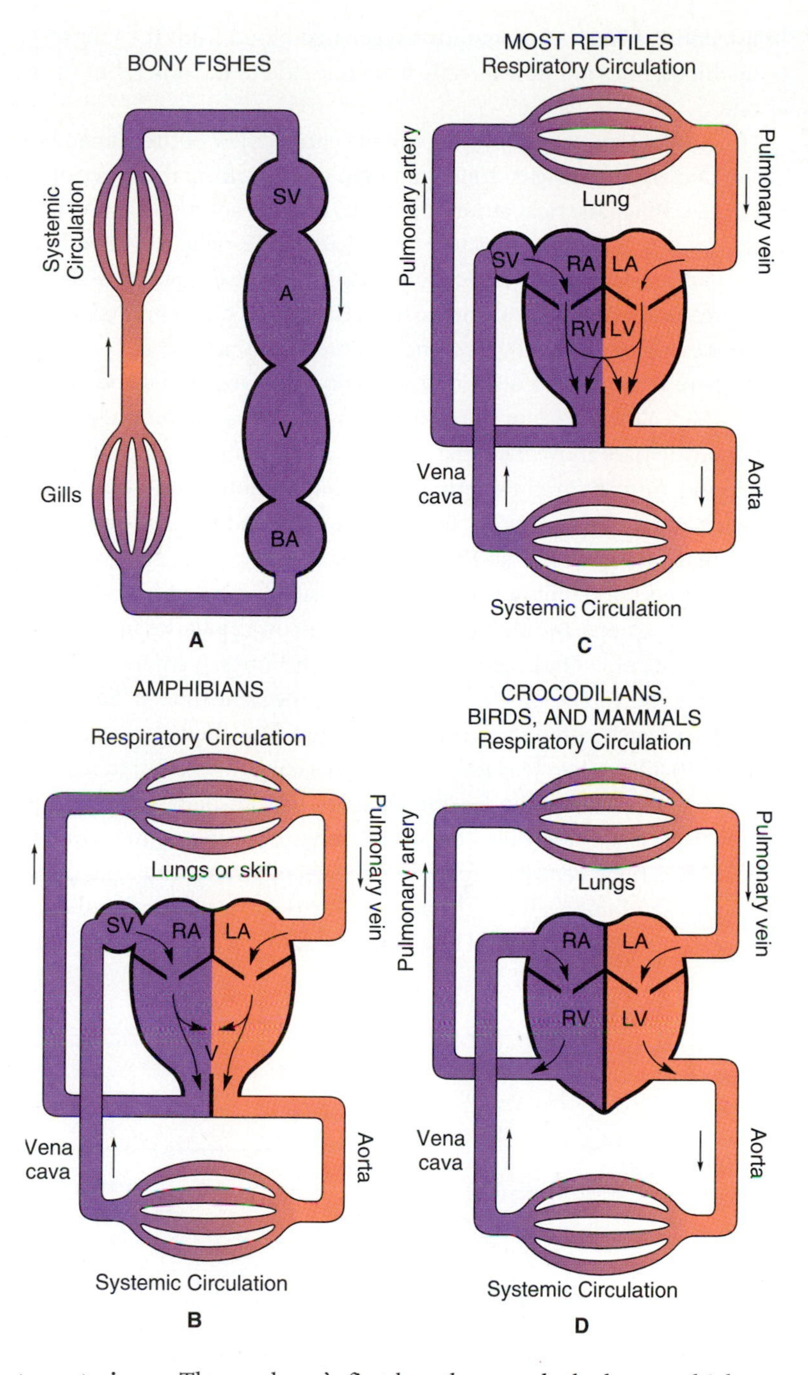

FIGURE 11.4

Circulatory patterns of vertebrates. (A) Bony fishes have four chambers in series: the sinus venosus (SV), the atrium (A), the ventricle (V), and the bulbus arteriosus (BA). There is no separation of respiratory and systematic circulation. (B) In amphibians the blood from the lungs and skin enters the left atrium (LA), and blood from the body enters through the sinus venosus and right atrium (RA). The blood from both atria empties into the one ventricle, which then pumps it into the respiratory and systemic circulations. (C) In reptiles there is a greater degree of anatomical division of the ventricle into two halves (RV and LV). (D) In crocodilians, birds, and mammals the ventricle is completely divided, forming a four-chambered heart, with the flow of blood through the lungs completely separated from the flow to other tissues.

called the **ductus arteriosus.** The newborn's first breath expands the lungs, which reduces the resistance to blood flow and allows the blood to flow through the pulmonary artery. Within hours the ductus arteriosus also closes.

Structure and Function of the Human Heart

More is known about the hearts of humans than about those of any other species, mainly because the prevalence of heart disease provides a great deal of incentive, as well as material, for research. Figure 11.4D outlines the circulatory "plumbing" of humans. As in the four-chambered hearts of all birds and mammals, there is complete separation of systemic from pulmonary circulation, so it is useful to consider each side a separate heart. The "right heart" pumps deoxygenated blood from the body to the lungs, and the "left heart" pumps oxygenated blood from the lungs to the rest of the body. Blood does not flow directly from one side to the other, but the two hearts beat as one.

Figure 11.5 gives a more anatomically correct view of the human heart. Deoxygenated blood enters the right atrium and right ventricle from the **superior** and **inferior venae cavae.** When the right atrium contracts, it forces additional blood into the right ventricle through the **atrioventricular** (AV) **valve.** The right AV valve has three cusps of tissue and is therefore also called the **tricuspid valve.** Contraction of the atrium produces little pressure because its walls are so thin. The atrium functions mainly as a reservoir that ensures complete filling of the ventricle before it contracts.

There is normally a delay of 0.07 seconds between atrial and ventricular contraction, which allows time for the ventricle to fill. When the ventricle contracts, the pressure quickly rises above the atrial pressure, forcing the AV valve shut. The only exit for the blood from the right ventricle is through another one-way valve into the pulmonary artery. This valve has flaps of tissue shaped like half-moons, so it is called a **semilunar valve.** When the right ventricle relaxes, the semilunar valve helps ensure that the blood does not flow back into it. Instead, the arteries, which are now inflated with blood, squeeze the blood through the narrow capillaries in the lungs.

After the blood has been oxygenated in the lungs, it enters the left atrium through pulmonary veins. The left atrium contracts at the same time as the right atrium, emptying the oxygenated blood under low pressure into the left ventricle through the left atrioventricular valve. This AV valve has two cusps of tissue and is called the **bicuspid valve.** It must have reminded the ancient anatomists of a bishop's pointed hat, or miter, because it is also called the **mitral valve.** The left ventricle contracts at the same time as the right ventricle, forcing the bicuspid valve shut and sending the blood through a semilunar valve in the aorta. The pressurized blood in the aorta then closes the aortic semilunar valve and flows into the systemic arteries and capillaries.

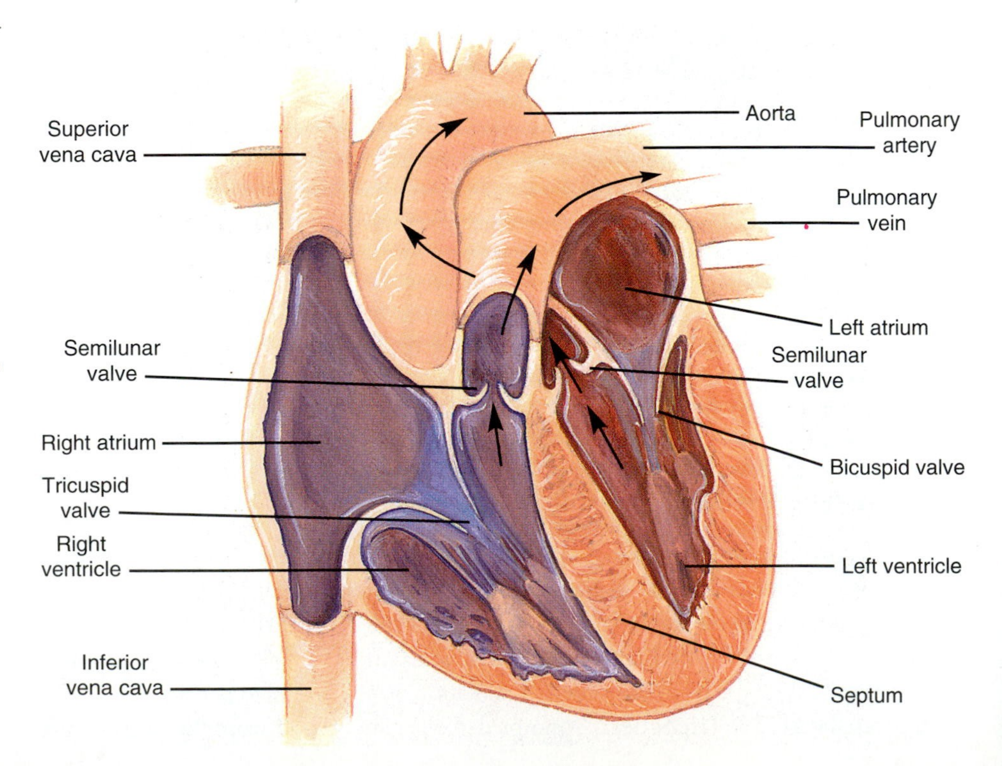

FIGURE 11.5

The human heart, showing the chambers and valves. Arrows indicate the direction of blood flow.

■ **Question:** If the normal left ventricle pumps 5 liters of blood per minute, how much does the right ventricle pump per minute? Why does failure of the left ventricle to pump blood adequately cause fluid to collect in the lung?

Electrical Activity of the Heart

The above description explains the mechanical events of the heart, but it does not tell us why the heart contracts. Contraction is, in fact, due to a flow of electrical potentials that is entirely separate from the flow of blood. Like all vertebrate hearts, the human heartbeat is **myogenic,** which means that the electrical stimulus for contraction originates in muscle cells of the heart itself. (In contrast, the hearts of many invertebrates are **neurogenic:** the beat is triggered by the nervous system.) The stimulus for each contraction is an action potential that occurs because the voltage across the plasma membrane slowly decreases until it reaches threshold. Many muscle cells are capable of producing action potentials, but those that produce them with the highest frequency act as the **pacemaker** for the entire heart. Ordinarily a cluster of cells in the **sinoatrial (SA) node** on the back of the right atrium is the pacemaker. The action potential travels rapidly from the pacemaker to adjacent cardiac muscle cells, which are electrically and mechanically joined to each other through **intercalated discs** (p. 209). As the action potential travels from one cell to the next, it triggers a new action potential and contraction.

An action potential from the SA node quickly spreads through the two atria. In mammals, however, atrial and ventricular muscle fibers are not electrically connected, so the action potentials have to take a special route to excite the ventricles. This pathway consists of the **atrioventricular (AV) node,** two **bundles of His** of conducting fibers, and **Purkinje fibers** going to the ventricular walls. The AV node receives the action potentials from the atria and delays them 0.07 seconds, allowing the ventricles to fill. After the delay, the AV node sends the action potentials through the right and left bundles of His, which conduct them through the Purkinje fibers. These fibers then activate all areas of the two ventricles simultaneously (Fig. 11.6).

The Control of Cardiac Output

Blood carries almost all the oxygen required by vertebrates. Consequently the **cardiac output**—the volume of blood pumped per minute—must be adjusted to match the demand for oxygen. Cardiac output equals the rate of contraction multiplied by the volume of blood pumped during each contraction of the left ventricle (the **stroke volume**). (Of course, the cardiac output of the right ventricle must be the same.)

$$\text{cardiac output} = \text{heart rate} \times \text{stroke volume}$$

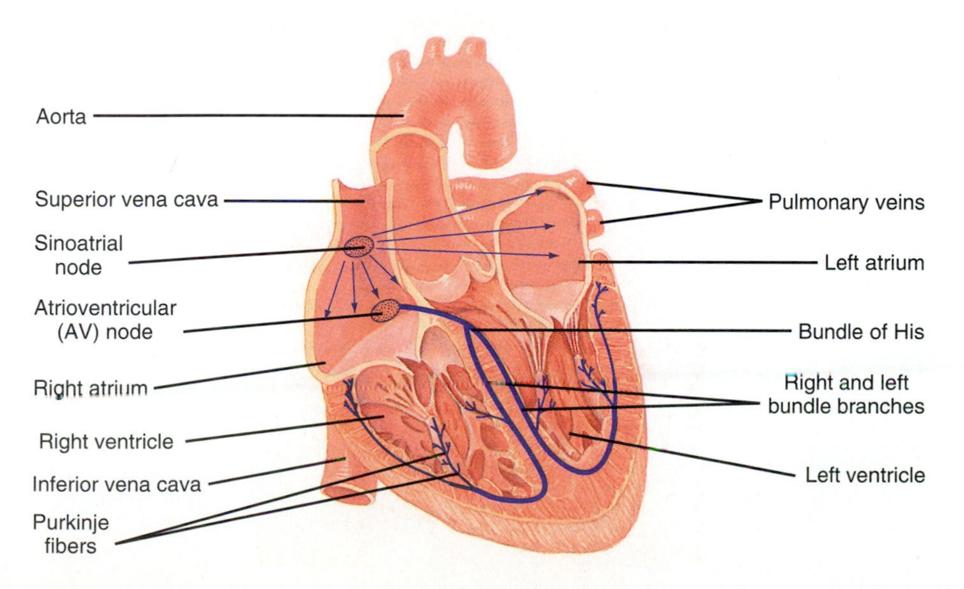

FIGURE 11.6
The structures of the human heart that trigger and coordinate its contraction. The sinoatrial (SA) node includes pacemaker cells that trigger action potentials (arrows) and therefore contraction in the atria. The atrioventricular (AV) node picks up action potentials from the atria, delays them while the ventricles fill with blood, then sends them to the ventricles via the two bundles of His and the Purkinje fibers.

Clinical Signs of Cardiac Functioning

The action potentials of cardiac cells spread throughout the body and can easily be recorded from the skin as the **electrocardiogram** (ECG, or EKG following the original German spelling) (Fig. 11.7). The first event in the ECG, the P wave, results from the combined action potentials when cells of the atria contract. Next comes the QRS complex, which results from the combined action potentials in the cells of the ventricles. Finally the T wave results from repolarization of the membranes of the ventricular muscles as they relax. Changes in the shape or interval between waves indicate a disturbance to electrical conduction in the heart. Such changes can be used to diagnose disorders such as heart attack or heart block (blockage of the bundles of His).

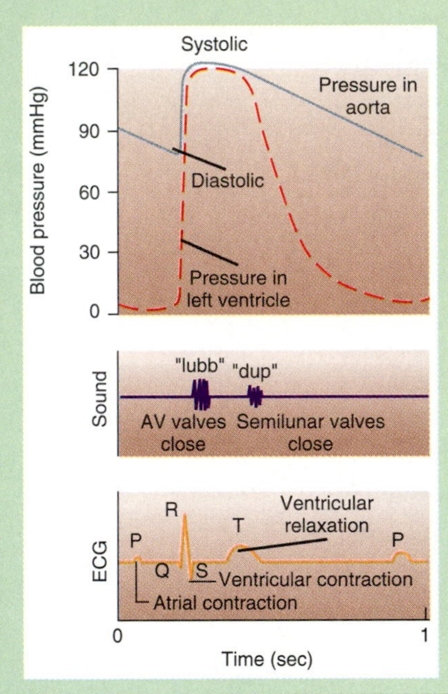

FIGURE 11.7

Arterial blood pressure, sounds, and the electrocardiogram (ECG) of the normal human heart. The P wave results from the electrical activity of atrial contraction. The QRS complex results from electrical activity of ventricular contraction. Note that at the time of the QRS complex the blood pressure shoots up from its lowest (diastolic) value of about 80 mmHg to its maximum (systolic) value of about 120 mmHg. The rise in ventricular pressure causes the atrioventricular valves to shut, producing the "lubb" heart sound. The T wave of the ECG results from repolarization of membranes when the ventricles relax. At this time the arterial pressure starts to exceed the ventricular pressure, which forces the two semilunar valves shut, producing the "dup" heart sound. Although ventricular pressure remains low until the next contraction, the arterial pressure declines gradually because the blood can only trickle through the fine capillaries.

The most familiar external signs of heart activity are the **heart sounds,** commonly designated "lubb" and "dup." The lubb sound occurs at about the time of ventricular contraction and is due to closing of the AV valves. The dup sound occurs during ventricular relaxation as the semilunar valves close. The heart sounds are therefore useful in diagnosing disorders of the valves. One such disorder is **heart murmur,** in which a blowing or roaring during the first or second heart sound indicates poor closing of an AV or semilunar valve, respectively. Even in this age of advanced medical technology, one of the most valuable talents of the physician is the ability to assess the functioning of the heart simply by listening to its sounds with a stethoscope.

Another time-honored part of a physical examination is the measurement of **arterial blood pressure.** The measurement of blood pressure uses the fact that the height to which a fluid rises in a tube is proportional to the pressure applied to the fluid at the base of the tube. The device used to measure blood pressure, called a **sphygmomanometer,** consists simply of a tube partly filled with mercury. The procedure for measuring blood pressure is to compress the brachial artery in the upper arm with an inflatable cuff attached to the sphygmomanometer. The physician then listens with a stethoscope for sounds of blood flowing past the cuff as the pressure is slowly released. A gauge on the cuff shows the pressure being applied to the artery. The blood will spurt audibly through the artery when the cuff pressure falls just below the highest pressure as the left ventricle contracts. At this point the blood pressure is read simply as that height in millimeters of mercury (mmHg or torr). This peak arterial pressure is called the **systolic pressure.** As the cuff pressure continues to fall, a final sound can be heard as it closes the artery only momentarily. This sound indicates when the minimum, or **diastolic pressure,** occurs, just before the ventricle contracts. Normal systolic pressure in humans is 120 mmHg, and normal diastolic pressure is 80 mmHg.

Systolic and diastolic pressures are usually indicated like a fraction, though they are never divided:

$$\frac{120}{80}$$

box continues

A systolic pressure of 120 mmHg is enough to raise a column of blood to a height of 1632 mm = 1.632 meters. For most people, that is greater than the distance between the feet and the heart, so even when you stand on your head your heart pumps blood to your feet. Excessive systolic or diastolic blood pressure often indicates **hypertension** (high blood pressure), which can put an added strain on the heart and circulatory system, possibly leading to weakened heart muscle (heart failure), heart attack, or stroke.

Thus there are two and only two ways by which cardiac output can be adjusted to match the oxygen demand: by changing heart rate and by changing stroke volume. At rest the average human heart has a rate of 72 beats per minute and a stroke volume of 70 ml per beat, for a cardiac output of 5.0 liters per minute. (About 60 barrels of blood per day!) Each of us has about 5.0 liters of blood, so the heart can pump all of it in about a minute. Even that prodigious amount is too little during exercise or unusual metabolic demand, so either rate, or stroke volume, or both must be increased.

Control of the cardiac output is mainly by neurosecretions from both divisions of the autonomic nervous system—the parasympathetic and sympathetic (pp. 178–179). At rest the parasympathetic nervous system is more active and slows the resting rate of the human heart to its normal average. The parasympathetic fibers responsible for slowing the heart are in the **vagus nerve,** so this slowing is called **vagal inhibition.** The vagus nerve originates in the part of the brainstem called the medulla, which controls vagal inhibition. The vagus nerve slows the heart by releasing acetylcholine, which increases the voltage across membranes of pacemaker cells, making them take longer to reach threshold for each action potential. The medulla also activates vagal inhibition in response to increased pressure in the aorta and carotid arteries. This response probably helps prevent cardiac output from becoming too high.

The sympathetic nervous system increases cardiac output by releasing norepinephrine onto cardiac muscles. Norepinephrine increases heart rate. It also increases the force of cardiac-muscle contraction, which increases the stroke volume. Circulating epinephrine from the adrenal medulla increases cardiac output in similar ways. Finally, the output of one ventricle compared with the other is automatically adjusted by the fact that ventricular muscle increases its force of contraction when it is stretched by accumulating blood. This phenomenon is referred to as **Starling's law of the heart.**

Peripheral Circulation

ARTERIES. The ceaseless beating of the heart is so dramatic that it tends to overshadow the silent streaming of blood through arteries, capillaries, and veins. Yet the functioning of these vessels is as important and complex as that of the heart. The arteries, which by definition conduct blood away from the heart, must have the resilience to inflate and deflate some 40 million times a year throughout your life. Figure 11.8A on page 226 shows the elastic tissue that makes this possible. In the more peripheral arteries and their small branches, the **arterioles,** the middle layer of tissue consists largely of smooth muscle. Hormones and the autonomic nervous system control the tension of these muscles, thereby regulating local blood flow and overall blood pressure. Contraction of these arterial smooth muscles causes **vasoconstriction,** which reduces blood flow through the arteries. If vasoconstriction occurs in many arteries simultaneously, the blood that is circulating through them is squeezed into a smaller volume, which increases the pressure. Thus an increase in vasoconstriction, as well as an in-

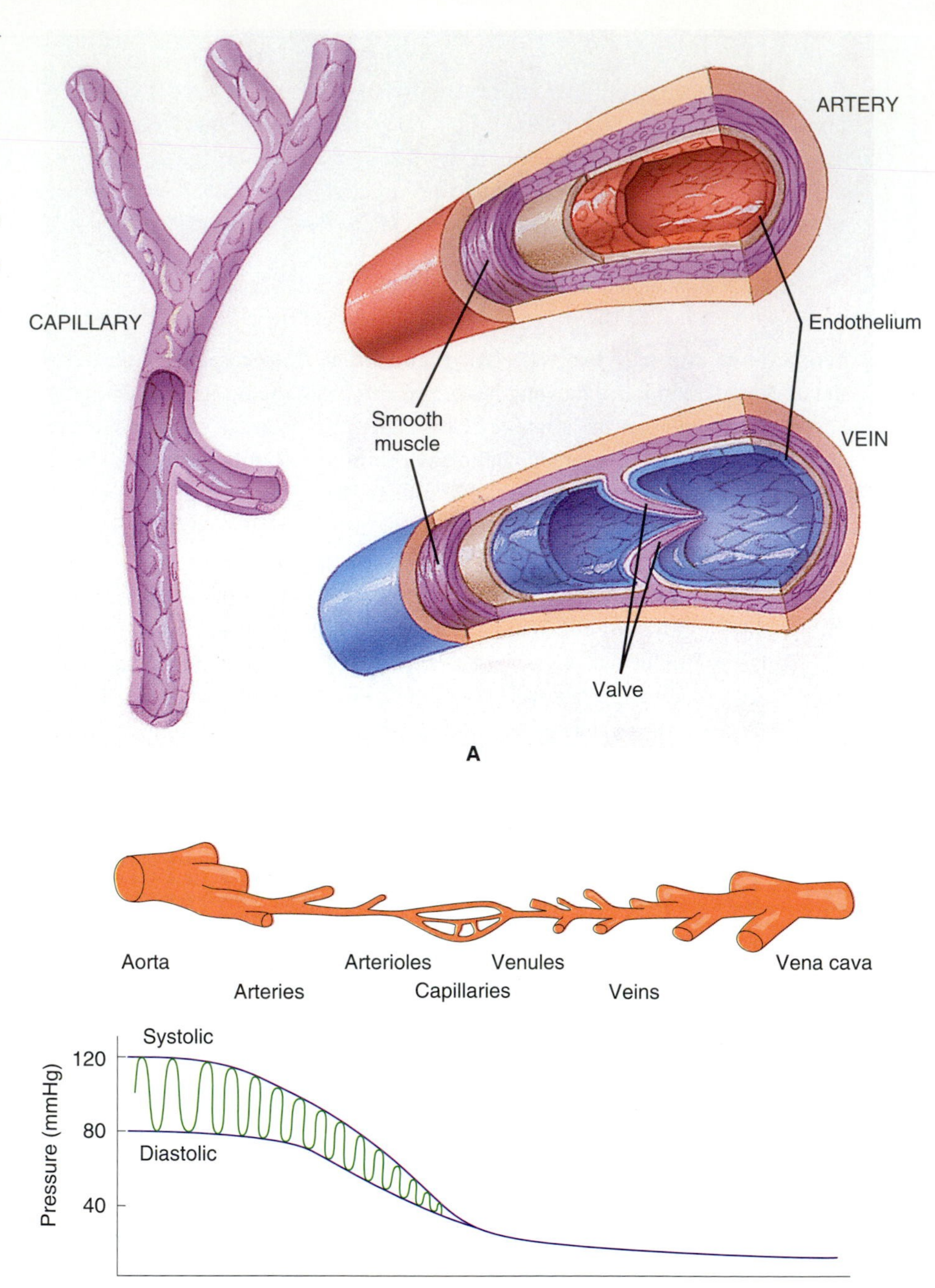

FIGURE 11.8

(A) The structure of arteries, capillaries, and veins. Smooth muscle in arteries regulates the local flow of blood and overall blood pressure. A valve is shown in the sectioned vein. (B) Blood pressures throughout the circulatory system. The arterial pressure varies between the limits shown (systolic and diastolic pressures). Most of the pressure is lost in the small arterioles and in the capillaries, which have an inner diameter just large enough for red blood cells to squeeze through.

crease in cardiac output, increases blood pressure. Relaxation of the smooth muscles causes **vasodilation,** which has the opposite effect as vasoconstriction on blood flow and blood pressure.

VEINS. Veins and their smaller branches, the **venules,** also have a layer of smooth muscles. However, contraction of these muscles, called **venoconstriction,** has a different effect from vasoconstriction. The veins usually contain about 70% of the blood, but during exercise and stress venoconstriction forces some of that reserve into circulation. The pressure in veins remains low even during venoconstriction. Generally venous blood pressure is about one-tenth as high as blood pressure in large arteries (Fig. 11.8B) and is due mainly to the hydrostatic pressure of gravity. The effect of this hydrostatic pressure can be seen by watching the veins of your hand swell as you lower your hand below the level of your heart.

With such low blood pressure, how do veins get the blood back up to the heart? The answer is the **venous pump,** which is not a separate organ itself, but the result of one-way valves and massaging by neighboring skeletal muscles. When skeletal muscles contract they compress nearby veins, forcing the blood in the only direction the valves permit it to flow: toward the heart.

CAPILLARIES. The formation of interstitial fluid, which is what circulation is all about, is due to the capillaries. In contrast to arteries and veins, capillaries have extremely thin walls with gaps between their cells. Blood pressure within the capillaries forces plasma through the gaps, leaving the blood cells and large proteins behind. The resulting filtrate of blood is the interstitial fluid. Why doesn't all the plasma get filtered out of the blood, drowning the tissues in interstitial fluid? One reason is that at any one time most capillaries are closed off by bands of smooth muscle at the arteriole end—the **precapillary sphincters.** Precapillary sphincters constrict when local conditions (low CO_2, normal temperature, and normal pH) indicate that there is adequate blood flow.

THE STARLING EFFECT. A second reason why excess interstitial fluid does not form is the Starling effect, named for E. H. Starling, who also proposed the law of the heart mentioned previously. The Starling effect refers to a balance between the blood pressure, which tends to force plasma out of the capillaries, and the **colloid osmotic pressure,** which tends to draw fluid back into the capillaries. The colloid osmotic pressure (also called oncotic pressure) is due to proteins (mainly serum **albumin**) that are too large to pass across the capillary walls and that are therefore in higher concentration in the blood than in the interstitial fluid. At the arterial end the blood pressure in the capillary is greater than the colloid osmotic pressure, so fluid leaves the capillary. As blood squeezes through the narrow capillary, however, it loses much of its remaining pressure. At the venule end, therefore, the blood pressure is lower than the colloid osmotic pressure, so fluid is drawn back into the blood (Fig. 11.9).

The Starling effect has many everyday consequences. One of these is **swelling** (= edema), which occurs when blood pressure is excessive or osmotic pressure is diminished. For example, standing for long periods without contracting the leg muscles (venous pumping) can cause blood to pool in the veins. This can increase pressure at the

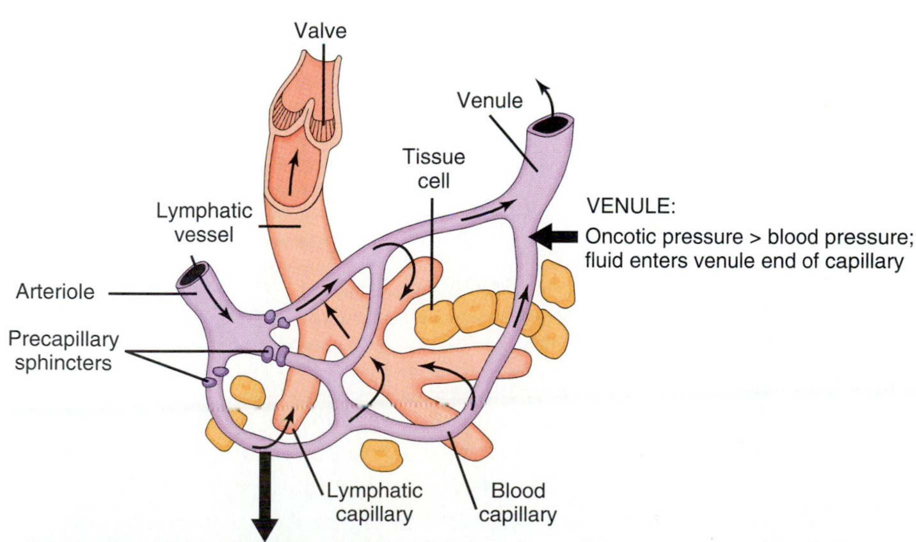

FIGURE 11.9

A capillary network. The precapillary sphincters remain closed much of the time in tissues with low oxygen demand, such as skin. Even though the blood pressure at the arteriole end tends to force fluid out of the thin and highly permeable capillary walls, the colloid osmotic pressure of the blood at the venule end draws the fluid back in.

Historical Aspects of Circulation

Nothing in physiology has quite the impact of blood and the beating heart. This and the fact that bleeding and the cessation of the heartbeat are so often associated with death probably account for ancient beliefs that life itself is located within the blood or the heart. According to Jewish belief that is literally as old as Moses, "the blood is the life." This is the basis for the kosher preparation of meats. Some religions still oppose blood transfusions because of the ancient association of life and blood.

One of the first to try to understand blood scientifically was Galen, a Roman physician of the second century. Galen based his ideas on what he could see in cadavers. Since the veins appear to the naked eye to be closed at their ends, he thought blood simply surged back and forth in them, carrying what he called "animal spirit." He knew that some blood also flowed from the right ventricle into the lungs, but since the diameter of the pulmonary artery in a cadaver is smaller than that of the vena cava, he concluded that not all the blood that entered the right ventricle from the vena cava went to the lungs. Some of it, he thought, must pass through invisible pores in the septum into the left ventricle, where it mixed with **pneuma** in the air from the lungs. Since the arteries of cadavers are usually empty of blood, Galen assumed that the mixture of pneuma and blood formed an invisible "vital spirit" that pulsed to and fro in the arteries.

Scientists in the early Renaissance tended to trust Galen more than their own eyes. When anatomy students tried in vain to poke straws through the pores between the two ventricles, many of their professors explained that the authority of Galen was superior to the efforts of students, or that they just weren't making hearts the way they did in Galen's day. In 1553 Michael Servetus's book *The Restoration of Christianity* poked holes of a different sort in this idea. Servetus correctly suggested that all the blood passes through the lungs and that none flows directly between the ventricles. Servetus failed to convince many people, perhaps because Calvin had him and all but three copies of his book burned for heresy.

In the 17th century, scientists began to realize the dangers of relying on the views of dead authorities based on dead animals. One of the first biologists to use the experimental approach with living animals was William Harvey, the physician to King Charles I. In 1628 Harvey published a book in which he argued that the heart was a pump, and that there is a **circulation** of blood in one direction only. He based his argument partly on simple experiments that showed that valves prevent the veins from carrying blood away from the heart. You can perform one of Harvey's experiments in a few seconds using your own forearm, as shown in Fig. 11.10. After doing this demonstration you may well wonder why it took 14 centuries for anyone to think of it. The probable answer is that experimental science was simply not an accepted approach to truth until the Renaissance.

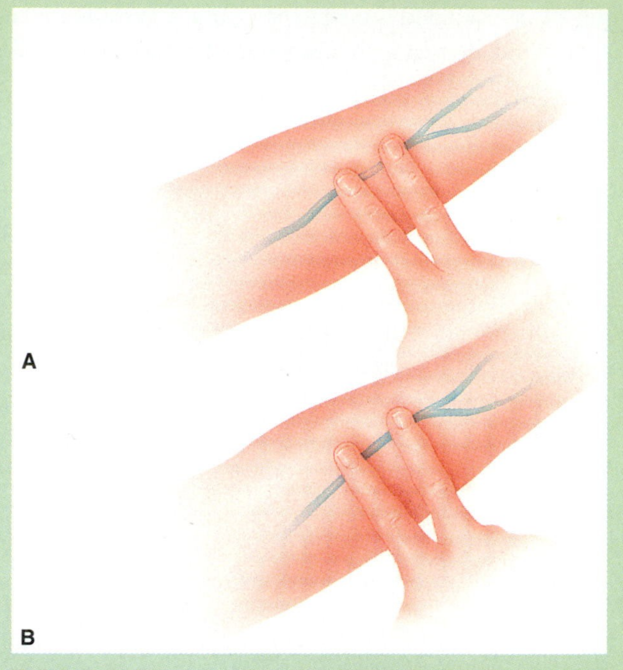

FIGURE 11.10
A simple experiment to demonstrate the one-way circulation of blood. (A) Press the fore and middle fingers over a prominent vein on the inner arm, then sweep the forefinger along the vein to squeeze the blood toward the heart. Then lift the forefinger while keeping the vein compressed with the middle finger. The cleared portion of the vein remains invisible because valves prevent blood returning from the direction of the heart. (B) Now reverse the experiment, *gently* sweeping the middle finger along the vein away from the heart, then lifting it. The vein remains visible because valves prevent the blood from being squeezed out in the direction away from the heart. (If done gently, there is little risk in this procedure. You probably do the same thing when you roll over in your sleep.)

The 17th-century microscopists Malpighi and Leeuwenhoek soon confirmed Harvey's theory of circulation by directly observing connections between arteries and veins. Numerous other experiments have confirmed that the heart is "merely" a pump, and few people still revere it or the blood as the seat of life and spirit. Yet, even in an age when having "a change of heart" is no longer just a figure of speech, the drama of circulation still holds its mysteries for scientists and nonscientists alike.

venous ends of capillaries to the point where it exceeds the osmotic pressure, thereby preventing the return of fluid from the tissues into the blood. Swelling also occurs in inflamed tissues, which release chemicals that make capillaries leakier. The leaky capillaries allow proteins to enter the interstitial fluid, eliminating the osmotic pressure difference.

Hemostasis

Since the blood in closed circulatory systems is pumped under high pressure, there is a constant danger that an injury to an artery will cause a fatal loss of blood. The pressure in large arteries is so high that nothing but an externally applied constriction will prevent death if the artery is cut. In veins and in smaller arteries, however, there are mechanisms that automatically limit blood loss. This limitation of blood loss is called **hemostasis.** (Note the difference between hemostasis and homeostasis). One mechanism of hemostasis is the contraction of smooth muscles, which may be sufficient to close the cut ends of arteries and veins. In addition, blood **platelets** are attracted to the site of injury by the exposed cell contents. Platelets, which are actually fragments of certain cells (**megakaryocytes**), adhere to the injured site and to each other and form a **platelet plug** (Fig. 11.11A).

For larger injuries a second mechanism of hemostasis comes into play: **clotting.** Clotting (coagulation) results from a chemical chain reaction that forms strands of a fibrous protein called **fibrin,** which entangle red blood cells at the site of injury. Approximately 16 proteins, ions, and other factors are involved in the series of reactions leading to the formation of fibrin. In the final steps, **prothrombin** changes to **thrombin,** and thrombin converts the globular protein **fibrinogen** into fibrin. A few hours after it forms, the clot begins to shrink, squeezing out serum. This shrinkage, called **clot retraction,** results from the contraction of the platelets. (Platelets contain more actin and myosin than any other cells except those of muscle.) Clot retraction pulls the injured tissue together and forms a more solid barrier to blood loss. Eventually the clot is removed by an enzyme that breaks up the fibrin.

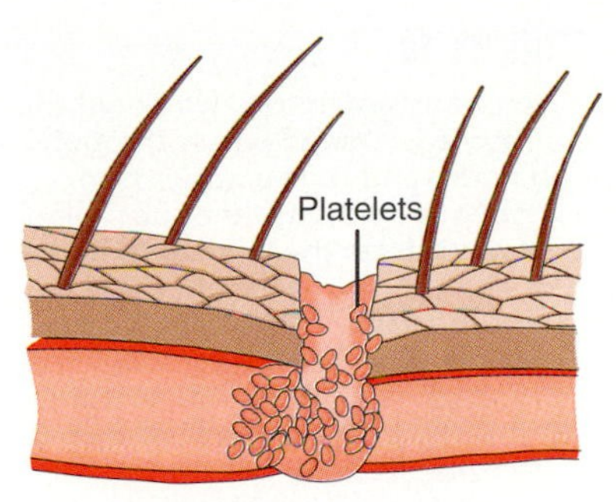

A Formation of platelet plug

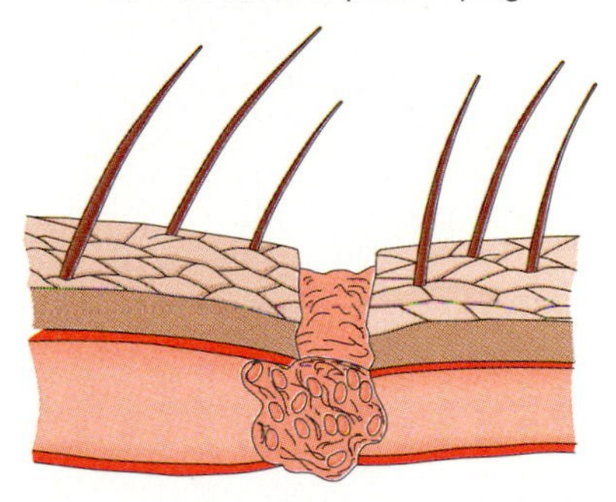
B Formation of fibrin

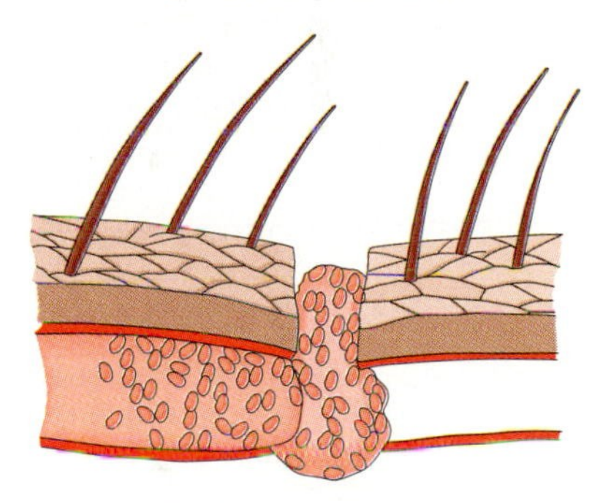
C Formation of clot, consisting largely of red blood cells

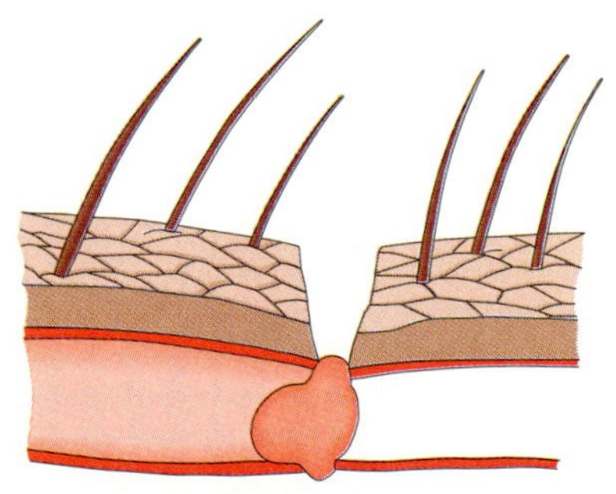
D Clot retraction

FIGURE 11.11

Key steps in hemostasis. (A) Collagen from damaged cells causes blood platelets to adhere to the damaged vessel, forming a platelet plug. (B) Thrombin from the platelets and a tissue factor from injured cells then trigger the formation of fibrin strands from the protein fibrinogen in the blood. (C) Red blood cells entangled in the fibrin make up the bulk of the clot. (D) Finally the platelets contract, and clot retraction pulls the tissue together and forms a solid plug.

The Lymphatic System

Between 1% and 4% of the interstitial fluid is not recycled back into capillaries by the Starling effect. In addition, some protein and cells escape from blood, and debris from dead cells also tends to accumulate in interstitial fluid. Why, then, aren't we all puffy from stuff that has collected in our tissues over the years? The answer is that there is a second circulatory system—the lymphatic system—that keeps the interstitial fluid free of such material. The fluid that circulates in the lymphatic system is called **lymph.** Lymph enters the lymphatic system through the closed but highly permeable ends of **lymphatic capillaries** (Fig. 11.12 on page 230). It then flows into **lymphatic vessels.** In mammals these lymphatic vessels have one-way valves like those in veins, so lymph is propelled by a mechanism similar to venous pumping. (Fishes, amphibians, and reptiles, however, have contractile vessels, called "lymph hearts," that pump lymph.) Lymphatic vessels all converge at one of two **thoracic ducts** near the collar bones, where the lymph empties into veins. If it were not for the lymphatic system, we would all eventually show symptoms like those of elephantiasis, which occurs when the nematode worm *Wuchereria bancrofti* clogs the lymph vessels. Look at Fig. 26.7 and be thankful if you have a healthy lymphatic system.

The lymphatic system also performs other essential functions. Some lymph vessels in the small intestine have small projections called **lacteals** that absorb digested fats (see Fig. 13.9). The lymphatic system is also associated with **lymphoid organs,** includ-

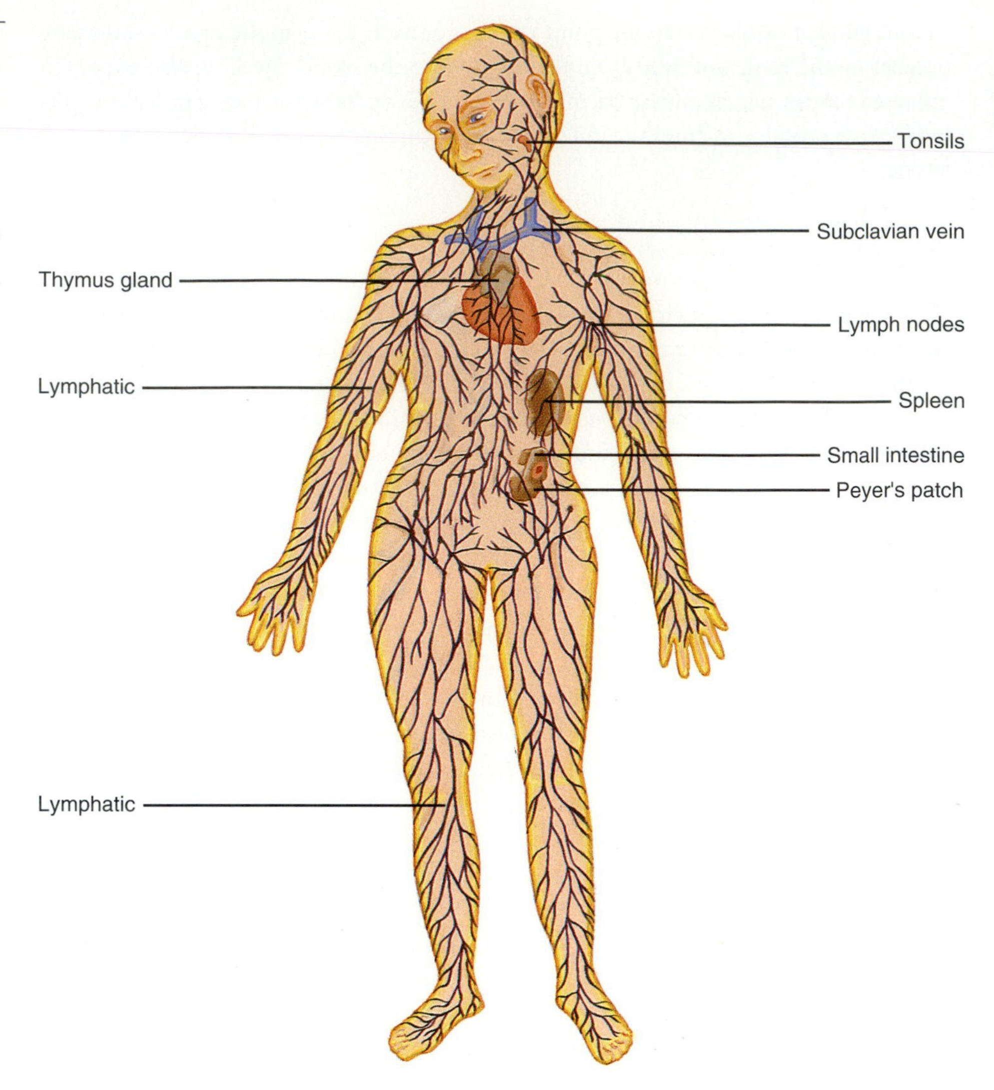

FIGURE 11.12

The lymphatic system and lymphoid tissue of humans. Lymph passes through lymph nodes and then empties from the lymphatic ducts into the subclavian veins. Lymph from the head and right arm drains into the lymphatic duct of the right subclavian vein. Lymph from the lower body and left arm drains into the lymphatic duct of the left subclavian vein. The thymus gland is degenerate in adults.

ing lymph nodes, tonsils, spleen, thymus, and Peyer's patches in the small intestine (Fig. 11.12). Lymphoid organs produce infection-fighting cells called **lymphocytes.** In addition, each organ has other specific functions. **Lymph nodes** contain phagocytic cells that engulf bacteria, the remains of white blood cells that have leaked into infected tissues, and other cellular debris. The lymph nodes also contain lymphocytes. Many lymph nodes are located in the groin, arm pits, and neck. These nodes, commonly referred to as "glands," are normally 1 to 25 mm in diameter, but they become noticeably swollen when they have to deal with an infection. The tonsils, likewise, become enlarged during certain infections. The spleen serves as a reservoir for red blood cells and as a site where lymphocytes mature. The thymus is another major site where lymphocytes develop the ability to combat infections, as will be discussed later.

General Defenses Against Invasion

An animal's interstitial fluid is a suitable habitat not only for its own cells, but also for a variety of viruses, bacteria, fungi, protozoans, and parasitic animals. Probably all animals have mechanisms that resist such invaders. Several types of molecules that punch holes in bacterial cell walls have been identified in various animals, including cecropins from insects, squalamines from sharks, and magainins from frogs. In humans and other mammals, **lysozyme** in sweat, tears, and other secretions kill bacteria before they enter the body. If bacteria do manage to get into the bloodstream, they face a system of proteins called **complement** that can destroy them. In addition, **natural killer cells** and **granulocytes** can destroy bacteria and many other kinds of invaders in the blood (Fig. 11.1).

A group of glycoproteins called **interferons** provides a general defense against viruses. Interferons are produced by infected cells that are already doomed from having their genetic machinery commandeered to produce new viruses. They interfere with the ability of other infected cells to produce a broad range of viruses, including those that cause certain cancers, herpes, viral pneumonia, hepatitis, influenza, and the common cold.

A general defense against all these invaders is **inflammation,** which is a reaction to any kind of damage to tissues. The signs of inflammation—warmth, pain, itching, reddening, and swelling—are familiar to anyone who has had a cold, a sunburn, or an allergic reaction. Inflammation is triggered by a variety of chemical messengers. **Histamine** and **kinins,** for example, dilate arterioles, thereby increasing the flow of blood to the area. This causes the warmth and reddening. They also make capillaries and blood vessel walls leaky. This leakiness causes swelling, since blood proteins leak into the interstitial fluid, upsetting the balance between blood pressure and colloid osmotic pressure (the Starling effect). More important, the leakiness allows certain white blood cells to enter the tissue spaces. First among these white blood cells are granulocytes, which are soon followed by **macrophages,** whose name literally means "big eater." Macrophages engulf and digest many of the invaders and damaged cells (Fig. 11.13).

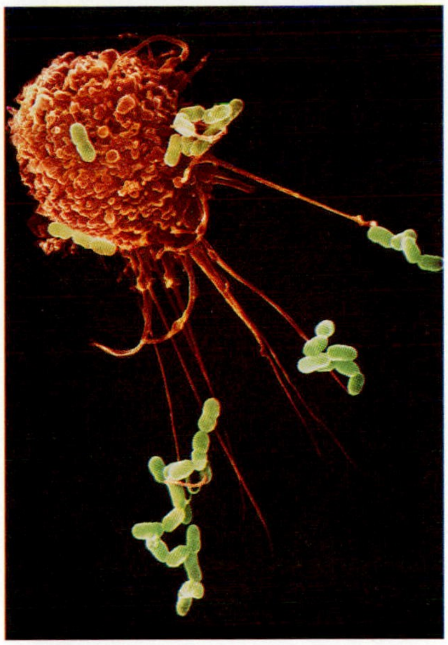

FIGURE 11.13
False-color scanning electron micrograph of a macrophage destroying bacteria by phagocytosis.

Immunity

Besides general defenses, there are specific defenses against invading viruses, foreign cells, complex foreign molecules, and parasites. These specific defenses are termed immunity. Immunity is a response to molecules called **antigens** that do not normally occur in a particular individual and therefore indicate infection. Macrophages and certain other **antigen-presenting cells** digest antigens into fragments that combine with proteins produced by a cluster of genes called the **major histocompatibility complex** (MHC). Antigens bound to MHC protein are then brought to the surface of the antigen-presenting cell, where they are recognized by receptor molecules on certain cells called **helper T lymphocytes** (= CD4 cells). (T is for thymus, where T cells mature in the embryo.) Each helper T cell has membrane receptor molecules that recognize a particular antigen. When that happens, the helper T cell divides and produces interferon and other **cytokines** (= lymphokines); which are substances that promote inflammation as well as defense against the specific invader that brought in the antigen.

There are two types of immune response, depending on the nature of the antigen. Antigens from bacteria, protozoans, cancerous or virus-infected cells, and foreign tissues, including transplanted hearts and other organs, trigger a **cell-mediated immune response.** This response involves a direct attack on cells by another kind of T lymphocyte: **cytotoxic T cells** (= CD8 cells). After exposure to the antigen and activation by cytokines from helper T cells, cytotoxic T cells divide into **clones** of identical cells with receptors for the antigen. They then destroy the plasma membrane of any cell bearing the antigen (Fig. 11.14 on page 232).

Foreign molecules and antigens from worms and other parasites trigger the second type of immune response: the **humoral immune response.** Humoral immunity involves blood-borne proteins called **antibodies** (= immunoglobulins), which are produced by clones of identical **plasma cells.** Plasma cells are **B lymphocytes** that have

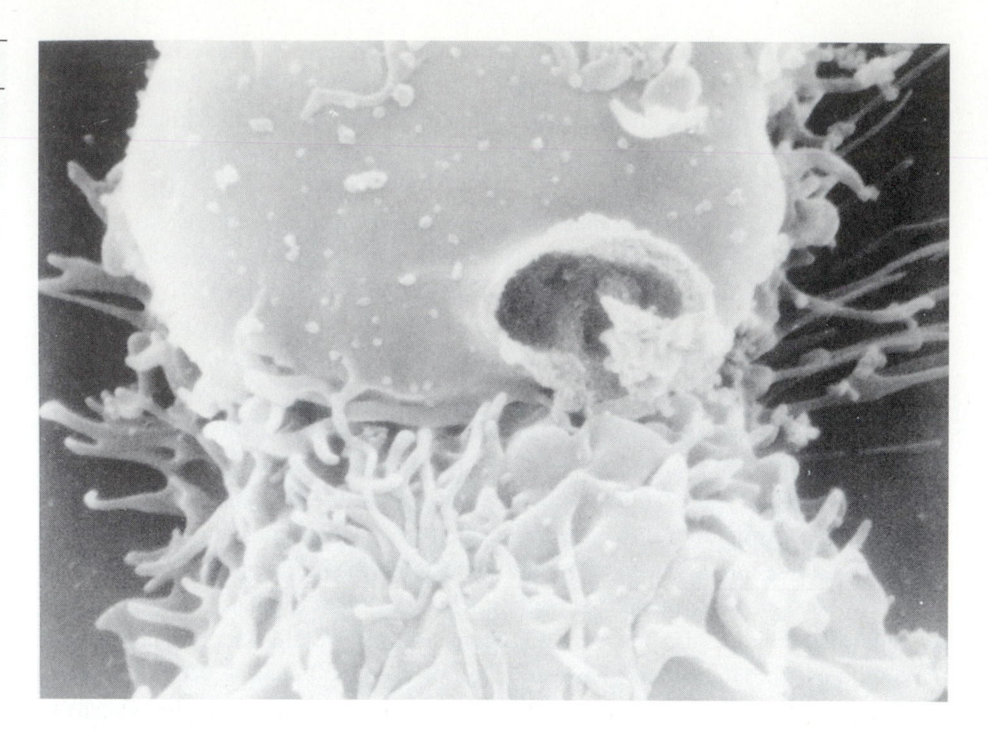

FIGURE 11.14

Scanning electron micrograph of a cytotoxic T cell (bottom) killing an invading cell. The T cell has secreted a protein called perforin that binds to the target cell's plasma membrane and makes holes in it. Natural killer cells and eosinophils work in a similar way. Shown 5700 times actual size.

■ **Question:** Would cytotoxic T cells attack viruses directly, or only cells invaded by viruses?

been exposed to antigen and activated by cytokines in lymphoid tissue, after which they enter the bloodstream. Antibodies **agglutinate** antigens into harmless clumps (Fig. 11.15). They also trigger destruction of foreign cells by natural killer cells and attract complement. The main interactions of the immune response are summarized in Fig. 11.16.

Immunological Memory

The immune response may take several days to build up the first time a particular antigen is encountered. If the same or a similar antigen is encountered later, however, the response can be quite rapid. This phenomenon is called immunological memory. After an invasion is successfully repelled by the B and T cells described previously, some of

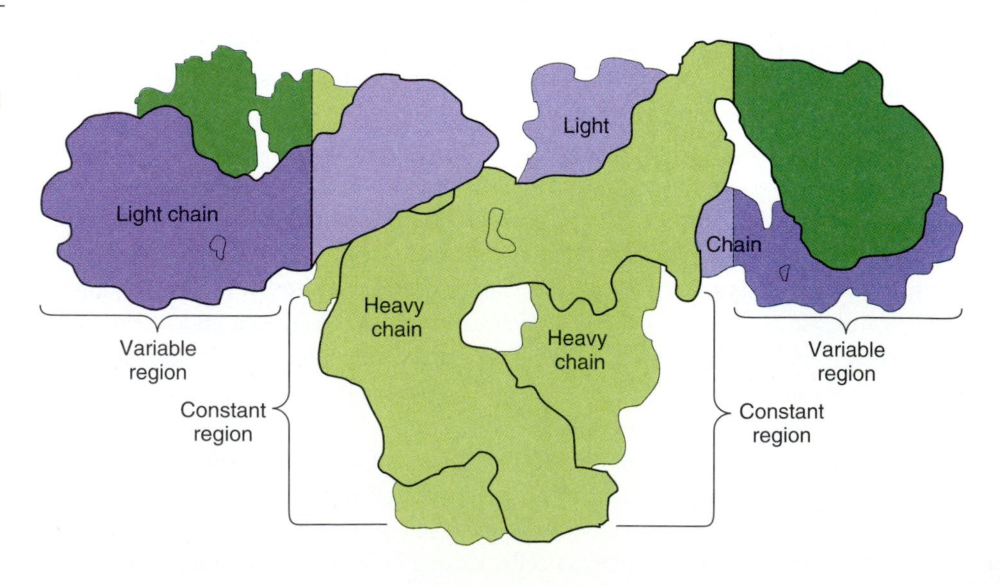

FIGURE 11.15

Structure of an antibody. Antibodies consist of two pairs of proteins: two identical heavy chains and two identical light chains. There are five types of heavy chain, giving rise to five types of immunoglobulin (IgG, IgA, IgM, IgE, IgD) with distinct functions. For example, IgE protects animals against parasites and is also involved in allergies. There are antibodies against virtually any potential antigen, because the variable regions can form more than 16 million kinds of attachment sites that bind to antigens.

■ **Question:** Since humans have only about 100,000 genes, how can we produce 16 million different variable regions?

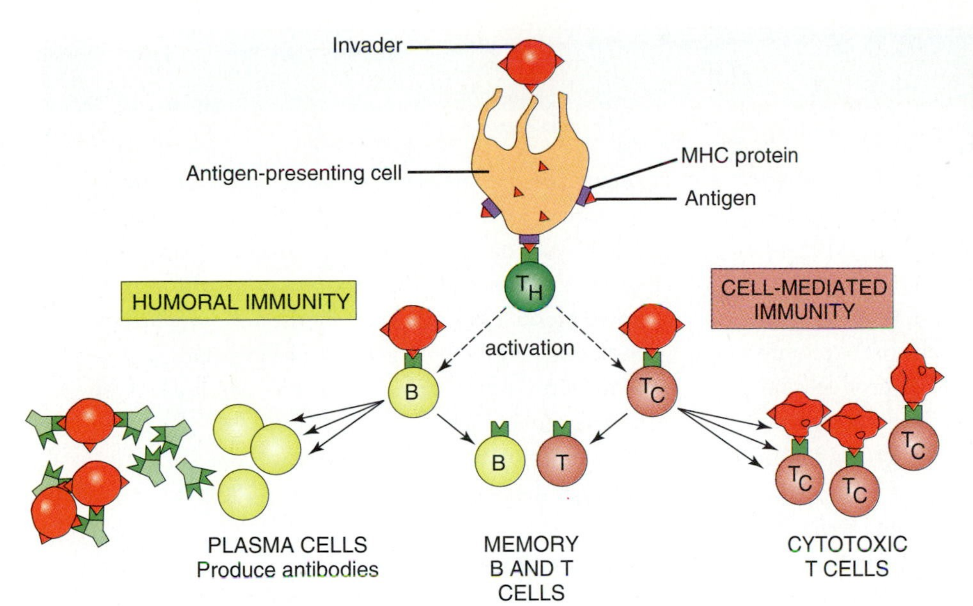

FIGURE 11.16

Some major events in the immune response. The immune response begins with a macrophage or other antigen-presenting cell engulfing an invader or damaged cell (pink) and presenting its antigens (red triangles) to helper T cells (T_H). Helper T cells then activate cytotoxic T cells that destroy cells or B cells that become antibody-producing plasma cells. These processes are coordinated by interferon, interleukins, and other cytokines.

■ **Question:** Why doesn't the immune system always work? How many ways can you think of for a virus, bacterium, or protozoan to avoid the immune system?

the cloned cells remain in reserve as **memory lymphocytes.** Memory lymphocytes can recognize the antigen and trigger an immediate and massive production of cytotoxic T cells and plasma cells upon a second exposure to it. Immunological memory therefore protects against repeated infection by the same invader. In some cases, however, immunological memory can be a nuisance or even a hazard. It is the basis for **allergies,** which are immune reactions against antigens that are not harmful in themselves. In some cases an immune reaction can be so sudden and massive that it causes a sudden vasodilation that drastically lowers blood pressure. Such **anaphylactic shock** is a serious hazard to people who are especially sensitive to certain antigens, such as those in bee stings.

Immunological memory makes it possible to **vaccinate** people and animals against infections by inoculating them with the antigens of infectious viruses or bacteria. Some vaccines, such as those for polio, are prepared from weakened virus. Other vaccines are prepared from viruses that are antigenically similar to the pathogenic virus, but less dangerous. Until recently, for example, vaccination with the cowpox virus was used to induce immunological memory against the smallpox virus. Thanks to vaccination, smallpox virus now survives only in laboratory cultures, and the only reminders of its existence are the round scars on the arms of people old enough to have been vaccinated against it.

People and other animals can also be immunized against toxins by several small injections of the toxin. Herpetologists and others who must handle poisonous snakes, for example, can be immunized in this way against venom. Some herpetologists simply wait for the snakes to make the injections. Another defense against toxin is **antiserum,** which is serum from animals (often horses) that contains antibodies to a toxin or other antigen. It is prepared by inoculating the animal with the antigen to trigger antibody production.

Just as important as immunological memory for certain antigens is the "memory" of the **self,** so that the immune system does not attack it. This memory usually results when T and B cells that would attack one's own cells and molecules are killed or inactivated in the embryo. Cells of another vertebrate are distinguished by their MHC proteins, which are unique and therefore antigenic except in an identical twin. Occasionally, however, lymphocytes attack self molecules such as receptors to synaptic

AIDS

There is no better way to appreciate the importance of the immune system than to consider what happens when it is destroyed, as it is in victims of AIDS—acquired immune deficiency syndrome. AIDS is caused by a human immunodeficiency virus (HIV) that attacks helper T cells, weakening the defense against other infections. Immediately after infection, HIV produces mild and temporary flulike symptoms. It then enters lymph nodes, where it reproduces for up to 10 years. Then symptoms of AIDS begin in the form of various infections that a normal immune system would easily have suppressed. Death usually occurs a few years later. The ultimate killers are often protozoans that cause pneumonia and toxoplasmosis (p. 469), tuberculosis, or a rare cancer of the blood vessels called Kaposi's sarcoma. HIV also damages brain cells.

HIV is spread by direct transfer of blood or through sexual intercourse. The most common routes of infection are therefore the following:

1. **Sexual intercourse, either anal or vaginal.** Anal intercourse between male homosexuals and bisexuals is the predominant route of transfer of HIV. Since this mode of transmission is now well known, most homosexuals and bisexuals take precautions to avoid transmission. The most rapidly increasing mode of infection is now vaginal intercourse between heterosexuals. Promiscuity vastly increases the danger of infection with HIV, because it exposes one not only to the body fluids of each sex partner, but, indirectly, to the fluids of the partner's former partners, and to all their former partners. Condoms reduce by tenfold the risk of infection by HIV, not to mention chlamydia, herpes, gonorrhea, syphilis, hepatitis B, and numerous other sexually transmitted diseases.
2. **Receipt (not donation) of blood or blood products by injection or transfusion.** Sharing of hypodermic syringes by intravenous drug users is the second most common mode of HIV transmission. Hemophiliacs were formerly at high risk, because they must take periodic injections of a blood-clotting factor that they lack. Now, however, the risk of acquiring AIDS in the developed parts of the world from blood has all but been eliminated by tests that detect the presence of antibodies against HIV in donated blood. (A positive test for the antibody merely indicates that the potential donor has been exposed to the virus, not that he or she is still carrying HIV or will develop AIDS.)
3. **Transmission from an infected mother to her fetus through the placenta.**

AIDS apparently originated in Africa from a virus that had previously infected only monkeys, and it was first identified in 1981 from a cluster of fewer than a hundred cases. It spread rapidly in the United States during the 1980s, and approximately 700,000 North Americans are now thought to be infected with HIV. In Africa, South America, and southern Asia more than 10 million people are infected, and the number is still growing rapidly. Health officials predict millions of deaths from AIDS in the coming decade.

In the United States the epidemic of AIDS appears to have leveled off at about 100,000 new cases per year, primarily among those who practice unsafe sex or use drugs intravenously. For others there is little chance of being infected with HIV, even if they have contact with those who are infected. There is no known case in which HIV was transmitted from an infected person to a family member who was not a sexual partner or newborn infant, even if the family member engaged in normal affectionate contact and shared dishes or even toothbrushes. No nurse who has administered mouth-to-mouth resuscitation to an AIDS victim has acquired the disease. In one study only 4 out of 870 health-care workers who accidentally jabbed themselves with HIV-contaminated needles subsequently tested positive for the virus. Out of 104 workers who were splashed with blood from AIDS patients, none became infected. Heterosexuals who are not extremely promiscuous and who do not share syringes with others are much more likely to die from overweight, smoking, or automobile accidents than from AIDS.

transmitters or hormones. The result is an **autoimmune disease.** Examples of autoimmune diseases include myasthenia gravis (a muscle degeneracy), one form of diabetes mellitus, rheumatoid arthritis, and multiple sclerosis.

Blood Typing

ABO GROUPING. For centuries physicians dreamed of transfusing blood from a healthy person into one who was bleeding to death. Early attempts, however, had variable success. Some patients were helped almost immediately, while others were killed just as quickly. Around the turn of the century, research on animals showed that the deaths resulted from the sudden agglutination (clumping) and destruction of the injected red blood cells. Karl Landsteiner (1868–1943) found that the results of transfusions depended on a proper matching of blood in four groups—A, B, AB, and O. Type A blood would usually not agglutinate in people with blood type A or AB, but would agglutinate in blood from people type B or O. Likewise, type B blood would not agglutinate in people with type B or AB blood, but would agglutinate in those with type A or O. Type O blood cells would usually not agglutinate in people with any blood type, but type AB blood cells would agglutinate in any type of blood other than AB.

We now know that the Landsteiner blood groups depend on whether red blood cells bear genetically inherited A and/or B antigens. If either of those antigens is not on the red blood cells of the recipient of the transfused blood, that molecule will be foreign, and he will have antibodies against it. These antibodies will agglutinate the transfused red blood cells, and the cytotoxic T cells will destroy them. To prevent this, blood typing is now routinely performed using antiserum for each antigen. For example, if the red blood cells in a drop of blood agglutinate when mixed with A antiserum, but not when mixed with B antiserum, the blood is type A. (What would be the blood type if the red cells agglutinate in both A and B antisera? Why wouldn't type O blood agglutinate in either serum?)

RH FACTOR. In 1940, 10 years after he received the Nobel Prize for discovering the ABO grouping, Landsteiner, together with A. S. Wiener, discovered another important and independent antigen: the Rh factor. The Rh factor, named for the rhesus monkey *Macaca mulata* in which it was discovered, is either present (Rh+) or absent (Rh–) on red blood cells. The presence or absence of the Rh factor is indicated by a + or – following the ABO type. For example, a common blood type in the United States is O+. Seventeen other blood groupings are also known, but they result from antigens that are much weaker than the A, B, and Rh antigens, and they can be ignored for purposes of transfusions.

Knowing the Rh factor is crucial in another way that clearly illustrates the importance of immunity. If an Rh– woman gives birth to an Rh+ baby, red blood cells from the baby can leak into the maternal circulation across the disintegrating placenta. These Rh+ red blood cells can then trigger the production of Rh antibodies by the mother, and these antibodies can enter the bloodstream of a later fetus. If this fetus is also Rh+, the antibodies from the mother will destroy the fetus' red blood cells. The result is called **erythroblastosis fetalis,** and it can kill the fetus or produce brain damage. Erythroblastosis fetalis is now rare in the United States because Rh– mothers who have just given birth to Rh+ babies are injected with anti-Rh antibodies. These antibodies destroy any Rh+ red blood cells in her circulation before they can trigger immunity.

Summary

Animals must create and maintain an internal environment—the interstitial fluid—in which their cells can live. Generally the immediate source of the interstitial fluid is blood, which usually consists of cells and plasma and is pumped through a circulatory system by a heart. Arthropods and some other invertebrates have open circulatory systems, while many other invertebrates and vertebrates have closed circulatory systems, in which the blood is always contained within the heart and blood vessels.

In vertebrates there is an increasing separation of oxygenated and deoxygenated blood from fishes to crocodilians, birds, and mammals. In the four-chambered hearts of humans and others there are two atria and two ventricles, with complete separation between the pulmonary circulation (which pumps deoxygenated blood to the lungs) and the systemic circulation (which pumps oxygenated blood from the lungs to the rest of the body).

Vertebrate hearts are myogenic: The action potentials that trigger contractions originate in pacemaker cells in the sinoatrial node of the heart. The action potentials are conducted over the atria, then delayed by the atrioventricular node before being transmitted by bundles of His and Purkinje fibers to the ventricles. One-way atrioventricular and semilunar valves prevent backflow of the blood. Cardiac output is regulated by effects of the autonomic nervous system and hormones on the heart rate and stroke volume.

Blood pumped through arteries under high pressure flows into capillaries and veins, losing pressure along the way. Because of the Starling effect, fluid tends to leave capillaries at the arterial end and to re-enter at the venous end. Fluid and cellular debris that accumulate in intercellular spaces return to the bloodstream through the lymphatic system, which is also important in immune responses. Loss of blood is limited by several processes of hemostasis, especially clotting.

There are several general defenses against invading organisms, including inflammation. Processes of inflammation also trigger immune responses, which are directed against specific invaders. Upon exposure to specific antigens, T lymphocytes attack cells that bear that antigen. B cells become plasma cells that release antibodies against the antigen. The response of antibodies to A, B, and Rh antigens on red blood cells is the basis for blood typing.

Key Terms

interstitial fluid
blood
plasma
serum
red blood cell
white blood cell
open and closed circulatory systems
atrium
ventricle
pulmonary circulation
systemic circulation
atrioventricular valve
semilunar valve
sinoatrial node
atrioventricular node
cardiac output
artery
vein
capillary
Starling effect
hemostasis
lymph
inflammation
immunity
T lymphocyte
B lymphocyte
antigen
antibody

Chapter Test

1. Why do most large animals require a circulatory system? (p. 217)
2. Explain how the following body fluids are related to each other, and how they differ: blood, plasma, serum, lymph, interstitial fluid. (pp. 217, 229)
3. What is the difference between open and closed circulatory systems? What are the advantages and disadvantages of each? Name one group of animals with open circulatory systems. (p. 217)
4. Briefly compare the hearts of various vertebrates. Explain what is meant by the following statement: "In the evolution of vertebrates there has been a trend toward greater separation of pulmonary and systemic circulation." (pp. 218–221)
5. Diagram the human heart from memory and show the pathway of blood flow through it. (pp. 221–222)
6. Explain what is meant by saying that the heart is myogenic. Briefly explain how contraction of the chambers of the heart is coordinated. (p. 223)
7. What role is served by nerves going to the human heart? (p. 225)
8. Explain how interstitial fluid is formed from blood. (p. 227)
9. What are the roles of the lymphatic system? (pp. 229–230)
10. Briefly discuss heart transplants. What property of the heart permits it to continue beating after its removal from the body? How does the recipient's body attempt to reject the transplant? (pp. 223, 231)
11. Describe as completely as you can how the body would defend itself against the venom of a rattlesnake. (pp. 230–232)

■ Answers to the Figure Questions

11.5 The right ventricle has to pump exactly the same volume of blood per minute that the left ventricle does, because all the blood that leaves one ventricle goes into the other. Weakening of the left ventricle causes blood to accumulate in the ventricle, which increases the pressure in the left atrium during its contraction. This increased pressure is transmitted through the pulmonary veins to the capillaries in the lungs and forces fluids across the capillary linings into the air spaces.

11.14 Viruses have no plasma membrane, so there is nothing for perforin to act upon except in cells.

11.15 Evidently, different combinations of genes gives rise to the enormous number of amino-acid sequences in different antibodies. (There are 4800 different heavy-chain genes and 400 light-chain genes in a human.)

11.16 As explained in Chapter 22, an invader can elude the immune system by frequently changing its surface proteins, by living within the host's cells, by encasing itself in a protective cyst, by reproducing so rapidly that it swamps the immune system, or by damaging the host's immune system.

Readings

Recommended Readings

Ada, G. L., and G. Nossal. 1987. The clonal-selection theory. *Sci. Am.* 257(2):62–69 (Aug).

Anderson, R. M., and R. M. May. 1992. Understanding the AIDS pandemic. *Sci. Am.* 266(5):58–66 (May).

Aral, S. O., and K. K. Holmes. 1991. Sexually transmitted diseases in the AIDS era. *Sci. Am.* 264(2):62–69 (Feb).

Atkinson, M. A., and N. K. Maclaren. 1990. What causes diabetes? *Sci. Am.* 263(1):62–71 (July). (*Insulin-dependent diabetes is an autoimmune disease.*)

Caren, L. D. 1991. Effects of exercise on the human immune system. *BioScience* 41:410–415.

Cooper, E. L. 1990. Immune diversity throughout the animal kingdom. *BioScience* 40:720–722. (*Introduction to several papers on this topic.*)

Doolittle, R. F. 1981. Fibrinogen and fibrin. *Sci. Am.* 245(6):126–135 (Dec).

Engelhard, V. H. 1994. How cells process antigens. *Sci. Am.* 271(2):54–61 (Aug).

Golde, D. W., and J. C. Gasson. 1988. Hormones that stimulate the growth of blood cells. *Sci. Am.* 259(1):62–70 (July).

Golde, D. W. 1991. The stem cell. *Sci. Am.* 265(6):86–93 (Dec).

Johnson, H. M., F. W. Bazer, B. S. Szente, and M. A. Jarpe. 1994. How interferons fight disease. *Sci. Am.* 270 (5):68–75 (May).

Johnson, H. M., J. K. Russell, and C. H. Pontzer. 1992. Superantigens in human disease. *Sci. Am.* 266(4):92–101 (Apr).

Nadel, E. R. 1985. Physiological adaptations to aerobic training. *Am. Sci.* 73:334–343.

Nossal, G. J. V. 1993. Life, death and the immune system. *Sci. Am.* 269(3):52–62 (Sept). (*Introduction to an entire issue with articles on the immune system, AIDS, allergy, autoimmunity, and related subjects.*)

Old, L. J. 1988. Tumor necrosis factor. *Sci. Am.* 258(5):59–75 (May).

Robinson, T. F., S. M. Factor, and E. H. Sonnenblick. 1986. The heart as a suction pump. *Sci. Am.* 254(6):84–91 (June).

Schwartz, R. H. 1993. *T* cell anergy. *Sci. Am.* 269(2):62–71 (Aug). (*How T cells are prevented from attacking self.*)

Vogel, S. 1992. *Vital Circuits.* New York: Oxford University Press. (*A light-hearted look at the fluid mechanics of circulation.*)

Wakelin, D. 1984. *Immunity to Parasites: How Animals Control Parasitic Infections.* Baltimore: Edward Arnold.

Young, J. D-E., and Z. A. Cohn. 1988. How killer cells kill. *Sci. Am.* 258(1):38–44 (Jan).

Zucker, M. B. 1980. The functioning of blood platelets. *Sci. Am.* 242(6):86–103 (June).

Additional References

American Zoologist 29: 367–480 (1989) has several papers on immune responses to parasites in a variety of animals.

CHAPTER

12

Oxygen

Aquatic spider (*Argyroneta aquatica*) with oxygen supply.

Origin and Importance of Oxygen

Having looked at the systems responsible for creating and regulating the interstitial fluid, we shall now begin considering the mechanisms by which the interstitial fluid obtains the necessities that make it a hospitable internal environment for animal cells. Oxygen is one of those necessities. The requirement for O_2 is a paradox, because O_2 is toxic, and there was essentially none on Earth for the first two billion years of life's history. All the O_2 now in the atmosphere originated from photosynthesis, beginning with cyanobacteria around two billion years ago (p. 385). Oxygen is a waste product—a pollutant, if you will—of photosynthesis, and many organisms must have become extinct when it began to accumulate in the air and water. In some, however, there evolved a way to detoxify oxygen by combining it with hydrogen to form water, and further evolution turned necessity into a virtue by extracting ATP in the process (pp. 56–60). Since then, the descendants of these organisms, including animals, have been not only tolerant of oxygen, but dependent on it for the production of most of their ATP. In fact, one of the major trends in the evolution of animals has been toward obtaining O_2 more easily, by living in air rather than in water (Fig. 12.1 on page 240).

Concepts of Gas Diffusion

No animal has yet evolved a mechanism for actively transporting oxygen, so O_2 has to enter body fluids simply by diffusion through moist, permeable membranes (pp. 38–39). Carbon dioxide, the waste produced in the oxidation of organic nutrients, must also be eliminated by diffusion. The cellular membranes across which O_2 and CO_2 diffuse are either totally part of the integument of the animal or mainly confined to respiratory organs, such as gills and lungs. If we consider the simple case of an aquatic organism in which O_2 and CO_2 are dissolved in water on the outside of the animal and in body fluids on the inside, then the direction and rate of diffusion of O_2 and CO_2 depend on their concentrations on each side of the membrane. Thus O_2 diffuses from seawater through the membranes of a shark's gills into its blood plasma as long as the concentration of O_2 in the water is higher than that in the plasma. The greater the difference in concentration on the outside compared with the inside, the higher the rate of diffusion.

For terrestrial animals the situation is different, because O_2 dissolves much more readily in air than it does in water. A terrestrial animal can easily suffocate in air that has a higher concentration of O_2 than the plasma has, because the O_2 is more soluble in air than in body fluids. Instead of considering concentration, therefore, we must think more generally in terms of **partial pressure.** Partial pressure is the contribution by one kind of molecule in a fluid to the total pressure of the fluid. For example, the partial pressure of O_2 (pO_2) in air is 21% of the atmospheric pressure, since 21% of a volume of air consists of O_2. At sea level the average atmospheric pressure is 760 torr (= mmHg, millimeters of mercury), so the pO_2 is 21% of 760 torr = 160 torr. The overall direction of diffusion is always from a high partial pressure to a low partial pressure. In effect, the partial pressures on each side of a membrane push O_2 in opposite directions, so the net rate of diffusion is proportional to the pO_2 on one side minus the pO_2 on the other side. The rate of diffusion therefore increases in direct proportion to the difference in partial pressure. If the partial pressures are equal, there is no net diffusion, even if the concentrations are different (Fig. 12.2 on page 240).

The rate of diffusion is also directly proportional to the area available for diffusion. Thus two lungs are twice as effective as one. Another factor affecting the rate of

FIGURE 12.1
Some animals and the levels of oxygen in their habitats. Fishes and the hydra, like the earliest animals, live in water, which has lower levels of oxygen. Near the middle of the figure the earthworm, snail, and frog represent animals that can breathe air but require moist habitats. At the top of the figure are completely terrestrial animals, which generally evolved most recently.

■ **Question:** Some animals whose ancestors were terrestrial, such as whales and sea snakes, have returned to water. Do they then evolve gills, or do they continue to breathe air with lungs?

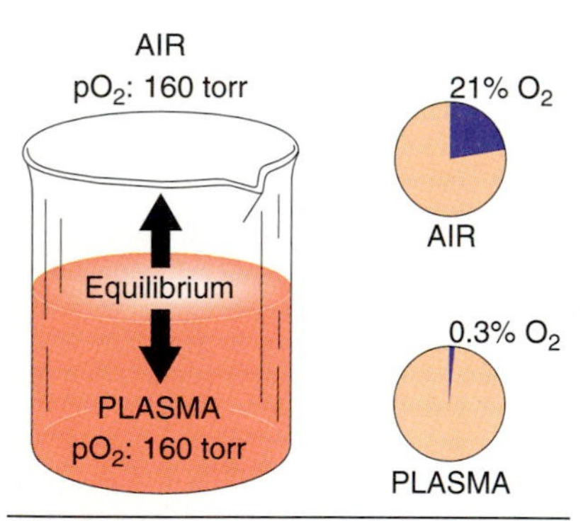

FIGURE 12.2
When considering the diffusion of gas from air to water or blood plasma, the partial pressure of the gas, rather than its concentration, determines the direction and rate of diffusion. Diffusion of O_2 from air to plasma stops when the partial pressures are the same, even though the amount of O_2 per milliter of plasma is much less than the amount per milliliter of air.

■ **Question:** What happens to the partial pressure of O_2 in water when the atmospheric pressure increases?

diffusion is membrane thickness. The thinner the membrane, the higher the rate of diffusion through it. To summarize, the rate of diffusion is proportional to

$$\frac{\text{partial pressure difference} \times \text{membrane area}}{\text{membrane thickness}}$$

(The algebraic equivalent of this statement is known as Fick's equation.) Animals have an enormous variety of respiratory organs by which they obtain oxygen, but all of them can be understood in terms of these concepts. No matter how diverse they are in structure, gills, lungs, and other such respiratory organs are all adapted to (1) maximize the difference between the pO_2 in the environment and that in the body fluids and (2) increase the area and minimize the thickness of tissue through which O_2 must diffuse.

One means of maintaining a large difference in pO_2 is by keeping the body fluids and external air or water moving, for example, by circulating blood and by breathing. Keeping the external fluid moving prevents it from being depleted of O_2, and keeping the plasma moving prevents it from becoming saturated with O_2. Examples will be discussed in various animal groups later.

Respiratory Pigments

Respiratory pigments in the blood plasma or in blood cells are also adaptations for achieving large differences in pO_2. Respiratory pigments bind O_2, thereby removing it from blood plasma. This keeps the pO_2 in the plasma low, thereby favoring continued diffusion of O_2 into the plasma. In the tissues, respiratory pigments unload the O_2 into plasma. This maintains a high pO_2 that favors diffusion to the interstitial fluid and cells. Respiratory pigments also transport CO_2 from tissues to the respiratory organs, and, incidentally, they give the blood of animals their various colors. They consist of a metal, usually iron, in an organic matrix. The metal binds the oxygen noncovalently. Respiratory pigments occur either suspended in the plasma or contained within blood cells, depending on the species. Several types of respiratory pigments occur in various animal groups (Table 12.1).

STRUCTURE OF HEMOGLOBIN. By far the best-known respiratory pigment is hemoglobin (Hb), primarily because it occurs in so many animals, including humans. Hemoglobin consists of **hemes** and protein. The hemes of all hemoglobins are identical and similar to the hemes in the cytochromes of mitochondria. Each heme contains an iron atom that binds an O_2 molecule. The proteins in hemoglobin vary widely, indicating that they evolved independently among different groups of animals. In vertebrates the protein is called **globin,** and it consists of two α (alpha) and two β (beta)

TABLE 12.1

Types of respiratory pigment, listed in order of frequency of occurrence.

	HEMOGLOBIN	HEMOCYANIN	HEMERYTHRIN	CHLOROCRUORIN
Animal Groups:	Vertebrates; some flatworms, nematodes, annelids, crustaceans, insects	Some molluscs; horseshoe crabs, crustaceans, some spiders	Mainly sipunculids; priapulids, brachiopods; some nematodes, polychaete annelids	Some polychaete annelids
Metal:	Iron	Copper	Iron	Iron
Location in Blood:	Cells or plasma	Plasma	Cells	Plasma
Color				
Oxygenated:	Red	Blue	Violet	Green
Deoxygenated:	Red-purple	Colorless	Colorless	Colorless

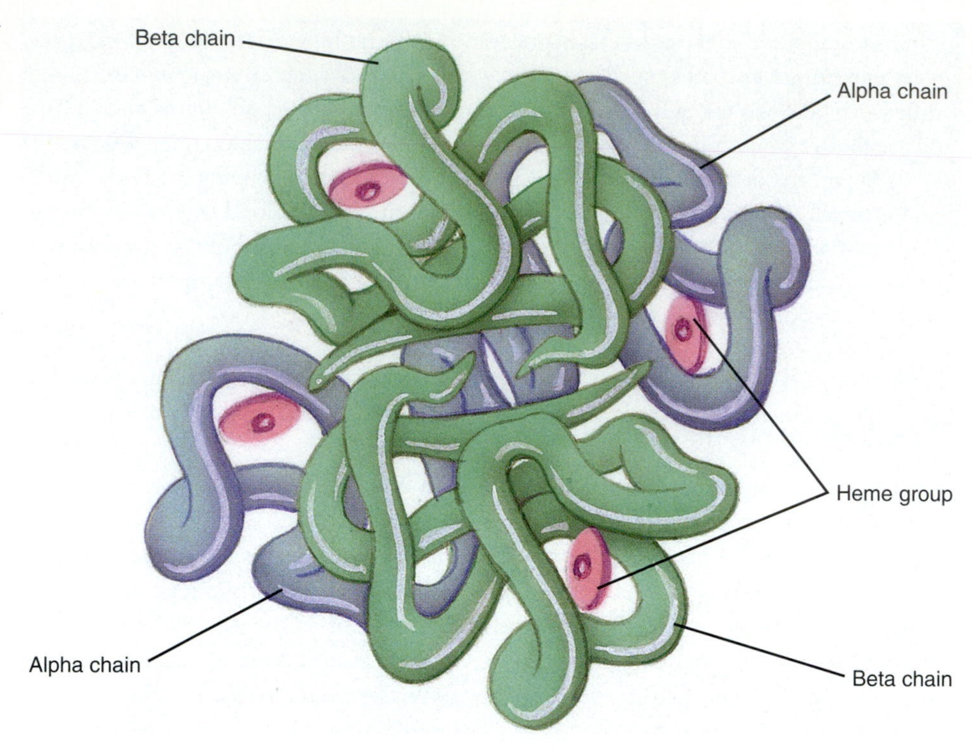

FIGURE 12.3
The structure of vertebrate hemoglobin (Hb). There are two α and two β protein chains (globins), each of which embraces a heme group (iron porphyrin). One O_2 binds to the iron atom in each of the heme groups.

chains. Each chain bears one heme group (Fig. 12.3). When O_2 binds to heme there is no exchange of electrons, so the iron is not oxidized by the oxygen. (In other words, hemoglobin does not rust.) Instead the iron simply holds the O_2 in place electrostatically. A hemoglobin molecule bearing O_2 is said to be **oxygenated,** rather than oxidized. Oxygenated hemoglobin, abbreviated HbO, has the familiar red color of blood that has been exposed to air. Hemoglobin that is deoxygenated is a darker red that looks blue in the veins when viewed through skin.

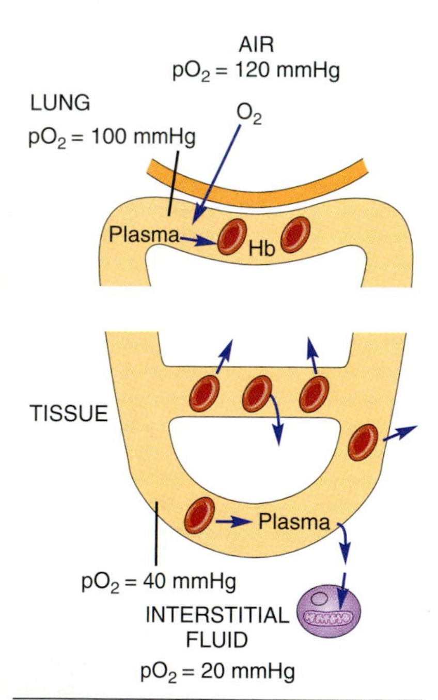

FIGURE 12.4
The diffusion of oxygen in mammalian lungs and in tissue. In the lungs, Hb removes O_2 from plasma, maintaining a lower pO_2 in plasma than in the air in the lungs. Oxygen therefore continues to diffuse into plasma. In other body tissues, Hb releases the O_2 into the plasma, maintaining a higher partial pressure in plasma than in interstitial fluid, which favors diffusion of O_2 into interstitial fluid.

Oxygen Binding and Release. Hemoglobin increases by 70 times the amount of O_2 that can be picked up by a given volume of blood in the respiratory organs. Just as important, hemoglobin gives up its oxygen to the tissues that need it (Fig. 12.4). How does hemoglobin "know" that it is supposed to bind O_2 in the capillaries of the respiratory organs and release it in the systemic capillaries? Of course it doesn't know in any conscious sense: The hemoglobin molecules simply respond to the chemical conditions around them. When O_2 binds to one heme, it increases the ability of other hemes in the same Hb molecule to bind to O_2. In other words, O_2 increases the **affinity** of Hb for O_2. Therefore when blood is in the respiratory organs, where the pO_2 is high, Hb binds O_2 more strongly. When blood is in the systemic capillaries, where the pO_2 is low, Hb has a lower affinity for O_2 and therefore releases it to the tissues.

The effect of O_2 on the affinity of Hb is shown in the **oxygen dissociation curve** (= oxygen equilibrium curve; Fig. 12.5). At high pO_2, virtually all the Hb is oxygenated (100% HbO). As the blood circulates to tissues and the pO_2 of the plasma falls, the Hb loses its affinity for O_2. Hb therefore gives up its oxygen. In the range of pO_2 normally found in tissues (20 to 40 mmHg) the oxygen dissociation curve is especially steep. Thus even a small drop in pO_2 in the tissue causes Hb to unload a lot of O_2 into the plasma. The steepness of the curve in the range of pO_2 found in tissues is not merely a stroke of good luck: The structure of the globin molecules, which determines the shape of the oxygen dissociation curve, has evidently undergone extensive evolution.

The globin molecules are also adapted to respond to CO_2 and other conditions in the plasma. CO_2, which is released by active tissues, reduces the affinity for O_2, causing

Hb to give up even more of its O_2 to those tissues. This adaptation, called the **Bohr effect,** results in a shift in the oxygen dissociation curve to the right (Fig. 12.6). High temperature and low pH, which also occur in active tissue, have similar effects on Hb. As a result, HbO unloads even more oxygen in the presence of highly active tissue than it would ordinarily.

Fetuses of humans and most other mammals produce a **fetal hemoglobin** that has a greater affinity for oxygen than does the Hb of their mothers. This enables the fetus to extract O_2 from the mother's blood, even though the pO_2 in the uterus is as low as that on Mt. Everest. Skeletal muscle, especially the slow-twitch kind, also has a substance that has a higher affinity for O_2 than does Hb (see p. 208; Fig. 12.6). This substance, **myoglobin,** is similar to the β chain of Hb. Since myoglobin binds O_2 more strongly than Hb does, it easily obtains the O_2 that is essential for the production of ATP.

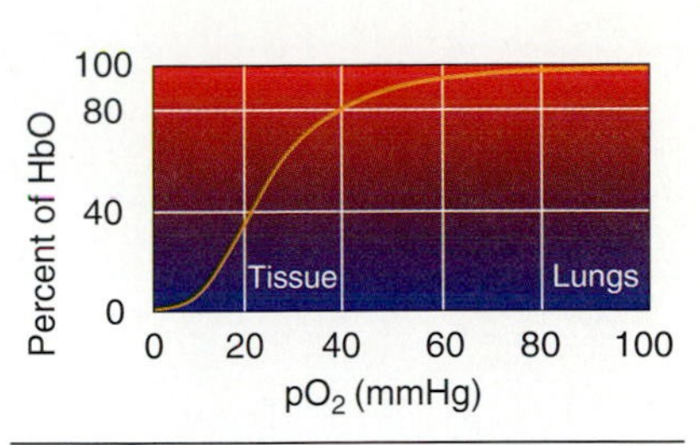

FIGURE 12.5

The oxygen dissociation (oxygen equilibrium) curve shows how the binding of O_2 by Hb depends on the partial pressure of O_2 (pO_2) in the surrounding plasma. In the lungs, where the pO_2 is high, Hb has a high affinity for O_2, and virtually all the Hb is in the oxygenated state (HbO). In the tissues, where pO_2 is about 30 to 40 mmHg, Hb loses its affinity for O_2 and gives it up to the plasma and interstitial fluid.

■ **Question:** Suppose the percentage of HbO was directly proportional to pO_2. Show on the graph what the oxygen dissociation curve would look like, and estimate how much O_2 the Hb would give up when pO_2 went from 40 to 20 mmHg.

CO_2 Transport

Red blood cells and hemoglobin also play important roles in the transport of carbon dioxide from tissues to the respiratory organs. CO_2 diffuses from cells into interstitial fluid and plasma, and from plasma into red blood cells. Some carbon dioxide reacts with water to form carbonic acid, which then dissociates into bicarbonate and hydrogen ion:

$$CO_2 + H_2O \longrightarrow H_2CO_3 \longrightarrow HCO_3^- + H^+$$

Ordinarily this reaction is too slow to carry off large amounts of CO_2, but red blood cells have an enzyme, **carbonic anhydrase,** that catalyzes the formation of the bicarbonate. Hemoglobin binds the H^+, accelerating the dissociation of the bicarbonate and preventing the acidification of blood. Some CO_2 is also transported by binding to the amine groups on the globin of hemoglobin, forming **carbamino compounds** ($HbCO_2$). These reactions reverse when the blood arrives in the respiratory organs.

Integumentary Exchange of Gases in Amphibians and Other Animals

In this chapter we shall focus on the three types of organs for exchange of respiratory gases: integument, gills, and lungs. As noted above, all have cellular membranes that must be thin and have a large surface area. Very small animals have such thin integuments and such large surface areas compared with body volume that they do not require gills or other special respiratory organs. Among these animals that exchange respiratory gases through the integument are round worms (phylum Nematoda) and annelids (phylum Annelida), such as earthworms and leeches. In many larger animals, all or part of the integument is specially adapted for respiration and can properly be called a respiratory organ. The phyla that use only integumentary exchange are those in Fig. 21.11 without a letter in the column labeled "Respiratory system." (Refer to the index for details on specific groups.) In many animals, such as land snails, eels, and frogs, integumentary exchange supplements gills or lungs.

Among the most familiar animals with integumentary exchange are amphibians. Certain salamanders satisfy all their requirements for O_2 by integumentary respiration. Frogs and most other amphibians have lungs, but even when breathing air they eliminate as much CO_2 through their skin as from their lungs. When submerged, of course, the integument is the only functioning respiratory organ in the frog. Unlike breathing through lungs, integumentary exchange requires no neural control. One can therefore destroy the nervous systems of a frog by "pithing" it with a probe, and the other organs

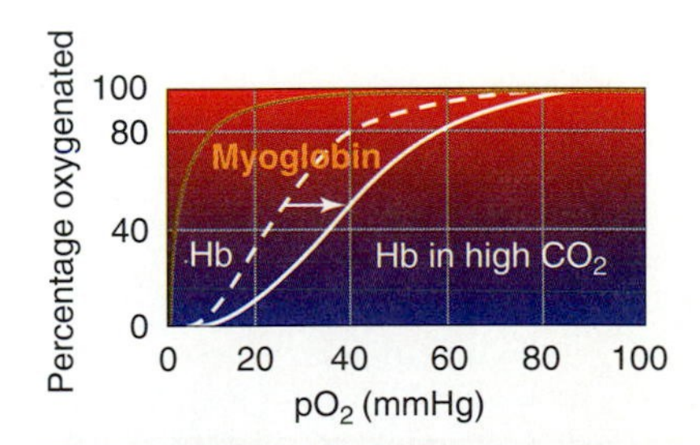

FIGURE 12.6

Carbon dioxide shifts the oxygen dissociation curve of Hb to the right, reducing the affinity for O_2 in tissues even if pO_2 remains the same. Thus Hb gives up more O_2 to active tissues. Myoglobin (left curve) has a higher affinity for O_2 than does Hb, so it takes O_2 away from blood to be used by muscle mitochondria.

will not deteriorate from anoxia. This is one reason why frogs are used in so many experiments.

Special adaptations for integumentary respiration include some or all of the following in various species:

1. **Reduced thickness of the integument.** For example, the thin skin of amphibians permits it to act as a respiratory organ under water.
2. **Increased area.** Some amphibians have folds of skin that provide additional area.
3. **Increased vascularization, with capillaries lying close beneath the skin.** This has the effect of increasing area, as well as quickly carrying off the oxygen so that pO_2 stays low in the blood.
4. **Means of keeping the integument moist and permeable to O_2.** Earthworms, amphibians, and many other animals with integumentary respiration continually secrete mucus for this function, and they generally have to stay in damp habitats.

Gills of Fishes and Other Aquatic Animals

Larger and more active animals require more specialized respiratory organs. In aquatic species these organs are called gills. Gills have evolved independently in many groups, including molluscs, crustaceans, and fishes. They vary greatly in their structure, but all have large areas and thin membranes, usually with numerous sheets of tissue (Fig. 12.7). Because of this structure, gills are usually delicate and require the buoyant support of water. Most animals with gills cannot breathe air directly, since the gills collapse

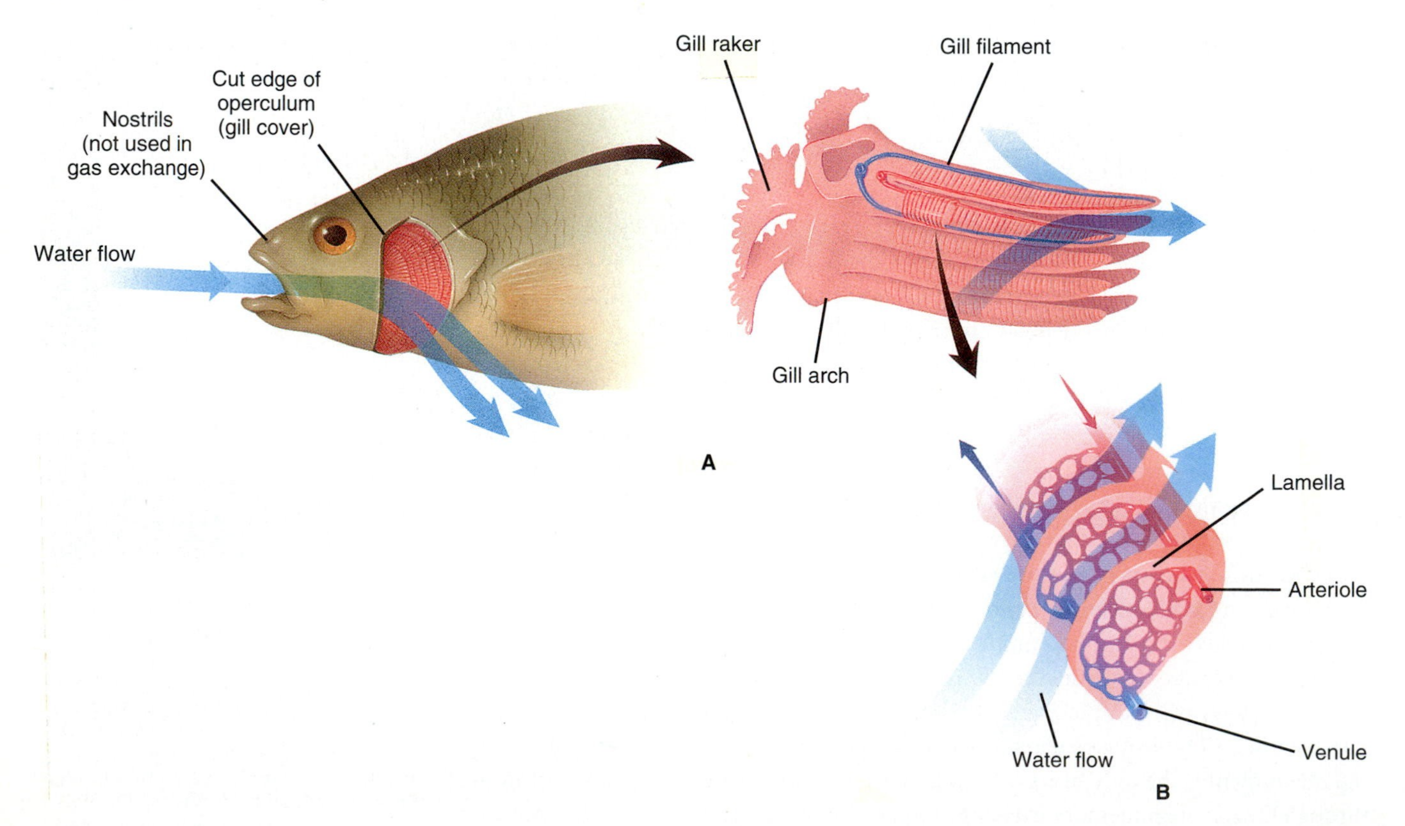

FIGURE 12.7
The structure of fish gills. (A) The location of the gills in a teleost (bony) fish. (B) One gill. Gill arches contain the blood vessels and bear the gill filaments. Gill rakers are used for feeding on small prey (p. 766). Numerous lamellae are superposed on each filament, increasing the surface area. The direction of water flow is opposite to the direction of blood flow. This constitutes a passive countercurrent mechanism for the exchange of O_2, as described in the box on p. 245.

when out of water. Some crustaceans and fishes, however, can breathe air, using reinforced gills that do not collapse when out of water.

Animals with gills usually have some mechanism that keeps oxygenated water continually flowing past them so that the external pO_2 stays high. Bivalve molluscs, such as clams, maintain the water flow by means of cilia on the gill surfaces. Sedentary animals often pump water past the gills by body contractions. Fishes such as bass, catfish, and aquarium fishes that swim slowly in calm water take water into the mouth (**buccal cavity**), then force it past the gills in the **opercular chamber.** This process is called **buccal pumping** (see Fig. 35.19). Fishes that swim rapidly or live in swift water, such as tuna and trout, can maintain the flow of water simply by opening their mouths into the current. The circulation of blood through the gills of fishes is in the direction opposite to that of the water. Fish gills therefore take advantage of **countercurrent exchange,** which increases the amount of O_2 they extract from the water. (See box below.)

Countercurrent Exchange

Countercurrent exchange is an elegant adaptation for extracting more oxygen from water, and it also occurs in numerous other organs adapted to maximize the exchange of molecules or heat. The following explanation of countercurrent exchange is for diffusion of molecules, but it applies as well for conduction of heat. The basic arrangement for a **passive countercurrent system** is that two fluids with different concentrations (or partial pressures, or temperature) flow past each other in opposite directions, separated by a permeable membrane. If the two fluids were flowing in the same direction side-by-side, the concentrations in both fluids would soon become equal, and diffusion would stop (Fig. 12.8A). By having the fluids flow in opposite directions, however, one fluid is always beside another fluid with a different concentration. Diffusion therefore continues, and more molecules are exchanged (Fig. 12.8B).

Much higher concentrations can be produced in an **active countercurrent exchange system.** In an active system the fluid flows through a tube that doubles back so that the direction of flow reverses (Fig. 12.8C). As fluid leaves the system, molecules are actively transported from it into the fluid that enters the system. This process repeats over and over, generating extremely large concentrations of the molecule. Examples of active countercurrent exchange in the kidney and in the swim bladder of fishes will be described later (see Chapters 14 and 35).

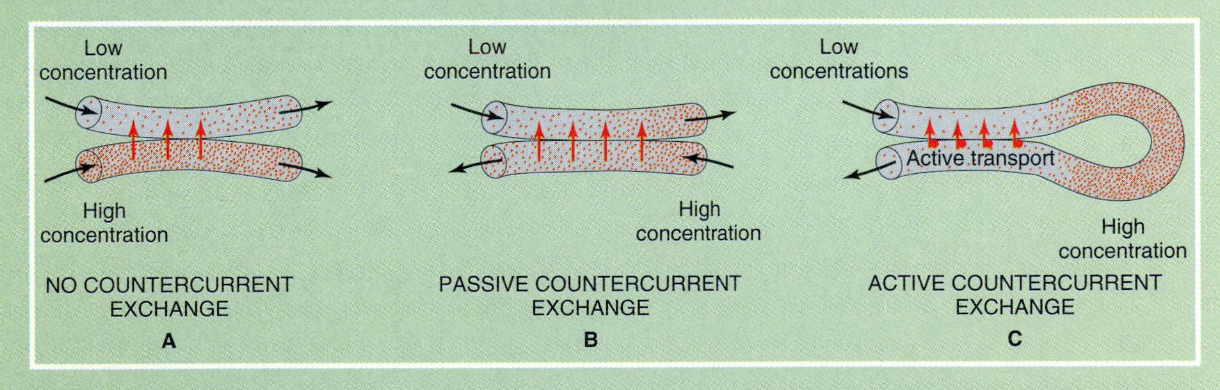

FIGURE 12.8
Countercurrent exchange. (A) A fluid with high concentration flowing past a fluid with low concentration in the same direction loses molecules by diffusion, but at a declining rate as the concentrations approach each other. (B) If the fluids move in opposite directions, however, the fluid with the higher concentration is always adjacent to fluid with a lower concentration. It therefore loses many more molecules to the other fluid by diffusion. (C) In an active countercurrent exchange system, molecules are continually recycled back into a fluid by active transport, resulting in a very large concentration.

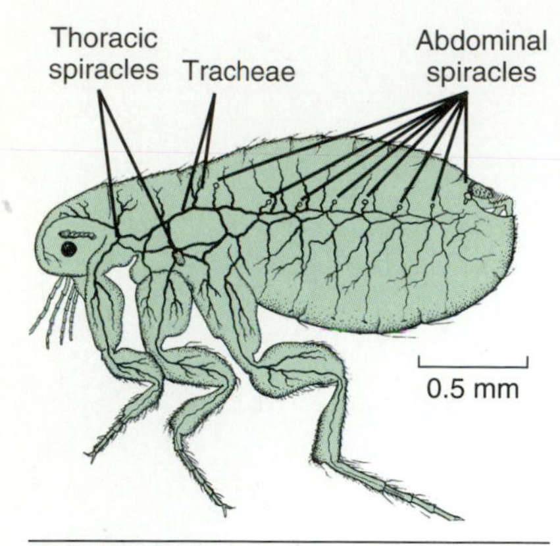

FIGURE 12.9

The tracheal system of an insect (a flea). Air enters through the spiracles and diffuses through tracheae and tracheoles.

Tracheae of Insects

The intricate surfaces of gills present such a high resistance to water flow that in some species 20% of the oxygen obtained by gills is used just to produce enough energy to keep water flowing past them. Little wonder that many animals evolved mechanisms for breathing air, which not only contains more oxygen, but also flows more easily. The invertebrates that are best adapted to air-breathing are spiders and insects. Spiders depend mainly on book lungs, which will be described later. In addition, hollow tubes called tracheae (TRAY-key-ee) contribute to respiration in some spiders, and they are the main respiratory organs in insects. The principle of tracheal respiration is straightforward: Tracheae simply provide a pathway from air to tissues (Figs. 12.9 and 32.7). In some cases the smallest tracheae, called **tracheoles,** connect directly to the plasma membranes of individual cells. The outer end of a trachea, the **spiracle,** is guarded by a valve that opens when internal CO_2 levels are high, but that otherwise stays closed, reducing the loss of water from the body. The diffusion of air through the tracheae is often passive, but some insects can pump air through the tracheae by contracting their bodies.

One advantage of tracheal respiration is that the circulation of blood does not have to be efficient, since it does not have to supply oxygen. Insects therefore get by with an open circulatory system with very low blood pressure. On the other hand, tracheal respiration does limit the size of insects, since O_2 does not diffuse readily through long, narrow tubes. Most insects are therefore only a few millimeters wide, and the largest insects are only(!) about 20 cm wide.

Lungs

The other major type of organ for breathing air is the lung, which usually consists of one or more internal air-filled cavities. Lungs evolved independently among three major groups: pulmonate molluscs, spiders, and vertebrates. Pulmonate molluscs (Latin *pulmo* lung) are the land snails, slugs, and some freshwater species that evolved from them. The lung of a pulmonate is a cavity that takes in air through an opening on one side (see Fig. 28.14). The lungs of spiders are called **book lungs** because they consist of flat pouches stacked like pages in a book (see Fig. 30.14).

The lungs of vertebrates originated as simple sacs branching from the digestive tracts of fishes during a dry spell some 400 million years ago. These sacs originally served as air reservoirs that allowed fishes to burrow into mud and escape drought. In most fishes the sacs later evolved into swim bladders that maintain buoyancy, but in others, such as lung fishes, the sacs became adapted for exchange of respiratory gases. Some of these fishes are thought to have been ancestors of amphibians, reptiles, birds, and mammals.

RESPIRATION IN AMPHIBIANS AND REPTILES. Although all vertebrate lungs had the same evolutionary origins, they diverged greatly in structure. The lungs of amphibians and most reptiles are essentially sacs. Folds in the walls of the lungs, especially in reptiles, increase the surface area for gaseous exchange (Figs. 12.10A and 37.13). All vertebrates with lungs have to breathe to maintain a high pO_2 and a low pCO_2 in the air within the lungs. The patterns of breathing in amphibians and reptiles differ from each other and from that of mammals. Amphibians draw fresh air into the buccal (mouth) cavity, exhale stale air from the lungs past the fresh air, then force the fresh air from the buccal cavity into the lungs (Fig. 12.10B). Reptiles, like mammals, inhale by expanding the body, which enlarges the lungs and allows air to enter. The major difference be-

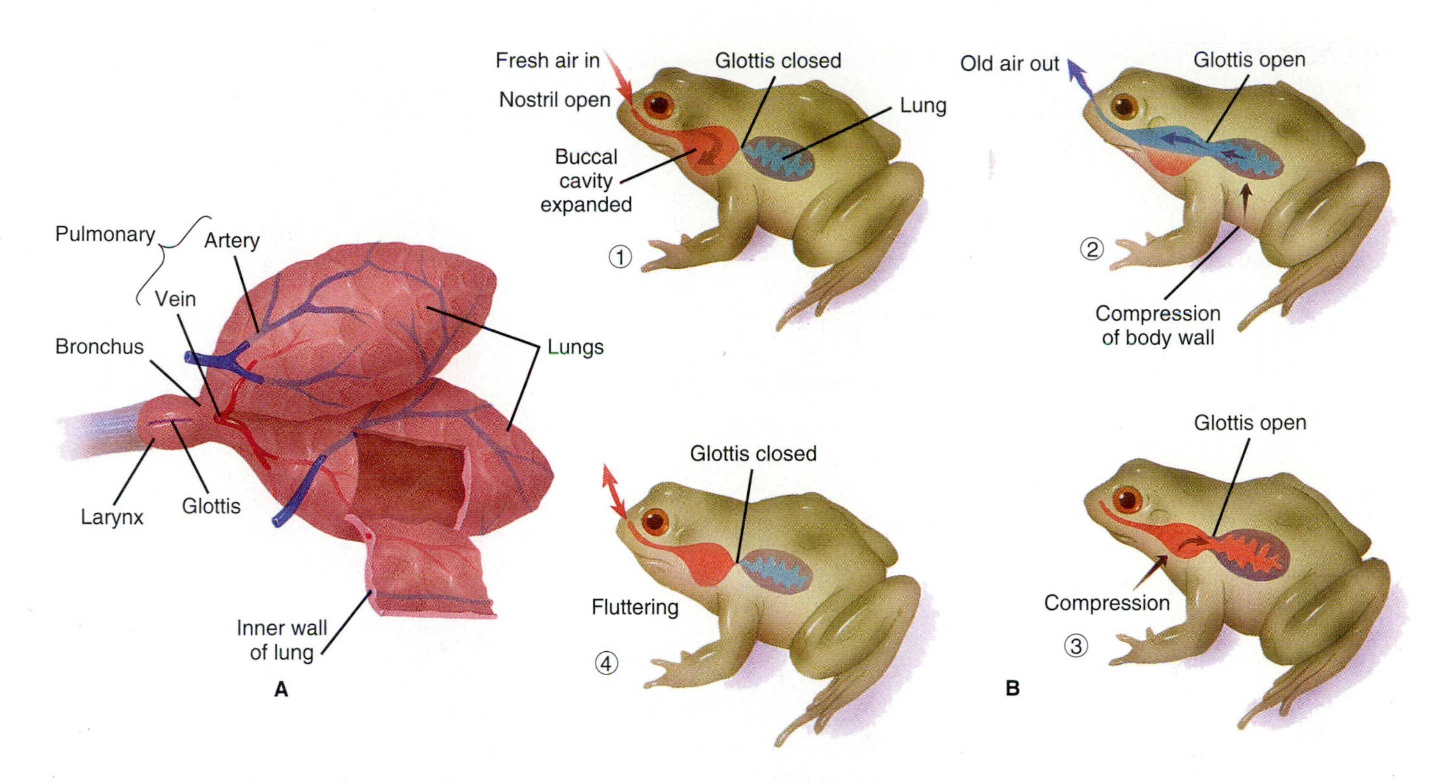

FIGURE 12.10

(A) The lungs of a frog. The lungs of reptiles are similar except that they usually have more extensive folding of the walls. (B) Schematic representation of breathing in the bullfrog *Rana catesbeiana.* ① With the nostrils open and the glottis closed, the buccal cavity expands, allowing atmospheric pressure to force air in. ② The frog then opens its glottis, and the elasticity of the lungs and contraction of the body force out the air already in the lungs. There is little mixing with air in the buccal cavity. ③ The frog then closes its nostrils and contracts the throat, forcing fresh air into the lungs. ④ With the glottis closed, the frog then flutters the throat to flush out the buccal cavity before beginning the cycle again.

tween breathing in reptiles and mammals is that reptiles hold their breaths for as long as several minutes after each inhalation. Reptiles can hold their breaths for so long because their metabolic rates are so much lower than those of mammals.

Respiration in Birds. Unlike the lungs of other vertebrates, those of birds are essentially solid except for numerous parallel tubes called **parabronchi** (Fig. 12.11 on page 248). Each parabronchus is divided lengthwise into many **air capillaries,** which increase the surface area for exchange of gases with the blood capillaries surrounding the parabronchi. In contrast to the sacs in the lungs of amphibians, reptiles, and mammals, in which air stagnates, the parabronchi permit air to flow completely through the lungs in only one direction during both inhalation and exhalation. **Air sacs** play an essential role in this one-way air flow. During inhalation the posterior air sacs expand, drawing in fresh air (see Fig. 38.13). The anterior air sacs also expand, and the lungs contract, so air flows out of the parabronchi into the anterior air sacs during inhalation. During exhalation the air sacs are compressed while the lungs expand slightly. Fresh air in the posterior air sacs is therefore forced through the parabronchi, and the used air in the anterior air sacs is exhaled.

Respiration in Humans and Other Mammals. The lungs and associated structures in mammals are more like those of reptiles than of birds (Fig. 12.12 on page 249). Air inhaled through the nasal passages or mouth enters the **trachea** (windpipe) then passes into each lung through a **bronchus.** The trachea is built of various connective tissues, including C-shaped rings of cartilage that bend rather than break when compressed. The trachea lies in front of the esophagus in the neck, which means that food has to pass over the opening of the trachea, called the **glottis** or **larynx,** when it is swallowed. This "design defect" presents a danger of choking while eating. This is es-

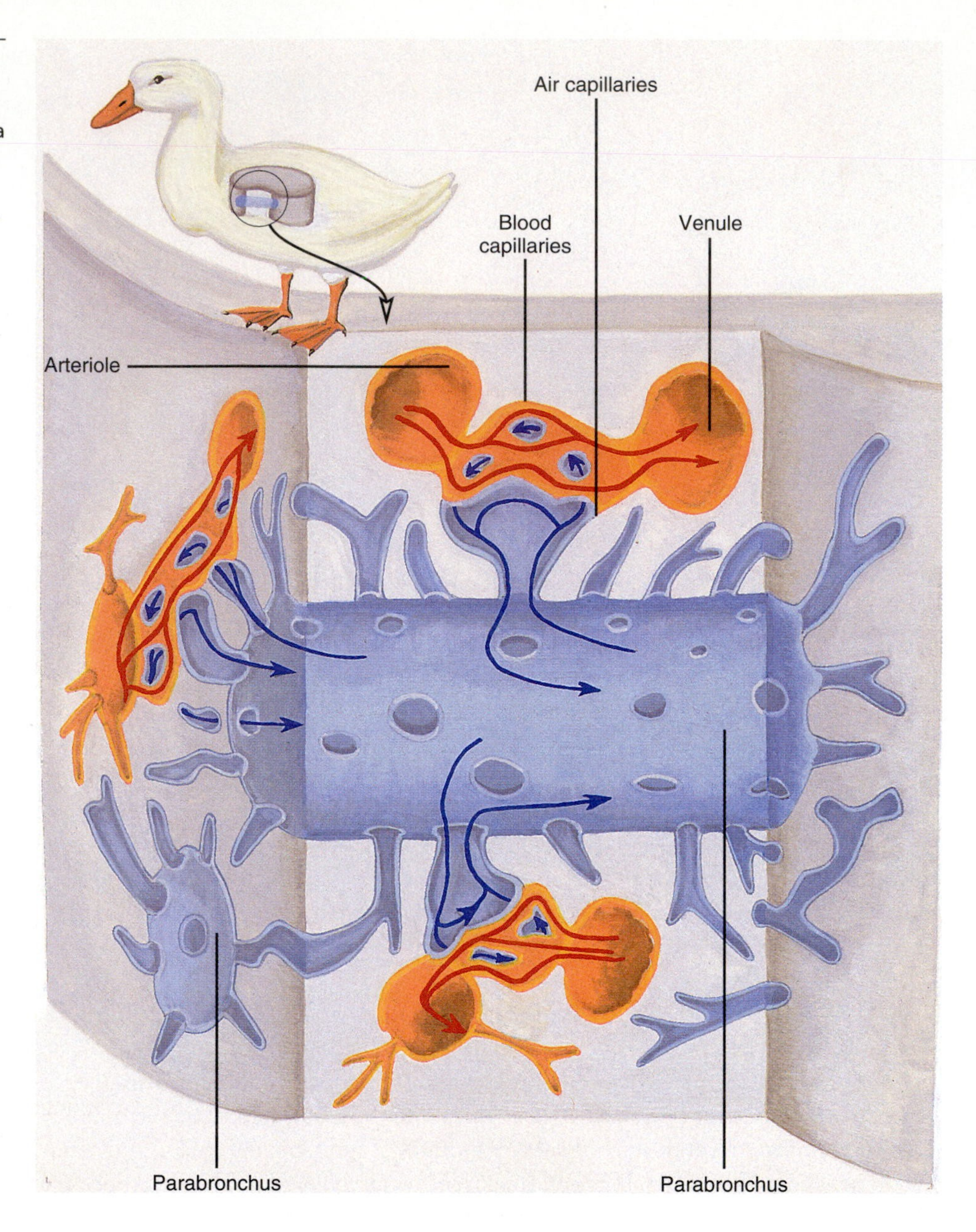

FIGURE 12.11

Detail of bird lung showing parabronchi. Blood and air move in opposite directions (arrows), establishing a countercurrent exchange.

■ **Question:** In order for this to function as a countercurrent exchange mechanism, the air must always flow in the same direction, opposite to that of blood. Does that occur?

pecially the case in humans, because the larger space above the larynx, which is necessary for speech, can accumulate food. Most of the time, however, the **epiglottis** blocks the glottis during swallowing.

Each bronchus divides within the lung into a network of numerous branches like an inverted tree. The smallest twigs in this bronchial tree are called **bronchioles.** Each bronchiole terminates with a cluster of hollow sacs called **alveoli** (plural of alveolus), where the actual exchange of oxygen and carbon dioxide occurs (Fig. 12.13 on page 250). Breathing ventilates the alveoli, bringing in fresh air that is rich in O_2 and removing the air that is laden with CO_2. As deoxygenated blood from the pulmonary artery flows through capillaries surrounding the alveoli, CO_2 diffuses out of the plasma while O_2 diffuses in. As one would expect from the requirements for diffusion, the alveolar walls are thin. Also the presence of numerous small alveoli provides a much larger surface area than would a few large alveoli with the same volume.

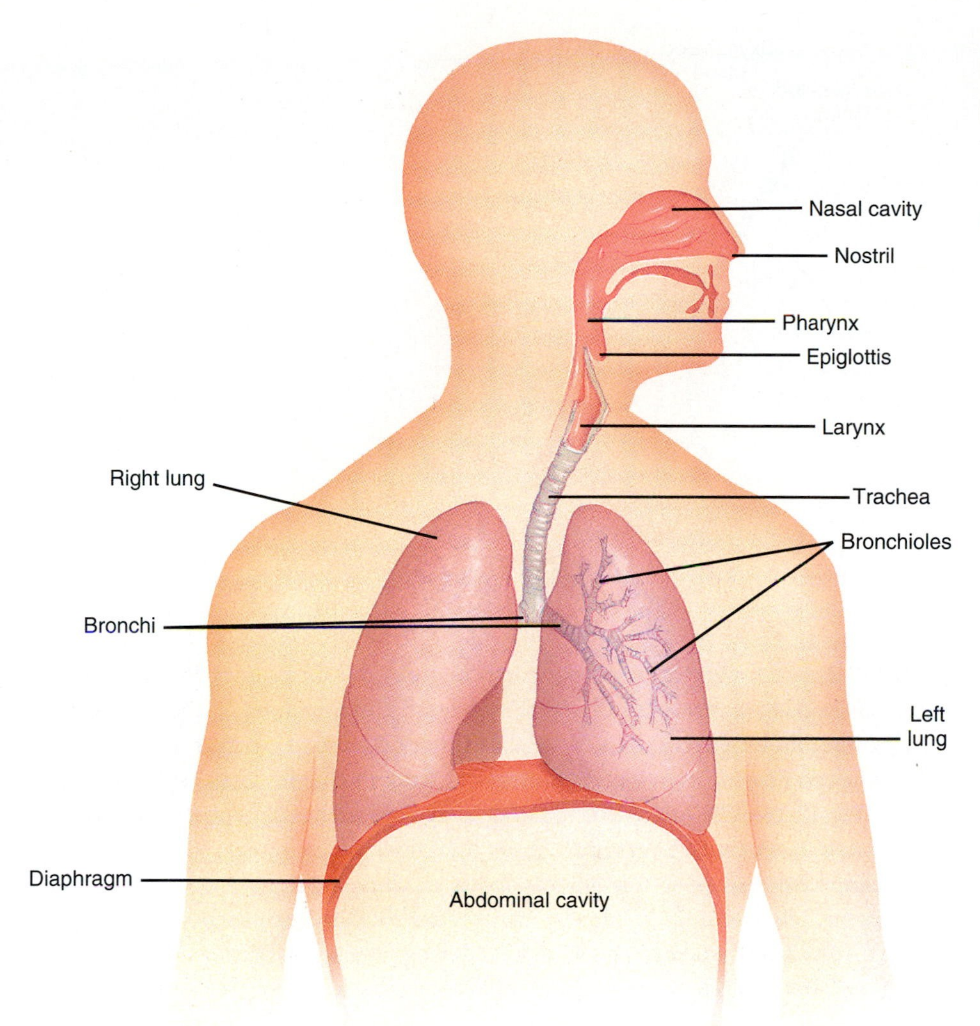

FIGURE 12.12

The lungs and air passages in a human. Air enters the lungs through the trachea and two bronchi. The larynx, also called the glottis, is part of the trachea.

■ **Question:** What happens to the lungs if the chest wall is penetrated by a wound (pneumothorax)? What would you have to do to restore breathing?

It takes many additional adaptations to keep the alveoli functioning. Mucus lining the respiratory passages protects them from the drying effects of air, and it traps dust and bacteria. Cilia lining the bronchi and bronchioles continually propel the mucous layer toward the mouth where it may be swallowed or spit out, depending on how you were raised. The importance of bronchial cilia is evident in those who have destroyed theirs with cigarette smoke. "Smoker's cough" is the only way they can bring up the mucus. The alveoli are also protected by antibiotic secretions and phagocytes that discourage bacteria, fungi, and other invaders, as well as by enzymes that break down any mucus that enters.

Breathing in Mammals

MECHANICS. In mammals, **inhalation** (= inspiration) is normally an active process, requiring contraction of skeletal muscles that enlarge the thoracic cavity. This enlargement reduces the pressure within the lungs by as much as 30 mmHg below atmospheric pressure, so air enters the alveoli. As the thorax expands, the lung is protected from abrasion by a surrounding double membrane called the **pleura.** The pleura also absorbs water or air that may become trapped between the lungs and the chest wall.

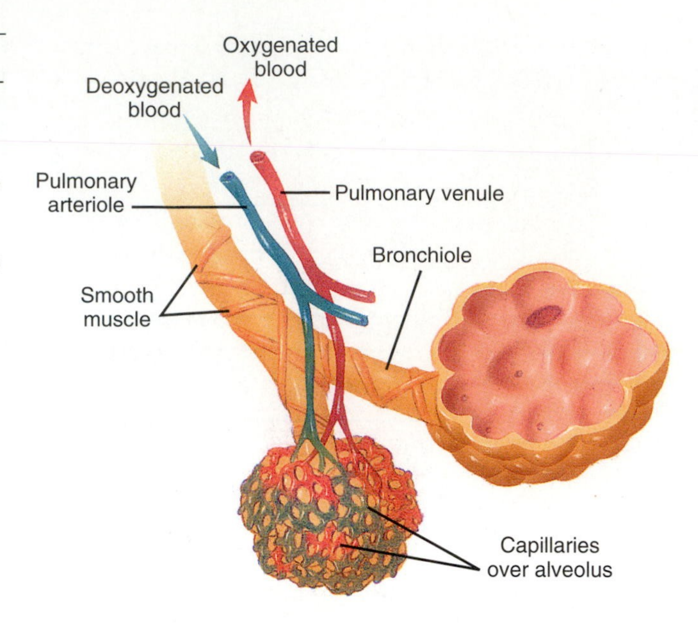

FIGURE 12.13

A bronchiole and several alveoli. The diameter of each alveolus is approximately 0.1 mm.

■ **Question:** From which chamber of the heart does the deoxygenated blood come? To which chamber does the oxygenated blood go?

Inhalation is triggered by neural impulses from the medulla of the brainstem (shown in Fig. 9.8) to two sets of skeletal muscles: the diaphragm and the intercostal muscles. The **diaphragm** is a sheet of skeletal muscle between the thoracic and abdominal cavities. When the diaphragm contracts, it flattens, thereby increasing thoracic volume (Fig. 12.14). The **intercostal muscles** lie in diagonal bands between the ribs (as you can see the next time you eat ribs). The contraction of these muscles causes the ribs to swivel upward from their attachments on the vertebrae, and this enlarges the chest cavity. From their anatomical arrangement, it is thought that the outer layer, consisting of the **external intercostal muscles,** contracts during inhalation. During normal **exhalation** (= expiration) the diaphragm and intercostal muscles simply return to their relaxed positions. Air can also be forcefully exhaled, during coughing and speaking, for example. The **internal intercostal muscles** may be partly responsible for forced exhalation. Forced exhalation also involves contractions of the abdominal muscles, which press on the abdominal contents and force the diaphragm upward.

REGULATION. Like the other essential requirements of cells, oxygen is maintained in appropriate levels in the interstitial fluid. This homeostasis is accomplished partly by changing the flow of air through bronchi, and partly by changing the rate and depth of breathing. The flow of air depends on bronchial diameter, which is controlled by smooth muscles (Fig. 12.13) that respond to neural and hormonal signals. During exercise or stress the sympathetic nervous system releases norepinephrine, which causes the smooth muscles to relax. The hormone epinephrine (adrenaline) from the adrenal medulla has a similar effect. Relaxation of the bronchial muscles allows the bronchi to dilate, letting air pass more easily into the alveoli. In the absence of stress or exercise, the parasympathetic nervous system releases acetylcholine onto these smooth muscles, which causes them to contract.

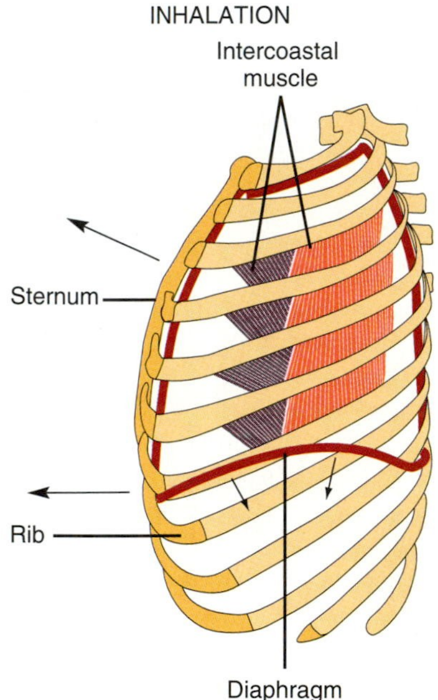

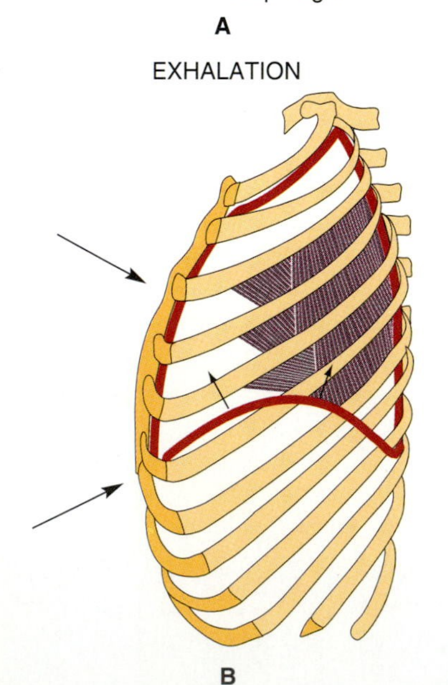

FIGURE 12.14

The mechanisms of inhalation and exhalation in humans. (A) Bursts of action potentials trigger inhalation by stimulating the diaphragm and intercostal muscles to contract. The diaphragm flattens, and the ribs move upward and outward. Both actions increase thoracic volume and thereby reduce the pressure within the lungs. (B) During normal exhalation the ribs and diaphragm return to their relaxed positions, thereby decreasing thoracic volume.

■ **Question:** What parts of the brain must interact to enable you to talk?

Breathing is controlled by bursts of action potentials to the diaphragm and intercostal muscles from a central pattern generator in the medulla. Each burst triggers an inhalation, so the number of bursts per minute equals the rate of inhalation. The frequency of action potentials within each burst controls the force of muscle contraction and therefore the depth of breathing. The medulla increases the rate and depth of breathing in response to signals indicating either high levels of CO_2 or low levels of O_2. High levels of CO_2 in the cerebrospinal fluid directly stimulate the medulla to increase breathing rate and depth. In addition, the medulla increases rate and depth of breathing in response to action potentials from **carotid and aortic bodies** in the carotid arteries to the head and in the aorta where it arches above the heart. The carotid and aortic bodies are sensitive to pH, pO_2, and especially to pCO_2. There are also other chemoreceptors in the linings of the airways that may stimulate the medulla in response to low levels of O_2 (Youngson et al. 1993).

Summary

The adaptations by which various animals obtain oxygen reflect the principles of diffusion. Oxygen diffuses at a rate that is proportional to the difference between external and internal partial pressures, proportional to the surface area for diffusion, and inversely proportional to the thickness of the diffusion surface. Animals therefore must maintain a sufficient difference in pO_2 and have a large, thin surface across which O_2 can diffuse. Many animals keep the external and internal fluids moving to maintain large differences in pO_2. Respiratory pigments such as hemoglobin serve the same function by (1) removing O_2 from blood plasma in respiratory tissues where pO_2 is high and (2) releasing the O_2 in other tissues where pO_2 is low.

In small animals the integument is thin enough and large enough in relation to body volume that no special respiratory organ is needed to obtain sufficient O_2. In many larger animals the integument is specially adapted as a respiratory organ. The adaptations include reduced thickness, increased area, increased blood circulation, and maintenance of a moist surface. The same adaptations occur in the gills of aquatic animals and the lungs of terrestrial animals. The lungs of amphibians, reptiles, and mammals are essentially sacs subdivided to various degrees. The lungs of birds contain numerous parabronchi through which air flows into and out of air sacs. The lungs of mammals contain numerous alveoli. Lungs are ventilated during breathing, which results from contractions of intercostal muscles and the diaphragm controlled by the medulla.

Key Terms

partial pressure
respiratory pigment
hemoglobin
integumentary respiration
gill
buccal pumping
countercurrent exchange
trachea
lung
parabronchus
bronchus
alveolus
diaphragm
intercostal muscle

Chapter Test

1. Why must partial pressure rather than concentration be used in figuring the direction and rate of diffusion of oxygen in terrestrial animals? (p. 239)
2. What respiratory pigment occurs in vertebrates? Describe its structure and function. How is its function modified in response to O_2 and CO_2? (pp. 241–243)
3. Write a biography of a hemoglobin molecule as it makes a complete journey through the circulatory system. (pp. 241–243)
4. Name four types of respiratory organ. For each type give the common name of an animal with that organ. Why is that type of respiratory organ appropriate for that animal? (pp. 243–246)
5. Explain why respiratory organs are so delicate. (p. 244)
6. How do the lungs of birds differ in structure and function from those of amphibians, reptiles, and mammals? (pp. 239–241)
7. In mammals, which muscles are involved in breathing? What causes them to contract? How are the rate and depth of breathing adjusted to meet the requirements for O_2? (pp. 249–251)

Answers to the Figure Questions

12.1 The probability of an animal reversing its course of evolution and thereby acquiring respiratory organs like those of its ancestors is extremely small. Animals like whales and sea snakes, whose ancestors apparently became aquatic relatively recently (within 100 million years), still breathe air using lungs. Some aquatic animals with terrestrial ancestors, such as insects, do have gills, however, although they are not at all like the gills of the insects' more remote aquatic ancestors.

12.2 The partial pressure of a gas is the total pressure multiplied by the fraction of the atmosphere due to that gas. Thus as long as the percentage of the atmosphere due to O_2 is constant, the partial pressure of O_2 will increase in direct proportion to the atmospheric pressure.

12.5 Draw a straight line from the bottom left corner of the graph to the top right corner. The Hb would give up only 20% of its O_2 (approximately 40% at 40 mmHg minus approximately 20% at 20 mmHg), rather than 50% (approximately 80% at 40 mmHg minus approximately 30% at 20 mmHg).

12.11 Yes, the air always flows in the same direction.

12.12 The lungs of humans and other mammals remain inflated only if the internal pressure is greater than that outside the lungs. Pneumothorax makes the pressure outside the lungs equal that inside, so the lungs collapse. First aid for pneumothorax consists of reinflating the lungs by blowing into the mouth while holding the nose shut, then blocking the wound to the chest.

12.13 Deoxygenated blood comes from the right ventricle; oxygenated blood goes to the left atrium.

12.14 While speaking, the brain must modify the pattern of breathing. The part of the medulla that triggers inhalation has to be inhibited while speaking, presumably by the motor cortex (see Figs. 9.9, 9.11, and 9.14). At the same time, the contraction of the internal intercostal muscles and the abdominal muscles is controlled to regulate the loudness of speech. The pattern of speech must be regulated to allow for periodic inhalation between phrases and sentences; otherwise the medulla will eventually override the speech areas of the brain, forcing an inhalation in the manner of someone speaking while out of breath.

Readings

Recommended Readings

Feder, M. E., and W. W. Burggren. 1985. Skin breathing in vertebrates. *Sci. Am.* 253(5):126–142 (Nov).

Kanwisher, J. W., and S. H. Ridgway. 1983. The physiological ecology of whales and porpoises. *Sci. Am.* 248(6):111–120 (June).

Schmidt-Nielsen, K. 1971. How birds breathe. *Sci. Am.* 225(6):72–79 (Dec).

Sebel, P., et al. 1985. *Respiration: The Breath of Life*. New York: Torstar. (*A lavish presentation of the physiology and sociology of breathing.*)

Additional References

Graham, J. B. 1988. Ecological and evolutionary aspects of integumentary respiration: body size, diffusion, and the invertebrata. *Am. Zool.* 28:1031–1045.

Liem, K. F. 1988. Form and function of lungs: the evolution of air breathing mechanisms. *Am. Zool.* 28:739–759.

Youngson, C., et al. 1993. Oxygen sensing in airway chemoreceptors. *Nature* 365:153–155.

CHAPTER

13

Nutrients

Raccoon, *Procyon lotor.*

Many children theorize that the food they eat simply falls into their hollow legs, and when the legs get full they have to go to the bathroom. As we get older we realize that this theory leaves a number of questions unanswered. Why does the food that goes in look so different from what comes out? Why does one feel the need to eat? And what is the point of it all, anyway? The objective of this chapter is to answer those questions and others. A concise answer is that the interstitial fluid of humans and of all animals must contain nutrients if it is to serve as a suitable internal environment for cells. These nutrients are released from food and absorbed into the body during the process of digestion.

Inorganic Nutrients

Like all molecules, nutrients are either inorganic or organic. The inorganic nutrients include ions and metals such as the following: sodium and potassium, which produce membrane potentials of neurons and other cells; calcium, which is an intracellular messenger and major component of shells and bone; and iron, which occurs in hemoglobin and cytochromes (Table 13.1). Many other elements must be present in animal diets in small amounts. The requirement for some of these **trace elements** is so small that it is often difficult to determine the consequences of a deficiency, since they cannot be totally excluded from experimental diets. One of the trace elements is **copper,** which mammals require in extremely small amounts for hemoglobin synthesis and mitochondrial functioning. Copper also occurs in the respiratory pigment hemocyanin in some

TABLE 13.1

Some important minerals required in the human diet.

Sodium (Na^+) and **Chloride** (Cl^-).	These are the most common ions in the interstitial fluid and in blood. Their net concentrations are largely responsible for maintaining the osmotic pressure in cells, and they are important in producing resting and action potentials across cell membranes. They must be present in the diet in rather high levels (several grams per day for humans, primarily as NaCl).
Potassium (K^+).	This is mainly an intracellular ion, important in maintaining the membrane voltage. Several milligrams are required each day.
Calcium (Ca^{2+}) and **Magnesium** (Mg^{2+}).	These divalent cations are among the most important in coordinating cellular functions. Ca^{2+} is a second messenger that triggers the release of synaptic transmitter, hormones, enzymes, and other secretions. It also triggers the contraction of muscles. It is important in calcium phosphate, the major mineral in bone. Mg^{2+} is required as a cofactor for enzymes such as ATPases.
Iron (Fe^{2+}).	This metal occurs in hemoglobin, myoglobin, and the cytochromes. Relatively small amounts are needed in the diet because Fe^{2+} is recycled. When the spleen breaks down dead red blood cells, it transfers the iron to the liver, where it is stored on the protein ferritin until bone marrow needs it to produce more hemoglobin.

invertebrates. Other trace elements are **cobalt** (in vitamin B_{12}), **iodine** (in the hormones thyroxine and triiodothyronine), **arsenic, chromium, fluoride, manganese, molybdenum, nickel, selenium, silicon, tin, vanadium, zinc,** and probably others. While essential in trace amounts, most of these elements are toxic in large concentrations.

Organic Nutrients

The organic nutrients include carbohydrates, fats, proteins, and vitamins. Cells need organic nutrients for two main reasons: for energy and for materials to make new cells. Organic nutrients consist mainly of six elements, easily remembered by the abbreviation "CHNOPS." By definition, **carbon** (C) is an element of all organic molecules, representing 9.5% of all the atoms in organic nutrients. The other elements and their proportions in organic nutrients are **hydrogen** (63%), **nitrogen** (1.4%), **oxygen** (25.5%), **phosphorus** (less than 1%), and **sulfur** (less than 1%).

CARBOHYDRATES. Carbohydrates serve primarily as immediate energy sources during cellular metabolism (pp. 53–60). Metabolism of 1 gram of carbohydrate produces about 4 kilocalories of energy (4 kcal/g), most of it heat, and a little less than half ATP. [One kilocalorie is enough energy to raise the temperature of 1 kilogram of water 1 degree Celsius. It is the same as the more familiar Calorie (capitalized) used in nontechnical literature such as diet books.] An average adult human ordinarily requires between 2000 and 3000 kcal per day, depending on body size, sex, and level of activity. Most of these calories come from carbohydrates—mainly starches and simple sugars, such as glucose, sucrose, and lactose (Table 13.2). Carbohydrates not immediately needed are stored in cells as the starchlike substance glycogen, or they are converted to fat.

LIPIDS. Lipids, also called fats, include several kinds of molecules, including neutral fats, phospholipids, and cholesterol. Fats that are liquid at body temperature are called

TABLE 13.2

Some important carbohydrates.

Glucose	The major blood sugar of vertebrates and most other animals (see Fig. 2.9).
Sucrose (common table sugar)	Combines two simple sugars and is therefore a **disaccharide** (see Fig. 2.10A). It consists of glucose linked to the fruit sugar, fructose.
Lactose	A disaccharide consisting of glucose linked to the similar sugar, galactose. Lactose occurs in milk, so it is mainly a nutrient for infant mammals.
Starch	The most common source of carbohydrate for herbivores (plant-eating animals) and most omnivores (animals such as humans that eat both plants and animals). Starch consists of many molecules of glucose, so it is a **polysaccharide.**
Glycogen	A starch-like polymer of glucose that is synthesized as a between-meal energy reserve (see Fig. 2.10B and p. 61). In humans the skeletal muscles store about 480 kcal of glycogen, and the liver stores about 280 kcal.

oils. Neutral fats are **triglycerides** (= triacylglycerols), which consist of a three-carbon glycerol backbone to which are linked three fatty acids (see Fig. 2.11). Fats are stored mainly in **adipose tissues** as long-term energy reserves, providing about 9 kcal/g. The average American has about 150,000 kcal of fat reserve (about 17 kg), which is enough to sustain life for up to 75 days. Animals can synthesize most fatty acids, but for most species there are some **essential fatty acids** that cannot be synthesized and that must therefore be present in the diet. For humans there are three essential fatty acids (arachidonic, linoleic, and linolenic). Phospholipid is similar to triglyceride except that a phosphate-linked group substitutes for one fatty acid (see Fig. 2.12). Phospholipids make up much of the substance of cell membranes. Cholesterol is also a major component of biological membranes, and it is the raw material for the synthesis of steroid hormones (see Fig. 7.2). Cholesterol is produced by the liver from fatty acids, and as everyone knows by now, excessive levels increase the risk of heart attack and stroke.

PROTEINS. Proteins serve as enzymes, hormones, and antibodies. Proteins also form collagen, actin, myosin, and other structural components of cells. Proteins are also the energy source of last resort, supplying 4 kcal/g, for a total of up to 24,000 kcal in a starving human. Proteins consist of precise sequences of amino acids (pp. 30–32). Most of the 20 amino acids in protein can be synthesized by all animals, but there are also some **essential amino acids** that must be obtained from the diet. For most animals, including humans, there are nine essential amino acids (cysteine, isoleucine, leucine, lysine, methionine, phenylalanine, threonine, tryptophan, and valine). The lack of just one of these amino acids is as damaging to growth and health as the lack of all of them, because the synthesis of a protein halts when mRNA calls for an amino acid that is not available. Unlike fatty acids, amino acids cannot be stored for use later, so proper growth and functioning require that all the essential amino acids be taken in frequently and at about the same time.

The "quality" of a protein is a measure of how closely the proportions of essential amino acids match the proportions required by the animal. For each animal the food with the highest quality of protein is obviously another animal of the same species. Most animal proteins are similar enough, however, that carnivores and omnivores can get enough of the essential amino acids without resorting to cannibalism. However, herbivores, including human vegetarians, may find it difficult to get the essential amino acids from plants. Many herbivores compensate by eating a variety of plants simultaneously, with each plant food making up for the deficiencies of others. Many cultures have evolved the same strategy. For example, in Mexico a popular dish is beans with rice. Beans make up for the deficiency of lysine in the rice, and the rice makes up for the deficiencies in methionine and cysteine in beans. Similarly, Jamaicans have a taste for rice and peas, and Indians mix wheat and pulses (seeds of legumes).

VITAMINS. Vitamins are organic molecules that are required in the diet but that are not essential fatty acids or amino acids. Vitamins serve various functions (Table 13.3), but unlike other organic nutrients they are not broken down, so they provide no energy. (This fact surprises most Americans, who have been misled by generations of advertising to believe that vitamins produce energy.) The same vitamins are not required by all species. For example, ascorbic acid (vitamin C) is not a vitamin for amphibians, reptiles, and many birds and mammals, which can synthesize it. It is a vitamin, however, for invertebrates, fishes, some birds, and primates (including humans). There are two groups of vitamins: the lipid-soluble vitamins that can be stored in body fat, and the water-soluble vitamins that cannot be stored and must therefore be ingested frequently. While small amounts of all vitamins are required, too much can be harmful. This is especially so for fat-soluble vitamins, which accumulate in the body.

TABLE 13.3

Vitamins required by humans.

FAT-SOLUBLE

A, formed from beta carotene, is a precursor for retinal in the visual pigments, and it is required for the maintenance of epithelial tissue and bones. It is also an antioxidant that helps protect membranes by neutralizing molecules with highly reactive unpaired electrons, called free radicals. (Free radicals are produced during normal functioning of mitochondria, and they can cause cancer by binding to DNA and membranes.) Deficiency of vitamin A causes night blindness and retardation of growth; excess causes neurological disturbances and arthritis-like symptoms. Sources include green leafy and orange-yellow vegetables, milk, egg yolks, liver, and fish-liver oil.

D Vitamin D_3 is synthesized by skin exposed to the ultraviolet rays of sunlight, so it does not normally fit the definition of a vitamin. The liver and kidneys convert vitamin D_3 into 1,25-dihydroxycalciferol (= calcitriol), which is required for absorption of Ca^{2+} by the gut. Deficiency causes bone malformations (rickets) in children; excess causes abnormally high levels of calcium, which cause neurological and kidney disorders. Dietary sources include egg yolks, butter, liver, and fish-liver oils.

E (tocopherol) is an antioxidant like vitamin A that helps protect membranes from free radicals. Vitamin E is so abundant, and so little is needed, that deficiency symptoms are never observed. Sources include vegetable oils, wheat germ, leafy vegetables, egg yolk, margarine, and legumes.

K (phylloquinone) is required for the synthesis of prothrombin, which is required for blood clotting. In many mammals sufficient vitamin K is produced by the bacteria in the large intestine. Other sources include leafy vegetables, vegetable oils, and liver.

WATER-SOLUBLE

B_1 (thiamine) is part of an enzyme required for the production of ATP by mitochondria. Deficiency usually occurs only in people subsisting on polished rice, or in alcoholics, malnourished pregnant women, or others with inadequate intake of the vitamin or inability to absorb it. Mild deficiency causes neurological disturbances, such as tiredness, irritability, insomnia, and pain. Severe deficiency causes beriberi, with severe neurological and circulatory disruptions that may be fatal. Sources of thiamine include dried yeast, whole grains, meat, nuts, legumes, and potatoes.

B_2 (riboflavin) is part of FAD, which is involved in the production of ATP by mitochondria (p. 58). Deficiencies, which are rarely seen in humans, appear mainly as skin disorders. Sources include milk, cheese, liver, meat, eggs, and enriched cereals.

Niacin (nicotinic acid) is part of NAD and NADP, which help produce ATP in the mitochondria (p. 54). (Nicotinic acid is not chemically related to nicotine.) The main symptom of deficiency is pellagra, which damages the nervous system, skin, and digestive tract. Sources include dried yeast, liver, meat, fish, legumes, and enriched cereals.

Pantothenic acid is part of coenzyme A in the Krebs cycle in mitochondria. It is present in so many foods that deficiency symptoms, mainly neurological, occur only in animals on experimental diets.

B_6 (pyridoxine and related molecules) are coenzymes for metabolism of amino acids and fat. Deficiencies are rare, resulting in neurological and skin disorders. Sources include dried yeast, liver, whole-grain cereals, fish, and legumes.

Table continues

TABLE 13.3, *continued*

Folacin (folic acid) is a coenzyme for metabolism of amino acids and nucleic acids, and in the maturation of red blood cells. Deficiency results in anemia. Sources include green leafy vegetables, meat, liver, and dried yeast.

B_{12} (cyanocobalamin) is required for formation of red blood cells and the metabolism of nucleic acid. Unlike most vitamins, which are synthesized by plants, vitamin B_{12} is produced only by certain bacteria. Humans obtain it mainly in meat, liver, eggs, and milk. Absorption depends on an "intrinsic factor" from the stomach. Failure of this factor results in pernicious anemia. Deficiencies can also result from extreme vegetarianism or as the result of a fish tapeworm (p. 528) or excessive *E. coli* bacteria in the gut, either of which absorb large amounts of the vitamin.

Biotin is a coenzyme in fat and glycogen synthesis and in amino acid metabolism. Deficiency, which is rare in humans, produces skin irritation. Sources include meat, liver, and egg yolks.

C (ascorbic acid) is required for collagen synthesis and maintenance of the intracellular matrix of bone and cartilage cells. It is also an antioxidant that protects the cytoplasm from damage by free radicals. The name "ascorbic" refers to the ability to prevent scurvy, a degeneration of the skin and gums. Sources include citrus fruits, tomatoes, potatoes, cabbage, and green peppers.

Dissolved Nutrients

The first cells probably lived in their food and simply absorbed nutrients from the "organic soup" across their cell membranes. Some animals still obtain at least a portion of their nutrients by absorption of **dissolved organic material** (DOM) across the body surface. Animals that depend mainly on DOM require a large surface area relative to body volume. They are therefore either long and thin or quite small. Such animals often do not even have digestive organs, since the nutrients are either already broken down to a form suitable for metabolism, or they can be broken down by intracellular digestion using **lysosomes** (pp. 40–41). Many aquatic invertebrates, such as sponges, various larvae, and small worms, obtain at least part of their nutrition in this way. Dissolved organic material amounts to only about 1 milligram per liter of seawater, so quite a lot of water must pass over such an animal to provide significant nutrition. **Endoparasites,** which live within the gut or tissues of other animals, also commonly absorb dissolved nutrients across the body surface. Among these are tapeworms and some flukes (phylum Platyhelminthes; see Chapter 25).

Many other animals acquire some of their nutrients from **endosymbionts** such as algae or bacteria that live within the animal. Some corals, sponges, and clams derive a large portion of their nutrients from photosynthesis by endosymbiotic protists called **zooxanthellae.** Some sponges benefit from photosynthetic cyanobacteria. The beard worm *Riftia pachyptilia* (phylum Pogonophora) obtains nutrients from sulfur-metabolizing bacteria that it cultivates within a special chamber. Some mammals, perhaps including humans, obtain significant amounts of vitamins B_{12} and K from bacteria in the large intestine.

Feeding

To the extent that animals can acquire nutrients by absorption across the body surface or from endosymbionts, they are relieved from the necessity of feeding. Most animals, however, must find, ingest, and digest liquid or solid food to create an internal environment with suitable amounts of nutrients. Except for reproduction, probably no other activity requires as much of an animal's attention and ingenuity as nutrition. One can understand much of an animal's structure and behavior simply by knowing what it eats. There are numerous modes of feeding that depend on whether the food is liquid or solid and, if solid, the size of the food. A sampling of feeding mechanisms follows.

SUSPENSION FEEDING. When food occurs as large numbers of small particles, suspension feeding is common. In suspension feeding, food particles suspended in liquid are trapped by some anatomical structure, secretion, or device constructed by the suspension-feeding animal. Each food particle can be as small as 1 micrometer (the fine detritus consumed by the small crustacean *Daphnia*) or as large as several centimeters (the crustacean krill consumed by baleen whales). One type of suspension feeding is filter feeding, in which a filtering structure acts like a sieve, trapping anything above a certain size. Other suspension feeders use mucus, electrical charge, and other surface properties to collect food. Obviously the ratio of food to nonfood particles must be high for suspension feeding to be efficient, so the presence of suspension feeders is often a good indication that water has not been polluted by inorganic debris. Figure 13.1 illustrates some of the varieties of suspension feeding, and others will be described in Unit 4.

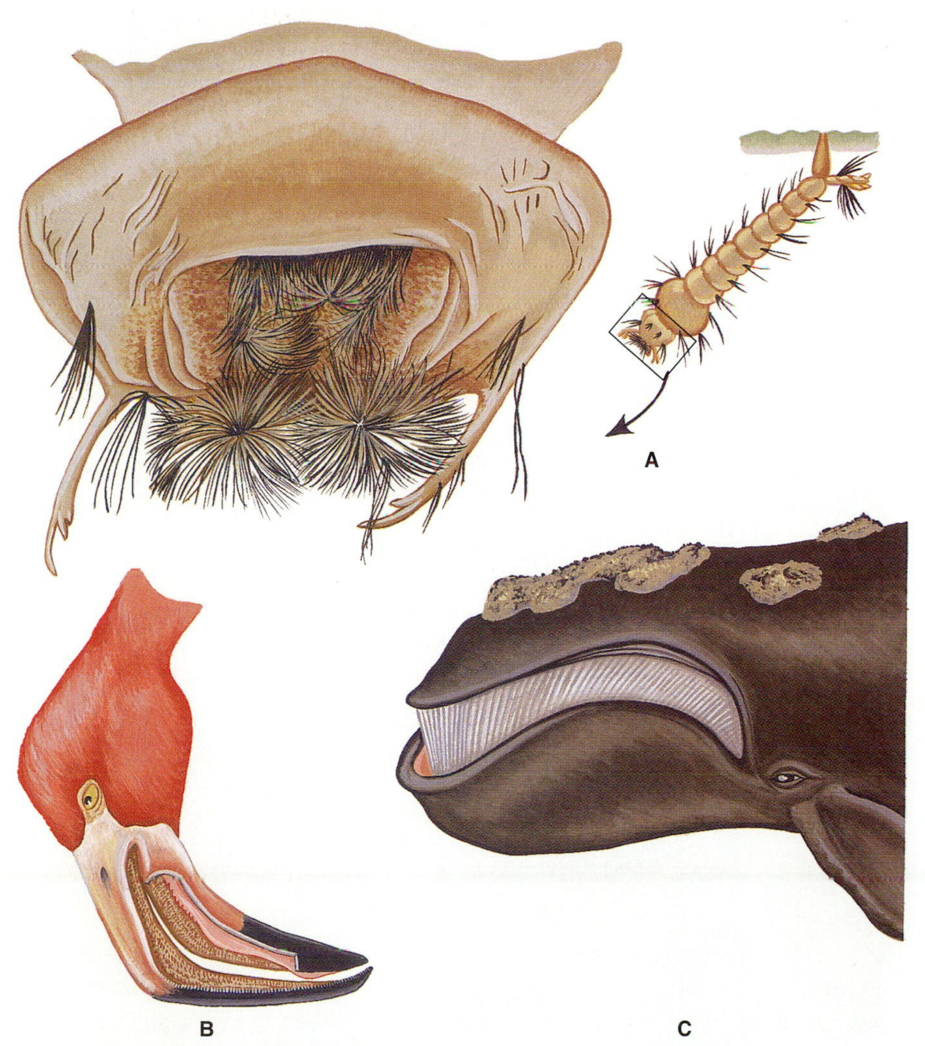

FIGURE 13.1

Adaptations for suspension feeding. See also the fish gill rakers in Fig. 12.7. (A) The mouthparts of the larva of a mosquito (*Culiseta*), which feeds while hanging from the surface of still water. (B) The head of the lesser flamingo, *Phoeniconaias minor,* in its normal feeding position with the head under shallow water. The beak is shown partly cut away to reveal comblike structures that filter cyanobacteria from brackish water. (C) The black right whale, *Eubalaena glacialis,* one of the largest of the suspension feeders, uses parallel filaments of baleen to filter out small crustaceans (krill). See also Fig. 39.15B.

Biting and Swallowing. Methods other than suspension feeding must be employed for relatively large food masses that occur in isolated lumps. There are three general ways of obtaining large foods, depending on the size of the food and whether it is a living animal. (1) If each food particle is smaller than the feeding animal and is sedentary (single cells, small plant parts, or small carrion), the feeder can ingest it in one or a few bites. Numerous examples will come to mind, such as birds eating berries. (2) If the food is larger, the animal can chew, burrow, or scrape its way through it. Carrion beetles, termites, and many endoparasites feed in this manner. (3) If the food is another animal capable of fleeing, the feeding animal(s) must first capture it and then either swallow it whole or bite off parts. A few of the adaptations for capture and ingestion are illustrated in Fig. 13.2, and many others will be described in Unit 4.

Piercing and Sucking. A different method is needed for fluid food, such as the sap of plants or the body fluids of other animals. Animals that feed on fluid often have piercing and sucking mouthparts. For example, aphids can pierce the tough cuticle of plants to get to the sap, and mosquitoes can penetrate the epidermis of mammals to tap the bloodstream (Fig. 13.3).

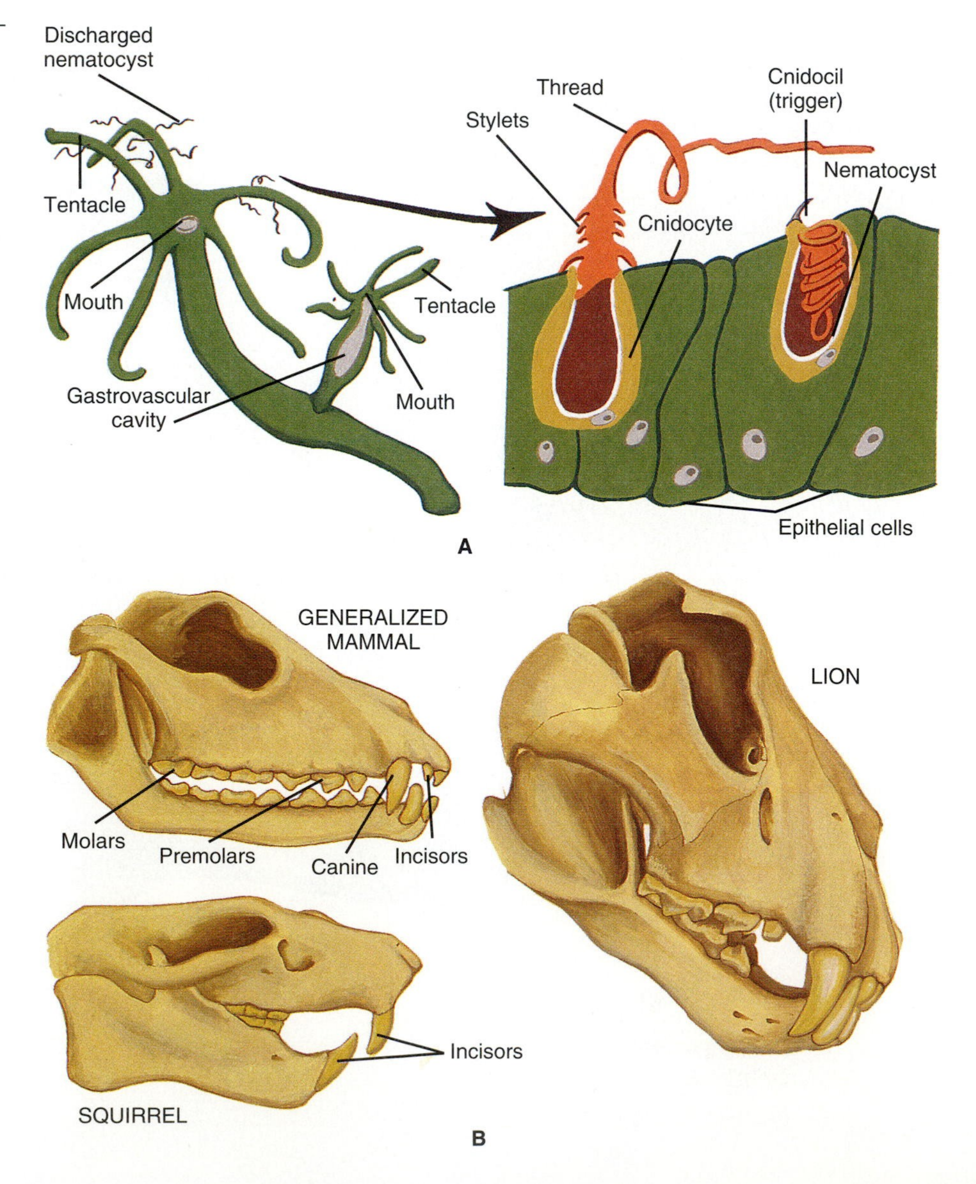

FIGURE 13.2

Some mechanisms for the capture and ingestion of food. (A) *Hydra* (phylum cnidaria) has nematocysts that discharge when prey contact the trigger (cnidocil). The barb and thread poison the prey, immobilizing it for capture by the tentacles. The prey is then taken through the mouth into the gastrovascular cavity where enzymes from gland cells digest it. Nutrients are absorbed into cells by endocytosis. (B) The dentition of mammals. Incisors are used for biting, canines for tearing of meat, and premolars and molars for shearing (lion) or crushing (squirrel). The numbers and sizes of each kind of tooth vary with the diet of the mammal. Carnivores have more pronounced canines than do herbivores.

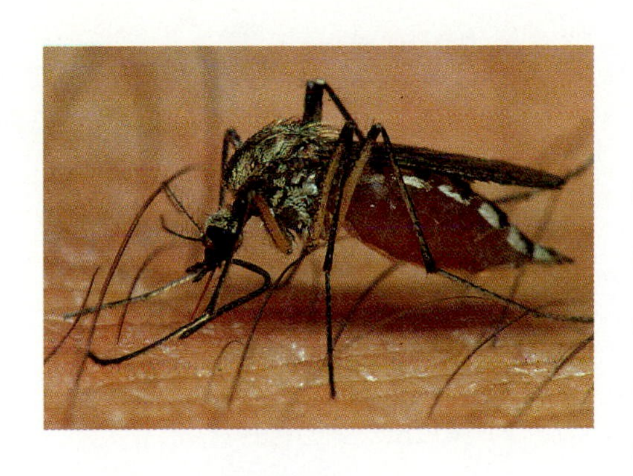

A

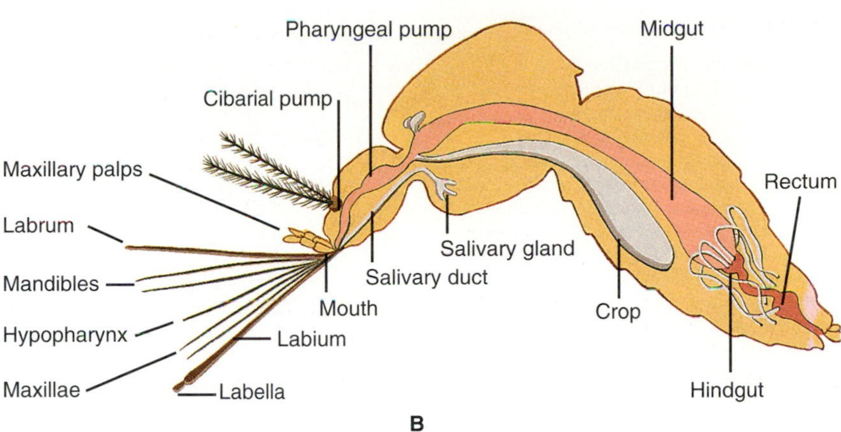

B

FIGURE 13.3

(A) A mosquito (family Culicidae) piercing human skin and sucking blood. Only females feed on blood. Note the change from the suspension feeding larva (Fig. 13.1A). (B) Side view of the mosquito with the mouthparts separated. During piercing of skin the sheath-like labium folds out of the way. Blood is drawn through the labrum into the midgut, where it is digested. All insects have the same mouthparts, but they are greatly modified for different kinds of foods (Fig. 32.10).

Intracellular Versus Extracellular Digestion

A few animals can absorb small bits of food directly into cells by **endocytosis** (pp. 40–42). Endocytosis encompasses both **phagocytosis** (cellular uptake of solids) and **pinocytosis** (uptake of liquids). Food is enclosed within a vacuole within the cell, then broken down by enzymes in the lysosome into molecules that can be metabolized (Fig. 13.4A). This mechanism of intracellular digestion provides at least some of the nutrients in protozoans, sponges, cnidarians, flatworms, rotifers, bivalve molluscs, and invertebrate chordates.

Digestion of larger masses of food must be extracellular (Fig. 13.4B). Extracellular digestion occurs not only outside cells, but anatomically outside the body, because the inside of the digestive tract (the **lumen**) is really an extension of the external environment. Many starfish, in fact, do not swallow food, but evert their stomachs over it. Theoretically you could do the same, but your mother would probably slap your hand.

■ **Question:** Adult male mosquitoes have similar mouthparts, but they do not consume blood. What could they eat?

Digestive Systems

Extracellular digestion requires digestive systems, which vary widely among different groups of animals. The simplest system is merely a cavity in which digestive enzymes convert food into nutrients that can be absorbed. The **gastrovascular cavity** of cnidarians such as *Hydra* is an example of such a simple system (Fig. 13.2A). More complex systems have additional organs for the secretion of enzymes and the absorption of nutrients (Fig. 13.5 on page 262). The gastrovascular cavity is an example of an **incomplete digestive system.** This does not mean that it is unfinished, but that there is only one opening that serves as both mouth and anus. Most animals (see Fig. 21.11 for survey) have **complete digestive systems.** The advantage of having an anus separate from the mouth probably does not have to be explained.

FIGURE 13.4

A comparison of intracellular (A) and extracellular digestion (B). In intracellular digestion, digestive enzymes from lysosomes combine with food in a vacuole. In extracellular digestion the enzymes are released not only outside the cells but morphologically outside the body. The nutrients are then absorbed into the body fluids.

■ **Question:** Could an animal feed by secreting digestive enzymes onto prey and slurping up the nutrients? (Hint: Use the index to look up feeding in spiders.)

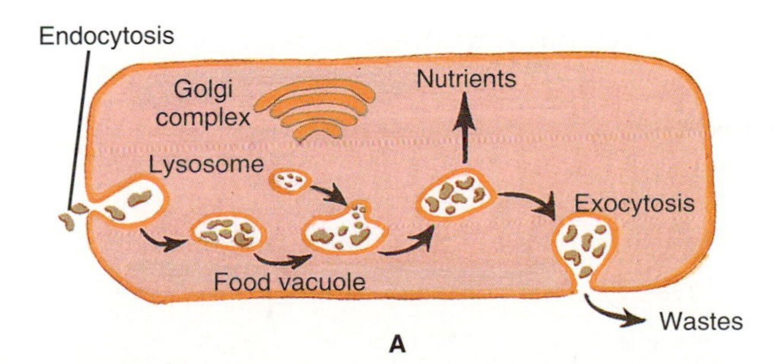

A

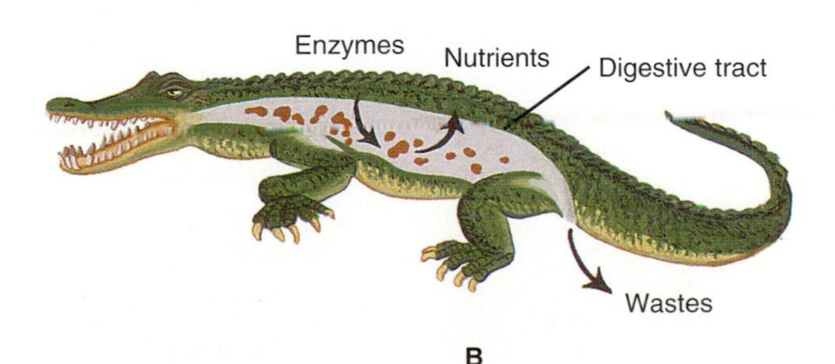

B

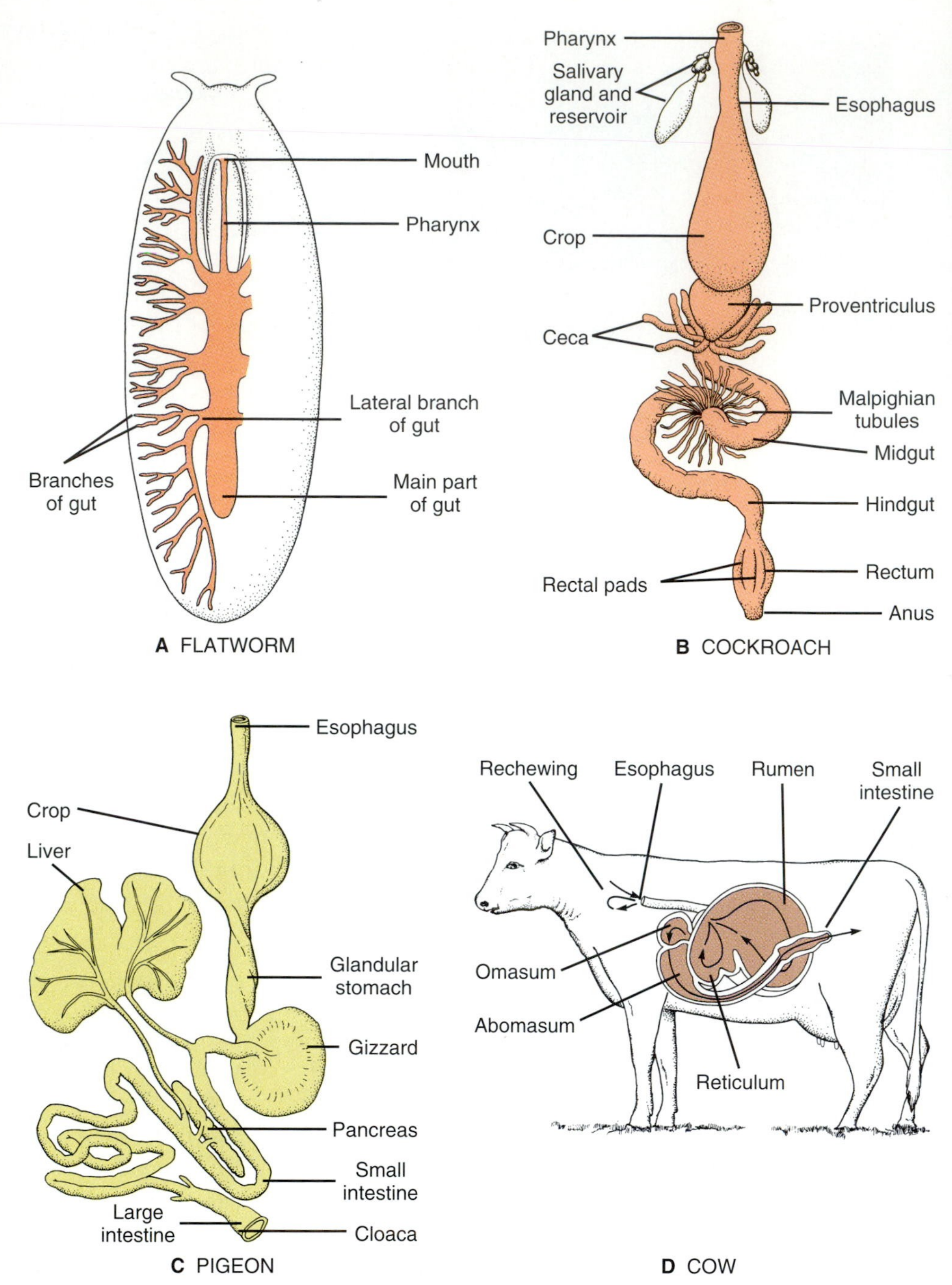

FIGURE 13.5

Various types of digestive organs. (See also Fig. 13.2A.) (A) Flatworms (phylum Platyhelminthes) typically have a branched gut with few specialized digestive organs. There is a mouth but no anus, so this is an incomplete digestive system. (B) Insects have highly developed digestive systems. (Compare Fig. 13.3B.) In the omnivorous cockroach the salivary gland secretes an enzyme (amylase) that digests starch. The crop serves primarily as a food reservoir, holding up to two months' supply. The proventriculus acts as a gizzard, breaking up solids. Digestion and absorption occur primarily in the midgut and ceca, which secrete a variety of enzymes. The rectum reabsorbs water from the feces. The Malpighian tubules are not part of the digestive tract, but excrete metabolic wastes and regulate ion concentrations. (C) The digestive tract of the pigeon *Columba livia*. The crop stores food, and in pigeons it also secretes a milklike fluid for the young. (D) *Bos*, a device for converting grass into milk. The digestive system, especially the four-chambered stomach, occupies most of the abdominal cavity. Grass is fermented by microbes in the rumen, then regurgitated for a second chewing before being swallowed again. Following this rumination the grass enters the reticulum.

RUMINANTS. Among the most elaborate digestive systems are those of ruminants such as cattle, sheep, and deer (order Artiodactyla). These mammals have a large forestomach—the **rumen**—in which the initial breakdown of plant material occurs (Fig. 13.5D). The partially digested material (the cud) is then regurgitated for chewing, reswallowing, and further digestion in other parts of the digestive tract. This process of regurgitation and chewing is called **rumination.** (Regurgitation by camels and llamas is also called rumination, but these animals are not considered to be ruminants, because they do not have rumens.) The rumen serves as a huge culture flask in which bacteria and protozoa partially digest cellulose and starch. The microorganisms obtain their energy by fermenting the sugars released, but the anaerobic environment of the digestive tract prevents them from completely oxidizing the sugars to CO_2 and H_2O. The fermentation products are therefore available to the host animal as nutrients.

Other animals in which the stomach is adapted for microbial fermentation include a South American bird called the hoatzin (pronounced WAT-sin; *Opisthocomus hoazin*), some kangaroos, and colobine monkeys of Africa (genus *Colobus*). Bacteria in the colobines release gaseous fatty acids that are absorbed across the stomach lining (Milton 1993). Howler monkeys of South America (*Alouatta*) have bacteria that play a similar role in the large intestine and the **cecum** (SEE-come) at the junction of the small and large intestines. Humans may also absorb significant amounts of volatile fatty acids from bacteria in the large intestine. Many rodents and rabbits also house microbial fermenters in their ceca, but they eat their own feces to recover the released nutrients. The guts of termites and some cockroaches contain bacteria and protozoa that are often said to be essential for the digestion of the cellulose in the cell walls of wood. In fact, however, termites and cockroaches can secrete their own cellulose-digesting enzyme, and the role of microbes in their guts is not known (Slaytor 1992).

Absorption of Nutrients

The function of digestion is to break food down into nutrient molecules that can be absorbed into blood and interstitial fluid. Absorption is therefore certainly as important as feeding and digestion. The mechanisms by which nutrients are absorbed across the cells lining the digestive tract are similar to those by which other molecules enter cells—namely, diffusion and active transport (pp. 38–40). **Diffusion** is always from high to low concentration, usually from the lumen into the blood. The nutrients absorbed by simple diffusion are small ions that can enter through channels in cell membranes, and hydrophobic molecules (lipids), which can pass through the phospholipid bilayer. Therefore the following nutrients are probably the ones absorbed by diffusion: K^+, Cl^-, fats, steroids, and lipid-soluble vitamins. Water is also absorbed by simple diffusion.

All other nutrients are probably absorbed by either **facilitated diffusion** through transporters or **active transport** by pumps. Sodium and calcium are absorbed into the bloodstream by pumps in the membranes of cells in the intestinal lining. There are also transporters or pumps for glucose, amino acids, peptides, and some vitamins.

Digestion in Humans

CHEWING. Numerous adaptations for digestion in other animals will be described in Unit 4. For now we can illustrate the concepts with our own species. Humans are omnivores. That does not mean that we literally eat everything, but it does mean that our digestive systems are capable of handling a wide variety of foods from both plants and animals. Our 32 **teeth,** for example, have a variety of shapes adapted for different foods (Fig. 13.6 on page 264). As you can verify with your tongue, each half of each jaw has two chisel-shaped **incisors** in front that are adapted for biting, one **canine** adapted for tearing meat, and two **premolars** and three **molars** adapted for grinding vegetable material and shearing meat. (You may be missing one or more of these, especially the hindmost molars, called "wisdom teeth.") As food changes consistency during chewing, the tongue manipulates it to the appropriate teeth.

The tongue and teeth also mix the food with **saliva** from the salivary glands. Normally, human salivary glands secrete about a liter per day, and most of it is water that gets recycled by swallowing. Saliva also contains mucus, which lubricates the

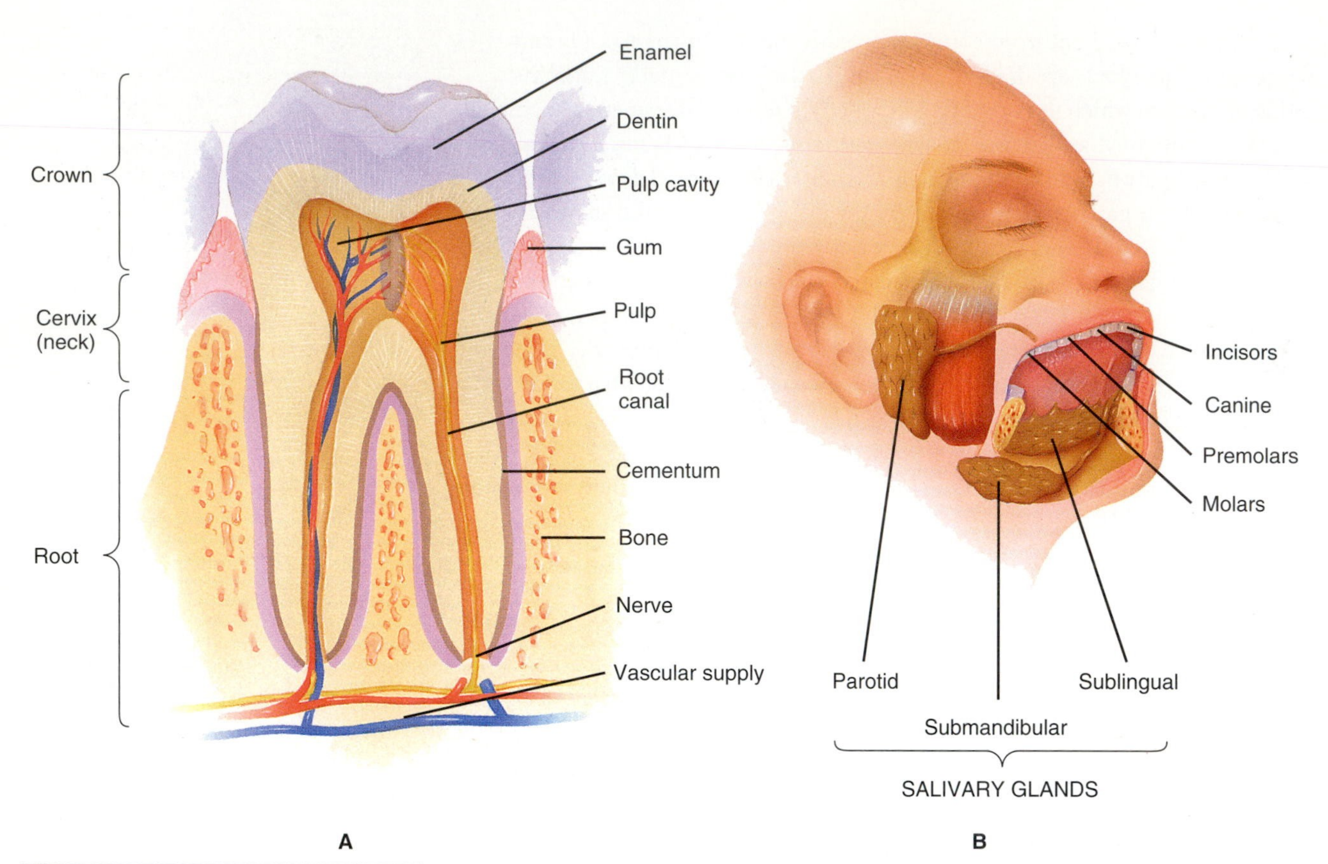

FIGURE 13.6

Digestive structures in the human mouth. (A) A longitudinal section of a molar. (B) Salivary glands and teeth.

chewed food, and an **amylase** enzyme that helps digest starch into the disaccharide **maltose.** In addition, the salivary glands and tongue secrete **lingual lipase,** an enzyme that helps digest triglycerides into monoglycerides.

SWALLOWING. After food has been chewed to a certain consistency, it forms into a **bolus** that the tongue pushes to the back of the mouth. Pressure of the bolus against the soft palate closes off the nasal passages and pushes the epiglottis down to block the glottis (the opening of the trachea). The bolus also stimulates touch receptors at the opening of the pharynx, at the back of the mouth. These receptors activate a swallowing center in the brainstem, which triggers contractions of numerous skeletal muscles in the pharynx. These muscles force the bolus into the tubular **esophagus** that connects to the stomach. Smooth muscles in the esophagus produce waves of contractions, called **peristalsis,** that propel the bolus to the stomach.

THE STOMACH. The stomach is a storage area that eliminates the need to eat continuously. It is separated from the esophagus and the small intestine by the **cardiac** and **pyloric sphincters.** In mammals the stomach secretes **hydrochloric acid** (HCl), which helps destroy contaminants, and **pepsin,** an enzyme that helps digest protein. The reduction in pH by the HCl inactivates salivary amylase but not lingual lipase, and it activates the pepsin. Contractions of the smooth muscle in the stomach churn the food and secretions into a watery, acidic mixture called **chyme** (pronounced kyme).

You might wonder how cells that secrete HCl and protein-digesting enzymes avoid digesting themselves. For HCl the answer is that the acid is secreted in two different processes. **Parietal cells** (= oxyntic cells) secrete H^+ and by a separate process secrete Cl^-, so the strong acid HCl is never in the cell (Fig. 13.7). The pepsin is secreted in

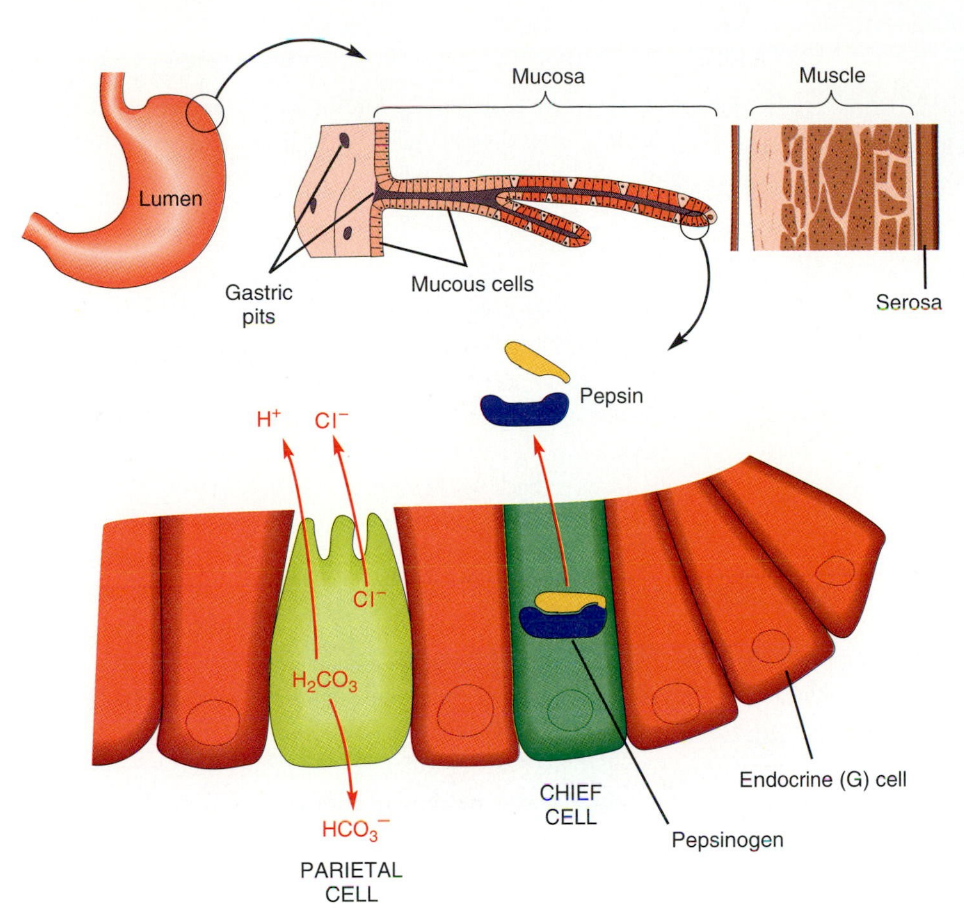

FIGURE 13.7

Hydrogen ion comes from the weak carbonic acid in parietal (= oxyntic) cells and is actively transported into the stomach by proton (H^+) pumps. Pepsin is secreted as inactive pepsinogen by the chief cells.

■ **Question:** What is the advantage of having the secretory cells located in pits?

the inactive form called **pepsinogen.** Pepsinogen gets activated in the stomach when HCl and other pepsin molecules remove a portion of the protein that blocks the active site on the pepsinogen. This explains how the secretory cells keep from destroying themselves, but what keeps the acid and protease from digesting the stomach? The answer to this question is not entirely known, but the layer of mucus that normally covers the lining, or **mucosa** of the stomach, is probably important. A similar mucosa also protects the rest of the digestive tract from digestive secretions and abrasion. Even with the protection of mucus, however, the acid and digestive enzymes destroy some 10 billion mucosal cells a day.

THE SMALL INTESTINE AND ASSOCIATED DIGESTIVE ORGANS. The stomach is important, but people can live without one, because most digestion occurs in the small intestine. When chyme enters the first section of small intestine (the **duodenum**) it mixes with enzymes and other secretions from several sources. Cells lining the small intestine release several digestive enzymes, but the main source is the **pancreas,** which is located just below the stomach in humans. Pancreatic enzymes enter the small intestine through a duct. (They are therefore **exocrine** secretions of the pancreas, not to be confused with the *endocrine* secretions, insulin and glucagon, which go into the bloodstream.) The pancreatic enzymes include lipase, several proteases, and an amylase similar to salivary amylase. As with pepsin, the pancreatic proteases are secreted in inactive form, so they do not digest the exocrine cells that secrete them. The pancreatic enzymes cannot function in an acidic environment, but the pancreas also releases bicarbonate (HCO_3^-), which neutralizes the acidity of the chyme. **Goblet cells** in the mucosa secrete mucus, which protects the intestinal lining from di-

William Beaumont

William Beaumont
Born: 21 November 1785 in Lebanon, Connecticut
Died: 25 April 1853 in St. Louis, Missouri
Education: Studied medicine by reading and apprenticeship; Army surgeon; pioneer in the study of stomach physiology

Considering its importance, it is surprising how recently people began to understand the stomach. Long after the heart was known to be a pump, the human stomach was described by some authorities as a mill, a fermenting vat, or a stew pan. William Beaumont provided the first indication that it was none of these. Beaumont got his medical training the old-fashioned way, by reading and apprenticing to a physician. He was licensed in 1812 and immediately joined the Army in Plattsburgh, New York for an ill-fated invasion of Canada at the start of the War of 1812. Beaumont was also present at the Battle of Plattsburgh in 1814, which turned out much better for the United States. According to local accounts, Beaumont's interest in digestive functions was apparent even then. It is said that he had the intestines of dead soldiers strung about his office. After the war Beaumont was released from the Army and settled in Plattsburgh, where he practiced medicine and sold pharmaceuticals, groceries, tobacco, and liquor.

In 1819 Beaumont accepted a commission in the Army and was posted at Fort Mackinac (MAC-kin-aw) in the remote Upper Peninsula of Michigan. At that time this was the northwest frontier of the United States, and Beaumont was the only physician within 300 miles. Beaumont's name might have passed into oblivion had it not been for an accident in 1822. A 19-year-old French-Canadian fur trader, Alexis St. Martin, was accidentally hit in the left side by a shotgun blast at close range. Beaumont was summoned and quickly pronounced the case hopeless. Fortunately, Beaumont was mistaken, because over the next several years St. Martin slowly recovered. St. Martin's survival was largely due to Beaumont, who took him into his home and personally cared for him after the Army and his employers cut off financial support.

The healing was not complete, however. A hole remained in St. Martin's side and stomach, providing Beaumont with a view of the inner workings of the stomach. In 1825 Beaumont began to take advantage of the situation to study the role of the stomach in digestion. At first the grateful St. Martin cooperated, but soon he ran away to Canada, having recovered sufficiently to canoe his family all the way back to Montréal from Prairie du Chien, Wisconsin! Four years later Beaumont found St. Martin and convinced him to resume experimentation. Further attempts to run away were thwarted by having St. Martin enlist in the Army. It is said, however, that when Beaumont looked in on St. Martin's stomach he often found it inflamed with alcohol. Supposedly, St. Martin took the most direct means to achieve that end, emptying the bottle directly into the opening in his side. Experiments continued until 1833, when Beaumont published his findings. St. Martin then managed to return to Canada and refused all further efforts of Beaumont and other scientists to continue the experiments. After St. Martin died in 1883 at age 80, his widow, afraid of grave-robbing scientists, allowed the body to decay before burying it at twice the normal depth in an unmarked grave.

Beaumont's work quickly attracted the attention of physiologists. European researchers, especially, were taken with the achievements of the "Frontier Doctor." While many of Beaumont's findings had already been suggested from studies of regurgitated gastric juice and partly digested food, Beaumont confirmed the following points by direct observation:

1. Gastric digestion results from a secreted fluid.
2. This secretion occurs at numerous points on the stomach lining.
3. Eating stimulates gastric secretion, and emotion influences secretion.
4. The fluid is acid as soon as it is secreted. (It had been thought that the acidity was due to decomposition.)
5. The stomach empties rather quickly.
6. Foods differ in their digestibility, and alcohol and other substances affect digestion.
7. Gastric digestion is due not only to the action of acid, but to some other substance. (Enzymes were unknown at the time.)

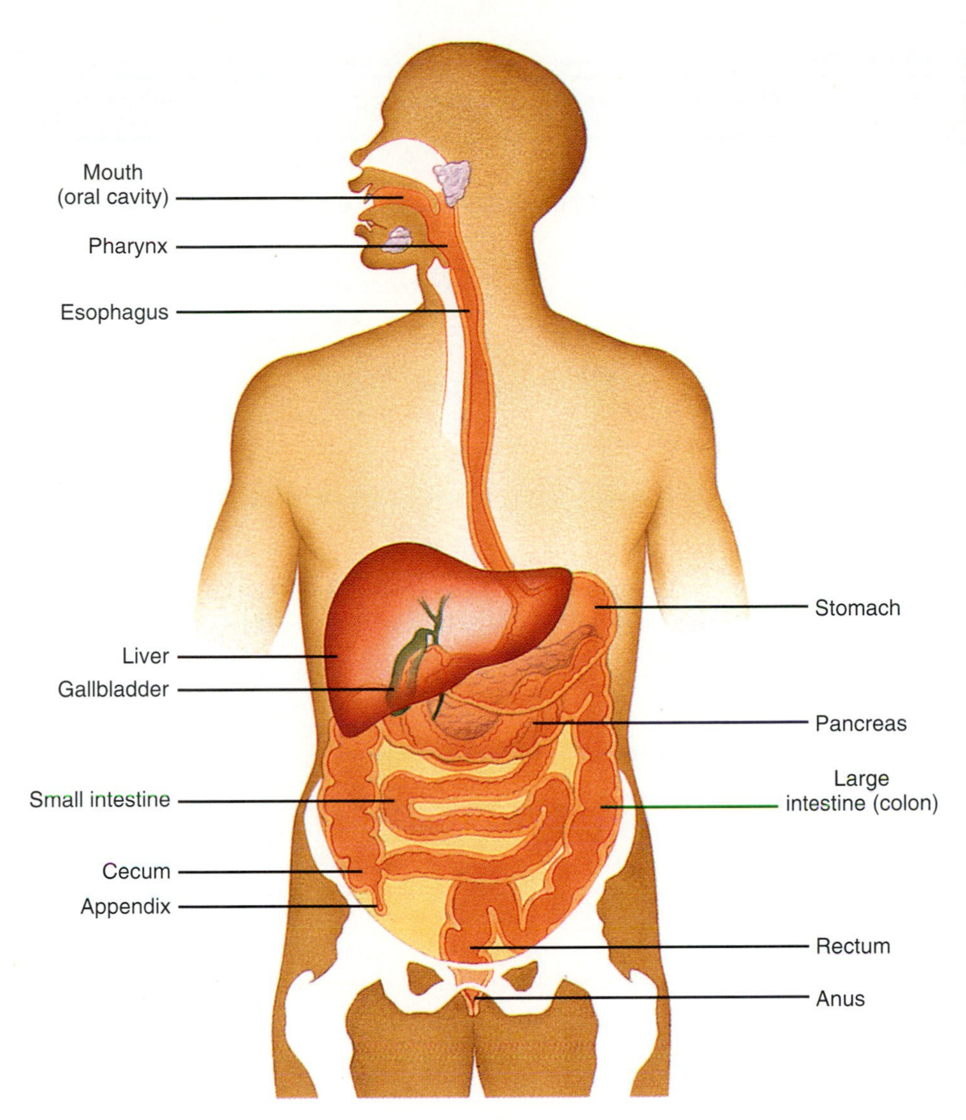

FIGURE 13.8

The human digestive tract consists of the mouth, pharynx, esophagus, stomach, small intestine, large intestine, and rectum. The part below the esophagus is often called the gastrointestinal (GI) tract. Food does not enter the liver, gallbladder, or pancreas. Rather, these organs secrete bile salts, enzymes, and bicarbonate into the small intestine where they aid digestion.

gestion. Figure 13.8 shows the arrangement of the major digestive organs, and Table 13.4 summarizes the digestive secretions.

BILE. Lipids are usually the most difficult nutrients to digest, because they form large globules rather than dissolving in the watery chyme. The digestive system solves this problem the same way dish-washing detergents do: by **emulsifying** the large fat globules into smaller ones. Small droplets present a greater surface area for lipases to work on. Some emulsification results from the movements of the stomach, and further emulsification results from substances in bile, which is a brownish, yellow, or olive-colored fluid produced by the liver and stored in the gallbladder. **Bile salts** and phospholipids in bile achieve emulsification by mixing with both lipids and water. Bile also contains substances not directly related to digestion, including cholesterol and bilirubin. **Bilirubin** is a breakdown product of hemoglobin and gives feces their characteristic color. One symptom of a disorder of the liver or gallbladder, such as jaundice or gallstones, is gray feces and yellowish skin, due to the retention of bilirubin.

THE LARGE INTESTINE. By the time material has passed through the three or so meters of the human small intestine, virtually all the absorbable nutrients have been removed, leaving a brown watery fluid containing indigestible material such as cellulose.

TABLE 13.4

Enzymes and other digestive secretions in humans (and most other vertebrates).

MOUTH

Salivary amylase breaks down starch to the disaccharide maltose by hydrolyzing every other bond between the glucose subunits.

Lingual lipase breaks down triglyceride to monoglyceride and two fatty acids by hydrolyzing ester bonds.

STOMACH

Pepsin, released as inactive pepsinogen by chief cells, hydrolyzes particular peptide bonds, breaking up proteins into peptide fragments.

HCl, released by parietal (oxyntic) cells, serves mainly to kill bacteria and activate pepsin.

LIVER

Bile salts and **phospholipids** in bile emulsify large fat droplets into small ones, which present more surface area for lipases to work on in the small intestine. Bile is stored in the gallbladder.

PANCREAS

Pancreatic amylase breaks down starch, like salivary amylase.

Other **carbohydrases** digest different carbohydrates. For example, **sucrase** (= invertase) digests sucrose, and **lactase** digests lactose (milk sugar). (Lactase occurs in infants, but not always in adults. Therefore many adults cannot digest most dairy products.)

Several proteases hydrolyze different peptide bonds. The most important proteases are **trypsin** (pronounced TRIP-sin), **chymotrypsin**, and **carboxypeptidase**. Carboxypeptidases hydrolyze the last peptide bond at the carboxyl end of a peptide. Like pepsin, these proteases are released in inactive form (trypsinogen, chymotrypsinogen, procarboxypeptidase).

Pancreatic lipase converts triglycerides to monoglycerides and fatty acids.

Bicarbonate (HCO_3^-) neutralizes acid from the stomach.

This material, which is essentially dilute feces, enters the large intestine, where it first passes a pouch called the **cecum.** In humans the cecum is small and ends with an **appendix,** which is of no known benefit except to surgeons. The rest of the large intestine reabsorbs water into the body, contains bacteria, and stores feces until it is convenient to eliminate them. (Note that "feces" is always plural.) The reabsorption of water is due to sodium pumps, which set up an osmotic gradient that draws water out of the feces and into the bloodstream. Feces are eliminated through the anus by the process of defecation, which combines voluntary and reflex contractions of skeletal and smooth muscles.

The bacteria in the large intestine constitute a complex and poorly understood community called the "intestinal flora." These bacteria ferment cellulose into gases, including volatile fatty acids that may be nutritionally significant (Milton 1993). They also produce vitamins B_{12} and K, and they contribute one-fourth to one-half of the bulk of feces. *Escherichia coli,* the research bacterium familiar as *E. coli,* is the best known member of the intestinal flora, although it represents less than 1% of the cells in this community. Another bacterium in the gut is *Sarcina ventriculi,* whose daily fermentation produces up to 30 grams of ethyl alcohol—the equivalent of more than a pint of beer.

Control of Digestion. The control of digestion is conveniently divided into three phases. The **cephalic phase** refers to stimulation or inhibition of secretion and motility by the nervous system before and during eating. The cephalic phase is responsible for the fact that merely thinking about food can stimulate digestive activities such as stomach movements (often audibly and at the most solemn moments). The autonomic nervous system (p. 178) plays a major role in the cephalic phase. Generally the **parasympathetic branch** stimulates digestive activities, and the **sympathetic branch** inhibits them. The autonomic nervous system controls motility by stimulating or inhibiting networks of nerve cells within the digestive tract that coordinate its movements. These **myenteric** and **submucosal plexuses** have been referred to as a "little brain," because they seem to function independently of the central nervous system. The autonomic nervous system controls secretion by acting on the salivary glands, parietal cells, and so on.

The **gastric phase** refers to control mechanisms that are active once food enters the stomach. Distension of the stomach, as well as stimulation of chemoreceptors, activates contraction of smooth muscles in the stomach lining. In addition, the hormones **gastrin** and **oxyntin** from G cells of the stomach (Fig. 13.7) stimulate stomach motility and the secretion of pepsin and HCl (Livingston and Guth 1993). Like all hormones, gastrin and oxyntin must go into the bloodstream to reach their targets, the chief and parietal cells. The stimulus for gastrin and oxyntin secretion are distension and peptides in the stomach, respectively.

Finally, the **intestinal phase** begins once chyme empties into the small intestine. The lining of the first portion of the small intestine secretes two families of hormones that help coordinate digestion. One family includes **cholecystokinin** (CCK), which is also called **pancreozymin** (PZ). The two names of CCK–PZ indicate their two roles in coordinating digestion: (1) stimulating the gallbladder to contract and release bile and (2) stimulating the pancreas to release digestive enzymes. Chole-cysto-kinin means gall-bladder-mover, and pancreo-zymin alludes to the pancreatic enzymes. The second family of hormones includes **secretin** (se-KREE-tin), which stimulates the secretion of bicarbonate by the pancreas and the synthesis of bile by the liver. The stimulus for secretion of CCK-PZ and secretin is the presence of chyme in the small intestine. Thus these two families of hormones are involved in negative feedback loops. They trigger the release of substances to digest the material that was the original stimulus for hormone secretion.

Absorption of Nutrients in Humans. Nutrient molecules released from food by digestion are absorbed across the mucosal lining of the small intestine. Both diffusion and active transport are involved, as described previously, and both rely on the large surface area of the mucosa. The large surface area is due to extensive folding and to **villi,** which are finger-like projections up to 1 mm long that give the mucosa a velvety texture. Individual cells on each villus have further submicroscopic projections called **microvilli,** which form a **brush border** that provides even more surface area (Fig. 13.9 on page 270). The villi and microvilli contribute so much area that if the intestinal mucosa were spread out it could serve as a tennis court, though a rather slippery one.

The large surface area allows virtually complete diffusion of the fatty acids and monoglycerides that result from the digestion of triglycerides. In addition, steroids, vitamins A, D, E, and K, thyroxine and triiodothyronine, alcohol, and other lipid-soluble molecules easily diffuse across the mucosa. These substances do not enter the bloodstream directly, but first go into **lacteals** within the villi. The lacteals, named for the milky appearance of the fat-laden lymph within them, are projections of the lym-

FIGURE 13.9
The structure of the mucosal lining of the small intestine.

■ **Question:** What adaptations would an intestinal parasite require in order to live here?

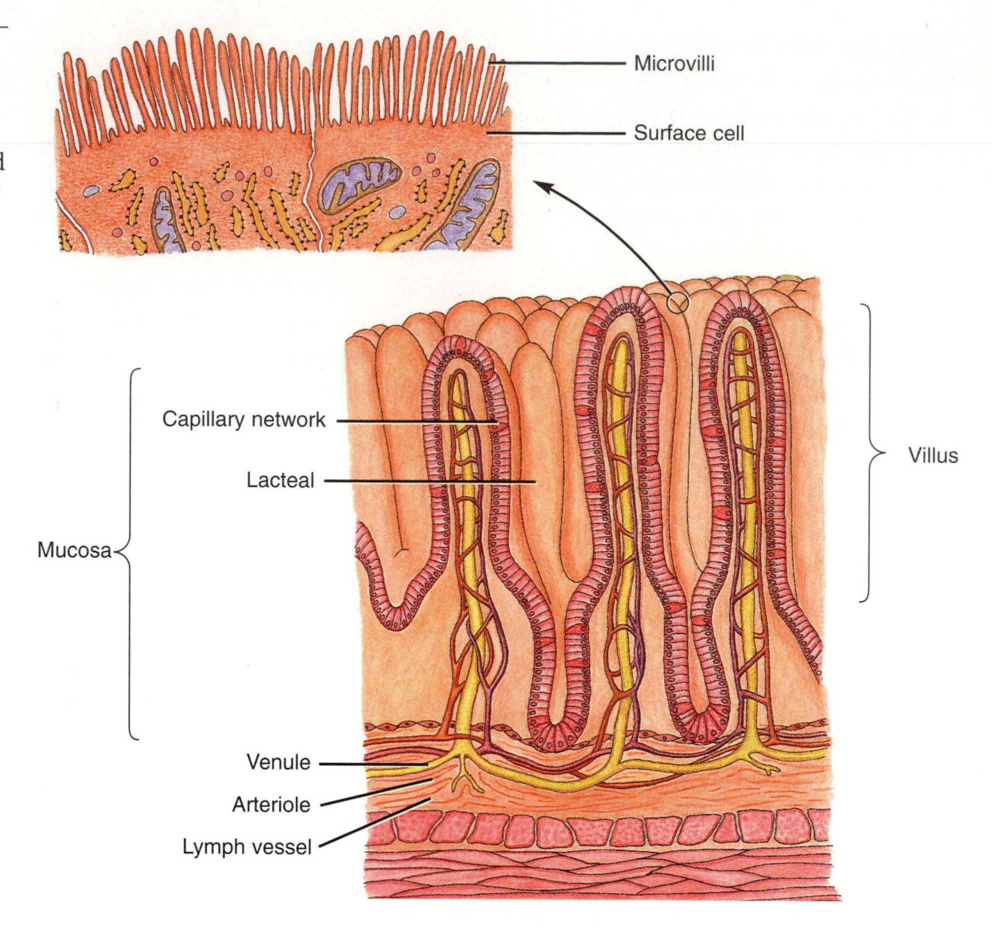

phatic system (pp. 229–230). The nutrients in the lacteals circulate in the lymph until they enter the bloodstream at a vein beneath the left collar bone.

Simple sugars, amino acids, dipeptides, tripeptides, water-soluble vitamins, most ions, and most other nutrients are absorbed by facilitated diffusion or active transport. Transporters or pumps in the plasma membranes transport the nutrients across the outer cells of the villi. The nutrients then diffuse into capillaries within the villi. The blood from the small intestine flows first through a portal vein to the liver. The liver then "decides" whether to store the glucose as glycogen, whether to metabolize the amino acids for energy or to use them for protein synthesis, and so on.

Regulation of Blood-Glucose Concentration

The levels of a nutrient in the blood and interstitial fluid naturally tend to vary with the diet, and the resulting fluctuations could raise havoc among many cells. This is especially true in the brain, because the activity of nerve cells varies in direct proportion to levels of glucose in the cerebrospinal fluid. Normally, such fluctuations in glucose are limited by the hormones insulin and glucagon, which are secreted into the bloodstream by endocrine cells located in the islets of Langerhans of the pancreas. Excessive levels of blood glucose (hyperglycemia) trigger the secretion of insulin by the β (beta) cells of the pancreatic islets. Insulin then circulates through the blood and stimulates cells, especially in muscle and adipose tissue, to take up more glucose. Muscle and liver cells then convert the glucose into protein and the storage form glycogen, while adipose tissue converts it to fat. This negative-feedback mechanism prevents glucose levels from getting too high immediately following the ingestion of carbohydrates. When glucose levels fall below normal (hypoglycemia), the α (alpha) cells of the pancreatic islets secrete **glucagon.** Glucagon stimulates liver cells to convert their glycogen reserves into glucose, thereby restoring blood-glucose concentrations.

Diabetes mellitus (dye-uh-BEE-teez MEL-it-us) is a common disorder that illustrates the importance of this homeostatic regulation of blood-glucose levels. In diabetes mellitus, excess glucose in the blood spills over into the urine, osmotically bringing water with it and increasing urine volume. **Type I,** or **insulin-dependent diabetes mellitus** (IDDM), results from the inability of the pancreas to secrete insulin. The cause of Type I diabetes is destruction of the β cells by a person's own immune system, which may be targeted against the β cells by a viral infection in those who are genetically predisposed. The more common form, afflicting approximately 2% of American adults, is called **Type II,** or **non-insulin-dependent diabetes mellitus** (NIDDM). It develops gradually in those genetically predisposed to it—primarily overweight people with "pear-shaped" obesity, rather than those with body fat distributed higher on the torso. The cause of Type II diabetes is the inability of cells to respond to insulin.

Regulation of Food Intake

Most animals maintain constant weights throughout their adult lives as long as sufficient food is available. They must therefore have mechanisms that regulate food intake. In house flies a simple stretch receptor in the abdomen indicates when enough food has been ingested (p. 403). The mechanisms in most other animals, including humans, are more complicated and poorly understood in spite of considerable research. The **hypothalamus** appears to be involved in the regulation of caloric intake in vertebrates, but how is still controversial. In the past, theories have centered on a hypothetical "feeding center" and "satiety center," but these ideas are greatly oversimplified.

Whether or not such neural centers exist, it is likely that hunger and satiety are evoked by some kind of signal associated with caloric intake. Candidates for a satiety signal include stretch receptors in the stomach, increased blood-glucose levels, cholecystokinin (CCK), and secretions from adipose tissue.

1. Stretch receptors do signal when the stomach is full, but most animals stop eating long before these receptors are activated.
2. Glucose is a logical choice as a "satiety signal," and glucose-responding nerve cells do occur in the hypothalamus. High levels of glucose could stimulate satiety, and low glucose levels could stimulate appetite. However, glucose levels fluctuate too much to keep body weight constant year after year.
3. Several studies indicate that cholecystokinin secreted by the small intestine during digestion suppresses appetite for a short time.
4. A more logical choice for long-term regulation would be some signal indicating the amount of body fat. Recently, a protein named **leptin,** which is produced by adipose cells, has been shown to cause a dramatic loss of fat in mice. Tests are in progress to determine whether leptin has the same effect in humans.

There is ample evidence all around us that in societies where food is plentiful and appetizing, the mechanisms that regulate food intake either do not function properly or are overridden by other factors affecting the balance between caloric intake and expenditure. The consequence is usually excessive body weight, which is called obesity if it is greater than 20% above what is considered normal for body height. Obesity can contribute to heart failure, diabetes, and other health-threatening disorders.

The First Law of Thermodynamics implies that there is only one cause of excess body weight: taking in more calories than are expended. One cause of excess body weight may therefore be inadequate expenditure of calories. In many mammals, excess calories are consumed by **brown adipose tissue** (= brown fat). Brown adipose tissue accounts for up to 6% of body weight in infants and mammals well adapted to cold. In the average human adult, brown adipose tissue is only about 1% of the body weight, concentrated in the neck, upper back, chest, and around the kidneys. Brown fat has a large number of mitochondria that do not produce much ATP, but break down fats into heat, thereby burning off excess calories. Many humans may be obese because this brown adipose tissue is inactive.

The other possible cause of excessive body weight is excessive caloric intake. Overeating may result from a variety of behavioral or physiological causes. Perhaps it is due to the simple fact that for most of us, food tastes better now than it has throughout almost all of our evolution. Most animals, and most humans for that matter, have a bland, monotonous diet, and avoiding overeating may simply be a matter of getting tired of the food. On the other hand, even rats become obese if they are allowed to eat the tasty foods most of us consume.

Pharmaceutical firms are naturally interested in understanding and controlling the mechanisms that regulate caloric intake. Americans spend 17 billions dollars per year on diets, low-calorie foods, weight-loss programs, and weight-control pills that seldom provide lasting weight reduction, so one can imagine how much money could be made from a method that does work. The only thing better than a pill to curb the appetite would be one that would allow you to eat all you wanted without gaining weight. Hundreds of such pills have actually been claimed. For example, a nationally circulated weekly television guide carried an ad for pills that would "burn fat" and let you "melt away pounds without diets or exercise." The pills were, in fact, nothing but benzocaine to anesthetize the taste buds before eating. They came with a little booklet of diets and exercises.

Summary

Nutrients are either inorganic ions and metals or organic molecules. Among the latter are carbohydrates (the chief sources of energy), fats (the long-term stores of energy), proteins, and vitamins. Some animals obtain nutrients by absorbing them in dissolved form, but most get them by feeding. Major modes of feeding include suspension feeding, biting and swallowing, and piercing and sucking.

Once food is ingested, it must be digested to break it into molecules small enough to be absorbed into the body fluid. Some animals rely mainly on intracellular digestion, but most use extracellular digestion, in which digestive enzymes are secreted into the lumen of a digestive tract. Lipid-soluble nutrients such as fats and certain vitamins then diffuse across the lining of the digestive tract into the body fluids. Most other nutrients, such as sugars and amino acids, enter the body fluids by facilitated diffusion or active transport.

In humans the digestive enzymes include salivary amylase and lingual lipase (which digest starch and fats), pepsin (which partially digests proteins in the stomach), and various carbohydrases, proteases, and lipases that are secreted by the pancreas into the small intestine. Other important digestive secretions include hydrochloric acid from the stomach, bicarbonate from the pancreas, and bile from the liver. Nutrients are absorbed mainly across the lining of the small intestine, which has a large surface area by virtue of its folds, villi, and microvilli. Lipid-soluble molecules diffuse into lacteals in the villi, and carbohydrates and amino acids are transported into capillaries in the villi.

Digestion is controlled in three phases—cephalic, gastric, and intestinal—by the autonomic nervous system and by hormones such as gastrin, oxyntin, cholecystokinin, and secretin. The concentration of glucose in the blood is regulated by the hormones insulin and glucagon. Appetite is thought to be controlled by areas of the brain associated with the hypothalamus.

Key Terms

vitamin
suspension feeding
intracellular digestion
extracellular digestion
ruminant
amylase
lipase
stomach

pepsin
chyme
pancreas
bile
trypsin
chymotrypsin
carboxypeptidase
gastrin
oxyntin
cholecystokinin
secretin
villi
microvilli
lacteal
α and β cells
diabetes mellitus

Chapter Test

1. List as many categories of nutrients as you can (e.g., carbohydrates) and explain how each category is appropriate for its general function. (pp. 254–256)
2. Describe the feeding methods that are appropriate for the following types of food: food particles in large numbers but much smaller than the feeding animal; large food masses; fluid food. Give an example of an animal that uses each feeding method. (pp. 259–260)
3. Digestion is said to occur outside the body in mammals. Explain what that means. (p. 261)
4. Describe the "strategy" of extracellular digestion. (p. 261)
5. Which kinds of nutrients can be absorbed by diffusion, and which ones must be absorbed by active transport? (p. 263)
6. Some animals feed on plants that contain chemicals that are potentially toxic, carnivores ingest their prey's hormones, and humans ingest laxatives and other drugs that would be harmful if they entered the bloodstream. Many such molecules are safe because they are inactivated during digestion. What kinds of molecules would be inactivated by digestion? What kinds of molecules would not be destroyed during digestion? Of these, what kinds could get into the bloodstream? (pp. 263, 268)
7. Describe what happens to food as it passes through the following parts of the human digestive tract: mouth, stomach, small intestine, large intestine. (pp. 263–268)
8. Explain the differences between digestive enzymes and digestive hormones. Give an example of each, and explain how it functions. (pp. 268–269)
9. Diagram a villus from memory and indicate the routes followed by fatty acids and by amino acids during absorption. (pp. 269–270)
10. Explain why glucose concentrations in the blood remain fairly constant even though the ingestion of glucose is periodic. (pp. 270–271)
11. Explain what is meant by a satiety signal. What are some of the possible candidates for satiety signals? How could a knowledge of satiety signals be exploited for weight control? (p. 271)

■ Answers to the Figure Questions

13.3 It would have to be a fluid; in fact, it is usually the nectar of flowers.

13.4 Because digestion essentially occurs outside the body in most animals, there is no reason why food necessarily has to be swallowed first. Spiders feed in this way.

13.7 Secretory cells within pits are protected from exposure to stomach contents. Unlike most cells of the stomach lining, secretory cells therefore do not have to be replaced as often.

13.9 An intestinal parasite would have to be protected by a body covering that resists digestive enzymes. It would also need some means of attachment in order to keep from being swept downstream in the digestive traffic. See Chapters 25 and 26 on tapeworm and intestinal round worms for specific examples.

Readings

Recommended Readings

Dethier, V. G. 1962. *To Know a Fly.* San Francisco: Holden-Day. (*An entertaining account of Dethier's research on feeding in flies.*)

Grajal, A., and S. D. Strah. 1991. A bird with the guts to eat leaves. *Nat. Hist.* 100:48–55 (Aug). (*The hoatzin.*)

Livingston, E. H., and P. H. Guth. 1993. Peptic ulcer disease. *Am. Sci.* 80:592–598.

Martin, R. J., B. D. White, and M. G. Hulsey. 1991. The regulation of body weight. *Am. Sci.* 79:528–541.

Merritt, R. W., and J. B. Wallace. 1981. Filter-feeding insects. *Sci. Am.* 244(4):132–144 (Apr).

Mertz, W. 1981. The essential trace elements. *Science* 213:1332–1338.

Milton, K. 1993. Diet and primate evolution. *Sci. Am.* 269(2):86–93 (Aug).

Moog, F. 1981. The lining of the small intestine. *Sci. Am.* 245(5):154–176 (Nov).

Owen, J. 1980. *Feeding Strategy.* Chicago: University of Chicago Press.

Sanderson, S. L. and R. Wassersug. 1990. Suspension-feeding vertebrates. *Sci. Am.* 262(3):96–101 (Mar).

Additional References

Cohen, I. B. (Ed.). 1980. *The Career of William Beaumont and the Reception of His Discovery.* New York: Arno Press. (*Includes two of Beaumont's early notebooks, an account of his work, and an evaluation of his influence.*)

Myer, J. S. 1939. *Life and Letters of Dr. William Beaumont.* St. Louis: Mosby.

Slaytor, M. 1992. Cellulose digestion in termites and cockroaches: what role do symbionts play? *Comp. Biochem. Physiol.* 103B:775–784.

CHAPTER

14

Water and Ions

Lioness, *Panthera leo.*

The Importance of Water and Ions

For virtually all of their history, cells have evolved in an environment of seawater. Not surprisingly, therefore, the cells of animals still require an environment similar to seawater, even when the animals, themselves, live in fresh water or on land. Water has many unique properties that make it necessary for life (pp. 19–24). One of these essential properties is its ability to dissolve a wide range of substances, including ions. A few animals, such as some nematodes, tardigrades, and rotifers, can survive almost total drying in an inactive state called **cryptobiosis,** but most animals die if only partially dehydrated. Humans, for example, are about 75% water, and a loss of only 15% to 20% of that water is fatal. Even a loss of 10% (only 5 kg of body weight) is life-threatening.

Ions are as essential for life as is water. In many animals the concentrations of ions required in the interstitial fluid do not match the ions available in the food or external environment. For example, herbivores ingest large amounts of potassium from the cellular fluid of plants, yet the potassium concentration in their interstitial fluids must remain low to avoid depolarizing the membranes of neurons and other cells. In such cases, homeostatic systems must regulate the concentration of the ion.

The Importance of Concentrations

Not only the concentrations of individual ions, but also the total ionic concentration of the interstitial fluid, must be regulated. In addition, nonionic solutes contribute to the total concentration of the interstitial fluid. This total concentration is called the **osmolarity.** Incorrect osmolarity can affect protein structure and function, and it can cause water to move into or out of cells or organisms by osmosis. Such osmotic movement of water into or out of a cell can lead to swelling or shrinking of cells and tissues, which can be fatal. A solution with an osmotic concentration higher than that of a cell or organisms will tend to draw water out of the cell or organism by osmosis. Such a solution with a high osmotic concentration is said to be **hyperosmotic** (Fig. 14.1). A solution that has an osmotic concentration that is lower than that of a cell or organism, such that the cell or organism tends to take in water from the solution, is said to be **hypo-osmotic.** A solution in which there is neither a net loss nor a net gain of water by the cell or organism is termed **iso-osmotic.** In most cases an animal must have an interstitial fluid that is iso-osmotic with respect to its cells. Either the animal must live in an environment that naturally maintains an iso-osmotic interstitial fluid, or the animal must have mechanisms of regulating the osmotic concentration of that fluid.

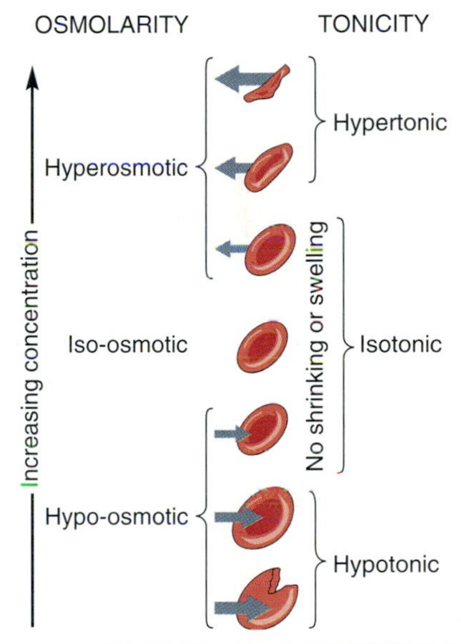

FIGURE 14.1
The effects of varying concentrations on a red blood cell. The direction and width of the arrows indicate the net direction and rate of water movement. The left side of the figure represents the effects of concentration on osmolarity. A red blood cell (or any other cell or animal) in an iso-osmotic fluid neither gains nor loses water due to osmotic pressure. Increasing the concentration makes the solution hyperosmotic, and the cell tends to lose water. Decreasing the concentration below the iso-osmotic level makes the solution hypo-osmotic, and the cell tends to gain water. Tonicity refers to the effect of a concentration on shrinking or swelling. Whatever its concentration, a solution is isotonic to a particular cell or animal if there is no shrinking or swelling. If the solution makes the cell or animal shrink, it is hypertonic. If it makes the cell or animal swell, it is hypotonic.

Although all animals have interstitial fluids that resemble seawater, there are wide variations in osmolarity and in the concentrations of individual ions (Table 14.1 on page 276). There is some order in these variations, however, that depends on the ionic concentrations in the environment. For example, a fish that lives in seawater (SW) is likely to have a higher sodium concentration in its interstitial fluid than a fish that lives in fresh water (FW). In general, the relationship of ion concentrations should be as follows:

$$\text{SW} \geq \text{marine animal} > \text{terrestrial animal} \geq \text{freshwater animal} > \text{FW}$$

The basis for this relationship is that less energy is required to maintain the difference between the internal and external concentrations if that difference is small. A major exception to this generalization is that marine vertebrates, such as bony fishes (for example sculpin), reptiles (sea snakes and turtles), and mammals (whales and seals), generally maintain internal concentrations more like those of freshwater animals. The explanation is that these and other vertebrates evolved from freshwater fishes.

TABLE 14.1

The concentrations of ions in seawater, fresh water, and in the body fluids of animals representative of different habitats.

Numbers below each ion are concentrations in millimolar (mM).

	Na^+	K^+	Ca^{2+}	Mg^{2+}	Cl^-
Seawater					
Average ocean	470	10	10	54	548
Marine Invertebrate					
Starfish *Asterias*	428	10	12	49	487
Marine Vertebrate					
Sculpin *Myoxocephalus*	194	4	3	2	177
Terrestrial Invertebrate					
Cockroach *Periplaneta*	161	8	4	6	144
Terrestrial Vertebrate					
Human *Homo*	142	4	5	2	104
Freshwater Invertebrate					
Crayfish *Cambarus*	146	4	8	4	139
Freshwater Vertebrate					
Salmon *Oncorhynchus*	147	9	3	1	117
Fresh Water					
(average North American river)	0.39	0.04	0.52	0.21	0.23

Osmoconformers

Many marine animals do not have to regulate their ionic concentrations, because seawater is unvarying and similar in ionic composition to their interstitial fluids. For small marine animals it would also take too much energy to maintain a concentration difference across their relatively large body surfaces. Many marine animals, especially small ones, therefore lack mechanisms that would regulate internal ionic concentrations if they happened to be placed in an environment that differed from seawater. Such animals are said to be osmoconformers.

Many osmoconformers die if placed in water that is more concentrated or less concentrated than seawater. The reason is that their cells either lose or gain water osmotically, and the shrinking or swelling destroys membranes and organelles. Some osmoconformers, however, can survive being placed in dilute or concentrated solutions because their cells resist shrinking or swelling. One mechanism for avoiding shrinking and swelling is to change intracellular concentrations in proportion to changes in extracellular concentration. Even if the cells of an osmoconformer do not shrink or swell, the entire animal may do so as the interstitial fluid loses or absorbs water. Such animals are sometimes referred to as "living osmometers," because one could theoretically use them to measure the osmolarity of a solution by weighing them before and after they had been in the solution.

Speaking of Concentrations

The total **concentration** of dissolved material depends on the number of individual particles in a given volume of solution. Even in a few milliliters of fresh water, the number of dissolved particles is unimaginably large (billions or trillions). Rather than enumerate particles, therefore, it is more convenient to express concentrations in moles per liter. A mole equals 6.022×10^{23} particles. A concentration of *X* moles/liter can also be expressed as an *X* molar concentration. Millimolar (mM = 10^{-3} M) concentrations are more common in living systems.

As long as the concentration is small enough that particles do not interact with each other, the osmotic effect—the osmolarity—will be directly proportional to the sum of the molar concentrations of all the dissolved particles. Thus a 1 mM solution of NaCl will have about twice the osmolarity of a 1 mM solution of sucrose, since the NaCl will dissociate into 1 millimole of Na^+ plus 1 millimole of Cl^-. $CaCl_2$ will have three times the osmolarity of an equal concentration of sucrose. At the concentrations normally found in body fluids, many salts do not completely dissociate into ions, so the osmolarity is somewhat lower than would be predicted simply from adding concentrations. Osmolarity is measured in osmoles/liter, or osmolar (Osm). Average seawater is 1 osmolar. Human interstitial fluid has an osmolarity of approximately 300 mOsm.

Osmolarity is one of several properties, called **colligative properties,** that depend on the concentration of particles and not on the nature of those particles. Osmolarity can be determined by measuring any of those other colligative properties. The **freezing point depression**—the number of degrees below 0°C at which a solution freezes—is one such property. Many **osmometers** determine the osmolarity of a solution by measuring the temperature at which it freezes. While pure water freezes at 0°C under standard conditions, average seawater, for example, freezes at −1.86°C. Seawater therefore has a freezing point depression of 1.86°C. The measurement in degrees Celsius can be converted to osmoles, but often it is not. One often sees osmolarities expressed in degrees Celsius.

Salinity is another way of expressing osmotic concentration, especially in marine biology. Salinity is essentially the number of grams of solids remaining after the water has evaporated from 1 kilogram of solution. Salinity therefore indicates the number of parts per thousand of dissolved material. Since percentage, (parts per hundred) is indicated by the symbol %, salinity is indicated by the symbol ‰. The dissolved solids in average seawater makes up 3.5% of its weight, so average seawater has a salinity of 35 ‰.

The scale in Fig. 14.2 may prove useful in converting among the various units of measurement. As a point of reference for comparison to other solutions, it would not be a bad idea to remember that average seawater has an osmolarity of 1000 mOsm, a freezing point depression of 1.86°C, and a salinity of 35 ‰.

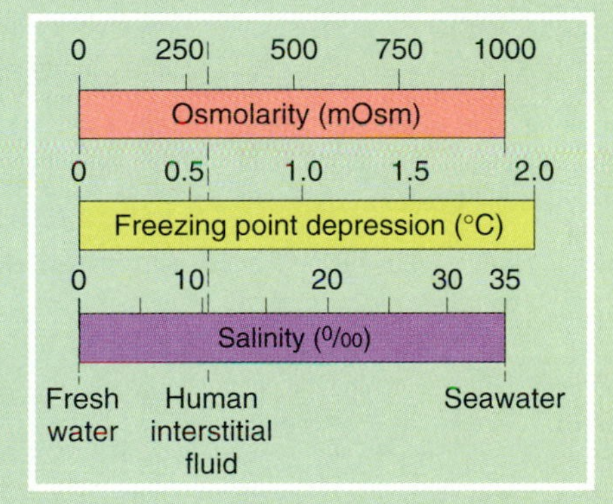

FIGURE 14.2
Three different measures of osmotic concentration.

■ **Question:** At what temperature does human blood freeze?

Osmoregulators

Many marine animals maintain internal levels of ions that are lower than those of seawater. Such animals are osmoregulators. Freshwater and terrestrial animals are also osmoregulators, since they have to maintain ionic concentrations in the interstitial fluid that are higher than those in their environments. The most adept osmoregulators live in environments with the greatest variability in ionic composition, such as **estuaries,**

Ringer's Solution to a Big Problem

It is doubtful whether much progress in physiology would have been possible without the ability to keep tissues functioning outside the body. Early physiologists were handicapped because dehydration destroyed isolated organs. Adding water destroyed tissues even more quickly than dehydration. Eventually someone found out that the correct amount of sodium chloride in the water would allow many organs to continue functioning for longer periods. These isotonic NaCl solutions, called **physiological salines,** are useful because they provide the correct osmotic pressure for cells and because they provide the two most prevalent ions (Na^+ and Cl^-) found in most extracellular fluids. Even so, few tissues can continue functioning for long in simple salines.

In 1883 a combination of error, insight, and research provided the solution to this problem—in both senses of the word "solution." Sidney Ringer was conducting research on isolated frog hearts at London University College Hospital when he noticed that the hearts were continuing to beat in the physiological saline for much longer than the usual half hour. At first he thought this was just one of many seasonal variations in frog physiology, and he actually published this incorrect explanation. Then Ringer found out that a laboratory technician had erroneously made up the physiological saline using ordinary tap water instead of laboratory distilled water. Ringer quickly realized that ions in tap water were required for the normal functioning of frog hearts, and by trial and error he found the optimum concentrations of calcium and potassium salts. The result was an improved solution that is still known as **Ringer's solution.** It consists of 6 grams of NaCl, 0.22 gram of KCl, and 0.29 gram of $CaCl_2$ per liter of solution. Bicarbonate (0.2 gram of $NaHCO_3$) is added to buffer the pH. Ringer's solution works well for most amphibians, but for other animals many other physiological solutions, known generally as Ringer's solutions, have been developed. For examples, Locke's, Tyrode's, and Krebs's solutions are for various mammalian tissues. The major ions in Ringer's solutions are Na^+, K^+, Ca^{2+}, Mg^{2+}, H^+ (pH), and Cl^-. Often glucose is added as an energy source.

where fresh water empties into salt water. The blue crab *Callinectes sapidus,* which inhabits estuaries, can maintain a fairly constant internal osmotic concentration no matter how dilute the external water is. Animals that can tolerate wide ranges of osmolarity are said to be **euryhaline** (Greek *eurys* broad + *halin* of salt). Those that can tolerate only a limited range are said to be **stenohaline** (Greek *stenos* short) (Fig. 14.3).

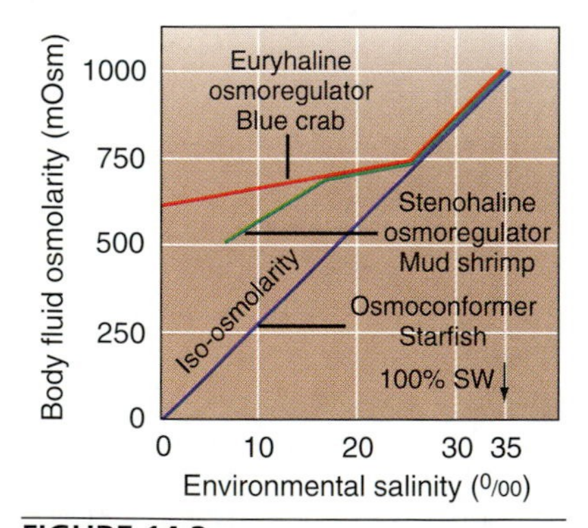

FIGURE 14.3
Internal osmolarity versus external osmolarity for three representative animals: The osmoconforming starfish *Asterias,* The euryhaline osmoregulating blue crab *Callinectes sapidus,* and the stenohaline osmoregulating mud shrimp *Upogebia pugettensis.* The *x* and *y* axes are labeled in different but equivalent units. The internal osmolarity (*y* axis) for the starfish is the same at all points as the external osmolarity (*x* axis).

Osmoregulatory Mechanisms. There are only a few things an osmoregulator can do to prevent or compensate for the loss or gain of water due to osmosis or the loss or gain of ions by diffusion.

1. **The osmoregulator can actively transport ions in the direction opposite from that of diffusion.** The mechanism for such active transport is generally an **ionic pump** located in the plasma membranes of certain cells (pp. 39–40). Such pumps enable animals living in a hyperosmotic environment to pump out excess ions to compensate for diffusion and intake with food. Ionic pumps can also be used to set up a concentration difference that will transport water osmotically. Using this arrangement, an animal in a hyperosmotic environment can pump in ions to bring in water, then pump the ions back out. Conversely, animals living in hypo-osmotic environments can pump ions in to compensate for loss, or use ion pumps to eliminate excess water.
2. **The osmoregulator can vary its permeability to water or to ions.** This approach allows for control of the amounts of water and ions entering or leaving the body.

Different organisms employ these principles in a variety of ways. Freshwater protozoans and sponges osmoregulate by means of **contractile vacuoles** (= water expulsion vesicles) in each cell (Fig. 14.4). The vacuole slowly swells with the osmotically absorbed water until it reaches a certain size, then it quickly expels its contents. Contractile vacuoles are surrounded by mitochondria, but it is not known how energy is harnessed to transport water into the vacuoles. The expulsion of water is apparently due to protein filaments similar to those in muscle.

Most animals have kidneys or various types of **nephridia** for osmoregulation (summarized in Fig. 21.11). Osmoregulatory functions in the nephridia and other organs of invertebrates will be described briefly in the next section, and in more detail in the chapters dealing with group. The kidneys of vertebrates will be described more thoroughly later in this chapter. All these organs generally work in two stages: **filtration** and active transport. First blood or another body fluid is filtered into the osmoregulatory organ, leaving cells, proteins, and other large molecules behind. The filtrate thus consists mainly of water, ions, and small dissolved molecules such as glucose and amino acids. Pumps also actively transport excess ions and molecules into the filtrate. This active transport from body fluid to filtrate is called **secretion.** Ions and molecules needed by the animal are then pumped out of the filtrate and back into the body fluid. Active transport from filtrate into body fluid is called **reabsorption.**

Excretion of Nitrogen Wastes

Organs that regulate ion concentrations in interstitial fluids are often called excretory organs, because in addition to regulating ion concentrations, they excrete certain metabolic wastes. The most important of these wastes is the nitrogen from amino acids. There are three major forms of nitrogenous waste: ammonia, urea, and uric acid (Fig. 14.5 on page 280). **Ammonia** is extremely toxic (as you could guess from inhaling its fumes), but it is easily diluted in water. It is the major nitrogenous waste of most aquatic animals, since they can simply excrete it into the environment as fast as it is produced. Many terrestrial animals, including all mammals, most amphibians, and some turtles, produce **urea** as their major nitrogenous waste. Urea denatures proteins, but is less toxic than ammonia, so it can be tolerated in the bloodstream until eliminated by the kidneys. Egg-laying terrestrial animals, including most reptiles and birds, eliminate nitrogenous waste as **uric acid.** Unlike ammonia and urea, which are soluble and would accumulate to toxic levels

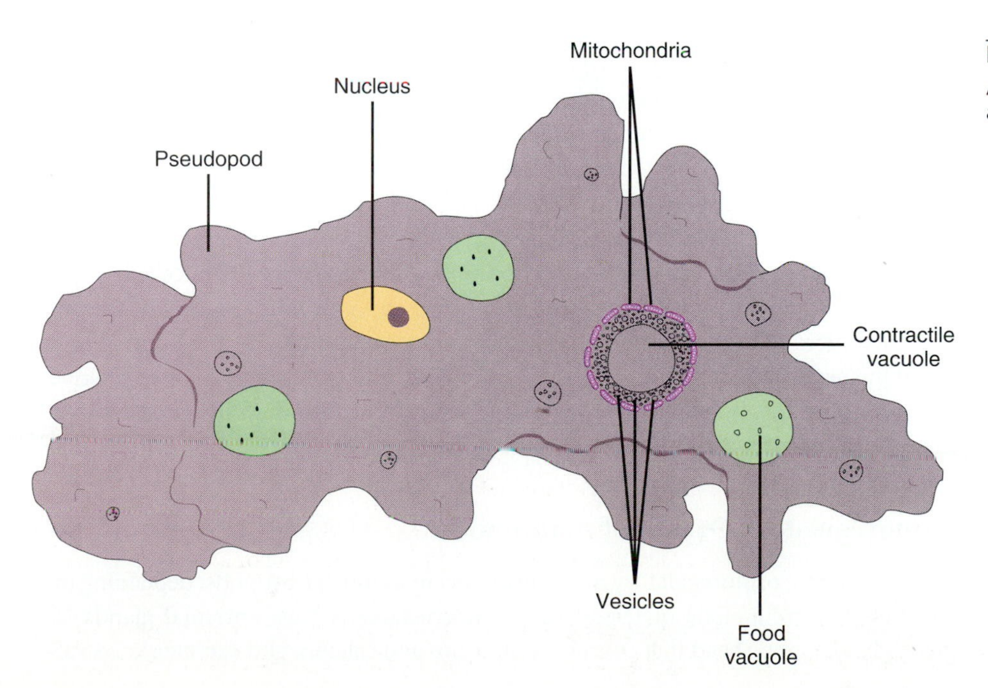

FIGURE 14.4
A contractile vacuole in a freshwater ameba.

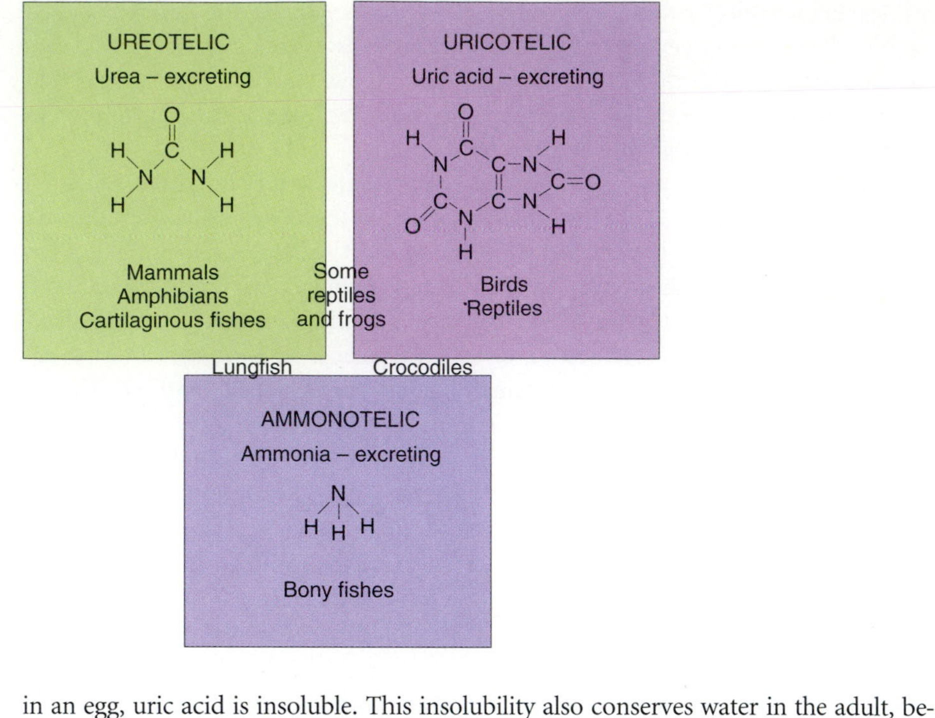

FIGURE 14.5
The most common nitrogenous wastes of various groups of vertebrates.

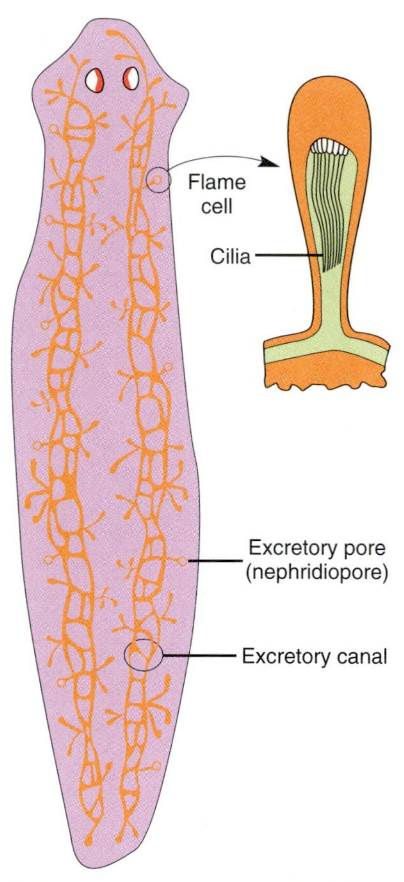

FIGURE 14.6
The protonephridia of a planarian (phylum Platyhelminthes). Protonephridia are branched and closed at the internal ends, where there are often flame cells (inset). Protonephridia of some animals have solenocytes, which differ from flame cells in having only one flagellum.

■ **Question:** In a freshwater planarian, would the protonephridia secrete or reabsorb water?

in an egg, uric acid is insoluble. This insolubility also conserves water in the adult, because almost all the water needed to eliminate uric acid can be reabsorbed into the body fluids, leaving the white paste that is familiar in bird droppings.

Nephridia

The tubular osmoregulatory organs of most invertebrates, called nephridia, use the same mechanisms of filtration and active transport as our kidneys. Unlike our kidneys, however, nephridia form a filtrate of some body fluid other than blood. Also unlike vertebrate kidneys, the pressure that drives the body fluid into the nephridium is not fluid pressure, but usually the activity of either cilia or flagella. After the filtrate is produced, the nephridium secretes excess ions into it and reabsorbs needed ions, nutrients, or water. The resulting fluid containing excess ions, metabolic wastes, and water is then excreted out of the body through a **nephridiopore** at the external end of the nephridium.

Nephridia with closed internal ends are called protonephridia, and those with open internal ends are called metanephridia. **Protonephridia** usually have branched tubes. A striking feature of many protonephridia are **flame cells** located at the closed internal ends of the tubes (Fig. 14.6). Flame cells are named for the flickering tuft of cilia that projects into the nephridial tube. The cilia may be responsible for producing the pressure that filters the body fluids into the nephridia. **Metanephridia** (often called simply nephridia) occur only in invertebrates with true coeloms. Metanephridia are unbranched tubes that open not only at the nephridiopore, but also at the internal end, called the **nephrostome,** where coelomic fluid enters (Fig. 14.7). In addition to filtering the coelomic fluid and actively transporting material into or out of it, metanephridia may also exchange material with blood in surrounding capillaries.

Osmoregulatory Organs of Arthropods

Several different osmoregulatory mechanisms occur among arthropods, depending on whether they are aquatic or terrestrial. Some crustaceans have **antennal glands** (= green glands) in the head that conserve potassium and calcium and excrete excess sul-

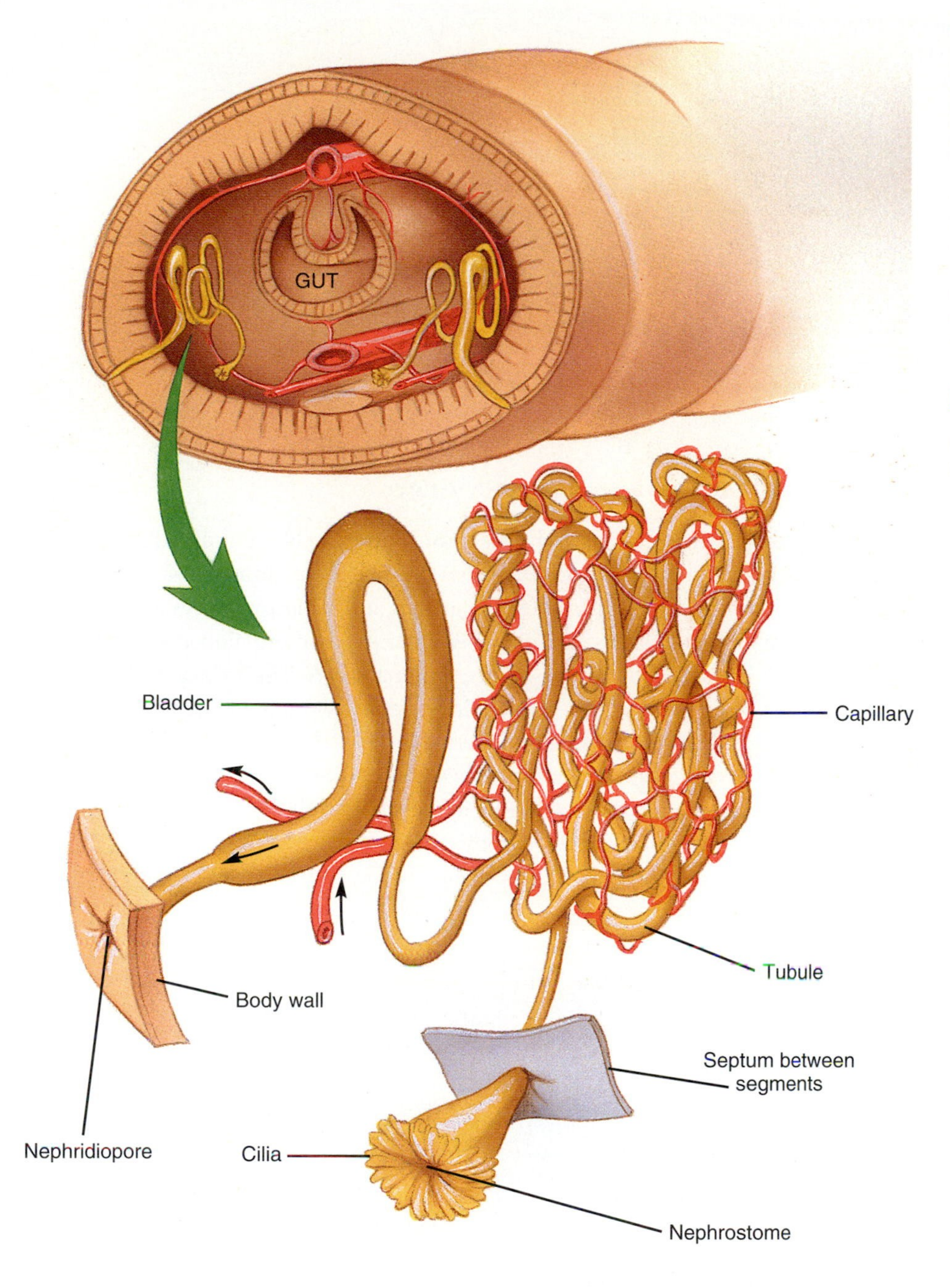

FIGURE 14.7

Metanephridia of the earthworm *Lumbricus terrestris.* The upper part of the figure shows a pair of nephridia in each segment and some of the major blood vessels. The lower part of the figure shows the capillary surrounding the unbranched tubule. Coelomic fluid from the next anterior segment is drawn into the tubule by a ciliated nephrostome. Fluid is excreted from the tubule at the nephridiopore. There is also exchange between tubular fluid and blood in the capillary.

■ **Question:** Would the metanephridia secrete or reabsorb water in this species?

fate and magnesium. In freshwater crustaceans, such as crayfish, the antennal glands also eliminate excess water. A filtrate of blood enters an **end sac,** then usually a **labyrinth,** and finally a coiled **tubule.** Reabsorption and secretion occur in all three structures, forming urine that is stored in a bladder until it is emptied through a pore at the base of the antenna (Fig. 14.8 on page 282).

Most insects and spiders are terrestrial arthropods whose main problem is to conserve water while eliminating metabolic wastes and excess ions. These functions are served by a combination of **Malpighian tubules** and the rectum. Two to several hundred Malpighian tubules are attached to the digestive tract usually between the midgut and hindgut, and they wave about within the abdominal cavity (see Figs. 13.5B and 32.9). These tubules transport ions, especially potassium, out of the blood and into the gut for elimination with the feces. Body fluids are filtered into the Malpighian tubules osmotically. As the fluid passes through the gut into the rectum, water is reabsorbed by **rectal pads.**

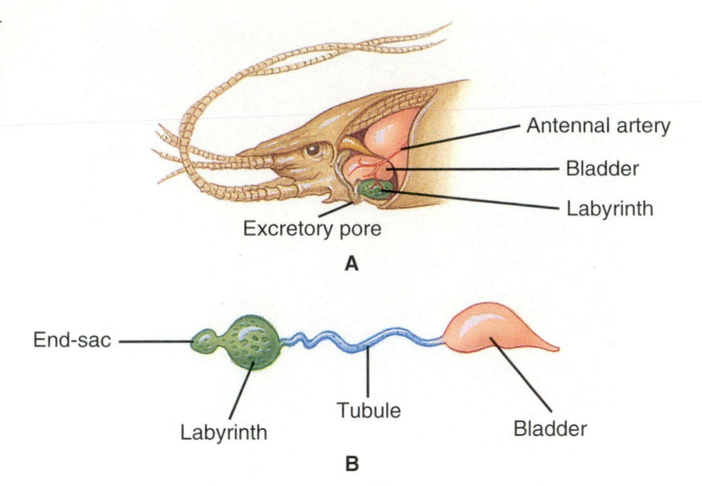

FIGURE 14.8

The antennal gland (green gland) of the crayfish *Astacus pallipes* in its natural configuration (A) and extended (B).

Structure and Function of Human Kidneys

Unlike nephridia, the kidneys of vertebrates form a filtrate of blood rather than that of another body fluid, and the filtration pressure is due mainly to blood pressure. Several kinds of kidneys occur in different vertebrate groups and at various stages of development, as will be described later. The following description of the human kidney also applies to the kidneys of adult reptiles, birds, and mammals. The kidney of the adult human is a fist-sized organ that is . . . well, kidney-shaped. (Mammals are the only animals in which the kidney has a definite shape.)

The two kidneys are located in the lower back on each side of the aorta and vena cava (Fig. 14.9). At any given time, about one-fifth of the blood from the aorta is entering the kidneys through the **renal arteries.** A slightly smaller volume exits through the **renal veins** into the vena cava. In general, the flow of fluid is from the outer **cortex** of the kidney

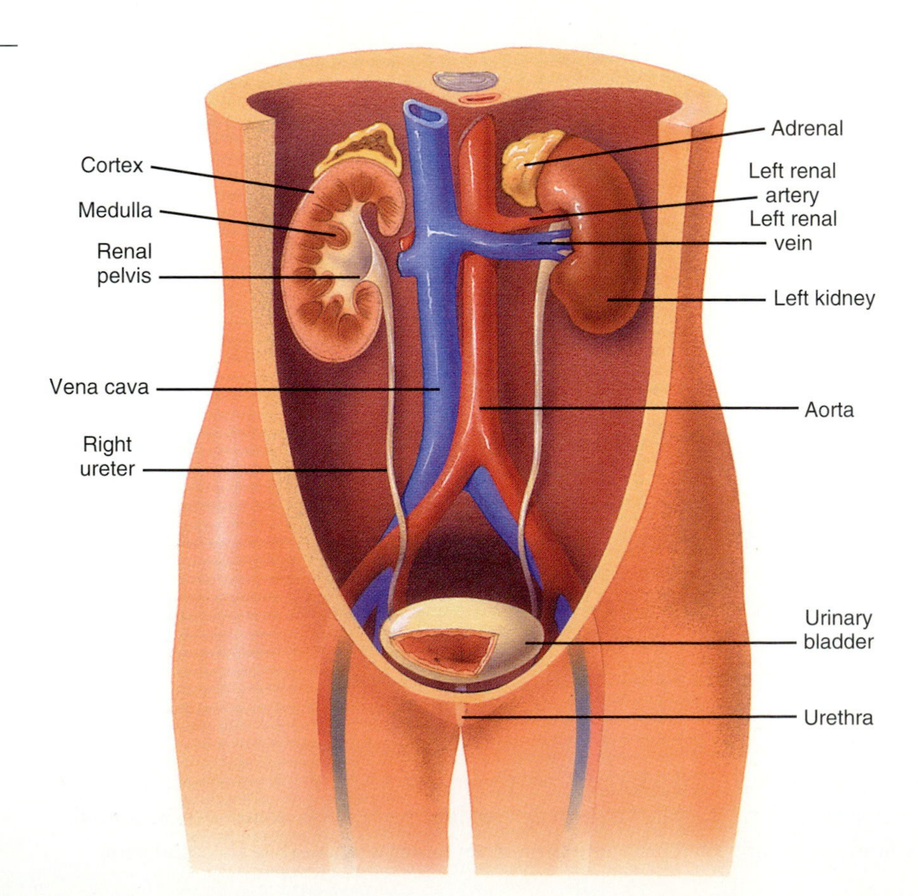

FIGURE 14.9

The structure of the human renal system. The ureters transport urine, containing excess ions and metabolic wastes removed from the blood, into the urinary bladder.

toward the inner **medulla.** The difference between arterial and venous blood volumes equals the 1.5 liters of urine that daily flows into the **urinary bladder** through the **ureters.** This description of the plumbing suggests that the overall function of the kidneys is to remove water, blood-borne wastes, and excess ions from blood. To understand how kidneys perform this task, we have to look at the microscopic and submicroscopic levels.

The Nephron

Each human kidney contains about a million subunits called nephrons that do the actual work (Fig. 14.10). In each nephron, arterial blood enters a tuft of capillaries called the **glomerulus.** Here the blood pressure drives the first stage of renal function, **filtra-**

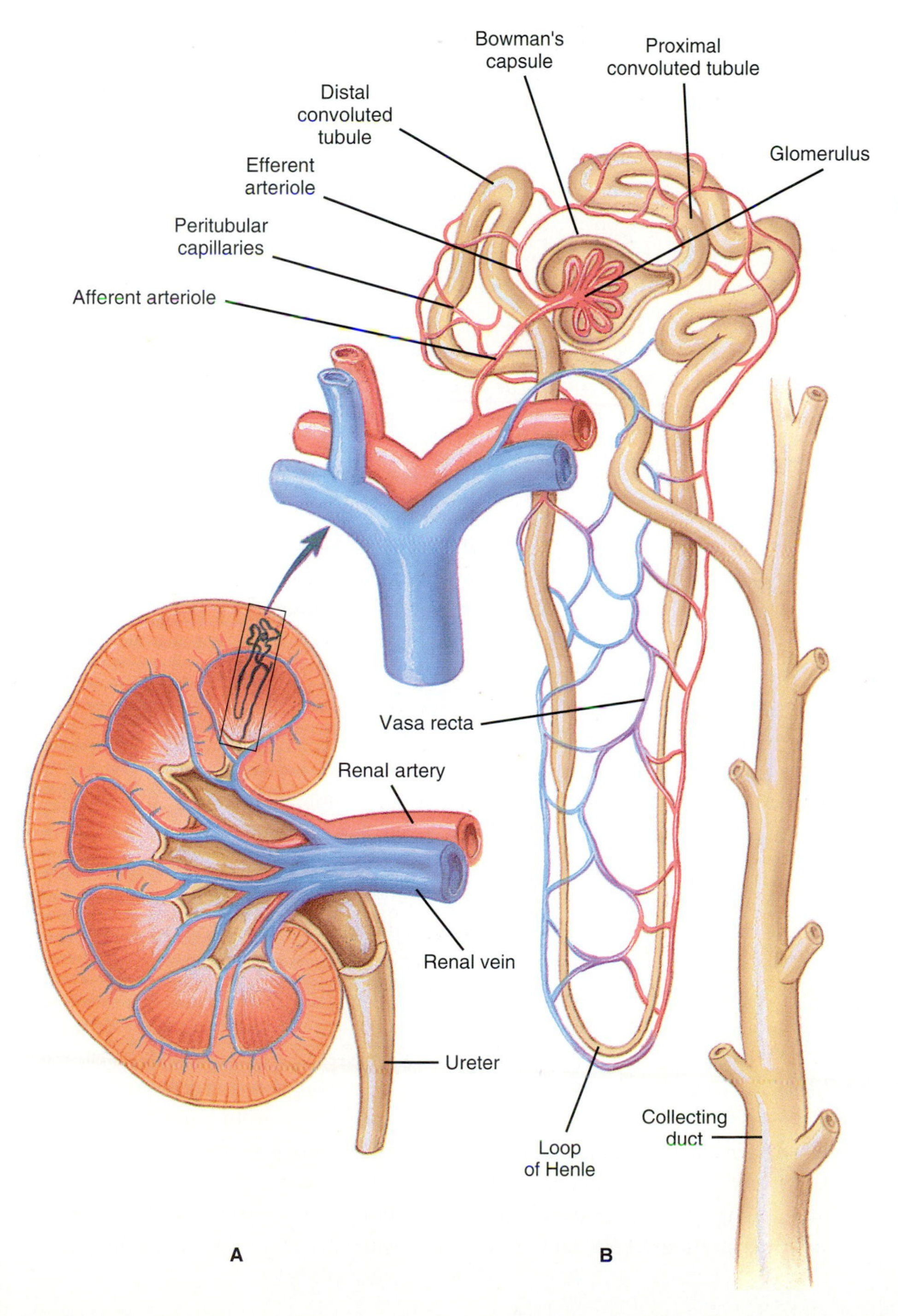

FIGURE 14.10

(A) A sectioned human kidney. The location of one nephron is indicated. (B) A nephron. The total length of an outstretched nephron is 2.0 to 4.4 cm.

■ **Question:** When kidney stones form from salts in the urine, where are they located?

tion. (Filtration in glomeruli and other capillaries is often called ultrafiltration.) As in other capillaries, the fluid portion of the blood is forced across the walls, leaving behind the cells and proteins. The resulting filtrate (or ultrafiltrate) enters the surrounding **Bowman's capsule,** which conducts it into the **proximal convoluted tubule.** In humans approximately 180 liters per day make it this far. Fortunately, not all of the filtrate goes into urine. In the proximal tubule, much of the inorganic ions and almost all of the glucose and amino acids are pumped out of the filtrate. These ions and molecules then diffuse back into the blood in the **peritubular capillary,** which surrounds the tubule. At the same time, most of the water in the filtrate is reabsorbed osmotically.

The tubular fluid next passes through a hairpin turn called the **loop of Henle.** The functioning of the loop of Henle is a complicated subject best left for later. For now it is enough to note that it aids in the reabsorption of water. In the **distal convoluted tubule** there is further active transport of inorganic ions. Depending on whether their concentration in the blood is too high or too low, there will be secretion or reabsorption of H^+, K^+, or other ions. Herbivores, for example, are apt to have an excess of potassium in the bloodstream, so pumps in the distal tubule secrete K^+ out of the peritubular capillaries into the distal tubule. H^+ is either reabsorbed or secreted to adjust the pH of blood. Urea, the nitrogenous waste that always comes to mind when we think of urine, is generally thought to simply diffuse from blood into filtrate. This assumption has recently been called into question, however, by the isolation of a urea transporter in the kidney (You et al. 1993). By the time the tubular fluid enters the **collecting duct,** it is essentially dilute urine, called pre-urine. Humans produce about 9 liters of pre-urine each day. Under normal circumstances most of the water in the pre-urine is reabsorbed, leaving concentrated, industrial-strength urine.

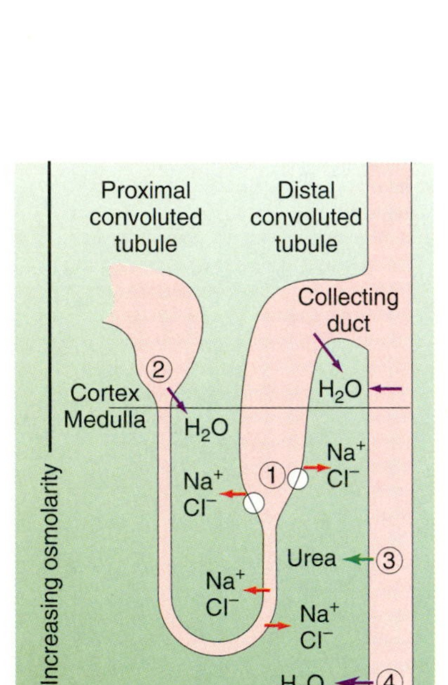

FIGURE 14.11

How the loop of Henle increases water reabsorption. At the ascending thick portion of the loop ①, Cl^- and Na^+ are actively transported from tubular fluid into the surrounding space. The increased concentrations of Na^+ and Cl^- draw water out of the distal tubule and the descending loop ②. As a result of the loss of both ions and water from the fluid in the loop, urea becomes concentrated in the fluid, but it can leave only in later parts of the collecting duct ③. The increased concentration of urea in the inner medulla draws more water out of the loop of Henle, producing extremely high Na^+ and Cl^- concentrations at the bend. Some of these ions therefore diffuse out, adding to what was actively transported at ①. The resulting concentration gradient increases from cortex to medulla, drawing increasing amounts of water out of the collecting duct ④.

Conservation of Water

The ability to produce concentrated urine depends on the loops of Henle, as was first suspected from the fact that vertebrates that have to conserve water generally have more and longer loops of Henle. Fishes, amphibians, and reptiles lack loops of Henle, and they are incapable of producing urine that is hyperosmotic to plasma. Birds have short loops of Henle and can produce urine that is slightly hyperosmotic. In mammals the loops of Henle occur in both short and long forms. The short forms are contained entirely within the outer cortex. The long forms extend into the medulla. The greater the proportion of long to short Henle's loops, the more concentrated the urine and the better the mammal can survive water stress. In the mountain beaver *Aplodontia rufa,* which lives in wet areas of the American Northwest, there are no long loops of Henle. This rodent's urine is only 2.4 times as concentrated as its blood plasma. In contrast, the loops of Henle of the kangaroo rat *Dipodomys merriami* are all long, and its urine is from 10 to 14 times as concentrated as its plasma. Approximately 14% of the loops of Henle in humans are long, and our urine averages about 4 times the osmolarity of plasma. Depending on circumstances, human urine can be from one-fifth to five times as concentrated as plasma.

After many years of research, physiologists now understand that the loops of Henle are an elaborate version of an active countercurrent-exchange mechanism (p. 245). Briefly, they create a large osmotic gradient of Na^+, Cl^-, and especially urea that increases from cortex to medulla. As pre-urine passes through the collecting ducts, it is exposed to this increasing osmotic concentration, which draws out most of the water. This water then enters the blood in the **vasa recta:** the capillaries surrounding the loop.

The system is driven by active transport of Cl^- and Na^+ from the tubular fluid into the surrounding space at the ascending thick portion of the loop (Fig. 14.11). The increased concentrations of Na^+ and Cl^- then draw water out of the descending loop and distal tubule. Urea therefore becomes highly concentrated in the fluid and leaves the collecting duct in the inner medulla. The increased concentration of urea draws water out of the loop of Henle, producing extremely high Na^+ and Cl^- concentrations at the bend. Some of these ions therefore diffuse out, adding to what is actively transported. The result is a gradient of ions and especially of urea that increases from cortex to medulla. As pre-urine flows through this gradient in the collecting duct, the increasing concentrations draw out more and more water.

Hormonal Control of Kidney Function

Antidiuretic Hormone. The amount of water reabsorbed depends on how permeable the collecting ducts are to water. This permeability is increased by antidiuretic hormone (ADH). The term *antidiuretic* refers to the ability of ADH to reduce the amount of water lost as urine. The posterior pituitary secretes ADH whenever the hypothalamus detects high osmolarity in the cerebrospinal fluid. Since ADH helps conserve water, it is part of a negative-feedback homeostatic mechanism. ADH also helps dilute body fluids by promoting thirst. Another name for ADH is vasopressin, which alludes to its ability to increase blood pressure.

The importance of ADH is clear from the disorder **diabetes insipidus.** Like the more common form of diabetes, diabetes mellitus, diabetes insipidus is characterized by excessive urine production and thirst. It results from either a reduced ability to secrete ADH or a lack of responsiveness in the collecting ducts. The result is that untreated individuals excrete up to 10 times the normal volume of urine—up to 15 liters per day. Before synthetic ADH became available, victims of diabetes insipidus spent much of their lives in the bathroom. Alcohol also temporarily inhibits ADH secretion, which explains the excessive urination and subsequent thirst that typically accompany drinking.

Renin–Angiotensin. Two hormones, renin (REE-nin) and angiotensin, together help ensure that fluid is filtered by the glomerulus into the tubules at a rate sufficient for the tubules to function. An adequate **glomerular filtration rate** (GFR) is essential for sufficient reabsorption and secretion. Renin is secreted into the bloodstream by the **juxtaglomerular apparatus** (JGA) near the glomerulus when the blood pressure and therefore GFR are low. The JGA also secretes renin if the level of NaCl in the tubule is too low, or in response to stimulation by the sympathetic nervous system. Renin is a hormone and also an enzyme that converts **angiotensinogen** from the liver into angiotensin I. Angiotensin converting enzyme (ACE) in the lungs then converts angiotensin I to angiotensin II. Angiotensin II has several effects, all of which help counteract the low blood pressure. (1) It increases blood pressure by causing vasoconstriction. (2) It stimulates the secretion of ADH. (3) It stimulates thirst. (4) It stimulates secretion of aldosterone, which will be described in the next section. **Hypertension** (high blood pressure) is often caused be excessive levels of angiotensin, and drugs that inhibit ACE are often prescribed as a treatment.

Aldosterone and Atrial Natriuretic Peptide. Aldosterone is a steroid hormone (mineralocorticoid) secreted by the adrenal cortex. Aldosterone stimulates the Na–K pumps in the loops of Henle and the collecting ducts, increasing the reabsorption of Na^+ and therefore the amount of water reabsorbed into the blood. The overall effect is to increase the blood volume and pressure.

Aldosterone generally works for only a matter of weeks. Eventually the increase in blood volume causes the heart to release atrial natriuretic peptide. Atrial natriuretic peptide counteracts the effects of aldosterone by inhibiting its secretion, inhibiting sodium reabsorption, inhibiting the renin–angiotensin system, and causing dilation of arteries.

PARATHYROID HORMONE. Finally, parathyroid hormone stimulates the reabsorption of Ca^{2+} by the loop of Henle and the distal convoluted tubule. The net effect is to increase the level of Ca^{2+} in the body fluids. Parathyroid hormone also promotes the release of Ca^{2+} from bones.

Types of Kidneys in Vertebrates

The kidney just described is a type called the **metanephros.** Two other types—the **pronephros** and the **mesonephros**—occur among different groups of vertebrates and within each vertebrate during embryonic development. The pronephros, which lies anteriorly in the abdomen, appears only briefly in many vertebrates, and not at all in the human embryo. In some vertebrates, however, the pronephros is the first osmoregulatory and excretory organ of the embryo, and it remains as the functioning kidney of tadpoles and other amphibian larvae. During embryonic development or during metamorphosis in amphibians, the pronephros is replaced by the mesonephros, which is elongated and posterior to the pronephros. The mesonephros is the functioning embryonic kidney of humans and many other vertebrates, and it remains as the permanent kidney of adult fishes and amphibians (see Figs. 6.24, 35.10, 35.22, and 36.7). In reptiles, birds, and mammals the mesonephros becomes part of the reproductive system (see Fig. 6.24), and the metanephros develops as the functional kidney.

Adaptations of Vertebrate Kidneys for Different Habitats

BONY FISHES. Whether mesonephric or metanephric, the kidneys of adult vertebrates all work on principles similar to those described for humans, but with adaptations suitable for their own environments. In freshwater fishes the nephrons of the mesonephric kidneys have glomeruli that produce large amounts of filtrate. Their nephrons, however, lack loops of Henle, and they are therefore incapable of producing urine more concentrated than blood plasma. Normally, of course, they would not need to, since these fishes are not in any danger of dehydration. Their problem is to eliminate excess water and to absorb and retain ions.

For marine bony fishes the problem is just the opposite. Like all vertebrates, they evolved from freshwater fishes, and they retain low internal ionic concentrations (Table 14.1). Their osmotic problem is to conserve water and eliminate ions. The nephrons of their kidneys have reduced glomeruli or none at all, which prevents water loss by filtration. Only a small amount of urine forms in the tubules to eliminate Mg^{2+} and SO_4^{2-}. More important are the gills, which actively transport Cl^- and Na^+ out of the body fluids.

CARTILAGINOUS FISHES. Sharks and rays (cartilaginous fishes) are also marine vertebrates that evolved from freshwater ancestors, and they too have ion concentrations lower than those in seawater. Unlike marine bony fishes, however, their body fluids are iso-osmotic or even hyperosmotic to seawater, thanks to special molecules called **osmolytes.** One osmolyte in cartilaginous fishes is the nitrogenous waste urea. The level of urea in the blood of a shark or ray is many times the concentration that would kill us and most other animals. Sharks and rays not only tolerate the urea, they require it for muscle function. The tolerance to urea was a deep mystery for many years, until it was found that sharks and rays have a second group of nitrogenous osmolytes, **methylamines,** that counteract the detrimental effects of urea. Because of the osmolytes, water loss is not a problem. The kidneys of cartilaginous fishes have glomeruli that produce copious urine. There is still a tendency for sodium, chloride, and other ions to accumulate in the body fluids by diffusion, however. These ions are eliminated by the **rectal gland** (see Fig. 35.10), which excretes a concentrated solution into the posterior end of the digestive tract.

AMPHIBIANS AND REPTILES. Most adult amphibians are adapted to living on land, but their permeable skin makes them highly dependent on a moist environment. Like freshwater fishes, the nephrons of their mesonephric kidneys have glomeruli, and they are incapable of producing urine more concentrated than blood plasma. In many amphibians, however, the urinary bladder can reabsorb water from the urine. The metanephric kidneys of reptiles also have glomeruli and are incapable of forming hyperosmotic urine. Reptiles are less dependent than amphibians on a moist environment, however, because they have impermeable skin. As noted previously, the metanephric kidneys of birds and mammals are more adapted to terrestrial life, since most have nephrons with loops of Henle that produce concentrated urine.

MAMMALS. Marine mammals have kidneys that require little water to eliminate the salts ingested with their food. It takes only 650 ml of water for a whale's kidneys to eliminate the salt in 1 liter of seawater, providing a net gain of 350 ml of pure water. In contrast, the less-efficient kidneys of humans require 1.35 liters of water to eliminate the salt from 1 liter of seawater, for a net *loss* of 350 ml. This explains why shipwrecked sailors survive longer if they drink nothing than if they give in to the urge to drink seawater.

Mammals that live in the desert also osmoregulate primarily by means of efficient kidneys that can excrete concentrated urine. The kangaroo rat of the American Southwest has such efficient kidneys that it never has to drink water. As long as the kangaroo rat remains in its burrow during the daytime to avoid excessive evaporation, the water produced in mitochondria (p. 60) makes up for losses (Fig. 14.12). Extraordinary as this seems, the common house mouse *Mus musculus* is almost as good at doing without water, unless it gets into our protein-rich food and needs extra water to eliminate the urea.

FIGURE 14.12

The daily water budget for the kangaroo rat *Dipodomys merriami,* which drinks no water. A daily water budget for humans is shown for comparison. About half our water is contained in food, and a little less than that comes from drinking. Sixty percent of the water taken in is lost in the urine, and most of the rest is lost by evaporation.

KANGAROO RAT

Gain (mil)		Loss	
Present in food	6.0	Urine	13.5
Drinking water	0.0	Evaporation	43.9
Metabolic product	54.0	Feces	2.6
	60.0		60.0

HUMAN

Gain (ml)		Loss(ml)	
Present in food	1200	Urine	1500
Drinking water	1000	Evaporation	950
Metabolic product	350	Feces	100
	2550		2550

Salt Glands

Reptiles, birds, and mammals that live in or near the sea face special osmoregulatory challenges. Even if their kidneys can produce urine more concentrated than blood plasma, they are incapable of eliminating all the salt taken in with food and water. Most marine reptiles and birds therefore depend on **salt glands** that secrete a highly concentrated solution. Salt glands have probably evolved independently in various groups of marine reptiles. In the marine iguana *Amblyrhynchus cristatus* of the Galápagos Islands the salt gland empties through the nostril. In sea turtles the salt gland is located in the orbits of the eye, and tears carry away the hyperosmotic secretion. In sea snakes the salt gland is located beneath the tongue. Saltwater crocodiles such as *Crocodylus acutus* (see Fig. 37.6A), which lives around the tip of Florida, have salt glands in their tongues. Almost any bird fed salty food will develop enlarged nasal glands above the eyes that can serve as salt glands.

Salt Intake

All animals require an adequate intake of salt, especially NaCl. For marine animals this obviously presents no problem. Most terrestrial and freshwater carnivores also get sufficient salt, because their prey are likely to have optimal ion concentrations in their tissues. Herbivores inhabiting seashores or former sea floors also get sufficient salts from eating plants that have absorbed the ions. Many freshwater and terrestrial herbivores, however, have to supplement the dietary intake of ions. Many freshwater animals obtain ions by actively absorbing them from water through the gills, intestine, and skin. Terrestrial herbivores generally have to find deposits of salt, especially NaCl. These animals usually have a tremendous hunger for salt, and any source is a valuable commodity. Porcupines will consume leather gloves, tool handles, and even the seats of outdoor toilets to get the salt left by human sweat. Many groundhogs risk being hit by cars to eat salt that has been spread on highways to melt snow.

In human cultures in which meat, milk, or blood are the main foods, supplementary salt is seldom needed. In other cultures, however, salt is as essential to people as to most other terrestrial animals. In the United States, for example, the ability to eat just one salted peanut is considered proof of superior will power. Roman soldiers gladly accepted salt as payment for their services. (The word "salary" is derived from the Latin word *salarius,* meaning "of salt.") Ancient rulers waged wars for control of salt deposits, and Jesus praised his disciples as "the salt of the earth." Now that salt is plentiful, many of us may be getting too much of a good thing. The average American consumes 4 to 6 grams of NaCl per day, which is several times the recommended intake. Increased sodium concentration in body fluid is believed to increase blood pressure in many people. Since hypertension leads to heart failure and other circulatory diseases, the salt-hunger for these individuals may be as risky as it is for groundhogs on the side of the road.

Summary

The concentrations of several ions are crucial for the proper functioning of cells. In addition, the total concentration of ions and other molecules in the body fluids has a profound osmotic effect on cells. Some marine animals, which live in environments where ion concentrations are constantly suitable, do not regulate concentrations in their body fluids: They are osmoconformers. Other animals, especially marine vertebrates and freshwater and terrestrial animals, must be osmoregulators, maintaining internal concentrations different from external concentrations. Osmoregulation utilizes at least one of the following two operations: (1) active transport of ions, possibly coupled to osmotic transport of water, and (2) changing the permeability to water or ions. These operations are generally carried out in two stages: (1) filtration of blood or other body fluids and (2) reabsorption of needed substances back into the blood and secretion of excess substances out of the blood into the urine. A variety of organs carry out these operations. Freshwater protozoans and sponges use contractile vacuoles. Many invertebrates have nephridia. Some crustaceans have antennal glands, and insects have Malpighian tubules. All vertebrates have kidneys.

In mammals the nephrons in the kidneys form a filtrate of blood in the glomerulus and surrounding Bowman's capsule, and the tubule carries out reabsorption and secretion. The filtrate enters the proximal convoluted tubule, where most of the water and almost all the glucose and amino acids are reabsorbed. It then passes through the loop of Henle to the distal convoluted tubule, where reabsorption and secretion of ions occur. The pre-urine then enters the collecting duct. Active transport of Na^+ and Cl^- in the loop of Henle sets up a concentration gradient that draws urea out of the collecting duct. The resulting concentration gradient of urea and ions draws most of the water out of the pre-urine, forming concentrated urine. Active transport and permeability are adjusted by several hormones. The structure of the kidney varies in different groups of vertebrates. In addition, various vertebrates have special adaptations, such as salt glands, for living in marine and desert habitats.

Key Terms

osmolarity
osmoconformer
osmoregulator
contractile vacuole
filtration
secretion
reabsorption
nitrogenous waste
urea
nephridium
antennal gland
Malpighian tubule
nephron
glomerulus
Bowman's capsule
proximal convoluted tubule
distal convoluted tubule
loop of Henle
collecting duct
salt gland

Chapter Test

1. Define osmolarity. Why is it important? (p. 275)
2. How does the osmolarity of body fluids generally compare with the osmolarity of the environment for the following groups of animals: marine invertebrates; marine vertebrates; freshwater vertebrates and invertebrates? (Use the terms "iso-osmotic," "hypo-osmotic," and "hyperosmotic" in your answers.) What problems in osmoregulation must animals in each group solve? (pp. 275–276)
3. Graph the internal osmolarity versus external osmolarity for an osmoconformer, a stenohaline osmoregulator, and a euryhaline osmoregulator. In labeling the axes of your graph, what units of measure can you use? (pp. 276–278)
4. Describe the following types of osmoregulatory organs: contractile vacuoles, nephridial organs, antennal glands, Malpighian tubules, kidneys. Which kinds of animals have each type of organ? (pp. 278–283)
5. Sharks, marine bony fishes, sea snakes, sea birds, and whales have body fluids with lower ionic concentrations than those in seawater. How do members of each of these groups maintain this state? (pp. 286–288)
6. Diagram a nephron and label its parts. Indicate where filtration occurs. What molecules are reabsorbed, and where? Which ions are secreted, and where? (pp. 283–284)
7. Give a general explanation for how the loops of Henle help the kangaroo rat survive without drinking water. (pp. 284–285)
8. List four hormones involved in regulating kidney function. Briefly state the function of each hormone. (pp. 285–286)

■ Answers to the Figure Questions

14.2 Human blood has an osmolarity similar to that of interstitial fluids, since they are formed as a filtrate of blood. Thus blood freezes at about between –0.5° C and –0.6° C.

14.6 Any freshwater animal would require the osmoregulatory organs to secrete excess water accumulated osmotically by the body.

14.7 The metanephridia of an earthworm would secrete water when the soil was more moist than the body fluids and would reabsorb water when the soil was less moist than body fluids.

14.10 Probably the medullary portion of the collecting duct, since fluids in the nephrons are dilute elsewhere in the kidney.

Readings

Recommended Readings

Beauchamp, G. K. 1987. The human preference for excess salt. *Am. Sci.* 75:27–33.

Cantin, M., and J. Genest. 1986. The heart as an endocrine gland. *Sci. Am.* 254(2):76–80 (Feb).

Crowe, J. H., and A. F. Cooper, Jr. 1971. Cryptobiosis. *Sci. Am.* 225(6):30–36 (Dec).

Dantzler, W. H. 1982. Renal adaptations of desert vertebrates. *BioScience.* 32:108–113.

Fertig, D. S., and V. W. Edmonds. 1969. The physiology of the house mouse. *Sci. Am.* 221(4):103–110 (Oct). (*The mouse's ability to conserve water is one key to its success.*)

Schmidt-Nielsen, K. 1981. Countercurrent systems in animals. *Sci. Am.* 244(5):118–128 (May). (*Among other functions, these help conserve water in the kidneys and in the noses of many mammals.*)

Additional Reference

You, G., et al. 1993. Cloning and characterization of the vasopressin-regulated urea transporter. *Nature* 365:844–847.

CHAPTER

15

Temperature

Japanese snow monkeys, *Macaca fuscata,* in hot spring.

__Heat__ is energy due to molecular motion. __Temperature__ is a measure of the average of this energy per molecule. Heat is commonly measured in calories or kilocalories (cal or kcal), and temperature is measured in degrees. It takes 1 calorie of heat to increase the temperature of 1 gram of water by 1 degree Celsius. Since animals consist mainly of water, it also takes about 1 calorie to warm 1 gram of tissue by 1 degree. Thus for a person who weighs 50 kg it takes 50,000 cal (50 kcal) to raise the body temperature by 1°C, from the normal of 37°C to 38°C. This heat can be absorbed from the environment or produced as a byproduct of metabolism.

Why Temperature Matters

Few people would doubt that the subjects of the previous three chapters—oxygen, nutrients, water, and ions—are necessities that animals actively seek out and regulate within themselves. It may not be as obvious to people that suitable temperature is also a necessity, but many animals behave as if they know it (Fig. 15.1). Many animals seek favorable environmental temperatures and have elaborate mechanisms for regulating the temperatures of their internal environments. This chapter describes why temperature is so important and the ways in which animals adapt to temperature.

The temperature range for animal life extends from below –50°C (–58°F) for some Arctic insects up to 50°C (122°F) for a certain fish (*Barbus thermalis*) in hot springs of Sri Lanka. This upper limit is well above the temperature of 45°C that is painful to humans. Most animals live within a narrower range of environmental temperatures, usually between 0°C and 35°C. Life is restricted to particular environmental temperatures because body temperature tends to become like the environmental temperature, and body temperature has a profound effect on the rates of all physiological processes. The reason is that almost all physiological processes depend on chemical reactions, and body temperature provides the **activation energy** for these reactions (p. 52). An increase of 10°C in body temperature typically doubles or triples the rate of a biochemical process (Fig. 15.2). A change in body temperature can therefore seriously affect the rates of activities. Low body temperature can kill an animal by reducing its rates of activity, even if the temperature is above freezing. High temperatures can be just as lethal by destroying enzymes and other proteins, even if the temperatures are far below the temperatures for boiling or combustion.

Ectotherms and Endotherms

Small animals, especially aquatic ones, have body temperatures that are essentially the same as their environmental temperatures. The reason is that a small animal has a large heat-conducting surface relative to its body mass, and water conducts heat readily.

FIGURE 15.1
Painted turtles *Chrysemys picta* crowd onto a log to bask in the sun after spending a winter buried beneath a frozen pond.

■ **Question:** What is the advantage of raising the body temperature by basking? What are the potential disadvantages for the turtles?

Such animals, in which the body temperature is determined primarily by the environment rather than by internally produced heat, are referred to as ectotherms (Greek *ecto-* outside). Ectotherms are generally also **poikilotherms,** meaning that their body temperatures are variable (Greek *poikilo-* varied). This need not be the case, however. Ectotherms that live in the constant temperature of a cave will have a constant body temperature, and lizards, though ectothermic, can maintain a constant body temperature by moving to cooler or warmer places.

Many animals, especially large, terrestrial ones, have body temperatures that depend more on their own heat production than on environmental temperature. Such animals are called endotherms (Greek *endo-* within). Endothermy enables some animals, primarily birds and mammals, to maintain fairly constant body temperatures in spite of varying environmental temperatures. These animals are therefore homeotherms (Greek *homeo-* same). Not all endotherms are homeotherms, however. For example, tunas, mackerels, swordfish, marlins, mackerel sharks, and thresher sharks maintain body temperatures above environmental temperatures by conserving body heat, but the body temperature is not constant.

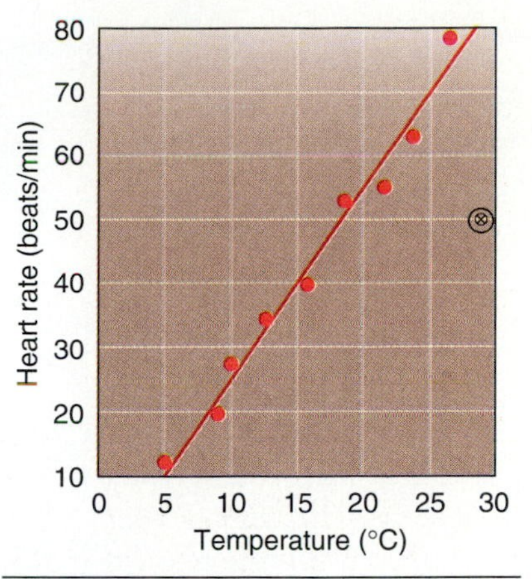

FIGURE 15.2

The rate of contraction of the isolated heart of a grass frog *Rana pipiens* in Ringer's solution at various temperatures. Irreversible damage, indicated by the fall in heart rate (the circled X), occurs at a little above 27°C. Physiologists express the dependence of a physiological rate on body temperature by the Q_{10}, which is the rate of a process at one temperature, divided by the rate at a temperature 10°C lower. In this example, the Q_{10} over the temperature range 10°C to 20°C was approximately 52/25 = 2.08. (Data from Harris 1976.)

■ **Question:** Purely physical processes such as diffusion vary in proportion to absolute temperature, which equals 273 + the temperature in °C. What would be the Q_{10} for a purely physical process?

Adaptations of Poikilotherms to Severe Cold

Torpor and Diapause. When any animal's body temperature drops so low that physiological activities slow down drastically, the animal is said to be in a state of **torpor.** An animal in torpor may appear to be dead, but its normal activities return as soon as the body is warmed. During spring and fall an ectotherm may become torpid each night, then recover during the day. Torpor can also last throughout the winter, as with amphibians, reptiles, and some insects. Torpor is simply a natural response to the reduction in enzyme activities, and it requires no special adaptation. The advantage of seasonal torpor is that it allows an ectotherm to survive winters without feeding. A disadvantage of torpor is that a sudden rise in temperature will immediately arouse the animal, perhaps at a time when there is no food available. This hazard is avoided by an adaptation called **diapause,** which occurs in some ectotherms, such as the house fly *Musca domestica.* Diapause is like torpor except that some physiological mechanism prevents arousal during brief warm periods. Torpor and diapause are often called "hibernation." It is advisable, however, to reserve the term hibernation for a quite different process that occurs only in a few birds and mammals, as will be described later.

Protection Against Freezing. While torpor and diapause allow animals to remain alive at low temperatures, they will not protect an animal from destruction of body tissues by ice crystals. One common protection in ectotherms exposed to cold is lowering the freezing point by increasing the concentration of solute in body fluids. For freshwater animals, such as fishes and frogs, the normal concentrations of body fluids are sufficient to prevent freezing as long as they are in liquid water, since the water freezes at a higher temperature than the body fluids. Some animals that are exposed to freezing temperatures, especially terrestrial ectotherms, lower their freezing points by increasing the concentrations of certain molecules, such as glucose and glycerol.

Besides lowering the freezing point by increasing concentration, glycerol is also an **antifreeze.** Antifreezes lower the freezing point by changing the arrangement of water molecules. Glycerol is a major antifreeze in insects and some frogs (and at one time in automobile radiators). The larva of the wasp *Bracon* has a 5 molar concentration of glycerol that enables it to survive –45°C winters in Canada and Siberia. Another type of antifreeze in insects and in certain spiders and mites is **thermal hysteresis protein** (THP). THPs are somewhat mysterious in that they lower the freezing point without affecting the temperature at which body fluid melts once it is frozen. Some Antarctic fishes, such as *Trematomus*, produce **antifreeze proteins** (AFPs) that bind to ice crystals and prevent their growth.

Some insects, deep-water Antarctic fishes, frogs, and turtles actually live at temperatures below their freezing points in a state of **supercooling.** Supercooled fluids are below the freezing temperature, but they do not crystallize because there is no site—or **nucleator**—on which crystals can get started. One disadvantage of supercooling is that a supercooled animal that contacts a nucleator freezes instantly. A school of Antarctic fish becomes flash-frozen if one of them bumps into a piece of ice.

Some ectotherms actually survive with up to 65% of their body water frozen. So far the animals known to survive partial freezing include barnacles and molluscs exposed at low tide, some frogs, garter snakes (*Thamnophis sirtalis*), painted turtle hatchlings (*Chrysemys picta*), and some insect larvae. Instead of avoiding freezing, these animals actually produce nucleators that encourage the freezing of water in extracellular spaces. At the same time, however, they increase the levels of antifreeze compounds that keep the ice crystals small. As the extracellular water freezes, the concentration of ions in the liquid water naturally increases, and this increased concentration draws water out of the cells osmotically. The intracellular concentration therefore increases, lowering the freezing point. This reduced freezing point protects the cells from damage due to freezing, but the cells would be damaged from shrinkage if it were not for certain **cryoprotectant** molecules, such as the disaccharide trehalose and the amino acid proline.

Acclimation of Ectotherms to Low Temperature

Acclimation is a physiological adjustment to an environmental change that occurs over a period of days or longer. Acclimation to low temperature is quite common in ectotherms that live more than a few months in temperate regions (Fig. 15.3). The effect of temperature on the rates of biochemical reactions can be compensated during acclimation by changing enzymes that catalyze the reactions. One approach is simply to increase the amount of enzyme in the cold and decrease it when the temperature is high. A second approach is to change the amount of thermal activation energy required, by producing a different form of an enzyme. The different enzyme, called an **isoenzyme** (= isozyme), requires different genes to direct its synthesis. One example of this method of acclimation occurs in the brains of trout and involves acetylcholinesterase, the enzyme that breaks down the synaptic transmitter acetylcholine. In the winter and spring, trout brains produce an isoenzyme of acetylcholinesterase that works best at low temperature. In summer and autumn the gene for that isoenzyme is turned off, and another gene produces an isoenzyme of acetylcholinesterase that works best and resists denaturation at higher temperatures.

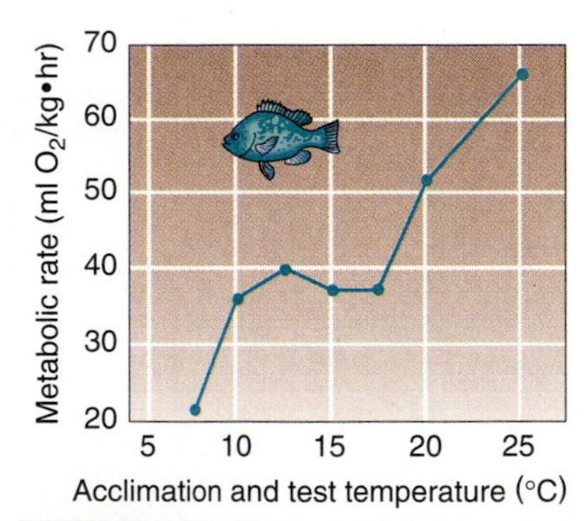

FIGURE 15.3
Compensation to environmental temperature by the common pumpkinseed sunfish *Lepomis gibbosus.* After allowing the fish to acclimate to each of the temperatures indicated by a datum point, its metabolic rate (rate of oxygen consumption) was measured at that temperature. Over the range of 10°C to 18°C the fish had about the same metabolic rate. (Data by J. L. Roberts, adapted from Cossins and Bowler 1987.)

Compensation for Environmental Temperature by Changing Body Temperature

BEHAVIORAL MECHANISMS. Although poikilotherms cannot maintain a constant body temperature, they often have behavioral mechanisms that enable them to control it. Being animals, they can seek out more favorable **microclimates**: areas where the local temperature differs from the general air or water temperature. Numerous studies on flies, fishes, and other animals show that many select the temperature at which their growth and reproduction are best. Another behavioral mechanism, **basking** in the infrared radiation from the sun, can elevate the body temperature above the air temperature (Fig. 15.1). Insects and reptiles can often be found on cool, sunny mornings basking with their bodies oriented so as to absorb the most solar radiation.

PARTIAL ENDOTHERMY. Many animals that are usually ectothermic can temporarily compensate for low environmental temperature by producing and conserving more body heat, which raises the body temperature. This phenomenon is called partial endothermy. The rise in body temperature can be due to basking, as mentioned above, or to **thermogenesis.** Thermogenesis is any kind of physiological activity whose major effect is the production of heat. Partial endotherms generally also have adaptations for conserving the heat.

Bernd Heinrich of the University of Vermont has studied many examples of partial endothermy among insects. His best-known study is on bumble bees *Bombus* spp. Heinrich grabbed bumble bees while they were foraging for nectar and inserted tiny thermocouples to measure the temperatures in various parts of their bodies. He found that even when the air temperature was only 5°C, the temperature of the flight muscles of these "cold-blooded" animals was at least 36°C, and sometimes as high as 45°C—equal to or above that of many "warm-blooded" animals. At the same time, the abdomen of the bumble bee was often 20°C cooler than the thorax. Heinrich and the late Ann E. Kammer found that bumblebees first shiver to warm up their flight muscles to the minimum temperature required for hovering, and then they maintain that temperature by conserving heat while flying. Bumble bees conserve heat from the flight muscles with their fuzzy **pile** and by countercurrent heat exchange (p. 245), which prevents the heat from circulating into the abdomen along with the blood (Fig. 15.4). This allows bumble bees to keep their flight muscles at working temperatures as they forage among widely separated flowers. If flowers are close together on a plant, however, bumble bees allow the flight muscles to cool, and they simply walk from flower to flower.

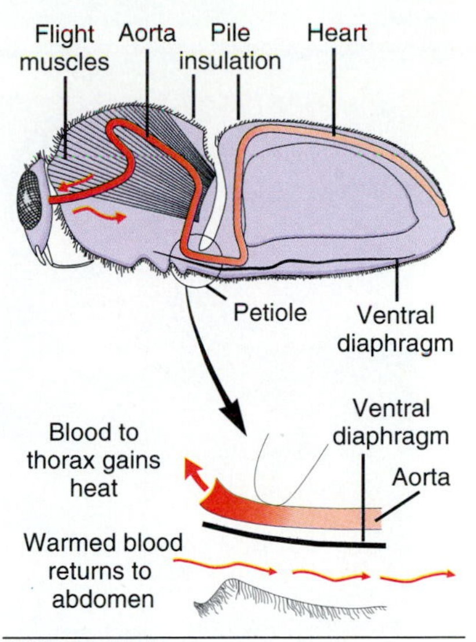

FIGURE 15.4
Heat conservation in the bumble bee *Bombus* sp. As cool blood passes from the abdomen anteriorly through the aorta, it picks up heat (represented by darker red) from the flight muscles. The ventral diaphragm pushes the warmed blood back into the abdomen through the narrow "waist" (petiole). At the same time it shuts off flow of the cooler blood from the heart, thereby reducing exchange of heat from hot to cooler blood.

Thermogenesis

Partial and complete endotherms raise their body temperatures above environmental temperatures by thermogenesis. Usually thermogenesis is divided into two categories: shivering and nonshivering thermogenesis. **Shivering** is a contraction of muscles that converts the energy from ATP into heat rather than useful mechanical work. Shivering and increased muscle tension are important mechanisms of thermogenesis during the first minutes of exposure to cold.

Nonshivering thermogenesis refers to metabolic processes with low efficiencies in which most of the released energy goes into heat. In mammals, nonshivering thermogenesis occurs mainly in **brown adipose tissue** (= brown fat), which takes its name and color from the large amount of cytochromes in the mitochondria. Brown adipose tissue constitutes up to 6% of a mammal's body weight, especially in infants, small mammals, and mammals acclimated to cold. Brown adipose tissue is unusual in having mitochondria that can be made to convert energy into heat rather than ATP.

Control of Heat Exchange by Endotherms

Endothermy implies that the body temperature can differ from the environmental temperature. That, in turn, implies that the endotherm can control the tendency of heat to be lost from the body in the cold or to be gained from a warmer environment. Physicists know of only four ways by which heat is exchanged—conduction, convection, radiation, and vaporization—so these are the only processes by which an endotherm can control heat exchange.

Bernd Heinrich

Bernd Heinrich
Born: 19 April 1940, in Germany
Education: University of Maine, B.A., M.S.: UCLA, Ph.D. (Zoology)
Professor of Zoology, University of Vermont. Ultramarathon runner: United States record for distance in 24 hours (157 miles); United States record for 100 km on road; 100 mile record on track (12 hr 27 min).
Author of: Bumblebee Economics, *a popular account of thermoregulation and energetics in bumblebees;* In a Patch of Fireweed, *an autobiographical account of his life and research;* One Man's Owl; Ravens in Winter; The Hot-Blooded Insects; A Year in the Maine Woods; *and numerous research articles.*

Excerpts from an interview a few days after Heinrich won the Lake Waramaug Ultramarathon (100 km) and a few days before he set the 100-mile track record:

You start your book In a Patch of Fireweed *with your early childhood in postwar Germany, in the first few years of your life, when your main concern seems to have been foraging [in the Hahnheide Forest] to get enough to eat. I wonder if that affected your later research.... I don't want to get too psychoanalytical....*

It sounds that way, but no, because I went into biology at first just being interested in all kinds of mechanisms of living things. I got really interested in cell biology at University of Maine. I was also very much interested in field biology, but at that time all the big advances were made in molecular biology, so I thought the really significant stuff was going to be in molecular biology. Eventually as I got more secure, I suppose, I decided that, "Well, I'll do what's more fun for me."

So you switched from cell biology into field studies for the fun of it. You use that word in your book over and over again: "fun."

Yes. I just enjoy it.

The switch from molecular biology to field biology was quite a decision for a young graduate student to make.

Oh it was, it was. A really tough decision, and I agonized over it for a long, long time. I said, "Well, I've got a long life ahead, and if I'm going to be 30, 40, 50 years, or whatever on a lab bench, and if I don't really enjoy it, I'm not going to make any progress." I got much satisfaction out of it [cell biology]. I was really enthusiastic. I got several papers. There was intellectual stimulation, but I was getting results very slowly, and I felt there was an awful lot of competition, and you have to have access to tremendous resources when you get into cellular biology. You're tied to huge machines that cost thousands and thousands of dollars, so you're tied into getting money. You're tied into being at a place where you can get that, so I felt that I'd lose a lot of freedom. I liked the intellectual challenge, but there were too many costs associated with it.

It seems to me that your running is almost as great a passion as your science. I mean who would run 157 miles in 24 hours? Is there any connection between that and your science? Do you ever get any insights from running that you couldn't have gotten from reading and from experiments on animals?

I doubt it.

That's a real disappointment. I thought you were going to say something about bioenergetics. But maybe something your running and your research do have in common is a high threshold for pain.

That is true. I talk about its being fun all the time, but that might be a little bit misleading, because it's in fun that you exert yourself the most and are willing to suffer the most. But you get satisfaction back from it. So when I look back on the 24-hour run, I had to suffer pretty hard for 24

box continues

hours, and actually for a couple of months beforehand getting in training. But I still look at it in terms of fun, because now for years later I'll be having that satisfaction. And the same with the work with the bees. I might have to really extend myself at times and really go through a lot of pain and effort. It's sort of like a long-distance run to go through a research project. You have to work on it, because if you slow down and stop, you essentially have destroyed the project. But then afterwards you get all the benefits.

There are some aspects of it that are more painful than others I imagine. What parts of research do you find the least fun?

Well, for example, right now I was doing some stuff on heat transfer in carpenter bees, and it's not much fun to have to get this animal set up with these little thermocouples—and you have to have a whole bunch of them inside—they are extremely fine things like hairs. They have to be in exactly the right place, and they pull out if you look at them sideways. It's frustrating to get them in just the right place and to get it set up when the animal with its legs flailing is destroying it all the time. You have to do it over and over again. It really takes a lot of persistence sometimes.

There seems to be a growing interest in the kinds of studies you're doing. A sort of . . . I don't know whether to call it a counterrevolution. Maybe that's just my impression, but do you sense that people are coming back to the classical biology studies of the kind you're doing?

Well, there seems to be an awful lot of work that way. It seems like in the 50s and 60s there was sort of a downgrading of this type of biology, but I think in the meantime studies of behavior, ecology, et cetera have matured more and have gained a lot more interest. If you look at the journals, the outpouring of work is just phenomenal since then, so I think there is a lot of work being done.

I'm wondering how much of your present research and things like the Bumblebee Economics *book depend on your formal education, and how much depended on your self-education. Is there any way you could divide that up? If you had to go back to college again would you have taken the same courses and curriculum?*

Well, I think it's always pretty hard to say exactly what you're going to need and precisely how a course you might take is going to feed into what you're going to do. I'm sure that a broad range of biology courses really helped me a lot in getting a feel for what the different types of things were in biology. In that way I could choose better and relate better by having the whole picture. So it's a matter of giving me more confidence in doing what I'm doing. Basically most of it is probably learned on my own, but if I wasn't sure about what I had to learn, then I might spend most of my time reading. Certainly there are many potential places for input from courses directly to, let's say, *Bumblebee Economics.* But in that particular case I think it would mostly be in my graduate education, in actually working in the lab and actually seeing what other graduate students were doing, and in going out in the field and observing for myself and taking the measurements.

At one point in In a Patch of Fireweed *you describe a feeling of being trained to fill a niche in the job market, and your dissatisfaction with that.*

Yeah, that's right. I went to college essentially knowing I wanted to be a farmer or cabinet-maker or something practical. I wanted to work with my hands and wanted to be outside and be my own boss. I had absolutely no concept at all that education was correlated with a job. I thought you went to college to learn something, simply for the sake of knowledge and for no other reason whatsoever. That it's a means to earning a living never even occurred to me.

I imagine there are quite a few students who will read your book on Bumblebee Economics *and kind of resent the fact that you've taken the problem from them. That problem is no longer available for them to work on. Do you ever wonder yourself whether you're going to run out of projects—use them all up?*

I sometimes think about it. When I first finished the study with the moth thermoregulation [for the Ph.D.] I said, "Well, here now I've found *the* solution to how insects thermoregulate—at least the ones which are important. I've skimmed the cream." I said, "Now there's no place to go but down, and there's just little details to pick up." And then I'd pick up bumblebees, and I'd see something entirely new that I didn't even imagine, and I'd work on them a little bit more, and I'd see something else that I didn't imagine. I keep thinking I'm going to run out, but in actuality I just get overwhelmed with new problems.

Do you deliberately go out looking for projects?

No, I don't do it deliberately, ever. I like being outdoors, so I might spend a day out in the woods just going out looking at birds and watching things. Just being outdoors. It's like . . . if you really like a woman, then you see a lot of beautiful things.

CONDUCTION. Conduction is the transfer of thermal agitation of molecules in one body to the molecules in another body with which it is in contact. Heat loss by conduction is proportional to the surface area of the animal (S), the **thermal conductivity** of the surface (C), and the difference between the core body temperature (T_b) and the ambient (environmental) temperature (T_a): $SC(T_b - T_a)$. Since S, C, T_b, and T_a are the only variables in this equation that the animal controls, endotherms can control conduction only by changing the exposed surface area, changing the conductivity of the surface, changing the body temperature, or seeking an area with a different temperature. Some animals change the core body temperature T_b, as will be described later, and most do seek out more favorable environmental temperatures T_a.

Animals also control conduction by changing the exposed surface area S and the conductivity C. One of the most familiar ways of changing S is folding your arms when it is cool, thereby preventing heat loss from the inner surface of the arms and part of the torso. Dogs stretch out to increase their exposed surface area when they are warm. The importance of surface area to homeothermy is suggested by the large size of birds and mammals compared with the size of most poikilotherms. Even the smallest mammal, Savi's pygmy shrew (*Suncus etruscus*) of the Mediterranean area, has a mass of 2 grams, which is much larger than the mass of most poikilotherms. The smallest bird, the bee hummingbird (*Mellisuga helenae*) of the Caribbean, has a mass of 3 grams. Because homeotherms are larger, they have less surface area per gram of tissue through which heat can conduct (and also radiate).

Conductivity C in bumble bees and some other partial endotherms is minimized by the hairlike pile that traps warmed air. In terrestrial homeotherms, hair and feathers serve a similar function. Figure 15.5 shows the importance of hair in thermoregulation by mice. The effectiveness of hair can be increased by **piloerection:** the elevation of hair due to contractions of smooth muscles (the arrector pili muscles in Fig. 10.1B). Birds have a similar mechanism to fluff the feathers. Humans have lost most of their fur, so piloerection now produces only ineffectual "goose bumps." Many mammals and birds also have a layer of low-conductance fat beneath the skin. Seals, whales, and other marine mammals have the most impressive layers of such **subcutaneous fat** in the form of blubber.

CONVECTION. Convection is like conduction except that heat is conducting to a moving fluid that carries off the heat. As air or water is warmed by contact with an animal, it becomes less dense and rises, carrying heat with it. This warm air or water is

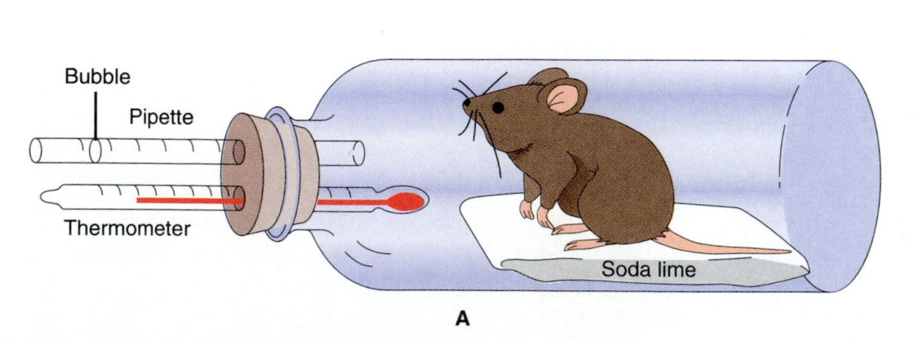

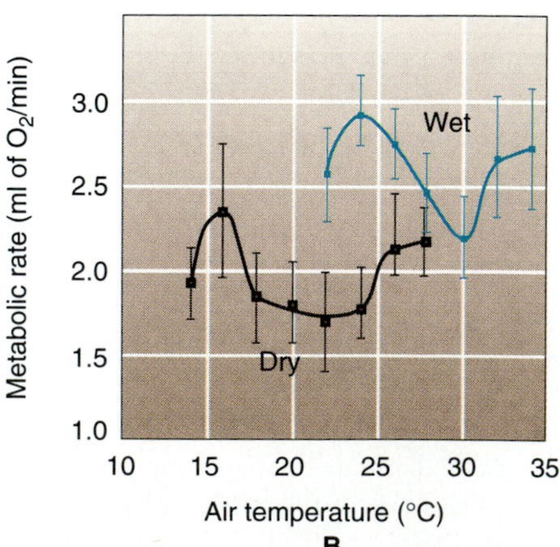

FIGURE 15.5

The metabolic rates of mice at various temperatures, before and after wetting the fur. (A) A laboratory mouse is placed in a small bottle with a packet of soda lime to absorb CO_2 and water vapor. Any change in the volume of air in the bottle results from the consumption of oxygen. Movement of a bubble in a graduated pipette indicates the rate of O_2 consumption. A thermometer measures air temperature in the bottle, which can be varied by covering the bottle with ice. (B) Metabolic rate. The vertical bars show the variation (standard error of the mean). Wetting the fur of the mouse shifts the curve upward and to the right, showing that higher metabolic rates occur at correspondingly higher environmental temperatures because of the loss of insulation and because of vaporization of the water.

then replaced by cooler air or water, which absorbs more heat from the animal. The resulting currents are called **natural convection.** There can also be **forced convection** due to wind and water currents. The rate of heat loss depends on the surface area and on the air or water speed. Changing the position of the body can increase or decrease the speed of natural or forced convection currents, thereby changing the rate of heat loss. Pile, hair, and feathers that slow the movement of convection currents can also aid endotherms in adjusting heat loss by convection.

RADIATION. Radiation is the loss of thermal energy by the emission of electromagnetic radiation in the infrared range. As with conduction and convection, the rate of heat loss by radiation is proportional to the surface area. If the difference between the effective environmental temperature (T_a) and the surface temperature of the animal (T_s) is small, the rate of heat loss by radiation is also roughly proportional to the difference between T_a and T_s. The rate of radiant heat loss is therefore proportional to $S(T_s - T_a)$. This means that an animal can control the rate of heat loss by changing its exposed surface area as described above, by changing its skin temperature (which is not necessarily the same as core body temperature T_b), or by seeking an area with a different effective temperature T_a.

T_a is not necessarily the temperature you would measure with a thermometer. The air temperature on a clear night may be above freezing, but an animal will be radiating heat to deep space, which has an effective temperature far below freezing. On a clear night an animal is therefore much better off under a tree, rock, or other kind of shelter that will reflect infrared. On a clear day, of course, animals can gain radiant heat by basking in the infrared radiation from the sun. Seeking shelter and basking can therefore be viewed as two behavioral means of controlling radiation by changing T_a.

Some partial endotherms also reduce heat loss due to radiation by reducing the surface temperature T_s. Tuna and mackerel sharks, for example, reduce surface temperature by means of a **rete mirabile** ("wonderful net") of intertwined veins and arteries that serves as a countercurrent exchange mechanism for heat (p. 245; Fig. 15.6 on page 300). Homeotherms also alter the skin temperature, by vasodilation and vasoconstriction. **Vasoconstriction**—the constriction of arteries—reduces the flow of blood to the skin, so that in an environment with low T_a the skin temperature will be less than the core body temperature. **Vasodilation** increases the flow of blood—and therefore heat—to the skin, so that more heat is radiated from the body. Vasodilation and vasoconstriction are easily seen in the pink or pale color of the skin.

Homeotherms also use vasoconstriction and vasodilation to regulate countercurrent exchange of heat from venous blood that has been warmed by muscle activity to cooler arterial blood (Fig. 15.7 on page 301). Because of vasoconstriction and countercurrent heat exchange, the feet, noses, and other extremities of sled dogs, ducks, and other animals exposed to cold for long periods often have temperatures barely above freezing. One reason the cells are able to function at such temperatures is that lipids in their cell membranes remain oily at temperatures that would congeal the lipids in other cells.

VAPORIZATION. Vaporization removes heat that is needed to convert water from a fluid to a gaseous phase. The rate of heat loss by vaporization is directly proportional to the amount of water lost, which depends on how fast the animal can evaporate water. In general, birds and hairy mammals evaporate water from the lungs, mouth, and nasal passages by **panting.** Many hairy mammals, such as mice, also lick themselves, which vaporizes water and also reduces the insulation of fur. Mammals with sparse

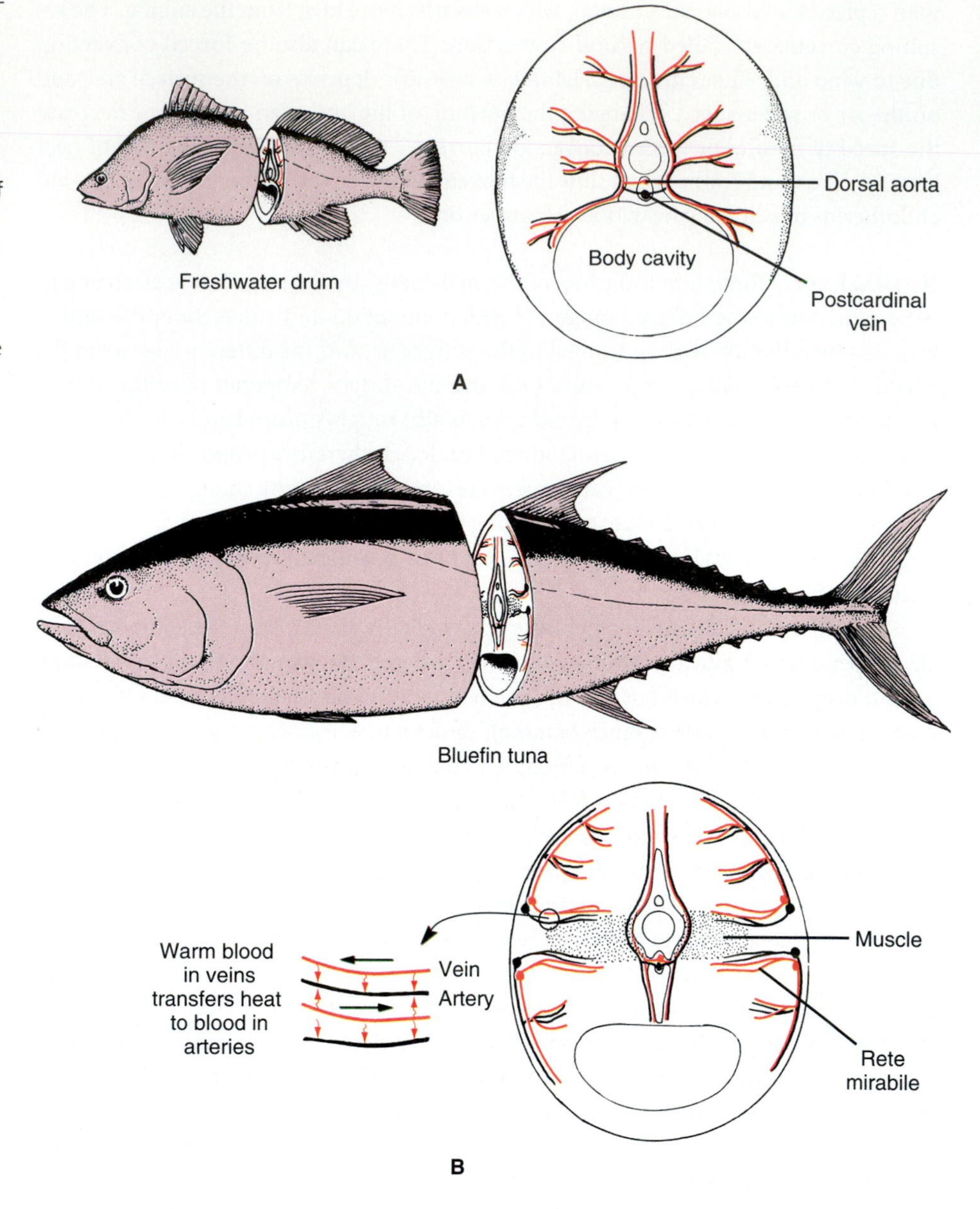

FIGURE 15.6
(A) In most fishes, such as the freshwater drum *Aplodinotus grunniens,* the major arteries (color) and veins radiate directly to the body surface, where much heat is lost to the water. (B) In bluefin *Thunnus thynnus* and other tunas a rete mirabile ("wonderful net") of intertwined veins and arteries recycles much of the heat from venous blood to arterial blood by countercurrent exchange.

■ **Question:** Would a rete mirabile be effective in a fish much smaller than tuna?

hair vaporize water by **sweating** (perspiring). A human can sweat as much as 10 liters per day. Even at room temperature (20°C) there is some **insensible perspiration.**

The more water vapor already in the air, the more difficult it is to vaporize more. Humans who have adapted to tropical climates where the relative humidity is high do not waste water in futile sweating. It would also be useless for aquatic animals to sweat, since the water obviously cannot evaporate. In relatively dry climates where there is plenty of drinking water, however, vaporization is of major importance in endothermy. Of all the channels for heat loss, vaporization is the only one that can cool the body below the environmental temperature.

Homeothermy

Among living animals, only birds and mammals maintain body temperature within a few degrees of a set point for extended periods. Placental mammals usually have core body temperatures of around 37°C over a wide range of environmental temperature,

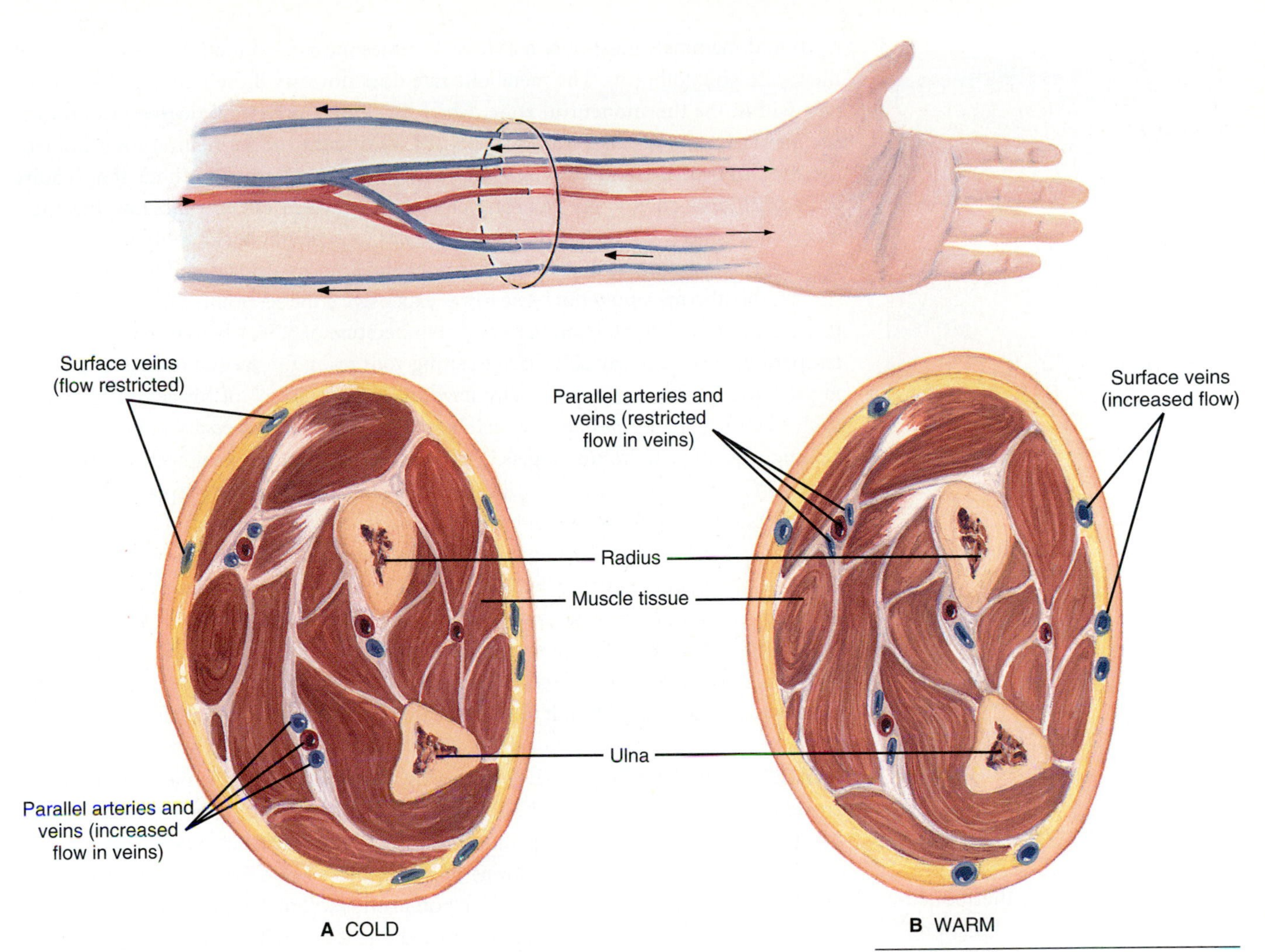

FIGURE 15.7

Countercurrent heat exchange in the human arm. (A) In cold weather the superficial veins of the arm constrict, reducing heat loss from the surface. The deep veins dilate, so heat from the venous blood recycles to the nearby arteries. (B) In warm temperatures the surface veins dilate and eliminate body heat, and the deep veins constrict.

and birds maintain a body temperature of about 41°C (Fig. 15.8A on page 302). Birds and mammals can therefore remain active at environmental temperatures in which they would freeze or enter torpor if they were poikilotherms, and they do not need to adjust enzyme amounts or produce alternative isoenzymes to compensate for changes in body temperature. Only in extreme cold or heat do the homeothermic mechanisms break down. At extremely low temperatures the animal is not able to produce or conserve enough heat to thermoregulate, and the body temperature drops (**hypothermia**). Hypothermia will cause enzyme activities to slow down, which will make it even more difficult to thermoregulate. Death is inevitable unless heat is generated by exercise or applied from an external source. At extremely high temperatures the heat produced as a byproduct of thermoregulation may actually exceed the amount of heat lost. When that happens the body temperature starts to rise (**hyperthermia**), the enzymes speed up their rates, and even more heat is produced. In a matter of seconds the body temperature can rise sufficiently to denature proteins, resulting in death.

The Cost of Homeothermy. Homeotherms pay dearly for the advantages of homeothermy. Even at the temperature to which it is acclimated, the metabolic rate of a homeotherm is many times higher than that for a poikilotherm of the same size. According to Robert Bakker (1975), "the total energy budget per year of a population of endothermic birds or mammals is from 10 to 30 times higher than the energy budget of an ectothermic population of the same size and adult body weight." This means that

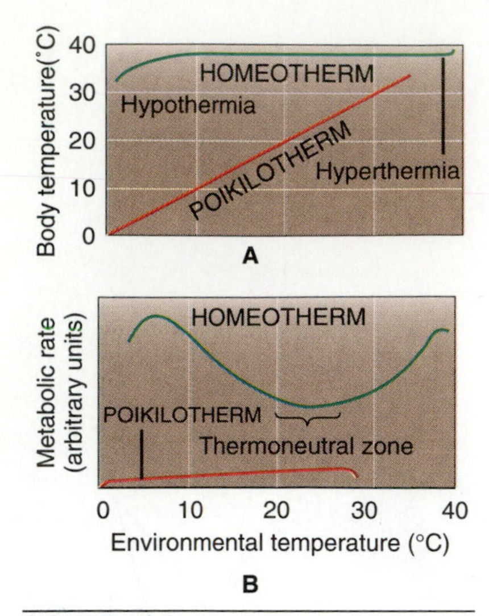

FIGURE 15.8

(A) The idealized body temperatures of a homeotherm over a range of environmental temperatures, compared with a poikilotherm, both acclimated to 22°C. (B) For the poikilotherm the metabolic rate increases continuously with body temperature until a few degrees above the acclimation temperature, where enzymes denature and the animal dies. The metabolic rate of the homeotherm varies with a U-shaped relationship to environmental temperatures over the range in which the body temperature can be regulated.

■ **Question:** Could the thermoneutral zone shift during acclimation? What physiological processes might account for such a shift?

birds and mammals must obtain at least 10 times more food than they would need if they were poikilotherms. The metabolic rate does not vary if the temperature rises or falls within the **thermoneutral zone,** which is a range of several degrees around the acclimation temperature (Figs. 15.5B, 15.8B). Within the thermoneutral zone, homeothermy can be maintained by changing body posture and other methods that require little additional energy. At environmental temperatures above or below the thermoneutral zone, however, the homeotherm has to expend considerably more energy on shivering, panting, and other heating or cooling mechanisms described earlier.

Homeothermy would not be nearly as expensive if the set points for body temperature were not so high. Maintaining a body temperature of 37°C when the environmental temperature averages, say 20°C, is like setting your room thermostat to 98°F and opening the windows to keep cool. Why have the "thermostats" of birds and mammals evolved to be so high? Perhaps the high body temperature is necessary to prevent overheating when the environment gets unusually warm. If our normal body temperature were 25°C, we would save many calories that would not have to be replaced by eating, but whenever the environmental temperature rose above 25°C it would be difficult to keep from overheating. The only way to avoid overheating would be vaporization of sweat, which might lead to dehydration. By keeping the body temperature near the maximum encountered in the environment, we reduce the threat of overheating and of dehydration. Another possible reason for high body temperature is that we cool much faster the hotter we are compared with the environment. During exercise, therefore, it is easier to eliminate excess heat if the body temperature is high to begin with.

HOMEOSTATIC CONTROL. If you destroy a thermostat, the furnace or air conditioner will respond erratically, if at all. If you apply ice or a lighted candle to a thermostat, the system will react as if the temperature of the entire room had changed. Similar effects occur if portions of the **hypothalamus** of a mammal's brain are damaged, cooled, or heated. Destroying one area of a dog's hypothalamus will make it unresponsive to heat: In a hot room the dog will not pant and would die of hyperthermia. Heating an area of the dog's hypothalamus will trigger panting even when the body and environmental temperatures are normal. Cooling a different area of the hypothalamus will trigger shivering and piloerection. These experiments suggest that homeothermy in mammals is coordinated in the hypothalamus. Apparently the hypothalamus compares actual temperature with what the body temperature should be, and it corrects any difference by triggering appropriate actions to regulate production and conservation of body heat.

The hypothalamus apparently gets information about actual temperature from two sets of **thermoreceptors.** Peripheral thermoreceptors in the skin provide an early warning of environmental temperature, and central thermoreceptors in the spinal cord and in the hypothalamus itself sense the core body temperature. Information from these thermoreceptors is compared with the **set point** (37°C), which is somehow programmed into the hypothalamus. The hypothalamus then triggers the appropriate behavior and responses of the autonomic nervous system, which controls piloerection, vasodilation, vasoconstriction, and sweating.

Resetting the Thermostat

HETEROTHERMY. The term *homeo*therm must be taken rather loosely, because no bird or mammal maintains exactly the same core body temperature throughout its life. Body temperature normally varies over a cycle of about 24 hours, even if environmental temperature is constant. Human body temperatures are generally a degree or so lower at 2 A.M. than at 2 P.M. (This is an example of a **circadian rhythm;** see Chapter

20.) In some species the daily variation in body temperature is even greater, and these animals are referred to as heterotherms. A camel (*Camelus*) that is deprived of water reduces panting and lets its daytime temperature rise to 41°C. At night the camel's hypothalamus lets body temperature drop to 35°C, thereby reducing conduction, convection, and the radiative heat loss to the clear desert sky. In other words, camels have automatic set-back thermostats like those found in many homes.

Some hummingbirds conserve energy by allowing the body temperature to drop to around 14°C each night. Many small mammals also let their body temperatures drop for several hours during cold weather. During these periods of reduced body temperature the animal is inactive, in a state of **daily torpor.** In addition to daily cycles of body temperature, some homeotherms also have seasonal variations. For example, squirrels and bears allow the body temperature to drop several degrees while they sleep through periods of severe cold. This **winter sleep** reduces the amount of heat lost by conduction and radiation.

HIBERNATION. Winter sleep, as well as the torpor and diapause of poikilotherms, is sometimes called hibernation. It would be best, however, to reserve the term for a quite different phenomenon that occurs only in one species of bird, the common poorwill (*Phalaenoptilus nuttallii*), and a few species of mammals, including bats, ground squirrels, and woodchucks. Hibernation differs from the torpor and diapause of poikilotherms in that the hibernating animal is still a homeotherm, but with a set point as low as 2°C. In the autumn, as if they knew what lay ahead, mammals that hibernate put on large amounts of fat, which will later act as insulation and nutrient reserves. They then den up and start making "test drops" of the body temperature as they become dormant. Eventually they enter a period of reduced body temperature that lasts from hours to weeks. Because of the reduced body temperature, enzyme activity is slowed, and the hibernator is dormant. Hibernators often arouse briefly during the winter, especially in response to disturbance or life-threatening cold. Arousal is perhaps the trickiest part of hibernation, because the body temperature must be raised using enzymes that are working at far below the normal body temperature. The rapid warming may be aided by thermogenesis in large amounts of brown adipose tissue that most hibernators have.

FEVER. Homeotherms and even some poikilotherms, such as lizards, can turn up the thermostat in response to bacterial infections. The resulting fever has been shown to inhibit reproduction in many disease-causing bacteria. Fever is triggered by proteins called **pyrogens** (Greek *pyro-* fire), especially interleukin-1 from macrophages (p. 231). Pyrogens apparently turn up the set point in the hypothalamus, making it "think" the body temperature is too low. The hypothalamus then triggers vasoconstriction, shivering, and heat-conserving behaviors until the body temperature equals the new set point. After the infection is over, the set point falls back to 37°C. The hypothalamus then senses that the body temperature is too high, so it triggers panting or sweating, vasodilation, and cooling behaviors until the body temperature returns to normal. People with infections often go through cycles of chill and fever, with the temperature rising and falling as if the hypothalamus had two roommates who could not agree on a setting for the thermostat.

Acclimation in Homeotherms

It is a familiar fact that one can get used to different seasons and climates. In New England the first 40°F day in autumn can seem unbearably cold, but in spring the same temperature can induce euphoria. In mammals a major mechanism for acclimation to cold is increased **thermogenesis.** Another major mechanism is **delayed responsiveness.** Animals acclimated to cold start shivering at lower temperatures than do nonacclimated animals, and those acclimated to heat begin sweating at higher temperatures. Acclimated animals also show **increased tolerance for discomfort.** Many mammals also grow extra hair and fat in the winter, but this is not an example of acclimation since it occurs even if they are kept warm.

In order to study the ability of humans to acclimate, P. F. Scholander (1958) somehow persuaded eight Norwegians to camp out for six weeks, sleeping at below-freezing temperatures in one-blanket sleeping bags. (Perhaps the "volunteers" were his students.) At the end of a six-week acclimation period, Scholander measured the metabolic rates and foot temperatures of his subjects as they slept, and he compared the measurements with those of Norwegians who had never before attempted to sleep under such conditions. As one would expect, the unacclimated subjects never managed to get to sleep, and their metabolic rates were higher than their daytime resting rates. Also the metabolic rates were extremely irregular. Looking at the jagged line on the graph of metabolic rate, one can almost see the subjects stomping and swearing throughout the night (Fig. 15.9). In spite of that activity, they were not very successful homeotherms. The temperature of their feet dropped to as low as 18°C. In contrast, the acclimated Norwegians kept their feet at a comfortable sleeping temperature by maintaining a steady, high metabolic rate.

Scholander then went to Australia to make comparable measurements on aborigines, who preferred to sleep naked at subfreezing temperatures. For his studies, however, Scholander had the Australians use sleeping bags like those of the Norwegians. In order to ascertain that the conditions were comparable, Scholander himself made an unsuccessful attempt to sleep next to them. "We ourselves provided the control material. It turned out, that while we shivered and thrashed about all night, wishing that the sun would rise, the Australians lay motionless, snoring their way through the night, usually on a subbasal metabolic rate, cooling off even more than we did on the surface." That is, the Australian aborigines managed to sleep without the elevation in metabolic rate of the Norwegians, by abandoning total homeothermy. They allowed their foot temperatures to drop below 15°C and simply ignored the discomfort.

It is interesting that the Norwegians' and the Australians' hypothalami chose different but appropriate strategies for acclimation. Assuming that Scholander was not so heartless as to starve as well as freeze his Norwegian volunteers, it would make sense

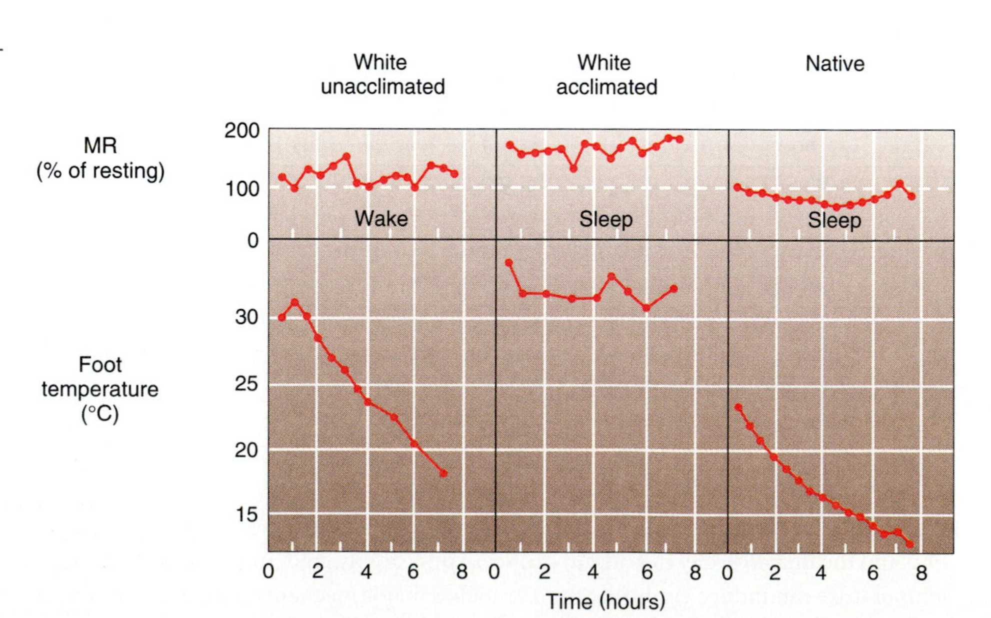

FIGURE 15.9

The metabolic rates and foot temperatures of three men sleeping or trying to sleep in the cold. The unacclimated Caucasian had an irregular, elevated metabolic rate, could not keep his feet warm, and could not sleep. After acclimation a Caucasian (middle part) could sleep and keep his feet warm by maintaining a high metabolic rate. The Australian native could also sleep, even without additional metabolic expenditure to keep his feet warm.

for their hypothalami to maintain homeothermy of the feet regardless of the metabolic cost. On the other hand, the Australian aborigines and their ancestors had probably experienced frequent periods of inadequate caloric intake. It would make more sense for their hypothalami to conserve energy by not attempting to keep their feet warm. We should not be too surprised that the hypothalamus can learn to balance the need for homeothermy against the need to conserve energy. The hypothalamus is, after all, a part of the brain.

Summary

The rates of all the activities of an animal depend on the rates of biochemical reactions, which depend on body temperature. In ectotherms the body temperature depends on the environmental temperature. Most ectotherms are therefore poikilotherms—with varying body temperatures. Ectotherms can, however, compensate for low environmental temperatures by producing more enzymes or a different isoenzyme. Many avoid the destructive effects of freezing by such means as increasing the concentration of solute in body fluids, by antifreezes, or by supercooling. Some poikilotherms are partial endotherms: They have periods in which the body temperature is determined by their production and conservation of heat.

In complete endotherms the body temperature is almost always determined by the production and conservation of body heat, and most complete endotherms are homeotherms. The production of heat is either shivering or nonshivering thermogenesis. Heat is conserved or eliminated by changing the area, temperature, and conductivity of the body surface, which affects conduction, convection, and radiation of heat. Vaporization of water by panting or sweating is another means by which endotherms eliminate body heat.

Homeothermy allows most physiological functions to occur at a constant rate regardless of environmental temperature. The major exception to this rule is the metabolic rate, which is high even at the acclimation temperature and is even higher at colder or hotter temperatures. Some homeotherms conserve energy at extreme temperatures by allowing the body temperature to deviate from normal.

Key Terms

heat
temperature
ectotherm
endotherm
poikilotherm
homeotherm
torpor
diapause
acclimation
partial endothermy
thermogenesis
brown adipose tissue
conduction
convection
radiation
vaporization
piloerection
vasoconstriction
vasodilation
thermoneutral zone
hibernation
pyrogen

Chapter Test

1. Explain how temperature affects an animal's activities. (p. 292)
2. What two properties of enzymes can be modified to compensate for a reduction in body temperature? (p. 294)
3. Define the following: poikilotherm, homeotherm, ectotherm, endotherm. In general, what kinds of animals are found in each category? Name some specific examples. (pp. 292–293)
4. Describe three ways by which poikilotherms avoid freezing. (pp. 293–294)
5. What are the four processes by which heat is exchanged between an animal and its environment? For each of the four, give one method by which an endotherm can modify the rate of heat loss by that process. (pp. 298–300)
6. In the experiment illustrated in Fig. 15.5, students typically observe the following responses in dry mice as the temperature falls: The mouse's fur fluffs out, the mouse curls up, the ears and tail become pale, and the mouse shivers. As the temperature rises, they observe that the mouse licks its fur, and its ears and tail become red. Explain the significance of each of these observations. Which of these actions are behavioral, and which ones are under hypothalamic control? (p. 302)
7. How do torpor, diapause, winter sleep, and mammalian hibernation differ from each other? (pp. 293, 303)

Answers to the Figure Questions

15.1 The advantage of a higher body temperature is an increase in the rates of chemical reactions responsible for growth, development of gametes, and other biological activities. The disadvantage is greater exposure to potential predators.

15.2 For a range of temperature from 10°C to 20°C the range of absolute temperatures is 283 K to 293 K. (K stands for Kelvin, the unit of measure for absolute temperature.) Because the rate of a physical process is directly proportional to absolute temperature, the Q_{10} equals the ratio of the rates at 293 K and 283 K, which would equal 293/283 = 1.035. Thus the rate of a physical process would increase by less than 4% for a 10°C (or 10 K) increase.

15.6 A rete mirabile in a small fish would probably not be able to conserve enough heat to overcome that lost to water by the large body surface relative to the body volume.

15.8 The thermoneutral zone could shift to the left or right on a graph like part B of this figure. Mechanisms involved could include (1) changes in conduction, convection, and radiation by changing the amounts of fat, fur, or feathers or (2) changes in the basal metabolic rate to increase or decrease heat production.

Readings

Recommended Readings

Bakker, R. T. 1975. Dinosaur renaissance. *Sci. Am.* 232(4):58–78 (Apr).

Carey, F. G. 1973. Fishes with warm bodies. *Sci. Am.* 228(2):36–44 (Feb).

Coutant, C. C. 1986. Thermal niches of striped bass. *Sci. Am.* 255(2):98–104 (Aug).

Eastman, J. T., and A. L. DeVries. 1986. Antarctic fishes. *Sci. Am.* 255(5):106–114 (Nov).

French, A. R. 1988. The patterns of mammalian hibernation. *Am. Sci.* 76:568–575.

Heinrich, B. 1979. *Bumblebee Economics.* Cambridge, MA: Harvard University Press.

Heinrich, B. 1981. The regulation of temperature in the honeybee swarm. *Sci. Am.* 244(6):146–160 (June).

Heinrich, B. 1987. Thermoregulation in winter moths. *Sci. Am.* 256(3):104–111 (Mar).

Heinrich, B., and G. A. Bartholomew. 1979. The ecology of the African dung beetle. *Sci. Am.* 241(5):146–156 (Nov).

Heinrich, B., and H. Esch. 1994. Thermoregulation in bees. *Am. Sci.* 82:164–170.

Kanwisher, J. W., and S. H. Ridgway. 1983. The physiological ecology of whales and porpoises. *Sci. Am.* 248(6):110–120 (June).

Karow, A. M. 1991. Chemical cryoprotection of metazoan cells. *BioScience* 41:155–160.

Lee, R. E., Jr. 1989. Insect cold-hardiness: to freeze or not to freeze. *BioScience* 39:308–313.

Schmidt-Nielsen, K. 1981. Countercurrent systems in animals. *Sci. Am.* 244(5):118–128 (May).

Southwick, E. E., and G. Heldmaier. 1987. Temperature control in honey bee colonies. *BioScience* 37:395–399.

Storey, K. B., and J. M. Storey. 1990. Frozen and alive. *Sci. Am.* 263(3):92–97 (Dec).

Additional References

Cossins, A. R., and K. Bowler. 1987. *Temperature Biology of Animals.* London: Chapman and Hall.

Harris, C. L. 1976. Temperature vs rate of the isolated frog heart: three hypotheses. *Physiol. Teacher* 5(3):1–3.

Heinrich, B. 1993. *The Hot-Blooded Insects.* Cambridge MA: Harvard University Press.

Schmid, W. D. 1988. Supercooling and freezing in winter dormant animals. In: R. W. Peifer (Ed.). *Tested Studies for Laboratory Teaching: Proceedings of the Ninth Workshop/Conference of the Association for Biology Laboratory Education (ABLE).* Dubuque, IA: Kendall/Hunt, Chapter 11.

Scholander, P. F. 1958. Studies of man exposed to cold. *Fed. Proc.* 17:1054–1057.

CHAPTER

16

Physiology of Reproduction

Green mamba *Dendroaspis angusticeps* hatching.

Most animals decline and die soon after they lose the ability to reproduce. For the female octopus, in fact, there is literally no life after sex. A "suicide gland" (the optic gland) does her in soon after she ovulates. Reproduction is also the end of life in another sense of the word: It is the end for which physiology is the means to that end. No matter how impressive an animal's nervous system, however mighty its muscles and intricate its hormonal and circulatory systems, they are merely the means for creating an internal environment in which the reproductive cells can function. It is therefore appropriate that this unit on physiology ends with reproduction. Reproduction is so important, however, that it involves numerous aspects of zoology besides physiology. The developmental processes by which gametes are produced and fertilized and develop into new individuals were described in Chapter 6. Chapter 20 describes the behavior that enables animals to breed and to nurture their offspring.

The Birds and the Bees

Asexual Reproduction. Human reproduction is traditionally compared with "the birds and the bees," but about the only thing these and other species of animals have in common is that they *can* reproduce sexually. Many animals, however, frequently do not do so. Asexual reproduction is especially common in sedentery animals that would have difficulty finding a mate and in parasites that must produce large numbers of offspring to compensate for the poor chances that any one of them will find a new host. Sponges, sea anemones, flatworms, ribbonworms (phylum Nemertea), and some echinoderms such as starfishes can reproduce asexually by **fragmentation.** (See Fig. 21.11 for a taxonomic summary of modes of reproduction.) In fragmentation an individual divides into two or more parts, each of which develops into a new individual. Another mode of asexual reproduction is **budding,** in which a new individual develops from a bud of cells in another individual. In *Hydra* and related cnidarians, and also in many sponges, new individuals develop by **external budding** of cells on the body surface (see Fig. 24.4). Many freshwater sponges and ectoprocts produce **internal buds** that enable the species to survive harsh winters and other environmental stresses in a dormant state (see Fig. 23.7).

A problem with asexual reproduction is that the new individual is genetically identical to the individual from which it developed. Thus an entire population could accumulate genetic defects or become so genetically homogeneous that it would be wiped out by an environmental change. In sexual reproduction, on the other hand, each new individual generally combines genes from two individuals. This genetic recombination increases the likelihood that in some individuals adaptations will evolve that will enable the species to survive environmental changes. This may well be the only benefit of sex, from an evolutionary point of view.

Sex also has its drawbacks, however. One problem with sex is that the formation of gametes by meiosis produces about as many males as females, yet only a few males are needed to fertilize many females. The surplus males merely use food and space that would better serve females and their offspring. Another problem with sexual reproduction is that finding mates of the same species and opposite gender requires a great deal of energy and time, and it exposes animals to predators. Some animals have modified forms of sexual reproduction that avoid some of these disadvantages.

Hermaphroditism. In many species of animals, reproduction is sexual, but each individual functions as both male and female. Such species are said to be **hermaphroditic,** after the mythological character Hermaphroditus, who had his body fused with that of a nymph. Hermaphroditic species are also called **monoecious** (mon-EE-shus), from the Greek for "one house." Like asexual reproduction, hermaphroditism is most common in sedentary animals that would have trouble finding mates, although self-fertilization is usually a last resort. Hermaphroditic individuals serve as both sexes either simultaneously or sequentially. **Simultaneous hermaphroditism,** in which each individual has both types of reproductive organs at the same time, is common in sponges, flatworms, snails (see Fig. 28.19), earthworms (see Figs. 29.17 and 29.18), corals, barnacles (see Fig. 31.16B), and certain fishes.

In **sequential hermaphroditism** each individual is first one gender and then the other (Fig. 16.1). Sequential hermaphroditism in which females change into males is called **protogyny** (Greek *proto-* first + *gyne* female). If the change is from male to female, it is called **protandry** (Greek *andros* male). Sequential hermaphroditism ensures a supply of both sexes. In some species the sex change occurs if there are too few of one gender. In schools of the seabass *Anthias squamipinnis,* for example, removing a certain number of males causes an equal number of females to change to males. These protogynous fishes then have a greater chance of breeding. Female saddleback wrasse *Thalassoma duperrey* change to males if they see that their school consists of a large proportion of small (usually female) fish. In species in which there are harems of females dominated by a single male, the largest female may improve its chances of mating by changing into a male.

FIGURE 16.1
Sequential hermaphroditism in the fairy basslet *Anthias* sp. The males at the bottom were once females like those above.

Most animal species consist of two sexes and are therefore said to be **dioecious** (die-EE-shus; from the Greek for "two houses"). The term **gonochoristic** is also frequently applied to animals in which there are separate sexes. Dioecious species avoid the genetic homogeneity of asexual reproduction and self-fertilization, but they may have problems in ensuring an appropriate number of each gender and in allowing mates to find each other. Being monoecious solves those problems, but another solution is to be dioecious but to keep mates attached to each other. The female anglerfish *Ceratias hobolli* keeps her tiny mate permanently on the outside of her body. *Enteroxenus,* a molluscan parasite of sea cucumbers, keeps her mate inside her body. *Enteroxenus* was thought to be hermaphroditic for many years until it was discovered that her "testis" was actually the degenerate male.

PARTHENOGENESIS. Another solution to the mate-supply problem in dioecious species is for the female to reproduce without having her eggs fertilized. This arrangement is called parthenogenesis (Greek *parthenos* virgin + *genesis* birth). Parthenogenesis is fairly common among sedentary and sparsely distributed animals, such as rotifers and certain insects. In some insects, parthenogenesis is a mechanism that produces different types of individuals adapted for different functions. In honey bees (*Apis mellifera*), male drones are produced parthenogenetically from unfertilized eggs, and female workers and queens are produced from fertilized eggs. In aphids, parthenogenetic and sexual generations alternate and are quite different from each other. Among vertebrates, parthenogenesis occurs in several species of fishes, some desert lizards, and domestic turkeys.

In animals except butterflies, moths, birds, and some reptiles, the females have two identical sex chromosomes (p. 78), so the parthenogenetic females produce only daughters. In such species the males are genetically superfluous and may even disappear, resulting in **unisexual species.** Some unisexual species of fishes and amphibians practice **hybridogenesis,** in which males of related species fertilize the eggs, but their chromosomes are then discarded. In whiptail lizards, parthenogenic females take turns acting like males, thereby inducing ovulation in each other (see Fig. 37.15B). In some species in which parthenogenesis is common, however, a genuine male of the same species is required for reproduction, even if its genes are not. In some fishes, for example, the eggs must be activated by sperm, but the male's chromosomes are not used. This phenomenon is called **gynogenesis.**

Sexual Variations

External Fertilization. In spite of the disadvantages, the majority of animals reproduce using two separate sexes. Evidently, sex has some advantage that outweighs all the disadvantages, but there is considerable controversy about what the advantages may be (pp. 371–372). Pleasure is usually not one of the advantages. In many dioecious species, individuals simply release sperm or ova into the environment, often without even seeing each other (Fig. 16.2). The sperm may be chemically attracted to ova of the same species, or they may simply swim or drift until they encounter them. Fertilization then usually occurs outside the body and is called external fertilization. External fertilization exposes gametes to damage from the environment, especially in fresh water or air. Exposed gametes are also likely to become food for other species. In many external fertilizers the risk is reduced by having many individuals, often of many different species, releasing sperm and eggs at the same time in **mass spawnings.** Corals inhabiting extensive areas of reef, for example, simultaneously release their gametes during the same hour. The advantage of mass spawning is that predators become satiated before they can completely consume the gametes. The means of synchronizing the release of gametes may be a chemical (**pheromone**), or a visual, tactile, or auditory signal during **courtship.** Courtship will be described in more detail later (pp. 418–420).

A

B

FIGURE 16.2

Two examples of external fertilization. (A) A basket sponge releases sperm into seawater, causing divers to report "smoking sponges." (B) Golden toads *Bufo periglenes* in amplexus, the breeding embrace. The smaller male behind releases sperm as the female releases a strand of white eggs. (This Costa Rican species is now apparently extinct.)

■ **Question:** How could the egg of one of these species avoid fertilization by sperm of another species?

Internal Fertilization. Internal fertilization is less hazardous for gametes and prevails among flatworms, earthworms, molluscs, insects, reptiles, birds, and mammals. Internal fertilization often requires some type of courtship behavior to ensure that the object of an animal's attentions belongs to the correct gender and species and is capable of producing gametes. Among predators, courtship is also important to prevent a mate from becoming a meal. Courtship climaxes with sperm transfer, usually through **copulation.** The sperm may be transferred in fluid **semen** or in a packet called a **spermatophore.** The copulatory and sperm-transfer organ is usually a **penis,** although other appendages may be adapted to that purpose. Octopuses use an arm (Fig. 16.3), spiders use a pedipalp on the head (Fig. 30.16A), and sharks use special claspers (Fig. 35.7). Some animals achieve internal fertilization without copulation. In scorpions, some insects, and salamanders, the male releases a spermatophore that the female then takes into her oviduct.

Oviparity. In almost all external fertilizers the young develop outside the females. Even in many species with internal fertilization, most of the embryonic development occurs outside the female, usually in an **egg** that is protected by a shell or some other structure. Such species are said to be oviparous (Latin *ovum* egg + *parere* to bring forth). Oviparity is common in many invertebrates and in birds, most fishes, amphibians, and reptiles, as well as in primitive mammals (monotremes) such as the duck-billed platypus. Following release of the egg from the mother (**oviposition**), the developing embryo obtains its nourishment from **yolk.** Although the parents cannot provide oxygen, nutrients, or other physiological support directly to the embryo inside the egg, many oviparous species do provide other kinds of support. Many **brood** their eggs, guarding or warming them during development. Some species of fish even brood eggs within the mouth or gill cavity. The rare (perhaps extinct) Australian frog *Rheobatrachus silus* broods eggs within its stomach (see Fig. 36.14).

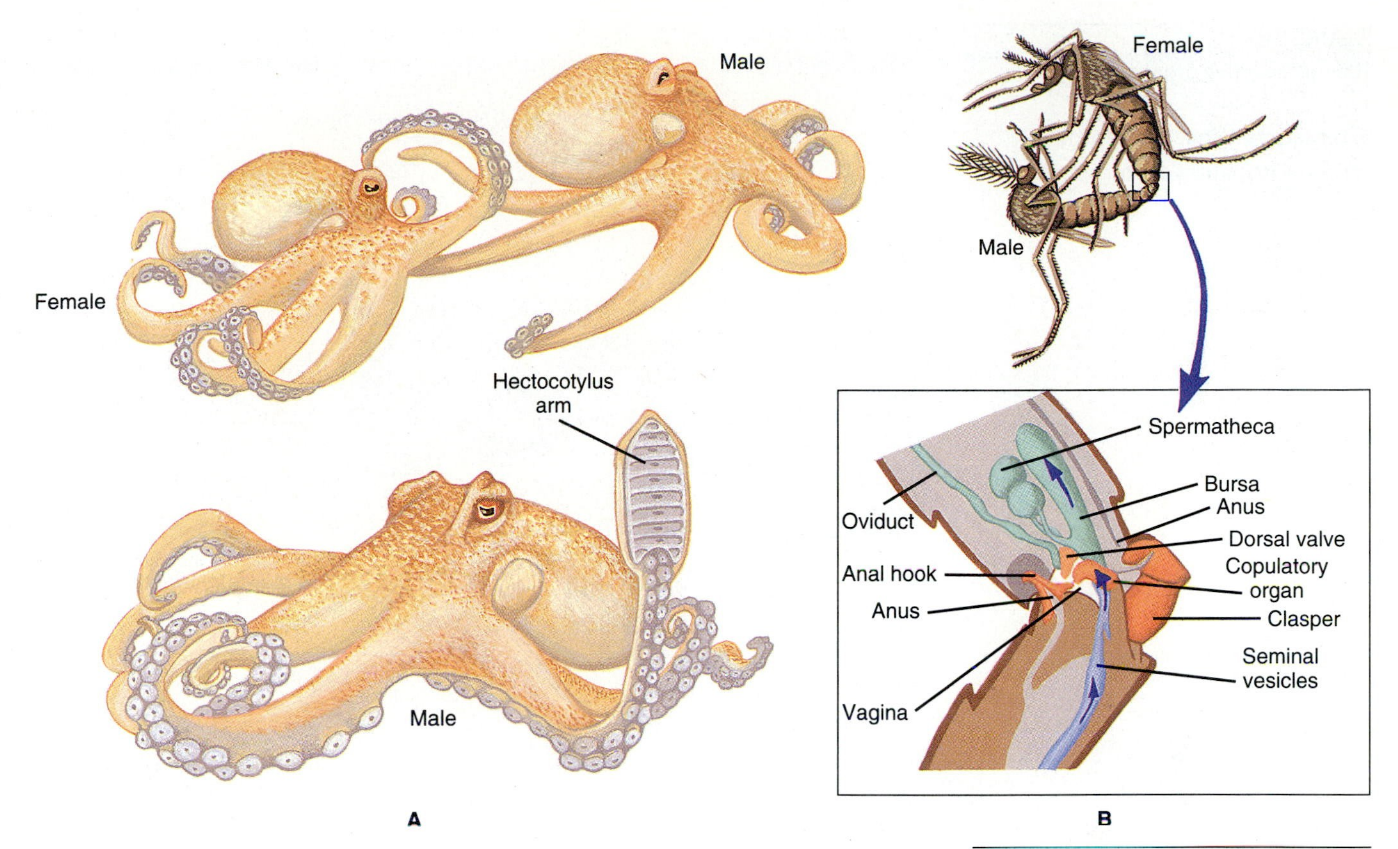

FIGURE 16.3

Two examples of intromittent organs for the introduction of sperm during internal fertilization. (See Fig. 16.4 for another example.) (A) The male octopus has one arm specialized as a "copulatory arm" with which he inserts spermatophores into the female. After mating, the end of the copulatory arm detaches, rendering him incapable of reproducing again. Some eminent scientists of the 19th century mistook the detached end of the arm for a parasite in the female. They named the "parasite" *Hectocotylus octopodis,* and the copulatory arm is still often called the hectocotylus arm. (B) Internal fertilization in the yellow-fever mosquito *Aedes aegypti* involves an intricate interlocking of genitals. The male's clasper and anal hook stabilize and widen the female's vagina. The copulatory organ is then engaged, widening the opening through which semen enters the bursa. All this takes about 20 seconds. Afterward the sperm, about a thousand of them, swim out of the bursa and into the spermathecae to fertilize ova as they are deposited.

VIVIPARITY. Some females retain shelled eggs within their bodies during embryonic development, so the young hatch while still in the oviduct. This phenomenon is called **ovoviviparity** if the mothers do not provide any nourishment or oxygen to the embryo. There is considerable doubt that this can ever happen, however (Guilette 1993). Embryos must obtain oxygen from somewhere, and in many species that are said to be ovoviviparous the shell is reduced and the oviduct has some adaptation for increasing exchange of respiratory gases and nutrients. These adaptations constitute at least a rudimentary **placenta,** and these species should be termed viviparous (Latin *vivus* living).

Viviparity occurs in all placental mammals (mammals except monotremes) and in every other major group of vertebrate except jawless fishes (Agnatha) and birds. Females of viviparous species give birth to living young after nurturing them within their bodies during embryonic development. Some of the adaptations for viviparity will be discussed later in this chapter and in the chapter on mammals (Chapter 39).

Reproductive Functioning of Male Mammals

All mammals are internal fertilizers, and all but the monotremes are viviparous. Mammals are accordingly equipped with quite elaborate organs for fertilization and for bearing young. The functioning and coordination of these organs are similar in most mammals and can be represented by humans. Fig. 16.4 on page 312 shows the reproductive organs in the adult human male. The external genitalia include both the beginning and the end of the male's direct role in reproduction. The **seminiferous tubules** in the **testes** produce sperm in the process called spermatogenesis (p. 107). The sperm mature in the **epididymis** on each testis, and they are stored there until ejaculation. A tube called the **vas deferens** transports sperm to the **urethra** of the **penis** during ejaculation. The

FIGURE 16.4
The reproductive organs of a man.

■ **Question:** What organ probably prevents urination during erection?

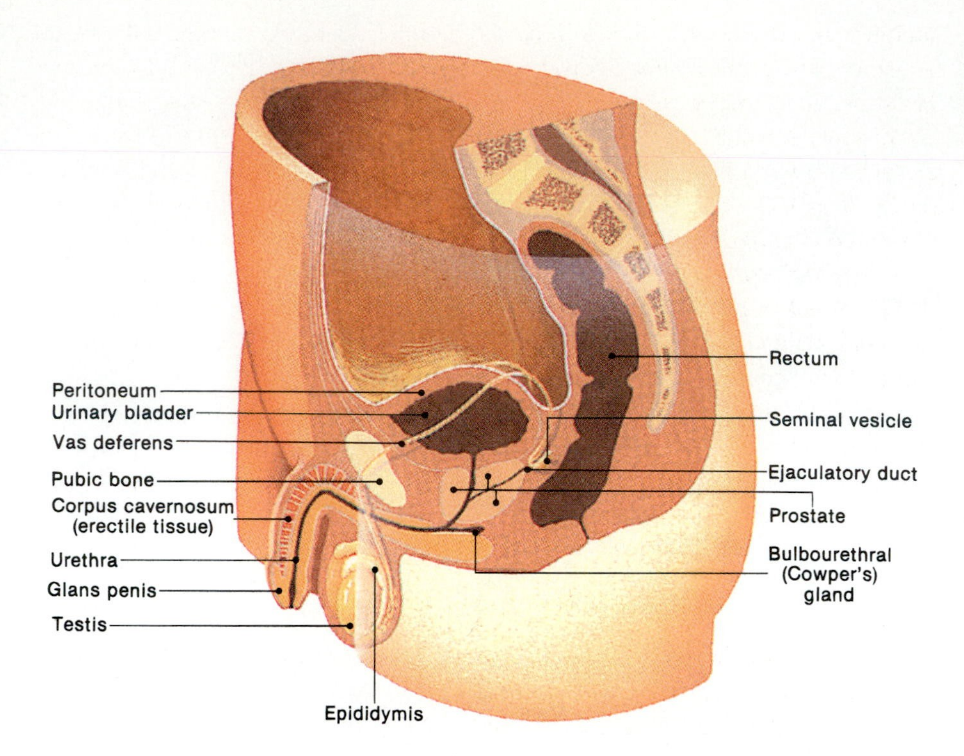

sperm are propelled by contractions of smooth muscles in the epididymides (ep-ee-DID-ee-MY-deez, plural of epididymis) and the **vasa deferentia** (plural of vas deferens).

During ejaculation a man releases about 200 million sperm in 3 milliliters of semen. (This number is about 40% lower than the average sperm count recorded in 1940. The cause of the decline is not known, but many suspect DDT and other chemical pollutants [Stone 1994].) Semen consists of secretions from three sets of glands:

1. Two **bulbourethral glands** (= Cowper's glands) secrete a clear, mucus-containing solution for lubrication.
2. Two **seminal vesicles** produce a viscous fluid containing fructose as an energy source for the sperm.
3. The **prostate gland** produces a milky fluid that maintains a suitably alkaline environment for the sperm.

Intromission of the normally flaccid penis is made possible by **erection.** Unlike most mammals, men do not have a bone to assist in erection, but depend on three cylinders of erectile tissue that fill with blood. The circulatory changes responsible for erection are controlled by the autonomic nervous system. The parasympathetic nervous system stimulates erection in response to pleasurable stimulation of mechanoreceptors in the penis, as well as olfactory, visual, and other types of stimulation. The sympathetic nervous system inhibits erection in response to fear or aggression. The sympathetic nervous system is, however, responsible for ejaculation.

Hormonal Control of Reproduction in Males

Reproductive physiology in the male is also controlled by hormones, especially two **gonadotropic hormones** from the anterior pituitary. One such gonadotropin is **follicle-stimulating hormone** (FSH). FSH stimulates production of sperm in the **seminiferous tubules** of the testes. The second gonadotropin from the anterior pituitary is

luteinizing hormone (LH). LH stimulates the interstitial cells (Leydig cells) of the testes to produce **androgens.** Androgens, such as **testosterone,** are steroid hormones that stimulate development and functioning of the reproductive organs. Androgens also promote **secondary sex characteristics.** These are traits that are useful but not essential for reproduction. In men they include hair on the face and body, deep voice, and behavioral characteristics such as aggression (Rubin et al. 1981). Androgens are also **anabolic steroids,** which means that they promote growth of muscle tissue.

Androgens also have a negative effect on their own release, by inhibiting the secretion of LH and possibly FSH. Androgens may directly inhibit the anterior pituitary, or they may act through the hypothalamus by inhibiting secretion of **gonadotropin releasing hormone** (GnRH), the peptide that stimulates release of both FSH and LH (pp. 141, 144). This inhibition tends to keep the reproductive state of men from fluctuating as much as in females, although androgens do surge during sexual arousal. The testes also produce **inhibin,** which inhibits GnRH and FSH secretion. These hormonal interactions are illustrated in Fig. 16.5.

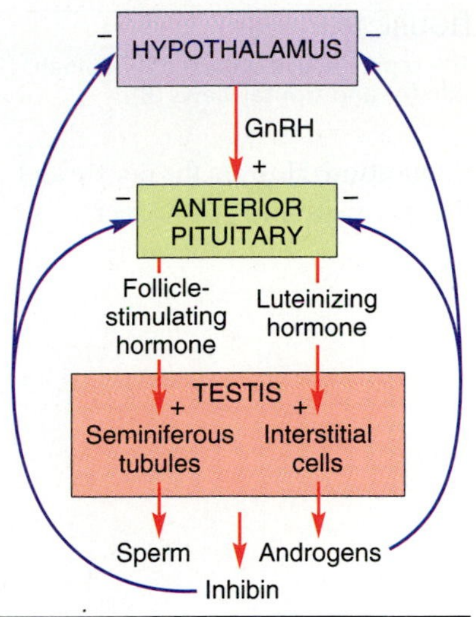

FIGURE 16.5

Stimulatory (+) and inhibitory (–) interactions among the male reproductive hormones.

■ **Question:** Can you suggest a chemical contraceptive that would interfere with one of the hormones in men?

Reproductive Functioning of Female Mammals

The reproductive organs of a woman are illustrated in Fig. 16.6 on page 314. The **vagina** is a receptacle for the penis, and it, as well as the clitoris and labia, has mechanoreceptors that produce the sensation of pleasure in women. Oogenesis, the production of eggs, occurs in the two **ovaries** (p. 107). In mammals the egg is called an **oocyte** (OH-ah-site), and release of the oocyte is called **ovulation.** After ovulation the oocyte enters one of the two **oviducts** (= Fallopian tubes) and is propelled toward the **uterus** (= womb) by cilia. Fertilization, if it occurs, usually happens in the oviduct, and the fertilized egg (= **zygote**) begins developing into an embryo before arrival in the uterus (see Fig. 6.9). In women and other primates the uterus has a single chamber that is adapted for the development of one offspring at a time. Most other mammals have a Y-shaped **bicornate uterus,** with two horns in which litters of offspring develop.

Girls are born with more than a million primary oocytes, of which about 400,000 survive past puberty. A few hundred of these develop, usually one at a time, within **follicles** in the ovaries (Fig. 16.7 on page 315). Ovulation usually occurs approximately every 28 days for three decades, first in one ovary and then in the other. During ovulation the oocyte is ejected into the abdominal cavity, but it is usually picked up by the ciliated **fimbriae** at the end of the oviduct. Part of the follicle remains in the ovary and forms a **corpus luteum** (Latin for "yellow body").

Hormonal Control of Reproduction in Females

Coordination of reproductive physiology and behavior in the female mammal is orchestrated by an impressive array of mechanisms, many of them hormonal. As in males, the anterior pituitary releases two gonadotropins, follicle-stimulating hormone (FSH) and luteinizing hormone (LH). FSH in women stimulates development of a follicle. LH triggers ovulation and stimulates the development of the corpus luteum. The secretions of both FSH and LH are due to gonadotropin releasing hormone (GnRH) from the hypothalamus. The follicle and the corpus luteum are also endocrine organs involved in regulating the ovarian cycle. The follicle produces steroid hormones called **estrogens,** including estradiol. Estrogens stimulate the development of the uterine lining—the **endometrium**—preparing it for implantation. Estrogens also promote secondary sexual characteristics, such as absence of hair on the face and body, high-pitched voice, more subcutaneous fat, and appropriate reproductive behaviors. (Male

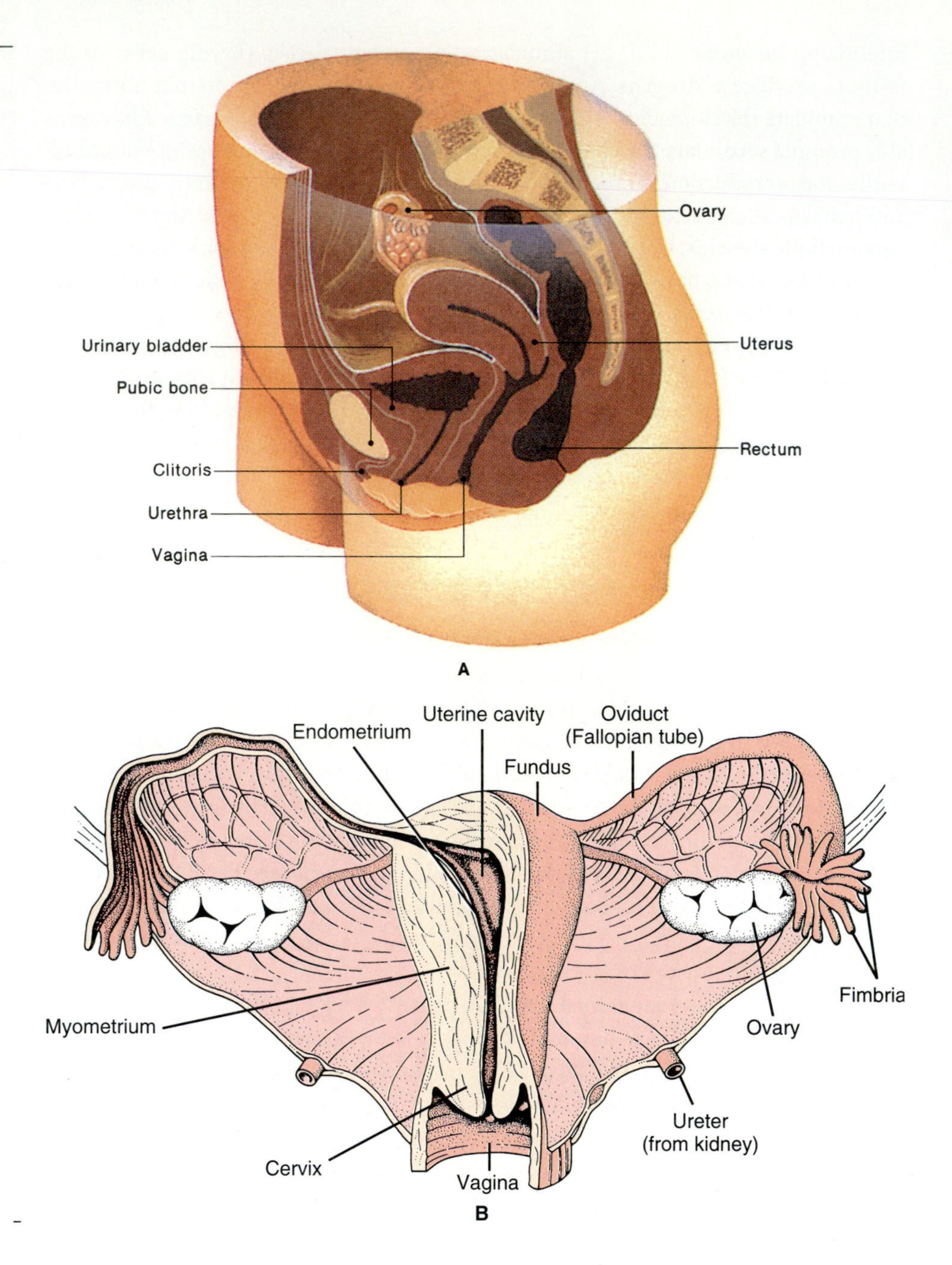

FIGURE 16.6
The reproductive organs of a woman: side (A) and frontal views (B).

■ **Question:** How do the oocyte and the sperm get into the oviduct?

rats that have been castrated and injected with estrogens build nests, something no self-respecting male rat would ordinarily do.) The corpus luteum also produces estrogens, as well as another steroid hormone, **progesterone.** Like estrogens, progesterone stimulates the endometrium in preparation for implantation.

In addition, estrogens and progesterone have feedback effects on the hypothalamus and anterior pituitary. Estrogens from the ripening follicle and from the corpus luteum inhibit FSH secretion, preventing the development of other oocytes. Estrogens also stimulate LH secretion, triggering ovulation. Progesterone from the corpus luteum then turns off LH secretion. The ovary also produces inhibin, which inhibits production of FSH. Besides acting directly on the anterior pituitary, estrogens and inhibin may inhibit FSH secretion by inhibiting release of GnRH from the hypothalamus. These interactions are illustrated in Fig. 16.8.

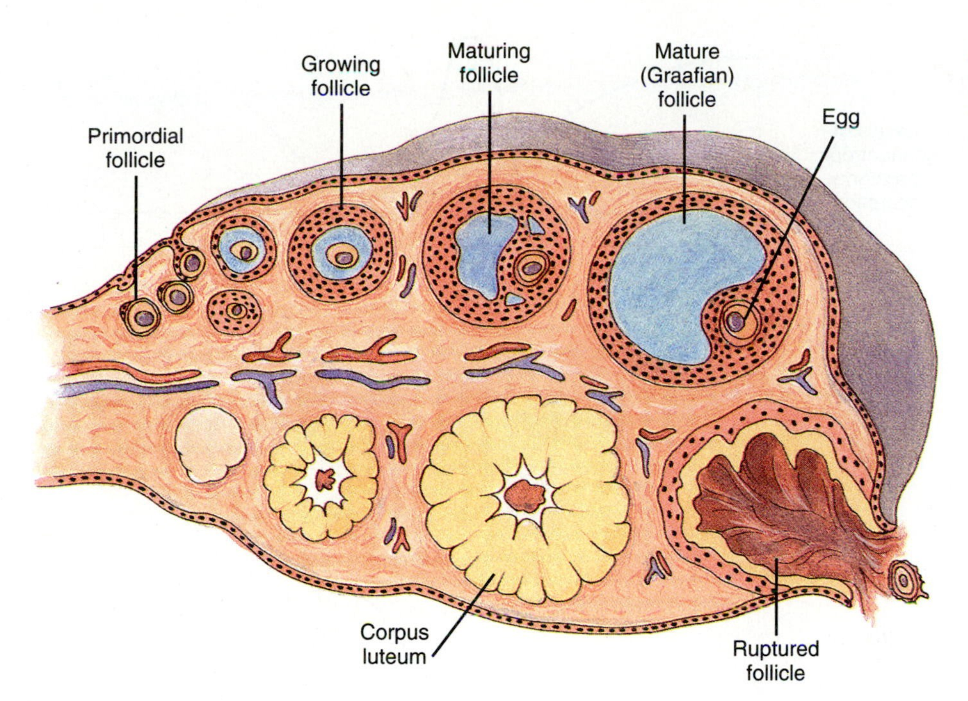

FIGURE 16.7

An ovary, showing stages in the maturation of a follicle (clockwise from upper left).

Estrous and Menstrual Cycles

In many female mammals, including women, the hypothalamus releases GnRH according to a regular cycle. By controlling the release of FSH and LH, this cycle also controls the development and ovulation of the oocyte. In most mammals the females become sexually receptive at the time of ovulation, and they are then said to be "in heat," or, more politely, in **estrus.** In these mammals the reproductive cycle is called the **estrous cycle.** (Note the difference in spelling of the noun "estrus" and the adjective "estrous.") In humans and gorillas, the females are sexually receptive without regard to whether there is an oocyte ready to be fertilized. In these species, sex serves not only for the conception of offspring, but also to keep both parents together during the long gestation and upbringing of the offspring. In these and other primates the decline in levels of estrogens and progesterone as the corpus luteum degenerates causes some of the endometrium to slough off, forming a **menstrual discharge.** In primates, therefore, the cycle is called the **menstrual cycle** rather than the estrous cycle.

The changes in hormone levels and in the ovary and endometrium are represented in Fig. 16.9 on page 316. The first day of menstruation is arbitrarily taken as day 1 of the cycle. At this time, estrogen and progesterone levels are low, so secretion of FSH is not inhibited. As FSH levels rise, a new follicle ripens, and an oocyte develops within it. The ripening follicle then produces estrogens, which trigger a surge in LH secretion. The LH surge triggers ovulation and formation of a corpus luteum. The corpus luteum continues to produce estrogens and begins producing progesterone, which terminates LH secretion. Both estrogens and progesterone stimulate development of the endometrium. If conception does not occur, the corpus luteum degenerates, levels of estrogens and progesterone drop, and a new cycle begins.

Many female mammals do not have estrous or menstrual cycles, but depend on external stimuli to trigger ovulation. The external stimulation may be seasonal, so that ovulation and estrus occur only at a particular time of the year. This ensures that young will be born when temperature, food supply, and other conditions are more favorable to their survival. In some species of cats, rabbits, and other mammals the females ovu-

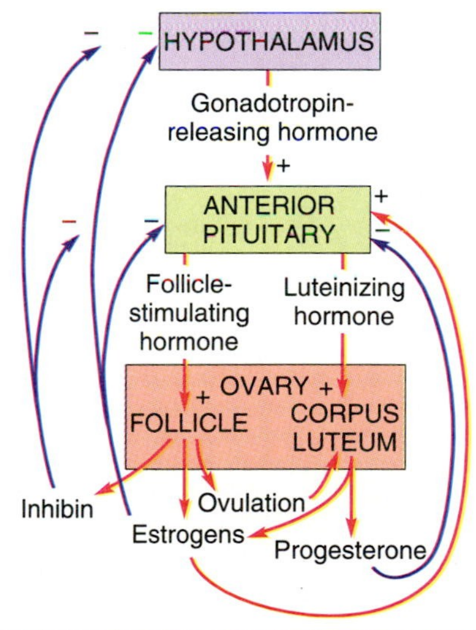

FIGURE 16.8

Stimulatory (+) and inhibitory (–) interactions among the female reproductive hormones.

■ **Question:** Why does increasing the level of progesterone prevent fertility?

FIGURE 16.9
Changes in hormone levels, the follicle, and the endometrium during a menstrual cycle.

■ **Question:** During pregnancy the corpus luteum continues to function. How would the levels of these hormones be affected?

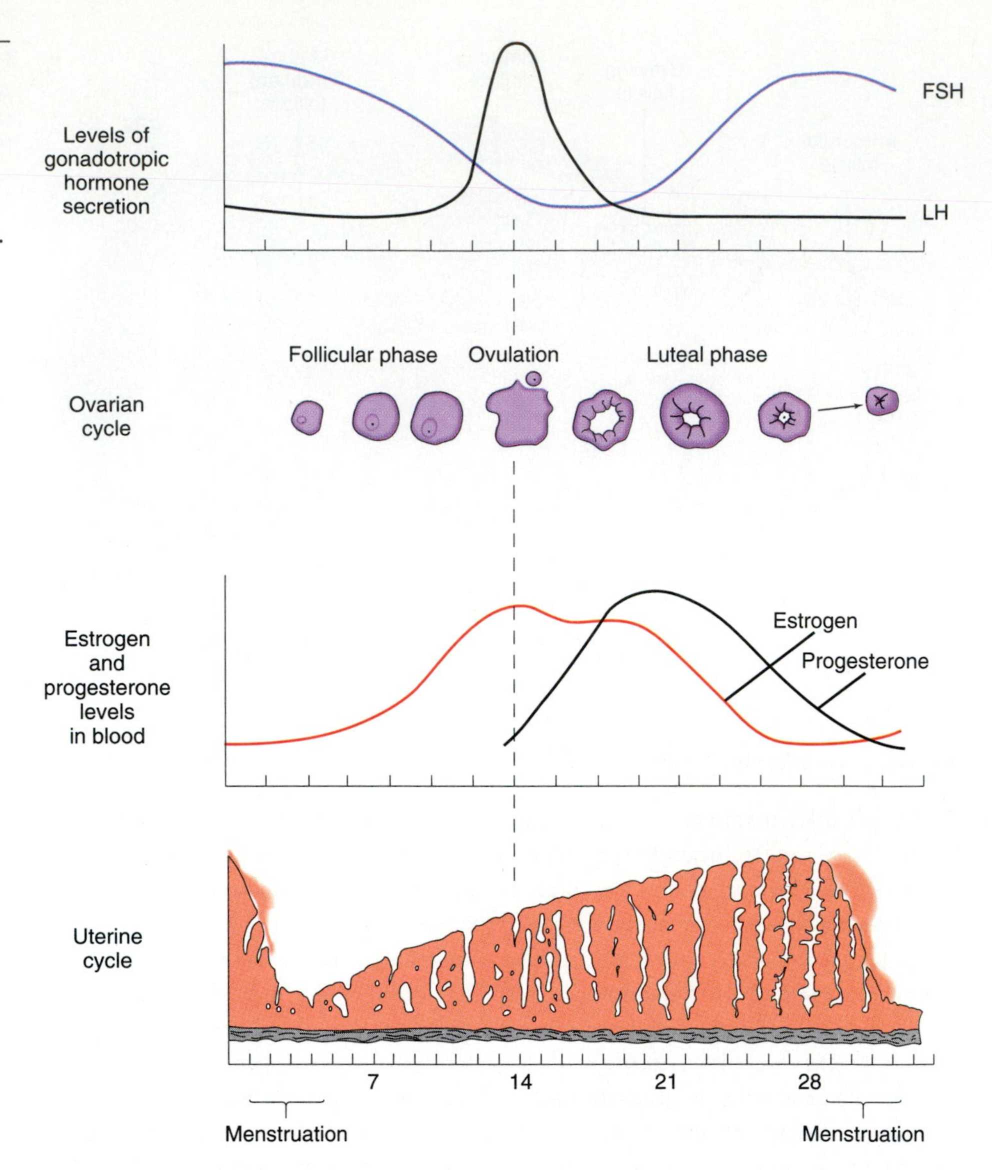

late only when stimulated during copulation. This **reflex ovulation** avoids wasting eggs and time.

Pregnancy

If fertilization occurs in a female placental mammal, the reproductive system must be shifted out of the estrous cycle and into preparation for pregnancy. This radical change in the physiology of the mother-to-be is triggered by the **blastocyst:** the cluster of cells that develops from the fertilized ovum (pp. 110–111). The outer layer of cells of the blastocyst, called the **trophoblast**, forms a **chorion.** One of the functions of the chorion is to secrete a protein hormone called **human chorionic gonadotropin** (hCG). hCG (or simply **chorionic gonadotropin** in nonhumans) enters into the circulatory system of the mother and prevents the corpus luteum from degenerating. Consequently, instead of the levels of estrogens and progesterone declining, the levels of these steroid hormones increase. Estrogens and progesterone continue to stimulate development of the endometrium, and menstruation does not occur. hCG is therefore the basis for the traditional first hint of pregnancy, the "missed period." hCG is also the basis for

Birth Control

Surgical Methods Birth control is becoming increasingly important in managing populations not only of humans, but also of wild and captive animals. One approach to birth control is **castration:** the surgical removal of ovaries or testes. Because castration also removes the sources of sex steroids, it has not enjoyed much success lately among humans. It is most often used by farmers and pet owners to prevent breeding and to make animals more docile.

A less radical surgical procedure is **vasectomy:** cutting and tying the vasa deferentia to prevent sperm from being ejaculated. Vasectomy may now be the most common surgery in the United States, and it is probably the most common form of contraception in married men. It is performed through the scrotum under local anesthetic during a short visit in the surgeon's office. Vasectomy is also becoming increasingly common in the management of animal populations in zoos. In some parts of the United States, beaver populations are controlled by vasectomy. The analogous procedure in females is **tubal ligation:** tying the oviducts. Blocking the oviducts by tubal ligation is more complicated than vasectomy, since it requires entry into the abdominal cavity.

Another surgical approach to birth control is **abortion,** in which the fetus is removed from the uterus by scraping or sucking it from the endometrium or by injecting hypertonic salt solution. Abortion is generally more dangerous to the woman than is contraception, although it is less risky than pregnancy itself.

Hormonal Interference In Europe a drug called **RU 486** (= mifepristone) has gained wide acceptance as a safe and effective means of inducing abortion. RU 486 works by blocking the receptor molecules for progesterone in the uterus. It is usually used in conjunction with a prostaglandin (p. 138) that induces uterine contractions. In the United States, hormonal interference is used only to prevent conception. The popularity and social impact of chemical contraception can be judged by the fact that of all the pills available in the United States, only one is instantly recognized as ***the* pill.** *The* pill usually consists of a combination of synthetic estrogens and progesterone that inhibits FSH and/or LH production. If used according to directions, the pill is 98% effective in preventing conception. The advantage of using steroids as oral contraceptives is that they can be swallowed and absorbed without being destroyed by digestive enzymes. The disadvantage with the pill is that it must be ingested every day. Many women find it more convenient to have progesterone-releasing sticks implanted under the arm once every five years.

Mechanical and Other Several other techniques are widely used, even though they are less effective in humans and generally impractical in other animals. One approach to contraception is spermicidal jelly or foam that kills sperm. These are about 80% effective when used alone, and more so when used with a diaphragm: a flexible device that fits over the cervix (neck) of the uterus and prevents entry by sperm. The condom also mechanically prevents sperm from fertilizing eggs. One advantage of the condom over all other forms of contraception is that it reduces the spread of AIDS and other sexually transmitted diseases (p. 234). However, because of slipping and leaking, condoms fail up to 14% of the time.

In China and some other developing countries, the main method of birth control is the **intrauterine device** (IUD). IUDs come in a variety of forms and are inserted into the uterus. Like any foreign object in the uterus, an IUD blocks implantation. IUDs are approximately 95% effective. Because of some risk of uterine damage, however, physicians seldom prescribe them in the United States, and many manufacturers have stopped selling them to avoid lawsuits.

One family-planning method that does not rely on technology is the **rhythm method,** in which intercourse is avoided for three days before and after ovulation. (Eggs and sperm can each survive for only about three days in the female.) The rhythm method reduces the likelihood of conception from the normal probability of 25%, *if* the time of ovulation can be accurately determined. Often it cannot. Out of 100 women using the rhythm method exclusively, between 13 and 24 will get pregnant within a year. The rhythm method is therefore used mainly because it is the only one permitted by the Catholic Church.

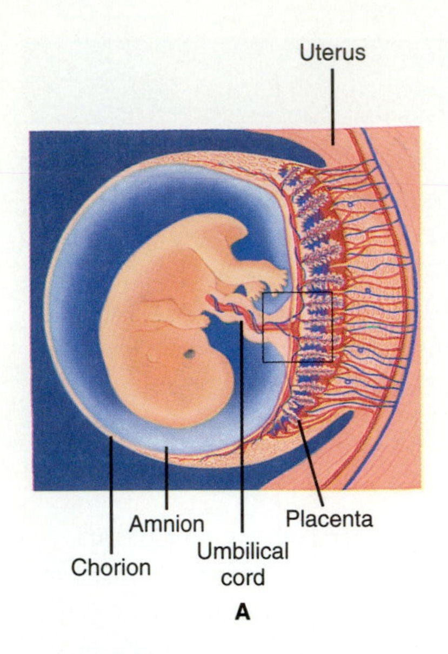

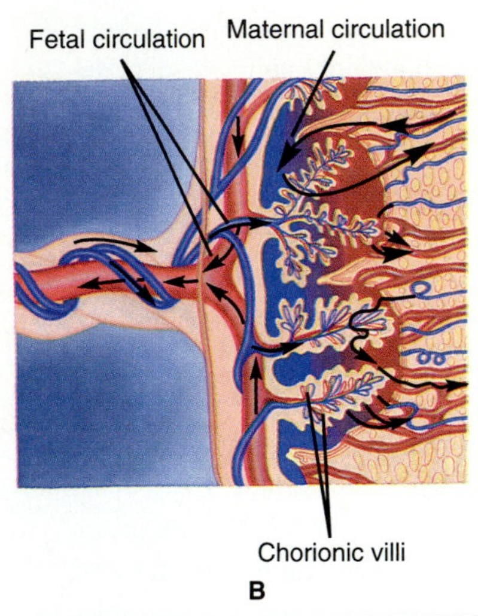

FIGURE 16.10

(A) A human fetus at about the eighth week of gestation (compare Fig. 6.25), showing the placenta. (B) Detail of the placenta showing some of the chorionic villi surrounded by pockets of circulating maternal blood. Blood flows away from and toward the fetus through vessels in the umbilical cord.

modern tests for pregnancy, which use antibodies to detect hCG in the urine of a pregnant woman.

Besides secreting hCG, the chorion also forms finger-like processes called **chorionic villi,** which embed themselves into the endometrium around five or six days after fertilization. During the next week these chorionic villi develop into the fetal portion of the **placenta** (Fig. 16.10). The main function of the placenta is to establish an exchange of nutrients, metabolic wastes, respiratory gases, and other materials between the fetus and the blood flowing in the maternal portion of the placenta. Nutrients, oxygen, antibodies, and other materials in the blood of the mother diffuse or are transported across the membranes of the villi and into the circulatory system of the fetus. Carbon dioxide and other metabolic wastes from the fetus move in the opposite direction into the maternal circulation.

At around the third month after conception the level of hCG declines, and the placenta takes over most of the production of estrogens and progesterone (Fig. 16.11). In addition to stimulating uterine development, these hormones promote development of the mammary glands, preparing them for **lactation**—the production of milk. At the same time, progesterone inhibits lactation until after birth. Estrogens and progesterone, as well as other placental hormones, also suppress the T cells in the mother's immune system, preventing the rejection of the foreign embryonic tissue developing within her.

Toward the end of the gestation period the uterine contractions of **labor** begin, gradually becoming stronger and more frequent. These contractions are due to increased secretion of **oxytocin** by the uterus and increased numbers of oxytocin receptors on smooth muscle (**myometrium**) of the uterus (Lefebvre et al. 1992). **Relaxin** from the corpus luteum and placenta also relaxes the pubic ligaments and the cervix. Eventually **cortisol** from the adrenal cortex of the fetus circulates into the mother's blood and alters enzyme activity in the uterus. This triggers **parturition**—the delivery of the child.

Lactation

The mother's levels of estrogens and progesterone drop sharply as soon as the placenta is delivered after the birth of the baby. This allows **prolactin** from the anterior pituitary to initiate the production of milk—lactation. Prolactin continues to be secreted as long as suckling occurs regularly, because nerve impulses from the nipple prevent the hypothalamus from secreting a hormone (dopamine) that inhibits prolactin secretion. These nerve impulses also trigger the release from the posterior pituitary of oxytocin, which causes the actual ejection of the milk. This discharge, which goes by the unromantic name **milk let-down,** is due to contractions of smooth muscles around the sacs (**alveoli**) that contain the milk (Fig. 16.12).

As long as the nipple is stimulated regularly, lactation can continue for several years. If suckling occurs several times every hour, as it does in some cultures, the release of FSH and LH are blocked, and the mother remains infertile. Lactation thus serves as a natural means of spacing births. As the young are **weaned,** however, the level of prolactin drops, lactation ceases, and the mother resumes ovulation.

In addition to its nutritional and family-planning benefits, human milk protects infants from one of their greatest threats, intestinal infections. Aggregated lymphatic follicles (Peyer's patches) in the small intestine of the mother (see Fig. 11.12) contain plasma cells that produce antibodies to bacteria ingested by the mother. These plasma cells migrate to the breasts and enter the milk, thereby providing infants with a source of antibodies to bacteria that they, too, are likely to encounter. Ironically, in develop-

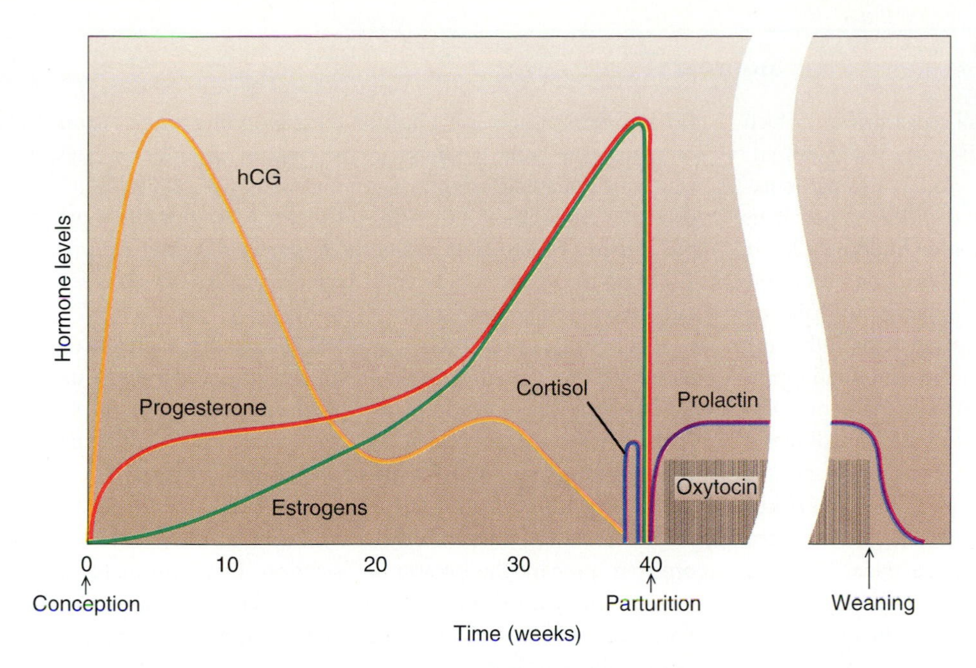

FIGURE 16.11

Hormonal changes in the mother during pregnancy, parturition, and lactation. Cortisol and hCG are from the embryo.

■ **Question:** Suppose the action of hCG were blocked by a drug. What would be the consequences?

ing countries, where children would benefit most from improved resistance to disease, many mothers give up nursing in favor of bottle-feeding, so that they can be like women in developed countries. Meanwhile, increasing numbers of women in developed countries are realizing the benefits of nursing.

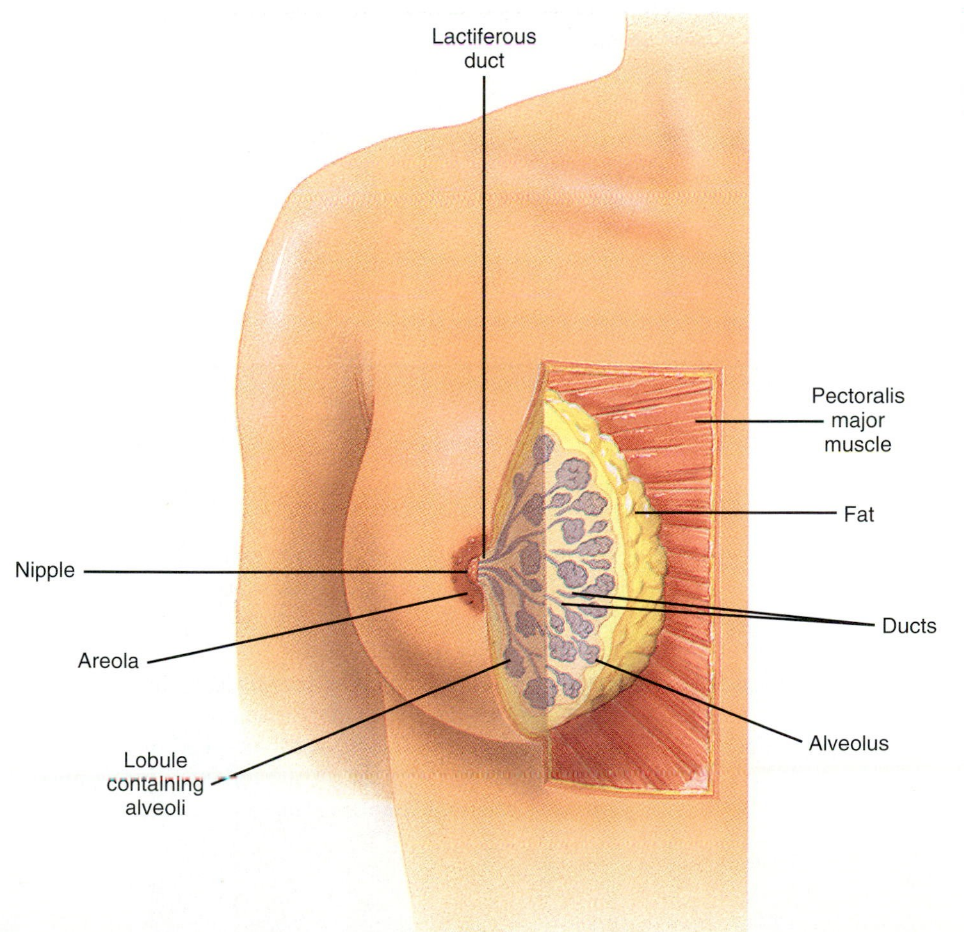

FIGURE 16.12

The structure of the human mammary gland. Milk is secreted by cells of the alveoli.

Summary

All species of animals are capable of sexual reproduction, which has the advantage of increasing genetic diversity but the disadvantage of producing superfluous males and requiring sometimes-complicated matings. Some animals avoid these problems by reproducing asexually, by fragmentation or budding. Others are simultaneous or sequential hermaphrodites, combining the functions of both sexes in one animal. Other animals are parthenogenetic: The female produces offspring without the eggs being fertilized. Most animals, however, are dioecious and reproduce only sexually. Fertilization may occur externally or internally, and the females may be oviparous or viviparous.All mammals have internal fertilization, and all but a few are viviparous.

The testes of male mammals produce sperm, which is ejaculated with semen. Reproductive functions are coordinated largely by hormones, especially two gonadotropins, FSH and LH, from the anterior pituitary. Androgens from the testes stimulate development and functioning of the reproductive organs, as well as secondary sexual characteristics. In females, FSH and LH are released in a cyclic fashion corresponding to the estrous or menstrual cycle. Follicles in the ovaries produce oocytes, which are released during ovulation into an oviduct. The follicle produces estrogens and then after ovulation develops into a corpus luteum, which produces both estrogens and progesterone. These hormones stimulate reproductive functions as well as secondary sexual characteristics.

Following fertilization, chorionic gonadotropin from the blastocyst prevents the corpus luteum from degenerating, and the blastocyst becomes implanted in the endometrium. A placenta then forms and exchanges oxygen, nutrients, and other necessities from the mother's blood to the fetus. Estrogens and progesterone, first from the corpus luteum and then from the placenta, maintain the endometrium. These hormones and also prolactin prepare the breasts for lactation. Oxytocin increases uterine contractions, and cortisol from the fetus triggers parturition. After birth, prolactin stimulates lactation, and oxytocin causes milk let-down.

Key Terms

fragmentation
budding
hermaphrodite
monoecious
dioecious
parthenogenesis
external fertilization
internal fertilization
oviparity
viviparity
testis
vas deferens
seminal vesicle
prostate gland
follicle-stimulating hormone
luteinizing hormone
androgen
vagina
ovary
ovulation
oviduct
uterus
follicle
corpus luteum
estrogens
progesterone
estrus
estrous cycle
menstrual cycle
chorionic gonadotropin
prolactin
oxytocin
lactation

Chapter Test

1. What are the disadvantages of sexual reproduction? How do fragmentation and budding avoid those disadvantages? How do hermaphroditism and parthenogenesis avoid those disadvantages? (pp. 308–310)
2. What are the advantages of internal fertilization? (p. 310)
3. Define oviparity and viviparity. In what kinds of animals does each occur? What are the advantages and disadvantages of each? (pp. 310–311)
4. Describe the career of a human sperm cell from the time it leaves the testis until it fertilizes an ovum. (pp. 311–312)
5. Diagram a sequence of neural and hormonal events that could account for the reflex ovulation of a cat or rabbit after copulation. (pp. 313–314)
6. From your understanding of hormonal effects, sketch a graph showing changes in hormones, the follicle, and the uterus throughout a complete menstrual cycle. Compare your graph with Fig. 16.9.
7. In your graph for the preceding question, show how the levels of hormones would change after ovulation if the egg is fertilized. What triggers these changes in the mother? (pp. 316, 318)
8. What are the roles of prolactin and oxytocin during pregnancy and lactation? (pp. 316, 318)

Answers to the Figure Questions

16.2 Presumably the gametes of the same species recognize each other by a ligand–receptor interaction like that between a hormone and a receptor molecule or between an enzyme and its substrate. In addition, different species may release sperm at different times, and the release of sperm and eggs may be synchronized by courtship, as in the frog.

16.4 Judging from its position at the junction of the urethra and the outlet of the urinary bladder, the prostate serves this function. Indeed, difficulty in controlling urination is a common symptom of prostate disease.

16.5 There are many possibilities if you don't care about side effects. A synthetic androgen (an anabolic steroid, for example) would inhibit GnRH release, thereby causing sterility by preventing the secretion of FSH.

16.6 Oocytes are released into the abdominal cavity and are taken into the oviduct by the fimbriae. Sperm propel themselves into the oviduct by flagella.

16.8 Progesterone inhibits LH secretion, thereby blocking ovulation.

16.9 Essentially the first week of the luteal phase (at about day 21) is prolonged. Levels of estrogens and progesterone remain high, and levels of FSH and LH remain low.

16.11 Blocking the action of hCG would allow the corpus luteum to degenerate as it normally does in the absence of conception (see Fig. 16.9). Because the endometrium would degenerate, implantation would probably not occur.

Readings

Recommended Readings

Beaconsfield, P., G. Birdwood, and R. Beaconsfield. 1980. The placenta. *Sci. Am.* 243(2):94–102 (Aug).

Cole, C. J. 1984. Unisexual lizards. *Sci. Am.* 250(1):94–100 (Jan).

Crews, D. 1994. Animal sexuality. *Sci. Am.* 270(1):108–114 (Jan).

Eberhard, W. G. 1990. Animal genitalia and female choice. *Am. Sci.* 78:134–141.

Frisch, R. E. 1988. Fatness and fertility. *Sci. Am.* 258(3):88–95 (Mar).

Kirkpatrick, J. F., and J. W. Turner, Jr. 1985. Chemical fertility control and wildlife management. *BioScience* 35:485–491.

Short, R. V. 1984. Breast feeding. *Sci. Am.* 250(4):35–41 (Apr).

Ulmann, A., G. Teutsch, and D. Philbert. 1990. RU 486. *Sci. Am.* 262(6):42–48 (June).

Warner, R. R. 1984. Mating behavior and hermaphroditism in coral reef fishes. *Am. Sci.* 72:128–136.

Additional References

Conn, D. B. (Ed.). 1991. *Atlas of Invertebrate Reproduction and Development.* New York: Wiley-Liss.

Guilette, L. J., Jr. 1993. The evolution of viviparity in lizards. *BioScience* 43:742–751.

Lefebvre, D. L. et al. 1992. Oxytocin gene expression in rat uterus. *Science* 256:1553–1555.

Rubin, R. T., J. M. Reinisch, and R. F. Haskett. 1981. Postnatal gonadal steroid effects on human behavior. *Science* 211:1318–1324.

Stone, R. 1994. Environmental estrogens stir debate. *Science* 265:308–310.

OVERVIEW

UNIT

Interactions of Animals with Their Environments and with Each Other

The topics in the previous units are essential for a complete understanding of how animals function. Understanding how cells and physiological systems work, however, cannot completely explain complex interactions, such as the cleaning symbiosis shown in this figure. Even with a complete understanding of cells and physiology, who could have predicted that the Spanish hogfish (*Bodianus rufus*) would feed on parasites and dead tissues of another fish, and that the yellowmouth grouper (*Mycteroperca interstitialis*) would let it?

Because of the incompleteness of cellular and physiological explanations, many zoologists specialize in other approaches to understanding the interactions of animals with each other and their environments. A zoologist specializing in ecology might study cleaning symbiosis in terms of the energy saved to both cleaner and client. A zoologist interested in evolution might wonder how such an interaction ever evolved and might look for antecedents in related species or in the fossil record. A zoologist specializing in behavior would carefully observe the behavior under various conditions to learn how the client signals that it will not eat the cleaner. The chapters in this unit describe some of the questions, assumptions, and techniques of zoologists with each of these interests.

CHAPTER

17

Ecology

Oxpeckers (*Buphagus* sp.) eating parasites on black rhinoceros (*Diceros bicornis*).

Abiotic Interactions

Animals get the requirements for life—oxygen, nutrients, water, ions, suitable temperature—from the physical environment and from other organisms. Every animal therefore depends on at least some parts of the physical environment and on at least one other organism. In fact, it is probably near the truth to say that every animal interacts with everything else, at least indirectly. The study of the interactions of organisms with each other and their environment is called **ecology,** a word derived from the Greek word *oikos,* meaning "house." The word evokes an image of all living and nonliving things thrown into a house—our planet—and left to work out their relationships. "Ecology" also calls to mind "economics": the study of the production, distribution, and consumption of goods. For living things, goods are the inorganic and organic—the abiotic and biotic—requirements for life. It is impractical to study the interaction of everything on Earth with everything else, so ecologists generally focus on one particular ecological system at a time. An ecological system, or **ecosystem,** is an association of all the organisms in an area, together with their physical environment. This definition of ecosystem is rather imprecise, but one does not have to be a trained ecologist to see that there are differences between such ecosystems as grassland and open ocean. Such differences arise partly from differences in abiotic factors, such as oxygen, water, ions, and temperature.

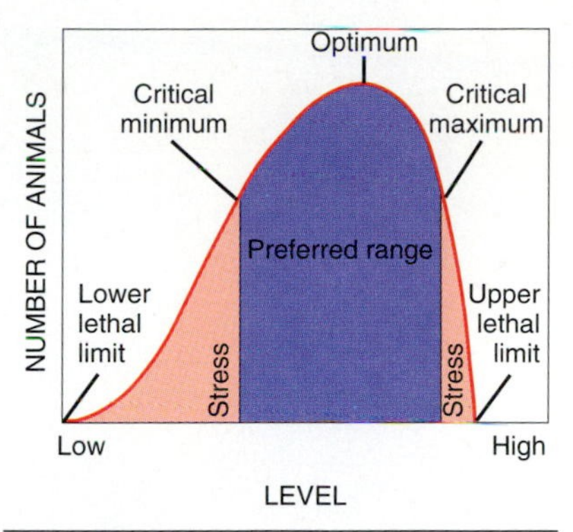

FIGURE 17.1

A hypothetical tolerance curve, showing the expected numbers of animals at various levels of an abiotic factor such as oxygen, water, ions, or temperature.

■ **Question:** Can you think of one situation that might skew the maximum population away from its optimum habitat?

It is a familiar fact that different species live in different ecosystems. Each species can live only in an ecosystem in which ranges of temperatures, oxygen levels, ionic concentrations, and other abiotic factors are suitable for that species. In other words, a species can live only within its **limits of tolerance.** The upper and lower extremes of the limit of tolerance are **lethal limits,** where the level of an abiotic factor is either too high or too low to permit the organism to live (Fig. 17.1).

Even within the limits of tolerance, conditions may not be optimal for particular organisms, and they may show signs of stress such as hypothermia, gasping, or dehydration. Although organisms may be able to survive such stressful conditions, they are less likely to be found there, because they generally do not reproduce well in those conditions. Organisms are more likely to be found where abiotic factors are within their **preferred range,** and especially near the **optimum.** Theoretically, the preferred habitat for a given species should simply be one that combines the preferred ranges for each abiotic factor. In reality, however, one physical factor can affect the preferred range of another. For example, the optimal temperature for a fish might be 20°C, but in a real environment the fish might suffocate in such warm water because warm water holds less oxygen and high temperature increases oxygen demand.

A further complication is that the levels of all these physical factors usually change over time. There are annual cycles of temperature and precipitation, for example, and these cycles differ in various parts of the world. Temperate zones have large seasonal variations in temperature and smaller variations in precipitation. The most stressful time of the year is generally the winter, when many animals die from cold and lack of food. In contrast, the temperature in tropical regions remains fairly constant through the year, but in many areas a season of heavy rains alternates with a dry season. The dry season is the most stressful to animals, when vegetation is scarce and watering places dry up.

Biotic Factors

Biotic factors are those associated with other organisms, primarily exchange of energy and nutrients from one organism to another. Because of these biotic factors, an animal may be excluded from a particular habitat even if all of its own abiotic factors are in their preferred ranges. The organisms on which the animal depends for food must also have its abiotic needs satisfied. Animals depend on other organisms because they, unlike plants, are incapable of using raw solar energy and most inorganic matter directly. The energy and matter must first be **fixed**—incorporated into organic molecules by **autotrophs.** The term "autotroph" comes from the Greek words for "self" and "food," and it refers primarily to photosynthesizers—certain bacteria, algae, and green plants. Autotrophs are called **primary producers,** and their fixation of matter and energy in forms that can be used by animals and other organisms is called **primary production.** Primary production supports not only the autotrophs themselves, but also the animals and other organisms that feed on them and on each other. Such nonautotrophic organisms are called **heterotrophs.** Heterotrophs are also called **consumers,** because they obtain energy by consuming organic material synthesized by other organisms.

Ecologists are very much interested in primary production, because it sets the limit on how many organisms can live in a given ecosystem. Primary production varies in different areas and at different times, because photosynthesis shuts down when there are inadequate levels of light, temperature, moisture, or certain minerals. Much of the primary production of summer days may be used by producers themselves to sustain life during the night and winter. Whatever is left over is the **net primary production,** which can sustain consumers. The annual net primary production is the yearly increase in organic matter or energy within an area. As Fig. 17.2 shows, ecosystems vary greatly in annual net primary production.

Nutrient Cycles

THE CARBON CYCLE. Biotic and abiotic factors are not strictly separated from each other, but interchange repeatedly in **biogeochemical cycles,** which are also called nutrient cycles. Carbon, nitrogen, and many other chemical elements in the environment go from the inorganic to the organic states in nutrient cycles by being fixed into organic molecules by autotrophs. **Carbon fixation** occurs primarily in green plants and algae when they convert carbon dioxide and water into oxygen and glucose through photosynthesis. During photosynthesis, red and blue light supplies the energy that links the carbon atoms from CO_2 into chemical bonds that make glucose ($C_6H_{12}O_6$). Cells then break down the glucose and extract the energy in a form they can use. The breakdown of glucose through cellular respiration releases the CO_2 back into the environment (Chapter 3). Cellular respiration is therefore essentially the opposite of photosynthesis, as represented in the following summary equation:

$$6\,CO_2 + 6\,H_2O + \begin{matrix}\text{solar energy} \\ \text{ATP}\end{matrix} \; \underset{\text{respiration}}{\overset{\text{photosynthesis}}{\rightleftarrows}} \; C_6H_{12}O_6 + 6\,O_2$$

In photosynthesis, cellular respiration, and other metabolic processes, carbon cycles between an inorganic state as CO_2 and an organic state as glucose and other organic molecules. The alternation between CO_2 and organic molecules forms a major loop of the carbon cycle (Fig. 17.3 on page 328). Carbon also drops out of this loop for long periods as fossil fuels (peat, coal, and oil) until liberated as CO_2 during combustion. Carbon also becomes locked up for long periods as calcium carbonate ($CaCO_3$) and other minerals in the shells and other hard parts of molluscs, corals, and other marine animals. For example, carbon in the shells of molluscs that lived 400 million years ago became limestone, which is eventually released through the slow process of weathering or the even slower process of volcanic action.

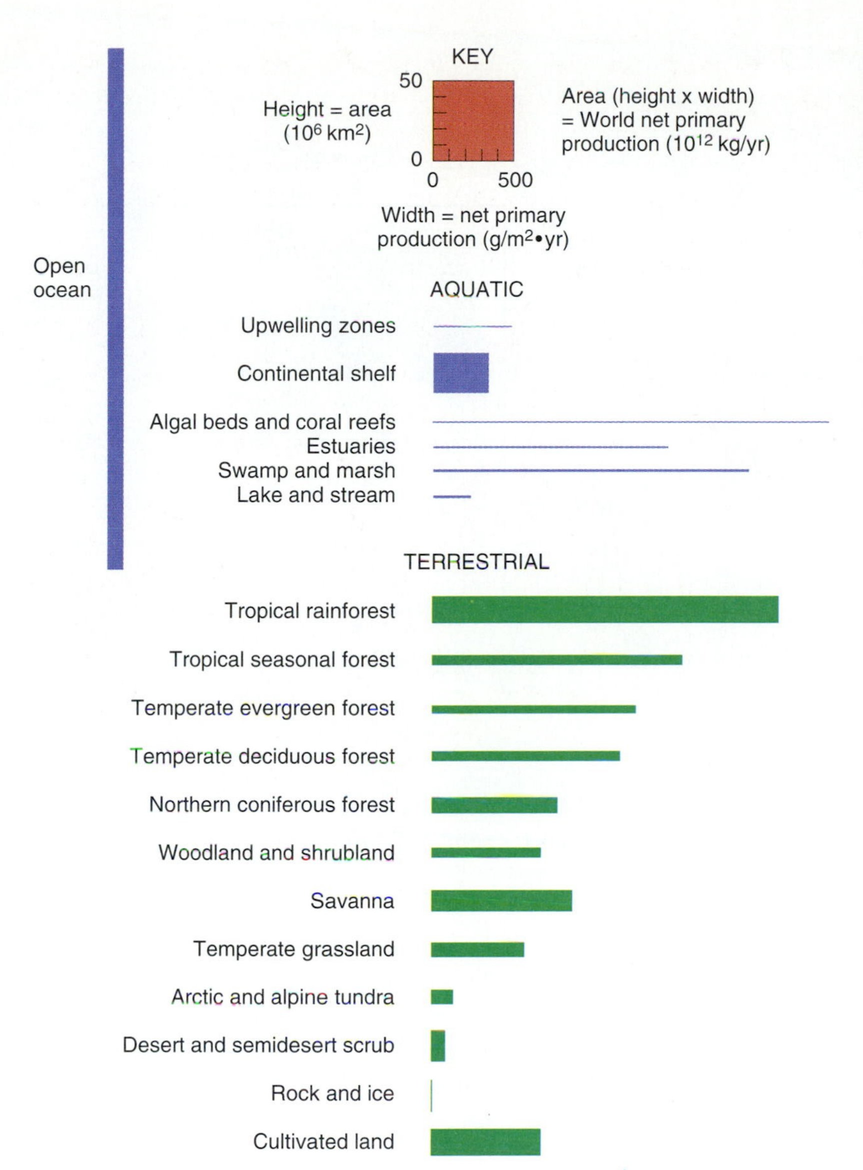

FIGURE 17.2
Primary production in various aquatic and terrestrial ecosystems. The height of each rectangle represents the total area of the ecosystem type on Earth. The width of each rectangle represents the average annual net primary production per unit area, estimated by comparing the dry weight of organic material at the beginning and end of one year. The area of each rectangle therefore represents the world net primary production for that type of ecosystem. (Data from R. H. Whittaker. *Communities and Ecosystems,* 2nd ed. New York: Macmillan, 1975.)

Decomposers, which are usually bacteria or fungi, are essential links in the biogeochemical cycles of carbon and other nutrients, because they break down complex organic molecules into simple organic and inorganic molecules. Without decomposers, much carbon and other nutrients in dead organisms and their waste products would not recycle. Animals that feed on excrement, dead organisms, and other nonliving organic matter—the **detritus feeders**—are also important, because they provide decomposers with access to dead plants and animals. Detritus feeders include numerous small animals of the soil (Fig. 17.4 on page 328). Most people would rather not think about the activities of dung beetles, carrion beetles, and other detritus feeders, but without them life would be even more unthinkable. Fields and forests would be nightmarish landscapes of fallen and undecayed trees, dead animals, and excrement.

Oxygen and Water Cycles. Coupled to the carbon cycle is an oxygen cycle. Oxygen is produced during photosynthesis and is consumed during respiration and the combustion of fossil fuels. The oxygen cycle is also coupled to another important biogeochemical cycle: the water cycle (= **hydrologic cycle**). Almost all the oxygen

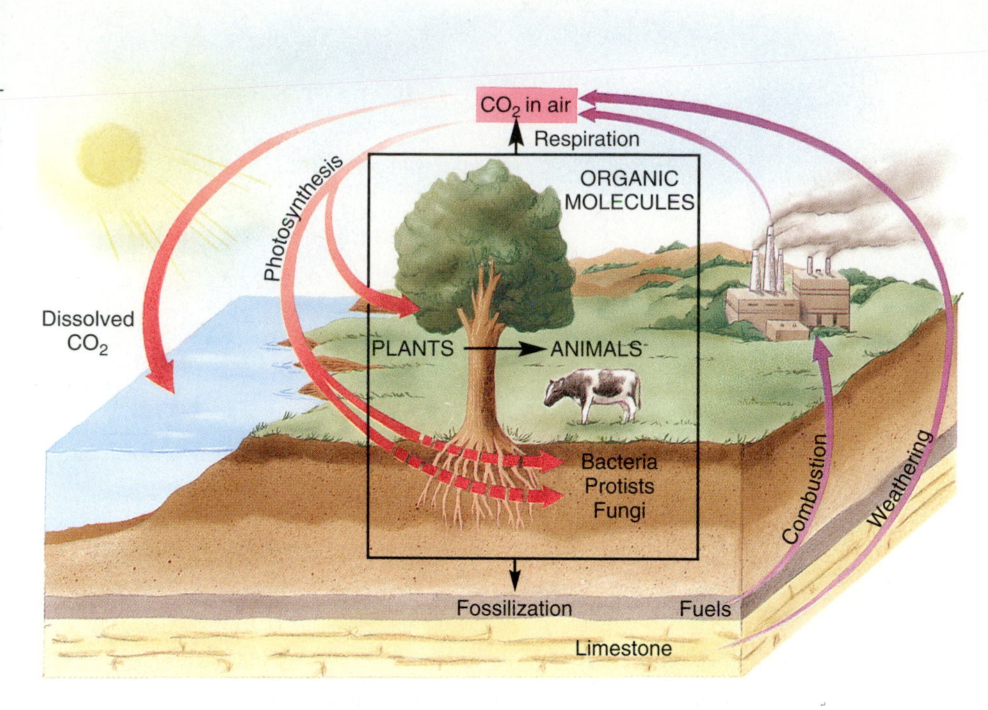

FIGURE 17.3

The carbon cycle. During photosynthesis, carbon in CO_2 dissolved in air and water is used to synthesize organic molecules. A small portion is soon converted back to CO_2 through respiration, but almost 2000 times as much carbon enters a slower loop as fossil fuels, animal shells, and other biogenic minerals. Weathering and combustion eventually complete the cycle.

available to animals and other organisms comes from water broken down during photosynthesis. The water cycle is completed during cellular respiration, when O_2 and H^+ combine into H_2O.

The Nitrogen Cycle. Because nitrogen is an essential element in proteins and many other organic molecules, the nitrogen cycle is also important (Fig. 17.6 on page 330). N_2 makes up 78% of the atmosphere, but before it can be used in the synthesis of organic molecules it must be converted into water-soluble nitrate NO_3^- or ammonia NH_3. This conversion is called **nitrogen fixation.** Some nitrogen is fixed as nitrate NO_3^- by lightning and by combustion in gasoline engines, and some is introduced as

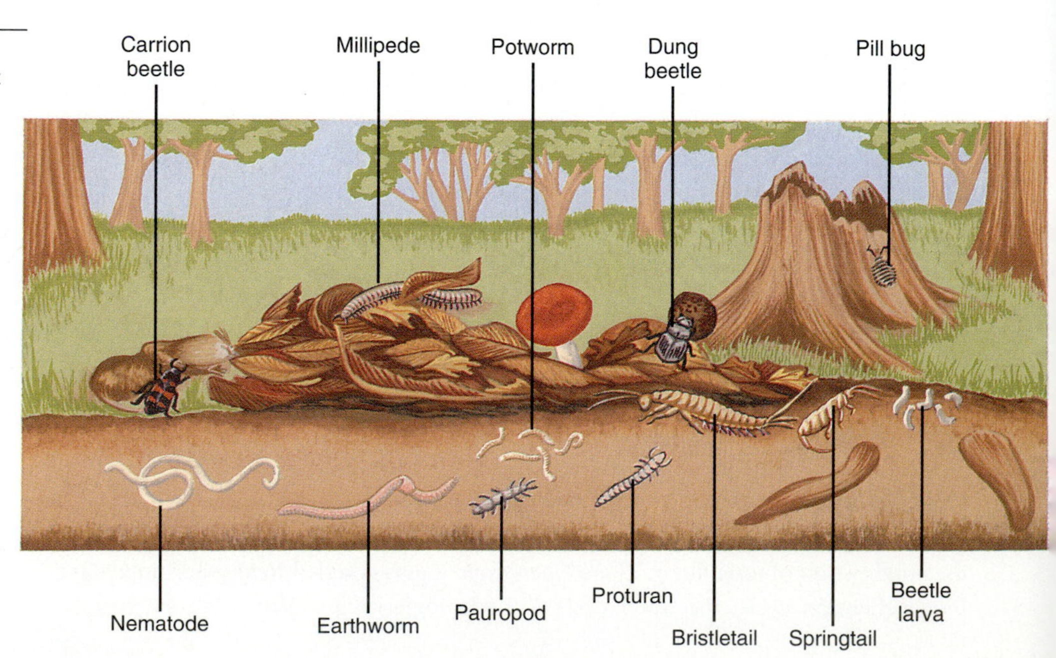

FIGURE 17.4

Many animals help complete nutrient cycles by feeding on detritus.

The Greenhouse Effect

The inputs and outputs of the carbon cycle resemble the incomes and expenditures of a government, a household, or a corporation, so ecologists speak of the "carbon budget." The carbon budget would be balanced if CO_2 production by respiration and combustion equaled CO_2 uptake by photosynthesis and absorption in oceans. Like so many other budgets, however, the carbon budget is unbalanced. Since 1958 the amount of CO_2 in the atmosphere has increased at least 12%, as measured in the relatively clean air of Hawaii (Fig. 17.5). This increase is attributed mainly to burning of fossil fuels, destruction of forests, and reduction in CO_2 absorption by the oceans. There is a lively controversy about which of these factors is most important.

The debate over CO_2 is not merely of academic interest, because CO_2, along with H_2O and many synthetic gases, reflects back to Earth much of the heat that would otherwise be radiated into space. Atmospheric CO_2 therefore has a greenhouse effect, increasing temperature on the Earth's surface. There is no doubt that the greenhouse effect occurs: Without it, Earth would be too cold to support life. The question is whether increased levels of greenhouse gases are raising temperatures. The level of CO_2 is expected to double by the year 2050, and various computer models predict a resulting increase in global temperature of from 1.5°C to 4.5°C. By the time we know what the precise effect is, it will be too late to do anything about it. The increased CO_2 and warming could have a variety of consequences. Increased CO_2 and temperature might increase plant production, but there could also be ecological and economic disruptions. Many areas that are now productive, such as the farm belt of the United States, could become deserts. The increased temperature could also melt the polar ice caps, flooding coastal areas, including many large cities.

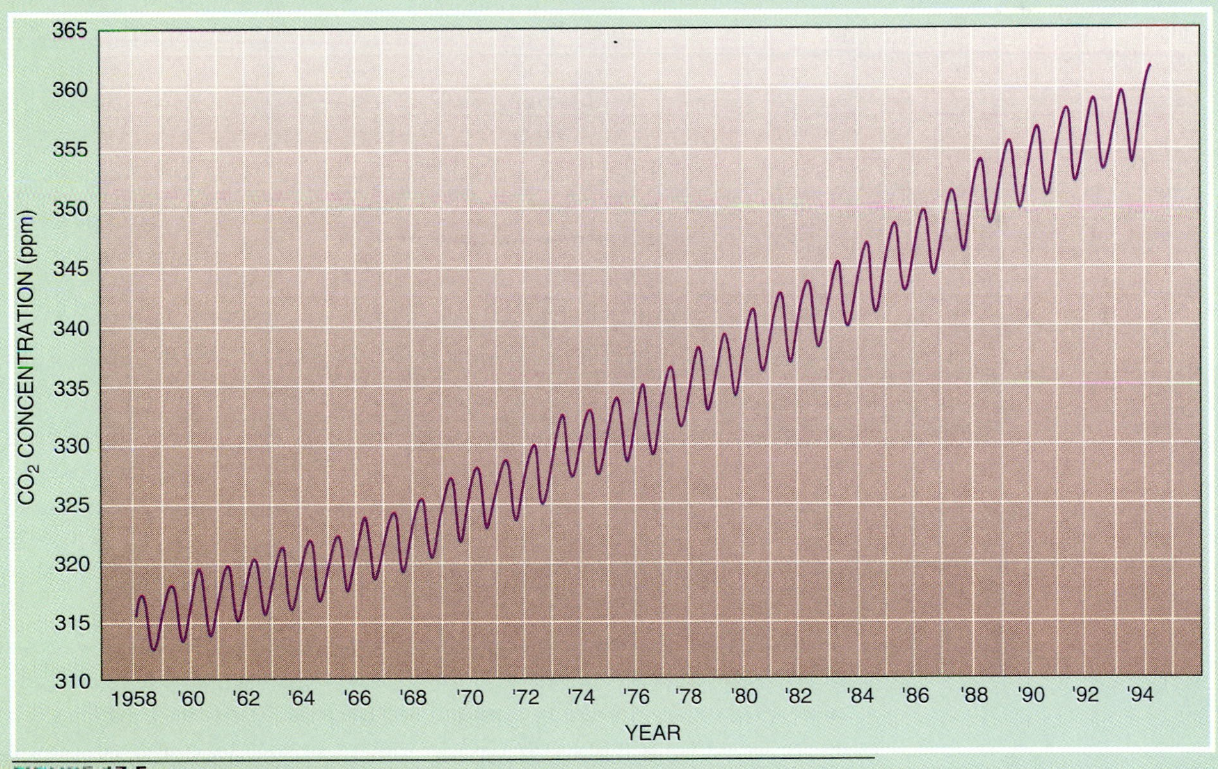

FIGURE 17.5

Levels of CO_2 in parts per million (ppm) measured at Mauna Loa on the island of Hawaii since 1958. Seasonal variations are due to increased photosynthesis in summer and increased plant respiration in winter, as well as changes in fossil-fuel emissions and absorption by oceans. Data are from Keeling et al. (1989) and personal communication with C. D. Keeling and T. P. Whorf.

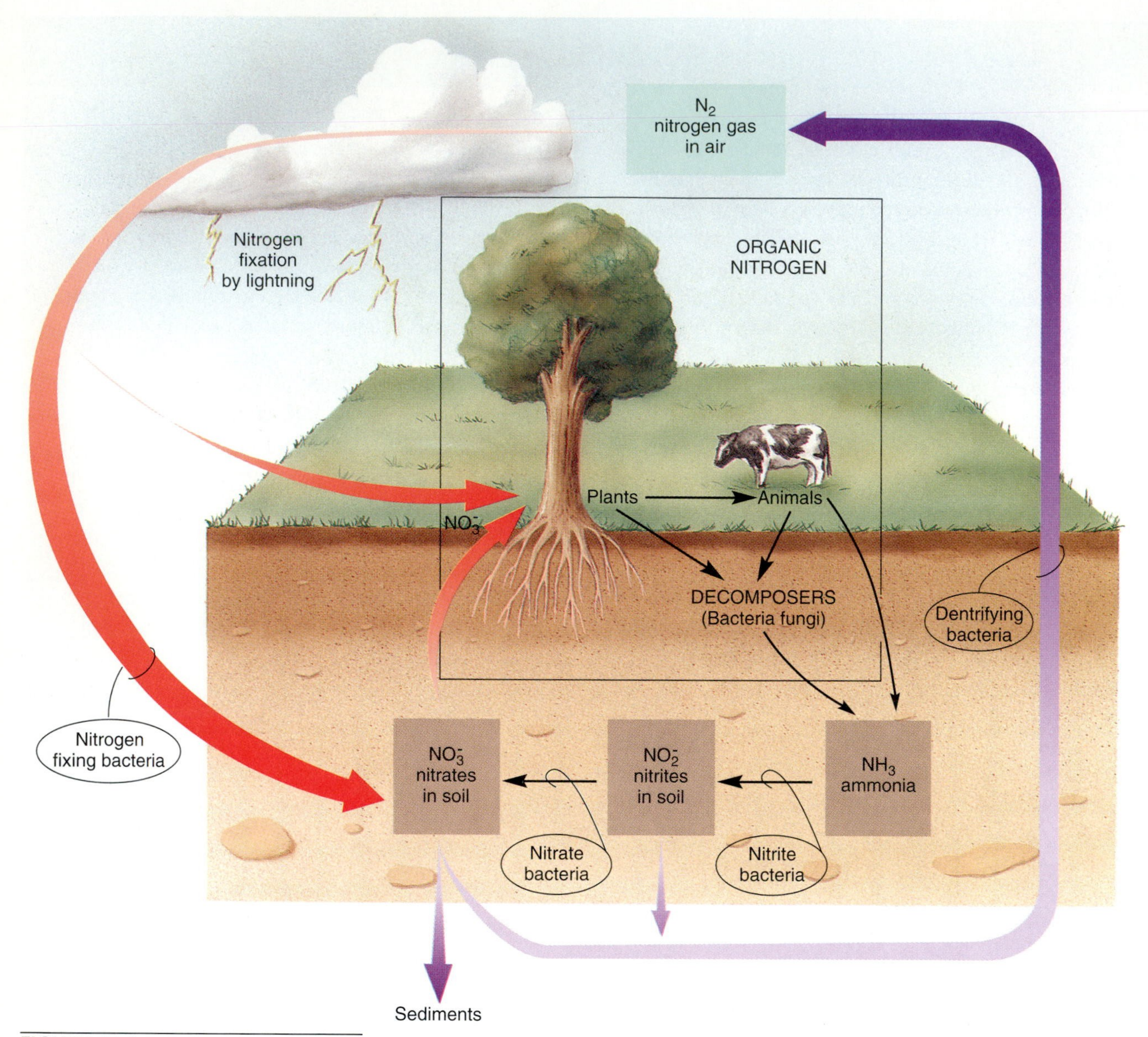

FIGURE 17.6

The nitrogen cycle for terrestrial organisms.

■ **Question:** How would this figure be modified to represent an aquatic ecosystem?

ammonia in synthetic fertilizers. The major source of fixed nitrogen, however, is nitrate from nitrogen-fixing bacteria. Many nitrogen-fixing bacteria are free-living in soil and water, but bacteria in the genus *Rhizobium* are symbionts within root nodules of legumes (such as peas, alfalfa, soybeans) and provide nitrates directly to their hosts. Nitrogen is usually the least-abundant nutrient in soils, so plant growth is often limited by how many nitrogen-fixing bacteria live in an area.

Another biological source of fixed nitrogen for plants are nitrogenous wastes, such as urea and the proteins of dead organisms. These are converted to ammonia and nitrate by decomposers, including **ammonifying bacteria. Nitrite bacteria** then convert some of the ammonia into nitrite (NO_2^-), which may then be converted into nitrate (NO_3^-) by **nitrate bacteria.** The nitrate can then be used by plants. Much of the nitrite and nitrate, however, are removed from biological utility when **denitrifying bacteria** convert them back into gaseous nitrogen (N_2). Denitrifying bacteria are the main reason why modern agriculture is so dependent on artificial fertilizers to replenish the nitrogen in soils. If denitrifying bacteria were destroyed, food production could presumably be greatly expanded in countries that cannot afford artificial fertilizers. So little is known about these bacteria, however, that destroying them would be foolhardy.

Energy

Biological processes require energy, most of which comes from breaking chemical bonds. Therefore, the flow of energy through an ecosystem generally follows the same path as carbon during photosynthesis and cellular respiration. This connection between organic molecules and energy explains why primary production can be expressed in either units of energy or units of mass. The movements of carbon and of energy differ in one major respect, however. Carbon cycles repeatedly between CO_2 and organic molecules, but energy is dissipated as it is transformed from sunlight to ATP and finally to heat. Eugene P. Odum expresses it succinctly: "Matter circulates; energy dissipates." The dissipation of energy is a corollary of the Second Law of Thermodynamics (see pp. 24–25), and it explains why life on Earth requires the Sun to continuously replenish the lost energy.

A few ecologists have actually undertaken the enormous task of accounting for all the energy transfers within a particular ecosystem. One classic example of such an **ecosystem analysis** was conducted at Silver Springs, Florida by Howard T. Odum (brother of Eugene P. Odum) in the 1950s. A summary representation of Odum's findings is shown in Fig. 17.7.

TROPHIC LEVELS. Because energy dissipates as heat each time it is transformed by an organism, the amount of energy available to an animal depends on how many transformations have occurred before energy reaches the animal. The position of an animal in the sequence of transformations is called its **trophic level,** from the Greek word *trophe,* meaning food. The lowest trophic level of animals is that of **primary consumer**—an animal that eats plants. Primary consumers include **herbivores,** which eat plants only. **Omnivores,** which eat both plants and animals, function as primary consumers when eating plants.

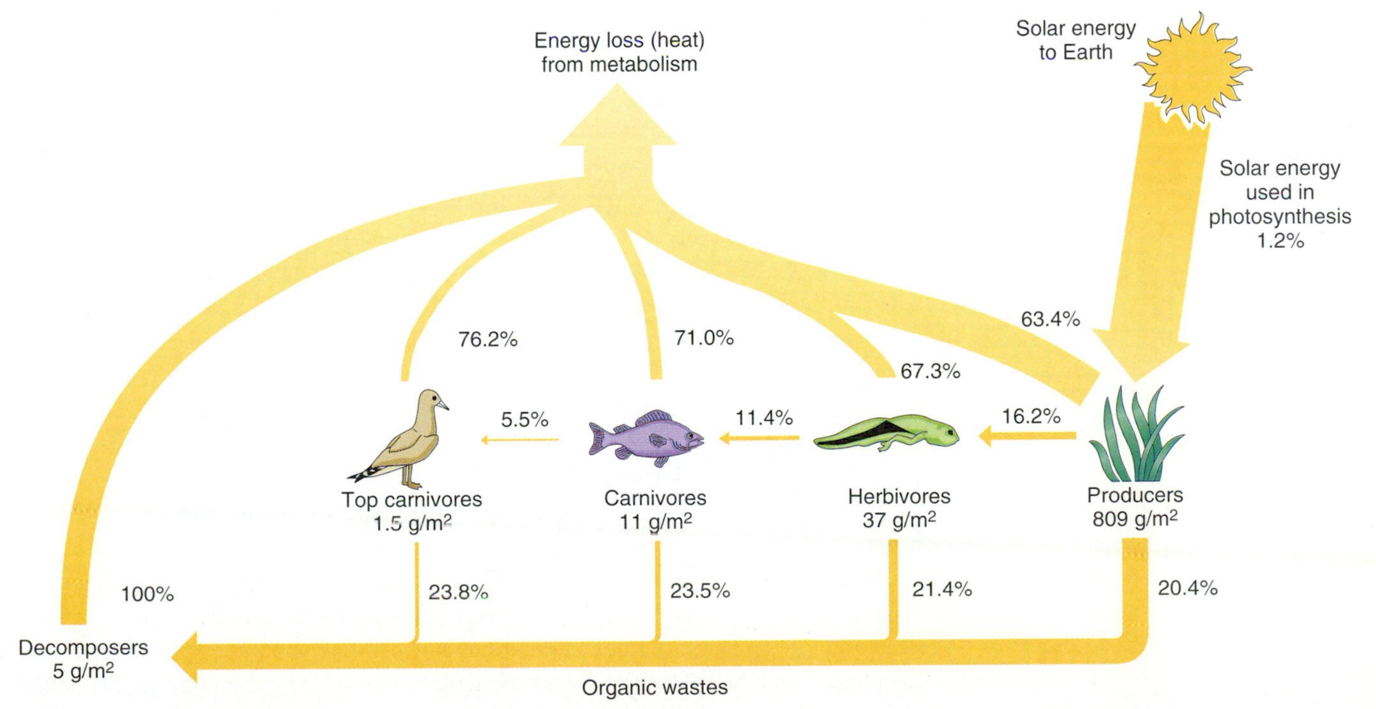

FIGURE 17.7

Annual energy flow through an ecosystem, based on a study in Silver Springs, Florida. All the incident solar energy is eventually radiated back to space as heat, but 1.2% of it is first used by producers in photosynthesis (arrow from Sun). This energy is then available to producers and to consumers. Numbers beneath the producers and the consumers represent the biomass in grams per square meter. During metabolism, producers convert 63.4% of the energy available from photosynthesis directly into heat (upward arrow). Of the energy made available in photosynthesis, 20.4% goes into organic wastes and is eventually converted to heat during metabolism by decomposers (arrow down and to the left). The remaining energy from photosynthesis (16.2%) is consumed by herbivores (arrow to left). Arrows from the herbivore and other consumers likewise represent the proportions of the energy input that go into heat and organic wastes and to the next consumer. (Data from Odum, H. T. 1957. Trophic structure and productivity of Silver Springs, Florida. *Ecol. Monogr.* 27:55–112.)

Some of the energy ingested by a primary consumer is lost as indigestible plant fiber in feces. The rest is **assimilated** into the animal. Part of the assimilated energy is lost as dissipated heat or is used immediately by the primary consumer. Whatever energy is left is available to produce new tissues, through fat storage, growth, and reproduction.

In the long run, only this new biomass produced by part of the assimilated energy is available as an energy source at the next trophic level, that of the **secondary consumer.** A secondary consumer is an animal that eats a primary consumer. Secondary consumers can be omnivores or, if they eat only other animals, **carnivores.** Once again, the secondary consumer loses some of the ingested energy (hair, bones, and other indigestible wastes) in feces. The secondary consumer loses some of the remaining assimilated energy as heat, and whatever is left is either used or stored in new biomass. Again, only this biomass can benefit an animal at the next trophic level that eats the secondary consumer. The process of energy dissipation continues in this way from trophic level to trophic level, up to the **top carnivore.**

ECOLOGICAL GROWTH EFFICIENCY. The energy that is converted into new biomass divided by the ingested energy is defined as the ecological growth efficiency. The ecological growth efficiency varies with the organisms involved. One study on Isle Royale in Lake Superior, for example, showed that for every 100 kcal of plant energy that moose consumed, only about 1 kcal eventually produced new tissues in the wolves that preyed on moose. Thus the ecological growth efficiency may have been approximately 10% for the moose and 10% for the wolves. Much assimilated energy is not converted into new tissues, but is lost as feces or used for immediate needs. In birds and mammals (homeotherms), much of the assimilated energy is used to keep the body warm (Fig. 17.8). Poikilotherms, in contrast, can generally store more of the assimilated energy.

Ecological Pyramids

Ecosystem analyses invariably show that there is less available energy at higher trophic levels than at lower ones. This relationship can be expressed by a **pyramid of energy,** in which the amount of energy at each trophic level is represented by the area of each level of the pyramid (Fig. 17.9 on page 334). Because higher trophic levels must get all their energy from lower trophic levels and because much of the energy is lost with each energy transfer, upper levels have less area than lower levels. Hence the pyramid shape. Because the energy is stored in an organism's tissues, the pyramid of energy also implies a **pyramid of biomass.** In the pyramid of biomass the area of each level of the pyramid represents the "standing crop": the total mass of living organisms at that level at a given time.

Pyramids of energy and of biomass can temporarily become partly inverted, as when top carnivores survive the temporary depletion of their prey. When averaged over a long time, however, the Second Law of Thermodynamics dictates that the pyramids will taper steadily from lower to higher trophic levels.

If the average size of individual organisms is the same at each trophic level, then the pyramids of energy and biomass also imply a **pyramid of numbers** of individual organisms at each trophic level. Usually there are indeed fewer carnivores than herbivores in an area, and fewer herbivores than individual food plants. However, the exact numerical relationship between individuals in each trophic level depends on their relative body sizes. For example, even though ants are at a higher trophic level than an acacia tree, thousands of ants can live on one tree because each ant is so small. If the body sizes of the organisms are known, the number of animals at each trophic level can be calculated from the amount of energy available to them. Using this concept, Schneider and Wallis (1973) were able to calculate the maximum number of Loch Ness monsters.

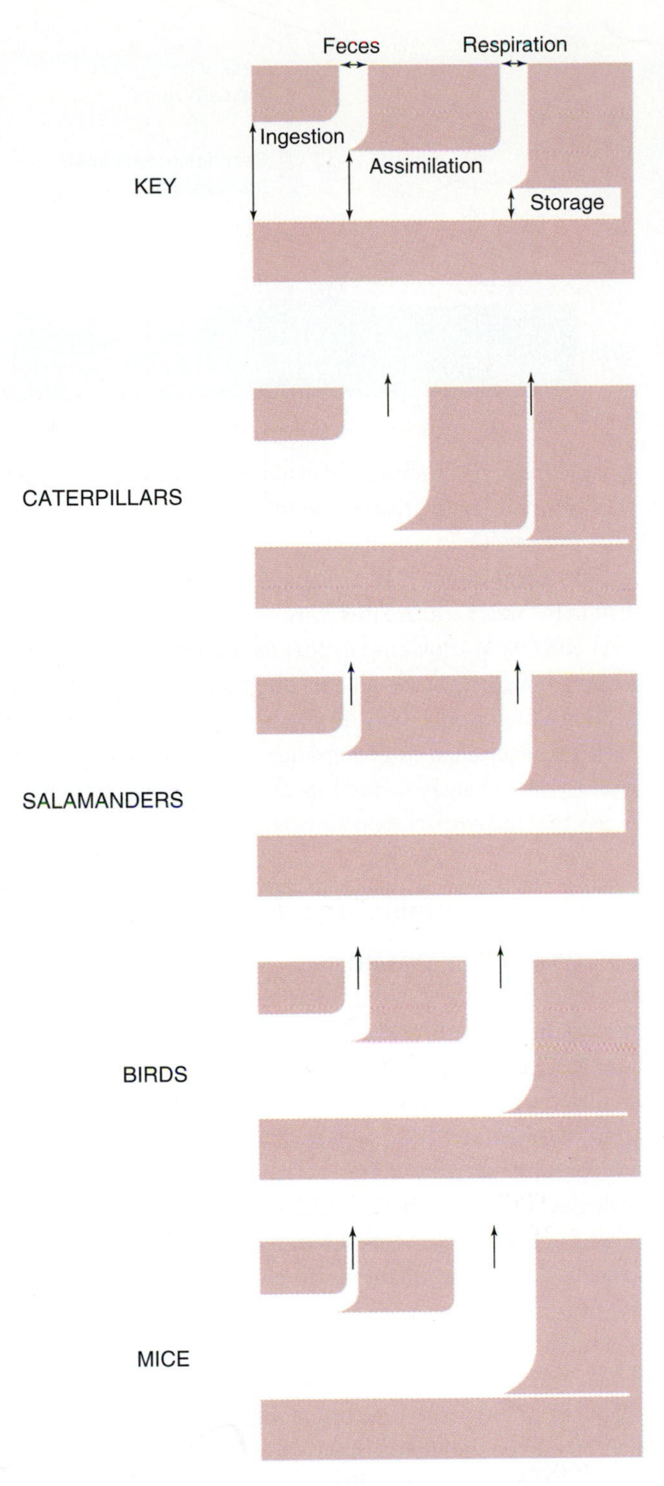

FIGURE 17.8

Energy inputs and outputs for four kinds of animals. Ingested energy is either assimilated or lost as feces (see key). Of the assimilated energy, part is consumed in mitochondrial respiration to satisfy the energy needs of the animal, and the remainder is stored as new tissue. Caterpillars lose much of the ingested energy as feces. Birds and mammals use much of their energy in cellular respiration to keep warm. Salamanders, which do not maintain a high body temperature, store the most energy as tissue. [Based on a study from Hubbard Brook Experimental forest in New Hampshire. See J. R. Gosz et al. 1978. The flow of energy in a forest ecosystem. *Sci. Am.* 238(3):92–102 (Mar).]

They found that the energy available in Loch Ness could sustain 157 small Nessies (average body mass 100 kg) or 10 large ones (average mass 1500 kg).

Food Chains and Webs

The transfer of energy from organism to organism can be represented as a food chain: a linear sequence in which one species is food for the next species in the sequence. Each food chain generally consists of producers plus two levels of consumers in terrestrial

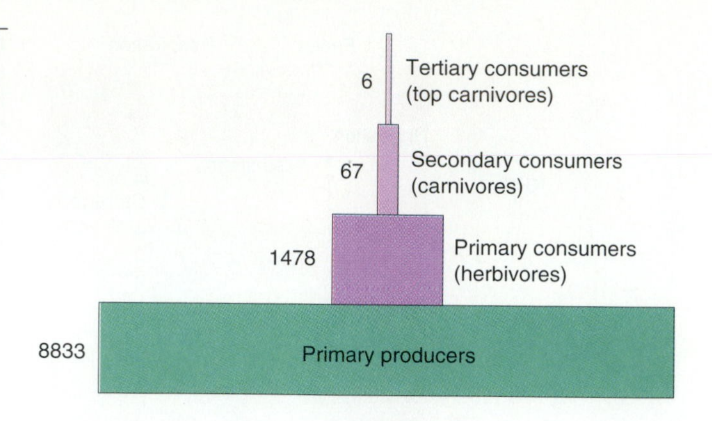

FIGURE 17.9

A pyramid of energy for the Silver Springs ecosystem shown in Fig. 17.7. The area of each level of the pyramid represents the amount of energy (kcal/m^2) in each trophic level.

■ **Question:** How would this pyramid be affected if the tertiary consumers were eliminated?

ecosystems or three levels of consumers in aquatic ecosystems. The difference in number of trophic levels arises from differences in the sizes of producers. In terrestrial systems the producers are mainly plants, which are large enough that large animals can feed on them directly. In aquatic ecosystems the producers are mainly algae, which drift in the water and are therefore by definition part of the **plankton.** The extra trophic level consists of tiny animals that feed on the algae, thereby concentrating them and making the energy accessible to larger animals. Usually several food chains intertwine to form a **food web.** Fig. 17.10 shows some of the members of a food web in Antarctica. Even this simplified food web includes several parallel food chains leading from the photosynthetic algae to the top carnivore, the killer whale *Orcinus orca.* This figure shows that many species occupy more than one trophic level.

Interactions Among Organisms

PREDATION AND SYMBIOSIS. Because organisms depend on each other for food and other biotic factors, they inevitably interact with each other. These interactions can be classified into several categories, the major ones being predation, symbiosis, and competition (Table 17.1 on page 336). Predation refers to the capture and eating of one animal, the **prey,** by another animal, the **predator.** Symbiosis includes any close association of organisms belonging to different species (not necessarily animals). In everyday speech the term symbiosis often refers to a mutually beneficial association, which biologists call **mutualism.** Besides mutualism, however, biologists also refer to two other kinds of associations as symbiotic. One of these is **parasitism,** in which one organism (the **parasite**) benefits from living on or in another living organism (the **host**), which is harmed but not killed outright. The second is **commensalism,** in which one organism (the **commensal**) benefits from associating with a host, which is neither helped nor harmed. In practice it is often difficult to distinguish predation from parasitism, or mutualism from commensalism. Most of the organisms that are called parasites cannot be shown to harm their hosts.

COMPETITION. Competition can occur whenever two or more organisms in the same habitat depend on the same limited resource, such as food, space, or nesting sites. The more similar organisms are to each other, the more likely they are to depend on the same limited resources, so competition is likely to be most intense among members of the same species and to decrease as evolutionary relatedness decreases. Many ecologists, but not all, find it useful to think of an organism as competing for a **niche** that is defined by the organism's interactions with other organisms and the environment. The niche is not a geographic location, but instead the organism's place in an

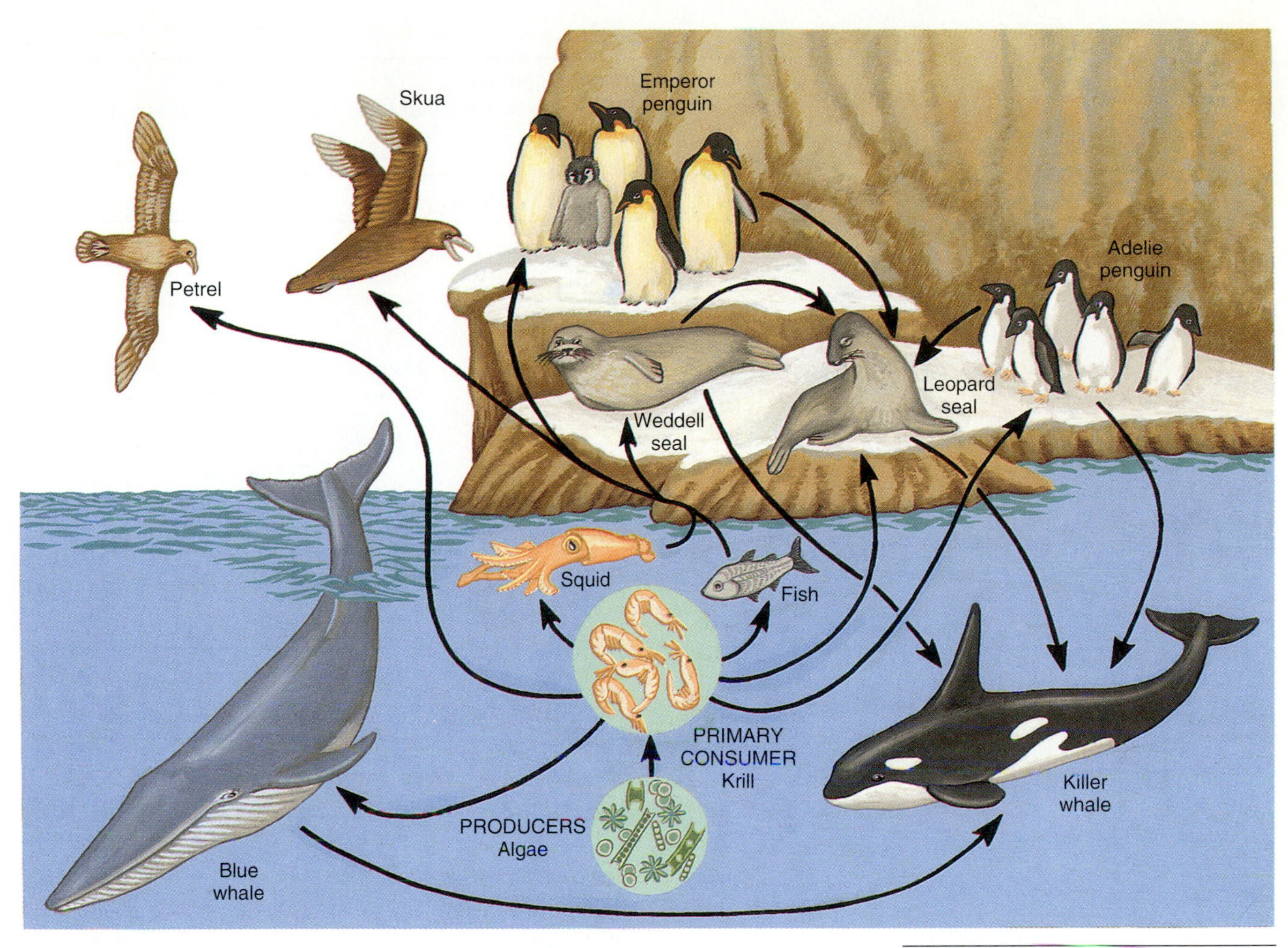

FIGURE 17.10

A simplified representation of an Antarctic food web. See Laws (1985) for a more nearly complete representation and discussion of this food web.

■ **Question:** What is the trophic level of the blue whale? The killer whale? Which whale do you think would be more abundant?

ecosystem as consumer of algae, predator on sheep, or whatever. A niche can be thought of as a job that can be filled only by those with the exact qualifications. Just as there are often many applicants for one job, two or more species may be able to compete for the same niche.

It is generally assumed that where there is competition for a niche, only one species will be successful and the others will be excluded from the habitat. This concept is called **the principle of competitive exclusion.** It is also called "Gause's principle," after G. F. Gause who published a laboratory study of it in 1934. Gause found that *Paramecium aurelia* and *P. caudatum* thrived when cultured separately in a particular medium, but when the two protozoans were cultured together, *P. aurelia* suppressed the population of *P. caudatum.* Similar results have been demonstrated with flour beetles (*Tribolium*) and with *Drosophila.*

Although competitive exclusion has clearly been demonstrated in the laboratory and in some field studies, it may not be a common occurrence in nature. One reason is that different species would seldom compete for precisely the same niche in the same habitat (Mares 1993). In seeming defiance of Gause's principle, for example, vast herds of wildebeest, zebra, topi, and Thomson's gazelles graze peacefully together on the grasses of the Serengeti. They apparently avoid competition by feeding on different parts of the grasses. Zebra tend to feed first, taking the tall grass stems. Topi and wildebeest then follow, taking the leaves and lower portions of the grasses that were exposed by the zebra. Finally Thomson's gazelles forage primarily on new grass shoots and on

TABLE 17.1

Types of interactions among organisms.
The symbols +, –, and 0 indicate whether the individual organism is benefited, harmed, or unaffected by the interaction.

INTERACTION	EFFECT		EXAMPLE
Predation	Predator +	Prey –	Coyote/rabbit
Symbiosis			
Mutualism	Mutualist +	Mutualist +	Cleaner fish/grouper
Parasitism	Parasite +	Host –	Tapeworm/human
Commensalism	Commensal +	Host 0	Some bacteria/human
Competition	Competitor –	Competitor –	Coyote/fox

broad-leaved plants and their fruits (Fig. 17.11). (See p. 374 for another example of how species avoid competition.)

A second reason why competitive exclusion may be unusual is that predation, disease, and accidents such as fire and storms can keep populations so low that competition does not occur. The role of predation in preventing competition was shown in a classic study by Robert T. Paine on the Pacific coast of Washington state and published in 1966. Paine found that the starfish *Pisaster ochraceous* normally depressed the populations of molluscs, barnacles, and other organisms clinging to the rocks. When he removed the starfish from a small area, populations of their prey increased, and they began to compete for space. Eventually one competitor, the mussel, *Mytilus californianus,* covered almost all the area.

Succession

Within each ecosystem are numerous associations among distinct groups of species. These associations, called **communities,** can be found repeatedly in many different areas. For example, the assortment of bacteria, fungi, and insects in rotting logs is so typ-

FIGURE 17.11

Resource partitioning in four species of grazing ungulates in Africa. At the close of the wet season, each species specializes on different parts of plants, thereby avoiding competition.

■ **Question:** How would the extinction of zebras affect the other three species?

ical that ecologists around the world recognize the phrase "rotting-log community." Although ecological communities are somewhat stereotyped, they do undergo changes, just as human communities do. Often one ecological community replaces another in a regular sequence called succession.

Ecologists recognize two major kinds of succession. Succession that begins with a complete absence of life, including soil organisms, is called **primary succession.** On freshly exposed rock, for example, primary succession often begins with lichens, which are mutualistic associations of fungi and algae. The lichens trap wind-blown soil and slowly erode the rock, creating a substratum that facilitates the growth of mosses. Dead mosses accumulating as a layer of organic humus will, in turn, facilitate the growth of grasses and other shallow-rooted plants. These grasses will shade out the mosses and contribute to the humus, which will eventually become deep enough to support larger shrubs and then trees. All these stages of primary succession are often visible on different parts of the same rock.

Successional change that begins with some life already present is called **secondary succession.** Agriculture is the most common initiator of secondary succession in North America. In the eastern United States the most familiar example of secondary succession is **old-field succession** on abandoned farms. After crop lands and pastures are abandoned and no longer mowed or grazed, they revert to tall grasses, then to weeds and shrubs. These "pioneer plants" inhibit the growth of other species, but they are generally short-lived and intolerant of shade. Eventually they are replaced by trees such as aspens and pines. These trees in turn will be replaced by tree species whose seedlings can tolerate shade. Succession ends when the area is covered in shade-tolerant trees whose seedlings can grow up in the shade of their parents. Such stable plant communities are called **climax** stages of succession. Oak and hickory dominate climax forests in the mid-Atlantic states. Farther north, beech, maple, and birch are the dominant species in deciduous climax forests.

As the plants in a community change during succession, so must the animals that depend on the plants for food and habitat. Such change in animal species is called **faunal succession** (Fig. 17.12). In addition to passively following plant succession, ani-

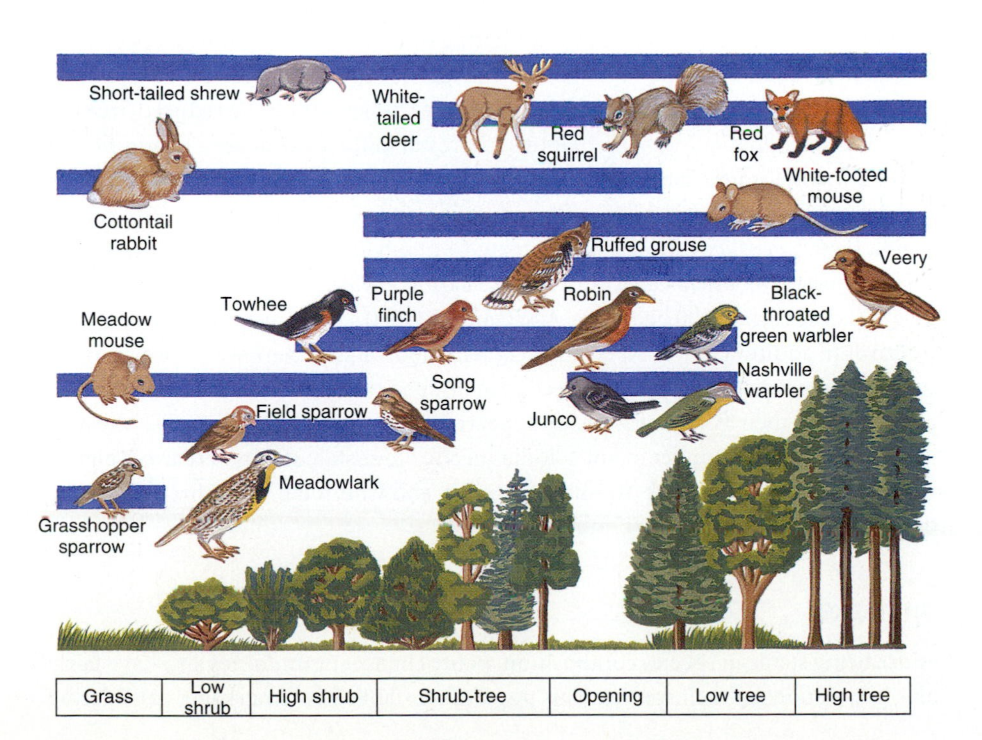

FIGURE 17.12

Faunal succession in the northeastern United States. The bars beneath each species show its range of habitats. As fields gradually succeed to climax forest (left to right), the animal community also succeeds.

■ **Question:** What species would be lost as a field succeeds to forest? What species would be gained? What species would be unaffected?

mals also contribute to it. Insects and other herbivores retard old-field succession by cropping the shoots of weeds and trees. The animal with perhaps the most obvious impact on succession is the beaver *Castor canadensis,* whose dams convert wooded valleys into ponds. The ponds then gradually fill with organic matter, forming meadows that eventually succeed to level woods.

Biodiversity and Stability

The species within a climax community are often remarkably stable from year to year. Many ecologists assume that the stability of a community increases with biodiversity (the number of species), because each species is less likely to be totally dependent on another. For example, one would not expect killer whales to leave an Antarctic community if penguins become scarce, because they can switch to preying on leopard seals. The relationship of biodiversity to stability seems self-evident and has been verified in at least one study (Tilman and Downing 1994). It is also possible that in addition to biodiversity leading to stability, stability may contribute to biodiversity, by providing time for species to become established in a community.

There is one situation in which increasing biodiversity leads to instability rather than to stability. If an **exotic species** enters an ecosystem where it has no competitors, predators, or parasites, its population can soar while other species are extinguished. During the last century at least 4500 exotic species have entered the United States, usually with the help of humans. Besides the ecological consequences, the cost to agriculture, industry, and health is estimated to be $97 billion. Numerous examples of exotic species will be presented in Unit 4, but the accidental introduction of gypsy moths *Lymantria dispar* is a good example. In 1869 a naturalist brought gypsy moths from Europe into a suburb of Boston in hopes of hybridizing them with the silkworm moth *Bombyx mori* to get the silk-making ability of *Bombyx* without its finicky taste for mulberry leaves. A few of the gypsy moths escaped, however, and now each summer throughout the eastern United States their descendants turn thousands of acres of foliage into a deluge of feces.

In addition to biodiversity, stability also depends on other factors, such as area. For example, isolating half a forest by lumbering or by fragmenting it with roads does not simply produce a smaller forest, but it changes the species within the community. The reason appears to be that populations of some species within the reduced area become so reduced that random fluctuations can result in their extinction. In addition, it is more difficult for new species to colonize a smaller area. For a given ecosystem type, the number of species is proportional to area$^{0.27}$. As a rule of thumb, therefore, reducing area by 90% reduces species biodiversity to one-half. The effect of reducing area has been studied in a tropical rainforest by clearing a wide perimeter to isolate tracts with areas of 1, 10, and 100 hectares. (A hectare is 10,000 square meters: approximately 2.5 acres.) In addition to the expected decline in biodiversity, there were species losses that were not predictable. In the 10-hectare area, species of birds that follow army ants and prey upon their fleeing victims disappeared within a few years, because the area was too small for the number of ant colonies needed to sustain them. Peccaries (piglike animals) abandoned even the 100-hectare tracts, and with them went three species of frogs that breed in the peccaries' mud-wallows.

Population

Besides being stable in species composition, many climax communities also have fairly stable populations of each species. This population stability is remarkable considering that most breeding pairs of animals can produce many offspring in a year. Many insects, for example, produce hundreds of fertilized eggs per year, many pairs of birds commonly produce more than eight eggs per year, and many mammals produce large litters of offspring. Even a couple of members of our own slow-breeding species could generate 100,000 new humans in a century if each couple had 10 children. One would therefore expect the population of every species to increase exponentially. If the population of rabbits increased tenfold each year, then 10 rabbits one year could become 10^2 rabbits in the second year, 10^3 the next, 10^4 the next, and so on.

Exponential population growth does, in fact, occur when populations are allowed to reproduce unchecked by crowding, competition, disease, or predation. These conditions commonly occur during the first few years after an exotic species is introduced (Fig. 17.13A). Exponential population growth is also occurring among humans because the natural checks on population have been temporarily lessened by advances in agriculture and medicine (see pp. 888–889). Exponential population growth cannot continue indefinitely, however. There are only 10^{72} electrons in the entire universe, so a species with a tenfold rate of population growth would use up all the electrons in the universe in less than 70 generations. Long before that, of course, the species would have killed off its food species, either directly or by damaging the environment. Once that happens the population necessarily "crashes" (Fig. 17.13B).

Exponential growth and crashes are temporary phenomena. Over the long term many natural populations fluctuate at levels around the maximum number that the ecosystem can support. This maximum sustainable population level is called the **carrying capacity** (Fig. 17.14). A population may rise toward or above the carrying capacity temporarily, but then the depletion of food, the spread of communicable diseases and parasites, or simply the stress of overcrowding increases the death rate and decreases the birth rate. The population then falls below the carrying capacity, reducing disease and stress and allowing food sources to recover. The population may then start to increase once more.

POPULATION CYCLES. Fluctuations in population around the carrying capacity often occur in regular patterns called population cycles. For example, the sheep population shown in Fig. 17.14 peaked in about 1854, 1880, and 1910, suggesting a population cycle with a period between 26 and 30 years. Such cycles are believed to result from the delayed effects of overpopulation. For example, it may take several years for food resources to be depleted or for the populations of parasites and predators to increase. A population cycle in one species can induce a cycle in another species in the same ecosystem. Figure 17.15 on page 340 shows how the population cycle of the

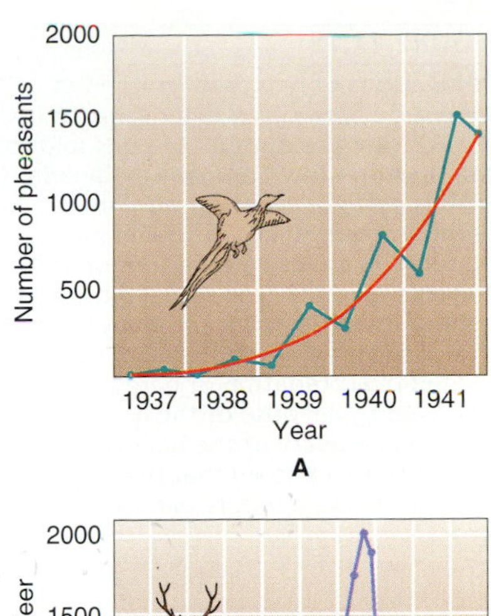

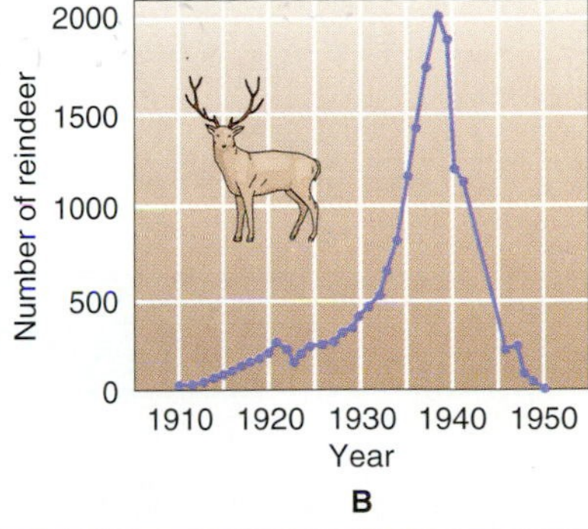

FIGURE 17.13

(A) Exponential growth in the population of ring-necked pheasants *Phasianus colchicus* following introduction on Protection Island, Washington. The smooth curve averages the fluctuations due to spring hatching and winter kill. The population approximately doubled each year following introduction. (B) After 18 reindeer *Rangifer rangifer* were introduced onto the Pribilof Islands, Alaska, their population grew exponentially for 30 years. Then, presumably because of overgrazing, the population crashed. A decade later there were only eight survivors.

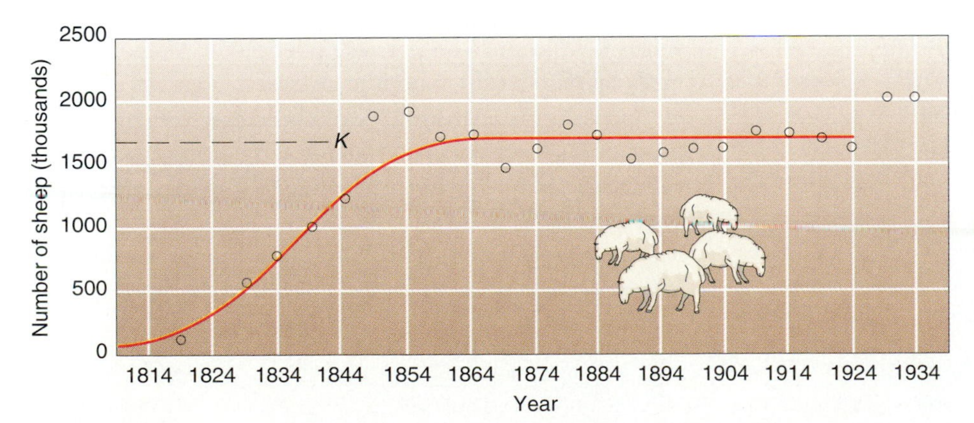

FIGURE 17.14

The population of domestic sheep *Ovis aries* in Tasmania. The data points show populations averaged over five years. The dashed line indicates the presumed carrying capacity (*K*).

■ **Question:** These data were collected before many of the modern sheep farming methods were introduced. How could some of those methods affect the carrying capacity?

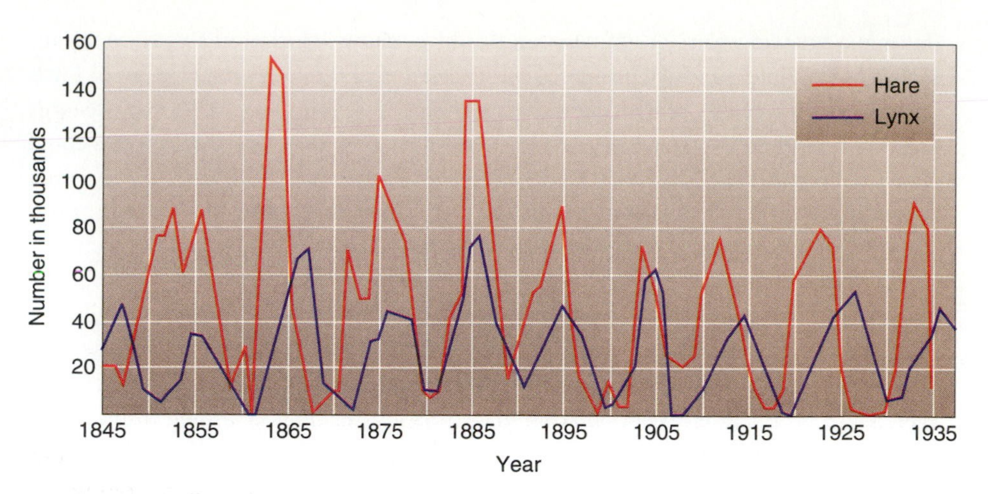

FIGURE 17.15

Populations of lynx *Felis lynx* and of snowshoe hare *Lepus americanus* as inferred from the number of pelts sold to the Hudson's Bay Company in Canada. (This is perhaps the most published graph in biology.) Both species have population cycles with periods of about 10 years, and the lynx cycle tends to lag behind the hare cycle. For many years this cycle was thought to show repeated overpredation of hares by lynx, followed by collapse of the lynx population and recovery of the hare population. It is now known that the hare population cycle occurs even without lynx.

r Selection, *K* Selection, and Life-History Traits

If certain assumptions are made, exponential population growth, followed by a leveling off near the carrying capacity, can be modeled mathematically by a **logistic equation:**

$$\frac{dN}{dt} = \frac{rN(K-N)}{K}$$

where dN/dt is the rate of population growth (number of new individuals per year); r is the natural rate of population growth if predation, starvation, and so on, do not limit it; N is the population at a given time; K is the carrying capacity.

Note that when N is much smaller than K, as when a species is newly introduced into an ecosystem, $dN/dt = rN$ approximately. This equation describes an exponential curve. When N equals K, dN/dt is zero: The population stops increasing. $(K-N)/K$ represents the unused proportion of the carrying capacity. For example, if the carrying capacity is 100, but the population is only 90, then 10% of the carrying capacity is still available for population growth. In words, this logistic equation says that the rate of population growth is proportional to the natural rate of growth, times the population, times the proportion of the carrying capacity still available.

This logistic equation suggests that high reproductive success can result from two kinds of adaptations: those that increase the reproductive rate r and those that maintain the population as high as possible relative to the carrying capacity K. These adaptations affect **life-history traits.** Life-history traits that increase r include early age of first reproduction, large number of offspring per mating, and large number of matings per season. Life-history traits that increase K include increased parental care, large body size, and long life. Life-history traits that increase r tend to be incompatible with the traits that increase K and vise versa, so many ecologists once thought that organisms tended to have either r-selected traits or K-selected traits. Examples of r-selected animals would be gypsy moths, most flies, and many small rodents. r-selected organisms would tend to increase their populations by reproducing rapidly whenever the chance arose, even if the population could not be sustained for long. Many birds and mammals, including humans, were considered to be K-selected. K-selected animals would tend to maintain high stable populations.

Most ecologists now consider the dichotomy of r- versus K-selection to be an oversimplification. Many animals have life-history traits that increase r at one stage of life and traits that increase K at another stage. Moreover, the genetic traits that affect r and K can quickly change within species. Rather than categorize each species as being r-selected or K-selected, ecologists now consider it necessary to study how each life-history trait individually affects population.

snowshoe hare *Lepus americanus* in Canada induced a cycle in the population of one of its predators, the lynx *Felis lynx*. Note that the lynx cycle lagged the hare cycle, because it took a year or so for the rise and fall of the hare population to influence the fertility and death rates of lynx.

Biomes

Communities are often similar to each other in widely separated places that share similar climates and other abiotic characteristics. It is therefore useful to refer to such communities as belonging to different ecosystem types, such as prairie, tropical rainforest, lake, desert, and so on. Aquatic ecosystems are mainly divided into freshwater or marine, depending on ionic concentration. Several types of freshwater and marine ecosystems are listed in Fig. 17.2 and will be described later. Terrestrial ecosystem types are called biomes. The two most important abiotic factors defining a biome are temperature and moisture (Fig. 17.16 on page 342 and Fig. 17.17 on page 343). These two factors determine which kinds of plants grow in a biome, and the plants largely determine the kinds of animals there. The major kinds of biomes are described below, and tropical rainforests are described in greater detail because so many different species of animals live in them.

TUNDRA. On the polar ice caps no plant grows. The few animals, such as polar bears in the Arctic and penguins in the Antarctic, obtain food from the seas and are best considered members of an aquatic ecosystem rather than a biome. The northernmost biome is the belt of **arctic tundra,** where extreme cold and extended darkness are the two major facts of life. There is little precipitation, but so little water evaporates that the upper layer of soil is saturated in summer. Beneath this soggy layer is permanently frozen **permafrost.** Trees cannot sink their roots deeply, so the vegetation consists mainly of short grasses and other nonwoody plants. Similar plants are often found on mountain tops above tree line, in **alpine tundra,** where the soil is too shallow to anchor trees against the wind. The few animals that survive year-round in Arctic tundra are mainly mammals such as the arctic fox, snowshoe hare, lemming and other rodents, caribou, polar bear, and musk-ox. One of the few birds that live year-round in the arctic tundra is the ptarmigan, but many migratory birds arrive in the summer. The eggs and larvae of mosquitoes and some other insects can also survive the winters, and in summer their populations often reach plague proportions.

NORTHERN CONIFEROUS FOREST. South of the arctic tundra in North America and Eurasia is the **taiga** (pronounced TYE-guh), which is dominated by vast forests of cone-bearing evergreen trees, such as spruces and fir. Similar coniferous forests grow at higher elevations farther south. These forests often seem dark and forbidding, because the needle-shaped leaves of conifers do not transmit sunlight. Little vegetation grows on the forest floor, not only because of the darkness but also because the soil is acidic and poor in nutrients. Most animals, including squirrels, birds, and hordes of insects, therefore forage high in the trees. Broad-leaved shrubs and trees do grow along streams, lakes, and other clearings, however. Deer, bear, beaver, moose, and hare can often be found feeding there.

TEMPERATE DECIDUOUS FOREST. In moist temperate regions, such as in the eastern United States, conifers give way to broad-leaved trees that lose their leaves in autumn. The fallen leaves of autumn form a nutrient-rich **humus** on the forest floor. Even in spring and summer, deciduous forests appear lighter than coniferous forests,

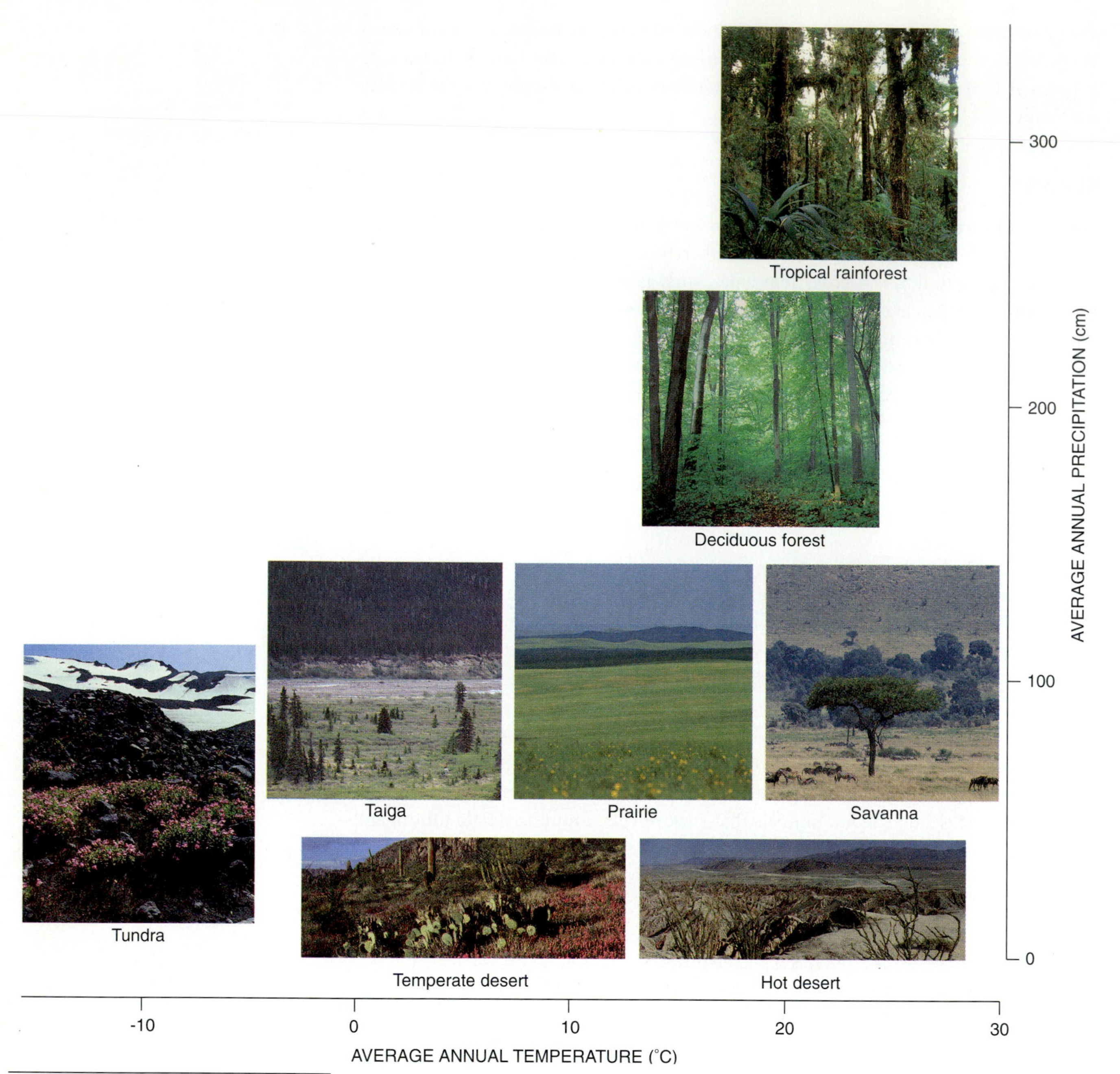

FIGURE 17.16

Temperature and precipitation determine which plants and animals can survive in a biome.

■ **Question:** Why are there no biomes in the upper left of the figure?

because the leaves transmit more light. In contrast to coniferous forests, therefore, the floors of deciduous forests are often crowded with flowering plants, shrubs, and small trees. Animals are diverse and abundant at all levels. Among the most conspicuous are deer, black bears, squirrels, chipmunks, rabbits, and birds of all kinds.

Temperate Grassland (Prairie). In central North America and in parts of Africa, South America, and Asia, ground water is too scarce to support trees. In addition, fires are often important in maintaining and extending prairies. Grasses and other nonwoody plants are dominant plants, and grazers such as antelopes and bison are (or were) the dominant animals. Natural prairies are now quite rare, because the deep, fertile soil proved too tempting as farmland, and the absence of forest cover made the animals easy targets for hunters. Smaller vertebrates, such as prairie dogs, rabbits, foxes, coyotes, and prairie chickens, still thrive on the remaining grasslands.

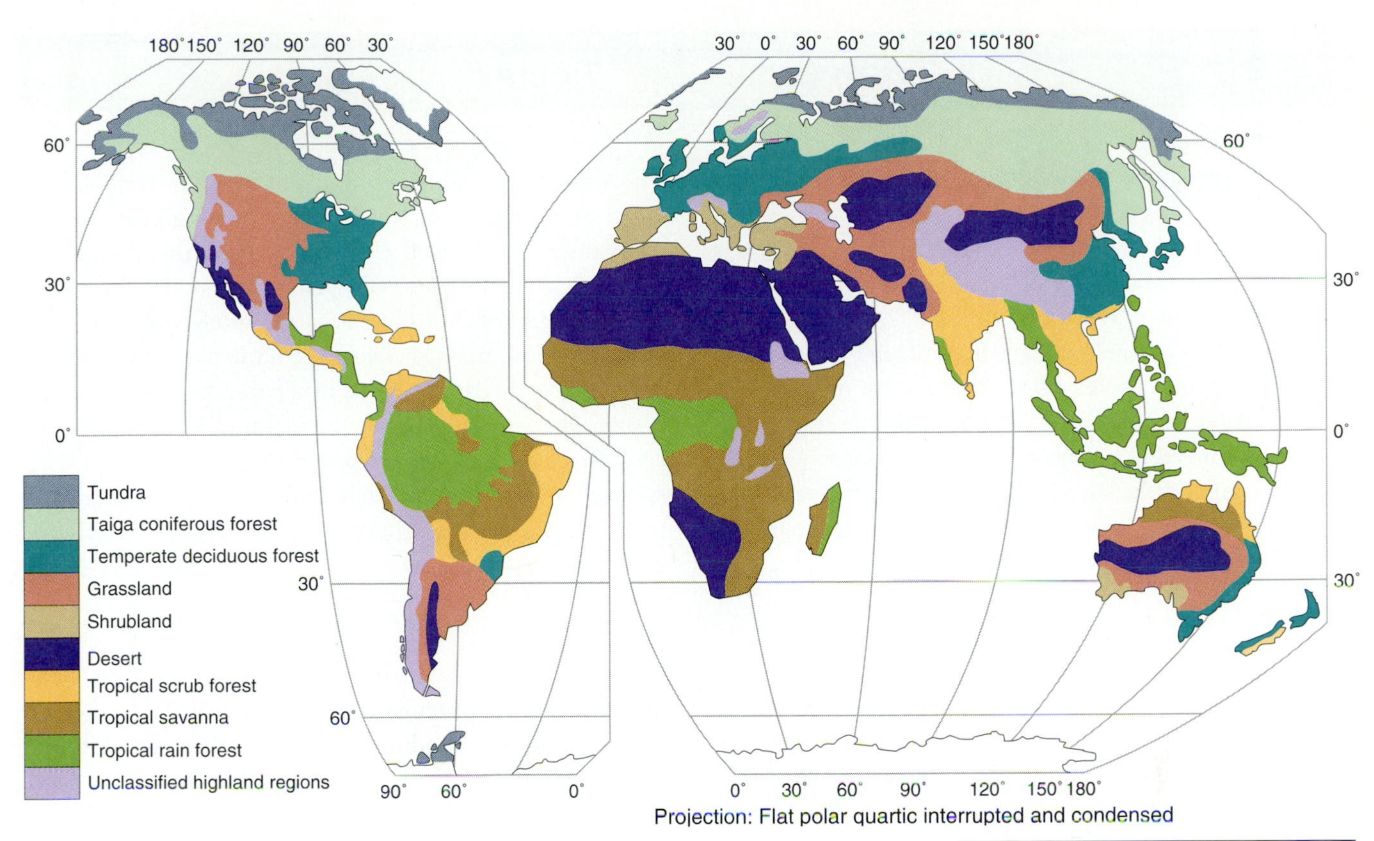

FIGURE 17.17
Distribution of major biomes of the world.

■ **Question:** Based on this map, in what part of Fig. 17.16 would shrublands and tropical scrub forest appear?

TROPICAL GRASSLAND (SAVANNA). In the savannas of Africa and South America the average annual rainfall is abundant, but periodic dry seasons and fires limit trees to isolated clumps. These trees, such as acacia and baobabs in Africa, are centers of intense animal activity. Ants nest within thorns on the acacias and pay their rent by protecting the trees from other insects. Only giraffes browse among the branches. The baobab trees provide nest sites for birds, at least until they are knocked over and eaten by elephants. In Africa the grasses support the largest diversity of hoofed animals on earth, including zebras, giraffes, and many species of antelope, such as wildebeest, topi, and gazelle. They graze peacefully but watchfully in view of lions, hyenas, and other predators that wait to take advantage of the weak or lame.

DESERT. Deserts occur wherever the potential for evaporation exceeds rainfall. In the **hot deserts** of the Sahara and the American Southwest the high evaporation is due to high daytime temperatures. **Cool deserts** occur in areas of extremely low precipitation, such as in Washington and Oregon, where the Cascade Mountain range intercepts moisture from the prevailing winds. Desert plants and animals must be adapted to quickly absorb and hold whatever moisture is available. Many desert animals obtain much of their water from metabolism, and their excretory organs are adapted to conserve water. In hot deserts they are often nocturnal and remain underground during the heat of the day. Among these are numerous reptiles, insects, scorpions, and rodents. Grasslands can become deserts if they are stripped of vegetation, which normally shades the soil and holds its moisture. **Desertification** is occurring in much of Africa because of overgrazing by cattle.

TROPICAL RAINFOREST. Tropical rainforests occur in equatorial regions with constant high temperature and moisture. A related biome, the **semi-evergreen, seasonal tropical forest,** occurs in equatorial areas subject to alternating dry and rainy

An Endangered Biome

You might expect an ecosystem as diverse as tropical rainforests to be stable, yet they lose several species per average day, with 4000 to 6000 species already lost. This is about 10,000 times the natural rate of extinction. The cause of this instability is an exotic species—*Homo sapiens.* It is now well documented in many tropical rainforests that the arrival of humans was followed by a wave of extinction, due mainly to overhunting. Now the major cause of extinction is lumbering and agriculture. Approximately 154,000 km^2—an area the size of Florida—are cleared each year. Half the tropical rainforests have already been destroyed, and at the present rate there will be none left in 50 years.

Among the casualties of tropical rainforest destruction may be many North American songbirds that overwinter there, as well as many other species of beautiful and fascinating animals, plants, and other organisms. In addition, we may also lose many organisms that could be of practical benefit to humans. About half the available pharmaceuticals originated from tropical plants. Pharmaceutical firms have found it more efficient to isolate new drugs from tropical plants than to synthesize them from scratch, and more than a thousand anticancer drugs have been discovered in this way.

seasons. Tropical rainforests now cover approximately 3% of the planet, but they account for more than 20% of its net primary production and more than half of its biomass. Perhaps more than half the world's species live in tropical rainforests: the figure is uncertain because only 1.4 million of the 10 to 100 million species on Earth have been described and named. The diversity of plants is so great that one must often search for hours to find two individuals of the same species. In an 800-km^2 area of Central American rainforest, 500 to 600 resident species of birds have been counted. Five hundred species of insects were caught in 2000 sweeps of a net near the ground, and many more species live in the canopy. In the early 1980s Terry L. Erwin of the National Museum of Natural History exploded insecticide bombs in the canopy and found that the majority of insects brought down were previously unknown. Among these, E. O. Wilson of Harvard University identified 43 species of ants living on one tree in Peru—as many as live throughout Canada or the British Isles. With so many different species, no single one is numerous enough to be called dominant.

The immense biodiversity of tropical rainforests is due largely to **vertical stratification** of light, humidity, temperature, and wind speed, which gives each layer of forest its own microclimate (Fig. 17.18). Light is only about 4% as bright on the ground as in the canopy. Consequently, most of the primary production occurs in the canopy, and there is little vegetation near the ground. ("Jungles" of thick undergrowth usually occur only along rivers or in abandoned clearings.) Vertical stratification of vegetation also leads to vertical stratification of animals. Many species spend their entire lives in the canopy without ever descending to the ground. These species include insects, frogs, lizards, birds, bats, and monkeys that feed on leaves, fruit, and nectar. The male birds are often brightly colored and engage in showy courtship displays that make them visible to potential mates in the thick foliage. Many mammals and insects in the canopy contribute to the ever-present chorus of courtship and other sounds.

Many tree-climbing species range between the canopy and the ground. These include frogs, many of which breed in water that collects in the leaves of **epiphytes**—plants that grow on the limbs and trunks of trees. Some snakes can also climb tree trunks in pursuit of prey. Most of all there are the ants and termites that range up and down the trees. These two groups alone account for one-third of the animal biomass in the rainforest. Biting ants defend many trees against both epiphytes and tree-climbing animals.

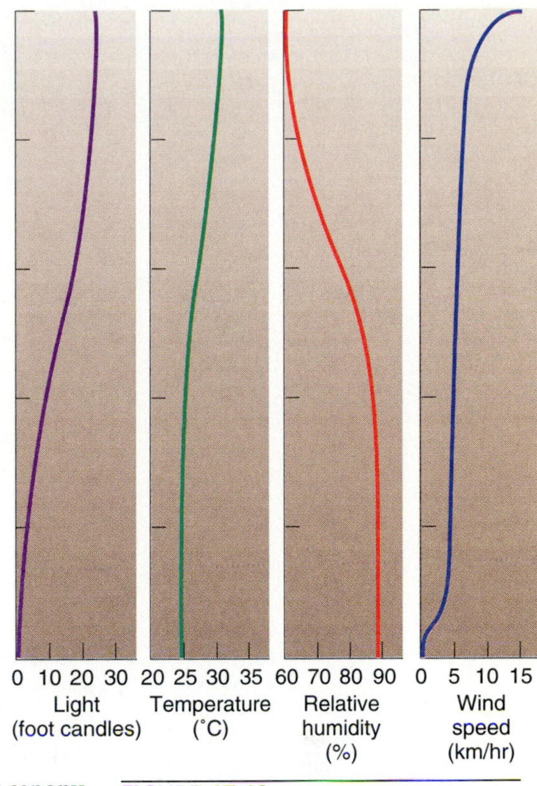

FIGURE 17.18

Vertical stratification in a tropical rainforest. From the canopy to the forest floor, light, temperature and wind velocity all decrease, while humidity increases. For scale, see figure of zoologist dangling from tall tree at right.

These overly protective ants once prevented ecologists from studying the forest canopy, but now special climbing gear, platforms, and other devices have solved that problem.

Ants and termites are as important on the ground as in the trees. As detritus feeders they aid bacteria and fungi in decomposing fallen leaves, fruits, and dead organisms. So rapid is decomposition in the warm, moist tropics that almost no organic matter accumulates in the soil. The soil is also poor in minerals, because of leaching by the heavy rain and rapid absorption by the shallow roots of trees. The only type of agriculture that is practical in such soil is the "slash and burn" system. Burning the vegetation adds some minerals to the bare clay, but after a few plantings these become depleted, and the field must be left to succeed to jungle.

■ **Question:** At what height would a frog find conditions most suitable? Why would a frog not necessarily be found at that height?

Freshwater Ecosystem Types

PONDS AND LAKES. Bodies of water with little or no current are called **lentic** ecosystems. Whether a lentic body of water is called a pond or a lake is often a matter of local custom, but, in general, ponds are more shallow than lakes. Natural ponds and lakes occur most often within moist biomes, especially northern coniferous forests. Beavers frequently create ponds, and Ice Age glaciers have hollowed out deep basins that now hold lakes. Lakes can be divided into three zones, depending on how much sunlight penetrates the water and supports photosynthesis (Fig. 17.19 on page 346). The **littoral zone** includes shallower water, usually around the lake margin, where sunlight penetrates to the bottom. Rooted plants, such as pond lilies, grow there, and a variety of animals take advantage of the plants for food and shelter. These animals include leeches, snails, and numerous insects that may spend all or part of their lives under water. Small fishes and amphibians are often numerous in the littoral zone, and larger vertebrates, such as moose and raccoon, come from the forests to feed.

Enclosed by the littoral zone is open water where sunlight does not penetrate deeply enough to support rooted plants. In the upper portions of the open water is the **limnetic zone,** where sunlight is sufficient to support photosynthesis by algae. Algae

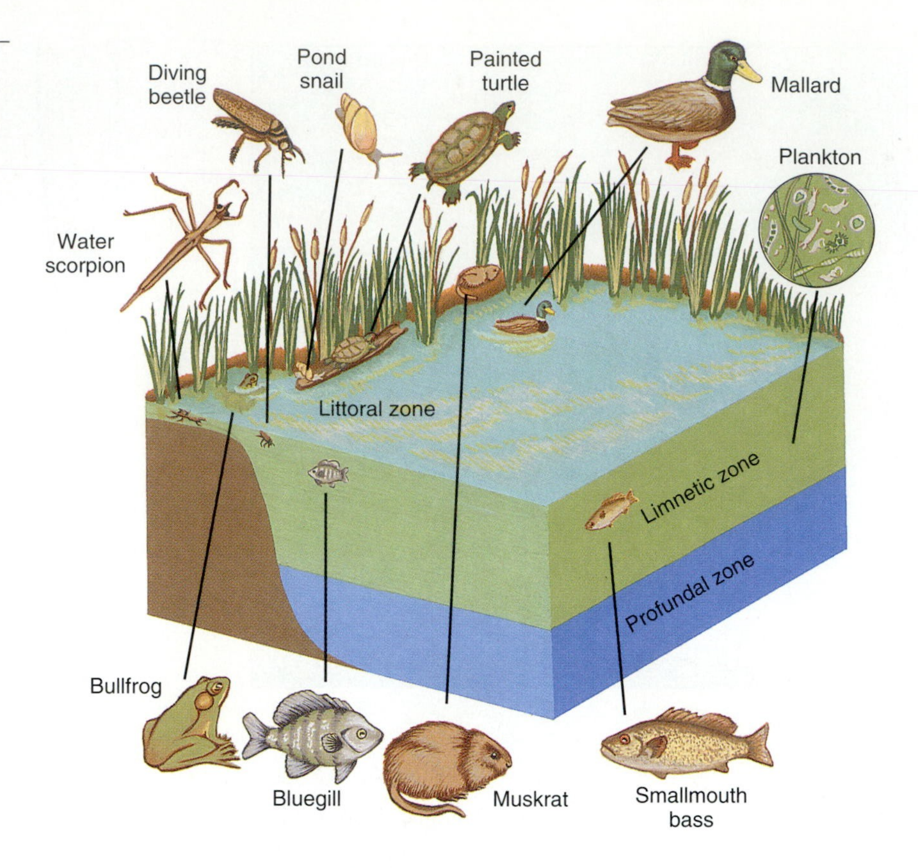

FIGURE 17.19
Zones within a lake and animals that live in them.

and other drifting, photosynthesizing organisms make up the **phytoplankton.** The tiny animals that feed on phytoplankton are called **zooplankton.** The zooplankton are eaten by fishes and other small animals, which are, in turn, eaten by larger animals. Beneath the limnetic zone is the **profundal zone,** which is devoid of plants. Because of the low oxygen concentration, few animals are found there except for fishes escaping the heat of summer. Often the bottoms of lakes are covered with muck in which anaerobic bacteria break down organic material that drifts down from the productive limnetic zone.

Lakes in temperate areas often show **thermal stratification** because the density of water varies with temperature (Fig. 17.20). Water is most dense at 4°C (39° F), so both warmer and colder water will float above 4°C water. In winter, therefore, 4°C water settles at the bottom, with colder water above it and ice on top. In summer, warmer water remains above the colder, denser water on the bottom. This thermal stratification also leads to **oxygen stratification,** partly because of stagnation and partly because there is no photosynthesis on the bottom. Water at the bottom therefore lacks oxygen, while water at the surface is continually replenished from the air. In the spring and fall there is generally a circulation of water, which redistributes the temperature and oxygen in lakes. The **spring overturn** occurs when the surface water warms to 4°C and wind or some other disturbance mixes it with the bottom water at the same temperature. **Fall overturn** occurs when the surface water cools to 4°C and sinks below the warmer, less-dense water on the bottom. Spring and fall overturns often bring up bottom sediments, including nutrients that can be used by photosynthesizers.

Rivers and Streams. Rivers, streams, and other moving freshwater are **lotic** ecosystems. The differences between lotic and lentic ecosystems are mainly due to the force of moving water. In a slow-moving river one might find the same species as in a lake. In a swift brook, however, one would find animals such as trout and specialized insect larvae that are able to withstand the current. Many rivers and streams consist of rapids alternating with slow-moving pools. Fishes and other animals will often be

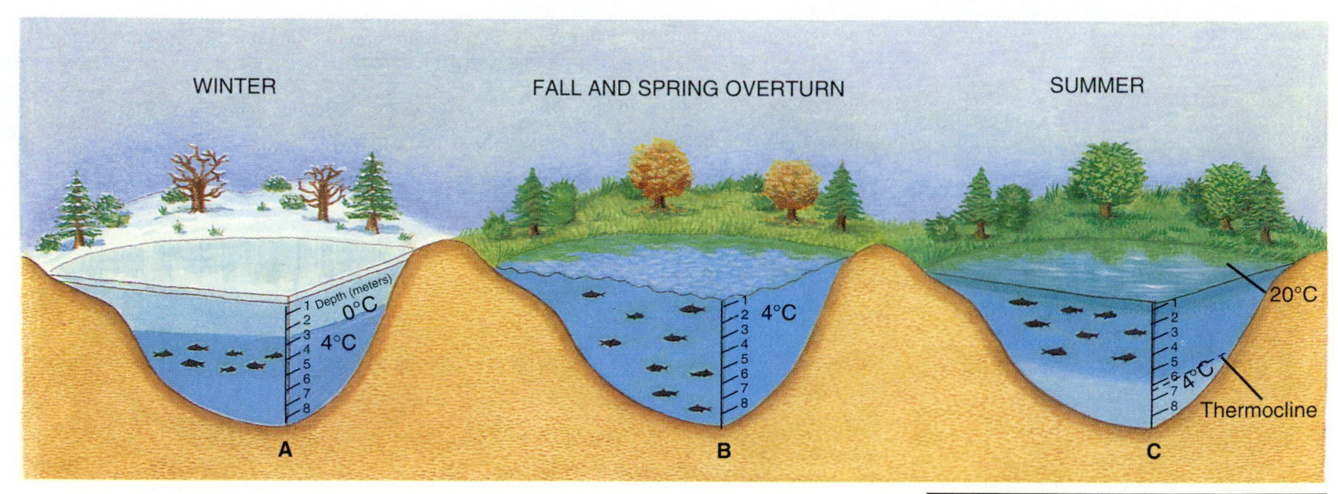

FIGURE 17.20

Seasonal changes in the distribution of temperature and oxygen in a temperate-zone pond and the effect on the distribution of fish. (A) In winter the water at the surface is coldest, and denser water at 4°C settles to the bottom. (Colder water is represented by darker blue on the right side of each figure.) Oxygen cannot diffuse through snow-covered ice into the water, and more O_2 is dissolved in warmer water (darker blue on the left side of each figure). (B) In spring the surface water warms to 4°C, and wind causes an overturn that eliminates both thermal and oxygen stratification. A similar overturn occurs in the fall when surface waters cool to 4°C. (C) In summer, warm, oxygenated water at the surface may be sharply separated from cold, oxygen-poor water by a layer called the thermocline.

found in the pools, where they do not have to continually struggle against the current. At the same time they can benefit from oxygen absorbed by water in the rapids.

MARSHES, BOGS, AND OTHER WETLANDS. Wetlands occur wherever shallow water covers land for at least part of the year. They can result from periodic flooding from rivers or lakes or from poor drainage. The type of wetland is defined mainly by its plant species, which depends largely on the periodicity of flooding. A wooded river bank that is briefly flooded each spring may differ little from nearby upland areas. A marsh, in contrast, is characterized by grassy vegetation, such as cattails and sedges. Bogs support acid-tolerant plants such as sphagnum moss, which often forms a floating mat.

Even more than in other ecosystem types, conditions in a wetland can change rapidly over both time and space. In other words, wetlands are "patchy." For that reason, ecologists since the 1970s have adopted a flexible approach to analyzing wetlands, called **patch dynamics.** Using patch dynamics, a wetland would not have to be permanently flooded in order to be classified as a marsh. The important criterion would be whether the wetland's ecology is overall like that of a marsh. This difference in categorization has greatly expanded the area covered by government regulations designed to preserve wetlands as nesting and feeding sites for ducks, geese, and many other birds. Wetlands also support many invertebrates, amphibians, reptiles, and small mammals. In addition, they can remove toxins and particles from streams before they can enter lakes and oceans, and for that reason some communities have created artificial wetlands for sewage treatment.

Marine Ecosystem Types

ESTUARIES. Estuaries occur at river mouths and bays where fresh water mixes with seawater. Most estuaries are surrounded by tidal marshes that are flooded at high tides. Because the salinity of the water varies with the tide and with the distance from the source of fresh water, estuaries provide a complex variety of habitats. They are also highly productive, because rivers continually supply nutrients. Mammals generally inhabit only the edges of estuaries, but many birds can take advantage of the entire surface area. The water hosts a variety of fishes, crustaceans, and molluscs capable of tolerating the changing salinity (Chapter 14).

CONTINENTAL-SHELF WATERS (INSHORE WATERS). The greatest diversity of marine animals occurs near shore in waters above the gently sloping continental shelves. The part of the continental shelf that is exposed at low tide, called the **intertidal zone,** hosts numerous species of seabirds, crustaceans, and other animals familiar to beachcombers. Rocky intertidal zones are especially rich in "seaweeds" and animals that find a foothold against the waves. These species often occur in distinct bands, with those more tolerant of exposure occurring higher up. Tide pools that collect in the intertidal zone often trap a fascinating variety of marine animals that normally occur in deeper water. Most flounder, cod, salmon, tuna, and other commercial fishes are netted in deeper continental-shelf waters, which have a high productivity because of the continual supply of nutrients from land and from sediments stirred up by wave action. Inshore waters are so turbid with sediments that sunlight seldom penetrates more than 30 meters.

Open Ocean. Beyond the margin where the continental shelf drops off sharply to depths of 7 kilometers or more is the open-ocean (**pelagic**) ecosystem. Here the water is clearer, and sunlight penetrates to 100 or 200 meters. In spite of the increased potential for photosynthesis by phytoplankton, the absence of nutrients makes open ocean one of the least productive ecosystems per square kilometer. Because two-thirds of the Earth's surface is open ocean, however, it is the habitat of an enormous variety of animals and has the largest primary productivity (Fig. 17.2).

Like tropical rainforests, oceans are vertically stratified. Near the surface is a "canopy" of **diatoms** and other photosynthesizers. This phytoplankton is fed upon by primary consumers, mainly zooplankton, that swim up to the surface at night. At dawn they reverse the **vertical migration,** swimming down below 500 meters, where they are less likely to be preyed upon. Most zooplankters (the term for individual species of zooplankton) are only a few millimeters long, but they make up for their small sizes by being the most numerous animals on Earth. Among the most important are two groups of crustaceans, **krill** and **copepods** (numbers 21 and 35 in Fig. 17.21; pp. 669 and 672). Krill

FIGURE 17.21

Plankton and nekton of the open ocean. The upper oval shows zooplankton, and the lower oval shows smaller plankton, mainly phytoplankton. The fishes shown at the bottom are members of the benthos—inhabitants of the ocean floor.

1 Dolphins, *Delphinus*
2 Tropicbirds, *Phaethon*
3 Paper nautilus, *Argonauta*
4 Anchovies, *Engraulis*
5 Mackerel, *Pneumatophorus* and sardines, *Sardinops*
6 Squid, *Onykia*
7 *Sargassum*
8 Sargassum fish
9 Pilotfish
10 White-tipped shark
11 Pompano, *Palometa*
12 Ocean sunfish, *Mola mola*
13 Squids, *Loligo*
14 Rabbitfish, *Chimaera*
15 Eel larva
16 Deep sea fish
17 Deep sea angler, *Melanocetus*
18 Lanternfish, *Diaphus*
19 Hatchetfish, *Polyipnus*
20 "Widemouth", *Malacosteus*
21 Krill, *Nematoscelis*
22 Arrowworm, *Sagitta*
23 Amphipod, *Hyperoche*
24 Sole larva, *Solea*
25 Sunfish larva, *Mola mola*
26 Mullet larva, *Mullus*
27 Sea butterfly, *Clione*
28 Copepods, *Calanus*
29 Assorted fish eggs
30 Stomatopod larva
31 Hydromedusa, *Hybocodon*
32 Hydromedusa, *Bougainvillia*
33 Salp (pelagic tunicate), *Doliolum*
34 Brittle star larva
35 Copepod, *Calocalanus*
36 Cladoceran, *Podon*
37 Foraminifer, *Hastigerina*
38 Luminescent dinoflagellates, *Noctiluca*
39 Dinoflagellates, *Ceratium*
40 Diatom, *Coscinodiscus*
41 Diatoms, *Chaetoceras*
42 Diatoms, *Cerautulus*
43 Diatom, *Fragilaria*
44 Diatom, *Melosira*
45 Dinoflagellate, *Dinophysis*
46 Diatoms, *Biddulphia regia*
47 Diatoms, *B. arctica*
48 Dinoflagellate, *Gonyaulax*
49 Diatom, *Thalassiosira*
50 Diatom, *B. vesiculosa*

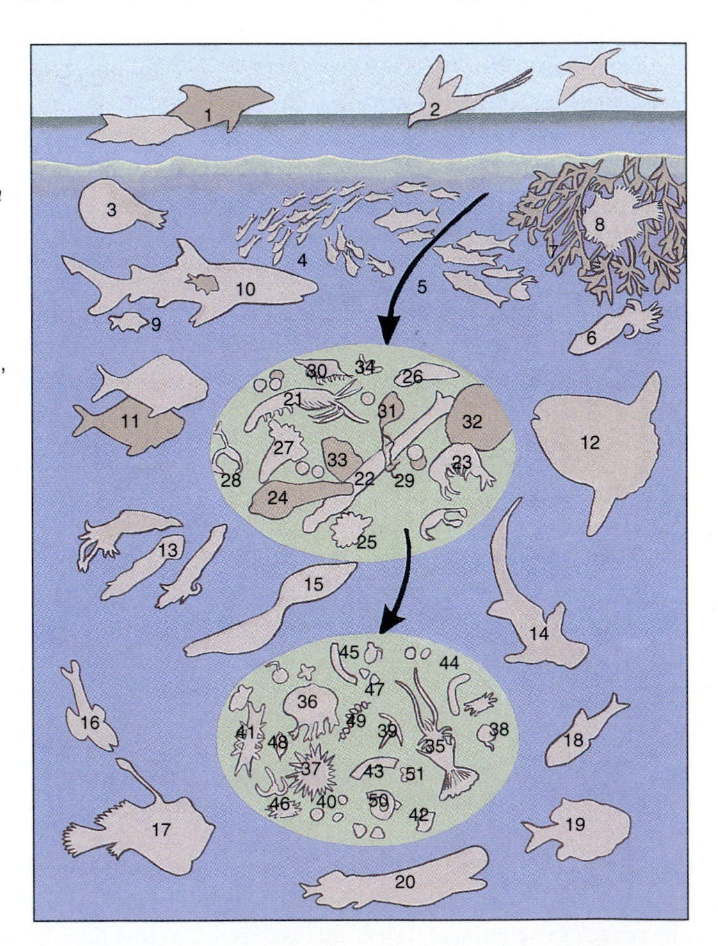

35 mm
5 mm

refers to *Euphausia superba* and approximately 85 other species of shrimplike crustaceans that are the major source of food for filter-feeding whales. As is evident in Fig. 17.10, krill are an essential link in food chains of polar regions. In temperate and tropical oceans, 2000 to 3000 species of copepods, especially of the genus *Calanus,* dominate the zooplankton.

Fishes and other large animals make up **nekton:** animals that swim actively rather than drift with currents. Nekton occur all the way down to the ocean floor, an average of 4 km below the surface. Nutrients for animals of the deep rain down as detritus from the productive regions above. Many of the animals that live in the perpetual darkness of the ocean depths have light-emitting organs by which they lure prey and communicate (see Fig. 35.30). In some fishes, mutualistic bacteria generate the light.

Upwelling Zones. Upwelling regions occur where nutrient-rich cold water rises from the ocean bottom to the surface. There photosynthesizing phytoplankton incorporate the nutrients into organic molecules, which then become available to the zooplankton and to larger animals that consume plankton. Upwelling zones are about four times as productive per square kilometer as open-ocean areas. The importance of upwelling regions becomes evident every few years off the coast of Peru and Ecuador when an invasion of warm Pacific water coincides with weak trade winds. This combination produces a thermal stratification that prevents the cold water from rising to the surface for 6 to 18 months. When extreme, such an event is called **El Niño**—the boy—because it often arrives around Christmas. (El Niño is also called El Niño—Southern Oscillation event, or ENSO.) El Niño of 1982–1983 reduced production by 95%, resulting in a sharp decline in fish populations and the closing of many fisheries. On the Galápagos Islands all the newborn fur seals (*Arctocephalus galapagoensis*) died during that year, no seabirds bred, and almost all the reef-building animals died.

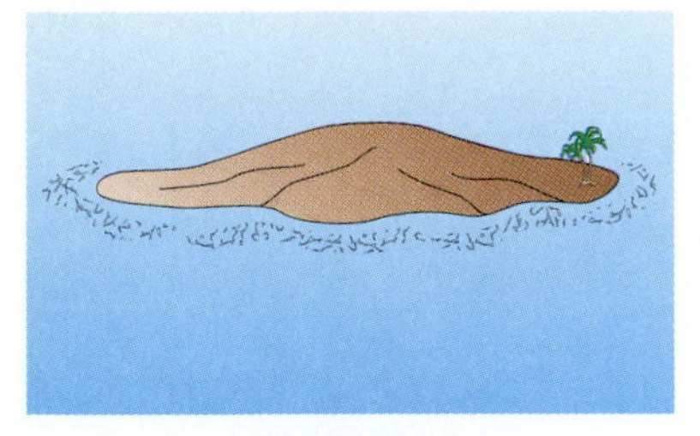

Fringing reef

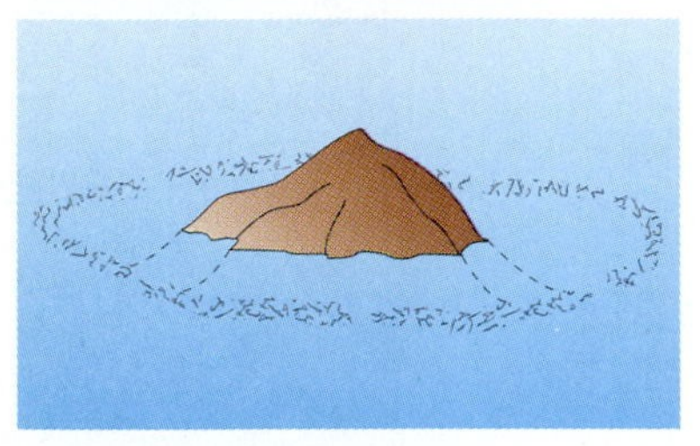

Barrier reef

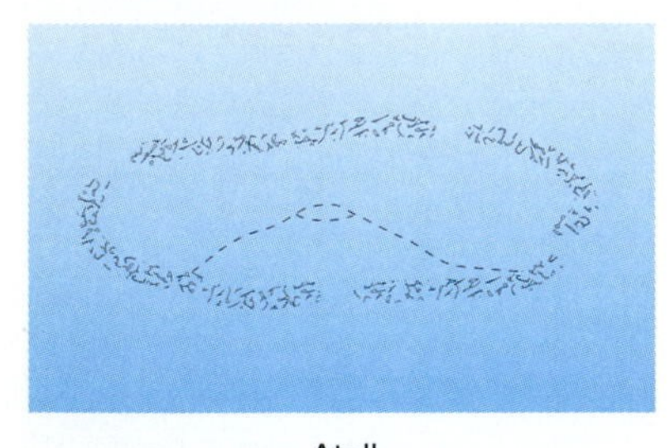

Atoll

FIGURE 17.22
Types of coral reef.

■ **Question:** Could an atoll have once been a barrier reef? Is there a sharp distinction between fringing and barrier reefs?

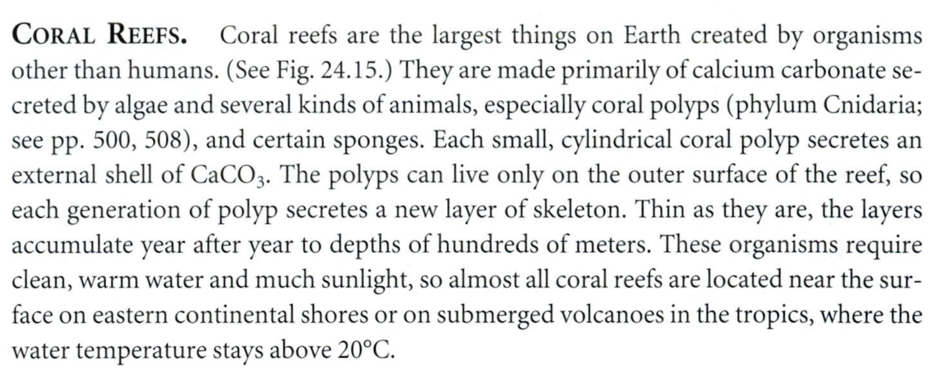

Coral Reefs. Coral reefs are the largest things on Earth created by organisms other than humans. (See Fig. 24.15.) They are made primarily of calcium carbonate secreted by algae and several kinds of animals, especially coral polyps (phylum Cnidaria; see pp. 500, 508), and certain sponges. Each small, cylindrical coral polyp secretes an external shell of $CaCO_3$. The polyps can live only on the outer surface of the reef, so each generation of polyp secretes a new layer of skeleton. Thin as they are, the layers accumulate year after year to depths of hundreds of meters. These organisms require clean, warm water and much sunlight, so almost all coral reefs are located near the surface on eastern continental shores or on submerged volcanoes in the tropics, where the water temperature stays above 20°C.

Charles Darwin identified three kinds of coral reefs:

1. A **fringing reef** grows in a narrow band parallel to a coast in shallow water (Figs. 17.22 [top] and 17.23). This is the common type of reef along the Florida Keys and the Caribbean.
2. A **barrier reef** forms a broad band separated from the coast by a lagoon. The most spectacular example is the Great Barrier Reef off the northeast coast of Australia, which is 2000 km long and 145 km wide. Actually the Great Barrier Reef is so large that it also includes fringing reefs within it.
3. An **atoll** is a ring of coral that encloses a lagoon thousands of meters across and only tens of meters deep. Atolls form on sinking volcanoes, as Charles Darwin first proposed during his voyage on the *Beagle* and as drilling confirmed in the 1950s.

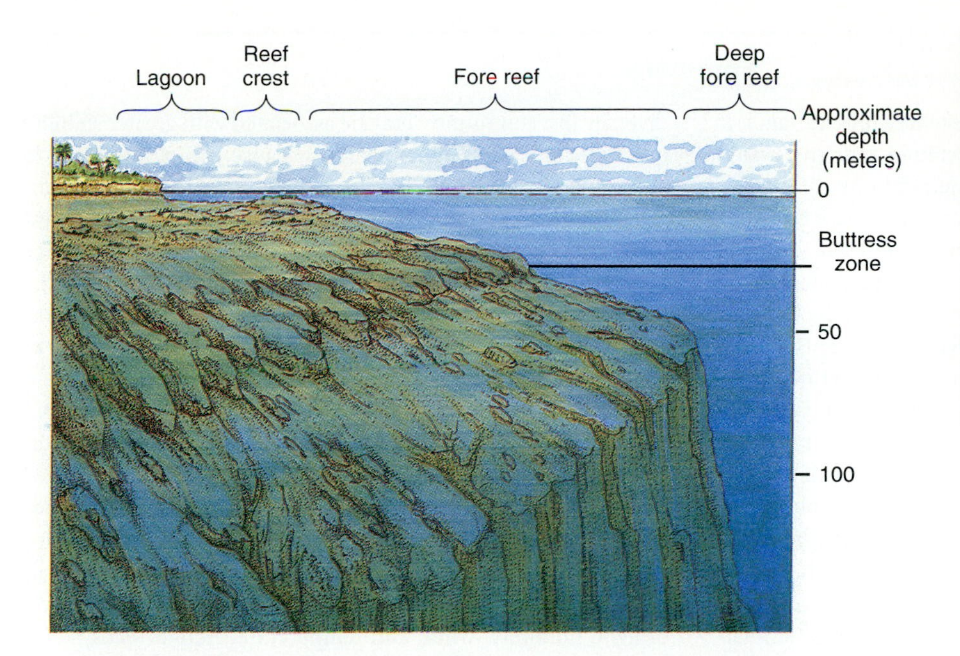

FIGURE 17.23

Structure of a fringing reef. The reef crest is a shallow band of growing coral approximately parallel to the coast. Often the reef crest encloses a shallow lagoon. Water in the lagoon is calm because the surf breaks against the reef crest. Most of the wave action is absorbed by the buttress zone in the sloping fore reef. Channels in the fore reef allow debris from eroding coral to fall into the deep fore reef instead of clogging the coral polyps. Few corals live in the deep fore reef, where light drops from 25% to 5% of the intensity at the surface.

The reason reef-building corals require sunlight is that the polyps obtain many of their nutrients from photosynthesizing protists (**zooxanthellae**) within their cells. (See p. 465.) The rate at which corals build reefs is closely coupled to the photosynthetic activity of these **endosymbionts** (intracellular mutualists). Coral polyps are also secondary consumers. At night their tentacles emerge from the skeletons and prey upon zooplankton, stunning them with poisonous barbs (**nematocysts**). The prey are mainly small crustaceans and worms that graze on algae. Polyps also absorb small amounts of **dissolved organic material** from excrement and dead organisms. Coral reefs have the highest primary production of any ecosystem (Fig. 17.2). This high productivity is rather paradoxical, because the clear waters above coral reefs are among the poorest in nutrients. The high productivity supports an enormous biomass and biodiversity of animals. Among the Caribbean reefs the common fishes alone number some 300 species.

The variety of habitats present in coral reefs partly accounts for the great diversity of animals that live among them (Fig. 17.23). Near shore the reef protects shallow lagoons where animals such as sea urchins and sea cucumbers (phylum Echinodermata) graze upon algae. The summit of the reef, called the **reef crest, coral rampart,** or **reef flat,** is exposed at low tide. It is dominated by large rounded and branching corals that form a network in which more delicate corals and algae live. In this highly productive zone, space seems to be the most limited resource. Many species of coral use their stinging nematocysts against potential competitors, and sponges secrete chemicals that inhibit crowding. Invertebrates and fishes also compete for places to hide during the day. Among these animals are many that tend to erode the reef. Parrot fish chew up the coral skeletons with their beak-like mouths, the boring sponge *Cliona* takes out small flakes, and sea worms bore through it.

On the seaward side of the crest is the **fore reef** or **reef front.** Near the top of the fore reef, corals form a wall deeply scarred with channels that provide numerous and varied habitats for animals. More important, these channels protect the reef by dissipating wave action and by carrying away the fine, white sand of decomposed corals that would otherwise smother the coral polyps. The reef front then drops off almost vertically for approximately 100 meters into the darkness. As the depth increases, the light-requiring algae and symbiotic corals give way to sponges and other animals that feed on the detritus raining down from the productive zone above.

Summary

Animals and other organisms in an ecosystem require certain levels of such abiotic factors as oxygen, temperature, nutrients, water, and ions. Primary producers are also required to fix carbon and nitrogen into forms that can be used by animals. These elements, as well as oxygen and water, are involved in nutrient cycles in which they are transferred from organic to inorganic forms and back. In addition, all animals depend on primary producers to fix solar energy into usable chemical bond energy in nutrients. Energy tends to be transferred along with carbon, except that instead of cycling, energy tends to dissipate. Consequently, there is less energy available at each higher trophic level, from primary producer, to primary consumer, to secondary consumer, and so on. This results in ecological pyramids of energy, biomass, and numbers. The transfer of energy from one organism to another can be represented as a food chain or, more realistically, as a food web.

Animals in an ecosystem inevitably interact with each other through competition, predation, and symbiosis. The animals that succeed in such interactions make up part of a community. Many ecologists refer to the role of each species in a community as its niche and believe that competition for niches is the main factor that determines community structure. There is often little overlap in niches, however, and predation and chance events may suppress populations and prevent competition. A community is first established in the process of primary succession, and it may be replaced during secondary succession. Increased biodiversity in a community may be associated with greater stability in species and populations, either because complexity leads to stability or because stability allows biodiversity to become established. Biodiversity also increases with area. In natural ecosystems, populations are often stable even though most organisms have the intrinsic capability of reproducing at an exponential rate. When the population reaches the carrying capacity, deaths from hunger, disease, and predation tend to equal the rate of birth. Often there are population cycles in which the population of a species fluctuates periodically around the carrying capacity.

There are various ecosystem types that vary with temperature and other abiotic factors. Terrestrial ecosystem types, called biomes, include tundra, temperate deciduous forests, and grasslands. Tropical rainforest is an extremely important biome to zoologists, because it includes possibly half of all species of animals. This biodiversity is due largely to the vertical stratification of light, humidity, wind, and temperature. Aquatic ecosystems are either freshwater (such as lakes, rivers, and wetlands) or marine (including estuaries, inshore waters, open ocean, and coral reefs). Because of its huge volume, open ocean supports a large number of species of zooplankton and larger animals. Coral reefs, which are created partly by small animals called polyps, are also host to a tremendous species diversity that depends on primary production by endosymbiotic zooxanthellae and the variety of habitats at various depths.

Key Terms

ecology
ecosystem
limit of tolerance
autotroph
primary production
heterotroph
consumer
nutrient cycle
fixation
decomposer
detritus feeder
trophic level
herbivore
omnivore
carnivore
ecological pyramid
food web
symbiosis
predation
mutualism
parasitism
commensalism
competition
niche
competitive exclusion
succession
community
climax community
biodiversity
carrying capacity
biome
stratification

Chapter Test

1. In your own words explain what ecology is, and why the study of ecology is essential for a complete understanding of animals. (p. 325)
2. What are some of the reasons for the differences in primary productivity shown in Fig. 17.2. For example, why is primary productivity higher in a temperate deciduous forest than in a northern coniferous forest? (pp. 341–345)
3. Figure 17.3 shows some terrestrial plants and animals involved in the carbon cycle. What kinds of marine organisms would play corresponding roles in each part of the carbon cycle? (pp. 348–350)
4. Diagram the oxygen cycle. (*Hint:* Use Fig. 17.3 as a model.) (pp. 327–328)
5. Explain the meaning of the following phrase: "Matter circulates, energy dissipates." (p. 331)
6. Identify three trophic levels represented in the opening figure on p. 324. Is the pyramid of biomass satisfied here? What about the pyramid of numbers? (pp. 331–333)
7. Define "secondary succession" and describe an example of it, preferably from your own experience. (pp. 336–338)
8. What are krill and copepods? Explain why they are so important to marine ecosystems. (pp. 348–350)

9. Explain in your own words why the stability of a community may increase with its biodiversity. Under what circumstances can increasing biodiversity reduce the stability? Give an example. (p. 338)
10. Tropical rainforests and open oceans are both vertically stratified. Explain what this means. What effect does vertical stratification have on the diversity and distribution of various organisms at various depths in both ecosystems? (Where are the primary producers located? At what trophic level are the animals on the bottom? What kinds of animals are the major primary consumers in each ecosystem?) (pp. 343–345, 348–350)
11. Using the concept of limits of tolerance, explain why different species of animals are found in different biomes. Name four biomes; for each one, name an animal species that can live there and one that cannot. For the latter species, explain why it is excluded. (pp. 340–345)
12. Name four types of aquatic ecosystem. For each one, name an animal that lives in it. (pp. 345–351)

■ Answers to the Figure Questions

17.1 The optimum habit for a species might also be the optimum for its predators or parasites, or it might not sustain the organisms on which the species depends.

17.6 Replace N_2 in air with N_2 dissolved in water; replace plants with algae; replace nitrates and nitrites in soil with nitrates and nitrites dissolved in water.

17.9 Eliminating the tertiary consumers would be expected to allow for more secondary consumers, which might reduce the number of primary consumers, which might allow for more primary producers.

17.10 Blue whales feed on krill, so they are secondary consumers. Killer whales are top carnivores: either tertiary or quaternary consumers. Because blue whales are lower on the food chain, one would expect them to have a larger total biomass. Each individual blue whale is much larger than a killer whale, however, so there could actually be fewer blue whales.

17.11 One could suppose that the elimination of zebras would allow more food for the other three grazers. It is not clear, however, that the other three species could take advantage of the parts of the plants normally consumed by zebras. They might, in fact, fare worse.

17.12 As fields succeed to forests, one would expect the grasshopper sparrow to go first. These would be replaced temporarily by species that live in shrubs and forest margins (the other sparrows, meadowlark, towhee, purple finch, robin, junco, black-throated green warbler, nashville warbler, and cottontail rabbit). These and the meadow mouse would be lost as the forest took over. The species that would finally be gained would be the veery. Ruffed grouse, white-footed mouse, white-tailed deer, red squirrel, red fox, and short-tailed shrew would remain throughout the succession.

17.14 The effect of modern agricultural practice is to increase the number of organisms per unit area of land, which is the carrying capacity. Removal of predators, prevention of contagious disease, and supplemental feeding would all increase carrying capacity for sheep.

17.16 Where the average annual temperature is near freezing, and there is a large amount of precipitation, one would expect the ground to be covered in snow or ice for most of the year, preventing the growth of plants for primary production.

17.17 Shrublands occur in areas bordering the Mediterranean Sea and southern Australia and are bordered by temperate grasslands, deserts, and temperate deciduous forests. One therefore expects them to be moderately dry and temperate to hot. In Fig. 17.16, shrublands might therefore be squeezed in just beneath deciduous forests. Tropical scrub forests occur in South America, Southeast Asia, and northern Australia generally adjacent to tropical savanna and tropical rainforest. The temperature would be tropical and the precipitation moderate, so in Fig. 17.16 tropical scrub forests might lie just to the right of deciduous forests.

17.18 Reproduction in frogs depends on water, so one would expect to find them near the ground, where the humidity is high. When not reproducing, however, they might spend most of their time in the canopy, where there are more insect prey.

17.22 Yes, an atoll is essentially a barrier reef in which the central island has been submerged. The difference between a fringing and a barrier reef is one of degree: how narrow or broad the reef is.

Readings

Recommended Readings

Ausubel, J. H. 1991. A second look at the impacts of climate change. *Am. Sci.* 79:210–220.

Bierregard, R. O., et al. 1992. The biological dynamics of tropical rainforest fragments. *BioScience* 42:859–866.

Cronon, W. 1983. *Changes in the Land.* New York: Hill and Wang. (*A fascinating account of how Indians and colonists altered the ecology of New England.*)

Grassle, J. F. 1991. Deep-sea benthic biodiversity. *BioScience* 41:464–469.

Holloway, M. 1993. Sustaining the Amazon. *Sci. Am.* 269(1):90–99 (July).

Houghton, R. A., and G. M. Woodwell. 1989. Global climatic change. *Sci. Am.* 260(4):36–44 (Oct).

Jackson, J. B. C., and T. P. Hughes. 1985. Adaptive strategies of coral-reef invertebrates. *Am. Sci.* 73:265–274.

Kusler, J. A., W. J. Mitsch, and J. S. Larson. 1994. Wetlands. *Sci. Am.* 270(1):64–70 (Jan).

Laws, R. M. 1985. The ecology of the southern ocean. *Am. Sci.* 73:26–40.

May, R. M., and J. Seger. 1986. Ideas in ecology. *Am. Sci.* 74:256–267.

Nicol, S., and W. de la Mare. 1993. Ecosystem management and the Antarctic krill. *Am. Sci.* 81:36–47.

Perry, D. R. 1984. The canopy of the tropical rain forest. *Sci. Am.* 251(5):138–147 (Nov).

Peterson, C. H. 1991. Intertidal zonation of marine invertebrates in sand and mud. *Am. Sci.* 79:236–248.

Post, W. M., et al. 1990. The global carbon cycle. *Am. Sci.* 78:310—326.

Ramage, C. S. 1986. El Niño. *Sci. Am.* 254(6):77–82 (June).

Reice, S. R. 1994. Nonequilibrium determinants of biological community structure. *Am. Sci.* 82:424–435.

Robison, B. H. 1995. Light in the ocean's midwaters. *Sci. Am.* 273(1):60–64 (July). (*Bioluminescent animals.*)

Schneider, S. H. 1989. The changing climate. *Sci. Am.* 261(3):70–79 (Sept).

Sebens, K. P. 1985. The ecology of the rocky subtidal zone. *Am. Sci.* 73:548–557.

Terbogh, J. 1992. *Diversity and the Tropical Rain Forest.* New York: Scientific American Books.

White, R. M. 1990. The great climate debate. *Sci. Am.* 263(1):36–43 (July).

Wilson, E. O. 1992. *The Diversity of Life.* Cambridge MA: Harvard University Press.

Additional References

Keeling, C. D., et al. 1989. A three-dimensional model of atmospheric CO_2 transport based on observed winds: Observational data and preliminary analysis. *Geophys. Monogr. Am. Geophys. Union* 55: 165–236.

Mares, M. A. 1993. Desert rodents, seed consumption, and convergence. *BioScience* 43:372–379.

Menard, H. W. 1986. *Islands.* New York: Scientific American Books. (*See Chapter 7 on coral reefs.*)

Naiman, R. J. 1988. Animal influences on ecosystem dynamics. *BioScience* 38:750–752. (*Introduction to several papers on beaver, moose, prairie dogs, and other animals.*)

Schneider, W., and P. Wallis. 1973. An alternative method of calculating the population density of monsters in Loch Ness. *Limnol. Oceanogr.* 18:343.

Tilman, D., and J. A. Downing. 1994. Biodiversity and stability in grasslands. *Nature* 367:363–365.

CHAPTER

18

Causes of Evolution

Male blue-footed booby, *Sula nebouxii,* displaying on the Galápagos Islands.

The Main Concepts of Evolution

Fossils speak as eloquently of past evolution as the ruins of temples and monuments speak of past civilizations. The anatomy, physiology, and genetics of living organisms likewise reveal their evolutionary heritage as surely as nations bear the imprint of their cultural histories. For these reasons there are as few biologists who doubt that evolution occurred as there are historians who doubt that ancient Rome existed and influenced our own culture.

This is not to say that all biologists agree on every aspect of evolution. Even the definition of evolution has been argued. In the past evolution usually referred to the transformation of one species into another. Now, however, biologists think of evolution as any **change in the gene frequency in a population of organisms**. Thus, evolution is not something that happens to an entire species, but to **populations** (groups of interbreeding organisms) within a species. By **gene frequency** we mean the proportion of each form of a gene at a given locus relative to the total number of genes at that locus in a group of organisms. Gene frequency is also known as allele frequency, because each form of a given gene is called an **allele**. For example, each person carries two alleles at the locus for the gene that determines blood type. Therefore, in a group of 50,000 people there are 100,000 alleles at that locus. If 25,000 of those 100,000 alleles produced the antigen for type A blood, then the gene frequency for that allele would be 25%. One example of evolution would be a change in that gene frequency from 25% to 26%. In this case, as in most evolution, there would be no visible change in the population.

Many theories have been proposed to explain how and why evolution occurs, but the most widely accepted is the **synthetic** or **neo-Darwinian theory.** As the two names suggest, the theory synthesizes Charles Darwin's concept of **natural selection** with modern genetics. The synthetic theory is based on three elementary facts of life:

1. Within populations of organisms there are differences in the genetic makeup of individuals. These differences arise from genetic **mutations** and from **recombinations** of these mutations during sexual reproduction (Chapter 4).
2. Some individuals inherit combinations of genes that increase **fitness,** which is defined as the ability of individuals with a given combination of genes to reproduce relative to others in the population. Therefore, individuals with greater fitness have more offspring, and those offspring tend to inherit the genes that contribute to fitness.
3. Therefore there tends to be an increase in the frequency of genes that contribute to fitness.

Evolution is the change in gene frequency resulting from changes in fitness. These changes in fitness can result in different **adaptations** of the organism for survival and reproduction in its environment. Such changes can occur within a population without resulting in a new species (**microevolution**). It is generally assumed, however, that the accumulation of such changes can lead to the evolution of new species and to higher taxa (**macroevolution**). The synthetic theory itself is evolving as concepts mutate and recombine and scientists select the ones with the greatest fitness. This chapter describes the outcome of that evolution up to the present.

The Evolution of Darwinism

Like most of the really important concepts in science, evolution is simple now that some extraordinarily gifted people have explained it to us. What surprises us now is that no one discovered it until two and a half centuries ago. One reason for the failure to appreciate evolution is that it is usually not apparent within the lifetime of a human or even within the history of civilization. Another reason was the overwhelming influence of Aristotle in Western thought. In the 4th century B.C. Aristotle created an all-embracing view of science and metaphysics that emphasized design and purpose. It would have made no sense to him for a species that had been designed for one purpose to change into another species designed for an entirely different purpose. In the 13th century Thomas Aquinas incorporated Aristotelian philosophy into Christian theology, and Aristotle's concepts became woven into the fabric of Christian culture. So ingrained are the concepts of design and purpose that even evolutionists often resort to the word "design" when describing adaptations, and they have to fight the temptation to think of adaptations as having a purpose rather than simply a function.

Early Evidence and Theories of Evolution. The concept that organisms could slowly change by natural processes first arose in France in the middle of the 18th century during a period called the Enlightenment. This period just prior to the French Revolution was a time of general discontent with all established authority, including the Church, the King, and Aristotle. Philosopher-scientists of the Enlightenment argued that human progress is possible under natural law (not under the laws of the Church and the King), and they seized upon any evidence that humans and other species can naturally progress from lower to higher forms. The two earliest philosopher-scientists to clearly state this evolutionary view were **Pierre-Louis Moreau de Maupertuis** and **Denis Diderot**. Maupertuis (pronounced mo-per-TWEE) proposed in 1745 that each organism was formed from particles coming from all parts of its parents (**pangenesis**), and that new types of organisms arose when unique combinations of such particles were maintained for several generations. Diderot (deed-ROW) suggested in 1754 that entire species could be born, develop, and die just like individuals. Although Maupertuis and Diderot were politically motivated, they based their arguments on scientific evidence such as the following:

1. **Comparative anatomy.** It had recently been found that in spite of their outward differences, horse hooves and human hands were similar in the number and position of their bones. This suggested that hooves and hands were not specially designed to serve the needs of each species, but had come from a shared origin (Fig. 18.1 on page 358). Such structures that have a shared evolutionary origin are now said to be **homologous.** Structures such as the wings of bats and of butterflies are, in contrast, **analogous,** because the most recent common ancestor of bats and birds did not have wings. The most striking examples of homologous structures are **vestigial organs,** which serve no function, but resemble a functional organ in a presumably ancestral species. Examples of vestigial organs include pelvic bones in some whales and snakes, and the remaining tail bones and muscles in humans.
2. **Comparative embryology.** Anatomical similarities are even clearer in the early embryos of related species, as Karl von Baer first noted for vertebrates (pp. 130–131).
3. **Fossils.** Fossils had been known for many centuries, but they were generally treated as rare curiosities rather than as relics of extinct species. With increased digging for ores and fuels during the Industrial Revolution, however, fossils could no longer be ignored. Evolution was directly suggested by the fact that fossils were similar to, yet different from, living species. Fossils also raised the

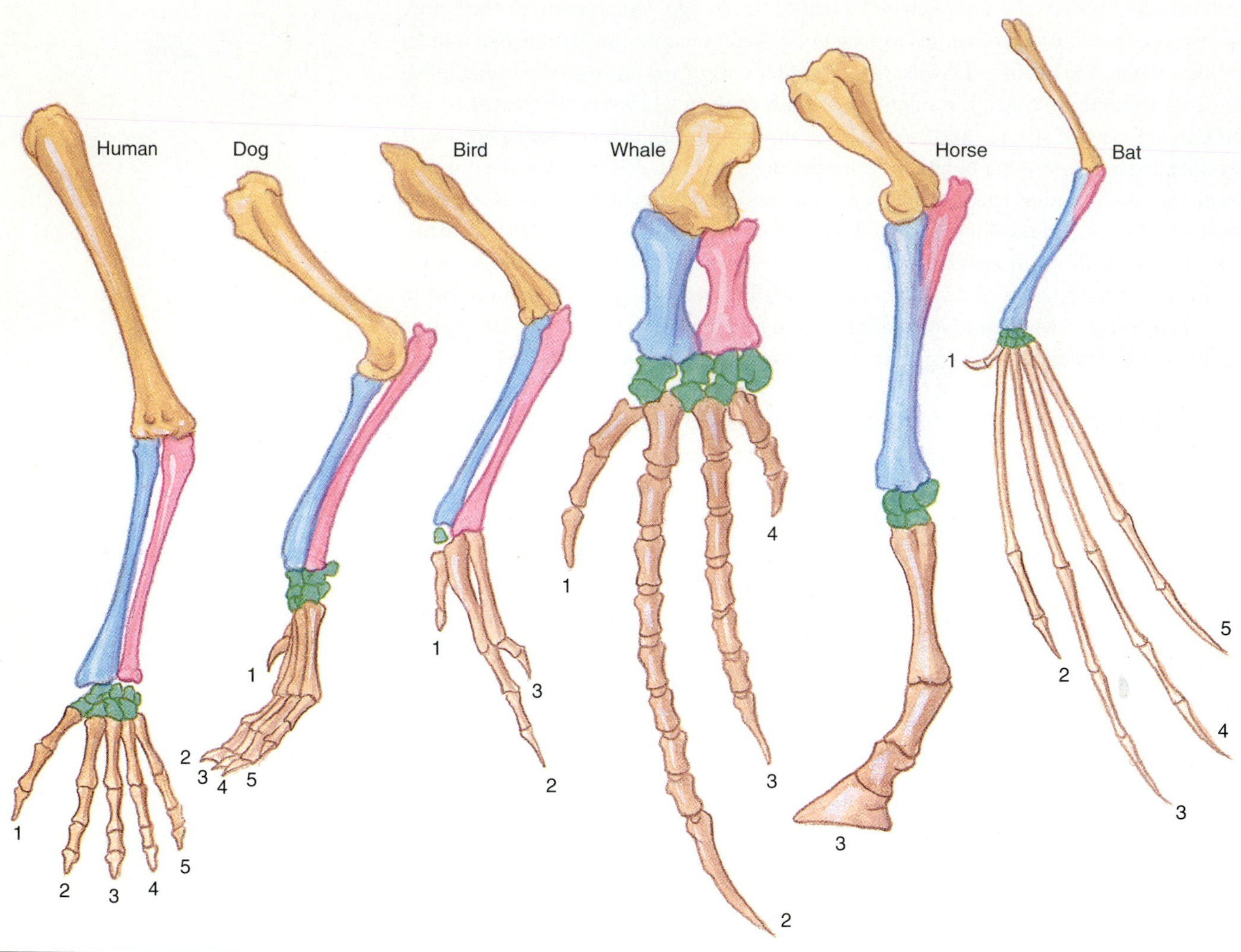

FIGURE 18.1

Homologous structure of the limbs of vertebrates. Even such diverse structures as the bat's wing, a horse's leg, and a human arm have similar numbers and sequences of bones. The number of bones can vary, however, if two or more of them fuse together during evolution, as in the horse.

■ **Question:** Are the wings of birds and bats homologous or analogous?

question of whether God would have designed species only to let them become extinct (Fig. 18.2).

4. **The age of the Earth.** According to biblical chronology the world was created only a little more than 6000 years ago—too recently for significant evolution to have occurred. Scientists gradually became convinced that the Earth was much older, providing more time for species to change.
5. **Artificial selection.** Entirely new kinds of plants and animals had been produced in relatively short times by selective breeding. It seemed likely that what humans could do in a short time, nature could certainly do in the much longer time available (Fig. 18.3 on page 360).

In order to avoid the censors, Maupertuis and Diderot had to write briefly and obscurely. In 1809, however, the French Revolution gave **Jean Pierre Antoine de Monet de Lamarck** the freedom to write a long book on the transformation of species and its causes. Lamarck was a gifted scientist who revolutionized the classification of animals by distinguishing between vertebrates and invertebrates. He was also one of the first to use the word *biologie,* thereby recognizing the study of life as a distinct science. Lamarck believed there were two major causes of the transformation of species: (1) a natural tendency for each species to progress toward a "higher" form, and (2) the **inheritance of ac-**

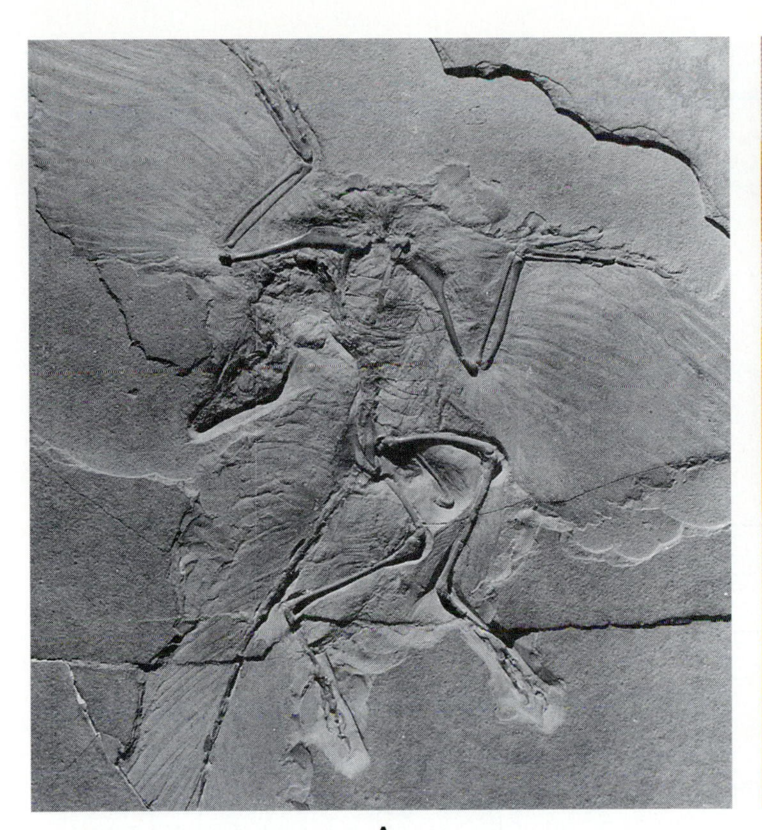

A

B

FIGURE 18.2

(A) *Archaeopteryx,* one of the best examples of fossil evidence for evolution. Like reptiles, it had teeth and vertebrae in its tail. (B) *Archaeopteryx* as it might have appeared in life, as a crow-sized predator. (See also Fig. 38.1.)

quired characters. The latter principle means that bodily changes that occur during the life of an organism can be inherited by its offspring. The concept of inheritance of acquired characters is nowadays referred to as **Lamarckism,** but it was popular thousands of years before Lamarck (p. 67). Lamarck argued that the struggle of individual animals to survive causes their nerves or other tissues to secrete a fluid that enlarges the organs involved in the struggle, and that the effects of this fluid are passed on to succeeding generations. In this way, for example, giraffes evolved long necks through their attempts, generation after generation, to browse on higher and higher leaves.

Lamarck's theory was quickly abandoned, mainly because of criticism by Baron Georges Léopold Chrétien Frédérick Dagobert Cuvier, one of the biggest names in science. Cuvier (pronounced KEW-vyay) was a pioneer in paleontology (the study of fossils) and in comparative anatomy. Although he rejected the idea of transformation of species, he certainly could not ignore the fossil evidence that different species had occupied the Earth in different ages in the past. His explanation for these changes in species was that life had been extinguished periodically by catastrophes, then replaced by different species. This theory is called **catastrophism.**

OTHER EVOLUTIONISTS BEFORE CHARLES DARWIN. In spite of Cuvier, evolution still intrigued scientists, especially in England. **Erasmus Darwin,** the most famous English poet and physician of his time and the grandfather-to-be of Charles Darwin, independently proposed a theory similar to Lamarck's in a poem (1789) and in a medical book (1794). Dr. Darwin's theory was attacked and then forgotten in England. In 1813 **William Charles Wells,** who had grown up in Charleston, South Carolina and gone to England during the American Revolution, argued that the black and white races of humankind had diverged because those with darker skins survived better in

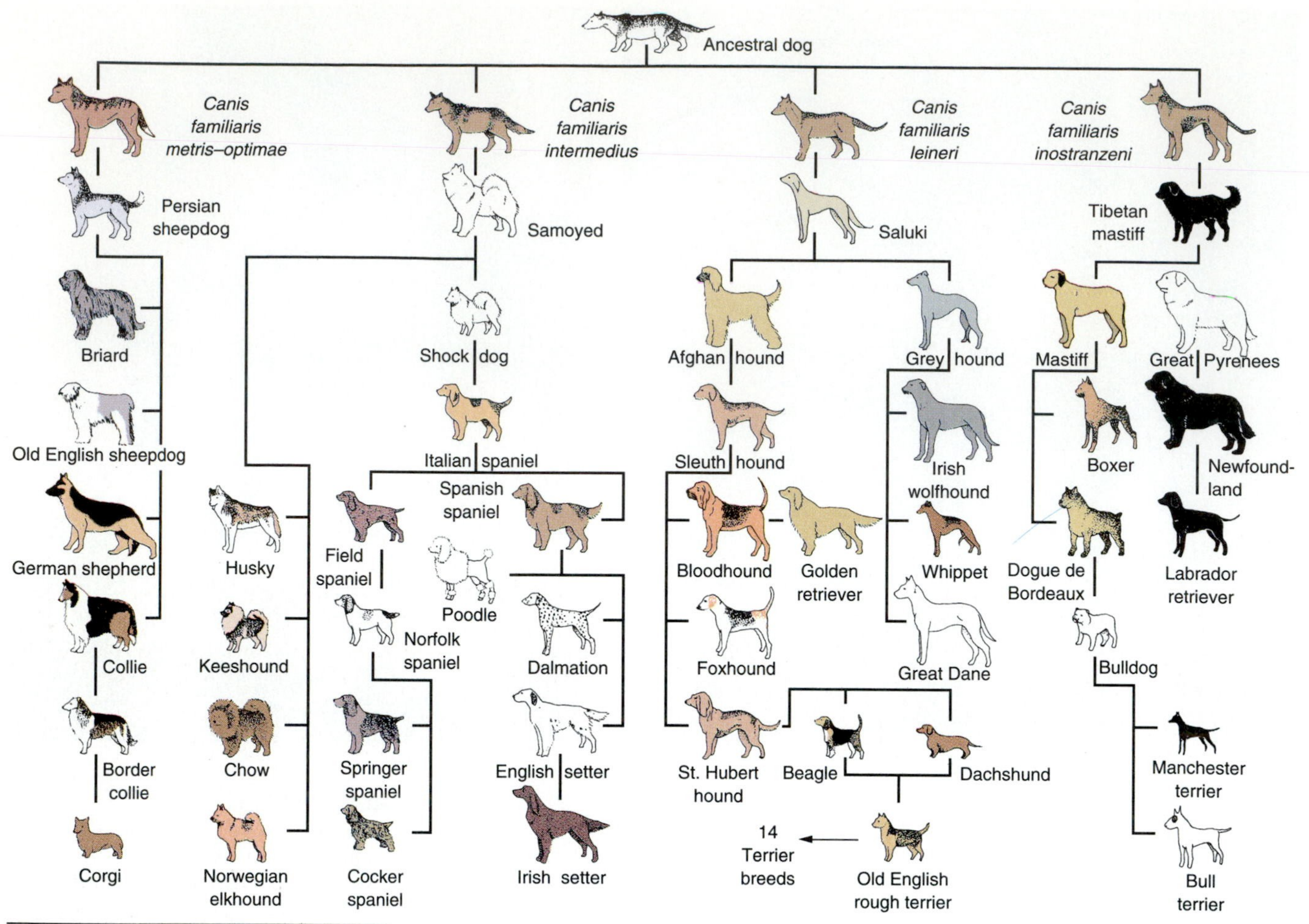

FIGURE 18.3

One of several pedigrees that have been suggested for the major breeds of dogs. All breeds are classified within the species *Canis familiaris* because they are known to have diverged from a common wolf ancestor within recent times. The anatomical differences among many of the breeds are far greater, however, than the differences between many pairs of different but related species, such as wolves *Canis lupus* and coyotes *C. latrans.*

the tropics. Wells did not suggest that one species could change into another, but by today's definition of evolution we would have to call him an evolutionist. Moreover, he grasped the concept of natural selection. **Patrick Matthew,** a wealthy orchardist in Scotland, actually published the main features of Darwin's theory of natural selection in 1831, in a book on the growing of timber for ships.

Between 1836 and 1856 at least 19 others suggested that evolution had occurred, but none provided convincing evidence or adequate explanations for evolution. With few exceptions their writings attracted little attention until the publication of Darwin's *The Origin of Species* in 1859. One reason why Charles Darwin succeeded where so many before him had failed was that he was one of the first to abandon the catastrophism of Cuvier and accept the new geological concepts of **Charles Lyell.** A lawyer by profession, Lyell in the 1830s published a three-volume book on geology that, starting with Darwin and continuing to this day, replaced catastrophism with a more gradual view of geological change. Lyell's view, called **uniformitarianism,** is that geological changes in the past were due mainly to erosion and other processes that are still occurring. Although Lyell criticized evolution in his book, it provided Darwin with a model for how small changes over long periods can lead to large changes.

The Origin of the Synthetic Theory

By the turn of the century, *The Origin of Species* had convinced almost all biologists that evolution occurred, but it was not so successful in convincing them that natural

Creation "Science"

At the same time that Enlightenment philosophers were starting to doubt a recent creation of species, biblical scholars began to doubt the literal truth of the two differing accounts of creation in *Genesis.* Since then it has become increasingly rare to find either theologians or scientists who seriously believe that all species were created individually during a short period within the past few thousand years. Curiously, however, since 1970 there have arisen vocal groups of "creation scientists" who insist on just such a view. Even more curious, many school boards, some state legislatures, and even a President of the United States have taken the claims of creation scientists seriously enough to advocate teaching creationism in science courses.

Two groups largely responsible for this situation are the Institute for Creation Research and the Creation Research Society. Although they claim to represent science, their activities are not scientific in the usual sense. Most of their research is done in libraries, where they scour the scientific literature for disagreements about *how* evolution occurred and then misrepresent them as disagreements about *whether* evolution occurred. Contrary to their public claims, these groups admit in private that their motive is to promote religion. Typical is a July 1993 letter from Henry M. Morris, President of the Institute for Creation Research, which includes the following: "We believe that America urgently needs the message of creation, for creation is the foundational component of the gospel (Revelation 14:6,7) and is essential to a real understanding of the person and work of Jesus Christ (Colossians 1:16), while evolutionism is the pseudo-scientific rationale of all that is false and harmful in our country."

Further evidence of the real nature of creation "science" is the oath that all members of the Creation Research Society must sign, which includes the following: (1) "The Bible is the word of God, and . . . all of its assertions are historically and scientifically true; (2) All basic types of living things, including man, were made by direct creative acts of God during Creation Week as described by Genesis; (3) The great Flood described by Genesis . . . was an historical event . . . ; and (4) Finally, we are an organization of Christian men of science. . . ."

Creation scientists have succeeded mainly because of skilled debaters who effectively present their case to a public that is poorly educated in the theories and methods of science. Creationists have also taken advantage of the fact that at first most evolutionists were too busy with current scientific controversies to spend much time debating 18th-century controversies. Lately, however, evolutionists have been better prepared for debates, and many have taken the trouble to write books and articles answering all of the creationist arguments. More important in the long run, many science educators are teaching science in a more realistic manner—as a continuing effort to create testable theories about phenomena of the knowable world, rather than as a collection of immutable facts. Only in this way can we hope to avoid future situations in which people cannot tell the difference between dogma and science.

selection was the cause. One reason for the reluctance to accept natural selection was that many theories of inheritance, including Darwin's own theory (a form of pangenesis), allowed for the possibility that the environment or the organism itself directly altered the heredity of the organism. The work of Gregor Mendel (pp. 67–71) disproved this idea, but it was unknown to Darwin and to most others until 1900. In fact, Mendel's work was first taken as an argument against natural selection, because Mendelians emphasized the importance of mutations rather than natural selection. They believed that evolution resulted from large genetic changes occurring within one generation, rather than from selection of small genetic variations occurring over many generations. During the first two decades of this century, therefore, many of the leading biologists were confidently telling their students that Darwinism was extinct.

Contributions of Theoretical Population Genetics. The neo-Darwinian or synthetic theory of evolution emerged gradually as a combination of Darwin's theory of natural selection and modern genetics. One aspect of modern genetics

The Convergent Evolution of Darwin and Wallace

Charles Robert Darwin
Born: 12 February 1809 in Shrewsbury, England
Died: 19 April 1882 in Down, England
Education: Undergraduate medical student for 1½ years at Edinburgh University. B.A. (in divinity) Christ's College, Cambridge University.
Naturalist. Author of The Voyage of the Beagle, The Origin of Species, The Descent of Man, and Selection in Relation to Sex, The Expression of the Emotions in Man and Animals, *and numerous other books and articles.*

FIGURE 18.4
Charles Darwin at age 45, four years before receiving Wallace's manuscript.

Alfred Russel Wallace
Born: 8 January 1823
Died: 7 November 1913
Education: Six years of elementary school
Collector and naturalist. Author of Darwinism *and several books of travel and natural history.*

FIGURE 18.5
Alfred Russel Wallace around 1865, after his return to England from the Malay Archipelago.

FIGURE 18.6
Charles Lyell, who warned Darwin that Alfred Russel Wallace might also be working on the theory of natural selection. Lyell had previously played a crucial role in evolution by convincing most geologists that changes in the Earth had come about gradually over long periods of time.

Late in the spring of 1858 Charles Darwin (Fig. 18.4) sat in his large home at Down, 20 miles outside London, following his routine of reading the morning mail before returning to his study for another hour and a half of work. His eyes gazed wearily beneath his overhanging brows. He was not yet 50, but poor health and the labor of writing several books on geology and biology had left their marks. Now Darwin was struggling to complete a book summarizing 20 years of evidence that species had evolved through a process he called natural selection.

Darwin picked up an envelope and recognized it as being from young Alfred Russel Wallace (Fig. 18.5), the gifted naturalist who was making his living in the Malay Archipelago by selling specimens he collected in the tropical forests. Darwin eyed the envelope warily. Less than two years before, his friend Charles Lyell (Fig. 18.6) had warned him that Wallace seemed on the verge of proposing a theory much like his own. Heeding Lyell's warning, Darwin had begun to write a brief summary for early publication, but the "brief summary" had already grown into a large book, and it was still growing. Darwin opened Wallace's envelope and found a cover letter asking Darwin's opinion on an enclosed manuscript. Darwin scanned the manuscript and was stunned. In a few pages it summarized the theory over which he had labored for two decades.

It may have crossed Darwin's mind to destroy Wallace's paper and pretend he had never received it. Mail took several weeks between England and the Malay Archipelago, so Darwin could easily have put together his own short paper before Wallace discovered that his manuscript was missing. Instead Darwin wrote to Lyell: "Your words have come true with a vengeance. . . . Please return me the [manuscript] which he does not say he wishes me to publish, but I shall of course, at once write and offer to send to any journal. So all my originality, whatever it may amount to, will be smashed, though my book, if it will ever have any value, will not be deteriorated; as all the labour consists in the application of the theory." A week later Darwin and Lyell decided it would not be dishonorable if some of Darwin's previous sketches of the theory were read before the Linnaean

box continues

Society at the same time they presented Wallace's manuscript. In the meantime, Darwin set to work condensing the big book he had started into an "abstract" that we now know as *The Origin of Species.*

The convergence of Charles Darwin and Alfred Russel Wallace on the same theory of evolution seems a remarkable coincidence, especially considering the differences in their backgrounds. Charles Darwin's father was a prominent physician and son of Erasmus Darwin, and his mother was of the famous Wedgwood family of potters. As a member of the landed gentry, Charles Darwin had few practical worries. In fact he gave up studying medicine partly because he realized he would inherit a fortune anyway. His father convinced young Darwin to study theology as an alternative to becoming an idle sporting man. Charles Darwin agreed partly because ministering to the parishioners of some rural church would not interfere with his main interest, collecting beetles. Darwin was religious in a polite sort of way, and had no objection to any of the doctrines of the Church of England, including the belief that God had created each species individually.

Wallace, on the other hand, was brought up without religion, or much of anything else. He had to begin earning a living after only six years of schooling. Through his own efforts he acquired enough training to become a teacher. Many hours in libraries, museums, and the field later made him a leading authority on natural history. These intellectual pursuits were, by necessity, mingled with the need to earn a living.

From the perspective of science the differences between Darwin and Wallace are less important than the similarities in their backgrounds. Chief among these similarities were: (1) collecting, which provided an appreciation of the variations within species; (2) traveling, which provided an appreciation of the distribution of species; and (3) reading the *Essay on Population* by Thomas Malthus, which provided an appreciation of the struggle for survival within species.

FIGURE 18.7
Variation within a single species of beetle, the Asiatic lady beetle *Harmonia axyridis.* Each variant is geographically isolated from other variants.

■ **Question:** The dogs in Fig. 18.3 are all considered to be members of the same species because their histories are known. By what criteria would these beetles be judged to be members of one species? That is, what is a species? What is implied by the fact that this question is so hard to answer?

Collecting As a boy, Darwin collected shells, pebbles, eggs, and other natural miscellany, but at Cambridge in the late 1820s he focused on beetles. The intensity of Darwin's passion for beetle collecting can be judged from the following episode described in his *Autobiography.*

> One day, on tearing off some old bark, I saw two rare beetles and seized one in each hand; then I saw a third and new kind, which I could not bear to lose, so that I popped the one which I held in my right hand into my mouth.

In trying to classify his treasures, Darwin could hardly have missed the differences among individuals within the same species (Fig. 18.7). Variability is essential to natural selection, because if there are no inherited differences among individuals, then there is nothing to select.

box continues

FIGURE 18.8

The *Beagle* in the Straits of Magellan, between South America and Antarctica. Darwin was one of 74 people crammed on board this vessel, which was 90 feet long and 24 feet wide. In the foreground are natives of Tierra del Fuego, whose wildness made a profound impression on Darwin.

Wallace became interested in natural history at age 13 during his first job helping his brother, a surveyor. A decade later, Wallace met Henry Walter Bates, who interested him in insects. Together in 1848 they went to South America, planning to finance the voyage by selling rare insects they collected. Collecting was to be Wallace's major source of income for the next 14 years. As with Darwin, the variations he observed within species played an important role in his conception of natural selection.

Travels Soon after graduating in 1831, Darwin received a letter from a former teacher offering a position as naturalist on a round-the-world surveying voyage aboard the *Beagle* (Figs. 18.8 and 18.9). As the letter noted, Darwin was not a "finished naturalist," but he was "amply qualified for collecting, observing, and noting, anything worthy to be noted in Natural History." In addition, Darwin was of the correct social class to serve as a companion to the 26-year-old captain. The previous captain of the *Beagle* had committed suicide, probably from loneliness. So at age 22 Darwin bade farewell to his family and the woman to whom he was practically engaged and set off on a hazardous voyage that was supposed to last three years, but which lasted five.

As a going-away present, Darwin was given a copy of the first volume of *Principles of Geology,* which had

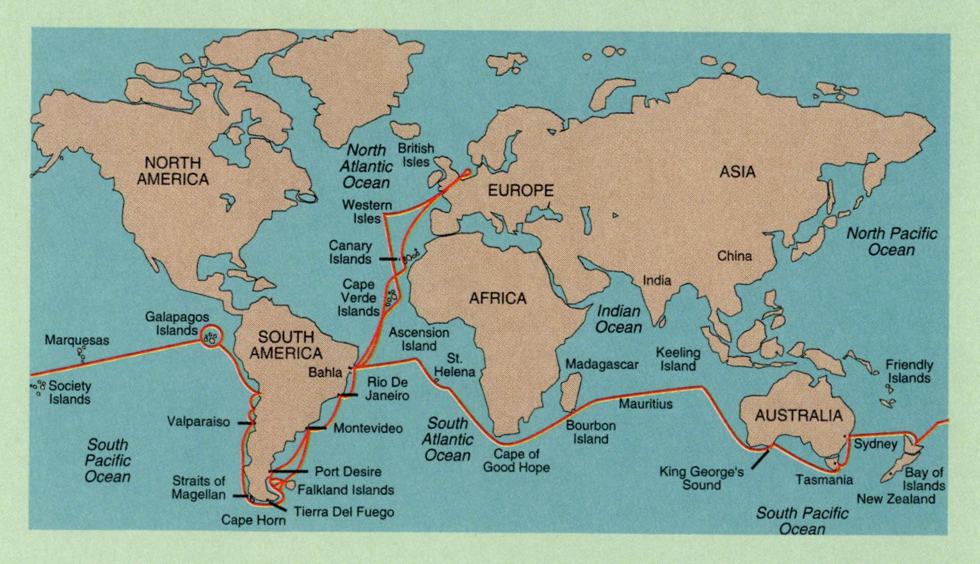

FIGURE 18.9

The route of Darwin's voyage around the world from 1831 to 1836. Observations made on the Galápagos Islands were crucial in eventually convincing Darwin that evolution had occurred.

box continues

just been published by a lawyer named Lyell. Darwin thereby became the first naturalist to see the New World through uniformitarian eyes. With this perspective he made several important contributions to geology, including the correct explanation for the origin of coral atolls. Lyell's book also focused Darwin's attention on the fossil record and the geographical distribution of species. Observations in these two areas eventually convinced Darwin of evolution. This conversion did not occur until March 1837, however, at least five months after Darwin's return to England. During the voyage, Darwin continued to believe the theory he had been taught by those who compromised between *Genesis* and the fossil record. This theory was similar to Cuvier's: Species had been periodically extinguished by catastrophes, then replaced by newly created species.

The following observations eventually made Darwin doubt that theory:

1. In South America, Darwin found the fossils of several animal species that were different from, but related to, the species still living in the area. Such discoveries had been made before, but catastrophists explained them as the result of God's having used similar species to replace the ones just destroyed. Darwin, however, realized that the evolution of the fossil species into the present species was more consistent with uniformitarianism.
2. Darwin found that similar animals on the Galápagos Islands, such as the birds now known as Darwin's finches, differed from one island to the next, even though the islands were physically similar and close to each other. It seemed unreasonable that a creator would have made different species for each island. At first, Darwin assumed that all the similar animals were simply varieties within the same species, but on returning to England he learned from the experts that his "varieties" were in fact distinct species. This made him wonder why it was so difficult to tell the difference between species and varieties if species had been created individually.

Wallace's voyage to South America was less propitious: On the return trip in 1852 the ship with many of his specimens burned. Undaunted, Wallace scraped together enough money for an expedition to the Malay Archipelago between 1854 and 1862. There his discoveries paralleled those of Darwin. In 1855 he published a paper in which he, too, observed that the fossils of extinct species are found in areas where similar species now live. This was the paper that prompted Lyell to warn Darwin of the possibility of being "scooped" by Wallace. Wallace also discovered that the marsupials and other unique groups on Australia and neighboring islands were separated from the Oriental groups on other islands of the Malay Archipelago by a sharp boundary, now called Wallace's Line (see Fig. 19.9). As with the Galápagos species, the simplest explanation was that the species on both sides of Wallace's Line evolved in different directions after they became isolated from each other.

Malthus and the Struggle for Survival Although Darwin realized several months after the *Beagle* voyage that species had evolved, he merely hinted at his ideas in public. He was too busy arranging for his collections to be analyzed by experts, writing an account of the voyage (which subsequently became *The Voyage of the Beagle*), and making important contributions in geology. He must also have been distracted by his first cousin, Emma Wedgwood, because he married her early in 1839. Darwin also began to suffer from heart palpitations, vomiting, and other disorders that did not abate until *The Origin of Species* was finally published. Even if Darwin had had the time and good health to publish his idea of evolution right away, he must have known that doing so would have jeopardized his standing as a geologist. Lyell had demolished "Lamarckism," and most biologists, including Darwin, had a low opinion of other theories of evolution. Darwin knew that he needed more evidence to support his claim that evolution had occurred, and he also needed a plausible explanation for how it occurred.

For clues to how nature altered species, Darwin began to study artificial selection, by which humans had altered numerous domesticated species. His major problem was to figure out how nature, lacking the consciousness of the breeder, could select which individuals would reproduce. Darwin read farm journals, quizzed breeders, and became a pigeon breeder, himself, in hopes of learning the secret, but in the end the clue

box continues

came from an essay on social theory, which Darwin read "for amusement" late in 1838. It was the *Essay on Population,* in which the Reverend Thomas Malthus pointed out that the ability of humans to reproduce far exceeds the capacity of the land to feed the resulting population. Malthus concluded that the human "struggle for survival" through famine, war, and social injustice was therefore natural and unavoidable. (Malthus mentioned the alternative of birth control only to dismiss it as a "vice" worse than starvation, war, and disease.)

Malthus intended his essay as an argument against French-Enlightenment dreams of human progress, but Darwin saw it as a mechanism for biological change. If more individuals are produced than can survive, then competition among members of each species is inevitable. Rather than stifle change, as Malthus thought, competition could cause evolutionary change. The best-adapted individuals would be the best competitors, and they would contribute their heritable adaptations in greater proportion to later generations. Darwin might have published his theory in 1838, only to have it forgotten. Instead he continued to collect evidence and did other research, including an eight-year study of the anatomy and classification of barnacles. As we have seen, it was Wallace's manuscript proposing the same theory that finally forced Darwin to publish.

By a remarkable coincidence, Wallace's insight was also triggered by Malthus's *Essay on Population,* which he had read 12 years before in England. While lying down in his forest shack during an attack of malaria, his fevered brain happened to remember Malthus's essay and relate it to the problem of evolution. In the next two days Wallace wrote out the main ideas of the theory of evolution by natural selection, unaware that the man he intended to send the manuscript to had already devoted 20 years to the same theory.

that played a crucial role in neo-Darwinism was **population genetics.** In contrast to Mendel with his controlled breeding experiments, population geneticists study patterns of inheritance in *natural* populations, usually by means of mathematical models. The term "population" in this context has a precise meaning: It is a group of sexually interbreeding organisms.

G. H. Hardy and W. Weinberg made a major contribution to population genetics in 1908 by independently disproving one argument against natural selection. It had been assumed that even if a new allele did improve the fitness of an individual, then in spite of natural selection it would quickly be diluted with each mating over succeeding generations. If such blending did occur then evolution by natural selection would hardly have a chance to get started. What is now called the **Hardy–Weinberg Principle** shows, however, that allele frequencies do not change if the population satisfies the following conditions:

1. The size of the population is practically infinite, so random variations in the reproduction of genes can be ignored.
2. Individuals mate at random; That is, the presence of a particular allele does not affect which individuals mate with each other.
3. The alleles do not increase or decrease the ability of organisms to reproduce. (In other words, natural selection does not occur in the population.)
4. There is no gain or loss of new alleles from another source; that is, there is no immigration or emigration.
5. There is no new mutation.

A population that satisfies all of these conditions is defined as an **equilibrium population.** Obviously, equilibrium populations exist only as models. Nonetheless, the

Hardy–Weinberg Principle is useful because it demonstrates that a single new allele does not automatically disappear by blending, even in an infinitely large population. Moreover, the Hardy–Weinberg Principle shows that because gene frequencies do not change in an equilibrium population, an equilibrium population is by definition a nonevolving one. Thus the equilibrium population serves as a baseline for measuring rates of evolution in real populations. The Hardy–Weinberg Principle also suggests that any departure from the five equilibrium conditions listed above can cause a change in gene frequencies and can therefore be a mechanism of evolution. Thus the possible causes of evolution are random changes in small populations (genetic drift), nonrandom mating (sexual selection), natural selection, emigration or immigration, and mutation. Each of these possible causes will be discussed later in this chapter.

Another major theoretical support for the synthetic theory came in the 1920s and 1930s when **R. A. Fisher, J. B. S. Haldane,** and **Sewall Wright** showed by three different mathematical proofs that natural selection is a more effective mechanism of evolution than is mutation. In order for random mutations to be the agent of evolution they would have to occur so frequently that harmful mutations would undo any previous evolution that had occurred. Mutations are, of course, necessary for evolution, because they produce genetic variation. These variations in **genotype** then contribute to the variations in **phenotype** (the expressed characteristics of an organism) on which selection acts. By selecting one phenotype over another, nature indirectly selects one genotype over another, thereby altering gene frequencies.

EXPERIMENTAL CONTRIBUTIONS. In addition to theoretical population genetics, numerous field studies during the 1930s and 1940s produced results that agreed more with the predictions of "selectionists" than with those of "mutationists." Among the major contributors to these studies were the following:

Theodosius Dobzhansky was a *Drosophila* geneticist who took his knowledge into the field. He discovered, among many other phenomena, that in natural populations there is a great deal of genetic variation. Consequently, evolution does not depend on just the right mutation occurring at just the right time.

Ernst Mayr is, among many other distinctions, a taxonomist who clarified the concept of species. Most biologists had thought of species as fixed types, rigidly defined by their appearance. This "typological" concept made it difficult to understand how one species could evolve into another. In contrast, Mayr defined species as **"groups of actually or potentially interbreeding natural populations, which are reproductively isolated from other such groups."** This **biological species concept** greatly clarifies evolution, since it invites consideration of species as groups of genetically variable populations rather than as fixed types.

Bernhard Rensch drew attention to the way populations of a given species tend to vary with environmental conditions in an adaptive way, as predicted from natural selection. For example, a mammal in a cold climate tends to have shorter ears, tail, and feet than a member of the same species in a warmer climate (Allen's Rule). Rensch pointed out that this phenomenon could be explained by natural selection, because shorter appendages lose heat less readily.

George Gaylord Simpson was a paleontologist who showed that species do not evolve from "lower" to "higher" forms in a straight line of progress. Instead, as expected from random mutation and natural selection, the pattern of evolution is like a tree, with many branches leading only to extinction (Fig. 18.10 on page 368).

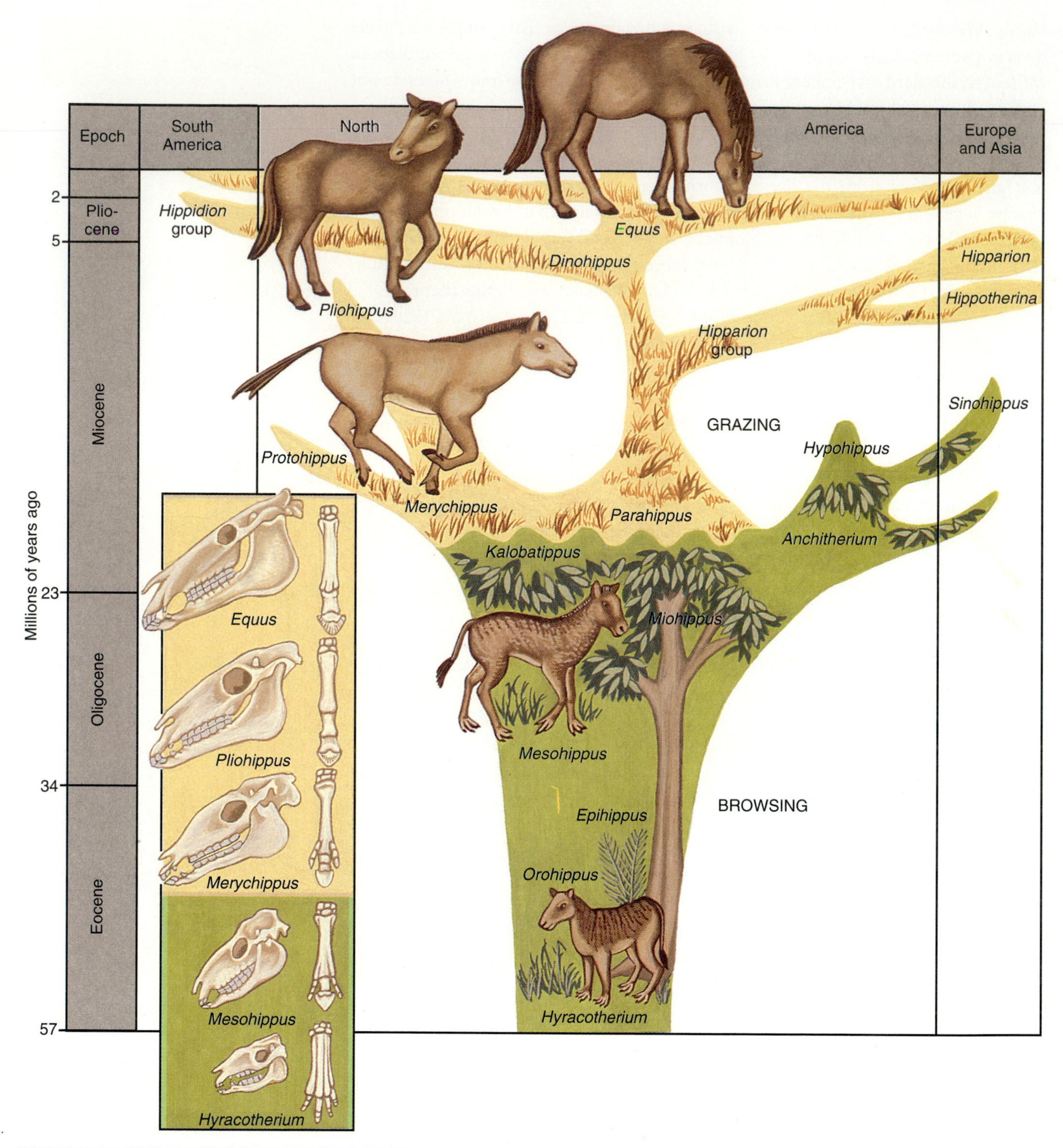

FIGURE 18.10

George Gaylord Simpson used the evolution of horses to argue against a linear progression, which might have been inferred mistakenly from the skulls and hooves shown in the inset. The left vertical axis shows the time scale, and the top horizontal axis shows geographical distribution. The change from browsing in the collie-sized *Hyracotherium* to grazing in the modern *Equus* paralleled the decreased humidity in North America and the change from tropical forest to prairie and savanna. During the last ice age, horses crossed into Europe and Asia, then into South America. They then became extinct in the Americas until the Spanish reintroduced them. This phylogeny, showing only the major lineages, is based on Prothero and Schoch (1989).

Speciation

According to the biological species concept, a new species begins when a group of organisms can no longer interbreed with the population from which it originated. How does this come about? One possible mechanism is that part of the population accumulates genetic changes that make it unable to reproduce with others in the population, even though they continue to live in the same areas. This pattern, called **sympatric speciation** (Greek *sym* together + *patris* native land), has been documented in some species. One cause of sympatric speciation, especially in plants, is polyploidy (p. 94): a sudden multiplication of chromosome number. Sympatric speciation appears to be less common than **allopatric speciation** (Greek *allo* other), in which a subpopulation first becomes physically separated from its parent population, then evolves differences that would prevent the subpopulation from interbreeding with the parent population.

Physical separation leading to allopatric speciation can occur in two ways. First, in an extremely large population, individuals on the fringes may not be able to mix well with the majority of the population, and the gene frequencies of this subpopulation may be different from the population as a whole. The subpopulation can then give rise to a genetically different population. This result is called the **founder effect,** because the descendants of the subpopulation will resemble the founders rather than the entire population, just as the inhabitants of a small town often resemble the founders of the town. The second mechanism of physical separation is called **vicariance.** Vicariance can result from a geological event, such as the rise of a mountain range, from the death of individuals living in the middle of the population, or from the straying of part of the population from the rest. The different species of Darwin's finches are thought to have arisen by vicariance, as a subpopulation wandered or was blown from the mainland to a Galápagos island and from there to other islands.

The Logic of Natural Selection

There are three kinds of arguments for natural selection as a mechanism of evolution: It is intuitively obvious, it is mathematically feasible, and it has been observed in fast-breeding organisms, such as antibiotic-resistant bacteria and pesticide-resistant insects. Each of these arguments shows that if there is genetic variability in a population and if conditions favor the reproduction of some individuals more than others, then the frequencies of alleles in the fitter individuals will increase at the expense of those in the less fit. Almost no evolutionist doubts, therefore, that natural selection is one mechanism—perhaps the only one—for the evolution of adaptations. This synthetic theory of evolution is so widely accepted that it could be called the orthodox view. Like most orthodoxies, it has given rise to several fallacies of logic that should be critically examined.

Evolution and Progress. One of the most prevalent fallacies of evolution is the notion that it is progressive—that is, that organisms can be ranked on a scale according to how far they have progressed toward the most highly evolved, most advanced, and most complex species (ourselves, naturally). In Darwin's time, many European scientists even fancied themselves on the top rung on such an evolutionary scale, above "less-evolved" human races. As Simpson and others showed, however, there is no evolutionary scale, but an evolutionary tree. Many biologists still use the terms "lower" and "higher," but only to indicate relative position on the tree, which depends on whether a group arose earlier or later than another. Because mammals evolved before birds, humans are out on a limb of the evolutionary tree with birds roosting over our heads. That does not necessarily make us inferior to birds.

Survival of the Fittest. The social theorist Herbert Spencer coined the phrase "survival of the fittest," and Alfred Russel Wallace persuaded Darwin that it would be a good summary of their theory of natural selection. "Survival of the fittest" is an unfortunate statement of a scientific theory, however, because it is not experimentally testable. Fitness is defined as the ability to produce viable offspring, so "survival of the fittest" means "survival of the populations that are best able to survive." No one ever bothers to do an "experiment" to test whether the fittest survive, because the outcome is obvious.

Lately the concept of survival of the fittest has also been weakened by a new appreciation of the role of chance. Asteroid impacts, changes in climate, and other random catastrophes have wiped out thousands of species that were apparently just as fit as the species that survived. As we shall see in the next chapter, one such mass extinction eliminated the mighty dinosaurs that had thrived for 140 million years, while it spared the lowly mammals of the time.

The Panglossian Paradigm. Many biologists assume, often without being aware of it, that every feature of an organism is optimally adapted for its function. Stephen Jay Gould and Richard C. Lewontin referred to this notion as the "Panglossian paradigm," after Dr. Pangloss, a character in a novel by Voltaire, who kept insisting through earthquakes and other catastrophes that this is the best of all possible worlds. The concept that every feature of an organism is an adaptation that contributes to fitness is useful when it stimulates research to discover how the presumed adaptation functions. Often, however, instead of creating a testable hypothesis, one is tempted just to make up a plausible story of how the adaptation evolved. Before breakfast any good biologist can come up with six explanations for why bats have wings, and half a dozen equally good explanations for why mice do not. Such "explanations" are often embarrassingly like Rudyard Kipling's *Just So Stories*, which explain how the whale got its throat, how the camel got its hump, how the rhinoceros got its skin, and so on.

"For the Good of the Species." A common fallacy in thinking about natural selection is that it is capable of producing whatever adaptation benefits a species. Natural selection is not, however, a conscious force that cares about the good of a species. The problem with "for the good of the species" thinking is that it often obscures more interesting concepts of evolution. For example, it is tempting to accept the following explanation uncritically: "Male deer charge each other with their potentially lethal antlers during contests for mating rights, but for the good of the species they have evolved behavioral mechanisms that make them avoid inflicting injury." It may be useful to the species for its members not to destroy each other, but a moment's thought reveals that this is not an adequate mechanism for the evolution of this behavior. There could have been natural selection for the behavior only if it allowed the *individuals* displaying the behavior to leave more of their genes in the next generation. Yet failing to kill an opponent, thereby perhaps losing the privilege of reproducing, would *decrease* the likelihood of a buck transmitting its genes into the next generation.

Genetic Drift

Besides natural selection, the Hardy–Weinberg Principle suggests that there are four other possible causes of evolution: genetic drift, sexual selection, immigration or emigration, and mutation. Changes in gene frequency (evolution) due to immigration and emigration are fairly obvious and need no further discussion. The role of mutation in evolution was dealt with by Fisher, Haldane, and Wright. Genetic drift and sexual selection remain as two potentially important factors in evolution.

The Hardy–Weinberg Principle indicates that equilibrium could occur in an infinite population. Real populations are always finite, however, and in a small population random events can have large effects on allele frequency. For example, the death of one organism with a particular allele will have a much greater effect on the frequency of that allele in a population of ten than it would have in a population of hundreds or thousands. Such random fluctuations in allele frequencies are referred to as **genetic drift** (Fig. 18.11).

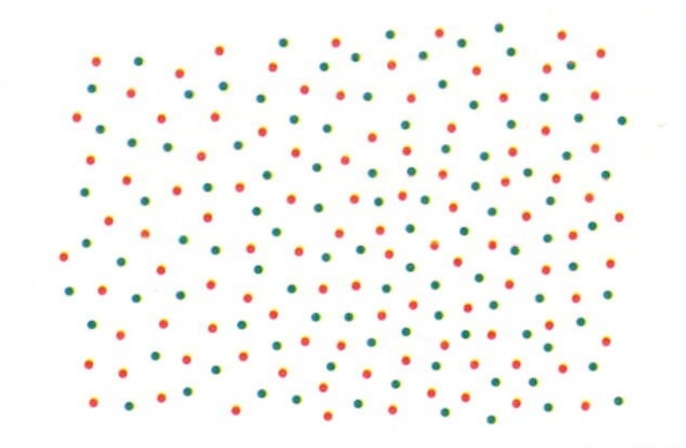

FIGURE 18.11

A model of genetic drift. This random distribution of equal numbers of red and green dots represents all the individuals of one species. The "allele frequencies" for red and green dots are 50% each. Place the blunt end of a pencil anywhere on the figure to select a random sample of organisms that will reproduce. Estimate the "allele frequencies" for the next generation by counting the number of red dots under the pencil end and dividing by the total of red and green dots under the pencil. Is the "allele frequency" still 50%? Now repeat the exercise using a small coin instead of a pencil to isolate a larger sample. Is the allele frequency closer to 50%?

Most evolutionists recognize genetic drift as a major factor in evolution, and some argue that it is even more important than natural selection. Their **neutral theory of evolution** proposes that most mutations are not selected for or against because they have no effect on the phenotype or the fitness of the organism. Selection is "blind" to such neutral mutations, so they arise unchecked and become established in populations through genetic drift. Neutralists, as advocates of this theory are called, contend that if evolution is defined as changes in gene frequencies, then evolution is due more to neutral mutations than to natural selection. Many adherents to the synthetic theory counter by saying that because neutral mutations are by definition nonadaptive, they cannot account for the important changes that occur during evolution. This controversy is not likely to be resolved in the near future. In the meantime, it is safe to say that natural selection is the major cause of adaptive evolution, but that it may not be the cause of all evolutionary changes.

Sex and Evolution

SEXUAL SELECTION. In an equilibrium (nonevolving) population, mating is random. In many species of animals, however, nonrandom mating can result from either mate choice, or competition. In mate choice members of one sex, usually females, choose mates based on behavior and appearance. Competition, which usually occurs among males, often results in only the strongest reproducing. Mate choice has apparently resulted in the evolution of behaviors and structures, mainly in males, that attract mates. One example of such sexually selected traits is the elaborate tail of the peacock (see Fig. 10.2B). Such displays apparently make no other contribution to fitness, because they generally do not occur after the breeding season. In fact, they may actually endanger the animal by attracting predators and interfering with locomotion. Competition has also resulted in structures and behaviors such as the antlers of deer and the head-butting of rams. Generally mate choice and competition occur in different species. In deer and sheep, for example, females are attracted to the sounds of antlers rattling and heads butting, but which male they breed with is decided not by them, but by the outcome of the competition.

It is often assumed that courtship displays, large antlers, and head-butting work against natural selection by making an animal vulnerable to predators and consuming energy. Some evolutionists point out, however, that the survival of an animal with such sexually selected traits may indicate that it has other traits that contribute to survival. For example, a peacock with a handsome, symmetric tail must be able to ward off parasites, and competition among males for mates may select for strength and other traits that contribute to survival. In such cases where sexually selected traits honestly represent overall fitness, sexual selection may be merely a special case of natural selection.

IS SEX WORTH THE TROUBLE? According to the synthetic theory, natural selection operates as if it were the purpose of each organism to get as many of its alleles as possible into the next generation. Yet most animals and many other organisms seemingly defy natural selection by using sex to reproduce. Not only does sexual reproduction make use of only half of each organism's alleles to make one offspring, but it has further disadvantages: (1) It requires energy to produce gametes, (2) it may take time and energy to acquire mates, (3) it renders many animals vulnerable to parasites and predators during reproduction, (4) it requires elaborate mechanisms to ensure that gametes are not wasted on a member of the wrong species, (5) it requires the cumbersome mechanisms of meiosis and fertilization, (6) it requires the existence of males that often contribute nothing to the species except sperm, and (7) in many species, including humans, the genes contributed by males have more mutations than those of females because they have gone through more cycles of mitosis. Every few years an evolutionist proposes an explanation for why sex evolved in spite of these disadvantages, but of the 20 or so theories, none has yet gained a consensus. Biologists are as uncertain as most 12-year-olds about whether sex is a good idea.

Most biologists assume that the advantage of sexual reproduction is that it increases genetic variability in two ways. First, the two pairs of chromosomes common in sexual species allow for **recombination:** the shuffling of alleles between homologous pairs of chromosomes during meiosis (pp. 74–77). Second, sexual reproduction can increase genetic variability through **outcrossing:** the mixing of alleles from two different individuals. The advantage of increased genetic variability could be twofold. First, organisms with many variable offspring are likely to have at least *some* that are well-adapted. Second, sexual species are less likely to become extinct, because the increased variability allows them to adapt more quickly to environmental change. The first advantage applies only to species with high fertility, however. In mammals and others with relatively few offspring, high genetic variability due to sexual reproduction might cause all the offspring to be defective rather than result in a few being better-adapted. In addition, there is no evidence that asexual organisms are less able to adapt than sexual ones.

Some Recent Studies of Evolution

COMPETITION IN DARWIN'S FINCHES. According to the synthetic theory, natural selection is based on competition among individuals within populations. Competition therefore plays a crucial role in evolution, but not necessarily the obvious role pictured in Alfred, Lord Tennyson's phrase, "Nature, red in tooth and claw." In most animals, and certainly in plants and other organisms, competition is usually a quiet, invisible struggle to avoid predators and obtain energy, mates, and the other necessities for survival and reproduction. The struggle is so quiet and invisible that most people are surprised when it is pointed out that, on average, just as many organisms die as are born. Because competition is so subtle, scientists require care and patience to observe it. The Galápagos Island birds now called Darwin's finches have proved quite useful in such studies.

There are 14 different but related species of Darwin's finches in the world: 13 on the Galápagos Islands and one on the Cocos Islands, 1000 km northeast (Fig. 18.12). All apparently evolved from a single unidentified and now-extinct species after the islands erupted from the sea several million years ago. All the species are drab colored with short tails; all the males court by presenting nest material and food to the females; and all build large, roofed nests and lay pink-spotted white eggs. The major external differences among the species are in their beaks and feeding habits. They also differ in distribution among the islands; not all species are found on all islands.

Because all the birds can easily fly from one island to another, why aren't all the species found on all the islands? How did the different species evolve in the first place? The synthetic theory of evolution suggests that competition is the answer to both ques-

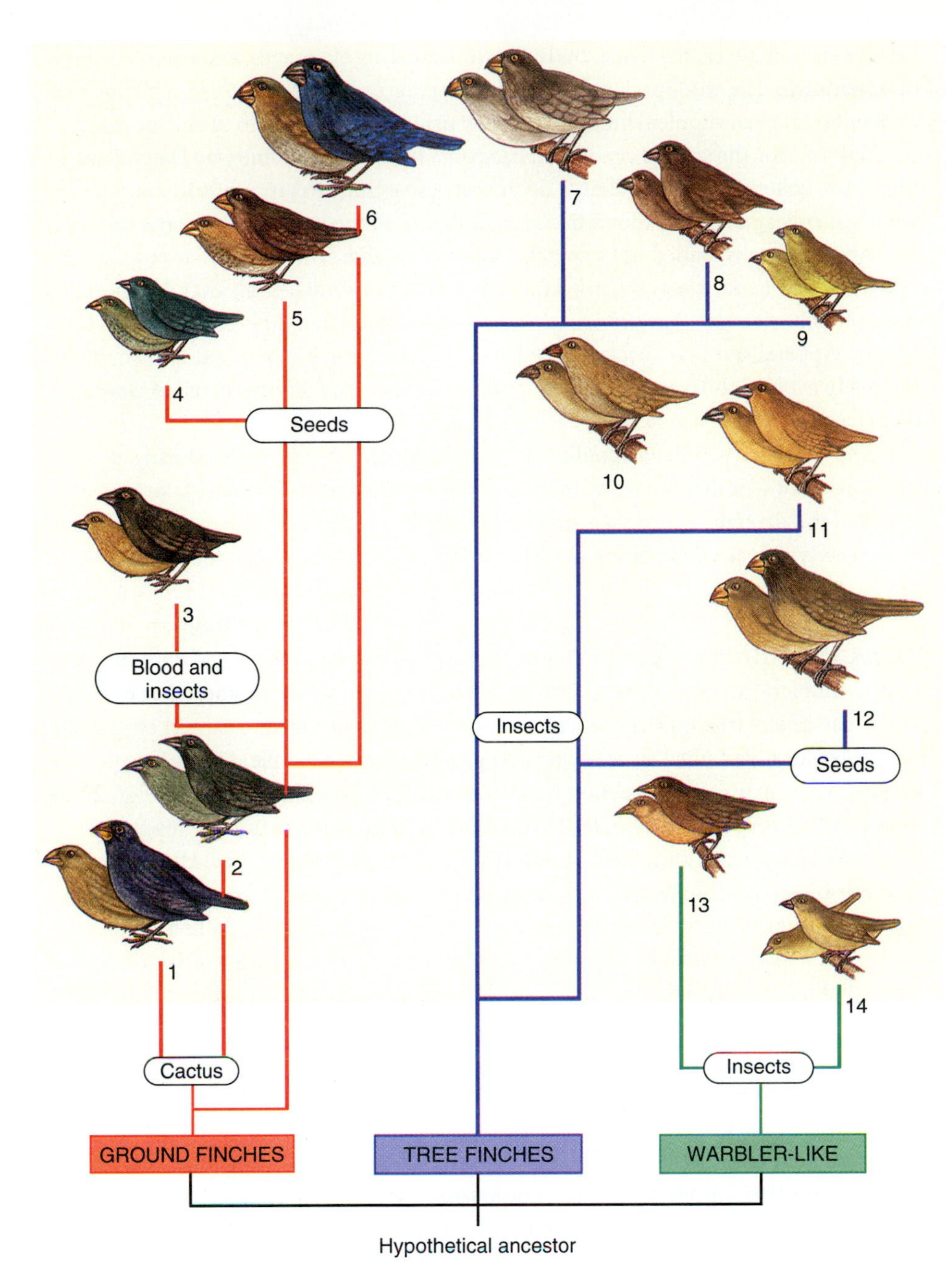

FIGURE 18.12

"Darwin's finches" of the Galápagos and Cocos Islands occur in three major groups—ground finches, tree finches, and warbler finches—that differ primarily in their beaks and feeding habits. Lines showing evolutionary relationships are inferred from comparisons of protein structures. 1. Large cactus ground finch *Geospiza conirostris.* 2. Cactus ground finch *G. scandens.* 3. Sharp-beaked ground finch *G. difficilis.* 4. Small ground finch *G. fuliginosa.* 5. Medium ground finch *G. fortis.* 6. Large ground finch *G. magnirostris.* 7. Large tree finch *Camarhynchus psittacula.* 8. Medium tree finch *C. pauper.* 9. Small tree finch *C. parvulus.* 10. Woodpecker finch *Cactospiza pallida.* 11. Mangrove finch *C. heliobates.* 12. Vegetarian finch *Platyspiza crassirostris.* 13. Cocos finch *Pinaroloxias inornata.* 14. Warbler finch *Certhidea olivacea.*

tions, with beaks and feeding habits being the most likely bases of competition. If this hypothesis is correct, then the following testable predictions should be true:

1. The beaks and feeding habits of species on each island are adapted for the kinds of food on the island.
2. Species with similar beaks and feeding habits do not occur on the same island.
3. A change in the food available on an island will favor species with certain beak characteristics and hurt those with other beak characteristics.

Because Darwin's finches haven't starved to extinction, the first prediction appears obvious. The interesting point, however, is that the finches have beaks adapted in quite specific ways for virtually every potential food source—insects, ticks, seeds of

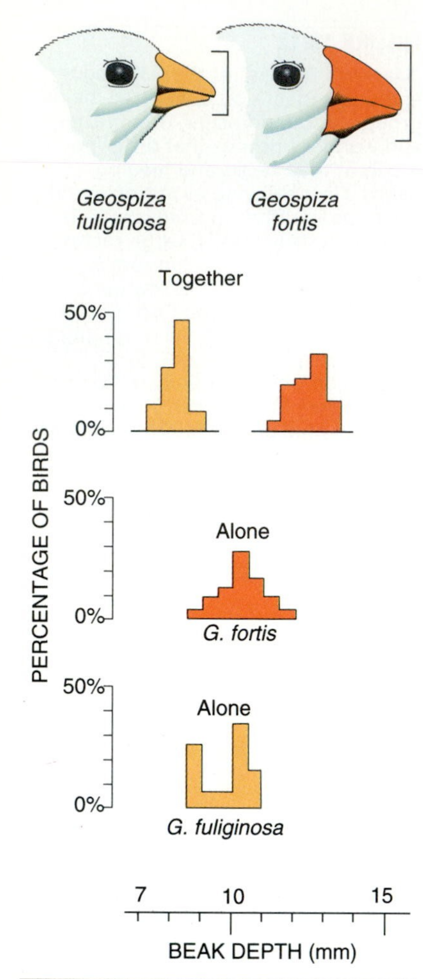

FIGURE 18.13

Character displacement in two species of Darwin's finches, the small and medium ground finches (numbers 4 and 5 in Fig. 18.12). On four islands where the two species compete with each other for seeds, the depth of the beaks is less in all of the small finches than in any of the medium finches (top graph). On islands where each species occurs alone, beak depth is similar in the two species (middle and bottom graphs).

■ **Question:** What is the connection between beak size and fitness?

various sizes, fruit, cactus tissue, buds, blossoms, mangrove leaves, and even the blood of other birds. The woodpecker finch and the mangrove finch (*Cactospiza pallida* and *C. heliobates*) even supplement their beaks by using cactus spines to probe for insects.

Support for the second prediction has come from a classic study by David Lack in the 1940s, followed by more than a decade of research by Peter and Rosemary Grant and their colleagues. They found that species that used the same food because they had similar-sized beaks tended not to inhabit the same island. In several cases, populations of two different species occupying different islands had similar beak sizes, but populations of those same two species occupying the same island had different beak sizes (Fig. 18.13). Apparently, two species forced to compete because they had similar beaks gradually evolved different beaks. The resulting divergence of form is called **character displacement.**

The third prediction was confirmed by P. T. Boag and Peter Grant during a severe drought on one of the islands in 1977. They found that larger birds with larger beaks survived the drought better than smaller birds with smaller beaks because only large, tough seeds withstood the drought, and only larger birds could crack the shells of these seeds.

MIMICRY IN BUTTERFLIES. Among the most striking products of natural selection is mimicry, in which one species (the **mimic**) resembles another species or an inanimate object (the **model**). In **aggressive mimicry** an animal, or part of an animal, resembles an object that attracts prey. Anglerfishes, for example, "fish" for prey by means of a structure near the mouth that works like a lure (Fig. 18.14A). (Figs. 28.18 and 37.18B show other examples.) In **defensive mimicry** the animal resembles an inedible or dangerous organism or object (Fig. 18.14B). (See Fig. 32.14B for another example of defensive mimicry.) The occurrence of mimicry must be totally mystifying to anyone unfamiliar with natural selection. Under Lamarckism and other theories of evolution, one must imagine that a potential mimic recognizes the potential model, realizes that it is not preyed upon, then somehow alters its genetic makeup so that its offspring resemble the model. Natural selection offers a more plausible explanation: An individual that happens to resemble an unpalatable model is likely to have more offspring, which will also be likely to have more offspring the more closely they resemble the model.

If the model is unpalatable, it may "advertise" the fact with **warning coloration.** Predators quickly learn to avoid prey with warning coloration, and they may also avoid mimics that are similarly colored (see Figs. 32.13 and 37.5C, D). If the mimic is palatable, the situation is called **Batesian mimicry,** after its discoverer, Henry Walter Bates (Wallace's companion in South America). The viceroy butterfly *Limenitis archippus* was for many years assumed to be a palatable Batesian mimic of the monarch butterfly, *Danaus plexippus*, which is generally unpalatable because it feeds on milkweeds. It now appears, however, that viceroys are also often unpalatable (Fig. 18.15). The two unpalatable species may therefore reinforce each other's protection from predation. Such mimicry of one unpalatable species by another is called **Müllerian mimicry,** named for Fritz Müller.

CAMOUFLAGE IN MOTHS. Camouflage may be thought of as mimicry of the surroundings. (See Figs. 32.14C, 35.21B, 36.6B, and 38.6A for examples.) One of the most celebrated examples of camouflage occurs in the peppered moth *Biston betularia* in England. Like many other insects, many peppered moths have genes that produce the dark pigment melanin. If such darkening is more common in areas where soot covers the vegetation on which the insects rest during the day, it is called **industrial**

FIGURE 18.14

(A) Aggressive mimicry in a fish of the genus *Antennarius* from the Philippines. Many anglerfish (order Lophiiformes) have dorsal fins modified into lures that they wave about in front of their mouths to attract prey. The bait on this species is particularly convincing. (B) Defensive mimicry by *Calloplesiops altivelis,* also from the South Pacific. When threatened by a predator, *Calloplesiops* hides its head and exposes its rear (bottom), which resembles the head of the dangerous moray eel *Gymnothorax meleagris* (top).

melanism. Prior to the mid-19th century, industrial melanism was rare in peppered moths. Almost all had speckled light-gray wings that were hard to see against lichen-covered tree trunks. The first **melanic** (dark-winged) moth was reported in 1849 near Manchester, where industries burned large amounts of coal. As pollution increased in England, so did the proportion of melanic moths. By 1886 up to 98% of peppered moths in some industrial areas were dark-winged.

Some biologists suggested that industrial melanism was due to the inheritance of acquired characters or to the direct induction of mutations by the pollution. In the 1930s E. B. Ford, one of the early contributors to the synthetic theory, suggested that natural selection was the cause. He hypothesized that air pollution destroyed lichens and exposed the dark tree trunks on which the moths rested during the day. The camouflage of the light-winged moths was therefore destroyed, while that of the dark-

FIGURE 18.15

Three types of mimicry in two species: the monarch butterfly *Danaus plexippus* and the viceroy butterfly *Limenitis archippus.* Most monarch butterflies become unpalatable (left) from feeding on toxic species of milkweed. Viceroys are also unpalatable, at least in some areas (Ritland and Brower 1991). Their similarity to monarchs reinforces the lesson to potential predators: Orange and black butterflies are not good to eat. Such mimicry of two or more generally unpalatable species by each other is called Müllerian mimicry. Viceroys were formerly thought to be palatable but avoided by predators because of their resemblance to unpalatable monarchs. This type of mimicry would be Batesian mimicry. Some monarchs are also palatable and may be protected by automimicry of unpalatable members of their own species.

A

B

FIGURE 18.16

Camouflage in light- and dark-winged peppered moths *Biston betularia.* The dark wings are due to increased deposits of melanin, which result from a single dominant allele. (A) One dark moth and one light moth resting on the lichen-covered bark of a tree. (B) Dark and light moths resting on the bark of a tree in an area in which industrial pollution has destroyed the lichen.

winged moths was improved (Fig. 18.16). Predatory birds then acted as the agents of natural selection, eating a greater proportion of the easily seen light-winged moths.

H. B. D. Kettlewell of Oxford University tested Ford's hypothesis by tagging hundreds of light and dark moths, releasing them in polluted and nonpolluted areas, and estimating the survivability of each form from the numbers recaptured. The results agreed with Ford's hypothesis. Near the industrial city of Birmingham, Kettlewell recaptured 53% of the dark moths and only 25% of the light moths. Near unpolluted Dorset, Kettlewell recaptured only 4.7% of the dark forms and 13.7% of the light forms.

As Kettlewell has pointed out, there are numerous factors that could complicate these results, including the fact that the moths tend to migrate to areas where they are well camouflaged. The small proportion of light-winged moths recaptured in Birmingham and of dark-winged moths recaptured in Dorset could therefore have been due to migration out of each area. In order to see whether predation really was a factor, Kettlewell posed dead moths of both varieties on light or darkened tree trunks. He found that, indeed, birds did consume more light-winged moths on dark trunks, and more melanic moths on lichen-covered trunks. In the past few decades governments have conducted their own "experiments" by tightening controls on air pollution. As expected, the proportion of melanic moths has declined as the air has become cleaner.

BIOCHEMICAL STUDIES OF EVOLUTION. Until the 1960s, evolutionists could compare the anatomical effects of genetic differences among organisms, but now they can examine the genetic differences directly. The first technique of molecular comparison to contribute significantly to evolutionary theory was **electrophoresis.** In this procedure, proteins are placed into a gel, and an applied voltage makes each protein move at a rate that depends on its net charge, size, and shape. By comparing the rate of movement of homologous proteins from two different species, one can infer how closely related the species are.

Electrophoresis was instrumental in solving one puzzle of evolution: If natural selection eliminates the unfit and if mutations are rare, how can organisms ever adapt to changing conditions? Electrophoresis showed that natural populations have a high degree of **polymorphism,** or genetic variety within a single trait. Thus mutations are not rigorously eliminated, but remain as the raw materials for future adaptations. In *Drosophila,* 53% of proteins examined electrophoretically were found to occur in more than one form within a single population; in the average fly, 15% of the genes were found to have two different copies (alleles). Polymorphism tends to be less for vertebrates, but even in humans more than 25 out of approximately 100 enzymes studied were found to be polymorphic (Lewontin 1982, Chapter 3).

Because of redundancy in the genetic code, some genetic differences do not affect protein structure. In addition, proteins are generally more difficult to handle than the nucleic acids (DNA and RNA) that code for them. For these reasons **nucleic acid sequencing**—determining the sequence of bases in DNA or RNA—is increasingly common as a tool of molecular comparison. The choice of which nucleic acid to use depends on the taxonomic scale at which comparisons are made. For comparisons among lower taxa (orders, families, genera, species), one must use a nucleic acid that accumulates mutations rapidly, such as mitochondrial DNA (mtDNA). If one is interested in comparing different kingdoms or phyla, one must use a highly conserved nucleic acid (one that mutates slowly), such as one of the RNAs in ribosomes (rRNA).

These and other techniques of molecular comparison have made possible **molecular phylogenetics,** which is a relatively new approach to determining the evolutionary

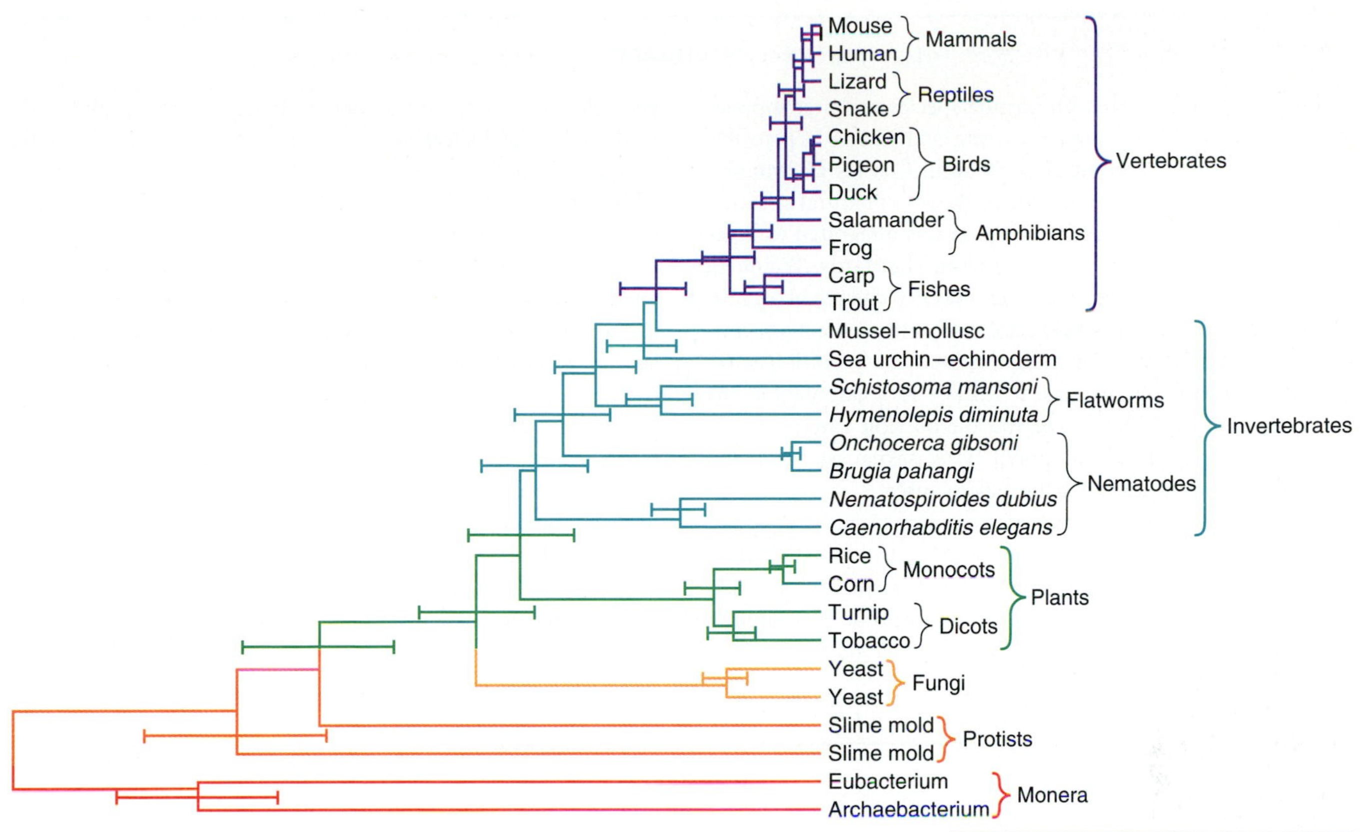

FIGURE 18.17

The phylogeny of 19 animals and other organisms as inferred from comparisons of a ribosomal RNA. The number of nucleotide substitutions between two organisms, and therefore the evolutionary distance between them, is proportional to the total length along the horizontal lines connecting the two. Thus the farther left the vertical line connecting two horizontal lines, the earlier the inferred evolutionary divergence of the organisms joined by the two lines. Each light horizontal line represents the range of possible error of measurement. Several alternative patterns are possible within these ranges of error. This figure suggests, among other patterns, that flatworms diverged from other invertebrates before mussels and sea urchins did. (Adapted from Qu, L-H., M. Nicoloso, and J-P. Bachellerie. 1988. Phylogenetic calibration of the 5′ terminal domain of large rRNA achieved by determining twenty eucaryotic sequences. *J. Mol. Evol.* 28:113–124.)

history (phylogeny) of organisms. Molecular phylogeny is subject to the same errors and differences in interpretation that plague comparative anatomy and other traditional methods of phylogeny, but it has the advantage that each molecular difference is an independent datum, so each molecule contains a large amount of data. Computerized statistical tests are used to determine the phylogeny most consistent with the mass of data. Figure 18.17 shows the phylogeny deduced by comparing the base sequences in a ribosomal RNA from 11 vertebrates, 8 invertebrates, and some plants, fungi, protists, and bacteria. It shows that, as expected, the ribosomal RNA of a human is more like that of another mammal (the mouse) than that of a vertebrate in any other class (bird, reptile, amphibian, or fish). It also shows that human ribosomal RNA differs from that of any of the invertebrates more than it differs from any of the vertebrates. This result, together with independent research using different ribosomal RNAs, supports the traditional classification based on comparative anatomy.

Research in molecular phylogenetics also provides an argument that comes as close to a proof of evolution as anyone could hope for. This argument is based on the unlikelihood that exactly the same errors will occur twice independently. Compilers of street guides and trivia games routinely use this concept by making up erroneous addresses and "facts" to trap plagiarists. A plagiarist can always argue that the correct facts in his work came from the same source that another author used; he has a more difficult time explaining to a jury how the same fictitious address or fact came to be in his work. Many related species contain identical useless sections of DNA called **pseudogenes** that can be thought of as plagiarized errors. Several such pseudogenes occur in humans and apes, but not in other mammals (Max 1990). The most likely explanation for the same pseudogene in two or more species is that one species "plagiarized" it from the other or that both got it from the same ancestor.

Summary

That species evolved from preexisting species was first appreciated in the 18th century, but convincing evidence and a plausible explanation for it were not available until Charles Darwin and Alfred Russel Wallace proposed their theory of natural selection in 1858. Subsequently, natural selection was integrated with genetics in the synthetic theory of evolution. The Hardy–Weinberg Principle shows that gene frequencies do not change in an infinite population without sexual selection, natural selection, emigration or immigration, or mutation. Because evolution is regarded as any change in gene frequency in a population, the Hardy–Weinberg Principle suggests the possible causes of evolution. Genetic drift in small populations, sexual selection, and natural selection appear to be the major causes.

Small changes in gene frequency can accumulate to the point that organisms can no longer interbreed in the population from which they originated. They then become members of a new species according to the biological species concept. Speciation is generally allopatric, with a subpopulation becoming physically separated by geological or other means (vicariance) or by being isolated on the fringe of the parent population (founder effect). Although sex is thought to contribute to evolution by increasing genetic variety, how or why it evolved is a mystery.

Some common notions of evolution are not valid. Evolution and natural selection do not result in progress, and they do not necessarily happen just because they would be good for the species. Not all features that evolve increase fitness, and the idea of survival of the fittest is untestable. Evolution and natural selection are supported by many experimental studies, such as those demonstrating predictable changes resulting from competition in Darwin's finches, and studies of mimicry of the monarch butterfly and industrial melanism in peppered moths. Molecular biological studies also demonstrate that patterns of differences in proteins and nucleic acids are generally consistent with phylogenies deduced by anatomical comparisons.

Key Terms

population
gene frequency
synthetic (neo-Darwinian) theory
natural selection
fitness
adaptation
homologous
analogous
uniformitarianism
Hardy–Weinberg Principle
biological species concept
sympatric speciation
allopatric speciation
founder effect
vicariance
genetic drift
sexual selection
Darwin's finches
molecular phylogenetics

Chapter Test

1. Describe the modern concept of evolution. (p. 358)
2. Define the following terms and relate them to the concept of evolution: population, mutation, species, fitness, natural selection. (pp. 356, 367)
3. Describe three kinds of evidence that evolution has occurred. (pp. 357–359)
4. Briefly describe the theory of evolution proposed by Charles Darwin and Alfred Russel Wallace. How does this theory differ from Lamarck's? (pp. 358, 366)
5. Briefly explain the synthetic theory of evolution in your own words. How does it differ from Darwin's original theory of evolution? (pp. 356, 366)
6. Describe some of the main contributions to the synthetic theory of evolution. (pp. 366–368)
7. What is the relevance of the Hardy–Weinberg Principle to evolution? (pp. 366–367)
8. Many creationists have charged that evolution is mathematically impossible because it depends entirely on chance. Does evolution depend only on chance? Which aspects of evolution do depend on chance, and which ones do not? (pp. 370–371)
9. Creationists also charge that evolution cannot be observed, and it is therefore not a fit subject for science. Is this a valid argument? Give an example of evolution research that satisfies your definition of science. (pp. 372–377)
10. Prepare an outline for an essay on the relationship of geology to evolution. (pp. 357–360)
11. Prepare an outline for an essay on the relationship of genetics (both population genetics and molecular genetics) to evolution. (pp. 366–367, 376–377)
12. Define genetic drift. How can it cause evolution? (pp. 370–371)
13. Describe the most likely way by which a new species arises. (p. 369)

■ Answers to the Figure Questions

18.1 Even though the wings of birds and bats are both forelimbs of vertebrates, they are analogous rather than homologous, because the most recent common ancestor of birds and bats did not have wings.

18.7 These beetles are judged to be members of the same species because each population is capable of interbreeding with other populations where their ranges overlap. It may be, however, that populations from widely separated areas may not interbreed and would therefore not satisfy this criterion for defining a species. The very fact that many species are so hard to define implies that they arise by evolution from shared ancestors.

18.13 Fitness is the ability to reproduce, which in birds depends to a great extent on having a healthy appearance that mates find attractive, and also on having the ability to provide food to nestlings. These requirements, in turn, depend upon having a break with a size and shape adapted for the food that is available.

Readings

Recommended Readings

Darwin, C. (N. Barlow, Ed.) 1958. *The Autobiography of Charles Darwin.* New York: W. W. Norton.

Darwin, C. 1859. *The Origin of Species.* Cambridge, MA: Harvard University. Press. (*Reprint of the first edition. Also in numerous other editions. Still worth the effort.*)

Darwin, C. 1962. *The Voyage of the Beagle.* Garden City, NY: Doubleday. (*One of the all time great travel books. Widely available in paperback.*)

Gilbert, L. E. 1982. The coevolution of a butterfly and a vine. *Sci. Am.* 247(2):110–121 (Aug).

Gould, S. J. 1977. *Ever Since Darwin.* 1980. *The Panda's Thumb;* 1983. *Hen's Teeth and Horse's Toes;* 1985. *The Flamingo's Smile;* 1991. *Bully for Brontosaurus;* 1993. *Eight Little Piggies.* New York: W. W. Norton. (*Essays, mostly related to evolution, collected from Gould's "This View of Life" column in Natural History Magazine.*)

Grant, P. R. 1991. Natural selection and Darwin's finches. *Sci. Am.* 265(4):82–87 (Oct).

Herbert, S. 1986. Darwin as a geologist. *Sci. Am.* 254(5):116–123 (May).

Lewin, R. 1982. *The Thread of Life.* Washington, D.C.: Smithsonian Books. (*A beautifully illustrated introduction to evolution.*)

Mossman, D. J., and W. A. S. Sarjeant. 1983. The footprints of extinct animals. *Sci. Am.* 248(1):74–85 (Jan).

Owen, D. 1980. *Camouflage and Mimicry.* Chicago: University. of Chicago Press.

Pietsch, T. W., and D. B. Grobecker. 1990. Frogfishes. *Sci. Am.* 262(6):96–103 (June). (*The anglerfish and related mimics.*)

Stebbins, G. L., and F. J. Ayala. 1985. The evolution of Darwinism. *Sci. Am.* 253(1):72–82 (July).

Wilson, A. C. 1985. The molecular basis of evolution. *Sci. Am.* 253(4):164–173 (Oct).

Additional References

Harris, C. L. 1981. *Evolution: Genesis and Revelations.* Albany, NY: SUNY Press. (*The history and philosophy of evolutionism.*)

Lewontin, R. 1982. *Human Diversity.* New York: Scientific American Library.

MacFadden, B. J. 1992. *Fossil Horses.* New York: Cambridge University Press.

Max, E. E. 1990. Letter. *Creation/Evolution* 10(1):45–49.

Mayr, E., and W. B. Provine (Eds.). 1980. *The Evolutionary Synthesis.* Cambridge, MA: Harvard University Press. (*Accounts of the origin of the synthetic theory of evolution by many of its founders.*)

Moore, J. A. 1984. Science as a way of knowing—evolutionary biology. *Am. Zool.* 24:467–534. (*A concise summary of evolutionary biology, with sound advice on teaching it.*)

Prothero, D. R., and R. M. Schoch (Eds.). 1989. *The Evolution of Perissodactyls.* New York: Oxford University Press.

Ritland, D. B., and L. P. Brower. 1991. The viceroy butterfly is not a Batesian mimic. *Nature* 350:497–498.

CHAPTER

19

History of Evolution

Bonobo (formerly pygmy chimpanzee) *Pan paniscus.*

Before Life on Earth

In the previous chapter we considered general concepts of evolution that could apply to life on any planet. Now it is time to consider how these concepts have shaped the course of evolution on the one planet whose life we vaguely know. We must begin the history of life on Earth before its beginning, because although the origin of life was different from its evolution, the particular way life originated has affected all subsequent evolution.

THE PREBIOTIC WORLD. There is considerable uncertainty about the conditions that gave rise to life on Earth. A variety of measures (see box on pp. 383–384) indicate that the Earth was formed about 4.5 billion years ago (Table 19.1 on page 382). For at least the first 700 million years there was no atmosphere to protect the Earth from icy comets and stony asteroids, and it probably looked something like the moon. Unlike the moon, however, the Earth had enough gravitational attraction to hold the water vapor, nitrogen, carbon dioxide, and other gases ejected by impacts and spewed from volcanoes. Thus was formed the early atmosphere, which brought weather, probably in the form of severe thunderstorms. Water from comets and condensed from the cooling atmosphere formed oceans, which became salty with minerals leached from rocks.

For biologists, one of the most important questions about the prebiotic Earth is the composition of its atmosphere, because atmospheric gases reacting in shallow pools may have formed the first organic molecules from which life arose. Many geophysicists believe that when life began almost four billion years ago the atmosphere consisted mainly of molecular nitrogen (N_2), with smaller amounts of carbon dioxide, methane, and ammonia. Unlike our present atmosphere, it lacked O_2. The absence of oxygen is important, because the organic molecules needed to synthesize the first life would have been oxidized immediately if O_2 had been present. Oxygen is one reason why organic molecules do not now occur on Earth unless they are synthesized within the protected environment within cells.

PREBIOTIC SYNTHESIS OF ORGANIC MOLECULES. How did the first organic molecules come about in the absence of organisms? Charles Darwin speculated in 1871 that organic molecules might have been synthesized in "some warm little pond." Although many scientists now believe that life could have originated on clays or in hydrothermal vents on the ocean floor, most of the research on the prebiotic synthesis of organic molecules has elaborated on Darwin's concept that it occurred in shallow water well mixed with the atmosphere. In the 1920s J. B. S. Haldane and A. I. Oparin independently suggested mechanisms by which simple molecules containing carbon and nitrogen might have reacted in water to form a "primordial soup" of organic molecules. In the early 1950s Stanley L. Miller and Harold C. Urey tested this idea by discharging a spark in a synthetic atmosphere containing the gases then thought to have been abundant in the prebiotic atmosphere (Fig. 19.2 on page 384). After a week, several amino acids common to life (as well as some that are not) had accumulated in the apparatus. Many variations on this experiment, using different gases and energy sources, have demonstrated that not only amino acids, but also nucleic acid bases and other biologically important molecules could easily have been formed before there was life. In fact, it is hard to see how the formation of a primordial soup could have been avoided.

TABLE 19.1

Milestones in the history of Earth.

Dates, in millions of years before present, based mainly on Harland et al. (1990).

DATE	EON	MAJOR EVENTS
		Ice ages. Evolution of humans.
	Phanerozoic	Birds, mammals, flowering plants.
		Dinosaurs.
		First vertebrates.
544		Sudden appearance of most animal phyla.
	Proterozoic	Oldest fossils of animals.
1000		
1500		
2000		Atmospheric O_2 at present level.
		Oldest fossil eukaryotes: algae.
2500		
3000	Archean	
		Oldest fossils—bacteria
3500		"Primordial soup."
		Oceans form.
		Atmosphere of N_2, CO_2, but no O_2.
4000	Priscoan	Oldest datable rocks.
		Intense impact and volcanic activity.
		No atmosphere.
4500		Formation of Earth's crust.
		Origin of solar system.

Measuring the Ages of Rocks

At various times and places, sand, mud, shells, and other sediments have slowly accumulated in layer upon layer for millions of years. Preserved in many of these strata, like photos in the pages of an album, are the fossils of organisms that lived when the deposits were formed. The fact that the oldest fossils are naturally in the deepest strata allows for **relative dating.** Relative dating is not what happens when you go out with your cousin; it is simply a method of ranking various forms of life from oldest to youngest (Fig. 19.1). Relative dating allows geologists to divide the history of the Earth into eons, eras, periods, epochs, and ages, much as taxonomists classify organisms into kingdoms, phyla, classes, and so on.

Until this century, few scientists imagined the world to be more than a million years old, and there was no method for determining the absolute ages of various fossils. Since the 1950s, however, geologists have had several methods for the **radioactive dating** of rocks and the fossils in them. Radioactive dating relies on the fact that radioactive isotopes decay into stable products at a rate that is not affected by temperature, pressure, or other variables. By comparing the amount of product with the amount of radioactive isotope in a rock, one can determine how long the isotope has been decaying, and therefore how old the rock is.

For example, half the atoms of potassium 40 (the isotope of potassium with a total of 40 protons and neutrons

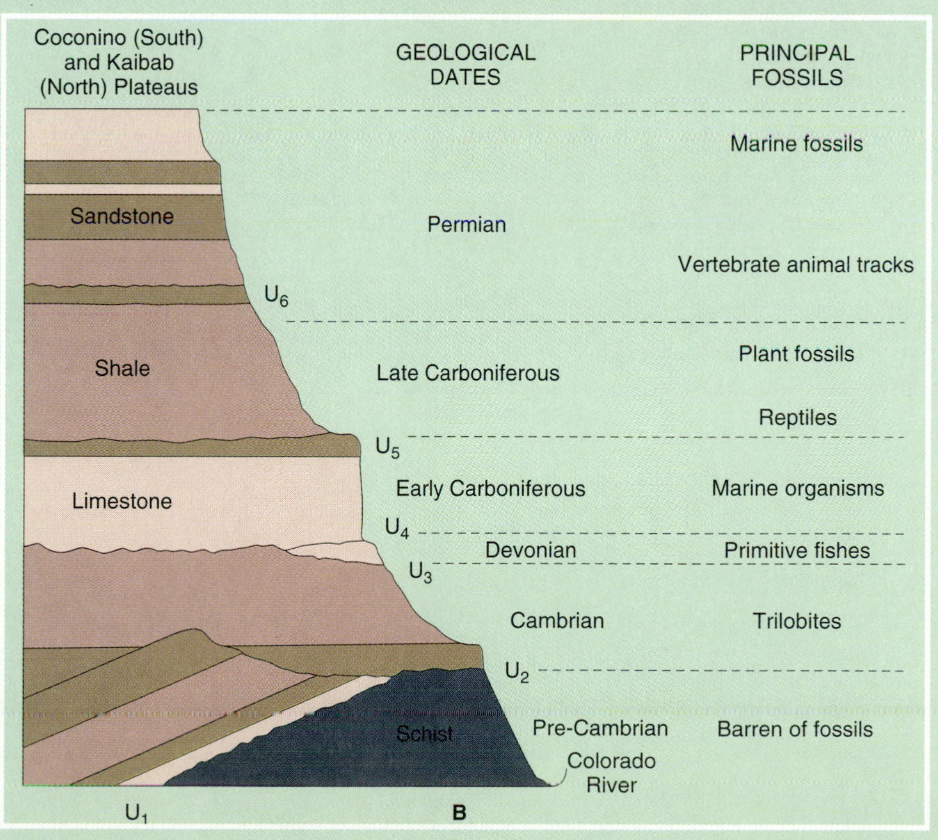

FIGURE 19.1
(A) The north rim of the Grand Canyon at its greatest depth, more than one mile. Erosion by the Colorado River has revealed distinctive strata formed by sediments of different materials deposited at different times. (B) The geological periods corresponding to the strata. The Grand Canyon exposes one of the most complete series of strata known, with most of the Paleozoic era represented. There are a few unconformities (U_1 through U_6) due to erosion or to the absence of sedimentation when the land rose above sea level. Unconformity U_3 represents a gap of approximately 100 million years.

■ **Question:** If the Grand Canyon had been deposited during the biblical flood, as creationists claim, what sequence of fossils would you expect?

box continues

in its nucleus) decays into either argon 40 or calcium 40 every 1.28 billion years. In other words, potassium 40 has a half-life of 1.28 billion years. Unlike calcium, argon is a gas at normal temperatures and does not react chemically; therefore all the argon found inside a rock must have gotten there by decay of potassium 40 within the rock. The greater the ratio of argon 40 to potassium 40 inside a rock, the older the rock must be. In order to be useful for dating, both the radioactive isotope and its decay product must be present in measurable amounts. Carbon 14 is useful mainly for dating organic materials less than 60,000 years old because its half-life is only 5,730 years.

The Origin of Genetic Control. Although the experiments of Miller and others indicate that organic molecules could have been synthesized prebiotically, there was still a major hurdle before those molecules could form life. Somehow a mechanism had to arise by which the sequence of nucleic-acid bases translated into the sequence of amino acids in proteins. At the same time some method had to arise to replicate the nucleic acids for reproduction. Because the replication of DNA now requires proteinaceous enzymes, and protein synthesis requires DNA, we have a classic chicken-and-egg problem. One possible solution would be for a single molecule, perhaps RNA, to have served the functions of both genetic material and enzymes. This idea has seemed more reasonable since the discovery that some RNA can catalyze reactions on itself. After life had gotten a start by using RNA as both gene and enzyme, the machinery might have evolved by which the RNA directed the synthesis of both DNA for genes

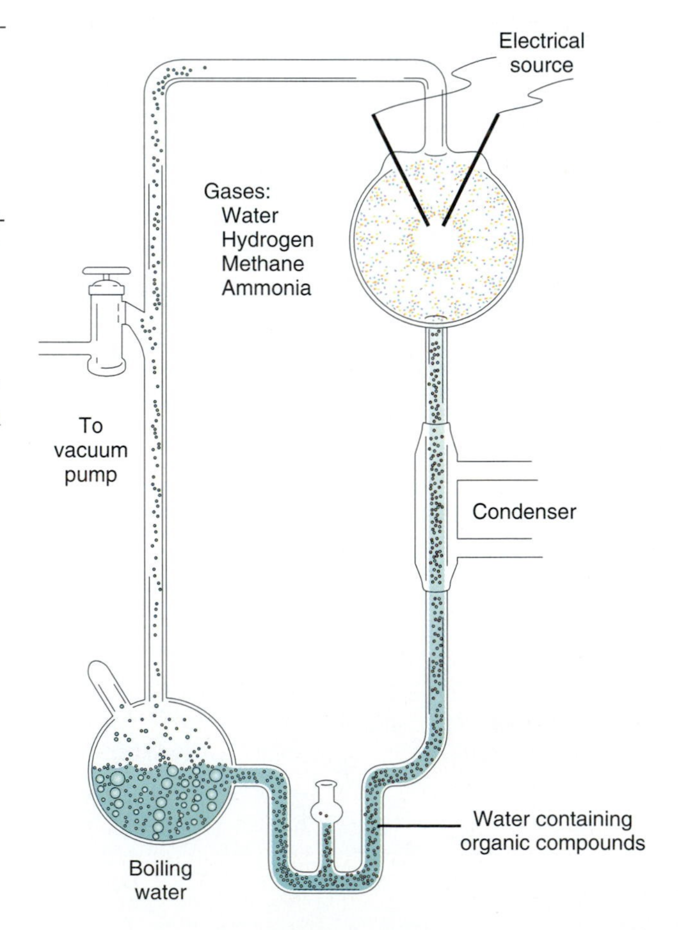

FIGURE 19.2

The apparatus used by Stanley L. Miller in 1953, which demonstrated how organic molecules could have been synthesized on the prebiotic Earth. Water vapor, methane, ammonia, and hydrogen (the gases then thought to have been most abundant in the prebiotic atmosphere) were exposed to sparks simulating lightning. Amino acids and other organic compounds were collected from the condensed fluid.

■ **Question:** Why would this apparatus have to be connected to a vacuum pump prior to boiling the water?

and proteins for enzymes. Eventually, RNA might have been relegated to its present role of translating the information from DNA to make protein.

Early Evolution: The Proterozoic Eon

THE FIRST ORGANISMS. The problems of genetic control and replication appear to have been solved in a relatively short time. Life is believed to have originated within a few hundred million years after the atmosphere began to provide a shield against asteroids and comets around 3.8 billion years ago. The first organisms may have been **prokaryotes**—organisms like existing bacteria and archaebacteria that lacked nuclei and other membrane-bound organelles (see pp. 7–8). The oldest known fossils are of microbes similar to cyanobacteria, which lived approximately 3.5 billion years ago in what is now Australia (Schopf 1992). (Of course, the continents have drifted so much that it is hard to say where in the world Australia was then.) These photosynthesizing bacteria may have released appreciable amounts of oxygen into the atmosphere. Atmospheric O_2 may have reached present levels by two billion years ago, when there were large numbers of algae and stromatolites (pillow-shaped colonies of photosynthesizing cyanobacteria).

The oxygen-enrichment of the atmosphere, although essential for the eventual evolution of animals, must have been an ecological catastrophe at the time. Many organisms exposed to the oxygen were probably killed when their organic molecules were oxidized. The survivors must have been confined to anoxic environments, or they must have evolved mechanisms for removing the oxygen.

THE ORIGIN OF EUKARYOTIC CELLS. Animals, plants, fungi, protozoa, and algae are all **eukaryotes,** with cells that have membrane-bound nuclei, mitochondria, and other organelles. Algae are known to have existed for 2.1 billion years, and there is evidence that other eukaryotes are almost as old as prokaryotes (Han and Runnegar 1992; Knoll 1992). For many years the origin of eukaryotic cells was an enigma to scientists. A few biologists had suggested that the organelles of eukaryotes could have originated as prokaryotes that were incorporated into other cells symbiotically, but few biologists took this concept seriously. Then in the early 1970s, Lynn Margulis drew attention to similarities of certain bacteria to mitochondria, chloroplasts, and other organelles of eukaryotes. Like bacteria, mitochondria and chloroplasts have their own nucleic acids, which enable them to make some of their own enzymes and reproduce themselves. These nucleic acids are more like those of bacteria than those in nuclei. The internal membranes of mitochondria and chloroplasts are also more like bacterial membranes than eukaryotic membranes.

Margulis refers to her scenario for the evolution of eukaryotes as the Serial Endosymbiotic Theory. According to this theory, eukaryotes evolved from a primordial eukaryote by acquiring mitochondria, chloroplasts, and cilia or flagella in a series of steps. The primordial eukaryote (protoeukaryote) may already have had a nucleus from an infolding of its plasma membrane. It may have acquired mitochondria by associating with aerobic prokaryotes, taking advantage of their ability to remove oxygen from their surroundings. Eventually the association between anaerobic and aerobic prokaryotes could have become intimate enough that the aerobe became an **endosymbiont** inside the host, and then a mitochondrion. This may sound unlikely, but even now some species of bacteria live inside other bacteria. The endosymbiont could then have evolved the ability to produce ATP using nutrients from the host, just as mitochondria now produce ATP from the pyruvate provided by cytoplasm. Chloroplasts, which are now responsible for photosynthesis in algae and plants, may have been acquired as cyanobacteria after the mitochondria were present to remove the oxygen they produced.

Flagella and cilia may have originated as motile prokaryotes like spirochaete bacteria. Similar associations occur now in which spirochetes propel other bacteria about. Centrioles, which move chromosomes during mitosis and meiosis and are structurally related to flagella and cilia, may have been acquired by a similar process. The transition from binary fission to mitosis, as well as the evolution of sexual reproduction, would therefore have occurred after the acquisition of flagella and cilia. Fig. 19.3 summarizes how various groups of organisms may have arisen according to the Serial Endosymbiotic Theory.

ORIGIN OF ANIMALS. The fossil record is silent on the origins of animals (**metazoa**), but several lines of indirect evidence suggest that it occurred around a billion years ago (Conway Morris 1992). By 600 million years ago many animal types were distributed worldwide (Fig. 19.4). These animals of the late Proterozoic eon (Greek *protero-* first + *zoon* animal) were first collected from the Ediacara (eddy-A-kara) Hills of South Australia, so they are referred to as **Ediacaran fauna.** Many of the Ediacaran animals appear to have been leaf-shaped or quilted organisms that lacked mouths and perhaps internal organs. Their flattened shapes may have given each cell access to nutrients and oxygen in the water. Many have been classified in such diverse groups as

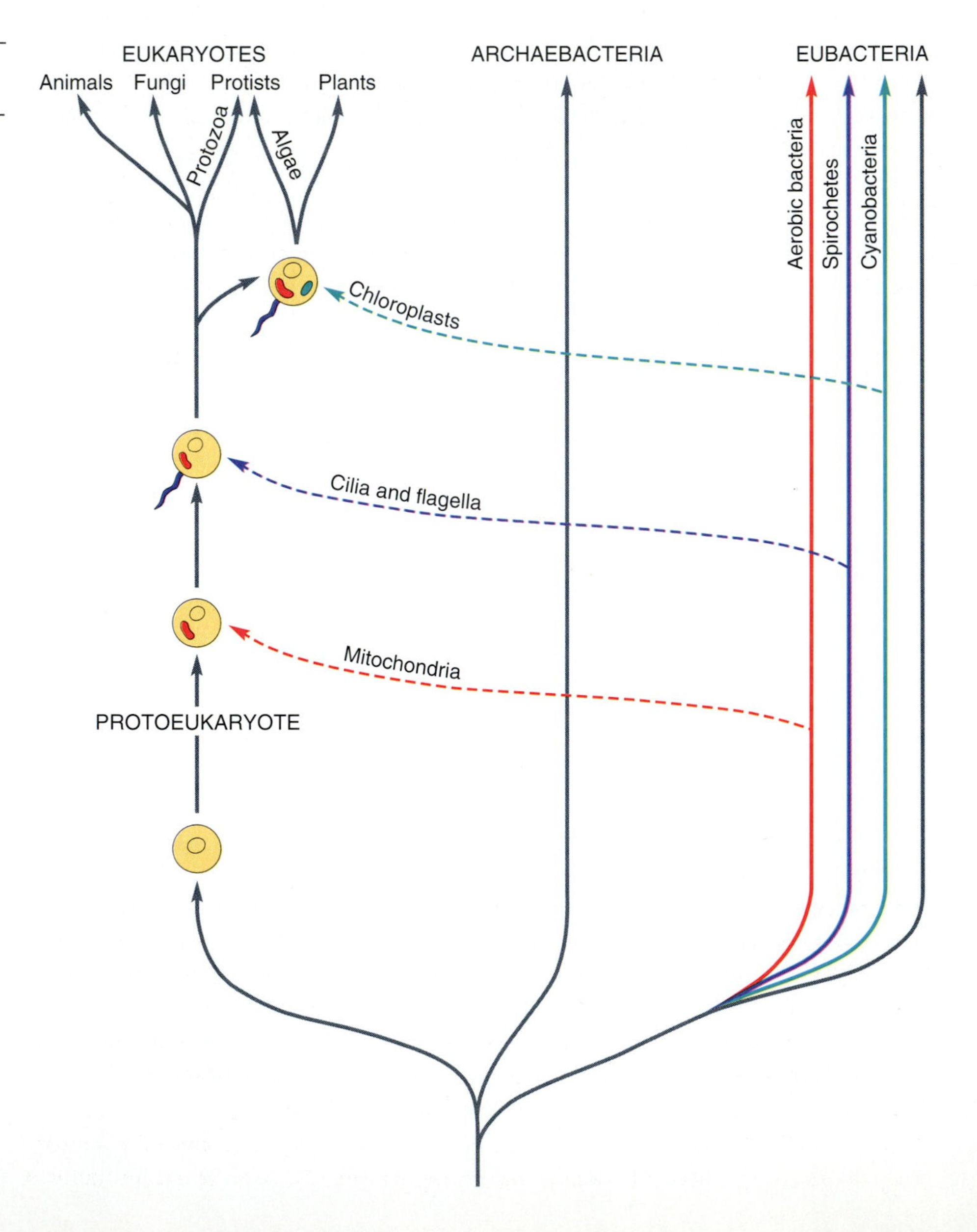

FIGURE 19.3
The main lines of early evolution showing the acquisition of organelles according to the serial endosymbiotic theory.

■ **Question:** Why is it likely that mitochondria were acquired before chloroplasts?

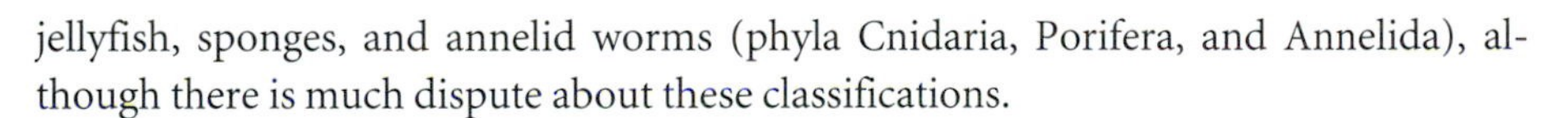

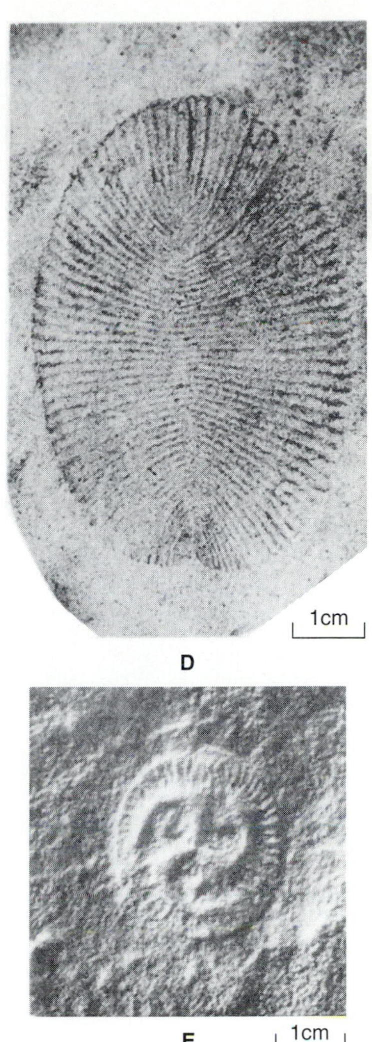

FIGURE 19.4

Fossils of animals that lived in the Ediacaran period between 600 and 590 million years ago. (A) *Charniodiscus arboreus,* which resembled modern sea pens (phylum Cnidaria, class Anthozoa). (B) *Cyclomedusa radiata,* which resembled jellyfish (phylum Cnidaria, class Scyphozoa). (C) *Spriggina floundersi,* which resembled a polychaete (phylum Annelida, class Polychaeta). (D) Another annelid-like animal, *Dickinsonia costata,* which was up to a meter long but only 3 mm thick. (E) *Tribrachidium heraldicum* is unlike any known animal.

jellyfish, sponges, and annelid worms (phyla Cnidaria, Porifera, and Annelida), although there is much dispute about these classifications.

The Paleozoic Era

Animal fossils become suddenly abundant in sedimentary rocks formed 544 million years ago, so paleontologists consider that time as the transition from the Proterozoic to the Phanerozoic eon (Bowring et al. 1993). (Greek *phanero-* visible.) It is also the dawn of the Paleozoic era (Greek *paleo-* old) and the Cambrian period (Table 19.2 on page 388). During the Cambrian period, within the brief span of 530 to 525 million years ago, most of the animal phyla that still exist became established (Figs. 19.5 and 19.6 on page 389). At least 20 explanations have been proposed for this "Cambrian explosion," but none has attracted a consensus. Among the phyla making their debut during the Cambrian explosion was our own phylum Chordata (see Fig. 34.5). The first vertebrates came soon after in the form of jawless fishes. The Cambrian animals were all marine, but within a hundred million years, during the Silurian period, some plants and animals declared their independence from their ancestral oceans (see pp. 634–635).

Punctuated Equilibrium. The Cambrian explosion in animal diversity was for many decades a problem to biologists and geologists who had come to accept unifor-

TABLE 19.2

Major evolutionary events of the Phanerozoic eon.

Dates (in millions of years before present) are mainly from Harland et al. (1990). See Fig. 34.17 for major evolutionary changes in chordates.

DATE	ERA	PERIOD	MAJOR EVENTS
2		Quaternary	Humans. Ice Ages.
	Cenozoic	Tertiary	Continents in modern positions. Flourishing of mammals and birds. Mass extinction.
65	Mesozoic	Cretaceous	First primates. Early placental mammals. First flowering plants.
146		Jurassic	First birds. Age of dinosaurs, other reptiles. Pangaea breaks up.
208		Triassic	First mammals. First dinosaurs
251	Paleozoic	Permian	Mountain-building. Marine extinctions. Continents aggregated into Pangaea. Decline of amphibians; rise of reptiles.
290		Carboniferous	First winged insects. First reptiles. Extensive forests. First amphibians.
363		Devonian	Diverse fishes, trilobites. First insects.
409		Silurian	First jawed fishes. First land animals and plants.
439		Ordovician	
510		Cambrian	First vertebrates—agnathans. "Cambrian explosion" of animal diversity. Marine invertebrates more abundant.
544		"Precambrian"	

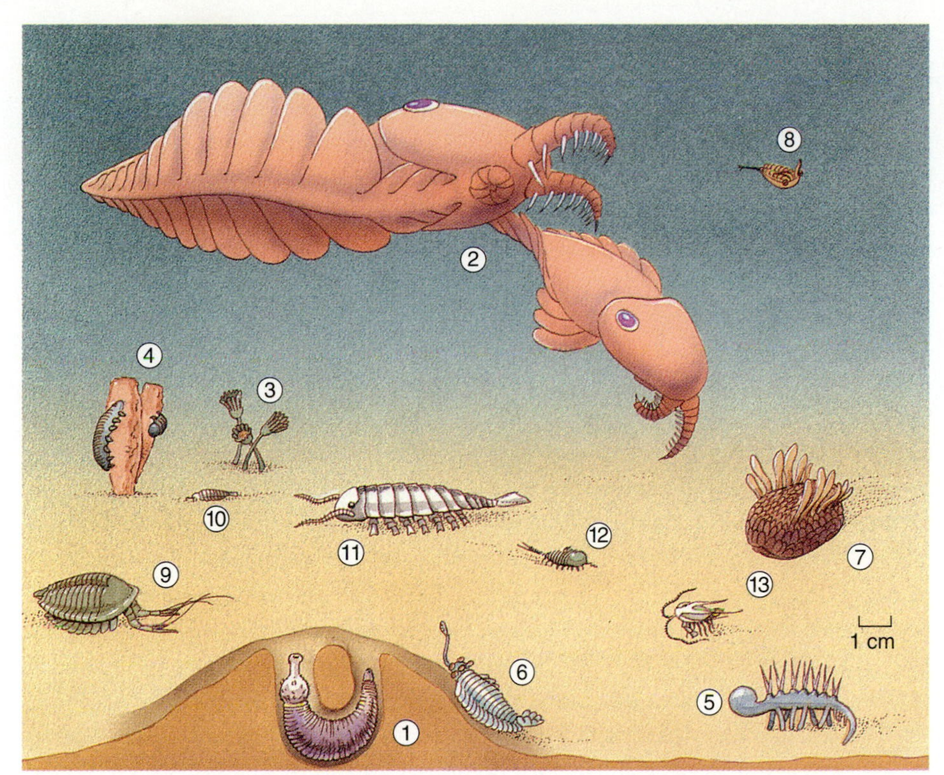

FIGURE 19.5

Some animals of the middle Cambrian period. The habitat was the base of a tropical reef a hundred meters below sea level, and it now forms the Burgess Shale at more than two kilometers altitude in the Rocky Mountains of British Columbia. The animals were apparently washed down from the reef above. In the foreground is a specimen of *Ottoia* ① (phylum Priapulida or Hemichordata) in its burrow. Members of two species of *Anomalocaris* ② loom overhead. These fearsome predators were the largest animals in this habitat. The mouth was once thought to be an animal itself, and the claws were thought to be shrimp. *Dinomischus* ③ was apparently sessile, like the sponges on which the onychophoran *Aysheaia* ④ (phylum Arthropoda) fed. The aptly named *Hallucigenia* ⑤, another onychophoran, was protected by seven pairs of spines that until recently were thought to have been the legs. *Opabinia* ⑥ was unique in having five eyes and one tentacle. *Wiwaxia* ⑦ had toothed jaws with which it fed as it crept along. Others shown here were arthropods, perhaps related to crustaceans. *Sarotocercus* ⑧ may have swum upside down. *Leanchoilia, Yohoia, Sidneyia, Habelia,* and the elegant *Marrella* (⑨ through ⑬) were adapted for walking on the bottom. See Fig. 34.5 for a chordate from the Burgess Shale.

mitarianism and a gradual pace of evolution without question. This **gradualism** is so ingrained that it took many years for scientists to accept that the Cambrian explosion was real. The Cambrian explosion is only one of many examples of a sudden change in the fossil record. There are numerous instances in which fossils appear unchanged throughout a stratum that was deposited over millions of years, but differ in the stratum immediately above. Because of the acceptance of uniformitarianism and gradualism, such apparently sudden changes in the fossil record have usually been attributed to interruptions, or **gaps,** in the deposition of fossils.

From time to time, some biologists and geologists pointed out that such gaps might actually result from a sudden quickening in the pace of evolution, but this concept did not really become popular until 1972, when Niles Eldredge of the American Museum of Natural History and Stephen Jay Gould of Harvard University gave it the name punctuated equilibrium. The main idea of punctuated equilibrium is that species remain largely unchanged—

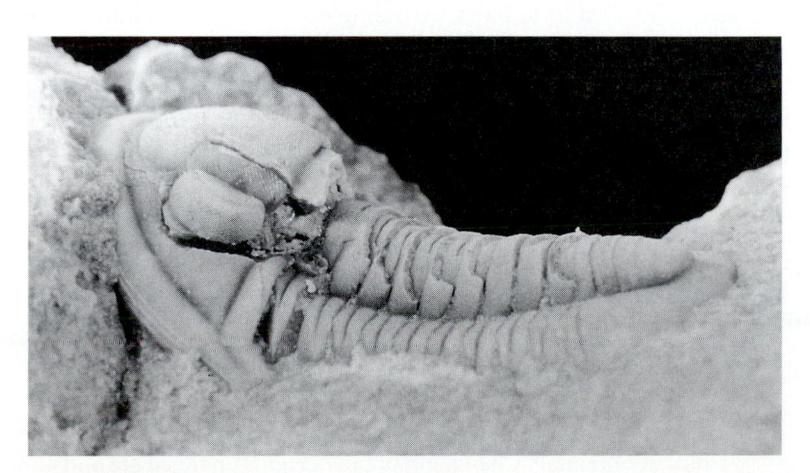

FIGURE 19.6

Cornuproetus sculptus, one of the 4000 known species of trilobites that flourished during the Paleozoic era. This individual died during the Devonian period about 400 million years ago in seas that covered what is now eastern Europe. Length approximately 1 cm. See pp. 636–638 for further description of trilobites.

■ **Question:** Do you see any flaw in the "design" of trilobites that could explain why they became extinct?

in equilibrium—for millions of years, and this lack of evolution is periodically punctuated by brief spurts of evolutionary change. (To a paleontologist, of course, "brief" means within tens of thousands of years.) Eldredge and Gould argue that most species arise during these punctuations. The concept that evolution usually occurs in spurts is now accepted by most biologists, and it could even be said to have replaced gradualism as a new orthodoxy. As Gould and Eldredge (1993) have acknowledged, however, there are many undisturbed fossil sequences that do show gradual evolution. Moreover, their claim that speciation coincides with punctuation remains controversial, because speciation does not necessarily involve morphological change, and morphological change can occur without speciation.

The Mesozoic Era

MASS EXTINCTIONS. The causes of most evolutionary punctuations are unknown, but some were due to worldwide catastrophes that caused mass extinctions. At least 10 catastrophic mass extinctions have occurred (not counting the one now being caused by humans). The most devastating one so far marked the end of the Paleozoic and the beginning of the Mesozoic era around 251 million years ago. This **Permian mass extinction** is believed to have resulted from a drop in sea level and extensive volcanic activity in what is now Siberia. The resulting marine and atmospheric changes led to the extinction of 4 of every 5 species of marine invertebrate within a few million years. Marine communities that had been dominated by soft-bodied-sedentary animals untroubled by predators were replaced by heavily shelled snails and molluscs that burrowed deeply to escape predators such as cephalopod molluscs and reptiles. This mass extinction also affected terrestrial evolution, possibly clearing the way for the Mesozoic era to become "the Age of Reptiles" (see pp. 811–813).

Another mass extinction marks the end of the Mesozoic era, between the Cretaceous (K) and the Tertiary (T) periods 65 million years ago. During this **K/T mass extinction** approximately 70% of all species vanished, including dinosaurs. The K/T mass extinction has aroused considerable interest since the discovery that it was probably caused, at least in part, by the impact of one or more comets or asteroids. A 300-kilometer buried crater from that time has been located near the Yucatan Peninsula of Mexico. The energy of the impact would have vaporized much of the ocean, sent up a cloud that darkened the skies for at least several months, started a global fire storm, poisoned the air and water with acid and metals, lowered temperatures, and shut down photosynthesis.

CONTINENTAL DRIFT. Another unpredictable influence on the diversity and distribution of animals was continental drift. Contrary to the almost universal belief prior to the 1960s, continents are not firmly anchored to Earth's core, but are plates floating on a layer of molten rock. India, for example, is now plowing into Asia and forcing up the Himalayas, and South America and Africa are getting farther apart. According to theories of **plate tectonics,** molten lava pushing up from **mid-ocean ridges** causes the sea floor to spread at a speed of several centimeters per year. When the spreading sea floor encounters the margin of a continental plate, it either pushes it away or sinks beneath it, pushing up mountains in the process (Fig. 19.7).

Continental drift and mountain building have affected the evolution of animals in two major ways: by changing dispersal patterns and by altering climate. Continental drift affected climate in several ways:

1. As continents drifted toward the poles or toward the equator, their climates became colder or warmer.
2. The uplift of mountains changed wind patterns and therefore altered the distribution of rainfall. Changes in ocean currents changed sea and air temperatures as well as atmospheric moisture.

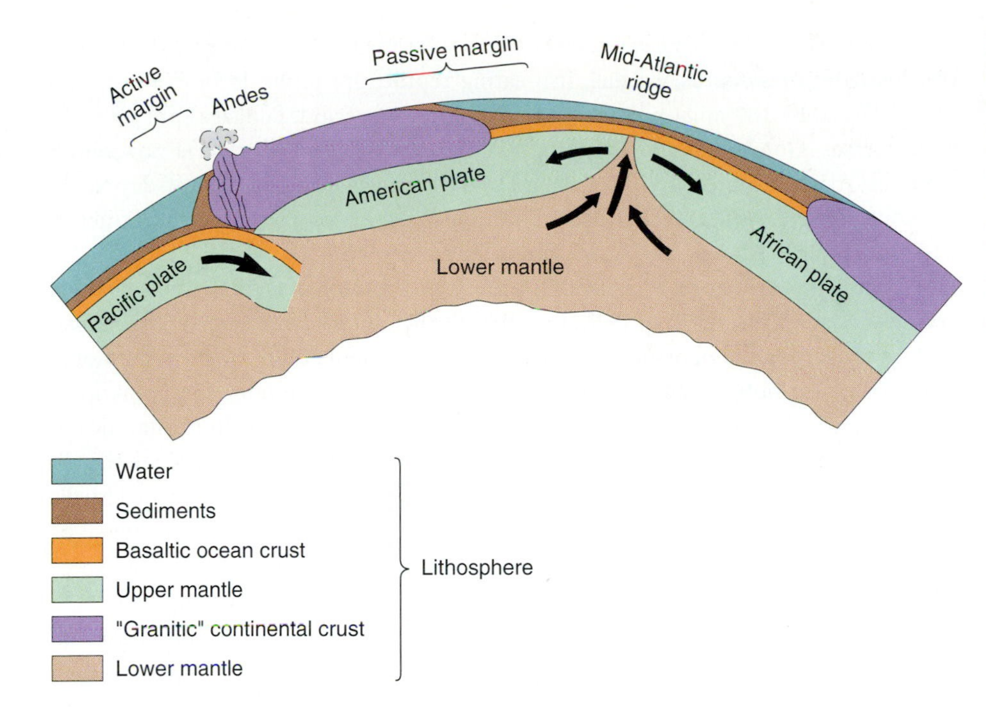

FIGURE 19.7

How plate tectonics builds mountains and moves continents. On the west coast of the Americas the Pacific plate beneath the floor of the ocean pushes under the lighter rocks of the continents. The resulting heat forms volcanoes and mountain ranges inland. On the east coast the American plate is being pushed away from the African and Eurasian plates by sea-floor spreading at the Mid-Atlantic Ridge.

3. As islands aggregated into large land masses, their temperatures became more diverse and variable and their humidities declined. The greater diversity of climates on large continents gave rise to a greater diversity of organisms.

Continental drift has probably been occurring since the Earth's crust formed, but the patterns of drift are clearest in the Mesozoic era. Early in the Mesozoic era, until approximately 200 million years ago, all the present continents were aggregated into one land mass, called **Pangaea** (Fig. 19.8). Continents that are now widely separated proba-

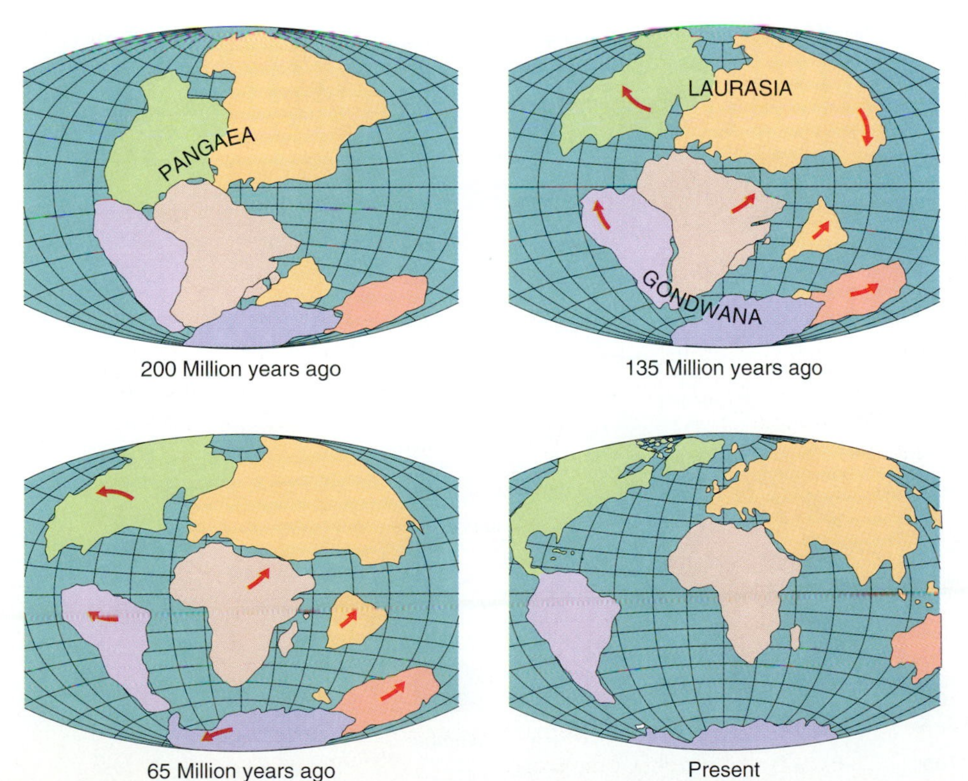

FIGURE 19.8

Continental drift. Approximately 200 million years ago the continents were aggregated into one land mass, Pangaea. By 135 million years ago Pangaea was divided into Laurasia and Gondwana. Sixty-five million years ago South America and India were completely separate island continents. Subsequently, North America separated from Eurasia, a land bridge emerged between North and South America, and Australia separated from Antarctica.

bly shared many of the same species when they were in contact as Pangaea. For example, fossils of *Mesosaurus*, a small, fish-eating reptile, are found in both Brazil and South Africa. By 180 million years ago, in the Jurassic period, Pangaea split into two large masses, **Gondwana** and **Laurasia.** Gondwana consisted of what is now South America, Africa, Antarctica, Australia, and India. Laurasia included present-day North America, Europe, Asia, and Greenland. By the end of the Mesozoic era, South America had split from Gondwana, and Africa and India were joining Asia.

Continental drift helps to explain the present distribution of animals into six **faunal regions** (= faunal realms) (Fig. 19.9). Consider the marsupials, for example. Marsupials such as kangaroos and opossums are native to Australia and South America, but not to Asia. This distribution suggests that they evolved some time after the breakup of Pangaea, but before Gondwana split into Australia, South America, and Antarctica. If so, marsupials would also have lived on what is now Antarctica, at a time when the climate was much warmer there than it is now. This prediction has been confirmed by the discovery of a fossilized marsupial jaw in Antarctica. Until it was colonized by Europeans, Australia remained isolated from Asia at Wallace's Line, and with few exceptions that island continent retained a uniquely marsupial assemblage of mammals.

Other faunal regions are less distinctive because land bridges have periodically allowed certain species to expand their ranges into new continents. For example, less than three million years ago continental drift built the Panama land bridge that reunited the Americas, allowing an interchange of mammals between the two continents. This event, which was another discovery by Alfred Russel Wallace, is called the Great American Interchange. Prior to the interchange, South America had been free of carnivores, but as cats, bears, dogs, and other placental mammals came south, many South American mammals became extinct. Half the genera of land mammals now in South America are descendants of those who came south across the land bridge. South American mam-

FIGURE 19.9

The six faunal regions of the world are determined largely by present climatic conditions, but also by past climates and dispersal pathways. Regions that were separated from each other during the Mesozoic era, during the early diversification of mammals, usually have different native types of mammals. The Australian fauna differ from the Oriental fauna, for example, because these regions have always been separated (by Wallace's Line).

■ **Question:** Nonhuman primates are native in four faunal regions. What are those regions? Assuming that primates originated in the Ethiopian region, how did they disperse?

A

B

FIGURE 19.10

Some large mammals of the Quaternary period (within the past two million years). (A) The Irish elk *Megaloceros* had antlers with a spread of three meters. (B) The sabertooth cat *Smilodon* lurked around present-day Los Angeles and also spread to South America.

mals that came north did not fare as well. Only three North American animals—the porcupine, opossum, and armadillo—are survivors of the northern migration.

The Cenozoic Era

The extinction of the big reptiles at the end of the Mesozoic era opened opportunities for birds and mammals. Prior to the start of the present Cenozoic era, some 65 million years ago, the few mammals had been scarcely larger than rats. Most of them remained small throughout the early Cenozoic era, during the Tertiary period. Later, during the Quaternary period, however, many mammals evolved truly impressive proportions (Fig. 19.10). As will be described later, a medium-sized (but no less interesting) species, *Homo sapiens,* also evolved during the Quaternary period.

ICE AGES. One of the factors that probably contributed to the success of birds and mammals during the Cenozoic era was **homeothermy**—the ability to maintain a constant body temperature. Homeothermy seems to have evolved just in time, because the Cenozoic era has witnessed one of only seven **ice eras** that have occurred on Earth. An ice era is a period lasting about 50 million years during which ice ages occur. During an ice age the average global temperature may drop only several degrees, but that is sufficient to allow large glaciers to expand over the land. The most recent ice age lasted from about 120,000 until about 10,000 years ago. Its ice sheet was up to 3 km (2 miles!) thick and covered all of Canada, the Great Lakes, and the northeastern United States as far south as New York. (Long Island was, in fact, formed from debris scraped up by the ice.) Three "little ice ages" have occurred in the past 10,000 years. The last one, between the 15th and 19th centuries, produced crop failures, famine, and wandering bands of marauders.

Ice ages have affected the evolution of animals not only through direct thermal effects, but also by reducing and modifying habitats. It is obvious that the ice sheet deprived many terrestrial animals of habitat, but marine organisms were similarly affected. So much water was locked up in ice that the seas dropped 90 meters below their present level, exposing coral reefs and shallow sea margins. The lowering of the seas also exposed land bridges that allowed the migration of animals between faunal regions. Between 30,000 and 10,000 years ago the Bering Strait was a major route for the spread of mammals. Horses migrated from America to Asia, and other large animals spread in the reverse direction, with human hunters in pursuit.

Human Evolution

Differences Between Humans and Apes. One of the most interesting events in the Cenozoic era—at least to us—was the origin of our own species. Because of this interest, much more is known about the evolution of our own species than about the evolution of any other group of animals. Nevertheless, there are still major gaps in knowledge of our evolutionary history. For the past century there has been little doubt that humans are most closely related to apes, and that both apes and humans, collectively called **hominoids,** are related to monkeys. Anatomical and molecular evidence support this view, although the fossil record is sketchy (Fig. 19.11). More is known about the times and steps by which **hominids** (humans and their two-legged relatives) diverged from apes. The fossil record and molecular comparisons indicate that the closest living relative to humans is either the common chimpanzee *Pan troglodytes* or the bonobo (formerly called pygmy chimpanzee) *P. paniscus.* In the surprisingly brief time of 6 million years, we came to differ from them in the following four ways, listed in their probable order of appearance:

1. **Bipedal locomotion,** rather than **brachiation** (swinging by the arms) or the **knuckle-walking** of chimpanzees (Fig. 19.12).
2. **Tool making and tool use,** which occur only sporadically in chimps.
3. **Increased brain size,** characterized by a brain mass that is about 75% larger in humans than it would be in chimpanzees if they were the same size as humans.
4. **Language,** which has communication abilities far exceeding those of the vocalizations and facial expressions of chimps.

Prior to the 1970s the acquisition of these human traits was commonly seen as the outcome of an epic struggle of the ape to acquire a large brain and language and to master tools and fire in order to become a man. The term "man" is appropriate here because the role of females in evolution was generally ignored. During the Vietnam

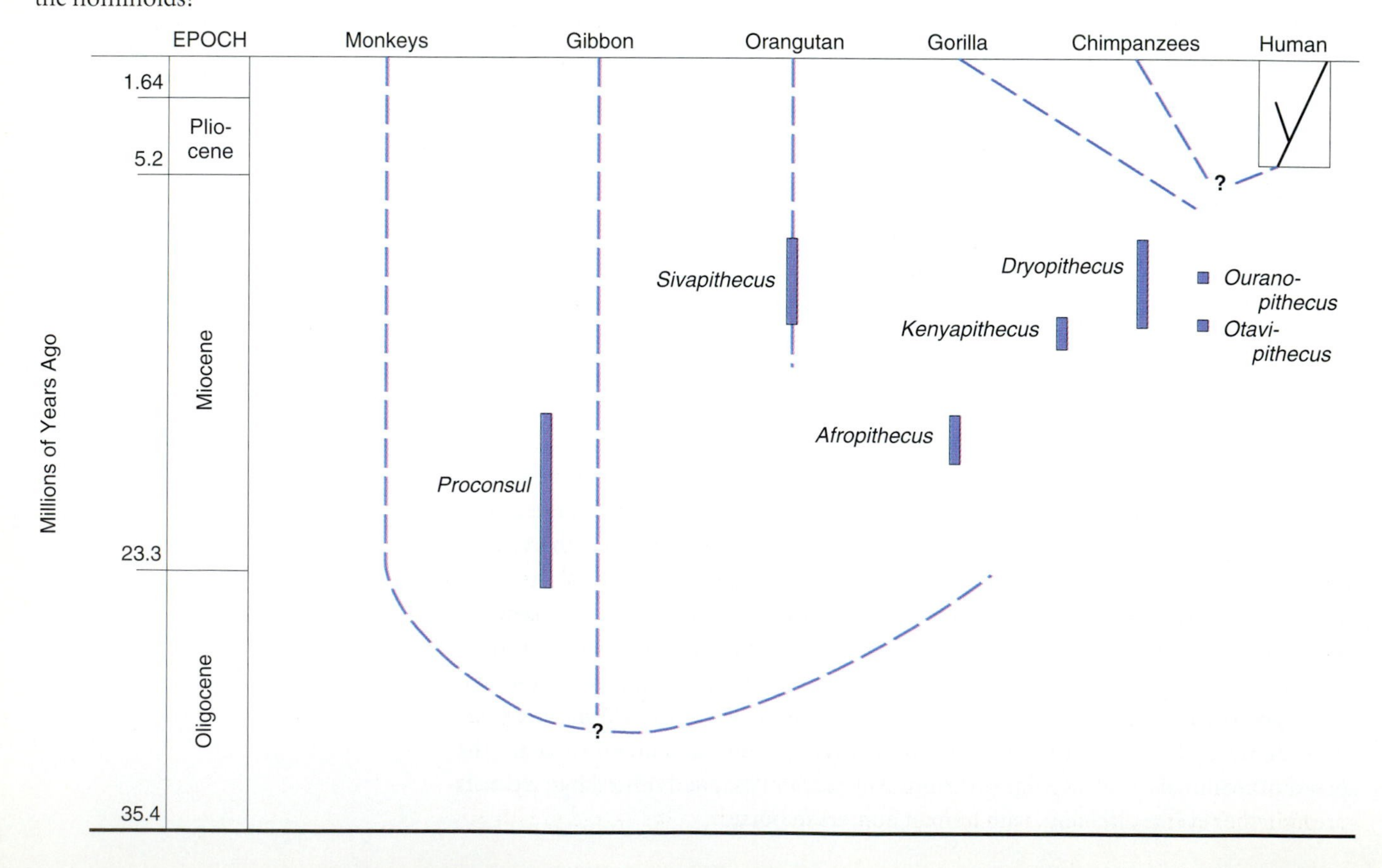

FIGURE 19.11
What little is known about the evolution of anthropoids (monkeys plus hominoids; see pp. 881–884). Dashed lines show inferred relationships with few fossils. See Fig. 19.13 for details in the box at upper right.

■ **Question:** What groups constitute the hominoids?

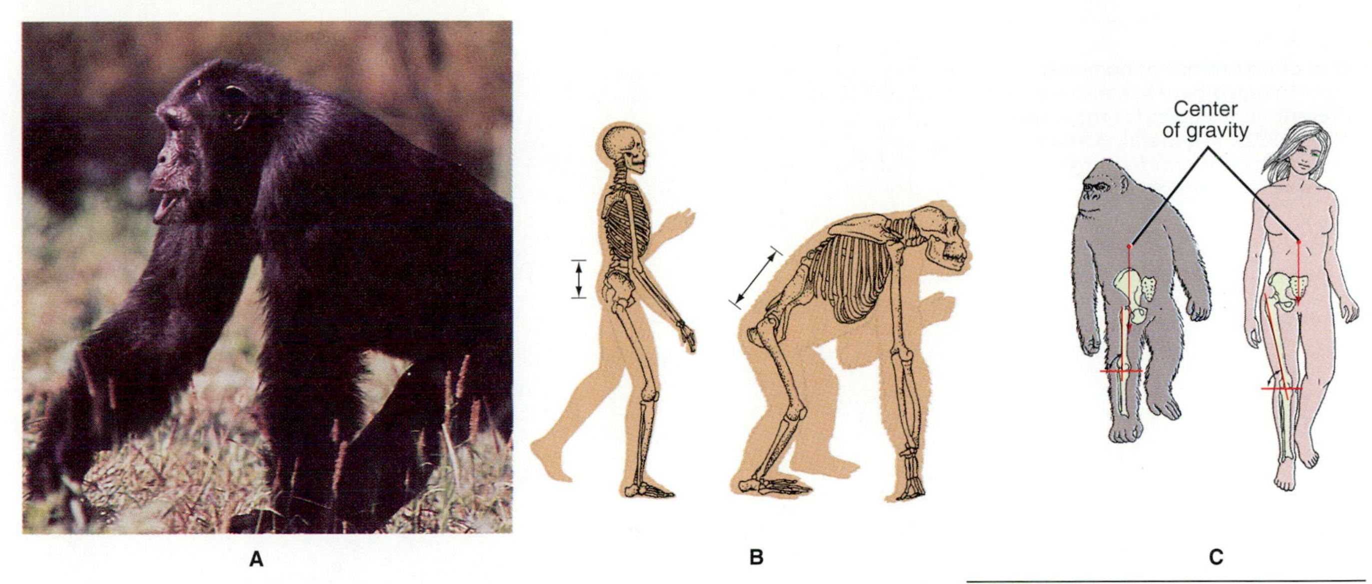

FIGURE 19.12

(A) The chimpanzee *Pan troglodytes,* like most apes, can climb well, but is likely to be found knuckle-walking on the ground. It can walk upright only briefly. (B) In apes there is a greater distance between the attachment of the vertebral column and the sockets of the femurs in the pelvis, making the animal unstable when upright. Note also that the vertebral column in the ape joins the skull at an angle, while that of the human is vertical and directly beneath the skull. (C) In apes the angle of the femur at the knee is approximately 90°. Therefore, when the ape lifts one foot during bipedal walking, it must shift its body to keep the center of gravity over the other foot. Note also that in the ape the first digit of the foot is angled like a thumb rather than a big toe.

War this scenario took a backward turn, and the image of man as Killer Ape became fashionable. To some extent, popular culture still influences paleoanthropology, as can be seen in some recent evolutionary scenarios that give the female star billing in human evolution and relegate the male to supporting cast.

Most paleoanthropologists now approach human evolution more scientifically, by formulating testable hypotheses. Instead of asking how apes managed to improve and become more like us, they investigate the ecological conditions that made it advantageous to be more like a human than like an ape. For example, under what conditions is it energetically more advantageous to walk upright than to knuckle-walk; what ecological conditions favor eating meat rather than only fruits and leaves; and when is scavenging more advantageous than hunting? To answer such questions, paleoanthropologists have to study not only the fossils of our ancestors, but also the climate, plants, and animals on which they depended. It is much more difficult and tedious work than making up stories of heroes, but it is much more scientifically rewarding.

Viewed in this way, the divergence of humans from apes is not a matter of humans leaving the apes behind in the evolutionary race, but simply a matter of each group making different uses of the available resources. The ancestors of chimps and gorillas remained in the trees, while our own ancestors became less arboreal. Although our current understanding of human evolution is now on surer footing, there is still much disagreement. The number of fossil species is relatively small, so each new discovery calls previous interpretations into question. What follows is therefore a tentative description of some of the major events in our recent evolution.

AUSTRALOPITHECINES. Hominids are divided into australopithecines and humans (genus *Homo*). Australopithecines generally had larger jaws and canine teeth, suggesting a diet of coarse plant material and greater dependence on teeth for defense and as tools. Australopithecine brains also tended to be more ape-like in size. The oldest known australopithecine is the recently discovered *Australopithecus ramidus,* which lived more than 4 millon years ago (Fig. 19.13 on page 396; White et al. 1994). *Australopithecus ramidus* was about the size of chimpanzees, and its tooth structure suggests that it was related to them, though clearly hominid. The next known species thought to be ancestral to humans was *Australopithecus afarensis,* which lived between 3.9 and 3.0 million years ago in eastern Africa (Fig. 19.14 on page 397). *Australopithecus afarensis* was at least partly adapted for bipedal locomotion. Bipedal locomotion is much slower than running on four legs, but it would have enabled hominids to see over tall grasses and greatly re-

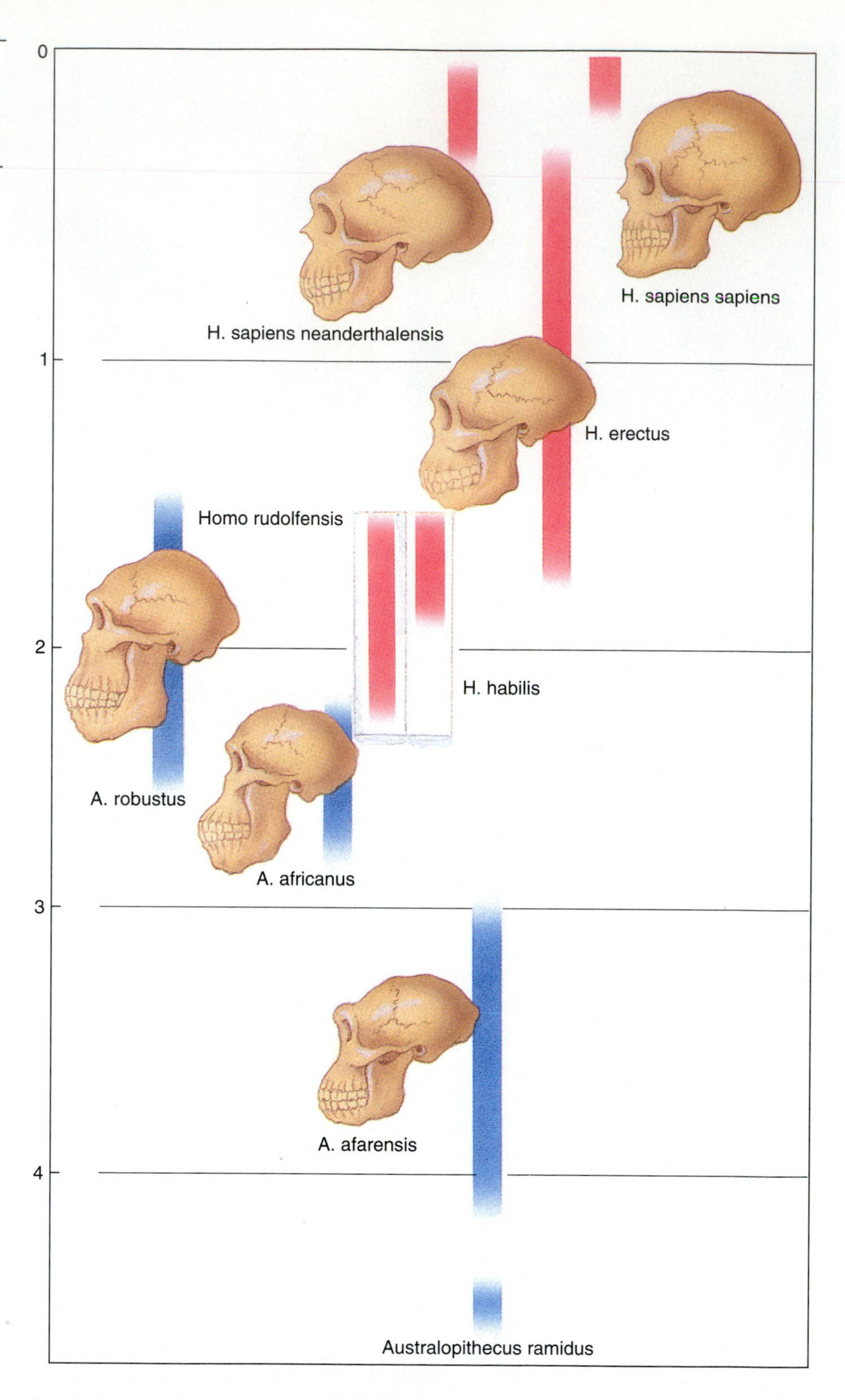

FIGURE 19.13

Times of occurrence of hominids. Paleontologists have suggested several different genealogies for these species (Wood 1992). In general, *Homo* is thought to have split from the australopithecines at *A. afarensis.*

duced the absorption of solar heat. Judging from wear marks on bones and horns found with its fossils, *A. afarensis* apparently used them as digging tools. The most surprising thing about *A. afarensis* was that its brain was so small. Until the discovery of *A. afarensis,* most anthropologists had assumed that a large brain came before the ability to use tools, which then forced the evolution of upright posture to free the hands. *Australopithecus afarensis* shows that bipedal locomotion and tool use came before a large brain.

Australopithecus afarensis may have been an ancestor of other australopithecines, including the larger and more widespread hominid *Australopithecus africanus.* Raymond

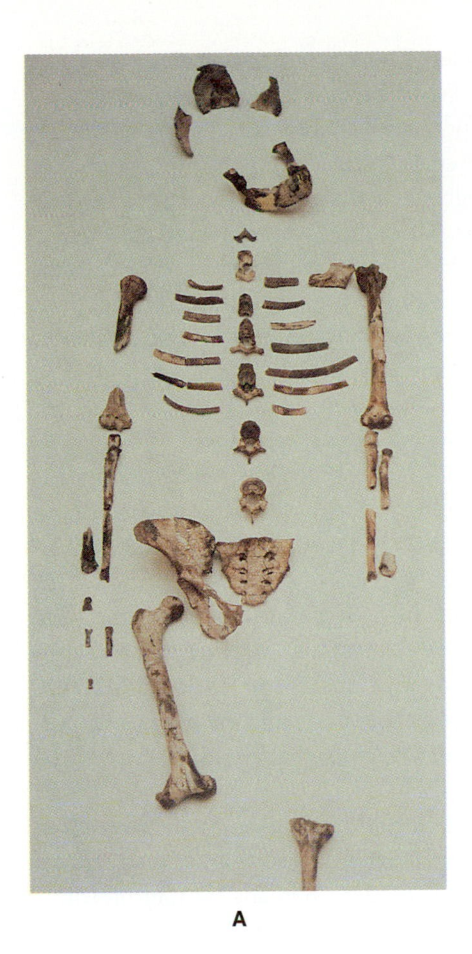

FIGURE 19.14

(A) The mortal remains of "Lucy," the first specimen of *A. afarensis* discovered by Donald Johanson in 1974. She stood approximately a meter tall and would have weighed less than 30 kilograms. The small pelvis and inward angle of the femur in this and other specimens indicates bipedality. The arms were relatively long, however, suggesting that brachiation was still used. (B) Artist's reconstruction of an australopithecine couple (*A. afarensis*) making the tracks shown in Fig. 20.4B. The skin color, amount of hair, shape of ears, and facial expression are speculative.

Dart discovered the first specimen of *A. africanus*, the famous Taung Child's skull, in South Africa in 1924, long before anthropologists were ready to accept such a small-brain primate as anything but an ape. *Australopithecus africanus* apparently became extinct some two million years ago and was replaced by several other species. One was a larger australopithecine, *A. robustus* (= *Paranthropus robustus*), which was almost as large as a modern human and had a stout, humanlike thumb suggesting that it could use tools (Susman 1994).

The First Humans. Two other apparent descendants of *A. afarensis* were the first species recognized as human: *Homo rudolfensis* and *Homo habilis. Homo rudolfensis* is poorly defined and often lumped with *H. habilis. Homo habilis*—literally, "handy man"—gets its name from the large numbers of sharp stone flakes and other tools often found with its fossils. *Homo habilis* males were approximately 1.3 meters tall and weighed 40 kg. Females were only half as large. Such differences in the sexes generally occurs when a dominant male controls a harem of females, as in gorillas. *Homo habilis* was smaller than *A. robustus* and lived at the same time and places in southern and eastern Africa and perhaps elsewhere. In some cases the skeletons of the two species have been found in the same deposits and caves, inspiring tales about the struggle between the small but skillful human and the large, clumsy australopithecine. They may not have competed with each other, however. Analysis of tooth wear suggests that *A. robustus* ate fruits while *H. habilis* scavenged for meat. Broken animal bones found with fossils of *H. habilis* suggest that they used their tools to extract the marrow. In this way they could have had a protein-rich diet without having to hunt or compete with much stronger and faster scavengers. Enlargement of the skull over Broca's speech area (see p. 187) suggests that *H. Habilis* could also use language.

The australopithecines and *Homo habilis* were eventually replaced by *H. erectus* (Fig. 19.15). (The name was given before it was realized that erect posture was nothing

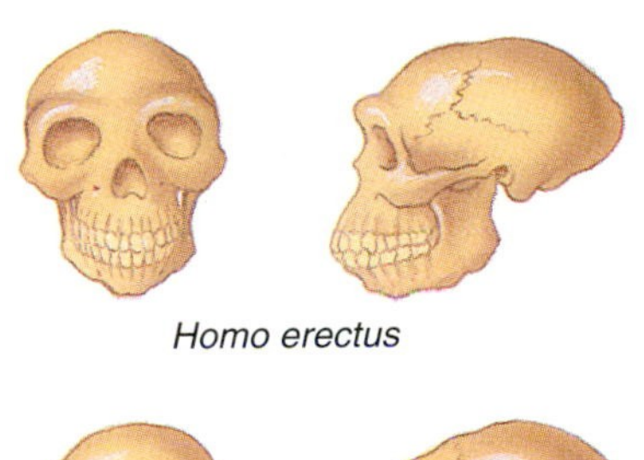

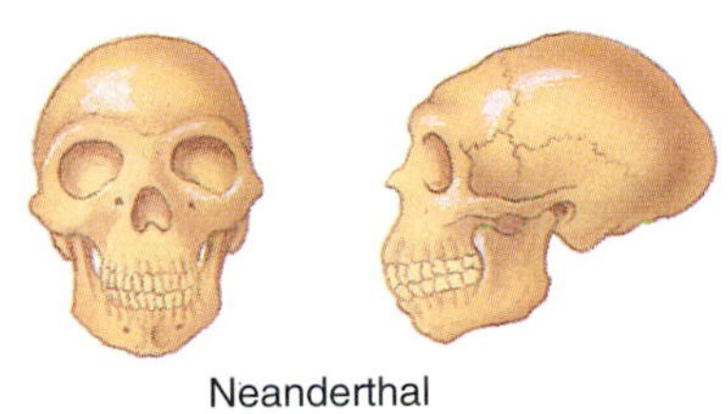

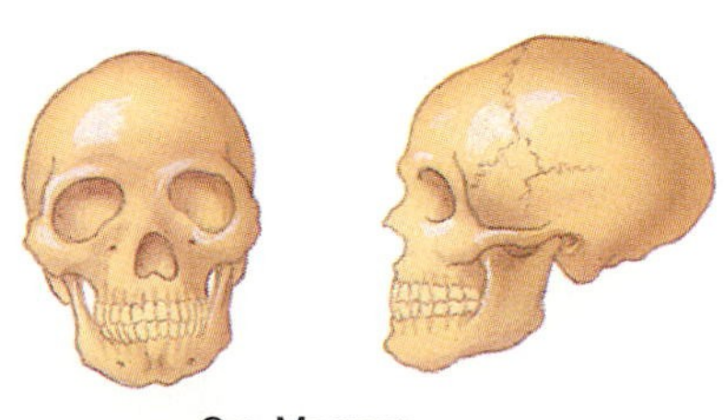

FIGURE 19.15

The skulls of *Homo erectus, H. sapiens neanderthalensis,* and *H. s. sapiens.*

■ **Question:** What does the reduction in jaw length in *H. sapiens sapiens* suggest?

new.) *Homo erectus* was almost as large as modern humans, and the females were closer in size to the males. The reduced difference in size suggests a change in social organization to one of monogamy and more cooperation between the sexes. The brain size enlarged dramatically, sometimes reaching 1100 cm^3, which is within the normal range of modern human brains. Perhaps not surprisingly, the tools produced by *H. erectus* were more sophisticated than those of its ancestors. These tools included hand axes, which may have enabled *H. erectus* to butcher carcasses quickly before other scavengers arrived. Starting around a million years ago, *H. erectus* also controlled fire. By that time, *H. erectus* had begun to migrate out of Africa into Europe and Asia, leaving numerous fossils including those now named Peking Man and Java Man.

Homo sapiens. *Homo erectus* evolved into our own species more than 100,000 years ago, but how it did so is the subject of raging controversy. Most evidence suggests that *H. sapiens* evolved in only one place, probably Africa, then spread throughout Europe and Asia, diverging into various races and replacing *H. erectus*. Some anthropologists oppose this **out of Africa model** with a **multiregional model** in which *H. sapiens* evolved from interbreeding populations of *H. erectus* throughout Africa and Asia. Human origins are further complicated by the fact that there were at least two major early groups: Neanderthal people and modern humans. Some paleoanthropologists consider Neanderthals to be a distinct species (*Homo neanderthalensis*), but most regard them as a subspecies, *H. sapiens neanderthalensis,* that could have interbred with our own subspecies, *Homo sapiens sapiens.*

Neanderthal people lived in the Near East, Europe, and western Asia from about 300,000 until 36,000 years ago. Some paleoanthropologists believe that Neanderthals never became extinct, but simply merged with modern humans by interbreeding. Neanderthals are popularly considered to have been an inferior grade of human and to have lost out in the competition with modern humans. This reputation of Neanderthals as stooped and stupid may have arisen because the first Neanderthal skeleton happened to be arthritic and was found at a time when scientists expected to see an "ape-man." Even normal Neanderthals would have been intimidating, however, with their short but powerful builds, prominent brows, and receding foreheads and chins (Fig. 19.15). Nevertheless, they may have been quite nice folks. Their brains were actually larger on average than our own. They buried their dead, sometimes with flowers, and they may have been able to talk. Members of our own subspecies appear to have tolerated them as neighbors and perhaps mates for thousands of years.

The oldest known remains of **modern humans,** *Homo sapiens sapiens*—"wise wise man"—were deposited in Africa and the Middle East by 100,000 years ago. By 35,000 years ago some had reached Europe. They are called **Cro-Magnon people,** after the place in France where their bones and tools were first discovered in a cave. As if to justify its Latin name, *Homo sapiens sapiens* survived ice ages, learned to spear mammoths, and even took the time to create cave art. Driven by climate changes and the pursuit of game, they spread throughout the world, even sailing to Australia and crossing the Bering Land Bridge to America. Some even became zoologists.

Summary

Life is believed to have begun with bacteria-like organisms almost four billion years ago, at a time when the atmosphere lacked oxygen. Eventually, however, some of these microorganisms altered the atmosphere by producing oxygen as a byproduct of photosynthesis. Some organisms acquired the ability to use the oxygen to make ATP, and these are believed to have become endosymbionts and then mitochondria within the ancestors of eukaryotic cells. Eukaryotic cells may also have acquired chloroplasts, flagella, and cilia by endosymbiosis. Some of these eukaryotic organisms became multicellular and evolved into animals before the Cambrian period some 600 million years ago.

Almost all the major groups of animals alive today, as well as many that became extinct, were present during the Cambrian period. The Cambrian and other periods of the Paleozoic era were dominated by marine invertebrates. The most significant animals of the Mesozoic era were dinosaurs and other reptiles. There were also mammals, but they did not become dominant until the present Cenozoic era. Many dinosaurs and many other groups of animals became extinct at the start of the Cenozoic era, during a mass extinction caused by the impact of a comet or asteroid.

Geological events have profoundly influenced evolution. The movement of continents during the Mesozoic era separated various groups of mammals from each other, dividing them into faunal realms. Ice ages during the Cenozoic further altered the distribution of animals by lowering sea levels, creating land bridges, and changing climates. Continental drift and ice ages also directly affected evolution by their effects on climate.

Humans evolved from a primate that was also ancestral to chimpanzees. The main differences acquired during human evolution were bipedal locomotion, the ability to make and use tools, increased brain size, and language. The oldest known hominoid that may have been ancestral to humans was *Australopithecus ramidus. Australopithecus afarensis* was at least partly bipedal more than three million years ago. The oldest ancestor considered to be human was *Homo habilis,* which made tools more than two million years ago. *Homo erectus,* which lived after *H. habilis,* had a brain almost as large as that of modern humans. Our own species, *H. sapiens,* dates back to 100,000 years ago, and is represented by Neanderthal people and modern humans, *H. sapiens sapiens,* including Cro-Magnon people.

Key Terms

endosymbiotic theory
Paleozoic era
Mesozoic era
Cenozoic era
Cambrian period
mass extinction
continental drift
plate tectonics
faunal region
australopithecine
Homo habilis
Homo erectus
Neanderthal
Cro-Magnon

Chapter Test

1. List the major ways that biological and geological events interacted with each other during the first four billion years of the Earth's existence. Estimate how long ago each event occurred. (pp. 381–382)
2. Briefly summarize the current state of knowledge regarding the origin of life. (pp. 384–385)
3. Describe the Serial Endosymbiotic Theory of the origin of eukaryotes. (pp. 385–386)
4. Briefly describe life in the Paleozoic era. (pp. 387–389)
5. What effects did geological events, especially asteroid collisions and continental drift, have on evolution during the Mesozoic era? (pp. 390–393)
6. How would you account for the success of mammals during the Cenozoic era? (p. 393)
7. What are the main distinctions between humans and apes? Briefly outline the sequence by which each of these distinctions could have evolved. What is the nature of the evidence for this sequence? (pp. 394–398)

■ Answers to the Figure Questions

19.1 Any flood turbulent enough to bring trilobites or other marine organisms to the American Southwest would have left no orderly sequence of fossils, but a random jumble. If the sediments were deposited gradually over the course of a year or so, the remains of reptiles and other terrestrial vertebrates that sink rapidly would be found in the bottom layers, while the remains of plants and fishes would be in the top layers. There would be no vertebrate animal tracks in sediments deposited by a flood that killed all animals except those on Noah's Ark in the Middle East.

19.2 Atmospheric gases, particularly O_2, would have to be eliminated to prevent oxidation of the synthesized molecules.

19.3 Mitochondria utilize the O_2 that is produced by chloroplasts.

19.6 No defect in trilobites has ever been blamed for their extinction. Except for a few spectacular catastrophes, such as asteroid impacts, the causes of past extinctions are largely unknown.

19.9 Primates are native to the Neotropical, Ethiopian, Oriental, and Palearctic faunal regions. (A trick question. Not shown in the figure is the Barbary ape, which is actually the monkey *Macacca sylvanus,* a native of Northern Africa.)

19.11 Hominoids are apes and humans. All but the monkeys in this figure are hominoids.

19.15 The reduction in jaw size suggests less dependence on jaws as tools and weapons.

Readings

Recommended Readings

Allègre, C. J., and S. H. Schneider. 1994. The evolution of the Earth. *Sci. Am.* 271(4):66–75 (Oct).

Alvarez, W., F. Asaro, and V. E. Courtillot. 1990. What caused the mass extinction? *Sci. Am.* 263(4):76–92 (Oct).

Bar-Yosef, O., and B. Vandermeersch. 1993. Modern humans in the Levant. *Sci. Am.* 268(4):94–100 (Apr).

Blumenschine, R. J., and J. A. Cavallo. 1992. Scavenging and human evolution. *Sci. Am.* 267(4):90–96 (Oct).

Briggs, D. E. G. 1991. Extraordinary fossils. *Am. Sci.* 79:130–141.

Coppens, Y. 1994. East Side Story: The origin of humankind. *Sci. Am.* 270(5):88–95 (May).

Gould, S. J. 1994. The evolution of life on the Earth. *Sci. Am.* 271(4):84–91 (Oct).

Gove, R. 1993. Explosion of life: The Cambrian period. *Natl. Geogr.* 184:120–136 (Oct).

Horgan, J. 1991. In the beginning. . . . *Sci. Am.* 264(2):116–125 (Feb).

Jones, S., R. D. Martin, and D. R. Pilbeam (Eds.). 1992. *The Cambridge Encyclopedia of Human Evolution.* New York: Cambridge Unversity Press.

Knoll, A. H. 1991. End of the Proterozoic eon. *Sci. Am.* 265(4):64–73 (Oct).

Levinton, J. S. 1992. The big bang of animal evolution. *Sci. Am.* 267(5):84–91 (Nov). (*The Cambrian explosion.*)

Lewin, R. 1993. *The Origin of Modern Humans.* New York: Scientific American Library.

Lovejoy, C. O. 1988. Evolution of human walking. *Sci. Am.* 259(5):118–125 (Nov).

Marshall, L. G. 1988. Land mammals and the great American interchange. *Am. Sci.* 76:380–388.

McMenamin, M. A. S. 1987. The emergence of animals. *Sci. Am.* 256(4):94–102 (Apr).

Orgel, L. E. 1994. The origin of life on the Earth. *Sci. Am.* 271(4):76–83 (Oct).

Pääbo, S. 1993. Ancient DNA. *Sci. Am.* 269(5):86–92 (Nov).

Schopf, J. W. 1992. *Major Events in the History of Life.* Boston: Jones and Bartlett.

Simpson, G. G. 1983. *Fossils and the History of Life.* New York: Scientific American Library. (*An attractive summary of the significance of fossils by one of the founders of modern paleontology.*)

Stanley, S. M. 1984. Mass extinctions in the oceans. *Sci. Am.* 250(6):64–72 (June).

Stanley, S. M. 1987. *Extinction.* New York: Scientific American Library.

Storch, G. 1992. The mammals of island Europe. *Sci. Am.* 266(2):64–69 (Feb).

Vidal, G. 1984. The oldest eukaryotic cells. *Sci. Am.* 250(2):48–57 (Feb).

Walker, A., and M. Teaford. 1989. The hunt for *Proconsul. Sci. Am.* 260(1):76–82 (Jan).

Weinberg, S. 1994. Life in the universe. *Sci. Am.* 271(4):44–49 (Oct). (*Introduction to an issue on the evolution of Earth and its life.*)

Wilson, A. C., R. L. Cann, A. G. Thorne, and M. H. Wolpoff. 1992. Where did modern humans originate? *Sci. Am.* 266(4):66–83 (Apr).

Ward, P. D. 1992. *On Methuselah's Trail: Living Fossils and the Great Extinctions.* New York: W. H. Freeman.

York, D. 1993. The earliest history of the Earth. *Sci. Am.* 268(1):90–96 (Jan).

Additional References

Andrews, P. 1992. Evolution and environment in the Hominoidea. *Nature* 360:641–646.

Bowring, S. A., et al. 1993. Calibrating rates of Early Cambrian evolution. *Science* 261:1293–1298.

Briggs, D. E., and P. R. Crowther (Eds.). 1990. *Palaeobiology: A Synthesis.* Boston: Blackwell Scientific Publications. (*Concise reviews of all aspects of evolution.*)

Conway Morris, S. 1992. The fossil record and the early evolution of the Metazoa. *Nature* 361:219–225.

Conway Morris, S., and H. B. Whittington. 1985. Fossils of the Burgess shale. Geological Survey of Canada, Miscellaneous Report 43.

Gould, S. J., and N. Eldredge. 1993. Punctuated equilibrium comes of age. *Nature* 366:223–227.

Glaessner, M. F. 1984. *The Dawn of Animal Life.* New York: Cambridge University Press.

Han, T-M., and B. Runnegar. 1992. Megascopic eukaryotic algae from the 2.1-billion-year-old Negaunee Iron-Formation, Michigan. *Science* 257:232–235.

Harland, W. B., et al. 1990. *A Geologic Time Scale 1989.* New York: Cambridge University Press.

Knoll, A. H. 1992. The early evolution of eukaryotes: a geological perspective. *Science* 256: 622–627.

Schopf, J. W. 1993. Microfossils of the Early Archean Apex chert: new evidence of the antiquity of life. *Science* 260:640–646.

Simons, E. L. 1989. Human origins. *Science* 245:1343–1350.

Susman, R. L. 1994. Fossil evidence for early hominid tool use. *Science* 265:1570–1573.

Tattersall, I., E. Delsen, and J. V. Couvering (Eds.). 1988. *Encyclopedia of Human Evolution and Prehistory.* New York: Garland.

White, T. D., G. Suwa, and B. Asfaw. 1994. *Australopithecus ramidus,* a new species of early hominid from Aramis, Ethiopia. *Nature* 371:306–312.

Wood, B. 1992. Origin and evolution of the genus *Homo. Nature* 355:783–790.

CHAPTER

20

Behavior

Male and female three-spined sticklebacks (*Gasterosteus aculeatus*).

The Scientific Study of Animal Behavior

Many people who are impressed when a computer generates a complex and beautiful pattern on its screen will not stop to watch a spider constructing a web. Perhaps they take the spider for granted, imagining they could do the same thing if they could make the silk. For them the spider merely builds its web in anticipation of catching a meal, just like the fisherman repairing his nets. It takes a scientist to turn a "simple" phenomenon like the spider's web into a difficult problem. Scientists realize that the mechanisms and motivations of the spider are not those humans would have under similar circumstances, and that our own behaviors are not even as simple as we think they are.

The error of ascribing to animals the thoughts, emotions, and motivations of humans is now rejected by most behavioral scientists as **anthropomorphic** (from the Greek words meaning "man" and 'form"). The banishing of anthropomorphism was due largely to the behaviorist school of psychology, which argued that one could only observe animal behavior, not try to infer the mental states underlying it. Later behaviorists, like B. F. Skinner, went even further, arguing that mental states, such as free will, did not occur even in humans. The behaviorist approach to animal behavior has produced more experimentally testable theories than the anthropomorphic approach, but most of these theories have been based on studies of the albino laboratory rat under extremely artificial conditions. Behaviorists have therefore been accused of replacing the anthropomorphic view of animals with a "ratomorphic" view of humans. For zoologists, the behavior of rats under laboratory conditions is not only a poor model for the behavior of humans, it also a poor model for the natural behavior of rats.

Neural Events in Behavior

Reflexes. Zoologists are more likely than psychologists to be interested in the behavior of animals in their natural contexts and for their own sake, rather than primarily as models for human behavior. As biologists they are more likely to accept that neural, hormonal, and other physiological mechanisms cause behaviors. This assumption has, in fact, been verified for some of the simplest behaviors. One such simple type of behavior for which the mechanism is completely known is the reflex. As explained in Chapter 9 (p. 175), a reflex is a stereotyped response to a stimulus that is controlled by simple chains of nerve cells that link sensory input directly to muscular or glandular output.

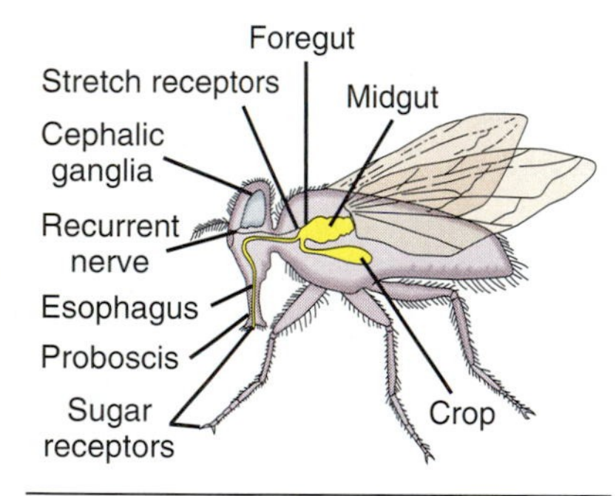

FIGURE 20.1

The feeding system of the blow fly *Phormia regina.* Sugar receptors on the feet trigger the reflex lowering of the proboscis. Sugar receptors on the proboscis then trigger the reflex ingestion of food. After a certain amount of food has been ingested, stretch receptors in the foregut and body wall activate the retraction of the proboscis by means of action potentials in the recurrent nerve.

■ **Question:** What would happen if the recurrent nerve were cut?

Reflexes can also account for some complex behaviors, such as a fly's uncanny ability to arrive just when and where dinner is being served and to select the tastiest dishes. Vincent Dethier (1915–1993; pronounced de-TEER) studied these abilities in the black blow fly *Phormia regina,* which normally lives on nectar, sap, and decaying plants or animals. The fly locates its food by flying upwind when the chemoreceptors on its antennae detect a suitable odor. The fly then lands on the food and tastes it with sugar and salt receptors on its feet. The fly feeds only if these receptors signal that the food is sweet enough and not too salty. If the salt receptors in the feet are not too active, action potentials from the sugar receptors travel to the head ganglia and trigger the reflex lowering of the fly's proboscis (Fig. 20.1). Sugar receptors on the proboscis then activate muscles that pump food into the crop. Feeding continues for a minute or so until the sugar receptors become less responsive, or until stretch receptors in the foregut or body wall signal that the fly has eaten enough.

Taxes. Another kind of elementary behavior is a **taxis,** in which an animal orients its movement with respect to the direction of a stimulus. One example of a taxis is the tendency of many animals to move toward or away from the source of an odor. This would be an example of a **chemotaxis.** A taxis toward or away from light is a positive or negative **phototaxis.**

KINESES. Another example of an elemental behavior is a **kinesis,** which is a change in the rate of movement rather than the direction of movement. Many animals display **chemokinesis** when they encounter molecules indicating the presence of food. The increased rate of movement, though not directed toward the source of food, increases the probability that they will encounter it by chance. Other animals become restless in the presence of light (**photokinesis**), heat (**thermokinesis**), or other adverse conditions, making it more likely that they will escape from those conditions.

OTHER INNATE BEHAVIORS. Reflexes, taxes, and kineses tend to produce stereotyped behaviors that are characteristic of the species. These neural subsystems are "wired" into the nervous system during development, presumably according to genetic instructions. There are also more complex behaviors that appear to be inborn. For example, the female brown-headed cowbird *Molothrus ater* may never have heard the song of a male brown-headed cowbird before, because cowbirds are raised in the nests of other species. Yet the first time a mature female hears a male's song, she responds with a stereotyped posture that signals sexual receptivity (Fig. 20.2). Such apparently inborn behaviors are said to be **instinctive** or **innate.**

FIGURE 20.2
The copulatory posture of the female cowbird *Molothrus ater.* Even though the female has been raised in the nest of another species and may never before have heard the courtship song of a male cowbird, it responds with this posture the first time it hears the song as an adult.

■ **Question:** Can you suggest a mechanism by which a gene directs this behavior?

LEARNING. Innate behaviors can be thought of as lying at one end of a spectrum, with completely learned behaviors at the other end. In between the extremes are innate behaviors that are modified by learning, as well as behaviors that can be learned only if there is an innate ability. Most experimental psychologists recognize two major kinds of learning: classical conditioning and operant conditioning. The Russian physiologist Ivan Pavlov (1849–1936) was the first to study **classical conditioning,** which is sometimes called Pavlovian or Type I conditioning. Classical conditioning occurs when an animal learns to respond to an irrelevant stimulus, the **conditioned stimulus,** in the same way that it normally responds to a relevant stimulus, the **unconditioned stimulus.** The most famous example of classical conditioning was Pavlov's dogs, which began to salivate whenever a bell signaled feeding.

The American psychologist B. F. Skinner was the major student of **operant conditioning,** which is also called Skinnerian, instrumental, or Type II conditioning. Operant conditioning occurs when an animal gradually develops a behavior after the rewarding (reinforcement) of closer and closer approximations to that behavior. In a typical demonstration, a rat learns to press a button if it is rewarded with food every time it hits the button by chance.

Theories of classical and operant conditioning have had little influence on studies of behavior in nature, perhaps because these theories were developed under artificial laboratory conditions. However, both types of learning probably do occur in nature. Animals probably become either tame or shy of humans by associating our presence with benefit or danger (classical conditioning). Honey bees probably learn by operant conditioning when to arrive at flowers that open at a certain time of day.

Genes and Behavior

Because innate behaviors do not have to be learned, they must be assumed to result from neural mechanisms that are inherited through genetic processes like those described in Chapters 4 and 5. In addition, there are patterns of behavior that are characteristic of

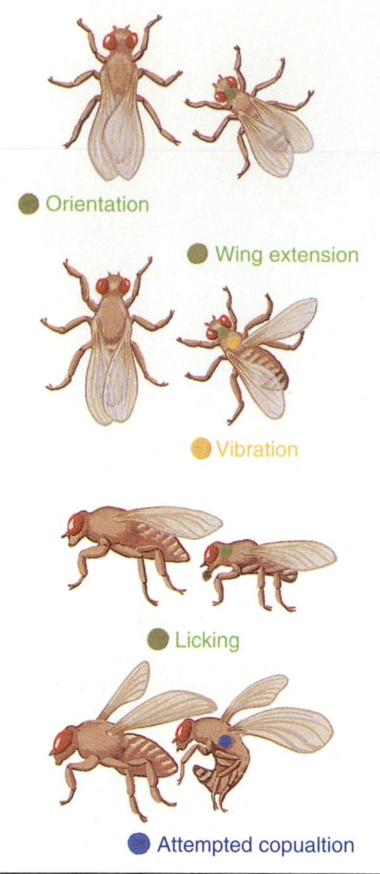

FIGURE 20.3

Male courtship behavior in *Drosophila melanogaster* requires that certain parts of the central nervous system be genetically male. An individual in which certain parts of the cephalic ganglion are genetically male will initiate courtship of a female (left) by orienting toward her and extending a wing. If the individual also has certain parts of the thoracic ganglion that are male, it will vibrate a wing. It will then lick the female's genitals. The individual will attempt copulation only if parts of the abdominal ganglion are male.

each species and appear to be genetically controlled. House cats, for example, behave like their lone, stealthy-hunting kin, while dogs are more noisy and social, like their pack-hunting relatives. These species differences persist even though domesticated cats and dogs have been raised under similar circumstances for thousands of years.

For a few behaviors the evidence for a genetic basis is as strong as Mendel's evidence for genetic factors in pea plants. For example, Fischer's lovebird *Agapornis personata* carries its nest material (bark or shredded paper) to the nest site in its beak, while the peach-faced lovebird *A. roseicollis* carries its nest material tucked in its feathers. Hybrids of the two species attempt to carry nest material by both methods. They repeatedly tuck the material into their feathers, then grasp it in their beaks, then tuck it into their feathers again. Finally, they fly off with the material either in their beaks or in their feathers. Evidently, the hybrids inherit alleles for both types of behavior.

The genetic basis of behavior has also been well established in fruit flies, ever since drosophila geneticists early in this century noticed that some of their mutants behaved differently. Since then, *Drosophila* **behavioral mutants** have provided much material for studying how genes affect behavior. Some *Drosophila* mutants, such as those named *dunce* and *amnesiac,* cannot remember whether to turn right or left in a simple Y maze. Biochemical analysis shows that the normal allele of the *dunce* gene codes for an enzyme that breaks down cyclic AMP. Mutations of this gene result in abnormally high levels of cAMP that somehow interfere with learning.

Genetic mosaics of *Drosophila* have also provided insights into the role of genes in behavior. Genetic mosaics are individuals with anatomical parts that differ genetically because of a mutation during development. Some of the most interesting genetic mosaics combine both male and female parts. These **gynandromorphs** (from the Greek words meaning "female," "male," and "form") are produced when a mutation in a female embryo causes one X chromosome of one cell to become lost during mitosis. The cells that descend from that mutant cell are genetically male. Which part of the adult becomes male depends on where and at what stage of development the X chromosome was lost. These sex mosaics display some confused courtship behaviors. Normally a male *Drosophila* initiates courtship of a female by **orienting** toward her (Fig. 20.3). If she has already mated, she sticks out her genitalia, which discourages(!) him. Otherwise the male taps her abdomen with a foreleg. She may then avoid him, but the male will follow and eventually **extend** and **vibrate** one wing in a courtship "song." The male also **licks** the female's genitalia, after which he usually attempts copulation. The female may eventually spread her wings and extend her genitalia to be mated.

Orienting, following, wing-extension, and licking occur only if a particular cluster of nerve cells in the dorsal part of the head ganglion is genetically male. This is true even if the rest of the gynandromorph is female. Even though wing-extension may occur in gynandromorphs with male head ganglia, singing (wing vibration) occurs only if particular cells in the thoracic ganglion are genetically male. The greater the number of male cells in the abdominal ganglia, the longer the gynandromorph persists in its attempts to mate.

The Evolution of Behavior

Behavioral Fossils. If some behaviors are genetically inherited, and if they can affect the ability to reproduce, then it follows that those behaviors can evolve. Unfortunately, behavior does not fossilize, but in many cases evidence of behavior

Genes for Behavior?

Many who do not hesitate to speak of the inheritance of wrinkling in peas or the inheritance of color blindness in humans are discouraged from speaking of the inheritance of human behaviors. Yet the phrase "gene for behavior" has essentially the same meaning as "gene for wrinkling in peas." It means simply that there is a gene that directs the synthesis of a protein (a growth factor, enzyme, synaptic transmitter, or hormone) that affects the development or functioning of the nervous system in such a way that a particular behavior is modified. For most behaviors, there are likely to be many genes that contribute, and each gene that affects behavior is likely to have other effects as well. Of course, as with other inherited traits, environment also plays a role.

The reluctance to speak about genes being involved in behavior no doubt arises from a fear that it will encourage the stupid and evil perversion of science that occurred in Nazi Germany and elsewhere. Even in the United States in fairly recent times, some judges, physicians, and psychologists evidently thought insanity, idiocy, criminality, and poverty were genetic, and they prescribed forced sterilization as a remedy. (For some reason they never recommended sterilizing everyone with a Y chromosome, even though that would eliminate virtually all violent crime.) Such absurdities occurred long before any knowledge of genetics, however, and bringing back the good old days before Mendel probably would not eliminate them. It seems more likely, in fact, that knowledge is a better antidote than ignorance. Thanks to genetics, we now know that most genetic traits could not be eliminated by genocide or forced sterilization in a reasonable time even if it were desirable to do so. One reason is that in humans even inherited behavioral traits are highly variable because of environmental differences. Many of those with a gene for a particular behavior would not display the behavior, and many of those without the gene would display the behavior.

Although it seems likely that human behaviors are influenced by genetics, it is generally difficult to say which behaviors are so influenced and to what extent. This will probably remain the case as long as it is impractical and unethical to perform the kinds of controlled-breeding studies on humans that can be done in peas and fruit flies. However, in some studies of families in which certain behaviors occur repeatedly, the behavior has been shown to be highly correlated with the inheritance of particular genes. In one family in which many of the boys committed impulsive crimes of aggression, arson, attempted rape, and lewd behavior, those who showed the behavior had inherited a mutated gene on the X chromosome that codes for an enzyme that normally breaks down the neurotransmitter serotonin (Brunner et al. 1993). Studies of inheritance patterns also suggest that alcoholism and some forms of schizophrenia have a genetic component. Contrary to the fears of some, an understanding of the genetic basis for such behaviors appears to encourage treatment that is more humane and effective than the old methods of shunning, imprisonment, or worse.

Other evidence for genetic influences on human behavior comes from studies of monozygotic twins reared apart (MZAs). Because monozygotic twins are genetically identical, any unusual similarities in the behaviors of such twins reared separately may be due to genetics. Such similarities do, indeed, occur. One striking case involved monozygotic brothers who had been separated shortly after birth, with one being raised by his father as a Jew in the Caribbean, and the other by his grandmother as a Catholic Nazi in Czechoslovakia. When the two were reunited after 47 years, they naturally discovered many differences between themselves, but there were also many odd similarities. Both wore similar glasses and moustaches, liked spicy foods and sweet liqueurs, thought it was funny to sneeze in public, and saved rubber bands on their wrists. It is not clear what to make of such curiosities. Is there a gene for saving rubber bands on the wrist, or is that just one way in which particular gene combinations happen to be expressed? Could it be that any two people chosen at random would show an equal number of similarities?

A more scientifically rigorous approach uses personality tests to determine the degree to which behavior differs between MZAs compared with monozygotic twins reared together (MZT). If the differences in scores of MZAs is similar to the differences in scores of MZTs, then those differences must be due more to heredity than to environment. Bouchard and his colleagues (1990) at the University of Minnesota have conducted such studies

box continues

on 128 MZAs. They have found that for several personality traits the variations in scores are due more to genetics than to environment. One of these traits is the intelligence quotient (IQ). It should be noted, however, that this research has been done mainly on middle-class Caucasians, so there is no basis for inferences about other groups. It does not follow, for example, that those with low IQ scores cannot benefit from education, or that differences in average IQ scores of different races are due to genes rather than to environment. It should also be remembered that IQ may not measure anything except the ability to take IQ tests.

does (Fig. 20.4). For example, we know that human ancestors walked upright as early as four million years ago because of fossilized footprints that were discovered in 1978 by colleagues of Mary Leakey during a game of elephant-dung tossing.

THE COMPARATIVE METHOD. In some cases evolution provides the only apparent justification for a behavior. Consider, for example, the female balloon fly *Hilara sartor,* which mates only with males carrying empty silken balloons. The females have absolutely no use for such balloons, and one would expect that they would soon be as wise as a woman whose suitor kept giving her empty candy boxes. A possible explanation of the female balloon fly's seeming gullibility comes from a comparison with the courtship behavior of other flies in the same family (Empididae). An early stage may be represented by certain species of the empidid family in which the male presents the fe-

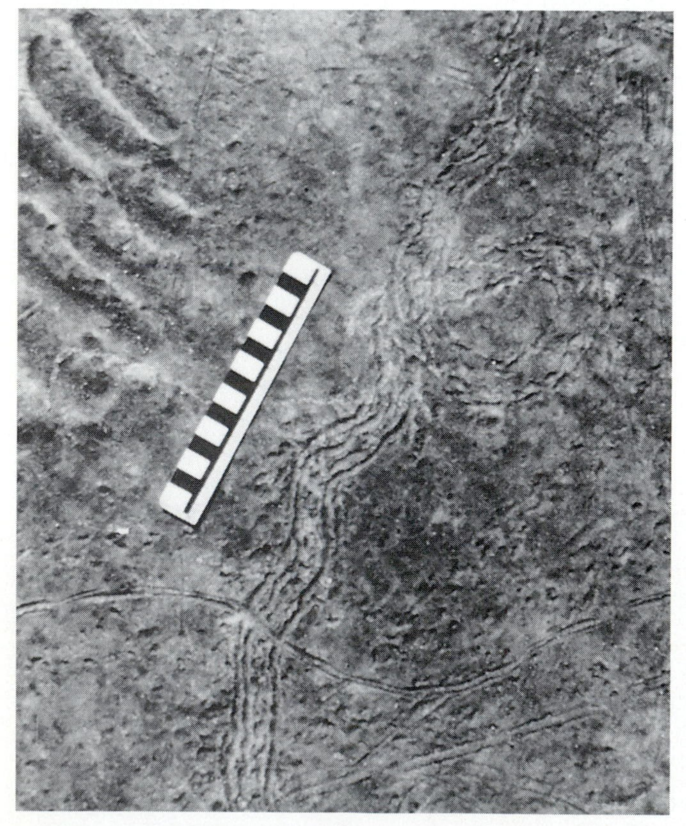

A

B

FIGURE 20.4

Behavioral fossils. (A) Tracks of trilobites (?) from 500-million-year-old rocks on what is now the shore of Lake Champlain. The ripple marks (upper left) indicate that these animals lived in shallow water. Scale in centimeters. (B) Fossil footprints made by a small and a large hominid who walked through volcanic ash 3.5 million years ago. See Fig. 19.14B for a recreation of the event. (Tracks going to the right were made by the extinct three-toed horse *Hipparion*.)

male with dead prey during courtship. Three factors may have favored the evolution of this behavior.

1. Males with demonstrated prey-catching abilities would be more likely to pass to succeeding generations genes contributing to that ability.
2. Males that ensured that their mates were well fed were more likely to get their genes into the next generation.
3. Males that provided the females with something other than themselves to eat were more likely to avoid becoming prey of the female during copulation.

The next stages of the sequence may be represented by species in which the male wraps the prey in silk before presenting it to the female. Possibly this gift-wrapping gives the male more time to copulate without becoming prey to the female, thereby assuring more nearly complete fertilization. In other species of empidid flies the male first sucks all the juices out of the wrapped fly, depriving the female and the eggs he fertilizes of nutrients, but still giving evidence of prey-catching ability and buying time for copulation. Still other species of empidids are nectar-feeding, so there is no need for the male to prove his predatory skill or to evade being eaten by the female. Yet the males of these species find insect fragments, wrap them up and present them to the females. At this stage the behavior appears to be merely an evolutionary relic of a once-functional behavior. *Hilara sartor* represents the final step in the sequence, presenting only the wrapping without even a hint of prey.

Communication

The interaction of male and female balloon flies, like so many other behaviors, depends on communication. There are several channels for communication, each with particular advantages and disadvantages.

VISUAL COMMUNICATION. Visual displays can transmit large amounts of information, but usually over short distances. Many, such as the courtship displays of birds, also carry the risk of attracting predators. Mammals are more subtle, using mainly facial expressions and body postures. Charles Darwin argued that such gestures evolved as ritualized symbols for behaviors. For example, when wolves expose their teeth in a snarl, they are hinting at a behavior that may be forthcoming (Fig. 20.5). Many arthro-

FIGURE 20.5

A dominant wolf (left) and a subordinate visually communicating. Note how each element of the communication is opposite in each animal. (1) The dominant's body is erect, while the subordinant is crouching. (2) The dominant's ears are forward, while the subordinate's are flattened. (3) The dominant's hackles (the hair on the back) are raised, while the subordinate's are not.

pods also use visual communication, but their rigid exoskeletons do not permit facial expressions. Fireflies more than compensate for this deficiency with their remarkable **bioluminescent organs.** Flying males solicit females by emitting pulses of light in a pattern that is specific for their species (Fig. 20.6). Females reply with another species-specific pattern. The male then approaches the female, they exchange several more flashes, and then they copulate.

CHEMICAL COMMUNICATION. Many animals communicate by chemical signals, which have the unique advantage of persisting for some time after the messenger has left an area. They also have the advantage that they will be detected only by those with receptors that respond to the chemical, so they are less likely to attract predators. Some chemical messages have a hormone-like ability to induce specific behavioral responses in recipients in the same species. Such chemical messages are called **pheromones.** The best-known pheromones are insect sex attractants, many of which have been isolated and chemically analyzed. The first such pheromone to be studied was bombykol, which is produced in minute amounts by glands near the anus of the female silk moth *Bombyx mori.* The glands from half a million females had to be processed to yield 12 mg of pheromone. (One lab worker was reportedly overheard complaining, "The end is always in sight, but the work is never done.") A single molecule of bombykol is

FIGURE 20.6
The patterns of light flashing by males of nine species of fireflies. The females do not fly, but answer the males by flashing from the ground (not shown).

■ **Question:** What role, if any, could learning play in flashing by the male?

enough to evoke an action potential from the antenna of a male silk moth, and several hundred molecules are enough to make the male fly upwind, toward the female.

Other functions served by pheromones include alarming members of the same species, causing members of a group to aggregate, identifying members of a group, marking territorial boundaries, marking food trails, and indicating social rank. Many of these functions are performed by various pheromones in social insects (termites, bees, and ants) (Fig. 20.7). As with humans, the ability to communicate gives rise to the ability to deceive. The larvae of certain beetles induce ants to feed and care for them by secreting a pheromone that mimics that of the ant larvae.

Whether humans produce pheromones has long been a question of some interest. There are numerous anecdotal observations of male mammals in zoos becoming attentive to female workers during their menstrual periods. Good scientific evidence began with studies by Martha McClintock based on observations she made as an undergraduate at Wellesley College in the late 1960s. McClintock found that the menstrual periods tend to become slightly longer and synchronized in women who are exposed to other women and few men, such as in dormitories at women's colleges. Subsequent research pointed to a substance in underarm sweat. Men apparently produce a different substance in their underarm sweat that counteracts the effect of the female substance.

AUDITORY COMMUNICATION. Auditory communication, such as the courtship songs of birds, has the advantage of allowing animals to convey large amounts of information over long distances. In a typical area during the breeding season, for example, male birds of dozens of species sing their distinctive songs that attract females. Each song must encode at least enough information to identify the species. A disadvantage of this and other forms of auditory communication is that the sound may also attract potential predators and parasites. There are certain flies, for example, that home in on the chirping of crickets as hosts for their parasitic larvae.

One of the most remarkable examples of communication is the **dance language of honey bees.** Beekeepers have known for thousands of years that once one honey bee discovers a source of food, many honey bees soon converge upon it. Aristotle (*History of Animals,* Book IX, Chapter 40) thought that the first honey bee led the others to the food source, but the real explanation turns out to be much more interesting. When a worker honey bee, the **scout,** first discovers a rich source of food, she returns to the

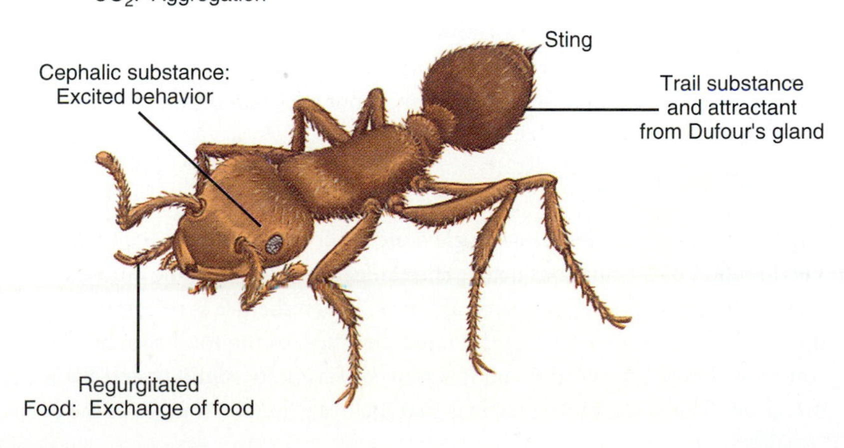

FIGURE 20.7

Like many social insects, the fire ant *Solenopsis* produces a variety of pheromones. Some pheromones are not released by a specific gland. For example, high concentration of CO_2 produced by a crowd of ants will attract stray ants and stimulate digging. Ants also pick up odors that are unique to each nest, and they attack ants in the nest that have a different nest odor. Secretions from the body surface also prompt ants to groom each other. A gland in the head of an attacked *Solenopsis* worker produces an alarm substance that evokes excited behavior in other workers. At the same time, Dufour's gland emits a substance that attracts the workers. When workers are foraging, this or another chemical is deposited by the stinger as a trail substance that other workers follow to the food source. The stinger is used not only to deposit trail substance, but also for attack and defense, which has made this aggressive species a nuisance since its accidental importation into the southern United States from South America.

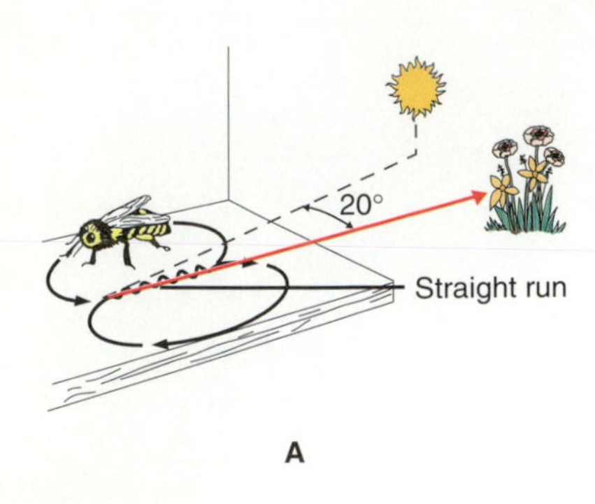

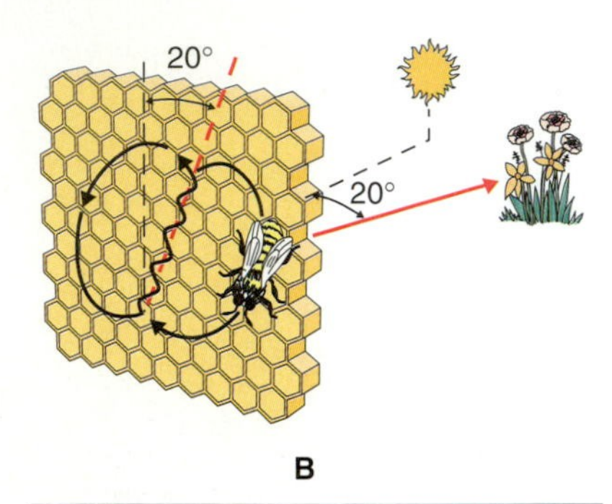

FIGURE 20.8

The waggle dance of a scout honey bee indicates the direction of a food source with reference to the sun. (A) When the scout dances on a horizontal surface, the straight run points to the food source. (B) Usually the scout dances on the vertical combs within the hive, and the angle of the straight run with reference to gravity equals the direction of the food source with respect to the sun.

hive and performs a kind of song and dance to which her sister workers pay keen attention. Evidently these workers, called **recruits,** learn from the scout's dance what and where the food is, because within minutes they fly out and begin foraging on it.

In 1945, after more than 20 years of research, Karl von Frisch managed to decipher the basic vocabulary of the dance language, and research since then has revealed the essential role that sound plays. If the food is located close by (between 5 and 85 meters, depending on the race of bee), the scout performs a **round dance** in a broad figure-eight while vibrating its wings. Workers in the hive detect the sound of the wings and follow the movements of the dancer with receptors in their antennae. The round dance apparently contains no information about the exact distance or direction of the food. The other workers also make a sound by pressing the throrax against the comb. This signals the scout to stop dancing and regurgitate a sample of the food. The recruits then simply fly out in the vicinity of the hive until they locate the food by its odor.

For distant food sources the scout performs an even more remarkable **waggle dance** by running in a straight line while waggling its abdomen some 14 times per second and vibrating her wings. She then circles back and repeats the performance. As described before, recruits follow the sound and movements with antennal receptors, and they emit their own sound to make the scout regurgitate. Evidently the waggle dance communicates information about the direction of the food, because instead of searching randomly within an area, the recruits fly more-or-less directly to the source. Recruits can also learn the distance of the source, because they take off with only enough fuel to fly the correct distance and back. Most of the recruits land within 20% of the exact distance and within 25° of the correct direction.

Von Frisch found that the scout communicates the distance and direction essentially by re-enacting the trip out to the food source. If the food is very distant or requires flying into the wind, then the duration of the straight run (waggle portion) of the dance increases. In one type of honey bee each second of the straight run translates as approximately 1 kilometer to the food source. If the waggle dance is performed on a horizontal surface, the straight run is aimed in the direction of the food (Fig. 20.8). On the vertical combs that honey bees usually build, however, the scout makes a remarkable translation in its dance language. It performs the waggle portion of the dance at an angle with respect to vertical that equals the angle between the food source and the direction of the Sun. For example, if the food source is 20° to the right of the Sun, the scout orients its waggle dance 20° to the right of vertical.

Understandably, there have been doubts that any insect could perform such calculations of distance and angles and then translate them into an abstract dance language that could be understood by other bees. Some doubts arose from the fact that recruits do not necessarily have to rely on the information of the dance. Nevertheless, several studies, such as one by James L. Gould in 1975 while he was a graduate student at Rockefeller University, have shown that honey bees can use the information in the waggle dance (Fig. 20.9). More recently, researchers found that they could send recruits to any area they chose by using tiny robot bees that performed the waggle dance.

Biological Clocks

Suppose a honey bee senses a waggle dance that indicates that a source of nectar is in the direction of the Sun, but before she can begin foraging some curious zoologist imprisons her in a dark box for several hours. When the bee is released will it fly toward the Sun, which is no longer in the same direction as the food source? Or will the bee somehow know how far the Sun has moved during its captivity and fly in the correct direction? The remarkable answer is that the bee generally forages in the correct direc-

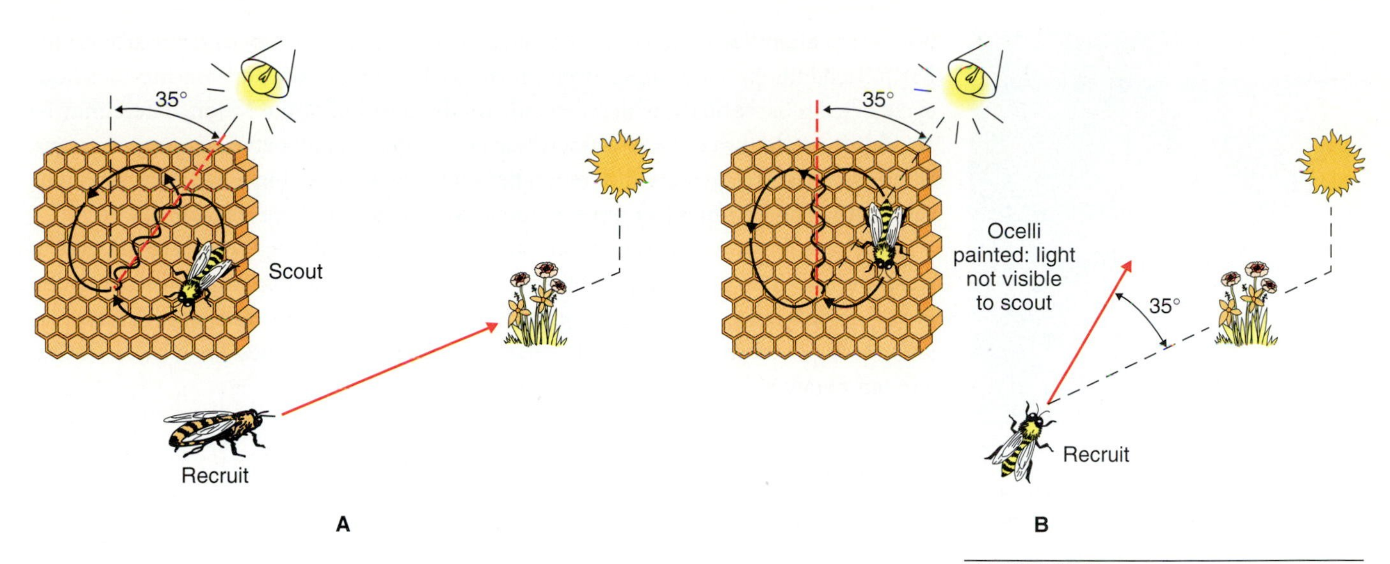

FIGURE 20.9

J. L. Gould's experiment demonstrating that the waggle dance does convey directional information. (A) A bright light inside the hive causes the scout to dance with reference to it rather than with reference to the vertical. If the food source is in the direction of the sun (angle = 0°), the scout will orient its dance toward the light, at 35° from vertical in this example. (B) Bees see the light with their simple eyes, the ocelli. If the ocelli of the scout is painted over so that it cannot see the light, it dances with reference to the vertical. Because the recruits still see with their ocelli, however, they interpret the dance with reference to the light, and they fly off in the wrong direction.

■ **Question:** In what direction would a recruit fly if its ocelli, but not those of the scout, were painted over?

tion. Apparently, bees and many other animals can, in effect, tell the time of day by means of **endogenous clocks.**

Animals use endogenous clocks not only for navigation, but also to regulate feeding, body temperature, sleeping, and other activities. These activities often follow a daily rhythm that is synchronized with the rhythms of Earth—the cycles of day and night, heat and cold. If such rhythms have a periodicity of approximately one day they are called **circadian rhythms** (Latin *circa* around + *dies* day). Many endogenous clocks maintain a circadian rhythm even without external cues, although the clocks tend to run a little fast or slow. In the spring of 1972 a French cave explorer volunteered to live alone in constant light and temperature in a Texas cave so that his endogenous rhythms could be studied. He went to sleep when he felt sleepy and ate when he felt hungry. Generally he maintained a normal sequence of daily activities, but his self-imposed days usually lasted a little longer than 24 hours. By the time the experiment ended on August 10, the volunteer thought it was only mid-July.

Determining what and where the endogenous clock is has proved extremely difficult, perhaps because, like a real clock, it has many parts, only one of which regulates time keeping. In some birds and other vertebrates the **pineal gland** near the brainstem appears to be an important component of a circadian clock. The pineal gland (which is actually the vestige of a primitive third eye) responds directly to light by ceasing its production of the hormone melatonin. Consequently, there is normally a 24-hour rhythm of melatonin synthesis, with melatonin production ending at sunrise. Sue Binkley and others have found that the circadian rhythm of melatonin synthesis persisted for several days even in a pineal gland that is removed from a chick's brain and kept in the dark.

In mammals the pineal gland also plays a role in circadian rhythms, but it is not directly responsive to light and does not have the central role it has in birds. Instead, the rhythm of melatonin synthesis by the pineal gland is driven by two **suprachiasmatic nuclei** (SCN), which are just above the optic chiasm, where the optic nerves cross. In bright light a special neural pathway from the retinas inhibits the SCN, thereby inhibiting melatonin synthesis. Removing the two SCN from mammals (but not from birds) permanently abolishes some of their daily rhythms. The rhythms can be restored by implanting the SCN from another animal into their brains.

In addition to circadian rhythms, many animals display **lunar cycles** that are cued by the position of the Moon. First there may be a rhythm with a periodicity of 24.8

FIGURE 20.10

At certain times of the year, shortly after high tide when the Moon is either full or new, California grunion *Leuresthes tenuis* come onto the beaches in huge numbers to deposit and fertilize their eggs in the sand. The next maximum high tide washes over the eggs, triggering hatching, uncovering the fry, and sweeping them out to sea. Length approximately 18 cm.

hours (**circalunidian**) that is cued by the position of the Moon in its orbit around the Earth. In addition, many shore animals have a rhythm that depends on the 12.4-hour cycle of tides (**circatidal**), which depends on the position of the Moon. There may be an additional rhythm of 14.75 days, which equals the interval between quarter moons, when there is the greatest difference between sea level at high and low tide. The California grunion times its reproduction according to this cycle (Fig. 20.10).

Perhaps most widespread are **circannual rhythms,** although they are difficult to study because it takes so long to collect sufficient data. Circannual rhythms ensure that many animals reproduce, migrate, grow long hair, or hibernate in the appropriate seasons. Many of these rhythms are triggered by environmental cues, such as day length and temperature.

Navigation

Biological Compasses. In addition to a biological clock, many animals also have biological compasses that enable them to navigate. Navigation has been most extensively studied in homing pigeons *Columba livia,* which have been bred and trained for centuries to return to the home loft from vast distances. Homing pigeons usually combine information about the time of day from their endogenous clocks with the position of the Sun as inferred from the pattern of polarized light in the sky. (This is the pattern of lighter and darker blue areas that you see when looking at different areas of the sky through polarized sunglasses.) Pigeons do not have to see the area over which they are flying, because they home just as well if they are wearing frosted contact lenses.

After decades of skepticism, it is now clear that migrating birds can also use Earth's magnetic field to find direction. Evidence for the existence of magnetic compasses in birds is due largely to the late William T. Keeton of Cornell University, who found that attaching magnets near the heads of homing pigeons disrupted their navigation on cloudy days. Birds that migrate at night can also use the stars as a compass. Apparently, many birds in the Northern Hemisphere learn which way is north by observing the stars as they rotate around the North Star. Afterward they can orient toward the north simply by viewing a few of these stars. If birds are raised in a planetarium in which the star field rotates around a star other than the North Star, they will later migrate in a direction that would be appropriate if that star were in the north. Birds apparently learn the star pattern during the weeks just prior to their first fall migration. If they cannot see the stars during that sensitive period, they will never learn to migrate by means of the star pattern.

Migration of Monarch Butterflies. No migration is more remarkable for distance covered per unit of body weight than that of the monarch butterfly *Danaus plexippus.* With four orange and black wings as flimsy as paper, these intrepid insects fly hundreds of kilometers to roost in dense clusters during the winter. Monarchs in the western United States overwinter in approximately 40 sites along the California coast that have so far escaped destruction by developers (Brower and Malcolm 1991). The wintering sites for monarchs from the eastern United States was a mystery for many decades. From 1937 to 1995, however, Fred Urquhart of the Scarborough Campus of the University of Toronto, his wife, Nora, and a total of more than 4000 associate members of the Insect Migration Association attached hundreds of thousands of small, numbered tags to the wings of living monarchs in order to track their migration (Fig. 20.11). Through their efforts the first of 10 known sites was discovered in 1975 at an altitude of 3000 meters in the Sierra Madre Mountains, about 400 km northeast of Mexico City. After several decades these patient and intelligent scientists had located a spot that each of the monarchs had found in less than a month.

A

B

FIGURE 20.11

(A) A monarch butterfly *Danaus plexippus* just after receiving a numbered tag from a member of the Insect Migration Association. (B) Professor Fred Urquhart in the Sierra Madre Mountains of Mexico surrounded by overwintering monarchs. In spring the females will mate and set out on northeasterly journeys, depositing fertilized eggs on milkweed plants along the way. The eggs hatch into yellow, black, and white-striped caterpillars (see Fig. 32.11) that feed on the milkweed plants, then metamorphose into adults that carry on the migration and reproduction. In late autumn the survivors reverse direction toward the wintering grounds. As many as five generations can elapse between the spring and fall migrations.

Ethology

The kinds of behaviors discussed so far have generally been studied using methods that are common among zoologists. The rest of this chapter will survey more complex behaviors that have given rise to approaches that are unique to behavioral scientists. One of the earliest of these approaches to the biological study of behavior was ethology, which became popular in the 1950s mainly as the result of research by Konrad Lorenz and Niko Tinbergen. The term "ethology" is often applied to any biological study of behavior, but its meaning is perhaps best explained by examining some of the best-known studies by these and other ethologists. Many of these studies focus on behaviors that are characteristic of a species and assumed to have evolved under the influence of natural selection. One example of such a behavior is the stereotyped movements of the beak by which a goose retrieves an egg that has rolled out of the nest. Once a goose sees the egg out of the nest, it goes through the movements of retrieving it even if someone first takes the egg away. Ethologists originally call such behaviors **fixed action patterns** (FAPs), but the term **motor program** is now more common. They call the stimuli that evoke motor programs **sign stimuli** or **releasers.** Motor programs and releasers occur frequently in courtship, intraspecific aggression, and territorial behavior, all of which have been studied extensively by ethologists.

Courtship, aggression, and territoriality all occur in the common freshwater fish called the three-spined stickleback (*Gasterosteus aculeatus*), which was the subject of a classical ethological study by Tinbergen. During the breeding season, the male stickleback establishes a territory from which it excludes other males. Within its territory the male excavates a shallow depression in the sand, then builds a tunnel-shaped nest of algae glued together with a secretion from the kidneys. Immediately after it completes the nest, the male stickleback develops brilliant colors. The eyes turn bright blue, the back goes from dull brown to shiny blue, and the belly becomes red.

If one brightly colored male strays into the territory of another, the resident male attempts to drive out the intruder by darting toward it with mouth open and spines erect on its back. If the trespasser does not immediately flee, the resident escalates the threat by making jerky movements while "standing on its head." The red belly is the main releaser of this territorial aggression. In fact, Tinbergen found that a male stickleback reacted more aggressively toward an artificial fish with a red underside than toward a real male stickleback without a red belly. The use of such models is characteristic of ethological research.

While the males are preparing their nests and changing colors, the females develop a silvery gloss, and their bellies swell with ripe eggs. If a female enters a male's territory, he repeatedly swims toward then away from her in what is called a zigzag dance (Fig. 20.12 on page 414). A model of a fish with a swollen belly is an equally effective releaser of the zigzag dance by the male. A female that is not intimidated turns toward the zigzagging male, which then goes straight to the nest. If the nest appears suitable she follows, and the male leads her into the tunnel-like nest, where her swollen belly entraps her. The male then prods the base of her tail to stimulate spawning. After the eggs have been deposited, the female easily escapes from the tunnel. Having lost her figure, the male chases her away. He then enters the nest and fertilizes the eggs. Afterwards the male spends several weeks guarding the eggs and fanning water over them with his fins.

FIGURE 20.12

Courtship and mating in the three-spined stickleback *Gasterosteus aculeatus.* The male zigzags toward a female whose belly is swollen with eggs, then he guides her to the nest. The male leads the female through the tunnel-shaped nest, but she is unable to exit because of her swollen belly. Below, the male is shown prodding the female to lay her eggs. Once this occurs she is able to leave. The male then enters the nest and fertilizes the eggs.

■ **Question:** Suggest an ethological experiment to test whether the red belly of the male serves as a releaser for the behavior of the female.

Sociobiology

Another productive approach to understanding animal behavior is sociobiology. Like ethologists, sociobiologists assume that many behaviors are largely dependent on genetics and have evolved in response to natural selection. As their title suggests, however, sociobiologists generally confine their interests to social behaviors, and they are less interested in the mechanisms of such behaviors than in the contributions of those behaviors to evolutionary fitness. Unfortunately, many people believe that sociobiology is nothing more than the naive assumption that all human behavior is genetically determined. This misconception is due largely to politically motivated attacks on E. O. Wilson and his 1975 book *Sociobiology: The New Synthesis.* In the first 546 pages of the book, Wilson surveyed the biology of social behavior in nonhumans, and in the final 29 pages he attempted to extrapolate to explanations of how such human behaviors as homosexuality, entrepreneurship, and religion could have evolved. Attacks on those

29 pages came mainly from biologists in the United States who followed the Marxist doctrine that human social behavior is determined by the economic system, not by biology.

FIGURE 20.13
Belding's ground squirrel *Spermophilus beldingi.*

ALTRUISM. The controversy has often obscured the real contributions of sociobiology. Among these is one possible explanation for altruism, which is defined as behavior of an animal that benefits others of the species while being harmful or of no benefit to the altruist. Altruism has plagued evolutionists since Darwin, because natural selection stresses competition and the struggle for survival, yet Darwin and everyone else knows that many animals cooperate with other members of their species. For example, it is well known that in many species parents will risk their lives to ensure the survival of their offspring. This kind of altruism presents no difficulties for Darwinism, because it directly favors an increase in the number of individuals bearing the genes that encourage parental altruism. Altruism toward those that are not offspring, however, was difficult for Darwinians to explain. Why, for example, do worker honey bees sacrifice their labor and even their lives to defend the hive, even though they usually do not reproduce? How could the genes responsible for altruism be selected, because the workers do not reproduce?

A mathematical solution to this problem was provided by William D. Hamilton when he was a graduate student in 1964. He showed that an organism's fitness—its potential for transmitting its genes into the next generation—includes not merely its individual fitness, but also its contribution to the fitness of others carrying the same genes. This total is called the **inclusive fitness** of an organism. Because kin are likely to carry genes identical to one's own, inclusive fitness is increased by altruism toward relatives. Giving your life so that two siblings or eight cousins can reproduce, for example, would result in your own individual fitness becoming zero, but it would not decrease your inclusive fitness as measured by the number of your genes in the next generation. For sterile worker honey bees the only way to increase inclusive fitness is to ensure the fitness of the queen, with whom they share genes. Of course, honey bee workers do not mathematically calculate their inclusive fitness and the advantages of altruism any more than most humans calculate how many siblings or cousins are worth dying for.

One of the contributions of sociobiology is that it has prompted animal behaviorists to look for evidence of **kin selection** and **kin recognition.** As a result, many surprising examples of kin recognition and selection have been discovered. Female Belding's ground squirrels *Spermophilus beldingi* of the American West, for example, evidently distinguish between their full sisters and half sisters, because they are less likely to fight with full sisters, with whom they share more genes (Fig. 20.13).

RECIPROCITY. This line of argument helps explain some of the controversy over human sociobiology. Because people of different races are easily recognized as being unrelated, it could be argued that racism has a genetic basis that evolved as a means of favoring one's own kin by discriminating against unrelated competitors. If one insists on deriving moral lessons from nature, however, then there are also species that show that aiding non-kin can be beneficial to all. This kind of altruism, in which unrelated individuals aid each other, is called reciprocity. An impressive example of reciprocity occurs in the vampire bat *Desmodus rotundus,* which feeds on the blood of mammals (including humans) in Central and South America. A vampire bat that has recently fed will regurgitate blood to an unrelated vampire bat that would otherwise starve to death. This and other examples of reciprocity occur when the benefit to the recipient is greater than the cost to the donor, and when every individual has a good chance of being a recipient.

Behavioral Ecology

Behavioral ecology, one of the most recent approaches to animal behavior, shares with ethology and sociobiology the assumptions that behavior is largely inherited and has evolved due to natural selection. Unlike ethologists and sociobiologists, however, behavioral ecologists are most interested in the ecological costs and benefits associated with particular behaviors. Behavioral ecologists often begin with the hypothesis that behaviors are **optimal** for efficiency of energy use. This approach is illustrated in a study by Geoff A. Parker on reproduction in yellow dung flies (*Scatophaga stercoraria*), which mate on fresh cow pats. It takes about 100 minutes for a male to fertilize all of one female's eggs, so one might think that the optimal duration of copulation would be 100 minutes. This would indeed be the case if each male copulated with only one female. After copulating with one female and guarding her from other males while she lays her eggs, however, the male flies off to find another female.

Parker found that guarding and searching took an average of 2.5 hours. The male is therefore faced with a dilemma: Should he copulate for the full 100 minutes to fertilize all the eggs of each female, or should he stop copulating sooner so that he can get an early start on finding the next mate? Somewhere between 0 minutes and 100 minutes there is an optimal duration of copulation that will result in the male's leaving the maximum number of offspring. Behavioral ecologist start with the hypothesis that the actual duration of copulation will be near that optimum. By actually measuring the proportion of eggs fertilized for various durations of copulation, Parker determined that the optimal duration was 41 minutes. The actual average duration of copulation was 36 minutes. The 5-minute difference between the theoretical optimum and the actual duration of copulation is probably within the normal range of variation that one expects in biology. Of course it is also possible that some other approach to studying the behavior would provide a better prediction and explanation.

Living in Groups

Having examined some of the elements of behavior and some approaches used in studying it, we can now turn our attention to some of the more complex types of animal behavior. One such behavior is group living. "Why," you might ask, "do some animals live in groups, while others get together barely long enough to reproduce? What are the advantages of group living that compensate for the increased competition for food and mates, the likelihood that predators will be attracted to a group, and the increased spread of disease and parasites? Why, to pick one example, did buffalo (*Bison bison*) roam in vast herds that extended across the American prairies as far as the eye could see, and why did wolves (*Canis lupus*) form packs that lurked along the fringes of those herds?"

Unfortunately it is too late to study these herds and packs in detail, but it should not be hard for you to imagine several advantages that could have led to the evolution of group living in these species. One obvious advantage is that the large number of bison provided some security from attacks by wolves. Not only were the bison intimidating in their size and numbers, but also the large number of eyes made it unlikely that a pack of wolves could approach the herd unnoticed. Each individual bison could therefore spend more time grazing with its head down and less time looking out for wolves. Another advantage for prey living in groups is the **saturation effect,** which results from the fact that a predator can eat only so much prey at one time. With all these advantages for a herd of bison, a single wolf would not have had much chance preying on them. Only by hunting in packs could they isolate a small or weakened bison.

Territoriality

Many animals that live in groups share a **home range.** In winter, for example, many birds flock together in a home range. After the spring migration to their new habitats, however, most male birds undergo a Jekyll-to-Hyde transformation and establish a **territory** that they vigorously defend against other males of their species. As noted before, sticklebacks, as well as numerous other animals, also establish territories. The essential difference between a home range and a territory is that the territory is, by definition, defended against encroachment by others of the same species. Male birds advertise possession of their territories by songs and visual displays. An intruding male's song or plumage is the releaser for **territorial aggression** by the resident. Resident males will often attack a tape recorder playing another male's song or a tuft of feathers the same color as the male's breeding plumage. The only way for a male bird to lurk about within the territory of another is to keep quiet and out of sight. Such "lurkers" are less likely to breed, however, because the songs and display are also required to attract females.

In some species a territory may be occupied by two or more males and their mates. Wolves are the classic illustration of a species that defends a group territory. The average wolf pack is an extended family of from five to eight individuals with a territory of a few hundred square kilometers. An **alpha male** and an **alpha female** lead the pack. They define the pack's territory by releasing a pheromone during a characteristic **raised-leg urination** about every 450 meters as they patrol its perimeter. To wolves from neighboring territories these olfactory boundary markers are, in one sense at least, nothing to sniff at. Packs of wolves have been seen abandoning a deer chase rather than cross into another pack's territory.

A

B

C

Aggression

Aggression often plays a major role in maintaining order among members sharing a territory. This intraspecific aggression should not be confused with defense or predation, which are sometimes referred to as aggression (Fig. 20.14). Aggression or threats of aggression are often used in establishing a **dominance hierarchy**—often called a pecking order—within the group. The dominance hierarchy determines which individuals have priority in mating, feeding, or other activities. Generally the largest and fittest individuals dominate a hierarchy. In wolf packs the alpha male and the alpha female are dominant. The alpha male dominates a hierarchy of subordinate males, and

FIGURE 20.14

Three types of aggressive behavior in the rattlesnake *Crotalus.* (A) During dominance displays, rattlesnakes neck-wrestle until one snake is pinned by the other. The defeated snake then assumes a submissive posture and slinks away unharmed. They do not use the venom, which is as toxic to rattlesnakes as to other animals. Nor do they use the rattlers, because snakes are deaf to air-borne sounds. (B) In defensive aggression the rattlesnake uses its rattlers as a warning and bites if necessary. (C) In predation the snake silently stalks its prey, then seizes and swallows it whole.

■ **Question:** Can you think of another example of how giving the same name to different behaviors can lead to erroneous inferences?

the alpha female dominates the subordinate females. Probably only the alpha pair breeds within most packs. The alpha male directs the hunting, brings food to the female and the pups, and helps tend the pups. The alpha female determines the location of the den where she will bear her pups, and she sometimes directs the hunting.

Subordinate wolves sometimes challenge the alpha male or female, but unless the alpha is old or wounded the insurrection is usually put down without bloodshed. Visual communication is the most important means by which alpha wolves demonstrate and maintain their dominance (Fig. 20.6). The above description should not be taken as the rule for all wolf packs, and certainly not for any other species. Wolves are extremely adaptable animals, and the pattern of dominance can change under different circumstances.

In some species, subordinate animals suffer considerably because of continual harassment from dominant individuals. Subordinate male baboons (*Papio*), for example, are often plagued by parasites and infected wounds because the glucocorticoid hormones secreted in response to social stress inhibit immune responses. In other species, subordinate animals appear to suffer little stress as long as each individual knows its place in the hierarchy. When two roosters (*Gallus gallus*) are first introduced to each other, both their heart rates increase from a resting level of about 200 beats per minute to approximately 350 beats per minute. Within a minute, however, one of the roosters establishes its dominance by crowing or threatening the other, and the heart rates of both roosters quickly decline to resting levels. Evidently the stress to roosters of not knowing their place on the pecking order is greater than the stress of being subordinate.

Courtship

Functions of Courtship. Not every species of animal courts. Sponges, for example, simply emit sperm in vast clouds and let the currents carry them to eggs of the same species. Other animals, such as some insects and snakes, mate with little or no courtship. For many species, however, courtship is essential. There are many advantages to courting:

1. In predaceous animals such as balloon flies, courtship keeps one mate from attacking the other.
2. Courtship ensures that an animal will not waste its gametes on a member of a different species. Probably only females of the same species would be seduced by the zigzag dance of a male three-spined stickleback, for example.
3. Courtship also ensures that the partner-to-be is physiologically capable of reproduction. For example, female sticklebacks whose bellies are not swollen with eggs will not attract a male.
4. In many species, courtship appears to synchronize reproductive activities. For example, a female whooping crane *Grus americana* does not ovulate until a male has danced about her, jumping in the air and flapping his wings. (Whooping cranes are now so scarce that human volunteers perform the dance. The eggs are then artificially inseminated with whooping crane semen, however.)
5. Some aspects of courtship appear to prolong copulation, helping to ensure that as many eggs as possible are fertilized. For example, the empty silk balloon that male empidid flies present to females may allow the male more time to copulate.
6. Courtship may also be a mechanism of natural selection that prevents individuals with undesirable traits from passing them on (see pp. 371–372). The restricting of courtship to the dominant members of a hierarchy or to those with territory also helps ensure that individuals too weak to maintain dominance and territory will not pass genes into the next generation.

Leks. In certain species of insects, amphibians, birds, and mammals, males gather and display for females. These gathering places are called leks (Fig. 20.15). Because mass courtship at a lek is so intense and easily observed, leks have been studied extensively. Nevertheless, there remains some disagreement over the function of a lek. It is usually assumed that the males select the site of the lek, and that the females are drawn to the lek, where they exercise their preference in males. It is also possible, however, that the site of the lek is one that is normally frequented by females.

FIGURE 20.15

A male sage grouse *Centrocercus urophasianus* struts at a Wyoming lek, using its showy display of plumage and the loud booming sound made by forcing air out of its esophageal sac.

Mating Systems. The successful males in a lek may copulate with many females but form no social bond with any of them. Such **promiscuity** also occurs among most other species. At first glance, promiscuity appears to be the most efficient way to produce offspring, especially for males. In fact, however, the offspring in a promiscuous species may find it hard to survive, because their own male parents are competing with them for resources. Often more offspring can be produced if males and females remain socially bonded as mates and both contribute to the offspring as parents. In many species, one male and one female remain together as mates at least for one breeding period. This arrangement, or mating system, is called **monogamy** (Greek *mono-* single, *gamos* marriage). Such social monogamy does not necessarily imply sexual monogamy.

Although monogamy seems most natural to people brought up in Western cultures, **polygamy** is the norm in many animals and in some human cultures. There are two kinds of polygamy. **Polygyny** (pronounced puh-LIJ-uh-nee; Greek *poly* many, *gyne* lady) is a mating system in which one male has more than one female mate. Polygyny is most common in environments in which resources are distributed in patches. Females that depend on the resources of the patch to raise their young mate with the male that controls the patch, even though they have to share the male and the patch with other females. In human cultures, polygyny also appears to be most frequent in patchy environments, such as deserts.

In **polyandry** (Greek *andro-* male), one female has more than one male mate. Polyandry generally occurs in species in which the female is larger than the male or for some other reason controls resources on which the male depends. Honey bees, for example, are polyandric. The drones, which lack stingers and cannot forage, are totally dependent on the hive, which in turn depends on the queen (pp. 697–699). Polyandry is also common among plovers and related birds (order Charadriiformes), in which the female is larger than the male. In phalaropes (genus *Phalaropus*) of the subarctic, the females have showier plumage and establish leks that attract males. The males perform many of the parental duties. In the related American jacana *Jacana spinosa,* which lives on lily-covered lakes in Central America, the female dominates a number of smaller males in her territory (Fig. 20.16). The males are largely responsible for incubating her many clutches of eggs.

FIGURE 20.16

American jacana *Jacana spinosa.*

Reproductive Investment. It is evident that males and females of many species make a considerable investment of time and energy in securing a suitable mate through courtship. The female can be certain that her courtship efforts will not be wasted, because her eggs are certainly bearing her genes. Males, on the other hand, run the risk that their reproductive investment will be diluted or nullified by other males. In many species there are mechanisms that reduce this risk. Internal fertilization is one such mechanism that avoids fertilization by extraneous sperm. Even with internal fertilization, however, a male may find his reproductive efforts undone by a later male. Some male insects avoid cuckoldry by releasing an antiaphrodisiac pheromone onto just-mated females. A male wolf or other canid often cannot disengage its penis for as long as a half hour after ejaculation. This ensures that the male's sperm will have a

FIGURE 20.17
When the female African mouthbrooder *Haplochromis burtoni* (left) tries to take into her mouth the egglike spots on the male's anal fin, she will also take in some sperm that will fertilize the eggs that are already in her mouth.

■ **Question:** Can you suggest a sequence of stages by which this behavior evolved?

FIGURE 20.18
A reed warbler *Acrocephalus scirpaceus* with a cuckoo *Cuculus carorus* nestling. It is easy for humans to see that something is amiss when the young dwarfs the parent, but the gaping mouth in the nest is an irresistible releaser to this surrogate parent. The absence of young warblers in the nest suggests that the cuckoo, with its larger mouth, outcompeted them for parental attention.

chance to fertilize ova before another male comes along. In snakes the male secretes a **copulatory plug** that mechanically prevents a later mating of the female. Male spiny-headed worms (phylum Acanthocephala) cement the females' genitals shut.

In spite of such stratagems, zoologists have found that in virtually every species studied the offspring in a brood are fathered by more than one male. This is true even for monogamous species. In the black-capped chickadee *Parus atricapillus,* for example, females mated with low-ranking males will sneak away from the nest and mate with males of higher rank. Using the technique of DNA fingerprinting (pp, 100–101), zoologists have found that up to 70% of the offspring in birds' nests are not fathered by the male that helps raise them. The only "faithful" couples found so far occur in the California mouse *Peromyscus californicus.*

Parental Behavior

Parental Investment. Another way of protecting a reproductive investment is to help ensure the survival of eggs and offspring through parental behavior. Sociobiologists have created a considerable body of **parental investment theory** to account for the circumstances in which parental behavior is a good investment of time and energy. Parental behavior is clearly not advantageous when an animal cannot be sure that at least some of the eggs or offspring it is tending are bearing its own genes. It is not surprising, therefore, that parental behavior is uncommon in species with external fertilization. The exceptions occur mainly when the parent keeps track of the eggs from the moment of fertilization, as is the case with the male three-spined stickleback. The male banded jawfish *Opisthognathus macrognathus* can also be fairly sure of the eggs he has fertilized, because he quickly takes them into his mouth, where they develop. In other **mouthbrooder** fishes, such as those in the genus *Tilapia,* it is the female that incubates the eggs in her mouth. In the African mouthbrooder *Haplochromis burtoni* and others, fertilization actually occurs within the mouth (Fig. 20.17).

Cooperative Breeding. In some species, individuals defy the logic of the preceding paragraph by helping to raise the offspring of others. In addition to honey bees, which have been discussed before, such cooperative breeding is best known in several species of birds, such as the acorn woodpecker *Melanerpes formicivorus* of Central America and the American Southwest, the green woodhoopoe *Phoeniculus purpureus* of Africa, and the Florida scrub jay *Aphelocoma coerulescens.* These birds often live in flocks within territories that can support only one breeding pair each, either because food or nest sites are scarce or because of competition from neighboring flocks. Presumably, cooperative breeding evolved because the nonreproducing helpers benefited through either kin selection or reciprocity.

Care of Young. The extent to which parents will sacrifice for their offspring is familiar to anyone who has watched either a robin tirelessly bringing worms to a nest full of gaping mouths or a human parent writing checks to pay for tuition. Humans are apt to interpret such selflessness as a manifestation of love, but many ethological studies suggest that in birds, at least, the behavior is a motor program. In many species of birds, any gaping mouth will evoke feeding by parents, whether the mouth belongs to their own young or not. European cuckoos and other **brood parasites** exploit this motor program to their own advantage (Fig. 20.18). A further sign that parental behavior in birds is automatic can be seen when the chicks of gulls stray outside the nest. The parents apparently do not recognize a chick outside the nest, because they stand idly by while they starve to death or are eaten by predators.

Reductionism

Many people will object to the direction this chapter is leading us—to the concept that even parental love can be reduced to an automatic behavior or physiological state and presumably to the actions of genes and nerve cells. Critics of such concepts often brand them with the label reductionism, ignoring the fact that it is the role of all science to try to reduce complex and poorly understood phenomena to more elementary and better-understood phenomena.

A more valid criticism of reductionism may be that it is not the most efficient approach to understanding behavior. An example from physics will illustrate why this may be. Theoretically, the temperature and pressure of a volume of gas could be calculated from the velocities of each molecule in the gas, using the laws of momentum and energy. In practice, however, such an approach is hopelessly difficult. The principles of thermodynamics were, in fact, derived from the behavior of large volumes of gas, and no one has ever reduced them to the laws of physics that apply to interactions of individual molecules. It seems just as unlikely that anyone will ever reduce a complex animal behavior to the actions of genes or nerve cells. The complete reduction of behavior in a human seems even more remote. Any brain capable of understanding itself will probably be so complicated that it can never be understood. Nevertheless, while scientific reductionism may never totally succeed in explaining animal behavior, it has so far been more fruitful than any of the alternatives.

The attachment that young birds and some young mammals show to parents is also somewhat automatic. These young can become **imprinted** to a parent during a sensitive period shortly after hatching. Thereafter the animal will tend to follow the imprinting stimulus and as an adult may be socially bonded to such stimuli. If imprinting to a parent does not occur, the young can be imprinted to a milk bottle, an electric train, or some other object (Fig. 20.19).

Humans and other primates do not become imprinted in the same sense, but strong social bonds do form between parents and offspring. For want of a more precise term, we can call this bond in primates **love.** Ethologists, sociobiologists, and behavioral ecologists have proposed reasonable explanations of the adaptive advantages that

FIGURE 20.19

Like many birds, goslings are imprinted to the first active object they see or hear during a critical period shortly after hatching. Usually that imprinting stimulus is the mother, but eminent ethologists will also do. This particular imprinting stimulus is Konrad Lorenz, who was one of the first ethologists to study imprinting.

■ **Question:** Is it likely that all birds imprint? What about cowbirds and European cuckoos?

FIGURE 20.20

An infant rhesus monkey *Macaca mulatta* clings to a surrogate mother in a study pioneered by Harry F. Harlow. If deprived of even this minimal substitute for parental behavior, monkeys fail to develop appropriate social behaviors.

led to the evolution of love. In primates it is physiologically as essential an adaptation as mother's milk. Even if infant monkeys are provided with all their other physiological needs, they develop pitifully abnormal behaviors without the care that, in humans, at least, comes only with love (Fig. 20.20).

Summary

Although much animal behavior is learned, much of it results from neural and other physiological processes that are partly controlled by genes and have evolved through natural selection. Some simple behaviors, such as reflexes, can be completely explained in such terms. Another facet of behavior that directly depends on physiological mechanisms is communication. Communication may be auditory, as in bird songs, visual, as with fireflies, or chemical, as with pheromones. Physiological mechanisms are also responsible for endogenous clocks, which regulate circadian rhythms, such as those of sleeping and waking, and circannual rhythms, such as those that regulate reproduction and migration at different times of the year. Endogenous clocks and biological compasses work together to allow many animals to navigate using the Sun, stars, and magnetic fields.

The assumption that many behaviors have evolved through natural selection is the basis for several approaches to the study of more complex behaviors. One of these approaches is ethology, which focuses on the roles of releasers and motor programs in courtship, aggression, and territoriality. Sociobiology deals with social interactions such as altruism and the benefits that may have led to their evolution. Behavioral ecologists start with the hypothesis that behavior is optimal, and they attempt to test that hypothesis by measuring the time and energy expenditures involved in behavior.

These and other approaches provide insight into the evolution of many types of complex behaviors. Living in groups, for example, can be advantageous by providing better defense against predators. Many animals that live in groups defend a territory, and they maintain order within the group by a dominance hierarchy imposed by aggression. Another complex behavior that can be understood from a biological perspective is courtship, which ensures that a potential mate is of the appropriate species and sex, is fertile, and is not likely to turn predatory. Courtship may also synchronize the physiological processes of mating and can serve as a means of sexual selection. Various animals have different mating systems, such as promiscuity, polygamy, and monogamy, which provide different advantages under different circumstances. Monogamy is often associated with intense parental behavior, which helps protect the reproductive investment of both parents.

Key Terms

reflex
taxis
kinesis
classical conditioning
operant conditioning
pheromone
waggle dance
endogenous clock
circadian rhythm
ethology
motor program
sign stimulus
sociobiology
altruism
inclusive fitness
kin selection
reciprocity
behavioral ecology
territory
aggression
dominance hierarchy
mating system
monogamy
polygyny
polyandry

Chapter Test

1. Name three people who have contributed to the scientific study of behavior and briefly describe their contribution. (pp. 402–422)
2. Describe a reflex and give an example of one. (p. 402)
3. Explain how classical conditioning differs from operant conditioning. Give an example from your own experience of each type of learning. (p. 403)
4. Describe one kind of experimental evidence that some behaviors can be genetically inherited. Exactly what do we mean when we say that a behavior is genetically inherited? (pp. 403–406)
5. What is the evidence that many animals have endogenous clocks? (pp. 410–412)
6. Birds can use several alternative mechanisms to determine direction during migration. Describe two of those mechanisms. What kind of evidence suggests that those mechanisms are available? (p. 412)
7. Compare the following approaches to the scientific study of animal behavior: ethology, sociobiology, and behavioral ecology. What premises do these three approaches share? How do they differ? Give an example of a study characteristic of each approach. (pp. 413–416)
8. What are the advantages to predators and to prey of living in groups? (pp. 416–417)
9. Describe a specific example of the role of communication in each of the following: predation, defense, courtship, parental behavior. (pp. 417–421)
10. The term "aggression" is sometimes applied to several distinct types of behavior: territorial aggression, dominance, predation, and defense. Explain the differences among these, and give an example of each. (pp. 417–418)
11. Explain three advantages of courtship. Which of these three, if any, apply to human courtship? (pp. 418–419)
12. Describe the circumstances under which parental behavior is a sound investment. What type of return do the parents get for their investment? (pp. 419–420)

■ Answers to the Figure Questions

20.1 Without feedback from the recurrent nerve, the fly continues to eat until its foregut bursts.

20.2 It is difficult to imagine how one gene could affect the brain of a female cowbird in such a way that the particular sound of the male triggers this particular behavior. Like most inherited traits, this behavior probably depends on the interaction of many genes and environmental factors.

20.6 It is conceivable that the pattern of flashing by males of each species is entirely learned; they could flash randomly until they get a response from the female of the same species, which would serve as a reinforcement for perfecting the flash pattern. There is no evidence that this happens, however. This explanation would explain flashing by males entirely by means of conditioning, but it is difficult to see how recognition of the correct pattern by females and recognition of the female's response by males could be entirely learned.

20.9 The scout would dance in the correct direction of 35°, which the recruit would perceive as indicating food 35° to the right of the sun. However, if recruits perceive the direction of the sun with their ocelli, which are painted over, they would presumably not know in which direction to fly.

20.12 The classical ethological technique used by Tinbergen is to see if an artificial model with the feature under investigation releases the behavior. Females do, in fact, follow artificial zigzagging models as long as they have red bellies.

20.14 One example is the term "homosexual." In humans the term refers to sexual preference for members of the same sex, but it has sometimes been applied to humans and other animals of the same sex that engage in sexual activity under circumstances that do not suggest preference. Another example is "rape," which has been applied to forced copulation among mallard ducks and other animals with the unlikely implication that the cause is the same as in rape by humans. The term "territory" is likewise often misappropriated from humans to other animals, or vice versa. Likewise, the term "slavery" is applied to ants that force ants of a different species to work in their colonies. A better term for the situation in ants would be "domestication," since slavery in humans involves members of the same species. However, the term "slavery" is now too well established in entomology to change.

20.17 Any spots on the male's fin that may have been mistaken for the female mouthbrooder's eggs could have increased the odds that she would take in sperm that would fertilize the eggs. This would have increased the chances that genes for those spots would be passed on to the offspring. Through successive generations, the behavior of the female mouthbrooder would have selected for spots that increasingly resemble eggs.

20.19 Nest parasites would not imprint. If they did, they would be socialized toward the wrong species.

Readings

Recommended Readings

Agosta, W. C. 1992. *Chemical Communication.* New York: Scientific American Library.

Alkon, D. L. 1983. Learning in a marine snail. *Sci. Am.* 249(1):70–84 (July).

Batra, S. W. T. 1984. Solitary bees. *Sci. Am.* 250(2):120–127 (Feb).

Bekoff, M. 1984. Social play behavior. *BioScience* 34:228–233.

Blaustein, A. R., and R. K. O'Hara. 1986. Kin recognition in tadpoles. *Sci. Am.* 254(1):108–116 (Jan).

Borgia, G. 1986. Sexual selection in bowerbirds. *Sci. Am.* 254(6):92–100 (June).

Carter, C. S., and L. L. Getz. 1993. Monogamy and the prairie vole. *Sci. Am.* 268(6):100–106 (June). (*Hormones and sexual behavior.*)

Clutton-Brock, T. H. 1985. Reproductive success in red deer. *Sci. Am.* 252(2):86–92 (Feb).

Crews, D., and W. R. Garstka. 1982. The ecological physiology of a garter snake. *Sci. Am.* 247(5):158–168 (Nov).

Dyer, F. C., and J. L. Gould. 1983. Honey bee navigation. *Am. Sci.* 71:587–597.

Fitzgerald, G. J. 1993. The reproductive behavior of the stickleback. *Sci. Am.* 268(4):80–85 (Apr).

Gould, J. L., and C. G. Gould. 1989. *Sexual Selection.* New York: Scientific American Library.

Gould, J. L., and C. G. Gould. 1994. *The Animal Mind.* New York: Scientific American Library.

Gould, J. L., and P. Marler. 1987. Learning by instinct. *Sci. Am.* 256(1):74–85 (Jan).

Greenspan, R. J. 1995. Understanding the genetic construction of behavior. *Sci. Am.* 272(4):72–78 (Apr).(*Courtship in drosophila.*)

Gwinner, E. 1986. Internal rhythms in bird migration. *Sci. Am.* 254(4):84–92 (Apr).

Halliday, T. R., and P. J. B. Slater (Eds.). 1983. *Animal Behaviour,* Three Volumes. Salt Lake City, UT: W. H. Freeman. (*A beautifully illustrated presentation of reproductive behaviors.*)

Hölldobler, B., and E. O. Wilson. 1983. The evolution of communal nest-weaving in ants. *Am. Sci.* 71:490–499

Holmes, W. G., and P. W. Sherman. 1983. Kin recognition in animals. *Am. Sci.* 71:46–55.

Johnson, C. H., and J. W. Hastings. 1986. The elusive mechanism of the circadian clock. *Am. Sci.* 74:29–36.

Kirchner, W. H., and W. F. Towne. 1994. The sensory basis of the honeybee's dance language. *Sci. Am.* 270(6):74–80 (June).

LeVay, S., D. H. Hamer, and W. Byne. 1994. Is homosexuality biologically influenced? *Sci. Am.* 270(5):43–55 (May). (*Debate on the relative importance of neurophysiology, genetics, and environment.*)

Pfennig, D. W., and P. W. Sherman. 1995. Kin recognition. *Sci. Am.* 272(6):98–103 (June).

Preston-Mafham, R., and K. Preston-Mafham. 1993. *The Encyclopedia of Land Invertebrate Behaviour.* Cambridge: MIT Press.

Prestwich, G. D. 1983. The chemical defenses of termites. *Sci. Am.* 249(2):78–87 (Aug).

Ryker, L. C. 1984. Acoustic and chemical signals in the life cycle of a beetle. *Sci. Am.* 250(6):112–123 (June).

Scheller, R. H., and R. Axel. 1984. How genes control an innate behavior. *Sci. Am.* 250(3):54–62 (Mar).

Seeley, T. D. 1983. The ecology of temperate and tropical honeybee societies. *Am. Sci.* 71:264–272.

Shettleworth, S. J. 1983. Memory in food-hoarding birds. *Sci. Am.* 248(3):102–110 (Mar).

Stacey, P. B., and W. D. Koenig. 1984. Cooperative breeding in the acorn woodpecker. *Sci. Am.* 251(2):114–121 (Aug).

Sulzman, F. M. 1983. Primate circadian rhythms. *BioScience* 33:445–450.

Sweeney, B. M., et al. 1983. Biological clocks. *BioScience* 33:424–450.

Topoff, H. 1990. Slave-making ants. *Am. Sci.* 78:520–528.

Turek, F. W. 1983. Neurobiology of circadian rhythms in mammals. *BioScience* 33: 439–444.

Waterman, T. H. 1989. *Animal Navigation.* New York: Scientific American Books.

Wilkinson, G. S. 1990. Food sharing in vampire bats. *Sci. Am.* 262(2):76–82 (Feb).

Winfree, A. T. 1987. *The Timing of Biological Clocks.* New York: Scientific American Library.

Winston, M. L., and K. N. Slessor. 1992. The essence of royalty: honey bee queen pheromone. *Am. Sci.* 80:374–385.

Additional References

Barinaga, M. 1994. From fruit flies, rats, mice: evidence of genetic influence. *Science* 264:1690–1693.

Bouchard, T. J., Jr., et al. 1990. Sources of human psychological differences: the Minnesota study of twins reared apart. *Science* 250:223–228.

Brower, L. P., and S. B. Malcolm. 1991. Animal migrations: endangered phenomena. *Am. Zool.* 31:265–276.

Brunner, H. G., et al. 1993. Abnormal behavior associated with a point mutation in the structural gene for monoamine oxidase A. *Science* 262: 578–580.

Hall, J. C. 1994. The mating of a fly. *Science* 264:1702–1714.

Holden, C. 1987. The genetics of personality. *Science* 237:598–601.

Mann, C. C. 1994. Behavioral genetics in transition. *Science* 264:1686–1689.

OVERVIEW

UNIT

The Diversity of Animals

The organisms shown here represent a small sample of the enormous diversity of animals. The solid structure of the coral reef was itself created by animals called polyps, which are classified in the phylum Cnidaria. Polyps are minute, but so numerous that the calcium carbonate they secrete over the decades accumulates into structures that dwarf any creation of humankind. The many nooks and crannies within the reef are habitats for a wide variety of other animals. Corals are so still and so simple in organization that they were not even recognized as animals until relatively recently. In contrast, the fishes shown here (phylum Chordata, subphylum Vertebrata) move rapidly and have a complex organization that makes them immediately recognizable as animals.

The other vertebrate shown here belongs to one of the most interesting of all species. By various contrivances, this one species has managed to adapt to almost every type of environment on earth. It is also the only species that systematically tries to study every other species of animal. The growing impact of our species makes such studies imperative for the survival of all animals, including ourselves.

CHAPTER

21

Systematics

Frog (*Rana* sp.) with dragonfly (*Plathemis* sp.)

The Importance of Systematics

More than a million living species of animals have been named, and some zoologists believe that at least 30 times that many remain to be discovered (May 1992). Although around 13,000 new species of animals are named and described each year, most animals will remain forever unknown. Many are so tiny and elusive that only the greatest good fortune would lead to their being noticed by someone who can appreciate their significance. Many other species are becoming extinct faster than they can be discovered, especially in tropical rainforests, where perhaps half the world's species are endangered by human activity. Even in the United States the government has only recently begun a National Biological Survey to inventory all its species. Biology, medicine, and agriculture are constantly handicapped because researchers cannot identify the organisms they are using.

Naming and describing species is one concern of the field of systematics, also called biosystematics. **Systematics** as a whole has been defined as the study of the diversity and relationships among organisms. In addition to naming and describing species, systematists are increasingly concerned with determining the **phylogeny** (evolutionary history) of organisms and classifying them into groups accordingly. Classification is what makes it possible for you to learn so much about animals even though this and other books can describe only a relatively few species. From studies of a few vertebrates, or a few mammals, or a few fishes, we can infer what features to expect in other vertebrates or mammals or fishes. Without systematics, zoology would be a disorganized collection of facts about apparently unrelated species, and this text would be (even more) enormous and chaotic.

What Is a Species?

It is obviously important for zoologists to know what they are referring to when they use the term "species." Many adhere to the **biological species concept** (p. 367), which defines a species as a group of organisms in which individuals are capable of interbreeding within the group, but not with individuals in other such groups. In essence, the organisms, themselves, define the species to which they belong when they find some organisms to be appropriate as mates, but not others. Obviously the criterion that each species be reproductively isolated from all other species is of little use for organisms that reproduce asexually, or for fossils. In addition, it is often impractical to study whether groups are reproductively isolated from each other. For that reason, many biologists consider a group of organisms to be a species if they are similar to each other and sufficiently different in some way from all other such groups of organisms. This criterion is often consistent with the biological species concept, because major differences often indicate that a group does not interbreed with another group.

It is not always true, however, that two reproductively isolated species differ significantly from each other. For example, the alder flycatcher and the willow flycatcher are virtually indistinguishable to human eyes and were once lumped together in one species. They were split into two species after it was discovered that slight differences in song and habitat make the males of one species unacceptable to the females of the other species (Fig. 21.1 on page 430). The opposite situation also occurs. Among birds approximately 9% of the groups generally regarded as separate species hybridize at least occasionally (Grant and Grant 1992). In some cases, two "species" are found to hybridize so often that they are reclassified as a single species. The Baltimore oriole of the eastern United States and Bullock's oriole of the West were once classified as different species because they look so different (Fig. 21.2 on page 430). After they were caught interbreeding where their ranges overlap, they were lumped together as the northern oriole, much to the dismay of amateur ornithologists, who had to drop a species from their life lists.

FIGURE 21.1

The willow flycatcher (left) and alder flycatcher look virtually identical to us, and they were once considered the same species. They do not interbreed, however, so according to the biological species concept they are classified as separate species. Reproductive isolation is due to differences in habitat and song. The willow flycatcher *Empidonax traillii* is more likely to be found in old orchards and willows, and the male's song sounds like "fitz-bew." The alder flycatcher *Empidonax alnorum* is more likely to be found in alder thickets near water, and the male's song sounds like "fee-bee-o."

Nomenclature

One essential area of systematics, called nomenclature, deals with the naming of species. The rules of nomenclature result in every species having a scientific name consisting of two words, such as the weird italicized words you have seen sprinkled throughout this book. If you don't happen to be fluent in Latin, these names may seem strange, and, in fact, most zoologists remember only a few of them. Nevertheless, they are a vast improvement over the names once used. In the Middle Ages, naturalists gave each organism a descriptive name in New Latin (the universal language of scholars in Europe), and as increasing numbers of organisms were discovered they had to add more and more descriptive words to the names to distinguish similar species from each other. If you had been a student in the 18th century, you would have had to learn that the honey bee is *Apis pubescens, thorace subgriseo, abdominae fusco, pedibus posticis glabris utrinque margine ciliatis*—the fuzzy bee with the grayish thorax, dusky-brown abdomen, hairless hind feet bordered with small hairs on both sides.

In the 18th century, Carolus Linnaeus (pronounced lin-NEE-us) simplified the lives of biologists considerably by assigning to each species a name consisting of only two words from the New Latin spoken in his day, which borrowed heavily from Greek. We still use his **binominal system.** (The system was formerly called the "binomial sys-

FIGURE 21.2

In spite of differences in appearance, the Baltimore oriole (left) and Bullock's oriole are now considered to be merely races within one species, the northern oriole *Icterus galbula.*

tem," and many zoologists still refer to it by that name.) The honey bee then became simply *Apis mellifera:* the bee that brings honey. The Linnaean (lin-NEE-an) system was quickly adopted throughout Europe. Unfortunately, however, there was no international body to ensure that species names introduced in one country would be accepted in other countries. English-speaking biologists often had one Latin name for a new species, and French-speaking biologists had another Latin name for the same species. They might as well have used their native tongues. In the early 1900s, however, an international organization was established to ensure that scientific names are the same around the world, no two animals have the same name, and the names do not change arbitrarily. These goals of **universality, uniqueness,** and **stability** are safeguarded by the International Commission on Zoological Nomenclature. Generally the Commission selects the name that was proposed by the first person to publish a clear description of the species after 1758, when the 10th and last edition of Linnaeus' *Systema Naturae* was published. The Commission's *International Code of Zoological Nomenclature* (Ride et al. 1985) includes guidelines for assigning names to species. The Commission explicitly states, however, that the guidelines must not restrict "the freedom of taxonomic thought or action." Some of the guidelines that are most helpful in becoming familiar with nomenclature are summarized below. Additional comments are in brackets.

1. The name of a species consists of two words (binomen) and that of a subspecies of three words (trinomen). In each case the first word is the name of the genus, the second word is the species epithet, and the third word, when used, is the name of the subspecies group. [Examples of binomens and trinomens: *Homo sapiens; Homo sapiens sapiens.* A subgenus may be given in parentheses after the generic name but is seldom used. Unfortunately, first names of species frequently have to be changed as a result of reclassification into a different genus, and the older name is commonly, but incorrectly, given in parenthesis after the new one; for example, the spring peeper, *Pseudacris (Hyla) crucifer,* moved from genus *Hyla* to genus *Pseudacris.* The generic name may be used alone. In order to refer to a particular species when only its generic name is known, the genus is followed by "sp."; for example, a frog in genus *Rana* would be *Rana* sp. To indicate more than one species in a genus, use "spp."; for example, *Rana* spp.]
2. The name must be either Latin or latinized. [The name should be written in a different typeface, usually italics, from the body of text in which it occurs. If different typefaces are not available, the name is underlined.]
3. The name of the genus and subgenus must be a noun in the nominative singular.
4. The species epithet must be an adjective agreeing in gender (masculine, feminine, or neuter) with the generic name [*Homo sapiens,* wise man]; or a noun in the nominative singular (like the genus) [*Felis catus,* cat cat]; or a noun in the genitive case [*Charadrius wilsonia,* Wilson's plover]; or an adjective derived from the name of an organism with which the animal is associated [*Eurosta solidaginis,* a fruit fly that develops in the goldenrod *Solidago*]. [Exceptions to these rules are sometimes allowed. Classification of one species of the beetle *Agra* was so troublesome that the name *Agra vation* was accepted.]
5. Genus and subgenus names are always capitalized; species epithets never are, even if derived from a proper name. [This differs from the rule in botany, which allows the species epithet to be capitalized if derived from a proper name. A generic name may also be used as the common name, in which case it is neither capitalized nor italicized. For example, drosophila.]
6. The name of the author who first described the species, or an abbreviation of his or her name, may be appended to the species names, not italicized. If the species has been transferred to a new genus, the name or abbreviation is en-

closed in parentheses. [Example: *Periplaneta americana* (L.), the American cockroach, originally described by Linnaeus (L.) in another genus.]

Classification

The Fallacious Chain of Being. Another essential function of systematics is to classify the named species so that we can organize our knowledge about them and make generalizations. Most people in Western cultures have an almost irresistible urge to deal with diversity by ranking groups on a single scale from the lowest to the highest. The inability to see groups as not merely different from each other, but as lower or higher, goes back at least to the medieval notion that everything from minerals, to plants, to animals, to humans, and then the celestial spirits could be arranged on a Great Chain of Being. To most people brought up in that tradition, animals presumed to be simpler and less successful are lower. Often they are also considered to be less worthy of existence.

Such an approach inevitably leads to contradictions. For example, many of the "lowest" forms are the most successful, if we measure success by numbers of species or individuals. Round worms (phylum Nematoda), many of which are "lowly," simplified parasites, may have more species than any other group of animals, although zoologists have gotten around to describing only 12,000 of them (Fig 21.3). There is also an enormous number of individual nematodes: 100 billion in a square meter of soil, for example. By comparison, vertebrates and most other "higher" animals have failed miserably. "Lowliness," simplicity, and success are clearly worthless criteria in systematics.

Taxonomy. Linnaeus was largely responsible for establishing a more logical approach to classification. He may thus be regarded as the founder of **taxonomy,** which is the branch of systematics that deals with the principles of classification. The Linnaean taxonomic system is a hierarchical one in which similar species are assigned to one genus, similar genera are assigned to the same higher category, and so on. Today there are six taxonomic categories that a zoologist must specify when describing a newly discovered species. These required categories are **kingdom, phylum** (plural phyla), **class, order, family,** and **genus** (plural genera). The kingdom level distinguishes among a few major groups, such as animals, plants, or bacteria. Phyla distinguish among groups in the same kingdom that have major differences. Animals in different phyla are supposed to have different "body plans," as will be discussed later in this chapter. Each phylum generally includes several classes, which differ greatly but share similar body plans. Each class may include several

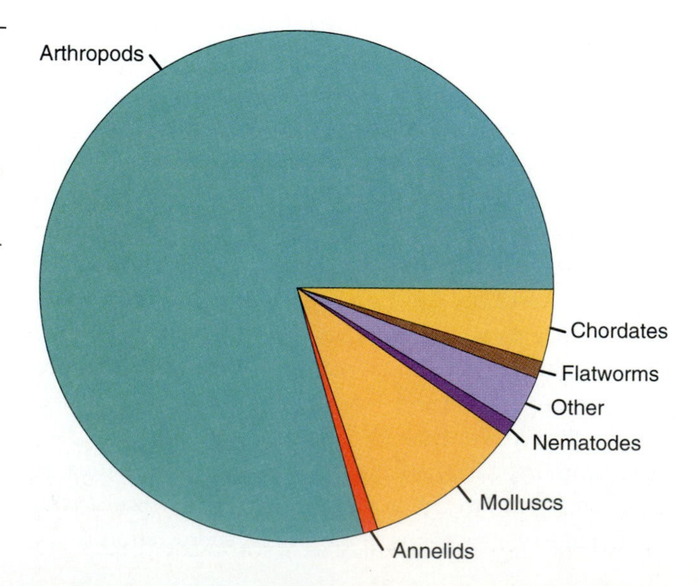

FIGURE 21.3
The proportion of the million named animals in various phyla. Only Chordata, Arthropoda, Mollusca, Annelida, Nematoda, and Platyhelminthes have more than 10,000 species each. One of the delightful implications of this chart is that anyone familiar with these six phyla is automatically familiar with the vast majority of animals.

orders, with lesser differences, each order may include several families with even lesser differences, and each genus may include several species with still smaller differences.

Unlike species, which can (sometimes) be determined by reproductive isolation, these taxonomic levels are not based on rigorous biological criteria. Taxonomists spend a great deal of time debating whether two groups of organisms differ sufficiently to be, for example, in different families or merely in different genera in the same family. Even at the phylum level, there are numerous debates about what constitutes a sufficiently different body plan.

The names of groups within each category are called **taxa** (plural of taxon). The required taxa for our own species are:

kingdom Animalia
phylum Chordata
class Mammalia
order Primates
family Hominidae
genus Homo
species *Homo sapiens*

In addition, each of these required levels may have higher or lower categories, often designated with prefixes and suffixes such as super-, sub-, and infra-. For example, we classify ourselves in the subphylum Vertebrata, the suborder Haplorhini, the infraorder Catarrhini, and subfamily Homininae. The International Commission on Zoological Nomenclature has not standardized terminology for these categories except to some extent at the family level. The name of the family group is formed by adding the appropriate suffix to the generic name of the "type genus" chosen by the author as being representative of the new family group. For superfamily the suffix is -OIDEA, for family it is -IDAE, for subfamily it is -INAE, and for tribe it is -INI.

MOLECULAR PHYLOGENETICS. Organisms are grouped within the same or different taxa on the basis of similarities and differences in anatomy, molecules, behavior, and other traits called characters. Simply put, a **character** is anything that can assist in classification. Since the 1950s a growing number of taxonomists have used homologous molecules in different groups of animals as sources of characters (pp. 376–377). For example, the sequences of bases in a ribosomal RNA from humans, lizards, chickens, sea urchins, and so on, have been used to infer the phylogeny of these groups of animals (see Fig. 18.17). This method, called molecular phylogenetics, has the advantage that each molecule provides a large number of characters. For example, each base in a nucleic acid, or each amino acid in a protein, can be considered a separate character. Thus the number of similarities or differences in the base sequence or amino acid sequence can be used as a measure of the evolutionary difference between two groups. Systematists then use computers to select the most statistically probable phylogeny out of the vast number of possibilities.

The advantage of molecular phylogenetics has been demonstrated with simulations and laboratory organisms (Hillis et al. 1994). In one study the molecular phylogeny of 10 strains of laboratory mice was inferred from chromosomal differences and compared with the phylogeny known from breeding records (Fitch and Atchley 1987). The molecular phylogeny was exactly correct, even when five different statistical methods of analyzing the data were used. In contrast, phylogenies based on morphology (lower jaw structure) or life-history traits (litter size, body mass at different ages, etc.) gave different phylogenies, none of which was correct.

Although molecular phylogenetics provides taxonomists with a powerful tool, it is still essential that morphological and other characters be studied. After all, what one sees in the field are shells, bones, and skins—not molecules. Unfortunately, zoos, museums, and other traditional sources of information on such characters have been neglected in the rush of enthusiasm for the new molecular tools.

Evolutionary Systematics

In *The Origin of Species,* Charles Darwin predicted that an evolutionary point of view would provide a natural basis for taxonomy. Since then, taxonomists have indeed tried to make classification of each group reflect its evolutionary history, or **phylogeny.** Until a few decades ago there was no name for this method of systematics, because there was no alternative, but now it is often called traditional systematics or evolutionary systematics. The goal of evolutionary systematics is to group each species into taxa that are related to each other through evolution. The results of evolutionary systematics are published in verbal descriptions of taxa and also in the form of **dendrograms** or **phylogenetic trees** that graphically show the relationships among taxa. (See, for example, Figs. 18.11 and 21.4).

Taxonomies are based on characters, but not all characters are useful in taxonomy. In order to determine how useful characters are likely to be, taxonomists first perform a **character analysis.** In evolutionary systematics, character analysis involves the following:

1. **Determining whether the character is homologous or analogous. Homologous** characters, like the bones in birds' wings and human arms (see Fig. 18.1), are so similar that they were most likely inherited from the same ancestral species. If, on the other hand, two groups have evolved the same character independently, the character is **analogous.** For example, the wings of bats and butterflies are analogous characters, since they clearly differ in structure. Only homologous characters provide any information about phylogeny.
2. **Determining whether a character is primitive or derived. Primitive** characters are those that are retained from earlier in an evolutionary lineage. For example, the four limbs of reptiles, birds, and mammals constitute a primitive character inherited from their amphibian ancestors. In this context the term "primitive" does not

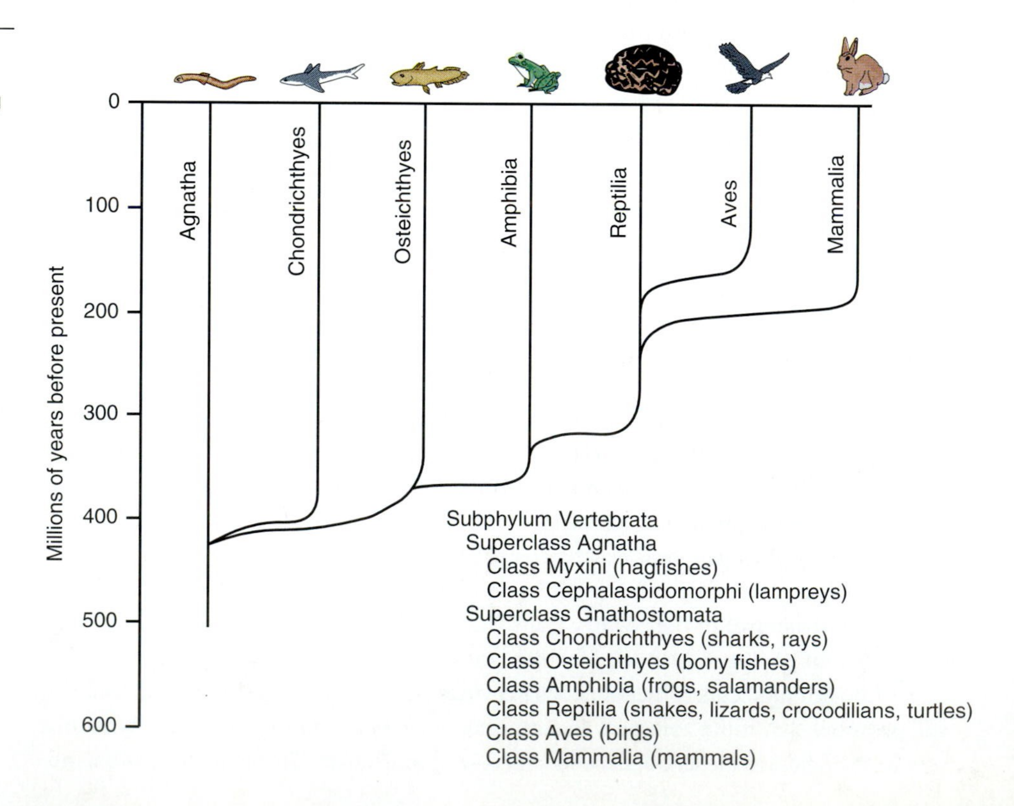

FIGURE 21.4
A dendrogram and list of taxa for the major groups of vertebrates, according to evolutionary systematics.

mean crude or poorly adapted. **Derived** characters, on the other hand, are later evolutionary modifications. The modification of the forelimbs of bats into wings is a derived character. Determining whether a character is primitive or derived helps establish whether a group is ancestral to or descended from another group.

3. **Estimating the degree of evolutionary difference between the characters in two groups.** The degree of evolutionary difference, along with the fossil record, provides information on the times of evolution. The degree of evolution also helps determine taxonomic levels; that is, it helps determine whether two groups will be classified in the same genus, or family, or order, and so on.
4. Finally, character analysis involves **determining how much weight to give to different sets of characters.** Characters based on developmental patterns might suggest a totally different phylogeny from characters based on behavior, for example, so the evolutionary systematist must decide how much importance to assign to each type of character.

Often character analysis suggests several different ways in which a group of animals can be classified. Choosing among the alternatives often depends on what could be called the personality or style of the individual systematist. Knowledge, experience, and judgment combine in a complex way to shape this scientific personality or style. One systematist may give more weight to morphological characters, while another may weigh molecular comparisons more heavily. One may be a "splitter" who tends to break up large groups of animals, while another, for equally valid reasons, may be a "lumper" who tends to combine several groups into one. Because of such individual differences, you will often find different classifications for the same group of organisms. One is free to choose the classification that appears to be most convenient and most consistent with phylogeny.

Numerical Taxonomy

During the 1960s some taxonomists became dissatisfied with the subjectivity involved in character analysis and sought more objective approaches to classification. One approach was numerical taxonomy, also called **phenetics.** Pheneticists avoid subjective aspects of character analysis simply by not doing any character analysis. They include as many characters as possible, whether homologous or analogous, and they give them all equal weight. The assumption is that with a large number of characters, the effects of incorrect ones will cancel each other. Each character in each species is assigned a number, and all the numbers are entered into a computer. The computer then groups the organisms into clusters based on similarity. There is no attempt to infer phylogeny from the result. Numerical phylogeny has been largely abandoned as a method of taxonomy, but its statistical and computer methods are still widely used.

Cladistics

Cladistics, also called **phylogenetic systematics,** is an increasingly important method of systematics first proposed by the German entomologist, Willi Hennig, in 1950. Its acceptance by zoologists has increased rapidly in recent years, even though its methods and results seem alien to those accustomed to traditional methods. Cladists, who often refer to themselves as phylogeneticists, work as follows:

1. In order to avoid the subjectivities of character weighting and judging degrees of evolution, phylogeneticists base their classifications on only the number of derived, homologous characters. One approach to determining whether a homologous character is primitive or derived is to see whether it occurs in an **outgroup,**

which is a group of organisms thought to be closely related to the groups being classified, but not part of them. A homologous character that occurs among the groups being classified, but not in the outgroup, is assumed to be derived. In the jargon of cladistics, such shared, derived homologies are called **synapomorphies.**

2. Cladists judge the phylogenetic relationships among groups only from the number of synapomorphies they share. Two or more groups with the greatest number of synapomorphies in common are assumed to have had the same recent ancestor. These groups plus their most recent common ancestor form a **clade.** A clade is the only valid taxonomic group in cladistics.
3. Phylogeneticists represent phylogenetic relationships as a pattern of branching lines, called a **cladogram** (Fig. 21.5). A cladogram is constructed by representing the most recent common ancestor of each clade as a point called a **node,** and all of the descendants as the ends of lines branching from the node. Each clade is joined to related clades and lines, forming larger clades. (Actually computers do this work, because the number of possibilities can be enormous. With only 10 groups there are more than 280,000,000 possible cladograms!) Cladograms represent phylogeny but not times or degrees of evolution.

Many zoologists have embraced cladistics because cladograms can reflect phylogeny more accurately than does evolutionary systematics. On the other hand, clades have some properties that many zoologists find hard to accept:

1. Two or more clades that diverge from the same node are **sister groups** with the same taxonomic rank, regardless how different they are. Thus, because crocodilians and birds share a recent common ancestor, they are sister groups at a different level from lizards, even though lizards are anatomically and physiologically more like crocodilians than birds (Fig. 21.6).
2. Clades do not lend themselves to classification in the traditional categories because one quickly runs out of names for different levels. In addition, groups that occupy the same taxonomic rank traditionally, such as class Mammalia and class Aves, are seldom at the same level cladistically. Many cladists indicate taxonomic levels on indented lists.
3. Clades must always be **monophyletic,** because they originate at a single node representing one ancestral group. A group that does not include its node is not

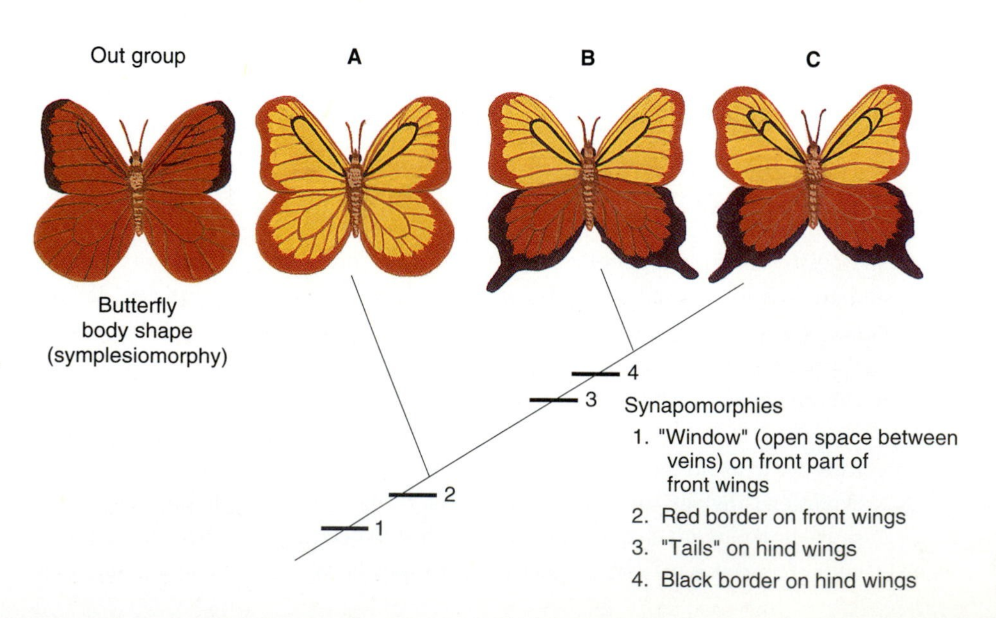

FIGURE 21.5

A simple, hypothetical cladogram. A, B, and C represent three different groups of insects being classified by comparison with an outgroup. The three groups A, B, and C share homologous characters 1 and 2 that are not found in the outgroup and that are therefore assumed to be derived (synapomorphies). Groups B and C also share synapomorphies 3 and 4 that are not found in A. B and C are therefore assumed to be more closely related to each other than either is to A. B and C, together with the node where their lines join, represent one clade. A, B, and C, together with the other node, represent a clade that includes the B-plus-C clade.

■ **Question:** How would the cladogram be affected by including a newly discovered species with "windows" and red borders on the front wings, and tails but not black borders on the hind wings? (Answers to this and other brainteasers in Unit 4 are at the end of each chapter.)

- Vertebrata
  - Agnatha
  - Gnathostomata
    - Chondrichthyes (sharks, rays)
    - Osteichthyes (bony fishes and their descendants)
      - Actinopterygii (ray-finned fishes)
      - Sarcopterygii
        - Dipnoi (lungfishes)
        - Choanata (tetrapods)
          - Batrachomorpha (amphibians)
          - Reptilomorpha (reptiles, including birds, and mammals)
            - Mammalia (mammals)
            - Sauropsida (reptiles including birds)
              - Chelonia (turtles)
              - Diapsida
                - Lepidosauria (snakes, lizards, tuataras)
                - Archosauria
                  - Crocodilia (crocodiles, alligators, etc.)
                  - Aves (birds)

FIGURE 21.6

A cladogram showing the major vertebrate clades and a corresponding indented list of clades. The names listed at the top are the groups that are not included in each succeeding clade. The level of the clade is shown by the amount of indentation in the list.

■ **Question:** Is this cladogram consistent with the dendrogram in Fig. 21.4?

a clade and is not taxonomically valid. Such a group is said to be **polyphyletic,** because its members appear to have descended from different recent ancestors. The most recent common ancestor of all birds must therefore be classified as a bird, even if it turns out to have been a featherless dinosaur incapable of flight. Phrases such as "birds evolved from dinosaurs" are gibberish to a cladist.

4. A group that includes the node but not all its branches is also invalid. It is said to be **paraphyletic.** Thus the traditional group "reptiles" is paraphyletic, and therefore nonexistent, unless one defines birds and mammals as reptiles.

Neither traditional evolutionary systematics nor cladistics is either right or wrong. One is simply better for showing similarities, and the other is better for showing evolutionary history. Zoologists must therefore be familiar with both methods.

Body Plans

Although there is no consensus on which method of systematics is better, it is safe to say that no method can be any better than the characters on which it is based. Unfortunately, at the phylum level of classification there are few good characters from which systematists can deduce phylogeny. Most of the present phyla appeared during the brief "Cambrian Explosion" without leaving a good fossil record (see pp. 387–389), and molecular comparisons have only begun to suggest phylogenetic relationships. The best characters available so far are the major features of body plans. A **body plan,** often referred to by the German term *Bauplan,* is the overall structural and functional organization of an animal. In contrast to individual organs, body plans are established early in embryonic development and appear to be conserved during evolution.

One of the major features of a body plan is its type of **symmetry,** if any. Some animals, including many sponges, are **asymmetric** (Fig. 21.7). That is, there is no way to cut one of these animals and get two similar halves. Other animals, such as sea anemones, are **radially or biradially symmetric.** Radially symmetric animals have approximately cylindrical bodies that can be equally divided like a cake by cutting through the central axis at any angle. Biradially symmetric animals are essentially cylindrical in shape, but with some paired features distributed on opposite sides of the body. A birthday cake with two candles is biradially symmetric. Animals that are radially or biradially symmetric tend to be sessile (attached to substratum). Most animals that move about actively are **bilaterally symmetric:** There is only one way in which

FIGURE 21.7

Types of symmetry in animals. The asymmetry of snails and the radial symmetry of echinoderms are derived, in the sense that these animals develop from larvae that are bilaterally symmetric. Biradially symmetric animals are radially symmetric in most features, but they have some paired structures on opposite sides of the body.

■ **Question:** Is any animal exactly symmetric? What about humans? (Consider the heart and liver.)

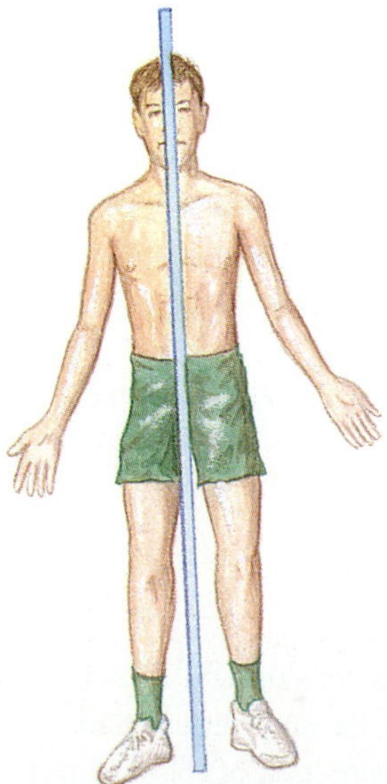

they can be cut down the midline into two approximately equal halves. When using type of symmetry in classification, one must be careful to determine whether the type is primitive or derived. For example, the asymmetry of adult snails is considered to be a derived trait, for two reasons. (1) Other molluscs, and presumably the ancestors of snails, are bilaterally symmetric; (2) snails develop from bilaterally symmetric larvae.

Bilaterally symmetric animals also differ in that their tissues develop from three **germ layers** (endoderm, mesoderm, and ectoderm; p. 117). They are therefore often referred to as being **triploblastic.** Sponges, cnidarians, and other animals that are primitively asymmetric, radially symmetric, or biradially symmetric, on the other hand, have traditionally been considered to have only two germ layers. They are therefore often referred to as **diploblastic.** In fact, however, the number of tissue layers in these animals is difficult to determine and may range from two to four. Regardless of the exact number, the germ layers in "diploblasts" are probably not homologous with germ layers in triploblasts. It is apparently valid to distinguish "diploblasts" from triploblasts, because they appear to have diverged very early (Christen et al. 1991).

Another useful feature is whether there is a fluid-filled, completely enclosed body cavity. Animals that are primitively asymmetric, radially symmetric, or biradially symmetric do not have such a body cavity. Nor do certain bilaterally symmetric animals, called **acoelomates** (ay-SEAL-oh-maytz). Some bilaterally symmetric animals have a body cavity called a **pseudocoel** (SOO-doe-seal; = pseudocoelom) and are called **pseudocoelomates.** In many cases the pseudocoel is a blastocoel that persists into adulthood and is often located between the endoderm and the mesoderm (p. 111; Fig. 21.8).

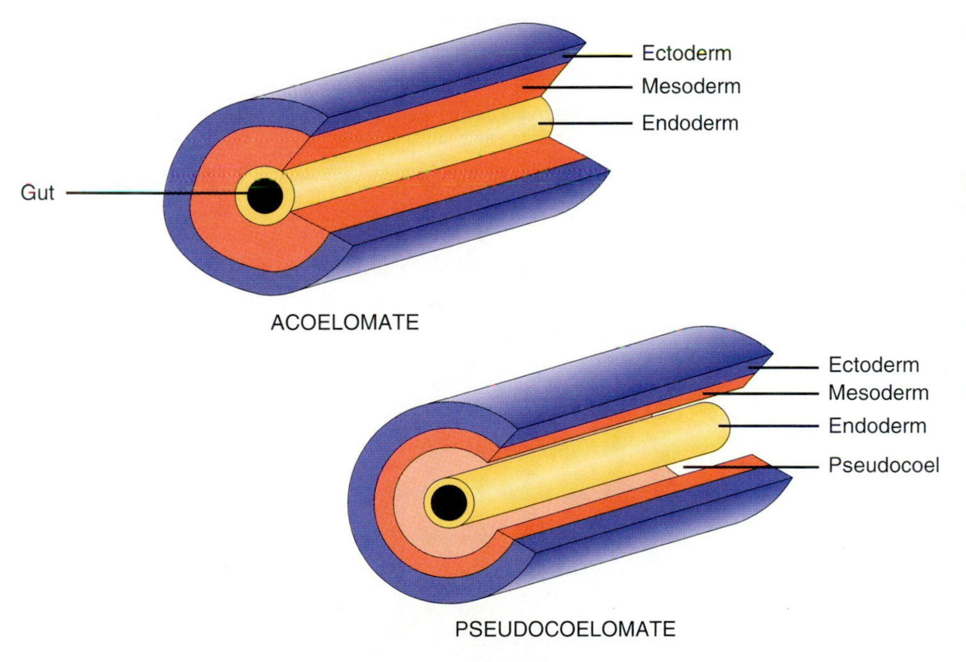

FIGURE 21.8

Schematic representation of the differences among acoelomates, pseudocoelomates, and coelomates.

■ **Question:** Can you think of more than one way in which a pseudocoelomate could have evolved from an acoelomate? If pseudocoels evolved in more than one way, would they be a good character for classification?

Other bilaterally symmetric animals, called **coelomates,** have a body cavity called a **coelom** (SEAL-um). A coelom is enclosed within mesoderm and is lined, at least in part, with a cellular **peritoneal membrane** (= peritoneum). The functions of these body cavities are discussed in Chapters 26 and 27.

Zoologists generally classify coelomates as either **protostomes** or **deuterostomes.** In animals in the division Protostomia the mouth (*stoma*) originates from the **blastopore,** which is the first (*proto*) opening to appear in the embryo (pp. 116–117). Arthropods and most other invertebrates are protostomes. In animals in the division Deuterostomia the mouth originates from a second (*deutero*) opening, and the blastopore usually becomes the anus. Echinoderms and chordates are the two major groups of deuterostomes. The validity of distinguishing between protostomes and deuterostomes is supported by molecular phylogenetic studies (Lake 1990; Telford and Holland 1993).

Protostomes and deuterostomes also frequently differ in the pattern of cell movement during cleavage and in the origin of the coelom. This **cleavage pattern** in most protostome groups is spiral (Fig. 21.9). (Protostomes are therefore often referred to as the **spiralia.**) Other cleavage patterns occur in some protostomes, however, especially in arthropods and some molluscs. The cleavage pattern is often assumed to be primitively radial in deuteros-

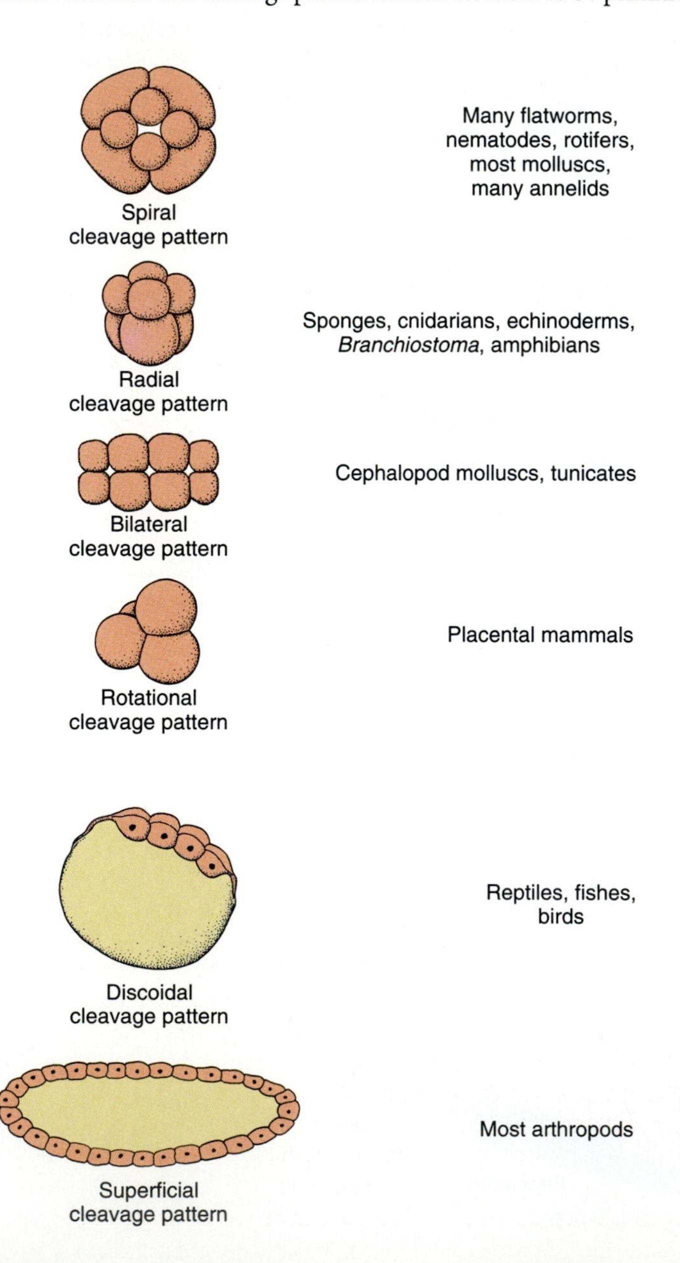

FIGURE 21.9

Patterns of cell movement during cleavage in various groups of animals. Spiral: new cells move into the furrows between older ones. Radial: dividing cells remain aligned. Bilateral: cells form two rows. Rotational: one pair of cells lies at right angles to the other in the four-cell stage. Discoidal: yolk pushes dividing cells into a disc at one side. Superficial: yolk pushes dividing cells to the surface.

tomes, because it is radial in echinoderms, in the invertebrate chordate *Branchiostoma,* and in amphibians. However, different cleavage patterns occur in fishes, reptiles, birds, and placental mammals. Most protostomes and some chordates are **schizocoelous,** meaning that the coelom results from a splitting of the mesoderm. Echinoderms and some chordates are **enterocoelous:** The coelom originates as a pouch formed by the mesoderm (pp. 112, 114).

The Animal Phyla

Between 30 and 35 groups of animals are generally considered to have body plans sufficiently distinct to justify classifying them into different phyla. In the following chapters I recognize 31 phyla (Table 21.1). Zoologists do not agree on how many phyla there

TABLE 21.1

The 31 phyla of animals recognized in this book.

Phylum	Common names
ASYMMETRIC, RADIALLY SYMMETRIC, OR BIRADIALLY SYMMETRIC	
Porifera (pore-IF-er-uh)	sponges
Placozoa (plak-oh-ZO-uh)	
Mesozoa (mess-oh-ZO-uh)	
Cnidaria (nigh-DARE-ee-uh)	hydra, jellyfish, sea anemones, corals, and others
Ctenophora (ten-OFF-or-uh)	comb jellies
BILATERALLY SYMMETRIC	
Acoelomates	
Platyhelminthes (plat-ee-hell-MINTH-eez)	flatworms
Gnathostomulida (NATH-oh-stom-YOU-lid-uh)	jaw worms
Gastrotricha (gas-tro-TRICK-uh)	
Pseudocoelomates	
Nematoda (NEE-ma-TOAD-uh)	round worms
Nematomorpha (nee-MAT-oh-MORE-fuh)	horsehair worms
Kinorhyncha (KIN-oh-RING-kuh)	
Loricifera (lore-i-SIFF-er-uh)	
Priapulida (PRE-ah-PEW-lid-uh)	
Rotifera (row-TIFF-er-uh)	rotifers
Acanthocephala (a-KANTH-oh-SEF-full-uh)	
Entoprocta (EN-toe-PROK-tuh)	
Coelomates	
Ectoprocta (ek-toe-PROK-tuh)	
Phoronida (for-OH-nid-uh)	
Brachiopoda (brack-ee-OP-uh-duh)	
Nemertea (NEM-ur-TEE-uh)	
Sipuncula (sy-PUNK-you-luh)	peanut worms
Echiura (ek-ee-YOUR-uh)	spoon worms
Tardigrada (tar-di-GRADE-uh)	water bears
Pogonophora (PA-gone-OFF-or-uh)	beard worms
Mollusca (ma-LUS-kuh)	snails, octopus, squid, shellfish, etc.
Annelida (an-EL-lid-uh)	segmented worms, leeches, etc.
Arthropoda (are-THROP-uh-duh)	insects, spiders, crustacea, etc.
Chaetognatha (key-TOG-nah-thuh)	arrowworms
Echinodermata (e-KINE-oh-DER-mah-tuh)	starfish, sea urchins, etc.
Hemichordata (HEM-i-core-DAY-tuh)	acorn worms. etc.
Chordata (core-DAY-tuh)	fishes, amphibians, reptiles, birds, mammals

are, or on how they are related to each other. One can find numerous cladograms and dendrograms that purport to show such relationships, but it is rare to find two that agree. Figure 21.10 is my attempt at showing relationships among phyla. It may be a good idea to memorize the positions of the major ones (those in boldface) as landmarks for navigating among the phyla. Figure 21.11 is a pictorial summary of the phyla, intended as a concise reference for the numbers of species, habitats, sizes, and physiological features. See also the inside cover.

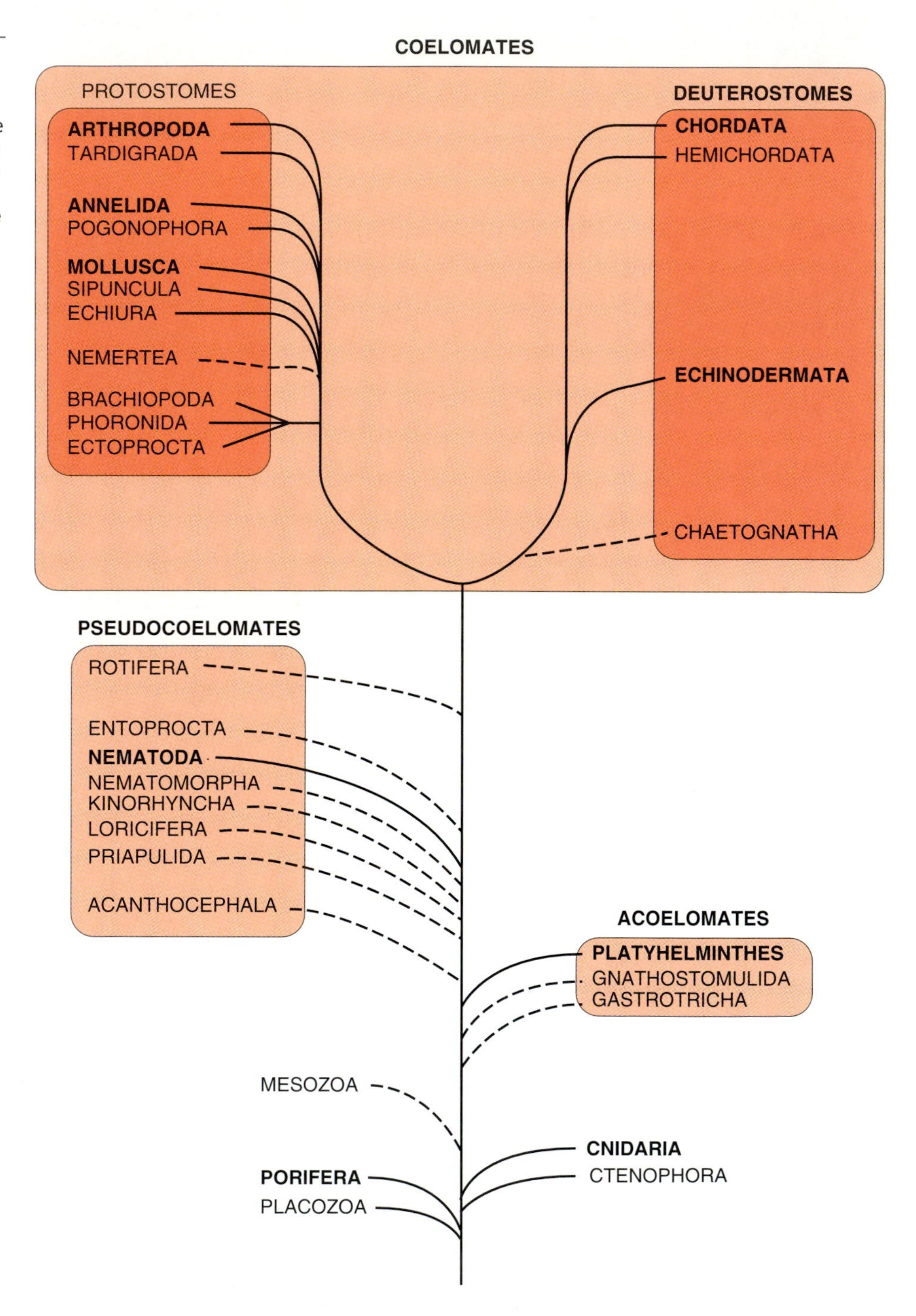

FIGURE 21.10

Possible phylogenetic relationships among 31 phyla of animals. Relationships shown by dashed lines are even more uncertain than others. Phyla with more than 5000 species are shown in bold type. See Eernisse et al. (1992) for cladograms representing alternative phylogenies from 12 other textbooks.

FIGURE 21.11

A summation of animal phyla. Parentheses indicate that a character occurs in relatively few members of the phylum.

KEY:

SIZE

< 1 mm   1 mm to 1 cm   > 1 cm

HABITAT

Terrestrial (T), Freshwater (F), Parasitic within other animals (P), Marine (M)

Dot indicates few species in a habitat

CIRCULATORY SYSTEM

Blank	No heart or vessels
	Open system (blood not restricted to heart and vessels)
	Closed system
	Closed, no heart

RESPIRATORY SYSTEM

Blank	Through integument only
G	Gills
T	Tracheal tubes
L	Lungs

OSMOREGULATORY SYSTEM

C	Contractile vacuoles (Porifera only)
P	Protonephridia (closed tubes)
M	Metanephridia (open tubes)
MT	Malpighian tubules (Arthropoda only)
K	Kidneys (Chordata only)

DIGESTIVE SYSTEM

Incomplete (no anus)

Complete

SKELETAL SUPPORT

Exoskeleton

Endoskeleton

Hydroskeleton

REPRODUCTION

Asexual budding or fission

Parthenogenesis

Hermaphroditic (monoecious)

Separate males and females (dioecious)

Phylum		Size	# Named species	Habitat	Physiology: Circulatory system	Respiratory system	Osmo-regulatory system	Digestive system	Skeletal support	Reproduction
PORIFERA	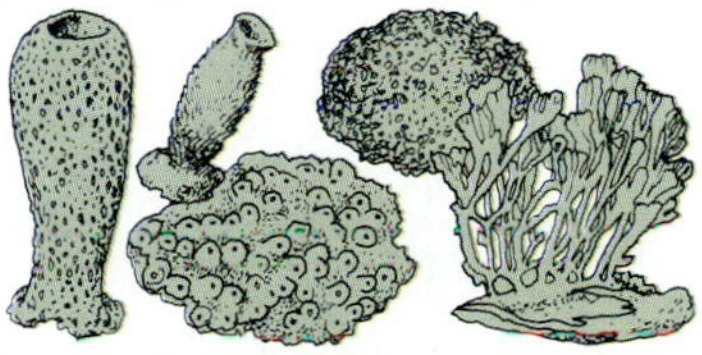	> 1 cm	5000	F (dot), M			C		Endoskeleton	Asexual budding or fission; Hermaphroditic; Separate males and females
PLACOZOA	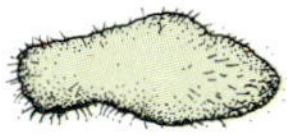	1 mm to 1 cm	1	M						Asexual budding or fission; Separate males and females
MESOZOA	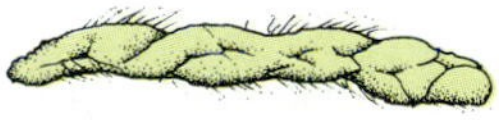	< 1 mm; 1 mm to 1 cm	100	P						Hermaphroditic; Asexual budding or fission; Separate males and females
CNIDARIA	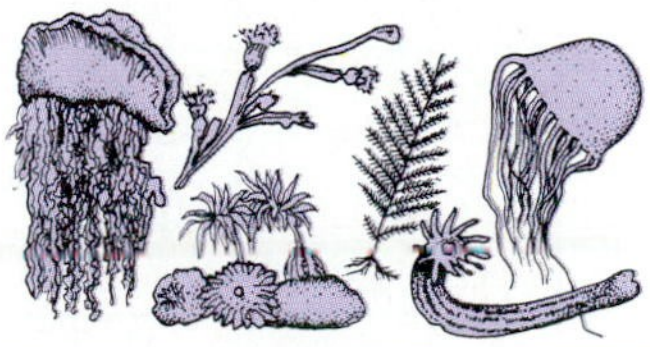	< 1 mm; 1 mm to 1 cm; > 1 cm	9000	F (dot), M				Incomplete	Endoskeleton; Hydroskeleton	Asexual budding or fission; Hermaphroditic; Separate males and females
CTENOPHORA	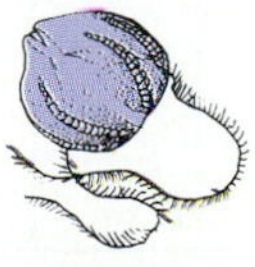	> 1 cm	100	M				Incomplete	Hydroskeleton	Hermaphroditic

ACOELOMATES										
					Physiology					
Phylum	Size	# Named species	Habitat		Circulatory system	Respiratory system	Osmo-regulatory system	Digestive system	Skeletal support	Reproduction
PLATYHELMINTHES		12,200	F P M				P	or None		
GNATHOSTOMULIDA		89	M				P	( )		
GASTROTRICHA		430	T F P M				None or P			
PSEUDOCOELOMATES										
NEMATODA		12,000	T F P M							
NEMATOMORPHA		320	F P M							
KINORHYNCHA		150	M				P			
LORICIFERA		10	M				P			
PRIAPULIDA		17	M				P			
ROTIFERA		1800	F				P			
ACANTHOCEPHALA		1100	P				None or P	None		
ENTOPROCTA		150	M				P			

COELOMATES – Protostomes										
					Physiology					
Phylum	Size	# Named species	Habitat		Circulatory system	Respiratory system	Osmo-regulatory system	Digestive system	Skeletal support	Reproduction
ECTOPROCTA		4500	M							
PHORONIDA		15	M				M			
BRACHIOPODA		335	M				M			
NEMERTEA		900	M				P			
SIPUNCULA		320	M				M			
ECHIURA		150	M				M			
TARDIGRADA		600	F M				MT?			
POGONOPHORA		140	M				M			
MOLLUSCA		110,000	T F M		or	G (L)	M			

COELOMATES – Protostomes										
				Physiology						
Phylum	Size	# Named species	Habitat	Circulatory system	Respiratory system	Osmo-regulatory system	Digestive system	Skeletal support	Reproduction
ANNELIDA 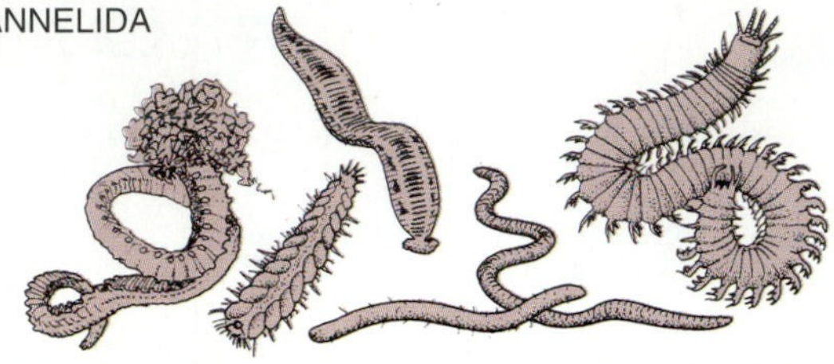		12,000	T F M			P or M			(○)
ARTHROPODA 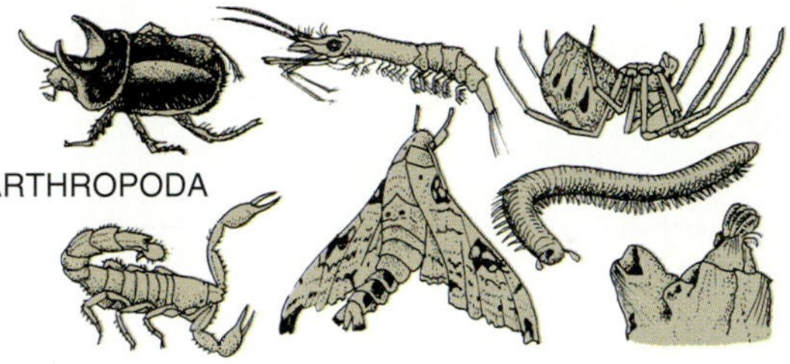		874,000	T F P M		T L G	MT and other			
Deuterostomes									
CHAETOGNATHA 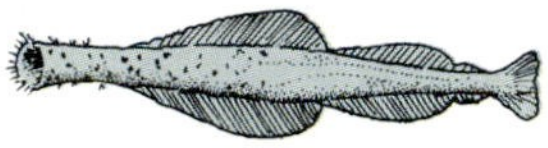		100	M						
ECHINODERMATA 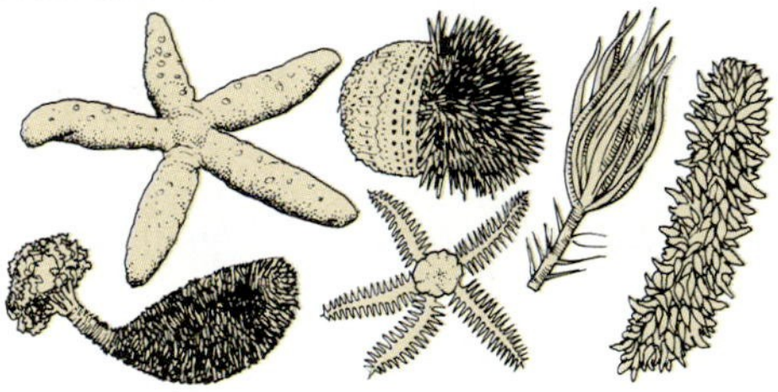		6100	M				or		
HEMICHORDATA 		85	M						
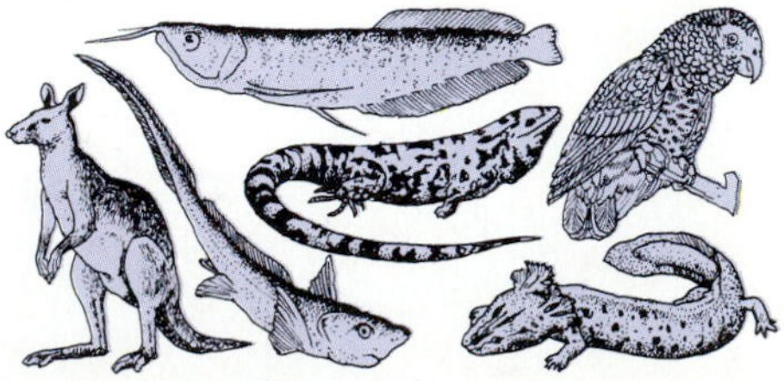		48,000	T F M		G L	K			

Summary

Systematics is the study of all diversity and relationships among organisms. It includes the field of nomenclature, which tries to establish universal, unique, and stable names for each species, using the binominal system in which each species is assigned a two-word name. When possible, a species may be defined as a group of interbreeding organisms that is reproductively isolated from other such groups.

Systematics also includes taxonomy, which establishes methods for classifying groups of organisms based on their phylogeny, so that one can generalize about them. Classification is based on anatomical, behavioral, and other characters that are assumed to have been inherited through evolutionary descent. Molecular phylogenetics, which compares molecules from different groups of organisms, has become an important source of characters in recent decades.

There are two major approaches to systematics: evolutionary systematics and cladistics. The first step in evolutionary systematics is character analysis, by which the systematist selects homologous characters, determines whether they are primitive or derived, estimates their degree of evolution, and determines how much weight to give them. The results are represented on dendrograms, which show the times of divergence of taxonomic groups, and in descriptions of the taxonomic categories (genus, family, order, etc.) to which the groups belongs. Cladistics, an increasingly important alternative to evolutionary systematics, uses only synapomorphies (derived homologies) to relate groups to each other in clades. The relationships among clades are shown in cladograms, which represent phylogenies but not the times or degree of evolutionary divergence.

Between 30 and 35 phyla of animals are now generally recognized. These phyla are often grouped according to similarities in body plans, such as whether they are asymmetric, radially symmetric, biradially symmetric, or bilaterally symmetric; whether they are triploblastic; and whether they are acoelomates, pseudocoelomates, or coelomates. Coelomates may also be distinguished by whether they are protostomes or deuterostomes, as well as by their cleavage patterns and the origin of their coeloms.

Key Terms

systematics
phylogeny
biological species concept
nomenclature
binominal system
taxon
kingdom
phylum
class
order
family
genus
molecular phylogenetics
character
evolutionary systematics
homologous
analogous
primitive
derived
cladistics
synapomorphy
clade
cladogram
sister group
body plan
asymmetric
radially symmetric
biradially symmetric
bilaterally symmetric
germ layer
triploblastic
acoelomate
pseudocoelomate
coelomate
protostome
deuterostome
cleavage pattern
schizocoelous
enterocoelous

Chapter Test

1. Most people conceive of animals as ranked linearly from lowest to highest. Explain why this is wrong. Suggest a more useful approach to classifying animals. (p. 432)
2. Define "character." What kinds of features of animals can be characters? What kinds of considerations enter into character analysis? (pp. 433–435)
3. What are the objectives of evolutionary systematics? (p. 434)
4. List the required taxonomic levels in descending sequence. What, if any, biological basis is there for each level? (p. 433)
5. Describe the main differences between evolutionary systematics and cladistics. What are the advantages and disadvantages of each? (pp. 434–438)
6. Briefly explain the differences among the following body plan characters (pp. 438–440):
   asymmetric/radially symmetric/biradially
   symmetric/bilaterally symmetric
   acoelomate/pseudocoelomate/coelomate
   protostome/deuterostome
7. For each character listed in the previous question, name an animal with that character. (pp. 438–440)

Answers to the Figure Questions

21.5 The new species, with synapomorphies 1, 2, and 3, but not 4, would go between species A and B.

21.6 The sequence of branching in the cladogram is not inconsistent with that in the dendrogram.

21.7 No animal is exactly symmetrical. In humans the heart is to the left, and the liver is on the right. Even the face is somewhat asymmetric, which is why your image in a photograph is different from your image in a mirror.

21.8 A comparison of the diagrams for the acoelomate and pseudocoelomate body plans suggests that the pseudocoel evolved by separation of mesoderm from endoderm. It is also possible, however, that in some pseudocoelomates the mesoderm could have separated from ectoderm. It is also possible that the pseudocoel originated as a cavity within mesoderm, and that one wall of the mesoderm was subseqently lost. Pseodocoels that originated in these different ways would not be homologous with each other, and the pseudocoel would therefore not be a good character.

Readings

Recommended Readings

Blunt, W. 1971. *The Compleat Naturalist: A Life of Linnaeus.* New York: Viking.

Duellman, W. E. 1985. Systematic zoology: slicing the Gordian Knot with Ockham's Razor. *Am. Zool.* 25:751–762.

Goto, H. E. 1982. *Animal Taxonomy.* London: Edward Arnold.

Gould, S. J. 1980. The telltale wishbone. In: *The Panda's Thumb.* New York: W. W. Norton, pp. 267–277. (*Some implications of cladistics.*)

Gould, S. J. 1983. What, if anything, is a zebra? In: *Hen's Teeth and Horse's Toes.* New York: W. W. Norton. pp. 355–365. (*Further implications of cladistics.*)

Jeffrey, C. 1973. *Biological Nomenclature.* London: Edward Arnold.

May, R. M. 1992. How many species inhabit the earth? *Sci. Am.* 267(4):42–48 (Oct).

Stafleu, F. A. 1971. *Linnaeus and the Linnaeans: The Spreading of Their Ideas in Systematic Botany, 1735–1789.* Utrecht: A. Oosthoek's Uitgeversmaatschappij.

Additional References

Christen, R., et al. 1991. An analysis of the origin of metazoans, using comparisons of partial sequences of the 28S RNA, reveals an early emergence of triploblasts. *EMBO J.* 10:499–503.

Eernisse, D. J., J. S. Albert, and F. E. Anderson. 1992. Annelida and Arthropoda are not sister taxa: A phylogenetic analysis of spiralian metazoan morphology. *Syst. Biol.* 41:305–330.

Eldredge, N., and J. Cracraft. 1980. *Phylogenetic Patterns and the Evolutionary Process.* New York: Columbia University Press.

Fitch, W. M., and W. R. Atchley. 1987. Divergence in inbred strains of mice: a comparison of three different types of data. In: C. Patterson (Ed.). *Molecules and Morphology in Evolution: Conflict or Compromise?* New York: Cambridge University Press, pp. 203–216.

Forey, P. L., et al. 1992. *Cladistics: A Practical Course in Systematics.* New York: Oxford University Press.

Grant, P. R., and B. R. Grant. 1992. Hybridization of bird species. *Science* 256:193–197.

Hillis, D. M., J. P. Huelsenbeck, and C. W. Cunningham. 1994. Application and accuracy of molecular phylogenies. *Science* 264:671–677.

Lake, J. A. 1990. Origin of the Metazoa. *Proc. Natl. Acad. Sci. USA* 87:763–766.

Minelli, A. 1993. *Biological Systematics: The State of the Art.* New York: Chapman and Hall.

Ride, W. D. L., et al. (Eds.). 1985. *International Code of Zoological Nomenclature,* 3rd ed. Berkeley: University of California Press.

Telford, M. J., and P. W. H. Holland. 1993. The phylogenetic affinities of the chaetognaths: a molecular analysis. *Mol. Biol. Evol.* 10:660–676.

Willmer, P. G. 1990. *Invertebrate Relationships: Patterns in Animal Evolution.* New York: Cambridge University Press.

Wilson, E. O. 1988. The current state of biological diversity. In: E. O. Wilson (Ed.). *BioDiversity.* Washington, DC: National Academy Press, Chapter 1.

General References

Alexander, R. M. 1990. *Animals.* New York: Cambridge University Press.

Allaby, M. (Ed.). 1992. *The Oxford Dictionary of Zoology.* New York: Oxford University Press.

Ayers, D. M. 1972. *Bioscientific Terminology: Words from Latin and Greek Stems.* Tucson: University of Arizona Press.

Barnes, R. S. K., P. Calow, and P. J. W. Olive. 1993. *The Invertebrates: A New Synthesis,* 2nd ed. Palo Alto, CA: Blackwell Scientific Publications.

Borror, D. J. 1960. *Dictionary of Word Roots and Combining Forms.* Palo Alto, CA: Mayfield.

Brusca, R. C., and G. J. Brusca. 1990. *Invertebrates.* Sunderland, MA: Sinauer Assoc.

Buchsbaum, R. 1987. *Animals Without Backbones,* 3rd ed. Chicago: University of Chicago Press.

Dorit, R., W. F. Walker, Jr., and R. D. Barnes. 1991. *Zoology.* Philadelphia: Saunders College Publishing.

Grzimek, B. (Ed.). 1968. *Grzimek's Animal Life Encyclopedia* (13 volumes). New York: Van Nostrand Reinhold. (*Available in relatively inexpensive paperback edition. Somewhat dated.*)

Harrison, F. W., et al. (Eds.). 1991–. *Microscopic Anatomy of Invertebrates* (15 volumes). New York: Wiley–Liss.

Hickman, C. P., Jr., L. S. Roberts, and A. Larson. 1993. *Integrated Principles of Zoology,* 9th ed. St. Louis, MO: Mosby.

Hyman, L. H. 1940–1967. *The Invertebrates.* (6 vols.) New York: McGraw–Hill. (*This monumental work by Libbie Hyman (1888–1969) is the starting point for invertebrate zoology.*)

Laverack, M. S., and J. Dando. 1987. *Lecture Notes on Invertebrate Zoology,* 3rd ed. Palo Alto, CA: Blackwell Scientific Publications.

Meglitsch, P. A., and F. R. Schram. 1991. *Invertebrate Zoology,* 3rd ed. New York: Oxford University Press.

Margulis, L., and K. V. Schwartz. 1987. *Five Kingdoms: An Illustrated Guide to the Phyla of Life on Earth,* 2nd ed. San Francisco: W. H. Freeman.

Miller, S. A., and J. P. Harley. 1996. *Zoology,* 3rd ed. Dubuque, IA: Wm. C. Brown.

Orr, R. T. 1982. *Vertebrate Biology,* 5th ed. Philadelphia: Saunders.

Parker, S. P. (Ed.). 1982. *Synopsis and Classification of Living Organisms.* New York: McGraw–Hill.

Pearse, V., J. Pearse, R. Buchsbaum, and M. Buchsbaum. 1986. *Living Invertebrates.* Palo Alto, CA: Blackwell Scientific Publications.

Pennak, R. W. 1989. *Fresh-Water Invertebrates of the United States,* 3rd ed. New York: Wiley.

Pierce, S. K., T. K. Maugel, and L. Reid. 1987. *Illustrated Invertebrate Anatomy: A Laboratory Guide.* New York: Oxford University Press.

Ruppert, E. E., and R. D. Barnes. 1994. *Invertebrate Biology,* 6th ed. Philadelphia: Saunders College Publishing.

Stachowitsch, M. 1992. *The Invertebrates: An Illustrated Glossary.* New York: Wiley–Liss.

Thorp, J. H., and A. P. Covich (Eds.). 1991. *Ecology and Classification of North American Freshwater Invertebrates.* New York: Academic Press.

CHAPTER

22

Prelude to Animals: Protozoa

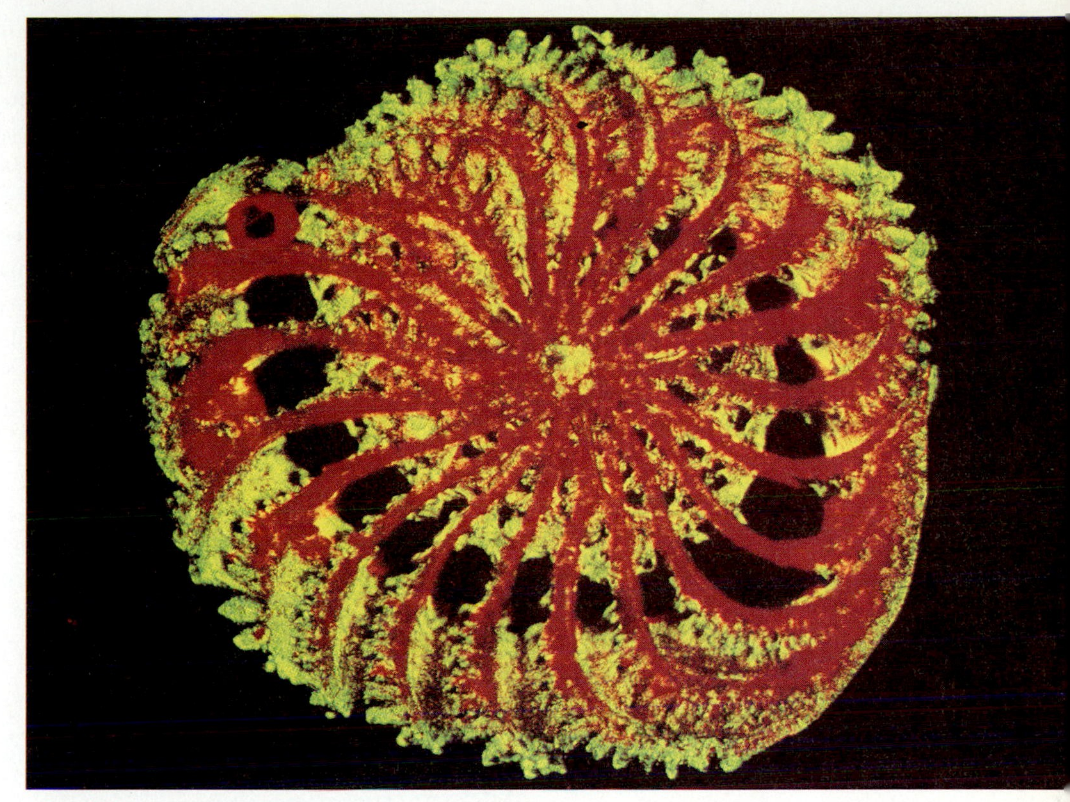

A foraminiferan, *Eliphidium crispum.*

There is a little-known planet where millions of bizarre organisms pursue an ancient struggle for survival. Formless beasts flow like jelly, surrounding and engulfing their prey. They in turn are sucked by whirling fans into gaping jaws. Others consume their victim's blood from within. Still others drift placidly in glassy shells, sparkling in the sunlight. The planet is our own—or a part of it as seen through a microscope. The organisms are protozoans (Fig. 22.1).

When Antony van Leeuwenhoek first observed protozoa in 1674 he called them "animalcules"—little animals—because all organisms were supposed to be animals if they moved, and plants if they didn't. Protozoa are no longer considered to be animals, because protozoa are generally single-celled, while animals are now defined as being multicellular (metazoa). Nevertheless, the proper study of zoology still includes protozoa. Protozoans make excellent models for the study of many phenomena that occur

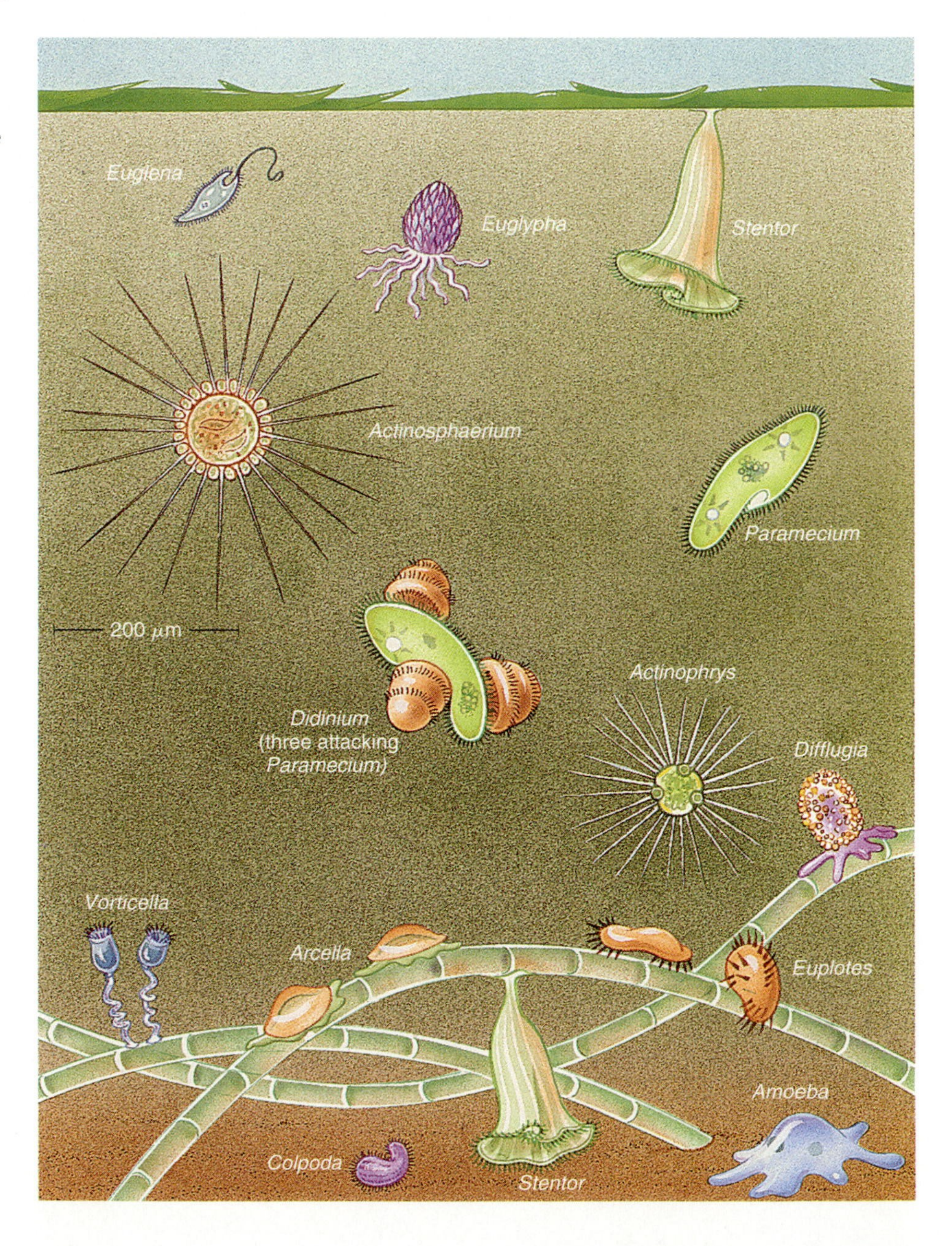

FIGURE 22.1

Common protozoa of a freshwater pond. Many will be discussed in this chapter. The green strands are algae. All are drawn to the same scale. The length of *Paramedium* is about half the diameter of the period at the end of this sentence.

in animals, and many of them have an enormous impact on animals, including ourselves.

The Unicellular Way of Life

The impressive thing about a protozoan, and the thing that is hardest to believe while watching one under a microscope, is that only one cell performs all the tasks essential to its life. Some tasks that are essential for animals, however, are not necessary in a unicellular organism. Protozoa do not need neural and hormonal systems, since there is seldom a need for intercellular coordination. Respiratory, excretory, and circulatory systems are also unnecessary, because the small size of the protozoan creates an enormous ratio of surface to volume, which allows oxygen, carbon dioxide, and wastes to diffuse readily across the plasma membrane. About all that remains for a protozoan is to move to a favorable habitat, to feed, and to reproduce (Table 22.1).

Many protozoans do not actively move. Some, such as *Vorticella,* are **sessile** (attached to substratum), and others, such as *Actinosphaerium,* drift with water currents (Fig. 22.1). Parasitic forms are transported within their hosts. Many other protozoans, however, move actively. *Amoeba* and many others move over surfaces by forming flexible extensions called **pseudopodia,** then flowing into them (p. 210). Many other protozoans, including *Paramecium,* move by means of **cilia,** and others, such as *Euglena,*

TABLE 22.1

Characteristics of protozoa

Protozoa PRO-toe-ZO-uh (Greek *proto* first + *zoon* animal)	
Morphology	**Unicellular eukaryotes**. Cells of colonial forms show little, if any, differentiation. Various symmetries: none, radial, bilateral, spherical. Either one nucleus, many similar nuclei, or nuclei of two different sizes. Usually lacking both endo- and exoskeletons; some species secrete protective shells.
Physiology	No organs; organelles carry out most functions. Nutrition of various types: ingesting other cells, absorbing organic material, or photosynthesis.
Locomotion	**Eukaryotic flagella**, **cilia**, **pseudopodia**, or cytoplasmic streaming without pseudopodia. Some forms do not actively move.
Reproduction and Development	Reproduction either **asexual** by **fission**, or **sexual** by fusion of nuclei within an individual, or between two individuals or gametes. No embryonic development.
Habitat, Size, and Diversity	Free-living in fresh water, seawater, or moist soil; or parasitic within other organisms. Mostly **microscopic** (5 to 250 μm), but some flagellates as small as 1 μm, and some foraminiferans up to 6.5 cm. Approximately **30,800 living species** described, with an average of more than one new species per day being identified. Approximately the same number of extinct species known from fossilized shells.

use **flagella** (pp. 210–211). Cilia and flagella are often embedded in a protein coat called the **pellicle,** which provides a semirigid structure that transmits the force of the cilia or flagella to the entire body of the protozoan.

Protozoans in the mature, feeding stage of life (**trophozoites**) obtain food in a variety of ways. Many consume organic detritus, bacteria, algae, other protozoa, and even small animals such as rotifers. Others live within the bodies of animals as endosymbionts, either absorbing nutrients from body fluids or intercepting nutrients in the digestive tract. A few protozoans, such as *Euglena,* produce their own nutrients through photosynthesis.

Protozoa reproduce by almost every method conceivable. The usual method is **binary fission,** with one protozoan dividing into two asexually. In some protozoans, such as *Paramecium,* reproduction involves the combining of nuclei when two individuals merge temporarily (**conjugation**). Conjugation is said to be a sexual process, because it results in recombination of genetic material. In contrast to sexual reproduction in animals, however, it does not involve gametes and does not produce a completely new individual. Some protozoans do form a new individual from the fusion (**syngamy**) of gametes. These protozoan sexual practices approach those of animals and suggest possible solutions to the puzzle of how and why sex evolved. Unlike animals, however, not even these protozoans develop from embryos, and there is no visible difference between the sexes. The life cycles of many species involve a stage called a **cyst** or **spore** that is capable of surviving environmental stress. Cyst production is especially common in the parasitic species, which must survive the harsh environment when passing from one host to another.

A Bewildering Diversity

Protozoans are usually classified in the kingdom Protista (called Protoctista by some), which includes all eukaryotes that are not fungi, plants, or animals. As might be expected from such a catch-all definition, the kingdom is not a monophyletic group (clade). In general, the members of kingdom Protista are considered to be algae if they photosynthesize and protozoans if they do not. This distinction is not strictly valid, however, because some protists photosynthesize even though quite similar species do not. *Euglena,* for example, photosynthesizes, but its close relatives do not. Botanists therefore consider *Euglena* and its relatives to be algae, and zoologists count them as protozoans (Fig. 22.2).

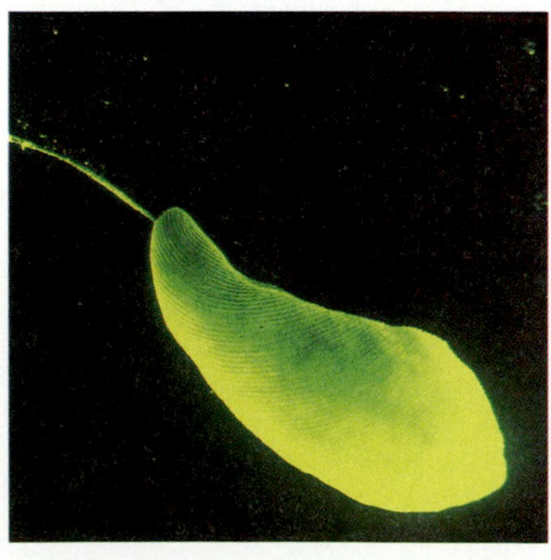

A

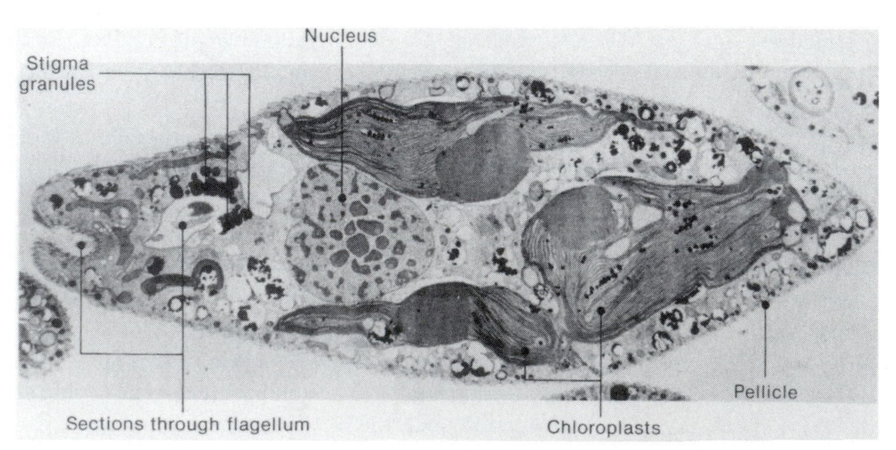

B

FIGURE 22.2

(A) Scanning electron micrograph (SEM) of *Euglena.* Beating by the flagellum pulls the organism through water. Note the outer covering, called the pellicle, which consists of spiral strands of proteins. The pellicle is flexible enough to permit *Euglena* to change shape and squirm through tight places (euglenoid movement). Length approximately 60 μm. (The color in this and other SEMs in this chapter is added.) (B) Transmission electron micrograph of *Euglena.* The chloroplast enables *Euglena* to photosynthesize and gives it a green color. The stigma (eyespot) directs its movements toward light. The stigma is not a photoreceptor but is instead a pigment that controls the amount of light hitting the base of the flagellum, and it therefore controls the direction of swimming.

The protists usually considered to be protozoans are informally divided into four major groups, depending on how or whether they move. **Flagellates** use flagella in locomotion, **sarcodines** form pseudopodia, and **ciliates** use cilia. **Sporozoans** generally do not actively move, and many form spores (= cysts). This way of classifying protozoans is unnatural, but it is still convenient as an informal way of referring to broad categories. Protozoologists have not arrived at a consensus on a formal classification, because protozoans are so diverse and discoveries about them are being made so fast. Research in molecular phylogenetics, for example, has suggested the following surprising conclusions:

1. Most flagellates are not closely related to other protozoans (Baroin et al. 1988, Sogin et al. 1986). Several groups of flagellates should be in other phyla, if not in other kingdoms. For example, dinoflagellates are more closely related to ciliates and certain sporozoans (apicomplexans) than to other flagellates, and *Giardia* is as closely related to bacteria as it is to flagellates (Sogin et al. 1989, Walters 1991). One group of flagellates, the choanoflagellates, are more closely related to animals than to other protozoa (Wainright et al. 1993).
2. Ciliates are genetically as diverse as all animals combined (Sogin and Elwood 1986).
3. Cellular slime molds, which are often considered to be sarcodines, are not closely related to sarcodines or to other protozoans (Qu et al. 1988, Field et al. 1988).
4. Myxozoa, which are usually classified as a phylum of sporozoans, may be animals more closely related to nematodes (Smothers et al. 1994).

Because of these difficulties, no formal classification is presented here. Instead, some of the major groups that are of most interest to zoologists are briefly described in the box on pp. 454–455.

Moving and Feeding

THE TROUBLE WITH BEING SMALL. We can scarcely imagine the problems protozoans have in moving, because the forces they encounter are so different from those we face. Like other large animals, we must contend mainly against gravity and inertia. Even the viscosity of water (its resistance to flow) does not keep us from diving and swimming. The situation is quite different for protozoans. Because of their buoyancy and small mass, they are not affected by gravity and inertia but are greatly affected by the viscosity of water. As a result, protozoans do not so much swim as crawl through water. Perhaps the only way for us to really understand protozoans would be to dive into a pool of molasses. Considering the problem of viscosity, protozoans move surprisingly rapidly, especially when you are trying to keep one in view under a microscope. Paramecia have been clocked at 1 millimeter per second, which is about 10,000 body-lengths per hour. For us that would be the equivalent of swimming more than 10 miles an hour.

CILIA. Paramecia and many other protozoans move by means of cilia (plural of **cilium**), which are short, numerous, hair-shaped organelles. Often the cilia are arranged in rows, called **kineties.** Like the cilia of vertebrates, which propel mucus along the bronchial tubes and ova through the oviducts, the cilia of protozoans generate a force parallel to the membrane surface. The beating of cilia all over the plasma membrane is

Some Groups of Protozoans Important to Zoologists

Each group is considered by most protozoologists to be either a phylum or class. Genera mentioned elsewhere in this chapter are noted.

Flagellates

Zoomastigina ZO-oh-mass-ti-JY-nuh (Greek *zoon* animal + *mastix* whip). At least one flagellum in mature organisms. One type of nucleus. Asexual reproduction by binary fission; rare syngamy (fusion of haploid gametes). Many parasitic. *Codosiga, Giardia, Leishmania, Opalina, Trichomonas, Trichonympha, Trypanosoma* (Figs. 22.5, 22.15).

Euglenida yew-GLEN-id-uh (Greek *eu-* true + *glini* eye). Either with chloroplasts or similar to species with chloroplasts. Enclosed by a pellicle. *Euglena* (Figs. 22.1, 22.2).

Dinoflagellata DY-no-flaj-el-LAH-tuh (Greek *dinos* a whirl + Latin *flagellum* small whip). With two flagella. Photosynthesizing or not. Mostly planktonic, but some endosymbionts. *Gambierdiscus, Protogonyaulax, Symbiodinium* (Fig. 22.13).

Sarcodines

Rhizopoda rye-ZOP-ah-duh (Greek *rhiza* root + *podos* foot). Pseudopodia lobe-shaped (lobopodia) or thread-shaped (filopodia). Amebas. *Amoeba, Arcella, Difflugia, Entamoeba, Euglypha* (Figs. 10.16, 14.4, 22.1, 22.6, 22.7).

Granuloreticulosea GRAN-you-low-re-TICK-you-LOW-see-uh (Latin *granulum* little grain + *reticulum* small net). With reticulopodia: thin pseudopodia, usually interconnecting as a net. In foraminiferans (Latin *foramen* opening + *ferre* to carry) reticulopodia extend through openings in an external skeleton. Sexually reproducing generation alternates with asexual generation. *Elphidium, Globigerina* (Chapter opener; Figs. 22.7, 22.12).

Actinopoda ACT-in-OP-ah-duh (Greek *actinos* rays). With axopodia: pseudopodia reinforced by microtubules. Usually spherical and planktonic, often with external skeleton. Marine forms are called radiolarians. *Actinophrys, Actinosphaerium* (Figs. 22.1, 22.7).

Sporozoans

Apicomplexa A-pi-kom-PLEX-uh (Latin *apex* + *complex* combination). With submicroscopical apical complex of variously shaped organelles

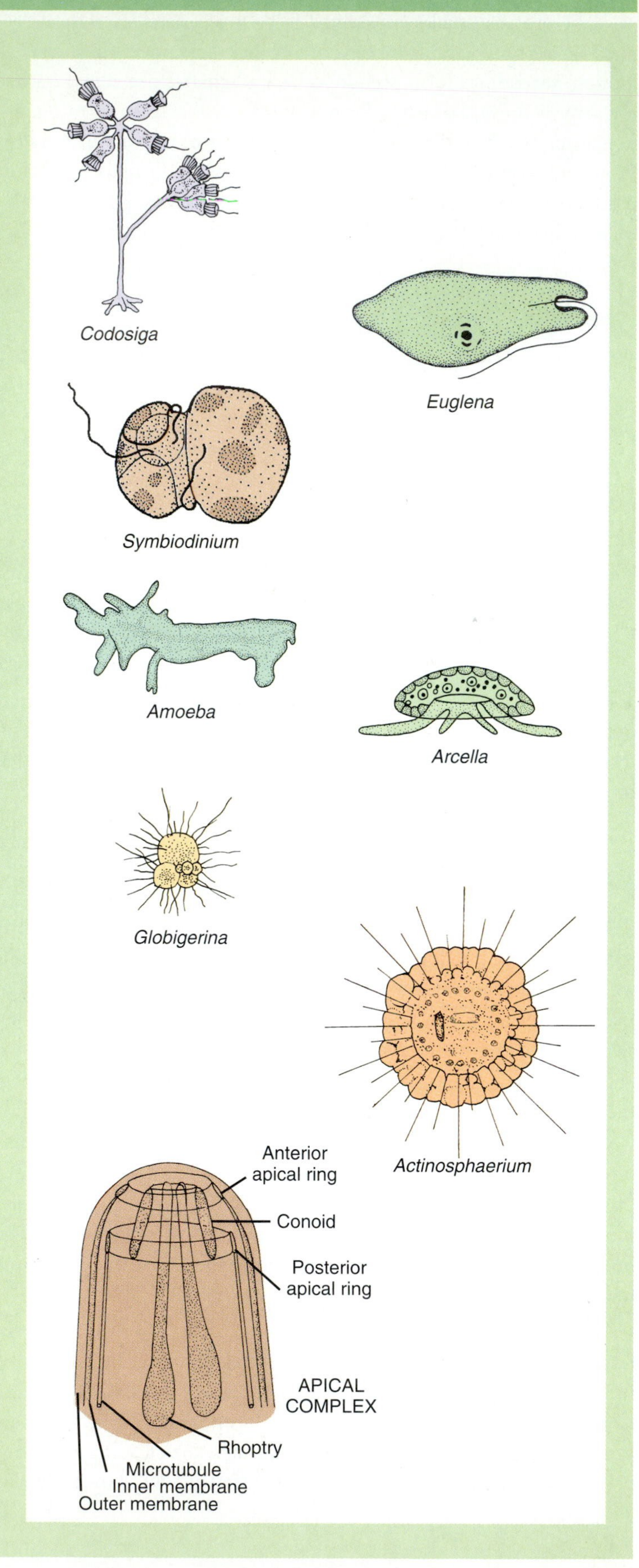

that enhance entry into host cells and may protect it from expulsion by the cell. No cilia. Sexual reproduction in some species. All parasitic. *Plasmodium, Toxoplasma.*

Microspora my-CROSS-spore-uh (Greek *micros* small + *sporos* seed). Spore-forming parasites inside cells in nearly all major animal groups. Spore has coiled filament through which cytoplasm enters host cell. No mitochondria.

Myxozoa mix-oh-ZO-uh (Greek *myx* slime). Parasites, mainly of fishes and annelid worms, that form multicellular spores containing structures resembling the nematocysts of phylum Cnidaria. The longstanding suspicion that they are degenerate animals is supported by molecular comparisons (Smothers et al. 1994).

Ciliates

Ciliophora sill-ee-OFF-or-uh (Latin *cilium* eyelash). Cilia, or at least the membrane structures of cilia, present in at least one stage. Two types of nucleus in almost all species. Sexual reproduction of several types. Contractile vacuole usually present. Most species free-living. Classification within this phylum is controversial. *Colpoda, Didinium, Euplotes, Heliophrya, Paramecium, Stentor, Stylonychia, Tetrahymena, Tokophrya, Vorticella* (Figs. 22.1, 22.3, 22.4).

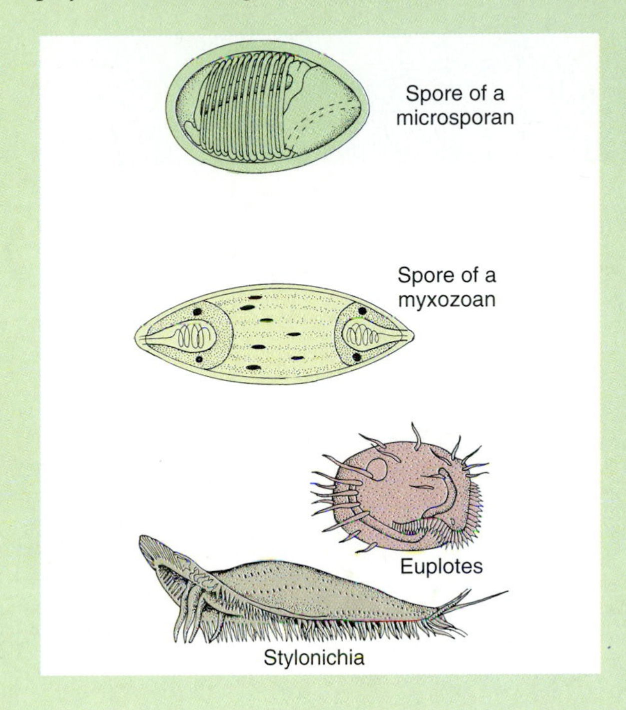

somehow coordinated, as can be seen from the synchronized waves in Figs. 10.17B and 22.3.

In addition to propelling protozoans, cilia also bring food into their **cytostomes** (literally the "cell mouths"). In many ciliates, such as *Paramecium, Vorticella,* and *Stentor,* cilia continually draw in water from which food is filtered into the cytostome. Once ingested, the food enters a food vacuole that fuses with organelles called **lysosomes,** which contain digestive enzymes. Nutrients released by digestion are then transported across the vacuole membrane into the cytoplasm. In some species the cilia combine near the cytostome as specialized buccal ciliature. Buccal ciliature may be in the form of long, fin-shaped **undulating membranes** or smaller, plate-like **membranelles.** Cilia can also fuse into **cirri,** which some ciliates use for "walking" (Fig. 22.4 on page 457).

Some ciliates do not have cilia in the mature stage. They are sedentary and obviously feed by some mechanism other than cilia. Suctorians, for example, lie in wait for other protozoans to pass, then snatch them by means of tentacles (Fig. 22.4). The tentacles usually paralyze prey within seconds, then begin drawing in their cytoplasm. Feeding by suctorians probably does not involve suction, in spite of their name. *Paramecium* often escapes suctorians, either by twisting free or by discharging numerous hair-shaped **trichocysts** from the pellicle. It is usually assumed that the trichocysts of ciliates work like arrows, but numerous other functions have been suggested. Spoon and his colleagues (1976) have observed that the discharge of trichocysts loosens the cilia, perhaps making predators lose their grip. This would explain why trichocysts are ineffective against *Didinium,* which grabs *Paramecium* bodily and swallows them whole (Fig. 22.1).

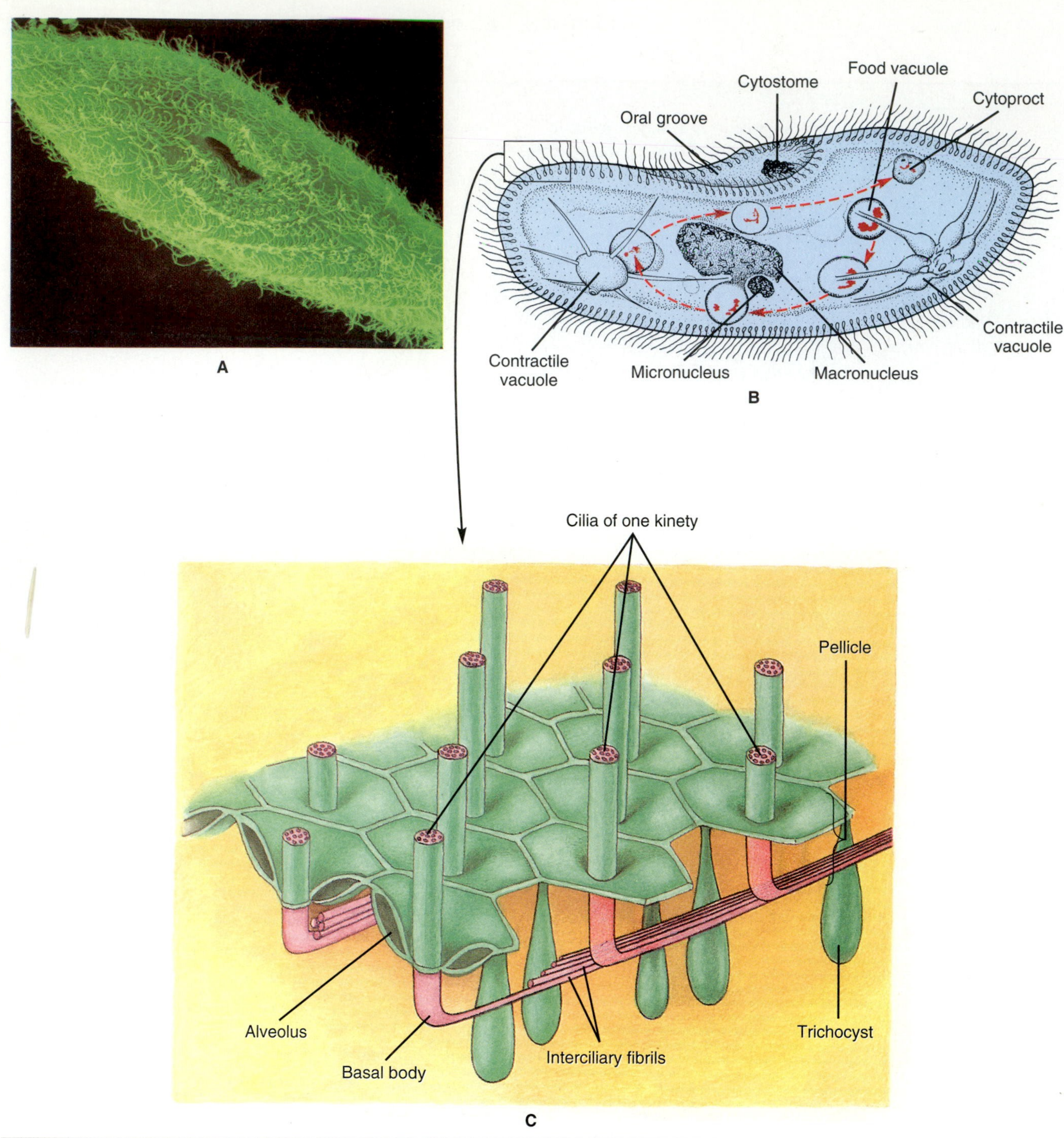

FIGURE 22.3

(A) SEM of *Paramecium,* the most familiar ciliate. (Approximate length 200 μm.) Approximately 10,000 cilia are evenly distributed in rows over the plasma membrane. The cilia are coordinated, as is apparent from the metachronal waves. (See also Fig. 10.17B.) The opening is the cytostome (cell mouth). (B) Schematic diagram of the organization of *Paramecium.* Cilia bring food into the cytostome, where it enters a food vacuole for digestion. (See Fig. 22.8B.) While food is being digested in the vacuole it migrates in a loop (shown by arrows) to the cytoproct where wastes are eliminated. The star-shaped organelles are contractile vacuoles (water expulsion vesicles) that eliminate water, an essential task in this freshwater genus (pp. 278–279). (C) The pellicle of *Paramecium* consists of double-walled chambers (alveoli) formed by plasma membrane and arranged in rows called kineties. Each alveolus is penetrated by a cilium, which originates from a basal body (= kinetosome). Kinetosomes in each kinety are linked by interciliary fibrils (= kinetodesmata). These fibrils have been called "nervelike," but it is unlikely that they conduct action potentials. Trichocysts, which are not part of the pellicle, contain filaments that are ejected when *Paramecium* is attacked by a predator.

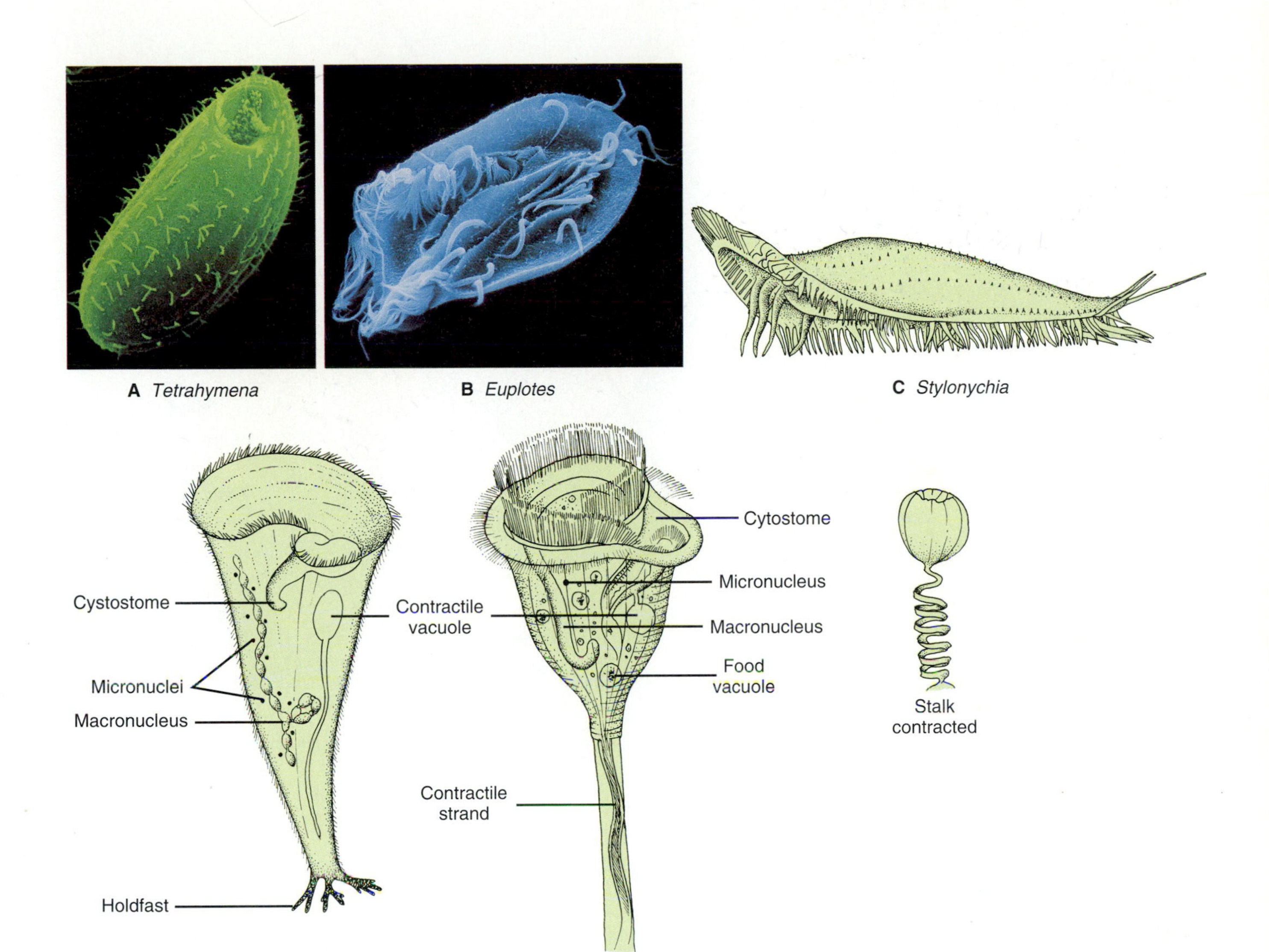

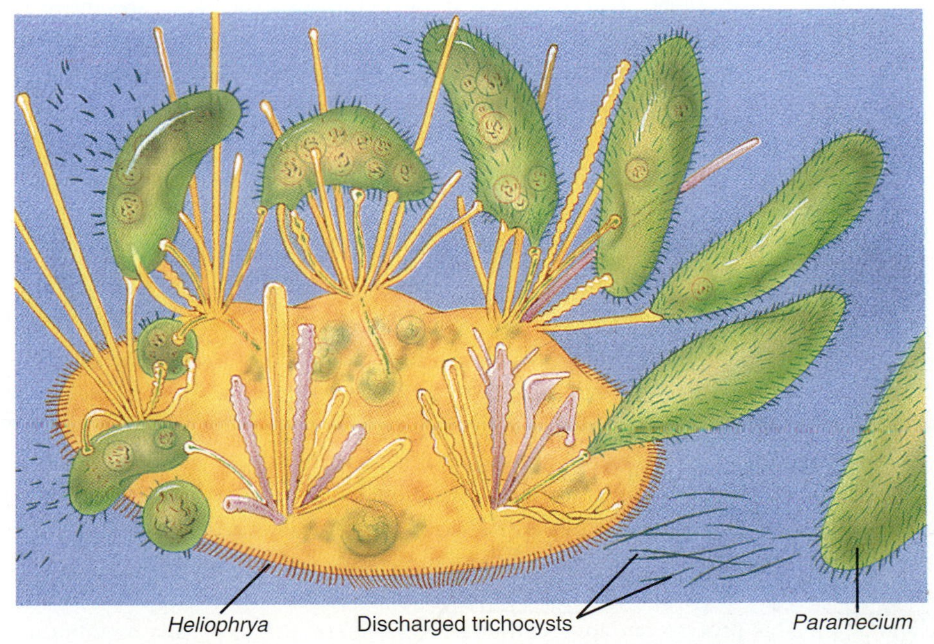

FIGURE 22.4

Varieties of ciliates. (A) In *Tetrahymena* most of the cilia are arranged in slanted rows, as in *Paramecium.* Near the mouth, however, cilia fuse into an undulating membrane. (B) In *Euplotes* the cilia are fused into tufts called cirri, which are used for "walking." Other cilia fuse into membranelles used in feeding. (C) *Stylonychia* also "walks" on cirri. (D) *Stentor,* a sessile protozoan anchored by its holdfast, uses its cilia exclusively to draw food-bearing water into the cytostome. Note the unusual beads-on-a-string structure of the macronucleus. (E) *Vorticella* is another sessile protozoan that uses its cilia to produce a feeding current. Periodically it contracts its stalk and appears to gulp. (F) The suctorian *Heliophrya* feeding on *Paramecium.* Various stages of feeding are shown, advancing counterclockwise from the lower right. Note the progressive loss of coordination of the prey's cilia. In the lower far right a *Paramecium* escapes after discharging trichocysts, which may loosen the cilia. (See Spoon et al. 1976.)

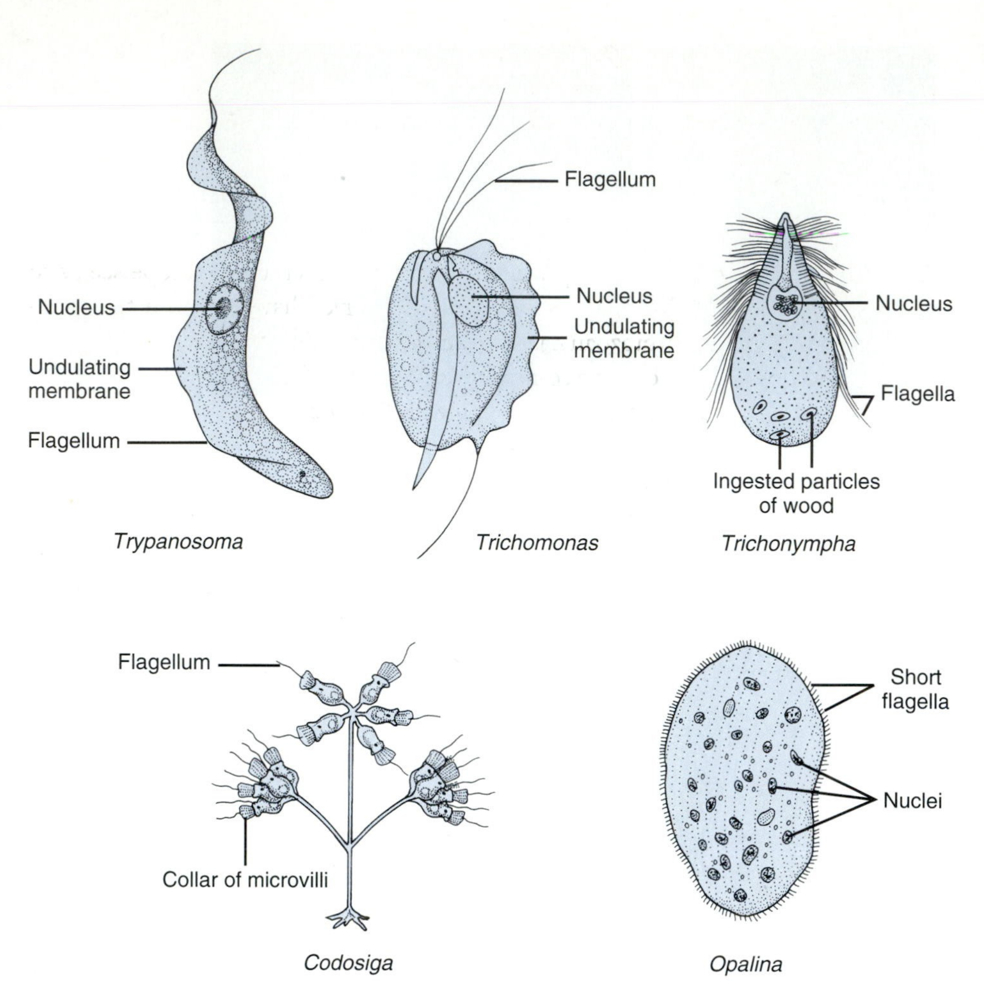

FIGURE 22.5

Some representative flagellates. *Trypanosoma,* with its single flagellum, is a parasite of animals, including humans. *Trichomonas vaginalis* infects the reproductive tracts of humans. Other species of *Trichomonas,* as well as *Trichonympha,* are symbiotic in the guts of wood-eating insects. *Codosiga* is colonial and sessile. The flagellum of each individual draws water through the sieve-like collar, which filters out food. Because it has cilia, *Opalina* was long considered a ciliate. However, its many nuclei are all of one size, unlike those of ciliates. *Euglena* (Figs. 22.1 and 22.2) is also a flagellate.

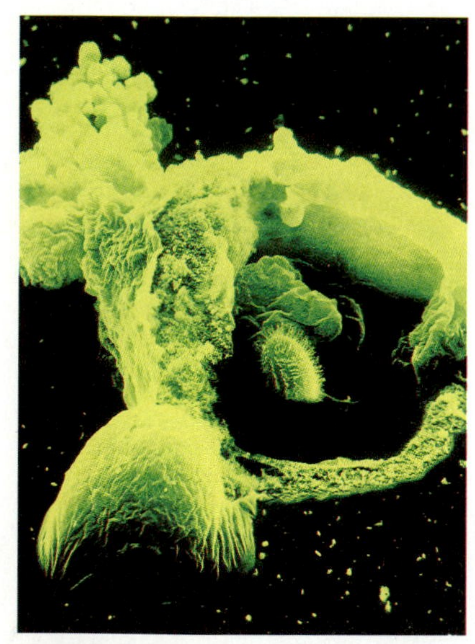

FIGURE 22.6

SEM of an *Amoeba* engulfing a ciliate (on the right).

■ **Question:** How do you suppose *Amoeba* detects its prey? Because ciliates can "out run" sarcodines, how does *Amoeba* catch them?

FLAGELLA. The beating of a flagellum generates a force parallel to the long axis of the flagellum (Fig. 22.5). In most flagellates, such as *Euglena,* the force pulls the protozoan through water, rather than pushing it, as occurs with sperm. Flagellates move at about 0.2 mm/sec, or about one-fifth as fast as ciliates. In some protozoans, such as *Trypanosoma,* the force generated by the flagellum is increased by an attached **undulating membrane.** In *Codosiga* and some other flagellates the base of the flagellum is surrounded by a sievelike collar of microvilli that filter out particles of food. This mechanism is similar to that in the **choanocytes** (collar cells) that assist in suspension-feeding in sponges and some other animals (see Fig. 23.1C).

PSEUDOPODIA. Pseudopodia are mobile extensions into which cytoplasm flows (see Fig. 10.16). In *Amoeba* and other sarcodines that use pseudopodia in locomotion on surfaces, pseudopodia bend downward, elevating the rest of the body from the substratum. Thus, when viewed from the side, such a sarcodine appears to creep about "on tiptoes." In a race against a typical ciliate and a typical flagellate, a sarcodine would come in a distant third, creeping along at about 0.005 to 0.02 mm/sec. Like cilia and flagella, pseudopodia are used not only for locomotion, but also for feeding. The process, called **phagocytosis,** has been studied extensively in *Amoeba* in an attempt to understand how vertebrate white blood cells ingest bacteria and other invaders. During phagocytosis, pseudopodia surround a food particle and enclose it in a **food vacuole** (Fig. 22.6). This process can apparently occur anywhere on the plasma membrane.

(Ameboid cells do not have cytostomes like those in paramecia.) Once the food vacuole is formed by phagocytosis, lysosomes fuse with it and digest the food. Four types of pseudopodia can be distinguished on the basis of shape (Fig. 22.7).

Behavior

Phototaxis in *Euglena*. A protozoan observed under a microscope seems to move about with as much purpose in life as we, ourselves, have. One must keep reminding oneself that the protozoan has no brain, or even a single nerve cell. Even without a microscope you can observe some impressive protozoan behavior. If you find a shaded, stagnant pond whose surface is green with *Euglena* and shine a bright light at one side, soon all the green will be at that end. The euglenoids swim to the bright side (**phototaxis**), leaving the other flagellates and ciliates swimming throughout the water and leaving the ameboids on the bottom. The response of *Euglena* to light is due to the light-sensitivity of the base of the flagellum. When *Euglena* is swimming directly toward or away from light, the base of the flagellum is continuously illuminated. When *Euglena* is swimming at an angle with respect to light, however, pigments in the **stigma**

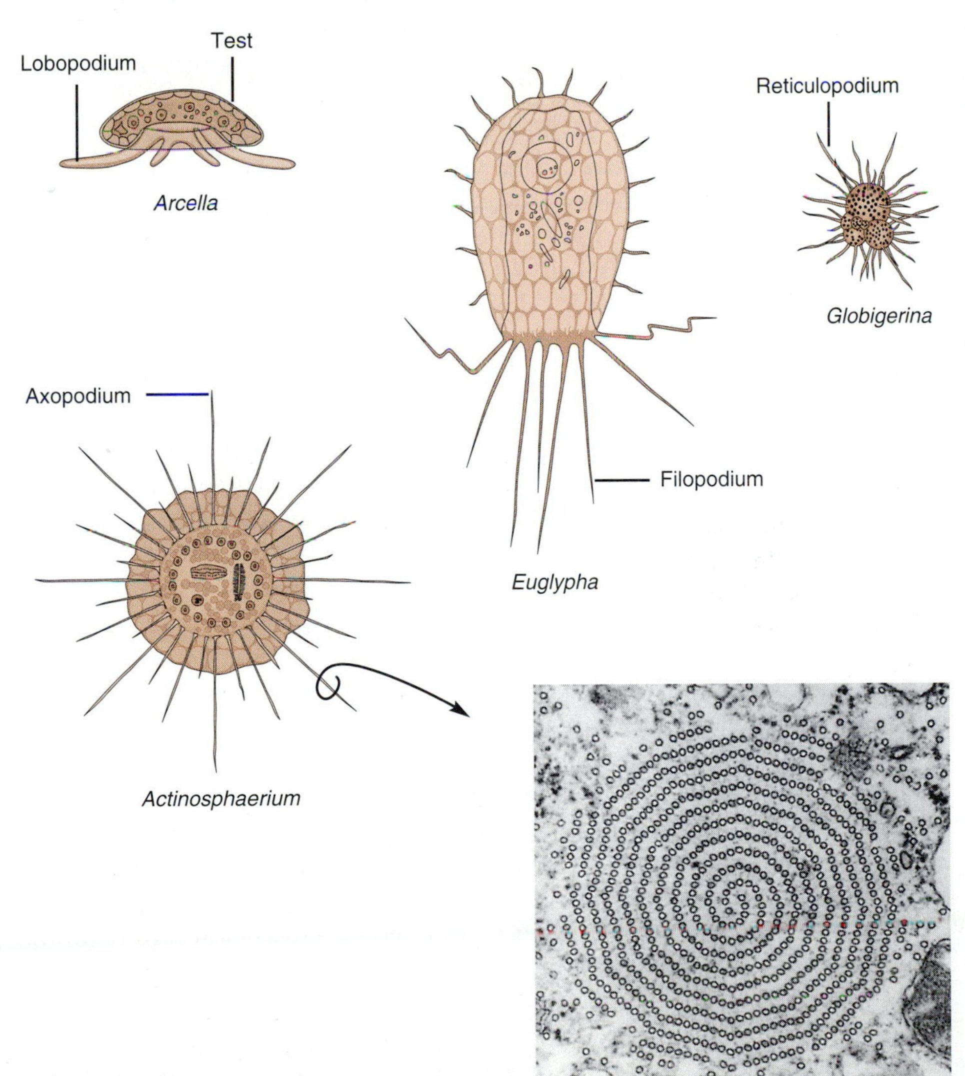

FIGURE 22.7
Four varieties of pseudopodia. Like *Amoeba, Arcella* has the lobopodium form of pseudopodium. The lobopodia are restricted to the portion of membrane not covered by a shell (test). See also *Difflugia* in Fig. 22.1. *Euglypha* has long, thin filopodia. *Globigerina* has reticulopodia, which are similar to filipodia except that they interconnect. The reticulopodia of foraminifera project through openings in the test. *Actinosphaerium* has axopodia, which are reinforced by microtubules. Inset: Electron micrograph of an axopod of *Actinosphaerium* in cross section.

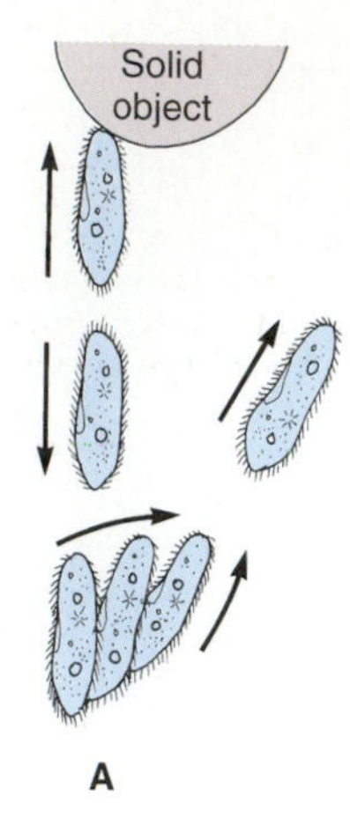

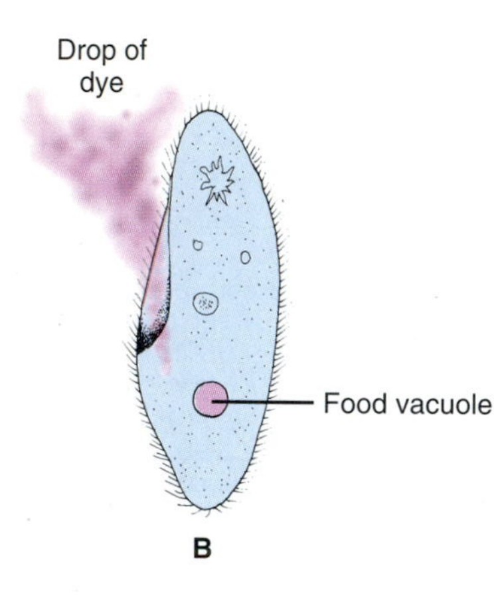

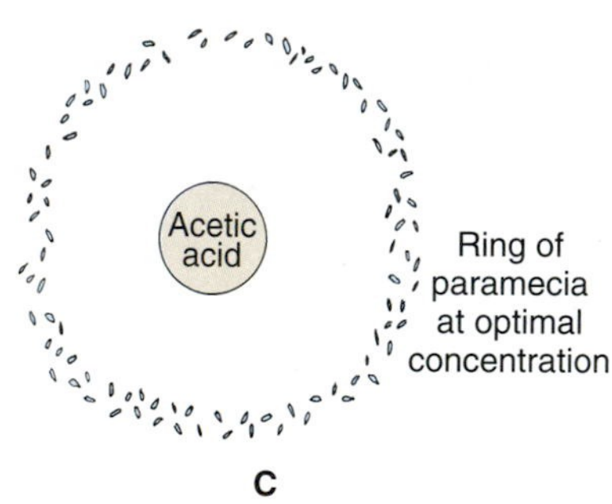

FIGURE 22.8

Behavior in *Paramecium.* (A) The avoiding response of *Paramecium* when it swims into an obstacle. (B) *Paramecium* samples the environment ahead by drawing in a stream of water, as can be demonstrated by injecting a drop of dye in front of it. (C) A drop of acetic acid in water diffuses outwardly in a decreasing concentration. Because of chemokinesis, paramecia aggregate in a zone around the drop, avoiding the high acidity in the center while taking advantage of the nutrient.

(= "eyespot;" Fig. 22.2) shade the base of the flagellum periodically as the body rotates. Each time a shadow of the stigma falls on the base of the flagellum, the long axis of the flagellum turns a few degrees in the direction of the stigma and therefore toward the light. Thus the flagellum pulls the body of *Euglena* toward the light.

BEHAVIOR OF *PARAMECIUM*. One of the most studied protozoan behaviors is the **avoiding reaction** of *Paramecium*. When *Paramecium* bumps into an object, it does not persist in hammering at it, like a fly against a window. Instead the slipper-shaped organism backs up a little, turns on its heel, as it were, and swims off to the side (Fig. 22.8A). The intriguing question about the avoiding reaction is how the *Paramecium* "perceives" and responds to contact without sensory organs or a nervous system. The cilia of *Paramecium* and other ciliates are interconnected by fibrils (Fig. 22.3) that have been called "nervelike" and "coordinating," but it is unlikely that these fibrils conduct action potentials or that they are involved in ciliary coordination. Coordination is more likely due to changes in the intracellular Ca^{2+} concentration triggered by changes in the voltage across the plasma membrane. Contact at the anterior end of *Paramecium* triggers an action potential during which calcium flows into the cell. The Ca^{2+} causes the cilia to reverse their direction of beating, backing the paramecium out of trouble. After the action potential, Ca^{2+} is pumped out, and the cilia resume their normal beating. Similar mechanisms are responsible for **chemokinesis:** changes in the speed of movement in response to chemicals. The cilia increase their rate of movement in repellent substances and slow down in attractive ones (Fig. 22.8C).

As in animals, the behavior of *Paramecium* is under some degree of genetic control. Mutants called "pawns" have smaller action potentials and less influx of Ca^{2+} in response to contact. Therefore these "pawns," like those in the game of chess, can only go forward. Another mutant of *Paramecium* is called "paranoiac," because its action potentials last so long that it has an exaggerated avoiding reaction.

Reproduction

ASEXUAL REPRODUCTION. Reproduction in protozoans is almost always asexual and usually by fission. Fission is preceded by mitosis, which differs in many protozoans from that of animal cells (pp. 72–74). In some protozoans the spindle apparatus occurs within the nucleus, centrioles are not present, or the nuclear envelope may remain intact throughout mitosis. Mitosis must get quite complicated in *Amoeba* and in some of the other sarcodines that have more than 500 chromosomes. The most common type of fission is **binary fission,** with one protozoan dividing into two roughly equal "daughter cells" (Fig. 22.9). Binary fission in the Ciliophora is especially interesting because there are two kinds of nuclei. In general the larger type, the **macronucleus,** directs metabolism, development, and the physical traits of the cell. The **micronucleus** transmits genetic information during reproduction.

Another means of asexual reproduction is **multiple fission,** in which numerous daughter cells form at the same time. Multiple fission is especially common in the sarcodines and sporozoans, and it will be described later in the discussion of malaria. In many ciliates, especially suctorians, a daughter cell forms from a small outgrowth of the parent, in the process called **budding.** Usually the bud forms outside the "parent," but *Tokophrya* and some other suctorians "give birth" to young that develop from internal buds.

SEXUAL REPRODUCTION. There are several forms of reproduction that involve the recombination of genetic material from two individuals, and are therefore said to be

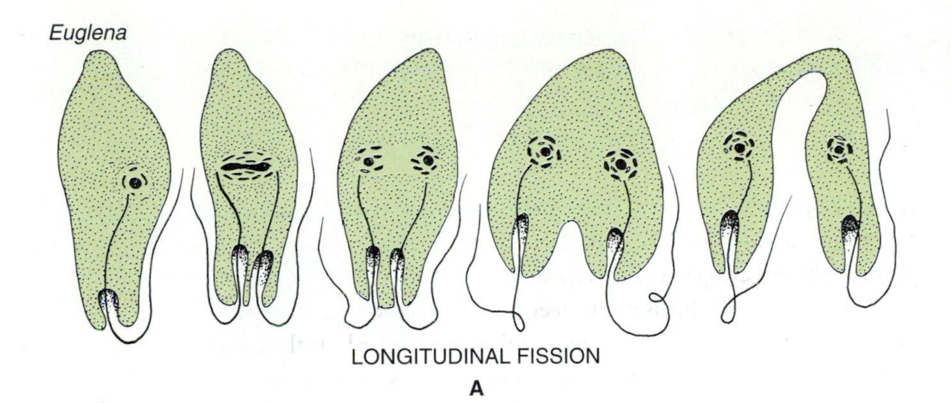

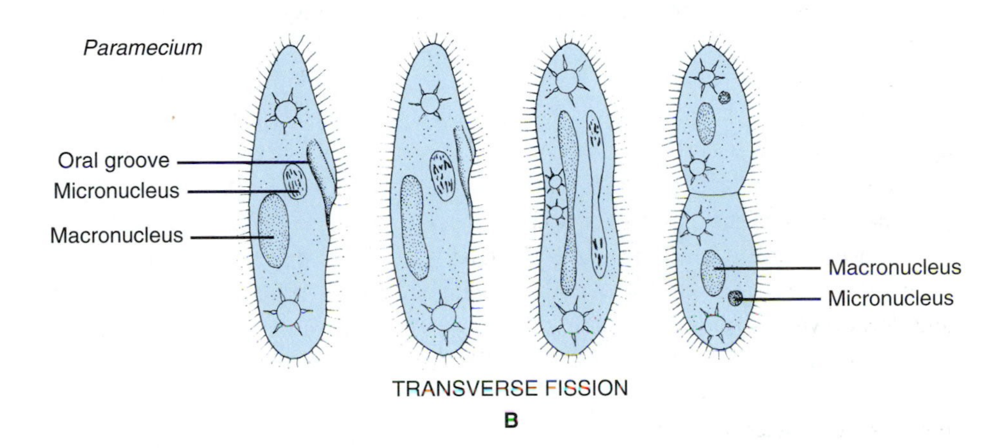

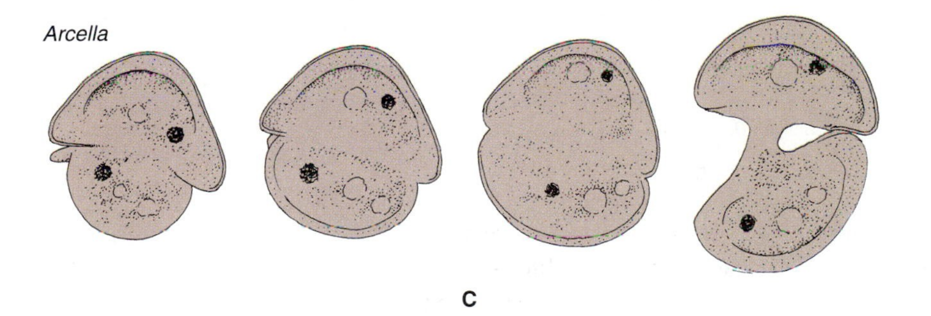

FIGURE 22.9

Binary fission in the flagellate *Euglena,* in the ciliate *Paramecium,* and in the sarcodine *Arcella.* Fission is lengthwise in flagellates and transverse in ciliates. In *Arcella* the shell does not divide, but new shell material is extruded along with the new daughter cell.

sexual. **Conjugation** in ciliophora is one of these. In *Paramecium,* conjugation begins with two ciliates attaching to each other by their cytostomes. The micronucleus of each conjugant then undergoes meiosis, producing four haploid **pronuclei** (Fig. 22.10 on page 462). The conjugants give each other one pronucleus, which fuses with one already present to form a diploid micronucleus. The two conjugants then separate, and the new micronucleus in each one divides several times to form new micronuclei. The old macronucleus degenerates, but some of the new micronuclei enlarge and become new macronuclei. Finally, two divisions of the cell and micronuclei result in four cells with one micro- and one macronucleus each.

Pairs of ciliates in the same species will conjugate with each other only if they belong to the same **variety.** In addition, they must belong to different **mating types** in order to conjugate, just as two animals must belong to different sexes in order to copu-

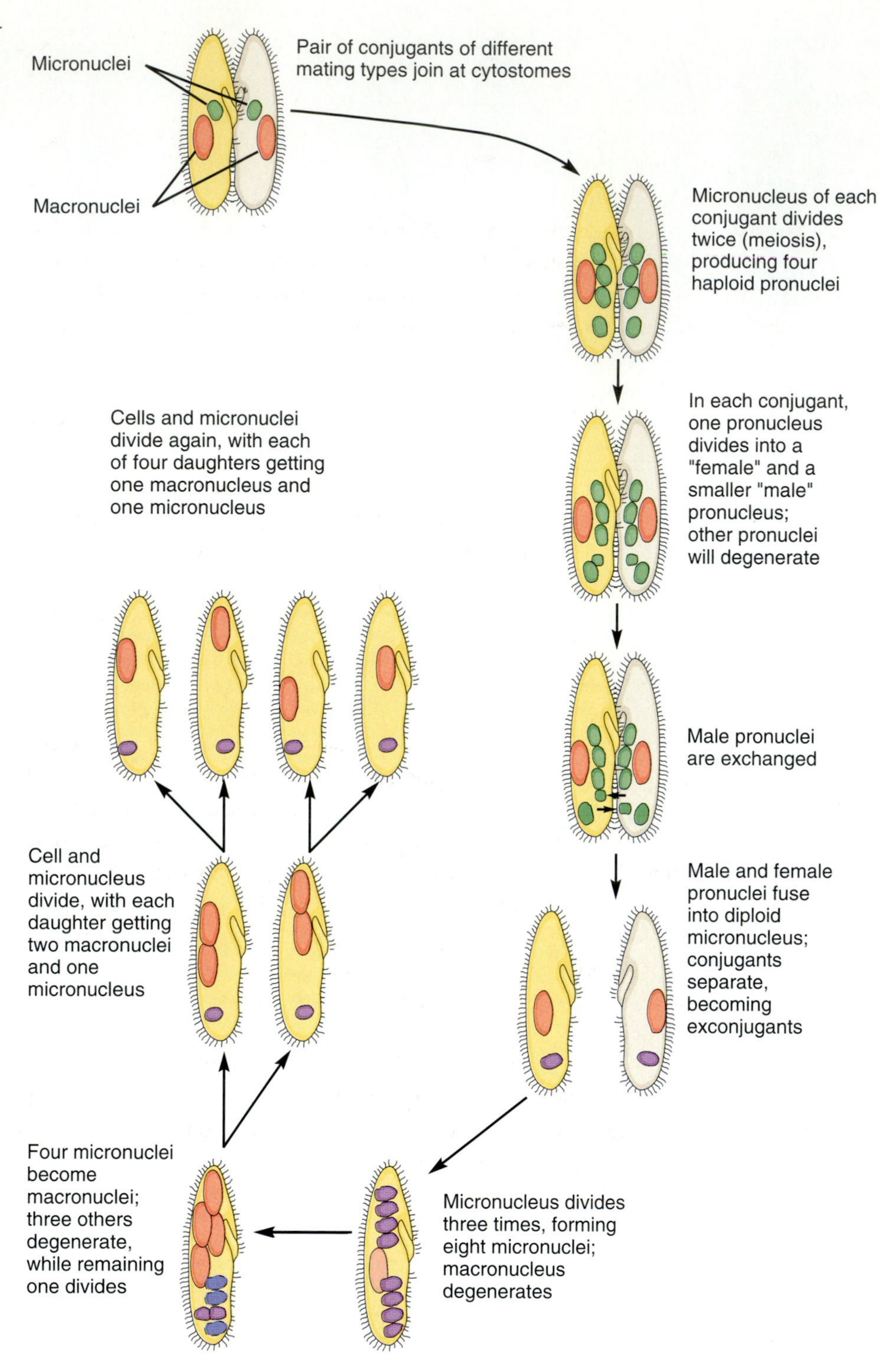

FIGURE 22.10
Conjugation in *Paramecium caudatum.* Conjugation produces eight daughter cells for each pair of conjugants, and all of them are likely to be genetically different.

■ **Question:** What happens to the original conjugants? How does their fate differ from that of animal parents?

late. In one species of *Paramecium* there are 14 varieties, each with up to five mating types. (It broadens the mind to consider what human society might be like if animals had evolved from conjugating ciliates.)

Another form of sexual reproduction is **autogamy,** which occurs in *Actinophrys* and some other protozoans. Autogamy is similar to conjugation except that it occurs entirely within a single individual. The haploid pronuclei that result from meiosis fuse without an exchange of genetic material between individuals.

In some protozoans (but not ciliates), reproduction is similar to that in animals, in that one gamete fertilizes another and forms a zygote. This process is called **syngamy** in both protozoans and animals. Meiosis always occurs, preventing the chromosome number from doubling. As in humans, it is completed in the zygote *after* fertilization, rather than during the formation of gametes.

ENCYSTMENT. In some species the zygote surrounds itself with a protective coating that forms a **cyst.** Several generations of binary fission may then occur within the cyst. Protozoans may also encyst prior to dispersal or in response to environmental stress. Such cysts are often called **spores.** Like the seeds of plants, they allow the protozoan to survive in a dormant state until environmental conditions are favorable. During encystment the protozoan loses any cilia or flagella, pumps out water, becomes spherical, and secretes a protective coating around itself. In this form, protozoans typically survive winters, dry seasons, and other severe conditions. *Colpoda* cysts have survived in the laboratory for many hours at temperatures of 180°C below freezing and for several hours in boiling water. Excystment occurs upon return to favorable conditions. Several species of protozoans were found cheerfully swimming about in museum soil samples that had been moistened for the first time in 49 years!

Speculations on the Origin of Animals

Because of the similarity of certain protozoans to certain animals, many zoologists assume that animals (metazoans) evolved from a protozoan. There are three major theories that are intended to explain how this evolution occurred. The **syncytial ciliate theory** proposes that a multinucleate protozoan—probably a ciliate—became multicellular as each of its nuclei became partitioned from the others by internal membranes. The eventual result was an animal somewhat like a planarian (phylum Platyhelminthes; class Turbellaria; Chapter 25), which, in turn, evolved into the majority of other animals. One criticism of this theory is that the proposed process resembles embryonic development only in insects, not in flatworms or other animals where one would expect it. Another problem is that no known ciliate reproduces in the way that animals do—by syngamy. Finally, the syncytial ciliate theory does not account for the evolution of sponges, cnidarians, and several other groups of animals that evolved before flatworms.

A second theory, called the **blastea theory,** was first proposed in 1874 by Ernst Haeckel. Haeckel suggested that animals evolved from hollow colonies of flagellates ("blastea") resembling the blastula stage of animal embryos (see Fig. 6.2). Haeckel and many zoologists since him have studied the flagellated protist *Volvox* as a model of the ancestor of all animals. Unfortunately, *Volvox* and all other protists that form hollow colonies are algae rather than protozoans, and they are more closely related to plants than to animals (Rausch et al. 1989). The second problem with the blastea theory is that Haeckel based it on his notion that development of the individual animal goes through all the stages of the evolution of its species, which is now discredited.

Although Haeckel's theory is undoubtedly wrong, it happens that recent evidence does support the concept that animals evolved from a colonial flagellate. Instead of *Volvox,* however, the protists that are most closely related to animals appear to be choanoflagellates such as *Codosiga* (Fig. 22.5). The **choanoflagellate theory** was first proposed because of the similarity between choanoflagellate cells and the choanocytes (collar cells) found in certain animals such as sponges (see Fig. 23.1). It is supported by molecular phylogenetic studies suggesting that animals shared an ancestor with choanoflagellates more recently than they did with other protozoans that have been studied (Wainright et al. 1993) (Fig. 22.11 on page 464).

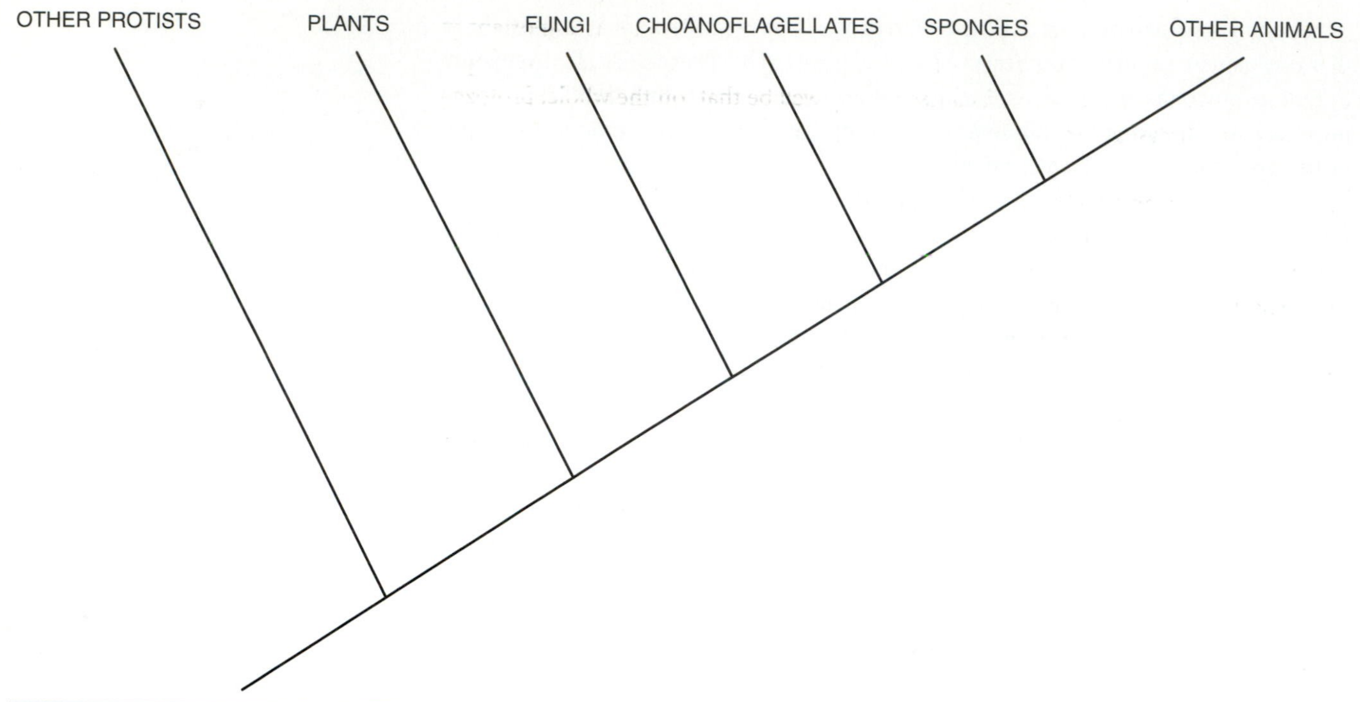

FIGURE 22.11

Cladogram showing the phylogeny of animals and other groups as inferred from comparisons of rRNA by Wainright et al. (1993). Animals shared a more recent common ancestor with choanoflagellates than with the other protozoa studied, including ciliates, apicomplexans, and dinoflagellates. The relationship of animals to fungi and plants differs in various studies (see Figs. 1.4 and 18.17).

FIGURE 22.12

A sample of foraminiferan tests.

■ **Question:** Some of these tests look like the shells of molluscs. Does that indicate a direct evolutionary connection, or simply convergent evolution because the functions of the shell were similar in the two groups? What are those functions?

Interactions with Humans and Other Animals

FORAMINIFERANS. Most people are not aware how much of their everyday surroundings are due to protozoa. Much of the chalk we write with and the limestone we walk upon was created by **foraminiferans** (Granuloreticulosea), which once made up a large portion of the plankton of many oceans. As foraminiferans died, their shells, up to 10 cm in diameter, sank and formed deep sediments on the ocean floor (Fig. 22.12). The shells changed into chalk deposits, such as the famous White Cliffs of Dover in England, and limestone, such as that found in the quarries of Indiana and in the Egyptian Pyramids. One esteemed scientist early in this century was so impressed with foraminiferans that he thought all rocks, including meteorites, had been created by them (Gould 1980).

Indirectly, foraminiferans must also be credited with many of our petroleum products. The different shells produced by species of foraminiferans that lived in different ages have formed "indicator fossils" that petroleum geologists use in searching for oil. Because of foraminiferans, one is as likely to find protozoologists in the petroleum industry and in geology departments as in zoology departments.

Almost all the foraminiferans were wiped out in the mass extinction at the end of the Mesozoic era 65 million years ago (p. 390), but the survivors, especially *Globigerina,* still produce a steady rain of shells to the ocean floor. "Globigerina ooze" is now accumulating at a rate of about 60 cm per century, and it covers approximately half the ocean floor. Pressure at depths greater than 4 km dissolves the calcium-based shells of these foraminiferans, but not the silicon- or strontium-based shells of radiolarians (Actinopoda). Instead of globigerina ooze, one therefore finds "radiolarian ooze" at the bottom of the deepest oceans. Radiolarian ooze has made its own contribution to civilization, having formed the flint that went into early tools.

ECOLOGY. The immense numbers of protozoa imply that they must have an enormous day-to-day impact on the environment. Protozoa form essential links in food chains by consuming other protozoans, algae, and bacteria and then being consumed by a variety of small animals. Their decomposing of organic matter is important in breaking down pollutants, and their consumption of bacteria is essential for the proper functioning of sewage-treatment facilities. It may well be that, on the whole, protozoa prevent more diseases than they cause.

SYMBIOSES. Many protozoans are symbiotic with animals, acting as harmless commensals, beneficial mutualists, or parasites. These symbioses are often highly specialized. Some species of suctorians, for example, have never been found anywhere except on the backs of turtles, and members of *Opalina* and its relatives are almost entirely restricted to the rectums of amphibians.

Especially important are dinoflagellates called **zooxanthellae,** which are mutualistic in reef-building corals, sea anemones, giant clams, and a few other animals (see Figs. 24.10A and 28.22). The most common species of zooxanthella is *Symbiodinium microadriaticum,* a brown species that occurs in several varieties in corals around the world. Zooxanthellae release 40%, and perhaps up to 90%, of their photosynthetic product as glycerol, with smaller amounts of sugars and amino acids. This nutritional contribution is essential to the formation of coral reefs, so many other reef inhabitants are indirectly dependent on zooxanthellae. High temperature and other stresses can make corals evict their zooxanthellae. The consequent "bleaching" is currently a matter of great concern to ecologists (p. 508).

Less beneficial are the dinoflagellates whose populations bloom to the point of creating "red tides" (which may actually be red, green, brown, colorless, or luminescent). Some dinoflagellates that produce "red tides," such as *Protogonyaulax* (originally classified in *Gonyaulax*), contain saxitoxin, a nerve blocker that can accumulate in clams and other molluscs (Fig. 22.13). Saxitoxin does not usually harm the molluscs, but it causes **paralytic shellfish poisoning** in humans who eat them. Another neurotoxic species of dinoflagellate is *Gambierdiscus toxicus,* which is responsible for **ciguatera poisoning,** a severe and prolonged neural disorder in humans. Ciguatera poisoning is due to ciguatoxin, which accumulates in large, carnivorous fishes that inhabit coral reefs. Most cases of ciguatera poisoning in the United States were due to eating large barracuda (*Sphyraena* spp.) from southern Florida until that state outlawed its sale.

Several species of *Trichomonas* and *Trichonympha* (Fig. 22.5) inhabit the guts of termites and wood-eating cockroaches. Their combined weight may equal one-third the weight of the host. These flagellates are often assumed to aid the host insect in di-

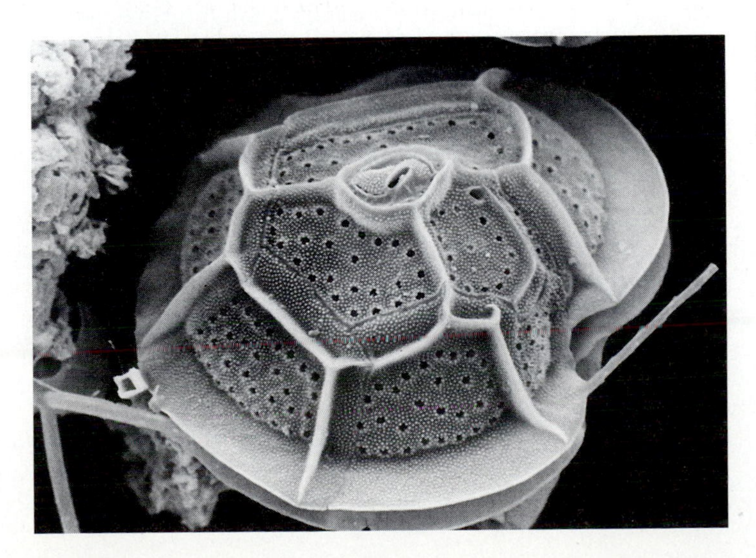

FIGURE 22.13

Scanning electron micrograph of the dinoflagellate *Protogonyaulax tamarensis.*

gesting the cellulose in plant cell walls, but there is little evidence that they do (Slaytor 1992). Ruminant mammals such as cattle also contain in their stomachs large numbers of ciliates that are assumed to play a similar role. Cattle fare just as well without their ciliates, however, so the symbiosis is probably commensal rather than mutualistic.

DISEASES. Most people in less-developed parts of the world are constantly aware of the importance of protozoa, because they and their animals are plagued by parasitic species. The box on p. 467 summarizes some of the reasons for the success of parasites, and Table 22.2 summarizes some of these diseases. A detailed treatment of them is possible only in a course on parasitology, but the later discussion of malaria will give some idea of the biological, economic, political, and sociological complexities of protozoan infections in general.

The Biology of Malaria

THE DISEASE. Most people in developed countries think of malaria as a conquered disease, if they think of it at all. Yet almost 300 million people in the world have malaria, and the number is increasing by about 5% per year. Each year more than two million people, mostly children, die from the direct effects of malaria. Millions of others are so weakened that they cannot support themselves. Malaria therefore ranks as one of the most serious diseases of humankind. It is also one of the most difficult diseases to control, having defeated all the scientific breakthroughs that have regularly promised to eliminate it. At present, malaria is locked in battle against the latest techniques of biological engineering, and the outcome is uncertain. The difficulty of eliminating malaria is partly due to social factors (see box on pp. 471–472), but also to the intricate biology of the causative agent, the sporozoan *Plasmodium.* There are more than 50 species of *Plasmodium* that attack a variety of birds and mammals, many of which act as reservoirs for species that cause malaria in humans.

The first symptoms of malaria—irregular chills and fever, malaise, and headache—are easily confused with influenza. These soon disappear, only to be followed by anemia and by debilitating chills and fever every few days. As the fever develops, shivering may literally rattle the bed. Fever occurs at two- or three-day intervals and is referred to as **tertian** or **quartan,** respectively. (These terms are from the Latin for "three" and "four," following the Roman custom of starting a count from one rather than zero.) The most serious form of the disease in humans is malignant tertian fever, which is caused by *Plasmodium falciparum.* This form of malaria is often fatal because it makes red blood cells attach to the inner surface of blood vessels, blocking circulation to the brain, kidney, heart, or other organs. *Plasmodium falciparum* causes about half the cases of malaria and almost all the deaths.

Until the turn of the century it was thought that malaria was caused by something in warm, humid air. (*Mal'aria* is Italian for "bad air"). The "something" turned out to be mosquitoes of the genus *Anopheles* (see Fig. 13.3). Only a few of the 390 known species of *Anopheles* transmit the disease to humans. These are the species that occur in the same habitat as humans and prefer their blood.

LIFE CYCLE OF *PLASMODIUM.* Like many parasites, *Plasmodium* completes different stages of its life cycle in two different hosts, both of which require unique adaptations. Reproduction occurs mainly in the human final host (steps 1 through 6 below

To Be a Parasite

Many protozoans and animals described in this and later chapters will strike you as being rather nasty characters, because they live in or on other organisms, including you and me. Many use their hosts mainly for transportation or shelter, or feed on their dead tissues or wastes. These harmless species might better be called commensals rather than parasites (see p. 334). Others, however, are frankly parasitic. They steal nutrients from within the host's guts, or they feed upon the host's own living tissues. Because the host provides transportation, shelter, and food, many parasites do not even have their own adaptations for locomotion, protection, and feeding.

A parasite's life might be idyllic if it were not for two major problems. The first is that the host has an immune system that tries to eliminate its unwanted guests. In this and some later chapters we will see that parasitic protozoans and animals can evade the immune system in a variety of ways:

1. A parasite may form a protective cyst, although this means that it must remain inactive.
2. It may cover itself with molecules that are frequently shed or changed, preventing the immune system from mounting an attack.
3. It may invade a cell of the host, where the immune system cannot recognize it as an invader.
4. It may reproduce so rapidly that it overwhelms the immune system.
5. It may attack or interfere with crucial components of the immune system.

The second major problem for the parasite is that living on or in the body of another organism may limit the opportunities for **transmission** of offspring to new hosts. One potential solution might be to produce massive numbers of offspring to compensate for the poor odds of successful transmission. Such increased reproduction, however, might increase the parasite's **virulence** (tendency to cause disease) so much that it would kill or severely weaken the host. A dead or disabled host would endanger the parasite's own life, and it would be a poor vehicle for the transmission of offspring to new hosts. Until recently it was widely assumed that parasites and hosts coevolve in ways that limit virulence, and that parasites cause severe disease only during early encounters with a particular host. Parasitologists now realize that there are several ways in which a parasite can become virulent without endangering its transmission:

1. In organisms that move over long distances—notably humans in recent times—symptoms of infection may be delayed, enabling the host to travel and infect new hosts. Many epidemiologists believe that this accounts for the recent resurgence and increased virulence of many human diseases.
2. Many parasites have two or more stages in their life cycles, and each stage may be specialized for either reproduction or transmission. Transmission can occur either by air or water, or it can be facilitated by a different host organism. The organism in which the last stage of the parasite occurs is called the **final** (= primary or definitive) **host.** An organism in which an earlier stage of the parasite occurs is an **intermediate host.** Each host requires that the parasite have a new set of adaptations for transmission and resistance to immune mechanisms, but these disadvantages may be overcome by the advantages of separating reproduction from transmission. A parasite may have a low reproductive rate and low virulence in the final host, which facilitates transmission, but it may asexually reproduce massive numbers of offspring and be virulent in an intermediate host. Alternatively, a parasite may reproduce and be extremely virulent in the final host and use an unharmed intermediate host mainly for transmission. Hosts that mainly facilitate the transmission of parasites are called **vectors.**
3. The parasite may have several alternative hosts, each of which can act as a **reservoir** in the event that other hosts are unavailable.
4. The parasite may modify the behavior of the host in such a way that the chances of transmission are increased. For example, a parasite may make its intermediate host more conspicuous in appearance and behavior, increasing the chances that it will be ingested by the final host.
5. The parasite may synchronize its life cycle with that of the host in a way that increases the probability of transmission without increasing total numbers of offspring. For example, reproduction by the parasite may be triggered by reproduction of the host, ensuring that new hosts are available for the parasite's offspring.

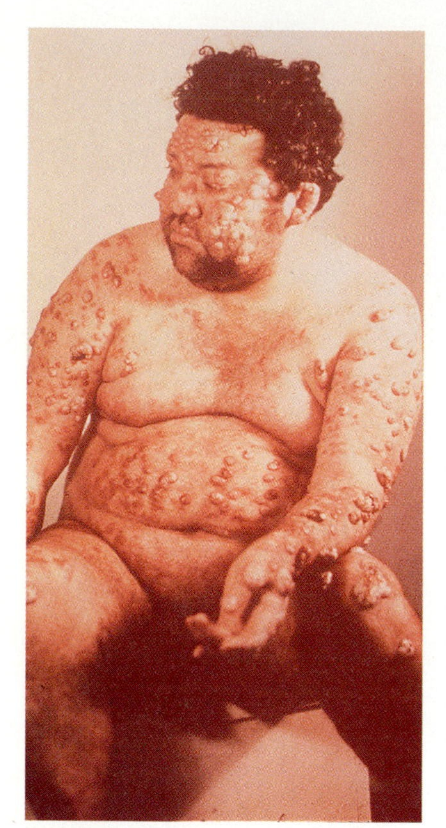

FIGURE 22.14

A victim of cutaneous leishmaniasis. Some deformities are even more ghastly.

TABLE 22.2

Some important protozoan diseases.

Leishmaniasis refers to three distinct tropical diseases caused by different species of the flagellate *Leishmania*, which is spread among humans and other mammals by blood-sucking sand flies. There are 12 million victims in tropical Africa and South America. *Leishmania* interferes with the immune system in several ways, and it can even survive and reproduce after being phagocytized by macrophages. Two forms of leishmaniasis damage the skin and mucous membranes (Fig. 22.14). Another form, called kala-azar, attacks the circulatory system and damages the liver and spleen. Although kala-azar can usually be cured with drugs, many children in developing countries die from it because their parents cannot afford treatment (less than $20).

Trypanosomiasis (African sleeping sickness), caused by *Trypanosoma brucei rhodesiense* and *T.b. gambiense* (Fig. 22.5). Transmitted to cattle and people throughout central Africa by the tsetse fly *Glossina*. Hundreds of thousands of victims, with 25,000 new cases per year. For the first two years the symptoms of sleeping sickness are fever, lethargy, and malaise (just not feeling well). Gradually the face swells and the lethargy gets worse as the trypanosomes attack the central nervous system. Eventually the desire to sleep overpowers the desire to eat, and the patient becomes malnourished. Finally the victim enters a coma from which he never awakens. Unlike most parasites that invade host cells to escape the immune system, *Trypanosoma* lives in the bloodstream exposed to antibodies. It survives by frequently changing its antigens (**variable surface glycoproteins**).

Chagas' disease, caused by *Trypanosoma cruzi*. Affects 12 million people in Central and South America. First invades macrophages, then attacks other tissues, especially nerve and muscle. Damage to the heart can cause death within a few years or after many years. Trypanosomes from the feces of the "kissing bug" *Triatoma* invade the wounds or mucous membranes. (It was once thought that Charles Darwin's mysterious illness was Chagas' disease, but his symptoms are now known to have been quite different.)

Trichomoniasis, a common infection of the genital tract caused by *Trichomonas vaginalis* (Fig 22.5). Sometimes produces inflammation of the vagina, but is often symptomless.

Giardiasis, caused by *Giardia lamblia* (Fig. 22.15). Leeuwenhoek, curious about the cause of his own diarrhea, was the first to observe *Giardia*. It is common in the human intestine, but it does not always produce symptoms because cells of the gut lining are replaced every three or four days. In severe infections the entire lining of the intestine may be covered with the trophozoites. Symptoms then include abdominal pain, gas, and "explosive" diarrhea. Once considered an affliction of travelers to developing countries, it is now common throughout the United States. Since the 1970s, careless defecation by increasing numbers of backpackers has brought *Giardia* cysts to even the remotest streams. Many animals are reservoirs for *Giardia* , continually reinfecting water. *Giardia* is unusual among eukaryotes in that it lacks mitochondria, endoplasmic reticulum, and Golgi bodies. These might have been lost as an adjunct of parasitism, but molecular studies suggest that *Giardia* is intermediate between prokaryotes and eukaryotes.

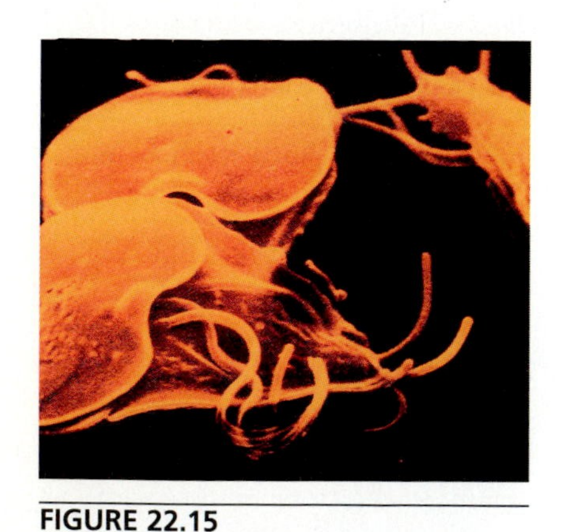

FIGURE 22.15

Giardia lamblia trophozoites attach to intestinal cells by means of the adhesive discs. Although these discs are sometimes called "suckers," it is doubtful that they adhere by suction. Each trophozoite is approximately 15 μm wide.

Amebic dysentery, caused by *Entamoeba histolytica*. Unlike several harmless species of amebas that normally live in the large intestine, *E. histolytica* sometimes attacks the tissues of the host. The infection may then spread to the liver or other organs, and damage to the colon may allow lethal bacteria to invade the abdominal cavity. More commonly the symptoms are diarrhea, vomiting, cramps, and malaise, which may be fatal to children. Approximately 10% of humans harbor the parasite, with about one-fourth of these showing symptoms. Transmission of cysts is through food and water, with flies and cockroaches often acting as vectors between exposed feces and food. The use of human feces as fertilizer is a common route of infection in much of the world.

Toxoplasmosis of cats, humans, and many other mammals, caused by *Toxoplasma gondii*. Cysts are transmitted through feces and undercooked meat. By means of the apical complex the trophozoite invades a cell, where it remains within a vesicle, unmolested by the immune system. Usually *Toxoplasma* lives in cells lining the gut, which are replaced rapidly enough that symptoms seldom occur. Severe cases can, however, kill kittens within a few weeks. Cats are especially liable to infection, which they acquire by eating rodents. Humans with immune deficiencies such as AIDS are also severely affected. Pregnant mammals, including women, can pass the infection to fetuses, with the resulting **congenital toxoplasmosis** causing miscarriages, stillbirths, and birth defects. Pregnant women should therefore avoid contact with cats.

and in Fig. 22.16), and transmission is by the mosquito intermediate host (steps 7 through 9).

1. Humans acquire the *Plasmodium* parasite from *Anopheles* females that have previously fed on the blood of a malaria victim. (Male mosquitoes do not feed on blood.) While drawing the victim's blood, the mosquito ejects saliva containing several hundred *Plasmodium* in the **sporozoite** stage.

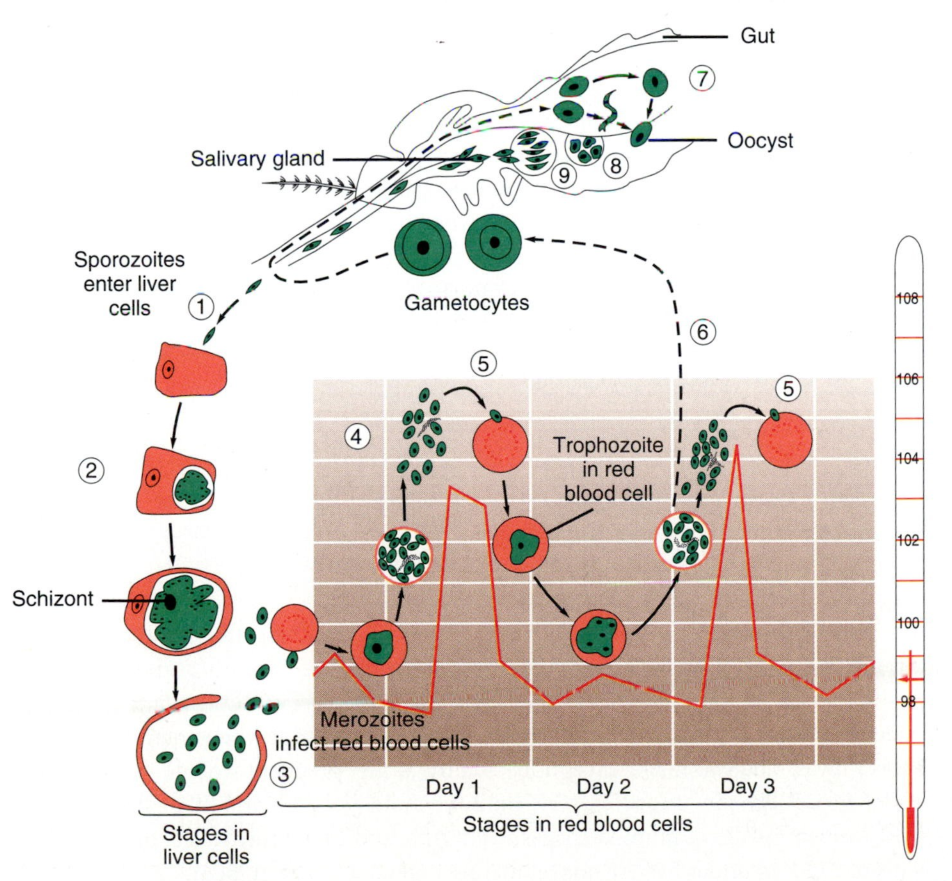

FIGURE 22.16

The life cycle of *Plasmodium*. The stages that occur in red blood cells are superposed on a graph of the body temperature of a person with tertian fever.

■ **Question:** Which of the adaptations listed on p. 467 does *Plasmodium* have?

2. Within an hour each sporozoite that survives the immune system enters a liver cell. During the next 6 to 12 days, depending on the species, the sporozoite develops into a large, multinucleate **schizont,** which splits into 5,000 to 10,000 individual cells by multiple fission (**schizogony**). The schizont eventually ruptures the liver cell, releasing thousands of **merozoites** into the blood.
3. Within seconds each merozoite invades a red blood cell and forms an ameboid **trophozoite.** The trophozoite feeds on the hemoglobin, turning it into an insoluble waste called **hemozoin.** The combination of erythrocyte destruction and the inability to recover the iron in hemoglobin produces anemia in the victim.
4. Each trophozoite undergoes schizogony once more, splitting into between 10 and 20 new merozoites that are released from what is left of the red blood cell. Each merozoite infects another red blood cell.
5. The release of the merozoites and/or their metabolic wastes triggers fever in the victim. At first the release is spread over time, producing a mild fever, but later it becomes synchronized with the daily rhythm in the host's body temperature, causing bed-rattling fever every 48 or 72 hours. (The synchronized release of merozoites may be an adaptation that swamps the victim's immune system with more invaders than it can handle at one time.)
6. After a few cycles some of the trophozoites develop inside blood cells into male and female **gametocytes** rather than merozoites. If another female *Anopheles* attacks the victim, it can ingest some of these gametocytes. As they are being sucked in, the increase in pH resulting from the elimination of CO_2 from the surrounding blood triggers the gametocytes to develop into gametes. Several flagella sprout out of each male gametocyte in a process called **exflagellation.**
7. In the gut of the mosquito, male gametes fertilize female gametes. The zygote then burrows through the gut lining and forms an **oocyst** (OH-oh-sist).
8. A thousand or more sporozoites develop within the oocyst. This process, by which a zygote divides into many cells, is called **sporogony.** (Compare schizogony in step 2 above.) Sporozoites are not spores. Because *Plasmodium* is always inside one or the other host, it never needs a spore stage to resist environmental stress. *Plasmodium* is therefore a sporozoan without spores.
9. The sporozoites migrate through the body cavity of the mosquito and enter the salivary glands to await the next human victim. The mosquito stage of the *Plasmodium* life cycle takes from 7 to 18 days, or longer if the temperature is low.

NATURAL DEFENSES. Adult humans mount an **immune response** against *Plasmodium* just as they do against any other foreign cell. Generally, however, the immune system is ineffective in preventing malaria. White blood cells have less than an hour to attack sporozoites, and it takes only one surviving sporozoite to cause the disease. Antibodies are produced too late to combat the first invasion of sporozoites, although they do respond to subsequent infections by attaching to the **circumsporozoite (CS) protein** that coats the sporozoites. Antibodies against CS protein are only partly effective, however, because the sporozoite simply sheds its CS coat and synthesizes a new one. Moreover, each CS protein has multiple copies of the antibody-attaching site, forcing the victim to use up copious amounts of antibody.

Although the immune system is largely ineffective in preventing malaria, it can reduce the symptoms of some forms of the disease to the point where victims can lead normal lives. The two major exceptions are, tragically, pregnant women and children under five. Pregnancy suppresses the immune system, so malaria often strikes pregnant women, resulting in miscarriages, stillbirths, and prematurity. The immune systems of children under five are not yet mature enough to protect them.

The Social Dimension of Malaria Research

All we know about the biology of malaria was accumulated bit by bit by knowledgeable scientists motivated by the challenge of a difficult problem and the wish to make a lasting and favorable impression on humanity. Students should know, however, that knowledge, curiosity, and good intentions do not ensure the smooth progress of science. In any kind of research with potentially important applications, social forces almost always intervene. The history of malaria research provides numerous examples of why scientists need to be aware of the political, economic, and cultural environment in which they work.

Malaria was described by Egyptians as early as 1500 B.C., so *Plasmodium* and *Anopheles* have had many centuries to adapt to the ways of *Homo*. One of the characteristics of humanity that *Plasmodium* has made good use of is the desire of some people to dominate others. Infected slave traders and Crusaders probably brought malaria into Europe from Africa and the Middle East. The Spanish Conquistadores then apparently brought it to America. Until the germ theory of disease was established, many Europeans probably thought malaria was God's punishment and racked their brains (and each other) trying to figure out what sin they had committed.

After Robert Koch and Louis Pasteur established that microbes cause diseases and that those diseases can be cured and prevented, the search began for a biological cause of malaria. Just as colonialism had been a major factor in spreading the disease, it became a major motive for stopping it. Malaria was a nuisance in European countries and the United States, but in tropical colonies it was an absolute plague, decimating not only natives, but also legions of white soldiers, merchants, and administrators. It is no coincidence that much of the research on malaria was by physicians attached to colonial armies.

In 1880 Louis Alphonse Laveran, a physician with the French army in Algeria, observed exflagellation in blood from a malaria patient and concluded that a protozoan caused malaria. Acceptance of Laveran's discovery was slow in coming, however. Koch and Pasteur had emphasized that bacteria cause disease; the claim by an unknown military physician that protozoans can cause infection did not count for much in comparison. After seven years, however, most physicians were convinced, and Laveran was awarded the Nobel Prize in 1907.

The next major step—the discovery of how malaria is transmitted—was due to Surgeon-Major Ronald Ross, a physician with the British army in India. Ross was passionately fond of poetry and mathematics, and he studied medicine only to satisfy his father. In India, however, he eventually became interested in malaria, which he studied mainly in his spare time. After pursuing many false leads, Ross discovered that a few species out of the swarms of mosquitoes transmitted the protozoan. He then succeeded in tracing the development of *Plasmodium* within the mosquito. Ross' superiors rewarded this crucial breakthrough in malaria research by assigning him to work on leishmaniasis as well!

Ross's discovery of how *Anopheles* transmits malaria contributed to the control of the disease, but probably not as effectively as economic development. In the United States many breeding areas for *Anopheles* were drained, but not so much to eliminate malaria as to increase the value of real estate. *Anopheles* was also inhibited by better housing and window screens and by mechanization, which allowed farm communities to spread out. Another economic factor was the cultivation of the tree *Cinchona ledgeriana,* from which the antimalarial drug quinine was derived. Quinine had been introduced to Europeans by South American natives in the 17th century, but it was too expensive for most people until *Cinchona* plantations were established in Southeast Asia. Quinine somehow disrupts the formation of merozoites in red blood cells and was the only treatment for malaria until the 1930s.

Americans had largely ceased to worry about malaria until Japan occupied the cinchona plantations during World War II. Scientists were then put to work developing alternatives to quinine, such as chloroquine. At about the same time, DDT was developed, and the eradication of mosquitoes from large areas became feasible. After the war the World Health Organization began a massive campaign to exterminate *Anopheles,* and therefore malaria, with DDT. Parts of the world did not have the political or cultural organization for such an effort, but in other places the campaign was an overwhelming success. By the mid-1960s, after spending $6 billion (a lot of money in those days), malaria had been eliminated from 80% of its former range, and the number of cases dropped from 350 million in 1948 to 100 million in 1965. Most governments were so confident that malaria was doomed that they saw no rea-

box continues

son for further research, and young scientists avoided choosing malaria as a research topic.

It soon became clear, however, that *Anopheles*, like many other insects, was evolving resistance to DDT and that *Plasmodium* was evolving resistance to the antimalarial drugs. In the Vietnam War probably more soldiers were put out of action by chloroquine-resistant malaria than by bullets. Since then the number of malaria victims has skyrocketed and is now increasing at a rate of about 5% annually. Developed countries without large numbers of troops at risk were unconcerned until recently. The total spent in the United States was reduced in the late 1980s to $31 million per year. Malaria, along with amebic dysentery, trypanosomiasis, and hookworm, has therefore become one of the Great Neglected Diseases of Mankind. In recent years, however, there has been a renewed interest in controlling malaria, with growing optimism that a vaccine can be developed.

Another type of natural defense against malaria is **sickle cell trait,** often found in those whose ancestors came from malaria-infested parts of Africa. A single base substitution in the gene for the alpha chain of hemoglobin causes the red blood cell to become sickle-shaped in low oxygen. The change in shape causes potassium to leak out of the red blood cell, and the low K^+ concentration kills the trophozoites. Presumably sickle-cell trait evolved because the advantage of resistance to malaria outweighed the disadvantages of sickle-cell anemia (see p. 100). **Thalassemia,** a genetic variant common around the Mediterranean and in Southeast Asia, also confers resistance to malaria. In thalassemic red blood cells, the membrane is damaged by hydrogen peroxide released by trophozoites. This allows K^+ to escape, killing the parasite.

Summary

Although most protozoans consist of only one cell, they nevertheless carry out all the functions of life. The large ratio of surface to volume eliminates the need for respiratory and circulatory systems, leaving locomotion, feeding, and reproduction as the main tasks. Protozoans were formerly classified into four groups—sarcodines, ciliates, flagellates, and parasitic sporozoans—on the basis of their modes of locomotion. Many sarcodines move on surfaces by forming pseudopodia into which their cytoplasm flows. Ciliates and flagellates propel themselves through the relatively enormous viscosity of water. Sporozoans generally drift passively or are transported within hosts. Classification based on mode of locomotion is no longer accepted, although the informal names for those groups are still useful.

Pseudopodia, cilia, and flagella are also used in feeding. Sarcodines use their pseudopodia to trap and absorb food by phagocytosis. Ciliates and flagellates set up water currents from which they filter food particles into their cytostomes. Many protozoans perform intricate behaviors, such as phototaxis and chemotaxis, in spite of the absence of a nervous system.

Most protozoans reproduce asexually by fission. Some, however, recombine genetic material from two individuals and are therefore said to be sexual. Conjugation is a common form of sexual reproduction in ciliates. Many protozoans, especially parasites, form protective cysts at some stage in their life cycles. It is generally assumed that animals evolved from a protozoan. The syncytial ciliate and the blastea theories of animal origins are now mainly of historical interest. It now appears more likely that animals shared a common ancestor with choanoflagellates.

Foraminiferans were responsible for massive stone and chalk deposits and serve as indicator fossils for petroleum geologists. Protozoans are also important links in food chains and as decomposers. They also form important symbioses with some animals, such as reef-building corals. Many protozoans cause diseases in humans and other animals. One of the most serious is malaria, which is caused by *Plasmodium* and is transmitted by *Anopheles* mosquitoes.

Key Terms

pseudopodium
trophozoite
binary fission
conjugation
syngamy
cyst
spore
flagellate
sarcodine
ciliate
cytostome
phagocytosis

food vacuole
contractile vacuole
macronucleus
micronucleus
zooxanthella
final host
intermediate host
sporozoite
merozoite
gametocyte

Chapter Test

1. Based on its appearance, characterize each of the protozoans in Fig. 22.1 as flagellate, ciliate, or sarcodine. (Why aren't sporozoans represented here?) (p. 453)
2. In a pool of stagnant water, where would you expect to find the following types of Protozoa, and why? Photosynthesizing flagellates, other flagellates, amebas. (p. 458)
3. Describe a process of asexual reproduction in a particular protozoan. Describe a process of sexual reproduction in a particular protozoan. (pp. 460–463)
4. Sketch a *Paramecium* from memory, making sure to include the following organelles: cilia, contractile vacuole, food vacuole, cytostome, cytoproct, macronucleus, micronucleus. State the function of each of these organelles. (Fig. 22.3)
5. Describe two theories suggesting that animals evolved from a colonial flagellate. What kind of evidence is there for and against each theory? (p. 463)
6. What are foraminiferans and radiolarians? In what ways are they similar? How do they differ from each other? Why are geologists so interested in them? (pp. 454, 464)
7. Describe three diseases of humans or other animals caused by protozoans. Explain how the protozoan produces the disease symptoms, and how it is transmitted. (pp. 466–470)
8. Starting with the sporozoite, arrange in correct sequence the following stages of *Plasmodium*. State where each stage occurs—which host and which organ. Sporozoite, gametocyte, oocyst, schizont, trophozoite, merozoite, gamete. (pp. 469–470)

Answers to the Figure Questions

22.6 In fact, *Amoeba* probably does not often ingest ciliates unless electron microscopists arrange for them to do so. Occasionally, however, a ciliate might blunder into *Amoeba* and stick.

22.10 The original conjugants never really die, but parts of them remain in the daughter cells of succeeding generations. This differs from the situation in animals, in which each offspring is a totally new organism separate from the parents.

22.12 The similarity of these tests to the valves of molluscs is undoubtedly an example of evolutionary convergence. For example, the spiral shape of some of these tests resembles that of *Nautilus* (see Fig. 28.24B) probably because this allows the test to remain compact even as the organism is growing.

22.16 (1)*Plasmodium* does form a protective cyst in the mosquito. (2) The sporozoite does frequently shed its circumsporozoite proteins. (3) *Plasmodium* does live primarily within cells of the human host. (4) *Plasmodium* does reproduce rapidly within red blood cells and overwhelms the immmune system when the red blood cells rupture. (5) *Plasmodium* does have a two-host life cycle in which virulence is low in the mosquito vector and high in the human final host. (6) *Plasmodium* can have reservoir species. (7) Sending the human host to bed in a debilitated, feverish state may make the host more vulnerable to mosquitoes.

Readings

Recommended Readings

Anderson, D. M. 1994. Red tides. *Sci. Am.* 271(2):62–68 (Aug).
Desowitz, R. S. 1991. *The Malaria Capers.* New York: W. W. Norton. (*Entertaining account of biological and sociological factors in kala-azar and malaria.*)
Donelson, J. E., and M. J. Turner. 1985. How the trypanosome changes its coat. *Sci. Am.* 252(2):44–51 (Feb).
Ewald, P. W. 1993. The evolution of virulence. *Sci. Am.* 268(4):86–92 (Apr).
Friedman, M. J., and W. Trager. 1981. The biochemistry of resistance to malaria. *Sci. Am.* 244(3):154–164 (Mar). (*How sickle cell trait and thalassemia confer resistance to malaria.*)
Godson, G. N. 1985. Molecular approaches to malaria vaccines. *Sci. Am.* 252(5):52–59 (May).
Harrison, G. 1978. *Mosquitoes, Malaria and Man: A History of the Hostilities Since 1880.* New York: E. P. Dutton. (*An entertaining and instructive history of attempts to control malaria.*)
Jahn, T. L., E. C. Bovee, and F. F. Jahn. 1979. *How to Know the Protozoa,* 2nd ed. Dubuque, IA: Wm. C. Brown. (*A guide to identification.*)
Kabnick, K. S., and D. A. Peattie. 1991. *Giardia:* A missing link between prokaryotes and eukaryotes. *Am. Sci.* 79:34–43.
Kolata, G. 1984. Scrutinizing sleeping sickness. *Science* 226:956–959.
Lewin, R. 1982. Nairobi laboratory fights more than disease. *Science* 216:500–503.
Rennie, J. 1992. Living together. *Sci. Am.* 266(1):122–133 (Jan).
Wakelin, D. 1984. *Immunity to Parasites: How Animals Control Parasitic Infections.* Baltimore: Edward Arnold.
See also relevant selections of General References listed at the end of Chapter 21.

Additional References

Baroin, A., et al. 1988. Partial phylogeny of the unicellular eukaryotes based on rapid sequencing of a portion of 28S ribosomal RNA. *Proc. Natl. Acad. Sci. USA* 85:3474–3478.

Ewald, P. W. 1994. *Evolution of Infectious Diseases.* New York: Oxford University Press.

Fenchel, T. 1986. *Ecology of Protozoa: The Biology of Free-Living Phagotrophic Protists.* Madison, WI: Science Tech/New York: Springer-Verlag.

Field, K. G., et al. 1988. Molecular phylogeny of the animal kingdom. *Science* 239:748–753.

Gould, S. J. 1980. Crazy old Randolph Kirkpatrick. In: *The Panda's Thumb.* New York: W. W. Norton, Chapter 22.

Harrison, F. W., and J. O. Corliss (Eds.). 1991. *Microscopic Anatomy of Invertebrates. Vol. 1: Protozoa.* New York: Wiley-Liss.

Jeon, K. W. (Ed.). 1961. *The Biology of Amoeba.* New York: Academic Press.

Lee, J. J., S. H. Hutner, and E. C. Bovee (Eds.). 1985. *The Illustrated Guide to the Protozoa.* Lawrence, KA: Society of Protozoologists.

Margulis, L., et al. (Eds.). 1990. *Handbook of Protoctista.* Boston: Jones and Bartlett.

Margulis, L., H. I. McKhann, and L. Olendzenski (Eds.). 1993. *Illustrated Glossary of Protoctista.* Boston: Jones and Bartlett.

Marshall, E. 1990. Malaria research—what next? *Science* 247:399–400.

Qu, L-H., M. Nicoloso, and J-P. Bachellerie. 1988. Phylogenetic calibration of the 5′ terminal domain of large rRNA achieved by determining twenty eucaryotic sequences. *J. Mol. Evol.* 28:113–124.

Rausch, H., N. Larsen, and R. Schmitt. 1989. Phylogenetic relationships of the green alga *Volvox carteri* deduced from small-subunit ribosomal RNA comparisons. *J. Mol. Evol.* 29:255–265.

Slaytor, M. 1992. Cellulose digestion in termites and cockroaches: what role do symbionts play? *Comp. Biochem. Physiol.* 103B:775–784.

Sleigh, M. A. 1989. *Protozoa and Other Protists.* New York: Routledge, Chapman and Hall.

Smothers, J. F., C. D. von Dohlen, L. H. Smith, Jr., and R. D. Spall. 1994. Molecular evidence that the myxozoan protists are metazoans. *Science* 265:1719–1721.

Sogin, M. L., and H. J. Elwood. 1986. Primary structure of the *Paramecium tetraurelia* small-subunit ribosomal RNA coding region: phylogenetic relationships within the ciliophora. *J. Mol. Evol.* 23:53–60.

Sogin, M. L., H. J. Elwood, and J. H. Gunderson. 1986. Evolutionary diversity of eukaryotic small-subunit rRNA genes. *Proc. Natl. Acad. Sci. USA* 83:1383–1387.

Sogin, M. L., et al. 1989. Phylogenetic meaning of the kingdon concept: An unusual ribosomal RNA from *Giardia lamblia. Science* 243:75–77.

Spoon, D. M., et al. 1976. Observations on the behavior and feeding mechanisms of the suctorian *Heliophrya erhardi* (Rieder) Matthes preying on *Paramecium. Trans. Am. Microsc. Soc.* 95:443–462.

Wainright, P. O., G. Hinkle, M. L. Sogin, and S. K. Stickel. 1993. Monophyletic origins of the Metazoa: an evolutionary link with fungi. *Science* 260:340–342.

Walters, J. 1991. The troublesome parasites—molecular and morphological evidence that Apicomplexa belong to the dinoflagellate-ciliate clade. *BioSystems* 25: 75–83.

CHAPTER

23

Sponges, Placozoa, and Mesozoa

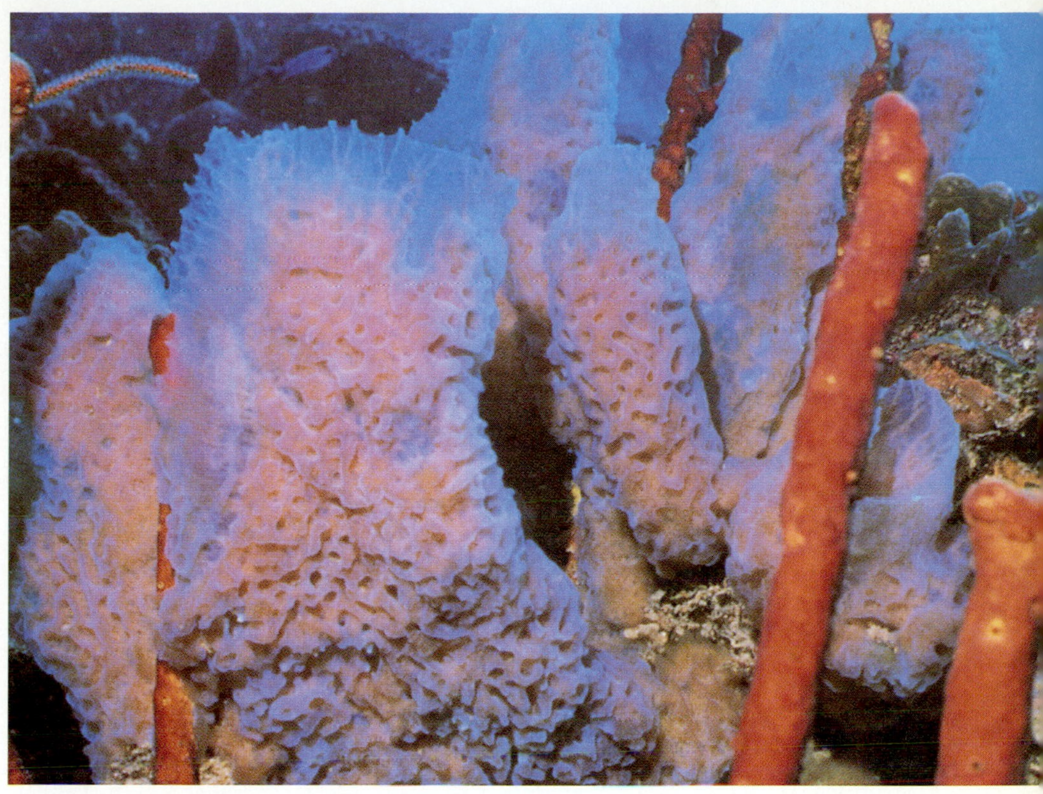

Azure vase sponge *Callyspongia*.

Are Sponges Animals?

The boundaries separating the kingdoms of life are as uncertain as those that separated the kingdoms of humans before there were accurate surveys and maps. Some of the colonial protozoans seem on the verge of invading the animal kingdom, while the animals discussed in this chapter seem to owe their allegiance to the Protista. Sponges, for example, resemble certain colonial protozoans (see pp. 458, 463). Aristotle considered sponges to be animals in spite of the fact that they do not actively move. It was not until 1766 that Aristotle's belief was confirmed by the discovery that sponges expel water. They were then thought to be related to corals (phylum Cnidaria) until one of Darwin's teachers, R. E. Grant, proposed for them the separate phylum Porifera.

Systematists now avoid such boundary disputes by defining animals as organisms that:

1. Are multicellular, with more than one type of cell;
2. Are heterotrophic (cannot produce their own food);
3. Reproduce sexually (at least sometimes) from a zygote formed from two different haploid gametes;
4. Go through a blastula stage during embryonic development.

Although sponges resemble colonial protozoans in some ways, having many cells of different types gives them and all other animals distinct advantages:

1. Animals are generally larger than protozoans. Greater body mass enables animals to move actively rather than drift with currents, and it provides an advantage in predation and avoidance of predators.
2. Having many cells allows each cell to specialize. Thus, unlike a protozoan, whose structure is a compromise for all the different roles it must play, each cell of an animal can be adapted for one particular role.
3. Having many specialized cells enables them to form organs that maintain an internal environment—the interstitial fluid between cells. Unlike the environment of protozoans, the internal environment of animal cells can be regulated for optimal levels of nutrients, oxygen, ions, and sometimes temperature, as discussed in Unit 2.

Of course, with each advantage there are also disadvantages. Having many specialized cells in a large body generally requires neural and hormonal mechanisms to coordination their activities. However, the organisms described in this chapter—Porifera, Placozoa, and Mesozoa—manage to do without such coordination. It is sometimes said that they have a cellular grade of organization rather than an organ grade of organization. These animals are so different that they are often considered to have evolved early and parallel to other metazoans, as represented in Fig. 21.10. The phyla Porifera and Placozoa are therefore often referred to as the Parazoa (Greek *para* beside + *zoon* animal). This view that Porifera and Placozoa separated early from the main line of animal evolution is supported by molecular phylogenetics (Christen et al. 1991, Wainright et al. 1993). As its name implies, the phylum Mesozoa (Greek *mesos* middle) has been thought for more than a century to be intermediate between protozoa and metazoa. This conjecture has also found support in molecular phylogeny (Hori and Osawa 1987).

SPONGES: PHYLUM PORIFERA

Introduction

Having a relatively large body with actively moving cells enables sponges to move water through internal channels and to extract nutrients and oxygen from it. The most significant features of sponges are all related to this basic plan (Table 23.1). These features are (1) the canals or spaces through which water circulates in the body, (2) the cells that pump the water, and (3) the endoskeleton that keeps the body from collapsing (Fig. 23.1 on page 478). The canal system is lined with **choanocytes,** also called collar cells, which resemble choanoflagellate protozoans such as *Codosiga* (Fig. 22.5). Indeed, it is not hard to imagine a sponge as a colony of *Codosiga* that has proliferated into a solid mass. The beating of the flagella of the choanocytes draws a continuous stream of water through the canal system, bringing food and oxygen to the sponge and removing wastes. Surrounding the flagellum of each choanocyte is a fringe of microvilli, the collar, which traps the food that the choanocyte then phagocytizes. A unique skeleton of protein fibers and mineral **spicules** (little spikes) keeps the whole thing from collapsing.

TABLE 23.1

Characteristics of sponges.

Phylum Porifera pore-IF-er-uh (Latin *porus* pore + *ferre* to bear)

Morphology	Multicellular animals without organs. Asymmetric or with radial symmetry. Numerous microscopic **ostia** by which water enters the **canal system** or spaces through the body. One or few **oscula** from which water exits. Water propelled by **choanocytes**. **Endoskeleton** of protein fibers and/or mineral spicules. No germ layers in embryo.
Physiology	A **cellular grade of organization**: many physiological functions performed by individual cells that are not coordinated by neural or hormonal systems. Osmoregulation in freshwater species by contractile vacuoles.
Locomotion	Adults **sessile**. Larvae swim by means of flagella.
Reproduction	**Asexual** reproduction by buds, and **sexual** reproduction. Most species hermaphroditic. Fertilization external or internal.
Development	Radial cleavage pattern. Development indirect (with a juvenile stage, called a larva, that does not resemble the adult).
Habitat, Size, and Diversity	All species aquatic; most marine. Distribution worldwide, including Arctic and Antarctic. Sizes range from 1 mm to 2 meters. Dull white or gray, or with red, orange, yellow, or purple pigments. Some freshwater species green from symbiotic algae. Approximately **5000 living species** described.

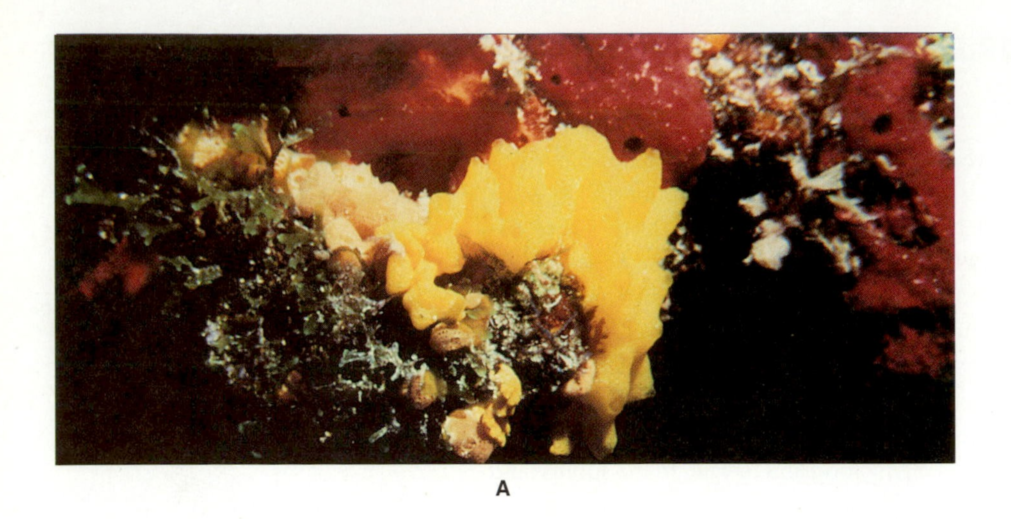

A

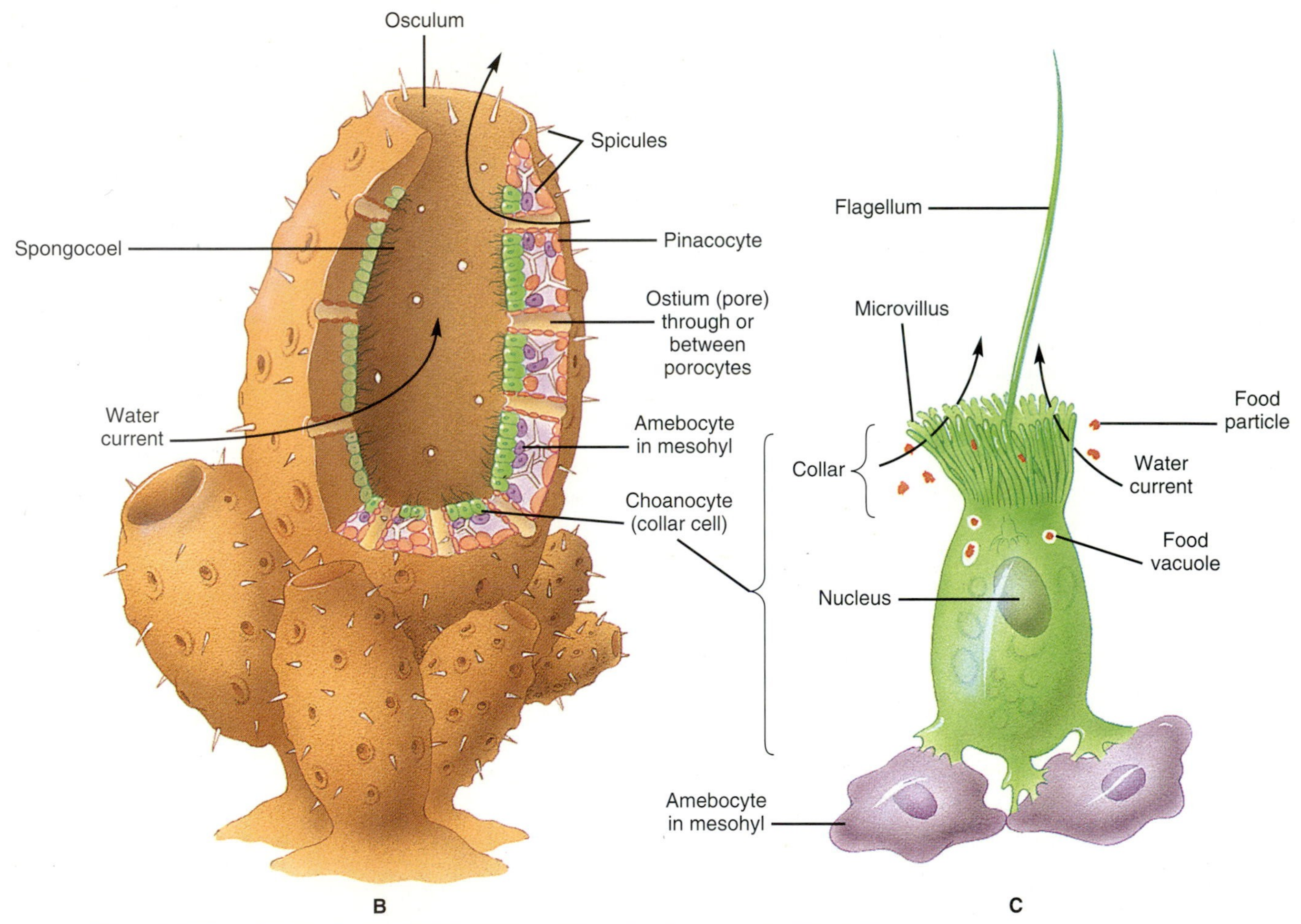

FIGURE 23.1

(A) The yellow sponge *Leucosolenia canariensis* has a simple organization. The red sponge is *Haliclona rubens.* (B) A section through *Leucosolenia,* showing the canal system formed by ostia, the central chamber (spongocoel), and the osculum. (C) The mechanism of suspension feeding. Water is drawn through the canal system by the beating of flagella on the choanocytes. Food particles are trapped in the collars of the choanocytes, which are formed by 30 to 40 microvilli. The choanocytes then phagocytize the food.

The outer layer of a sponge consists of cells called **pinacocytes** (Table 23.2). The life-giving water is drawn through the pinacocyte layer through numerous microscopic **ostia** (the pores that give poriferans their name). It exits by one large **osculum** or several oscula. Sandwiched between the pinacocytes and canal system is a jellylike **mesohyl** (often called mesenchyme or mesoglea). The cells of the mesohyl do not form a tissue, since they are not interconnected. Moreover, no sponge cell is coordinated with others by hormones or nerves, and many continue to function even after isolation from the body. Other cells of the mesohyl include several kinds of **amebocytes,** which, among other functions, ferry food from the choanocytes to other cells.

"Mesohyl," "mesoglea," and "mesenchyme" are three of the confusing terms applied to the gelatinous material between inner and outer cell layers in a variety of invertebrates. Some of these terms are defined below. This text uses the term most common for each group of animals.

Mesenchyme: *Middle layer consisting of cells and cell products in a jellylike mesoglea. Called mesohyl in sponges.*

Mesoglea: *Jellylike portion of mesenchyme. Synonymous with mesenchyme that contains no cellular material, as in many cnidarians and ctenophores.*

Collenchyme: *Mesenchyme with little cellular material, as in some cnidarians and ctenophores.*

Parenchyme: *Mesenchyme in which cells are densely packed, especially in acoelomates. Often spelled parenchyma.*

TABLE 23.2

Cells of sponges.

CELLS OF INNER AND OUTER LAYERS

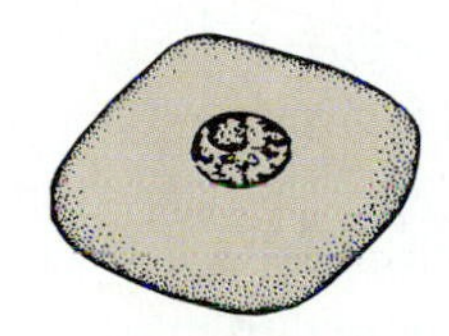

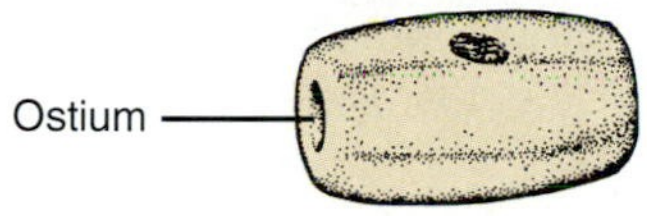

Choanocytes (= collar cells; Fig. 23.1) form a flagellated cell layer (= choanoderm). They:

1. Propel water through the sponge with their flagella.
2. Trap food particles with the collar, phagocytize them, and transfer them to amebocytes.
3. Give rise to cells that form sperm and ova in some species.

Pinacocytes line the outer surface (= epidermis, pinacoderm) and the nonflagellated channels in most sponges.

Porocytes are tubular pinacocytes that surround some pores (except in hexactinellids). They can regulate water flow by contracting.

CELLS IN MESOHYL

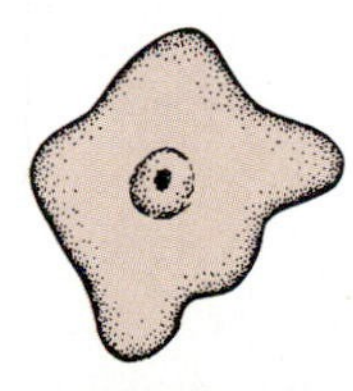

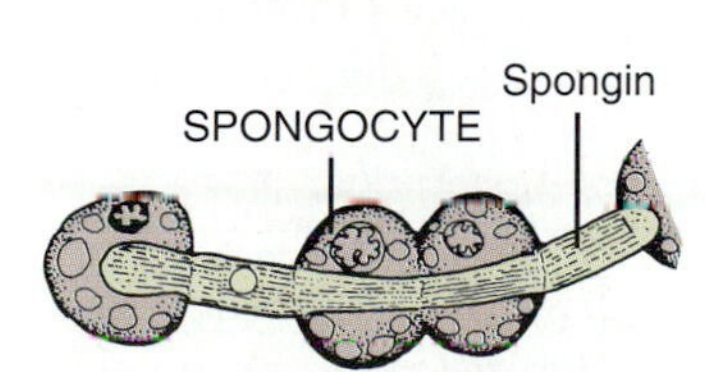

Myocytes contract around oscula and channels, thereby regulating water flow (in some demosponges).

Amebocytes

Archaeocytes:

1. Undergo differentiation into any other cell type.
2. Transport food vacuoles from choanocytes into mesohyl.
3. Form sperm and ova in some species.

Sclerocytes (= scleroblasts) form spicules. (Founder and thickener cells; Fig. 23.3.)

Collencytes, **lophocytes**, and **spongocytes** form fibers of the protein collagen. The spongocytes secrete a form of collagen called **spongin**, which is unique to sponges.

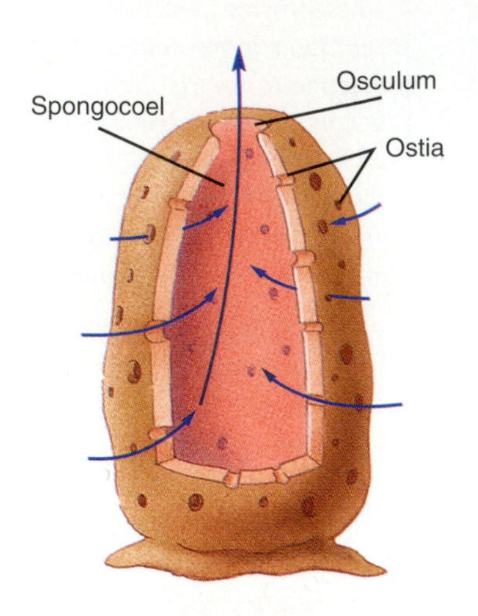

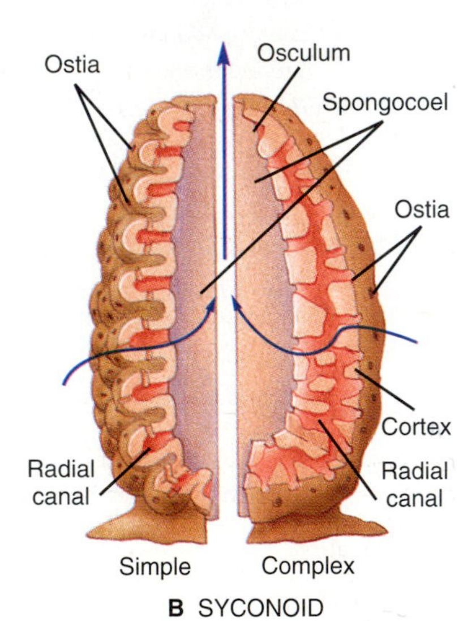

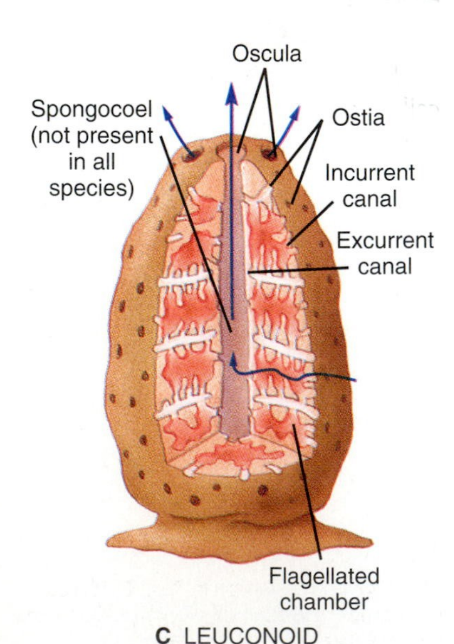

Canal Systems

As noted above, the basic body plan of most sponges is a system of canals through which water brings in food and oxygen. Because there is no other circulatory system, each cell has to be close to the water, and because water often has little organic material in it, quite a lot of it has to be pumped through. According to one estimate, a marine sponge has to pump 30,000 grams of water to add 1 gram of cells to its mass. As a sponge grows, however, the volume of cells that have to be supplied increases as the cube of body length, while the area available for choanocytes increases only as the square of body length. This fundamental concept therefore limits the sizes of sponges. Sponges with the simplest body plans have an **asconoid** (bag-shaped) canal system (Greek *askos* bag). In asconoid systems the choanocytes occur only on the lining of the central chamber called the **spongocoel** (pronounced SPUN-jo-seal) (Figs. 23.1A and 23.2A). Because of the limited surface area for choanocytes, asconoid sponges never grow larger than 10 centimeters.

Other sponges can grow larger because they have more elaborate canal systems that provide more area for choanocytes. In the **syconoid** canal system (Fig. 23.2B) the choanocytes do not line the spongocoel but are found within numerous **radial canals.** The spongocoel of syconoid sponges serves merely as an exhaust leading to the osculum. Still more area is available for choanocytes in the majority of sponges, which have **leuconoid** canal systems. In most leuconoids there is no spongocoel at all. The choanocytes are located within numerous **flagellated chambers,** which draw in water from **incurrent canals** and expel it by an **excurrent canal** directly to the osculum. One 7-cm leuconoid sponge was estimated to have several million flagellated chambers. The large number of choanocytes enables some leuconoid sponges, such as the loggerhead, *Spheciospongia vesperia,* to reach a height of 2 meters and to pump approximately 2 liters of water every minute of every day.

The Endoskeleton

The canal system will function only if the body can be prevented from collapsing. That is the function of the endoskeleton, which consists of a network of protein fibers and (usually) mineral spicules. The fibers are made of collagen, including a unique form called **spongin**. The spicules are either **calcareous** or **siliceous** and range in length from 10 μm to 40 cm, depending on species. Calcareous spicules are composed mainly of calcium carbonate (in the form of calcite or aragonite), and siliceous spicules are made of a glassy mineral made with silicon. Contrary to the general assumption, recent evidence suggests that spicules do not deter predators (Pawlik 1993).

Some siliceous spicules are constructed on a strand of protein that serves as a kind of scaffolding. Cells called **sclerocytes** secrete the silica onto the strand, and the silica then crystallizes into a spicule with the required shape. Some calcareous spicules are constructed without such a scaffolding. In order to construct a three-rayed spicule, three sclerocytes congregate at the construction site, and each divides into a **founder**

FIGURE 23.2
The types of canal system found in sponges. Darkened areas show the locations of choanocytes. (A) In asconoid sponges the choanocytes are restricted to the spongocoel. (B) In syconoid sponges, named for the genus *Sycon* (Fig. 23.4A), the choanocytes occupy the sides of radial canals. The left and right halves of this figure show simple and complex versions of the syconoid arrangement. In the simple syconoid each ostium leads into only one radial canal. In the complex system each ostium conducts water into more than one radial canal. (C) Leuconoid sponges (named for *Leuconia,* now *Leucandra*) have flagellated chambers distributed throughout.

cell and a **thickener cell** (Fig. 23.3). Each founder cell moves away, laying down a ray of calcite, and the thickener cells then move along these rays adding more $CaCO_3$. When the spicule is completed, all six cells slide off the rays and disintegrate. Virtually nothing is known about the factors that guide sclerocytes in constructing the spicules appropriate for their species.

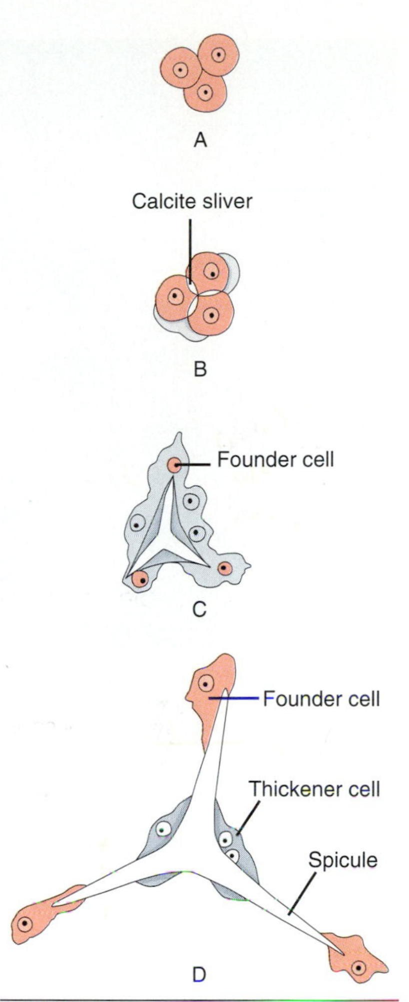

FIGURE 23.3

Formation of a spicule. (A) Three sclerocytes assemble. (B) They develop into three thickener cells and three founder cells; the latter lay down a sliver of calcite. (C) The founder cells move apart, elongating the calcite rays. (D) The thickener cells add to the spicule thickness.

■ **Question:** How do you think these cells coordinate their activities without nerves or hormones?

Classification

The spongy nature of poriferans makes it difficult to classify them on the basis of their shapes. Taxonomists have therefore had to rely largely on their only solid and regular parts, the spicules. Three classes of sponges are now generally recognized, based mainly on the composition and shape of spicules.

Class Calcarea (= class Calcispongiae) is characterized by calcite spicules with three or four rays (Fig. 23.4). Calcarea are commonly called calcareous, limy, or chalky sponges. Their bodies are bristly with spicules, and they are usually dull-colored and less than 4 cm long. Calcareous sponges are all marine and are more common in shallow than in deep water.

Sponges in **class Hexactinellida** (= class Hyalospongiae), commonly known as the glass sponges, are named for the six-rayed siliceous spicules. Often their spicules are cemented together into a roughly cylindrical or conical skeleton 10 to 30 cm tall, as in the beautiful Venus' flower basket *Euplectella aspergillum* (Fig. 23.5 on page 482). This class is one of the least understood, owing to their inaccessibility in deep oceans. They are never found in depths less than 23 meters, and most live between 200 and 2000 meters. Some have been dredged up from depths greater than 6 km. Structural peculiarities include the lack of mesohyl and epidermis (pinacoderm). Glass sponges also lack true canal systems, since the water passes through spaces rather than canals on its way to the flagellated chambers. Almost all the cells are syncytial (fused). This unusual feature has led some zoologists to classify the Hexactinellida in their own subphylum, or even in a separate phylum Symplasma.

FIGURE 23.4

(A) *Sycon,* a member of class Calcarea. (B) Three- and four-rayed spicules from various calcareous sponges.

FIGURE 23.5

(A) The skeleton of Venus' flower basket *Euplectella aspergillum,* a hexactinellid. Note the fusion of spicules and the spun-glass appearance at the end. Approximately 30 cm long. (B) Spicules of hexactinellids are generally six-rayed.

Species in **class Demospongiae** also have siliceous spicules (or none at all), but the spicules are not six-rayed (Fig. 23.6). This class includes 90% of all species, mainly because taxonomists tend to include within it any sponge they cannot fit into the other classes. Demosponges can be found at all depths of the ocean down to 9 km, although each species has its preferred depth. All 150 of the freshwater species of Porifera belong to this class. Also included in this class are the bath sponges (*Spongia* and *Hippospongia*), which (obviously) do not have spicules.

Classes Calcarea and Demospongiae also include the **coralline sponges,** which are so remarkable that for about two decades they were given their own class (Sclerospongiae). The skeletons of these reef-building sponges consist of a large base of calcium carbonate, in addition to siliceous spicules. Cells grow in a thin layer within

Extant Classes of Phylum Porifera

Genera mentioned elsewhere in this chapter are noted.

Class Calcarea kal-sa-REE-uh (Latin *calcis* limestone). Skeleton composed of spicules of calcium carbonate, needle-shaped or with three or four rays. Includes species with asconoid, syconoid, and leuconoid canal systems. All marine. *Leucandra, Leucosolenia, Sycon* (Figs. 23.1, 23.4).

Class Hexactinellida hex-ACT-in-ELL-id-uh (Greek *hex* six + *aktis* ray). Siliceous six-rayed spicules, often fused into networks. Most cells syncytial. All marine. *Euplectella* (Fig. 23.5).

Class Demospongiae dim-oh-SPONGE-ee-ee (Greek *demos* people + *spongia* sponge) Skeleton of siliceous spicules and/or spongin fibers, or (rarely) neither. Spicules are not six-rayed as in the hexactinellids. All leuconoid. *Callyspongia, Cliona, Cryptotethya, Halichondria, Haliclona, Hippospongia, Microciona, Spheciospongia, Spongia, Spongilla, Verongia* (Chapter opener; Figs. 23.6, 23.7, 23.8).

A

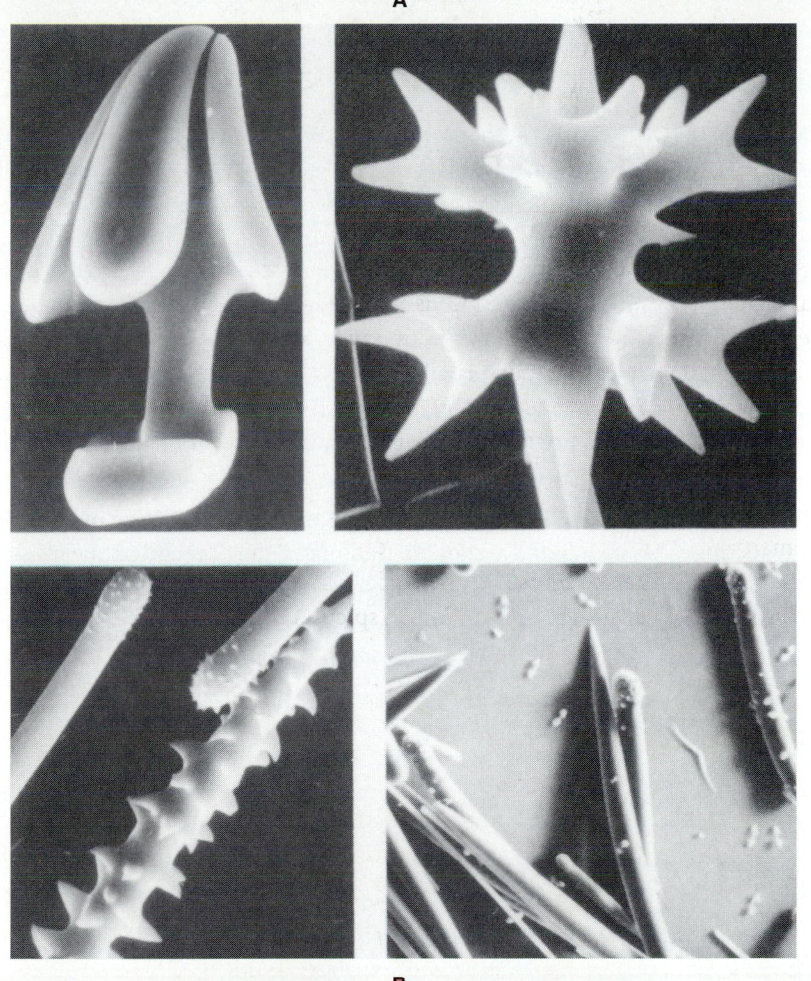

B

FIGURE 23.6

(A) Crumb-o'-bread sponge *Halichondria panicea,* so called because dried specimens resemble moldy bread. This demosponge encrusts rocks and seaweed and can often be seen at low tide. (B) Scanning electron micrographs of spicules from various demosponges.

and over the base. Also distinctive are bulging excurrent canals that form star-shaped patterns (astrorhizae) radiating from each osculum. Fossil coralline sponges were long known, but all were thought to have become extinct until the 1960s. Some living species had been misclassified as corals. Many others had remained undiscovered on steep, dark reef banks where they were missed by dredging equipment.

Reaggregation

One of the most remarkable displays of a cellular grade of organization occurs when cells of sponges are separated from each other. Not only can many survive by themselves for a considerable time, but in some species they can actually reassemble into a working sponge. This phenomenon, called reaggregation, was discovered by H. V. Wilson around the turn of the century. It was long forgotten until some scientists began to use it as a model system for the study of organ transplantation.

Wilson forced a redbeard sponge *Microciona prolifera* through a cloth to separate its cells. When allowed to stand in seawater for a few weeks, the cells reaggregated into a functioning sponge again. Wilson also mixed cells from the red *M. prolifera* with cells from the purple *Haliclona oculata,* and within a few weeks they somehow sorted themselves out and reaggregated into one red and one purple sponge again. Some sponges not only refuse to aggregate with cells of another species, but they also will not reaggregate with cells of another individual of the same species. Generally, however, sponges of the same species readily merge with each other. Reaggregated cells link together by means of a complex of proteins and carbohydrates called **aggregation factor.**

Reproduction

Asexual Reproduction. Sponges can reproduce asexually by **budding.** In some species, mainly the hexactinellids, archaeocytes form external buds, usually on long stalks. The buds then drift away and form new sponges. Freshwater sponges and some marine species, such as the boring sponge *Cliona lampa,* can reproduce from internal buds, called **gemmules.** Gemmules are small (0.3 mm), spherical structures capable of surviving environmental stresses, such as would be faced in a pond in winter (Fig. 23.7). Before winter some unknown stimulus, perhaps reduced light or temperature, triggers gemmule formation. Archaeocytes group together, and a tough coat of spongin and spicules surrounds them. In spring the archaeocytes in the gemmules develop into choanocytes and all the other cell types required to start a new sponge.

Sexual Reproduction. Like all animals, sponges are also capable of reproducing sexually. Because they cannot move about to find mates, they depend on the vicissitudes of water currents to ensure that sperm find an ovum of the correct species. Most individuals are hermaphroditic (monoecious), but each one usually produces ova and sperm at different times, thereby avoiding self-fertilization. The sexes look alike in dioecious

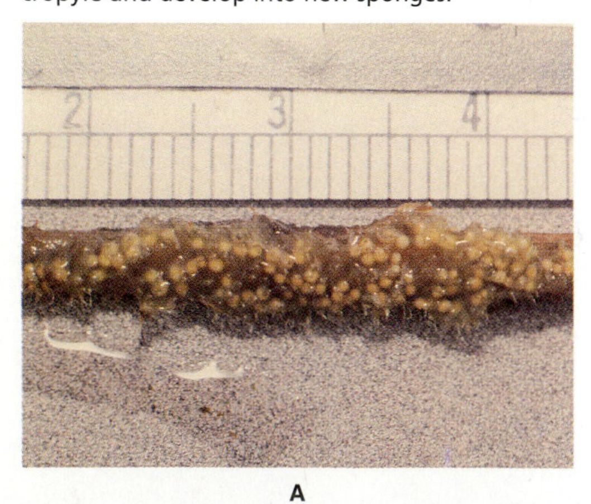

A

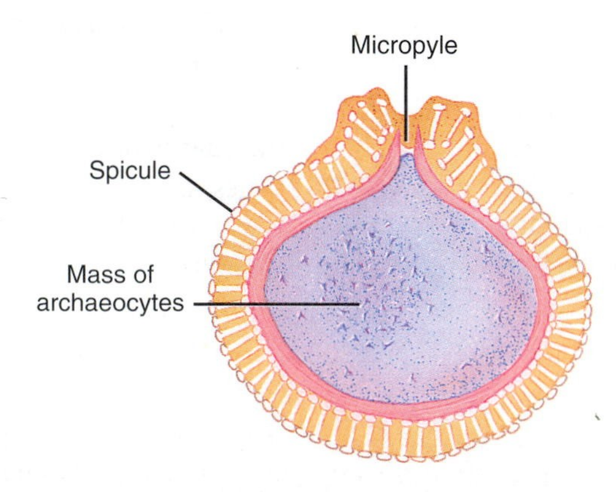

B

FIGURE 23.7

(A) The freshwater sponge *Spongilla* in autumn, filled with gemmules. This sponge is frequently encountered on submerged vegetation in sunlit ponds and streams. Symbiotic algae give it a green color in spring and summer. Scale in millimeters. (B) Diagram of a sectioned gemmule of *Spongilla.* In spring the archaeocytes will spill out of the micropyle and develop into new sponges.

species. By examining the gametes produced by each individual, it has been found that females usually outnumber males. Sperm are often released in such prodigious numbers that divers report "smoking sponges" (see Fig. 16.2A). The sperm drift downstream and are taken into the ostia of other sponges, where the choanocytes phagocytize them as if they were food. If the sperm are of the same species, however, the choanocytes do not pass them on to amebocytes for distribution as food. Instead, the choanocytes lose their collars and flagella, and they themselves carry the sperm to egg cells.

Development

Development in sponges is indirect, meaning that there is a larval form that does not resemble the mature sponge. The larvae have flagella by which they move about, which they must do to disperse to new habitats. Some species are oviparous, with the eggs being released soon after fertilization. Most sponges, however, are viviparous, with some larval development occurring in the mesohyl of the parent. Often **nurse cells** attend the developing larva. In one study a sponge gave birth to an average of one larva every 15 seconds for three or four days.

Most demosponges, and probably some species in other classes, produce larvae called **parenchymulae.** The parenchymula larva is essentially a solid blastula covered with flagella (Fig. 23.8A). Using the flagella, the parenchymulae swim about for up to two days. If they are lucky enough not to become food for some other organism and to find suitable substratum, they **settle** and begin developing into a sponge. In most calcareous sponges and some demosponges there is a different type of larva, the **amphiblastula** (Fig. 23.8B). The amphiblastula larva develops from a hollow blastula in which flagella are pointed inward within the blastocoel. The larva then undergoes **inversion,** which pushes the flagellated cells to one end (the animal pole) of the outer surface, allowing the larva to swim. The amphiblastula is then released from the parent's mesohyl and soon settles. After settling, the flagellated cells of the amphiblastula turn inward once more and become choanocytes.

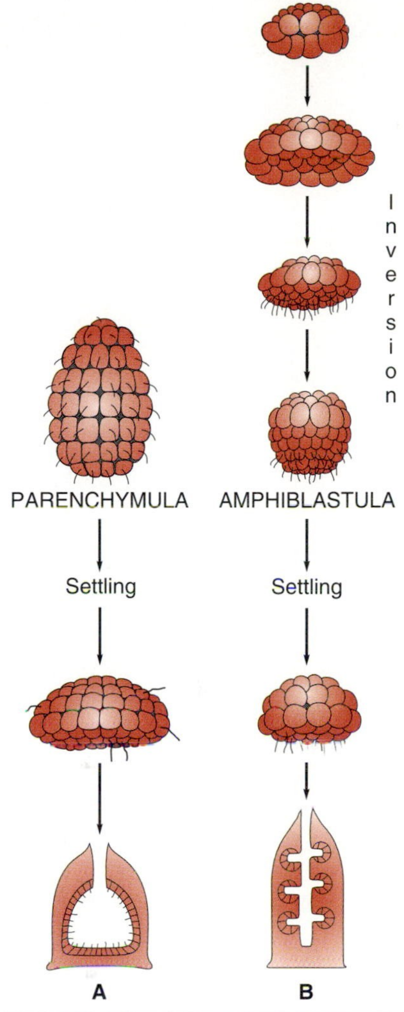

FIGURE 23.8
Development of two types of sponge larvae. (A) A parenchymula larva. It was formerly believed that the external flagellated layer inverted inward and produced the choanocyte chamber after settling, but this does not appear to be the case (Bergquist and Glasgow 1986, Misevic et al. 1990). (B) An amphiblastula larva.

Interactions with Humans and Other Animals

Considering that sponges lack teeth and claws and are not even capable of fleeing, they are remarkably free of predators. Apparently few animals relish a mouth full of spongin. However, a few bony fishes and the hawksbill turtle (*Eretmochelys imbricata*) feed exclusively on sponges (Meylan 1988). Some molluscs also prey on marine sponges, and the larvae of spongillaflies (order Neuroptera, family Sysyridae) feed on freshwater sponges. Some sponges are toxic to fishes, bacteria, and other organisms. Many other sponges, however, play host to large numbers of mutualistic bacteria. In some species, in fact, the volume of bacterial cells is twice as great as the volume of sponge cells. Many coral-reef sponges harbor photosynthesizing cyanobacteria inside them—something no other animal does.

Numerous animals live inside the channels of sponges, taking advantage of the protection and continuous flow of water. One specimen of the loggerhead sponge *Spheciospongia vesparia* from Florida was found to harbor 16,000 snapping shrimp (*Synalpheus brooksi*)—impressive hospitality even for a sponge that can grow 2 meters tall. Rachel Carson in *The Edge of the Sea* comments on the considerable racket produced by so many snapping claws. Venus' flower basket *Euplectella* (Fig. 23.5) often contains a male and a female shrimp (*Spongicola*) that have set up housekeeping and then grow too large to escape. Dried specimens were once popular gifts to newlyweds in Japan, symbolizing the wish that the couple remain together until death.

The greatest problem faced by sponges is finding suitable substratum for attachment. Their chief competitors in this contest are corals and other sponges. Many sponges avoid crowding by secreting toxins that kill or inhibit the growth of other organisms. Boring sponges, such as *Cliona lampa,* create their own habitat by chemically etching holes in corals and shells. Their work can be found on just about any shell on a beach. Boring sponges are so active around Bermuda and some other places that they have changed the reef structure. Some sponges are provided a substratum by the sponge crab (*Dromia*), which attaches sponges to itself for camouflage and protection.

Humans have valued bath sponges (*Spongia* or *Hippospongia*) since the Bronze Age, nearly 4000 years ago. What is so valuable is not the entire sponge, but the skeleton that remains after the sponge is killed, dried, and kneaded, leaving the spicule-free spongin skeleton with its useful ability to hold up to 35 times its weight in liquid. As the demand for sponges has grown, divers have had to go as deep as 30 meters for them. Until recently, their only diving equipment was a heavy rock. Many divers off Tunisia still use only a lead weight and a rope. Kalymnos, Greece was a major port for sponge divers for thousands of years until a blight wiped out their sponges, and other Mediterranean countries closed their territorial waters to foreign divers. Many of the Greek divers have now gone to Tarpon Springs, Florida, whose sponge industry has recently recovered from killer fungi and red tides (population explosions of dinoflagellates).

A wild sponge takes about five years to reach marketable size (12.5 cm), and it sells for around 10 dollars. Many people are willing to pay that price for a bath sponge that is so much better than the synthetic kind. For many applications, such as metal polishing, there is no good substitute for a real sponge. Small pieces of elephant ear sponge *Spongia officinalis lamella* are also preferred for pottery making. Sponges may someday prove even more useful as a source of medicines. Cytosine arabinoside, from the Caribbean sponge *Cryptotethya crypta,* is a major anticancer drug that blocks DNA synthesis in tumors (Ruggieri 1976, Tucker 1985). Several other potentially useful drugs from sponges are currently being studied (Flam 1994).

PHYLUM PLACOZOA

Phylum Placozoa is represented so far by only one described species, *Trichoplax adhaerens* (Fig. 23.9). This organism has been known since 1883, but was mistaken for the larva of a cnidarian or for a mesozoan until 1971. Until the late 1980s the only place *Trichoplax* had been found was on the walls of marine aquaria. Vicki B. Pearse has since found it swimming freely at various depths throughout the Pacific. It appears to be somehow protected from predation. It has a few thousand cells and is barely visible to the unaided eye (up to 3 mm). *Trichoplax* reproduces asexually by binary fission and budding, and also sexually. There are no organs and little differentiation of cells, as shown by its ability to regenerate a complete animal from any excised portion. Further testimony to the simplicity of this organism is the fact that it has the least DNA per cell of any animal.

Trichoplax moves in any direction, so there is no front, rear, or sides, and there is no symmetry. There are, however, differences between the dorsal and ventral surfaces. The dorsal surface consists of flat ciliated cells and shiny spherical cells ("bright spheres"), and the ventral surface has columnar ciliated cells and gland cells. Between the dorsal and ventral cells is a cavity containing fluid and fibrous cells. The fibrous cells contract the body in a way that makes *Trichoplax* resemble an ameba. Locomotion is due, however, to the ventral cilia. The animal probably feeds by covering algae or other food, secreting digestive enzymes, and absorbing the nutrient molecules (Table 23.3).

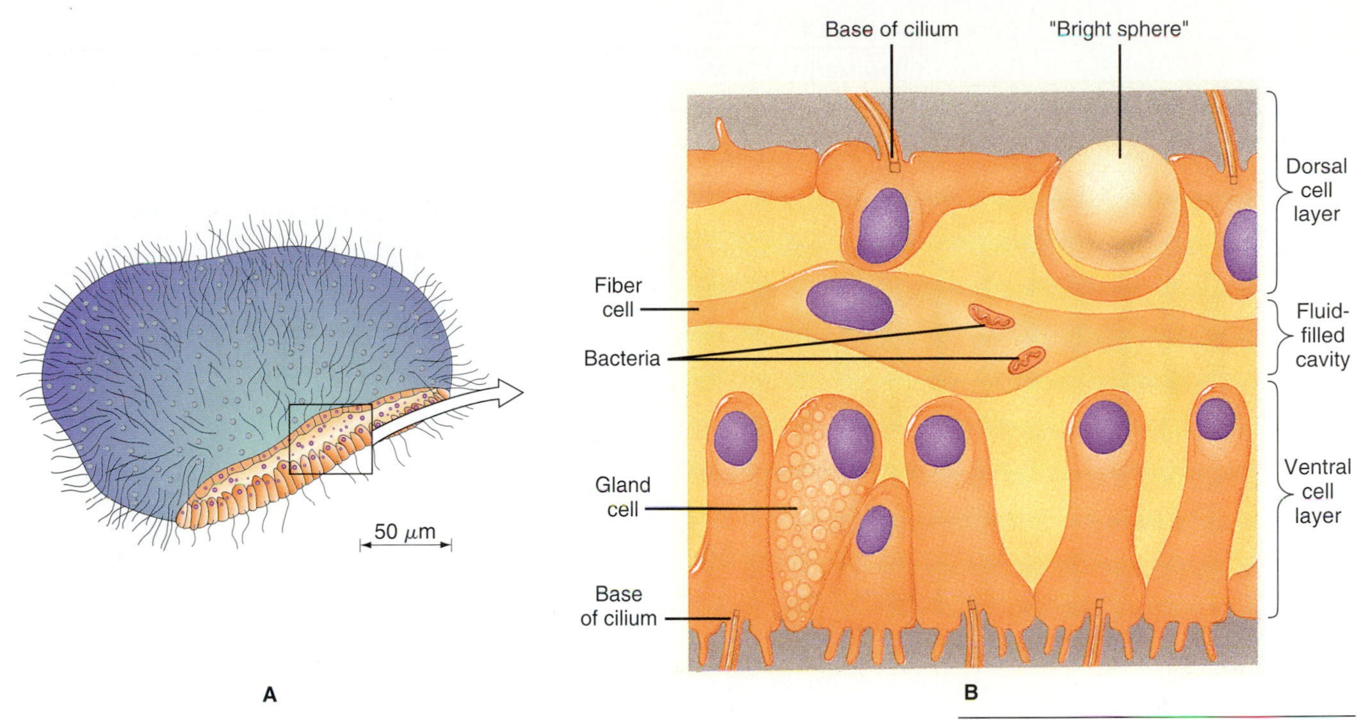

FIGURE 23.9

(A) *Trichoplax adhaerens.* (B) Cross section showing the four cell types.

■ **Question:** What is the difference between this animal and a colonial protozoan?

TABLE 23.3

Characteristics of Placozoa (based on one species).

Phylum Placozoa plak-oh-ZO-uh (Greek *plax* plate)

Morphology	Multicellular animals without organs. Outer layer of ciliated cells. Inner cavity contains fluid and fibrous cells. Asymmetric.
Physiology	No organ systems. External digestion.
Locomotion	Swimming and creeping on ventral cilia.
Reproduction	Sexually, as well as by binary fission and budding.
Development	Unknown.
Habitat, Size, and Diversity	Marine. Natural history unknown. Size up to 3 mm. **One species** described. An unnamed species of *Trichoplax* is also known, and another possible species (*Treptoplax reptans*) was reported in 1893, but not observed since.

PHYLUM MESOZOA

The Mesozoa are mysterious, small, parasitic, worm-shaped (vermiform) animals generally consisting of 20 to 30 cells. At present the phylum is generally divided into two classes: Rhombozoa (or Dicyemida) and Orthonectida. Many zoologists, however, think these should be two different phyla (Table 23.4 on page 488).

We are less ignorant about *Dicyema* than about any other mesozoan. *Dicyema* is a vermiform animal less than 7 mm long usually found in the kidney of octopus and other cephalopod molluscs. Virtually nothing is known about what the dicyemid finds

TABLE 23.4

Characteristics of Mesozoa.

Phylum Mesozoa mess-oh-ZO-uh (Greek *mesos* middle)

Morphology	**Two layers of cells** (not homologous to germ layers of metazoan embryos). Asymmetric.
Physiology	Lacking circulatory, respiratory, osmoregulatory, digestive, and nervous systems.
Locomotion	Cilia.
Reproduction	Asexual and sexual.
Development	Indirect.
Habitat, Size, and Diversity	Endoparasites of marine invertebrates. Sizes range from less than 1 mm to 7 mm. **One hundred known species.**

so attractive about cephalopod urine, or why the cephalopod tolerates dicyemids in its kidneys. *Dicyema* most often reproduces asexually within the cephalopod kidney by releasing small vermiform larvae. The adults that produce vermiform larvae are called **nematogens** (Greek *nema* thread + *genes* born) (Fig. 23.10). When the population grows to the point that virtually the entire inner surface of the kidney is covered, some

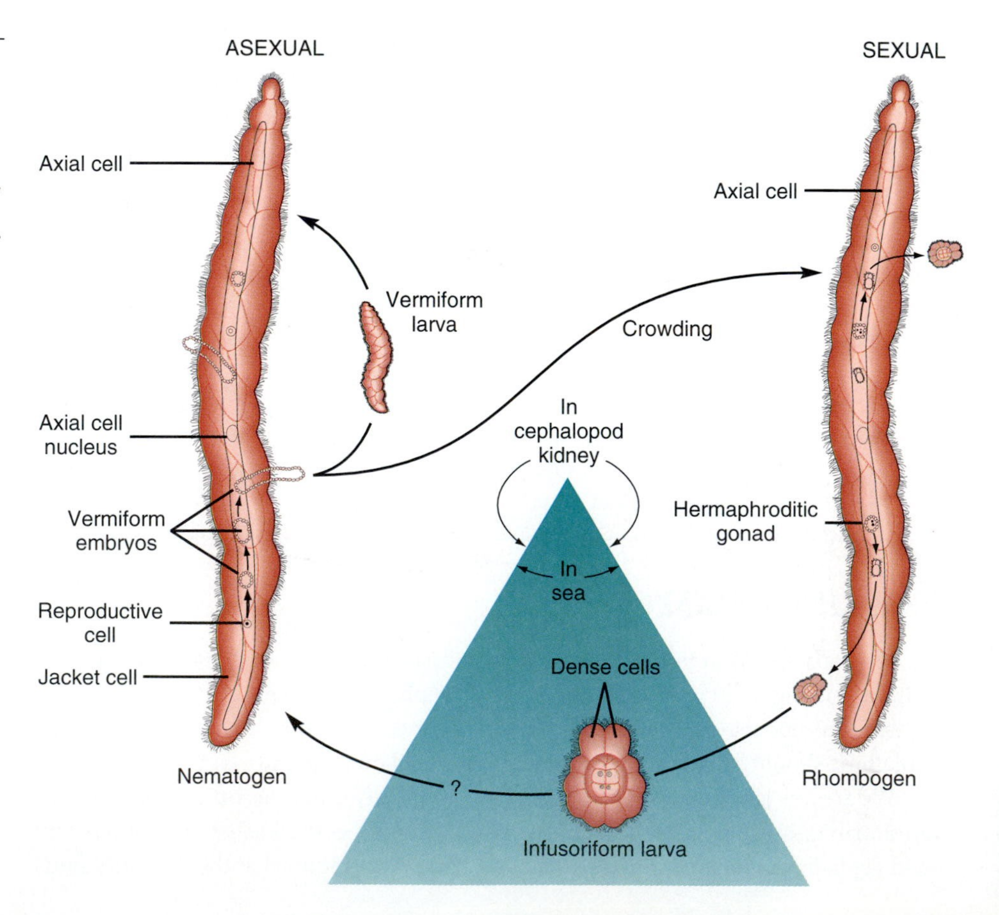

FIGURE 23.10

The life cycle of the dicyemid *Dicyema truncatum.* Generally the nematogen forms reproduce asexually in the kidney of a cephalopod mollusc. When the lining of the kidney becomes too crowded, hermaphroditic rhombogen forms are produced, and these reproduce sexually. The resulting infusoriform larvae are then released into water.

factor triggers the production of sexually reproducing **rhombogens.** The larvae released by rhombogens are shaped like toy tops and covered with cilia. They are called **infusoriform larvae,** because they resemble ciliated protozoa (infusoria). After being released when the cephalopod urinates, infusoriform larvae sink to the bottom by virtue of two dense cells. Infusoriform larvae are not immediately infective to another cephalopod at this stage, but must undergo some unknown steps of maturation, perhaps in a secondary host. Both the vermiform and infusoriform larvae are produced within an **axial cell** that runs up the center of the adult. Within this cell are numerous other cells(!) that produce vermiform larvae in nematogens. In rhombogens the axial cell contains hermaphroditic gonads that produce self-fertilizing gametes.

Adults in class Orthonectida are free-swimming and bisexual (Fig. 23.11). The dimorphic males and females are essentially containers of gametes. After internal fertilization the female gives birth to ciliated larvae. The larvae swim to a host and enter, then lose their cilia and begin parasitizing a variety of tissues. The orthonectid larva then undergoes repeated mitosis without cell division, resulting in a syncytial **plasmodium.** Some of the nuclei of the plasmodium then develop into male or female free-swimming adults, which leave the host and begin the next cycle.

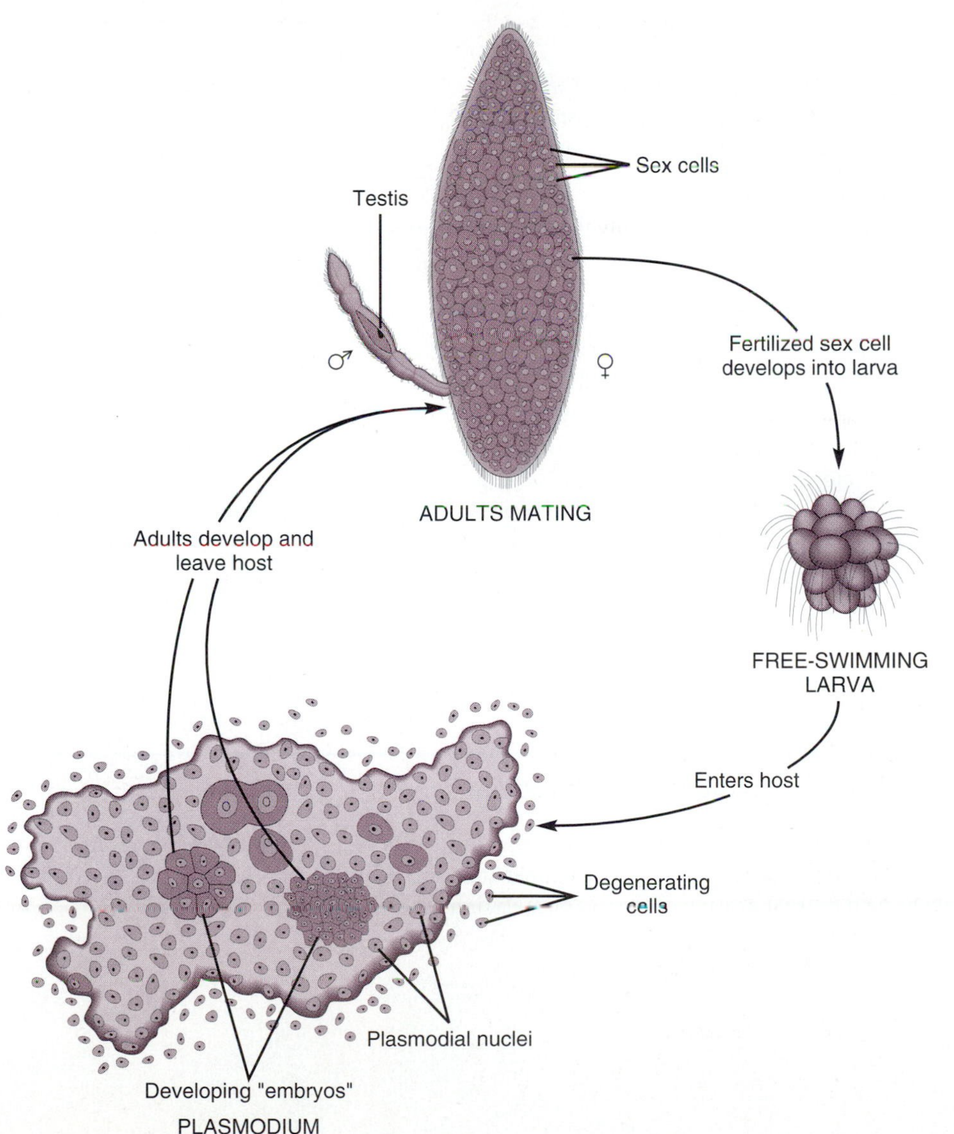

FIGURE 23.11
Life cycle of the orthonectid *Rhopalura ophiocomae.* Length of the adult female is approximately 0.25 mm.

Summary

The phyla Porifera, Placozoa, and Mesozoa comprise animals, since they are multicellular, heterotrophic, and sexual and have a blastula stage of development. Unlike most animals, however, there is a cellular grade of organization, in which there are no organs and no neural or hormonal coordination. These animals are also asymmetric.

Most members of phylum Porifera—sponges—are characterized by a canal system through which choanocytes pump water and extract food. The body is supported by an endoskeleton of collagenous fibers and spicules. Canal systems are classified as asconoid, syconoid, or leuconoid, in order of increasing complexity and increasing surface area for choanocytes. Three classes of sponges are distinguished, mainly from differences in the spicules. Most species belong to class Demospongiae, which has siliceous spicules (if any) that are not six-rayed. Sponges reproduce asexually by budding, or sexually by releasing gametes. Larvae are flagellated and responsible for dispersal to new habitats. The cellular grade of organization is manifest in the remarkable ability of sponges to reaggreate from dispersed cells.

Phylum Placozoa has just one named species, *Trichoplax adhaerens,* which is a flattened mass of cells that swims and crawls with cilia. Phylum Mesozoa includes small, worm-shaped parasites with only 20 to 30 cells.

Key Terms

choanocyte
spongocoel
mesohyl
ostium
osculum
spicule
asconoid
syconoid
radial canal
leuconoid
flagellated chamber
archaeocyte
reaggregation
gemmule

Chapter Test

1. Explain why choanocytes are essential to the life of sponges. How do ostia, oscula, and skeleton relate to the functioning of choanocytes? (pp. 477–480)
2. Describe three types of sponge canal systems, using the terms asconoid, syconoid, and leuconoid. What problem is solved by the development of the syconoid and leuconoid types? (p. 480)
3. Explain what is meant by the phrase "cellular grade of organization." Give one example illustrating how the cells of sponges behave independently. Give an example of how the cells of sponges interact. (pp. 476, 480–484)
4. What are the main criteria by which the three classes of sponge are distinguished? Name the three classes and describe the type of skeleton each has. (pp. 481–482)
5. Explain how sponges manage to reproduce sexually. What types of cells enable them to reproduce asexually. What is a gemmule? (pp. 484–485)
6. Both Porifera and Placozoa are often classified in the subkingdom Parazoa. What similarities do these phyla share that would support this classification? What are some of the important ways in which Porifera and Placozoa differ? (pp. 477, 487)
7. What is the significance of the term "Mesozoa?" (p. 476)

■ Answers to the Figure Questions

23.3 Good question. Perhaps each founder cell secretes a chemical that repels the others, leading to the formation of rays on the spicule. Thickener cells might be held near the center by either attraction or repulsion by the founder cells.

23.9 The cells are specialized for different functions. Another possible difference may be that placozoans develop from a blastula, but that is not known.

Readings

Recommended Readings

Lapan, E. A., and H. J. Morowitz. 1972. The mesozoa. *Sci. Am.* 227(6):94–101 (Dec).

Wood, R. 1990. Reef-building sponges. *Am. Sci.* 78:224–235.

See also relevant selections in General References listed at the end of Chapter 21.

Additional References

Bergquist, P. R. 1978. *Sponges.* Berkeley: University of California Press. (*A detailed but readable account of recent research.*)

Bergquist, P. R. 1985. Poriferan relationships. In: S. Conway Morris et al. (Eds.). *The Origins and Relationships of Lower Invertebrates.* New York: Oxford University Press, pp. 14–27.

Bergquist, P. R., and K. Glasgow. 1986. Developmental potential of ciliated cells of ceractinomorph sponge larvae. *Exp. Biol.* 45:111–122.

Christen, R., et al. 1991. An analysis of the origin of metazoans, using comparisons of partial sequences of the 28S RNA, reveals an early emergence of triploblasts. *EMBO J.* 10:499–503.

De Vos, L. et al. 1991. *Atlas of Sponge Morphology.* Washington, DC: Smithsonian Institution Press.

Flam, F. 1994. Chemical prospectors scour the seas for promising drugs. *Science* 266:1324–1325.

Fry, W. G. (Ed.). 1970. *The Biology of the Porifera.* Zoological Society of London Symposium 25. New York: Academic Press.

Grell, K. G., and A. Ruthmann. 1991. Placozoa. In: F. W. Harrison and J. A. Westfall (Eds.). *Microscopic Anatomy of the Invertebrates,* Vol. 2. New York: Wiley–Liss, pp. 13–27.

Harrison, F. W., and Louis De Vos. 1991. Porifera. In: F. W. Harrison and J. A. Westfall (Eds.). *Microscopic Anatomy of the Invertebrates,* Vol. 2. New York: Wiley–Liss, pp. 29–89.

Hori, H., and S. Osawa. 1987. Origin and evolution of organisms as deduced from 5S ribosomal RNA sequences. *Mol. Biol. Evol.* 4:445–472.

Meylan, A. 1988. Spongivory in hawksbill turtles: a diet of glass. *Science* 239:393–395.

Misevic, G. N., V. Schlup, and M. M. Burger. 1990. Larval metamorphosis of *Microciona prolifera:* evidence against the reversal of layers. In: K. Rützler (Ed.). *New Perspectives of Sponge Biology.* Washington, DC: Smithsonian Institution Press, p. 182.

Pawlik, J. R. 1993. The role of spicules in defending Caribbean sponges from predation: glass just doesn't cut it. *Am. Zool.* 33(5): abstract 92.

Ruggieri, G. D. 1976. Drugs from the sea. *Science* 194:491–497.

Rützler, K. (Ed.). 1990. *New Perspectives in Sponge Biology.* Washington, DC: Smithsonian Institution Press.

Simpson, T. L. 1984. *The Cell Biology of Sponges.* New York: Springer-Verlag. (*Far more comprehensive than the title suggests.*)

Tucker, J. B. 1985. Drugs from the sea spark renewed interest. *BioScience* 35:541–545.

Vacelet, J. 1985. Coralline sponges and the evolution of Porifera. In: S. Conway Morris et al. (Eds.). *The Origins and Relationships of Lower Invertebrates.* New York: Oxford University Press, pp. 1–13.

Wainright, P. O., G. Hinkle, M. L. Sogin, and S. K. Stickel. 1993. Monophyletic origins of the metazoa: an evolutionary link with fungi. *Science* 260:340–342.

CHAPTER

24

Cnidaria and Ctenophora

Strawberry anemone *Corynactis californica.*

Introduction

As we saw in the previous chapter, sponges manage to grow to large size by bringing the necessities of life to their cells in through a canal system. A different solution to life's problems has evolved in the phylum Cnidaria, which includes jellyfish, freshwater polyps, corals, and sea anemones, and also in the phylum Ctenophora (Fig. 24.1). Both cnidarians and ctenophores consist of two or three layers of cells organized around a central chamber, so no cell is far from the source of oxygen, nutrients, water, and ions. The cells of each layer are interconnected and function together, and they therefore form tissues. This **tissue grade of organization** distinguishes cnidarians and ctenophores from the sponges, placozoans, and mesozoans considered in the previous chapter. The fact that their tissues do not combine into organs distinguishes them from the animals to be considered in later chapters. Another feature of both cnidarians and ctenophores is that most have **biradial symmetry.** That is, the body tends to be radially symmetric, but some paired structures are positioned on opposite sides (p. 438).

Because of their similarities, cnidarians and ctenophores are often referred to together as "radiata," and they were once combined in the phylum Coelenterata. Many zoologists assume that because of their relatively simple organization these "coelenterates" must have evolved earlier than most other animals. That suspicion is supported

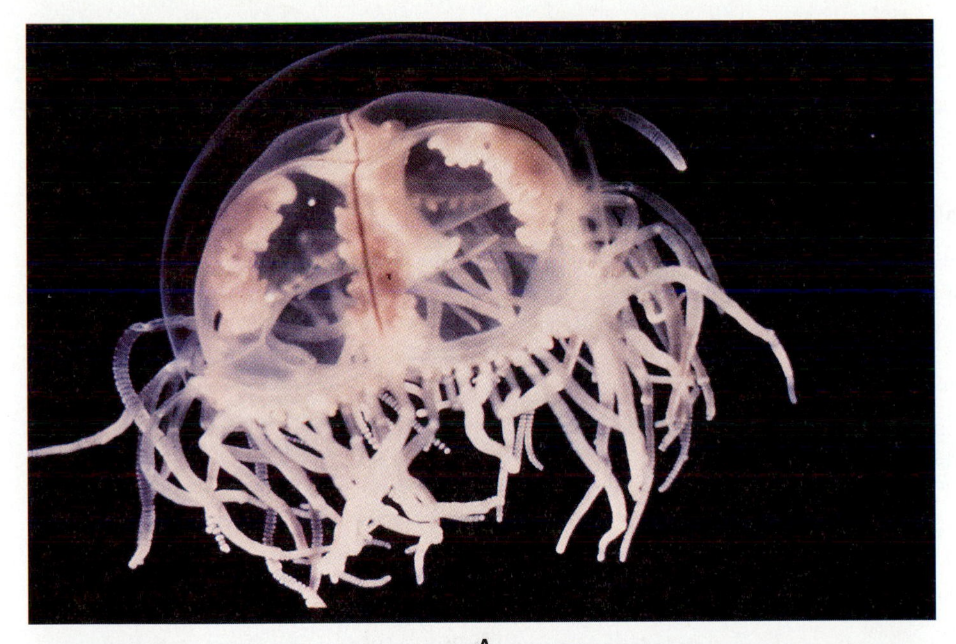

A

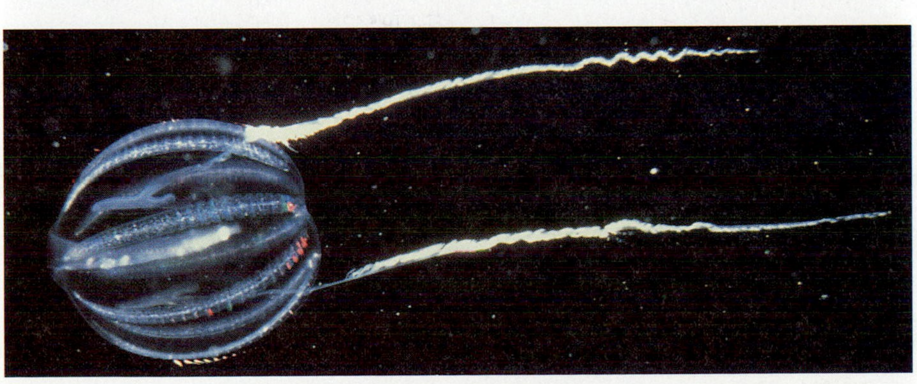

B

FIGURE 24.1

Members of the phyla Cnidaria and Ctenophora have thin tissues oriented radially or biradially around a hollow cavity. (A) The jellyfish *Gonionemus* (phylum Cnidaria). The four white structures beneath the transparent bell are gonads. Diameter of bell 2 cm. (B) The comb jelly commonly called sea gooseberry, *Pleurobrachia pileus* (phylum Ctenophora). Diameter approximately 2 cm.

by comparisons of ribosomal RNA sequences, which also support the classification of Cnidaria and Ctenophora in different but related phyla (Christen et al. 1991, Wainright et al. 1993).

CNIDARIANS

General Organization

GASTROVASCULAR CAVITY. The gastrovascular cavity (= coelenteron) is central both structurally and figuratively in the life of a cnidarian (Table 24.1, Fig. 24.2). As its name implies, the gastrovascular cavity serves the same functions as the digestive and circulatory organs of other animals, and most cnidarians spend much of their lives putting food into it. Cnidarians usually eat zooplankton, such as copepods (Crustacea), and large individuals can even ingest fish. The main tools for obtaining fish and other large prey are unique subcellular structures called **cnidae** (NY-dee; singular cnida), which spear, poison, and lasso prey. After cnidae capture a prey, the **tentacles** pull it toward the mouth, which draws it into the gastrovascular cavity. The mouth is the only opening into the gastrovascular cavity, so it also serves as an anus. (In other words, the digestive tract is **incomplete.**)

TABLE 24.1

Characteristics of cnidarians.

Phylum Cnidaria nigh-DARE-ee-uh (Greek *knide* nettle [referring to the stinging cnidae])

Morphology	**Biradial symmetry** (essentially radial in hydroids). Either bell-shaped medusae or cylindrical polyps. With hollow tentacles. Three (or two) thin layers of tissue around a central gastrovascular cavity. No coelom. Either hydrostatic skeleton or elastic skeleton of mesoglea. Some also secrete chitinous sheaths or calcareous exoskeletons.
Physiology	Diffuse **nerve net**; no central nervous system. Cnidae used in predation and defense. Mostly carnivorous, but some contribution from dissolved organic material and photosynthesizing endosymbionts. Few organs; individual cells perform most physiological functions.
Locomotion	Planula larvae swim by means of cilia. Medusae swim by slow undulations due to epitheliomuscle and nutritive muscle cells. Polyps usually sessile (attached to substratum). Some cnidae used in locomotion.
Reproduction	Asexual fission or budding; also sexual. Usually dioecious. Fertilization usually external.
Development	Radial cleavage pattern. Cleavage indeterminate (removal of a cell during cleavage does not result in an incomplete embryo.). Ciliated planula larvae.
Habitat, Size, and Diversity	All **aquatic**; almost all marine. Sizes of adult individuals range from less than 1 mm to 70 meters long, including tentacles. Approximately **9000 living species** described.

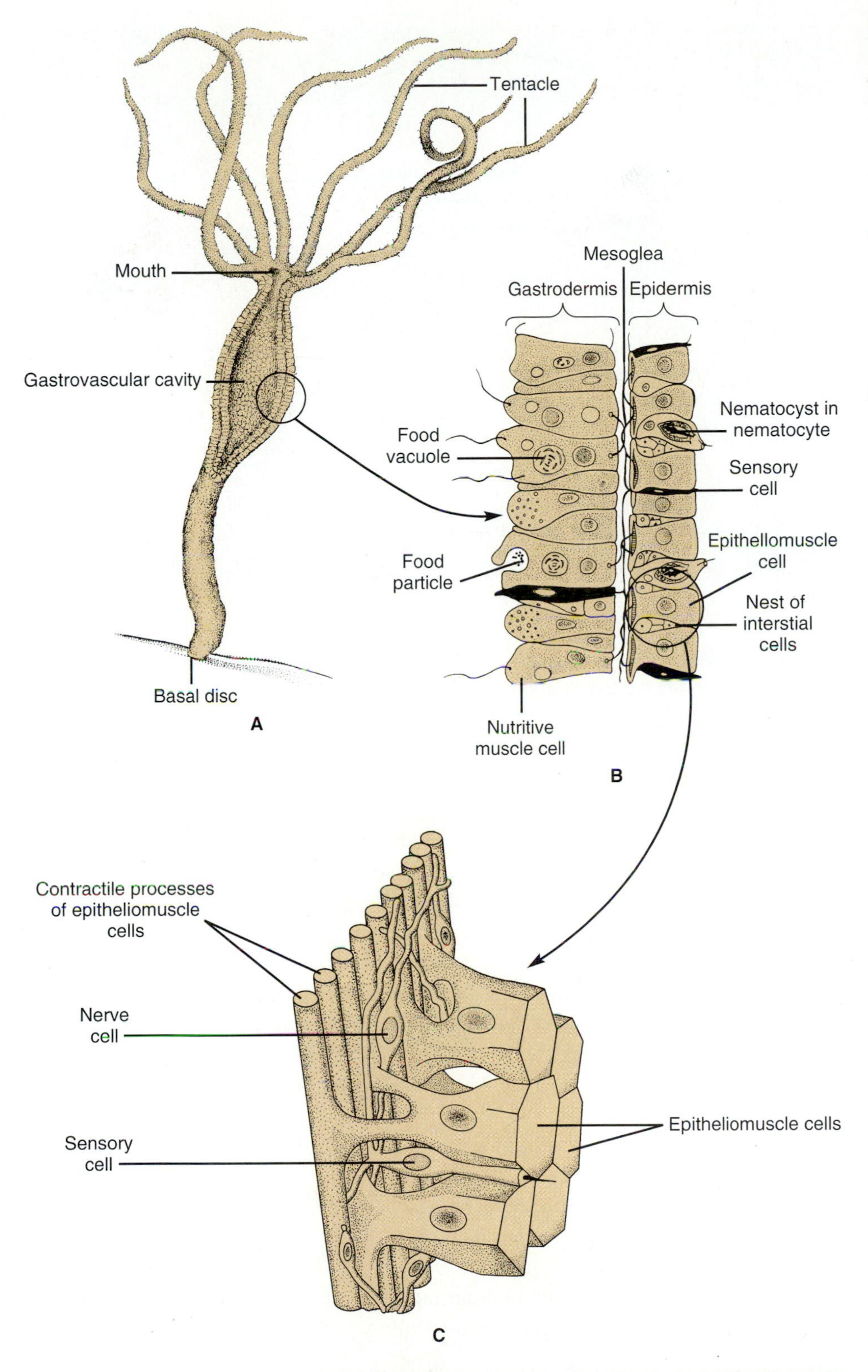

FIGURE 24.2

(A) Schematic longitudinal section of *Hydra.* (B) Representation of a section through the body wall of *Hydra,* showing the epidermis and the gastrodermis lining the gastrovascular cavity. The majority of epidermal and gastrodermal cells are contractile cells (epitheliomuscle and nutritive muscle cells). The latter absorb food in vacuoles and bear pairs of flagella. The interstitial cells differentiate into other cell types. Nematocytes produce the stinging structures called nematocysts. (C) Portion of the epidermis showing epitheliomuscle cells.

■ **Question:** How do you think the sensory cells, nerve cells, and epitheliomuscle cells interact? What kind of behavior might they produce in response to a predator?

TISSUE LAYERS. The lining of the gastrovascular cavity is a tissue layer called the **gastrodermis.** Gastrodermal cells secrete enzymes that partially digest prey in the gastrovascular cavity, then they phagocytize food particles for further digestion. Digestion is therefore both extracellular and intracellular. Flagella on certain gastrodermal cells draw water into the gastrovascular cavity, thereby producing a pressure that maintains a

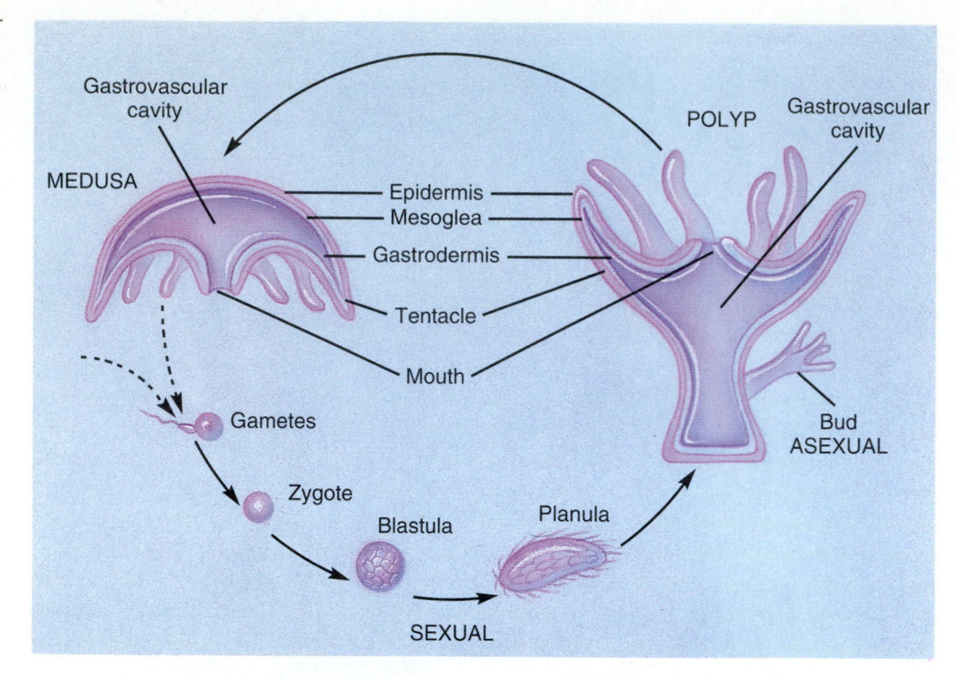

FIGURE 24.3

Many Cnidaria alternate between a sexual, swimming medusa (left) and an asexual, sessile polyp. Although they look quite different externally, this schematic diagram shows that both the medusa and the polyp have a similar organization of tissues enclosing a gastrovascular cavity.

hydrostatic skeleton. The flagella also apparently mix food and enzymes. The external covering of the body is a tissue layer called the **epidermis.** Cells of the epidermis apparently obtain nutrients from the gastrodermis by diffusion. Epidermal cells may also absorb dissolved organic material directly. In many cnidarians, such as corals, the epidermal cells also harbor photosynthetic protists (**zooxanthellae**) that produce nutrients.

The epidermis and gastrodermis are often considered to be the only two tissue layers in cnidarians, which are therefore often said to be **diploblastic.** There is an increasing tendency now to regard as a third tissue layer the **mesoglea** (= mesenchyme; p. 479), which lies between the gastrodermis and the epidermis. Cnidarians may therefore be **triploblastic,** like most animals, although the three tissue layers of cnidarians are probably not homologous with those of other animals. The mesoglea (Greek *mesos* middle, *gloia* glue) is a translucent, gelatinous material that is often invaded by neural and other kinds of cells. In some cnidarians, such as *Hydra,* it is merely an adhesive layer that holds the epidermis and gastrodermis together. In others, especially jellyfishes, the mesoglea constitutes most of the body. The mesoglea performs several functions:

1. It is an elastic skeleton, returning the animal to its normal shape after contraction.
2. It maintains **neutral buoyancy,** because it contains less sulfate (SO_4^{2-}) than seawater, and it is therefore less dense.
3. It may be a reservoir of nutrients.
4. It increases body size without requiring a corresponding increase in the number of cells requiring oxygen and nutrients.

FIGURE 24.4

A *Hydra* polyp with a bud forming on one side and an ovary opposite. In some species of *Hydra* budding occurs only in the presence of a secretion from bacteria or prey. High levels of CO_2 in crowded colonies trigger the formation of gonads.

■ **Question:** What is the advantage of being able to reproduce both sexually and asexually? Under what circumstances is each mode of reproduction more advantageous that the other?

Life Cycles

In many species of Cnidaria each individual changes from a **polyp** stage to a **medusa** stage, or vice versa (Figs. 24.3). This change is often considered to be a primitive state of cnidarians, but in most species only the polyp or the medusa stage occurs. Polyps are usually approximately cylindrical and attached to substratum, as in sea anemones,

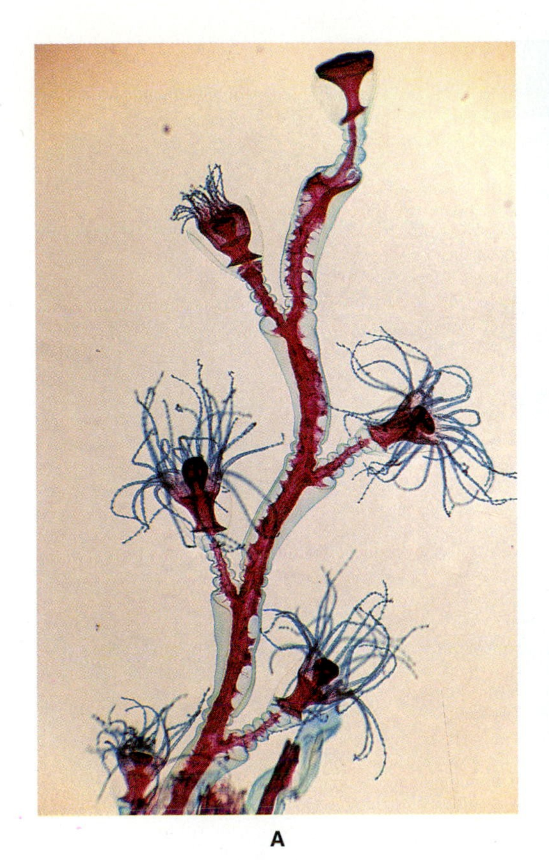

A

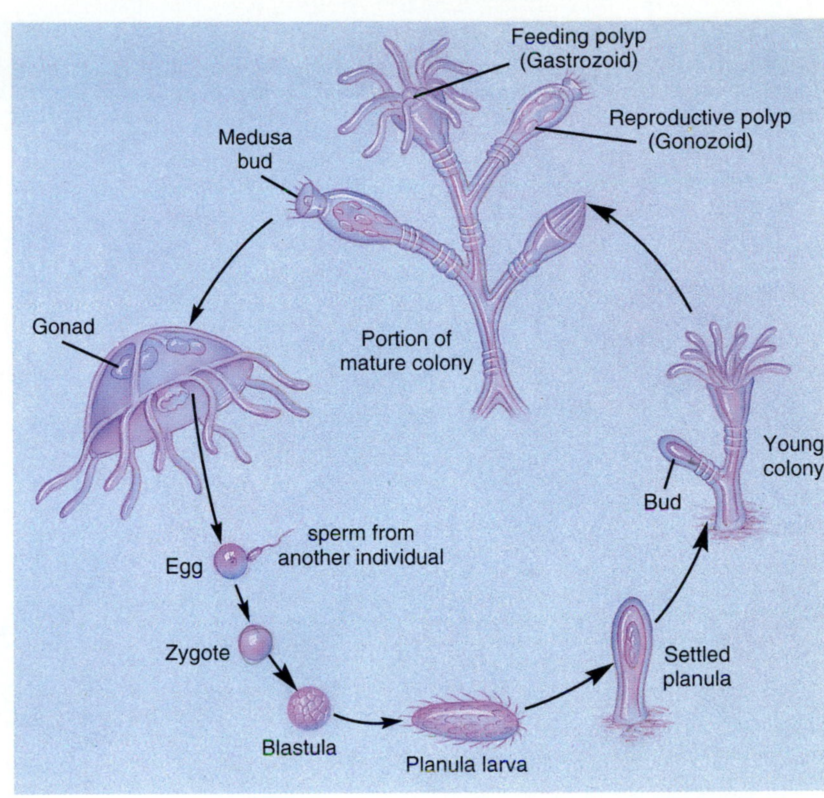

B

FIGURE 24.5

(A) The colonial hydrozoan *Obelia.* Width of each member of the colony approximately 1 cm. (B) The life cycle of *Obelia.* Note the similarity to Fig. 24.3.

■ **Question:** What is an advantage of being colonial? What is a disadvantage?

corals, and hydras (Fig. 24.4). If the medusa form is absent, as in hydras and corals, polyps reproduce new polyps either sexually or by budding. Polyps that produce medusae do so by asexual budding.

Medusae are shaped like bells or umbrellas, and they can swim with a kind of jet propulsion by contracting the body. The familiar jellyfishes (Fig. 24.1A) are in the medusa stage. Medusae usually have separate sexes, although they are difficult to tell apart. Fertilization usually occurs externally, and the zygote develops into a **planula** larva that swims by means of cilia. In most species the planula **settles** and forms a polyp. The polyp then buds asexually, with each bud developing into either a new medusa or a new polyp. The life cycle of the colonial cnidarian *Obelia* is similar to that sketched in Fig. 24.3, and it is commonly used as a typical example (Fig. 24.5).

A Survey of the Major Groups

CLASS HYDROZOA. The number of classes of Cnidaria ranges from three to six in recent taxonomies, but four is a common number. One class, the hydrozoans, includes all the freshwater cnidarians, such as *Hydra,* as well as the fire corals. Both of these forms occur only as polyps, but many other hydrozoans, such as *Obelia,* have both medusa and polyp stages. The medusae of hydrozoans resemble jellyfishes (class Scyphozoa), but many can be distinguished from jellyfishes by a shelf-like **velum** around the inner margin of the bell, which concentrates the force of water when the bell contracts. Other hydrozoans that can be confused with jellyfishes are siphonophores such as the familiar Portuguese man-of-war *Physalia* (Fig. 24.6 on page 499). Siphonophores are actually colonies of polyps and medusae, each of which is called a **zooid** (ZO-oyd). All the zooids of a siphonophore develop from the same ovum, but each subsequently becomes adapted as a float, as a tentacle for predation, as a digestive organ, or as a reproductive

Major Extant Groups in Phylum Cnidaria

Genera mentioned elsewhere in this chapter are noted.

Class Hydrozoa hy-dro-ZO-uh (Greek *Hydra* a many-headed water serpent in Greek mythology + *zoon* animal). Asexual polyps and/or sexual medusae. Medusae, when present, may have a velum (Latin for veil)—an inward fold of the edge of the bell. Gastrovascular cavity not divided by septa. Hydroids, fire corals, and siphonophores. *Chlorohydra, Hydra, Obelia, Physalia, Gonionemus* (Figs. 24.4, 24.5, 24.6).

Class Scyphozoa sy-foe-ZO-uh (Greek *skyphos* cup). Free-swimming medusae with four-part radial symmetry. Polyp reduced or absent. Prominent gelatinous mesoglea. Margin of bell with rhopalia. No velum. Jellyfish. *Aequorea, Aurelia, Cyanea* (Figs. 24.1A, 24.7).

Class Cubozoa cube-oh-ZO-uh (Greek *kubos* cube). Bell cubical and bent inward at margin to form a velarium (not a true velum). Tentacles suspended from four flat pedalia at corners of bell. Cubomedusans. *Carybdea, Chironex* (Fig. 24.8).

Class Anthozoa an-tho-ZO-uh (Greek *anthos* flower). No medusae; all sessile polyps. Pharynx leads into gastrovascular cavity, which is divided by six or more septa (mesenteries) bearing cnidae. Mesoglea contains cells.

Subclass Octocorallia (= Alcyonaria) OK-toe-core-AL-ee-uh (Greek *octo* eight + *korallion* pebble). Eightfold biradial symmetry: eight tentacles; gastrovascular cavity divided by eight septa. Endoskeleton of calcareous spicules. All colonial. Includes soft corals, sea whips, sea fans, sea pens, and sea pansies. *Corallium, Gorgonia, Heliopora, Plexaura* (Fig. 24.9).

Subclass Hexacorallia (= Zoantharia) HEKS-a-core-AL- ee uh (Greek *hex* six). Often with sixfold biradial symmetry: gastrovascular cavity divided by six septa. Includes sea anemones, stony (true) corals, tube corals, black (thorny) corals. *Adamsia, Anthopleura, Antipathes, Calliactis, Corynactis, Heteractis, Metridium, Palythoa, Platygyra, Pocillopora, Stomphia*

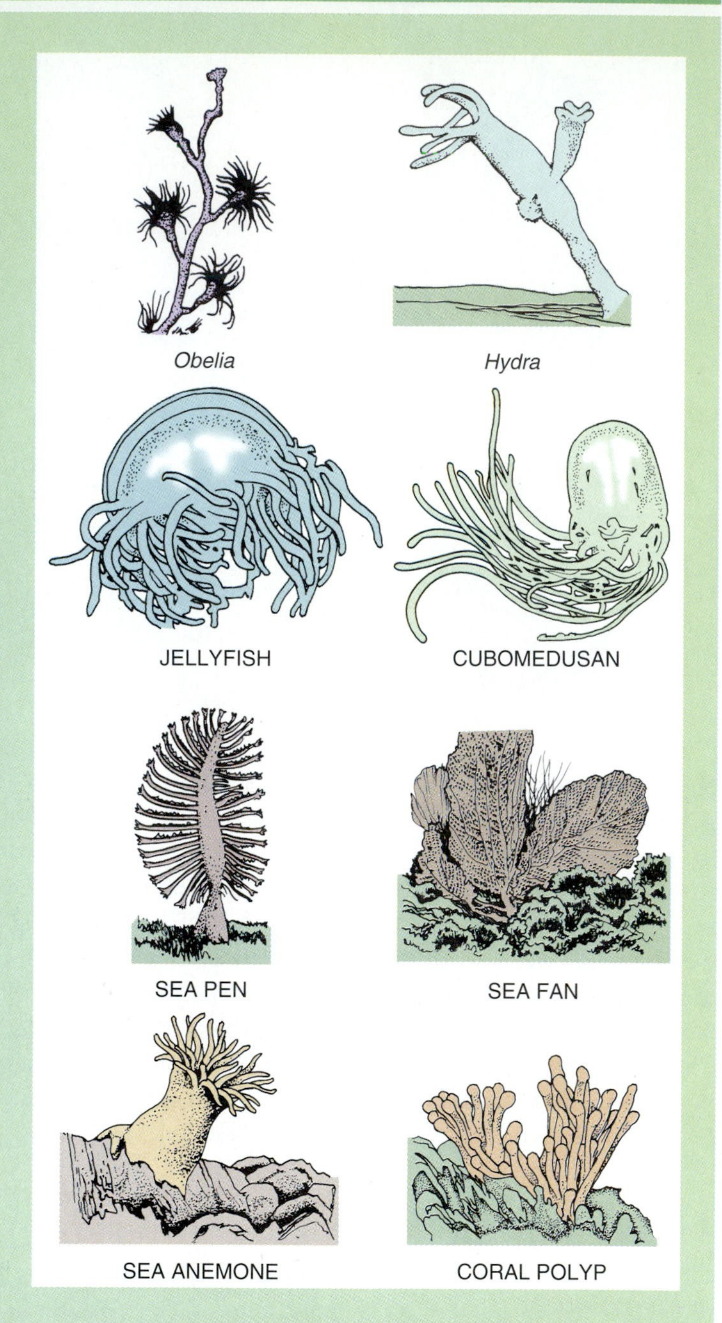

FIGURE 24.6

The Portuguese man-of-war *Physalia* resembles a jellyfish in appearance, as well as in the effect of its nematocysts. It is not a single medusa with dangling tentacles, however, but a floating colony of many polyps (zooids). A highly modified polyp forms the float (pneumatophore), which contains a high concentration of carbon monoxide. The fishing polyps (dactylozoids) can extend to 13 meters. Prey—usually small fishes—are ingested by different polyps (gastrozoids, near the dactylozoids). Other polyps are responsible for sexual reproduction.

■ Question: Are these tentacles homologous with the tentacles of an individual polyp?

organ. For more than a century zoologists have debated whether the zooid or the entire siphonophore should be considered an individual animal (Gould 1985).

CLASS SCYPHOZOA. Class Scyphozoa includes most of the cnidarians commonly called jellyfishes. (The term jellyfish is also a general term for a medusa in any class of Cnidaria.) The polyps of scyphozoans are reduced or absent, but the medusae are usually prominent, owing to the large, jellylike mesoglea. The largest, *Cyanea,* has a bell up to 2 meters in diameter and has tentacles up to 70 meters long. The common jellyfish of the seashore, *Aurelia aurita,* is more typical in size, with a diameter of about 10 cm and short tentacles (Fig. 24.7 on page 500). This jellyfish is distributed from the equator to both poles, and it is a major predator of larval herring and other fishes.

The structure of *Aurelia* is representative of the class. There is a prominent four-part, radial symmetry. Four **oral arms** bearing cnidae lead into the mouth and gastrovascular cavity, and the gonad is shaped like a four-leaf clover (Fig. 24.7A). The umbrella is scalloped in eight sectors (four or sixteen in other scyphozoans). On the margin of the umbrella where two sectors meet, there is a sensory organ called the **rhopalium.** The rhopalium includes a statocyst organ for balance, as will be discussed later. In some scyphozoans the rhopalia also bear simple eyes. Four **radial canals** branch out from the center like the ribs of an umbrella and join the **ring canal** that runs around the margin.

Unlike some jellyfishes, *Aurelia* has a polyp stage (Fig. 24.7C). *Aurelia* is also unusual in having internal fertilization. The males emit sperm from their mouths, and females draw it through their mouths into the gastrovascular cavity where fertilization occurs. Development of the ciliated planula larvae occurs in special pouches near the mouth. Eventually the larvae escape, settle on the sea floor, lose their cilia, and become polyps.

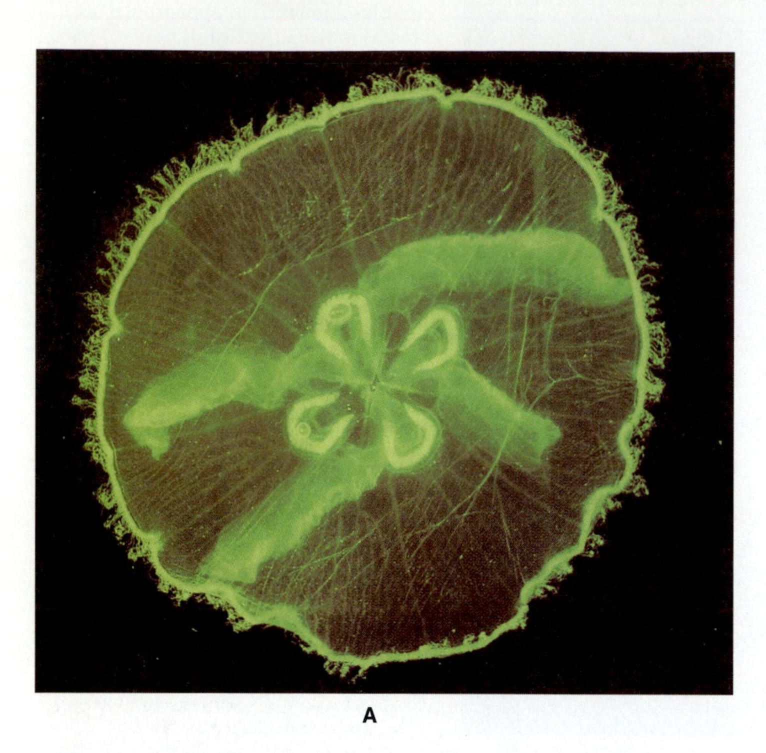

A

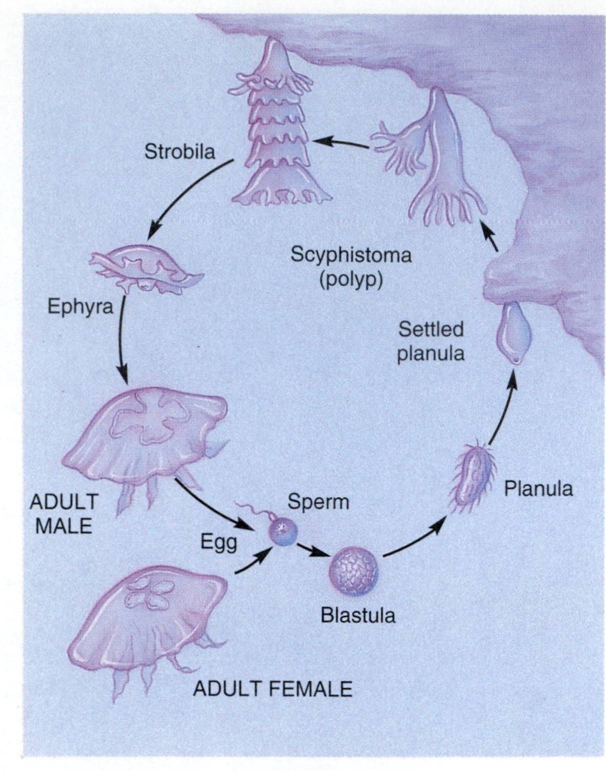

B

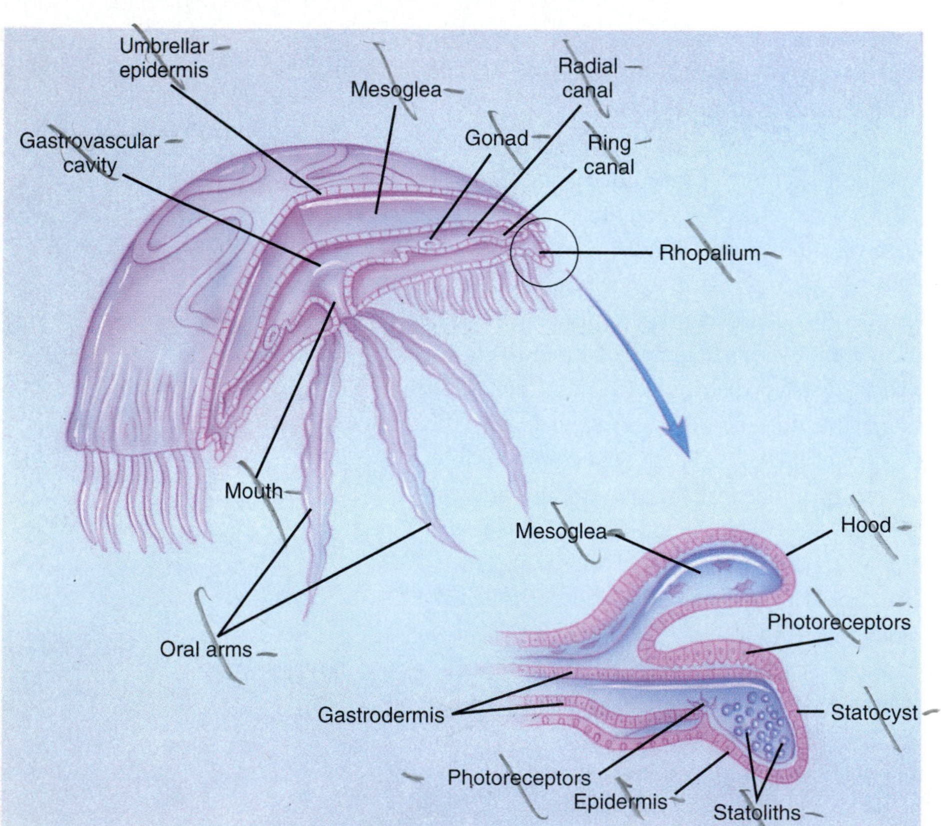

C

FIGURE 24.7

(A) The jellyfish *Aurelia aurita,* commonly called a moon jelly. The clover-leaf pattern on the umbrella is the "gonad" (testis or ovary). Diameter of bell approximately 40 cm. (B) The life cycle of *Aurelia.* Unlike some jellyfishes, *Aurelia* has a polyp stage, called the scyphistoma. The scyphistoma develops asexually into a strobila, which produces immature medusae, called ephyrae. Each ephyra then matures into an adult medusa. (C) Structure of *Aurelia.* .

Question: Of what use would the photoreceptors and statocyst be to a jellyfish?

CLASS CUBOZOA. The 16 or so species of cubozoans, or cubomedusans, resemble true jellyfishes, and they are commonly called "box jellyfishes" because of their cubical umbrellas. Cubomedusans differ from jellyfishes, however, in having four **pedalia** with which they swim rapidly. Some species come into frequent and sometimes lethal contact with people, because they are attracted to prey in shallow water and by pollution and lights (Fig. 24.8).

CLASS ANTHOZOA. Members of class Anthozoa have no medusae and differ so much in other respects from other Cnidaria that some taxonomists propose elevating Anthozoa to a subphylum. Anthozoans are generally divided into two subclasses. Subclass Octocorallia includes soft corals, sea whips, sea fans, sea pens, sea pansies, and others (Fig. 24.9 on page 502). All are colonial polyps, each of which has eight tentacles. Subclass Hexacorallia is more diverse and includes sea anemones, stony (true) corals, tube anemones, and black (thorny) corals (Fig. 24.10 on page 502). Tube anemones and black corals are so different from the others that many taxonomists put them into their own subclasses (Ceriantheria and Antipatheria).

The most prominent individuals in class Anthozoa are the sea anemones (Fig. 24.11 on page 503). They are much more massive than the polyps of hydrozoans, ranging in diameter from less than 5 mm to more than a meter. Many form colorful beds of "flower animals" in warm, shallow waters. The hydrostatic pressure that helps support

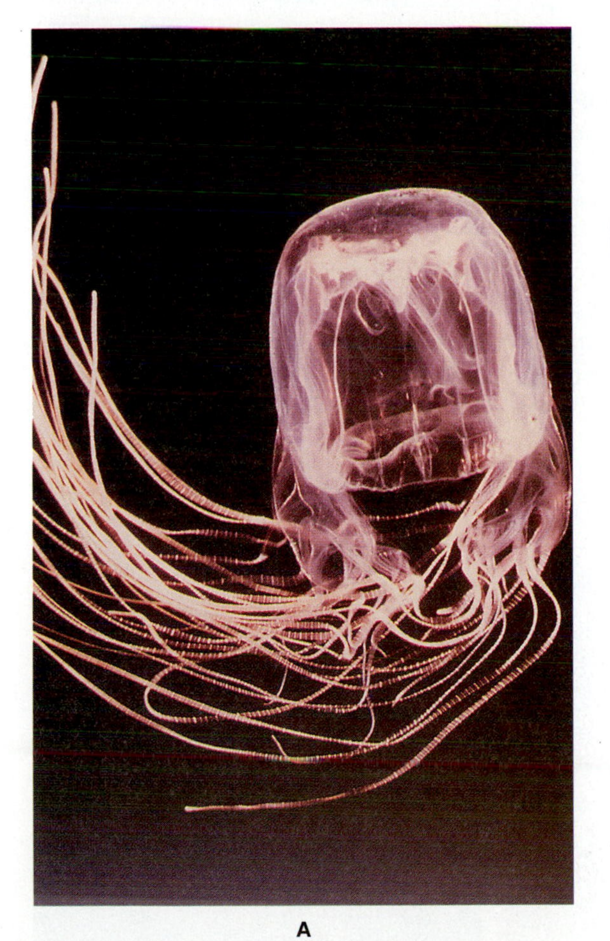

A

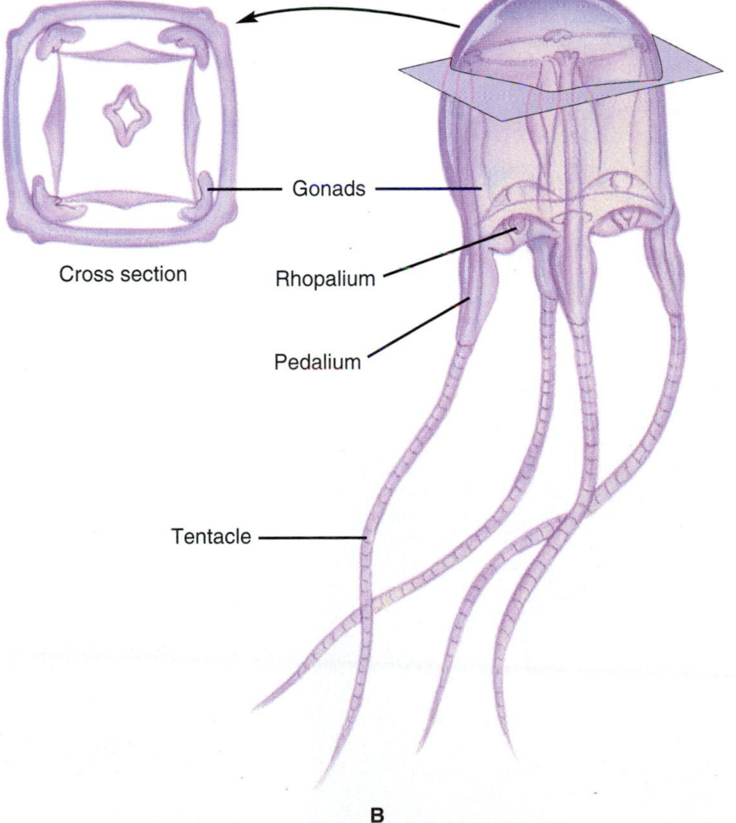

B

FIGURE 24.8

Since 1980 more than 70 swimmers in Australia stumbled ashore in severe pain, then dropped dead within minutes. (A) The cause—nematocysts of the sea wasp *Chironex fleckeri,* a cubozoan. Width of bell approximately 25 cm. (B) Structure of another cubozoan *Carybdea marsupialis.*

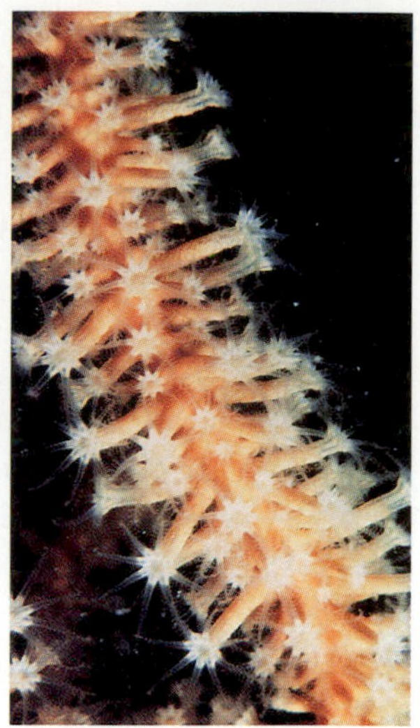

A B

FIGURE 24.9

Octocoral anthozoans. (A) Colony of sea fans, *Gorgonia adamsi.* (B) Closeup showing several extended polyps of a different species.

such a large body is generated by the action of cilia at one or both ends of the long slit-shaped mouth, in a groove called the **siphonoglyph.** The siphonoglyph draws a steady current into the pharynx and gastrovascular cavity, and cilia elsewhere in the pharynx pump out the water. Support is also provided by **septa** (= mesenteries) that run vertically through the body. Six pairs of **complete septa** (= perfect mesenteries) connect the pharynx to the outer wall, dividing the upper part of the gastrovascular cavity radially into six chambers. Pairs of **incomplete septa** (= imperfect mesenteries) partially divide these chambers. In addition to reinforcing the body, the septa also increase the surface area for digestion and absorption and are the sites for development of gametes.

On the edges of incomplete septa, below the gonads, are **filaments** with cnidae and gland cells that help subdue and digest prey. In some species the filaments continue into **acontia** that bear cnidae and can be extended through pores or the mouth to augment the tentacles in subduing prey. The numerous tentacles are arrayed in two rings on the **oral disc** surrounding the mouth. Some sea anemones have **catch tentacles** that are longer than the others and are used in intraspecific competition (Fig. 24.12).

FIGURE 24.10

Hexacoral anthozoans. (A) Polyps of the stony coral *Pocillopora* sp. The brown spots are due to the symbiotic photosynthetic protist (zooxanthella) *Symbiodinium microadriaticum.* (B) Schematic diagram of a polyp within its calcium carbonate skeleton. (C) A brain coral *Platygyra sinensis.* The tentacles retracted when illuminated by the photographer's light.

■ **Question:** What kind of interactions among polyps could lead to the regular pattern seen in part C?

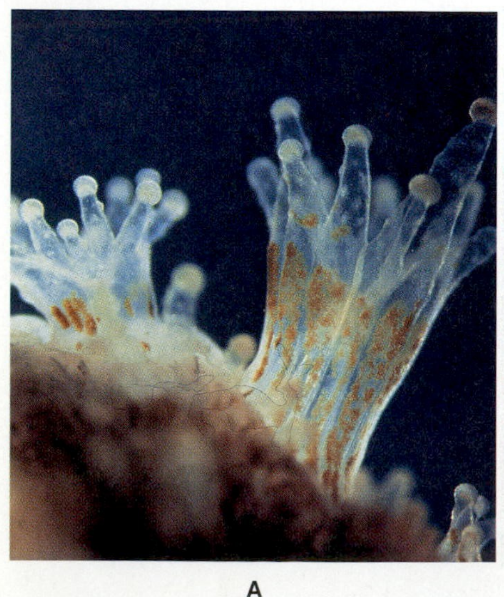

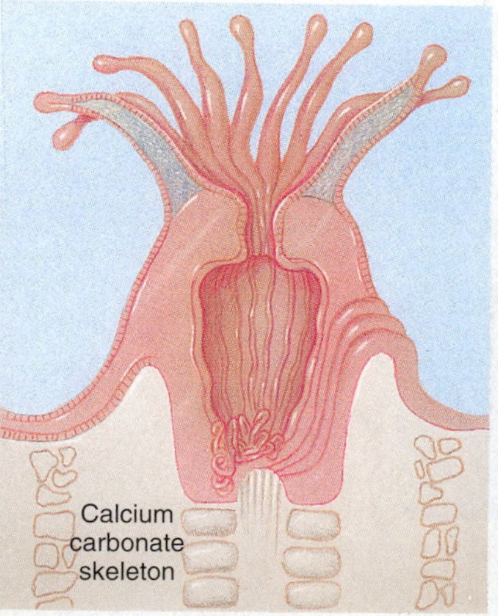

A B C

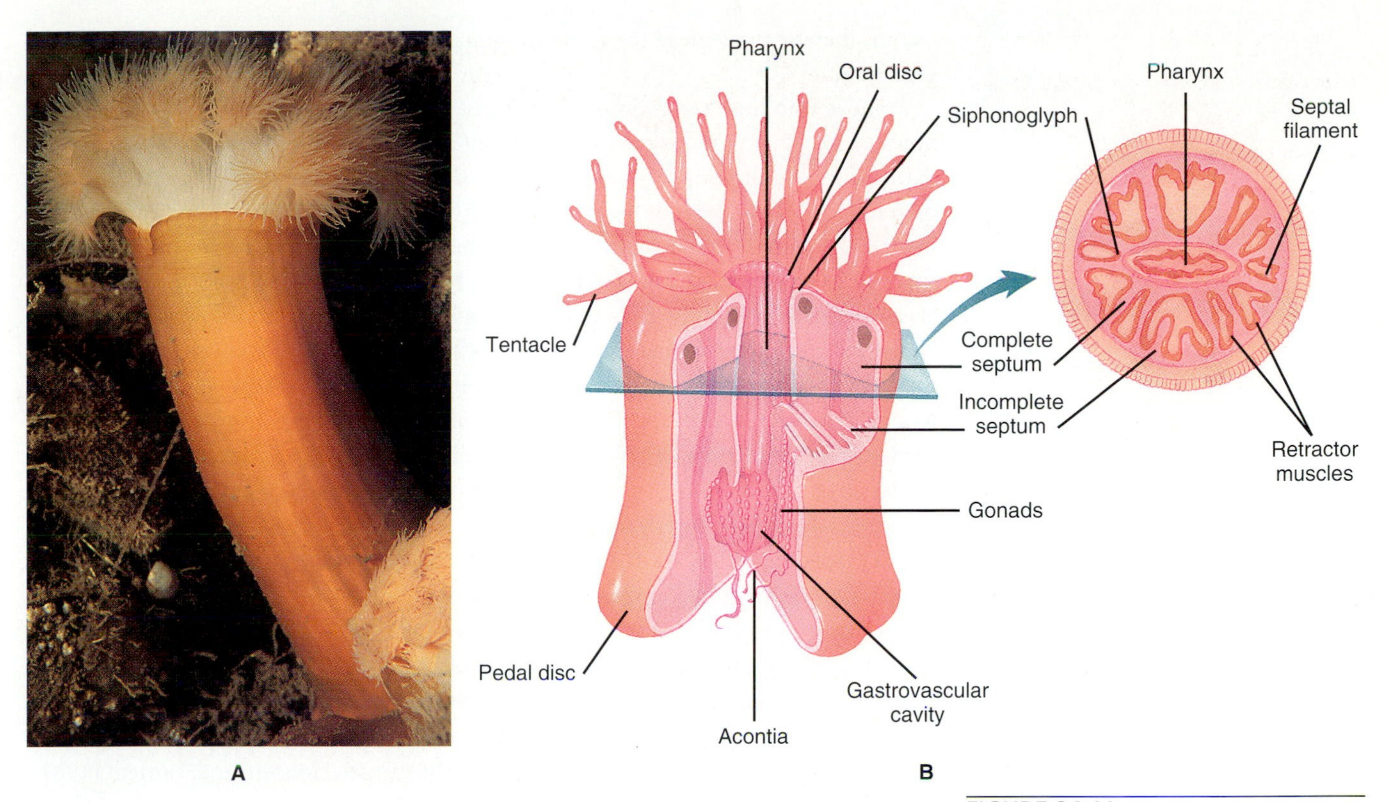

FIGURE 24.11

The sea anemone *Metridium.* (A) Living specimens. Height 45 cm. (B) Diagram of sectioned anemone.

Cnidae

FUNCTIONS. Cnidarians are the only animals that produce intracellular structures called cnidae, which are used in predation, defense, and locomotion. There are more than 25 types of cnidae. Some assist in locomotion by sticking to solid objects. Others entangle prey before capture by the tentacles. Most cnidae are of the spearing variety, called **nematocysts,** which occur in the type of cell known as a **nematocyte.** A nematocyst punches a hole into a prey or predator by means of **stylets,** then launches a hollow **thread** into the wound (see Fig. 13.2A). The thread is from 0.5 mm to several millimeters long and serves as a syringe that injects a toxin that destroys cell membranes. The nematocyst is usually discharged by contact with a triggerlike structure called the **cnidocil.** Usually chemoreceptors must first have detected the presence of a suitable

FIGURE 24.12

The sea anemone *Metridium senile* on the left attacking a member of its own species with a catch tentacle (thick tentacle in center). In response, the anemone on the right has inflated several of its own catch tentacles.

victim, thereby preventing the waste of nematocysts if the cnidocil contacts an inorganic object.

Those who have swum into a jellyfish or Portuguese man-of-war will appreciate how effective these nematocysts can be. The toxin can sting for hours and produce inflammation. Attempting to wipe off or wash off the tentacles merely triggers more nematocysts. In some parts of the world, many beachgoers take the precaution of carrying meat tenderizer or vinegar to inactivate the toxin. In severe cases the toxin destroys tissue and can produce a fatal cardiac arrest. The stings of the sea wasp *Chironex fleckeri* and some others in the class Cubozoa can kill a person within minutes or scar survivors for life. Not all animals are bothered by nematocysts. The sea slug *Glaucus* not only eats the Portuguese man-of-war, but somehow manages to get the undischarged nematocysts into its skin for its own protection. The flatworm *Microstomum* uses the nematocysts from its prey, *Hydra,* in a similar way. Such nematocysts, taken from cnidarians by their predators, are called **kleptocnidae** (Greek *kleptein* to steal).

Mechanism of Discharge. The small size (0.1 mm) of the nematocyst and the speed of firing (within 3 msec) have made it difficult to learn what force ejects the thread, but various plausible candidates have been suggested. The **osmotic theory** holds that water enters the nematocyst capsule osmotically, swelling the capsule and ejecting the thread. The **tension theory** suggests that the thread is fired by tension built up springlike during synthesis. The **contractile theory** proposes that contracting structures surrounding the nematocyst squeeze out the thread. Holstein and Tardent (1984) in Zurich have managed to obtain evidence bearing on the mechanism by filming the discharge with ultrahigh-speed microcinematography. They found that in *Hydra attenuata,* nematocysts are discharged in four steps, shown in Fig. 24.13:

1. First a pressure of 150 atmospheres swells the nematocyst by about 10%.
2. The nematocyst's cover flips open, and three stylets emerge and punch a hole in the prey or predator. (This phase takes less than 10 μsec, so the acceleration of the stylets is approximately 40,000 G!)
3. The stylets back out of the hole within about 150 μsec. This means that the initial force that opens the cover must have been expended.
4. The thread shoots into the hole in the prey or predator, turning inside-out as it extends.

Holstein and Tardent suggest that the initial swelling of the capsule is due to osmosis (osmotic theory), and that the discharge of the thread is due to stored mechanical energy from the capsule wall (contractile theory).

Nerve Cells and Movement

Nerve Nets. Compared with nematocyst discharge, most behaviors of cnidarians are slow, simple, and nonspecific. This is to be expected from the diffuse nervous system, the general absence of complex receptor organs, and the poorly differentiated contractile cells. The nerve cells do not connect specific receptors with specific muscles via well-defined neural pathways, but form a nerve net that sends a wave of excitation radiating from the area of stimulation (see Fig. 9.3). In many cnidarians, nerve cells can produce two types of action potential: (1) sodium spikes that trigger fast tentacle retraction and other escape movements and (2) calcium spikes that trigger slower feeding movements. In jellyfish the epithelial cells also conduct electrically and mediate very slow movements. Many of the synapses have transmitter vesicles on both sides of the cleft, which enables them to transmit in two directions.

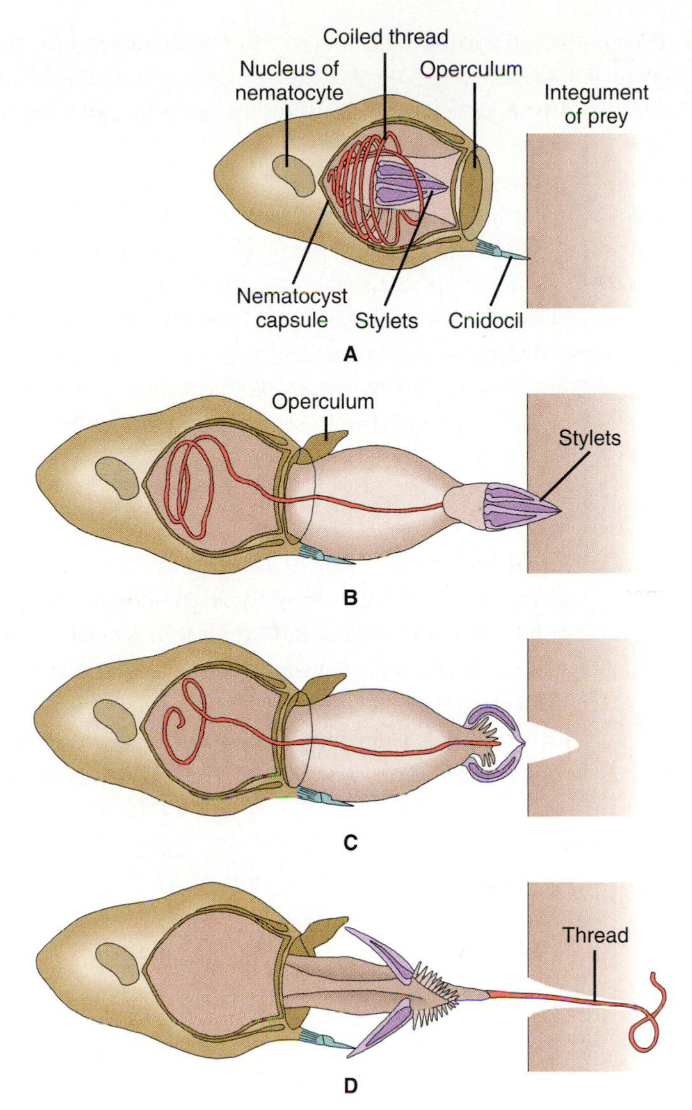

FIGURE 24.13

Discharge of a nematocyst. (A) The capsule swells. (B) The operculum opens, and the stylets emerge and puncture the prey or predator. (C) Stylets start to withdraw. (D) Thread emerges and enters hole in prey or predator.

Contractile Cells. In many Cnidaria most of the cells of the epidermis and gastrodermis are capable of contracting. These **epitheliomuscle cells** and **nutritive muscle cells** (Fig. 24.2) are not considered to be true muscle cells because they are not derived from a middle tissue layer, as in other animals. The contractile portion of epitheliomuscle cells runs lengthwise along the body and tentacles, while the contractile portion of nutritive muscle cells runs circularly around the gastrovascular cavity. These contractile cells do not attach to a hard skeleton, but work against the elastic mesoglea and hydroskeleton.

Receptors. Judging from their behavior, Cnidaria must have chemoreceptors and touch receptors on the epidermis, although they are not well-defined structurally. Perhaps the best-known evidence for chemoreceptors is the mouth-opening and tentacle movements of *Hydra* in the presence of prey. The substance that triggers this behavior is glutathione, a tripeptide that leaks from wounded animals. Many jellyfishes and cubomedusans also have combined receptor organs called rhopalia distributed around the margin of the bell (Fig. 24.7B). Included in each rhopalium is a balance organ, the **statocyst,** and often several simple eyes called ocelli. The statocyst works like those found in other aquatic invertebrates. Dense structures called **statoliths** rest on mechanoreceptor hairs, deflecting some of them, depending on the orientation of the organ with respect to gravity. Activation of a statocyst inhibits contraction on the same side of the medusa, so contractions on the opposite side tend to keep the animal oriented correctly.

Cubozoa have relatively elaborate ocelli, each with a lens and sensory layer reminiscent of those in the eyes of cephalopod molluscs and vertebrates. The ocelli connect directly to the nerve net without going through a ganglion, so it is doubtful that these animals can analyze much visual information. Nor is it clear what they would do with the information. The ocelli might detect luminescent prey or, because Cubozoa copulate, help in finding mates. Apparently, ocelli are not the only means of detecting light. Even without ocelli the green hydra *Chlorohydra* can move into sunlight, somersaulting end over end to where its photosynthesizing symbionts can function.

Behavior

Many cnidarians are sessile or slow-swimming, and they tend to behave as individuals. Large shoals of jellyfishes that wash onto beaches are not due to social schooling or a gang assault on human swimmers, but simply to the vagaries of wind and water. Social behavior does occur in corals, however, during **mass spawning** in which many different species in a reef release their gametes at the same hour. The most complex behaviors are displayed by sea anemones. Some sea anemones, such as *Stomphia* and *Calliactis,* detach and swim away in the presence of chemicals released by certain predatory starfishes or sea slugs. When other sea anemones, such as *Anthopleura elegantissima,* detect a predator, they secrete a chemical alarm pheromone that causes other anemones in the colony to pull in their tentacles. Many sea anemones actively compete for space with others of the same or a different species (Fig. 24.12).

Regeneration in *Hydra*

Hydras were first seen by Antony van Leeuwenhoek in 1702, but these retiring organisms were unable to compete for attention among all the other marvels being revealed by the microscope. In the 1740s, however, Abraham Trembley, a Swiss naturalist working in The Netherlands, made hydras superstars among organisms by showing (he thought) that they bridge the gap between plants and animals. This created such a sensation because it seemed to confirm the assertion of philosophers that all organisms are linked from lowest to highest on a Chain of Being.

Trembley first noticed hydras as green, flower-shaped structures attached to plants in ditch water. Because they were green and apparently rooted, he first thought they were plants. He found, however, that they contracted when disturbed, and even moved into sunlit parts of their containers. This puzzled Trembley, because only animals were supposed to move. Trembley decided to test whether his organisms were plants or animals by cutting them in two; only plants were supposed to be able to survive this procedure. To his surprise these organisms not only survived, but the flower-like head regenerated a new body, and the decapitated body sprouted a new head. This phenomenon later inspired Linnaeus to name the organism *Hydra,* after the sea serpent that grew two heads for every one that was chopped off. Since then the genus that Trembley studied has been renamed *Chlorohydra:* green hydra. Trembley concluded that hydras must be plants, but then changed his mind for the last time after observing that they captured prey and that the green color was due to symbiotic algae.

Trembley also discovered that the head from one hydra could be transplanted onto the body of another. (This experiment in 1742 was the first permanent transplant of animal tissue.) Since then more than 2000 published reports have followed Trembley's work on regeneration in hydras. Many such studies are now aimed at determining how the cells of an embryo "know" where they belong in the body and what they are supposed to become when they get there. Hydras are useful models in such studies of development for two reasons. First, like some sponges, hydras are able to reaggregate into functioning animals after being separated into individual cells. Second, like some vertebrates they display **polarity** during regeneration. If the head (mouth + tentacles) and the foot (basal disc + basal stalk) are removed from a polyp, the remaining **gastric column** always regenerates a new head at the oral end and a new foot at the other end (Fig. 24.14A). Any section removed from the gastric column also regenerates a head and foot at the correct ends. Polarity can be reversed by grafting the head and foot onto the wrong ends of the gastric column and leaving them for a day.

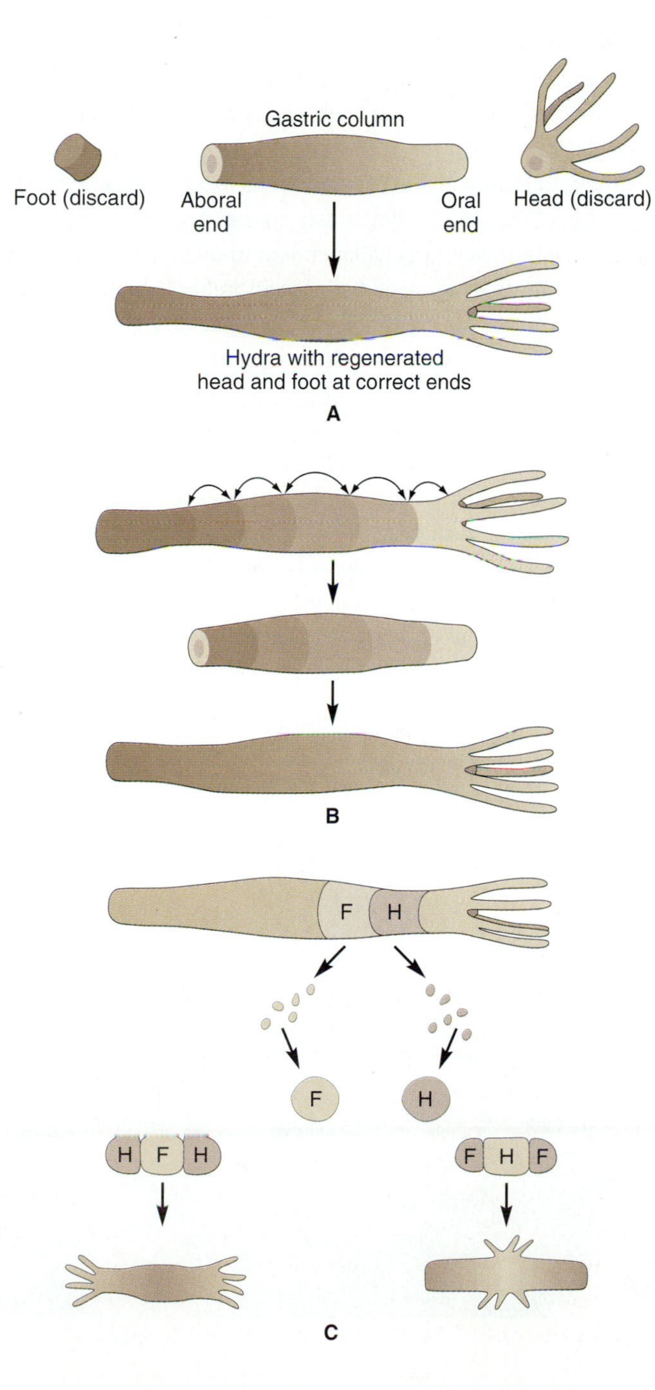

FIGURE 24.14

(A) Polarity in the regeneration of a new head and foot from the gastric column of *Hydra.* (B) Polarity is preserved even if the gastric column is cut into many sections, each of which is reversed. This indicates that polarity is due to a chemical gradient. (C) Polarity is preserved even in tissue clumps that have reaggregated from isolated cells.

■ **Question:** What do you think would happen if the discarded head in part A were applied to one clump of H cells on the lower left in part C?

There are at least two possible explanations for polarity: (1) Either each section of gastric column includes oriented cells that mark the polarity, or (2) there is a chemical gradient due to different concentrations of substances at the oral and foot ends. If the first theory were correct, then cutting the gastric columns into several sections and reversing each section while maintaining them in the same sequence should cause a reversal of polarity (Fig. 24.14B). This does not occur, however. There are, in fact, gradients of four different peptides involved in polarity. The regeneration of a head is due to a high concentration of **head activator.** Once the head is regenerated, it produces a **head inhibitor** that prevents another head from regenerating. **Foot activator** and **foot inhibitor** have similar effects at the opposite end.

Interactions with Humans and Other Animals

Coral Reefs. Of all the interactions of cnidarians with other animals, none is more impressive than those of a coral reef. Reef-building (**hermatypic**) corals are undoubtedly among the most ecologically important animals (pp. 350–351). Coral reefs provide substratum for a multitude of other sessile marine animals and also provide resting and hiding places for numerous mobile ones. Moreover, by their location and sheer bulk (Fig. 24.15) they buffer coastal organisms from the effects of storms. Reef-building by corals is largely dependent upon glycerol, amino acids, and other nutrients leaked by mutualistic protists (**zooxanthellae**) sheltered within their polyps (p. 465). These zooxanthellae are dinoflagellates that vary considerably, although most are classified in one species, *Symbiodinium microadriaticum.* Environmental stresses such as increased water temperature cause coral polyps to expel their zooxanthellae, a phenomenon called **bleaching.** Widespread bleaching in the Caribbean has caused a great deal of concern since the late 1980s, and the cause is not fully established.

Thanks to their cnidae, cnidarians have few predators. Two important exceptions are parrotfishes (family Scaridae) and certain starfish. Parrotfish chew up coral reefs to extract the polyps. The crown-of-thorns starfish (*Acanthaster planci*) is one of the most voracious destroyers of corals (see Fig. 33.18). During the 1960s and 1970s these starfish appeared to endanger even the enormous Great Barrier Reef of Australia, but since then the threat has subsided as mysteriously as it began.

FIGURE 24.15

A coral reef formed in the Devonian period more than 350 million years ago rises above the now-arid landscape of Western Australia.

SYMBIOSES. Zooxanthellae are only one of many kinds of organisms involved in symbiosis with cnidarians. The mutualistic green alga of *Chlorohydra* is another example, already noted. While the green hydra benefits from nutrients supplied by the algae, the algae benefit from the hydra's ability to move into sunlight. Another example of mutualism is the association between the Portuguese man-of-war and the fish *Nomeus*. *Nomeus* swims among the tentacles of the Portuguese man-of-war unharmed and safe from predators. In return, *Nomeus* probably serves as bait, luring predatory fishes into the tentacles of the Portuguese man-of-war. A similar mutualism occurs between certain species of sea anemones and certain fishes, especially damsel fishes (also called anemone fishes and clown fishes; family Pomacentridae). These sea anemones secrete fish attractants that may differ for each species of anemone and attract only a particular species of damsel fish. After a brief period of acclimation, the fish can swim unharmed among the tentacles, feeding on scraps from the prey captured by the anemone, which the fish, themselves, may have lured (Fig. 24.16). Until recently it was thought that the damsel fish coated itself with the anemone's mucus as protection from the nematocysts. It now appears more likely that protection involves a change in the fish's own mucus. (See Shick 1991, pp. 300–307, for a review.)

Perhaps the most bizarre example of mutualism occurs between some sea anemones and certain hermit crabs. While it is young, some sea anemones, such as *Adamsia palliata,* attach to mollusc shells that are occupied by young hermit crabs. They then spend the rest of their lives riding about with their mouths just above and behind those of the crabs, feeding on scraps. The anemone gradually surrounds and absorbs the shell, replacing it with its own body and growing at the same rate as the crab. The crab is not only defended and camouflaged, but also, unlike other hermit crabs, it does not have to find a new shell when it outgrows the old one.

HUMANS. Only occasionally do humans prey on cnidarians for food. In the Orient some jellyfishes, as well as the highly dangerous *Chironex*, are eaten. They are probably prized for their rubbery texture more than their nutritional value. The most important interaction of humans with cnidarians is our impact on coral reefs. Pollution from seaside development, agricultural runoff, and oil spills clogs the feeding systems of coral polyps, shades their zooxanthellae, and promotes the growth of algae and other competitors (Burke 1994). In the Philippines, fishermen have destroyed 95% of the coral reefs by illegally using cyanide and dynamite to catch fish for food and for unscrupulous pet dealers. A recent surge of interest in marine aquaria has resulted in hundreds of tons of coral reef ("live rock") being sold to more than 10 million hobbyists. Virtually all this coral dies within two years. Around Jamaica, overfishing, combined with hurricane damage in 1980, has led to a 90% reduction of coral-reef area (Hughes 1994). The removal of predaceous fishes caused the overpopulation and subsequent collapse of the sea urchin *Diadema antillarum,* which removed the main control on algal growth.

FIGURE 24.16
The skunk clown fish *Amphiprion akallopisos* with its host anemone, *Heteractis magnifica.*

Increasing numbers of boats and divers damage coral reefs in shallow waters, and collectors destroy large blocks of dead reef to sell as coffee-table curios. Precious corals, especially red precious coral *Corallium rubrum,* have been prized for thousands of years for jewelry. Divers with modern equipment have made them rare in places. In parts of the Mediterranean the tiniest fragment of precious coral is quickly seized. Black corals (*Antipathes*) and blue corals (*Heliopora*) are also used for jewelry. Coral jewelry is not only attractive, but was once believed to protect against the "evil eye." The incidence of "evil eye" has declined drastically in recent times, so precious corals are no longer prescribed much.

■ **Question:** What do you suppose is the adaptive value of the bright colors of anemone fishes? Would they warn potential predators away from the fish or from the anemone?

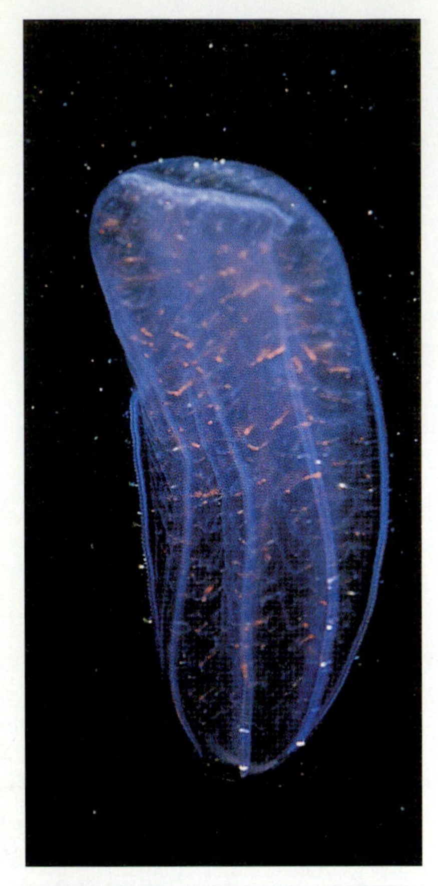

FIGURE 24.17
Beroe cucumis (class Nuda) preys on other ctenophores. It is about 1 cm in diameter. (*Beroe* is often spelled *Beroë*, in violation of a rule of nomenclature, to indicate that the *e* is pronounced, and not part of the diphthong *oe*.)

Other cnidarians may soon find a place in medicine, however (Ruggieri 1976, Tucker 1985). Palytoxin, from the Hawaiian soft coral *Palythoa toxica,* is one of the most potent toxins known and was once used as a spear poison. It is now used as an antitumor drug. After eight years of effort, organic chemists have succeeded in synthesizing palytoxin, which is one of the most complex organic molecules known, with 129 carbon atoms. Some other corals, such as *Plexaura,* produce both antitumor and antimicrobial substances. Surgeons are also experimenting with coral for bone replacement. Aequorin, from the jellyfish *Aequorea aequoria,* has been used to study intracellular processes in both normal and abnormal cells, because it emits light in proportion to calcium concentration.

CTENOPHORES

General Features

Members of phylum Ctenophora, commonly called "comb jellies," are delicate marine animals sometimes encountered swimming or drifting harmlessly near the surface (Figs. 24.1B and 24.17). Often they are found along beaches as translucent, gelatinous lumps called "sea walnuts" or "sea gooseberries." At night they may make their presence known by brilliant flashes of light. The use of submersible research vessels in recent years has shown that ctenophores are much more numerous in deep water than was previously suspected. The evolutionary origin of the ctenophores is unclear, partly because of a virtually total lack of fossils. Like cnidarians, ctenophores have thin tissues oriented around a gastrovascular cavity (Table 24.2). They differ from cnidarians in the following ways:

1. Ctenophores have no alternation of polyp and medusa generations.
2. They do not produce cnidae, but capture food by adhesive cells (colloblasts).

TABLE 24.2
Characteristics of ctenophores

Phylum Ctenophora ten-OFF-or-uh (Greek *kteis* a comb + *phora* bearing)

Morphology	**Biradial symmetry**: opposite tentacles (if present) and internal canals on a radially symmetric body. Tentacles (if present) consist of solid epidermis. Some species compressed into a flattened shape. **Monomorphic** (no alternating polyp and medusa). Always solitary and mobile. Thin tissues around a central gastrovascular cavity. No coelom.
Physiology	Few organs. **Statocyst** and **nerve net** control locomotion and feeding. Adhesive **colloblasts** trap food. **Hydrostatic skeleton** only.
Locomotion	By eight rows of **comb plates** (fused cilia), or by undulations of the body using true muscle cells.
Reproduction	Sexual reproduction only; most hermaphroditic.
Development	Cleavage determinate. Development indirect.
Habitat, Size, and Diversity	All **marine**. Size of adult from about 1 cm to 1.5 m. Approximately **100 living species** described.

3. Most swim by means of rows of fused cilia called comb plates.
4. Their cleavage is determinate, as in most other invertebrates.
5. Their tentacles (when present) are solid and consist only of epidermis.
6. They have true muscle cells derived from mesenchyme.

Ctenophores are the largest individual animals that swim by means of cilia. They are named for the eight rows of fused cilia called **comb plates** (Greek *kteis* comb) that most species use in swimming (Fig. 24.18). The cilia of the combs move in sequence from the anus toward the mouth, making luminescent species flash like neon signs. A **statocyst** near the anal opening coordinates the activity of the comb plates. This coordination allows the animal to swim in a preferred direction, usually horizontally with its mouth in front. The operation of this statocyst is quite elegant. The supporting cilia on which the **statolith** rests normally beat with an intrinsic rhythm, and these cilia impose the same rhythm on all the cilia in the comb plates that radiate from the statocyst. When the statolith presses against a particular group of supporting cilia, because of gravity or acceleration, it changes their rate of beating, thereby changing the rate of beating in the cilia of particular comb plates.

Feeding

Ctenophores are all predatory, although nutrition in some species is supplemented by photosynthetic algae. In ctenophores with tentacles, the **colloblasts** (glue cells) are the

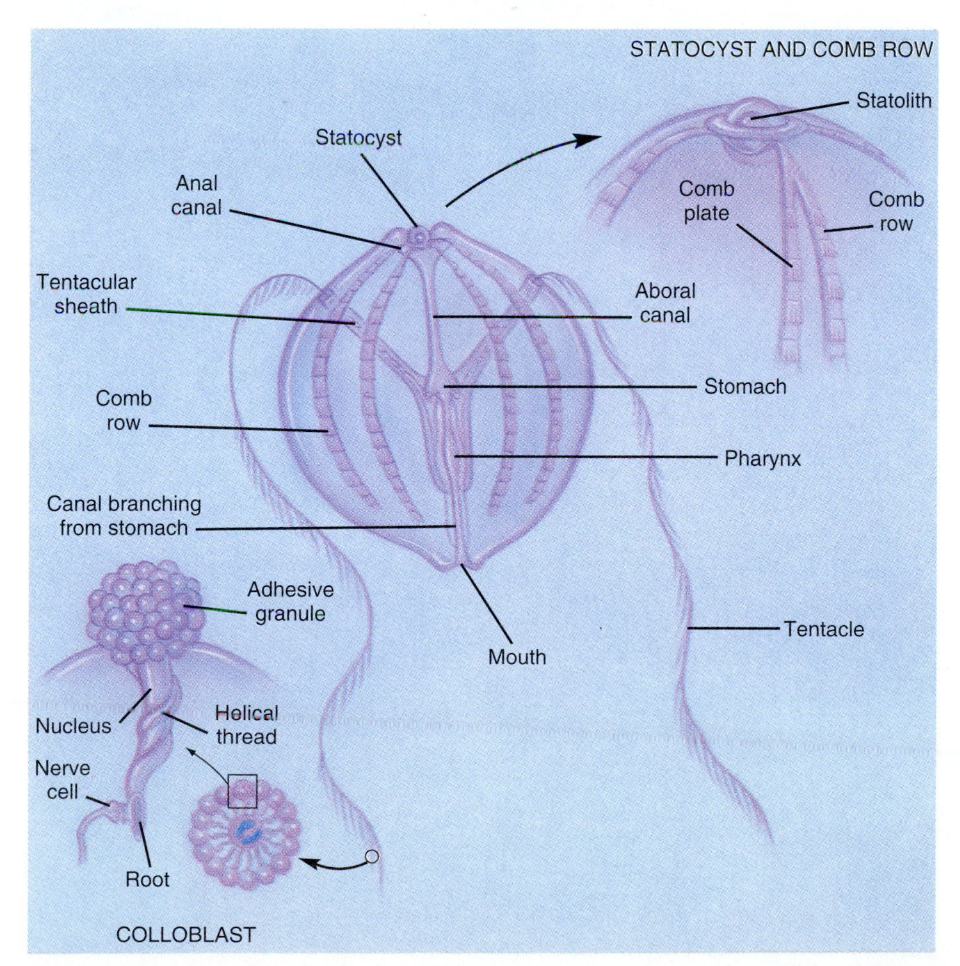

FIGURE 24.18
Pleurobrachia. (Compare Fig. 24.1B.) Insets show detail of colloblast and of statocyst and comb rows.

major organs of predation. (One species, *Haeckelia rubra,* has nematocysts that are apparently kleptocnidae acquired from cnidarian prey.) The colloblasts are not ejected, like nematocyst threads, but adhere to small organisms as the tentacles gracefully sweep the water. The tentacles then pass across the mouth, as if the ctenophore were licking its fingers. In the pharynx, food particles adhere to a mucous layer, which tufts of fused cilia (**membranelles**) propel into the stomach. Species with reduced or no tentacles feed by swimming with their mouths agape, trapping plankton on the mucous surfaces of the enlarged **oral lobes.** Some species simply swallow large prey whole. *Beroe* (Fig. 24.17), which has no tentacles, aggressively pursues and swallows only other ctenophores, including those larger than itself.

Digestion begins in the pharynx and continues in the stomach. The gastrovascular cavities of ctenophores branch into canals that distribute the partially digested food for intracellular digestion. The ctenophore gastrovascular cavity differs from that of cnidarians in having two **anal canals** through which only a little of the wastes are excreted. (The digestive tract is considered to be incomplete.)

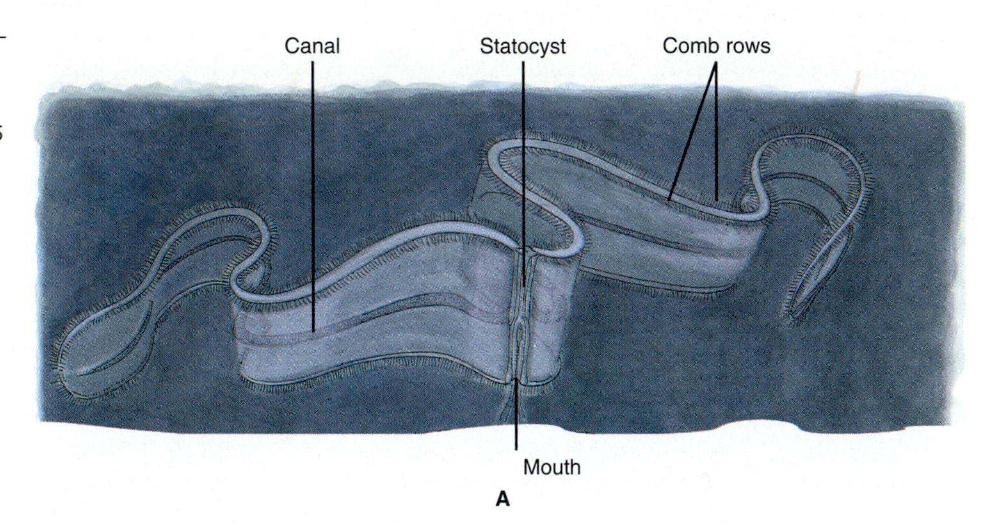

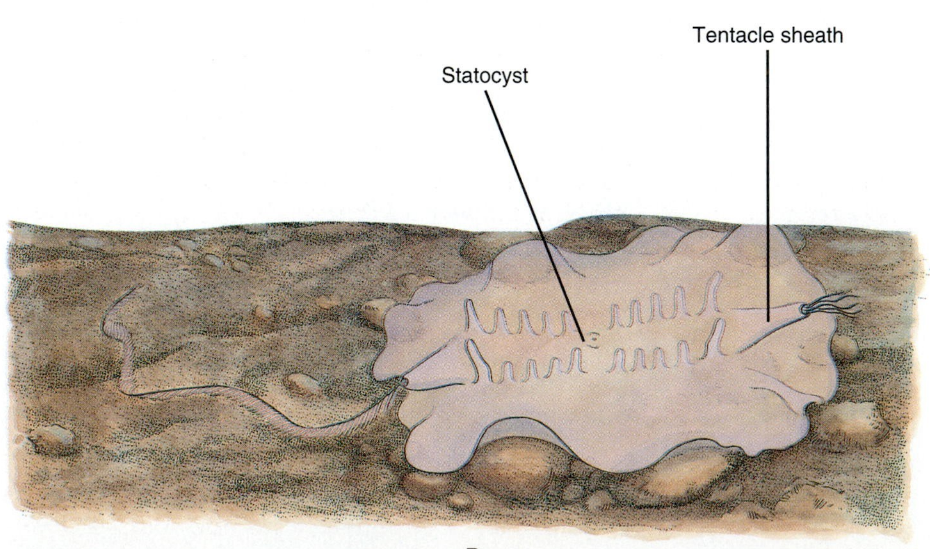

FIGURE 24.19

Two bilaterally symmetric ctenophores. (A) *Cestum,* commonly called a Venus girdle. Total length is approximately 1.5 meters. (B) *Coeloplana* differs from *Cestum* in the direction of flattening. The mouth forms the bottom surface. Length approximately 1 cm.

■ **Question:** How do you imagine Coeloplana gets food to its mouth?

Reproduction

Ctenophores are generally hermaphroditic and can reproduce only sexually, rather than by budding. Gametes generally exit through the mouth. Development occurs externally and produces a cydippid larva, which looks like a miniature adult.

Classification

Two classes of ctenophores, Tentaculata and Nuda, are distinguished by the presence or absence of tentacles. The 50 species of *Beroe* (Fig. 24.17) belong to class Nuda. *Pleurobrachia,* whose name means "side arms," is often considered representative of the tentaculates (Figs. 24.1B and 24.18). The body of *Pleurobrachia* is from 1.5 to 2.0 cm in diameter, but the tentacles can extend from their protective **tentacle sheaths** to a length of approximately 15 cm. Most of the body is composed of mesoglea (often called **mesenchyme** or **collenchyme** in ctenophores; p. 479). Within the class Tentaculata there is a wide divergence of shapes, with two genera having lost almost all semblance of biradial symmetry (Fig. 24.19). *Cestum,* commonly called Venus's girdle, is flattened along the oral–aboral axis, while *Coeloplana* is compressed with the oral and aboral portions close to each other on opposite sides. *Coeloplana* lacks comb rows; it creeps about on substratum.

Summary

Members of the phyla Cnidaria and Ctenophora generally have biradial symmetry and three (or two) thin layers of tissues enclosing a gastrovascular cavity. This body plan allows for a large size without a circulatory system or other complex organs to supply the necessities of life to all the cells. In spite of these similarities, major differences justify classifying cnidarians and ctenophores in different phyla. One difference is that the life cycle of cnidarians is thought to be primitively bimorphic, with a sessile, asexual polyp alternating with a swimming, sexual medusa. No such alternation of generations occurs in ctenophores. Second, cnidarians prey using nematocysts, while ctenophores use colloblasts. Third, the tentacles of cnidarians include gastrodermis, epidermis, and mesoglea (if present), while the tentacles of ctenophores are solid and consist only of epidermis. Fourth, cnidarians move by means of nutritive muscle cells and epitheliomuscle cells, while ctenophores have true muscle cells. Another difference is that ctenophores swim by means of comb plates that are unique to the phylum.

Four classes of cnidarians are generally recognized: Hydrozoa, Scyphozoa, Cubozoa, and Anthozoa. Hydrozoans include freshwater polyps, such as *Hydra,* the Portuguese man-of-war, and fire corals. Scyphozoans are the true jellyfish, in which the medusa is the most prominent, or only, stage. Cubozoans are similar to jellyfish, but are box-shaped. Anthozoans are extremely diverse polyps, including true corals, sea fans, sea whips, sea pansies, and sea anemones.

Cnidarians are virtually all carnivorous, using their cnidae to spear, poison, or entangle prey. Nematocysts first punch holes in prey, then shoot threads into the holes. Discharge of nematocysts is triggered by a combination of chemical and mechanical stimulation, without the involvement of the nervous system. The nervous system consists of a diffuse nerve net that causes slow movements of the body and tentacles during feeding, or fast movement during escape. There are few other organs. Parts of hydras can regenerate a new head or foot at the correct ends, because of chemical gradients that establish polarity.

Key Terms

biradial symmetry
gastrovascular cavity
gastrodermis
epidermis
mesoglea
polyp
medusa
planula
nerve net
radial canal
cnida
nematocyst
statocyst
mass spawning
polarity
zooxanthella
comb plate
colloblast

Chapter Test

1. Describe the ways in which Cnidaria and Ctenophora resemble each other. What differences account for their being classified in different phyla? (pp. 493, 510–511)
2. Explain how the cnidarians and ctenophores are able to maintain such large bodies without a circulating body fluid. (p. 493)
3. Some cnidarians alternate between polyp and medusa generations. What are the characteristics of each form? What advantages might the alternation of forms confer on a species? (pp. 496–497)
4. Describe the life cycle of *Obelia.* (p. 497)
5. Describe two examples of mutualism involving members of class Anthozoa. (pp. 508–509)
6. At the beach you find a biradially symmetric, gelatinous animal with tentacles. How can you tell from its appearance whether it is a scyphozoan, a cubozoan, or a tentaculate? Which of these would be safe to touch? (pp. 498–500, 511–512)
7. What are the differences between cnidae and colloblasts? (pp. 502–504, 511–512)
8. Describe the organization and functioning of the nervous systems of cnidarians. (pp. 504–506)
9. Describe one experiment supporting the chemical-gradient theory of polarity in *Hydra.* (pp. 506–508)

■ Answers to the Figure Questions

24.2 The nerve cells are in a position to coordinate interactions of sensory cells and epitheliomuscle cells. Detection of predators by sensory cells could trigger contraction by epitheliomuscle cells.

24.4 Asexual reproduction can compensate for a scarcity of mates and would be important for a sessile animal like *Hydra.* Sexual reproduction, however, has the advantage of producing offspring that are not genetically uniform.

24.5 The advantage is that individual zooids can specialize for different functions, and offspring produced asexually need not find new habitat. A disadvantage might be that predators would find the entire colony more tempting than many small individuals.

24.6 A fishing tentacle probably is homologous with a tentacle of an individual polyp. A feeding tentacle, however, is most likely the elongated body of a polyp, including the gastrovascular cavity.

24.7 Photoreceptors and statocysts would be necessary to enable the jellyfish to swim to the correct depth in water.

24.10 The even spacing suggests that each polyp competes for space with its neighbors.

24.14 The head might keep the H cells from forming a new head.

24.16 It seems unlikely that the bright colors of anemone fishes are warning coloration, since the anemone itself is already brightly colored, and fishes are not toxic. More likely the bright colors of anemone fishes lure potential predators, which then become the prey of anemones.

24.19 *Coeloplana* feeds by covering its prey, perhaps after subduing it with the tentacles.

Readings

Recommended Readings

Derr, M. 1992. Raiders of the reef. *Audubon* March/April, pp. 48–56. (*Destruction of coral reefs to supply marine aquaria.*)

Faulkner, D., and R. Chesher. 1979. *Living Corals.* New York: Clarkson N. Potter. (*Beautifully illustrated.*)

Fautin, D. G. 1992. A shell with a new twist. *Nat. Hist.* April, pp. 51–57. (*Symbioses between anemones and hermit crabs.*)

Gierer, A. 1974. Hydra as a model for the development of biological form. *Sci. Am.* 231(6):44–54 (Dec).

Goreau, T. F., N. I. Goreau, and T. J. Goreau. 1979. Corals and coral reefs. *Sci. Am.* 241(2):124–136 (Aug).

Gould, S. J. 1985. A most ingenious paradox. In: *The Flamingo's Smile.* New York: W. W. Norton, Chapter 5. (*On the question of individuality in the Portuguese-man-of-war.*)

Jackson, J. B. C., and T. P. Hughes. 1985. Adaptive strategies of coral-reef invertebrates. *Am. Sci.* 73:265–274.

Kaplan, E. H. 1982. *A Field Guide to Coral Reefs.* Boston: Houghton Mifflin. (*And their inhabitants.*)

Koehl, M. A. R. 1982. The interaction of moving water and sessile organisms. *Sci. Am.* 247(6):124–134 (Dec). (*How anemones adapt to ocean currents.*)

Lenhoff, H. M., and S. G. Lenhoff. 1988. Trembley's polyps. *Sci. Am.* 258(4):108–113 (Apr).

See also relevant selections in General References listed at the end of Chapter 21.

Additional References

Burke, M. 1994. Phosphorus fingered as coral killer. *Science* 263:1086.

Christen, R., et al. 1991. An analysis of the origin of metazoans, using comparisons of partial sequences of the 28S RNA, reveals an early emergence of triploblasts. *EMBO J.* 10:499–503.

Harrison, F. W., and J. A. Westfall (Eds.). 1991. *Microscopic Anatomy of the Invertebrates. Vol. 2: Placozoa, Porifera, Cnidaria, and Ctenophora.* New York: Wiley–Liss.

Holstein, T., and P. Tardent. 1984. An ultrahigh-speed analysis of exocytosis: nematocyst discharge. *Science* 223:830–833.

Hughes, T. P. 1994. Catastrophes, phase shifts, and large-scale degradation of a Caribbean coral reef. *Science* 265:1547–1551.

Murata, M., et al. 1986. Characterization of compounds that induce symbiosis between sea anemone and anemone fish. *Science* 234:585–587.

Ruggieri, G. D. 1976. Drugs from the sea. *Science* 194: 491–497.

Sebens, K. P. 1994. Biodiversity of coral reefs: What are we losing and why? *Am. Sci.* 34:115–133.

Shick, J. M. 1991. *A Functional Biology of Sea Anemones.* New York: Chapman & Hall.

Spencer, A. N. 1989. Neuropeptides in the Cnidaria. *Am. Zool.* 29:1213–1225.

Tucker, J. B. 1985. Drugs from the sea spark renewed interest. *BioScience* 35:541–545.

Wainright, P. O., G. Hinkle, M. L. Sogin, and S. K. Stickel. 1993. Monophyletic origins of the metazoa: an evolutionary link with fungi. *Science* 260:340–342.

CHAPTER

25

Flatworms and Other Acoelomates

Unidentified marine flatworm.

Introduction

If the evolution of animals can be compared to the growth of a tree, then the animals we have studied so far appear to have grown out from the roots. In this chapter we begin to study the limbs of the tree. Unlike the asymmetric, radially symmetric, or biradially symmetric animals described in the previous two chapters, the animals that will be discussed from now on are **bilaterally symmetric** and are often called the **bilateria.** In addition, in these animals, unlike those described in previous chapters, there is no doubt that their tissues are derived from three germ layers: ectoderm, mesoderm, and endoderm. That is, they are **triploblastic.** In addition, their tissues are organized into definite organs that perform specific functions. In other words, they have an **organ-system grade of organization,** rather than a cellular or tissue level of organization.

In this chapter we start with three groups generally thought to be on the lowest branches (see Fig. 21.10). These are the acoelomate phyla **Platyhelminthes, Gnathostomulida,** and **Gastrotricha,** commonly known as flatworms, jaw worms, and gastrotrichs. (Nemertea, covered in Chapter 27, are also thought by some to be acoelomates.) All are elongated, legless invertebrates known classically as **worms** or helminths. The three phyla described in this chapter are considered together for several reasons:

1. Unlike animals to be considered in later chapters, they lack a body cavity (pseudocoel or coelom; see p. 439). Thus they are called acoelomates (a-SEAL-oh-maytz). Instead of a body cavity, the tissue between ectoderm and endoderm—the mesenchyme—consists of densely packed mesodermal cells and is called **parenchyme** (or parenchyma; see p. 479).
2. These animals have similar solutions to the problem of getting the necessities of life to all their cells. The main solution is that they are flattened dorsoventrally, which provides a large ratio of surface to volume and ensures that no cell is far from a source of oxygen, nutrients, water, and ions, even though there is no circulatory system or body cavity.
3. Species in each phylum have ciliated epidermal cells at some stage.
4. All three phyla include species with protonephridia (osmoregulatory tubules open only at the external end).
5. All have a central nervous system consisting of cerebral ganglia and longitudinal nerve cords.

FLATWORMS: PHYLUM PLATYHELMINTHES

Characteristics and Major Kinds of Flatworms

Flatworms vary considerably, mainly because some are free-living while others are parasitic within other animals. The major characteristics that most have in common are summarized above and in Table 25.1.

Classification within the phylum Platyhelminthes is unsettled. Traditional classes are defined according to the degree to which the group is free-living or parasitic. Members of class **Turbellaria,** which are almost all free-living, tend to have the best-developed organ systems, since they must search for and secure food, and they must tolerate a variety of external environments (Fig. 25.1 on page 518). Turbellarians can creep about on cilia in a film of secreted mucus. Turbellarians can also move by alternately contracting longitudinal and circular muscles. On encountering suitable prey—usually a small animal—turbellarians cover it with their bodies and begin feeding through a ventral mouth.

A

B

FIGURE 25.1

(A) An unidentified marine flatworm. Many marine flatworms swim gracefully and are brightly colored. (B) *Bipalium kewense,* a terrestrial turbellarian native to the tropics of Asia. Having been inadvertently imported with ornamental plants, it is now common in gardens and greenhouses in the southern United States. It preys on earthworms and has become a serious pest in some worm farms. Length is up to 25 cm.

■ **Question:** What is the most likely function of the bright coloration of these flatworms: courtship, camouflage, or warning coloration?

TABLE 25.1

Characteristics of flatworms.

Phylum Platyhelminthes plat-ee-hell-MINTH-eez (Greek *platys* flat + *helminthos* worm)

Morphology	Acoelomate: no body cavity. **Parenchyme** consisting of mesodermal cells separates ectodermal and endodermal tissues. **Bilateral symmetry**. Body usually flattened dorsoventrally.
Physiology	**Central nervous system**, typically with a pair of anterior ganglia and up to four pairs of longitudinal nerve cords joined by branches. Also a diffuse nerve plexus beneath the epidermis. Mainly carnivorous (predatory, scavenging, or parasitic); some with mutualistic algae in epidermis. Absorption of dissolved organic material across the epidermis in endoparasites, and probably others. Pharynx and often highly branched digestive cavity. Digestive tract (if present) incomplete (without a separate anus). Digestion partly intracellular. Osmoregulation by protonephridia. No circulatory, respiratory, or skeletal system.
Locomotion	Creeping on substratum, swimming, or passive within host. Often with duogland adhesive system for temporary anchorage to substratum.
Reproduction	Usually **monoecious**, but self-fertilization usually avoided. Asexual reproduction by transverse fission in some forms.
Development	Spiral cleavage pattern (when not distorted by yolk). Determinate. Blastospore becomes mouth. Development direct, or indirect with one or more larval stages.
Habitat, Size, and Diversity	Almost all **aquatic** or **parasitic**. Sizes of adults from less than 1 mm to more than 10 meters long. Approximately **12,200 living species** described.

The parasitic flatworms called **flukes** were formerly included in a single class Trematoda. Now, however, they are divided into two classes, depending on whether they have one host species (**monogenetic**) or two or more hosts (**digenetic**). Like most animals, monogenetic flukes, now placed in **class Monogenea,** reproduce only during the final, adult stage of development. Their eggs develop directly into nonparasitic juveniles that resemble the adults. Most are external parasites on the skin or gills of fish, to which they cling by suckers, hooks, or clamps. Other flukes remain in **class Trematoda,** and most of these belong to **subclass Digenea.** Digenetic flukes have at least five distinct stages in their life cycles, and they can reproduce asexually during at least one of them. At least one stage parasitizes an **intermediate host,** usually a mollusc. The final stage is usually an internal parasite of a vertebrate **final** (= definitive or primary) **host,** in whose tissues it attaches by means of suckers.

Tapeworms, class Cestoda, are parasites of the vertebrate digestive tract, to which they attach by suckers and/or hooks. Tapeworms do not have digestive systems of their own. In fact, only the reproductive system is well developed. Development is indirect, and most have larval forms that parasitize intermediate host species different from the final host of the adult.

Major Extant Groups of Phylum Platyhelminthes

Genera mentioned elsewhere in this chapter are noted.

Class Turbellaria TUR-bell-AIR-ee-uh (Latin *turbellae* a turbulence). Usually free-living, with ventral mouth from which the pharynx can be extended and with a branched or lobed intestine. Locomotion by creeping on cilia in a layer of secreted mucus. Epidermis secretes rod-like rhabdites in mucus that may release chemical repellents. Mostly hermaphroditic; some also reproduce by transverse fission. *Bipalium, Dugesia, Mesostoma* (Chapter opener; Figs. 25.1, 25.3).

Class Monogenea MON-oh-gen-EE-uh (Greek *monos* single + *genea* generation). Flukes with one host per life cycle. Mainly leaf-shaped ectoparasites of fish. Attachment to host by hooks, clamps, and/or suckers on posterior. Gut well developed. Adult covered by nonciliated, syncytial tegument. Juveniles ciliated and free-swimming. Monoecious. *Gyrodactylus, Oculotrema* (Fig. 25.6).

Class Trematoda TRIM-a-TOAD-uh (Greek *trematodes* with holes [referring to suckers]). Flukes with five or more developmental stages and two or more hosts. Some early stages reproduce asexually. Final stage mainly round or leaf-shaped endoparasites of vertebrates. Final stage attaches to host's tissues by suckers. Gut with two branches. Final stage covered with nonciliated tegument. Mostly monoecious. *Opisthorchis, Fasciola, Schistosoma* (Figs. 25.7, 25.8).

Class Cestoda (= Cestoidea) ses-TOW-duh (Greek *kestos* belt). Adults elongated, flattened endoparasites in intestines of vertebrates. Usually one or more larval stages and intermediate hosts. Adult attaches to host's intestine by hooks and/or suckers. No mouth or digestive organ. Adult covered with nonciliated tegument. Tapeworms. *Diphyllobothrium, Dipylidium, Echinococcus, Taenia, Taeniarhynchus, Vampirolepis* (Fig. 25.9).

Turbellarians

All free-living flatworms are assigned to class Turbellaria, although the class also includes some species that are symbiotic in marine organisms. Because free-living forms cannot depend on a host to provide a homeostatically regulated environment, most turbellarians are fully equipped with their own organs. These organ systems enable them to inhabit a wide range of habitats: marine, freshwater, and terrestrial.

Feeding and Digestion. In most turbellarians the most prominent organ is the **intestine.** It is often highly branched into the thickest layer of tissue, the parenchyme. The intestine partly substitutes for the absence of a circulatory system and body cavity. Most turbellarians have a pharynx that can be everted through the mouth. Digestive enzymes are secreted through the pharynx into the food, and the partially digested food can then be drawn through the pharynx into the intestine. Partially digested food is then phagocytized into parenchymal cells for intracellular digestion (Fig. 25.2 on page 520).

Turbellarians are primarily predators and scavengers. Typically they glide about with head slightly raised until they are over small prey such as nematodes, rotifers, copepods, and insects. The turbellarian then grips the prey and immobilizes it with mucus, which is secreted by the epidermis. The mucus also contains rod-shaped structures called **rhabdites,** which are produced in the parenchyme and are thought to release toxic substances. Mere contact with the mucus of some turbellarians paralyzes and kills prey (Fig. 25.3 on page 521). The mucus also appears to be toxic to potential predators. This may explain why many marine turbellarians display beautiful warning

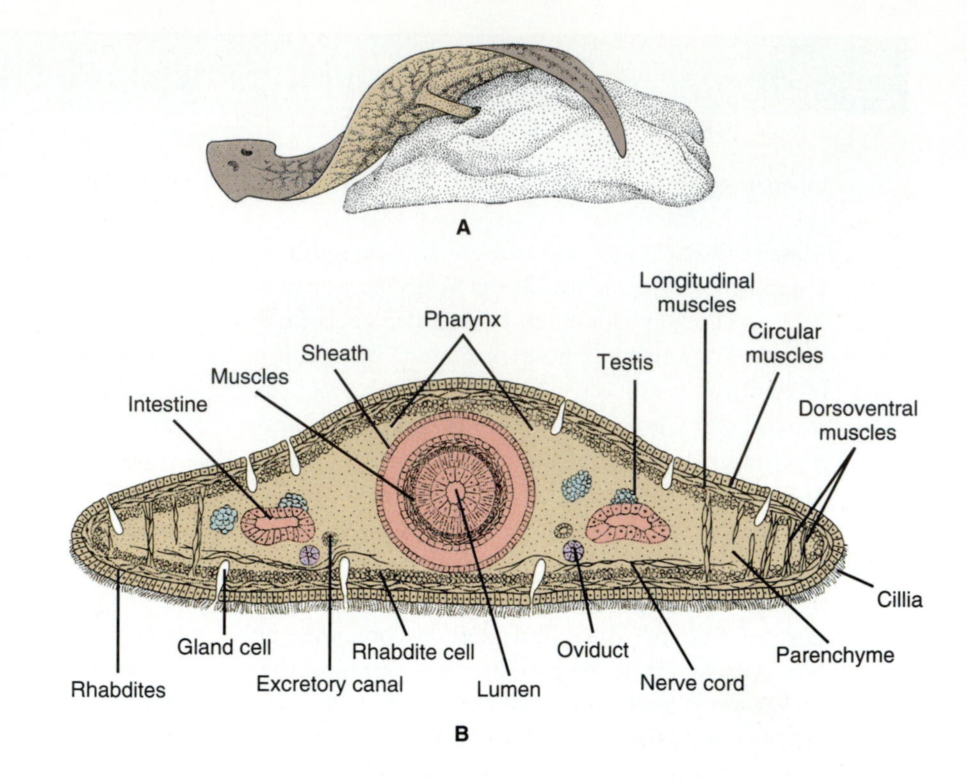

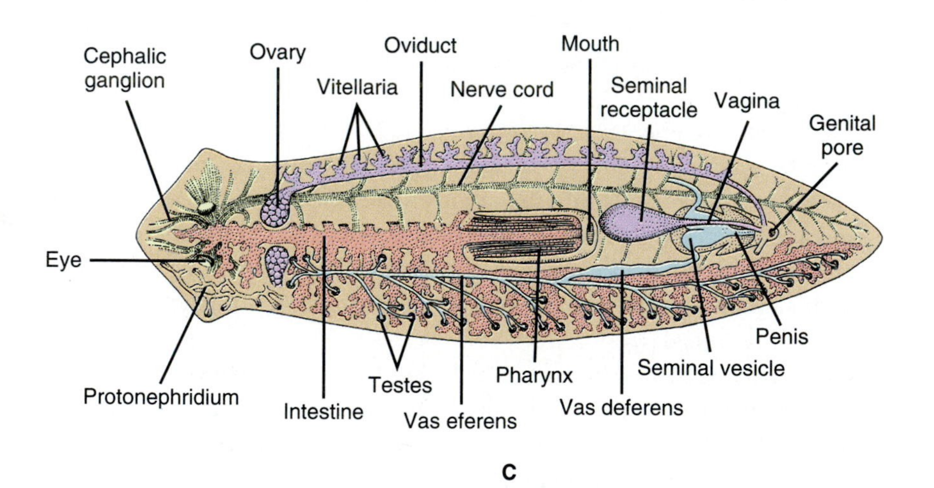

FIGURE 25.2

A free-living flatworm, the planarian *Dugesia,* from 0.5 to 2.5 cm long. (A) Feeding on a mass of food, using the pharynx that has been everted through the mouth. (B) Cross section of the body through the pharynx. (C) Schematic diagram of the organ systems. The upper part of the diagram shows the nervous system and the female reproductive organs. The medial and left part shows the digestive tract and male reproductive system. Part of the protonephridial system that courses throughout the parenchyme is shown in the left anterior portion.

coloration. Many turbellarians also secrete a kind of glue that may help keep them anchored while struggling with prey, as well as during locomotion. This adhesive comes from a **viscid gland** cell on the ventral surface, and it goes onto microvilli of **anchor cells** for temporary anchorage. A **releasing gland** later secretes a substance that breaks the bond. The combination of viscid and releasing glands is called a **duogland adhesive system.**

Excretion and Osmoregulation. Ammonia and other metabolic wastes are eliminated partly through the epidermis and partly through **protonephridia.** Protonephridia are also responsible for osmoregulation. They are especially important in freshwater turbellarians, which would swell up without them. Protonephridia are, by definition, excretory tubules that are closed internally and open to the outside. The inter-

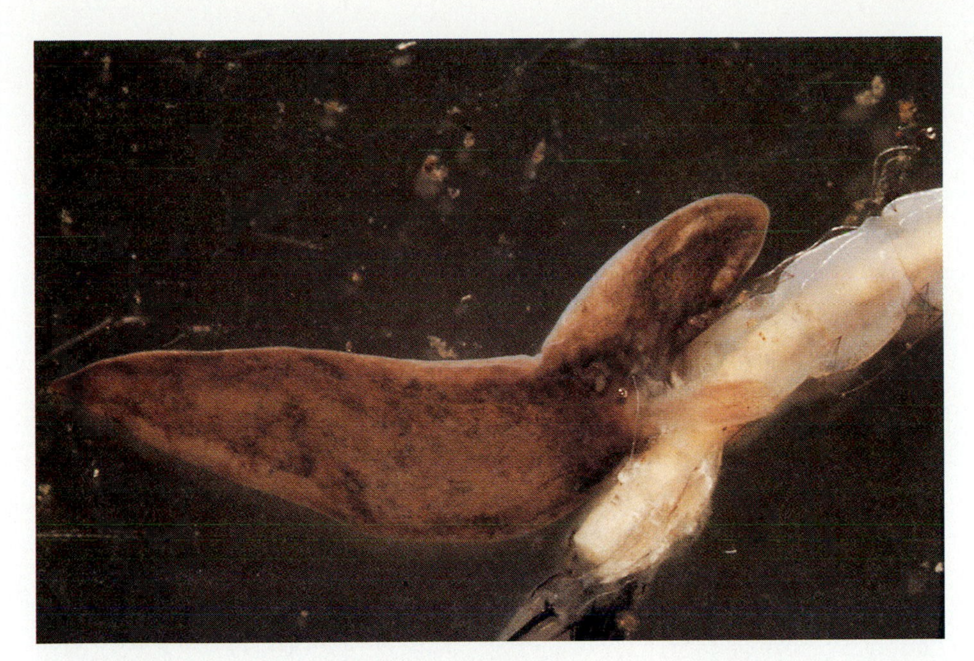

FIGURE 25.3

The turbellarian *Mesostoma* (left) with its pharynx around a mosquito larva *Culex tarsalis.* Turbellarians may significantly reduce mosquito populations.

nal endings of the tubules terminate in **flame cells,** so called because each cell has one tuft of cilia inside that flickers like a flame (see Fig. 14.6). In some flatworms each cell at the end of the protonephridia has numerous ciliated **flame bulbs.** The cilia of flame cells and flame bulbs are believed to pull fluid through the membranes of the flame cell, producing a filtrate of water and small molecules, but leaving cells and large molecules within the interstitial fluid. As the filtrate moves along the protonephridial tubules, active transport or diffusion may eliminate wastes and reabsorb nutrients into the parenchyme.

NERVOUS SYSTEM. Turbellarians are the first animals encountered so far in this unit that have central nervous systems. Because bilateral symmetry is associated with movement of an animal in one direction, it is natural for sensory receptors to be concentrated at the anterior end, which encounters the environment first, and for nerve cells to be concentrated at the same end. As in most bilaterally symmetric invertebrates, concentrations of nerve cells occur in two **cephalic ganglia** (= cerebral ganglia). Cephalic ganglia are sometimes called "brains," but unlike the brains of vertebrates, they have little control of the worm's overall activities. Most behavior is coordinated in the longitudinal nerve cords (**connectives**) and their ladder-like branches (see Fig. 9.3). In addition, there is a **subepidermal nerve plexus** that functions independently of the cephalic ganglia, in much the same way as the cnidarian nerve net. The cephalic ganglia serve primarily as centers for the integration of sensory information from head receptors, including the two **eyespots** (ocelli). The dark areas of these eyespots look like the irises of human eyes, but they are actually pigment cups that block the light from the medial direction (Fig. 25.4). Without this pigment the animal could not determine the direction of the light, because the ocelli have no lenses that could form an image. The epidermis is equipped with a variety of other receptors, especially in the lateral lobes (**auricles**) on the head. Behavioral studies indicate that such receptors enable flatworms to react appropriately to a variety of chemical and mechanical stimuli. Some turbellarians also have a **statocyst** on top of the head that detects the direction of gravity.

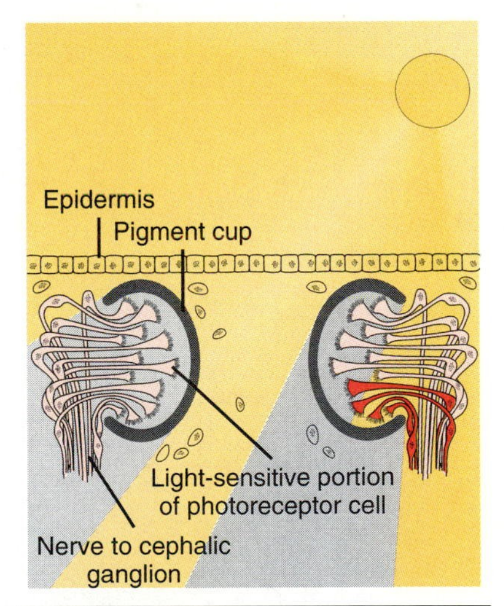

FIGURE 25.4

Ocelli of *Dugesia.* The pigment cups block light from the opposite side, enabling the flatworm to determine the direction of light.

■ **Question:** Why might a turbellarian need to determine the direction of light?

REPRODUCTION AND DEVELOPMENT. Asexual reproduction by transverse fission (lengthwise division) is common in terrestrial and freshwater turbellarians. Most turbellarians and other flatworms, however, reproduce sexually. Like most flatworms, turbellarians are hermaphroditic and apparently favor copulation over self-fertilization. Copulation increases genetic diversity and may avoid the waste of gametes that occurs with external fertilization. All flatworms have a penis (also called a **cirrus**), and most also have a vagina. Those without vaginas nevertheless manage internal fertilization by **hypodermic impregnation,** in which one flatworm stabs the other with its penis. The internal reproductive systems are highly varied, but the following description and terminology apply to many flatworms and to many other invertebrates. Sperm leave each testis through a **vas efferens,** and these vasa efferentia collect in a **vas deferens** (= sperm duct) on each side (Fig. 25.2C). During copulation, sperm enter the **seminal receptacle** of the partner, where they remain until ovulation. During ovulation the ova travel down an **oviduct** on each side and are joined along the way by yolk cells from **vitellaria** (= yolk glands).

In turbellarians some ova and thousands of yolk cells are packaged into a cocoon that may be attached to substratum. The embryos usually develop directly into young, which resemble their parents except that they lack reproductive organs. Some marine species first hatch into larvae that swim by means of cilia.

LABORATORY STUDIES OF PLANARIANS. The freshwater turbellarians called planarians are often used in laboratory studies. *Dugesia,* with its appealing cross-eyed look, is especially popular (Fig. 25.2). It is easily collected from beneath leaves, rocks, and other objects in ponds and streams by using a bit of raw meat for bait, and cultures are easy to maintain in the laboratory. *Dugesia* is often used in studies of **regeneration.** Like *Hydra* (pp. 506–508) they demonstrate polarity. That is, a section removed from the middle of a planarian will grow a new head and tail at the correct ends. If the head is split lengthwise and the two halves are kept apart, each half will regenerate the missing parts to form a two-headed worm. This can be repeated indefinitely to produce planaria with 10 or more heads. Flatworms also readily accept grafts of heads and other body parts, even from other species.

In most animals, regeneration serves the obvious function of restoring lost limbs. In flatworms it serves at least two other functions. First, starved flatworms often metabolize some of their own body parts, so regeneration allows them to recover from prolonged food shortages. Second, regeneration enables them to reproduce by transverse fission. In warm water, well-fed turbellarians often appear to pinch themselves in two, and each half regenerates into a new and complete individual. In some species, transverse fission is incomplete; the flatworm produces additional body parts, but actual separation is delayed. The result is a chain of attached **zooids** that superficially resembles a segmented worm.

During the 1960s many psychologists used *Dugesia* as a simple model to study **learning.** They even had their own journal with the whimsical title *Worm Runner's Digest.* Numerous reports in this journal claimed that *Dugesia* could not only learn various types of mazes and conditioned responses, but could acquire such learning by cannibalizing trained planaria! Other laboratories were unable to repeat the experiments, however, and they are now seldom mentioned. One problem with this type of research using planaria is that they "run" the mazes (actually they glide through them) on a layer of mucus secreted from the ventral surface. Care must be taken to ensure that the planarian is actually learning to run the maze correctly, rather than simply following an old mucous trail.

Monogenetic Flukes

The main differences between turbellarians and flukes are adaptations of the latter to parasitism. One such adaptation is the **tegument,** a body covering that protects flukes from host defenses and from the external environment. The tegument was formerly

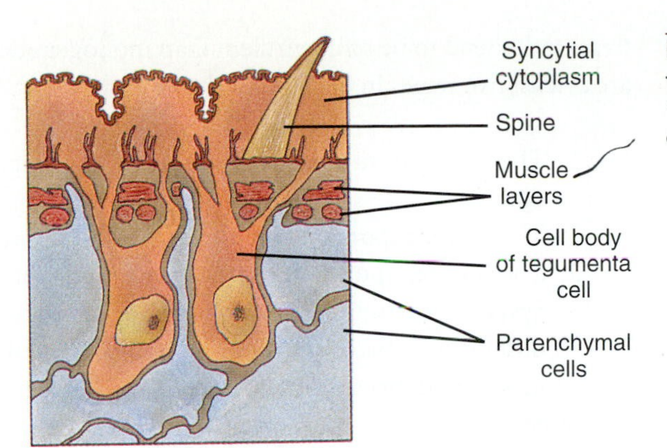

FIGURE 25.5

The tegument of *Fasciola hepatica,* the "liver rot" trematode of sheep and cattle.

considered to be cuticle—a protective layer of dead cells or secretions, such as occurs in insects. Electron microscopy has revealed, however, that the tegument actually consists of inactive portions of a syncytium whose active portions are submerged within the parenchyme (Fig. 25.5). Other adaptations of flukes—and of parasites in general—compensate for the low probability that a particular parasite will find a proper host (see p. 467).

Flukes are usually flattened, oval, and a few centimeters long. All flukes used to be placed in class Trematoda, but most monogenetic forms—those that have only one host species—are now assigned their own class Monogenea. They usually have direct development with a single juvenile form called an **oncomiracidium** (ON-ko-MERE-a-SID-ee-um). Most oncomiracidia spend some time swimming by means of their cilia before they find a suitable host, and many perish before they succeed. Enormous numbers of oncomiracidia are produced, thereby compensating for the poor odds. Once an oncomiracidium attaches to a host, it develops into an adult fluke, which remains attached by means of a posterior **opisthaptor** equipped with various hooks, clamps, and suckers (Fig. 25.6).

Most adults of class Monogenea are ectoparasites on the skin or gills of fishes, but some parasitize other vertebrates that spend a lot of time in water. *Oculotrema hippopotami,* for example, lives on the eyes of hippos, as its name implies. Like many parasites, monogeneans are highly selective with regard to host species and even host tissues. A species may be found on one gill arch of a fish, but never on a different gill arch, or on a gill filament near the base, but never near the end.

Because monogenetic flukes must depend on a single host for both reproduction and transmission, they have evolved mechanisms that usually ensure that the parasites do not endanger the lives of the hosts. This low virulence is not due to the kindness of the parasite, but is simply a consequence of natural selection. Only when transmission to hosts becomes especially easy, as in crowded fish farms, do monogenetic flukes produce disease symptoms. The parasites may then become so numerous that the host is literally smothered by flukes covering the gills, or by scar tissue caused by excessive damage. *Gyrodactylus,* for example, is a serious pest of trout, bluegills, and goldfish in ponds (Fig. 25.6). One of the keys to the success of *Gyrodactylus* is that the larva develops within a parent's uterus, with another larva inside it, another inside that one, and perhaps even a fourth-generation larva inside that one!

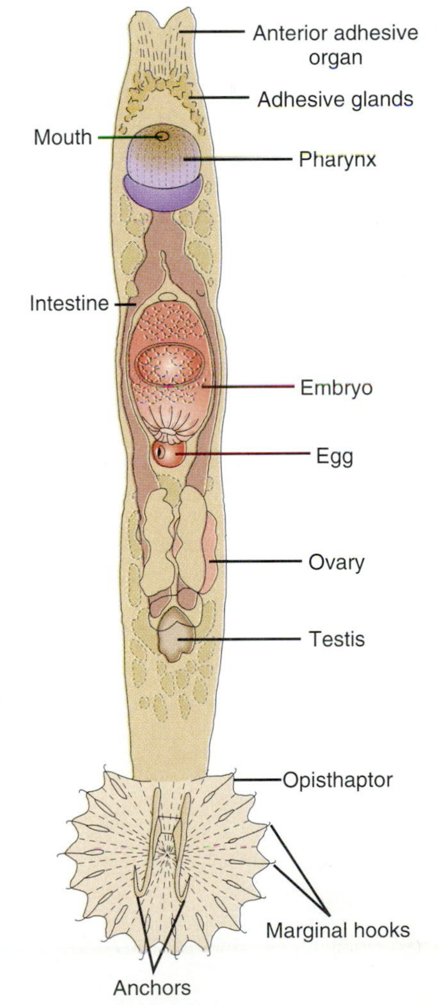

FIGURE 25.6

The monogenean fluke *Gyrodactylus cylindriformis* showing the adhesive glands and opisthaptor by which it clings to the gills of fish. Length approximately 0.4 mm.

Trematodes

The Digenea differ from the Monogenea in having at least two host species. Because one host can be used mainly for transmission and another mainly for reproduction, trematodes tend to be more virulent than monogenetic flukes (see p. 467). Often there are at least five stages in the trematode life cycle:

1. The **miracidium** hatches from the egg and usually infects a mollusc as intermediate host.
2. The **mother sporocyst** develops from the miracidium.
3. The mother sporocyst asexually produces either numerous **daughter sporocysts** or one to three generations of **rediae,** depending on species. The major difference between the sporocyst and the redia is that the sporocyst lacks a mouth and digestive tract and obtains nutrients by absorption across the tegument, while the redia actively consumes tissues of the intermediate host.
4. In species that do not form rediae, cells within each daughter sporocyst develop into **cercariae,** which leave the intermediate host and swim freely to infect vertebrate final host. Inside the final host, they develop into sexually reproducing flukes.
5. In species that produce rediae, the cercaria infects a second intermediate host in which it forms an additional larval stage, the **metacercaria.** The metacercaria then parasitizes the vertebrate final host.

LIFE CYCLE OF THE LIVER FLUKE. One of the most complex life cycles of digenea is that of the liver fluke *Opisthorchis sinensis* (formerly *Clonorchis sinensis*). This fluke is a serious problem in the Orient, where custom, a lack of cooking fuel, and poor sanitation encourage the eating of raw fish from polluted water. Infection of the human final host begins with the ingestion of metacercariae encysted in the muscles of fish (Fig. 25.7). After developing into mature flukes in the human intestine, they migrate into the bile duct, which connects the gallbladder to the liver. Light infections may not produce any symptoms, but heavy infections, with up to 20,000 flukes in the bile duct, can cause damage to the liver and other abdominal organs. Each hermaphroditic fluke can live up to 50 years, reach a length of several centimeters, and produce thousands of eggs that are released with the feces of the host. If an egg is fortunate enough to land in water and be ingested by a snail, a miracidium larva hatches out. The snail then becomes the first intermediate host. Inside the snail's tissues the miracidium develops into a sporocyst, which then asexually produces thousands of rediae. Each redia asexually produces up to 50 cercariae, which emerge from the snail's epidermis. Cercariae swim freely by means of their tails, then bore into the muscles or under the scales of a fish, which becomes the second intermediate host. There each cercaria becomes encysted as a metacercaria. The life cycle is completed when the final host, another vertebrate, eats the fish containing the metacercariae.

SCHISTOSOMIASIS (= bilharziasis). The blood fluke *Schistosoma* (SHISS-toe-SO-muh; Greek *schistos* split + *soma* body) is another serious cause of disease, affecting approximately 200 million people (Fig. 25.8). Schistosomiasis can be caused by three species of *Schistosoma,* each with a different intermediate snail host, geographic distribution, and symptoms. *Schistosoma mansoni* is common in Africa, South America, and the Caribbean. *Schistosoma haematobium* is common in Africa and the Middle East. *Schistosoma japonicum* occurs in the Far East. (See box on p. 527.)

Schistosoma differs from the liver fluke in that it skips the redia and metacercaria stages. Miracidia produce sporocysts within snails, and the sporocysts produce more sporocysts. The second-generation sporocysts then produce cercariae that penetrate the skin of humans, without a second intermediate host. Penetration of the skin by cercariae (Fig. 25.8B) triggers an immune response that results in itching and a rash.

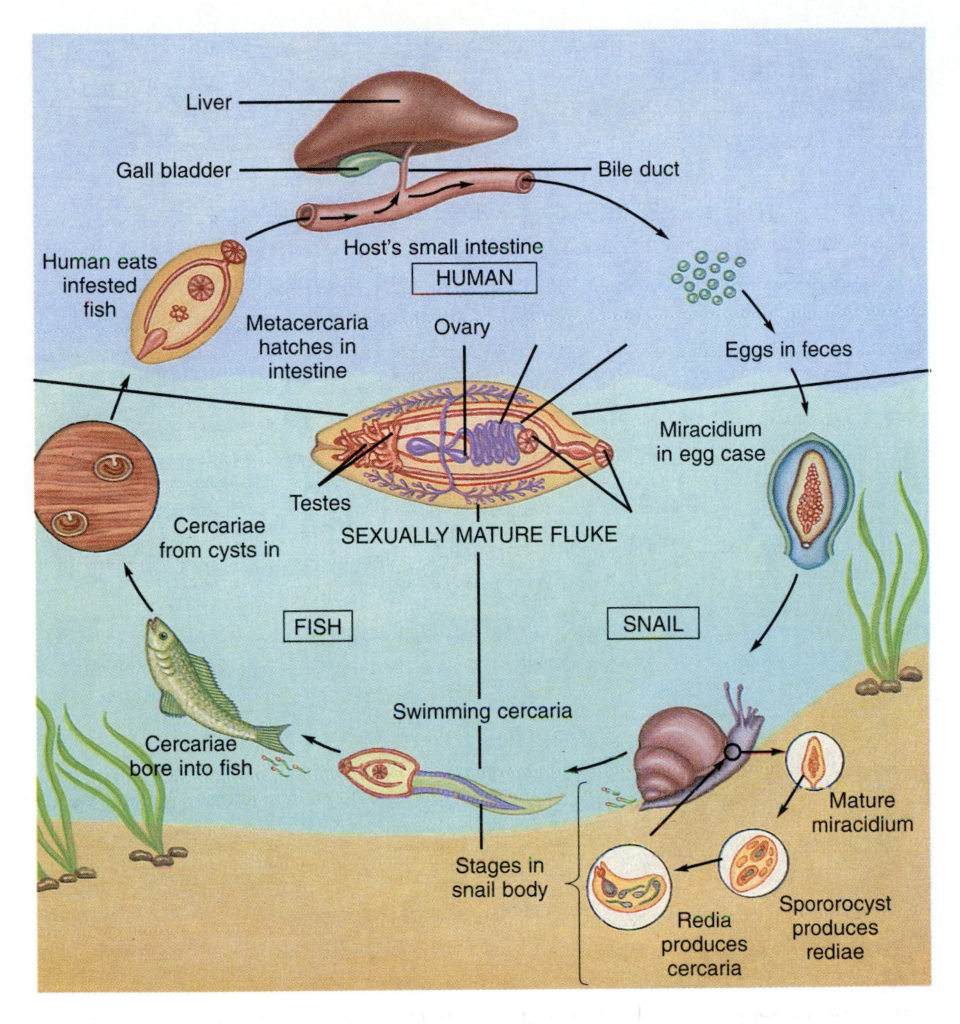

FIGURE 25.7
The life cycle of the human liver fluke *Opisthorchis.*

■ **Question:** Which of the adaptations listed on p. 467 does *Opisthorchis* have?

(These immediate symptoms are similar to "swimmer's itch," which results when a species of *Schistosoma* that is common in North American lakes enters the skin. This species cannot use humans as a host, however.)

The immediate symptoms are followed by 4 to 10 weeks of fever, abdominal pain, throat irritation, diarrhea, and enlargement of the liver and spleen. During that time

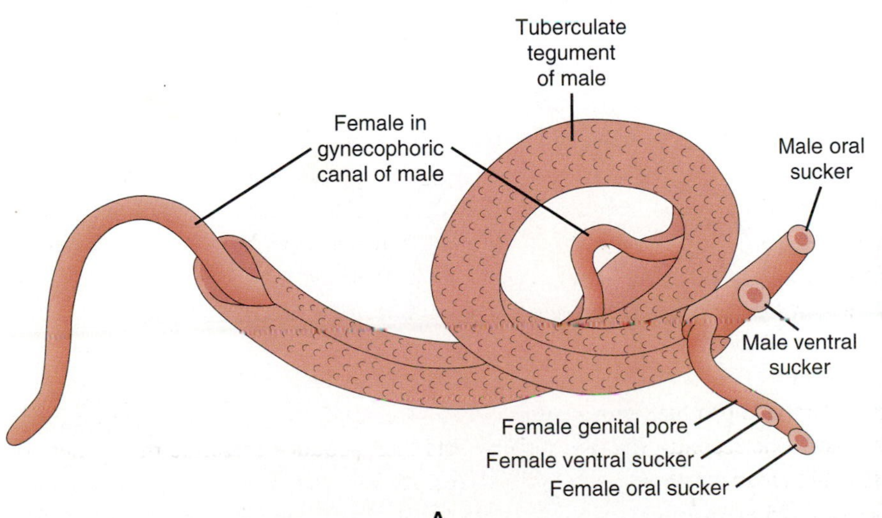

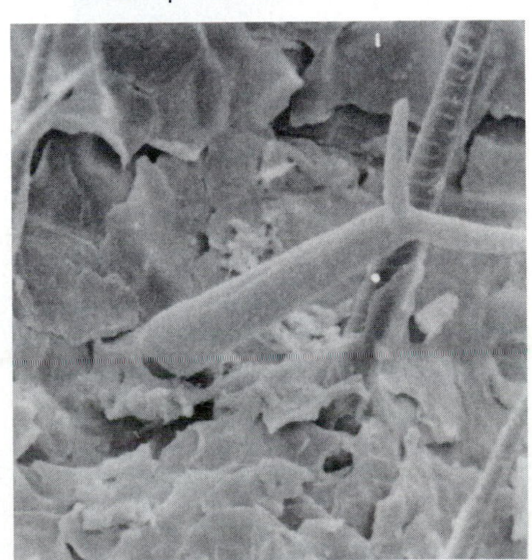

FIGURE 25.8
(A) Male and female *Schistosoma mansoni* copulating, as they are often found in their hosts. Note the gynecophoric canal of the male, which embraces the female and gives the genus its name. During copulation the female receives carbohydrates as well as sperm from the male. (B) Scanning electron micrograph of a cercaria of *S. mansoni* entering the skin of a mouse. The cercaria secretes a protein-digesting enzyme that enables penetration.

the cercariae mature into adults and take up residence in the portal vein, which carries nutrient-rich blood from the intestine to the liver. *Schistosoma* apparently settles in the portal vein because tumor necrosis factor, a cytokine involved in the host's immune system, is more concentrated there. Once in the portal vein, a female snuggles into the groove of a male's body, and the tumor necrosis factor stimulates her to lay hundreds of eggs every day. Some eggs emerge through damaged vessels in the intestine or urinary bladder and are excreted in either the feces or urine, often accompanied by blood. Many of the eggs, however, lodge in the liver. There tumor necrosis factor plays a second vital role in the life of *Schistosoma:* It triggers the host tissue to encase the eggs in pinhead-sized growths (granulomas). These granulomas protect the host from the eggs, but they also protect the eggs from the host's immune system. Over the next few years the granulomas and secretions from the eggs damage the liver and other organs. Eventually the victim stops passing eggs, but affected organs may already have been damaged permanently, and often fatally.

Tapeworms

Of all Platyhelminthes, those of class Cestoda are most dedicated to a life of parasitism (Fig. 25.9). They have no choice, because they cannot synthesize certain essential nutrients and they lack a digestive tract. The adults get their food already digested in the intestine of a vertebrate host. The adaptations of cestodes to parasitism reveal a terrible kind of beauty. The tegument of the tapeworm must be permeable to nutrients, yet resist damage by digestive enzymes and alkaline secretions of the host. The teguments of tapeworms are adapted for these conflicting requirements in two ways. First, they are covered with **microtriches** (singular = microthrix), which resemble the microvilli on the vertebrate intestinal lining. Like the microvilli, the microtriches increase the surface area for the absorption of nutrients. Second, many tapeworms secrete substances that inhibit digestive enzymes. Tapeworms also reduce the pH to a level at which they, but not the digestive enzymes, can function. The reduction in pH may be due to acidic products of anaerobic glycolysis, which is the worm's major source of ATP in the oxygen-deficient environment of the gut. Because glycolysis is so inefficient, tapeworms must consume a lot of the host's food. Hence the phrase "he must have a worm" is applied to someone with an excessive appetite.

FIGURE 25.9

(A) A fish tapeworm *Diphyllobothrium latum.* (B) Scanning electron micrograph of the scolex of a tapeworm with a circlet of hooks and the four suckers. (C) A mature proglottid of a tapeworm filled by gonads of both sexes. The exact function of Mehlis' gland is unknown, but it is thought to be involved in the formation of egg shell. (Width 5 mm.)

■ **Question:** What organs visible in part C are not associated with reproduction?

A

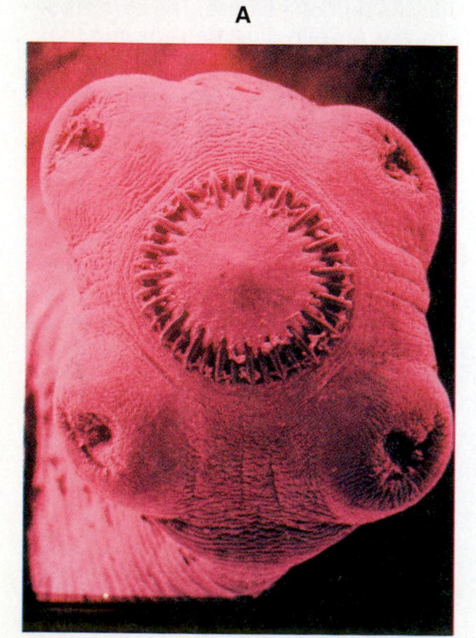

B

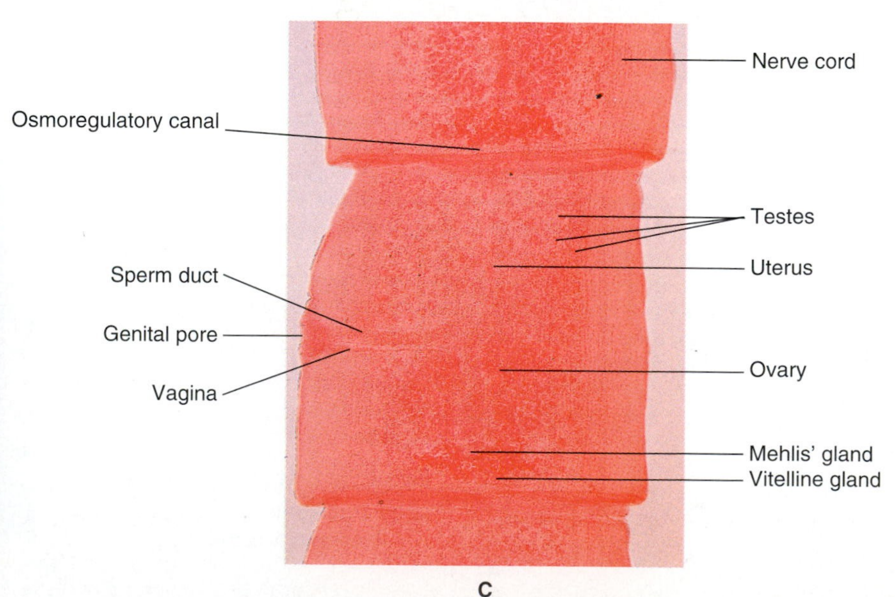

C

Schistosoma and Society

Schistosoma and *Homo* have probably had a long relationship. The bloody urine characteristic of *S. haematobium* infection is noted in 4000-year-old Egyptian papyri, and the eggs of this species have been found mummified along with their hosts of 3000 years ago. Conditions for *Schistosoma* are probably more favorable now than they have ever been, however. The combination of high population density, poor sanitation, and irrigated fields where snails thrive seems made-to-order for the disease. Poor sanitation and the need to live close to both drinking water and farm fields ensure that eggs will be excreted into the water, where host snails abound. The cercariae that develop in the snails readily find a human host. The victim can be a fisherman, a peasant working in a rice paddy, a woman washing clothes, or a child playing in the water.

To those in developed countries it would appear to be a simple matter to eliminate schistosomiasis. All one need do is interrupt the life cycle of *Schistosoma* long enough. This could be done in three ways:

1. Destroy the flukes in their human victims. There are now several drugs that will rid human victims of *Schistosoma,* and new drugs and vaccines are being developed.
2. Destroy the snail hosts. There are poisons capable of ridding streams of snails without endangering vertebrates. Since 1955 in the People's Republic of China, workers have systematically dug up the banks of streams and irrigation ditches to destroy the snails.
3. Improve sanitation and provide alternatives to fishing, planting, swimming, and washing in contaminated water.

In most countries none of these solutions is affordable. In Gambia, for example, the amount of money available per year to treat *all* health problems is less than 50 cents per person. This is not enough for drugs or poisons, and certainly not enough for sewage systems, water treatment, or swimming pools. In some countries, people are so debilitated by schistosomiasis and other diseases that they are incapable of earning even 50 cents for health care.

In principle, an improvement in the economy of impoverished countries should help eliminate the conditions that favor parasites. In practice, however, attempts to improve the economies of less-fortunate countries sometimes do more harm than good, because of ignorance of the biological situation. One example is the Aswan High Dam in Egypt, which was largely financed and planned by Great Britain, the former Soviet Union, and the United States. This dam was intended to provide flood control, irrigation, and power, but it has also provided an enlarged habitat for snails. As was predicted, the resulting increase in schistosomiasis has deprived many Egyptians of the hoped-for benefits from the dam. Before the dam was completed in 1971, approximately 5% of people in the 500-mile valley between Aswan and Cairo were victims of schistosomiasis. Four years later the prevalence was approximately 35% in people living below the dam and 76% in people living above the dam.

Hope for the eradication of schistosomiasis has been raised by the development of an inexpensive chemical (niclosamide) that can be applied to the skin to block penetration by the cercariae (Cherfas 1989). The United States Army has begun tests of the antipenetrant, but there does not appear to be much interest except to protect the armies of developed nations when they invade less-developed ones. The focus has now shifted to development of a vaccine (Cherfas 1991).

In most cestodes (those in the subclass Eucestoda) a headlike **scolex** (= holdfast) with a combination of suckers and hooks keeps the tapeworm from being swept away by traffic in the gut (Fig. 25.9B). Most drugs that are effective against cestodes work by making the gut so inhospitable or the worm so sick that the scolex lets go. Behind the scolex a **germinative zone** asexually produces a steady stream of subunits called **proglottids.** In some species an individual may produce a dozen proglottids daily, for as long as 20 years. Hundreds of proglottids, collectively referred to as a **strobila,** may form a chain more than 10 meters long in some species. The proglottids are united by the nerve cords, protonephridia, and tegument, but reproductively each proglottid is

FIGURE 25.10
Hydatid cysts of *Echinococcus granulosus* in the liver of a 41-year-old man who had been suffering abdominal pains for six years before a portion of his liver was removed. Hydatid cysts can also form in other organs, sometimes accumulating several liters of fluid each. Humans become infected from unhygienic association with dogs. Herbivores such as moose and sheep are more likely than humans to be intermediate hosts, because they pick up the eggs from the feces of canines. The canines then complete the life cycle when they prey upon the moose or sheep.

an individual (Fig. 25.9C). Any two proglottids, either on the same or on different tapeworms, are capable of exchanging sperm. Inside the uterus of the proglottid, fertilized ova develop into larvae, called **oncospheres,** which remain enclosed within shelled eggs. Eggs then leave the proglottid, or the oldest proglottids at the end of the strobila break off with up to 100,000 eggs inside. **Gravid** (egg-bearing) proglottids often crawl out of the anus. Proglottids of *Dipylidium caninum,* glistening white and several millimeters square, can sometimes be seen crawling on an infected dog, cat, or child. Proglottids can also be found crawling about in the bed or clothing of an infected person. Once the eggs of a tapeworm are released, they must usually be ingested by an intermediate host in order to hatch. Humans are seldom intermediate hosts of *D. caninum,* since most of us avoid eating food contaminated by feces. However, the unsanitary habits of dogs, along with humans kissing them, sometimes leads to people becoming intermediate hosts (Fig. 25.10). From the worm's point of view, humans make poor intermediate hosts, since we seldom get eaten by another vertebrate that could serve as a final host.

Usually the intermediate host is a vertebrate prey of the final host, or an ectoparasite (such as a flea or louse) that is commonly eaten by the final host. After the intermediate host ingests the eggs, its digestive secretions hatch them, and an oncosphere emerges from each. The oncospheres then bore through the intestinal wall and into the blood or lymphatic circulatory system of the intermediate host. The blood or lymph then takes them to the skeletal muscle, heart, or some other organ. The oncospheres can survive in many different tissues, since they quickly secrete a protective cyst around themselves. They eventually develop into a variety of forms, depending on species. For example, in the beef tapeworm and the pork tapeworm (*Taeniarhynchus saginatus* and *Taenia solium*) the cyst develops into a juvenile called a **bladder worm** or **cysticercus** (SIS-tee-SIR-kus). Each of the cysticerci (SIS-tee-SIR-sy) is essentially an inside-out scolex that everts after the infected tissue of the intermediate host is eaten by the final host. The scolex then attaches to the lining of the intestine by means of suckers and hooks.

Table 25.2 lists some common tapeworms and their hosts. Because the intestine of one mammal is much like that of another, the species that serves as final host depends mainly on whether its diet includes the intermediate host. Fig. 25.11 illustrates the life cycle for the beef tapeworm *Taeniarhynchus saginatus* (formerly *Taenia saginata*).

TABLE 25.2

Some important tapeworms of humans and other hosts.

SPECIES	INTERMEDIATE HOSTS	FINAL HOST
Fish tapeworm *Diphyllobothrium latum*	Copepod, fish	Human, other fish-eating mammals
Dog tapeworm *Dipylidium caninum*	Flea	Dog, cat, human
Beef tapeworm *Taeniarhynchus saginatus*	Cattle	Human
Pork tapeworm *Taenia solium*	Pig, human	Human
Dwarf tapeworm *Vampirolepis nana*	Flour beetle (optional)	Mouse, human

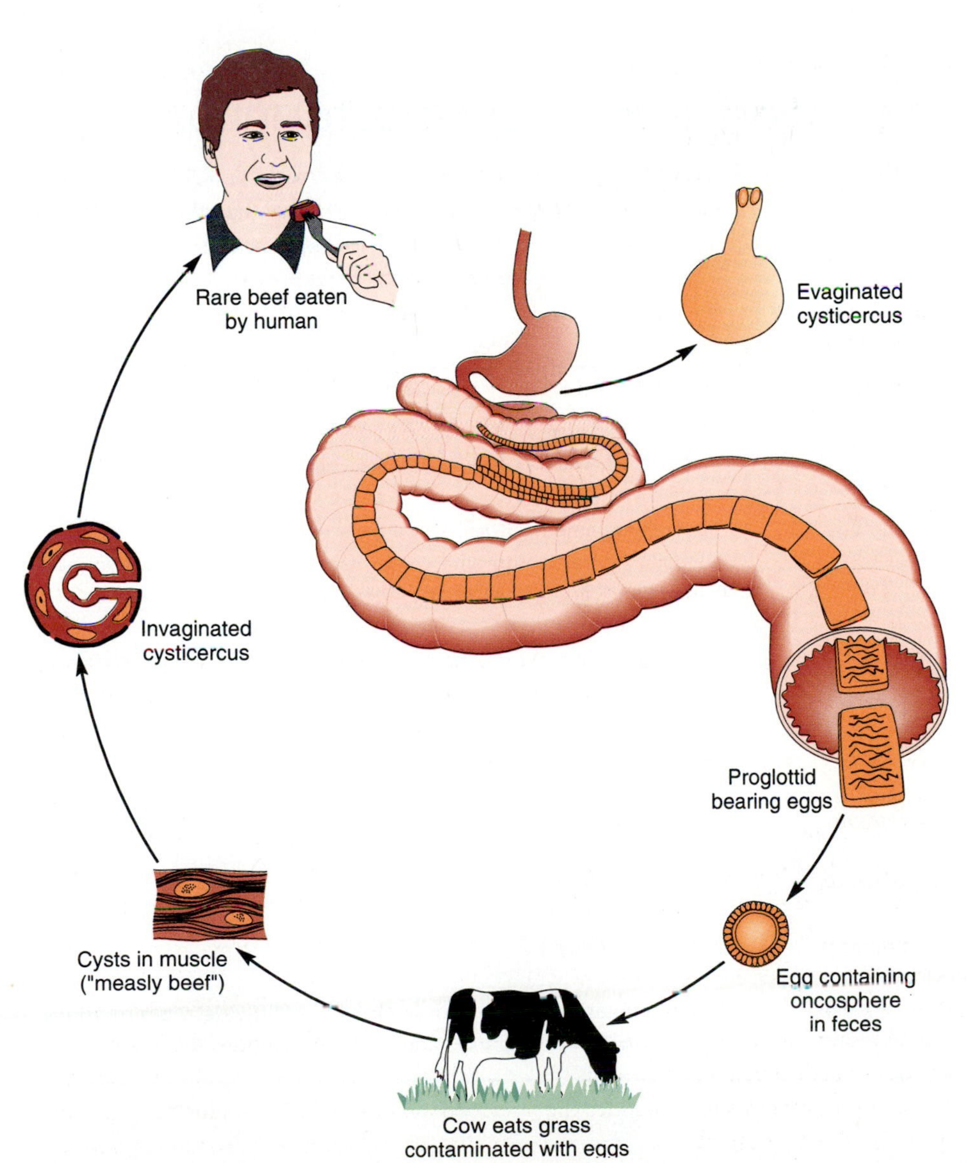

FIGURE 25.11

The life cycle of the beef tapeworm *Taeniarhynchus saginatus.*

■ **Question:** Based on this life cycle, what cultural differences would account for differences in rates of infection among various groups of people?

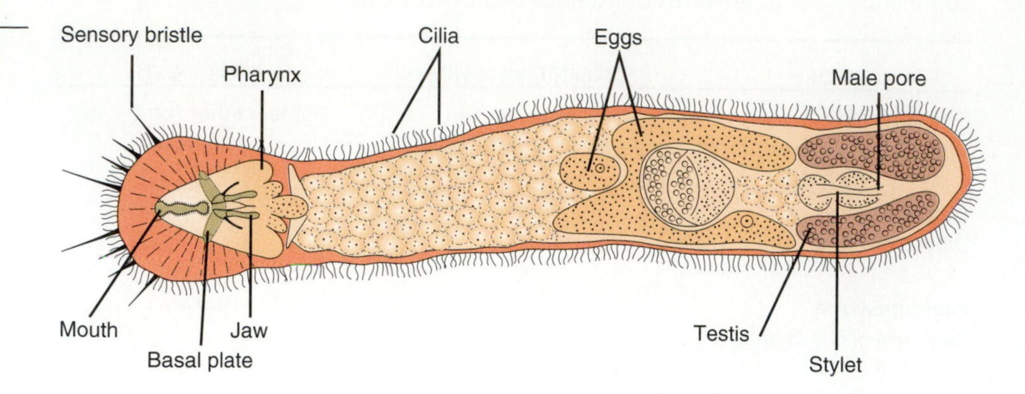

FIGURE 25.12
Structure of the gnathostomulid *Problognathia minima.*

This worm often infected humans in the United States until meat inspection became widespread. Even now infections can occur. Only 80% of domestic beef is inspected, one out of four samples with cysts are missed during inspection, and beef is often served so rare that cysticerci are not killed.

JAW WORMS: PHYLUM GNATHOSTOMULIDA

Gnathostomulids (Fig. 25.12) were discovered in 1956, but were considered to be tiny turbellarians until 1969. Among the numerous characteristics that they share with turbellarians in addition to the flattened shape and acoelomate condition are the lack

TABLE 25.3

Characteristics of gnathostomulids.

Phylum Gnathostomulida NATH-oh-stom-YOU-lid-uh (Greek *gnathos* jaw + *stoma* mouth

Morphology	**Bilaterally symmetric**. Dorsoventrally flattened. Parenchyme poorly developed. Ventral mouth with comb-shaped **basal plate** and two **jaws** made of hard organic material.
Physiology	Central nervous system consists of cephalic ganglia and longitudinal nerve cords. Protnephridia with solenocytes (= cyrtocytes). No circulatory, respiratory, or skeletal system. Digestive tract usually incomplete, but some have functional anal pore.
Locomotion	Cilia; one per epidermal cell.
Reproduction	Monoecious. Fertilization internal.
Development	Development direct.
Habitat, Size, and Diversity	Found on marine plants or in sediments worldwide. Sizes of adults from 0.3 to 3.5 mm. Approximately **89 living species** described.

of a circulatory system, hermaphroditism, protonephridia, ciliated epithelium, and a pharynx (Table 25.3). However, Gnathostomulida differ from Platyhelminthes in the following ways:

1. Instead of flame bulbs in the protonephridia, there are **solenocytes,** which have flagella rather than tufts of cilia. (In Gnathostomulida, solenocytes are also called **cyrtocytes**—curved cells.)
2. The epithelium differs from that of turbellarians in having only one cilium per cell.
3. Instead of extending the pharynx out of the mouth when feeding, gnathostomulids scrape food up with a comb-shaped **basal plate** and a pair of hard **jaws.**

Jaw worms are less than 1 mm long and feed on bacteria, protists, and fungi. None is known to be parasitic. They are usually found on marine algae and plants or in shallow sediments. Many marine forms live in the black, anaerobic layer produced by sulfur-metabolizing bacteria. Many others live among sand grains, where they share their **interstitial** habitat with other microscopic animals, such as gastrotrichs and nematodes. Six thousand gnathostomulids were found in a 1-liter sample of such sediment. In spite of their vast numbers, probably fewer than one-tenth of the living gnathostomulid species are known. Because they are transparent and small, they are easily overlooked. Also it requires special technique to free them from sand without damage.

PHYLUM GASTROTRICHA

Gastrotrichs (Fig. 25.13 on page 532) are often considered to be pseudocoelomates, because a small space can be seen between the ectoderm and mesoderm in specimens prepared for microscopy. This space is largely filled with mesodermal cells, however, and W. D. Hummon (1982 and personal communication) has shown that, in fact, it is an artifact from fixing the tissue for microscopy. For these reasons, and because of several similarities to flatworms, specialists now consider gastrotrichs to be acoelomates. Like turbellarians, gastrotrichs glide on cilia, have duogland adhesive systems for temporary attachment, and are hermaphroditic. Freshwater species have protonephridia with solenocytes. As in gnathostomulids, there is only one cilium per epidermal cell (Table 25.4).

Gastrotrichs are all free-living and use their ventral cilia to draw organic debris and protists into the mouth. Marine forms occur on coral and in subtidal or intertidal sediments. Freshwater species occur in bogs and marshes—seldom in clear streams. With a hand lens one can often find them creeping about on lily pads and duckweed. They are easily mistaken for ciliated protozoa. Some are parthenogenic, but most are sequentially hermaphroditic (protandric). That is, each individual functions first as a male, then as a female. Fertilization is internal, with an acting male transferring a spermatophore (sperm packet) to an acting female. Gastrotrichs produce the largest ova in relation to adult body size of any animal. Two types of egg can be formed by freshwater gastrotrichs. One has a thick shell and must undergo drying, freezing, or heating before it will develop. The other has a thin shell that hatches immediately. Development is direct in both cases.

FIGURE 25.13

(A) External view of the freshwater gastrotrich *Chaetonotus* sp. (B) Internal structure.

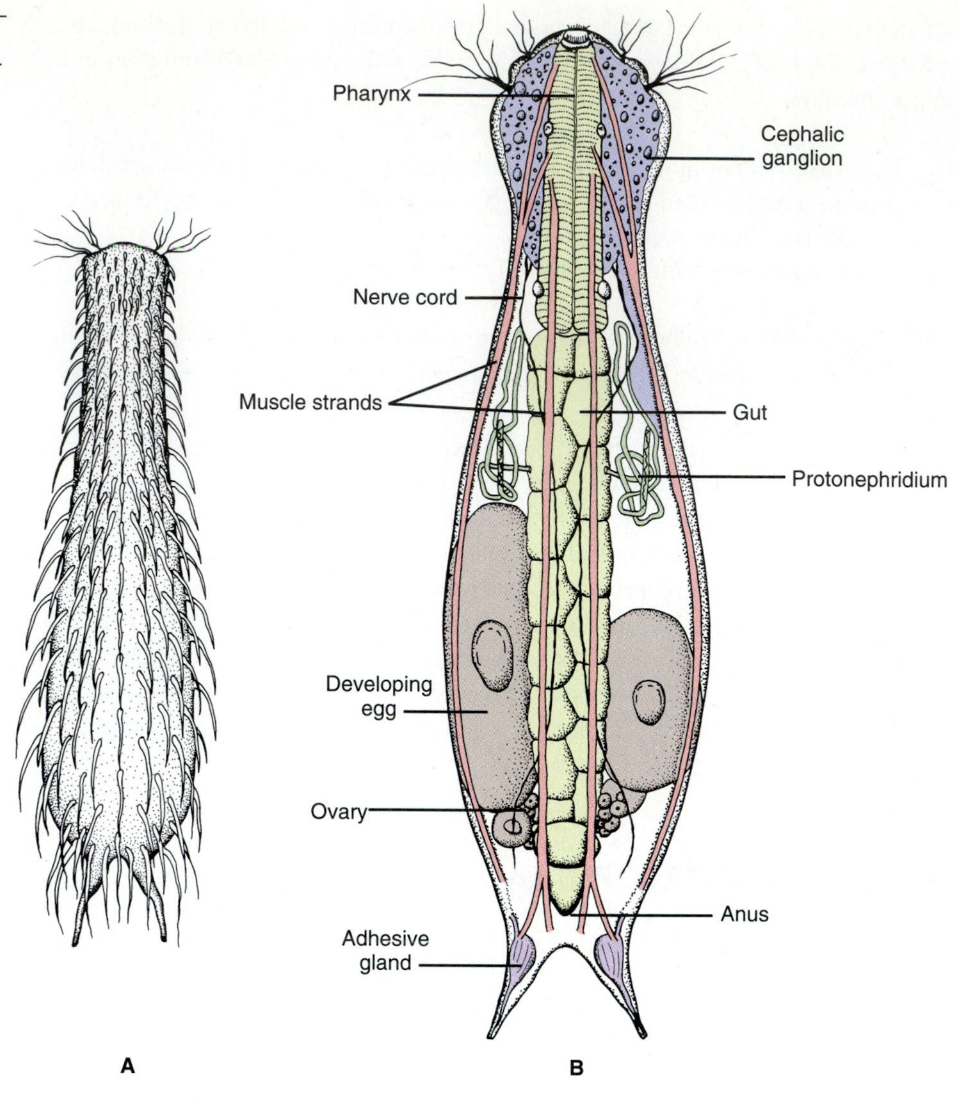

TABLE 25.4

Characteristics of gastrotrichs.

Phylum Gastrotricha GAS-tro-TRICK-uh (Greek *gaster* belly + *trichos* hairs)

Morphology	**Bilaterally symmetric**.. Dorsoventrally flattened.
Physiology	Feeding by means of ventral cilia. Protonephridia with solenocytes bearing one or two flagella. No circulatory, respiratory, or skeletal system. Central nervous system consists of cephalic ganglia and longitudinal nerve cords. Digestive tract complete.
Locomotion	By muscular creeping on ventral cilia. With duogland adhesive system for temporary attachment.
Reproduction	Monoecious. Fertilization internal.
Development	Development direct.
Habitat, Size, and Diversity	Living in sediments and on plants; about half the species are marine and half freshwater. Sizes of adults from 0.1 to 4 mm. Approximately **430 living species** described.

Summary

Flatworms, jaw worms, and gastrotrichs (phyla Platyhelminthes, Gnathostomulida, and Gastrotricha) are acoelomate. Access of oxygen, nutrients, water, and ions to cells is provided by a flattened body, which gives a large surface area per volume. In the larger flatworms a highly branched intestine also aids in the distribution of nutrients. There are also protonephridia and other organ systems that help maintain an interstitial fluid. All are bilaterally symmetric. Other features shared by the three phyla include (1)locomotion by creeping on ciliated cells and (2) the presence of a central nervous system with cephalic ganglia.

Flatworms include turbellarians, monogenetic flukes, digenetic flukes (trematodes), and parasitic tapeworms (cestodes). Turbellarians generally prey on and scavenge smaller animals by everting the pharynx and secreting digestive enzymes. Monogenetic flukes are usually ectoparasites on the gills of fish, to which they attach by the opisthaptor. Trematodes parasitize vertebrates as adults and parasitize molluscs as larvae. Fertilized eggs leave the vertebrate primary host in the feces or urine, then hatch into miracidia. The miracidia then enter a mollusc intermediate host, in which they develop into a sporocyst and then into daughter sporocysts or rediae. They then form cercariae that leave the mollusc and enter a new final host either directly or after further development. Liver flukes and schistosomes are two trematodes that cause diseases in humans.

Tapeworms inhabit the intestines of vertebrates, to which they attach by hooks and suckers on the scolex. They produce a chain of proglottids, each of which sexually produces eggs that contain oncospheres. The oncospheres are ingested by an intermediate host, usually an insect or vertebrate, in which they encyst as cysticerci. The life cycle is completed when a new final host eats the intermediate host.

Key Terms

bilateral symmetry
triploblastic
parenchyme
duogland adhesive system
protonephridium
flame cell
tegument
oncomiracidium
opisthaptor
miracidium
sporocyst
redia
cercaria
metacercaria
scolex
proglottid
oncosphere
cysticercus

Chapter Test

1. Explain the term "acoelomate." (p. 517)
2. Acoelomates are said to have an organ-system grade of organization. Illustrate what this means by describing an organ system of an acoelomate. (pp. 519–522)
3. Name the four classes of Platyhelminthes, and describe the major features that distinguish them from each other. (pp. 517–519)
4. Explain how the differences among turbellarians, flukes, and tapeworms are correlated with their degrees of parasitism. (pp. 519–528)
5. Explain how irrigation and poor sanitation favor schistosomiasis. What role does the host's immune system play in the disease? (pp. 524–526)
6. Describe the life cycle of any tapeworm, using the terms oncosphere, proglottid, and cysticercus. (pp. 526–529)
7. Gnathostomulids and gastrotrichs are both small acoelomates. List the major similarities and differences between them. (pp. 530–532)

■ Answers to the Figure Questions

25.1 The colors of flatworms probably would not function in courtship, because these animals do not have good vision. Nor is there any background against which these colors would camouflage them. The most likely explanation is warning coloration.

24.4 Perhaps being able to determine the direction of light helps avoid hot, dry habitats exposed to direct sunlight.

25.7 (1) The cercariae of *Opisthorchis* do form cysts in the fish host. (2) *Opisthorchis* has a three-host life cycle in which two of the hosts, the snail and the fish, serve as vectors for transmission to the human final host.

25.9 None.

25.11 The beef tapeworm thrives where people consume rare beef and allow feces to contaminate pastures.

Readings

Recommended Readings

Desowitz, R. S. 1981. *New Guinea Tapeworms and Jewish Grandmothers: Tales of Parasites and People.* New York: W. W. Norton. (*Informative and entertaining.*)

Rennie, J. 1992. Living together. *Sci. Am.* 266(1):122–133 (Jan).

See also relevant selections in General References listed at the end of Chapter 21.

Additional References

Boaden, P. J. S. 1985. Why is a gastrotrich? In: S. Conway Morris et al. (Eds.). *The Origins and Relationships of Lower Invertebrates.* New York: Oxford University Press, pp. 249–260.

Cherfas, J. 1989. New weapon in the war against schistosomiasis. *Science* 246:1242–1243.

Cherfas, J. 1991. New hope for vaccine against schistosomiasis. *Science* 251:630–631.

Harrison, F. W., and B. J. Bogitsh (Eds.). 1991. *Microscopic Anatomy of Invertebrates, Vol. 3: Platyhelminthes and Nemertinea.* New York: Wiley–Liss.

Hummon, W. D. 1982. Gastrotricha. In: S. P. Parker (Ed.). *Synopsis and Classification of Living Organisms*, Vol. 1. New York: McGraw–Hill. pp. 857–863.

Lammert, V. 1991. Gnathostomulida. In: F. W. Harrison and E. E. Ruppert (Eds.). *Microscopic Anatomy of Invertebrates,* Vol. 4: *Aschelminthes.* New York: Wiley–Liss, pp. 19–39.

Ruppert, E. E. 1991. Gastrotricha. In: F. W. Harrison and E. E. Ruppert (Eds.). *Microscopic Anatomy of Invertebrates,* Vol. 4: *Aschelminthes.* New York: Wiley–Liss, pp. 41–109.

Sterrer, W., M. Mainitz, and R. M. Rieger. 1985. Gnathostomulida: enigmatic as ever. In: S. Conway Morris et al. (Eds.). *The Origins and Relationships of Lower Invertebrates.* New York: Oxford University Press, pp. 181–199.

CHAPTER

26

Nematodes and Other Pseudo-coelomates

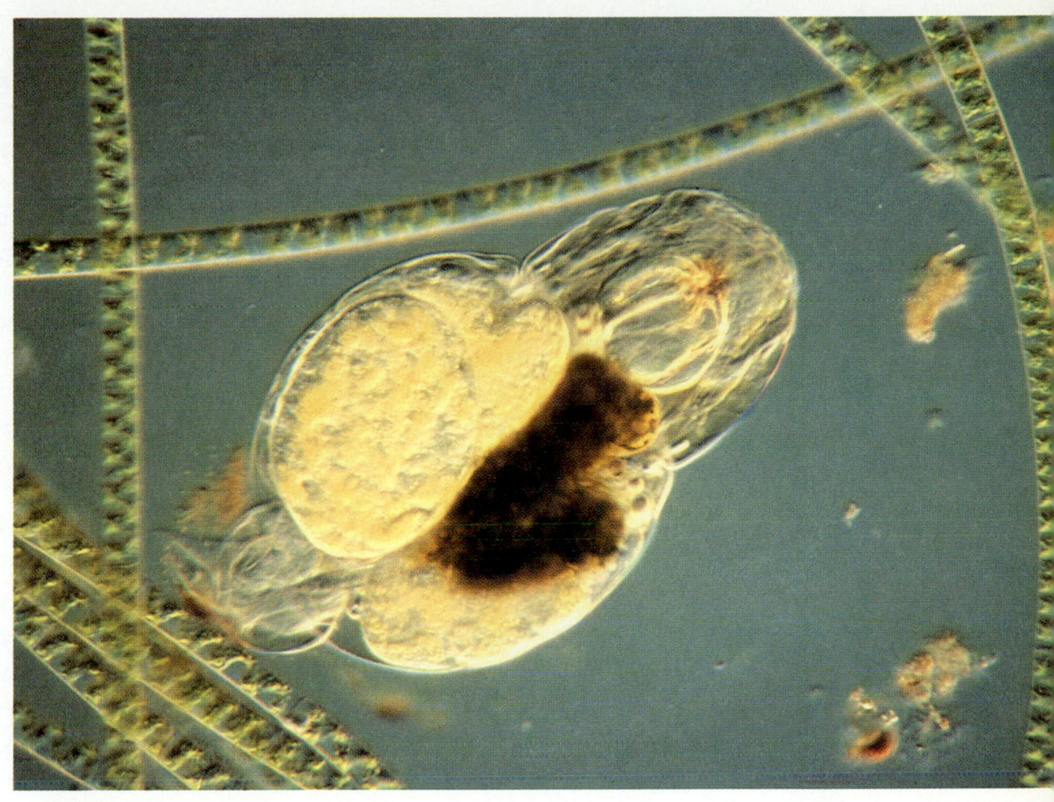

A rotifer (Notomata cerberus).

Introduction

The eight phyla considered in this chapter—**Nematoda, Nematomorpha, Kinorhyncha, Loricifera, Priapulida, Rotifera, Acanthocephala,** and **Entoprocta**—are extremely diverse. The reason they are often considered together is that unlike the acoelomates considered in the previous chapter, the digestive tract is not merely hollowed out of a more-or-less solid body. Instead the digestive tract lies within a **body cavity,** producing a "tube-within-a-tube" organization. This body cavity encloses other organs as well and is filled with fluid that may contain unattached cells of various kinds. There are two major kinds of body cavities: the **coelom** and the **pseudocoel.** Most animals, including ourselves, have a coelom (SEAL-um), which is by definition enclosed in mesoderm and is at least partly lined with a cellular membrane. Nematodes and other pseudocoelomates have a body cavity called the pseudocoel (SOO-doe-seal; also called pseudocoelom) that lacks a mesodermal peritoneal membrane (see Fig. 21.8).

It was formerly believed that all pseudocoels are simply blastocoels that persist after embryological development (pp. 110–111). It now appears that pseudocoels can arise in several ways in different phyla, and even in different groups within a phylum. Thus the pseudocoelomates are not a natural taxonomic group. Despite the prefix "pseudo-," the pseudocoel is just as real and just as functional as a coelom. Even some coelomates (molluscs, arthropods, and three other groups described in the next chapter) have what is technically a pseudocoel, although it is called a hemocoel in those animals. The functions of pseudocoels (and of coeloms) include the following:

1. The pseudocoel (or coelom) provides room for the digestive tract to grow, enabling the surface area for the absorption of nutrients to grow at the same pace as body volume. It also allows the digestive tract to move within the body, enabling food to be propelled through it.
2. The body cavity (pseudocoel or coelom) may also provide greater flexibility in the development of other organs so that they can grow as needed. This may be especially important for gonads, allowing them to enlarge seasonally and to produce large eggs.
3. The body cavity may circulate nutrients, oxygen, water, and ions, thereby substituting for or complementing a blood vascular system.
4. The body cavity may hold metabolic wastes and excess water or ions. Evidence for this function comes from the fact that many osmoregulatory and excretory organs filter the fluid in the body cavity.
5. The body cavity may function as a hydrostatic skeleton, providing a semirigid body structure against which muscles can contract.

Five of the pseudocoelomate phyla—Nematoda, Nematomorpha, Rotifera, Kinorhyncha, and Priapulida—as well as the acoelomate phylum Gastrotricha (often considered pseudocoelomate) were formerly treated as classes in the phylum Aschelminthes (pronounced ask-hel-MIN-theez). The phylum Aschelminthes is obsolete, although the name is still sometimes used informally. Further taxonomic revisions can be expected, because we know little about the relationship of the pseudocoelomates to each other and to other groups. Our ignorance about their origin, and therefore about the early evolution of animals, is due to the lack of a fossil record for these soft-bodied animals and to the relative neglect by zoologists. Except for the economically important parasitic species, even the huge phylum Nematoda has received inadequate attention.

ROUNDWORMS: PHYLUM NEMATODA

The best known phylum of pseudocoelomates is Nematoda, mainly because so many of them are parasitic in people. Like flatworms, they appear to have arisen early in the evolution of bilaterally symmetric animals (see Fig. 18.17). They may rival insects as the most numerous and diverse animals on Earth. Only about 12,000 species have been described, but there are probably hundreds of thousands awaiting discovery and description. Virtually every sample of soil, water, animals, or vegetation in an unstudied area will yield new species. The number of individual roundworms is as impressive as the number of species. A square meter of bottom sediment off the coast of the Netherlands was found to harbor more than 4 million nematodes; in one rotting apple 90,000 nematodes of several species were counted; and 10^{11} nematodes have been estimated to live in an acre of farmland. As N. A. Cobb remarked in 1915, "If all the matter in the universe except the nematodes were swept away, our world would still be dimly recognisable [as] a thin film of nematodes. . . ."

Organization of the Body

The tube-within-a-tube construction is apparent on dissecting a large nematode, such as the intestinal parasite *Ascaris,* which may reach a length of 30 cm (Fig. 26.1 on page 538). The inner tube is the complete digestive tract, which includes the mouth, pharynx, intestine, rectum, and anus (Table 26.1). Together with the reproductive organs,

TABLE 26.1

Characteristics of nematodes.

Phylum Nematoda NEE-ma-TOAD-uh (Greek *nema* thread)

Morphology	**Vermiform** (worm-shaped): long, round, bilaterally symmetric, without appendages, tapered at both ends. With a **pseudocoel. Cuticle** containing collagen. Outer layer often decorated with grooves or structures shaped like hairs, rods, or bumps. Epidermis usually syncytial. Pseudocoel under pressure forms a hydroskeleton, and cuticle serves as exoskeleton.
Physiology	Central nervous system consists of nerve ring around pharynx, ventral and caudal ganglia, and several nerve cords. Complete, unbranched digestive tract. No circulatory or respiratory system. No protonephridia. Some species have **gland cells** or **renette cells** that are assumed to be osmoregulatory because they are located in the pseudocoel and have ducts to the outside.
Locomotion	Longitudinal muscles contract against cuticle and hydroskeleton, producing dorsoventral waves. No appendages. No cilia or flagella except for cilia in sensory receptors.
Reproduction	Usually **dioecious** and dimorphic. Sperm ameboid.
Development	**Eutelic**: number and position of somatic cells constant in all adults of same species. Usually four juvenile stages, often called larvae, resemble the adult.
Habitat, Size, and Diversity	**Aquatic** (in fresh water or seawater), terrestrial (in moist soil), or endoparasitic. Lengths of adults from a few millimeters to approximately 1 meter. Approximately **12,000 living species** described.

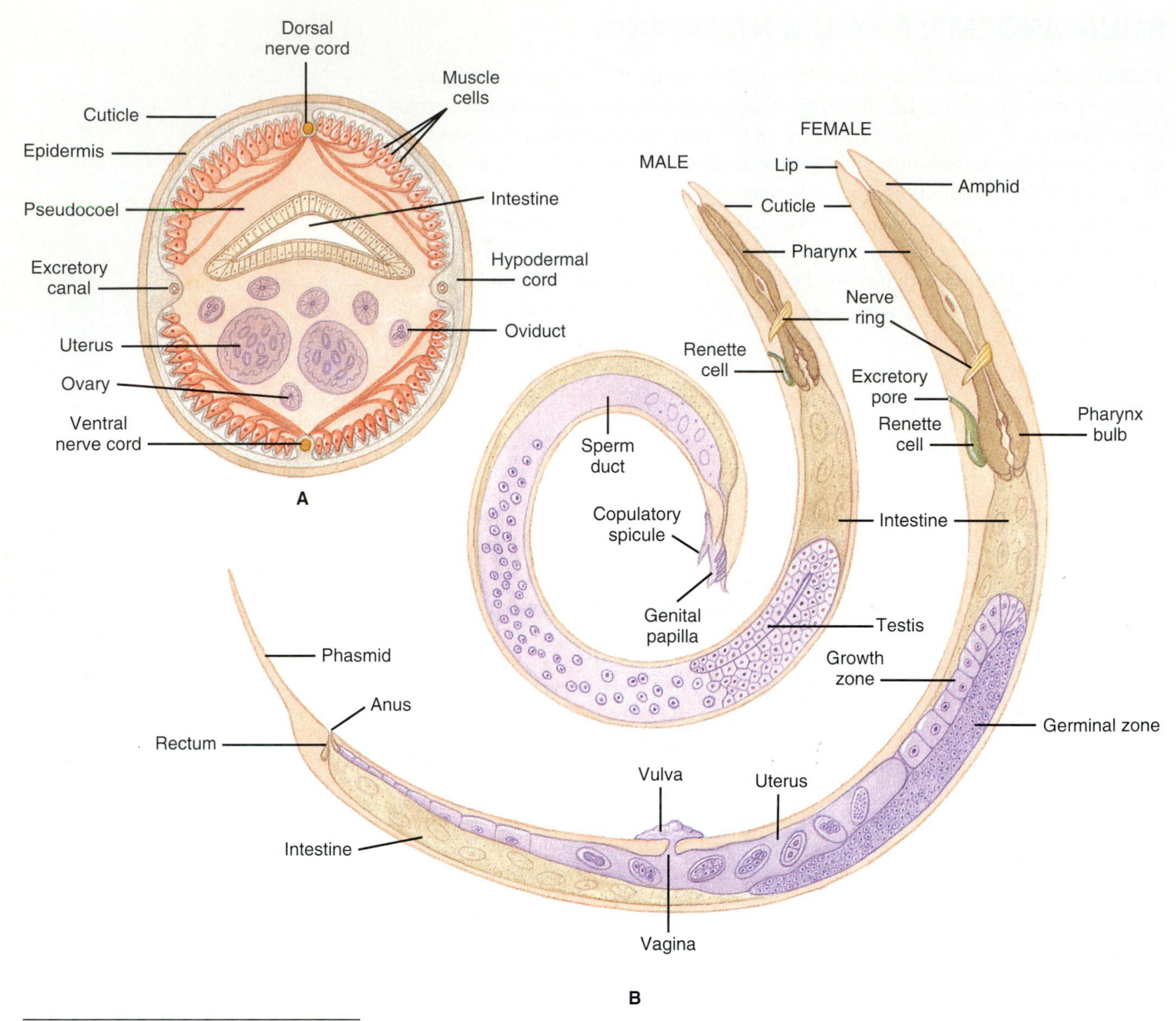

FIGURE 26.1

(A) Cross section of the parasitic nematode *Ascaris* showing the tube-within-a-tube organization. Note that muscle fibers branch to nerve rather than vice versa. (B) Structure of the female and male free-living, freshwater nematode *Rhabditis.* Ameboid sperm are shown maturing within the testis of the smaller male. Eggs and embryos are shown developing within the uterus of the female.

the digestive tract is suspended within the pseudocoel. (These tissues shrink in alcohol and other preservatives, so the pseudocoel is not really as spacious as it appears in Fig. 26.1.) The outer tube consists of the muscular body wall that is usually enclosed by a syncytial epidermis (= hypodermis). The epidermis secretes a tough **cuticle** with several layers containing the fibrous protein collagen. The cuticle protects parasitic forms from digestive enzymes and protects free-living forms from abrasions and other hazards. Many zoologists attempting to kill and preserve nematodes in alcohol have been surprised to find them still thrashing about hours later.

Locomotion

Unlike most worms, which have muscles arranged both lengthwise along the body and circularly around the body, nematodes have only the longitudinal muscles. They do not have circular muscles that would maintain an internal pressure and keep the body from collapsing when the longitudinal muscles contract. Instead, high pressure in the pseudocoel serves as a **hydroskeleton,** and the tough cuticle serves as an **exoskeleton.**

Classification of Nematoda

The classification of Nematoda is satisfactory only for lower taxa. Two classes are tentatively recognized.

Class Adenophorea (= Aphasmidia) ad-DEAN-oh-FOR-ee-uh (Greek *aden* gland + *phoros* bearing). Mainly free-living in marine sediments. No phasmids (posterior sensory organs). Amphids (anterior sensory organs) large, variously shaped. Epidermis not syncytial in two species that have been examined. Excretory canals rudimentary or lacking. *Trichinella* (Fig. 26.8).

Class Secernentea (= Phasmidia) sess-ur-NIN-tee-uh (Latin *secernens* secreting). Mainly parasitic, or microbivorous in soil. Rarely aquatic. Phasmids present. Amphids pore-shaped. Excretory system with lateral canals. *Ancylostoma, Ascaris, Caenorhabditis, Dracunculus, Enterobius, Necator, Onchocerca, Rhabditis, Toxocara, Wuchereria* (Chapter opener; Figs. 26.1, 26.5, 26.6B)

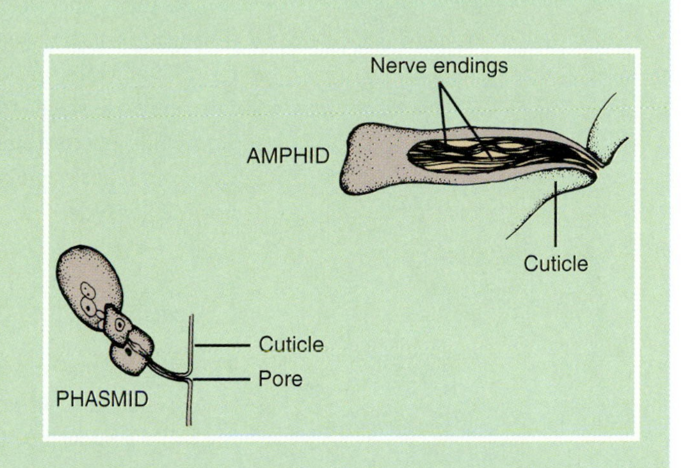

The alternating contraction of the dorsal and ventral longitudinal muscles produce dorsoventral waving of the body that propels the worm slowly among soil particles or host cells. The dorsal and ventral muscles send out branches to the nervous system, rather than the other way around. (This pattern was once thought to be unique to nematodes, but it has been found also in flatworms, gastrotrichs, and other groups.) The muscle branches are interconnected by electrical synapses, so all the dorsal muscles contract in synchrony, as do all the ventral muscles.

Nervous System

The dorsal and ventral muscles are controlled by a dorsal and a ventral nerve cord, respectively. The rest of the central nervous system consists of a nerve ring around the pharynx, with several ganglia connected to it. There are various sensory organs, including the relatively complex **amphids** recessed on each side of the head. Parasitic species generally have **phasmids,** which are similar to amphids but are located near the tail. These organs apparently include receptors for chemicals released by hosts and prey and for sex attractants.

Reproduction and Development

Reproduction in nematodes is conventional in many respects, but unusual in others. Most species have separate, dimorphic sexes, and fertilization is internal. In some species the female apparently crawls along the curl of the male's tail to effect copulation. Because the hydrostatic pressure in the pseudocoel is quite large, however, a male probably could not insert a penis into the female. Instead of a penis, therefore, males have one or two **copulatory spicules** that hold the female's **vulva** open and allow sperm to enter the vagina. The sperm are unusual in lacking acrosomes and in being ameboid rather than flagellated (see Fig. 6.4). Evidently they work well enough: Some females lay hundreds of thousands of shelled eggs every day. There are usually four juvenile stages, L1 through L4, which are often called larvae even though they resemble the adults (Fig. 26.2). The juveniles must shed the cuticle (**molt**) before developing into the next, larger stage. In some species one or two molts may occur within the egg before hatching.

In many species, adverse environmental conditions cause development to stop during a particular juvenile stage, producing a juvenile that is capable of surviving for months or years until conditions improve. In many species this resistant form is the third juvenile stage, which is then called a **dauer larva** (German *dauer* enduring). A "crowding pheromone" or the absence of food triggers formation of dauer larvae in *Caenorhabditis elegans,* and secretions from food organisms trigger resumption of the normal life cycle.

Cryptobiosis

The phenomenal success of roundworms must depend in part on their ability to survive and succeed in adverse conditions. Anywhere there is a film of water, there are probably nematodes. In fact, in many places where there is no water there are nematodes, although in a state of arrested activity called cryptobiosis. Cryptobiosis also occurs in several other groups of animals, including rotifers (Crowe and Cooper 1971, Rensberger 1980). Leeuwenhoek discovered cryptobiosis in rotifers in 1702. Its discovery in nematodes in 1743 created a sensation in France. Soon all the best people were entertaining their guests by bringing nematodes back from the dead with a drop of wa-

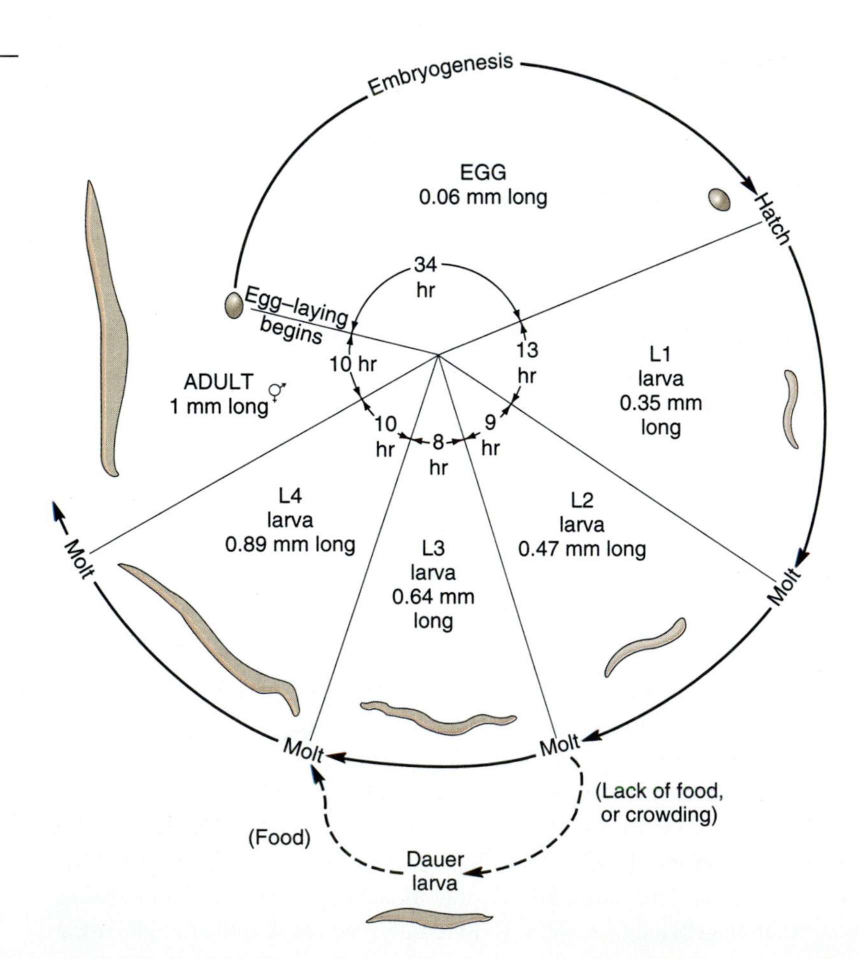

FIGURE 26.2
The 3 ½-day life cycle of a hermaphrodite of the free-living nematode *Caenorhabditis elegans.*

Caenorhabditis elegans: The New Drosophila?

Nematodes have contributed to some of the most important concepts in biology. The discovery that fertilization involves the fusion of two pronuclei in the egg was first made in nematodes by O. Bütschli. E. van Beneden used nematodes to show that the chromosome number is halved during meiosis and restored during fertilization. Some of the first studies of sex chromosomes and of cleavage were also made using nematodes. Around the turn of the century the study of development in *Ascaris* eggs led to important discoveries about how cells "decide" what they are going to be when they grow up. These studies had to use fixed, whole mounts of eggs, however. Because the cells migrate and the embryo moves during development, it was impossible with those methods to keep track of development beyond the 102-cell stage of embryogenesis. After a lapse of more than half a century, dozens of laboratories around the world have once again resumed the study of development in a nematode, and once again a humble roundworm is yielding fundamental biological discoveries.

The modern era of nematode developmental biology began in 1963 with the realization by Sidney Brenner that molecular biology had progressed to a point where the genetic control of development could be completely understood if one could find a suitable species to study. A suitable species would have to have a short life cycle so that complete development could be followed in a reasonable time. The cells would also have to be few so that every one could be traced. The animal would have to be small and easily cultivated in large numbers for controlled breeding experiments. The species would also have to have a suitable genetic system. All these advantages and more were provided by soil-inhabiting nematode *Caenorhabditis elegans* (SEE-no-rab-DYE-tiss ELL-i-ganz).

1. The life cycle of *C. elegans* takes only 3.5 days (Fig. 26.2). The adult has only 959 somatic cells (excluding gonads). Moreover, as in all nematodes, the fate of each cell is identical in every individual of the same species (see Fig. 6.13). That is, the animal has the property of **eutely** (YEW-tell-ee).
2. The length of the adults is only about 1 mm. They are easily cultured on agar plates provided with a lawn of *E. coli* bacteria on which to feed. Each hermaphroditic parent produces an average of 300 offspring.
3. There are only about 80 million pairs of DNA nucleotides comprising 2000 to 4000 genes. These are distributed on five pairs of autosomes, plus one X chromosome in the male and two X chromosomes in the hermaphrodite. (The male and hermaphrodite resemble, respectively, the male and female *Rhabditis* shown in Fig. 26.1.) The hermaphrodites are protandric, which means they first produce sperm, then ova. The hermaphrodite stores the sperm and can use them to fertilize its own eggs. If the hermaphrodite mates with a male, however, that male's sperm are used first. Whether the sperm are acquired by copulation or self-fertilization, half the offspring are males and half are hermaphrodites.
4. Self-fertilizing hermaphrodites are advantageous for research, because they tend to have homozygous offspring in which recessive mutations are easily detected. Another advantage of self-fertilization is that even if mutant hermaphrodites cannot mate, they can still produce offspring.
5. An additional advantage of *C. elegans* is that they survive freezing in liquid nitrogen (−196°C!), so mutants can be stored indefinitely for later study.
6. Finally, the eggs, juveniles, and adults are all transparent. Therefore it is not necessary to kill, fix, and stain them for microscopy. Cells can be observed dividing and migrating in living eggs and juveniles using Nomarski optics (see Fig. 2.20C).

In spite of doubts from many biologists that the project was feasible, or even worthwhile, Brenner published his first paper on the genetics of *C. elegans* only 10 years after his initial proposal. Within another decade, by 1983, the complete lineage of the 959 somatic cells had been worked out by Einhard Schierenberg at the Max-Planck Institute for Experimental Medicine in Göttingen, Germany, H. R. Horvitz at the Massachusetts Institute of Technology, John Sulston at the MRC Laboratory of Molecular Biology, and their many colleagues. Judith Kimball was mainly responsible for tracing the lineages of the 55 cells that make up the male gonads and the 143 cells of the hermaphrodite's gonads. John White and his colleagues pieced together a thousand electron micrographs to trace the formation of all 302 nerve cells and every one of their 8000 synapses.

box continues

Some of the conclusions from this study were completely unexpected. First, commonly accepted generalizations about the fate of the three germ layers are not absolute (p. 117). Some muscle cells do not come from mesoderm, and some nerve cells do not come from ectoderm, for example. Second, bilateral symmetry in the adult does not result from bilaterally symmetric development in the embryo. Homologous cells on each side of the adult may come about in quite different ways. Third, development is uneconomical: A large proportion of neurons die soon after they develop, apparently when only one daughter cell formed by mitosis is needed. Finally, the development of this simple worm cannot be summarized in a few simple statements, but must be described cell-by-cell. The organism results from numerous biological interactions among individual cells, none of which has any "conception" of what the final structure will be like.

ter. The phenomenon also ignited a century-long debate between science and the church, provoked by scientists who insisted on calling the phenomenon "resurrection." Most scientists believed that the nematodes actually returned to life after being dead. Clerics, however, insisted that life after death was a miracle that God did not lightly bestow upon worms. They argued that the nematodes were not actually dead, but merely in a state of extremely low metabolism. For once the clerics were right.

Interactions with Humans and Other Animals

Nematodes are generally divided between free-living and parasitic forms. Free-living species feed on all manner of organic debris, bacteria, fungi, algae, protozoans, and small animals such as other nematodes. Of course they are also food for other organisms (Fig. 26.3). Their ecological importance must be enormous, though largely unknown. In agriculture we know how important nematodes are. Some parasitic forms cause billions of dollars per year in damage to livestock and crops. Some of those billions of dollars are spent on nematocides. However, these poisons also kill beneficial ne-

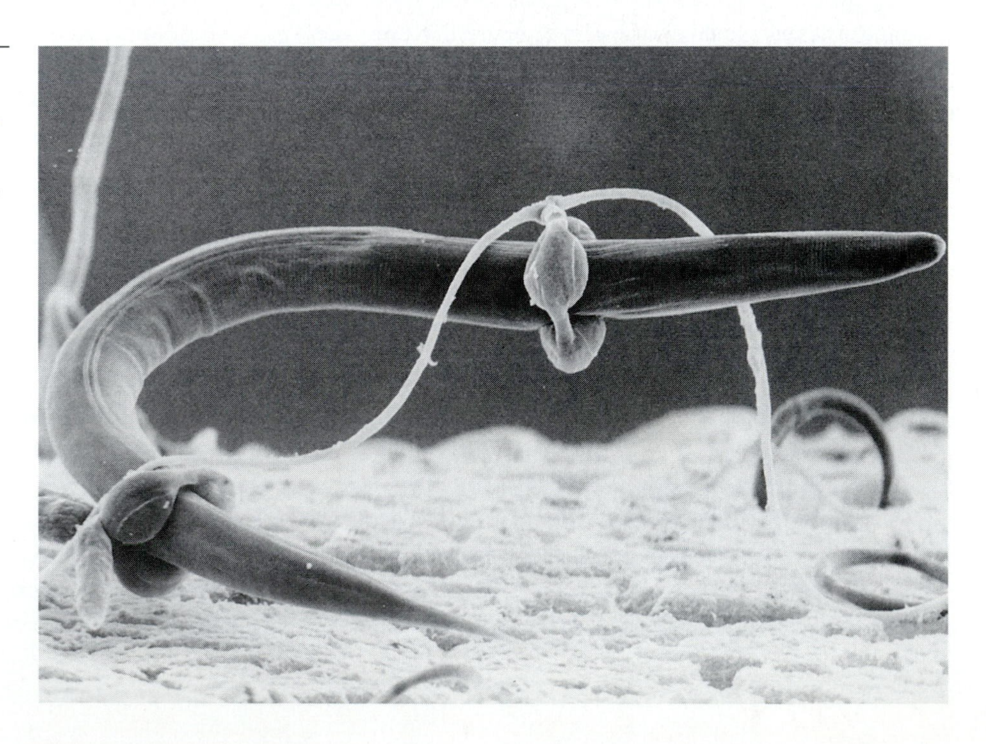

FIGURE 26.3

Scanning electron micrograph of a nematode trapped by a predatory fungus (*Arthrobotrys anchonia*). Three cells in the fungal loops expand on contact, trapping the prey. The fungus then enters the worm and digests it. At least 150 other species of fungi also prey on nematodes.

matodes that prey on insect pests, consume bacteria and fungi, decompose biological toxins and wastes, and aerate soil. Nematodes that infect insects and other agricultural pests are now sold in dehydrated form (just add water!) as an alternative to chemicals. Parasitic nematodes are equally important to humans in much of the world, where virtually everyone is host to one or more forms. Like all parasites, these nematodes are adapted in various ways for life in other organisms (see p. 467). Some of the most important nematode parasites in humans are described in the following sections.

Large Intestinal Roundworms

Ascaris is one of the largest nematode parasites, and it is therefore a common subject for nematode research. In *Ascaris lumbricoides,* the species that infects humans, adult females are up to 30 cm long, and the males are about 20 cm long. They occur in the intestines of approximately a billion people around the world, mainly where sanitation is poor and feces are used as fertilizer. Even in parts of the United States, ascariasis is not uncommon. The major cause of infection is fecally contaminated food. The infective second-stage (L2) juvenile can survive up to seven years in its egg, much longer than other traces of fecal contamination. Once in the small intestine, digestive secretions hatch the egg and the L2 juvenile emerges. Curiously, even though the L2 juvenile is already in the intestine, it does not directly develop into an intestinal parasite. It first must pass through the mucosa into the lymph or blood veins, then circulate through the right heart and into a lung (Fig. 26.4). The L2 juvenile then has to break into an alveolus (air sac), where it remains for 10 days while molting twice into an L4 juvenile. The L4 then ascends the trachea and finally gets back into the digestive tract by being swallowed. The L4 arrives in the lower small intestine about two months after having been swallowed the first time as an egg. The juvenile then molts into a fertile adult and begins feeding on the intestinal contents. If another worm of the opposite gender is available, mating occurs. Each female then begins releasing up to 200,000 eggs per day into the feces to complete the cycle.

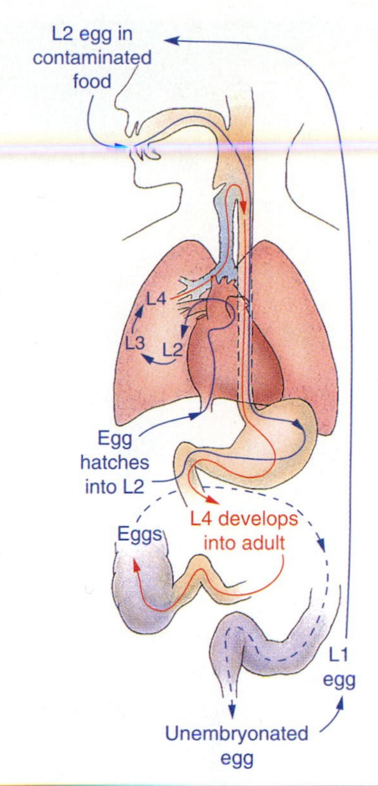

FIGURE 26.4
The life cycle of the large intestinal roundworm *Ascaris lumbricoides.*

■ **Question:** How and why do you think the complicated route of *Ascaris* evolved? Hint: Read the following description of the life cycle of hookworm.

Infection by *Ascaris* produces a variety of symptoms, depending on the site and stage of infection and on the number of worms. When the L2 juveniles penetrate the intestine they usually do not cause much damage, presumably because there is such a rapid turnover of the intestinal lining. However, L2 juveniles that go astray into the wrong organ can cause local inflammation. A great deal of damage, including a severe form of pneumonia, can be produced when large numbers of L2 juveniles break through the alveoli. Adults in the small intestine usually cause only minor symptoms as long as there are few of them and they remain in the intestine. Sometimes females, perhaps following the urge to pass through the curled tail of a male for mating, wander into the appendix, bile duct, or pancreatic duct, where they can cause serious blockage. They may also get into the lungs and other organs by wandering up the esophagus. They can also exit through the mouth or anus. While not physically damaging, the psychological effect of a foot-long worm coming out the anus or mouth should not be underestimated.

Heavy infections by adults can block the intestine or penetrate its lining, with fatal consequences. *Ascaris* can also cause malnutrition and underdevelopment by robbing the human host of its food and by releasing toxic metabolic wastes. The resulting poor health and increased demand for food make it even less likely that a victim will be able to work his way out of the poverty that is often associated with eating fecally contaminated food. A victim's children will probably also grow up having to eat food contaminated by feces. As usual, the life cycle of the parasite is entwined in the economic cycle of the host.

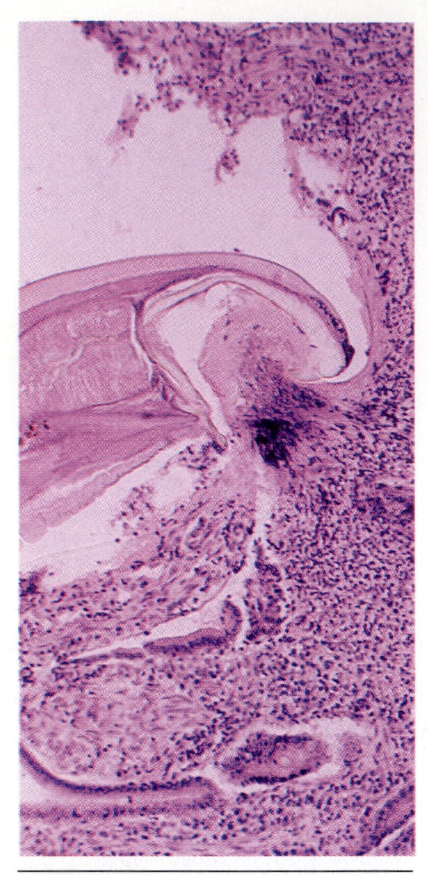

FIGURE 26.5

Microscopic section showing the attachment of a hookworm to the intestinal lining. A "tooth" deep in the mouth of the worm penetrates the mucosa and makes it bleed.

The common roundworms of puppies and kittens, *Toxocara canis* and *T. cati,* are similar to *Ascaris,* in appearance and life cycle, but they are smaller. All puppies and 20% of kittens are infected with *Toxocara* until they are "wormed." People, especially children, can also become infected with *Toxocara,* even though *Toxocara* cannot complete its life cycle in a human. The juveniles, however, wander about through a variety of organs, producing an inflammation.

Hookworms

Hookworms are also parasitic nematodes of the intestine, but they are quite different from *Ascaris* in their life cycle and mode of parasitism. *Necator americanus* and *Ancylostoma duodenale* are the major hookworms of humans, infecting a billion people in warmer regions of the world, including several million in the southeastern United States. *Necator* adult females are approximately 1 cm long, and males are about 0.7 cm long. The life cycle of *Necator* begins when an egg is released with feces onto the soil. Within a week the free-living L1 juvenile hatches out and molts twice into an infective L3 juvenile, which can survive for several weeks in warm, moist soil if protected from sunlight. In order to complete the life cycle, however, it must burrow into the skin of a human, often a barefoot child. The juvenile then circulates with the blood into a lung, where it breaks through into the respiratory tract and is swallowed. On arriving in the upper portion of the small intestine, L3 molts twice into an adult. The adult then begins moving about on the intestinal lining, where it finally attaches to villi with its "teeth" and begins feeding on blood (Fig. 26.5). Each worm consumes 0.03 ml of blood per day, and infection by many worms can use up to 200 ml of blood per day. The anemia, malnutrition, and retarded development that result tend to perpetuate poverty.

Pinworms

One of the most common nematodes in humans is the centimeter-long pinworm *Enterobius vermicularis.* It seldom constitutes a health problem, however, because it apparently feeds on wastes and bacteria in the cecum and appendix of the large intestine, rather than on the host's nutrients or tissues. Usually the most troublesome symptom of infection is embarrassment. In the United States perhaps one out of three children and one out of ten adults are victims. It is highly contagious and therefore tends to be shared by every member of a family. *Enterobius* thrives in spite of excellent sanitation because it does not depend on fecal contamination of food or soil, but on the spread of eggs or juvenile directly from the anus.

The life cycle of *Enterobius* is triggered by the drop in body temperature of the sleeping host. After copulating in the large intestine, the males die and the females make a one-way journey to the anus. Each female then lays approximately 1500 eggs on the anus. Four to seven hours later the L1 juveniles hatch out. If bathing does not occur, the juveniles can migrate back into the anus. A juvenile must go all the way up to the small intestine to complete its development before returning to the large intestine as an adult. The L1 juveniles can also get into the small intestine in eggs transferred directly from the anus to the mouth by the host. Transmission by this route is more likely because in some people the females cause anal itching when they lay eggs. The tiny eggs can also be picked up from bed clothes and furniture and can even be inhaled. These eggs provide the most convenient method of diagnosis, which is done by dabbing the anus with cellophane tape and inspecting it with a microscope.

Guinea Worm

Of all animals, *Dracunculus medinensis* may well be the least lovable. The female of this species is up to a meter long and spends its adult life eating its way under human skin. The female retains the eggs in her body during early development. Eventually pressure, irritation, or aging of the female causes L1 juveniles to erupt from the mother. The human host reacts by forming a blister in the skin; when the blister breaks, millions of juveniles spill out. Because the blisters often occur on the leg or foot, and because victims often try to find relief from the burning pain by bathing, the juveniles often emerge directly into water. This is essential for completion of the life cycle, because the L1 juvenile must be ingested by a particular freshwater crustacean within about a week in order to complete its development. This crustacean is the copepod *Cyclops,* a frequent inhabitant of drinking water in parts of Africa, India, and the Middle East (Fig. 26.6A). The juvenile invades the body cavity of this intermediate host, then molts twice within approximately two weeks into an infective L3 juvenile.

The L3 juvenile in the copepod enters the final host, *Homo,* with drinking water. For the next three weeks the juvenile burrows through a variety of host tissues until it reaches the subcutaneous connective tissue in the pelvic region. Then the third and fourth molts occur, forming an adult approximately 43 days after the contaminated water was drunk. By the third month of infection, mating occurs. Males then die, but the female continues to eat and grow beneath the skin.

A number of drugs are effective against guinea worms, but for economic reasons the treatment that is still widely used is to wind out the female a little at a time on a stick (Fig. 26.6B). This must be done over a period of several weeks. If the worm breaks, it will retreat, introducing bacteria beneath the skin. The timing of the cycle is often such that 30% to 40% of adults and children are incapacitated at the time of planting or harvesting. To most of us it would appear that instead of treatment, the potential victims should try prevention, by keeping either *Cyclops* or people out of the drinking water. It is not easy, however, to get this message into remote villages, or to change habits that have persisted for thousands of years. (Even in the United States, thousands of people get AIDS each year in spite of extensive publicity on how to prevent it.) The World Health Organization has targeted the guinea worm for extinction by distributing inexpensive cloth filters to remove *Cyclops* from water. Through such efforts the number of cases has declined from more than 10 million to less than 250,000 within a decade.

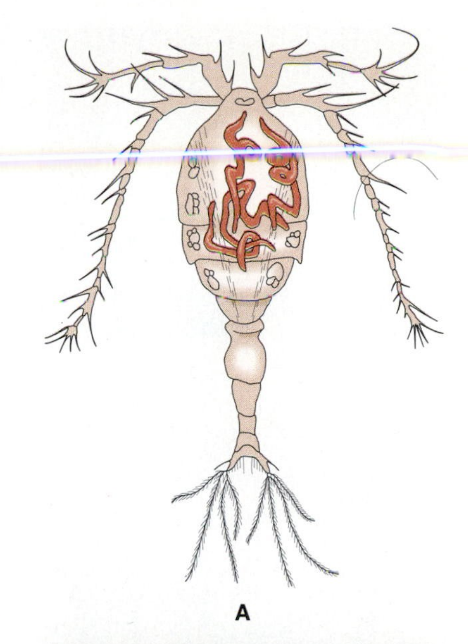

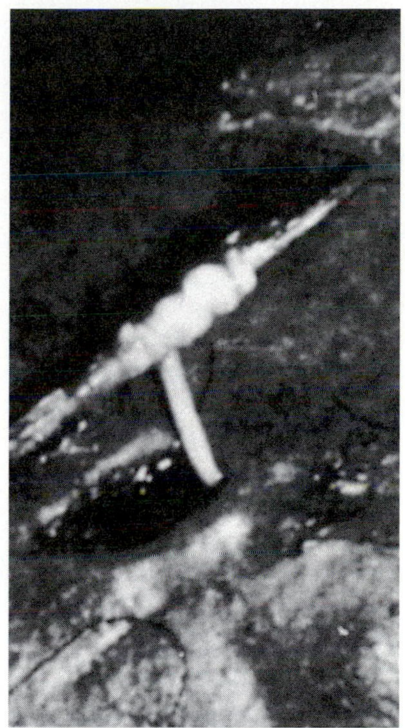

FIGURE 26.6
(A) Guinea worm juveniles in the copepod *Cyclops,* the intermediate host. *Cyclops* is about the size of the head of a pin. (B) The traditional way of removing *Dracunculus,* by winding it a little each day onto a stick.

Filarial Worms

There are eight species of filarial worms parasitic in human tissues, and almost all have blood-sucking insects as intermediate hosts. *Wuchereria bancrofti* is the best-known filarial worm, because of the gross disfigurement of the legs and genitals that it can cause (Fig. 26.7 on page 546). This condition is commonly called "elephantiasis." **Filariasis** is the appropriate name, however, because the disease is caused by filarial worms, not by elephants. Filariasis occurs when lymphatic vessels are blocked by female and male worms, which are 10 and 4 cm long, respectively. The disease can occur anywhere in the world that is warm enough for the nematodes to develop within insect intermediate hosts. There are now approximately 400 million victims in tropical regions around the world, and some experts believe it is the world's fastest-spreading disease.

Adult filarial worms feed on lymph and mate within the lymph vessels, and the females give birth to juveniles called **microfilariae.** The microfilariae pass through the thoracic lymph duct into the bloodstream, where they can be ingested by a blood-

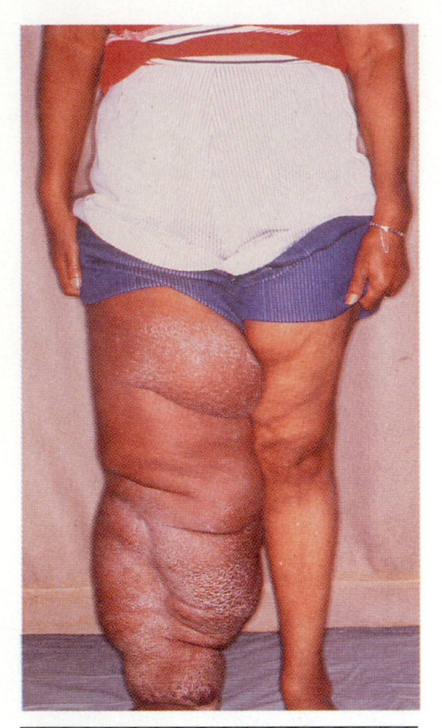

FIGURE 26.7
A victim of elephantiasis.

sucking insect. The insect is the intermediate host and the vector for the transmission to humans. Most microfilariae are produced while the host is asleep, so the intermediate hosts are often nocturnal mosquitoes. Inside the insect the microfilariae penetrate the gut and molt twice into infective L3 juveniles. When the insect feeds on a human again, these juveniles emerge from the insect's mouth and enter the wound in the human's skin. Once in the lymphatics, the L3 molts twice into an adult.

Another filarial worm, *Onchocerca volvulus,* has a similar life cycle except that its intermediate hosts are black flies (*Simulium* spp.). Although the adults may reach lengths up to 40 cm, it is the young microfilariae that cause the troublesome disease **onchocerciasis.** Large numbers infecting the skin cause an itching that sometimes drives the victim to suicide. The worms are so dense that positive diagnosis is made simply by snipping off a small piece of skin and seeing the worms in it. Infections of the cornea cause blindness. Onchocerciasis is often called **river blindness,** because it is more prevalent around rivers, in which black fly juveniles live. In some villages in Africa and South America, almost everyone over 50 is blind from onchocerciasis, and some of the most productive land is uninhabitable. River blindness may soon be history, however. The Onchocerciasis Control Program of the World Health Organization is eliminating black fly juveniles from rivers, and the pharmaceutical firm Merck & Company is distributing without cost the drug Ivermectin, which kills the microfilariae before they can cause blindness.

Trichina Worms

Trichinella was discovered in 1835 by a first-year medical student, James Paget, in London. Paget noticed that calcified particles in the cadaver he was dissecting kept dulling his scalpels. Other students and faculty had previously dismissed such particles as bone fragments, but under the microscope Paget saw that they were actually encysted worms. Later they were named *Trichinella spiralis,* and still later these trichina worms were found to cause the disease now called trichinosis or trichinellosis. Humans and other animals usually get trichinosis by eating cysts in raw or undercooked meat. In the small intestine the cysts are digested away, freeing the juveniles to molt into adults. The adults then burrow into the mucosa lining the gut, where they mate. Human victims generally suffer nausea and abdominal pain at this time. During the next 4 to 16 weeks each female may give birth to 1500 juveniles before she dies. Some juveniles enter the digestive tract and may spread through feces to a new host. Other juveniles enter the blood circulation and eventually arrive in muscle. There the juveniles burrow into muscle fibers, which then serve as "nurse cells." Protected from the immune system by these nurse cells, the juveniles curl up and become encysted until ingested by a carnivore or omnivore. In the encysted state, juveniles can live for 10 months, absorbing nutrients and growing to a diameter of 1 mm (Fig. 26.8).

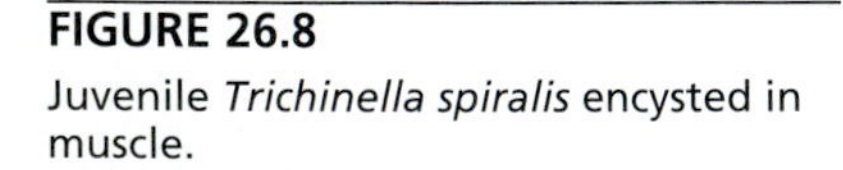

FIGURE 26.8
Juvenile *Trichinella spiralis* encysted in muscle.

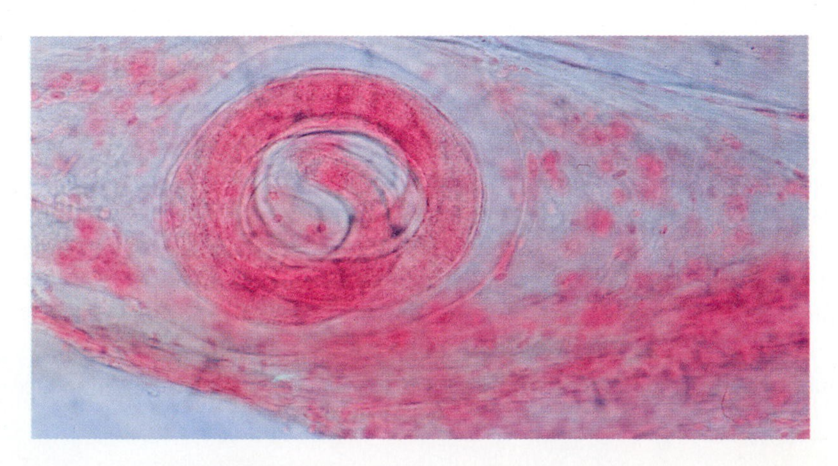

In humans the cysts of *Trichinella* merely produce vague muscle aches often mistaken for flu. The adults and juveniles can, however, produce severe pain, inflammation, fever, and fatal damage as they burrow through tissues. Humans can avoid trichinosis by not consuming the meat of any carnivore or omnivore, especially pork, without cooking it thoroughly. Pork and wild game should be cooked until it is no longer pink, or, in the case of pork sausages, until all the meat reaches 170°F. These precautions would appear to be simple in wealthier countries, yet approximately 1 in 25 Americans has *Trichinella* worms in his muscles (Kolata 1985). Most cases are discovered only during autopsy. The incidence was much higher until feeding pigs table scraps including pork was made illegal. Even now, however, approximately one in every thousand pigs in the United States gets trichinosis by eating rats or other pigs' tails.

HORSEHAIR WORMS: PHYLUM NEMATOMORPHA

Adult nematomorphs are long and thin and can often be found in puddles, contorted into knots. They are commonly called horsehair worms, because they were so common in watering troughs that many people thought they came from horses' hair. They are also known as gordian worms, after King Gordius, who tied an intricate knot and declared that whoever untied it would rule the world. Zoologists once considered them to be nematodes because of the external similarity of the adults, and because they also have a cuticle, longitudinal muscles only, and a nerve ring around the pharynx (Table 26.2). However, the larvae of nematomorphs, which are parasitic in arthropods or leeches, are quite different from juvenile nematodes (Fig. 26.9). The larvae in fact resemble priapulid worms and kinorhynchs.

Nectonema agile, the one species of marine nematomorph that lives in seas bordering the United States, has larvae that parasitize crabs. The free-living adults can sometimes be found along the beach on moonless summer nights. Other nematomorphs spend their brief adult lives reproducing in fresh water. They do not—in fact,

TABLE 26.2

Characteristics of horsehair worms.

Phylum Nematomorpha (= Gordiacea)
NEE-mat-oh-MORE-fuh (Greek *nema* thread + *morphe* form)

Morphology	**Vermiform. Pseudocoel** usually filled with parenchyme.
Physiology	**Nerve ring** around pharynx. Digestive tract degenerate. Juveniles, endoparasitic in arthropods, absorb nutrients across **cuticle**. No respiratory, circulatory, or osmoregulatory system.
Locomotion	Longitudinal muscles produce undulations of body.
Reproduction	**Dioecious**. Fertilization external.
Development	Indirect. Blastopore becomes mouth.
Habitat, Size, and Diversity	Adults free-living, mainly in fresh water. **Larvae parasitic** in arthropods or leeches. Lengths of adults up to 70 cm. Approximately **320 living species** described.

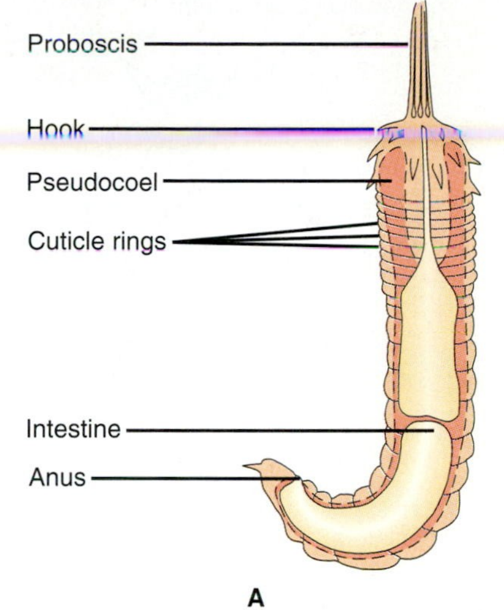

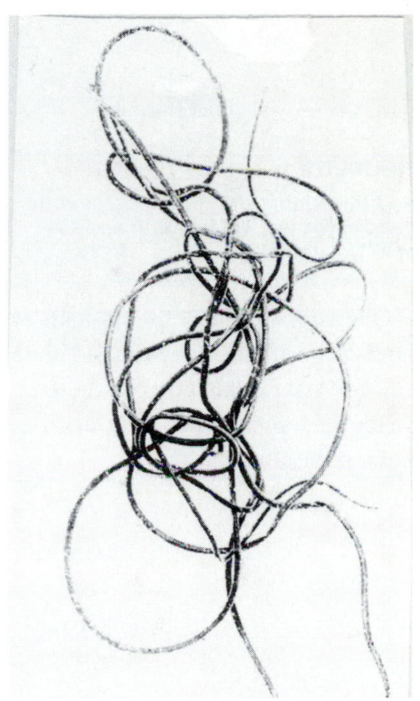

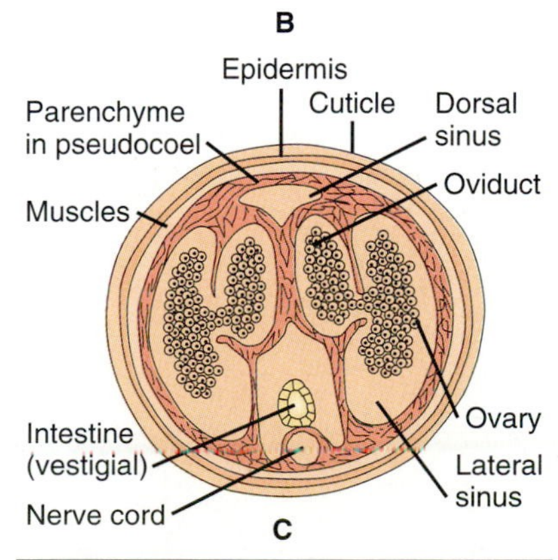

FIGURE 26.9

(A) Larva of a nematomorph with proboscis extended. Length approximately 250 μm. (B) An adult *Gordius.* (C) Cross section of a female *Gordius.*

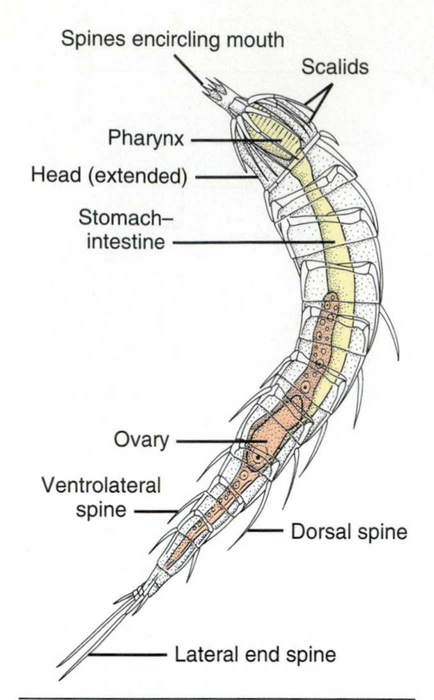

FIGURE 26.10
An adult kinorhynch *Echinoderes* with head extended. Total length approximately 0.2 mm.

■ **Question:** Compare the adult kinorhynch with the priapulid in Figure 26.12. What explanation could there be for the resemblance to the larval but not the adult priapulid?

cannot—feed, because the digestive tract is degenerate. The anus serves only as an exit for gametes. The males crawl and swim actively until they find a female, deposit sperm near her anus, then die. The ova are fertilized as they emerge, often in strands more than 2 meters long. Several weeks later the eggs hatch into larvae that are ingested by the host. The larvae absorb nutrients across the cuticle and develop into adults inside the host. If the host is a terrestrial arthropod, the adult worms emerge only when the host returns to water, which the nematomorph larvae somehow induce it to do. If the adult worm develops in autumn, it overwinters as a cyst.

PHYLUM KINORHYNCHA

Kinorhynchs are tiny, free-living worms that burrow through marine sediments, feeding on diatoms and organic material. The organ for burrowing is an extrusible proboscis and circlet of spines (**scalids**) similar to those of nematomorph larvae. The kinorhynch burrows by forcing its head into the sediment, anchoring the spines, then pulling the body forward by retracting the head. Unlike nematodes and nematomorphs, kinorhynchs have not only longitudinal but also circular and diagonal muscles. Recent examinations show that the "large body cavity" described in older literature is an artifact of fixation, as in Gastrotricha. Further investigation may therefore show that kinorhynchs are acoelomates. There are six juvenile stages, each followed by a molting of the chitinous cuticle. The cuticle of the trunk is divided into 11 overlapping joints, but the body is generally not considered to be segmented (Fig. 26.10, Table 26.3).

TABLE 26.3
Characteristics of kinorhynchs.

Phylum Kinorhyncha (= Echinodera)
KYN-oh-RING-kuh (Greek *kinein* to move + *phynchos* beak)

Morphology	"Body cavity" largely filled with amebocytes. Retractable head with scalids. Vermiform. Body flattened ventrally and domed dorsally. Body divided into proboscis, head, neck, and trunk.
Physiology	**Nerve ring** around pharynx. Digestive tract complete in adults. Pair of **protonephridia** with solenocytes (as in the acoelomates Gnathostomulida and Gastrotricha). No respiratory or circulatory system.
Locomotion	Burrowing with **retractile head**.
Reproduction	**Dioecious**. Sexes similar. Fertilization never observed, but presumed to be internal.
Development	Direct.
Habitat, Size, and Diversity	Free-living in marine sediments worldwide, to depths of 6 km. Length of adults less than 2 mm. More than **150 living species** described.

PHYLUM LORICIFERA

The newest phylum of animals was erected in 1983 by Reinhardt Kristensen of the University of Copenhagen. Kristensen came across Loricifera in sand and gravel from Denmark, Greenland, and the Coral Sea beginning in 1975, usually while looking for gnathostomulids and other interstitial animals (**meiofauna**). Some of the delicate specimens of Loricifera were lost in microscopic preparation, however, and others were larvae that could not be used as a basis for a new phylum. In 1982 Kristensen was spending his last day on a project in France when he received a large sample of bottom gravel. Because he did not have time to extract the small animals by the standard method (bathing in $MgCl_2$), he decided to use fresh water to osmotically loosen them from the gravel. He then preserved all the meiofauna in formalin for later study. To his delight, this sample yielded a definitive series of larvae and adults that he named Loricifera.

More than 50 species of loriciferans have been found, but their numerous scalids make them difficult to describe. So far only 10 species have been formally described and named. The first one formally defined was *Nanaloricus mysticus* (Latin *nano* dwarf + *lorica* corset; *mysticus* secret). Its retractable head and thorax ringed by scalids suggest a kinship with kinorhynchs (Fig. 26.11). Also like kinorhynchs, they have mouths on cones with stylets (Table 26.4). The **lorica** that girdles the abdomens of adults and larvae resembles that of some rotifers. The adults are sedentary, but the larvae swim by means of two propeller-like **toes.**

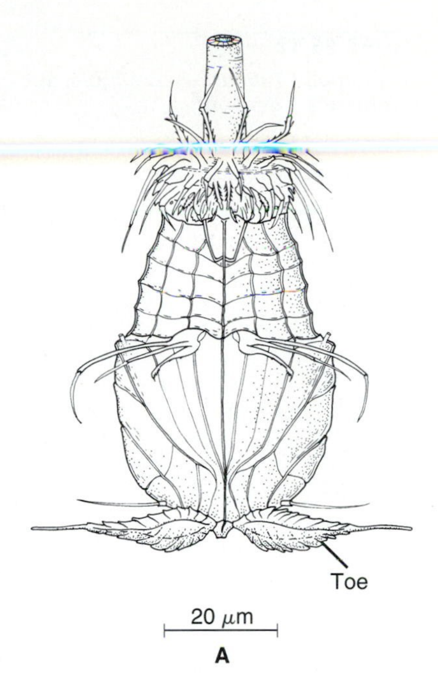

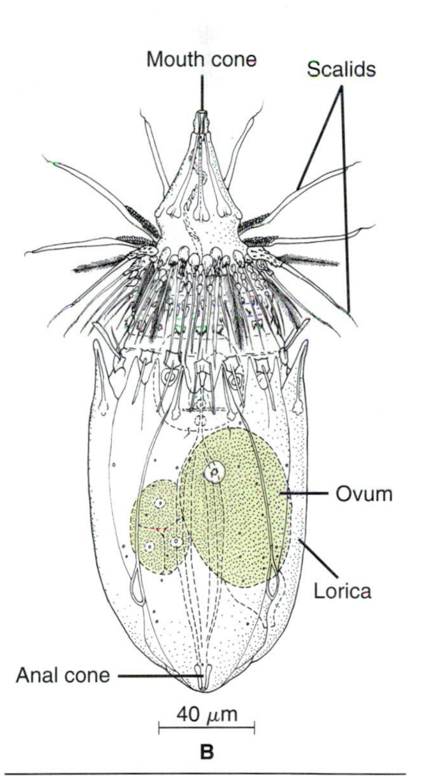

FIGURE 26.11
(A) A larva (Higgins-larva) of *Nanaloricus mysticus,* the first loriciferan discovered. (B) Adult female *N. mysticus.*

■ **Question:** What do you think the functions of the scalids would be in a burrowing animal?

TABLE 26.4
Characteristics of loriciferans.

Phylum Loricifera lore-i-SIFF-er-uh (Latin *loricus* corset + *fero* to bear)

Morphology	Head, neck, and thorax retract into abdomen. **Lorica** of six cuticular plates covers abdomen.
Physiology	Nervous system consists of cephalic ganglion, a ring of ganglia around the pharynx, and at least one abdominal ganglion. Pair of protonephridia. Digestive tract complete.
Locomotion	Adults believed to be sedentary; larvae propel themselves by two caudal toes.
Reproduction	Dioecious; sexes differ. Fertilization never observed, but presumed to be internal.
Development	Indirect, with perhaps three to five larval stages.
Habitat, Size, and Diversity	Free-living in marine sediments worldwide, to depths of several kilometers. Length of adults up to 383 μm. **Ten species** out of more than 50 formally described.

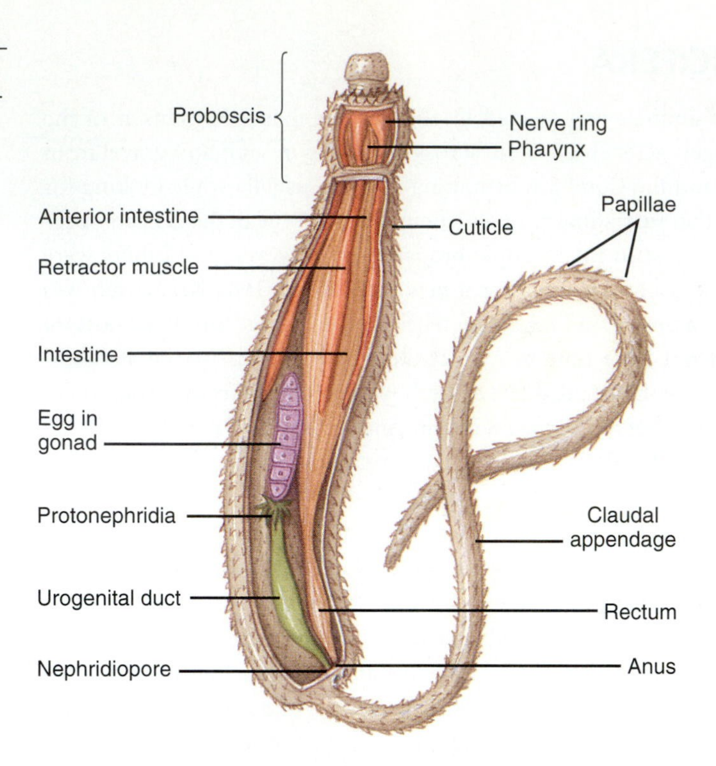

FIGURE 26.12
The priapulid *Tubiluchus corallicola*, approximately 1 mm long.

PHYLUM PRIAPULIDA

Priapulida are often considered to be coelomates, but the body cavity is lined with muscles rather than a peritoneum (Malakhov 1994, p. 240; Land and Nϕrrevang 1985). They are treated as pseudocoelomates here for that reason, and also because of their similarity to some other phyla in this chapter. Priapulids live partly buried in marine sediments where they feed by grabbing passing polychaete worms (phylum Annelida) with a kinorhynch-like proboscis. The body of the adult resembles a nematomorph larva, except that most priapulids have one or two unique caudal appendages (Fig. 26.12). Like loriciferans and some rotifers, the larvae have a lorica of cuticular plates. Like nematodes, priapulids have a nerve ring around the pharynx, and like kinorhynchs they have protonephridia with solenocytes (Table 26.5).

PHYLUM ROTIFERA

Except for nematodes, rotifers make up the largest and most diverse group once included in the phylum Aschelminthes. Molecular phylogenetics suggests that they are more closely related to coelomates than to nematodes (Hori and Osawa 1987), and they may have become pseudocoelomates by losing the peritoneum. It is easy to find rotifers with a hand lens in virtually any freshwater habitat: rivers, lakes, bogs, sandy beaches, ditches, moisture that collects in mosses and tree bark, and even bird baths. They are so numerous that they constitute a major link in aquatic food webs. A few species contribute to marine plankton, and some are symbiotic on or in other animals. Some species are colonial. Like many nematodes, rotifers are capable of withstanding almost complete desiccation in a state of cryptobiosis (pp. 540, 542), and they can also survive freezing in liquid nitrogen.

Most species attach to a substratum by means of a **foot** with one or more **toes,** or swim by means of a ciliated **corona** (Table 26.6). The constantly whirling cilia make ro-

TABLE 26.5

Characteristics of priapulids.

Phylum Priapulida PRE-ah-PEW-lid-uh (Greek *priapos* penis)

Morphology	Arguably pseudocoelomate. Vermiform. Bilateral symmetry, tending toward radial symmetry. Cuticle with numerous projections in rows that give the appearance of segmentation. Yellow or brown.
Physiology	**Predatory** by means of eversible proboscis. Nerve ring around pharynx. Digestive tract complete in adults. **Protonephridia** with solenocytes. No respiratory or circulatory system (although cells in body cavity of some species contains the respiratory pigment hemerythrin.)
Locomotion	Burrowing by contraction of longitudinal and circular muscles.
Reproduction	Dioecious. Fertilization external. One species lacks males and is presumably parthenogenetic.
Development	Indirect. Larvae enclosed in cuticular lorica that is molted between stages.
Habitat, Size, and Diversity	Free-living in marine sediments worldwide, to depths of several kilometers. Lengths of adults 2 mm to 8 cm. **Seventeen living species** described.

tifers, also called "wheel animals," fascinating animals to watch under the microscope. Their transparent bodies provide a window to their inner workings (Fig. 26.13 and Fig. 26.14 on page 552). Behind the mouth can be seen the uniquely modified pharynx, called the **mastax,** with its chitinous jaws (**trophi**) rapidly grinding detritus, algae, and

TABLE 26.6

Characteristics of rotifers.

Phylum Rotifera row-TIFF-er-uh (Latin *rota* wheel + *ferre* to bear)

Morphology	With anterior ciliated **corona** and posterior **toes**. Cuticle, forming a lorica in some species, but skeletal support due mainly to intracellular structures. Many tissues **syncytial**. Transparent; often taking on color of food.
Physiology	Nervous system consisting of dorsal ganglion and longitudinal nerves. Digestive tract complete in females; usually degenerates in males. **Protonephridia** with flame cells. No respiratory of circulatory system.
Locomotion	Free-swimming forms use corona. Sessile forms attach by a foot.
Reproduction	Dioecious. Most species entirely or partly parthenogenetic. Many species mainly parthenogenetic, but periodically reproducing sexually.
Development	Direct. Spiral cleavage. **Eutelic.**
Habitat, Size, and Diversity	Mainly free-living in fresh water. Lengths of adult females 50 μm to 2 mm. Approximately **1800 living species** described.

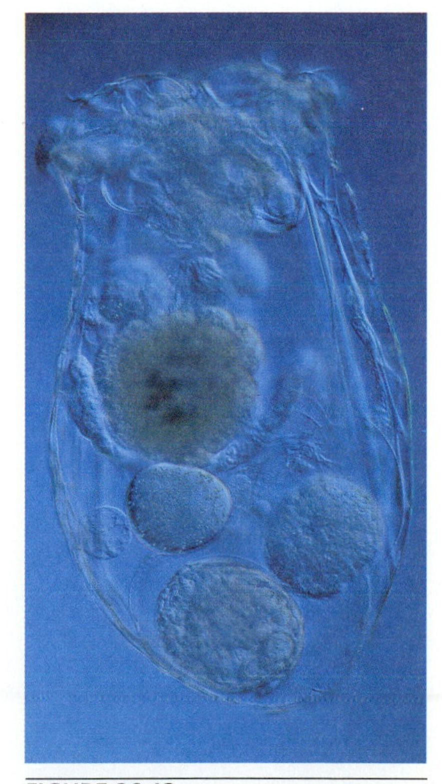

FIGURE 26.13

A female *Asplanchna sieboldi,* one of the largest rotifers (up to 2 mm long). Near the middle is the stomach with two round digestive glands on top.

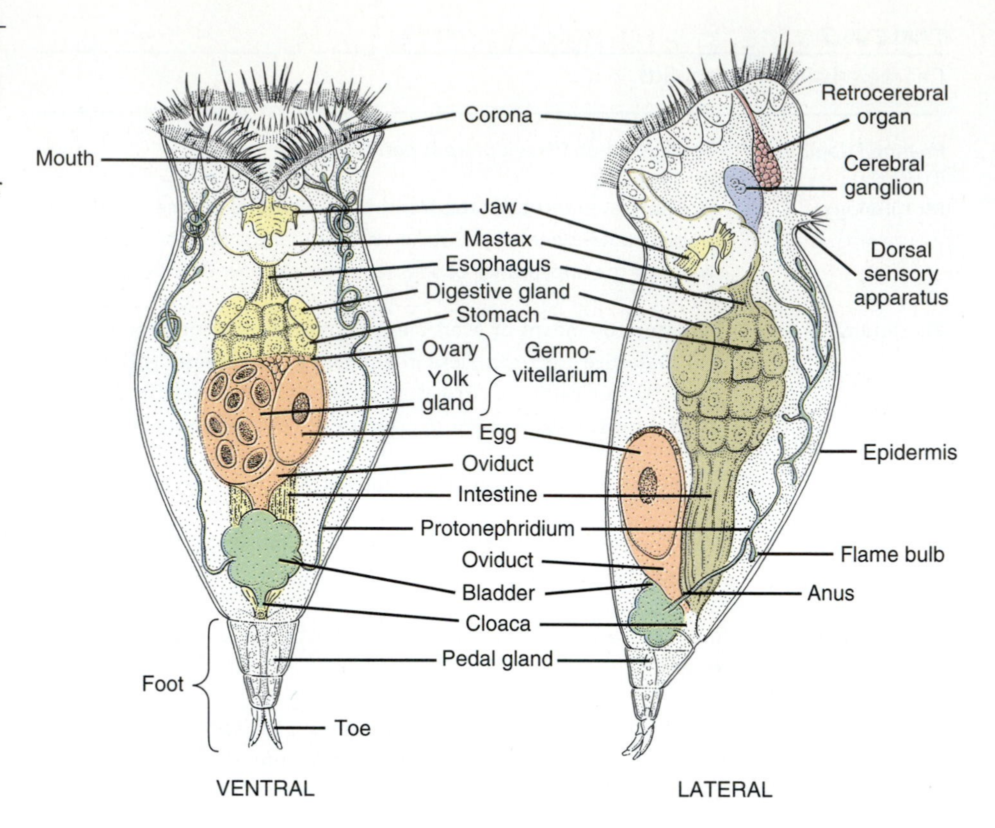

FIGURE 26.14

Ventral and lateral views of female rotifers *Epiphanes*. Length approximately 0.5 mm.

■ **Question:** Is this a monogonont or a digonont?

other small organisms. Also unique in many species is the **retrocerebral sac,** which is a mucus-secreting organ. The digestive tract is complete with an anus. In females the anus empties into a single receptacle called the **cloaca** (klo-A-kuh), which is a Latin word meaning "sewer." By definition, a cloaca is also the terminus of the oviduct and of the excretory system. In freshwater rotifers the osmoregulatory system is quite important and consists of a pair of protonephridia that empty into a bladder. The phylum is divided into two classes, Monogononta and Digononta, depending on whether the gonad is unpaired or paired, respectively.

Chances are good that any rotifer you see will be a female that reproduces parthenogenetically. The only two species that have a conventional sex life are parasitic on the gills of a particular marine crustacean. Rotifers in the class Digononta apparently don't produce males, and those in the class Monogononta produce them only when they are needed. During most of the summer the females of *Asplanchna* and other species of monogononts are parthenogenetic, producing only diploid, thin-shelled eggs that hatch into females only. Such females are said to be **amictic** (Greek *a* not + *miktos* mixed). Late in autumn, however, unknown stimuli cause the eggs to develop into females that are said to be **mictic.** Mictic females lay haploid eggs that hatch into degenerate males. The males then hypodermically impregnate the mictic females (including perhaps their own mothers). The females then lay fertilized diploid eggs with thick shells. These "winter eggs" are actually encysted females. The thick shells enable "winter eggs" to survive drying, cold, and other adverse conditions. They can also be distributed widely by wind or by adhering to birds and other animals. Upon reaching favorable, moist conditions, these eggs hatch into the next generation of parthenogenetic (amictic) females.

SPINY-HEADED WORMS: PHYLUM ACANTHOCEPHALA

You are not likely to see an acanthocephalan unless you look inside the body cavity of an insect or crustacean intermediate host, or inside the small intestine of a fish, bird, or mammalian final host (Fig. 26.15). The adults attach to the intestine of the vertebrate host by means of the spiny proboscis that gives the phylum its name. Many acanthocephalans modify the behavior of arthropod intermediate hosts in ways that make them more susceptible to predation by vertebrate final hosts. Janice Moore (1984) at Colorado State University has found, for example, that when the pill bug *Armadillidium vulgare* (actually a terrestrial crustacean) is infected by the acanthocephalan *Plagiorhynchus cylindraceus,* it is more likely to come out of its normal dark, damp habitat into the open, where it is easy prey for birds.

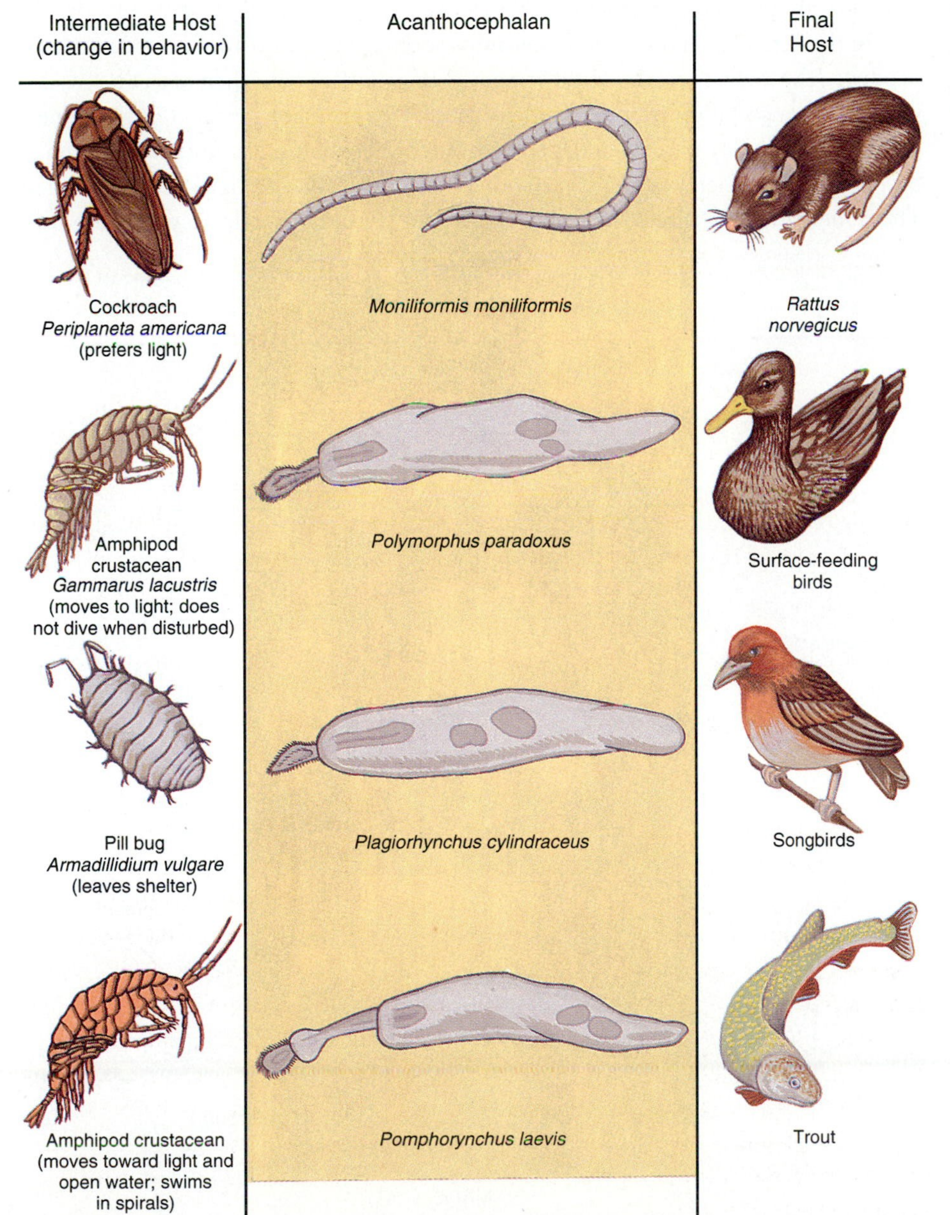

FIGURE 26.15
Some acanthocephalans and their intermediate and final hosts. As noted in parentheses, many acanthocephalans alter the behavior of the intermediate hosts in ways that make them more likely to be preyed upon by the final hosts.

After copulation within the final host, females release eggs containing embryos, called **acanthors,** into the host's feces. If an appropriate species of insect or crustacean ingests the egg while eating feces, the egg hatches, and the acanthor bores through its gut. Once in the intermediate host's body cavity, the acanthor develops into another embryonic form, the **acanthella.** The acanthella eventually develops into an encysted **cystacanth.** If the intermediate host is then eaten by the appropriate species of final host, the cystacanth develops into an adult. Because humans do not intentionally eat arthropods that eat feces, they are seldom infected by acanthocephalans. Sometimes the cystacanth passes through the digestive tracts of several species before it reaches the intestine of a suitable final host. Up until this time the proboscis has been withdrawn into its receptacle. On entering the small intestine of an appropriate vertebrate, however, the proboscis everts (Fig. 26.16B). The recurved spines then penetrate the intestinal wall. A heavy infection can cause massive and apparently painful destruction of the mucosa. Acanthocephalans have no mouth or digestive tract, but absorb the host's food directly across the tegument (Table 26.7).

In the vertebrate's intestine, cystacanths soon develop into adults and start to reproduce. The male impregnates the female by means of a penis, then seals the female's vagina with a secretion from his **cement gland.** The primary function of the cement gland is presumably to prevent copulation by a subsequent male, but males have also been known to cement the genitalia of other males, which keeps them from copulating. The cement dissipates from the female in time for her to begin releasing eggs. Another

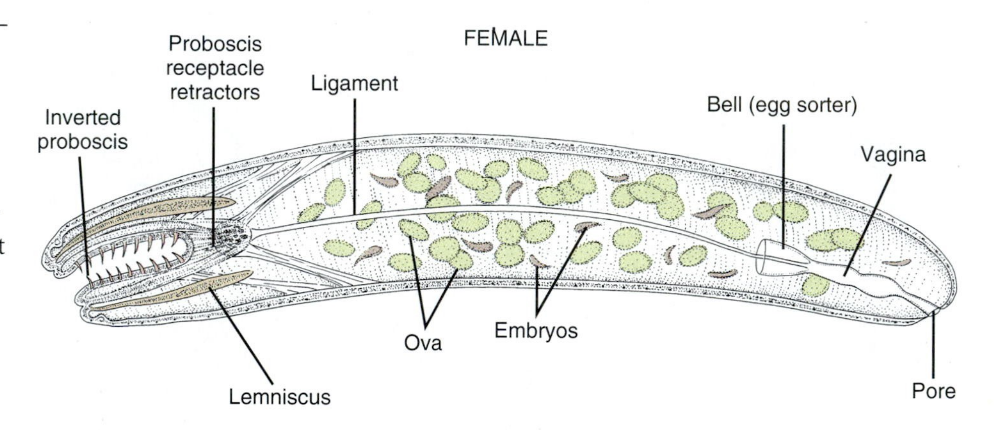

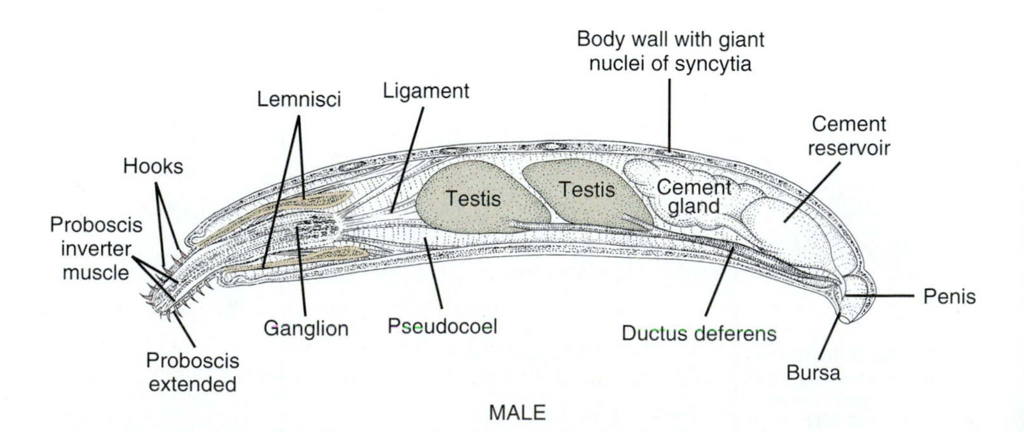

FIGURE 26.16

Structure of a male and female acanthocephalan. The lemnisci are thought to function in digestion, metabolism, excretion, or eversion of the proboscis.

■ **Question:** What is the adaptive advantage of the male's having a cement gland with which to seal the reproductive tract of a mated female?

TABLE 26.7

Characteristics of spiny-headed worms.

Phylum Acanthocephala a-KANTH-oh-SEF-full-uh (Greek *akantho* thorn + *kephale* head)

Morphology	With spiny, retractable proboscis. Body of adult usually flattened in life, but often cylindrical following preparation for microscopy. Body covered by syncytial tegument.
Physiology	Parasitic; no digestive system. Nutrients absorbed across tegument. Nervous system with ganglion in proboscis receptacle. Protonephridia, when present, with flame cells. No respiratory system. No circulatory system; lacunae within the tegument may distribute nutrients.
Locomotion	Passive in host.
Reproduction	**Dioecious**. Fertilization internal.
Development	Indirect. Cells of nervous system and gut eutelic.
Habitat, Size, and Diversity	White or cream colored: seldom seen because they are internal parasites. Distributed worldwide. Lengths of adults 1 mm to 70 cm. More than **1100 living species** described.

curious feature of acanthocephalan reproduction is that the female has an "egg sorter" that releases the mature eggs, but withholds the immature ones for further development (Fig. 26.16).

PHYLUM ENTOPROCTA

Entoprocts are an obscure group with uncertain relationships to the other animals. Some zoologists consider them to be acoelomates, because the body cavity is filled with gelatinous material. Some form colonies that resemble those of hydroids (compare Figs. 24.5A and 26.17). As in colonial hydroids, each individual **zooid** (zo-oyd) in a colony has a mouth surrounded by a ring of tentacles (Table 26.8 on page 556). Each zooid also has the capacity to regenerate and to reproduce by budding and is mainly sessile, although some can move by somersaulting. Instead of a gastrovascular cavity, however, each zooid has a complete, U-shaped digestive tract, with the mouth and anus both enclosed by tentacles (Fig. 26.18 on page 556). Also the functioning of the tentacles differs greatly from that of hydroids. Instead of capturing prey with nematocysts and pulling them to the mouth, the tentacles of entoprocts have cilia that draw water past the tentacles, trapping food in a layer of mucus. Entoprocta superficially resemble Ectoprocta (next chapter), and the two phyla have often been confused. Ectoprocts are coelomates, however, and their anal opening lies outside the ring of tentacles. Nearly all entoprocts are marine, but one freshwater genus, *Urnatella*, is oddly distributed in lakes in the eastern and midwestern United States, India, and a few other isolated places.

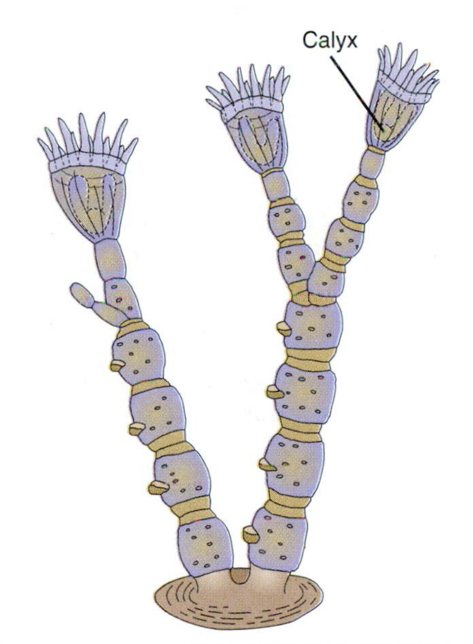

FIGURE 26.17
Part of a colony of the freshwater entoproct *Urnatella*. The calyx dies each autumn, but a new one is regenerated from the stalk in the following spring. Height of each zooid approximately 3 mm.

■ **Question:** Compare the shape of this colony with that of *Obelia* (Fig. 24.5). Does the similarity necessarily indicate a phylogenetic relation? What could be another explanation?

TABLE 26.8

Characteristics of entoprocts.

Phylum Entoprocta (= Kamptozoa) IN-toe-PROK-tuh (Greek *entos* within + *proktos* anus)

Morphology	Bilateral symmetry, tending toward radial symmetry. Solitary or forming colonial "animal mats." Each individual or zooid with a **calyx** (body) atop a **stalk** with adhesive base. Exoskeleton present.
Physiology	Digestive system complete and U-shaped, with both anus and mouth within circle of from 6 to 36 ciliated **tentacles**. Cilia draw water past tentacles, trapping food particles on mucus. Nervous system with ganglia. **Protonephridia** with flame cells. No circulatory or respiratory system.
Locomotion	Adults sessile; larvae creep by means of cilia.
Reproduction	Mainly monoecious. Fertilization internal. Reproduction also by budding in colonial species.
Development	Indirect. Spiral, determinate cleavage. Blastopore becomes anus.
Habitat, Size, and Diversity	Almost all **marine**. Length of zooids 1 to 5 mm; colonies greater than 1 cm. Approximately **150 living species** described.

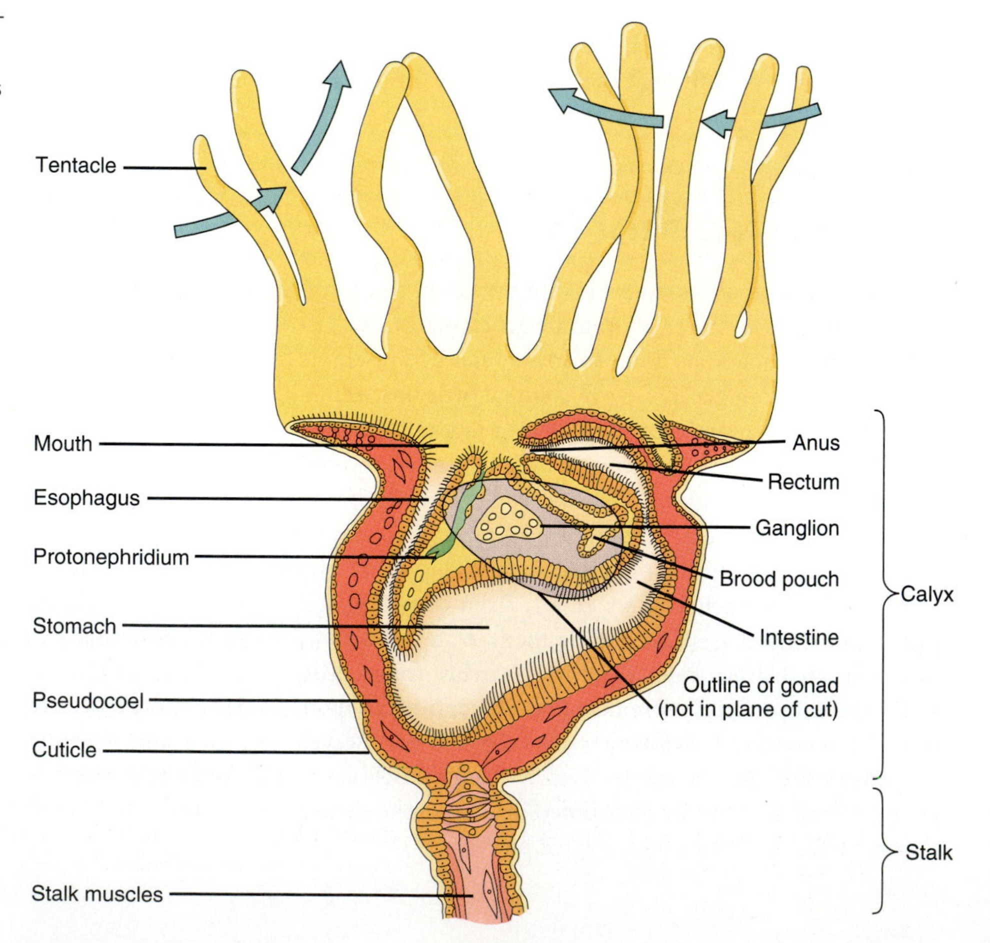

FIGURE 26.18

Structure of the calyx and part of the stalk of the entoproct *Barentsia*. Arrows show the direction of water movement.

Summary

Nematodes, nematomorphs, rotifers, kinorhynchs, loriciferans, priapulids, acanthocephalans, and entoprocts generally have little in common except a pseudocoel: a body cavity without a cellular lining. The digestive tract, if present, lies within the pseudocoel, giving a tube-within-a-tube organization. The pseudocoel can provide room for the digestive system and other organs to enlarge, substitute for a circulatory system, and function as a hydroskeleton.

The most numerous and best known pseudocoelomates are roundworms (nematodes). They move about within soil or organisms by contracting the longitudinal muscles against the hydroskeleton of the pseudocoel and the stiffness of the cuticle. Many are free-living, but others are parasitic. One of the most important roundworms is the free-living *Caenorhabditis elegans*, which is used extensively in studies of developmental genetics. Among the important parasitic forms in humans are large intestinal roundworms, hookworms, filarial worms, guinea worms, pinworms, and trichinella worms.

Other pseudocoelomates are diverse in appearance and life styles and are of uncertain relationship to each other and to other animals. Most adult horsehair worms (Nematomorpha), which resemble very long nematodes, are free-living in fresh water, but the larvae are parasitic in arthropods. Kinorhynchs use retractile heads to burrow in marine sediments. Loricifera, the newest animal phylum, also live in marine sediments, and they bear a girdlelike lorica. Priapulid worms also inhabit marine sediments, where they prey by means of an eversible proboscis. Rotifers are common inhabitants in fresh water, where they use their ciliated coronas to filter out small prey. Adult spiny-headed worms (Acanthocephala) are parasitic in the guts of vertebrates, where they attach by means of an eversible proboscis. Their larvae are parasitic in arthropods. Entoprocts are solitary or colonial hydralike animals that feed by means of ciliated tentacles that enclose both mouth and anus.

Key Terms

pseudocoel
cuticle
copulatory spicule
cryptobiosis
eutely
amictic
mictic

Chapter Test

1. Explain what a pseudocoel is. How does it differ from a coelom? Give three possible functions of the pseudocoel. (p. 536)
2. There are many uncertainties in the classification of pseudocoelomates. Give two examples of uncertain phyla, and explain the source of the uncertainty. (pp. 547–556)
3. Nematodes are said to be successful. Explain the basis for this statement. (p. 537)
4. *Caenorhabditis elegans* has made major contributions to our understanding of development. Briefly describe the nature of these contributions. State four reasons why this species is so suitable for this research. (pp. 541–542)
5. Describe the life cycles of two species of parasitic nematodes. (pp. 543–547)
6. Among the eight phyla in this chapter there is a great diversity of feeding mechanisms. Describe four of these mechanisms and name one phylum in which the mechanism occurs. (pp. 547–556)

■ Answers to the Figure Questions

26.4 Most likely *Ascaris lumbricoides* evolved from a species that got into the lung through some means other than by being swallowed; perhaps the ancestral species bored through skin.

26.10 Kinorhynchs and priapulids could have shared an ancestor, and kinorhynchs could have retained juvenile features in the adult (paedomorphosis).

26.11 Scalids would prevent the worm from slipping backward during burrowing.

26.14 There is only one ovary, so this must be a monogonont.

26.16 Sealing the reproductive tract would prevent the female from mating with other males.

26.17 The similar branching patterns of *Urnatella* and *Obelia* probably result from convergent evolution of a shape that minimizes competition of zooids with each other.

Readings

Recommended Readings

Crowe, J. H., and A. F. Cooper, Jr. 1971. Cryptobiosis. *Sci. Am.* 225(6):30–36 (Dec).

Gilbert, J. J. 1980. Developmental polymorphism in the rotifer *Asplanchna sieboldi. Am. Sci.* 68:636–646. (*How vitamin E alters the morphology of the female.*)

Hotez, P. J., and D. I. Pritchard. 1995. Hookworm infection. *Sci. Am.* 272(6):68–74 (June).

Moore, J. 1984. Parasites that change the behavior of their hosts. *Sci. Am.* 250(5):108–115 (May).

Rensberger, B. 1980. Life in Limbo. *Science 80*:36–43 (Nov). (*On cryptobiosis.*)

See also relevant selections in General References listed at the end of Chapter 21.

Additional References

Bird, A. F., and J. Bird. 1991. *The Structure of Nematodes,* 2nd ed. San Diego CA: Academic Press.

Cobb, N. A. 1915. Nematodes and their relationships. U. S. Department of Agriculture Yearbook, Washington, DC: U.S. Government Printing Office, pp. 457–490.

Croll, N. A. (Ed.). 1976. *The Organization of Nematodes.* New York: Academic Press.

Crompton, D. W. T., and B. B. Nickol (Eds.). 1985. *Biology of the Acanthocephala.* New York: Cambridge University Press.

Hori, H., and S. Osawa. 1987. Origin and evolution of organisms as deduced from 5S ribosomal RNA sequences. *Mol. Biol. Evol.* 4:445–472.

Kenyon, C. 1988. The nematode *Caenorhabditis elegans. Science* 240:1448–1453.

Kolata, G. 1985. Testing for trichinosis. *Science* 227:621, 624.

Land, J. van der, and A. Nϕrrevang. 1985. Affinities and intraphyletic relationships of the Priapulida. In: S. Conway Morris et al. (Eds.). *The Origins and Relationships of Lower Invertebrates.* New York: Oxford University Press, pp. 261–273.

Levine, N. D. 1980. *Nematode Parasites of Domestic Animals and of Man,* 2nd ed. Minneapolis: Burgess.

Lorenzen, S. 1985. Phylogenetic aspects of pseudocoelomate evolution. In: S. Conway Morris et al. (Eds.). *The Origins and Relationships of Lower Invertebrates.* New York: Oxford University Press, pp. 210–223.

Malakhov, V. V. 1994. *Nematodes.* Washington, DC: Smithsonian Institution Press.

Nicholas, W. L. 1984. *The Biology of Free-Living Nematodes,* 2nd ed. New York: Oxford University Press.

Poinar, G. O. 1983. *The Natural History of Nematodes.* Englewood Cliffs, NJ: Prentice–Hall.

Roberts, L. 1990. The worm project. *Science* 248:1310–1313.

Wharton, D. A. 1986. *A Functional Biology of Nematodes.* Baltimore: Johns Hopkins University Press.

Wood, W. A., et al. (Eds.). 1988. *The Nematode Caenorhabditis elegans.* Cold Spring Harbor, NY: Cold Spring Harbor Laboratory.

Zuckerman, B. M. (Ed.). 1980. *Nematodes as Biological Models* (2 volumes). New York: Academic Press.

CHAPTER

27

Miscellaneous Coelomates

The brachiopod *Lingula*.

FIGURE 27.1

Part of a colony of *Bugula neritina* ectoprocts. This species is common on rocks and pilings and is easily confused with algal mats.

Introduction

The animals to be discussed in the rest of this book are all generally considered to be coelomates. That is, all have a body cavity—the **coelom**—that develops within the mesoderm and is at least partly lined with a cellular membrane called the **peritoneum** (see Fig. 21.8). Often the peritoneal membrane forms **mesentery** membrane, which holds the digestive tract and other organs suspended within the coelom. The coelom performs the same functions as a pseudocoel (p. 536), and it is sometimes difficult to tell them apart.

The coelomates are traditionally divided into two groups, depending on the fate of the blastopore (pp. 439–441). In arthropods, annelids, molluscs, and most of the animals discussed in this chapter, the blastopore (the first opening of the blastula) becomes the mouth. These animals are therefore **protostomes.** Many protostomes also share other similarities in development. In chordates, echinoderms, hemichordates, and arrowworms the mouth forms from an opening other than the blastopore, so these animals are **deuterostomes.** Many deuterostomes also share other similarities in development. Although the protostome/deuterostome dichotomy appears to be valid, there are numerous variations in developmental patterns (Willmer 1990). The nine phyla described in this chapter—**Ectoprocta, Phoronida, Brachiopoda, Nemertea, Sipuncula, Echiura, Tardigrada, Pogonophora,** and **Chaetognatha**—comprise few species compared with those to follow. Phylogenetic relationships among them is obscure at best, but Fig. 21.10 provides a tentative representation.

PHYLUM ECTOPROCTA

The Ectoprocta were once joined with the Entoprocta in phylum Bryozoa (Greek *bryon* moss), because many members of both phyla resemble each other in forming mosslike colonies (Fig. 27.1). Many zoologists still prefer the name Bryozoa for the phylum Ectoprocta, and they refer to ectoprocts as bryozoans or "moss animals." Unlike entoprocts, ectoprocts are coelomates, and the anus of each member of the colony (zooid) lies outside its ring of tentacles. The latter difference explains the term ectoproct (Greek *ektos* outside + *proktos* anus). The tentacles constitute a **lophophore,** which is defined as a ring of tentacles enclosing the mouth but not the anus, with each tentacle containing a branch of the coelom (Fig. 27.2). The fact that the anus lies outside the lophophore is significant to taxonomists, but even more so to the ectoprocts themselves. This arrangement allows the cilia on the tentacles to maintain a stream of

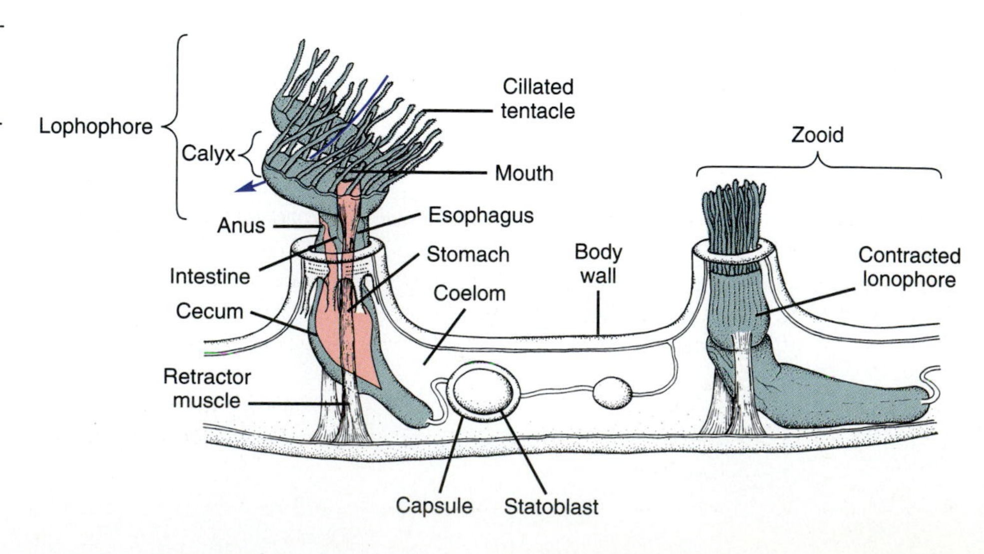

FIGURE 27.2

The ectoproct *Plumatella,* which occurs on the undersides of rocks in fresh water. Arrow shows direction of water current. The statoblast is a resistant structure that asexually regenerates a new colony after the old one dies in winter.

■ **Question:** How would you define an individual ectoproct?

food-bearing water directly toward the mouth without drawing in the animal's own feces. (Compare Fig. 26.18.)

Because the phyla Phoronida and Brachiopoda also have lophophores, Ectoprocta are usually thought to be more closely related to them than to Entoprocta, even though they do not look like phoronids or brachiopods. The three phyla—Ectoprocta, Phoronida, and Brachiopoda—are referred to as **lophophorates.** Classification of the lophophorates is confused by the fact that the pattern of development in many of their species is more like that of deuterostomes than that of protostomes. In all three phyla cleavage is radial, as in many deuterostomes. Also in many ectoprocts and brachiopods the mouth does not develop from the blastopore. In fact, many animals in these two phyla do not form blastopores in the embryo. Many zoologists resolve these taxonomic uncertainties by placing the lophophorates in a group separate from both the protostomes and the deuterostomes, while others consider them to be deuterostomes. Comparisons of molecular structures, however, indicate that lophophorates are related to protostomes such as molluscs and annelids (Halanych et al. 1995, Hori and Osawa 1987, Lake 1990).

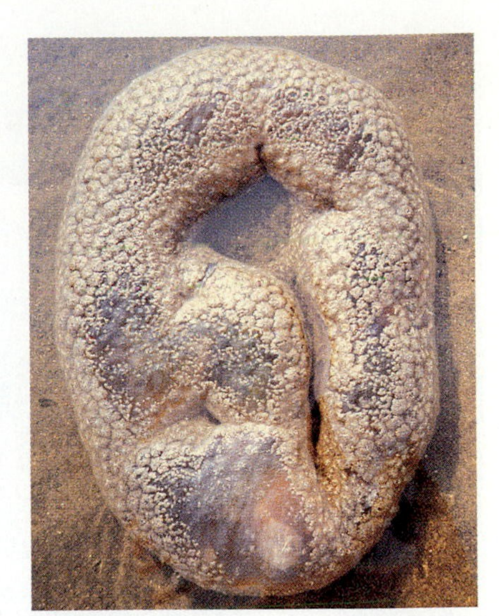

FIGURE 27.3
A large ectoproct colony *Pectinatella magnifica* found in Lake Champlain. Diameter approximately one meter.

Ectoprocts are colonial, often consisting of millions of zooids with their coeloms interconnected. Each zooid consists of a fleshy **polypide** living within a secreted chamber called the **zoecium** (zo-EE-she-um; from the Greek for "animal house"). The polypide includes the lophophore, digestive tract, and the nervous system and muscles. The zoecia may be cubic, conical, oval, or cylindrical, depending on the species. The colonies may be mossy, globular (Fig. 27.3), brushy, or crusty. Many colonies are protected by an external covering that includes various proportions of gelatin, **chitin** (the substance in insect exoskeletons), calcium carbonate, and sometimes sand (Table 27.1).

TABLE 27.1
Characteristics of ectoprocts.

Phylum Ectoprocta (= Bryozoa) ek-toe-PROK-tuh (Greek *ektos* outside + *proktos* anus)

Morphology	Each zooid is bilaterally symmetric. **Lophophorate**: circular or crescent-shaped lophophore encloses mouth but not anus. **Exoskeleton** of protein, chitin, $CaCO_3$ and/or sand.
Physiology	Ciliated tentacles of lophophore draw water toward mouth and past tentacles. Lophophore retracted by muscles; extended by muscles, deformation of exoskeleton, or coelomic pressure. Digestive system complete, U-shaped. Nervous system with nerve ring around esophagus and ganglion between mouth and anus. No excretory, respiratory, or circulatory system.
Locomotion	Adults usually **sessile** (attached) and colonial. Some freshwater colonies creep.
Reproduction	**Monoecious**. Fertilization internal. Colonies grow by asexual budding.
Development	Protostomate, although cleavage tends to be radial, and the mouth does not develop from the blastopore. Neither schizocoelous nor enterocoelous, coelom forms during metamorphosis of larva.
Habitat, Size, and Diversity	Most species marine. Found worldwide. Length of each zooid less than 1 mm; colonies greater than 1 cm. Approximately **4500 living species** described.

FIGURE 27.4

(A) A "garden" of phoronids, *Phoronis vancouverensis.* (B) Longitudinal section of *P. australis.* The length of the trunk is approximately 5 mm. Arrow shows direction of water current.

■ **Question:** How are the gametes released for external fertilization?

Ectoprocts can also defend themselves by retracting their delicate lophophores into the zoecia, using some of the fastest muscles known. At the same time, the disturbed zooids alert others in the colony by means of nerve impulses. Still others sprout thorny growths when preyed upon. In some ectoprocts, such as *Bugula,* some zooids are specialized as **avicularia,** which defend the colony by means of sharp, snapping beaks like those of birds. Other ectoproct colonies have zooids equipped with a long bristle that sweeps off sediments and the larvae of sponges, coral, and other animals that compete for space.

Because adult zooids are confined within zoecia, their sex lives are somewhat constrained. Colonies generally grow by asexual budding of the zooids, and new colonies start from eggs produced and fertilized in the same hermaphroditic zooid. A zooid broods the eggs within the coelom or in a special zoecium. The larvae are mobile and thus able to disperse the species to new habitats. Each larva selects a suitable substratum, then starts a new colony by reproducing asexually. Many freshwater ectoprocts can also reproduce asexually by means of internal **statoblasts,** which have tough capsules that can survive winter. Statoblasts of some species aid in dispersal by floating or attaching to other animals.

Evidently this kind of life has been successful; ectoproct fossils have been dated as far back as the early Ordovician, half a billion years ago. Petroleum explorers value ectoprocts as "indicator fossils." Someday living ectoprocts may also be valued as an ally in the battle against cancer. Seventeen antitumor substances have been isolated from *Bugula neritina* alone (Tucker 1985, Flam 1994). Now, however, *Bugula* and some other marine ectoprocts are known mainly for their ability to foul up ship bottoms and pilings.

PHYLUM PHORONIDA

Phoronids are a small phylum of lophophorates that live in shallow seas, either singly in sediments or in tangled groups on pilings and rocks. Each phoronid is confined in a leathery or chitinous tube that it secretes. Many are colored bright orange, pink, green, or yellow. A group of them with extended lophophores can make the sea floor resemble a flower bed (Fig. 27.4A). The lophophore has up to 50 ciliated tentacles in two spirals. Cilia on the tentacles propel water downward, and food particles are trapped in mucus in a food groove between the spirals. Cilia then propel the food-bearing mucus to the mouth (Fig. 27.4B).

Phoronids have some interesting physiological innovations that are probably related to their half-buried existence (Table 27.2). First, they have **closed circulatory systems,** with blood entirely confined within vessels. There is no heart; blood is pumped by contracting vessels. Like vertebrates, they have red blood cells containing the oxygen-transporting pigment hemoglobin. Second, adult phoronids have a pair of **nephridia,** which are excretory tubules open both to the outside of the body and at the coelom. Such nephridia are called **metanephridia** to distinguish them from the protonephridia of some acoelomates and pseudocoelomates, which are closed on the internal end by flame cells or solenocytes. Larval phoronids also have protonephridia. Most phoronids are hermaphroditic, but fertilization is generally by sperm from another individual. Eggs are fertilized internally and released through the metanephridia, which therefore serve as **gonoducts.** This is a common arrangement in coelomates with metanephridia.

The eggs exit among the tentacles and are brooded there in some species. The larvae are free-swimming members of the planktonic community. (They are called actinotroch larvae, because they were once thought to be adults of a different kind of animal with the generic name *Actinotrocha.*) These actinotrochs must eventually settle on appropriate substratum before metamorphosis into an adult.

TABLE 27.2

Characteristics of phoronids.

Phylum Phoronida for-OH-nid-uh (Phoronis: surname of the goddesses Io in Greek mythology)

Morphology	**Lophophorate**: spiral lophophore with 20 to 50 tentacles enclosing mouth but not anus. Adults **vermiform**, inhabiting a tube secreted by the epidermis. Not colonial.
Physiology	Digestive system complete, U-shaped. Probably some contribution to nutrition by absorption of dissolved organic material. Nervous system diffuse, mainly in epidermis. Adults have pair of **metanephridia**, which also serve as **gonoducts**. Closed circulatory system with red blood cells; no heart.
Locomotion	None in adult.
Reproduction	Monoecious or dioecious. Fertilization internal. Reproduction also by transverse fission or budding in some species.
Development	Protostomate even though some are enterocoelous, and cleavage is radial and indeterminate. Development indirect; larvae free-swimming.
Habitat, Size, and Diversity	**Marine**, to depths of 400 meters. Distributed worldwide. Length from 1mm to 50 cm. Approximately **15 living species** described.

PHYLUM BRACHIOPODA

Brachiopods are easily confused with bivalve molluscs such as clams and mussels and were once included with them in the phylum Mollusca. Unlike molluscs, however, brachiopods have **valves** (shells) on the ventral and dorsal sides rather than on the left and right (Fig. 27.5). In addition, brachiopods have shells that are either without hinges (class Inarticulata) or unequal in size (class Articulata). (Many brachiopods in the latter group

FIGURE 27.5

(A) Unidentified brachiopods, one with gape open showing the lophophore, and another showing the pedicel.
(B) Side view of the brachiopod *Magellania.* This is a typical position, with the ventral valve above.
(C) Vertical view into one valve of a brachiopod, showing the lophophores and the direction of water (blue arrows) and food movement (red). Food particles trapped on the tentacles are transported to the mouth along a ciliated food groove.

A

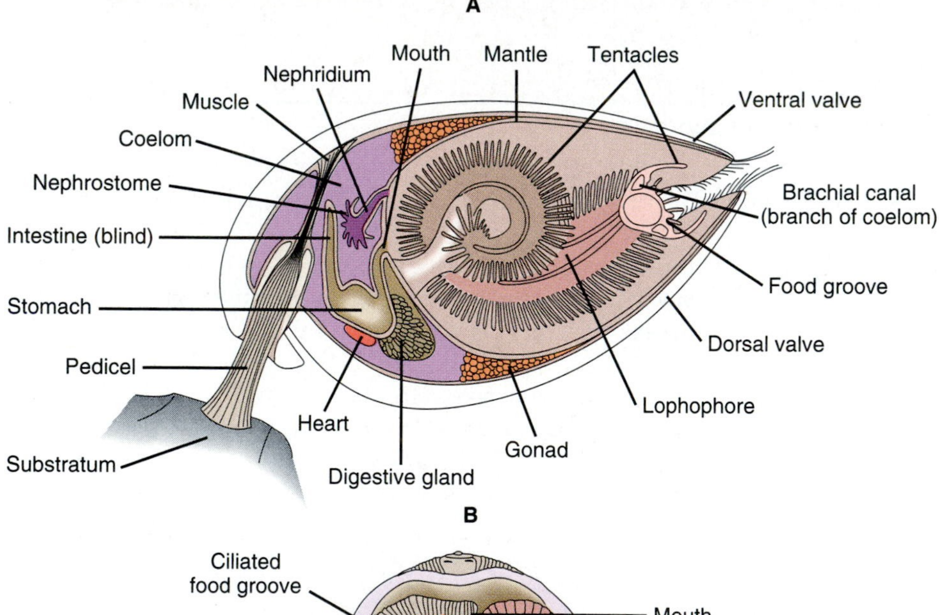

B

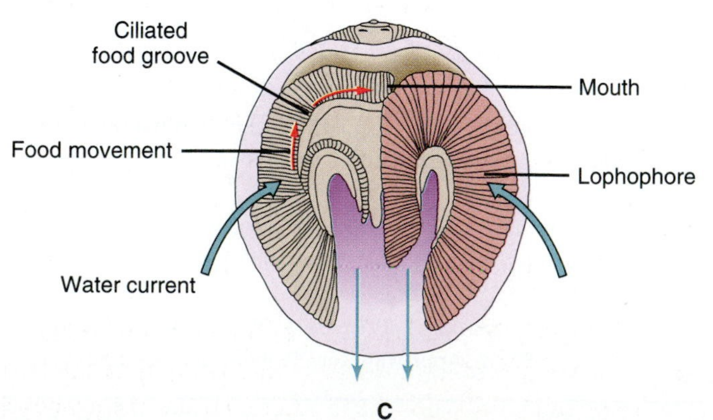

C

TABLE 27.3

Characteristics of brachiopods.

Phylum Brachiopoda brack-ee-OP-oh-duh (Greek *brachion* arm + *podos* foot)

Morphology	**Lophophorate**: lophophore with two circular or spiral arms within the shell. Shell of two **calcareous valves**, usually unequal in size.
Physiology	Digestive system either complete or without anus. Nervous system with several ganglia connected to a nerve ring around intestine. With **metanephridia** that function as gonoducts as well as excretory organs. Open circulatory system with contractile **heart**.
Locomotion	Adults usually **sessile**, attached by pedicel. A few species use the pedicel for locomotion. Larvae free-swimming.
Reproduction	**Dioecious**. Fertilization usually external.
Development	Molecular phylogenetics suggests kinship with protostomes, but mouth does not develop from blastopore. Cleavage also radial as in deuterostomes, and some species are enterocoelous. Development indirect with free-swimming larvae.
Habitat, Size, and Diversity	**Marine**: at all depths from intertidal to abyssal. Length from 1 mm to more than 9 cm. Approximately **335 living species** described.

are called lamp shells, because the larger ventral valve resembles a Roman oil lamp.) Also unlike molluscs, which have fleshy bodies, brachiopods have large, curled lophophores as their most prominent organs (Table 27.3).

Another difference is that most brachiopods attach to substratum by a **pedicel** on the ventral valve, usually with the ventral valve above the dorsal valve. The pedicel serves as an anchor against the turbulence of shallow water. Muscles attached to the pedicel can also contract to shake sediment off the brachiopod. Some brachiopods do not form a pedicel, but cement themselves to rock or are simply so dense that they do not need an attachment. Others, such as *Lingula* (see opening figure for this chapter), burrow into sediment.

Brachiopods have among the lowest rates of metabolic activity of all animals, and many can survive long periods without oxygen. Their minimal food requirements are satisfied mainly by organic debris and algae, which they filter out of water currents generated by cilia on the lophophore. Brachiopods have apparently fallen on hard times in the past 250 million years. They are among the most common fossils from the Paleozoic, but there are only about 1% as many species alive now. The cause of their decline may have been competition from bivalve molluscs. Some brachiopods have survived quite well, however. *Lingula* has remained virtually unchanged for more than half a billion years. It may be the oldest "living fossil."

PHYLUM NEMERTEA

Members of phylum Nemertea (= Rhynchocoela, Nemertina, Nemertinea) are often considered to be acoelomates, because the coelom consists only of the **rhynchocoel** (RING-ko-SEAL). The rhynchocoel is a chamber that houses the proboscis (Greek

FIGURE 27.6
The nemertean *Lineus* sp. on soft coral.

■ **Question:** What do you suppose is the function of the bright colors?

rhynchos snout + *koilos* cavity; Fig. 27.7). Also like some other acoelomates, nemerteans creep over substratum by muscle contractions or by means of cilia and a mucous secretion, and they have nervous systems and protonephridia like those of flatworms. On the other hand, the rhynchocoel is technically a coelom, since it forms within mesoderm and is lined with a cellular membrane (Turbeville and Ruppert 1985). Molecular phylogeny also suggests that nemerteans are more closely related to lophophorates, molluscs, and other coelomates than to flatworms (Hori and Osawa 1987). In addition, ne-

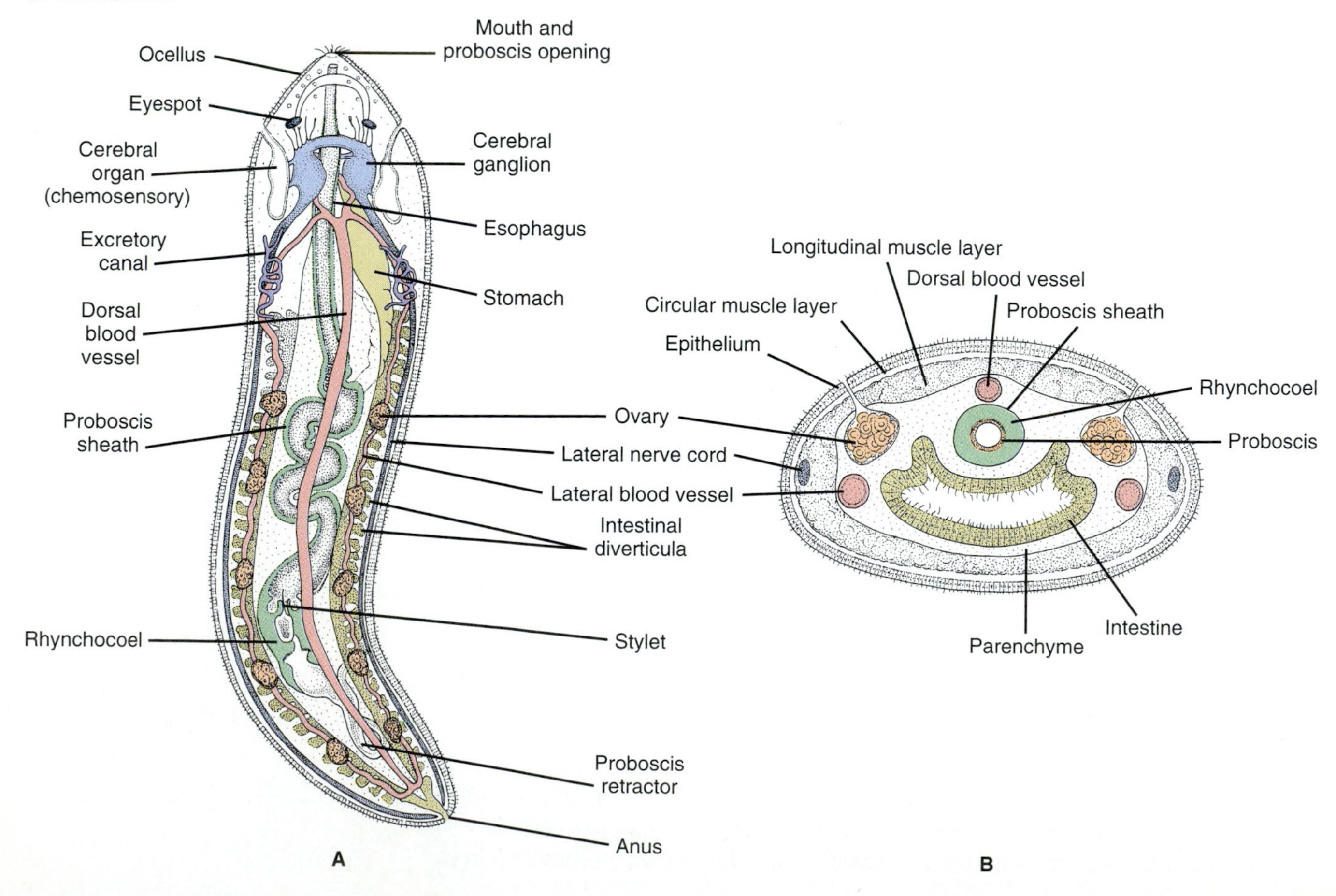

FIGURE 27.7
The anatomy of the nemertean *Amphiporus pulcher.* (A) Longitudinal section. (B) Cross section.

■ **Question:** Compare this figure with Fig. 25.2 and with several others in this chapter. Do you think nemerteans are more like acoelomates or coelomates?

merteans differ from acoelomates in being mostly dioecious and in having a complete digestive tract and a vascular system for the circulation of blood. There is no heart: Blood is propelled by contractions of large vessels, first in one direction, then the other.

Most nemerteans live among seaweed and gravel, where they scavenge and prey mainly upon annelid worms and other marine organisms, including diatoms, flatworms, nematodes, molluscs, crustaceans, and small fishes. Their main tool of predation, as well as locomotion, is the distinctive **proboscis,** which resides in the rhynchocoel when it is not being used. By contracting the body, the nemertean shoots out its proboscis, often with great accuracy and for a distance up to three times its body length. The proboscis turns inside out as it extends. It coils around the target, trapping it in sticky and sometimes toxic mucus. In some species, nail-shaped **stylets** also stab the prey repeatedly. The proboscis is not connected to the digestive tract. Contraction of a muscle pulls prey to the mouth, then withdraws the proboscis back into the rhynchocoel.

Nemerteans have impressive powers of regeneration. If a proboscis is damaged or hopelessly entangled, the nemertine simply sheds it and grows a new one. This regenerative ability also enables some nemerteans to reproduce asexually by **fragmentation.** An adult can spontaneously break up, and each fragment can develop into a new individual. Nemerteans also reproduce sexually, with external fertilization and usually with separate sexes (Table 27.4).

Nemertean worms are common along the sea coast, and many are brightly colored and large. The long, round ones are often called boot lace worms, and the long, thin

TABLE 27.4

Characteristics of nemerteans.

Phylum Nemertea (= Rhynchocoela) NIM-ur-TEE-uh (Greek *Nemertes* a sea nymph, unerring)

Morphology	Adults with an eversible **proboscis** normally sheathed within the **rhynchocoel**, which is technically a coelom. Most of the body filled with parenchyme.
Physiology	Some organ systems similar to those of Platyhelminthes: protonephridia with flame bulbs; no respiratory or skeletal system; a central nervous system consisting of cephalic ganglia and longitudinal nerve cords. Up to several hundred planarian-type eyespots in some species. A **closed circulatory system**, but no heart. **Complete digestive tract**; digestion partly intracellular. Mainly carnivorous. Probable absorption of dissolved organic material.
Locomotion	By muscular contraction in adults. Proboscis sometimes used for gripping and burrowing.
Reproduction	Usually **dioecious**. Fertilization external. Asexual reproduction by fragmentation.
Development	Spiral, determinate cleavage. Rhynchocoel is schizocoelous. Development generally direct.
Habitat, Size, and Diversity	Almost all marine. Some symbiotic in mantle cavity of molluscs, or in gills or egg masses of crabs. Adults range from less than 1mm to almost 60 meters long. Approximately **900 living species** described.

A

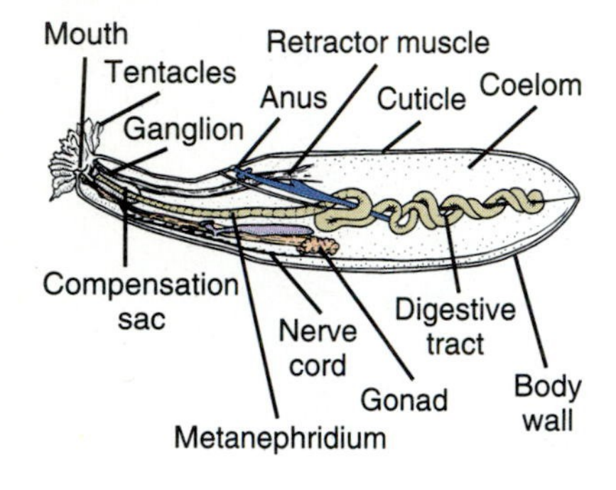

B

FIGURE 27.8
(A) The sipunculan *Dendrostomum pyroides* with tentacles everted.
(B) Longitudinal section with introvert extended.

■ **Question:** Why do you suppose the anus is located anteriorly?

ones are often called ribbon worms. The longest animal in existence is a nemertean, *Lineus longissimus.* One specimen was almost 60 meters long. It is difficult to measure the body length accurately, because many nemertines can stretch to 10 times their resting length. In spite of their size and appearance, nemertines are not often seen because they do not attack humans, and they burrow into sand or mud during the day and at low tide. Fishermen who know how to find them, however, often use nemerteans for bait. Several genera, such as *Prostoma,* are common in freshwater ponds. Several species have made the transition from marine to terrestrial life and are sometimes accidentally imported into greenhouses. A few are symbiotic. *Carcinonemertes errans* lives on the egg masses of crabs, including the commercially important Dungeness crab (*Cancer magister*).

PEANUT WORMS: PHYLUM SIPUNCULA

Like lophophorates, sipunculans also have ciliated tentacles that enclose the mouth but not the anus. These tentacles are not part of a lophophore, however, because the space within each tentacle is not part of the coelom. Instead, the tentacles are part of what is called the **introvert:** the retractable anterior portion of the worm (Fig. 27.8). The introvert is extended by pressure in the coelom due to contraction of the body wall, and the tentacles are extended by fluid pressure in the **compensation sac.** Sipunculans use their tentacles to feed on detritus on the ocean floor, where they live in burrows, in mollusc shells, in crevices of coral reefs, or among roots of plants. Some can burrow into coral and cause considerable damage to reefs. Although many sipunculans are colorful and large (up to a meter long), they are seldom seen because of their reclusiveness, scarcity, and inactivity. When disturbed they slowly pull in the introvert. When thus contracted, some sipunculans resemble peanuts—hence the common name peanut worm.

Most sipunculans have separate sexes, although they cannot be distinguished by external appearance. Gametes exit from the gonads in the coelom through the metanephridia, which are therefore gonoducts. Fertilization occurs externally. Reproduction is synchronized by some substance in sperm that induces females to release their ova. Development in some species is direct, with the egg immediately forming a worm. In other species the egg first develops into a trochophore larva, which has a ring of cilia around its body for locomotion (Table 27.5). Similar larvae also occur in molluscs and annelids (Figs. 28.7A and 29.5A). This similarity, among others, leads many zoologists to believe that sipunculans are related to molluscs (Hori and Osawa 1987, Lake 1990, Rice 1985).

TABLE 27.5

Characteristics of sipunculans.

Phylum Sipuncula sy-PUNK-you-luh (Latin *siphunculus* a small pipe)

Morphology	Retractable **introvert** with tentacles.
Physiology	Digestive system complete and U-shaped. Nervous system with ganglion and connectives around esophagus. Pair of **metanephridia**, which function as gonoducts. No circulatory or respiratory system, but some species have hemerythrin in coelomic fluid.
Locomotion	Usually none in adults.
Reproduction	Almost all **dioecious**. Fertilization external. Some reproduction by transverse fission.
Development	Spiral cleavage. Development either direct or with one or two larval stages including a trochophore.
Habitat, Size, and Diversity	**Marine**; worldwide, especially in the tropics. Adults usually **sedentary**, often burrowed in sediment or coral. At all depths from intertidal to abyssal. One species, *Golfingia procera*, parasitic in the annelid *Aphrodite*. Length from several centimeters to almost a meter. Approximately **320 living species** described.

SPOON WORMS: PHYLUM ECHIURA

Echiurans are also burrowing marine animals, but unlike most phyla discussed previously in this chapter, they do not have tentacles. Instead, most echiurans burrow and feed with a ciliated, mucus-coated proboscis, which has a **gutter** that brings detritus into the mouth (Fig. 27.9 on page 570). The elastic proboscis can extend more than 150 cm in an echiuran with a body only 40 cm long. It does not retract into the body, but is simply shortened by muscle contraction. Often the proboscis is spoon-shaped, giving the phylum its common name of "spoon worms." Echiurans are often fairly large, and those that are not transparent may be colored red, yellow, brown, or gray. However, they are seldom seen except by clam diggers, because they spend much of their lives within burrows, or in crevices or shells abandoned by other species. Although fossils of spoon worms are not found, remnants of their burrows appear in 450-million-year-old rocks.

Almost all spoon worms are dioecious and release their gametes through the pore of the metanephridium (gonoduct) for external fertilization (Table 27.6 on page 571). The larva is a trochophore, suggesting to some taxonomists that echiurans, like sipunculans, are related to molluscs and annelids. This finding is also consistent with molecular phylogeny (Hori and Osawa 1987). In most echiurans the sexes are similar, but *Bonellia* and other members of the family Bonelliidae show one of the most bizarre examples of sexual dimorphism of any animal. The females are up to 2 meters long, including the proboscis, and they resemble other echiurans. The males, however, are only about 1 to 3 mm long and live inside the females. Whether a larva develops into a female or male depends on whether it settles near the proboscis of a female. A substance from the female's proboscis causes a larva to migrate into her nephridium and to develop into a male. Inside the nephridium the degenerate male functions essentially as a testis, fertilizing the female's eggs, which develop within the nephridium. The nephridium has no excretory function in this genus, but serves only as a uterus.

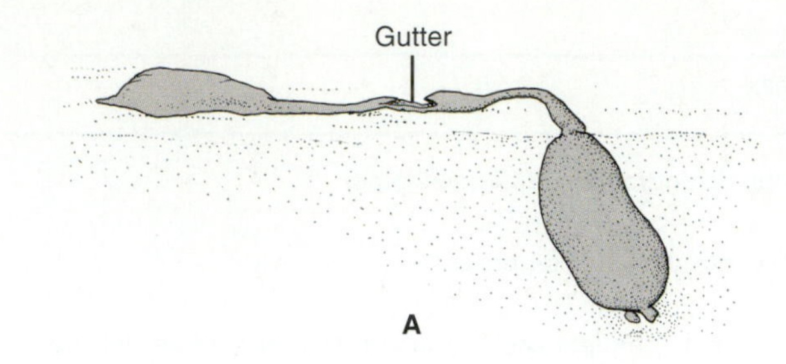

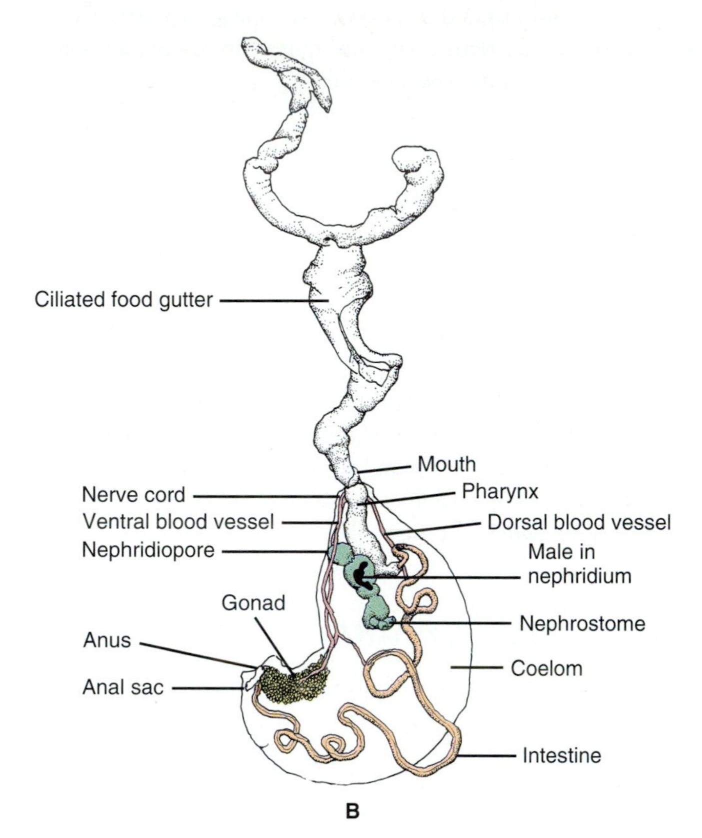

FIGURE 27.9

(A) The echiurid *Tatjanellia grandis* feeding with trunk buried in sand. (B) The female *Bonellia tasmanica* from the coast of southeastern Australia. This species is green from a pigment in the algae it eats. The genus is unusual in its sexual dimorphism. The tiny male is located within the nephridium of a female. Total length approximately 10 cm.

WATER BEARS: PHYLUM TARDIGRADA

The name "water bears" was given to the tardigrades by an 18th-century zoologist who was apparently reminded of bears as he watched them lumber along on his microscope slide (Fig. 27.10). Tardigrades are somewhat smaller than bears—less than a millimeter

FIGURE 27.10

Scanning electron micrograph of the tardigrade *Echiniscus.*

TABLE 27.6

Characteristics of spoon worms.

Phylum Echiura ek-ee-YOUR-uh (Greek *echis* snake + *oura* tail)

Morphology	Ciliated, mucus-coated proboscis, which is highly extensible but cannot be retracted into the body. **Vermiform**.
Physiology	Digestive system complete. Nervous system without ganglia. **Metanephridia** function mainly as gonoducts; excretion of metabolic wastes due largely to **anal sac**. Closed circulatory system; no heart. No respiratory system, but proboscis and anus may function like gills. Cells in coelom carry hemoglobin.
Locomotion	Adults **sedentary**, often burrowed in sediment or living in abandoned shells.
Reproduction	Almost all **dioecious**. Fertilization external in most species.
Development	Indirect, with free-swimming **trochophore** larva.
Habitat, Size, and Diversity	**Marine**; worldwide, especially on the bottoms of warm oceans. At all depths from intertidal to abyssal. Length of body from a few millimeters to 40 cm. Proboscis up to 1.5 meters long when extended. Approximately **150 living species** described.

long—and they usually live in pond sediments, in the water film in soil or plants, or in marine sediments. They feed on plants, nematodes, rotifers, and other animals (including other tardigrades) by piercing them with two **stylets** that protrude from the mouth, then sucking out their juices. Tardigrades are probably close relatives of arthropods. As in arthropods, the coelom is restricted to a space around the gonad, and the main body cavity is technically a pseudocoel (Table 27.7 on page 572). Other evidence of this relationship is a periodically molted cuticle, the resemblance to mites, and branches from the gut that resemble Malpighian tubules (Fig. 27.11).

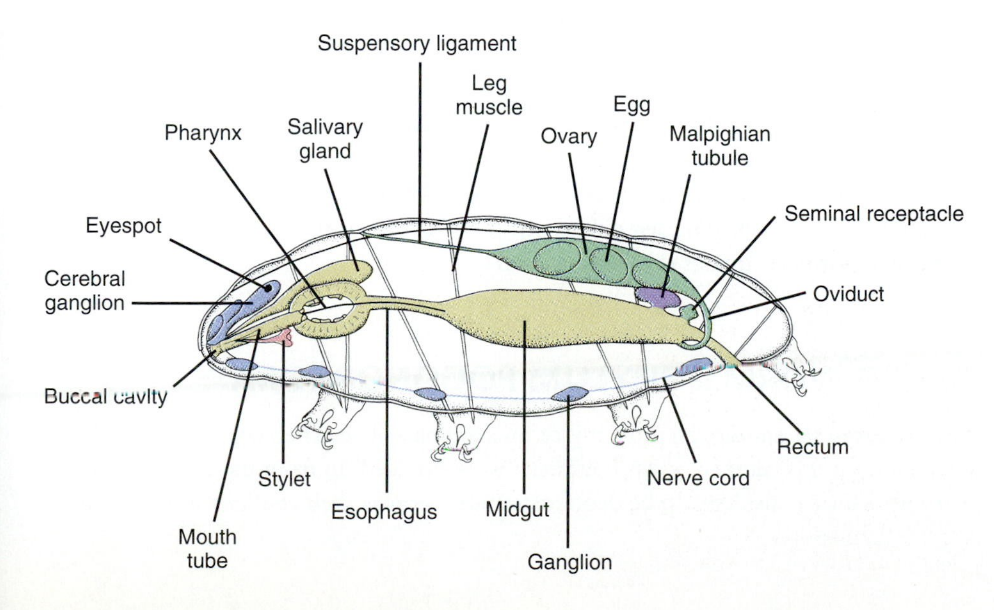

FIGURE 27.11

Longitudinal section of a generalized tardigrade. Note the bands of muscles that move the legs.

TABLE 27.7

Characteristics of water bears.

Phylum Tardigrada tar-di-GRADE-uh (Latin *tardus* slow + *gradus* step)

Morphology	Coelom reduced to cavity around gonad. Main body cavity a hemocoel. Body not divided into head, thorax, abdomen. Proteinaceous **cuticle** divided, but opinion differs on whether body is segmented. Four pairs of **legs** with four to eight claws each.
Physiology	Nervous system with cephalic ganglion around pharynx, and four ventral ganglia. Usually with pair of eyespots. Digestive tract complete. No circulatory or respiratory system. Branches from gut considered by some to be excretory organs: Malpighian tubules like those of insects.
Locomotion	Slow creeping on legs.
Reproduction	Mainly **dioecious**. Sexes similar. Some species parthenogenetic.
Development	Enterocoelous. Development direct, with periodic molting of cuticle. **Eutelic.**
Habitat, Size, and Diversity	**Aquatic**, mainly in freshwater ponds or in water film in soil or on plants. Some interstitial in marine sands. Some parasitic on crustaceans, mussels, and sea cucumbers. Worldwide, and at elevations ranging from the Himalayas to the ocean abyss. Body length from 0.1 to 1.7 mm. Approximately **600 living species** described.

Tardigrades are generally dioecious, but in some species males have never been identified, and the females are presumed to be parthenogenetic. Apparently females release eggs only when about to molt. In some species the copulating males inject sperm between the old and new cuticle, and the fertilized eggs are shed with the old cuticle, in which they develop. In other species, copulation does not occur; males deposit sperm on the molted cuticle containing the eggs. Tardigrade eggs hatch into miniatures of the adults. The juveniles grow by increasing the size of cells, rather than by increasing the number of cells through mitosis. Like nematodes and some other pseudocoelomates, tardigrades have a constant number of cells (**eutely**).

Ordinarily, tardigrades live for only a few months, but if they are gradually dried out they can live for decades or possibly centuries in a state of **cryptobiosis.** During cryptobiosis the body is barrel-shaped, inactive, and light enough to be dispersed by wind. In this condition it can survive temperatures near absolute zero (–270°C) and as high as 150°C, as well as ionizing radiation a thousand times more intense than the lethal dose for humans. On being returned to water at normal temperature they resume their normal activities within a few hours.

BEARD WORMS: PHYLUM POGONOPHORA

Beard worms are among the most mysterious animals alive. They were not even discovered until this century, when fragments were dredged up from the ocean depths. They were once considered to be deuterostomate, because their coeloms seemed to be divided by septa into three chambers, as in some other deuterostomes. Then in 1963 it was discovered that this conclusion was based on samples that had lost their segmented tails (**opisthosomes**), which have several coelomic compartments. Some zoologists now think beard worms belong in the same phylum with other segmented worms (Annelida), but the majority think they are only close relatives of annelids. Molecular phylogeny supports the latter conclusion (Lake 1990, Kojima et al. 1993).

A pogonophoran lives within an upright tube of chitin and protein that it secretes around itself, with from one to many thousand tentacle-like **branchiae** protruding from the upper end of the tube. The branchiae are the "beard" of beard worms. Without their tubes many beard worms would look like threads that are badly frayed at one end (Fig. 27.12 on page 574). The branchiae are often called tentacles, but they absorb gases and are not used to capture solid food. In their typical cylindrical arrangement the branchiae resemble an intestine, complete with microvilli that increase the surface area.

Adult pogonophorans mystified zoologists for many years because they have no mouths, intestines, or digestive glands (Table 27.8). It is now known that they obtain most of their nutrients from bacteria that live within cells in the **trophosome,** a long cylindrical organ in the posterior part of the trunk. Apparently beard worms harvest the bacteria or their metabolic products. These bacteria are of a type that synthesize carbohydrate $[CH_2O]$ using the energy released from the oxidation of hydrogen sulfide, according to the following reaction:

$$CO_2 + H_2S + O_2 + H_2O \longrightarrow H_2SO_4 + [CH_2O]$$

TABLE 27.8

Characteristics of beard worms.

Phylum Pogonophora PA-gone-OFF-or-uh (Greek *pogon* beard + *phora* bearing)

Morphology	Several body cavities, at least one of which is a coelom. Segmented. **Vermiform.** From one to hundreds of thousands of **branchiae.** Living within a **chitinous tube.**
Physiology	No digestive tract in adults. Dissolved nutrients absorbed by branchiae and from mutualistic chemoautotrophic bacteria. Closed circulatory system with heart; hemoglobin dissolved in blood. No special respiratory organ. Nervous system without ganglia. Metanephridia.
Locomotion	None.
Reproduction	Almost all **dioecious.** Males release water-borne spermatophores that may be caught by females' branchiae. Fertilization external, in tube of female.
Development	Trochophore larvae are brooded within the maternal tube in some species.
Habitat, Size, and Diversity	**Marine,** world-wide on ocean floors from 20 meters to 10 km deep. Many in sulfur-rich waters, especially near hydrothermal vents. Length up to 3 meters. Approximately **140 living species** described.

FIGURE 27.12

(A) Schematic diagram of a pogonophoran removed from its tube. The body has four divisions: cephalic lobe, forepart, trunk, and opisthosome. The thickness of the body relative to length is greatly exaggerated here. (B) Upper portion of the pogonophoran *Siboglinum* retracted in its tube. This species has only one branchia. (C) Detail of cephalic lobe of *Siboglinum*.

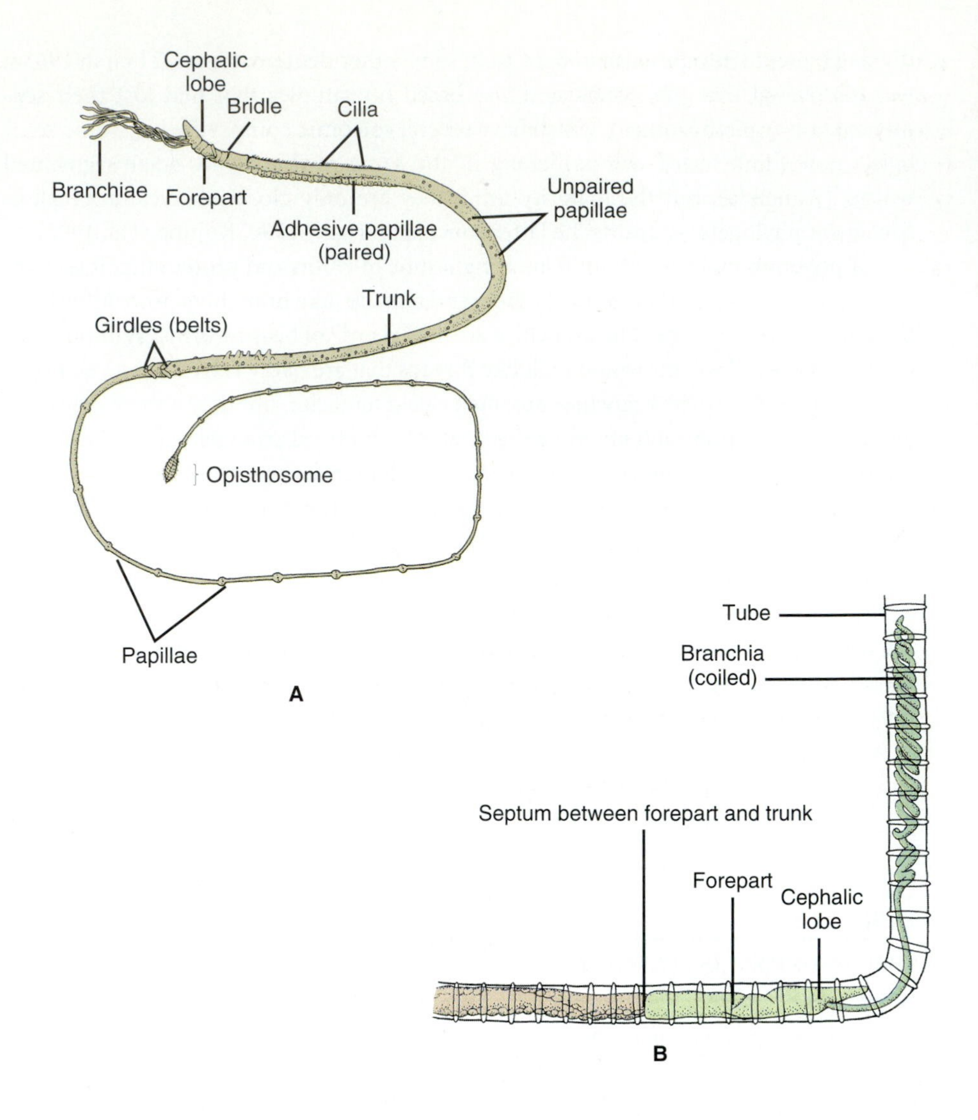

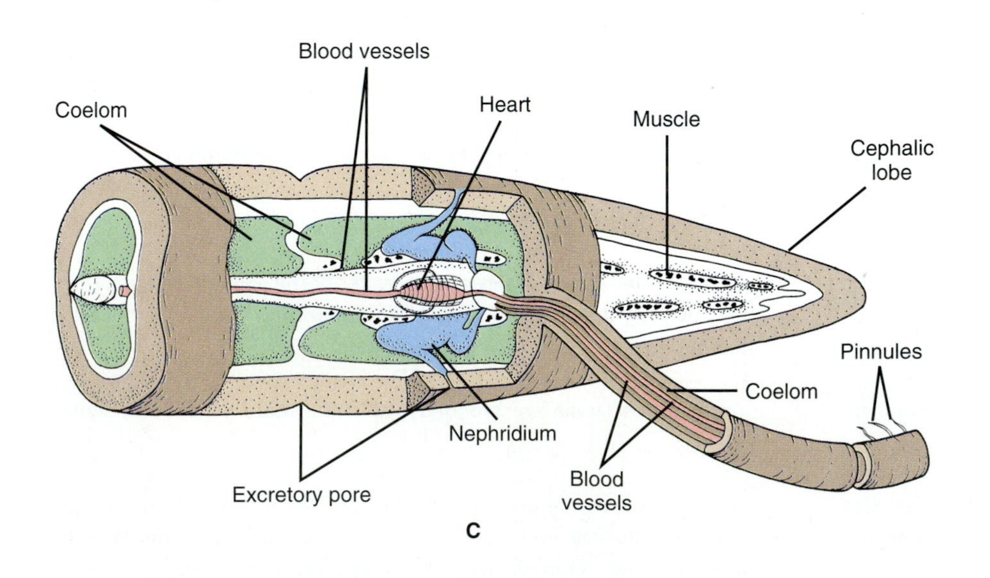

Hemoglobin dissolved in the blood is crucial for this process. It not only carries oxygen, like other hemoglobins, but also binds the hydrogen sulfide, which would otherwise kill the animal. Thus the hemoglobin transports both H_2S and the O_2 to oxidize it directly to the bacteria. Many beard worms live in sulfur-rich habitats, such as hydrothermal vent areas, which are essentially undersea volcanoes. The first beard worm discovered in such a seemingly inhospitable place was *Riftia pachyptila,* which, together with 11 other species, is classified in the group Vestimentifera. (Some advocate making this a separate phylum.) These **vestimentiferan** tube worms have more than 200,000 branchiae that stick out like a rolled tongue, bright red with the hemoglobin in the blood (Fig. 27.13).

FIGURE 27.13
Vestimentiferan tube worms *Riftia pachyptila* just centimeters away from a scalding death at a Galápagos hydrothermal vent. Other members of the hydrothermal vent community shown here are crabs and mussels. *Riftia pachyptila* is up to 3 meters long and unusually thick—2 to 3 cm compared with less than a millimeter for most beard worms.

■ Question: How did this species get to the Galápagos hydrothermal vent without a means of locomotion, and apparently with nonmotile larvae?

ARROWWORMS: PHYLUM CHAETOGNATHA

Arrowworms, named for their streamlined appearance, are among the most enigmatic of animals. Zoologists for more than a century have debated their relationship to other phyla, and proposals for their next of kin range from nematodes to chordates. The body cavity of the adult is a pseudocoel; arrowworms are considered to be coelomates only because the juvenile has a temporary body cavity lined with a mesodermal cellular membrane. They are deuterostomate, because their mouths develop from an opening other than the blastopore, but they have little else in common with other deuterostomes. Molecular phylogeny suggests that they are not closely related to either protostomes or other deuterostomes, but may belong to a separate lineage that originated early in coelomate evolution (Telford and Holland 1993, Wada and Satoh 1994).

The daily lives of chaetognaths are equally obscure, although most are large (about 4 cm long) and make up a large proportion of marine biomass. If you have been in the ocean, you have probably been surrounded by them, but they are virtually impossible to see because they are transparent. As many as a thousand arrowworms have been found in a cubic meter of seawater.

Chaetognaths are the most important predators of copepods, and they also consume fish and each other. They capture prey in bristles (Greek *chaite*) around the jaws (*gnathos*) (Fig. 27.14, Table 27.9 on page 576). They have a hood that snaps over the head and jaws to entrap prey and presumably to enhance streamlining as they swim. Some chaetognaths paralyze prey with the neurotoxin tetrodotoxin (p. 156), which is

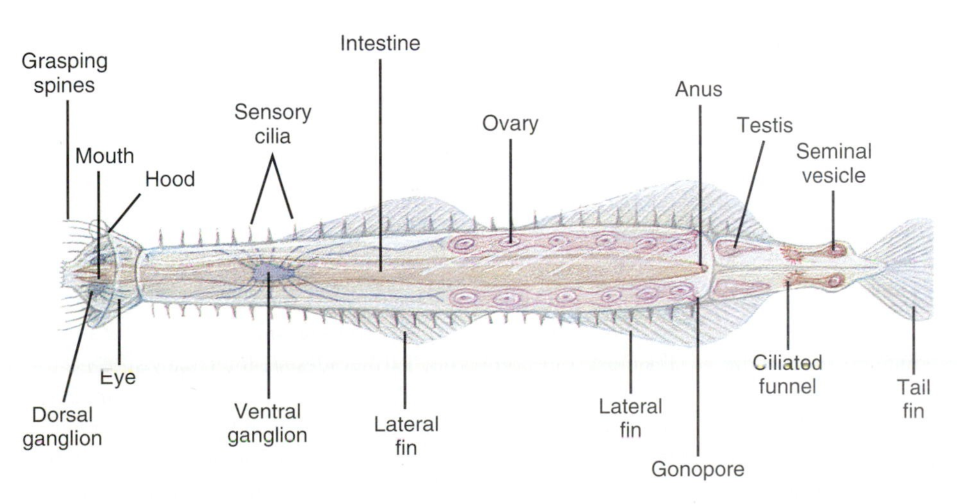

FIGURE 27.14
The chaetognath *Sagitta* in ventral view.

■ Question: What is the function of the ciliated funnel? How would self-fertilization occur.

TABLE 27.9

Characteristics of arrowworms.

Phylum Chaetognatha key-TOG-nah-thuh (Greek *chaite* bristles + *gnathos* jaws)	
Morphology	Coelom in juveniles; body cavity in adult lacks peritoneum. Body **dart-shaped**, with **bristles around jaws** and retractable hood over head. Bilateral symmetry. Not segmented. Epidermis with **cuticle**.
Physiology	Straight, complete digestive tract. Nervous system with ganglia and specialized receptors. No excretory, circulatory, or respiratory organ.
Locomotion	Most species swim by dorsoventral undulation of the finned body.
Reproduction	Monoecious. Fertilization internal.
Development	Deuterostomate. Radial cleavage. Not enterocoelous.
Habitat, Size, and Diversity	All species marine, planktonic. Length of adult ranges from 0.5 to 12 cm. Body generally transparent. More than **100 living species** described.

produced by symbiotic bacteria. The fluid-filled body cavity serves as a hydroskeleton. Swimming is due to dorsoventral oscillations of the body, with propulsion due to fins fixed to the sides of the trunk. The only internal organs are a straight digestive tract and gonads. The nervous system includes a ventral ganglion in the trunk wall and a dorsal ganglion and several smaller ganglia in the head. Two dorsal eyes presumably detect light and control vertical migration at dawn and dusk.

Arrowworms are all protandrous hermaphrodites, meaning that they function first as males, then as females. Each individual has two ovaries and two testes. Sperm circulate within the body cavity as they mature, then enter the seminal vesicles. During spawning the sperm break through the epidermis and migrate into the gonopore to the ovary of the same or a different individual. Fertilization occurs internally. The eggs are then released, and in several days they hatch into immature chaetognaths that reach maturity within another week or so.

Summary

Animals classified in the nine phyla in this chapter all have a coelom, which is a chamber formed within mesoderm and lined with a cellular peritoneum. Eight of these phyla—Ectoprocta, Phoronida, Brachiopoda, Nemertea, Sipuncula, Echiura, Tardigrada, and Pogonophora—are considered to be protostomate, which means that the blastopore usually becomes the mouth. Chaetognatha are deuterostomate, which means that the mouth originates from an opening other than the blastopore. Ectoprocta, Phoronida, and Brachiopoda are lophophorates: They have lophophores, which are circles of tentacles enclosing the mouth but not the anus. Nemerteans are often considered to be acoelomates related to flatworms, but they have a coelom into which the proboscis retracts. Tardigrades share many similarities with arthropods, and pogonophorans appear to be related to annelids.

These are minor phyla considering the number of species, but several are important for other reasons. Brachiopods were extremely numerous in the Paleozoic era, and their shells are among the most common fossils. Tardigrades are interesting subjects for the study of cryptobiosis. Beardworms are ecologically interesting, because many of them live on the products of bacteria in the total darkness of hydrothermal vents. Chaetognaths are among the most numerous marine animals, and they are ecologically important predators.

Key Terms

coelom
peritoneum
protostome
deuterostome
lophophore
closed circulatory system
metanephridium
gonoduct
valve
rhynchocoel
trochophore larva
branchis

Chapter Test

1. Explain what a coelom is and why it is important. (p. 560)
2. What is a lophophore? How does it function? What is its taxonomic significance? (pp. 560–561)
3. Some phyla in this chapter have tentacles but are not lophophorates. Name the phyla. Why are they not considered to be lophophorates? (pp. 568, 572)
4. Name two phyla discussed in this chapter that do not feed by tentacles. Describe how they do feed. (pp. 565–576)
5. For each of the nine phyla give a characteristic or combination of characteristics that is not found in any of the other eight phyla. (pp. 560–576)

■ Answers to the Figure Questions

27.2 This is a question that has often been debated. From an evolutionary perspective, the individual would be a genetically unique zooid. The zooids that develop asexually from statoblasts would not be individuals by this definition.

27.4 The anatomy suggests that the metanephridium is the route by which gametes are released. The metanephridia are, in fact, gonoducts.

27.6 Because there are only eyespots rather than image-forming eyes, the colors are probably "intended" for other animals rather than members of the same species. It does not appear likely that the colors provide camouflage, so warning coloration seems the most likely explanation.

27.7 The cross section in part B shows that most of the body is filled with parenchyme, as in flat worms and other acoelomates. Functionally, therefore, nemerteans are more like acoelomates than coelomates. The rhynchocoel is a coelom, however, so technically and perhaps phylogenetically, nemerteans are coelomates.

27.8 Having the anus anterior is a common and sensible arrangement for burrowing animals.

27.13 Presumably *Riftia* would have slowly spread generation by generation from one source of sulfur to another. Because hydrothermal vents in fact move across the ocean floor, *Riftia* could have moved with them. It has recently been proposed that the carcasses of whales could have provided an alternative source of sulfur.

27.14 The ciliated funnel leads to the seminal vesicle, and its opening is close to the testis. The logical inference is that the ciliated funnel transfers sperm from the testis to the seminal vesicle. It is there, in fact, that sperm are packaged into "sperm balls." During normal spawning, sperm break through the epidermis around the seminal vesicle and enter the gonopore of a mate. There is no reason why they cannot also enter the gonopore of the same individual. Usually, however, this would not lead to self-fertilization, because the individual is not a simultaneous hermaphrodite.

Readings

Recommended Readings

Crowe, J. H., and A. F. Cooper, Jr. 1971. Cryptobiosis. *Sci. Am.* 225(6):30–36 (Dec).

Nelson, D. R. 1975. The hundred-year hibernation of the water bear. *Nat. Hist.* 84(7):62–65.

Rensberger, B. 1980. Life in Limbo. *Science 80*:36–43 (Nov).

Richardson, J. R. 1986. Brachiopods. *Sci. Am.* 255(3):100–106 (Sept).

Tucker, J. B. 1985. Drugs from the sea spark renewed interest. *BioScience* 35:541–545.

See also relevant selections in General References at the end of Chapter 21.

Additional References

Bieri, R., and E. V. Thuesen. 1990. The strange worm *Bathybelos*. *Am. Sci.* 78:542–549.

Bone, Q., A. Pierrot-Bults, and H. Kapps (Eds.). 1991. *The Biology of Chaetognatha*. New York: Oxford University Press.

Flam, F. 1994. Chemical prospectors scour the seas for promising drugs. *Science* 266:1324–1325.

Halanych, K. M., et al. 1995. Evidence from 18S ribosomal DNA that the lophophorates are protostome animals. *Science* 267:1641–1643.

Harrison, F. W., and M. E. Rice (Eds.). 1993. *Microscopic Anatomy of Invertebrates,* Vol. 12: *Onychophora, Chilopoda, and Lesser Protostomata*. New York: Wiley–Liss. (*Includes chapters on Tardigrada, Echiura, Sipuncula, Pogonophora, and Vestimentifera.*)

Hori, H., and S. Osawa. 1987. Origin and evolution of organisms as deduced from 5S ribosomal RNA sequences. *Mol. Biol. Evol.* 4:445–472.

Kinchin, I. M. 1994. *The Biology of Tardigrades.* London: Portland.

Kojima, S., et al. 1993. Close phylogenetic relationship between Vestimentifera (tube worms) and Annelida revealed by the amino acid sequence of elongation factor-1α. *J. Mol. Evol.* 37:66–70.

Lake, J. A. 1990. Origin of the Metazoa. *Proc. Natl. Acad. Sci. USA* 87:763–766.

Rice, M. E. 1985. Sipuncula: developmental evidence for phylogenetic inference. In: S. Conway Morris, et al. (Eds.). *The Origins and Relationships of Lower Invertebrates.* New York: Oxford University Press, pp. 274–296.

Ross, J. R. P. (Ed.). 1987. *Bryozoa: Present and Past.* Bellingham, WA: Western Washington University.

Telford, M. J., and P. W. H. Holland. 1993. The phylogenetic affinities of the chaetognaths: a molecular analysis. *Mol. Biol. Evol.* 10:660–676.

Turbeville, J. M. 1991. Nemertinea. In: F. W. Harrison, and B. J. Bogitsh (Eds.), *Microscopic Anatomy of Invertebrates,* Vol. 3: *Platyhelminthes and Nemertinea.* New York: Wiley–Liss.

Turbeville, J. M., and E. E. Ruppert. 1985. Comparative ultrastructure and the evolution of nemertines. *Am. Zool.* 25:53–71.

Wada, H., and N. Satoh. 1994. Details of the evolutionary history from invertebrates to vertebrates, as deduced from the sequences of 18S rDNA. *Proc. Nat'l. Acad. Sci. USA* 91:1801–1804.

Willmer, P. 1990. *Invertebrate Relationships.* New York: Cambridge University Press.

Woollacott, R. M., and R. L. Zimmer (Eds.). 1977. *The Biology of Bryozoa.* New York: Academic.

CHAPTER

28

Molluscs

Spanish shawl nudibranch (*Flabellina iodinea*).

The Significance of Molluscs

Before the arrival of Europeans in what is now northern California, a Yurok man could get a bride from a wealthy family for 10 strings of the tusk-shaped shells of the mollusc *Dentalium pretiosum* (Fig. 28.1A). The fine for adultery was five strings of tusk shells. European aristocrats considered seashells too valuable for such transactions; they preferred to use gold.

Until the 18th century, Europeans collected shells only for their beauty, with little thought to the molluscs (also spelled mollusks) that made them. Lamarck (later to be reviled for his theory of evolution) was one of the first to realize that if dead shells are so interesting and beautiful, then the living molluscs must be even more so. During the French Revolution he convinced the government to employ in the Museum of Natural

FIGURE 28.1

A variety of molluscs. (A) The scaphopod *Dentalium vulgare.* (B) The chiton *Mopalia muscosa.* (C) The common land snail *Helix.* Note the two sets of head appendages, with eyes on the upper, longer pair. (D) The razor clam *Tagelus plebeius,* which normally stays buried in marine sediment, pumping water through its long siphons. (E) In some molluscs, such as the giant octopus *Octopus dofleini,* the shell is reduced or absent.

FIGURE 28.2

Fossils of ammonites (*Dactylioceras*) from about 200 million years ago.

History a specialist in shells and molluscs, as well as experts on other groups of animals. Thus in France and elsewhere, even before the terms "zoologist" and "scientist" had been invented, there were people whose chief occupation was collecting and studying molluscs and their shells. We would now call them malacologists and conchologists. These were perhaps the first people actually paid for studying animals.

In spite of two centuries of such studies, molluscs are still mysterious. Their body plans are so different from those of other animals that some zoologists do not consider them coelomates, and some suggest that they are not directly related to other protostomes. Evidence from molecular phylogenetics, however, supports the majority view that molluscs are closely related to annelids and other coelomate protostomes (see Fig. 21.10) (Field et al. 1988, Hori and Osawa 1987). The fossil record of molluscs goes back more than half a billion years, to the Cambrian period. By then they had already diversified into all the major groups now recognized. At least 35,000 extinct species are known from their fossilized shells. The grandest of these were the ammonoids (Fig. 28.2), with spiral shells up to 2 meters in diameter. They were quite common throughout the Paleozoic and Mesozoic eras, then became extinct rather suddenly. Possibly they were caught in the Cretaceous/Tertiary mass extinction (p. 390), or perhaps predators such as marine reptiles and crabs finally evolved the ability to crush the ammonoids' shells. Approximately 110,000 species of molluscs have survived such catastrophes, making the phylum Mollusca one of the most diverse. Another measure of molluscan success is the wide range of habitats molluscs have invaded. Most are marine organisms, but many live in fresh water and moist terrestrial habitats, and a few are parasitic. About the only thing no mollusc can do is fly.

General Organization

The great diversity of molluscs makes it difficult to discuss them as a single group. Many zoologists attempt to remedy this difficulty by describing "the hypothetical ancestral mollusc." Unfortunately, this animal seems to have existed only in textbooks, and most malacologists have abandoned it. Instead of describing such a hypothetical

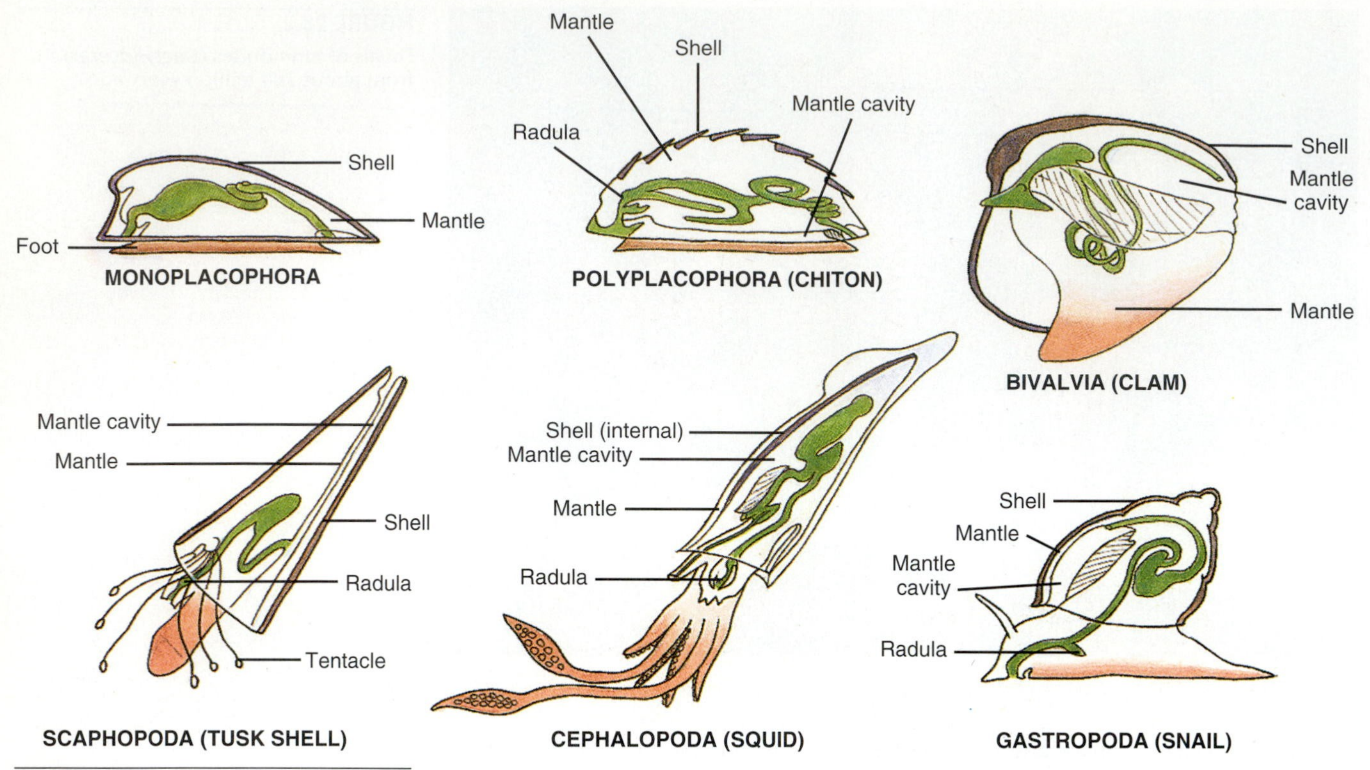

FIGURE 28.3

Representations of six classes of molluscs showing the shell, foot (red), and visceral mass. The latter is indicated by the digestive tract and includes the mantle. The position of the radula, where present, is also indicated.

■ **Question:** What are the key differences between Monoplacophora and Polyplacophora? Between Cephalopoda and Scaphopoda?

ancestor, it is just as well to describe the features shared by most molluscs and discuss the variations later.

The body of a mollusc can generally be divided into the shell and the fleshy, living part. The fleshy parts of a mollusc can be further divided into the **foot** and the **visceral mass** (= visceral hump) (Fig. 28.3). The foot is adapted in a variety of ways for locomotion. The visceral mass includes the organs for digestion, circulation, reproduction, and respiration.

The visceral mass also includes two external flaps of tissue called the **mantle** (= pallium). The mantle secretes the shell and encloses a **mantle cavity.** The mantle cavity performs many of the functions that the coelom performs in other animals. The fluid in the mantle cavity, which in aquatic molluscs is continually replaced with water from the outside, carries away excess water, ions, and wastes, and helps circulate nutrients and oxygen. One final structure that is unique to molluscs and is found in all groups except bivalves and a few others is the **radula** (RAD-jul-uh). In most forms the radula is a rasping organ near the mouth that is used in scraping up algae and other food.

Shells

The shell is absent in one class of molluscs (Aplacophora), and it is either absent or vestigial in octopus, squid, and cuttlefish (class Cephalopoda). In most molluscs, however, the shell is obvious and important. It generally supports the soft body and protects it from many predators. Little else about the shell is obvious. We still do not know the function, if any, of many of the ornate shapes and colors that make shells so attractive and valuable. Nor do we understand the developmental processes that direct molluscs in the designing of their shells. Shells generally consist of a chalky **prismatic layer,** with the outside often covered with an organic layer called the **periostracum** and the inside usually covered by a smooth **nacreous layer** (Fig. 28.4).

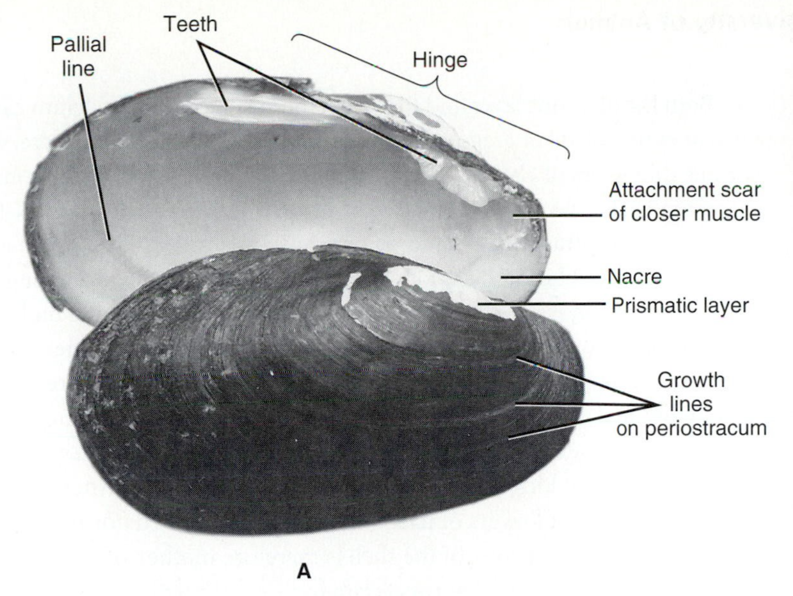

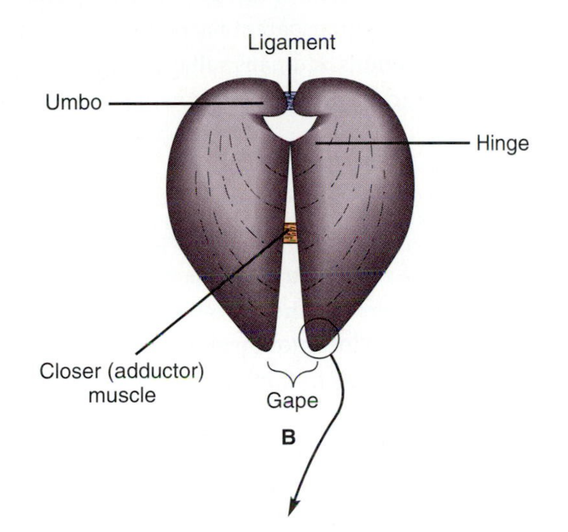

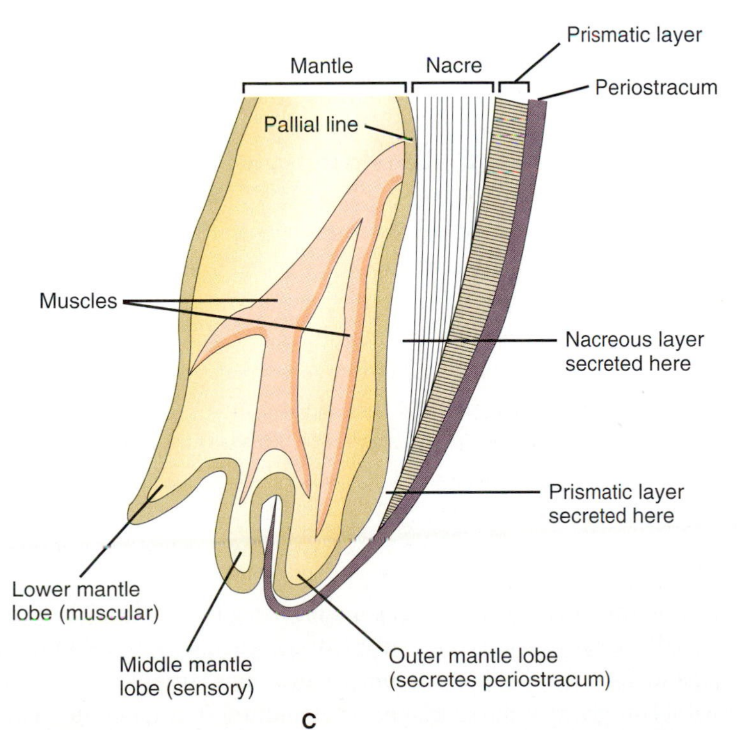

FIGURE 28.4

(A) Valves of the freshwater mussel *Margaritifera margaritifera* showing the internal nacre (top) and outer periostracum. Part of the periostracum has eroded from the oldest part of the shell (the umbo), exposing the chalky prismatic layer. Attachment sites (scars) for the muscles that open and close the shells are visible on the inner surface. The teeth prevent the shells from slipping sideways. The annual growth rings indicate that this 8-cm specimen was approximately 6 years old. (B) Schematic representation of the mechanism that opens and closes the shell. The elastic ligament at the umbo opens the shell. The adductor muscles are a type known as catch muscles (see p. 210) and are among the most powerful known. (C) Cross section of the edge of the shell and mantle. The mantle attaches to the inner surface of the shell at the pallial line (visible in part A), forming a chamber where new nacre is secreted.

■ **Question:** What kinds of receptors do you suppose the sensory lobe bears, and what behaviors would they evoke?

Both the nacreous layer and the prismatic layer consist of calcium carbonate in the form of either calcite or aragonite. The nacreous layer, also called **nacre,** consists of numerous thin layers of $CaCO_3$. The prismatic layer of the shell usually consists of prism-shaped crystals held within a matrix of protein called **conchin** (KON-kin; = conchiolin). The periostracum is made of conchin and protects the prismatic layer from abrasion and dissolution by acids. Protection from naturally occurring acids is especially important in freshwater and terrestrial species. Some marine molluscs secrete little or no periostracum. The rate of shell growth varies with water temperature and chemistry, resulting in annual growth rings and smaller rings that are thought to record lunar, daily, and tidal cycles. The shell may also shrink (Downing and Downing 1993).

Many molluscs, such as oysters, secrete nacre around particles of foreign tissue or debris that become lodged between the mantle and the shell. After several years, hundreds or thousands of layers of nacre will have accumulated, forming a **pearl.** Another name for the nacreous layer of the shell is therefore mother-of-pearl. Mother-of-pearl is often iridescent, because light at certain wavelengths (colors) is canceled by its reflection through the partially transparent nacreous layers, while other wavelengths are reinforced by their reflections. Humans value this iridescence in mother-of-pearl, but, of course, it is invisible and therefore useless in the living mollusc.

The Foot

As the name implies, the foot is usually responsible for locomotion. In many species the foot secretes a layer of mucus, like the familiar slime trails left by snails and slugs. The mollusc glides upon this mucus track by waves of ciliary movement or muscle contraction. This form of locomotion also occurs in many flatworms (see Chapter 25), which leads some zoologists to think flatworms were the direct ancestors of molluscs. Some freshwater snails can also move upside down beneath the water surface (see Fig. 2.3A). In bivalves (clams, oysters, etc.) the foot is used for locomotion in another way, by rooting into sand or mud and pulling the animal along. In squid, octopus, and other cephalopods, the head and foot combine into a tentacled mass with a funnel for jet propulsion.

The Radula

In most molluscs the radula bears teeth on a membrane that is bent around a cartilaginous support (the **odontophore**) (Fig. 28.5). Ordinarily the radula is kept within a **radula sac** beneath the mouth. By contracting certain muscles to the odontophore and radula membrane, the radula can be extended and made to grind off food particles, much as a belt sander grinds off bits of wood. The radula also serves as a conveyor belt, carrying food particles into the mouth. The teeth are made of **chitin,** the same substance in the cuticle of insects, and number from a few to hundreds of thousands, depending on the species. In some species the radula is used to scrape algae off rocks or to bore through the shells of other molluscs. This is rather hard on the teeth, but they can be replaced.

Respiration

In most molluscs the respiratory organs are **gills** (Table 28.1 on page 586). As discussed in Chapter 12, gills are adapted to exchange O_2 and CO_2 in water by having a large surface area and thin membranes. In molluscs the gill also has unique adaptations that justify giving it the special name **ctenidium** (ten-ID-ee-um; Greek *kteis* a comb).

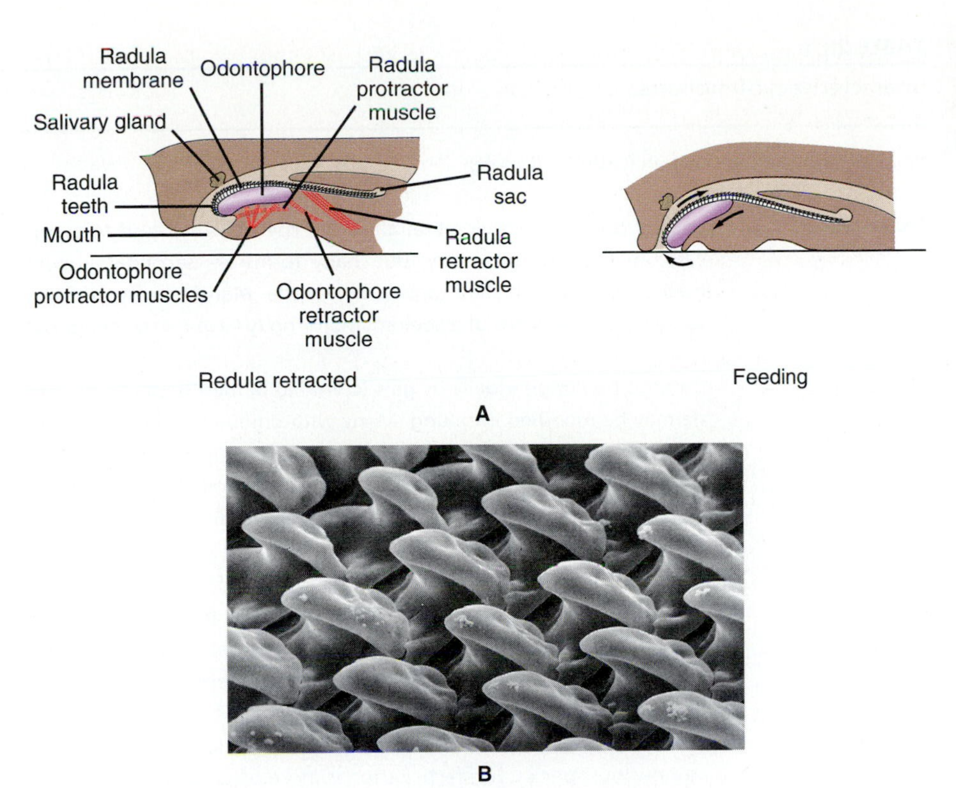

FIGURE 28.5

(A) Side view of a generalized radula. When the mollusc is not feeding, the radula is retracted by the odontophore retractor muscles. During feeding, the odontophore protractor muscles move the odontophore forward, pressing the radula against substratum. The radula protractor and retractor muscles then slide the radula back and forth.
(B) Scanning electron micrograph of the radula teeth of the snail *Cupedora*. Each tooth is approximately 35 μm long.

Ctenidia consist of sets of **filaments** (= lamellae) covered with cilia (Fig. 28.6). As the cilia propel water across the surface of the filaments, oxygen diffuses across the membrane into the blood, and carbon dioxide diffuses out. In some molluscs, such as clams and other bivalves, the cilia also sort out particulate matter, sending food particles on a string of mucus to the mouth. After passing the gills, the water stream usually goes past the anus and the outlets of the kidneys, carrying off wastes. In many molluscs, water enters and leaves through **incurrent and excurrent siphons.** Before it reaches the gills, the incoming stream of water is sampled by a sense organ, the **osphradium,** which may detect silt, food, or predators.

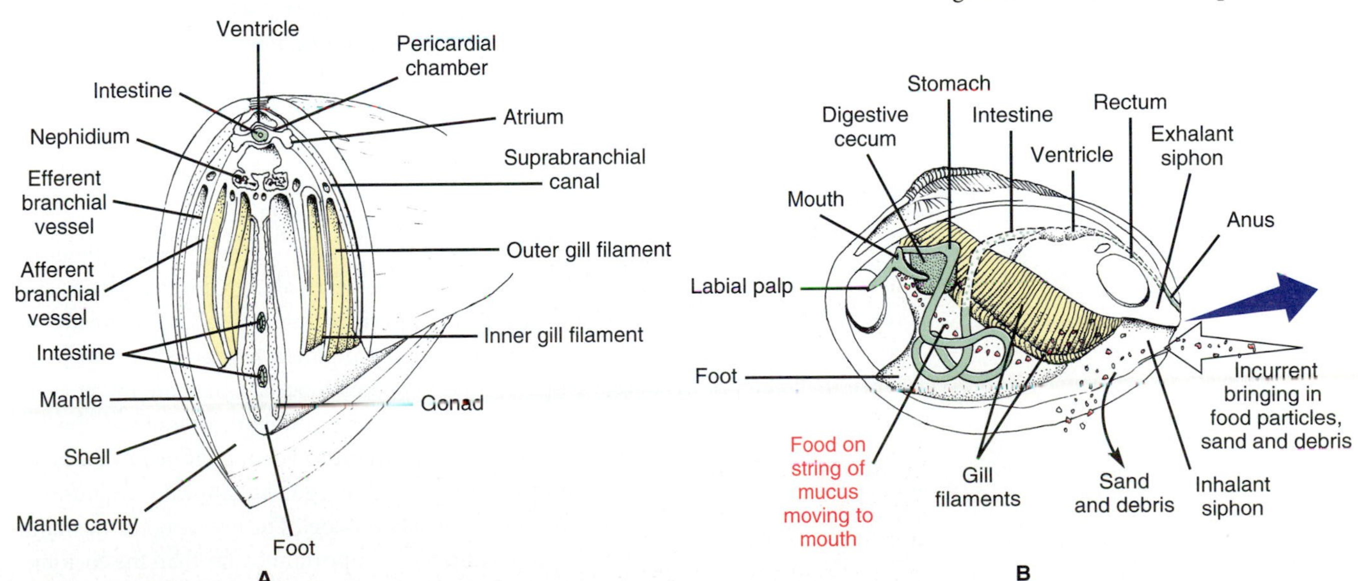

FIGURE 28.6

Function of the gills (ctenidia) of a typical bivalve. (A) Cross section showing parts involved in respiration. Cilia on the gills draw water into the filaments through pores, then through suprabranchial canals, and out through the exhalant siphon. O_2 and CO_2 are exchanged with blood circulating through the filaments. (B) Longitudinal section showing parts involved in feeding in *Mercenaria.* As water passes through the pores in the filaments, particles of food, sand, and debris are filtered out. Food collects on a mucous string, which is then propelled by cilia toward the mouth, and from there to the stomach and intestine within the foot. (See Fig. 28.21.) Sand and debris settle out.

■ **Question:** Why would bivalves be good indicators of water pollution?

TABLE 28.1

Characteristics of molluscs.

Phylum Mollusca ma-LUS-kuh (Latin *mollis* soft)

Morphology	Usually considered coelomate, although coelom is reduced to a pericardium. Bilateral symmetry, but many forms are secondarily asymmetric. Usually with **shell** formed by **mantle. Mantle cavity** performs some of the functions of a coelom. Major body cavity is the **hemocoel.**
Physiology	Gaseous exchange usually by **gills** (ctenidia) in mantle cavity, or mantle may be modified into lung. Many with unique **radula** for scraping up food particles. Digestive tract complete, but much digestion intracellular. Nervous system with several pairs of ganglia linked by nerves. Circulatory system open or closed; with heart. Osmoregulation and excretion by metanephridia.
Locomotion	Muscular **foot**, which may be highly modified for functions other than locomotion. **Cilia** and **mucus** often important in locomotion and other organ functions. Many forms sessile or sedentary.
Reproduction	Mostly dioecious, but some monoecious. Copulation common even in monoecious species, but fertilization is more often external.
Development	Spiral cleavage pattern, except bilateral in cephalopods. Protostomate. Development usually indirect, with **trochophore larva**, and a veliger larva in some species.
Habitat, Size, and Diversity	Primarily marine, but many freshwater and terrestrial species, and a few parasitic. Adult body length from less than 1 mm to more than 20 meters (giant squid *Architeuthis*, including tentacles). Approximately **110,000 living species** described.

Some molluscs do not have gills, but exchange respiratory gases directly across the mantle surface. Many terrestrial snails have no gills, but have part of the mantle modified into a **lung** for air-breathing. Some of these **pulmonate** snails have reinvaded aquatic habitats, but still retain the lung. Often they can be seen in ponds rising to the surface for air.

Circulation

The circulatory system consists of a heart and blood vessels. In most species the heart has three chambers: two atria and one ventricle. The heart, along with the gonads and kidneys, is enclosed within a chamber called the **pericardium,** which is the greatly reduced coelom. In most forms the copper-containing pigment **hemocyanin** transports oxygen in the blood. In squid, octopus, and other cephalopods, which are much more active than other molluscs, the circulatory system is closed, but in most molluscs the circulatory system is open. That is, the blood (= **hemolymph**) is not confined to the heart and vessels, but percolates under low pressure through irregular channels and sinuses in the tissues. These channels and sinuses make up a **hemocoel.** The hemocoel is the major body cavity in these molluscs and is larger and more important by far than the coelom.

Nervous System

The central nervous system of molluscs typically consists of a ring of ganglia (see Fig. 9.3). A pair of **pedal ganglia** usually control the foot, **cerebral ganglia** integrate sensory information, and other ganglia control the functions of other parts of the body. The nervous system of many molluscs also secretes hormones that regulate such functions as egg laying and growth (see Table 7.2). Most molluscs move at the proverbial snail's pace, and they are generally regarded as rather dull, but this may be because few zoologists or psychologists have had the patience to test their intellectual acuity. Some molluscs, especially octopus, squid, and other cephalopods, are active predators that can learn to recognize prey with their sharp eyes and sensitive touch receptors. An octopus can even learn a task by watching another octopus perform the task (Fiorito and Scotto 1992).

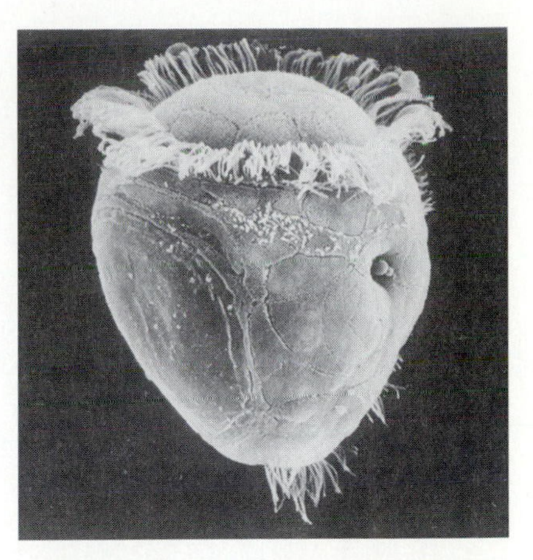

A

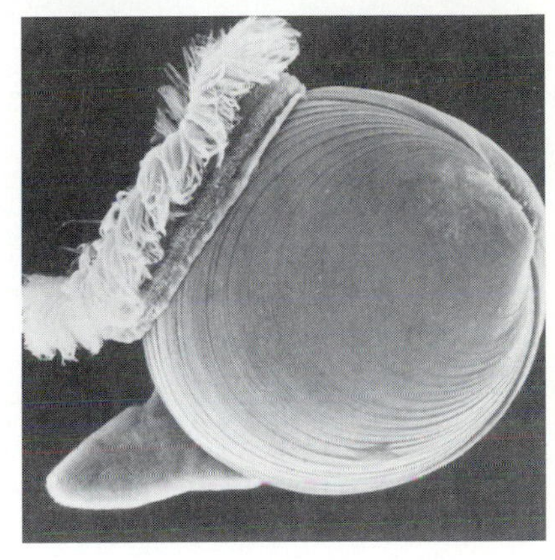

B

FIGURE 28.7

(A) Scanning electron micrograph of the trochophore larva of the shipworm *Teredo navalis.* Trochophores are defined by the circle of cilia (prototroch). See Fig. 29.5A for anatomical details of a trochophore, and see Fig. 28.20A for an adult *Teredo.* Diameter approximately 25 μm. (B) SEM of the veliger larva of a different shipworm, *Lyrodus pedicellatus.* The foot (lower left) and valves are evident. Locomotion is due to the ring of cilia. Diameter of the valves approximately 175 μm.

Osmoregulation and Excretion

Molluscs have one or more pairs of **nephridia** that remove excess water, ions, and metabolic wastes from the coelomic fluid and transport them into the mantle cavity for excretion. The mollusc nephridium is of a type called a **metanephridium,** since its tubule has both an external pore (**nephridiopore**) and an internal opening (**nephrostome**). The nephridia of molluscs are often called kidneys, although they function quite differently from the kidneys of most vertebrates, since they filter coelomic fluid rather than blood. Also unlike the kidneys of vertebrates, the nephridia of many molluscs serve as gonoducts that transport gametes from the gonads into the mantle cavity.

Reproduction and Development

Most molluscs are either one sex or the other, although one would never know it by looking at them. Some are hermaphroditic. **Development** is direct in cephalopods, many freshwater snails, and some bivalves. In many marine molluscs there is a **trochophore** larva (Fig. 28.7A), as in some flatworms, annelids (see Fig. 29.5), and other protostomes. Trochophores are characterized by a ring of cilia that helps compensate for the adult mollusc's limited ability to travel to new habitats. In gastropods and bivalves, and in no other phylum, another larval form, the **veliger,** occurs after the trochophore (Fig. 28.7B). The veliger also has cilia for motility and, unlike the trochophore, also has the beginnings of the shell and most of the adult organs.

Classification

Most taxonomists now recognize seven or eight classes of molluscs, based mainly on differences in the foot and shell. Usually these differences are quite apparent, making it easy to identify on sight the class to which a mollusc belongs. The first three or four classes to be described are interesting but minor in terms of the number of species.

APLACOPHORANS. Fewer than 300 species of wormlike, shell-less molluscs constitute **class Aplacophora** (Greek *a* without + *plax* plate). Most aplacophorans are about 2.5 cm long and have calcareous scales and spicules instead of shells. Most creep on deep ocean floors and are often found entwined in colonial hydroids or soft corals (phylum Cnidaria) on which they feed. The foot is modified into a groove up the ventral midline (Fig. 28.8). Many taxonomists recognize as a separate class (Caudofoveata) about 70 species that spend most of their lives burrowed vertically in the sea floor, feeding on microbes and detritus with the posterior end uppermost.

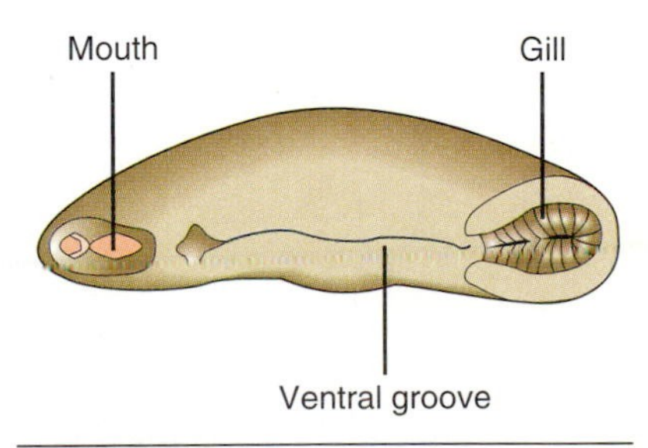

FIGURE 28.8

An aplacophoran, *Neomenia carinata.*

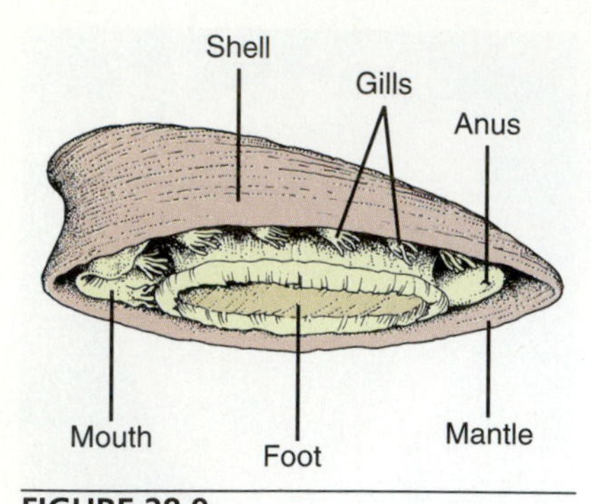

FIGURE 28.9

A monoplacophoran, *Neopilina*. The longest dimension is up to 3.7 cm.

MONOPLACOPHORANS. Class Monoplacophora was known from fossils and had been considered extinct until the 1950s. Then while examining some preserved specimens that had been collected deep off the west coast of Mexico, Henning Lemche of Copenhagen came across one labeled as a limpet. Turning it over, he saw that the animal had five pairs of gills and was therefore not only not a limpet, but also not a member of any other class of molluscs thought to be extant. Thus the first recent monoplacophoran, *Neopilina galatheae,* was discovered. Since then about a dozen living species of Monoplacophora have been collected from around the world at depths of 2 to 7 km. As the name implies, monoplacophorans have a one-piece shell (Fig. 28.9). The stomach contents suggest that *Neopilina* feeds mainly on radiolarian protozoans. One of the most interesting features of Monoplacophora is the serial repetition of organs, with five or six pairs of gills, six pairs of nephridia, eight pairs of pedal-retractor muscles, and so on. Some zoologists consider this to be evidence that molluscs are related to segmented animals such as annelids.

CHITONS. Members of **class Polyplacophora** are commonly referred to as **chitons,** because their dorsal shells of eight overlapping plates resemble Greek tunics with that name (Fig. 28.1B). No other mollusc has such a shell. The shells of chitons are also different in other ways. Chiton plates have only two layers: (1) an outer **tegmentum** made of conchin and calcium carbonate and (2) an inner **articulamentum** that is entirely calcareous. The plates are embedded in a part of the mantle called the **girdle.** Often the girdle is defended with spines, scales, and bristles. The plates also bear many smaller sensory projections called **esthetes.** An esthete may have a simple tactile or visual receptor or, in some species, a well-developed eye, complete with a lens. Esthetes allow the chiton to detect predators and wave action even though its head is shielded under the shell. The mantle cavity is a groove between the foot and the edge of the mantle. There are from 6 to 88 pairs of gills (Fig. 28.10).

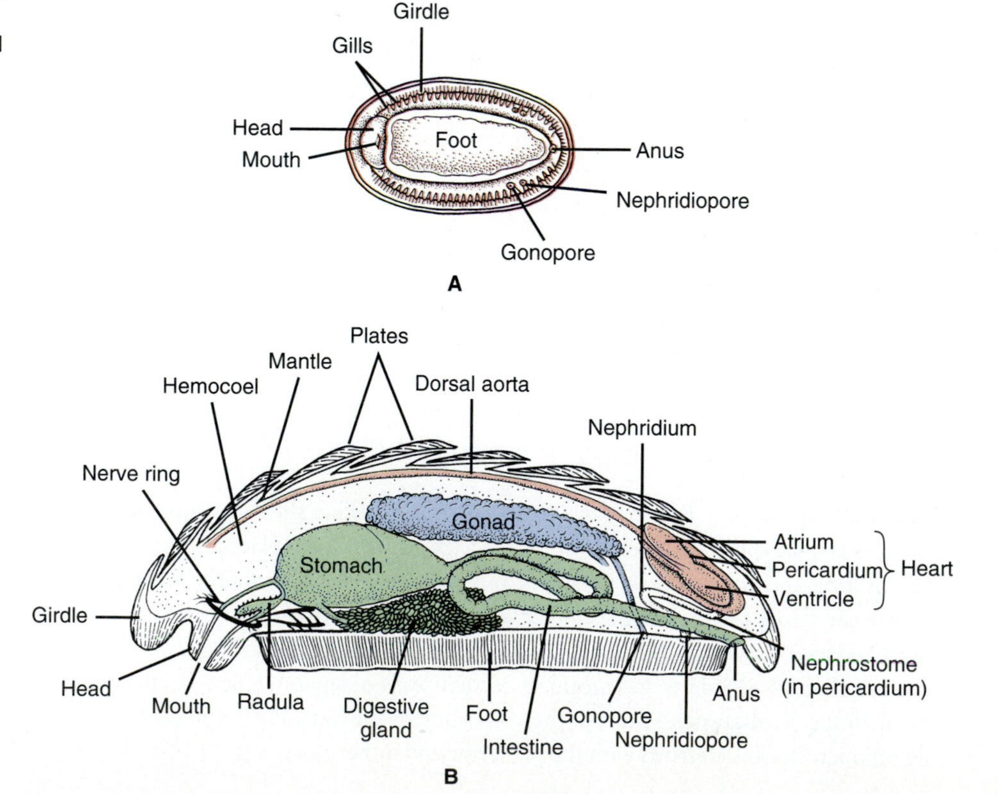

FIGURE 28.10

(A) Ventral view of a chiton. (B) Internal structure of a chiton.

Major Extant Groups in Phylum Mollusca

Genera mentioned elsewhere in this chapter are noted.

Class Aplacophora AYE-plak-OFF-for-uh (Greek *a* without + *plakos* plate + *phoros* bearing). Vermiform, up to 30 cm long. No shell but with calcareous scales and spicules. *Neomenia* (Fig. 28.8).

Class Monoplacophora MON-oh-plak-OF-for-uh (Greek *monos* single). With single cap- or cone-shaped dorsal shell. Nephridia, gills, and other organs serially repeated. Benthic. Dioecious; fertilization internal. *Neopilina* (Fig. 28.9).

Class Polyplacophora POL-ee-plak-OF-for-uh (Greek *poly* many). Shell of eight overlapping plates. Flattened. Head reduced; foot broad and flat. Benthic. Dioecious. Chitons. *Mopalia* (Fig. 28.1B).

Class Scaphopoda skaf-FOP-oh-duh (Greek *skaphe* trough + *podos* foot). Tooth- or tusk-shaped shell open at both ends. Burrowing in marine sediment, head down. With tentacles. No gills; perhaps no circulatory system, except for blood sinuses. Dioecious; fertilization external. Tusk shells. *Dentalium* (Fig. 28.1A).

Class Gastropoda gas-TROP-oh-duh (Greek *gaster* belly). Shell of one piece, often spirally coiled; absent in slugs. Body undergoes torsion during development. Secondarily asymmetric. Both trochophore and veliger larvae may occur, or development may be direct.

Subclass Prosobranchia pro-so-BRANK-ee-uh (Greek *pros* forward + *branchia* gills). Gills anterior to heart as result of torsion. One pair of tentacles. Mainly dioecious. Limpets, abalone, periwinkles, conchs, cowries, slipper shells, whelks, cone shells. *Busycon, Conus, Cypraea, Diodora, Haliotis, Lambis, Littorina, Murex, Thais, Truncularioposis* (Fig. 28.12).

Subclass Opisthobranchia oh-PISTH-oh-BRANK-ee-uh (Greek *opisthe* behind). Gill, if present, posterior, as result of detorsion. One to four pairs of tentacles. Shell, gill, and/or mantle cavity reduced or absent. Marine. Monoecious. Bubble shells, sea hares, pteropods (sea butterflies), sea slugs (order Nudibranchia). *Aplysia, Coryphella, Flabellina, Glaucus, Phidiana* (Chapter opener; Figs. 20.4, 28.13).

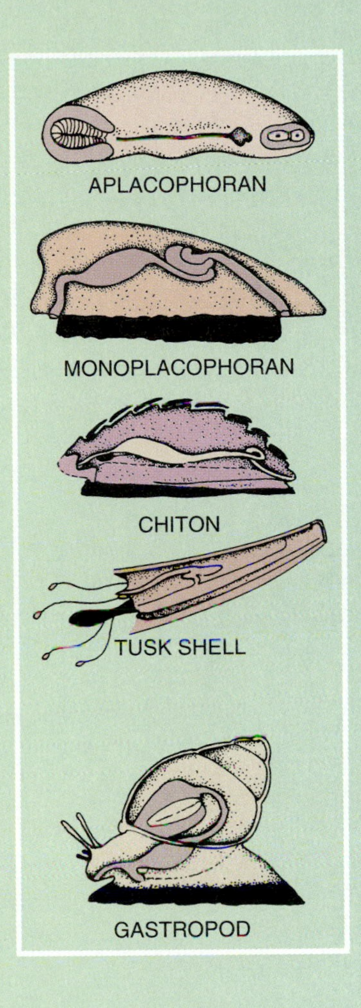

box continues

Subclass Pulmonata pull-mon-ATE-uh (Latin *pulmo* lung). Mantle cavity modified into lung. Degree of detorsion variable. Shell, if present, a spiral. Monoecious. Development direct. Terrestrial, freshwater, and a few marine snails and slugs. *Ariolimax, Cupedora, Helix, Limax* (Figs. 2.3A, 28.1C).

Class Bivalvia (= Pelecypoda) bi-VALVE-ee-uh (Latin *bi* two + *valva* folding door). Shell of two lateral valves. Foot usually wedge-shaped. No radula. Gill often large and used for suspension feeding. Usually dioecious, with external fertilization. Often with both trochophore and veliger larvae. Benthic, in either fresh or salt water. Clams, mussels, oysters, scallops, cockles, shipworms. *Bankia, Chlamys, Crassostrea, Dreissena, Lyrodus, Margaritifera, Meleagrina, Mercenaria, Mytilus, Ostrea, Pecten, Tagelus, Teredo, Tridacna* (Figs. 28.1D, 28.20, 28.22).

Class Cephalopoda SEF-a-LOP-oh-duh (Greek *kephale* head). Head well developed, with prominent arms or tentacles, eyes, and jaws. Funnel for jet propulsion. Dioecious. Development direct. All marine.

Subclass Nautiloidea naw-til-OY-dee-uh (Greek *nautilos* sailor). External coiled shell, divided into chambers. Many (80 to 90) tentacles. One genus, *Nautilus* (Fig. 28.24).

Subclass Coleoidea col-ee-OY-dee-uh (Greek *koleon* sheath). Shell reduced, internal, or absent. Eight or ten arms with suckers. Squid, cuttlefish, octopus. *Loligo, Octopus, Sepia, Todarodes* (Chapter 8 opener; Figs. 28.1E, 28.26).

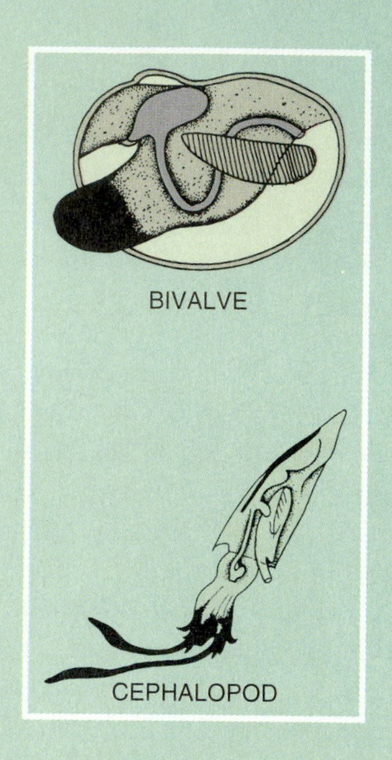

The 800 or so species of chitons are fairly common in the intertidal zone and in deeper waters, and some species are up to 40 cm long. Often they can be found in shallow water, creeping along on the broad ventral foot and scraping up algae and invertebrates with the radula. Getting a look at the foot and the radula can be difficult, however. When disturbed, the foot and mantle edge create suction against the substratum, making it hard for waves and predators, as well as curious zoologists, to dislodge the animal.

TUSK SHELLS. Tusk shells, such as the species mentioned at the start of this chapter, belong to **class Scaphopoda** (Fig. 28.1A). They are also called tooth shells. The shell is open at both ends. Scaphopods spend most of their lives with the posterior end sticking up from the sand or mud in water as deep as 6 km. Cilia in the mantle cavity draw water in through the buried anterior end, and sudden contractions of the foot periodically squirt the water out the posterior opening of the shell (Fig. 28.11). Cilia and mucus on the foot and tentacles (= **captacula**) trap protozoa and detritus and transfer them to the mouth and radula. There is no gill; oxygen from the water diffuses directly into the mantle. The coelom, heart, and blood vessels are either greatly reduced or nonexistent.

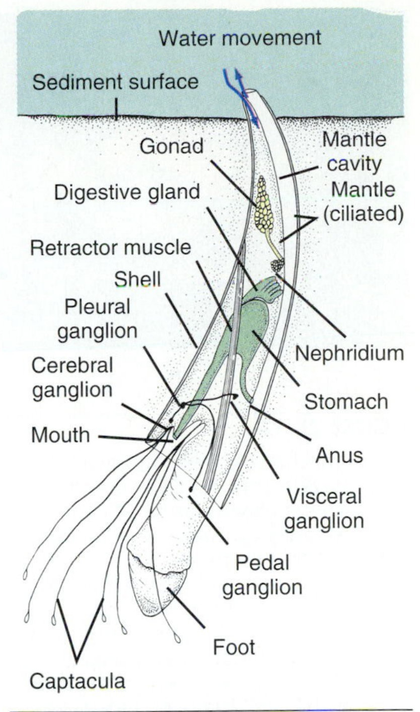

FIGURE 28.11
Structure of the scaphopod *Dentalium*.

MAJOR GROUPS: GASTROPODS, BIVALVES, AND CEPHALOPODS. The next three classes—Gastropoda, Bivalvia, and Cephalopoda—comprise the majority of molluscs (Fig. 28.1C, D, and E). Each of these classes will be discussed later in a separate section. **Class Gastropoda** includes forms with a one-piece shell, such as snails, limpets, conchs, and whelks, as well as slugs and other relatives, in which the shell is greatly reduced. **Class Bivalvia** (= Pelecypoda) includes oysters, mussels, clams, and other molluscs in which the shell consists of two **valves**. **Class Cephalopoda** includes squids, octopuses, nautiluses, and other molluscs with tentacles.

Gastropods

More than one-third of all mollusc species are gastropods. They are so diverse that it is convenient to divide the class into three subclasses. Members of subclass Prosobranchia produce many of the shells familiar on beaches—the periwinkles, conchs, cowries, whelks, and cone shells. Another prosobranch is the abalone (AB-a-LOW-nee), which, if you live near the Pacific, may be familiar on your dinner plate (Fig. 28.12 on page 592). The subclass Opisthobranchia includes sea hares such as *Aplysia* and *Phidiana* (= *Hermissenda*), which are often used in studies on the mechanisms of learning (Fig. 28.13 on page 592). Subclass Pulmonata includes the air-breathing snails and slugs, familiar scourges of the garden (see Figs. 28.1C and 28.14 on page 593).

Figure 28.15, on page 593, is a schematic representation of the internal structure of gastropods. The dominant features are the coiled shell and the foot. The shell encloses the visceral mass and continues to enlarge at the anterior edge as the animal grows. When threatened, many gastropods retreat into the mantle cavity and seal the entrance with a tough **operculum** on the back of the foot. This is a bit of a bother, because the leading edge of the mantle has to be repositioned precisely in order to resume shell growth.

TORSION. One of the peculiarities of gastropod organization evident in Fig. 28.15 is the way the digestive tract doubles back on itself, with the mouth and anus at the same end. This is due to torsion, which is one of the few features found in all living gastropods and in no other group. Torsion begins in the veliger larva (except in snails and others that do not have distinct larval stages) as the rear of the visceral mass makes two 90° twists to the right, bringing the mantle cavity behind and above the head (Fig. 28.16 on page 594). Depending on the species, torsion can occur within a few minutes or over a much longer period. The immediate cause of torsion is a muscle on the right side that is stronger than the corresponding muscle on the left. The evolutionary cause is not as well understood. Numerous advantages of torsion have been proposed, but each has weaknesses. One obvious advantage is that torsion enables the gastropod to protect its delicate head and foot by pulling them into the mantle cavity. The same advantage could be provided, however, simply by making the shell a little larger.

Having the anus hanging over the gills is a definite hygienic concern. Many gastropods avoid fouling their gills by tying up the feces in strings of mucus. Others with

A

B

C

D

E

FIGURE 28.12

Prosobranchia. (A) Yellow periwinkle *Littorina obtusata.* (B) Overturned shell of the spider conch *Lambis.* (C) The Chinese cowry *Cypraea chinensis.* The pink mantle is partly separated, exposing the shell. (D) Shell of the lightning whelk *Busycon contrarium,* along with egg cases. The elongated portion of the shell is the siphonal canal, which encloses the siphon. Whelks use the sharp edge of the shell to pry open bivalves. This species is unusual among gastropods in that it is left-handed (sinistral). That is, the shell coils counterclockwise away from the apex, which is at the opposite end from the siphonal canal. (E) The abalone *Haliotis* showing the series of holes for the escape of respiratory gases. As the abalone grows, it closes off older holes and develops new ones.

■ **Question:** What are the sources of the colors in the molluscs shown in parts A, C, and E? What are the functions of the colors, if any?

coiled shells avoid fouling by having just one gill and eliminating feces toward the side without a gill. Gastropods with two gills, such as the abalone *Haliotis* and the keyhole limpet *Diodora,* void the feces through one or more holes in the shell. Gastropods in subclass Opisthobranchia avoid the danger of fouling by **detorsion,** in which the gut is straightened out to some degree. The sea hare *Aplysia* shows the largest degree of detorsion, with the anus almost completely at the rear of the animal.

COILING. In addition to torsion, many gastropods also undergo the separate process of coiling in the development of the visceral mass and shell. All living gastropods, including those without shells, show signs of having descended from species with coiled shells. Coiling is one way that a shell can grow along with the animal with-

FIGURE 28.13

The sea slug *Phidiana,* a popular subject for the study of the neural mechanisms of learning.

FIGURE 28.14

The banana slug *Ariolimax columbianus* of the American Northwest, showing the large respiratory pore (pneumostome). This bright yellow mollusc is the mascot of the University of California at Santa Cruz.

out becoming so long that it impedes locomotion. (The other way is to have roughly circular shells, as in bivalves.) Without coiling, gastropods would have shells like scaphopods and would probably have to be sedentary like them.

The simplest form of coiling is one in which all whorls are in the same plane (planospiral). In other words, the shell will lie flat on either side. Such shells are satisfactory for some aquatic snails, but they are too bulky for terrestrial gastropods. A

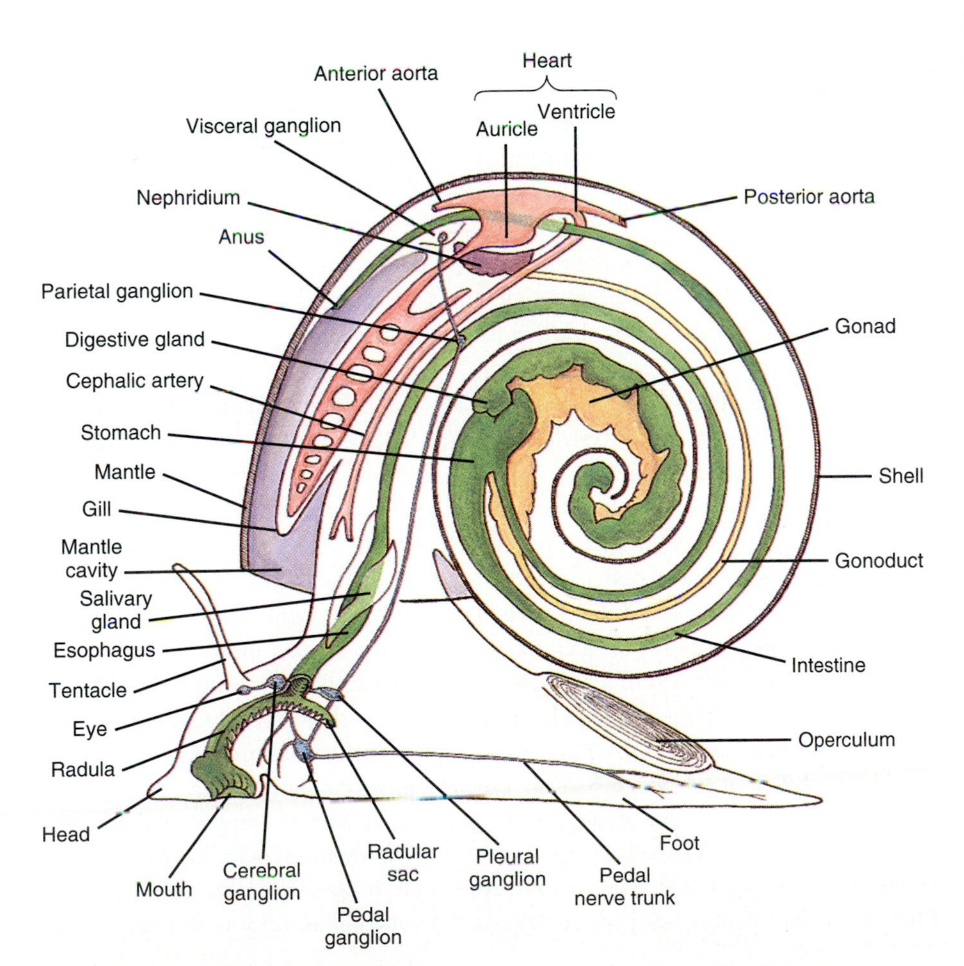

FIGURE 28.15

Internal structure of a generalized gastropod.

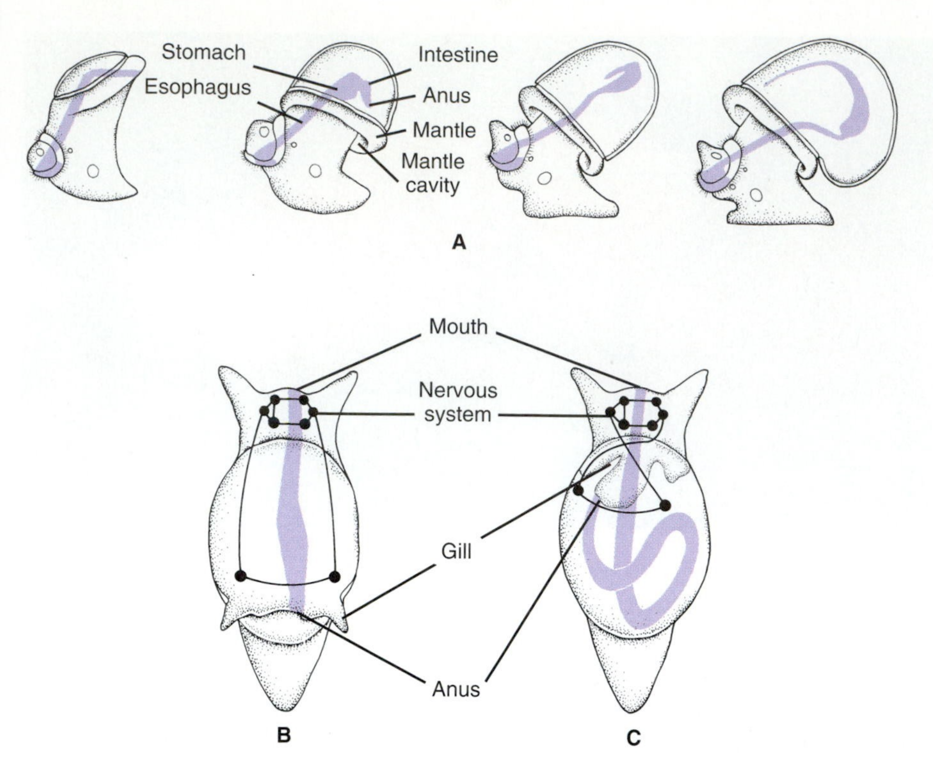

FIGURE 28.16

(A) Torsion in the veliger larva of a gastropod. The digestive tract is shown in color to indicate movement of the visceral mass. (B) An adult gastropod as it would appear without torsion. The nervous system and digestive tract are indicated. (C) A torted adult. The result of torsion is that the mantle cavity enclosing the gills and anus lies above and behind the head, and the nervous system is twisted into a figure eight.

■ **Question:** After undergoing torsion, some gastropods undergo detorsion and end up looking like B. Which gastropods are they? What does this indicate about their evolution?

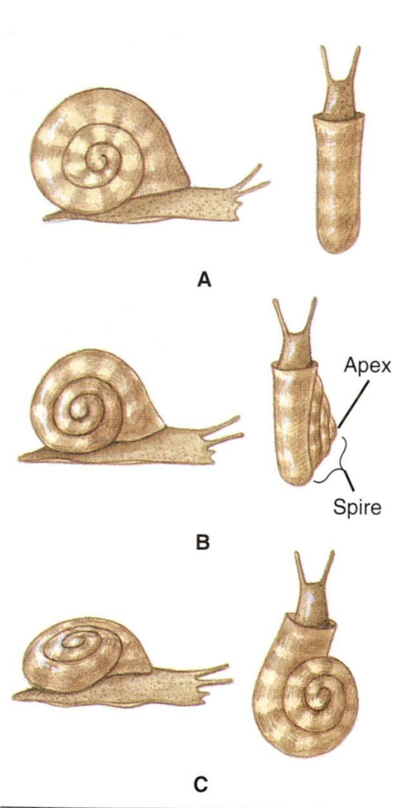

FIGURE 28.17

Side and top views of coiled gastropod shells. (A) A planospiral coil is appropriate for a buoyant mollusc, but its high profile would be easily upset by gravity or water currents. (B) A more compact (conispiral) shell results from offsetting each new turn in the coil to one side. The weight is now unbalanced, however. (C) Balance is restored by turning and tilting the shell so that its weight is over the foot.

■ **Question:** What type of symmetry does each of the three animals have?

more compact form that is less likely to be tipped over by gravity results when each whorl develops a little to one side of the preceding whorl (conispiral). This pattern leaves the older and smaller whorls—those forming the **apex** and **spire**—sticking out to one side. (Whether the apex is on the right or left is determined early in development by the same gene that determines the direction of the spiral movement of cells during cleavage.) This arrangement leaves the shell off balance (Fig. 28.17B). This problem is solved by a shell that is partially turned to one side, with the spire tilted over the back (Fig. 28.17C). The evolution of this adaptation must have presented an additional problem: The weight of the shell would have rested on one of the delicate gills, nephridia, and auricles of the heart. In modern gastropods this final problem is solved by omitting those organs on the side that is beneath the shell.

Defense and Predation. The shell is a constant refuge against predation, drying, and other threats, but it restricts habitat to places where calcium is available. Snails, for example, are plentiful only in fresh water or soils near limestone, which consists of fossilized shells (including perhaps those of the snails' ancestors). Where calcium is lacking, one often finds slugs but few snails. Slugs make up for the lack of a protective shell by being active only when it is damp and by secreting mucus that is sticky and noxious. (One zoologist has reported that slug mucus numbs the tongue.)

Although there is plenty of calcium in the oceans, sea slugs and sea hares also lack shells. Like land slugs, most of these opisthobranchs also defend themselves with toxic mucus, which many of them obtain from sponges and other prey. Opisthobranchs often advertise their toxicity with showy **warning coloration** that probably deters many would-be predators. John Steinbeck in *Cannery Row* refers to "orange and speckled nudibranchs [that] slide gracefully over the rocks, their skirts waving like the dresses of Spanish dancers." Other shell-less gastropods protect themselves in more novel ways. *Glaucus* shoots predators with nematocysts that it gets from its own prey, the Portuguese man-of-war (see p. 497). *Aplysia* distracts predators by secreting a cloud of

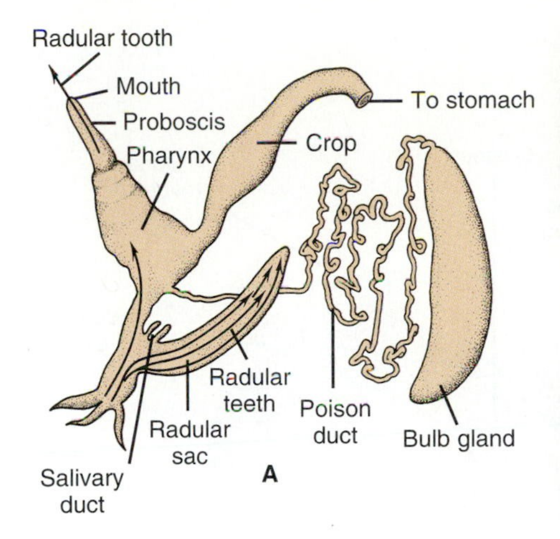

B

FIGURE 28.18

(A) The feeding mechanism of the marine snail *Conus.* The radula is highly modified into a poison "harpoon" gun. (B) Aggressive mimicry by *Conus purpurascens* of the eastern Pacific. The sequence begins when the cone shell senses a passing fish and sticks out its wormlike proboscis. The fish nibbles at the lure, is stung in the mouth, and is ingested.

ink when disturbed. Most gastropods graze and browse placidly, but a few are as innovative in predation as they are in defense. Cone shells (*Conus*) have modified radula teeth that they use like harpoons to inject neurotoxins into passing prey (Barinaga 1990). (The toxin can also be lethal to a human who steps on certain species of *Conus.*) *Conus purpurascens* enhances its hunting abilities by **aggressive mimicry,** using its red, worm-shaped proboscis to lure prey (Fig. 28.18).

REPRODUCTION. Reproduction is another area in which gastropods reveal their inventiveness. Some slugs will court for several days before copulating. A pair of slugs *Limax maximus* (Fig. 28.19 on page 596), which may be more than 10 cm long, sometimes display their commitment to each other by dangling from a tree limb by a 40-cm strand of mucus while they exchange sperm packets. One is tempted to applaud such exuberance, but it may be only a way to avoid predation during this vulnerable time. No less strange is the courtship of many snails, such as the common *Helix.* A pair of these snails will display their mutual interest by stabbing each other in the foot with a **dart.** The calcareous or chitinous dart, sometimes called a "love arrow," is a centimeter long in some snails. Fortunately, unlike Cupid's arrows, these darts are not aimed for the heart. In the 18th century the French philosopher-scientist Maupertuis speculated that the function of the dart was to sexually excite these lethargic animals. However, recent studies suggest that the dart actually injects a pheromone (Adams and Chase 1990).

LOCOMOTION. Some gastropods, such as sea hares, swim by undulatory movement of the body, and a few, such as conchs, drag themselves along with a proboscis. In the majority of gastropods, however, the foot is responsible for locomotion. Snails, for example, move by ripplelike contractions of the foot on a mucus trail. This is an inefficient form of locomotion, since the speed is limited by the rate of mucus secretion, and the secretion alone uses up to one-fourth of the gastropod's total energy budget. The efficiency is not as low as it might be, however, because of a peculiar property of the mucus. It sets up like glue beneath stationary portions of the foot, but turns liquid un-

A

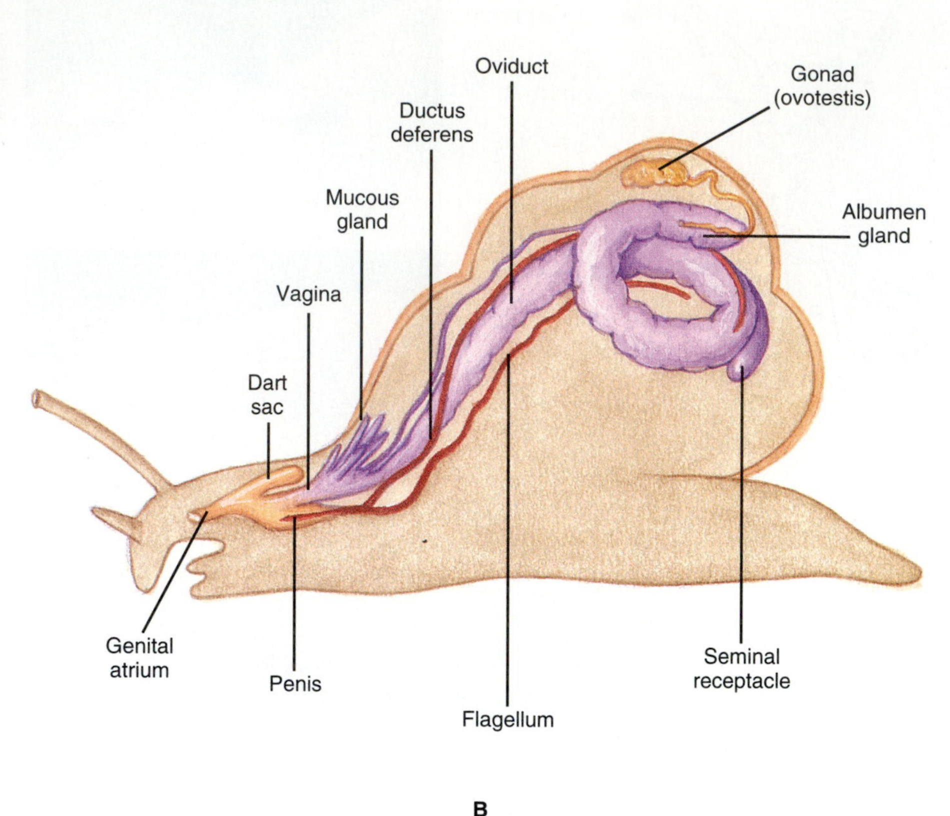

B

FIGURE 28.19

(A) Two hermaphroditic slugs *Limax maximus* mating by entwining their reproductive organs, which emerge from the genital atrium on the right side of the head. They contain both male and female parts and exchange sperm.
(B) Reproductive system of a hermaphroditic pulmonate snail. Organs with a primarily male function are shown in red; those that are primarily female are in blue; those with shared functions are in yellow. During copulation the penis of each hermaphroditic snail enters the genital atrium in the head of the other and inserts a sperm packet (= spermatophore) into the vagina. The flagellum is believed to form the sperm packet.

der a part of the foot that starts to move. In some limpets the mucus also recovers some energy by trapping and stimulating the growth of algae, which the limpets eat as they retrace their mucus trails during tidal migrations.

Bivalves

Members of class Bivalvia include the familiar clams, mussels, oysters, and other animals with shells consisting of two valves (Fig. 28.20). Unlike the valves of brachiopods (see pp. 564–565), those of bivalves are positioned laterally and are symmetrical to each other. Normally the valves are held open by an elastic ligament near the hinge, but when threatened the bivalve can "clam up" using its **adductor muscles** (Fig. 28.4B). (The adductor muscle is what you eat when you have bay scallops.) Another distinctive feature of bivalves is the wedge-shaped foot—hence the former name of the class, Pelecypoda (Greek *pelekys* axe + *podos* foot).

LOCOMOTION. Most bivalves that move about do so by inserting the foot into soft substratum, inflating it with blood, then using it as an anchor to pull the body along. Other bivalves, such as the razor clam *Tagelus* (Fig. 28.1D), use the foot to burrow as deep as half a meter. The scallop *Pecten* claps the two valves together, forcing water out around the hinge and jet propelling the scallop in the direction of the gape. The ship worms *Teredo* and *Bankia* use their tiny valves to bore into sea grasses and wooden pilings or boat hulls. Many other bivalves are sessile as adults. The oyster *Ostrea* cements

FIGURE 28.20

(A) Two shipworms *Teredo* in wood. Note the highly modified valves around the suckerlike foot. Shipworms do considerable damage by digesting the cellulose in wooden structures. *Teredo,* a native of the Caribbean, nearly left Columbus stranded. (B) The scallop *Chlamys* showing numerous eyes bordering the gape. Bivalves show little cephalization; not even the eyes are concentrated in a head.

its left valve to substratum, and sea mussels such as *Mytilus* attach themselves by **byssal threads** secreted from the rudimentary foot. These byssal threads contain a substance that rivals epoxy as an adhesive and has the additional advantage of working under water. Researchers are studying ways to use the adhesive in dentistry or medicine.

Feeding and Digestion. Bivalves are unusual among molluscs in that they lack radulae. Most use the gills (ctenidia) for filter feeding. As cilia draw water across the surface of the ctenidia, they also bring in particles of food and inorganic debris (Fig. 28.6). The particles cannot pass through the pores into the gills, and they therefore become trapped in a continuously secreted layer of mucus. Cilia separate the debris from the food, which they propel to the mouth in a mucous string. The mucous string is reeled into the stomach by the **crystalline style,** which is rotated by cilia lining the style sac (Fig. 28.21 on page 598). The crystalline style also releases digestive enzymes, such as amylase, as it grinds against the **gastric shield** on the stomach lining. Nutrient molecules released during digestion in the stomach enter the **digestive glands,** where further digestion occurs intracellularly. Larger indigestible particles go into the intestine and are eliminated.

One important variation in this pattern of feeding should be mentioned. Six species of **giant clams** and one closely related cockle do not move about and feed in detritus-rich sediments, but stay anchored in nutrient-poor coral reefs with the valves gaping upward. Some of these giant clams (*Tridacna gigas*) are a meter long and have a mass of 250 kg. This prodigious size naturally raises the question of what they eat. Contrary to old sailors' tales, giant clams probably do not subsist on human flesh, since their digestive systems are no better equipped to handle meat than are those of other clams. Much of their nutrition comes from photosynthetic protists called zooxanthellae (see p. 465). These zooxanthellae *(Symbiodinium microadriaticum)* reside in the

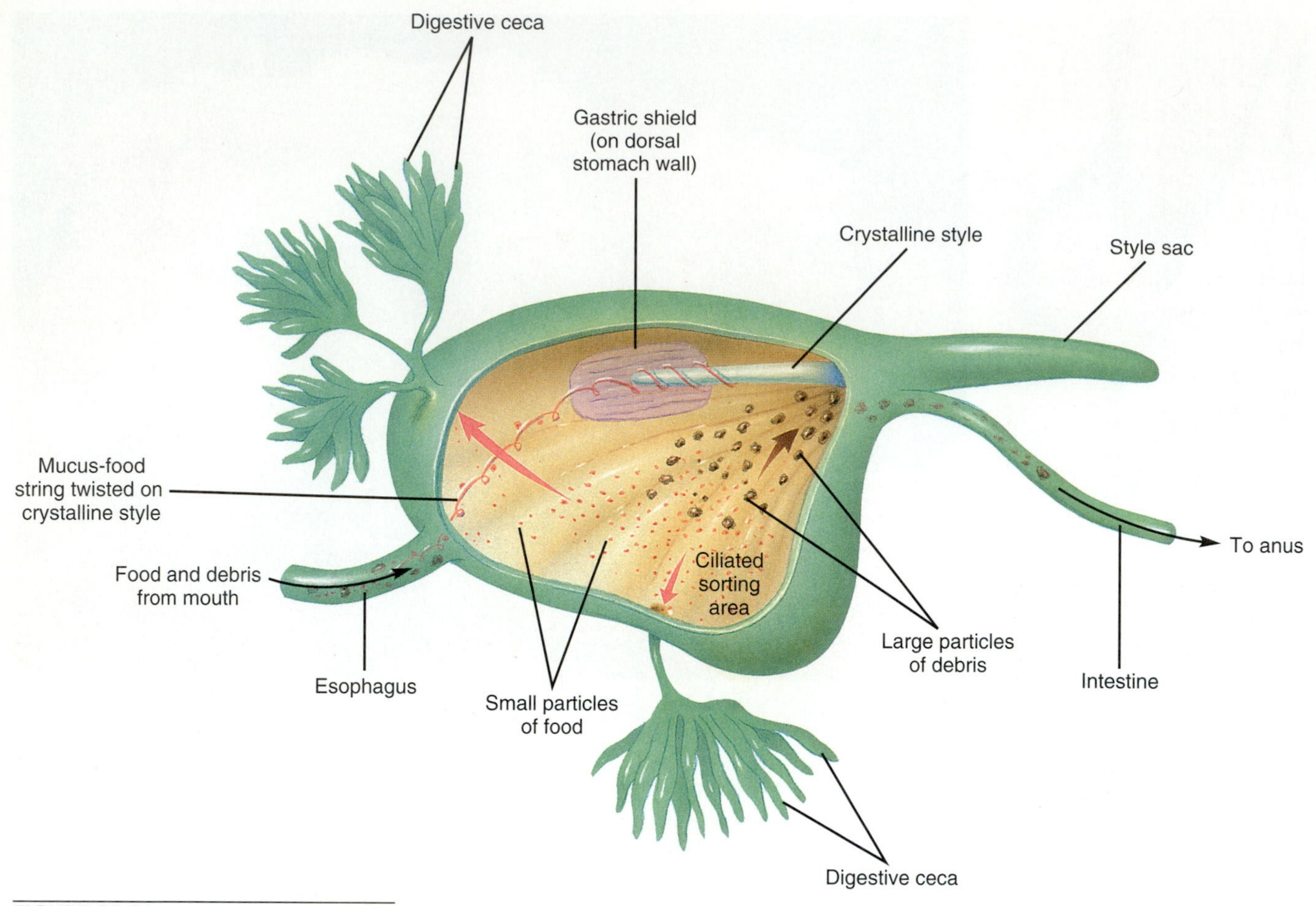

FIGURE 28.21

Digestive system of a bivalve. Cutaway view of the stomach showing the crystalline style reeling in the mucous string bearing food particles. The crystalline style releases digestive enzymes as it grinds against the gastric shield. The ciliated sorting area separates small food particles from larger, indigestible bits. Arrows indicate the movements of larger particles to the intestine and of small particles to the digestive ceca, where intracellular digestion occurs.

mantle just inside the edges of the valve and produce glycerol and alanine (Fig. 28.22). Often the mantle of giant clams has bright pigments that block the wavelengths of sunlight that would damage the clam's tissues.

REPRODUCTION AND DEVELOPMENT. Bivalves appear to have used up their share of evolutionary innovation on the ctenidia and foot, since the other organs are fairly simple. There is not even a concentration of receptors and neural tissue to indicate where a head would be. This simplicity also extends to reproduction. As a rule, bivalves show little of the courtship or other social behaviors displayed by gastropods and some other molluscs. Most simply discharge gametes into the water, where fertilization occurs, although some brood the embryos. The majority of gametes and larvae become food for other species, but most females release tens of millions of eggs per season.

Various behavioral adaptations increase the odds of successful reproduction. Some female bivalves inhale sperm, thereby effecting internal fertilization. The female Pacific Coast oyster *Ostrea lurida* also broods her larvae within the mantle cavity until they become **spats**—veligers ready to attach to substratum. Some freshwater clams have specialized larvae that attach to the gills or skin of fish (Fig. 28.23). These parasitic larvae are called **glochidia** in one group. The glochidia encyst in the fish tissue, where they are nourished and protected until able to live independently. The female clam *Lampsilis ventricosa* lures fish that serve as adoptive parents for her glochidia. She has part of the mantle shaped remarkably like a small fish, even down to an eyespot. When a larger fish approaches the lure for a meal, the clam releases her glochidia, which then hitch a ride in the fish's gills.

Cephalopods

Cephalopods are named for the foot, which is integrated into the head in the form of arms, tentacles, and/or a **funnel** (= siphon). Cephalopods expel water through the funnel for locomotion, and they use their tentacles and arms mainly in predation. Cephalopods are divided into two quite different subclasses that differ in the number and form of the tentacles. Subclass Coleoidea comprises squids, cuttlefishes, and octopuses, all of which have eight arms, with suckers lining the inner surfaces. Squids and cuttlefishes resemble each other, and both have in addition to the eight arms two longer tentacles with cup-shaped suckers at the ends. Chitinous teeth or hooks around the edges of the suckers augment suction in grasping prey. Subclass Nautiloidea includes six species in the genus *Nautilus,* which have more than 90 sticky tentacles without suckers, each retractable in its own sheath.

BODY SUPPORT AND BUOYANCY. Unlike most molluscs, octopuses have no shell, and their bodies are bulbous and flaccid when relaxed. Squids and cuttlefishes have only a vestigial shell inside the mantle. In squids it is a long **pen** (= gladius) running down the back, and in cuttlefishes it is the flat **cuttlebone** often seen in bird cages. The cuttlebone has gas-filled chambers that make the cuttlefish buoyant. The giant squid *Architeuthis,* which has the largest mass of any invertebrate (up to 450 kg), is also buoyant. Its buoyancy is due to high levels of low-density ammonium ions in its tissues. (This was discovered by Clyde Roper and two others from the aroma of some *Architeuthis* steaks they were cooking up to celebrate completion of a doctoral examination.)

Unlike most cephalopods, *Nautilus* has a massive shell that holds gases that make the animal buoyant. As the living part of *Nautilus* grows, it enlarges its shell and moves into the newest, widest area. It seals off the old chamber by secreting a calcareous sep-

FIGURE 28.22
The giant clam *Tridacna* on a coral reef. The mantle edges contain symbiotic zooxanthellae.

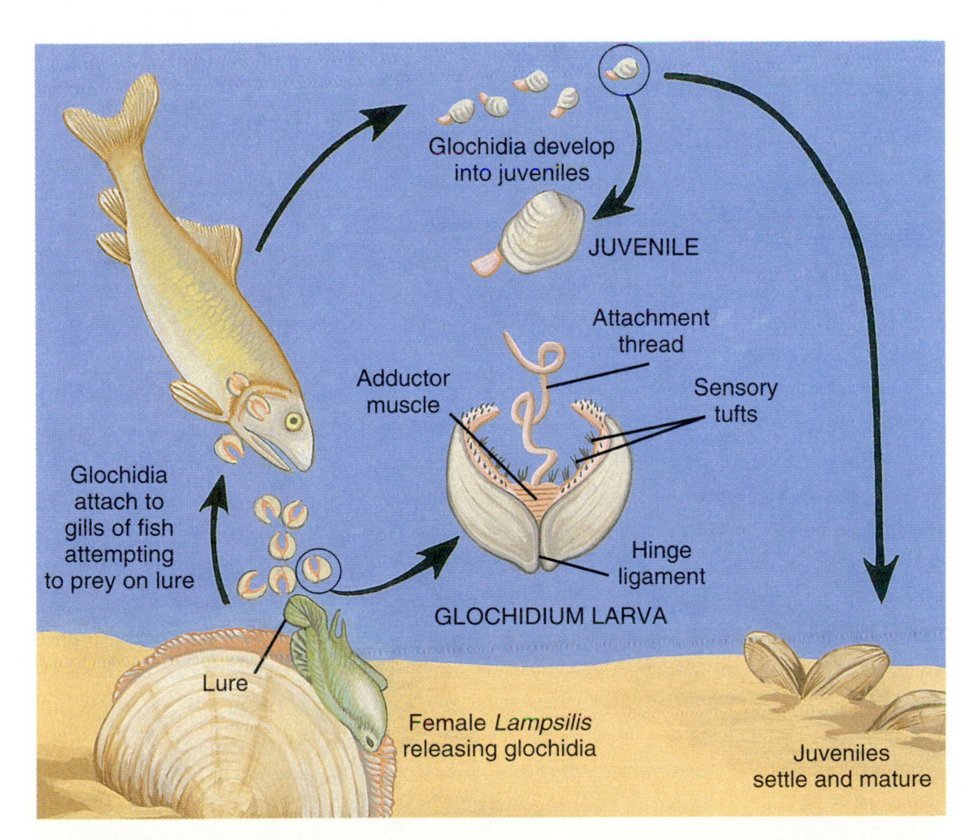

FIGURE 28.23
Life cycle of the freshwater clam *Lampsilis ventricosa,* including details of the glochidium larva.

■ **Question:** Lampsilis has no eyes and never sees its lure or a fish. What does this imply about the evolution of the lure, and about evolution in general?

A

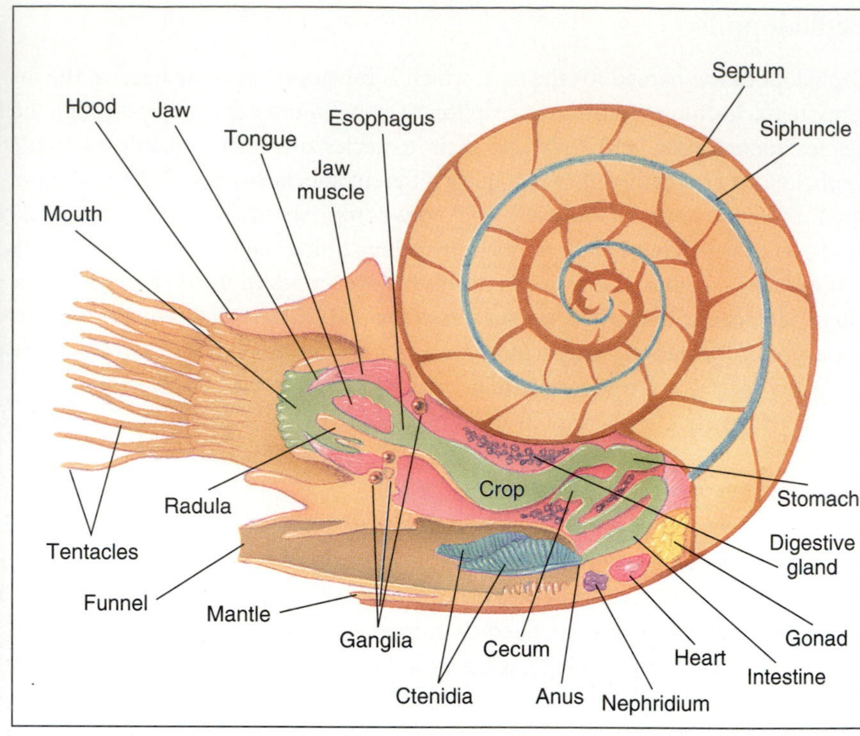

B

FIGURE 28.24

(A) *Nautilus* feeding. (B) Internal structure of *Nautilus.*

tum behind it (Fig. 28.24). The animal cannot simply leave an empty chamber, however, because it lives at depths of more than a hundred meters, where the pressure would crush it. Instead, a **cameral fluid** that is ionically similar to nautilus blood (as well as to seawater) fills the abandoned cavity until the new septum is formed. Gradually, cameral fluid is removed osmotically and replaced by a gas—essentially air with the oxygen removed. The low density of the gas compensates for the added mass of shell and body, thereby providing buoyancy.

Nautilus replaces the cameral fluid with gas even at depths greater than 500 meters, where the pressure is more than 50 atmospheres. This is equivalent to blowing up a balloon with a 25-ton boulder on it! This feat is performed by the **siphuncle,** the strand of tissue that runs through all the chambers. The siphuncle actively transports ions out of the cameral fluid, creating a local osmotic gradient that draws the water out of the chambers. The gas then infiltrates the chambers, presumably by diffusion.

LOCOMOTION. Octopuses usually prowl the ocean floors at a typical molluscan pace, and *Nautilus* drifts lazily up and down in a daily migration in pursuit of prey. They can also propel themselves rapidly by expelling water from the mantle cavity through the funnel during the escape response, which is triggered by the famous **giant axons** (see pp. 153–155). Escape from predators is often aided by another distinctive cephalopod trait, the release of a blob of mucous **ink,** which may startle or confuse the predator. Jet propulsion is too inefficient to be used for routine locomotion by octopus and *Nautilus,* however. Only cuttlefish and squids, with their streamlined bodies and fins, use jet propulsion commonly. Squids are, in fact, the fastest swimmers of all invertebrates. The squid *Todarodes pacificus* reaches speeds of 11 km/hr during 2000-km migrations.

Jet propulsion is due to bands of radial and circular muscles enclosed in a collagenous **tunic** (Fig. 28.25 inset). The tunic maintains the overall shape of the mantle and

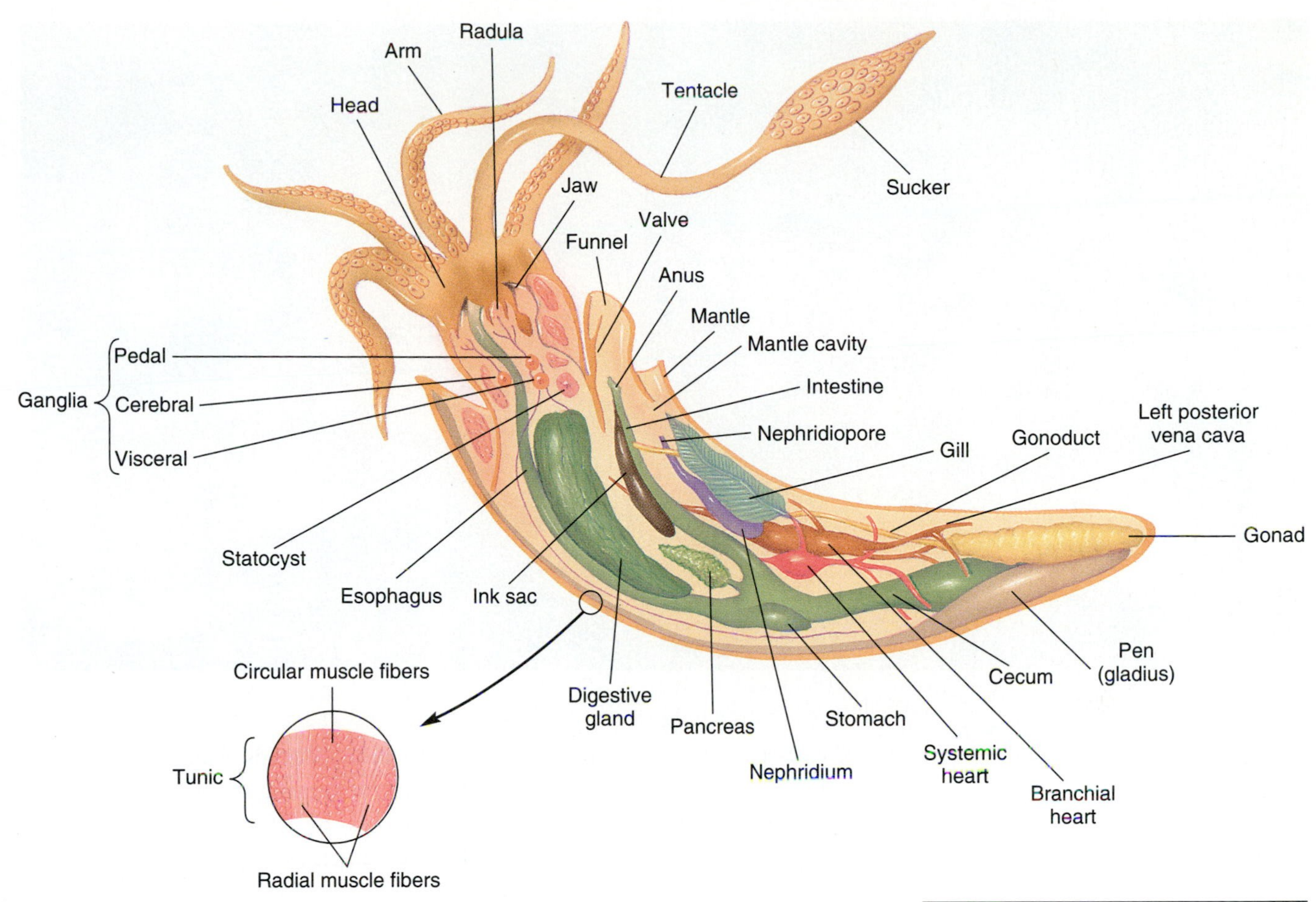

FIGURE 28.25

Internal structure of the squid *Loligo*. The inset shows a section of the mantle, with radial and circular bands of fibers enclosed by collagenous sheaths called the tunic. Radial fibers contract lengthwise in the plane of the paper, thickening as they do so and thereby expanding the diameter of the body. Circular muscles compress the body when they contract.

keeps it from expanding lengthwise when the muscles contract. The mantle is filled with water when the radial muscles contract. As they contract they thicken, thereby expanding the diameter of the body and drawing in water past the head. Expulsion of the water through the funnel is due to contraction of circular muscles.

INTERNAL ORGANIZATION. Unlike most molluscs, cephalopods have a closed circulatory system capable of sustaining high metabolic levels (Fig. 28.25). In many cephalopods there is not only a **systemic heart** that pumps oxygenated blood from gills to body tissues, but also two accessory **branchial hearts** that return deoxygenated blood from the body tissues to the gills. The digestive system begins with a pair of chitinous jaws shaped like those of a parrot. There is a radula, but it is used mainly in swallowing the chunks of prey bitten off by the jaws. Digestive secretions are produced by **salivary glands,** a **liver,** and a **pancreas,** which are, of course, not homologous with vertebrate organs of the same names. The anus is near the gills, but strong contractions of the mantle prevent fouling.

■ **Question:** Describe the escape reflex of the squid. (Refer to Fig. 8.4.) What is the function of the valve in the funnel?

BEHAVIOR. Cephalopods display a variety of complex behaviors. Octopuses can quickly learn to approach or avoid various visual, chemical, and tactile stimuli, and they retain memory for long periods. Some of the most interesting behaviors of coleoids derive from their ability to change their appearances. This feat is accomplished by means of pigment-containing **chromatophore** cells, which are expanded by surrounding muscles. Depending on the species, yellow, orange, red, blue, and black chromatophores may be present in three layers, allowing the coleoid to assume a vari-

FIGURE 28.26

Two cuttlefish *Sepia* copulating. Note the differences in appearance due to the chromatophores.

ety of patterns or to flash from one color to the next. Coleoids change their appearances during courtship and other social behaviors, presumably as a form of communication (Fig. 28.26). The chromatophores give squids, cuttlefishes, and octopuses an uncanny talent for camouflage. Some squids and cuttlefishes also have **photophores:** luminescent organs they can turn on and off. Photophores may be used for communication, but they are also known to serve in camouflage. By using the photophores on the ventral surface to match the amount of overhead illumination, squids make themselves invisible to prey and predators below them.

As would be expected from such behaviors, cephalopods have good vision. The eyes of coleoids show striking similarities to those of vertebrates. (Compare Figs. 8.12A and 28.27D.) Like our own eyes, they have a focusing lens, an iris that regulates the amount of light, and a retina that converts the image into action potentials. That these are examples of evolutionary convergence, rather than products of shared ancestry, is shown by developmental differences and the fact that the photoreceptors of the molluscan retina aim toward the lens. Figure 28.27 suggests a possible sequence by which the octopus eye could have evolved.

REPRODUCTION. Cephalopods have separate sexes, and they copulate during a head-to-head mating embrace. Male coleoids do not use a penis as the intromittent organ but have the bottom left arm modified for that function. This arm plucks a sperm packet (**spermatophore**) from the male's mantle cavity and inserts it into the female's mantle cavity (see Fig. 16.3A). The mating embrace is frequently violent in octopuses, and the arm often detaches inside the female and is never regenerated. Some 19th-century zoologists thought the detached arms found in females were parasites, which they named *Hectocotylus octopodis.* The copulatory arm of the male is still often called the **hectocotylus arm.** Each female is limited to one mating, because a secretion of the **optic gland** kills her within a few weeks while it promotes development of the fertilized eggs. If the optic gland is surgically removed, the female survives mating but is sterile. Most coleoids live for only one or two years, but nautiloids, which do not self-destruct in mating, may live for more than 20 years.

Interactions with Humans and Other Animals

Molluscs are important to humans as well as to other animals as food. Snail shells are a major source of calcium for some birds, and the decline in some bird species has been attributed to the decline of snails due to acid precipitation (Graveland et al. 1994). According to one study, sperm whales alone consume as large a mass of squid as humans do of all species of fish combined. Squid is, of course, also an item in the human diet. Fishing boats from Korea often use drift nets up to 30 miles long to catch squid. These nets also catch and kill thousands of sea birds, turtles, and other nonfood animals. Because of pressure from environmental groups, Japan and Taiwan have dropped the practice. Other molluscan foods include clams, oysters, mussels, abalone, snails (escargot), and scallops. Giant clams (*Tridacna*) have long been a staple in the diet of Pacific Islanders, so much so that the species is now endangered. Scientists are now studying ways to domesticate giant clams for food (Heslinga and Fitt 1987).

The consumption of molluscs goes back centuries. In the Mediterranean area, fishermen for thousands of years have been dropping pots into the water to take advantage of the tendency of octopuses to crawl into dark refuges. Archeologists frequently find ancient mounds of discarded shells (middens) indicating that bivalves have long been important in the human diet. Oysters were first cultivated for food in the first century B.C. by Romans. Humans have also found a way to use oysters to increase the food supply indirectly: The crushed shells attract microorganisms that kill nematodes that are agricultural pests.

Molluscs also nourish humans culturally. Rare and beautiful shells have been prized throughout history, and many are still extremely valuable to collectors. In some cultures, mollusc shells served for money. The *wampum* of East-Coast Indians was made from the purple or white shells of the edible clam, or quahog, *Mercenaria mercenaria.* Until colonists taught Indians the mercenary uses of *wampum,* it was merely a ceremonial item. Shells and other parts of molluscs have long been used in bodily adornment. Pearls immediately come to mind. Many bivalves form pearls when a foreign object lodges between the shell and the mantle, but the pearls of most species lack the desired luster. Pearl oysters are still sometimes gathered by divers who go as deep as 40 meters, equipped only with a clamp over their noses. Now, however, most pearls are cultured in Japan. They are produced by the oyster *Meleagrina* sp., usually in response to a piece of nacre from a different species inserted beneath their shells. Demand for the clam shell used as a seed for cultured pearls has created an economic boom in the upper Mississippi region and has also endangered the species.

Bivalve shells also provide the material from which cameos are carved. Finally, the ink of *Sepia* is still used as a source of the artists' ink called sepia. During the first 2000 years B.C. three species of marine snails—*Murex brandaris, Thais haemastoma,* and *Truncilariopsis trunculus*—made their contribution to decoration as the sources of a permanent purple dye for cloth. The dye was laboriously collected from hypobranchial glands in the mantles of vast numbers of snails, then processed by secret recipes. In Rome only a select few were permitted to make or use the "royal purple." According to *Numbers* 15:38, however, all the Hebrews of the Exodus sewed a single band of the purple into their garments to remind them of the commandments. Jews no longer follow the practice, because the art of making the authentic purple has been lost.

Royalty also found less benign uses for molluscs. Agrippina, the mother of Emperor Nero, is said to have poisoned her son's rivals with secretions from sea slugs. Other molluscs that have detrimental impacts on humans include snails, which are essential intermediate hosts for the fluke *Schistosoma* (see pp. 524–526). Shipworms cause extensive damage to wooden pilings and boat hulls and are becoming increasingly troublesome as improved water quality allows more of them to survive. Since

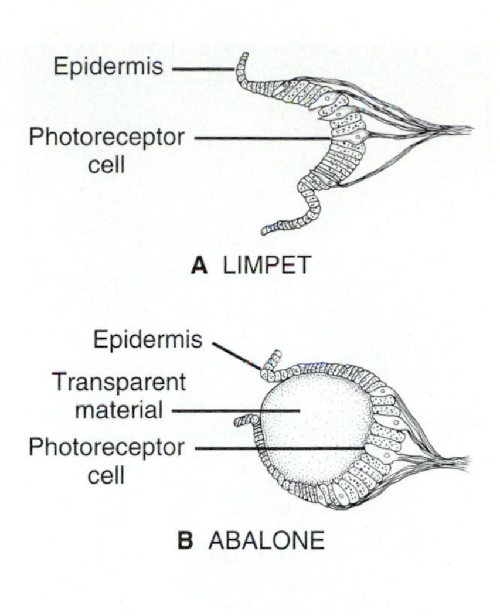

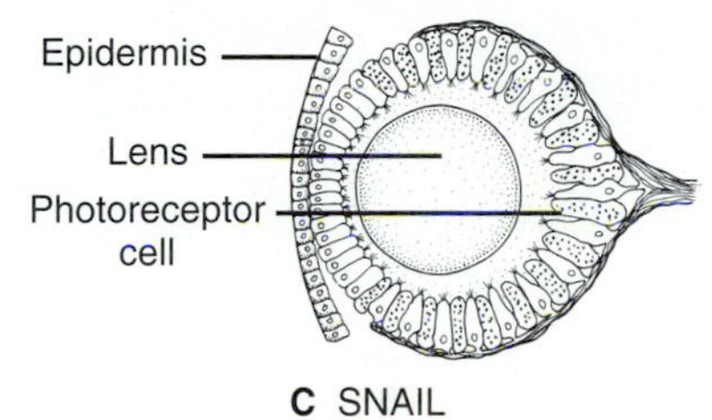

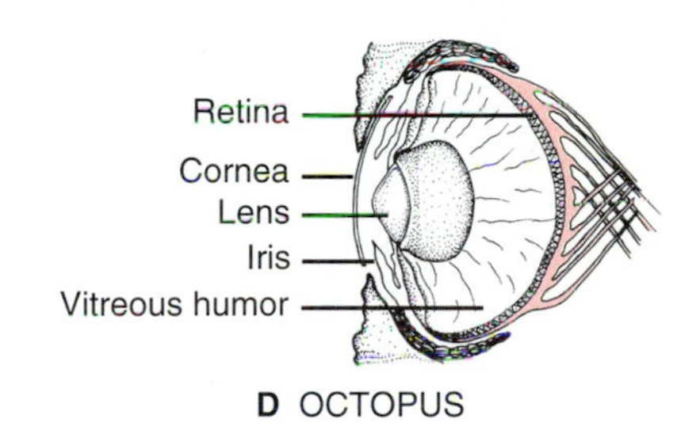

FIGURE 28.27

Photoreceptor organs of molluscs, suggesting a possible sequence by which the octopus eye could have evolved. (A) The eyespot of a limpet. (B) The abalone eye, which contains a transparent substance that may have been the precursor of a lens. (C) The eye of a pulmonate. (D) The eye of an octopus.

FIGURE 28.28
Each zebra mussel *Dreissena polymorpha* uses up to 2000 byssal threads to attach to pipes, rocks, and the shells of other bivalves.

1988 another mollusc, the zebra mussel *Dreissena polymorpha,* has become an even bigger nuisance in freshwater lakes and streams in the northeastern United States. Zebra mussels cling to substratum in huge masses, clogging municipal water intakes and cooling intakes in boat motors, fouling beaches with their sharp shells and rotting corpses, and outcompeting other bivalves and sport fishes (Fig. 28.28). The mussel is believed to have been introduced in 1986 when a ship emptied the ballast water it had carried across the Atlantic into the Great Lakes. Although scientists in 1981 had warned of just such a threat from dumping ballast water, Congress has still not outlawed the practice. Control of this species is expected to cost more than $5 billion by the end of the century. A second, unnamed species of *Dreissena,* the quagga mussel, was discovered in 1992 and may also cause problems.

Humans are undoubtedly more harmful to molluscs than vice versa. Because bivalves are filter feeders, they tend to accumulate pollutants, and in many places they are collected and analyzed as a means of monitoring water pollution. Agricultural runoff, sewage, and toxic wastes have severely reduced the catch of Eastern oysters *Crassostrea virginica.* In spite of massive efforts to clean up the pollution, oysters have continued to decline because of overfishing, a mysterious blood disease, and the protozoan parasites MSX and Dermo (*Haplosporidium nelsoni* and *Perkinsus marinus*). In the past century the yield of Eastern oysters from Chesapeake Bay has dropped from 15 million bushels per year to less than half a million bushels. There is now a controversial plan to replace the Eastern oyster in Chesapeake Bay with the Pacific oyster *Ostrea lurida* or the Asian species *Crassostrea gigas.*

Summary

Molluscs are extremely diverse, but most have a similar body plan. The living tissue is usually divided into a foot and a visceral mass. The foot is adapted in a variety of ways for locomotion. The visceral mass includes the mantle, which secretes the shell and which forms a mantle cavity enclosing the respiratory organs. Another unique feature of most molluscs is the radula, which is used to scrape up food or bore into shells. Although molluscs have a small coelom enclosing the heart, the main body cavity in most forms is the hemocoel through which the open circulatory system pumps blood. Classification is based largely on the form of the shell. There are three major groups. Gastropods include snails, conchs, whelks, and others with a one-piece, coiled shell, as well as slugs and related forms in which the shell is reduced or absent. Mussels, clams, oysters, and others with a shell of two valves are classified as bivalves. In cephalopods—octopus, squid, cuttlefish, and nautiloids—the foot is modified into tentacles.

Many gastropods, such as snails and slugs, creep about on a mucous secretion, which is toxic in some species. The larvae of gastropods undergo a process of torsion by which the rear portion of the body bends forward over the anterior end. In most gastropods a separate process of coiling makes the shell more compact.

Bivalves generally use the foot as an anchor during locomotion, although some swim by clapping the two valves together. Most feed by filtering food particles out of the water current that flows over the gills (ctenidia). The particles are trapped in a mucous strand that is reeled in by the crystalline style. Food is digested intracellularly in the digestive gland with the aid of enzymes from the crystalline style.

Cephalopods are generally more active than other molluscs, and they can swim rapidly by compressing the body wall to produce a jet of water through the funnel. Squids, cuttlefishes, and octopuses have eight tentacles with suckers, and they possess either a reduced shell or none at all. Nautiloids have numerous sticky tentacles and an elaborate spiral shell that maintains buoyancy.

Key Terms

foot
visceral mass
mantle
mantle cavity
radula
nacre
prismatic layer
periostracum
ctenidium
hemocoel
trochophore larva
veliger
torsion
adductor muscle
crystalline style
cameral fluid
siphuncle

Chapter Test

1. Identify the class to which each mollusc in Fig. 28.1 belongs. (pp. 587–591)
2. For each of the classes listed in the previous question, describe the foot and explain its function. (pp. 587–599)
3. Mucus plays several roles in the lives of molluscs. Describe three such roles and give an example of a mollusc in which mucus serves that function. (pp. 594–598)
4. Describe the mantle and relate its functions to key features of the molluscan body plan. (p. 582)
5. What are the functions of the shell in molluscs? Explain how the shell grows to keep pace with the growth of the mollusc. How do some molluscs adjust the shapes of their shells to keep pace with body size? (pp. 582–584)
6. Describe the radula and the gill of a mollusc. What roles do they play? (pp. 584–585)
7. Describe the process of torsion. In what kinds of molluscs does it occur? (pp. 591–592)

■ Answers to the Figure Questions

28.3 As their names imply, monoplacophorans have a "one-plate" shell, while polyplacophorans have a "many-plate" shell. One obvious difference between scaphopods and cephalopods is that the shell is internal and much reduced in cephalopods.

28.4 A good guess would be that the sensory lobe would respond to external threats, triggering retraction of the mantle from the edge and closure of the valves. Another likelihood is that increased exposure of the sensory lobe as the mantle grows would trigger synthesis of new shell to cover the mantle.

28.6 Because bivalves filter large amounts of water during respiration and feeding, they tend to accumulate toxins.

28.12 See Table 10.1. The yellow periwinkle probably gets its color from pigments in the periostracum. The pink of the Chinese cowry is mainly pigment in the exposed mantle, with smaller amounts in the periostracum. The iridescence of the abalone is a structural color. The function of the yellow of periwinkle is unknown. The cowrie's pink may be warning coloration. The abalones iridescence may be a byproduct of the process of formation of the shell.

28.16 Opisthobranchs and some pulmonates undergo detorsion. The fact that all gastropods undergo torsion even if the torsion is subsequently reversed indicates that Gastropoda is probably a monophyletic group that evolved from an ancestor that underwent torsion.

28.17 The planospiral snail in part A is bilaterally symmetric. The other two are asymmetric. (This asymmetry is derived, since it develops from a bilaterally symmetric larva.)

28.23 The evolution of the lure clearly was not guided by visual information in the clam, but by visual information in the fish, which was the agent of natural selection. What this implies in general is that evolution is better explained by natural selection than by some Lamarckian mechanism in which the individual organism guides the evolution of its own species.

28.25 In response to a threat, action potentials in giant axons trigger contraction of the circular muscles, which squeezes water out of the mantle cavity through the funnel. This propels the squid in a direction opposite to the direction in which the funnel is aimed. The valve in the funnel would prevent water from being sucked back in through the funnel when the mantle expands, which would draw the squid back toward danger.

Readings

Recommended Readings

Gosline, J. M., and M. E. DeMont. 1985. Jet-propelled swimming in squids. *Sci. Am.* 252(1):96–103 (Jan).

Gould, S. J. 1985. *The Flamingo's Smile.* New York: W. W. Norton. (*Chapters 3, 9, 11, and 16 discuss various aspects of molluscs, including Gould's own research on snails.*)

Hayes, B. 1995. Space-time on a seashell. *Amer. Sci.* 83:214–218. (*Computer modeling of mollusc shells*.)

Hoffmann, R. 1990. Blue as the sea. *Am. Sci.* 78:308–309. (*On royal purple.*)

Linsley, R. M. 1978. Shell form and evolution of gastropods. *Am. Sci.* 66:432–441.

Ludyanskiy, M., D. McDonald, and D. MacNeill. 1993. Impact of the zebra mussel, a bivalve invader. *BioScience* 43:533–545.

Morse, A. N. C. 1991. How do planktonic larvae know where to settle? *Am. Sci.* 79:154–167. (*Mainly on abalone.*)

Rehder, H. A. 1981. *The Audubon Society Field Guide to North American Seashells.* New York: Alfred A. Knopf.

Roper, C. F. E., and K. J. Boss. 1982. The giant squid. *Sci. Am.* 246(4):96–105 (Apr).

Safer, J. F., and F. M. Gill. 1982. *Spirals from the Sea: An Anthropological Look at Shells.* New York: Clarkson N. Potter.

Scheller, R. H., and R. Axel. 1984. How genes control an innate behavior. *Sci. Am.* 250(3): 54–62 (Mar). (*The egg-laying hormone of the marine snail* Aplysia.)

Solem, A. 1974. *The Shell Makers: Introducing Mollusks.* New York: Wiley.

Ward, P. 1983. The extinction of the ammonites. *Sci. Am.* 249(4):136–147 (Oct).

Ward, P. D. 1988. *In Search of Nautilus: Three Centuries of Scientific Adventures in the Deep Pacific to Capture a Prehistoric-*

Living-Fossil. New York: Simon & Schuster.

Ward, P., L. Greenwald, and O. E. Greenwald. 1980. The buoyancy of the chambered nautilus. *Sci. Am.* 243(4):190–203 (Oct).

Yonge, C. M. 1975. Giant clams. *Sci. Am.* 232(4):96–105 (Apr).

There are numerous field guides and other books on seashells from around the world. See also relevant selections in General References at the end of Chapter 21.

Additional References

Adams, S. A., and R. Chase, 1990. The "love dart" of the snail *Helix aspera* injects a pheromone that decreases courtship duration. *J. Exp. Zool.* 255:80–87.

Baker, B. 1992. Botcher of the bay or economic boon? *BioScience* 42:744–747. (*On plans to replace the depleted Eastern oyster with the Pacific oyster in Chesapeake Bay.*)

Barinaga, M. 1990. Science digests the secrets of voracious killer snails. *Science* 249:250–251.

Downing, W. L., and J. A. Downing. 1993. Molluscan shell growth and loss. *Nature* 362:506.

Field, K. G., et al. 1988. Molecular phylogeny of the animal kingdom. *Science* 239:748–753.

Fiorito, G., and P. Scotto. 1992. Observational learning in *Octopus vulgaris. Science* 256:545–547.

Graveland, J., R. van der Wal, J. H. van Balen, and A. J. van Noordwijk. 1994. Poor reproduction in forest passerines from decline of snail abundance on acidified soils. *Nature* 368:446–448.

Harrison, F. W., and A. J. Kohn (Eds.). *Microscopic Anatomy of Invertebrates.* Vol. 5: *Mollusca I.* New York: Wiley–Liss.

Heslinga, G. A., and W. K. Fitt. 1987. The domestication of reef-dwelling clams. *BioScience* 37:332–339.

Hori, H., and S. Osawa. 1987. Origin and evolution of organisms as deduced from 5S ribosomal RNA sequences. *Mol. Biol. Evol.* 4:445–472.

Ward, P. D. 1987. *The Natural History of Nautilus.* Boston: Allen & Unwin.

CHAPTER

29

Annelids

Christmas tree tube worm *Spirobranchus giganteus*.

The Significance of Segmentation

A year before he died in 1882, Charles Darwin charmed thousands with a book entitled *The Formation of Vegetable Mould through the Action of Worms, with Observations on their Habits.* What apparently captivated Darwin's readers was the revelation of how the unseen labors of the humble earthworm contributed to the fertility of the land, and thereby to their own existence. Earthworms belong to the phylum Annelida, along with other oligochaetes and with two other major groups. One group, the polychaetes, are not quite as familiar as earthworms, and the other group, leeches, are not quite as charming.

Oligochaetes, polychaetes, and leeches look very different from each other, but they share certain similarities that lead taxonomists to group them in the same phylum (Table 29.1 on page 610). One of these similarities is that their bodies are divided into **segments** (Fig. 29.1A). Annelids are therefore commonly called segmented worms. The segments, also called **somites** or **metameres,** are not produced by repeated division and subdivision of the body, as might be expected. Instead they are generated sequentially from the rear, the way a machine turns out sausage links. (See pp. 126–127 for more on segmentation in development.) True segmentation (= metamerism) occurs unequivocally in only three other extant phyla: Pogonophora, which some taxonomists regard as annelids or at least close relatives (pp. 572–574); Arthropoda (next three chapters); and our own phylum Chordata. The essential features of segmentation are that the coelom and the muscles of the body wall show serial repetition, at least in the embryo. That is, the coelom and muscles are homologous from one metamere to the next. In addition, the nervous system, excretory organs, and other organs may also be serially repeated.

Most of the signs of segmentation are obscure in chordates, and to some extent in arthropods. In these phyla, segments fuse and differentiate into distinctive body regions, such as the head, thorax, and abdomen. Each of these regions is called a **tagma** (plural tagmata). Annelids show little, if any, of this **tagmosis** (= tagmatization). Earthworms, for example, are essentially alike from front to rear. In most annelids, segmentation remains apparent throughout life, with each segment evident externally as a little ring (*annellus* in Latin). Segmentation also remains evident inside most annelids. Each somite is usually separated from its neighbors by **septa** and has its own set of muscles, its own coelomic cavity, its own ganglion on the ventral nerve cord, its own nephridia, and many other organs (Fig. 29.1B).

Zoologists have proposed several hypotheses to explain the adaptive advantages of segmentation. One of the most plausible hypotheses is that the segmentation of the coelom and muscles was an adaptation for burrowing, as explained in Chapter 10 (p. 195). Another possibility is that segmentation is simply a developmentally convenient way of achieving large body size. Rather than increase the size of every organ in the body, the embryo simply repeats parts. Segmentation has undoubtedly evolved independently in chordates and perhaps arthropods, so there may have been different advantages in each of these groups.

Origins and Classification

While the origins of most phyla are uncertain because of a paucity of evidence, the origin of annelids is obscured by too many clues, many of them contradictory. The segmentation and body shape of annelids suggest a kinship to Pogonophora (see Fig. 21.10). The trochophore larvae of some Annelida are like those of some members of the phyla Echiura, Sipuncula, and Mollusca. Except for fossilized worm tracks and burrows, there are few paleontological clues to assist in unraveling annelid evolution. Many taxonomists once considered a group of small, simple annelids called Archiannelida to be relics

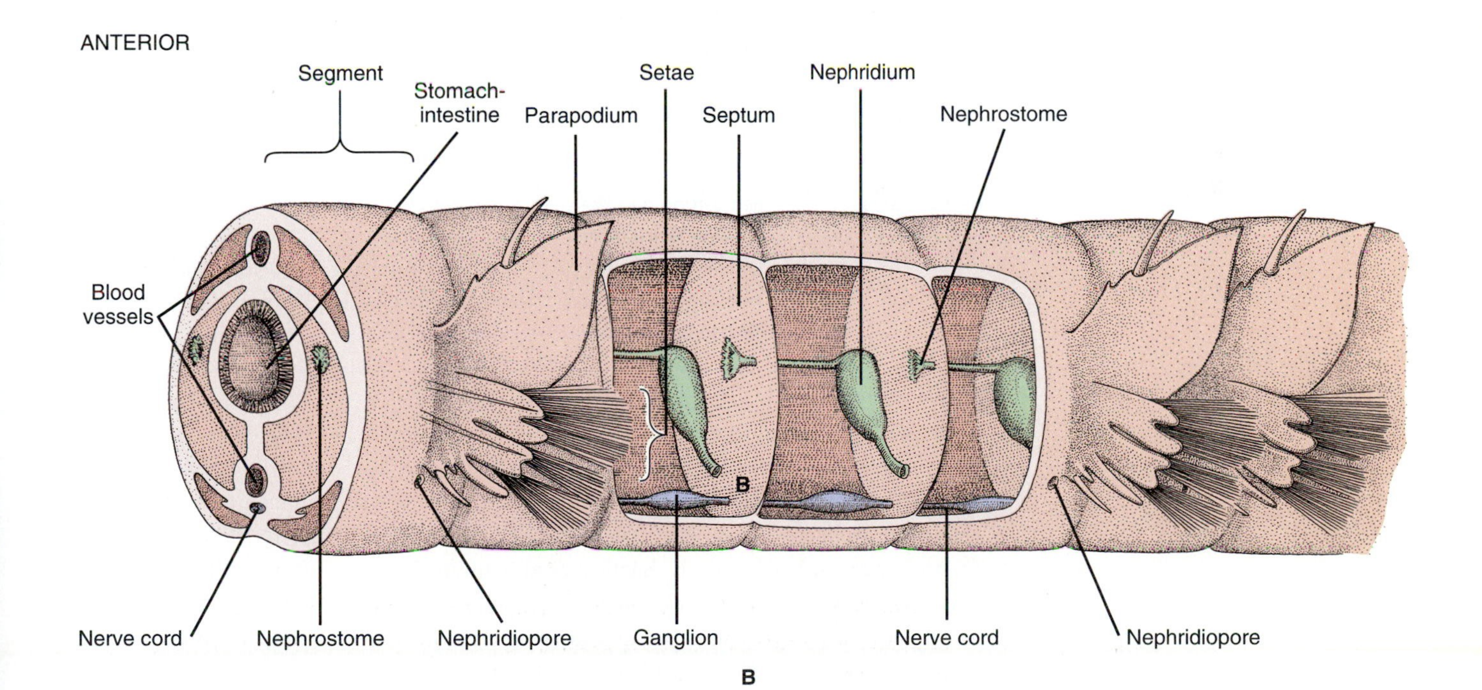

FIGURE 29.1

(A) The clam worm, also called sand worm, *Nereis virens,* a polychaete annelid. The segmentation of the body is evident. (B) A few of the 200 or more somites of *Nereis,* emphasizing segmentation. The cross section at the left side of the figure is at one septum at the front of a segment.

TABLE 29.1

Characteristics of annelids.

Phylum Annelida an-EL-lid-uh (Latin *annellus* little ring)

Morphology	Body **segmented**. Integument covered with thin, nonchitinous cuticle secreted by epidermis. Often bearing **setae** made of chitin. Coelom divided into a cavity on each side of each segment, except in most leeches, in which the coelom is reduced to channels and spaces.
Physiology	Gaseous exchange usually through integument; sometimes through specialized respiratory structures. Digestive tract complete. Central nervous system a ladderlike array of ganglia on a pair of ventral nerves (which may be fused into a single cord). Cephalic ganglion dorsal to esophagus. Except in leeches, circulatory system closed, with ventral and dorsal longitudinal vessel. Respiratory pigments often present in blood cells or plasma, and/or in coelomic fluid. Many segments with pairs of nephridia for osmoregulation. A layer of circular muscle surrounds bundles of longitudinal muscles; diagonal and dorsoventral muscles also present in some groups.
Locomotion	Creeping or swimming with parapodia (many polychaetes), peristaltic burrowing (earthworms), or undulatory swimming (leeches). Coelomic pressure produces hydrostatic skeleton; important for support as well as locomotion.
Reproduction	Polychaetes mainly dioecious; fertilization external. Oligochaetes hermaphroditic; fertilization external following copulation. Leeches hermaphroditic; fertilization internal following copulation. Asexual budding or fragmentation in some polychaetes.
Development	Cleavage pattern often spiral. **Trochophore larvae** in most polychaetes; development direct in oligochaetes and leeches.
Habitat, Size, and Diversity	More than half the species are marine polychaetes, but there are many freshwater and terrestrial species, especially oligochaetes and leeches. Distributed worldwide except on Antarctica and Madagascar. Body length from less than 1 mm to approximately 3 meters. Approximately **12,000 living species** described.

of the early evolution of annelids. Now, however, archiannelids are regarded as being polychaetes that have become simplified as an adaptation to living within sediment.

Within the phylum, classification is a little clearer. Most zoologists recognize three classes: polychaetes (class Polychaeta), oligochaetes (class Oligochaeta), and leeches (class Hirudinea). Class Polychaeta includes mainly marine worms with numerous bristles. (*Polychaites* in Greek means "many hairs.") These bristles, or **setae,** occur chiefly on appendages called **parapodia,** which function in locomotion and respiration (Fig. 29.1). Setae are fewer on oligochaetes (Greek *oligo-* few.) Because of the setae on polychaetes and oligochaetes, annelids are often called "bristle worms."

The leeches, class Hirudinea, lack setae. They live mostly in fresh water, but a few are terrestrial or marine. Both the oligochaetes and leeches have a swelling called a **clitellum** (Latin *clitella* pack saddle) on the back of certain segments. The clitellum is a permanent and familiar feature on some earthworms (Fig. 29.14A). In most oligochaetes and

Major Extant Groups of Phylum Annelida

Genera mentioned elsewhere in this chapter are noted.

Class Polychaeta POL-lee-KEY-tuh (Greek *polys* many + *chaite* long, flowing hair). Most segments with parapodia bearing numerous setae. Coelom well developed and divided into two lateral compartments per segment. Head distinct, with eyes and tentacles. No clitellum. Usually dioecious; asexual budding or fragmentation in some. Some with trochophore larvae. Mostly marine. *Amphitrite, Arenicola, Autolytus, Chaetopterus, Glycera, Nereis, Palola, Sabella, Spirobranchus, Syllis, Vanadis* (Chapter opener; Figs. 29.1A, 29.6, 29.8, 29.9, 29.10, 29.11).

Class Oligochaeta OH-lig-oh-KEY-tuh (Greek *oligos* few). Body divided into obvious segments, each with few setae. New segments added throughout life. Coelom well developed and compartmented. With clitellum. No parapodia or distinct head. Hermaphroditic; asexual budding in some. No larval stage. Usually terrestrial, burrowing in moist soil. Some burrow in freshwater or marine sediments. Earthworms and other oligochaetes. *Lumbricus, Megascolides, Tubifex* (Figs. 29.12, 29.13, 29.14, 29.17).

Class Hirudinea HEAR-oo-DIN-ee-uh (Latin *hirudo* leech). Same number of segments in all species (30 to 34, depending on the authority followed). Each segment divided externally into as many as 14 transverse ridges (annuli). Body usually flattened dorsoventrally. No setae (except on anterior segments of the "living fossil" *Acanthobdella*). Coelom not divided into segmental compartments; filled with spongy connective tissue and muscles. With clitellum. Usually with anterior and posterior suckers for attachment to host and for locomotion. No parapodia or tentacles. Hermaphroditic. No larval stage. Predators, ectoparasites, or scavengers. Usually in fresh water; some marine or terrestrial. Leeches. *Haemadipsa, Haementeria, Hirudo, Limnotis, Macrobdella* (Figs. 29.19 and 29.20).

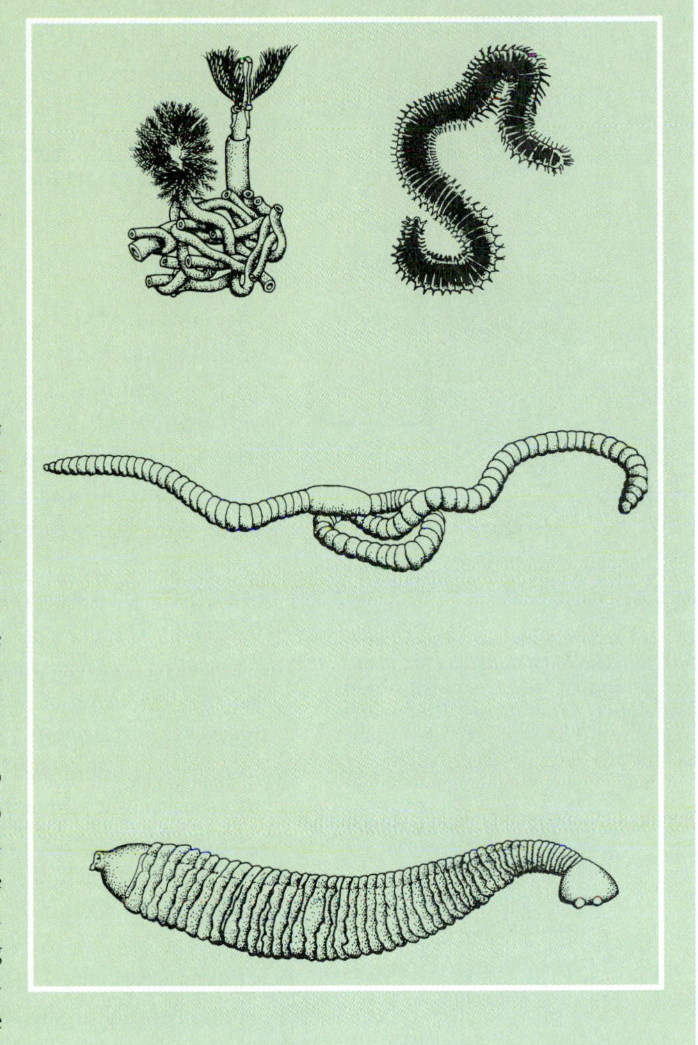

leeches, however, the clitellum appears only during the period of reproduction, when it secretes a cocoon in which embryos develop. Oligochaetes and leeches are often referred to as clitellates.

Polychaeta is the largest and most diverse of the three classes, and it is often considered to be most like the ancestral annelids. Most polychaetes live within the upper 50

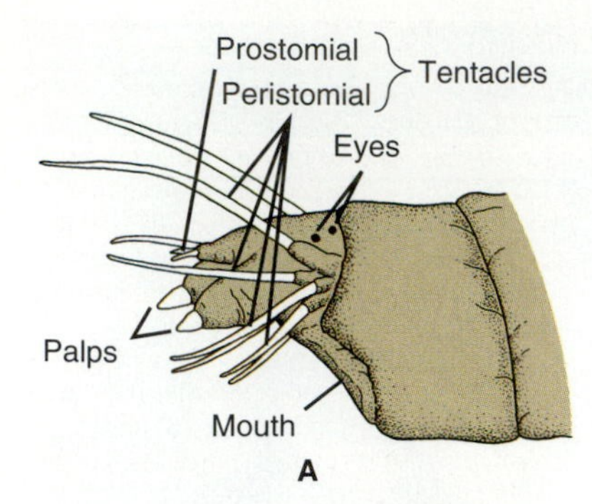

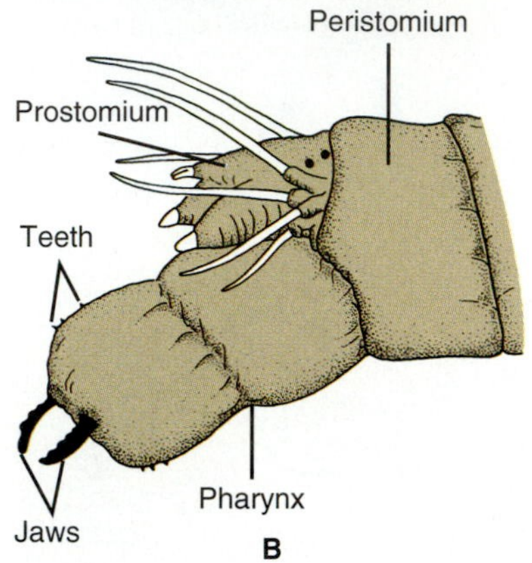

FIGURE 29.2

The head of *Nereis virens* consists of the prostomium (Greek *pro* in front of + *stoma* mouth), and the peristomium (Greek *peri*-around). (A) Pharynx retracted. (B) Pharynx extended out of the mouth.

■ **Question:** What is the advantage of an eversible proboscis?

meters of oceans, but some have been found as deep as 5 km and some have adapted to fresh water. Polychaetes are typically 5 to 10 cm long, but they range in length from less than 1 mm to several meters. Many are dull in appearance, but others are iridescent, luminescent, or brightly pigmented. Many are red where hemoglobin shows through the integument. Polychaetes are either freely moving (errant) predators or sedentary in tubes or burrows where they feed on detritus or plankton. On this basis they were formerly classified into the two subclasses Errantia and Sedentaria. This taxonomy is no longer considered valid, since it does not reflect phylogeny. Still, however, it is convenient to refer to polychaetes as either errant and sedentary. In errant species all the segments tend to be similar; that is, there is little, if any, tagmosis. Sedentary forms have some degree of tagmosis, with the anterior metameres specialized for feeding.

Polychaetes

The clam worm *Nereis,* named for a type of sea nymph in Greek mythology, is often considered representative of polychaetes, especially the errant forms (Fig. 29.1A). One member of the genus, *Nereis virens,* is common along the Atlantic coast. It is shiny greenish with red appendages, and its more than 200 somites extend for up to 90 cm. It lives as a juvenile for two or more years along the low-tide line. Adults spend their days in mucus-lined burrows or under rocks with the head protruding. At night they swim or crawls about, preying on other invertebrates or feeding on carrion and algae.

FEEDING. To seize prey, *Nereis* quickly shoots out its proboscis, sometimes with an audible pop (Fig. 29.2). The proboscis is in actuality the **pharynx,** which is everted by contracting circular muscles in the anterior somites, thereby increasing the pressure in the coelomic cavities of the head. (This is one advantage of having the coelom segmented into chambers.) As the pharynx retracts, it turns outside in, closing the chitinous jaws and **denticles** ("teeth") on the prey. Food passes through a short esophagus into the straight **stomach-intestine,** which extends from about the 12th somite to the anus in the last somite. The gut secretes a variety of enzymes in this omnivore, including cellulase, which digests the cell walls of algae. Polychaetes, as well as earthworms, are among the few animals that can produce this enzyme.

NERVOUS SYSTEM. *Nereis* detects prey by receptors on the head, especially chemoreceptors (**nuchal organs**) on the prostomial palps and on a pair of prostomial tentacles. Touch receptors on the prostomial tentacles and on the four pairs of peristomial tentacles are probably also important for locating food and also shelter. *Nereis* also has two pairs of eyes. These are relatively simple pigment-cup organs, with a retina of **rhabdomeric** receptor cells (p. 666) surrounding a spherical lens (Fig. 29.3A). In *Nereis* the lens is too close to the retina to focus an image on it, so the eyes probably function primarily as light detectors. However, in at least two planktonic, predaceous species, the lens can focus images on the retina (Fig. 29.3B). The general structure of these eyes shows remarkable evolutionary convergence with the eyes of vertebrates and cephalopod molluscs.

Information from the receptors on the head is integrated in the most anterior ganglion of the central nervous system, the cephalic ganglion, located dorsally in the prostomium. Two **circumpharyngeal connectives** connect the cephalic ganglion to the **ventral nerve cord.** As is typical in annelids, the ventral nerve cord consists of a series of ganglia joined by a pair of longitudinal **connectives** (Fig. 29.1B). In *Nereis* and many other annelids the pair of connectives is fused into a single connective. There are more than 200 ganglia, each of which controls one somite. Overall behavior is coordinated by impulse traffic in nerve cells in the connective, with the cephalic ganglion playing a major

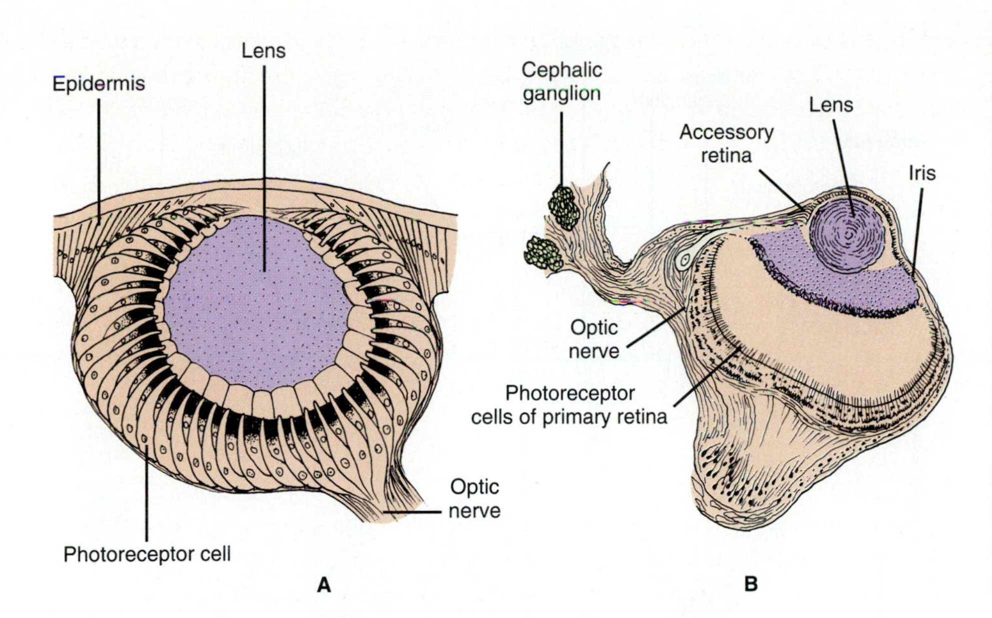

FIGURE 29.3

(A) One of the four eyes of *Nereis.* Note that the photoreceptor cells are too close to the lens to resolve an image. (B) The eye of the deep-sea polychaete *Vanadis* is like that of vertebrates and cephalopods in having its retina in the focal plane of the lens. Also like some deep-sea fishes and cephalopods, *Vanadis* has an accessory retina that responds to wavelengths different from those of the primary retina. Because different wavelengths penetrate to different depths in the ocean, this could enable the eye to be used as a depth gauge.

■ **Question:** What is the advantage of having a lens in Nereis if it does not focus an image? (Hint: Why are there receptor cells all around the lens?)

role. Polychaete neurophysiology is not known in detail except for the five cells called **giant nerve fibers** that trigger escape responses. The single median giant fiber is quite sensitive to mechanical stimulation in the first 60 somites and triggers rapid contraction of the longitudinal muscles of posterior somites. The behavioral response is therefore a rapid withdrawal of the head from a potential threat in front of the animal. Two paramedial giant fibers similarly trigger a withdrawal reflex of the anterior somites when the posterior is threatened. In addition, there are two lateral giant fibers that trigger shortening of the entire body in response to strong stimulation anywhere on the body.

MOVEMENT AND SUPPORT. The longitudinal muscles responsible for the withdrawal reflexes lie in four thick bands outside the peritoneum that surrounds the coelom (Fig. 29.4 on page 614). The longitudinal muscles are also responsible for rapid crawling. Those on opposite sides of the segments contract alternately, producing a wavelike wriggling. As in all annelids, a layer of circular muscle surrounds the longitudinal muscles. The circular muscles are weak in polychaetes, but they are important in creating the coelomic pressure for the hydrostatic skeleton that makes locomotion possible. In addition, oblique muscles to the parapodia assist in crawling.

FUNCTIONS OF PARAPODIA. Each parapodium has a dorsal lobe, the **notopodium,** and a ventral lobe, the **neuropodium.** A stiff, chitinous **aciculum** reinforces each lobe. Each lobe also bears numerous setae and a **cirrus** that is richly endowed with sensory receptors. The notopodia of *Nereis* serve as **gills.** This function is possible because of the large number of capillaries in each notopodium. Hemoglobin, which is dissolved in the plasma, shows red through the thin integument of these dorsal lobes. (Polychaetes that have the respiratory pigment chlorocruorin have green gills.)

CIRCULATION. The circulatory system is closed, as it must be to transport blood between segments separated by septa. The dorsal vessel pumps the blood anteriorly, and branches in each somite transport the blood to capillaries in the organs. Blood then collects in the ventral vessel, which carries it posteriorly back to the dorsal vessel. At rest, blood pressure is due primarily to peristaltic contraction of the dorsal vessel (Figs. 29.1B and 29.4). The blood pressure is less than 10 mmHg in resting *Nereis,* but it can double in active individuals, because of increased coelomic pressure.

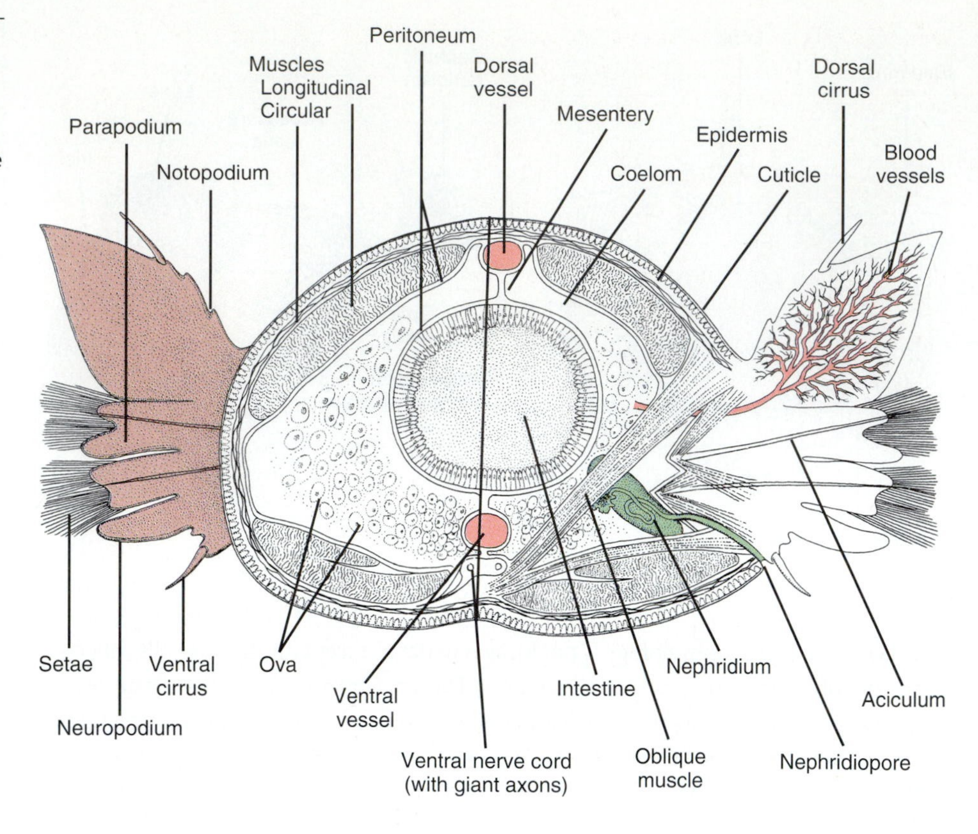

FIGURE 29.4

A cross section of *Nereis,* emphasizing the muscles and parapodia. The right half of the figure represents a cut through a parapodium. In the coelom are ova, which are released through the nephridia.

As in most annelids, the coelomic fluid of *Nereis* carries some of the burden of transporting oxygen, nutrients, and wastes within each segment. In some polychaetes the coelomic fluid carries the entire burden, because there is no blood. In these worms the coelomic fluid may contain hemoglobin, and it can circulate between segments because the septa are incomplete. One example is *Glycera,* which is commonly and paradoxically known as the bloodworm. The coelom of *Nereis* and other polychaetes is divided along the midline of each segment into a pair of coelomic cavities surrounded by the peritoneum. Where the peritonea from the two coelomic cavities in one segment meet they form a **mesentery** along the midline (Fig. 29.4). In most annelids the mesenteries and/or septa are incomplete between some coelomic cavities. In *Nereis,* for example, the septa are absent from anterior segments, forming the large chamber that everts the pharynx.

EXCRETION. In addition to maintaining the hydrostatic skeleton and transporting oxygen and nutrients, the coelomic fluid is evidently important in eliminating metabolic wastes through the **metanephridia,** which are often called simply nephridia. The nephrostome receives the coelomic fluid of the segmental cavity anterior to the nephridium and also exchanges material with blood in surrounding capillaries. Each nephridium expels these wastes through the nephridiopore in the body wall (Fig. 29.1B). The nephridia have little to do in regulating ion and water levels in *Nereis* and most other marine polychaetes, since these animals are osmoconformers, with body fluids similar to the seawater that normally surrounds them. In estuarine and freshwater polychaetes, however, the nephridia are more highly developed and maintain a higher internal ionic concentration than occurs in the environment.

In many annelids, part of the peritoneum around the gut differentiates into **chloragogue cells** (= chloragog or chloragogen cells), which assist the nephridia in eliminating nitrogenous wastes. These greenish or brownish cells remove the amine groups from amino acids and convert them into either ammonia or urea. Bits of chloragogue tissue containing these nitrogenous wastes periodically break off and either leave through the nephridia or form deposits in the body wall. Chloragogue cells also store glycogen and fats as energy reserves, and they gather around maturing ova and provide them with nutrients.

SEXUAL REPRODUCTION. The peritoneum and nephridia are also important in reproduction. *Nereis* and other polychaetes have no distinct gonads or genital organs. Instead the peritonea produce the gametes, and the nephridia often serve as routes for their emission. Reproduction in *Nereis* and many other polychaetes occurs during a remarkable transformation called **epitoky** (Greek *epi-* on + *tokos* birth). After living two or more years as a crawling, benthic juvenile (**atoke**), in which reproduction is inhibited by a neurosecretion from the cephalic ganglia, *Nereis* changes to a swimming, pelagic adult (**epitoke**). The body enlarges, the eyes increase in relative size, the head appendages shrink, the parapodia change shape, and muscles change from slow-twitch to fast-twitch types. The epitokes leave the shore and head for open water. Then on a night determined partly by the phase of the moon, the female epitokes release a pheromone that excites the males to swim about them in circles. As the sun rises the epitokes of both sexes release vast numbers of gametes. This synchronous mating improves the chances of fertilization. It also ensures that some of the gametes will survive predation by fishes and other animals, since there are too many gametes to be eaten at one time. (This is an example of **predator saturation;** pp. 416–417.) As in many flatworms, molluscs, and some other protostomes, the zygote of *Nereis* and other polychaetes develops into a trochophore larva, which is characterized by a girdle of cilia (Fig. 29.5).

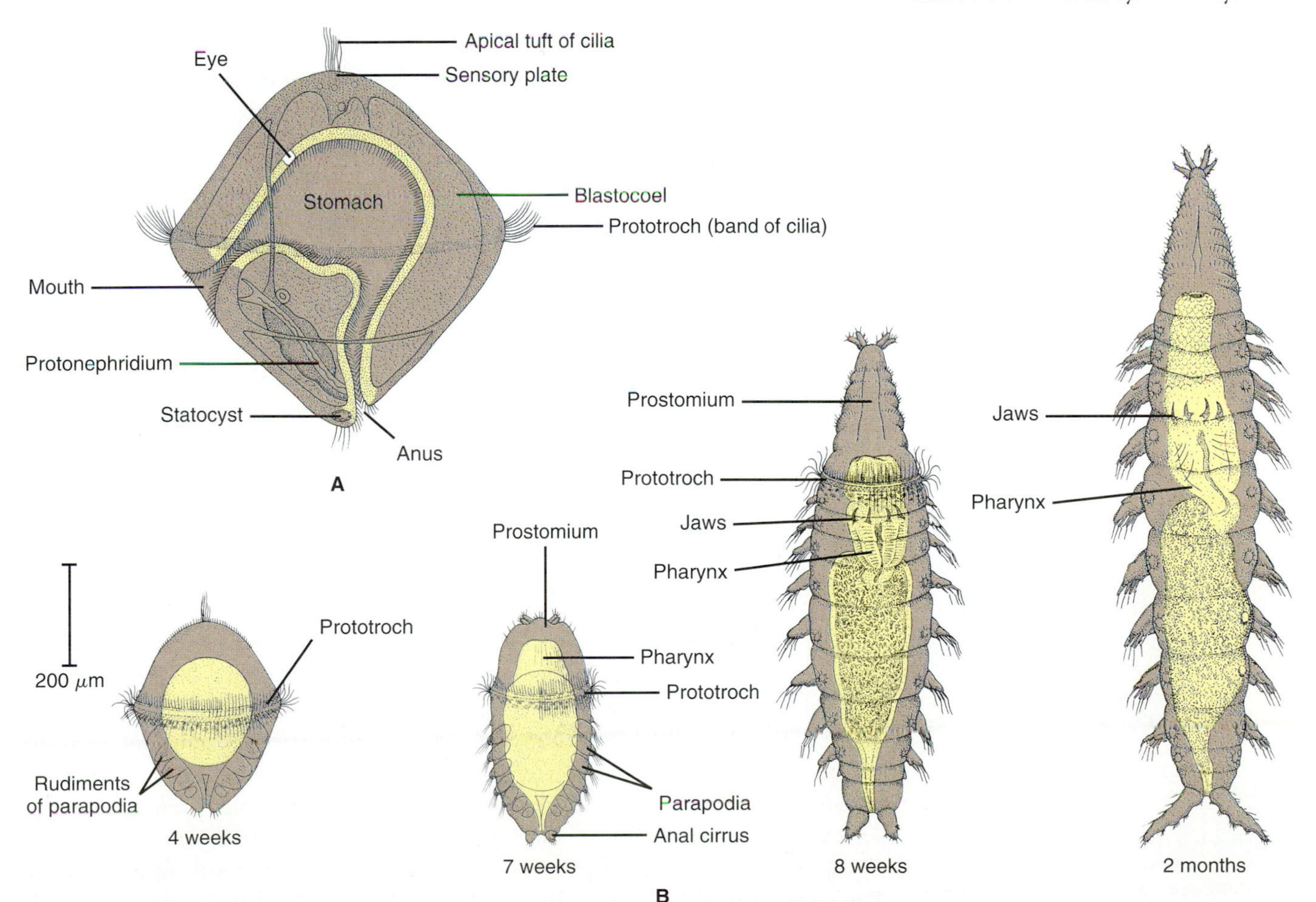

FIGURE 29.5

(A) A polychaete trochophore larva. (B) Development in the bloodworm *Glycera.* At four weeks the parapodia are starting to develop, but the larva still uses the ring of cilia to swim. At seven weeks the larva still swims by means of its cilia. The digestive system is not developed sufficiently to permit feeding. At eight weeks *Glycera* has become a post-larva. It can now crawl with its parapodia and catch prey with its eversible pharynx armed with jaws.

■ **Question:** What would be the likely function of the statocyst? The eye?

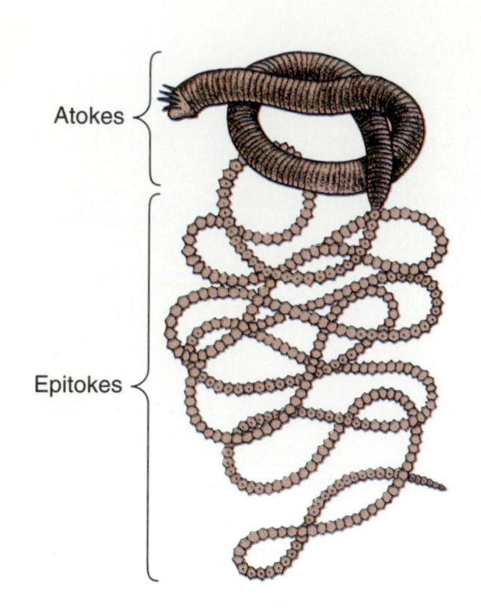

FIGURE 29.6
The Pacific palolo worm *Palola viridis* during epitoky. Only the posterior segments become epitokes, which break off and release the gametes within them. Each epitoke has an eyespot that guides its swimming toward the surface.

Epitoky is different and equally notable in the palolo worm *Palola viridis* (formerly *Eunice viridis*). In this errant polychaete each posterior segment becomes an individual epitoke that detaches from the anterior portion of the worm (the atoke; Fig. 29.6). After detaching, the epitokes swim to the surface and release their gametes in such profusion that they cloud the water. The atokes return to burrow in coral reefs where many live to breed the following year. The swarming of the palolo worm occurs so predictably, in October or November at the start of the moon's last quarter, that native Samoans schedule it as a holiday on which to gather and feast on epitokes.

Asexual Reproduction and Regeneration. Asexual reproduction occurs in some errant polychaetes by either budding or fragmentation. In budding, new individuals develop on a segment, then detach. The segment that serves as the **growth zone** for buds is genetically determined, although that segment differs among individuals of the same species. In *Autolytus,* buds themselves undergo budding, resulting in a chain of zooids (Fig. 29.7). Some polychaetes reproduce by spontaneous fragmentation. They divide into two parts, and each fragment regenerates the parts it lacks. *Syllis,* for example, spontaneously breaks at a predetermined site between two somites. The anterior fragment grows a new tail, and the posterior fragment grows a new head. The ability to reproduce by spontaneous fragmentation is perhaps not surprising in a phylum noted for the ability to regenerate lost parts. Many annelids can regenerate a new animal from a few somites that remain after others have been lost by accident or human experimentation. In experimental fragmentation the head and tail always regenerate at the anterior and posterior ends of the fragment, respectively. That is, annelid regeneration shows **polarity,** as well as many other similarities to regeneration in *Hydra* (pp. 506–508).

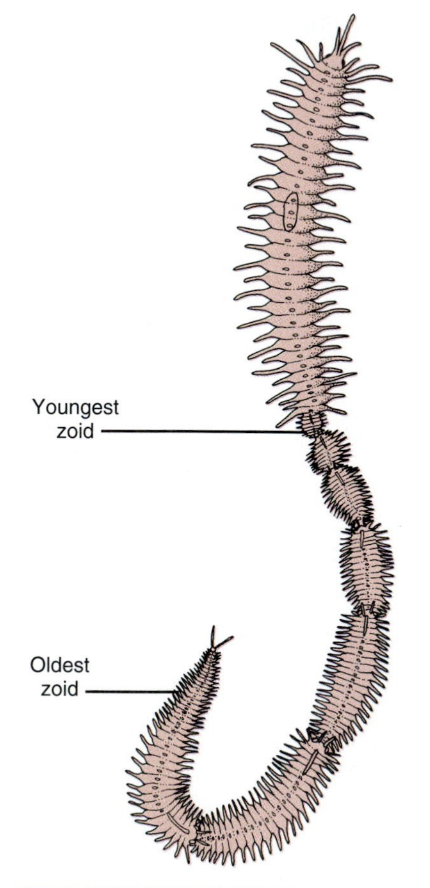

FIGURE 29.7
Budding in the errant polychaete *Autolytus* forms a chain of zooids. Budding proceeds posteriorly, so the oldest zooids are at the rear. (Compare the formation of new segments, which proceeds anteriorly, with the oldest segments in front.)

Sedentary Polychaetes. Many polychaetes are adapted to burrowing or living in tubes. The most apparent adaptations involve the parapodia and the mechanisms of feeding. Generally the parapodia are reduced or modified for some function other than locomotion, and various food-gathering structures compensate for the inability to go out to eat. These two types of modification have led to a greater degree of tagmosis than in errant polychaetes.

Probably the most beautiful adaptations for food gathering are the plumes of tentacles in the **annelid tubeworms** (families Sabellidae and Serpulidae). *Sabella,* for example, has a funnel-shaped plume of tentacles that gives the worm its common name "feather duster" (Fig. 29.8A). Each tentacle, or **radiole,** bears numerous branches lined with cilia (Fig. 29.8B). The cilia produce an upward current of water that draws particles into food grooves in the radiole. Cilia then carry the particles to the mouth. Along the way the particles are sorted according to size. Plankton are carried to the mouth and ingested, the largest particles are rejected, and medium particles of sand are stored for future tube construction.

Other sedentary polychaetes live either partially or entirely within burrows, and they feed on detritus rather than plankton. *Amphitrite* burrows near the low-tide line of quiet bays, but it feeds with its long tentacles extended over the sediment (Fig. 29.9 on page 618). Cilia lining the gutter-shaped tentacles carry food particles to the mouth.

The **parchment worm** *Chaetopterus* lives entirely within its U-shaped burrow (Fig. 29.10 on page 618). The burrow is lined with collagen (the "parchment") and lies buried in the ocean bottom down to 30 meters. The burrow has a diameter of approximately 2 cm in the middle and is often shared by crabs and other commensals. Both ends taper and extend above the sediment and thereby keep out debris. Although this living arrangement provides protection, it presents *Chaetopterus* with problems of obtaining food and oxygen and of eliminating wastes. These problems are solved largely by maintaining a flow of water from anterior to posterior. Fans, consisting of the mod-

A

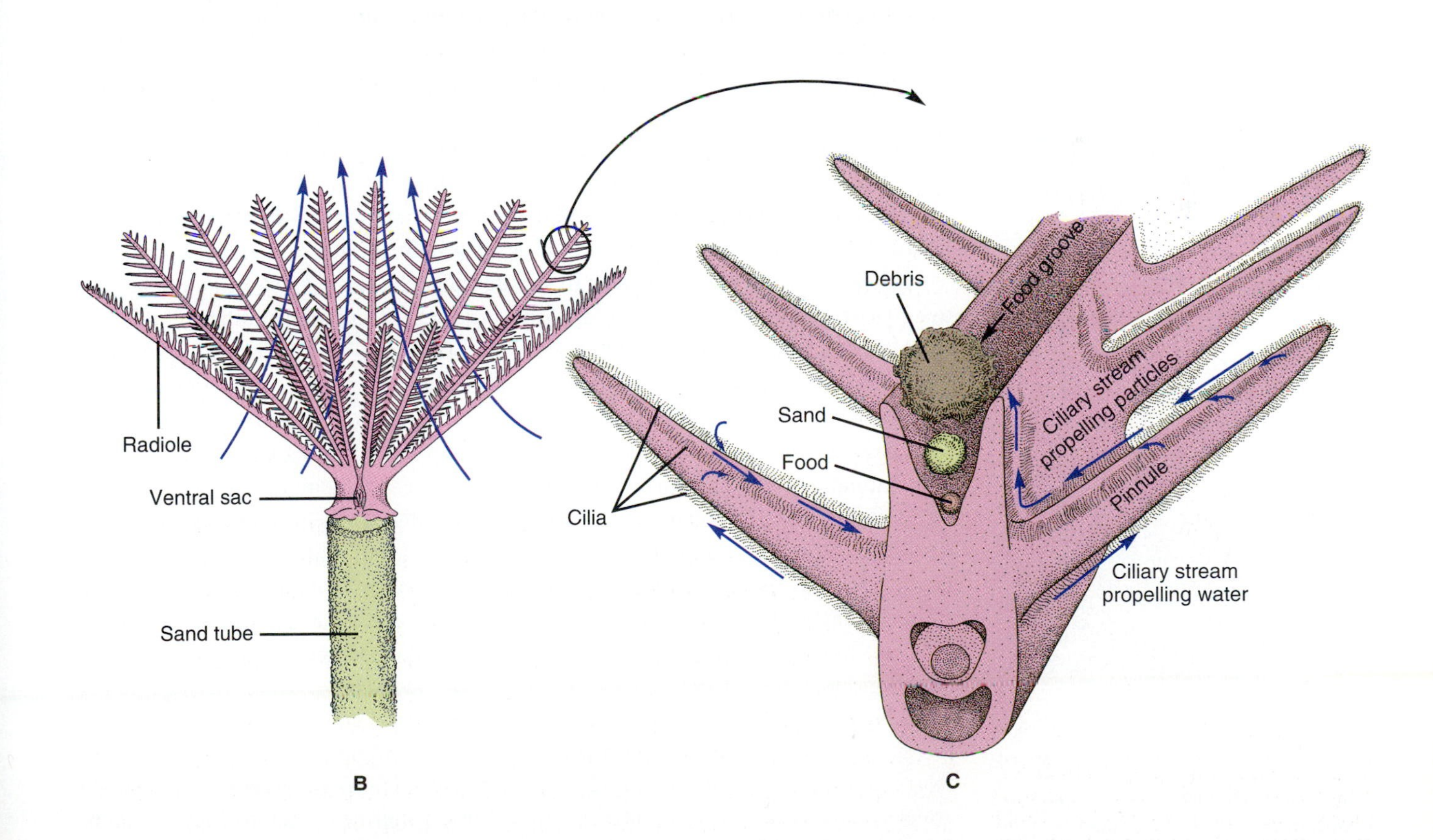

FIGURE 29.8

(A) Feather dusters (sabellids). Each worm is approximately 5 cm long and lives in a tube 10 cm long, into which it withdraws when endangered. (B) Filter feeding by *Sabella.* Arrows show the direction of water currents produced by cilia on the numerous pinnules on each radiole. (C) Movement and sorting of particles in the food groove of a radiole

■ **Question:** How does the mechanism of feeding in a sabellid differ from that of an entoproct (Fig. 26.18) or ectoproct (Fig. 27.2)?

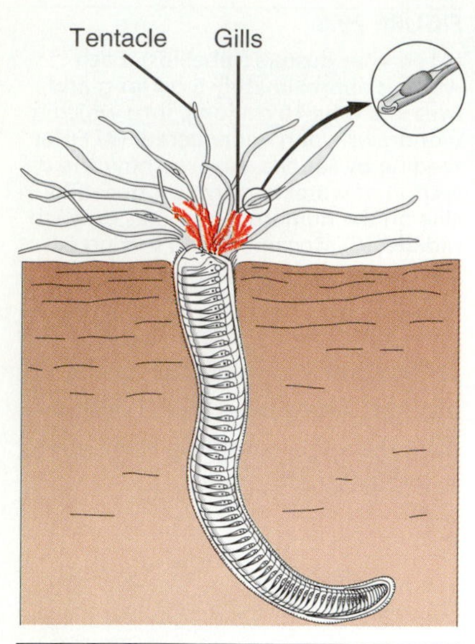

FIGURE 29.9
Amphitrite feeding with its long tentacles. The inset shows food being propelled down the groove of the tentacle by means of cilia. The smaller appendages on the head are gills, which are blood-red with hemoglobin. The worm is approximately 35 cm long.

ified parapodia of segments 14 through 16, generate the flow. Particulate food drawn in by the current is trapped by a web of mucus continuously secreted by the enlarged notopodia of segment 12. The food enters a small pouch in the web, which lies within a food cup on the back of the worm, just ahead of the fans. When the mass of food within the mucous pouch reaches a diameter of about 3 mm, it is pinched off and pushed into a groove along the middle of the back. The fans then stop beating while cilia in the groove move the mucus-wrapped food to the mouth. *Chaetopterus* is clearly one of the most tagmatized of annelids.

The **lugworm** *Arenicola*, which is common in intertidal mud flats, has a less elaborate solution to the problems of living in a burrow (Fig. 29.11). Peristaltic contractions of the lugworm's body draw water into the posterior end of its L-shaped burrow. The unlined burrow is closed near the worm's head, and water percolating upward causes the sand or mud to collapse. *Arenicola* feeds by continually ingesting sand or mud with its eversible proboscis. Periodically the worm backs out of the burrow to defecate. The caving in of sand or mud above the head forms a disc-shaped depression that, together with the mound of fecal castings around the nearby opening of the burrow, is easily recognized at low tide. By continually cycling sediments through its body, the lugworm plays a role in shallow marine environments similar to that played by earthworms in terrestrial environments. In areas where lugworms are abundant, they bring to the surface some 691,000 kg per hectare of sand (1900 tons per acre) each year and bury an equal amount of sediment. On average, all the material within the upper 50 cm of the bottom recycles through lugworms every two years.

Earthworms and Other Oligochaetes

The origin of oligochaetes is uncertain, but it is not difficult to imagine sedentary polychaetes similar to lugworms burrowing generation after generation from the bottoms of shallow bays into the beds of freshwater streams and eventually into moist soil far from flowing water. Such worms would have become increasingly adapted to fresh water, and their delicate tentacles, gills, and parapodia would have become reduced in the abrasive soil. In this way *Arenicola*-like polychaetes may have evolved into the roughly 3000 species of oligochaetes. A few species of oligochaetes have apparently retraced this evolutionary history and now live in burrows or tubes in freshwater or marine sediments.

One of the most common of the freshwater oligochaetes is *Tubifex tubifex*, familiar to aquarists as a fish food. These red worms are up to 10 cm long and live on the bottoms of lakes, duck ponds, and polluted streams. They keep their heads within tubes on the bottom while waving their bright red tails in the water (Fig. 29.12). In certain polluted tidal rivers, tubifex worms are so numerous that the banks appear bright red at low tide. *Tubifex* is also common at sewage outflows and settling tanks. Often *Tubifex* is the only animal seen in its habitat, because few others are so tolerant of low oxygen levels. The ability of *Tubifex* to extract oxygen is due to the high concentration of hemoglobin dissolved in its blood and to the corkscrew motion of its tail, which draws aerated water downward. Surprisingly, *Tubifex* does not absorb oxygen through its integument, but from the lining of its intestine. To make the paradox complete, *Tubifex* obtains all its nutrients by absorbing dissolved organic material across its integument.

FIGURE 29.10
A model of the parchment worm *Chaetopterus*. Parchment worms are often three times as long as this 7 cm specimen. The smaller worm with *Chaetopterus* is the sipunculan *Phascolosoma*.

The Night Crawler. The majority of oligochaetes are not aquatic like *Tubifex*, but spend most of their lives burrowing through soil. Some of these terrestrial earthworms are over 3 meters long (Fig. 29.13). More typical in length, however, is the night crawler *Lumbricus terrestris*, favorite of robins, fishermen, and biology teachers. *Lumbricus* (Fig. 29.14 on page 620) is up to 30 cm long when mature. It usually bur-

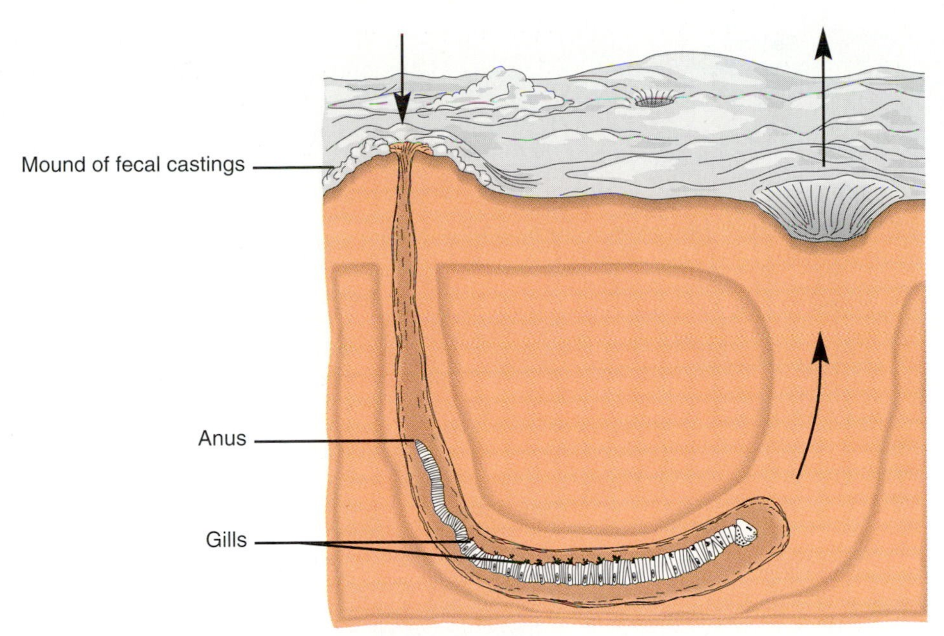

FIGURE 29.11
The lugworm *Arenicola* in its burrow. Arrows show the direction of water movement. The worm is approximately 30 cm long.

■ **Question:** Do you see a possible disadvantage in this direction of water flow?

rows within the upper 30 cm of moist soils rich in organic material, but it can descend to 2 meters to escape drying or cold. On warm, damp nights *Lumbricus* extends the anterior portion of its body out of the burrow to forage for leaves and stems or to mate. They can often be seen on the surface under these conditions, although most of them quickly retreat into the burrow if a light is shone on them. Except after a heavy rain, *Lumbricus* spends its days within the burrow. There it feeds on organic material in the soil or on the vegetable matter it has previously pulled into the burrow. The fecal castings near the entrances of the burrows are often the only clues most people have of the vast numbers of earthworms living beneath their feet. According to one estimate, 54,000 earthworms might inhabit a single acre in England. *Lumbricus* and other terrestrial earthworms amply repay the soil for shelter. Charles Darwin estimated that the dry weight of soil turned over annually by earthworms was as much as 18.1 tons per acre. At the same time, earthworms buried a corresponding amount of vegetable matter, forming up to 0.2 inches of humus per year on an acre of English soil. Like Darwin's previous studies of evolution and the formation of coral reefs, his studies of earthworms show how much change can be effected by small actions multiplied countless times.

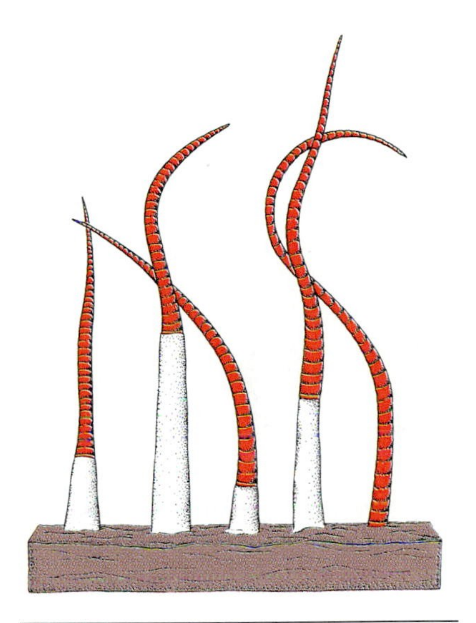

FIGURE 29.12
The oligochaete *Tubifex tubifex,* with its head inside its tube. A spiraling motion of the tail brings oxygen-bearing water downward, and the oxygen is absorbed by the gut.

BURROWING. Earthworms have two methods of burrowing. In hard soil they literally eat their way through. In soft soil and on surfaces, earthworms move by means of peristaltic contractions (see Fig. 10.3). This mode of locomotion uses circular and longitudinal muscles to alternately extend and thicken each segment. Four pairs of setae on each somite (except the first and last) dig into the soil and prevent back-sliding. First the earthworm contracts the circular muscles of the anterior segments. This narrows and lengthens the first segments, which then probe into the soil ahead. The longitudinal muscles of these segments then contract, thickening the segments and widening the channel. In the meantime the circular muscles contract in the segments farther back, narrowing and lengthening those segments so that they are pulled forward when the anterior segments shorten. The longitudinal muscles in those segments then contract and pull them forward, while circular muscles farther back contract. This sequence of alternating contractions of circular and longitudinal muscles moves rearward like a wave, pulling the worm smoothly through the soil.

FIGURE 29.13
An Australian giant earthworm *Megascolides australis* being extracted from its burrow. The two men located it by hearing the gurgling sounds of burrowing.

FIGURE 29.14

Structure of the earthworm *Lumbricus terrestris.* (A) External structure. Some of the 115 to 200 segments are numbered here and will be mentioned later. The ventral surface is somewhat flattened and is not as dark red as the dorsal surface. (B) Cross section showing the muscles, setae, and gut, all used in burrowing. See Fig. 14.7 for the nephridia. See also Fig. 29.16.

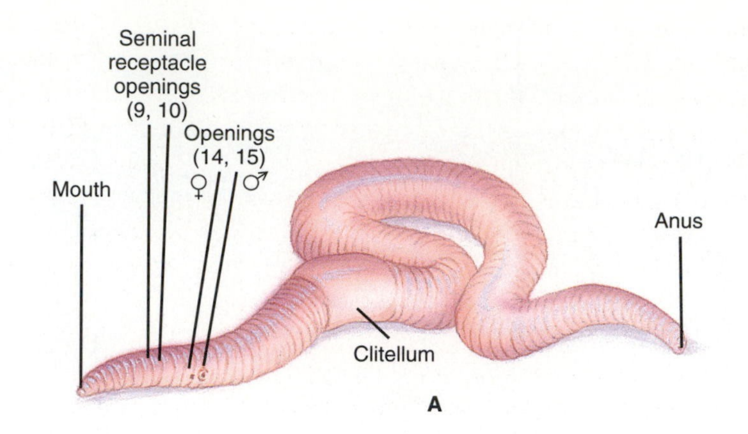

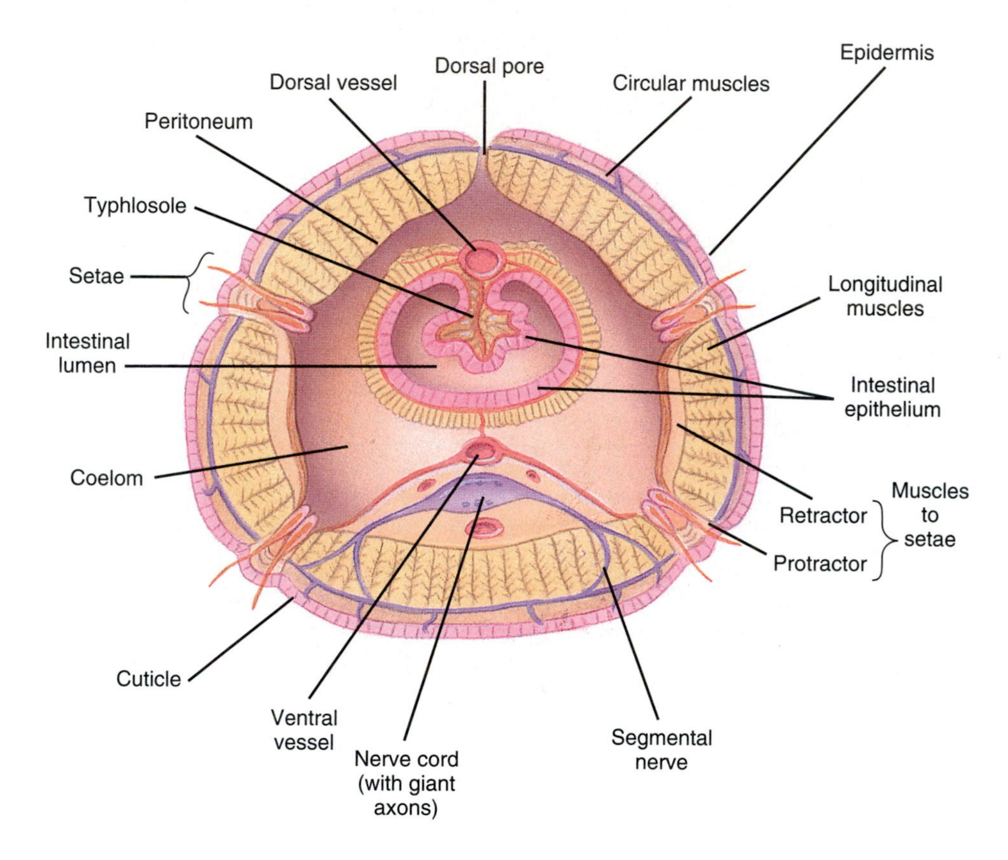

Peristaltic burrowing works only because septa completely separate the coelomic cavities of most of the segments (except between segments 1 through 4, and 17 and 18). Without the septa, contraction of the circular muscles in one segment would narrow that segment but would not lengthen it, because the pressure would be dissipated over the entire length of the animal. Likewise, contraction of the longitudinal muscles of one segment would not thicken the segment to widen the burrow and anchor the setae. Segmentation is therefore essential for this type of locomotion. The setae, in spite of small sizes and numbers, contribute significantly to burrowing. The force they generate is evident from the vigorous tugging robins must use to extract an earthworm from its burrow.

INTEGUMENT. Burrowing through soil is feasible only in an animal with an integument that can resist abrasion by soil particles. The integument of earthworms has a protective outer layer of thin, collagenous cuticle secreted by the epidermis (Figs. 29.14 and 29.15). Along the midline of each segment is a dorsal pore that allows coelomic fluid to leak out slowly and moisten the cuticle. The epidermis also includes numerous glands that secrete mucus that lubricates the cuticle and protects it from drying. In cuticle that has been stripped off and allowed to dry on a microscope slide, it is easy to see the pores leading to the mucous glands, as well as pores leading to numerous sensory cells. These sensory cells presumably include chemoreceptors and mechanoreceptors, which would be of obvious use in a burrowing animal. Earthworms have no eyes, but there are numerous photoreceptor cells in the epidermis of each segment.

In spite of the lubricating and moisturizing secretions, the delicate integument is certainly vulnerable to injury. A more armorlike cuticle such as that in insects would, however, make the segments inflexible and would therefore interfere with burrowing. Moreover, a thicker cuticle would interfere with **integumentary respiration.** Because parapodia and gills would be even more of a liability in burrowing, earthworms must rely exclusively on integumentary respiration. The uptake of oxygen is aided by an extensive system of capillaries beneath the cuticle, by the presence of hemoglobin dissolved in the blood plasma, and by the efficient closed circulatory system with its five aortic arches ("hearts") that aid the dorsal vessel in pumping (Fig. 29.16).

DIGESTION. Considering the amounts of food and soil that pass through them, earthworms' digestive systems are surprisingly simple. The digestive tract of *Lumbricus* starts with the mouth, which lies beneath the prostomium in the first segment. Food is pumped by the pharynx into the esophagus and is stored in the crop. (On each side of the esophagus are calciferous glands, which are thought to excrete excess calcium from the body fluids.) Following the crop in *Lumbricus* is a muscular gizzard lined on the inside with cuticle, where food is ground up. Food then enters the straight intestine, where most enzymatic digestion and absorption occur. Among the enzymes secreted into the intestine is cellulase, which digests plant cell walls.

A feature characteristic of many oligochaetes is an inward fold along the length of the intestine, the **typhlosole,** which increases the area for absorption of nutrients (Fig. 29.14B). Surrounding the outside of the intestine are chloragogue cells, which store

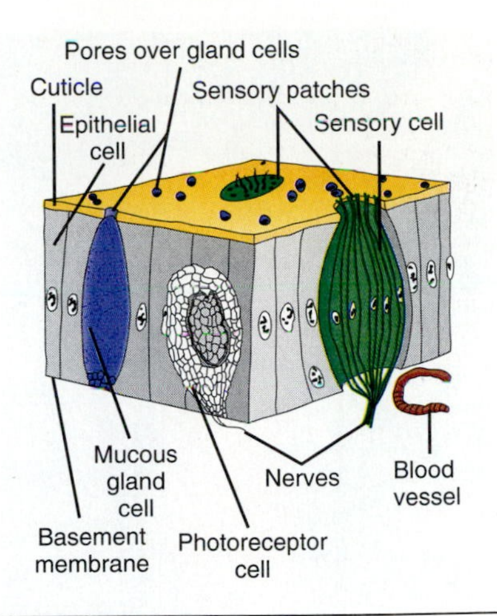

FIGURE 29.15
The integument of *Lumbricus.*

■ **Question:** Which sense organs would be involved in determining which direction is up during burrowing?

FIGURE 29.16
Cutaway view of the internal organs of *Lumbricus.* Note the five aortic arches ("hearts") joining the dorsal and ventral vessels in segments 7 through 11. Reproductive organs are shown in greater detail in Fig. 29.18A.

■ **Question:** What is the main role of the cephalic ganglion in an earthworm?

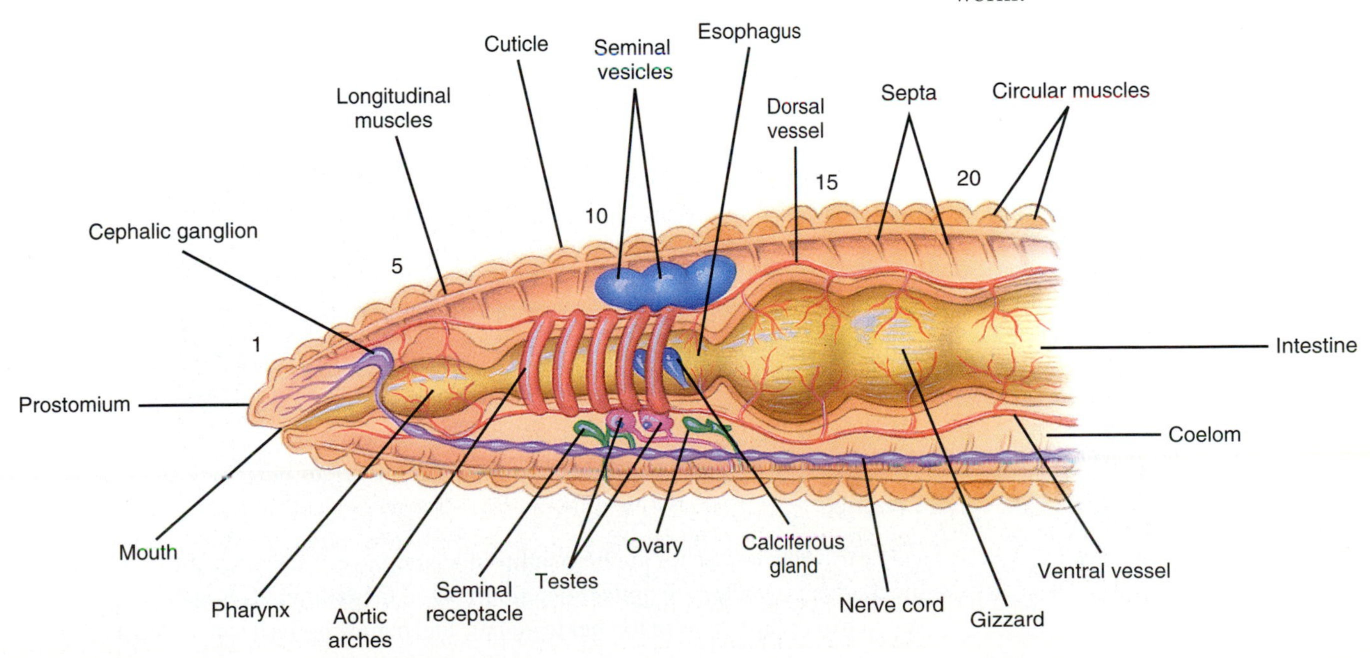

FIGURE 29.17
Two earthworms, *Lumbricus terrestris,* copulating.

nutrients and eliminate nitrogenous wastes, as described previously for polychaetes. The nephridia of freshwater and terrestrial oligochaetes are much like those of marine polychaetes except that they eliminate larger amounts of water during osmoregulation (see Fig. 14.7).

REPRODUCTION. Oligochaetes clearly must have a method of reproduction different from that of their aquatic ancestors. Releasing large numbers of gametes into the environment and trusting to chance simply will not work in soil. Unlike polychaetes, which rely on the coelom and nephridia for reproduction, each oligochaete has specialized reproductive organs of both sexes. Although hermaphroditic, oligochaetes usually mate with each other rather than with themselves. Pairs of *Lumbricus* mate by extending the anterior portions of their bodies from their neighboring burrows and exchanging sperm. Copulation lasts two to three hours, and an individual can copulate every three or four nights if the air is sufficiently warm and damp to allow them to stay out. During copulation the two worms are held in mutual embrace by mucus secreted from the clitellum, which lies on somites 32 through 37 (Fig. 29.17). The clitellum also has specialized **genital setae** that clasp the mate by penetrating its integument. The clitellum is conspicuous and permanent in sexually mature *Lumbricus,* but is temporary in earthworms that breed only during particular times of the year.

Sperm are produced by two pairs of small testes, which are in segments 10 and 11 of *Lumbricus* (Figs. 29.16 and 29.18A). After maturing within the seminal vesicles (= sperm sacs), sperm pass through funnel-shaped openings into sperm ducts. During copulation, cilia lining the sperm ducts propel the sperm through the male pore on the ventral surface of somite 15. The sperm from each worm are then deposited in its mate's **seminal receptacles** in somites 9 and 10. It might appear that the easiest way to exchange sperm would be for two earthworms to lie beside each other with each male pore releasing sperm directly into the mate's seminal receptacles. Many earthworms do, in fact, mate in this position. *Lumbricus,* however, copulates with the male pore opposite the clitellum of its mate, so the sperm must journey some 20 segments to reach the seminal receptacles of the mate (Fig. 29.18B). The sperm pass through a **seminal groove:** a wrinkle on the ventral surface formed by contractions of certain muscles. Rhythmic contractions of these muscles propel the sperm through the groove. The groove is enclosed by a tube of slime surrounding each worm, so sperm from each mate do not mix as they pass each other.

After copulating, the worms separate and return to their burrows. Fertilization and egg-laying occur a few days later, in one of the most remarkable reproductive processes among animals. First each worm secretes a sheath of mucus around its clitellum and anterior segments (Fig. 29.18C). The clitellum then secretes albumen as nourishment for the eggs, and it envelops the mucus and albumen in a tough chitinlike material that will form a **cocoon** for the eggs. The worm then backs out of the cocoon. As the cocoon slips over segment 14, it receives an ovum from a pore leading to the oviduct. (The ovum was previously produced by an ovary in segment 13, and it matured in the coelom.) As the cocoon continues to slide, it picks up sperm from the sperm receptacles in segments 9 and 10. Once the cocoon has completely slipped off, its ends close, forming a yellowish, lemon-shaped body approximately 7 mm long. Fertilization occurs within the cocoon. Two to three weeks later, one new worm emerges.

Leeches

Like earthworms, leeches are hermaphroditic and have clitella that form cocoons. In most other ways they are quite different. Many of these differences appear to have resulted from adaptations of leeches to aquatic life, mainly in fresh water. Peristaltic bur-

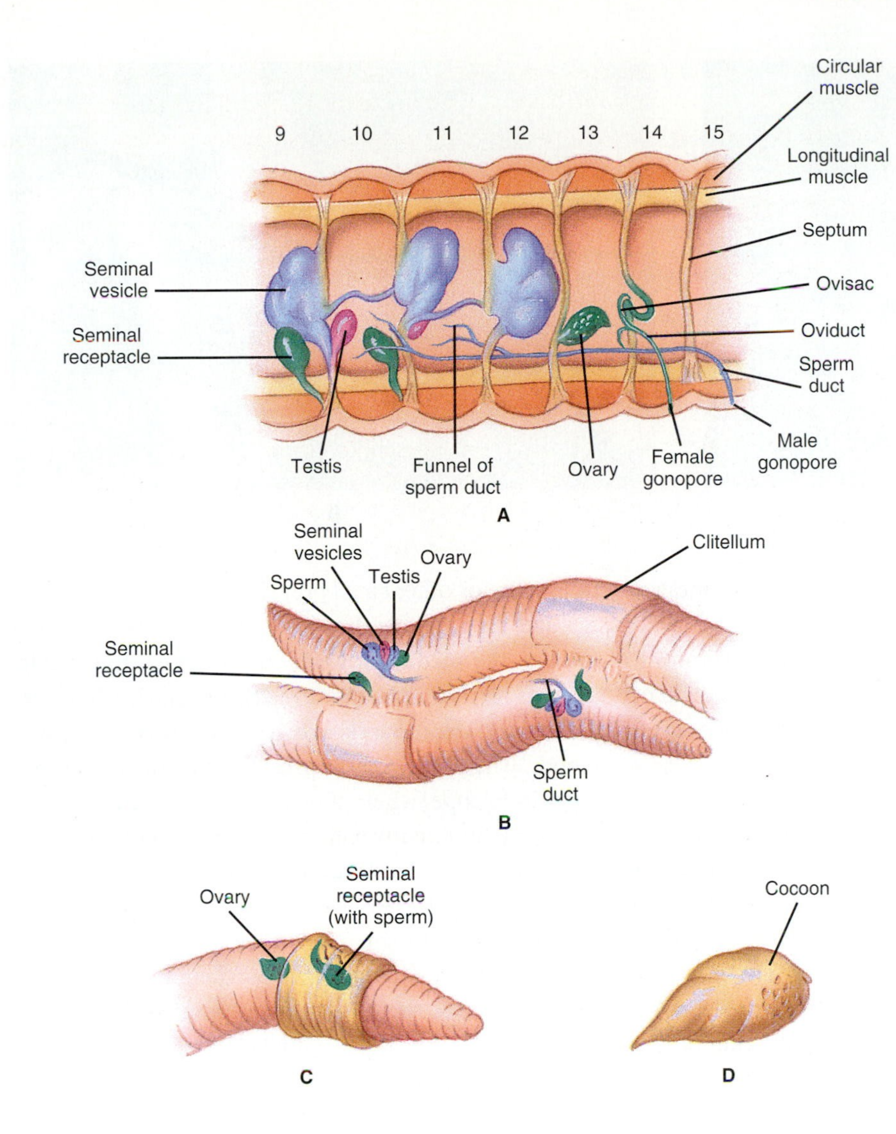

FIGURE 29.18

Reproduction in *Lumbricus.* Male and female organs are shown in different colors. (A) During copulation the fertile male somites (10 and 11) discharge sperm through a pore on somite 15. Later the fertile female somite (13) releases ova that mature in the coelom, then exit through a pore in somite 14. (B) Copulation. Sperm must travel outside the body to reach the seminal receptacle of the mate. (C) Fertilization occurs at the time the cocoon forms. As the cocoon slides anteriorly from the clitellum, it picks up ova, then sperm from the seminal receptacles. (D) A cocoon.

rowing is no longer appropriate, and most leeches lose their setae and intersegmental septa during embryonic development. The coelom functions as a single large chamber, which is occupied mainly by spongy tissue and dorsoventral muscles. The latter muscles, in addition to diagonal, circular, and longitudinal muscles, allow leeches to swim quite gracefully by means of dorsoventral undulations of the body (Fig. 29.19 on page 624). Some leeches are adapted to terrestrial living, at least part of the time. On solid substratum they move in inchworm fashion, alternately attaching and detaching anterior and posterior suckers. Leeches also differ from oligochaetes in that the segments are constant in number and are often subdivided externally into numerous **annuli.** Because the fissures between annuli look like those between segments, it is hard to determine from appearance where the boundaries of each segment are. Internally, however, each segment can be defined by the fact that it is innervated by one ganglion of the ventral nerve cord.

FEEDING. Because of their reputations as blood suckers, leeches do not enjoy much esteem among humans. However, few of the 500 or so species of leeches are ectoparasites on humans. Many leeches are highly specific for their hosts, and they feed on the body fluids of invertebrates (including leeches of other species) or fishes, but not on mammals. Many lack cutting or probing mouthparts and must eat worms, invertebrate

A

B

FIGURE 29.19

Locomotion in the American medicinal leech *Macrobdella decora.* (A) Using the anterior and posterior suckers to creep up the side of an aquarium. (B) Swimming by dorsoventral undulations of the body. Length approximately 10 cm.

larvae, the eggs of fishes and amphibians, or other small prey. Many also scavenge carrion. Some leeches, however, can and do consume the blood of mammals, including humans. The best known of these is the medicinal leech, *Hirudo medicinalis* (Fig. 29.20). The term "medicinal" comes from the fact that until a century ago leeches were used to treat a variety of human ailments. The theory was that somehow leeches could suck out disease-causing "bad blood." This ability was so prized that medicinal leeches were collected almost to extinction in Europe, where they are now protected by law. They were introduced into America, but apparently without lasting success. Although relatively rare in nature, they are widely cultured in laboratories for research, and they still have some medical uses.

The ability of some leeches to parasitize humans, other mammals, and occasionally other vertebrates, is due to a variety of adaptations that allow them to pierce or cut into the skin without notice. Leeches attach to their hosts by means of suckers. Some then pierce the skin of the host with a sharp proboscis. Others cut into the skin with sharp jaws. *Hirudo* uses three sharp-toothed jaws like circular-saw blades to make a Y-shaped incision in skin (Fig. 29.21A). While cutting the skin, it secretes an unidentified local anesthetic, as well as a histamine-like substance that keeps the blood vessels open. The blood is then sucked in by the pharynx. While the blood is being swallowed, it

FIGURE 29.20

The medicinal leech *Hirudo medicinalis* removing accumulated blood (a hematoma). This European species is up to 20 cm long.

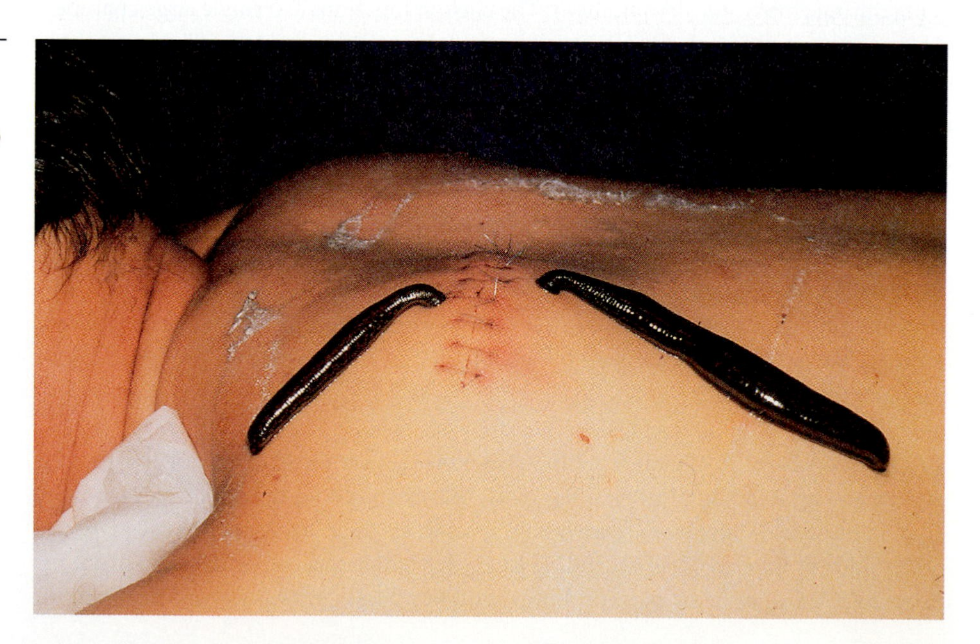

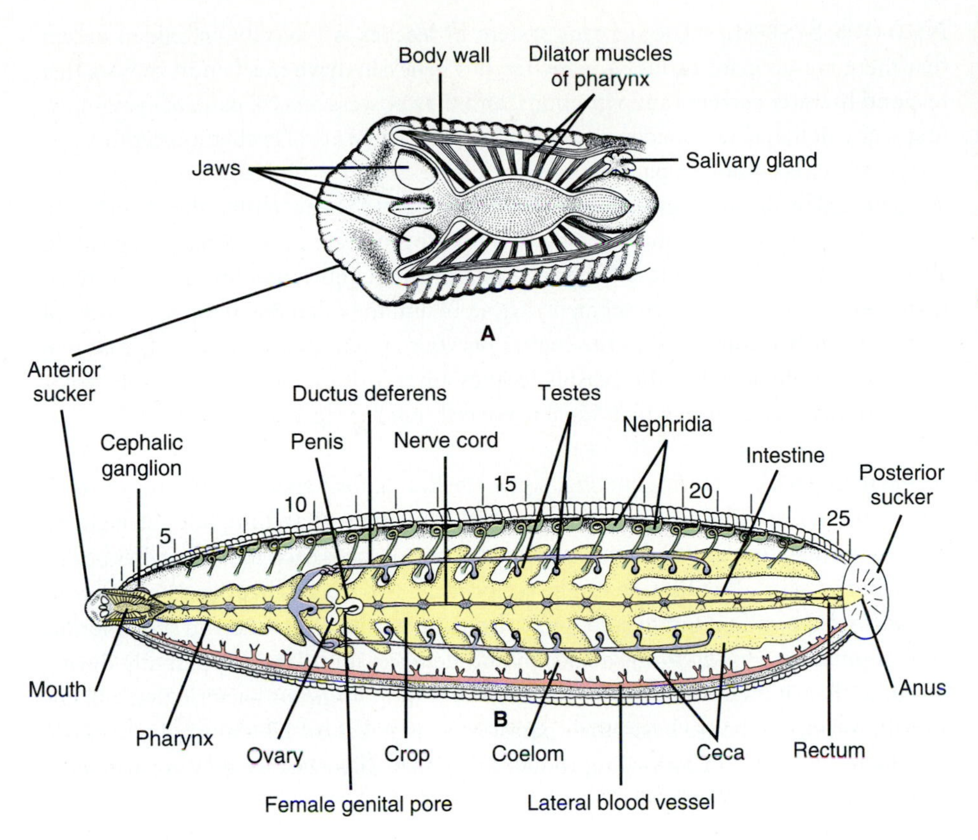

FIGURE 29.21

(A) The feeding apparatus of the medicinal leech *Hirudo medicinalis.* The three toothed jaws slice through skin. The dilator muscles of the pharynx then contract to suck blood. (B) Ventral view of the internal structure of *H. medicinalis.* Vertical lines and numbers (top) indicate segments.

■ **Question:** Like a flatworm, the body of a leech is largely occupied by spongy tissue. Why, then, is a leech considered to be a coelomate? What essential differences distinguish leeches from flatworms? (Compare Fig. 25.2.)

mixes with a protein called **hirudin,** which prevents clotting by inhibiting the action of thrombin (p. 229). Without this anticoagulant a blood clot in the leech would interfere with locomotion.

A medicinal leech takes up to 15 ml of blood during a single feeding. Much of this volume is essentially salt water, which the 17 pairs of nephridia soon eliminate. The concentrated blood is then stored within numerous diverticula of the crop. The proteins of hemoglobin and other components of the blood are then digested extremely slowly. In one study *Hirudo* took 200 days to digest a single meal, and it continued to live for another 100 days on the energy stored from that feeding. One unexpected feature of digestion in leeches is that their guts do not secrete any proteolytic enzymes. Instead they rely on bacteria in the gut for enzymatic digestion. In *H. medicinalis* only one species of bacterium, *Pseudomonas hirudinis,* digests the blood. *Pseudomonas hirudinis* also produces an unidentified antibiotic that eliminates any other bacteria from the digestive tract, allowing red blood cells to remain unspoiled within the leech intestine for more than a year. Because of the antibacterial action of *P. hirudinis,* one need not worry about getting a bacterial infection from *Hirudo.* Some disease-causing protozoans and tapeworm cysts, however, are transmitted by leeches.

CIRCULATION. The ability of *Hirudo* to live for many months on a single meal clearly indicates a low metabolic rate. Its circulatory system is reduced accordingly. In many leeches, including *Hirudo,* there are not even any blood vessels. The coelomic fluid, which pulsates slowly through sinuses of the tissue-filled coelom, performs the functions of blood. (Because the circulatory system is open, the coelom is also considered to be a **hemocoel.**) In *Hirudo* and many other leeches, the coelomic fluid contains hemoglobin. (It is produced by the leech rather than obtained from the host.) A few leeches possess gills, but *Hirudo* and most others exchange gases through the integument.

Nervous System. The nervous system of leeches is typically annelidan except that there are no giant axons. Aquatic leeches generally have mechanoreceptors that respond to water currents and vibrations, and they possess several pairs of eyes on the first segment that detect shadows of potential prey. There are also chemoreceptors and temperature receptors. Apparently hungry leeches attach to virtually any moving object detected by the mechano- and photoreceptors, but they drop off if the chemo- and thermoreceptors indicate that the object is not appropriate prey. *Hirudo medicinalis* also has well-defined receptors for touch, pressure, and injurious stimuli concentrated in the central annulus of each segment. These 14 and the other 340 or so nerve cells in each segment are connected in the central nervous system in the same way from one individual to the next. For this reason, leeches are favorite subjects of neurophysiologists. Another reason is that individual nerve cells readily regenerate.

Reproduction. During mating, some leeches deposit **spermatophores** on each other's bodies, and the sperm bore through the integument and migrate through the coelom to reach the eggs. In *Hirudo* and many other leeches each individual has both a penis and a vagina. Fertilization can be reciprocal if two leeches happen to copulate in a head-to-tail position, but this does not always occur. Between one and nine months after copulation, *Hirudo* forms a cocoon and deposits it on land. Apparently during this time the parent also inoculates the cocoon with *P. hirudinis* bacteria that will enable the young leeches to begin their parasitism immediately. Like most oligochaetes, leeches are incapable of asexual reproduction. Unlike oligochaetes, they are also incapable of regenerating lost body parts.

Interactions with Humans and Other Animals

Annelids attract little notice from most people, perhaps because so few of them are edible to us, and vice versa. Indirectly, however, annelids certainly have a profound effect on humans and other animals. Even if we ignore the polychaetes, about which there is little ecological knowledge, the role of annelids is impressive. The effect of earthworms in fertilizing and aerating soil has already been noted. They are so beneficial that in some areas where they do not occur naturally, or where they have been eliminated by overuse of poisons, farmers pay to have them brought in. Earthworms have also been put to work reclaiming mined soils. The night crawler *Lumbricus,* the clam worms *Nereis,* and the blood worm *Glycera* are also the bases for a bait-worm industry worth tens of millions of dollars. This figure does not include the thousands of entrepreneurs who dig worms in the backyard and sell them at roadside.

Although earthworms are not a major item in the human diet, they do feed a large number of animals that are important to us. *Turdus,* a genus that includes the American robin and the European blackbird, comes immediately to mind. Less familiar as predators on earthworms are burrowing mammals such as moles and shrews. In many areas, foxes also get most of their protein from earthworms during part of the year.

Leeches have less human impact, which suits most people. In North America they are occasionally bothersome; in other areas they can be life-threatening. In areas around the Mediterranean, *Limnotis nilotica* invades the tracheae and esophagi of horses, cattle, and even humans who drink from springs and ponds that the leeches inhabit. Large numbers can cause asphyxiation. Even more distressing are land leeches in the genus *Haemadipsa,* which live in moist tropical areas. They detect groups of mammals by means of odors, vibrations, and warm air currents and actually creep toward them. They can penetrate small openings in clothing and attach to victims undetected. The worst comes after the leeches fall off, because the anticoagulant causes the wounds to bleed severely.

Even leeches have their virtues, however. Although the curative powers of the medicinal leech fell into disrepute because of excessive claims in the previous century, *Hirudo medicinalis* still has its valid medical uses. They are sometimes applied to remove the discolorations of bruises and to improve local circulation after severed limbs are reattached. Leeches would be of even greater medical service if hirudin could be extracted in large enough quantities to prevent blood clots, but it is not practical to extract the anticoagulant from such small animals. The gene for hirudin has been cloned, however, so this problem may be solved. In the meantime, Roy T. Sawyer went to French Guiana in 1977 in pursuit of the Amazon leech *Haementeria ghilianii,* which would be large enough to supply large amounts of hirudin. Sawyer found his leeches, and eventually he and his colleagues discovered that this species produces a previously unknown anticoagulant. This enzyme, called hementin, breaks down fibrin strands. Unlike hirudin, therefore, it is effective treatment for heart attacks even after clots have already formed. Sawyer has started a leech farm in Wales that markets several useful leech products.

Summary

Annelids include such diverse animals as polychaetes, oligochaetes, and leeches. In spite of their differences, all share the feature of segmentation, which may have evolved as an adaptation for burrowing or as a means of achieving larger size without all the organs growing larger. Polychaetes are either errant or sedentary. The errant polychaetes can wriggle and can swim by means of parapodia. Many prey by means of an eversible proboscis. The sedentary polychaetes, such as annelid tube worms, generally live in burrows or secreted tubes, and many feed on particulate matter by means of tentacles.

Earthworms (oligochaetes) burrow in soil either by eating their way through or by alternately contracting circular and round muscles, using their few setae to grip the sides of the burrow. Because the coelom is segmented into compartments, each segment can narrow or shorten independently of others during burrowing. Earthworms feed on vegetable material. They are hermaphroditic but generally exchange sperm. Leeches are mainly aquatic and either creep about with their suckers or swim. Many prey on smaller animals, but some suck the blood of vertebrates.

Key Terms

segment
tagma
septum
seta
parapodium
clitellum
ventral nerve cord
mesentery
epitoky
radiole

Chapter Test

1. Explain what segmentation is. What is the importance of segmentation in annelids? How is segmentation reflected in the structure of the nervous system, metanephridia, and the coelom? (p. 608)
2. What are the three classes of annelids? Briefly state the differences among them. (pp. 608–612)
3. Describe how one member of each class obtains food. Sketch the feeding apparatus for each one. (pp. 612–625)
4. Describe the means of locomotion in errant polychaetes, in earthworms, and in leeches. Explain how the coelom is adapted for locomotion in errant polychaetes, earthworms, and leeches. (pp. 613–623)
5. What are some of the advantages and disadvantages of a sedentary life in polychaetes? How are the disadvantages overcome? (pp. 616–618)
6. Compare sexual reproduction in a marine polychaete with that in a terrestrial earthworm. Explain how each method of reproduction is appropriate to the habitat. Why do you think leeches reproduce in a manner similar to that of earthworms? (pp. 615, 622)

■ Answers to the Figure Questions

29.2 Everting the proboscis to snare prey is faster than lunging after them and is less likely to startle the prey.

29.3 The lens of *Nereis* would concentrate light entering the small opening of the eye onto a few receptor cells, enabling the eye to determine the direction of light.

29.5 The statocyst would enable the trochophore to sense the direction of gravity, which would be necessary for controlling position and direction of swimming. The eye would also aid in directing vertical movement toward or away from the water surface.

29.8 In entoprocts and ectoprocts there is no food groove that sorts food from sand and debris.

29.11 An obvious disadvantage is that the water that brings in oxygen must pass the mound of fecal castings and the anus.

29.15 Unless the burrow is very deep, the photoreceptor cells could be used to distinguish up from down. In addition, mechanoreceptive cells in the integument could determine which part of the worm is being pressed by gravity against the wall of the burrow.

29.16 The anterior segments are much like the others except that feeding and burrowing begin there. It seems likely, then, that the cephalic ganglion is largely involved in coordinating those activities.

29.21 Leeches are considered to be coelomates mainly because they share characteristics of other annelids, which are clearly coelomates. Unlike flatworms, leeches are segmented, they move by means of undulations rather than by cilia and mucus, and they have metanephridia, a ventral nerve cord with segmental ganglia, a complete digestive tract, and clitella.

Readings

Recommended Readings

Darwin, C. 1881. *The Formation of Vegetable Mould through the Action of Worms, with Observations on their Habits.*

Lent, C. M., and M. H. Dickinson. 1988. The neurobiology of feeding in leeches. *Sci. Am.* 258(6):98–103 (June).

Nicholls, J. G., and D. Van Essen. 1974. The nervous system of the leech. *Sci. Am.* 230(1):38–48 (Jan).

Stent, G. S., and D. A. Weisblat. 1982. The development of a simple nervous system. *Sci. Am.* 246(1):136–146 (Jan). (*In the leech.*)

See also relevant selections in General References at the end of Chapter 21.

Additional References

Lee, K. E. 1985. *Earthworms: Their Ecology and Relationships with Soils and Land Use.* Orlando: Academic Press.

Mill, P. J. (Ed.). 1978. *Physiology of Annelids.* New York: Academic Press.

Pennak, R. W. 1989. *Freshwater Invertebrates of the United States,* Vol. 1, 3rd ed. New York: Wiley, Chapter 13.

Satchell, J. E. (Ed.). 1983. *Earthworm Ecology.* New York: Chapman and Hall.

Sawyer, R. T. 1986. *Leech Biology and Behaviour* (three volumes) New York: Oxford University Press.

CHAPTER

30

Introduction to Arthropods: Spiders and Minor Groups

Crab spider *Misumena vatia.*

General Features of Arthropods

There are so many spiders, crustaceans, insects, and other arthropods that it will require three chapters to introduce them (Fig. 30.1). This great variety seems almost like an outburst of evolutionary exuberance, as if nature were celebrating having at last perfected and completed the protostome division. (See Fig. 21.10.) It is more likely, however, that the enormous diversity and numbers of arthropods are due to their success in invading land and adapting to the great variety of terrestrial habitats.

A major key to the success of arthropods is the integument, which has a hard **cuticle.** The integument reduces the loss of water from the body and serves as a sup-

FIGURE 30.1

Some representative arthropods. (A) The centipede *Lithobius.* Approximately 3 cm long. (B) The onychophoran *Peripatus.* Note the absence of joints in the legs (lobopodia). (C) The six-spotted fishing spider *Dolomedes triton* on water. Leg spread approximately 3 cm. (D) A crayfish. Length approximately 10 cm. (E) The southeastern lubber grasshopper *Romalea microptera.* Length 6 cm.

■ **Question:** Construct a cladogram (see pp. 436–437) for these arthropods based on the characters visible here. Compare your result with Fig. 30.5.

portive exoskeleton and protective armor. The rigid cuticle of most arthropods imposes a number of other characteristic features. One of these is the jointed appendages of most forms, which give the phylum its name. Because the cuticle serves as an exoskeleton, a pressurized hydroskeleton is unnecessary, and the coelom is greatly reduced. The main body cavity is the **hemocoel,** which consists of tissue spaces through which blood circulates. Some other features of arthropods are found also in annelids. Annelids also have a cuticle, although it is not composed of the same material as the cuticle of arthropods (chitin). Arthropods are also segmented like annelids, although the segments are not separated internally by septa, and many are tagmatized into distinct regions such as head, thorax, and abdomen. As in annelids, the central nervous system is primitively a ventral nerve cord with segmental ganglia, although many of these fuse during development. Most arthropods also have compound eyes (each with numerous lenses), which occur in a few annelids but in no other group. Unlike annelids, however, arthropods have open circulatory systems (Table 30.1).

TABLE 30.1

Characteristics of arthropods.

Phylum Arthropoda are-THROP-uh-duh (Greek *arthron* joint + *pod-* pertaining to the foot)

Morphology	Coelomate, although coelom reduced to portions of reproductive and excretory systems. Major body cavity is the **hemocoel** consisting of blood-filled spaces among tissues. Integument with **chitinous cuticle** that serves as **exoskeleton**. Entire integument **molts** periodically. Body **segmented**, usually with many segments altered as **tagmata** (head, thorax, cephalothorax, abdomen).
Physiology	Gaseous exchange through **tracheal tubes, gills, book lungs**, or **body surface**. Digestive tract complete. Central nervous system usually a chain of ganglia on a ventral nerve cord. Cerebral ganglion located dorsally, usually with inputs from various combinations of compound and simple eyes. Circulatory system open. Heart consisting of a pulsatile vessel along the dorsal midline, with blood entering through one-way **ostia**. Osmoregulation mainly by **Malpighian tubules** and/or glands.
Locomotion	Legs usually jointed and moved by striated muscles attached to cuticle. Motile cilia absent except in sperm of a few groups.
Reproduction	Generally dioecious. Sexual dimorphism common. Fertilization internal. Usually oviparous.
Development	Superficial cleavage pattern (Fig. 21.9). Protostomate. Development direct or with characteristic larval stages.
Habitat, Size, and Diversity	Crustaceans generally aquatic; others generally terrestrial. Distributed worldwide. Body length (excluding legs) from less than 0.1 mm to 60 cm. Approximately **874,000 living species** described; many more undescribed.

The Cuticle of Arthropods

Structure. As previously noted, the cuticle is a major distinction of arthropods. The arthropod cuticle consists largely of **chitin** and proteins, which are secreted by the epidermis. In crustaceans the cuticle also incorporates calcium phosphate and calcium carbonate. Chitin consists of polysaccharide chains that line up to form crystalline, hexagonal rods. Bundles of these rods held together in a protein matrix account for the shape and strength of cuticle (Fig. 30.2A, B).

Chitin occurs only in the two inner layers of cuticle, the **endocuticle** and **exocuticle** (Fig. 30.2C). The difference between the two layers is that endocuticle remains pliable, while exocuticle becomes hardened, or **sclerotized,** in most areas. Sclerotization is also called **tanning.** Like the preparation of leather, the tanning of arthropod cuticle results from the cross-linking of proteins. Covering the exocuticle is a thin, sclerotized **epicuticle,** which consists of proteins and waxes (fatty acids linked to alcohols). The waxes in the epicuticle retard water loss. Generally, the more arid the environment, the greater the amount of wax in the epicuticle.

The cuticle of arthropods is also responsible for their often spectacular colors that serve in camouflage, recognition, and warning coloration. These colors arise in two major ways: from pigments and from the fine structure of the cuticle. Pigments in the exocuticle or in the wax are generally responsible for long-wavelength colors: brown, red, orange, and yellow. Short-wavelength colors (green, blue, violet) are generally **structural colors** (Table 10.1, p. 194).

Running through the exocuticle and endocuticle are **pore canals** and **wax canals** that are thought to be paths by which the epidermis secretes waxes onto the epicuticle. The pore canals are also thought to be the routes by which calcium salts become incorporated into the endocuticle in the exoskeletons of crustaceans. There are also canals leading from **dermal glands,** whose function is unknown. Parts of the cuticle are modified into a variety of sensory receptors that enable the arthropod to know what is going on outside. Pores through the cuticle allow detection of chemicals by chemorecep-

FIGURE 30.2

The structure of chitin and the cuticle. (A) Chitin is made of chains of *N*-acetylglucosamine. (B) Crystalline rods of chitin held together by protein form the main structural elements of cuticle. The cuticle is reinforced by calcium carbonate in most crustaceans. (C) Cuticle, secreted by the epidermis, has three layers: the chitinous endocuticle and exocuticle and the lipid-rich epicuticle.

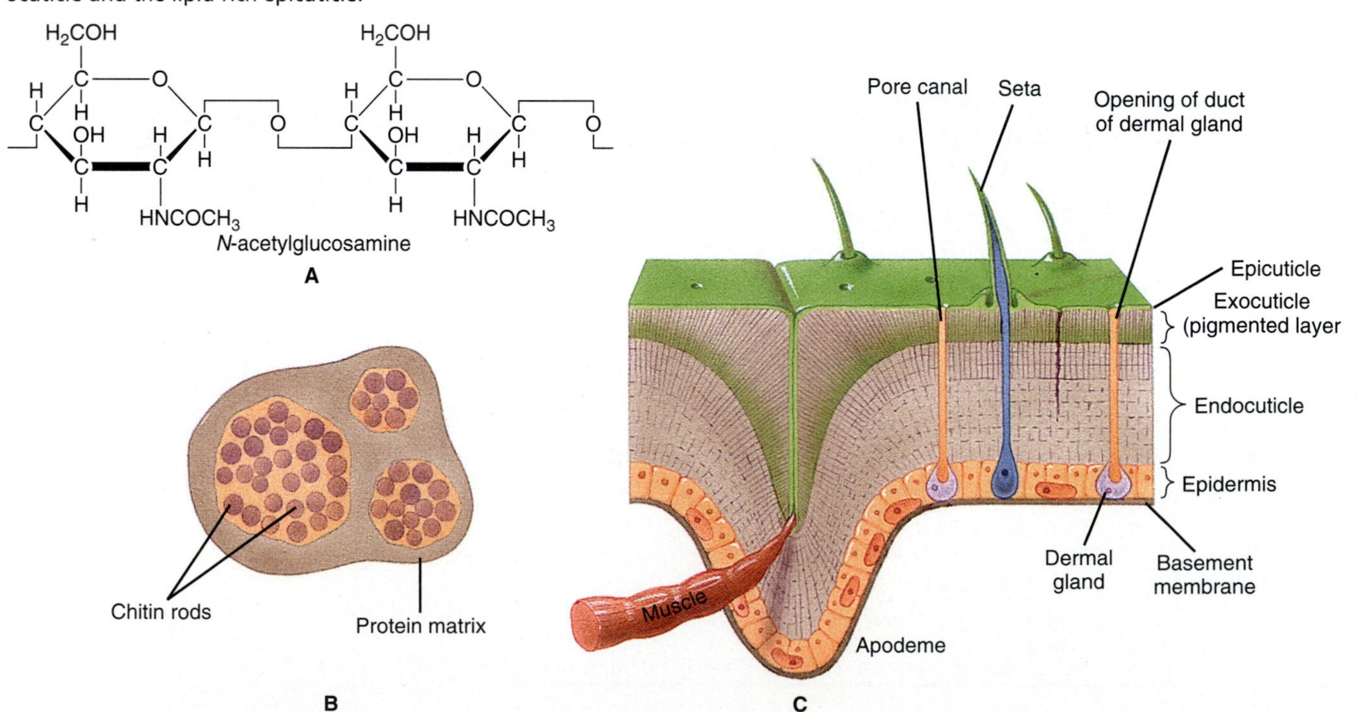

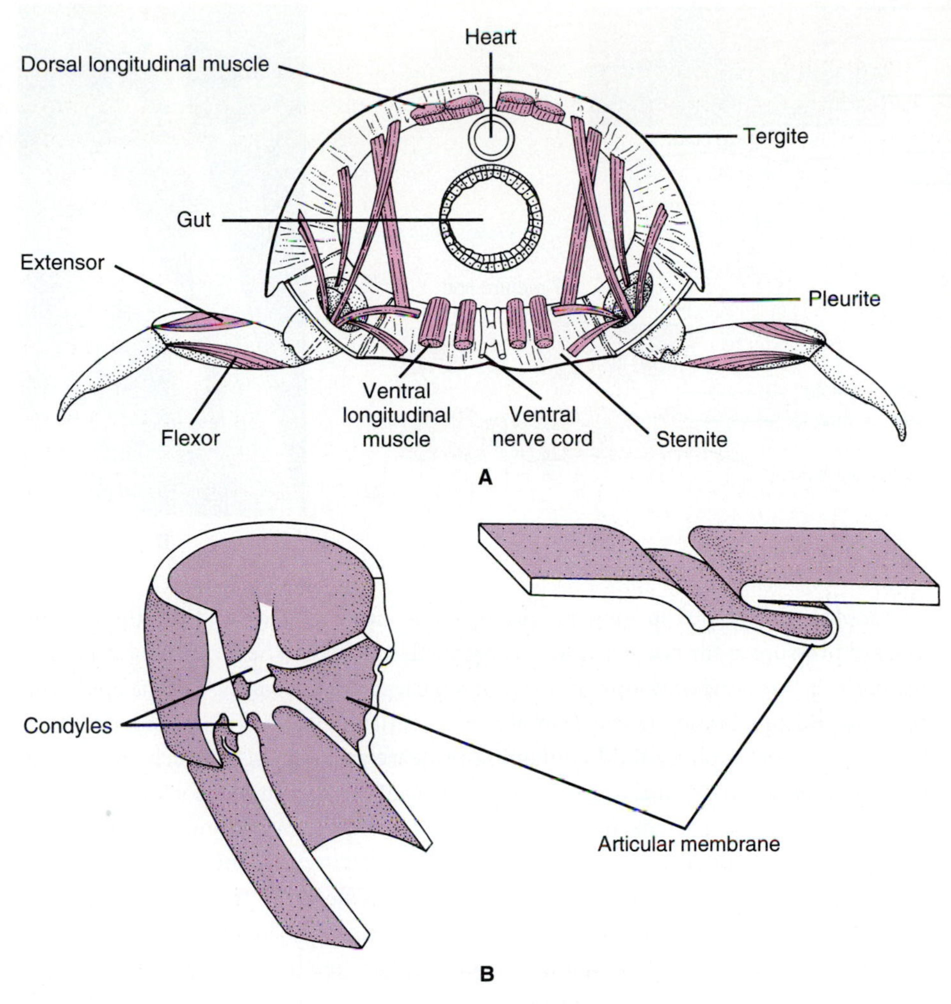

FIGURE 30.3
(A) Schematic representation of an arthropod in cross section, showing cuticular plates (sclerites) and tubes and the attachment of some muscles. Each segment primitively has four cuticular plates: a dorsal tergite, a ventral sternite, and two lateral pleurites. Usually some of these plates combine or divide during development. Note the absence of circular muscles, which are not needed in an animal without a hydroskeleton. Appendages are enclosed in cuticular tubes. (B) Articular membranes occur between cuticular plates and at the joints of limbs. Condyles at joints compensate for the weakness of articular membranes.

tor cells, and various deformable plates, slits, and hairlike setae connect to mechanoreceptor cells.

THE CUTICLE AS EXOSKELETON. Like pieces of armor, sclerotized cuticle is linked together as plates (= **sclerites**) over the body surface and as tubes around appendages (Fig. 30.3A). The combination of sclerotized plates and tubes makes up the exoskeleton. Projecting internally from the exoskeleton are **apodemes** to which muscles attach (Fig. 30.2C). Movement of the body is possible because the sclerites and tubes are hinged together by a type of cuticle called **articular membrane** (= arthrodial membrane) (Fig. 30.3B). Articular membrane is flexible and transparent because it lacks exocuticle.

MOLTING. The arthropod exoskeleton is so useful for support and protection that one might wonder why it doesn't occur in all animals. One explanation may be that it is difficult for an animal to grow while encased in a rigid exoskeleton. Molluscs solve this problem by continually enlarging their shells as they grow. Arthropods, however, periodically replace the cuticle when they outgrow it. The process of shedding the old cuticle is called molting or **ecdysis** (EK-dis-sis; Greek *ekdysis* an escape). Insects go through a fixed number of molts until they develop into adults, which do not molt. Other arthropods, such as spiders, molt an indefinite number of times, both as juveniles and adults.

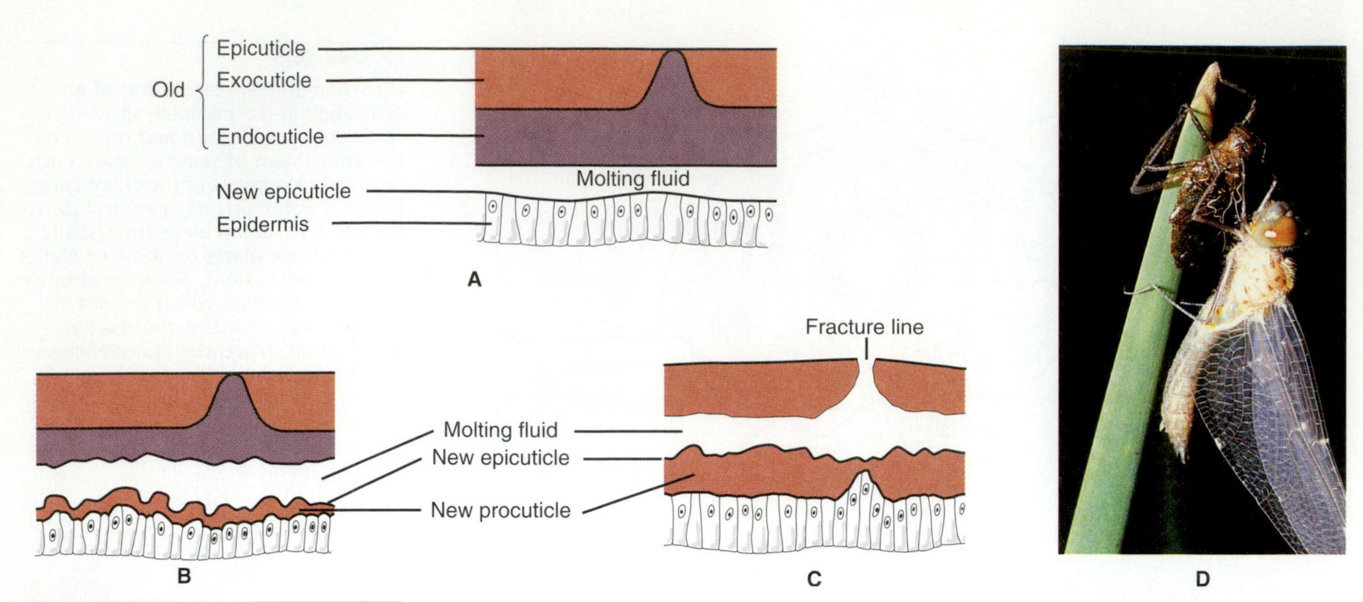

FIGURE 30.4

(A) Molting begins when the epidermis separates from the old cuticle, forms a new epicuticle, and secretes molting fluid. (B) Molting fluid eats away the old endocuticle. The epidermis then forms new procuticle, which will become new exocuticle and endocuticle. (C) Finally the old cuticle ruptures. (D) A just-molted adult dragonfly clings to its old cuticle. The white strands are molted cuticle of tracheae.

■ **Question:** How did the large dragonfly in D fit into its old cuticle?

Molting is initiated in some species by environmental cues, and in others by increased pressure in the body, due to tissue growth or a meal. These cause the release of the molting hormone **ecdysone** (p. 138), which triggers the separation of the epidermis from the old cuticle and the secretion of new epicuticle (Fig. 30.4). The epidermis then begins to secrete **molting fluid** containing protease and chitinase, which eat away at the old, nonsclerotized endocuticle. The products of this digestion are used in producing the new cuticle. The enzymes do not attack tanned exocuticle and new epicuticle, or the nerve and muscle connections, so the animal retains its mobility and protection for as long as possible. After the old endocuticle separates from the new epicuticle the epidermis secretes a new **procuticle,** which will later differentiate into the new exocuticle and endocuticle. The old exocuticle is finally shed when the arthropod inflates itself with air or fluid, which ruptures the old cuticle at predetermined fracture lines. Cuticle that lines the gut and tracheae molts at the same time (Fig. 30.4D).

It may require many minutes for the arthropod to extricate itself from its old exoskeleton. During that time it is especially vulnerable. Not only does the new unsclerotized cuticle lack protective hardness, but the arthropod must maintain an inflated posture during sclerotization in order for the cuticle to become large enough for future growth. The weakness of the exoskeleton just after molting is probably one factor that has limited the size of arthropods, especially terrestrial arthropods that do not have the benefit of buoyancy to help support their weight. The largest insect is only(!) about 30 cm long, and most are much smaller.

Terrestrial Pioneers

Phylum Arthropoda includes the first animals discussed so far in this text that are truly adapted to a life on dry land. The oldest known fossils of terrestrial animals are arthropods that lived on what is now Australia during the Silurian period about 420 million years ago. Only a few million years younger are fossils of tiny spiderlike arthropods found in Scotland and New York. Because we are terrestrial animals ourselves, we are apt to take living on land for granted. The transition from water to land was by no means simple, however. Several major adaptations were required (Table 30.2). Two of these adaptations were already present in the aquatic arthropods: (1) an exoskeleton that can retard water loss and (2) the appendages that had been used for crawling or burrowing on the ocean floor.

TABLE 30.2

Adaptations of arachnids and other animals for terrestrial life.
See Little (1990) for further discussion.

1. **The ability to resist drying,** primarily by means of an impermeable integument.
2. **Improved organs for maintenance of ion and water balance.**
3. **Modification of metabolic pathways with nitrogenous wastes eliminated as urea or uric acid, rather than as ammonia.** (See pp. 279–280.)
4. **The ability to obtain oxygen directly from air.**
5. **Improved ability to compensate for changes in temperature,** which occur more rapidly in air than in water. (See the discussion of poikilotherms in Chapter 15.)
6. **Modification of appendages for terrestrial locomotion, and strengthening of muscles and skeleton to compensate for the loss of buoyancy.**
7. **Modification of sensory receptors,** with less dependence on water-borne chemical information and more dependence on airborne chemicals, vision, and vibration of substratum and air.
8. **Internal fertilization,** so that gametes remain moist. This in turn implies the development of sophisticated neural mechanisms for courtship and anatomical arrangements for copulation.
9. **Either viviparity or eggs that retain water,** which provide a moist environment for development.
10. **Large supply of nutrients to the embryo, either by nutrient-rich eggs or viviparity,** since development must often be prolonged until the immature animal is capable of terrestrial living.

Classification

The classification of arthropods has a long, tortured history that is still not settled. It is worth reviewing it here as an illustration of the kinds of considerations that enter into classification. For many years, living arthropods were divided into two subphyla based mainly on differences in mouthparts (Fig. 30.5A). The subphylum Chelicerata (ke-LIS-ur-AH-tuh) included the spiders, scorpions, mites, and other arthropods with a pair of oral appendages called **chelicerae.** Lacking mandibles for chewing food, such chelicerates generally predigest their food outside the body and suck it up as a semiliquid. Most crustaceans, insects, and myriapods (such as millipedes and centipedes) have **mandibles** adapted for chewing solid food, and for many decades taxonomists combined them all in the subphylum Mandibulata. Although crustaceans differ greatly from insects and myriapods, there did not appear to be any fundamental difference that could be used to separate them taxonomically.

In the 1960s Sidnie Manton seemed to offer a way out of the dilemma. From detailed anatomical studies, she concluded that the mandibles of crustaceans develop from the bases of appendages, while those of insects and myriapods develop from entire appendages. She also concluded that the appendages of crustaceans are primitively two-branched (biramous), while those of insects and myriapods are unbranched (uniramous) (Fig. 30.6 on page 636). Based on these differences, Manton proposed splitting arthropods into three phyla: phylum Chelicerata for the chelicerates; phylum Crustacea for mandibulates in which the mandible develops from part of an appendage, and the appendages are biramous; and phylum Uniramia, which would include insects and myriapods, as well as onychophorans (Fig. 30.5B). Most zoologists have rejected the suggestion that arthropods

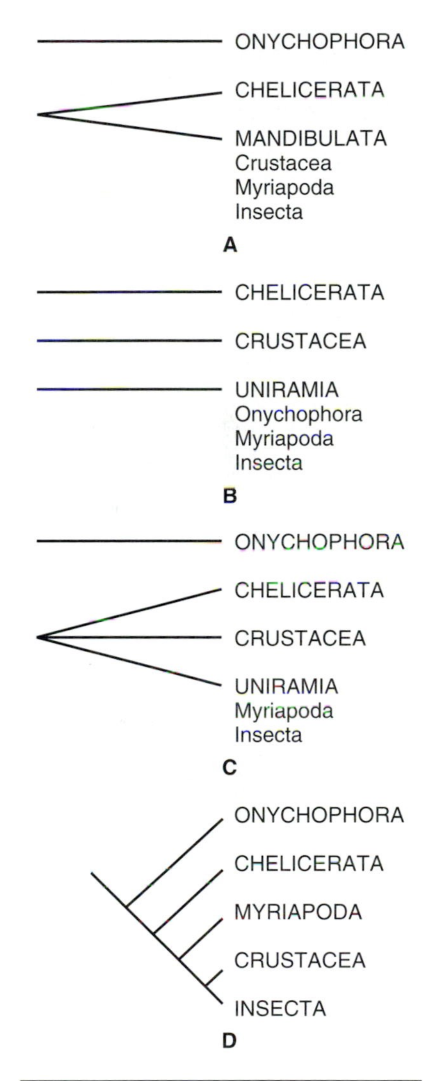

FIGURE 30.5

The evolution of arthropod classification. (A) Before Manton's work was accepted, most zoologists divided arthropods into two subphyla: Chelicerata and Mandibulata. Onychophora was a separate phylum. (B) Manton proposed that the arthropods comprised three phyla, one of which included Onychophora. (C) Most zoologists retained Onychophora and Arthropoda as two separate phyla, but accepted division of Arthropoda into three subphyla. (D) Phylogeny assumed in this edition. The position of Myriapoda is especially uncertain.

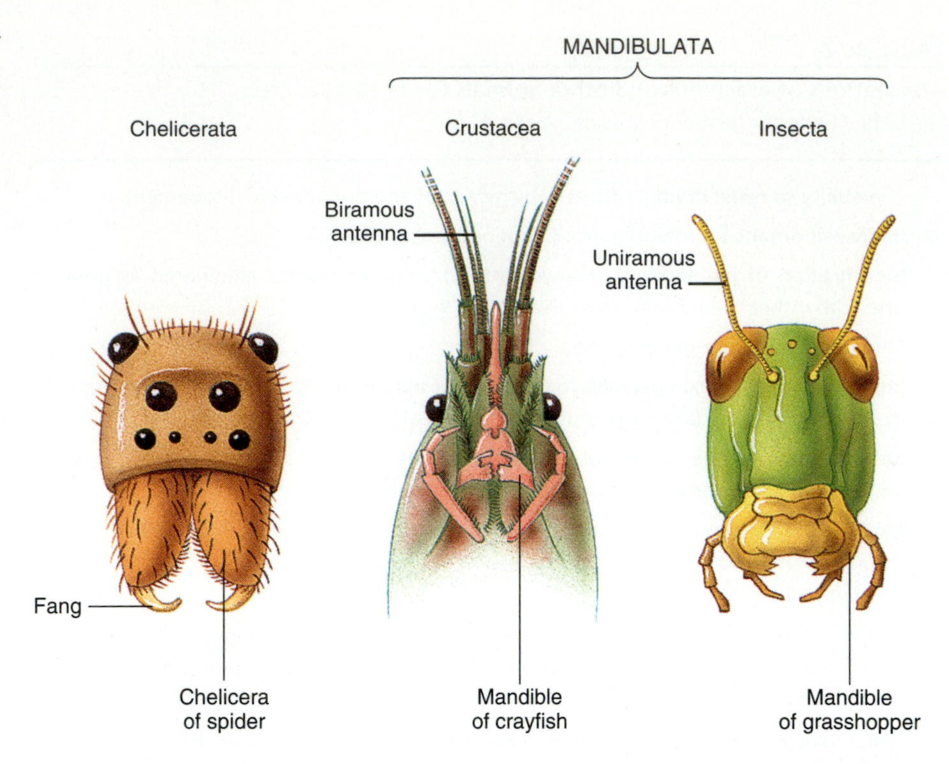

FIGURE 30.6
The morphological basis for arthropod classification, according to S. M. Manton. Spiders and other chelicerates are distinguished from crustaceans and insects by their mouthparts, which include chelicerae instead of mandibles. Manton thought that the mandibles of crustaceans develop from only the proximal portions of appendages, and that the mandibles of insects develop from entire appendages. She also thought that the appendages of insects and myriapods were primitively uniramous, while those of crustaceans were biramous.

should be divided among three phyla, since they all have similar body plans, and there is considerable fossil and molecular evidence that they are a natural (monophyletic) group (Briggs and Fortey 1989, Turbeville et al. 1991). Many zoologists, however, welcome Manton's rationale for separating Crustacea from the other mandibulates, and they therefore divide the arthropods into three subphyla: Chelicerata, Crustacea, and Uniramia (Fig. 30.5C). Most were reluctant to include the onychophorans, however, because they appeared to be related as closely to annelids as to uniramians.

This system was challenged in 1990 by Emerson and Shram, who found fossil evidence that the biramous appendages of crustaceans resulted from the fusion of pairs of body segments in a uniramous arthropod. This finding suggested a closer relationship between crustacea and uniramia. Further challenge to a three-way split of arthropods came from the paleontologist Jarmila Kukalová-Peck (1992). She pointed out that Manton's evidence for "whole-leg mandibles" in Uniramia was based on studies of myriapods and onychophorans but not insects, and that numerous fossil and living crustaceans and insects have polyramous (many-branched) appendages. She therefore essentially recommended going back to the classification accepted prior to Manton's work.

Recent molecular and morphological studies confirm that Arthropoda is monophyletic and may include Onychophora (Ballard et al. 1992, Boore et al. 1995, Friedrich and Tautz 1995, Wheeler et al. 1993). We have probably not heard the last word on arthropod phylogeny and classification. Until we do, it may be convenient simply to regard all five groups as having equal taxonomic rank (Fig 30.5D). The classification of Arthropoda listed in the box on p. 637 therefore recognizes as subphyla the Onychophora, Chelicerata, Myriapoda Crustacea, and Insecta. All but Crustacea and Insecta are described in this chapter.

Trilobites

This text is concerned mainly with living animals, but one extinct group of arthropods is too important to overlook. Trilobites, usually classified in their own subphylum Trilobita, flourished during the 85 million years of the Cambrian period, then slowly declined. They

A Classification of Phylum Arthropoda

Subphylum Onychophora on-i-KOFF-or-uh (Greek *onyx* claw + *phora* bearing). Segmentation not apparent. Legs not jointed. With mandibles and no compound eye. Velvet worms.

Subphylum Chelicerata ke-LIS-ur-AH-tuh (Greek *khele* claw + *keras* horn). With chelicerae. Spiders, scorpions, mites, horseshoe crabs, sea spiders.

Subphylum Myriapoda meer-ee-AP-oh-duh (Greek *murios* countless + *podos* foot). Body divided into head and long trunk, with many trunk segments bearing jointed legs. With mandibles and no compound eye. Centipedes and millipedes.

Subphylum Crustacea krust-AY-see-uh (Latin *crustacea* shelled). With mandibles and compound eyes. Generally aquatic. Crayfish, crabs, shrimp, barnacles, and others. (See Chapter 31.)

Subphylum Insecta in-SEK-tuh (Latin *insectus* cut into [referring to the segmentation]. With mandibles and compound eyes. Generally terrestrial, usually with wings. Hexapods. (See Chapter 32.)

finally vanished during the mass extinction at the end of the Paleozoic era, after more than 300 million years of existence. There were so many species and numbers of trilobites that they were certainly the dominant animals during much of the Paleozoic era.

Trilobites are named for the division of the chitinous carapace into three longitudinal lobes (Figs. 19.6 and 30.7). The known fossils range in length from 3 mm to almost 70 cm, but 2 to 7 cm is typical. The bodies of trilobites were clearly segmented and tag-

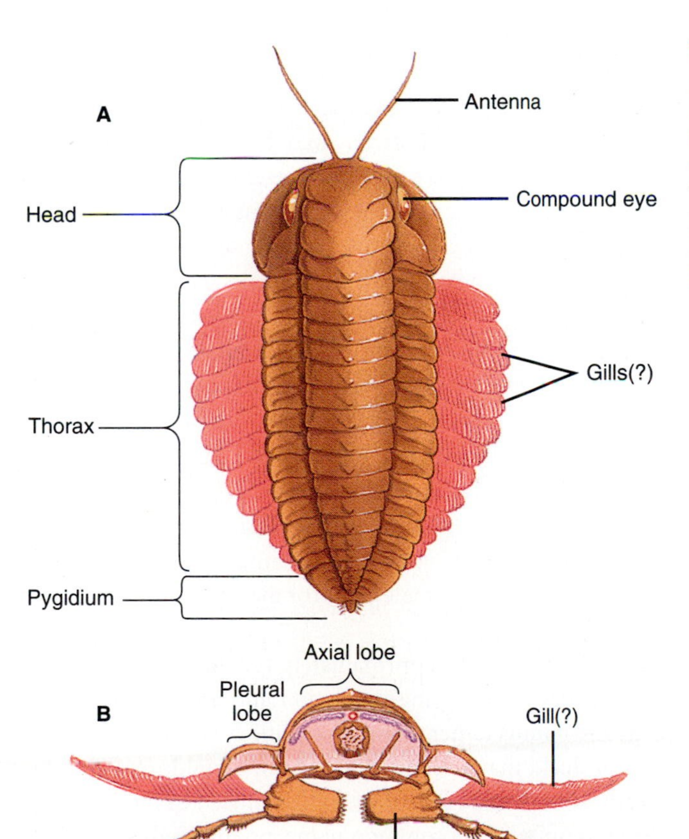

FIGURE 30.7

Triarthrus eatoni, a trilobite that lived in the ocean above what is now Rome, New York, during the Ordovician period some 450 million years ago. Of the 4000 known species of trilobites, painstaking reconstructions from fossils have made this the best known. (A) Dorsal view. (B) Cross section. Body length approximately 4 cm.

matized into three distinct regions: the head (= **cephalon**), the thorax, and the **pygidium.** The head incorporated several fused segments and bore a pair of antennae, compound eyes, mouth, and legs. The heads of many trilobites were armed with long, sharp spines, suggesting that they were subject to predation. Many fossils are found with the thorax curled in a defensive posture. Fossilized burrows and the shovel shape of the heads of some species suggest that some trilobites burrowed in marine sediments for food or protection. On the other hand, the dorsal position of the eyes, the adaptation of the appendages for walking, and numerous fossilized tracks indicate that most trilobites crawled. A few smaller species probably swam. The thorax consisted of 2 to 29 segments, with the number increasing as the animal grew. Each segment except the last bore a pair of biramous legs, with the ventral branch apparently adapted for locomotion. The dorsal branch was filamentous and may have served as an oar, a gill, or a net.

Myriapods

Among the living arthropods, the Myriapoda may be the most primitive group, judging from the relative lack of tagmatization. Often, however, they are considered to be closely allied with insects, because they are also terrestrial mandibulates and have internal organs similar to those of insects. Unlike insects, however, myriapods have legs on most of their body segments. Consequently, instead of having three tagmata as in insects (the head, leg-bearing thorax, and abdomen), the bodies of myriapods are divided into a head and a long, leg-bearing trunk. There are four major kinds of myriapods: centipedes, millipedes, pauropods, and symphylans.

CENTIPEDES. Most of the roughly 3000 species of **centipedes** (Chilopoda) are nocturnal predators, well adapted to chasing down insects and other small prey (see Fig. 30.1A). You will seldom see one unless you are curious about what lives under logs and stones. In houses where cockroaches and other insects are available as prey, however, the house centipede *Scutigera* may be encountered in bathrooms and other damp places. Centipedes have the appendages on the first trunk segment modified as poisonous fangs. *Scutigera* is harmless to humans, but some tropical centipedes that grow up to 25 cm long can be dangerous, though seldom lethal.

On each of the other trunk segments except the last two is a pair of walking legs that extend laterally. The total number of walking legs ranges from 15 to 191 pairs, depending on the species. (The actual number of legs is never a hundred, as the name "centipede" implies, since the number of pairs is always odd.) The legs propel the flexible, flattened body rapidly in leaf litter. Prey are detected by the single pair of antennae on the head. In most species the eyes are clusters of simple **ocelli** that are incapable of forming an image. Like many predaceous animals, centipedes engage in prolonged courtship. After several hours the male deposits a sperm-filled sac called a **spermatophore,** which he then transfers into the female's genital opening with his mouth. In many species the female cares for the eggs and young in underground burrows.

MILLIPEDES. The approximately 10,000 named species of **millipedes** (Diplopoda) feed mainly on decaying vegetation and are adapted for burrowing through rotting logs and humus rather than for predation. When burrowing, the head is bent beneath a dorsal shield that acts like the blade of a bulldozer. The short legs are positioned beneath the robust body and are synchronized to move in a slow, wavelike motion (Fig. 30.8). In spite of the common name, no millipede has a thousand legs. The maximum number is 752, and most have fewer than 50. The body is usually cylindrical in cross section, and the cuticle of most species is reinforced with calcium salts, as in crus-

Classes in Subphylum Myriapoda

Class **Chilopoda** ky-LOP-oh-duh [Greek *chilo* lip (referring to the poison "fangs," which are actually legs)]. Body usually flattened; from 18 to 184 segments in the trunk, depending on species. Most segments with one pair of walking legs. Centipedes. *Lithobius, Scutigera* (Fig. 30.1A).

Class Diplopoda dip-LOP-oh-duh (Greek *diplos* double). Body usually cylindrical. Trunk of 9 to more than 100 diplosegments. Most diplosegments with two pairs of legs each; anterior segments have only one pair of legs or none. Millipedes. *Glomeris, Pachydesmus* (Fig. 30.8).

Class Pauropoda paw-ROP-oh-duh (Greek *pauros* few). Less than 2 mm long, with 11 or 12 segments. Nine or ten pairs of legs. No eye; antennae three-branched.

Class Symphyla SIM-fy-luh (Greek *sim* same + *phyli* tribe). Less than 6 mm long. Twelve trunk segments, each with one pair of legs. No eye. *Scutigerella.*

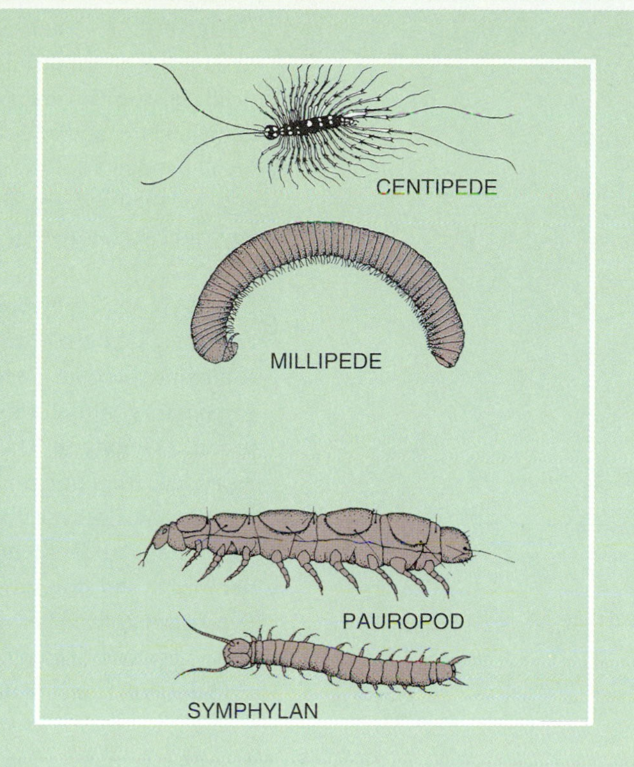

taceans. These appear to be adaptations that prevent the body from folding as the legs push it forward. The number of places where such folding can occur has also apparently been reduced by the fusion of pairs of trunk segments into **diplosegments.** Evidence for such fusion is that each diplosegment has two pairs of openings to the heart and tracheae, two ganglia on the ventral nerve cord, and two pairs of legs (hence the name Diplopoda). The first four trunk segments, often called the thorax, remain unfused and bear either one pair of legs each or none at all.

Like centipedes, millipedes are most often restricted to moist areas under logs, stones, and leaf litter. When disturbed, some of them, called pill millipedes, curl up like pill bugs. Some spray or secrete toxic or irritating defensive substances, including hydrogen cyanide. The European millipede *Glomeris marginata* secretes a substance that is chemically similar to the tranquilizer Quaalude. Within a few hours after eating this

FIGURE 30.8
The millipede *Pachydesmus.* Approximately 5 cm long.

millipede, a spider becomes totally relaxed for several days. It is not clear what good this does the millipede.

PAUROPODS. There are only about 400 species of **pauropods**, but the number of individuals is enormous. Five million may live in a hectare of forest litter, largely unnoticed because they are colorless and less than 2 mm long. They feed on dead plant and animal matter and on fungi. The trunk consists of 11 or 12 segments, with a pair of legs on all segments but the first and last. The antennae are branched. What appears to be eyes on the head are sensory organs of some other, unknown function. Pauropods have no circulatory or respiratory organ.

SYMPHYLANS. Symphylans number only a little over a hundred species, and they are generally confined to moist areas where they feed on living or dead vegetation. They sometimes become greenhouse pests. Although they are less than 1 cm long, they are sometimes confused with centipedes. The common "garden centipede" *Scutigerella* is, in fact, a symphylan. There are 12 trunk segments, each with a pair of legs. Many species reproduce from unfertilized eggs (parthenogenetically), and others reproduce sexually in a way that does not even require that the two sexes meet. Males simply leave a stalked spermatophore for a female to find by chance. The female takes the spermatophore into her mouth and stores the sperm in special pouches. To reproduce, she uses her mouth to take an egg from her genital opening, which is located on the ventral surface of an anterior segment. She deposits the egg on substratum such as moss, then smears it with stored sperm from her mouth.

Onychophorans

Approximately 110 species of onychophorans, commonly called velvet worms or walking worms, have been named, and up to 50 others are known but not yet described. They are easily mistaken for shiny green, blue-black, orange, or whitish caterpillars as they creep over the forest floor on their 13 to 43 pairs of clawed appendages (Fig. 30.1B). Although they are up to 15 cm long, they are seldom seen because they avoid light and live in rainforests in India, Africa, South America, and Australia. They are apparently ancient animals. Fossils half a billion years old in the Burgess Shale of British Columbia have been interpreted as those of aquatic onychophorans (see number 4 in Fig. 19.5). Onychophorans, together with tardigrades (see pp. 569–572), have long been considered close relatives of insects and other arthropods because of their chitinous cuticle, dorsal tubular heart, open circulatory system, hemocoel as the major body cavity, and breathing tubes (**tracheae**) (Fig. 30.9). Traditionally, however, they were placed in their

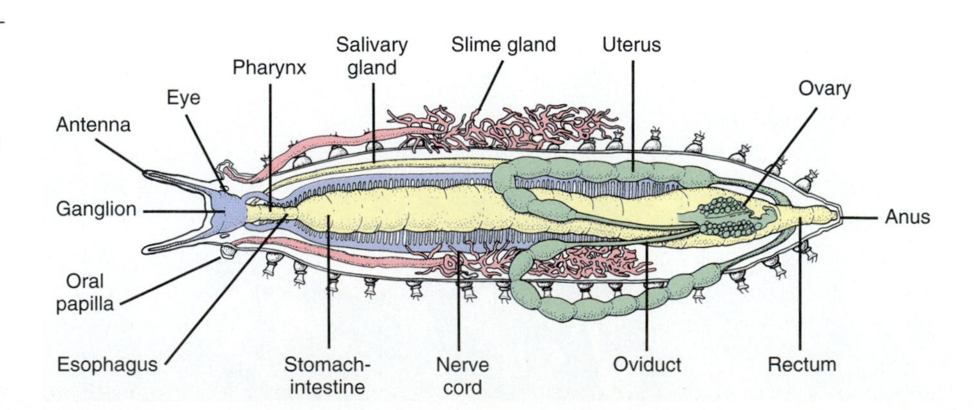

FIGURE 30.9

Longitudinal section of a female *Peripatus* sp., dorsal view. Note the presence of a uterus in this viviparous species. For simplicity, the tubular heart along the dorsal midline is not shown, nor are the tracheal tubes nor nephridia near each leg. Length approximately 7 cm.

own phylum Onychophora because of several differences. Their cuticle does not harden, their appendages are not jointed, there are no thoracic or abdominal ganglia, they have nephridia like those of annelids, and openings of the tracheae cannot close to limit the loss of body moisture. The mode of predation also differs. Onychophorans capture small prey by squirting them with jets of milky fluid from two **oral papillae** at distances up to 30 cm. The fluid congeals into slime that entangles the prey, which is then digested externally by juices secreted by the onychophoran.

Chelicerates

Subphylum Chelicerata includes approximately 65,000 named species of horseshoe crabs, sea spiders, true spiders, scorpions, ticks, mites, and other arthropods with chelicerae. Unlike mandibles that usually function as jaws for chewing solid food, chelicerae are used to seize, pierce, or tear prey, which are then partially digested externally before being consumed as a liquid. Often feeding is aided by a pair of appendages called **pedipalps** near the mouth. Antennae do not occur in any chelicerate. Although this combination of anterior appendages defines the Chelicerata, most people do not get close enough to see them. Instead they recognize most chelicerates by the four pairs of walking legs and the head and thorax fused into a **cephalothorax.**

Taxonomists usually divide extant chelicerates among three classes. Class Merostomata includes the horseshoe crabs, which are, of course, not really crabs at all. They are frequently encountered along Atlantic beaches when they come ashore to mate. Class Pycnogonida includes so-called sea spiders, which are seldom encountered except by divers. To humans and other terrestrial animals the most familiar chelicerates are members of class Arachnida: the spiders, harvestmen, scorpions, ticks, mites, and others.

Horseshoe Crabs

Class Merostomata is an ancient group of aquatic chelicerates, with some fossils nearly half a billion years old. One major group, subclass Eurypterida, has been extinct for about 250 million years. The only living merostomes belong to three genera of horseshoe crabs (subclass Xiphosura). Three of the four species of horseshoe crabs occur in the western Pacific. The fourth species, *Limulus polyphemus,* is brown, up to 60 cm long, and familiar along the east coast of North America. In early summer so many *Limulus* arrive on Atlantic beaches that they look like an invasion by midget marines, with only their helmets showing above water. The objective of this invasion is reproduction. At high tide the female comes ashore with one or more males clinging to her with their modified front walking legs (Fig. 30.10). The female digs a hole into which she deposits several hundred greenish eggs, and the males deposit sperm on them before the female covers them. Several weeks later larvae hatch and catch the next high tide out to sea. The larvae are called "trilobite larvae," because of their superficial resemblance to their presumed ancestors.

FIGURE 30.10
A female horseshoe crab *Limulus polyphemus* comes ashore to lay her eggs, dragging a smaller male. The females are up to half a meter long.

As any eastern beachcomber knows, many horseshoe crabs become stranded and die on beaches in their efforts to reproduce, and millions of the eggs become food for migrating birds. In spite of these perils, *Limulus* remains abundant, protected from predators by the hard carapace over the cephalothorax. The carapace bears a compound eye on each side and two simple eyes near the middle. In addition, there are five light-sensitive organs beneath the carapace. The abdomen can be folded beneath the animal to protect the delicate ventral appendages. The backward-pointing spines on the abdomen not only deter predators, but dig into the sand and prevent waves from

Major Extant Groups in Subphylum Chelicerata

Genera mentioned elsewhere in this chapter are noted.

Class Merostomata MER-oh-STOW-ma-tuh (Greek *meros* thigh + *stoma* mouth). Aquatic. Body divided into cephalothorax and abdomen, joined by thick "waist." Cephalothorax bears a pair of compound eyes and a pair of simple eyes or ocelli. Abdomen with paired appendages bearing gills, and a long tail spine (**telson**). *Limulus* (Fig. 30.10).

Class Pycnogonida PICK-no-GO-nid-uh (Greek *pyknos* crowded + *gony-* knee). Cephalic somite only partly fused with thorax; abdomen vestigial. Four (sometimes five or six) pairs of walking legs with eight or nine joints each. Sucking mouth on long proboscis. Four simple eyes. Sea spiders. *Nymphon* (Fig. 30.12).

Class Arachnida a-RACK-nid-uh (Greek *arachne* spider). Four pairs of legs. Abdomen usually lacking locomotory appendages. No compound eye. Mainly terrestrial. Spiders, harvestmen, scorpions, pseudoscorpions, ticks, mites, and others. *Acarapis, Argiope, Boophilus, Brachypelma, Centruroides, Chelifer, Corythalia, Demodex, Dermacentor, Dicrostichus, Dolomedes, Hogna, Ixodes, Latrodectus, Leiobunum, Limnochares, Loxosceles, Lycosa, Mastophora, Misumena, Paruroctonus, Pandinus, Sarcoptes, Steatoda, Trombicula, Vachonium, Varroa* (Chapter opener; Figs. 30.1C, 30.13, 30.16, 30.19, 30.20, 30.21, 30.22, 30.23, 30.24, 30.25).

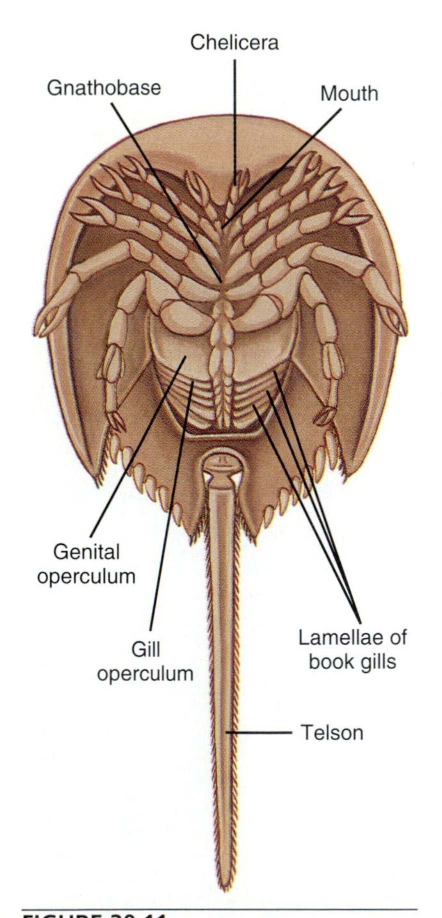

FIGURE 30.11
Ventral view of the horseshoe crab *Limulus.*

washing the animals out to sea as they attempt to go ashore. The tail spine, or **telson,** may also deter large predators, but it functions mainly as a lever during burrowing and for righting the animal if it is flipped over by a wave. *Limulus* can swim by flapping its gills and opercula (gill covers), but most of the time it crawls and burrows along the bottom on five pairs of legs. Each of the front four pairs of legs is tipped with a pincer-like **chela** (KEE-luh).

Horseshoe crabs eat molluscs and polychaete worms. This diet is more appropriate for mandibulates than for chelicerates, but horseshoe crabs make up for the lack of jaws by chewing food with their legs. Chelae on the chelicerae seize and partly macerate the food, then pass it back to the first four pairs of legs (Fig. 30.11). These legs have proximal segments, called **gnathobases,** with tooth-shaped projections that enable them to function as jaws. After the gnathobases chew up the food they transfer it forward to the mouth. The fifth pair of legs do not engage in feeding, and they lack chelae. They provide most of the force for burrowing, and they also sweep the gills clean. These gills are called **book gills,** because the modified appendages that form them resemble pages. Oxygen uptake is encouraged by movement of the gills. There is also a respiratory pigment called hemocyanin, which makes *Limulus* blood blue.

Sea Spiders

Sea spiders get their common name from the fact that most of them have eight legs, though some have 10 or 12. The origin of their formal names, Pycnogonida (crowded knees) or Pantopoda (everywhere feet), is obvious (Fig. 30.12). The legs extend up to 70 cm in one species, but 1 cm is more typical. In many species there is an extra pair of legs, called **ovigerous legs** (= ovigers). Sea spiders use the ovigerous legs to clean themselves, and the males also use them to collect and brood the eggs they have fertilized. Approximately 1000 species of pycnogonids are common in all oceans, especially cold ones. They can often be found on the soft tissues of sponges, hydroids, soft corals, anemones, ectoprocts, and clams, on which they feed by means of a sucking mouth at the end of the moveable proboscis. Some species lack chelicerae. Digestion occurs in

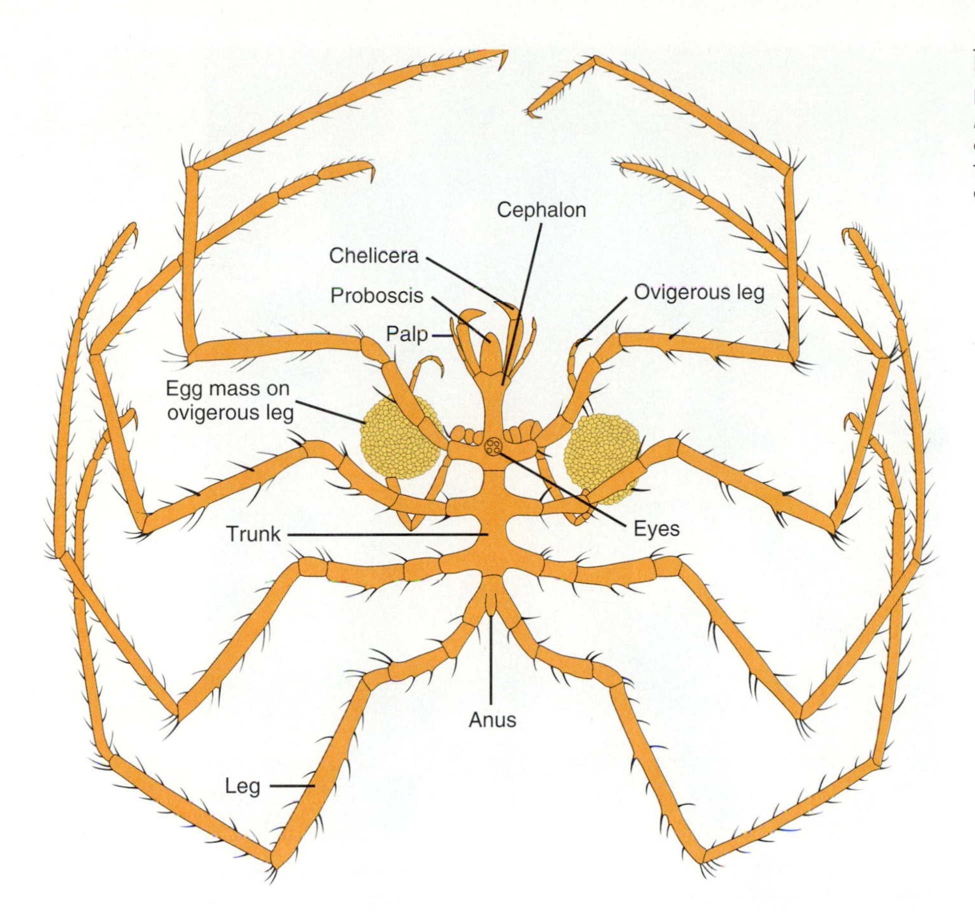

FIGURE 30.12
Dorsal view of a male pycnogonid, *Nymphon rubrum.* Note the unusual location of the eyes. Most sea spiders are transparent or yellowish-brown to cream colored.

mucosal cells of the gut, which has branches radiating almost to the tips of the legs. The gonads also extend into the legs. Thanks to the enormous surface area relative to body mass, pycnogonids get by without any respiratory or excretory system.

Spiders

Most chelicerates belong to class Arachnida, which includes more than 73,000 known species of true spiders, harvestmen, scorpions, ticks, mites, and a few other groups. The most familiar arachnids are spiders (order Araneae), with approximately 34,000 known species. Most spiders are from 2 to 10 mm long, excluding the legs. Some female tarantulas have bodies up to 90 mm long (Fig. 30.13 on page 644). Male spiders are always smaller than females of the same species. The cuticle-covered body, divided into cephalothorax and abdomen, and the four pairs of legs are features by which most people recognize spiders quickly, and often with revulsion. The revulsion may come from an abhorrence of long, hairy legs. (If that is the case, however, why are Irish setters so popular?) The revulsion to spiders might also be due to fear of their fangs, even though only a few spiders can harm humans. The majority could not pierce human skin even if they tried.

External Structure. The external structure of spiders arises by a process that is typical for arthropods. Each metamere (= segment) of the annelid-like embryo has a pair of appendages, and there is extensive tagmosis. In spiders the anterior six metameres fuse during development to form the cephalothorax (= **prosoma**), while the posterior metameres form the abdomen (= **opisthosoma**). These two tagmata are linked by a narrow **pedicel.** The six pairs of anterior appendages become the two che-

FIGURE 30.13

The Mexican red-legged tarantula *Brachypelma smithi* showing its chelicerae. Two leg-shaped pedipalps occur on each side of the chelicerae. This and other tarantulas defend themselves by rubbing off barblike hairs that irritate the throat and skin of vertebrates. (The name "tarantula" was erroneously applied to several large genera of the American tropics by colonists, who apparently mistook them for the true tarantulas, *Lycosa* or *Hogna,* of the Mediterranean. Both true and American tarantulas are feared, but in fact their bites, though lethal to insects and mice, are no worse for humans than wasp stings.)

■ **Question:** Why do you think this picture evokes revulsion in some people?

licerae, the two pedipalps, and the four pairs of legs. These three kinds of structures are therefore **serially homologous** to each other.

CHELICERAE. In most spiders each chelicera has a fang that is normally kept within a groove, like the blade of a folding knife (Fig. 30.6). During predation or defense, the fangs extend and inject venom from a poison gland into the prey or predator. Depending on species, the venom includes various neurotoxins, protein-digesting enzymes, and pain-inducing amines. After biting its prey, a spider generally backs off while the toxins kill or paralyze it. Many spiders further subdue prey by wrapping it with silk, either before or after biting. After the prey has been overcome, some spiders mash it with their chelicerae while regurgitating digestive enzymes from the gut onto it. Other spiders regurgitate enzymes into the prey through the fang holes. After allowing the enzymes to work for a few seconds, spiders suck up the semiliquid food into their mouths. Some spiders use their chelicerae not only for feeding but for carrying prey or egg cocoons, for grasping objects, or for digging burrows. In addition to the chelicerae, spiders also have a pair of pedipalps (also called simply palps) that help manipulate prey and substitute to some extent for chewing mandibles. In adult male spiders the pedipalps are specialized as copulatory organs, as will be described.

DIGESTION. Spiders suck up the externally digested food by expanding the pharynx and **pumping stomach** (Fig. 30.14). Most of the internal digestion and absorption of nutrients occurs in the midgut, which has numerous diverticula branching anteriorly into the bases of the legs. Digestive enzymes for both internal and external digestion are

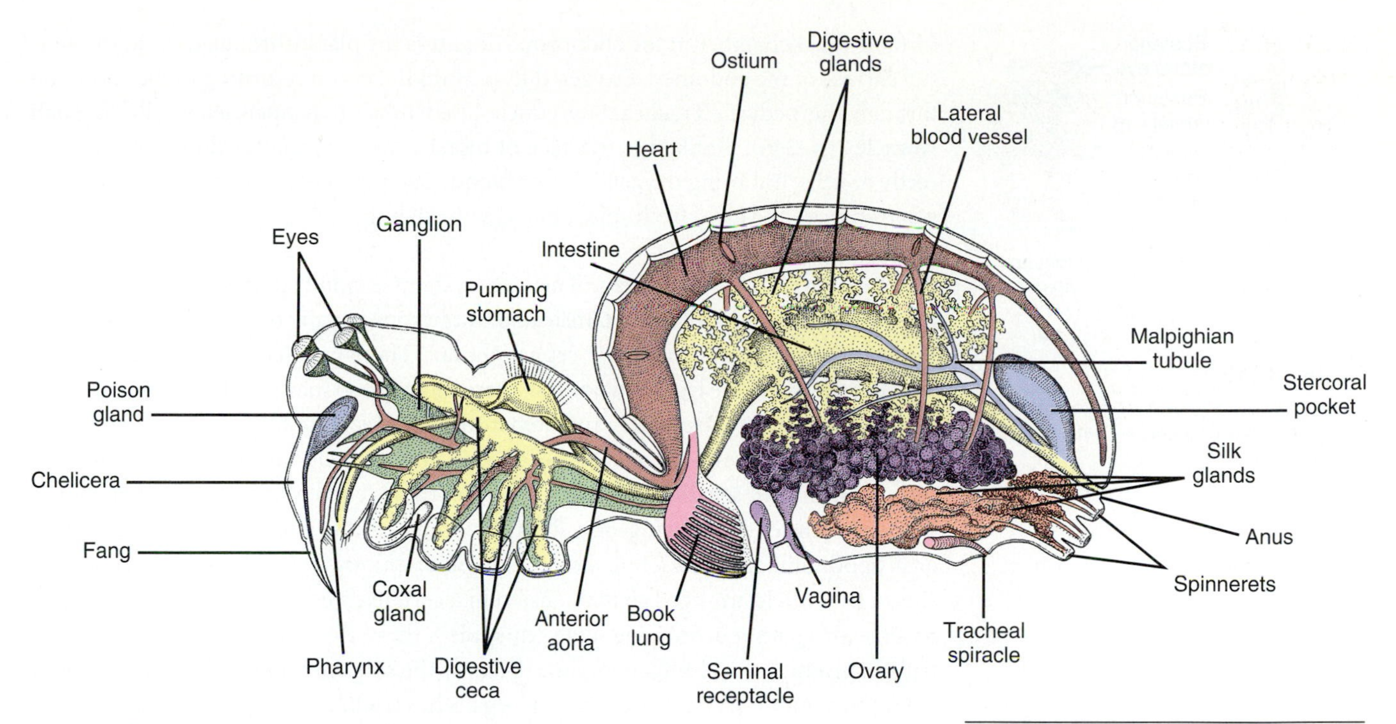

FIGURE 30.14
Internal structure of a female spider.

■ **Question:** After a spider eats its web the silk can be used again. What route does the silk take during this recycling?

usually secreted by a **digestive gland,** which is unfortunately often called a "liver." Because of the highly developed digestive system and low rate of metabolism, most spiders can live for long periods without eating. Tarantulas can go for several months between meals, and adult black widows (*Latrodectus mactans*) can live for 200 days without eating. The abdomens of spiders can expand enormously after a meal, thanks to the arthrodial membrane linking the sclerites. Often this expansion triggers molting.

EXCRETION. The main excretory organs of spiders are **Malpighian tubules** (p. 281). Malpighian tubules actively transport excess ions and metabolic wastes out of the blood in the abdomen and into fluid in the **stercoral pocket,** which is just dorsal to the rectum. As the fluid is excreted, rectal glands reabsorb much of the water. In addition, spiders have **coxal glands** that eliminate wastes through pores near some of the coxae (the proximal joints of the legs).

CIRCULATION. The coelom of spiders, and of all arthropods, is even more reduced than in molluscs. The only traces of the coelom in spiders occur in the coxal glands, gonads, and a few other places. The main body cavity is the hemocoel: the blood-filled space around the internal organs. The circulatory system is typically arthropodan, with a dorsal tubular heart that pumps blood (= **hemolymph**) anteriorly through an aorta. The blood percolates through the hemocoel of the thorax, returns to the abdomen through the pedicel, and is taken back into the heart through one-way valves called **ostia.** Although the circulatory system is open, spiders are capable of producing high blood pressures (up to 480 mmHg in tarantulas). This high blood pressure serves two functions. First, two joints in each leg lack extensor muscles, so complete extension of spider legs depends on blood pressure. (This explains why a spider's legs fold when it dies.) Second, the high pressure enables the blood to transport oxygen efficiently.

RESPIRATION. Respiration is due to a pair of **book lungs** and one or two pairs of tracheae. The book lungs are similar in structure to the book gills of *Limulus* and are serially homologous with the legs. Each lung consists of 15 to 20 air-filled plates within

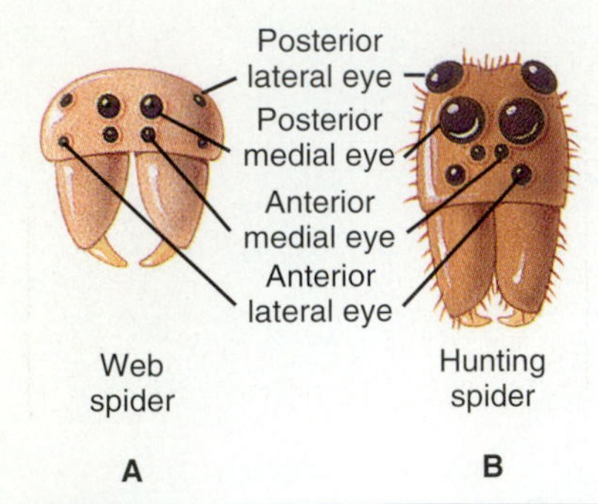

FIGURE 30.15

A comparison of the eyes of orb-weaving spiders (A) and those that stalk their prey (B).

a blood-filled chamber in the abdomen. Air enters the plates through a slit in the ventral surface of the abdomen. Oxygen diffuses into the blood returning to the abdomen through the pedicel. Tracheae are cuticle-lined tubes that open externally through **spiracles** (p. 246). Unlike the tracheae of insects, those of spiders do not branch directly to cells, but bring oxygen into the blood. The pigment hemocyanin aids in oxygen transport and gives fresh spider blood a bluish hue.

NERVOUS SYSTEM. The central nervous systems of adult spiders are different from those of most other arthropods. Instead of numerous ganglia on a ventral nerve cord, there are two large ganglia in the cephalothorax. These ganglia are largely concerned with coordinating the legs (a major task since there are so many of them) and integrating the sensory input from eyes and numerous other receptors.

Spiders have several types of mechanoreceptors, including "hairs," receptors at the joints, and **slit sensilla.** The hair-shaped receptors enable spiders to hear air-borne sounds, such as the buzzing of a fly. (The ability to hear is common in terrestrial arthropods and chordates, but absent in all other animals.) The slit sensilla consist of slits in the cuticle arranged so that tension in a certain direction causes the slit to open or close, triggering action potentials. In spiders these slits often occur on the legs in parallel groups called **lyriform organs** (because the slits are arranged like the strings of a lyre). A spider with damaged lyriform organs has trouble finding its way back to prey once it has left it.

Typically there are six or eight simple eyes (ocelli). Arachnids, unlike trilobites, crustaceans, and insects, do not have compound eyes. Spider eyes are arranged about the head in various patterns that are useful to taxonomists for classification (Fig. 30.15). The eyes cover overlapping fields of view, with virtually no blind spot. Spiders that capture prey with webs generally rely more on mechanical information from the web than on vision. Most web spiders are active at night, and blinding them by painting their eyes has no apparent effect on their ability to make webs or to capture prey trapped in them. Presumably their eyes function mainly to detect movements by daytime predators. On the other hand, hunting spiders, such as jumping spiders (family Salticidae), hunt during the day by pursuing and pouncing on prey.

Salticids commonly have a pair of large **primary eyes** with excellent resolution over a narrow angle. If the smaller secondary eyes detect possible prey, the salticid orients its body so that the primary eyes are aimed at it. The eyes, of course, cannot move. The retinas, however, can move within the eyes. If potential prey moves within about 20 cm of the spider, the spider scans it by moving its retinas rather than its eyes or body. At a distance of less than 10 cm the retinas scan it rapidly to determine whether it is prey or a member of its own species. In the latter case the salticid may respond with courtship or some other appropriate social behavior.

REPRODUCTION. Reproduction in spiders is rather conventional except that the males use their pedipalps as copulatory organs. After the final molt to adulthood, a male spider ceases to feed and devotes the rest of its abbreviated life to reproduction. It constructs a special **sperm web** on which it deposits a drop of semen, then dips the bulbs of its pedipalps into the semen to fill them with sperm (Fig. 30.16A). It then goes in search of a mate. The search is often aided by pheromones that females with unfertilized ova release into the air. As in many predators, copulation is usually preceded by an elaborate courtship that helps prevent any fatal mistakes. In web spiders the male carefully approaches the female on her web, avoiding vibrations that the female could misinterpret as those of struggling prey. In some species, the male identifies himself by plucking on the web in a particular way. Hunting spiders (salticids and others) generally use visi-

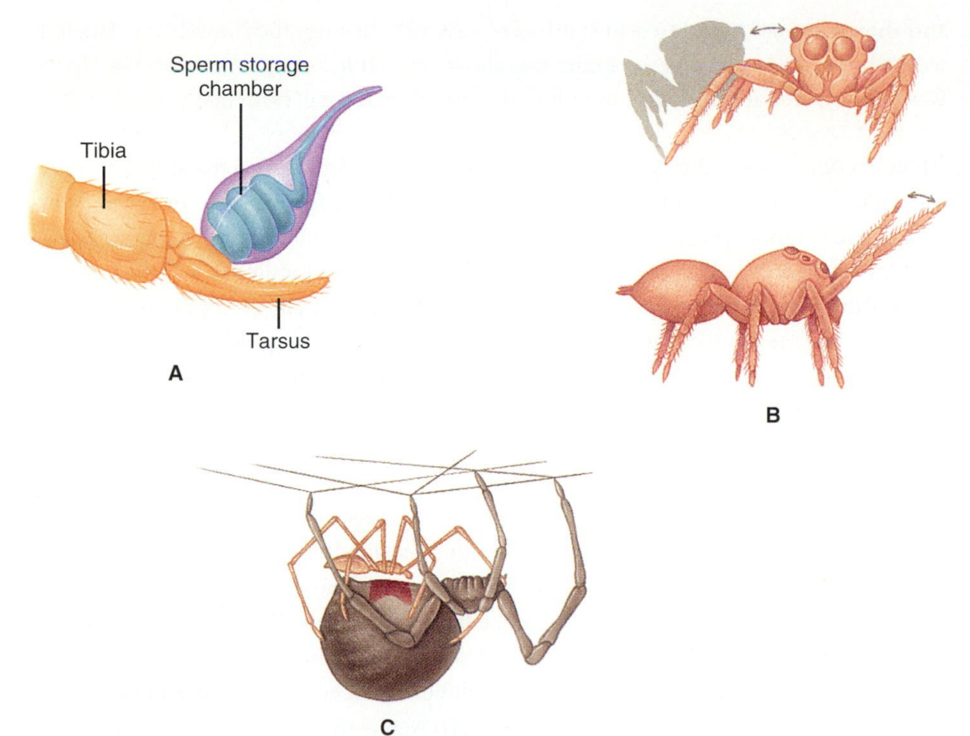

FIGURE 30.16
(A) A pedipalp of a male spider. These copulatory organs are extremely variable and useful in classification. (B) The jumping spider *Corythalia* courts by rocking sideways and by waving its forelegs. (C) Sperm transfer by black widow spiders *Latrodectus mactans*. Contrary to the myth that gives the black widow her name, the male (shown as brown) usually survives the encounter.

■ **Question:** Why wouldn't orb-weaving spiders be likely to use displays like that in part B?

ble or audible signals in courtship (Fig. 30.16B). Such communication is usually successful in saving the male's life from the much larger female. Contrary to popular myth, female spiders in most species do not routinely eat males during or after copulation. Following courtship, insemination is achieved when the male inserts the pointed tip of a pedipalp into a pore leading to the female's seminal receptacle (Fig. 30.16C).

After mating, the male goes in search of another mate, while the female remains at her web. Soon she begins depositing hundreds or even thousands of fertilized eggs into silken cocoons. Female spiders often show a touching degree of care for these cocoons and for the spiderlings that hatch from them. Many people who have been about to step on a wolf spider have been startled and distracted by curiosity upon seeing it divide into dozens of tiny spiderlings that the mother had been carrying on her back. Eventually all young spiders leave their mothers and disperse. A common means of dispersal is "ballooning." The young spider stands exposed to the wind, playing out a string of silk that eventually becomes long enough to carry it aloft. Ballooning spiders have been reported at altitudes up to 5 km. Along with tiny insects they constitute an "aerial zooplankton."

Spider Silk

Composition. Spider silk consists of proteins (keratins) synthesized as an aqueous solution by glands in the abdomen (Fig. 30.14). Generally there are six glands, each of which produces a different type of silk for each function: sperm web, drag line, cocoon, and various parts of web. The silk is extruded from one of three pairs of **spinnerets,** and each fiber consists of two strands. As fibers are forced out by muscle contraction or the pull of a hind leg, tension changes the proteins from a dissolved alpha-helical form to an insoluble beta sheet (see pp. 32–33). Spider silk is only 0.01 µm to a few micrometers in diameter, yet its density is so low and its strength so high that a strand would have to be 80 km long to break under its own weight. Spiders do not waste such a valuable substance. After the silk has served its function it is eaten, and the amino acids are used to synthesize new silk. In one study a web was labeled with radioactive tracer, and the spider was allowed to eat it, as it normally would. Up to 90% of the tracer appeared in a web the spider made just a half hour later.

FUNCTIONS. Virtually every aspect of spider biology depends on the ability to produce silk. As previously noted, males produce sperm webs, females weave cocoons out of silk, and juvenile spiders use silk to "balloon" to new habitats. In addition, most spiders lay down a **drag line** of silk as they move about. The drag line is attached to the substratum and allows the spider to recover if it falls accidentally or deliberately drops to evade a predator. Of course the most familiar use of silk is in spider webs.

Approximately half the species of spiders construct webs. Some use the web as a lining for their burrows. Trapdoor spiders (family Ctenizidae) live within burrows covered with a silk trap door. When they feel prey passing overhead, they spring out and drag the prey into the burrow. Other spiders detect prey with trip lines radiating from the mouth of the burrow. The most common type of web is the **sheet web,** which is approximately horizontal and for many species leads to a funnel-shaped retreat in which the spider awaits prey. These webs are conspicuous on lawns after a heavy dew. **Cobwebs** are more loosely woven than sheet webs, and they depend on sticky threads to snare prey walking beneath them (Fig. 30.17). The third type of web is the familiar **orb web** that usually hangs vertically and snares flying insects. It will be discussed in more detail in the next section. Because of its intricate geometry, the orb web is thought to be the most evolutionarily advanced. However, a few spiders that belong to groups that generally weave orb webs have evolved interesting alternatives. Bolas spiders (*Dicrostichus* in Australia and *Mastophora* in the Americas) produce a single, sticky strand of silk that they throw at flying insects. The usual prey of *Mastophora* are male moths, which the spiders attract with secretions that mimic the sex attractants of female moths. Many spiders live in the webs of other species, eating the host's prey or the host, itself, which may have been lured to its demise by movements mimicking those of struggling prey.

ORB WEBS. The orb web is among the most fascinating of all animal constructions. Few of us would know how to begin constructing a web between one structure and an-

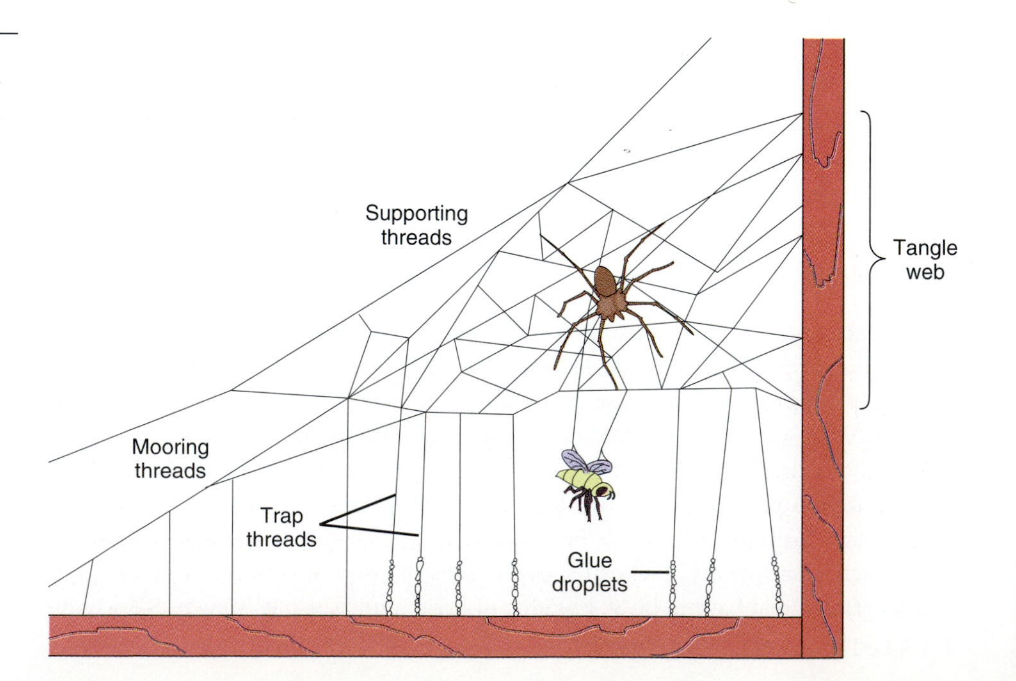

FIGURE 30.17
A cobweb spider *Steatoda borealis* approaching a fly that has walked into a trap thread.

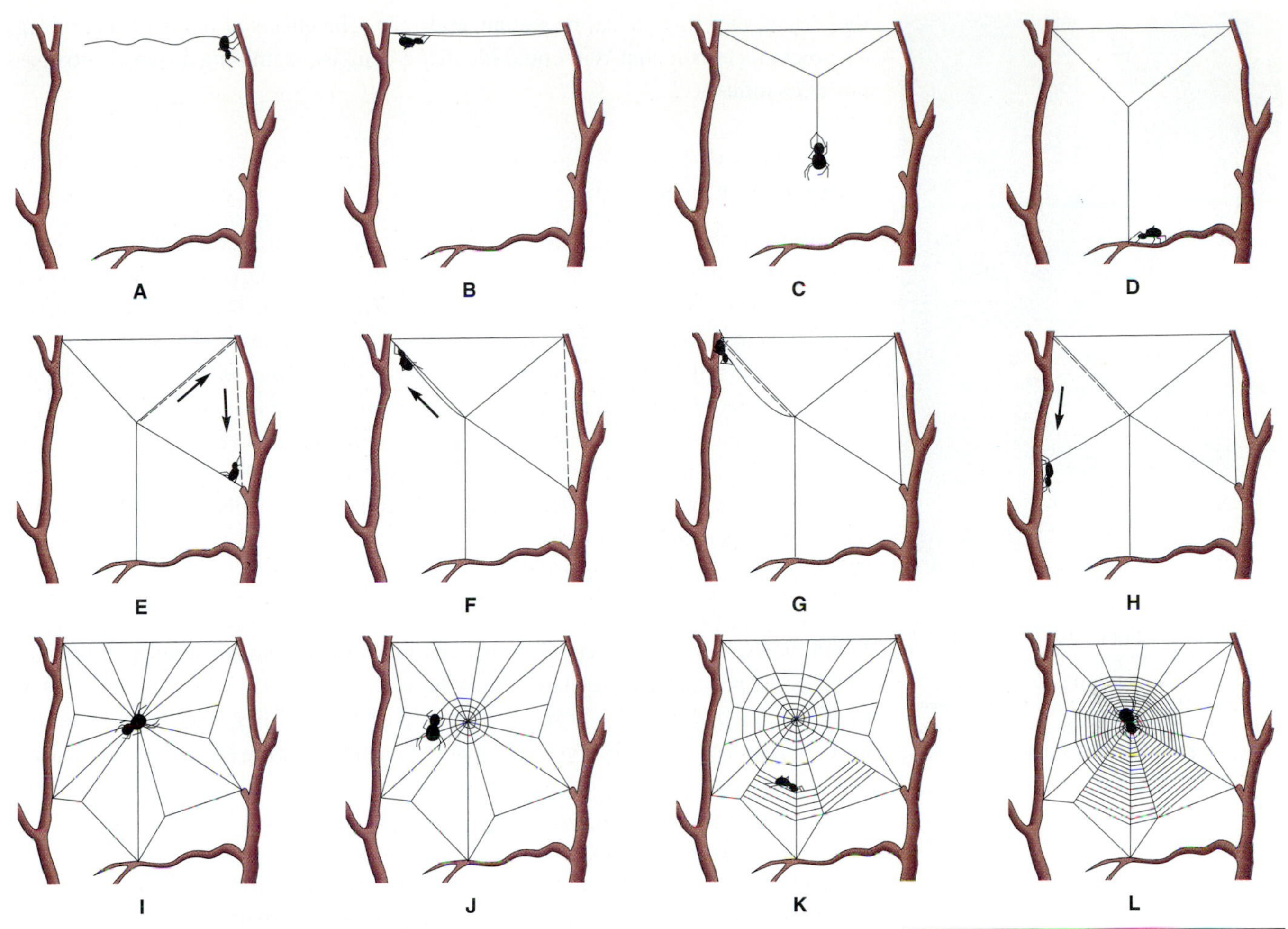

FIGURE 30.18

Steps in the construction of an orb web by one species of spider. (A) The spider plays out a strand in the wind to form a support. (B) The spider follows the support and lays down a parallel strand and then (C) pulls down at the middle, forming three radii shaped like a Y (D). (E) The spider then climbs up the stem of the Y to one of the supports, descends, and attaches a frame and another radial strand. (F–H) Additional radial and frame threads are attached. (I, J) The spider constructs an auxiliary spiral from the hub that serves as a temporary scaffolding. (K, L) It adds the catching spiral, then removes the auxiliary spiral.

■ **Question:** Do you suppose the spider sees ahead of time that there is a lower branch for attachment, as in (D)? What does it do if there is none?

other separated by a distance of tens or hundreds of body lengths. Yet many orb-weaving spiders perform this feat within a few minutes virtually every night, and they do it without vision (Fig. 30.18). The portion of the web that snares prey is the **catching spiral,** which works in one of two ways, depending on the species of spider. Many spiders deposit droplets of glue on threads of the catching spiral, while others (**cribellate** spiders) construct the catching spiral of wooly threads that entangle prey. How the spider avoids getting entangled in its own catching spiral is still a largely unexplored mystery. After constructing the web many spiders go to the hub or to a special refuge area until prey is detected. (See Fig. 30.18.) When the spider senses vibrations induced by trapped prey it rushes to the hub, if it is not already there. It then plucks the radial threads and uses their vibrations to determine the direction of the prey (Fig. 30.19 on page 650).

The general pattern of orb weaving varies among different species of web spiders. There are even small differences among individual spiders. These individual variations may be due in part to variations in the sizes of the front legs, which spiders use to measure lengths and angles. Webs can also be affected by many of the same drugs that affect human behavior. Peter N. Witt, a physician formerly at Duke University, first discovered this fact when asked by a zoologist to prescribe a drug that would make spiders construct webs during the daytime, when it was more convenient to observe them. Witt first tried amphetamine, which did not affect the time the web was built, but made the spiders place the radial threads and catching spiral abnormally. Intrigued by this result, Witt spent decades studying webs and the effects of psychoac-

FIGURE 30.19

The orange argiope *Argiope aurantia* reaping the harvest of its web. The dense vertical band of silk, called the stabilimentum, is thought to deter accidental destruction by birds, camouflage the spider, or attract insect prey by reflecting ultraviolet.

■ **Question:** Which of these explanations for the stabilimentum seems unlikely for this colorful spider?

FIGURE 30.20

The harvestman *Leiobunum* on a milkweed seed pod.

tive drugs, such as caffeine, mescaline, and LSD. The effects of many of these drugs are so characteristic that Witt could identify a drug by examining a web constructed under its influence.

Some Other Arachnids

HARVESTMEN. Besides spiders (order Araneae) there are several other important groups of arachnids. Harvestmen (= daddy longlegs; order Opiliones) are often conspicuous in late summer, at about harvest time (Fig. 30.20). They are readily distinguished from spiders by the fused cephalothorax and abdomen, and usually by legs that are extremely long and thin in relation to the body. They feed and scavenge on a variety of invertebrates and plants. A few species are among the few chelicerates capable of ingesting solid food. Harvestmen lack poison fangs. Perhaps for that reason most people take more kindly to them than to spiders. Harvestmen defend themselves from predators by an abdominal "stink gland" and by readily disconnecting a leg that has been grasped. Such **autotomy,** the detachment of a leg or other appendage to save a life, is quite common among arthropods.

SCORPIONS. Another group of arachnids includes scorpions (order Scorpionida), which most people have heard of but few have seen (Fig. 30.21). Scorpions are found mainly in the tropics and subtropics, especially in deserts. They usually stay in underground burrows during the day and come out at night to prey on spiders and insects. Desert scorpions detect prey by sensing vibrations in sand with slit sensilla and sensory hairs on their legs. The sand scorpion *Paruroctonus mesaensis* of the Mojave Desert uses differences in the times that vibrations arrive at its eight legs to determine the direction of prey (Brownell 1984). In this way a sand scorpion can quickly and accurately home in on a cockroach up to 50 cm away. After detecting prey, a scorpion chases it down, then seizes it in the chelae of the two pedipalps.

The scorpion dismembers prey with its chelicerae and digests it externally. If the prey is large the scorpion first subdues it by injecting a paralytic venom. While holding the prey in its pedipalps it quickly arches the tail over its head, directing the stinger with great accuracy. The venoms of different species contain neurotoxins of two major types that work on the sodium channels responsible for generating action potentials (see p. 156). Although most scorpions can inject only enough venom to cause pain and swelling in humans, the stings of some species can be fatal unless treated with antibodies to the neurotoxin.

Like spiders, scorpions engage in elaborate courtship and parental behaviors. During courtship the male and female grasp each other's chelae and promenade back and forth while the male kneads the female's chelicerae with his own, in the scorpion equivalent of smooching. Then he rather unromantically stings her. The sting may subdue the female's aggression, but does no permanent harm. The promenade usually lasts 30 to 60 minutes, but may go on for days until the male locates a suitable place on which to deposit a spermatophore. He then maneuvers the female until her external genitals flip open a lid on the spermatophore, which releases the sperm. In some species the male then flees, but in many species he makes an involuntary contribution to the nutrition of the young by being cannibalized by the female. Scorpions are viviparous, with the embryos absorbing nutrients from the mother's digestive tract. Up to 90 ova may be fertilized in one mating, and development can take more than a year in some species. Immediately after birth the young scorpions climb onto the mother's back and remain there until they become independent, about a week later.

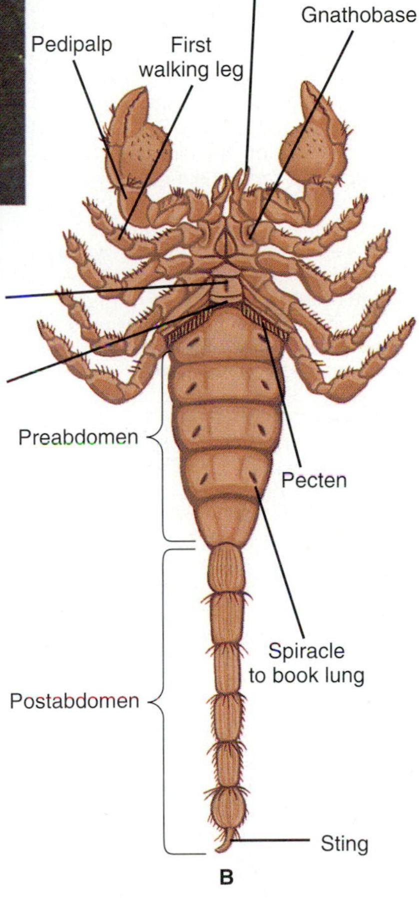

FIGURE 30.21

(A) Sand scorpion *Paruroctonus mesaensis* feeding on a burrowing cockroach. (Photographed at night under ultraviolet.) (B) Ventral view of the scorpion *Pandinus imperator,* the giant African scorpion (length up to 20 cm). The scorpion cephalothorax, to which the pedipalps and legs attach, is relatively short and is broadly joined to the abdomen. The abdomen has two distinct regions: the preabdomen and postabdomen. On the second abdominal segment are comb-shaped pectines (plural of pecten), which are mechanoreceptive and perhaps chemoreceptive organs unique to scorpions.

PSEUDOSCORPIONS. Approximately 2000 species of pseudoscorpions (order Pseudoscorpionida) are so named because of their prominent chelae (Fig. 30.22 on page 652). They lack the elongated postabdomen and stinger of scorpions, however, and never exceed 7 mm in length. They are common inhabitants in soil, under stones, and in animal nests. One species, *Chelifer cancroides,* is a harmless inhabitant of libraries and homes, which it enters by hitching onto house flies!

TICKS AND MITES. Ticks and mites (order Acari = Acarina) differ from the other arachnids in having lost all external signs of segmentation (Fig. 30.23 on page 652). They also differ in having a projecting mouth region, the **capitulum,** visible in Fig. 30.23A. Mites are usually less than 1 mm long, and ticks differ from them mainly in being much larger. Approximately 30,000 species of mites have been described, but many authorities believe the unknown species far outnumber all the known ones. Mites are often so small that they can be identified only with the scanning electron microscope. Close examination of a small sample of leaf litter from virtually anywhere in the world will usually reveal hundreds of mites of several species. Many mites parasitize various tissues of terrestrial vertebrates, while others prey on invertebrates or feed on plants.

There are approximately 850 species of ticks, all of which are parasitic on reptiles, birds, or mammals. Most species have a **three-host life cycle,** with a different host for each of three active life stages—larva ("seed tick"), nymph, and adult. In spring the six-legged larva ascends vegetation and stays until a host passes. It then hooks its pedipalps into the skin, then pierces the skin with sucking mouthparts. The larva feeds on blood until enormously swollen, then drops off the host and molts into an eight-legged nymph. The nymph then attaches to a new host, feeds, and molts into an adult. The adult then feeds on a third host before mating.

FIGURE 30.22
The cave blind pseudoscorpion *Vachonium* sp. As in many cave-dwelling animals, the eyes have degenerated in this species.

Interactions with Humans and Other Animals

Benefits. Spiders, scorpions, ticks, mites, and other chelicerates evoke negative feelings in most people. We know so little about many of these animals, however, that they may turn out to be beneficial in ways that we can scarcely imagine. Only in the 1960s was it discovered that the blood of horseshoe crabs clots when exposed to certain bacterial toxins. Now the blood of horseshoe crabs is widely used to screen substances for potential toxicity in humans. Even more recently, horseshoe crabs were found to be essential for the survival of numerous migrating shore birds, such as the red knot (*Calidris canutus*). In early May these robin-sized sandpipers leave their wintering sites in Brazil, having fattened themselves on snails. By the time they arrive at Delaware Bay, however, they are too thin and exhausted to reach their nesting grounds in the Arctic. Fortunately, horseshoe crabs are depositing millions of eggs in the sand at this time, and red knots and other migrating birds can replenish their energy supplies by feasting on them.

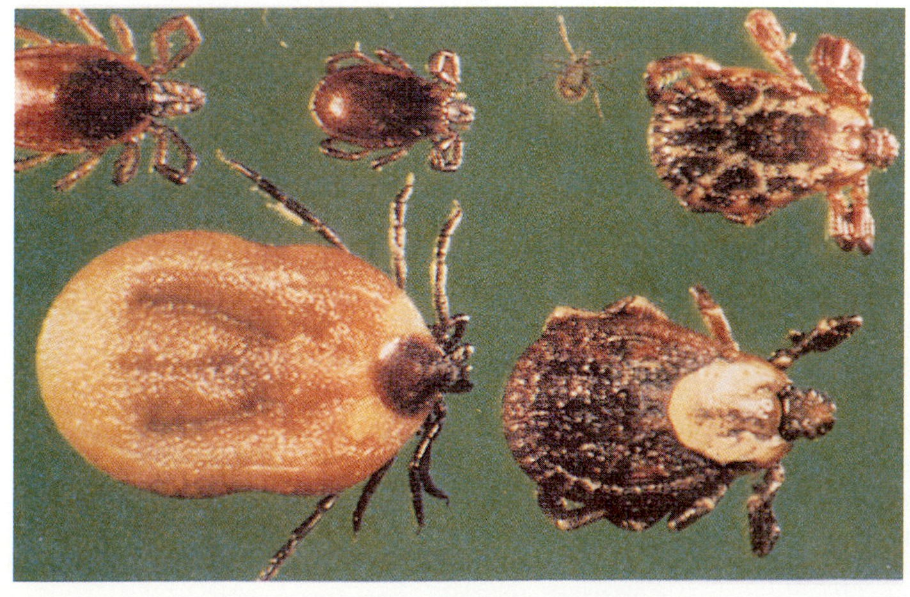

A

B

FIGURE 30.23
(A) The deer tick *Ixodes scapularis* (= *I. dammini*) at various stages of development on the left, and two dog ticks *Dermacentor variabilis* on the right (male and female, top and bottom, respectively). At top, starting from the upper left, are the female, male, and nymph of the deer tick. The year-old nymph, which transmits diseases to humans, is only a little larger than the period at the end of this sentence. On the lower left is a female deer tick engorged with blood. During its two-year life cycle each tick feeds just three times. As a larva and then as a nymph it usually feeds on deer mice and other small mammals; as an adult it feeds mainly on deer. If a larva feeds on a mammal carrying the bacterium responsible for Lyme disease, and then feeds as a nymph on a person, it may transmit the disease to that person. This species of tick is also responsible for a growing number of cases of human babesiosis in the United States. (B) The red freshwater mite *Limnochares americana.* This relatively large species (3 mm) can be found in ponds throughout North America. Water mites are among the few chelicerates that have returned to the aquatic habitat. Immature stages frequently live as parasites on the backs of aquatic insects. (See Fig. 32.15.)

VENOMOUS SPECIES. There is no denying that for many animals, including humans, interactions with chelicerates often have unpleasant consequences. A few species of spiders and scorpions can produce painful and even fatal bites and stings in humans. The scorpion *Centruroides* sp. kills hundreds and perhaps thousands of people each year in Mexico alone. The most dangerous spider in North America is the female black widow *Latrodectus mactans* (Fig. 30.24A). Her venom is 15 times more toxic than that of the prairie rattlesnake, although there is so little of it that only 1% of bites are fatal. The venom causes neuronal synapses to release the transmitter acetylcholine, resulting in muscle spasms, abdominal rigidity and cramps, sweating, salivation, high blood pressure, and sometimes convulsions. Another dangerous spider is the brown recluse *Loxosceles reclusa,* which is common in the south-central United States (Fig. 30.24B). It is brown with a dark violin-shaped marking on the dorsal cephalothorax. Its venom contains enzymes that destroy blood cells and induce white blood cells to attack surrounding tissues. Its bite can be fatal to children, and the craterlike wounds in adults may require months to heal. Both spiders have an unfortunate tendency to live in homes, outhouses, and other buildings. In the future the venoms of spiders, scorpions, and mites may prove more beneficial than harmful. Recently, scientists attempting to develop new approaches to insecticides have begun inserting the genes for these venoms into viruses that infect insects.

A

B

FIGURE 30.24

(A) The female black widow *Latrodectus mactans* with a cocoon. (B) The brown recluse *Loxosceles reclusa,* which can be identified by a violin-shaped pattern on the cephalothorax. Body approximately 1.5 cm long.

VECTORS OF DISEASE. Even more devastating to humans and other animals are some ticks and mites. Not only can ticks weaken terrestrial vertebrates by feeding on their blood, but they may also transmit disease-causing organisms. The Rocky Mountain wood tick *Dermacentor andersoni* is a vector for several diseases of humans, including **Rocky Mountain spotted fever.** In spite of its name, Rocky Mountain spotted fever occurs most commonly in the eastern United States. Wild rodents serve as a reservoir for the rickettsia bacterium that causes the disease. Deer, raccoons, squirrels, mice, and other mammals serve as reservoirs for other diseases for which ticks are vectors. A recent surge in such diseases may be due to the rise in numbers of these suburban animals.

The incidence of another tick-borne disease has increased even more dramatically since the 1970s. **Lyme disease** is caused by a bacterium (*Borrelia burgdorferi*) that is transmitted by deer ticks, *Ixodes* spp. (Fig. 30.23A). Thousands of people and untold numbers of dogs, cows, and horses have been infected all over the world. The first stage of the disease is often marked by a distinctive bull's-eye-shaped rash and flu-like symptoms. Diagnosis is difficult, but most people recover without treatment, and treatment with antibiotics is highly effective during this stage. Many untreated people, however, develop second and third stages of the disease that may last for years. In the second stage there may be nerve damage, meningitis, partial paralysis, and heart irregularities. In the third stage there may be arthritis-like inflammation of the joints. The best treatment is prevention, by wearing long pants and sleeves outdoors in the summer in areas where the disease occurs, by using a repellant containing DEET, and by examining every square centimeter of skin and removing the tiny nymphs that are the main culprits. Another approach is to kill the ticks while they are on alternative hosts, especially wild mice. Cotton laced with the poison Damminix can be left outdoors so that when mice use it in their nests the Damminix will kill the ticks. Controlling deer populations will also limit the number of ticks, but trying to kill all the mammals and birds that are reservoirs of the bacterium is futile.

Ticks can be just as devastating by transmitting diseases to cattle and wild animals on which people depend for food. Among the most severe tick-borne diseases of ruminants is **babesiosis,** caused by the protozoan *Babesia.* The cattle tick *Boophilus* is a vector for the species of *Babesia* that causes Texas cattle fever, also called red-water fever because of the bloody urine of its victims. Researchers have recently developed a vac-

FIGURE 30.25

Symbiotic mites. (A) The follicle mite *Demodex folliculorum* isolated and in its normal habitat. This species is usually a harmless commensal in hair follicles on the human face. A related species, *D. brevis,* lives in sebaceous glands on the face. The genus *Demodex* includes the smallest of all known arthropods (less than 0.1 mm long). (B) Parasitic mites *Acarapis woodi* in the trachea of a honey bee.

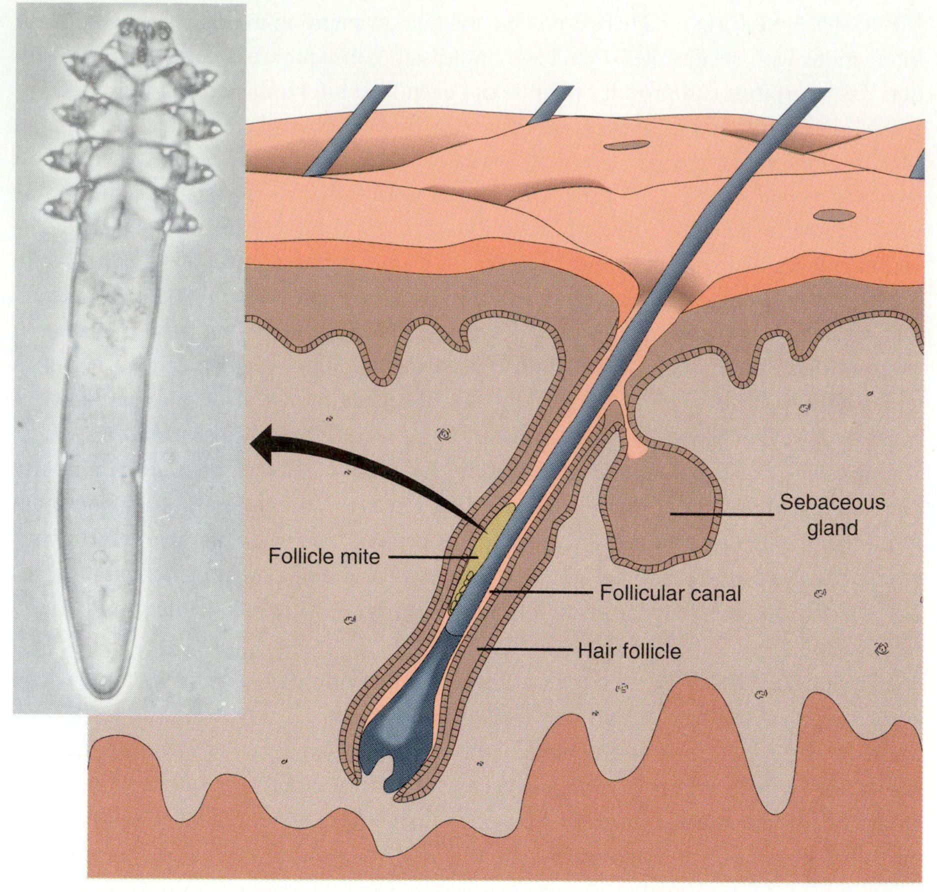

A

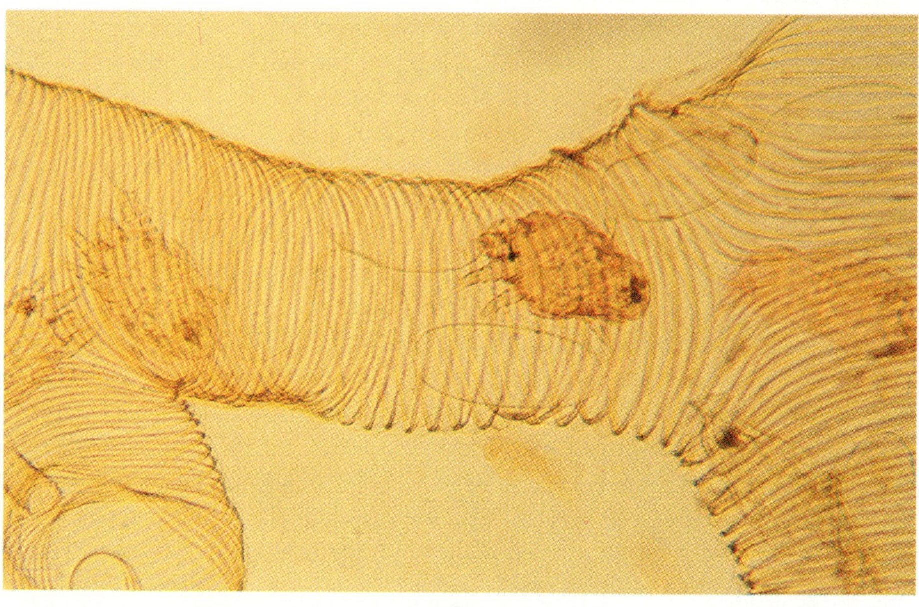

B

cine that protects cattle against babesiosis and other tick-borne diseases in a most unusual way. It causes the cattle to produce antibodies against the gut lining of the tick, preventing females from absorbing the cattle hormones they require for reproduction.

Many **mites** are harmless commensals on animals (Fig. 30.25A), but others cause severe economic losses by transmitting diseases and by parasitism. For example, several species of follicle mites (genus *Demodex*) cause mange in dogs, and they cause other skin eruptions in cattle, horses, and hogs. The tracheal bee mite *Acarapis woodi,* which inhabits the tracheae of honey bees, has recently spread over much of the United States, threatening millions of dollars worth of honey and crops that depend on bees for pollination (Fig. 30.25B). Even more recently the mite *Varroa jacobsoni* has become a serious threat to honey bees. Mites can also be a health nuisance for humans. Larvae of the chigger mite *Trombicula* cause a maddening itch when they secrete digestive enzymes into the skin and feed on the tissues. Female itch mites *Sarcoptes scabiei* tunnel through human skin, producing the disorder commonly called scabies or "the seven-year itch." Finally, feces and other products from mites that inhabit house dust and pillows have been identified as major cause of allergies and childhood asthma.

Summary

The phylum Arthropoda is an extremely diverse group that includes the Chelicerata (spiders, scorpions, and others), Crustacea, Insecta, Myriapoda (centipedes, millipedes, pauropods, and symphylans), and perhaps Onychophora. Classification is still controversial. All these forms have an integument with a cuticle consisting largely of chitin and protein. The cuticle is divided into endocuticle, exocuticle, and epicuticle. The exocuticle and epicuticle are usually sclerotized in large areas, forming an exoskeleton of plates and tubes hinged together by arthrodial membrane. The epicuticle contains waterproof waxes, which, with numerous other adaptations, enabled arthropods to be among the first terrestrial animals. In order for arthropods to grow they have to molt periodically. The rigid cuticle of most arthropods requires that the legs be jointed. In addition, all have segmented bodies and other similarities.

Onychophorans are an exceptional group of caterpillarlike animals in which the cuticle is flexible, and the legs are not jointed. They are restricted to moist habitats and trap prey by squirting a sticky substance at them. Myriapods and most other arthropods have rigid exoskeletons and jointed legs. Centipedes are predaceous, with one pair of legs modified as poisonous fangs. Millipedes, pauropods, and symphylans feed mainly on decaying vegetation and animals. Chelicerates are mainly carnivorous, and most feed by means of chelicerae.

Chelicerates include three major groups: horseshoe crabs, sea spiders, and arachnids (true spiders, harvestmen, scorpions, pseudoscorpions, ticks, and mites). Horseshoe crabs and sea spiders are aquatic chelicerates that are quite distinct from each other and from arachnids. Horseshoe crabs have gills and dome-shaped carapaces, and sea spiders use integumentary respiration through the enormous surface area of their long legs. Arachnids are mainly terrestrial and have book lungs. Spiders make up the majority of arachnids. Their bodies are divided into two tagmata: the cephalothorax and abdomen. There are four pairs of legs.

Most spiders attack prey with the fangs on their chelicerae, then feed by regurgitating digestive enzymes and ingesting the semiliquid. Spiders have three or four pairs of simple eyes. Web-building spiders have small eyes and rely mainly on mechanoreceptors; hunting spiders have larger and more acute eyes. Many spiders use silk to construct webs with which they trap prey. There are three main kinds of webs: sheet webs, cobwebs, and orb webs. Orb webs have a catching spiral that traps prey in glue or in wooly threads. Silk is also used by males for sperm webs, by females for cocoons, by juveniles for ballooning to new locations, and by all for drag lines.

Key Terms

cuticle
hemocoel
chitin
endocuticle
exocuticle
epicuticle
sclerotization
articular membrane
molting
trachea
chelicera
mandible
cephalothorax
telson
gnathobase
book gill
Malpighian tubules
spinneret
sheet web
cobweb
orb web

Chapter Test

1. Explain the argument in favor of dividing arthropods into three phyla. What is the argument in favor of retaining the phylum Arthropoda? (pp. 635–636)
2. How do Myriapoda and Onychophora differ from each other and from other arthropods? (pp. 638–641)
3. Explain how arthropods are able to move while inside an exoskeleton. What features of the exoskeleton account for its rigidity? How does the exoskeleton retard the loss of body water? (pp. 632–633)
4. List three major groups of chelicerates by their common names, and briefly describe each group. What features do they have in common, and what are the differences among them? (pp. 641–643)
5. Describe four ways in which spiders are adapted to terrestrial living. (pp. 643–647)
6. Describe four ways spiders use silk. (pp. 647–650)
7. Describe how hunting spiders differ physiologically and behaviorally from web spiders. (pp. 646–647)
8. For each of the following, state one way in which they differ from spiders: harvestmen, scorpions, ticks. (pp. 650–651)

■ Answers to the Figure Questions

30.1 Answers will be somewhat subjective, based on the synapomorphies chosen. Some possible synapomorphies are: (1) legs jointed, (2) abdomen separate and clearly segmented, (3) head clearly distinct from leg-bearing segments, (4) with antennae, (5) fewer than eight walking legs. The arthropods and the synapomorphies they display are: centipede (Myriapoda) 1, 4; Onychophora 4; spider (Chelicerata) 1; crayfish (Crustacea) 1, 2, 4; grasshopper (Insecta) 1, 2, 3, 4, 5. Crustacea and Insecta share the most synapomorphies (three), so they could be sister groups. The clade consisting of Crustacea plus Insecta shares two synapomorphies with Myriapoda and only one with Onychophora and Chelicerata, so Myriapoda might be the sister group of Crustacea plus Insecta. The relationship to Chelicerata and Onychophora would be difficult to resolve with these synapomorphies.

30.4 The dragonfly was not as large when it was inside the old cuticle. It inflated itself after emerging.

30.13 Good question.

30.14 After passing through the pharynx, stomach, intestine, and digestive glands, it somehow has to get to the silk glands. Because there does not appear to be a direct connection to the silk glands, the amino acids released from the digested silk must travel through the hemolymph to the silk glands, where they are resynthesized into silk.

30.16 This kind of courtship display would probably not be appreciated by an orb-weaving female, since her eyes are so poor. Moreover, it might be mistaken for the struggle of prey.

30.18 An orb-weaving spider in the dark probably cannot see if there is a branch on which the silk will attach. If the silk does not attach, the spider probably reels it in and tries again.

30.19 This spider would probably appear more conspicuous to a bird than the stabilimentum does, so the first explanation is unlikely. Whether the second explanation is likely depends on what the spider would look like to insect prey that perceive UV. If both the stabilimentum and the spider reflect UV, the spider might well be camouflaged by the stabilimentum. The last explanation is also possible.

Readings

Recommended Readings

Barlow, R. B., Jr. 1990. What the brain tells the eye. *Sci. Am.* 262(4):90–95 (Apr). (*On vision in the horseshoe crab.*)

Brownell, P. H. 1984. Prey detection by the sand scorpion. *Sci. Am.* 251(6):86–97 (Oct).

Burgess, J. W. 1976. Social spiders. *Sci. Am.* 234(3):100–106 (Mar).

Foelix, R. 1982. *Biology of Spiders.* Cambridge, MA: Harvard University Press.

Forster, L. 1982. Vision and prey-catching strategies in jumping spiders. *Am. Sci.* 70:165–175.

Gertsch, W. J. 1979. *American Spiders,* 2nd ed. New York: Van Nostrand Reinhold. (*Beautifully illustrated.*)

Gray, J., and W. Shear. 1992. Early life on land. *Am. Sci.* 80:444–456.

Habicht, G. S., G. Beck, and J. L. Benach. 1987. Lyme disease. *Sci. Am.* 257(1):78–83 (July).

Hadley, N. F. 1986. The arthropod cuticle. *Sci. Am.* 255(1):104–112 (July).

Jackson, R. R. 1985. A web-building jumping spider. *Sci. Am.* 253(3):102–115 (Sept).

Jackson, R. R. 1992. Eight-legged tricksters. *BioScience* 42:590–598. (*Spiders that prey on other spiders.*)

Kantor, F. S. 1994. Disarming Lyme disease. *Sci. Am.* 271(3):34–39 (Sept).

Kaston, B. J. 1978. *How to Know the Spiders,* 3rd ed. Dubuque, IA: Wm. C. Brown.

Levi, H. W., 1978. Orb-weaving spiders and their webs. *Am. Sci.* 66:734–742.

Levi, H. W., and L. R. Levi. 1968. *A Guide to Spiders and Their Kin.* New York: Golden Press.

Levi-Setti, R. 1975. *Trilobites: A Photographic Atlas.* Chicago: University of Chicago Press.

Marshall, S. D. 1992. The importance of being hairy. *Nat. Hist.* 101:41–47 (Sept). (*On tarantulas.*)

McDaniel, B. 1979. *How to Know the Ticks and Mites.* Dubuque, IA: Wm. C. Brown.

Miller, J. A. 1987. Ecology of a new disease. *BioScience* 37:11–15. (*On Lyme disease.*)

Milne, L., and M. Milne. 1980. *The Audubon Society Field Guide to North American Insects and Spiders.* New York: Alfred A. Knopf.

Myers, J. P. 1986. Sex and gluttony on Delaware Bay. *Nat. Hist.* 95:68–77 (May). (*On horseshoe crab eggs as food for migrating birds.*)

Preston-Mafham, R., and K. Preston-Mafham. 1984. *Spiders of the World.* New York: Facts on File. (*Beautifully illustrated.*)

Rudloe, A., and J. Rudloe. 1981. The changeless horseshoe crab. *Natl. Geogr.* pp. 562–572 (Apr).

Shear, W. A. 1993. One small step for an arthropod. *Nat. Hist.* 102:46–51. (*On terrestrial colonization.*)

Shear, W. A. 1994. Intangling the evolution of the web. *Am. Sci.* 82:256–266.

Vollrath, F. 1992. Spider webs and silks. *Sci. Am.* 266(3):70–76 (Mar).

See also relevant selections in General References at the end of Chapter 21.

Additional References

Arnaud, F., and R. N. Bamber. 1987. The biology of Pycnogonida. *Adv. Marine Biol.* 24:1–96.

Ballard, J. W. O., et al. 1992. Evidence from 12S ribosomal RNA sequences that onychophorans are modified arthropods. *Science* 246:241–243.

Barbour, A., and D. Fish. 1993. The biological and social phenomenon of Lyme disease. *Science* 260:1610–1616.

Boore, J. L., et al. 1995. Deducing the pattern of arthropod phylogeny from mitochondrial DNA rearrangements. *Nature* 376:163–165.

Briggs, D. E. G., and R. A. Fortey. 1989. The early radiation and relationships of the major arthropod groups. *Science* 246:241–243.

Eberhard, W. G. 1990. Functions and phylogeny of spider webs. *Annu. Rev. Ecol. Syst.* 21:341–372.

Emerson, M. J., and F. R. Shram. 1990. The origin of crustacean biramous appendages and the evolution of Arthropoda. *Science* 250:667–669.

Friedrich, M., and D. Tautz. 1995. Ribosomal DNA phylogeny of the major extant arthropod classes and the evolution of myriapods. *Nature* 376:165–167.

Harrison, F. W. and M. E. Rice (Eds.). 1993. *Microscopic Anatomy of the Invertebrates,* Vol. 12: *Onychophora, Chilopoda, and Lesser Protostomata.* New York: Wiley–Liss.

Hopkin, S. P., and H. J. Read. 1992. *The Biology of Millipedes.* New York: Oxford University Press.

Kukalová-Peck, J. 1992. The "Uniramia" do not exist: the ground plan of the Pterygota as revealed by Permian Diaphanopterodea from Russia (Insecta: Paleodictyopteroidea). *Can. J. Zool.* 70:236–255.

Little, C. 1990. *The Terrestrial Invasion: An Ecophysiological Approach to the Origin of Land Animals.* New York: Cambridge University Press.

Manton, S. M. 1972. *The Arthropoda: Habits, Functional Morphology and Evolution.* New York: Clarendon.

Polis, G. A. (Ed.). 1990. *The Biology of Scorpions.* Stanford, CA: Stanford University Press.

Shear, W. A. (Ed.). 1986. *Spiders: Webs, Behavior, and Evolution.* Palo Alto, CA: Stanford University Press.

Sonenshine, D. E. 1991. *Biology of Ticks.* New York: Oxford University Press.

Turbeville, J. M., et al. 1991. The phylogenetic status of arthropods, as inferred from 18S rRNA sequences. *Mol. Biol. Evol.* 8:669–686.

Wheeler, W. C., P. Cartwright, and C. Y. Hayashi. 1993. Arthropod phylogeny: A combined approach. *Cladistics* 9:1–39.

Witt, P. N., and J. S. Rovner. 1982. *Spider Communication: Mechanisms and Ecological Significance.* Princeton, NJ: Princeton University Press.

CHAPTER

31

Crustaceans

Scarlet crab *Grapsus.*

General Features of Crustaceans

Although chelicerates and crustaceans are both usually classified in phylum Arthropoda, they are so different from each other that even nonzoologists easily tell them apart. Some people who drool at the sight of lobster, crab, or shrimp on their dinner plate would faint at the sight of a spider or scorpion, even if it were boiled and served with melted butter. The reasons for this prejudice are not clear.

Zoologically there are differences between chelicerates and crustaceans that are not merely matters of taste. Unlike members of the subphylum Chelicerata, which have chelicerae, crustaceans have mandibles. Crustaceans are often referred to as the aquatic mandibulates, because few have adapted to terrestrial life. In spite of these differences, most of the approximately 40,000 species of crustaceans share the important arthropod features (see Table 30.1), including segmentation, jointed appendages, and chitinous exoskeleton.

SEGMENTATION. Segmentation is apparent in the crayfish, which we shall take as a representative crustacean (Fig. 31.1). Segmentation is most pronounced in the abdomen, where each of six somites is covered by two plates of cuticle: a dorsal **tergite** and a ventral **sternite.** In crayfishes and many other crustaceans, segmentation is less obvious anteriorly, because somites are tagmatized as a single cephalothorax that is shielded dorsally by a **carapace.** The 13 pairs of appendages on the cephalothorax reveal, however, that this tagma originates from 13 embryonic segments.

JOINTED APPENDAGES. The 13 pairs of jointed appendages on the cephalothorax, along with the 6 pairs of appendages on the abdomen, are all serially homologous with each other. Each appendage is often considered to develop from a generalized appendage with two branches: a lateral branch called the **exopod** and a medial branch called the **endopod.** These two branches are attached to a **protopod.** (See top panel in Fig. 31.2 on page 660.) During development the exopod may be lost from some appendages, but additional branches may grow out of protopods, endopods, or remaining exopods with each molt. In this way the appendages become modified for a variety of functions, some of which are surprising even for an arthropod.

In crayfishes and many other crustaceans, the first two pairs of segmental appendages develop into two pairs of antennae that bear receptors for touch and chemoreception. The appendages on the third through eighth segments make up a

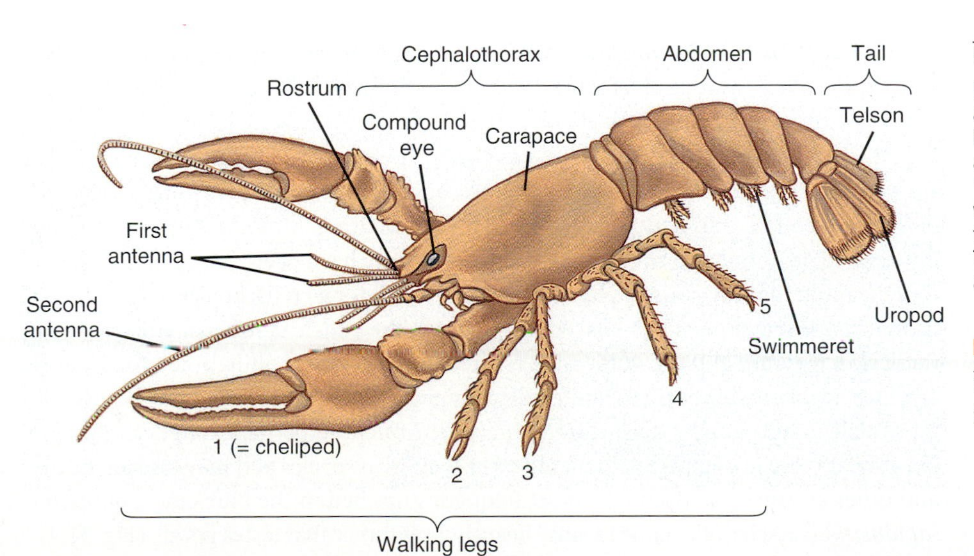

FIGURE 31.1
External features of a crayfish. There are several genera and approximately a hundred species of crayfishes. In some areas they are known as "crawfish" or "crawdads," and they are eaten. The walking legs are numbered. Four of the five swimmerets on one side are shown. The first pair of swimmerets are adapted as copulatory organs in males.

■ **Question:** Why aren't crayfish off balance with their walking legs so far forward?

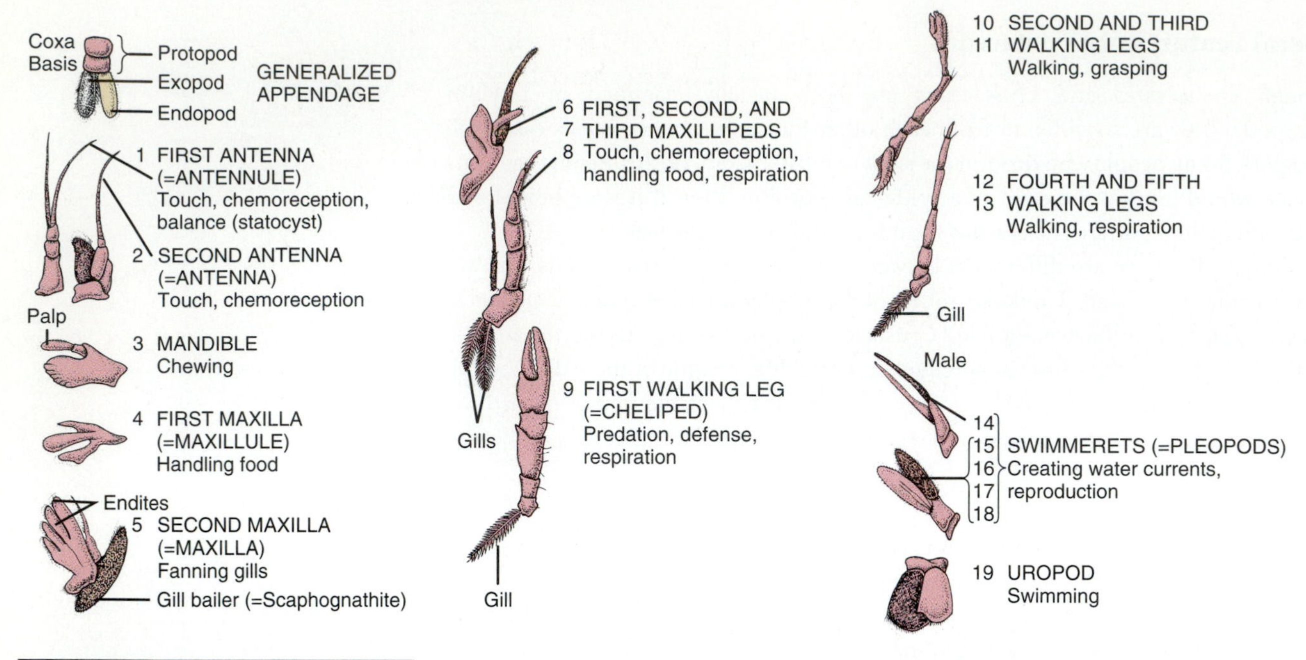

FIGURE 31.2

Types of appendages of a crayfish and their functions. Numbers indicate the segment. All the appendages are thought to originate from a primitive biramous appendage schematically represented in the top panel. The protopod of this appendage has two joints, the coxa and basis. Attached to the basis is a lateral exopod and a medial endopod. During development the exopod is commonly lost, as in the mandibles and cheliped, but additional branches may sprout. A medial branch is called an endite (see the second maxilla), and a lateral branch is called an exite. An exite on a protopod is called an epipod. All the gills are epipods. Many zoologists add the suffix -ite to most of these structures—hence exopodite, endopodite, and so on.

complicated assortment of **mandibles, maxillae,** and **maxillipeds** generally involved with feeding (Fig. 31.3). The two mandibles crush food; the two pairs of maxillae shred food and pass it forward to the mandibles. The second pair of maxillae also have "gill bailers" that draw water anteriorly past the gills. The gills are epipods on the last pair of maxillipeds (segment 8) and on some walking legs (segments 9, 12, and 13), and they are hidden beneath the carapace.

The first pair of walking legs, called **chelipeds** (segment 9), bear **chelae** (pincers) used in grasping food, predation, and defense. They are generally busy exploring the substratum. The second and third pairs of walking legs, but not the fourth and fifth, are also chelate. The appendages on abdominal segments 14 through 18 are called **swimmerets** (= pleopods), although they are not used for swimming in crayfish and many other crustaceans. Females use their swimmerets to carry eggs, and males have the first pair of swimmerets modified as copulatory organs. The 19th and last pair of segmental appendages are two **uropods,** which combine with the telson to make a large fin used for swimming.

EXOSKELETON. Like other arthropods, crustaceans have an exoskeleton with a cuticle reinforced by chitin. Most large crustaceans also have substantial deposits of calcium carbonate in the three layers of cuticle (epicuticle, exocuticle, endocuticle; see Fig. 30.2C). The blue crab *Callinectes sapidus,* for example, has the equivalent of four sticks of chalk deposited in its claws, carapace, and other hard areas of cuticle. Because of the rigid cuticle, crustaceans can grow only after molting. Molting is an opportunity not only to grow but also to regenerate limbs that may have been damaged or lost before the molt. Molting is also a physiological and behavioral crisis, however. Chitin and protein are lost and must be replaced. Even worse, the cuticle is soft just after molting, rendering the animal more vulnerable to predators. Before molting, therefore, many crustaceans must stop feeding and other activities to seek a protective shelter. While the cuticle is still soft, the lobster *Homarus americanus* and perhaps other crustaceans are also less able to compete with members of their own species and may lose territorial and other advantages. Molting is most frequently studied in the blue crab *Callinectes sapidus,* whose scientific name means "beautiful swimmer that tastes good" (Fig. 31.4).

Soft-shell crabs, which you may have eaten, are just-molted members of this species. One reason molting is so often studied in the blue crab is that commercial crabbers already have a great deal of knowledge and interest in the subject (Warner 1976).

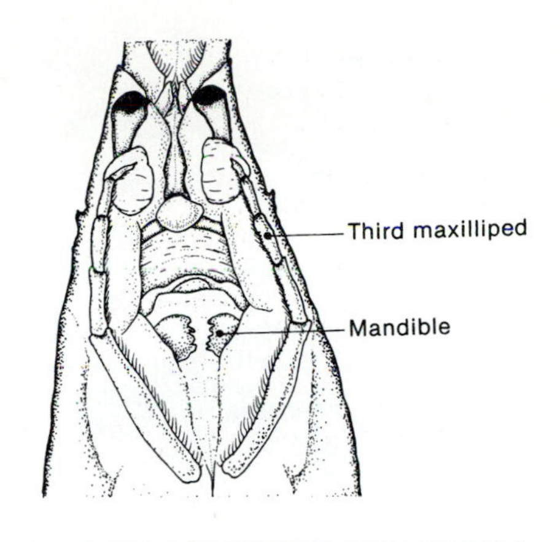

FIGURE 31.3
Mouthparts of the crayfish *Cambarus.*

■ **Question:** For what kinds of food would these mouthparts be adapted?

Reproduction

The blue crab illustrates how major events in a crustacean's life—even reproduction—must be organized around molting. Like most crustaceans, *Callinectes sapidus* has separate sexes and reproduces by copulation. The female can mate only after her final, or "nuptial," molt as an adult. There are several reasons for this. First, the molting hormone ecdysone is required for the production of ova. Second, copulation is physically impossible unless the female's cuticle is still soft from molting. Third, during the nuptial molt the female exchanges her narrow **apron,** which is actually the abdomen folded beneath the cephalothorax, for the broad apron required for copulation and for carrying eggs.

The nuptial molt usually occurs in autumn, in the shallow, dilute seawater of bays and estuaries. A week or so before the nuptial molt the female becomes attractive to males, probably because of a pheromone. A male (known to crabbers as a "Jimmy") that finds a female about to undergo the nuptial molt courts her by standing on the tips of his walking legs, extending his claws sideways, waving his swimming legs over his rear, and kicking up sand (Fig. 31.5A on page 662). The female responds by waving her claws in a beckoning motion. She may impatiently back under the male. If not, the male makes a sudden grab for her and cages her beneath him within his walking legs. The seizure of the female by the male triggers her production of ecdysone. With both crabs facing forward, the male carries the female around for up to a week, looking for a refuge in dense grass and waiting for the female to molt (Fig. 31.5B). Crabbers call these pairs "doublers" and scoop them up to get the soon-to-molt females.

If not interrupted by crabbers, the male continues to cradle the female during the two to three hours required to molt, as well as during the post-molt period in which she inflates herself with water to stretch the cuticle for future growth. The male then turns the female onto her back. Then the female unfolds her broad abdomen to embrace the male and to expose two **genital pores** that lead to **sperm receptacles.** The male pulls

FIGURE 31.4
The edible blue crab *Callinectes sapidus,* which is found all along the Atlantic coast. In this and other adult crabs the abdomen is tucked beneath the cephalothorax to form an apron that encloses the swimmerets. The rear pair of legs are modified as paddles that allow the blue crab to swim rapidly in any direction. Width 23 cm.

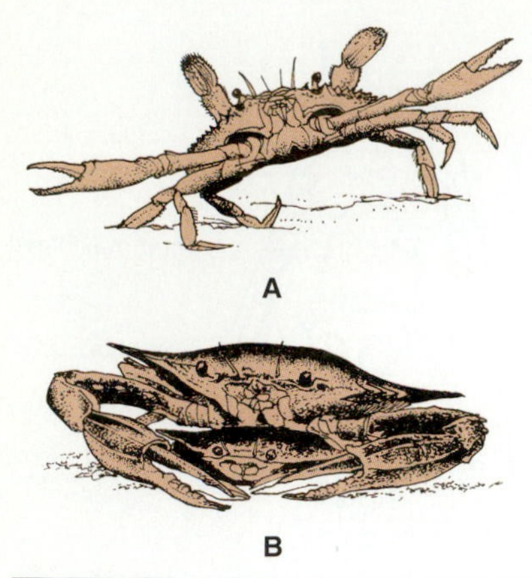

FIGURE 31.5

Mating in blue crabs. (A) A male courting a female. (B) The male cradling the female until she molts.

two hollow swimmerets from beneath his narrow abdominal apron and inserts these copulatory organs into her genital pores. For the next 5 to 12 hours the male transfers two packets of sperm released from the bases of these swimmerets into the sperm receptacles of the female. This injection is essential because the sperm of crabs and most other crustaceans lack flagella and are immotile.

After copulation the male turns the female upright again and continues to cradle her for at least two more days. In this way he protects his reproductive investment until her shell hardens. The male then lets go of the female, which soon starts to swim toward the saltier waters at the mouth of the bay or estuary. Although blue crabs copulate, fertilization of the eggs occurs externally and after a long delay. In the spring following mating, the female releases a sticky mass containing some two million eggs and mixes it with one of the packets of sperm. The female broods the fertilized eggs beneath her abdominal apron. The eggs are orange at first, then turn brown as the larvae within them develop.

Development

In some crustaceans, including crayfish, development is direct. That is, the just-hatched young are miniatures of the adult. In blue crabs, however, the egg hatches into a larva, called a **zoea** (pronounced ZO-ee-uh; Fig. 31.6A). The zoea is barely a millimeter long, has no claws, cannot actively swim, and does not feed. The zoea usually molts six times in as many weeks, developing in size and structure with each molt. The sixth-stage zoea then molts again into a **postlarva,** which, by definition, has the same complement of functional legs as the adult. In crabs the postlarva is called a **megalops** and looks more like a crayfish or shrimp than a crab (Fig. 31.6B). Hordes of these megalops continue the migration to sea. (So many pass Virginia Beach on their way out of the Chesapeake Bay that the irritating nips from their tiny claws chase out the bathers, who complain of "water fleas.") In the ocean the megalops join the countless other kinds of zooplankton that form the base of most food webs. The odds of any one megalops surviving to adulthood are perhaps one in a million, but there are so many of them that blue crab populations remain stable in spite of heavy predation.

After two weeks a megalops molts into a pinhead-sized version of the adult, which migrates toward shores and into estuaries and bays. By November of the first year of its life, this adult form has molted seven or eight times and has grown to a few centimeters long. It then burrows into the bottom, where it is protected from winter cold. In spring it resumes its movements and molting. A blue crab may live approximately three years, grow to approximately 12 cm long, and molt more than 20 times until it is ready to complete the life cycle by mating.

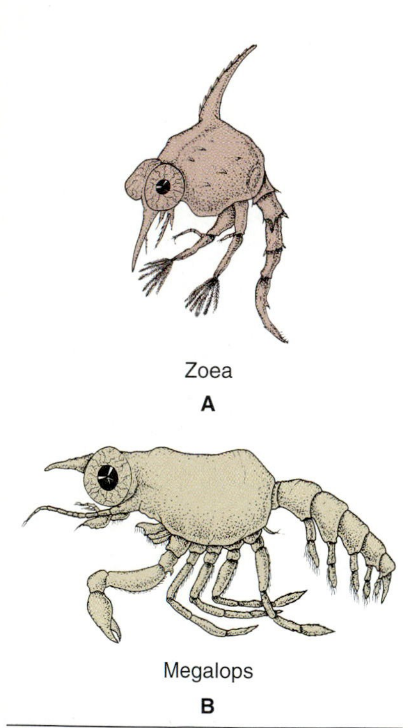

FIGURE 31.6

Stages in the life of the blue crab. (A) The egg hatches into a zoea, which usually molts seven times, increasing in size with each molt. (B) The seventh molt produces the megalops, which now has claws and five pairs of legs like the adult. The megalops larva then molts into the adult form.

■ **Question:** What could be the function of the dorsal spine on the zoea?

Although the first free-swimming larval form in the blue crab is the zoea, it is assumed that another larval form, the **nauplius,** occurs within the egg, simply because the nauplius occurs as the first larval form within every major group of Crustacea. The nauplius (NAW-plee-us), with its oval, unsegmented, microscopic body, is unique to crustaceans (Fig. 31.7). Perhaps the oddest feature of the nauplius larva is the one median eye, called the **naupliar eye.** Its three pairs of appendages develop into the first and second antennae and the mandibles.

Internal Structure and Function

DIGESTION. The digestive system of crayfish and many other crustaceans includes a short esophagus connecting the mouth to the stomach, a long, straight intestine terminating at the anus, and the **digestive gland** (Fig. 31.8). The final stages of digestion occur within the digestive gland (= midgut gland). The digestive gland is also called the hepatopancreas, because it stores fats and glycogen like the vertebrate liver and secretes digestive enzymes like the vertebrate pancreas. The stomach is divided into two chambers in crayfishes. The anterior **cardiac** chamber is used primarily for grinding and storage. Chitinous teeth attached to the muscular lining of the cardiac stomach make up a **gastric mill** that crushes and tears large pieces of food. Food that has gone through the gastric mill passes into the smaller **pyloric stomach,** which is lined with **setae** that sort out particles small enough to pass into the intestine, and still smaller particles that go to the digestive gland.

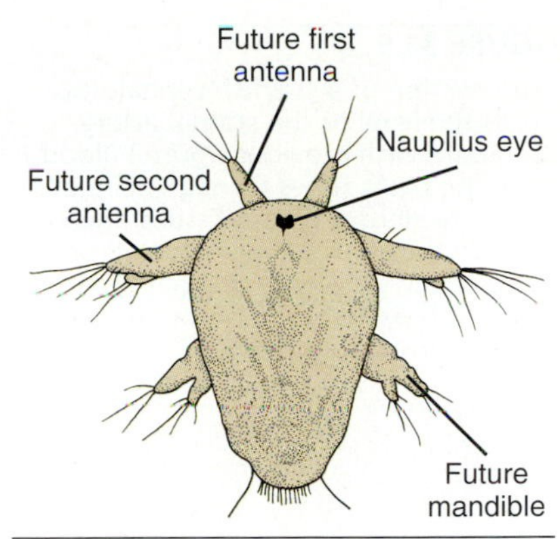

FIGURE 31.7

The planktonic nauplius larva of the copepod *Cyclops fuscus.* The first antenna is uniramous. The biramous second antennae and mandibles are used in swimming. The nauplius does not feed.

■ **Question:** What could be the function of the naupliar eye?

EXCRETION AND OSMOREGULATION. The crustacean organs for excretion and osmoregulation are unlike those of other arthropods. Excretion of metabolic wastes, primarily ammonia, is due to diffusion across the gills into the environment, even in terrestrial crustaceans. Little osmoregulation is required in marine crustaceans, since the body fluids have ion concentrations similar to those of seawater. In freshwater crustaceans, such as crayfishes, and in estuarine crustaceans, such as the blue crab, osmoregulation is due mainly to active transport of ions across the gill surface into the blood. Crustaceans also have either **antennal glands** (= green glands) or **maxillary glands** that conserve potassium and calcium and excrete excess sulfate and magnesium in most species. The two types of gland are similar to each other; the names depend on whether the excretory pores (nephridiopores) are at the bases of the second antennae or the second maxillae. Crustacean embryos have both antennal and maxillary glands, but one or the other is lost during development in most species.

An antennal or maxillary gland consists of an **end sac** (which is a rudiment of the coelom) and a **tubule** that leads into a urinary **bladder** (see Fig. 14.8). The antennal gland works by a combination of filtration and active transport. Pressure in the hemocoel forces hemolymph into the end sac. The filtrate then enters the green **labyrinth,**

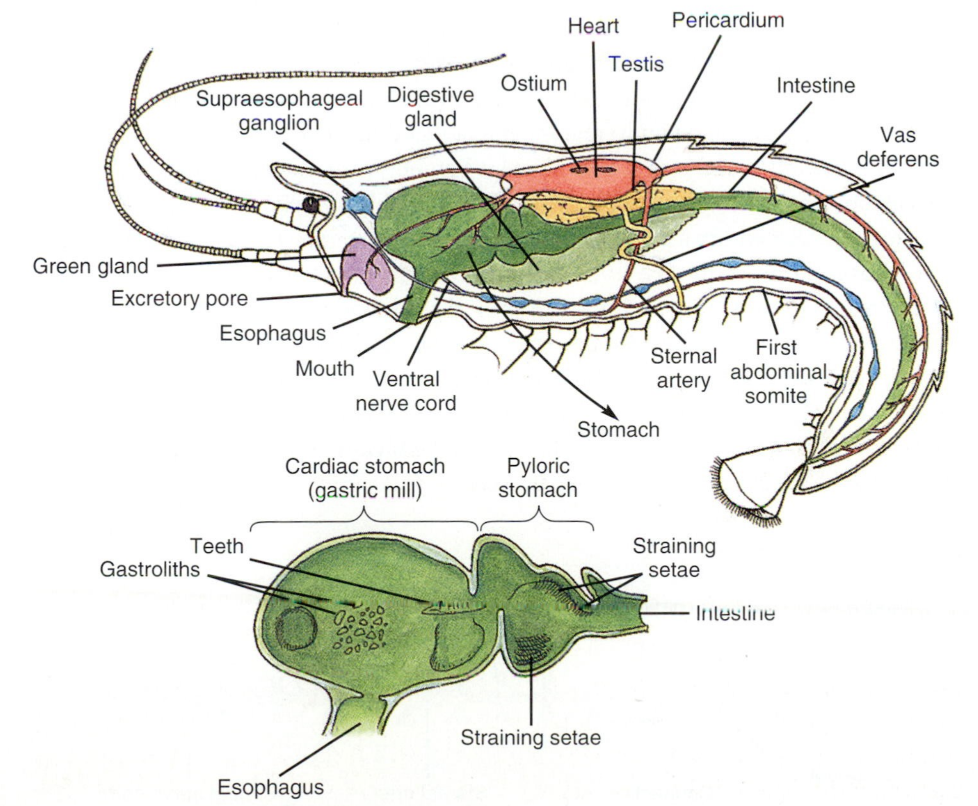

FIGURE 31.8

Longitudinal section of a male crayfish, showing the internal organs. The inset shows details of the stomach, which is divided into cardiac and pyloric sections. Gastroliths are deposits of calcium that are reabsorbed by the epidermis during molting.

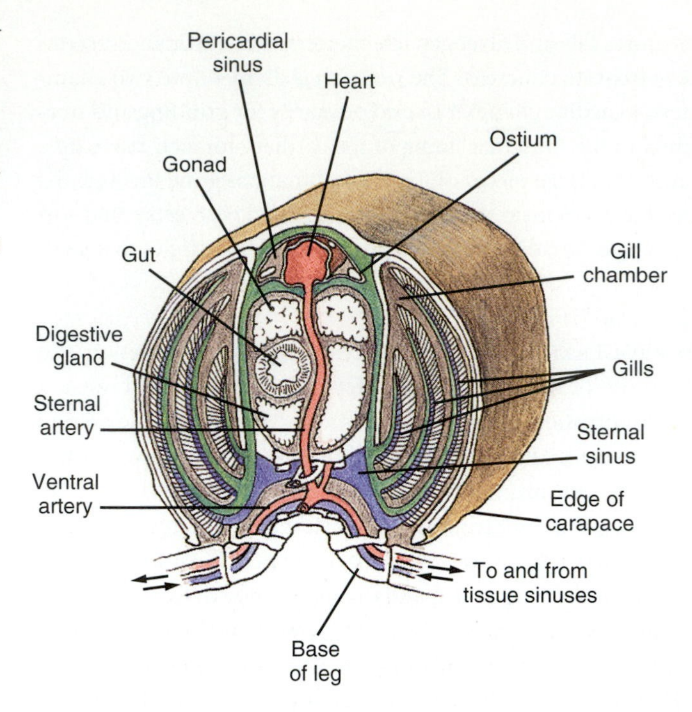

FIGURE 31.9
Cross section of a crayfish cephalothorax at the level of the sternal artery. (Compare with previous figure.) Blood from the heart flows through arteries (red) into sinuses within tissues, then returns to the sternal sinus (blue). From there it flows through the gills (green), where it is oxygenated before returning to the heart through ostia.

the function of which is unknown. As the filtrate passes through the tubule, nutrients and ions are actively reabsorbed across the lining into the hemolymph, and water is actively secreted into the urine. The urine then enters the bladder and is eventually excreted through the excretory pore (Fig. 31.3).

Circulation and Respiration. Crustaceans, like all arthropods, are coelomates, but the coelom is rudimentary. In crayfish the coelom occurs only in the excretory organ and as a chamber surrounding the gonads. The main body cavity is the **hemocoel,** which consists of blood-filled spaces inside and around major organs. As in other arthropods, the circulatory system is open, and there is no clear distinction between the blood (= hemolymph) and the interstitial fluid. The almost colorless blood contains some dissolved respiratory pigment **hemocyanin,** as well as ameboid blood cells involved in clotting and phagocytosis. The blood is pumped under low pressure by a heart that lies within a **pericardium** along the dorsal midline (Figs. 31.8 and 31.9). Blood leaves the heart through several arteries, including a **sternal artery** that connects the heart to a ventral artery. From these arteries the blood passes through the hemocoel, then collects in the **sternal sinus,** where it enters the gills. Oxygenated blood from the gills then enters the pericardium, where it returns to the heart through one-way valves called **ostia.** The gills lie on both sides of the thorax within gill chambers formed by the carapace. As noted previously, the gills are branches of the legs, and they are irrigated by the gill bailers on the second maxillae.

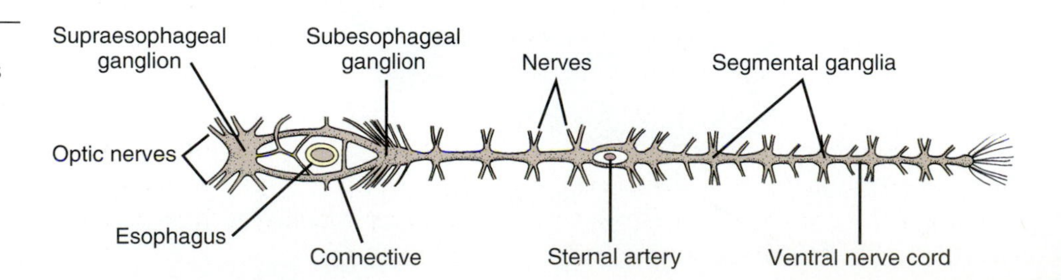

FIGURE 31.10
Outline of the crayfish central nervous system.

Central Nervous System. The crustacean nervous system is typically arthropodan, with ganglia linked in series along a ventral nerve cord. In primitive crustaceans, such as fairy shrimp, each segment has a pair of ganglia side by side. The ganglia are joined to each other by connectives, forming a ladderlike array similar to that in annelids. In most crustaceans the pair of ganglia in each segment fuse into one ganglion. Ganglia from adjacent segments also tend to fuse, especially in the cephalothorax (Fig. 31.10). In some crustaceans the connectives include giant axons that trigger the rapid backward movement that is familiar to anyone who has ever tried to catch a crayfish. These large-diameter giant axons are activated by mechanoreceptors that respond to movements of potential predators. The giant axons then trigger flexion of the abdomen, which causes the rapid flip of the telson that propels the crayfish backward. The speed of the reflex is due to rapid conduction of action potentials not only by giant axons but also by electrical synapses that link the sensory nerve cells to the giant axons and link the giant axons to the motor neurons.

The largest of the fused ganglia are the **subesophageal** and **supraesophageal ganglia** in the head. The subesophageal ganglia are mainly involved in coordinating feeding movements. An important part of this coordination is the integration of information from chemoreceptors and the numerous mechanoreceptive **tactile hairs** on the mouthparts. The supraesophageal ganglia integrate information from mechanoreceptors and chemoreceptors on the antennae, from the statocyst organ, and from the eyes. The supraesophageal ganglia also exercise dominance over many of the other ganglia, though not to the extent implied by the term "brain" that is often used.

Vision. Most crustaceans have good vision, often with eyes on long, mobile stalks. The main eyes of crustaceans, horseshoe crabs, and insects are **compound eyes.** A compound eye consists of many subunits called **ommatidia,** each of which accepts light from a small area of the visual field (Fig. 31.11A). Photoreceptor cells in each ommatidium send action potentials to the optic and supraesophageal ganglia, which process information about the intensity, color, and angle of polarization of the light received by the ommatidium. In this way the arthropod compound eye analyzes a visual stimulus bit by bit in the same way a computer builds an image on the screen from many individual pixels. The supraesophageal ganglia presumably integrate the information from all the ommatidia into some kind of picture of the visual stimulus, but it is impossible to say what such a picture looks like to an arthropod. One often sees photographs taken through compound eyes that purportedly show an arthropod's view of the world. All such photos really show, of course, is what the world would look like to us if we used arthropod eyes as contact lenses.

Each eye is covered by transparent cuticle called the **cornea,** which may be divided into numerous **facets,** with one facet covering each ommatidium (Fig. 31.11A). Ommatidia are square in cross section in crayfish and many other crustaceans, and they tend to be hexagonal in insects. Beneath each facet is a **crystalline cone.** Compound eyes focus light in a variety of ways among different arthropods. In terrestrial arthropods it is mainly the change in refractive index from air to cornea that focuses the light. In most aquatic arthropods the crystalline cone acts as a lens that focuses the light. In crustaceans with square facets the light is focused by reflection from the walls of the ommatidia.

In crustaceans and insects that are active in bright light, each ommatidium is shielded from its neighbors by pigment, so it samples only a small region of the visual field. Compound eyes of this type are called **apposition eyes.** In crayfish and other arthropods that are active at night, in shade, or in dark water, light spreads among several neighboring ommatidia. Their eyes are termed **superposition eyes.** Superposition eyes are hundreds of times more sensitive than apposition eyes.

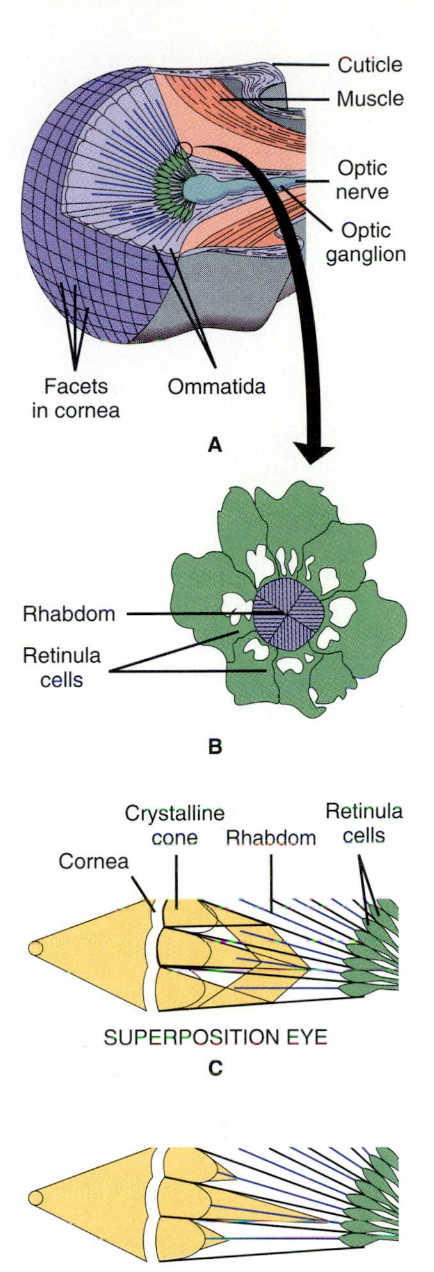

FIGURE 31.11

(A) The compound eye of a crayfish. (B) Cross section of a retinula. The retinula cells form the rhabdom, where light is transduced into an electrical response. (C) Crayfish have superposition compound eyes, which are more sensitive because each rhabdom receives light from several ommatidia. (D) Many other crustaceans have apposition compound eyes in which each rhabdom receives only light from its own ommatidium, which is screened from adjacent ones by pigments.

The photoreceptive region of the ommatidium is the **retinula** ("little retina"), which usually consists of seven or eight **retinula cells** (Fig. 31.11B). The retinula cells have numerous parallel microvilli projecting toward the central axis of the ommatidium and constituting the **rhabdom.** The rhabdom contains the photopigments and is believed to be the site where light energy is transduced into voltage changes that evoke action potentials. Many arthropods have good color vision, owing to several different kinds of pigments that are sensitive to different wavelengths. Some mantis shrimps, in fact, have ten different photopigments, compared with only three in humans. Many arthropods can also perceive ultraviolet wavelengths that are invisible to humans. The parallel array of microvilli in the rhabdoms also permits some arthropods to detect the angle of polarization of light. This ability allows the animal to determine the position of the sun for navigation even when the sun is hidden by clouds, the shoreline, treeline, or other objects.

THE STATOCYST ORGAN. Within a chamber at the base of each first antenna in crayfishes and some related crustaceans is a mechanoreceptor organ, the statocyst, that allows a crayfish to tell which way is up. The statocyst organ works much like the utricle and saccule of the human inner ear (see Fig. 8.15). A **statolith,** consisting of a dried secretion or sand grains stuck together, rests on hair cells that send action potentials to the central nervous system. As the orientation of the body changes with respect to gravity, the statolith rests on different hair cells, allowing the crustacean to determine its orientation. The statocyst is lined with cuticle, so the animal must replace its statolith after each molt. A crayfish cannot tell up from down until fresh sand enters the statocyst. Mischievous zoology students have been known to replace the sand in a crayfish's aquarium with iron filings, so that after molting the crayfish will flip onto its back when a magnet is held above it.

ENDOCRINE SYSTEM. Hormones play major roles in coordinating the physiology of crustaceans. The most important endocrine organ is the **X organ–sinus gland (XOSG) complex,** which is located near the optic nerve. Another important endocrine organ is the **Y organ,** which is located at the base of each maxilla. Among the hormones from the XOSG system is **molt-inhibiting hormone** (MIH). MIH blocks molting by inhibiting the secretion of **ecdysone** from the Y organs. Because of MIH, molting occurs only when certain environmental cues, such as changes in temperature or day length, inhibit the X organs, allowing the Y organ to secrete ecdysone.

Other secretions from the XOSG complex control **chromatophores,** enabling the integument to change color. One hormone causes the pigment to become more concentrated inside the red chromatophores, thereby making the integument less red. Another secretion from the XOSG system, the **crustacean hyperglycemic hormone,** is analogous to adrenaline and glucagon in vertebrates, increasing the conversion of glycogen stores into glucose. Still another secretion, the **distal retinal–pigment hormone,** aids in the adaptation of the compound eyes to dim light.

Malacostracans

Crustaceans are extremely adaptable in morphology, so their taxonomy is unsettled. It does not take much searching to come up with half a dozen different schemes for classifying them. Most taxonomists, however, agree in placing the majority of crustaceans, including crayfishes, hermit crabs, true crabs, lobsters, shrimp, and terrestrial wood lice, in the same taxon, considered here to be the class Malacostraca. Most of these seemingly unrelated animals share a combination of traits referred to as the "caridoid facies." The major features of the caridoid facies appear in crayfishes (Fig. 31.1). These

Classes of Crustacea According to Abele (1982)

Genera mentioned elsewhere in this chapter are noted.

Class Remipedia rim-i-PEE-dee-uh (Latin *remipedes* oar-footed). Primitive inhabitants of marine caves. Thirty-two trunk segments. Biramous, oar-shaped trunk appendages are all similar. *Speleonectes* (Fig. 31.21A).

Class Cephalocarida SEF-a-low-CARE-i-duh (Greek *kephale* head + *karis* shrimp). Primitive. Horseshoe-shaped head, 19-segment trunk, uniform trunk appendages. No carapace or eyes. Benthic; marine. Less than 4 mm long. *Hutchinsoniella* (Fig. 31.21B).

Class Branchiopoda BRAN-key-OP-oh-duh (Greek *branchia* gills + *podos* foot). Thoracic appendages leaf-shaped. Water fleas, fairy shrimp, and brine shrimp. *Artemia, Daphnia* (Fig. 31.20).

Class Maxillopoda max-ill-OP-oh-duh (Latin *maxilla* jaw). Large maxillae, used for feeding. Trunk reduced, usually to 11 segments. Fish lice, copepods, barnacles, rhizocephalans, tongue worms. *Cyclops, Lepas, Linguatula, Sacculina, Semibalanus* (Figs. 31.16, 31.17, 31.18).

Class Ostracoda os-TRACK-oh-duh (Greek *ostracodes* having a shell). Body enclosed in hinged, bivalved carapace. Swimming by second antennae and up to two pairs of trunk appendages. Seed shrimp. *Entocytheria* (Fig. 31.19).

Class Malacostraca mal-a-KOS-tra-kuh (Greek *malakos* soft + *ostrakon* shell). Typically 19 somites (5 head, 8 thorax, 6 abdominal). Head and some thoracic somites fused; usually covered with carapace. Abdomen with appendages. Usually with stalked, compound eyes. Sand fleas, krill, true shrimps, crayfishes, lobsters, spiny lobsters, hermit crabs, true crabs, wood lice. *Armadillidium, Callinectes, Cambarus, Euphausia, Grapsus, Lysmata, Homarus, Limnoria, Mithrax, Pagurus, Porcellio, Talorchestia, Uca* (Chapter opener; Figs. 30.1B, 31.4, 31.12, 31.13, 31.14, 31.15, 31.22).

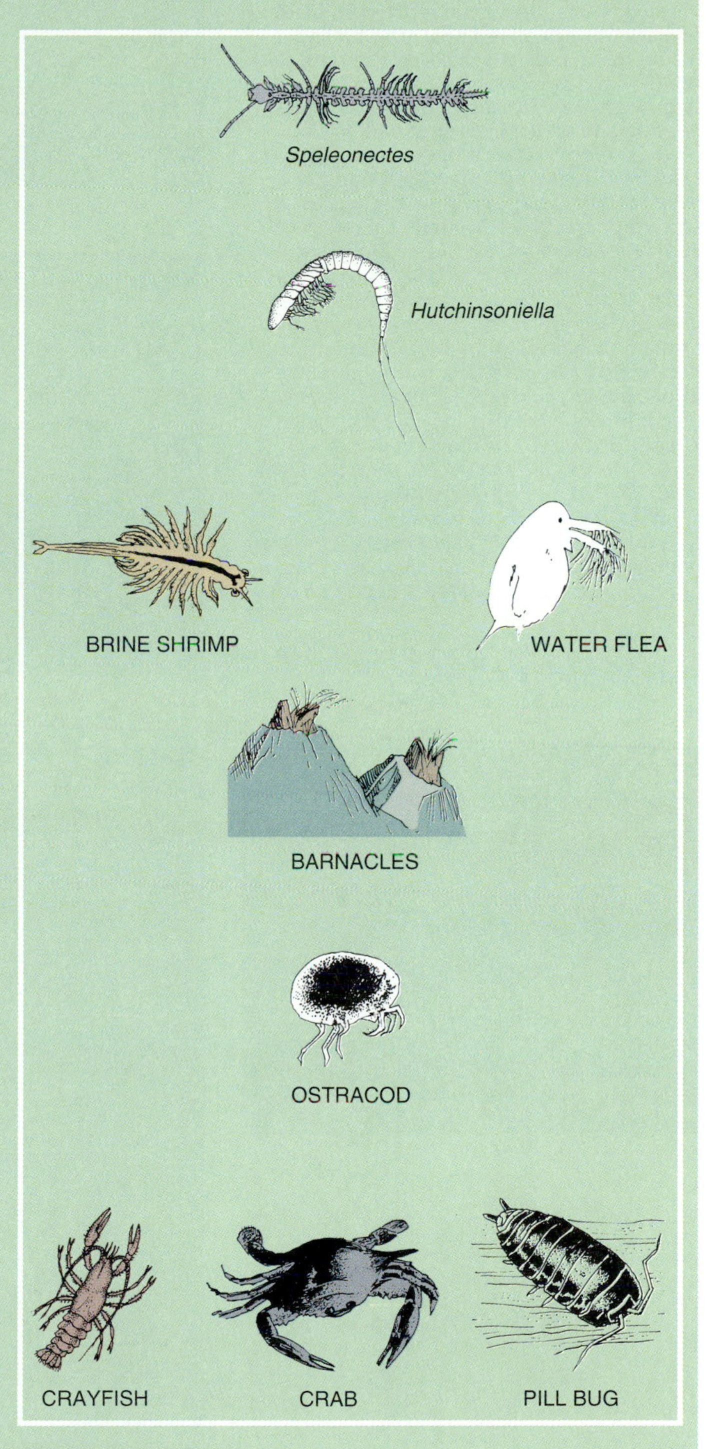

features include the particular number of somites in the cephalothorax (13) and abdomen (6), the carapace covering the cephalothorax and terminating anteriorly in a rostrum, the telson at the end of the abdomen, and the particular assortment of appendages (stalked eyes, biramous first antennae, walking legs, and swimmerets).

FIGURE 31.12

Some decapod crustaceans. (A) The common edible lobster *Homarus americanus,* found along the Atlantic coast from Labrador to Virginia. In life the lobster is greenish black on back; it turns all red only after boiling breaks down pigment molecules. Length 86 cm. (B) The fiddler crab *Uca minax.* The enlarged claw of the male is used in territorial defense and sexual display, as well as to block its burrow after it has retreated into it. Fiddler crabs and other land crabs do not have rear legs adapted for swimming, as do blue crabs and other swimming crabs. Width approximately 4 cm. (C) Hermit crabs, such as *Pagurus,* are distinct from blue crabs, fiddler crabs, and other true crabs. The hermit crab does not secrete a shell, but uses one from a deceased mollusc. As the hermit crab grows, it must find a larger shell, which it does by inspecting any object that releases calcium but not substances that indicate the shell is already occupied. The drawing shows *Pagurus* without a shell. Approximately 4 cm long. (D) Red-backed cleaner shrimps *Lysmata grabhami* on a moray eel. Like certain fishes (see p. 322), these colorful shrimp remove parasites from animals that would ordinarily be their predators. Length 2.5 cm.

A

B

C

D

DECAPODS. Among the many groups of malacostracans, the group Decapoda is the best known. Approximately 8500 species of crayfishes, lobsters, true crabs, hermit crabs, true shrimps, and many other forms are decapods (Fig. 31.12). As the name implies, decapods have five pairs of walking legs. The anterior pair is usually specialized with large chelae. Decapods are also characterized by three pairs of maxillipeds. Crayfishes and lobsters with large claws are similar to each other, except that lobsters live in seawater and are generally larger. Spiny lobsters are differentiated from other lobsters by the absence of enlarged chelae on the first walking legs (Fig. 31.21) and by their unique **phyllosoma** larvae, which are so thin that they are transparent. True crabs have flattened carapaces, large chelae, and reduced abdomens folded beneath the cephalothorax. Hermit crabs have coiled abdomens that allow them to back into empty mollusc shells. True shrimps, called prawns in Britain, are usually compressed laterally, and they are smaller than other decapods but larger than the crustaceans mentioned below that are also called shrimp.

KRILL. Another important malacostracan group is Euphausiacea, which includes about 85 small, shrimplike species known as krill (Fig. 31.13). These abundant marine planktonic forms are a major item in the diets of whales, seals, penguins, cephalopods, and other consumers (see Fig. 17.10). Often they occur in swarms with up to 30,000 individuals per cubic meter, churning the water and coloring it reddish brown. They are often more visible at night than in daylight because of light from a **photophore.** Unlike decapods, euphausids lack maxillipeds and chelae. Instead, all the thoracic appendages are legs used to trap food. The carapace does not cover the gills.

AMPHIPODS. Another group of malacostracans is the Amphipoda, which differ from decapods and euphausids in several respects. Amphipods lack a carapace, and the abdomen is not sharply different from the thorax. The compound eyes lie flat on the sides of the head, rather than on stalks. There are two or more types of leg adapted for different functions. In most species the body is compressed laterally. Like decapods, female amphipods brood the eggs and young. Overlapping flaps from the thoracic legs form the ventral brood chamber (**marsupium**). Most amphipods live in marine or freshwater habitats, and a few are parasitic. The most familiar forms are the "beach fleas" or "sand hoppers" that are often glimpsed burrowing into sand after being uncovered at the shore line (Fig. 31.14 on page 370).

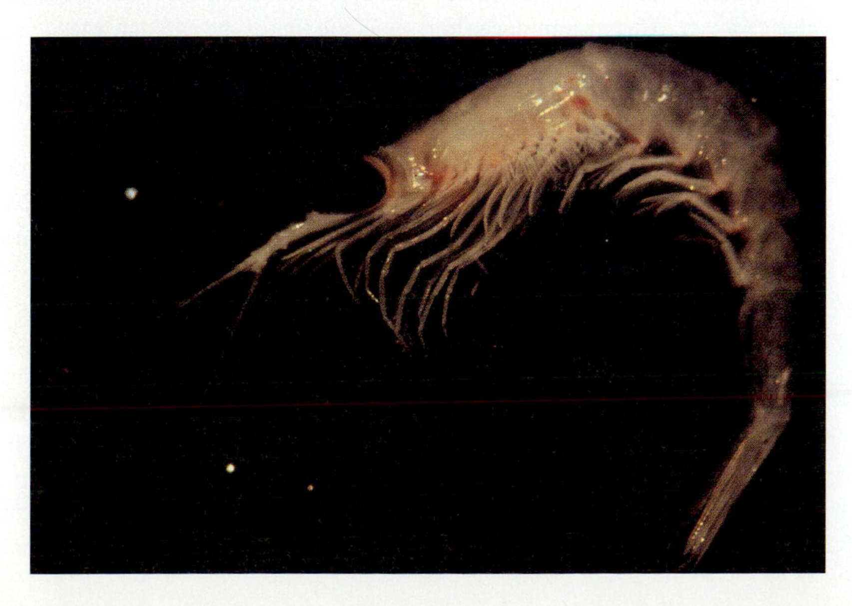

FIGURE 31.13
Euphausia superba, one of the most important species among krill. The legs form a thoracic basket that traps smaller plankton. Length up to 6 cm.

FIGURE 31.14
An amphipod: the beach flea *Talorchestia* burrowing in sand.

ISOPODS. Isopods resemble amphipods in the absence of a carapace and the use of a marsupium for brooding young (Fig. 31.15). They differ from amphipods in being flattened dorsoventrally and in having legs of only one type (hence the name). The group Isopoda includes not only marine, freshwater, and parasitic forms, but also terrestrial species. The most familiar isopods are wood lice, such as the sow bug *Porcellio* and the pill bug *Armadillidium.* (The difference between sow bugs and pill bugs is that the latter roll into a ball when threatened.) These crustaceans breathe through gills, and are therefore confined to moist environments, such as basements and beneath stones.

Maxillopoda

BARNACLES. Another very diverse group of crustaceans is the class Maxillopoda, which includes forms that hardly resemble crustaceans at all. Barnacles, for example, were considered to be molluscs until 1830 because many have thick calcareous shells and sessile habits (Fig. 31.16). The clue to their true identity is the nauplius larva, which allows for dispersal of the species. The nauplius develops into a **cypris larva,** named after an ostracod it resembles. The cypris larva has compound eyes and a bivalve carapace. After swimming a short time, the tips of the antenna attach to substratum by secreting a polysaccharide cement that is the strongest adhesive known. The carapace then develops into a **mantle** that secretes the calcareous plates of the shell in which the adult develops. Inside the mantle cavity of the adult most of the typical arthropod structures are present, although inverted and greatly modified (Fig. 31.16C). In nonparasitic species the legs form **cirri** (thus the name of the group, Cirripedia). The cirri draw food into the mouth except when the plates are tightly closed for protection from predators or exposure at low tide.

FIGURE 31.15
The common pill bug *Armadillidium.*

RHIZOCEPHALANS. The group Cirripedia also includes some even more bizarre animals, the Rhizocephala, that illustrate so well the delicate adaptations of parasites to hosts. The best known rhizocephalan is *Sacculina,* which begins parasitism when its female cypris larva attaches to a crab and injects a mass of undifferentiated cells. These cells migrate to the intestine of the host and, like germinating seeds, develop rootlike growths that spread throughout the host's body. The cells then multiply and differentiate into the adult, reproductive form, which erupts as an **external mass** beneath the

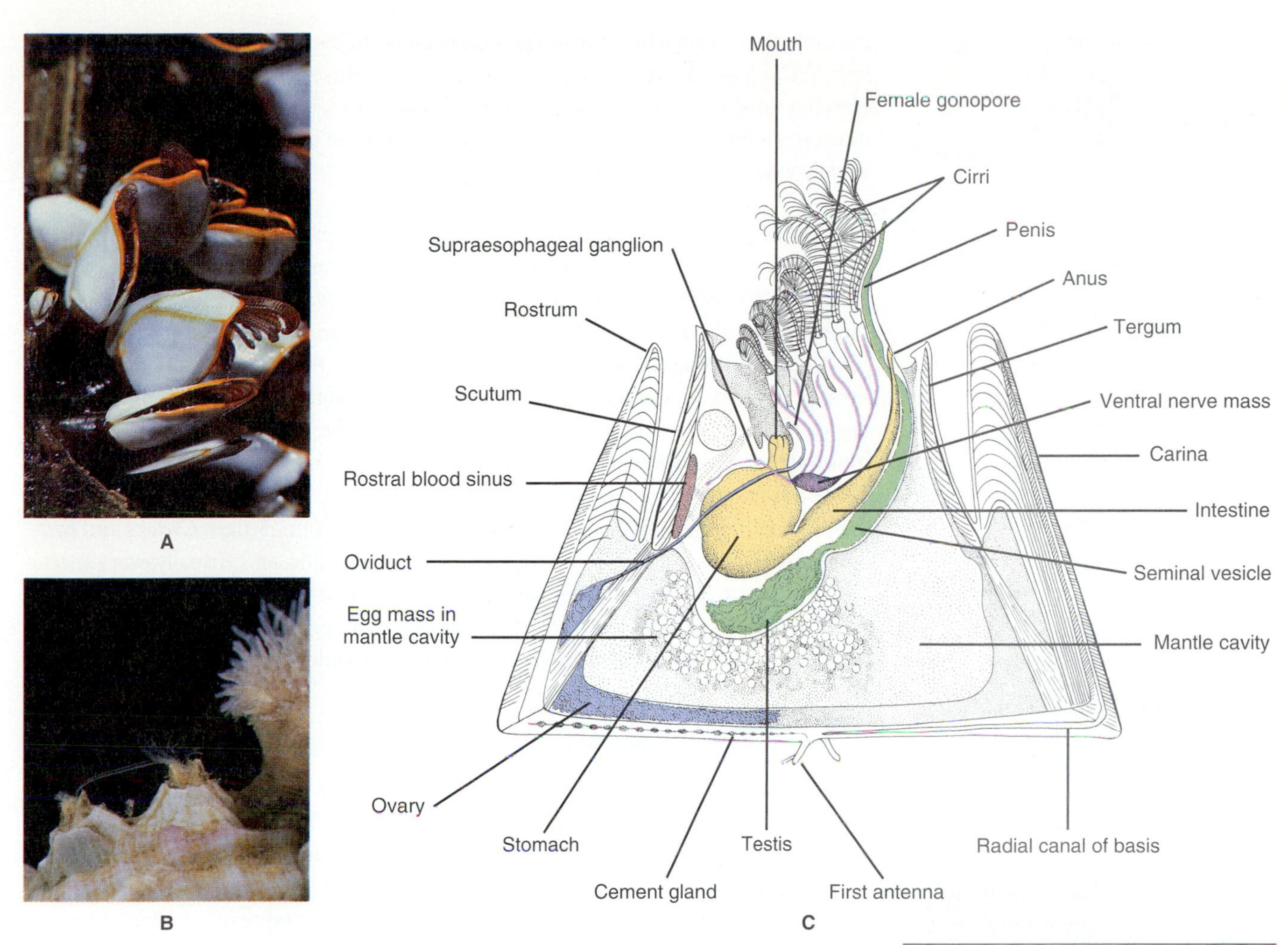

FIGURE 31.16

Barnacles. (A) The goose barnacle *Lepas anatifera.* The common and Latin names (*anatifera* = "goose-bearing") come from the resemblance of the shell and stalk to the body and neck of a goose. In the Middle Ages it was commonly believed that goose barnacles developed into geese. Note the cirri. Length 15 cm. (B) A barnacle has the longest penis relative to body size of any animal. Here two northern rock barnacles *Semibalanus balanoides* mate by extending their penes to each other. Barnacles are hermaphroditic, but avoid self-fertilization. The other structures extending from the shells are cirri. Barnacles often encrust ships' hulls, but also choose other animals, such as whales, molluscs, and crabs, as substratum. Height approximately 2.5 cm. (C) Internal organization of *Semibalanus.* Comparison with Fig. 31.8 shows that the barnacle has typical crustacean features, but is extremely modified in organization.

crab's apron. The term "external mass" is about as descriptive as one can get; the only resemblance of rhizocephalans to other crustaceans occurs in the larvae. The position of the external mass beneath the apron is crucial. In any other position it would be wiped off, but while it is beneath the apron it is treated by a female host as if it were her own egg mass. In fact, survival of the parasite depends on the host ventilating, grooming, and protecting the external mass. If the host crab happens to be a male, the parasite castrates it, and it develops the broad apron and brooding behavior of females. The external mass has her own brood chamber, into which a male cypris larva releases cells that become a testis. Sperm from this testis then fertilizes the parasite's ova, which develop into free-swimming nauplius larvae and then new cypris larvae.

Fish Lice and Tongue Worms. Another group of maxillopods, the Branchiura, includes two forms of ectoparasites that live on vertebrates. The first form, commonly called fish lice, live on marine and freshwater fishes. They have flattened bodies, compound eyes, and sucker-shaped maxillae by which they attach to hosts. Fish lice are often a problem in goldfish aquaria. The second form are the tongue worms, which are ectoparasites in the respiratory tracts of vertebrates. As might be expected from this odd habitat, the adults are unlike other crustaceans or any other kind of animal. They are so different, in fact, that until recently they were classified in their own phylum Pentastomida (Fig. 31.17 on page 672). Pentastomids are up to 16 cm long, but the males are usually much smaller. The mouth lies between two pairs of re-

■ **Question:** What could the function of a barnacle's antennae be?

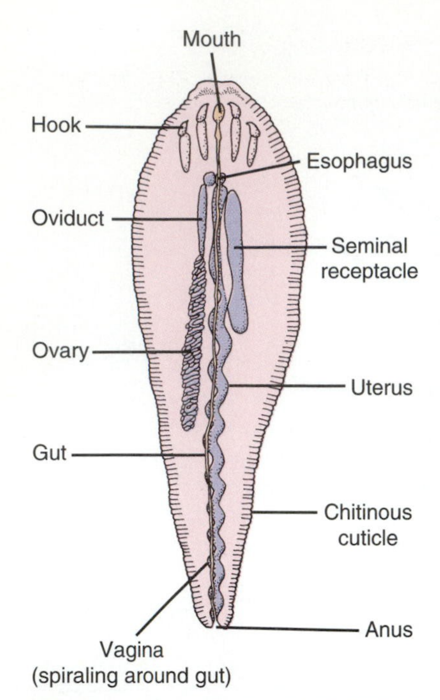

FIGURE 31.17

A female pentastomid, *Linguatula serrata.* Note the simple digestive system and the absence of respiratory, circulatory, and excretory organs. Approximately 10 cm long.

FIGURE 31.18

Cyclops, a common freshwater copepod. This mature female is carrying two egg sacs.

tractable claws, with which the tongue worm clings to the respiratory epithelium of the host, feeding on blood, mucus, lymph, and tissue. Most pentastomids parasitize reptiles, but some live in the air sacs of sea birds, and others live in the nasal passages of canines and felines. Several species occasionally infect the nasal passages of humans, but usually without causing symptoms.

COPEPODS. The group Copepoda includes many small but important planktonic crustaceans (Fig. 31.18). Some zoologists estimate that copepods are the most numerous animals on the planet. Copepods vary considerably in form, but in general they lack a carapace and abdominal appendages. Usually there are four pairs of biramous appendages and a single pair of uniramous maxillipeds. Often the first antennae are the most prominent appendages, and they aid in locomotion. There is no compound eye, but many species retain the single median eye from the nauplius larva. The common freshwater genus *Cyclops* was named for the one-eyed monster of Greek mythology. Copepods are major food sources for whales and fishes. Like the much larger animals that feed on them, many free-living copepods are suspension feeders. *Cyclops* and some others are predatory. Many other copepods are ectoparasitic, and virtually every species of fish is host to at least one species of copepod. Each species of copepod is selective about its host species, and even about the site of attachment. Some copepods attach only to the nostrils of cod, some to the gills of salmon, and some to the spiracles of skates.

Some Other Crustaceans

OSTRACODS. Several other major groups of Crustacea are generally recognized. The class Ostracoda includes the seed shrimps, which look more like tiny clams than shrimps (Fig. 31.19). Ostracods live either in seawater or fresh water, on the bottom, on plants, or as plankton. Most ostracods are scavengers, but many are predators or parasites. Although small, they are quite numerous and widespread, and they comprise an important link in food webs between producers (algae) and consumers, such as fishes.

BRANCHIOPODS. Members of the group Branchiopoda include the tiny, but rather familiar, "water fleas," brine shrimp, and several other forms of so-called shrimp (Fig. 31.20). The class takes its name from the leaf-shaped appendages, which exchange respiratory gases. "Water fleas," also known as cladocerans (order Anomopoda), get their name from their jumpy movements as they swim with their enlarged second antennae. *Daphnia* and other cladocerans make up a large portion of zooplankton in many lakes and ponds. They are efficient feeders on algae, and they can significantly improve the clarity of water. One of the secrets of their success is their mode of reproduction, which is remarkably like that of rotifers (p. 552). In the summer when conditions are good the females parthenogenetically produce more females every two or three days. When conditions become unfavorable, the females produce some eggs that have to be fertilized and some males to fertilize them. The encysted embryos, which are commonly called "eggs," are extremely resistant to cold and drying, and they can be transported to new habitats by wind or in mud that clings to other animals.

Other branchiopods are fairy shrimp and brine shrimp (order Anostraca), which live in pools that are often temporary and extremely saline. After the pools dry up, the "eggs" (encysted embryos) commonly survive for decades until the rains come again or the wind carries them to a new pond. The "eggs" then hatch into nauplius larvae within a day or two. Even "eggs" shown by carbon-dating to be 10,000 years old have hatched successfully. "Eggs" also survive shipment from San Francisco Bay or the

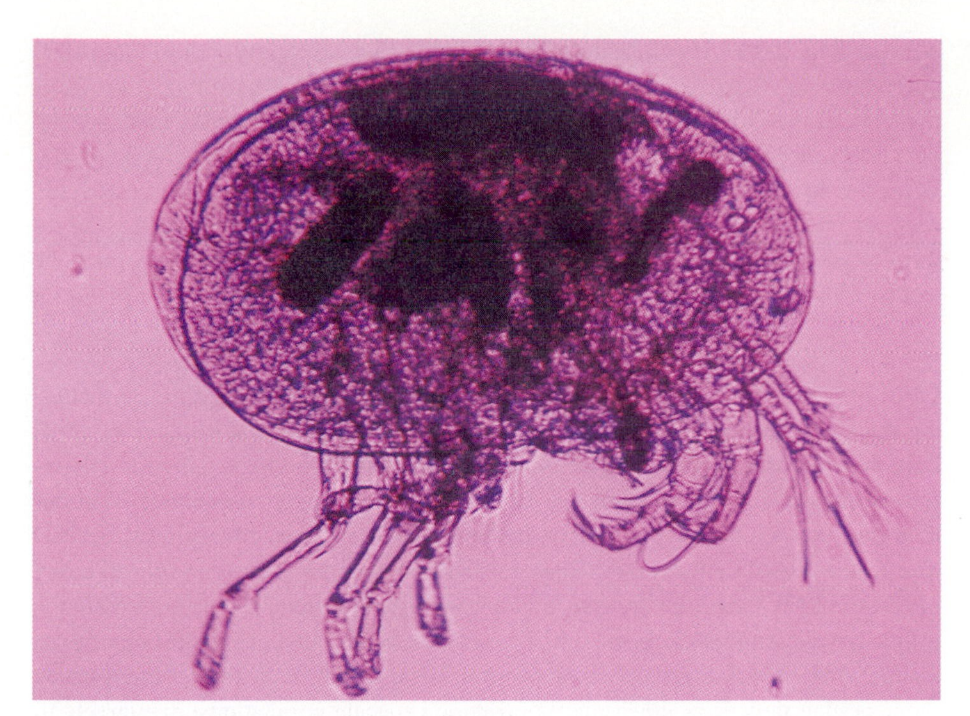

FIGURE 31.19
The freshwater ostracod *Entocytheria*. Diameter 1.1 mm.

Great Salt Lake to pet stores throughout the country, where they are sold as a convenient way to get live food for aquarium fish. Brine shrimp "eggs" are also marketed to children as "sea monkeys."

Remipedians and Cephalocaridans. Two other groups deserve mention here because they are usually given the taxonomic rank of classes (Fig. 31.21 on page 674). Class Remipedia was discovered by Jill Yager in 1981 in submarine caves in the Bahamas. Fewer than a dozen species of remipedes are now known, all from submerged caves. Little is known about them because of their inaccessibility. Class Cephalocarida is also thought to be primitive on account of its uniform legs.

Interactions with Humans and Other Animals

Detrimental Effects on Humans. Crustaceans have a much better reputation among humans than do most other arthropods. Many crustaceans are good to eat, and few of them have poison fangs or spread diseases. One of the few crustaceans that

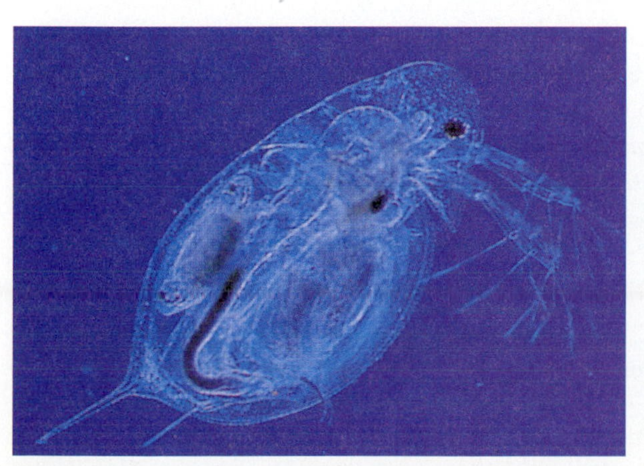

A

B

FIGURE 31.20
Branchiopods. (A) The female water flea *Daphnia*, common in freshwater ponds. Two juveniles can be seen dorsally developing in the brood chamber. Every few days the females molt, releasing the young and reloading the brood chamber with eggs. (B) The brine shrimp *Artemia salina* can live in ponds several times more saline than the sea. The female above is carrying eggs.

is a vector for disease is a copepod of the genus *Cyclops,* which serves as an intermediate host for the dreaded Guinea worm (*Dracunculus*) (p. 545). Some other crustaceans present mainly economic problems. The marine isopods called gribbles, *Limnoria lignorum,* cause considerable damage to wooden pilings and boat hulls (Fig. 31.22). Barnacles have long been a nuisance to shippers. A ship with a badly encrusted hull may have its top speed reduced by half, and the cost of fuel greatly increased. Until recently the only solution was to beach the ship and scrape the hull. Hulls can now be coated with a plastic to which the barnacle glue cannot adhere.

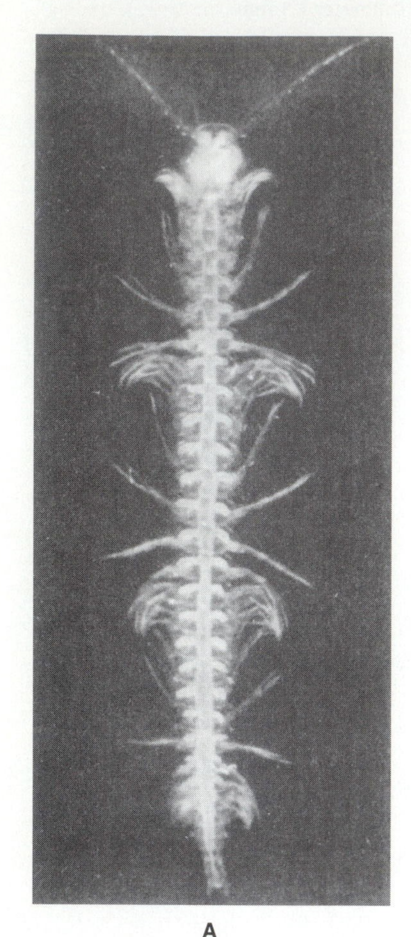

A

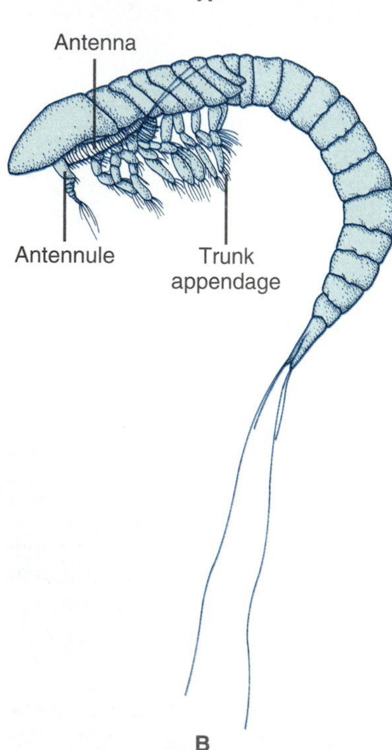

B

CRUSTACEANS AS FOOD. Crustaceans feed vast numbers of other animals, including humans. Tiny crustaceans, especially copepods, ostracods, and krill, are essential links between producers and larger consumers in natural food webs. These zooplankton eat phytoplankton, and they are eaten by fishes and other consumers. Without them, animal populations in aquatic habitats would collapse. By reproducing efficiently, these small crustaceans manage to maintain high populations in spite of heavy predation. Krill, *Euphausia superba,* is being harvested for human consumption in the waters around Antarctica. As many as 12 tons have been netted in a single hour, but so far the total annual catch by humans is minuscule compared with consumption by other predators. It is assumed that the decline in whale populations, due largely to commercial whaling, has left a surplus of krill available for fishing.

Larger crustaceans, such as barnacles and decapods, tend to be protected from predators by their heavily calcified cuticles. They are, nevertheless, important food sources for a few predators. In the future, large crustaceans will be even more important components in the human diets, as techniques for ocean farming and ranching are developed. Already the harvest of spiny lobsters (= rock lobsters) on the Caribbean coast of Mexico is being increased by providing rough platforms under which they gather. There is some question, however, whether this technique increases production or merely increases the catch temporarily by attracting lobsters to the platforms.

Walter H. Adey (1987) and his staff at the Marine Systems Laboratory of the Smithsonian Institution have created artificial coral reef microcosms in the laboratory to help study the potential for increasing the productivity of edible crustaceans. They have also constructed rafts in the Caribbean as artificial reefs, providing substratum for algal growth. They suggest that edible animals such as the West Indian spider crab *Mithrax spinosissimus* would graze on the algal turfs and could be the basis for oceanic ranches that are similar in concept to cattle ranches.

DETRIMENTAL EFFECTS OF HUMANS. Ironically, just as we are becoming aware of the potential of crustaceans as sources of protein, we are also becoming aware of how some of our other activities are endangering that potential resource. Toxic chemicals deliberately and accidentally dumped into the oceans may be harming the copepods, ostracods, and numerous other organisms on which fishes and other animals de-

FIGURE 31.21

(A) The remipedian *Speleonectes.* Approximately 2 cm long. (B) The cephalocaridan *Hutchinsoniella.* Length approximately 4 mm.

■ **Question:** Why are uniform appendages considered primitive?

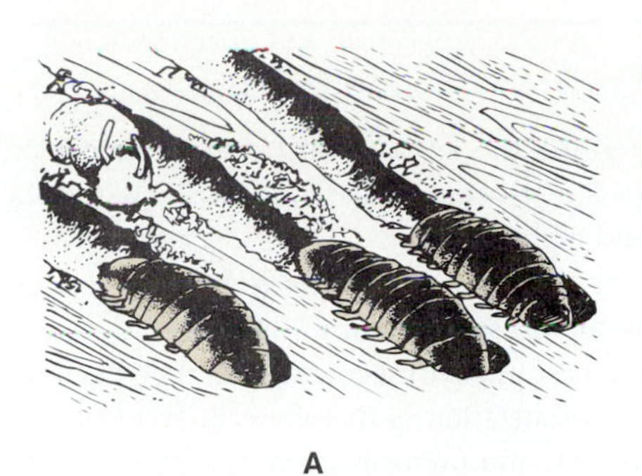
A

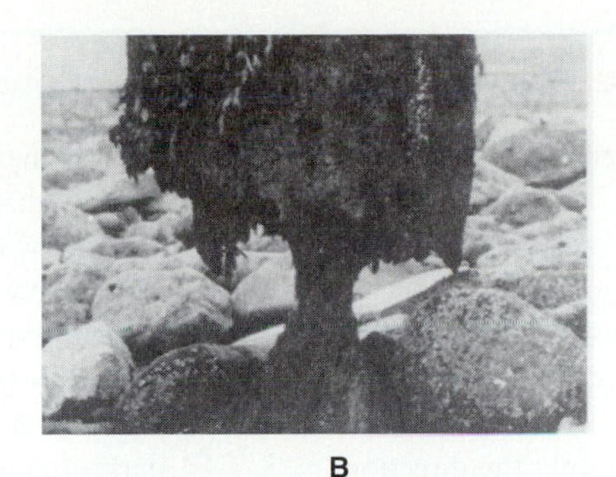
B

FIGURE 31.22
(A) Wood-eating marine isopods *Limnoria lignorum,* commonly called gribbles. (B) A wooden piling heavily damaged by gribbles.

pend. Many of the toxic chemicals include residues of poisons that were intended to kill insects and which are likely to be just as toxic to crustaceans. Deliberate and accidental oil spills may be equally devastating.

Summary

Like most arthropods, crustaceans are segmented animals with jointed appendages. Also like most other arthropods, the coelom is much reduced, and the main body cavity is a hemocoel. Crustaceans are distinguished from chelicerates in having mandibles and from insects mainly by aquatic adaptations. Most crustaceans retain many more appendages than do other arthropods, and these appendages are modified into a great variety of serially homologous structures, including antennae, mouthparts, walking legs, and swimmerets.

Other distinguishing features, such as gills and appendages adapted for swimming, are due largely to the retention of an aquatic mode of life. Some crustaceans also have unique organs, such as the statocyst organ for determining the direction of gravity, the antennal gland for osmoregulation and excretion, and the X organ–sinus gland complex that coordinates molting. With some notable exceptions, such as barnacles, crustaceans are dioecious. Development is direct in many species, but many others produce a distinctive one-eyed nauplius larva.

Most familiar crustaceans, including crayfishes, hermit crabs, true crabs, lobsters, true shrimps, and wood lice, are classified in the group Malacostraca. Other important malacostracans are krill, amphipods, and ostracods. Other major groups include such important and familiar forms as copepods, barnacles, and brine shrimp.

Key Terms

carapace
exopod
endopod
protopod
mandible
maxilla
maxilliped
cheliped
swimmeret
nauplius
antennal gland
compound eye
rhabdom
statocyst
X organ–sinus gland complex

Chapter Test

1. Describe three features that crustaceans share with other arthropods. (pp. 659–661)
2. Describe three features that distinguish crustaceans from chelicerates. (pp. 662–666)
3. Name five different types of crustacean appendage, and describe the function of each. (pp. 659–660)
4. Diagram a compound eye, showing the location of the cornea, the retinula cells, and the rhabdoms. (pp. 665–666)
5. Give the common names of five kinds of malacostracans, and briefly describe each kind. (pp. 666–670)
6. Give the common names of four kinds of nonmalacostracan crustaceans, and briefly describe each kind. (pp. 670–673)

■ Answers to the Figure Questions

31.1 In water the abdomen does not weigh as much as it might appear to.

31.3 Crayfish mouthparts are evidently adapted for hard foods such as aquatic invertebrates and plants.

31.6 The dorsal spine is presumably defensive. The animal is so small that the spine is not likely to serve as a rudder in swimming.

31.7 The naupliar eye most likely detects only the direction and intensity of light as an aid in orientation. This information may guide the larva toward richer sources of food. Perhaps it also helps regulate the level of activities during day and night.

31.16 Most likely the antennae serve only to help the cypris larva locate a suitable substratum.

31.21 Embryos start with uniform segments and then become tagmatized. Tagmatization is therefore likely to be the derived condition, and uniform segments the primitive character state.

Readings

Recommended Readings

Benson, A. A., and R. F. Lee. 1975. The role of wax in oceanic food chains. *Sci. Am.* 232(3):76–86 (Mar). (*Copepods, even more than other marine animals, store much of their energy as wax.*)

Browne, R. A. 1993. Sex and the single brine shrimp. *Nat. Hist.* 102:34–39 (May).

Caldwell, R. L., and H. Dingle. 1976. Stomatopods. *Sci. Am.* 234(1):81–89 (Jan).

Cameron, J. N. 1985. Molting in the blue crab. *Sci. Am.* 252(3):102–109 (May).

Cronin, T. W., N. J. Marshall, and M. F. Land. 1994. The unique visual system of the mantis shrimp. *Am. Sci.* 82:356–365.

Fitzpatrick, J. F., Jr. 1983. *How to Know the Freshwater Crustacea.* Dubuque, IA: W. C. Brown.

Nicol, S., and W. de la Mare. 1993. Ecosystem management and the Antarctic krill. *Am. Sci.* 81:36–47.

Nilsson, D-E. 1989. Vision optics and evolution. *BioScience* 39:298–307. (*Includes excellent description of various kinds of compound eyes.*)

Warner, W. W. 1976. *Beautiful Swimmers: Watermen, Crabs and the Chesapeake Bay.* Boston: Atlantic–Little, Brown. (*A popular account of the lives of crabs and crabbers.*)

Wicksten, M. K. 1980. Decorator crabs. *Sci. Am.* 242(2):146–154 (Feb).

See also relevant selections in General References at the end of Chapter 21.

Additional References

Abele, L. G. (Ed.). 1982. *The Biology of Crustacea,* Vol. 1. New York: Academic Press.

Abele, L. G., W. Kim, and B. E. Felgenhauer. 1989. Molecular evidence for inclusion of the phylum Pentastomida in the Crustacea. *Mol. Biol. Evol.* 6:685–691.

Adey, W. H. 1987. Food production in low-nutrient seas. *BioScience* 37:340–348.

Harrison, F. W., and A. G. Humes (Eds.). 1992. *Microscopic Anatomy of Invertebrates,* Vol. 9: *Crustacea.* New York: Wiley–Liss.

Harrison, F. W., and A. G. Humes (Eds.). 1992. *Microscopic Anatomy of Invertebrates,* Vol. 10: *Decapod Crustacea.* New York: Wiley–Liss.

Questin, L. B., and R. M. Ross. 1991. Behavioral and physiological characteristics of the Antarctic krill, *Euphausia superba. Am. Zool.* 31:49–63.

Ross, R. M., and L. B. Quetin. 1986. How productive are Antarctic krill. *BioScience* 36:264–269.

Tangley, L. 1987. Enhancing coastal production. *BioScience* 37:309–312.

CHAPTER

32

Insects

Caterpillar of the tiger swallowtail *Papilio.*

Diversity

If a probe from another planet arrived on Earth to collect random samples of four species of living organisms, chances are that three of those four species would be insects. Those insects would probably convince whoever sent the probe that our little planet is well worth a visit. Insects have mastered virtually every habitat except the ocean, and they share the skies with only a few vertebrates. Their impact on humans and other animals is incalculable. If it were not for a few accidents of physiology that kept them small, insects would be even more dominant over the world, and we might not be here to read about them.

Insects, like the spiders, crustaceans, and other animals discussed in the previous two chapters, are members of the phylum Arthropoda. They are considered in this text to constitute the subphylum Insecta. As noted previously (pp. 635–636), the relationship of insects to other arthropods is unclear, although there is good reason to believe they are closely related to crustaceans. Like crustaceans, insects are mandibulates. The main difference between insects and crustaceans is that most insects are secondarily adapted for terrestrial living, and most crustaceans retained aquatic adaptations. Myriapods (centipedes and millipedes) are also terrestrial mandibulates, but they differ from insects in having more than six legs. Insects can therefore be defined as terrestrial mandibulates with six legs.

Insects have bodies divided into three tagmata: the head; the thorax, which bears the legs and wings, if any; and the abdomen, which retains obvious segmentation and bears the external genitalia on the terminal segments (Fig. 32.1). With the demise of the subphylum Uniramia as a valid taxon, the group Insecta is here elevated to the level of a subphylum that comprises two classes. The **class Apterygota** includes the insects that lack wings: orders Collembola, Protura, Diplura, Archaeognatha, and Thysanura. These apterygotes tend to have degenerate compound eyes and Malpighian tubules (excretory organs), and there is little, if any, metamorphosis. Collembola are commonly called "spring tails" or "snow fleas" because they have springlike forked "tails" that enable them to leap (Fig. 32.2). They are extremely numerous, but are so small that they are seldom noticed. Close examination will reveal hundreds of them jumping on the surface of a pond or snow in the spring. Proturans and diplurans are more difficult to see, because they inhabit soil and decaying vegetation. The orders Archaeognatha and Thysanura includes bristletails and silverfish. Silverfish are familiar inhabitants of many homes.

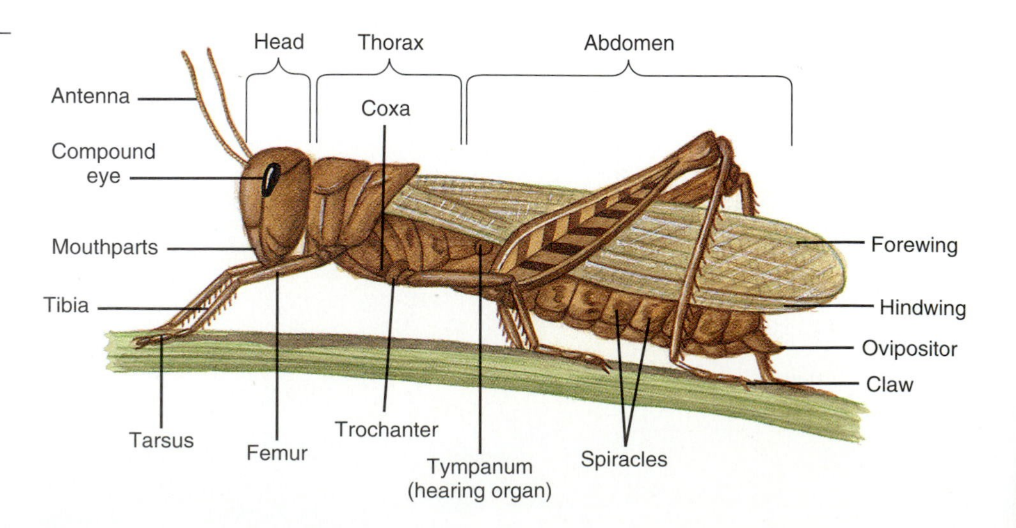

FIGURE 32.1
The external structure of a female grasshopper, showing the body parts commonly found in insects.

Insects that are primitively winged (meaning they either have wings or have lost them secondarily) make up the **class Pterygota.** This includes almost all insects, so the class is further divided into those that show some trace of wing before reaching adulthood and those that show wings only as adults. These two groups, here considered to be subclasses, also differ in the degree of metamorphosis. The **subclass Exopterygota** includes insects that change their basic form gradually or not at all as they mature. Exopterygotes are said to be **hemimetabolous.** The egg hatches into an immature stage called a **nymph.** (Aquatic nymphs are often called **naiads.**) The nymph molts a definite number of times, becoming a different **instar** each time. The wing buds begin to appear externally during the later instars before adulthood—hence the name Exopterygota (Figs. 32.3). The insect becomes a winged adult during the final molt.

Members of the **subclass Endopterygota** are **holometabolous,** changing form radically usually twice in their lives (Fig. 32.4 on page 681). Commonly they hatch as wormlike larvae, called caterpillars, maggots, or grubs, depending on the order to which they belong. After several molts the larva becomes a **pupa** (= chrysalis). The winged adult develops during pupation, and it emerges by the process of **eclosion.** Metamorphosis in both Exopterygota and Endopterygota is regulated by several hormones, as described on pages 145–147.

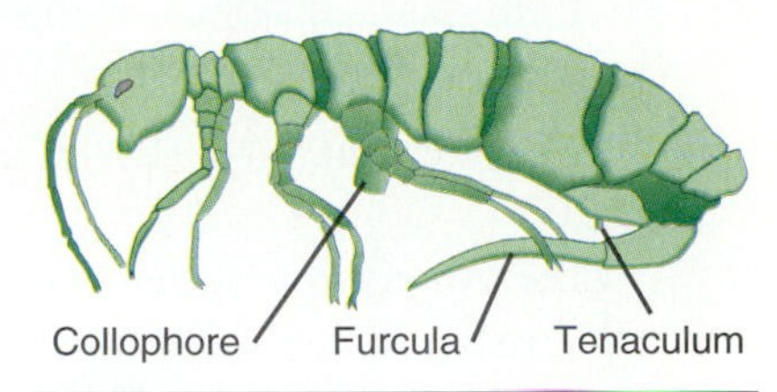

FIGURE 32.2

The springtail *Isotomurus,* which lives on the water surface. Other collembolans are frequently seen hopping on the surface of snow and are therefore called "snow fleas." The tenaculum is thought to act as a catch, holding the furcula while muscular or hydrostatic tension builds up. When the tenaculum releases the furcula, it launches the animal up to 30 cm. The collophore is a tube of unknown function. Approximately 2.5 mm long.

■ **Question:** Why would hopping be more advantageous than walking in this and other insects?

Insect Success

In numbers of individuals, numbers of species, and range of habitats, insects are enormously successful. More than three-quarters of a million insect species have been named so far. If this figure is too high now, it won't be for long, because thousands of new species are described each year. Terry Erwin, using a biodegradable insecticide to knock insects out of the forest canopy in the Amazon Basin, found so many new species that he estimated the total number of insect species to be 50 million. This is approximately 35 times the number of described species of *all* living organisms. Probably only a minuscule portion of these insects will ever be described, because there are not enough taxonomists and because species are now becoming extinct so rapidly.

The great diversity of insects may be due to their having evolved before most other terrestrial animals, and to a low rate of extinction (Labandeira and Sepkoski 1993). Insects evolved during the Devonian period at least 390 million years ago, when there were few other terrestrial animals to compete with them. They had about 40 million years to find empty habitats before reptiles came along, and at least 200 million years

FIGURE 32.3

Gradual metamorphosis in a typical exopterygote, the American cockroach *Periplaneta americana.* Shown clockwise from the bottom left are four nymphs representing 4 of the 13 instars of this species. Note the wingbuds in the last nymph. Adult male and female, with wings. The female is carrying an egg case. Another egg case is shown in the center.

Major Groups in Subphylum Insecta

Genera mentioned elsewhere in this chapter are noted.

Class Apterygota ap-ter-i-GO-tuh (Greek *a-* not + *pterygotos* winged). Primitively wingless. Little or no metamorphosis. Up to 30 mm long. Includes orders Collembola (springtails), Diplura, Protura, Archaeognatha (bristletails), Thysanura (silverfish). *Isotomurus* (Fig. 32.2).

Class Pterygota ter-i-GO-tuh. Adults with wings, or wings secondarily lost. At least some metamorphosis. See Fig. 32.5 for a visual synopsis of the major pterygote orders.

Subclass Exopterygota (= Hemimetabola) ex-OP-ter-i-GO-tuh (Greek *ex-* external). Metamorphosis gradual, with wings appearing as buds on nymphs. Nymphs with compound eyes. True bugs,* mayflies, dragonflies, damselflies, stoneflies, grasshoppers, crickets, cockroaches, mantids, termites, earwigs, lice, thrips, aphids, leafhoppers, periodic cicadas, scale insects. *Abedus, Belostoma, Coenagrion, Gerris, Microtermes, Nasutitermes, Periplaneta, Phymata, Podisus, Prociphilus, Schistocerca* (Figs. 7.13B, 30.1E, 30.4D, 32.3, 32.11, 32.12A, 32.14, 32.15, 32.17A, 32.18C, 32.19).

Subclass Endopterygota (= Holometabola) in-DOP-ter-i-GO-tuh (Greek *endo-* internal). Complete metamorphosis; larva does not resemble adult, and wings appear suddenly in adult. Larvae lack compound eyes. True flies,* ant lions, dobsonflies, beetles, scorpionflies, hangingflies, caddisflies, butterflies, moths, fleas, sawflies, ants, bees, and wasps. *Actias, Aedes, Amphibolips, Anopheles, Apis, Bibio, Cactoblastis, Canthon, Chrysolina, Cryptoses, Danaus, Drosophila, Epilachna, Glossina, Hyalophora, Hyles, Hylobittacus, Lymantria, Megarhyssa, Metasyrphus, Papilio, Pediobius, Photuris, Pieris, Scarabaeus, Solenopsis, Vespula* (Chapter opener; Figs. 1.7, 12.9, 18.7, 18.15, 18.16, 20.11A, 32.4, 32.12B, 32.13, 32.17B, 32.18A,B, 32.20, 32.21, 32.22, 32.23.)

*The terms "bug" and "fly" are used in the common names of many crawling or flying insects, but only insects in orders Heteroptera and Diptera, respectively, are true bugs and flies. To avoid taxonomic confusion, many entomologists indicate that an insect is not a true bug or a true fly by making "bug" or "fly" part of a longer common name. For example, the flying beetles that glow in the dark should be termed "lightningbugs" or "fireflies." However, bed bugs are true bugs, and house flies are true flies.

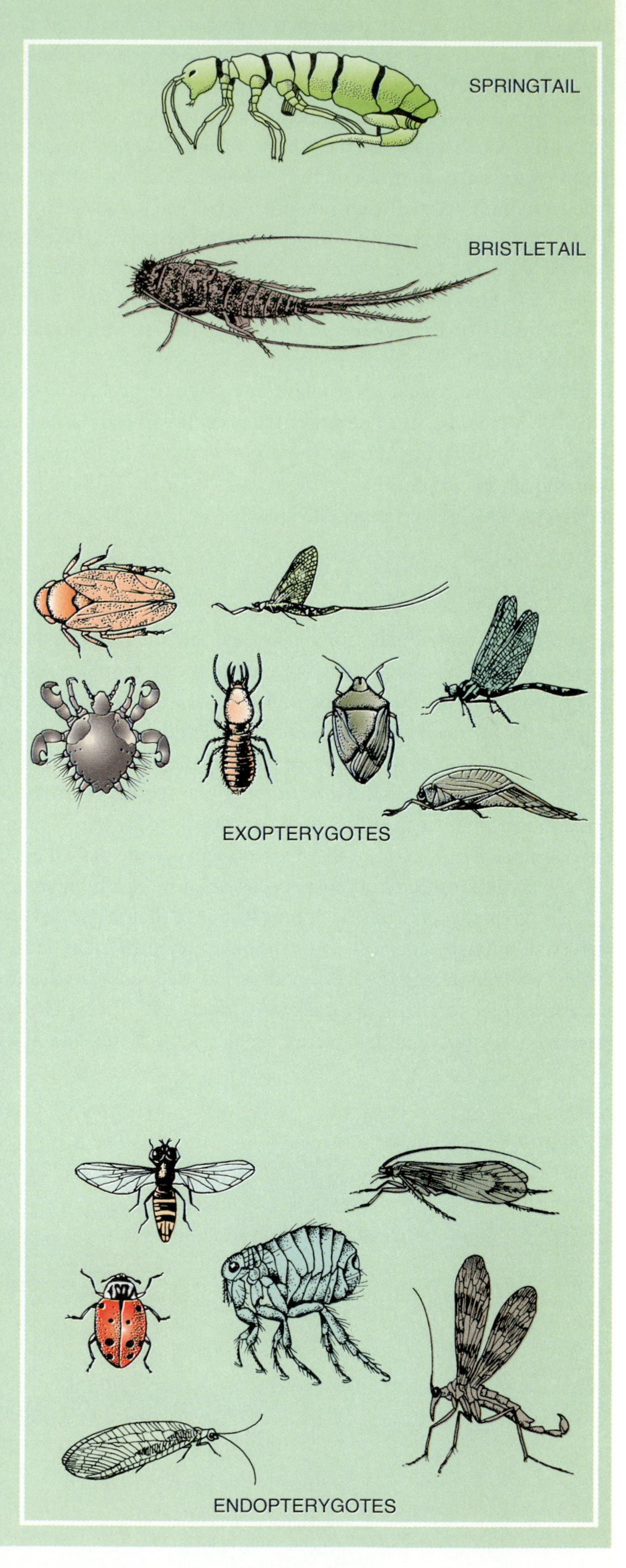

A

B

C

FIGURE 32.4

Complete metamorphosis in an endopterygote, the sphinx moth (= hawk moth) *Hyles gallii.* (A) Caterpillar. Length approximately 10 cm. (B) Pupa. Length 5 cm. (C) Adult male. Length 4 cm.

■ **Question:** What is the advantage of complete metamorphosis?

before the first birds and mammals. (Refer to Table 19.2.) The adaptations that allowed insects to take advantage of the many possibilities of life on land included wings that aided in dispersal, resistance to dehydration, improved organs for the maintenance of water and ion balance, the ability to obtain oxygen from air, the ability to adjust to extreme changes in air temperature, and sensory receptors for airborne chemicals and sound. (See Table 30.2.)

Another key to their diversity is that a variety of adaptations for feeding evolved very early. The basic insect mouthparts proved to be highly modifiable for chewing, sucking, or sponging nutrients from plants and later from animals. The ability to fly to new habitats must also have been important. From the equatorial regions of the supercontinent Pangea (see Fig. 19.8), where fossil remains suggest they originated, insects spread to every terrestrial region except the poles.

Survival on land also required adaptations in reproduction. Fertilization had to be internal to prevent drying of gametes, and embryos had to develop within nutrient-rich eggs that let in oxygen without losing water. The diversification of insects also benefited from rapid reproduction and short life cycles that enabled new habitats to be colonized even if only a small proportion of the insect pioneers were able to survive in them. The following sections of this chapter focus on these adaptations for terrestrial success.

Osmoregulation and Excretion

The impermeable integuments of their aquatic ancestors gave insects, along with other terrestrial arthropods, a head start toward life on land. Terrestrial life requires much more, however, than merely preventing the loss of body water. Unlike crustaceans, insects cannot simply rely on a marine environment to supply ions and water in correct proportions, and they cannot simply allow metabolic wastes to diffuse into the environment. Instead, insects depend on **Malpighian tubules** and the rectum for excretion and osmoregulation (Fig. 32.6 on page 684). Each insect has from two to several hundred thin Malpighian

FIGURE 32.5

A visual guide to the familiar orders of insects. Main features useful in identification are outlined in color. Horizontal bars show a typical body length. The orders Homoptera and Heteroptera are often combined as order Hemiptera.

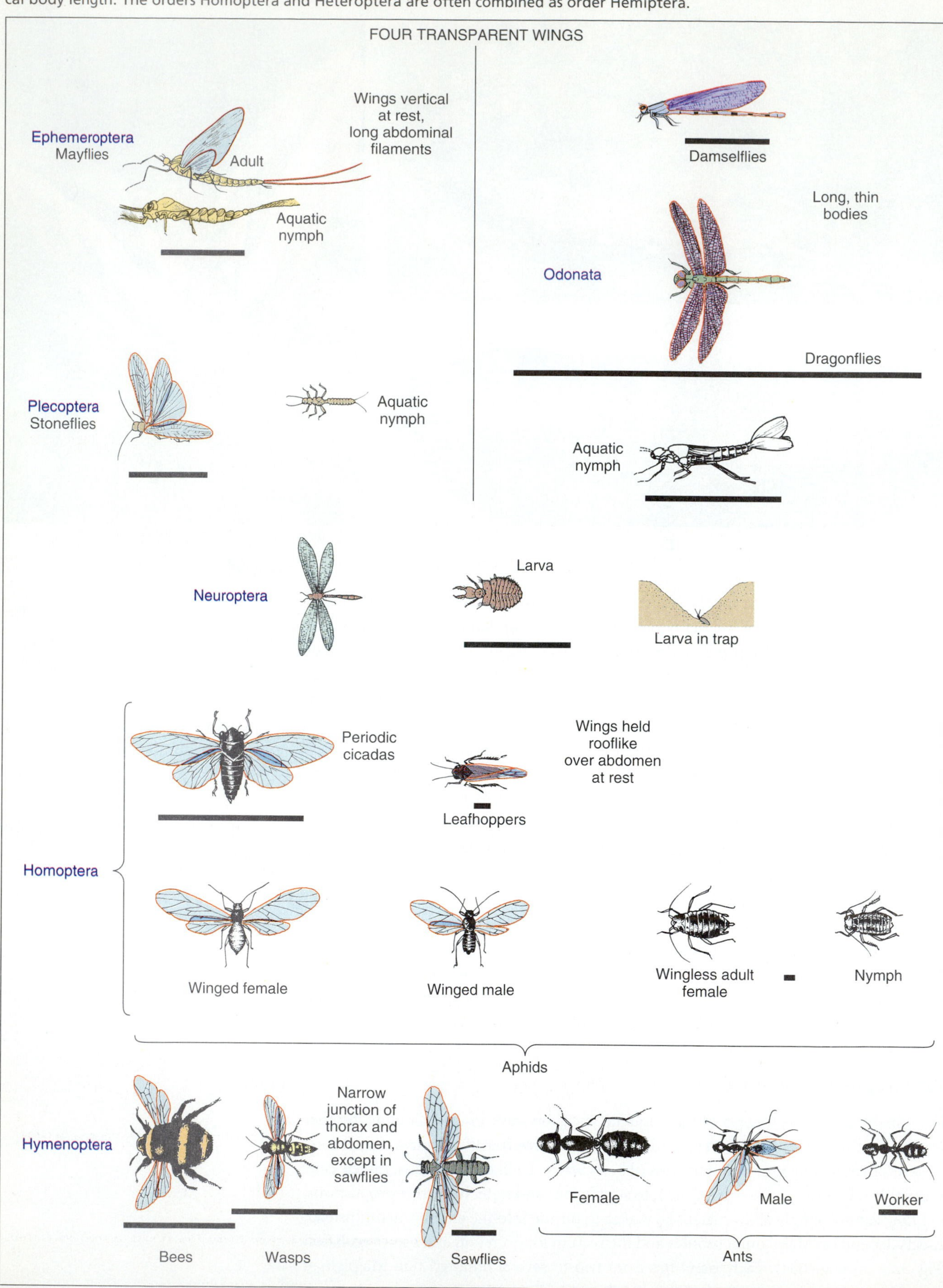

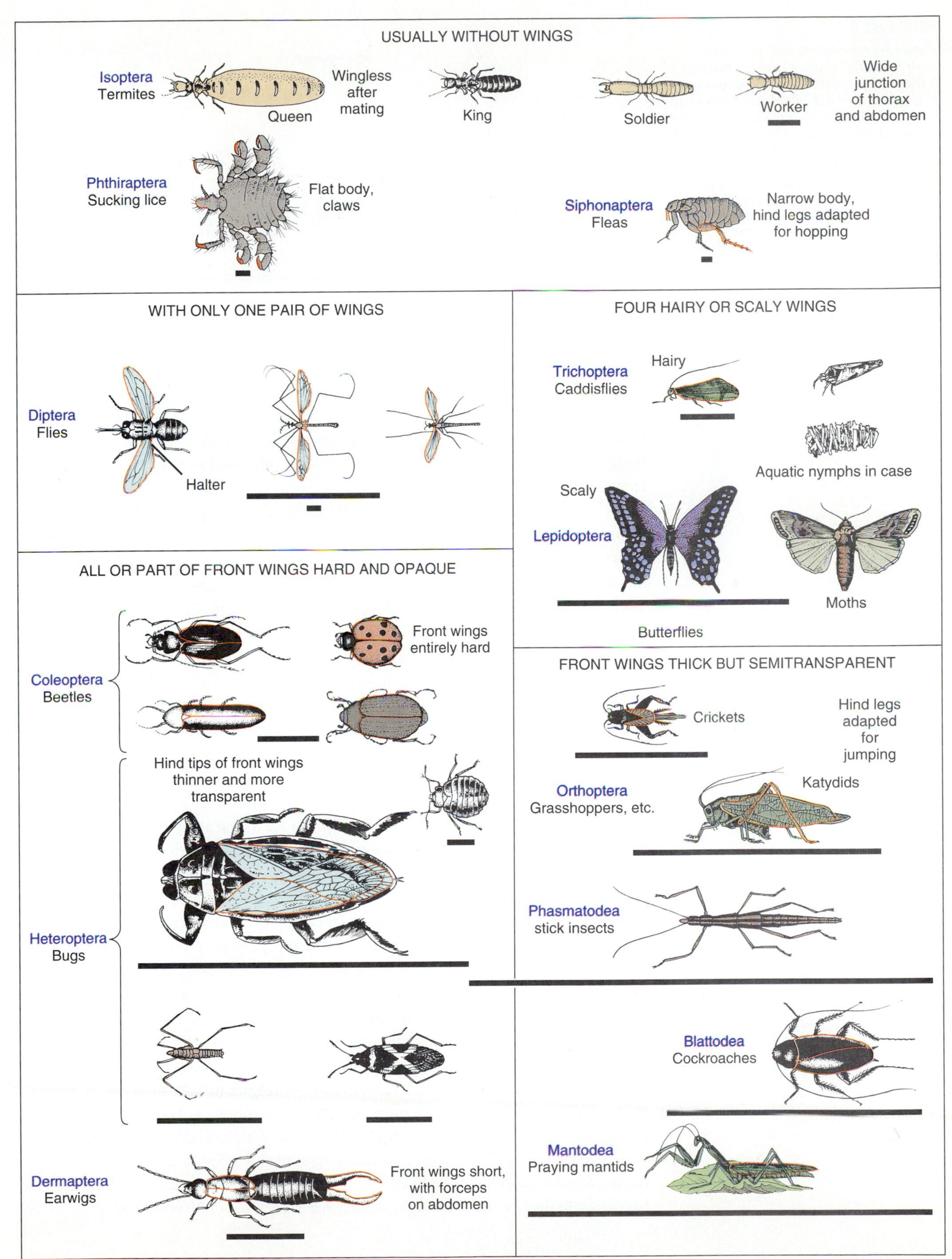

USUALLY WITHOUT WINGS
Isoptera
Termites
Queen
Wingless after mating
King
Soldier
Worker
Wide junction of thorax and abdomen
Phthiraptera
Sucking lice
Flat body, claws
Siphonaptera
Fleas
Narrow body, hind legs adapted for hopping
WITH ONLY ONE PAIR OF WINGS
Diptera
Flies
Halter
FOUR HAIRY OR SCALY WINGS
Trichoptera
Caddisflies
Hairy
Aquatic nymphs in case
Scaly
Lepidoptera
Moths
Butterflies
ALL OR PART OF FRONT WINGS HARD AND OPAQUE
Coleoptera
Beetles
Front wings entirely hard
Hind tips of front wings thinner and more transparent
Heteroptera
Bugs
FRONT WINGS THICK BUT SEMITRANSPARENT
Crickets
Hind legs adapted for jumping
Katydids
Orthoptera
Grasshoppers, etc.
Phasmatodea
stick insects
Blattodea
Cockroaches
Dermaptera
Earwigs
Front wings short, with forceps on abdomen
Mantodea
Praying mantids

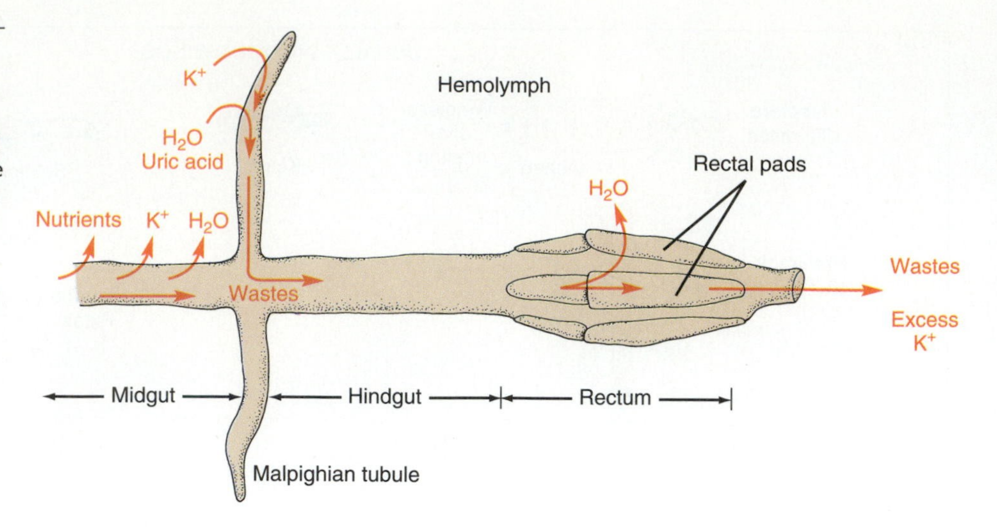

FIGURE 32.6

Schematic representation of the functioning of the insect excretory system. In the midgut nutrients, potassium ions (K^+), and water are absorbed from food into the hemolymph. Potassium ions are actively pumped out of the hemolymph into the Malpighian tubules. Water follows osmotically, carrying dissolved wastes such as uric acid that join the digestive wastes in the hindgut. Six rectal pads reabsorb the water and needed ions back into the hemolymph.

tubules, which are often yellow and have muscle cells that keep them moving within the hemocoel in the abdomen. Each Malpighian tubule attaches at one end between the midgut and the hindgut (Figs. 13.6 and 32.9). The other end is either unattached or attached to the rectum.

The blood pressure is so low that Malpighian tubules require an alternative means of producing filtrate. They actively absorb ions, especially potassium, from the hemolymph, and water follows the potassium osmotically. Along with the water come dissolved molecules, including uric acid, the major nitrogenous waste of insects (p. 280). The fluid then enters the hindgut, where it mixes with digestive wastes. Much of the ions may be reabsorbed into the hemolymph across the gut lining. Plants are extremely rich in potassium, and herbivorous insects excrete most of the K^+ in the feces. As the feces pass through the rectum, **rectal pads** recycle much of the water back into the hemolymph. In some insects, such as the desert locust *Schistocerca gregaria,* virtually all the water that enters the Malpighian tubules is recovered by the rectal pads.

Respiration

TRACHEAE. Another adaptation for terrestrial life is the system of breathing tubes, called tracheae, that allow insect tissues to obtain oxygen directly from the air. These tracheae are up to several millimeters in diameter in large insects. They are lined with glistening white cuticle that molts along with the exoskeleton (see Fig. 30.4D), and they are reinforced by a spiral strand of chitin. Each trachea opens to the surface through a **spiracle** (Fig. 32.7). There are two pairs of spiracles in the thorax and eight pairs in the abdomen. Each spiracle has a valve that reduces the loss of water from body fluids and helps keep out parasites, particles, and liquid water. The valve opens in response to neural excitation signaling a buildup of CO_2 in the hemolymph. The large tracheae branch off into smaller and smaller tracheae, forming a pattern that varies with the species. (Compare Figs. 12.9 and 32.7B.) The thinnest branches, the **tracheoles,** are generally less than 0.1 μm in diameter. Unlike the tracheae of spiders, those of insects bring air directly to tissues rather than to blood. Some tracheoles, in fact, indent the plasma membranes of cells, bringing oxygen directly to the mitochondria that require it. Because the blood does not have to supply oxygen, insects get by with open circulatory systems that operate under low pressure. A dorsal, tubular heart simply pumps blood anteriorly, then eventually draws the blood back in from the hemocoel through one-way ostia (see Figs. 11.3B and 32.9).

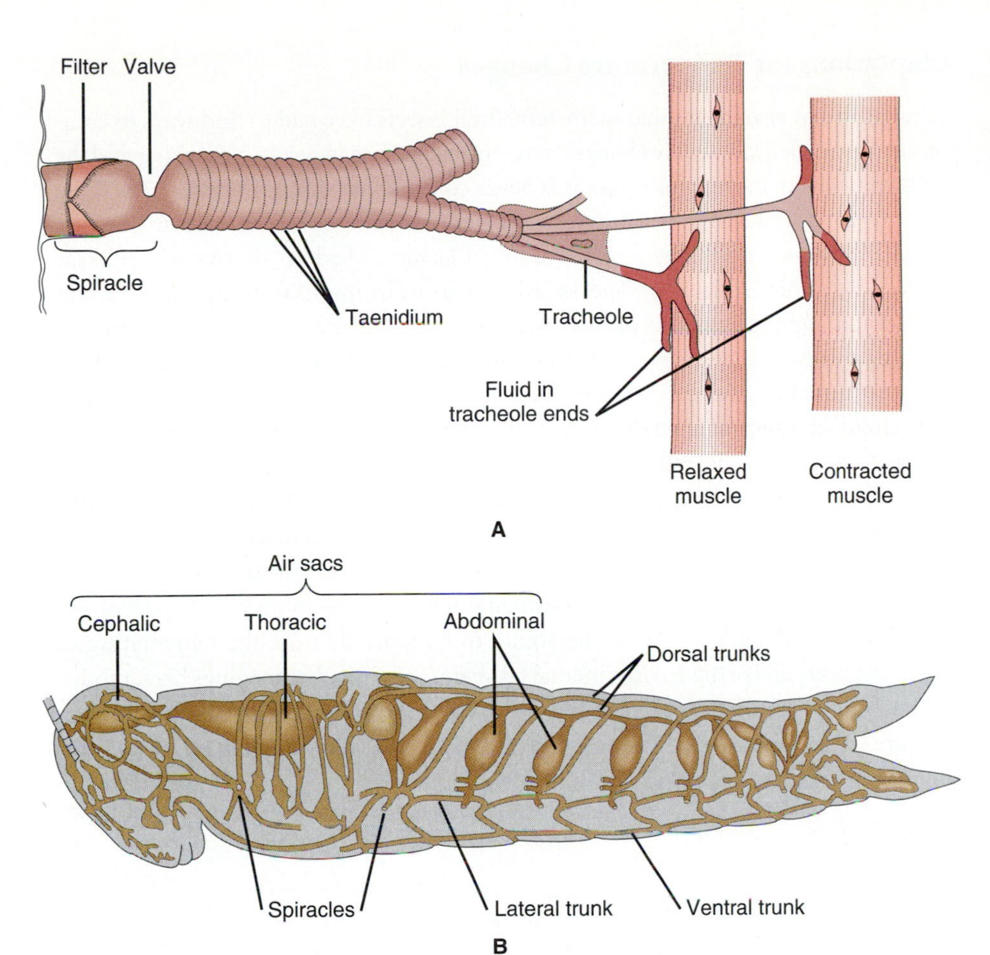

FIGURE 32.7

The tracheal system. (A) Schematic representation of a trachea branching to tracheoles. The filter and valve in the spiracle keep out particles, parasites, and liquid water and prevent the evaporation of body water. The tracheae are reinforced by a spiral strand of chitin called the taenidium. Tracheoles bring air to within a few micrometers of most cells, and they actually invaginate some cells. They are usually filled with fluid at the ends, especially if the cell is not active. The largest tracheae may be several millimeters in diameter; the tracheoles are often less than 0.1 μm in diameter. (B) The tracheal system of a grasshopper. Longitudinal tracheae and air sacs allow grasshoppers, unlike many other insects, to ventilate the tracheae by breathing.

■ **Question:** Tracheae are a straightforward means of delivering oxygen to cells. Why do you think they have not evolved in other animals, such as humans?

The movement of the O_2 along the tracheal tubes is largely by diffusion. Some insects, however, aerate the tracheae by means of breathing movements that tend to take air into the thoracic spiracles and expel it through the abdominal spiracles. Still, the movement of gases in any long, thin tube is so slow that it may be one of the factors that has kept insects relatively small. In the largest insects, which are approximately 30 cm long, O_2 has to travel perhaps 10 cm from spiracles to the innermost tissues.

AQUATIC INSECTS. Although insects are fundamentally terrestrial animals, some have returned to the water as either larvae or adults, and obviously they have some other way of obtaining oxygen. These aquatic insects have not simply "reinvented" the gills of their ancestors. Some insect larvae, such as mosquitoes, hang from the water surface, breathing air through a tube (see Fig. 13.1A). Other insect larvae absorb oxygen from the water by means of **tracheal gills,** which are areas rich in tracheae that are especially adapted for absorbing O_2. In dragonfly larvae the tracheal gills are in the rectum, which moves water in and out much as lungs inhale and exhale air.

Other aquatic insects, such as the true bugs called water boatmen (family Corixidae) and backswimmers (family Notonectidae), have **physical gills.** These are not anatomical structures, but bubbles of air the insects carry with them beneath the water surface. These bubbles are not merely oxygen reserves. In fact, they work much better if filled with air than if filled with oxygen. As the bugs absorb oxygen from the bubbles, the partial pressure of O_2 (pO_2) in the bubble falls below that of the water. As a result, O_2 diffuses from the water into the bubble (pp. 239–241).

Adaptations for Temperature Changes

Living in air has also meant that many terrestrial insects have had to find ways to compensate for sudden and severe changes in temperature. The majority of insects avoid the problem by living in climates where it is never cold. Others simply die after producing cold-tolerant eggs that start an entirely new generation with the return of warm weather. However, a surprisingly large number of insect larvae and adults survive winter as far north as the Arctic Circle. Some species avoid freezing by means of compounds such as glycerol and thermal hysteresis proteins, others become supercooled, and still others actually survive partial freezing. (Refer to Chapter 15 for details and references on these and other topics in this and the following paragraph.) Protected against freezing, they may remain in a state of inactivity called **torpor** for as long as the body temperature is too low to sustain normal rates of enzyme function. Other species enter a phase of inactivity called **diapause** when winter approaches. Diapause, in contrast to torpor, is not a passive response to cold, but a physiological state that the insect induces.

Torpor and diapause permit survival in an inactive state, but some insects can remain active at surprisingly low environmental temperatures by means of **partial endothermy.** Partial endothermy is the ability to temporarily raise the temperature of part of the body above the environmental temperature. On cold mornings bees, moths, and many other insects bask in sunlight or shiver their flight muscles to raise the temperature of the thorax enough to permit flight. The heat is often conserved by **pile,** which is an insulating coat of hairlike structures.

Sensory Organs

MECHANORECEPTION. Like other arthropods, insects have a variety of mechanoreceptors that inform them of the world outside the exoskeleton. These include hairlike projections that respond to deflections, as well as a variety of structures that respond to deformation of the cuticle. Many insects also have mechanoreceptors that detect airborne sounds. The ability to hear has evolved several times among insects, judging from the variety of auditory organs. Many insects can also detect vibrations of the substratum using mechanoreceptors in or on the legs. These mechanoreceptors are associated with various cuticular structures, such as hair-shaped **setae.** Setae may also respond to wind or air movements produced by predators.

FIGURE 32.8
The antenna of a male luna moth *Actias luna* showing numerous branches that increase surface area for receptor cells.

CHEMORECEPTION. Insects can also detect airborne molecules by a process comparable to olfaction in humans. Molecules pass through small pores in the cuticle, especially on antennae, and then bind to neuronal receptor cells (Fig. 32.8). Among the important molecules detected are **pheromones,** which are substances emitted by one animal that induces a social behavior in another member of the same species (see Fig. 20.7). Many insects also have chemoreceptors for substances emitted by the plants or animals they feed on. For example, mosquitoes and some other "biting" flies have chemoreceptors in their antennae that respond to carbon dioxide, to humid convection currents, and to other volatile substances from mammals. Detection of the CO_2 increases the tendency of mosquitoes to fly. They are then apt to encounter convection currents and volatile chemicals, which they follow to the host. Repellents work by blocking one or more of the receptors, thereby preventing the mosquitoes from locating a host.

VISION. Most insects have two compound eyes that resemble those in crustaceans, except that they tend to have hexagonal rather than rectangular facets (pp. 665–666).

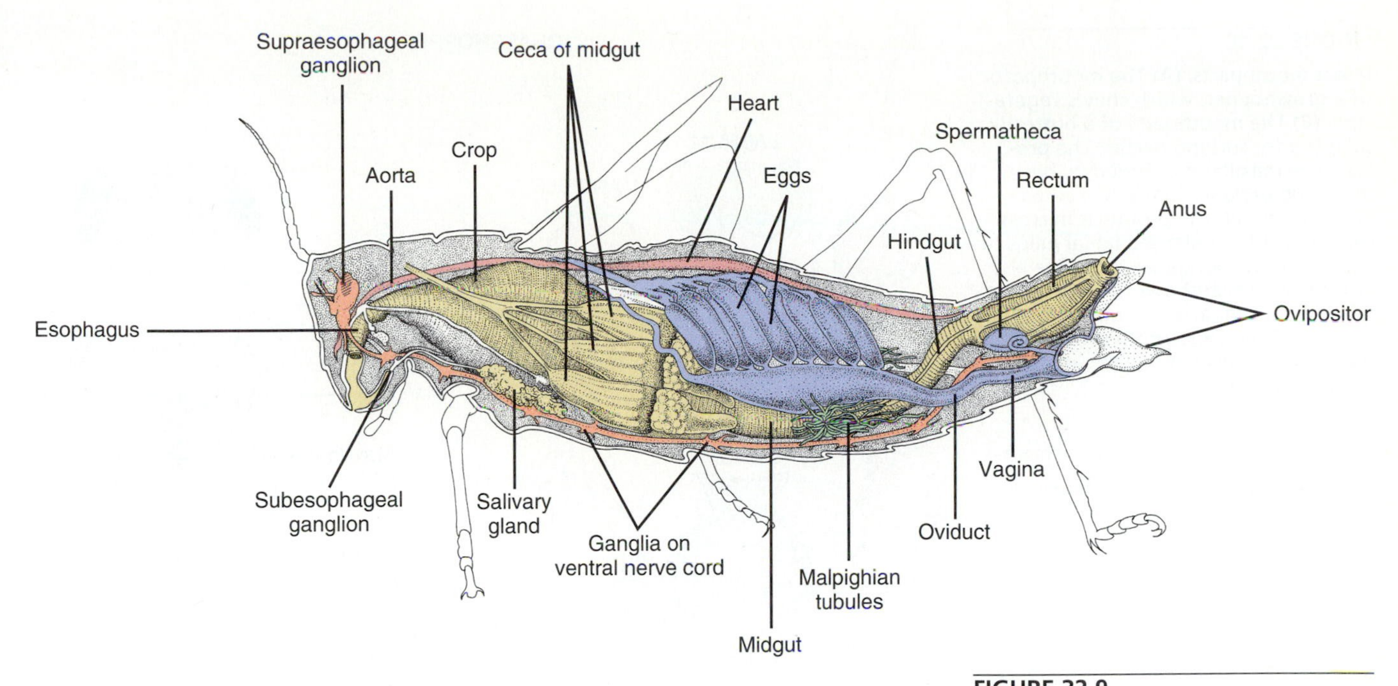

FIGURE 32.9

Internal structure of a female grasshopper.

Usually there are also up to three simple ocelli. Insects are generally blind to red light, but they can perceive ultraviolet and also the direction of polarization.

The Central Nervous System

Other aspects of insect physiology have required less modification for terrestrial life, and they are generally similar to those described in the previous chapter for crustaceans. (Compare Fig. 32.9 with Fig. 31.8.) This applies to the central nervous system, which consists of a chain of ganglia on a ventral nerve cord. Each ganglion is largely autonomous and controls local activities. The thoracic ganglia generally control walking and flight. The ganglia in the head (the supra- and subesophageal ganglia) are mainly involved with the integration of sensory information from the eyes, antennae, and mouthparts and with control of feeding. The supra- and subesophageal ganglia also exercise overall control over many functions coordinated by other ganglia. For example, the subesophageal ganglion of the male mantis usually inhibits copulatory movements that are coordinated by the last abdominal ganglion. When the antennae detect pheromones from a sexually receptive female, however, the subesophageal ganglion stops inhibiting copulatory activity. Decapitating a male mantis also removes the inhibition and causes the male mantis to attempt to copulate with a finger or other object of about the same size and shape as a female mantis. Contrary to popular belief, however, female mantids do not have to bite the heads off males in order to get them to mate.

Feeding and Digestion

MOUTHPARTS. Insect mouthparts have diversified for a wide range of foods, yet they are all fundamentally the same. The mouthparts include two **mandibles,** a **hypopharynx,** a **labrum** and **labium,** and a pair of **maxillae** with sensory palps. In grasshoppers and other insects that chew food, the mandibles function like jaws, the hypopharynx works like a tongue, and the labrum and labium are analogous to upper and lower lips.

FIGURE 32.10

Insect mouthparts. (A) The mouthparts of a grasshopper, which chews vegetation. (B) The mouthparts of a butterfly, adapted for sucking nectar. The proboscis (= maxilla) is unfurled by hemolymph pressure in the two galeae. The inset shows the proboscis in cross section. (C) The enlarged labial palps of the house fly sponge up food. See also Fig. 13.1A, which shows the suspension-feeding mouthparts of a mosquito larva, and Fig. 13.3, which shows the piercing and sucking mouthparts of an adult mosquito.

■ **Question:** Which mouthparts appear to be homologous with legs?

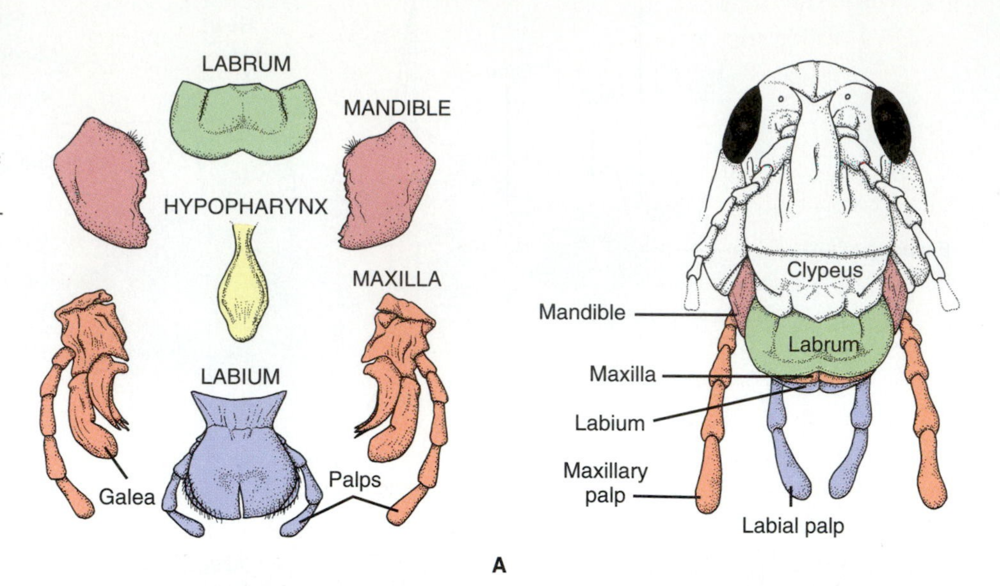

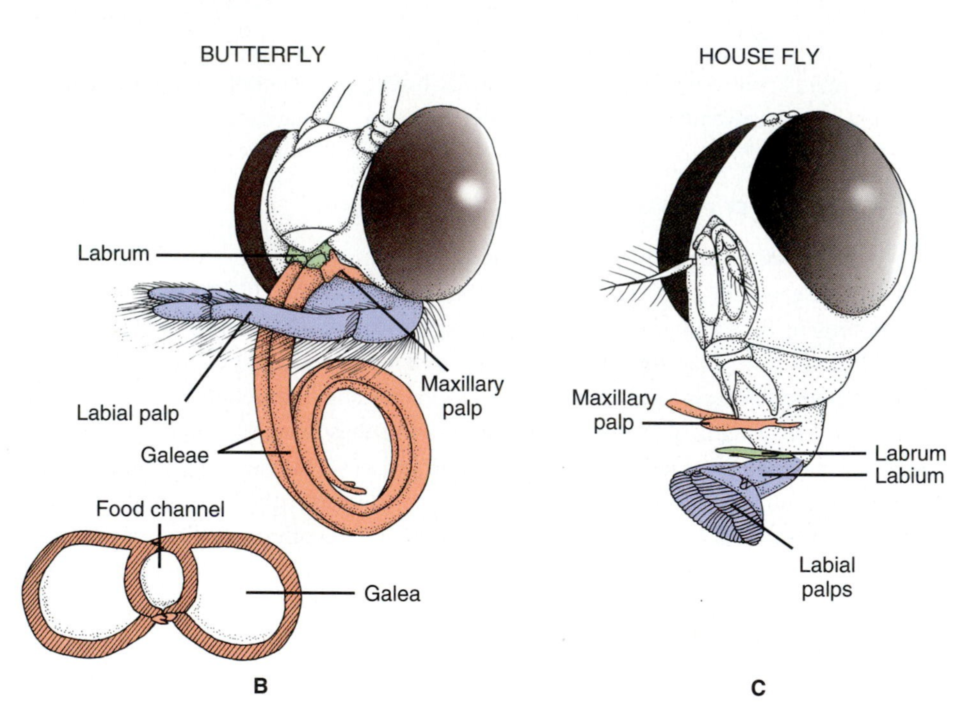

FIGURE 32.11

A spined soldier bug *Podisus* demonstrating its piercing and sucking ability on the caterpillar of a monarch butterfly. Toxins from the milkweed leaf on which the caterpillar was feeding do not deter the bug.

In other insects, however, these structures vary considerably in structure and function (Figs. 32.10 and 32.11).

DIGESTION. The digestive tract is generally a tube divided into three parts: **foregut, midgut,** and **hindgut** (Figs. 13.3, 13.5B, 20.1, and 32.9). The foregut is often modified as a **crop** in which food is stored. Some digestion may also occur in the crop from enzymes secreted by the **salivary glands** and regurgitated from the midgut. The midgut often has pouches called **ceca.** The midgut and ceca are the main sites for secretion of enzymes and absorption of nutrients into the surrounding hemocoel. The gut secretes a variety of digestive enzymes, including cellulase in wood-eating termites and cockroaches. (See pp. 465–466.) The hindgut and rectum are where water and nutrients from the Malpighian tubules are reabsorbed.

INSECT AGRICULTURE. As a final topic in the subject of insect feeding, it is worth noting that many insects have gone through the equivalent of the Neolithic Revolution, giving up the hunter–gatherer life in favor of agriculture. Many ants and termites cultivate fungi within their burrows (see Fig. 1.7). Some ants guard aphids, in exchange for which the aphids let the ants "milk" them of a nectarlike secretion from the anus. Daniel Janzen has found that in tropical forests ant colonies inhabit the swollen thorns of acacia trees and feed on the nectar. The ants keep the acacias free of other herbivores and prune vegetation that would otherwise shade and kill the acacias. Acacias die without their ant colonies, and ants die without the acacia thorns.

Predation and Defense

Insects, of course, not only eat but are often eaten. Many, however, have elaborate defenses against predation. The bombardier beetle puts up one of the most dramatic chemical defenses, directing a hot, explosive secretion toward potential predators (Fig. 32.12). Many insects that are toxic either from their own secretions or those of host plants adver-

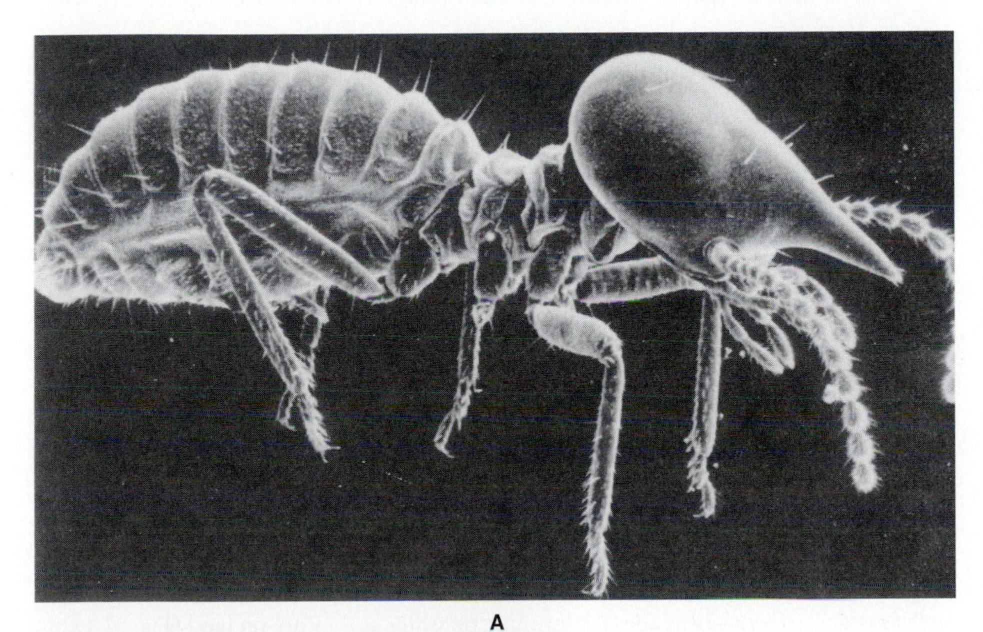

A

B

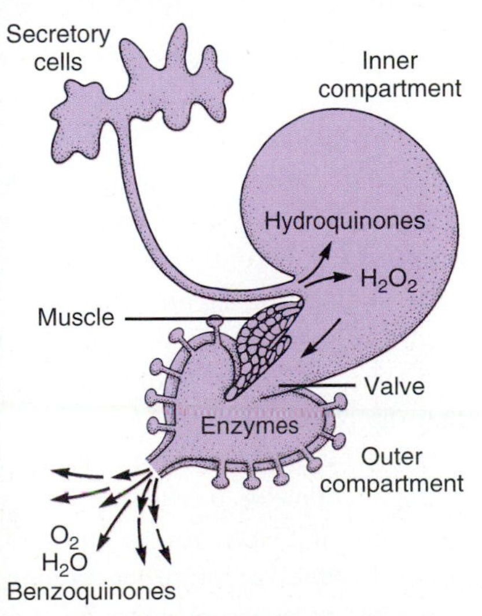

C

FIGURE 32.12

(A) A soldier termite *Nasutitermes corniger* defends its colony with a head modified as a glue gun. Soldiers of other species secrete various other adhesives and irritants. (B) Self-defense by the bombardier beetle (family Carabidae). The beetle (held in place by wax) directs a 100°C stream of benzoquinone toward a potential predator—in this case forceps pinching the left front leg. (C) This behavior has been known for a long time, but its mechanism, along with the defense of numerous other insects, was discovered by Thomas Eisner and his colleagues at Cornell University. The beetle stores a mixture of hydroquinones and hydrogen peroxide (H_2O_2) in an inner compartment in the abdomen. When alarmed, the beetle opens a valve that allows the mixture to enter an outer compartment containing enzymes that catalyze the conversion of the H_2O_2 into H_2O and O_2 and the exothermic oxidation of hydroquinones into benzoquinones. The O_2 propels the hot benzoquinones with an audible pop.

A

B

FIGURE 32.13

The harmless hover fly *Metasyrphus americanus* (A) resembles a yellowjacket or hornet such as *Vespula* (B). Many other dipterans, called bee flies, also derive protection from predators by mimicking bees, wasps, and yellowjackets.

tise the fact with **warning coloration.** Many other insects have appearances that **mimic** such coloration and provide protection from predation (Figs. 18.15 and 32.13). Another approach to defense against predation is **camouflage** (pp. 374–376; Fig. 32.14). The sources of these colors are described on page 194.

A

B

C

FIGURE 32.14

Defensive and aggressive camouflage. (A) The nymph of a spittlebug (Homoptera) partly exposed from its bubbly defensive secretion. (The nymph is upside down and facing left. Note the curled wing buds of this exopterygote.) (B) The wooly alder aphid *Prociphilus tessellatus* covers itself in a waxy secretion that not only hides it, but also sticks to the mouthparts of predators. These aphids are also guarded by ants. Thomas Eisner and colleagues (1978) have found that some predaceous lacewing larvae (order Neuroptera) disguise themselves in the "wool," which enables them to approach the aphids undetected by ants that guard them. (C) This ambush bug *Phymata* (left) was also apparently well camouflaged when the honey bee arrived to feed on goldenrod.

Locomotion

RUNNING AND JUMPING. Living on land enables insects to add running, jumping, and flying to the possible modes of locomotion. The majority crawl on all sixes using an **alternating tripod gait.** While the fore and hind legs on one side and the middle leg on the other side are fixed on substratum, the other three legs advance. Insects readily vary this gait, however, if they have lost a leg or if the substratum is irregular. Some insects, notably fleas, grasshoppers, and crickets, have hind legs adapted for jumping. In many cases, jumping is due to the sudden release of stored energy. Among the many other fascinating discoveries made by the self-taught entomologist Miriam Rothschild is the fact that the legs of fleas are cocked in a sort of catch while pressed against a rubbery protein called **resilin.** The flea then releases the catch, and the resilin pushes the legs back, sending the flea tumbling through the air. Grasshoppers use a different technique for cocking and releasing the hind legs. They first build up tension in both the extensor and flexor muscles of the hind legs, then suddenly relax the flexor muscles. Many insects, such as water striders, are as adept at jumping on water as they are on land (Fig. 32.15).

FLYING. Impressive as these feats of locomotion are, it is unlikely that they could have gotten insects very far evolutionarily. The success of insects probably owes much more to the ability to fly. Theories about the origin of flight have long been controversial, but it is worth describing a few of them to show how zoologists apply concepts to generate hypotheses. The oldest theory is that wings originated as small flaps (paranota) that first permitted insects to glide or parachute, and that the paranota later became adapted for flapping flight. The concept is usually referred to as the paranotal theory, or sometimes as the flying-squirrel theory. Joel Kingsolver of Brown University and Mimi A. R. Koehl of the University of California at Berkeley, however, have shown that the paranotal theory is unlikely. Using models of insects with wings of various lengths, they found that wings provide little help in gliding until they reach a length of about a centimeter. It is fundamental to the concept of natural selection that a structure can evolve if it improves the fitness of an animal, but not merely because it might improve the fitness of future generations. Therefore the structures that eventually became wings had to serve some other function initially.

FIGURE 32.15

A common water strider *Gerris* (Hemiptera) on a pond. Fine, water-repellant hairs on the legs, and the location of the sharp tarsi above the ends of the legs keep the legs from breaking through the surface tension. The middle pair of legs do most of the rowing; the hind legs act as rudders, and the short front legs are used mainly for seizing prey. The water strider can also jump to seize prey or to escape. Prey are detected by the ripples they generate. Male water striders can also determine the sex of other water striders by the difference in ripple frequency. A population of *Gerris* generally contains some winged adults that can also fly to other habitats. (The large red spheres on the thorax are immature red mites that parasitize water striders.)

Kingsolver and Koehl proposed that the precursors of wings could first have been used as solar collectors to raise body temperature, and later for gliding and flight as they became larger. A serious problem with this theory is that in solitary adult insects flight is virtually the only advantage of an elevated temperature. Thus the use of wings for flight should have preceded their use in thermoregulation (Heinrich 1993, pp. 104–115). Another theory, by the paleontologist Jarmila Kukalová-Peck of Carleton University in Ottawa, is that wings originated in aquatic nymphs, perhaps as gills and then as fins before they became adapted in the terrestrial adults for gliding. More recently, James Marden and Melissa Kramer (1994) at Penn State suggested that wings could first have been used in skimming across the surface of water, as stoneflies still use them.

For insects smaller than dragonflies and butterflies the viscosity of air is high relative to inertia. (In other words, the Reynolds number is low. See pp. 212–213). Therefore most insects cannot fly the way airplanes do, by slicing through the air with smooth wings. Flying for most insects is like swimming is for us. A common pattern of wing movement, in fact, resembles the butterfly stroke of human swimmers (Fig. 32.16). The wings do not flap straight up and down, but are flung forward during the downstroke and backward during the upstroke. Also during downstroke the rear portions of the wings tilt up and provide forward thrust, and during the upstroke the rear portions of the wings tilt down and reduce air resistance.

The tilting of the rear portions of the wings is due to the attachment of the wings near the front edge, and to the greater rigidity of the wing near the front. This rigidity is due to a greater concentration of **veins** in front. There are numerous other adaptations for flying that occur in insects but not in any other animal or machine. For example, in damselflies, butterflies, and certain moths (families Saturniidae and Noctuidae) the wings also perform a maneuver in which they clap together at the end of the upstroke, then fling apart at the start of the downstroke. This **clap and fling** pattern generates air currents that provide additional lift.

The complex movements of the wings dictate some intricate attachment of muscles. In dragonflies the flight muscles (elevators and depressors) are attached directly to the bases of the wings. (See left portion of Fig. 10.4B.) A burst of action potentials from the thoracic ganglia triggers each contraction of the flight muscles. These are therefore called **synchronous muscles.** Synchronous muscles directly attached to the

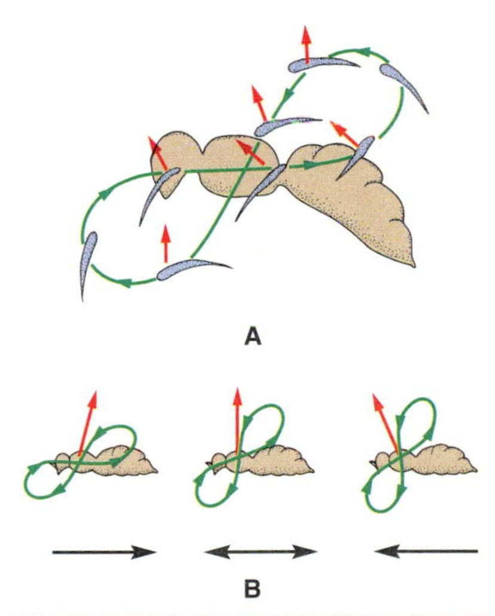

FIGURE 32.16

Wing movements during flight. (A) During the downstroke the wing (dark profile) is angled downward, providing forward thrust as well as lift. (Arrows show direction of force on body.) During the return stroke the wing (light profile) moves backward with respect to the body and is angled to reduce drag and provide additional forward thrust. (B) By changing the orientation of this figure-eight pattern, many insects can hover and fly either forward or backward.

wings are said to be primitive, although they are good enough to permit dragonflies to hover, capture prey, and even mate on the wing. Insects considered more advanced, such as flies, beetles, bugs, and bees, have flight muscles attached to cuticular plates in the exoskeleton. (Right part of Fig. 10.4B.) Deformation of these plates "clicks" the wings up and down. The flight muscles are also different, in that their contraction is not triggered by a synchronous burst of action potentials. They are therefore called **asynchronous muscles.** Asynchronous muscles are triggered to contract by being stretched, as described elsewhere (p. 210).

The fastest flying insect is said to be a sphinx (= hawk) moth (family Sphingidae), which has been clocked at 15 m/sec—roughly 30 mph. This is about 1000 body lengths per second, compared with only 100 lengths per second for a jet fighter at Mach 3. The distances covered by insects are also impressive, especially in units of body length. Monarch butterflies have been known to cover 425 km in a day during migration (pp. 412–413). From the perspective of an insect, even a few meters is a long distance to navigate. Insects use polarized light and probably other cues to help them find their way over such vast distances relative to their body sizes.

Reproduction

EGGS. Another key to insect success on land is the ability to reproduce vast numbers of offspring under the hostile conditions of terrestrial life. One of the major adaptations for terrestrial reproduction is the self-sufficient, or **cleidoic,** egg. In contrast to the eggs of most other invertebrates, those of insects do not depend on an aqueous medium to provide water and dissolve metabolic wastes. The insect egg consists of an ovum and a nutrient yolk enclosed in a shell made of proteins and waxes. The shell offers mechanical protection and is permeable to oxygen but not to water.

After the final molt into adulthood, most female insects begin egg production, **oogenesis.** Oogenesis is soon followed by **vitellogenesis,** which is the production in the **fat body** of proteins that become part of the yolk. The ovary then constructs a shell around each ovum and yolk. The release of the egg, called **oviposition,** occurs after mating. Pores (**micropyles**) allow sperm to penetrate the shell for fertilization of the ovum as it passes down the oviduct during ovulation. Other pores (**aeropyles**) that are permeable to O_2 but not to H_2O allow the embryo to respire even if the egg is soaked in rain or in a puddle. Early development of the egg is described on page 114.

COURTSHIP. Many insects achieve a high reproductive rate through adaptations in behavior, including courtship. Courtship helps ensure that the prospective mates belong to the same species, are of opposite sex, and are physiologically ready to reproduce. Insects employ almost every imaginable sensory channel when locating mates fulfilling these requirements. The most common channel of courtship communication is by sex-attractant pheromones released by one sex, usually the female, and attractive to members of the other sex in the same species. A few insects, such as cicadas, mosquitoes, bark beetles, katydids, grasshoppers, and crickets, use acoustic signals in courtship. A male cricket chirps by rubbing the medial edge of a front wing (the **scraper**) across a row of ridges (the **file**) on the inner surface of the other front wing. Chirps emitted in a particular pattern and rate attract females of the same species to the male's territory. A different pattern of chirps repels males of the same species.

Insects generally use vision only to locate and orient toward the mate at close range. (See Fig. 20.3 for the use of vision by courting drosophila.) Fireflies are a striking exception. Males court in flight by flashing in specific patterns that sexually receptive females answer from the ground (see Fig. 20.6). The light is produced by a molecule called **luciferin** in the presence of ATP and the enzyme **luciferase.**

In some insects, courtship also allows one member of the pair to assess the fitness of the other (sexual selection). Courtship may also prolong copulation, thereby increasing the number of sperm transferred. In predaceous species, courtship also helps keep one mate from being eaten by the other. Randy Thornhill (1980) of the University of New Mexico has demonstrated all these functions in the black-tipped hangingfly (order Mecoptera). The male black-tipped hangingfly *Hylobittacus apicalis* begins courting by capturing another arthropod as prey, then emitting a pheromone while hanging by its forelegs. A female attracted by the pheromone and "satisfied" with the size and palatability of the prey will lower her wings as a signal that she is receptive. The female hangingfly will then hang around to feed and mate. The larger the prey, the longer copulation will last, and the more sperm will be transferred. After copulating, the male and female fight over what is left of the prey, with the male usually winning the prey and using it as bait for another female. A male may also steal prey from another male, sometimes after tricking the male by lowering its wings like a receptive female.

FERTILIZATION. Fertilization occurs internally in all insects, and it is achieved through copulation in all but a few (Fig. 32.17). The genitals of males and females generally couple by a complicated "lock and key" mechanism (see Fig. 16.3B). Fertilization usually does not occur during copulation. Instead the sperm are stored in a **spermatheca** and are released during oviposition, sometimes months or years after copulation. As the eggs move down the oviduct, the sperm enter them through the micropyles. In some species, several fertilized eggs are then enclosed within a cuticular case, or **ootheca** (Fig. 32.3).

OVIPOSITION. Most insects oviposit where moisture and other physical conditions are suitable for development, or on particular plant or animal hosts that will provide food for the young. Many flies and wasps, for example, lay their eggs in or near feces, decaying meat, or living animals (Fig. 32.18A on page 694 and Fig. 32.22). Many wasps inject other insects or spiders with a paralyzing but nonlethal venom, then place the helpless prey into burrows with their eggs. This behavior assures that the larvae will

FIGURE 32.17
(A) Sperm transfer in the damselfly *Coenagrion puella*. The male (above) is grasping the female with claspers near his anus. At the same time the female is bending her abdomen forward to receive sperm that the male has previously placed in accessory genitals on his second and third abdominal segments. (B) Almost all other insects transfer sperm directly. Here a pair of March flies (probably *Bibio*) copulate. Unfortunately the male (below) has not noticed that a crab spider has killed the female.

■ **Question:** The male damselfly continues to grip the female while she oviposits. What is the significance of that behavior?

A

B

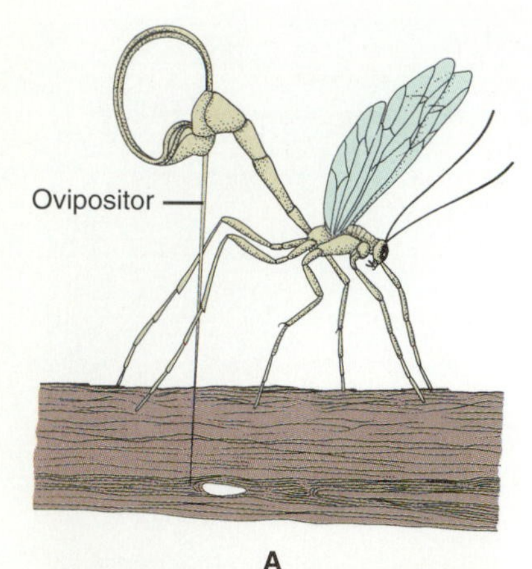

A

B

C

FIGURE 32.18

Reproductive behavior. (A) A female ichneumon wasp *Megarhyssa macrurus* bores into wood with her long ovipositor and deposits an egg near the larva of a wood-boring insect. (B) Oak apple galls. One is cut open to show the pupa of the gall wasp *Amphibolips* sp. whose mother caused the oak to form the gall. (C) A male giant water bug *Belostoma* with eggs on its back.

■ **Question:** How does the ichneumon wasp locate a host?

have fresh food when they hatch. Such larvae, and other parasites that ultimately kill the host, are called **parasitoids.**

Many insects that oviposit in plant tissues somehow induce the plants to form abnormal growths, called **galls,** around the egg. Galls protect the egg and larva and may provide food (Fig. 32.18B). It is not clear whether gall-forming insects inject a growth hormone or some other substance that induces the growth, or whether the gall is an adaptation of the plant that protects its tissues from the egg. Because suitable hosts are apt to be scarce, there is a danger that so many eggs will be deposited on a single host that the nutrients will be depleted before the larvae fully develop. Many species reduce this risk by laying conspicuous eggs or by marking the host with a pheromone that deters other ovipositing females.

The majority of insects are egg-laying, or **oviparous.** Some females, however, provide even further protection of the embryos from drying and predation by retaining the eggs within themselves until they hatch into larvae. If such an insect provides the embryo with nourishment, it is said to be **viviparous,** like us. In the female tsetse fly *Glossina,* not only an egg but also the larva develops within a uterus. Other insects may retain the eggs until hatching but may not provide any nourishment. Such insects are said to be **ovoviviparous.** In aphids there is an alternation of generations, in which a generation of females with wings reproduces sexually and oviparously, followed by several generations of wingless, ovoviviparous females produced from unfertilized eggs (parthenogenetically).

PARENTAL BEHAVIOR. Insects generally provide little parental care but produce large numbers of offspring that ensure that at least a few survive. (In contrast, humans and some other mammals produce only a few offspring and devote a great deal of parental care to ensure their survival.) A few species of insects do, however, provide some care for offspring, generally by brooding eggs. Almost always it is the female that provides parental care, presumably because she has already made the greater reproductive investment in the form of the egg and because it is more certain that the offspring she cares for are her own (p. 420). Certain giant water bugs are notable exceptions, since the male carries eggs on its back (Fig. 32.18C). In *Abedus herberti* the male allows a female to oviposit on his back, but only if she interrupts oviposition to copulate with him several times, thereby assuring his paternity. (See Tallamy 1984 for discussion.)

EUSOCIALITY. The most elaborate reproductive behaviors are performed by **eusocial insects.** By definition, eusociality occurs when there is cooperative rearing of the young, when infertile individuals work on behalf of individuals that reproduce, and when there is an overlap of at least two generations that contribute to labor in the colony. Eusociality has evolved independently in four orders of insects: Isoptera (termites), Hymenoptera (ants and certain wasps and bees), Coleoptera (at least one species of beetle), and Homoptera (certain aphids). In these groups there are **castes** of sterile workers that differ morphologically from the reproductive individuals. (See Fig. 32.5 for representatives of the various castes of termites and ants.) The sterile workers generally far outnumber the reproductives and focus much of their activity on them. This arrangement can be compared to an individual organism, in which the functions of most cells are directed toward maintaining optimal conditions that enable the reproductive cells to function. In some eusocial species the reproductive individuals are so specialized that they are, in fact, little more than gonads. The female reproductives, called **queens,** do little but lay eggs. They cannot even feed themselves (Fig. 32.19). The reproductive males are also useless except to fertilize the queens. The fact that eusociality evolved several times suggests that it increases reproductive success even though

FIGURE 32.19

The queen termite *Microtermes* has her abdomen enormously swollen with eggs. Her head and thorax are on the right in this picture. The dark dorsal stripes are terga. The queen is attended by numerous workers.

■ **Question:** What kind of cuticle expands so much to accommodate the eggs of the female? (See p. 633.)

the great majority of individuals lose their ability to reproduce. A major triumph of sociobiology is that it offers a solution to this paradox. (See pp. 414–415 and below.)

Social organization varies considerably among different species of eusocial insects. Colonies of **termites** generally include one queen and one king that remain together during the life of the colony. The king periodically fertilizes the queen, who spends almost all her time laying eggs. The other members of the colony are male and female workers. They do not reproduce as long as they remain in a colony with a functioning queen. The largest workers are soldiers that defend the colony (Fig. 32.12A). One or more castes of smaller workers are specialized for foraging, brooding eggs and nymphs, building nests, and maintaining proper temperature and humidity in the nest. The nests consist of elaborate tunnels and chambers in wood, or in mounds constructed of mud. Whether an individual nymph develops into a reproductive termite or one of the worker castes is determined by pheromones released by workers attending the nymphs. Periodically, workers produce new queens and kings. These are the only individuals that ever have wings. After their final molt they emerge from the colony in a **nuptial flight,** pair up as queen and king, break off their wings, then start a new colony.

Social structure among many eusocial **ants** is similar to that in termites. A major difference in ants and other eusocial hymenopterans is that males never become workers. All adult males, called **drones,** are fertile, and their only function is to copulate once with a queen. The drones either die during copulation, or the workers sting them to death or exile them from the colony and let them starve. Drones cannot defend themselves, since they lack stingers (which are modified ovipositors). The queen "decides" whether to produce females or males by opening or closing a valve from the spermatheca during oviposition. If the valve is closed the eggs will not be fertilized, and they will develop into males. More often, however, the queen allows the eggs to be fertilized, and they develop into females. Because the eggs that develop into drones receive only one set of chromosomes from the queen, they are haploid. The females are all diploid. This system of sex determination is therefore called **haplodiploidy.**

Whether a female will develop into a new worker or a new queen depends on how well it is fed as a larva. This, in turn, depends on pheromones from the queen that inhibits workers from feeding female larvae enough to make them develop into queens. Only if the queen pheromones are in low concentrations will workers feed certain of the female larvae enough for them to develop into new queens. This can happen if the queen weakens or dies, or if the colony becomes large enough for it to divide into new colonies. In many species of ants the production of various castes of workers also depends on how well the larvae are fed. Poorly fed larvae develop into members of a smaller caste of workers. The best-fed larvae develop especially large heads and mandibles and become soldiers.

Social structure is as complex and diverse among other eusocial hymenopterans—certain wasps and bees—as among ants. Honey bees have been the most thoroughly studied, and several aspects of their organization have been discussed throughout this text. The accompanying box brings together and summarizes these and other aspects of honey bee life.

Interactions with Humans and Other Animals

THE IMPORTANCE OF INSECTS. The number of individual insects has been estimated at 10^{18}—a billion billions. It is not surprising, therefore, that virtually every terrestrial animal, including ourselves, interacts in some way with some insect. Numerous fishes, birds, and mammals use insects as a major source of nutrients. Humans in most cultures also eat fried grasshoppers, beetle grubs, and assorted larvae and pupae. People in some parts of Africa get most of their protein from insects, and in Colombia movie-goers munch roasted ants instead of popcorn. Westerners are virtually alone in shunning insects as food, although anyone who eats sausages, peanut butter, or foods made with stored grains inadvertently eats some insects. Insects contribute to our diets mainly as pollinators. Approximately 75% of all commercial crops in the United States, worth approximately $10 billion per year, depend on pollination by honey bees. (These honey bees were all imported from Europe after North American bees proved unsuited to pollinating European crops.)

On the other hand, insects eat a great deal of food intended for humans. Approximately 13% of crops are lost to insects each year, and there are approximately 800 insect species considered to be agricultural pests in the United States alone. Farmers—and ultimately consumers and taxpayers—spend billions of dollars each year trying to eliminate these pests. Comparable assaults are made by foresters against insects that damage trees, and by home-owners against insects that damage homes and gardens. Numerous isolated skirmishes are fought against biting flies, cockroaches that spread disease and trigger allergies, and other insects that are merely annoying. Elsewhere in the world the major enemies are flies that spread malaria, yellow fever, river blindness (p. 546), and bacterial contamination. In parts of Africa, plagues of locusts are still as common as in biblical times. If this were a battle of people against people instead of people against insects, it would rank as World War III. So far the insects appear to be winning the war. Each year the mosquito *Anopheles* brings death to more people through malaria than were killed in combat in World War II. Our weapons have failed to wipe out a single species of pest insect, the cost of insect control continues to mount, and the number of victims of malaria is increasing (p. 466).

INSECTICIDES. Few people would have bet on the insects in the 1940s and 1950s, when DDT was coming into wide use. DDT (dichloro-diphenyl-trichlorethane) was the first poison that could be used practically to kill virtually any insect. In World War II it eliminated infestations of lice and fleas, two camp followers that had brought discomfort and disease to soldiers and civilians in all previous wars. Soon DDT was being used to kill mosquitoes, farm and forest pests, and even household flies and cockroaches. The very success of DDT was largely responsible for our subsequent disappointment in it and later poisons, because it raised our expectations. Before DDT we would have been content to **suppress** insects; now we expect total **eradication.** Consumers who once accepted the risk of biting into an apple and finding half a larva now reject any apple that has the slightest cosmetic damage from insects. Farmers who once had to accept that insects were going to get a portion of their crop, and in some years all of it, can no longer afford to lose part of the crop to insects. Malaria, once only

Honey Bees

Although the goal of zoology is to provide a complete understanding of an organism, necessity dictates that most zoologists specialize in only one discipline, such as physiology, behavior, genetics, or ecology, and that they pursue their discipline in only a few model organisms. Consequently, we may know a great deal about the genetics of drosophila or the ecology of termites, but seldom have an integrated view of all aspects of the biology of one species. With the honeybee *Apis mellifera,* however, we are as close to achieving that goal as we are for any species, even our own. Several aspects of honey bee biology are discussed elsewhere in this text, but it may be useful to summarize these and other topics in one place to gain some appreciation of how various aspects of biology work together in one species.

A Queen Is Born

The life of a honey bee makes sense only in the context of the life of its colony. A mature colony consists of 20,000 to 80,000 workers and one queen. **Queen mandibular pheromone,** a combination of five secretions from the queen's mandibular glands, inhibits the workers from raising a new queen. In late spring the queen reduces her secretion of the pheromone, and the workers respond by constructing 20 or more special queen cells near the bottom of the comb. Unlike the classic hexagonal cells in which workers store honey and pollen and raise new workers, queen cells are long and tapered (Fig. 32.20). The queen deposits eggs in these cells. After hatching, the resulting larvae are fed a sugary substance called **royal jelly,** which is secreted from the hypopharyngeal glands in the workers' heads. Royal jelly is similar to the **worker jelly** that is fed to other larvae, but it is much richer in sugar. Sixteen days after the eggs were laid, eclosion occurs, and the new queens emerge from the royal cells.

Swarming

In the meantime the workers have been preparing the existing mother queen to emigrate from the hive with a swarm of workers. Like trainers in a weight-loss program, the workers feed the queen less and less and jostle her more and more. This combination of enforced dieting and exercise causes the queen to lose about one-fourth of her body weight and to become restless. Finally, while the new queens are pupating, the mother queen flies out of the hive, taking about half the workers in a swarm. The swarm clusters on a limb or other substratum not far away. Worker scouts leave the swarm to find a new nest cavity, which may be a cave, hollow tree, or artificial hive. After they find a suitable site, the scouts return and communicate its location using the waggle dance (pp. 409–410). The scouts also communicate the suitability of the site by how long they persist in dancing. "Enthusiastic" scouts will gradually convince less enthusiastic ones to inspect their sites. Over a period of several days a consensus will be reached, and the scouts will lead the entire swarm to the chosen site.

Mating

The workers in the old colony remain queenless for about eight days, until a new queen emerges. This virgin queen then takes off on a "nuptial flight" to a special "drone congregation area" where drones from nests within several kilometers

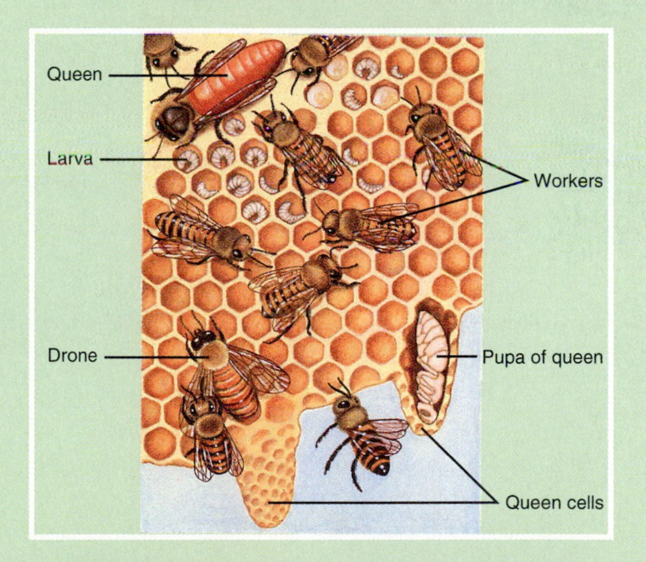

FIGURE 32.20
Portion of a honey bee colony. The queen (upper left) is larger than the workers. Workers continually crowd around her, feeding her and spreading chemical messages throughout the hive. Other workers tend eggs and larvae and place honey and pollen into open cells. Other open cells hold eggs and larvae. Capped cells enclose honey and pupae. A drone (lower left) is being dragged by its wing and will be killed or evicted. New queens are raised in special queen cells (lower right), one of which is shown opened to reveal the pupa inside.

box continues

gather to mate. Using her queen mandibular pheromone as a sex attractant, the queen attracts a hundred or more of these drones, which form a comet-shaped mass as they fly after the queen. Approximately 10 or 12 drones succeed in mating with the queen, dying in the process when their genitals explode during copulation. Unsuccessful drones will either be stung to death by their sister workers or exiled from the colony and left to die. During the nuptial flight the queen accumulates more than enough sperm in her spermatheca to fertilize up to 2500 eggs per day for several years. Almost all the fertilized eggs will develop into workers.

Winter Activities For the rest of the summer the workers forage for pollen and nectar, which they feed to the larvae that will develop into new workers. Workers also gather plant resins, called **propolis,** with which they seal crevices in the nest. They also bring back water with which to cool the nest by evaporation. Much of the nectar is allowed to partially evaporate to make honey. The high concentration of sugar in the honey, as well as some antibiotics, permits it to be stored in capped cells through the winter without contamination by bacteria or fungi. Stored honey and pollen allow the colony to be active through the winter. This is a major difference between honey bees and other eusocial wasps and bees. In bumble bees, for example, only the queen lives through winter, and even then she is in the inactive state of torpor. After the spring warmth arouses her, she must find a nest and raise up a brood of new workers before she can make new queens. In honey bees the workers use energy from the honey and pollen to generate heat, which allows the queen to begin laying eggs at the start of winter. The pollen provides the queen with proteins for eggs. A new generation of workers will therefore be available to take advantage of the earliest flowers of spring. By late spring the colony is already producing new queens and beginning a new annual cycle.

This year-round activity of the colony is thought to have originated in the tropics, but it now allows honey bees to live in climates where the temperature drops to as low as –30°C. The survival of honey bees depends on their maintaining relatively high temperatures within the nest. Workers on the periphery of the hive produce heat by shivering with their flight muscles, and they conserve the heat with their hairlike pile and by huddling together. The mass of workers can produce as much heat as a light bulb (40 watts or more). By clustering around the larvae the bees can warm them to around 35°C even when the outside temperature is as low as –30°C. In the northeastern United States it takes approximately 20 kg of honey to sustain a hive at this level of thermoregulation through winter.

Division of Labor Unlike many termites and ants, in which there are several specialized worker castes, honey bees have only one worker caste. During the spring and summer each worker gradually progresses through various tasks, such as tending to larvae, cleaning cells in the comb, capping cells containing pupae and honey, building comb, and maintaining the comb temperature and humidity. Finally each worker becomes a forager and will remain one until she dies, usually less than two months after having become an adult. The division of labor is not rigidly scheduled, but varies with environmental and genetic factors that affect the levels of juvenile hormone.

Foraging A worker is well equipped anatomically for the task of foraging for nectar and pollen. She uses the proboscis to suck up nectar, which consists mainly of water and sucrose. The nectar is then stored in the crop, or "honey stomach," during the flight back to the nest. A valve, called the "honey stopper," prevents the nectar from entering the midgut unless the bee needs the sugars for energy. Upon returning to the nest the forager looks for an empty cell. If none is available she keeps the nectar in her crop, which causes the wax glands to redevelop, enabling her to build new cells. If she finds an empty cell she regurgitates the nectar and begins the process of making honey out of it. Along with the nectar she secretes some sucrase enzyme that digests the sucrose into the simple sugars glucose and fructose. She also secretes an enzyme that converts some of the glucose to an acid, which helps prevent bacterial contamination. Workers then fan their wings, thereby speeding up evaporation of water from the nectar, which reduces its volume and helps preserve it from microbial contamination.

Pollen accumulates on the pile, and the worker removes it with **pollen brushes** on the fore and middle legs (Fig. 32.21). She then moistens the pollen with nectar and removes it from the brushes with the **pollen rake** on the opposite hind legs. Finally she uses the **pollen packer** on the hind leg to pack the mixture of pollen and nectar into the pollen basket, which is a depression bordered by hairs. (A full pollen basket can be seen in Fig. 32.14C.) When the pollen baskets are full the worker flies back to the nest. Before then, however, some pollen gets transferred to other flowers, thereby effecting pollination. Pollination by honey bees is necessary for the survival of many plant species and ultimately, therefore, for the survival of honey bees as well.

Sources of nectar and pollen are usually located by workers sent out as **scouts.** Once a scout locates a new source of food she returns to the nest and communicates the location to other workers (pp. 409–410). If the flowers are nearby, the scout performs a round dance while workers crowd around. The scent on the scout's body tells the **recruits** what flowers to search for. If the flowers are far away the scout performs the remarkable waggle dance mentioned previously.

FIGURE 32.21

A worker honey bee, showing the structures used in foraging. Nectar is sucked into the crop by the proboscis. Pollen is scraped off the body with the pollen brushes, mixed with nectar, transferred to the pollen rakes, then packed into the pollen baskets by the pollen packers.

The Advantages of Altruism Although many questions remain, this brief description indicates that zoologists understand quite a lot about *how* worker honey bees perform their tasks. They have even been able to suggest a satisfactory answer to the most fundamental question of all: *Why* do the workers perform those tasks? According to the theory of natural selection, a behavior should evolve only if it increases the fitness of the individual performing the behavior. That is, only those behaviors should evolve that increase the transmission of genes for that behavior into future generations. How, then, could worker altruism have evolved, since workers do not normally reproduce? Why haven't there evolved lazy workers that refuse to work, or selfish workers that eat the nectar themselves rather than donate it to the colony? Because workers retain their ovaries and can, in fact, lay eggs if neither queen substance nor larvae are present in the nest, why don't they reproduce their own young?

So far the most satisfactory answer to these questions has come from W. D. Hamilton and other sociobiologists (pp. 414–415). Hamilton, while still a graduate student in 1964, pointed out that workers carry many of the same genes as their mother and sister queens, so by helping the queens reproduce, the workers are also helping reproduce their own genes. Of course, honey bees know nothing of genetics. The "reason" the workers work is that they have inherited the appropriate genes. If they inherited a gene for laziness or selfishness, the queen would presumably be less successful in reproducing that gene. Such a gene would therefore disappear from the species. Likewise, if a worker inherited a gene that made her try to reproduce on her own, she would have less time to devote to maintaining the colony. Both the worker and the queen would suffer, and they would be less able to reproduce that gene.

one of many common diseases, seems much more serious now that antibiotics and other wonder drugs control many of those other diseases.

Another reason for disappointment in DDT and other poisons is that insects evolve **resistance** to them. In any population of insects there are a few with genes whose products break down the chemicals or block their entry into their systems. Even if all the other insects of the species are wiped out, these few can reproduce rapidly enough to restore the population, which will inherit the resistance. Since 1970 the number of such resistant insect pests has more than doubled to approximately 500. Because of resistance, poisons have to be used in ever-increasing concentrations. On Long Island, New York potato farmers must now spray up to 10 times per season at a cost of $300 per acre, and the heavy dose of poisons has polluted many water wells.

According to one estimate, pesticides (including herbicides, insecticides, and fungicides) poison a million people per year worldwide and kill 20,000 (Pimental et al. 1992). When the insects get so resistant that safe and affordable levels of poisons are ineffective, new poisons must be developed. To find an effective new poison an average of 10,000 chemicals must be screened at a cost that now exceeds $45 million. This cost, of course, is passed on from the manufacturer to the farmer, and ultimately to the person who buys groceries.

In many cases chemical agents have proved not only ineffective but counterproductive. Insect pests actually increase following some applications of poisons. In one experiment, cabbage plants sprayed with DDT had twice as many cabbage worms (caterpillars of the moth *Pieris rapae*) as did unsprayed cabbages. The reason was that the DDT wiped out beetles that had preyed on the caterpillars. Furthermore, improper spraying often directly damages crops or kill honey bees that are needed to pollinate them.

Many poisons intended for insect pests harm not only beneficial insects but also a wide range of other animals, as Rachel Carson pointed out in 1962. Ten years after her book *Silent Spring* was published, the Environmental Protection Agency finally banned the use of DDT in the United States. Since then many people have been lulled into believing that all government-approved poisons are safe. In fact, however, only a few of the poisons in use before 1972 have even been tested to determine whether they cause cancer, birth defects, or mutations in mammals, and many poisons developed since then are carcinogenic, teratogenic, or mutagenic in experimental animals. In spite of regulations and inspections, approximately 35% of the food consumed in the United States is contaminated by pesticides. Not only humans, but also farm animals, pets, fishes, and wild birds, may be affected. The term "poison" is therefore more accurate than "insecticide" and "pesticide," since a chemical cannot tell the difference between an insect, a human, or another animal, or the difference between a beneficial insect and one that we consider a pest.

INTEGRATED PEST MANAGEMENT. Because of the ineffectiveness, expense, and dangers of poisons for insect control, entomologists have moved toward a more sophisticated approach called Integrated Pest Management (IPM). IPM employs a growing arsenal of biological weapons against selected insects. The goal is not total eradication, but suppression of insects to the point where their damage costs less than further control would. Rather than saturating an area with chemicals as soon as insects appear, IPM stresses a thorough understanding of the target insect so that a safe, effective control can be applied at the right time. Obviously such thorough understanding requires a great deal of research by many entomologists. In just a few decades since IPM has become widespread, however, such research has contributed several weapons to the arsenal against insect pests:

1. **Biological control: the use of predators, parasitoids, and pathogens.** Among the most promising weapons are parasitoid wasps that lay their eggs in the insect host (Fig. 32.22). Also effective in many cases is the bacterium *Bacillus thuringiensis* (BT), which produces a protein crystal that fatally destroys the midguts of insects that ingest the bacteria. BT has been used successfully against tent caterpillars and aquatic larvae of mosquitoes and black flies. Through genetic engineering the gene for the protein crystal has also been introduced into soil bacteria for potential use against insects that attack plant roots. Some insects have evolved resistance to BT.
2. **Saturating the area with male insects sterilized by radiation.** This method works for species in which the female mates only once, and is therefore unable

FIGURE 32.22
A parasitoid wasp *Pediobius foveolatus* oviposits on the larva of the Mexican bean beetle *Epilachna varivestis* in spite of the larva's defensive spines.

to reproduce after mating with a sterile male. The most successful use has been against several species of blow flies whose larvae, called screw worms, feed on wounds in cattle.

3. **Modifying farm methods.** Changing the spacing between rows, alternating rows of different crops (intercropping), crop rotation, and changing the time of planting are among the modifications that make it harder for insect populations to reach pest levels.
4. **Use of hormones and other substances to disrupt metabolism or behavior.** Synthetic juvenile hormone (JH) and chemicals that disrupt JH functioning have been applied successfully in particular cases (pp. 145–147). Recently azadirachtin, a drug from the neem tree that blocks the action of the molting hormone ecdysone, has generated considerable optimism. Around 250 synthetic sex-attractant pheromones are sold to disrupt reproduction in various species, but so far their main use has been as bait in insect traps.
5. **Limited use of poisons.** Poisons are still used as part of IPM, but at low enough levels to avoid resistance, and at times when the target insect is most vulnerable and beneficial insects are least endangered.

EXOTIC INSECTS. The most troublesome insects are those introduced into an area where plants have no defenses against them, and where the insects have no natural enemies. The classic example of such an exotic insect is the **gypsy moth** *Lymantria dispar*. In 1869 caterpillars of this moth crawled from a house in Medford, Massachusetts, where they had been brought by a French entomologist hoping to cross them with oriental silk worms to get a hybrid silk producer that would not have to feed on mulberry leaves. Twenty years later the citizens of Medford were complaining of caterpillars denuding the trees, falling down their collars, and making the sidewalks slippery with their crushed bodies. Efforts to control the outbreak were just about to succeed when state legislators decided there was no further need to appropriate money. As a result the gypsy moth is now entrenched over most of the Northeast. Its caterpillars denude one to two million acres in an average year and ten times that area in a bad year. Massive sprayings of DDT and other poisons have proved powerless to stop them. Nor have any of 47 imported enemies been able to keep up with the gypsy moths' prodigious ability to reproduce and the ability of the caterpillars to spread by riding the wind on strands of silk. Other tools of integrated pest management, such as synthetic sex-attractant (Disparlure) and BT, are economically feasible only in limited areas. Foresters in the Northwest are now concerned about an invasion of an Asian strain of the gypsy moth that was discovered in May 1991.

One might think gypsy moths would have taught a valuable lesson, yet exotic insect pests continue to enter the country. Since its arrival in Birmingham, Alabama in the late 1930s, the Brazilian red fire ant, *Solenopsis invicta,* has become a major pest in the southeastern United States, eliminating many native insects and inflicting painful bites to humans and animals. In 1956 a geneticist brought **Africanized honey bees,** *Apis mellifera scutellata,* into Brazil in hopes that it would be a better honey producer than the European bee. A year later, a visitor allowed 26 colonies to escape, and the remaining queens were then given to bee keepers. Eventually it was discovered that these Africanized bees, popularly known as "killer bees," sting with little provocation. Because they were better adapted to the tropics than European races, they spread throughout South and Central America in only three decades, and they have killed approximately 1000 people. In 1990 they arrived in the United States, where they have killed one person. Many bee keepers have adapted to the Africanized bees, but in the United States increased liability-insurance premiums could eliminate bee keeping as a hobby and threaten this $140 million segment of the economy.

Potentially more dangerous than killer bees are Asian tiger mosquitoes, *Aedes albopictus,* which have recently entered the United States in water that collected in old tires. This mosquito feeds on a wide range of mammals and has the potential to spread numerous diseases, including encephalitis and dengue fever. This exotic is now established throughout the Southeast and Midwest.

FIGURE 32.23
The elephant dung beetle *Heliocopris dominus* of Thailand. Females oviposit in balls of dung, then both females and males roll the balls to suitable places where they bury them. The developing larvae then feed on the dung. In the process, dung beetles play an important role in eliminating feces and in recycling nutrients. The ancient Egyptians revered another dung beetle, the scarab *Scarabaeus sacer,* which they visualized as wheeling the Sun about the heavens.

In some carefully managed situations, exotic insects have proved beneficial, especially in controlling other exotic species. Thousands of acres of rangeland in the American West were cleared of klamath weed by two species of *Chrysolina* beetles, and Australian rangeland was rescued from the prickly pear cactus by the moth *Cactoblastis cactorum.* Also in Australia, dung beetles of several species have been introduced to deal with the cow dung that would otherwise choke the grasses and provide breeding ground for parasites. (See Fig. 32.23.) Scientists are currently experimenting with certain parasitic flies that may be able to control the fire ant (Grisham 1994).

Although humans have never succeeded in deliberately extinguishing a species of insect, they have unintentionally brought about the extinction of at least 33 North American species. In the tropics the rate of extinction due to human activities is incalculably large. We will never know which of these extinct species would have controlled Africanized bees or disease-bearing mosquitoes, or which species would have brought as much pleasure as honey bees and silk worms, or which would have contributed as much to science as *Drosophila,* or which would have brightened the darkness and excited the imagination like fireflies.

Summary

Insects are divided into the wingless apterygotes, such as silverfish, and the pterygotes. The pterygotes are further divided into the exopterygotes, in which the juveniles (nymphs) resemble the adults and have wing buds, and the endopterygotes, in which there is complete metamorphosis from larva to pupa to adult.

Approximately three out of every four named animals is an insect. Their enormous diversity may be due to adaptations that enable them to survive in a wide variety of terrestrial habitats. The impermeable integument, as well as Malpighian tubules and other organs for osmoregulation and excretion, enable them to conserve body water. The tracheal system enables their tissues to obtain oxygen directly from air. Many have special adaptations that enable them to survive cold. These include antifreeze compounds, partial endothermy, torpor, and diapause. Their sensory organs are adapted mainly for use in air, enabling them to hear and detect airborne chemicals. The mouthparts, though consisting of essentially the same structures in all species, are highly specialized for different modes of feeding, such as chewing, sucking, and sponging. Insects also have several modes of locomotion, including running, jumping, and especially flying. Finally, insects are enormously successful at reproduction, largely because of the cleidoic eggs that exchange respiratory gases but conserve water. In eusocial species of termites, ants, and bees, colonies may function as superorganisms, with only a few reproducing individuals served by one or more castes of workers.

The success of insects makes some of them pests for our own species. Some insects are vectors of disease, and many feed on forests and crops. Until a few decades ago, the main approach to controlling insects was ever-increasing applications of poisons such as DDT. Increasingly, insects are being controlled by Integrated Pest Management, which employs moderate doses of poisons together with biological controls and manipulation of crops to reduce infestation.

Key Terms

hemimetabolous
nymph
holometabolous
pupa
Malpighian tubule
trachea
spiracle
tracheole
torpor
diapause
partial endothermy
pheromone
mouthparts
mandible
hypopharynx
labrum
labium
maxilla
foregut
midgut
hindgut
alternating tripod gait
synchronous muscle
asynchronous muscle
cleidoic egg
vitellogenesis
fat body
oviposition
spermatheca
ootheca
gall
ovoviviparous
eusocial
haplodiploidy
queen mandibular pheromone
royal jelly
altruism
integrated pest management

Chapter Test

1. Describe the differences between insects and other arthropods. (p. 678)
2. Describe the major difference between Apterygota and Pterygota, and name one example of each. (pp. 678–680)
3. Describe the differences between endopterygote and exopterygote insects. Which ones are hemimetabolous, and which are holometabolous? Which ones have nymphal stages, and which ones have pupae? Name an example of an endopterygote and an exopterygote. (pp. 679–680)
4. Describe two adaptations to terrestrial life in insects, and explain how they may have contributed to the success of insects. (pp. 681–694)
5. Describe how Malpighian tubules and tracheae are adapted for terrestrial living. (pp. 681–685)
6. Explain how the mouthparts of insects are adapted for chewing, sucking, or sponging up food. For each of these three modes of feeding, name a kind of insect in which it occurs. (pp. 687–688)
7. Briefly describe two competing theories about how flight evolved in insects. (pp. 690–691)
8. Explain the functioning of an insect egg. How is it adapted for terrestrial life? What is the role of vitellogenesis in its formation? How does the shell allow fertilization? (pp. 692–693)
9. A colony of honey bees has sometimes been compared to a "superorganism," with queens, drones, and workers performing tasks analogous to those of different organs. Elaborate on this concept. What "organs" do the queens, drones, and various workers represent in the "superorganism?" What are some differences between a honey bee colony and a superorganism? (pp. 697–699)
10. Describe an example of each of the following: parasitoid, pheromone, haplodiploidy, castes, metamorphosis, courtship. (pp. 679–700)
11. Explain why simply poisoning insect pests with chemicals has proved unsatisfactory. Describe two alternatives used in integrated pest management. (pp. 696–701)

■ Answers to the Figure Questions

32.2 Hopping is faster and therefore more effective in evading predators. In fleas and grasshoppers it is probably also easier than scrambling through hair or grass. In such tiny animals as springtails, hopping might also be less hindered by adhesion to water on the surface of ponds and snow.

32.4 Perhaps metamorphosis evolved as an adaptation permitting the juvenile stages to feed and grow rapidly while permitting the adult stages to reproduce and disperse effectively. Insects that do not undergo complete metamorphosis, on the other hand, have essentially the same body shape throughout life, and this is probably a compromise among the needs for feeding and growth in the juveniles versus reproduction and dispersal in the adults.

32.7 It is always difficult to try to reconstruct evolutionary history, but one of the guiding concepts in speculating on evolutionary history is that natural selection works with the present rather than looking forward to the future. That is, while humans and many other animals might have been better off with tracheae, the fact that they survive and reproduce well enough with lungs or other respiratory organs means that there was little selection pressure favoring the evolution of alternatives. A second guiding concept is that natural selection is only capable of modifying structures that already exist, not of originating completely new organs. Thus in humans and other terrestrial vertebrates, lungs evolved as a means of breathing air because the precursors of lungs were already present in their ancestors (lungfish). In insect ancestors, on the other hand, there was no structure that could be modified into a lung, and the open circulatory system would not have been suitable for the transport of oxygen anyway. Tracheae was the appropriate solution. Unfortunately we do not know what structure served as the precursor.

32.10 The labial and maxillary palps look most like legs, and in fact they are modified appendages of segments that combine to form the head. The antennae are also homologous with legs (see Fig. 6.27).

32.17 The male's behavior is an adaptation that protects his reproductive investment by preventing a subsequent male from mating. In fact, many male dragonflies have genitals that can scoop out the sperm deposited by previous males, so this guarding behavior is highly advantageous.

32.18 A good guess would be that ichneumon wasps have chemoreceptors on the ovipositor that detect chemicals given off by the larvae of wood-boring insects.

32.19 This would be arthrodial membrane.

Readings

Recommended Readings

Batra, S. W. T. 1984. Solitary bees. *Sci. Am.* 250(2):120–127 (Feb).

Berenbaum, M. R. 1994. *Bugs in the System: Insects and Their Impact on Human Affairs.* Reading MA: Addison-Wesley.

Camazine, S., and R. A. Morse. 1988. The Africanized honeybee. *Am. Sci.* 76:464–471.

Dethier, V. G. 1984. *The World of the Tent-makers: A Natural History of the Eastern Tent Caterpillar.* Amherst: University of Massachusetts Press.

DeVries, P. J. 1992. Singing caterpillars, ants and symbiosis. *Sci. Am.* 267(4):76–82 (Oct).

Duplaix, N. 1988. Fleas: the lethal leapers. *Natl. Geogr.* 173(5):672–694 (May).

Eberhard, W. G. 1980. Horned beetles. *Sci. Am.* 242(3):166–182 (Mar).

Evans, H. E. 1985. *The Pleasures of Entomology: Portraits of Insects and the People Who Study Them.* Washington, DC: Smithsonian Institution Press.

Evans, H. E., and K. M. O'Neill. 1991. Beewolves. *Sci. Am.* 265(2):70–76 (Aug). (*Beewolves are wasps that prey on bees.*)

Fabre, J. H. (Edited by E. W. Teale). 1949. *The Insect World of J. Henri Fabre.* New York: Dodd, Mead. (*One of several collections of the enjoyable writings of the 19th-century observer of insect behavior.*)

Franks, N. R. 1989. Army ants: a collective intelligence. *Am. Sci.* 77:139–145.

Funk, D. H. 1989. The mating of tree crickets. *Sci. Am.* 261(2):50–59 (Aug).

Gilbert, L. E. 1982. The coevolution of a butterfly and a vine. *Sci. Am.* 247(2):110–122 (Aug).

Gordon, D. M. 1995. The development of organization in an ant colony. *Am. Sci.* 83:50–57.

Gould, J. L., and C. G. Gould. 1988. *The Honey Bee.* New York: Scientific American Library.

Handel, S. N., and A. J. Beattie. 1990. Seed dispersal by ants. *Sci. Am.* 263(2):76–83 (Aug).

Heinrich, B. 1981. The regulation of temperature in the honeybee swarm. *Sci. Am.* 244(6):146–160 (June).

Hölldobler, B., and E. O. Wilson. 1983. The evolution of communal nest-weaving in ants. *Am. Sci.* 71:490–499.

Huber, F., and J. Thorson. 1985. Cricket auditory communication. *Sci. Am.* 253(6):60–68 (Dec).

Klowden, M. J. 1995. Blood, sex, and the mosquito. *BioScience* 45:326–331 *(How mosquitos feed.)*

May, M. 1991. Aerial defense tactics of flying insects. *Am. Sci.* 79:316–328. (*Mainly on crickets' response to bat sonar.*)

Merritt, R. W., and J. B. Wallace. 1981. Filter-feeding insects. *Sci. Am.* 244(4):132–144 (Apr).

Morse, D. H. 1985. Milkweeds and their visitors. *Sci. Am.* 253(1):112–119 (July).

Nijhout, H. F. 1981. The color patterns of butterflies and moths. *Sci. Am.* 245(5):140–151 (Nov).

Pasteur, G. 1994. Jean Henri Fabre. *Sci. Am.* 271(1):74–80 (July). (*Life of the great 19th century observer and writer on insects.*)

Prestwich, G. D. 1983. The chemical defenses of termites. *Sci. Am.* 249(2):78–87 (Aug).

Rinderer, T. E., B. P. Oldroyd, and W. S. Sheppard. 1993. Africanized bees in the U. S. *Sci. Am.* 269(6):84–89 (Dec).

Rosenthal, G. A. 1983. A seed-eating beetle's adaptations to a poisonous seed. *Sci. Am.* 249(5):164–171 (Nov).

Rosenthal, G. A. 1985. The chemical defenses of higher plants. *Sci. Am.* 254(1):94–99 (Jan).

Ryker, L. C. 1984. Acoustic and chemical signals in the life cycle of a beetle. *Sci. Am.* 250(6):112–123 (June).

Seeley, T. D. 1989. The honey bee colony as a superorganism. *Am. Sci.* 77:546–553.

Southwick, E. E., and G. Heldmaier. 1987. Temperature control in honey bee colonies. *BioScience* 37:395–399.

Stokes, D. W. 1983. *A Guide to Observing Insect Lives.* Boston: Little, Brown.

Thornhill, R. 1980. Sexual selection in the black-tipped hangingfly. *Sci. Am.* 242(6):162–172 (June).

Topoff, H. 1990. Slave-making ants. *Am. Sci.* 78:520–528.

Tumlinson, J. H., W. J. Lewis, and E. M. Vet. 1993. How parasitic wasps find their hosts. *Sci. Am.* 266(3):100–106 (Mar).

Winston, M. L., and K. N. Slessor. 1992. The essence of royalty: honey bee queen pheromone. *Am. Sci.* 80:374–385.

Wooton, R. J. 1990. The mechanical design of insect wings. *Sci. Am.* 263(5):114–120. (Nov).

See also relevant selections in General References at the end of Chapter 21.

Additional References

Arnett, R. H., Jr. 1986. *American Insects: A Handbook of the Insects of America North of Mexico.* New York: Van Nostrand Reinhold.

Breed, M. D., C. D. Michener, and H. E. Evans (Eds.). 1982. *The Biology of Social Insects.* Boulder CO: Westview Press.

Dover, M. J., and B. A. Croft. 1986. Pesticide resistance and public policy. *BioScience* 36:78–85.

Eisner, T., et al. 1978. "Wolf-in-sheep's-clothing" strategy of a predaceous insect larva. *Science* 199:790–794.

Greany, P. D., S. B. Vinson, and W. J. Lewis. 1984. Insect parasitoids: finding new opportunities for biological control. *BioScience* 34:690–696.

Grisham, J. 1994. Attack of the fire ant. *BioScience* 587–590.

Harris, P. 1988. Environmental impact of weed-control insects. *BioScience* 38:542–548.

Heinrich, B. 1993. *The Hot-Blooded Insects.* Cambridge MA: Harvard University Press.

Hölldobler, B., and E. O. Wilson. 1990. *The Ants.* Cambridge MA: Belknap.

Hölldobler, B., and E. O. Wilson. 1994. *Journey to the Ant: A Story of Scientific Exploration.* Cambridge MA: Belknap.

Labandeira, C. C., and J. J. Sepkoski, Jr. 1993. Insect diversity in the fossil record. *Science* 261:310–315.

Lambert, B., and M. Peferoen. 1992. Insecticidal promise of *Bacillus thuringiensis. BioScience* 42:112–122.

Lewin, R. 1985. On the origin of insect wings. *Science* 230:428–429.

Marden, J. H., and M. G. Kramer. 1994. Surface-skimming stoneflies: A possible intermediate stage in insect flight evolution. *Science* 266:427–430.

Pimental, D., et al. 1991. Environmental and economic effects of reducing pesticide use. *BioScience* 41:402–408.

Pimental, D., et al. 1992. Environmental and economic costs of pesticide use. *BioScience* 42:750–760.

Roitberg, B. D., and R. J. Prokopy. 1987. Insects that mark host plants. *BioScience* 37:400–406.

Seeley, T. D. 1985. *Honeybee Ecology.* Princeton, NJ: Princeton University Press.

Tallamy, D. W. 1984. Insect parental care. *BioScience* 34:20–24.

Tangley, L. 1987. Regulating pesticides in food. *BioScience* 37:452–456.

CHAPTER

33

Echinoderms

Gulf star *Pentaceraster cumingi.*

Diversity

Echinoderms witnessed the dawn of the Paleozoic era more than half a billion years ago, and they contributed more than their share to the "Cambrian explosion" of animal diversity (pp. 387–388). More than 20 classes once flourished, but only six widely diverse classes are extant. The most familiar are starfish, also known as sea stars or asteroids, which make up the class Asteroidea (Fig. 33.1). Other common souvenirs from trips to the beach include sand dollars and sea urchins, which belong to the class Echinoidea. Three other classes—Ophiuroidea (brittle stars), Crinoidea (feather stars and sea lilies), and Holothuroidea (sea cucumbers)—are familiar to divers. Members of the sixth class, Concentricycloidea, are rarely seen, and they were not even discovered until 1986.

These classes differ greatly in such major features as the presence of arms (rays) and the position of the mouth and anus (Fig. 33.2 on page 708). These differences make classification difficult, and no classification based only on morphology has earned a consensus. The rich fossil record of echinoderms (except for Concentricycloidea), as well as molecular phylogeny, suggests that crinoids are the oldest extant class of echinoderms. Starfish and brittle stars are closely related to each other and arose at about the same time, followed by sea urchins and sea cucumbers (Smith et al. 1993, Wada and Satoh 1994). Relationships to other phyla are even more uncertain. Echinodermata are assumed to be distantly related to our own phylum Chordata, because both are **deuteros-**

FIGURE 33.1

Representative echinoderms. (A) Northern starfish *Asterias vulgaris* are common on rocky, northeastern shores of the United States. Diameter 20 cm. (B) Brittle stars. (C) The common edible sea urchin *Echinus esculentus.* Diameter 10 cm. (D) A group of feather stars on the Great Barrier Reef off the coast of Australia. Those at the upper center (white) and right (dark) are *Oxycomanthus bennetti.* The red one at the lower center and the yellow one to the left are *Himerometra robustipinna.* (E) The sea cucumber *Pseudocolochinus axiolagus.* Note the feeding tentacle being withdrawn from the mouth (upper left) and the warning coloration.

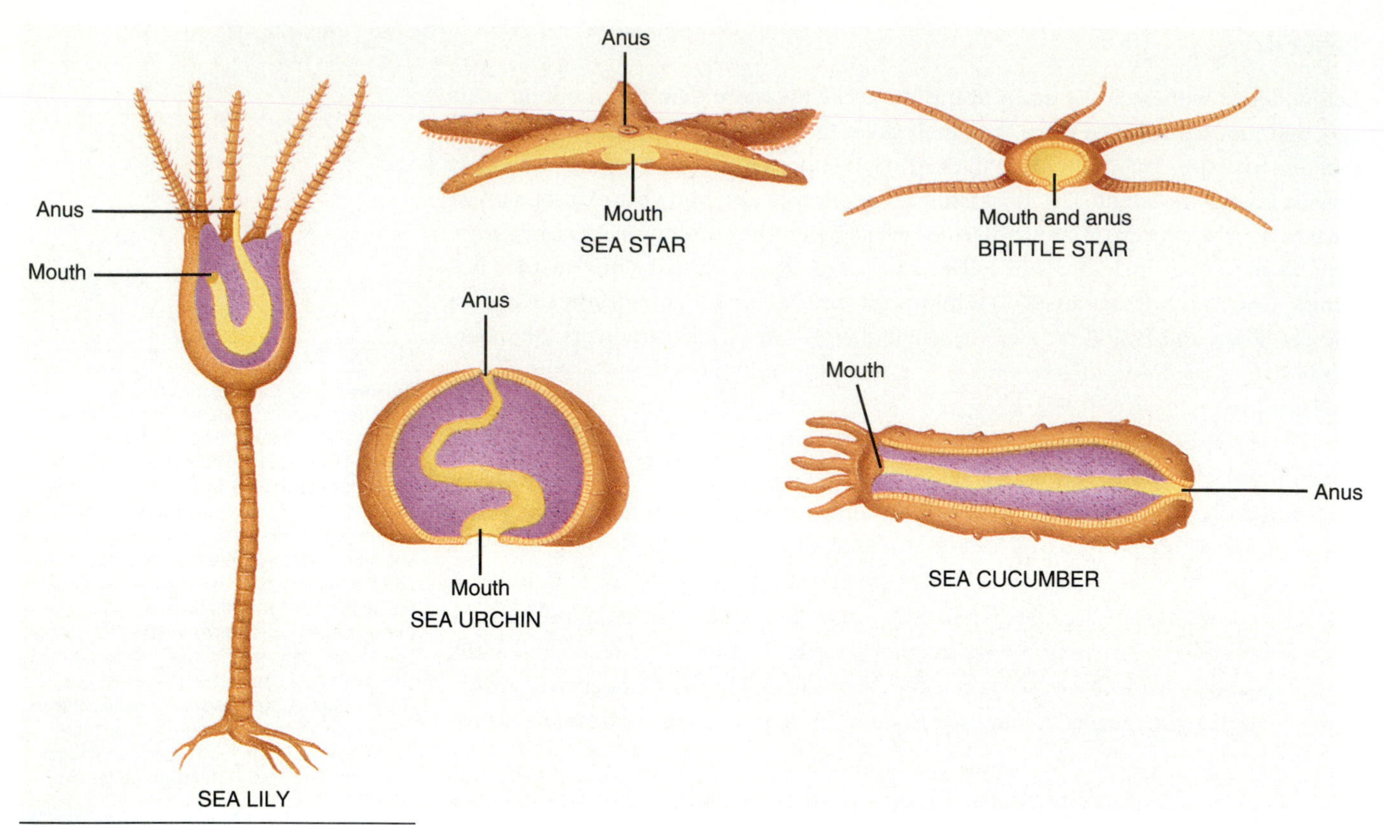

FIGURE 33.2

Basic organization of the major kinds of echinoderms. The surface bearing the mouth is called the oral surface. The opposite side is the aboral surface. Sea cucumbers lie on their sides with the mouth at one end, and they have evolved bilateral symmetry.

■ **Question:** Based on the characters visible in this and the preceding figure, construct a cladogram (pp. 435–437) showing the relationships among these five groups.

tomes (see Fig. 21.10). Deuterostomia are defined as coelomates in which the mouth develops from a second or later opening in the embryo rather than from the blastopore (p. 440). (The term "deuterostome" comes from the Greek words for "second" and "mouth.") Two other deuterostomate phyla are Chaetognatha (arrowworms), which are described on pp. 575–576, and Hemichordata, which will be described in the next chapter. In addition to the origin of the mouth, all deuterostomes are often assumed to share two other developmental similarities: cleavage of the earliest embryonic cells in a **radial** rather than a spiral pattern, and development of the coelom from pouches in the gut (**enterocoely**). (See Chapter 21.) In fact, however, the combination of radial cleavage and enterocoely is characteristic only of echinoderms.

Sea Stars as Representative Echinoderms

In spite of the great differences in appearance of the classes of echinoderms, all have an organization that is fundamentally like that of starfish (Table 33.1 on page 710). Starfish will therefore be used as the major example in describing the features of echinoderms.

Radial Symmetry. The most evident feature of starfish and most other echinoderms is that they tend to be radially symmetric. It is clear, however, that this is a derived rather than a primitive feature. There is a continuous fossil record of echinoderms indicating that their ancestors were bilaterally symmetric, and the larvae of echinoderms are still bilaterally symmetric. Apparently the radial symmetry of adult echinoderms is a secondary adaptation to their sedentary lives on the ocean floor. In most living species of echinoderms the radial symmetry is **pentamerous,** as is evident in the five arms (or rays) of most sea stars. In other words, the bodies of most adult echinoderms can be divided into five (Greek *penta-*) similar parts (*meros*).

Extant Classes of Phylum Echinodermata

Genera mentioned elsewhere in this chapter are noted.

Class Asteroidea AS-tur-OY-dee-uh (Greek *aster* star). Usually star-shaped, with five (or more) arms thickly joined to the central disc. Open ambulacral grooves and tube feet on oral (bottom) surface. Podia, usually with suckers, used in locomotion. Madreporite and anus (if present) on aboral surface. Starfish. *Acanthaster, Asterias, Linckia, Luidia, Pentaceraster* (Chapter opener; Figs. 33.1A, 33.11A, 33.18).

Class Ophiuroidea OFF-ee-your-OY-dee-uh (Greek *ophis* snake + *oura* tail). Five (or more) flexible arms clearly distinct from central disc. Closed ambulacra. Tube feet without suckers; not used in locomotion. Stomach, but no gut or anus. Madreporite on oral (lower) surface. Brittle stars. (Fig. 33.1B).

Class Crinoidea krin-OY-dee-uh (Greek *krinon* a lily). Flower-shaped, with cup-shaped body and five branched arms. Body of adult on long, attached stalk (sea lilies) or without stalk (feather stars). Feeding is by means of ciliated ambulacral grooves and podia. Anus on oral (upper) surface. No spines or madreporite. *Antedon, Himerometra, Oxycomanthus* (Figs. 33.1D, 33.8, 33.9).

Class Echinoidea EK-in-OY-dee-uh (Greek *echinos* a sea urchin or hedgehog). No arms; body hemispherical (sea urchins and heart urchins) or disc-shaped (sand dollars). Heart urchins and sand dollars tend toward bilateral symmetry (irregular). Ossicles form a rigid test. Mouth on bottom surface, usually with special chewing apparatus (Aristotle's lantern). Locomotion by movable spines and (in sea urchins) tube feet with suckers. Closed ambulacral grooves. *Diadema, Echinus, Mellita, Psammechinus, Strongylocentrotus* (Figs. 33.1C, 33.10).

Class Holothuroidea HA-low-thur-OY-dee-uh (Greek *holothourion* a term Aristotle originally applied to a sea polyp). Body cucumber-shaped. Central axis elongated, with mouth and anus at opposite ends. Bilateral symmetry. No arms or spines. Podia at oral (anterior) end elongated into tentacles. Remaining tube feet (if any) have suckers used for locomotion. Sea cucumbers. *Cucumaria, Pseudocolochinus, Thyone* (Figs. 33.1E, 33.13, 33.14).

Class Concentricycloidea con-SEN-tree-sy-KLOY-dee-uh (Latin *concentricus* concentric + *cyclus* ring). Water vascular system with two ring canals concentric on ventral surface. Disc-shaped body without arms, mouth, or anus. Ventral surface covered by velum. *Xyloplax* (Fig. 33.15).

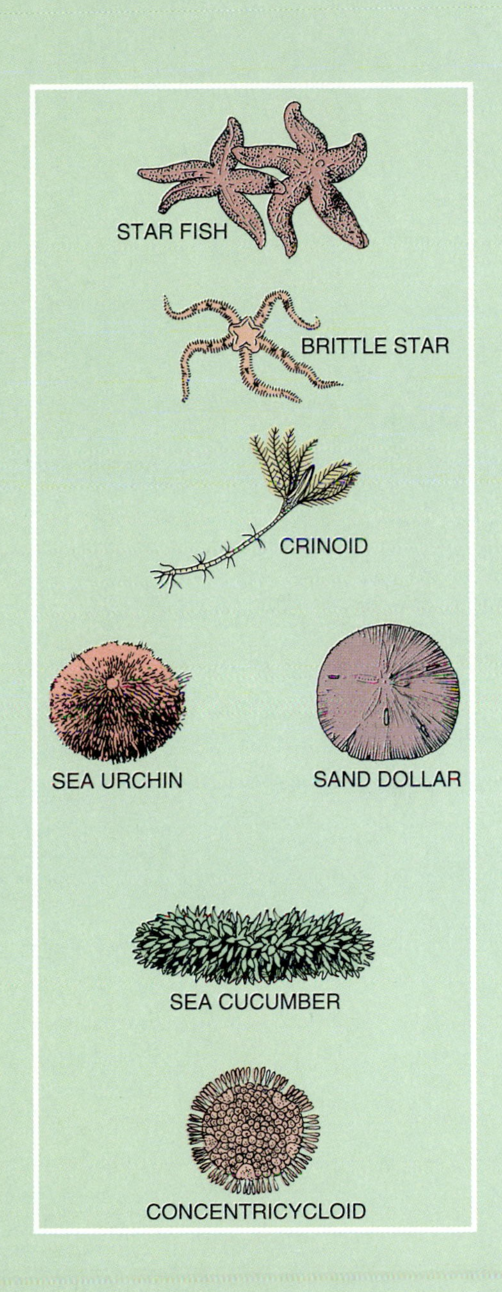

TABLE 33.1

Characteristics of echinoderms.

Phylum Echinodermata e-KINE-oh-DER-mah-tuh (Greek *echinos* a sea urchin or hedgehog)

Morphology	Adults with derived, **pentamerous radial symmetry**, tending toward bilateral symmetry in some forms. Starfish, brittle stars, sea lilies, and feather stars have five (or more) arms. No head. With endoskeleton of calcareous ossicles. Integument of many forms bears spines and/or beaklike **pedicellariae**.
Physiology	Coelomic fluid resembling seawater circulated by ciliated peritoneum. Nervous system consists mainly of a plexus beneath the epithelium, together with a circumoral ring and radial nerves. No brain or other ganglia; few specialized receptor organs. No osmoregulatory organ. Exchange of respiratory gases and elimination of metabolic wastes generally through integument, especially **papulae**. Digestion by means of a simple stomach and intestine; with or without an anus.
Locomotion	Many forms move by means of **tube feet**, which may have suckers on the ends. Tube feet extend and retract by **water vascular system** and muscles. Crinoids sessile; some crawl or swim by waving arms; some burrow by peristaltic contraction or movable spines.
Reproduction	Most forms dioecious, but with little external difference between sexes. Fertilization external. In at least one species of starfish, *Luidia* sp., the larvae can also reproduce asexually by fission.
Development	Radial, indeterminate cleavage. Deuterostomate. Enterocoelous. Development usually indirect, with a variety of **bilaterally symmetric larvae**. Lost body parts readily regenerated in adults.
Habitat, Size, and Diversity	Exclusively marine. Many forms primarily intertidal or subtidal; others range in depth down to ocean trenches. None parasitic. Largest dimension of adults ranges from a few millimeters to 2 meters. Approximately **6100 living species** described.

TUBE FEET. Another distinction of echinoderms is the tube foot, or **podium** (Fig. 33.3). Podia, which are not found in any other phylum, cling to substratum and food. Starfish that live on rocky shores or coral reefs would be unable to move and would be tossed around by waves if they did not have tube feet. In these starfish the end of each tube foot is shaped like a suction cup and secretes mucus that forms a tight contact with solid substratum. Each tube foot of the common starfish *Asterias* can lift 29 grams, so hundreds of them can exert an impressive force. Sea stars can cling to rocky shores in spite of the buffeting of heavy waves, and they can pull open the shells of clams and other prey. Some tube feet in starfish also serve as feelers, and in some other echinoderms they function as tentacles.

A starfish uses its tube feet during locomotion by slightly lifting the tip of whichever arm points in the direction it is going, then extending the tube feet at the tip of that arm until they make contact. Muscles in those tube feet then contract to apply suction and to shorten the tube feet, thereby pulling the asteroid a short distance.

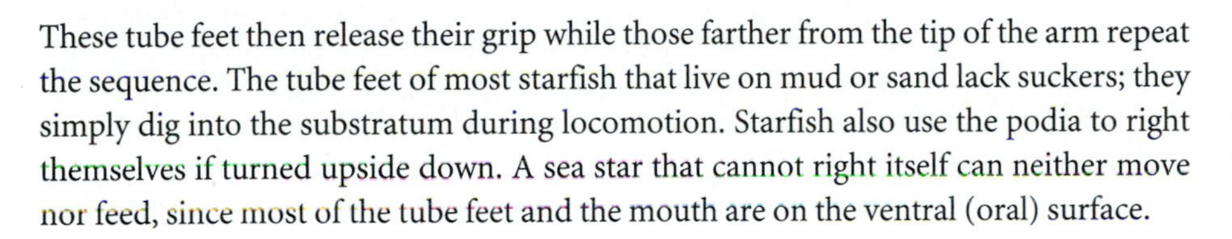

A

These tube feet then release their grip while those farther from the tip of the arm repeat the sequence. The tube feet of most starfish that live on mud or sand lack suckers; they simply dig into the substratum during locomotion. Starfish also use the podia to right themselves if turned upside down. A sea star that cannot right itself can neither move nor feed, since most of the tube feet and the mouth are on the ventral (oral) surface.

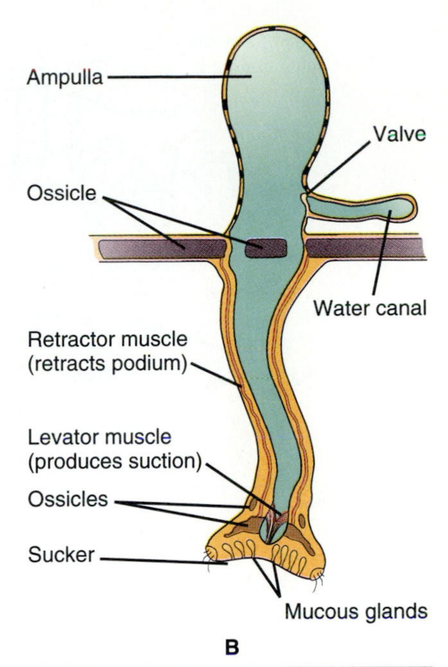

B

FIGURE 33.3

(A) Side view of a starfish arm showing tube feet. (B) The hollow tube foot is filled with seawater, which enters from the water canal through the one-way valve. The tube foot is extended by contraction of muscles surrounding the ampulla. Selective contraction of retractor muscles allows the tube foot to be extended in any direction. Contraction of levator muscles causes the tube foot to apply suction. After the suckerlike endings attach to solid substratum or prey, contraction of all the retractor muscles retracts the tube foot.

AMBULACRAL GROOVES. The tube feet of starfish are concentrated along ambulacral grooves, or **ambulacra,** that radiate from the mouth (Fig. 33.3A). Usually there is one ambulacral groove on the oral surface of each arm, and the podia are arranged in rows on both sides of each groove. In sea stars the ambulacral grooves are open, exposing the radial nerves, which help coordinate movements of the arms and tube feet. Each radial nerve also conducts some visual and tactile information from the ocellus and tentacle at the end of the arm, as well as information from chemoreceptors all over the integument (Fig. 33.6). This information is apparently used mainly in local reflexes responsible for defense, locomotion, righting, and feeding. There is no brain or other ganglion for overall coordination of neural activity.

THE WATER VASCULAR SYSTEM. Extension of the tube feet is due to another system unique to echinoderms: the water vascular system (Fig. 33.4 on page 712). As its name suggests, the water vascular system compensates for the absence of a blood circulatory system. It comprises a network of canals filled with seawater containing certain cells (**coelomocytes**), proteins, and concentrated potassium ions. The water enters through a sievelike **madreporite** on the aboral (upper) surface and passes through the **stone canal.** Cilia on the lining (peritoneum) of the canals draw the water into a **ring canal** that encircles the mouth. Five (or more) **radial canals** branch from the ring canal into the arms, where they bring water into the tube feet through valves. The water is stored in bulblike **ampullae,** which extend the tube feet when they contract.

The water vascular system is one of four major parts of the coelom in adult asteroids. The largest part, the **perivisceral coelom,** surrounds the visceral organs and branches into the arms. It is primarily responsible for transporting nutrients and perhaps respiratory gasses.

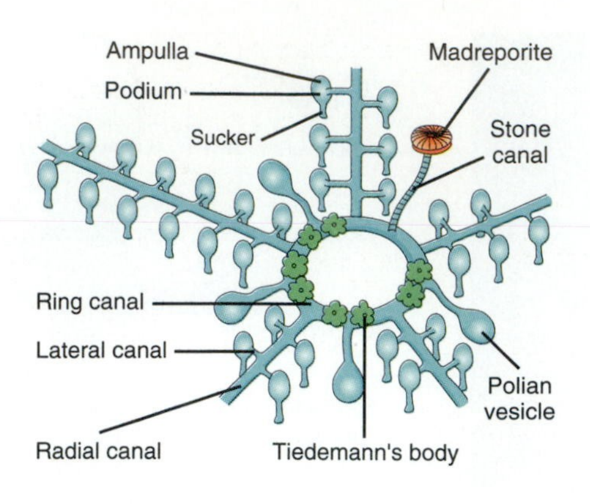

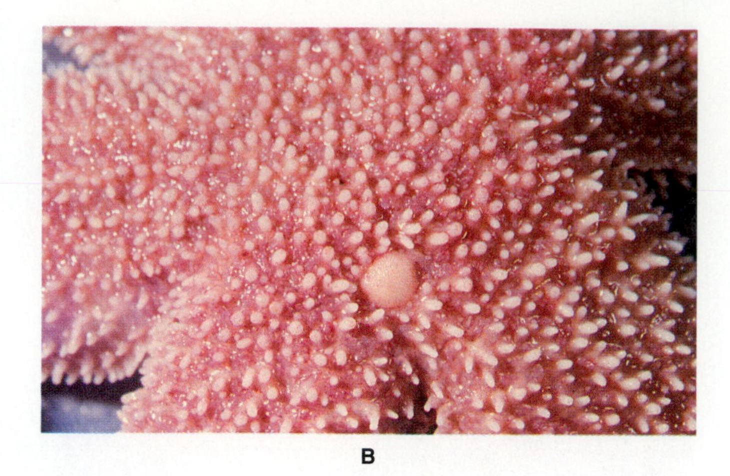

FIGURE 33.4
(A) The water vascular system of a starfish. Sea water enters the madreporite on the aboral surface and passes through the stone canal into the ring canal and radial canals. The water then enters the ampullae and tube feet, as described in Fig. 33.3. Polian vesicles and ampullae serve as reservoirs. Tiedemann's body is believed to produce cells (coelomocytes) in the coelom. (B) The madreporite of the sea star *Asterias.*

■ **Question:** What do you suppose would happen if the madreporite were blocked and the sea star sank or rose to the surface?

THE INTEGUMENT. The integument contains bony plates called **ossicles,** each of which is a single crystal of calcium carbonate with some magnesium carbonate. Because these ossicles are enclosed in tissue, they constitute an **endoskeleton.** The ossicles are secreted by the dermis, which lies beneath the thin epithelium that covers the animal. In sea stars, and also in brittle stars and sea urchins, the ossicles have spiny projections on the aboral surface that may deter predators. Starfish also have a unique **catch connective tissue** that stiffens the integument and helps defend against predation. The aboral surfaces of starfish and many other echinoderms also bear numerous beaklike structures called **pedicellariae** (Figs. 33.5 and 33.11). They apparently keep the integument free of sponges, corals, and other encrusting organisms, and they are used in defense and feeding. Pedicellariae are so remarkable that when first discovered in the late 18th century they were thought to be parasites. Also important in defense may be the striking pigmentation of many echinoderms. Because echinoderms have only rudimentary eyes at best, these pigments may serve as camouflage or warning coloration rather than as a means of identifying members of the same species.

The aboral surface is also responsible for the elimination of wastes and the exchange of respiratory gases between the coelomic fluid and seawater. These functions do not require elaborate mechanisms, because of the low metabolic rates of these slug-

FIGURE 33.5
(A) Detail of the aboral surface of a starfish, showing pedicellariae, spines, and papulae. (B) Structure of a pedicellaria. The two jaw ossicles (often called valves) pivot on the basal piece.

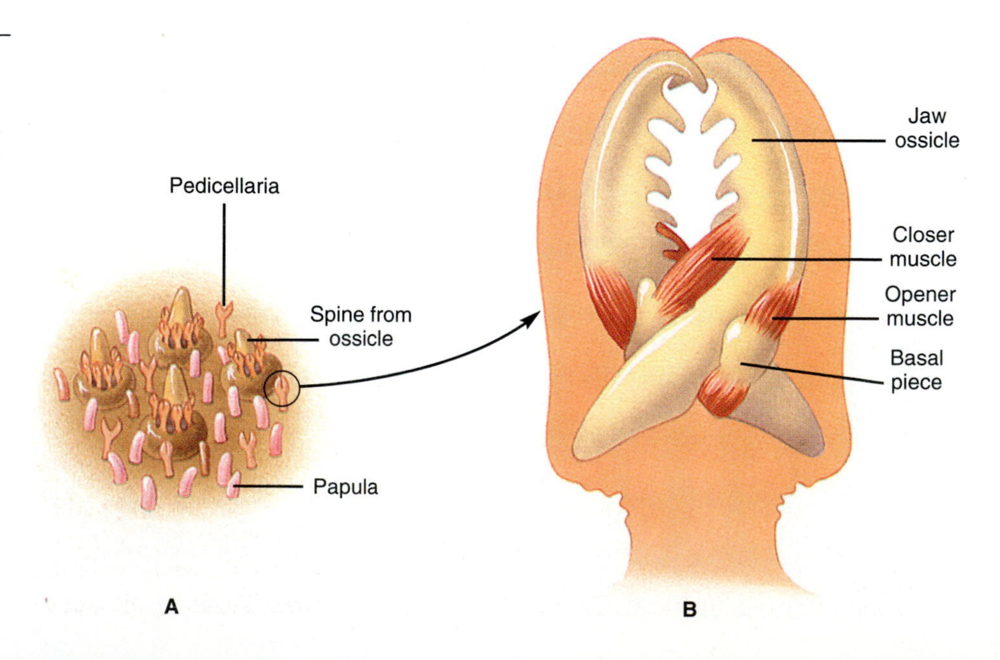

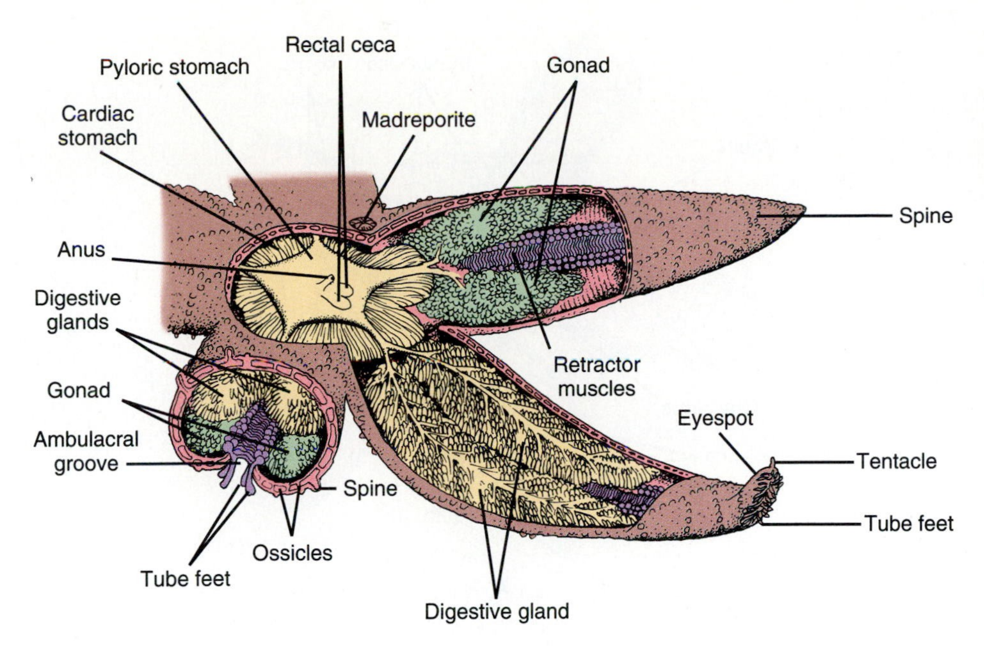

FIGURE 33.6

Internal structure of a starfish viewed from the aboral surface. Note the two chambers of the stomach. The function of the rectal ceca is unknown. The gonads release either sperm or ova externally through gonopores between the arms.

■ **Question:** Where are the respiratory and excretory organs?

gish animals. Tiny projections called **papulae** provide enough surface area for the diffusion of ammonia, O_2, and CO_2. Diffusion across the papulae also maintains ionic concentrations in the coelomic fluid similar to those in seawater. Echinoderms lack organs for osmoregulation and can therefore live only in marine habitats.

DIGESTION. Starfish have fairly simple digestive systems (Fig. 33.6). The stomach is the main organ of digestion and fills most of the central disc. It has two chambers—the **cardiac stomach** and the **pyloric stomach.** Digestive enzymes enter the stomach from a pair of **pyloric ceca** (= digestive glands) in each arm. The pyloric ceca are also the major sites for storage of food and absorption of nutrients. The intestine and anus are either absent or too small to be of much use, so the digestive tract is essentially incomplete. Most sea stars feed on externally digested material, so they generate little solid waste. Those that feed on small particles use the pedicellariae to pass the food into the mouth and directly to the stomach.

Many starfish prey on surprisingly large and well-protected animals, such as molluscs, crustaceans, and coral polyps. While gripping the prey with their podia, these sea stars evert their cardiac stomachs out through their mouths and partially digest the prey externally. To feed on crustaceans and coral polyps they simply cover the prey with their cardiac stomachs. *Asterias* and some other asteroids can insinuate the cardiac stomach through tiny gaps between the valves of oysters and other bivalve molluscs. A space of only 0.1 mm is all *Asterias* needs to digest an oyster inside its own shell. After partially digesting the prey the cardiac stomach retracts to bring in the nutrients.

Other Echinoderms

BRITTLE STARS. With their five or more arms, brittle stars (ophiuroids) resemble sea stars in general appearance. In contrast to the lumbering movements of sea stars, however, brittle stars can scuttle rapidly across substratum, and they sometimes swim. Unlike the arms of starfish, those of brittle stars are slender and more flexibly attached to the central disc, enabling them to make rapid, snakelike movements. A common name for brittle stars is, in fact, serpent stars. Also contributing to the agility of brittle stars are the closely articulating ossicles that surround the arms (including the ambu-

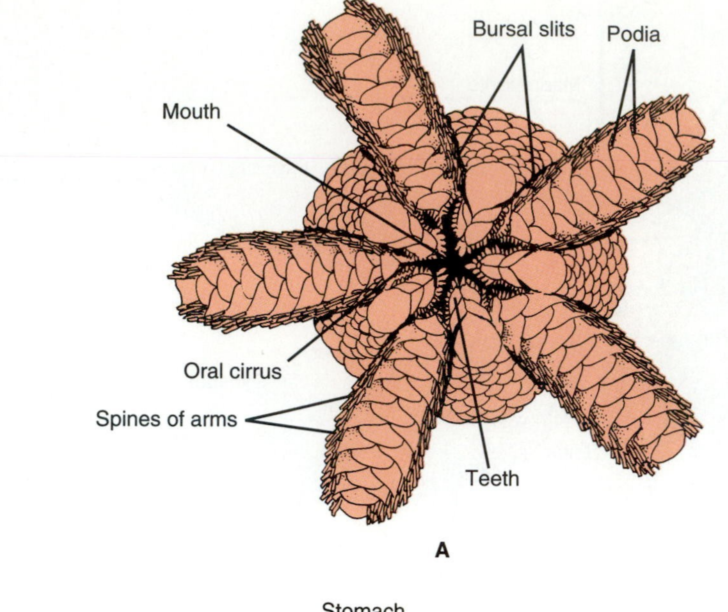

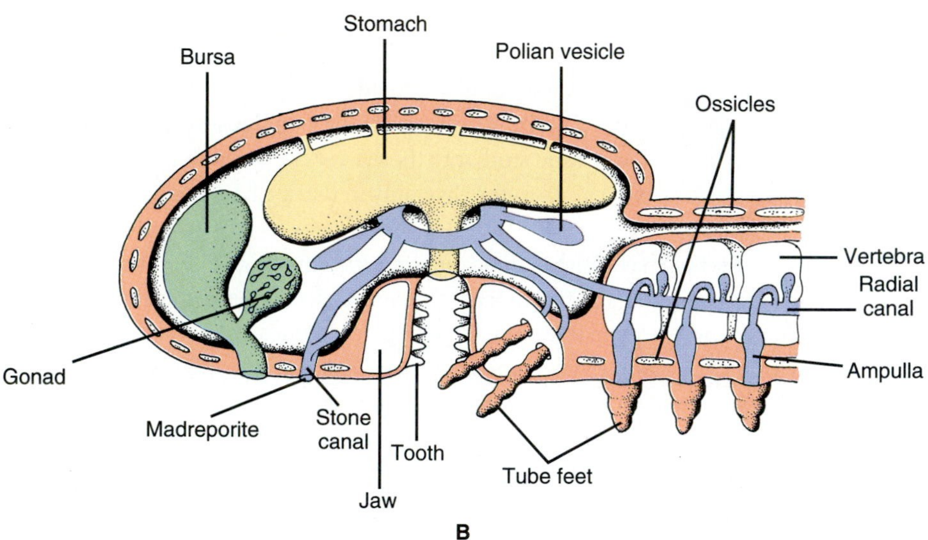

FIGURE 33.7

The structure of brittle stars. (A) Oral surface of the central disc. (See Fig. 33.1B.) (B) A section through the central disc and part of one arm.

lacra) (Fig. 33.7). These spiny ossicles enable the arms to twist and bend much like the spinal column of vertebrates; in fact the ossicles are often called vertebrae.

Brittle stars lack suckers on their tube feet, so they are generally found in deeper water, away from wave action. Most of the time they stay hidden in dark crevices, feeding on small particles that they capture with the tube feet and in mucus on the arms. Brittle stars have no pedicellariae. The mouth is armed with five-toothed jaws (Fig. 33.7). There is a stomach but no intestine or anus. Brittle stars also lack papulae. They exchange respiratory gases by pumping water through unique slits into five **bursae** (sacs). In addition to serving as respiratory organs, the bursae also serve as attachment sites for the gonads. Either sperm or ova break through the bursae and escape through the bursal slits for external fertilization.

CRINOIDS. Sea lilies and feather stars are relics of the ancient class Crinoidea, and they are considered the most primitive echinoderms (Fig. 33.8). Crinoids have multiples of five featherlike arms that are extensively branched and covered with numerous **pinnules.** The arms wave above the cup-shaped body, which is called the **calyx** (plural = calyces). In sea lilies the calyx is at the end of a long, flexible stalk whose other end is

A

B

FIGURE 33.8

(A) A "meadow" of crinoids (sea lilies) as they appeared about 350 million years ago in the shallow sea that covered what is now the central United States. Some extinct crinoids were more than 20 meters long, but no living form is more than a meter long. (B) Ossicles from the stalks and bodies of crinoids that lived approximately 400 million years ago in what is now New York State. Such ossicles are among the most common fossils in Paleozoic limestone.

usually attached to substratum. Feather stars also have long stalks in their late larval stages, but on becoming adults the calyces break off the stalks and swim away by waving their arms. It appears, therefore, that feather stars evolved from sea lilies. Feather stars attach to substratum by means of clawlike **cirri** (visible in Fig. 33.1D).

At first glance, crinoids bear little resemblance to starfish. If one imagines a crinoid feeding upside down on the substratum, however, it is not so different from a starfish. Crinoids use tube feet on their arms to trap small organisms, which are passed to the mouth by the open ambulacral grooves (Fig. 33.9 on page 716). The pinnules greatly increase the area available for this suspension feeding. Gametes develop within the arms and pinnules, and they break through the epithelium for external fertilization.

ECHINOIDS. Sea urchins, heart urchins, and sand dollars, collectively known as echinoids, can be visualized as starfish with their arms raised and fused above them, so that the five ambulacra bend up toward the anus. Sea urchins are hemispherical in shape (if one ignores the spines), and they are said to be **regular** because they are radially symmetric. Heart urchins resemble sea urchins except that they have shorter spines and have the mouth toward one edge of the oral surface. They therefore tend to be bilaterally symmetric and are said to be **irregular.** Sand dollars are disc-shaped and also irregular (Fig. 33.10 on page 716). In all echinoids the ossicles are close-fitting, forming tests that are often found along beaches. Living sea urchins are covered with long spines that pivot in sockets at the base. By flexing these spines, sea urchins can use them like numerous legs to creep across the ocean floor, and they can use them like pincers to aid the pedicellariae in capturing prey and deterring predators (Fig. 33.11 on page 717). Each spine consists of a single crystal of $CaCO_3$ that would be as brittle as chalk if it were not for certain glycoproteins in it. Heart urchins and sand dollars have shorter spines that they use instead of their tube feet for burrowing through sand as they feed on detritus.

Sea urchins generally feed on algae and encrusting animals, which they scrape up and chew with remarkable toothed jaws that make up **Aristotle's lantern** (Fig. 33.12 on page 718). The digestive tract is more elaborate than that of asteroids. The esophagus, stomach, and intestine make up a long tube that spirals inside the perivisceral coelom.

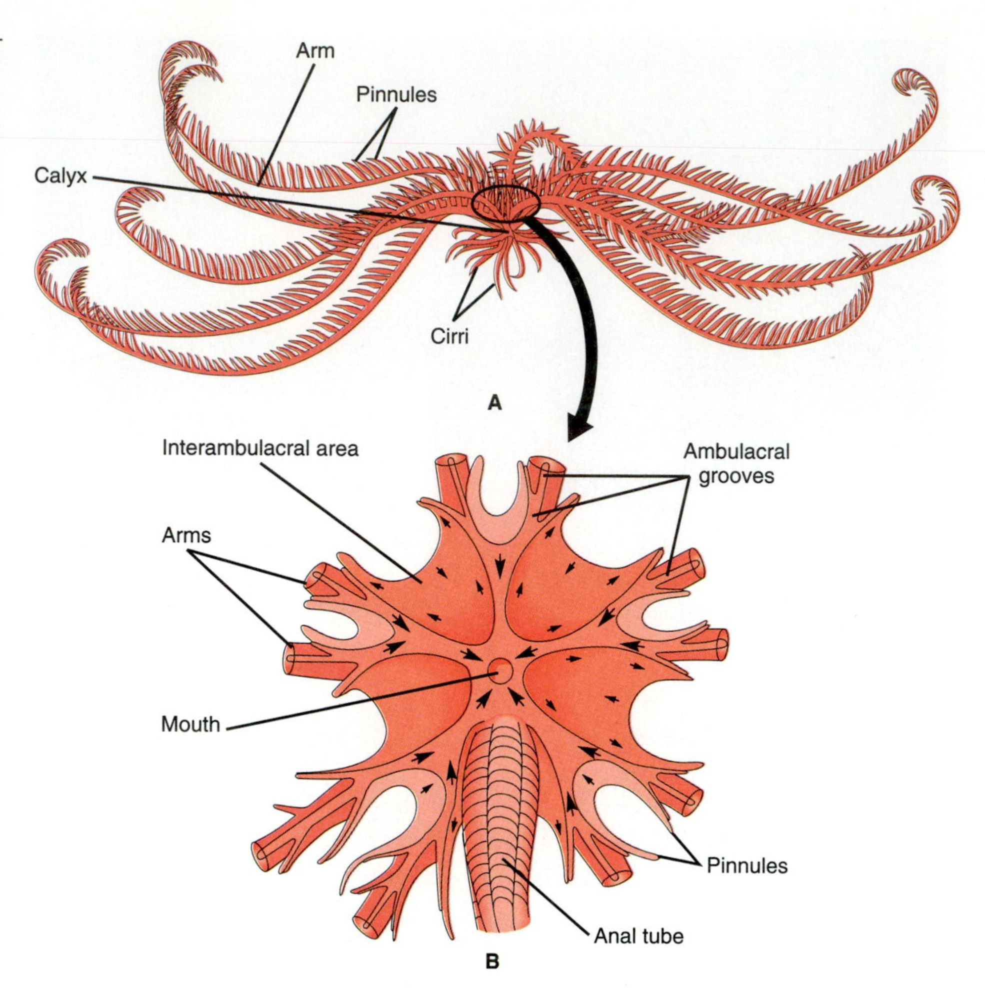

FIGURE 33.9

Structure of the feather star *Antedon.* (A) *Antedon* normally attaches to substratum by its cirri, but it can walk with its arms. (B) Prey's-eye view into the calyx. Tube feet on the pinnules thrash about and toss small organisms into the open ambulacral grooves, where they are trapped in mucus and moved by cilia toward the mouth (broad arrows). Other cilia keep the interambulacral area clear of debris (thin arrows). Fouling is avoided by the position of the anus on the anal tube.

■ **Question:** If you were sent to dive for a sample of a crinoid, how would you be sure not to bring back a sabellid worm (Fig. 29.8) by mistake?

Water that is swallowed with food bypasses the stomach and much of the intestine through a ciliated tube called the **siphon.** Heart urchins feed on organic detritus, which they pick up with tube feet as they burrow. Sand dollars generally feed by sifting sand through their short spines and passing organic material toward the mouth on cilia. In

FIGURE 33.10

The keyhole urchin *Mellita quinquiesperforata,* a common sand dollar. The "keyholes" (lunules) in this species are believed to help prevent the sand dollar from sailing in water currents and to enable food particles to be passed from the aboral to the oral surface. The petal-shaped pattern (petaloid) shows the location of tube feet specialized for exchange of respiratory gases.

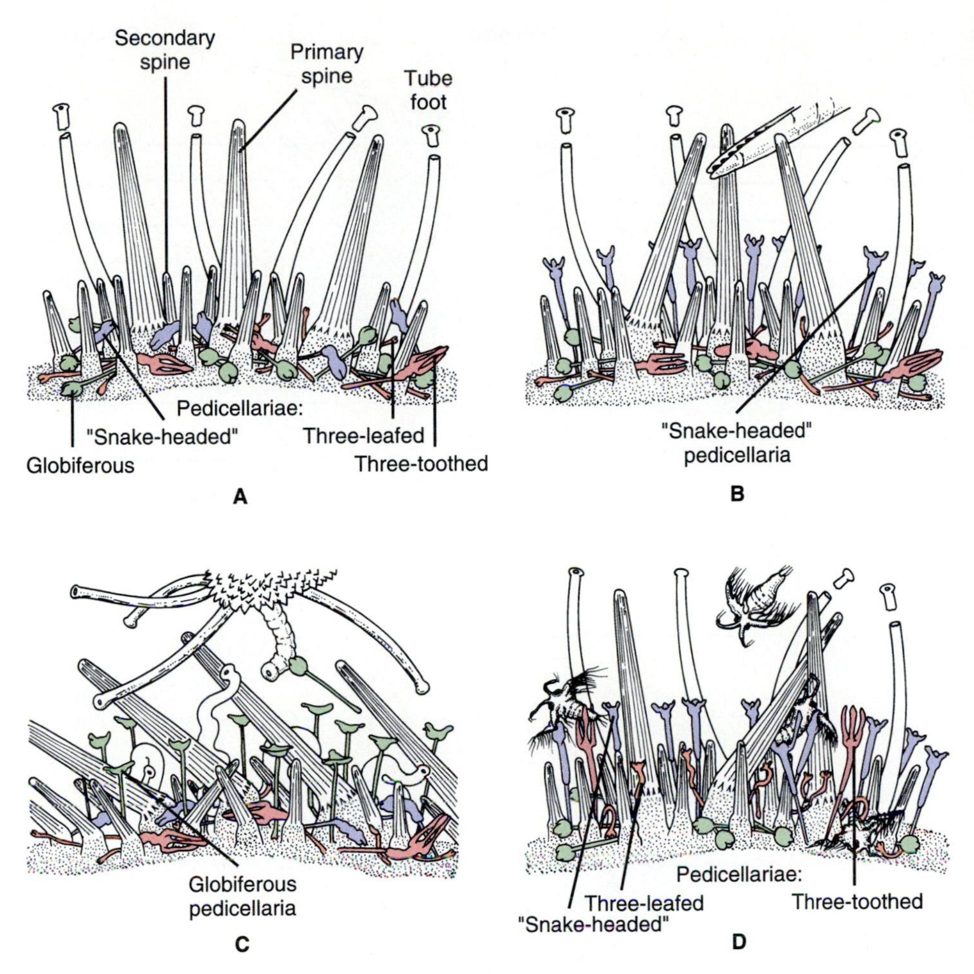

FIGURE 33.11
Defense and predation by spines and pedicellariae on the aboral surface of the sea urchin *Psammechinus miliaris.* (A) At rest the primary and secondary spines and tube feet are extended, while the four types of pedicellariae are relaxed. The primary spines are about 5 mm long. Only parts of the much longer tube feet are shown. (B) In response to mechanical disturbance (forceps) the spines press against the object, and ophiocephalous (snake-headed) pedicellariae prepare to bite. (C) If a predaceous sea star threatens, the spines bend away, exposing poisonous, toothed globiferous pedicellariae that attack the starfish's tube feet. (D) Two types of pedicellariae, ophiocephalous and tridentate (three-toothed), seize plankton, such as these brine shrimp larvae (*Artemia*) , while primary spines crush them. Small triphyllous (three-leafed) pedicellariae seize injured plankton that fall onto the test. (Adapted from Jensen 1966.)

moderate currents, however, some species dig their anterior edges into sediment with the oral end facing the current to capture plankton. Like sea urchins, sand dollars have Aristotle's lanterns. Echinoids also absorb significant amounts of dissolved organic matter across the body surface.

SEA CUCUMBERS. Sea cucumbers resemble their vegetable namesake in overall shape, and they are so unlike other echinoderms that they were once classified in the phylum Sipuncula. They lack arms and are elongated along the oral–aboral axis of the body. Some species lack tube feet for locomotion; they burrow by contracting circular and longitudinal body muscles peristaltically. Burrowing sea cucumbers feed by trapping organic material on sticky tentacles (modified tube feet), then stuffing the tentacles one at a time into the mouth, like someone licking his fingers. Other sea cucumbers creep along substratum on tube feet equipped with suckers, grazing with the tentacles. In these species, tube feet tend to be best developed on three of the five ambulacra that are usually on the bottom. These species therefore have a ventral surface (called the sole) and a dorsal surface, and they are therefore bilaterally symmetric. Podia on the dorsal side lack suckers, and they probably serve as feelers (Fig. 33.13 on page 719). The ossicles are microscopic, and there are no spines, so the integument ranges in texture from soft to leathery.

Sea cucumbers are as strange inside as they are externally. Instead of the madreporite being on the external surface, it hangs freely within the perivisceral coelom. The digestive tract is long like that of sea urchins, but it terminates in a **cloaca** before reaching the anus. The muscular cloaca serves as a respiratory organ, pumping water into and out of the res-

FIGURE 33.12

(A) Internal structure of a sea urchin. (B) Internal view of Aristotle's lantern, so-called because Aristotle described it as looking like a lantern with the panes left out. (C) External view of the teeth of the purple sea urchin *Strongylocentrotus purpuratus.*

■ **Question:** What sequence of muscles contractions would allow the jaws to open, then extend, then close.

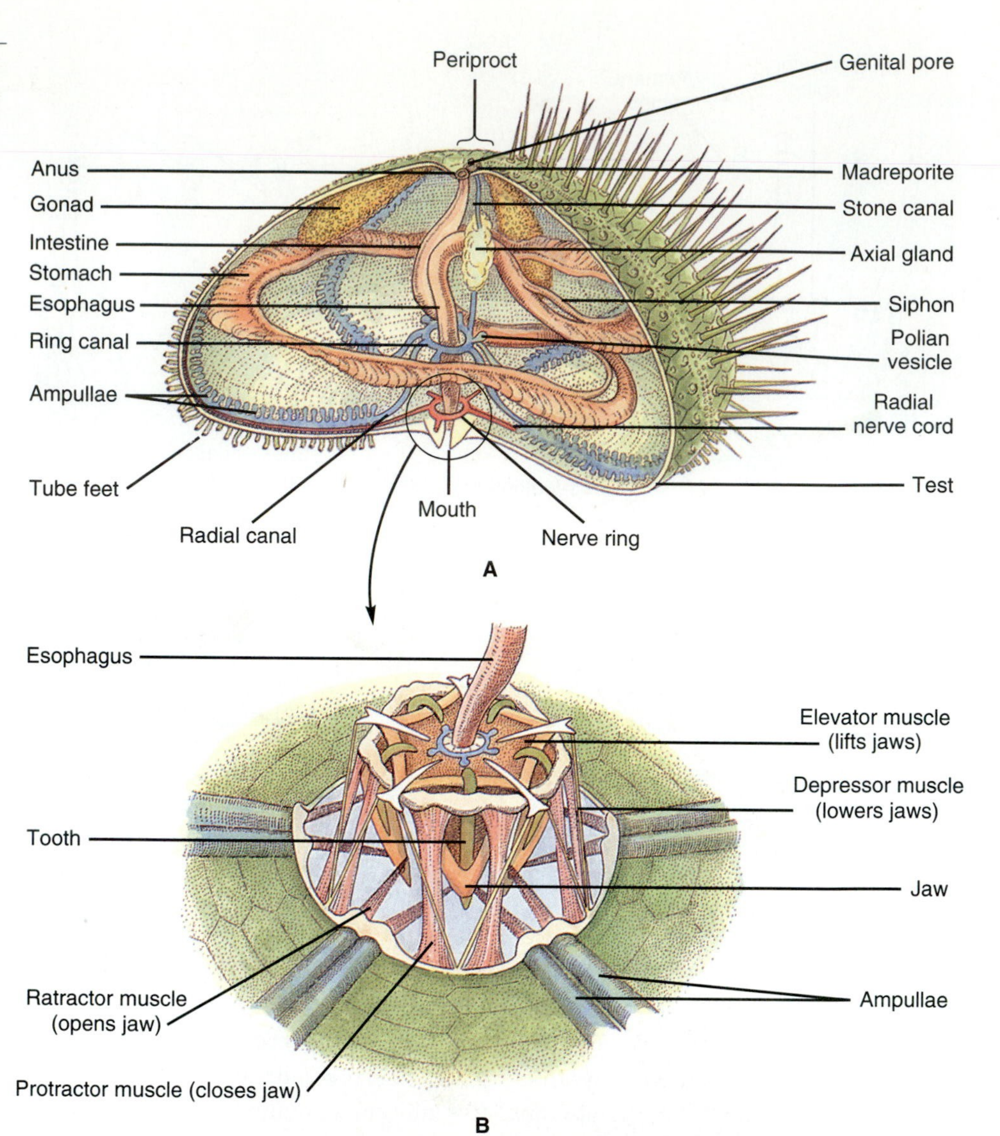

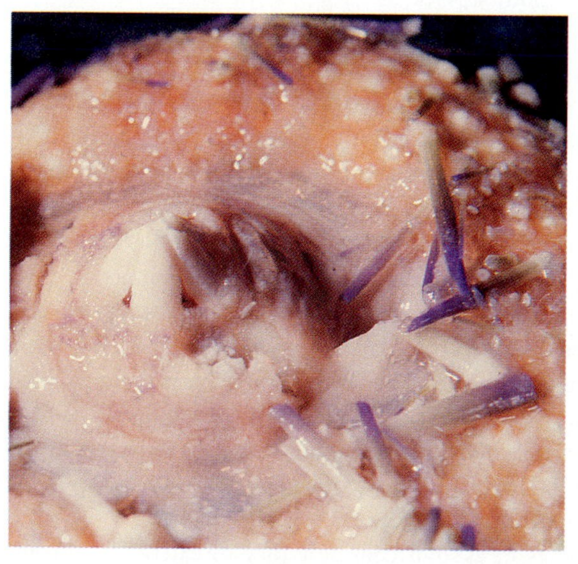

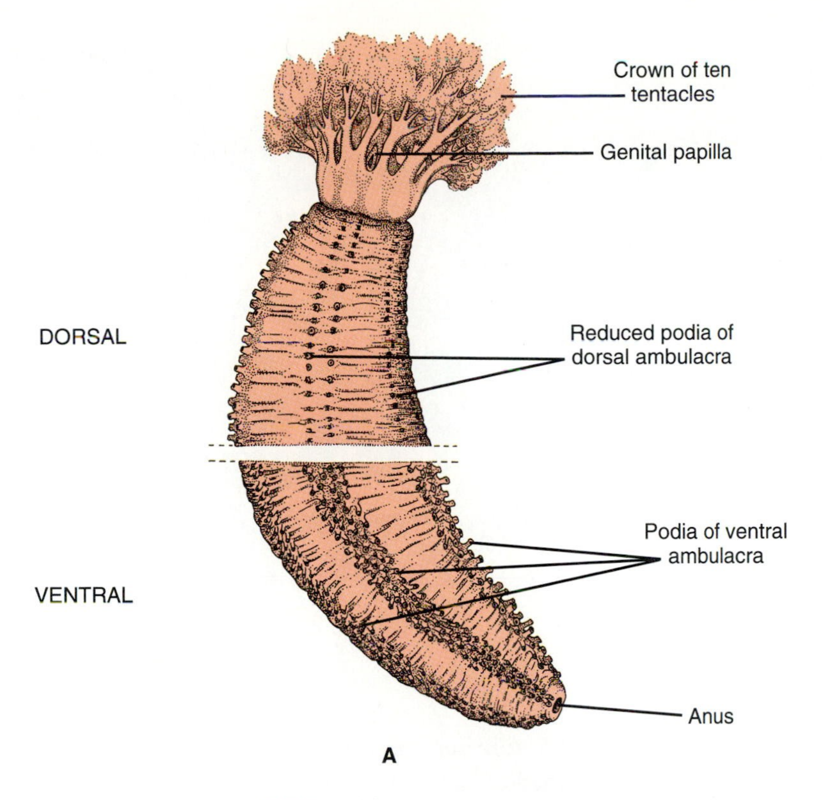

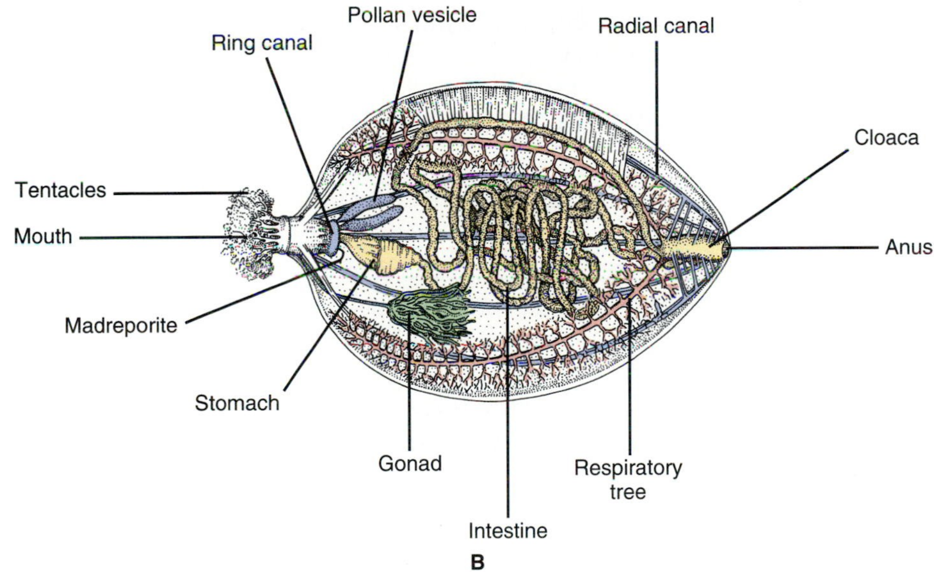

FIGURE 33.13
(A) External structure of the sea cucumber *Cucumaria frondosa* showing the differences in tube feet on the ventral and dorsal surfaces. (B) Internal organs of *Thyone.*

piratory tree that branches into the coelom. Inside the anus of some sea cucumbers are white, pink, or red **tubules of Cuvier.** These sticky tubules are shot out at crabs, fishes, and other potential predators, which can become fatally entangled in them (Fig. 33.14). The sea cucumber readily regenerates new Cuvierian tubules for the next predator. As if this were not impressive enough, sea cucumbers also squeeze out their own digestive tracts, gonads, and respiratory trees under certain circumstances that apparently have nothing to do with defense. Following this self-evisceration they regenerate the lost organs.

CONCENTRICYCLOIDS. One species of echinoderm not belonging to any of the above groups was discovered in 1986 inside wood collected more than a kilometer deep off the coast of New Zealand. Class Concentricycloidea now includes two species,

FIGURE 33.14
A sea cucumber has expelled its Cuvierian tubules from its anus toward the photographer. The Cuvierian tubules will regenerate.

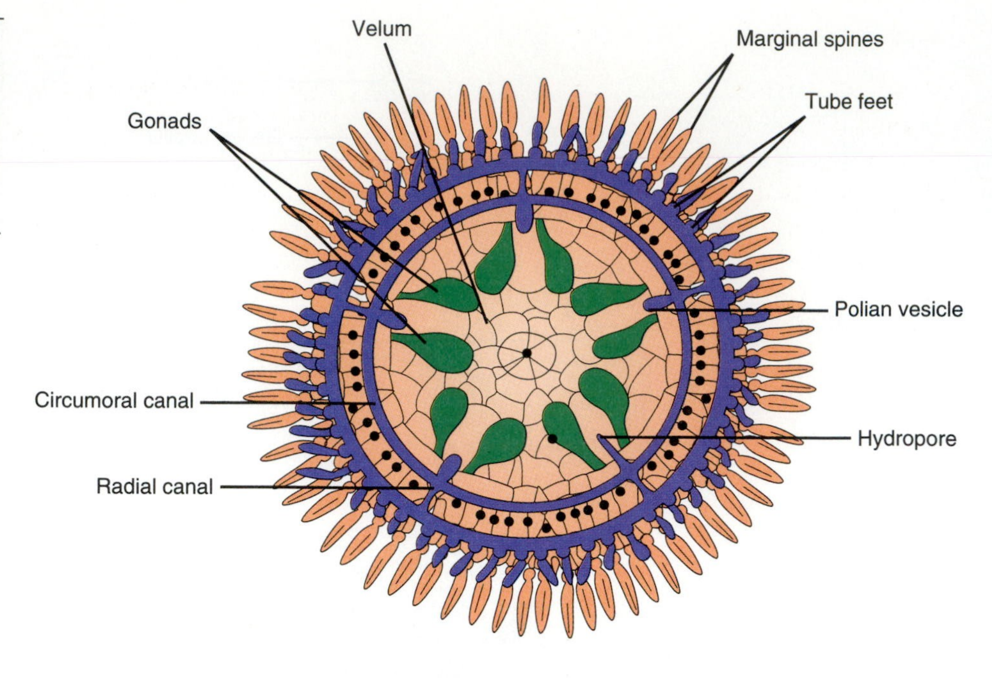

FIGURE 33.15

Xyloplax medusiformis, the first named species of the new class Concentricycloidea. Ventral view showing the gonads and the water vascular system with its two concentric canals. Embryos and larvae in all stages of development were found within the gonads. The hydropore opens into the water vascular system and is homologous with the madreporite.

Xyloplax medusiformis and *X. turnerae.* These animals have been called "sea daisies," but it remains to be seen whether they will ever become familiar enough to require a common name. Individuals are disc-shaped with a fringe of short spines around the edge (Fig. 33.15). They are up to 9 mm in diameter and lack arms. They also lack a mouth, gut, or anus. The ventral surface (**velum**) of the animal may actually be the lining of the stomach, which digests food externally. The water vascular system has two concentric ring canals.

Development

Echinoderms, especially sea urchins and starfish, have long been favorite subjects of developmental biologists. A major reason is that adults can be made to release millions of ova or sperm by electrically stimulating them or by injecting them with potassium chloride or acetylcholine solution. In most species the ova are relatively large (0.1 to 0.2 mm in diameter), and they have little yolk to interfere with observation of intracellular processes. Fertilization and development can therefore be watched under low magnification. (See Conway et al. 1984 for procedures.)

The early stages of development (pp. 116–117) are similar in almost all echinoderms. The shapes of the **larvae** vary, however, depending on the class (Fig. 33.16). In some species in which there is internal fertilization and brooding, the larval stage may be abbreviated. In all other echinoderms the larvae are independent organisms competent at every function except reproduction. The larvae are better adapted to disperse the species than are adult echinoderms. After several weeks of planktonic existence, the bilaterally symmetric larva begins the remarkable metamorphosis into a radially symmetric adult. First the larva swims downward and attaches to substratum, usually with the left side down. Contrary to what one might expect, the mouth and anus of the larva usually do not migrate to their new locations on the oral and aboral sides, but simply disappear. The stumps of the foregut and hindgut then reorient to the oral and aboral sides, and a new mouth and anus develop near them. Likewise, the arms of larvae do not develop into the arms of the adult, but simply disappear.

STARFISH

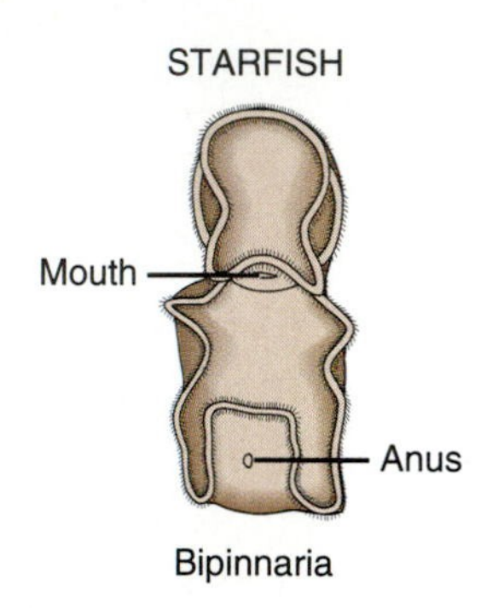

Bipinnaria

SEA CUCUMBERS

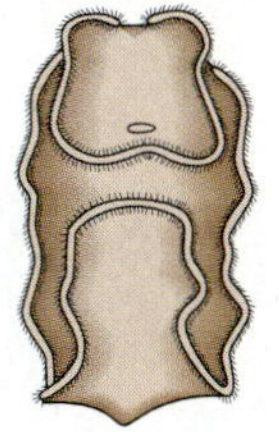

Auricularia

BRITTLE STARS

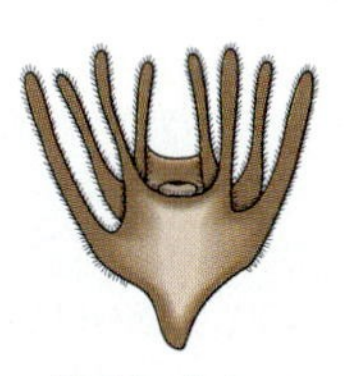

Ophiopluteus

SEA URCHINS

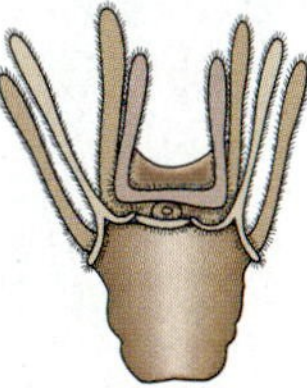

Echinopluteus

SEA LILIES

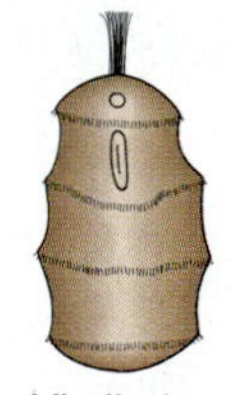

Vitellaria

FIGURE 33.16

Types of echinoderm larvae. Note that the forms that most resemble each other as larvae do not necessarily resemble each other as adults. Compare the ophiopluteus and the echinopluteus, for example. This is one of the sources of difficulty in echinoderm classification.

■ **Question:** Construct a cladogram based on the morphology of these larval types. Compare this cladogram with the one you constructed from Fig. 32.2.

Regeneration

Many adult echinoderms have remarkable abilities to regenerate lost body parts. Asteroids often cast off an arm that has been seized by a predator, then regenerate a new one. The sea star *Asterias vulgaris* can, in fact, recover its original form from only one-fifth of the central disc attached to one arm. Some starfish can even recover their original form from only a single arm (Fig. 33.17). Some species of starfish use their regenerative abilities to reproduce asexually, by splitting in half across the central disc, with each half regenerating the missing parts. Before their regenerative abilities became widely known, oyster fishermen used to try to kill starfish by cutting them in half and throwing them back into the water—a worse than useless strategy. Brittle stars and crinoids have less regenerative ability and require an intact central disc or calyx in order to regenerate a lost arm. Sea cucumbers lack arms, but regenerate their tubules of Cuvier and viscera, as described before.

FIGURE 33.17
The starfish *Linckia multifora* regenerating from a single arm. At this stage of regeneration the starfish is aptly called a "comet." Members of this genus are especially adept at regeneration and often reproduce asexually in this way.

Interactions with Humans and Other Animals

Crown-of-Thorns Starfish. Echinoderms almost never attack humans or transmit diseases, and only rarely does a careless diver die from handling sea urchins that have especially poisonous spines or pedicellariae. On the other hand, echinoderms can indirectly have devastating effects on humans and other animals. In the late 1960s, for example, the crown-of-thorns starfish *Acanthaster planci* created near panic among people of the western Pacific, and they alarmed ecologists around the world (Fig. 33.18). This sea star, which feeds on coral polyps, underwent a population explosion that appeared to threaten the existence of even the vast and priceless Great Barrier Reef of Australia. Numerous other coral reefs, along with the shores and islands they protect and the fish on which the islanders depend, were also endangered. The crown-of-thorns starfish has, in fact, damaged approximately 500 km of the Great Barrier Reef and a large area of the reef off Guam. Tens of thousands of the starfish have been killed by injecting formaldehyde in an attempt to curb the damage. Although the sea stars

FIGURE 33.18
The crown-of-thorns sea star feeding on coral. Each year an individual approximately half a meter in diameter reduces over five square meters of coral to bare, white skeleton. In some areas of the South Pacific more than 90% of the coral has been destroyed. The starfish itself is safe from most predators because of the spines and irritating mucus that cover it. One of the few major predators is a mollusc, the giant triton (*Charonia tritonis*). It was originally thought that removal of hundreds of thousands of these predators by collectors was responsible for the rise in the sea star's population, but this now appears to be too simple an explanation.

continue to destroy large areas, it now appears that the coral can recover from the damage, and the threat no longer seems as grave as it once did.

SEA URCHINS. Predation by sea urchins is another cause of ecological and economic concern. Sea urchins have destroyed several valuable kelp forests of the west coast of the United States (see Fig. 1.6). Populations of the sea urchins there have swelled due to depletion of predaceous sea otters by fur trappers and the dumping of raw sewage into the ocean. To stop this destruction, sea urchins have been dredged up and killed and the area "reforested" with kelp embryos raised in the laboratory. In the Atlantic, excessive populations of sea urchins have become a nuisance by robbing lobster traps. Many former lobstermen have now become sea-urchin fishermen, and they sell the roe to Japan for $100 to $150 a pound.

Meanwhile, throughout the Caribbean the population of the black sea urchin *Diadema antillarum* also exploded because of depletion of predatory fishes. The population of *Diadema* then suddenly collapsed from an infectious disease, leaving the coral reefs without any effective control of algae. The resulting overgrowth of algae contributed to a 90% reduction in coral reef area since 1980 (Hughes 1994). The complexity and unpredictability of these population fluctuations and their effects stand as evidence of how little we understand about the role of echinoderms in marine ecology.

SEA CUCUMBERS. In China and the Pacific islands boiled and dried sea cucumbers are a delicacy, and in some areas they are believed to be aphrodisiacs. Demand is so great that much of the known Pacific sea cucumber beds have been depleted, and fishermen have begun to threaten beds off South America and the Galápagos Islands. In the future sea cucumbers may be collected as medicine. Pacific islanders have long known that cut-up sea cucumbers can be used to poison fish in tidal pools. It now turns out that the poison, holothurin, has various effects on nerve and muscle and also suppresses the growth of certain tumors (Ruggieri 1976).

Summary

Echinoderms are deuterostomes, since they are coelomates in which the mouths of larvae develop from an embryonic opening other than the blastopore. In addition, cleavage is radial and indeterminate, and they are enterocoelous. All are marine animals. Most echinoderms tend to have pentamerous radial symmetry as adults. Unlike any other animal, they have tube feet arranged along ambulacral grooves, a water vascular system, and an endoskeleton of ossicles. The tube feet are used to anchor the animal to substratum and for locomotion and feeding. They are extended by pressure from the water vascular system, which circulates seawater. Echinoderms are named for spiny protrusions on the ossicles. In addition, the integument may bear pedicellariae, which are used in predation and defense, and papulae, which exchange respiratory gases and metabolic wastes.

There are six major groups of echinoderms: starfish, brittle stars, crinoids (sea lilies and feather stars), echinoids (sea urchins, sand dollars, and others), sea cucumbers, and the little-known concentricycloids. Starfish generally have five stiff arms and feed by everting the stomach. Brittle stars resemble starfish, but they have longer and more flexible arms. In crinoids the arms branch from a calyx, which sits atop a long stalk in sea lilies. Echinoids have close-fitting ossicles that form a solid test, and they feed by means of Aristotle's lantern. Sea cucumbers are tube-shaped and have tube feet formed into tentacles.

Key Terms

deuterostome
radial cleavage
enterocoelous
pentamerous radial symmetry
podium
ambulacral groove
water vascular system
madreporite
ring canal
radial canal
ossicle
pedicellaria
papula
pinnule
calyx
Aristotle's lantern
respiratory tree

Chapter Test

1. Describe the features that define a deuterostome. (pp. 707–708)
2. Give the common names for the five major kinds of echinoderms. Describe each of the five briefly. What shared features justify placing them all in the same phylum? (pp. 707–708)
3. Explain how podia work. What is the role of the water vascular system in the functioning of podia? What is the main function of podia in sea stars? (pp. 710–711)
4. Describe the role each of the following structures plays in the life of an echinoderm: ambulacral grooves, papulae, pedicellariae, ossicles. (pp. 711–713)

■ Answers to the Figure Questions

33.2 The cladogram will depend on the characters chosen. Some possible ones are: (1) five arms, (2) radial canals extending into arms, (3) complete digestive tract, (4) mouth on ventral surface, (5) sessile, (6) bilateral symmetry. The groups and their synapomorphies are: sea stars, 1, 2, 3, 4; brittle stars 1, 4; sea lily 1, 3; sea urchin 3, 4; sea cucumber 1, 3, 6. A cladogram is not easily resolved from these synapomorphies, and this reflects the current situation in echinoderm taxonomy.

33.4 Normally the madreporite would appear to allow water to enter or leave the water vascular system to compensate for changes in pressure. If the madreporite were blocked, one could expect the water vascular system to shrink or swell as the sea star sank or rose.

33.6 No distinctive respiratory or excretory organs are shown, so one might surmise that the integument played those roles. In fact, the papulae do.

33.9 The sabellid would retreat into its tube when you reached for it; the crinoid has no tube into which it can withdraw.

33.12 First retractor muscles would open the jaws, then depressor muscles would lower them, then protractor muscles would close the jaws, and finally elevator muscles would raise them.

33.16 Possible synapomorphies are: (1) with arms, (2) with longitudinal rather than circular bands of cilia, (3) with tuft of cilia at one end. The groups and their synapomorphies are: starfish 2; sea cucumbers 2; brittle stars 1; sea urchins 1; sea lilies 3. The cladogram would link starfish with sea cucumbers and brittle stars with sea urchins. The relationship to sea lilies could not be determined. This cladogram is at variance with current thinking on echinoderm phylogeny.

Readings

Recommended Readings

Birkeland, C. 1989. The Faustian traits of the crown-of-thorns starfish. *Am. Sci.* 77:154–163.

Feder, H. M. 1972. Escape responses in marine invertebrates. *Sci. Am.* 227(1):92–100 (July). (*How sea urchins and other prey escape sea stars.*)

Inoué, S., and K. Okazaki. 1977. Biocrystals. *Sci. Am.* 236(4):82-92 (Apr). (*How a sea urchin shapes its ossicles.*)

Macurda, D. B., Jr., and D. L. Meyer. 1983. Sea lilies and feather stars. *Am. Sci.* 71:354–365.

See also relevant selections in General References at the end of Chapter 21.

Additional References

Conway, C. M., D. Igelsrud, and A. F. Conway. 1984. Sea urchin development. In C. L. Harris (Ed.). *Tested Studies for Laboratory Teaching: Proceedings of the Third Workshop/Conference of the Association for Biology Laboratory Education (ABLE).* Dubuque, IA: Kendall/Hunt, Chapter 4.

Harrison, F. W., and F-S. Chia (Eds.). *Microscopic Anatomy of Invertebrates,* Vol. 14: *Echinodermata.* New York: Wiley–Liss.

Hughes, T. P. 1994. Catastrophes, phase shifts, and large-scale degradation of a Caribbean coral reef. *Science* 265:1547–1551.

Jensen, M. 1966. The response of two sea-urchins to the sea-star *Marthasterias glacialis* (L.) and other stimuli. *Ophelia* 3:209–219.

Lawrence, J. 1987. *A Functional Biology of Echinoderms.* Baltimore: Johns Hopkins University Press.

Ruggieri, G. D. 1976. Drugs from the sea. *Science* 194:491–497.

Smith, M. J., A. Arndt, S. Gorski, and E. Fajber. 1993. The phylogeny of echinoderm classes based on mitochondrial gene arrangements. *J. Mol. Evol.* 36:545–554.

Wada, H., and N. Satoh. 1994. Phylogenetic relationships among extant classes of echinoderms, as inferred from sequences of 18S rDNA, coincide with relationships deduced from the fossil record. *J. Mol. Evol.* 38:41–49.

CHAPTER

34

Introduction to Chordates

Tunicates, *Polycarpa* sp.

PHYLUM HEMICHORDATA

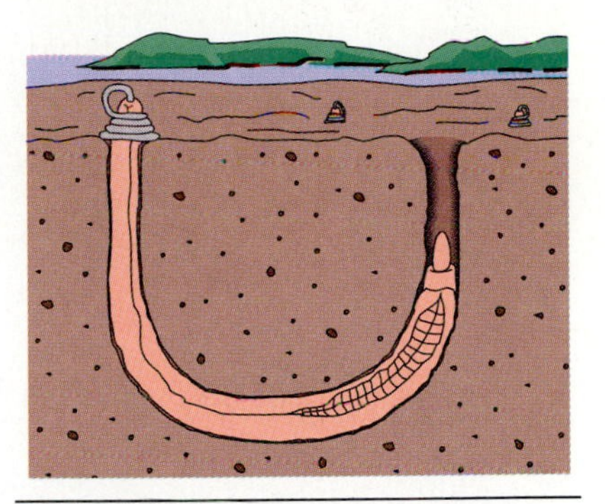

FIGURE 34.1

The acorn worm *Balanoglossus* (phylum Hemichordata) defecating while in its U-shaped burrow in shallow seawater. The anal end of the burrows are marked by distinctive coils of castings formed as the worm passes sediment through the gut. This species is beige or pink in color. The folds on the trunk are genital wings near the area of the gonads. Some members of this genus are up to 2.5 meters long.

Fewer than 5% of the named species of animals belong to the phylum Chordata, yet these relatively few species receive the most attention from zoologists and from people in general. When most people use the word "animal" they mean a chordate of some kind—a fish, amphibian, reptile, bird, or mammal. People are chordates, so this interest in one phylum is probably due in part to "phylocentrism." A zoology book written by a snail would probably be mainly about molluscs. On the other hand, there are also good, objective reasons for paying attention to chordates. Except for arthropods there is no other phylum whose members have so successfully adapted to so many modes of living. Regardless where an animal lives, it is difficult for it to get through life without encountering chordates. That is as true for snails and other molluscs as it is for people.

One of the things most people wish to know about their own phylum is where it came from. As we shall see later, there is little fossil evidence bearing on the origins of chordates, but there is a group of extant animals—phylum Hemichordata—that may provide some clues to our humble origins (see Figs. 21.10 and 34.1). Not only are chordates and hemichordates both deuterostomes, but hemichordates share with chordates two significant features: **pharyngeal slits** and **dorsal nerve cord** (Table 34.1). It was once thought that hemichordates also shared the feature that defines chordates: the dorsal skeletal rod called the notochord. For that reason, hemichordates were formerly considered to be a type of chordate. Now, however, it is recognized that what was thought to be a notochord in hemichordates is actually an extension of the buccal cavity called the **buccal diverticulum** (= stomochord), which is not homologous with the notochord.

TABLE 34.1

Characteristics of hemichordates

Phylum Hemichordata HEM-i-core-DAY-tuh (Greek *hemi* half + *chorda* cord)

Morphology	Coelomate (although peritoneum may be secondarily lost in some species). Body divided into proboscis, collar, and trunk. Not segmented. With **dorsal nerve cord**, **pharyngeal slits**, and **buccal diverticulum**, but no notochord.
Physiology	Circulatory system essentially open, with contracting dorsal and ventral vessels and heart vesicle. A network of epidermal nerve cells, concentrating in some areas to form middorsal and midventral nerve cords. Respiration through integument. Digestive system complete.
Locomotion	Peristaltic contraction of the proboscis. Larvae ciliated.
Reproduction	Sexes usually separate but similar. Some pterobranchs monoecious. Fertilization external. Some asexual reproduction by fragmentation or budding.
Development	Deuterostomate. Coelom (when present) does not always arise in classic deuterostomate fashion, enterocoelously. Development either direct or with tornaria larvae.
Habitat, Size, and Diversity	All species marine, benthic. Length of adult enteropneusts ranges from 2.5 cm to 2.5 meters; pterobranchs generally less than 1 cm long. Approximately **85 living species** described.

Phylum Hemichordata is divided into two classes that differ from each other greatly as adults, although they resemble each other embryologically. Class Enteropneusta includes most of the approximately 85 extant species of the phylum. The enteropneusts, commonly called acorn worms, burrow in marine sediments and have wormlike bodies divided into a proboscis, collar, and trunk. Members of class Pterobranchia have tentacles and live in colonies inside secreted tubes. Because the tentacles contain branches of the coelom, they are technically **lophophores** (pp. 560–561).

Acorn Worms

GENERAL FEATURES. Members of class Enteropneusta are commonly called acorn worms because the head and collar resemble the nut and cup of an acorn. Behind the collar is a long, slimy trunk that is more than 2 meters long in some species. Each of the three main body sections (head, collar, and trunk) has a separate coelomic compartment filled with spongy tissue that provides some mechanical support, but the body is so soft that it is easily broken by handling. Acorn worms live in U-shaped burrows in shallow water (Fig. 34.1), or they burrow through marine sediments or live under rocks or seaweed. The proboscis is the main organ of locomotion, and it pulls the trunk along behind it. Many hemichordates feed by ingesting large amounts of mud or sand from which the gut extracts organic debris. Others feed by means of cilia on the proboscis, which pass food particles back toward the mouth (Fig. 34.2). The food particles are bound on a mucus string and swallowed along with water.

INTERNAL STRUCTURE. Water swallowed during feeding exits through **pharyngeal slits** and **gill pores.** The pharyngeal slits are U-shaped openings in the pharynx that carry swallowed water into **pharyngeal pouches** and out through the gill pores (Fig. 34.3). Pharyngeal slits are sometimes called gill slits, although they are much

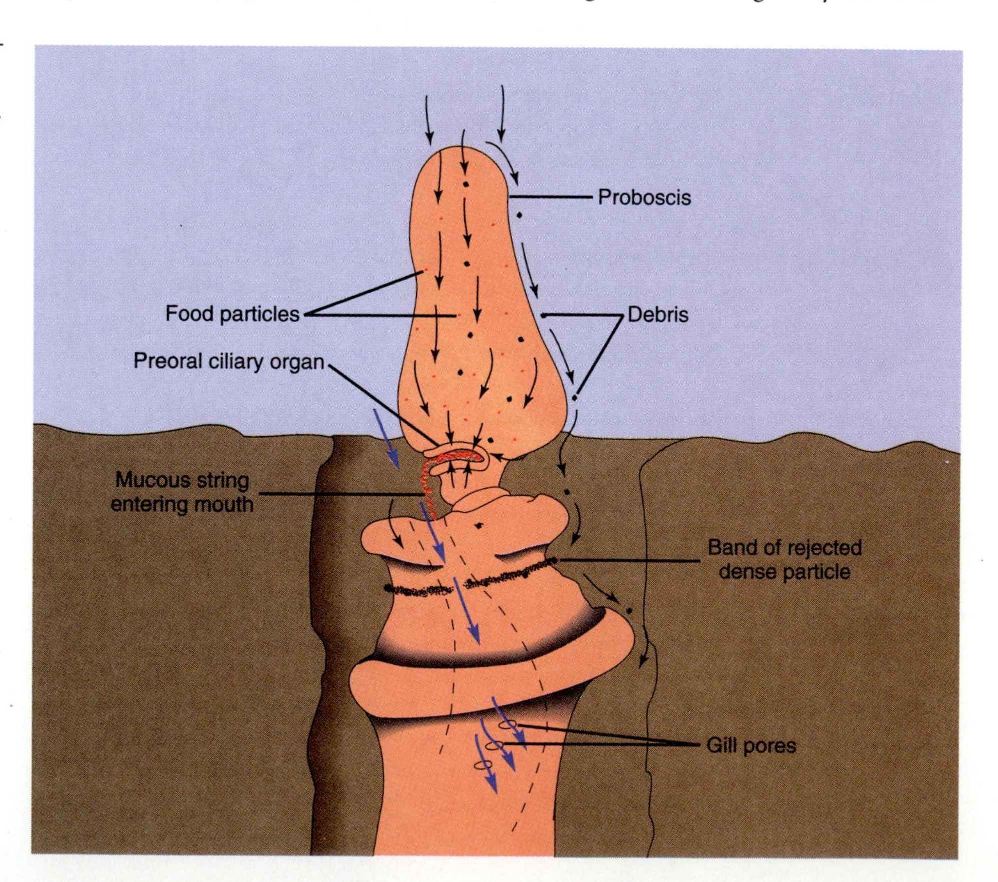

FIGURE 34.2

Method of feeding in some acorn worms. Cilia on the proboscis pass particles toward the mouth (thin arrows). Sand and other dense, inorganic materials fall to the collar, while less-dense organic particles are carried into the mouth on a mucous string. Water entering the mouth during feeding exits through the pharyngeal slits and gill pores (thick arrows). The collar can be pulled forward to close the mouth.

■ **Question:** What other kind of animal feeds by trapping particles on a mucus string and reeling it in on cilia?

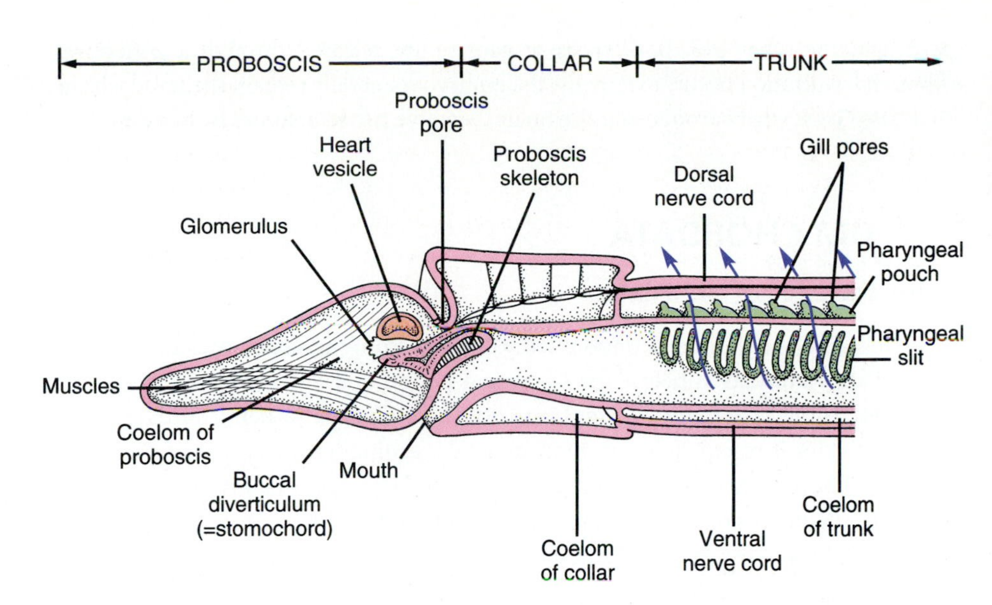

FIGURE 34.3
The internal structure of the head, collar, and anterior part of the trunk of an enteropneust. Arrows show movements of water through pharyngeal slits, pharyngeal pouches, and gill pores. The buccal diverticulum was originally thought to be a notochord.

more important in feeding than in exchange of respiratory gases. The body surface is the main route of respiratory exchange. A contracting **heart vesicle** draws colorless blood anteriorly through a dorsal vessel into the proboscis, then pumps it posteriorly through a ventral vessel. The blood circulates through sinuses, making it an essentially open circulatory system. One such network of sinuses, the **glomerulus,** is assumed from its structure to be an excretory organ. The nervous system consists largely of a diffuse network in the base of the epidermis. Along the dorsal and ventral midlines this plexus is concentrated into dorsal and ventral nerve cords, which lack ganglia. In some places the dorsal nerve cord is often hollow, like the dorsal nerve cords of chordates. Sensory receptors are scattered over the integument, especially on the proboscis.

REPRODUCTION AND DEVELOPMENT. Acorn worms have separate sexes, although there is little evident difference between males and females. The gonads are distributed in rows, often forming finlike bulges on the trunk (Fig. 34.1). Each gonad has a separate gonopore, and fertilization occurs externally, apparently during mass spawning initiated by the females. Early development is typically deuterostomate: Cleavage is radial, the blastopore becomes the anus, and the coelom forms as an outpocketing of the archenteron. Development is direct in some species, but others produce **tornaria larvae,** so called because ciliary contraction causes them to spin. (Latin *tornare* to turn on a lathe.) Tornarias look much like the bipinnarias of sea stars (see Fig. 33.16).

Pterobranchs

The 20 or so species in class Pterobranchia are organized much like the Enteropneusta, with similar development and with bodies divided into proboscis, collar, and trunk. As adults, however, they differ greatly in external appearance (Fig. 34.4 on page 728). They look more like ectoprocts (pp. 560–562) than other hemichordates, presumably because they have evolved similar adaptations for living colonially in secreted tubes. The collar expands dorsally into tentacled arms that look like the lophophores of ectoprocts. Cilia on these tentacles direct food into ciliated grooves that carry it to the mouth. Also as in ectoprocts, the alimentary canal is U-shaped, with the anus outside the fringe of tentacles. In most species there is one pair of pharyngeal slits, but some have none.

Some pterobranchs are dioecious, but as might be expected for animals confined to a tube, many are hermaphroditic. In the monoecious zooids, one gonad produces

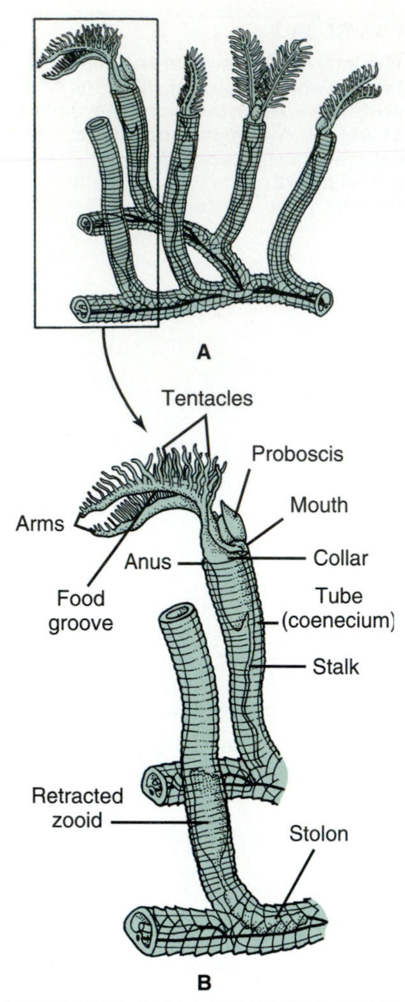

FIGURE 34.4
(A) A colony of *Rhabdopleura*. Any disturbance causes each zooid to contract the stalk and withdraw into its tube (= coenecium), as shown on the left. Afterward the ciliated proboscis climbs back up the stalk to feed. The tube is secreted by the proboscis. (B) Detail showing the proboscis, tentacled collar, and trunk. Unlike most pterobranchs, members of this genus lack pharyngeal slits. Each zooid is about 1 mm long, not including the stalk.

■ **Question:** Compare this figure with Fig. 27.2. What key differences are there between adult ectoprocts and enteropneusts? Are these sufficient to differentiate them at the phylum level?

sperm and the other ova. Both types of gamete are released through a gonopore. Although fertilization occurs externally, the embryos generally remain sheltered within the tubes. The sexually produced individuals then give rise to colonies by budding.

PHYLUM CHORDATA

Major Features of Chordates

Chordates and hemichordates undoubtedly evolved from the same ancestor, but that ancestor apparently lived before the Cambrian period and did not leave any fossils. The oldest known chordate, *Pikaia,* bears little resemblance to hemichordates (Fig. 34.5). Nor does it resemble arthropods, annelids, echinoderms, or any of the other groups that have sometimes been proposed as chordate ancestors. *Pikaia* already has most of the hallmarks of chordates: pharyngeal slits, a dorsal nerve cord, a notochord, a post-anal tail, and myomeres (Table 34.2).

The Notochord. The notochord is a flexible rod of cells along the dorsal midline that all chordates have, at least as embryos (Fig. 34.6). The notochord is responsible for differences in body plans that justify placing hemichordates and chordates in different phyla. Besides giving the phylum Chordata its name, the notochord is important in embryonic development and body support. The notochord is a major organizer during embryonic development (pp. 119, 123). It establishes the axis of the body and

TABLE 34.2
Characteristics of chordates

Phylum Chordata core-DAY-tuh (Latin *chorda* cord, referring to the notochord)

Morphology	With **notochord, pharyngeal slits, dorsal hollow nerve cord,** and a **post-anal tail,** at least in embryo. **Myomeres** (segmented muscle blocks) in most species. Segmented.
Physiology	Circulatory system closed in most forms, usually with ventral heart. Central nervous system originating as dorsal, tubular nerve cord (becoming brain and spinal cord in vertebrates). Digestive system complete. Osmoregulation and excretion largely by kidneys. Respiration through gills, lungs, and/or integument. Birds and mammals homeothermic.
Locomotion	Swimming, creeping, running, or flying, using skeletal muscles attached to endoskeleton in vertebrates.
Reproduction	Sexes almost always separate, often with marked sexual dimorphism. Some species hermaphroditic; some parthenogenetic. Some invertebrate forms reproduce by budding. Fertilization generally external in aquatic forms; internal in terrestrial forms. Oviparous or viviparous.
Development	Cleavage pattern radial (lancelet and amphibians), bilateral (tunicates), discoidal (fishes, reptiles, birds), or rotational (placental mammals). Deuterostomate. Primitively enterocoelous; mammals and some other vertebrates schizocoelous. Development indirect or direct.
Habitat, Size, and Diversity	Marine, freshwater, amphibious, or terrestrial. From a few millimeters to 32 meters long. Approximately **48,000 living species** described.

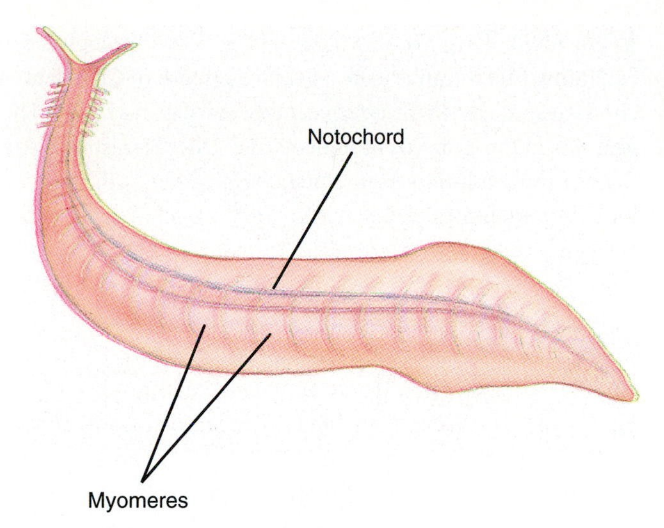

FIGURE 34.5

The notochord and myomeres identify *Pikaia* as one of the earliest known chordates. Its fossils are found in the Burgess Shale formation, which is more than half a billion years old (pp. 387–390).

guides development of the nerve cord. In most vertebrates the notochord guides the formation of the vertebral column, which becomes the flexible support for locomotion. In the invertebrate chordates that swim or burrow the notochord persists in adults, maintaining flexibility while keeping the body from collapsing.

DORSAL, TUBULAR NERVE CORD. A second characteristic of chordates is the dorsal, hollow nerve cord. This nerve cord lies along the midline, dorsal to the notochord. In vertebrates the spinal cord forms from the dorsal nerve cord and becomes enclosed by vertebrae. The brain forms at the anterior end of the nerve cord. Many other animals, such as annelids and arthropods, also have nerve cords, but they are ventral except in hemichordates. The chordate nerve cord differs from those in other phyla in being hollow, although the channel often becomes obscure during development.

PHARYNGEAL SLITS. A third distinguishing feature of chordates is the possession of pharyngeal slits in the embryo. These slits form as pockets of ectoderm grow inward and fuse with pockets of endoderm lining the pharynx. In some of the chordates discussed in this chapter, as well as in fishes and some amphibians, these slits are the precursors of gills. In many terrestrial vertebrates the pharyngeal slits never completely perforate the neck, and they usually disappear before birth or hatching.

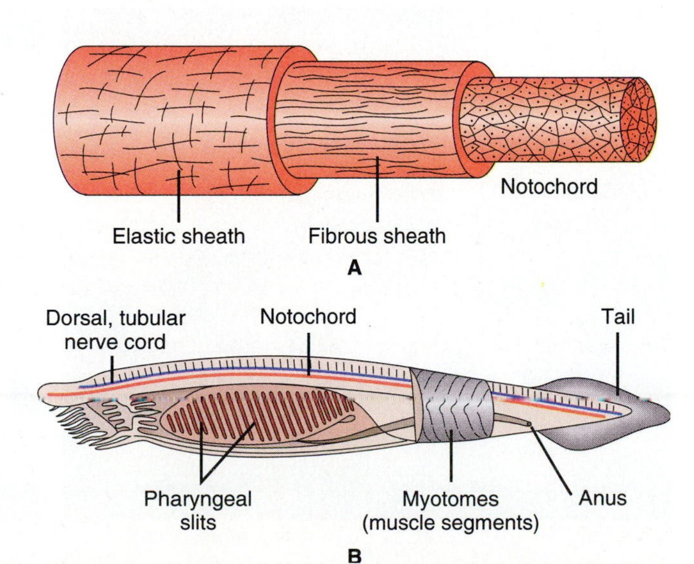

FIGURE 34.6

(A) Structure of the notochord. Its stiffness is due to fluid pressure in the closely-packed cells, and to the sheaths. (B) The amphioxus (*Branchiostoma*), showing the features considered diagnostic of chordates: notochord, dorsal tubular nerve cord, pharyngeal slits, and post-anal tail. Segmental muscles are not unique to chordates but are a common feature.

Post-anal Tail and Myomeres. Some zoologists consider the presence of a tail extending past the anus to be a fourth hallmark of Chordata. A post-anal tail occurs in all chordates, at least in the embryo. (Yes, you also had one.) The post-anal tail and pharyngeal slits in the embryos of humans and other terrestrial vertebrates are regarded as vestiges of our evolution from aquatic vertebrates with tails. Another feature that is common in vertebrates and some invertebrate chordates is bands of segmental muscles, called **myomeres** (= myotomes).

Classification

Chordates are generally divided into two major groups: Craniata and Protochordata (see Fig. 34.16). The group Craniata (Greek *kranion* skull) comprises the subphylum Vertebrata, in which the head is well developed. The protochordates, also called Acrania, are the invertebrate chordates. Protochordates are divided into two subphyla, Urochordata and Cephalochordata, which will be described in more detail shortly. Members of subphylum Urochordata have tadpolelike larvae with typical chordate features. As adults, however, members of the largest group of urochordates resemble sponges more than chordates at first glance. Most are sessile on the ocean floor, lack a coelom, and pump water through their hollow bodies through two siphons (Chapter opener; Fig. 34.7). A common name for them is sea squirts. Urochordates are also called tunicates, because their bodies are reinforced with a tough **tunic.** Members of the second protochordate subphylum, the Cephalochordata, retain chordate features throughout life. The adults look fishy, but they have no fins, scales, or bones (Fig. 34.8 on page 732). *Pikaia* was probably a cephalochordate. Because of their resemblance to a surgeon's lancet, cephalochordates are often called lancelets.

Tunicates

Ascidians. Of the approximately 2000 species in the subphylum Urochordata, most belong to the class Ascidiacea. Ascidiaceans are also the urochordates you are most likely to see, because their bag-shaped bodies are often brightly or delicately colored, or starkly black or white. The showy appearance may serve as warning coloration that deters predators, since many ascidians are highly poisonous to eat. (Julius Caesar's wife is said to have eliminated his rivals by serving them ascidians.) Individuals are up to 30 cm long,

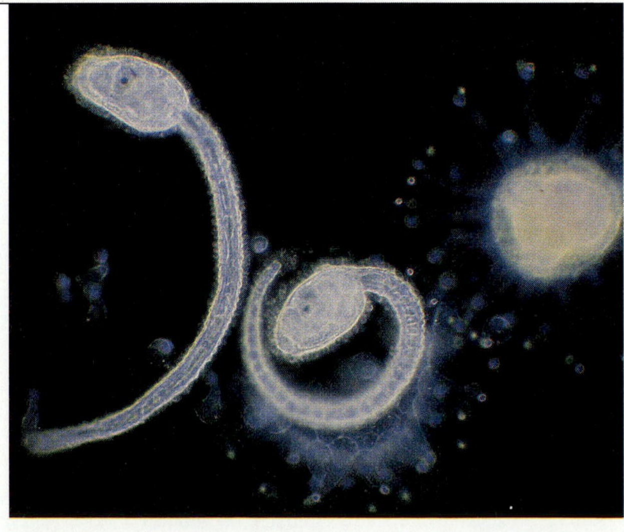

A

B

FIGURE 34.7

The urochordate *Ciona intestinalis.* (A) From right to left, an egg and two just-hatched tadpole larvae. Note the eyespot, and the anterior adhesive structures by which the tadpole attaches to substratum before metamorphosis into a bag-shaped, sessile adult. (Compare Fig. 34.10.) (B) An adult. Water enters and leaves by the incurrent and excurrent siphons at the center and right, respectively.

Major Groups of Phylum Chordata

Genera mentioned elsewhere in this chapter are noted.

Subphylum Urochordata (= Tunicata) YOUR-oh-core-DAY-tuh (Greek *oura* tail + *chordi* cord). Tadpole larva with notochord extending into tail. Adult usually lacks notochord, dorsal hollow nerve cord, coelom, and segmentation.

Class Ascidiacea a-sid-ee-ACE-ee-uh (Greek *askidion* little wineskin). Sac-shaped as adults. Sessile; enclosed in well-developed fibrous tunic. With two siphons near each other. Incurrent siphon draws water into pharynx for filter-feeding and gas exchange. Water exits through dorsal excurrent siphon. Sea squirts (= ascidians = tunicates). *Botryllus, Ciona, Halocynthia, Polycarpa* (Chapter opener; Figs. 34.6, 34.8).

Class Larvacea (= Appendicularia) lar-VASE-ee-uh (Latin *larva* ghost). Adults retain tadpole shape of larvae, including persistent notochord and nerve cord. Tunic not persistent. Pelagic (open-ocean) forms often enclosed in gelatinous "house" that traps prey. Larvaceans. *Oikopleura* (Fig. 34.12).

Class Thaliacea thal-ee-ACE-ee-uh (Greek *thalia* plenty). Adults barrel-shaped, enclosed in tunic, with siphons at opposite ends. Either drifting in currents or swimming by jet propulsion of water through siphons. Colonial forms swim entirely by ciliary action. Pyrosomes, salps, and doliolids. *Doliolum, Pyrosoma* (Figs. 34.12, 34.13).

Subphylum Cephalochordata SEF-a-low-core-DAY-tuh (Greek *kephale* head). Notochord and nerve cord extend full length of body and persist in adult. Body of adult slender, fishlike; without scales or tunic. Lancelets. *Branchiostoma* (Fig. 34.8).

Subphylum Vertebrata VER-te-BRAY-tuh (Latin *vertebratus* backboned). Vertebrae surround spinal cord. Anterior part of nerve cord becomes brain, which is enclosed in cranium.

Superclass Agnatha (= Cyclostomata) AG-na-thuh (Greek *a-* without + *gnathos* jaws). Without jaws or paired appendages. Living forms have neither scales nor bones. Notochord persists in adult. Lampreys and hagfishes. (See Chapter 35.)

Superclass Gnathostomata NATH-oh-STOW-ma-tuh (Greek *stoma* mouth). Jaws present. Notochord usually lacking in adult. Usually with paired appendages.

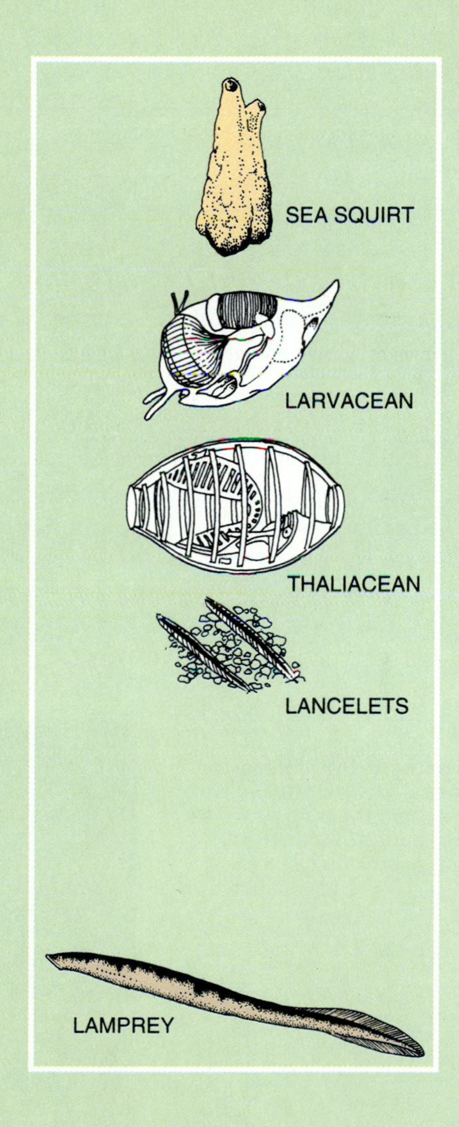

box continues

Class Chondrichthyes kon-DRIK-the-eez (Greek *chondros* cartilage + *ichthys* a fish). Fishes with cartilaginous skeletons. Scaly skin. Five to seven pairs of gills with separate, uncovered openings. Sharks, rays, and chimaeras. (See Chapter 35.)

Class Osteichthyes OS-tee-IK-the-eez (Greek *osteon* bone). Fishes generally with bony skeletons. Scaly skin. Pairs of gills, with gill covers. (See Chapter 35.)

Class Amphibia am-FIB-ee-uh (Greek *amphi-* double + *bios* life). Respiration by gills, integument, or lungs. Tetrapods (primitively with paired, lateral appendages). Frogs, toads, newts, salamanders, and caecilians. (See Chapter 36.)

Class Reptilia rep-TIL-lee-uh (Latin *repere* to creep). Horny scales. Breathing through lungs. Without larval stages. Tetrapods, although legs may be absent. Snakes, lizards, sphenodonts, turtles, and crocodilians. (See Chapter 37.)

Class Aves AY-veez (Latin *aves* birds). Scales on feet only; feathers elsewhere. Tetrapod, with forelimbs modified as wings. Breathing through lungs. No larval stage. Birds. (See Chapter 38.)

Class Mammalia ma-MALE-ee-uh (Latin *mamma* breast). With mammary glands. Neither scales nor feathers; hairy skin. Tetrapod, with forelimbs modified as flippers or wings in some forms. With lungs. No larval stages. Mammals. (See Chapter 39.)

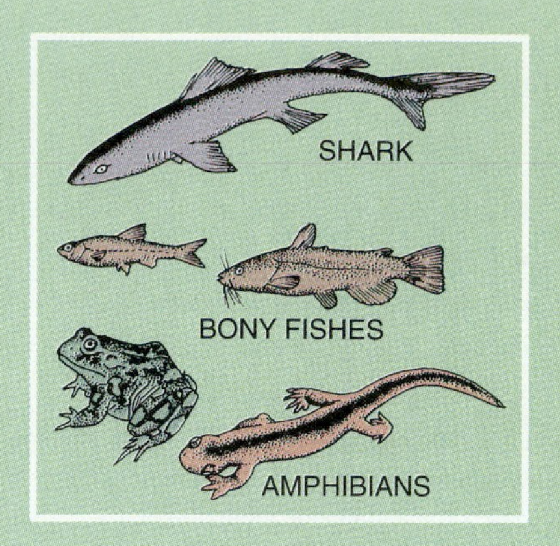

FIGURE 34.8

Adult cephalochordates, amphioxus (*Branchiostoma*). *Branchiostoma,* also called the lancelet, spends much of its time partially buried in coarse sediment. Length approximately 6 cm.

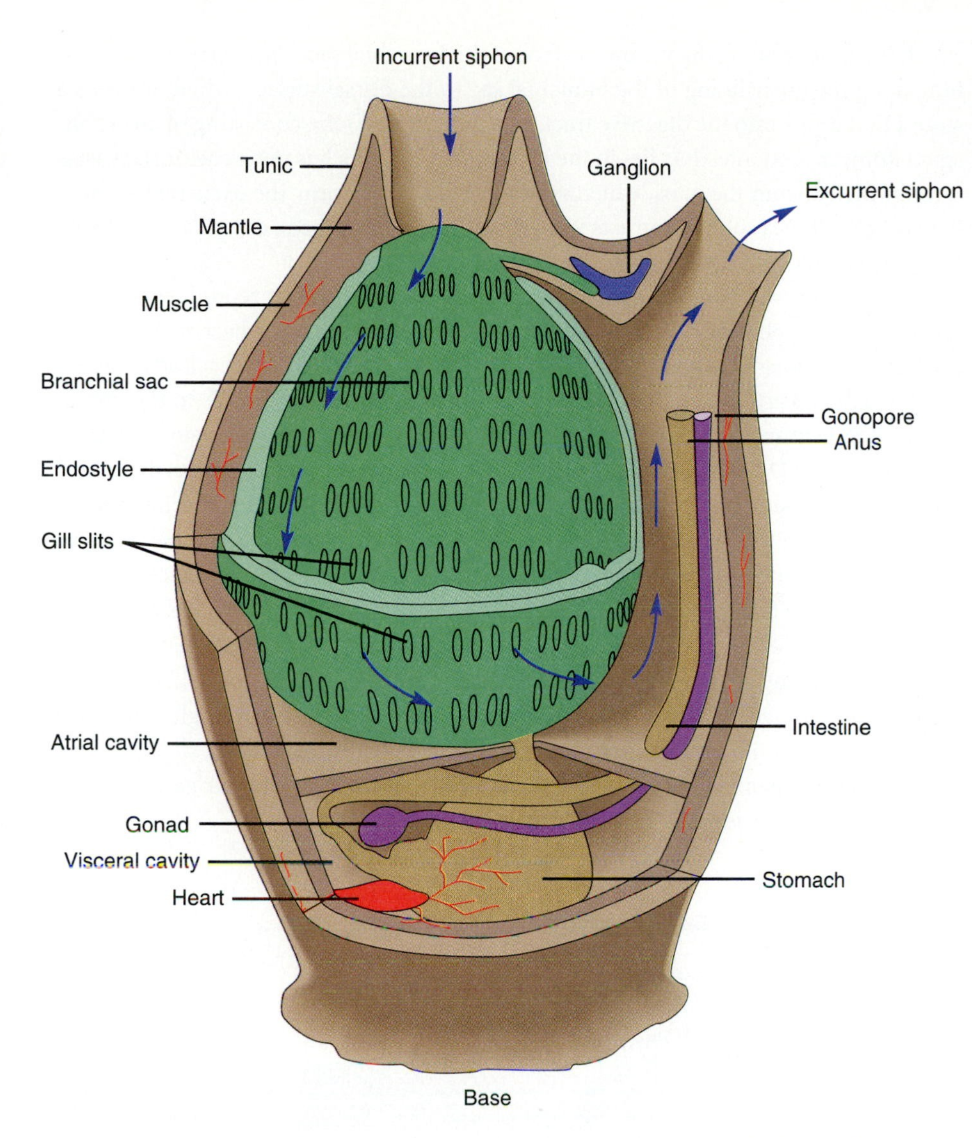

FIGURE 34.9
Internal structure of a solitary ascidian. Dorsal (defined by the location of the ganglion) is on the right, and anterior is up. Arrows show the direction of water movement. The endostyle collects iodine and is thought by some to be homologous with the thyroid gland.

and many species form larger colonies. Along rocky shores at low tide they are hard to ignore. If you try to pick one up you may be startled by a jet of water from the excurrent siphon. This defensive reaction gives ascidians the common name sea squirts.

The current of water that flows through a sea squirt's body is not only for defense, but is also necessary for most of its other vital functions. Water passes into the incurrent siphon and enters the **branchial sac,** which fills most of the upper cavity (the **atrium**) of the body (Fig. 34.9). Cilia around numerous **pharyngeal slits** (often called gill slits) in the branchial sac are responsible for maintaining this flow of water. In a large ascidian the cilia can pump several liters per hour. One function of the water current is to keep the soft body from collapsing, by producing an internal pressure that maintains a **hydroskeleton.** There is no endoskeleton, and the coelom, if any, consists of a pair of small sacs. Outward bulging of the body wall (**mantle**) is prevented by the tunic, which is strengthened by fibrous molecules similar to cellulose.

The branchial sac is also essential for **respiration.** Water passing through the branchial sac exchanges O_2 and CO_2 with the blood circulating in the lining. The **circulatory system** consists of a rudimentary heart and sinuses that form a hemocoel. One of the curious and unexplained features of ascidians is that the direction of circulation reverses every few minutes, even if the heart is removed from the body. The flow of water through the branchial sac is also necessary for **filter feeding.** Small food particles in the water current become trapped in a film of mucus that is secreted by the **endostyle,** which is a groove along the ventral surface of the branchial sac. Cilia carry the mucous film along the inner lining of the branchial sac to the dorsal surface, where it forms a strand that drops into the digestive tract. The digestive system, consisting of an esophagus, stomach, and intestine, lies in the **visceral cavity** (which is not a coelom). Wastes are eliminated from the anus, which is in the atrial cavity near the excurrent siphon. Periodically the sea squirt contracts its body to clear out the feces, as well as any debris that may have accumulated in the branchial sac.

The flow of water out the excurrent siphon is also essential for reproduction. Most species are hermaphroditic, with both an ovary and a testis near the stomach in the visceral cavity. Contraction of the body forces ova and sperm out through ducts that lead to the excurrent siphon. Fertilization occurs externally in solitary species and leads to the formation of free-swimming tadpole larvae (Fig. 34.7A). A tadpole larva generally swims for only a few hours. During this time it must find a new habitat, which it does with the help of a simple eye, a statocyst (gravity receptor), and receptors for chemical and mechanical stimuli. After locating a suitable substratum, the larva attaches by means of three **adhesive papillae.** It then undergoes **metamorphosis** (Fig. 34.10). During metamorphosis the larva loses some of the features that identify ascidians as chordates: the notochord, the dorsal nerve cord, and the tail. The internal organs also undergo a rotation of about 90°.

Ascidians may also occur as a colony of individuals much like the solitary ascidians described above, or a colony may share one circulatory system, a single tunic, and one excurrent siphon. Colonial ascidians reproduce sexually like solitary ascidians, except that all the members of the colony synchronize the release of their gametes (Fig. 34.11). Colonial species also form buds that reproduce asexually.

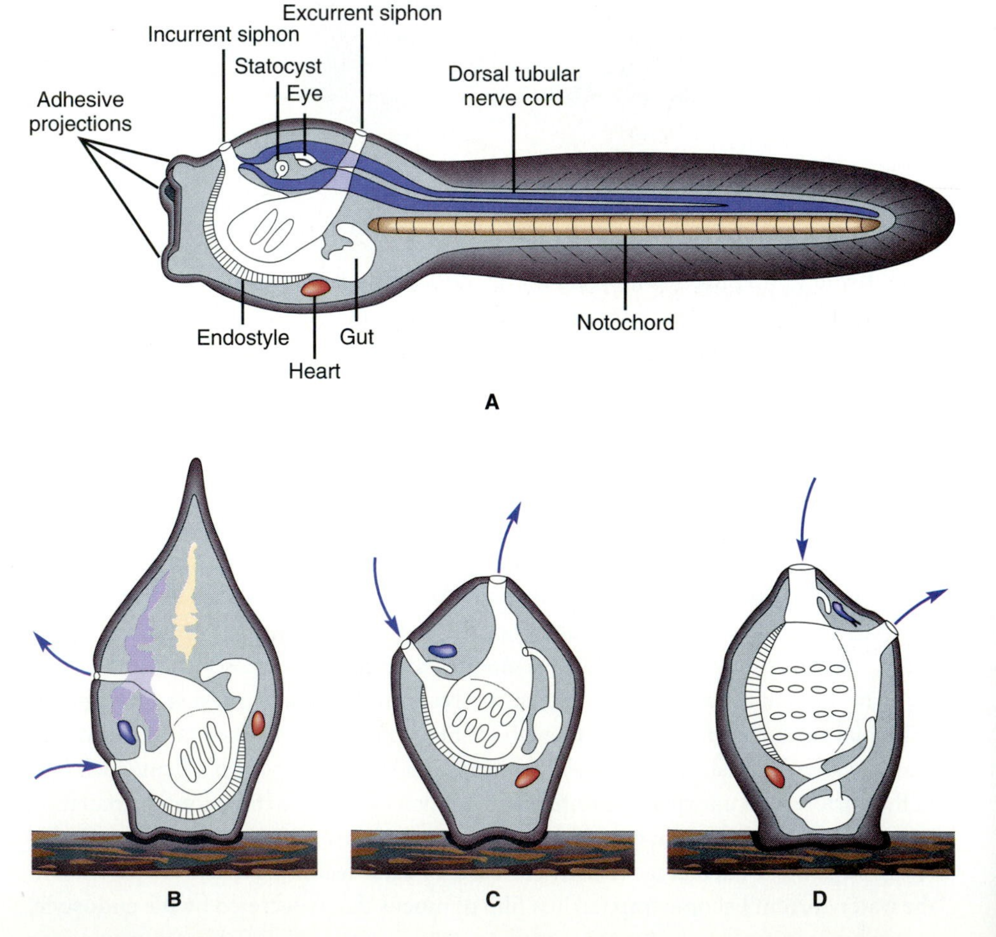

FIGURE 34.10
Metamorphosis in a solitary ascidian. (A) The free-swimming tadpole larva. (B) Immediately after settling and attaching to substratum. (C) Later in metamorphosis. (D) The adult.

■ **Question:** How does the locomotor apparatus of the ascidian larva compare with that of a larva of sponge, jellyfish, ectoproct, echinoderm, hemichordate, or other invertebrate with motile larvae and sessile adults?

FIGURE 34.11

The compound colonial ascidian *Botryllus schlosseri.* There is one star-shaped pattern, approximately 1.8 mm long, for each colony. Each ray of a star is the incurrent siphon of one ascidian. A single excurrent siphon in the center of the rays is shared by all the colony. Two or more colonies can combine, but only if they recognize each other as being kin. (See Pfennig and Sherman 1995.)

LARVACEANS. There are fewer species in the class Larvacea than in the class Ascidiacea, but in numbers of individuals larvaceans may be the most common planktonic forms in many seas. They are seldom noticed because they are only a few millimeters long and virtually transparent (Fig. 34.12). Many secrete **houses** several centimeters in diameter, but these are also seldom noticed because they are made of

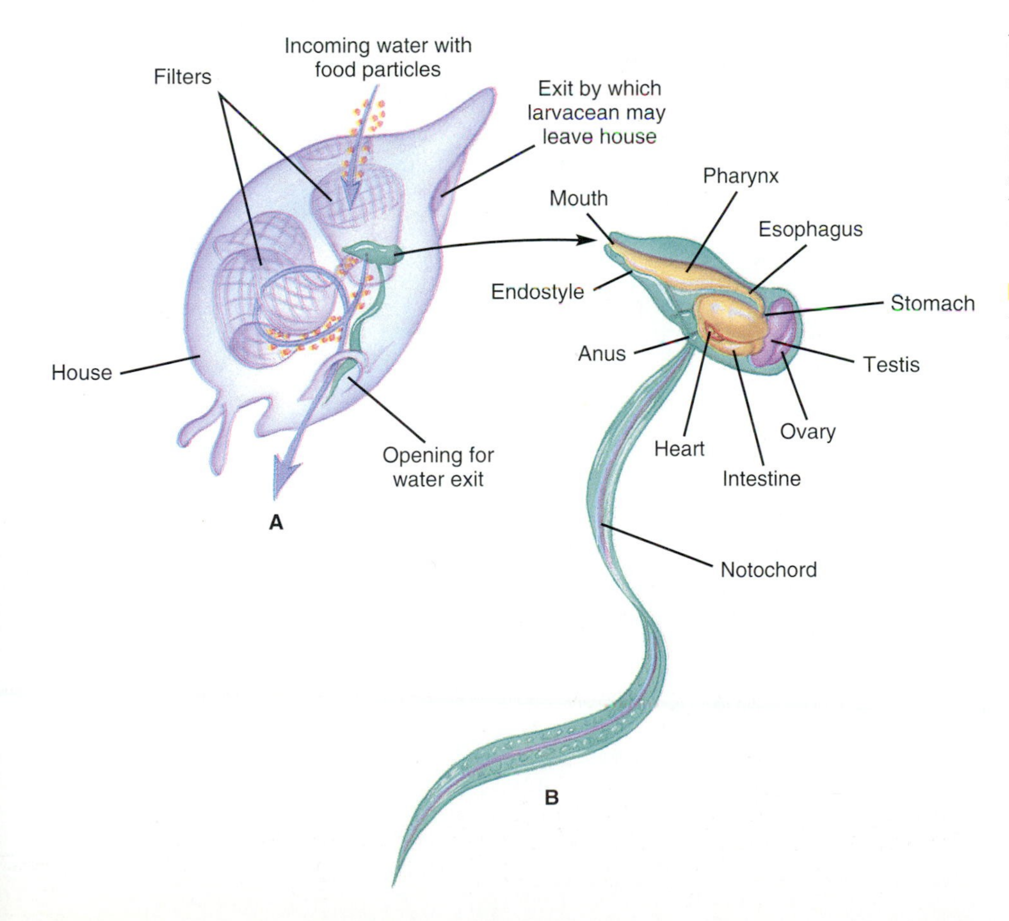

FIGURE 34.12

(A) The larvacean *Oikopleura albicans* "at home." Arrows show the direction of the water current generated by the beating of the tail. The wing-shaped structures are feeding filters that trap tiny plankton. (B) *Oikopleura* outside its house.

■ **Question:** What is the function of having two sets of filters?

transparent mucus that disintegrates in nets. Many larvaceans and their houses glow at night, but they are often mistaken for luminescent algae. Only since submersible research vessels have come into wide use have zoologists begun to appreciate the enormous numbers and sizes of houses and their ecological impact.

The house is used for filter feeding. When the larvacean is "at home" it continually waves its tail, drawing water through the house. The notochord, which persists in the adult, provides the reinforcement required for the tail to function as a pump. Water enters at the anterior end through two **incurrent filters,** which have a mesh so fine that only plankton smaller than a few micrometers can enter. Once inside the house the plankton become trapped in a wing-shaped **feeding filter.** Houses have no opening to eliminate feces, so several times each day the larvacean abandons its home and builds a new one. Reconstruction takes only a few minutes. A larvacean will also abandon its home if threatened by a predator. Such abandoned houses, with their accumulation of plankton, may represent a significant contribution to the diets of many marine animals, which would not otherwise be able to feed on such tiny plankton.

THALIACEANS. The class Thaliacea comprises three groups: doliolids, which are generally about a centimeter long; salps, in which individuals are up to 20 cm long and may form chainlike colonies; and pyrosomes, which form colonies up to 20 meters long. Doliolids and salps are shaped like barrels with rings of muscles that look like the hoops (Fig. 34.13). These muscles force water through the body, producing a current that is used for swimming, filter feeding, and exchange of respiratory gases. The most unusual thing about doliolids and salps is the alternation of sexual and asexual generations. In some species the hermaphroditic adults in the sexual generation produce larvae externally, like ascidians. In other species the hermaphroditic adult broods a solitary egg that develops directly. In either case the adults in the next generation are asexual. They produce a string of buds, each one of which develops into a hermaphroditic adult. Pyrosomes are colonies that are propelled entirely by ciliary action of the individuals (Fig. 34.14). They get their name (Greek *pyr* fire + *soma* body) from the

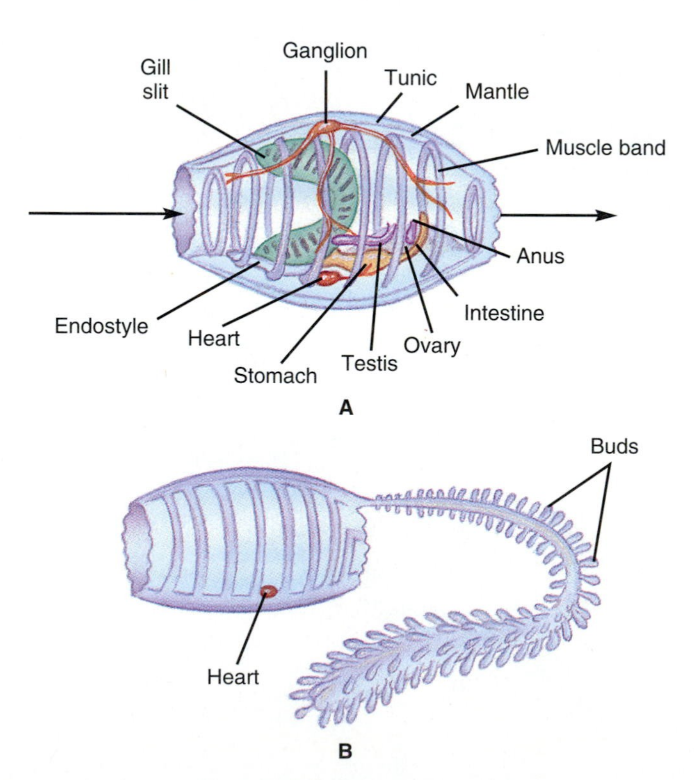

FIGURE 34.13

The thaliacean *Doliolum.* (A) Sexual adult. Arrows show the direction of water movement. (B) An adult in the asexual ("nurse") stage.

■ **Question:** What is the advantage of alternating sexual and asexual generations? What other animals have such alternating generations?

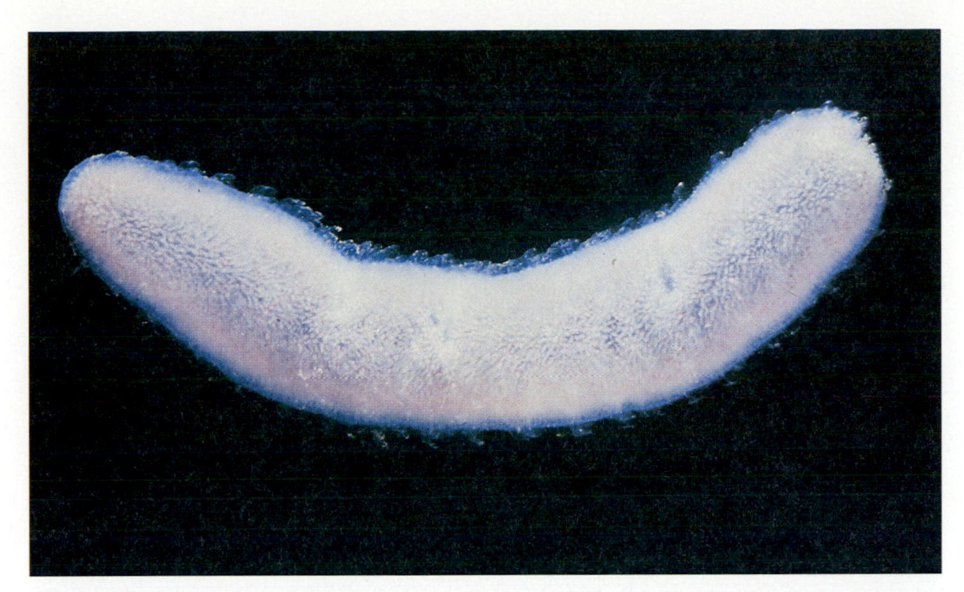

FIGURE 34.14
A pyrosome of genus *Pyrosoma.*

brilliant waves of light that are triggered when individuals in the colony encounter an obstacle or unsuitable conditions. This luminescence inhibits ciliary movement in neighboring individuals in the colony, which then produce their own light and stop propelling the pyrosome forward.

Lancelets

Subphylum Cephalochordata, which comprises lancelets, derives its name from the notochord that extends to the front of the body. There is no actual head or brain: A dorsal, hollow nerve cord extends to the front in both larvae and adults. Other chordate characteristics of lancelets include the numerous pharyngeal (gill) slits and the tail that projects beyond the anus (Fig. 34.15). Lancelets, also called amphioxus, even have several features reminiscent of vertebrates, and it is often used to introduce essential features of vertebrates. In fact, lancelets appear to be the closest living relatives of vertebrates (Gee 1994, Wada and Satoh 1994). The vertebrate-like features of cephalochordates include the following: blocks of segmental muscles (**myomeres**); a ventral contracting blood vessel that may be homologous to the vertebrate heart; a diverticulum of the intestine that resembles the precursor of the liver in vertebrates; separate ventral and dorsal roots of the nerve cord; and rudimentary olfactory and optic receptor organs.

The 25 or so species of lancelets are found along shallow marine shores around the world. Although they can swim by side-to-side contractions of the fin-shaped body,

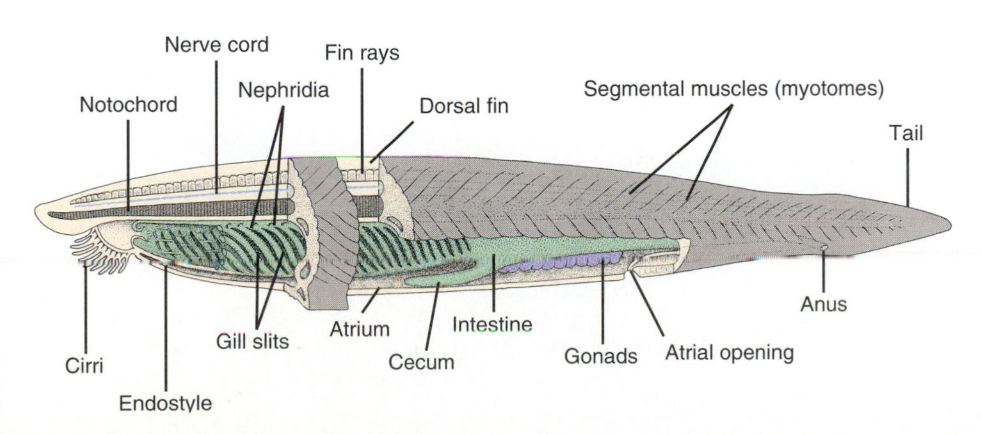

FIGURE 34.15
Schematic representation of a lancelet. This is one of the few chordates in which the notochord persists in the adult.

they are usually found with their tails buried in sand (Fig. 34.8). The springy notochord prevents the body from collapsing during both swimming and burrowing. Lancelets feed and respire in much the same way as adult ascidians: Cilia on the pharyngeal slits drive a current of water through the pharynx. Food trapped on a sheet of mucus produced by the endostyle is passed back to the cecum (the gut diverticulum mentioned before) for digestion. From the pharynx water passes through the atrium and out through the atrial opening.

Unlike tunicates, lancelets have separate sexes. As gametes ripen in the more than two dozen gonads, they burst into the atrial cavity and out through the atrial opening for external fertilization. The larvae resemble the adults but are covered by cilia that are used for swimming and propelling food toward the mouth. As the larva grows, the area of the ciliated body surface fails to keep pace with the body mass. As a result, the larva tends to sink, and it eventually adopts the burrowing habit of the adult.

Vertebrates

Traditional taxonomists divide the living members of subphylum Vertebrata into eight major groups comprising agnathans, fishes, amphibians, reptiles, birds, and mammals (see Figs. 21.4, 21.6, and 34.16). The large number of major groups may indicate the success of vertebrates in diversifying, or it may simply reflect the human tendency to see differences more easily in the familiar. Groups of vertebrates are often placed in clades according to criteria of major evolutionary significance. The jawless fishes are

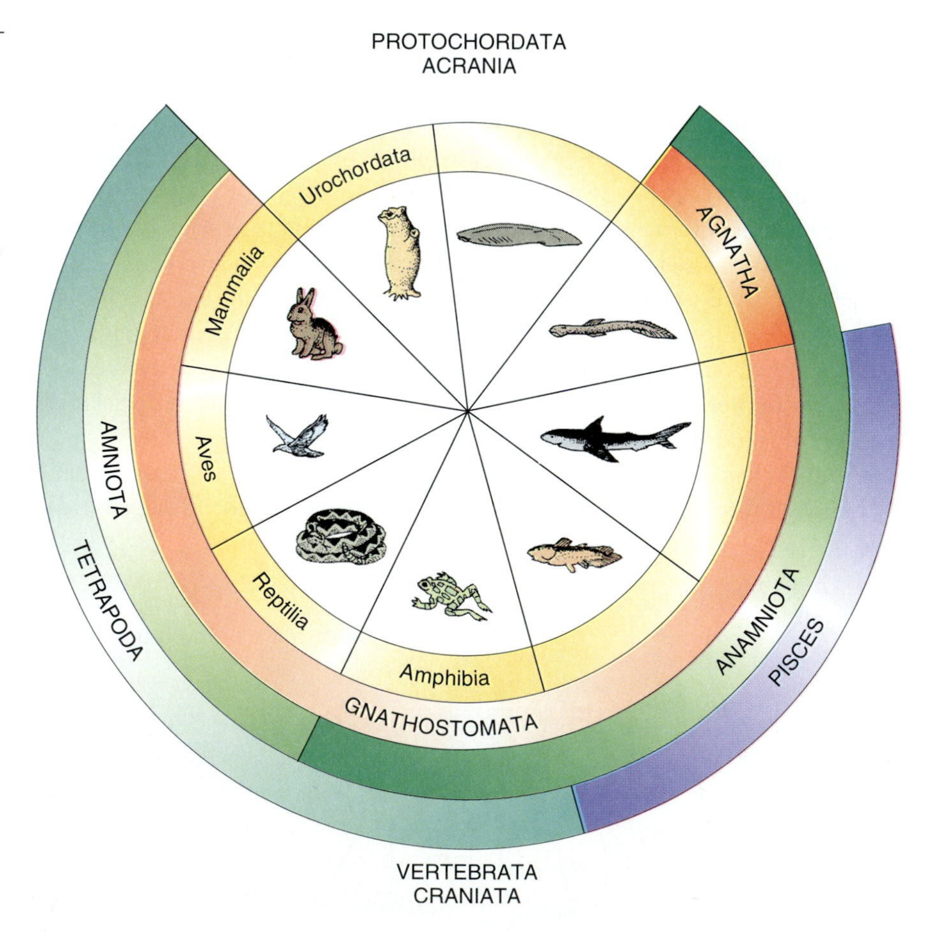

FIGURE 34.16
The major groups of chordates.

Question: From Fig. 21.6, what is the name of the clade that is synonymous with tetrapods? What is the name of the clade that is synonymous with amniotes? Why are Pisces and Anamniota not valid clades?

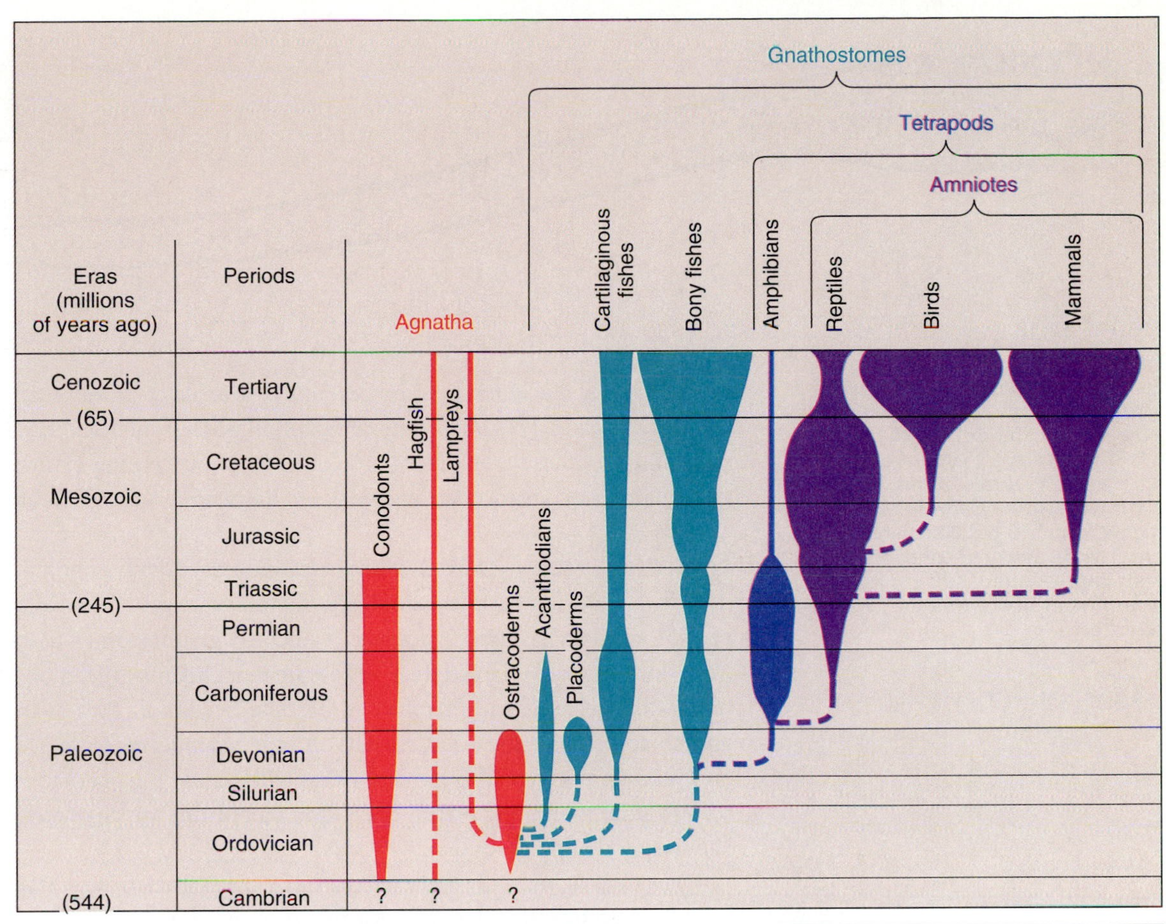

FIGURE 34.17

Evolution of the major groups of vertebrates. The width of each group suggests its diversity. Dashed lines represent hypothetical relationships.

referred to as Agnatha; all other vertebrates have mouths with jaws and are therefore called Gnathostomata. The gnathostomes are divided into Pisces (PI-seez), which are the jawed fishes with paired fins, and Tetrapoda (TET-ra-PO-duh), which generally have two pairs of limbs primitively adapted for terrestrial locomotion. Fishes and amphibians, which do not develop within a fluid-filled sac (the **amnion**), are called Anamniota. As will be described shortly, reptiles, birds, and mammals develop within an amnion, and they are therefore referred to collectively as Amniota.

Milestones in Vertebrate Evolution

AGNATHANS. The evolutionary origins of major groups of vertebrates are uncertain and controversial, but we can at least sketch out some of the major transitions. The oldest known vertebrates were jawless forms called agnathans. Fossils of agnathans are found in North America and Asia in marine deposits more than 500 million years old (Fig. 34.17). The oldest group of agnathans may have been **conodonts,** which were known for many years only by abundant fossils of their teeth (Fig. 34.18 on page 740). Conodonts appear to have been related to **hagfish,** species of which are still extant and will be described in the next chapter. Another extant group of agnathans are **lampreys.** Lampreys may have evolved from **ostracoderms,** so-called for the bony plates covering them (Greek *ostrakon* shell + *derma* skin). Judging from their tiny mouths and flat bellies, ostracoderms ate small food particles on the bottoms of streams and ponds (Fig. 34.19 on page 740). Ostracoderms apparently had cartilaginous rather than bony skeletons, and many lacked

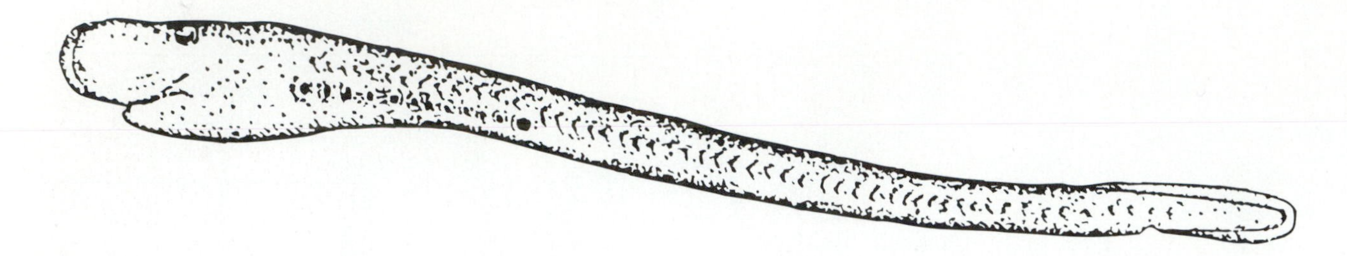

FIGURE 34.18

Reconstruction of an early conodont. Length approximately 4 cm. Until the 1980s, conodonts were known only by their conical "teeth," which had been attributed to every kind of organism from algae to arrowworms. These teeth are important indicator fossils for sediments from 515 to 200 million years old. The discovery of conodont fossils with impressions of soft tissues provided the evidence that they were vertebrates.

■ **Question:** What evidence visible here suggests that conodonts were chordates?

paired fins. By the Silurian period millions of years later, most radiated into freshwater streams and lakes. Presumably the competition was less intense there, with few large animals and no other vertebrate of any kind. Before becoming extinct at the end of the Devonian period, ostracoderms gave rise to the ancestors of vertebrates with jaws.

GNATHOSTOMES. The oldest known jawed vertebrates, or gnathostomes, appeared during the Silurian period around 420 million years ago. The evolution of jaws was a pivotal event in evolution, because it enabled gnathostomes to become predatory. More important, it enabled their vertebrate descendants to feed on plants and on animals with skeletons and therefore to become terrestrial. Two of the earliest kinds of gnathostomes were the **acanthodians** and the **placoderms** (Fig. 34.20). Acanthodians and placoderms were also the first fishes with paired fins, and acanthodians also had scales rather than bony plates. Scales and paired fins are characteristic of most cartilaginous and bony fishes that live today.

Jaws are commonly believed to have originated from cartilaginous supports called **gill arches.** (See Fig. 12.7 for the location of gill arches in modern fishes.) On each side of the head of the gnathostome ancestor a pair of gill arches may have evolved into part of the upper and lower jaws (Fig. 34.21). No fossil has been found showing an intermediate stage of this evolution, but there is indirect evidence that it occurred: (1) In certain primitive fishes the jaws resemble gill arches; (2) in sharks the jaws develop from gill arches; and (3) the branching of the cranial nerve to the jaws resembles the branching of cranial nerves to the gill arches in cartilaginous fishes.

FIGURE 34.19

Reconstruction of a Devonian ostracoderm, *Hemicyclaspis.* The two flaps behind the gills are not paired fins. Length approximately 20 cm.

A

B

FIGURE 34.20

Two early gnathostomes. (A) An acanthodian, *Euthacanthus.* Acanthodians are also called spiny sharks because of a spine on the leading edge of each fin. Length approximately 20 cm. (B) Cast of the head of a placoderm, *Dunkleosteus terrelli,* in which the jaws were developed with a vengeance. With a body length of more than 10 meters, this must have been a fearsome predator.

Initially the jaws may have served only to close the mouth. Later they became armed with scales that evolved into teeth.

TETRAPODS. An innovation that was equally important in the evolution of terrestrial vertebrates occurred during the Devonian period approximately 370 million years ago, when some bony fishes developed reinforcements in their fins that enabled them to support their weight in shallow water and on land. They are believed to have been the first tetrapods, which later diverged into amphibians, reptiles, birds, and mammals. The type of fish that evolved into tetrapods is still a matter of debate, as will be discussed in the next chapter. The relationship of tetrapods to fishes is clear in *Ichthyostega,* the earliest tetrapod for which there is a complete reconstruction. Although it shares many features with fishes, *Ichthyostega* is considered to have been an amphibian because it had a skeleton adapted to support its weight on land, digits at the ends of its limbs, and a middle ear that could hear airborne sounds (Fig. 34.22 on page 742). Its fossils are found in what is now eastern Greenland, in deposits about 360 million years old.

AMNIOTES. Although amphibians are adapted for living on land at least part of the time, they must have water to reproduce. The water is required to keep the embryo moist and as a place into which metabolic wastes from the embryo can diffuse. The complete independence from water required a self-contained egg capable of storing water and wastes, as well as nutrients for the embryo. A self-contained egg is called a **cleidoic egg** (kly-DOH-ik). The cleidoic egg of a vertebrate is also an **amniote egg,** because an internal water supply bathing the embryo is contained in an extraembryonic membrane called the **amnion.** Another extraembryonic membrane, the **allantois,** holds the metabolic wastes and helps a third extraembryonic membrane, the **chorion,** exchange respiratory gases (pp. 117–118). The evolution of the amniote egg was a major milestone that allowed the evolution of reptiles, birds, and mammals starting in the early Carboniferous period some 340 million years ago. The evolution of these groups will be described in more detail in later chapters.

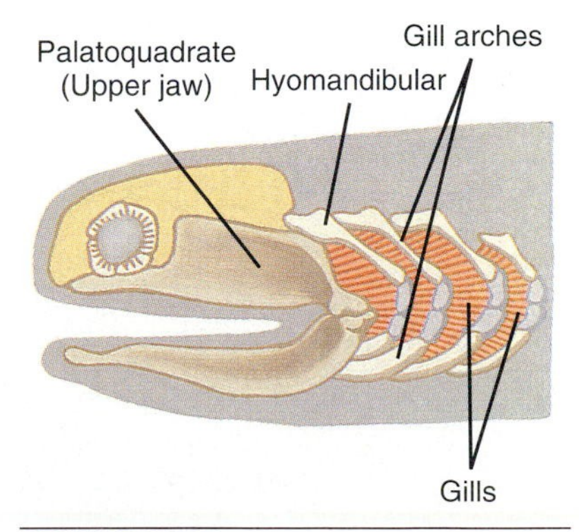

FIGURE 34.21

Possible evolution of a cartilaginous gill arch into the left half of an upper and lower jaw.

FIGURE 34.22

(A) Similarities of the lobe-finned fish *Eusthenopteron* and the amphibian *Ichthyostega,* both of which were about a meter long and lived during the Devonian period. Neither animal is thought to have been ancestral to extant tetrapods, but they are chosen because their reconstruction from fossils is so nearly complete. The skulls (B and C) and teeth (D) of both animals had similar structures. The middle-ear bone (columella) of the amphibian evolved from a jaw bone of the fish. The pineal eye was a third eye that still occurs in some vertebrates. *Ichthyostega* belonged to an extinct group of amphibians called labyrinthodonts because of the labyrinth-like folding of the dentine in the teeth. (E) The lobed fins of the fish show homology with the limbs of the tetrapod.

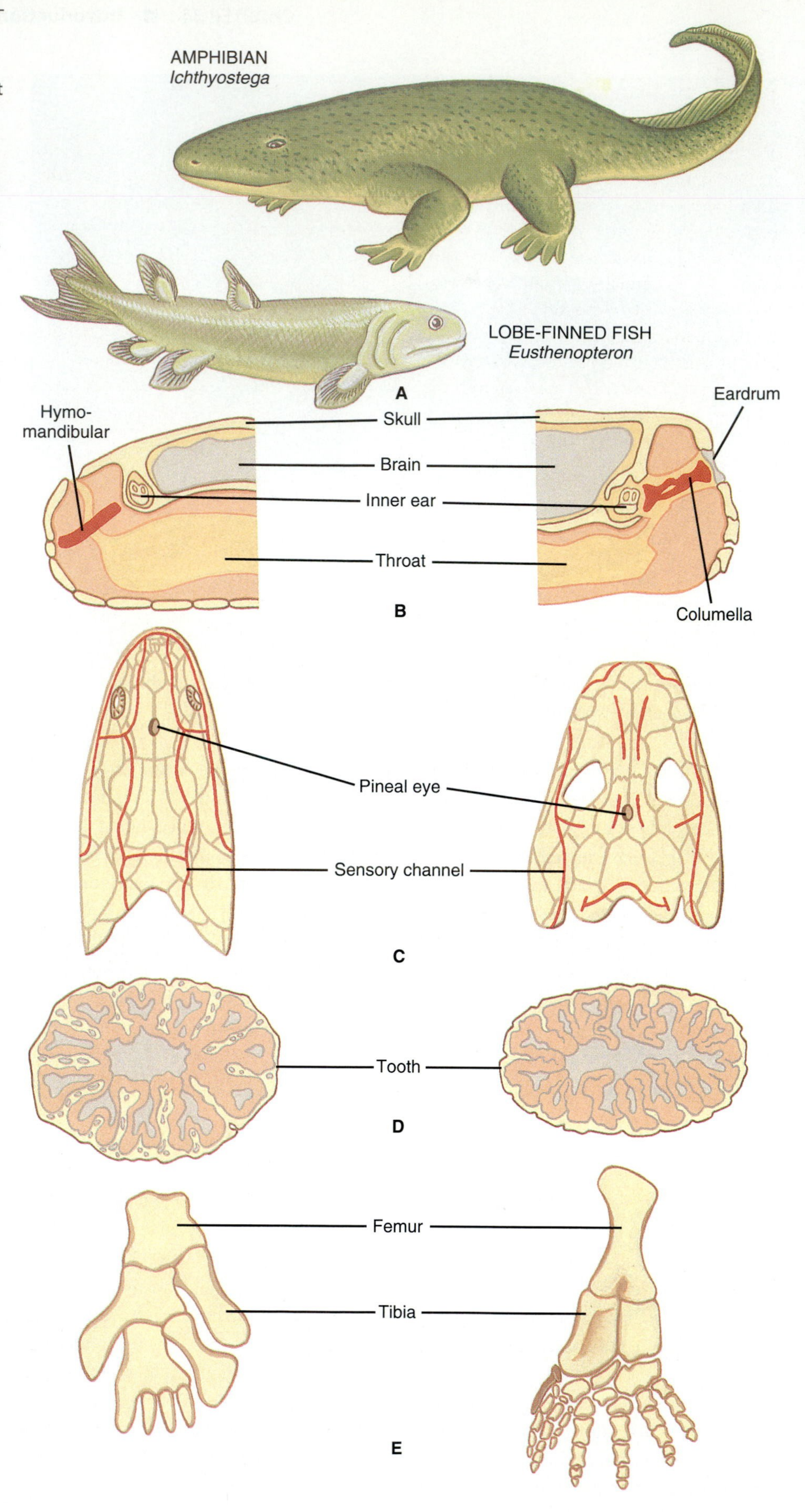

Summary

The phyla Hemichordata and Chordata are apparently closely related. Both are deuterostomate and have pharyngeal slits and a dorsal nerve cord. Phylum Hemichordata comprises enteropneusts (acorn worms), which burrow in marine sediments, and pterobranchs, which are usually colonial and live in secreted tubes. The essential differences between hemichordates and chordates are due to the notochord. The notochord guides embryonic development, and it is replaced by vertebrae in most vertebrates. Also unique to chordates are a post-anal tail and myomeres. Chordates are divided into protochordates and vertebrates. Protochordates include urochordates (sea squirts, larvaceans, and thaliaceans) and cephalochordates (lancelets). Vertebrates are divided into the agnathans, which are jawless fishlike animals, and gnathostomes. The gnathostomes are vertebrates with jaws: fishes, amphibians, reptiles, birds, and mammals. All gnathostomes except fishes are tetrapods, and all tetrapods except amphibians are amniotes. The evolution of amniotic eggs enabled reptiles, birds, and mammals to reproduce on land, since the eggs were self-contained. Four legs support their bodies on land, and jaws enable these animals to feed on land plants and animals.

Key Terms

pharyngeal slit
dorsal nerve cord
notochord
post-anal tail
myomere
branchial sac
agnathan
gnathostome
gill arch
tetrapod
amniote
cleidoic egg

Chapter Test

1. Describe the two classes of hemichordates. Why are these dissimilar animals in the same phylum? (pp. 726–728)
2. What distinguishing characters occur in both hemichordates and chordates? How does the notochord provide for the different body plans of the two phyla? (p. 725)
3. Give the common names for the three subphyla of chordates. Name a representative example in each subphylum. (pp. 730–738)
4. Describe the functions of the branchial sac in adult ascidians. (pp. 733–734)
5. Explain why the lancelet is so often studied in biology courses. (pp. 737–738)
6. Name a common example of each of the following kinds of vertebrates: agnatha, gnathostome, pisces, tetrapod, anamniote, amniote. To which of these groups do humans belong? (pp. 738–741)
7. Explain why each of the following is considered a milestone in vertebrate evolution: jaws; tetrapody; the amniote egg. (pp. 740-741)

■ Answers to the Figure Questions

34.2 Bivalve molluscs. See Fig. 28.6B

34.4 The body plans of adult ectoprocts and pterobranchs are remarkably similar, and some taxonomists have suggested that they are more closely related to each other than pterobranchs are to enteropneusts.

34.10 The other invertebrates listed here propel themselves by means of cilia; the ascidian larva propels itself with a tail, which is possible only because of reinforcement by the notochord.

34.12 The incurrent filters would keep out sand and other debris; the feeding filters traps smaller food particles.

34.13 Perhaps the asexual generation is adapted for producing large numbers of offspring, while the sexual generation maintains genetic variety. Aphids also have alternating sexual and asexual generations.

34.16 Tetrapods are referred to as Choanata in Fig. 21.6, and amniotes are the Reptilomorpha. Pisces would be paraphyletic, since it would not include all the descendants of the most recent common ancestor of Chondrichthyes and Osteichthyes. Anamniota would likewise be paraphyletic.

34.18 Myomeres, pharyngeal slits, and possibly post-anal tail.

Readings

Recommended Readings

Alldredge, A. 1976. Appendicularians. *Sci. Am.* 235(1):94–102 (July).

Forey, P., and P. Janvier. 1994. Evolution of the early vertebrates. *Am. Sci.* 82:554–565.

Pfennig, D. W., and P. W. Sherman. 1995. Kin recognition. *Sci. Am.* 272(6):98–103 (June).

Radinsky, L. B. 1987. *The Evolution of Vertebrate Design.* Chicago: University of Chicago Press.

See also relevant selections in General References at the end of Chapter 21.

Additional References

Barrington, E. J. W., and R. P. S. Jefferies (Eds.). 1975. *Protochordates.* Symposium of the Zoological Society of London, No. 36. New York: Academic Press.

Briggs, D. E. G. 1992. Conodonts: a major extinct group added to the vertebrates. *Science* 256:1285–1286.

Carroll, R. L. 1988. *Vertebrate Paleontology and Evolution.* New York: W. H. Freeman.

Conway Morris, S., and H. B. Whittington. 1979. The animals of the Burgess Shale. *Sci. Am.* 241(1):122–133 (July).

Gee, H. 1994. Return of the amphioxus. *Nature* 370:504–505.

Gould, S. J. 1989. *Wonderful Life: The Burgess Shale and the Nature of History.* New York: W. W. Norton.

Sansom, I. J., et al. 1992. Presence of the earliest vertebrate hard tissues in conodonts. *Science* 256:1308–1311.

Wada, H., and N. Satoh. 1994. Details of the evolutionary history from invertebrates to vertebrates, as deduced from the sequences of 18S rDNA. *Proc. Nat'l. Acad. Sci. USA* 91:1801–1804.

CHAPTER

35

Fishes

Spine cheek clown fish *Premnas biaculeatus.*

The Success of Fishes

From the sunlit surface to the deepest ocean trench, and from the clearest mountain stream to the bottom of the murkiest pond, fishes abound. More than 21,000 named species—nearly half of all vertebrates—are fishes. Fishermen, zoologists, and amateur divers who invade their habitat with nets, submersible research vessels, and diving gear discover approximately 100 new species of fish each year. The number of individual fishes is also greater than that of all other vertebrates. The evident success of fishes may be due to their having been among the first to take full advantage of the chordate features summarized in the previous chapter. These features are adapted in fishes in such a way that most can swim actively throughout a three-dimensional aquatic habitat, rather than settling to substratum, like many aquatic invertebrates.

One chordate distinction, the notochord, is replaced during development in most fishes by a more supportive yet flexible **vertebral column.** This part of the skeleton provides a framework for the attachment of other skeletal structures and muscles that enable fishes to swim rapidly. A second feature of chordates, the post-anal tail, gives rise to the **caudal fin** that supplies most of the propulsive force in swimming. The pharyngeal slits become openings to the efficient **gills** that extract the oxygen needed to sustain a high level of activity. Finally, the dorsal hollow nerve cord is greatly enlarged at the anterior end, forming a **brain** that coordinates activity in relation to information about the environment.

An Overview of the Kinds of Fishes

AGNATHANS. The term "fishes" is an informal one referring to primitively aquatic vertebrates: agnathans, cartilaginous fishes, and bony fishes. Agnathans are often considered not to be fishes because they do not have jaws, scales, or paired fins on the sides. In fact, they lack vertebrae, even though they are considered to be vertebrates. Agnathans were among the first to take advantage of the possibilities of vertebrate life (pp. 739–740). Their jawless descendants constitute the clade or superclass Agnatha (see Figs. 21.4, 21.6, 34.16, 34.17). Agnatha now includes approximately 75 species of hagfishes and lampreys (class Myxini and class Cephalaspidomorphi) (Fig. 35.1).

CARTILAGINOUS FISHES. Jawed fishes and all other extant vertebrates make up the clade or superclass Gnathostomata. One of the two major groups of jawed fishes,

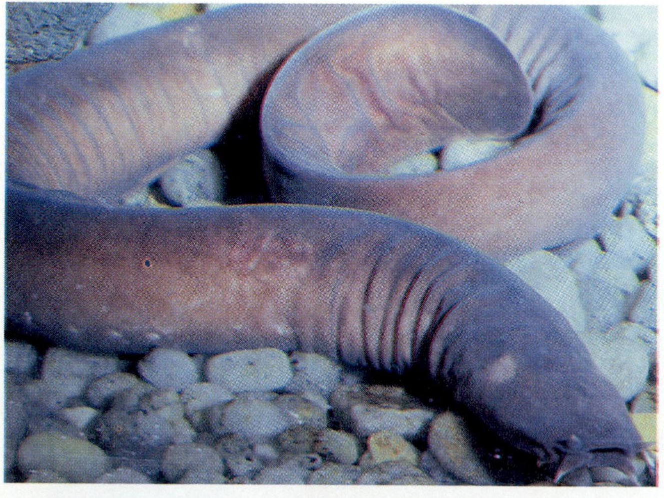

A

B

FIGURE 35.1
Jawless fishes. (A) The Pacific hagfish *Eptatretus stoutii.* Length approximately 65 cm. (B) Sea lampreys *Petromyzon marinus* parasitizing a carp, *Cyprinus carpio.* Length of each lamprey approximately 80 cm.

the clade or class Chondrichthyes, comprises the jawed fishes with cartilaginous skeletons. Class Chondrichthyes is divided into two subclasses, the Elasmobranchii and the Holocephali. The elasmobranchs include more than 300 species of sharks and more than 400 species of rays (including skates). Sharks are the most familiar cartilaginous fishes. Like most fishes, they have scales and two pairs of fins on the sides, as well as a caudal fin and unpaired fins along the dorsal and ventral midline. Rays have few scales, and their fins are highly modified in keeping with their overall flattened shape. Most elasmobranchs are predatory, and their ability to engulf prey is aided by an upper jaw that swings open like the lower jaw. The second major group of cartilaginous fishes, the subclass Holocephali, comprises approximately 30 species of chimaeras (pronounced ky-MERE-uhs; Fig. 35.2). Chimaeras, commonly called ratfishes, are distinguished from elasmobranchs by an upper jaw that is fused to the skull. Their bodies are intermediate in shape between those of most sharks and most rays.

BONY FISHES. The other major group of jawed fishes is the traditional class Osteichthyes. In these fishes bone replaces much of the cartilage in the skeleton during

A

B

C

FIGURE 35.2

Cartilaginous fishes. (A) The spiny dogfish shark *Squalus acanthias,* which is often used as a laboratory subject. (Length approximately 1.5 meters.) (B) The blue-spotted ray *Taeniura lymma.* Stingrays such as this one defend themselves with two venomous spines about midway along the tail, which can cause excruciating pain and even death to a person. Length approximately 2.4 meters. (C) Spotted ratfish (chimaera) *Hydrolagus colliei* of the Pacific. (Length approximately 1 meter.)

FIGURE 35.3

Representative bony fishes. (A) A lobe-finned fish, the Australian lungfish *Neoceratodus forsteri.* Of the six living species of lungfishes, this one is most like the lungfishes of the Paleozoic era. (Length approximately 1.5 meters.) (B) The yellow perch *Perca flavescens,* which is often studied in zoology laboratories as a representative ray-finned fish. (Length approximately 38 cm.)

■ **Question:** Why is the perch called a ray-finned fish?

A

B

development. There are two main groups of bony fishes: lobe-finned fishes and ray-finned fishes. Lobe-finned fishes are named for their fan-shaped lateral fins with fleshy lobes at the base (Fig. 35.3A). These lobes are reinforced with bones that are homologous to the limb bones of tetrapods (see Fig. 34.22E). Because of this homology, and because many have lungs and can breathe air, a lobe-finned fish of some kind is generally thought to share its most recent ancestor with amphibians and other tetrapods. Most bony fishes have fins that are not lobed but reinforced by bony rays. Among these ray-finned fishes (clade or subclass Actinopterygii) are almost all the familiar species that come to mind when one thinks of fish, including trout, tuna, bass, goldfish, and perch (Fig. 35.3B). Most fishes are ray-finned, so they will be discussed extensively later in this chapter.

Agnathans

HAGFISHES. One group of jawless fishes is the class Myxini, which comprises more than 30 living species of hagfishes (Figs. 35.1A and 35.4 on page 751). Hagfishes inhabit temperate sea floors at depths greater than 25 meters, where they eat small invertebrates and scavenge dead and dying fishes and larger invertebrates. Using their tongues, which are toothed like snail radulas, they burrow into carcasses and feed from within. This habit and their shape may account for Linnaeus having mistaken them for intestinal worms.

Hagfishes are sometimes a nuisance to deep-sea fishermen, since they swim into nets to feed on the catch. The worst part comes when a fisherman removes the hagfish

A Classification of Fishes, According to Nelson (1994)

Genera mentioned elsewhere in this chapter are noted.

Superclass Agnatha (= Cyclostomata) AG-na-thuh (Greek *a*-without + *gnathos* jaws). Jawless vertebrates. Vertebrae rudimentary or absent. Modern forms without true appendages (paired fins), scales, or bones. Notochord persists in adult. Gill pouches empty through uncovered pores rather than slits.

Class Myxini mix-EYE-nye (Greek *myxa* mucus). Terminal mouth with four pairs of sensory barbels. Nasal sac connecting to pharynx. Five to 15 pairs of gill pouches with one shared opening on each side. Hagfishes. *Eptatretus, Myxine* (Fig. 35.1A).

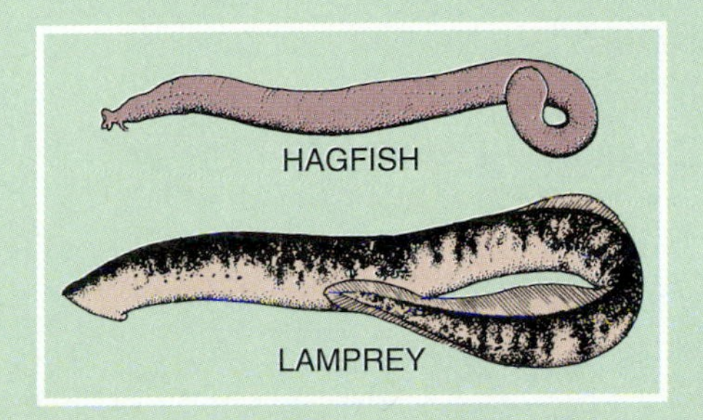

Class Cephalaspidomorphi sef-a-LASS-pid-oh-MORE-fy (Greek *kephale* head + *aspido-* shield + *morphe* shape). Sucking mouth. Nasal sac not connected to pharynx. Seven pairs of gill pouches. Lampreys. *Petromyzon* (Fig. 35.1B).

Superclass Gnathostomata NATH-oh-STOW-ma-tuh (Greek *stoma* mouth). With jaws. Notochord usually absent in adult; replaced by vertebrae. Most with paired appendages. Jawed fishes, amphibians, reptiles, birds, and mammals.

Class Chondrichthyes kon-DRIK-the-eez (Greek *chondros* cartilage + *ichthys* fish). Fishes with cartilaginous skeletons. No swim bladder or lung. Spiral valve in intestine. Male with pelvic claspers for sperm transfer. Teeth not fused to jaws; replaced continually.

Subclass Elasmobranchii e-LAZ-mo-BRAN-kee-eye (Greek *elasmos* plate + *branchia* gills). Five to seven pairs of gill slits without covers, and usually a pair of gill openings modified as spiracles. Upper jaw movable. Placoid scales usually present. Rigid dorsal fin. Sharks and rays (including skates). *Carcharodon, Cephaloscyllium, Cetorhinus, Isistius, Odontaspis, Pristis, Raja, Rhincodon, Squalus, Taeniura, Torpedo* (Figs. 35.2A, 35.2B, 35.9B, 35.12).

Subclass Holocephali HO-low-SEF-al-lye (Greek *holos* whole). Upper jaw fused to skull. Four gill slits with a single cover on each side. No spiracles. Without scales. Males with club-shaped clasping organ in front of the eyes (in addition to pelvic claspers). Chimaeras. *Hydrolagus* (Fig. 35.2C).

box continues

Class Osteichthyes OS-tee-IK-the-eez (Greek *osteon* bone). Fishes with bony skeletons. Scaly skins. Several gills on each side with one gill cover. Usually with air sacs that function either as lungs or as swim bladders for buoyancy.

Subclass Dipneusti (= Dipnoi) dip-NEW-sty (Greek *di-* two + *pneustikos* of breathing). One or two lungs. Lobed pectoral fins. All median fins fused together. Spiral valve in intestine. Lungfishes. *Neoceratodus, Protopterus* (Fig. 35.3A).

Subclass Crossopterygii cross-op-ter-RIJ-ee-eye (Greek *krossoi* fringe + *pteryx* wing). With lobed pectoral fins. Two lungs in extinct species; one fat-filled lung in the living species. Spiral valve in intestine. The only known living species is the coelacanth *Latimeria chalumnae*. (Fig. 35.13).

Subclass Brachiopterygii (= Cladistia) brake-ee-op-ter-RIJ-ee-eye (Greek *brachion* arm). Body elongated or eel-like. With lungs and lobed pectoral fins. Dorsal fin divided into 5 to 8 finlets, each with a spine. Spiral valve in intestine. (Often grouped in subclass Actinopterygii, superorder Chondrostei.) Bichirs and reedfish.

Subclass Actinopterygii ACT-in-op-ter-RIJ-ee-eye (Greek *aktis* ray). Fins supported by rays. Swim bladders instead of lungs. Ray-finned fishes. (See Fig. 35.14 for examples.)

Superorder Chondrostei kon-DROSS-te-eye. Primitive ray-finned fishes. Skeleton mostly cartilaginous. Notochords in adults. Ganoid scales. Spiral valve in intestine. Sturgeons and paddlefishes. (Reedfish and bichirs often included.) *Acipenser* (Fig. 35.15).

Superorder Holostei ho-LOS-te-eye. "Intermediate" ray-finned fishes. Skeleton mainly bony. Swim bladder adapted as a lung. Scales ganoid or cycloid. Spiral valve in intestine. Bowfin and gars. *Amia, Lepisosteus* (Fig. 35.16).

Superorder Teleostei TEL-lee-OS-te-eye (Greek *teleos* complete). Skeleton almost completely bony. Scales cycloid, ctenoid, or absent. No spiral valve in intestine. Vestigial notochord. Swim bladder maintains buoyancy. Nelson lists 35 extant orders; some of the more important ones are represented in Fig. 35.14. *Alosa, Anguilla, Anomalops, Astroscopus, Bothus, Cyprinus, Danio, Electrophorus, Gnathonemus, Gymnarchus, Gymnotus, Helena, Hippocampus, Holacanthus, Malaptererus, Monocirrhus, Oncorhynchus, Pendaka, Perca, Poecilia, Premnas, Pterois, Salvelinus* (Opening photos for Unit 3, for Chapters 2 and 20, and for this chapter; Figs. 16.1, 18.14, 20.10, 20.17, 24.16, 35.3B, 35.21, 35.24A, 35.29, 35.30).

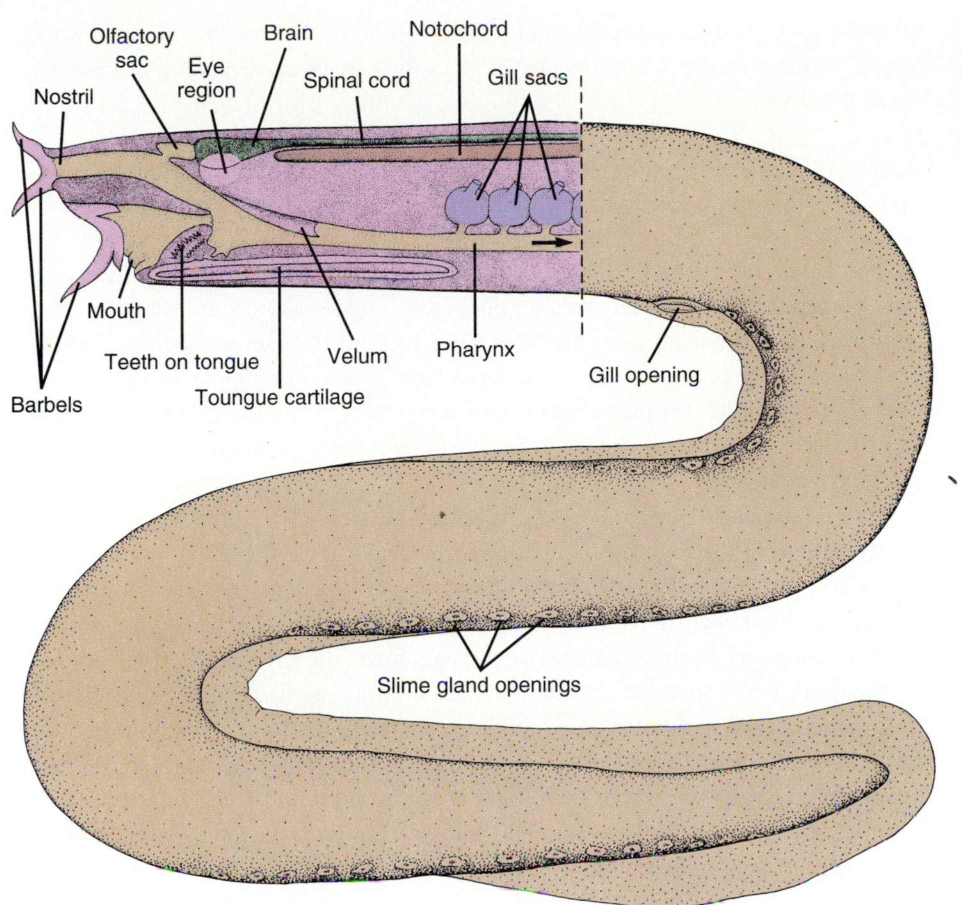

FIGURE 35.4

The Atlantic hagfish *Myxine glutinosa.* The eyes are degenerate and functionless, but there are light-sensitive organs in the skin. The velum pumps water from the single nostril through the gill sacs. Note the absence of vertebrae and the presence of the notochord in the adult. The "teeth" are not homologous with those of most vertebrates.

■ **Question:** What features of the hagfish are also found in conodonts (Fig. 34.18)?

and ends up covered in **slime.** The slime comes from unique glands with 90 openings distributed along the sides. Out of these openings come entire cells that break upon release. There are two kinds of slime gland cells; one kind releases a glycoprotein that forms clear mucus in seawater, and the other kind releases protein threads up to 60 cm long that make the mucus sticky and viscous. The resulting slime effectively protects hagfishes from predators but could endanger the hagfish itself by plugging the nostril through which it takes in water. The hagfish rids itself of slime by tying itself into an overhand knot at the rear, then sliding the knot forward over its head. The hagfish uses the same maneuver to slip out of the grasp of a predator, or to pull off chunks of flesh when feeding (Fig. 35.5 on page 752).

In spite of the slime, many zoologists like hagfishes. They are the closest living relatives of conodonts (p. 739) and may provide a glimpse of the physiology of the earliest vertebrates. Like most marine invertebrates, but unlike other vertebrates, hagfishes have ion concentrations in their body fluids that are approximately as high as those in seawater. Thus they do not require the special osmoregulatory adaptations of most other fishes. Hagfishes are also the only vertebrates that retain in the adult both a **pronephros** and a **mesonephros:** two kinds of kidney that occur only in the embryos of most vertebrates (p. 286). Perhaps the strangest thing about hagfishes is that they have four different hearts that boost the blood pressure through gills and other areas. What little we know about the reproduction of hagfishes also appears unusual. Each individual has one long gonad, the front part of which becomes an ovary in females, and the back part of which becomes a testis in males. Fewer than 1% of netted hagfish are males. The female lays approximately two dozen large eggs (approximately 2 cm) with tufts at one

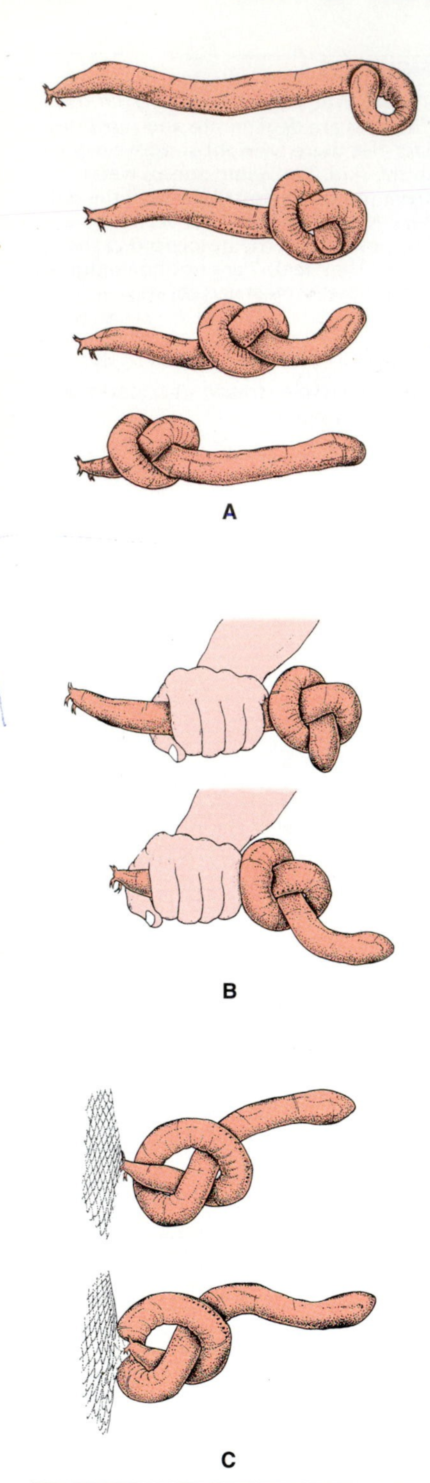

FIGURE 35.5
The cartilaginous body enables the hagfish to tie itself into a sliding knot to (A) clear itself of slime, (B) escape a predator, or (C) tear flesh.

end that anchor them to each other and to substratum. A prize for describing how the eggs get fertilized has been unclaimed ever since the Copenhagen Academy of Science offered it in 1864.

LAMPREYS. The remaining 41 living species of agnathans are lampreys (class Cephalaspidomorphi). Like hagfishes, lampreys lack jaws and paired fins, and they have eel-like bodies with cartilaginous skeletons. Lampreys differ from hagfishes in many ways, several of which reflect their different ways of feeding. Many species, such as the sea lamprey *Petromyzon marinus,* parasitize (or prey on) fishes, using the keratinized "teeth" on their tongues and **oral discs** to bore through the skin to feed on blood (Figs. 35.1B and 35.6). Other lampreys have poorly developed "teeth" and are not parasitic. Parasitic lampreys locate hosts with eyes that are better developed than those of hagfishes. Lampreys may also be able to detect electric fields generated by the muscle activity of hosts.

Also unlike hagfishes, lampreys spend much of their lives in fresh water, where they reproduce. Using its mouth, a male lamprey begins moving pebbles to excavate a shallow nest and is soon joined in the work by one or more females. Males then spawn repeatedly for more than a week, fertilizing numerous eggs (up to a quarter of a million for the sea lamprey). Each egg hatches into a larval form, the **ammocoete** (pronounced AM-mo-seat), that is so unlike the adult that it was long classified as a different species. The ammocoete drifts out of the nest and burrows tail first into a muddy bottom with its oral hood sticking out for suspension feeding. Generally, after four to seven years the ammocoete metamorphoses into an adult. In nonparasitic species the adults remain in the stream without feeding, and they live only long enough to reproduce. Some parasitic species migrate downstream to the ocean, but paradoxically the sea lamprey migrates into lakes without ever entering the sea. After a period of feeding, the parasitic species return to the streams to breed, and they die soon afterward.

The sea lamprey is now familiar to, and despised by, those who fish the Great Lakes, Lake Champlain, and many other large lakes in North America. An average sea lamprey deprives sport fishermen of approximately 20 kg of fish during the year and a half of its life as an ectoparasite. In some lakes virtually every lake trout, whitefish, salmon, carp, or other game fish has at least one lamprey on it, or the scar of a lamprey. This was not always the case. In Lake Ontario, where sea lampreys were native, they had little effect on fisheries. The construction of the Welland Canal around Niagara Falls in 1829, however, provided the sea lamprey with access to Lake Erie, and from there they migrated into Lakes Huron, Michigan, and Superior. By the 1940s and 1950s they had virtually eliminated fisheries in these lakes. Those fisheries recovered after the compound TFM (3-trifluoromethyl-4-nitrophenol) was found to be toxic to ammocoetes when applied to the streams in which they live. (TFM damages gills, but it is not known how.) In the past decade, however, lampreys have begun to breed in streams where this compound cannot be applied economically or safely, and they are now threatening a $2 billion economy based on sport fishing.

Sharks

It seems appropriate to begin the description of gnathostomes with sharks: They immediately come to mind when one hears the word "jaws." Most sharks are too shy to live up to their reputations as vicious maneaters, but there is no denying that many are highly adapted to swiftly pursue and quickly dismember prey, including humans if the opportunity arises.

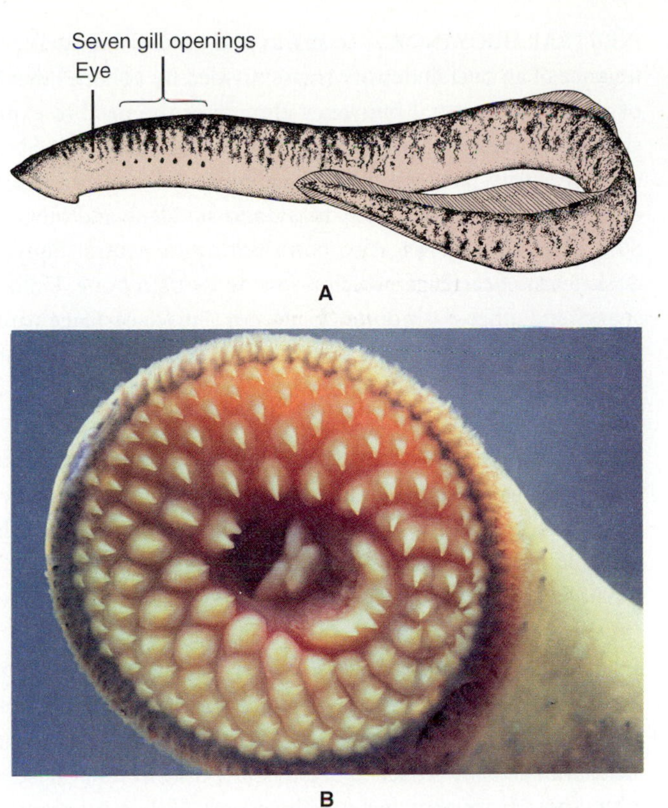

FIGURE 35.6

(A) The sea lamprey *Petromyzon marinus.* (B) The oral disc of a sea lamprey. The sea lamprey uses its oral disc not only for feeding, but also to move stones for nest building and to anchor itself to stones. (Hence the name *Petromyzon,* from the Greek words for "stone" and "suck.") Lampreys also use the oral disc to attach to other fishes for free transportation. The mouth is free to perform these functions because water flows into and out of each one of the seven pairs of gills through an individual opening.

FINS. The predatory abilities of sharks are due largely to their ability to swim with speed and agility, thanks to their characteristic fins. The caudal fin, which is attached to a narrow **caudal peduncle,** provides most of the forward propulsion in sharks and other fishes. In most sharks the vertebral column extends into the dorsal lobe of the caudal fin, which is larger than the ventral lobe (Fig. 35.7). Such fins with lobes of different sizes are termed **heterocercal.** For several decades it has been thought that a heterocercal tail tends to spin a fish head downward about its center of mass, and that this downward pitch has to be overcome by lift from the pectoral fins. Careful analysis by K. S. Thomson (1990), however, indicates that this is not necessarily the case. Sharks can apparently make small changes in the shape of the heterocercal tail to change the direction of thrust for diving, climbing, or turning.

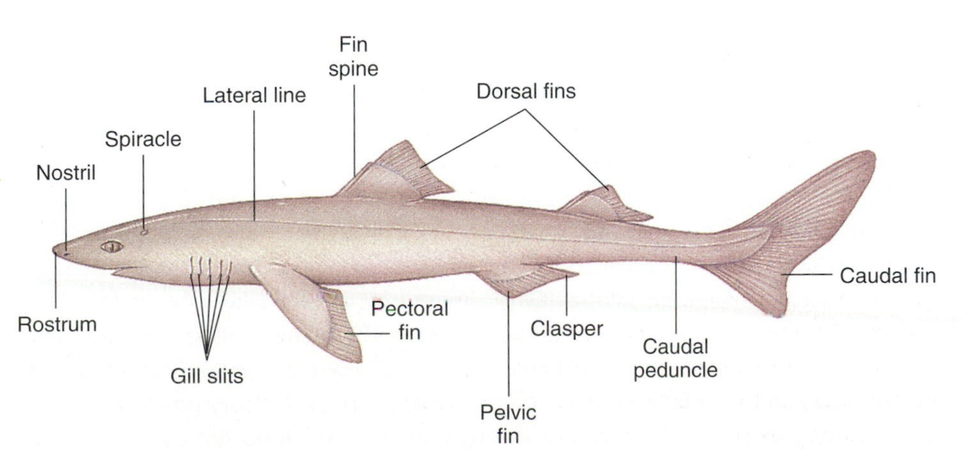

FIGURE 35.7

External features of a male dogfish. (Compare Fig. 35.2A.)

■ **Question:** How would a female dogfish differ?

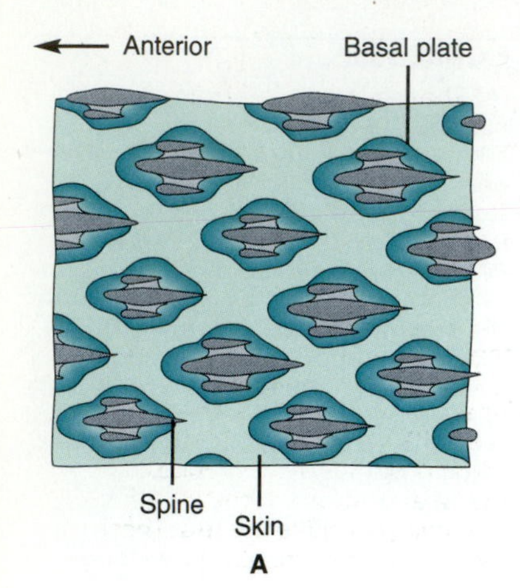

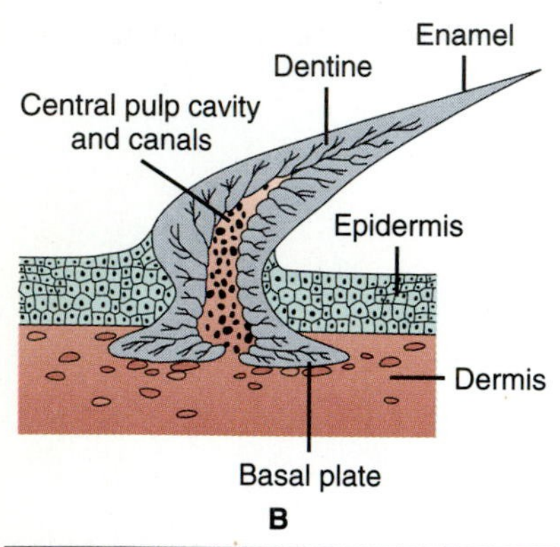

FIGURE 35.8

Placoid scales. (A) Surface view of shark skin. (B) Section through the spine of a single scale. The scales are homologous with the teeth of vertebrates. (Compare Fig. 13.6A.)

NEUTRAL BUOYANCY. In sharks and other fishes, swimming is aided by the maintenance of an overall density (mass divided by volume) that is nearly the same as that of water. This neutral buoyancy eliminates the need to expend energy to keep from sinking or floating to the surface. Sharks maintain neutral buoyancy mainly by means of low-density fats, especially **squalene,** in their large livers. The liver typically accounts for 25% of total body mass, and squalene and other low-density oils make up 80% of the liver mass. Also contributing to neutral buoyancy in sharks is the endoskeleton of cartilage, which is less dense than bone. Unlike most other vertebrates, sharks and other elasmobranchs do not replace cartilage with bone during embryonic development, except for a few places around the teeth. This is a derived character, since the ancestors of elasmobranchs were bony.

SKIN. The tough skin of sharks aids the cartilaginous skeleton in supporting the body and in swimming. Unlike skeletal muscles of most vertebrates, which apply their force to the skeleton, the muscles of sharks pull on the skin, which transmits the force to the caudal fin during swimming. The muscles used in swimming account for much of the body mass, especially in the tail. These muscles are arranged in segmental blocks (**myomeres**), each of which is shaped like a W on its side. The significance of this arrangement will be discussed in a later section on locomotion. Another adaptation of the skin to swimming are the numerous scales with plate-shaped bases embedded in the skin. These **placoid scales** have backward-pointing spines that feel rough to the human touch, but actually reduce hydrodynamic drag in water. Placoid scales are homologous with teeth. They have a central pulp, dentin, and an outer covering of enamel (Fig. 35.8).

FEEDING AND DIGESTION. The structural similarity of a placoid scale to a vertebrate tooth (see Fig. 13.6A) is not coincidental. The teeth of sharks are merely placoid scales that become enlarged as they migrate over the jaw (Fig. 35.9). These teeth, which are generally numerous, sharp, and curved into the mouth, are all the more formidable because the loose attachment of both jaws to the cranium allows predaceous sharks to open their jaws widely to engulf larger prey. The loose attachment of the jaws also enables many sharks to extend the jaws forward so that the protruding snout does not interfere with feeding. Many predaceous sharks, such as the great white shark *Carcharodon carcharias,* bite out a chunk of flesh and let prey bleed to death before feeding on it. Their preferred prey are mammals with lots of body fat, such as dolphins and whales. There is some indication that they attack humans by mistake and find us rather unappetizing (Klimley 1994). Not all sharks use their teeth in feeding. The largest sharks feed on plankton, which they filter out of the water with their gills. These include basking sharks (*Cetorhinus*), which are up to 15 meters longs, and whale sharks (*Rhincodon*), which are up to 18 meters long. The digestive tracts of sharks are similar to those of other vertebrates (see Chapter 13), except for the **spiral valve** in the intestine, which slows the passage of food and increases the surface area for digestion and absorption (Fig. 35.10 on page 756).

SENSORY RECEPTORS. Sharks locate prey using a variety of sensory receptors. The small eyes are usually not considered to be of much use in predation, although there is some debate on this point. Like all fishes, amphibians, and snakes, sharks focus by moving the lens back and forth in the eye rather than by changing its shape. The most important receptors appear to be chemoreceptors. There are chemoreceptors on the body surface and mouth, but the most sensitive ones are in olfactory pits in the head. Water circulates to the olfactory pits through nostrils, which do not connect to the pharynx, and therefore have no function in respiration.

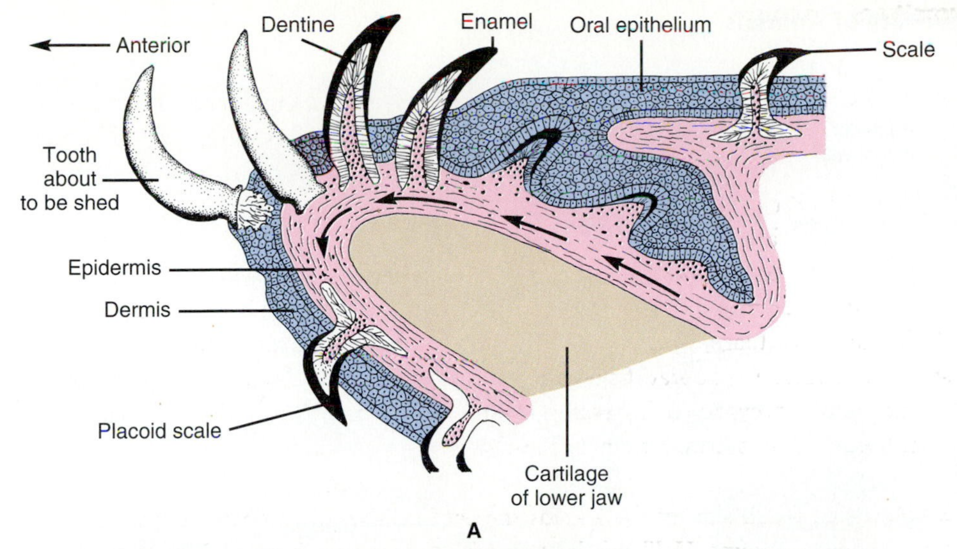

A

B

C

FIGURE 35.9

(A) The lower jaw of a shark, showing the teeth, which are placoid scales. Like all placoid scales, the teeth of sharks are continually lost and replaced. Arrows indicate the continual outward movement of the teeth as they mature and are shed. Unlike the teeth of other vertebrates, sharks' teeth are not attached to the skeleton. (B) Head of the sand tiger shark *Odontaspis taurus,* showing the teeth. This species feeds on fishes, squids, and the occasional human. Total length is over four meters. (C) Cookie-cutter sharks, *Isistius plutodus* and *I. brasiliensis,* have a few large teeth with which they slice out pieces of flesh from whales, porpoises, and other sharks after attaching to them with their suckerlike mouths. These sharks are approximately 40 cm long.

Question: Do sharks chew their food?

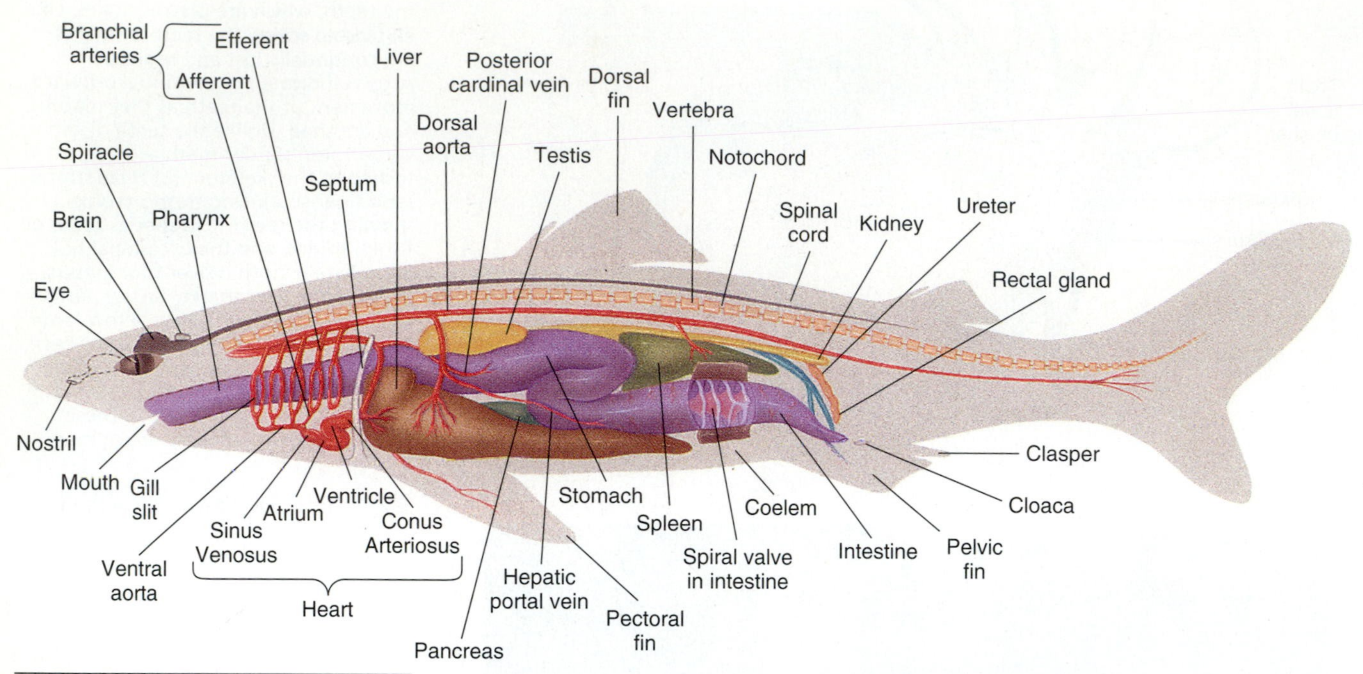

FIGURE 35.10

Internal structure of a male spiny dogfish.

■ **Question:** Why is the liver so large?

Several kinds of receptors are concentrated in a canal called the **lateral line organ,** part of which lies just beneath the **lateral line** visible in Fig. 35.7. The lateral line organ also curves around the head. Among the kinds of receptors in the lateral line organ are electroreceptors that are so sensitive they can detect the electrical activity of muscles in potential prey buried in sand. Electroreceptors may also enable sharks to detect ocean currents and to sense Earth's magnetic field for navigation. Also in the lateral line organ and elsewhere on the body surface are mechanoreceptors called **neuromasts,** which sharks may use to detect water currents generated by prey, predators, other sharks, and obstacles. The neuromasts are similar to the mechanoreceptors of the **inner ears** (see Fig. 8.15C). The inner ears of sharks consist of the **semicircular canals,** which sense acceleration of the body. Associated with the semicircular canals are **otolith organs** that respond to gravity and perhaps also to low-frequency sound. A fish has no need for external ears because the tissues have a density similar to that of water, and they are therefore transparent to sound. Additional mechanoreceptors possibly involved in prey detection are located in **spiracles,** which are modified gill slits unique to elasmobranchs.

RESPIRATION AND CIRCULATION. Gills supply the oxygen needed to sustain predation and other activities of sharks. There are generally five pairs of gills, although a few species have six or seven pairs. Each gill lies in its own gill pouch and has a separate gill slit. The structure and functioning of elasmobranch gills are essentially the same as for other fishes, as described elsewhere (pp. 244–246). Slow-swimming and bottom-dwelling species pump water past the gills by expanding and contracting the gill pouches. Pelagic sharks, which swim in the open ocean almost constantly, use **ram ventilation,** in which water is forced into the open mouth and past the gills. When the mouth is busy with other shark business, water enters through the spiracles. Blood circulates through the gills via branchial arteries and enters the dorsal aorta after becoming oxygenated. As in most vertebrates, red blood cells contain the respiratory pigment hemoglobin. After the blood circulates through the body tissues and becomes deoxygenated, it is pumped back to the gills by the heart, which has four chambers in series: the **sinus venosus,** the **atrium,** the **ventricle,** and the **conus arteriosus.**

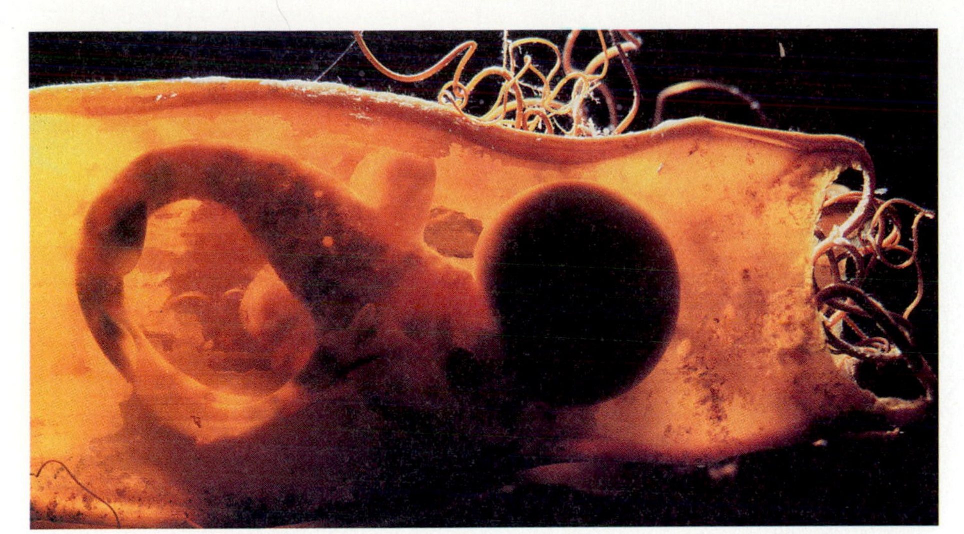

FIGURE 35.11

Egg capsule of the swell shark *Cephaloscyllium ventriosum,* containing an embryo. The tendrils at the ends may anchor the capsule to seaweed. Note the large, spherical yolk sac.

EXCRETION AND OSMOREGULATION. The gills of sharks and other fishes also play a role in excretion and osmoregulation. Ammonia, a nitrogenous waste from the metabolism of amino acids, diffuses across the gills into water. Sharks also produce other nitrogenous wastes, urea and methylamines, that do not diffuse across the gills. These are retained in high concentrations in the body fluids, where they serve as **osmolytes** that help maintain an osmotic concentration comparable to that of the surrounding seawater (p. 286). Although the total osmolarity of the body fluids is similar to that of seawater, the concentration of salts in the body fluids is lower than that of seawater, suggesting that sharks evolved from freshwater fishes. The ions that tend to accumulate in the body fluids from the environment are excreted mainly by the **rectal gland.** The rectal gland empties into the **cloaca,** which also receives feces and gametes. The **mesonephric kidney** is of secondary importance (p. 286).

REPRODUCTION. Fertilization is internal in most sharks. In males sperm are transported to the cloaca from the testes via the **mesonephric ducts** (= Wolffian ducts), which also drain the kidneys. A **clasper** behind the pelvic fins transfers the sperm from the cloaca into the oviduct (Müllerian duct) of the female. Some species are definitely viviparous, with a placenta-like structure connected to the embryo by an umbilical cord. In some sharks the embryo develops within the oviduct, and the mother gives birth to live young. These sharks are often said to be **ovoviviparous,** which implies that the egg is simply retained inside the oviduct during hatching, without any other support from the mother. Usually, however, there is little or no shell around the egg, and an area of the oviduct modified as a uterus provides some support to the embryo. It is therefore likely that even these live-bearing sharks are **viviparous.** A few species are indisputably **oviparous.** Their eggs are enclosed within a tough capsule in which the embryo develops externally (Fig. 35.11). These embryos depend on a generous supply of yolk, which makes the egg quite large (up to 10 cm).

Rays

About half the species of elasmobranchs are rays, including skates. Rays differ from most sharks in having few scales and in being adapted for feeding on bottom-dwelling animals, such as molluscs and crustaceans. Rays have a few platelike teeth adapted for crushing such prey. Their bodies are flattened dorsoventrally, enabling them to glide slowly over the bottom (Fig. 35.2B). The eyes and spiracles are on top of the head.

FIGURE 35.12
The smalltooth sawfish *Pristis pectinatus,* which occurs in estuaries and shallow coastal waters from the Chesapeake Bay south. (Length approximately 5 meters.)

Because the mouth is often buried in bottom deposits, water for the gills enters through the spiracles on the dorsal surface and exits through gill slits on the ventral surface. The pectoral fins are greatly enlarged, and they undulate gracefully during swimming. The tail fins are reduced or absent. Some rays, notably the electric rays, have electric organs used for predation, defense, and perhaps communication (Fig. 35.28). These organs consist of noncontracting muscle cells whose action potentials summate to as much as 220 volts. Most rays are viviparous, but those commonly called skates (family Rajidae) release eggs within rectangular cases—the "mermaids' purses" familiar to beachcombers. Another distinctive group of rays is the sawfishes (family Pristidae), which have sawlike snouts with which they thrash at schools of fish and sift through sediments for small prey (Fig. 35.12).

Chimaeras

Chimaeras are not as well known as elasmobranchs, since they are generally smaller (less than 2 meters long), are never fished, and do not eat people. Someone must have been impressed by them, however, to have named them after the chimaera of Greek myth, which was said to have a lion's head, a goat's body, and a serpent's tail. This description fits their overall shape (Fig. 35.2C). Chimaeras were numerous during the Devonian period some 400 million years ago, but the surviving 30 species are now restricted to deep offshore waters, where they feed mainly on molluscs, crustaceans, and other invertebrates. Like rays, chimaeras have a few large, platelike teeth. Unlike elasmobranchs, chimaeras have the upper jaw fused to the skull (hence the name of the subclass: Holocephali).

Lobe-Finned Fishes

Among the bony fishes (class Osteichthyes), three extant groups are often referred to as lobe-fins, because most have lateral fins with fleshy lobes at the base. These lobes are reinforced by bones that articulate with the rest of the skeleton and are homologous with the limb bones of tetrapods. In these fishes the caudal fins typically taper posteriorly to a point; this shape is called **diphycercal** (DIF-i-SIR-kal).

LUNGFISHES. Members of one group of lobe-fins, the lungfishes (subclass Dipneusti or Dipnoi), have one or two lungs with which they can breathe air. These lungs connect to the esophagus and are inflated by swallowing air. Six species of lungfishes survive from better times hundreds of millions of years ago. Four species in the genus *Protopterus* occur in freshwater swamps in Africa, one species lives in South America, and one inhabits shallow rivers in Queensland, Australia (Fig. 35.3A). The

FIGURE 35.13

A living coelacanth photographed from a submersible research vessel in its natural habitat 100 to 200 meters deep in the western Indian Ocean. Note the diphycercal tail. The fish is approximately 1.5 meters long.

■ **Question:** Based on appearance alone, which group of lobe-finned fishes seem most closely related to amphibians and other tetrapods: *Eusthenopteron* (Fig. 34.21), lungfishes (Fig. 35.3A), or the coelacanth?

South American and African species use their lungs to hold air reserves while burrowed into mud for up to two years to escape drought. In addition, while buried, the African lungfishes enter a state called **estivation,** in which they surround themselves in a cocoon of mucus and drastically reduce their metabolic activities. Considerable evidence points to lungfishes as the closest living relatives of tetrapods, although some zoologists contend that the following group, coelacanths, have that distinction. [The extinct group Panderichthyidae is now considered to have been more closely related to the tetrapods than either lungfishes or coelacanths (Ahlberg and Milner 1994).]

COELACANTHS. Coelacanths (SEAL-uh-kanth; subclass Crossopterygii) flourished at the same time as lungfishes, and they also had lungs. Rather than burrow into mud, however, they are more likely to have crawled away from drying ponds on their stout fins. Coelacanths are interesting primarily because they were long thought to have been the ancestors of tetrapods and to have become extinct 60 million years ago. Then in 1938 fishermen near the Comoro Islands in the western Indian Ocean caught a living coelacanth that found its way to Marjorie Courtenay-Latimer, curator of a small museum in South Africa, who informed J. L. B. Smith, a chemist and ichthyologist. Headlines around the world bannered the discovery. It was as if a living Neanderthal person had been spotted on the streets of Hollywood. In honor of Miss Courtenay-Latimer and the mouth of the river near where it was caught, the coelacanth was named *Latimeria chalumnae* (Fig. 35.13).

Zoologists were as excited as anyone by the discovery of *Latimeria,* because a living coelacanth would provide a rare opportunity to test their speculations about the evolution of tetrapods. They had to wait until 1952 for a second coelacanth, but since then they have acquired approximately 200 specimens, enough to raise alarm about endangering the species. Subsequent research on the coelacanth has taken some of the edge off the excitement. In some ways *Latimeria* is more like an elasmobranch than an ancestor of tetrapods. It produces urea and other osmolytes that are used in osmoregulation, and it has a spiral valve in the intestine. *Latimeria* has only the vestige of one lung, which is filled with low-density fat that helps maintain buoyancy. One feature not found either in elasmobranchs or modern tetrapods is a joint over the brain that allows the front part of the skull to flip upward when the mouth opens. The female gives birth to live young from eggs the size of tennis balls. How these eggs get fertilized is a mystery, since the males have no apparent copulatory organs. Unfortunately, coelacanths live for less than a day after being brought to the surface, so many details of physiology are unclear.

Bichirs and Reedfishes. Eleven extant species of bichirs and reedfishes (subclass Brachiopterygii) constitute the third major group of lobe-fins. Bichirs and reedfishes share many features with ray-finned fishes, and they are sometimes classified with them. All inhabit stagnant ponds and streams in Africa, where they get most of their oxygen by gulping air into lungs rather than by using their gills. They propel themselves along the bottom and through vegetation with their lobed fins, and they can also feed on land.

Ray-Finned Fishes

Paddlefishes and Sturgeons. Members of the largest subclass of bony fishes, the Actinopterygii, are characterized by fins reinforced by bony rays that are not attached to the rest of the skeleton. The ray-finned fishes include most fishes, which have adapted in virtually every imaginable way to almost every aquatic habitat. Some of the orders of ray-finned fishes are represented in Fig. 35.14. One distinctive group is the superorder Chondrostei, which comprises the paddlefishes and sturgeons (Fig. 35.15 on page 764). (Many taxonomists also include the bichirs and reedfishes in this group.) As a group, chondrosteans are quite ancient and retain many features that are considered primitive. Most of the 25 living members have sharklike features such as ventral mouths, spiracles, spiral valves in the intestine, unfused upper jaws, and largely cartilaginous skeletons. Unlike most sharks, however, adult chondrosteans have small teeth or none at all. Sturgeons suck up invertebrates from the bottom, and paddlefishes filter feed by trapping plankton with **gill rakers,** which will be described later. Chondrosteans also differ from sharks in having a cover (**operculum**) over the gill openings.

Bowfin and Gars. Gars and the bowfin, *Amia calva,* both have spiral valves in their intestines and can breathe air (Fig. 35.16 on page 764). Because they share these structures they are often combined in the same superorder Holostei. The bowfin and gars differ, however, in the kinds of scales they have. The bowfin has thin, round, overlapping **cycloid scales,** and gars have thick, nonoverlapping **ganoid scales** (Fig. 35.17 on page 765).

Teleosts

The remaining ray-finned fishes are extremely diverse and undoubtedly polyphyletic, but they are commonly united under the name teleosts. There is only enough space here for a pictorial survey of some of the familiar teleosts (Fig. 35.14), a more detailed discussion of the yellow perch *Perca flavescens* as a representative teleost, and comments on some of the most striking departures from the perciform model. Special adaptations of teleosts will be noted in later sections.

Fins. Like most ray-finned fishes, perches have fins supported by calcified **fin rays** (Figs. 35.3B and 35.18 on page 765). In the perch the fin rays are either stiff and unjointed spines, as in the anterior dorsal fin, or flexible and branching, as in the posterior dorsal fin. These dorsal fins, as well as the single anal fin, have limited mobility, and they mainly help keep the fish from rocking and turning when the caudal fin moves.

The pectoral and pelvic fins are more mobile and help control vertical movement, turning, and sudden stopping. In perches the pelvic fins are close to the pectoral fins, an arrangement that is considered more advanced than in most other teleosts, especially lobe-fins, chondrosteans, and holosteans. The pectoral fins also provide most of the for-

FIGURE 35.14

Some common ray-finned fishes, representing 14 of the 36 extant orders recognized by Nelson (1994). Numbers of species worldwide are according to Nelson.

■ **Question:** Which of these fishes have you eaten?

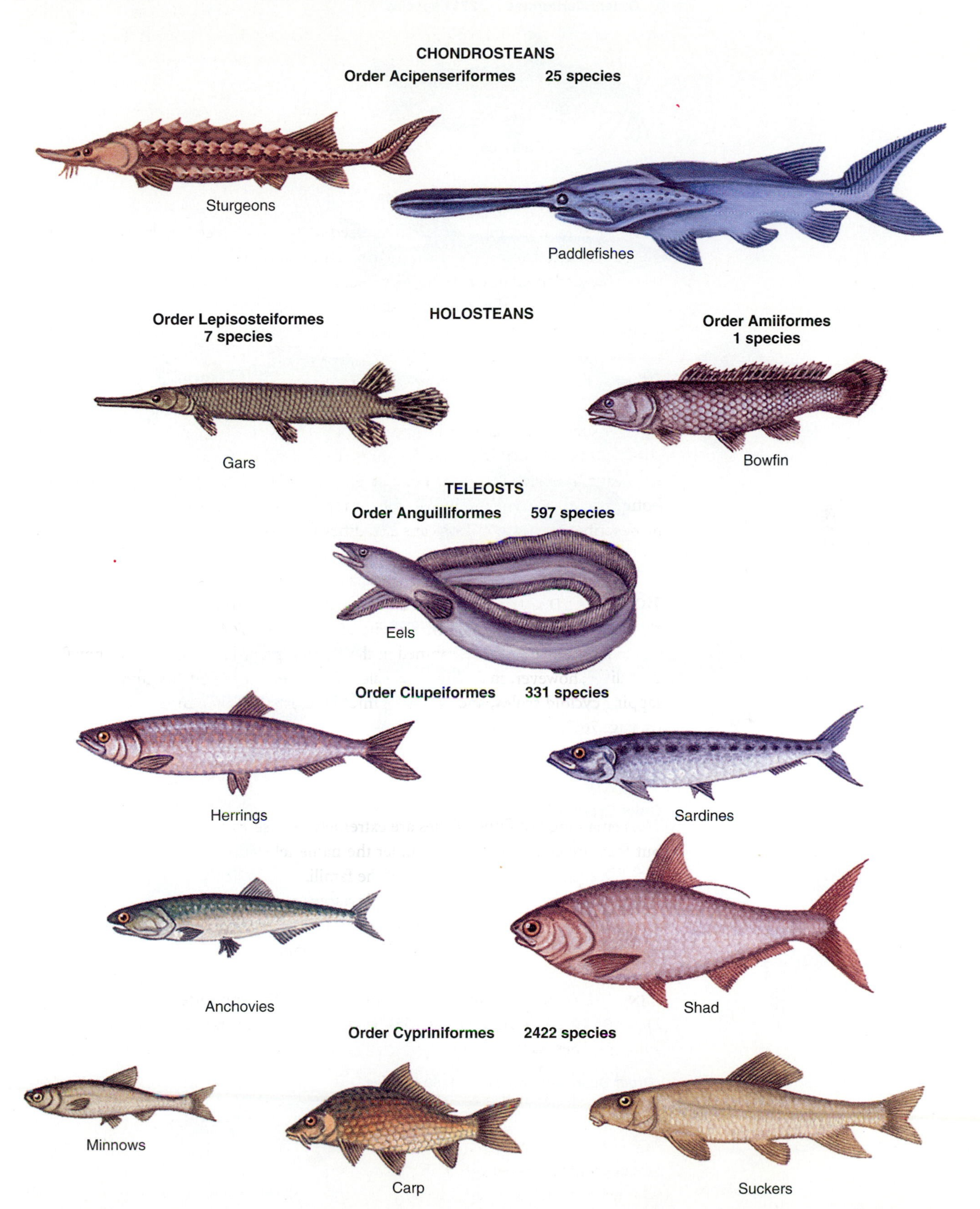

Figure continues

Order Siluriformes 2211 species
Catfishes
Order Salmoniformes 320 species
Pikes
Trout, salmon, whitefishes, and other salmonids
Order Gadiformes 414 species
Cods
Hakes
Order Cyprinodontiformes 845 species
Flyingfishes
Killifishes
Order Scorpaeniformes 1160 species
Scorpionfishes
Sculpins

Order Perciformes 7791 species
Perches
Basses
Seabasses
Sunfishes
Drums
Barracudas
Tunas
Mackerels
Swordfish
Order Pleuronectiformes
538 species
Order Tetraodontiformes
329 species
Halibut, flounders, soles, and
other flat fishes
Puffers

FIGURE 35.15

A lake sturgeon *Acipenser fulvescens.* These fish attains lengths of up to 2.4 meters. Schools of them may account for persistent stories of monsters in Lake Champlain and other freshwater lakes.

■ **Question:** What would be the function of the barbels on the mouth?

ward thrust during slow swimming; the caudal fin is used for acceleration and cruising. As in most teleosts, the dorsal and ventral lobes of the caudal fin are similar in size, and the vertebral column does not extend into either lobe. Such symmetric caudal fins are termed **homocercal,** in contrast to the heterocercal caudal fins of sharks (Fig. 35.7) and chondrosteans and the diphycercal caudal fins of lobe-finned fishes (Fig. 35.13).

GILLS. Just anterior to the pectoral fin on each side is the operculum, which covers the four gills (see Fig. 12.7). When swimming rapidly, fishes simply open the mouth and allow **ram ventilation** to drive water past the gills and out the operculi. Tunas and mackerels are, in fact, obliged to use ram ventilation, and they must cruise continually

FIGURE 35.16

A Florida gar *Lepisosteus platyrhincus* hangs motionless in a sluggish stream. Neutral buoyancy, due to the swim bladders, is essential for the "lie-in-wait" style by which gars prey on other fishes. Note the diamond-shaped ganoid scales.

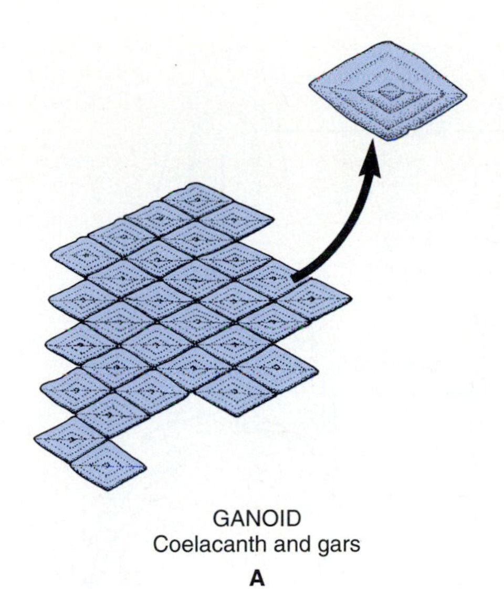

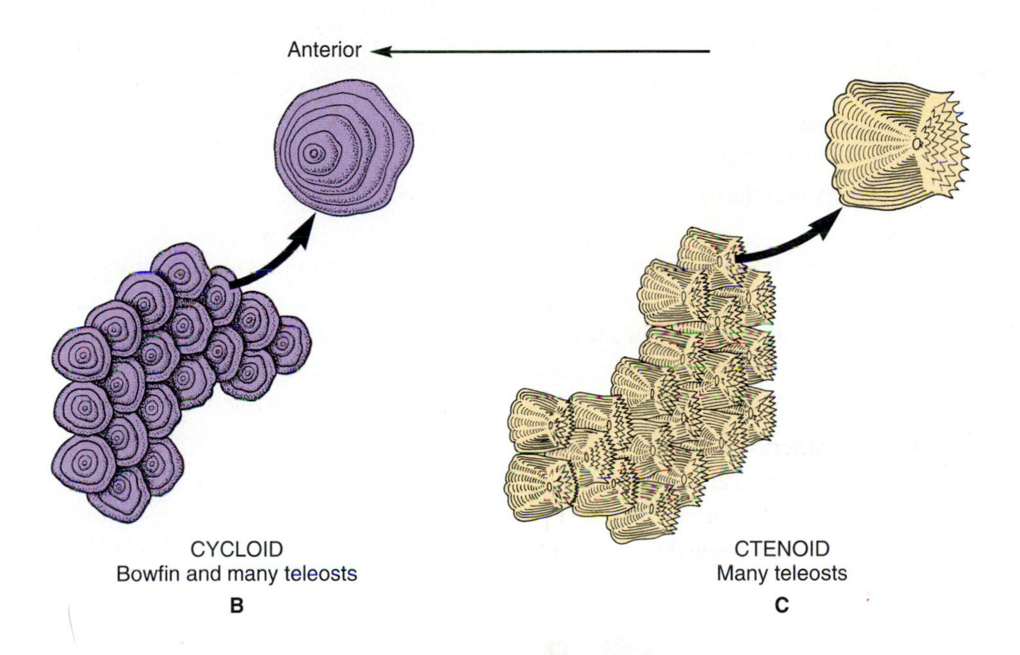

FIGURE 35.17

Types of scales in ray-finned fishes. (A) Ganoid scales are thick and nonoverlapping and are composed of bone overlaid with an enamel-like substance called ganoin. (B) Cycloid scales are thin and overlapping, permitting more flexibility. Unlike ganoid scales, cycloid scales grow as the fish grows, and in some species they show annual growth rings. (C) Ctenoid scales are essentially cycloid scales with teeth at the posterior edges. The teeth are believed to reduce hydrodynamic drag during swimming. Paddlefishes, catfishes, eels, and some other fishes sacrifice the protection afforded by scales in favor of the increased flexibility of having none.

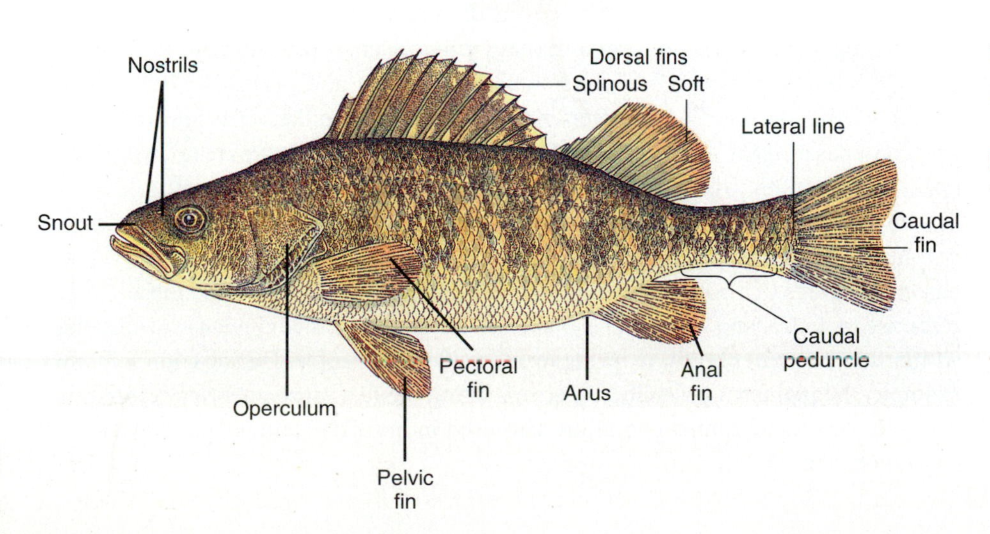

FIGURE 35.18

External structure of the yellow perch.

■ **Question:** Do you think the term "pelvic fin" is appropriate?

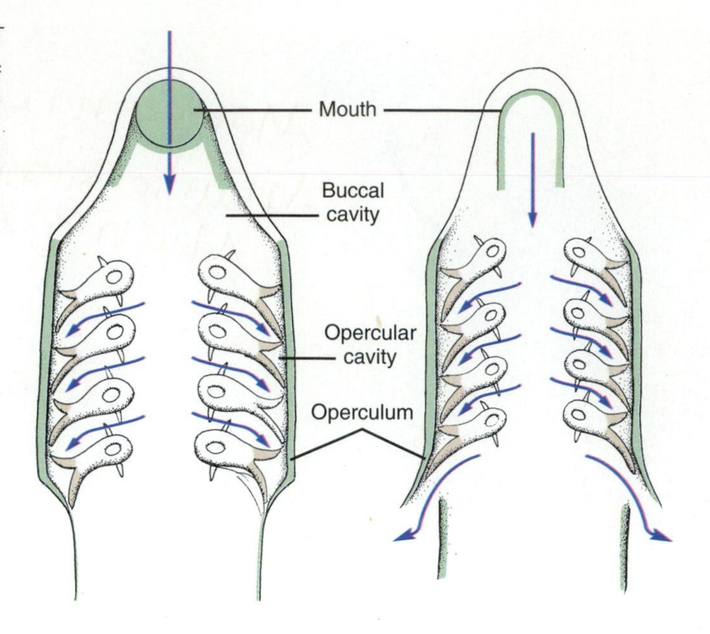

FIGURE 35.19

Buccal pumping viewed from the top of a fish. Water is first drawn into the buccal cavity (left). The mouth then closes, and the buccal cavity contracts, forcing water into the opercular cavity (right). After a short delay the operculum opens, emptying the opercular cavity.

in order to obtain oxygen. When stationary or swimming slowly, perches and other teleosts ventilate their gills by **buccal pumping,** which involves compressing the **buccal** (mouth) **cavity** and the **opercular** (gill) **cavity** in sequence (Fig. 35.19).

FEEDING. Perches and other teleosts also use buccal and opercular pumping to "slurp" up plankton and small invertebrates and fishes. This **suction feeding** is aided by a complicated arrangement of jaw bones that enables both the upper and lower jaws to swing open and forward with incredible speed (Fig. 35.20). **Gill rakers** on the gill arches aid in trapping small food particles, either by filtering them out of the water or by directing them to the mucus-covered roof of the oral cavity (Sanderson et al. 1991). (See Fig. 12.7.) Larger bits of food are caught in numerous small teeth fused to the bones of the jaw and on the pharynx. Teleosts that prey on large fishes, eat coral, or scrape algae and other food off rocks generally have a firm, nonprotrusible upper jaw, which is considered to be a primitive character.

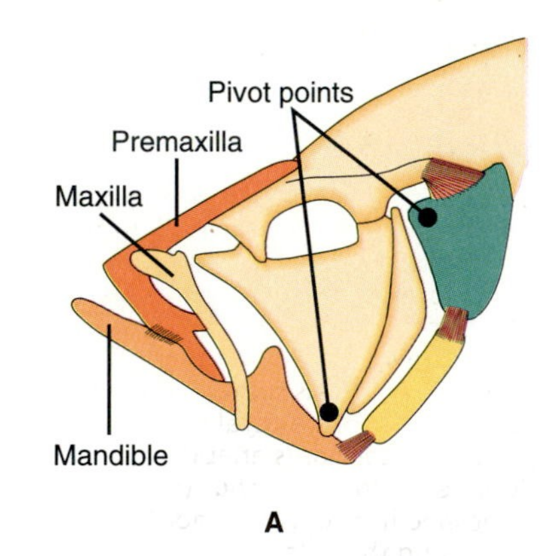

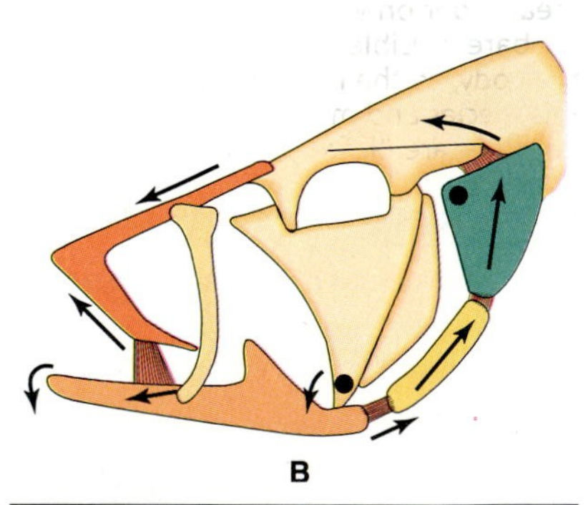

FIGURE 35.20

Mechanism of jaw opening and protrusion in the South American leaffish *Monocirrhus polyacanthus* (order Perciformes). (A) Jaws closed. (B) Jaw opening and protruding. Arrows show directions of movement. This is one of four mechanisms of jaw protrusion (Motta 1984).

INTEGUMENT. The entire body of the perch is covered with mucus-secreting epidermis. The mucus protects against abrasion and infection and probably reduces hydrodynamic drag. The epidermis also covers the **scales,** which grow out of the dermis, overlapping like shingles. Perches and many other teleosts have **ctenoid** (TEN-oyd) **scales,** while other teleosts have cycloid scales or none at all (Fig. 35.17C). Unlike sharks, teleosts do not shed and replace their scales. Instead the scales grow continually, compensating for wear. The **colors** that make fishes so attractive are not within the scales, but are mainly due to special skin cells called **chromatophores** (see Table 10.1). The silvery and iridescent colors common among fishes are due to reflection from chromatophores that contain guanine. The major colors of the yellow perch are due to chromatophores that contain yellow and black pigments (carotenoid and melanin). In response to neural and hormonal signals, these pigments either condense or disperse within the chromatophores, reducing or intensifying the color. During courtship, for example, the pigments generally disperse, making the fish more intensely colored. In many teleosts the chromatophores are also used in mimicry, camouflage, or warning coloration (Fig. 35.21).

FIGURE 35.21

Uses of skin pigmentation in teleosts. (A) The lion fish *Pterois volitans* (order Scorpaeniformes) is gaudy at night. Like many coral-reef fishes, lion fishes are dull-colored and camouflaged during the day. Lion fishes use the feather-shaped, venomous spines to herd prey fish into crevices in coral reefs. (B) Chromatophores enable the eyed flounder *Bothus manchus* (order Pleuronectiformes) to camouflage itself by copying the color and pattern of the ocean floor on which it rests. Both eyes are barely visible on the upper side of the body, to the left. Most members of this species and most other species of flatfishes are "lefteyed" like this. Young flatfishes are born bilaterally symmetric like most fishes, but gradually the mouth and eyes migrate to one side or the other, and the fish settles to the bottom.

■ **Question:** How do you think the flounder sees the bottom in order to camouflage itself?

Receptor Organs. One of the most important receptor organs is the lateral line organ whose position is shown by the lateral line faintly visible on the side. As in sharks, the lateral line organ of bony fishes includes mechanoreceptors that are sensitive to water movements. In many bony fishes the lateral line organ also has electroreceptors, as will be described later in this chapter. The **eyes** of bony fishes are relatively large and sensitive not only to color, but also to ultraviolet, which penetrates farther through water than do the colors that are visible to humans. The ultraviolet receptors also enhance the perception of prey and enable the fishes to detect the polarization of sunlight for navigation (Hawryshyn 1992). Teleosts also have **olfactory sacs** with chemoreceptors that are used in detecting prey, and also for chemical communication by means of **pheromones.** As will be discussed later, salmon also use olfaction to navigate back to their native streams for spawning. One or two nostrils on each side of the head allow water to flow into and out of olfactory sacs. In a few teleosts the nostrils connect to the pharynx, enabling the fishes to breathe while most of the body is buried in mud.

Bony fishes also have **inner ears** consisting of the otolith organs and semicircular canals, as in sharks. There is no outer ear or eardrum, but in many fishes the **swim bladder** (Fig. 35.22 on page 768) may serve the same function by intercepting low-frequency vibrations that would otherwise pass through the body. In many teleosts the swim bladder is close to the inner ears, and in herrings a thin extension of the swim bladder contacts them. In carps, catfishes, and some other teleosts certain bones called **Weberian ossicles** transmit vibrations from the swim bladder to the inner ears.

The Swim Bladder. Flounders, certain coral reef species, and a few other fishes specialize in resting or scrambling about on the substratum, but most fishes would be severely limited if they had to continually struggle to keep from sinking. Like most

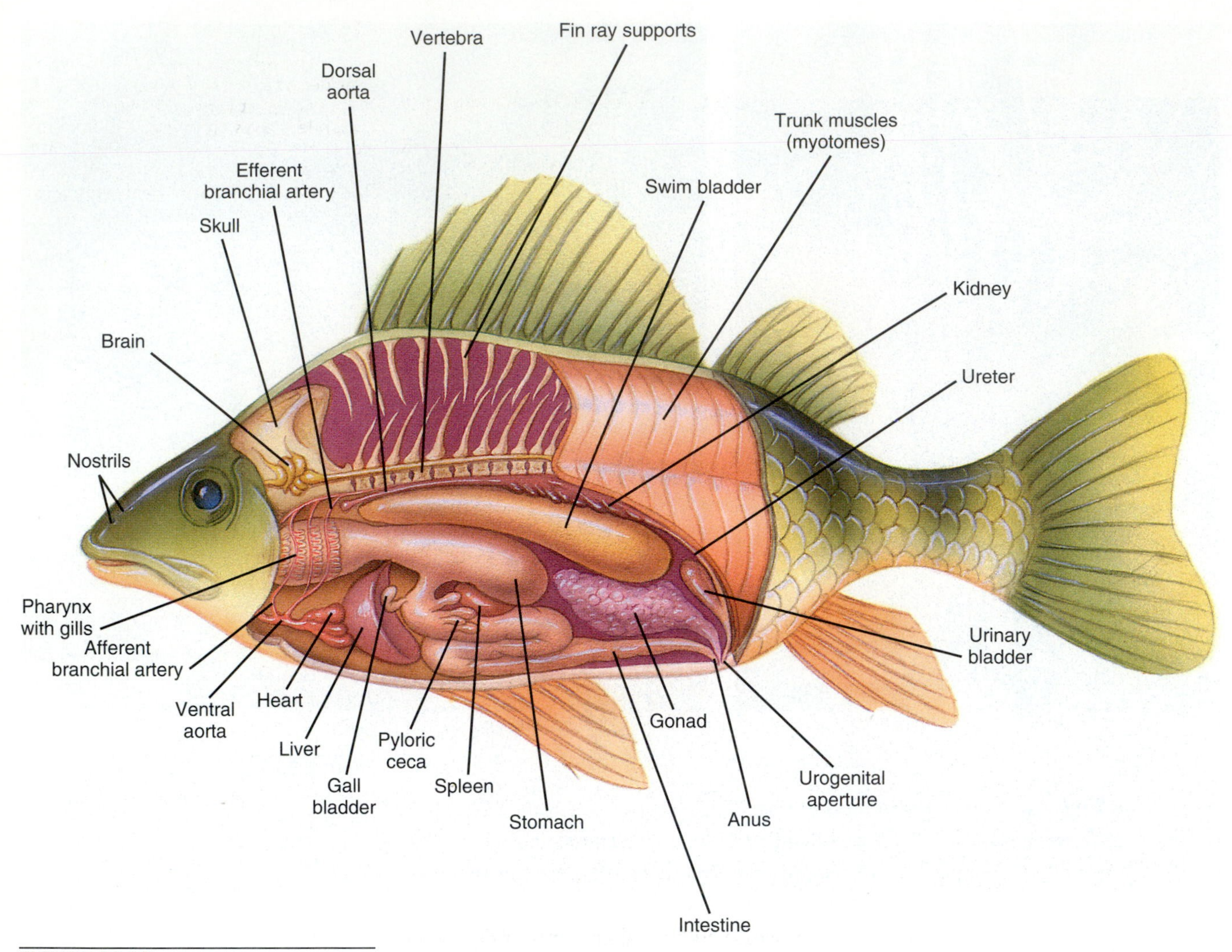

FIGURE 35.22
Internal structure of the yellow perch.

■ **Question:** Why is the swim bladder located near the center of the fish?

teleosts, perches have a gas-filled swim bladder that maintains neutral buoyancy. The swim bladder contains gases that make the overall density of the fish approximately the same as that of water. It occupies about 7% of the body volume in the perch and other freshwater fishes, and about 5% in fishes that live in the more-buoyant salt water. Swim bladders probably originated in early fishes as pouches on the esophagus that served both as air reservoirs and as aids to buoyancy. Subsequently these esophageal pouches specialized for one function or the other. In the evolutionary line that led to the lobe-finned fishes and to tetrapods, the pouches evolved into lungs, with inner surfaces specialized for transferring oxygen to the bloodstream. In teleosts the pouches evolved into swim bladders with linings that retain gases.

In some teleosts with access to air, such as trout, the swim bladder remains connected to the esophagus by a **pneumatic duct** and can be inflated by swallowing air. In other teleosts the pneumatic duct closes during embryonic development. How do these fishes inflate their swim bladders? How do deep-sea teleosts, especially, inflate their swim bladders under pressures of several tons per square centimeter? The answer lies in the **gas glands,** which facilitate the diffusion of gas, mainly oxygen, into the swim bladder (Fig. 35.23). Each gas gland is supplied by a network of capillaries called a **rete mirabile** (pronounced REE-tee meer-AH-bill-lay). By means of a countercurrent mechanism (p. 245), these retia concentrate blood gases, which the gas glands then release into the swim bladder. The gas glands increase the release of oxygen by secreting

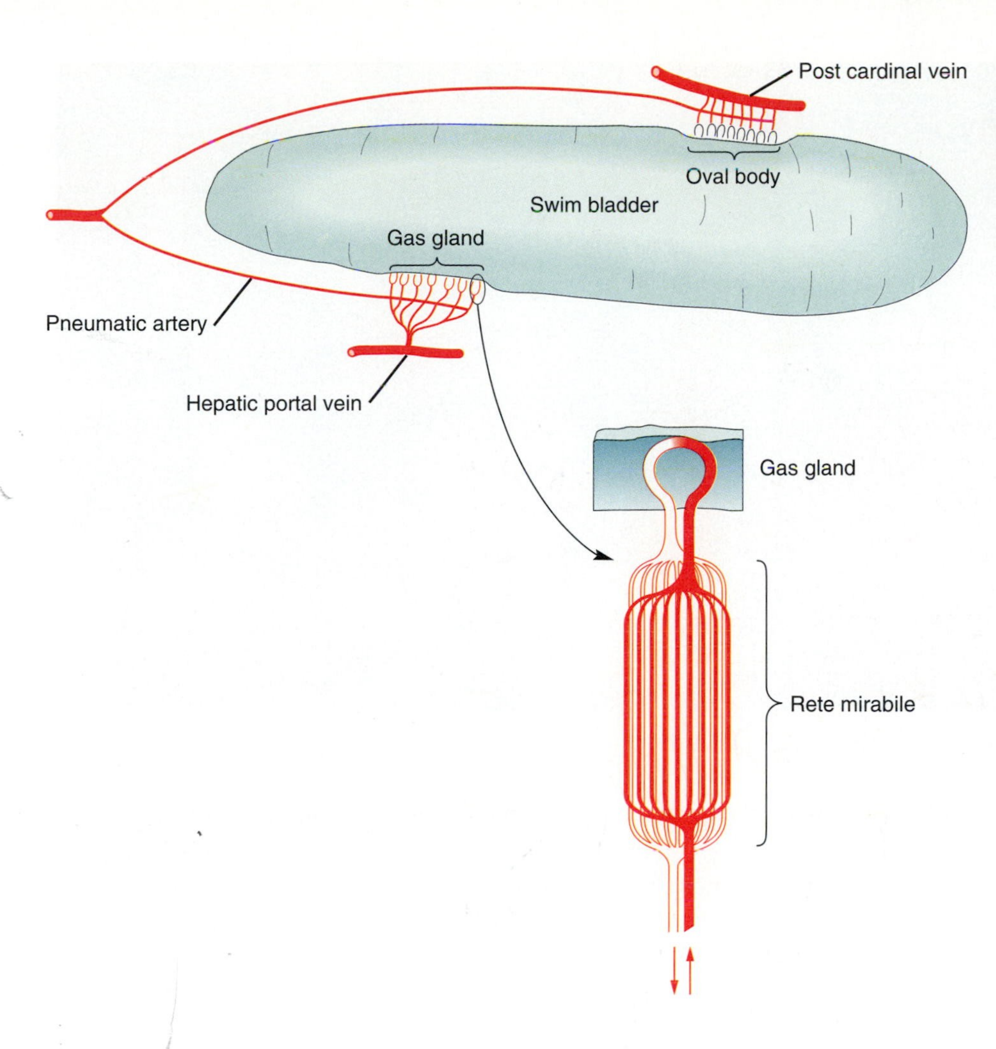

FIGURE 35.23

Schematic representation of the swim bladder of a perch and mechanisms of filling and emptying it. The rete mirabile is a system of capillaries that concentrates gases in the blood, which are released into the swim bladder by the gas gland. The oval body serves as a release valve for excess pressure.

lactic acid, which causes the hemoglobin to release additional amounts of O_2. When fishes rise toward the surface the pressure on the swim bladder decreases, causing it to expand. Rupture of the swim bladder is avoided by an **oval body** that safely leaks excess gas back into the blood.

Many fishes (orders Batrachoidiformes, Perciformes, and Scorpaeniformes) also use the swim bladder to produce sounds. Drums and croakers have muscles that vibrate the swim bladder to produce sound, and grunts use the swim bladder to amplify sounds made by grinding teeth in the pharynx. These sounds are produced mainly by males during courtship and aggressive displays.

OTHER INTERNAL ORGANS. The other internal organs of teleosts are typical for vertebrates (Fig. 35.22). (Refer to the index for information from Unit 2 on various aspects of vertebrate physiology.) Here I note several differences between perches and the sharks that were previously described. Perches and other teleosts differ from sharks in lacking spiral valves in the intestine. The yellow perch does, however, have three **pyloric ceca** that may serve the same function of increasing the duration of digestion and the area for absorption of nutrients. The heart of teleosts has four chambers in series (sinus venosus, atrium, ventricle, and bulbus arteriosus; see Fig. 11.4A). Like 80% of fishes, the yellow perch is a freshwater species with osmoregulatory problems that differ from those of sharks and other marine fishes. To compensate for ions that diffuse out of the body, the gills actively transport ions from the water into the blood. The mesonephric kidneys eliminate most of the excess water that enters through the gills

A

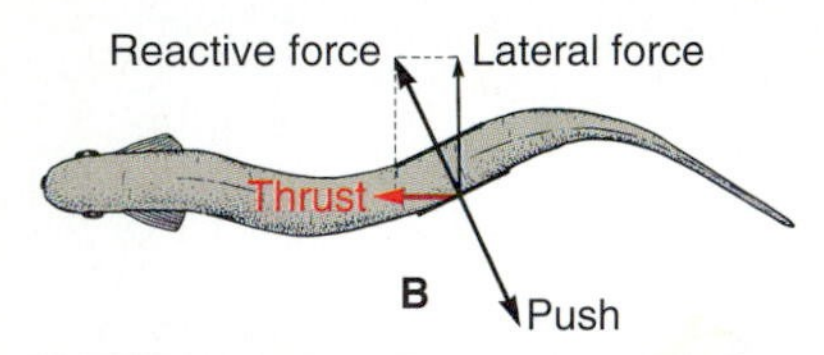

FIGURE 35.24

(A) Moray eel *Helena muraena* demonstrating undulatory (anguilliform) swimming in a coral reef. Note the absence of pectoral fins, reduction of other fins, and the lateral compression of the body. Length approximately 1 meter. (B) Analysis of forces in anguilliform swimming. As the middle portion of the body swings left, it pushes against the water, which generates an equal and opposite reactive force on the fish. The reactive force has a lateral component that tends to push the fish sideways, and it also has a forward component—thrust—that propels it forward.

■ **Question:** Diagram a shark or perch from above, showing the caudal fin during rapid swimming and the forces generated.

and digestive tract by producing copious, dilute urine. Ammonia and other metabolic wastes are also eliminated through the urine, as well as by diffusion through the gills. (For osmoregulation among marine teleosts see p. 286.) Urine is kept in the urinary bladder until excreted through the **urogenital aperture** just anterior to the anal fin. This aperture is also used for the release of gametes. The yellow perch has separate sexes and external fertilization, but other teleosts have numerous interesting variations in reproduction, some of which will be described later.

Locomotion

SWIMMING. Swimming requires enough momentum that the viscosity of water does not stop forward movement between strokes. Some immature fishes are too short to sustain this forward momentum and are therefore either immobile or planktonic. However, even the smallest adult fish, the centimeter-long pygmy goby from the Philippines (*Pendaka pygmaea*), is long enough to swim. (See the discussion of Reynolds number, pp. 211–212.) Fishes swim in a variety of ways, and the variety of shapes of fins and bodies reflects adaptations to different styles of swimming, among other requirements for life. Eels and eel-shaped fishes use the entire body for propulsion, bending it in a series of tight undulations (Fig. 35.24). Generally the fins of such fishes are small, but the body is compressed from side to side and serves as one large fin.

The fins and bodies of many other fishes are adapted for a particular type of swimming, such as maneuvering, cruising, or accelerating (Fig. 35.25).

1. **Maneuvering** is important for fishes that feed on coral and other sessile organism in reefs, weedy bottoms, and other restricted areas. In these fishes, speed is less important than maneuvering, so their bodies are generally short. Forward thrust and turning are due largely to oarlike pectoral fins rather than to the caudal fin, which is usually short. At slow speeds, drag due to water pressure is negligible, so streamlining is not required. The main source of drag is the fric-

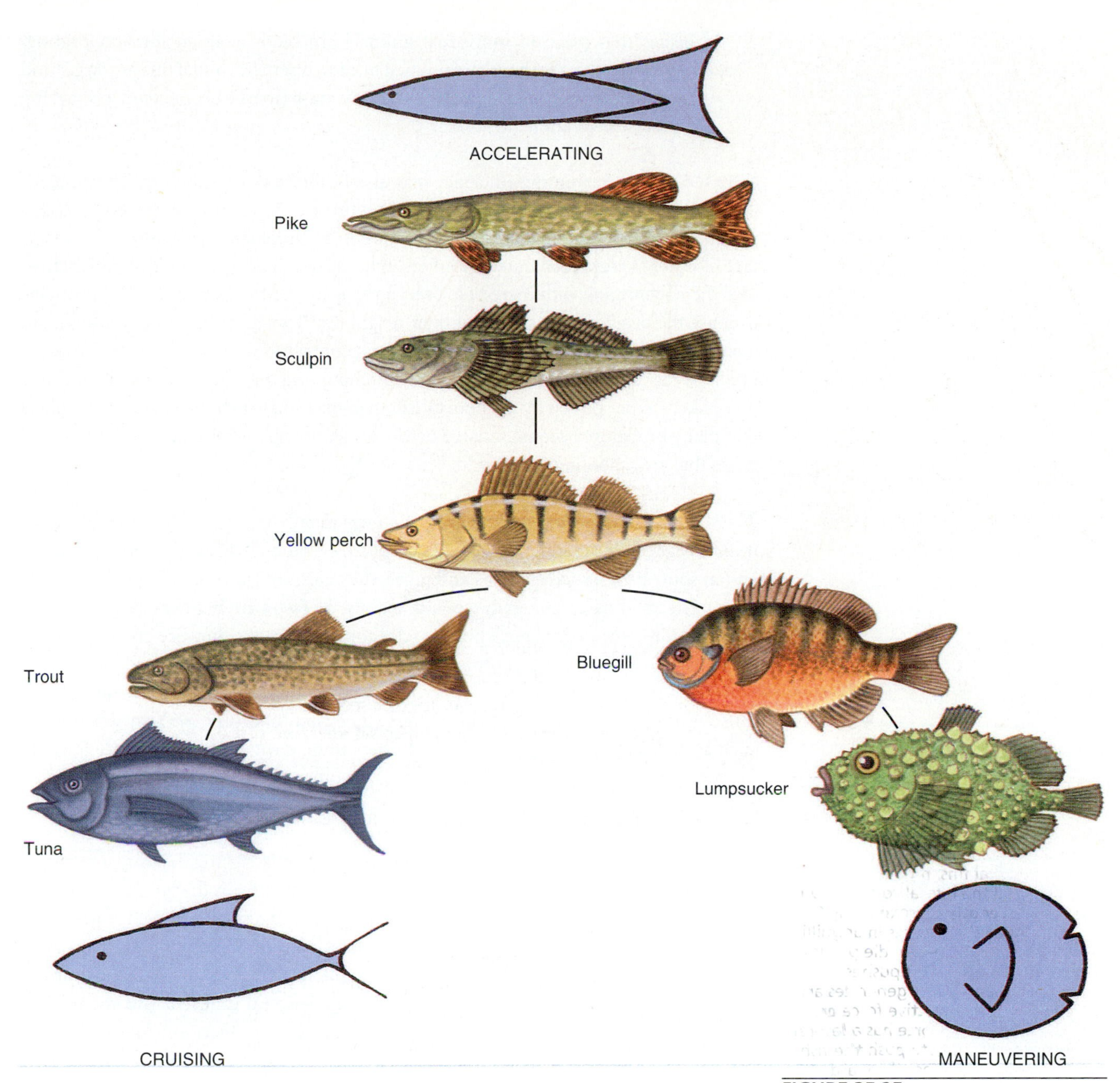

FIGURE 35.25
Specializations of fishes for maneuvering, cruising, and accelerating. Ideal shapes for each mode are shown at the outer edges of the figure. Most fishes, such as the yellow perch and others shown near the center of the figure, have bodies and fins that compromise among the requirement for all three types of swimming, as well as for other functions.

tion of water against the body surface. This friction is minimized by a spherical body shape, which has the least surface area for a given volume.

2. Fishes that feed on plankton and schools of smaller fishes are often adapted for **cruising** at steady, moderate speeds in the open ocean or lakes. Lateral movement of the caudal fin provides most of the forward thrust, and it and the caudal peduncle have to be thin to minimize turbulence and reduce side-to-side movement of the body. These fishes generally have massive, elongated bodies and large median fins that limit sideways movement during swimming. The larger surface areas of the body and fins increase drag due to friction, but friction drag during fast swimming is negligible compared to the drag due to water pressure against the front of the body. This pressure drag is reduced by streamlining.

3. Fishes that prey on individual animals are often adapted for **accelerating.** Generally they also have streamlined bodies, but their caudal fins are larger and can therefore generate more thrust. Body mass (inertia) is reduced, permitting greater acceleration.

SCHOOLING. Swimming efficiency may also be affected by schooling. The slime released from all the fish in a school appears to lubricate the water in some way. It is also possible that maintaining a particular position in a school allows a fish to take advantage of water currents generated by others in the school. Fish in schools often synchronize their swimming with incredible precision, using their eyes and lateral line organs to sense the position and speeds of their neighbors. The contribution of schooling to locomotion is probably less important than its contribution to avoiding predation. A predator has a harder time selecting and catching one fish out of a school, and it is more likely to be spotted by a school of fish than by a single fish. In addition, it is likely that many predators become satiated before they can wipe out an entire population of fishes that are schooling together.

MUSCLES. The power for acceleration comes mainly from the segmental blocks of muscles called myomeres (Figs. 35.22 and 35.26). These muscles envelop the viscera, and in some fishes, such as tunas and pikes, they account for more than half the body mass. The myomeres generally consist mainly of **fast-twitch fibers** (type II white fibers; p. 208). Although these white muscles contract rapidly, they tire quickly because they produce ATP by the anaerobic breakdown of glycogen into lactic acid. Most of the power for cruising comes from darker muscles that produce ATP aerobically. These muscles are made up of **slow-twitch fibers** (type I red fibers; p. 208). In some fishes, such as mackerel, salmon, and carp, red fibers mingle with anaerobic white fibers in the myomeres. In tunas, striped bass, and some others, red fibers occur in lengthwise bundles outside the myomeres (see Fig. 15.6).

In bony fishes the myomeres are anchored to spines and ribs on the vertebrae. When the myomeres contract, the vertebral column bends, flexing the caudal peduncle and caudal fin. A horizontal septum separates the dorsal from the ventral myomeres on each side. Adjacent myomeres nest against each other and are held together by septa. (These septa dissolve when a fish is cooked, causing the myomeres to separate into flakes.) Each myomere has a complex shape, folding forward and backward in such a way that a transverse section cuts through several different myomeres. The overlapping arrangement of myomeres may allow the contraction to be distributed smoothly over a greater length of body. Within each myomere, individual muscle fibers are not aligned lengthwise along the body. In fact, their orientation is so complex that it is not clear how their contraction bends the vertebral column.

Migration and Homing

EELS. One of the most important and fascinating uses of locomotion in fishes is in migration. Many species of fishes, such as shad, salmon, trout, and eels, are as impressive as birds in the distances over which they migrate and the precision with which they navigate. Every American eel (*Anguilla rostrata*) and European eel (*A. anguilla*) begins its life in the calm, clear Sargasso Sea south and southeast of Bermuda, migrates into freshwater streams in either America or Europe, then returns (**homes**) to the Sargasso Sea up to 15 years later to reproduce (Fig. 35.27 on page 774). Hundreds of meters beneath the floating sargassum seaweed that accumulates there, each female deposits approximately a million and a half eggs, males fertilize them, and then the

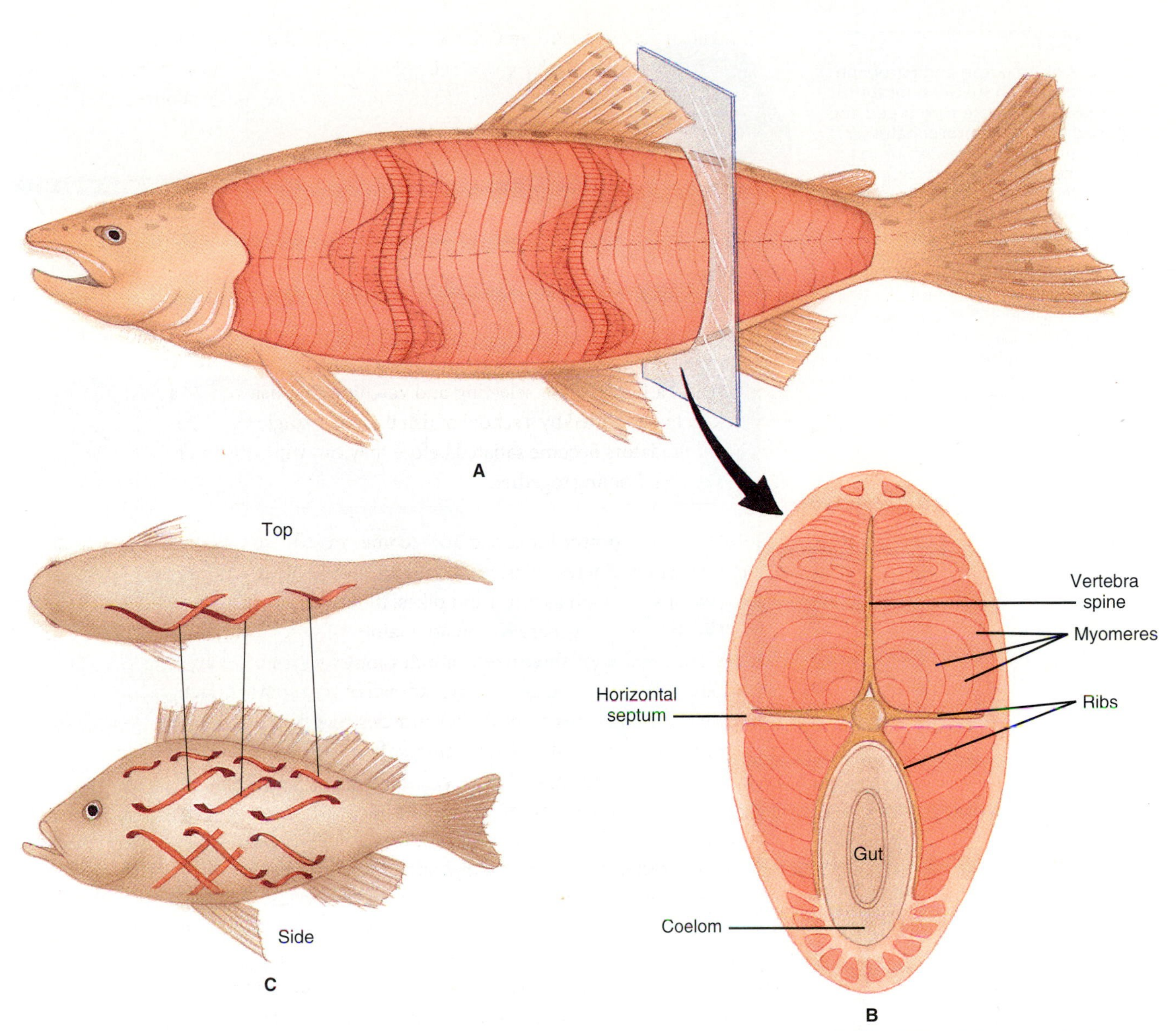

FIGURE 35.26

(A) Myomeres (= myotomes) of a salmon. (B) Cross section of the salmon, showing the division of myomeres by septa and the relation to the vertebrae. (C) Schematic representation of the orientation of muscle fibers within the myomeres of a typical teleost.

adults die. After two days each egg hatches into a hermaphroditic larva only a few millimeters long. The larva is so unlike an adult eel that it was once classified in a completely different genus, *Leptocephalus,* and it still goes by that name. Too short to swim, the leptocephalus larvae drift into currents. The American and European species drift together at first, yet in less than a year the European eels somehow find their way to European estuaries, and the American eels arrive at American estuaries (Lecomte-Finiger 1994).

During migration the leptocephali metamorphose into more eel-like juveniles called **elvers.** Upon arrival in estuaries the osmoregulatory systems of the elvers gradually adapt to less saline waters. Some elvers remain in the estuaries, and after several years they change from hermaphrodites to males. Other elvers, although apparently identical to those that remain in the estuaries, migrate upstream in the spring. Often they travel for several hundred kilometers, sometimes over land on moist nights, breathing through their gills and skin. Once they find a suitable location, these elvers will spend up to 40 years there, eating frogs and invertebrates by day in the warm

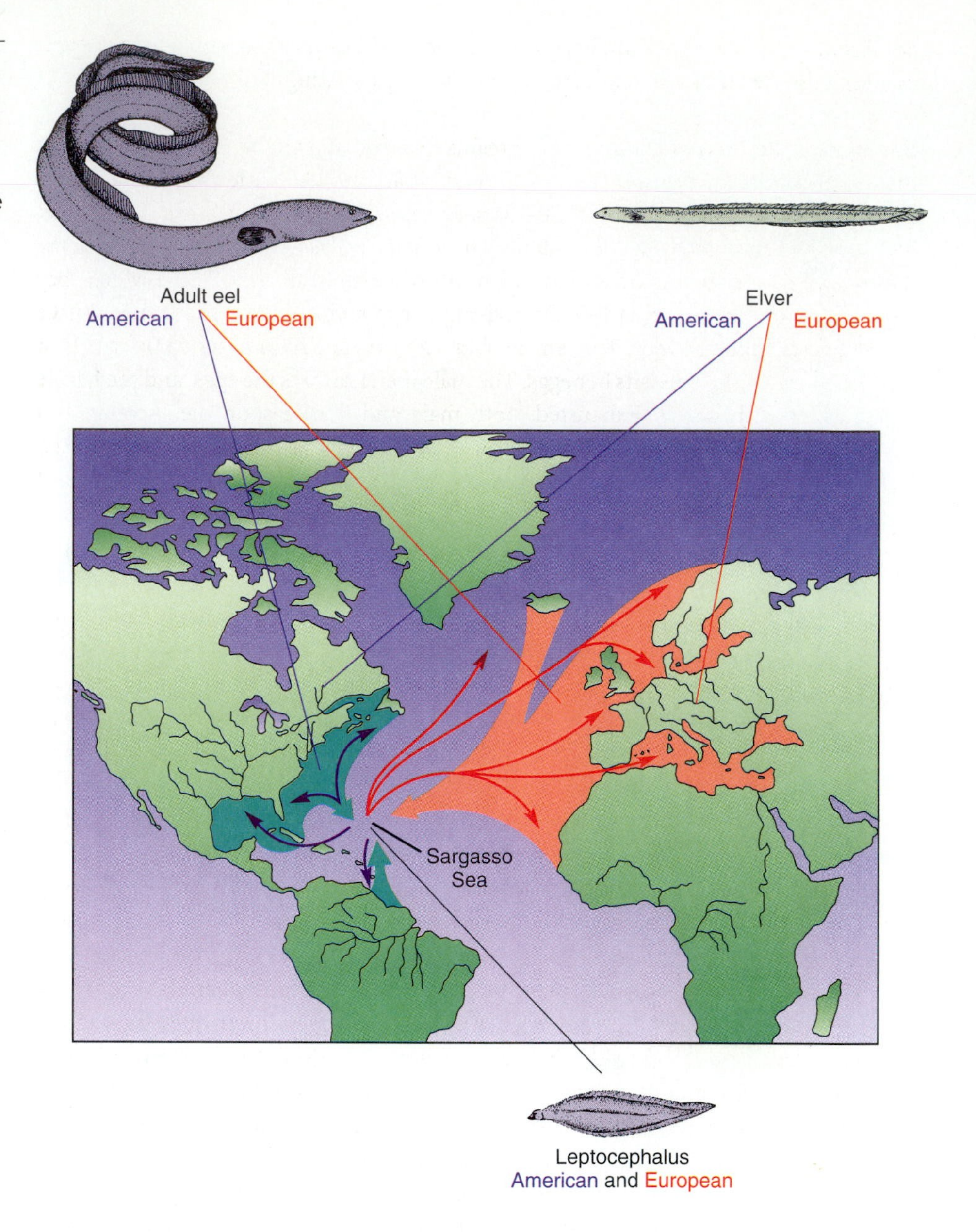

FIGURE 35.27

Migrations of American and European eels. Adults of both species migrate to the Sargasso Sea where they breed and die (broad arrows). Leptocephalus larvae then migrate to American or to European estuaries (thin arrows), where they become elvers. Female elvers migrate up streams. After several years the elvers mature into adults, which complete the cycle. The only morphological difference between the two species is in the average number of vertebrae (average of 107 in the American eel versus 114 in the European). Comparisons of mitochondrial DNA, however, show that the two species do not interbreed, even though their breeding areas in the Sargasso Sea overlap.

months and remaining buried in mud during the cold. Finally they all mature into females and in the autumn return downstream to the estuary where males join them for the return voyage to the Sargasso Sea. How eels navigate thousands of kilometers of open ocean back to the Sargasso Sea is a mystery. Some fishes can apparently determine direction from the position of the Sun, but adult eels generally migrate in the total darkness of deep water. Perhaps like some other fishes, eels can detect Earth's magnetic field.

Eels and approximately 40 other species are said to be **catadromous,** which means that they hatch in seawater and migrate up freshwater streams to feed. For catadromous species the greatest navigational challenge is to find an area in the vast ocean where they can successfully reproduce. Eighty-seven other species, such as some lampreys, shad, salmon, and trout, are **anadromous:** They hatch in freshwater streams and migrate downstream to feed in the ocean. For anadromous species, migration would appear to be easier, since all they have to do to find a breeding area is to enter the mouth of a stream and keep going against the current. This approach is risky, however, because many streams dry up, become laden with silt, or prove unsuitable for reproduction in some other way that a fish has no way of predicting.

Salmon. Perhaps because some streams become unsuitable for reproduction, many anadromous fishes have evolved a homing ability that guides them to a stream that has proved suitable in the past—namely the stream in which they, themselves, hatched. For example, adult coho salmon *Oncorhynchus kisutch* return to spawn in the same rapid, gravel-bottomed streams of northwestern North America in which they hatched two or three years earlier. On arriving in November a male and female engage in a ritualized mating dance. The female then uses her caudal fin to scoop out a nest (a **redd**) into which she deposits her eggs. The male then fertilizes the eggs, and the female covers them with gravel. Exhausted, both male and female soon die. Around two months later the eggs hatch. The hatchlings, or **alevins,** remain under gravel for several months, until the nutrient supply in their yolk sac is depleted. Then the fish, now called **fry,** emerge from the nest and establish a territory in which they feed on plankton and insects that drift downstream. During a new moon in the spring of the second year of life, a surge in the levels of the thyroid hormones thyroxine and triiodothyronine triggers metamorphosis into **smolts.** During the smoltification process the osmoregulatory systems prepare for seawater. The smolts swim en masse downstream to the ocean, where most will feed for a year and a half. At the end of that period they stop feeding and begin to migrate back to their native streams, perhaps using geomagnetic cues to navigate. They remain at the mouth of the correct river for about a month while their osmoregulatory systems readjust to fresh water and their gonads develop. At this point they acquire a red mating coloration and distinctive hooked jaws. They then continue migrating upstream to the breeding area where they began life.

How do salmon find their way up rivers and tributaries to the correct breeding area? An answer to this question occurred to Arthur D. Hasler of the University of Wisconsin during a visit to a mountain slope he had known as a boy. "As I hiked along a mountain trail in the Wasatch Range of the Rocky Mountains where I grew up, my reflections about the migratory behavior of salmon were soon interrupted by wonderful scents that I had not smelled since I was a boy. . . . [S]o impressive was this odor that it evoked a flood of memories of boyhood chums and deeds long since vanished from conscious memory. The association was so strong that I immediately applied it to the problem of salmon homing" (Hasler and Scholz 1983, p. xii).

Hasler began to wonder whether salmon might be **imprinted** to the odor of their native stream while they are still smolts, then home in on that odor as adults. For several decades Hasler and his colleagues and students conducted experiments to test the theory. First they showed that coho salmon could indeed be conditioned to distinguish between water from two different streams, and that they lost this ability if their olfactory sacs were destroyed or if organic molecules were filtered from the water. Hasler and his coworkers also trapped and tagged migrating salmon in two branches of a Y-shaped stream, plugged the nostrils of some of them, then released the salmon downstream below the junction of the two branches. Nearly all the fishes with unplugged nostrils were later recovered in the same branch in which they had been trapped originally, but the salmon with plugged nostrils were much less likely to make it back to the correct branch of the stream. Conclusive evidence came from studies in which smolts from hatcheries were exposed to one of two artificial substances. When these fishes were ready to migrate, one of these substances was added to one stream, and the other substance was added to another stream. Although the salmon could have entered any one of more than a dozen streams, the great majority migrated up the stream containing the substance to which they had been imprinted as smolts.

Electroreceptors and Electric Organs

Many fishes apparently use sensitive electroreceptors in the lateral line to navigate over long distances. Some electroreceptors can respond to a voltage gradient as small as 0.005 microvolt per centimeter, which is equivalent to connecting one terminal of an ordinary 1.5-volt battery to San Francisco Bay and the other at Victoria, British Columbia. Such sensitivity would enable the fish to navigate by detecting the voltage induced as its body moves through Earth's magnetic field. Electroreceptors could also enable a fish to detect the voltage generated by ocean currents. In addition, experiments show that electroreceptors can detect the muscle potentials generated by prey.

Many fishes not only detect electrical potentials, but also generate them. These **electric fishes** produce voltages using electric organs that are generally derived from muscles that have lost their ability to contract. The voltage produced by each cell is an action potential no larger than that in an ordinary muscle cell (approximately 0.1 volt), but hundreds or thousands of cells in the electric organ are arranged in a series that allows the voltages to summate. In strongly electric fishes the electric organ discharge is up to several hundred volts, which is sufficient to stun or kill prey (Fig. 35.28). It is not clear how these fishes avoid electrocuting themselves. Electric organs generate much lower voltages in weakly electric fishes, such as knifefishes (family Gymnotidae) and

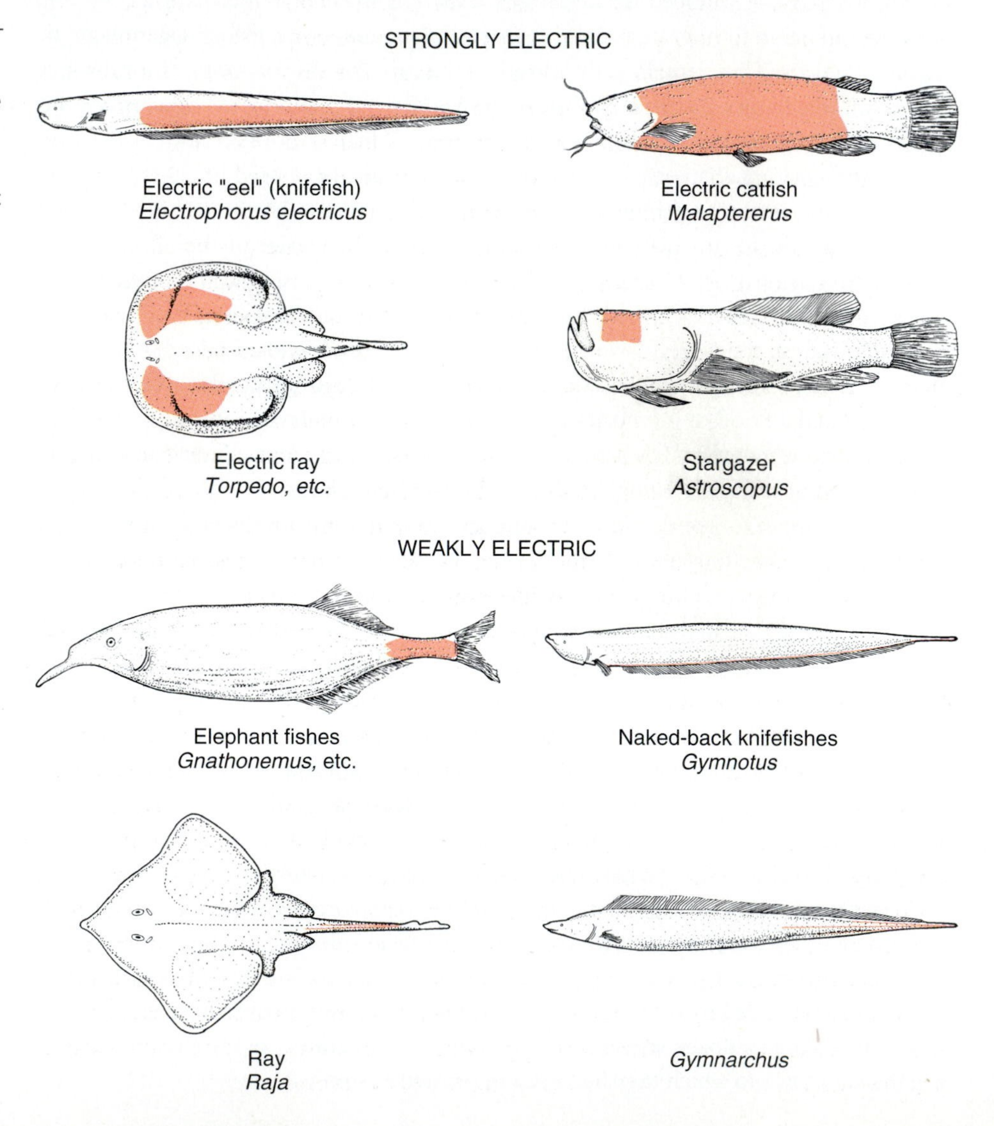

FIGURE 35.28

Electric fishes. Strongly electric fishes have large electric organs that generate tens of volts used to stun or kill prey. Weakly electric fishes use their smaller electric organs for communicating and locating objects. The diversity of electric organs in several distinct groups of fishes indicates that they evolved independently several times.

elephantfishes (family Mormyridae). Gymnotids and mormyrids are nocturnal and live in murky streams in South America and Africa, respectively. They use their electric organ discharges to locate nearby objects, by detecting the distortions in the electric fields produced by the objects. These gymnotids and mormyrids also use the electric organ discharges to communicate with other members of their own species. Although many related species may inhabit the same streams, they can distinguish one species from another, and many can discriminate age, sex, and even particular individuals on the basis of the electric organ discharge.

Reproduction

One can find among fishes virtually every reproductive adaptation imaginable, and a few that are almost unimaginable. Like the yellow perch, most species have separate sexes, although it is seldom easy to distinguish them visually. Fertilization is external in most teleosts. Some, such as tuna, sardines, whitefish, and certain shads (*Alosa*), are mass spawners, releasing their gametes within large schools. Others choose particular mates, which they recognize by visual cues such as the belly of the female swollen with eggs or the bright coloration due to chromatophores. Others recognize potential mates by pheromones, and, as noted previously, males and females of some species literally electrify each other with electric organ discharges. Once potential mates recognize each other, courtship may ensue. Courtship often involves a stereotyped swimming display, such as that of the three-spined stickleback (p. 413). The display often climaxes in a nest, where the female releases her eggs and the male fertilizes them. In other species the eggs may be attached to vegetation or, as in salmon, hidden under gravel.

Following external fertilization most fishes simply swim off, but others provide some degree of support and protection for the eggs and hatchlings. In some species one or both parents remain near the nest, fending off the numerous potential predators that feed on fish eggs and fry. Some aerate the eggs by fanning water currents over them with their fins. The greatest support is provided by various species that brood the eggs and sometimes the fry in or on their bodies. **Mouthbrooders** take the eggs into their mouths (p. 420), and some allow the fry to take refuge from danger within their mouths. Others brood eggs in their gill chambers or on their skin. Perhaps the most unexpected method of brooding occurs in seahorses and some pipefishes and catfishes. In these species the sex roles seem to be reversed. The female has a penislike organ with which she deposits eggs into a pouch (**marsupium**) on the belly of the male, which then fertilizes and broods them (Fig. 35.29). In some species the male pouch even has a placenta with which he provides nourishment to the eggs.

FIGURE 35.29
A male seahorse *Hippocampus* (order Syngathiformes) "giving birth."

■ **Question:** Why is this reproductive arrangement so rare among animals?

Other teleosts depart radically from the dioecious, externally fertilizing "norm." Like sharks and coelacanth that were mentioned previously, some are truly viviparous, with the eggs being retained in the ovary or oviduct of the female and supplied with nutrients and oxygen. Among these live-bearing fishes are surfperches and the famous aquarium guppy *Poecilia reticulata.* In these species, fertilization is, of course, internal. Some live-bearing fishes have eliminated the males altogether and have become parthenogenetic and unisexual. Other teleosts are hermaphroditic. Some, such as most sea basses, are **simultaneous hermaphrodites.** They can release both eggs and sperm, and they alternate between serving as male and female several times during a single mating. Other teleosts are **sequential hermaphrodites,** which change permanently from one sex to another. The sex change is generally triggered by some environmental cue, such as a low number of males in a school (p. 309). These fishes have generated considerable research in behavioral ecology because they provide a means of studying the circumstances under which it is reproductively advantageous to be one sex or the other.

Interactions with Humans and Other Animals

Ecology. It is probably safe to say that removing all the fishes will affect every other organism in an aquatic ecosystem. Many fishes serve vital roles in food webs by eating plankton and channeling the energy and nutrients to higher trophic levels, often other species of fishes. Many birds, mammals, and other animals that are not members of aquatic communities also depend heavily on fishes for food. Fishes also interact symbiotically with other species. Two beautiful examples have been noted elsewhere: that of anemone fishes with sea anemones (see Fig. 24.16), and that of cleaner fishes with various other species of fishes (p. 323). Another fascinating symbiosis occurs with luminescent bacteria in the light organs of certain species of anglerfishes and flashlight fishes (see numbers 17 and 18 in Fig. 17.21, and Fig. 35.30).

Benefits to Humans. Fishes benefit humans in a variety of ways, not least of which is the pleasure of watching and learning about them. Hobbyists spend millions of dollars and countless hours annually on diving equipment and on aquarium supplies and exotic fishes. Unfortunately the latter hobby can also exact a high cost to the fishes as well, since some collectors of marine species use cyanide, which kills nine fish for every one collected. A threatened boycott by American dealers has greatly reduced the use of cyanide. Sport fishermen derive as much pleasure and spend even more money on their hobby (approximately $8 billion per year in the United States), again often at high cost to the fish. Of course the most common benefit from fishes is food. The average human consumes 20 kg of fish per year. Fish is becoming more popular in the United States as people become increasingly concerned about the fats and cholesterol in red meats. Besides the tuna, cod, and numerous other fishes enjoyed in the United States, eels are popular in Europe, and caviar (sturgeon eggs) are considered the ultimate luxury food.

Fishes also benefit humans with unique medical products. One of these is tetrodotoxin from puffer fish, which is widely used in neurophysiological research and as a local anesthetic (p. 156). Researchers in the United States and the Middle East are also studying the healing properties of slime from certain catfishes. Arab fishermen

FIGURE 35.30
The flashlight fish *Anomalops katoptron* (order Beryciformes), a nocturnal species of Pacific coral reefs. Every few seconds the fish flashes the light organ by lowering a flap. Flashlight fishes are generally benthic or nocturnal, and they use the luminescent organs for communication. They may also use the light organs to attract and detect prey. The light organs are usually located near the eyes, which is the optimal location for detecting prey by means of reflected light.

have long been in the habit of rubbing this slime on cuts, and scientists have now extracted from it a variety of components that stop bleeding and promote tissue growth. Researchers have also recently isolated a unique antimicrobial steroid (squalamine) from the dogfish shark *Squalus acanthias.* The zebra fish *Danio rerio,* a common inhabitant of home aquaria, promises even greater contributions to science and medicine. It is replacing drosophila in many laboratories because it can be manipulated to produce offspring that contain only genes from the mother, and it has transparent embryos that permit observation of developmental processes.

IMPACT OF PEOPLE ON FISHES. Until recently, fish were as plentiful as . . . well, the fish in the sea. The 15th-century explorer John Cabot reported that the fishes were so numerous off the Atlantic coast of Canada that his men had trouble rowing ashore. Fishes are now so scarce there that fishing for cod, haddock, and flounder in the Georges Bank fishery had to be stopped in 1994, and 12 of the 16 other major fisheries of the world are on the verge of collapse. The main culprit appears to be modern fishing methods (Holloway 1994). Huge nets now "strip mine" the ocean floor, engulfing not only fishes of marketable size, but also the catch of future years as well as immense numbers of "useless" animals whose dead bodies are thrown overboard. Some species of sharks are also threatened by the practice of "finning," in which the fins, which contain a noodle-like cartilage much prized in Asia, are cut off, and the rest of the shark is dumped overboard to bleed to death.

Fishes are also threatened by destruction of habitat. When settlers arrived in America, alewives, menhaden, and other fishes were so plentiful that a thousand fish cost only a dollar, and they were used as fertilizer. Mill dams then stopped many of the spawning runs. Then came sawdust and silt from mills and mines, erratic water levels due to clearing of forests, and fertilizers and pesticides from farming. Canals compounded these problems by introducing exotic species such as the sea lamprey. More recently, air pollution has made many lakes and streams too acidic to support fish, and polychlorinated biphenyls (PCBs) and other toxic wastes have either killed the fish or made them unsafe to eat. In many parts of the country, people are warned not to eat more than one meal per month of locally caught fish. Even large bodies of water like the Chesapeake Bay and offshore areas once considered bottomless dumps for garbage, toxic wastes, and sewage sludge, are now yielding reduced catches and fishes that are unhealthy and unhealthful.

Many government and other groups have attempted to alleviate some of these problems by reducing pollution, by building fish ladders to allow migration past dams, and by artificially manipulating ecosystems. Sometimes the results are disastrous. In Montana between 1968 and 1975, for example, the exotic opossum shrimp (*Mysis relicta*) was introduced into the Flathead River-Lake system with the idea of increasing the food supply for salmon (*Oncorhynchus nerka*). Instead of feeding the salmon, however, the shrimp depleted the zooplankton. As a result the salmon essentially disappeared by 1987, and along with them went the population of lake trout (*Salvelinus namaycush*), bald eagles, and numerous other animals that had fed on the salmon (Spencer et al. 1991).

In many cases it has proved cheaper and more politically acceptable to simply replace wild fishes with domestic fishes. Most sport fishes are raised in hatcheries and bear little resemblance to native species. Thanks to biotechnology, they are selected, hybridized, and genetically manipulated in the attempt to increase size, tolerance to deteriorated habitats, and "return" (meaning they are easier to catch). Generally the eggs are heat-shocked or otherwise treated to interfere with meiosis, so that they are diploid and produce triploid offspring that grow up sterile. Thus if stocking turns out to damage a habitat, the damage will be reversible. Techniques of genetic engineering promise further "improvements" of this kind.

Fish farming, or **aquaculture,** has also become common. Virtually all rainbow trout and catfish sold in the United States now come from fish farms. Aquaculture also depends heavily on biotechnology. Some trout farms produce only females, which look and taste better than males, since they don't waste energy in aggressive encounters. In order to get sperm for reproduction, some eggs are treated with testosterone, which causes the embryos to develop into trout that are physiologically male, though still genetically female. Because the sperm have only X chromosomes, all the fertilized eggs develop into females.

Summary

Three groups of primitively aquatic vertebrates are referred to as fishes. One group is Agnatha, which includes the jawless hagfishes and lampreys. Class Chondrichthyes comprises sharks, rays, and chimaeras, which have cartilaginous skeletons. Sharks have uncovered gill slits and placoid scales, some of which continuously form teeth. Class Osteichthyes comprises the bony fishes, in which bone replaces much of the cartilage during development, and the gills are covered by an operculum. Bony fishes include lobe-finned and ray-finned species. Most fishes—and indeed most vertebrates—are ray-finned fishes called teleosts.

Fishes owe their success largely to being efficient swimmers. The force used in swimming comes mainly from the caudal fin, which is heterocercal in sharks, diphycercal in lobe-finned fishes, and usually homocercal in ray-finned fishes. The pectoral fins provide some forward thrust as well, and dorsal and pelvic fins provide directional stability. The shapes of the fins and body are adapted in different fishes for cruising, accelerating, or maneuvering. Sharks and tuna tend to have narrow caudal peduncles and fins for efficient cruising. Fishes that lunge after prey, such as pike, tend to have broader tails. Fishes that maneuver among coral reefs or grasses, such as butterfly fish, tend to be spherical or disc-shaped.

The shapes of scales, whether placoid, ganoid, cycloid, or ctenoid, influence the amount of friction during swimming, as does the secretion of mucus. Neutral buoyancy is also important in swimming, since the fish does not have to expend energy to keep from sinking. Sharks maintain low body density by means of the cartilaginous skeleton and squalene, and most ray-finned fishes use a swim bladder that is inflated by a gas gland.

Fish detect prey, communicate, and navigate by a variety of receptors, including chemoreceptors on the body surface and in olfactory pits, mechanoreceptors and electroreceptors in the lateral line organ, and the eyes. Some fishes can generate electricity for communication, navigation, and predation. Some, such as eels and salmon, can use Earth's magnetic field, electric currents, chemicals, or other cues to migrate long distances, and even to home to their native waters.

Most fishes are dioecious, fertilize the eggs externally, and provide little care for eggs and young. In some, however, fertilization is internal, and the young may be born alive. Some tend the eggs and young in the nest, or even in the mouth. Some are sequential or simultaneous hermaphrodites, and some are parthenogenetic.

Key Terms

vertebral column
caudal fin
caudal peduncle
heterocercal
neutral buoyancy
myomere
placoid scale
spiral valve
lateral line organ
diphycercal
operculum
cycloid scale
ganoid scale
homocercal
ram ventilation
buccal pumping
swim bladder
ctenoid scale
gas gland
catadromous
anadromous

Chapter Test

1. Describe a major distinction of each of the following groups, and give the common name of one member of each group: Agnatha, Gnathostomata, Chondrichthyes, Osteichthyes, Elasmobranchii, Holocephali, Crossopterygii, Actinopterygii, chondrosteans, holosteans, teleosts. (pp. 746–750)
2. For each of the following groups, describe one trait that is considered to be primitive: hagfishes, sharks, lobe-finned fishes. (pp. 748–760)
3. Describe the following kind of scales and name a fish that has each kind: placoid, ganoid, cycloid, ctenoid. What advantage does each kind of scale confer to the fish? (pp. 754, 765)
4. Describe some major ways in which the physiology and life cycle of the sea lamprey differ from that of the American eel. (pp. 752, 772–774)
5. Why is it important for a fish to maintain neutral buoyancy? Explain how neutral buoyancy is maintained in a shark and in a bony fish. (pp. 754, 767–769)
6. For each of the fishes shown in photographs in this chapter, state on the basis of its fins and body whether it specializes in maneuvering, cruising, or accelerating. (pp. 770–772)
7. Describe the differences between osmoregulation in the dogfish shark and that in the yellow perch. (pp. 757, 769–770)
8. Describe the functions of lateral line organs, electroreceptors, and olfactory receptors in fishes. (pp. 754–756, 767)

■ Answers to the Figure Questions

35.3 Note the bony rays in the fins.

35.4 Eel-shaped body, jawless mouth, pharyngeal openings.

35.7 A female shark would not have a clasper.

35.9 Sharks do not have molars, and the teeth are not firmly anchored to enable them to apply much grinding force. They are clearly adapted for slicing off bits or food that are then swallowed whole.

35.10 One likely explanation is that the liver, with its high content of squalene, maintains neutral byoyancy. Another explanation is that the liver produces the large amount of osmolytes needed to maintain osmotic balance.

35.13 A close call. All these look more like the fishes they are than like salamanders or other amphibians.

35.14 Chances are good that a surprisingly large number of these will appear on your list.

35.15 In the murky depths of many lakes, the barbels would be better able than the eyes to detect bottom-dwelling prey.

35.18 No fish has a pelvis, and in perch and many other "advanced" species the pelvic fin is not far behind the pectoral fin (which is also misnamed).

35.21 This is something of a mystery, since the eyes are on the upper part of the body. Presumably it gets a look at the bottom before it settles.

35.22 If the swim bladder were below the midline, the fish would tend to flip upside down.

35.24 The arrows showing the forces produced should look like those in part B, except that they would be coming from the caudal fin of a shark or perch.

35.29 Generally the female has already made a larger reproductive investment by producing large, yolky eggs, so there is more advantage to her in brooding eggs and young. In addition, it is more likely for females than for males that the offspring they are caring for are carrying their genes.

Readings

Recommended Readings

Bass, A. H. 1990. Sounds from the intertidal zone: vocalizing fish. *BioScience* 40:249–258.

Boschung, H. T., Jr., et al. 1983. *The Audubon Society Field Guide to North American Fishes, Whales, and Dolphins.* New York: Alfred A. Knopf.

Coutant, C. C. 1986. Thermal niches of striped bass. *Sci. Am.* 255(2):98–104 (Aug).

Donaldson, L. R., and T. Joyner. 1983. The salmonid fishes as a natural livestock. *Sci. Am.* 249(1):51–58 (July).

Eastman, J. T., and A. L. DeVries. 1986. Antarctic fishes. *Sci. Am.* 255(5):106–114 (Nov).

Fricke, H. 1988. Coelacanths: the fish that time forgot. *Natl. Geogr.* 173(6):824–838 (June).

Gould, S. J. 1993. Full of hot air. In: *Eight Little Piggies.* New York: Norton, Chapter 7. (*On the evolution of lungs and swim bladders.*)

Hawryshyn, C. W. 1992. Polarization vision in fish. *Am. Sci.* 80:164–175.

Klimley, A. P. 1994. The predatory behavior of the white shark. *Am. Sci.* 82:122–133.

Leggett, W. C. 1990. The spawning of the capelin. *Sci. Am.* 62(5):102–107 (May).

Levine, J. S., and E. F. MacNichol, Jr. 1982. Color vision in fishes. *Sci. Am.* 246(2):140–149 (Feb).

Moyle, P. B. 1993. *Fish: An Enthusiast's Guide.* Berkeley: University of California Press.

Partridge, B. L. 1982. The structure and function of fish schools. *Sci. Am.* 246(6):114–123 (June).

Pietsch, T. W., and D. B. Grobecker. 1990. Frogfishes. *Sci. Am.* 262(6):96–103 (June).

Policansky, D. 1982. The asymmetry of flounders. *Sci. Am.* 246(5):116–122 (May).

Robison, B. H. 1995. Light in the ocean's midwaters. *Sci. Am.* 273(1):60–64 (July). *(Bioluminescent fishes.)*

Triantafyllou, M. S., and G. S. Triantafyllou. 1995. An efficient swimming machine. *Sci. Am.* 272(3): 64–70 (Mar). *(Efforts to model the mechanics of fish swimming.)*

Warner, R. R. 1984. Mating behavior and hermaphroditism in coral reef fishes. *Am. Sci.* 72:128–136.

Webb, P. W. 1984. Form and function in fish swimming. *Sci. Am.* 251(1):72–82 (July).

Wu, C. H. 1984. Electric fish and the discovery of animal electricity. *Am. Sci.* 72:598–607.

See also relevant selections in General References at the end of Chapter 21.

Additional References

Ahlberg, P. E., and A. R. Milner. 1994. The origin and early diversification of tetrapods. *Nature* 368:507–514.

Bone, Q., and N. B. Marshall. 1982. *Biology of Fishes.* London: Blackie.

Evans, D. H. (Ed.). 1993. *The Physiology of Fishes.* Boca Raton LA: CRC Press.

Hasler, A., and A. Scholz. 1983. *Olfactory Imprinting and Homing in Salmon.* New York: Springer-Verlag.

Holloway, M. 1994. "Diversity blues." *Sci. Am.* 271(2):16, 18.

Lecomte-Finiger, R. 1994. The early life of the European eel. *Nature* 370:424.

Motta, P. J. 1984. Mechanics and functions of jaw protrusion in teleost fishes: a review. *Copeia* 1984:1–18.

Moyle, P. B., and J. J. Cech, Jr. 1982. *Fishes: An Introduction to Ichthyology.* Englewood Cliffs, NJ: Prentice-Hall.

Nelson, J. S. 1994. *Fishes of the World,* 3rd ed. New York: Wiley.

Sanderson, S. L., J. J. Cech, Jr., and M. R. Patterson. 1991. Fluid dynamics in suspension-feeding blackfish. *Science* 251:1346–1348.

Shapiro, D. Y. 1987. Differentiation and evolution of sex changes in fishes. *BioScience* 37:490–497.

Spencer, C. N., B. R. McClelland, and J. A. Stanford. 1991. Shrimp stocking, salmon collapse, and eagle displacement. *BioScience* 41:14–21.

Thomson, K. S. 1990. The shape of a shark's tail. *Am. Sci.* 78:499–501.

Thomson, K. S. 1990. *Living Fossil: The Story of the Coelacanth.* New York: W. W. Norton.

Webb, P. W. 1988. Simple physical principles and vertebrate aquatic locomotion. *Am. Zool.* 28:709–725.

CHAPTER

36

Amphibians

The green frog *Rana clamitans.*

An Overview of Amphibians

Most people think about amphibians—if they think about them at all—as cold, slimy, and not very bright. There was a time, however, when amphibians were the most exciting animals on land. They were, in fact, the *only* animals on land except for some arthropods and an occasional lobe-finned fish. As terrestrial animals ourselves, we are tempted to look back on the colonization of land as a simple and natural step, but there was nothing simple about it. Numerous adaptations were required before amphibians could spend an appreciable length of time on land, and they never succeeded in making a complete break from water. (See Table 30.2 for the adaptations required for independence from water.)

Class Amphibia or clade Batrachomorpha (see Figs. 21.4 and 21.6) now comprises more than 4300 species divided among three distinct groups: the wormlike caecilians (se-SIL-yens), the tailed salamanders, and the tailless frogs (including toads) (Fig. 36.1). The most obvious sign that salamanders and frogs are adapted to terrestrial living is their four legs. (The absence of legs in caecilians is a derived adaptation for burrowing. All amphibians are technically tetrapods.) The two pairs of legs are strongly attached to pectoral and pelvic girdles on the vertebral column (Fig. 36.2). The force generated by the legs pushing against the substratum is therefore transferred directly to the vertebral column. The bones of amphibians are also generally denser and stronger than those of fishes, providing support without the buoyancy of water. Another major terrestrial adaptation are the lungs that occur in most adult amphibians.

In spite of these adaptations, few amphibians have become completely independent of water. Immature forms are almost always aquatic, and most adults must either live in moist habitats or frequently return to water. One reason is that amphibians depend on water for reproduction, since their eggs lack an impermeable shell that would prevent water loss or an amnion that would enclose the embryo in fluid (p. 118). Most amphibians also have an integument that is highly permeable to water, and they must therefore have access to fresh water to replace the body water lost by evaporation.

A

B

C

FIGURE 36.1

Representatives of the three major extant groups of amphibians.
(A) Caecilians, such as this yellow-striped caecilian *Ichthyophis kohtaoensis,* are rarely seen because they are worm-like burrowing animals of the tropics.
(B) Salamanders have legs adapted for walking, and a tail. Most remain in or close to water, but the subadult Eastern newt, *Notophthalmus viridescens,* known as a red eft, spends several years wandering over land. (C) Because of their enlarged hind legs, frogs walk poorly but can hop and swim well. The bullfrog *Rana catesbeiana* is one of the largest and most common frogs of North America. It spends much of its time resting in water with its green head camouflaged among algae and vegetation. The nostrils and upper halves of the eyes are usually above water, while the lower halves of the eyes watch beneath the surface.

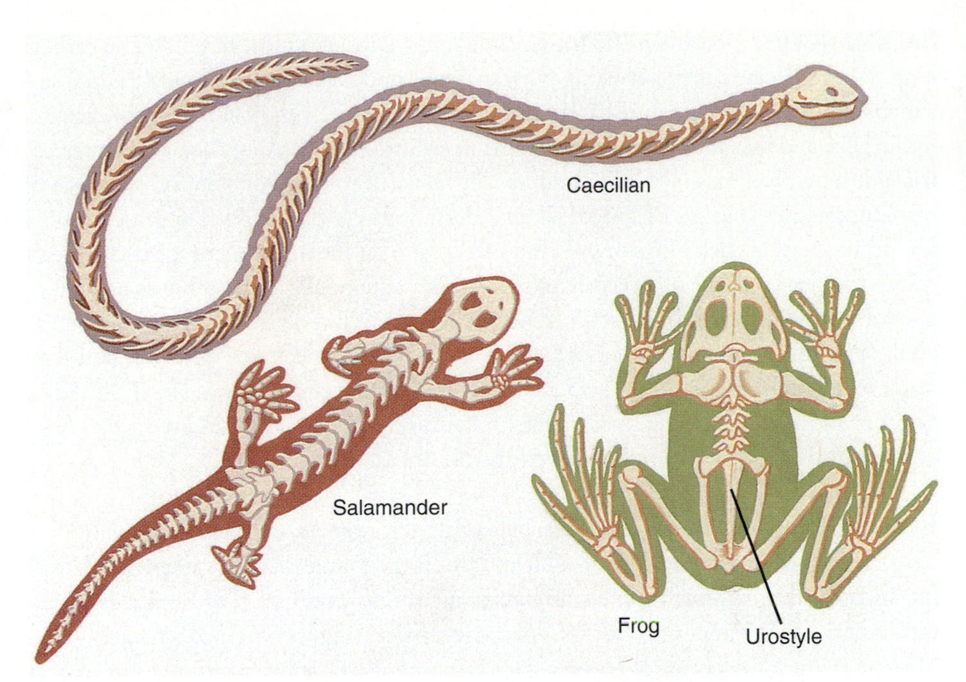

FIGURE 36.2

The skeletons of caecilians, salamanders, and frogs are adapted for their different modes of locomotion. In salamanders the back-and-forth movement of the legs is assisted by flexion of the vertebral column. In frogs the urostyle transmits the force of the hind legs to the rest of the body during jumping.

■ **Question:** Which of these three types of skeleton appears to be most primitive? What is meant by primitive?

The colonization of land by vertebrates was begun late in the Devonian period, almost 400 million years ago, by lobe-finned fishes that were already equipped with lungs and fleshy fins as adaptations for escaping the frequent droughts of that warm, dry period (p. 741). Escape from the recently evolved jawed fishes and the lure of edible insects may also have selected for the ability to lead double lives in water and on land (Greek *amphi-* double + *bios* life). Early amphibians, such as *Ichthyostega,* had four legs and a prominent tail (see Fig. 34.22). A variety of other amphibians evolved between 360 and 250 million years ago—a period sometimes called the Age of Amphibians. Reptiles, mammals, and birds shared the same ancestry. They are much better adapted to land and probably contributed to the decline of amphibians during the Mesozoic era. The fossil record of extant amphibians cannot be traced back any farther than about 200 million years, so relationships among them and to earlier amphibians are unclear.

Kinds of Extant Amphibians

CAECILIANS. Caecilians (order Gymnophiona) are little known because they are burrowing animals, and even when noticed they are easily mistaken for earthworms. As a result of evolutionary convergence, they have several superficial similarities with snakes. Early caecilians had limbs, but all have been lost as adaptations for burrowing (Jenkins and Walsh 1993). Some species have scales that arise from the dermis and protect against abrasion. In addition, the lungs are long and thin, and the left one is often much smaller than the right. Although caecilians appear to be all tail, they in fact have little or no tail, since the anus is near the tip of the body. They range in length from 10 cm to more than 1.5 meters, and they have from 60 to 285 vertebrae. They prey on invertebrates, which they locate by protrusible sensory tentacles between the eyes and nostrils. There are some 160 species that live in tropical forests of Central and South America, Africa, India, and the Seychelles islands in the Indian Ocean. Because no amphibian can survive for long in seawater, this far-flung distribution was a great mystery prior to the discovery of continental drift. Now it is understood that caecilians originated on Gondwana before it broke up (see Fig. 19.8).

SALAMANDERS. Salamanders (order Caudata = Urodela) most resemble the earliest amphibians, because they have long post-anal tails and legs of approximately equal size suitable for walking. Immature caudates are aquatic larvae, with pharyngeal gill slits, external gills, and tails that serve as caudal fins in swimming. Metamorphosis into a terrestrial adult involves loss of the gills and usually a switch to lungs or skin for exchange of respiratory gases (Fig. 36.1B). Most salamanders, however (family Plethodontidae and a few other species), lack lungs or gills and rely totally on the integument for exchange of respiratory gases. Some other salamanders remain aquatic throughout life as a result of **paedomorphosis,** which is the occurrence of external gills and other larval traits in the adult (Figs. 6.32 and 36.3). There are approximately 380 species of urodeles, and they range in length from less than 15 cm to 1.5 meters. They occur mainly in North America, Europe, and Asia, reflecting their origins in Laurasia (see Fig. 19.8). Many species have crossed land bridges into South America and Southeast Asia.

FROGS. Most amphibians are frogs, belonging to order Anura (Greek *an-* without + *oura* tail). The dominant feature of adult anurans is their large hind legs adapted for leaping and hopping, although some entirely aquatic frogs never hop. The hind toes are almost as long as the upper two sections of the leg, increasing the leverage during leaping. The toes of the hind feet are also webbed for swimming. The short front legs serve mainly as props when resting, as shock absorbers when landing after a hop, and as detectors of substrate vibration. Males also use the front legs to clasp the females during mating. Frogs have fewer vertebrae than do salamanders, and several are fused into a long, flat bone (the **urostyle**) that transmits the force of the hind legs to the body (Fig. 36.2).

A

B

FIGURE 36.3

Paedomorphosis in salamanders. (A) The adult mudpuppy *Necturus maculosus,* which is common in the Mississippi River drainage. The gills, red with blood, grow larger than shown here if the mudpuppy lives in stagnant, muddy water. Length approximately 40 cm. (B) An adult female two-toed amphiuma (*Amphiuma means*) guarding her clutch of eggs. The legs, not visible here, are reduced to useless prongs. Amphiumas inhabit swamps and canals in the southeastern coastal plain of the United States. They may live for more than two decades and grow to a length of more than a meter. They are voracious predators of crayfish, frogs, and fish, and they can inflict a painful bite.

Orders Within Class Amphibia

Genera mentioned elsewhere in this chapter are noted.

Order Gymnophiona JIM-no-FEE-oh-nuh (Greek *gymno* naked (that is, without scales) + *ophioneos* snakelike). Limbless and wormlike. Little or no post-anal tail. Eyes degenerate. *Ichthyophis* (Fig. 36.1A).

Order Caudata (= Urodela) caw-DAH-tuh (Latin *caudatus* tailed). Long tail. Separate head and trunk. Usually two pairs of limbs approximately equal in size. Salamanders. *Ambystoma, Amphiuma, Necturus, Notophthalmus, Taricha* (Figs. 36.1B, 36.3, 36.16).

Order Anura un-YOUR-uh (Greek *an-* without + *oura* tail.) Adult has no tail. Head and trunk fused. Hind legs enlarged for jumping. Frogs (including toads). *Agalychnis, Ascaphus, Bufo, Dendrobates, Epipedobates, Hyla, Leptodactylus, Nectophrynoides, Rana, Rheobatrachus, Xenopus* (Opening figures for Chapters 11, 21, and this chapter; Figs. 16.2B, 36.1C, 36.4, 36.6, 36.13, 36.14).

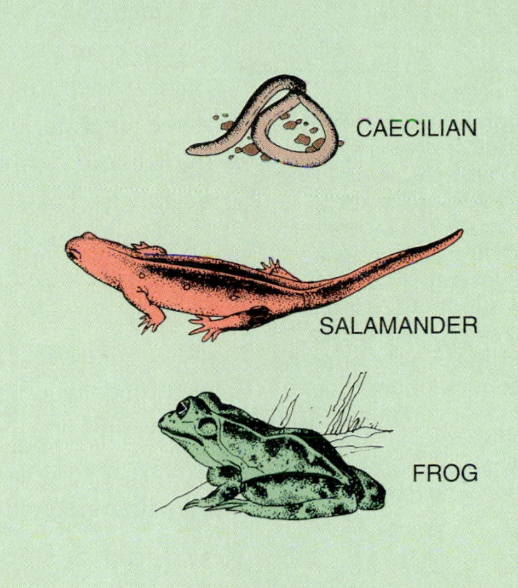

Adults never have tails, and the head is essentially fused to the trunk. Adult frogs rely on a combination of lungs and integument for exchange of O_2 and CO_2. Certain frogs, commonly known as toads, have thicker skins and are better adapted for life away from water. Adult frogs range from 1 to 30 cm long, and they feed mainly on insects and other invertebrates. In contrast to salamanders, most frogs undergo a dramatic metamorphosis. The terrestrial, carnivorous adults develop from tadpole larvae that swim with their tails, have internal gills, and feed on algae, plants, and detritus (Fig. 36.4).

FIGURE 36.4
Tadpole larvae of the common European frog *Rana temporaria.* These two siblings are both eight weeks old, but the one on the right has already developed legs.

■ **Question:** What might account for the different rate of development in these two tadpoles?

Skin

Integumentary Respiration. Although children are attracted to amphibians, most grownups regard them as ugly, largely on account of their cold, slimy skin. The exceptions are herpetologists (specialists in amphibians and reptiles), who consider the integument to be one of the most beautiful features of amphibians. The skin of an amphibian is adapted as a respiratory organ that is often more important than the gills or lungs, especially in adult frogs immersed in water. Even on land the skin absorbs almost as much O_2 as the lungs do, and it eliminates most of the CO_2. Frogs can continue to respire through the skin even if breathing ceases to ventilate the lungs. A major reason why frogs are frequently used in studies of vertebrate physiology is that they go on respiring even after the central nervous system is completely destroyed (pithed) to stop movement and pain.

The respiratory function of the skin depends to a large extent on its thinness (Fig. 36.5A). The **epidermis** is only a few micrometers thick, permitting easy diffusion of gases to and from the blood capillaries in the underlying dermis. There is an outer **cornified layer** containing the protein keratin, but it is usually extremely thin. During the months when they are active, frogs shed the cornified layer periodically (every 3 to 19 days in the toad *Bufo*). The frog splits the cornified layer starting at the head, peels it off with its legs, then usually eats it.

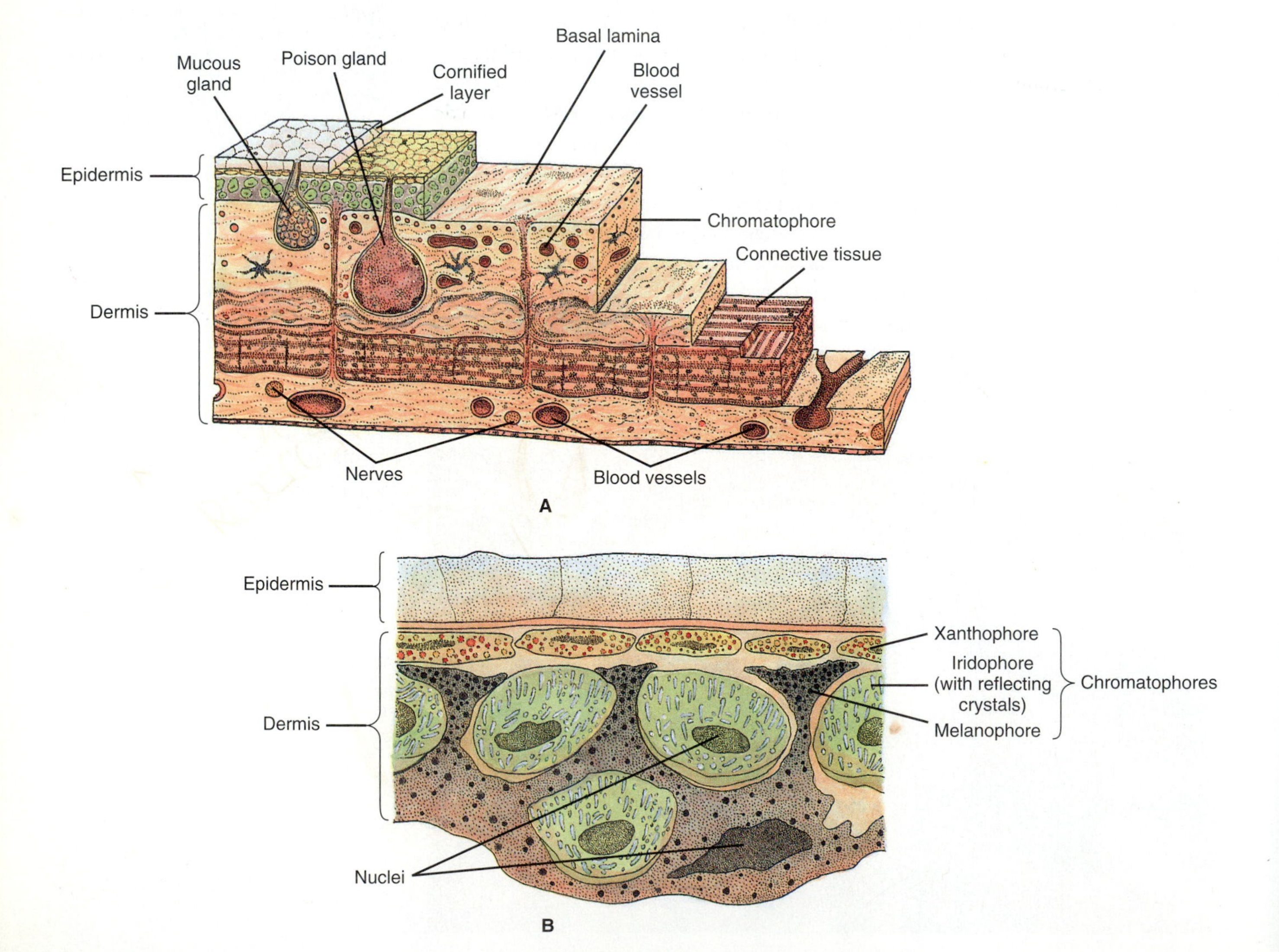

FIGURE 36.5

(A) Schematic representation of frog skin. Total thickness is generally less than 50 μm, although the skin of many frogs, especially toads, is thicker in some places ("warts"). (B) Section of skin from the green tree frog *Hyla cinerea* showing arrangement of chromatophores. Xanthophores contain yellow, orange, or red pigments. Blue structural color reflecting through the xanthophores from iridophores produces the green color common in frogs. Dark melanin pigments can migrate into branches of melanophores, covering the iridophores and darkening the skin.

DEFENSE. Frog skin is too thin to provide much protection against abrasion, dehydration, or predators. Because it is attached to underlying tissues in only a few places, however, predators may find it difficult to get a good grip on the body. A clear, slippery secretion from **mucous glands** in the dermis may also help frogs slip away from predators. The main function of the mucous glands is to keep the skin moist and therefore permeable to respiratory gases. The skins of all amphibians also contain **poison glands** (= granular or serous glands). These glands are often concentrated in thickenings of the skin or in bulges behind the eyes, called **parotoid glands.** (Parotoid glands must not be confused with the parotid salivary glands. Parotoid glands can be seen in Fig. 36.13B.)

When an amphibian is stressed, the poison glands secrete a frothy material containing one or more noxious substances. The secretions of some amphibians are only distasteful, but others contain toxins that are among the most potent among animals and are dangerous even to touch. Some induce paralysis by blocking ion channels or pumps in the membranes of nerve and muscle cells. Examples of these toxins are tetrodotoxin from certain newts (*Taricha*), which blocks sodium channels, and batrachotoxin from some poison-dart frogs of Central and South America, which holds the sodium channels open (p. 156). Toxin from the giant toad *Bufo marinus* works like the neurotransmitter serotonin. At one time there was a wave of toad-licking as rumors circulated that this secretion was hallucinogenic.

COLORS. Many toxic amphibians are brightly colored, and potential predators soon learn to avoid them (Fig. 36.6A). Red, orange, and yellow warning colors are due to pigments in certain **chromatophores,** and blue is a structural color reflected from other kinds of chromatophores (see Table 10.1; Figs. 36.5B and 36.6A). Less toxic species tend to be camouflaged. Amphibians that live among green algae and plants are often green (Fig. 36.1C). The green color comes from structural blue reflecting through yellow pigment. Many amphibians that live on the forest floor or in trees are camouflaged with brown or gray melanin pigments in chromatophores (Fig. 36.6B). These chromatophores, called **melanophores** or **melanocytes,** are also partly responsible for the darkening of the skin that often occurs when amphibians are on a dark substratum or in darkness. Control of the melanocytes is partly due to **melanocyte stimulating**

FIGURE 36.6

(A) Warning coloration in the red-backed dart-poison frog *Epipedobates cainarachi* of Peru. Dart-poison frogs (family Dendrobatidiae) get their name from three species in the genus *Phyllobates,* which natives of western Colombia use to poison their blow-gun darts. (B) The wood frog *Rana sylvatica* of the northeastern United States and southeastern Canada, camouflaged against a background of leaves.

■ **Question:** Why would it be more important for the wood frog to be brown rather than green, as camouflage against green vegetation?

A

B

hormone (MSH) from the pituitary. Secretion of MSH is in turn directed by the **pineal body** in the brain, which is connected by a nerve to a light-sensitive **pineal eye** (= median eye) in the top of the skull. (See Fig. 34.22C for location.) The pineal body also apparently controls daily (circadian) rhythms of physiology and behavior.

Internal Organization

LUNGS. The beauty of amphibians is not merely skin deep, but extends to the adaptations of their internal organs (Fig. 36.7). Many of the organs are similar in structure and function to those of other vertebrates, as described in Unit 2. (See index for details on particular organs.) The focus here is on the adaptations for terrestrial life, especially in frogs. The lungs of amphibians are essentially sacs, although there may be some internal walls that increase the surface area for diffusion of gases. Lungs often account for more than half the oxygen taken into the blood when a frog is on land and for virtually all the oxygen when it is floating in stagnant water with only the eyes and nostrils above the surface. Small amounts of respiratory gases are also exchanged across the lining of the buccal cavity. The structure of the lungs and the pattern of breathing are described in Fig. 12.11.

CIRCULATION. The evolution of air breathing entailed changes in the circulatory system from that of fishes. The amphibian heart has a second atrium that allows deoxygenated blood from the body tissues to be separated from oxygenated blood coming from the lungs and skin (Figs. 11.4B and 36.8). Deoxygenated blood enters a chamber on the back of the heart called the **sinus venosus,** which pumps the blood to the right atrium. At the same time, oxygenated blood from the skin and lungs enters the left atrium. In response to action potentials from the pacemaker region between the sinus venosus and the atria (the SA node), the two atria contract simultaneously, forcing both deoxygenated and oxygenated blood through atrioventricular valves into the one ventricle.

In some salamanders and frogs, such as the African clawed frog *Xenopus laevis,* the deoxygenated and oxygenated blood mix in the ventricle. In many salamanders

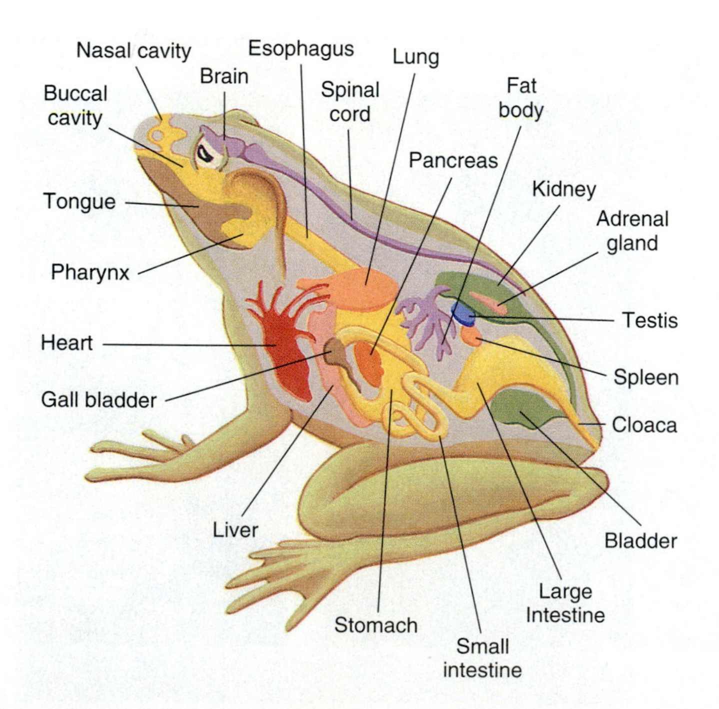

FIGURE 36.7
Internal organization of a male frog.

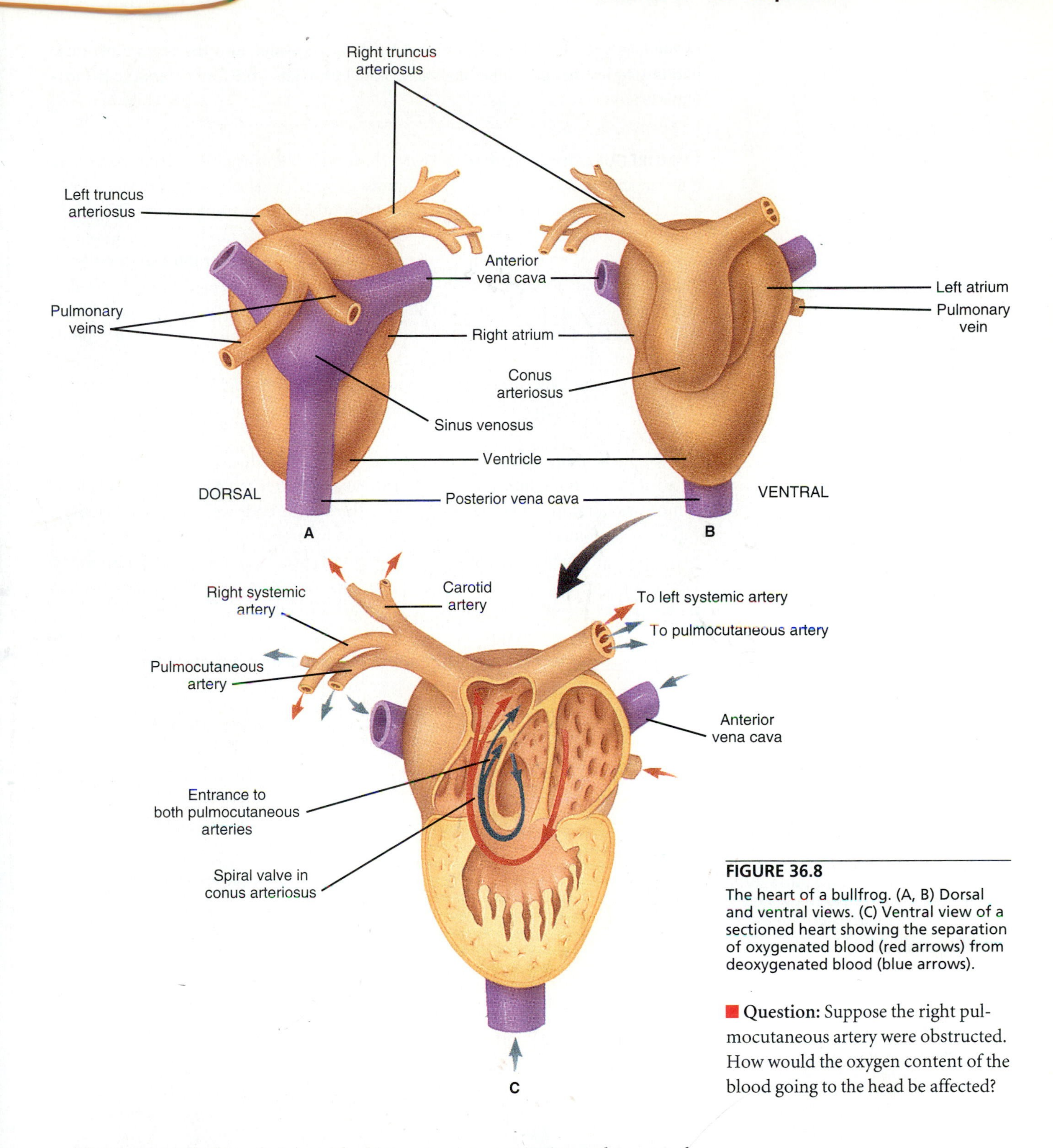

FIGURE 36.8
The heart of a bullfrog. (A, B) Dorsal and ventral views. (C) Ventral view of a sectioned heart showing the separation of oxygenated blood (red arrows) from deoxygenated blood (blue arrows).

■ **Question:** Suppose the right pulmocutaneous artery were obstructed. How would the oxygen content of the blood going to the head be affected?

and frogs, especially those that depend on integumentary respiration under water, the oxygenated blood in the left side of the ventricle remains separate from deoxygenated blood in the right side of the ventricle. When the ventricle contracts, it forces the blood into a chamber called the **conus arteriosus.** A **spiral valve** in this conus diverts the oxygenated blood into the **carotid artery** to the head and into the **right systemic artery.** Some oxygenated blood also enters the **left systemic artery.** The two systemic arteries merge at the dorsal aorta, which supplies most of the blood to the body. Most of the deoxygenated blood is diverted by the spiral valve into the two **pulmocutaneous arteries** that go to the lungs and skin, although some also enters the left systemic artery.

OSMOREGULATION AND EXCRETION. Most amphibians do not drink water, and the skin and the kidneys are the main routes by which ions and water are gained and lost. The kidneys conserve ions by reabsorbing them from urine into the blood. In fresh water the skin can actively transport ions into the body fluids. (Frog skin is a favorite model system for research on ionic transport.) When an amphibian is on land, the thin skin can also lose a great deal of water, and when it is submerged, the skin tends to absorb water osmotically. The kidneys compensate for this loss or gain of water through the skin by either reabsorbing water from the urine or producing copious amounts of dilute urine. Like fishes, adult salamanders and frogs have mesonephric kidneys (p. 286). The urine passes from the kidneys through the **ureters** into the **cloaca,** and it may then be excreted immediately or stored in the bladder (Figs. 36.7 and 36.9).

Amphibians are incapable of conserving water by producing urine more concentrated than the body fluids, but they can reabsorb water from the urinary bladder. Some frogs can store urine equivalent to one-third of the body weight in their bladders, and most amphibians can store large amounts of water in lymph sacs beneath the skin. No amphibian can live in seawater, although some species can tolerate brackish water. The crab-eating frog *Rana cancrivora* of Southeast Asia lives in mangrove swamps that are about 80% seawater. Like sharks and some bony fishes, these frogs have high concentrations of **osmolytes** (urea and methylamines) that help prevent loss of body water by osmosis. Some desert frogs conserve water by eliminating nitrogenous wastes in the form of uric acid, which requires less water from the kidneys. This adaptation is like that of birds and reptiles, but unlike that of most amphibians. Generally the kidneys of

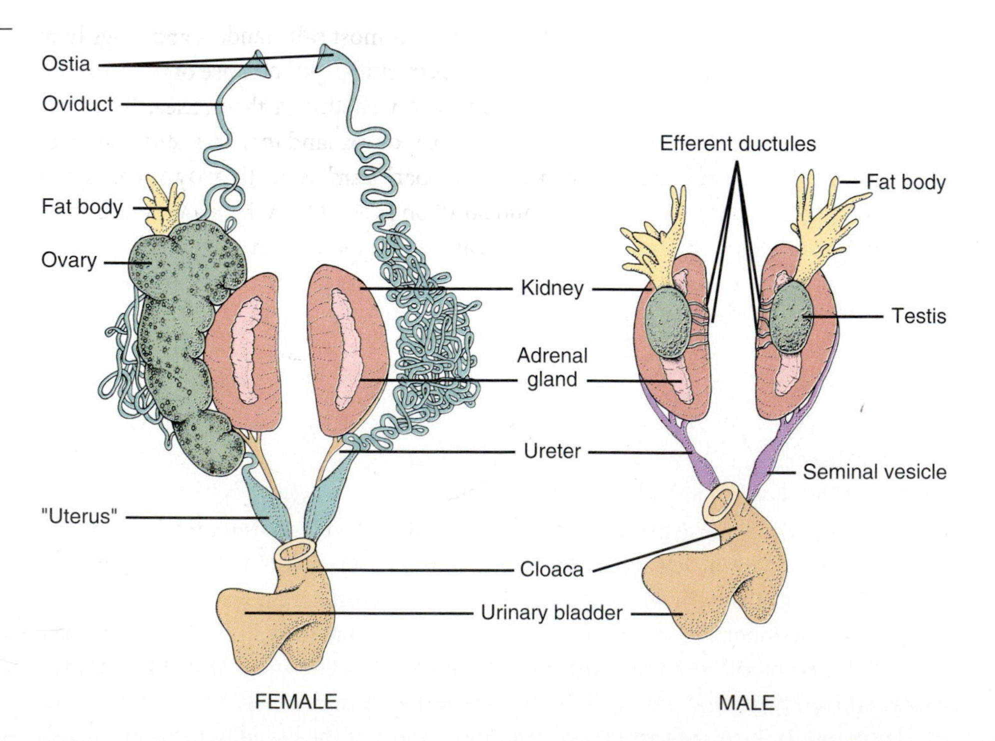

FIGURE 36.9
Ventral view of the urogenital systems of female and male frogs. Only the right ovary is shown. Sperm enter the kidney through efferent ductules, and then are stored in the ureters. Ova enter the oviducts and are stored in the "uterus."

■ **Question:** What does the "uterus" have in common with the uterus of a mammal? Why is "uterus" a poor choice of terms for an amphibian?

terrestrial amphibians eliminate nitrogenous wastes as urea, and larvae and aquatic adults eliminate it as ammonia.

Sensory Receptors

MECHANORECEPTION AND OTHER SENSORY MODALITIES. In the larvae and purely aquatic adults of many salamanders and frogs there are lateral line organs with mechanoreceptors and electroreceptors like those of fishes (pp. 754, 756). In terrestrial adults, however, most sensory receptors are adapted for detecting differences in the propagation of pressure, chemicals, and light in air compared with water. For example, air does not have sufficient density or electrical conductance to activate mechanoreceptors and electroreceptors like those in the lateral line organ of fishes. In terrestrial adult amphibians, touch, pressure, temperature, and injury are sensed mainly by free nerve endings in the skin. Terrestrial amphibians sense airborne molecules with olfactory epithelia in the nasal cavities and with **Jacobson's organs** (= vomeronasal organs) in the roof of the mouth.

Like other vertebrates, amphibians have an **inner ear** that detects position and acceleration of the head, as well as auditory vibrations. As in fishes, the inner ear consists of **semicircular canals** and **otolith organs.** An outgrowth of one of the otolith organs, called the **lagena,** is most sensitive to sound, and it is homologous with the cochlea of reptiles, mammals, and birds. While in water, amphibians hear in the same way that many fishes do, by detecting vibrations that pass through body tissues to the inner ear. On land, frogs have two other auditory channels. The first is by way of the external **eardrums** (= tympanic membranes; visible in Fig. 36.1C) that respond just like our own eardrums to airborne vibrations. A small bone, the **columella** (= stapes) transmits the vibrations from the eardrum to the inner ear. (See Fig. 34.22B.) The second, which is available to both frogs and salamanders, is through the forelimbs and muscles leading toward the inner ears. Low-frequency "seismic" vibrations conducting through these firm tissues may warn of large predators.

VISION. The eyes of caecilians are degenerate, but most salamanders and frogs have good vision. On land the cornea aids the lens in refracting light, because of the large differences in the index of refraction of air compared with that of the cornea. Frogs can therefore accommodate (focus on) much closer objects on land than in water. Amphibians accommodate by moving the lens forward or backward rather than by changing the shape of the lens. (Compare Figs. 8.12A and 36.10 on page 794.) As in mammals, **lachrymal (tear) glands** and eyelids reduce abrasion and drying of the cornea in air. Frogs blink by retracting the eyes into the orbit, which causes a translucent part of the lower lid, called the **nictitating membrane,** to sweep over the eye.

The **retina** of the amphibian eye has several kinds of rods and cones, but it is not clear whether it is capable of discriminating different colors. The retina also contains other kinds of cells, and it apparently does most of the processing of visual information before transmitting it to the brain. For example, in leopard frogs (*Rana pipiens*) one group of retinal cells responds to small moving objects that are darker than the background. These cells are called **net convexity detectors** or, more descriptively, "bug detectors." The retina does not respond to small, dark objects that don't move, and, in fact, an insect is perfectly safe in the presence of a leopard frog as long as it keeps still. The retinas of other frogs are apparently wired differently, however. The giant toad *Bufo marinus* and the African clawed frog *Xenopus* do see and eat food that is not moving, which makes them much easier to raise in the laboratory.

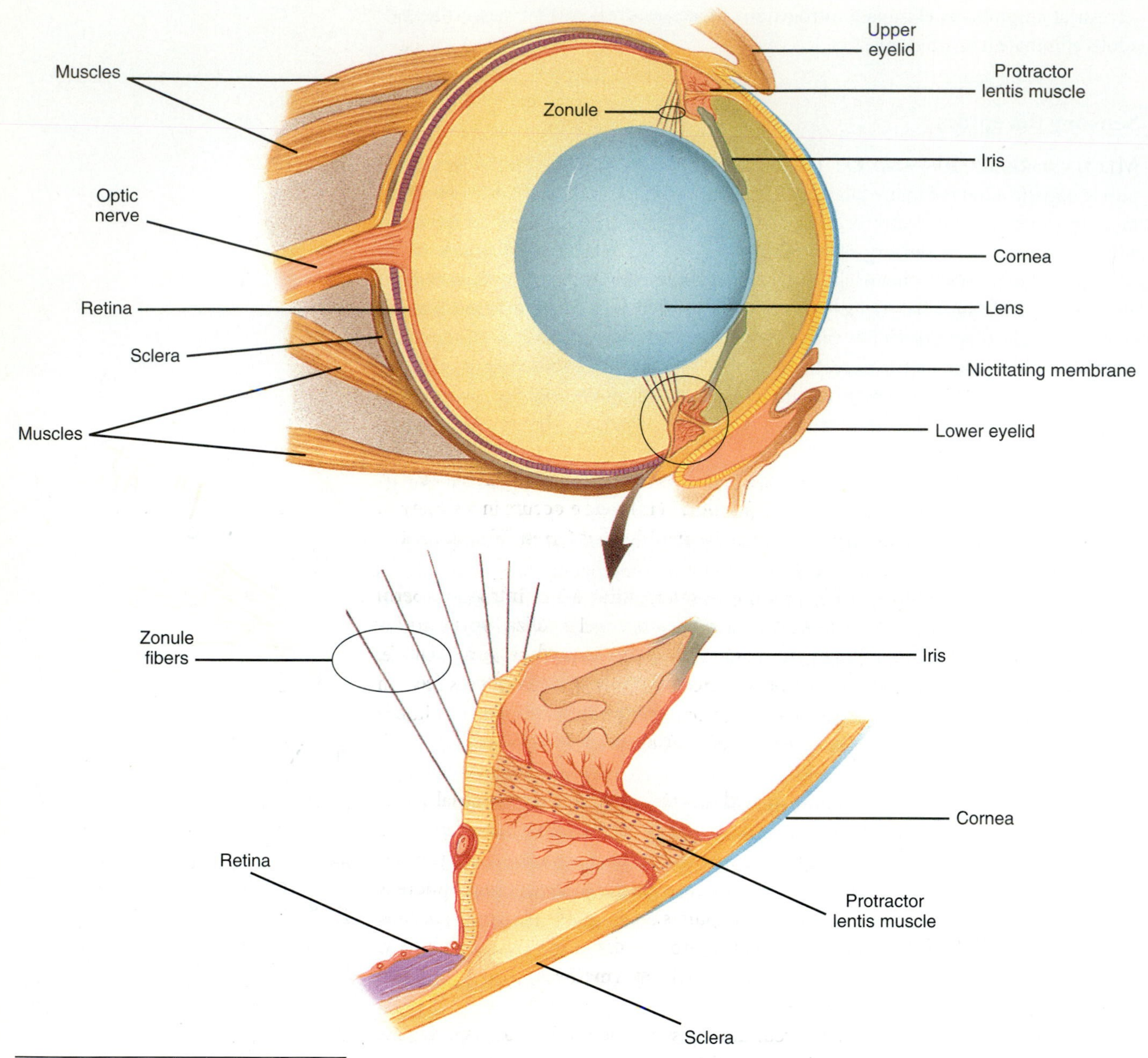

FIGURE 36.10

The eye of a leopard frog *Rana pipiens.* Inset shows detail of the mechanism for accommodation. Contraction of the protractor lentis muscle moves the lens forward, focusing nearby objects on the retina.

Feeding

Adult amphibians are carnivorous. Probably the most familiar means of feeding in adult amphibians is capturing prey on the sticky tongue and drawing it into the mouth to be swallowed whole without being chewed. Generally amphibians can capture prey on protrusible tongues, except for caecilians and permanently aquatic forms, such as *Xenopus.* The mechanism for protruding the tongue varies. In bullfrogs and true toads (families Ranidae and Bufonidae) the tongue is attached to the front of the mouth and flipped out like a whip (Fig. 36.11). Adult lungless salamanders (family Plethodontidae) extend their tongues by contracting lateral muscles that squeeze the tongue forward in much the same way as in humans. Few humans are as good at catching prey with their tongues as salamanders are, however. Some salamanders can catch an insect as far away as 80% of their body length within 50 milliseconds.

Amphibians do not necessarily limit their diet to what they can catch with their tongues. Bullfrogs and giant toads, for example, will eat small mammals, birds, snakes, fishes, and other frogs if the chance arises. The African clawed frog *Xenopus* and other species in the family Pipidae have no tongues at all. These purely aquatic species usually filter zooplankton by pumping water into and out of the mouth, and they use their fingers to stuff larger food into their mouths.

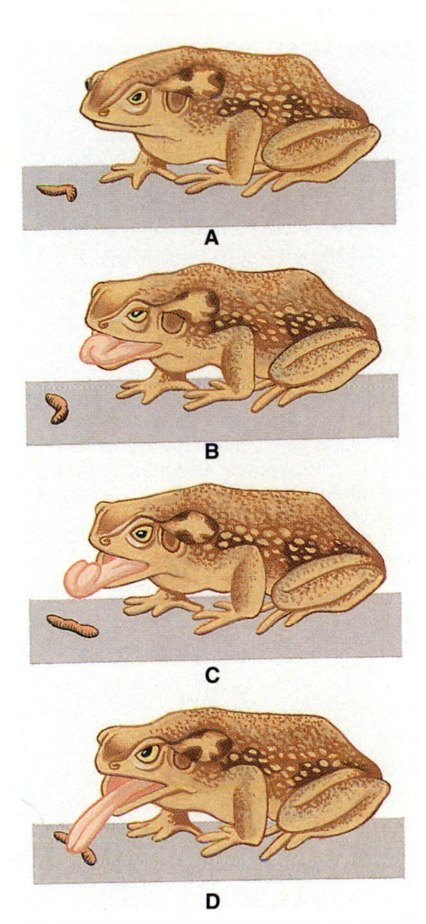

FIGURE 36.11

The lingual flip demonstrated by the giant toad *Bufo marinus,* which is native to fresh water in South America. It is one of the rare anurans that will feed on immobile prey. (A–C) The mouth opens and muscles contract, stiffening the base of the tongue and flipping it forward. (D) The flexible end of the tongue has catapulted forward and struck the prey. The entire sequence, including retrieval of prey, takes less than 0.15 seconds. Based on tracings of high-speed films by Carl Gans and G. C. Gorniak (1982).

Reproduction

The reproductive organs of amphibians are closely associated with the kidneys, ureters, and cloaca, all of which make up the **urogenital system** (Fig. 36.9). Sperm from the testes pass through fine tubules into the kidneys and ureters, where they are stored until released through the cloaca during mating. Fat bodies apparently store nutrients for the breeding season, when feeding generally does not occur. During the breeding season the ovaries of females produce hundreds or thousands of eggs. These eggs enter the body cavity and are taken into the two oviducts. The oviducts surround the eggs with a jelly coat as they propel them toward storage chambers near the cloaca. The jelly swells in water, producing shapeless blobs or long strands. Fertilization occurs in a variety of ways, depending largely on the degree to which the amphibian is terrestrial or aquatic.

CAECILIANS. As a group, caecilians are the most terrestrial. Males introduce sperm into the female by means of a protrusible copulatory organ, and fertilization occurs internally. The gametes therefore avoid desiccation and other hazards of terrestrial life. Most caecilians are viviparous, so the embryos are also protected. The fetuses feed on secretions and on tissues they scrape from the lining of the mother's oviduct. Little is known about courtship or many other aspects of reproduction in caecilians.

SALAMANDERS. In about 90% of salamanders, whether aquatic or terrestrial, fertilization is internal, although copulation does not occur. Instead the male deposits a **spermatophore,** which consists of a mass of sperm capping a conical blob of jelly secreted by a gland on the lining of the male's cloaca. Transfer of the spermatophore is achieved during courtship, in which the male captures or blocks the path of the female, then maneuvers or leads her until the spermatophore is drawn into her cloaca (Fig. 36.12 on page 796). A pouch in the female's cloaca (the **spermatheca**) stores the sperm until the eggs pass through the cloaca.

Various modes of reproduction occur among salamanders. In many species the eggs, larvae, and adults are all aquatic. Many others are terrestrial as adults, and these often produce terrestrial eggs. In some species with terrestrial eggs the larvae must find water, but in some the larval stage occurs entirely in the egg. Some terrestrial salamanders are viviparous.

Terrestrial salamanders that return to their native habitats to breed often do so by **homing** from rather long distances. Victor C. Twitty (1966) found many tagged red-bellied newts (*Taricha rivularis*) back in their native streams within a year after he had transported them across 8 km of rugged California hills. Although many amphibians are known to navigate using the position of the Sun, blinding these newts did not interfere with homing. Cutting their olfactory nerves did, however. Adult Eastern newts *Notophthalmus* also home to their native ponds, and they do so under circumstances that rule out olfaction. In the spring they orient in the direction of their home ponds even after being brought into a laboratory up to 30 km away. The directional cue appears to be geomagnetic, since their orientation can be disrupted with artificial mag-

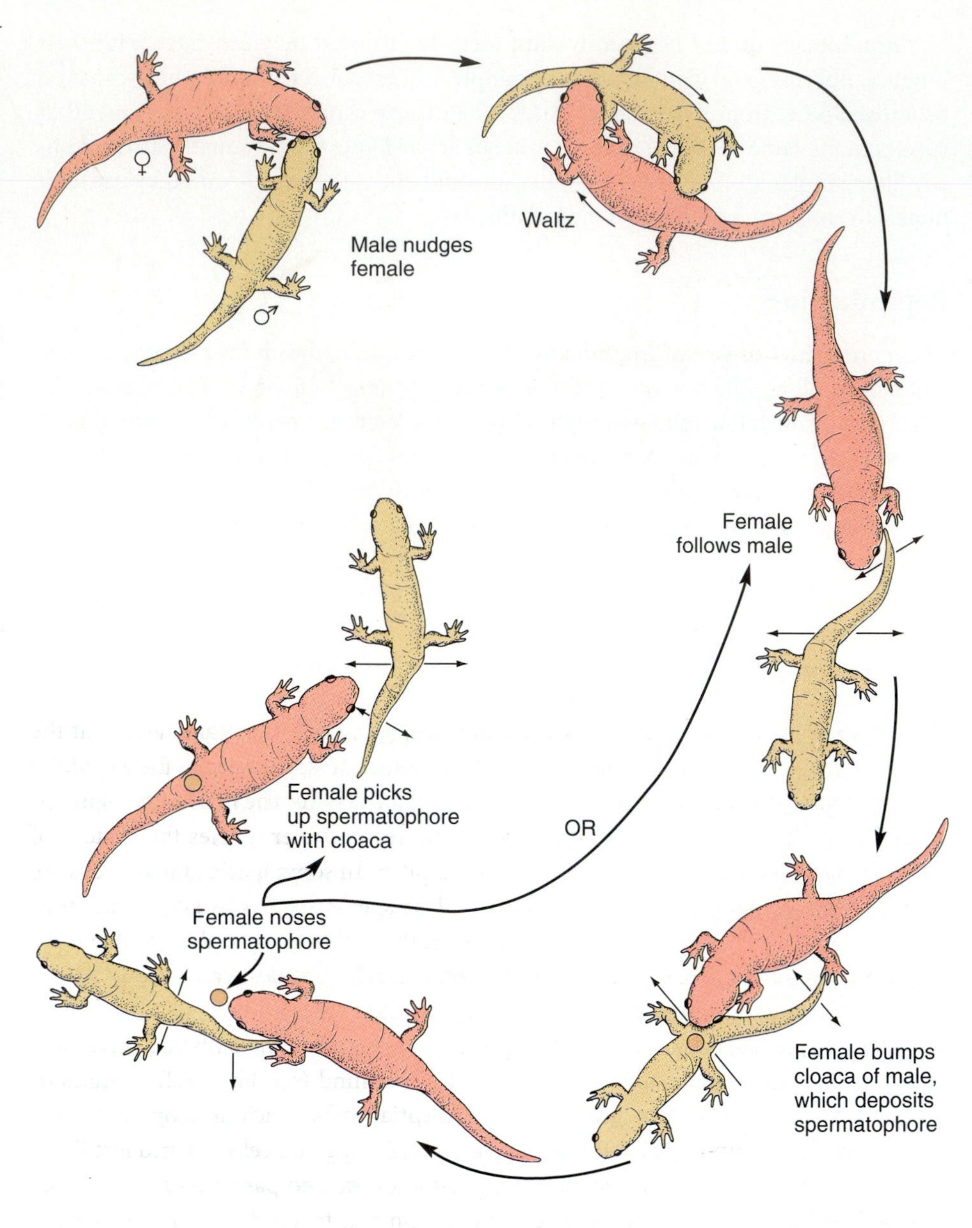

FIGURE 36.12
Courtship in the mole salamander *Ambystoma talpoideum,* which breeds in shallow pools in the Gulf states. The male approaches and nudges a larger female, then each pushes against the other's cloaca in a circular "waltz." After one or two rounds the male pulls himself forward with his front legs while shuffling his rear from side to side and fanning his tail across the female's snout. The female follows for several minutes, then nudges the male's cloaca. He then pauses to deposit a spermatophore, then resumes his movements. After nosing the spermatophore the female walks over it. If her cloaca picks it up she leaves the male. If not, she follows the male again for another try. Modified from Shoop (1960).

netic fields. Like birds (p. 412), amphibians apparently use more than one kind of directional cue in navigation.

FROGS. Almost all frogs are oviparous, and fertilization is usually external. This means that reproduction has to occur in water, or at least in a damp situation. Male frogs generally attract females during the breeding season by means of an **advertisement call** that is produced by exhaling air through the larynx. In most species, males have a prominent **vocal sac** that enlarges during the breeding season and inflates during calling (Fig. 36.13). The vocal sac has often been assumed to amplify the call by acting as a resonator. Because filling the sac with helium has no effect on the sound, however, this assumption appears to be incorrect (Rand and Dudley 1993). In male white-lipped frogs (*Leptodactylus albilabris*) of Puerto Rico, the vocal sac may be used in seismic communication, because it thumps the ground during calling. Advertisement calls can be risky. In the tropics there are bats that prey on frogs by homing in on their calls (see Fig. 39.17A).

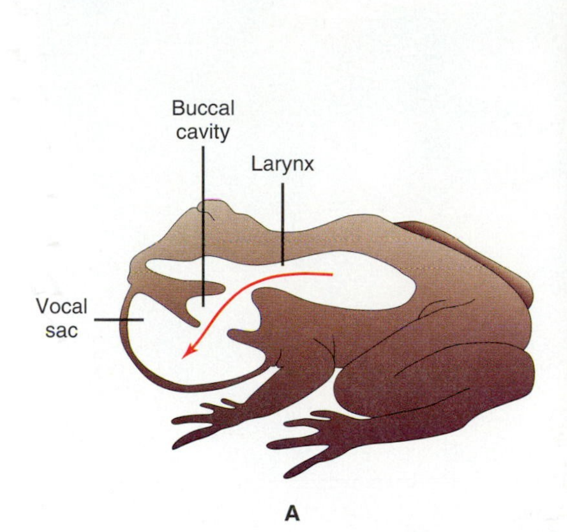

A

B

FIGURE 36.13

Advertising calling by male frogs. (A) Frogs produce sound by blowing air from the lungs across the vocal cords in the larynx and into the vocal sac. Females can also vocalize, but they do not have vocal sacs. (B) A singing American toad *Bufo americanus.* The song is a musical trill up to half a minute long. Body length approximately 10 cm.

In choice habitats there are usually many species calling simultaneously, but the calls are extremely species-specific in pattern and pitch. In some species the ear of the female is narrowly tuned to the pitch of the advertising call for the male of her species, and she may not even hear the calls of other species. In other species the pattern of notes within the call is more important than the pitch. In some species the females are most attracted to the lowest-pitched calls for their species, which usually come from larger males. Males not only attract females with their advertising calls, but also repel other males—usually. In some species, however, **satellite males** wait silently near a calling male and often intercept females he has attracted.

In most species, eggs and sperm are released during a mating embrace called **amplexus,** in which the male grasps the female from behind (see Fig. 16.2B). Males of many species grip the females with the help of **nuptial pads,** which develop on the fingers, arms, or chest in response to hormones. After the eggs are released and fertilized, they are usually left to hatch into free-living tadpoles with no parental care. The eggs and tadpoles are extremely vulnerable to predation and to accidents such as drying, but usually a female produces hundreds or thousands of eggs in a season.

Although this is the general pattern of mating, there are many interesting variations that afford more protection for the offspring and allow some independence from water. Frogs with such adaptations generally produce fewer offspring, but a greater proportion survive. One such adaptation is internal fertilization. In at least one species, *Ascaphus truei,* the male has a tail-like extension of the cloaca that serves as an intromittent organ during amplexus. In several species of African toads in the genus *Nectophrynoides,* the male and female achieve internal fertilization simply by pressing their cloacae together during amplexus. Five species in this genus are also viviparous. Many other frogs mate in the usual fashion but brood the eggs or tadpoles to protect them from drying and predation. Some construct nests, some guard the eggs, and some carry the eggs and tadpoles on their backs. In 60 species of "marsupial frogs" the females even have pouches or pits in the skin on their backs in which the eggs and tadpoles develop. Perhaps the oddest example of parental care is gastric brooding, in which a parent takes eggs or tadpoles into its stomach for brooding (Fig. 36.14 on page 798).

FIGURE 36.14
A frog is "born." The froglet *Rheobatrachus silus* from Queensland, Australia, emerges from the mouth of its mother after having developed in her stomach for 37 days. During this time the mother had not eaten, and secretion of acid and digestive enzyme from the stomach was suppressed. Similar gastric brooding occurred in a related species, *R. vitellinus*. Neither species has been seen for several years, and they and the phenomenon are presumed to be extinct.

■ **Question:** What medical value might research on these frogs have had?

Development

DEVELOPMENT OF THE TADPOLE. In some frogs, development is direct. In most frogs, however, a free-living tadpole stage precedes metamorphosis to the adult. The pattern in the family Ranidae is best known. The eggs generally begin developing immediately after fertilization. First the jelly coat surrounding the eggs swells with water, forming masses or strands of spawn. Within a few minutes to a day, depending on species and temperature, the egg cell begins dividing repeatedly in the process called **cleavage** (Fig. 36.15). Cleavage results in the formation of a hollow mass of cells called the **blastula** several hours later. The blastula then undergoes gastrulation during the next several hours. For the next several days the **gastrula** elongates and a tail bud forms. The egg then hatches into a tadpole, typically five to ten days after fertilization. At first the tadpole clings to vegetation with its sucker, obtaining nutrients from the remains of its yolk sac. Soon, however, the tadpole begins swimming with its tail fin as it feeds on algae and organic particles with its horny mouth. The gills are external at first, but soon an **operculum** grows over each gill, and a **spiracle** forms as an exit for water drawn through the mouth and past the gills. The next noticeable change in the tadpole is the development of the hind legs, which are usually complete two or three weeks after fertilization. Several days later the forelimbs emerge from the gill chambers in which they have been developing. Development generally takes longer in large species in cold climates: Bullfrogs in the northern United States, for example, usually spend two years as tadpoles.

METAMORPHOSIS. The emergence of the forelimbs signals one of the most remarkable processes in animals: the metamorphosis of the aquatic, herbivorous tadpole into a semiterrestrial, carnivorous frog. These changes are apparently triggered by a rise in the levels of thyroid hormones (thyroxine and triiodothyronine) and a fall in the level of prolactin. Over a period of a week or so, the tadpole's mouth softens and becomes wider, jaws and a tongue develop, the gut shortens, and the tail regresses. The gills also regress, and the lungs develop. Eventually, from one to several months after fertilization, the froglet is able to emerge onto land. It may live as an amphibious juvenile for several years before it is able to reproduce.

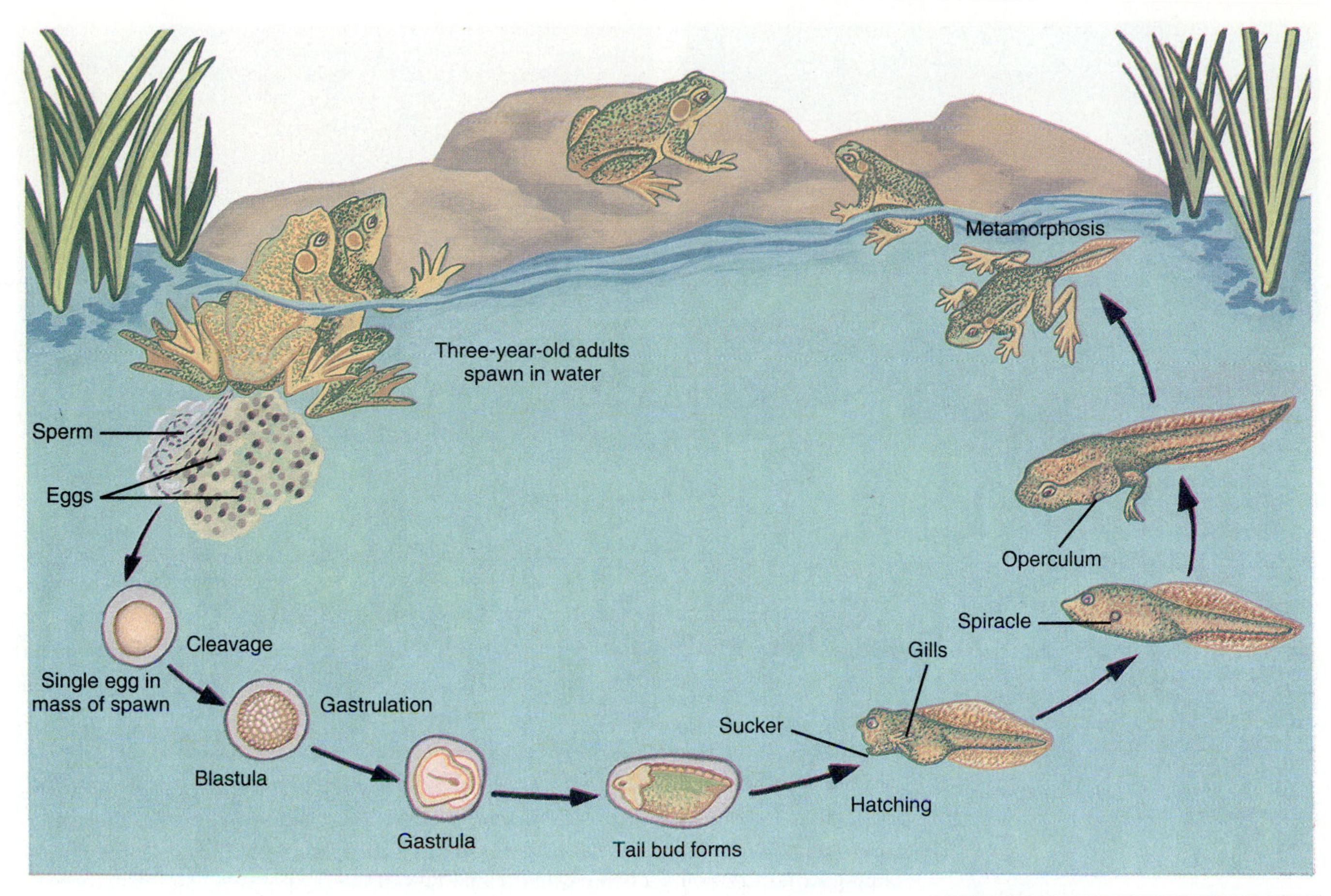

FIGURE 36.15
Development in anurans. For details of cleavage and gastrulation see Figs. 6.12 and 6.16. See Fig. 36.4 for photographs of tadpoles.

Question: According to von Baer (pp. 130–131), the development of an individual vertebrate recapitulates the evolution of its species. To what extent does this seem to be true for the metamorphosis of frogs?

Development in Salamanders. Both larval and adult salamanders have legs and tails, so their metamorphosis is not as dramatic as the transition from tadpole to frog. In fact, as a result of paedomorphosis many salamanders become sexually mature with few external changes from the larval form (see p. 131). In some species, such as the mudpuppy and the eel-like amphiuma, paedomorphosis occurs regardless of environmental conditions (Fig. 36.3). In other salamanders, including subspecies of tiger salamanders in the western United States and Mexico, paedomorphosis occurs only if there is plenty of water. If their habitats dry up they lose their gills, develop lungs, and take to land. Subspecies of tiger salamanders that live in the eastern United States are seldom paedomorphic (Fig. 36.16 on page 800).

Interactions with Humans and Other Animals

Except for an occasional meal of frog legs, amphibians are not a major item in the human diet, and humans are never on an amphibian menu. Perhaps for that reason about the only people who pay much attention to amphibians are children and zoologists. It is hard to imagine a group of animals that has made a greater contribution to human welfare, however. Much of what we know about medicine originated in studies on frogs, many going back a century and more. Many of these studies are now repeated by young biologists like you as essential steps in their training. An estimated 20 million frogs per year are used in education in the United States alone. Frogs also continue to contribute to new knowledge. For example, the oocytes of the African clawed frog *Xenopus laevis* are favorite cells for testing the functions of nucleic acids. RNA isolated from other species and injected in the oocyte will often be translated into protein, enabling its func-

A B

FIGURE 36.16

Adults of two subspecies of the tiger salamander *Ambystoma tigrinum,* one paedomorphic (A) and the other not (B). Like many amphibians, tiger salamanders occur throughout the United States in several different forms that may interbreed where their ranges overlap. Subspecies from separated areas will often not interbreed, however. The colors of various subspecies of *A. tigrinum* range from gray to yellow to olive. The forms that undergo paedomorphosis are common in the Western United States. Length in part B approximately 30 cm.

tion in a living cell to be studied. *Xenopus* has also joined other frogs and salamanders in a long tradition of amphibian contributions to embryology. Researchers have also isolated substances from frog skin that may provide medical benefits. One substance, epibatidine from the dart-poison frog *Epipedobates,* stopped pain 200 times more effectively than morphine in tests on mice. Antibiotics called magainins have also been isolated from frog skin and are now undergoing clinical tests.

During the 1980s many zoologists began to report extinctions and alarming declines in various amphibians from many different habitats around the world, including the United States. Acid precipitation, deforestation, and changing climate may be contributing to these declines. A recent study shows a correlation between declining populations and increased exposure to ultraviolet radiation (see Blaustein and Wake 1995). Another study suggests that the "declines" are normal, random population fluctuations that are being noticed now only because zoologists are paying more attention to amphibians (Pechmann et al. 1991). One reason for the recent interest in amphibians is that they are extremely sensitive to environmental changes, and a decline may be regarded as a forewarning of future declines for animals that are not as sensitive. Moreover, a decline in amphibians may directly affect other species. A glimpse of what might happen can be seen in Bangladesh, where frogs were hunted out for sale to European restaurants. Since then the government has had to spend more on insect control than they earned from sale of the frogs. While most areas are suffering a decline in amphibians, there is a literal plague of frogs in the sugar cane fields of northern Australia. The self-inflicted problem is with the giant toad *Bufo marinus,* which cane-growers imported from Central America to eliminate beetles. Unfortunately, they ignored the warnings of herpetologists that the frogs would probably eat everything except the beetles, since they are nocturnal and the beetles are diurnal. They have expanded their range by several kilometers per year, consuming other amphibians and reptiles that get in their way.

Summary

Amphibians are tetrapods that are partly adapted to terrestrial living. There are three extant groups: caecilians, salamanders, and frogs. Extant caecilians have no legs, and their skeletons are adapted for burrowing. In salamanders there are four approximately equal-sized legs firmly attached to the rest of the skeleton, and the vertebral column extends beyond the anus into the tail. In frogs the hind legs are greatly enlarged for hopping, and there is no tail in the adult. Most amphibians are capable of living on land as adults, although they are still dependent on fresh water for reproduction. Terrestrial forms generally have lungs, aquatic forms generally have gills, and all have thin skins capable of integumentary respiration. The skin also contains mucous glands that help protect it and keep it moist, chromatophores that provide camouflage or warning coloration, and poison glands.

Caecilians generally reproduce by copulating, salamanders generally reproduce by transferring spermatophores, and most frogs reproduce by amplexus, in which the eggs are usually fertilized externally. Most male frogs attract females with advertising calls. Sounds, as well as vibrations of water or substratum, are detected by the inner ears. In frogs the fertilized eggs develop into tadpoles, which are aquatic and herbivorous. Changes in hormone levels trigger their metamorphosis into terrestrial carnivores, beginning with the development of the hind legs, followed by shrinkage of the tail and the emergence of the front legs from the gill openings. Many adult salamanders remain aquatic and retain the gills or other juvenile features as a result of paedomorphosis.

Key Terms

urostyle
parotoid gland
chromatophore
sinus venosus
conus arteriosus
pulmocutaneous artery
cloaca
Jacobson's organ
columella
nictitating membrane
spermatophore
vocal sac
amplexus
nuptial pad
metamorphosis

Chapter Test

1. Discuss three of the major adaptations of amphibians that enable them to survive on land. Describe two features of amphibians that prevent their being totally independent of water. (p. 784)
2. Name the three modern groups of amphibians, and describe the major differences among them. (pp. 785–787)
3. Describe how the amphibian skin is adapted for life in water and on land. (pp. 788–790)
4. Describe the roles of poison glands and chromatophores in defending amphibians against predation. (pp. 789–790)
5. Describe the structure of the frog's heart and the pathway of blood through it. (pp. 790–792)
6. Describe two ways that sensory receptors in amphibians are adapted to air rather than water. (p. 793)
7. Compare the courtship and reproduction of a typical salamander with that of a typical frog. (pp. 795–797)
8. Compare development in a salamander with complete metamorphosis with that of a typical frog. (pp. 798–799)

■ Answers to the Figure Questions

36.2 The salamander skeleton is most primitive because it is closer to the original tetrapod condition, whereas the loss of the legs and the adaptation of hind legs for hopping are derived conditions. Primitive does not mean poorly adapted.

36.4 Even though the eggs were presumably fertilized at about the same time, one may have been in warmer water than the other. Or the difference in rate of development could have been due to genetic variation.

36.6 Like many frogs, wood frogs breed in spring, before green leaves emerge.

36.8 Obstructing one of the two pulmocutaneous arteries would reduce the respiratory functioning of the skin or lungs by about half, so one could expect a corresponding drop in oxygen levels to the brain.

36.9 In contrast to the uterus of placental mammals, the "uterus" of the amphibian merely stores ova without providing any nutritive or respiratory support.

36.14 These frogs might have provided a new approach to the suppression of stomach acidity in the treatment of ulcers. (In fact, there has been research along those lines.)

36.15 The tadpole does not resemble a fish ancestor. As von Baer noted, however, the tadpole does resemble a stage in the larval development of fishes. (See Fig. 6.31.)

Readings

Recommended Readings

Blaustein, A. R., and R. K. O'Hara. 1986. Kin recognition in tadpoles. *Sci. Am.* 254(1):108–116 (Jan).

Blaustein, A. R., and D. B. Wake. 1995. The puzzle of declining amphibian populations. *Sci. Am.* 272(4):52–57 (Apr).

Del Pino, E. M. 1989. Marsupial frogs. *Sci. Am.* 260(5):110–118 (May).

Duellman, W. E. 1992. Reproductive strategies in frogs. *Sci. Am.* 267(1):80–87 (July).

McClanahan, L. L., R. Ruibal, and V. H. Shoemaker. 1994. Frogs and toads in deserts. *Sci. Am.* 270(3):82–88 (Mar).

Myers, C. W., and J. W. Daly. 1983. Dart-poison frogs. *Sci. Am.* 248(2):120–133 (Feb).

Narins, P. M. 1995. Frog communication. *Sci. Am.* 273(2):78–83 (Aug).

Phillips, K. 1990. Where have all the frogs and toads gone? *BioScience* 40:422–424.

See also relevant selections in General References at the end of Chapter 21.

Additional References

Behler, J. L., and F. W. King. 1979. *The Audubon Society Field Guide to North American Reptiles and Amphibians.* New York: Alfred A. Knopf.

Dawid, I. B., and T. D. Sargent. 1988. *Xenopus laevis* in developmental and molecular biology. *Science* 240:1443–1453.

Dudley, R., and A. S. Rand. 1993. Frogs in helium: the anuran vocal sac is not a cavity resonator. *Am. Zool.* 33: abstract 588.

Duellman, W. E., and L. Trueb. 1994. *Biology of Amphibians,* 2nd ed. Baltimore Johns Hopkins University Press.

Gans, C., and G. C. Gorniak. 1982. How does the toad flip its tongue? Test of two hypotheses. *Science* 216:1335–1337.

Halliday, T. R., and K. Adler (Eds.). 1986. *Encyclopedia of Reptiles and Amphibians.* New York: Facts on File.

Hankin, J. 1989. Development and evolution in amphibians. *Am. Sci.* 77:336–343.

Jenkins, F. A., Jr., and D. M. Walsh. 1993. An Early Jurassic caecilian with limbs. *Nature* 365:246–250.

Pechmann, J. H. K., et al. 1991. Declining amphibian populations: the problem of separating human impacts from natural fluctuations. *Science* 253:892–895.

Rand, A. S., and R. Dudley. 1993. Frogs in helium: the anuran vocal sac is not a cavity resonator. *Physiol. Zool.* 66:793–806.

Shoop, C. R. 1960. The breeding habits of the mole salamander in southeastern Louisiana. *Tulane Stud. Zool.* 8:65–82.

Twitty, V. C. 1966. *Of Scientists and Salamanders.* San Francisco: W. H. Freeman. (*Amusing accounts of research.*)

Zug, G. R. 1993. *Herpetology: An Introductory Biology of Amphibians and Reptiles.* San Diego: Academic Press.

CHAPTER

37

Reptiles

The bush viper *Atheris squamiger*.

What Is a Reptile?

Few animals inspire as much curiosity, awe, and terror as reptiles. Even though most of the 6300 living species of turtles, snakes, lizards, and crocodilians are harmless, the thought of encountering a snake is enough to keep many people out of the woods, and even dinosaurs that have been safely extinct for more than 65 million years can still make a child sleep with the lights on. Living reptiles are also interesting enough to keep mature herpetologists up past bedtime. Taxonomists also lose sleep because of reptiles.

Although everyone can immediately tell the difference between a scaly reptile and a feathered bird or hairy mammal, defining reptiles zoologically has proved difficult. Extant reptiles, birds, and mammals are all descended from the same ancestor. All are tetrapods with adaptations for completely terrestrial living, including eggs with an amnion that encloses the embryo in fluid, and skin that is impermeable to water and respiratory gasses. Taxonomists traditionally define the class Reptilia to include amniotes with scales rather than feathers or hair, but this distinction is relatively minor compared with the many similarities between, for example, birds and crocodiles. It is also seldom helpful to paleontologists, since feathers and hair don't often fossilize. Moreover, cladists require that a natural group include all the descendants of the most recent ancestor, so they do not recognize a clade that includes reptiles but not birds and mammals (see pp. 435–437). Perhaps future zoology texts will not even have a chapter on reptiles. For now, however, it is convenient to follow tradition as long as we remember that the term "reptile" refers to a diverse assemblage of extinct and living forms, some of which are closer to mammals and birds than to turtles and snakes.

The oldest known reptilian fossil dates to the Carboniferous period around 340 million years ago (see Table 19.2, Fig. 34.17, and Fig. 37.1). Reptiles therefore originated only some 30 million years after amphibians. Within 100 million years after their first appearance, reptiles became so numerous, diverse, and often large that we refer to the entire Mesozoic era as the Age of Reptiles. Paleontologists divide reptiles into three major lineages (Fig. 37.2). Two relatively insignificant lines of descent led to turtles and to mammals. The third line of descent was the most diverse and spectacular. It included the ancestors of living snakes, lizards, and crocodiles, as well as extinct marine reptiles (ichthyosaurs and plesiosaurs), flying reptiles (pterosaurs), dinosaurs, and birds.

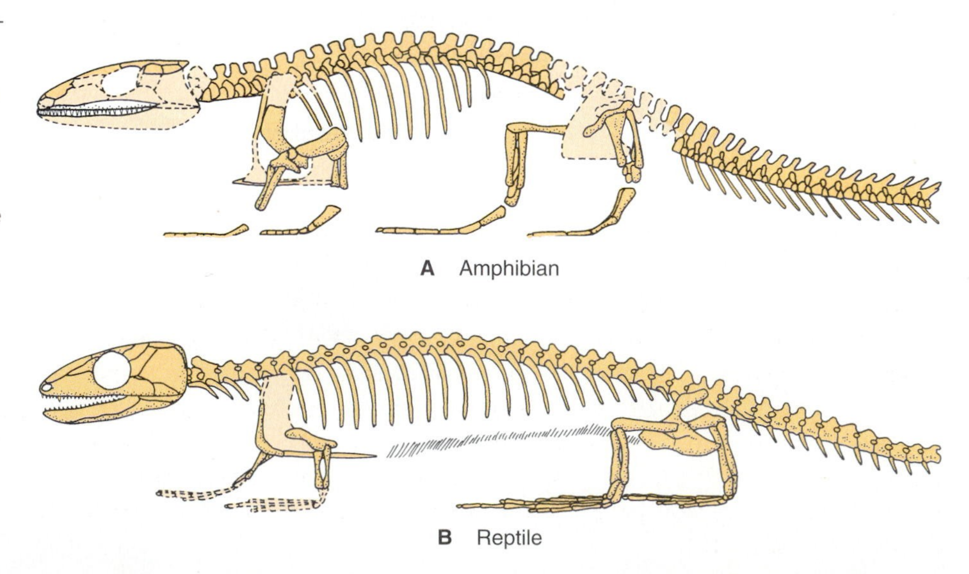

FIGURE 37.1

Reconstructed skeletons of a fossilized amphibian and a reptile, both of which lived during the Carboniferous period, and both of which were approximately 25 cm long. (A) *Bruktererpeton fiebigi* belonged to a group of amphibians that probably descended from the same group as reptiles. (B) *Hylonomus lyelli,* judged to be a reptile from details of the skeleton. Note the gastral ribs on the belly. This fossil was found in what is now Nova Scotia, inside a fossilized hollow tree stump where the animal is presumed to have been trapped.

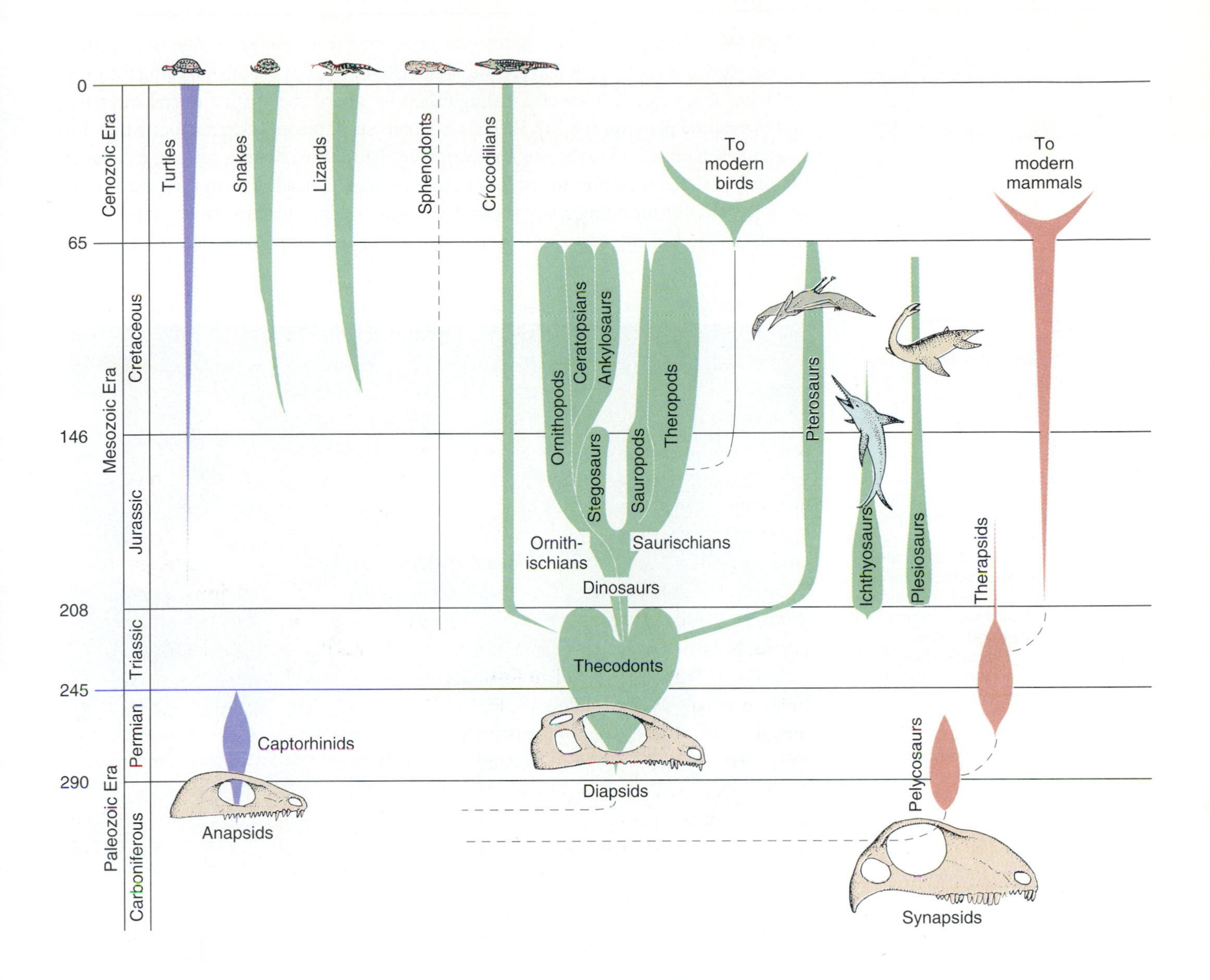

FIGURE 37.2

The evolution of major groups of reptiles. Numbers on the left are millions of years before present. The three major lineages are recognized by the number of temporal openings in the skull behind the orbit of the eye. Captorhinids and living turtles have no temporal openings and are called anapsids. Synapsids had one temporal opening on each side and were ancestral to mammals. Diapsids had two temporal openings on each side.

■ **Question:** Construct a cladogram showing the branching sequence for turtles, crocodilians, birds, dinosaurs, and mammals. Compare with Fig. 21.6.

Extant Reptiles

Paleontologists recognize these three lineages largely from the number of **temporal openings** in the skull behind the orbits of the eyes. The function of these openings is not certain, but they may relate to the attachment of jaw muscles. Temporal openings often occur in the skull near bony arches where muscles attach. (The cheek bone of humans is an example of such an arch.) Turtles lack these arches and temporal openings, and they are therefore referred to as **anapsids** (from the prefix *an,* meaning without, and "apse," the arch-shaped wing that houses the altar of a cathedral). The ancestors of mammals had one pair of temporal openings, and they are called **synapsids** (Greek *syn* with). The remaining early reptiles, with two pairs of temporal openings, are called **diapsids** (Greek *di* two).

TURTLES. Today only a few hardy orders survive from the great Age of Reptiles. Turtles (order or clade Chelonia) are the most distinctive from the others, being the only surviving anapsids. Most are also distinguished by a hard shell with a dorsal **carapace** and the ventral **plastron** (Fig. 37.3). The shell consists of pieces of dermal bone that are fused to each other and to the ribs and vertebrae. The dermal bone is covered by a layer of horny material similar to the scales of other reptiles. Instead of teeth they use keratinous jaw sheaths for biting. Only about 300 species of chelonians survive. Some occur mainly on land, many occur only near fresh water, and some live in seawater except

A

FIGURE 37.3

Turtles. (A) The common eastern box turtle *Terrapene carolina,* showing the carapace. This individual is a male; females have yellow-brown eyes. The turtle can close its shell tightly enough to retain moisture and keep out even the smallest predator. It is usually harmless, although people have died from eating box turtles that have consumed poisonous mushrooms. This and other species of turtles can live many decades and have been found with dates of more than a century ago carved on their carapaces. (Length of carapace up to 22 cm.) (B) The Midland painted turtle *Chrysemys picta marginata,* showing the plastron. This species is common in and around ponds in eastern and northern states. (See also Fig. 15.1.) Faint annual growth rings on each area of the plastron allow the age to be determined. (C) The leatherback turtle *Dermochelys coriacea* does not have a hard shell, but is covered with a leathery skin with dermal plates imbedded. This is the largest living turtle, reaching lengths of more than 2 meters. It wanders widely in the Atlantic and Pacific Oceans and can dive to 1.2 kilometers for more than half an hour. Like many sea turtles, this species is endangered. This female is digging a hole in which to deposit eggs that were fertilized at sea.

B

C

when breeding. The British call the terrestrial chelonians tortoises, the freshwater forms terrapins, and the marine species turtles. Americans usually just call them all turtles.

LIZARDS, WORM LIZARDS, AND SNAKES. Other living reptiles are diapsids, although the number of temporal openings has often changed from the original two. The order Squamata (clade Lepidosauria) includes approximately 6000 species about evenly divided between lizards and snakes. Most extant lizards (suborder Lacertilia) look like their ancestors, with four sprawling legs and long tails (Fig. 37.4). Worm lizards and snakes evolved from lizard ancestors, and cladistically they are specialized clades of lizards adapted for burrowing. There are about 140 species of worm lizards that are classified in either the suborder Lacertilia with other lizards or in the suborder Amphisbaenia. Worm lizards have no legs, except for *Bipes,* which has two. Snakes are generally classified in the suborder Serpentes. Most snakes have lost all external traces of legs, although some retain vestiges of the pelvic girdle (Fig. 37.5 on page 808).

CROCODILIANS. Order or clade Crocodilia includes 21 species of crocodiles, alligators, caimans, and a gavial (Fig. 37.6 on page 809). Crocodilians are more closely related to birds than to other extant reptiles, but they look like overgrown lizards, with

A

B

C

FIGURE 37.4

Lizards. (A) The common iguana, *Iguana iguana,* is green when young, but it turns brown or black with age. Some individuals grow to 2 meters long. They are nonvenomous but will defend themselves by biting, scratching, and lashing with their tails. Generally they are found in trees overhanging water, to which they escape when threatened. Common iguanas are native to Central America, but they have been introduced into Florida. (B) The two-legged worm lizard *Bipes* sp., sometimes classified in the suborder Amphisbaenia. (C) A real dragon. Komodo monitors, *Varanus komodoensis,* are up to 3 meters long and feed on deer, pigs, carrion, and sometimes each other. Only a few thousand wild Komodo dragons survive on Komodo and a few neighboring islands in Indonesia, but many zoos have succeeded in breeding them.

A

B

FIGURE 37.5

Snakes. (A) The common garter snake *Thamnophis sirtalis* lives throughout the United States and well into Canada. There are numerous subspecies that vary in color and pattern. When disturbed, it gives off a foul musk. It bites but is nonpoisonous. Females are up to 60 cm long and give birth to as many as 85 live young each year. Males are about one-third as long. (B) The timber rattlesnake *Crotalus horridus* of the eastern United States lives for up to 30 years and reaches lengths up to 1.9 meters. Rattlesnakes and their allies, copperheads and cottonmouths, are called pit vipers for the pit between the nostril and the eye, which leads to an infrared receptor that detects warm prey. Each time the rattlesnake molts, usually two to four times per year, it adds a new segment to its rattle. (C) The highly venomous Arizona coral snake *Micruroides euryxanthus.* Another species of coral snake occurs in southeastern coastal areas of the United States, and there are more than 200 other species mainly in the tropics. The bright colors are warning coloration that predators learn to avoid. (Length up to 53 cm.) (D) In areas where there are coral snakes, harmless snakes with similar coloration are presumed to be mimics. The milk snake *Lampropeltis triangulum* may be an example. Milk snakes were named for the common belief that they drank milk from cows. This is only one of many superstitions about snakes. (Length up to 63 cm.)

C

D

A

B

C

FIGURE 37.6

Crocodilians. (A) The American crocodile *Crocodylus acutus* lives at the extreme tip of Florida, in Central America, and on the brink of extinction. Adults may reach lengths of 7 meters, but large individuals are now rare. (B) Alligators have broader, rounder snouts than crocodiles. They commonly rest with all but their nostrils submerged. American alligators *Alligator mississippiensis* were hunted for their hides to the verge of extinction and were also sold by pet stores, but they are now common sights in the Gulf States. They reach lengths of up to 6 meters and occasionally attack humans in water or on land. (C) After trade in alligators was prohibited, spectacled caimans *Caiman crocodilus,* named for the bony ridges around the eyes, were imported from Central and South America for sale as pets. Many were released, and some now inhabit drainage ditches in southern Florida. (Length up to 2.6 meters.)

four sprawling legs and a tail. These largest of extant reptiles live in or near water, and they use their webbed hind feet and tails with great effectiveness for swimming.

TUATARAS. One final group of extant reptiles (order Sphenodonta) is also diapsid, and the two extant species in it deserve separate discussion because they retain many of the most primitive reptilian characters. These lizardlike sphenodonts, commonly called tuataras (TOO-ah-TAH-ruh), cling to survival on approximately 30 small islands of New Zealand, thanks to government protection (Fig. 37.7 on page 811). Until recently it was thought that there was only one species, *Sphenodon punctatus*, simply because a classification more than a century old had been ignored. Comparisons of enzyme structures and morphology have confirmed a second species, *S. guntheri.* This species is close to extinction because it was not recognized and protected on the one island where it occurs.

Tuataras can live more than 75 years as individuals, and they are also long-lived in evolutionary terms. They are descendants of a group that is more than 200 million

Extant Groups of Class Reptilia

Genera mentioned elsewhere in this chapter are noted.

Order Chelonia (= Testudines) kee-LOW-nee-uh (Greek *chelone* tortoise). No temporal openings (anapsid). Body enclosed in a two-part shell of dermal plates fused to vertebrae and ribs. Jaws without teeth. Turtles. *Caretta, Chelonia, Chrysemys, Dermochelys, Eretmochelys, Lepidochelys, Macroclemys, Pseudemys, Terrapene, Trachemys* (Figs. 15.1, 37.3, 37.18B.)

Order Squamata skwah-MAH-tuh (Latin *squama* scale). Body covered with epidermal scales that are shed with skin periodically. Paired copulatory organs. Jaws usually with teeth.

Suborder Lacertilia (= Sauria) lay-sir-TILL-ee-uh (Latin *lacerta* lizard). Two pairs of limbs usually present. Slender body. Most with external ear opening and movable eyelids. Lizards and worm lizards (family Amphisbaenidae or separate suborder Amphisbaenia). *Amblyrhynchus, Anolis, Basiliscus, Bipes, Chamaeleo, Cnemidophorus, Draco, Heloderma, Iguana, Ptyodactylus, Tupinambis, Varanus* (Chapter opener; Figs. 37.4, 37.15, 37.18A and C, 37.21)

Suborder Serpentes sir-PEN-teez (Latin *serpens* serpent). Elongated, legless body. No external ear opening. Lidless eyes covered by transparent scale. No urinary bladder. Left lung usually vestigial. Jaws extremely flexible. Snakes. *Agkistrodon, Atheris, Boiga, Crotalus, Dispholidus, Hydrophis, Lampropeltis, Micruroides, Naja, Python, Thamnophis* (Figs. 37.5, 37.17, 37.19, 37.20).

Order Crocodilia KROK-oh-DILL-ee-uh (Latin *crocodilus* crocodile). Stout body with two pairs of limbs and webbed, clawed toes. Tail laterally compressed, used as a fin in swimming. Four-chambered heart. Crocodiles, alligators, caimans, and gavial. *Alligator, Caiman, Crocodylus* (Fig. 37.6).

Order Sphenodonta (= Rhynchocephalia) SFEE-no-DONT-uh (Greek *spheno-* wedge + *odontos* tooth). Lizardlike. Pineal (median) eye apparent. Tuatara. Two extant species. *Sphenodon* (Fig. 37.7).

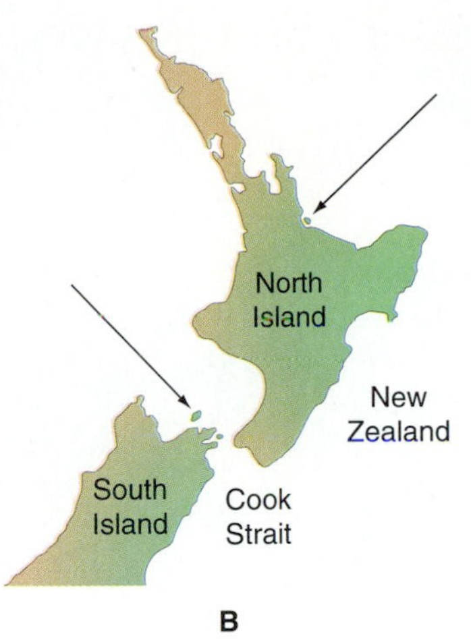

A B

FIGURE 37.7

(A) Two species of sphenodonts, or tuataras, survive thanks to protection by the government of New Zealand. Length is approximately 70 cm. (B) Present range of tuataras.

years old and was once thought to have become extinct 100 million years ago. One of the primitive features of these living fossils is the **pineal eye** (= median eye) complete with lens and retina. The pineal eye lies in the middle of the skull and regulates the activity of the pineal gland in the brain. Its ultimate function is not understood.

The Age of Reptiles

ICHTHYOSAURS AND PLESIOSAURS. This book is mainly about the living, but extinct reptiles have drawn so many people to zoology that we owe them some attention. Ichthyosaurs and plesiosaurs are among the most fascinating, especially for those who fantasize that they may survive in Loch Ness and elsewhere. In their prime, they appear to have been exclusively marine. Ichthyosaurs ("fish-lizards") were shaped somewhat like sharks except that the vertebral column bent downward rather than upward in the caudal fin (Fig. 37.8). Plesiosaurs had legs modified as flippers. There were two forms: long-necked and short-necked (Fig. 37.8).

FIGURE 37.8

Portion of the mural "Into the Swim Again" at the National Museum of Natural History. On the left is the Late Cretaceous turtle *Protostega* being pursued by three plesiosaurs, *Dolichorhynchops.* On the lower right are two Late Jurassic ichthyosaurs, *Stenopterygius.*

FIGURE 37.9

Part of the Age of Reptiles mural painted in 1947 by Rudolph Zallinger at the Peabody Museum of Natural History at Yale University. The mural is shown reversed here so that the earliest forms occur on the left. All these species lived in North America during the Mesozoic era at different times and places. Colors are conjectural. (1) *Dimetrodon* and (2) *Edaphosaurus* were pelycosaurs ("finbacks"). Their dorsal fins may have absorbed or dissipated heat, and they may have been used in social displays. These mammal-like synapsids gave rise to the even-more mammal-like therapsids such as (3) *Cynognathus.* The cat-sized *Saltoposuchus* (4) was a thecodont *Coelophysis* (5), with a length of up to 2.5 meters, was a rather small dinosaur. *Plateosaurus* (6) and *Camptosaurus* (7), up to 7 meters long, approached the sizes we associate with dinosaurs. *Allosaurus* (8), *Stegosaurus* (9), and *Apatosaurus* (10), with lengths up to 12, 9, and 21 meters, respectively, fully live up to our expectations of size. The triangular plates on *Stegosaurus* probably absorbed and radiated heat. *Archaeopteryx* (11) is shown soaring above. Most of the large dinosaurs up to the Cretaceous were herbivorous, including *Edmontosaurus* (12), a duck-billed dinosaur 13 meters long. *Tyrannosaurus* (13), with a length up to 12 meters, was one of the few that could take on another large reptile. *Ankylosaurus* (14), up to 11 meters long, may have defended itself against such large predators with the bony armor and club at the end of its tail. *Triceratops* (15), up to 9 meters long, protected its head and neck with horns and a bony shield. *Pteranodon* (16), a pterosaur with a wingspan of 7 meters, took to the air.

PTEROSAURS. The sizes and shapes of ichthyosaurs and plesiosaurs conform to what we would expect of marine predators, but no aeronautical engineer would have designed a flying reptile like *Pteranodon* (Fig. 37.9, number 16). For many years it was assumed that *Pteranodon,* with a wingspan of 7 meters and a mass of 17 kg, could only have glided from one perch to another. The discovery of fossilized fish within the fossilized ribs of pterosaurs, however, indicated that they must have flown over water. It is hard to see how they kept from crumpling their wings if they splashed into the water after prey, and even more difficult to understand how such large animals regained altitude. Perhaps they soared on ocean breezes. Even so, the lack of a stabilizing tail and the position of the wings behind the center of gravity made them aerodynamically unstable. The rudderlike head may have provided some lateral stability, but some of the other 100 pterosaur species, such as the huge *Quetzalcoatlus* (wingspan 12 meters), were even more unbalanced and lacked such rudders. Flight in *Quetzalcoatlus* has been compared to shooting an arrow backward. Nevertheless, in 1986 a half-scale plastic model of *Quetzalcoatlus* flew successfully, thanks to an on-board computer to correct its attitude.

DINOSAURS. Of course the best known and most impressive extinct reptiles were the dinosaurs, loosely defined as Mesozoic terrestrial reptiles with legs more upright than sprawling. Until a few decades ago about the only people interested in dinosaurs were children and some paleontologists who were themselves sometimes suspected of being fossils. Dinosaurs were widely thought of as slow and stupid and of having become extinct from their own mass. (The term "dinosaur" is still applied disparagingly to large and cumbersome structures destined to become extinct, such as textbooks overburdened with parenthetical information.) Today dinosaurs are usually depicted as active and intelligent, and the study of them has also become more active and intelligent.

About 350 genera of dinosaurs have been discovered, about half of them in the last two decades. Dinosaurs originated as **thecodonts,** which were relatively small reptiles mainly from the Triassic period. Some thecodonts, such as the one numbered 4 in Figure 37.9, were bipedal, and some resembled early crocodilians. There were two main lines of dinosaurs, saurischians and ornithischians, that are differentiated by pelvic structure. In the **saurischian**—"lizard-hipped"—dinosaurs the pubis projected forward and with the ischium made a fork pointing downward between the hind legs (Fig. 37.10A on page 814). In **ornithischian**—"bird-hipped"—dinosaurs the pubis bent back along the ischium, as in modern birds (Fig. 37.10B; compare Fig. 38.12B). (Oddly enough, birds are not closely related to bird-hipped dinosaurs, but to lizard-hipped di-

nosaurs similar to the species in Fig. 37.10A.) All the bird-hipped dinosaurs were herbivorous and had a beaklike structure on the lower jaw.

Saurischians are generally divided into two main groups: sauropods and theropods (Fig. 37.2). Most **sauropods** were enormous, plant-eating quadrupeds. *Apatosaurus* (once known as *Brontosaurus;* number 10 in Fig. 37.9) was one of the largest, with a body length of up to 21 meters. *Plateosaurus* (number 6) is considered to be an early relative of sauropods. **Theropods** were the only carnivorous dinosaurs. They ranged in length from less than a meter (Fig. 37.10A) to 12 meters for *Tyrannosaurus* (13 in Fig. 37.9), and all were bipedal. Other theropods in Fig. 37.9 are *Coelophysis* (5), *Allosaurus* (8), and *Archaeopteryx* (11), if, like most cladists, you consider birds to be dinosaurs.

There are four main groups of ornithischians. **Ornithopods,** such as *Camptosaurus* and *Edmontosaurus* (7 and 12 in Fig. 37.9), could walk on two legs, but probably preferred four. **Stegosaurs,** such as *Stegosaurus* (9), are distinguished by dermal plates and spikes on the tail. **Ankylosaurs,** such as *Ankylosaurus* (14), had heavy armor and a heavy cudgel at the end of the tail. **Ceratopsians,** such as *Triceratops* (15), had horns.

Fossilized tracks show that even large dinosaurs could run rapidly, and many apparently migrated in vast herds. John R. Horner and others have also discovered fossilized dinosaur eggs and young in nest colonies, indicating a high degree of social behavior. Such discoveries refute the idea of dinosaurs having become extinct from lethargy and stupidity. But why *did* dinosaurs, after an enormous success lasting more than 150 million years, die out at the end of the Mesozoic era? There have been as many theories about the dinosaurs' demise as there have been about the decline and fall of Rome, but the discovery that a large asteroid or comet struck Earth at the same time has convinced most paleontologists that the impact was the major cause (p. 390).

Amniote Eggs

Although reptiles ultimately failed to maintain dominance, they did hand down a legacy or adaptations that enabled surviving reptiles, birds, and mammals to succeed them. One of these was the amniote egg. Unlike the eggs of amphibians, the eggs of reptiles and other amniotes have **extraembryonic membranes** that provide water and oxygen for the embryo and isolate metabolic wastes. These extraembryonic membranes include the **yolk sac,** the **allantois,** the **chorion,** and the **amnion** (pp. 117–118). The allantois stores metabolic wastes and assists the chorion in exchanging respiratory gases. The amnion holds amniotic fluid in which the embryo is bathed. In many

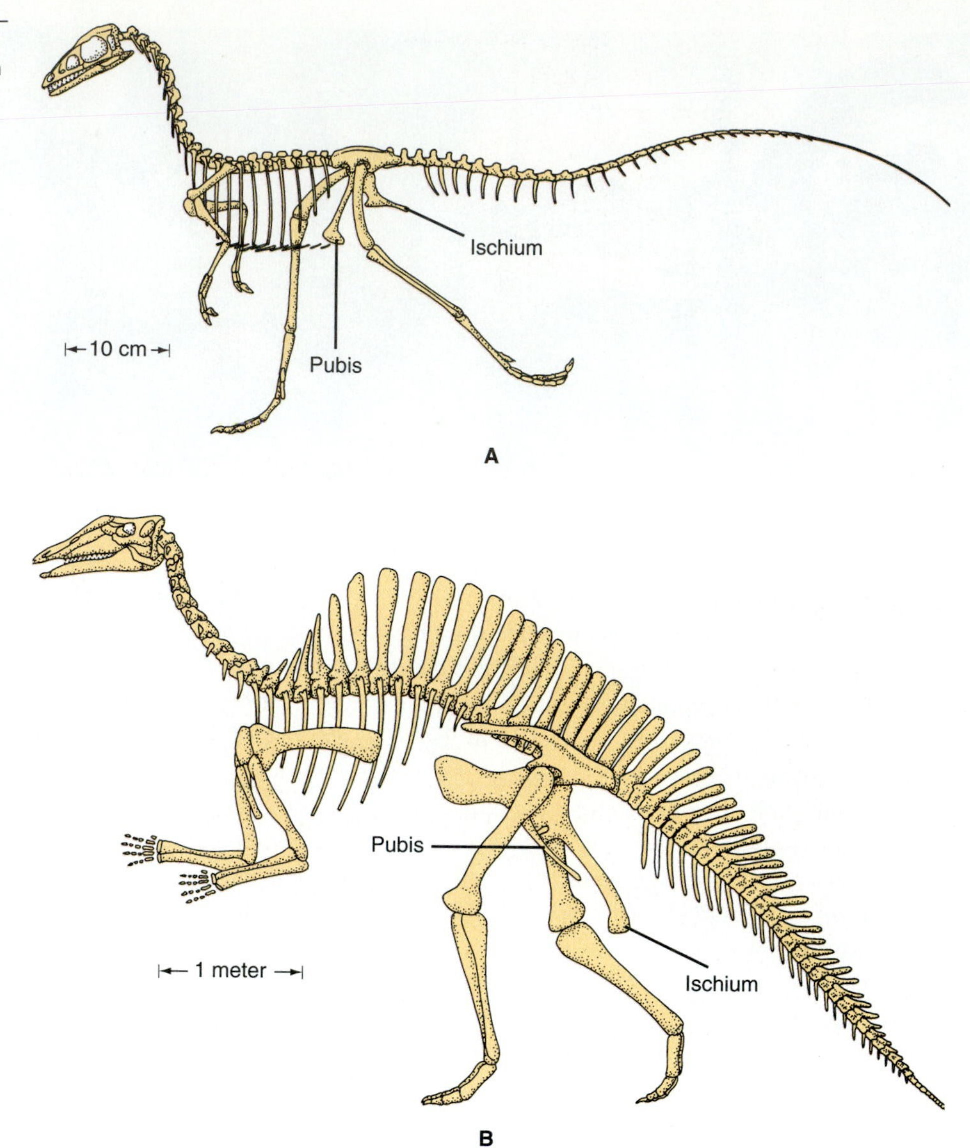

FIGURE 37.10

(A) A saurischian, *Compsognathus*, with the pubis and ischium indicated. *Compsognathus* was a theropod of the late Jurassic period. (B) *Ouranosaurus*, an ornithischian (ornithopod) from the early Cretaceous, showing the parallel pubis and ischium.

■ **Question:** What features of these two reconstructions reveal the paleontologist's preconceptions?

species the **albumen** ("white") of the egg also serves as a water store. The eggs of amniotes are generally larger than those of fishes and amphibians, and they hatch into miniatures of the adults rather than into larvae.

Although the amnion does hold a reserve of fluid, the eggs of most reptiles must absorb additional moisture from the environment. Most reptiles therefore deposit eggs in moist soil. Most reptile eggs have a leathery "parchment shell" that is more permeable to O_2 and CO_2 than the hard shells of birds and some reptiles. In some snakes and lizards the eggs are totally independent of environmental supplies of water, because they are incubated within the oviduct. These eggs have little or no shell, and the oviducts provide water, respiratory gases, and perhaps nutrients to the embryo. These reptiles are therefore viviparous. Viviparity is thought to have evolved approximately 100 times in reptiles.

Scales

Another adaptation that enables most extant reptiles to survive on land are horny (keratinized) scales that protect the skin from abrasion. Because scales occur in all living species of reptiles, it is reasonable to assume that they evolved early. Unlike the scales

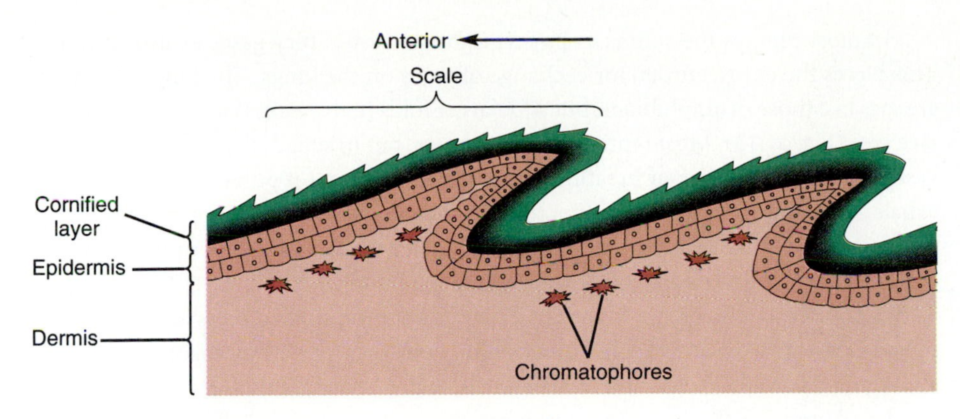

FIGURE 37.11
Epithelium of a reptile. Compare the epidermal scales with dermal scales of a fish, Fig. 35.8. In snakes and lizards new scales form beneath the old cornified layer of epidermis, which is shed periodically. Chromatophores often produce striking colors, to which may be added iridescent structural colors due to the fine sculpting of scales.

■ **Question:** Why do most scales have the free edge pointing posteriorly?

of fishes and amphibians, those of reptiles arise from epidermis rather than dermis (Fig. 37.11). Contrary to what one would naturally assume, the scales do not appear to be major barriers to the loss of water from the body, since snakes that have had their scales removed do not dehydrate. Lipids in the skin are apparently more important in preventing dehydration.

Osmoregulation, Respiration, and Circulation

The maintenance of water and ionic balance is due mainly to the kidneys. Instead of mesonephric kidneys, as in fishes and amphibians, reptiles have metanephric kidneys like those of birds and mammals (p. 286; Fig. 37.12). Like birds, most reptiles excrete a thick, white urine with uric acid as the main nitrogenous waste. Uric acid is insoluble and therefore requires little water to eliminate. The urine empties through the cloaca, which is also the exit for feces and gametes. The external opening of the cloaca, called the **vent,** is a slit that runs lengthwise in crocodilians and turtles, and crosswise in lizards, snakes, and tuataras. Marine reptiles such as sea turtles, the marine iguana (Fig. 37.18A), and sea snakes (Fig. 37.17) also have **salt glands** in the head that excrete excess salt (p. 288).

FIGURE 37.12
Internal structure of a male crocodile. The anatomy and physiology are similar to those of mammals, as described in Unit 2. Refer to the index for specific topics.

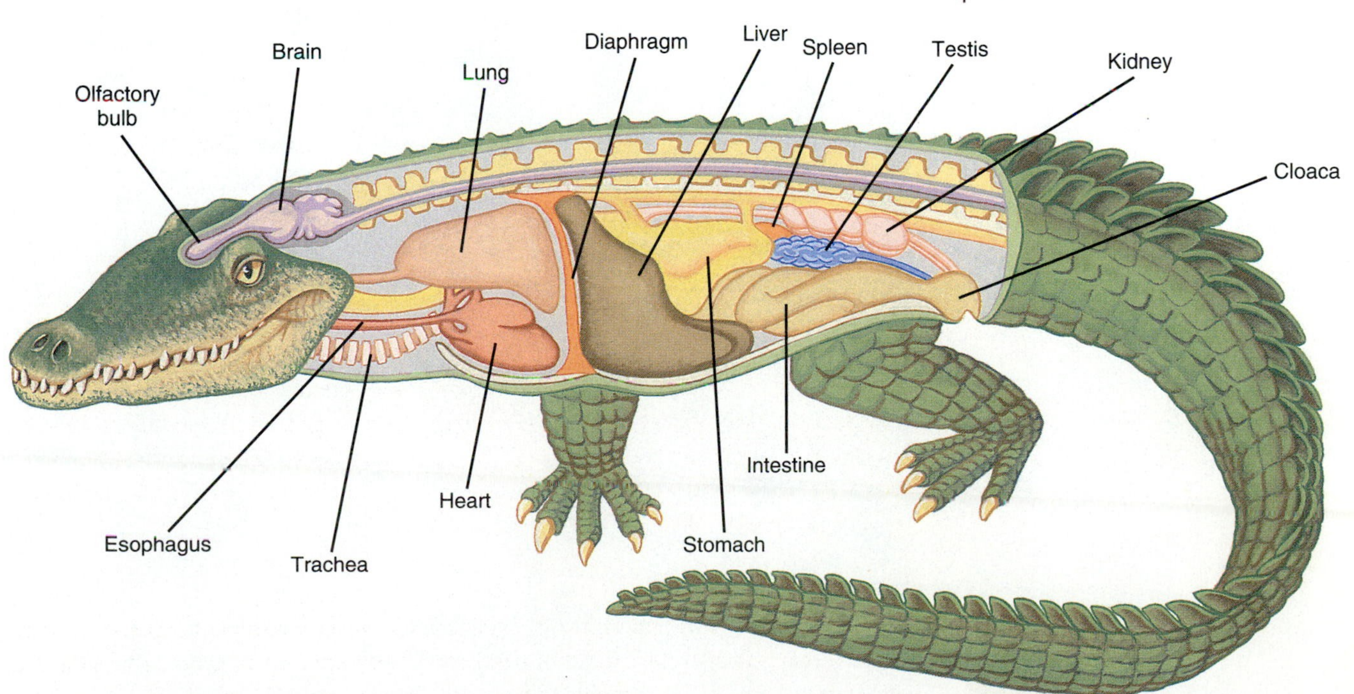

In most reptiles the skin is as impermeable to respiratory gases as it is to water. This places the entire burden for exchange of gases on the lungs. The lungs of reptiles are sacs like those of amphibians, but with larger folds in the walls that increase the surface area (Fig. 37.13). Most reptiles have two lungs, but in snakes the left lung is usually vestigial. The mechanics of breathing varies widely among reptiles. Most inhale and exhale by expanding and contracting the chest cavity using the intercostal muscles between the ribs, as in humans and other mammals (pp. 249–250). Because the intercostal muscles of lizards are also involved with locomotion, they cannot breathe while running. In crocodilians the lungs are expanded during inhalation when the liver is pulled backward by a muscular **diaphragm** (which is not homologous to the diaphragm of mammals). Because of their shells, turtles must use their leg muscles to expand and contract the body cavity for breathing. Reptiles generally have metabolic rates less than a tenth as high as those of birds or mammals of similar sizes, so their breathing rates are correspondingly low.

A few reptiles exchange respiratory gases through the skin. These include the sea snakes (family Hydrophiidae) and soft-shelled turtles (family Trionychidae). In addition, aquatic turtles can obtain oxygen by pumping water into and out of the mouth and cloaca, the linings of which function like gills. The respiratory tracts of most reptiles do not include a vocal apparatus, although many can make hissing noises by exhaling. In crocodilians and some lizards, however, there is a **larynx** that produces chirps, grunts, or roars.

Like amphibians, most reptiles have hearts with two atria and one ventricle (see Fig. 11.4), but a partial septum in the ventricle generally separates oxygenated from deoxygenated blood. The septum is complete in crocodilians, so they have four-chambered hearts like birds and mammals. The respiratory and circulatory systems of reptiles can provide more oxygen to tissues than can those of amphibians, but they are still less effi-

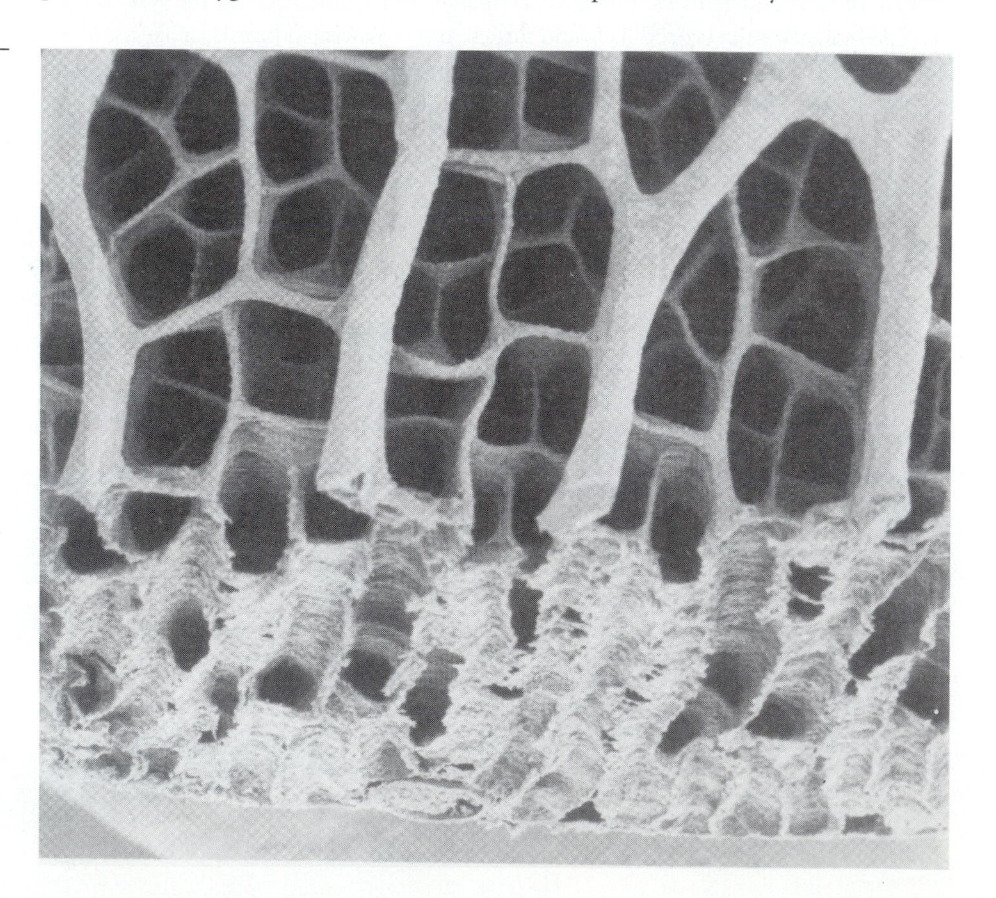

FIGURE 37.13

Scanning electron micrograph of the wall of the lung of the tegu lizard *Tupinambis nigropunctatus,* magnified approximately 35 times.

cient than those of birds and mammals. Nevertheless, many reptiles can move quite rapidly, using their anaerobic fast-twitch muscles (Type II, p. 208). These muscles tire quickly, however, and an exhausted reptile takes hundreds or thousands of times longer than a mammal does to recover. This explains why predaceous reptiles do not usually chase prey over long distances, but lunge at them when they wander into range.

Nervous System

The nervous system of reptiles is similar to that of mammals in its organization, but generally smaller than that of a mammal (or bird) of the same body size (see Figs. 9.6 and 9.7). The reptilian brain also lacks a cerebral cortex, which is responsible for the most sophisticated behaviors in mammals. Vision is the most important sensory modality. Except for snakes, most reptiles that are active during the day have retinas with numerous cones, and they presumably have good color vision. In the cones of most reptiles are colored oil droplets that narrow the range of colors to which each cone responds. Most reptiles focus their eyes by changing the shape of the lens in a manner similar to that in mammals, but snakes move the lens back and forth. Most reptiles have paired eyelids, and also translucent **nictitating membranes** that sweep sideways and clear the cornea. Snakes, however, have no eyelids, which accounts for their steely gaze. Instead their eyes are protected by transparent scales—probably an adaptation for burrowing. In *Sphenodon* and many lizards there is also a **pineal eye** on top of the head that probably detects only light intensity and may control biological rhythms. Some snakes can "see" warm prey in the dark by means of infrared receptors. Pit vipers have an infrared detector on each side of the head in a pit between the nose and eye (Fig. 37.19). Pythons have as many as 13 pairs of infrared receptors around their mouths, and the related boa constrictor has infrared-sensitive scales around the mouth.

Reptiles have well-developed olfactory bulbs with inputs from the olfactory epithelia in the nostrils and from **Jacobson's organs** in a pair of depressions in the roof of the mouth. Jacobson's organs, also called **vomeronasal organs,** are chemoreceptors used in detecting prey or members of the same species. For many decades it has been assumed that the forked tongues found in many squamates were adaptations for transferring molecules to the two Jacobson's organs. Recently, however, Kurt Schwenk (1994) of the University of Connecticut presented compelling evidence that the tongue does not contact Jacobson's organs, and that they are forked as an adaptation for following odor trails.

Most reptiles have hearing mechanisms similar to those of amphibians, with one middle-ear bone (the **columella** or **stapes**) that transfers sound from the tympanum (eardrum) to the inner ear (see Fig. 39.13A). In most turtles and crocodilians the tympanum is external, as in frogs, but in most lizards the tympanum is recessed in a canal. Some other lizards, as well as all snakes, lack tympani altogether. Instead the columella touches a jawbone that picks up vibrations. This arrangement is probably responsive to vibrations of the substratum rather than to airborne sound.

Reproduction

SEX DETERMINATION. Reptiles have separate sexes, although generally only they can tell the difference by casual inspection. Whether a reptile develops into a male or a female is determined genetically, but only certain squamates have different sex chromosomes. In many lizards the female has two X chromosomes, and the male has one X and one Y chromosome, as in mammals. In other lizards and in many snakes, it is the female that is **heterogametic** (has differing sex chromosomes). In female heterogamety, which also occurs in birds, the sex chromosomes are designated ZW in females and ZZ in males. In some lizards, many turtles, and probably tuataras (Cree et al. 1995) and all crocodilians, the activation of genes that control sex depends on temperature during a critical period of incubation. In some species embryos at high temperatures become males, in others they become males only at low temperatures, and in still others they become males only at moderate temperatures. The advantage of this temperature-dependent sex determination is hotly debated. It may ensure that when the population is endangered by thermal stress the more-valuable sex will be produced. It may also prevent inbreeding, since siblings tend to be of the same sex.

Copulation and Fertilization. Several dozen species of lizards reproduce without the eggs being fertilized and without the existence of males. Except in these fortunate parthenogenetic and unisexual species, however, reptile eggs have to be fertilized. Because most reptiles lay shelled eggs, internal fertilization is virtually dictated. Male tuataras have no copulatory organs; mating is achieved simply by pressing the cloacae together. All other male reptiles have at least one penis; snakes and lizards have two, called **hemipenes** (Fig. 37.14). The male uses only one hemipenis at a time, depending on sensory information relating to which testis has the most mature sperm. The reptile penis or

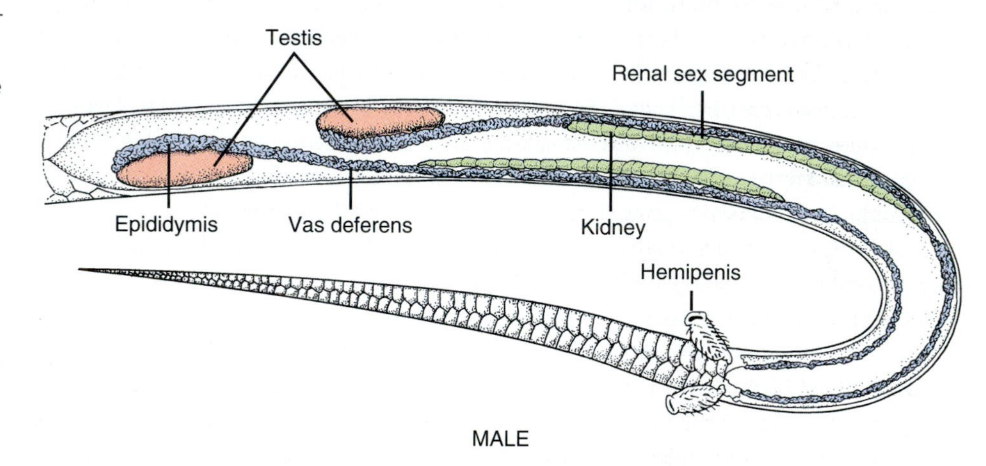

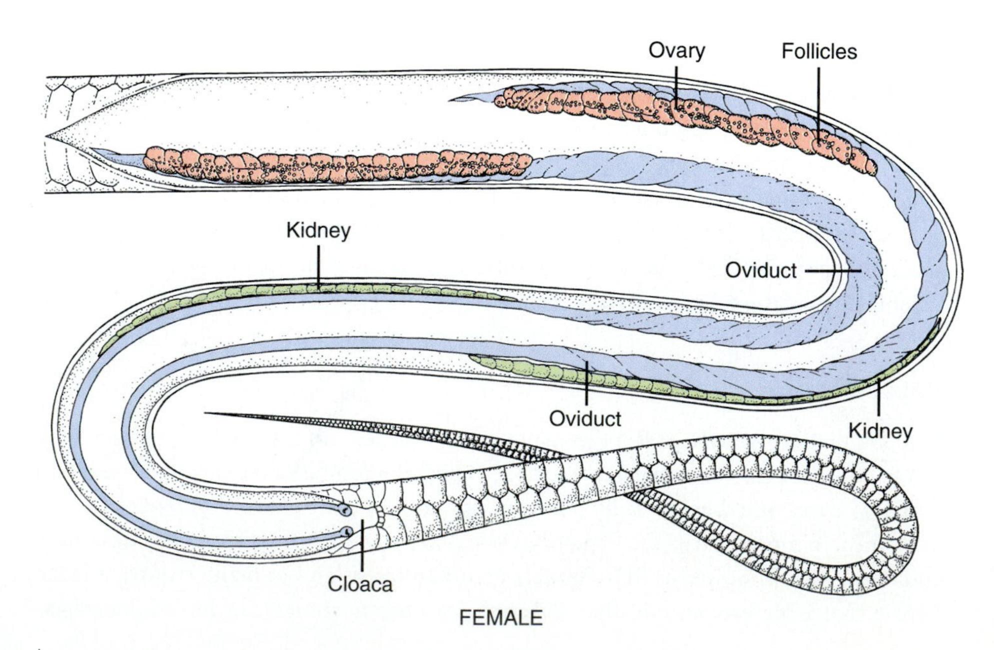

FIGURE 37.14

Reproductive systems of male and female garter snakes. During erection one hemipenis of the male everts from the cloaca like the finger of a glove turned inside-out. The sex segments of the male's kidney produce seminal fluid and pheromones. In this viviparous species the oviduct also serves as a uterus.

■ **Question:** Are the hemipenes of a snake homologous with the penis of a mammal?

hemipenis has a groove that guides the sperm into the female's cloaca during copulation. The sperm travel up the oviduct and may become embedded in the lining for months or years before they fertilize the ova as they ripen. After fertilization the eggs pass down the oviduct, acquiring yolk, albumen (in many species), and finally the shell.

The reproductive hormones of reptiles include steroids (androgens and estrogens) from the testes and ovaries, as in mammals, and one or two gonadotropic hormones from the anterior pituitary (pp. 312–314). In most reptiles the females probably become sexually receptive while estrogen levels are high, around the time of ovulation. Reproductive behavior has been studied for relatively few species. In garter snakes, rattlesnakes, and perhaps other snakes, the skin of ovulating females secretes a pheromone that attracts males. Male garter snakes do not have to travel far to find females, since thousands of both sexes den up together during the winter. The males emerge first during the spring, and dozens crowd around each female, forming a "mating ball" as she emerges. The successful male copulates by aligning his body along hers and opening her cloaca with the spines on one hemipenis. After copulation, part of the ejaculate forms a plug in her cloaca that emits an **antiaphrodisiac** that discourages other males. One of the odd features of mating in garter snakes is that it occurs at a time when the gonads are regressed and sex steroid levels are low in both sexes. Apparently males fertilize eggs with sperm produced in the previous summer.

In crocodilians and lizards it is generally the male that attracts females. Male crocodiles seduce females with loud bellowing and a pheromone, while lizards rely mainly on visual displays (Fig. 37.15A). Perhaps the most surprising courtship behavior occurs in many lizards, including 15 of the 45 species of whiptail lizards (family Teiidae). What is so surprising is that courtship occurs at all, since these species consist only of parthenogenetic females. An individual behaves like a female when its estrogen levels are high and ovulation is imminent. After ovulation, when estrogen levels are low and progesterone levels are high, the same individual behaves like males do in bisexual species of whiptails (Fig. 37.15B). Each female alternates between female and male behaviors several times per breeding season.

PARENTAL CARE. Most reptiles apparently consider their parental duties done once they have laid fertilized eggs or, in the case of viviparous species, once the female has given birth. Crocodilians, however, are unexpectedly tender in their parental be-

A

B

FIGURE 37.15

(A) Green anoles *Anolis carolinensis* of the American Southeast can change color in seconds and are sold by pet stores as chameleons. Males are usually brown, especially when basking. When fighting or courting, they turn green, extend the pink dewlap on the throat, and bob the head up and down. (Length up to 21 cm.) (B) Pseudocopulation in desert-grassland whiptails *Cnemidophorus uniparens,* a unisexual species consisting entirely of parthenogenetic females. The upper female behaves like males of closely related bisexual species of whiptail lizards. (Length approximately 20 cm.)

■ **Question:** What physiological function could be served by pseudocopulation?

havior, and most dig nests for the 25 to 60 eggs laid by a female. The American alligator provides the nest with wet vegetation that may generate heat by fermentation. The nest may also favor the growth of acid-secreting bacteria that will help etch away the hard shell and make hatching easier. Young crocodilians often begin chirping like chicks at the time of hatching, and this encourages the mother to uncover the nest. Then, using jaws that could just as easily tear off a person's leg, she gently picks up the hatchlings and carries them to the water. Not only the mother but also the father and other adults respond to distress calls from young, but as soon as the young reach subadulthood, adult males no longer tolerate their presence.

Locomotion

TURTLES. The basic pattern of reptilian locomotion is that of lizards, on four sprawling legs that are aided in their movements by the flexible body. Turtles retain this mode of locomotion except that they are unable to move the legs forward by bending the body. A land turtle may spend its long life plodding no more than a hundred meters from where it hatched. At the other extreme, however, are **sea turtles,** especially the leatherback (Fig. 37.3C), green turtles (*Chelonia*), hawksbills (*Eretmochelys*), loggerheads (*Caretta*), and ridleys (*Lepidochelys*). Many sea turtles migrate thousands of kilometers, and the females return periodically to one beach to bury their fertilized eggs in the sand. It is assumed that hatchlings become imprinted to the smell of their native beach, and that upon reaching adulthood the females home to that beach first by magnetic and celestial cues and then by smell.

Newly hatched sea turtles scramble directly toward the ocean after a month or more of incubation in the sand. They are apparently attracted toward the brightest part of the horizon, over the water. Once in the water they appear to be guided by mechanical stimulation from waves, and later by magnetic fields. They swim straight out for many kilometers, drawing power from the yolk located within their guts. For many years no one knew where they went at sea. Finally in 1984 after three decades of study, Archie Carr (1909–1987) and others solved the mystery by tracking tagged turtles. They discovered that sea turtles that hatch at Florida swim out to the Gulf Stream then float on rafts of sargassum sea weed that drift clockwise around the Sargasso Sea in the North Atlantic. Here the turtles find rest and shelter from predators while they themselves prey on shrimps, crabs, jellyfish, and other animals also attracted to the rafts. After as many as 50 years they reach sexual maturity, mate at sea, and return to their native beaches to start a new cycle.

SNAKES. The locomotion of snakes is radically altered from that of their lizard ancestors, and it is quite adaptable to circumstances. During terrestrial locomotion, snakes anchor their bodies at two or three points, then advance the portions of the body that are not anchored (Fig. 37.16). At the same time the anchor points shift backward at the same speed as the snake moves forward, giving the impression that the snake is flowing effortlessly like a stream of water. Four basic styles of terrestrial locomotion can be recognized:

1. The most common is **lateral undulation,** which is used by smaller snakes on firm and irregular ground. In lateral undulation the snake pushes sideways against rocks, plants, or other elevated points (the optimum number is three).
2. Many large snakes use **rectilinear movement,** in which the body moves straight forward without bending, using two or three points of contact between the sub-

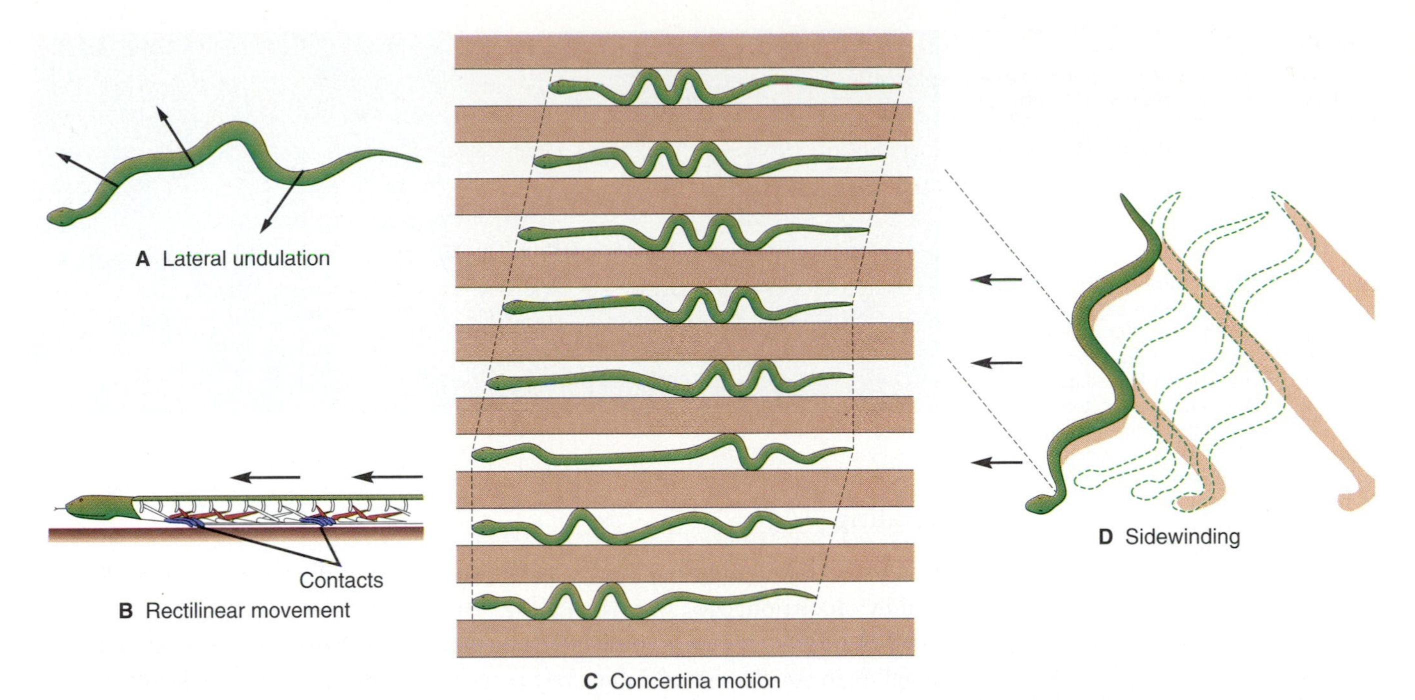

FIGURE 37.16
Modes of terrestrial locomotion in snakes. (A) In lateral undulation the sides of the body push against points of contact. Arrows show the direction of forces, which are not parallel to the body axis but have a net direction that propels the snake forward. The forward movement does not depend on friction against the points of contact. In fact, lateral undulation is fastest if the contacts can rotate. (B) In rectilinear movement the body is aligned in the direction of movement, and ventral scales are dug in at several points (shown in color). Contractions of muscles posterior to these points pull the body forward (arrows). New points of contact are continually formed posterior to the previous ones. (C) In concertina movement, successive portions of the body are bent into waves that brace-against the sides of a burrow. (D) Sidewinding is used in loose sand and produces characteristic J-shaped tracks. The hook of the J is formed by the head, which is flung forward, followed by posterior portions of the body. Dashed outlines show previous positions. The body is at right angles to the direction of travel, and the tracks are diagonal to it.

stratum and ventral scales. Rectilinear movement is less visible to prey, and it is often used by smaller snakes while stalking.

3. Burrowing snakes use what is called **concertina motion,** in which one part of the body bends into an S shape that is braced against the walls of the burrow. The front part of the bend then straightens out to extend the body, while the bend is continually reformed at the rear. Before the bend reaches the end of the snake a new one forms behind the head. Concertina motion is also used by snakes climbing rough tree trunks.
4. Finally, many snakes on loose sand use a unique **sidewinding** motion. When sidewinding, a snake twists its body into a spiral at right angles to the direction of overall motion, with only two parts of the body contacting the substratum. The head is then flung forward, followed by posterior parts of the body. The advantage of sidewinding is that the body tends to dig in for better traction, and contact with hot sand is minimized.

MISCELLANEOUS FORMS. Many squamates have interesting modifications of these locomotory patterns. Sea snakes, for example, have laterally compressed tails that adapt lateral undulation for swimming (Fig. 37.17 on page 822). Sea snakes can also dive effectively and stay submerged for up to eight hours. Apparently they do not inflate the lungs before diving, since that would make them too buoyant. Instead they obtain oxygen through the integument. "Flying snakes" of the tropics leap great distances between trees. Peculiar adaptations for locomotion can also be found among lizards. Some geckos, such as the house gecko *Ptyodactylus hasselquistii* of northern Africa, have loose folds of skin on their toes that enable them to walk on walls and ceilings. The flying dragons (*Draco* spp.) of the Asian tropics have elongated ribs that can open up to form a parachutes, enabling these lizards to glide between trees. No less impressive are basilisks of the American tropics (*Basiliscus* spp.), which are sometimes called "Jesus Christ lizards" because they can run on their hind feet across water at speeds up to 12 km/hr.

FIGURE 37.17

The banded sea snake *Hydrophis fasciatus* has a laterally compressed tail that it uses like a fin when swimming. On shore it is virtually helpless because its scales are so smooth.

Feeding

Except for some turtles and lizards, almost all extant reptiles are carnivorous, feeding mainly on carrion, eggs, worms, molluscs, and insect larvae that can be caught without a chase (Fig. 37.18A). The alligator snapping turtle hardly moves at all except for wiggling the worm-shaped fishing lure in its mouth (Fig. 37.18B). On the other hand, many predaceous reptiles are quite agile hunters. Crocodilians run and swim rapidly after prey, then thrash violently to tear off large chunks. Like all living reptiles, they swallow without chewing, since they lack the variety of cutting, tearing, and grinding teeth found in mammals. The teeth of reptiles are generally all sharp and recurved. In

A

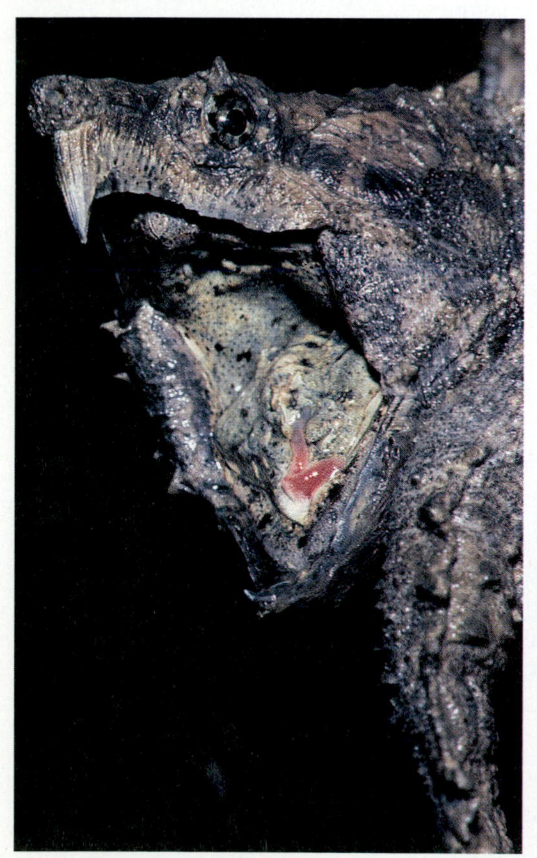

B

C

FIGURE 37.18

(A) Marine iguanas *Amblyrhynchus cristatus* bask after feeding on algae in the chilling waters off the Galápagos Islands. These harmless vegetarians reach lengths of up to 1.8 meters. (B) The alligator snapping turtle *Macroclemys temminckii* lives in the Gulf states and Mississippi area, where it lies in water with its mouth open, wiggling the wormlike fishing lure in its toothless beak. (C) The European chameleon *Chamaeleo chamaeleon* demonstrates its hunting prowess. After anchoring itself to a branch with its tail, it suddenly inflates its tongue with blood and launches the end at the target, which may be as far away as its own body length (up to 28 cm). The chameleon then hauls the prey into its mouth and coils up its tongue. This form of predation obviously requires good vision. Each of the chameleon's eyes peeps out of fused eyelids and can be moved independently of the other. Chameleons are famous for their ability to instantly change color, which they do for social communication and camouflage.

addition, many reptiles have teeth on the palate (**vomerine teeth**) that hold struggling victims.

Most carnivorous reptiles swallow their prey whole, which means they can't eat anything larger than their heads. In snakes, however, the jaw bones are loosely hinged to each other in such a way that the mouth can accommodate prey several times wider than the head (Fig. 37.19 on page 824). Flexible portions of epidermis between the scales also allow the skin to stretch. Snakes do not choke on large prey because the tracheal opening is located near the front of the mouth.

Venoms

A few squamates subdue or kill prey before swallowing it. Boas and pythons (family Boidae) first suffocate their prey by wrapping themselves around the body and tightening the coils each time the prey exhales (Fig. 37.20 on page 825). Some squamates subdue prey with venom. The venoms of rattlesnakes and most other venomous species are essentially saliva containing mixtures of digestive enzymes such as proteases, collagenases, and phospholipases. In addition to paralyzing and killing prey, venom starts the digestive process from inside the prey. Whether these enzymes work mainly on blood, heart, muscles, nerves, or some other tissue depends on where the venom is injected and the size of the prey. The venoms of cobras contain toxins that work specifically on the nervous system, usually causing death by paralyzing the respiratory muscles.

There are two species of venomous lizards, including the gila (HE-lah) monster (Fig. 37.21 on page 825). There are at least four families of venomous snakes.

1. Coral snakes and cobras (family Elapidae) are the most dangerous. Indian cobras (*Naja naja*) by themselves cause an estimated 10,000 human deaths each year. These snakes have two permanently erect fangs in front, each with an external groove through which venom flows into the bite. The spitting cobra *Naja nigricollis* can also blind a victim by spitting venom into its eyes.
2. Sea snakes (family Hydrophiidae) are similar to cobras except for their marine adaptations.
3. Puff adders and other Old World vipers (family Viperidae) comprise a third group of venomous snakes. New World pit vipers—rattlesnakes, copperheads, and water moccasins—are also often included in this family, or in a separate family Crotalidae. Viperids and crotalids have front fangs that fold back when not in use, and each fang works like a hypodermic needle (Fig. 37.19B).
4. Family Colubridae includes most of the common nonvenomous snakes such as garter snakes and milk snakes, but also some species, such as the African boomslang *Dispholidus typus*, with saliva so toxic that it is wise to consider it a venom. Venomous colubrids have grooved fangs in the rear of the mouth.

Interactions with Humans and Other Animals

Most people's images of reptiles are colored by childhood phobias and myths. There are people in the United States who kill any snake they see because a serpent supposedly tempted Eve to eat the forbidden fruit in the Garden of Eden. (Yet they eat apples.) Many others are mistrustful of all reptiles because of the few that are dangerous. While it is true that venomous snakes are extremely dangerous in many parts of the world, the average American is more likely to be killed by another person than to be

FIGURE 37.19

Feeding mechanisms of snakes. (A) A copperhead *Agkistrodon contortrix* swallowing a deer mouse (*Peromyscus*). Note that the opening of the trachea (the glottis) is at the front of the mouth, preventing suffocation during the long process of swallowing. (B) Jaw bones of a rattlesnake enable the mouth to open widely. In addition, the jaw bones on one side can separate from those on the other side. (C) The mouth of a rattlesnake.

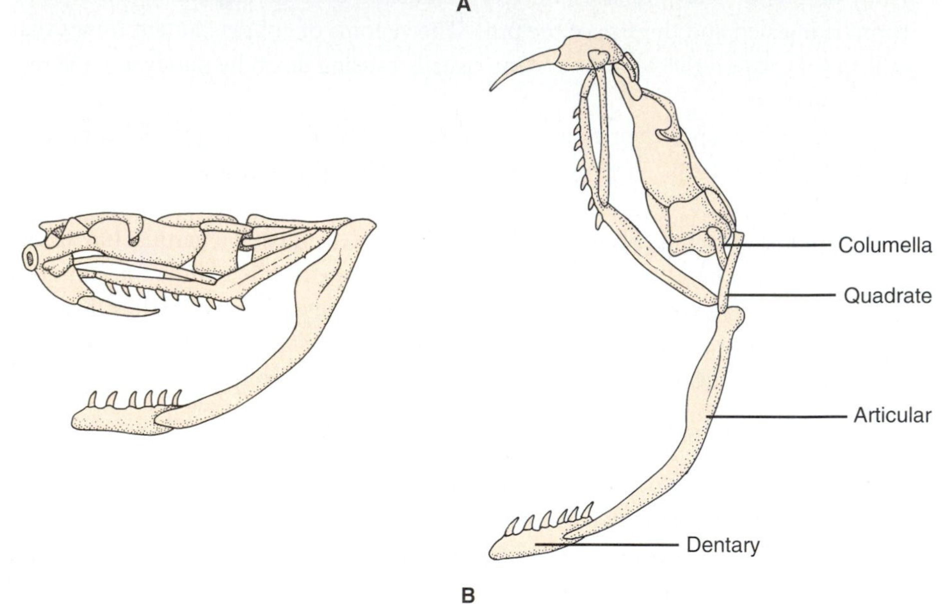

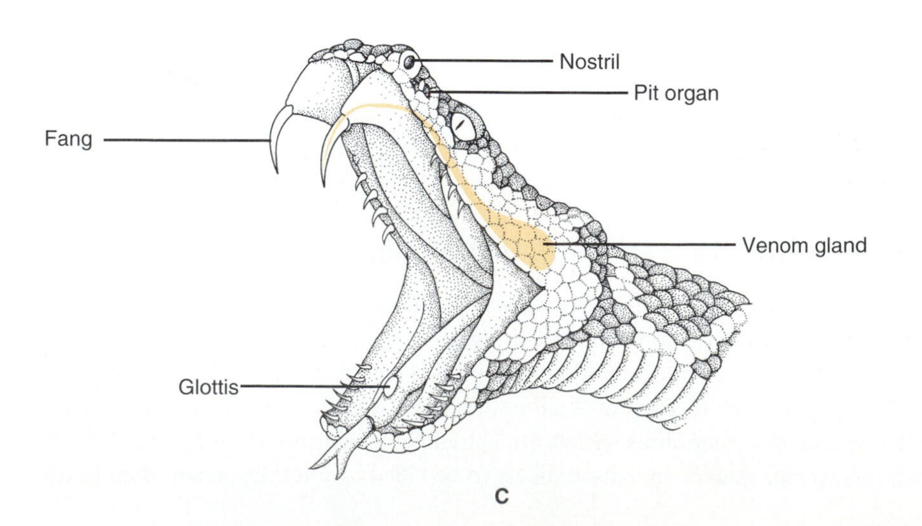

bitten by a venomous snake. Even if a person is one of the 8000 Americans bitten each year, there is a 99.8% chance of surviving.

Probably reptiles are a net benefit to human welfare by keeping populations of rodents in check. Of course many reptiles also prey on desired species, such as birds. Such predation can be especially troublesome with exotic reptiles. The mildly venomous brown tree snake (*Boiga* sp.) has wiped out most of the forest birds in Guam since its arrival in 1960 in the landing gear of airplanes. These 1.7-meter-long snakes are now causing anxiety in Hawaii. In Europe, conservationists are worried that the American red-eared turtle *Trachemys scripta elegans,* which is imported for pets, threatens to outcompete native turtles.

Reptiles are much less a threat to humans than humans are to reptiles. Besides deliberately killing them, humans also destroy their habitats. The situation is especially critical for sea turtles, which have the misfortune of depending on the same sunny beaches preferred by tourists and developers. Sea turtles are also threatened by human garbage and spilled oil, which accumulate in the same ocean convergences as the sargassum rafts where sea turtles spend their first year. Sea turtles are often found with their digestive tracts blocked by plastic bags presumably mistaken for jellyfish, or their mouths glued shut by tar balls from oil spills.

Several species of sea turtles are so close to extinction that many people have begun to take notice. The Mexican government now has armed troops protecting the only nesting site of Kemp's ridley turtle (*Lepidochelys kempi*) from poachers, who sell the eggs for their supposed aphrodisiacal powers. After years of lobbying by conservationists, United States lawmakers have finally taken steps to protect Kemp's ridley and other sea turtles by requiring escape hatches (turtle excluding devices, or TEDs) on shrimp nets. These measures appear to be succeeding, because the number of nesting Kemp's ridleys was up to 580 in 1994, almost three times the 1985 number, though still far below the 40,000 of 1947 (Louma 1994).

While sea turtles have attracted the most attention, many other reptiles are also endangered by humans. Many are killed for their hides, which are often imported illegally into the United States to make shoes and handbags. Three species of sea snakes in the Philippines have become extinct from such hunting. Some reptiles, such as turtles and rattlesnakes, are hunted for food. Many other reptiles are traded legally or illegally

FIGURE 37.20

The reticulated python *Python reticulatus* of Southeast Asia killing a rat by constriction. The python does not exert much muscle contraction, but simply locks its coils to keep the victim from inhaling. This is one of the longest snakes, reaching lengths of up to 10 meters.

FIGURE 37.21

The gila monster *Heloderma suspectum* of the southwestern United States and Mexico is one of only two venomous lizards. The other is the Mexican beaded lizard *H. horridum.* The venom comes from a gland in the lower jaw and is not injected but simply flows into the bite. The venom is seldom fatal to humans; but once a gila monster has bitten, it hangs on tenaciously. Gila monsters feed mainly on eggs, young birds, small rodents, and other lizards. They store nutrients in the stout tail.

as exotic pets. Approximately 25,000 box turtles are exported to Europe as pets each year, with an estimated 90% dying during shipment. Approximately 100,000 iguanas are imported into the United States each year from Central America as pets, and most soon die from improper care. Some populations are being restored by a program of captive breeding and restoration of habitat. The American alligator is an example of what can be accomplished with timely, serious efforts. Once nearly extinguished by poachers, it is now a common tourist attraction and a $30 million business for hunters who can now sell the hides legally.

Summary

The three major lineages of reptiles—anapsid, synapsid, and diapsid—are distinguished mainly by the number of temporal openings in the skull. Turtles are the only extant anapsids. Synapsids are extinct except for their descendants, mammals. Extant diapsid reptiles are the tuataras, lizards, worm lizards, snakes, and crocodilians. Reptiles were much more numerous and diverse during the Mesozoic era and included marine ichthyosaurs and plesiosaurs, flying pterosaurs, and terrestrial dinosaurs.

One of the keys to the ability of reptiles to survive on land is the amniote egg, with its extraembryonic membranes that hold water and perform other functions. Reptiles are generally oviparous, but some snakes and lizards are viviparous. Because the eggs of viviparous species become covered with a shell that is usually parchmentlike, fertilization is necessarily internal. There is little parental care except in crocodilians. Reptiles conserve water by means of lipids in their scaly skins. The skeletons are adapted for terrestrial locomotion, but snakes and worm-lizards have secondarily lost the legs as an adaptation for burrowing. Extant reptiles have uniform teeth and jaws not adapted for chewing. Some snakes and lizards immobilize prey with venoms, which are essentially toxic forms of saliva.

Key Terms

temporal openings
anapsid
synapsid
diapsid
carapace
plastron
amniote egg
extraembryonic membranes
allantois
chorion
amnion
albumen
salt gland
heterogametic
hemipenis
lateral undulation
rectilinear movement
concertina motion
sidewinding

Chapter Test

1. Describe in simple terms the features that distinguish turtles, lizards, snakes, and crocodilians from each other. (pp. 805–811)
2. For each of the major groups in the previous question, briefly describe their evolutionary origins. In what way is each related to dinosaurs? (pp. 805–811)
3. Describe the amniote egg and explain how it enables most reptiles to reproduce independently of water. (pp. 813–814)
4. Describe the integument of snakes and the functions of scales. (pp. 814–815)
5. Why are the circulatory and respiratory systems of most reptiles less effective than those of humans? How can reptiles live with these less-efficient systems? (pp. 815–817)
6. Describe reproduction in one species of reptile. (pp. 817–820)
7. Describe the special adaptations of a rattlesnake for predation. (p. 823)
8. Explain how a snake is able to move on land without legs. (pp. 820–821)

Answers to the Figure Questions

37.2 This requires some extrapolation from the figure, but extending the lines leading to turtles and to mammals should give the same branching sequence as in Fig. 21.6. (Dinosaurs would be in the place of Lepidosauria.)

37.10 The saurischian is represented as being an active runner; the ornithischian is shown in a less dynamic stance, with its tail dragging.

37.11 In a snake the raised posterior edge would prevent backsliding, and in any reptile it would be less likely to snag and accumulate debris.

37.14 One would guess not. Not only are the hemipenes double, but they erect in a totally different manner.

37.15 Like copulation in their bisexual ancestors, pseudocopulation may be required to trigger ovulation.

Readings

Recommended Readings

Buffetaut, E. 1979. The evolution of the crocodilians. *Sci. Am.* 241(4):130–144 (Oct).

Carr, A. 1986. Rips, FADS, and little loggerheads. *BioScience* 36:92–102.

Cole, C. J. 1984. Unisexual lizards. *Sci. Am.* 250(1):94–100 (Jan).

Crews, D. 1979. The hormonal control of behavior in a lizard. *Sci. Am.* 241(2):180–187 (Aug).

Crews, D. 1987. Courtship in unisexual lizards: a model for brain evolution. *Sci. Am.* 257(6):116–121 (Dec).

Crews, D., and W. R. Garstka. 1982. The ecological physiology of a garter snake. *Sci. Am.* 247(5):158–168 (Nov).

Eckert, S. A. 1992. Bound for deep water. *Nat. Hist.* 101:29–35 (Mar). (*On the leatherback turtle.*)

Gans, C. 1970. How snakes move. *Sci. Am.* 222(6):82–96 (June).

Heatwole, H. 1978. Adaptations of marine snakes. *Am. Sci.* 66:594–604.

Horner, J. R. 1984. The nesting behavior of dinosaurs. *Sci. Am.* 250(4):130–137 (Apr).

Langston, W., Jr. 1981. Pterosaurs. *Sci. Am.* 244(2):122–136 (Feb).

Lillywhite, H. B. 1988. Snakes, blood circulation and gravity. *Sci. Am.* 259(6):92–98 (Dec).

Lohmann, K. J. 1992. How sea turtles navigate. *Sci. Am.* 266(1):100–106 (Jan).

Newman, E. A., and P. H. Hartline. 1982. The infrared "vision" of snakes. *Sci. Am.* 246(3):116–127 (Mar).

Pooley, A. C., and C. Gans. 1976. The nile crocodile. *Sci. Am.* 234(4):114–124 (Apr).

Shine, R. 1994. Young lizards can be bearable. *Nat. Hist.* 103:34–39 (Jan). (*On viviparous skinks.*)

Vickers-Rich, P., and T. H. Rich. 1993. Australia's polar dinosaurs. *Sci. Am.* 269(1):50–55 (July).

See also relevant selections in General References at the end of Chapter 21.

Additional References

Bakker, R. T. 1986. *The Dinosaur Heresies.* New York: William Morrow. (*A controversial book that brings the dinosaurs back to life.*)

Behler, J. L., and F. W. King. 1979. *The Audubon Society Field Guide to North American Reptiles and Amphibians.* New York: Alfred A. Knopf.

Carroll, R. L. 1988. *Vertebrate Paleontology and Evolution.* New York: W. H. Freeman.

Cohn, J. P. 1989. Iguana conservation and economic development. *BioScience* 39:359–363.

Cree, A., M. B. Thompson, and C. H. Daugherty. 1995. Tuatara sex determination. *Nature (London)* 375:543.

Daugherty, C. H., et al. 1990. Neglected taxonomy and continuing extinctions of tuatara (*Sphenodon*). *Nature* (*London*) 347:177–179.

Ferguson, M. W. (Ed.). 1984. *The Structure, Development and Evolution of Reptiles.* New York: Academic Press.

Guillette, L. J., Jr. 1993. The evolution of viviparity in lizards. *BioScience* 43:742–751.

Halliday, T. R., and K. Adler (Eds.). 1986. *Encyclopedia of Reptiles and Amphibians.* New York: Facts on File.

Louma, J. R. 1994. Endangered ridley turtle makes a comeback on Mexican coast. *New York Times,* Nov. 29, C4.

Russell, F. E. 1984. Snake venoms. In: M. W. Ferguson (Ed.). *The Structure, Development and Evolution of Reptiles.* New York: Academic Press, pp. 469–480.

Schwenk, K. 1994. Why snakes have forked tongues. *Science* 263:1573–1577.

CHAPTER

38

Birds

King bird of paradise *Cincinnurus regius.*

The Origin of Birds

It doesn't take a zoologist to tell the difference between a living reptile and a bird. Yet there is really only one fundamental difference between the two groups: Birds have feathers. Feathers, in turn, enable birds to fly and to maintain a constant, high body temperature, unlike extant reptiles. Flight and homeothermy led in turn to other physiological and anatomical differences from reptiles. The farther one goes back in the fossil record, however, the harder it is to see a difference between birds and reptiles. The oldest known bird, *Archaeopteryx,* had a skeleton similar to that of a small theropod dinosaur, like *Compsognathus* (see Fig. 37.10A). If impressions of its feathers had not been preserved, the fossils of *Archaeopteryx* would be collecting dust in museum basements, labeled as just another unimpressive dinosaur (Fig. 38.1). Three of the seven known fossils of *Archaeopteryx* were, in fact, discovered in museum collections in just that way. Because *Archaeopteryx* had feathers, however, zoologists place it in class Aves (pronounced AY-veez).

Archaeopteryx was discovered in 1861, only two years after the publication of Darwin's *Origin of Species.* Since then it has been considered to be the best example of the evolution of one major kind from another. Because of the similarities in skeletons, most early Darwinians, such as Thomas Henry Huxley, considered birds to have

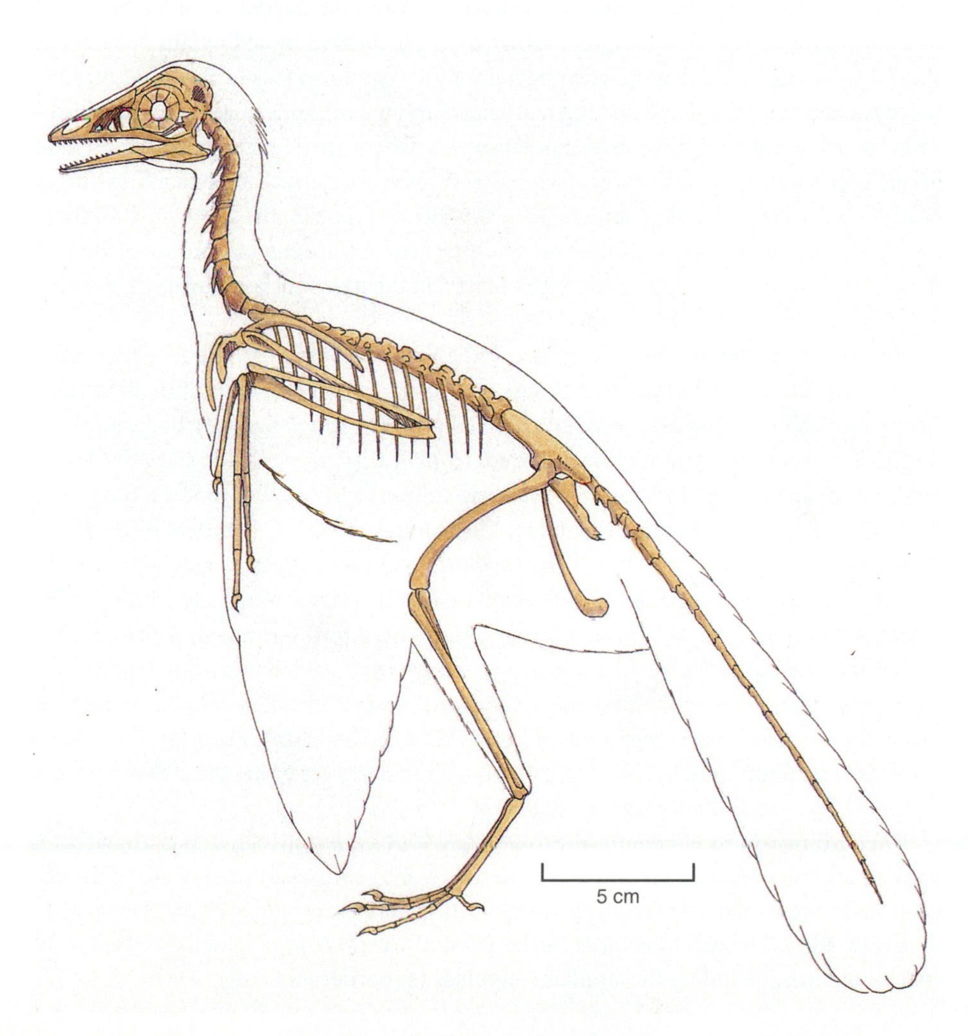

FIGURE 38.1
Reconstructed skeleton of *Archaeopteryx lithographica.* See also Fig. 18.2. Like reptiles, *Archaeopteryx lithographica* had heavy scales on its breast and gastral ribs along the belly, which may have substituted for the lack of a bony sternum to which flight muscles could attach. A recently discovered new species, *Archaeopteryx bavarica,* did have a sternum, however. Both species also had teeth and a bony tail. Unlike reptiles, however, *Archaeopteryx* had a furcula ("wishbone"). The curved hind toe indicates that *Archaeopteryx* perched in trees (Feduccia 1993).

■ **Question:** Ostrom represented *Archaeopteryx* as running along the ground with the wrist of the wing straight out instead of bent downward as shown here. What function could the wings have served in that position?

FIGURE 38.2

Representatives of 28 commonly recognized orders of birds, with distributions and estimated numbers of species. Many ornithologists combine orders Phoenicopteriformes and Ciconiiformes.

evolved from a dinosaur. Then in 1927 a book by the Dutch lawyer and paleontologist, Gerhard Heilmann, persuaded most zoologists that birds evolved much earlier from a thecodont (see Fig. 37.2). Since the 1970s, however, research by John H. Ostrom of Yale has converted the majority of zoologists back to the view that birds evolved from a theropod dinosaur—or, as cladists would say, birds *are* theropod dinosaurs. Some paleontologists interpret the evidence differently. Larry D. Martin of the University of Kansas and Samuel Tarsitano of Southwest Texas State University still consider it more likely that birds diverged from a thecodont or an intermediate between thecodonts and crocodilians.

Diversity

We know *Archaeopteryx* only because the sediment in which it was buried was fine enough to form clear impressions of the feathers, and its bones were thicker than those of modern birds. Most sediments are too coarse to preserve the details of feathers, and most birds are so light that they float rather than get covered by sediments. Most birds also have hollow bones that disintegrate before they can form fossils. For these reasons there are relatively few fossils between *Archaeopteryx* and modern birds, although recent discoveries of fossil birds in Mongolia and elsewhere are filling in many gaps. These fossils combine reptilian features such as teeth with modern features such as finger bones fused together as wing bones. *Archaeopteryx* was much more reptilian than these birds. Alan Feduccia (1995) of the University of North Carolina has proposed that all these extinct birds with reptile features be placed in the new subclass Sauriurae. All others would be in subclass Ornithurae.

Beginning with the Cenozoic era around 65 million years ago, the fossil record is more nearly complete, but modern orders appear so abruptly that it is difficult to trace the relationships among extant orders. Molecular phylogenetics suggests relationships that differ somewhat from those inferred from morphology (Sibley and Ahlquist 1991). Both anatomical and molecular criteria support a division of modern birds into two major groups: **ratites** and **carinates.** Most ratites are large, flightless birds of the Southern Hemisphere (Fig. 38.2). They include ostriches, rheas, cassowaries, emus, and kiwis, as well as the giant moa and elephant bird that were hunted to extinction by humans. Many live on islands or in other areas where they did not need to fly from predators until humans arrived. Ratites have sternums shaped like flat-bottomed boats (*ratis,* in Latin), without keels for the attachment of flight muscles. The other modern birds, the carinates, have deep **sternal keels** (Latin *carina* ship's keel) for the attachment of large flight muscles (see Fig. 38.10). All carinates are either good fliers or, like penguins, have secondarily become flightless.

According to a recent count, approximately 9700 extant species of birds have been named, and on average four new ones are discovered each year, mainly along the remote headwaters of the Amazon. Ornithologists generally recognize 27 or 28 extant orders (Fig. 38.2). The perching birds (order Passeriformes) comprise well over half of all species of birds, including the familiar songbirds (suborder Passeres).

Figure continues

Order Anseriformes
150 species (worldwide)
Ducks
Swans
Geese
Order Falconiformes
288 species (worldwide)
Vultures
Eagles
Hawks
Order Galliformes
268 species (worldwide)
Grouse
(Winter)
Ptarmigans
♂
♀
Pheasants
Order Gruiformes
209 species (worldwide)
Rails
Coots
Cranes
Order Charadriiformes
331 species (worldwide)
Plovers
Sandpipers
Snipes
♀
♂
Phalaropes
Terns
Gulls
Puffins
Order Columbiformes
303 species (worldwide)
Doves
Pigeons
Order Psittaciformes
340 species (tropical)
Parrots, parakeets

Order Strigiformes
146 species (worldwide)
Order Caprimulgiformes
105 species (worldwide)
Order Apodiformes
428 species (worldwide)
Swifts
Barn owls
Other owls
Nighthawks
Hummingbirds
Order Coliiformes
6 species (southern Africa)
Order Trogoniformes
37 species (tropical)
Order Coraciiformes
200 species (worldwide)
Order Piciformes
383 species (worldwide)
Mousebirds
Trogons
Kingfishers
Woodpeckers
Order Passeriformes
more than 5000 species (worldwide)
Swallows
Wrens
Thrashers
Thrushes
Nuthatches
Buntings
Vireos
Blackbirds
Finches
Starlings
Crows
Jays
Warblers
Sparrows

Major Groups of Class Aves

Genera mentioned elsewhere in this chapter are noted.

Subclass Sauriurae saw-ree-YOUR-ee (Greek *sauro-* lizard + *oura-* tail). Birds with reptilian features such as bony tails, unfused finger bones, true teeth. *Archaeopteryx* (Fig. 18.2).

Subclass Ornithurae orn-ith-UR-ee (Greek *ornitho-* bird). Birds without reptilian features: no vertebrae in tails, finger bones fused in wings, no true teeth. *Acrocephalus, Alcedo, Aptenodytes, Apteryx, Archilochus, Athene, Bonasa, Branta, Brotogeris, Bubo, Cardinalis, Chlamydera, Cincinnurus, Collocalia, Cuculus, Cyanocitta, Cygnus, Delichon, Ectopistes, Erithacus, Falco, Gallus, Grus, Gymnogyps, Haliaeetus, Lobipes, Melanerpes, Meleagris, Melopsittacus, Menura, Mimus, Otis, Parus, Passer, Pavo, Phalaropus, Phalacrocorax, Pharomachrus, Phasianus, Philetairus, Phoeniculus, Pitohui, Ploceus, Psittacus, Scolopax, Steganopus, Sterna, Struthio, Sturnus, Troglodytes, Turdus, Tyto* (Openers for Chapters 7, 9, and 18 and for this chapter; Figs. 20.2, 20.15, 20.16, 20.18, 38.2, 38.5B and C, 38.6, 38.11, 38.17, 38.19A, 38.20, 38.21, 38.22, 38.23, 38.25, 38.26, 38.27).

Feathers

Development. T. H. Huxley called birds "glorified reptiles." He might also have added that feathers are glorified scales. Feathers are in fact homologous with reptilian scales. The development of a feather begins with a **papilla** that elongates and sinks into a **follicle** in the dermis (Fig. 38.3). Blood vessels and nerves from the dermis then enter the papilla, forming a **pulp.** The pulp supplies nutrients for the surrounding epidermis to grow and produce keratin, which forms the feather as well as the sheath that protects the developing feather.

In the development of the most familiar feathers, called **vane feathers,** a **shaft** develops within the papilla, and numerous **barbs** branch out at an angle. These barbs are curled up within the sheath, but once the feather is fully developed the sheath bursts and the barbs unfurl into two **vanes,** one on each side of the shaft. The pulp then dries up, leaving a hollow **quill.** The portion of the shaft extending from the quill is the **rachis** (RAY-kis). Extending from each barb are numerous tiny **barbules.** In vane feathers the barbules have hooked ends that interlock with adjacent barbs like Velcro, causing the vanes to form a single flexible surface (Fig. 38.4 on page 836).

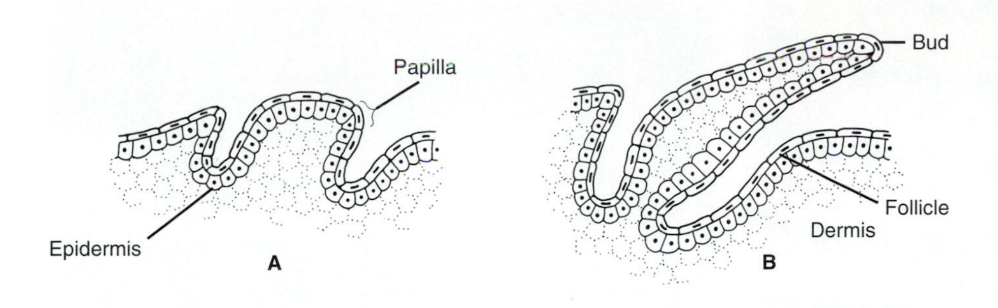

FIGURE 38.3

Development of a vane feather. (A) The feather starts as a papilla similar to a reptilian scale (compare Fig. 37.11). (B) The papilla forms a bud that sinks into a follicle in the dermis. (C) The shaft and attached barbs of the feather develop between a protective sheath and the nutritive pulp. (D) When the feather is mature the sheath bursts. The pulp dries up, leaving a hollow quill.

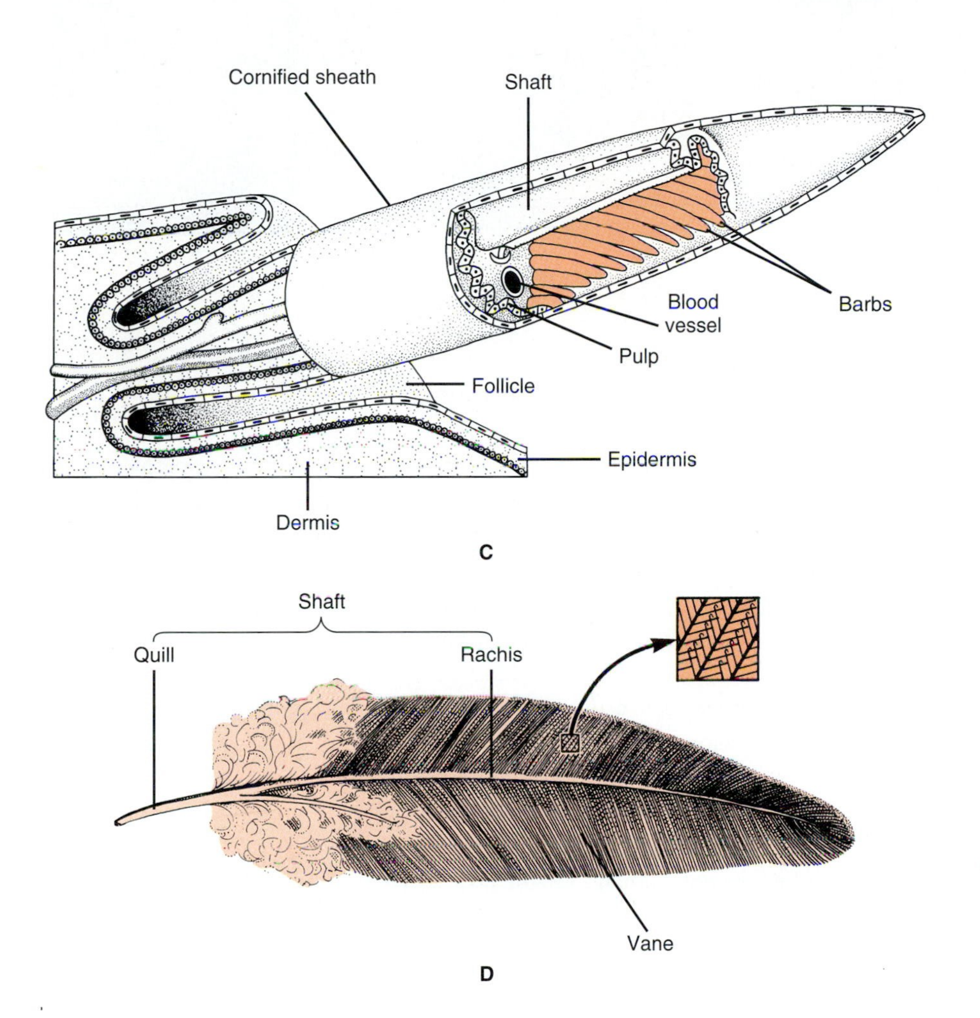

According to the convention adopted by the American Ornithological Union, each word in the common names of bird species should be capitalized. Most ornithologists would therefore write "Ruby-throated Hummingbird" and "Tundra Swan." For consistency with common names in the rest of this book, however, the names of birds are not capitalized.

DISTRIBUTION AND TYPES. The number and types of feathers depend on species, body size, and other factors. One ruby-throated hummingbird *Archilochus colubris* was found to have 940 feathers, and a tundra swan *Cygnus columbianus* had 25,216. Papillae are not distributed randomly all over the skin, but lie on a few **feather tracts** (= pterylae). In most birds, vane feathers, also called **contour feathers,** cover most of the body. (A notable exception is the lower legs, which are covered with scales like those of reptiles.) Vane feathers also extend from the wings and tail as **flight feathers.** Flight feathers are asymmetric, with one vane narrower than the other. This allows flight

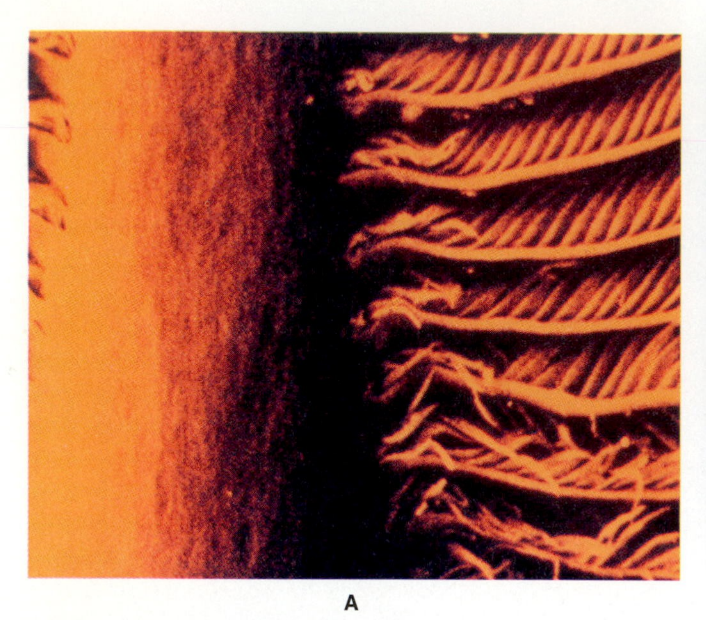

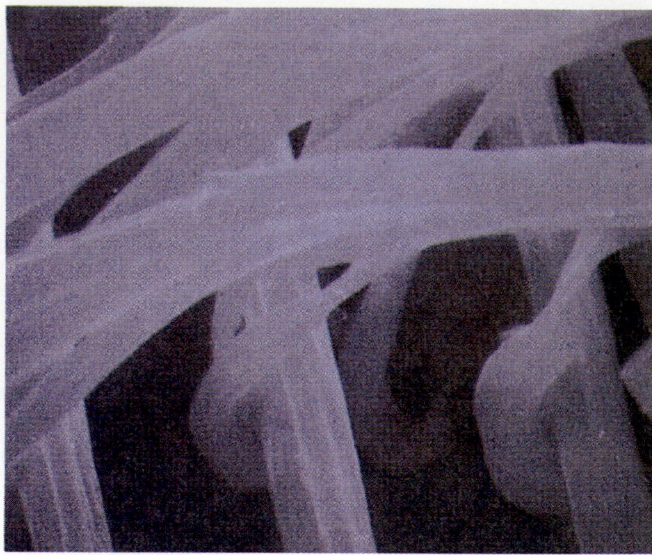

FIGURE 38.4

Scanning electron micrographs of vane feathers of an American robin (*Turdus migratorius;* order Passeriformes). (A) The shaft is to the left. (B) Higher magnification shows the hooked barbules by which barbs of adjacent feathers interlock. Such interlocking allows the feathers to form a single flexible surface.

feathers on the wings to pivot during the upstroke, opening spaces that reduce air resistance. There are also other types of feathers (Fig. 38.5).

COLOR. A feather is naturally white unless it acquires color, either as pigment from chromatophores during development or as structural color (pp. 193–194). Black, brown, dull yellow, and dull red colors are due to melanin pigments. Carotenoids account for bright yellow, orange, and in many cases, red feathers. In other birds, bright red or green is due to porphyrin pigment. Blue color in birds is Tyndall blue, caused by the scattering of light by particles or overlapping keratin layers. The iridescence of peacocks and starlings, which changes with the viewing angle, is a different kind of structural color due to bundles of tubules in the feathers. The activation of different combinations of chromatophores during development allows a single feather to acquire a pattern of more than one color.

Color generally functions either in camouflage or social communication. In many species, juveniles and females are well camouflaged with melanin-pigmented feathers, while adult males in breeding plumage are brightly colored (Fig. 38.6 on page 838). In a few toxic species, bright plumage might also serve as warning coloration. Several brightly colored species of *Pitohui* in New Guinea produce batrachotoxin, a neurotoxin similar to those produced by certain frogs (p. 789; Dumbacher et al. 1992).

MAINTENANCE. With so many functions depending on the feathers, it is not surprising that birds spend so much time maintaining them. One familiar maintenance behavior is **preening,** which consists of running the feathers through the beak to zip the barbules together. While preening, most birds also oil the feathers with a water-repellant secretion from the **oil gland** (= uropygial gland) at the base of the tail. Birds also care for their feathers by bathing, both in water and in dust. Bathing apparently eliminates ectoparasites such as ticks and lice. Many birds also rid themselves of parasites by the curious practice of "anting," in which they wallow in anthills and induce the soldier ants to attack the parasites. Although the ants undoubtedly bite the bird as well, birds go into such a frenzy of excitement during anting that it is hard to believe they don't enjoy it.

MOLTING. In spite of maintenance, feathers do become damaged and have to be replaced. In addition, feather replacement occurs during regular periods of molting. Most

FIGURE 38.5

(A) Varieties of feathers. Some vane feathers have aftershafts. A down feather lacks hooks on the barbules and forms an insulating undercoat in geese and many other waterfowl. Filoplumes have barbs only at the tips and are used mainly to detect movement of air and feathers. Bristles have a few barbs concentrated at the base and also serve in mechanoreception. (B) Filoplumes employed as decoration on the head of a peacock *Pavo cristatus.* Most of the other feathers are vane feathers. (C) Bristles form the eyelashes of the ostrich *Struthio camelus.* Most of the other feathers of the ostriches have hairlike shafts without hooked barbules.

■ **Question:** Which of these types of feathers would the kiwi (Fig. 38.2) have?

birds undergo a gradual molt, in which they lose only a few feathers at a time. Usually one matched pair of flight feathers is lost at a time, so balance is not upset during flight. However, many large water birds, such as ducks and geese, fly poorly, if at all, following the loss of a pair of flight feathers. Instead of remaining flightless throughout a long, gradual molt, these species replace all the flight feathers during a shorter period of several weeks. In most species, molting occurs once a year after the breeding season. Molting may occur more than once a year in birds that depend on their feathers for long migrations and in birds whose feathers are subject to much damage. In addition, many birds, especially males, undergo partial molts that uncover their breeding plumage.

Flight

PERFORMANCE. Birds commonly cruise at speeds of about 40 km/hr, and some, such as the peregrine falcon *Falco peregrinus,* can reach 190 km/hr in a power dive. Just as remarkably, many birds can hover at zero speed. Climbers on Mount Everest, gasping for enough oxygen to take the next step, have reported birds flying overhead, apparently un-

A

B

FIGURE 38.6

Functions of feather color. (A) An American woodcock, *Scolopax minor* (formerly *Philohela minor*), camouflaged on its nest. (B) A male resplendent quetzal *Pharomachrus mocino* sacrifices camouflage to attract females in the dark, thick foliage of Central American rainforests. The quetzal was revered as god of the air by the Mayas and Aztecs, who associated it with the god Quetzalcoatl. In the past, many Europeans doubted its existence; because of destruction of rainforests there is now reason to doubt its future existence. The length including the tail feathers is more than a meter.

■ **Question:** What behavioral adaptation is required for the camouflage of the woodcock to be effective?

perturbed by the thin atmosphere. Many birds make aerial migrations of thousands of kilometers twice a year (see p. 412), and certain juvenile terns are thought to spend months or even years on the wing, not landing until summoned by the urge to reproduce. Birds make these feats look so easy that it is surprising how hard their flight is to understand.

EVOLUTION. One poorly understood question about bird flight is its origin, which is closely related to the controversy over the origin of birds, themselves. One point that is clear is that birds did not inherit the ability to fly from a pterosaur or a bat (Fig. 38.7). But how did flight originate? Scales or feathers do not enlarge and form wings in the "hope" that they will someday enable a descendant to fly. Changes in the forelimbs would have had to be advantageous at every stage in their evolution as wings. For more than a century there have been two major theoretical scenarios for how this evolution could have occurred: the arboreal theory and the cursorial theory. According to the **arboreal theory,** also known as the "trees-down theory," wings evolved in reptiles that climbed up trees hunting insects, then glided to the base of the next tree. This is a highly efficient form of hunting that is widely used by some birds, such as woodpeckers. The **cursorial,** or "ground-up theory," proposes that wings evolved in a running (cursorial) reptile, perhaps as rudders that helped the reptile lunge after prey. The arboreal theory tends to be favored by those who argue that birds evolved from thecodonts, while the cursorial theory is favored by those who favor a theropod origin of birds. The earliest known bird, *Archaeopteryx,* was almost certainly arboreal, judging from the curve of its claws and other anatomical features. Whether *Archaeopteryx* could fly or not is still very much up in the air (Speakman and Thomson 1994).

BODY SIZE. In addition to questions about its origins, there are also many questions about the mechanisms of bird flight, even though the basic principles have been known since the Wright brothers. One of the elementary principles was discovered the

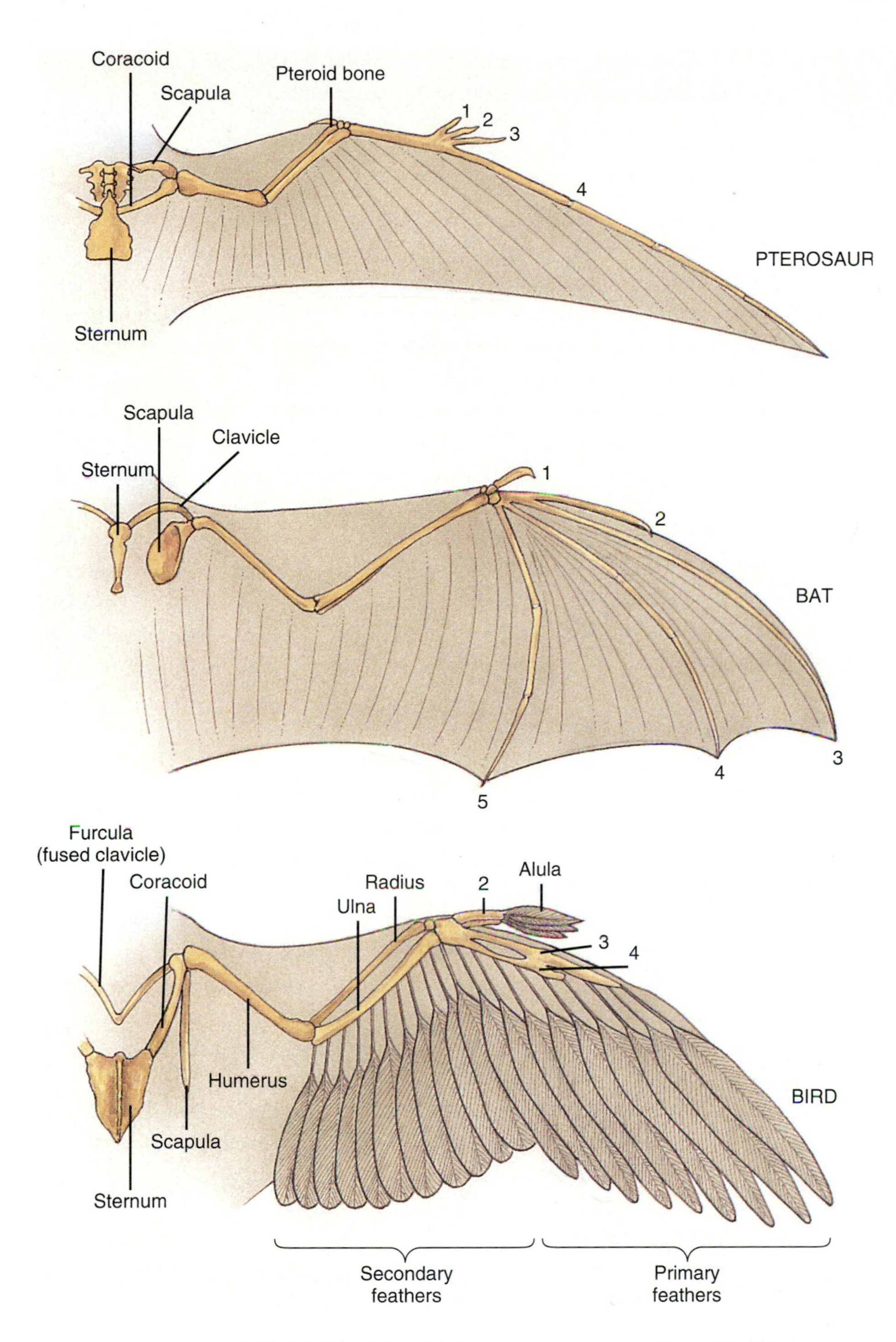

FIGURE 38.7

Comparison of the wings of pterosaurs, bats, and birds shows that flight evolved independently in reptiles, mammals, and birds. In pterosaurs and bats the main wing structure is skin supported by the finger bones (the fourth finger in pterosaur, and the second through fifth in bats). In birds the flight feathers form the wing surface. The primaries are attached to the fused third and fourth digits, and the secondaries are attached to the ulna of the lower arm. An auxiliary wing, the alula, consists of feathers attached to the second digit. There are also differences in the pectoral bones supporting the wings. Pterosaurs lacked clavicles (collar bones), but unlike mammals, they had a coracoid that helped support the wings. In birds the clavicles are fused into a furcula ("wishbone"), and the coracoid provides additional support.

■ **Question:** How does this figure illustrate the concept of analogy? What parts of the three wings are homologous?

hard way by the early pioneers of human flight: Wings have to be large enough to generate enough **lift** to support the weight of the body. In other words, the **wing loading** (body mass divided by wing area) must be below a certain value. Small wing loading requires large wings, light bodies, or, as in most birds, both. Birds have a variety of adaptations that reduce body weight, including respiratory air sacs and hollow bones, which will be described later. In spite of these adaptations, wing loading inevitably increases as the body gets larger, because mass increases more rapidly than area. The largest bird that can fly is the great bustard *Otis tarda*, with a mass up to 22 kg. There are compensations for being large, however, including greater ability to escape predators and to conserve body heat. Ratites, penguins, and some others have given up flight

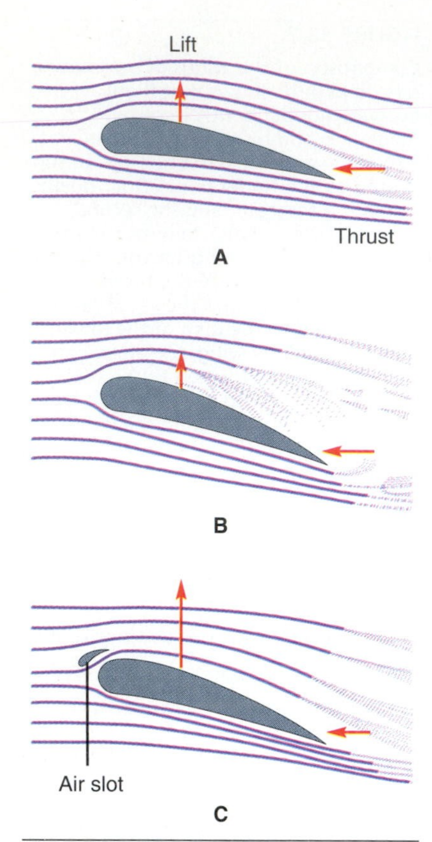

FIGURE 38.8
The pattern of air movement over an idealized wing viewed from the end. Arrows show the direction and magnitude of lift and thrust. (A) Air forced to move more rapidly over the top has a lower pressure, which produces most of the lift. Additional lift is produced by the pressure of air against the bottom surface of the wing. This high-pressure air moving past the rear generates thrust. Increasing the angle of attack decreases thrust, and it increases lift up to a point. (B) Increasing the angle of attack too much induces turbulence, which causes a loss of lift (stalling). (C) Stalling is prevented by adding a wing slot to channel a fast stream of air across the upper surface. Slotting is achieved in birds by spreading the alula and primary feathers.

in favor of these advantages. There is also a minimum size required for flight (pp. 212–213), and all birds are, in fact, larger than the minimum.

GLIDING. Gliding, in which the wings are held out while the bird slowly falls through the air, is relatively easy to understand. Air escaping past the hind part of the wing generates **thrust,** which pushes the wing and the bird forward. This forward movement generates lift because the profile of a bird's wing is like that of an airplane: The front of the wing is thicker than the rear, and the rear curves downward. This forces air to flow over the wing faster than it flows beneath the wing. The faster the velocity of air, the lower its pressure, so there is reduced pressure on the upper surface of the wing, which lifts the wing up (Fig. 38.8A). Some lift is also produced by the pressure of air against the bottom surface of the wing. **Soaring** on thermal updrafts or into the wind uses the same principles as gliding.

Increasing the angle of attack of the wing (angling the front of the wing upward) increases the lift but decreases the thrust. Birds commonly increase the angle of attack to maintain altitude at slow speed or to slow down their forward movement while maintaining enough lift for a gentle landing. If the angle of attack is too steep, however, the flow of air over the wing becomes turbulent, resulting in **stalling** (Fig. 38.8B). To prevent stalling, birds do what airplane pilots do when landing: They open wing slots. In birds these wing slots result from the spreading of primary feathers and **alulae** (Fig. 38.8C).

FLAPPING FLIGHT. Flapping flight is much more complicated than gliding because the wing movements vary in so many ways, depending on the size of the bird, the direction and speed of flight, air temperature, and other factors. However, some of the lift and thrust are produced in the same way as in gliding, except that the wing is moving actively rather than passively through air. Birds do not flap their wings straight up and down, but in a figure-eight pattern (Fig. 38.9). During the downstroke the **pectoralis muscles** pull the wings downward and forward, reducing pressure on top of the wings and increasing it beneath (Fig. 38.10). During the upstroke the **supracoracoideus muscles** pull the wings upward and backward. As noted previously, the asymmetry of the primary feathers causes them to pivot during the upstroke, opening spaces for air to flow between them. The wing also partly folds during the upstroke.

ADAPTATIONS OF WING SHAPE. Each kind of flight makes conflicting demands on wings. Most birds have wings that compromise among all these demands, but some specialize in satisfying the requirements for one type of flight.

1. Optimal **gliding and soaring** require the largest wing surface that the bones and muscles can support. (See Fig. 38.26.)

FIGURE 38.9
Wing movements during flapping flight in the little owl *Athene noctua.* The wing sweeps forward during the downstroke and backward in the upstroke. Note the pivoting of the flight feathers due to their asymmetry, which reduces resistance during the upstroke.

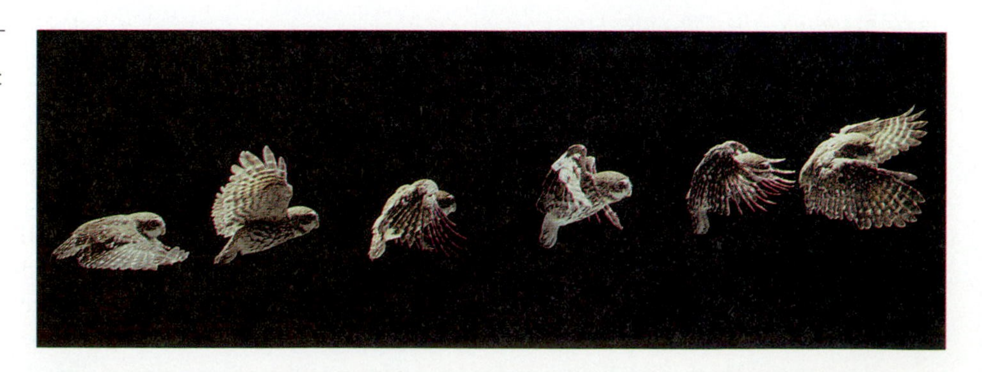

2. Efficient **flapping flight** requires a longer wing than for gliding, since much of the thrust of flapping flight comes from the tips, which move the most. Longer wings are also more efficient, because they have proportionately less area at the tips, where there is the most turbulence. Swept-back wings with pointed tips improve efficiency by reducing turbulence and also drag. (See Fig. 38.27C.)
3. **Maneuvering** among tree limbs requires that wings be shorter than would be optimal for flapping flight. The consequent reduction in lift and thrust can be compensated by flying faster.

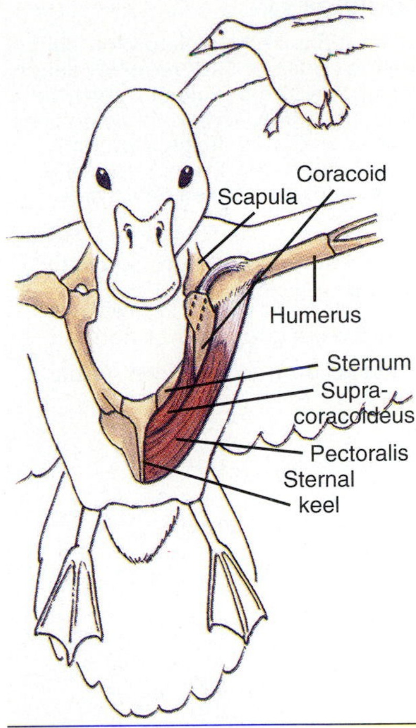

FIGURE 38.10
Frontal view of the pectoral girdle and attached wing muscles. The downstroke of flapping flight is due to contraction of the pectoralis muscles, which are attached to the sternal keel and to the wishbone (not shown). The upstroke is due to contraction of the supracoracoideus muscle, which is also attached to the sternal keel. The contraction of the supracoracoideus is redirected by the bending of its tendon over the scapula, like a pulley. See also Fig. 38.7 for a ventral view, and see Fig. 38.11B for a lateral view. In chickens and most other birds that fly weakly or not at all, the flight muscles are type II white and produce ATP anaerobically. In good fliers the muscles are red and aerobic.

ADAPTATIONS OF THE SKELETON. Flight makes enormous mechanical demands on a bird. The skeleton has to be strong to withstand the stresses from the flight muscles, but it must also be light. This combination of strength and lightness is achieved by the resorption of bone marrow early in life, leaving most of the bones hollow except for internal reinforcing struts (Fig. 38.11A on page 842). Contraction of the flight muscles imposes large stresses on the ribs, which avoid fracturing by folding like a knife (Fig. 38.11B). The **furcula** serves as a spring that helps the wings rebound during the upstroke.

The specialization of the forelimbs as wings means that locomotion on land is bipedal rather than quadrupedal. Like humans, therefore, birds are slower and less stable than most terrestrial vertebrates when walking and running. Birds have feet that are specialized for particular kinds of locomotion and for other functions (Fig. 38.12 on page 843). Bipedality also requires that the pelvic girdle support the entire body weight when the bird is standing. Consequently the pelvic girdle of birds consists of large bones fused to the vertebrae of the back. In contrast, the vertebrae of the neck are quite flexible. This flexibility is essential for several reasons: (1) It enables the bird to use its head to maintain balance while flying and (quite noticeably in the pigeon) walking. (2) It compensates for the limited ability of birds to rotate their eyes. (3) It allows the bird's beak to reach the oil gland at the base of the tail and all the feathers for preening.

Physiology

THERMOREGULATION. Most birds maintain body temperatures between 40°C and 42°C, which is almost always well above the environmental temperature. In other words, birds are endothermic homeotherms (pp. 292–293). The ability to regulate body temperature enables birds to be active in a wide range of seasons and climates, but it exacts an enormous cost. In general, birds have metabolic rates at least 10 times higher than they would need if they did not regulate body temperature. This, in turn, means that they must have at least 10 times the food intake. Endothermic homeothermy is possible because the heat produced by this high metabolic activity is conserved by feathers, which reduce thermal conductivity and radiation.

RESPIRATION. Although the physiological systems of birds are, in general, similar to those of reptiles and mammals (as described in Unit 2), the demands of flight and homeothermy have apparently selected for several unique adaptations. This is especially so for the respiratory system. Unlike the lungs of reptiles and mammals, which consist of sacs in which air can stagnate, birds' lungs consist of numerous tubules called **parabronchi** through which air flows in only one direction (pp. 247–248). The lungs are ventilated to a small degree by their own expansion and contraction, but mainly by the

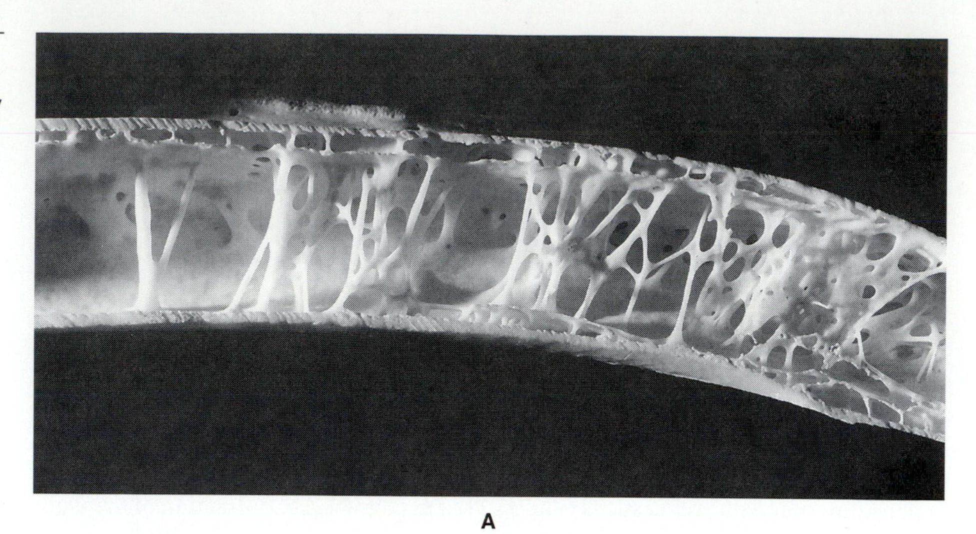

FIGURE 38.11

Adaptations of the skeleton for flight. (A) The bodies of birds have low density (weight divided by volume), partly because the bones, such as this wing bone from an eagle, are hollow. Strength is largely due to the internal struts. (B) Skeleton of the chicken *Gallus gallus.*

■ **Question:** If you didn't know the species, would you say the bird in part B was a good flyer or not? What could account for the deep sternal keel?

expansion and contraction of **air sacs** (Fig. 38.13 on page 844). Inhalation and exhalation are aided by movement of a nonmuscular **diaphragm** between the abdominal and thoracic cavities, and breathing may also be coupled to movements of the sternum and furcula during flight.

The air sacs of some species branch into the bones of the wing. In fact, if one of these bones is opened to air, the trachea can be shut off completely without suffocating

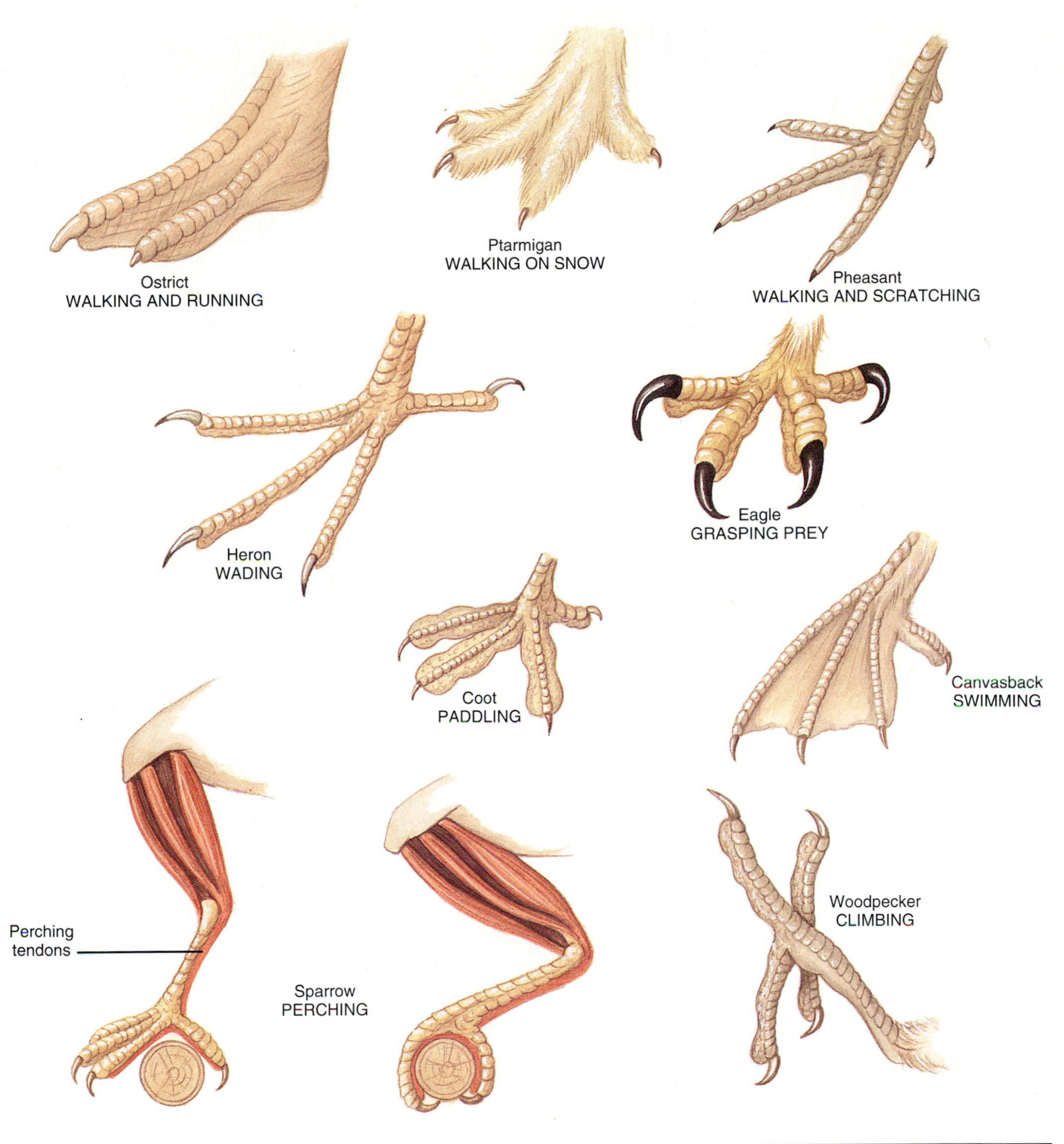

FIGURE 38.12

Adaptations of birds' feet. The two figures on the bottom right show perching tendons. When the bird's legs fold the perching tendon automatically flexes the claws, closing them around the perch. This mechanism enables birds to sleep without falling.

the bird. Normally, of course, air enters and exits through the nasal passages, trachea, and bronchi (Fig. 38.14 on page 844). At the upper end of the trachea is a **larynx** that, unlike the larynges (= larynxes) of mammals and some reptiles, lacks vocal cords. Instead, voice is produced by the **syrinx** at the other end of the trachea, near the bronchi. In chickens and many other birds there is one syrinx in the trachea near its junction with the bronchi. The syrinx has a pair of membranes (**tympani**) that vibrate as air blows past them. The pitch of the sound is controlled by muscles that adjust the tension on the tympani. Songbirds have essentially two syringes (= syrinxes), which enable them to sing with two voices at the same time.

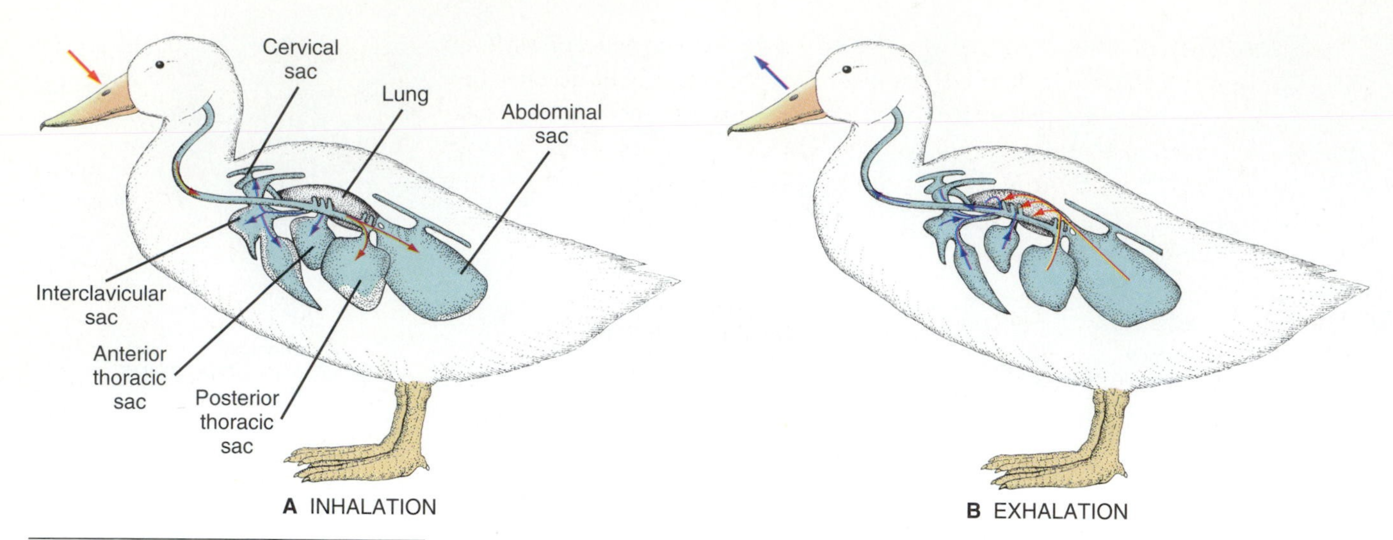

FIGURE 38.13

Breathing in birds. (A) During inhalation the lungs contract somewhat, and relaxation of body muscles allows the air sacs to expand. These movements draw fresh air (red) through a central tube (the mesobronchus) in each lung into the posterior air sacs and force stale air (blue) from the lungs into the anterior air sacs. (B) During exhalation the lungs expand and the air sacs contract, moving fresh air from the posterior air sacs through the lungs and emptying stale air from the anterior air sacs.

■ **Question:** In what order would the following parts of the respiratory system be encountered by an O_2 molecule during breathing: lung, anterior thoracic sac, abdominal sac. Are any parts encountered twice?

CIRCULATION. The high metabolic rates of birds also place demands on the circulatory system. Like crocodilians and mammals, birds have four-chambered hearts with complete separation of oxygenated and deoxygenated blood (see Fig. 11.4D). The heart rate increases with metabolic demand and is generally higher in smaller birds. The heart of a flying hummingbird beats so fast (1200 beats per minute) that it hums. The blood is similar to mammalian blood except that mature red blood cells retain their nuclei.

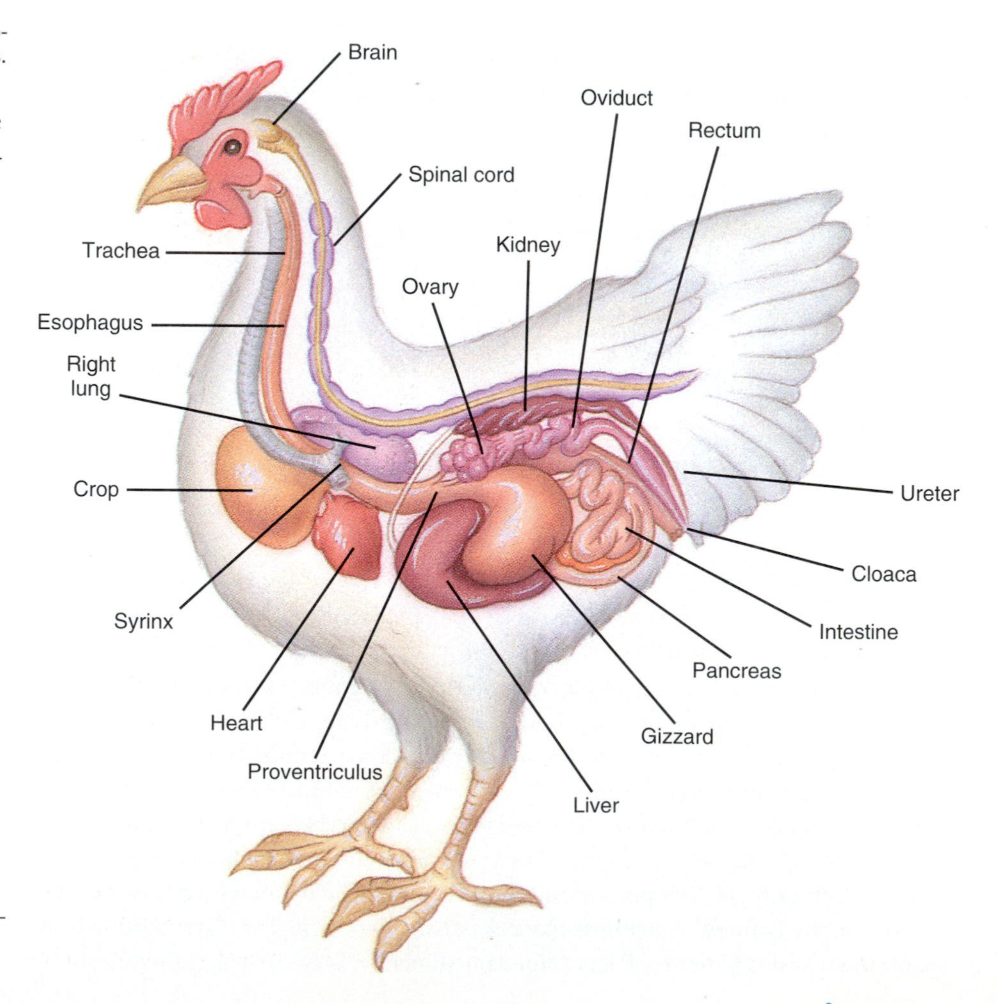

FIGURE 38.14

Internal organs of a domestic hen.

OSMOREGULATION AND EXCRETION. Other physiological systems of birds are generally similar to those in reptiles and mammals. The kidneys are of the metanephric type. In marine birds, **salt glands** that empty through the nostrils assist in osmoregulation (p. 288). Birds have nephrons with short loops of Henle that help conserve water (pp. 284–285). Water is also conserved by the excretion of insoluble uric acid as the main nitrogenous waste (pp. 279–280). The white, pastelike uric acid is eliminated through the cloaca, along with feces.

THE NERVOUS SYSTEM. The **brain** is highly developed, although along somewhat different lines from that of mammals. It is really not much of an insult to be called a "bird-brain," since the brains of birds are about as large as those of most mammals of the same body size (see Fig. 9.7). As one might expect in animals that fly so skillfully, the cerebellum is disproportionately large compared with that of a reptile or mammal (see Fig. 9.6). Bulging between the cerebellum and the cerebral hemispheres are the **optic lobes.** The cerebral cortex is not as well developed as in most mammals. Instead the core of the cerebrum, the large **corpus striatum,** appears to handle most of the complex behaviors.

Many avian behaviors seem to depend only on **instinct** and can often create the impression that birds are stupid. For example, a male domestic turkey *Meleagris gallopavo* will try to copulate with a human hand if the hand is holding the decapitated head of a female turkey. The male Eurasian robin *Erithacus rubecula* will become enraged at a tuft of red feathers in its territory, apparently "thinking" it is a rival male. Many other male birds will attack their own images in a mirror placed within their territories. (This is a useful trick for determining the boundaries of territories.) Most bird behaviors, however, reveal an interaction of instinct and **learning.** For example, many male songbirds can make a fair approximation of their mating song without having heard it before, but to sing correctly they must have learned the song for their species during a certain sensitive period soon after they hatch.

The vocal abilities of parrots, starlings, mockingbirds, crows, and some other birds are impressive but have usually been dismissed as being simple mimicry. Recently, however, Irene Pepperberg at the University of Arizona has succeeded in training African grey parrots *Psittacus erithacus* to vocalize in ways that go well beyond mere "parroting" and may qualify as language. One of the parrots, Alex, can name the colors and shapes of objects, ask for specific foods, and state whether two objects differ in shape, color, or size. Perhaps just as impressive are more natural feats of learning. The marsh tit *Parus palustris,* a chickadee-like bird of Europe, can remember dozens of places where it has hidden seeds. Some birds can learn the pattern of stars in a few days, remember it for years, and can use that memory and other cues to navigate over enormous distances (p. 412).

The **eyes** (Fig. 38.15A) are about as large as the cerebrum, and in hawks they are so large that they touch in the middle of the head. In most birds the muscles that control eye movements are virtually nonexistent, presumably having been sacrificed to reduce weight and make room for larger eyes. Birds therefore move the head in order to change the direction of vision. Birds that are preyed upon tend to have the eyes on the sides of the head, allowing them to see in all direction. Predatory birds tend to have eyes toward the front of the head, with some overlap in the visual field permitting visual perception of depth. Each eye is protected by two ordinary lids and by a transparent **nictitating membrane.** In ducks, loons, and some other diving birds the nictitating membrane has an area shaped like a contact lens that helps the bird focus under water.

Most birds are diurnal and have good perception of colors, with numerous cones. In nocturnal birds, rods predominate. Because birds have smaller eyes than do hu-

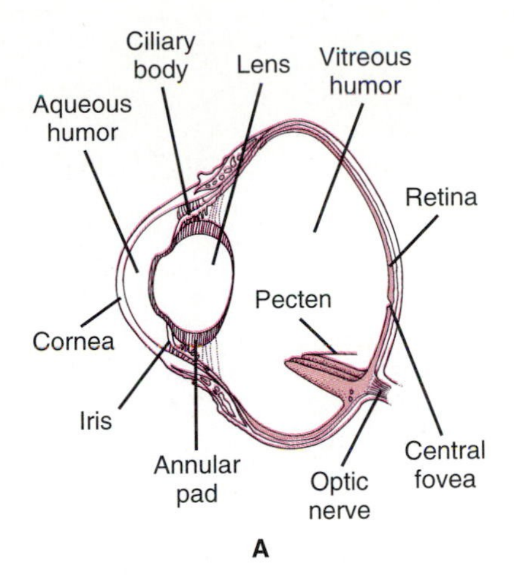

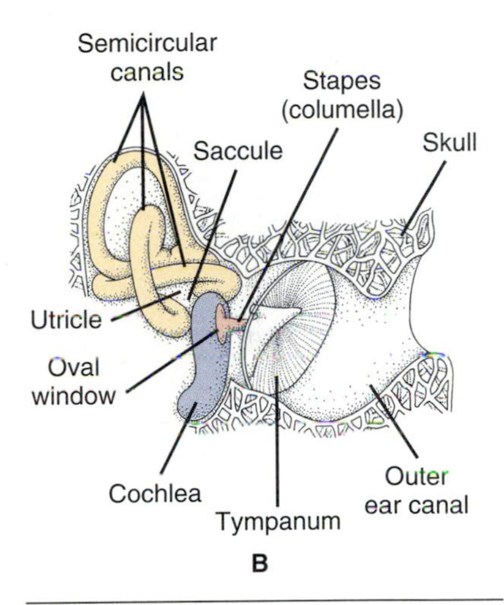

FIGURE 38.15
Sensory receptors of birds. Compare Figs. 8.12A and 8.15A. (A) The eye. The lens is shown focused close. To accommodate for distant vision the cornea and lens are flattened in front. The pecten is a series of triangular ridges believed to supply oxygen and nutrients, compensating for the absence of blood vessels in the retina. The fovea is the center of focus, with the sharpest vision. Some birds have two or three foveae. (B) Frontal section of an inner ear.

mans, their vision would be expected to be poorer, because the image is compressed onto a smaller retina. In many species, however, the rods and cones in parts of the retina are packed more densely than they are in humans. Many larger birds, in fact, have a visual acuity several times better than that of humans. A vulture can resolve two closely spaced lines from three times farther away than a person can.

Like other vertebrates, birds have an **inner ear** with semicircular canals and otolith organs that respond to acceleration and static position, respectively (Fig. 38.15B). In addition, birds have a well-developed **cochlea** specialized for hearing. Unlike the snail-shaped cochlea of a mammal, that of a bird is short and straight. Sound vibrations are picked up by the eardrum and transmitted to the cochlea by a single middle-ear bone, the stapes (columella). Most birds can hear only sounds with frequencies below 12 kHz (12,000 cycles per second), compared with an upper limit of 20 kHz for humans. Within that range, however, some birds have an acuteness of hearing that astounds humans. Pigeons and many other birds can hear subsonic vibrations (below 20 Hz), such as those generated by wind in mountains, from distances of hundreds or thousands of kilometers. They may use such "sounds" to navigate. The barn owl *Tyto alba* can catch mice just as well when blindfolded as not. The ear openings of owls are located beside the eyes in the disc-shaped ruff of feathers around the face, and the left opening is aimed downward while the right is aimed upward. By comparing the time of arrival and the intensity of a sound in the two ears, barn owls get a fix on the source of the sound in both the horizontal and vertical direction.

Olfaction in birds is generally much less acute than in most mammals. Vultures are exceptional in that they can locate carrion by smell alone. In most birds there is only a small olfactory epithelium and no Jacobson's organ. There are also fewer taste receptors than in mammals, and little need for any, since birds bolt down their food without chewing it. Behavioral studies show that pigeons and some other birds can detect Earth's magnetic field and changes in air pressure, although the receptors have not been identified.

DIGESTION AND FEEDING. Birds, especially small ones, need large amounts of nutrients to supply the high rate of metabolism. The 11-gram blue tit *Parus caeruleus* needs about 30% of its own weight in seeds every day. The great horned owl *Bubo virginianus* eats an average of 5% of its body weight in mice per day. (For a human that would be the equivalent of eating about 2 pounds of mice for breakfast, lunch, and dinner.) Adaptations for handling these prodigious diets differ in several ways from those in other vertebrates. One important difference is that extant birds have no teeth, having apparently jettisoned these heavy structures to save weight. The inability to chew food is compensated by the **gizzard** (= muscular stomach), which is lined with horny plates that crush and grind hard foods. Breaking food into small pieces allows better mixing with digestive enzymes and greatly increases the rate at which the digestive system can supply nutrients. Many seed-eating birds swallow grit that aids the grinding action of the gizzard. In owls and many other birds that eat vertebrates whole, the gizzard crushes the bones, which are then regurgitated in a pellet that also contains hair and other indigestible parts.

Anterior to the gizzard is the **proventriculus** (= glandular stomach), which secretes protein-digesting enzymes. In many birds the proventriculus and gizzard together are too small to store much food. Instead, an enlargement of the esophagus, called the **crop**, enables birds to store food and transport it to the young. In pigeons, doves, and flamingos with young the crop also secretes a "pigeon's milk" analogous to mammalian milk, which they feed to the young. Male penguins produce a similar secretion from the esophagus.

Just as the teeth of mammals are adapted for the kinds of foods they eat, the beaks of birds are similarly adapted (Fig. 38.16). Even among some closely related species,

FIGURE 38.16

Adaptations of beaks for different ways of feeding.

■ **Question:** Which birds in the order Passeriformes shown in Fig. 38.2 are probably granivorous (seed-eaters)? Which ones are likely to be insectivorous?

such as Darwin's finches, the beaks often differ in shape and size (see Fig. 18.12). Birds that feed on seeds tend to have heavier bills than those that feed on insects. The size of the bill is quite variable, so natural selection enables a population to adapt to changing food supplies within a few generations (pp. 372–374). Feeding behavior is also diverse, specialized, and adaptable to circumstances (Fig. 38.17 on page 848).

Reproduction: Genetics and Physiology

As in many snakes and lizards, the sex of a bird is determined by whether it inherits two Z chromosomes or, alternatively, a Z and a W. In contrast to mammals, it is the female that is **heterogametic** (has two different sex chromosomes, Z and W). In many adult birds the two sexes differ greatly in coloration during the breeding season, but

FIGURE 38.17

The Eurasian kingfisher *Alcedo atthis* catching fish. (A) Diving for the fish, using the wings and tail for guidance. (B) The fragile wings fold just before impact. (C) Success. The kingfisher beats the fish against a branch to kill it, then swallows it head first.

they are generally indistinguishable at other times and as juveniles. In many other species the two sexes resemble each other year round, except that they may differ in size. Fertilization is necessarily internal, since females lay eggs that are covered with shells. Male ratites, chickens, ducks, storks, and some other groups have an erectile penis on the ventral wall of the cloaca, which is considered to be a primitive trait. As in reptiles, an external groove on the penis guides sperm into the cloaca of the female. In most birds, however, internal fertilization is effected simply by the male pressing his cloaca to that of the female while treading upon her back. Generally the testes are small except during the mating season, probably as an adaptation that reduces body weight. Females of most species save weight by having only the left ovary.

Reproductive activities are coordinated by hormones similar to those in mammals (pp. 312–315). The anterior pituitary secretes gonadotropins (FSH and LH) that trigger gamete production and ovulation. Prolactin, which stimulates the production of milk by female mammals, also stimulates the production of "pigeon's milk" by pigeons and doves. Estrogens and androgens, steroid hormones from the ovaries and testes, respectively, stimulate appropriate reproductive behaviors. The secretion of these hormones generally follows an annual rhythm synchronized by day length. Many of the same hormones are also involved in coordinating other behaviors that have to be synchronized with reproduction. For example, testosterone triggers the partial molt to breeding plumage, and it promotes aggression and territorial defense. Prolactin, together with cortisol from the adrenal cortex, helps coordinate the timing and direction of migration.

Eggs

FORMATION. All birds are oviparous. It is a long-standing mystery why viviparity has never evolved in birds as it has several times in fishes and amphibians, approximately a hundred times in reptiles, and once in mammals. The commonly offered answer is that it would be difficult for a pregnant bird to fly, but that does not seem to bother bats. Egg formation begins at ovulation, when the egg or ovum is essentially a

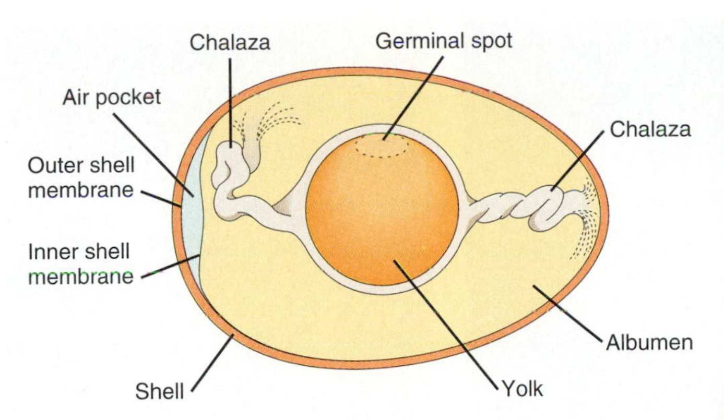

FIGURE 38.18

A just-laid, fertilized egg. The embryo develops from the germinal spot. (Compare the later embryo in Fig. 6.17.) The chalaza, which consists of thick strands of albumen, keeps the ovum near the center of the egg and twists as the egg is turned during incubation, allowing the denser yolk (vegetal pole) to stay below the embryo (animal pole).

giant cell consisting of the yolk and the **germinal spot** from which the embryo will form. The ovum enters the oviduct through the funnel-shaped **infundibulum.** Fertilization usually occurs at this point. As the ovum passes down the oviduct, it is covered with **albumen** (egg white), a process that takes about 3 hours in the domestic hen. Farther down the oviduct the albumen becomes enclosed in two keratin membranes, which later become separated at one end of the egg to form an air pocket.

The egg then enters an enlarged area of the oviduct (the "uterus"), where a protein matrix is deposited. On this matrix a shell consisting mainly of calcium carbonate crystallizes. The calcium comes from the female's bones, and much of it is gradually absorbed by the developing chick to make its bones. Formation of the shell takes about 20 hours in a hen. The final product released from the cloaca is a self-contained (cleidoic) egg (Fig. 38.18). Pores in the shell allow for the exchange of respiratory gases, assisted later by the chorion and allantois (pp. 117–118). The air pocket expands to fill the space left by the lost water. In the last hours of development the chick breaks into the air pocket and gets its first breath of air. Finally the chick breaks through the egg shell using an **egg tooth** on the outside of the upper beak.

FIGURE 38.19

(A) The egg of the least tern *Sterna antillarum* is hardly visible even though exposed on the beach. (B) A nest of the great reed warbler *Acrocephalus arundinaceus* of Europe containing one cuckoo egg. Many cuckoos specialize in laying their eggs in the nests of one host species. Presumably through natural selection the cuckoo eggs come to resemble those of the host species. See also Fig. 20.18.

SIZE AND COLOR. Each egg is relatively large, generally from 2% to 12% of the female's body weight. In the kiwi *Apteryx* the egg is about 25% as massive as the adult. The egg of the extinct "elephant bird" of Madagascar held 8 liters, and aborigines used the shells as dishes.

The eggs of birds that nest in cavities are generally white. In species whose nests are exposed to potential predators and to the Sun, however, the shells are often exquisitely spotted and colored with pigments. The pigments are breakdown products of hemoglobin that are usually brown, olive, green, or blue. Unlike other kinds of pigments, these reflect infrared radiation from the Sun and thereby help prevent overheating. Experiments have established that the colors of many birds' eggs camouflage them from potential predators (Fig. 38.19A). "Robin's-egg blue" and other conspicuous colors are more difficult to explain. Like the colors of pearls, they might simply be incidental to some other function. On the other hand, they may enable the adult birds to detect and eject the eggs of European cuckoos, cowbirds, and other brood parasites that lay their eggs in the nests of other species. European cuckoos (*Cuculus* spp.) foil this adaptation by laying eggs that mimic those of the host species (Fig. 38.19B).

Reproductive Behavior

COURTSHIP. Many of the bird songs, colors, and behaviors that people find so interesting are even more interesting to the birds, because they are involved in finding mates. Courtship songs and displays enable individuals to select mates that are mem-

A

B

FIGURE 38.20

Two male birds, both belonging to order Passeriformes and living in Australia, attract females with unusual visual and vocal techniques. Both birds are excellent singers, often mimicking other birds and even cats, dogs, sheep, and human noises. (A) The male superb lyrebird *Menura novaehollandiae* has a fragile tail that probably hinders its movements through the eucalyptus forests. The ability to survive with such a tail demonstrates its fitness. Females prefer such tails to more practical ones. (B) The male spotted bowerbird *Chlamydera maculata* is plain looking except for a purple ruff on the back of its neck. It compensates for its drabness by constructing a bower and decorating it with bright objects, many of which it pilfers from people. Males also steal from rivals and attempt to destroy their bowers. The ability to maintain an attractive bower is therefore evidence of the male's fitness.

bers of the same species, of opposite sex, and physiologically able to reproduce. In most species the males play the more active role of displaying their plumage and singing, and the females simply signal rejection or receptivity by means of a particular posture (see Fig. 20.2). However, in phalaropes (*Steganopus, Lobipes,* and *Phalaropus;* Fig. 38.3, order Charadriiformes), it is the female that is more brightly colored and attempts to attract the male. This reversal of sex roles is associated with a reversal of the usual pattern of hormone levels. Female phalaropes secrete more androgens, and the males secrete more estrogens than in most birds.

A bird generally selects from a number of potential mates, resulting in **sexual selection** (p. 371). The mate choice is based on differences in song and plumage that are seldom noticed by humans. It was only recently that researchers discovered that the symmetry on each side of a bird is as important as its showiness. Birds with colored leg bands have been found to be more attractive to mates if the bands match on the two legs. Mate choice may be simply a matter of individual preference with no other relationship to fitness, but there is growing evidence that attractiveness to mates depends on having other traits that also contribute to fitness. Songs and attractive courtship displays may be such a handicap that the ability of a bird to survive in spite of them could indicate superior abilities to avoid parasites and predators (Fig. 38.20).

Many birds use displays similar to those used in courtship to defend **territories** from rivals during the breeding season (p. 417). Defense of the territory may persist only until mating is completed, or it may continue until the young are raised. In the latter case the territory must be large enough to provide all the resources needed by the family.

MATING SYSTEMS. In some species the male and female separate immediately after copulation, and either or both then breed again with different partners. Such **promiscuity** occurs in bowerbirds and some other species. Most species, however, are **monogamous** in the sense that they remain with each other during each breeding period, usually until the young are able to leave the nest. Even in monogamous pairs, however, one or both individuals usually engage in **extra-pair copulations.** Using DNA fingerprinting (pp. 100–101), ornithologists have determined that even in monogamous species, from 10% to 70% of the chicks in a nest were not genetically related to the male that was helping to care for them.

Monogamy is common in songbirds and other **altricial** species, in which the young are born helpless and without feathers. Experiments have shown that in altricial species the brood fails or the chicks develop slowly if the male parent is removed. Monogamy also occurs in ducks and some other **precocial** species, in which the young have feathers and can feed themselves soon after hatching. In these species the male parent plays little role as a parent, except perhaps to defend the territory. **Cooperative breeding** occurs in some monogamous birds, such as the acorn woodpecker *Melanerpes formicivorus* and the green woodhoopoe *Phoeniculus purpureus.* In these species, only one monogamous pair in a flock mates, but other members of the flock help in rearing their young. Cooperative breeding occurs mainly in poor environments that can support only a limited number of breeding pairs (p. 420).

In most monogamous species the pair remains together for only one breeding season, but in some species a pair may stay together for many years and perhaps for life. Long-term data are difficult to gather, but there is good evidence that albatrosses, petrels, and some other birds mate for life. Low "divorce rates" are also reported for swans and geese in captivity. Long-term monogamy is most common in long-lived species and has the obvious advantage of avoiding having to find a new mate each year.

Like some human cultures, some species of birds practice **polygamy,** in which each bird of a particular gender typically has more than one mate at the same time. Polygamy in which each male breeds with more than one female at the same time is called **polygyny.** Males in most polygynous species do not contribute to rearing the young except by defending territory. Polygyny is especially common in precocial birds such as grouse, pheasants, peafowl, and chickens. It also occurs most often in habitats where the distribution of resources is patchy and controlled by males. In contrast to the usual situation in which the females select which males they will mate with, in most polygynous species previous competition among males determines whether a male gets to breed. In some polygynous species, however, the males form **leks** where they compete for the attention of females (p. 419). Each successful male breeds with several females and contributes nothing to rearing the young.

Polygamy in which each female has more than one male mate is called **polyandry.** Polyandry also occurs in patchy habitats, but the resources are dominated by the females, which are usually larger than the males. Polyandry is much less common than polygyny and occurs mainly among plovers, jacanas, phalaropes, and others in order Charadriiformes. Each male incubates the eggs it has fertilized, and it rears the young with littie help from the female. Among red phalaropes (*Phalaropus fulicarius*) it is the female that has the showy plumage and establishes a lek.

NESTS. In many species, reproductive behavior includes the construction of a nest in which eggs will be deposited and incubated. Nests may be built by one or both sexes and range from a mere depression in the ground to a structure that seems far more elaborate than required (Figs. 38.19 and 38.21 on page 852). Each species appears to follow an instinctive "blueprint," constructing nests so precisely that one can identify many species from their nests alone. Even though each bird was raised in a nest like the one it builds, it is nevertheless amazing that it can construct a nest like that of its parents. After all, even though we have all grown up in houses, few of us are carpenters. The major function of a nest is to hide the eggs or make them inaccessible to predators. Nests also help protect the eggs and young from cold, rain, and exposure to the Sun. Many birds line their nests with their own down or feathers, or they use moss, bark, or other insulating material.

INCUBATION. As with most biological processes, temperature has a profound effect on the rate of development of bird embryos. Eggs of the house wren *Troglodytes aedon,*

FIGURE 38.21

(A) Among the most skillful nest builders are weavers of Africa, which weave oval or spherical nests out of grass and plant fibers. The black-headed weaver *Ploceus cucullatus* is shown here working on its nest. (B) Stages in the construction of a weaver's nest. The male adds the entrance tube only after he has attracted a mate. (C) Many weavers construct nests near each other. In social weavers *Philetairus socius,* many pairs cooperate in building one huge canopy, beneath which each pair later constructs a separate nest. (D) House martins *Delichon urbica* formerly built their mud-and-pebble nests beneath cliffs. Now they use the eaves of houses.

for example, require 13 days to hatch at the normal temperature of 35°C, but five days longer if the temperature of the eggs is 32°C. During those extra five days the eggs and parents are subject to predation and other hazards. Not surprisingly therefore, most birds incubate their eggs night and day, maintaining them at a high temperature (usually between 35°C and 40°C). In most monogamous species both parents help incubate the eggs, or one remains on the eggs in the nest while the other brings it food. Many birds lose feathers on the belly, forming one or more **brood patches** that transfer body heat more effectively to the eggs. The brood patch also allows the bird to sense the temperature of the egg. The urge to incubate eggs is overwhelming in many species. The male emperor penguin even stands in temperatures down to –60°C (–77°F) for eight to nine weeks during the dark Antarctic winter, incubating its one egg between its feet and abdomen (Fig. 38.22). It will even try to incubate a rock if it has lost its egg.

FIGURE 38.22

Male emperor penguins *Aptenodytes forsteri* gorge on fish during the Antarctic summer, then march up to 120 km to a rookery where they fast for 115 days during the winter breeding season. Afterward the male exchanges places with its mate, which has been fishing.

Parental Care. Incubation is a burden on parents, but it is nothing compared with the demands placed on them after the young hatch. In altricial species the parents have to provide not only for their own high metabolic demands, but also for those of their rapidly growing chicks. The body mass of a typical altricial chick increases approximately tenfold between the time it hatches and the time it leaves the nest a week or two later. This mass and more has to be provided in food brought by the parents. The parents also have to remove the considerable amount of feces produced by the young. Usually the feces are enclosed in a mucous coat, and the parents swallow it or carry it away in their beaks. In addition, altricial birds are generally not capable of thermoregulation for several days, so the parents must continue to provide warmth and protection from exposure.

Even after the young leave the nest, most birds continue to feed them for some time. In addition, while young birds are developing the feathers, muscles, and behavioral skills needed for flying and feeding, most adults protect them from predators (Fig. 38.23). Protection takes many forms, ranging from alarm calls to the mobbing

FIGURE 38.23

Wild turkey *Meleagris gallopavo* defending her chicks.

and physical attack of predators. Many birds, such as the ruffed grouse *Bonasa umbellus,* divert potential predators from the chicks to themselves by behaving as if they had broken wings. Such parental solicitude might be interpreted as love for the young, but other evidence suggests that these behaviors are instinctive. Many birds devote as much or more attention to **brood parasites**—offspring of different species, such as European cuckoos and cowbirds—as they do to their own offspring (see Fig. 20.18). Some gulls apparently do not recognize their own chicks if they stray from the nest by as little as a centimeter, and they will let them starve or be killed. And many of the same birds that valiantly defend a chick against a predator will do nothing to prevent the chick from being killed by a nest mate. Chicks also form a social bond to their parents, but this too appears to be an instinctive process, called **imprinting** (p. 421). Young birds can just as easily become imprinted to humans, electric trains, or to anything else that moves and makes noise during the first days of their lives.

Interactions with Humans and Other Animals

ECOLOGY. Because birds are such active and mobile consumers, they have enormous ecological impacts on a wide range of other animals. A small titmouse (*Parus* sp.) consumes an average of one insect every 2.5 seconds during a winter day, and a barn owl (*Tyto alba*) eats several mice in an average night. Even herbivorous birds affect other animals indirectly by altering the distribution of plants. Birds are major vectors for dispersing seeds that cling to their bodies or pass intact through their digestive tracts. Many tropical plants have flowers that can be pollinated only by hummingbirds or honeycreepers, which are more reliable in rainy weather than are insect pollinators. Without hummingbirds, half the plants in the Andes would become extinct within a year.

On the other hand, there are species of birds that are ecologically detrimental, especially when introduced into an alien habitat. Two familiar examples of such troublesome **exotic species** are the starling *Sturnus vulgaris* and the house sparrow *Passer domesticus.* Both were deliberately released in New York City in the 19th century, allegedly by people who thought it would be neat to have all the birds mentioned in Shakespeare. They now range throughout North America, competing with native songbirds, spreading diseases, and damaging crops. Other introduced species have been less troublesome. The ring-necked pheasant *Phasianus colchicus* is a valued game bird. The canary-winged parakeet *Brotogeris versicolorus* and the budgerigar *Melopsittacus undulatus* have become established around New York City and in Florida as escaped pets with apparently no adverse consequences.

BIRDS AND PEOPLE. The relationships between birds and humans are varied and complex (Fig. 38.24). Many early archeological artifacts reveal wonder, envy, and fascination for the colors, songs, and flight of birds. Cave paintings, totems, and hieroglyphics frequently depict birds, often in the forms of gods. In numerous cultures even today, people decorate themselves with feathers as if to borrow some of the bird's powers, or at least some of its beauty. In developed countries, where values are measured with money, our fascination with birds is calculated in the billions of dollars spent by millions of people for travel, field guides, binoculars, bird feed, nest boxes, and other ornithological paraphernalia. Birds account for approximately 12% of the billions of dollars in sales by pet stores, compared with 19% for dogs and 5% for cats. Economically, bird watching is more important than bird hunting.

FIGURE 38.24

One of the most unusual interactions of birds and humans: a Chinese fisherman and a cormorant (*Phalacrocorax*). In many parts of the Orient, such tame cormorants dive for fish that are lured by lantern to the fisherman's boat. A loop around the cormorant's neck keeps it from swallowing the catch.

The increased availability of food and nest sites provided by humans have enabled many species to extend their ranges northward. The northern mockingbird *Mimus polyglottos* and the northern cardinal *Cardinalis cardinalis* have become northern birds only in the last few decades, apparently because of bird feeders and certain cultivated plants. In many areas Canada geese *Branta canadensis* and other birds have stopped migrating south, apparently because farmlands inadvertently provide food for them throughout the winter.

Birds also provide food for people, though not in the same way. Chickens and their eggs probably contribute as much protein to the human diet as any other single species except cows. Turkey, duck, goose, pheasant, quail, and pâté de foie gras (a paste made from the livers of force-fed geese) provide rarer treats. For those who fancy the truly exotic, there is bird's-nest soup, the main ingredient of which is the dried saliva that oriental cave swiftlets (*Collocalia* spp.) use to construct their nests. Other products include goose down for clothing and comforters, as well as feathers for hat decorations.

EXTINCT AND ENDANGERED SPECIES. On the whole the impact of people on birds has been detrimental to the birds. Approximately two-thirds of the world's bird species are declining in numbers, and 10% are threatened with the fate of the passenger pigeon. Passenger pigeons were once so plentiful in America that Alfred Russel Wallace suggested in his famous evolution paper of 1858 that they were indestructible. Wallace clearly underestimated the abilities of sport and market hunters, because the last passenger pigeon died just a year after Wallace did (Fig. 38.25A,B on page 856). Most bird hunters in the United States now work as hard as anyone to protect the species on which they depend for sport. Nevertheless, their activities sometimes have unforeseeable consequences. Many ducks and other birds that feed on the bottoms of ponds ingest shot gun pellets and die from lead poisoning. Federal law now forbids lead shot in some areas, but hunters have successfully blocked efforts to extend the prohibition, because steel shot costs more, may damage gun barrels, and is thought to be less effective. Eagles and other birds that feed on wounded animals also succumb to lead poisoning. One of

A

B

FIGURE 38.25

(A) John James Audubon's painting of passenger pigeons, *Ectopistes migratorius.* Audubon once reported a flock of passenger pigeons that took three days to pass overhead. (B) Shooting passenger pigeons for sport in Iowa in 1867. Not much skill was required. Another favorite technique was to tie one pigeon to a stool and throw a net over the hundreds that would land to investigate. This is said to be the origin of the term "stool pigeon."

the victims may be the California condor, a species that had to be saved from extinction by captive breeding (Fig. 38.26).

The destruction of habitat is even more harmful to birds than shooting them. Millions of acres of wetlands are drained for agriculture, development, and mosquito control, depriving waterfowl and many other birds of the habitat they require for breeding. Countless species of birds are also endangered by destruction of tropical rainforests. Many of these birds are "our" familiar summer songbirds that spend their winters in the rainforests. Brood parasites such as cowbirds, as well as nest predators such as blue jays (*Cyanocitta cristata*), raccoons, and opossums, also reduce songbird populations. The fragmentation of forests by suburban development makes it easier for these parasites and predators to reach woodland songbird nests.

Other ways that birds die from human-altered environments include collisions with electric lines, broadcast towers, and skyscrapers in fog, as well as with windows that reflect the sky. Many birds also collide with automobiles and airplanes. The latter can be as fatal to humans as to birds. In September 1988 a pelican brought down one of the most sophisticated and expensive planes in the world, a B-1B bomber, killing its crew of four. To prevent such a tragedy at some commercial airports, a falcon is sent aloft to scare off birds before each landing and takeoff. Another hazard to birds is getting covered with some of the oil it takes to fuel all these human activities. Thousands of seabirds regularly meet this fate. In spite of the efforts of volunteers to cleanse the birds, few survive. Many birds are also still damaged by pesticides. DDT was banned in the United States in 1972, after it was linked to the decline in reproduction of eagles and other birds due to egg-shell thinning. Since the ban the populations of eagles and other affected birds have started to recover, but DDT is still used in other countries. Many other pesticides undoubtedly get passed up through the food web and concentrated in top carnivores in the same way that DDT did, but most pesticides are not tested for toxicity in humans, much less in birds.

Many species of birds are threatened by trade. More than six million live birds were imported into the United States in the 1980s, and probably two to three times that many were traded illegally or died during capture and shipment. This number includes a large number of parrots sold on the streets of Mexico and smuggled into the United States in

FIGURE 38.26

By 1987 so many California condors *Gymnogyps californianus* had died of lead poisoning from eating wounded deer that all 27 survivors were captured for breeding. This controversial effort succeeded well enough that some were returned to the wild in 1992.

A

C

B

FIGURE 38.27

Back from the brink of extinction. (A) Male and female whooping cranes *Grus americana.* (B) The bald eagle *Haliaeetus leucocephalus,* which was virtually extinct in the lower 48 states. Its populations have recovered so well that in 1993 the species was moved from the federal list of endangered species to the list of threatened species. (C) The peregrine falcon *Falco peregrinus* feeding on a pheasant. The peregrine falcon was removed from the federal list of endangered species in 1994.

tourists' luggage. Such trade, combined with deforestation, endangers approximately a third of the 140 parrot species found in the Western Hemisphere (Beissinger and Bucher 1992). Unlike legally imported birds, which must be quarantined before sale, some illegally sold birds have contagious diseases that can wipe out a chicken or turkey farm. Some also carry psittacosis, which can be fatal to native birds and to humans. The smuggling of some birds out of the United States is also a problem. Especially at risk are falcons, which are shipped to the Middle East and trained to hunt for sport.

CONSERVATION. Many people have become sufficiently alarmed to demand more conservation laws and tougher punishment for violating them. In 1992 the United States Congress banned the importation of ten endangered species, including some lovebirds, cockatoos, and parakeets, and it restricted the importation of some other birds. Education in the "conservation ethic" has also been effective. One now seldom hears of hunters shooting hundreds of hawks for target practice, or killing eagles in the belief that they compete for game or kill livestock. Populations of some endangered species are now recovering partly because of such efforts, but also partly because of artificial rearing (Fig. 38.27). The whooping crane, which was down to 21 individuals in 1941, now numbers nearly 300, thanks largely to a program of artificial insemination and using the related sand-hill crane (*Grus canadensis*) as foster parents. The peregrine falcon has also made a comeback thanks to artificial rearing (**hacking**). The technique, pioneered by Tom Cade and the Cornell University Laboratory of Ornithology, has led to the reintroduction of falcons into their native habitats and also into cities, where skyscrapers and bridges make fair substitutes for the cliffs preferred as nest sites. The falcons return the favor by controlling the populations of pigeons, starlings, and house sparrows. Hacking and reintroduction are now being used to restore other species of hawks, as well as eagles, owls, and other birds.

Summary

Archaeopteryx and other birds are thought by most paleontologists to have shared ancestry with theropod dinosaurs. Feathers enabled *Archaeopteryx* and other birds to maintain high body temperatures, and they apparently led to the other differences now seen between extant birds and reptiles. Feathers develop from papillae that are homologous with reptilian scales. There are several kinds of feathers. The most familiar is the vane feather which has a shaft with barbs growing outward on two sides. The barbs have hooked barbules that interlock and form a continuous surface. These feathers form the area of wings used in flying, and they insulate the body thermally. They generally are colored in specific patterns that function in camouflage or courtship and territorial defense.

During flight, air flows more rapidly over the top of wings than beneath them, generating the lift that keeps the bird aloft. Thrust results from the flow of air past the rear of the wings. The angle of the wings can be changed to increase lift, but the primary feathers and alulae must then open to prevent stalling. During flapping flight, the wings are depressed and raised by strong muscles attached to the sternal keel. Wings can have a variety of shapes depending on whether they are specialized for gliding, speed, or maneuvering. The body mass is minimized by means of air sacs and hollow bones. The air sacs pump air completely through the parabronchi of the lungs, efficiently exchanging respiratory gases, as required by the high metabolic requirements of thermoregulation and flight. Birds must also eat and digest prodigious amounts of food. The beaks of birds are generally specialized for particular foods.

All birds are oviparous and produce cleidoic eggs. Fertilization is internal and usually follows an elaborate courtship behavior involving singing and displays of plumage. In most monogamous species both males and females incubate eggs and care for young, especially if they are altricial. Most promiscuous, polygynous, or polyandrous species tend to be altricial, and only one parent incubates the egg and cares for young. Birds are extremely important ecologically, and many are threatened by human activities, either directly or by destruction of habitat.

Key Terms

ratite
carinate
sternal keel
vane feather
shaft
barb
rachis
barbule
lift
wing loading
thrust
alula
parabronchi
air sac
syrinx
promiscuity
monogamy
altricial
precocial
polygamy
polygyny
lek
polyandry
brood patch

Chapter Test

1. How do ratites differ from carinates? Give the common names for two kinds of ratites and several carinates. (p. 830)
2. Sketch a vane feather, labeling the quill, rachis, and barbs. Describe the structure and function of barbules. (pp. 834–836)
3. Describe three adaptations of carinates that enable them to fly. (pp. 838–841)
4. Explain how wings produce lift and thrust during flight. (p. 840)
5. Explain how alulae and wing slotting are used in slow flight. (p. 840)
6. Describe how the wings of different species of birds are specialized for gliding, speed, or maneuverability. (pp. 840–841)
7. Sketch four different kinds of bird feet specialized for such functions as perching, climbing, swimming, grasping prey, and walking. (Fig. 38.12)
8. Explain why the metabolic rate of a bird is so much higher than that of a reptile of the same size. (p. 841)
9. Briefly explain how birds sing. Why do they sing? (pp. 843, 850)
10. Bird behaviors depend on complex interactions of instinct and learning. Describe an example of a bird behavior that appears to be mainly instinctive, one that appears to be

mainly learned, and one that appears to be instinctive but modified by learning. (pp. 845–854)

11. Describe the production and development of a bird egg. Describe two special adaptations of the shell. (pp. 848–849)
12. Describe three mating systems in birds: monogamy, polygyny, and polyandry. Give the common name of a bird with each mating system. What is the relationship between each mating system and the degree to which the young are developed at hatching? (pp. 850–851)

Answers to the Figure Questions

38.1 Ostrom originally believed that the dinosaur ancestors of birds ran along the ground catching insects with their forelimbs, which subsequently evolved into wings used as stabilizers during running, and later into wings for flight.

38.5 The hairlike feathers of the kiwi appear to be filoplumes.

38.6 Woodcock remain perfectly still even when a potential predator is only centimeters away.

38.7 These three wings are themselves analogous in the sense that none of the ancestors shared by pterosaurs, birds, and bats had wings. On the other hand (or other wing), the bones do show homology, since all the shared ancestors had similar bones.

38.11 The relatively small wing and large legs suggest that the chicken is not a good flyer, yet the deep sternal keel indicates strong pectoral muscles. This trait is probably due to artificial selection for frying rather than flying.

38.13 Lung, abdominal air sac, then lung again, then anterior thoracic sac. During the first passage through the lungs the air goes through mesobronchi; during the second passage they pass through parabronchi, where oxygen is exchanged.

38.16 Primarily granivorous passeriforms would have robust beaks: buntings, finches, sparrows. Primarily insectivorous passeriforms would have more pointed beaks: swallows, wrens, thrashers, nuthatches, vireos, warblers. Many passeriforms, such as thrushes and crows, are omnivorous and have beaks of intermediate shape.

Readings

Recommended Readings

Beehler, B. M. 1989. The birds of paradise. *Sci. Am.* 261(6):116–123 (Dec).

Borgia, G. 1986. Sexual selection in bowerbirds. *Sci. Am.* 254(6):92–100 (June).

Calder, W. A., III. 1978. The kiwi. *Sci. Am.* 239(1):132–142 (July).

Davies, N. B., and M. Brooke. 1991. Coevolution of the cuckoo and its hosts. *Sci. Am.* 264(1):92–98 (Jan).

Ehrlich, P. R., D. S. Dobkin, and D. Wheye. 1988. *The Birder's Handbook.* New York: Simon & Schuster/Fireside. (*Concise source of information on United States birds.*)

Feduccia, A. 1980. *The Age of Birds.* Cambridge, MA: Harvard University Press. (*A beautifully illustrated guide to the evolution of birds.*)

Grajal, A., and S. D. Strahl. 1991. A bird with the guts to eat leaves. *Nat. Hist.* 100:48–55 (Aug). (*The hoatzin.*)

Heinrich, B. 1986. Why is a robin's egg blue? *Audubon* 88(4):64–71 (July).

Knudsen, E. I. 1981. The hearing of the barn owl. *Sci. Am.* 245(6):112–125 (Dec).

Ligon, J. D., and S. H. Ligon. 1982. The cooperative breeding behavior of the green woodhoopoe. *Sci. Am.* 247(1):126–134 (July).

Marshall, L. G. 1994. The terror birds of South America. *Sci. Am.* 270(2):90–95 (Feb).

Mock, D. W., H. Drummond, and C. H. Stinson. 1990. Avian siblicide. *Am. Sci.* 78:438–449.

Myers, J. P., et al. 1987. Conservation strategies for migratory species. *Am. Sci.* 75:18–26.

Nottebohm, F. 1989. From bird song to neurogenesis. *Sci. Am.* 260(2):74–79 (Feb). (*The brains of canaries develop new nerve cells as they learn new songs.*)

Proctor, N. S., and P. J. Lynch. 1993. *Manual of Ornithology.* New Haven CT: Yale University Press.

Rahn, H., A. Ar, and C. V. Paganelli. 1979. How bird eggs breathe. *Sci. Am.* 240(2):46–55 (Feb).

Seymour, R. S. 1991. The brush turkey. *Sci. Am.* 265(6):108–114 (Dec).

Shettleworth, S. J. 1983. Memory in food-hoarding birds. *Sci. Am.* 248(3):102–110 (Mar).

Simons, T., et al. 1988. Restoring the bald eagle. *Am. Sci.* 76:253–260.

Smith, P., and C. Daniel. 1982. *The Chicken Book.* San Francisco: North Point Press. (*Everything about the chicken.*)

Stacey, P. B., and W. D. Koenig. 1984. Cooperative breeding in the acorn woodpecker. *Sci. Am.* 251(2):114–121 (Aug).

Taborsky, M., and B. Taborsky. 1993. The kiwi's parental burden. *Nat. Hist.* 102:50–57 (Dec).

Terborgh, J. 1992. Why American songbirds are vanishing. *Sci. Am.* 266(5):98–104 (May).

Waldvogel, J. A. 1990. The bird's eye view. *Am. Sci.* 78:342–353.

Wellnhofer, P. 1990. Archaeopteryx. *Sci. Am.* 262(5):70–77 (May).

Wiley, R. H., Jr. 1978. The lek mating system of the sage grouse. *Sci. Am.* 238(5):114–125 (May).

Zimmer, C. 1992. Ruffled feathers. *Discover* pp. 45–54 (May). (*Controversy surrounding the claim that* Protoavis *was a bird predating the dinosaurs.*)

See also any of the numerous field guides to birds and relevant selections in General References at the end of Chapter 21.

Additional References

Beissinger, S. R., and E. H. Bucher. 1992. Can parrots be conserved through sustainable harvesting? *BioScience* 42:164–173.

Dumbacher, J. P., et al. 1992. Homobatrachotoxin in the genus *Pitohui:* chemical defense in birds? *Science* 258:799–801.

Feduccia, A. 1993. Evidence from claw geometry indicating arboreal habits of *Archaeopteryx. Science* 259:790–793.

Feduccia, A. 1995. Explosive evolution of Tertiary birds and mammals. *Science* 267:637–638.

Schultze, H-P., and L. Trueb. 1991. *Origin of the Higher Groups of Tetrapods.* Ithaca, NY: Comstock Publishing Associates. (*See chapters by Witmer, Ostrom, Martin, and Tarsitano.*)

Sibley, C. G., and J. E. Ahlquist. 1991. *Phylogeny and Classification of Birds: A Study in Molecular Evolution.* New Haven, CT: Yale University Press.

Speakman, J. R., and S. C. Thomson. 1994. Flight capabilities of *Archaeopteryx. Nature* 370:514.

CHAPTER

39

Mammals

Chimpanzee *Pan troglodytes.*

Major Features of Mammals

Any vertebrate with hair is immediately recognizable as a mammal, and if there is any doubt, closer inspection will confirm the classification by revealing the presence of mammary glands. Hair is one of many adaptations for **homeothermy**—the maintenance of relatively constant body temperature—which, in turn, accounts for many other crucial differences between mammals and most other animals. Like birds, mammals expend approximately 10 times as much energy as reptiles of the same body size, and they therefore require about 10 times as much food. Many mammals can obtain the large amounts of food required by homeothermy only because homeothermy enables them to feed at times and in places where low environmental temperatures eliminate poikilotherms from competition. In temperate zones, mammals rule the night, and they share winters only with some birds. Moreover, homeothermy enables mammals to exploit wide ranges of habitat. The mice that live in the tropics are fundamentally like those of the Arctic, and temperature is no barrier to the wanderings of whales from the poles to the equator.

The second major distinction of Mammalia, and the one for which the class was named, are the mammary glands (pp. 318–319). All mammals have them (even males, though lactation occurs only rarely in men and some other male mammals, and regularly only in a species of fruit bat). The presence of mammary glands may also be related to homeothermy. Many zoologists believe they evolved from sweat glands.

Origins

SYNAPSIDS. Like birds, mammals shared ancestry with reptiles and gained ascendancy over them during the present Cenozoic era—the era we presume to call the Age of Mammals. Mammals originated before birds, during the Triassic period approximately 200 million years ago. Our ancestors belonged to the group called synapsids (see Fig. 37.2). The earliest synapsids were **pelycosaurs,** such as *Dimetrodon* and *Edaphosaurus* (numbers 1 and 2 in Fig. 37.9). These finback reptiles may have used their fins to help regulate body temperature by absorbing solar radiation. They were replaced by another group of synapsids called **therapsids.** Most therapsids, such as *Cynognathus* (number 3 in Fig. 37.9), fled the evolutionary scene almost as soon as the dinosaurs made their appearance, but their close relatives the mammals were eventually more successful.

Although the fossil record yields no direct link between them, there is little doubt that mammals shared their ancestry with therapsids (Fig. 39.1 on page 862). Like dinosaurs and mammals, therapsids had legs positioned beneath the body rather than sprawling, as in other reptiles. This posture provides greater support for large body weight. Most therapsids were only as large as domestic dogs, however, so this added support makes sense only if they ran for long periods at a time. Paleontologists infer, therefore, that unlike reptiles, which can run only briefly before becoming exhausted (p. 817), therapsids could sustain prolonged running, like mammals. Supporting this deduction is the fact that the teeth and jaws of therapsids were better adapted to process large amounts of food, which would have provided both the means and the necessity for sustained locomotion. Instead of the uniform, peg-shaped teeth of most reptiles, which are useful only for gripping prey or tearing off leaves before swallowing them whole, therapsids had cheek teeth adapted for chewing. Chewing breaks up food, allowing digestive enzymes to process nutrients faster. The lower jaws of therapsids were also stronger, consisting of one strong bone with a simple joint, rather than several bones attached by a complex joint.

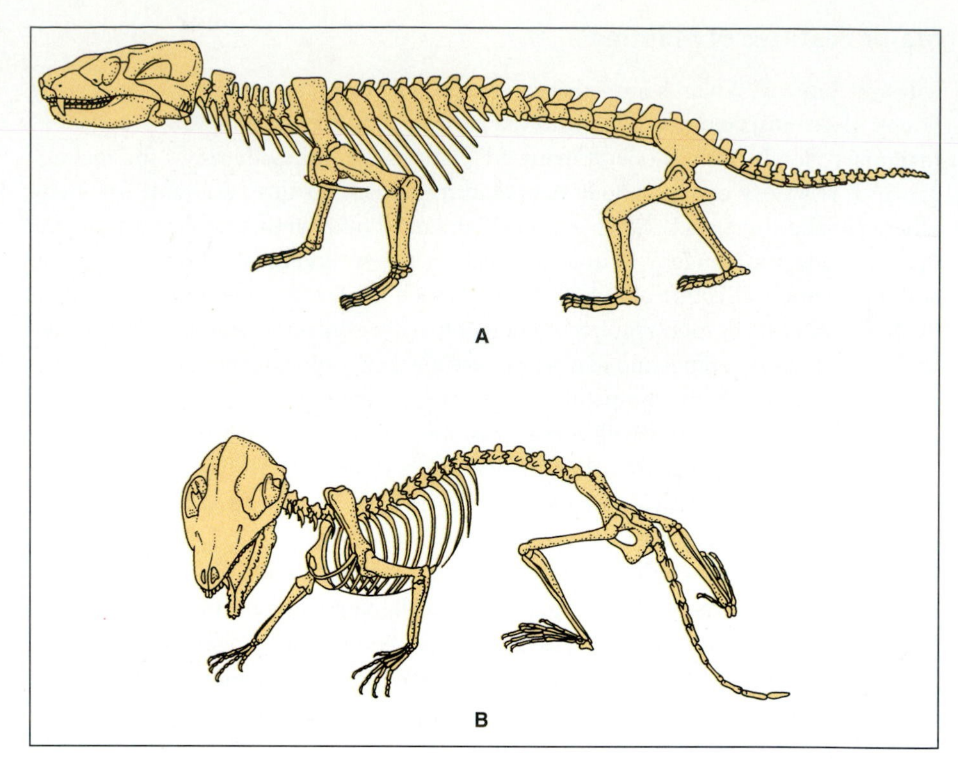

FIGURE 39.1

Comparisons of the skeletons of a therapsid and an early mammal. (A) The skeleton of *Thrinaxodon,* a therapsid reptile of the cynodont suborder that is closest to mammals. *Thrinaxodon* fossils are approximately a half meter long and found in Lower Triassic deposits of South Africa and Antarctica. (B) Skeleton reconstructed from bones of two early mammals (mainly *Megazostrodon*) that lived at the same time as *Thrinaxodon,* some 200 million years ago. Note the large size of the skull relative to the rest of the body, indicating a relatively large brain that is characteristic of mammals (Fig. 9.7). This animal would have been approximately 10 cm long, not counting its tail. A primitive mammal with these teeth would have been carnivorous, probably feeding on insects. It probably resembled a mole or shrew in appearance and behavior (order Insectivora).

EARLY MAMMALS. Because hair, mammary glands, and other definitive features of mammals do not often fossilize, paleontologists rely on differences in bones, especially those of the jaws and middle ears. In mammals, two pairs of jaw bones have become bones of the middle ear. As will be described later, the loss of these bones from the jaw provides more strength, enabling the mammal to chew food rather than bolt it down whole. The larger brains of these early mammals also suggest higher levels of behavioral activity. Their skeletons and teeth are similar to those of modern moles and shrews, suggesting that early mammals lived much like moles and shrews, feeding voraciously on insects, probably at night when they were safe from reptilian predators.

Diversity

MONOTREMES. Class Mammalia now comprises more than 4600 named species, and approximately 16 new species are discovered each year. Taxonomists generally divide mammals into two subclasses: Prototheria and Theria. The subclass Prototheria comprises one order, Monotremata, and includes the egg-laying mammals, called monotremes. Monotremes originated more than 130 million years ago and perhaps provide a glimpse of what early mammals were like. Only three species survive, all restricted to the Australian faunal region (see Fig. 19.9), having apparently been isolated there some 50 million years ago by the breakup of Gondwana (pp. 390–392). Two species, *Zaglossus bruijni* and *Tachyglossus aculeatus,* are echidnas, also called spiny anteaters (Fig. 39.2A). The most famous monotreme is the duck-bill or platypus, *Ornithorhynchus anatinus* (Fig. 39.2B).

At first glance, echidnas and the platypus seem unlikely relatives. Echidnas are terrestrial and burrowing animals, while the platypus is largely aquatic, living in the banks

A

B

FIGURE 39.2

Monotremes. (A) The short-beaked echidna *Tachyglossus aculeatus.* This species has electroreceptors in its snout with which it presumably detects ants, termites, and worms on which it feeds with its long, sticky tongue (Andres et al. 1991). (B) A mother platypus with young in their burrow. A platypus swims with its beaverlike tail and otterlike webbed feet, feeding on bottom-dwelling invertebrates with its ducklike bill. The first specimen of a platypus sent to Europe was at first dismissed as a hoax concocted by sewing together the parts of different animals. While under water the platypus keeps its eyes, nostrils, and ears closed. It rapidly scans the bottom with its bill, which contains touch receptors and electroreceptors. The latter detect the action potentials generated by the muscles of prey. The sexes are similar except that the body length is approximately 30 cm for females and 45 cm for males.

of streams in which it feeds. The horny beak, or **rostrum,** of the echidna is tubular and adapted for feeding on ants and termites. The rostrum of the duck-bill is adapted for feeding on crustaceans, insect larvae, and other aquatic invertebrates. The hair of echidnas is also quite different from that of the platypus. Many hairs of spiny anteaters are modified like those of porcupines as defensive weapons, while the hair of the platypus is so luxuriant that the species was nearly hunted to extinction for its fur. In spite of these differences, however, several similarities justify classifying the three species in the same order. In both echidnas and the platypus the males have spurs on their hind legs that they use during aggressive encounters with other males. In the platypus the spur injects a venom that is variously described as being "mildly toxic" or as being capable of killing a dog and causing excruciating pain in a human. Both the platypus and echidnas also have electroreceptors in their snouts with which they detect prey. Monotremes also are unique among mammals in having a common outlet, the **cloaca,** for feces, urine, and eggs. (Hence the name monotreme, from the Greek for "one hole.") Most important, both echidnas and the platypus share the distinction of being the only living egg-laying mammals, having apparently retained that mode of reproduction from the earliest mammals. Oviparity in monotremes will be described in greater detail later.

MARSUPIALS. The second subclass of mammals is Theria, which includes marsupials and placental mammals. In marsupials (metatherians), most of the development of the embryo occurs outside the uterus, usually in a pouch (**marsupium**) on the mother's belly. There are now approximately 270 species of marsupials, but they were once far more numerous (Fig. 39.3 on page 864). Most extant marsupials, including koalas, kangaroos, bandicoots, and wombats, are unique to Australia, where they are thought to have been generally free from competition with placental mammals.

Many marsupials of Australia have become adapted to habitats that would otherwise be occupied by placental mammals, and some are remarkably like them. Such in-

FIGURE 39.3

Marsupials. (A) A female opossum *Didelphis virginiana* with young. The opossum is the only extant marsupial in the United States. Following a tradition that can be traced back to an 18th-century artist, baby opossums are often portrayed in neat rows with their tails wrapped around their mother's. As shown here, however, they are much less organized than that. Opossums thrive in spite of being hunted for fur, sport, and food. (B) The Tasmanian devil *Sarcophilus harrisii* has a reputation for viciousness, based largely on its behavior when confined in poor conditions. In the wild, Tasmanian devils prey along streams for a variety of small vertebrates and invertebrates, and their powerful jaws enable them to take an occasional sheep. Destruction of "the bush" eliminated them from eastern Australia, but they still live on the island of Tasmania. (C) The koala *Phascolarctos cinereus* feeding on its only food, eucalyptus leaves. Because the eucalyptus leaves are toxic to other animals, the koala has few competitors. Its malodorous flesh also renders it safe from most predators. The species was once endangered by hunting for amusement and fur, and it is now threatened by destruction of eucalyptus forests by developers.

dependent evolution of similar adaptations is referred to as **convergent evolution.** Kangaroos, for example, live on semiarid grasslands like those inhabited by antelope and buffalo elsewhere in the world, and like these ruminants, kangaroos have multichambered stomachs (pp. 262–263). Many marsupials resemble their placental counterparts externally as well (Fig. 39.4).

Placental Mammals. Placental mammals, also called eutherians, differ from marsupials mainly in that placental support in the uterus is prolonged. The oldest clearly identified placentals lived around 70 million years ago in what is now Mongolia, and they expanded their ranges to virtually everywhere except Antarctica and remote islands. During the Cenozoic era the diversity of placental mammals mushroomed (see Fig. 34.17). They may have eliminated many of the marsupials, many other nonplacental mammals, and numerous other kinds of animals. The diversity of placental mammals has shrunk somewhat in the last few thousand years, with the loss of some of the most spectacular species (see Fig. 19.10). There are now more than 4000 known species of eutherians, usually divided among 19 orders. Representatives of all the orders of eutherians are shown in Fig. 39.5 on pages 866–868. The structure and functioning of most mammals are similar to those of humans, as reviewed in Unit 2. (See the index for particular subjects.) Much of this chapter focuses on the adaptations of mammals that set them apart from their reptilian ancestors. Finally we look more closely at our own species and our place among the other animals.

FIGURE 39.4

Evolutionary convergences of marsupial and placental mammals.

■ **Question:** What kinds of characters would *not* be useful in classifying these animals?

FIGURE 39.5

Representatives of the orders of Eutheria. The total comes to 4629 (not counting the three species of monotremes and 272 species of marsupials). Numbers of species are according to Wilson and Reeder 1993.

■ **Question:** Think of the last 10 mammals you have seen. To which orders do they belong?

Figure continues

Order Edentata (= Xenarthra) 29 species
Anteaters
Armadillos
Sloths
Order Tubulidentata 1 species
Aardvark
Order Pholidota 7 species
Pangolins
(= scaly anteaters)
Order Cetacea 78 species
Whales
Porpoises
Order Rodentia 2021 species
Mice
Rats
Voles
Squirrels
Chipmunks
(Winter)
(Summer)
Lemmings
Marmots
Prairie dogs
Order Lagomorpha 80 species
Rabbits
Hares

Major Extant Groups of Class Mammalia

Genera mentioned elsewhere in this chapter are noted.

Subclass Prototheria PRO-toe-THEER-ee-uh (Greek *proto-* first + *ther* wild beast). Egg-laying mammals (monotremes). Adults lack teeth. Reproductive tract opens into cloaca. Duck-billed platypus and echidnas. *Ornithorhynchus, Tachyglossus, Zaglossus* (Fig. 39.2).

Subclass Theria THEER-ee-uh. Exit of reproductive tract separate from that of alimentary tract (no cloaca).

Infraclass Metatheria MAY-tah-THEER-ee-uh (Greek *meta-* after). After an abbreviated gestation, young complete development outside the uterus, usually in a marsupium. Opossums, marsupial mice, Tasmanian devil, numbat, bandicoots, possums, wombats, kangaroos, wallabies. Marsupials. *Didelphis, Macropus, Phascolarctos, Sarcophilus* (Chapter 5 opening photograph; Figs. 39.3, 39.4, 39.6).

Infraclass Eutheria yew-THEER-ee-uh (Greek *eu-* true). Placental mammals. *Alces, Alopex, Balaena, Balaenoptera, Bubalus, Canis, Castor, Cervus, Dasypus, Diceros, Equus, Eschrichtius, Felis, Giraffa, Heterocephalus, Lepus, Loxodonta, Manis, Monodon, Mustela, Odobenus, Orcinus, Ovis, Pan, Phoca, Rangifer, Rhinolophus, Trachops, Tursiops, Ursus* (Opening photographs for Chapters 3, 4, 10, 13, 14, 15, 17, and 19, this chapter, and Unit 2; Figs. 1.8, 20.6, 20.14, 39.5, 39.7, 39.8, 39.9, 39.10, 39.11, 39.14, 39.15, 39.16, 39.17, 39.19, 39.20, 39.21, 39.22, 39.23).

Reproduction

OVIPARITY IN MONOTREMES. The major groups of mammals differ primarily in their modes of reproduction. Monotremes are thought to have diverged quite early, and it is possible that their reproduction is similar to that of therapsids and the earliest mammals. The female platypus produces from one to three eggs in a season, and these are fertilized by one male following an elaborate courtship and copulation. Fertilization occurs in the left oviduct. (The right ovary and oviduct are nonfunctional, as in birds, coincidentally.) As the fertilized eggs pass down the oviduct they acquire the first proteinaceous layer of shell. The eggs then arrive in an enlarged area of the oviduct (the uterus), where a second layer of shell is applied. While in the uterus the eggs grow within the soft shell, nourished by glandular secretions from the mother. Finally the eggs acquire their third and last soft shell layer before entering the cloaca. Each ripe egg is oval, measuring approximately 14 mm across the middle and 17 mm from end to end.

Around 20 days after mating, the female retires to her burrow to lay her eggs. No one has observed oviposition, but it is supposed that the female bends so that the eggs stick to her abdomen as soon as they leave her cloaca. She then holds the eggs against her abdomen with her broad tail, incubating them for perhaps 11 days with her body heat. The young hatch out by tearing the rubbery shell with an egg tooth similar to that in birds and many reptiles. The female continues to hold the young to her abdomen as they suckle milk from the nipple (Fig. 39.2B). Weaning occurs some four months after hatching.

Oviparity in the platypus, and presumably in early mammals, has several disadvantages. The female has to invest a large supply of protein from her own body to form large, yolky eggs. The necessity of incubating the eggs hinders her feeding, even though she must soon sacrifice large amounts of fat and proteins in her milk. The incubation period is necessarily brief, with the result that young are generally **altricial** (not well enough developed to feed or defend themselves). Because the young are born so early developmentally, the mother must invest even more time (four months in the platypus) before they can be weaned.

VIVIPARITY IN MARSUPIALS AND PLACENTALS. In contrast to oviparous mammals, viviparous mothers can avoid predators and continue to feed themselves through almost the entire gestation period, which can therefore be prolonged. Viviparity requires many physiological adjustments, however, to redirect the physiology of the mother from fertilizing the eggs toward sustaining the embryo. The adaptations for viviparity in marsupials differ markedly from those of placental mammals (pp. 316, 318). Unlike the embryos of placental mammals, those of marsupials do not implant into the uterine lining, and they do not form a close and lasting placental connection with the mother's blood. Instead, the embryos of most marsupials spend a brief time in a depression of the uterine lining, and the placenta provides limited nutrition by diffusion from the mother's tissues. The brevity and weakness of the placental connection may be necessary to keep the embryo from triggering an immune reaction against it by the mother. Otherwise, the immune system would have to be inhibited as it is in placental mothers.

Because the placenta is in place so briefly in marsupials, there is little opportunity or need for the embryo to secrete chorionic gonadotropin or any other signal to redirect the mother's reproductive system from fertilization to pregnancy (pp. 316, 318). In the Virginia opossum *Didelphis virginiana,* and perhaps in some other marsupials, the mother's estrous cycle continues just as it would have if conception had not occurred. Its intrauterine gestation period lasts only 12.5 days and coincides with the luteal phase of the estrus cycle (pp. 315–316). In large kangaroos the gestation period lasts 33 days—two days less than the estrous cycle—and the mother generally conceives again shortly after giving birth.

Because the gestation period of marsupials is so brief, the neonates are still essentially embryos, comparable in developmental stage with that of a human embryo only a few weeks old (Fig. 39.6A). Development must therefore continue outside the uterus, yet in a protected place where nourishment can be provided. In many marsupials this place is a pouch in the skin of the abdomen, which contains nipples. In a sense the marsupium serves the same function as a uterus, and the teat serves the function of a placenta. The neonate, helpless as it is, must make its own way into the marsupium without any help from the mother.

In the case of the red kangaroo, it is just as well that the mother does not offer to help, since she weighs 30,000 times more than the 1-gram neonate. The forelimbs of the neonate marsupial are more developed than the rest of its body, and on leaving the birth canal it struggles hand-over-hand through the hair on the mother's belly. The neonate is blind, but odor, gravity, or perhaps some unknown cue guides it to the marsupium.

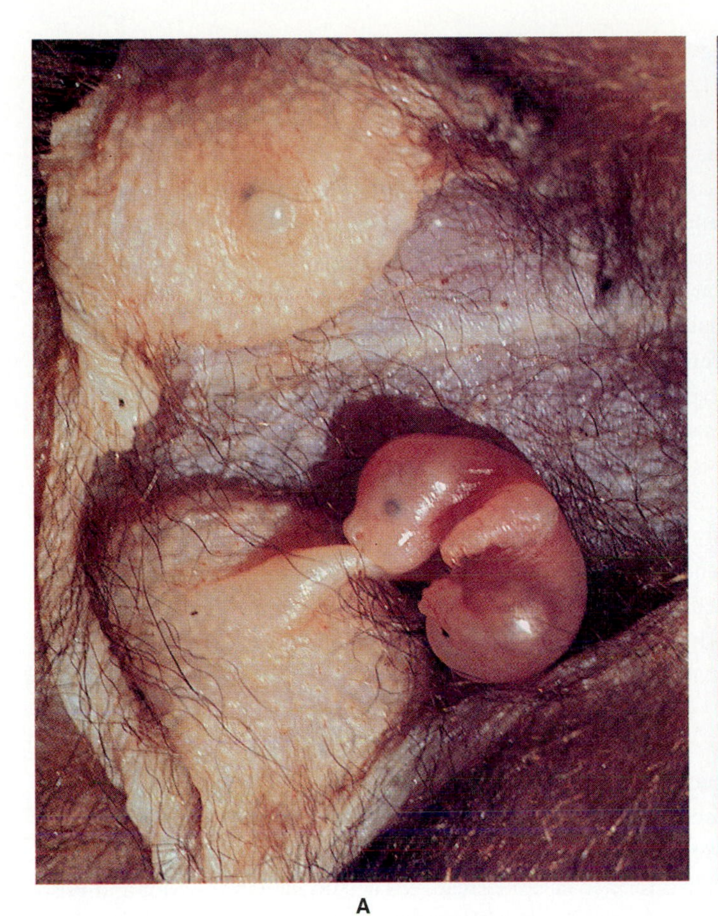
A

B

FIGURE 39.6

(A) A neonate marsupial attached to a nipple in the mother's pouch. Its developmental stage is comparable to that of a human embryo only six or seven weeks old (Fig. 6.23B). (B) Mother red kangaroo *Macropus rufus* (= *Megaleia rufa*) with joey in her pouch (more or less).

Once inside, the neonate finds a nipple and begins suckling. The nipple swells in the neonate's mouth, preventing it from becoming dislodged while the mother hops or runs about. Development then continues in the pouch. The young red kangaroo opens its eyes around 144 days after entering the pouch, and a few days later it pokes its head out for the first time. After approximately 190 days in the marsupium the young kangaroo, now called a joey, leaves the pouch for the first time. The joey will return periodically for another 50 days for protection and to suckle from the same teat (Fig. 39.6B). It will remain near the mother for several months after weaning, until it is almost a year old.

While the young kangaroo has been developing in the marsupium, the mother may well have conceived again. This second embryo remains in the uterus in a dormant stage called **embryonic diapause** as long as there is already one offspring in the pouch. After about 200 days in embryonic diapause, however, the mother has evicted the joey from her pouch, and the second embryo resumes development. Thirty days later this neonate climbs into the marsupium just as the first one did, and it attaches to a second teat. The joey can still nurse from its own teat, however. Curiously the two teats produce different kinds of milk. The neonate gets a high-protein, low-fat milk, while the joey gets a high-fat, low-protein milk. Unlike most mammals, in which lactation blocks the resumption of the estrous cycle (p. 318), the mother kangaroo might well get pregnant again while nursing both a neonate and a joey. In this way the red kangaroo can produce a steady supply of young at approximately 18-month intervals.

Reproductive Behavior. All female mammals can produce milk, and this fact necessarily implies parental behavior, at least on the part of the mother. In the kangaroo, dolphins, chimpanzees, and many other mammals, the males make no parental investment, and they generally have little to do with their offspring. These animals are often **promiscuous**: The males mate with any receptive females they encounter. A few species of mammals, such as wolves and other canids, are **monogamous.** The male mates with only one female at a time and may make a significant parental investment. **Polygyny** is the most common mating system of mammals. It occurs among lions, gorillas, sea lions, deer, and many others. The common arrangement with polygyny is for a group occupying a territory to be dominated by one or several males that have exclusive mating privileges with all the receptive females in the group. Often, as in deer, the males compete in ritual or actual combat that establishes a **dominance hierarchy** for mating and other preferences.

Polygyny is one type of polygamy. The other type, **polyandry,** in which a female has several male mates simultaneously, is rare among mammals. One species in which polyandry does occur is the naked mole rat *Heterocephalus glaber,* a burrowing rodent of eastern Africa (Fig. 39.7). Jennifer Jarvis of the University of Cape Town in South Africa discovered that the social organization of naked mole rats shows remarkable evolutionary convergence with termites (p. 695). A typical colony of approximately 75 mole rats has one breeding female, one to three reproductive males, and two or three castes that differ in body size and the degree to which they help dig burrows.

Hair

FUNCTIONS. The next several sections describe some special adaptations of mammals, many of which are associated with the skin. One of these is hair. Many insects and some other kinds of animals produce hairlike body coverings (pile), but only mammals produce real hair. A hair is formed as epidermal cells accumulate in the root and push older cells out from the follicle (see Fig. 10.1B). Like the scales of reptiles and the feathers of birds, hair therefore consists of dead, keratinized epidermal cells. Hair is not homologous with scales or feathers, but appears to have evolved independently, perhaps in the synapsids. The major function of hair is to reduce the conduction of body heat to the environment. The resistance to heat conduction is increased by contraction of **arrector pili** muscles, which lever the hair outward in a process called **piloerection.** Piloerection thickens the fur coat (**pelage**), usually in response to cold. In many mammals, such as canids and apes, piloerection also occurs during aggressive encounters and has the effect of making the animal appear larger (see Fig. 20.5). Most humans don't have enough body hair either to conserve heat or to put on an imposing display, but the arrector pili muscles still respond under similar circumstances, producing "goose bumps."

Hair also serves other functions. Hair holds and concentrates pheromones produced by the **apocrine sweat glands,** which are used in territorial marking, courtship, and other social communication. (Most cooling is due to eccrine sweat glands.) In adult humans the apocrine sweat glands are concentrated around the groin and armpits, which may explain our peculiar distribution of body hair. Often sensory neurons are attached to the hair root, allowing the hair to function as a touch receptor. **Vibrissae** ("whiskers") are especially adapted as touch receptors. Some mammals, such as spiny anteaters, hedgehogs, and porcupines, have thick, sharp hairs (**quills**) that deter predators. Contrary to popular myth, a porcupine cannot fling its quills, but it will slap at a predator or a nosy dog with its spiny tail. The quills easily detach from the skin of the porcupine, but tiny hooks at the other ends make them difficult to remove from the skin of the victim.

COLOR. As hairs grow they generally acquire black, brown, or yellowish-red pigments called **melanins** from epidermal cells called melanocytes (pp. 193–194). Many

A

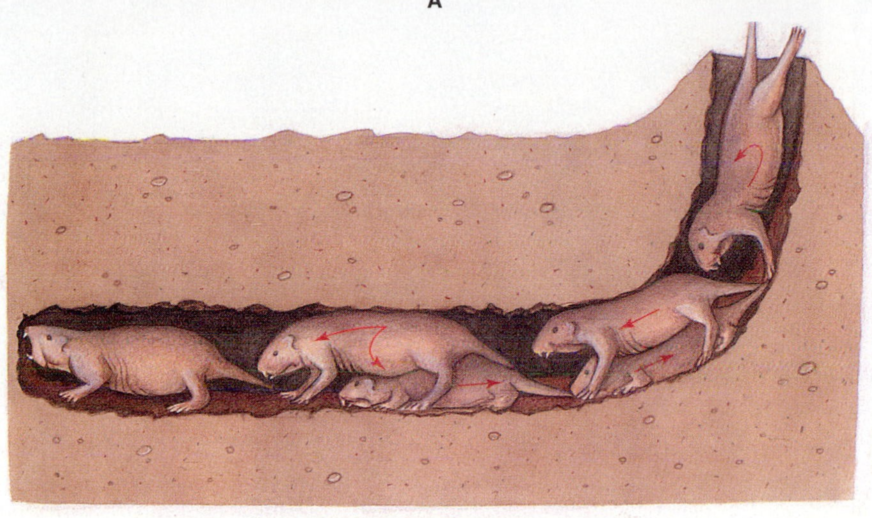

B

FIGURE 39.7

(A) A naked mole rat *Heterocephalus glaber,* one of 12 species of African mole rats (family Bathyergidae). Frequent workers use the incisors for gathering roots and for burrowing tunnel systems several kilometers long. The lips close behind the incisors to keep dirt out of the mouth, and a flap of skin protects the nostrils. Being hairless, they regulate body temperature by basking and huddling. Length approximately 5 cm. (B) A digging chain of frequent workers. The digger kicks the excavated soil to the second mole rat on the bottom, which usually pushes it backwards as it backs out beneath other mole rats to the mouth of the tunnel. This mole rat, descriptively called a "volcanoer," then kicks out the dirt. The volcanoer then crawls over the backs of other mole rats, eventually taking its turn once more behind the digger. The digger is replaced periodically.

melanocytes do not produce melanins, however, resulting in the white hair of albino and elderly mammals (see opening figure for Chapter 5). Many Arctic mammals also have white hair, which presumably camouflages them against the snowy landscape (Fig. 39.8A on page 874). Black and brown hair provides good camouflage against soil, rocks, tree bark, dry grasses, and forest shadows. Some temperate-zone animals, such as the snowshoe hare, molt twice a year, changing their color to match the seasons (Fig. 39.8B). In contrast, some mammals are conspicuously colored. The colors of some mammals, especially around the face, serve for social communication and recognition. In skunks the white stripe on the black background is undoubtedly **warning coloration.** The contrasting stripes of the zebra serve as **disruptive coloration.** Disruptive coloration makes it difficult for a predator to pick out an individual from a herd and anticipate its speed and direction (Fig. 39.8C). (Early in World War II the British Navy painted some of its ships with garish stripes for the same reason. It proved effective until the Germans acquired radar.)

FIGURE 39.8

Color patterns in mammals. (A) The Arctic fox *Alopex lagopus*. The thick, white fur conserves heat and provides camouflage, which is essential for predation in open terrain. In accordance with Allen's Rule, the ears and nose are shorter than in southern foxes, which reduces the loss of body heat. (B, C) The ermine, or short-tailed weasel, *Mustela erminea* in summer and winter. Weasels are voracious predators and usually kill their prey by biting through the skull. (D) Burchell's zebra *Equus burchelli* are clearly not camouflaged, but one can see how confusing the stripes would be in a stampeding herd.

Claws, Antlers, and Horns

Mammalian skin also produces a variety of hard structures that are used in predation, defense, and social aggression. In some mammals the skin produces epidermal scales that serve as armor (Fig. 39.9). In rhinoceroses and some other animals the skin itself is thick enough to serve as armor. In most mammals skin also produces **claws,** which are used in defense, capturing prey, climbing, digging, and grooming (Fig. 39.10). Claws are keratinized sheaths that cover the pointed ends of the distal phalanges (toe bones). In primates at least some of the claws are modified as flat **nails** that do not enclose bone. In **ungulates** the claws are modified as hooves. Ungulates are classified in two orders, depending on whether each hoof has an odd number of toes (order Perissodactyla) or an even number (order Artiodactyla).

Ungulates are now the only mammals that have **horns** and **antlers.** These two forms of head ornamentation are often confused with each other, but they differ in several respects:

1. An antler consists of dermal bone that is initially covered by furry skin, called velvet. A horn, on the other hand, has a core of bone that projects partly into a hollow keratin sheath that grows out continually from patches of epidermal cells.

A

B

FIGURE 39.9

Scaly mammals. (A) The cape pangolin, also called scaly anteater (*Manis temmincki*) of the African savannas. The scales are made of fused modified hairs and are shed periodically. They protect the pangolin not only from predators, but also from the stings of ants and termites while it raids their nests. It scoops up insects with a long, sticky tongue that retracts into a sheath that extends down the abdomen as far as the pelvis. (B) The long-nosed (= nine-banded) armadillo *Dasypus novemcinctus.* The carapace consists of plates of dermal bone covered with keratinized epidermis. Armadillos curl into well-protected balls when threatened.

2. Except in caribou, antlers occur only in mature males, while horns generally occur in both sexes.
3. Antlers are always shed and regrown annually, while horns are generally permanent and irreplaceable. Although antlers are shed each year, they tend to get larger with age as the animal produces higher levels of sex hormones and is able to mobilize more nutrients to produce the antlers.

Cattle, sheep, goats, and antelopes (order Artiodactyla; family Bovidae) have a pair of hollow, symmetric horns that cover bone cores near the top of the head (Fig. 39.11A on page 876). In most of these even-toed ungulates the horn sheath is never shed and, if lost, cannot be replaced. The American pronghorn *Antilocapra americana,* however, grows new horn sheaths inside the old ones, which are shed each year. (There is some debate over whether these should be called horns.) The horn of rhinoceroses, which are odd-toed ungulates, is very different (Fig. 39.11B). A rhinoceros horn grows along the midline of the snout and consists of agglutinated fibers of keratin (sometimes erroneously said to be hair). Antlers occur in deer, caribou, and elk (order Artiodactyla; family Cervidae; Fig. 39.11C). During development in the summer the antlers are sheathed in skin velvet, which is shed prior to the breeding season, generally in late autumn (Fig. 39.23B).

Adaptations for Feeding

JAWS. Because homeotherms have to process much more food than do reptiles, their feeding adaptations differ greatly. The chief difference is that most mammals speed up di-

FIGURE 39.10

A bobcat *Felis rufus* demonstrates the value of claws on a snowshoe hare *Lepus americanus.*

■ **Question:** For what function is the hind foot of the snowshoe hare adapted?

A

B

C

D

FIGURE 39.11

Horns and antlers. (A) Like other bovids, Dall's sheep (*Ovis dalli*) have a pair of symmetric horns. (B) Horns seem like overkill in a 2000-kg black rhino (*Diceros bicornis*). Ironically, the horns, which are valued in North Yemen for knife handles and in parts of Asia for "medicines," may well bring about the extinction of the species through illegal poaching. A rhino horn sells for approximately $24,000. (C) The elaborate antlers of the caribou *Rangifer rangifer*. (D) The bony protrusions on the head of the giraffe (*Giraffa camelopardalis*) are neither horns nor antlers. Their function, if any, is unknown.

■ **Question:** In what way are the horns of the caribou like the tail of the lyre bird (Fig. 38.20A)?

gestion by chewing food, thereby breaking it into smaller pieces and mixing it with enzymes. Birds achieve this with a gizzard, but most mammals use their teeth. Mastication places severe mechanical strain on jaws and teeth, requiring a stronger lower jaw and jaw joint. Most reptiles have a complex jaw with several small bones (quadrate, articular, and dentary; Fig. 39.12A). The complex jaw joint provides little resistance to strain, but is adequate in animals that swallow their food without chewing. In mammals, however, the only bone of the lower jaw is the large dentary in which the teeth are rooted (Fig. 39.12B). The enlarged dentary is stronger and provides more area for the attachment of jaw muscles.

The jaw muscles attach at the other end to the **zygomatic arch** (cheek bone) and top of the skull, which may have a **sagittal crest** that provides more space for attachment. The jaw joint usually involves only the dentary and the squamosal bone of the skull. In spite of the strength of the joint, it is still mobile enough in herbivores and many omnivores such as ourselves that the teeth can grind against each other in all directions, as in a cow chewing its cud. In an animal that has to breathe as often as a mammal does, prolonged chewing also requires that the nasal passages not open into the mouth, as they do in most reptiles. Thus mammals have a shelf of bone called the **secondary palate** that separates the nasal passages from the mouth, as well as a **soft palate** that closes the nasal passages during swallowing.

The quadrate and articular bones were not simply discarded during the evolution of a simpler jaw joint, but were modified as middle-ear bones. Fossil evidence suggests that the quadrate and articular bones of mammal-like reptiles had already played a role in transmitting sound vibrations from the external tympanum to the inner ear. As they became smaller and lost their functions in the jaw joint, they assumed an increasing role in audition. Gradually the quadrate and articular joined the stapes (= columella) and became the **incus** and **malleus** (Fig. 39.13 on page 878). The presence of three middle-ear bones in mammals, rather than only the stapes, as in amphibians, reptiles, and birds, is thought to increase the efficiency of energy transfer from the air to the liquid of the inner ear (impedance matching). It may account for the fact that mammals can generally hear higher pitches than can most other terrestrial vertebrates.

TEETH. Increased efficiency of chewing in mammals also owes a great deal to adaptations in the teeth. In contrast to the uniform, peg-shaped teeth of most reptiles, which serve mainly to grip food, most mammals have a diversity of teeth that serve a variety of functions (see Figs. 13.2B and 13.6). In other words, mammals are **heterodont** rather than **homodont.** In front are the **incisors,** which are often chisel-shaped and used for biting off pieces of food. Lateral to the incisors are the **canines,** which are used by carnivores and omnivores for holding, piercing, and tearing flesh, not only in feeding but also in aggressive encounters. Finally there are the **premolars** and **molars.** In carnivores and many omnivores, including humans, the premolars and molars have sharp ridges for cutting meat. In herbivores they are flat for crushing and grinding vegetable matter. The adaptation of premolars and molars for herbivory undoubtedly accounts for much of the success of mammals, making them virtually the only vertebates that can feed on leaves and other coarse vegetable matter. The number, size, and form of each kind of tooth vary with species and reflect diet. Because jaws and teeth are often preserved long after the rest of the skeleton has been scattered and decayed, dentition is useful in identifying carcasses. Numbers of teeth are commonly expressed by **dental formulae** (Table 39.1).

TABLE 39.1

Dental formulae for some mammals.

Each pair of numbers gives the number of incisors, canines, premolars, or molars on each side of the upper and lower jaw (I/I, C/C, P/P, M/M). Thus the formula 2/1, 0/0, 3/2, 3/3 indicates the following: two incisors on each side of the upper jaw and one incisor on each side of the lower jaw; no canine; three premolars on each side of the upper jaw and two on each side of the lower jaw; and three molars on each side of the upper and the lower jaw. The total number of teeth in this species would be 28.

DENTAL FORMULA	FEEDING	SPECIES
1/1, 0/0, 1/1, 3/3	Eating tree bark, nuts, roots	Beavers, porcupines, many squirrels, pocket mice
2/1, 0/0, 3/2, 3/3	Browsing	Hares, rabbits
0/3, 0/1, 3/3, 3/3	Grazing	Deer, bison, sheep, caribou
2/2, 1/1, 2/2, 3/3	Omnivory	Humans
3/3, 1/1, 4/4, 2/3	Predation	Dogs, foxes, wolves, bears

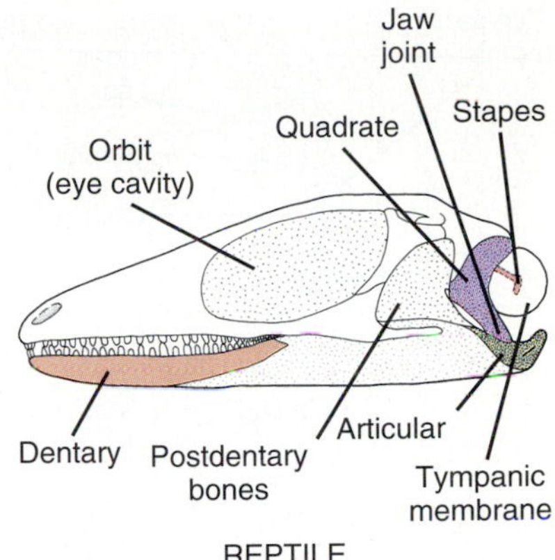

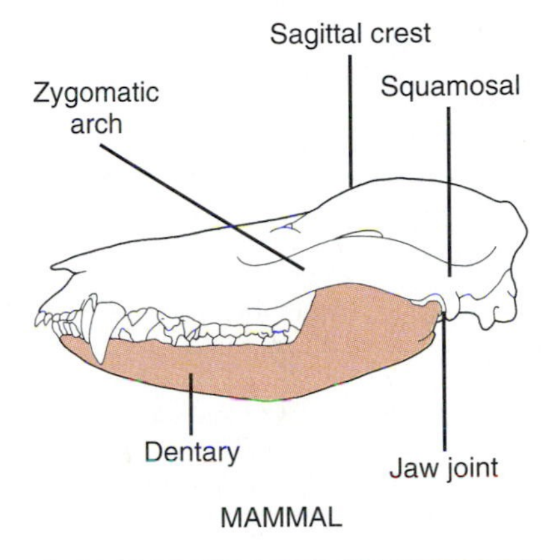

FIGURE 39.12

The jaw of the early Triassic reptile *Prolacerta* compared with that of the American opossum. See also Fig. 37.19B. The dentary of the opossum and other mammals is enlarged and stronger than the combined quadrate, articular, and other bones of the reptilian jaw. Large jaw muscles of mammals attach to the sagittal crest and zygomatic arch. (The sagittal crest is clearer in Fig. 39.1B.) Note also the diversity of teeth in the opossum.

■ **Question:** Do you see a good place for the attachment of jaw muscles in part A? Why not?

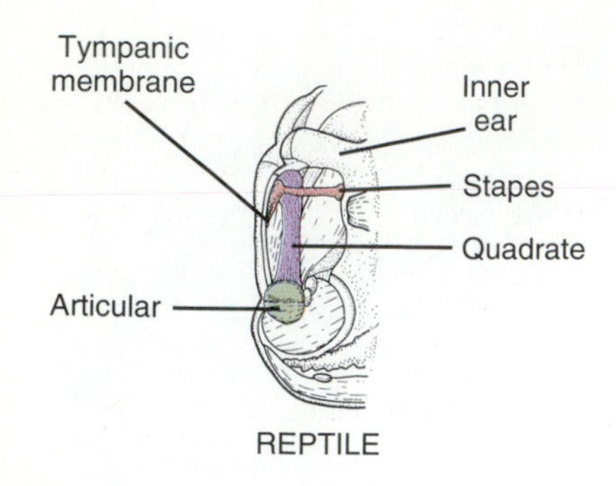

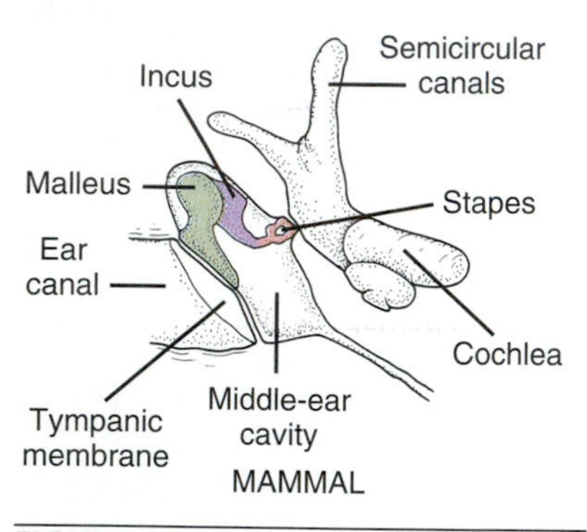

FIGURE 39.13

The auditory organs of the same reptile and mammal as in Fig.39.12, viewed from the rear. The quadrate and articular of the reptile have become the incus and malleus of the mammal.

FIGURE 39.14

Tusks. (A) The walrus *Odobenus rosmarus* has enormous upper canines, which it may use to skid along the ocean floor or to rake up clams. Body length approximately 3 meters. (B) Narwhals *Monodon monoceros* have only two teeth. In males usually the left one, rarely both, forms a tusk. Females seldom have tusks. Total body length is up to 5 meters.

Reptiles' teeth do have one advantage over those of mammals: They are continually replaced. Most mammals get only two sets of natural teeth. The first set consists of **deciduous teeth** ("milk teeth"), which do not include molars. Deciduous teeth are replaced early in life by **permanent teeth.** The permanent teeth have to last the mammal for the rest of its life, even though they are subject to considerable wear, especially in herbivores. Rodents and some other mammals compensate for such wear by having incisors that grow continually.

Some mammals have teeth adapted for functions other than biting, tearing, and chewing. In elephants the upper incisors are enlarged as **tusks** that are used to manipulate trees and other large objects. The walrus has upper canines enlarged as tusks, which it uses to haul itself onto ice and perhaps as skids or rakes as it feeds on shellfish on the ocean floor (Fig. 39.14A). Most unusual is the single, spirally grooved tusk of the narwhal, which develops from an upper tooth and is up to 2.7 meters long in males (Fig. 39.14B). In all three of these species the tusks may be as important in competition among males and in defense as they are in feeding.

Whales are traditionally divided into two suborders, Odontoceti and Mysticeti, depending on whether they do or do not have teeth, respectively (Fig. 39.15). (Molecular phylogeny suggests that this division does not accurately reflect phylogeny and that whales should be included within Artiodactyla rather than in their own order cetacea [Graur and Higgins 1994, Milinkovich et al. 1994]). The toothed whales are narwhals, sperm whales, and dolphins (which are called porpoises if they have rounded rather than bottle-shaped noses). Like reptiles, most of these whales have large numbers of similar teeth, which they use for grasping prey rather than for chewing. Crushing of food occurs in a three- or four-chambered stomach, especially in the first chamber, which contains stones, bones, and shells. Other whales do not have teeth except as fetuses. Instead they comb food out of water or sediments with rows of keratin plates called **baleen,** which hangs from the upper jaw like a moustache. For example, the gray whale (family Eschrichtidae) scoops up bottom sediments from which it filters out amphipod crustaceans (Fig. 39.15B). Right whales (family Balaenidae) have baleen only on the sides of the mouth (see Fig. 13.1C). They swim slowly through schools of plankton (mainly copepods), engulfing huge masses of them. They then use their large tongues to squeeze water out the sides of the mouth, through the baleen, which traps the plankton. Rorquals (family Balaenopteridae), which include the humpback, blue, and minke whales, feed on krill or fish.

A

B

A

B

FIGURE 39.15

Toothed and toothless cetaceans. (A) Part of a group of approximately 30 killer whales (actually a type of dolphin) *Orcinus orca* attacking a young blue whale *Balaenoptera musculus.* Of the odontocetes, killer whales are among the few that take on such large prey. They also attack seals and penguins. This blue whale was approximately 20 meters long and might have become one of the largest animals ever to live (up to 30 meters). Killer whales are up to 9 meters long. (B) The gray whale *Eschrichtius robustus* and other baleen whales have a comb of baleen on the upper jaw. Most baleen whales spend about half the year in polar waters feeding on abundant food, then migrate to warmer waters to mate and give birth. The gray whale filters crustaceans out of mouthfuls of sediment. The stirring up of sediments by the gray whale and perhaps by the walrus is believed to return minerals from the ocean floor back into nutrient cycles. Only about one-eighth of this 14-meter-long whale is shown.

ECHOLOCATION. Sailors at least as far back as Aristotle's time have reported wails, creaks, groans, and screams from both baleen and toothed whales, and recordings have made these haunting "songs" familiar. Whales apparently use them to communicate with each other, and in a few cases zoologists have been able to interpret the social context in which different songs occur. Each dolphin has a unique "signature whistle" that may serve as its name. Dolphins and some other toothed whales also produce explosive sounds that may stun or even kill fish within a few meters. Besides these sounds, toothed whales emit pulses of ultrasound ("sound" with a frequency above the upper limit of human hearing) that are used in echolocation. Echolocation is similar to radar except that instead of radio waves it uses ultrasound. An ultrasonic beam is focused like a search light, and the reflected beam provides information about the direction, distance, speed, and shape of objects it strikes. Using echolocation, whales can find prey in the dark ocean depths and in water clouded with silt or plankton. The echolocation pulses from toothed whales do not come from the larynx. In dolphins they apparently originate near the blow hole. The skull, air sacs, and perhaps the fatty

FIGURE 39.16

(A) A bottlenosed dolphin *Tursiops truncatus*. Like other odontocetes, dolphins have only one nostril, the blow hole. (Mysticetes have two blow holes.) (B) Schematic representation of the head of a dolphin. While underwater, the nasal plug shuts the blow hole. Ultrasonic vibrations appear to be produced by shuttling air between air sacs in the head, which causes them and the nasal plug to vibrate. The skull, air sacs, and "melon" focus the vibrations into an echolocating beam. Returning echoes are thought to be conducted to the inner ear by the lower jaw.

A

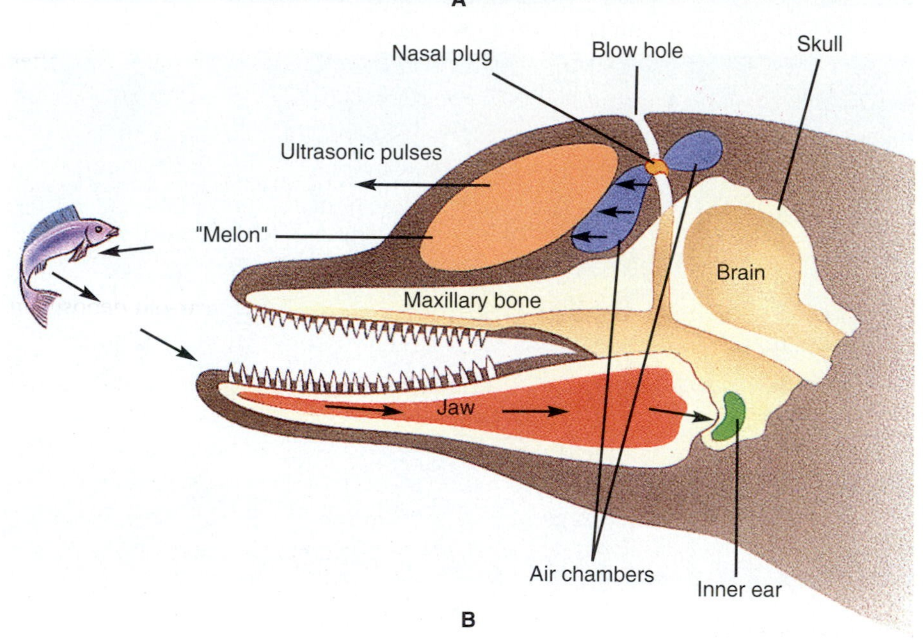

B

"melon" in the head focus the vibrations into a beam (Fig. 39.16). The lower jaw picks up the returning echo and conducts it to the inner ear.

It is well known that many **bats** use echolocation to navigate and avoid collisions with each other as they fly to and from their roosts. While many bats feed on fruit and nectar, and some feed on small vertebrates, most use echolocation to prey on nocturnal flying insects, availing themselves of a source of food that virtually no other kind of animal can exploit (Fig. 39.17). Bats use the larynx to produce ultrasound with frequencies exceeding 100,000 hertz in some species. The ultrasound emerges from either the mouth or the nose. The latter is advantageous, since a bat that has just caught prey could otherwise have difficulty navigating with its mouth full. Bats detect the echoes with their ears, as Lazaro Spallanzani demonstrated two centuries ago by plugging them with wax. Many bats have relatively huge external ears, which may help collect the echoes. Many others have bizarre ornamentation around the nose that may act like a megaphone that directs the ultrasound (Fig. 39.17).

A

B

FIGURE 39.17

(A) A fringe-lipped bat *Trachops cirrhosus* of Panama about to capture a frog. Most bats that prey on large, immobile animals emit low-intensity echolocation pulses and are therefore called "whispering bats." This particular species also homes in on the vocalizations of its prey. (B) The greater horseshoe bat *Rhinolophus ferrumequinum* of Europe, Asia, and northern Africa closing in on a moth. Note the complex nose of the bat, which presumably is adapted for focusing high-intensity sonar pulses. The moth has heard the pulses and is taking evasive action.

Primates

ORIGINS. Now that we are approaching the end of this book in which we have considered so many kinds of animals with scientific detachment, it might be interesting to consider our own order Primates with the same objectivity. There is no reason, after all, why understanding humanity should be any less a goal for zoology than it is for anthropology, sociology, psychology, history, theology, or art. In fact, none of these disciplines can provide a complete understanding of humanity if it is divorced from a consideration of ourselves as animals. It should already be apparent that much of what we are and what we do relates to the fact that we are animals—in particular mammals, and most especially primates.

How did we get started? A single tooth found in 70-million-year-old deposits in Montana provides the earliest trace of the origins of primates. This molar has the squarish shape and blunt cusps that paleontologists generally associate with primates, though not all agree that this particular tooth belonged to a primate. Animals of this kind flourished early in the Cenozoic era. They were small, tree-climbing animals, perhaps somewhat like modern tree shrews (order Scandentia; Fig. 39.5). The earliest undisputed primate fossils belong to two distinct groups dating to approximately 55 million years ago in what is now Europe and North America. In addition to molars typical of primates, they had other characteristic skull features. These include eye orbits that face forward for stereoscopic vision and a brain case that is large even for mammals. Both groups of early primates became extinct, but not before producing ancestors of the modern extant primates. One group is believed to have given rise to lemurs, lorises, and the other primates commonly called prosimians (Figs. 39.18 and 39.19 on page 882). The second group is thought to have given rise to tarsiers (Fig. 39.20 on page 883), and then to monkeys, apes, and humans, which are commonly referred to as anthropoids.

MONKEYS. The anthropoids (humanlike animals) are the monkeys, apes, and, of course, humans. Monkeys fall into two distinctive groups: the New World monkeys of South and Central America, and the Old World monkeys of Africa and Asia (Fig. 39.21 on page 883). Among the New World monkeys are the capuchin monkeys, howler monkeys, squirrel monkeys, spider monkeys, marmosets, and tamarins. New World monkeys lack opposable thumbs: Their thumbs bend around branches on the same side as the fingers. These monkeys, especially those with **prehensile tails,** are nevertheless quite at home in the trees. Old World monkeys have opposable thumbs and big toes that allow them to grip tree limbs, and they use their tails for balance rather than

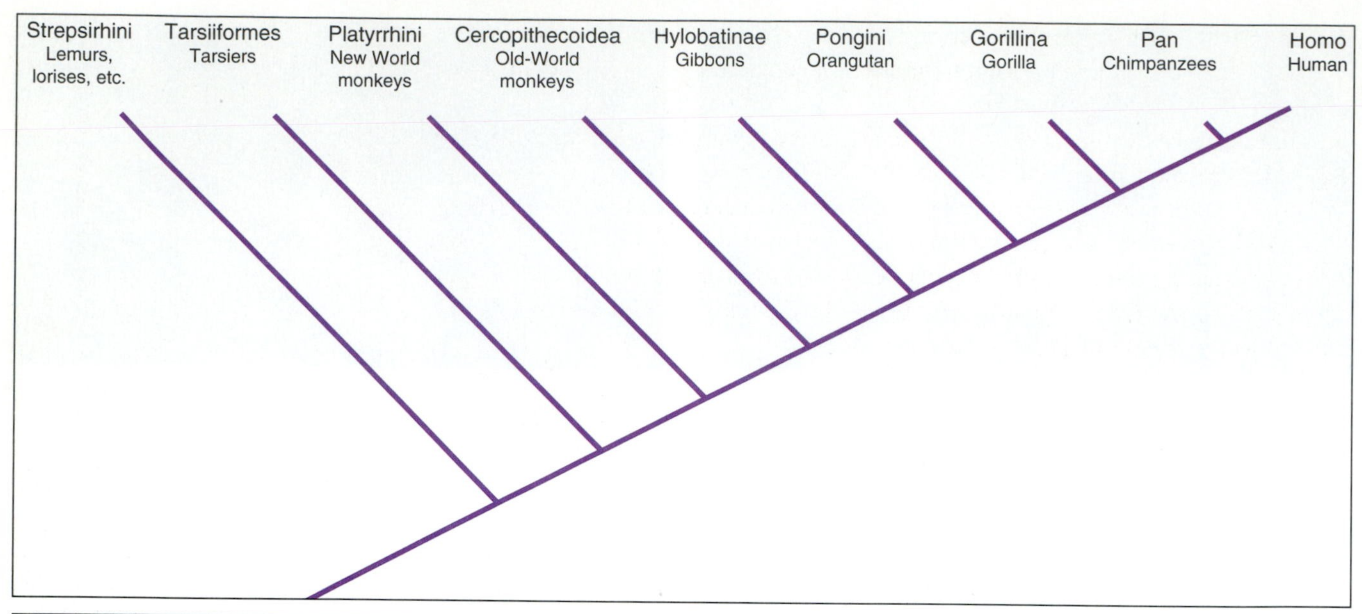

FIGURE 39.18

Cladogram showing the relationships among primates. Names of clades are from Miyamoto and Goodman (1992).

for grasping. Among the Old World monkeys are several species of macaques (including the rhesus), baboons, and mandrills.

APES. Apes and humans appear to differ from all monkeys, since they lack tails and tend to be larger and more terrestrial. In the kinds of characters that matter to taxonomists, however, apes and humans are more closely related to Old World monkeys than these monkeys are to New World monkeys. All three groups—Old World monkeys, apes, and humans—differ from New World monkeys in having opposable thumbs, noses with high bridges, and nostrils that point to the front rather than to the sides. Old World monkeys, apes, and humans also share the dental formula 2/2, 1/1, 2/2, 3/3 (two incisors, one canine, two premolars, three molars on each half of both jaws), which dif-

FIGURE 39.19

Prosimians (suborder Strepsirhini). (A) A ring-tailed lemur *Lemur catta* with young. In its native Madagascar it is sometimes kept as a house pet. Length, including tail, 1.2 meters. (B) The slender loris *Loris tardigradus.* Lorises have no tails. They hunt insects and other small animals at night by carefully stalking them.

■ **Question:** In what ways are these and other primates adapted for arboreal life? Which of those adaptations have we inherited?

fers from that of New World monkeys. Molecular comparisons also show that Old World monkeys, apes, and humans are more closely related to each other than any of them is to New World monkeys.

The apes include three species of African Great Apes: the gorilla *Gorilla gorilla* (see Fig. 1.8), the chimpanzee *Pan troglodytes* (see the opening photograph and Fig. 19.12A), and the bonobo (formerly pygmy chimpanzee) *Pan paniscus* (see opening photo for Chapter 19). The bonobo is the most like humans in appearance and behavior, often walking upright and mating in a face-to-face position. Other apes live in southeastern Asia: the orangutan *Pongo pygmaeus* (opening photograph for Chapter 10) and several species of gibbons *Hylobates* spp.

FIGURE 39.20

Tarsiers (*Tarsius* sp.) are extremely agile in trees, making tremendous leaps as they feed on insects, lizards, and other small animals. All are nocturnal, as might be guessed from the owlish eyes. Adhesive pads on the fingertips help maintain grip. The body is about 15 cm long. The hairless tail probably assists in maintaining balance.

RELATIONSHIP OF HUMANS TO OTHER APES. Like most people, traditional taxonomists believe there is an enormous difference between humans and apes. They therefore classify humans in the exclusive family Hominidae, while either relegating all the apes to the family Pongidae or placing the gibbons in a third family Hylobatidae. Separating humans from apes may have less to do with science, however, than with tradition and other considerations. When Linnaeus named *Homo sapiens* he saw little scientific basis for separating humans from apes. To avoid controversy he simply omitted taxonomic criteria from the description of humans, which reads in its entirety "*Homo nosce Te ipsum*"—Know yourself, man. He confessed what he had done in a private letter in 1747 (quoted in Frängsmyr 1983, p. 172.):

> I ask you and the whole world for a generic differentia between man and ape which conforms to the principles of natural history. I certainly know of none. . . . If I were to call man ape or vice versa, I would bring down all the theologians on my head. But perhaps I should still do it according to the rules of science.

Recent studies based on molecular comparisons and on morphology support Linnaeus. Segments of human DNA are identical to corresponding segments of DNA

A

B

FIGURE 39.21

Monkeys. (A) A New World monkey, the spider monkey *Ateles geoffroyi*, demonstrates how handy a prehensile tail can be. (B) An Old World monkey, the rhesus monkey *Macaca mulatta*, is one of the most familiar primates. It is commonly used in medical research. Like these two, most monkeys and apes spend considerable time grooming each other. Grooming not only removes external parasites but is essential in maintaining normal social bonds.

from bonobo, chimpanzee, or gorilla at more than 98% of their bases. These findings, and many others based on molecular phylogenetics, suggest that humans are more closely related to bonobos, chimps, and gorillas than any either of these African apes is to the orangutan. In fact, several studies show that humans are more closely related to chimps than chimps are to gorillas (Miyamoto and Goodman 1990; Begun 1992). Therefore if bonobos, chimpanzees, gorillas, and orangutans belong in the same family, then humans belong in that family with them. In addition, any taxon that includes apes but not humans is paraphyletic and therefore cladistically unacceptable. In short, we are all apes.

Interactions with Humans and Other Animals

ECOLOGY. Many groups of animals far outnumber mammals in numbers of individuals and species. However, because of their high levels of activity, around the clock and throughout the year, and their ability to adapt to such a broad range of habitats, mammals have a disproportionate influence on other animals. Many mammalian herbivores are so large and graze in such vast herds that they create and maintain entire biomes—prairies and savannas. Large carnivorous mammals dominate many food webs and thereby control the populations of other kinds of animals. Even mice, moles, shrews, and bats, although small as individuals, collectively consume huge amounts of plant material, insects, and other foods. They, in turn, are food for birds and larger carnivorous mammals. Many other examples of the ecological roles of mammals have been noted in this chapter and elsewhere.

PEOPLE AND OTHER MAMMALS. In our mechanized world we tend to overlook the importance of other mammals in human affairs, but it was not so long ago that much of the labor of farming and transportation was done by horses, oxen, mules, and other domesticated mammals. In many parts of the world, that is still the case. Millions of people still depend on the labor of an estimated 75 million domesticated water buffalos *Bubalus bubalis.* Cattle and numerous other domesticated mammals also contribute more directly to the human food supply, as meat. In many parts of the world, hunted mammals provide a major portion of the protein in the diet. In the United States and other developed countries the nutritional value of hunting is secondary to the recreational importance. Other economic contributions of mammals include furs and leather. Each year millions of mammals also contribute to human welfare as research animals. Mammals have been indispensable in finding treatments and cures for diseases in the past, and they continue to be essential in research on cancer, AIDS, and other diseases.

The relationship between people and other mammals is not always one of unilateral exploitation. Many people form close attachments with pets and vice versa. The importance of pets to people must be enormous considering the prices of dogs and cats and the quality, variety, and cost of pet food. This is especially obvious when we consider how many of our own species we allow to go hungry. People also value seeing mammals that are not pets but are not quite wild either. Deer, raccoons, chipmunks, squirrels, and other mammals contribute immeasurably to rural and suburban life, and captive mammals in zoos are major cultural assets in many cities. In addition, many people cherish wild animals, even if they have little hope of ever seeing them. People contribute millions of dollars to organizations devoted to the preservation of such mammals (Fig. 39.22).

CONSERVATION. As with so many things, we value wild mammals most when we are on the verge of losing them. As a result, conservation is now a well-established part of the culture of developed countries, at least until it threatens someone's livelihood.

FIGURE 39.22

Mother and baby harp seals, *Phoca groenlandica.* In the 1960s up to 300,000 harp seals were killed each year, including pups that were clubbed to death for their white fur. Public protest forced the Canadian government to set quotas and to ban the clubbing of pups in 1985. However, the penises of 50,000 harp seals per year are still sent to Hong Kong for use as supposed aphrodisiacs. The mother is up to 2 meters long.

Conservation is not yet established in many parts of the world, however. In Africa, for example, ivory poachers with AK-47 assault rifles and chain saws have killed more than half the 1.3 million African elephants (*Loxodonta africana*) that were alive in 1979. Poachers have killed all but 3000 of the 60,000 black rhinoceroses (*Diceros bicornis*) that were alive in 1970. In Namibia, wildlife authorities must cut off the horns of surviving black rhinos each year to save their lives. This procedure recently became controversial after researchers found evidence that the dehorned rhinos could no longer protect their calves from predators. In many parts of Africa and in much of South America the greatest threat to mammals is the destruction of habitat, especially rainforests. Many developing countries are so burdened by poverty and foreign debt that even if there are conservation laws, there is no money to enforce them. China, for example, has outlawed importation of rhinoceros horns, which are sold in the Orient as aphrodisiacs, but it does nothing to enforce the laws. Often poachers and illegal developers bribe the underpaid or greedy officials who are supposed to enforce the laws.

In many places, conservationists have succeeded in changing attitudes and laws regarding mammals—often by mobilizing public pressure, and increasingly by making conservation more profitable than extinction. In the United States and Canada, for example, wildlife is estimated to be worth $70 million per year. The state of Wyoming alone earns a billion dollars per year from people who come to hunt and view its animals (Geist 1994). Several African countries now get a major part of their revenue from tourists who come to see large mammals. Some of these countries have set aside large game reserves and have cracked down on poaching and illegal trading in ivory, horns, and furs. The President of Kenya has brought elephant poaching to a virtual halt by threatening to shoot poachers on sight and by burning 12 tons of ivory to dry up the trade. In order to win support from the growing human populations in these countries, imaginative schemes have been devised so that some of the profits of tourism reach the people who would otherwise be farming the land or profiting from poaching. In South America, tourism in rainforests is just beginning, so conservationists have worked out other incentives, such as "debt for nature" exchanges, in which governments sell off huge foreign debts that they have no hope of repaying in exchange for setting aside areas to be preserved.

Many countries, both developed and not, have also signed treaties aimed at conserving species. One example is the Convention on International Trade in Endangered Species (CITES), which outlaws trade in endangered animals or parts of them. Another is a ban on commercial whaling agreed to by members of the International Whaling

A

B

C

FIGURE 39.23

Recovered, recovering, and doubtful American mammals. (A) A Florida panther *Felis concolor,* subspecies of the mountain lion (= cougar or puma). Only 30 to 50 individuals survive, all in the southern tip of Florida. The survival of this subspecies is thought to be threatened by citrus farming, deer hunting (which deprives the cats of prey), disturbance due to hunting with dogs, automobile collisions, and severe inbreeding. Florida spent $13 million for highway underpasses for the panther, and zoologists are considering test-tube fertilization as a means of restoring the population. Body approximately 2.5 meters long. (B) Bison once grazed from the Rockies to as far east as Buffalo, New York, and often in herds that stretched as far as one could see. Countless numbers were killed to deprive hostile Indians of their food source, and many others were killed by market hunters who often took only the tongue. By 1900 fewer than 1000 remained. Today there are more than 30,000. Shoulder height approximately 2 meters. (C) The largest deer in the world, the moose, was nearly hunted out in the United States by 1900, but its population is now increasing throughout the northern tier of states. This bull, shedding its velvet, has a total height of approximately 2.5 meters.

Commission in 1982 to prevent the imminent extinction of several species of whales, including the right whale *Balaena glacialis* and the blue whale *Balaenoptera musculus.* Japan and Norway, however, take advantage of a loophole that allows them to kill several hundred of the abundant minke whales *Balaenoptera acutorostrata* each year for "research." (A typical "discovery" from such "research" is that pregnant female whales have higher levels of progesterone than nonpregnant females.) Studies of the DNA of whale meat sold in Japan suggests that illegal hunting of other whales may be continuing (Baker and Palumbi 1994). Norway has recently announced its intention to defy the commission and resume full-scale whaling, but adverse publicity and a boycott of Norwegian fish may change that plan.

The populations of many species will return to healthy levels spontaneously if their habitats are preserved and they are no longer killed by humans. Some of the North American species that have recovered from near extinction are beaver (*Castor canadensis*), which were decimated to satisfy the fashion for beaver hats, and bison (*Bison bison*), which were hunted nearly to extinction for meat and to deprive Indians of food (Fig. 39.23B). Other species, including the lynx (*Felis lynx*), the red wolf (*Canis rufus*), and the moose (*Alces alces*; Fig. 39.23C), have had to be reintroduced and protected. Many other native American mammals, ranging in diversity from the grizzly bear *Ursus arctos horribilis* to the Indiana bat *Myotis sodalis,* require considerably more help, and their survival is in doubt (Fig. 39.23A). Many species are so depleted or their habitats so precarious that captive breeding in zoos and preserves is the only alternative to extinction.

Conserving and restoring species requires the efforts of virtually every type of biologist, including ecologists, physiologists, and behaviorists, as well as nonbiologists such as economists and politicians. In the past decade many individuals from these various disciplines have come together under the banner of **conservation biology.** Conservation biology is now a distinct and growing field with its own organization and journal. Like any new field, this one is still the subject of doubts. Some biologists wonder whether conservation biologists are abandoning their scientific objectivity. Conservation biologists are apt to reply that without their efforts there may someday be no organisms to be objective about.

The Future of Humans and Other Animals

In the preceding section, as in most of the concluding sections in this unit, I have tried to be objective in discussing the ways in which humans are affecting other animals. Often I have had to struggle to keep my sadness from reaching these pages as I considered the tragic results of our actions on this planet—tragic not only for other organisms, but also for ourselves. Now that this book is near its close, you will, I hope, forgive a brief departure from cool, scientific objectivity. Zoologists are, after all, human, and it is characteristic of humans to try to foresee the future consequences of present activities, even if it cannot be done with much confidence. What follows, then, is my own vision of our dismal future if we continue rushing blindly into it, stampeding the other animals ahead of us.

EXTINCTION. As noted in virtually every chapter in this unit, the success of humanity has often been at the expense of other species. Of all the consequences of our success, the worst undoubtedly is extinction. It will be humanity's most enduring legacy, because unlike our dams, highways, cities, and other monuments to human civilization, extinction is forever. Our species extinguishes another species every day on average, threatening the continuity of three billion years of life. In the United States alone, approximately 3600 species are officially listed as endangered. This should strike most people as fundamentally contrary to the ethics and principles of fairness that we like to think separates us from apes and other "lower" forms of life. Many people, however, see nothing wrong with one species, and a relatively new one at that, destroying another. In the view of some, other species have to justify their existence by being useful to humans. One United States Senator argued in favor of weakening the Endangered Species Act on the grounds that "people should have dominion over fish, wildlife, and plants. Only where lower species are of benefit to mankind are they important." (Congressional Record, 18 July 1978, pp. S11040 and S11043. Note the Old Testament word "dominion," from the King James translation of *Genesis* 1:28.)

Few people express themselves so crudely these days. Most are now willing to let a "useless" species live as long as it doesn't interfere with some profitable venture. In fact, many people now consider the preservation of even noncommercial species to be important even if it causes some economic hardship for others. A recent example is the public outcry over attempts to waive the Endangered Species Act and allow the logging of old-growth forests in the Pacific Northwest, which is the habitat of the endangered spotted owl (*Strix occidentalis*). One reason for the change in favor of species may be that people increasingly see their own environments becoming unpleasant as a result

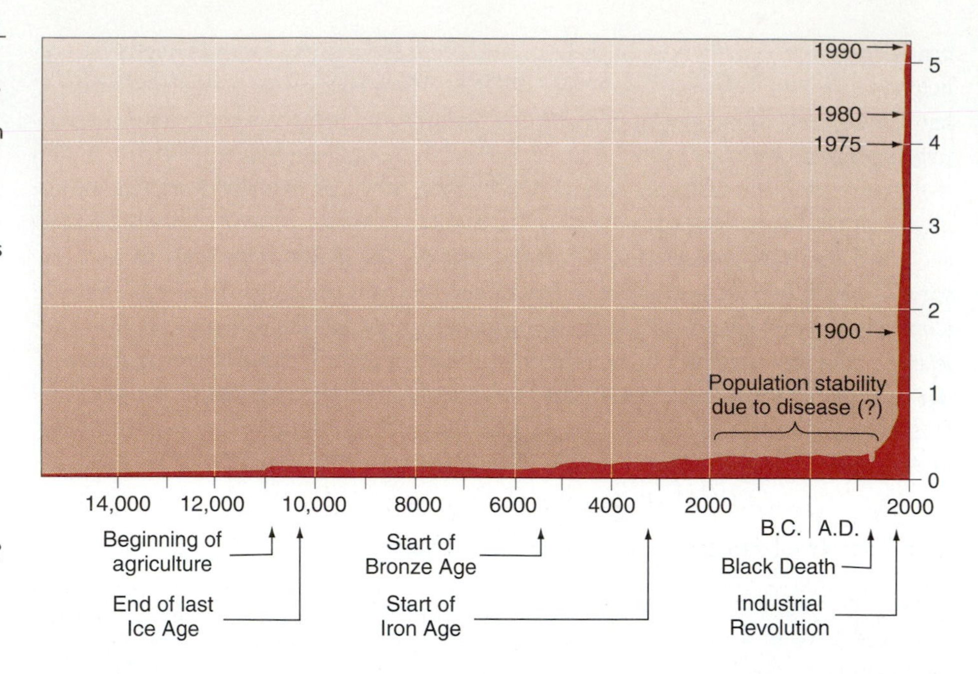

FIGURE 39.24

Estimated world population of humans. Population is believed to have jumped at about the same time as the invention of agriculture (Neolithic Revolution) and to have increased gradually thereafter. Contagious diseases may have played a large role in holding populations in check. The population began its current surge around 1700, following the Industrial Revolution, improvements in sanitation and medicine, and more productive agricultural methods. In spite of recent reductions in fertility due to contraception, the population is expected to reach 8 billion by 2050.

■ **Question:** What is your most optimistic estimate of the maximum human population the Earth can sustain? From the graph, estimate how long it will take to reach that population at the present rate of growth. What happens then?

of development, while the benefits go mainly to the developers. Another reason appears to be an increasing appreciation of the value of **biodiversity** for its own sake. It is now widely perceived that an ecosystem is not simply a collection of isolated species, but a system that ceases to function if too many of its parts are removed. More and more people are realizing that humans are part of that system, and that they cannot go on destroying other species without endangering themselves.

OVERPOPULATION. Unfortunately, well-intended efforts to preserve species appear destined to fail because the human population is growing so rapidly. Most of you will live to see twice as many of us competing with each other and other species for space and resources. A graph of human population since the Industrial Revolution looks like one for bacteria, weeds, rats, cockroaches, and other pests that have escaped from natural controls (Fig. 39.24). An objective observer from another planet might well diagnose Earth as suffering from an infestation or infection of us. The graph also looks like the early part of the curve for a species about to crash (see Fig. 17.13B), so we, ourselves, might be among the victims.

Incredibly, there are many people who, apparently carried away by theology or politics, deny that human overpopulation is a problem. An opponent of birth control has argued that the Earth can't be overpopulated, since all 5.5 billion of us could easily fit in the state of Rhode Island. (That is certainly true, but only if no one needs to go to the bathroom.) In the People's Republic of China the Marxist government of the 1970s discouraged birth control, arguing that the problem was not too many people, but inequitable distribution of goods due to capitalism (Ding Chen 1980). In 1984 a spokesman for the United States government argued in Mexico City, of all places, that increasing population was good for business. In 1987 a television evangelist and candidate for President urged United States citizens to reproduce to keep from being outnumbered by other countries. At the 1992 "Earth Summit" in Rio de Janeiro, overpopulation was virtually ignored because of pressure from the Vatican.

At last in 1994 the threat of overpopulation has become so obvious that nations meeting at the United Nations International Conference on Population and Development in Cairo resolved to do something about it (Roush 1994). Even with their best efforts, however, the world population will probably reach 8 billion by the year 2050. In Africa the population will double in the next 24 years, in spite of 13 million deaths due to AIDS. The next generation of leaders—that is, many of you—will be the ones who will either have to double the food supply to Africa or deal with the consequences such as those recently witnessed in Rwanda. Even in the United States the population is expected to double in the next century, and even sooner unless some way is found to keep out those fleeing their own overcrowded countries. Your children and grandchildren will have to find some way to cut per capita pollution and demand for resources in half just to maintain the present quality of life. One can only hope they will also find the time and energy to provide for the needs of other species that will be competing for space with twice as many Americans.

THE IMPORTANCE OF ZOOLOGISTS. Zoologists are uniquely qualified to bring such concerns to the attention of other people, because they can observe what consequences our actions have on other animals. Like the miners' canaries that stop singing when the air is poisoned, animal sentinels can provide early warnings that we are endangering their environments and therefore our own. Zoologists will be the first to observe when other animals have, so to speak, stopped singing. Zoologists will also be increasingly called upon to advise how to preserve species and their habitats. In many cases this preservation will have to take place in zoos, or perhaps someday even in petri dishes, until habitats recover sufficiently for reintroduction. Finally, for those species that are doomed to extinction, zoologists will be the only ones competent to memorialize them through research and to describe the many unique animals our children and grandchildren will never get to know.

Summary

The most significant adaptation in the evolution of mammals appears to have been homeothermy. Homeothermy was tied to the evolution of hair, which, together with mammary glands, virtually defines mammals. Homeothermy enabled mammals to spread to a wide range of climates and to remain active year round and at night. The cost for this increased activity was an extremely high metabolic rate and therefore a large demand for food. Mammals have stronger jaws and different kinds of teeth for chewing food and increasing the rate of digestion. Two of the jaw bones found in reptiles evolved into middle ear bones in mammals. Instead of uniform, peg-shaped teeth, mammals have various combinations of incisors, canines, premolars, and molars that vary with diet.

Hair reduces the conduction of body heat and in many mammals is colored in different patterns for camouflage, disruptive coloration, warning coloration, or social display. Skin also produces claws, hooves, horns, and antlers. Horns are permanent keratinized sheaths over bony cores, while antlers are temporary formations of dermal bone.

A few mammals, the monotremes, are oviparous. Most mammals are viviparous, meaning that the young develop at least partially within a uterus and are supported by means of a placenta. The advantage of viviparity is that development can be prolonged, and the young may be able to fend for themselves soon after birth. In marsupials the period in the uterus is brief, but development continues in a marsupium. Most mammals are polygynous.

Primates are generally divided into two suborders, with the prosimians (lemurs, lorises, and some others) in one, and tarsiers, monkeys, apes, and humans in the other. Old World monkeys, including baboons, are quite different from New World monkeys and are more closely related to apes and humans. The chimpanzee, the bonobo, and the gorilla are in turn more closely related to humans than to the orangutan or gibbons.

Mammals have a great impact on virtually every ecosystem in which they occur. This is especially true for our own species because of our large and growing population. Human activities have already extinguished numerous species of mammals and other organisms, and the rate of extinction will undoubtedly increase as our population doubles in the next 50 years.

Key Terms

homeothermy
marsupium
convergent evolution
piloerection
pelage
vibrissa
horn
antler
zygomatic arch
sagittal crest
secondary palate
soft palate
incus
malleus
incisor
canine
premolar
molar
dental formula
deciduous teeth
tusk
baleen
echolocation

Chapter Test

1. Describe the essential differences between mammals and reptiles. (pp. 861–862)
2. Explain how the evolution of homeothermy is related to the evolution of stronger jaws and more diverse teeth compared with those of reptiles. (p. 877)
3. Explain how the evolution of stronger jaws is related to the evolution of the middle ear bones in mammals. (pp. 875–877)
4. Describe the major features that distinguish monotremes, marsupials, and placental mammals from each other. (pp. 862–864)
5. Explain why nonflying placental mammals are native on every large land mass except Australia and Antarctica. (p. 862)
6. What physiological obstacles had to be overcome in the evolution of viviparity in mammals? How have marsupials overcome those obstacles? How have placental mammals overcome them? (pp. 870–871)
7. Explain the differences among the following: horns, antlers, tusks. (pp. 874–875, 878)
8. In Table 39.1 relate each dental formula to the diet of the mammal. (p. 877)
9. Describe the similarities in echolocation in toothed whales and bats. What differences are there between echolocation in these two groups? (pp. 879–880)
10. What is the relationship of humans and other apes to Old World and New World monkeys? What is the significance of the phrase "humans and other apes?" (pp. 881–884)

■ Answers to the Figure Questions

39.4 Any character shared by both marsupials and placentals would be misleading in classification below the category of class. This would include overall morphology and color pattern.

39.5 The answer will vary, of course, but most likely the order Primates will be on the list.

39.10 The name gives this away.

39.11 Both are cumbersome structures with no function other than to secure mates. In the lyre bird the tail influences mate selection by the female; in the caribou the horns are used in contests among males. Oddly enough, the female caribou also has horns.

39.12 Reptiles generally do not chew and therefore do not require strong jaw muscles.

39.19 The most obvious adaptations visible here are the grasping hands and stereoscopic vision needed to judge the distance of limbs. We have inherited both of these.

39.24 Regardless how high your estimate is, continued population growth at the 20th-century rate will reach that estimate probably in less time than you think. When the population reaches the maximum that the Earth can sustain, the only possible consequence is death by starvation, disease, or warfare of at least as many people as are born.

Readings

Recommended Readings

Austad, S. N. 1988. The adaptable opossum. *Sci. Am.* 258(2):98–104 (Feb).

Bekoff, M., and M. C. Wells. 1980. The social ecology of coyotes. *Sci. Am.* 242(4):130–148 (Apr).

Bongaarts, J. 1994. Can the growing human population feed itself? *Sci. Am.* 270(3):36–42 (Mar).

Bronson, F. H. 1984. The adaptability of the house mouse. *Sci. Am.* 250(3):116–125 (Mar).

Burney, D. A. 1993. Recent animal extinctions: recipes for disaster. *Am. Sci.* 81:530–541.

Clark, W. C. 1989. Managing planet Earth. *Sci. Am.* 261(3):46–54 (Sept). (*Introduction to an entire issue on the global effects of humans.*)

Clutton-Brock, T. H. 1985. Reproductive success in red deer. *Sci. Am.* 252(2):86–92 (Feb).

Dawson, T. J. 1977. Kangaroos. *Sci. Am.* 237(2):78–89 (Aug).

Degabriele, R. 1980. The physiology of the koala. *Sci. Am.* 243(1):110–117 (July).

de Waal, F. M. B. 1995. Bonobo sex and society. *Sci. Am.* 272(3):82–88 (Mar).

Fenton, M. B., and J. H. Fullard. 1981. Moth hearing and the feeding strategies of bats. *Am. Sci.* 69:266–275.

Fleming, T. H. 1993. Plant-visiting bats. *Am. Sci.* 81:460–467.

Ghiglieri, M. P. 1985. The social ecology of chimpanzees. *Sci. Am.* 252(6):102–113 (June).

Griffiths, M. 1988. The platypus. *Sci. Am.* 258(5):84–91 (May).

Hoage, R. J. (Ed.). 1985. *Animal Extinctions: What Everyone Should Know.* Washington, DC: Smithsonian Institution Press.

Honeycutt, R. L. 1992. Naked mole-rats. *Am. Sci.* 80:43–53.

Kanwisher, J. W., and S. H. Ridgway. 1983. The physiological ecology of whales and porpoises. *Sci. Am.* 248(6):111–119 (June).

Marshall, L. G. 1988. Land mammals and the great American interchange. *Am. Sci.* 76:380–388.

Moehlman, P. D. 1987. Social organization in jackals. *Am. Sci.* 75:366–375.

Moore, J. A., et al. 1985. Science as a way of knowing—human ecology. *Am. Zool.* 25:375–641.

Murray, J. D. 1988. How the leopard gets its spots. *Sci. Am.* 258(3):80–87 (Mar).

Nelson, C. H., and K. R. Johnson. 1987. Whales and walruses as tillers of the sea floor. *Sci. Am.* 256(2):112–117 (Feb).

O'Brien, S. J. 1987. The ancestry of the giant panda. *Sci. Am.* 257(5):102–107 (Nov).

O'Brien, S. J., D. E. Wildt, and M. Bush. 1986. The cheetah in genetic peril. *Sci. Am.* 254(5):84–92 (May).

O'Shea, T. J. 1994. Manatees. *Sci. Am.* 271(1):66–72 (July).

Reeves, B. O. K. 1983. Six millenniums of buffalo kills. *Sci. Am.* 249(4):120–131 (Oct).

Renouf, D. 1989. Sensory function in the harbor seal. *Sci. Am.* 260(4):90–95 (Apr).

Seyfarth, R. M., and D. L. Cheney. 1992. Meaning and mind in monkeys. *Sci. Am.* 267(6):122–128 (Dec).

Sherman, P. W., J. U. M. Jarvis, and S. H. Braude. 1992. Naked mole rats. *Sci. Am.* 267(2):72–78 (Aug).

Suga, N. 1990. Biosonar and neural computation in bats. *Sci. Am.* 262(6):60–68 (June).

Tattersall, I. 1993. Madagasgar's lemurs. *Sci. Am.* 268(1):110–117 (Jan).

Terborgh, J., and M. Stern. 1987. The surreptitious life of the saddle-backed tamarin. *Am. Sci.* 75:260–269.

Thewissen, J. G. M., and S. K. Babcock. 1992. The origin of flight in bats. *BioScience* 42:340–345.

Tuttle, R. H. 1990. Apes of the world. *Am. Sci.* 78:115–125.

Whitehead, H. 1985. Why whales leap. Sci. Am. 252(3):84–93 (Mar).

Wilkinson, G. S. 1990. Food sharing in vampire bats. *Sci. Am.* 262(3):76–82 (Feb).

Wilson, E. O. (Ed.). 1988. *Biodiversity.* Washington, DC: National Academy Press.

Wilson, E. O. 1989. Threats to biodiversity. *Sci. Am.* 261(3):108–116 (Sept).

Würsig, B. 1979. Dolphins. *Sci. Am.* 240(3):136–148 (Mar).

Würsig, B. 1988. The behavior of baleen whales. *Sci. Am.* 258(4):102–107 (Apr).

Zapol. W. M. 1987. Diving adaptations of the Weddell seal. *Sci. Am.* 256(6):100–105 (June).

See also relevant selections in References at the end of Chapter 21.

Additional References

Andres, K. H., M. von Düring, A. Iggo, and U. Proske. 1991. The anatomy and fine structure of the echidna *Tachyglossus aculeatus* snout with respect to its different trigeminal sensory receptors including electroreceptors. *Anat. Embryol.* 184:371–394.

Baker, C. S., and S. R. Palumbi. 1994. Which whales are hunted? A molecular genetic approach to monitoring whaling. *Science* 265:1538–1539.

Begun, D. R. 1992. Miocene fossil hominids and the chimp–human clade. *Science* 257:1929–1933.

Carroll, R. L. 1988. *Vertebrate Paleontology and Evolution.* New York: W. H. Freeman.

Cohn, J. P. 1988. Halting the rhino's demise. *BioScience* 38:740–744.

Committee on Animals as Monitors of Environmental Hazards, National Research Council. 1991. *Animals as Sentinels of Environmental Health Hazards.* Washington, DC: National Academy Press.

Crompton, A. W., and P. Parker. 1978. Evolution of the mammalian masticatory apparatus. *Am. Sci.* 66:192–201.

Daily, G. C., and P. R. Ehrlich. 1992. Population, sustainability, and Earth's carrying capacity. *BioScience* 42:761–771.

Ding Chen. 1980. The economic development of China. *Sci. Am.* 243:152–165 (Sept).

Frängsmyr, T. (Ed.). 1983. *Linnaeus: The Man and His Work.* Berkeley: University of California Press.

Geist, V. 1994. Wildlife conservation as wealth. *Nature* 368:491–492.

Gibbons, A. 1992. Conservation biology in the fast lane. *Science* 255:20–22.

Graur, D., and D. G. Higgins. 1994. Molecular evidence for the inclusion of cetaceans within the order Artiodactyla. *Mol. Biol. Evol.* 11:357–364.

Grzimek, B. (Ed.). 1989. *Grzimek's Encyclopedia of Mammals.* New York: McGraw-Hill.

Kennedy, P. 1993. *Preparing for the Twenty-First Century.* New York: Random House. (*On the consequences of human overpopulation.*)

King, F. A., et al. 1988. Primates. *Science* 240:1475–1482. (*Uses of primates in research.*)

Macdonald, D. W. (Ed.). 1984. *The Encyclopedia of Mammals.* New York: Facts on File.

Milinkovich, M.C., A. Meyer, and V. R. Powell. 1994. Phylogeny of all major groups of cetaceans based on DNA sequences from three mitochondrial genes. *Mol. Biol. Evol.* 11:939–948.

Miyamoto, M., and M. Goodman. 1990. DNA systematics and evolution of primates. *Annu. Rev. Ecol. Syst.* 21:197–220.

The Multimedia Encyclopedia of Mammalian Biology. New York: McGraw-Hill. (*A CD-ROM with Grzimek's Encyclopedia of Mammals and more.*)

Novacek, M. J. 1992. Mammalian phylogeny: shaking the tree. *Nature* (*London*) 356:121–125.

Nowak, R. M. 1991. *Walker's Mammals of the World,* 5th ed. Baltimore: Johns Hopkins University Press.

Robey, B., S. O. Rutstein, and L. Morris. 1993. The fertility decline in developing countries. *Sci. Am.* 269(6):60–67 (Dec).

Roush, W. 1994. Population: the view from Cairo. *Science* 265:1164–1167.

Vaughan, T. A. 1986. *Mammalogy,* 3rd ed. Philadelphia: W. B. Saunders.

Wilson, D. E., and D. A. Reeder (Eds.). 1993. *Mammal Species of the World,* 2nd ed. Washington, DC: Smithsonian Institution Press.

Würsig, B. 1989. Cetaceans. *Science* 244:1550–1557.

α. (*See under* **alpha entries.**)

A band. Middle band of a sarcomere in skeletal muscle, formed by thick and thin protein filaments.

abiotic. Without life. For example, abiotic synthesis of organic molecules, or abiotic factors in ecology, such as oxygen, water, ions, and temperature.

aboral. On the side opposite the mouth, especially in an echinoderm.

acanthella. Second larval stage of acanthocephalans, which develops from an acanthor, then becomes an encysted cystacanth in the intermediate host.

acanthor. Egg containing an embryonic acanthocephalan, released into the feces of the final host.

acclimation. Gradual development of tolerance to a change in an environmental condition, such as temperature.

accommodation. Focussing of the lens of the eye.

aciculum. Stiff, chitinous reinforcement of the parapodium of a polychaete annelid.

acidity. Relative concentration of hydrogen ions in a solution.

acoelomate. Having no body cavity (neither pseudocoel nor coelom).

acontium. Tentaclelike structure attached to mesentery that extends through the mouth of a sea anemone.

acousticolateralis system. Lateral line system and inner ear of a fish or amphibian.

acrosome (= acrosomal vesicle). Lysosome in sperm that contains enzymes that digest a pathway through the covering of eggs.

actin. Protein that forms intracellular microfilaments and thin filaments in muscle.

action potential. Brief, all-or-none change in the voltage across an excitable membrane of a nerve or muscle cell.

activation energy. Free energy that must be absorbed by a molecule to make it react chemically.

active site. Part of an enzyme that binds to substrate.

active transport. Energy-dependent movement of ions or molecules across a biological membrane.

adaptation. (1) Beneficial trait of an organism. (2) Reduction in sensitivity of a sensory receptor due to use.

adductor. (1) Muscle that brings a body part toward the body axis. (2) Muscle that closes the valves of a bivalve mollusc.

adenine. Nitrogenous base in nucleic acids and ATP.

adenohypophysis. Part of the pituitary gland consisting of the anterior pituitary and the pars intermedia.

adenosine. Nucleotide consisting of adenine and ribose.

adenosine triphosphate (ATP). Adenosine with three phosphate groups, which couples energy-producing processes with energy-consuming processes.

adhesion. Binding of different kinds of molecules to each other.

adhesive papilla. One of three organs by which an ascidian larva attaches to substrate prior to metamorphosis.

adipose. Pertaining to cells or tissue in which fats are deposited.

adipose fin. Finlike fatty lobe behind the dorsal fin of certain male salmonid fishes.

adrenal medulla (= chromaffin tissue). Central part of the adrenal glands, which secretes norepinephrine and epinephrine (adrenaline) in response to stress.

aeropyle. Pore in insect egg that is permeable to O_2 but not to H_2O.

aestivation (= estivation). Dormant condition during summer or dry period.

afferent. Conducting toward the center, such as a nerve cell conducting action potentials from sensory receptors into central nervous system, or a vein conducting blood to the heart. Compare efferent.

affinity. Ability of one molecule to bind another—for example, an enzyme to its substrate.

aftershaft. Secondary feather growing from the main shaft.

agglutination. Sticking together, especially of blood cells.

aggressive mimicry. Resemblance of a predator to a harmless animal or object.

air sac. Air-filled sac in the respiratory system of birds.

albinism. Inability to produce melanin, resulting in pale skin, white hair, and pink pupils.

albumen. Fluid of an amniote egg that contains large amounts of albumin. Egg white.

albumin. One of a group of soluble proteins common in blood, albumen, and other tissues.

all or none. Principle that action potentials either occur fully or not at all.

allantois. Cavity formed by extraembryonic membrane, which stores metabolic wastes and helps exchange oxygen and carbon dioxide through the egg shell in reptiles and birds and becomes the lining of the urinary bladder in mammals.

allele. One form of a gene occurring at the same locus on a chromosome as another allele, but producing a different variety of the same trait.

allele frequency (= gene frequency). Proportion of a given allele relative to all the alleles at the same locus in a group of organisms.

allergy. Immune response to an otherwise harmless antigen.

allometry (= scaling). Differential effect of changing linear, area, and volume dimensions of organisms.

allopatric. From different areas. Mainly referring to a mode of evolution in which members of a population become geographically separated and diverge into two different species.

allosteric effect. Stimulation or inhibition of an enzyme due to binding of a nonsubstrate to a site on the enzyme different from the active site.

alpha cell. Cell of the pancreas that secretes glucagon.

alpha helix. Coiled secondary structure of a protein.

alpha motor neuron. Efferent neuron with large-diameter axon that directly excites skeletal muscles.

alternating tripod gait. Walking pattern in insects, in which three alternating legs support the body at any instant.

altricial. Referring to a species, especially bird or mammal, in which the very young are incapable of feeding or caring for themselves. Compare precocial.

altruism. Behavior of an animal that is disadvantageous or without benefit to the animal, but beneficial to another of the same species.

alula. First digit on a bird's wing.

alveolus (plural **alveoli**). (1) Air sac in the mammalian lung. (2) Milk reservoir in the mammary gland of a female mammal.

ambulacral groove (= ambulacrum). One of five grooves radiating from the mouth in an echinoderm, which contains canals of the water vascular system.

amebocyte. Cell capable of locomotion by formation of pseudopodia.

ameboid movement. Movement of certain cells, such as *Amoeba* and white blood cells, in which a pseudopodium slowly advances and the rest of the body flows into it.

amictic. Referring to parthenogenetic female rotifers that lay diploid, thin-shelled eggs that hatch into females. Compare mictic.

amino acid. Molecule with an amine group and an acidic carboxyl group. Subunit of proteins.

ammocoete. Larva of a lamprey.

ammonifying bacterium. Bacterium that converts nitrogen-containing molecules into ammonia.

ammonotelic. Excreting ammonia as the main nitrogenous waste.

amnion. Fluid-filled cavity formed by extraembryonic membrane, which encloses the embryo of reptiles, birds, and mammals (amniotes).

amniote. (1) Having an amnion, as in an amniote egg. (2) A vertebrate (reptile, bird, or mammal) in which the embryo has an amnion.

amphiblastula. Free-swimming larva in certain marine sponges (most calcareous sponges and some demosponges), which develops from a blastula in which flagella are pointed inward within the blastocoel.

amphid. Complex sensory organ near the heads of certain nematodes.

amplexus. Mating embrace of frogs, in which male grasps female from behind as both release gametes.

ampulla (AM-pew-luh). (1) Bulge at the end of each semicircular canal containing hair cells that sense acceleration of the head. (2) Reservoir in the water vascular system of an echinoderm. (3) Low-frequency electroreceptor of certain fishes (ampulla of Lorenzini or ampulla of teleosts).

amylase. Enzyme that digests starch into the disaccharide maltose.

anabolism. Metabolism in which organic molecules are synthesized.

anadromous. Migrating from oceans to freshwater for spawning, like lampreys, shad, trout, and salmon.

anaerobic glycolysis. Fermentation of glucose in the absence of oxygen.

analogous. Similar because of convergent evolution rather than shared ancestry.

anaphase. Stage of mitosis or meiosis when sister chromatids move to opposite poles and become chromosomes.

anaphylactic shock. Drop in blood pressure due to excessive immune response.

anapsid. A vertebrate, especially a reptile, without temporal openings in the skull.

androgen. Steroid hormone, such as testosterone, with masculinizing effects. Compare estrogen.

androgenic gland. Masculinizing gland in crayfish and related crustaceans.

aneuploidy. Difference of at least one in the normal number of chromosomes in a set.

animal pole. Part of an egg containing less yolk.

anion. Negatively charged ion such as chloride.

annulus. External subdivision of a body segment of a leech.

antennal gland (= green gland). Excretory organ in the head of certain crustaceans.

anterior pituitary. Part of the adenohypophysis that secretes several hormones in response to other hormones from the hypothalamus.

anthropomorphism. Attribution of human traits to nonhuman animals.

antiaphrodisiac. Pheromone that discourages mating, usually produced after a female has already mated.

antibody (= immunoglobulin). Protein from plasma cells that binds to a specific foreign molecule as part of the immune response.

antibody labeling. Technique in which radioactive or fluorescent antibodies bind to particular molecules and reveal their presence.

anticodon. Triplet of three bases on transfer RNA that is complementary to a codon on messenger RNA.

antidiuretic hormone (ADH; = vasopressin). Hormone from the posterior pituitary that increases the reabsorption of water by the kidneys.

antifreeze. Substance that lowers the freezing point of a solution.

antigen. Molecule, usually foreign, that evokes an immune response.

antioxidant. Molecule such as vitamin A, C, or E that neutralizes free radicals.

antiserum. Serum containing antibodies to a toxin or other antigen.

antivenin. Antiserum to venom.

antler. Bony, usually temporary, growth on head of deer, antelope, and some other even-toed ungulates.

aorta (a-OR-tuh). (1) Largest artery of vertebrates, which conducts blood from the left ventricle to body tissues. (2) Anterior part of the heart of arthropods.

aortic body. Receptor in the arch of the aorta, which senses the level of CO_2 and O_2 in the blood.

apical complex. Group of organelles characteristic of protozoans in the group Apicomplexa.

apocrine sweat gland. Gland in the groin or arm pit that produces odiferous sweat for social communication.

apodeme. Internal projection of cuticle to which muscle attaches in arthropods.

apopyle. Opening of a radial canal into the spongocoel of a syconoid sponge, or into a flagellated chamber of a leuconoid sponge.

appendicular skeleton. Bones of the limbs, and the pectoral and pelvic girdles to which they are attached.

apron. Abdomen of certain crabs, which is folded beneath the cephalothorax.

arachnoid membrane. Membrane covering the brain, which has villi through which cerebrospinal fluid drains into blood.

arboreal. Living in trees.

archenteron. Cavity in embryo formed during gastrulation; the future gut.

archeocyte. Type of amebocyte in sponges that wanders through the intercellular substance (mesohyl) transporting food and differentiating into other cell types.

Aristotle's lantern. Beaklike chewing apparatus of a sea urchin.

arteriole. Smallest-diameter artery, which conducts blood into a capillary.

artery. Blood vessel that conducts blood away from the heart.

arthrodial membrane (= articular membrane). Flexible, transparent cuticle that hinges sclerotized cuticle in arthropods.

articulamentum. Inner calcareous layer of the plates of a chiton.

articular membrane. Arthrodial membrane.

artificial selection. Breeding controlled by humans selecting for particular hereditary traits.

asconoid. Type of canal system in simple sponges in which the choanocytes line a central chamber (spongocoel). Compare syconoid and leuconoid.

asexual reproduction. Reproduction by fission, budding, or some other method not involving the fertilization of gametes.

assimilated energy. Portion of ingested energy that is available to an organism for use or storage.

association cortex. Area of the cerebral cortex that integrates information from two or more sensory modalities.

astrocyte. Type of glial cell that helps form blood-brain barrier.

asymmetric. Incapable of being bisected into similar halves.

asynchronous muscle (= fibrillar muscle). Type of flight muscle in certain insects in which contraction is not synchronized with action potentials from central nervous system.

atoke. Creeping, benthic, juvenile stage of many polychaete annelids, or the anterior, nonreproductive segments of some other polychaetes.

ATP. Adenosine triphosphate.

atrial natriuretic hormone (= cardionatrin). Hormone from atrium of the heart that opposes aldosterone by increasing secretion of sodium from kidneys.

atrioventricular bundles (of His). Fibers in the heart that conduct action potentials from the atrioventricular node to the ventricles.

atrioventricular (AV) node. Tissue of the heart which is excited by action potentials in the atria and which then, after a delay, sends action potentials through atrioventricular bundles to ventricles.

atrioventricular (AV) valve. One-way valve that permits blood to flow from an atrium into a ventricle.

atrium. (1) Central chamber (spongocoel) of a sponge. (2) Upper chamber in the body of an adult ascidian, which contains the branchial sac. (3) Chamber of the heart that receives blood from the veins.

auricle. (1) External ear (pinna) or any similarly shaped structure. (2) Obsolete term for an atrium of the heart.

auricularia. Larva of a sea cucumber.

autogamy. Fusion of haploid nuclei inside the cell (usually a protozoan) that produced them.

autoimmune disease. Immune reaction against an organism's own molecules.

automimicry. Protection conferred on nontoxic members of a species by their resemblance to toxic members of the species.

autonomic nervous system. Part of the peripheral nervous system that involuntarily controls circulation, digestion, and so on. Consists of sympathetic and parasympathetic divisions.

autosome. Chromosome that occurs in the same number in both males and females. Any chromosome that is not a sex chromosome.

autotomy. Self-amputation to escape a predator.

autotroph (= primary producer). Organism that obtains its energy from inorganic sources, as in photosynthesis.

avicularium. Beak-shaped zooid of ectoprocts used in defense of the colony.

avoiding reaction. Behavior in which a *Paramecium* or other ciliate backs away from a barrier, turns, then moves around it.

axial. Located along the axis.

axial skeleton. Bones of the body axis, including the skull, vertebral column, and rib cage.

axon. Part of a nerve cell that conducts action potentials.

axoneme. Central core in a cilium or eukaryotic flagellum, usually consisting of a "9 + 2" arrangement of microtubules.

axopodium. Type of pseudopodium in protozoans in Actinopoda that is reinforced by microtubules and that shortens to draw food toward the body for phagocytosis.

β. (*See under* **beta entries.**)

B cell. Lymphocyte that transforms into a plasma cell that produces antibodies in the immune response.

baleen. Comb-shaped structure on upper jaw of certain whales that filters food out of water or sediments.

barb. One of the thin processes from the shaft of a feather.

barbule. Hooked structure at the end of the barb of a feather that connects the barb to other barbs.

basal body (= kinetosome). Organelle to which cilia and eukaryotic flagella attach. Identical to the centriole.

basal comb. Device used by gnathostomulids to scrape up food.

basal lamina (= basement membrane). Sheet of fibrous proteins underlying epithelial tissues.

base. (1) Substance that combines with H^+ in solution. (2) Nitrogenous part of nucleic acids.

basement membrane. Basal lamina.

basilar membrane. Membrane in the cochlea that supports the organ of Corti and vibrates in response to sound.

basking. Exposing the body to solar radiation.

Batesian mimicry. Protection from predation conferred on a palatable species by its resemblance to a toxic species.

Bauplan (German; plural Baupläne). Body plan of an animal.

behavioral ecology. Approach to animal behavior that bases hypotheses on the assumption that behavior is optimized.

behaviorism. School of psychology that focuses on behavior in the laboratory and avoids inferences about internal psychological states.

benthic. At the bottom of a lake or ocean.

benthos. Benthic organisms.

beta cell. Cell of the pancreas that secretes insulin.

bicuspid valve (= mitral valve). Left atrioventricular valve of the heart.

bilateral cleavage pattern. Pattern of cleavage in which cells move symmetrically with respect to the embryonic body axis. Characteristic of cephalopod molluscs and tunicates.

bilaterally symmetric. Capable of being bisected into two similar, mirror-image halves.

bile. Secretion from the liver that is stored in the gallbladder until released into the small intestine.

bile salt. Molecule in bile that emulsifies fats.

binary fission. Type of asexual reproduction in which the organism divides into two roughly equal parts.

bindin. Species-specific molecules on echinoderm sperm that bind to receptor molecules on the vitelline envelope of the egg of the same species.

binominal system. System of nomenclature popularized by Carl Linnaeus in which each species is uniquely named by two words: one for the genus group and one for the species group.

biogenetic law. Statement by Haeckel, now discounted, which states that "the history of the fetus is a recapitulation of the history of the race [species]." Ontogeny recapitulates phylogeny.

biogeochemical cycle. Nutrient cycle, such as the carbon cycle, that involves both biological and geological processes.

biological species concept. Ernst Mayr's definition of species as "groups of actually or potentially interbreeding natural populations, which are reproductively isolated from other such groups."

bioluminescent organ. Organ that produces light.

biomass. Total mass of living tissue in an area.

biome. Type of ecosystem, especially terrestrial.

bipedal. On two feet.

bipinnaria. Ciliated, bilaterally symmetric larva of a starfish.

biradial symmetry. Having a generally radial symmetry but with some structures on opposite sides.

biramous. With two branches.

bladder worm (= cysticercus). Juvenile tapeworm, consisting of an invaginated or introverted scolex in a fluid-filled cyst.

blastema. Mass of dedifferentiated cells of mesodermal origin from which a limb regenerates.

blastocoel. Cavity in the blastula.

blastocyst. Preferred term for blastula in mammals.

blastomere. Embryonic cell formed during cleavage.

blastopore. Opening of the archenteron of the gastrula, which usually becomes either the anus or the mouth.

blastula. Early embryonic stage consisting of a mass of cells (blastomeres), generally enclosing a cavity (blastocoel).

blood. Fluid that moves through a circulatory system consisting of tubular vessels.

blood group. System of antigens on red blood cells that partly determines whether a donor's blood will be compatible with that of a recipient.

blood–brain barrier. Blood capillaries and certain glial cells of the brain that allow glucose and other nutrients to enter the cerebrospinal fluid but prevent the entry of many unwanted molecules.

blood–CSF barrier. Choroid plexus and arachnoid membrane, which control the concentrations of ions and some other materials in the cerebrospinal fluid.

Bohr effect. Decrease in the affinity of hemoglobin for O_2 in the presence of CO_2.

book gill. Respiratory organ of horseshoe crabs, with membranes arranged like pages of a book.

book lung. Respiratory organ of spiders, with membranes arranged like pages of a book.

bolus. Rounded mass, such as swallowed food.

Bowman's capsule (= glomerular capsule). Spherical structure of each nephron of the kidney enclosing the glomerulus.

brachial. Pertaining to the arm.

brainstem. part of brain left after removing cerebrum and cerebellum.

branchia. Thin, tentacle-like structure at the anterior end of a beard worm.

branchial. Pertaining to gills.

branchial heart. Accessory heart that pumps deoxygenated blood from the body tissues to the gills in certain cephalopod molluscs.

branchial sac. Chamber in adult ascidians that produces the hydroskeleton and exchanges respiratory gases.

breathing. Ventilation of the respiratory organs.

bristle. (1) Type of feather with barbs only at the base of the shaft. (2) Short, hairlike structure.

bronchiole. Tubule in a lung that terminates in air sacs.

bronchus (BRON-kus; plural **bronchi,** pronounced BRONK-eye). Tube connecting the trachea to a lung.

brood. (1) Eggs, larvae, or young. (2) To guard or warm eggs, larvae, or young.

brood parasite. Species such as cuckoo that depends on another species to brood its young.

brood patch. Bare spot on the belly of a bird that increases heat transfer between the bird and its egg during incubation.

brown adipose tissue (BAT = brown fat). Fatty tissue adapted for generating body heat.

brush border. Group of microvilli on a plasma membrane.

buccal cavity. Mouth cavity.

buccal ciliature. Cilia near the cytostome (mouth) that certain protozoa use in feeding.

buccal diverticulum (= stomochord). Anterior projection of the buccal cavity of a hemichordate, formerly considered a notochord.

buccal pumping. In fishes, ventilation of gills by contracting the buccal and opercular cavities.

budding. Type of asexual reproduction in which offspring form from small growths (buds) in or on the parent.

buffer. Molecule or group of molecules that resists changes in the hydrogen-ion concentration (pH) of a solution.

bulbourethral gland (= Cowper's gland). Gland that secretes clear, mucous fluid for lubrication during copulation.

bulbus arteriosus. Conus arteriosus in a fish. Chamber into which blood is pumped from the ventricle of the heart.

bundles (of His). Fibers in the heart that conduct action potentials from the atrioventricular node to the ventricles.

bursa. Fluid-filled sac, such as one of five sacs that exchange respiratory gases in brittle stars.

byssus (= byssal threads). Filaments secreted by certain bivalve molluscs for attachment to substrate.

calcareous (kal-KARE-ee-us). Mineral structure based on calcium, especially calcium carbonate.

calmodulin. Protein that mediates the action of Ca^{2+} in cells.

calorie. Amount of heat required to warm a gram of water by 1°C.

calyx (plural calyces). Cup-shaped structure, such as the body of an entoproct or crinoid.

cancellous (= spongy = trabecular) **bone.** Bone with spaces containing marrow.

cancer (= malignancy). Tumor from which cells spread to other parts of the body.

canine. (1) Referring to a dog or other canid. (2) Sharp tooth adapted for tearing meat.

capillary. Thin-walled tubule between arteriole and venule, responsible for circulating blood in tissues.

capitulum. Projecting mouth region, as in ticks and mites.
captaculum. Ciliated tentacle of a scaphopod mollusc.
carapace. (1) Shieldlike dorsal plate over the cephalothorax of a crustacean. (2) Dorsal shell of a turtle.
carbamino compounds ($HbCO_2$). Hemoglobin with CO_2 bound to its amine groups.
carbohydrate. Organic molecule with aldehyde or ketone group and consisting of carbon, hydrogen, and oxygen in the ratio 1:2:1. Simple sugars and more complex molecules made of sugars.
carbon cycle. Biogeochemical movements of carbon among various organic and inorganic states.
carbonic anhydrase. Enzyme that catalyzes the formation of bicarbonate from carbonic acid.
carcinogen. Cancer-causing substance.
carcinoma. Cancer that begins in epithelium, especially in lungs, breasts, and intestines.
cardiac muscle. Type of muscle that forms the heart in chordates.
cardiac output. Volume of blood pumped by heart per unit of time; equals heart rate multiplied by stroke volume.
cardiac stomach. First of two chambers in the stomach of an echinoderm.
carina (= sternal keel). Long, thin process on the sternum of a bird, to which flight muscles are attached.
carnivore. Organism that feeds on animals.
carotenoid. Straight-chain hydrocarbon related to vitamin A and often found as a yellow, orange, or red pigment.
carotid body. Receptor in the sinus of the carotid artery that senses the level of CO_2 in the blood.
carrying capacity. Maximum number of individuals of a species that can be sustained in a habitat indefinitely.
cartilage. Flexible connective tissue in the skeleton of a vertebrate.
cartilaginous. Consisting of cartilage.
caste. Group of individuals within a species of eusocial insects (ants, bees, termites) that is different in appearance and behavior from other such groups.
castration. Surgical removal of ovaries or testes.
catabolism. Metabolic breakdown of organic molecules.
catadromous. Migrating from freshwater to seawater to spawn, like eels.
catastrophism. Generally discredited view that past geological changes occurred suddenly, due to floods, volcanism, and so on.
catch connective tissue. Connective tissue of echinoderms that stiffens as defense against predation.
catch muscle. Type of muscle in bivalve molluscs that allows the shells to remain closed without use of energy.
catching spiral. Part of an orb web of spiders in which prey are caught.
cation. Positively charged ion, such as sodium and calcium.
caudal. Pertaining to or toward the tail.
cecum (SEE-kum; plural **ceca**). Blind pouch, for example, at the beginning of the large intestine of vertebrates, or the midgut of insects.
cell cycle. Cycle of cell activity beginning with interphase in a new daughter cell at the end of a previous mitosis and continuing through the end of the following mitosis.
cell theory. Proposal by Schwann in 1839 "... that there exists one general principle for the formation of all organic productions, and that this principle is the formation of cells. . . ."
cell-adhesion molecule (CAM). Glycoprotein on cell surface responsible for adhesion of particular cells to each other.
cellular respiration (= respiration). Complete catabolism of glucose requiring oxygen and releasing carbon dioxide.
central canal. (1) Haversian canal in dense bone that contains vessels and nerves. (2) Channel through the spinal cord through which cerebrospinal fluid circulates.
central pattern generator. Network of nerve cells that produces a rhythmic pattern of action potentials.
centriole. One of a pair of microtubular structures that gives rise to mitotic spindles and to cilia or flagella. Identical with basal body.
centrolecithal. Egg with yolk concentrated in the center, as in arthropods.
centromere. Region of a chromosome where kinetochore is located, and which contains genes for kinetochore.
cephalic ganglion. Enlargement of the central nervous system in the head, often called the brain.
cephalon. Head of a trilobite.
cephalothorax. Fused head and thorax, as in spiders and many crustaceans.
cercaria. Larva of *Schistosoma* and other flukes that swims from molluscan intermediate host to final host or second intermediate host.
cercus (SIR-kus; plural **cerci** SIR-sy). Posterior antenna-like structures on certain insects.
cerebellum. Part of the vertebrate brain that coordinates learned movements.
cerebral commissure. Band of nerve fibers connecting the two hemispheres of the cerebrum.
cerebrospinal fluid (CSF). Fluid in the ventricles of the vertebrate brain and the central canal of the spinal cord.
cerebrum. Major part of the vertebrate brain responsible for perception and voluntary movement.
channel. Glycoprotein in plasma membrane that permits the passage of certain ions.
character. Any anatomical, molecular, or behavioral difference that provides a basis for classification.
character analysis. Determination of the nature and potential usefulness of a character in classification.
character displacement. Condition in which individuals with similar traits that cause them to compete with each other become less similar over succeeding generations, due to evolution.
chela (KEE-luh). Pincers on an appendage of an arthropod.
chelicera (ke-LIS-ur-uh). Paired mouthpart of spiders and other chelicerates.
chelipeds (KEL-i-pedz). First pair of walking legs, with pincers, in certain crayfish, crabs, and other decapod crustaceans.
chemical synapse. Structure on a nerve cell that releases chemical transmitter onto another nerve cell or muscle cell.
chemiosmotic theory. Proposal by Peter Mitchell in 1961 that diffusion of H^+ across inner membrane of mitochondria produces ATP.
chemoautotrophic. Referring to bacteria that produce energy for synthesis of organic molecules by oxidizing inorganic molecules.
chitin (KY-tin). Polysaccharide that is a major component of cuticle in arthropods and that occurs in integuments of many other invertebrates.

chloragogue cells (= chloragog = chloragogen). Greenish or brownish tissue in annelids that assists nephridia in eliminating nitrogenous wastes, stores glycogen and fats, and nourishes maturing ova.

chlorocruorin. Green respiratory pigment in some annelid worms.

choanocyte (ko-AN-oh-site) (= collar cell). Flagellated cell of sponges that propels water and traps food particles in its collar.

choanoderm (ko-AN-oh-derm). Lining of flagellated chambers of sponge.

chondrocyte (KON-dro-site). Cell in cartilage that secretes collagen and proteoglycans.

chorion. Extraembryonic membrane of vertebrates that exchanges respiratory gases in the eggs of reptiles and birds and develops into part of the placenta in most mammals.

chorionic villus. Fingerlike projection of the mammalian placenta into the endometrium of the uterus.

choroid plexus. Structure that helps produce cerebrospinal fluid by actively transporting ions and certain molecules from blood.

chromatid. One of two structures formed by a dividing chromosome which becomes a new chromosome following mitosis or meiosis.

chromatin. Molecular substance (DNA and protein) composing chromosomes.

chromatophore. Cell with pigment that can be covered or uncovered, or concentrated or dispersed to change color of integument.

chromosomal aberration. Visible abnormality in structure or number of chromosomes.

chromosome. Elongated or rounded structure consisting of DNA and proteins which divides during meiosis and mitosis to carry genetic information to gametes and to dividing cells.

chrysalis. Pupa of a butterfly or moth.

chyme (pronounced kyme). Partly digested food and acid in the stomach.

chymotrypsin. Pancreatic enzyme that digests proteins.

ciliary loop. Band of mechano- and/or chemoreceptors beneath the head of an arrowworm.

ciliary photoreceptor. Type of photoreceptor in vertebrates that develops around a cilium.

ciliate. Ciliated protozoan, or any protozoan in the group Ciliophora.

cilium. Short, hair-shaped motile structure on a cell surface that usually occurs in large numbers. (Compare flagellum.)

circadian rhythm. Biological rhythm with a duration of approximately one day.

circannual rhythm. Biological rhythm with a period of approximately one year.

circulatory system. Heart and vessels that transport blood to tissues and back.

cirrus. (1) Tuft of fused cilia used for locomotion by certain protozoans. (2) Copulatory organ of male flatworms and certain other invertebrates. (3) Tuft of setae on lobes of polychaete annelids and some other invertebrates. (4) Legs of barnacle adapted as brush-like structure for feeding. (5) Clawlike structures that hold feather stars to substratum.

clade. In cladistics, all the descendants of one ancestor, as represented by all the branches on a cladogram that can be traced to one node.

cladistics (= phylogenetic systematics). Method of systematics in which relationships among organisms are deduced from the number of derived homologous characters (synapomorphies) and represented in a cladogram.

cladogram. Pattern of branching lines representing the sequence, but not the times, of evolutionary descent of groups of organisms.

clap-fling. Pattern of wing movement in some insects in which the wings come together during upstroke then peel apart to create lift.

clasper. Modified fin used as an intromittent organ in male sharks and certain other cartilaginous fishes.

class. Taxonomic level below phylum and above order.

classical conditioning (= Pavlovian = Type I conditioning). Type of learning in which a normal response to one stimulus (the unconditioned stimulus) becomes associated with a new stimulus (conditioned stimulus).

classification. Assignment of organisms to taxonomic groups based on relationships among them.

cleavage pattern. Pattern of cell movement during embryonic cleavage. See also particular types: bilateral, discoidal, radial, rotational, spiral, superficial.

cleidoic (kly-DOH-ik). Referring to a self-contained egg of an insect or amniote (reptile, bird, or egg-laying mammal).

climax. (1) Final stage of succession after which species in a community remain stable. (2) Orgasm.

clitellum. External swelling on certain segments of oligochaetes and leeches that secretes mucus to form a cocoon.

cloaca (klo-A-kuh) (plural **cloacae** [klo-A-see]). Chamber into which gametes and excretory wastes empty.

clone. Group of genetically identical cells or organisms.

closed circulatory system. Circulatory system in which blood is confined to heart or vessels, as in vertebrates.

clot retraction. Shrinking of blood clot due to contraction of platelets.

clotting (= hemostasis = coagulation). Congealing of blood that stops bleeding.

cnida (NY-duh) (= cnidocyst). Adhesive or stinging organelle in cnidarians.

cnidocil (NY-doe-sill). Trigger for release of nematocyst of cnidarians.

cnidocyst (NY-doe-sist) (= cnida). Adhesive or stinging organelle in cnidarians.

cnidocyte (NY-doe-site). Cell that produces a cnida.

coated pit. Depression in a plasma membrane where receptor-mediated endocytosis occurs.

cobweb. Loosely woven spider web that uses sticky threads to snare prey walking beneath it.

cochlea. Tubular auditory organ of mammals, birds, and some reptiles.

cocoon. Hollow structure in which developing or resting stages occur.

codon. Triplet on messenger RNA that specifies addition of a particular amino acid or that signals the stop or start of synthesis of a protein.

coelenteron (seal-EN-ter-on) (= gastrovascular cavity). Cavity of an incomplete gut serving in both digestion and circulation in cnidarians and some other invertebrates.

coelom (SEAL-um). Body cavity that develops in mesoderm and is enclosed within a peritoneal membrane.

coelomocyte. Amoebocyte in coelomic fluid, or in vascular fluid of an echinoderm.

coenzyme. Organic cofactor that transfers a functional group to or from the substrate during an enzymatic reaction.
coevolution. Interaction of two species such that each evolves adaptations to the other.
cohesion. Binding of identical molecules to each other.
collagen. Fibrous protein reinforcing skin and other flexible tissues.
collar cell. Choanocyte of a sponge.
collenchyme. Mesenchyme (= mesoglea) of jellyfishes.
collencyte. Collagen-secreting amebocyte in sponges.
colligative. Property of a solution, such as freezing point and osmolarity, that depends on its concentration rather than on the nature of the solute.
colloid. Mixture of water and molecules that do not dissolve but are kept from precipitating by attraction with the water.
colloid osmotic pressure (= oncotic pressure). Hydrostatic pressure due to osmotic effect of colloid.
columella (1) Stapes. Sole middle-ear bone in an amphibian, reptile, or bird. (2) Central stalk of a gastropod shell.
columnar. Column-shaped, as in cells of columnar epithelium.
commensalism. Association between two species with no apparent benefit or harm to either species.
community. All interacting species within a habitat.
competitive exclusion. Exclusion of one species by another in a habitat in which both compete for the same resource.
complement. Group of proteins that attract phagocytic cells and aids antibodies in destroying bacteria.
complementary DNA (cDNA). Segment of DNA with base sequence complementary to another segment of DNA.
complete digestive tract. Digestive tract with a mouth and a separate anus.
compound eye. Arthropod eye consisting of numerous subunits (ommatidia).
concertina motion. Type of locomotion used by burrowing snakes, in which a portion of the body is bent and braced against the sides of the burrow.
conchin (= conchiolin). Proteinaceous substance that forms the periostracum that covers the shells of molluscs.
conditioned stimulus. Stimulus to which an animal learns to respond through classical conditioning.
conduction. Movement, as of action potentials, heat, or electric charge.
conductivity. Ability to conduct heat, charge, and so on.
cone. Photoreceptor cell in vertebrate eye that is responsible for color vision.
congenital defect. Nonhereditary developmental defect.
conjugation. Temporary coupling and exchange of genetic material in certain protozoans (ciliates) or between bacteria.
connective. Nerve linking two ganglia in a nervous system.
connective tissue. Tissue of mesodermal origin, including cartilage, bone, blood, and connective tissue proper (ligament, tendons, fat).
consumer (= heterotroph). Organism that obtains its energy from organic materials synthesized by other organisms.
continental drift. Movement of continents on a layer of molten rock.
contour feather (= vane feather). Major type of feather, in which barbs form two vanes on opposite sides of the shaft.
contractile vacuole. Osmoregulatory organelle in protozoa and freshwater sponges.
conus arteriosus. Chamber in the heart of a fish or amphibian into which blood is pumped from the ventricle.
convection. Transfer of heat through bulk movement of air or water.
convergence. Evolution of similar (analogous) traits independently in two different lines of descent.
cooperative breeding. Breeding in which nonreproductive individuals assist in raising offspring of others of the same species.
copulation. Transfer of sperm by a male into a female.
copulatory plug. Secretion by snakes and certain mammals that prevents subsequent males from copulating with a female, or prevents the semen from leaking from the vagina.
cornea. (1) External transparent coat of vertebrate eye. (2) Transparent cuticle covering the eye of an arthropod.
cornified layer. Dead, keratinized cells protecting epidermis from water and abrasion.
corona. Feeding apparatus of a rotifer, consisting of a ciliated disc surrounding the mouth.
corpus luteum. Remainder of an ovarian follicle after ovulation.
cortical reaction. Release of the contents of cortical granules beneath the plasma membrane of an egg during fertilization, which serves as a slow block to further fertilization.
countercurrent exchange. Exchange of heat, oxygen, and so on, between two fluids moving past each other in opposite directions.
courtship. Behavior by which male and female of same species prepare for sexual reproduction.
covalent bond. Chemical bond in which one or more pairs of electrons are shared by two atoms.
Cowper's gland (= bulbourethral gland.) Gland that secretes a clear, mucous fluid for lubrication during copulation.
coxal gland. Excretory organ of spider, which empties through a pore near the proximal joint (coxa) of the leg.
cramp. Rigidity of muscle caused by lack of ATP.
cribellate spider. Spider with cribellum that produces wooly silk for the catching spiral of its orb web.
crista. Infolding pocket of the inner membrane of a mitochondrion. Site of the respiratory chain.
crop. (1) Anterior part (= foregut) of insect digestive tract in which food is stored and partly digested. (2) Pouch in the esophagus of a bird, in which food is stored.
cross-bridge. Connection between thin and thick filaments in muscle.
crossing over. Intertwining of the four chromatids during prophase I of meiosis, during which they exchange genetic material.
cryoprotectant. Substance that protects tissues from damage due to formation of ice crystals.
cryptobiosis. Inactive state induced by dehydration.
crystalline cone. Structure in ommatidium of compound eye that, together with lens, concentrates light.
crystalline style. Rod-shaped structure in stomachs of bivalve molluscs that grinds against the gastric shield, releasing digestive enzymes.
ctenidium. Gill, especially of a mollusc.
ctenoid. With comb-shaped projections.
cuboidal. Cubical, like the cells of cuboidal epithelium.

cuticle. (1) Organic, noncellular body covering external to the epidermis, as in the integument of arthropods, annelids, and some other invertebrates. (2) Cornified layer of dead cells on vertebrate skin.

cuttlebone. Vestigial shell used for buoyancy in cuttlefish.

cyclic nucleotide. Nucleic acid base in which phosphate is attached at two places to the ribose. Usually cyclic adenosine monophosphate (cAMP) or cyclic guanosine monophosphate (cGMP).

cycloid. Disc-shaped.

cydippid larva. Free-swimming larva of a ctenophore.

cypris larva. Larval stage following the nauplius larva of a barnacle.

cyrtocyte. Curved solenocyte (flagellated cell in the protonephridium) in a gnathostomulid.

cyst. Capsule containing a protozoan or animal in an inactive state resistant to environmental stress.

cystacanth. Embryo of an acanthocephalan that is encysted in an intermediate host and develops into adult following ingestion by final host.

cysticercus (SIS-tee-SIR-kus) (= bladder worm). Juvenile tapeworm, consisting of an invaginated or introverted scolex in a fluid-filled cyst.

cytidine. A nitrogenous base in nucleic acids.

cytochrome. Protein with iron-containing heme group; part of electron transport chain of cellular respiration.

cytokine (= lymphokine). Interferon, interleukin, or other substance that coordinates inflammation and immunity.

cytokinesis. Division of cytoplasm during mitosis or meiosis.

cytoplasm. Cell contents, including organelles except the nucleus and plasma membrane.

cytoplasmic membrane system. System comprising endoplasmic reticulum, Golgi complexes, and lysosomes, responsible for formation of vesicles for synthesis of new plasma membrane, formation of lysosomes, and secretion of glycoproteins.

cytoplasmic streaming. Movement of endoplasm into a pseudopodium; believed responsible for ameboid movement.

cytosine. Nucleotide consisting of cytidine linked to ribose.

cytoskeleton. Network of microfilaments, microtubules, intermediate filaments, and other protein structures that reinforce cells.

cytostome. Opening through which food is ingested in many protozoa.

daily torpor. Hours-long drop in body temperature in a small bird or mammal in cold.

dart. Sharp structure with which certain snails stab each other during courtship.

dauer larva. Resistant larval form of some nematodes.

deciduous teeth (= milk teeth). Temporary teeth of a young mammal.

decomposer. Bacterium, fungus, or other organism that feeds on organic tissue or wastes and converts them to inorganic molecules.

defensive mimicry. Resemblance of a potential prey to a different animal or object that is not preyed upon.

definitive host (= primary or final host). Organism in which the mature form of a parasite develops.

dendrite. Branch of a nerve cell that conducts potentials toward the cell body.

dendrogram (= phylogenetic tree). Diagram representing the evolution of a group of organisms.

denitrifying bacterium. Bacterium that converts nitrite and nitrate, which can be used by plants, into gaseous nitrogen (N_2), which cannot be used.

dense bone (= compact bone). Bone that forms by replacement of cartilage and provides most structural support.

density-dependent factor. Ecological factor that depends on population density; for example, communicable diseases and food shortages.

density-independent factor. Ecological factor that does not depend on population density; for example, predation and natural catastrophes.

dental formula. Numerical key indicating numbers of each kind of tooth on each half of the upper and lower jaws of a mammal.

denticle. Tooth on jaw in pharynx of a polychaete annelid.

deoxyribonucleic acid (DNA). The material basis for genes. Nucleic acid with deoxyribose as its sugar. It generally occurs as a double helical strand.

depolarization. Reduction in the magnitude of a membrane potential.

derived. Referring to a trait that evolved later than a primitive trait.

dermal bone (= membranous bone). Bone that forms from layers of embryonic connective tissue rather than by replacement of cartilage (as in dense bone). Forms fish scales, turtle shells, antlers, and flat bones of vertebrate skulls.

dermal gland. Gland of unknown function in arthropod integument.

dermis. True skin underlying epidermis.

desmosome. Structure joining the plasma membranes of adjacent cells.

determinate cleavage (= mosaic development). Cleavage in which fates of cells are determined in early cleavage. Common among invertebrates.

determination. Permanent establishment of the fate of an embryonic cell.

detorsion. Partial reversal of torsion in certain snails.

detritus feeder (= detritivore). Animal that feeds on particles of decaying organisms.

deuterostome. Animal in which the mouth originates from an opening in the embryo other than the blastopore, and usually with other distinctive patterns of embryonic development.

diadromous. Migrating either from fresh water to sea water to spawn (catadromous) or from sea water to fresh water to spawn (anadromous).

diapause. Inactive state begun prior to adverse environmental condition, such as cold.

diaphragm. Thin sheet of tissue separating the thoracic and abdominal cavities of mammals, birds, and some reptiles.

diapsid. Reptile or other vertebrate with two temporal openings in the skull.

diastolic blood pressure. Minimum blood pressure just before ventricular contraction.

differentiation. Irreversible change in the structure and function of a cell during embryonic development.

diffraction colors. Structural colors formed by closely spaced grooves in integument, scales, or feathers.

diffusion. Passive movement of molecules from a region where they are highly concentrated to a region of lower concentration.

digestive gland. (1) Hepatopancreas in a crustacean, which secretes digestive enzymes and stores nutrient reserves. (2) Pyloric cecum in an echinoderm, which stores and absorbs nutrients.

dihybrid cross. Cross between two individuals heterozygous for two different genes.

dimorphism. Occurrence of two forms in the same species; for example, sexual dimorphism.

dioecious (die-EE-shus). Having separate sexes, as opposed to monoecious, in which each individual has both male and female reproductive organs.

diphycercal (DIF-i-SIR-kal). Pertaining to the caudal fins of lobe-finned and some other fishes, which taper symmetrically to a point at the rear.

diploblastic. Developing from two germ layers, as was once thought to occur in cnidarians and ctenophores.

diploid. Having chromosomes occurring in homologous pairs, rather than individually (haploid).

diplosegment. An apparent segment with two pairs of legs, which results from the fusion of a pair of true segments in a millipede.

direct calorimetry. Measurement of rate of body heat production to determine metabolic rate.

direct development. Development into an organism that resembles the adult form, rather than going through a larval stage.

disaccharide. Carbohydrate consisting of two simple sugars (monosaccharides) joined by a glycosidic link.

discoidal cleavage. Cleavage pattern is which embryo is compressed into a disc, characteristic of fishes, reptiles, and birds.

disruptive coloration. Pattern of coloration that masks outline of an animal; for example, a zebra.

distal convoluted tubule. Part of a nephron in the kidney that conducts tubular fluid to the collecting duct.

DNA hybridization. Technique for comparing DNA from two different species by combining single DNA strands from each species and determining the temperature at which they come apart.

dominance hierarchy (= pecking order). Social rank that determines which animals in a group has priority in mating, feeding, or attacking others.

dominant. Referring to an allele that is expressed in the presence of a different (recessive) allele for the same trait.

dorsal. Pertaining to the back or upper surface.

dorsal root. Dorsal branch of a nerve at the spinal cord, which contains axons from sensory receptors.

down feather. Feather in which barbs are not interconnected by barbules and are therefore fuzzy.

drag line. Silk strand played out by spiders as anchorage.

drone. Male honey bee or other hymenopteran.

ductus arteriosus. Tube connecting the pulmonary artery to the aorta in a mammalian fetus, permitting blood to bypass the lungs.

ductus deferens (plural **ductus deferentes**) (= sperm duct = vas deferens). Tube through which sperm are ejaculated.

duo-gland adhesive system. Combination of viscid and releasing glands used in locomotion and anchoring in flatworms.

eccrine sweat gland. Gland that produces sweat that cools the body by evaporation.

ecdysis (EK-duh-sis; = molting). Shedding of cuticle, skin, hair, or feathers.

ecdysone (= molting hormone). Steroid hormone of arthropods that triggers molting.

echinopluteus. Larva of a sea urchin.

echolocation. Location of prey or other objects by emission of ultrasound and reception of the echo, as in bats and whales.

eclosion. Emergence of an adult insect from the pupa.

ecological growth efficiency. Ratio of energy stored as new tissue to ingested energy over a period of one year.

ecosystem. Association of all the organisms in an area, together with the physical environment.

ectoderm. (1) Outer germ layer from which external epithelium and nerve cells develop. (2) Tissues derived from ectoderm.

ectolecithal egg. Egg with yolk outside plasma membrane of the ovum.

ectoplasm (= plasma gel). Outer, gelatinous portion of cytoplasm. Compare endoplasm.

ectotherm. Animal whose body temperature is determined primarily by the environmental temperature.

edema (uh-DEEM-uh). Swelling.

efferent. Conducting outward, such as a nerve cell conducting action potentials out of the central nervous system, or an artery conducting blood from the heart. Compare afferent.

egg. Female gamete, or the zygote, with associated membranes, yolk, and shell (if any).

egg tooth. Temporary projection on beak of hatching birds and certain reptiles, which enables them to break through the shell.

electrical synapse. Structure by which membrane potentials conduct directly from one nerve cell to another.

electrocardiogram (ECG or EKG). Record of voltages produced by action potentials of the heart.

electron acceptor. Molecule such as oxygen that accepts electrons from an electron transport chain.

electron transport chain. System of enzymes and cytochromes involved in ATP synthesis in mitochondria.

electrophoresis. Procedure for comparing proteins by their rate of movement in an electric field, which depends on their net charges, sizes, and shapes.

electroreceptor. Receptor in many fishes and some other vertebrates that responds to electric fields.

El Niño (= El Niño—Southern Oscillation event, or ENSO). Invasion of warm Pacific water that prevents normal upwelling of cold water off west coast of South America.

Eltonian pyramid. Diagram representing energy, biomass, or numbers of organisms at each trophic level.

elver. Juvenile eel.

embryo. Developing organism, especially in early stages. In humans the term is applied until the eighth week of gestation.

emulsify. To disperse large fat droplets into smaller ones.

endite. Medial minor branch of a crustacean appendage.

endocrine. Referring to a ductless gland that secretes its product (e.g., a hormone) into the blood.

endocuticle. Inner, nonsclerotized layer of arthropod cuticle.

endocytosis. Transport of material into a cell by enclosing it within a pit in the plasma membrane that becomes a vesicle.

endoderm. (1) Inner germ layer from which the lining of the gut, internal glands, and respiratory organs develop. (2) Tissues derived from endoderm.

endogenous clock. Hypothetical biological system that controls the timing of biological rhythms, or compensates for movement of Sun or stars during migration.

endogenous opiate. Secretion that induces sleep and relieves pain, mimicked by morphine, heroin, and other opiates.

endolecithal egg. Egg with yolk enclosed in the plasma membrane of the ovum.

endolymph. Fluid in cochlea and vestibular apparatus.

endometrium. Inner lining of the uterus.

endoparasite. Parasite that lives within the gut or tissues of host.

endoplasm (= plasma sol). Inner, fluid portion of cytoplasm. Compare ectoplasm.

endoplasmic reticulum (ER). Cytoplasmic membranes bearing ribosomes for production of proteins (rough ER) or lacking ribosomes (smooth ER).

endopod (= endopodite). Medial branch of a biramous appendage of a crustacean.

endopterygote. Referring to an insect in which the wings or wing buds are apparent only in the adult, which is therefore holometabolous.

endoskeleton. Skeleton surrounded by other body tissues, as in vertebrates.

endostyle. Groove in ventral surface of branchial sac of an ascidian, which secretes mucus that traps food particles.

endosymbiont. Organism that lives inside another organism.

endotherm. Animal whose body temperature is determined primarily by its own heat production and conservation.

endplate (= neuromuscular junction). Neuronal synapse on a skeletal muscle cell.

endplate potential (EPP). Change in voltage in plasma membrane of muscle cell caused by release of transmitter from endplate.

energy level. Energy associated with a particular position, such as distance of an electron from a nucleus.

enhancer. Portion of DNA that promotes transcription of a distant gene.

enterocoelous. Having a coelom that developed from pouches in the endoderm in the embryonic gut (archenteron).

entropy. Measure of the energy rendered unavailable for further work; equal to or greater than the amount of heat lost from a system divided by the temperature change of the system. A measure of disorder.

enzyme. Organic catalyst, usually a protein, which speeds up a chemical reaction without being changed by the reaction.

ephyra. Jellyfish larva that buds asexually from a strobila.

epiblast. In placental mammals part of the inner cell mass that forms the amnion and the embryo.

epicuticle. Waxy and proteinaceous covering of arthropod cuticle.

epidermis. Outer layer of surface tissue.

epididymis (EP-i-DID-i-mis; plural **epididymides**). Structure on the testis that stores sperm.

epigenesis. Embryonic development of new structures from those that did not previously exist as such. Compare preformation.

epiglottis. Flap of cartilage at the root of the tongue that closes the opening of the trachea (the glottis) during swallowing.

epipod (= epipodite). Lateral minor branch (exite) on a protopod of a crustacean appendage.

epithelial tissue (= **epithelium**). Tissue derived from ectoderm or endoderm that forms the body covering or the lining of body cavities.

epitheliomuscle cell. Contractile cell in the epidermis of a cnidarian.

epitoky. Transformation of certain polychaete annelids from juveniles (atokes) to reproductive forms (epitokes).

equilibrium. Balanced state in which no net change occurs.

equilibrium potential. Hypothetical membrane voltage that would be required to maintain a certain ratio of concentrations of an ion on each side of a membrane.

erythrocyte (= red blood cell). Cell that transports hemoglobin in blood of vertebrates.

esophagus. Tube connecting the mouth to the stomach.

essential. Required in the diet; for example, essential amino acid or essential fatty acid.

esthete. Compound receptor organ of a chiton.

estivation (= aestivation). Dormant condition during summer or dry period.

estrogen. Estradiol or related steroid hormone that promotes female behavior, morphology, and reproductive functions.

estrous cycle. Cycle of changes in the reproductive physiology of a female.

estrus. Sexual receptivity in a female vertebrate; "heat."

estuary. Area where a river empties into a sea.

ethology. Study of behavior that emphasizes mechanisms and evolution in the natural environment.

eukaryote (= eucaryote). Cell with nuclei and organelles, or an organism (protist, fungus, plant, or animal) with such cells. Compare prokayrote.

euryhaline. Able to tolerate wide shifts in environmental salinity. Compare stenohaline.

eutely. The property of having a fixed number and position of somatic cells from one individual to the next within a species.

evolution. Genetic change in a group of organisms.

evolutionarily stable strategy (ESS). In behavioral ecology, a behavior that persists because competing alternatives are less advantageous.

evolutionary systematics. Method of taxonomy in which times and patterns of evolution are deduced from comparative anatomy, paleontology, and other methods.

excitable cell. Cell capable of producing an action potential.

excurrent canal. In leuconoid sponges the canal that carries water from a flagellated chamber to the osculum.

exite. Lateral minor branch of a crustacean appendage.

exocrine. Secreted through a duct that does not empty into the bloodstream.

exocuticle. Sclerotized layer of arthropod cuticle beneath the epicuticle.

exocytosis. Transport of material out of a cell by enclosing it within a vesicle that fuses with the plasma membrane.

exon. Portion of messenger RNA or DNA that is used in translation of protein. Compare intron.

exopod (= exopodite). Lateral branch of a biramous appendage of a crustacean.

exopterygote. Referring to an insect in which the wings or wing buds appear in an immature stage, and which is therefore hemimetabolous.

exoskeleton. Supporting structure that encloses all or part of the body, such as in insects.

extant. Not extinct.

extensor. Muscle that extends a limb away from the body axis.

external fertilization. Fertilization of eggs outside the female's body.

extraembryonic membrane. Membrane outside the embryo of a reptile, bird, or mammal (amniote). Includes chorion, amnion, allantois, and yolk sac.

extracellular matrix (= ECM). Group of proteins outside the plasma membrane that helps maintain cell structure and adhesion.

eyespot. (1) Simple photoreceptor lacking lens or other structures found in true eyes. (2) Eye-shaped pattern that may confuse potential predators.

F_1 generation. Hybrid offspring resulting from an experimental cross of true-breeding parents.

F_2 generation. Offspring from interbreeding or self-fertilizing F_1 hybrids.

facet. Part of the cornea over one ommatidium of a compound eye of an arthropod.

facilitated diffusion. Diffusion aided by a transporter in the plasma membrane.

FAD. Flavine adenine dinucleotide.

Fallopian tube (= oviduct). Tube through which eggs of mammals travel after ovulation.

family. Taxonomic level below order and above genus.

fast-twitch fiber (= type II white fiber). Muscle cell that contracts rapidly but tires easily because of reliance on glycolysis.

fat. Lipid with up to three fatty acids linked to a glycerol. Mono-, di-, or triglyceride.

fat body. Fatty, diffuse organ in insects that stores energy reserves and produces yolk.

fate map. Diagram showing structures that will be formed by parts of an embryo.

fatty acid. Chain of carbon atoms (hydrocarbon chain) with an acidic carboxyl group at one end.

faunal region (= faunal realm). Six continent-sized areas characterized by their distinctive animals.

faunal succession. Changes in kinds of animals that inhabit an area as a result of plant succession.

feather tract (= pteryla). Linear region of bird skin that produces feathers.

feedback inhibition. Control mechanism in which the product of a process inhibits the process.

feral. Once-domesticated.

fermentation. Anaerobic catabolism, such as glycolysis.

fertilization. Fusion of a sperm with an egg, forming a zygote.

fertilization membrane. Membrane of egg after fertilization has triggered the cortical reaction that prevents further fertilization.

fetal hemoglobin. Form of hemoglobin in fetus and newborn mammal.

fetus. Unborn mammal. Developing human after the eighth week of gestation.

fever. Elevation of body temperature above normal, especially in response to infection.

fibrillar muscle (= asynchronous muscle). Muscle in certain insects in which contraction is not synchronized with action potentials.

fibrin. Blood protein that changes from globules (fibrinogen) to fibers to form a clot.

fibroblast. Cell in connective tissue proper that secretes collagen.

fibronectin. Glycoprotein that guides cell migration and makes cells adhere.

fibrous protein. Protein in the form of a long, insoluble strand.

Fick's equation. Equation stating that the rate of diffusion is directly proportional to the surface area of a membrane and the difference in concentrations or partial pressures on each side of the membrane, and inversely proportional to the thickness of the membrane.

filopodium. Long, thin pseudopodium of certain protozoans (Filosea).

filoplume. Type of feather with long, flexible shafts and barbs only at the tip.

filter feeding. Obtaining food by trapping small particles suspended in water or air.

fimbriae. Ciliated, finger-shaped structures that draw eggs into the end of the oviduct.

final host (= primary or definitive host). Organism in which the mature form of a parasite develops.

fission. Type of asexual reproduction in which an organism divides into two or more offspring.

fitness. (1) Average number of offspring per individual with a particular genotype. (2) Loosely, how well an organism or genotype is adapted.

5′ cap. Structure at 5′ end of messenger RNA.

fixed. (1) Adjective applied to energy or matter that has been incorporated into organic molecules. (2) Preserved, as a specimen in microscopy. (3) Adjective applied to an allele that occurs in every individual in a population.

fixed action pattern (FAP). Stereotyped behavioral response to a specific signal (releaser). Now more often called a motor program.

flagellate. (1) Bearing one or more flagella. (2) A flagellated protozoan.

flagellated chamber. In leuconoid sponges a chamber containing choanocytes.

flagellum (plural **flagella**). Long, hair-shaped motile structure on the cell surface that usually occurs individually or in small groups.

flame bulb. Hollow structure with cilia (the flame) that produces filtration pressure at closed end of certain excretory organs (protonephridia).

flame cell. Ciliated cell at closed end of an excretory organ (protonephridium) of certain animals.

flavine adenine dinucleotide (FAD). Electron carrier that is reduced to $FADH_2$ in Krebs cycle.

flexor. Muscle that brings limbs toward the body axis.

fluid mosaic model. Structure of plasma membrane in which proteins are dissolved within a bilayer of phospholipid.

fluke. (1) Parasitic flatworm in class Trematoda or Monogenea. (2) Flatfish in the group Pleuronectiformes. (3) One of two lateral extensions of a whale's tail.

follicle. (1) Structure in skin of mammal or bird that produces a hair or a feather. (2) Part of ovary in which an egg cell develops.

food chain. Linear sequence defining which organisms eat and are eaten by other organisms.

food vacuole. Vesicle formed as result of phagocytosis, which contains food for intracellular digestion.

food web. Entire trophic relationship in a community, consisting of all the intertwining food chains.

footprint pheromone. Substance distributed by honey bee queen that inhibits production of other queens.

forebrain (= prosencephalon). Anterior part of the vertebrate brain that develops into the cerebrum, olfactory bulbs, thalamus, hypothalamus, pituitary gland, and other structures.

foregut (= crop). Anterior part of insect digestive tract in which food is stored and partly digested.

founder cell. (1) Amebocyte (sclerocyte) in sponges that helps form mineral spicules. (2) Stem cell. Embryonic cell from which certain other cells develop.

founder effect. Tendency of reproductively isolated populations to reflect the atypical genetic composition of the founders of the population.

fovea. Depression on the retina where vision is sharpest.

fragmentation. Type of asexual reproduction in which an individual divides into two or more parts, each of which develops into a new individual.

free energy. Potentially available energy content of a system, minus the energy that is unavailable due to increase in entropy.

free radical. Molecule with a highly reactive unpaired electron.

freezing point depression. Reduction in freezing point of a solution due to concentration of solute.

fry. Recently hatched fish.

functional group. Group of atoms (hydroxyl, ketone, methyl, etc.) commonly behaving as a single unit in organic molecules.

furcula. (1) In birds, the fused collar bones (clavicles); the wishbone. (2) In springtails (Insecta, Collembola), the springing mechanism.

γ. (*See under* **gamma entries.**)

G phase. One of two phases (G_1 and G_2) in the cell cycle during which DNA is not synthesized.

G protein. Protein that binds guanosine triphosphate and helps control cellular responses to hormones or other stimuli.

gait. Pattern of leg movements corresponding to different speeds of locomotion.

gall. (1) Abnormal growth on plants often induced by a developing insect. (2) Bile.

gamete. Sperm or egg cell.

gametogenesis. Development of gametes from primitive germ cells.

gamma motor neuron. Small-diameter nerve cell that excites muscle portion of a spindle receptor organ.

ganglion. Part of a nervous system containing cell bodies and synapses.

ganoid scale. Thick, nonoverlapping scale of gars and certain other bony fishes.

gap junction. Group of protein channels through two adjacent plasma membranes that permits movement of ions, monosaccharides, amino acids, and electrical current between cells.

gap phase. One of two periods (G_1 and G_2) in interphase of the cell cycle when DNA synthesis does not occur.

gas gland. Gland that pumps gas into the swim bladder of a fish.

gastric mill. Chitinous teeth attached to the muscular lining of the cardiac stomach of crustaceans.

gastric shield. Part of stomach lining of bivalve molluscs against which the crystalline style grinds to release digestive enzymes.

gastrodermis. Inner cell layer of cnidarians.

gastrolith. Calcium deposit recovered from molting cuticle and stored in stomach of crustaceans.

gastrovascular cavity (= coelenteron). Cavity of an incomplete gut used for both digestion and circulation in cnidarians and some other invertebrates.

gastrula. Early embryo enclosed in two germ layers (ectoderm and endoderm) with proliferating mesoderm, and with a cavity (archenteron) opening through a blastopore.

gastrulation. Process by which a blastula develops into a gastrula.

gate. Portion of an ion channel that opens in response to a signal to increase permeability of the ion.

gemmule. Internal bud for asexual reproduction in freshwater and some marine sponges.

gene. Genetic unit responsible for one trait, or for producing one protein.

gene frequency (= allele frequency). Proportion of a given allele relative to all the alleles at the same locus in a group of organisms.

genetic drift. Change in gene frequency in a population due to chance.

genetic mosaic. Organism in which different parts of the body are genetically different.

genital pore. Opening to a sperm receptacle in a female crustacean.

genital setae. Bristles on the clitella by which earthworms and other oligochaetes clasp each other during mating.

genome. Genetic material in a haploid set of chromosomes.

genotype. Genetic makeup of an organism. Compare phenotype.

genus (plural **genera**). Taxonomic level above species and below family.

germ cell. Cell that develops into a gamete.

germ layer. One of three layers of cells (ectoderm, endoderm, and mesoderm) formed early in embryonic development.

germ plasm. Cells responsible for reproductive transmission of heredity. Compare somatoplasm.

germinative zone. Part of a tapeworm behind the scolex where proglottids are produced asexually.

gestation. Period in which offspring of mammals are in the uterus.

giant axon. Large-diameter axon that usually triggers an escape response.

Gibbs free energy (= free energy). Potentially available energy content of a system, minus the energy that is unavailable due to increase in entropy.

gill. Organ for exchange of respiratory gasses in aquatic organisms.

gill raker. Comblike portion of a fish gill in which plankton are trapped.

gizzard (= muscular stomach). Posterior portion of a bird's stomach, with a horny lining that crushes food.
gladius (= pen). Vestigial shell in the back of a squid.
gland. Secretory organ.
gland cell (= renette cell). Presumed osmoregulatory cell of a nematode.
glandular stomach (= proventriculus). Anterior portion of a bird's stomach that secretes digestive enzymes.
glial cell. Nonneural cell in a nervous system that forms myelin, provides support, or forms the blood–brain barrier.
globin. Protein portion of hemoglobin, myoglobin, or cytochrome.
globular protein. Rounded protein that is usually water-soluble.
glochidium. Parasitic larva of certain freshwater bivalves.
glomerular capsule (= Bowman's capsule). Spherical structure of each nephron of the kidney enclosing the glomerulus.
glomerular filtration rate (GFR). Rate (ml/min) at which fluid is filtered through the glomeruli of the kidneys.
glomerulus. (1) Capillary in each nephron of the kidney where filtration into Bowman's capsule and the tubule occurs. (2) Sinus network in an acorn worm that is assumed to be an excretory organ.
glottis. Folds of tissue in the larynx that close the trachea.
glyceride (= neutral fat). Lipid consisting of glycerol with one, two, or three fatty acids attached.
glycerol. Three-carbon molecule with three –OH groups.
glycogen. Branched polysaccharide of glucose used as short-term energy store in liver, muscle, and some other cells.
glycolipid. Lipid with carbohydrate attached.
glycolysis. Partial breakdown of glucose in the absence of oxygen.
glycoprotein. Protein with carbohydrate attached.
glycosidic link. Bond joining two simple sugars, formed by an oxygen atom.
gnathobase. Toothed proximal segment of a leg of a horseshoe crab that functions as a jaw for chewing food.
goblet cell. Cell in the mucosa of the digestive or respiratory tract that secretes protective mucus.
Golgi complex (= Golgi body or Golgi apparatus). Group of hollow, disc-shaped cytoplasmic membranes responsible for preparing vesicles for synthesis of plasma membrane, or containing glycoproteins for lysosomes or for secretion.
gonad. Organ—for example, ovary or testis—that produces gametes.
gonadotropin (= gonadotropic hormone). Hormone such as FSH, LH, and hCG that stimulates gonadal activity.
Gondwana. Southern supercontinent that split from Pangea before 180 million years ago, in the Jurassic Period. It later divided into South America, Antarctica, Australia, Africa, and India.
gonochoristic. Having either a male or a female gonad, but not both. Dioecious.
gonocoel. Reduced coelom enclosing the gonads in a tardigrade or mollusc.
gonoduct. Duct from a gonad through which gametes are released.
gonopore. Body opening through which gametes are released.
gram molecular mass (= gram molecular weight). Number of grams of a substance equal to the average number of protons and neutrons in a molecule of the substance. A mole.
granular gland (= poison gland = serous gland). Poison-secreting gland in skin of an amphibian.
gravid. Bearing ripe eggs. Pregnant.
gray matter. Parts of brain and spinal cord that contain nerve cell bodies and synapses. Gray because of absence of myelin.
green gland (= antennal gland). Excretory organ in the head of certain crustaceans.
growth cone. Growing part of a nerve cell during embryonic development.
growth factor. Hormonelike substance that stimulates mitosis and maintains particular cells.
GTP. Guanosine triphosphate.
guanine. A nitrogenous base in nucleic acids and GTP.
guanosine. Nucleotide consisting of guanine and ribose.
gustation. Taste.
gynandromorph. Genetic mosaic in which part of the body is male and the rest is female.
gynogenesis. Mode of reproduction in which sperm are required for egg activation, but genes from the sperm do not contribute to the zygote.

habituation. Type of learning in which response to a stimulus declines with repetition.
hair cell. Receptor cell of the cochlea and vestibular apparatus, as well as certain other mechanoreceptor organs.
halter (plural **halteres**). Club-shaped structure that occurs instead of a hind wing in a true fly.
haplodiploidy. Reproductive system in ants, bees, and other hymenopterans in which some eggs are fertilized and develop into diploid females, and others develop parthenogenetically into haploid males.
haploid. Having just one of each kind of chromosome.
Hardy–Weinberg principle. Mathematical derivation showing that gene frequencies do not change in a population under equilibrium conditions.
Haversian canal (= central canal). Channel in dense bone that contains vessels and nerves.
head group. Hydrophilic part of a phospholipid that is linked to the rest of the molecule through phosphate. A phosphatidyl group.
heart murmur. Blowing or roaring sound in the normal "lubb" or "dup" sound of the heart, indicating poor closure of the heart valves.
heart vesicle. Contracting vascular chamber in the proboscis of an acorn worm.
heat. Energy associated with molecular vibration.
heat capacity. The amount of heat required to change the temperature of a substance by a given amount. The heat capacity of water is 1 calorie per degree Celsius.
hectocotylus arm. Arm of a male octopus specialized as a copulatory organ.
hemal system. Diffuse spongy tissue that lies parallel to the water vascular system in an echinoderm.
heme. Iron-containing, nonprotein (porphyrin) portion of hemoglobin, myoglobin, and cytochromes.

hemerythrin. Violet, iron-containing respiratory pigment of some annelids and other invertebrates.

hemimetabolous. Referring to an insect in which the immature forms (nymphs) resemble the adult. Compare holometabolous.

hemipenis. One of the two copulatory organs of a snake or lizard.

hemocoel (HE-mo-seal). Body cavity in an animal with an open circulatory system, which contains a fluid that combines the functions of coelomic fluid and blood.

hemocyanin. Blue, copper-containing respiratory pigment of molluscs, crustaceans, and some spiders.

hemoglobin. Red, iron-containing respiratory pigment of vertebrates and many invertebrates.

hemolymph. Blood of animal with an open circulatory system.

hemostasis. Arrest of bleeding or blood flow.

Hensen's node. Site of cell movement during gastrulation in reptiles, birds, and mammals.

hepatic. Relating to the liver.

hepatopancreas (= digestive gland). Organ that secretes digestive enzymes and stores nutrient reserves, especially in crustaceans and molluscs.

herbivore. Animal that eats only plants.

heritability. Amount of variance that can be attributed to genetic differences.

hermaphroditic (= monoecious). Combining the reproductive functions of both genders in one individual.

heterocercal. Having a caudal fin with the dorsal lobe larger than the ventral, and with the vertebral column extending into the dorsal lobe, as in sharks.

heterochrony. Change in the timing of a developmental event; for example, paedomorphosis.

heterogametic. Having different sex chromosomes.

heterosis (= hybrid vigor). Increased fitness of an organism due to heterozygosity (having different alleles for the same gene).

heterotherm. Homeotherm in which body temperature makes large but controlled deviations.

heterotroph (= consumer). Organism that obtains energy from organic materials synthesized by other organisms.

heterozygous. Having two different alleles for the same gene.

hexose. Simple sugar with six carbons, such as glucose and fructose.

hibernation. (1) In certain mammals an adaptation to cold temperatures in which the body temperature is maintained a few degrees above freezing. (2) Loosely, torpor, diapause, or any other state of inactivity during cold.

hindbrain (= rhombencephalon). Part of the brain that forms the cerebellum, pons, and medulla.

hindgut. Part of insect digestive tract posterior to midgut.

histamine. Secretion from mast cells of injured tissues that triggers swelling and other signs of inflammation.

histone. Protein around which DNA is organized.

holdfast. (1) Structure for attachment of an animal to substratum. (2) The scolex of a tapeworm.

holoblastic. Complete cleavage of an egg, as opposed to meroblastic cleavage in which yolk prevents or delays cleavage of part of the egg.

holometabolous. Referring to an insect in which the immature forms do not resemble the adult.

holozoic nutrition. Feeding on particulate food or fluids.

home range. Area shared by a group of social animals of the same species.

homeo box. Sequence of 180 base pairs found in many homeotic genes of insects and many other animals.

homeostasis. Maintenance of certain conditions in a steady state.

homeothermy. Maintenance of stable body temperature.

homeotic gene. Gene that, when mutated, changes the identity of a segment, causing its appendages to develop into the homologous appendages of a different segment.

homocercal. Having a caudal fin with dorsal and ventral lobes of approximately equal size, and with the vertebral column not extending into either lobe, as in most bony fishes.

homologous. (1) Similar because of shared ancestry. (2) Referring to chromosomes that are similar in shape and carry identical genes. (3) Referring to different parts (e.g., legs and antennae) that develop from similar structures in different segments of segmented animals. Serially homologous.

homozygous. Having two identical alleles for a gene.

hormone. Secretion from a specific tissue that travels through the blood stream to evoke a response in distant cells.

horn. Keratinous growth, usually permanent, on the heads of cattle and certain other hoofed mammals.

Hox cluster genes. Homeotic genes arranged in clusters on a chromosome.

human chorionic gonadotropin (hCG). Hormone secreted by chorion of the human embryo and fetus that is essential for maintenance of pregnancy.

hybrid. Offspring of genetically different parents. A heterozygote.

hybrid vigor (= heterosis). Increased fitness of an organism due to heterozygosity (having different alleles for the same gene).

hybridogenesis. Fertilization in an all-female species by males of a related species with the male chromosomes subsequently discarded during oogenesis.

hydride ($H{:}^-$). Hydrogen atom with an additional electron.

hydrogen bond. Attraction of a strongly electronegative atom, especially O or N, for a hydrogen atom that is covalently bonded to a strongly electronegative atom.

hydroid. A cnidarian polyp, such as a hydra.

hydrologic cycle (= water cycle). Biogeochemical movement of water among various organic and inorganic states.

hydrolysis. Chemical reaction in which a molecule is decomposed by the addition of water.

hydrophilic. Having a high solubility in water, like most polar molecules.

hydrophobic. Having a low solubility in water, like fats.

hydrostatic skeleton (= hydroskeleton). Internal hydrostatic pressure that supports the body of certain animals.

hyperosmotic. Referring to a solution with a higher osmotic concentration than another solution.

hyperpolarization. Increase in the magnitude of a membrane voltage.

hypertension. High blood pressure.

hypertonic. Referring to a solution that causes a cell or organism to lose water and shrink.

hypnozoite. Dormant stage of malaria-causing *Plasmodium* in liver.

hypoblast. Part of the inner cell mass of a mammalian embryo that forms the yolk sac.
hypodermic impregnation. Insemination by inserting the penis through the integument, as occurs in certain flatworms.
hypodermis. Cuticle-secreting epidermis in animals such as nematodes, annelids, and arthropods.
hypopharynx. Tonguelike structure in mouthparts of an insect.
hypophysis (= pituitary gland). Compound gland that secretes several hormones in response to signals from the nearby hypothalamus in the brain.
hyposmotic (= hypo-osmotic). Referring to a solution with a lower osmotic concentration than another solution.
hypothalamus. Part of the brain that helps regulate feeding, water and ion balance, body temperature, and the secretion of hormones from the pituitary gland.
hypotonic. Referring to a solution that causes a cell or organism to gain water and swell.

Ia afferent. Large-diameter neuron that conducts action potentials from spindle receptors in muscles to the spinal cord.
I band. One of two bands at the end of a sarcomere in skeletal muscle, formed by thin protein filaments.
ice age. Period of reduced mean temperature when annual snow accumulation exceeds melting, causing the spread of glaciers.
ice era. Period lasting about 50 million years during which ice ages occur.
imaginal disc. Structure in insect larva that becomes a particular adult structure after metamorphosis.
imago (im-A-go). Adult form of an insect.
immunoglobulin (= antibody). Blood protein from plasma cells that binds to specific foreign molecules.
immunological memory. Ability of the immune system to recognize and quickly respond to a second exposure to a particular foreign molecule.
implantation. Process by which embryo enters the lining of the uterus.
imprinting. (1) Social bonding between a young animal (usually a bird) and a parent or other object. (2) Long-term memory of the odor of a native stream used by certain fishes to migrate to the same stream for spawning.
inbreeding. Breeding between closely related organisms.
incisor. Chisel-shaped tooth adapted for cutting.
inclusive fitness. Individual fitness (ability to reproduce) plus contributions to the fitness of kin multiplied by the probability that they are carrying genes identical to one's own.
incomplete digestive tract. Digestive tract with one opening that serves as mouth and anus.
incurrent canal. In leuconoid sponges a channel leading water into a flagellated chamber.
independent assortment. Inheritance of one allele regardless of whether another allele at a different locus is also inherited.
indeterminate cleavage (= regulative development). Cleavage in which fates of cells (blastomeres) are determined later in cleavage.
indicator fossil. Fossil that is characteristic of a particular age, and therefore useful in dating sediments.
indirect calorimetry. Determination of metabolic rate by measuring rate of O_2 consumption or CO_2 production.
indirect development. Development in which there is a larval form.
induction. During embryonic development, the influence of some cells (organizer) on the development of others.
industrial melanism. Darkening of the integument of insects in areas where soot has accumulated. Believed to be an evolved camouflage.
initiator. Mutagen or oncogenic virus that initiates cancer.
inflammation. Reaction to tissue damage involving swelling, pain, heat, redness, and itching.
infundibulum. Funnel-shaped opening through which eggs enter the oviduct of a bird.
ink. Dark or luminescent secretion released by cephalopods and some other molluscs in response to a threat.
inner cell mass. Cells in a developing reptile, bird, or mammal that form the embryo and extraembryonic membranes.
instar. Stage between molts in an immature arthropod.
integument. The body surface of an animal.
intercalated disc. Electrically conducting junctions between muscle cells in the heart.
intercostal muscle. Skeletal muscle between the ribs partly responsible for breathing.
interference color. Structural colors due to overlapping scales that cause incident and reflected rays of light to interfere with each other.
interferon. Secretion (a cytokine) from virus-infected cells that interferes with the ability of virus to reproduce in other cells.
interleukin. Chemical (a cytokine) that stimulates immune responses.
intermediate filament. Part of the cytoskeleton consisting of protein strands between microtubules and microfilaments in diameter (7 to 11 nm).
internal cellularization. Hypothetical process by which metazoans originated from a multinucleate protozoan in which the nuclei became separated by intracellular membranes.
internal fertilization. Fertilization of ova inside the mother's body.
interneuron. Neuron with both synaptic inputs and outputs to other neurons.
interphase. Period in the life cycle of a cell when mitosis and meiosis are not occurring, or between telophase I and prophase II of meiosis.
interstitial cell (= Leydig cell). Cell of the testes that produces androgens.
interstitial fluid. Fluid surrounding cells.
intertidal zone (= seashore). Part of the continental-shelf ecosystem lying between high and low tide levels.
intracellular digestion. Digestion of food inside cells, as in many protozoans, sponges, cnidarians, flatworms, rotifers, bivalve molluscs, and primitive chordates.
intron (= intervening sequence). Portion of messenger RNA that is removed during processing and does not code for a protein. Also the DNA coding for that portion of the mRNA.
introvert. Retractable anterior portion of a sipunculid.
inversion. Reversal of a portion of a chromosome.
ion. Atom or molecule with a net charge due to gain or loss of one or more electrons.

ionic bond. Chemical bond in which an electron from one atom is transferred to another. Characteristic of atoms that form ions.

iridophore. Type of chromatophore that contains guanine or other purine and produces silvery and iridescent colors by reflecting light.

isoenzyme (= isozyme). Form of an enzyme that differs from another form of the same enzyme.

isolecithal. Egg with sparse yolk that is evenly distributed.

isosmotic (= iso-osmotic). Having the same osmotic concentration as another solution.

isotonic. Referring to a solution that does not cause a cell or organism to shrink or swell.

isotope. Atom that differs in number of neutrons in the nucleus compared with another of the same element.

Jacobson's organ (= vomeronasal organ). Olfactory organ in the roof of the mouth of an amphibian, reptile, or mammal.

jelly layer. Covering of echinoderm egg.

jumping gene (= transposing element). Submicroscopic segment of a chromosome that moves to another chromosome or to a different position on the same chromosome.

juxtaglomerular apparatus (JGA). Structure near the glomerulus of the kidney that secretes renin.

karyotype. Chromosomes of an organism as they appear under the microscope.

keratin. Complex of globular and fibrous proteins held together by disulfide bonds. Replaces cytoplasm in epidermal cells as they become cornified, and forms nails, claws, and horns.

kidney. Osmoregulatory organ of vertebrates, or analogous organ in molluscs.

killer cell. Cell of the immune system that attacks foreign cells.

kinesis (plural **kineses**). Nondirectional movement in response to a stimulus. Compare taxis.

kinetochore. Protein on centromere of chromosome by which spindle fibers attach during mitosis and meiosis.

kin recognition. Recognition by one animal that another is related.

kin selection. Preferential behavior by one animal toward related animals.

kinetosome (= basal body). Organelle to which a cilium or eukaryotic flagellum attaches.

kinin. Peptide that relaxes arterial smooth muscle and increases the flow of blood out of capillaries.

kleptocnida (KLEP-to-NY-duh). Nematocyst of a cnidarian ingested by a predator and used for the predators' own defense.

knuckle-walking. Locomotion of gorillas and chimpanzees in which the weight of the upper body rests on the backs of the fingers.

Krebs citric acid cycle (= tricarboxylic acid or TCA cycle). Cycle of reactions in mitochondria by which pyruvate from glycolysis is oxidized to carbon dioxide during cellular respiration.

krill. Filtered food of toothless whales, especially small crustaceans in the group Euphausiacea.

labium. (1) Liplike fold of tissue, such as an inner or outer lip of the female genitals. (2) Lower lip of insect or crustacean mouthparts.

labrum. Upper lip of insect or crustacean mouthparts.

lachrymal gland. Tear gland.

lactate. Waste product of anaerobic glycolysis.

lacteal. Lymph vessel in the villus of the small intestine that absorbs digested fats.

lagena. Auditory receptor of an amphibian or reptile; homologous with the cochlea of birds and mammals.

Lamarckism. Inheritance of acquired characteristics, one component of Lamarck's theory of evolution.

lamella. Thin, layered structure, such as in dense bone.

larva. Juvenile that is sharply different in form from the adult.

larynx. Enlargement of the vertebrate trachea that contains the glottis; voice-producing organ of amphibians, reptiles, and mammals.

lateralization. Unequal distribution of functions on each side of the brain, with each side generally controlling functions of the other side of the body.

lateral line organ. Canal or line bearing mechanoreceptors on each side of a fish.

lateral undulation. Most common mode of locomotion in snakes, in which the body pushes sideways against several supports.

Laurasia. Northern supercontinent that split off from Pangea before 180 million years ago, in the Jurassic Period. Later divided into North America, Europe, and most of Asia.

Iek. Area where animals of one sex display to attract mates.

lentic. Referring to water with little or no current, as in a pond or swamp.

leptocephalus. Leaf-shaped larva of an eel.

lethal limit. Upper or lower extreme for a limit of tolerance. Maximum or minimum temperature, oxygen level, or other abiotic factor for an organism.

leuconoid. Type of canal system in most sponges, in which choanocytes are located in numerous flagellated chambers and there is no spongocoel.

leukemia. Cancer of white blood cells.

leukocyte (= white blood cell). One of several types of blood cells important in immunity.

life-history trait. Adaptation that affects reproductive success, such as age at sexual maturity and degree of parental investment.

lift. Force generated during flight that overcomes the force of gravity.

ligand. Molecule such as a hormone or synaptic transmitter that binds specifically with a receptor molecule without forming a covalent bond.

limax form. Type of locomotion in some amebas in which the whole cell creeps like a single pseudopodium.

limbic system. Part of the cerebral cortex that controls emotional behavior.

limit of tolerance. Range of temperature, oxygen level, ionic concentration, or other abiotic factor in which an organism can live.

limnetic zone. Surface waters away from shore.

lingual flip. Flip of tongue by which certain frogs capture prey.

linkage group. Group of genes that tend to be inherited together because all are found on the same chromosome.

linkage map. Diagram showing the locations of genes on a chromosome.

lipase. Enzyme that digests lipids.

lipid. Hydrophobic molecule; a neutral fat, oil, steroid, or wax.

lipid bilayer. Double layer formed spontaneously by phospholipids in water as hydrophobic tails point toward each other with hydrophilic head groups pointing outward.

littoral zone. Shallow water along a shore.

lobopodium. (1) Lobe-shaped pseudopodium of amebas and some other protozoans (Lobosea and Eumycetozoea). (2) Leglike appendage of an onychophoran.

locus (pronounced LOW-kus; plural **loci,** pronounced LOW-sy). Position of a gene on a chromosome.

logistic equation. An equation describing idealized exponential population growth, leveling off at the carrying capacity.

lophophore. Arch of hollow (coelomate) tentacles enclosing the mouth but not the anus.

lorica. Girdlelike enclosure of some protozoans, rotifers, and other invertebrates.

lotic. Referring to rivers, streams, and other moving freshwater.

loop of Henle. Bend in the tubule of a nephron of the kidney that is important in water reabsorption.

luciferase. ATPase that catalyzes emission of light in fireflies.

luciferin. Molecule in fireflies that emits light when activated by ATP and luciferase.

lumen. Cavity of an organ such as the gut or a blood vessel.

lunar cycle. Period of a biological rhythm that is determined by movement of Moon through its effect on tides.

lung. Organ for exchange of respiratory gases in terrestrial animals.

lymph. Fluid that accumulates in the lymphatic system from interstitial fluid.

lymph node. Lymphoid tissue, especially in the groin, arm pits, and neck, through which all lymph is filtered.

lymphatic capillary. End of a lymph vessel, where lymph collects from interstitial fluid.

lymphatic vessel. Tubular channel that transports lymph into the blood.

lymphocyte. Cell of the B or T type responsible for immunity.

lymphoid organ. Tissue that produces lymphocytes, including lymph nodes, tonsils, thymus, and spleen.

lymphoma. Cancer of solid tissues associated with the lymphoid tissue.

lyriform organ. Parallel array of mechanoreceptive slits in cuticle of spiders.

lysosome. Spherical, membrane-enclosed organelle containing enzymes that digest materials absorbed by endocytosis.

lysozyme. Enzyme in saliva, tears, and other secretions that breaks down bacterial cell walls.

macroevolution. Evolutionary change large enough to be recognizable at the level of species, genus, or higher taxon.

macromere. Blastomere that is larger than other blastomeres (micromeres).

macronucleus. Larger of two types of nuclei in certain protozoans (ciliates), which directs metabolism, development, and the physical traits of the cell, but not reproduction.

macrophage. Phagocytic white blood cell.

madreporite. Sievelike input of water vascular system of an echinoderm.

magnetoreceptor. Receptor in insects, salamanders, birds, and perhaps mammals that responds to magnetic fields.

major histocompatibility complex (MHC). Group of genes that direct synthesis of proteins important in recognition of foreign cells by the immune system.

Malpighian tubule. Excretory organ of insect or spider that is attached at one end between the midgut and hindgut.

mammary gland. Milk-producing gland of a mammal.

mandible. Jaw of a vertebrate, or jawlike part of arthropod mouthparts.

manometer. Fluid-filled tube for measuring pressure.

mantle (= pallium). (1) Part of the visceral mass of molluscs and brachiopods consisting of two flaps of skin on the dorsal surface of the body that secretes the shell and forms the mantle cavity. (2) Thin body wall of tunicates.

mantle cavity. Major body cavity of molluscs and brachiopods.

marsupium. In amphipods, marsupial mammals, and some fishes and other animals, a receptacle in which eggs and young are brooded.

mass extinction. Episode in which many species become extinct at about the same time.

mastax. Pharynx of a rotifer modified with crushing jaws (trophi).

maternal factor. Substance in egg that affects embryonic development.

mating system. Proportion of males to females involved simultaneously in reproduction. Monogamy, polyandry, or polygyny.

mating type. Group of ciliate protozoans capable of conjugating with each other, but not with members of other mating types.

matrix. Central space within mitochondrion, enclosed by the inner membrane.

maxilla. (1) Upper jaw bone of a vertebrate. (2) Part of arthropod mouthparts.

maxillary gland. Gland in crustacean that regulates ion concentrations.

maxilliped. Appendage associated with the mouth of a crustacean.

median eye (= pineal eye). Third eye of sharks and certain other vertebrates.

medulla. (1) Medulla oblongota. Part of the brainstem that regulates circulation, breathing, coughing, sneezing, and swallowing. (2) Inner part of an organ, especially the adrenal gland and kidney.

medusa. Jellyfish or any similar free-swimming cnidarian.

megalops. Larval stage of a true crab.

meiofauna. Small animals that live among gravel and sand grains.

meiosis. Process by which chromosomes in a diploid cell duplicate, and the cell divides twice into four haploid cells that form gametes.

melanin. Black, gray, or brown pigment in skin, hair, feathers, retina, and other structures.

melanophore (= melanocyte). Type of chromatophore containing melanin pigment.

membranelle. Plate-shaped buccal ciliature of certain protozoans (ciliates).

membrane potential. Voltage across a membrane.

membranous bone (= dermal bone). Bone that forms from layers of embryonic connective tissue rather than by replacement of cartilage. Forms fish scales, turtle shells, antlers, and flat bones of vertebrate skulls.

memory lymphocyte. Lymphocyte responsible for triggering immune response to second invasion by a particular foreign cell or molecule.

menstrual cycle. Cycle of changes in reproductive function (estrus cycle) of women and other primates that menstruate.

meridional. Cleavage furrow that is linear from pole to pole, like a meridian on a globe.

meroblastic. Term applied to an egg in which cleavage in one area is delayed or prevented by yolk.

merozoite. Stage of certain protozoans, such as *Plasmodium,* resulting from multiple fission (schizogony) by the schizont.

mesencephalon (= midbrain). Middle part of the brain between the cerebrum and brainstem (forebrain and hindbrain).

mesenchyme. Middle layer of connective tissue consisting of cells and cell products in a jellylike matrix (mesoglea).

mesentery. (1) Thin fold of peritoneum that holds viscera in place. (2) Septum of a sea anemone.

mesobronchus (plural **mesobronchi**) Central tube leading to parabronchi in the lung of a bird.

mesoderm. (1) Germ layer that gives rise to blood, bone, and other connective tissues, as well as muscle. (2) Tissue derived from mesoderm.

mesoglea. Jellylike substance in mesenchyme. Synonymous with mesenchyme in cnidarians and ctenophores.

mesohyl. Cellular layer between inner and outer layers of a sponge. Often called mesenchyme or mesoglea.

mesolecithal. Egg with moderate amount of yolk that is unevenly distributed. Moderately telolecithal.

mesonephros. Kidney in adult fishes and amphibians and in embryos of other vertebrates. Develops into epididymis, vas deferens, and seminal vesicle in male mammals.

messenger RNA (mRNA). Ribonucleic acid transcribed from DNA that determines the sequence of amino acids during synthesis of proteins.

metabolic rate (MR). Rate of overall metabolism; rate of production of body heat.

metabolism. Synthesis of new organic molecules from nutrients (anabolism) and breakdown of nutrient molecules to extract energy (catabolism)

metacercaria. Juvenile fluke that develops when the cercaria loses its tail and which encysts in second intermediate host.

metachronal wave. Coordinated wave of contraction of cilia.

metamere (= somite = segment). Linearly repeated subdivision of a segmented animal (Pogonophora, Annelida, Arthropoda, and Chordata).

metamerism (= segmentation). Division of body into segments (metameres).

metamorphosis. Development of a larva into the adult stage.

metanephridium. (Often called simply nephridium.) Excretory organ of many invertebrate coelomates, consisting of a tube open at both ends that drains coelomic fluid.

metanephros. Kidney of adult reptiles, birds, and mammals.

metaphase. Stage of mitosis or meiosis in which chromosomes are visible and line up between the cellular poles.

metazoon (plural **metazoa.** Adjective form **metazoan**). An animal (multicellular).

microclimate. Atmospheric condition in a restricted space.

microevolution. Evolutionary change in gene frequency that does not result in a change in species.

microfilament. Thinnest (6 nm) of three fibers in the cytoskeleton, consisting of the protein actin.

microfilaria. First larval stage of a filarial worm (phylum Nematoda).

micromere. Blastomere that is smaller than others (macromeres).

micronucleus. Smaller of two types of nuclei in certain protozoans (ciliates), which transmits genetic information during reproduction.

micropyle. Pore in insect egg that admits sperm for fertilization.

microthrix (plural **microtriches**). Microvillus-like projection on tegument of tapeworm that increases surface area for absorption of nutrients from gut of host.

microtubule. Thickest (22 nm) of the three fibers in the cytoskeleton.

microvillus. Submicroscopic projection of the plasma membrane.

mictic. Referring to female rotifers that lay haploid eggs that hatch into male rotifers. Compare amictic.

midbrain (= mesencephalon). Middle part of the brain between the cerebrum and brainstem (forebrain and hindbrain).

middle ear bones. Small bones (ossicles) that transmit vibrations from the ear drum to the cochlea.

middle piece. Part of sperm between head and flagellum that contains mitochondria.

midgut. Part of insect digestive tract between crop and hindgut, where most digestion and absorption occur.

milk let-down. Release of milk during suckling.

miracidium. Ciliated larva of a fluke, released in egg from final host and infectious to molluscan intermediate host.

mitochondrion (plural **mitochondria**). Organelle in which Krebs cycle and the respiratory chain occur. Major source of ATP.

mitosis. Process by which chromosomes duplicate and separate prior to cell division.

mitral valve (= bicuspid valve). Left atrioventricular valve.

molar. (1) Tooth with several cusps adapted for grinding. (2) Unit of measure of concentration. One mole per liter.

molarity. Measure of concentration. Number of moles of a substance dissolved in a liter of solution.

mole. 6.02×10^{23} molecules (Avogadro's number) of a substance; one gram molecular mass.

molecular phylogenetics. Determination of evolutionary relationships of organisms by comparing structures of proteins, nucleic acids, or other molecules.

molting (= ecdysis). Shedding of cuticle, skin, hair, or feathers.

molting fluid. Secretion of arthropod epidermis that digests endocuticle of old exoskeleton that is to be molted.

monoecious (mon-EE-she-us) (= hermaphroditic). Combining both male and female reproductive organs in one individual.

monogamy. Mating system in which one male pairs with one female at a time in reproduction.

monohybrid cross. Cross between two individuals with different alleles for a given gene.

monophyletic. Referring to a natural taxonomic group: one that has evolved from only one ancestral species. Compare polyphyletic.

monosaccharide. Simple sugar molecule, such as glucose.

monosynaptic reflex. Simple reflex in which there is only one synapse connecting sensory input to motor output.

morphogenesis. Developmental processes by which germ layers form and body pattern is established.

morphology. (1) Structure. (2) The study of structure.

morula. Solid mass of embryonic cells resulting from cleavage, prior to the blastula.

mosaic development (= determinate cleavage). Development in which fates of cells are determined in early cleavage. Common among invertebrates.

motor program. Stereotyped behavioral response to a specific signal (releaser). Formerly called fixed action pattern.

motor unit. One motor axon and all the muscle fibers it controls.

mouthbrooder. Fish that broods eggs or young in its mouth.

mucosa. Mucus-secreting tissue layer, such as the inner lining of the gut.

mucus. Glycoprotein solution that protects and lubricates tissues.

Müllerian duct. Embryonic duct that develops into the oviduct of a female.

Müllerian mimicry. Resemblance of two or more toxic species to each other, which presumably increases the deterrence to predation.

multiple fission. Type of asexual reproduction in which an organism divides into many offspring simultaneously.

muscular stomach (= gizzard). Posterior portion of a bird's stomach, with a horny lining that crushes food.

mutagen. Agent such as a chemical, ionizing radiation, or ultraviolet radiation that induces genetic mutations.

mutation. Hereditary change, especially in the sequence of bases in DNA.

mutualism. Type of symbiosis in which both symbionts benefit.

myelin. Layers of membrane surrounding an axon, periodically interrupted by nodes of Ranvier.

myenteric plexus. Network of nerve cells in the gut that coordinates motility.

myocyte. Contractile cell encircling an osculum or channel in a sponge.

myofibril. Longitudinal contractile subunit within a muscle fiber.

myogenic. Referring to hearts that contract spontaneously, without neural excitation.

myoglobin. Respiratory pigment that transports oxygen in muscle.

myomere (= myotome). Block of segmental muscles in a chordate.

myometrium. Smooth-muscle layer of the uterus.

myosin. Protein that forms thick filaments in muscle.

myotome. (1) Mesodermal segment of vertebrate embryo from which a block of muscles (myomere) develops. (2) A myomere.

nacre (= nacreous layer). Inner layer of mollusc shell.

NAD. Nicotinamide adenine dinucleotide.

naiad. Immature aquatic insect (nymph).

Na^+–K^+ ATPase (= sodium/potassium pump). Enzyme in plasma membranes responsible for coupled transport of sodium and potassium.

naris (plural **nares,** pronounced NARE-eez). Opening of a nasal cavity.

natural group. Taxonomic group that corresponds to an actual evolutionary group.

nauplius. Free-swimming microscopic larva of a copepod, barnacle, or other crustacean; with one median eye (naupliar eye).

nekton. Aquatic organisms that move actively through water, rather than drifting or floating passively like plankton.

nematocyst. Organelle in cnidarians that shoots out stinging thread for predation or defense. Most common type of cnida.

nematocyte. Cell that produces nematocysts.

neocortex. Outer few millimeters of the cerebral cortex in mammals, which contains most synapses.

neo-Darwinian theory (= synthetic theory). Current theory that combines Darwin's theory of natural selection with modern knowledge of genetics to explain most evolution.

neoteny. Retention of juvenile form in an adult.

nephridiopore. Opening through which a nephridium (protonephridium or metanephridium) excretes wastes and excess water.

nephridium. Tubular osmoregulatory and excretory organ of many invertebrates; either a protonephridium or a metanephridium.

nephron. Functional subunit responsible for osmoregulation and excretion in the vertebrate kidney, consisting of a glomerulus, Bowman's capsule, and a tubule.

nephrostome. Funnel-shaped, internal opening of a metanephridium through which cilia filter coelomic fluid for osmoregulation and excretion.

nerve net. Diffuse network of nerve cells forming the nervous system of cnidarians, ctenophores, and echinoderms.

nerve ring. Circular nerve around the pharynx that forms a major part of the nervous system of many small invertebrates.

net primary production. Increase in biomass or energy in an area at the end of a given period.

neural crest. Ectoderm that develops into sensory nerve cells, sympathetic ganglia, adrenal medulla, dentin of teeth, and connective tissues of head.

neural fold. One of two folds of ectoderm that form the embryonic neural tube from which the brain and spinal cord develop.

neural tube. Hollow cylinder of ectoderm from which the brain and spinal cord develop.

neurogenic. Referring to hearts that are stimulated to contract by the nervous system.

neurohypophysis (= posterior pituitary). Part of the pituitary gland that releases hormones from neurosecretory nerve endings.

neuroid cells. Syncytial amebocytes forming a nerve-like network in certain sponges.

neuromast. Mechanoreceptor cell in lateral line organ of fishes.

neuromuscular junction (= endplate). Junction between a nerve cell and a muscle cell.

neuron. Nerve cell.

neuropodium. Ventral lobe of a parapodium of a polychaete annelid.

neurosecretory cell. Nerve cell that secretes hormone or other substance rather than synaptic transmitter.

neurula. Stage of vertebrate development in which the neural tube and neural crest form.

neurulation. Formation of neural tube and neural crest in the embryo.

neutral. Having a pH of approximately 7, equal to that of pure water.

neutral buoyancy. Having a density (mass divided by volume) equal to that of the external medium and therefore being able to float.

neutral fat (= glyceride). Lipid consisting of glycerol with one, two, or three fatty acids attached.

neutral theory of evolution. Theory that most evolution results from mutations that have no immediate effect on the ability of organisms to survive and reproduce, and are therefore not subject to natural selection.

niche. Abiotic and biotic factors that define the role of a species in a community.

nicotinamide adenine dinucleotide (NAD^+). Coenzyme that is reduced to NADH during glycolysis and cellular respiration.

nictitating membrane. Translucent eyelid of many vertebrates.

nitrate bacterium. Bacterium that converts nitrite NO_2 into nitrate NO_3^-.

nitrite bacterium. Bacterium that converts ammonia into nitrite NO_2^-.

nitrogen cycle. Biogeochemical movements of nitrogen among various inorganic and organic states.

nitrogenous base. Adenine, guanine, cytosine, thymine, or uracil; the chemical structures that determine the genetic information in nucleic acids.

nitrogenous waste. Substance, usually ammonia, urea, or uric acid, by which an animal eliminates nitrogen.

node of Ranvier (RAHN-vee-ay). Gap between adjacent sections of myelin.

nondisjunction. Failure of chromatids to separate during meiosis, leading to gain or loss of chromosomes in gametes.

notochord. Stiff rod of tissue along the dorsal midline of lancelets and vertebrate embryos.

notopodium. Dorsal lobe of a parapodium of a polychaete annelid.

nuchal organ. Anterior chemoreceptor of polychaete annelid.

nuclear envelope. Porous, double membrane system that encloses the cell nucleus.

nucleic acid. DNA or RNA, each of which consists of chains of nucleotides.

nucleolus. Dark-staining part of cell nucleus that synthesizes ribosomes.

nucleosome. Fundamental subunit of chromatin consisting of a length of DNA partly wrapped around histones.

nucleotide. Nucleic acid base attached to ribose or deoxyribose, to which is attached a phosphate.

nucleus (plural **nuclei**). (1) Cluster of cell bodies and synapses in the brain. A ganglion. (2) Particle on which crystals of ice form. (3) Organelle in eukaryotic cells that contains the hereditary material.

numerical taxonomy (= numerical phenetics = phenetics). Method of classification in which numerical values are assigned to quantifiable differences as a basis for grouping organisms.

nuptial pad. Thickened skin on hands or other parts of male frog by which it grasps female during mating.

nuptial swarm. Aggregation of animals, especially insects, for breeding.

nurse cell. Cell that attends developing larvae in a sponge or other animal.

nutritive muscle cell. Contractile cell in the gastrodermis of a cnidarian.

nymph. (1) Immature insect that is hemimetabolous (gradually metamorphic, without a pupal stage). (2) Legless, immature stage of a pentastomid.

ocellus. Simple eye of an arthropod or other invertebrate.

odontophore. Cartilaginous support for the radula of a snail.

oil gland (= uropygial gland). Gland at the base of the tail of a bird that produces oil for preening.

olfaction (= smell). Chemoreception of airborne chemicals.

olfactory bulbs. Pair of brain structures just above the olfactory epithelium that are responsible for perception of odors.

olfactory epithelium. Epithelium lining the roof of the nasal cavity and bearing olfactory receptors.

ommatidium. Subunit of arthropod compound eye.

omnivore. Animal that eats both plants and other animals.

oncogene. Mutated gene or gene acquired by viral infection that initiates cancer.

oncomiracidium (ON-ko-MERE-a-SID-ee-um). Ciliated larva of flatworm in class Monogenea.

oncosphere. Egg-enclosed larva produced by proglottid of a tapeworm.

oncotic pressure (= colloid osmotic pressure). Hydrostatic pressure due to osmotic effect of a colloid.

ontogeny. Development of an individual.

oocyst (OH-uh-sist). Cyst enclosing zygote of malaria-causing *Plasmodium* and certain other protozoans.

oocyte (OH-uh-site). Cell in meiosis during development into an ovum.

oogenesis (oh-uh-GEN-uh-sis). Development of egg cells.

oogonium (oh-uh-GO-nee-um) (= primordial germ cell). Cell that will develop into an oocyte and then an ovum.

ootheca (oh-uh-THEEK-uh). Protective case containing fertilized eggs.

open circulatory system. Circulatory system found in insects and some other invertebrates, in which blood is not always confined within the heart or vessels.

operant conditioning (= Skinnerian, instrumental, or Type II conditioning). Learning of a behavior due to the rewarding (reinforcement) of closer and closer approximations to that behavior.

opercular chamber. Cavity enclosing gills of a fish.

operculum. Covering of a chamber, such as the gill chamber of a fish, the nematocyst of a cnidarian, or the shell of a snail.

ophiopluteus. Larva of a brittle star.

opisthaptor. Posterior adhesive organ of a flatworm in the class Monogenea.

opisthosoma. Segmented posterior portion of a pogonophoran or an arachnid.

optimum. Level of temperature, ion concentration, or other abiotic factors at which an organism lives best.

oral disc. Double ring of tentacles around the mouth of a sea anemone.

orb web. Spider web constructed in a vertical plane to snare flying insects.

order. Taxonomic level below class and above family.

organ of Corti (KOR-tee). Mechanoreceptive tissue in the cochlea responsible for hearing.

organelle. Discrete structure in a cell that performs a specific function.

organic molecule. Any carbon-based molecule.

organizer. Portion of an embryo that controls development of tissues around it.

organogenesis. Embryonic development of organs.

osculum. Opening by which water exits a sponge.

osmoconformer. Animal, generally a marine invertebrate, that does not regulate the solute concentration of its body fluids when the concentration in the environment changes.

osmolarity. Osmotic effect of a given solution. The concentration that would produce the same osmotic effect as that of the solution. Osmoles per liter.

osmolyte. Organic molecule, such as urea, that increases the osmolarity of body fluids to prevent loss of water due to osmosis.

osmometer. Device for measuring osmolarity.

osmoregulator. Animal that regulates the solute concentration of its body fluids as the concentration in the environment changes.

osmosis. Movement of water from a low solute concentration into a higher solute concentration.

osmotic pressure. Hydrostatic pressure due to osmosis.

osphradium. Sense organ that samples incoming water of bivalve molluscs and aquatic snails.

ossicle. (1) Small calcareous plate forming part of the endoskeleton of an echinoderm. (2) Any middle-ear bone.

osteoblast. Cell that forms bone by gradually replacing cartilage with calcareous minerals.

osteoclast. Cell that dissolves minerals in bone.

ostium. An opening, especially (1) a microscopic pore through which water enters a sponge, or (2) an opening by which blood enters the arthropod heart.

otolith (= **otoconium**). Mineral deposit that deflects hair cell receptors in the utricle and saccule (the otolith organs) of vertebrates or in the statocyst of certain invertebrates for perception of orientation.

outcrossing. Mixing of alleles due to interbreeding between two genetically different populations.

outgroup. Group of organisms used in cladistics as a reference in determining whether characters in other groups are primitive or derived.

oval body. Organ that leaks excess gas out of the swim bladder of a fish.

ovary. Organ of females that produces eggs.

ovicell. Zooid of an ectoproct colony specialized as an egg-brooding chamber.

oviduct. Tube through which eggs pass after release from the ovary.

oviger (= ovigerous leg). Leg of a male sea spider (Pycnogonida) that is specialized to gather and brood fertilized eggs.

oviparity. Egg laying.

oviposition. The act of egg laying, especially by an insect.

ovoviviparity. Retention of the fertilized egg in the oviduct until hatching, without nutritional support from the mother.

ovulation. Release of a female gamete from the ovary.

ovum. Mature female gamete.

oxidation. Loss of electrons by a molecule.

oxidation–reduction reaction (= redox reaction). Reaction involving loss of electrons from one molecule, which is oxidized, to another, which is reduced.

oxygen cycle. Biogeochemical movement of oxygen among various inorganic and organic states.

oxygen dissociation curve (= oxygen equilibrium curve). Graph showing the degree of oxygenation of a respiratory pigment at different levels of O_2 in blood.

oxyntic cell (= parietal cell). Cell of the stomach lining that secretes H^+.

pacemaker. Group of nerve or muscle cells that triggers activity in an organ, especially the heart.

paedomorphosis. Occurrence of juvenile features in an adult as a result of neoteny or progenesis.

pallium. Mantle of a mollusc or brachiopod.

palp. Small appendage usually involved in feeding.

pancreas. Organ that secretes insulin and glucagon into the blood, and digestive enzymes into the small intestine in vertebrates. In invertebrates, a digestive gland.

Pangea. Supercontinent comprising entire land mass of Earth approximately 250 million years ago.

pangenesis. Obsolete theory that genetic traits come from all parts of the bodies of parents.

papilla. Small nipple-shaped projection, such as a taste papilla on the tongue or a feather bud on a bird's skin.

papula. Projection on integument of echinoderms for exchange of respiratory gases and nitrogenous wastes.

parabiosis. Experimental procedure in which blood streams of two living animals are interconnected.

parabronchus (plural **parabronchi**). Tube for exchange of O_2 and CO_2 in a bird's lung.

parallel processing. Simultaneous analysis of different features of a stimulus, such as color and movement, by different parts of the nervous system.

paraphyletic. Referring to a group that does not include all descendants of its most recent common ancestor.

parapodium. Appendage used by polychaete annelids for locomotion and exchange of respiratory gases.

parasitism. Symbiosis in which one organism, the parasite, lives at the expense of the other (host).

parasitoid. Parasite, especially an insect larva, that consumes and ultimately kills the host.

parasympathetic division. Part of the autonomic nervous system active during rest.

parathyroid glands. Glands on the thyroid that secrete parathyroid hormone.

parenchyme (= parenchyma). Mesenchyme containing densely packed cells, especially in acoelomates.

parenchymula. Free-swimming larva of some sponges, which is solid and covered with flagella.

parental investment theory. Body of theory in sociobiology that attempts to explain the evolutionary advantages of parental behavior.

parietal cell (= oxyntic cell). Cell of the stomach lining that secretes H^+.

parotoid gland. Poisonous swelling behind eye of some frogs.

parthenogenesis. Development of an embryo from an unfertilized ovum.

partial endothermy. Temporary elevation of body temperature of a poikilotherm by production, absorption, and conservation of heat.

partial pressure. Portion of the pressure in a fluid due to one substance in the fluid.

parturition. Process of giving birth to young.

patch dynamics. Method of ecological analysis that does not assume uniformity of an ecosystem in space and time.

pecking order (= dominance hierarchy). Social rank that determines which animals in a group have priority in mating, feeding, or attacking others.

pectoral girdle. Bones that attach the bones of the forelimbs to the axial skeleton.
pedalium. Rudderlike appendage at each corner of a cubozoan.
pedicel. A small stalk, especially the narrow "waist" between the thorax and abdomen of a spider, ant, or other arthropod (also called a petiole); or the second segment of an insect antenna.
pedicellarium. One of many sharp or beaklike defensive structures on integument of a sea urchin or some other echinoderm.
pedicle (= peduncle). Stalk, such as that by which a brachiopod attaches to substratum, or the caudal fin attaches to a fish.
pedipalp. One of the second pair of appendages in arachnids, near the mouth.
peduncle. A pedicle.
pelage. Coat of hair on a mammal.
pelagic. Referring to open ocean.
pellicle. Plasma membrane and associated fibrous cytoplasm in ciliate and flagellate protozoans.
pelvic girdle. Bones by which bones of the hindlimbs attach to the axial skeleton.
pen (= gladius). Vestigial shell in the back of a squid.
penetrant. Type of nematocyst that penetrates tissue and injects a stinging or paralyzing toxin.
penis (plural **penes**). Male organ used for intromission and sperm transfer.
penis bulb. Organ in certain flatworms that stores sperm until copulation.
pentamerous radial symmetry. Five-part radial symmetry, characteristic of starfish and other echinoderms.
pentose. Simple sugar with five carbons, such as ribose or deoxyribose.
pentose shunt. Series of reactions responsible for metabolism of five-carbon sugars using three- and six-carbon sugars in glycolysis.
pepsin. Protein-digesting enzyme secreted by the stomach.
peptide. Molecule consisting of few amino acids.
peptide bond. Linkage between two amino acids, formed by removal of one hydrogen from the amino group of one amino acid and the OH from the carboxyl group of the other amino acid.
pericardial sinus. Sinus in body cavity of arthropod that contains the heart.
pericardium. Sac enclosing a heart.
periostracum. Outer protective layer of bivalve mollusc shell.
peristalsis. Wavelike contraction, such as occurs in the esophagus during swallowing.
peristaltic progression. Type of locomotion in earthworms and some other animals involving alternating contraction of body segments.
peritoneum (= peritoneal membrane). Membrane enclosing the coelom.
peritubular capillary. Capillary surrounding the tubule of a nephron in the kidney.
permeable. Permitting diffusion, as in a membrane that is permeable to ions.
pH. Measure of acidity, equal to the negative of the exponent of the H^+ concentration expressed as a power of 10.
phagocytosis. Transport (endocytosis) of solid material into a cell by formation of an enclosing vesicle.
pharyngeal pouch. Sac for filter-feeding by acorn worms.
pharyngeal slit. One of several grooves in the neck of vertebrate embryos.
pharynx (plural **pharynges**). Part of the digestive tract between the mouth and the esophagus. In birds and mammals, the place where the respiratory and digestive tracts cross.
phasmid. Posterior sensory organ in certain nematodes.
phenetics (= numerical phylogeny = numerical phenetics). Method of classification in which numerical values are assigned to quantifiable differences as a basis for grouping organisms.
phenotype. Expressed characteristics of an organism, especially those genetically influenced. Compare genotype.
pheromone. Airborne chemical secreted by one animal that influences the behavior of another in the same species.
phosphagen. Molecule, usually creatine phosphate or arginine phosphate, that stores phosphate bonds for ATP or that shuttles phosphate from mitochondria to make ATP.
phospholipid. Molecule consisting of glycerol to which are linked two fatty acids and one phosphate with a polar head attached.
phosphorylation. Chemical addition of a phosphate group.
photophore. Light-emitting organ.
phylogenetic systematics (= cladistics). Method of systematics in which outgroups are used to distinguish primitive from derived characters, and lines of descent but not times of divergence are deduced from the number of derived homologous characters (synapomorphies).
phylogenetic tree (= dendrogram). Diagram representing the evolution of a group of organisms.
phylogeny. Evolutionary history of an organism or taxon.
phylum (plural **phyla**). Taxonomic level below kingdom and above class.
physical gill. Bubbles of air that function as a gill in an aquatic insect.
physiological saline. NaCl solution osmotically balanced to substitute for a body fluid.
phytoplankton. Small algae and plants suspended in water.
pile. Hairlike covering on arthropod integument.
pilidium. Free-swimming larva of nemertine worms, enclosed in plates arranged like a cap with ear flaps.
piloerection. Erection of hair on a mammal due to contraction of the arrector pili muscles.
pinacocyte. Cell lining the external surface (pinacoderm) or a nonflagellated channel in a sponge.
pinacoderm. Outer cell layer (epidermis) of a sponge.
pineal eye (= median eye). Third eye of sharks, amphibians, and certain other vertebrates.
pineal gland (= pineal body). Gland in vertebrate brain derived from the pineal eye, which appears to be important in biological rhythms.
pinocytosis. Transport (endocytosis) of fluid and dissolved material into a cell by formation of an enclosing vesicle.
pituitary (= hypophysis). Compound gland that secretes several hormones in response to signals from the nearby hypothalamus in the brain.

placenta. Organ for exchange of oxygen, nutrients, wastes, and other substances between mother and fetus.

placoid. Plate-shaped.

plankton. Small aquatic organisms suspended in water. Includes phytoplankton (algae and small plants) and zooplankton (larvae and small animals). Compare nekton.

planula. Free-swimming ciliated larva of a cnidarian.

plasma. Fluid portion of blood that remains after cells are removed.

plasma cell. Cell formed from a B lymphocyte that makes antibodies.

plasma gel (= ectoplasm). Outer, gelatinous portion of cytoplasm. Compare plasma sol.

plasma membrane. Membrane enclosing a cell.

plasma sol (= endoplasm). Inner, fluid portion of cytoplasm. Compare plasma gel.

plasmid. Loop of DNA that bacteria exchange among themselves naturally.

plastron. Ventral shell of a turtle.

plate tectonics. Branch of geology relating to the movement of plates of Earth's crust due to sea floor spreading and subduction at continental margins.

platelet. Fragment of a white blood cell (megakaryocyte) in blood that initiates clotting.

pleopods (= swimmerets). Appendages on abdominal segments 14 through 18 of decapod crustaceans.

pleura. Membrane enclosing the lung.

pleurite. Cuticular plate (sclerite) on the side (pleuron) of an arthropod.

pneumatic duct. Connection from the esophagus to the swim bladder or lung of a bony fish.

podium (= tube foot). Hollow grasping organ of an echinoderm, controlled by the water vascular system and muscles.

poikilotherm. Animal that does not regulate its body temperature.

polar. (1) Property of certain molecules, such as water, in which positive and negative charges are not uniformly distributed. (2) Associated with a pole, such as the north or south pole, or the animal or vegetable pole of an egg.

polar body. One of three functionless haploid cells formed as a byproduct of meiosis in the development of an ovum.

polarized. Having different properties at opposite extremes. For example, regenerating different body parts from each end of excised tissue.

pole. One of a pair of opposites, such as the animal or vegetal pole of an egg.

pollen brush, comb, packer. Structures on legs of bees used to transport pollen.

poly-A tail. Sequence of 150 to 200 adenosine monophosphates attached to messenger RNA during processing.

polyandry. Mating system in which one female has more than one male mate at a time.

polygamy. Having more than one mate at a time. Either polygyny or polyandry.

polygyny (puh-LIJ-uh-nee). Mating system in which one male has more than one female mate at a time.

polymorphism. Variety in a genetically-determined trait within a population.

polyp. Sessile stage of a cnidarian.

polypeptide. Molecule consisting of many amino acids.

polyphosphoinositide lipid. Substance such as inositol triphosphate (IP_3) or diacylglycerol (DAG) that functions as a second messenger in a cellular response to a hormone or other signal.

polyphyletic. Referring to a group that combines two or more distinct lines of evolution. Compare monophyletic.

polypide. Fleshy part of an ectoproct or other zooid.

polyploidy. Occurrence of three or more sets of chromosomes.

polysaccharide. Carbohydrate consisting of many simple sugars joined by glycosidic links.

polysome. Messenger RNA with ribosomes attached during translation.

polyspermy. Fertilization of an egg by more than one sperm.

pons. Part of the brainstem with horizontal axons that bridge the two halves of the brain.

population. (1) Number of individuals in an area. (2) Group of interbreeding organisms.

population cycle. Periodic increase or decrease in population around the carrying capacity.

pore canal. Channel through arthropod cuticle thought to conduct waxes to the surface.

porocyte. Cell (pinacocyte) surrounding a pore (ostium) in a sponge.

portal vessel. Any blood vessel joining two capillary systems, such as the vessel connecting the hypothalamus with the pituitary, or the small intestine with the liver.

posterior pituitary (= neurohypophysis). Part of the pituitary that releases hormones as neurosecretions.

postsynaptic potential. Change in voltage of a nerve cell membrane in response to synaptic transmitter.

powder down. Feathers that continually grow and disintegrate into talcumlike powder in certain birds.

precapillary sphincter. Band of smooth muscle that closes to divert blood into a capillary system.

precocial. Referring to a species, especially bird or mammal, in which the young are able to feed and care for themselves soon after hatching or being born. Compare altricial.

preening. Running the feathers through the beak of a bird to restore the interconnections among barbs.

preferred range. Range of temperature, oxygen concentration, and other levels of abiotic factors in which the highest population density of a species is found.

preformationism. Obsolete view that each organism was already preformed within either the ovum or the sperm. Compare epigenesis.

prehensile. Adapted for grasping by wrapping around an object, as in the prehensile tails of New World monkeys.

premolar. Two-pointed mammalian tooth adapted for grinding.

pressure. Force divided by the area over which the force is distributed.

primary consumer. Animal that eats autotrophs.

primary host (= final or definitive host). Organism in which the adult form of a parasite develops.

primary producer (= autotroph). Organism that obtains its energy from inorganic sources, as in photosynthesis, and thereby makes it available to other organisms.

primary production. Fixation of matter and energy in organic molecules by primary producers.

primary succession. Establishment and subsequent changes in species composition in an area previously devoid of life.

primary structure. Sequence of amino acids in a protein.

primary transcript (= mRNA precursor). Messenger RNA before it has been processed by removing introns and attaching the poly-A tail and 5′ cap.

primitive. (1) Trait that is ancestral to some other (derived) trait. (2) Organism with many primitive traits, therefore resembling an ancestral organism.

primitive streak. Area along the dorsal midline of the vertebrate embryo where the development of the central nervous system occurs.

primordial germ cell. Cell that develops into a gamete.

prismatic layer. Chalky middle layer of mollusc shell.

probe. Radioactively labeled segment of DNA used to identify a portion of DNA to which it is complementary.

proboscis (plural **proboscides**). Snout or other projection on the head, especially associated with the nasal portion of vertebrates or with the mouth in planarians, leeches, and insects.

procuticle. Newly formed cuticle immediately after molting, before sclerotization.

profundal zone. Deepest water of a lake or pond, beneath the limnetic zone.

progenesis. Accelerated development of sexual maturity in an organism that otherwise retains its juvenile form.

proglottid. Reproductive subunit of a tapeworm.

prokaryote (= procaryote). Organism, mainly bacterial, without cell nuclei and other organelles.

promoter. (1) Sequence of DNA that initiates transcription of a nearby gene. (2) Anything that promotes proliferation of tumor cells after they have been initiated by an oncogene.

pronephros. Primitive kidney in the embryo of a vertebrate.

pronucleus. (1) Haploid nucleus of a gamete after fertilization but before fusion into the zygote nucleus. (2) Haploid product of meiosis of a micronucleus exchanged during conjugation by certain ciliate protozoans.

prophase. First stage of mitosis or meiosis, in which chromosomes become visible.

propolis. Plant resins with which honey bees seal the hive.

prosencephalon (= forebrain). Anterior part of the vertebrate brain including the cerebrum, olfactory bulbs, thalamus, hypothalamus, pituitary gland, and other structures.

prosoma (= cephalothorax). Fused head and thorax of a spider or other arthropod.

prosopyle. Opening of an incurrent canal into a radial canal of a syconoid sponge, or of a flagellated chamber into an excurrent canal in a leuconoid sponge.

prostaglandin. Hormonelike secretion that triggers contraction in uterine smooth muscle and affects blood pressure, blood clotting, inflammation, and numerous other processes.

prostate gland. Organ that produces milky fluid of semen.

prostomium. Part of the head projecting in front of the mouth, particularly in annelids and certain molluscs.

protandry. Sequential hermaphroditism in which gender changes from male to female.

protein. Molecule consisting of amino acids linked to each other by peptide bonds.

protein family. Group of similar proteins believed to have evolved by duplication and mutation of a single gene.

protein sequencing. Determining the sequence of amino acids in a protein.

proteoglycan. Complex of protein and carbohydrate that, together with collagen, makes up a fibrous network in cartilage.

prothrombin. Protein in blood that changes to thrombin as part of the clotting process.

protogyny. Sequential hermaphroditism in which gender changes from female to male.

proton pump. Membrane device responsible for active transport of hydrogen ions.

protonephridium. Organ for osmoregulation and excretion, consisting of branched tubule with closed ends internally. Occurs in some acoelomates and pseudocoelomate invertebrates and in the chordate amphioxus.

proto-oncogene. Normal gene that can mutate to form a cancer-initiating oncogene.

protopod (= protopodite). Base of a biramous appendage of a crustacean.

protostome. Animal in which the mouth originates from the blastopore, and usually with other characteristic patterns of embryonic development.

proventriculus. (1) Anterior portion of a bird's stomach (glandular stomach) that secretes digestive enzymes. (2) Gizzardlike structure between the crop and the midgut of an insect.

proximal convoluted tubule. Part of the tubule of a nephron between Bowman's capsule and the loop of Henle.

pseudocoel (= pseudocoelom). Major body cavity that is similar to a coelom but not derived from mesoderm or not enclosed by a peritoneum.

pseudocoelomate. Animal with a pseudocoel.

pseudogene. Segment of DNA that resembles a gene but is not transcribed.

pseudopodium (= pseudopod). Extension from a protozoan or ameboid cell used in locomotion and endocytosis.

pulmonary vein. Vein that carries blood from a lung to the left atrium of the heart.

pulp. Nerves, blood vessels, and other soft tissue in a tooth or feather.

pump. Structure in membranes responsible for active transport of ions and molecules.

punctuated equilibrium. Pattern of evolution in which long periods of little change are punctuated by brief periods (lasting tens of thousands of years) in which there is rapid evolution.

Punnett square. Diagrammatic device for determining the expected proportions of each genotype resulting from a cross.

pupa. Stage of development in certain insects (holometabolous) between the larva and the adult.

pupil. Opening in the iris of the eye that admits light.

Purkinje fiber. Fiber in the heart that conducts action potentials to the ventricle.

pygidium (py-JI-dee-um). Nonsegmented posterior of a trilobite or other segmented animal.

pyramid (of energy, mass, numbers) (= Eltonian pyramid). Diagram representing energy, biomass, or numbers of organisms at each trophic level.

pyrogen. Substance that induces fever.

pyrosome. Large, luminescent baglike colony of salps.

pyruvate. Three-carbon molecule produced in glycolysis and oxidized in the Krebs cycle.

Q_{10}. Measure of the effect of temperature on the rate of a process. Calculated by dividing the rate at one temperature by the rate at a temperature 10 degrees lower.

quadrupedal. On four feet.

quaternary structure. Combination of two or more protein subunits into a functional unit.

queen. Reproductive female termite, ant, or bee.

queen substance. Pheromone secreted from honey bee queen's mandibular glands that inhibit production of new queens.

quill. (1) Sharp, stiff defensive hair of a mammal such as a porcupine. (2) Hollow portion of the shaft of a feather.

R group. Part of an amino acid that differs from one amino acid to another.

rachis (RAY-kis). Portion of the shaft of a feather that is not hollow.

radial canal. Part of echinoderm water vascular system that branches into an arm.

radial cleavage pattern. Cleavage pattern in which newly divided cells are aligned with the central axis. Characteristic of echinoderms, amphibians, and some invertebrates.

radially symmetric. Forming two similar halves when bisected at any angle along the axis.

radiation. (1) Radial movement from a source. (2) Emission of electromagnetic or ionizing radiation.

radioactive dating. Technique for determining the age of materials by measuring quantities of decay products of radioactive isotopes.

radioimmunoassay (RIA). Technique for measuring small concentrations of hormones or other molecules.

radiole. Branched, ciliated tentacle of annelid tube worms.

radula (RAD-ju-luh). Rasping tonguelike organ of a snail or other mollusc.

ram ventilation. Method of ventilating the gills by swimming with the mouth open.

ratite. Having a breastbone without a sternal keel, as in flightless birds. Compare carinate.

reabsorption. Active transport from filtrate into body fluid by an organ of osmoregulation and excretion.

reaggregation. Movement of cells toward each other after they have been dispersed, as in sponges.

recapitulation. Summary repetition, especially the repetition of the pattern of evolution of an organism in its pattern of development.

receptive field. Group of receptors that directly affects a nerve cell's electrical activity.

receptor-mediated endocytosis. Type of endocytosis in which receptors binding a particular substance migrate to a coated pit on the plasma membrane, where a vesicle is formed that transports the material into the cell.

receptor molecule. Molecule in or on a cell that binds a particular hormone, synaptic transmitter, or other substance.

receptor potential. Change in voltage across membrane of a sensory receptor cell during stimulation.

recessive. Referring to an allele that is not expressed in the presence of a different homologous (dominant) allele.

reciprocity. Behavioral assistance by one organism that is returned by another.

recombinant DNA. DNA combined from two different chromosomes during meiosis, combined from two different individuals during fertilization, or combined by other natural or artificial means.

recruitment. Increasing the number of nerve cells activated, especially in increasing the force of contraction of a muscle.

rectal gland. Osmoregulatory organ of sharks and rays that secretes concentrated NaCl solution into the posterior intestine.

rectal pad. Structure on insect rectum that recycles water back into the hemolymph.

rectilinear movement. Locomotion used by snakes when stalking, in which the body lies in the direction of movement and creeps forward without twisting.

rectum. Terminal portion of a digestive tract.

red blood cell (= erythrocyte). Cell that transports hemoglobin through blood of vertebrates.

redd. Circular depression in gravel used as a nest by salmonid fishes.

redia. Larval form of a fluke that is produced by sporocysts and lives in an intermediate host.

redox reaction (oxidation–reduction reaction). Reaction involving loss of electrons from one molecule, which is oxidized, to another, which is reduced.

reduction. Gain of electrons by a molecule.

reductionism. Term applied, usually disparagingly, to attempts to explain complex biological phenomena in terms of simpler phenomena, especially physical and chemical.

reflex. Simple, stereotyped response to a stimulus.

reflex ovulation. Ovulation triggered by copulation.

regeneration. Development of a new limb or organ to replace a lost one.

regulative development (= indeterminate cleavage). Development in which fates of cells (blastomeres) are determined later in cleavage.

releaser. Sign stimulus that triggers a stereotyped response (motor program).

releasing gland. Part of the duo-gland system of a flatworm that dissolves the adhesive released by the viscid gland for attachment and locomotion.

releasing hormone. Peptide from hypothalamus that, together with release-inhibiting hormones, regulates secretion from the anterior pituitary.

renal. Pertaining to the kidney.

renette cell (= gland cell). Presumed osmoregulatory cell of a nematode.

rennin (REN-in). Stomach enzyme of infant mammals that coagulates milk.

replication. Duplication of a cell's DNA prior to mitosis.

reproductive isolation. Separation of two populations by geographic barriers or by differences in behavior or appearance so that they do not interbreed.

reservoir. Species in which a parasite that causes disease in another species lives without necessarily causing disease.

resilin. Rubbery protein that contributes to jumping in fleas and locusts.

respiration (= cellular respiration). Complete catabolism of glucose requiring oxygen and producing carbon dioxide.

respiratory chain. Oxygen-requiring sequence of reactions in mitochondria involving transfers of protons and electrons, resulting in production of ATP.

respiratory circulation. Circulation of blood through lungs, gills, or other organs for exchange of O_2 and CO_2.

respiratory pigment. Colored molecule such as hemoglobin that increases the capacity of blood to transport oxygen.

respiratory tree. (1) System of branching bronchioles in mammalian lungs. (2) Branched respiratory organ in the body cavity of a sea cucumber.

resting membrane potential. Voltage across the plasma membrane of a cell except when the potential is specifically changed, as by an action potential.

restriction enzyme. Enzyme produced by bacteria that cleaves DNA at particular base sequences.

restriction fragment length polymorphism (RFLP). Segment of DNA of varying length produced by restriction enzyme. Used as a marker in mapping chromosomes.

rete mirabile (REE-tee mir-AH-bi-lay; plural **retia mirabiles**). Intertwined veins and arteries conducting blood in opposite directions for the concentration of oxygen, heat, or ions.

reticular formation. Part of the brainstem that maintains wakefulness.

reticulopodium. Type of pseudopodium in certain protozoans (foraminiferans, Granuloreticulosea) that forms a food-trapping net.

retina. Structure of the eye that bears photoreceptors and other nerve cells for vision.

retinula. Photoreceptive region of an arthropod compound eye usually consisting of seven or eight retinula cells.

retrocerebral sac. Mucus-secreting structure of rotifers.

reuptake. Active transport of synaptic transmitter back into a synaptic knob.

reverse transcriptase. Enzyme produced by cells infected with retrovirus which causes the cell to transcribe the RNA of the viral genome into DNA.

Reynold's number (Re). Dimensionless number representing the ratio of inertial to viscous forces in swimming or flying. It increases with speed and body length.

Rh factor. Protein antigen on red blood cells that partly determines whether donated blood is compatible. Presence or absence designated by + or – following the ABO type.

rhabdite. Rod-shaped structure in epidermis of a turbellarian flatworm that helps produce mucus.

rhabdom. Microvilli in retinula of arthropod compound eye that contain the photopigment.

rhabdomeric photoreceptor. Type of photoreceptor cell in most invertebrates organized around microvilli or membrane lamellae, rather than cilia as in vertebrates.

rhombencephalon (= hindbrain). Part of the brain that forms the cerebellum, pons, and medulla.

rhopalium (= tentaculocyst). Rod-shaped structure enclosing a statocyst organ and sometimes ocelli in a jellyfish.

rhynchocoel (RING-ko-SEAL). Cavity that encloses the proboscis of a nemertean.

ribonucleic acid (RNA). Nucleic acid with ribose as its sugar. It is generally single-stranded and occurs as messenger RNA, ribosomal RNA, and transfer RNA.

ribose. Five-carbon sugar in ribonucleic acid, ATP, and cyclic AMP.

ribosome. Organelle consisting of protein and ribosomal RNA (rRNA) in the cytoplasm and on rough endoplasmic reticulum, which organizes the synthesis of proteins.

rigor. Cramping of muscle due to lack of ATP; especially rigor mortis, which occurs after death.

ring canal. Part of echinoderm water vascular system that encircles the mouth.

Ringer's solution. Solution containing balanced concentrations of Na^+, K^+, Ca^{2+}, and other ions for replacement of body fluids. Especially such a solution for use with amphibians.

rod. Vertebrate photoreceptor cell responsible for vision in dim light.

Root effect. Exaggerated Bohr effect in many fishes which enables the blood to release oxygen into the swim bladder upon exposure to acid.

rotational cleavage pattern. Cleavage pattern in which one pair of cells formed by the second cleavage lies at right angles to the other pair of cells. Characteristic of placental mammals.

rough endoplasmic reticulum (rough ER). Cytoplasmic membranes bearing ribosomes.

royal jelly. Sugary substance that worker honey bees feed to female larvae to induce them to develop into queens.

rumen. Large fore-stomach of a ruminant in which plant fibers are initially digested.

ruminant. Animal such as cow, sheep, and deer that has a rumen.

rumination. Regurgitation of food from the stomach for further chewing.

S phase. Part of interphase in the cell cycle during which DNA is replicated.

saccule. Mechanoreceptor in inner ear that responds to balance.

salinity. Measure of total solute concentration, especially in seawater.

saliva. Secretion from salivary glands into the mouth.

salt gland. Gland in marine vertebrates that secretes excess salt, usually from the eye sockets or nostrils.

saltatory conduction. Conduction in myelinated axons, in which action potentials "jump" from one node of Ranvier to the next.

saprozoic nutrition. Feeding on dissolved nutrients.

sarcodine. Protozoan that characteristically forms pseudopodia.

sarcoma. Cancer that arises from bone, cartilage, fat, muscle, or other tissues of mesodermal origin.

sarcomere. Subunit of a skeletal muscle fiber representing the structure between two Z disks.

sarcoplasmic reticulum. Membranous organelle of muscle that controls contraction by releasing and storing Ca^{2+}.

satellite male. Male, especially frog, that stays near a courting male and intercepts females he has attracted.

saturated fat. Neutral fat lacking double bonds so that two hydrogens are bonded to each carbon in the fatty acid chain.

saturation effect. Advantage of group living or mass hatching resulting from the inability of predators to eat more than a small proportion of the group at one time.

scalid. Circle of spines on the proboscis.

scaling (= allometry). Differential effects of changing linear, area, and volume dimensions in organisms.

schizocoelous. Having a coelom formed by division of mesoderm.

schizogony. Multiple fission, especially in certain protozoans.

schizont. Multinuclear form of *Plasmodium* or other protozoan that forms from a sporozoite and undergoes multiple fission (schizogony).

sclerite. Platelike piece of sclerotized cuticle.

sclerocyte (= scleroblast). Amebocyte that forms spicules in a sponge.

sclerotizing (= tanning). Hardening and darkening of arthropod cuticle.

scolex (= holdfast). Adhesive organ of a tapeworm.

scute. Horny or bony scale, such as that on a turtle or other reptile.

scyphistoma. Polyp of a jellyfish just after settling.

sebaceous gland. Skin gland that produces oily sebum.

second messenger. Intracellular chemical that mediates action of a hormone or other stimulus in a cell.

secondary consumer. Animal that consumes a primary consumer.

secondary structure. Conformation of part of a protein due to hydrogen bonding between adjacent protein strands. Secondary structure takes two common forms: α-helix and β-sheet.

secondary succession. Changes in species composition in an area already inhabited.

secretion. (1) Active transport of a substance from body fluid into an organ for excretion. (2) Movement of a substance out of a gland.

sedentary. Not moving.

segment (= somite = metamere). Linearly repeated subdivision of a body in Pogonophora, Annelida, Arthropoda, and Chordata.

segmentation (= metamerism). Division of body into segments.

segmentation mutation. Mutation in insects that affects the number or pattern of segments.

segregation. Independence of expression of each allele, as opposed to blending.

semen. Fluid containing sperm.

semicircular canals. Three canals in inner ear that respond to rotational acceleration of the head.

semiconservative replication. Duplication of DNA in such a way that each new molecule conserves one strand of the old DNA.

semilunar valve. Valve in the aorta or pulmonary artery that prevents regurgitation of blood into the ventricle.

seminal vesicle. Organ of male vertebrates that secretes fructose-rich fluid into semen.

seminiferous tubule. Tubule in testis that produces sperm.

semipermeable. Referring to a membrane that is permeable to some substances but not to others.

semiplume. Feather intermediate between a vane feather and down, lacking hooks on the barbules but with a rachis.

sensitive period. Period during which an organism is subject to a certain effect such as imprinting.

sensory neuron (= afferent neuron). Nerve cell that conducts action potentials from sensory receptors into central nervous system.

septum. Sheet of tissue dividing two chambers, such as two ventricles of the heart.

sequencing. Determining the sequence of amino acids in a protein or bases in a nucleic acid.

serial homology. Derivation of structures from corresponding parts of different segments in an animal. For example, the antenna of an insect is serially homologous to a leg.

serous gland (= poison gland = granular gland). Poison-secreting gland in skin of amphibians.

serum. Fluid part of clotted blood that remains after cells and the clot are removed.

serum albumin. Proteins in blood that are largely responsible for its osmotic pressure.

servomechanism. System that uses its own output for self-correction.

sessile. Attached to substratum.

set point. Hypothetical temperature, pH, or other level that is maintained by homeostasis.

seta (plural **setae**). Bristle, usually chitinous, on an annelid, arthropod, or other invertebrate.

settling. Process by which a free-swimming larva attaches to substratum prior to developing into a sedentary adult.

sex. (1) Reproductive process by which a new individual develops from gametes produced by two parent individuals. (2) Gender (male or female).

sex chromosome. Chromosome that occurs in different numbers in different sexes. Chromosome that is not an autosome.

sex linkage. Location of a gene on a sex chromosome, making it likely to be transmitted differently to different sexes.

sexual selection. Differential reproduction based on differences in sexual displays or physical competition among potential mates.

sheet web. Spider web built horizontally and often leading to a funnel-shaped retreat in which the spider awaits prey.

sickle cell. Deformed red blood cell resulting from a point mutation in a gene for hemoglobin. Causes sickle-cell trait in heterozygotes, and sickle-cell anemia in homozygotes.

sidewinding. Mode of locomotion used by snakes on sand, in which the body is formed into a spiral at right angles to the direction of locomotion.

sign stimulus. Signal that evokes a stereotyped behavioral response (motor program).

siliceous. Containing silica (silicon dioxide).

simple epithelium. Epithelium with one cell layer.

simple eye. Light detecting organ with a lens to concentrate light, but lacking the capability of resolving an image.

sinoatrial (SA) node. Cluster of cells in the right atrium that normally functions as pacemaker of the heart.

sinus venosus. Chamber between the vena cava and the right atrium of most vertebrates and in fetus of birds and mammals.

siphonoglyph. Ciliated groove in the mouth of sea anemones that generates hydrostatic pressure.

siphuncle. Strand of tissue that empties fluid from the chambers of *Nautilus.*

skin. (1) Integument of vertebrates, consisting of dermis and epidermis. (2) Soft integument of any animal.

sliding filament theory. Theory that thin and thick muscle filaments slide among each other, and that the force of muscle contraction comes from the interaction of the two types of filaments.

slit sensillum. Mechanoreceptor in the cuticle of arachnids arranged so that tension in a certain direction generates a response by opening or closing a slit.

slow-twitch fiber (= type I red fiber). Muscle fiber that contracts slowly and metabolizes aerobically.

smolt. Young salmonid fish at time of migration from native stream.

smooth endoplasmic reticulum (= smooth ER). ER without ribosomes, generally responsible for packaging glycoproteins from rough ER into vesicles for processing by Golgi complex.

smooth muscle. Involuntary muscle that lacks striations.

snRNP ("snurp"). Small nuclear ribonucleoprotein. Complex of RNA and protein that splices exons together during mRNA processing.

sociobiology. Biological study of animal behavior that emphasizes the evolution of social behavior.

solenocyte. Flagellated type of flame bulb that drives filtrate in certain excretory organs (protonephridia).

solubility. Amount of a substance that can be dissolved in a given volume of solvent.

solute. Substance dissolved in another substance.

solution. Liquid in which something is dissolved.

solvent. Liquid such as water that dissolves another substance.

somatoplasm. Body of an organism exclusive of the germ plasm that is responsible for transmission of hereditary information.

somite (= segment = metamere). (1) Linearly repeated subdivision of a body in a segmented animal (Pogonophora, Annelida, Arthropoda, or Chordata). (2) In vertebrates, one of 40 masses of embryonic mesoderm.

spat. Juvenile oyster with protective shells, ready to attach to substratum.

speciation. Evolution of a new species.

species (singular and plural). (1) For sexual organisms, "groups of actually or potentially interbreeding natural populations, which are reproductively isolated from other such groups" (Ernst Mayr). For sexual organisms, groups of organisms that are similar to each other but recognizably different from other such groups. (2) Taxonomic category below the genus or subgenus level, which may or may not correspond to the biological definition above.

sperm (singular and plural; = **spermatozoon**). One or more male gametes.

sperm receptacle. Reservoir for sperm in a female crab.

sperm web. Special web on which a male spider deposits a drop of semen to fill the bulbs of its pedipalps.

spermatheca. Sperm reservoir in the female.

spermatid. Haploid, immature sperm cell that has not acquired its flagellum.

spermatocyte. Sperm cell in the process of meiosis.

spermatogenesis. Development of sperm.

spermatogonium. Primordial germ cell that eventually develops into sperm.

spermatophore (sperm packet). Packet or globule of sperm produced by certain male invertebrates and salamanders.

spermiogenesis. Maturation of a spermatid into a sperm.

sphincter. Ring of muscle that controls flow through an aperture or tube.

spicule. Needle-shaped mineral particle that helps form the endoskeleton of a sponge or sea cucumber.

spindle apparatus. Arrangement of microtubules radiating from centrioles to chromatids during mitosis and meiosis.

spinneret. Tubule on abdomen of a spider from which silk is extruded.

spiracle. (1) External opening of an insect trachea. (2) Modified first gill opening of a shark. (3) Excurrent channel for tadpole gills.

spiral cleavage pattern. Cleavage pattern in which daughter cells of cleavage move spirally into the furrows between other cells. Characteristic of many invertebrates other than arthropods and echinoderms.

spiral valve. (1) Helical membrane in intestine of sharks and primitive fishes. (2) Membrane that separates oxygenated from deoxygenated blood in conus arteriosus of amphibian heart.

spliceosome. Body in nucleus that mediates mRNA processing by removing introns and splicing exons.

spongin (SPUN-jin). Fibrous protein that helps form the endoskeleton of a sponge.

spongocoel (SPUN-go-seal) (= atrium). Central chamber in an asconoid or syconoid sponge.

spongy (= cancellous = trabecular) **bone.** Bone with spaces containing marrow.

spontaneous generation. Supposed production of an organism without parents.

spore. Inactive, resistant form of a zygote. In certain protozoans, the product of an oocyst.

sporocyst. Larval form of *Schistosoma* and other flukes that lacks a mouth and digestive tract and absorbs nutrients from the intermediate host.

sporogony. Development of numerous spores or sporozoites within oocyst of malaria-causing *Plasmodium* and certain other protozoans.

sporozoan. Protozoan that reproduces from spores.

sporozoite. Form of malaria-causing *Plasmodium* and some other protozoans that forms from a spore and develops into a schizont.

squalene. Low-density fat that maintains buoyancy in sharks.

squamous epithelium. Epithelium with flattened cells.

stapes (stay-peez). A middle-ear bone.

Starling effect. Balance of blood pressure and colloid osmotic pressure in capillaries that prevents fluid from accumulating in tissues.

statoblast. Resistant form from which a freshwater ectoproct reproduces a new zooid asexually.

statocyst. In crustaceans and certain other aquatic animals a hollow mechanoreceptor organ containing granules that determines direction of gravity.

statolith. Dried secretion or sand grains in statocyst organs, which deflect hair cells in response to gravity.

stem cell (= founder cell). Embryonic cell from which certain other cells develop.

stenohaline. Able to tolerate slight changes in environmental osmolarity. (Compare euryhaline.)

stercoral pocket. Receptacle for metabolic wastes just dorsal to the rectum in a spider.

sternal keel (= carina). Long, thin process on the sternum of a bird, to which flight muscles are attached.

sternum. (1) Vertebrate breastbone. (2) Ventral surface of an arthropod.

sternite. Cuticular plate (sclerite) on ventral surface (sternum) of an arthropod.

steroid. Four-ringed lipid such as cholesterol, vitamin D, or a steroid hormone.

stigma (plural **stigmata**). (1) "Eyespot" of *Euglena* or other protozoan. (2) Type of spiracle on certain insects.

stoma (plural **stomata**). A mouth.

stomach. Enlargement of the digestive tract for storage and digestion of food.

stomochord (= buccal diverticulum). Anterior projection of the buccal cavity of a hemichordate, formerly considered a notochord.

stone canal. Tube that brings water from the madreporite into the rest of the water vascular system in an echinoderm.

stratified epithelium. Epithelium with more than one cell layer.

stress. Activation of the adrenal cortex and sympathetic nervous system.

stretch receptor. Mechanoreceptor that responds to stretch, especially in skeletal muscle.

stretch reflex. Self-correcting contraction of a skeletal muscle in response to sudden stretching.

striated muscle. Skeletal muscle, which has repeating subunits producing a striped pattern.

strobila. (1) Stage in development of jellyfish that develops into a series of free-swimming larvae (ephyrae). (2) Chain of tapeworm proglottids.

stroke volume. Volume of blood pumped by one contraction of the heart.

structural color. Color due to structure that alters the transmission or reflection of light, rather than to pigment.

structural white. White due to dust-sized, transparent particles that scatter all wavelengths equally.

stylet. Sharp projection, such as on the proboscis of a nemertean or tardigrade.

subepidermal nerve plexus. Portion of the nervous system of a flatworm that is independent of the cerebral ganglia.

subjunctional fold. Fold in the plasma membrane of a muscle cell opposite the neuromuscular junction.

submucosal plexus. Network of nerve cells in the gut that coordinates motility.

substrate. (1) Molecule that an enzyme binds to and catalyzes a reaction in. (2) Substratum.

substrate-adhesion molecule (SAM). Glycoprotein such as fibronectin that guides movement of cells during development.

substratum (= substrate). Surface on which something rests.

succession. Replacement of one community by another in the same area. See primary succession and secondary succession.

summation. Additive effect of multiple postsynaptic potentials or of muscle twitches.

supercooling. Reduction in temperature below the nominal freezing point without formation of ice crystals.

superficial cleavage pattern. Cleavage pattern in which blastomeres are confined to the periphery of the egg. Characteristics of arthropods.

suppressor gene. Gene whose product normally prevents abnormal growth of tissue (a tumor).

suprachiasmatic nucleus (SCN). Part of the hypothalamus above the optic chiasm that apparently controls the pineal gland's influence on biological rhythms in a mammal.

swelling. Accumulation of fluid in tissues due to inflammation, tissue damage, or other cause.

swim bladder. Gas-filled chamber that maintains buoyancy in many fishes.

swimmerets (= pleopods). Appendages on abdominal segments 14 through 18 of decapod crustaceans.

syconoid. Type of canal system in a sponge in which choanocytes do not line the spongocoel, but are found within radial canals.

symbiosis. Close association between members of different species. Includes mutualism, parasitism, and commensalism.

symmetry. Complementarity of shape such that an animal can be bisected into two similar halves.

sympathetic division. Part of the autonomic nervous system that is active during stress ("fight or flight") situations.

sympatric. In the same area. Referring especially to evolution in which a population becomes reproductively divided into two species even though they are not geographically separated.

synapomorphy. In cladistics a homologous trait that is assumed to be derived, because it occurs in two or more groups being classified but not in the outgroup. The only traits useful for classification in cladistics.

synapse. Structure by which a nerve cell excites or inhibits another nerve cell or a muscle cell.

synapsid. Reptile or other vertebrate with one temporal opening.

synapsis. Alignment of homologous chromosomes side by side during meiosis.

synaptic knob. Enlarged terminal of a nerve cell associated with a chemical synapse.

synaptic transmitter (= neurotransmitter). Chemical released from a synapse that tends to either excite or inhibit a nerve or muscle cell.

synchronous muscle. Muscle in insects in which contraction is synchronous with action potentials in the muscle. Compare asynchronous muscle.

syncytium (sin-SISH-ee-um). Multinucleate cell, usually resulting from the fusion of two or more cells.

syngamy. Fertilization of one gamete by another in sexual reproduction, especially in protozoans.

synthetic theory (= neo-Darwinian theory). Current theory that combines Darwin's theory of natural selection with modern knowledge of genetics to explain most evolution.

syrinx (plural **syringes** or **syrinxes**). Voice-producing structure in the trachea of a bird.
systematics. The scientific study of diversity and all relationships among organisms.
systemic circulation. Circulation of blood to metabolizing tissues, as opposed to respiratory circulation.
systolic blood pressure. Peak blood pressure, which follows ventricular contraction.

T cell. Type of lymphocyte responsible for destroying damaged cells.
t tubule. Transverse tubule in muscle that conducts action potentials into sarcomeres to trigger contraction.
taenidium. Spiral strand of chitin that reinforces insect trachea.
tagma (plural **tagmata**). Group of body segments forming a distinct region, such as head, thorax, or abdomen.
tagmosis (= **tagmatization**). Development of distinct body regions (tagmata) in segmented animals.
tanning (= sclerotizing). Hardening and darkening of arthropod cuticle.
tapetum. Reflecting layer in the eyes, especially in nocturnal animals.
taste. Chemoreception of substances not airborne.
taste bud. Structure on tongue bearing chemoreceptors for taste.
taxis (plural **taxes**). Orientation of an animal with respect to direction of a stimulus.
taxon (plural **taxa**). Category such as phylum, order, or species in which organisms are classified.
taxonomy. Study of the principles of classification.
tectorial membrane. Tissue in organ of Corti that deflects hair cells in response to sound.
tegument. Outer covering of parasitic flatworm (tapeworm or fluke) or acanthocephalan, consisting of hardened portions of living cells partially submerged within parenchyme.
teleology. Inappropriate attribution of a purpose or goal to a process.
telolecithal. Egg with abundant, unevenly distributed yolk.
telomere. Proteins and bases that protect chromosome tips.
telophase. Final stage of mitosis or meiosis in which chromosomes are separated to opposite poles and begin to disperse prior to cell division.
telson. (1) Posterior projections of a decapod crustacean that combines with the uropod to make a fin. (2) Spinelike posterior projection of a horseshoe crab.
temperature. Measure of the average energy associated with the thermal agitation of a molecule.
tendon. Connective tissue that attaches muscle to bone.
tentacle. Flexible, elongated, unsegmented extension, especially around the mouth.
tentaculocyst (= rhopalium). Rod-shaped structure enclosing a statocyst organ and sometimes ocelli in a jellyfish.
teratogen. Substance that causes congenital defects.
tergite. Cuticular plate (sclerite) on dorsal surface (tergum) of an arthropod.
tergum. Dorsal surface of an arthropod.
termination sequence. DNA base sequence that stops transcription.
territory. Area from which one animal excludes others of the same species.
tertiary structure. Shape of a protein determined by interactions among R groups.
test. A hard case, such as the shell of a sea urchin.
testis (plural **testes**). Organ that produce sperm.
testis determining factor (TDF). Substance produced by a gene on the Y chromosome that induces development of testes rather than ovaries.
tetrad. Four chromatids joined in synapsis during prophase I of meiosis.
tetrapod. Vertebrate belonging to a group that generally has four limbs (amphibian, reptile, bird, or mammal).
thalamus. Part of midbrain consisting of nuclei and neural tracts linking parts of the cerebrum with each other and with other parts of nervous system.
theca (THE-kuh; plural **thecae** THE-see). (1) A case, covering, receptacle, or sheath. (2) Overwintering cyst of a hydra.
thermal hysteresis protein (THP). Protein in insects that lowers the freezing point of tissues without affecting the melting point.
thermal stratification. Layering of temperature lakes resulting from differences in water density due to temperature.
thermogenesis. Production of body heat, especially to raise body temperature.
thick filament. Filament of the protein myosin in muscle.
thickener cell. Sclerocyte that helps form spicules in a sponge.
thin filament. Filament of the protein actin in muscle.
threshold. Minimum stimulus required for a response such as an action potential.
thrombin. Protein made from prothrombin during the process of blood clotting.
thrombosis. Blood clotting that interferes with circulation.
thrust. Force generated during flight that pushes and animal forward against friction and pressure drag.
thymine. Nitrogenous base in DNA.
tight junction. Band joining one cell to surrounding cells that blocks the passage of substances between them.
tissue. Group of cells structurally and functionally related to each other.
tissue culture. Technique of maintaining cells outside the body (*in vitro*).
tonofilament. Intermediate filament of keratin that anchors a desmosome to the cell contents.
top carnivore. Carnivorous animal that is not, itself, prey to another carnivore.
tormogen. Cell in arthropod cuticle that forms the socket of a hairlike structure.
tornaria larva. Larva characteristic of acorn worms, which spins as it swims.
torpor. State of inactivity due to low body temperature.
torsion. Twisting of veliger larva of a snail or other gastropod mollusc into a U shape.
toxin. Substance released by one organism that is harmful to another.
trabecular (= cancellous = spongy) **bone.** Bone with spaces containing marrow.
trace element. Inorganic nutrient required in minute amount.
trachea (plural **tracheae**). (1) Tube connecting mouth and nose with the bronchi in a vertebrate. The wind pipe. (2) In insects and spiders, a tube through which air diffuses to tissues.

tracheal gill. Area of aquatic insect that is rich in tracheae adapted for absorbing O_2 under water.
tracheole. Smallest branch of a trachea in tissue of an insect or spider.
transcription. Production of messenger RNA with a base sequence complementary to one strand of DNA in a gene.
transducin. Protein that removes cGMP during the response of rods and cones to light.
transfer RNA (tRNA). RNA that binds a particular amino acid and inserts it where called for by messenger RNA during protein synthesis.
translation. Synthesis of protein as directed by messenger RNA.
translocation. Change in location, such as movement of part of a chromosome to another location on the same or another chromosome.
transporter. Structure in the plasma membrane that is responsible for facilitated diffusion of molecules.
transposition. Movement of a submicroscopic segment of a chromosome (transposing element = jumping gene) to another chromosome or to a different position on the same chromosome.
triacylglycerol (= triglyceride = neutral fat). Lipid consisting of three fatty acids linked to a glycerol.
tricarboxylic acid cycle (= TCA cycle = Krebs citric acid cycle). Cycle of reactions in mitochondria by which pyruvate from glycolysis is oxidized to carbon dioxide during cellular respiration.
trichocyst. Hair-shaped defensive structure discharged from the pellicle of certain protozoans (ciliates).
trichogen. Cell in arthropod cuticle that forms a hairlike structure.
tricuspid valve. Right atrioventricular valve of the heart.
triglyceride (= triacylglycerol = neutral fat). Lipid consisting of three fatty acids linked to a glycerol.
triose. Simple sugar with three carbons.
triploblastic. Developing from three germ layers: ectoderm, mesoderm, and endoderm.
trochophore larva. Free-swimming larva with cilia encircling the mouth. Characteristic of molluscs and some marine flatworms, annelids, and brachiopods.
trophic level. Level of an animal in a food web or food chain that depends on the number of energy transfers between the animal and the primary producers. Examples: primary consumer, secondary consumer, and so on.
trophoblast. Cells surrounding the blastocyst that are responsible for implantation into the uterine lining.
trophosome. Chamber within certain pogonophorans containing symbiotic chemoautotrophic bacteria.
trophozoite. Mature, feeding stage of a parasitic protozoan.
tropomyosin. Protein on actin that blocks cross-bridge formation in vertebrate skeletal muscle.
troponin. Protein on actin that binds Ca^{2+} to trigger contraction in vertebrate skeletal muscle.
trypsin. Protein-digesting enzyme.
tube foot (= podium). Hollow grasping organ of an echinoderm, controlled by the water vascular system and muscles.
tuberous organ. High-frequency electroreceptor of certain fishes.
tubules of Cuvier. Sticky strands ejected by sea cucumbers when disturbed.
tumor. Mass of cells resulting from inappropriate cell division.
tun. Barrel-shaped body of a tardigrade in cryptobiosis.
tunic. (1) Collagenous sheath reinforcing radial and circular muscles of a cephalopod mollusc. (2) Sheath enclosing the body of a tunicate.
tusk. Greatly enlarged tooth of a mammal such as an elephant.
twitch. Response of a muscle to a single action potential.
tympanum. An ear drum or other vibrating membrane.
Tyndall blue. Structural color caused by scattering of light by submicroscopic particles.
typhlosole. Longitudinal infolding of the intestine of an earthworm.

ultrafiltrate. Fluid filtered across a capillary wall.
umbo (plural **umbones**). Bulge in each valve at the hinge of a bivalve mollusc, or the beak of a brachiopod.
unconditioned stimulus. In classical conditioning, a stimulus to which an animal normally responds.
undulating membrane. (1) Long, finlike group of cilia (buccal ciliature) near the cytostome (mouth) of a protozoan. (2) Finlike fold in pellicle and the flagellum of trypanosomes and related protozoa.
unequal crossing over. Failure of homologous chromosomes to break in exactly the same place during crossing over in meiosis, resulting in gain or loss of genes in gametes.
uniformitarianism. Concept that geological changes in the past occurred in the same way as those occurring now.
uniramous. Unbranched (one-branched).
unisexual species. Species in which only parthenogenetic females occur.
upwelling. Rise of cold, nutrient-rich water to the surface due to prevailing local winds or water currents.
urea. Nitrogenous waste excreted mainly by mammals and other viviparous animals.
ureotelic. Excreting urea as the major nitrogenous waste.
ureter. Tube conducting urine from the kidney to the urinary bladder.
urethra. Tube through which urine is voided from the bladder.
uric acid. Main nitrogenous waste in egg-laying terrestrial animals.
uricotelic. Excreting uric acid as the main nitrogenous waste.
urogenital aperture. Aperture through which urine and gametes are released by certain fishes.
urogenital system. Excretory and reproductive systems with shared parts.
uropods. Last pair of segmental appendages of a decapod crustacean, which combine with the telson to make a fin.
uropygial gland (= oil gland). Gland at the base of the tail of a bird that produces oil for preening.
urostyle. Fused vertebrae of a frog.
uterus (= womb). Cavity in which embryos develop in a viviparous animal.
utricle. Mechanoreceptor in inner ear that responds to balance.

vaccination (= inoculation). Stimulating immunity by exposing an animal to the antigen of a virus or bacterium.
vagal inhibition. Reduction in heart rate by the vagus nerve.
vagina. Receptacle for the penis during copulation.

vagus nerve. Major branch of the parasympathetic division of the autonomic nervous system.

valve. (1) One of two shells in a bivalve mollusc or brachiopod. (2) Device for controlling flow of a fluid, such as a valve in the heart.

vane feather (= contour feather). Major type of feather, in which barbs form two vanes on opposite sides of the shaft.

vaporization. Conversion of a substance from a liquid to a gas phase.

variety. Race or other genetically different group within a species.

vas deferens (plural **vasa deferentia**) (= sperm duct = ductus deferens). Tube through which sperm are ejaculated.

vas efferens (plural **vasa efferentia**). In flatworms and some other invertebrates, a tubule conducting sperm into the vas deferens.

vasa recta. Capillaries surrounding the loops of Henle in nephrons of the kidney.

vasectomy. Sterilization by cutting and tying the vasa deferentia.

vasoconstriction. Narrowing of arteries due to contraction of smooth muscles.

vasodilation. Widening of arteries due to relaxation of smooth muscles.

vector. (1) Organism that transmits a parasite or disease to another. (2) Virus or other organism used in recombinant DNA technology to transfer genetic material from one organism to another.

vegetal pole. Part of ovum containing more yolk than the animal pole.

vein. (1) Blood vessel that circulates blood from capillaries to heart. (2) Tubule in insect wing that circulates blood and contains nerves.

veliger. Larval form of certain molluscs that occurs between the trochophore larva and the adult.

velum. (1) Infolding flap on the edge of the bell of a hydrozoan medusa. (2) Ciliated swimming organ of a molluscan veliger larva. (3) Ventral surface of an echinoderm in class Concentricycloides. (4) Muscular membranes that pump water into gills of a jawless fish.

vena cava. Largest vein in a vertebrate, which returns blood to the heart.

venous pump. Forcing of blood toward heart due to contraction of skeletal muscle and one-way valves in veins.

vent. External opening of a cloaca.

ventral. Pertaining to the lower or front surface. Compare dorsal.

ventral root. Ventral branch of nerve from spinal cord, which contains motor neurons.

ventricle. (1) Chamber of the heart that pumps blood into an artery. (2) Chamber in the brain filled with cerebrospinal fluid.

venule. Small vein connected to a capillary.

vermiform. Worm-shaped.

vertical migration. Daily rhythm of vertical movements of zooplankton, usually down at dawn and to the surface at night.

vertical stratification. Systematic vertical differences in distribution of abiotic factors, such as light, humidity, temperature, and wind speed in tropical rainforests.

vesicle. Spherical, membrane-bound structure enclosing material being transported into a cell by endocytosis, or material (digestive enzyme, synaptic transmitter, or hormone) being transported out of the cell by exocytosis.

vestibular apparatus. Mechanoreceptive organs for detection of head motion and orientation, consisting of semicircular canals and otolith organs (utricle and saccule).

vestigial. Referring to a functionless structure that may have evolved from a useful structure in an ancestral species.

vibrissa (= whisker). Stiff facial hair adapted as a touch receptor.

vicariance. Reproductive isolation of two populations in a species by geological change or by extinction of members of species between the two populations.

villus. Projection of a tissue layer, especially of the small intestine or the placenta.

virulence. Potential to cause disease.

visceral mass. Fleshy part of a mollusc consisting of the foot and mantle.

viscid gland. Part of the duo-gland system of flatworms that secretes an adhesive for attachment and locomotion.

vitalism. Belief that organisms have powers unique to life.

vitamin. Organic molecule required in the diet for metabolism, but which is not itself metabolized.

vitellaria. (1) Larvae of sea lilies. (2) Plural of vitellarium.

vitellarium (= yolk gland). Gland that produces yolk, and in many flatworms, eggshell.

vitelline envelope. Covering of echinoderm egg between the plasma membrane and the jelly coat.

vitellogenesis. Production of proteins (vitellogenins) and other constituents of yolk.

viviparity. Birth to live young that have been nurtured during development within the reproductive tract of the mother. Compare ovoviviparity and oviparity.

vocal sac. Skin pouch on a male frog that amplifies the advertising call.

vomerine. Relating to the flat bone that forms part of the septum between the nostrils; for example, the vomerine teeth on the palate of some fishes, amphibians, and reptiles.

vomeronasal organ (= Jacobson's organ). Olfactory organ in the roof of the mouth of an amphibian, reptile, or mammal.

von Baer's law. Proposal that embryonic development in vertebrates goes from general forms common to all vertebrates to increasingly specialized forms characteristic of classes, orders, and lower taxonomic levels.

vulva. Outer chamber leading to female reproductive tract.

waggle dance. Form of tactile communication by which a scout honey bee communicates the distance and direction of food to other worker bees.

warning coloration. Highly visible coloration of toxic species that deters predation.

water cycle (= hydrologic cycle). Biogeochemical cycling of water among various organic and inorganic states.

water vascular system. Network of water-filled canals in an echinoderm.

wax. Long-chain fatty acid linked to a long-chain alcohol or carbon ring.

wax canal. Channel through arthropod cuticle believed to transport wax to the surface.

Weberian ossicle. Bone by which swim bladder is believed to transmit sound to semicircular canals and otolith organs of some fishes.

white blood cell (= leukocyte). Blood cell important in immunity.

white matter. Portions of the brain or spinal cord consisting largely of myelinated axons.

wing loading. Body mass divided by wing area.
withdrawal reflex. Simple reflex that pulls a limb away from injury.
Wollfian duct. Duct of the embryonic kidney (mesonephros) that develops into part of the reproductive system in males.
worker jelly. Sugary substance that worker honey bees feed to larvae.

X organ–sinus gland system (XOSG). Glands in crustacea that secrete molt-inhibiting hormone and several other hormones.
xanthophore. Type of chromatophore that contains red, orange, or yellow pigment.

Y organ. Gland in crustaceans that secretes molting hormone.
yolk. Nutrient deposit in an egg.
yolk sac. Membrane enclosing the yolk in developing vertebrates.

Z disk. Structure at each end of a sarcomere in skeletal muscle.
zoea (ZO-ee-uh). Larval stage of certain crabs.
zoecium (zo-EE-she-um). Cuticular chamber enclosing an individual zooid (polypide) of an ectoproct.
zona pellucida. Glycoprotein covering a mammalian egg.
zona reaction. Response to fertilization triggered by the cortical reaction, which blocks polyspermy by hardening the zona pellucida and preventing the binding of sperm.
zooid (ZO-oyd). Individual in a colony of animals or protozoans.
zooplankton (singular **zooplankter**). Aquatic larvae and animals suspended in water.
zooxanthella (ZO-oh-zan-THELL-uh). Mutualistic photosynthesizing protist (dinoflagellate) within a sponge, coral, clam, or other animal.
zygote. Fertilized ovum.

Line Drawings

Electronic Illustrations created by HRS Electronic Text Management.

Photographs

Page 2, Figure 1.1, American Museum of Natural History, no. 2373(4); *Page 3, Figure 1.2*, Lary Shaffer; *Page 3, Figure 1.3*, C. Leon Harris; *Page 8, Figure 1.5*, David Phillips/VU; *Page 8, Figure 1.6*, ©Bates Littlehales/Animals Animals, no. 442933F; *Page 9, Figure 1.7*, David Bray; *Page 9, Figure 1.8*, George Holton/Photo Researchers, no. 5T9850; *Page 21, Figure 2.3A*, C. Leon Harris; *Page 35, Figure 2.18A*, C. Leon Harris; *Page 35, Figure 2.18B*, C. Leon Harris; *Page 35, Figure 2.18C*, C. Leon Harris; *Page 35, Figure 2.18D*, C. Leon Harris; *Page 35, Figure 2.18E*, T. Fujita and D.W. Fawcett/VU; *Page 35, Figure 2.18F*, Reproduced from *A Textbook of Histology*, W. Bloom and D.W. Fawcett, 10th ed., ©1975 by permission of Chapman & Hall; *Page 35, Figure 2.18G*, Reproduced from A.L. Beyer, O.L. Miller, Jr., and S.L. Knight (1980). *Cell* 20:75-84 by copyright permission of Cell Press; *Page 35, Figure 2.18H*, Reproduced from A.L. Beyer, O.L. Miller, Jr., and S.L. Knight (1980). *Cell* 20:75-84 by copyright permission of Cell Press; *Page 37, Figure 2.20*, Robbin DeBiasio; *Page 43, Figure 2.26*, Reproduced from E. Lazarides and K. Weber (1974). *Proc. Natl. Acad. Sci. USA* 71:2268-2272, by courtesy of the National Academy of Sciences; *Page 44, Figure 2.27*, M. McGill; *Page 46, Figure 2.29A*, Reproduced from K. Tanaka, A. Iino, and T. Naguro (1976). *Arch. Histol. Jap*. 39:165-175 by permission of the Japan Society of Histological Documentation; *Page 46, Figure 2.29B*, Reproduced from *The Cell*, D.W. Fawcett, 2nd ed. ©1981 by permission of Chapman & Hall; *Page 46, Figure 2.29C*, Reproduced from K. Tanaka, A. Iino, and T. Naguro (1976). *Arch. Histol. Jap*. 39:165-175 by permission of the Japan Society of Histological Documentation; *Page 46, Figure 2.29D*, Reproduced from M. Bielinska, G. Rogers, T. Rucinsky, and I. Boine (1979). *Proc. Natl. Acad. Sci. USA* 76:6152-6156, by courtesy of the National Academy of Sciences; *Page 47, Figure 2.31*, Dr. Daniel Branton/Biological Laboratories, Harvard University; Page 57, *Figure 3.5A*, Courtesy Pierre and Nina Favard, Centre National de la Recherche Scientifique, France; *Page 67, Figure 4.1*, Granger; *Page 73, Figure 4.5A*, CNRI/Phototake NYC; *Page 73, Figure 4.5B*, CNRI/Phototake NYC; *Page 73, Figure 4.5C*, CNRI/Phototake NYC; *Page 73, Figure 4.5D*, CNRI/Phototake NYC; *Page 75, Figure 4.6*, Susanne M. Gollin and Wayne Wrap, Dept. of Human Genetics, University of Pittsburgh; *Page 78, Figure 4.9A*, American Philosphical Society; *Page 84, Figure 5.1*, Cold Spring Harbor Archives; *Page 86, Figure 5.3*, U.K. Laemmli; *Page 95, Figure 5.11*, Cytogenetics Laboratory, The Children's Hospital, Denver, CO; *Page 96, Figure 5.12*, Reprinted by permission of John Wiley & Sons Inc. from J. de Grouchey and C. Turleau, *Clinical Atlas of Human Chromosomes*, ©1977 by John Wiley & Sons, Inc., New York; *Page 102, Figure 5.15*, R.L. Brinster and R.E. Hammer, University of Pennsylvania; *Page 110, Figure 6.5A*, Lennart Nilsson; *Page 110, Figure 6.6*, Reproduced from H. Shatten and G. Shatten (1980). *Developmental Biology* 78:435-449, by copyright permission of Academic Press, Inc., Orlando, FL.; *Page 111, Figures 6.7A-D*, Reproduced from M.J. Tegner and D. Epel (1973). *Science* 129:685-688, by copyright permission of the AAAS; *Page 113, Figure 6.9A*, Reproduced with the permission of the author (Trounson et al. 1982. *Journal of Reproduction and Fertility* 64:285-294); *Page 115, Figure 6.12A*, CABISCO/Phototake NYC; *Page 115, Figure 6.12B*, CABISCO/Phototake NYC; *Page 115, Figure 6.12C*, CABISCO/Phototake NYC; *Page 115, Figure 6.12D*, CABISCO/Phototake NYC; *Page 115, Figure 6.12E*, CABISCO/Phototake NYC; *Page 115, Figure 6.12F*, CABISCO/Phototake NYC; *Page 116, Figure 6.13A*, Prof. Dr. E. Schierenberg/Zoologisches Institut der Universitat, Köln; *Page 116, Figure 6.13B*, Prof. Dr. E. Schierenberg/Zoologisches Institut der Universitat, Köln; *Page 119, Figure 6.16*, CABISCO/Phototake NYC; *Page 123, Figure 6.22B*, From R. O'Rahilly and F. Muller, 1987. *Developmental Stages in Human Embryos*. Washington D.C.: Carnegie Institution of Washington, Publication 637. Page 94, Figure 10-1A; *Page 124, Figure 6.23A*, From R. O'Rahilly and F. Muller, 1987. *Developmental Stages in Human Embryos*. Washington D.C.: Carnegie Institution of Washington, Publication 637. Page 158, Figure 14-1D; *Page 124, Figure 6.23B*, Carlo Bevilacqua/CEDRI; *Page 126, Figure 6.25*, Carlo Bevilacqua/CEDRI; *Page 127, Figure 6.26*, Jim Langeland, Steve Paddock, Sean Carroll, Howard Hughes Medical Institute, University of Wisconsin; *Page 127, Figure 6.27(1)*, Dr. F.R. Turner, Indiana University; *Page 127, Figure 6.27(2)*, Dr. F.R. Turner, Indiana University; *Page 128, Figure 6.28*, Courtesy of S.M. Rothman. From W.M. Cowan. (1979) *Scientific American* 241 (Sept.):113-130; *Page 146, Figure 7.12B*, C. Leon Harris; *Page 147, Figure 7.13*, W.S. Bowers/Dept. of Entomology, University of Arizona; *Page 152, Figure 8.1A*, Dr. Christine Gall, University of California at Irvine; *Page 157, Figure 8.6*, Dr. Cedric S. Raine/Albert Einstein College of Medicine; *Page 158, Figure 8.7A*, E.R. Lewis, Y.Y. Zeevi, T.E. Everhart, University of California/BPS; *Page 158, Figure 8.7B*, J.E. Heuser; *Page 185, Figure 9.11B*, W. Feindel, Montreal Neurological Institute; *Page 187, Figure 9.14*, M.E. Raichle, Washington University, St. Louis, MO; *Page 194, Figure 10.2A*, C. Leon Harris; *Page 194, Figure 10.2B*, C. Leon Harris; *Page 201, Figure 10.9C*, T.R. Resse and D.W. Fawcett/VU; *Page 203, Figure 10.11A*, C. Leon Harris; *Page 205, Figure 10.13A*, C. Leon Harris; *Page 205, Figure 10.13B*, Dr. H.E. Huxley; *Page 210, Figure 10.16*, R.D. Allen; *Page 211, Figure 10.17A*, Dr. William E. Barstow; *Page 211, Figure 10.17B*, Sidney L. Tamm, in Tamm, S.L., and G.A. Hsorridge. 1970. *Proc. Roy. Soc. London*; *Page 211, Figure 10.17C*, David M. Phillips/VU; *Page 231, Figure 11.13*, Boehringer Ingelheim International GMBH/Lennart Nilsson; *Page 232, Figure 11.14*, Dr. G. Kaplan/The Rockefeller University; *Page 244, Figure 12.7B*, Dr.Warren W. Burggren, Dept. of Zoology, University of Massachusetts, Amherst; *Page 266, Figure 13A*, Bettmann Archive; *Page 261, Figure 13.3A*, Rod Planck/Tom Stack & Associates; *Page 292, Figure 15.1*, C. Leon Harris; *Page 296, Figure 15A*, C. Leon Harris; *Page 309, Figure 16.1*, Warren Williams/Planet Earth Pictures; *Page 310, Figure 16.2*, Marty Snyderman/VU; *Page 310, Figure 16.2B*, Michael Fogden/Bruce Coleman; *Page 342, Figure 17.16A*, © Chip and Jill Eisenhart/Tom Stack & Associates; *Page 342, Figure 17.16B*, ©1985 John Sohlden/VU; *Page 342, Figure 17.16C*, © Doug Sokell/Tom Stack & Associates; *Page 342, Figure 17.16D*, ©1991 Ron Spomer/VU; *Page 342, Figure 17.16E*, © Joe McDonald/VU; *Page 342, Figure 17.16F*, © Steve McCutcheon/VU; *Page 342, Figure 17.16G*, © 1994, Lynn M. Stone; *Page 342, Figure 17.16H*, © L.L.T. Rhodes/Earth Scenes; *Page 359, Figure 18.2*, Carnegie Museum of Natural History; *Page 359, Figure 18.2A*, John Ostrom, Yale University; *Page 362, Figure 18.4*, The American Museum of Natural History; *Page 362, Figure 18.5*, Granger; *Page 362, Figure 18.6*, National Portrait Gallery; *Page 363, Figure 18.7*, © *Scientific American*; *Page 364, Figure 18.8*, American Philosophical Society; *Page 375, Figure 18.14A*, T.W. Pietsch and D.B. Grobecker; *Page 375, Figure 18.14B*, Steinhardt Aquarium, San Francisco; *Page 375, Figure 18.15*, C. Leon Harris;

Page 376, Figure 18.16, H.B.D. Kettlewell. Courtesy of David Kettlewell; *Page 383, Figure 19.1A*, Manfred Gottschalk/Tom Stack & Associates; *Page 387, Figure 19.4A-E*, P. Cloud & M.F. Glaessner, 1982, *The Ediacran Period and System: Metazoa Inherit the Earth, Science* 218:783-792; *Page 389, Figure 19.6*, E.N.K. Clarkson/University of Edinburgh, Grant Institute; *Page 393, Figure 19.10A*, The American Museum of Natural History; *Page 393, Figure 19.10B*, The American Museum of Natural History; *Page 395, Figure 19.12A*, Dr. Elmut Albrecht/Bruce Coleman; *Page 397, Figure 19.14A*, Science VU/VU; *Page 397, Figure 19.14B*, D. Finnin/C. Chesek, The American Museum of Natural History; *Page 403, Figure 20.2*, Andrew King; *Page 406, Figure 20.4A*, C. Leon Harris; *Page 406, Figure 20.4B*, John Reader; *Page 407, Figure 20.5*, ©1993 Charlie Paiek/Animals Animals; *Page 412, Figure 20.10*, Jeff Foott/Bruce Coleman; *Page 413, Figure 20.11A*, C. Leon Harris; *Page 413, Figure 20.11B*, Biana Lavies ©1978 Copyright, National Geographic Society; *Page 415, Figure 20.13*, John Gerlach/Tom Stack & Associates; *Page 419, Figure 20.15*, Diana Stratton/Tom Stack & Associates; *Page 419, Figure 20.16*, Tim Davis/Photo Researchers; *Page 420, Figure 20.17*, Jane Burton/Bruce Coleman; *Page 420, Figure 20.18*, Roger Wilmshurst/Bruce Coleman; *Page 421, Figure 20.19*, Nina Leen, *Life* Magazine ©Time Warner Inc.; *Page 422, Figure 20.20*, Martin Rogers/TSW; *Page 430, Figure 21.1A*, Mike Hoplak for the Cornell Lab of Ornithology; *Page 430, Figure 21.1B*, Robert C. Simpson/Tom Stack & Associates; *Page 430, Figure 21.2A*, Robert C. Simpson/Tom Stack & Associates; *Page 430, Figure 21.2B*, S. Maslowski/VU; *Page 452, Figure 22.2A*, David M. Phillips/VU; *Page 452, Figure 22.2B*, P.L Walne, and H.J. Arnott (1967) *Planta* 77:325-354, p. 465; *Page 456, Figure 22.3A*, Karl Aufderheide/VU; *Page 457, Figure 22.4A*, David M. Phillips/VU; *Page 457, Figure 22.4B*, Manfred Kage/Peter Arnold, Inc.; *Page 458, Figure 22.6*, K. Jeon/VU; *Page 459, Figure 22.7*, from L.E. Roth, D.J. Pihlaja and Y. Shigemaka (1970). *J. Ultrastruct. Res.* 30:7-37, by copyright permission of Academic Press, Inc., Orlando, FL; *Page 464, Figure 22.12*, Manfred Kage/Peter Arnold Inc.; *Page 465, Figure 22.13*, © Biophoto/Photo Researchers; *Page 468, Figure 22.14*, Armed Forces Institute of Pathology; *Page 468, Figure 22.15*, Jerome Paulin/VU; *Page 478, Figure 23.1A*, Larry Roberts/VU; *Page 481, Figure 23.4A*, Biological Photo Service; *Page 482, Figure 23.5A*, James Balodimas/The Field Museum of Natural History, Chicago; *Page 483, Figure 23.6A*, Glenn Oliver/VU; *Page 483, Figure 23.6B*, Patricia R. Bergquist, Dept. of Zoology, The University of Auckland; *Page 484, Figure 23.7A*, C. Leon Harris; *Page 493, Figure 24.1A*, John D. Cunningham/VU; *Page 493, Figure 24.1B*, Fred Bavendam/Peter Arnold; *Page 496, Figure 24.4*, CABISCO/Phototake NYC; *Page 497, Figure 24.5A*, Ed Reschke/Peter Arnold, Inc.; *Page 499, Figure 24.6*, ©1977 George Lower/Photo Researchers; *Page 500, Figure 24.7A*, John D. Cunningham/VU; *Page 501, Figure 24.8A*, Neville Coleman/VU; *Page 502, Figure 24.9A*, Daniel W. Gotshall/VU; *Page 502, Figure 24.9B*, James R. McCullagh/VU; *Page 502, Figure 24.10A*, Peter Parks/Oxford Sci/Animals Animals; *Page 502, Figure 24.10C*, Larry Tackett/Tom Stack & Associates; *Page 503, Figure 24.11A*, © Herb Segars/Animals Animals; *Page 503, Figure 24.12*, Jennifer Purcell; *Page 508, Figure 24.15*, Phillip Playford from P.E. Playford, 1980, Ordovician Great Barrier Reef of the Canning Basin, Western Australia. *Peteroleum Geologists* 64: 814-840, p. 518; *Page 509, Figure 24.16*, Don and Pat Valenti; *Page 510, Figure 24.17*, Laurence P. Madin/Woods Hole Oceanographic Institution; *Page 518, Figure 25.1A*, Dave Fleetham/Tom Stack & Associates; *Page 518, Figure 25.1B*, Biological Photo Service; *Page 521, Figure 25.3*, Dwight Kuhn; *Page 525, Figure 25.8B*, Harvey D. Blankespoor, Hope College, Holland, MI; *Page 526, Figure 25.9A*, Allan Roberts; *Page 526, Figure 25.9B*, S. Flegler/VU; *Page 526, Figure 25.9C*, John D. Cunningham/VU; *Page 528, Figure 25.10*, John Hart, M.D., University of Chicago (Department of Pathology); *Page 542, Figure 26.3*, N. Allin and G.L. Barron/Universty of Guelph, Canada; *Page 544, Figure 26.5*, R. Calentine/VU: *Page 545, Figure 26.6B*, Courtesy of Institute of Public Health Research, Teheran University School of Public Health; *Page 546, Figure 26.7*, Courtesy of Institute of Public Health Research, Teheran University School of Public Health; *Page 546, Figure 26.8*, Dwight Kuhn; *Page 547, Figure 26.9B*, Trustees of the British Museum (Natural History); *Page 551, Figure 26.13*, John Gilbert/Dartmouth College; *Page 560, Figure 27.1*, Kjell B. Sandved/VU; *Page 561, Figure 27.3*, C. Leon Harris; *Page 562, Figure 27.4A*, David J. Wrobel/BPS; *Page 564, Figure 27.5A*, Allan Roberts; *Page 566, Figure 27.6*, Kjell B. Sandved/VU; *Page 568, Figure 27.8A*, John D. Cunningham/VU; *Page 570, Figure 27.10*, Dr. Diane R. Nelson/Eastern Tennessee State University; *Page 575, Figure 27.13*, D. Foster/VU; *Page 575, Figure 27.14*, C. Leon Harris; *Page 580, Figure 28.1A*, Kjell B. Sandved/VU; *Page 580, Figure 28.1B*, Jeff Foott/Tom Stack & Associates; *Page 580, Figure 28.1C*, C. Leon Harris; *Page 580, Figure 28.1D*, C.R. Wyttenbach/University of Kansas/Biological Photo Service; *Page 580, Figure 28.1E*, Gary Milburn/Tom Stack & Associates; *Page 581, Figure 28.2*, James L. Amos/Photo Researchers; *Page 583, Figure 28.4A*, C. Leon Harris; *Page 585, Figure 28.5B*, The Field Museum of Natural History, Chicago; *Page 587, Figure 28.7A*, C.B. Calloway; *Page 587, Figure 28.7B*, C.B. Calloway; *Page 592, Figure 28.12A*, William H. Amos; *Page 592, Figure 28.12B*, James A. Carmichael/Bruce Coleman; *Page 592, Figure 28.12C*, Scott Johnson/Animals Animals; *Page 592, Figure 28.12D*, Matt Bradley; *Page 592, Figure 28.12E*, Brian Park/Tom Stack & Associates; *Page 592, Figure 28.13*, Kjell B. Sandved/VU; *Page 593, Figure 28.14*, Milton Rand/Tom Stack & Associates; *Page 595, Figure 28.18A*, A. Kerstitch/VU; *Page 595, Figure 28.18B*, A. Kerstitch/VU; *Page 595, Figure 28.18C*, A. Kerstitch/VU; *Page 596, Figure 28.19*, Geoff du Feu/Planet Earth Pictures; *Page 597, Figure 28.20B*, Dave B. Fleetham/Tom Stack & Associates; *Page 599, Figure 28.22*, Dave B. Fleetham/Tom Stack & Associates; *Page 600, Figure 28.24A*, Douglas Faulkner/Photo Researchers; *Page 601, Figure 28.25*, Douglas Faulkner; *Page 604, Figure 28.28*, ©1990 Scott Camazine/Photo Researchers; *Page 609, Figure 29.1A*, R. DeGoursey/VU; *Page 617, Figure 29.8*, D. Holden Bailey/Tom Stack & Associates; *Page 618, Figure 29.10*, American Museum of Natural History; *Page 619, Figure 29.13*, Globe Photo; *Page 622, Figure 29.17*, C. Leon Harris; *Page 624, Figure 29.19A*, C. Leon Harris; *Page 624, Figure 29.19B*, C. Leon Harris; *Page 624, Figure 29.20*, St. Bartholomew's Hospital/Science Photo/Photo Researchers; *Page 630, Figure 30.1A*, C. Leon Harris; *Page 630, Figure 30.1B*, Michael Fogden/Animals Animals; *Page 630, Figure 30.1C*, C. Leon Harris; *Page 630, Figure 30.1D*, ©Breck P. Kent/Animals Animals; *Page 630, Figure 30.1E*, C. Leon Harris; *Page 634, Figure 30.4D*, C. Leon Harris; *Page 639, Figure 30.8*, C. Leon Harris; *Page 641, Figure 30.10*, Lynn M. Stone; *Page 644, Figure 30.13*, Dick George/Tom Stack & Associates; *Page 650, Figure 30.19*, C. Leon Harris; *Page 650, Figure 30.20*, C. Leon Harris; *Page 651, Figure 30.21*, Phillip Brownell/Dept. of Zoology/ Oregon State University; *Page 652, Figure 30.22*, © Robert and Linda Mitchell; *Page 652, Figure 30.23A*, Ed Bosler Ph.D., State of New York Dept. of Health; *Page 652, Figure 30.23B*, William H. Amos; *Page 653, Figure 30.24A*, W.J. Weber/VU; *Page 653, Figure 30.24B*, Richard Thom/VU; *Page 654, Figure 30.25A*, Dr. C. Desch. Biology Dept., University of Connecticut; *Page 654,*

Figure 30.25B, Lilia Ibay de Guzman; *Page 661, Figure 31.4*, Lynn M. Stone; *Page 668, Figure 31.12A*, Fred Bavendam/Peter Arnold, Inc.; *Page 668, Figure 31.12B*, Joe McDonald/Tom Stack & Associates; *Page 668, Figure 31.12C*, Gerald & Buff Corsi/Tom Stack & Associates; *Page 668, Figure 31.12D*, Dave B. Fleetham/Tom Stack & Associates; *Page 669, Figure 31.13*, ©Adrian Wenner/VU; *Page 670, Figure 31.14*, William H. Amos; *Page 670, Figure 31.15*, C. Leon Harris; *Page 671, Figure 31.16A*, Tom Stack/Tom Stack & Associates; *Page 671, Figure 31.16B*, Heather Angel/ Biofotos; *Page 672, Figure 31.18*, Dwight Kuhn; *Page 673, Figure 31.19*, John D. Cunningham/VU; *Page 673, 31.20A*, Paulette Brunner/Tom Stack & Associates; *Page 673, Figure 31.20B*, Ed Degginger/Bruce Coleman; *Page 674, Figure 31.21A*, Jill Yager/Antioch University; *Page 675, Figure 31.22B*, Richard Walters/VU; *Page 679, Figure 32.3*, C. Leon Harris; *Page 681, Figure 32.4A*, C. Leon Harris; *Page 681, Figure 32.4B*, C. Leon Harris; *Page 681, Figure 32.4C*, C. Leon Harris; *Page 686, Figure 32.8*, Dwight Kuhn; *Page 688, Figure 32.11*, C. Leon Harris; *Page 689, Figure 32.12A*, Barbara L.Thorne; *Page 689, Figure 32.12B*, Thomas Eisner; *Page 690, Figure 32.13A*, C. Leon Harris; *Page 690, Figure 32.13B*, C. Leon Harris; *Page 690, Figure 32.14A*, C. Leon Harris; *Page 690, Figure 32.14B*, C. Leon Harris; *Page 690, Figure 32.14C*, C. Leon Harris; *Page 691, Figure 32.15*, C. Leon Harris; *Page 693, Figure 32.17A*, Hans Pfletschinger/Peter Arnold, Inc.; *Page 693, Figure 32.17B*, C. Leon Harris; *Page 694, Figure 32.18B*, C. Leon Harris; *Page 694, Figure 32.18C*, C. Leon Harris; *Page 695, Figure 32.19*, Dr. Edward S. Ross/CA Academy; *Page 701, Figure 32.22*, B.M. Shepard/G.R. Carner/J. Hudson; *Page 702, Figure 32.23*, E.R. Degginger/Animals Animals; *Page 707, Figure 33.1A*, Dwight Kuhn; *Page 707, Figure 33.1B*, Brian Parker/Tom Stack & Associates; *Page 707, Figure 33.1C*, Tom Stack/Tom Stack & Associates; *Page 707, Figure 33.1D*, David L. Meyer; *Page 707, Figure 33.1E*, Brian Parker/Tom Stack & Associates; *Page 711, Figure 33.3A*, Allan Roberts; *Page 712, Figure 33.4B*, C. Leon Harris; *Page 715, Figure 33.8A*, Ron Testa and Diane White/Field Museum; *Page 715, Figure 33.8B*, C. Leon Harris; *Page 716, Figure 33.10*, David S. Addison/VU; *Page 718, Figure 33.12C*, C. Leon Harris; *Page 719, Figure 33.14*, Ed Robinson/Tom Stack & Associates; *Page 721, Figure 33.17*, Ed Robinson/Tom Stack & Associates; *Page 721, Figure 33.18*, Ed Robinson/Tom Stack & Associates; *Page 729, Figure 34.5; Page 730, Figure 34.7A*, Robert Brons/Biological Photo Service; *Page 730, Figure 34.7B*, Charles R. Wyttenbach/BPS; *Page 732, Figure 34.8*, Heather Angel/Biophotos; *Page 735, Figure 34.11*, ©Jennifer Purcell; *Page 737, Figure 34.14*, Gary R. Robinson/VU; *Page 740, Figure 34.18*, slide #82670, The Field Museum of Natural History, Chicago; *Page 740, Figure 34.19*, slide #82670, The Field Museum of Natural History, Chicago; *Page 741, Figure 34.20A*, slide #82664/ The Field Museum of Natural History, Chicago; *Page 746, Figure 35.1A*, Ken Lucas/BPS; *Page 746, Figure 35.1B*, Tom Stack/Tom Stack & Associates; *Page 747, Figure 35.2A*, A.B. Joyce/Photo Researchers; *Page 747, Figure 35.2B*, Ed Robinson/Tom Stack & Associates; *Page 747, Figure 35.2C*, Dave Fleetham/Tom Stack & Associates; *Page 748, Figure 35.3A*, John D. Cunningham/VU; *Page 748, Figure 35.3B*, Patrice Ceisel/VU; *Page 753, Figure 35.6B*, Patrice Ceisel/ VU; *Page 755, Figure 35.9B*, William H. Amos; *Page 755, Figure 35.9C*, Leonard J. Compagno/Shark Research Center/ South African Museum; *Page 757, Figure 35.11*, David J. Wrobel/BPS; *Page 759, Figure 35.13*, Peter Scoones ©Planet Earth Pictures; *Page 764, Figure 35.15*, Tom Stack/Tom Stack & Associates; *Page 764, Figure 35.16*, Patrice Ceisel/VU; *Page 767, Figure 35.21A*, Denise Tackett/TomStack & Associates; *Page 767, Figure 35.21B*, Denise Tackett/VU; *Page 770, Figure 35.24A*, Tom McHugh/Photo Researchers; *Page 777, Figure 35.29*, ©Dr. Paul A. Zahl/Photo Researchers; *Page 778, Figure 35.30*, A. Kerstitch/VU; *Page 784, Figure 36.1A*, D.G. Barker/Tom Stack & Associates; *Page 784, Figure 36.1B*, C. Leon Harris; *Page 784, 36.1C*, C. Leon Harris; *Page 786, Figure 36.3A*, ©E.R. Degginger/Animals Animals; *Page 786, Figure 36.3B*, Allan Blank/Bruce Coleman; *Page 787, Figure 36.4*, Jane Burton/Bruce Coleman; *Page 789, Figure 36.6A*, D.G. Barker/Tom Stack & Associates; *Page 789, Figure 36.6B*, C. Leon Harris; *Page 797, Figure 36.13B*, C. Leon Harris; *Page 798, Figure 36.14*, M.J. Tyler, University of Adelaide, Australia; *Page 800, Figure 36.16A*, Victor Hutchinson/VU; *Page 800, Figure 36.16B*, Allan Roberts; *Page 806, Figure 37.3A*, Dwight R. Kuhn; *Page 806, Figure 37.3B*, Don and Pat Valenti; *Page 806, Figure 37.3C*, John C. Murphy/Tom Stack & Associates; *Page 807, Figure 37.4A*, Dwight R. Kuhn; *Page 807, Figure 37.4B*, David M. Denni/Tom Stack & Associates; *Page 807, Figure 37.4C*, John Nees/Animals Animals; *Page 808, Figure 37.5*, Lynn M. Stone; *Figure 37.5B*, Zig Leszczyminski/Animals Animals; *Page 808, Figure 37.5C*, Alan and Sandy Carey; *Page 808, Figure 37.5D*, Lynn M. Stone; *Page 809, Figure 37.6A*, Lynn M. Stone; *Page 809, Figure 37.6B*, David R. Frazier; *Page 809, Figure 37.6C*, Gary Milburn/Tom Stack & Associates; *Page 811, Figure 37.7A*, John Cancalor/Tom Stack & Associates; *Page 811, Figure 37.8*, Chip Clark/National Museum of Natural History/Smithsonian Institution; *Page 812, Figure 37.9*, Rudolph Zallinger/Peabody Museum of Natural History; *Page 816, Figure 37.13*, Daniel Luchtel/Michael Hlastala; *Page 819, Figure 37.15A*, J.H. Robinson/Animals Animals; *Page 822, Figure 37.18A*, William H. Amos; *Page 822, Figure 37.18B*, Mary Clay/Tom Stack & Associates; *Page 822, Figure 37.18C*, © Stephen Dalton/Photo Researchers; *Page 824, Figure 37.19*, Joe McDonald/VU; *Page 825, Figure 37.20*, C.B. Frith/Bruce Coleman; *Page 825, Figure 37.21*, Tom Brakefield/Bruce Coleman; *Page 836, Figure 38.4A*, David M. Phillips/VU; *Page 836, Figure 38.4B*, S.D. Carlson; *Page 837, Figure 38.5B*, C. Leon Harris; *Page 837, Figure 38.5C*, C. Leon Harris; *Page 838, Figure 38.6A*, C. Leon Harris; *Page 838, Figure 38.6B*, Michael Fogden/VU; *Page 840, Figure 38.9*, © Stephen Dalton/Photo Researchers; *Page 842, Figure 38.11A*, Andreas Feininger, *Life* Magazine © Time Warner, Inc.; *Page 848, Figure 38.17A*, © Sylvestris/Gross/Peter Arnold; *Page 848, Figure 38.17B*, Fritz Polking/Peter Arnold; *Page 848, Figure 38.17C*, © Silvestris/Gross/Peter Arnold; *Page 849, Figure 38.19A*, Lynn M. Stone; *Page 849, Figure 38.19B*, John Markham/Bruce Coleman; *Page 850, Figure 38.20A*, © Hans and Judy Beste/Animals Animals; *Page 850, Figure 38.20B*, Frithfoto/Bruce Coleman; *Page 852, Figure 38.21A*, John K. Nakata/Terraphotographics/BPS; *Page 852, Figure 38.21C*, David L. Pearson/VU; *Page 852, Figure 38.21D*, © Stephen Dalton/Photo Researchers; *Page 853, Figure 38.22*, Kjell B. Sandved/VU; *Page 853, Figure 38.23*, John Ebeling; *Page 855; Figure 38.24*, C. Leon Harris; *Page 856, Figure 38.25A*, Granger; *Page 856, Figure 38.25B*, Bettmann Archive; *Page 856, Figure 38.26*, John Borneman/Photo Researchers; *Page 857, Figure 38.27A*, Wendy Shattil/Bob Rozinski/Tom Stack & Associates*; Page 857, Figure 38.27B*, Don and Pat Valenti; *Page 857, Figure 38.27C*, Alan and Sandy Carey; *Page 863, Figure 39.2A*, J. Alcock/VU; *Page 863, Figure 39.2B*, Jean Phillipe Varin © Jacana Photo Researchers; *Page 864, Figure 39.3A*, Allan Roberts; *Page 864, Figure 39.3B*, Mark Newman/Tom Stack & Associates; *Page 864, Figure 39.3C*, Lynn M. Roberts; *Page 871, Figure 39.6A*, G.R. Roberts; *Page 871, Figure 39.6B*, Lynn M. Stone; *Page 873; Figure 39.7A*, John Visser/Bruce Coleman; *Page 874, Figure 39.8A*, Alan and Sandy Carey; *Page 874, Figure 39.8B*, Dwight R. Kuhn; *Page 874, Figure 39.8C*, Dwight R. Kuhn; *Page 874, Figure 39.8D*,

J.P. Scott/Planet Earth Pictures; *Page 875, Figure 39.9A*, ©Peter Pickford/DRK Photo; *Page 875, Figure 39.9B*, ©Phil Dotson/Photo Researchers; *Page 875, Figure 39.10*, Marty Stouffer Prod./Animals Animals; *Page 876, Figure 39.11A*, Alan and Sandy Carey; *Page 876, Figure 39.11B*, Gerald Corsi/Tom Stack and Associates; *Page 876, Figure 39.11C*, John Shaw/Tom Stack & Associates; *Page 876, Figure 39.11D*, Joe McDonald/Tom Stack & Associates; *Page 878, Figure 39.14A*, Thomas Kitchin/Tom Stack & Associates; *Page 878, Figure 39.14B*, Science VU/VU; *Page 879, Figure 39.15A*, Photo by Barbara L. Thorne, Museum of Comparative Zoology, Cambridge, Mass.; *Page 879, Figure 39.15B*, ©1983 Francois Gohier/Photo Researchers; *Page 880, Figure 39.16A*, Brian Parker/Tom Stack & Associates; *Page 881, Figure 39.17A*, Dr. Merlin D. Tuttle/Bat Conservation International; *Page 881, Figure 39.17B*, Steven Dalton/Oxford Scientific Films/Animals Animals; *Page 882, Figure 39.19A*, Lynn M. Stone; *Page 882, Figure 39.19B*, Zoological Society of San Diego; *Page 883, Figure 39.20*, Gary Milburn/Tom Stack & Associates; *Page 883, Figure 39.21A*, D.J. Chivers/Anthro-Photo; *Page 883, Figure 39.21B*, Norman Owen Tomalin/Bruce Coleman; *Page 885, Figure 39.22*, Bill Curtsinger; *Page 886, Figure 39.23A*, Don and Pat Valenti; *Page 886, Figure 39.23B*, T. Kitchin/Tom Stack & Associates; *Page 886, Figure 39.23C*, Lynn M. Stone; *Page 1, Chapter Opener 1*, Tom Stack and Associates; *Page 18, Chapter Opener 2*, ©W.Wirniewski/Okapia/Photo Researchers; *Page 50, Chapter Opener 3*, Francois Gohier/Photo Researchers; *Page 66, Chapter Opener 4*, J&B Photo/Animals Animals; *Page 83, Chapter Opener* 5, © Chris McHugh/Photo Researchers; *Page 105, Chapter Opener 6,* © Robert and Linda Mitchell; *Page 136, Chapter Opener 7,* © Lynn M. Stone; *Page 150, Chapter Opener 8,* © Tom McHugh/Photo Researchers; *Page 174, Chapter Opener 9,* John Shaw/Tom Stack & Associates; *Page 191, Chapter Opener 10,* © Kjell B. Sandved/VU; *Page 216, Chapter Opener 11,* Michael Fogden/Bruce Coleman; *Page 238, Chapter Opener 12,* Nuridsany et Perennou/Photo Researchers; *Page 253, Chapter Opener 13,* © Stephen J. Lang/VU; *Page 274, Chapter Opener 14,* © Mitch Reardon/Photo Researchers; *Page 291, Chapter Opener 15,* © Akira Uchiyama/Photo Researchers; *Page 307, Chapter Opener 16,* © Anthony Bannister/Animals Animals; *Page 324, Chapter Opener 17,* Joe McDonald/Tom Stack & Associates; *Page 355, Chapter Opener 18,* © Patti Murray/Animals Animals; *Page 380, Chapter Opener 19,* © John Giustina/The Wildlife Collection; *Page 401, Chapter Opener 20,* © 1975 Dwight R. Kuhn; *Page 428, Chapter Opener 21,* Allan Blank/Bruce Coleman; *Page 449, Chapter Opener 22,* © R. Oldfield/VU; *Page 475, Chapter Opener 23,* Nancy Sefton/Photo Researchers; *Page 492, Chapter Opener 24,* Brian Parker/Tom Stack & Associates; *Page 516, Chapter Opener 25,* © Dave Fleetham/VU; *Page 535, Chapter Opener 26,* R.B. Taylor/SPL/Photo Researchers; *Page 559, Chapter Opener 27,* A.J. Copley/VU; *Page 579, Chapter Opener 28,* ©Lawrence Naylor/Photo Researchers; *Page 601, Chapter Opener 29,* © Kjell Sandved/VU; *Page 629, Chapter Opener 30,* C. Leon Harris; *Page 658, Chapter Opener 31,* © George Holton/Photo Researchers; *Page 677, Chapter Opener 32,* © L.West/Photo Researchers; *Page 706, Chapter Opener 33,* © Chris Huss/The Wildlife Collection; *Page 724, Chapter Opener 34,* ©W. Gregory Brown/Animals Animals; *Page 745, Chapter Opener 35,* © Dave B. Fleetham/VU; *Page 783, Chapter Opener 36,* ©Zig Leszczynski/Animals Animals; *Page 803, Chapter Opener 37,* © Jim Merli/VU; *Page 828, Chapter Opener 38,* ©1994 Jack Fields/Photo Researchers; *Page 860, Chapter Opener 39,* © Robert and Linda Mitchell; *Page 16, Unit Opener 1,* Ken Eward/Photo Researchers; *Page 134, Unit Opener 2,* Tom J. Ulrich/Visuals Unlimited; *Page 322, Unit Opener 3,* © Charles V. Angelo/Photo Researchers; *Page 426, Unit Opener 4,* ©Al Grotell.

INDEX

Common roots of zoological terms

Each root is followed by its meaning and example(s). The root is from Greek unless noted by (L) as being from Latin.

A-, an- not, without (acoelomate, anaerobic)
ab- (L) off, from, away (aboral)
acanth thorn (acanthocephalan)
acti ray (Actinopterygii)
ad- (L) to, toward (adoral)
alb (L) white (albino)
allo- different, other (allele)
amphi around, on both sides, double (amphibian)
andr male (androgen)
aqua (L) water (aquatic)
arche ancient, first (archenteron)
astr, aster star (astrorhizae)
auto self (autotrophic)
axo, axi axis (axial)

Bi- (L) two, double (bilateral)
blast sprout (blastula, fibroblast)
brach (L) arm (brachiopod)
branch gill (nudibranch)

Card heart (cardiac)
caud (L) tail (caudal)
centr center (centromere)
cephal head (cephalopod)
chaet bristle, hair (polychaete)
chondr cartilage (Chondrichthyes)
chord string (notochord)
chori membrane (chorion)
chrom color (cytochrome)
clad branch (cladoceran)
coel hollow (coelom, schizocoel)
corn (L) horn (unicorn)
cyst bladder (statocyst)
cyt cell (cytoplasm, erythrocyte)

Dactyl finger or toe (pentadactyl)
de- (L) from, down, out (defecate)
dendr tree (dendrite)
derm skin (epidermis, placoderm)
di- two, double, separate (diploid)
dia- across, through (diaphragm)
diplo two, double (diploid)
dist (L) distant (distal)

Echin hedgehog, i.e. spiny (echinoderm)
ect outside (ectoparasite)
en-, end-, ent- inside (endoparasite)
enter intestine, gut (coelenterate)
entomo insect (entomology)
epi upon, over, beside (epidermis)
erythr red (erythrocyte)
eu- true, good (eukaryotic)
eury wide (euryhaline)
ex- (L) out, from (excrete)

Fer (L) bear (porifera)

Gam marriage (gamete)
gast stomach, belly (gastrula)
gen produce (carcinogen)
gene origin, birth (genetics)
gnath jaw (Agnatha)
gon seed, generation, offspring (gonad)
gyn female (gynandromorph)

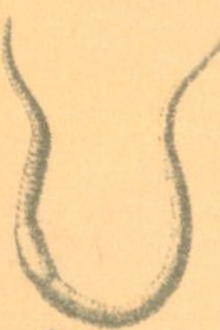

Helmin worm (helminthes)
hem blood (hemoglobin, uremia)
hemi- half (hemichordate)
hetero- other, different (heterozygote)
hex six (hexactinellid)
hol whole (holoblastic)
hom (L) human (hominid)
homeo-, homo- same, alike (homeostasis)
hyal glass (hyaline)
hydr water (hydrated)
hyper- over, excessive (hyperthermia)
hypo- under (hypotonic)

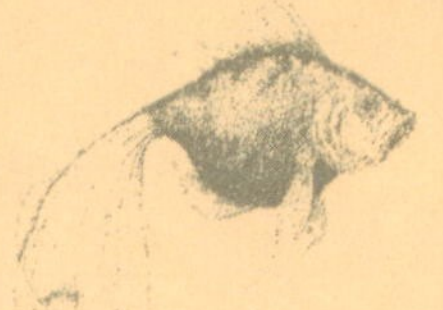

Ichthy fish (Osteichthyes)
inter- (L) between (intercellular)
intr- (L) inside (intracellular)
iso equal (isotonic)

Leuc, leuk white (leukocyte)
lob lobe (lobose)
loph crest, tuft (lophophorate)
lys, lyt loose (proteolytic)

Macro large (macromolecule)
melan black (melanin)
mer part (sarcomere, meroblastic)
meso middle (mesoderm)
meta after (metazoan)
micro small (microtubule)
mono one, single (monoecious)
morph form (metamorphosis)
myo muscle (myofibril)

Naut ship, sail (nautiloid)
nem thread (nematode)
neo new (neo-Darwinism)

Oo pronounced OH-oh; egg (oocyte)
ophi snake (Ophiuroidea)